ENCYCLOPEDIA OF APPLIED PLANT SCIENCES

ENCYCLOPEDIA OF APPLIED PLANT SCIENCES

Editor-in-Chief

BRIAN THOMAS

Editors

DENIS J MURPHY

BRIAN G MURRAY

ELSEVIER
ACADEMIC
PRESS

Amsterdam Boston Heidelberg London New York Oxford
Paris San Diego San Francisco Singapore Sydney Tokyo

Elsevier Ltd., The Boulevard, Langford Lane, Kidlington, Oxford, OX5 1GB, UK

Elsevier Inc., 525 B Street, Suite 1900, San Diego, CA 92101-4495, USA

First edition 2003

Library of Congress Control Number: 2003109025

A catalogue record for this book is available from the British Library

ISBN: 0-12-227050-9 (set)

This book is printed on acid-free paper

Printed in the United Kingdom

EDITORIAL ADVISORY BOARD

FOREWORD

I often tell visitors and policy makers who want to learn about current research in the plant sciences and agriculture that 'This is not your father's plant science'. As amply evidenced in these three volumes of the Encyclopedia of Applied Sciences, plant biology is undergoing a significant change in emphasis. The change engages far greater numbers of plant scientists in genetics, molecular biology, and biotechnology than ever before. This is not surprising given the growing amount of information on genome structure and functional genomics and the tools of genetic transformation of plants. These changes makes it possible for laboratory scientists to have a far greater impact on agriculture production and products than in any other period of time in the long history of applied plant sciences. The future promises to bring even more changes that bring sustainability to production agriculture and greater variability to its products.

These volumes contain chapters from world authorities in a range of topics and carry information that will inform the advanced reader, as well as the newcomer to the field, about the latest advances in the expanding fields of plant science. The reader will learn about the use of transgenic plants with altered metabolism that enhances accumulation of proteins and carbohydrates, alkaloids and terpenoids, and other components. Other chapters report the state of the science and recent breakthroughs in genetics and biochemistry of photosynthesis, plant growth and development, and plant nutrition. There is also a much needed update on the many experimental approaches to develop transgenic plants with novel characteristics, including enhanced resistance to pests and pathogens; biofortification and enhanced food safety; altering seed biology; reducing ripening times and post-harvest losses; bioremediation, and so on. The advances that have been made in the field are truly remarkable and are a testament to the many opportunities that modern genetics and biotechnology have brought to plant sciences.

However, while the advances have been great the informed reader will realize that very few of the discoveries reported in these chapters have yet found their way into commercialization. Furthermore, there has been a disturbing trend in recent years to slow the delivery of new biotechnology products to the marketplace. Governmental regulatory agencies in the US and the EU, as well as countries around the world, have shown reluctance to grant approval of new products, or to approve for sale products that have gained regulatory approval in another country.

At this juncture the products that make it to the marketplace are those for which significant profit can be accurately predicted: to date, the direct beneficiaries of the few approved products are the farmers. It remains to be seen whether subsequent products will be limited to plants whose traits benefit the farmer/producer, or whether products that have more obvious 'consumer benefit' will be approved for release. And, given the troubling trend of governments to expand regulatory oversight, and therefore cost of bringing a product to market, it also remains to be seen whether public sector institutions will or will not be a major player in developing genetically enhanced crops, or whether this opportunity will be left to major private sector companies.

Nevertheless, the quality of plant science research continues to improve, enhancing our knowledge of plant biology, and creating more and more opportunities to make contributions to sustainably produce sufficient amounts of nutritious and safe foods to feed an ever-growing population. And, to bring plant made

pharmaceuticals and other specialty products to market. The concerns aside, I remain optimistic and hope that that you share this optimism after reading these volumes.

Roger N. Beachy, Ph.D.
President
Donald Danforth Plant Science Center
St. Louis, Missouri, USA

INTRODUCTION

Plants have a huge impact upon human existence. All of our food and much of our other energy needs are ultimately derived from the products of plant photosynthesis. Plants also have their impact through the social and economic dimensions of human existence. All agriculture, which provides us with our food, is ultimately based on plant production, be it directly through arable crops such as cereals, horticultural crops including fruits and vegetables or indirectly through pasture, forage and plant derived animal feedstuffs. Industrial crops such as fiber crops or woodlands and forests are a vital source of renewable materials for clothing, building and recreation, that we take for granted in everyday life. Flowers and ornamental plants delight us with their beauty, fragrance and their seemingly infinite variety.

Societies and economies all over the world have been able to gain increasing benefits from plants as our scientific understanding of their properties and potential has increased rapidly over recent decades. The range and complexity of this new information about plants is now potentially bewildering. A real challenge is how can such accumulated knowledge be made available to those taking an interest in the area for the first time. Equally how can a specialist in one aspect of plant applications gain access to information from wider or complementary areas of the subject? This is where we hope that this work will be of the most value.

The Encyclopedia of Applied Plant Sciences provides a friendly and comprehensive portal to the huge amount of knowledge that is available on plants and how we make use of them. The Editors, assisted by a distinguished Editorial Board have commissioned articles from an unparalleled group of experts across the plant sciences to provide introductory overviews and summaries of the state-of-the-art for their areas. Each article is sufficient for the reader to gain an understanding of the essentials of the topic covered and also provides an entry to the substantial amount of literature that lies beneath the articles. Rigorous editing and comprehensive indexing by the publishers ensures that the articles are understandable and that information can be located as needed. Articles covering the breadth of plant science are grouped under clearly described entries that will help the reader in finding not only the article they require but also articles on closely related topics.

Wherever possible authors have tried to show how the underlying science and information links through to applications in practical situations. Plant science undoubtedly has an exciting future and in this Encyclopedia we signpost this through a range of entries on the potential of the new genetic technologies, both in addressing traditional challenges and in making novel contributions to agriculture and other applications of plant science.

Putting the encyclopedia together has been a long and involved project. We are grateful for the help and support from the staff at Elsevier, the expertise and time given by the editorial board and the huge amount of work by the authors of the articles. We believe that the end result is a unique assemblage of the key information in this important field and are privileged to have been involved in the project. The value in the work is down to the quality of the contributors. Omissions or lack of balance are entirely the responsibility of the Editors. We hope that the work proves to be not only of practical value but can act as a catalyst for readers to gain entry to the exciting world of plant science.

Brian Thomas
Denis Murphy
Brian Murray

GUIDE TO USE OF THE ENCYCLOPEDIA

Structure of the Encyclopedia

The material in the Encyclopedia is arranged as a series of entries in alphabetical order. Some entries comprise a single article, whilst the majority of entries are on more diverse subjects and consist of several articles that deal with various aspects of the topic. In the latter case the articles are arranged in a logical sequence within an entry.

To help you realize the full potential of the material in the Encyclopedia we have provided three features to help you find the topic of your choice.

1. Contents List

Your first point of reference will probably be the contents list. The complete contents list appearing in each volume will provide you with both the volume number and the page number of the entry. On the opening page of an entry a contents list is provided so that the full details of the articles within the entry are immediately available.

Alternatively you may choose to browse through a volume using the alphabetical order of the entries as your guide. To assist you in identifying your location within the Encyclopedia a running headline indicates the current entry and the current article within that entry.

You will find "dummy entries" where obvious synonyms exist for entries or where we have grouped together related topics. Dummy entries appear in both the contents list and the body of the text. For example, a dummy entry appears for Arable Crops which directs you to Production Systems and Agronomy: Agricultural Crops where the material is located.

Example
If you were attempting to locate material on Biotechnology via the contents list.

BIOTECHNOLOGY *see* DISEASES: Genetic Modification of Disease Resistance, Fungal Pathogens; Genetic Modification of Disease Resistance, Viral Pathogens. GENETIC MODIFICATION: Gene Cloning, General Principles; Insertional and Transposon Mutagenesis; Transformation, General Principles; Transformation in Dicotyledons; Transformation in Monocotyledons; Transformation in Plastids; Transgene Stability and Inheritance. GENETIC MODIFICATION, APPLICATIONS: Flower Color; Nitrogen Fixation; Oxidative Stress; Cell Factories; Molecular Farming; Plantibodies. GROWTH AND DEVELOPMENT: Control of Gene Expression, Posttranscriptional Regulation.

At the appropriate location in the contents list, the page numbers for these articles are given.

If you were trying to locate the material by browsing through the text and you looked up Biotechnology then the following information would be provided.

Biotechnology *see* **Diseases**: Genetic Modification of Disease Resistance, Fungal Pathogens; Genetic Modification of Disease Resistance, Viral Pathogens. **Genetic Modification**: Gene Cloning, General Principles; Insertional and Transposon Mutagenesis; Transformation, General Principles; Transformation in Dicotyledons; Transformation in Monocotyledons; Transformation in Plastids; Transgene Stability and Inheritance. **Genetic Modification, Applications**: Flower Color; Nitrogen Fixation; Oxidative Stress; Cell Factories; Molecular Farming; Plantibodies. **Growth and Development**: Control of Gene Expression, Posttranscriptional Regulation.

Alternatively if you were looking up Genetic Modification the following information would be provided.

GENETIC MODIFICATION

Contents

2. Cross References

All of the articles in the Encyclopedia have been extensively cross referenced. The cross references, which appear at the end of an article, have been provided at three levels:

i. To indicate if a topic is discussed in greater detail elsewhere

ROOT DEVELOPMENT/Root Growth and Development.
See also: **Growth and Development**: Cell Division and Differentiation Cell Growth. **Nutrition**: Ion Transport; Mineral Uptake; Nitrogen Fixation. **Root Development**: Genetics of Primary Root Development; Lateral Root Initiation; The Rhizosphere and its Microorganisms. **Water Relations of Plants**: Phloem; Xylem.

ii. To draw the reader's attention to parallel discussions in other articles

ROOT DEVELOPMENT/Root Growth and Development.
See also: **Growth and Development**: Cell Division and Differentiation; Cell Growth. **Nutrition**: Ion Transport; Mineral Uptake; Nitrogen Fixation. **Root Development**: Genetics of Primary Root Development; Lateral Root Initiation; The Rhizosphere and its Microorganisms. **Water Relations of Plants**: Phloem; Xylem.

iii. To indicate material that broadens the discussion

ROOT DEVELOPMENT/Root Growth and Development.
See also: **Growth and Development**: Cell Division and Differentiation; Cell Growth. **Nutrition**: Ion Transport; Mineral Uptake; Nitrogen Fixation. **Root Development**: Genetics of Primary Root Development; Lateral Root Initiation; The Rhizosphere and its Microorganisms. **Water Relations of Plants**: Phloem; Xylem.

3. Index

The index will provide you with the volume number and page number of where the material is to be located, and the index entries differentiate between material that is a whole article, is part of an article or is data presented in a table. Detailed notes are provided on the opening page of the index.

4. Color Plates

The color figures for each volume have been grouped together in a plate section. The location of this section is cited in the contents list. Color versions of black and white figures are cited in figure captions within individual articles.

5. Contributors

A full list of contributors appears at the beginning of each volume.

CONTRIBUTORS

W W Adams III
University of Colorado
Boulder, CO, USA

S R Adams
Horticulture Research International
Wellesbourne
UK

G N Agrios
University of Florida
Gainesville
FL, USA

B S Ahloowalia
International Atomic Energy Agency
Vienna, Austria

G Alibert
INP-ENSAT
Auzeville, France

T Altabella
Universitat de Barcelona
Barcelona, Spain

A Altman
The Hebrew University of Jerusalem
Rhovot, Israel

J L Araus
University of Barcelona
Barcelona, Spain

S Arcioni
Institute of Plant Genetics
Perugia
Italy

A E Arthur
John Innes Centre
Norwich, UK

S N Azam-Ali
University of Nottingham
Nottingham, UK

R A Azevedo
University of São Paulo
Piracicaba, Brazil

M Bacon
University of Lancaster
Lancaster, UK

M Båga
University of Saskatchewan
Saskatoon, SK, Canada

P Barcelo
DuPont Wheat Transformation Laboratory
Barcelona, Spain

H S Bariana
University of Sydney
Camden, NSW, Australia

A R Barr
University of Adelaide
Glen Osmond, SA, Australia

L Bastiaans
Wageningen University
Wageningen, The Netherlands

M H Behboudian
Massey University
Palmerston North, New Zealand

J N B Bell
Imperial College London
Ascot, UK

G D Bending
Horticulture Research International
Wellesbourne, UK

R L Benech-Arnold
IFEVA
Buenos Aires, Argentina

L R Benjamin
Rothamsted Research
Harpenden, UK

P Berjak
University of Natal
Durban, South Africa

A Bernadac
UMR 990 INRA-ENSAT
Toulouse, France

I J Bingham
Scottish Agricultural College
Aberdeen, UK

D Blakesley
Horticulture Research International
West Malling, UK

J M Bonga
Canadian Forest Service – Atlantic Forestry Center
Fredericton, NB, Canada

A-M Boudet
Université Paul Sabatier
Toulouse, France

M Bouzayen
UMR 990 INRA-ENSAT
Toulouse, France

J Brown
University of Idaho
Moscow, ID, USA

V Buchanan-Wollaston
Horticulture Research International
Wellesbourne, UK

M A S Burton
University of Hertfordshire
Hatfield, UK

P D S Caligari
Universidad de Talca
Talca, Chile

R W F Cameron
University of Reading
Reading, UK

S Cameron
Canadian Forestry Service
Fredericton, NB, Canada

M Cannell
Syngenta, Jealotts Hill International Research Centre
Bracknell, UK

M Cappadocia
Université de Montréal
Montréal, Canada

J S Carmichael
University of North Dakota
Grand Forks, ND, USA

A C Cassells
National University of Ireland
Cork, Ireland

K J Chalmers
University of Adelaide
Glen Osmond, SA, Australia

B Chanson
Université Montpellier II
Montpellier, France

D Chávez
University of Illinois
Chicago, IL, USA

S C Cheah
Malaysian Palm Oil Board
Kuala Lumpur, Malaysia

R N Chibbar
National Research Council Canada
Saskatoon, SK, Canada

V L Chopra
National Research Centre on Plant Biotechnology
New Delhi, India

L P Christensen
Danish Institute of Agricultural Sciences
Aarslev, Denmark

P Christou
Fraunhofer Institute of Molecular Biology and Applied Ecology
Schmallenberg, Germany

S D Clouse
North Carolina State University
Raleigh, NC, USA

K E Cockshull
Horticulture Research International
Wellesbourne, UK

J I Cooper
NERC Centre for Ecology and Hydrology
Oxford, UK

A Cordeiro
Universitat de Barcelona
Barcelona, Spain

C Darley
University of York
York, UK

E Dayan
Agricultural Research Organization
Hanegev, Israel

J Degenhardt
Max Planck Institute for Chemical Ecology
Jena, Germany

G-J de Klerk
Center for Plant Tissue Culture Research
Lisse, The Netherlands

J M DeLong
Agriculture and Agri-Food Canada
Kentville, Nova Scotia, Canada

B Demmig-Adams
University of Colorado
Boulder, CO, USA

N Dengler
University of Toronto
Toronto, ON, Canada

K Denyer
John Innes Centre
Norwich, UK

M Dow
National University of Ireland
Cork, Ireland

J G Dubrovsky
Universidad Nacional Autónoma de Mexico
Cuernavaca Morelos, Mexico

E M Duffy
National University of Ireland
Cork, Ireland

J M Dunwell
University of Reading
Reading, UK

G Dutfield
Queen Mary College
London, UK

P Elomaa
University of Helsinki
Helsinki, Finland

M R Emmett
University of Reading
Reading, UK

D E Evans
Oxford Brookes University
Oxford, UK

J D Everard
DuPont Experimental Station
Wilmington, DE, USA

W E Finch-Savage
Horticulture Research International
Wellesbourne, UK

K Fisher
Virginia Polytechnic Institute and State University
Blacksburg, VA, USA

M E Foley
USDA
Agricultural Research Service
Fargo, ND, USA

B V Ford-Lloyd
University of Birmingham
Birmingham, UK

C H Foyer
Rothamsted Research
Harpenden, UK

J B Free
University of Wales
Cardiff, UK

C Freeman
University of Wales
Bangor, UK

S C Fry
The University of Edinburgh
Edinburgh, UK

M D Gale
John Innes Centre
Norwich, UK

G Galiba
Agricultural Research Institute of the Hungarian Academy of Sciences
Martonvásár, Hungary

J Gershenzon
Max Planck Institute for Chemical Ecology
Jena, Germany

K G Gilbert
University of Bristol
Bristol, UK

T A Gorshkova
KIBB, RAS
Kazan, Russia

K S Gould
University of Auckland
Auckland
New Zealand

J Grace
University of Edinburgh
Edinburgh, UK

D J Gray
University of Florida
Apopka, FL, USA

J Gril
Université Montpellier II
Montpellier, France

K Groten
Rothamsted Research
Harpenden, UK

X Gu
North Dakota State University
Fargo, ND, USA

N D Hagan
CSIRO Plant Industry
Canberra, Australia

D L Hallahan
Central Research and Development
E I DuPont de Nemours & Company
Wilmington, DE, USA

J B Harborne[†]
University of Reading
Reading, UK

F Hay
Royal Botanic Gardens Kew
Ardingly, UK

P Hedden
Rothamsted Research
Harpenden, UK

J S Heslop-Harrison
University of Leicester
Leicester, UK

T J V Higgins
CSIRO Plant Industry
Canberra, Australia

D F Hildebrand
University of Kentucky
Lexington, KY, USA

H W M Hilhorst
Wageningen University
Wageningen
The Netherlands

S J Hiscock
University of Bristol
Bristol, UK

P J Holloway
The Long Ashton Research Station
Bristol, UK

R Hull
John Innes Centre
Norwich, UK

M O Humphreys
IGER
Aberystwyth, UK

R Hunt
University of Sheffield
Sheffield, UK

[†] Deceased

B S Jalani
Malaysian Palm Oil Board
Kuala Lumpur, Malaysia

P E Jameson
Massey University
Palmerston North, New Zealand

M F Jardinaud
Biotechnology and Plant Breeding, ENSAT
Auzeville, France

M C Jarvis
Glasgow University
Glasgow, UK

P P Jauhar
United States Department of Agriculture – Agricultural Research Service
Fargo, ND, USA

C E Jeffree
University of Edinburgh
Edinburgh, UK

R Johnson
Colby College
Waterville, ME, USA

D L Jones
University of Wales
Bangor, UK

H D Jones
Rothamsted Research
Harpenden, UK

L Jones
University of York
York, UK

V G Kakani
Mississippi State University
MS, USA

J Kallerhoff
INP-ENSAT
Auzeville, France

J Kang
University of Toronto
Toronto, ON, Canada

K J Kasha
University of Guelph
Guelph, ON, Canada

S P Kell
University of Birmingham
Birmingham, UK

R E Kendrick
Wageningen University
Wageningen, The Netherlands

A R Kermode
Simon Fraser University
Burnaby, British Columbia, Canada

A D Kinghorn
University of Illinois
Chicago, IL, USA

A J Kinney
DuPont Experimental Station
Wilmington, DE, USA

J D Klein
ARO-Volcani Center
Bet Dagan, Israel

R M D Koebner
John Innes Centre
Norwich, UK

A Kohli
John Innes Centre
Norwich, UK

M J Kropff
Wageningen University
Wageningen, The Netherlands

K Kunert
University of Pretoria
Pretoria, South Africa

A Kurian
Kerala Agricultural University
Vellanikkara, India

A Kushairi
Malaysian Palm Oil Board
Kuala Lumpur, Malaysia

M Labra
University of Milan
Milan, Italy

J Langridge[†]
Adelaide University
Glen Osmond, SA, Australia

P Langridge
Adelaide University
Glen Osmond, SA, Australia

E Larsen
Danish Institute of Agricultural Sciences
Aarslev, Denmark

A Latché
UMR 990 INRA-ENSAT
Toulouse, France

D W Lawlor
Rothamsted Research
Harpenden, UK

P A Lazzeri
DuPont Wheat Transformation Laboratory
Barcelona, Spain

P J Lea
Lancaster University
Lancaster, UK

R C Leegood
University of Sheffield
Sheffield, UK

R A Leigh
University of Cambridge
Cambridge, UK

Z Li
UMR 990 INRA-ENSAT
Toulouse, France

S E Lid
Agricultural University of Norway
Ås, Norway

A M Litterick
Crop and Soil Research Group, SAC
Aberdeen, UK

R E Litz
University of Florida
Homestead, FL, USA

[†]Deceased

G J Loake
University of Edinburgh
Edinburgh, UK

U Lüttge
Darmstadt University of Technology
Darmstadt, Germany

P Maliga
Rutgers University
Piscataway, NJ, USA

J Maluszynska
University of Silesia
Katowice, Poland

M Maluszynski
IAEA
Vienna, Austria

T Marks
Horticulture Research International
West Malling, UK

J M Martínez-Zapater
Centro Nacional de Biotechnología, CSIC
Madrid, Spain

G Matthews
Imperial College of Science, Technology and Medicine
Ascot, UK

N Maxted
University of Birmingham
Birmingham, UK

M R McAinsh
Lancaster University
Lancaster, UK

S McQueen-Mason
University of York
York, UK

J Memelink
Leiden University
Leiden, The Netherlands

M Mencuccini
University of Edinburgh
Edinburgh, UK

K Meyer
DuPont Central Research and Development
Wilmington, DE, USA

W B Miller
Cornell University
Ithaca, NY, USA

D J Miralles
University of Buenos Aires
Buenos Aires, Argentina

D J Murphy
University of Glamorgan
Cardiff, UK

B G Murray
University of Auckland
Auckland, New Zealand

D Nadella
North Dakota State University
Fargo, ND, USA

R M Napier
Horticulture Research International
Wellesbourne, UK

R E L Naylor
Trelareg Consultants
Banchory, Scotland, UK

P Newell
Formerly of The University of
the South Pacific
Suva, Fiji Islands

W E Newton
Virginia Polytechnic Institute
and State University
Blacksburg, VA, USA

P S Nobel
University of California
Los Angeles, CA, USA

P L Nurmberg
University of Edinburgh
Edinburgh, UK

A M Nuutila
VTT Biotechnology
Espoo, Finland

E O' Connor
Bio-Renewables Ltd
Ely, UK

C E Offler
University of Newcastle
New South Wales, Australia

K-M Oksman-Caldentey
VTT Biotechnology
Espoo, Finland

O-A Olsen
Pioneer Hi-Bred International
Johnston, IA, USA

J L Osborne
Rothamsted Research
Harpenden, UK

N W Pammenter
University of Natal
Durban, South Africa

Y S Park
Canadian Forest Service – Atlantic
Forestry Center
Fredericton, NB, Canada

J W Patrick
The University of Newcastle
New South Wales, Australia

M J Paul
Rothamsted Research
Harpenden, UK

J-C Pech
UMR 990 INRA-ENSAT
Toulouse, France

K V Peter
Kerala Agricultural University
Vellanikkara, India

M Petitprez
Biotechnology and Plant Breeding, ENSAT
Auzeville, France

P W B Phillips
University of Saskatchewan
Saskatoon, SK, Canada

M Piñeiro
Centro Nacional de Biotechnología, CSIC
Madrid, Spain

A H Pickering
Massey University
Palmerston North, New Zealand

M Pletsch
Universidade Federal de Alagoas
Maceió, Brazil

K M Plummer
University of Auckland
Auckland, New Zealand

Y Poirier
University of Lausanne
Lausanne, Switzerland

S Pond
Canadian Forestry Service
Fredericton, NB, Canada

S A Power
Imperial College London
Ascot, UK

R K Prange
Agriculture and Agri-Food Canada
Kentville, Nova Scotia, Canada

P V V Prasad
University of Florida
Gainesville, FL, USA

F Pupilli
Institute of Plant Genetics
Perugia, Italy

C B Purrington
Swarthmore College
PA, USA

A S Raghavendra
University of Hyderabad
Hyderabad, India

O Ramli
Malaysian Rubber Board
Selangor, Malaysia

K R Reddy
Mississippi State University
MS, USA

M J Reiss
Institute of Education
University of London, UK

P Roberts
University of York
York, UK

D Robinson
University of Aberdeen
Aberdeen, UK

T L Rost
University of California
Davis, CA, USA

E C Rowe
Wageningen University
Wageningen, The Netherlands

G Russell
The University of Edinburgh
Edinburgh, UK

P R Ryan
CSIRO Plant Industry
Canberra, Australia

F Sala
University of Milan
Milan, Italy

R A Sánchez
IFEVA
Buenos Aires, Argentina

S Sanogo
New Mexico State University
Las Cruces, NM, USA

G E Schaller
University of New Hampshire
Durham, NH, USA

P J Sharp
Plant Breeding Institute
University of Sydney, Australia

T Shepherd
Scottish Crop Research Institute
Dundee, UK

P R Shewry
Rothamsted Research
Harpenden, UK

K Shinozaki
RIKEN Tsukuba Institute
Tsukuba, Japan

N W Simmonds[†]
University of Edinburgh
Edinburgh, UK

L Skøt
Institute of Grassland and Environmental Research
Aberystwyth, UK

G A Slafer
University of Buenos Aires, Buenos Aires
Argentina

H Slater
Lancaster University
Lancaster, UK

S M Smith
Institute of Cell and Molecular Biology
University of Edinburgh, UK

D Sparkes
University of Nottingham
Loughborough, UK

E Stoger
RWTH Aachen
Aachen, Germany

R Straughan
University of Reading
Reading, UK

J Sutka
Agricultural Research Institute of the Hungarian Academy of Sciences
Martonvásár, Hungary

I Szarejko
University of Silesia
Katowice, Poland

D A Tabah
University of Bristol
Bristol, UK

A G Taylor
Cornell University
Geneva, NY, USA

[†]Deceased

T H Teeri
University of Helsinki
Helsinki, Finland

B Thibaut
Université Montpellier II
Montpellier, France

B Thomas
Horticulture Research International
Wellesbourne, UK

J M G Thomas
University of Florida
Gainesville, FL, USA

G A Thompson
University of Arkansas
Little Rock, AR, USA

A F Tiburcio
Universitat de Barcelona
Barcelona, Spain

K Tsunewaki
Fukui Prefectural University
Fukui, Japan

R M Twyman
University of York
York, UK

A J E van Bel
Justus-Liebig-University
Giessen, Germany

J E G van Dam
WUR, Wageningen
The Netherlands

B Veit
AgResearch
Palmerston North, New Zealand

R Verpoorte
Leiden University
Leiden, The Netherlands

J A Walsh
Horticulture Research International
Wellesbourne, UK

H L Wang
University of Arkansas
Little Rock, AR, USA

C B Watkins
Cornell University
Ithaca, NY, USA

C A Watson
Crop and Soil Research Group, SAC
Aberdeen, UK

A D Webster
Maidstone, Kent, UK

J L Weller
Univeristy of Tasmania
Hobart, Australia

P J White
Horticulture Research International
Wellesbourne, UK

E J Woltering
Wageningen University
Wageningen, The Netherlands

X B Yang
Iowa State University
Ames, IA, USA

K Yu
University of Kentucky
Lexington, KY, USA

J A D Zeevaart
Michigan State University
East Lansing, MI
USA

R L Zimdahl
Colorado State University
Fort Collins, CO
USA

M Ziv
The Hebrew University of Jerusalem
Rohovot
Israel

CONTENTS

VOLUME 1

C

D

E

F

A

ABIOTIC STRESSES

Contents

Cold Stress

J Sutka and G Galiba, Agricultural Research Institute of the Hungarian Academy of Sciences, Martonvásár, Hungary

Introduction

Apart from the availability of water, low temperature (chilling and frost) is the most important environmental factor that limits the productivity and geographical distribution of plants in large areas of the world. Cold temperatures can affect the development of plants in almost each phase, from germination through to seed set. The sensitivity of a plant species in a particular environment is determined by the limit to which its metabolic processes continue to function under low temperature stress or the point at which it may suffer permanent injuries that finally bring about death. The cold tolerance of a plant species is evolutionarily determined and depends on the climate of the area the plant originated from. According to their cold tolerance plants are divided into three categories:

1. Chill susceptible: damaged by temperatures below 12°C.
2. Chill tolerant but freezing susceptible; able to acclimate to temperatures below 12°C but unable to survive freezing.
3. Freezing tolerant (frost tolerant); able to acclimatize to survive temperatures significantly below freezing.

As shown in **Table 1**, tropical plants like banana (*Musa sapientum*), papaya (*Carica papaya*), etc. belong to the chill susceptible category. Several subtropical plant species (paprika (*Capsicum frutescens*), potato (*Solanum tuberosum*), tomato (*Lycopersicon esculentum*), etc.) which are cultivated in temperate regions are classified as chill tolerant but freezing susceptible. The herbaceous or woody plants characterized as freezing tolerant originated mostly from the temperate climates. There are substantial differences among the plant species classified as freezing tolerant. Most of the herbaceous plants survive only moderate freezing, between −7°C and −30°C. However, there are some much hardier woody species, which can tolerate temperatures below −80°C. The effect of cold on plants is not only determined by the magnitude of the drop of the temperature, but is to a large extent dependent upon the season, developmental stage, and for how long the low temperature persists.

Cold Acclimation of Plants

To achieve the full genetic potential of freezing tolerance the plants must have time to adapt (i.e., become hardened, or acclimated) gradually to the low temperatures. Under natural conditions the cold hardening (acclimation) takes place in autumn when the temperature gradually decreases to 0°C over several weeks. Cold hardening or acclimation can be defined as a nonheritable modification of structures and functions as a response to cold which minimizes damage and improves the fitness of an individual plant. The significance of the temperatures and/or daylength (photoperiod) decreases from autumn to winter. A cold acclimation or freezing tolerance inductor has been recognized since the nineteenth century; nonacclimated rye (*Secale cereale*), for instance, is killed by freezing at about −5°C, but after a period of exposure to low nonfreezing temperature can survive freezing down to about −30°C.

Table 1 Examples of cold-sensitive (tender) and freezing-tolerant (hardy) plants, including some important crop species

Family	*Species*	*Range of temperature causing injury (°C)*	
Musaceae	*Musa* spp. (banana)	+10	+12
Lauraceae	*Persea* spp. (avocado)	+6	+8
Caricaceae	*Persea* spp. (avocado)	+4	+10
Poaceae	*Oryza sativa* (rice)	+12	+15
	Zea mays (maize/corn)	+2	+12
	Avena sativa (oat)	−5	−10
	Hordeum vulgare (barley)	−7	−12
	Triticum aestivum (bread wheat)	−9	−18
	Secale cereale (rye)	−15	−30
Solanaceae	*Lycopersicon esculentum* (tomato)	+2	+5
	Capsicum annuum (paprika/pepper)	−2	+4
	Solanum tuberosum (potato)	−2.5	0
	Solanum acaule (wild potato species)	−6	−8.5
Brassicaceae	*Arabidopsis thaliana*	−9	−14
Rutaceae	*Citrus* spp. (orange and lemon)	−2.2	−10
Myrtaceae	*Eucalyptus* spp. (eucalyptus)	−8	−16
Cupressaceae	*Juniperus* spp. (juniper)	−25	−45
Pinaceae	*Pinus* spp. (pines)	−20	−60
Rosaceae	*Prunus* spp. (plum)	−20	−80

Table 2 Climatic program for freezing test in cereals

Time (days)	*Day temperature (°C)*	*Night temperature (°C)*	*Daylength (h)*	*Function*
3	20	20	16	Germination
7	15	10	12	Growth
14	10	5	12	Growth and hardening
14	5	0	8	Hardening
7	2	−2	8	Hardening
2	−4	−4	Dark	Second phase of hardening
1 or 2	Between −10 and −16		Dark	Freezing[a]
2	Between 0 and +2		Dark	Thawing
14	16	15	12	Recovery
1	16	15	16	Evaluation

[a] Freezing temperature depends on the genotype.

The capability of plants to harden is genetically determined, and individuals possessing higher freezing tolerance can be selected. In plants low positive temperature or freezing is the commonest inductive environmental factor for acclimation to cold. However, drought or the plant hormone abscisic acid (ABA) can substitute for cold at least in part. In woody species shortening daylength causing dormancy is often also a trigger for acclimation.

Farmers will grow winter cereals in preference to spring ones provided they can rely on the survival of the plants over winter. This preference is largely based on the superior yielding ability (the difference being roughly 30%) of winter-planted cereals. The capacity for overwintering in the temperate zone depends on several factors including low positive and freezing temperatures, waterlogging or ice encasement (both of which cause anaerobic stress), and wind (which enhances any tendency for shoot dehydration). In addition, overwintering plants can be attacked by diseases; these are known as snow molds since they develop under the snow. Among these factors freezing tolerance proved to be the most important, because an intact plant can better withstand the other damaging agents. One serious problem that plant breeders encounter in developing more hardy cultivars is the fact that very severe winters producing differential winter kill occur only erratically and infrequently in most locations. This necessitates the development of artificial testing procedures to help progress in the breeding programs.

The degree of freezing tolerance is highly dependent on the hardening conditions. Under controlled environmental conditions temperatures slightly above 0°C and photoperiods of 8–21 h are considered to be optimal for cold hardening of winter cereals (barley (*Hordeum vulgare*), wheat (*Triticum aestivum*), rye, etc.). **Table 2** shows one of the climatic hardening programs used at the Martonvásár (Hungary) plant research station. This program is

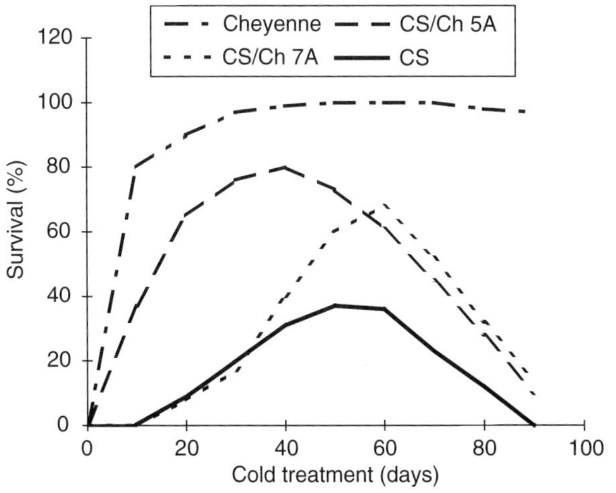

Figure 1 Time-dependent changes in frost tolerance during cold hardening at +2°C. Freezing tolerance was established as the percentage of survival. Plant material: Chinese Spring (CS), spring wheat; Cheyenne (Ch), winter wheat; CS/Ch 5A and CS/Ch7A, chromosome substitution lines.

based on the 30-year average autumn temperature characteristic for this area. There are two main characteristics of the program:

1. The temperature gradually decreases from +15°C to −2°C over 6 weeks.
2. At the end of the program the plants are kept at −4°C for 2 days in the dark.

It is considered that keeping the plants at freezing temperatures just below 0°C further increases their freezing tolerance. That phenomenon is known as the second phase of hardening.

Cold resistance is determined by exposing the plants to a temperature low enough to cause damage and determining either the temperature which causes death of 50% of the plants (LT_{50}) or the percentage of plants surviving. Since cold injury damages cell plasma membranes, injury may also be measured by electrical conductivity of the tissues or leakage of various compounds from the cells. Vital staining or comparing the capacity of the plants for regrowth after a recovery period are also widely used methods.

The cold hardiness of plants changes in characteristic ways with increase in the duration of the hardening process. In hardy wheats hardiness starts to increase after 2–3 weeks and rises to a maximum which usually persists for a few weeks, depending on the hardening conditions and the cultivar. Hardiness later falls (**Figure 1**).

Damage Caused by Freezing

The natures of chill and freezing stresses are different. Chill stress is a direct effect of low temperature on cell membranes. Freezing, however, often acts indirectly, damaging the cells by dehydration. As temperatures drop below 0°C, ice formation is generally initiated in the cellular spaces, due in part to the extracellular fluid having a higher freezing point (lower solute concentration) than the intracellular fluid. Ice nuclei may form spontaneously (homogenous nucleation), or around nonaqueous matter (heterogeneous nucleation). The probability of homogenous nucleation is very low until the temperature drops to −40°C. Heterogeneous nucleators include some biological debris or inorganic material. In addition, bacteria commonly found on plants (such as *Pseudomonas syringiae* and *Erwinia herbicola*) produce a protein able to nucleate freezing at temperatures as high as −2°C. Because the water potential of the extracellular ice is less than the water potential of liquid water within the cells, ice formation results in a drop in water potential outside the cell. Consequently, there is movement of unfrozen water down the water potential gradient from inside the cell to the intercellular spaces. At equilibrium, the extent of dehydration is a direct consequence of temperature: with progressively greater falls in temperature below freezing, the less intracellular liquid water remains, the more strongly collapsed are the cells, and the lower is the intracellular water potential. At −10°C, more than 90% of the osmotically active water typically moves out of the cells and the osmotic potential of the remaining fluid is greater than 5 osm.

Membranes

The targets of freezing-induced dehydration are the membranes, particularly the plasma membrane. This undergoes phase separation which destabilizes it, causing it to break up. A relationship between the phase transition of membrane lipids and the chilling sensitivity of plants was first postulated in the early 1970s by Raison and Lyons. They proposed that the formation of gel (or solid phase) by lipids in biological membranes at nonfreezing chilling temperatures induces damage to the same plant tissues that can lead to the death of the plant. Dehydration in the absence of cold has the same effect, confirming that damage is a consequence of the freezing-induced desiccation. Freeze-fracture electron microscopic studies reveal that the membrane phase separation which has been demonstrated to occur constitutes a removal of protein particles from areas of the membrane. Break-up of the membrane is mediated by local changes in phase from lamellar to nonlamellar, including hexagonal II. The precise sequence of events causing this is unclear. Most likely the loss of protein particles from areas of membrane could leave the nonlamellar lipids (such as phosphatidylethanolamine) in this area without stabile interactions. Other forms of membrane damage also can occur, e.g., expansion-induced lysis; or if intracellular ice forms adhesion with cell walls and membranes it can cause cell rupture. Thus, a key part of cold acclimation is to stabilize membranes against freezing induced injury.

Acclimation of bacteria, fungi, protozoa, plants, and animals to temperatures below their respective normal growth temperatures generally results in changes in the unsaturation of fatty acids in membrane lipids. The extent of unsaturation has a considerable effect on the fluidity of membrane lipids. When organisms are exposed to low temperatures, the fluidity of their membrane lipids decreases. Such exposure enhances the expression of genes for fatty acid desaturases, which introduce double bonds into the fatty acyl chains of membrane lipids, thereby compensating for the decrease in membrane fluidity. As a result, the original physical properties of the membranes are restored and can support the functions of the membrane associated proteins and their complexes. This hypothesis was proved experimentally by the inactivation of individual genes for fatty acid desaturases in the cyanobacterium *Synechocystis*. In higher plants, genetic manipulation of glycerol-3-phosphate acyltransferase and cyanobacterial $\Delta 9$ desaturase has allowed modulation of the level of unsaturation of phosphatidylglycerol in thylakoid membranes. Characterization of the photosynthetic functions of such genetically modified strains of cyanobacteria and higher plants has revealed that polyunsaturated fatty acids are essential for the protection of the photosynthetic machinery against photoinhibiton at low temperatures.

The importance of plant plasma membrane lipid composition in freezing tolerance was demonstrated by Steponkus and his colleagues using liposomes to modify the membrane lipid composition of protoplast isolated from nonacclimated rye leaves. They enriched the plasma membrane from acclimated rye leaves, or specific saturated or unsaturated phosphatidylcholine species. When the lipids from the acclimated membrane or unsaturated phosphatidylcholine molecules were supplied, these significantly increased the freezing tolerance of the protoplasts. These experiments well demonstrated that membrane lipid composition is an important factor in hardening to frost.

Oxidative Stress

Chilling and cold injury is mediated, in part, by oxygen free radicals (singlet oxygen, superoxide radicals, or hydrogen peroxide) as agents causing damage to the proteins, nucleic acids, and membranes. The activation of oxygen by the photosystems in the presence of excessive light is probably the major site of production of free radicals in leaves but other electron transport systems, including those on the mitochondria or plasmalemma, may also contribute especially in nonphotosynthetic tissues. The development of the symptoms of chilling injury is frequently coincident with peroxidation of fatty acids. In this way lipid peroxidation would alter the physical properties of membrane lipids, thereby inhibit the function of membrane-bound proteins contributing to the development of visual symptoms of injury caused by cold temperatures. Therefore, it is clear that the capability of the plants to enhance the free radical scavenging capacity by increasing the endogenous level of antioxidants e.g.: carotenoids, tocopherol, ascorbate, superoxidedismutase, glutathione, etc., is an important part of the plant's defense mechanism under cold stress conditions.

Genetics of Freezing Tolerance

Winter hardiness, including frost tolerance, has long been regarded as being under complex multigenic control. However, it would still be possible for major differences in cold adaptation between species or cultivars to depend on allelic differences in a small number of genes as suggested in cold acclimation in pea (*Pisum sativum*) and *Solanum*. Since common wheat is a hexaploid its vital genes are replicated. This has permitted the development of a series of chromosome-substitution lines. By comparing chromosome-substitution lines with the parental lines it was possible to determine which chromosomes carry the locus for freezing tolerance. The analysis of substitution lines showed that at least 10 of the 21 pairs of chromosomes are involved in the control of frost tolerance and winter hardiness (**Figure 1**). However, major genes influencing frost resistance (*Fr*) and vernalization requirement (*Vrn*) were localized on the long arm of chromosomes 5A and 5D. Of particular importance for adaptation to autumn showing are the genes for vernalization requirement. *Vrn* genes determine the needs for cold temperature required for flower development (*see* **Regulators of Growth:** Jasmonates). Recent studies indicated that the *Vrn1–Fr1* interval on chromosome 5A of wheat has a major effect on freezing tolerance (**Figure 2**). Conservation of gene order (synteny) in *Triticum* is well known and this is true for the *Vrn1–Fr1* interval studied in barley, rye, and *Triticum monococcum*, as well. For example a map of quantitative trait loci (QTL) in barley has resulted

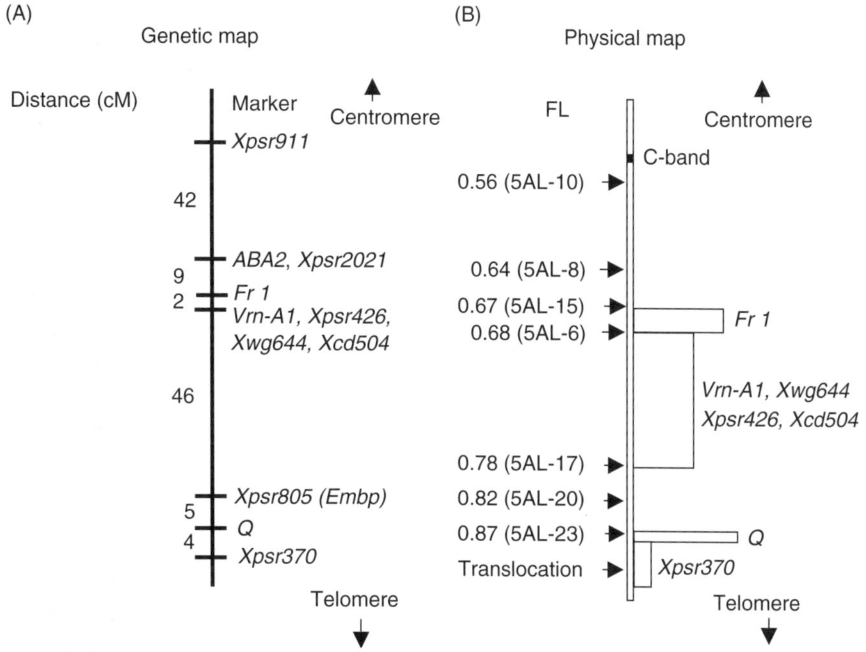

Figure 2 (A) Genetic and (B) cytogenetically based physical maps of the long arm of wheat chromosome 5A. Fraction lengths (FL) and deletion stock numbers are indicated on the left of the physical map. Marker names are indicated on the right of the chromosomes.

in the identification of a 21-cM region on chromosome 7 that has a major role in freezing tolerance. This region accounted for 32% of the variance in LT_{50} values and 39–79% of the variance in winter field survival. Chromosome 7 of barley (renamed recently as 5H) is homologous to the group 5 chromosomes of wheat. The 21-cM freezing tolerance interval in barley contains the *Xwg644* and *Xcdo504* molecular markers that are also linked to the *Vrn1–Fr1* interval in wheat.

The mechanism whereby the *Vrn1–Fr1* interval affects freezing tolerance remains to be determined. Regulatory genes affect cold-induced ABA, water-soluble carbohydrates, fructan and proline accumulation, more than do genes regulating the cold-induced expression of proteins in various functions that were mapped in this region. Thus the possibility is raised that the *Vrn1–Fr1* interval affects frost tolerance through encoding several proteins involved in regulating the expression of cold-inducible genes that have roles in freezing tolerance.

Cold-Induced Metabolic Responses

Role of ABA in Cold Acclimation

ABA is considered as a stress hormone, emphasizing its role in the adaptation of plants to different abiotic stress conditions such as drought, salinity, heat, and cold. At the beginning of the 1980s it was discovered that ABA levels increase transiently in response to low temperature in *Solanum commersonii*, a plant that cold acclimates, but not in *S. tuberosum*, a plant that does not cold acclimate. Moreover, the exogenous application of ABA increased the freezing tolerance of *S. commersonii* plants at warm temperatures. So it was hypothesized that cold acclimation is activated through the action of ABA. Subsequent studies extended this observation establishing that ABA levels increase, at least transiently, in a diverse group of plant species during cold acclimation and that exogenous application of ABA at warm nonacclimating temperatures enhances the freezing tolerance of several plant species that cold acclimate. Experiments with plants carrying mutations in ABA synthesis (*aba1*) or the ability to respond to ABA (*abi1*) showed less freezing tolerance than the wild-type. These results support the hypothesis that ABA has a key role in activating cold acclimation. Recent results, emerged from gene regulation studies on *Arabidopsis* and the experimental fact that during cold acclimation the increase in ABA levels peaks at 24 h and returns to essentially the normal level after 2 days (yet freezing tolerance increases for about 1 week and remains elevated for at least 3 weeks) adds a question mark to this hypothesis. Using only ABA as hardening agent the level of freezing tolerance is lower than can be achieved by cold treatment. We may elucidate this controversy approaching this question from gene regulation point of view. It is well established that some genes are strictly cold-inducible (such as barley *COR14b*), some are ABA

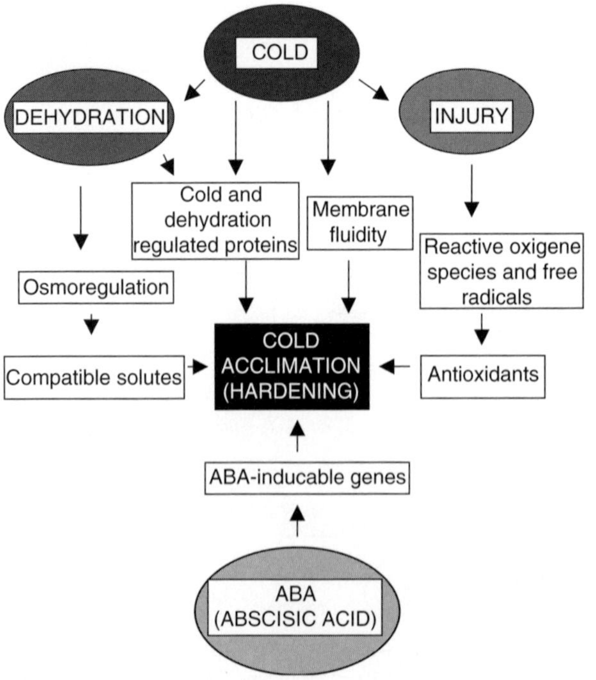

Figure 3 The metabolic changes that trigger cold acclimation.

responsive (*Arabidopsis RAB18* and *LTI65*), and others expressed are in response of both stimuli (*Arabidopsis COR78*). It is likely that to achieve full cold hardiness the coordinated effect of ABA-dependent and ABA-independent regulatory pathways is required (**Figure 3**).

Role of Cold-Inducible Genes in Cold Acclimation

In 1985 Guy with his colleagues established that changes in gene expression occur with cold acclimation. Since than, considerable effort has been directed at determining the nature of cold-inducible genes and establishing whether they have roles in freezing tolerance. This has resulted in the identification of many genes that are induced during cold acclimation. A large number of these genes encode proteins with known enzyme activities that potentially contribute to freezing tolerance. For instance, the *Arabidopsis FAD8* gene encodes a fatty acid desaturase that might contribute to freezing tolerance by altering lipid composition. Other genes encoding molecular chaperons including a spinach (*Spinacia oleracea*) *hsp70* gene and an oilseed rape (*Brassica napus*) *hsp90* gene might contribute to freezing tolerance by stabilizing proteins against freeze-induced denaturation. A notable feature of cold acclimation is the involvement of extracellular proteins. Several of the proteins extracted from the leaf apoplast of rye and related cereals have antifreeze activity *in vitro*. However, these so-called "antifreeze proteins" have strong sequence similarity to certain pathogenesis-related proteins. So, the key unresolved question about these proteins is whether or not their antifreeze property operates *in vivo*. Most likely they do not prevent freezing or defer it to lower temperatures, and instead it is suggested that they help control the sites at which ice forms and grows and the rate of growth. Also, cold-responsive genes encoding various signal transduction and regulatory proteins have been identified including a mitogen-activated protein (MAP) kinase, calmodulin-related proteins and 14-3-3 proteins. These proteins might contribute to freezing tolerance by controlling the expression of freezing tolerance genes or by regulating the activity of proteins involved in freezing tolerance. Whether these cold-responsive genes actually have important roles in freezing tolerance, however, remains to be determined.

The first direct evidence for a cold-induced gene having a role in freezing tolerance came as recently as 1996. The *COR15a* gene encodes a 15-kDa polypeptide that is targeted to the chloroplast. The constitutive expression of *COR15a* in nonacclimated transgenic *Arabidopsis* plants increases the freezing tolerance of both chloroplasts frozen *in situ* and isolated leaf protoplasts frozen *in vitro*. In many cases, the gene or protein itself is known to be cold responsive in a number of species, indicating at least a partial uniformity of response to cold amongst unrelated plants. Several cold-inducible genes are responsive to a variety of stresses, not just cold, or are also expressed during seed development in the nonstressed plant, such as late embryogenesis abundant (*LEA*) genes. Thus, though their expression is often related to low cytoplasmic water content, they are not uniquely expressed in stressed plants. Speculatively, the differences between tolerant and nontolerant species may be attributed to differences in the controls of stress-related gene expression, rather than differences in the types of functional genes present in the genome.

Signal Transduction Pathways

The prerequisite for the plants to acclimate to cold for surviving the approaching winter is the ability to sense the declining temperatures at the onset of winter. It is still an open question how plants sense cold. At least in the cyanobacterium *Synechocystis* it was demonstrated that membrane fluidity regulates the expression of low temperature-inducible genes. It can be postulated that the cold signal is first sensed by the plasma membrane by becoming more rigid. The cold signal is than transmitted to the Ca^{2+} channels via rearrangement of actin and tubulin cytoskeleton. Opening of the Ca^{2+} channels allows the frequently observed cold-triggered Ca^{2+} influx from the vacuole into the cytoplasm. The Ca^{2+} influx activates the calcium-dependent protein kinases which in turn will phosphorylate other proteins including other types of kinases. The kinases will activate the transcriptional factors which control the expression of stress-inducible genes. The significance of the transcriptional factors in cold acclimation was demonstrated by overexpression of the cold-related transcription factor *CBF1* in *Arabidopsis*. This caused enhanced expression of several cold-responsive genes and conferred freezing tolerance on plants growing in a warm noncold-acclimating environment.

Osmotically Active Solutes

The increased osmolarity of the cellular solute is an effective mechanism for avoiding intracellular ice formation and cellular dehydration. In most, if not all, hardy plants, cell solutes concentrate during the process of hardening. Numerous reports point out the accumulation of various osmotically active solutes, called compatible solutes, such as sucrose, fructan, quaternary ammonium compounds, and proline during hardening. They may function as cryoprotectants and many of these compounds may also stabilize membranes.

Proline Proline accumulation has often been shown to occur in plants as a consequence of environmental stress. The physiological significance of this accumulation is assumed to be associated with the ability of proline to act as osmoregulator, as protective agent for cytoplasmic enzymes and for membranes, or as storage compound for post stress growth. Significant positive correlations between proline level and frost tolerance have been found in a broad spectrum of plants. In breeding programs for frost tolerance in winter barley, the proline levels in cold-hardened leaves are indeed used as an early selection criterion for frost-tolerant lines. Further support for the involvement of proline in the adaptation mechanism for frost tolerance comes from *in vitro* selection studies. Isolated hydroxyproline-resistant (Hyp-resistant) cell lines of potato, barley, and wheat possess an increased amount of proline and of other amino acids as well as increased frost tolerance. The highest increase in tolerance of regenerated plants were at 2–3°C. These characteristics were shown to be heritable in wheat.

Glycine betaine Glycine betaine (hereafter referred to as betaine) is widely distributed in higher plants, animals, and bacteria. Betaine protects cells from salt stress by maintaining an osmotic balance with the environment and by stabilizing the quaternary structure of complex proteins. In photosynthetic systems, betaine stabilizes the oxygen-evolving photosystem II complex. Higher plants accumulate betaine in response to both salt and cold stress. Enhancement of either chilling or freezing tolerance by genetic engineering of betaine accumulation in plants has been demonstrated. Choline oxidase of the soil bacterium *Arthobacter globiformis* converts choline to betaine. Transformation of *A. thaliana* and rice with cloned choline oxidase genes enabled the plants to accumulate betaine and enhanced its tolerance to cold stress.

Carbohydrates Carbohydrate changes are of particular importance because of their direct relationship with such physiological processes as photosynthesis, translocation, and respiration. Among the soluble carbohydrates, sucrose and fructans have some potential role in adaptation to cold stress. Sucrose can act in water replacement to maintain membrane phospholipids in the liquid-crystalline phase and to prevent structural changes in soluble proteins. Next to the role as a plant carbohydrate reserve, fructans may have other functions, including involvement in drought and frost tolerance in many cereals, including wheat and barley. Studies on overwintering wheat and rye cultivars have shown that cold hardiness is strongly correlated with the capacity to increase soluble carbohydrate pools in field conditions. When the carbohydrate reserve is depleted (in March–April) cold hardiness drops considerably as well. Association between frost resistance and water-soluble carbohydrate content has been reported in barley and wheat. QTL controlling traits associated with winter hardiness (field survival, LT_{50}, vernalization response, and fructan and sucrose content) have been mapped to chromosome 7(5H) of barley and to 5A of wheat.

Applications of the Results of Cold Stress Research

The importance of cold stress research can be verified if we take account that globally there is an annual expenditure of US$100 million to minimize frost damage to crops, and annual losses of US$10–100 million or more from frost damage. Testing plants for their frost tolerance is standard practice in breeding programs which help to avoid the frost damage of winter crops. Recently, progress for winter hardiness improvement in cereals has been slow due to various reasons. As shown above, the genetic control of winter hardiness is extremely complex. Breeders have to work with small differences in large populations. Most importantly, the exploitable genetic variation for winter hardiness appears to be nearly exhausted. As a striking example for these statements, a survey of winter wheat cultivars available worldwide revealed than none of them is much more resistant to low temperatures than Kharkov 22MC, which was released for commercial production in Canada in 1912. Our better understanding of how plants are able to cope with cold stress and the usage of the recent tissue culture and molecular techniques certainly will help our breeders to improve freezing tolerance of cultivated plants.

Genetic Transformation

Manganese superoxide dismutase from *Nicotiana plumbaginifolia* was constituvely expressed in alfalfa and targeted to chloroplasts or mitochondria. The performances of transformant and control lines were compared in the field over 3 years, where the transformants showed a significantly higher winter survival and yield.

In Vitro Selection

Proline-overproducer somaclonal lines had also enhanced freezing tolerance. These characters were shown to be heritable in wheat, and in the third generation increased proline content and improved freezing tolerance of acclimated plants were correlated.

Marker-Assisted Selection

The segregation of *Vrn1–Fr1* interval affects the frost tolerance, followed in near-isogenic lines (NILs) by *Xwg644* restriction fragment length polymorphism (RFLP) marker. These lines derived from a cross between a freezing-sensitive spring wheat and a freezing-tolerant winter wheat. The NILs that inherited the *vrn1–fr1* interval from the winter wheat parent were about 4°C more freezing-tolerant than those carrying the spring locus (*Vrn1–Fr1*).

List of Technical Nomenclature

Abscisic acid (ABA)	A plant growth regulator.
Chilling	Suboptimal temperature stress above 0°C.
Cold acclimation	Nonheritable modification of structures and functions by cold temperatures in order to minimize cold-induced injuries.
Cold stress	The effect of suboptimal temperature resulting in structural and functional injuries.
COR proteins	Cold-regulated proteins.
Freezing	Temperature stress below 0°C.
Gene mapping	Determination of the position of a gene locus on a chromosome.
Oxidative stress	Accumulation of reactive oxygen species induced by various environmental effects.
Vernalization	Subjection of a plant to cold temperature in order to ensure the flower development.

See also: **Abiotic Stresses**: Free Radicals, Oxidative Stress and Antioxidants. **Crop Improvement**: Chromosome Engineering; Genetic Maps; Mutation Techniques. **Genetic Modification of Primary Metabolism**: Proteins. **Growth and Development**: Control of Gene Expression, Post Transcriptional Regulation; Control of Gene Expression, Regulation of Transcription. **Photosynthesis and Partitioning**: Photoinhibition; Primary Products of Photosynthesis, Sucrose and other Soluble Carbohydrates. **Regulators of Growth**: Abscisic Acid; Photomorphogenesis; Vernalization. **Tissue Culture and Plant Breeding**: Somaclonal Variation.

Further Reading

Dörffling K, Dörffling H, Lesselich G, *et al.* (1997) Heritable improvement of frost tolerance in winter wheat by *in vitro* selection of hydroxyproline-resistant proline overproducing mutants. *Euphytica* 93: 1–10.

Galiba G, Kerepesi I, Vágûjfalvi A, *et al.* (2001) Mapping of genes involved in glutathione, carbohydrate and *COR14b* cold-induced protein accumulation during cold hardening in wheat. *Euphytica* 119: 173–177.

Gombos Z and Murata N (1998) Genetic engineering of the unsaturation of membrane glycerolipid: effects on the ability of the photosynthetic machinery to tolerate temperature stress. In: Siegenthaler PA and Murata N (eds) *Lipids in Photosynthesis: Structure, Function and Genetics*, pp. 249–262. Dordrecht: Kluwer Academic.

Kocsy G, Galiba G, and Brunold C (2001) Role of glutathione in adaptation and signalling during chilling and cold acclimation in plants. *Physiologia Plantarum* 113: 158–164.

Li PH and Chen THH (1997) *Plant Cold Hardiness: Molecular Biology, Biochemistry, and Physiology*. New York: Plenum Press.

McKersie BD and Leshem YY (1994) *Stress and Stress Coping in Cultivated Plants*. London: Kluwer Academic Publishers.

Pearce RM (1999) Molecular analysis of acclimation to cold. *Plant Growth Regulation* 29: 47–76.

Sakamoto A and Murata N (2000) Genetic engineering of glycine betaine synthesis in plants: current status and implications for enhancement of stress tolerance. *Journal of Experimental Botany* 51: 81–88.

Steponkus PL and Web MS (1992) Freeze-induced dehydration and membrane destabilization in plants. In: Somero GN, Osmond CB, and Bolis CL (eds) *Water and*

Life: Comparative Analysis of Water Relationships at the Organismic, Cellular and Molecular Level, pp. 338–362. Berlin: Springer-Verlag.
Thomashov MF (1999) Plant cold acclimation: freezing tolerance genes and regulatory mechanisms. *Annual Review of Plant Physiology and Plant Molecular Biology* 50: 571–591.

Free Radicals, Oxidative Stress and Antioxidants

K S Gould, University of Auckland, Auckland, New Zealand

Introduction

Plants are continuously exposed to free radicals. These unstable and often highly reactive molecules present a formidable challenge to all plants, even under optimal growing conditions. If left unchecked, free radicals can cause oxidative injury by initiating chain reactions that disrupt membranes, denature proteins, fragment DNA, and ultimately precipitate cell death. The problems are exacerbated in plants that face additional stressors such as high light, temperature extremes, drought, or fungal infections. In these situations, protection from the effects of free radicals is not a luxury – it may be critical for survival! Fortunately, plants have evolved a sophisticated armory of antioxidant defense, a diverse assortment of enzymes, pigments, and secondary metabolites that serve to scavenge or quench the reactive molecules before they inflict injury. Efforts to enhance the antioxidant levels in plants have far-reaching implications both for crop productivity and human nutrition.

Free Radicals

A free radical is defined as any atom or molecule capable of independent existence that contains one or more unpaired electrons. An unpaired electron is one that occupies an atomic or molecular orbital by itself. Free radicals are usually denoted by a superscript dot after the chemical formula, such as $OH^{\bullet}$ for the hydroxyl radical. A related group of compounds, including hydrogen peroxide, singlet oxygen, and ozone, lack unpaired electrons, but, like the radicals, can participate in cellular redox reactions; these together with the oxygen containing radicals are collectively known as reactive (or active) oxygen species (ROS or AOS).

Table 1 Examples of free radicals and reactive oxygen species in plant cells

Name	*Formula*
Diatomic oxygen	O_2
Singlet oxygen	$^1O_2$
Superoxide	$O_2^{\bullet -}$
Hydroxyl	$OH^{\bullet}$
Hydrogen peroxide	H_2O_2
Transition metal atoms/ions	Fe, Cu
Thiyl	$RS^{\bullet}$
Peroxyl, alkoxyl	$RO_2^{\bullet}$, $RO^{\bullet}$
Nitric oxide	$NO^{\bullet}$
Peroxynitrite	$ONOO^-$

Free radicals can be generated in plants when a nonradical gains or loses an electron, or receives excitation energy from a photoactivated pigment. Most radicals are unstable and are therefore short-lived; half-lives in the order of a microsecond or a nanosecond are common. Stability is restored when two radicals meet and share their unpaired electrons in a covalent bond. However, when a free radical reacts with an organic molecule such as an unsaturated fatty acid in a membrane bilayer, a new radical results, and chain reactions are established. Such reactions can eventually lead to oxidative injury.

Free radicals and ROS are generated in the chloroplasts, mitochondria, endoplasmic reticulum, peroxisomes, glyoxysomes, plasma membrane, and apoplasm of plant cells. They are present in the roots and shoots of both vegetative and reproductive individuals. Many different types of radicals have been identified (**Table 1**). The oxygen-centered radicals have been most extensively characterized, possibly because of the pivotal roles of oxygen in the processes of respiration and photosynthesis. Indeed, diatomic oxygen is itself a free radical, which, although relatively stable, is toxic to plants at higher concentrations. However, it is becoming increasingly evident that the nitrogen radicals, primarily nitric oxide and the reactive nitrogen species peroxynitrite, also contribute significantly to a plant's oxidative load.

Sources of Reactive Oxygen and Nitrogen

Routine processes such as photosynthesis, respiration, and nitrogen metabolism generate free radicals and ROS. These are quickly scavenged by antioxidants, and are not usually harmful to plants. Indeed, some oxidants are beneficial at low concentrations; H_2O_2, for example, is required in the apoplast for lignin biosynthesis, and has been implicated as a signaling molecule in defense responses to fungal elicitors. Environmental and biotic stressors further increase the numbers of radicals produced, and can cause permanent oxidative injury.

Photosynthesis

Chloroplasts in leaves, stems, and other green tissues are the primary source of ROS in plants under light. In photosynthesis, chloroplasts harvest light energy, transform it into chemical energy and reducing equivalents, and generate diatomic oxygen from the oxidation of water. Electrons travel along the electron transport chains of photosystems II (PSII) and I (PSI), eventually to reduce the terminal accepter $NADP^+$ to NADPH. However, even under moderate light conditions, a proportion of the electron flux through PSI does not reach $NADP^+$. Instead, those electrons are diverted from the penultimate acceptor, ferredoxin (Fd) in its reduced state, to molecular oxygen. This results in the reduction of O_2 to superoxide radicals ($O_2^{\bullet -}$). The process is known as the Mehler reaction after its discoverer:

$$Fd_{red} + O_2 \rightarrow O_2^{\bullet -} + Fd_{ox}$$

Superoxide itself is a relatively unreactive free radical. It is capable of both oxidation and reduction and is known to react with some organic molecules, yet it diffuses only slowly and cannot penetrate cell membranes. However, chloroplasts attempt to remove all superoxide almost as soon as it is formed. An extremely efficient enzyme, superoxide dismutase (SOD), is used to disproportionate $O_2^{\bullet -}$ into H_2O_2 and O_2:

$$2\,O_2^{\bullet -} + 2H^+ \overset{SOD}{\rightarrow} O_2 + H_2O_2$$

Thus, in the process of removing $O_2^{\bullet -}$, SOD generates another ROS, hydrogen peroxide. H_2O_2 is a weak oxidizing agent that is toxic to most cells in the 10–100 $\mu mol\ l^{-1}$ range. It diffuses quickly within and between cells, and crosses the lipid bilayer of cell membranes faster than can be accounted for by diffusion alone. Described as only mildly reactive, H_2O_2 can nevertheless inactivate essential enzymes in the Calvin cycle and, by oxidizing thiol groups, disrupts the mechanism that couples carbon fixation in the stroma to photosynthetic electron flow. It must, therefore, be kept below micromolar concentrations in the chloroplasts. This is normally achieved by enzymatic antioxidants.

There is another very good reason why chloroplasts and other cellular components need to remove $O_2^{\bullet -}$ and H_2O_2. The two species can react together, in the presence of transition metal ions Cu^{2+} or Fe^{2+}, to form hydroxyl radicals ($OH^\bullet$). The reaction takes place in two steps:

(1) $H_2O_2 + Fe^{2+} \rightarrow OH^\bullet + OH^- + Fe^{3+}$
(the Fenton reaction)

(2) $O_2^{\bullet -} + Fe^{3+} \rightarrow O_2 + Fe^{2+}$
(ferrous ion recycling)

The net sum of reactions (1) and (2) is called the Haber–Weiss reaction:

$$H_2O_2 + O_2^{\bullet -} \rightarrow O_2 + OH^\bullet + OH^-$$

Hydroxyl radicals are potentially the most dangerous molecules in biological systems. In plants, animals, and humans, $OH^\bullet$ reacts nonspecifically with almost every type of molecule found in living cells: sugars, amino acids, phospholipids, nucleotides, and organic acids. Indeed, it is among the most reactive chemicals known. $OH^\bullet$ radicals diffuse rapidly, and react with any molecule in their vicinity to produce secondary radicals. No specific scavengers of $OH^\bullet$ are known. Because $OH^\bullet$ radicals are too reactive to be controlled directly, plants eliminate instead the less reactive precursors, H_2O_2 and $O_2^{\bullet -}$.

Respiration

The problems with ROS faced by mitochondria during respiration are essentially similar to those of the chloroplasts. In the process of oxidative phosphorylation, electrons from NADH are transferred to diatomic oxygen via an electron transport chain on the inner mitochondrial membrane. This transfer releases free energy, much of which is conserved through the synthesis of ATP from ADP and P_i. Although electron flow through the chain is tightly regulated, it has been estimated that 1–3% of the O_2 reduced in mitochondria may form $O_2^{\bullet -}$. Electrons are believed to leak directly onto O_2 from various sites (e.g., the cyanide insensitive alternative oxidase) early in the mitochondrial electron transport chain. Damage to mitochondrial membranes favors electron leakage, and thus increases $O_2^{\bullet -}$ production. As occurs in the chloroplasts, $O_2^{\bullet -}$ in the mitochondria is rapidly dismutated into H_2O_2 by SOD. In addition to the risk of producing $OH^\bullet$ radicals via Haber–Weiss reactions, H_2O_2 can also degrade proteins essential to mitochondrial function, such as cytochrome *c*. Mitochondrial H_2O_2 is scavenged predominantly by the enzyme catalase.

Nitrogen Metabolism

Shoots release into the atmosphere significant amounts of nitric oxide ($NO^\bullet$), a gaseous, unstable, and reactive free radical. $NO^\bullet$ emissions as high as 5×10^{-14} mol cm^{-2} s^{-1} have been reported for various species under strong light, and even greater spurts occur when the plants are returned to darkness.

The biochemical source of $NO^{\bullet}$ in plants is contentious, but there is growing evidence for enzymatic control. In mammalian cells, nitric oxide synthase generates $NO^{\bullet}$ in the conversions of the amino acid L-arginine to L-citrulline, a reaction that consumes oxygen and NADPH. Several, indirect lines of enquiry have indicated that a similar process occurs in plants.

An alternative explanation, recently supported by direct, *in vivo* evidence, holds that $NO^{\bullet}$ is formed during the metabolism of nitrate. Normally, nitrate is converted in the cytosol to nitrite by nitrate reductase. The nitrite is then reduced in the chloroplast to ammonium by nitrite reductase. The reducing power for this second step, the production of ammonium is by reduced nitrite reductase. However, at times when the supplies of Fd_{red} are low (at nightfall, or under environmental stress), nitrite is instead converted to $NO^{\bullet}$ by nitrate reductase (**Figure 1**). This mechanism accounts for the burst in $NO^{\bullet}$ evolution evident when plants are moved from the light to the dark, but it does not explain why $NO^{\bullet}$ can also be generated in the light, when Fd_{red} is plentiful. It is highly likely, therefore that two or more pathways operate for the biosynthesis of $NO^{\bullet}$ in plants.

$NO^{\bullet}$ has both cytotoxic and cytoprotective properties. $NO^{\bullet}$ applications at concentrations above $10^{-6}\,mol\,l^{-1}$ inhibit photosynthesis and retard shoot growth. In plant cell cultures, high $NO^{\bullet}$ levels have been associated with irreversible DNA fragmentation and cell death. In contrast, $NO^{\bullet}$ at lower concentrations promotes normal growth and development. It stimulates seed germination, root elongation, leaf expansion and phytoalexin production, inhibits etiolation, and retards the onset of senescence. $NO^{\bullet}$ is also involved as a signaling molecule in plant defense responses to fungal pathogens, and has been implicated as a promoter of tolerance to herbicides and drought.

Protective properties of $NO^{\bullet}$ have been tentatively ascribed to an antioxidant function of this radical in plant cells. $NO^{\bullet}$ diffuses freely, and can readily permeate cell and organelle membranes. It has a high affinity both for the transition metals and $O_2^{\bullet-}$, and may, therefore, prevent the formation of $OH^{\bullet}$ radicals from Haber–Weiss reactions. However, the reaction between $NO^{\bullet}$ and $O_2^{\bullet-}$ (**Figure 1**) generates the peroxynitrite ion ($ONOO^-$), which itself is considered to be a major cytotoxic agent of reactive nitrogen. Moreover, $ONOO^-$ can react with H_2O_2 to give singlet oxygen ($^1O_2$), which has the potential to damage photosynthetic membranes.

Stress Responses

Laboratories and controlled greenhouses are probably the only places where plants can experience constant environments. In more natural conditions, abrupt daily and seasonal changes are inevitable. Anthropogenic factors, such as the emission of pollutants, the deposition of heavy metals, and the impact of human activities on the climate, compound the stresses faced by plants. Stress has been defined as

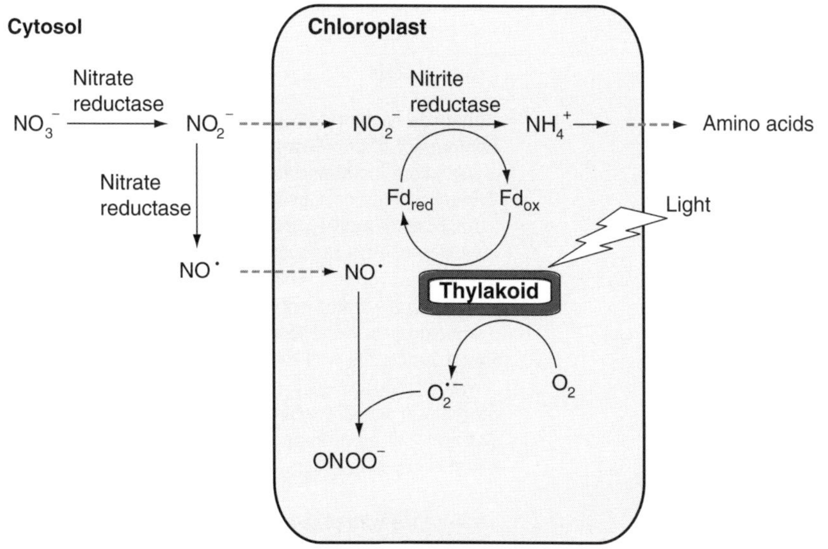

Figure 1 Tentative scheme for the biosynthesis of nitric oxide ($NO^{\bullet}$) and peroxynitrite ($ONOO^-$) in plant cells. Reproduced with permission from Yamasaki H (2000) Nitrite-dependent nitric oxide production pathway: implications for involvement of active nitrogen species in photoinhibition *in vivo*. *Philosophical Transactions of the Royal Society, London, Series B* 355: 1477–1488.

a state in which increasing demands made upon a plant lead to a destabilization of functions. Plants often recover from the effects of stress, and, indeed, subsequently show improved resistance. However, when the limits of tolerance are exceeded and the adaptive capacity is overworked, the result may be permanent damage or even death.

Almost all environmental and biotic stressors can lead to the overproduction of reactive oxygen and/or nitrogen in plants (**Table 2**). The experimental evidence is mostly indirect, based on observations of changes in levels of different antioxidants upon the application of a stressor. Recently, however, fluorescent probes have become available that are specific for a given reactive species. These probes have facilitated direct observations of $O_2^{\bullet-}$, H_2O_2, and $NO^{\bullet}$ inside plant cells under the microscope, and spectacular surges have been witnessed in response to fungal elicitors and mechanical injury (**Figure 2**).

Stressors can induce ROS formation both by upregulating the normal mechanisms for their biosynthesis, and by creating new routes (**Table 2**). Under hot, sunny conditions, for example, the Mehler reaction assumes increasing importance, accounting for up to 20% of the electron flux through PSI. These conditions can also trigger photorespiration (the C-2 oxidative photosynthetic cycle), which, in the process of converting 2-phosphoglycolate to 3-phosphoglycerate, generates H_2O_2 in the peroxisomes. Significant oxidative loads are generated even by those plants that are considered to be well adapted to arid conditions, such as the CAM (crassulacean acid metabolism) species of the desert. CAM plants close their stomata during the day to conserve water, and photosynthesize using CO_2 obtained via the decarboxylation of malic acid from vacuolar reserves. O_2 is formed, but because the stomata are closed, it cannot readily escape from the leaves. As a consequence, internal O_2 concentrations soar to around 30% (41.5% in the exceptional case of *Kalanchoë gastonis-bonnieri*), far in excess of the requirements for respiration. The excess oxygen leads to $O_2^{\bullet-}$ and H_2O_2 by photooxidation.

Plants rarely experience a single stressor in isolation. Combinations of stressors can be synergistically effective in promoting ROS formation. For example, the combination of high light and low temperatures can impair photosynthesis through the formation of singlet oxygen ($^1O_2$). Low temperatures abate the activities of Calvin cycle enzymes, but do not affect light capture. A sunny, cold climate thus leads to conditions where the absorbed light energy is in excess of the capacity to utilize ATP and NADPH in photosynthesis. The quanta excite ground-state chlorophyll first to its singlet state ($^1Chl^*$), and then to its overexcited triplet state ($^3Chl^*$). Triplet chlorophyll has a lifetime of around 1 ms, and releases its energy as heat, as fluorescence, or by reacting with O_2 to form $^1O_2$. $^1O_2$ is highly diffusible, but it is not a radical, and therefore does not initiate chain reactions directly. It can, however, oxidize the amino acids methionine, tryptophan, histidine, and cysteine, and $^1O_2$ leads to peroxidation of lipid membranes in the chloroplast envelope, thylakoids, tonoplast, and plasmalemma.

Table 2 Examples of primary reactive species, and putative mechanisms for their overproduction in response to environmental, anthropogenic, and biotic stressors

Stressor	*Reactive species*	*Mechanism(s)*
Strong light	$O_2^{\bullet-}$, H_2O_2, $^1O_2$	Enhanced Mehler activity; photorespiration; triplet chlorophyll excitation
Heat	$O_2^{\bullet-}$, H_2O_2, $NO^{\bullet}$	Impairment of photosynthetic and mitochondrial electron transport; enzyme inhibition; increased membrane permeability
Cold	$O_2^{\bullet-}$, H_2O_2	Enhanced Mehler activity; suppression of Calvin cycle enzymes; reduced antioxidant activity; decreased membrane fluidity
UV-B radiation	$OH^{\bullet}$, $O_2^{\bullet-}$, H_2O_2	Inhibition of PSII reaction center enzymes; possibly fission of H_2O_2
Drought	$O_2^{\bullet-}$, H_2O_2, $NO^{\bullet}$	Inhibition of rubisco; uncoupling of electron transport from ATP synthesis; enhanced Mehler activity; photoinhibition; inhibition of mitochondrial antioxidants; enhanced root respiration
Mechanical injury	$O_2^{\bullet-}$, H_2O_2, $NO^{\bullet}$	Elicitation by cell wall fragments; interference with redox systems on plasma membrane
Salinity	$O_2^{\bullet-}$, H_2O_2, $NO^{\bullet}$	Stomatal closure, causing $NADP^+$ deficit and O_2 reduction in mitochondria
Pathogens	$O_2^{\bullet-}$, H_2O_2, $NO^{\bullet}$	Activation of membrane bound NADPH oxidase or cell wall peroxidase
Pollutant gases	$OH^{\bullet}$, $O_2^{\bullet-}$, H_2O_2, $NO^{\bullet}$, $ONOO^-$	Metabolism of O_3, SO_2, NO_x entering through stomata
Herbicides	$O_2^{\bullet-}$, H_2O_2, $^1O_2$	Interference with photosynthetic electron transport; photoactivated herbicide interactions with O_2; inhibition of antioxidants
Heavy metals	Fe^{2+}, Cu^{2+}, $OH^{\bullet}$, $^1O_2$, $O_2^{\bullet-}$, H_2O_2	Direct uptake from contaminated soils; Haber–Weiss reactions; Fe dependent photosensitization

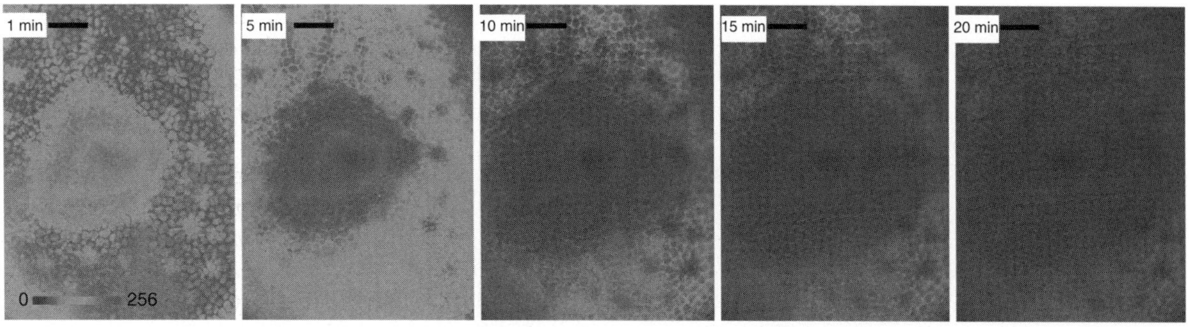

Figure 2 (see color plate 1) Oxidative burst in the epidermal cells of the leaves of *Pseudowintera colorata* at various times after puncturing with a needle. Hydrogen peroxide (red color), produced immediately after wounding, is gradually scavenged by antioxidants (increasing blue coloration). Scale bar = 200 μm. Reproduced with permission from Gould KS, McKelvie J, and Markham KR (2002) Do anthocyanins function as antioxidants in leaves? Imaging of H_2O_2 in red and green leaves after mechanical injury. *Plant, Cell & Environment* 25: 1261–1269.

Reactive species of oxygen and nitrogen are certainly not the only causes of cellular injury in plants under stress, yet they are evidently a significant contribution. Crops that have been genetically engineered to overproduce antioxidant enzymes often show enhanced cross-tolerance to several different stressors. Conversely, the underproduction of antioxidant enzymes is associated with an increased sensitivity to stress, indicating that adequate protection from the effects of ROS is essential.

Oxidative Injury

Lipid membranes, DNA, and proteins are all direct targets for attack by free radicals. Radicals also act on cell metabolism indirectly, by initiating signaling cascades involving the release of secondary messengers such as free Ca^{2+}. Oxidative injury manifests itself as chlorotic and/or necrotic lesions, reduced productivity, abnormal development, and ultimately the death of the plant.

Chlorophyll bleaching is often the first visible sign of oxidative injury. Lipids in the chloroplast envelope and thylakoid membranes contain a high percentage of polyunsaturated fatty acids, and are therefore particularly susceptible to peroxidation. Radical induced lipid peroxidation involves the initiation, propagation, and termination of a chain reaction. Initiation occurs when a radical (e.g., $OH^{\bullet}$) extracts a hydrogen atom from a methylene group in the fatty acid, creating a lipid alkyl radical ($R^{\bullet}$). During the propagation phase, the alkyl radical reacts with oxygen to form a lipid peroxyl radical ($RO_2^{\bullet}$), which can then remove a hydrogen from an adjacent fatty-acid side chain to form a new alkyl radical. Lipid peroxides are formed in ever-increasing numbers as the reaction propagates along the fatty-acid chains. Termination of the chain reaction may be achieved by dimerization of adjacent radicals. By disrupting lipid structure, peroxidation increases the leakiness of membrane bilayers, inactivates membrane bound enzymes, and generates cytotoxic aldehydes and ketones.

Oxidative stress greatly accelerates DNA damage. $OH^{\bullet}$, in particular, causes single- and double-strand breakages, fragmentation of the deoxyribose sugar, chemical alterations to bases, and cross-links between DNA and protein, which impair chromatin unfolding, DNA repair, and transcription. Proteins exposed to $OH^{\bullet}$ show damage to amino acid residues, polypeptide fragmentation, denaturation, and aggregation.

Antioxidants

An antioxidant has been defined as "any substance that, when present at low concentrations compared to those of an oxidizable substrate, significantly delays or inhibits oxidation of that substrate." The caveat that antioxidants be present at low concentrations is critical; because many antioxidants promote redox reactions, they can act as pro-oxidants if present in large amounts. Antioxidants function by: (1) preventing the production of free radicals; (2) scavenging the unpaired electron; (3) quenching the energy of excited molecules such as $^1O_2$; or (4) terminating chain reactions.

Plant cells normally contain a suite of antioxidants localized in different compartments (**Figure 3**). The antioxidant complement often changes as organs grow and develop, and is markedly affected by environmental stresses. Antioxidants are broadly categorized into two groups: the enzymatic antioxidants and the low molecular weight (nonenzymatic) antioxidants. Both groups are apparently required for the effective removal of reactive species. The enzymatic antioxidants are further subdivided into two types: those that deal directly with toxic

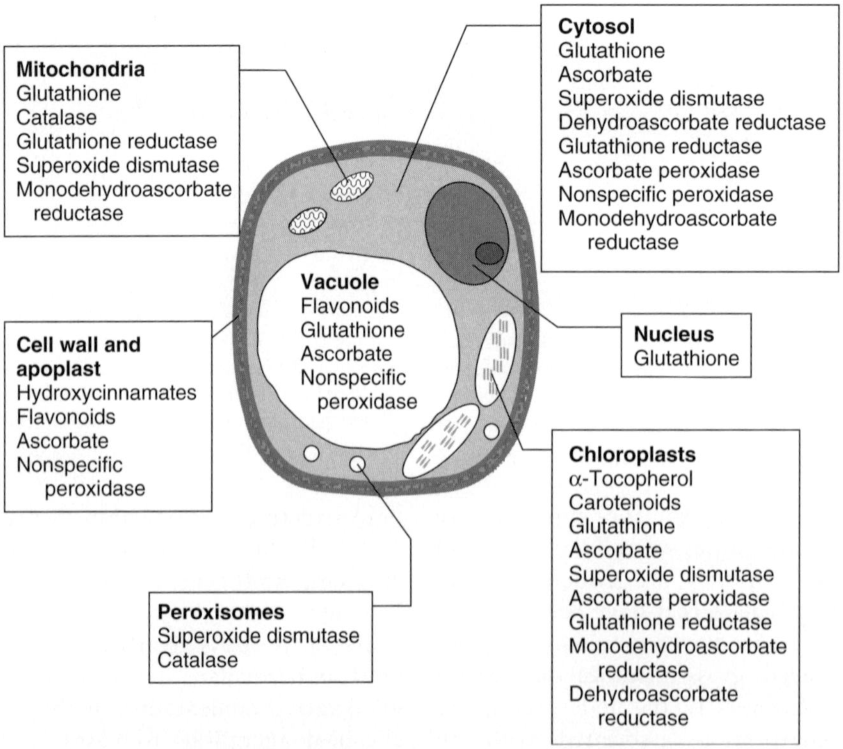

Figure 3 Localization of enzymatic and low-molecular-weight antioxidants in plant cells.

oxidants and those that help to restore the spent antioxidants to their reduced state.

Superoxide dismutase (SOD) is probably the best researched of the enzymatic antioxidants. It is extraordinarily efficient – one of the fastest enzymes known – scavenging $O_2^{\bullet -}$ produced both by the chloroplasts and the mitochondria. Three types of SOD isozymes occur in plants, distinguished by the type of metal cofactor required to catalyze the dismutation reaction: MnSOD (in mitochondria); FeSOD (in chloroplasts); and Cu/ZnSOD (in cytosol, peroxisomes, and chloroplasts). SOD activities increase appreciably in harsh environments, implicating a role in stress tolerance.

H_2O_2 can be scavenged either by catalase (CAT) or the peroxidases (PX). CAT occurs predominantly in the peroxisomes, where it scavenges H_2O_2 produced by photorespiration, but it may also be found in the mitochondria:

$$2H_2O_2 \overset{\text{CAT}}{\rightarrow} 2H_2O + O_2$$

In the chloroplasts, ascorbate peroxidase (APX) is the key enzyme involved in H_2O_2 scavenging. APX uses ascorbate (vitamin C) as a substrate in a redox reaction, producing monodehydroascorbate (MDA):

$$2\ \text{ascorbate} + H_2O_2 \overset{\text{APX}}{\rightarrow} 2\text{MDA} + 2H_2O$$

Although H_2O_2 has been eliminated, chloroplasts now need to recover ascorbate, a particularly useful molecule, from MDA. This can be achieved by various routes. In the thylakoids, MDA is reduced nonenzymatically by ferredoxin:

$$\text{MDA} + \text{Fd}_{\text{red}} \rightarrow \text{ascorbate} + \text{Fd}_{\text{ox}}$$

In the stroma, the enzyme MDA reductase (MDAR) is used:

$$\text{MDA} + \text{NADPH} \overset{\text{MDAR}}{\rightarrow} 2\ \text{ascorbate} + \text{NADP}^+$$

Further enzymes are needed because MDA can also spontaneously dissociate into ascorbate and dehydroascorbate (DHA). DHA is reduced to ascorbate by dehydroascorbate reductase (DHAR), using glutathione (GSH) as the reducing substrate:

$$\text{DHA} + 2\ \text{GSH} \overset{\text{DHAR}}{\rightarrow} \text{ascorbate} + \text{GSSG}$$

Glutathione reductase (GR) completes the "ascorbate–glutathione" pathway, regenerating GSH using NADPH from the light reaction:

$$\text{GSSG} + \text{NADPH} \overset{\text{GR}}{\rightarrow} 2\text{GSH} + \text{NADP}^+$$

In addition to the enzymatic antioxidants, there has been increasing recognition of the role of low-molecular-weight antioxidants (LMWAs) as protectants from oxidative stress. This has, in part, stemmed from research that shows that such compounds are important components of human nutrition, potentially protecting us from the development of diseases such as cancer. LMWAs are a diverse assortment of water and lipid soluble compounds, including various vitamins, glutathione, and several plant pigments. They occur in almost all compartments of the plant cell (**Figure 3**).

Vitamin E (α-tocopherol) is probably the most important of the mobile lipid soluble LMWAs. Found in thylakoid membranes, α-tocopherol effectively scavenges $O_2^{\bullet -}$, serves as a chain breaking agent to stop the propagation of lipid peroxides, and protects the membrane proteins from oxidation of S-H groups. In performing these tasks, α-tocopherol is itself oxidized into a radical, and is regenerated to its reduced state by ascorbate.

Ascorbate (vitamin C) is a particularly versatile antioxidant. Located in the cytosol, apoplast, and the chloroplasts, ascorbate can both quench $^1O_2$ and scavenge $O_2^{\bullet -}$. It also regenerates various antioxidants, such as α-tocopherol and the flavonoid and xanthophyll pigments, and serves as a substrate in enzyme catalyzed reactions for the detoxification of H_2O_2.

Plant pigments have unique, multiple roles in the protection from oxidative stress. First, they efficiently eliminate reactive oxygen species. β-carotene, an orange pigment that is a precursor of vitamin A, quenches $^1O_2$ produced by chloroplasts. The flavonoid pigments are particularly potent scavengers of most reactive species, including H_2O_2, $^1O_2$, $O_2^{\bullet -}$, $OH^{\bullet}$, $RO_2^{\bullet}$, and $ONOO^-$. Indeed, cyanidin-3-glucoside (an anthocyanin) and quercetin (a flavonol) have antioxidant capacities more than four times greater than those of vitamins C and E, as estimated using automated "oxygen radical absorbance capacity" (ORAC) assays. Anthocyanins, the pigments that cause red, purple, and blue colors in various plant organs, are predominantly located in the cell vacuole. Recent studies have shown that red leaves often have higher capacities to combat oxidative stress than do green leaves, and that cells with anthocyanic vacuoles remove H_2O_2 more rapidly than do colorless cells.

A second, important role of the plant pigments is to prevent the generation of ROS in the first place. Xanthophyll pigments (comprising three carotenoid pigments) achieve this by absorbing the excess energy in triplet chlorophyll before it can be transferred to molecular oxygen. Under strong light, violaxanthin is converted to antheraxanthin, and then to zeaxanthin, which dissipates the excess energy as heat. The anthocyanins in leaves function as a light filter, intercepting the energetic green wavelengths before they have an opportunity to cause free radicals in the chloroplasts. Similarly, the colorless flavonoids and hydroxycinnamates, which are especially abundant in cell walls of the upper epidermis, attenuate ultraviolet radiation before it can reach chloroplasts in the underlying cells.

Concluding Remarks

The potential for oxidative injury is a serious concern, particularly for crop plants, which are often grown in environments far-removed from those of their ancestral origins. Current changes to the global climate, the elevated temperatures and UV-B levels, and increasing incidence of severe droughts and floods are expected to increase the likelihood of oxidative stress even further. It is not surprising, therefore, that there has been a concerted effort by chemists, plant scientists, and medical researchers to understand and improve upon the natural antioxidant systems in plants. The emerging data are extremely encouraging. Cultivars of wheat (*Triticum* spp.), for example, have been identified that produce supernumerary levels of antioxidant enzymes in response to abiotic stressors; these cultivars have an enhanced tolerance to water stress, salinity, and heat. Similarly, in several species, plants that have been genetically engineered to overexpress the genes for SOD, APX, and/or CAT are demonstrably more tolerant to drought, salinity, chilling, freezing, and herbicides. Thus, the problems stemming from oxidative stress in cultivated plants, though significant, might not be insurmountable.

See also: **Abiotic Stresses**: Cold Stress; Mechanical Stress and Wind Damage. **Genetic Modification, Applications**: Oxidative Stress. **Photosynthesis and Partitioning**: Photoinhibition; C_3 Plants; C_4 Plants; CAM Plants. **Plants and the Environment**: Global Warming Effects; Ozone Depletion; Plants and Atmospheric Pollution. **Regulators of Growth**: Photoperiodism. **Water Relations of Plants**: Drought Stress; Salt Stress.

Further Reading

Foyer CH, Lelandais M, and Kunert KH (1994) Photooxidative stress in plants. *Physiologia Plantarum* 92: 696–717.

Gould KS and Lee DW (eds) (2002) *Anthocyanins in Leaves: Advances in Botanical Research*, vol. 37. London: Academic Press.

Inzé D, and Van Montagu M (eds) (2002) *Oxidative Stress in Plants*. London: Taylor & Francis.

Halliwell B and Gutteridge JM (1999) *Free Radicals in Biology and Medicine*, 3rd edn. Oxford: Oxford University Press.

Larson RA (1997) *Naturally Occurring Antioxidants*. Boca Raton: Lewis Publishers.

Leshem YY (2000) *Nitric Oxide in Plants: Occurrence, Function and Use*. Dordrecht: Kluwer Academic Publishers.

Polle A (1997) Defense against photooxidative damage in plants. In: Scandalios JG (ed.) *Oxidative Stress and the Molecular Biology of Antioxidant Defenses*, pp. 623–666. New York: Cold Spring Harbor Laboratory.
Wendehenne D, Pugin A, Klessig DF, and Durner J (2001) Nitric oxide: comparative synthesis and signaling in animal and plant cells. *Trends in Plant Science* 6: 177–183.

Mechanical Stress and Wind Damage

J Grace, University of Edinburgh, Edinburgh, UK

Introduction

Erasmus Darwin wrote about the sensitivity of plants to touch in 1794. Later, his grandson Charles Darwin carried out experiments on the touch-sensitivity of insectivorous and climbing plants. One of the most spectacularly sensitive plants, *Mimosa pudica*, folds its leaves completely within a few seconds of being touched. Many climbing plants have tendrils that respond to touch by curling, a process that requires differential growth on different sides of the tendril. Less obvious touch-sensitivity occurs in most other plants. They respond to mechanical stimulation by developing more structural tissue, enabling them to withstand the forces of gravity and wind. These structural tissues take various forms. They are most evident in the stems and branches of trees, which often develop mechanical tissues eccentrically in order to compensate for the tendency for directional displacement by the prevailing wind or by slope. However, similar growth responses are more or less ubiquitous in green plants, affecting leaves and roots as well as stems and branches, and can affect annual plants, grasses, and shrubs as well as trees.

Historically, the effects of wind were noted in the ecological and forestry literature, usually in the context of stress; but the observations almost always relate to situations where effects of other factors are involved. For example, the wind shaping and "flagging" of trees in coastal regions may be caused by the deposition of salt on the windward leaves rather than to the direct effect of wind. Likewise, the wind shaping of trees on mountains may be the consequence of small differences in the temperature of tissues on the windward versus leeward side of the crown, rather than the direct effect of mechanical stimulation on the tissues. Thus, to understand the effect of wind and mechanical stress on plants, it is usually necessary to carry out controlled experiments using wind tunnels, shaking machines, and mechanical rigs, which enable forces to be applied to plants. When plants are exposed to wind or to almost any mechanical challenge, their rate of growth declines and the allocation of new biomass shifts to achieve more mechanical tissue and less photosynthesizing tissue.

In trees, this mechanical tissue takes the form of thickened cells in the woody tissues: these tissues are found in particular parts of the tree where mechanical strengthening is required. The material formed is much denser than normal wood and is called reaction wood or compression wood, depending on whether it has been formed under tension or compression. Such structural changes have long been known to anatomists and wood technologists, but it took physiologists rather a long time to realize that there really was a direct response to mechanical stimulation by all plants. It was not until 1973 that the term thigmomorphogenesis was coined by Jaffe to describe the general phenomenon whereby plants perceive mechanical stimuli and undergo a growth response.

In this article, it is convenient to discuss the effects of wind and mechanical stresses under two headings: direct and indirect effects.

Direct Effects

Response to Movement of Plants in the Wind

In nature, terrestrial plants are more often in motion than not, as a result of air movement. There have been some notable experiments in which trees were prevented from moving. The result is very clear: when they are restrained, trees grow taller, but their stems are thinner (**Figure 1**). A lesser proportion of the water conducting cells, the tracheids, develop thick walls to confer mechanical strength. When supported at 45° and rotated, the thickening response was especially well developed (**Figure 1**). Somewhat similar results have been obtained from experiments on herbaceous plants such as corn (*Zea mays*; maize) and sunflower (*Helianthus* spp.). Many such observations point toward a self-regulating capacity whereby the plant builds its own supporting structure, enabling it to withstand the forces acting upon it. Thus, when external support is provided, the stems of the plant do not need to be as thick, and assimilates are directed to an increase in height growth at the expense of growth in diameter.

There are some interesting arboricultural implications of this phenomenon. When trees are planted in urban environments they are often provided with a

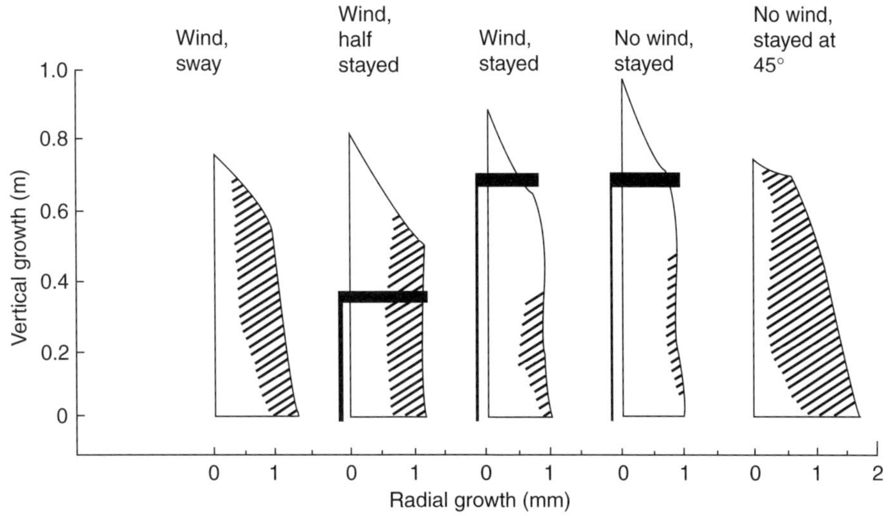

Figure 1 The influence of mechanical support on the distribution of wood increment in the stem of young trees (*Larix laricina*). The solid bars show how the plant was restrained using "stays." The shaded areas denote thick-walled (exceeding 5 μm) tracheids. Reproduced with permission from Larson PR (1965) Stem form of young *Larix* as influenced by wind and pruning. *Forest Science* 11: 412–424.

supporting stake, but if the stake is not removed when the root system has become established, the tree will grow tall and thin, with poor anchorage. When the stake is eventually removed (or when it rots) the tree will be too feeble to stay vertical, and it is likely to be blown over.

There appear to be similar mechanisms operating on the root system. Thus, when Stokes and coworkers removed the tap roots of young trees to mimic the death of the tap root, which happens in water logged soils, there was a 60% increase in the growth of roots on the windward side and a 45% increase in the growth of roots on the leeward side, relative to growth at right angles to the prevailing wind. When Goodman and Ennos grew corn and sunflower in soil of a high bulk density their roots systems developed in rather different ways, ensuring anchorage.

The formation of the buttresses of trees, especially rain forest trees, is presumed to be another example of the direct effect of mechanical forces. In 1991, the special case of trees was widely described by Mattheck. The base of the tree widens to form a huge plate-like structure, which is anchored into the ground by 'sinker' roots. Mattheck proposed that the development of the buttress and the sinker roots is a direct result of the mechanical tensions and compressions occurring when the tree is swaying in the wind. The biomechanical response has been called the "constant surface strain hypothesis," whereby the parts of the stem, buttress, and roots that have a high strain (deformation) develop strengthening tissues to reduce the surface strain.

Wood anatomists have described the special characteristics of the tissues that develop in relation to the force of gravity acting on branches. In coniferous species, the wood on the underside of branches is denser than elsewhere and less elastic. The tissue is called compression wood, and the constituent cells (tracheids) are shorter and thicker. In angiosperm species, the corresponding tissue is called tension wood, and develops on the upper side of the branches. This wood contains an increased number of thick-walled gelatinous fibers and few and smaller vessels.

Grasses show a somewhat similar response to mechanical stimulation. When they are either exposed to wind (using a wind tunnel) or gently shaken using a mechanical frame, there is a reduction in the length of the leaves, an increase in their thickness, and an increase in the extent of thick-walled mechanical tissues such as sclerenchyma and collenchyma, with an increase in the thickness of the cuticle. As a consequence, the stems become stiffer.

Touch sensitivity is already being exploited in horticulture. Seedlings of tomato, geranium (*Pelargonium* spp.), *Petunia* spp., and pansy (*Viola tricolor*) growing at densities of 1000–2000 plants m^{-2} are brushed using plastic netting. This results in sturdy plants that survive transplanting very well, and the final performance of the plant is not impaired by the treatment.

The response to mechanical stimulus is probably not linear. Some early work showed that brief mechanical stimulation each day elicited a morphogenic response. It may be that even plants in sheltered environments such as glasshouses respond to small air currents, the mechanical stimulation when they are irrigated, and the touch of the grower.

Possible Mechanism of Thigmomorphogenesis

The biochemical and molecular mechanisms involved in touch-sensitivity in plants is not well understood. Plants are thought to achieve touch sensitivity through detection of pressure changes within the cell, or by the statoliths. The statoliths are starch grains used by plants to detect gravity and orientation of roots and stems with respect to gravity. They are found in specialized cells, statocytes, often observed in roots but present in other parts of the plant as well. These grains rest against the endoplasmic reticulum (ER), and thus any displacement of the tissue (for example by wind) will cause a change in the location of the grains against the ER. This shift in contact is supposed to trigger biochemical events. However, some plants that lack starch and lack statoliths still retain some sensitivity to gravity, so this cannot be the only mechanism.

Touch and wind cause an increase in cytosolic calcium, and it is also known that synthesis of calmodulin mRNA occurs over a period of less than an hour. Calmodulin is a ubiquitous calcium binding protein, which regulates the activity of some enzymes, and binds calcium when the intracellular level becomes high. The Ca^{2+} ATPase pumps are situated in the plasma membrane and ER, and they cause transport of calcium from one cell to another, a process that has a well-known role in cell signaling, and so this is probably how other parts of the tissue "know" when one small part of the plant has been touched, rubbed, or damaged. Recently, Oufattole and coworkers showed that expression of genes for plasma membrane ATPase was enhanced by mechanical stress.

Finally, still less is known of the processes that follow these initial stages. However, mechanical stimulation of various kinds induces the production of ethylene, which acts like a growth hormone, and can initiate some of the morphogenetic changes that are typical of wind exposed plants, such as the development of sclerenchyma.

Mechanical Abrasion

Mechanical action may abrade the cuticle and disrupt the leaf surface. Leaf surfaces are made up of interlocking epidermal cells, coated with a more or less waterproof and gas-proof cuticle, composed of the lipid polyesters cutin and cutan. The cuticle proper is covered with a waxy epicuticle, which often gives a blue coloration to plant surfaces, for example, in many conifer leaves, in grapes (*Vitis* spp.), and in cabbages (*Brassica oleracea capitata*).

In strong wind, the leaves of plants are set in violent motion, and there are usually multiple collisions between leaves and twigs. Often there is a repeated pattern, whereby one leaf persistently rubs against another in more or less the same location on the lamina. Abrasive damage to leaf surfaces may be viewed using scanning electron micrography, which reveals that the epicuticular wax is polished, and that the cuticle is cracked. The damage is at its most intense on parts of the leaf that stand proud, such as on and near the veins, and at the leaf margins. Macroscopically, there may be no visible damage, but later the leaf margins turn brown and necrotic patches appear. These symptoms correspond to what horticulturists call "wind scorching." The necrotic regions are presumably desiccated as a result of the effective loss of the water impermeability in those areas of the leaf. Light microscopy with appropriate staining reveals that the borders of the necrotic regions become sealed off with corky tissue.

When this happens the resistance of the leaf to the diffusion of water and gases declines, and so water loss may increase. This may be demonstrated in wind tunnel experiments, where the rate of transpiration is seen to increase over the course of several days of wind treatment, both by day and by night (**Figure 2**). The wind exposed leaf may be detached and hung on a chemical balance, whilst weighing it periodically to determine the rate of water loss. In the early stages of dehydration, water is lost rapidly through the stomata and cuticle. Later, the stomata shut and the rate of dehydration is slower. This slower rate is 'cuticular water loss': in this phase the water is being lost through a much-damaged cuticle (**Figure 2**). The results from this sort of work suggest that storm damage increases cuticular water loss, and may also impair the capacity of the stomata to close or even increase the stomatal aperture. Repair of the cuticle can, however, occur. When leaf surfaces are wounded or penetrated, the exposed internal tissue takes on a new role as the impenetrable barrier. The synthesis of cutin, suberin, and lignin has been reported at such wound sites, and the suberized layer serves to isolate the damaged region from the surrounding foliar tissue, preventing excessive water loss.

Conifers and sclerophyllous plants, including many shrubs, have cuticles that are thick and relatively resistant to mechanical damage. They also often have a hypodermis below the epidermis, which may have a further protective role.

In nature, mechanical abrasion may also occur when particles of ice, sand, and soil impact upon the leaf or twigs. At high elevation sites, "wind-blasting" is used to describe the combination of these effects. In extreme cases, the bark is eroded from young stems and twigs.

Damage to the leaf surface may also predispose leaves to damage by atmospheric pollution. In one

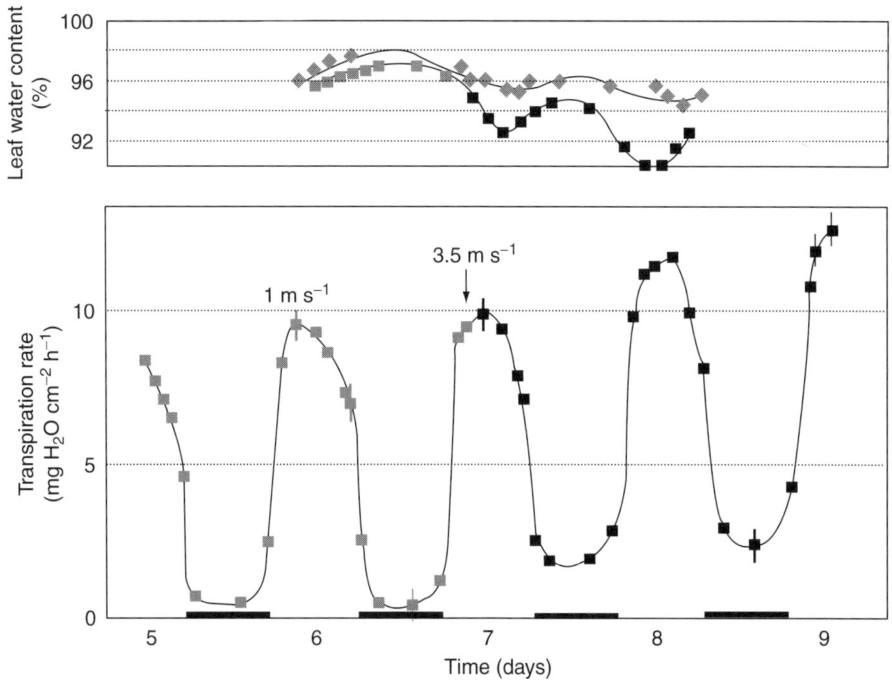

Figure 2 A wind tunnel experiment. The effect of increasing the wind speed from 1 m s^{-1} to 3.5 m s^{-1} on transpiration rate and water content of the leaves. Black squares denote plants were exposed to 3.5 m s^{-1}, the other symbols are low wind speed controls at 1 m s^{-1}. Note the upward trend from day 7 to 9. Reproduced with permission from Grace J (1974) The effect of wind on grasses 1. Cuticular and stomatal transpiration. *Journal of Experimental Botany* 25: 542–551.

study by Hoad and coworkers in 1992 on the leaves of beech (*Fagus sylvatica*), droplets of labeled sulfate (^{35}S-sulfate in an aqueous solution, applied in an "artificial acid rain" formulation) were placed on the leaf surface and allowed to dry. The leaves were then rinsed to remove the excess sulfate that had not been taken up. It was shown that exposed leaves absorbed much more sulfate than leaves taken from sheltered places (**Figure 3**).

Large-scale mechanical damage to trees, involving uprooting and breakage of mature trees, is an enormous problem in silviculture. Much research effort has gone into designing a silviculture that minimizes the risks of windthrow. The most important tactic in reducing risk is to avoid thinning the forest at a later stage. Nowadays, tree crops are often planted at a wider spacing than hitherto, to avoid the need for thinning. Moreover, the earlier practice of planting on the tops of a ploughed furrow is now less used, as it encourages an uneven development of the root system, and therefore predisposes the forest to windthrow when the wind is blowing perpendicular to the plough lines.

Indirect Effects

Wind and Transpiration

The stomatal pores are distributed over the leaf surface, and for most plants they are usually open for all or part of the day and close at night. It is through these stomata that carbon dioxide enters the leaves by diffusion from the atmosphere to the sites of photosynthesis. At the same time, water vapor inevitably diffuses from the substomatal cavities within the leaves to the outside, being driven by the difference in water vapor pressure between the inside and outside of the leaf. This vapor pressure difference is influenced by the humidity of the air and the thickness of the boundary layer surrounding the leaf. However, it is also influenced by the leaf temperature because the air in the substomatal cavity is water saturated, and the saturated vapor pressure of water in air is a steep function of temperature.

Early authors believed that one of the inevitable effects of wind was to increase the rate of transpiration. This belief would only be true if the leaf remained at a constant temperature, but it does not. In bright sunlight and light breezes, a leaf may be several degrees warmer than the air, but when wind speed is increased the leaf temperature is reduced to be closer to air temperature. This fall in leaf temperature makes the saturated vapor pressure of water inside the leaf also fall, and so the driving gradient for transpiration falls. Hence, the effect of wind speed on transpiration is rather complex, depending on factors that influence the thickness of the boundary layer (size of leaf and wind speed), and also on the energy balance of the leaf. In 1995, Van

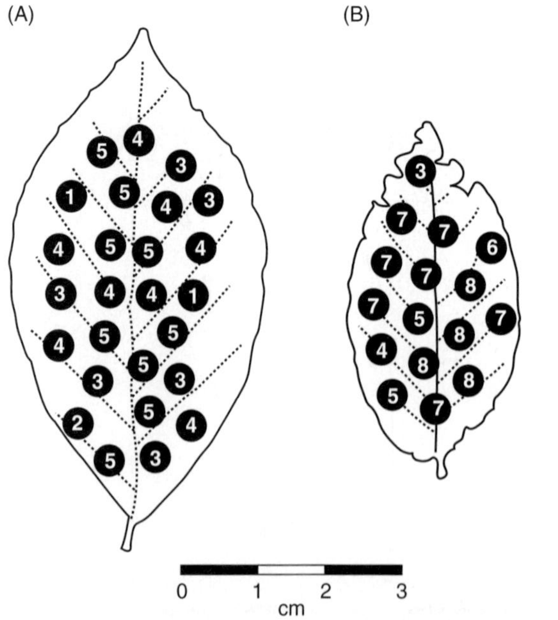

Figure 3 Droplets of $^{35}SO_4^-$ were placed on leaves from beech (*Fagus sylvatica*) to determine the permeability of the leaf surface to droplets of acid rain. (A) Results from sheltered leaves; (B) results from wind exposed leaves (Note the difference in size.) The uptake (pmol sulfur cm^{-2}) is shown by the numbers 1–8 as follows: 1 > 5; 2 = 5–10; 3 = 10–50; 4 = 50–100; 5 = 100–500; 6 = 500–1000; 7 = 1000–5000; 8 = 5000–10 000. Reproduced with permission from Hoad SP, Jeffree CE, and Grace J (1992) Effects of wind and abrasion on cuticular integrity in *Fagus sylvatica* and consequences for transfer of pollutants through leaf surfaces. *Agriculture, Ecosystems and Environment* 42: 275–289.

Gardingen and Grace carried out wind tunnel experiments and calculations based on the leaf energy balance, which showed that often the effect of moderate wind on transpiration is negligible, and sometimes wind actually decreases the transpiration rate (**Figure 4**). Here, the effect of wind on cooling the leaf may be substantial, especially in large leaves, so that the reduced vapor pressure of water inside the leaf causes a reduction in transpiration rate. There is also a possibility that an increase in wind speed may cause stomata to close, although the evidence for this phenomenon is rather slight.

The occasions where wind does cause water stress are those where there is the complication of mechanical abrasion of the cuticle, as discussed above. When the cuticle is damaged, the elevated transpiration continues after the wind has finished, and it may also continue at night (if there is a positive leaf–air gradient of water vapor pressure to drive it).

The realization that wind doesn't necessarily increase transpiration rate (and sometimes decreases it) generally leads us to abandon the idea that the dwarfing of plants in windy places (especially mountains) is something to do with water stress. It is much more likely to be the result of low temperatures and thigmomorphogenesis.

Wind, Energy Balance, and Surface Temperature

Plant organs The temperature of a plant organ in a particular environment may be calculated from simple energy balance considerations. The approach requires some knowledge of the effect of wind on the boundary layer resistance of the plant organ. If we are talking of flat leaves or spherical structures such as many fruits, this information is available from wind tunnel experiments. It is beyond the scope of this article to present the calculations in full, but the reader can refer to Van Gardingen and Grace's work for more details. Plant organs are intermediate between wet systems (such as open water) and dry systems. As evaporation is a major term in the energy balance, it is important that a value of the stomatal resistance can be estimated or assumed in the calculation. The result (**Figure 5**) shows that wind exerts a notable impact on the temperature of the plant organ. In bright sunshine and light wind, an increase in the wind speed cools the leaf, bringing it closer to the air temperature. This effect can be large (for large leaves that have thick boundary layers) or small (for small leaves with small boundary layers). As we saw above, there are corresponding influences on the rate of evaporation, because the falling temperature reduces the saturation vapor pressure inside the leaf.

Shelterbelts There have been many schemes, some of them enormous in scale, to improve agriculture and horticultural productivity by the creation of windbreaks and shelter belts. The earlier literature, of which there is an immense amount, suggests that crop growth is often increased by shelter, but not always so. The beneficial effect of windbreaks and shelterbelts was believed to have been caused by the creation of a favorable microclimate for plant growth. However, critical measurements of the temperature and humidity in the lee of shelterbelts showed that the effect is small in most cases, and distributed over a rather short distance from the belt. Thus, the overall benefit to the crop can only be very small. Moreover, land area is consumed by the belt itself, solar radiation is intercepted by the belt, and if the belt is living, water and nutrients are consumed also. Ripening of soft fruits, such as strawberry (*Fragaria x ananassa*), may be appreciably earlier in the lee of the belt.

The reduced mechanical stimulation in the lee of the belt is likely to be a more important consideration,

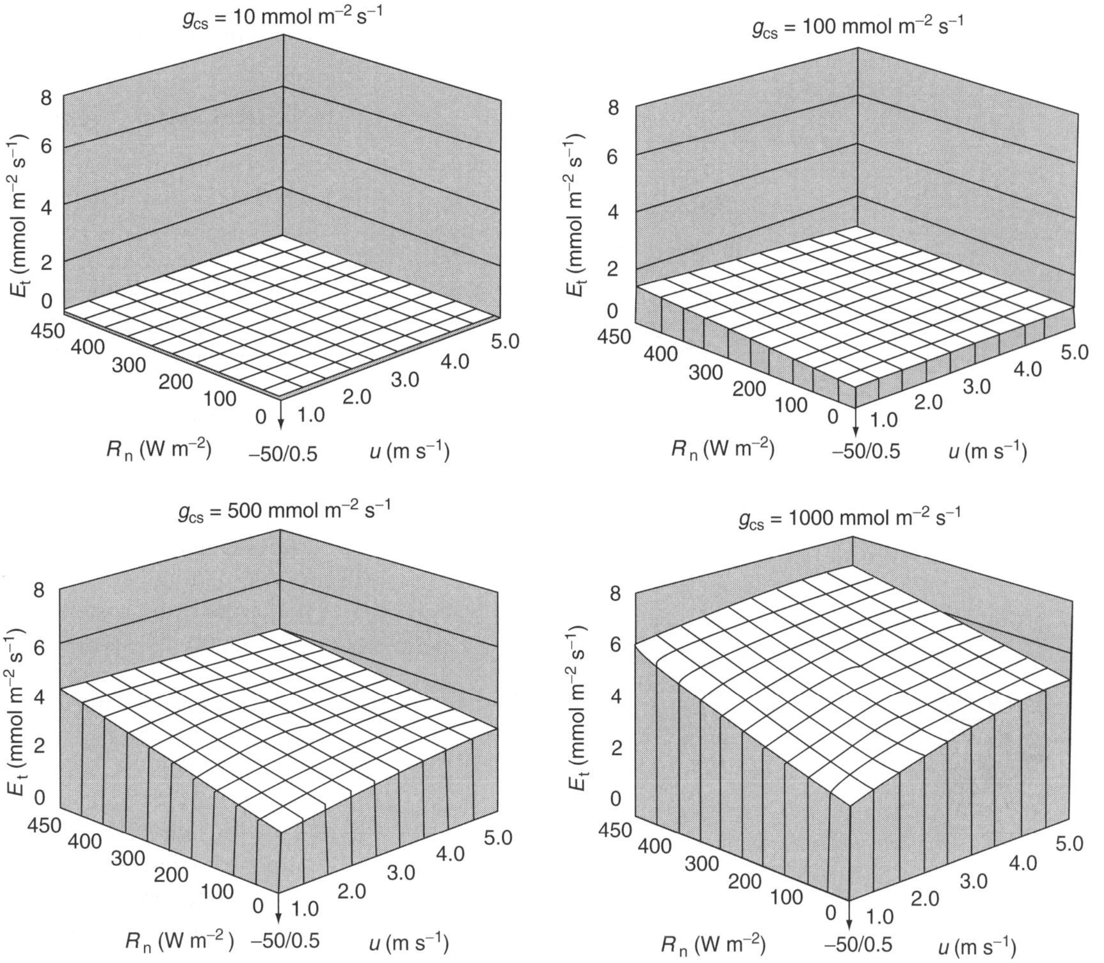

Figure 4 The effect of wind speed, u, and net radiation, R_n, on the rate of water loss by transpiration, E_t. The four diagrams show the result for leaves with four stomatal conductances, g_{cs}, from 10 mmol m^{-2} s^{-1} to 1000 mmol m^{-2} s^{-1}. Note that transpiration is not very sensitive to wind, somewhat more sensitive to the net radiation, and very sensitive to stomatal conductance. Reproduced with permission from van Gardingen P and Grace J (1991) Plants and wind. *Advances in Botanical Research* 18: 189–253.

especially in horticulture where the commercial quality of the produce is impaired by damage.

The aerodynamics of windbreaks and shelterbelts suggest that rather than provide shelter over a horizontal distance of 20 times their height, the shelter effect is spoilt by the tendency for a downward vertical movement of air at about 12 times the height. This feature may bring cooler air into contact with the crop, and it may cause mechanical agitation leading to damage of the leaf surfaces.

Mechanical Forces as a Stress

Mechanical forces have challenged plants throughout their evolutionary history, but especially since the time when plants colonized the land. It is not surprising that practically all vascular plants are able to alter their allocation patterns to cope with the "ordinary" mechanical forces associated with gravity and wind. As we have seen, this is achieved through the development of specialized tissues that strengthen the plant against gravity and against the buffeting of the wind. Mechanical failure does, of course, occur in exceptionally strong winds, but even then many plants have the capacity to recover by sprouting new shoots, utilizing their reserves of food stored in the root system.

Woody tissues have a dual function: they transport water and provide mechanical strength. Wood remains one of the strongest materials known to man. Leaf surfaces are also mechanically strong in many cases, providing the first line of defense against weathering, which may include sand-blasting, abrasion by ice particles, acid attack, attack by insects, and so forth. If this first line of defense is breached, the photosynthetic tissues are vulnerable to loss of water. However, if desiccation and death of leaves occurs, most plants have the capacity to regrow.

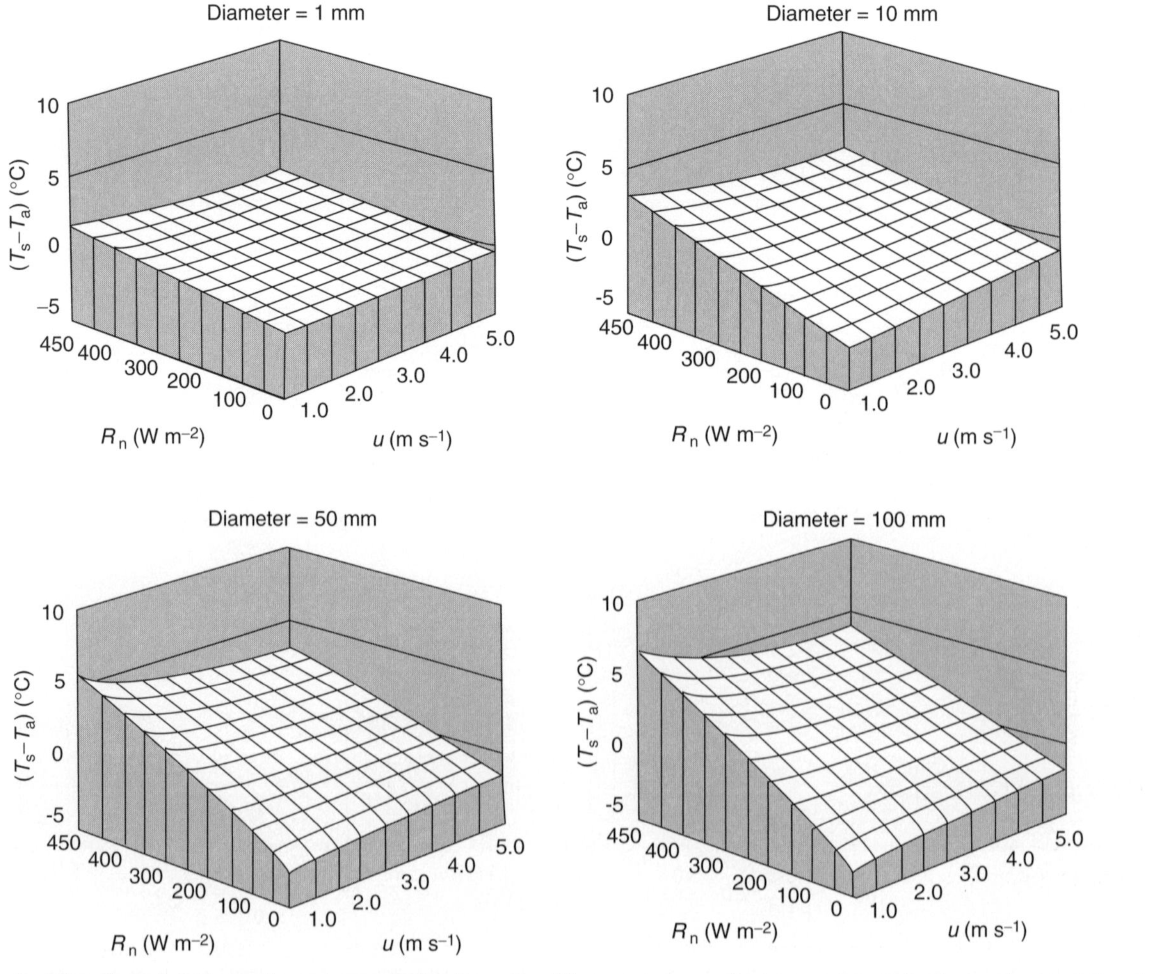

Figure 5 The effect of wind speed u and net radiation R_n on the difference between the temperature of the leaf surface and the air ($T_s - T_a$). The four diagrams show the result for leaves of four diameters: 1, 10, 50 and 100 mm. Note that wind cools the leaf appreciably when the leaf is large and strongly irradiated. Reproduced with permission from van Gardingen P and Grace J (1991) Plants and wind. *Advances in Botanical Research* 18: 189–253.

Wind damage to plants is economically very significant. Wind is perceived as a stress to plants by horticulturists because wind damage substantially reduces the appearance of the crop, and thus makes the crop less saleable: in the case of agriculture, wind may cause "lodging" or blowdown of entire fields of cereal crops, hence making the crop unharvestable; in the case of forestry, storms and hurricanes cause devastating damage to the timber crop, making the harvest very difficult to achieve.

List of Technical Nomenclature

ATPase An enzyme that releases phosphate and energy from adenosine triphosphate (ATP); crucial to the energetics of all cells.

Arboriculture The cultivation and care of trees.

Buttresses Protruding expansions of the lower part of the tree's stem, found especially in tropical trees.

Collenchyma Plant cells with thick (but not woody) walls, conferring mechanical strength.

Cuticle The waterproof and gas-proof layer of fatty substances on the epidermis of plants.

Cutin A mixture of fatty substances in the cuticles of plants.

Cytosol The soluble components of the cytoplasm.

Endoplasmic reticulum (ER) Extensive membranes in the cell, where lipids and some proteins are synthesized.

Lignin A complex cross-linked polymer giving wood its toughness and its resistance to decay.

Sclerenchyma	Plant cells having thick and woody walls, usually conferring mechanical strength.
Silviculture	Planting and care of forests.
Stomata	Adjustable pores on the leaf surface through which water vapor and gases diffuse between the leaf and the atmosphere.
Suberin	Waxy substances in the walls of some plant cells, that resist the flow of water.
Tap roots	Thick roots that penetrate deep into the soil, as opposed to superficial and fibrous roots.
Tracheids	Water conducting cells in plants, a main cell type in wood.
Transpiration	Loss of water vapor from plants via pores on the leaf surface (stomata).

See also: **Regulators of Growth**: Ethylene. **Water Relations of Plants**: Uptake, Loss and Control.

Further Reading

Coutts MP and Grace J (1995) *Wind and Trees*. Cambridge: Cambridge University Press.

Goodman AM and Ennos AR (1999) The effects of soil bulk density on the morphology and anchorage mechanics of the root systems of sunflower and maize. *Annals of Botany* 83: 293–302.

Grace J (1974) The effect of wind on grasses 1. Cuticular and stomatal transpiration. *Journal of Experimental Botany* 25: 542–551.

Grace J (1977) *Plant Response to Wind*. London: Academic Press.

Hoad SP, Jeffree CE, and Grace J (1992) Effects of wind and abrasion on cuticular integrity in *Fagus sylvatica* and consequences for transfer of pollutants through leaf surfaces. *Agriculture, Ecosystems and Environment* 42: 275–289.

Larson PR (1965) Stem form of young *Larix* as influenced by wind and pruning. *Forest Science* 11: 412–424.

Oufattole M, Arango M, and Boutry M (2000) Identification and expression of three new *Nicotiana plumbaginifolia* genes which encode isoforms of a plasma-membrane H^+-ATPase, and one of which is induced by mechanical stress. *Planta* 210: 715–722.

Sinclair W, Oliver I, Maher P, and Trewavas A (1996) The role of calmodulin in the gravitropic response of the *Arabidopsis thaliana* agr-3 mutant. *Planta* 199: 343–351.

Stokes A, Fitter AH, and Coutts MP (1995) Responses of young trees to wind: effects on root growth. In: Coutts MP and Grace J (eds) *Wind and Trees*, pp. 264–275. Cambridge: Cambridge University Press.

van Gardingen P and Grace J (1991) Plants and wind. *Advances in Botanical Research* 18: 189–253.

Agronomy *see* **Crop Improvement**: Plant Breeding, Principles; Plant Breeding, Practice; Hybridization and Plant Breeding; Marker Assisted Selection. **Growth and Development**: Field Crops. **Integrated Pest Management**: Principles; Practice; Disease Prediction Models. **Production Systems and Agronomy**: Agricultural Crops; Grassland; Horticultural Crops, Protected; Orchard Crops; Organic Farming; Commercial Flower Production Methodology; Nursery Stock and Houseplant Production; Plantation Crops and Plantations; Oil Palm and Coconut; Rubber; Multicropping.

Arable Crops *see* **Production Systems and Agronomy**: Agricultural Crops.

B

BIODIVERSITY AND CONSERVATION

Contents

Plant Diversity, Conservation and Use

N Maxted and S P Kell, University of Birmingham, Birmingham, UK

Plant Biodiversity

Introduction

One of the critical challenges facing the world today is the conservation of biological diversity, or biodiversity, and the sustainable use of its components for the benefit of humankind. Biodiversity is a finite world resource that has economic and ethical value to humankind, but this resource is being eroded or lost for perpetuity by careless, unsustainable practices. The exponential loss of biodiversity is well documented; what is most worrying is that this process shows no signs of slowing down and threatens to become even more severe in the future.

The fundamental importance of these issues was addressed at the United Nations Conference on Environment and Development (UNCED) held in Rio de Janeiro, Brazil in 1992, and has been enshrined in the Convention on Biological Diversity (CBD). Its objectives are:

> the conservation of biological diversity, the sustainable use of its components and the fair and equitable sharing of the benefits arising out of the utilization of genetic resources, including by appropriate access to genetic resources and by appropriate transfer of relevant technologies, taking into account all rights over those resources and to technologies, and by appropriate funding.

The Convention on Biological Diversity is the first comprehensive global treaty for the conservation of biodiversity and sustainable utilization of the earth's biological resources. However, it goes beyond these goals, to encompass the issues of access to genetic resources and the sharing of benefits arising from the exploitation of those resources, as well as recognizing the need to provide the biologically rich South with the support it needs to implement its provisions. The Convention represents a milestone in biodiversity conservation thinking, reflecting international acknowledgement of the loss of our biological resources and the urgent need for action. It is now recognized as the primary guiding framework for biodiversity conservation and sustainable use.

What Is Plant Biodiversity?

The widespread use of the term biodiversity began in the 1980s and was launched into popular use by Edward Wilson, the American conservationist, who defined it as:

> The variety of organisms considered at all levels, from genetic variants belonging to the same species through arrays of species to arrays of genera, families, and still higher taxonomic levels; including the variety of ecosystems, which comprise both communities of organisms within particular habitats and the physical conditions under which they live.

Plant biodiversity is that component of biodiversity that is included within the plant kingdom. There are many definitions of biodiversity, but the definition provided by the Convention on Biological Diversity can be adapted to refer to plant biodiversity alone:

> Plant biodiversity means the variability among living plants from all ecosystems (terrestrial, marine and other aquatic) and the ecological complexes of which they are part; this includes diversity within species, between species and of ecosystems.

Plant diversity exists at all levels of biological organization, as illustrated in **Figure 1**.

Therefore, plant biodiversity encompasses diversity at all levels of biological organization: communities, species, individuals, and genes. However, the most commonly used estimate of biodiversity is in terms of the numbers of species (see **Table 1**). Since many areas of high plant diversity remain unexplored, or unrecorded, estimates of the total number of plant species are inevitably inaccurate. However, based on the number of currently described species (i.e., those known to science), there is estimated to be roughly a total of 320 000. **Table 1** shows the approximate or known numbers of described species in each plant group, and an indication of how well each group is known.

Quantifying Plant Biodiversity

In order to conserve and use plant biodiversity we must first measure it, and this only becomes possible when quantitative values can be ascribed and compared; therefore, to disentangle the huge array of variation that biodiversity encompasses, we need to distinguish the key elements. It is customary to classify biodiversity into three main groups: genetic diversity, taxonomic (or organism) diversity, and ecological diversity. Within these groups are a number of elements or levels of biodiversity (**Table 2**), although these elements should not be thought of as being distinct from one another, but as being inextricably linked.

Genetic diversity is the heritable variation that is observed within and between populations of organisms. The basic genetic component is the gene. Genes

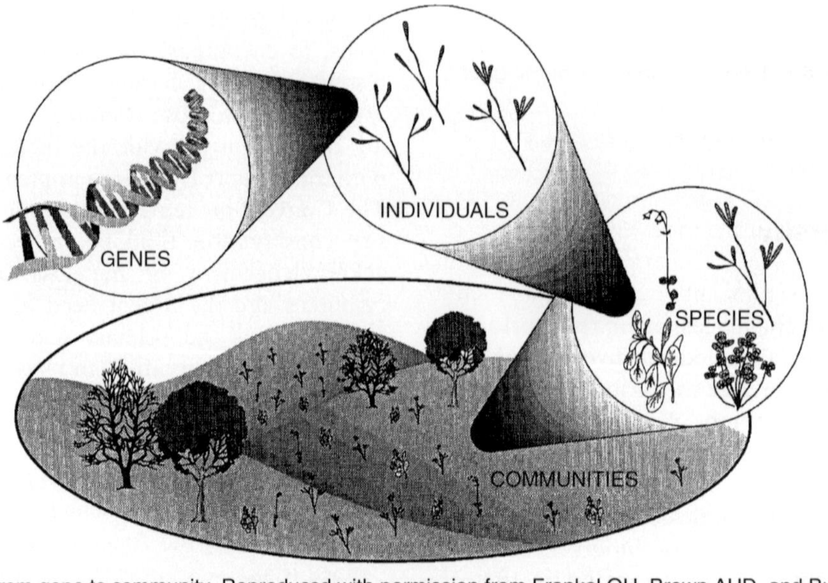

Figure 1 Diversity from gene to community. Reproduced with permission from Frankel OH, Brown AHD, and Burdon JJ (1995) *The Conservation of Plant Biodiversity.* Cambridge: Cambridge University Press.

Table 1 Species diversity for major plant groups

Taxon	*Common name*	*Number of described species*	*How well group is known*
Anthocerophyta	Horned liverworts	100	Moderate
Anthophyta	Flowering plants	270 000	High
Bryophyta	Mosses	10 000	Moderate/high
Coniferophyta	Conifers	630	High
Cycadophyta	Cycads	145	High
Filicinophyta	Ferns	12 000	High
Ginkgophyta	Ginkgo	1	High
Gnetophyta	Gnetophytes	70	High
Hepatophyta	Liverworts	6000	Moderate
Lycophyta	Club mosses	1000	Moderate
Psilophyta	Whisk ferns	10	Moderate/high
Sphenophyta	Horsetails	15	High
Total number of described plant species		299 971	

Data from Groombridge B and Jenkins MD (2002) *World Atlas of Biodiversity.* Berkeley: University of California Press.

Table 2 The composition and levels of biodiversity

Genetic diversity	*Taxonomic (or organism) diversity*	*Ecological diversity*
	Kingdoms	Biomes
	Phyla	Bioregions
	Families	Landscapes
	Genera	Ecosystems
	Species	Habitats
	Subspecies	Niches
Populations	Populations	Populations
Individuals	Individuals	
Chromosomes		
Genes		
Nucleotides		

are found in the nuclei of all cells of all organisms. Genes are made up of DNA, and are situated along chromosomes. Ultimately, it is the variation in the sequence of four nucleotide base pairs, which, as components of nucleic acid, constitute the genetic code, which is passed from generation to generation. New genetic variation arises in individuals by gene and chromosome mutations, and in organisms that reproduce sexually, by recombination. There can be various distinct forms of the same gene, referred to as alleles.

Individuals in a population or a species vary genetically for a range of characteristics or traits. Such genetically significant traits might include: height, fecundity, disease resistance, or tolerance to extreme environmental conditions such as drought. This variation, which may for instance be expressed morphologically, behaviorally, or physiologically, is referred to as the phenotype. The phenotype results from a combination of the individual's genotype (its genetic composition), interacting with the environment in which it is found.

The pool of genetic variation present within an interbreeding population is acted upon by selection. It is the genetic variation within and between individuals and populations of the same species, that ensures the species as a whole can adapt and change in response to natural (e.g., changing environment) and artificial (e.g., breeder's selection criteria) selection pressures. Therefore, genetic diversity enables evolution and adaptation of species within a changing environment, and is essential for the long-term survival of a species.

The diversity of organisms is classified using a taxonomic hierarchy, which varies depending upon the particular classification used. However, the taxonomic levels shown in **Table 2** are universally accepted, with interim levels occurring, making up a complex classification, e.g., below "family" level there may also be "subfamilies." Central to the concept of taxonomic diversity is the species, and for practical purposes, species are the most appropriate targets for biodiversity conservation and use. While genes provide the blueprint for the construction of organisms, they are only expressed through the form and function of species. Similarly, ecosystems are essentially manifestations of the interactions between organisms. It follows that neither genes nor ecosystems can be manipulated or managed without attention being given to the requirements of species; they are the entities in nature that adapt and evolve, occupy ecological space, and become extinct.

Ecological diversity describes biodiversity from population level and upwards through niches, habitats, ecosystems, landscapes and bioregions, to biomes (**Table 2**). At the largest scale, a "biome" describes any of a group of major regional terrestrial communities with its own type of climate, vegetation, and animal life. They are not sharply separated but merge gradually into one another. Examples include tundra, temperate deciduous forest, and desert. At the smallest scale, ecological diversity can describe "populations," which are local communities of potentially interbreeding organisms. In asexual organisms, the term usually refers to a local community of physiologically or morphologically similar individuals of the same species. Conservationists usually refer to habitat, ecosystem, or landscape conservation, but ecosystems have become the accepted unit of conservation within this group. An ecosystem is defined by the CBD as: "a dynamic complex of plant, animal and microorganism communities and their non-living environment interacting as a functional unit."

While it is convenient to think of biodiversity in terms of genetic, species, or ecosystem diversity, it is also important to note the significance of populations. Populations are local communities of potentially breeding organisms. It is within populations of species that we find the genetic diversity that is significant in terms of adaptation and survival. Therefore, while measurements of presence and absence of species may be useful for some studies, e.g., global checklists, we need to know how populations of species behave in order to put in place strategies for their conservation and sustainable use.

Where Plant Biodiversity is Found

Plant biodiversity is not evenly distributed across the surface of the earth. **Table 3** indicates the distribution of higher plants on a continental and regional basis, showing estimates of the total number of species found and the number of endemic species found on each continent and within each subregion. It can be clearly seen that plant species diversity decreases away from the equator towards the poles

Table 3 Regional distribution of higher plants

Continent	*Subregion*	*Number of species*		*Endemic species*	
		Subregion	*Continent*	*Number*	*%*
Europe			12 500	3500	28
Americas			133–138 000		
	North America	20 000		4198	21
	Central America	30–35 000		14–19 000	46–54
	South America	70 000		55 000	78.5
	Caribbean Islands	13 000		6555	50
Africa			40–45 000	35 000	77–87.5
	North Africa	10 000			
	Tropical Africa	21 000			
	Southern Africa	21 000			
Asia					
	Southwest Asia and Middle East	23 000	126–137 000	7100	31
	Central and North	17 500		2500	14
	Indian subcontinent	25 000		12 000	48
	Southeast Asia (Malesia)	42–50 000		29–40 000	70–80
	China and East Asia	45 000		18 650	41.5
Australasia	Australia and New Zealand		17 580	16 202	90
Oceania	Pacific Islands		11–12 000	7000	58–63
		Total number of endemic species		216–234 000	

Data from Davis S, Heywood UH and Hamiton AC (1994) *Centres of plant diversity: a guide and strategy for the conservation.* (3 volumes.) Gland: WWF and IUCN.

and that some ecosystems have higher levels of species diversity than others, even at similar latitudes. Even within habitats of the same type, some geographic areas are more diverse than others.

Various attempts have been made to pinpoint areas of particularly high levels of species diversity on a worldwide scale. For example, some 25 areas have been highlighted as having very high species numbers and endemicity, unusual combinations of community ecological characteristics and superspeciose taxa; these are the so-called plant "hotspots:"

- Mediterranean Basin
- Indo-Burma
- Brazilian Cerrado
- Sundaland
- Guinean forest of West Africa
- Tropical Andes
- Atlantic forest region
- Mesoamerica
- Mountains of South Central China
- Madagascar and Indian Ocean Islands
- Caucasus
- Wallacea
- California Floristic Province
- Philippines
- Central Chile
- New Zealand
- Caribbean
- Chocó–Darién–Western Ecuador
- Western Ghats and Sri Lanka
- Succulent Karoo
- Cape Floristic Province
- Polynesia/Micronesia
- Eastern Arc Mountains
- New Caledonia

Although not strictly hot spots, the Galápagos and Juan Fernandez Islands are particularly important in terms of plant diversity and should be regarded as additional hot spots. These 27 areas alone support about 131 339 endemic plant species (more than 40% of the world total), which with the nonendemic species means that about 70% of vascular plants are contained in 1.44% of the world's surface.

A similar picture of uneven geographic distribution also emerges for the geographical distribution of genetic diversity found within crop gene pools. The world-famous Russian geneticist and plant breeder N.I. Vavilov developed new concepts of centers of diversity for crop plants in the 1920s; these were generally mountainous areas situated in tropical or subtropical regions, and Vavilov concluded that there were eight of these primary centers throughout the world:

I **Chinese Center** – Western and Central China – millets, various beans, onion, radish, cabbage and fruit trees, as well as plants producing oils, spices, medicines, and fibers.

II **Indian Center** – India, Indo-Malaya, Indo-China, Burma, and Assam – rice, chickpea, various beans, many tropical fruits, including citrus fruits (lemons, tangerines, etc.), banana, mango, etc.; oil-producing species, fibers, spices, stimulants and dye plants; sugar.

III **Inner Asiatic Center** – Northwestern India, Afghanistan, Tajikistan, Uzbekistan, and western Tian Shan – wheat, peas, cabbage, lettuce, sesame, cotton, various vegetables and melon species, spice crops, etc.; fruit and nut trees.

IV **Asia Minor Center** – Transcaucasia, Iran, Turkmenistan, and Anatolia – wheat, rye, oats, chickpeas, lentil, vetches, peas; various forage plants such as alfalfa, clover, and sainfoin; melons; vegetables; fruit crops, including apple, pear, pomegranate, fig, grape, pistachio nut.

V **Mediterranean Center** – Mediterranean countries – Forage and vegetable species; various oil-producing plants and spices; olive, beets, cabbages, onion, asparagus, lettuce, parsnip; ethereal oil species.

VI **Ethiopian Center** – Ethiopia – wheat, barley, peas and beans, lupins, teff, finger millet, coffee, banana, and sorghum.

VII **South Mexican and Central American Center** – South Mexico and Central America – corn, various tropical beans, marrow, sweet potato, peppers, cotton, and tobacco.

VIII **South American Andean Center** – Peru, Ecuador, and Bolivia – potato and other tuberous crops, some fruit crops, lupins, beets, corn, and various beans.

VIIIa **Chiloan Center** – Chile – potato, oilweed, grasses, and strawberries.

VIIIb **Brazilian – Paraguayan Center** – Brazil and Paraguay – manioc, peanut, cocoa, rubber plant, and maté.

So far in the discussion of the geographical distribution of plant diversity, we have focused on the spatial distribution of species, but biodiversity shows patterns of distribution at all levels. If we now consider genetic diversity, we might also assume that infraspecific (i.e., within species) genetic diversity is evenly spread throughout the range of the species, but this is often not the case. Recent studies of genetic diversity in relation to wild lentils (*Lens* spp.), for example, have clearly shown that the genetic diversity is not partitioned evenly across its geographic range.

Whether at the species or population levels, some species have higher levels of genetic diversity than others, for example, corn (*Zea mays*; maize) has comparatively high levels and pine (*Pinus resinosa*) has much lower levels. The record for the highest number of alleles per gene locus, however, goes to red clover (*Trifolium pratense*), where self-incompatibility (i.e., the failure of gametes from the same plant to form a viable embryo) is controlled by a single, multiallelic gene expressed in the pollen, and it has been estimated that over 200 alleles exist for this one gene.

Why Plant Biodiversity Has Value

In a time of rapid human population growth (there will be 9.0 billion humans by 2025), our plant genetic resources are as valuable to us as gold and oil, if not more so! Loss of plant biodiversity at all levels has a fundamental detrimental impact on humans because of the direct and indirect benefits humans gain from their existence and use. The ways in which plants and plant products are used by humans are extremely diverse, including food, food additives, animal foods, materials, fuels, poisons, medicines (human and veterinary), ornamentals, recreational use, windbreaks, soil improvers, erosion control, pollution indicator species, and gene donors.

Of the approximate 300 000 known plant species, some 5000 provide food; however, 30 provide 95% of the world's calorie intake and three alone (corn, wheat, and rice) supply almost 50% of the calories and protein in the human diet. So there is clearly scope for diversification based on further domestication and the promotion of so-called neglected and underutilized crops. It has also been reported that approximately 80% of the world's people rely on traditional medicine for their primary healthcare needs and an estimated 85% of traditional remedies are derived from plant extracts.

So-called "keystone species" are also important in this context. These are the species that contribute most significantly to the structure of a community or its processes. The removal of a keystone species makes the other members of a community vulnerable to extinction, and therefore the community itself vulnerable to collapse. Plants also have cultural value, often playing a role as symbols in local cultural practices, religion, folklore, and art. Many countries have adopted plants as their national symbol; for example the cedar of Lebanon (*Cedrus libani*), valued over centuries for the quality of its timber.

Diversity at the genetic level ensures the short-term viability of individuals and populations and the evolutionary potential of populations and species. It allows adaptation to a changing environment and so facilitates survival. Genetic diversity has also been used directly by farmers for millennia; they selected and exchanged genotypes, as do contemporary plant breeders, who introduce new genes into crop varieties. Both farmers and plant breeders need access to genetic diversity if they are to keep their options open in their continual competition with evolving pests and diseases. The potential economic impact of disease resistance can be high. For example, the use of hessian fly resistance in wheat in the United States saved US$17 million in a single year. However, the consequences of lack of genetic diversity in crops can be devastating, as shown by the well-known case of potato blight in Ireland in 1845. No country is sufficiently wealthy in native genetic diversity to make it independent of other regions of the world. Take, for example, the botanically rich country of Brazil; two-thirds of its calorie consumption is based on crops originating from another continent (**Table 4**).

As well as their direct use to humans, plants in natural ecosystems provide a vital role in the effective functioning of the biosphere. Their loss affects the water cycle and may lead to an increase in the frequency of extreme conditions such as drought and flood. It can also lead to soil erosion and disruption of the nutrient cycle. It may also affect the climate, by altering the albedo (radiation reflective quality) of the earth's surface, air turbulence, levels of evapotranspiration, and the concentration of carbon dioxide in the atmosphere. Natural ecosystems also play an important role in the absorption and breakdown of pollutants. Furthermore, the aesthetic and recreational value of wild plants and habitats has always been an important impetus to conservation, since the establishment of the first national parks such as Abhayaranxyas (or "forest reserve") in India in the fourth century BC, to the Yellowstone National Park in the United States in 1872.

Threats to Plant Biodiversity

Plant biodiversity is increasingly threatened at the ecosystem, species, and genetic levels, largely as a result of human activities.

At the ecosystem level, climatic modeling studies suggest that since the beginning of agriculture, some 15% of forests had by 1970 been converted to cultivation. A study by the Food and Agriculture Organization of the United Nations (FAO) and United Nations Environment Program (UNEP) found that between 1980 and 1990, for some 87 countries, about 169 million hectares of forest were lost, or about 1% per year. Over 90% of natural wetlands in New Zealand have been lost since European settlement. More than 90% of other major ecosystems are either disappearing or being fragmented, and so losing their self-sustaining viability. Loss of mangrove swamps is rapid the world over, with figures of 40–80% being quoted for countries such as Thailand, the Philippines, and the Gambia. A UNEP study suggested that some 69.5% of the world's drylands have to some extent been degraded as a result of adverse human impacts.

Table 4 Source of plant-derived calories consumed in Brazil

Crop	*Share of plant-derived calories (%)*	*Center of origin*
Sugar	20.38	Indochina
Rice (paddy)	17.64	Asia
Wheat	15.29	West and Central Asia
Corn	12.20	Central America
Soybean	8.84	China–Japan
Cassava	7.10	Brazil–Paraguay
Beans	6.40	Andes
Bananas	2.22	Indochina

Data from Crucible Group (1994) *People, Plants and Patents.* Ottawa: International Development Research Center.

At the species level the 1997 IUCN *Red List of Threatened Plants* lists 751 described species from 192 families that are thought to be extinct in the wild and 12.5% of the world's flora as globally threatened. This is probably an underestimate as threat has only recently started to be estimated and it is difficult to prove that a species is extinct. IUCN have developed a system of categories of threat status, which requires detailed knowledge of the population dynamics of the species concerned, the so-called IUCN *Red List* categories (**Figure 2**). The 1997 IUCN study suggests that approximately 34 000 plant species are threatened to some degree, with some 6500 in the "Endangered" category, which suggests that they are likely to become extinct in the near future. There are numerous estimates of predicted extinction rates but the consensus view is that 15–20% of all plant species could become extinct by 2020.

It is more difficult to measure loss of genetic diversity or "genetic erosion," either within or between species. However, some examples of indirect measurement, in some cases anecdotal, involving crops, may be illustrative. The proportion of the wheat crop in Greece contributed by old, indigenous cultivars (landraces) declined from 80% in 1930 to less than 10% in 1970. Bread wheat landraces are also now difficult to find in Turkey, Iraq, Afghanistan, Pakistan, and the region where it was first cultivated, whereas 30 years ago they were common. In China, nearly 10 000 wheat varieties were in use in 1949, but only 1000 were still in use by the 1970s. In Cambodia, unique rice varieties were lost in the 1970s when war disrupted agricultural production. Stored seed in the national gene bank was eaten or rotted, and numerous landraces would therefore have died out were it not for the duplicates preserved in the International Rice Research Institute (IRRI) gene bank in the Philippines. In Mexico and

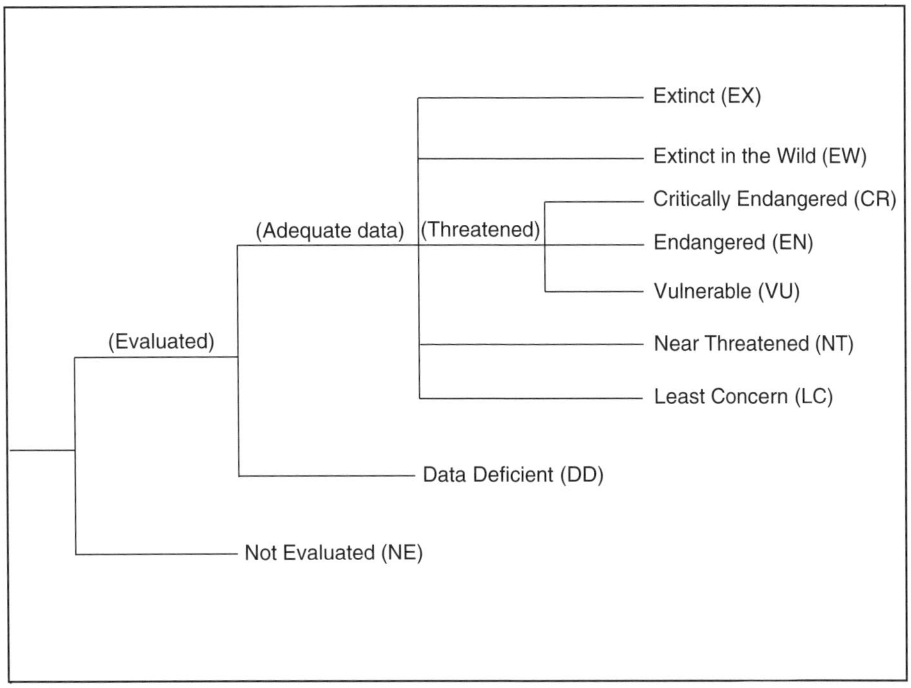

Figure 2 Structure of the IUCN *Red List* categories (IUCN (2001) *IUCN Red List Categories and Criteria: Version 3.1.* http://www.redlist.org).

Guatemala, urbanization has displaced some of the populations of teosinte (*Zea mexicana*), the closest relative of corn.

It has been estimated that the loss of rare wild plants represents a real and substantial economic loss; these endangered food crop relatives have a worth of about US$10 billion annually in wholesale farm values. Although genetic erosion cannot be quantified accurately, it seems likely that virtually all plant species are currently suffering loss of genetic variation to varying degrees. Therefore, assuming the consensus figure for loss of plant species by 2020 is correct, it can be further estimated that 25–35% of plant genetic diversity could possibly be lost by 2020.

What causes this loss of plant biodiversity? It is important to appreciate that both species extinction and genetic erosion can be natural events, just as specific and genetic evolution are natural; nature is, and it seems has always been, dynamic in this respect. However, the contemporary situation concerning species extinction and genetic erosion is quite different from that which existed in the past. Humankind now has the ability to alter drastically the world environment in ways not previously possible and it is these anthropogenic changes that have exponentially increased the speed of specific and genetic erosion and extinction. Therefore, the causes of loss of plant biodiversity are almost invariably anthropogenic in origin and may be broadly grouped under the general headings of:

- destruction, degradation, and fragmentation of natural habitats (e.g., road and reservoir building)
- overexploitation (e.g., medicinal plant extraction from the wild, fuelwood gathering, overgrazing)
- introduction of exotic species which compete with, prey on, or hybridize with native species (e.g., human introduction of alien species for exploitation that subsequently escape and become novel weeds or pests)
- human socioeconomic changes and upheaval (e.g., extinction of tribal cultures, urban sprawl, land clearances, food shortages)
- changes in agricultural practices and land use (e.g., displacement of landraces by modern varieties or shift to monoculture)
- calamities, both natural and manmade (e.g., floods, landslides, or wars).

So how can we avoid further deleterious loss of plant biodiversity? This diversity is vital for potential and actual human exploitation in the future. Therefore, we need to instigate active conservation programs and more sustainable exploitation policies to ensure future human development and well-being.

Plant Conservation

In order to develop practical techniques to achieve conservation and sustainable-use objectives, conservation managers must use their knowledge of genetics, ecology, geography, taxonomy, and many other disciplines to understand and manage the biodiversity they wish to conserve. Plant conservation is therefore a complex process involving the synthesis of a multitude of scientific disciplines and practical techniques.

An important element of conservation is the protection afforded by legal instruments, such as international treaties and national legislation. Such legislation can help provide the support required for practical conservation (e.g., the protection of a species from illegal collection from a nature reserve), or may even provide the justification for carrying out the conservation actions in the first place (e.g., species listed on the IUCN *Red List*).

Genetic and Ecological Conservation

It is important to stress that conservation and sustainable use are not just about individual plant species, but include all aspects of biodiversity from ecosystems (communities of organisms and their abiotic environment), through communities (collections of species found in a common environment or habitat), species and populations to genetic diversity within species. In recent years, a distinction has been made between conservation at ecosystem and genetic levels, and these may be referred to as ecological and genetic conservation respectively:

1. Ecological conservation focuses on the conservation and sustainable use of whole communities; individual survival and extinction are a major concern, but are seen in the context of overall community health. The UNESCO Man and the Biosphere Program (MAB), which established a network of biosphere reserves throughout the world, representing distinct biomes and ecosystems, exemplifies ecological conservation. The clear emphasis is on conservation of ecosystems: individual species should be conserved as components of ecosystems, rather than on a species-by-species basis.
2. Genetic conservation focuses more explicitly on particular taxa (usually species) and attempts to conserve the full range of genetic (allelic) variation within those taxa. The aim of genetic conservation is often linked intimately to human utilization. The raw materials of genetic conservation are genes within gene pools, the total diversity of genetic material of the particular taxon being conserved. The product of the gene pool is either preserved or utilized genetic diversity. The process linking the raw genetic matter and the preserved or utilized gene pool is conservation.

The distinction between the two basic forms of conservation is in practice blurred and may be viewed as artificial and of limited semantic importance, because the conservation of ecosystems and species are intimately linked. However, in practice conservation managers need to focus on something tangible that can be monitored and managed; therefore, the division into ecological and genetic conservation has practical implications.

Having made this point, it is important to stress that in order to undertake conservation, species interactions must be understood as far as possible. Even if the conservation target is a population of a species, no populations survive in isolation. They exist within a community or ecosystem, interacting with other species and the abiotic environment. Examples of interactions include pollinators, seed dispersers, microbial symbionts (i.e., organisms dependent on other organisms for their survival), herbivores (whether natural or introduced by humans), and pathogens. Thus, even when applying genetic *in situ* conservation, the maintenance of genetic diversity has to be considered within the context of whole ecosystems.

Genetic Conservation Strategies and Techniques

Plant genetic resources have been defined as the genetic material of plants, which is of value as a resource for present and future generations of people. Traditionally, this definition focused on crop plants and their wild relatives, but it is increasingly considered that all plant species are a potential resource for humanity. The ultimate goal of genetic resources conservation is to ensure that the maximum possible genetic diversity of a taxon is maintained and available for utilization. Plant genetic resource conservation acts as a link between the genetic diversity of a plant and its utilization or exploitation by humans (**Figure 3**).

Selection of Target Taxa

Conservation activities are always limited by the financial, temporal, and technical resources available; conservation has a cost and the effort expended is directly related to how much society values a species and is therefore willing to pay. It is impossible actively to conserve or monitor all species, so it is important to make the most efficient and effective selection of species upon which to focus conservation efforts. This choice should be objective, based on

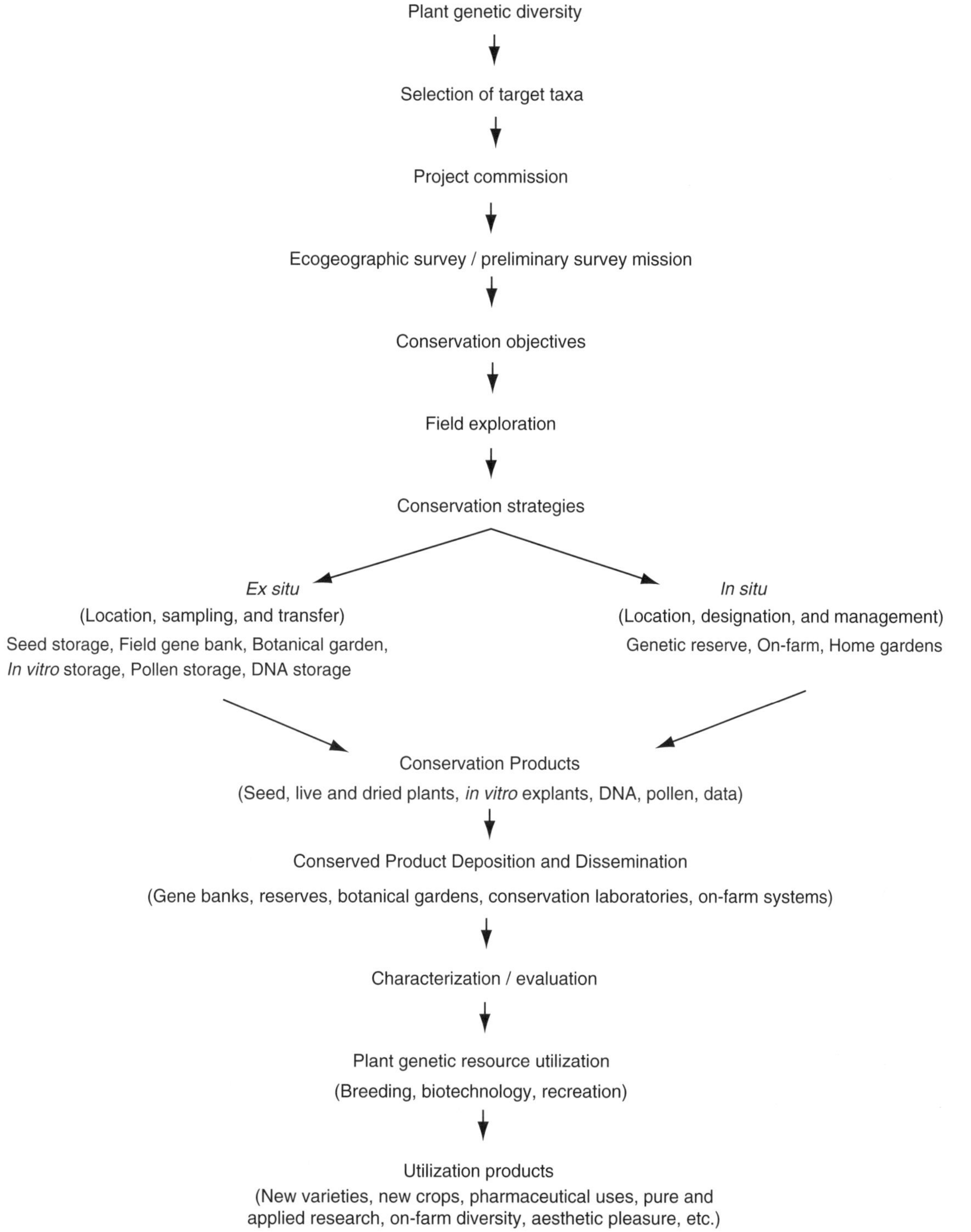

Figure 3 A model of plant genetic conservation. Reproduced from Maxted N, Ford-Lloyd BV, and Hawkes JG (1997) *Plant Genetic Conservation: The in Situ Approach* with kind permission of Kluwer Academic Publishers.

logical, scientific, and economic principles related to the perceived value of the species. In terms of conservation planning, the factors that provide a species with "value" include:

- current conservation status
- potential economic use
- threat of genetic erosion
- genetic distinction
- ecogeographic distinction
- national or conservation agency priorities
- biologically important species
- culturally important species
- relative cost of conservation

- conservation sustainability
- ethical and aesthetic considerations.

Rarely will one of the above factors on its own lead to a taxon being given conservation priority. More commonly, all or a range of these factors will be assessed for a particular taxon and then it will be assigned a certain level of national, regional, or world conservation priority. If the overall score passes a threshold level or is higher than competing taxa, then proposals will be made to conserve it; it will then be either conserved *in situ* in a genetic reserve or on-farm, or collected for *ex situ* conservation.

Having listed the factors that affect the selection of target taxa in terms of value to society, those related to potential economic use are commonly given higher comparative value, especially in poorer economies where income generation is of highest priority.

This anthropocentric and utilitarian view of the selection of conservation priorities may offend some conservationists. However, when financial resources for conservation are, and are likely to remain, limited and when people continue to suffer from malnutrition in many parts of the world, there appears no practical or ethical alternative to giving those species of most direct use to people the highest value and thus the highest conservation priority.

Project Commission

In practice, once taxa are selected for conservation, some form of commission statement necessarily precedes the actual conservation activities. The commission statement establishes the objectives of the conservation proposal, and specifies: the target taxa and target areas, how the material is to be utilized and where the conserved material is to be duplicated. It also gives an indication of which conservation techniques are to be employed. It is worthwhile taking the time to formulate a clear, concise commission statement, as it will help to focus subsequent conservation activities.

The commission may vary in taxonomic and geographic coverage, from a systematic collection program for a single species throughout its geographic range to a range of target taxa from a restricted location, e.g., onion (*Allium*) species of Central Asia, large cat species worldwide, or chickpeas (*Cicer*) from the western Tian Shan. In each case, however, a particular group of taxa from a defined geographical area must be considered currently insufficiently conserved (either *in situ* or *ex situ*), of sufficient actual or potential use, and/or endangered, to warrant active conservation.

Ecogeographic Survey and Preliminary Survey Mission

Once the target taxon or group of taxa has been selected and delimited, the conservationist begins to amass and synthesize fundamental biological data to help formulate an appropriate conservation strategy. The synthesis and analysis of these data enables the conservationist to make vital decisions concerning: which taxa to be included in the target group, where to find these taxa, which combination of *in situ* and *ex situ* conservation to use, what sampling strategy to adopt, where to site the reserve or store the germplasm, what collecting route should be followed, and when the collecting window (period when seed is ripe and suitable for collection) occurs.

For example, consider the close relative of the lentil, *Lens orientalis*. If the basic biological data for this species indicates that it has previously been found on stony slopes at the edge of the Fergana valley in Uzbekistan, then further material of this species is likely to be found under similar conditions, and is less likely to be found in different habitat types or far-distant regions.

The process of collating and analyzing geographical, ecological, taxonomic, and genetic data for use in designing conservation strategies is referred to as 'ecogeography', which can be defined as:

> 'An ecological, geographical, taxonomic, and genetic information-gathering and synthesis process for a particular taxon. The results are predictive and can be used to assist in the formulation of collection and conservation priorities.'

Ecogeographic studies involve the use of large and complex data sets obtained from the literature and from the compilation of passport data associated with herbarium specimens and germplasm accessions. The data compiled are essentially of four kinds: ecological, geographic, taxonomic, and genetic. These data are synthesized to produce three basic products: the database, which contains the raw data for each taxon, the conspectus, which summarizes the data for each taxon, and the report, which discusses the contents of the database and conspectus, as well as proposing appropriate conservation strategies.

Ecogeographic techniques enhance the efficiency of genetic conservation because they enable the conservationist to identify clearly the geographical regions and ecological niches that the taxon inhabits; therefore, not only can areas with high numbers of target taxa be identified, but also, areas that contain high taxonomic or genotypic diversity, uniqueness of habitat, and economic or breeding importance.

If the available ecogeographic data for the target taxon are limited, there will not be sufficient background biological evidence to formulate an effective conservation strategy. In this case, it would be necessary to undertake an initial survey mission to gather the novel ecogeographic data required on which to base the conservation strategy.

The survey mission may be in the form of "coarse grid sampling," which involves traveling throughout a likely target region and sampling sites at relatively wide intervals over the whole region. The precise size of the interval between sites depends on the level of environmental diversity across the region, but it may involve sampling every 1–50 km. The population samples and data collected during this mission can then be used to formulate further conservation priorities and develop an appropriate strategy, thus providing the same result as the ecogeographic survey for groups that are better biologically understood.

Conservation Objectives

The products of the ecogeographic survey or survey mission provide a basis to formulate future conservation priorities and strategies for the target taxon. Within the target area, zones of particular interest may be identified; for instance, areas with high concentrations of diverse taxa, very low or very high rainfall, and high frequency of saline soils, or extremes of altitude or exposure.

In general, areas with very distinctive characteristics are likely to contain plants with distinct genes or genotypes. If a taxon is found throughout a particular region then the conservation manager can use the ecogeographic data to select positively a series of diverse habitats to designate as reserves. If a taxon has been found at one location but not at another with similar ecogeographic conditions, then these similar locations should be searched.

Within the target taxon, specific variants may be identified which warrant conservation priority, for example: species that have previously unrecognized utilization potential, populations that are particularly in danger of genetic erosion, or those that had not previously been conserved.

If clear objectives have not already been established in the project commission, the conservationist should undertake to set out a clear, concise statement of the proposed conservation strategy for the target taxon and, if appropriate, prioritize actions. The statement should answer such questions as:

1. Which populations require conservation?
2. Can local farmers play a part in conservation activities?
3. Do population levels require close monitoring?
4. Should a national or international collecting team be directed to collect the priority target taxa?
5. What conservation strategy is appropriate?
6. What combination of conservation techniques is appropriate?
7. Is a more detailed study required before any of these questions can be answered?

Field Exploration

Once the conservation objectives have been clarified, whichever conservation strategy is to be applied, the ecogeographic information is used to locate and identify the general locality of the plant populations that are to be conserved; however, the ecogeographic data will rarely be sufficiently comprehensive to locate precisely actual populations. Therefore, the preparatory element of conservation activities will be followed by field exploration, during which actual populations are located.

Ideally, populations of the target taxon that contain the maximum amount of genetic diversity in the minimum number of populations will be identified, but how is this goal to be achieved? Commonly, there is too much diversity both in wild and domesticated species to conserve all their alleles; therefore, we must attempt to conserve the range of diversity that best reflects the total genetic diversity of the species. How many plants must be sampled, which specimens and what pattern of sampling is appropriate? Ideally, to answer these questions, the conservationist should know: the amount of genetic variation within and between populations, local population structure, breeding system, taxonomy, and ecogeographic requirements of the target taxon, as well as many other biological details. Some of this information will be supplied following the ecogeographic survey, but some will remain unavailable. Therefore, the practice of field exploration will be modified depending on the biological information on the target taxon and target area that is available.

Populations should be selected for conservation if they are found on the periphery of the target taxon's distribution or those containing morphological or ecological variants. Atypical populations or those existing under atypical conditions may possess genes or alleles that are unknown or extremely rare in the target taxon's center of diversity, and this material possibly contains genetic variation of special use to breeders, e.g., pest or disease resistance, or adaptation to climate.

In Situ Conservation Techniques

In situ conservation involves the maintenance of genetic variation at the location where it is encountered, either in the wild or in traditional farming systems. The three main *in situ* techniques may be broadly defined as: genetic reserve, on-farm, and home garden conservation.

Genetic reserve conservation of wild species involves the location, designation, management, and monitoring of genetic diversity in a particular, natural location. This technique is the most appropriate for the bulk of wild species, because it can, when the management regime is minimal, be relatively inexpensive. When dealing with plants, the object is for the reserve to contain the minimum number of individuals that can maintain genetic diversity within the species; too few and genetic diversity will decline over time, too many and resources may be wasted in managing the large population.

The genetic variation present in wild populations of plants is a requisite for evolutionary adaptation to a changing environment and hence the survival of species. Genetic variation is therefore essential for the maintenance of viable populations. In order to establish a genetic reserve for a target taxon, or a group of target taxa, the effective population size must be estimated.

Genetic reserves are appropriate for orthodox-seeded (i.e., species with seeds that can be maintained in long-term storage by a combination of reduction in moisture content and storage at low temperature) and nonorthodox-seeded (i.e., species with seeds that cannot be stored long-term) plant species. Genetic reserves also permit multiple taxon conservation in a single reserve and allow continued evolution of the species. However, the disadvantages are that the conserved material is not immediately available for human exploitation, and if the management regime is minimal, little characterization or evaluation data (i.e., information about a plant's habit and traits that are useful for breeding) may be available. In the latter case, the reserve manager may even be unaware of the complete specific composition of the reserve he or she is managing.

The process of domestication (the selection and adaptation of wild plants for use by humans), has taken place over thousands of years, and has led to the existence of an enormous number of different cultivars (plant domesticates). On traditional farms, the plants that are sown and harvested are generally known as 'landraces'. Each season the farmer keeps a proportion of harvested seed for resowing, and seed may be exchanged locally between villages. Thus, the landrace is highly adapted to the local environment and is likely to contain locally adapted alleles that may prove useful for specific breeding programs.

Modern-day breeding methods have largely taken the place of traditional farming and selection methods, which has led to the loss of many old landraces. There is a need to conserve the diversity found both in the wild relatives and in the old varieties of domesticated species to cater for changing environmental conditions and human needs. On-farm conservation involves the maintenance of traditional crop or cultivation systems, by farmers within traditional agricultural systems. The advantage of on-farm conservation is that it ensures the maintenance of ancient landraces and those wild species dependent on traditional agriculture. However, the landraces may yield less than their modern counterparts, and may be less popular with farmers. Therefore, some form of incentive, even subsidies, linked to regular monitoring may be required to ensure sustainability.

Home garden conservation is a variant of on-farm conservation, but on a smaller householder scale. It involves the cultivation of more species-diverse material in home, kitchen, backyard, or dooryard gardens and focuses on the cultivation of food crops, medicinal plants, herbs, and spices grown primarily for home consumption. Although individual population sizes are likely to be small, sufficient populations for genetic security can be maintained by agglomerating neighboring households. Home gardens are also often a source of traditional landraces lost from larger farming systems.

It should be noted that contemporary economic forces will tend to act against the continued maintenance of ancient landraces and they are undoubtedly currently suffering rapid genetic erosion, if not facing extinction. A backup system of *ex situ* conservation is therefore essential, as discussed below.

Ex Situ Conservation Techniques

Ex situ conservation involves the maintenance of genetic variation away from its original location. Samples of populations of species, subspecies, cultivars, breeds, or varieties are taken and conserved either as living collections of plants in field gene banks, botanic gardens, and arboreta, or as samples of seed, tubers, tissue explants, pollen, or DNA maintained under special artificial conditions.

In an ideal world it would be preferable to conserve all biodiversity *in situ* in "nature," rather than move it into an "artificial environment" to be conserved. However, due to the threat of genetic erosion in the original location and the need for easy access for exploitation, *ex situ* conservation techniques are necessary for many plant species. These techniques also provide a safety backup for *in situ* conservation techniques, where *in situ* conservation in a genetic reserve or on-farm alone cannot guarantee long-term security for a particular domesticated or wild species. One significant advantage of *ex situ* conservation is that the genetic material is always available to the breeder for evaluation. On the other hand, *in situ* conserved material may require time-consuming and costly expeditions to obtain samples of the material before the evaluation can take place. Thus, material conserved *ex situ* in a gene bank, once evaluated, will always be immediately available.

Ex situ collection and storage of seeds or embryos is the most convenient and widely used method of genetic conservation. They are the natural dispersal, storage, or generative organs for the majority of species. This technique involves samples being collected from individuals or populations and then transferred to a gene bank for storage under controlled conditions, usually at subzero temperatures.

In plants, seed and embryo storage is adopted for the bulk of orthodox-seeded species. After being dried to an appropriate moisture content (commonly 5–6%), the seed is frozen at $-20°C$. However, this method is only suitable for species that can be dried and stored at low temperature without losing viability. The advantages of this technique are that it is efficient and reproducible, and feasible for short-, medium-, and long-term secure storage. However, the disadvantages are that there are problems in storing recalcitrant-seeded species. The latter species cannot be dried and cooled in the way used for orthodox seeds; they often produce little seed and normally reproduce vegetatively (clonally).

In vitro conservation involves the maintenance of explants in a sterile, pathogen-free environment and is widely used for vegetatively propagated and recalcitrant-seeded species. *In vitro* conservation offers an alternative to field gene banks. It involves the establishment of tissue cultures of accessions on nutrient agar and their storage under controlled conditions of either slow or suspended growth. The main advantage is that it offers a solution to the long-term conservation problems of recalcitrant, sterile, or clonally propagated species. However, there are disadvantages: the risk of somaclonal variation (the genetic variation that arises during the period of dedifferentiated cell proliferation that takes place between the culture of an explant and production of regenerants), the need to develop individual maintenance protocols for the majority of species, and the relatively high-level technology and cost required.

The best answer for cheap, long-term *in vitro* conservation in the future may be cryopreservation, that is, the storage of frozen tissue cultures at very low temperatures, for example in liquid nitrogen at $-196°C$. If this technique can be perfected to reduce the damage caused by freezing and thawing, it may be possible to preserve materials indefinitely.

Historically, botanical gardens were often associated with physic gardens or displays of single specimens of botanical curiosities and as such, did not attempt to reflect the genetic diversity of the species. These gardens hold living collections of species that were collected in a particular location and moved into *ex situ* collections to be conserved. The advantage of this method of conservation is that gardens do not have the same constraints as many other conservation agencies; they have the freedom to focus on wild species that may otherwise not be given sufficient priority for conservation.

However, there are major disadvantages to this conservation technique. The number of species that can be genetically conserved in a botanical garden is limited because of the available space; the majority of these gardens are located in urban areas in temperate countries and site expansion would be prohibitively expensive. The greatest part of botanical diversity is located in tropical climates, yet the majority of botanical gardens are located in temperate countries; there is therefore a need to keep the species in expensive glasshouses or compounds, which also limit the space available. Also only very few individuals can be held of each species, thus neglecting the need to conserve the range of genetic diversity found in the wild; however, if the target species is very near extinction and only one or two specimens remain extant, this objection of course does not hold.

The conservation of germplasm in field gene banks involves the collection of plant specimens from one location and the transfer and conservation at a second site. Field gene banks have traditionally provided the answer for recalcitrant or sterile-seeded plant species, or for those species where it is preferable to store clonal material; they are commonly used for such species as cocoa (*Theobroma cacao*), rubber (*Hevea brasiliensis*), coconut (*Cocos nucifera*), mango (*Mangifera indica*), coffee (*Coffea arabica*), banana (*Musa sapientum*), cassava (*Manihot esculenta*), sweet potato (*Ipomaea batatas*), and yam (*Dioscorea* spp.). The advantages of field gene banks are that the species are easily accessible for utilization, and that evaluation can be undertaken while the material is being conserved. The disadvantages are that the material is restricted in terms of genetic diversity, is susceptible to pests, disease, and vandalism, and may involve large areas of land.

The storage of pollen grains is possible in appropriate conditions, and allows their subsequent use for crossing with living plant material. In the future, it may also be possible to regenerate haploid plants routinely from pollen cultures, but no generalized schedules have yet been developed. The disadvantage with pollen storage is that only paternal material is conserved.

The storage of DNA in appropriate conditions can be easily and inexpensively achieved given the appropriate level of technology; however, the regeneration of entire organisms from DNA cannot be envisaged at present, although single, or small numbers of genes could subsequently be utilized. The advantage of this technique is that it is low cost and simple, but the disadvantage is that there are problems with subsequent gene isolation, cloning, and transfer.

In practice, there may be cases where conservation cannot be accurately termed *in situ* or *ex situ*. Take, for example, the conservation of the legume tree genus *Leucaena*. In this case, the germplasm is often collected from native habitats and taken *ex situ* to more easily managed agroforestry areas for conservation by local communities. The trees are not conserved using standard applications of field gene bank or arboreta techniques, but within these recipient local communities the transplanted trees are to all intents and purposes managed by local people using traditional silvicultural techniques within an *in situ* on-farm system. In this particular case and cases like it, the term *circa situ* has been used to describe this form of conservation. The material is effectively managed within traditional systems by local farmers and may be considered directly comparable to the routine application of on-farm conservation of crops.

Conservation Products and Dissemination

The products of conservation activities are: conserved germplasm (seed, embryos), live plants, dried plants (herbarium specimens), cultures, and conservation data. As discussed above, *ex situ* conserved orthodox seed is commonly held in gene banks at subzero temperatures and low moisture content to prolong their life. Live plants are conserved in genetic reserves, field gene banks, botanical gardens, and research laboratories. Germplasm that is stored in a suspended form such as tissue, pollen, or DNA is stored as cultures in specialist laboratory facilities.

Dried plant voucher specimens are held in herbaria; they are linked to specific samples of germplasm and are as much as possible representative of the conserved populations. Conserved material is ideally associated with a range of passport data, which details the taxonomic, geographical, and ecological provenance of the material. All passport data associated with conserved material should be entered into a database and made available for the management of the material, the formulation of future conservation priorities and strategies, and any exploitation.

Whether the conservation products are conserved *in situ* or *ex situ*, they should ideally be duplicated in more than one location to ensure their safety. The distribution of duplicate sets of material avoids accidental loss of the material due to fire, economic or political difficulties, warfare or other unforeseen circumstances. Duplication of the passport data is relatively easy from the conservation database, and the commissioning agency, relevant host country institutes, and other interested parties should hold copies. The various conservation products, where they are stored, and where they should be duplicated are summarized in **Table 5**.

It should however be remembered that duplication of germplasm can have inherent problems, such as the casual infection of native diversity with alien pests, viruses, bacteria, fungi, or phytoplasma. Hence the need for special phytosanitary care (e.g., tests, quarantine, certification) when sending or receiving germplasm. Similar problems can result from the deliberate transfer of alien species around the world. Species transferred from one region to another for agricultural or forestry exploitation may result in escapes (either by cross-pollination with local relatives or by wind-, bird- or insect-borne propagules). The resulting invasive populations may outcompete or suppress local (and even worse, endemic) populations or become a noxious weed in farmers' fields.

Conservation Product Utilization

As already stressed, there is an intimate link between conservation and utilization. The products of conservation, whether they are "living" or "suspended," should be made available for utilization by humankind. Conservation can be seen as the safe-keeping of preserved material, so that the material is available for utilization at a future date. In certain cases, the material can be used directly, for instance, in the selection of forage accessions, where little breeding is undertaken. The conserved material may also be used in reintroduction programs where the primitive breed or landrace has been lost locally due to civil unrest or the application of perverse government incentives that encourage the alteration of traditional practices.

Commonly, the first stage of utilization involves recording genetically controlled characteristics (characterization). Genetic material may also be tested in diverse environmental conditions to evaluate and

Table 5 The storage and duplication sites of conservation products

Conservation product	*Storage site*	*Duplication site*
Plant germplasm (seed, vegetative organs, etc.)	Gene bank	National, regional, and international gene banks, duplication with other conservation techniques
Live plants	Field gene bank, botanical garden, genetic reserve, on-farm	Duplication with other conservation techniques, e.g., gene bank storage of seed
Dried plants	Herbarium or museum	National, regional, and international herbaria or museums
Explants or plantlets	Tissue culture	Duplication with other conservation techniques, e.g., gene bank storage of seed
DNA and pollen	Various cultures	Duplication with other conservation techniques, e.g., gene bank storage of seed
Conservation data	Conservation database	Duplication with other national, regional, and international conservation agencies

screen for tolerance to harsh conditions, or be deliberately infected with diseases or pests to screen for particular biotic resistance (evaluation). The biotechnologist screens for single genes, which once located may be transferred into a host organism, while the biochemist (bioprospector) screens for particular chemical products that may be of use to the pharmaceutical industry. The products of utilization are therefore numerous, including new varieties, new crops, improved breeds and pharmaceuticals, as well as a "good" environment for recreational activities.

Ecosystem Conservation

Much of the previous discussion is focused on conservation of species and genetic diversity within species, but just as genetic diversity can exist within species, species can only exist within ecosystems. No species exists in isolation, therefore if an attempt is being made to conserve a species *in situ* the community, habitat, or ecosystem in which it is found must also be conserved.

An ecosystem can be defined (e.g., by the Convention on Biological Diversity) as: "a dynamic complex of plant, animal and micro-organism communities and their non-living environment interacting as a functional unit." Significantly, the ecosystem approach has been adopted as a framework for the analysis and implementation of the objectives of the Convention, and in effect, it has become the fundamental paradigm for the Convention's activities.

The Convention on Biological Diversity (see below) states that "the ecosystem approach is a strategy for the integrated management of land, water and living resources that promotes conservation and sustainable use in an equitable way," and that it is

> based on the application of appropriate scientific methodologies focused on levels of biological organization, which encompass the essential structure, processes, functions, and interactions among organisms and their environment. It recognizes that humans, with their cultural diversity, are an integral component of many ecosystems.

Furthermore, "the ecosystem approach requires adaptive management to deal with the complex and dynamic nature of ecosystems and the absence of complete knowledge or understanding of their functioning." Therefore, we need to understand not only the biological components of diversity, but also the structure, processes, functions and interactions that comprise ecosystems. This involves the measurement of the spatial and temporal scales of biodiversity.

Just as species diversity at population and genetic levels can be quantified and monitored, so can ecosystems. But measuring and monitoring diversity at the ecosystem level presents a unique challenge. A key to understanding ecosystems is often through the study of keystone and indicator species. A keystone species is one that plays a pivotal role in ecosystems, e.g., species of palm or fig trees that maintain all the vertebrate frugivores in a Peruvian rainforest for 3 months of the year when other trees are not in fruit. An indicator species is one that flags changes in biotic or abiotic conditions. It is important to monitor these species, as they are important indicators of change. Species of amphibians, mollusks, birds, chironomid flies, fungi, lichens, corals, and other marine invertebrates may be useful indicators that reflect the quality and changes in environmental conditions and aspects of community composition.

Methods for large-scale measurement and monitoring of ecosystems revolve around the use of remote sensing techniques. The technique may be based on low-altitude aerial photography or high-altitude satellite imagery, the former providing high resolution in relatively small areas, and the latter providing low resolution over relatively large areas. The use of special films and filters allow different portions of the electromagnetic spectrum to be emphasized, permitting a focus on particular types of information. For instance, infrared film can show spatial variation in surface temperatures, and vegetation cover can be distinguished from rock, soil, or water.

When viewed from above, remotely sensed images are like maps spread over a table, with most elements of the landscape clearly drawn. Computer-based geographic information systems (GIS) are used to analyze spatially related data sets, such as topography, vegetation cover, rainfall, and temperature. Remote sensing can be particularly useful for detecting major changes in land use, such as deforestation and conversion of land to agricultural use; detecting large variations in species composition, which can be correlated with environmental and anthropogenic changes; monitoring inaccessible, large, and highly species-rich areas, such as tropical forest habitats in the Amazon basin, which present special challenges for traditional survey methods; and monitoring climate change (for instance, the encroachment of desertification may be assessed by studying rainfall maps over a prolonged period; understanding vegetation seasonality).

The ecosystem approach to conservation is exemplified by the Man and the Biosphere Program, which established a network of biosphere reserves throughout the world, representing distinct biomes and ecosystems. Biosphere reserves are areas of terrestrial and coastal/marine ecosystems, where, through appropriate zoning patterns and management mechanisms, the conservation of ecosystems and their biodiversity is combined with the sustainable use of natural resources for the benefit of local communities, including relevant research, monitoring, education, and training activities.

Complementary Conservation and Sustainability

In situ and *ex situ* conservation strategies should not be seen as alternatives or in opposition to one another, but rather as being complementary, as is stated in Article 9 of the CBD. Each conservation strategy or technique acts as a backup to another, the degree of emphasis placed on each depending on the conservation aims, the type of species being conserved, the resources available, and whether the species has utilization potential.

The efficacy of adopting a 'holistic' or integrated approach to conservation is now well established as the only sustainable option. Therefore, when formulating an overall genetic conservation strategy for a species, a combination of different techniques should be applied. The techniques may be thought of as slotting together like pieces in a jigsaw puzzle, to complete the overall conservation picture and thus ensure the maintenance of plant genetic diversity.

The adoption of an integrated approach requires the conservationist to address not only biological questions but also the practical and political ones; for instance:

1. What are the species storage characteristics?
2. What do we know about the species breeding in captivity?
3. Do we want to store the germplasm in the short, medium, or long term?
4. How important is the species?
5. Where is the species located, and how accessible is it/does it need to be?
6. Are there legal issues relating to access?
7. How good is the infrastructure of the established reserves?
8. What backup is necessary/desirable?
9. How does the species conservation strategy fit within the local community development program?

Given answers to these questions, the appropriate combination of techniques to conserve the gene pool can be applied in a pragmatic and balanced manner. The precise combination of techniques can be formulated for each species or group of species, demonstrating the flexibility of the integrated approach.

Sustainability in the sense of continuance is a fundamental concept for conservation. Whether seed or embryos are collected for *ex situ* conservation in a gene bank or a habitat is designated as a reserve, each option has a financial cost and it would be wasteful of limited conservation funds not to ensure the conservation project is sustainable at least in the medium term. If the material is lost from a reserve the resources expended on establishing the reserve would have been wasted and the cost of rehabilitating populations using materials stored *ex situ* would have to be considered. The latter option is commonly expensive and may require extensive research to ensure the reintroduced materials do not likewise go extinct. Unfortunately, many conservation projects are funded on a short-term basis, so it is essential that an effective project exit strategy be developed so the conservation program itself is sustainable.

Not only is it necessary to integrate the different conservation strategies and techniques, and local communities in a sustainable conservation project, it is also important to integrate the different potential agencies involved. This is particularly true for large *in situ* reserve projects where the project may naturally span national borders and professional disciplines.

Therefore, the project team must ensure that the local, provincial, national, regional, and international conservation agencies, as well as those from the different disciplines involved, such as environmentalists, foresters, agriculturalists, politicians, etc. work together to promote the success of the project.

Conservation Legislation

As already mentioned, legislation is an important component of biodiversity conservation. Conservation legislation governs our rights as individuals and states over the management of biological diversity. Without such legislation, there would be little protection for species and their habitats. There is a vast array of legislation relating to biodiversity, which exists at many levels, including customary laws, national policies, regional treaties, and international conventions.

The most fundamental principle in conservation legislation is that of national sovereignty over natural resources. This is a basic rule in international relations and means that, as sovereign entities, states may exercise their rights over all natural resources on their territory, subject to international law. In reality, this means they may exploit, destroy, or conserve their natural resources as they please. This principle is reflected in many international legal instruments relating to the conservation of species, including the CBD.

In many countries, customary and traditional legal measures for the preservation of biological diversity were in place long before national governments began to pass laws concerning the conservation and use of biodiversity. While many traditional cultures have disappeared or merged with dominant societies and economies, the survival of many communities still depends upon effective systems of managing biological resources. Such cultures can provide valuable information for the management of biodiversity and subsequently for the development of modern law.

Concerns about the loss of biodiversity and recognition of the need to conserve species and habitats have steadily grown as people have come to witness and understand the multitude of threats to natural resources, the rapidity of the loss in many cases, and the enormity of the problem. For whichever reason, widespread concern about environmental degradation has led to a global consensus to establish rules of international law to achieve a better balance between sovereign rights over natural resources and the need to preserve biological diversity on a global scale. Throughout the 1970s and 1980s, legislation to conserve biodiversity developed through a series of stages, one of the most significant of these being the Stockholm Declaration of 1972, which sets out 26 Principles placing emphasis on the need to protect both species and their habitats. A host of international environmental legal instruments have developed as a result of the Stockholm Declaration, but most of these could not have arisen without a large number of national laws already being in place. In turn, national legal instruments have built upon the objectives and minimum requirements of the international laws. A two-way dynamic interaction continues to take place in this way, leading to the strengthening of conservation legislation at both national and international level.

The culmination of efforts to develop legislation to protect biodiversity was the United Nations Conference on Environment and Development (UNCED or the Earth Summit) of 1992 in Rio de Janeiro, where the Rio Declaration on Environment and Development was adopted. This was a turning-point in international negotiations on issues of environment and development, and was the first time that governments worldwide treated biodiversity as a major concern. At the same time as the adoption of the Rio Declaration, UNCED adopted the Convention on Biological Diversity, to which there are now 187 parties.

The Convention on Biological Diversity is currently the most comprehensive global legal mechanism for biodiversity conservation. It provides an internationally recognized framework for the implementation of policy and programs related to biodiversity conservation and sustainable use. However, it goes beyond these goals, to encompass the issues of access to genetic resources and the sharing of benefits arising from the exploitation of those resources, as well as recognizing the need to provide the biologically rich South with the support it needs to implement its provisions. The Convention represents a milestone in biodiversity conservation thinking, reflecting international acknowledgement of the loss of our biological resources and the urgent need for action. It is now recognized as the primary guiding framework for biodiversity conservation and sustainable use.

Within the framework of the Convention on Biological Diversity, the Conference of the Parties, which meets every 3 years, has initiated several thematic work programs. As a result, a number of important declarations and additional treaties have been adopted. Most significantly for plant conservation is the Global Strategy for Plant Conservation, adopted in 2002. Other international treaties of major significance for plant conservation include the

Convention on Wetlands of International Importance especially as Waterfowl Habitat (Ramsar Convention), adopted in 1971, the Convention on International Trade in Endangered Species of Wild Fauna and Flora (CITES), adopted in 1975, and the International Treaty on Plant Genetic Resources for Food and Agriculture, adopted in 2001.

There are numerous other legal instruments of relevance to plant biodiversity conservation, both at national and regional levels. These include protected area legislation, endangered species acts, directives, such as the EU Habitats Directive, and other regional instruments such as the Agreement on the Conservation of Nature and Natural Resources (ASEAN Agreement), the Convention on Conservation of Nature in the South Pacific, and many more. Many countries have national legislation protecting specific areas and in some cases individual species. Species listed in the appendices of national, regional and international legal instruments are given high priority for conservation by the parties as they seek to implement their obligations. **Table 6** lists presents examples of legislation relevant to plant conservation.

Plant Utilization

How Plants Are Used

The ultimate reason for conserving plant biodiversity is to make it available for use by humankind, either now or in the future. Plants provide an extremely wide range of useful products, which are relied on by people in all regions of the world. A mixture of direct harvesting from the wild or cultivation, ranging from basic subsistence farming to sophisticated agricultural systems supplies food, medicines, and a wealth of raw materials. As well as the more obvious plant products such as food, medicines, and construction materials, plants provide a wide variety of resources used in industry and commerce. For instance, plant extracts are used in the manufacture of glues, soaps, cosmetics, dyes, plastics, lubricants, and polishes. In addition, wild plants are exploited for their aesthetic and medicinal qualities (e.g., wild orchids, sweet peas, evening primrose, and ginseng); the worldwide commercial trade in wild plants (with wild animals) has been valued at as much as US$5 billion per year. The main uses of plant genetic resources by humankind are as:

- food: crop species including beverages
- food additives: including processing agents and other additives
- animal foods: the fodder or forage species
- materials: such as wood, fibers, cork, cane, tannins, latex, resins, gums, waxes, oils, etc.
- poisons
- medicines: human and veterinary
- environmental: these will include species that are ornamentals, recreational, hedges, shade plants, windbreaks, soil improvers, plants for regeneration, erosion control, indicator species (e.g., pollution, underground water)
- gene donors: plants that contain desirable traits that can be transferred to other species to improve their use.

There are three ways in which we use these resources:

1. Consumptive use refers to any use of plant resources that literally "uses them up," e.g.,

Table 6 Examples of legislation relevant to plant conservation

Acronym	*Treaty*	*Date*	*Place*
—	Convention on Nature Protection and Wildlife Preservation in the Western Hemisphere	1940	Washington
IPPC	International Plant Protection Convention	1951	Rome
Ramsar	Convention on Wetlands of International Importance Especially as Waterfowl Habitat	1971	Ramsar
WHC	Convention concerning the Protection of the World Cultural and Natural Heritage	1972	Paris
CITES	Conventional on International Trade in Endangered Species of Wild Fauna and Flora	1973	Washington
Berne	Convention on the Conservation of European Wild Life and Natural Habitats (Europe)	1979	Berne
ITTA	International Tropical Timber Agreement	1983	Geneva
CBD	Convention on Biological Diversity	1992	Rio de Janeiro
—	Directive on the Conservation of Natural Habitats and of Wild Fauna and Flora (Europe)	1992	Brussels
ITPGRFA	International Treaty on Plant Genetic Resources for Food and Agriculture	2001	Rome
GSPC	Global Strategy for Plant Conservation	2002	The Hague

gathering plants from the wild for medicine or converting a forest to agricultural land.
2. Nonconsumptive use includes uses such as plant breeding, the use of sacred sites for cultural and religious practices, and recreational uses such as ecotourism.
3. Commercial consumptive use refers to any consumptive use of plant resources for economic gain, e.g., the international tropical timber trade, or the sale of locally harvested products in neighboring villages.

These are useful ways to categorize plant use but it should be noted that sometimes there is only a fine distinguishing line between them.

The beginnings of plant biodiversity use are as old as humankind; hunter–gatherers in preagricultural times exploited their environment not only for food, but also for fuel, medicines, building materials, tools, adornment, transport, and recreation. Before humankind learned to domesticate plants, and to sow and cultivate crops, it survived by the hunting and collection of resources. Early humans would have eaten berries, fruits, seeds, flowers, young succulent shoots, underground storage organs, and fleshy roots, all found within their local environment. These early forms of plant use were *in situ*, in the sense that plants were found and used within the hunter–gatherers' "home range". The origin of agriculture and the evolution of crop plants began between 5000 and 10 000 years ago, and it is now generally thought that agriculture originated more or less simultaneously in different parts of the world. These include the Fertile Crescent of the Near East (wheat, barley, pulses), the Huang He (Yellow River) region of China (millet, rice), Meso-America (maize, pulses, peppers, squashes) and the Central Andes of South America (potatoes and other tuberous crops).

From these early centers of agriculture, the spread of domesticated plants took place. Following conscious and unconscious selection pressures individual crops became increasingly diverse: in traditional agroecosystems newly domesticated plant types and primitive cultivars diverged from their wild ancestors, occasional crosses continued to occur between the early crops and their wild relatives and thus, useful traits were introduced to the crop, such as pest and disease resistance, or resistance to extreme conditions such as drought. With the beginning of human migration and trade, crop plants evolved in new environments and crossed with local wild relatives, leading to even greater genetic diversity. For instance, in West Africa, the indigenous cultivated rice (*Oryza glaberrima*) has hybridized with the introduced common Asian rice (*O. sativa*), enriching the rice gene pool in the region. These processes of natural selection in response to new ecological conditions, in conjunction with farmers selecting for particular characteristics over thousands of years of cultivation, have led to the vast array of crop diversity that we know today.

Today, however, relatively few botanical families account for the world's major domesticated plants; of the 511 plant families currently recognized, only 173 have domesticated representatives (**Table 7, Figure 4**), and of these, two families predominate: the Poaceae (wheat, barley, rice, maize, etc.) and the Fabaceae (faba bean, chick pea, lentil, etc.). Analysis of food energy supplies on a national basis shows that only 103 crop plants provide 90% of the per capita food plant supplies of all nation states.

Table 7 Cultivated species in the major botanical families

Plant family	*Number of species*
Poaceae	379
Fabaceae	337
Rosaceae	158
Solanaceae	115
Asteraceae	86
Cucurbitaceae	53
Lamiaceae	52
Rutaceae	44
Brassicaceae	43
Apiaceae	41
Chenopodiaceae	34
Zingiberaceae	31
Arecaceae	30

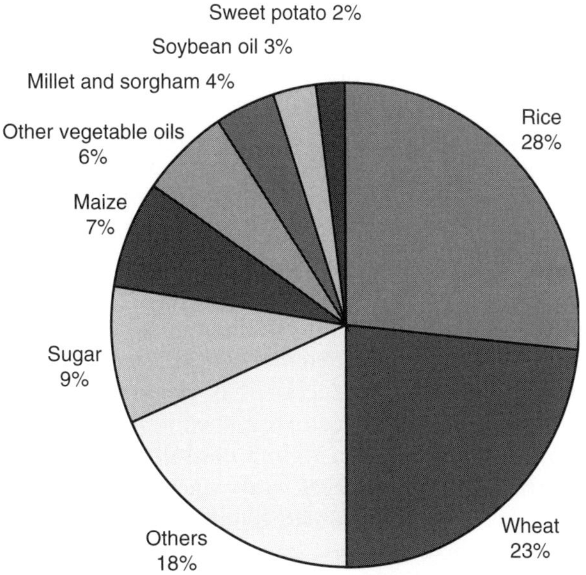

Figure 4 Most important crops for food energy supply. Reproduced with permission from FAO (1998) *The State of the World's Plant Genetic Resources for Food and Agriculture.* Rome: Food and Agriculture Organization of the United Nations.

However, while relatively few plants contribute to food production globally, at a local level plant resources provide a varied source of nutritional needs. For instance, in one region of Peru 193 fruits species are regularly consumed and of these, 120 species are exclusively wild-collected and a further 19 originate from both wild and cultivated species.

Wild species are also nutritionally important to many people, and form an integral part of the daily diets of many poor rural households. In many countries, people utilize wild plants during periods of famine, and especially during the hunger season, that precedes crop harvests. These include numerous root, tuber, leafy vegetable, and fruit-producing species. They are important because they are a source of important vitamins, minerals, and other nutrients, they complement the staple crops eaten by many of the most vulnerable people, including children and the elderly, and they represent ready sources of income for cash-poor households. For instance, in the United Republic of Tanzania in 1988, it was calculated that the value of all wild plant resources to rural communities, whether for subsistence consumption or sale, was more than US$120 million.

Plant Breeding

Farmers have carried out plant breeding for centuries. These years of breeding and selection have led to the production of many of the plant foods that we eat today, and have been fundamental to the development of human society. Modern methods in breeding and selection for new varieties have now largely taken over from traditional methods in most regions of the world, and inadvertently this has resulted in the loss of many old varieties or landraces. The production of new varieties can be crucial to farmers in both developing and developed countries. Extreme conditions such as drought, or susceptibility to pests and disease, can cause the loss of whole crops, economic ruin, and famine. New varieties can be bred to resist pests and disease, and harsh environmental conditions. Various studies, mostly conducted on cereals, have estimated that more than 50% of the increase in crop production has been due to the improvement of crop cultivars; such improvement is brought about by transferring desirable genes/traits to crops from landraces and other more distant germplasm sources. Maize hybrids in the tropical and subtropical areas have been improved by 25–40% during the past 10–15 years due to the introgression of a wider germplasm base.

Biotechnology has been defined as the application of biological organisms, systems and processes to the provision of goods and services. Biotechnology techniques can be divided into two categories: methods used before the discovery of the structure and function of DNA in the late 1950s and early 1960s, such as those used for millennia to make bread, cheese, beer, and wine; and the subsequently developed recombinant DNA technology, which modifies genes within cells and organisms by incorporating genetic material from other sources. This is also known as genetic modification or genetic engineering.

The biotechnology industries embrace many sectors, including pharmaceuticals, botanical medicines, agriculture, crop protection products, horticulture, cosmetics, perfumes, detergents, paper-making, and so on. Biotechnology companies apply enzymes and use biologically active compounds derived from genetic resources as an integral part of processes and products in almost every industry sector. The "access and benefit-sharing" provisions of the Convention on Biological Diversity are therefore particularly relevant to the biotechnology industries, but are extremely complex to implement.

Medicinal Plants and Pharmaceuticals

The use of plant biodiversity in medicine is significant. Today, approximately 80% of the world's population relies on traditional medicines derived from plants and animals as their primary source of medical treatment, while a large proportion of drugs in worldwide use are derived directly or indirectly from biological sources. Approximately 25% of prescriptions used in the United States contain products derived from plants while 119 pure chemical substances that have been extracted from 90 species of higher plants are used in medicines worldwide.

The early exploitation of crop species is paralleled by that of medicinal plants, when it was usually associated with older 'wiser' persons (wise women, witches, and later, herbalists). Historical records show that ancient Greeks, Persians, Chinese, Assyrians and Egyptians each used specific plants for their medicinal, benign, or malign, properties. Today, at a local level, an extremely wide range of plant species is used for medicinal purposes; the World Health Organization (WHO) has listed over 21000 plant names (including synonyms) that have reported medical uses around the world. China, for example, has a remarkably high number of genera and species that are used for medicinal purposes (**Table 8**).

There are many examples of plants yielding pharmaceutical compounds, such as: *Catharanthus roseus* of Madagascar, which was traditionally used

Table 8 Genera and species used in Chinese medicine

Plant family	*Total number of genera used*	*Percentage of genera used*	*Total number of species used*	*Percentage of species used*
Saxifragaceae	24	100	427	36
Asteraceae	227	68	2323	33
Ranunculaceae	41	83	737	57
Liliaceae	67	69	401	89
Rubiaceae	75	79	477	46

Data from Xiao PG and Peng Y (1998) Ethnopharmacology and research on medicinal plants in China. In: Prendesgast HDU, Etkin NL, Haris DR and Houghton PJ (eds) *Plants for Food and Medicine*, pp. 31–39. © The Board of Trustees of the Royal Botanic Gardens, Kew.

by various cultures for the treatment of diabetes, contains two important drugs that have been effective in the treatment of Hodgkin disease, childhood leukemia, and other blood cancers. The bark of the Pacific yew (*Taxus brevifolia*), used by indigenous peoples in North America for the treatment of some noncancerous conditions, has been found to contain high concentrations of paclitaxel (marketed as Taxol), a chemical that has proved very effective in preliminary tests for the treatment of ovarian and breast cancer. *Chondrodendron tomentosum* is used by indigenous peoples in South America in an arrow poison known as "curare"; the plant yields the compound D-tubocurarine, which is used as a model for a series of similar synthetic neuromuscular-blocking agents, some of which are now routinely used as anesthetics in surgical operations. *Rauwolfia serpentina*, which yields the antihypertensive alkaloid reserpine, known as the Indian snake root, has a long history of traditional medical use, including the treatment of mental disorder, snake bites, and as a tranquilizer. Reserpine revolutionized Western medical treatment of hypertension in the 1950s, and caused massive overharvesting of wild populations in India, leading to its inclusion in CITES Appendix II in 1990.

Medicinal plant species are still to a large extent harvested from the wild and relatively few are cultivated as crop plants. Plant breeding has only taken place with the commercially important plants such as *Papaver somniferum*, *Chamomilla recutita*, and *Mentha piperita*. Regions that are known to have important concentrations of major medicinal plants include Mexico and Central America, the west–central region of South America (Colombia, Ecuador, Peru), the Indian subcontinent, west Asia and parts of northeastern Africa. Overexploitation of medicinal plants extracted from the wild is leading to problems of genetic erosion in some of these regions. In India, for example, where traditional healers use 2500 plant species, species of *Aconitum*, *Dioscorea*, and *Ephedra* are some of the medicinal plants under threat in the wild. Some medicinal plants are subject to regulations such as CITES; however, as with any regulatory instrument, there are frequently problems in enforcing them efficiently and adequately, and the illegal trade in medicinal plants continues.

Plant-based pharmaceutical drugs were estimated to make up 25% of all pharmaceuticals sold and to generate sales of US$15.5 billion in the United States in 1990. This is a vast industry even when the cost of US$231 million to develop a marketable pharmaceutical drug is factored into the equation. Despite the historical and current prevalence of plants in the pharmaceutical industry, only between 5% and 15% of the approximately 300 000 described species of higher plants have been investigated for the presence of bioactive compounds. And of the 21 000 plant names that have reported medical uses around the world, about 5000 (predominantly temperate) species of higher plants have been properly investigated as potential sources of new drugs. However, of these only about 120 or so are regularly used around the world and these are derived from less than 100 species. There is thus a major effort by the pharmaceutical companies to identify new sources of medicinally useful compounds–"bioprospecting" as potential sources of new drugs.

Botanical medicines, as distinct from pharmaceuticals, are produced directly from whole plant material; they contain a large number of constituents and active ingredients working in conjunction with each other, rather than a single isolated active compound. The botanical medicine industry is experiencing rapid growth worldwide, with annual growth rates of between 10% and 20% in most countries. While botanical medicines are often scientifically poorly understood, they usually have long histories of traditional use. Traditional knowledge is widely used in the pharmaceutical industry as the basis for determining safety and efficacy, developing agronomic practices for cultivation of materials, and guiding the development if new products. Botanical medicines represent a range of product types; including raw herbs (dried or fresh), tinctures (infusions of herbs in alcohol), extracts (greater concentration of the original material from the plant with the aid of a solvent), and standardized extracts (plant material standardized to one or more chemical markers).

Other Uses of Plants

In addition to the use of plant biodiversity as the major source of food and medicine, there are numerous other uses for humankind. These include such diverse uses as construction and industrial materials, clothing, and tools, along with recreational and ethical usage.

A wide range of materials used for construction and industrial purposes are directly derived from plant biodiversity. These include wood, fibers, dyes, resins, gums, rubber, oil, and waxes. Wood in particular forms a vital commodity for many communities dependent on fuel for heating and cooking, as well as a highly valuable product in terms of lumber production. Wood is still predominantly harvested from the wild, and is one of the most significant commodities in international trade; in 1995 the annual value of fuelwood and wood-based products worldwide was estimated at US$418 billion, or nearly 2% of the world's gross domestic product. Wood provides the primary source of fuels in many developing nations, shelter in traditional home-building and sophisticated construction, and the basis for the international pulp and paper industry. In many tropical countries, timber products are being rapidly exported to fund industrialization and foreign debt repayments.

The bulk of wood in world trade comes from temperate sources, with the major exporters being the United States, the former USSR and Canada for logs and sawn wood, and the United States, the former USSR and Finland for plywood. The main tropical source countries are Malaysia, Papua New Guinea, and Gabon for logs, and Malaysia and Indonesia for sawn wood and plywood. Industrial timber plantations commonly consist of conifers, which lend themselves well to cultivation. Relatively few hardwoods have been cultivated as plantation timbers, and the majority of hardwoods in international trade are derived from natural forests, which are subject to varying degrees of management. Timber species-richness is particularly high in tropical regions. Ghana, for instance, has 674 tree species reaching timber size and about 60 of these have been traded on international markets.

After timber, rattans provide the second most important source of export earnings from tropical forests. Countries with major rattan industries include the Philippines, China, Indonesia, India, Sri Lanka, and Thailand. Local uses of rattans include the production of mats, baskets, fish traps, dyes, and medicines, while for the international market, rattans are mostly used in the production of cane furniture. The rattan industry relies almost entirely on wild stocks, with about 90% of the world's raw material supply extracted from the wild (mostly from Indonesia), and the remaining 10% from plantations in Central and South Kalimantan.

Plants also provide important utilitarian products for both domestic and commercial markets; they are used to make a variety of clothing and utensils. Another important use of plant biodiversity by humankind can be broadly classed as recreational use, and includes the exploitation of plants for ornamental purposes and the use of natural areas as parks and nature reserves. The cultivation of ornamental plants has a long history. For example, lilies have been cultivated in China for decorative purposes for around 2000 years. In Roman times, roses, lilies, violets, anemones, narcissi, and lavender were grown as garden plants in Europe. Today, the diversity of decorative plants established in cultivation around the world far surpasses the number of plants grown for food. The demand for good-quality ornamental plants has led to the existence of a lucrative market, and while the ornamental horticulture industry represents a modest market as compared to crop plants and vegetables, it is nonetheless significant in monetary terms. It has been estimated that the annual global market for ornamental horticultural products lies between US$16 billion and US$19 billion. The ornamental horticulture industry consists of five main areas: herbaceous plants, woody plants, cut flowers, foliage plants, and bulbs.

Orchids, bulbs, and cacti are amongst the most significant plant groups in the international horticultural trade. Although the majority of material traded is from artificially propagated stock, the demand for wild-collected plants still exists, and the effect of the industry on wild populations of orchids and cacti has led to the inclusion of the entire plant families on the CITES Appendices.

Many plants have social and cultural significance, and have influenced the philosophy, language, art, religion, and social structure of many societies. For example, symbolism: the cedar of Lebanon is a national symbol and appears on the Lebanese flag; the country has recognized this species as culturally important, and although the species distribution is now severely limited and populations depleted, the government is taking action to reintroduce the tree into the wild; religious ceremonies: flowers and plants with cultural significance are used in churches and religious ceremonies; mythology and art: in African cultures wild plants figure prominently in mythology and works of art such as carving and paintings and adornment, often for traditional dress to indicate status or hierarchy.

Perhaps the purest form of utilization of biodiversity by humankind is for its aesthetic values. This may be as simple as enjoying a walk in the country or nature reserve, weekend botanizing, ecotourism, or going on a trek. This basic requirement for human recreation is an important objective and attraction in establishing national parks and reserves, and in recent years, this concept has been exploited in the ecotourism industry. Humankind's need for recreational spaces can be a powerful tool in biodiversity conservation.

Sustainable Use

As already emphasized, plant conservation and use are inextricably linked; many believe our ultimate motive for conservation is to ensure humankind may continue to utilize the components of biodiversity, on which we rely. In fact, the survival of human life on earth depends on the continued use of wild fauna and flora, therefore non-use of plant diversity is not an option; however, we can no longer afford carelessly to eradicate species or genetic diversity through overexploitation.

The word "sustainability" has become increasingly widely used in recent years. The concept of sustainable utilization (or use) refers to the use of specific renewable natural resources; for instance, the sustainable use of timber. The sustainable use of natural resources is one of the three central themes of the Convention on Biological Diversity. Specifically, Article 10 of the Convention, "Sustainable Use of Components of Biological Diversity", states that: Each Contracting Party shall, as far as possible and as appropriate:

(a) Integrate consideration of the conservation and sustainable use of biological resources into national decision-making;
(b) Adopt measures relating to the use of biological resources to avoid or minimize adverse impacts on biological diversity;
(c) Protect and encourage customary use of biological resources in accordance with traditional cultural practices that are compatible with conservation or sustainable use requirements;
(d) Support local populations to develop and implement remedial action in degraded areas where biological diversity has been reduced; and
(e) Encourage cooperation between its governmental authorities and its private sector in developing methods for sustainable use of biological resources.

Sustainable use of biological resources is not a new concept. For centuries, many local communities have practiced customary use of biodiversity in accordance with traditional cultural systems. Many of these systems have involved sustainable harvest regimes in order to conserve the natural resources on which they rely for their survival. The mangrove swamps of the Sundarbans in Asia provide a good illustration; these form the single largest mass of contiguous mangrove forest in the world, and they have been under sustainable management for more than 120 years. They are important local sources of timber, fuelwood, poles, industrial raw materials, and a whole array of nontimber as well as nonforest products of great economic significance.

Based on the silvicultural requirement, local conditions, and the objective of fulfilling the demand for specific products, site and situation specific management regimes have been developed in these countries to guarantee the supply of desirable products, while ensuring the sustainability of the forests. These management practices have withstood the test of time and have proved that the management interventions have been appropriate because these forests still exist in good condition even though they have been providing a steady yield of products for several decades.

Another example of sustainable use of plant biodiversity is of the chaguar (*Bromelia hieronymi*), a plant traditionally used for food and fiber by the Wichí people, hunter–gatherers from the semiarid Chaco region in Argentina. It is also the basis of a range of handicrafts, the production of which provides stable employment for Wichí women. To harvest chaguar, Wichí women must travel many kilometers into the forest. A study analyzing the factors that affect chaguar availability and demand, and examining how interactions between social factors and the environment produce patterns of open access to the chaguar resource, has found that though there are no regulations governing chaguar use, social stigma appears to prevent misuse or waste. Within the particular context of the Wichí economy, open access does not lead to degradation of the resource.

Although many traditional cultural systems incorporate sustainable management practices, and much can be learned from these systems, it is important to note that there are also cases where use of resources by local communities is unsustainable. The reasons for this may be complex, and vary on a case-by-case basis. However, the broad underlying socioeconomic problems associated with unsustainable use of natural resources are often similar in many countries.

In many developing countries, these include reduction of tropical forests, marginalization of people due to war or political instability, lack of incentives for local people to manage state-owned land, and lack of resources and infrastructure within local government to monitor and control the use of resources.

Working towards sustainable use of wild resources is therefore a complex process involving many political, social, economic, and biological factors, which vary according to region and culture. A particular area of emphasis today is on initiation of community conservation projects, where in simplistic terms local people become the owners and stewards of the resources, and therefore reap the benefits of utilization. Such 'bottom–up' initiatives have resulted from the recognition that sustainable utilization can only be achieved with local community support, and through collaboration between farmers, local interest groups, nongovernmental organizations (NGOs), and government.

Another approach to increasing sustainability is certification and labeling systems. They have been targeted at providing credible information on, for example, the sustainability of forest management. In this case, independent audits are carried out to verify that (1) ground-level forest management complies with these criteria, and (2) the chain of custody is verified from the forest to the point of sale of end products to allow trustworthy communication on forest management as part of the environmental quality of the product. IUCN–The World Conservation Union has also launched the Sustainable Use Initiative, which is working through regional sustainable-use specialist groups, to evaluate management systems and identify those factors which influence, both in a negative and positive manner, the sustainability of biodiversity use.

Conclusion

Plant biodiversity is an essential prerequisite for human existence; therefore, the conservation of plant biodiversity is vital if humankind is to thrive. We currently can, by definition, have no conception of which plant species will be useful in the future. For instance, which species contain the pharmaceutical compounds required to fight diseases yet to evolve, which will provide options for feeding our livestock in a time of global climatic instability, and which are candidates for replacement of the currently major providers of calories? If we wish to maintain the current wealth of natural plant diversity, we need to classify that diversity, much of which still remains to be described, ensure that diversity is conserved, reduce the various human activities that threaten diversity, and promote its sustainable use.

See also: **Biodiversity and Conservation**: Germplasm Conservation. **Secondary Products**: Pharmaceuticals, Plant Drugs.

Further Reading

Convention on Biological Diversity (1992) *Convention on Biological Diversity: Text and Annexes*. Montreal: Secretariat of the Convention on Biological Diversity.

Crucible Group (1994) *People, Plants and Patents*. Ottawa: International Development Research Center.

FAO (1998) *The State of the World's Plant Genetic Resources for Food and Agriculture*. Rome: Food and Agriculture Organization of the United Nations.

Frankel OH, Brown AHD, and Burdon JJ (1995) *The Conservation of Plant Biodiversity*. Cambridge: Cambridge University Press.

Given D (1995) *Principles and Practice of Plant Conservation*. London: Chapman & Hall.

Global Strategy for Plant Conservation (2002) Text. www.biodiv.org/programmes/cross-cutting/plant/

Groombridge B and Jenkins MD (2002) *World Atlas of Biodiversity*. Berkeley: University of California Press.

Guarino L, Ramanatha Rao V, and Reid R (1995) *Collecting Plant Genetic Diversity: Technical Guidelines*. Wallingford: CAB International.

Hawkes JG, Maxted N, and Ford-Lloyd BV (2000) *The ex Situ Conservation of Plant Genetic Resources*. Dordrecht: Kluwer.

Heywood VH and Watson RT (1995) *Global Biodiversity Assessment*. Cambridge: Cambridge University Press.

International Treaty on Plant Genetic Resources for Food and Agriculture (2001) Text. ftp://ext-ftp.fao.org/waicent/pub/cgrfa8/iu/ITPGRe.pdf

IUCN (2001) *IUCN Red List Categories and Criteria: Version 3.1*. http://www.redlist.org

Maxted N, Ford-Lloyd BV, and Hawkes JG (1997) *Plant Genetic Conservation: The in Situ Approach*. London: Chapman & Hall.

Prendergast HDV, Etkin NL, Harris DR, and Houghton PJ (1998) *Plants for Food and Medicine: Proceedings of the Joint Conference of the Society for Economic Botany and the International Society for Ethnopharmacology*. Kew: Royal Botanic Gardens.

ten Kate K and Laird SA (1999) *The Commercial Use of Biodiversity: Access to Genetic Resources and Benefit Sharing*. London: Earthscan Publications.

Walters KS and Gillett HJ (1998) *1997 IUCN Red List of Threatened Plants*, compiled by the World Conservation Monitoring Center. Gland: IUCN.

World Conservation Monitoring Center (1992) *Global Biodiversity: Status of the Earth's Living Resources*. London: Chapman & Hall.

Germplasm Conservation

B V Ford-Lloyd, University of Birmingham, Birmingham, UK

Introduction and Brief History of Germplasm or Genetic Resources Conservation

In 1926, the Russian geneticist and plant breeder N. I. Vavilov proposed that crop improvement draw from wide genetic variability, and to this end he collected cultivated plants and their wild relatives from many parts of the world. It was Vavilov who first focused attention on the diversity to be found in crop plants and the fact that it was concentrated in what he termed "centers of diversity". He investigated the distribution of the major crop plants and determined areas where there were concentrations of botanical varieties, using detailed studies of their morphology, cytology, genetics and resistance to pests and diseases, and adaptation to climatic conditions.

Much later, in the 1960s, international genetic conservation of crops and crop relatives gained momentum. This was spearheaded by the Food and Agriculture Organization of the United Nations (FAO), and in the 1970s by the International Board for Plant Genetic Resources (IBPGR) and its successor organization the International Plant Genetic Resources Institute (IPGRI). Further impetus was provided by the Convention on Biological Diversity (CBD) in 1992, its objectives being stated as "…the conservation of biological diversity, the sustainable use of its components and the fair and equitable sharing of the benefits arising out of the utilisation of genetic resources…". Article 1 of the FAO International Undertaking on Plant Genetic Resources (adopted 2001) echoes the wording of the CBD for all three main aims and the CBD, together with FAO promotes the adoption of a complementary approach to conservation of plant genetic resources for agriculture (PGRFA) that incorporates both *ex situ* and *in situ* techniques.

Genetic resources are, therefore, the "total genetic diversity of cultivated species and their wild relatives". Specifically they will be made up of the diversity found in any modern cultivated, domesticated or semidomesticated plant species and its component cultivars (currently in use or obsolete), older traditionally grown varieties or "landraces", and the related wild species, which may be the immediate or more distant evolutionary relatives of the cultivated forms.

Current Status of Germplasm Conservation

There may be up to 500 000 species of vascular plants in existence, only 250 000 of which have been identified or described. Approximately 7000 have been collected or cultivated by humans for food and all of these should, therefore, be considered as important PGRFA. Most attention has focused upon the 30 crops that currently feed the world, and within those the nine crops that directly supply 75% of all humans' dietary energy, i.e., wheat (*Triticum aestivum*), rice (*Oryza sativa*), corn/maize (*Zea mays*), *Sorghum* species, millet (*Setaria italica*), potatoes (*Solanum tuberosum*), sweet potatoes (*Ipomoea batatas*), soybean (*Glycine max*), and sugar beet (*Beta vulgaris*)/sugar cane (*Saccharum officinarum*). However, minor crops and underutilized species include plants that serve a wide range of functions, and may be of considerable importance in specific regions of the world, e.g., yams (*Dioscorea* spp.), oca (*Oxalis* spp.), cañihua (*Chenopodium pallidicaule*), breadfruit (*Artocarpus altilis*), and buckwheat (*Fagopyrum* spp.). Vegetables, fruits and other species, including wild plants gathered for food, are often important for nutrition and dietary diversification, while multipurpose trees and other crops can contribute to agricultural diversification. A range of forage species where legumes and grasses are a main focus of attention are highly significant because of their importance in supporting livestock as a critical component of many farming systems and their indirect contribution to human dietary energy.

What of the quantity of germplasm that is being conserved (**Table 1**)? For the major crops, germplasm collections are substantial. For instance, according to the *State of the World's PGRFA* report of 1998, well over a million accessions of wheat, nearly half a million samples of rice and just over one quarter of a million accessions of corn are being conserved. Over 100 000 seed samples of rice have been collected and stored in one institute alone (International Rice Research Institute in the Philippines). However, there is extreme variation between crops and species in the numbers of samples that have been acquired and conserved, with the more underutilized crops and many wild species often being inadequately represented.

Numbers of samples do not necessarily give a reliable assessment of how much genetic diversity is being conserved, so in order to accurately identify

Table 1 Some examples of major seed collections of crops and wild species held in genebanks throughout the world (based on FAO, 1998)

Crop	*Genebank*	*Country*	*No. of accessions*	*% of world germplasm*
Wheat (*Triticum*)	VIR	Russia	35 959	5
Wheat (*Aegilops*)			3255	15
Rice (*Oryza*)	IRRI (CGIAR)	Philippines	80 634	19
Maize (*Zea*)	IARI	India	25 000	10
Bean (*Phaseolus*)	CIAT (CGIAR)	Colombia	41 061	15
Apple (*Malus*)	DFBBAL	Sweden	9012	9
Palm (*Elaeis*)	INERA	D.R. Congo	17 631	83
Medicago (*Medicago*)	SARDI	Australia	20 000	38
Cacao (*Theobroma*)	CENARGEN	Brazil	2286	24

Table 2 Examples of loss of genetic resources

Genetic vulnerability

- A new race of corn/maize leaf blight destroyed over 15% of the US crop in 1970 because the same cytoplasmic genes were used to produce all major varieties
- In 1972, the winter wheat cultivar 'Bezostaya' was grown on 15 million hectares in the Soviet Union. Its use had spread beyond its original area of cultivation into much colder regions and the crop was wiped out completely by one severe winter. It was uniformly cold susceptible
- In 1979–1980, one million tonnes of sugar was lost in Cuba because rust attacked the one variety of sugar cane being grown on 40% of the agricultural land
- All F_1 hybrid rice in China shares the same genes for male sterility (1990)
- Commercial production of bananas relies upon five varieties all derived from one parent variety, "Cavendish". These varieties are highly susceptible to a serious fungal disease called "Black Sigatoka".

Genetic erosion

- In Korea, 74% of varieties of 14 crops being grown on particular farms in 1985 had been replaced by 1993
- In China, nearly 10 000 wheat varieties were in use in 1949. Only 1000 were still in use by the 1970s
- The majority of varieties of agricultural plants documented historically in the USA can no longer be found. For example, documentation and descriptions exist for 7098 apple varieties (used between 1804 and 1904), but now, 86% of these have been lost, along with 95% of cabbage, 91% of field corn/maize, 94% of peas, and 81% of tomato
- In the UK, many older varieties of Brussels sprout (*Brassica oleracea gemmifera*) disappeared after the release of hybrid varieties in the 1960s
- Comparison of current data with an inventory taken in 1930 shows that in Mexico only 20% of the landraces are still grown; this is the result of a decrease in area of land planted with corn/maize, and its replacement with other more profitable crops

gaps in world collections of germplasm, the genetic diversity of existing collections must be ascertained. This can only be done once passport data have been analyzed and evaluation of germplasm has been carried out. Unfortunately, complete sets of passport data are often unavailable and germplasm evaluation is often inadequate. Information from ecogeographical surveying work and taxonomic studies is equally important, but again often inadequate, as are estimates of genetic diversity using molecular markers, all of which are important tools for conserving, managing and using PGRFA genes.

Why Conserve Germplasm: Genetic Erosion and Genetic Vulnerability?

Over the last 60 or 70 years, scientists throughout the world have been engaged in developing better and higher yielding cultivars of crop plants for growth on increasingly larger scales. This has inevitably involved the replacement of more variable, possibly lower yielding, locally adapted varieties (or landraces) grown in traditional agricultural systems (**Table 2**). As far back as 1969, fewer than five varieties of each of several major crops (common bean (*Phaseolus vulgaris*), cotton (*Gossypium* spp.), pea (*Pisum sativum*), potato, rice, sweet potato) occupied more than 50% of the planting area in the United States. Increasingly, every major crop has had a decreasing genetic base, and diversity within cultivated plants has been replaced by genetic uniformity, a situation that may give rise to uniform susceptibility to a pest or pathogen. This loss of diversity in the crops themselves has also been accompanied by further loss of genes found in wild and weedy species related to the crop plants, because of improved crop husbandry and the destruction of natural ecosystems by human development (see **Table 3**). The paradox is that in order to be able to develop new cultivars of crop plants in the future (**Figure 1**), plant breeders will need to have access to

Table 3 Important causes of genetic erosion

Direct	
• Replacement of local varieties	
• Overexploitation of species	
• Overgrazing	
• Reduced fallow	
• Changing agricultural systems	
Indirect	
• Land clearing	
• Population pressure	
• Environmental degradation	
• Legislation/policy change	
• Pests/weeds/diseases	
• Civil strife	
• Climate change	

the wealth of genes that are being lost: hence, the urgency for conservation.

Why Conserve PGRFA?

PGRFA need to be conserved because of the valuable genes they contain. PGRFA are made up of a diversity of genes combined into a vast array of different genotypes. This genetic variation is valuable because of the agronomic and other qualities that it can provide for future crop improvement, the direction of which we may not be able to predict at the present time. However, it is possible to identify many important examples where landraces or other PGRFA have led to improvements in crop production (Table 4).

Conservation Strategies

There are two basic conservation strategies, namely *ex situ* and *in situ*, each of which is composed of various techniques. *Ex situ* conservation involves the conservation of PGRFA outside their natural habitats. Samples are maintained as living collections of plants in field genebanks, botanic gardens, or arboreta, or samples of plants are stored as seeds, tubers, tissue cultured explants, or DNA using special artificial conditions. *In situ*, on the other hand, implies conservation of whole ecosystems and natural habitats and the genetic resources that exist in their natural surroundings, and in the case of domesticates or cultivated forms, in the surroundings where they have developed their distinctive properties.

Ex situ: Seed Storage

The most convenient way of maintaining most plant germplasm is by storing seeds. The main exceptions to this are plants that are normally vegetatively propagated and do not produce viable seeds (e.g. banana, *Musa sapientum*), and crops where the seeds produced are very short-lived (cacao, *Theobroma cacao*). The latter are often referred to as being "recalcitrant" because they will not withstand drying below some relatively high moisture content without loss of viability. There are also species with intermediate storage behavior, where the seeds can be dried to relatively low moisture levels (7–12%), low enough to qualify as orthodox, but that are sensitive to the low temperatures typically employed for storage of orthodox seeds. Many temperate and tropical tree species, including oil palm (*Elaeis guineensis*), rubber (*Hevea brasiliensis*), cocao, and coffee (*Coffea* spp.), fall into the recalcitrant category. Fortunately, the majority of crop plants that produce seeds show "orthodox" behavior, so allowing storage for relatively long periods.

The two most important factors for effective conservation of orthodox seeds are therefore seed moisture content and temperature. Roughly speaking, longevity is doubled for each 5°C reduction in temperature or for each 1% reduction in moisture content. This is now known as Harrington's "rule of thumb". Orthodox seeds stored in a seed genebank are dried to 5–6% moisture content and placed in sealed containers stored at temperatures below −18°C. The minimum requirement and standard for genebanks is that seeds should be stored under conditions "which would ensure that the accession's viability would remain above at least 85% for 10 to 20 years".

Ex situ: *In vitro* Conservation

There are genetic resources for which seed storage is not appropriate or even possible. Those crops that are normally vegetatively propagated (e.g., potatoes), or that do not produce viable seed (e.g., bananas), or that produce very short-lived seeds (recalcitrant, see above) must to be conserved by other means. For this reason, much interest has focused on the application of tissue culture or *in vitro* techniques. It is possible to store plants *in vitro* for short periods of time, or longer if subculturing is carried out after certain intervals. Such slow-growth storage has proved extremely successful for shoot cultures of potato, cassava (*Manihot* spp.), fruit crops, such as banana, apple (*Malus pumila*), pear (*Pyrus communis*) and strawberry (*Fragaria x ananassa*), and many other horticultural species. Strawberry has been stored for 6 years with occasional addition of water to the culture medium, and garlic shoots can be stored for short periods if subcultured every 18–24 months.

In vitro technology also opens up the possibility of ultra-low temperature storage or cryopreservation

Figure 1 (A) *Oryza rufipogon*, a wild rice used for rice improvement, particularly the introduction of tungro virus resistance into the 'New Plant Type' (NPT); and (B) breeding line of the NPT of rice in the process of development at the International Rice Research Institute.

Table 4 Selected examples of the value of PGRFA for crop improvement

- K20, a bean variety grown on 16% of sub-Saharan Africa, is derived from a cross between a local Ugandan variety and a Colombian landrace
- About 400 improved rice varieties have been released in India, which are the result of selection from landraces
- In Bangladesh, landraces have contributed directly to 20 out of 23 jute varieties and about 90% of all leafy vegetables
- In Venezuela, 30% of the material used to develop new varieties is native
- Resistance to brown planthopper, yellow stem borer, bacterial leaf blight, and tungro tolerance has been transferred to cultivated rice from a range of wild rice species
- The wild rice species *Oryza rufipogon* increased yield by 17% when introgressed into cultivated rice

at −196°C for indefinite storage of germplasm. Cell suspensions, callus cultures, meristems and somatic embryos of many species can now be cryopreserved in liquid nitrogen. Somatic embryos of species such as cassava can be desiccated and encapsulated to form "artificial seeds" for commercial purposes, and it is likely that these novel man-made propagules will be important units of conservation in the future.

Ex situ: DNA Storage

The storage of DNA, or of parts of plants such as leaves from which DNA can be extracted, can be achieved easily and cheaply. Genes can be stored virtually indefinitely in the form of DNA affording long-term conservation. However, it cannot be envisaged, with present technology at least, that a whole plant could be retrieved from storage. The current use of genes for PGRFA stored in this way would be limited to their isolation, cloning and transfer to a recipient plant, i.e., via the production of a transgenic plant. Nevertheless, DNA storage is being used for conservation of germplasm from tropical forests where loss of species is substantial and uncontrolled, and where there is no alternative for storing representative samples of the species under threat. DNA samples are also being prepared from some of the larger seed germplasm collections worldwide as an additional resource aimed at improving the efficiency of molecular marker studies.

Ex situ: Botanic Gardens, Arboreta and Field Gene Banks

The conservation of germplasm in field gene banks or plantations involves the collecting of material from one location and its transfer and planting in a second location. Field genebanks have traditionally provided the answer to the conservation of recalcitrant species and vegetatively propagated species and are used to conserve cacao, rubber, coconut (*Cocos nucifera*), mango (*Mangifera indica*), coffee, banana, cassava, sweet potato, and yam.

The maintenance of large living collections in field gene banks requires large inputs of labor and land. Likewise the conservation of fruit trees in the form of orchards is beset by many problems, including space considerations associated with the need to conserve an adequate sample of genetic variation.

There are two other *ex situ* conservation techniques that have been widely used to conserve the entire range of plant diversity: botanic gardens and arboreta. These are living collections of plants held for public display, educational benefit, economic exploitation and scientific enquiry. There are approximately 1600 botanic gardens worldwide, containing over 3.2 million living accessions of as many as 100 000 species. It has been estimated that between 12 000 and 15 000 of these species are threatened in the wild and about half of these botanic gardens have some form of conservation program. Some botanic gardens have specialist collections of economically important plants and their wild relatives, medicinal and forest species, and many have sections devoted to the flora of particular regions of the world.

In situ Conservation

In situ conservation involves the maintenance of the genetic variation making up PGRFA in the location where it is encountered naturally, either in the wild or within a traditional farming or domestic situation. While existing nature reserves or national parks are designed for the purpose of *in situ* conservation, this is normally targeted at animal species, habitats or ecosystems rather than specific PGRFA. Few have as their primary goal the conservation of plant species, let alone plant genetic resources. Furthermore, the conserved material is not immediately available for exploitation, and in reality little guarantee can be given as to the likely "longevity" of the germplasm being conserved. However, awareness of the importance of *in situ* conservation is increasing for a number of reasons (**Table 5**), and it is seen as an important component of "complementary conservation strategies".

Natural Reserves or Genetic Reserves

The objective of the conservation process should be to maximize the genetic diversity represented within a minimum number and size of genetic reserves. To achieve this, information is needed on the amount of genetic variation, population structure, breeding system, habitat requirements, and geographical distribution of the target taxa. Conservation of the wild species component of PGRFA should therefore involve the location, designation, management, and monitoring of genetic diversity in a particular natural location; a basic model for this is provided in **Table 6**.

Table 5 Some factors promoting the use of *in situ* conservation

- Greater public awareness of conservation resulting from the environmental movement
- Increasing importance of wild species as a source of genes for crop improvement
- The impracticality of conserving all PGRFA genes *ex situ*
- The need for PGRFA to continue to adapt and evolve under changing environmental conditions
- Increasing recognition by countries that plant diversity has political, social and economic use

A comprehensive example of setting and monitoring of a natural reserve is provided by the "Ammiad" experiment in Israel. This focused upon naturally occurring wild relative germplasm of emmer wheat (*Triticum turgidum*). A study of the natural dynamics of wild emmer populations has been undertaken to serve as a precursor to conservation of wild cereals in their native ecosystems. Biological aspects, demography, ecological affinities, and genetic and phenotypic variation of the species were examined. Numerous localities were surveyed initially, before the target site was selected near the settlement of Ammiad in Eastern Galilee. On the site, physical, edaphic, climatic, floristic, demographic, and phytopathological features were recorded. Detailed ecological data were recorded for each sampling point. Seedling and whole plant demography was assessed, and spatial and temporal genetic variation was measured using various molecular markers. Additionally, morphological characters were measured and fungal diseases recorded. The conclusion drawn after 10 years of study was that centrally located emmer populations were genetically stable. These would be suitable for *in situ* conservation with two provisos, namely, that major environmental catastrophes did not occur and that dynamic shifts in population structure did not take place over a much longer time period.

On-Farm and Home-Garden Conservation

These conservation techniques involve the maintenance of traditional crop varieties or cropping systems by farmers or gardeners within traditional agricultural systems. For example, landraces are sown and harvested and every so often the farmer keeps a portion of harvested seed for resowing in subsequent seasons. In this case, it is the farmer who will intentionally or unintentionally conserve germplasm. The conservationist can monitor the situation, but may have little to do with the actual conservation. While it is desirable that landraces be conserved in this way, it is precarious in the sense that farmers will always be in a position to move from growing landraces to modern cultivars, and may require economic incentives not to make that change, a situation which in itself may affect the germplasm being conserved. Studies of how traditional agricultural systems work and how farmers' choices may affect diversity in those landraces are very informative, but do not improve the reliability of on-farm conservation (**Table** 7).

Table 6 Basic model for natural reserve conservation (after Maxted N, Ford-Lloyd BV, and Hawkes JG (eds) (1997) *Plant Genetic Conservation: The* In-Situ *Approach*. London: Chapman & Hall)

Plan and establish the reserve
- Assessment of site
- Assessment of socioeconomic and political factors
- Design of reserve
- Assessment of taxon and reserve sustainability
- Management plan formulation

Manage and monitor reserve
- Initiation of reserve management plan
- Monitoring of reserve
- Assessment of community relationships

Utilize the reserve
- Use of reserve traditionally or professionally
- Linkage to *ex situ* conservation (complementary), research programs, educational organizations

The Future

Comparative genetics has already indicated that evolutionarily related taxa have highly similar genome organization; at the fundamental level of the gene, which in reality is what we are concerned with conserving, the nucleotide sequences are highly conserved. While acknowledging this, there is now a major impetus to research allelic diversity at the DNA sequence level. "Allele mining" of the conserved diverse genetic resources and at economically important gene loci uncovers the fact that very small changes in nucleotides can result in dramatic phenotypic changes. Understanding of the conserved nature of genomes and allelic diversity will increasingly enable us to utilize germplasm whatever its form, and wherever it comes from, taxonomically speaking, which in turn implies that PGRFA, not just of the relatively small number of major crops, but of more or less any plant species should be our focus of conservation. But for too long plant, animal, and microbial genetic resources have each been compartmentalized, and they now need to be considered as a unified source of genetic resources. This will however place a vastly increased burden upon the curators of germplasm, not only because they may be faced with managing rapidly increasing quantities of gene resources, but also because they may need to make choices about what

Table 7 Some of the factors on which farmers' choose rice landrace seed to save for future sowing

Agroecological	Use
Field adaptation	Yield
Maturity	Eating quality
Drought tolerance	Price
Lodging resistance	Volume expansion
Dependability	Texture (glutinous, vitreous, viscous)

germplasm to prioritize for conservation. It will also be necessary for them to utilize greater arrays of genomic tools if effective use of their genetic resources is to be made. Bioinformatics will become increasingly important in the context of germplasm management and exploitation. Effective data management or lack of it has been recognized as an important concern for some time. We shall need to draw on the massive bioinformatic advances that have been made over the last few years associated with genomics and proteomics to revolutionize the way in which genetic resources data are viewed if PGRFA are to be effectively conserved and utilized in the future.

List of Technical Nomenclature

Accession A sample of a crop variety collected at a specific location and time, i.e., sample of seed in a seed genebank.

Bioinformatics Computational biology.

Callus Undifferentiated mass of cells growing in tissue culture.

CBD Convention on Biological Diversity.

Cell suspension Single cells or small clumps of cells growing in liquid tissue culture medium.

CENARGEN Centro Nacional de Pesquisa de Recursos Geneticos e Biotecnologia (Brazil).

CGIAR Consultative Group on International Agricultural Research.

CIAT Centro Internacional de Agricultura Tropical.

Complementary conservation Combinations of techniques (such as on-farm plus seed storage) to maximize efficiency and safety of conservation.

DFBBAL Department of Horticulture and Fruit Breeding Balsgard, Sweden.

Ecogeographical study The gathering and synthesizing of taxonomic, geographical and ecological data for the formulation of collecting and conservation priorities.

Evaluation Undertaking field trials or experiments to measure different characteristics (e.g., seed yield, disease resistance, stress tolerance) of germplasm accessions.

FAO Food and Agriculture Organization of the United Nations.

Genetic diversity The range of differences among individual plants or organisms shown in particular DNA sequences (can be genes, or noncoding regions).

Germplasm All the genetic material that comprises the inherited characteristics of an organism or species.

IARI Indian Agricultural Research Institute.

IBPGR International Board for Plant Genetic Resources.

INERA Institut National pour l'Etude et la Recherche Agronomique (Democratic Republic of Congo).

***In vitro* culture** Plant organs, cells or tissues grown in glass or other culture vessels.

IPGRI International Plant Genetic Resources Institute (Formerly IBPGR).

IRRI International Rice Research Institute.

Meristem A growing point of a shoot or root made up of actively dividing cells, cells undergoing cell division, and one or two differentiated organs such as leaves (in the case of a shoot meristem).

Molecular marker A polymorphism in a DNA sequence or protein amino acid sequence that can be used to study genetic variation and its inheritance.

Passport data Data that are required to describe the time and place of origin and botanical identification of an accession (e.g., accession number, country of collection, date of collection).

PGRFA Plant Genetic Resources for Agriculture.

Somatic embryo A very young plant possessing a shoot and root axis that is derived from a single somatic cell (as opposed to a zygote).

Sub-culture The transfer of small groups of cells or organs growing *in vitro* to a new culture medium and vessel to allow for continued healthy growth.

VIR Vavilov Institute (Russian Federation).

See also: **Crop Improvement**: Comparative Genetics. **Weeds**: Weed Seed Biology; Weed Technology and Control.

Further Reading

Brown AHD, Clegg MT, Kahler AL, and Weir BS (eds) (1990) *Plant Population Genetics, Breeding, and Genetic Resources*. Sunderland, Mass: Sinauer Associates.

Callow JA, Ford-Lloyd BV, and Newbury HJ (eds) (1997) *Biotechnology and Plant Genetic Resources. Biotechnology in Agriculture Series, no. 19*. Oxford: CAB International.

Cooper HD, Spillane C, and Hodgkin T (eds) (2001) *Broadening the Genetic Base of Crops*. Wallingford: CABI Publishing.
Engels JMM, Rao VR, Brown AHD, and Jackson MT (eds) (2002) *Managing Plant Genetic Diversity*. Oxford: CABI Publishing.
Evenson RE, Gollin D, and Santaniello V (eds) (1998) *Agricultural Values of Plant Genetic Resources*. Wallingford: CABI Publishing.
FAO (1998) *The State of the World's Plant Genetic Resources for Food and Agriculture*. Rome: FAO.
Frankel OH, Brown AHD, and Burdon JJ (1995) *The Conservation of Plant Biodiversity*. Cambridge: Cambridge University Press.
Guarino L, Rao VR, and Reid R (eds) (1995) *Collecting Plant Genetic Diversity*. Wallingford: CAB International.
Hawkes JG (1983) *The Diversity of Crop Plants*. Cambridge, Mass: Harvard University Press.
Hawkes JG, Maxted N, and Ford-Lloyd BV (2000) *The* Ex Situ *Conservation of Plant Genetic Resources*. Dordrecht: Kluwer.
Maxted N, Ford-Lloyd BV, and Hawkes JG (eds) (1997) *Plant Genetic Conservation: The* In-Situ *Approach*. London: Chapman & Hall.
Vavilov, NI (1992) *Origin and Geography of Cultivated Plants* (English edn transl. by D Love). Cambridge: Cambridge University Press.

Recalcitrance

P Berjak and N W Pammenter, University of Natal, Durban, South Africa

What and How Widespread Is Seed Recalcitrance?

There is a common perception that mature seeds are shed as dry entities that do not undergo further development (usually germination) until water is provided. These dry seeds can be stored for long periods which are predictable from the initial seed water content and storage temperature. This type of storage behavior is termed "orthodox." Much of our information about seed structure, development, physiology, and germination has been derived from investigations on orthodox seeds. Most studies have been conducted on seeds of around 20 important crop and a few woody species – all of which are orthodox – and so most of the amassed information has been derived from the seeds of considerably fewer than 0.1% of all higher plants!

However, there is increasing evidence that orthodox seed behavior might not be the norm. The ever-growing list of species producing nonorthodox seeds is an extension of the compendium of recalcitrant crop seeds published some 20 years ago by Chin and Roberts. The term "recalcitrant" is applied to those seeds that are shed at high water contents and are desiccation-sensitive (i.e., are not able to tolerate removal of more than a small proportion of their tissue water), having been adopted because it describes seeds that cannot be stored at the low water contents and low temperatures used in the storage of orthodox seeds. In fact, recalcitrant seeds cannot be stored for any useful period.

Many seeds identified as being recalcitrant are produced by tree species of mesic tropical forests, where conditions continually favor germination and seedling establishment, although recalcitrant seeds are also produced by a few species of markedly seasonal temperate regions. Production of recalcitrant seeds is not exclusive to dicotyledonous tree species: there are gymnospermous species, dicotyledenous herbaceous species, and monocotyledonous species that produce recalcitrant seeds. The phenomenon of seed recalcitrance is widespread across families, and although there are families in which apparently no species produce recalcitrant seeds, there seems to be little pattern among families regarding the occurrence of seed recalcitrance. Nevertheless, recalcitrance has been suggested to be the ancestral seed condition, and hence cannot be considered to be a manifestation of deviant behavior. Rather, recalcitrant seeds might be thought to behave in an unexpected manner – but only because of entrenched perceptions that all seeds are similar to beans or wheat!

However, within each of the categories orthodox and recalcitrant, there are marked differences in response to desiccation. Even among orthodox species, not all can be as severely dehydrated (or survive as long) as others, and among recalcitrant types, there are notable differences in the proportion of water loss that can be tolerated under similar drying conditions, and the length of time for which hydrated seeds can be stored. A category of seeds that can withstand dehydration, but not to water contents as low as orthodox types, has been described as showing "intermediate" characteristics, but it is likely that when seed biology of a greater number of species has been studied, a continuum of seed behavior will emerge (**Figure 1**). This will be based not just on relative sensitivity or tolerance to desiccation, but also on a spectrum of other characteristics.

Storage of orthodox seeds at low relative humidity (RH) (to maintain low water content) and low temperature will generally conserve vigor and

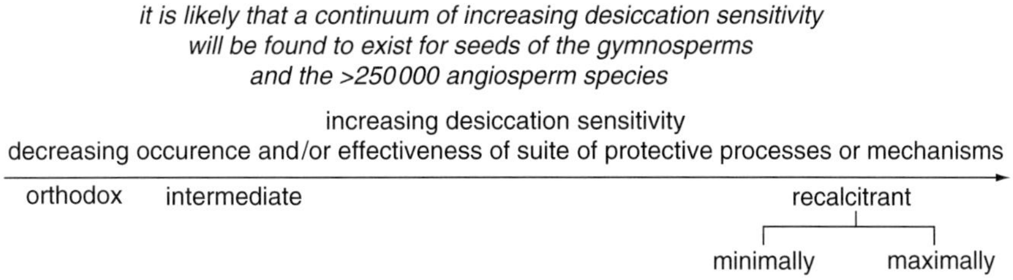

Figure 1 Schematic representation of the suggested continuum of increasing seed sensitivity to desiccation in relation to the decreasing occurrence and/or effectiveness of putatively protective phenomena.

viability for extended periods, provided the seeds are of good quality. Especially valuable seed (e.g., in genebank base collections) is conventionally stored at $-18\,^{\circ}\mathrm{C}$. However, similar approaches are precluded for recalcitrant seeds: they must remain hydrated to retain viability, and the high water contents preclude storage at below-zero temperatures. Additionally, recalcitrant seeds of some tropical species are chilling-sensitive. Recalcitrance thus poses major problems in terms of short- to medium-term storage for planting programs and the seed trade, and for long-term germplasm conservation.

Characterization of Recalcitrance

Preshedding development of orthodox seeds culminates in maturation drying, during which dry mass accumulation ceases, and seed water content and thus fresh weight decline significantly. This phase is essentially absent in recalcitrant types (**Figure 2**); a characteristic feature is that there is continued dry matter accumulation up to shedding, while the amount of water stays essentially constant. Consequently, water content may appear to decline to a limited extent. However, the water content of recalcitrant seeds at shedding is always high – from 0.3 to >4.0 g water (g dry mass)$^{-1}$ (g g^{-1}) (23% to 80% on a fresh mass basis (fmb)), depending on the species, the season, and the provenance. Mature orthodox seeds are generally shed at water contents <0.15 g g^{-1} (<13% fmb) and will survive subsequent desiccation to at least 0.05 g g^{-1} (5% fmb). On attempting to categorize unfamiliar seeds, water contents at shedding exceeding 0.3 g g^{-1} give the first clue to recalcitrance. (However, this is not a guarantee of recalcitrant behavior; some orthodox seeds, especially those from fleshy fruits, are shed at high water content, but they are tolerant of subsequent desiccation.)

Not only are there differences in desiccation sensitivity among recalcitrant seeds of different species, but even within seeds of individual species there is often notable inter- and intraseasonal variation (**Table 1**). Such differences are dependent on both intrinsic properties of the seeds and also the manner in which they are treated. For experimental purposes, recalcitrant seeds are generally hand-harvested from the parent plant. As there are generally few indications of the relative imminence of natural shedding, there could be considerable

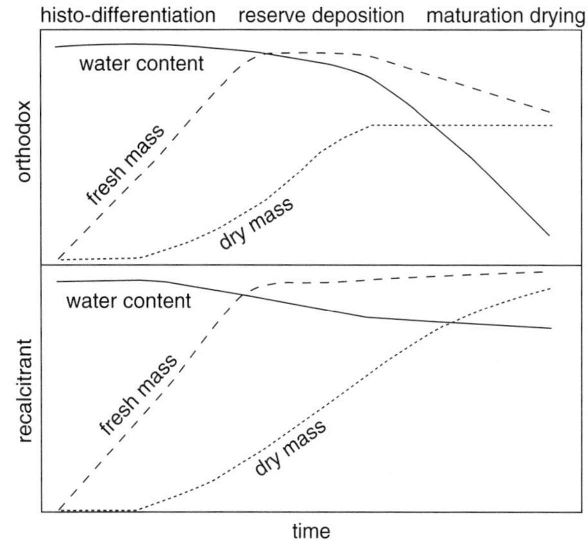

Figure 2 An overview, comparing development in orthodox and recalcitrant seed types. While dry mass accumulation ceases and water content declines sharply during the final developmental phase in orthodox seeds, in recalcitrant types dry matter continues to accumulate and there is very little change in seed water content.

Table 1 Variability in water content among different developmental stages and inter-seasonally for mature seeds of tea (*Camellia sinensis*)

Developmental status	*Water content (g g^{-1}, dmb)*
Immature	5.56±0.64[a]
Mature: June 1990	1.96±0.28[a]
June 1991	2.36±0.20[a]
July 1993	4.39±0.40
Mature: germinating	2.54±0.47[a]

[a] Data from Berjack P, Vertucci CW, and Pammenter NW (1993) Effects of developmental status and dehydration rate on characteristics of water and desiccation-sensitivity in recalcitrant seeds of *Camellia sinensis*. *Seed Science Research* 3: 155–166.

variation in the developmental status of the seeds that are harvested, contributing to this variation. If seeds are stored for a short while before being subjected to dehydration, they may become more sensitive. Another experimental treatment that affects the measured desiccation sensitivity is the rate of drying; seeds dried rapidly will lose viability at lower water contents than those dried slowly. Thus there is no unqualified "critical (threshold) water content" characteristic of a species, below which seed viability will be lost. (A misleading feature of much of the literature on individual recalcitrant seed species, before the importance of drying conditions was appreciated, is the identification of putative "critical water contents.")

The fundamental phenomenon underlying recalcitrant behavior is that the seeds are metabolically active when they are shed or harvested. When recalcitrant seeds are stored under conditions maintaining their water content (hydrated storage), developmental events grading into the onset of germination will occur. The longevity of the seeds in hydrated storage is suggested to be inversely related to the rate at which germinative metabolism proceeds: generally once cell division and extensive vacuolation occur during radicle extension, the seeds lose vigor and ultimately viability, because the requirement for additional water to support active growth cannot be met.

Two approaches to curtailing or delaying germinative metabolism in hydrated storage have suggested themselves. The first is the use of low (but above freezing) temperatures; this is possible in the case of seeds of temperate species, but does not provide a means for effective long-term storage, partly because of the complicating factor of seed-associated fungi. Cold storage is precluded for those (usually tropical) species that are chilling-sensitive. The second approach is the lowering of seed water content to a level permitting ongoing vital metabolism but curtailing germination (storage at "subimbibed" water content); this has been shown to be so injurious as to shorten seed lifespan relative to storage at the shedding water content. At subimbibed water content seeds are also overrun by proliferating fungi far more rapidly than are seeds stored without prior dehydration. Limited experimentation with metabolic inhibitors has similarly proved this approach to be ineffective.

When one considers that recalcitrant seeds are, in effect, germinating seeds or seedlings, then it is not surprising that they are no more amenable to these various manipulations than would be seedlings from orthodox seeds. In order to understand – and perhaps be able to manipulate – recalcitrant seeds, an appreciation of the factors conferring the unusual quality of desiccation tolerance in orthodox seeds is necessary.

Desiccation-Tolerance Mechanisms in Relation to Recalcitrance

Developmental Events: The Basis of Orthodoxy or Recalcitrance

During development there are no apparent differences between orthodox and recalcitrant seeds during histodifferentiation and reserve deposition (**Figure 2**). It is only after this that the patterns diverge. Nevertheless, in orthodox seeds there are profound changes that accompany the acquisition of desiccation tolerance and the ability to survive in the desiccated state. It is the lack and/or only partial expression of the suite of interacting processes and mechanisms operative in orthodox seeds that precludes recalcitrant types from entering the maturation drying phase. Recalcitrant behavior is thus inevitably a product of seed development, explaining at least partly why manipulations to extend their postharvest lifespan by imposing conditions to make them more like orthodox seeds have been such a dismal failure.

Intracellular physical characteristics Desiccation-tolerant plant cells must withstand the physical strains that accompany the volume reduction associated with the loss of considerable portions of cellular water. The water in fluid-filled spaces is generally replaced by space-occupying insoluble material, generally protein in the vacuoles, and starch and/or lipid external to the vacuole. The cytoskeleton has the ability to dissociate in an orderly manner. Although there is no direct evidence, it is also possible that the cell walls are plastic and can readily fold. Additionally, although direct data are lacking, the chromatin must assume a conformation that will protect the integrity of the genome in the desiccated condition. The nucleoskeleton, which determines nuclear architecture, must be modified in a strictly controlled manner.

These characteristics develop prior to or concomitant with maturation drying in orthodox seeds, but are lacking, or only partially manifested in recalcitrant seeds. Even from the relatively few studies that have been done, there appears to be a correlation between the degree of recalcitrance, and the manifestation of some of these features. For example, in *Avicennia marina*, *Ekebergia capensis*, and *Aesculus hippocastanum* there is a decreasing degree of vacuolation in embryo cells, and a decreasing sensitivity to dehydration under similar conditions. Nevertheless, *Ae. hippocastanum* seeds are desiccation sensitive, indicating that although protection against physical strains is necessary, it is not adequate in itself in conferring desiccation tolerance.

Intracellular dedifferentiation and metabolic "switch off" Intracellular dedifferentiation accompanies the onset of maturation drying in orthodox seeds. Mitochondria and plastids lose internal structure and endomembranes such as the rough endoplasmic reticulum become substantially reduced and the cisternae of Golgi bodies disassociated. These changes imply minimization of metabolic activity, including respiration and membrane synthesis and processing. The reduction in membrane surface area also reduces the sites that would undergo substantial, often deleterious, changes upon desiccation (**Figure 3**). A further indication of metabolic "switch off" is the cessation of DNA replication and the arrest of most embryo cells in the G1 phase (prereplication; DNA in the undoubled 2C form) with the onset of maturation drying. Dehydrated orthodox seeds are effectively ametabolic, not simply because there is no water available, but because of a controlled shutdown of activity and dismantling of structures preceding or accompanying maturation drying.

Marked dedifferentiation does not occur in any of the recalcitrant embryos of a variety of species examined to date, and respiration rates remain high. Interestingly, although the mitochondria of the recalcitrant embryos of both *Av. marina* and *Ae. hippocastanum* remain highly differentiated, the mitochondria occupy a greater proportion of cell area in the more desiccation sensitive species, *Av. marina*. There are insufficient data on cell cycling in recalcitrant embryos to comment unequivocally. However, in *Av. marina* seeds, only the most transient cessation of DNA replication occurs, with resumption of DNA synthesis (the S phase), resulting in its entering the more vulnerable, doubled state (4C), soon after shedding. In the temperate recalcitrant species *Acer pseudoplatanus*, on the other hand, >60% of the embryo cells have been reported to be arrested in the 2C state, although this might be associated with the dormancy of these seeds. The DNA of newly shed seeds of *Av. marina* is severely damaged by only a slight degree of dehydration, further indicating the vulnerability of actively metabolic subcellular components to water loss.

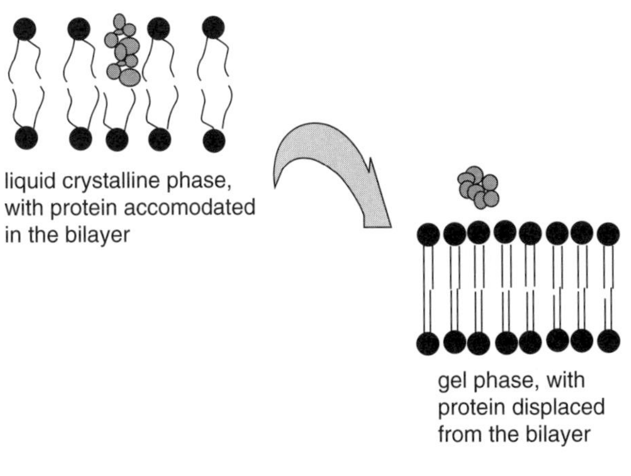

Figure 3 Lateral phase separation of membrane lipids from the functional liquid crystalline state (left) to the relatively rigid gel state is a potentially lethal consequence of dehydration in desiccation-sensitive tissues. Integral membrane proteins are likely to be displaced from the lipid bilayer, as shown (right). Uncontrolled free radical generation is thought to contribute to such phase changes.

Free radicals, reactive oxygen species, and anti oxidant systems Free radicals, which include reactive oxygen species (ROS), are strong oxidants and can cause, *inter alia*, peroxidation of membrane lipids leading to impairment of membrane structure and function (**Figure 3**). Reactive oxygen species are naturally produced during normal metabolism, but tissues contain a range of both enzymatic and nonenzymatic antioxidants which function to prevent injurious consequences of "escaped" ROS. However if metabolism is disturbed (by, for example, dehydration) there is the potential for unregulated ROS production with deleterious consequences such as oxidation of macromolecules and membrane deterioration. Hence, to prevent damage during the early stages of maturation drying, the presence and optimal operation of antioxidants is essential.

In developing recalcitrant seeds metabolism is not programmed to be "switched off": it continues, if not unabated, certainly at a relatively high level. It appears that, although recalcitrant seeds/embryos do possess antioxidants (and there are variations among species), these may become impaired or otherwise unable to cope with the level of ROS generation accompanying slow dehydration.

The presence and operation of putatively protective molecules

Sucrose with certain oligosaccharides or sugar alcohols Maturing orthodox seeds accumulate considerable amounts of sucrose and oligosaccharides (usually raffinose and/or stachyose), or sucrose coaccumulates with galactosyl cyclitols, depending on the species. As dehydration proceeds, these mixtures form a highly viscous supersaturated solution known as a glass. This glassy (vitrified) state can become so viscous as to curtail molecular diffusion, thus minimizing unregulated metabolism and its deleterious consequences. It is possible that the lifespan of mature orthodox seeds under defined storage conditions is influenced by the stability of the glassy state. Because they are metastable, glasses tend to break down, and this phenomenon may underlie the inevitable deterioration of orthodox seeds during storage.

Interestingly, the recalcitrant seeds of several species do accumulate sucrose and oligosaccharides in mass ratios conducive to glass formation. However, vitrification would occur only at water contents lower than those at which recalcitrant seeds lose viability on slow drying (whether "naturally" after being shed, or experimentally after harvest).

Late embryogenic accumulating/abundant proteins (LEAs) Synthesis of the set of robust, hydrophilic proteins termed LEAs (dehydrin-like proteins) precedes maturation drying during orthodox seed development. There is convincing evidence that LEAs are somehow involved in the acquisition and maintenance of desiccation tolerance in orthodox seeds, perhaps because their amphipathic nature facilitates interaction with a wide range of macromolecules and ions, thus preventing denaturation of the macromolecules under dehydrating conditions.

The situation with respect to the occurrence and possible role of LEAs in recalcitrant seeds is equivocal. Such proteins occur in recalcitrant seeds of a variety of species, from grasses to trees, from a range of habitats. On the other hand, LEAs were conspicuously absent from recalcitrant seeds of 10 tropical wetland species tested.

The evidence of the occurrence of both appropriate sugar/oligosaccharide combinations and LEAs in recalcitrant seeds underscores the contention that no one factor can be considered to be the key factor in either the acquisition, or the maintenance, of desiccation tolerance. The phenomenon must be the result of the interplay of a variety of mechanisms and processes which will surely emerge as being under multigenic control.

Amphipathic substances Endogenous amphipathic substances have been suggested to partition into membrane lipid bilayers during dehydration, preventing the formation of the gel phase in the desiccated state (**Figure 3**). On rehydration, the amphipaths have been shown to partition back into the cytoplasm. While this may be a further mechanism involved in desiccation tolerance in orthodox seeds, the status of amphipathic molecules in recalcitrant seeds has not yet been ascertained.

The Ability for Damage Repair on Rehydration

Storage of orthodox seeds at high temperatures and water contents causes damage which decreases vigor and brings about viability loss. However, before viability is reduced, the decreased vigor is manifested as an increasing time lag between seed imbibition and radicle extension. During this period, intracellular repair mechanisms become operational and repair must be effected before germination can occur. Repair during this lag phase in orthodox seeds occurs at the level of protein macromolecules, membranes, and nucleic acids, and restitution of free radical scavenging systems. In fact, the efficacy of osmopriming of low-vigor orthodox seeds arises from repair processes occurring while the seeds are held at water potentials that allow this metabolism, but preclude germination.

There are very few studies of repair by damaged recalcitrant seeds. However, following rehydration of the highly recalcitrant seeds of *Av. marina*, no DNA repair is possible once 22% of the originally present water has been lost (i.e., at water contents over $1.0\,g\,g^{-1}$ (about 53% wmb)), suggesting a very inadequate DNA repair system compared with orthodox seeds. In terms of free radical scavenging processes, there is evidence that antioxidant systems fail during dehydration of desiccation-sensitive seeds and seedlings, and are assumed to remain ineffective on rehydration. It appears that the repair mechanisms of recalcitrant seeds are as sensitive to water loss as all other processes.

The mechanisms and processes outlined above constitute some of a suite of protective mechanisms, probably all of which must be present, and which act together to confer tolerance to dehydration and the ability to survive for extended periods in the dry state. However, the list is probably by no means complete, with essential developmental phenomena remaining to be identified. In seeds that are not orthodox, the features are represented to differing extents, and some may not be present at all. This may be the underlying cause of the differing degrees of recalcitrance that are observed among species.

The Post-Shedding Response to Drying and Drying Rate

Biophysical studies have convincingly shown that in cells, water exists as at least five different states or types, the details of which are beyond the scope of this article. However, an appreciation that intracellular water up to around $0.28\,g\,g^{-1}$ will not freeze and is consequently different from "bulk," freezable water, is necessary. Nonfreezable water is either tightly associated with (bound to) intracellular surfaces, or is glassy – and, in fact, is considered as imposing ordered structure on macromolecules, membranes, and organelles. Freezable water, in contrast, may be considered as solution water.

During maturation drying of orthodox seeds, freezable water is progressively lost, in concert with the orderly shutdown of metabolic processes and the operation of the suite of mechanisms and processes discussed earlier. Below a water content of around $0.3\,g\,g^{-1}$, the water removed is from the nonfreezable fraction, being lost without deleterious consequences because the embryo cells are adequately protected against the strains of severe desiccation.

Recalcitrant seeds remain metabolic after shedding, and, because of their relatively large size and nature of the outer coverings they dry slowly in nature, as well as under usual dehydrating regimes used in the laboratory. As freezable water is lost metabolism will still proceed, but becomes unbalanced with continued dehydration over time. The more slowly water is lost, the greater is the time during which unbalanced metabolism can occur, reactive oxygen species can be produced, and the consequent damage accumulate. These processes will proceed so that the seeds generally will have lost viability by the time the water content has declined to around $0.8\,g\,g^{-1}$ (44% wmb). The lethal damage that has occurred is a direct consequence of unbalanced metabolism, so is termed metabolism-induced damage.

The more rapidly recalcitrant seeds are dried, the shorter is the time for which these damaging processes occur and the less the damage that accumulates on passing through a water content range. Consequently, material that is dried rapidly can be dried to lower water contents than material dried slowly. An appreciation of the importance of time as a factor in the desiccation responses of recalcitrant seeds removes some of the confusion concerning the idea of a species-specific critical water content.

However, because of their inherent desiccation sensitive nature, there is a lower limit below which recalcitrant seeds or embryos cannot be dried, no matter how rapid the dehydration. Drying below this point rapidly induces what is referred to as desiccation damage *sensu stricto*, which destroys cellular structures and organization, and is different, and has different causes, from metabolism-induced damage. Desiccation damage is a consequence of the absence of effective mechanisms that protect macromolecular structures and permit orthodox seeds to maintain viability when severely dehydrated. This lower limit to even the most rapid drying constitutes a true critical water content: it could well be the same (in terms of water potential, rather than water content) for all recalcitrant seeds, and it appears to coincide with the level of nonfreezable water.

"Conventional" Storage and Cryostorage

"Conventional" orthodox seed storage conditions of low temperature and RH are precluded for recalcitrant seeds which maintain viability only when they are stored hydrated. However, this option succeeds only in the short-term: recalcitrant seeds of tropical species generally lose viability within weeks to months, while those of temperate species may last up to a year or more, particularly if they can be cold-stored. Hydrated storage is thus of little use for long-term conservation of the genetic resources of species producing recalcitrant seeds. In addition, the storage period is normally significantly curtailed by the activity of the ubiquitous seed-associated fungi which survive on and in the seeds. The conditions of hydrated storage, particularly at room temperature, facilitate fungal proliferation which becomes rampant as the seed vigor inevitably declines. While concerted efforts are being made to overcome the major problems of hydrated storage, so as to extend postharvest vigor and viability, long-term storage, essential for the conservation of genetic resources, must be achieved by other means.

Cryostorage

Hydrated recalcitrant seeds cannot be stored at subzero temperatures in conventional freezers ($-20\,^{\circ}C$), as the slow cooling (freezing) results in lethal ice crystallization. The fact that material that is rapidly dried can survive (in the short term) to quite low water contents opened up a possible protocol for cryostorage. If material can be dried to the extent that freezable water is not present (without losing viability), then it may be possible to submit this material to freezing processes. Most seeds are too large to dry sufficiently rapidly to the required water contents while still retaining viability, and so the process of flash-drying was developed. In this technique the embryonic axes – which generally constitute only a minute fraction of the total seed volume/mass – are excised and exposed to moving dry air, so adequate dehydration can be achieved within minutes to hours, depending on the species. With such rapid drying the axes will tolerate dehydration to, or close to, the level where all remaining water is nonfreezable. Flash-dried axes supply the material for cryostorage, i.e. storage at temperatures below $-80\,^{\circ}C$, but conventionally in or over liquid nitrogen ($-196\,^{\circ}C$ or *c.* $-150\,^{\circ}C$, respectively).

However, successful cryostorage (resulting in the production of normal seedlings) is not straightforward. Whether or not ice crystal damage occurs depends on a combination of the tissue water content and the freezing rate. To minimize dehydration stress the axes should be dried as little as possible; to avoid freezing damage, the axes should be as dry as possible. Very rapid freezing rates (achieved by rapid plunging small tissue pieces into the cryogen (coolant)) will reduce the potential for freezing damage, but to achieve rapid freezing requires as low a tissue water content as feasible. The choice of cryogen is also important, as it must cause no biological injury but must also be a good conductor of heat away from the axis during its plunge. Additionally, thawing conditions and medium are also critical; not only can ice crystals form during warming, but rehydration damage must be avoided. A further complication is that germination of previously frozen axes must be achieved under *in vitro* conditions, which must be optimized before anything else is attempted. Additionally, all inoculum of the ubiquitous seed-associated microorganisms (generally fungi) must be totally eliminated, as these not only will overrun axes in culture, but survive flash-drying and cryostorage remarkably well.

Despite all these complications, it seems that there is no alternative to cryostorage for long-term conservation of the genetic resources of species producing recalcitrant seeds. Currently, the parameters required for successful cryopreservation have to be ascertained on a species basis, and using this approach some considerable success has been achieved. However, at the time of writing there are species which appear to be intractable to the necessary manipulations. Current research is directed at understanding the processes occurring during partial dehydration, rapid cooling, thawing, and regeneration, with the intention of progressing from the empirical approach to one that is more predictive of required conditions.

List of Technical Nomenclature

Free radicals	Molecular species containing an unpaired electron and therefore highly reactive.
Freezable water	Tissue water that undergoes a phase change from liquid to crystalline on cooling; the phase change can be detected by thermal analysis.
Glassy or vitrified state	A very concentrated "solution" that is so viscous as to inhibit molecular diffusion and hence chemical reactions.
Nonfreezable water	Water in tissue at low water contents where a phase change cannot be demonstrated by thermal analysis; this water does not have the properties of water in a solution.
Orthodox	A type of seed that can tolerate desiccation and can be stored at low temperature and water content for predictable periods.
Reactive oxygen species (ROS)	Free radical oxygen compounds.
Recalcitrant	A type of seed that cannot survive desiccation and cannot be stored at low water content.

See also: **Biodiversity and Conservation**: Seed Banks. **Postharvest Physiology**: Seed Storage. **Seed Development**: Onset of Desiccation Tolerance; Seed Production. **Tissue Culture**: Artificial Seeds. **Weeds**: Weed Seed Biology.

Further Reading

Black M and Pritchard H (eds) (2002) *Desiccation and Survival in Plants: Drying without Dying*. Wallingford: CAB International.

Chin HF and Roberts EH (eds) (1980) *Recalcitrant Crop Seeds*. Kuala Lumpur: Tropical Press.

Hong TD and Ellis RH (1996) *A Protocol to Determine Seed Storage Behaviour*. IPGRI Technical Bulletin No.1. Rome: International Plant Genetic Resources Institute.

Pammenter NW and Berjak P (1999) A review of recalcitrant seed physiology in relation to desiccation-tolerance mechanisms. *Seed Science Research* 9: 13–37.

Vertucci CW and Farrant JM (1995) Acquisition and loss of desiccation tolerance. In: Kigel J and Galili G (eds) *Seed Development and Germination*, pp. 237–271. New York: Marcel Dekker.

Seed Banks

F Hay, Royal Botanic Gardens Kew, Ardingly, UK

The History of Seed Banking

The term "seed bank" could be used to describe any artificial store of seeds – even a collection of homemade packets of seeds collected from the garden could be called a seed bank. However, the term is usually used to describe larger scale facilities where there is a more organized system of collection, cleaning, packaging, storage, and, perhaps, utilization and/or distribution of seeds. This manmade seed bank should not be confused with the bank or pool of seeds in the soil awaiting dormancy release and/or the right conditions for germination, and where seeds will experience a very different set of "storage" conditions to that of a seed bank.

Since the onset of agriculture, humans have stored seeds from one year to the next. The exchange of seeds and plants has also been practiced for thousands of years; at the most basic level, plant material may be distributed between members of a family or community. Exchange of material increased the desire for new crops and other plants, resulting in planned plant collecting expeditions. In the case of ornamental species, botanical gardens were probably the most influential in terms of mounting collecting trips and in distributing material. By the 1800s, many botanical institutes were regularly producing an *Index Seminum*, listing species for which seeds (and other propagules) were available for distribution to other botanical institutes around the world. These seed lists are still produced by many botanical gardens that do not have a seed bank facility *per se*. The seeds, collected from the living collections, are surplus to their own requirements for maintaining these living collections and are probably freshly harvested or, at most, just a few years old.

The first seed bank as a repository for germplasm was that developed by NI Vavilov at the All-Union Institute of Plant Introduction, now the Vavilov All-Union Scientific Research Institute of Plant Industry in St Petersburg, Russia. In the 1920s and 1930s, Vavilov made more than 100 collecting expeditions to different parts of the world, including Iran, the United States, Central and South America, the Mediterranean, and Ethiopia, amassing some 50 000 samples of crop plants.

In 1974, the Consultative Group on International Agricultural Research (CGIAR) set up the International Board for Plant Genetic Resources (IBPGR), which had the remit to coordinate the collection and distribution of plant genetic resources with emphasis on major crop species. The IBPGR was also instrumental in establishing seed banks in both developed and developing countries. By the early 1990s, the number of seed banks around the world had increased from ten in 1974 to 128. In 1991, the International Plant Genetic Resources Institute (IPGRI) based in Rome, Italy succeeded the IBPGR. The IPGRI continues to focus on the conservation and use of genetic resources, mainly of economically important species and largely in developing countries, not just through seed banking. It is also a key publisher of information and recommendations on the conservation of plant germplasm.

Thus, many existing seed banks focus on crop species and the need of plant breeders to have a pool of material that can be used in breeding programs. Such seed banks often contain only cultivars and races of the crop or crops of interest. For example, the International Rice Genebank of the International Rice Research Institute in the Philippines holds more than 100 000 samples of cultivated rice and related wild species that are used in research and breeding programs. Similarly, the Vegetable Gene Bank at Horticulture Research International in the UK, holds and distributes seeds of vegetables, such as onion (*Allium cepa*), cabbages (and other *Brassica*s), carrots (*Daucus carota*), and radishes (*Raphanus sativus*).

Over the last few decades, the use of seed banks as a means of *ex situ* conservation of wild plant species has increased, perhaps complementing *in situ* conservation programs. More and more countries are creating new or adding to existing seed banks so that, as well as relevant crop species, the wild flora is conserved away from threats such as desertification, over-grazing, and land clearance. The Millennium Seed Bank Project (MSBP) led by the Royal Botanic Gardens Kew is probably the largest undertaking that there has ever been in terms of banking seeds from wild plant species. At the beginning of 2002, the Millennium Seed Bank (MSB) held more than 14 000 samples of seeds from more than 5500 wild plant species, including 93% of the native UK flora. By 2010, the MSBP aims to have conserved 10% of the world flora (an estimated collecting target of ca. 25 000 species). An important aspect of this project is technology transfer and capacity building so that, as well as having duplicate samples stored in the MSB, collaborating organizations from around the world are able to conserve their own genetic diversity. In accordance with the Convention on Biological Diversity agreed at the 1992 Earth Summit in Rio de Janeiro, the partners of the MSBP all work under Access and Benefit Sharing agreements in order to control and protect the movement and use of the seeds stored in the MSB.

In an ideal world, there would be no need for active conservation of any sort. Second best would be to have vast areas of land that are minimally affected by humans. However, even the establishment of small nature reserves may not be possible in places where there are other more pressing problems involved in the realities of living from day to day. Any method of *ex situ* conservation, including living collections, tissue culture, seed banking, and DNA banking, can be seen as providing insurance against species extinction. Compared with other *ex situ* measures, seed banking is relatively inexpensive in terms of facilities, space, and input of manpower required to maintain the collection, especially since seeds from most species have the potential to remain viable for hundreds of years. Furthermore, seeds are propagules; releasing the genetic diversity that is held in a seed bank simply involves removing and germinating some seeds.

Can all Seeds be Stored in a Seed Bank?

In short, no, not all seeds can be stored in a seed bank – at least not in a conventional seed bank. The function of a seed bank is to ensure that the seeds remain viable for as long as possible. This is achieved by slowing down the rate at which seeds age by storing the seeds at low moisture contents and low temperatures. The IPGRI recommends that a seed bank should have an operating temperature of −18°C, and that the seeds should be dried to about 5% (i.e., 5% of the total weight of the seeds is water) and hermetically sealed in airtight containers before being placed at this temperature. Not all seeds can be banked because they do not tolerate drying to these low moisture contents, and if moist seeds are placed at subzero temperature, ice crystals can form and cause lethal cell damage.

Seed Storage Categories

Seeds that do not tolerate drying are described as desiccation intolerant, desiccation sensitive, or "recalcitrant," the last term being coined by Roberts in the 1970s. If a sample of recalcitrant seeds is dried, seeds will start to die as the moisture content of the sample falls below a certain high moisture content, typically in the region of 40–50% (the actual moisture content will vary between species). By the time the moisture content of the seed sample has reached 20%, all the seeds are likely to be dead.

A moisture sorption isotherm shows the equilibrium relationship between seed moisture content and relative humidity (RH) or relative vapor pressure (RVP = RH/100) (usually plotted as moisture content versus RVP) at a constant temperature, and can be used to indicate the level and tightness of water binding in the seeds. Recalcitrant seeds lose viability as phase III water (water that is effectively "bulk water") is removed (**Figure 1**).

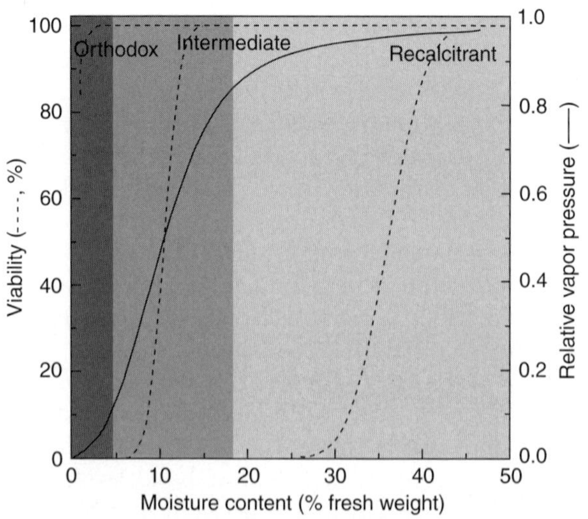

Figure 1 The relationship between moisture content and relative vapor pressure and the effect of drying on the germination of a sample of seeds of the three seed storage categories. Recalcitrant (desiccation sensitive or desiccation intolerant) seeds lose viability as bulk water (light shading) is removed; intermediate (partially desiccation tolerant) seeds tolerate removal of some of the weakly bound water (medium shading); orthodox (desiccation tolerant) seeds remain viable even when some of the strongly bound water is removed (dark shading), although longevity may be impaired.

Generally, it is thought that recalcitrant seeds are more prevalent in moist habitats, such as tropical rain forests, where seeds will be shed at high moisture contents without drying *in situ*. Examples of tropical species with recalcitrant seed storage behavior include cocoa (*Theobroma cacao*), lychee (*Litchi chinensis*), and rubber (*Hevea brasiliensis*). However, there are some temperate trees that have desiccation intolerant seeds, for example, horse chestnut (*Aesculus hippocastanum*), sycamore (*Acer pseudoplatanus*), and some oaks (*Quercus robur*, *Q. rubra*). In addition to desiccation sensitivity, recalcitrant seeds are often chilling sensitive. Unless the rates of cooling and warming are tightly controlled, recalcitrant seeds will not survive subzero temperatures: the water freezes resulting in cell death, both directly and indirectly (having a drying effect) by the formation of ice crystals. Tropical recalcitrant seeds may not survive exposure to temperatures lower than a critical temperature as high as 15°C; temperate recalcitrant seeds tend to tolerate much lower temperatures, perhaps even approaching 0°C. Although recalcitrant seeds cannot be stored in a conventional seed bank, it is generally possible to store them without complete loss of viability for a few months, under moist, aerated conditions at as low a temperature as is possible without incurring chilling damage.

Fortunately, the proportion of species known to have recalcitrant seeds is relatively small (7.4% of 6919 species reviewed to date, although it should be noted that those species that have been studied in any detail are biased toward crop and other species with economic value and to plants from temperate environments). Most species (88.6%) produce seeds that are described as "desiccation tolerant" or, after Roberts, "orthodox." These seeds survive drying to very low moisture contents, at least as low as ca. 5% (**Figure 1**), and subsequent storage at subzero temperatures. Furthermore, the longevity of orthodox seeds increases in a predictable manner in response to reduction in moisture content and temperature. At low temperatures and moisture contents, it is predicted that the seeds from many orthodox species will remain viable for hundreds of years. Such predictions are supported by reports of ancient seeds retaining the ability to germinate after many centuries. It is orthodox seeds that are routinely stored in a seed bank, and most of this article therefore relates to orthodox seeds.

There is a further category of seeds that do not conform to the definitions for either orthodox or recalcitrant seed storage behavior. Produced by seemingly few species (1.9%), the response of these seeds appears to be intermediate to the recalcitrant and orthodox response, i.e., they are partially desiccation tolerant, tolerating removal of at least some, if not most, of phase II type water, water that is weakly bound at nonionic, polar sites (**Figure 1**). In addition to this partial desiccation tolerance, and in contrast to orthodox seeds, the longevity of intermediate seeds, at least in the case of those from tropical species, is reduced at temperatures below ca. 10°C. Temperate intermediate seeds are likely to tolerate lower temperatures (at or just above 0°C), where their rate of metabolism will be reduced and they may remain viable for a number of years. However, with the limited longevity of both recalcitrant and intermediate seeds, other methods of conservation of seed tissues may be explored, in particular cryopreservation of embryonic axes.

Why Do Some Seeds Tolerate Drying and Others Do Not?

This is a question that has been asked by seed scientists for many years and has yet to be answered definitively. Common approaches to gain understanding in this area are to compare the composition, appearance, and/or response of orthodox seeds with that of recalcitrant seeds or to look at changes in composition, appearance, and/or response that occur in orthodox seeds during the acquisition of desiccation tolerance and/or subsequent loss of desiccation tolerance during germination. In some orthodox species, a desiccation tolerant state can be reinduced in germinating seeds (providing germination has not progressed too far), and these have also been studied for correlants of desiccation tolerance. Thus, a biochemical study might compare levels of a particular molecule or group of molecules: a "differential display" experiment would be carried out to look for genes that are only expressed in desiccation tolerant tissues; an ultrastructural study may compare and contrast the cellular structure of desiccation tolerant and intolerant tissues following drying. As a result of such studies, a number of mechanisms/strategies have been proposed that may contribute to a state of desiccation tolerance.

Most proposed desiccation tolerance mechanisms are based on the preservation of molecular structures that would otherwise be maintained by interactions with water molecules. In terms of changes in composition, a number of groups of molecules have been suggested to have this effect, including proteins and sugars. Proteins such as those that are switched on in the later stages of seed development, the so-called late embryogenesis abundant (LEA) proteins and/or heat shock proteins (HSPs) are classes of proteins that are thought to play a role in stabilizing the conformation of other proteins. However, it is more likely that the role they play is more critical in the early stages of water loss when phase III water is removed, and less important when weakly bound, phase II water is removed.

Sugars are believed to be a more serious contender in terms of conferring stability at low moisture contents. According to the "water replacement" theory, the hydroxyl groups of the sugar substitute for water, thereby providing the hydrophobic interactions required for preservation of the molecular structure of membranes and proteins. An alternative mechanism is that the sugar molecules, possibly in association with proteins, play a part in bringing about a glassy state, a semiequilibrium state where viscosity is high and thus molecules are less able to interact. Certainly the proportions of different types of sugar appear to be significant in terms of storage category; there is a general trend of increasing ratio of oligosaccharides (e.g., raffinose, stachyose) to sucrose with increasing desiccation tolerance in a broad range of species. High levels of oligosaccharides compared with monosaccharides may also reduce the availability of respiratory substrates, and thereby slow the rate of metabolic activity, leading to a quiescent state. In contrast to orthodox seeds, recalcitrant seeds do not appear to enter a dormant state, generally progressing from development straight to germination. Low levels of monosaccharides may also limit Maillard-type reactions that lead to disruption of protein structure.

However, although a glassy state may enhance stability, it is unlikely to confer desiccation tolerance *per se*. Furthermore, there is not a significant relationship between the amount of any one sugar and level of desiccation tolerance, and there are exceptions to this general trend. For example, seeds of the mangrove (*Avicennia marina*) are viviparous and highly recalcitrant and yet have a high oligosaccharide content.

Differences in lipid composition have also been observed between desiccation tolerant and desiccation intolerant tissues. Membranes are key to cell survival by defining cells, conferring organelle structure, and as the site for essential processes (for example, the electron transport chain of the mitochondrial membrane). A high level of unsaturation of the acyl chains of phospholipids may help to preserve membrane bilayer structure at low water contents, decreasing the likelihood of transition to a gel phase, which could lead to lateral phase separations and loss of membrane integrity. Lipid peroxidation leading to free radical production and further oxidative stress may also be a contributing factor in desiccation intolerance. Or rather, the ability to limit the damage caused by reactive oxygen species may contribute to the ability to withstand desiccation; free radical scavengers accumulate in orthodox seeds late in seed development and subsequently disappear during germination.

At the ultrastructural level, further differences between desiccation tolerant and desiccation intolerant tissues can be seen. In orthodox seeds, vacuoles are filled with storage protein during development. A lower level of vacuolation means there is less contraction of the cells as they are dehydrated, and membranes are less likely to shrink as a result of membrane vesicles breaking away from the cellular membrane. A consequence of this would be leakage of cellular constituents upon reimbibition. There also appears to be a reduction in internal membrane surface in orthodox seeds brought about by dedifferentiation of organelles. Again, this is likely to contribute to a quiescent state; dedifferentiation of mitochondria would result in lower rates of respiration. Lower rates of respiration would also reduce the rate of production of reactive oxygen species.

Clearly there are quite a number of significant changes/differences that have been identified. No single strategy identified so far appears to be the sole determinant of desiccation tolerance, and any number of these proposed mechanisms might contribute, to a greater or lesser extent, to the ability to withstand desiccation.

Seed Collection, Processing, and Storage

Seed Collection

Since the aim of seed banking is to conserve genetic diversity, seed collectors must consider how they should make their collections in order to ensure that the accession (the sample of seeds that is stored as a unit in the seed bank) contains as many of the genes that are present in the original population as possible and, ideally, in similar proportions. This may be relatively easy when the seeds being collected are from a crop cultivar where there is likely to be very little genetic variation from one plant to the next, particularly in the case of highly bred and selected grass crops such as wheat (*Triticum aestivum*). In some cases, plants may be grown with the specific purpose of regenerating a seed bank accession, in which case, all the seeds will be collected.

Whereas a crop seed bank will have multiple accessions of many genotypes, in the case of wild species conservation, the aim might be to simply conserve the species, in which case a single population can be sampled – collecting multiple populations of numerous species would be an enormous task. Some conservationists would, nonetheless, advocate sampling from a number of distinct populations in order to ensure that most of the genetic diversity within the species is conserved. The emerging field of conservation genetics can play an important role in identifying target populations with more genetic variation and/or that are more genetically distinct and in estimating sample sizes in order to ensure that the diversity sampled is optimized and that the resources available to a conservation program are well used. Where such tools are not available and the aim is simply to take a sample from a representative population of a species, it is important that collecting is carried out across the population and ideally at the time when the population as a whole is at peak seed maturity.

In terms of numbers of seeds that should be collected, this depends very much on the species being collected and how many seeds are produced by individuals and by the population as a whole. The IPGRI recommends minimum numbers of seeds of 4000 for genetically homogeneous accessions (e.g., crop cultivars) or 12 000 for heterogeneous accessions (e.g., wild species). However, whilst some wild species produce tens of thousands of seeds per plant each year, others produce relatively few. When the UK flora was being collected as part of the MSBP, a number of species proved very difficult to collect, in part because of their infrequent and limited flowering and/or fruiting. For example, some aquatic species such as *Utricularia australis* and *U. intermedia* only set seed sporadically, if at all, and four people spent more than 10 h collecting less than 30 seeds from an apparently healthy population of another aquatic plant, *Ceratophyllum demursum*. This can be compared with a plant such as foxglove (*Digitalis purpurea*), a biennial that has spikes of up to 80 flowers, which mature into dehiscent capsules that can each contain up to 1000 seeds; a single plant could have produced 80 000 seeds by the end of its 2-year cycle. Obviously, whatever is being collected, it is important that only a sample is taken so that there are plenty of seeds available to enter the pool of seeds in the soil, and the natural plant cycle and ecology of the area are not disturbed. The collecting protocol for the MSBP recommends that no more than 20% of what is ripe on the day of harvest be collected.

The size of a population may also be very small. When it is only possible to sample a very few plants, seeds from different plants might be stored separately so that if a sample is required, e.g., for a reintroduction program, it is possible to ensure that seeds from different plants will be included in the sample and the diversity of the new population is maximized (this is due to the statistics of sampling small numbers). For any small accession, it might be decided that it is necessary to do some bulking up – germinating a sample of the original seeds and allowing the plants to grow to maturity. This is, of course, unlikely to be a viable option for tree species.

Seed maturity is another important factor to consider when seeds are being collected for storage. Although orthodox seeds are defined as being desiccation tolerant, very immature seeds are not. The relative timing of the acquisition of desiccation tolerance during seed development varies between species. For a broad range of species, the ability to survive drying using seed banking methods has usually been acquired by or is acquired around the time that the accumulation of dry matter in the seeds is complete. This marker point during seed development is often referred to as mass maturity, although in the agricultural literature it is often termed physiological maturity. Collecting well-filled seeds should mean that they will be capable of germinating after drying and storage. However, seed quality traits – including measures such as levels of seedling establishment, rate of germination, and seed longevity – may continue to improve after the point of mass maturity (this is why this point in seed development may be termed mass maturity rather than physiological maturity). In some cases, the improvements in seed longevity between the point of mass maturity and the time of natural dispersal may be many-fold. For example, the relative longevity of a population of developing foxglove seeds increased nearly eightfold during this maturation phase of seed development.

In the very early stages of seed development, seed moisture content (as a proportion of the total weight of the seed) is very high. It will then start to decline gradually as dry weight accumulates, and then more rapidly following the formation of an abscission layer at the vascular connection that is present between the seed and the parent plant. Seed quality will increase throughout the early stages of seed development, and improvements may continue post-abscission provided that the moisture content of the seeds is high; once the seeds dry to a certain level, maturation processes will cease. The ideal time for collecting seeds is when they first dry down to this point when maturation processes stop; at this point, seed quality should be optimal and the seeds should be collected and dried following seed banking protocols.

The moisture level of the seeds at this time when quality is optimal will vary between species in terms of the amount of water as a proportion of the weight of the seeds; however, in terms of the water potential of the seeds it is relatively precise (due to the type of metabolic processes that can occur at different water potentials). It is somewhat difficult and time consuming to measure the water potential of seeds (since they are often small and/or encased in covering structures). However, it is relatively easy to measure the relative humidity (RH) or relative vapor pressure (RVP = RH/100) with which a sample of seeds is in equilibrium. RVP is the amount of water in the air at constant temperature compared with the amount of water that the air can hold. When seeds equilibrate to a certain air RVP at a given temperature, the seeds will have a precise moisture content; thus, at that moisture content, we can say that the seeds have that equilibrium RVP. The equilibrium RVP of a sample of seeds is approximately equal to the level of "water activity" in the seeds, and is thus a better indicator of water potential and the ability of the water in the seeds to facilitate metabolic activity than a measure such as moisture content. By the time RVP has fallen to approximately 0.85, respiration and maturation processes will have ceased and the seeds will start to age. Using a portable hygrometer in the field enables seed collectors to identify whether or not the seeds have reached this stage and should be collected.

This ideal time to harvest will be near the time at which seeds will naturally be dispersed, at least in the case of wild plant species – many major grass crops have been selected and bred for shatter resistance so that the seeds are retained on the plant and dry *in situ* without being dispersed. The farmer can then make a harvest at a convenient time. If collection is delayed beyond the time when RVP first falls below about 0.8, the moisture content (and water activity) of the seeds will fluctuate with changes in ambient RVP. The seeds may then start to age in the field resulting in losses in seed quality and ultimately in seed viability. For some crop species, if the moisture content of the seeds rises dramatically following rain, delaying collection may result in preharvest sprouting in the field. From a seed banking point of view, both preharvest sprouting and ageing *in situ* should be minimized. Crops should be collected promptly and seeds of wild plants remaining lodged after most of the seeds have been dispersed should be avoided. If a hygrometer is not available, other indicators may be used to determine whether or not seeds are at the point of dispersal. Although not foolproof, indicators such as fruit or seed color and signs of fruit dehiscence can be used to identify the optimum time for harvest for some species.

Of course, whilst we talk about seed collection, the actual dispersal unit that is collected will often be a fruit. Nonetheless, the above discussion still applies, at least for dry fruits. If a hygrometer is available and it is not possible to easily extract a few seeds, the equilibrium RVP of the fruit could be measured: if less than 0.85, the seeds are likely to have started drying, and it would be a suitable time to make a collection. In the case of fleshy fruits, the water activity of the seeds will still decline following the attainment of mass maturity and formation of the abscission layer. However the seeds will not be able to dry down to ambient RVP, and so fruit color might be a more useful indicator of whether or not the fruits should be harvested.

Seed Processing

Following collection, the first stage of processing is to dry the seeds. In dry tropical places, drying in the field may be achieved by spreading the seeds out in a thin layer in a place where there is likely to be good circulation of air, ideally not in direct sunlight. The seeds should be turned regularly in order to ensure all the seeds dry. In moister regions, using a desiccant, such as silica or other dry absorbent material (in an oven), may be a more reliable way of ensuring that the seeds dry to low moisture content. Once the seeds have dried as much as is likely in the field, they should be bagged up and sent to the seed bank.

In some cases, it will not be necessary (or perhaps possible) to dry the seeds in the field; the collections should be quickly transported back to the seed bank for drying. Most seed banks have a purpose-built dry room. The IPGRI recommends that seed bank dry rooms operate at about 15°C and a RVP of 0.10–0.15. Again, it is important that seeds are turned regularly, especially for large collections or collections of large seeds. Seed banks without dry rooms may have smaller drying cabinets or use desiccators containing silica gel.

Once the seeds have dried, at least partially, cleaning can commence. In some cases, it will be possible to do some of this in the field. This may aid the drying process by cutting down on bulk and removing covering structures. Different methods of cleaning may be used depending on the size of collection, fruit type, and the standards adopted by the processors. In some cases, for example species with single seeded, winged fruits, it may be acceptable and more cost effective to store the fruit rather than the actual seed.

Large collections of grass crops may be threshed as for normal harvests. For smaller collections, other machines, such as blowers, may also be used. Sieves can also be very effective for sorting the seed from the chaff. A stack of two sieves will sort the material into three fractions: large pieces of detritus will be retained in the top sieve (with large holes) and small pieces of detritus and dust will fall through the bottom sieve, hopefully leaving a relatively pure seed fraction in this bottom sieve. This process may have to be repeated a number of times, and rubbing the seeds against a sieve using bungs may help remove adhering fruit structures. Ultimately, for some small and/or fragile collections, hand sorting with dissecting equipment, perhaps under a microscope or a magnifying lens, will be the only effective method of cleaning the seeds. Once cleaned, the collections should be redried in case moisture has been taken up whilst the seeds are being cleaned, or in case covering structures restricted the drying process.

Once the seeds have dried back to equilibrium (or to ca. 5% moisture content if drying is achieved using a desiccator), they are ready for banking. (It is worth noting at this point that although the IPGRI recommends that seeds be stored at about 5% moisture content, where a dry room is in use the seeds will actually be dried to equilibrium with the RVP and running temperature of the dry room. For most species these conditions will give moisture contents in the range of 4–6%, but there will be slight variation depending on the species since the equilibrium relationship between seed moisture content and RVP varies depending on the chemical composition of the seeds.) At this banking stage of the processing, samples may be taken for testing purposes. It is obviously important to know the viability of a collection when it is placed into storage, but other details of the accession may also be required, for example, the number of seeds. An X-ray may also be carried out on some samples of seeds; if the X-ray indicates that there is a high proportion of empty, poorly filled, damaged, and/or insect infested seeds, further cleaning may be desirable. It may also be necessary to grow a plant in order to check the identity of the collection, particularly in the case of a collection from a wild plant species. It may not be possible to make two visits to a population of a target species or take a herbarium specimen at the time of seed collection (a herbarium pressing should normally be of a whole plant or sections of each part of a plant, including the flower, which may not be present when the fruits are being collected). A herbarium specimen is an essential reference for confirming the identity of an accession.

The seeds not required for this first stage of testing are sealed inside airtight containers; typical containers include steel cans, glass bottles or preserving jars, and heat-sealed, laminated aluminum foil bags. Double packing, i.e., sealing a smaller bottle or packet inside a larger bottle or packet, may be used to make sure that there can be no seepage of air into the container. The collection may be divided at this stage into an "active" and a "base" collection, the active collection being the one that may be accessed sooner or more regularly for testing or for research or breeding purposes. The size of the container chosen for the seeds is such that there is a relatively small volume of air compared with the volume of seeds. Obviously, it is necessary that the containers are accurately labeled; many seed banks now have computerized systems for recording and labeling accessions. Once sealed, the containers are ready to be placed in the actual seed bank. The IPGRI recommends a seed storage temperature of $-18°C$, either in purpose-built prefabricated cold rooms or in domestic deep freezers. Some seed banks may store the base collection at a lower temperature than the active collection, perhaps storing the active collection at just above $0°C$ and the base collection at $-18°C$, in order to reduce running costs. Other seed banks, such as the National Seed Storage Laboratory in the United States, routinely store seeds in the vapor phase above liquid nitrogen (a temperature of about $-160°C$), as well as at $-18°C$. These cryopreserved seeds will be held in polypropylene containers intended for cryogenic use.

It is important that the viability of the seeds in the seed bank is known and monitored. Different seed banks have different standards in terms of the level of viability that must be maintained in the collections. The viability standard used at the MSB is 85%. Because the MSB conserves seeds from wild plant species, the dormancy breaking requirements of different species may be complex and unknown. Thus, a series of samples may have to be taken soon after banking in order to ensure that the viability of each collection is at least 85% (with 95% confidence limits). Some wild species have complex and/or unknown requirements for breaking of dormancy and, as a consequence, a large proportion of the research work that is carried out at the MSB is on seed germination. For some species, a vital staining test is also carried out to make sure that samples of seeds that are failing to germinate have not already lost viability.

Once a collection has passed an initial viability test, subsequent samples may be taken at regular intervals (typically every 10 years) in order to check that viability is being maintained in the seed bank. If the viability of a collection is initially less than or falls below the set standard, it may be necessary to regenerate the collection or, if possible, re-collect it. However, regeneration raises issues relating to the maintenance of the genetic diversity of the original collection, particularly when it is carried out in an environment that is less favorable than the source environment. Re-collection, where possible, might be the preferable option provided the species still exists.

Seed Longevity

Once the seeds are in long-term seed bank storage, it is expected that they will age at a barely detectable rate, especially if the quality of the collection is high at the outset (with 99–100% viability). However, few seed banks have been in operation for long enough to determine the rates of ageing and viability loss at the recommended seed bank storage conditions. Most predictions of seed longevity, including predictions for seed bank storage, are derived using the viability equations that Ellis and Roberts defined in 1980. Ellis and Roberts stored seeds of barley (*Hordeum vulgare*) at a range of moisture contents in hermetically sealed packets at a range of temperatures, removing packets at regular intervals in order to determine viability through a germination test. For each set of storage conditions (moisture content $\times$ temperature), the pattern of viability loss through time followed a normal distribution. Using probit analysis (a statistical procedure that fits a generalized linear model to binomial data that follows a normal distribution), they were able to define an equation with parameters for the initial viability of the seed lot, K_i, and for the time, σ, for viability to fall by one normal equivalent deviate (or standard deviation) of the normal distribution of seed deaths over time. Since all the seeds were from a single population, K_i was the same for all the storage experiments, whilst σ depended on the moisture content, m (% fresh weight), and temperature, t (°C), during storage according to a second equation:

$$\log \sigma = K_E - C_W \log m - C_H t - C_Q t^2$$

where K_E and C_W are parameters that model the effect of moisture content on the longevity of orthodox seeds, and C_H and C_Q are parameters that model the effect of temperature on the longevity of orthodox seeds. Large increases in longevity are achieved by reducing the moisture content and temperature of storage due to the logarithmic linear σ-moisture content relationship and the semilogarithmic quadratic σ-temperature relationship (**Figure 2**).

With further experiments, Ellis and Roberts and other seed physiologists concluded that the parameters K_E and C_W are constant within species and that C_H and C_Q are constant within and between species. Consequently, if different seed lots of the same species are stored under the same conditions in terms of moisture content and temperature, the rate of viability loss in terms of normal equivalent deviates, σ, will be constant. It also means that once K_E and C_W have been determined from a particular set of storage experiments and using the universal values for C_H and C_Q (or determining these empirically as well), these equations can be used to predict the longevity of any seed lot of that species for any storage environment. This assumption that the parameters are constant between and/or within species can be applied broadly; however, seeds that are collected too early, before maturation processes have been completed, may lose viability at a faster

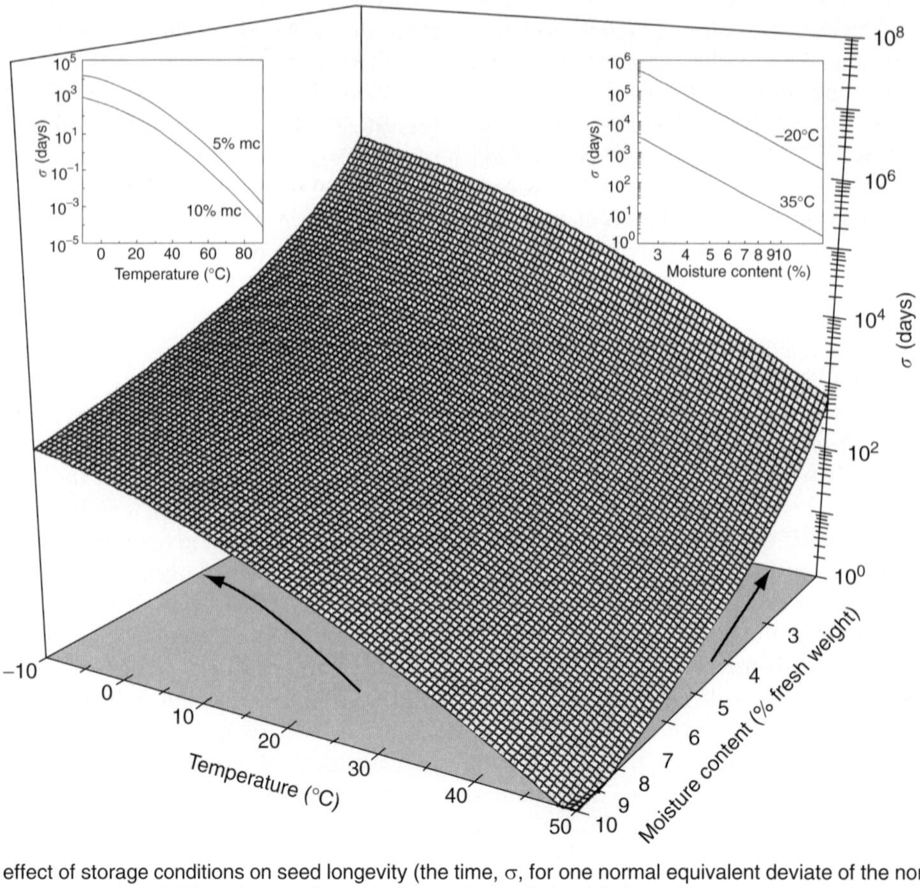

Figure 2 The effect of storage conditions on seed longevity (the time, σ, for one normal equivalent deviate of the normal distribution of seed deaths over time to elapse). There is a semilogarithmic quadratic relationship between temperature and σ at constant moisture content (inset top left and curved arrow on main graph) and a linear logarithmic relationship between moisture content and σ at constant temperature (inset top right and straight arrow on main graph). Data created using the Ellis and Roberts viability equations and viability constants for lettuce.

rate than seeds collected at the optimum point during seed development. There have also been some examples of species where there are differences between seed lots from different cultivars or ecotypes. Although these differences in estimates of parameters appear small, they are magnified when they are used to make predictions of longevity under seed bank conditions.

Storage experiments set up in order to estimate the parameters of the viability equations are usually conducted at relatively high moisture contents and temperatures (compared with seed bank conditions) so that the results are obtained within an acceptable experimental time frame. However, it has been argued that these estimated values (and indeed the equations themselves) should not be used to make predictions of longevity at moisture contents and/or temperatures significantly lower than the conditions used in the storage experiments. The equations do appear to apply over the approximately linear part of the moisture sorption isotherm (the medium-shaded region in **Figure 1**). This range corresponds to phase II of water binding, where all of the strong, phase I water binding sites are occupied and where, as the amount of water present increases (i.e., moving from left to right on the isotherm), more and more of the weak-binding sites become occupied. Theoretically, at a certain high moisture content, where water uptake is in phase III (water that is effectively bulk water), respiration, metabolic, and repair processes will resume resulting in increases in longevity at very high moisture levels. Seeds in the soil are likely to have this level of hydration, and this is why they are likely to remain viable in the soil, provided that they remain dormant; there have been numerous reports of seeds from temperate floras remaining viable after decades in the soil. However, the longevity of seeds in the soil is unlikely to match that of seeds stored at low moisture contents and temperatures.

Although the viability equations are used to predict seed death, they also apply to earlier measures of ageing: actual loss of viability is the final, catastrophic event in the ageing process. Early indicators that seeds are ageing include increased leakage of electrolytes when they are soaked in water (due to membrane damage), impairment of enzyme systems, a slower rate of germination, increased susceptibility to stresses, accumulation of chromosome abnormalities, and an increased proportion of abnormal seedlings. Thus, any ageing may reduce the likelihood of being able to produce healthy plants, and should be minimized as much as possible by ensuring that seeds are stored under optimum conditions. This is particularly important for accessions where there is a high level of genotypic heterogeneity and the loss of a few individuals will reduce the gene pool of the accession.

What Determines Seed Longevity?

This question could be asked either in relation to the effect of storage conditions and/or in relation to differences in seed longevity among species. Inter-species differences in seed longevity can be large. For example, if we consider storage at 5% moisture content and $-20^{\circ}C$, the predicted time for one normal equivalent deviate of the normal distribution of seed deaths over time (i.e., σ) to elapse is 30 years for foxglove seeds compared with 15 208 years for teff (*Eragrostis tef*) seeds.

Within a species, seed longevity increases as the amount of phase II water is reduced. There are two theories as to why removal of this water improves longevity. Firstly, quite simply, removal of water will limit its ability to facilitate and act as a substrate for reactions, thereby reducing metabolism and generally slowing down the rate at which ageing can occur. Alternatively, removal of phase II water increases longevity by reducing the molecular mobility in the cellular cytoplasm as the tissues move toward a glassy state, thereby reducing the likelihood of ageing reactions. Extending this idea that molecular mobility decreases as water is removed, the theoretical optimum moisture content for storage will be at the point when molecular mobility in the cytoplasm is at a minimum, when the cytoplasm is in a glassy state.

There certainly appears to be a low moisture content limit below which further drying does not improve longevity and may actually be detrimental to longevity. This detrimental effect may be due to loss of strongly bound, phase I water, which, even in orthodox seeds, is important for maintaining the integrity of molecular structure, particularly in the longer term. This is one reason why ultra-drying of seeds is not recommended by some seed physiologists. Moisture sorption isotherms can be used to indicate this optimum moisture content, just above the point where phase I type water binding is complete (**Figure 3**).

Since isotherms, which clearly indicate the type of water present, shift with temperature (**Figure 3**), a parallel can be drawn between the effects of changing temperature and the effects of changing moisture content. If the moisture content of the seeds is at a mid-point in the phase II region of water binding, lowering temperature will have a similar effect as lowering moisture content. A small decrease may only result in a reduction in the ability of water to facilitate reactions or in molecular mobility, and thus the observed reduction in longevity. However, a

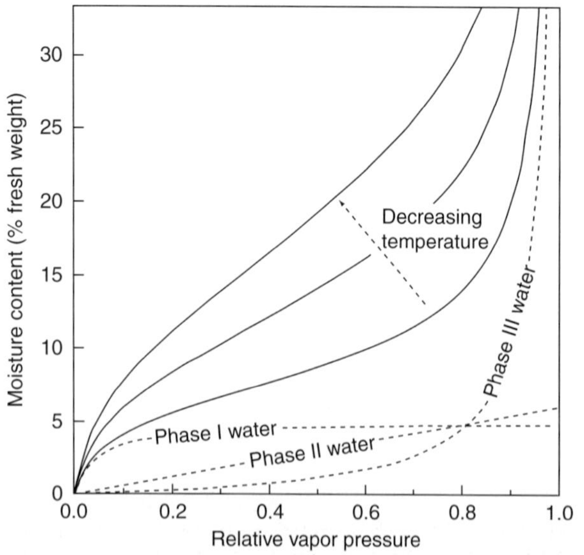

Figure 3 Hypothetical seed moisture sorption isotherms showing the effects of changes in temperature (solid curves). The shape of the isotherm indicates the phases of water binding that are present in the seed, and is the sum of the amount of phases I, II, and III water that is present (shown for the lowest isotherm (corresponding to the highest temperature) as the dashed curves).

larger reduction in temperature may result in the loss of phase I type water, which could then be detrimental to seed survival and longevity.

Seed moisture sorption isotherms also vary from species to species due to differences in composition. Thus, the region where the viability equations apply will also vary in terms of moisture content and temperature. In the case of lettuce (*Lactuca sativa*) seeds, work published in 1990 suggested that the species constants determined for this species were applicable between 2.5 and 15% moisture content and between −13 and +90°C. The limits might now be revised to higher moisture contents at lower temperatures and lower moisture contents at higher temperatures. There is some consistency across species in terms of the optimum RVP (and thus water activity or potential) for maximizing longevity. A consensus is forming that the optimum RVP for seed longevity is around 0.15 at the temperature at which the seeds are stored (not the temperature at which the seeds are dried).

Certainly, from the previous discussion, it appears that increasing molecular mobility as the tissues move away from the glassy state and/or increasing ability of water to facilitate reactions determines relative longevity within species. It may also be a major factor when we consider interspecies differences in longevity. Accepting that longevity is largely determined by molecular mobility, different compounds, in particular sugars and proteins, could influence the "strength" of the cytoplasmic glass, i.e., how quickly the glass will melt as water content and/or temperature increase, and may therefore be a determinant of relative seed longevity. However, there have been few studies of comparative longevity across a wide range of species, and most of the evidence to date has been derived from within species studies, some of which has given rise to conflicting evidence. For example, although primed seeds of *Impatiens walleriana* have reduced longevity and altered sugar contents, there was no increase in molecular mobility or change in the properties of the cytoplasmic glass. Thus, while sugars and proteins may play a role in the glassy state and hence in the stability of the system, other factors, such as the efficiency of antioxidant and free radical scavenging systems, may also be important.

Utilization

Clearly, utilization of seeds stored in seed banks will depend on the aims of the seed bank and its end-users. A seed bank aimed at conserving species is less likely to make large quantities of seed available to others unless required for reintroduction purposes. Some seed banks, like the MSB, will produce lists of seeds that are available for distribution. In the case of the MSB, the quantities of seeds involved are small, the main users will be other botanical, conservation, or nongovernment organizations, and samples will only be taken from collections where there are sufficient seeds and where the MSBP partner has agreed to allow their distribution for noncommercial use. In contrast, a seed bank set up for plant breeders is likely to regularly remove seeds for breeding programs, regularly regenerating the collection to ensure a steady supply.

Whilst the collection, processing, and storage of seeds are fundamental to the workings of a seed bank, the removal of samples should not be taken lightly. When the seeds are removed from cold storage and placed in a warmer environment, any water in the air may be able to condense on the surface of the seeds, increasing the likelihood of water uptake, which, combined with the higher temperature, could result in ageing. For this reason, the packets or bottles of seeds should be allowed to warm to ambient temperature before they are opened – ideally the containers will be opened in the dry room in order to minimize exposure of the bulk of the collection to higher humidities. The sampling procedure should ensure that each seed has an equal chance of being included in the sample, and should therefore involve some system of mixing and selecting. The sample may then be sealed in an appropriate container for distribution. The main collection should be allowed to re-equilibrate in the dry room in order to ensure that seed moisture content has not been raised, and may then be resealed and returned to the seed bank. Fortunately, there is no evidence that repeated sampling (i.e., repeated cooling and warming) is detrimental to long-term seed survival.

List of Technical Nomenclature

C_H, C_Q	Parameters in the viability equations of Ellis and Roberts that model the effect of temperature on the longevity of orthodox seeds. Values believed to be universal, i.e., constant for all species.
Glassy state	A nonequilibrium, amorphous solid phase where molecular structure is undefined (noncrystalline).
Intermediate seeds	Seeds that tolerate partial drying.
K_E, C_W	Parameters in the viability equations of Ellis and Roberts that model the effect of moisture content on the longevity of orthodox seeds. Values believed to be constant within species.
K_i	In the viability equations of Ellis and Roberts, an estimate of the initial viability of the seeds (measured as the number of normal equivalent deviates worth of viable seeds).
Moisture content	Usually calculated as the percentage of the total mass of the seeds that is water.
Optimum moisture content	The moisture content where seed longevity is maximized.
Orthodox seeds	Seeds that tolerate drying to low moisture contents (3–5% total weight) and whose longevity in dry storage is predictable.
Recalcitrant seeds	Seeds that do not tolerate drying below relatively high moisture content.
RVP	Relative vapor pressure. Proportion of water that is in the air as vapor, relative to the amount that the air could hold, at a given temperature.
Seed longevity	How long seeds will remain viable (mainly used in the context of dry storage of orthodox seeds).
σ	In the viability equations of Ellis and Roberts, refers to the time in days for one normal equivalent deviate of the normal distribution of seed deaths over time to elapse.
Sorption isotherm	Graphical plot showing the equilibrium relationship between seed moisture content and RVP at constant temperature.

See also: **Biodiversity and Conservation**: Germplasm Conservation; Recalcitrance. **Ethics and Biosafety**: Ownership of Plant Genetic Resources. **Seed Development**: Onset of Desiccation Tolerance; Physiology of Maturation; Seed Quality.

Further Reading

Black M and Pritchard HW (eds) (in press) *Desiccation and Survival in Plants.* Wallingford: CAB International.

FAO/IPGRI (1994) *Genebank Standards.* Rome: IPGRI.

Guarino L, Ramanatha Rao V, and Reid R (eds) (1995) *Collecting Plant Genetic Diversity. Technical Guidelines.* Wallingford: CAB International.

Hong TD, Linington S, and Ellis RH (1998) *Compendium of Information on Seed Storage Behaviour.* Kew, UK: Royal Botanic Gardens Kew.

Kigel J and Galili G (1995) *Seed Development and Germination.* New York: Marcel Dekker.

Probert RJ and Hay FR (2000) Keeping seeds alive. In: Black M and Bewley JD (eds) *Seed Technology and its Biological Basis*, pp. 375–410. Sheffield: Sheffield Academic Press.

Smith RD, Dickie JB, Linington SH, Pritchard HW, and Probert RJ (2003) *Seed Conservation: Turning Science into Practice.* Kew, UK: Royal Botanic Gardens Kew.

Tweddle JC, Turner RM, and Dickie JB (2002) *Seed Information Database (Release 2.0, Jan. 2002).* http://www.rbgkew.org.uk/data/sid

Biotechnology *see* **Diseases**: Genetic Modification of Disease Resistance, Fungal Pathogens; Genetic Modification of Disease Resistance, Viral Pathogens. **Genetic Modification**: Gene Cloning, General Principles; Insertional and Transposon Mutagenesis; Transformation, General Principles; Transformation in Dicotyledons; Transformation in Monocotyledons; Transformation in Plastids; Transgene Stability and Inheritance. **Genetic Modification, Applications**: Flower Color; Nitrogen Fixation; Oxidative Stress; Cell Factories; Molecular Farming; Plantibodies. **Growth and Development**: Control of Gene Expression, Post Transcriptional Regulation.

C

CELL WALLS AND FIBERS

Contents

Cell Walls

S C Fry, The University of Edinburgh, Edinburgh, UK

Introduction

Plant cell walls are metabolically active subcellular compartments. They govern the shape, size and growth rate of the cell, the texture and strength of the tissue, and the plant's resistance to microbial digestion. The best-known component, cellulose, usually constitutes only 25–40% of the wall's dry weight; the rest is mainly other polysaccharides (pectins and hemicelluloses). Structural (glyco)proteins and other structural polymers are also present, as well as enzymes and low-molecular-weight metabolites. Wall polysaccharides and glycoproteins are synthesized by the protoplast and released into the apoplast, where they are subjected to numerous modifications, including transglycosylation, covalent crosslinking, hydrolysis, and the making/breaking of hydrogen bonds. These processes contribute to the assembly of new wall material and to the tightening and loosening of existing wall material. Partial wall autolysis *in vivo* liberates oligosaccharides, some of which ("oligosaccharins") are active signaling molecules. Cellulose, as the most abundant organic chemical on earth, is a prime target for digestion – both in Nature (by pathogens and saprophytes) and in the biotechnology industry – but cellulose is often surprisingly resistant to digestion owing to the presence of other wall components.

Location and Roles: Cells, Walls, and Membranes

The cell wall – the 0.1–10-μm thick, outermost layer of the plant cell – is a physical and biological frontier zone between the plasma membrane and the "outside world," while still permitting intercellular communication. Mainly composed of polysaccharides, the wall differs fundamentally from the plasma membrane, which is composed of lipids and proteins and is much thinner (<0.01 μm). Bonds may exist between the plasma membrane and the wall. However, most such bonds are weak: the membrane quickly retracts from the wall when the cell is plasmolyzed (e.g. in $1\,\mathrm{mol\,l^{-1}}$ sucrose).

Conventionally, the wall is considered part of the cell. However, some workers now regard the wall as an "extracellular matrix," implying that it is outside the cell. Here, the cell excluding the wall is considered "a protoplast." The wall is thus unambiguously "extraprotoplasmic," whether or not it is considered "extracellular."

All the walls of a plant (plus the sap that permeates them) constitute the apoplast; all the protoplasts (interconnected by plasmodesmata) constitute the symplast.

The position of the cell wall as the outermost layer of the cell allows the wall to serve several important roles, including dictating the cell's shape and size, strengthening mechanical tissues, e.g., fibers, repulsing pathogens, sequestering toxic metal ions, attaching cells to their neighbors and producing intercellular messengers. Protoplasts and walls cooperate to produce a functioning plant.

Parts of the Cell Wall

The cell wall is biphasic: a scaffolding of rigid, inextensible, cellulosic microfibrils is embedded in a matrix of noncellulosic polymers. The matrix is more highly hydrated and exhibits greater molecular mobility than microfibrils.

A primary cell wall can grow in area (at the time of deposition of its microfibrils). Primary walls are usually ~0.1 μm thick, but are much thicker in specialized tissues such as collenchyma. Some primary walls later become altered (crosslinked or lignified), preventing further growth in area. The primary walls of adjoining cells are glued together by a pectin-rich middle lamella.

A secondary wall layer is one whose microfibrils are deposited, internal to the primary wall, after the cell has lost the ability to expand. Secondary walls may continue to thicken (often to ~10 μm) after the cessation of growth in area. Primary and secondary wall layers of the same cell often differ markedly in chemical composition.

Microscopy

Tissue-specific differences in wall composition can be demonstrated both microscopically and chemically. Microscopy provides more precise spatial information but lacks the ability to assign exact chemical structures. Methods for visualizing wall constituents include staining them with ruthenium red, calcofluor white, phloroglucinol–HCl, sudan III or toluidine blue O, direct observation of the auto-fluorescence of certain wall components, photographic documentation of UV absorbance, and the observation of wall components that specifically bind fluorescently labeled lectins or antibodies. Enzymes in the wall can be visualized using chromogenic or fluorescent substrates.

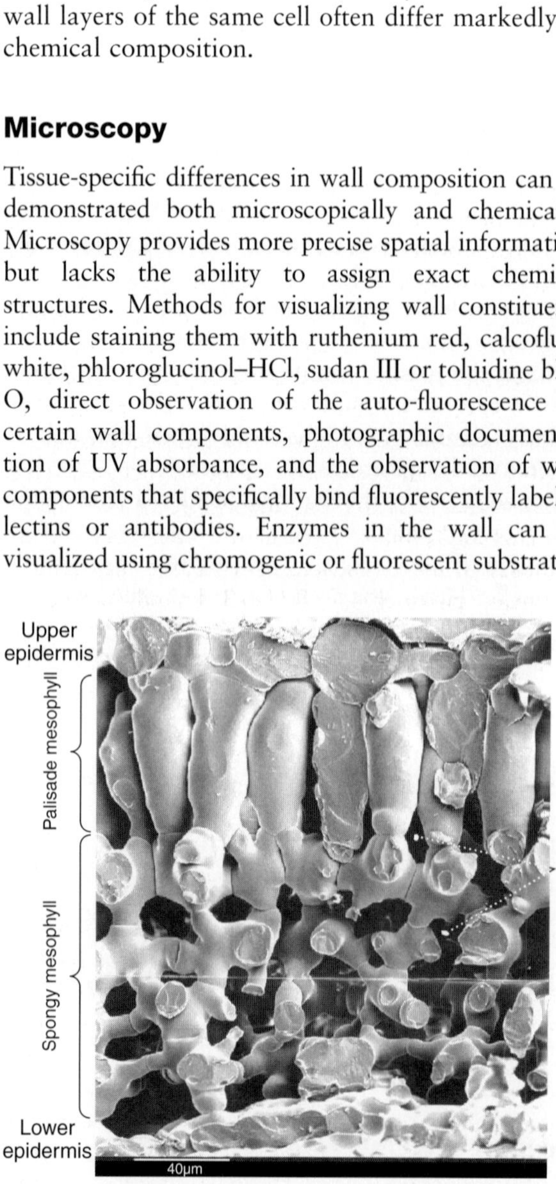

Figure 1 Scanning electron micrograph of the inside of a young bean leaf, *Phaseolus vulgaris*, 14 days after germination. The leaf was freeze-fractured at right angles to the surface and is viewed fully hydrated at −160 °C. The outer layer of each cell is a primary wall, which dictates the tissue-specific cell shape, cell size, and contact with neighboring cells. Bar = 40 μm. Electron micrograph supplied by Dr Christopher E. Jeffree, The University of Edinburgh.

Transmission electron microscopy (EM) indicates the thickness of individual walls, their attachment to neighboring walls, and the early stages of wall formation. Scanning EM provides surface views of wall textures and shapes (**Figure 1**).

Chemical Analysis

Isolation of Cell Walls

For chemical analysis, walls may first be separated from the rest of the cell: after vigorous tissue homogenization, the tough, dense wall fragments are pelleted by low-speed centrifugation in a dense medium, in which protoplasmic fragments remain suspended. Isolated wall fragments can be freed of starch and noncovalently associated proteins by washing in 90% (v/v) dimethylsulfoxide and phenol–acetic acid–water (2:1:1), respectively.

However, for many proposes, wall analysis can utilize "alcohol-insoluble residue" (AIR), obtained by grinding the tissue in 70% (v/v) ethanol, in which small, polar biochemicals (monosaccharides, oligosaccharides, amino acids, phenolics, sugar-phosphates, Krebs' cycle intermediates, etc.) dissolve and can thus be rejected. AIR includes most cellular polymers (walls, starch, protein, RNA, DNA, many lipids), among which walls often predominate.

Extraction of Wall Polymer Fractions

Isolated cell walls are often put through a sequence of extractants that solubilize successive classes of wall polymers. A commonly used sequence is:

1. a chelating agent (EDTA, CDTA, or EGTA), which, by removing Ca^{2+}, solubilizes some pectins (heating promotes solubilization but partially degrades pectins);
2. cold 0.05 mol l^{-1} Na_2CO_3 to solubilize additional pectins;
3. 0.5 mol l^{-1} NaOH to solubilize some hemicelluloses;
4. 6.0 mol l^{-1} NaOH (at 37 °C) to solubilize essentially all remaining hemicelluloses.

The NaOH-insoluble material is called the "α-cellulose" fraction. NaOH treatments are sometimes preceded by warm, acidified chlorite, which, by destroying lignin, may facilitate hemicellulose extraction. NaOH is usually supplemented with 1% (w/v) $NaBH_4$ to minimize oxidation of polysaccharides.

Cellulose, hemicelluloses, and pectins can be characterized, at a simple level, by chromatographic analysis of their constituent sugar residues (**Figure 5**). These sugars (monosaccharides) are released from the polysaccharide molecule by hydrolysis, catalyzed by hot acid or by suitable enzymes.

Chemistry of Cellulose

Cellulose, the main component of the "α-cellulose" fraction, is a water-insoluble, rigid, linear polymer of β-(1→4)-linked D-glucose residues (**Figure 2**). Microfibrils are bundles of parallel-lying cellulose molecules attached to each other via hydrogen bonds. The cellulose in a microfibril is largely crystalline and essentially free of internal water.

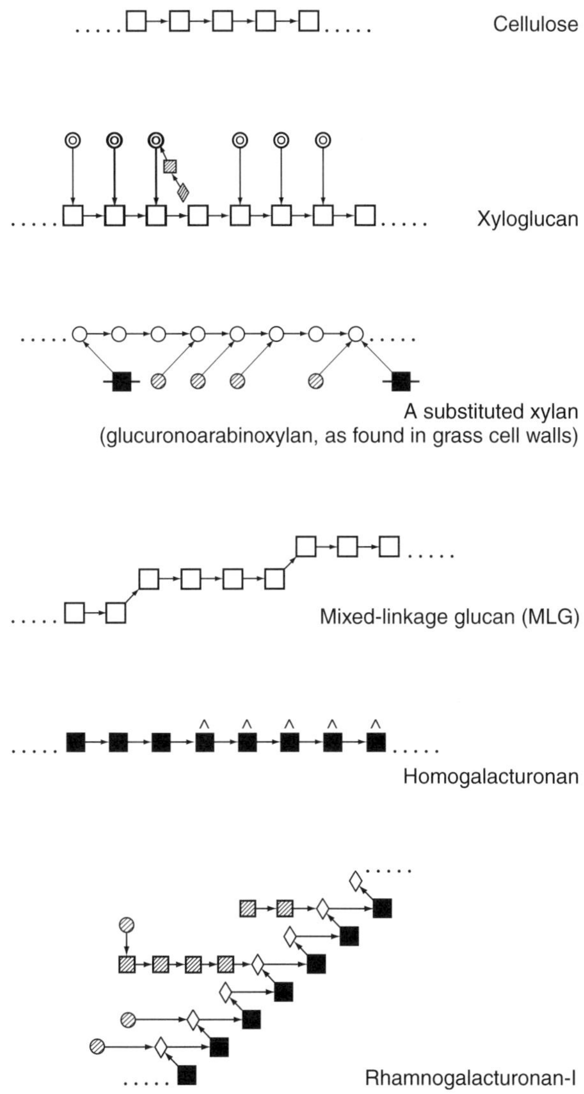

Figure 2 Schematic views of the construction of the major plant cell wall polysaccharides. Filling of symbols represents character of sugars: Solid fill = acidic sugars; Hatched = highly hydrophilic neutral sugars; No fill = more hydrophobic neutral sugars (relative to the corresponding sugar shown hatched). Shape of symbols represents molecular formulae of sugars: □ = hexoses ($C_6H_{12}O_6$ or $C_6H_{10}O_7$ if acidic); ◇ = 6-deoxyhexoses ($C_6H_{12}O_5$); ○ = pentoses ($C_5H_{10}O_5$). Bonds between neighboring sugar residues: ↖ = (1→2)-linkage; ↗ = (1→3)-linkage; → = (1→4)-linkage; ↓ = (1→6)-linkage. All the sugars shown here are in the pyranose ring form except arabinose, which is in the furanose ring form. Key: ■ α-D-Galacturonic acid; ■̂ Methyl α-D-galacturonate (neutral ester); -■- α-D-Glucuronic acid; ▨ β-D-Galactose; ◈ α-L-Fucose; ⊘ α-L-Arabinose; □ β-D-Glucose; ◇ α-L-Rhamnose; ◎ α-D-Xylose; ○ β-D-Xylose.

Chemistry of Hemicelluloses

Hemicelluloses and pectins are matrix (= nonmicrofibrillar) polysaccharides. Dilute and concentrated NaOH, applied sequentially, preferentially solubilize different hemicelluloses – arabinoxylans and xyloglucans, respectively. Xyloglucans have a backbone identical to cellulose; an arabinoxylan's backbone is similar but composed of β-(1→4)-linked D-xylose residues (**Figure 2**). Xyloglucans and arabinoxylans, unlike cellulose, have short side chains (predominantly α-D-xylose and α-L-arabinose, respectively), conferring hydrophilicity. Thus, after solubilization by NaOH, these hemicelluloses often remain water-soluble even when the alkali is removed or neutralized.

Owing to their cellulose-like backbones, hemicelluloses can often hydrogen bond to cellulose; this may be important for wall assembly, the hemicelluloses "tethering" microfibrils. The hydrophilic side chains prevent self-aggregation of hemicelluloses.

Other hemicelluloses with short side chains have mannose-rich backbones, including galactoglucomannans and glucuronomannans. Two other hemicelluloses of more restricted occurrence, i.e. callose (a linear polymer of β-(1→3)-linked D-glucose residues) and mixed-linkage glucan (MLG; **Figure 2**), lack side chains but nevertheless share many of the extractability and hydrogen-bonding properties of xyloglucans and arabinoxylans.

Chemistry of Pectins

Pectins are more hydrophilic than hemicelluloses and, once extracted, readily remain water-soluble. They have a high content of α-D-galacturonate residues. Some domains (homogalacturonan) of pectin consist predominantly of a linear chain of α-(1→4)-linked D-galacturonate residues (**Figure 2**). Some lengths of homogalacturonan are fully methyl-esterified and therefore neutral; others have free —COOH groups and are thus polyanions. Other recognizable domains of pectin include rhamnogalacturonan-I (**Figure 2**) and the extremely complicated rhamnogalacturonan-II. Homogalacturonan and the rhamnogalacturonans may be covalently linked as a single long pectic polysaccharide molecule.

Chemistry of Wall Proteins

The wall contains numerous diverse proteins, many of which are glycosylated. Wall proteins include enzymes and expansins (biological roles discussed below) and structural proteins, e.g., extensins (hydroxyproline-rich, basic glycoproteins) and proline-rich and glycine-rich proteins. Acidic, mucilaginous, arabinogalactan proteins may be attached to membrane lipids and possibly function in cell signaling.

Lignin, Cutin, Cutan, Suberin, Sporopollenin, Silica, and Tannins

Of these seven nonpolysaccharides, the first five are hydrophobic, indigestible polymers impregnated into certain specialized walls. Lignin occurs in the walls of tracheary elements and some fibers: lignification begins in their primary walls and only later spreads into the thick secondary walls. Cutin and suberin are largely aliphatic polyesters (composed of hydroxy fatty acids and fatty alcohols), but also contain tightly associated "lignin-like" aromatic material. Cutin occurs in the cuticle and epidermal wall, suberin in cork cell walls and the Casparian strip. Some plants possess cutan, a polythene-like polymer, instead of cutin. Sporopollenin, present in the exine of pollen grains, is an extremely resistant and incompletely characterized polymer built up of long-chain aliphatic and aromatic carbon skeletons. Some walls contain additional structural materials, e.g., silica (especially in grasses) or possibly condensed tannins (proanthocyanidins).

Variation in Wall Composition

Primary walls, secondary walls and the middle lamella differ markedly in composition (**Table 1**).

Different Cells: Different Walls

Specific cell types, adapted to particular biological roles, often have characteristic wall compositions: e.g., xylem elements (very rich in lignin), cork (suberin), epidermis (cutin), root cap cells (mucilaginous polysaccharides), nettle stinging cells (silica), and phloem sieve plates, and pollen tubes (callose).

During Response to Growth Hormones and Other Stimuli

When cell expansion rates are altered hormonally, little change occurs in wall polysaccharide composition, but specific polysaccharides may change in molecular weight. The cleavage of xyloglucans (in dicotyledons) and MLGs (in grasses) may be part of the wall-loosening mechanism by which auxin promotes growth. Callose can be deposited as an almost instant response to wounding.

Taxonomic Differences: Evolution of the Plant Cell Wall

Major shifts in primary wall composition have accompanied certain evolutionary leaps. Thus, xyloglucans occur in all embryophytes ("land plants," including bryophytes) but apparently not in algae, even those (Charophyceae) closely related to embryophytes. This implies that the first occurrence of xyloglucan coincided with the invasion of the land by plants. Similarly, bryophytes and pteridophytes have primary walls that are very rich in mannose, but this trait diminishes in the seed plants, especially the angiosperms. The origin of the lycopodiophytes (e.g., *Lycopodium* and *Selaginella*), regarded as an evolutionary side alley, was accompanied by the adoption of 3-O-methyl-D-galactose as a major wall

Table 1 Major polymers of the three principal layers of the extraprotoplasmic phase in dicots

Component	*Abundance of component in*			
	Middle lamella (dicots)	*Primary wall (dicots)*	*Primary wall (Gramineae)*	*Secondary wall (dicots)*
Cellulose	–	+++	+++	+++
Xyloglucan	–	+++	+	–
Xylans	–	+	++	+++
Feruloyl groups in xylans		–	+	–
Mannans	–	±	±?	++
Homogalacturonan	++++	++	+	–
Rhamnogalacturonans	+?	++	+	–
Mixed-linkage glucan	–	–	variable	–
Lignin	variable	variable	variable	variable
Extensin	–	+	±	–

The data are generalized; note that some exceptions occur.
–, absent.
±, trace.
++++, major component.
variable, amount strongly dependent on cell type.

component that is practically undetectable in non-lycopodiophytes. Finally, the emergence of the Poaceae as a separate family was apparently accompanied by the first occurrence of MLG and by several quantitative changes (**Table 1**).

Biosynthesis and Secretion of Wall Polymers

Cellulose

Cellulose is the most abundant organic chemical on earth but its biosynthesis is still poorly understood. This glucose polymer is produced by enzymic factories ("terminal complexes") in the plasma membrane, using cytoplasmic UDP-glucose as donor substrate. Each glucose residue is transferred from UDP-glucose, possibly via glycolipid and/or glycoprotein "carriers," most likely on to the nonreducing terminus of a growing cellulose chain. Each terminal complex simultaneously extrudes into the wall numerous cellulose chains, which almost instantly hydrogen bond to each other, forming one microfibril. As the microfibril grows, it pushes around within the plasma membrane the terminal complex that is synthesizing it. This movement, and hence the final orientation of the deposited microfibril, may be guided by protoplasmic microtubules. Microfibril orientation dictates the direction of cell expansion: transversely and longitudinally orientated microfibrils direct cell growth in length and girth, respectively.

Matrix Polysaccharides

Pectins and most hemicelluloses (excluding callose, which is made in the plasma membrane) are synthesized in Golgi bodies, using sugar nucleotides (especially UDP-glucose, UDP-galactose, UDP-glucuronate, UDP-galacturonate, UDP-xylose, UDP-arabinose, GDP-mannose, and GDP-fucose) as donor substrates, in transglycosylation reactions catalyzed by polysaccharide synthases. The polysaccharides are transported in vesicles from the Golgi bodies to the plasma membrane and secreted into the apoplast. Experiments in which radiolabeled monosaccharides are fed *in vivo* show that the transit time, from synthesis to secretion, is typically 15–25 min.

Polysaccharide synthase activity (rather than the supply of sugar nucleotides) is the main control point in wall polysaccharide biosynthesis during plant development. For example, pectin synthase and xylan synthase decrease and increase, respectively, during the transition from primary to secondary wall production in differentiating xylem.

Nonsugar Substituents on Matrix Polysaccharides

Before secretion, noncarbohydrate groups may be added to wall polysaccharides, especially methyl, acetyl, and feruloyl esters and methyl ethers (**Table 2**). Acetyl and methyl groups are added to polysaccharide chains from acetyl coenzyme A (acetyl-CoA) and *S*-adenosylmethionine, respectively. Feruloyl groups may be added from feruloyl coenzyme A (feruloyl-CoA).

Lignin, Cutin, and Suberin

Lignin, cutin, and suberin are synthesized in the cell wall itself and not by membranes. Lignin is formed by oxidative polymerization of small phenolic molecules (monolignols, e.g., coniferyl alcohol) catalyzed by wall-bound peroxidases and/or oxidases. Cutin and suberin appear to be formed by transacylation reactions using hydroxy-fatty-acyl-CoAs as donor substrates.

Wall Proteins

Proteins destined for the wall are synthesized on the rough endoplasmic reticulum and secreted via Golgi

Table 2 Presence and possible significance of noncarbohydrate substituents known to be attached to cell wall polysaccharides (for structures, see **Figure 5C**)

Noncarbohydrate substituent	*Location on wall polysaccharide molecule*	*Possible significance of the substituent's presence*
Methyl ester	Galacturonate residues (often several contiguous ones) of homogalacturonan	Abolishes the negative charge of galacturonate and thus prevents Ca^{2+} bridge formation; increases polymer's hydrophobicity
Methyl ether	Xylose and fucose residues in rhamnogalacturonan-II, glucuronate residues in glucuronoarabinoxylan; galactose residues in lycopod cell walls	Increases hydrophobicity, decreases enzymic digestibility?
Acetyl ester	Galacturonate residues in some pectins, galactose residues in xyloglucan, xylose residues in xylans	Increases hydrophobicity, decreases enzymic digestibility?
Feruloyl and *p*-coumaroyl esters[a]	Arabinose residues in glucuronoarabinoxylan (of Poaceae), arabinose and galactose residues in pectins (of Chenopodiaceae)	Site for covalent crosslinking by oxidative phenolic coupling; increases hydrophobicity, decreases enzymic digestibility

[a] Ferulic acid is 4-hydroxy-3-methoxycinnamic acid; *p*-coumaric acid is 4-hydroxycinnamic acid.

bodies and Golgi vesicles. Signal sequences at the proteins' N-termini direct them to the wall. The hydroxyproline residues of extensin are formed in the endoplasmic reticulum from initially incorporated proline residues; soon afterwards, in the Golgi bodies, many of the hydroxyproline residues acquire a tetra-arabinoside side-chain synthesized from UDP-Ara.

Assembly of the Cell Wall: Wall Tightening

When and Where?

Consider first the addition of new wall material to an existing wall (**Figure 3**). Matrix polysaccharides are secreted, a vesicle's worth at a time, into a hypothetical layer (the "periplasmic space"), presumed to exist between the plasma membrane and the wall proper. Since the turgid protoplast exerts pressure on the wall, each new vesicle-load of matrix polymers becomes squashed against the inner face of the wall and thus perhaps spread over a wide area. In addition, the newly secreted polysaccharides become concentrated by a sieving effect as water, but not polymers, is squeezed through the existing wall. Simultaneously, cellulose microfibrils are being extruded through this concentrated layer of new matrix polymers. Here, somehow, the components become entangled and cross-linked to form new, "tight" wall material. Pulse radiolabeling studies show that wall assembly is very rapid: xyloglucan becomes firmly wall-bound almost immediately (<1 min) after its secretion.

In some cells (e.g., young, rapidly expanding corn (*Zea mays*, maize) cultures), arabinoxylans are secreted into the "periplasmic space" not as a solution but as a large coagulum of molecules already covalently crosslinked via ferulate-based bridges (see below). Thus, in these cells, each new vesicle-load of arabinoxylans probably cannot spread to cover a large area of wall nor efficiently bind to microfibrils. Such a situation would lead to a relatively "loose" wall, which can expand very readily.

At the cell plate after mitosis, where a completely new wall is formed, vesicles first fuse to form the new opposing plasma membranes, and the vesicular contents form the middle lamella. Later, the membranes extrude microfibrils and additional vesicles arrive to deliver matrix polymers that entangle and crosslink with the microfibrils to form the new primary wall.

Crosslinks form not only during wall assembly but also postsynthetically, leading to wall tightening which may reduce the rate of cell expansion and increase resistance to microbial attack.

Hydrogen Bonding

One step in wall assembly is the nonenzymic hydrogen bonding of hemicelluloses (e.g., xyloglucans) to the

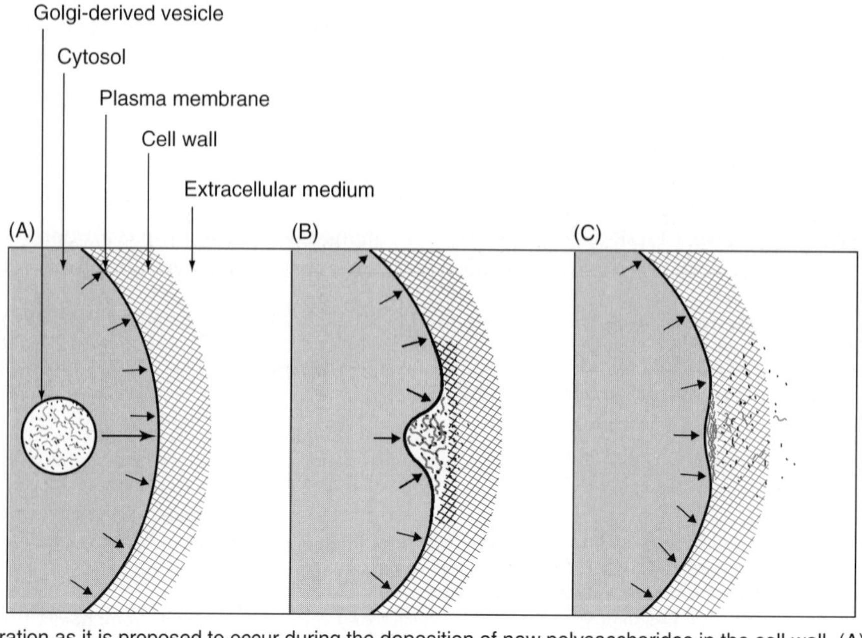

Figure 3 Ultrafiltration as it is proposed to occur during the deposition of new polysaccharides in the cell wall. (A) Vesicular traffic: a Golgi-derived vesicle, containing polysaccharides (~ ~ ~) and water (∴∵), moves towards the plasma membrane. (B) Exocytosis: the vesicle fuses with the plasma membrane, discharging its contents into the "periplasmic space." (C) Ultrafiltration: water from the vesicle is squeezed through the wall owing to the pressure exerted by the turgid protoplast. Most of the polysaccharide molecules remain trapped, at high concentration, in the "periplasmic space." Small arrows (→) represent turgor pressure (exerted by protoplast on cell wall); (——), membrane; (▩▩▩), existing cell wall.

naked surfaces of newly extruded lengths of microfibril. Hemicellulose–cellulose hydrogen bonding is readily demonstrated with purified polysaccharides *in vitro*. The creation of such bonds *in vivo* will be enhanced by the very high concentration of newly secreted hemicelluloses present in the periplasmic space owing to the "ultrafiltration" that probably occurs when water (but not large molecules such as hemicelluloses) from exocytosing vesicles is squeezed through the existing wall (**Figure 3**). Thus, any given length of newly extruded microfibril will soon find itself covered by a molecular monolayer of hydrogen-bonded hemicelluloses. Most hemicellulose molecules are much longer than the spacing between microfibrils, and it is proposed that individual hemicellulose molecules act as "tethers" between neighboring microfibrils. This arrangement may make a major contribution to wall assembly and strength.

Ionic Bonding

Many pectins and xylans possess acidic residues (galacturonate and glucuronate, respectively). Basic wall proteins, e.g., extensins and some enzymes, will be electrostatically attracted to acidic polysaccharides and in this way fixed in position within the wall. Such proteins (at least until additional, non-ionic bonds form) can be solubilized from the wall with concentrated salts, e.g., 0.2 mol l^{-1} $CaCl_2$, which competitively cleave polymer–polymer ionic bonds.

Calcium Bridges

The Ca^{2+} ion, being divalent, can bond to two homogalacturonan chains, thereby crosslinking them. For this reason, the crispness of pickled gherkins is enhanced by inclusion of $CaCl_2$ in the brine. Strong crosslinking depends on several contiguous galacturonate residues in both homogalacturonan molecules being acidic (lacking methylester groups):

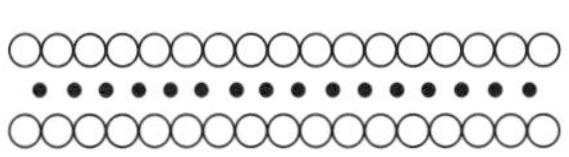

• Ca^{2+}
○ Galacturonate residue

Therefore, pectinesterase, which removes methyl ester groups from a run of contiguous galacturonate residues, is a "wall-tightening enzyme," enabling calcium bridging.

Borate Ester Bridges

Borate forms unusually strong ester linkages with specific apiose residues in rhamnogalacturonan-II. One borate can simultaneously bind to two such residues, covalently crosslinking two rhamnogalacturonan-II molecules. It is unclear when in the life of the polysaccharide molecule and where in the cell such crosslinks are formed.

Covalent "Grafting" by Transglycosylation

If cellulose synthesis is blocked by treatment with 2,6-dichlorobenzonitrile, the wall binding of xyloglucan continues for many hours. Thus, hydrogen bonding to new cellulose is not the only mechanism of xyloglucan integration. Another mechanism is the chemical "grafting" (transglycosylation) of a newly secreted xyloglucan to another that is already wall-bound. This is catalyzed in the wall by xyloglucan endotransglucosylase (XET) activity. XET is one of the two enzymic activities exhibited by proteins that are now termed XTHs (xyloglucan endotransglucosylase/hydrolases). The XET protein cuts a glycosidic linkage in the backbone of xyloglucan (○○○○○○) and transiently forms a xyloglucan–enzyme covalent complex:

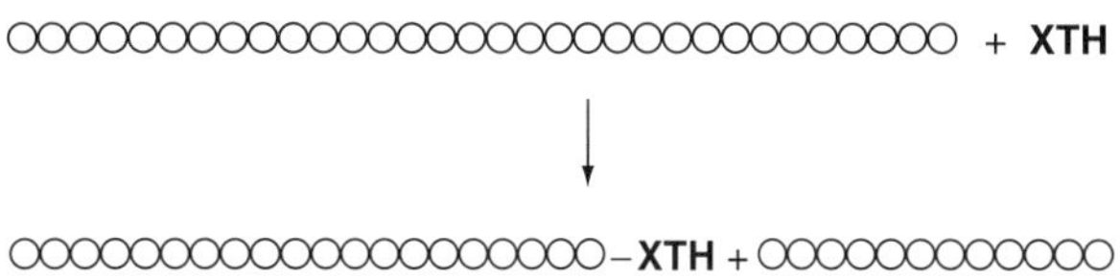

The xyloglucan moiety of the complex is then transferred on to the nonreducing end of another xyloglucan (⊙⊙⊙⊙⊙⊙):

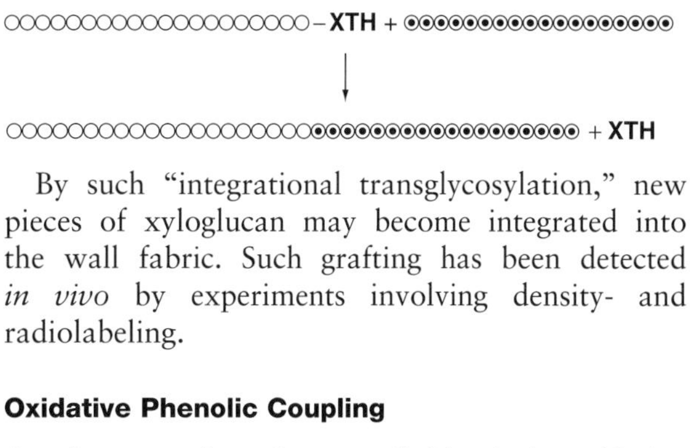

By such "integrational transglycosylation," new pieces of xyloglucan may become integrated into the wall fabric. Such grafting has been detected *in vivo* by experiments involving density- and radiolabeling.

Oxidative Phenolic Coupling

Another type of covalent crosslinking is the oxidative coupling of phenolic groups, probably catalyzed by peroxidases $+ H_2O_2$. Feruloyl groups couple to form various isomeric diferulates (plus larger quantities of incompletely characterized trimers, tetramers, etc.), probably crosslinking the arabinoxylans to which they are attached. Tyrosine residues also produce dimers, trimers, and tetramers (**Figure 4**), crosslinking the wall glycoproteins in which they occur. Oxidative phenolic coupling was long assumed to

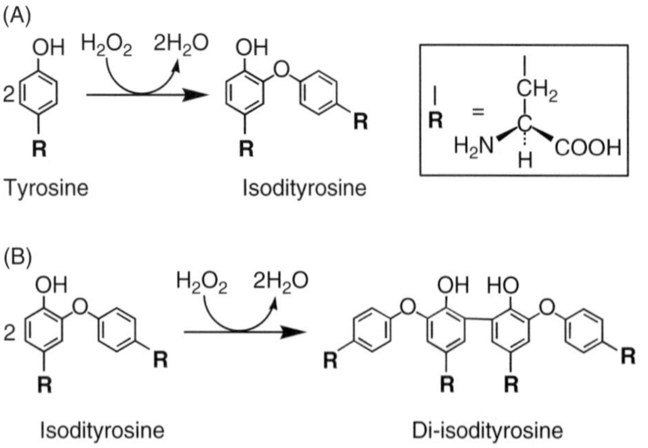

Figure 4 Oxidative phenolic coupling as a mechanism for crosslinking plant cell wall polymers, exemplified by the tyrosine residues of extensin, a cell wall glycoprotein. (A) Dimerization of tyrosine to form isodityrosine; (B) dimerization of isodityrosine to form di-isodityrosine. **R** represents the structure shown in the box. Peroxidase may be the enzyme responsible, with H_2O_2 as the oxidizing substrate. Other wall-bound phenolics, such as the ferulate groups that are ester-linked to arabinoxylans, also undergo a series of similar reactions. Reproduced with permission from Brady JD and Fry SC (1997) Formation of di-isotyrosine and loss of isodityrosine in the cell walls of tomato cell-suspension cultures treated with fungal elicitors or H_2O_2. *Plant Physiology* 115: 87–92.

happen only after secretion of the polysaccharides or glycoproteins; however, recent pulse-radiolabeling experiments with corn cells show that some feruloyl coupling occurs intraprotoplasmically, arabinoxylans being secreted into the apoplast already crosslinked.

Turnover of the Cell Wall: Wall Loosening

When and Where?

Once assembled, the cell wall is not a static structure serving only a passive role in plant architecture. Rather, walls may be modified postsynthetically, according to the changing needs of tissues. For example, the end walls of xylem vessel elements undergo controlled autolysis to enable these cells to function in sap translocation. Walls of parenchyma and epidermal cells may be loosened to promote cell expansion and to cause the tissue softening characteristic of abscission and fruit ripening.

Hydrolases in Autolysis

Primary walls contain numerous hydrolytic enzymes, many apparently able to attack the wall itself. There are endoglycanases, which cleave polysaccharide backbones, causing a rapid decrease in molecular weight; and there are exoglycanases, which remove sugar residues progressively from nonreducing end(s), causing only moderate erosion of the polysaccharide. In addition, there are esterases, and proteinases. Some wall glycanases and their possible biological roles are listed **Table 3**, but in many cases their precise biological roles remain uncertain (**Figure 5**).

Transglycosylation

Besides "integrational transglycosylation" (see above), XETs also catalyze transglycosylation between pairs of previously wall-bound xyloglucans ("restructuring transglycosylation"), as evinced by dual-labeling experiments *in vivo*. If, as believed, xyloglucans tether neighboring microfibrils, the transient cutting and rejoining of xyloglucans by XET activity may reversibly loosen the wall and permit cell expansion without permanent weakening. XET activity is elevated in many rapidly expanding cells (e.g., gibberellin-treated, or a certain distance from the apical meristem). By experiments using fluorescently labeled substrates *in vivo*, XET action (as opposed to "activity," assayed *in vitro* with extracted enzyme) was shown to peak in the root's elongation zone. In addition, XET activity rises during ripening, e.g., in kiwifruit and persimmons. Thus, XET often correlates with – and may cause – wall loosening.

Expansin Action

Expansins are wall-located proteins with little or no hydrolytic activity but which appear to catalyze the reversible breakage of hemicellulose–cellulose hydrogen bonds. Exogenous expansins thereby loosen plant cell walls, as shown by measurements of creep – the slow, irreversible extension of a wall specimen when stretched with a constant force. In model experiments, expansins even promoted the "creep" of stretched filter paper (which is almost pure cellulose), indicating cleavage of cellulose–cellulose hydrogen bonds. Expansins have acidic pH optima and their action may relate to the mechanism of "acid growth;" they may also contribute to the softening of ripening fruit. Properties of the two major classes of expansins are summarized in **Table 4**.

Action of Hydroxyl Free-Radicals: Nonenzymic Scission

Enzymes and expansins are not the only agents that could cause wall loosening *in vivo*. Hydroxyl free-radicals ($^{\bullet}OH$) are an additional possibility. These are produced *in vitro* simply upon mixing ascorbate, O_2 and traces of Cu^{2+}, via three nonenzymic reactions:

$$AH_2 + O_2 \rightarrow A + H_2O_2 \qquad [1]$$

$$\tfrac{1}{2}AH_2 + Cu^{2+} \rightarrow \tfrac{1}{2}A + Cu^{+} \qquad [2]$$

$$H_2O_2 + Cu^{+} \rightarrow {}^{\bullet}OH + OH^{-} + Cu^{2+} \quad \text{('Fenton reaction')} \qquad [3]$$

Table 3 Some hydrolytic enzymes present in the plant cell wall and their possible roles *in vivo*

Enzyme	*Possible substrates in cell wall*	*Chemical effect on substrates*	*Possible biological significance*
Endoglycanases			
Cellulase (endo-β-(1→4)-D-glucanase)	Xyloglucan, MLG, cellulose(?)	Mid-chain cleavage of glucan backbone	Wall loosening to enable cell expansion; fruit ripening in some species; abscission
Pectinase (endo-α-(1→4)-D-galacturonanase; 'endopolygalacturonase')	Homogalacturonan	Mid-chain cleavage of homogalacturonan backbone (acidic regions)	Fruit ripening in some species; abscission
Exoglycanases			
β-D-Glucosidase	Xyloglucan,[a] MLG	Removal of glucose residues one by one from nonreducing terminus	Degradation of MLG to recycle monosaccharide; role in xyloglucan unclear
β-D-Fructofuranosidase (invertase)	Sucrose (dissolved in apoplast)	Hydrolysis to glucose + fructose	Release of respiratory substrates from phloem-translocated sucrose
α-D-Galactosidase	Raffinose, verbascose, etc. (dissolved in apoplast)	Removal of galactose from oligosaccharide structure	Release of respiratory substrates from phloem-translocated oligosaccharides
β-D-Galactosidase	Xyloglucan, rhamnogalacturonan-I	Removal of galactose residues one by one from nonreducing termini of side chains	?
α-L-Fucosidase	Xyloglucan-derived oligosaccharides	Removal of fucose from oligosaccharide structure	Inactivation of oligosaccharin after perception of signal?
α-D-Xylosidase	Xyloglucan-derived oligosaccharides	Removal of xylose from nonreducing terminus of oligosaccharide structure	?
α-D-Galacturonidase	Homogalacturonan; oligogalacturonides	Removal of galacturonate residues one by one from nonreducing terminus until size of substrate is reduced to pentasaccharide	Inactivation of oligosaccharin after perception of signal?

[a] After initial attack by α-D-xylosidase.

where AH_2 = ascorbate and A = dehydroascorbate. Since ascorbate, O_2 and Cu^{2+} are all normally apoplastic components, •OH radical production *in vivo* seems plausible. •OH is the most reactive of all known molecules, readily causing nonenzymic scission of polysaccharides. Work is in progress to explore •OH production in the walls of living cells and whether such radicals cleave wall polysaccharides *in vivo* and thus loosen the cell wall during cell expansion or fruit ripening.

Biophysics of the Plant Cell Wall

Ion-Exchange Properties

Metal ions bind to cell walls and especially to pectins in the walls. Some bind reversibly (e.g., Na^+, K^+), while others bind much more durably (e.g., Cu^{2+}, Pb^{2+}). Histidine-rich proteins may also contribute to the binding of Cu^{2+}. Much of a plant's total Cu^{2+} may be wall-bound, and thus harmlessly sequestered if taken up excessively from polluted soils.

Permeability: Pore Size

Cell walls are freely permeable to small, uncharged molecules. However, wall "pore" dimensions exclude globular proteins above 5–7 nm diameter (≡36–67 kDa). Partial wall digestion with pectinase (but not cellulase or proteinase) enables permeation by larger molecules, suggesting that pectins govern pore size.

Despite these small pore size estimates, plant cell cultures secrete huge molecules (> 1000 kDa) into their medium (i.e., through the wall). Therefore, some larger pores must also exist. To date, all pore measurements have been made in dead or flaccid cells. In a turgid cell, the protoplast may widen some pores by stretching the wall.

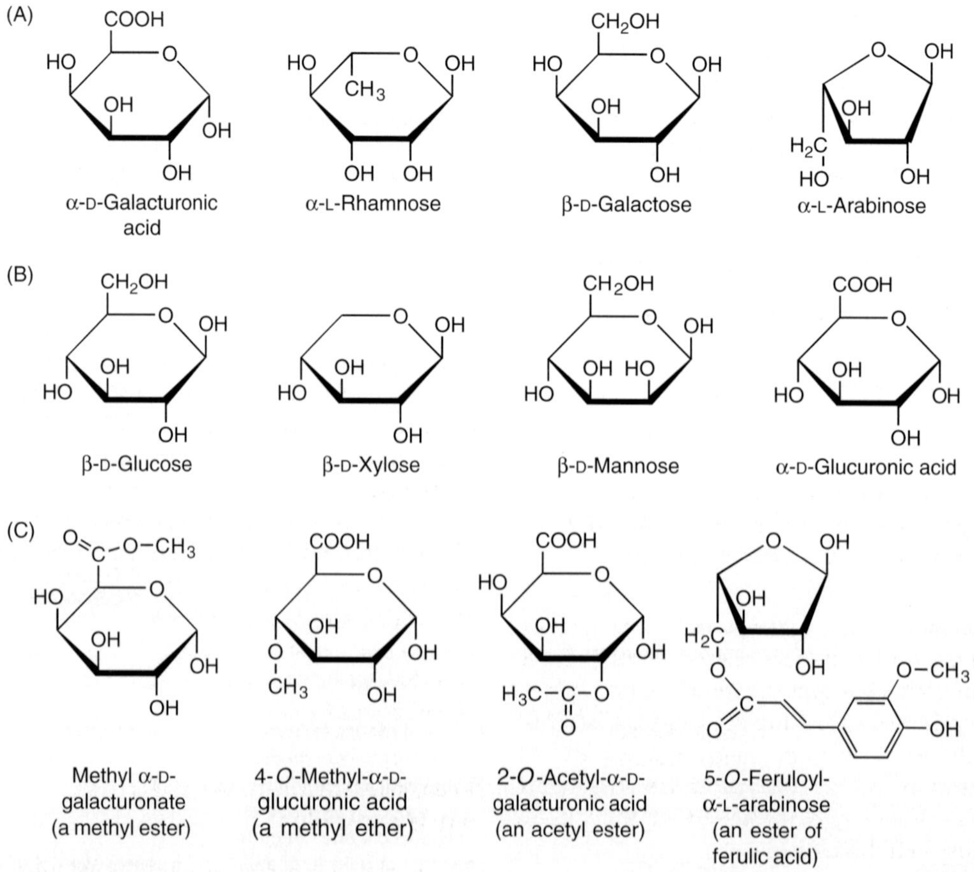

Figure 5 Chemical structures of representative sugar residues found in plant cell wall polysaccharides. (A) Sugar residues characteristic of pectins; (B) sugar residues characteristic of hemicelluloses; and (C) selected sugar residues with noncarbohydrate substituents. Note that the thumbnail classification into residues characteristic of pectins and of hemicelluloses is a simplification and that significant exceptions exist.

Table 4 Some properties of α- and β-expansins

Properties	*α-Expansins*	*β-Expansins*
Hydrophobicity	Hydrophobic	Hydrophilic
Glycosylation	Nonglycosylated	Glycosylated
Extraction from cell wall	Slow; requires high salt	Rapid, in water
Break cellulose–cellulose hydrogen bonds	Probably	Not tested
Break xyloglucan–cellulose hydrogen bonds	Probably	Not tested
Cause creep in dicot cell walls	Yes	Slightly
Cause creep in graminaceous cell walls	Slightly	Yes
Other comments	Numerous isoforms expressed, especially in dicots	Abundant protein on grass pollens; causes hay fever

Elasticity, Plasticity and Creep: Wall Loosening and Tightening

When mature walls are stretched (e.g., by increasing turgor pressure), some "elastic" (i.e., reversible) expansion occurs immediately; when the stretching force is removed, the wall springs back to its original size. Reversible expansion of guard cell walls causes the opening and closing of stomata.

"Plastic" expansion is slower but irreversible: when a young, primary wall is stretched and later relaxed, the wall does not quite return to its original size. Plastic expansion is "growth," i.e., an irreversible increase in cell volume. Some plastic expansion occurs immediately the force is applied, with an additional component (creep) occurring gradually. Walls with high and low plasticity are described as loose and tight, respectively (the terms "rigid" and "stiff", which are sometimes used instead of "tight," are unhelpful because they wrongly imply that the inextensible wall is difficult to bend). Wall loosening and tightening are the major factors regulating the growth rates of individual cells, at least when water is plentiful.

Breaching of the Cell Wall by Microbes

Fungal and Bacterial Strategies for Wall Digestion

Microbes attack plant cell walls in two main situations:

- as parasites or symbionts of living plants; and
- as saprophytes – either free (e.g., digesting leaf litter) or symbiotic (in animals' guts).

Such microbes produce endo- and exoglycanases that hydrolyze wall polymers. Often, pectinases are produced first, which digest the middle lamella; later, cellulases, xylanases, and other glycanases appear. To achieve cellulose digestion, microbes require the concerted action of an endoglycanase (cellulase, i.e., endo-(1→4)-β-D-glucanase) and an exoglycanase (cellulose cellobiohydrolase).

The Plant's Strategies for Bolstering its Walls

As though designed to enhance resistance to microbial digestion, plant cell walls often have features minimizing the action of microbial enzymes. Walls may be impregnated with relatively indigestible substances such as lignin, cutin, suberin, extensin, or silica. Sometimes, these substances are deposited rapidly as a defensive response to the first signs of microbial invasion. Plant cell walls may also constitutively possess substances that inhibit microbial enzymes, e.g., tannins or specific polygalacturonase inhibitor proteins (PGIPs).

The Counter-Attack

Clearly, these defensive strategies are surmountable. (So abundant are plant cell walls that if any minor wall component were not digestible, much of the world's carbon would by now be locked up in that form!) White rot fungi produce "ligninases" (which are oxidoreductases, not hydrolases), and some phytopathogens can penetrate the cuticle by secreting cutinase.

Agriculturally, vast quantities of cellulose produced annually are burnt or rotted. For example, straw is very rich in cellulose, but difficult to exploit because it is lignified. There would be great commercial benefit to be gained from an ability to digest straw cheaply to glucose.

Oligosaccharins

Cell walls can generate signaling molecules. Certain oligosaccharides, termed oligosaccharins, produced by partial hydrolysis of wall polysaccharides, possess potent biological activity. For example, homogalacturonan-oligosaccharins activate several defensive responses including phytoalexin or lignin biosynthesis and glycoprotein crosslinking. Certain fucose-containing xyloglucan-oligosaccharins inhibit growth at concentrations as low as $1\,\mathrm{nmol\,l^{-1}}$. Other possible oligosaccharins affect morphogenesis and protein synthesis.

Most experiments have concerned exogenous oligosaccharins produced *in vitro*. However, radiolabeling experiments confirm that traces of oligosaccharins are produced *in vivo* by partial cleavage of preformed polysaccharides. Thus, oligosaccharins can be added to the list of naturally occurring signaling molecules.

Wall Metabolism as a Target for Herbicides

Since the pectin–hemicellulose–cellulose wall of embryophytes is unique and indispensable, a specific inhibitor of any essential aspect of wall metabolism would be a potential herbicide. The few examples of wall-interfering herbicides reported to date, including 2,6-dichlorobenzonitrile (dichlobenil), isoxaben, quinclorac, triazofenamide, and a thiazolidinone, are all inhibitors of cellulose biosynthesis. Of these, quinclorac alone appears to inhibit hemicellulose biosynthesis also. Oxaziclomefone, a new herbicide particularly effective on the growth of grass roots, appears to act by specifically blocking cell wall expansion, without immediately affecting cellulose biosynthesis or cell division, and may therefore have a novel target in wall metabolism. There is an untapped resource of potential new targets for herbicide action in the numerous other processes in cell wall biochemistry, some of which are discussed here, while others no doubt have yet to be discovered. Of particular interest is the discovery of important taxonomic variation in wall biochemistry, raising the prospect of valuable new selective herbicides.

List of Technical Nomenclature

(1→3), (1→4), etc. Locants, designating the points of linkage between two neighboring sugar residues within a polysaccharide.

α-, β-	Prefixes to designate the chirality at carbon atom no. 1 of an aldose sugar (relative to that at its penultimate carbon atom). Starch is an α-glucan; cellulose is a β-glucan.
Apoplast	The aqueous solution that permeates the cell wall (includes xylem sap).
D-, L-	Prefixes to designate the chirality at the penultimate carbon atom of a sugar.
Endoglycanase	An enzyme that can split a polysaccharide molecule in mid-chain by hydrolyzing a glycosidic bond.
Exoglycanase	An enzyme that can release sugars from a polysaccharide by hydrolyzing glycosidic bonds at or near a nonreducing terminus.
GDP-mannose	Guanosine diphospho-mannose, a high-energy form of mannose used in mannan synthesis.
Glucan, xylan, mannan, galacturonan	Any polysaccharide with a backbone composed mainly of glucose, xylose, mannose, or galacturonate residues, respectively.
Glucuronoarabinoxylan	A xylan with side chains containing glucuronate and arabinose residues. (Further polysaccharide names can be constructed on a similar basis.)
Glycoprotein	A protein with mono-, oligo- or polysaccharides glycosidically linked to it.
Glycosidic bond	Covalent linkage formed between a sugar and some other molecule (which may itself be a sugar), with the loss of one molecule of H_2O.
Hemicellulose	Any of several nonpectic polysaccharides extracted from the cell wall with alkali.
Nonreducing terminus	A sugar residue that is glycosidically linked within a polysaccharide but has no other sugar residue glycosidically linked to it; a polysaccharide may have many nonreducing termini.
Oligosaccharide	As polysaccharide but with fewer sugar residues, e.g. 3–30.
Oligosaccharin	An oligosaccharide with biological signaling activity.
Pectin	Any of several galacturonate-rich polysaccharides of the cell wall.
Plasmolysis	Shrinkage of the protoplast away from the cell wall owing to loss of water.
Polysaccharide	A polymer composed of many sugar residues linked to each other.
Protoplast	The "living" part of the cell, bounded by the plasma membrane, i.e., excluding the cell wall.
Reducing terminus	The single sugar at one end of a polysaccharide that is not glycosidically linked to another sugar and will therefore reduce Fehling's solution.
Sugar residue	A glycosidically linked sugar molecule.
Transglycosylation	The process of splitting a glycosidic linkage and reattaching the former sugar residue to a different molecule.
UDP-glucose	Uridine diphospho-glucose, a high-energy form of glucose used in glucan synthesis.
Xyloglucan	A glucan with side chains containing xylose residues.

See also: **Diseases**: Bacterial Diseases; Fungal Diseases. **Growth and Development**: Cells. **Primary Products**: Cellulose; Proteins; Starch. **Secondary Products**: Lignin.

Further Reading

Aldington S and Fry SC (1993) Oligosaccharins. *Advances in Botanical Research* 19: 1–101.

Bolwell GP (1993) Dynamic aspects of the plant extracellular matrix. *International Review of Cytology* 146: 261–324.

Brett CT and Waldron K (1996) *Physiology and Biochemistry of Plant Cell Walls*, 2nd edn. London: Chapman & Hall.

Buchanan CJ, Fry SC, and Eastwood MA (1996) The use of ^{14}C-labelled substrates to study plant cell wall breakdown in the gastrointestinal tract. *Proceedings of the Nutrition Society* 55: 927–936.

Carpita NC and McCann MC (2000) The cell wall. In: Buchanan BB, Gruissem W, and Jones RL (eds) *Biochemistry and Molecular Biology of Plants*, pp. 52–105. Rockville, MA: American Society of Plant Physiologists.

Cosgrove DJ (1997) Assembly and enlargement of the primary cell wall in plants. *Annual Review of Cell and Developmental Biology* 13: 171–201.

Delmer DP (1999) Cellulose biosynthesis: exciting times for a difficult field of study. *Annual Review of Plant Physiology and Plant Molecular Biology* 50: 245–276.

Delmer DP and Stone BA (1988) Biosynthesis of plant cell walls. In: Preiss J (ed.) *Biochemistry of Plants: A Comprehensive Treatise*, vol. 14, pp. 373–419. New York: Academic Press.

Fry SC (1986) Cross-linking of matrix polymers in the growing cell walls of angiosperms. *Annual Review of Plant Physiology and Plant Molecular Biology* 37: 165–186.

Fry SC (1995) Polysaccharide-modifying enzymes in the plant cell wall. *Annual Review of Plant Physiology and Plant Molecular Biology* 46: 497–520.
Fry SC (2000) *The Growing Plant Cell Wall: Chemical and Metabolic Analysis*, pp. 1–333. Caldwell, NJ: The Blackburn Press.
Nishitani K (1998) Construction and restructuring of the cellulose–xyloglucan framework in the apoplast as mediated by the xyloglucan-related protein family – A hypothetical scheme. *Journal of Plant Research* 111: 159–166.
Popper ZA and Fry SC (2003) The primary cell wall composition of bryophytes and charophytes. *Annals of Botany* 91: 1–12.
Showalter AM (1993) Structure and function of plant cell wall proteins. *Plant Cell* 5: 9–23.
Staehelin LA and Moore I (1995) The plant Golgi apparatus – structure, functional organisation and trafficking mechanisms. *Annual Review of Plant Physiology and Plant Molecular Biology* 46: 261–288.
Wallace G and Fry SC (1994) Phenolic components of the plant cell wall. *International Review of Cytology* 151: 229–267.

Fiber Formation

J E G van Dam, WUR, Wageningen, The Netherlands
T A Gorshkova, KIBB, RAS, Kazan, Russia

Introduction

In all higher plants elongated fiber cells form the structural elements, giving strength and shape to the tissues of stems, branches, and roots, but also to softer tissues such as leaves and flowers or fruits. Plant fibers represent a renewable raw material of special technical importance. Their utilization is closely linked to the human history of technical development and exploration of natural resources. Millions of tons are consumed annually in processed forms for textiles, paper, chemical-grade cellulose, or for fuel and lumber.

Plant fibers derive their mechanical strength primarily from the special design of the cell architecture and the cellulose fibrils contained in the cell walls. Plant cell walls composed of polysaccharides, glycoproteins, and phenolic compounds form in the fibers thick layers (around 10 μm, as compared to 0.1–2.0 μm in the cells of other tissues) of structural material around the protoplast.

Because their extreme characteristics of cell length and wall thickness, plant fibers are obvious subjects to study biochemical and molecular biological mechanisms of plant cell elongation and cell wall formation. The intensity of these processes, the high turnover of specific metabolites in the dynamic stages of fiber formation, and the simultaneous changes in morphology of the cells all give clues to the highly organized process.

Definition

Plant fibers are found as structural elements in all higher plants. Botanically the fiber is considered to be an individual cell, which is part of sclerenchyma tissue and is characterized by a thick cell wall and a high length-to-diameter ratio (reaching 1000 and more). The basic function of sclerenchyma is providing mechanical integrity to the plant. Spindle-shaped cells with tapered ends are considered to be a characteristic feature of the fiber cell. According to this definition, cotton (*Gossypium* spp.) fiber – probably the most well-known plant fiber – is not a fiber, but a trichome, the extrusion of epidermal tissue.

Classification

Fiber cells are present in many different plant parts. Botanically fibers are divided according to their position within or outside the xylem. Libriform fibers and fiber tracheids, being a part of xylem, are called xylary fibers. Another type of xylary fiber, present in tension wood, is the gelatinous or mucilaginous fibers. Xylary fibers constitute an integral part of the xylem and develop from the same meristematic tissues as do the other xylem components. Extraxylary fibers (i.e., the fibers localized outside xylem) can be found within the cortex (cortical fibers), phloem (phloem or bast fibers), or at the periphery of the vascular bundles (perivascular fibers).

Possessing distinct tensile strength, elasticity, and flexibility, fiber cells can be discriminated from other plant tissues and be extracted for industrial purposes. The sclerenchyma fibers may occur singly as idioblasts, but more usually they occur in bundles. These fiber bundles include the dicotyledon bast fiber crops of commercial interest known as soft fibers: flax (*Linum usitatissimum*), hemp (*Cannabis sativa*), jute (*Corchorus* spp.), ramie (*Boehmeria nivea*), and kenaf (*Hibiscus cannabinus*). Bast fiber bundles run longitudinally along the stem from bottom to top and reach almost the full length of the plants, which may be 3 meters or more for hemp and kenaf (**Figure 1**). The leaf fibers, also known as hard fibers, are the agglomeration of fiber cells with vascular elements and are extracted from the fibrovascular system of leaves of monocotyledons: sisal (*Agave sisalana*), abaca (*Musa textilis*), yucca (*Yucca* spp.), henequen (*Agave fourcroydes*), *Phormium tenax*, some palms

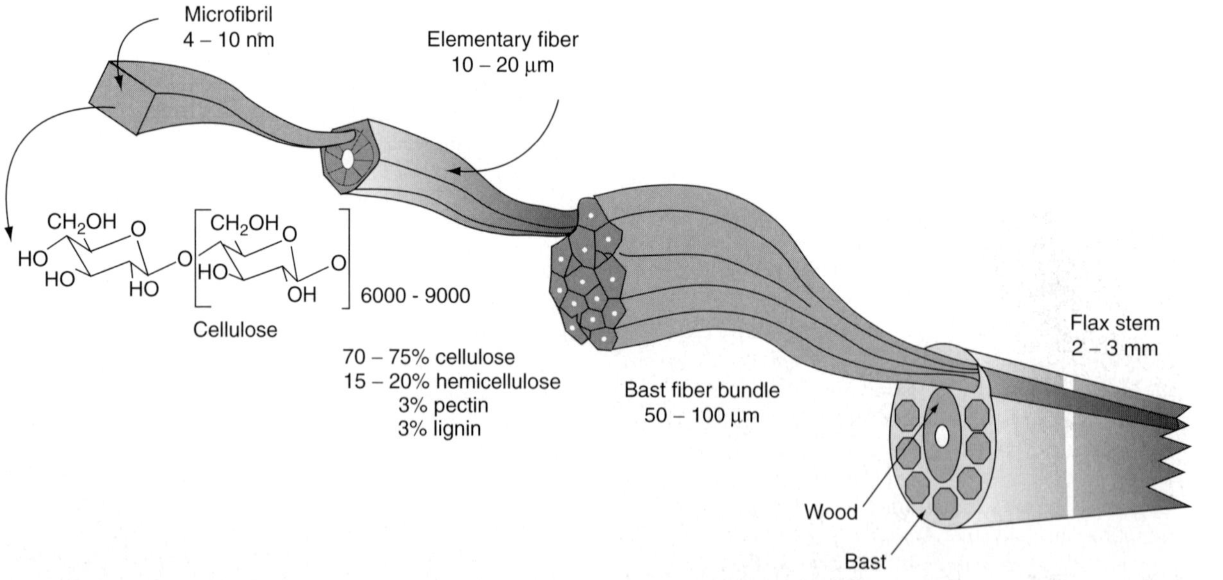

Figure 1 Schematic representation of different levels of magnification of flax bast fiber cells in a bundle.

(e.g., raffia (*Raphia ruffia*), etc). Fibrous tissues may also be obtained from other plant parts such as stalks and roots of grasses (sugar cane (*Saccharum officinarum*) bagasse, reeds, straws) or the mesocarp (coconut (*Cocos nucifera*) coir fiber).

Another class of fibers, that do not occur in bundles, are the lignified fiber cells in the fibrovascular woody tissues of trees and shrubs of both gymnosperm and angiosperm plants. Wood is a mixture of dead and living cells of many different cell types, including fiber-tracheids and libriform fibers. Libriform fibers (liber, inner bark) have extremely thick walls and simple pits. Fiber-tracheids are intermediate forms between tracheids and libriform fibers. Tracheids are not fibers, as their major function is conducting water and the cell shape is not typical of a fiber, though they have relatively thick cell walls. Both fiber-tracheids and libriform fibers have emerged from tracheids in the course of evolution and the intermediate forms can be easily found. This type of wood fiber is utilized in large quantities by the pulping industries for cellulose extraction and papermaking.

None of the plant fiber classifications is completely comprehensive, as there are always some forms, which do not fit any group or do fit more than one. The septate fibers for example cannot be considered as xylary or extraxylary. They can be found both within phloem and xylem, even of the same plant species. These fibers are characterized by the presence of internal septa, which are formed after thickening of the secondary cell wall.

Another classification of plant tissues, including fibers, is made according to their origin. The primary fibers are those formed from the primary (apex) meristem. The secondary fibers are produced by secondary meristem and are formed when the stem has reached its maximum length and secondary thickening takes place by tangential division of cambium cells. In general, primary fibers are thicker, with more compact cell walls and smaller lumen and therefore stronger. Though primary and secondary fibers are easily identified under the microscope, it is often technically difficult to separate them.

Plant fibers of commercial importance are classified according to their use in textile fibers (cotton, ramie, flax, jute, hemp), cordage fibers for making ropes and twines (sisal, coir, abaca, henequen), brush and mat fibers, stuffing and upholstery materials, papermaking fibers or wickerwork materials (raffia). The commercial value depends on the nature of the fiber with respect to length, fineness, strength, and stiffness. The embedding of fiber cells in the surrounding tissues affects the ease of extraction to a large extent, which dictates the commercial value and fiber quality characteristics (**Figure 2**; **Table 1**).

Composition

Plant fibers can be considered as composite structures of oriented cellulose fibrils embedded in a matrix of more or less rigid structures formed by the hemicellulose and lignin polymers. Important characteristics for fiber cells are determined by the amount and the distribution of the chemical constituents in various layers of the cell wall. By immunocytochemistry the cell wall components can be localized. General distribution of most fiber components is similar to that of any plant cell with a secondary cell wall. Pectins are localized mainly in middle lamellae

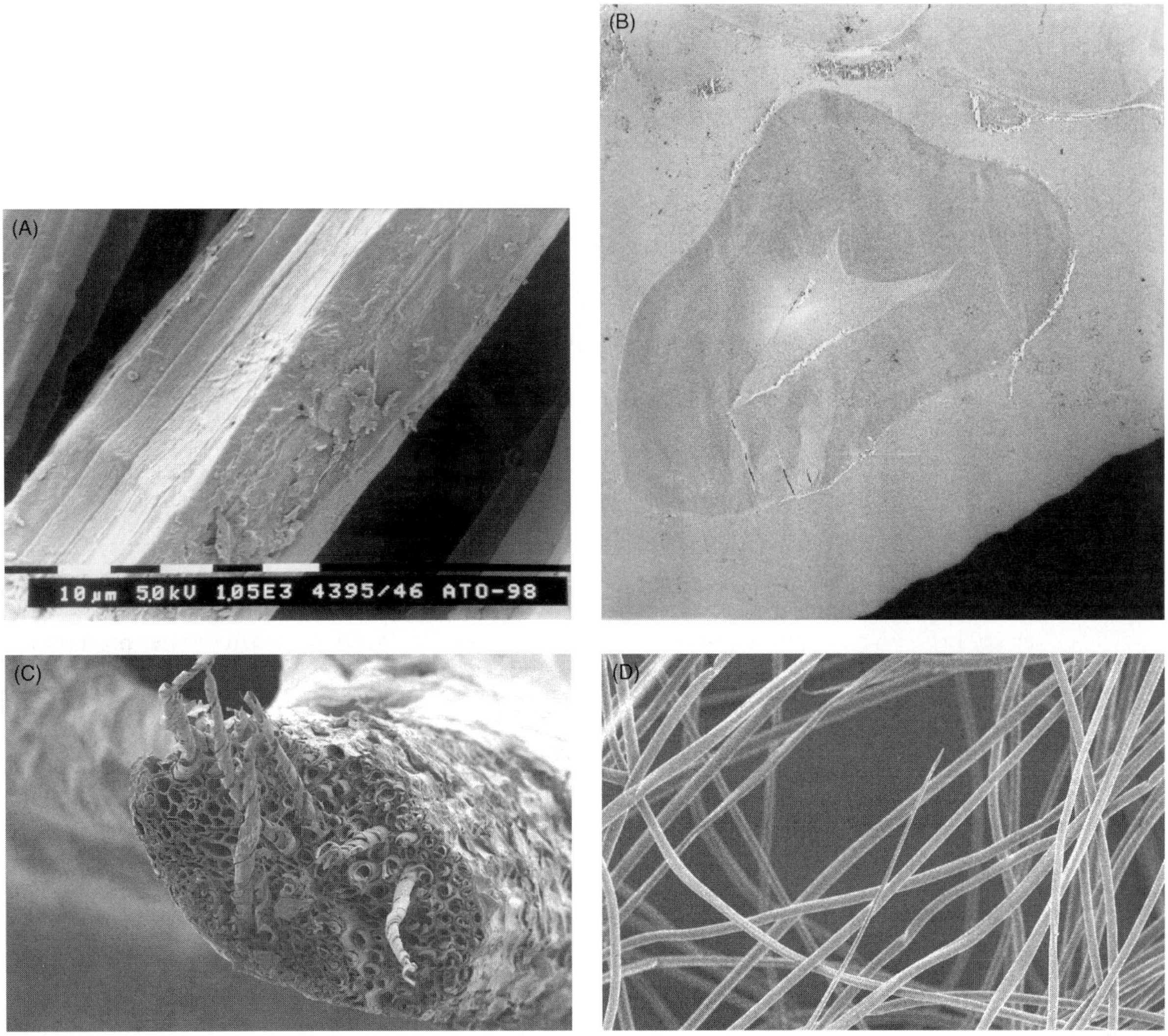

Figure 2 (A) Scanning electron micrograph of kenaf bast fiber (scale = 10 µm); (B) transmission electron micrograph of hemp bast fiber cross-section; (C) scanning electron micrograph of coir fiber (scale = 50 µm); (D) scanning electron micrograph of kapok fiber floss (scale = 100 µm).

and primary cell wall (though they may also be tightly associated to secondary cell wall cellulose). The secondary cell wall is highly enriched in cellulose.

Cellulose

The main stress-bearing compound of the plant cell wall is the crystalline cellulose microfibril, which typically consists of 10 nm wide crystallites of long and straight closely packed polymeric chains of exclusively β-(1→4)-linked D-glucopyranose units. Its insolubility in water or other solvents has been ascribed to its conformational features in crystalline form, in which the hydrogen bonding form strong intramolecular and interchain interactions. The hydrogen bonded sheets in the crystalline latice of native cellulose (cellulose I) is unique for biosynthetically produced fibrils and cannot (yet) be reproduced from chemically regenerated cellulose. The fibrillar orientation in the different cell wall layers and the high strength of the crystalline cellulose make structures strong and flexible.

Noncellulosic Cell Wall Polysaccharides

Though cellulose constitutes the major part of the wall, a number of other polysaccharides occur and are important for fiber properties. Quantification of cell wall polysaccharides is complex due to various types of bonds between the individual polysaccharides and/or other components. Indications of polysaccharide contents are obtained from the monosaccharide composition of the cell wall (**Figure 3**). However, the same monosaccharide can be the part of different polysaccharides: for example, glucose is a component of cellulose, glucomannan, and xyloglucan; xylose is a component of both xyloglucan and xylan, etc. The exact proportions and structures of

Table 1 Botanical description of some technical plant fibers

Common name	*Latin name*	*Family*	*Cell type*
Flax	*Linum usitatissimum*	Linaceae (Dicotyledonae)	Primary phloem fibers
Ramie	*Boehmeria nivea*	Urticaceae (Dicotyledonae)	Primary and secondary phloem fibers
Hemp	*Cannabis sativa*	Cannabaceae (Dicotyledonae)	Primary and secondary phloem fibers
Jute	*Corchorus capsularis*	Tiliaceae (Dicotyledonae)	Primary and secondary phloem fibers
Kenaf	*Hibiscus cannabinus*	Malvaceae (Dicotyledonae)	Primary and secondary phloem fibers
Sisal	*Agave sisalana*	Agavaceae (Monocotyledonae)	Fibers, associated to vascular bundles
Abaca	*Musa textilis*	Musaceae (Monocotyledonae)	Fibers, associated to vascular bundles
Coconut	*Cocos nucifera*	Arecaceae (Monocotyledonae)	Mesocarp fibers
Raffia	*Raphia ruffia*	Arecaceae (Monocotyledonae)	Fibers, associated to complex mixture of leaf base tissues
Cotton	*Gossypium hirsutum, G. barbadense*	Malvaceae (Dicotyledonae)	Trichome – extrusion of epidermis (seed)
Kapok	*Ceiba pentandra*	Bombaceae (Dicotyledonae)	Trichome – extrusion of epidermis (seed pod)

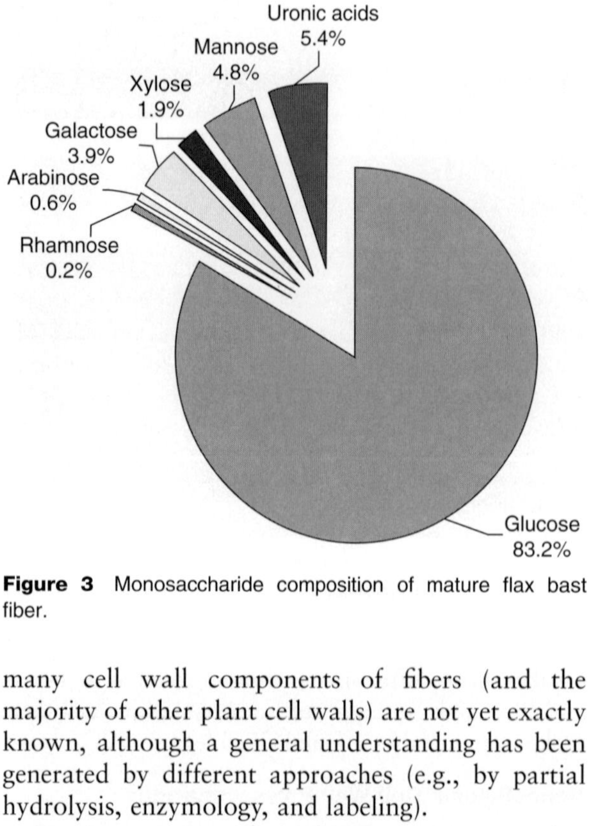

Figure 3 Monosaccharide composition of mature flax bast fiber.

many cell wall components of fibers (and the majority of other plant cell walls) are not yet exactly known, although a general understanding has been generated by different approaches (e.g., by partial hydrolysis, enzymology, and labeling).

An important polysaccharide type occurring mainly in the primary cell walls and middle lamellae of nonwoody plant tissues is pectin. Pectins are block polymers consisting either exclusively of α-(1→4)-D-galacturonic acid residues (homogalacturonan) or of these interrupted by α-(1→2)-L-rhamnose residues, which may be substituted by various branches (rhamnogalacturonans RG-I and RG-II). The galacturonic acid groups of pectins may be methylesterified or form Ca^{2+} bridges with a neighboring pectin molecule, forming a strong crosslink in a well-defined arrangement, the so called "egg-box" model. By applying chelating agents these ionic bonds can be disrupted leading to dissolving of the pectin and loosening of the bonds between plant cells. The same effect is achieved by applying pectolytic enzymes (pectinase, pectate lyase, and others), which are produced by microorganisms in the retting process to liberate the fibers.

Hemicelluloses or noncellulosic cell wall polysaccharides other than pectins are involved in the embedding of cellulose microfibrils in the cell walls. Through hydrogen bonding the often branched hemicellulosic polysaccharides form an amorphous gel, linking cellulose fibrils and other cell wall components. The hemicelluloses may be partially esterified (acetyl groups) affecting the hydrophilicity and solubility. Some covalent bonding may be present with lignin or ferulic acids (phenolic constituents). The major polysaccharides belonging to this group are xyloglucans, xylans, glucomannans, arabinogalactans, and callose. The occurrence of the hemicelluloses is not evenly distributed in different tissues or within a plant cell wall. Xyloglucans are mainly found in the primary walls, while xylan occurs predominantly in cell walls of woody (xylem) tissues. For some fibers the tissue-specific and growing stage-specific cell wall polysaccharides have been described, which occur at the onset of secondary wall formation and later disappear (e.g., callose in cotton fiber and galactan in flax fiber).

Lignin

A crucial component for the fiber properties is lignin, which occurs in primary walls as well as secondary walls and the middle lamellae of higher plants. Lignin is a complex polymer composed of phenyl propane units (e.g., *p*-hydroxyphenyl, guaiacyl, and syringyl units), which can be linked by means of numerous types of chemical bonds (ether, ester, carbon–carbon). The fibrous tissues of angiosperm predominantly contain guaiacyl structural elements (>95%), while

Table 2 Fiber morphological characteristics and average composition

Common name	*Fiber dimensions*			*Chemical composition*			
	Bundle length (cm)	*Length (mm)*	*Width (μm)*	*Cellulose (%)*	*Pectin (%)*	*Hemicellulose (%)*	*Lignin (%)*
Flax							
bast	30–90	13–60	12–30	75	3	15	<1
shives	—	0.1–0.5	10–30	37	3	25	30
Ramie	>150	50–200	15–80	76	2	15	1
Hemp							
bast	100–300	5–55	16–50	70	3	15	3
hurds	—	0.5–0.6	15–40	40	3	25	25
Jute, bast	15–36	0.8–7	5–25	62	1	22	13
Kenaf, bast	90–180	1.5–11	14–33	55	4	13	12
Sisal	60–100	0.8–8	10–40	73	1	13	11
Abaca	100–200	3–12	12–36	62	5	13	11
Coconut	5–20	0.3–1.0	12–24	33	5	13	33
Cotton	—	20–60	12–25	90	—	6	<1
Kapok	—	10–20	10–30	43	4	23	15

in the lignin of dicotyledonous xylem the syringyl : guaiacyl ratio shows considerable variation. In the lignin of monocotyledonous plants the presence of *p*-hydroxyphenyl groups is typical. Because of its difficult solubility and versatile chemical behavior there are no methods available to isolate the undisturbed polymer from plant tissues. Quantifying lignin contents often is unreliable. Lignin content and composition are extremely important for fiber properties and application. In general, soft fibers (flax, hemp, ramie, and cotton) are not much lignified, while the hard fibers (sisal, abaca, and coir) show higher lignin contents (**Table 2**).

Minor Components

The minor cell wall components, such as proteins, waxes, ash, and the noncell wall plant components (low molecular weight sugars, organic acids and secondary metabolites such as flavonoids, saponins, tannin, terpenes, etc.) do not show any specificity for fiber cells. This is also the case for the known modifiers of plant cell wall constituents, such as the hydroxycinnamic acids, methyl groups of uronic acids, and acetyl substituent groups of various polysaccharides.

Development

In the complex process of fiber formation from cell division to maturing and cell death different stages can be distinguished, each with its own characteristic anatomical, biochemical and molecular genetic changes. Extreme elongation and cell wall thickening processes are strikingly intensified in fibers. To study the biochemistry of fiber formation requires isolation from the surrounding tissues, which is difficult at early growing stages. A longitudinal section for cytological studies over the whole length of an individual fiber is practically impossible. Model systems to study fiber formation include the most important technical crops (e.g., cotton, flax) or the best characterized plants from a general biology point of view (e.g., *Arabidopsis*).

Formation of fiber cells in plants is a highly orchestrated process, involving the stages of fiber initiation, elongation, cell wall layer formation, and maturation. The regulation of gene expression and molecular mechanisms of genetic control of all these interdependent sequential or overlaid processes are still poorly understood.

Elongation

The elongation of sclerenchyme fibers occurs in two main stages: coordinated growth and intrusive growth. The initials of primary fibers appear early, before the organ in which they occur has elongated, and so they may grow in length symplastically together with the neighboring cells, which continue to divide. This type of growth is called coordinated. It involves cells of all the tissues of the organ and occurs without separation from the conjoined walls of the neighboring cells. It takes place at the very beginning of primary fiber formation, close to the meristem, where all tissues increase their volume. Coordinated growth allows some cells to become longer than others as they cease to divide earlier than neighboring cells.

When the rate of a cell's elongation exceeds the rate of the surrounding tissue growth, intrusive growth occurs. During intrusive growth cells have to force their way in between walls of other cells. There are few cell types able to grow in such a way. Intrusive growth leads to the increase of fiber cell numbers on the cross-section of the stem.

The elongation of a plant cell may occur by extension of the wall all over its surface (diffuse growth), or by localized extensions, for example, at the apex of the cell (tip growth). Root hairs, pollen tubes, and fungal hyphae grow exclusively at their tips. Such cells are characterized by a typical zonation of their organelles and a steep Ca^{2+} gradient in the tip region. Tip growth has not been observed in fiber cells yet.

Some plant fibers (e.g., flax, hemp, *Nicotiana*) are known to be multinuclear cells. Large numbers of nuclei can be considered as a prerequisite for the enormous elongation. However, other plant fibers (cotton) have a single nucleus.

The fiber cell volume of the fiber cell expands many thousand times during elongation. The expansion involves biosynthetic processes, leading to the formation of additional cell wall, plasmalemma, and tonoplast constituents, and requires extensibility of the cell wall and sufficient water availability to fill the space. Plant physiological studies have provided information on cell organelle involvement, transport of assimilate products, and energy supplies.

Of specific relevance for the formation of fiber cell walls is the process of cellulose deposition and the role of terminal complexes of rosettes involved in the biosynthesis and aggregation of linear glucan chains. Basically, the biosynthesis of cellulose occurs through polymerization of glucose residues donated by uridine diphosphate (UDP)-glucose, an activated form of glucose. The stiff and strong cellulose microfibrils are embedded in a matrix of noncrystalline polysaccharides, the hemicelluloses and pectins, which form a gel-like matrix by hydrogen bridges in which cellulose fibrils to a certain extent may slip over each other, allowing extension (in directions perpendicular to the fibril) and realignment.

The matrix polysaccharides of the cell wall are synthesized in the Golgi complex and by means of the vesicle transport system are delivered to the plasma membrane. Cellulose (and callose) are synthesized by enzymes in the plasma membrane. The cellulose molecules quickly aggregate to form crystalline fibrils. Cellulose content in the elongating fibers is only around 30% of the dry weight. The orientation of cellulose fibrils seems to be controlled by the orientation of cortical microtubules. In elongating fiber cells the newly formed cellulose microfibrils are initially laid transversely to the long axis of the cell, and reorient in the course of elongation, giving the impression that those microfibrils are not highly ordered. The cell wall thickness during fiber elongation is only 0.2–0.4 μm.

During elongation the fiber cell expands mainly by increase of vacuole volume. The volume of the cytoplasm does not increase much and the elongated fiber cell has a large central vacuole(s) and a thin layer of cytoplasm. The developing fiber cytoplasm has the features of the cell, specialized for the synthesis of cell wall: it is packed with active Golgi complexes and endoplasmic reticulum. Mitochondria are well developed, supplying energy for the highly energy-consuming process of cell wall formation, and some fibers have rather developed chloroplasts.

Thickening

The reinforcing of the full-grown fiber cell occurs through the deposition of a thick wall of highly oriented cellulose fibrils. The secondary cell wall is deposited inside the primary cell wall and pushes the layer of cytoplasm towards the interior of the cell at the expense of the vacuole. Elongation and cell wall thickening may partially overlap, so that the growing tips become thickened later than the "older" part of the cell wall. The thick secondary cell wall shows layering of cellulose microfibrils, which may be considered as a kind of "growth rings," as a result of the changes in growth conditions.

The very intensive process of cell wall polysaccharide deposition needs large-scale exocytosis of Golgi vesicles. Within a cell that is not increasing in volume, this leads to the excess of plasma membrane surface, which can be observed by its numerous invaginations. The volume of the vacuole decreases during secondary cell wall deposition, so the cell has also to reduce the tonoplast surface. At maturity the fiber almost completely fills the lumen with cell wall and actually squeezes the life out of itself.

Cell wall thickening proceeds in parallel with cell wall maturation, which involves postsynthetic modifications of the deposited polymers, forming or breaking of linkages, turnover of some polysaccharides, cation exchange, dehydration processes, etc. Phenolic components (lignin) may be introduced during maturing within the cell wall. Cytoplasm degradation, occurring sooner or later in the fiber cells can be important for fiber quality.

Flax

Flax bast fiber cells are an ideal example of sclerenchyma fibers, with extraordinary cell wall thickness and length. Their major function is to provide strength to the long (1 m) and thin (2 mm) flax stem. Flax bast fibers have been shown to be a useful model system for fundamental study of plant cell growth and cell wall formation. The fiber-enriched portion of flax stem can be peeled off from the woody part (xylem), making the sample less complex. Flax has the advantage that it has only primary fibers so is not complicated by the unknown interrelationship of primary and secondary fiber formation (as, for instance, in hemp). Moreover, the fiber elongation and cell wall thickening stages are largely separated and well characterized. The availability of the morphological reference point in the stem, where the transition in the fiber formation stages (e.g., elongation to thickening) takes place opens the possibility for functional genomics (cDNA microarrays for comparison of expression patterns in distinguished developmental stages, observation of upregulation and downregulation of enzymes involved in biosynthetic processes) specifically for cellulose fiber formation in plants.

Flax bast fibers are extraxylary fibers located along the whole stem. They are grouped in bundles, each with 10 to 30 fiber cells in the cross-section, giving totally around 600 fiber cells on the stem cross-section (**Figure 4**). Within the bundle the individual fiber cells end at different points and are slightly twisted around each other, forming a kind of thread. The length of flax fiber cells varies between 13 and 60 mm with an average usually between 20 and 30 mm (**Figure 5**). Up to 25–30 nuclei can be observed within one cell. Although bast fibers constitute around 6–7% of the total cell number seen in a stem cross-section, they contribute 30% of the stem's dry mass owing to the enormous development of their cell walls. An individual flax plant produces 10–15 000 fiber cells, yielding 0.3–0.5 g of fiber.

The anatomical characteristics, which influence the quality of fiber bundles, include the number of cells within a bundle, the length of individual fibers and the length uniformity of fiber cells, the degree of fiber helicoidal twisting within the bundle, the width of cell wall, and its ratio with lumen diameter. Biochemical fiber quality determinants are the content of cellulose and its characteristics (degree of crystallinity and molecular mass) but also the amount and types of noncellulosic cell wall polysaccharides and lignin content.

Flax bast fibers are primary phloem fibers and originate from procambium, close to the apical meristem. Bast fiber initiation is coordinated with other tissue formation, including xylem components and leaf primordia. Elongation of the individual fiber cell is very fast and takes place within 2 to 3 days at a rate of 1 to 2 mm day^{-1}. The cell diameter grows at most two- to threefold, while the length increases 3000–4000 times. For the first several hours a fiber cell enlarges by coordinated growth (top millimeters of the stem) and then by intrusive elongation, which presumably occurs by diffuse growth. During intrusive elongation the plasmodesmata (which are well pronounced during the coordinated elongation)

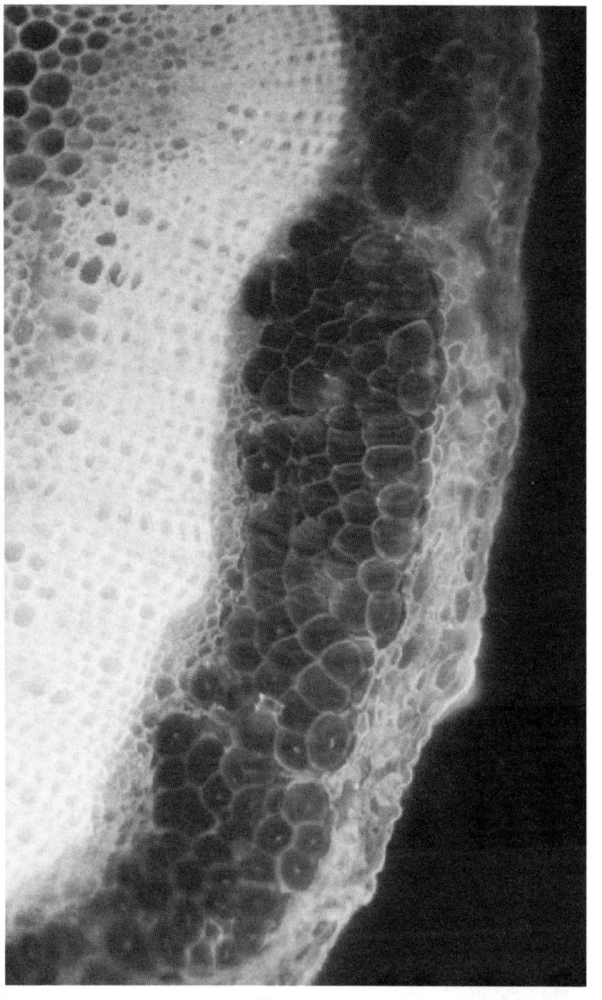

Figure 4 (see color plate 2) Cross-section of mature flax stem, stained with Cellofluor under UV.

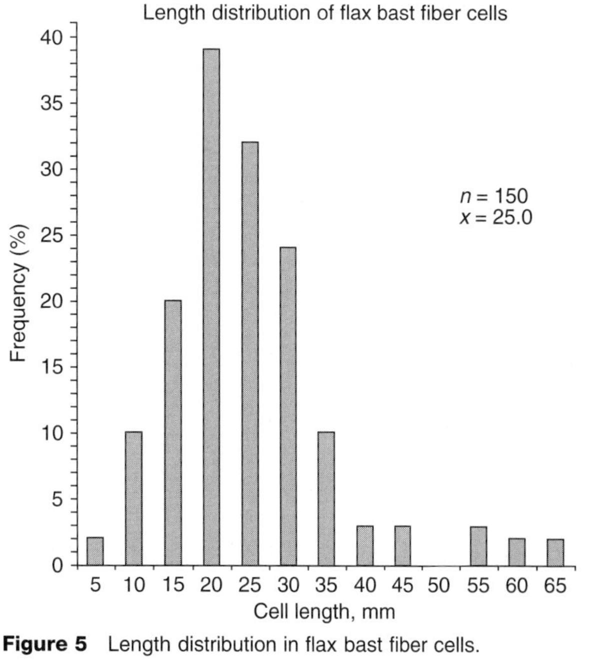

Figure 5 Length distribution in flax bast fiber cells.

have to be destroyed. The cytoplasm of a fiber cell gets isolated from the symplast and communicates with other cells only through apoplast. As a consequence, a self-supply of photosynthetic assimilates increases, which is obvious from the development of fiber cell chloroplast ultrastructure.

Bast fiber elongation is completed above the so-called "snap point" (6–8 cm from the top of the plant), where the stem changes its mechanical properties (**Figure 6**). Above this point the stem is fragile, while below it, considerable effort is needed to break the stem. The snap point is well pronounced during the fast growth stage (approximately 3 weeks in the middle of flax plant development), when the length of all bast fiber cells in the mature plant is determined and cannot be changed later. If unfavorable factors affect the plant at this crucial period, they not only decrease the fiber yield by reduced metabolism, but leads to a smaller amount of the fiber cells on the stem cross-section due to inhibition of fiber elongation above the snap point.

Below the snap point, cell wall thickening takes place in the fibers, with no further elongation. Thickening proceeds for around 2 months, almost to plant maturity. Besides the accumulation of cellulose fibril layers, postsynthetic modifications of the cell wall take place, as is obvious from structural changes in the inner cell wall layers. Though the majority of cell wall polymers of mature fibers are synthesized at the wall thickening stage, the rate of their synthesis during elongation can be even higher.

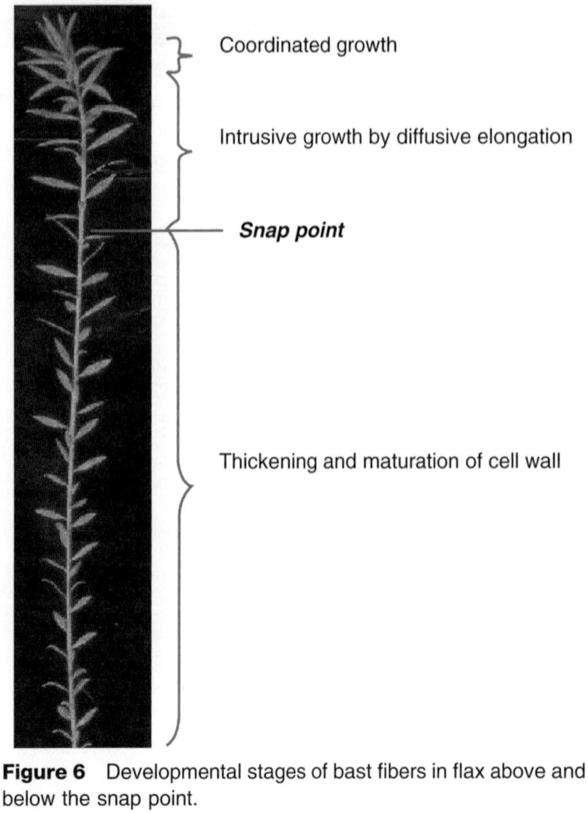

Figure 6 Developmental stages of bast fibers in flax above and below the snap point.

The major component of flax bast fiber is cellulose, which comprises around 75% and is known for high crystallinity. However, noncellulosic polysaccharides to a large extent determine the properties of flax fiber which is different from cotton, which is composed mainly of cellulose. Their presence influences water absorption, swelling, flexibility, etc. Xyloglucan, gluco(galacto)mannan, arabinan, rhamnogalacturonan I, and polygalacturonic acid have been described among the polysaccharides of mature flax fiber. Noncellulosic polysaccharides (including pectins) are localized not only within the middle lamellae and primary cell wall, but also within the layers of secondary cell wall, and are tightly bound to cellulose.

A prominent role in the process of flax bast fiber formation seems to be played by the fiber-specific galactan. It emerges at the fast growth stage of flax development immediately below the snap point, being absent during fiber elongation. Significant amounts of this polymer account for up to one-third of all noncellulosic cell wall polymers, but disappear in later stages and are not present in mature fibers. It was localized by immunochemistry within the inner cellulosic layers of the bast fiber cell wall. It seems to perform a specific role in flax fiber formation.

The lignin content of flax bast fibers is below 0.5%. The secondary cell wall of flax fiber is not lignified. At the end of growth some lignin is deposited in the middle lamellae and cell corners. Lignification does not take place in all fiber cells and it is not known, what determines the localization of phenolic compounds. Excessive lignification leads to a poorer quality of fibers and is one of the reasons to harvest the crop before the full maturity of the seed.

The protoplast of the fiber cell is long-lived but close to plant maturity it degrades. The process of protoplast degradation may influence the final fiber quality, and therefore relevant for the choice of the harvest moment.

Cotton

Cotton, kapok (*Ceiba pentandra*), and other floss form on the seeds (or seed pod), are extremely long and narrow hairs, which are not sclerenchyma elements.

Cotton production by far exceeds that of all other fiber crops taken together. Cotton fibers are almost pure (ca. 90%) cellulose. The fibers are easily detached from the seeds and do not need isolation from other tissues. After removal of the longer fibers (lint fibers), the seeds remain covered with very short fibers, so-called "fuzz."

Cotton fibers are single elongated cells formed from the seed coat protodermis (immature epidermis). Dozens of seeds are formed within a boll, each ovule producing from 10 000 to 20 000 fibers with a length of 20–60 mm. Isolation of the developing fibers makes analysis of the constituents and the rate of their synthesis possible. In elongating fibers the amount of cellulose per unit of fiber length remains constant at about $1\,ng\,mm^{-1}$, increasing sharply at the onset of secondary wall formation to reach about $130\,ng\,mm^{-1}$ at maturity.

Cotton fiber development is described taking as a reference point the day of anthesis in the flower. Fiber initials on the ovule are first observed immediately after anthesis. Between 10 and 25% of protodermal cells of the ovule develop a fiber. Fertilization is not a prerequisite for fiber initiation, but without it the fibers develop improperly. Fiber initials first develop as isodiametric bulbs on the surface of protodermal cells. Several waves of fiber initiation occur in the ovule, starting at the chalazal end of the ovule. Fiber initials that will develop into the lint fibers continue to be produced for 4 or 5 days. Fiber initials continue to appear for a week more, but those will develop into fuzz.

In fiber initials (but not in other epidermal cells) the nucleus enlarges, reaching a maximum at 6 to 10 days post anthesis (dpa), then declining slowly to a constant size at 20 dpa. Uridine, in particular, is intensively incorporated into nucleoli from 1 to 6 dpa, when large numbers of ribosomes are produced and used for rapid elongation. Thymidine is not incorporated during the critical period of fiber initiation, indicating that there is no endoreduplication.

Within 1 to 2 dpa the fibers start elongation, a process that lasts for several weeks. As fibers continue to elongate they begin to twist together, forming clusters. During the elongation phase the fiber produces a thin, flexible cell wall, capable of growing as the cell volume increases. The elongation proceeds by diffuse growth, when new material is deposited over the fiber length. Accompanying the increases in fiber length and diameter are increases in cell wall thickness. Starting about 15 to 20 dpa the fiber begins to deposit a thick, more rigid secondary cell wall, up to 10 μm thick at maturity. At the onset of secondary wall formation a thin layer of callose is formed at the interface between plasma membrane and cell wall. The polysaccharide is produced with high turnover and later disappears. Simultaneously, the rate of cellulose synthesis increases 100-fold. The ongoing diffuse growth overlaps at least for 1–2 weeks with the intensive cell wall thickening processes.

The direction of the twist changes at frequent intervals along the fibers. In between the primary and secondary cell wall there is an intermediary thin layer called the winding layer, which has a specific, steeply pitched helix orientation of cellulose microfibrils, essentially oriented transversely to the axis of the fiber. The diameter is larger and the helical gyration is oriented opposite to the secondary wall layers. Before the mature fiber dries it appears as a long cylinder which later partially collapses, producing a flattened, twisted ribbon structure. Daily cycling of temperatures during secondary cell wall formation may generate layering in the fiber wall. Swelling of the mature fiber in sodium hydroxide reveals a ringlike pattern under a microscope. At maturation, at 50 to 60 dpa, the fruit capsule opens, and the cylindrical fibers dehydrate and collapse to ribbon-like, twisted structures.

Having its metabolism largely oriented to cellulose formation, cotton was the first plant for which the genes for putative cellulose synthase were described. That was a major step for the development of plant cell wall research in general. Molecular genetic studies of fiber initiation, cell elongation, and callose metabolism have also been performed, using cotton as a model system.

See also: **Cell Walls and Fibers**: Cell Walls; Wood Growth and Development. **Growth and Development**: Cell Growth. **Primary Products**: Cellulose. **Secondary Products**: Lignin.

Further Reading

Basra AS (ed.) (1999) *Cotton Fibers: Developmental Biology, Quality Improvement, and Textile Processing.* Binghamton: Food Products Press.

Brown RM, Saxena IM, and Kudlicka K (1996) Cellulose biosynthesis in higher plants. *Trends in Plant Science* 1: 149–156.

Chaffey N (ed.) (2002) *Wood Formation in Trees: Cell and Molecular Biology Techniques.* London: Taylor & Francis.

Esau K (1977) *Anatomy of Seed Plants.* New York: John Wiley.

Fahn A (1990) *Plant Anatomy.* Oxford: Pergamon Press.

Fry SC (1988) *The Growing Plant Cell Wall: Chemical and Metabolic Analysis.* Harlow: Longman Science and Technology.

Gorshkova TA, Wyatt SE, Salnikov VV, *et al.* (1996) Cell-wall polysaccharides of developing flax plants. *Plant Physiology* 110: 721–729.

Gorshkova TA, Salnikov VV, Chemikosova SB, *et al.* (2003) Snap point: turning point in flax (*Linum usitatissimum* L.) bast fiber development. *Industrial Crops and Products* (in press).

Kim HJ and Triplett BA (2001) Cotton fiber growth *in planta* and *in vitro*: models for plant cell elongation and cell wall biogenesis. *Plant Physiology* 127: 1361–1366.

Kirby RH (1963) *Vegetable Fibers: Botany, Cultivation and Utilization*. New York: Wiley Interscience.

Marchessault RH and Sundarajan PR (1983) Cellulose. In: Aspinal GO (ed.) *The Polysaccharides*, vol. 2, pp. 12–95. New York: Academic Press.

McDougall GJ, Morrison IM, Stewart D, Weyers JDB, and Hillman JR (1993) Plant fibres: botany, chemistry and processing for industrial use. *Journal of the Sciences of Food and Agriculture* 62: 1–20.

Lin SL and Dence CW (1992) *Methods in Lignin Chemistry*. Berlin: Springer-Verlag.

Preston RD (1974) *The Physical Biology of Plant Cell Walls*. London: Chapman & Hall.

Seagull R and Alspaugh P (2001) *Cotton Fiber Development and Processing: An Illustrated Overview*. Lubbock, TX: International Textile Center, Texas Tech University and Cotton Inc.

Zhong R, Burk DH, and Ye Z-H (2001) Fibers: a model for studying cell differentiation, cell elongation, and cell wall biosynthesis. *Plant Physiology* 126: 477–479.

Wood Growth and Development

B Thibaut, J Gril and B Chanson, Université Montpellier II, Montpellier, France

Introduction

Everybody knows what wood is, but it is mainly associated with the decorative material used for handicrafts, furniture, and joinery. Obviously, everybody also knows that wood comes from trees, but how and why trees produce wood is less well known. Although pieces of furniture made from palm tree or bamboo flooring look like wood, these will not be included in this article. Here, the term “wood” is restricted to the material produced by gymnosperms and dicotyledonous angiosperms.

Wood: A Multifunctional Material for Tree Architecture

Wood was designed by nature millions of years ago as the supporting material of trees, the tallest and largest structures in the plant kingdom. These structures can be up to 100 m high and support a high above-ground complex structure (the crown), which can occupy a ground area equivalent to 10 houses (**Figure 1**), and can live for hundreds of years, surviving the action of rain, wind and snow.

Progressive Modular Building Design

As a tree is a living body, it must remain viable over the course of its life, which can span centuries. The mechanical function of tree structure is therefore somewhat accessory to characteristics such as having a large leaf surface area to capture solar energy and CO_2 from the atmosphere; at any stage in a tree's life, only a modest proportion of total energy is dedicated to mechanical function. The load bearing parts of a tree (trunk, branches) need to be just strong enough to support the developing crown, so that it can create more and more active leaf surfaces. This is made possible by the combined contribution of two separate growth mechanisms: primary and secondary growth. Primary growth results from the activity of the buds, which elongate along their own axis or divide into new ones, and leads to the architectural topology typical of the tree species; the terminal bud of the trunk (the “apex”) plays a more or less important role depending on the species in question or stage of development of the tree (**Figure 2**). Secondary growth follows on almost immediately from the creation and elongation of a stem axis. A cylindrical layer of dividing cells, the cambium, is generated just under the bark. It has the ability to produce wood material in successive cylind-

Figure 1 The Tree. A tall individual tree of the genus *Paninari* with its crown high above the canopy, in French Guiana.

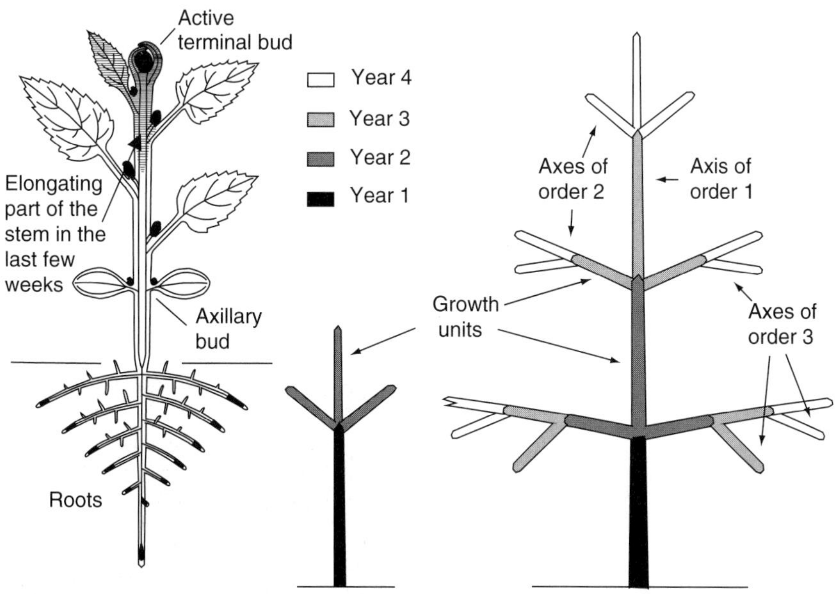

Figure 2 Primary growth of trees. A tree consists of two main parts: the aerial part (stem, branches, and leaves) and the root system. The aerial part is a branching system that grows by two simultaneous processes: axis elongation by cellular division in the terminal buds, and creation of a new axis by ramification from an existing one through cell division in axillary buds. From chronological and topological sequences, growth units should be characterized by their order and year of appearance. It is well known that the properties of wood are strongly linked to these "physiological" parameters. Left hand part of figure is reproduced from Troll W (1937) *Vergleichende Morphologie der höheren Pflanzen.* Berlin: Borntrager.

rical layers, as long as the stem axis (trunk or branch) remains alive, in a centripetal way (**Figure 3**). These two combined mechanisms allow maximization of the mechanical inertia and resistance of the tree against the direction of highest loading forces. Moreover, the simultaneous secondary growth of trunk and branches provides the tree with a very good assembly resistance (**Figure 4**). This is the origin of knots in lumber.

Multiple Functions of Wood

The woody part of the stem, or xylem, is not only the load bearing element, but also links the different parts of the tree by circulating fluids, food, and information, just like the veins and nerves in animals. It stores food for the periods in a tree's lifetime when it cannot produce any new food for growth, for example, when all the leaves have fallen.

During cell division in the cambium and further differentiation processes of the xylem, different types of cells can be specialized for each basic function. Gymnosperms (softwoods) contain essentially two cell types: tracheids, which are devoted both to water conduction and mechanical strength, and parenchyma, which is devoted to synthesis and storage of chemical compounds (fats, waxes, starch, tannins, etc.) usable as food by the tree. Angiosperms (hardwoods) contain, in addition, cells that are organized into vessels, which are devoted to water conduction, and long, thin cells called fibers, which impart mechanical strength. Vessels and fibers are always oriented in the axis direction of the stem; however, parenchyma cells are generally organized in radial files (medullar rays), but can also be arranged in axial patterns in the case of angiosperms (**Figure 5**).

Because water conduction is a critical function when there is climatic change (for example, early spring in temperate zones), many more conducting elements (vessels in angiosperms, thin-walled tracheids in gymnosperms) are produced during these periods. Typically, in temperate regions, wood is more porous (and softer) in spring than in summer leading, in many species, to visible successive annual rings (**Figure 6**).

Cells that are specialized for water conduction and mechanical strength die very quickly, usually within a few weeks. Parenchyma cells can live for several years and maintain their ability to produce toxic substances if the tree is attacked by fungi or insects. The external, living part of the stem is called the "sapwood." It is always whitish in color, and often noticeably paler than the inner part of the tree or "heartwood." In spruce (*Picea* spp.), fir (*Abies* spp.), or beech (*Fagus* spp.), for instance, it is hard to distinguish between sapwood and heartwood, whereas the heartwood of pine (*Pinus* spp.), oak (*Quercus* spp.), or chestnut (*Castanea* spp.) is more colorful than the corresponding sapwood. The transition from sapwood to heartwood, which is more or less progressive, is linked to the death of parenchyma cells and is accompanied by two phenomena: (1) the synthesis of organic molecules

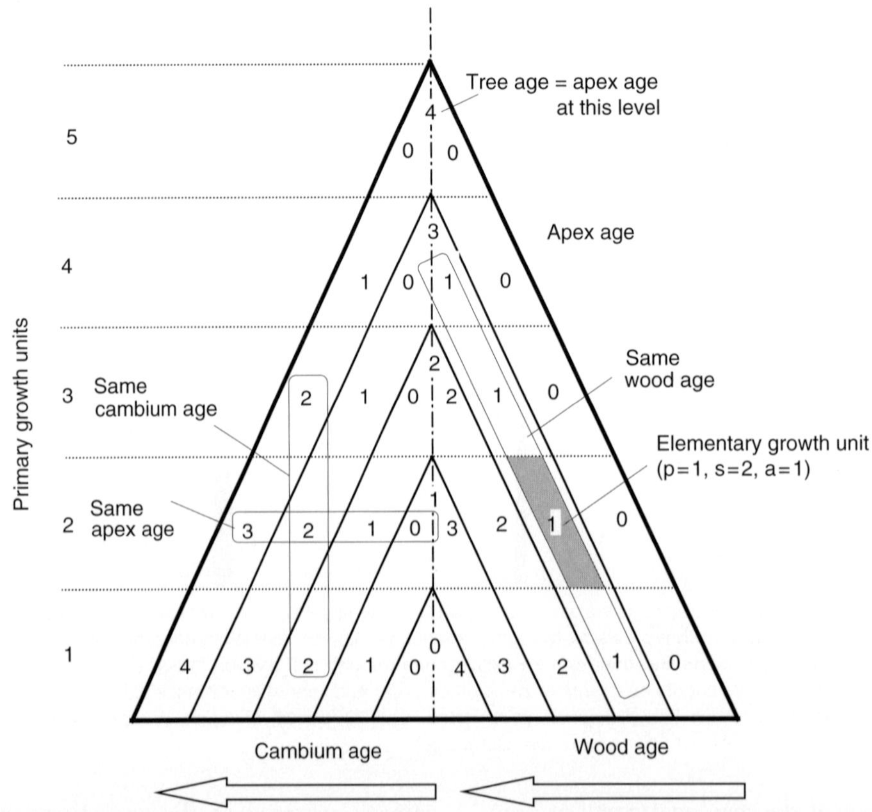

Figure 3 Secondary growth of stems. During primary growth, new annual growth units are created, which in turn will grow in diameter each year as a result of cell division at the periphery of the stem just below the bark. Thus, each primary growth unit consists of successive secondary growth units (cylindrical layers) visible as growth rings when looked at in cross-section. The number of growth rings in a given section can be used to measure time parameters in terms of cambium age, wood age, etc., provided that the age of the tree and the position of the stem in the tree topology are known (for branches). Rings with a low cambium age, usually called "juvenile wood," display significantly different mechanical behavior, i.e., lower stiffness and higher deformability in the elastic zone. This is considered to be an adaptation to changes in environmental conditions as the tree ages. Age (of tree, apex, cambium, wood) = number of complete time units (years for example) since the "birth" of the apex (equivalent to the age of the tree), cambium, and wood (at a given height level for a given elementary growth unit). A, present tree age; p, apex age (primary growth) at the moment of cambium birth; s, cambium age (secondary growth) at the moment of wood birth; a, wood age (ageing phenomena) since wood birth until present time; As a rule, $A = p + s + a$ in each elementary growth unit.

called secondary metabolites (or extractives), giving heartwood its distinctive color, smell, or natural resistance to biological decay; and (2) the obstruction of the vessels by thin membranes called tyloses. With the passing of time, however, these molecules lose some of their biocidical activity through some kind of chemical ageing, which explains why the most interior part of tree trunks are usually more rotten.

Wood and Biodiversity

Trees as individual living beings or as species must survive for very long periods of time, during which they may be subjected to many adverse hazards. By exhibiting a large diversity of structures and reaction strategies, trees are able to cope with this challenge. Biodiversity among species is apparent at each stage of wood formation. In the first phases of cell division and differentiation, each species displays a specific pattern of arrangements of fibers (or tracheids), vessels, axial and radial parenchyma, and other elements, such as resin ducts. This pattern is used by anatomists to identify a species from its wood (**Figure** 7). Furthermore, extracts from the wood of each species display a large cocktail of molecules (more than one hundred for oak), providing another method of botanical identification.

Wood as a Smart Material

Just like man-made concrete, wood is prestressed in the tree. In the first stages of cell development (division, elongation), the cell wall is very soft and cannot support high stress levels. During the maturation stage, the cell wall of fibers or tracheids thickens and hardens with the deposition of cellulose microfibrils embedded in a matrix of hemicelluloses and lignin. The polymerization process leading to a greater rigidity induces a small strain (change of

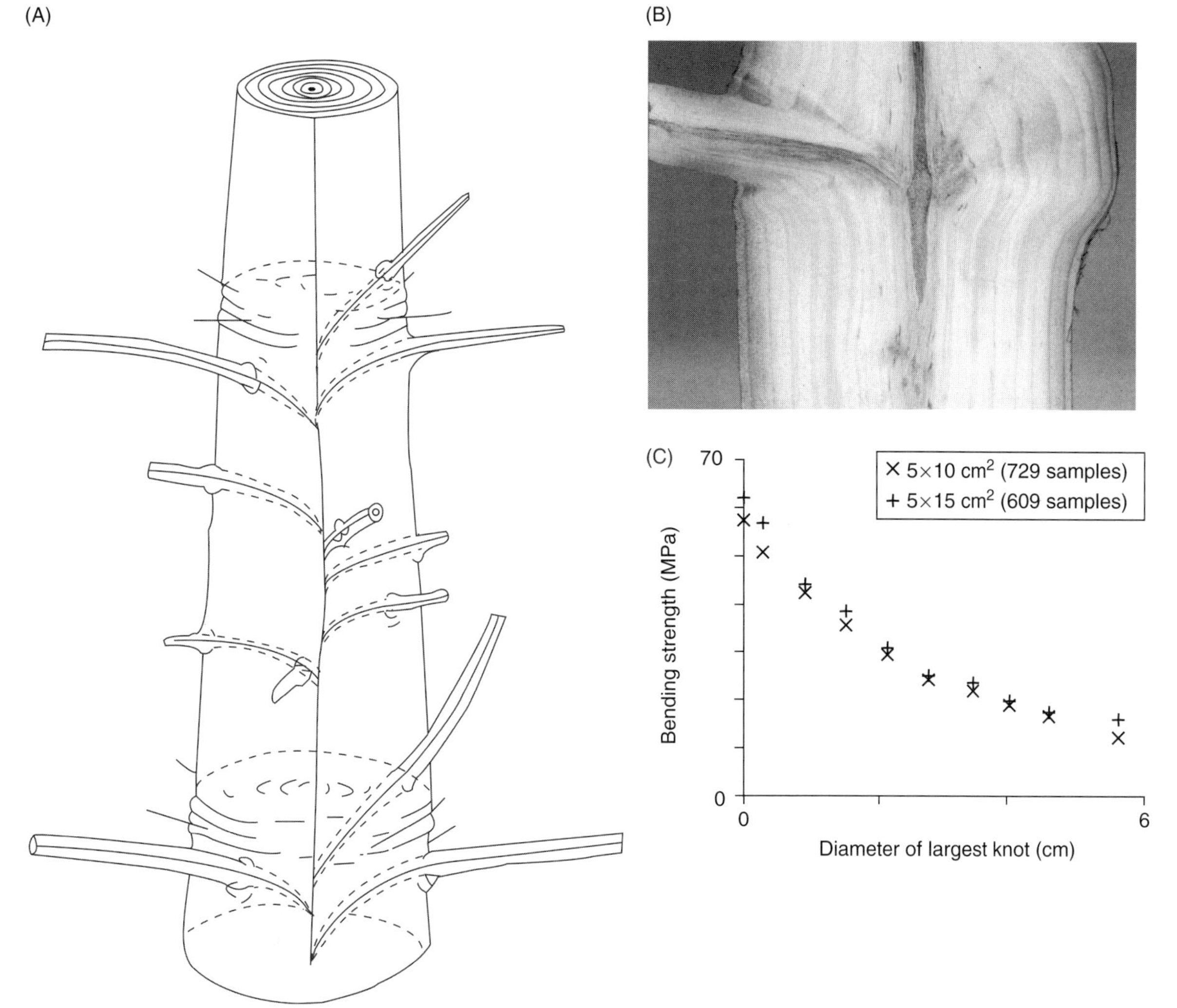

Figure 4 Degree of branching of the tree and degree of knottiness of the cut boards. (A) Owing to simultaneous growth in length and diameter of the two ramified axes, branches are progressively embedded inside the trunk (drawing by F. Colin) (B) This provides a great deal of strength at the point of connection. (C) Boards cut from the trunk of Scots pine (*Pinus sylvestris*) that contain more knots exhibit a lower mechanical strength. Data from Kollmann FFP and Côté WA (1968) *Principles of Wood Science and Technology*. New York: Springer-Verlag.

dimension divided by the initial length in a given direction). The order of magnitude of this maturation strain is typically 1 μm/1 mm to 10 μm/1 mm, to be compared to the levels of 1 mm/mm in the elongation phases occurring before maturation (**Figure 8**). On the whole, this maturation strain experienced by the newly formed wood layer is restrained by the rigid material already existing inside the stem. Just as a wet piece of paper glued to a wall would be subjected to tensile stress as it dried, the axial maturation shrinkage leads to a longitudinal tensile stress at the end of the process. Conversely, the restraint of the transverse expansion leads to a compressive tangential stress.

Prestressing, which is probably as old as life itself, allows natural structures to resist strong loads with minimal use of materials. Its existence in plants can easily be seen by observing the often spectacular deformations of plant portions resulting from cutting or splitting (**Figure 9**). Prestressing is beneficial to tree stems subjected to frequent bending loads, because it protects the external layers from being excessively compressed axially or pulled transversally to the fibers, which is the major mechanical weakness of wood.

Variable prestressing also became available at the same time as the evolution of prestressing by maturation strains, by small modifications in the maturation process leading to changes in maturation strains. Thus it was also possible to generate different levels of prestressing on both sides of a stem, the upper and lower side for example, by having different maturation strains on both sides. The bending moment resulting from this dissymmetric

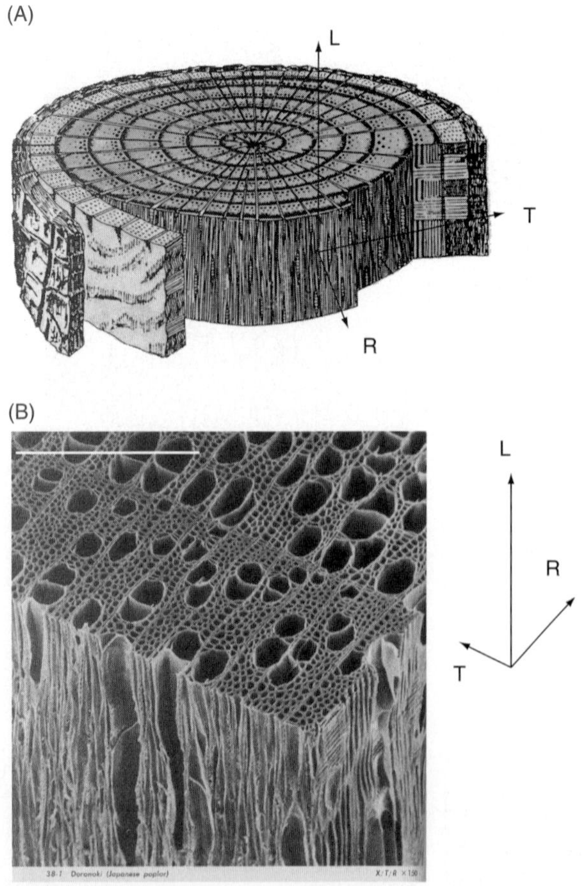

Figure 5 Variations in structure of wood. (A) Transverse section of a stem showing the successive growth rings with an initial porous zone produced in spring for high water conductivity and a final fibrous zone produced in summer imparting high mechanical strength. The radial thin bands are the ray parenchyma cells, which are specialized for the synthesis of food or protective chemicals. (B) A three-dimensional view of the cell pattern for a typical hardwood, with the vessels (the cells with the greatest diameter) being larger in the initial zone (back part of figure). L (longitudinal, parallel to stem axis), R (radial), and T (tangential) are the three main directions of anisotropy of wood. Most of the properties of wood have very different values when measured in the L, R, or T direction. Scale bar, 0.5 mm = 500 μm.

prestressing enables trees to restore their vertical habit after accidental leaning as a result of high winds, prevents the "weeping" of branches, and allows slow movement of the whole crown to search for light when the young tree is shadowed by a taller neighbor (**Figures 10** and **11**).

Wood Formation and Wood Properties

Engineers are always wary of the enormous variability of the wood properties even inside a tree. As suggested above, this variation in properties can be explained by the competing constraints that influenced the conditions of wood formation. Theoretically, for any given wood specimen, if we know the exact conditions of its formation, and the resulting structure and composition, we should be able to predict its properties. We are still a long way from being able to do this, but we can gain useful information by applying concepts of material design to wood.

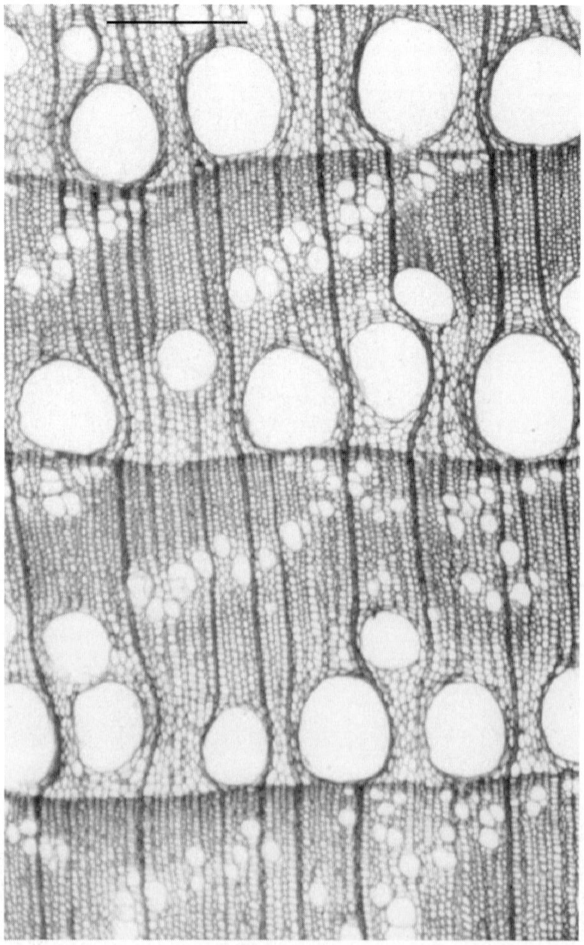

Figure 6 (see color plate 3) Water conduction and the multilayered nature of wood at the microscopic level. In temperate zones with their successions of warm and cold seasons, there is generally a dormancy period in winter when primary and secondary growth are stopped. During spring, there is a very active period of growth (new leaves, flowers, primary growth) during which water conduction from roots to leaves is of primary importance. "Conductive" cells (i.e., vessels or thin-walled tracheids) are produced in greater numbers than supporting cells (i.e., fibers or thick-walled tracheids). The reverse is true in summer, so each annual ring is composed of a very porous "spring wood" layer and a denser "summer wood" layer. For many woods, the mechanical properties of summer wood is two to four times higher than spring wood. In this cross-section of chestnut (*Castanea* spp.) wood, the spring wood has very large vessels, typical of ring porous species. Scale bar, 0.5 mm = 500 μm.

Cellular Structure

To the material scientist, wood resembles a honeycomb – a highly oriented foam with a great variety of cellular patterns. This conformation is known to be

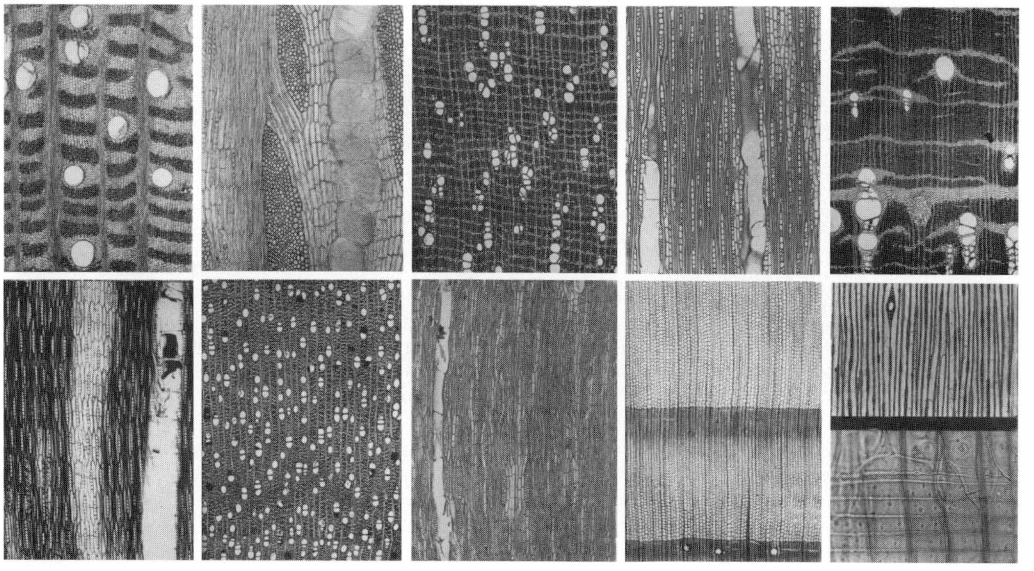

Figure 7 Transverse sections of a variety of woods showing variation in cellular organization and overall structure.

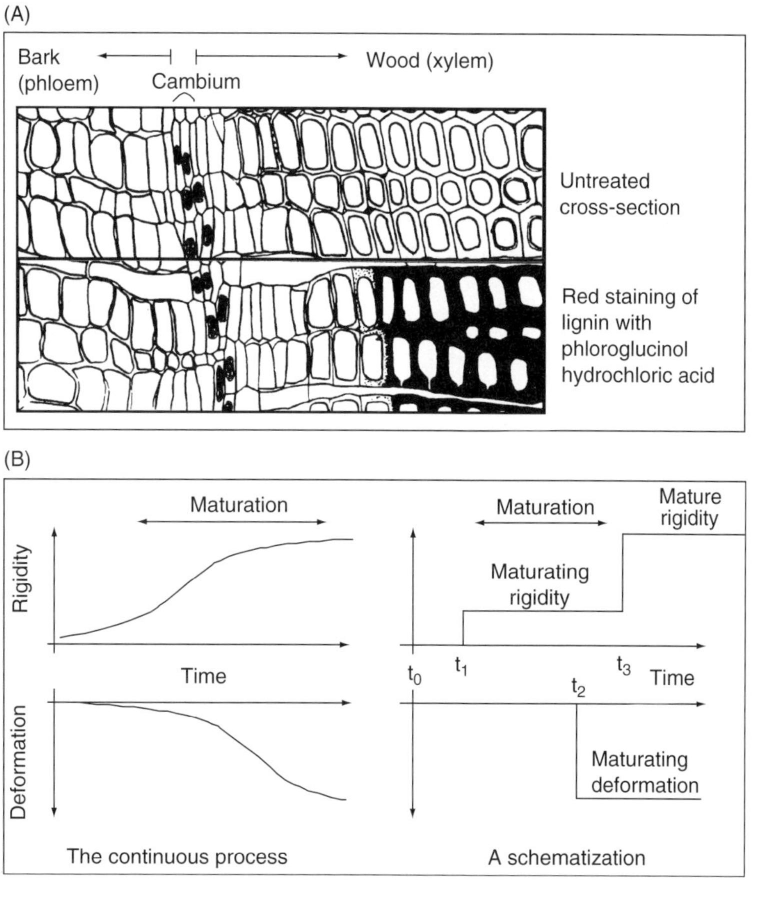

Figure 8 Wood cell differentiation, from early division to complete maturation. (A) A cross-section of wood including the cambium. The genesis of a wood cell is characterized by the following successive steps: cambial division leading to a new cell with a thin and loose wall, characterized by a very high deformability and low stiffness; large cell expansion (often more than 100%) in length and sometimes in diameter; and maturation characterized by cell wall thickening due to deposition of successive layers of microfibrils embedded in a lignin/hemicellulose matrix. (B) Stiffening and maturation strain. The cell wall becomes very stiff, and at the same time exhibits small maturation strains (around 0.1%) linked to the synthesis of polymers.

Figure 9 Evidence of prestressing in wood stems. The splitting that occurs after stems are cut into logs is an excellent way to demonstrate the existence of a residual stress field, characterized by peripheral longitudinal tension and central compression along the fiber. This phenomenon causes severe industrial loss of materials in the case of some species, such as beech. Photo kindly supplied by T. Beimgraben.

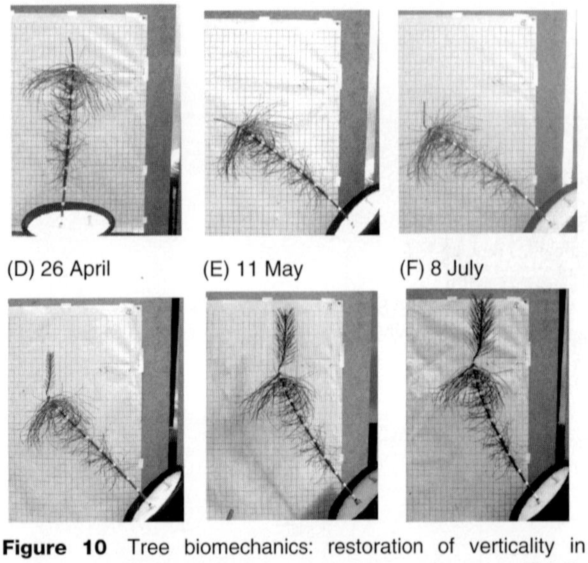

Figure 10 Tree biomechanics: restoration of verticality in young pine. Photographs of young pine specimens artificially inclined at 45° at the beginning of spring. During the 3-month growth period, an elastic downward bending is observed immediately (B); it is followed by further viscoelastic downward bending (C); although the top part containing the bud is already vertical; the trunk then experiences an upward bending over its whole length (D–E) until verticality is nearly restored.

cost effective in terms of quantity of matter needed to bear a given load. In this category of material, its porosity (the proportion of empty volume) is a key parameter that determines many properties e.g., mechanical resistance, sound and heat isolation, weight, etc. The specific gravity of wood cell walls being a constant 1.5 whatever the species, porosity (p) can be directly linked to specific gravity (ρ) of wood through the linear relation $p = 1 - 0.66\rho$. The

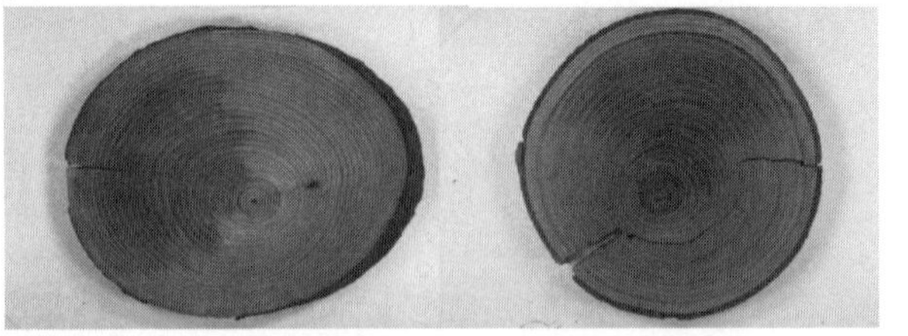

Figure 11 Angular distribution of reaction wood: the driving phenomena for long term movement of trees. Sections of inclined tree stems often exhibit whole angular sectors of wood that have a different aspect and strongly different properties. (A) Compression wood of conifers (dark brown sector on left part of the disk) is more lignified with more inclined microfibrils in the cell wall (high microfibril angle, MFA), resulting in a modulus of elasticity much lower than that of normal wood; the longitudinal maturation strain is positive (elongation) and leads to compressive stress in the new layer. (B) Tension wood of broadleaved trees (dark brown sector on top of disk) is less lignified, with more vertical microfibrils (low MFA) and a longitudinal modulus of elasticity a little higher than that of normal wood; the strongly negative maturation strains (shrinkage) lead to a high level of tensile stress in the new layer.

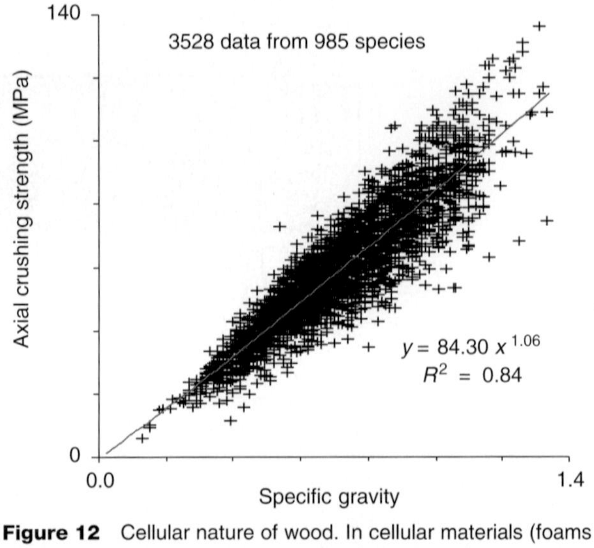

Figure 12 Cellular nature of wood. In cellular materials (foams and honeycombs), porosity or density are key factors because they are directly linked to the ratio between voids and solid matter. In wood, the cell wall density is about 1.5 g cm^3, whatever the species, so that specific gravity (ρ) and porosity (p) are linked by a very simple formula: $p = 1 - 2D/3$. All mechanical properties of wood specimens are strongly correlated to specific gravity (here being the axial crushing strength, i.e., the ability to support heavy loads in compression as a post).

value of ρ varies from 0.1 (balsa) to 1.3 (snake wood; *Brosimum guianense*). Most types of wood float on water, unless they are fully saturated with water, in which case only the most dense types would sink immediately. Specific gravity of wood is easily measured and can serve to evaluate other properties (**Figure 12**).

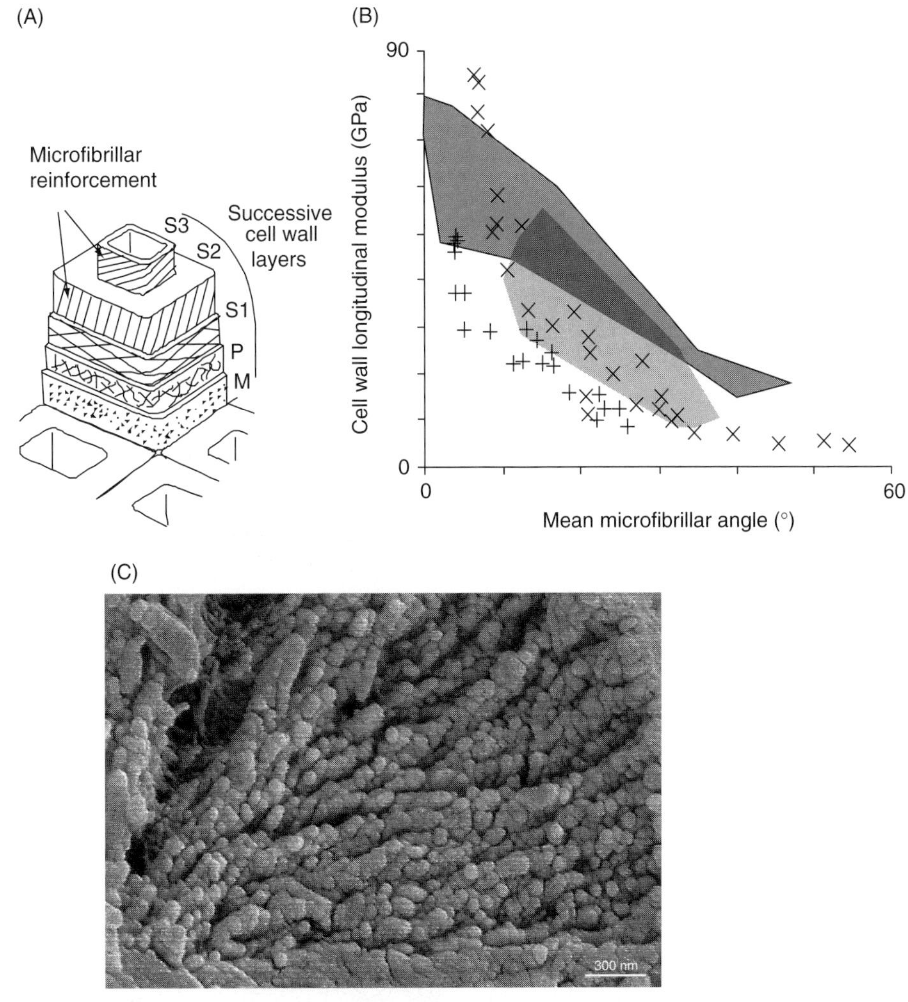

Figure 13 Fiber composite nature of cell wall. (A) The typical ultrastructure of tracheids or fibers. The cell wall is made up of successive layers, each one of them being characterized by a specific microfibril organization inside the hemicellulose + lignin matrix. Owing to its thickness, and more or less axial orientation of microfibrils, the S2 layer plays a major role in the longitudinal mechanical properties of wood. (B) The mean angle between microfibrils and cell axis (MFA) has a strong influence on the mechanical properties of wood. (C) The ultrastructure of the cell wall material. Scanning electron micrograph of a portion of the cell wall, showing elementary nanofibrils in the S2 layers of a broken oak cell. (Photo kindly provided by T. Zimmerman) × Cowdrey DR and Preston RD (1966) *Elasticity and mocrofibrillar angle in the wood of Sitka spruce*, Proceedings of the Royal Society of London-Series B, vol. 166, 1004, p. 245–272. + Norimoto M, Tanaka F, Ohogama T, and Ikimune R (1986) *Specific dynamic Young's modulus and internal friction of wood in the longitudinal direction*, Wood research and technical notes, vol. 22, p. 53–65. ■ Cave (1968) *The anisotropic elasticity of the plant cell wall*, Wood Science and Technology, vol. 2, 4, p. 268–278. ■ Salmen L and de Ruvo A (1985) *A model for the prediction of fiber elasticity*, Wood and Fiber Science, vol. 17, 3, p. 336–350.

Fiber Composite Technology in Cell Walls

In most man-made cellular materials (e.g., honeycombs, foams), the cell walls are homogeneous and isotropic. Nature uses a fiber composite technology to produce the cell walls, which is another cost-effective technique for managing a high number of mechanical properties. An increased mechanical strength in the fiber direction, for instance, can be obtained not only by producing thicker cell walls, but also by reducing the inclination of the microfibrillar reinforcements in the S2 layer (**Figure 13**). Conversely, a more oblique inclination, up to 45°, yields a higher compliance and compressibility along the fiber direction. The level of prestress generated by maturation, on the other hand, depends on this microfibrillar angle: exceptionally high tension is produced in the so-called tension wood of hardwoods, where it is almost parallel to the cell axis (**Figure 14**), but it becomes a compressive stress in the compression wood of softwoods.

Hydrocarbonate Polymers as Basic Components

More than 95% of wood is made from carbon, oxygen, and hydrogen, and 50% of the dry matter is carbon. In the cell wall, the matrix is a mixture of lignin (a complex and highly branched polyphenolic polymer) and hemicelluloses (polysaccharide polymers). The nanofibers (microfibrils) are made of cellulose and are mainly organized in crystal-like organelles. The respective proportions of these polymer families, often representing around 95% of wood dry matter, varies from species to species; for example, gymnosperms have a higher proportion of lignin as compared to angiosperms, explaining other variations in the basic properties of wood (**Table 1**).

Owing to parenchyma cell activities, secondary metabolites are always present in wood in proportions ranging from a few per cent to sometimes more than 20% of dry wood matter. The influence of these metabolites on wood properties can vary greatly depending on whether they are only located in cell lumens or inserted in the cell wall, on reactive sites of the primary polymers. Not only do these secondary metabolites provide a food source and biocidal properties against predators, such as fungi and insects, but they are also believed to play a role in the acoustic properties of musical instruments, for example, in Pernambucco wood used for violin bows. This can be explained by the rather small size of the molecules that can be inserted inside the microfibril/matrix architecture, affecting the rigidity of plasticity of the wood.

Finally, there are always some minerals (ash) present in small amounts, never exceeding a few per cent of dry wood matter, the most common being calcium or silica based minerals. A silica content in excess of 0.2% is known to have a serious blunting effect on tools, but also acts to repel marine borers.

Wood, a Natural and Renewable Material for Designers

Having been designed by nature as the ultimate building material, wood naturally became one of the first materials used by man. Today, in the context of sustainable management, existing world forests consume more than one billion tonnes of atmospheric carbon to produce more than two billion tonnes of wood each year. In a natural cycle, this wood will be destroyed by insects and fungi, and carbon will return to the atmosphere as CO_2. Wood, therefore, is an excellent source of renewable material that can help to improve the global carbon balance by being used in numerous products (buildings, furniture, toys, books, etc.), which can improve the quality of life of billions of people.

Living wood and the cut wood used by designers differ essentially in two ways:

1. Living wood always has a high moisture content; after being cut into boards it will dry out and then exhibit swelling or shrinking depending on changes in the air humidity.

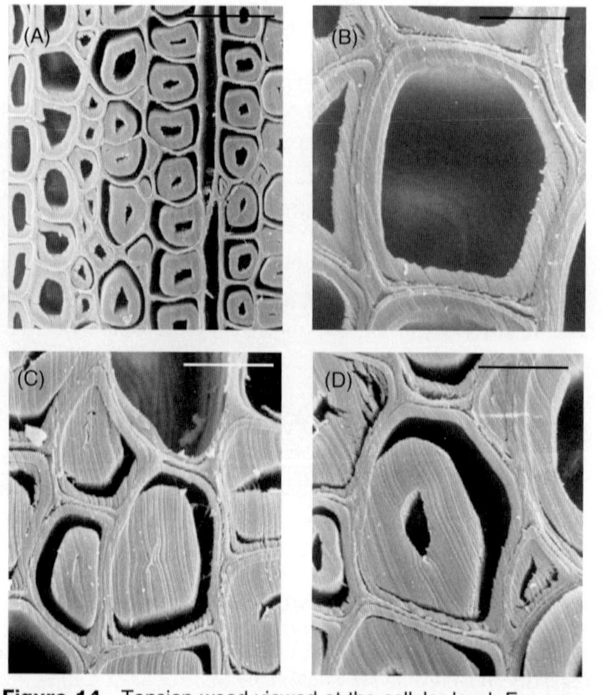

Figure 14 Tension wood viewed at the cellular level. For many broadleaved species (illustrated here by the sweet chestnut, *Castanea sativa*), the cell wall of tension wood fibers exhibits a specific pattern with its characteristic "G" layer, so named after its gelatinous aspect, which contains essentially cellulose with a very low MFA. Please note the section of "normal" wood in (B) for comparison. Scale bars: (A) 50 mm, (B) 10 mm, (C) 20 mm, (D) 20 mm.

Table 1 Chemical composition of some wood species (%)

Component	*Hardwoods*			*Softwoods*		
	Fagus grandifolia	*Populus tremuloides*	*Ulmus americana*	*Abies balsamea*	*Picea glauca*	*Pinus strobus*
Cellulose	45	48	51	42	41	41
Lignin	22	21	24	29	27	29
Xylan	26	24	19	9	13	9
Glucomannan	3	3	4	18	16	12
Pectin, starch, ash, etc.	4	4	2	3	3	2

After Kollmann FFP and Côté WA (1968) *Principles of Wood Science and Technology*. New York: Springer-Verlag.

2. The cells of living wood can prevent wood decay by producing metabolites *in situ*; cut wood, however, is completely dead. Designers must cope with potential biological risk by using naturally protected heartwood, or by using chemical treatments with biocides to protect wood in hazardous situations.

Conclusion

Wood is the material designed by plants for the building of large-scale structures, such as trees. Most of the properties of wood and their variability are linked to the tree's growth history. The huge biodiversity of plants is reflected in the great diversity of types of wood. There has always been a strong link between tree biology and wood technology. Certification or traceability, in a broad sense, could be more than just an indicator of good forestry practice; it could also be used to compile a comprehensive data bank of parameters, which could be employed by designers in their management of the variability of wood properties.

See also: **Cell Walls and Fibers**: Cell Walls. **Growth and Development**: Cell Division and Differentiation.

Further Reading

Anonymous (1987) *Wood Handbook: Wood as an Engineering Material.* Washington DC: US Department of Agriculture.

Archer RR (1986) *Growth Stresses and Strains in Trees.* Berlin, Heidelberg, New York, Tokyo: Springer Verlag.

Bodig J and Jayne BA (1982) *Mechanics of Wood and Wood Composites.* New York: Van Nostrand Reinhold Company.

Butterfield BG (1998) *Microfibril Angle in Wood.* Proceedings of the IAWA/IUFRO International Workshop on the Significance of Microfibril Angle to Wood Quality, Westport, New Zealand, 1997. Chicago: International Association of Wood Anatomists.

Fengel D and Wegener G (1984) *Wood: Chemistry, Ultrastructure, Reactions.* Berlin, New York: De Gruyter.

Hallé F, Oldeman RAA, and Tomlinson PB (1978) *Tropical Trees and Forests: An Architectural Analysis.* Berlin, New York: Springer-Verlag.

Higuchi T (1997) *Biochemistry and Molecular Biology of Wood.* Berlin, Heidelberg, New York: Springer-Verlag.

Kollmann FFP and Côté WA (1968) *Principles of Wood Science and Technology.* New York: Springer-Verlag.

Larson PR and Philip R (1994) *The Vascular Cambium. Development and Structure.* Berlin, Heidelberg, New York: Springer-Verlag.

Nair MNB (1998) *Wood Anatomy and Major Uses of Wood.* Selangor, Malaysia: University of Malaysia.

CROP IMPROVEMENT

Contents

Plant Breeding, Principles

N W Simmonds[†], University of Edinburgh, Edinburgh, UK
A E Arthur, John Innes Centre, Norwich, UK

[†] Deceased.

Plant Breeding as Applied Evolution

The essential principles of plant breeding are evolutionary as perceived by Charles Darwin. That is, they depend upon the generation of genetic variability by crossing unlike parents, followed by selection, not natural as in the wild, but artificial, in the hands of the plant breeder. Sometimes, indeed, truly natural selection in populations is useful but more often the breeder intervenes in bending complex populations to his/her interests. Procedures chosen depend upon the biology, mating system, and propagation of the crop under investigation. Essentially, the same principles apply to animal improvement and Darwin paid more attention to pigeons and dogs than he did to crops. But the fundamentals are constant: generate variability and select. Darwin had no real genetic understanding and it fell to his cousin, Francis Galton, to initiate the phase of genetics, namely quantitative genetics, which is of leading importance for crop improvement.

Objectives of Plant Breeding

Yield

This is nearly always the most important consideration in plant breeding. It is promoted, essentially, by noting that the plant produces so much biomass (dry matter per hectare per season) and the chosen crop variety converts this biomass to the desired product by partitioning it between product and "waste;" this waste is sometimes useful, e.g., cereal straw, where grain is the main product. The ratio of useful product to total dry matter is often called the "harvest index" and is a measure of efficiency. The term has been much used by crop physiologists and there is no doubt that enhanced partitioning has been a major feature of the great advances made in crop yields in recent years.

Many but not all plant diseases affect yield, but rarely by affecting partition. More commonly, they affect biomass and photosynthetic efficiency and thus reduce yield. A plant that is dead or defoliated can produce little or no yield.

Quality

Quality can be defined as fitness for purpose, and it has two components: inherent worth of the variety and condition. The breeder can and does work on the former but can do nothing about the latter, which lies in the hands of the grower, processor, and salesman. An excellent variety can produce a worthless product if mishandled. This is the area in which commercial maneuvers are commonest. There is often room for price adjustments to the grower's disadvantage, for example, when malting barley (*Hordeum vulgare*) is downgraded to feed status because the market is well supplied.

Socioeconomic Features

One might suppose that a grower's satisfaction is linearly related to yield, but it rarely is. Market price is usually a nonlinear function of both yield and quality, so the grower gets some return for crop quality, though possibly not much. The nonlinearity produces a curved relation between yield and price, despite being nearly a straight line over narrow ranges.

Genetics and Plant Breeding

There are fairly numerous examples of characters affected by the action of "major" genes or "oligogenes," as they may be called for contrast with the following category (i.e., polygenes giving rise to quantitative variation). Major genes affect characters in a fairly dramatic, all-or-nothing, qualitative fashion and they were studied extensively by Gregor Mendel (tall or short stature, smooth or wrinkled seed coats, etc.). The plant breeder often sees segregation of major genes in his material but they are often of no economic importance. Stature, colors, spininess, etc., are all affected and a full list of such traits for crop plants would be a long one. Examples of major genes that are of importance include some conferring disease resistance and plant stature. Short plants have been a prominent feature of selection over the past decades, primarily because of their improved partitioning (stature affects vegetable matter rather than grain or other floral products). Monogenic disease resistance is often hazardous because of the evolution of new races of the pathogen in question. Where major genes are of economic importance, they are generally easy to handle because they present straightforward segregations and simple Mendelian ratios. However, if several distinct major gene loci affect a single plant character, without dominance, F_2 segregations will approximate to 1:2:1, 1:3:3:1, 1:4:6:4:1, etc., and individual classes may be very hard to recognize. It should be noted that distributions quickly build up to a curve approximating the "normal," in which case the distinction between major genes and polygenes is blurred (see below).

In contrast to the above, there are "polygenes" and "polygenic characters," commonly called quantitative traits, that give rise to quantitative or continuous variation. Polygenes have small individual effects and, without critical evidence, are usually thought to be additive and open to a much simplified biometrical treatment. A great deal of crop variability is caused by polygenes and they are the major object of the plant breeder's activities. If the numerical argument of the preceding paragraph is extended to more numerous genes of smaller effects, then binomial distributions, defined by the marginal sums of a Pascal triangle, quickly emerge and approximate to normality. Normal statistics can be used to deal with these genes and this fact only became realized in about 1920, following Galton's pioneering work in the late nineteenth century. A normal distribution is characterized by its mean and variance, the latter being a measure of variability. Thus, denoting the mean as x bar ($\bar{x}$) the variance is the sum of squared deviations from it: $V = \sum(x - \bar{x})^2/n - 1$ and the standard deviation is the root of the variance. A crude variance contains both genetic and environmental elements, the latter being estimated from variation between replicates but called "error." The two sources of variation can, with appropriate designs, be separated, but details of the process will not be treated here. Note that "errors" are not mistakes, but are biological departures and interesting in their own right.

In assessing traits, breeders usually have to work with phenotypic (P) variation, which is the result of the combined effects of the influences of the genotype (G) and the environment (E), and the interaction between the two (G × E, discussed in more detail below). This can easily and most usefully be expressed in the simple equation:

$$P = G + E + (G \times E)$$

The role of genetic and biometrical analyses is to help determine the extent of the contributions of these parameters to the phenotype and, where possible and appropriate, attempt to partition the genetic effects further into additive and dominance contributions, along with more complex effects such as epistatic interactions.

However this is achieved, it is possible to derive equations of the general form:

$$h^2 = V_G/(V_G + V_E)$$

where h^2 is the heritability, V_G and V_E are the genetic and environmental variance, respectively, and V_P indicates phenotypic variation, that is total variation due both to genotype and environment, which is what the plant breeder nearly always has to deal with. Herein lies the difficulty; if V_E were always zero, V_G would be the same as V_P or unity and plant breeding would be simple. But heritabilities rarely approach unity and what the plant breeder actually sees is but a shadow of V_G.

The reader should be warned that this treatment is much simplified, but for practical plant and animal breeding purposes, the simple additive model serves quite well; indeed it has to, because it is the basis for prediction of response to selection (see below).

To summarize, then, the plant breeder may have to deal with major genes, which are easy to handle but relatively uncommon. Polygenic or quantitative variation, however, is much more important and selection for quantitative trait characters makes up the bulk of the breeder's work.

Genotype × Environment Interactions

We saw above that heritabilites are aimed at separating environmental from genetic effects in the P = G + E + (G × E) equation. However, frequently genotype × environmental (G × E) interactions occur; that is, different genotypes behave differently under different circumstances. Such effects are universal and can present great difficulties to the breeder. In particular, yield, a universal goal, is subject to such effects. Out of scores of examples that could have been cited, a neat example is that of the cocoa bean (*Theobroma cacao*) in Trinidad (**Figure 1**).

The three clones used in the study all increased in yield with increasing plant density but to difhyphen-qj;fering degrees; the planter when choosing a clone needs to take the intended planting density into account.

G × E interactions are well-nigh universal for quantitative characters and no general biometrical method of handling them has been offered, though many attempts have been made. It is a common experience that genotype × years interactions in annual crops are greater than genotype × sites interactions, but this hardly eases the breeder's problems. In practice, the only action that can be taken is to

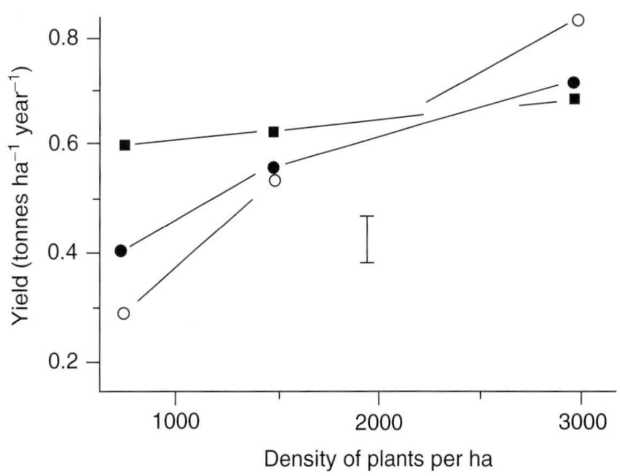

Figure 1 Genotype × environment (G × E) interactions among cocoa bean clones. Differential responses of three Trinidad (TSH) clones to density of stand. Yield of dry beans (t/ha year) at three densities of planting (density of plants per ha). ■, TSH1188; •, TSH730; ○, TSH919. The vertical bar is the standard error of a difference between two means. Reproduced with permission from Mooleedhar V and Lauckner FB (1990) Effect of spacing a yield in improved clones of *Theobroma cacao L. Tropical Agriculture, Trinidad* 67: 376–378.

conduct numerous well-run trials at diverse sites and over different seasons and to leave the growers to come to their own conclusions; a poor solution to the problem, maybe, but a realistic one. The breeder may have to capitalize on these G × E interactions by selecting genotypes that perform particularly well in certain environments; this restricts the sale of seed of single varieties, but can provide an acceptable solution to an otherwise difficult problem if G × E effects are very large.

Figure 1 not only shows G × E interactions, but also the effects of E, as such, and these are universal. No sensible agriculturist ever expects a single variety to give the same yield, for example, over different sites and seasons. E effects frequently exceed G effects and, indeed, G × E interactions may do so too. Together with the fact that heritabilities very rarely approach unity, G × E interactions rank as one of the leading problems facing the plant breeder.

Mating Systems in Crop Plants

All living organisms carry few to many deleterious recessive genes, generally in the heterozygous condition. Organisms that have been inbred (e.g., self-pollinated) for many generations have usually lost most of their recessive genes as a result of exposure to contrary selection. These organisms are called inbreeders and they are well represented among crop plants by many of the cereals (e.g., wheat, *Triticum* spp., barley, *Hordeum vulgare*, rice, *Oryza sativa*), many of the annual legumes, and a few perennials, such as *Coffea arabica* (coffee tree). By contrast, other plants have remained sufficiently outbred to retain the recessives, which are revealed by enforced inbreeding. Among the crops, they are represented by corn (*Zea mays*; maize), some millets (*Setaria italica*), brassicas, and the great majority of tree crops, such as rubber and most fruits. Outbreeding is maintained by various botanical devices such as sex, self-incompatibility, wind pollination, insect pollination, etc. There is a clear relation between mating system and life cycle, such that annuals tend to be inbreeders and perennials tend to be outbreeders. The rule is clear, though there are marginal cases and outright exceptions. *C. arabica* is an exception to the rule and some perennials can be bred as excellent inbreeders. Marginal cases include *sorghum* spp. and cottons (*Gossypium* spp.), which can be treated either way and a few wild cocoa bean plants are naturally somewhat inbred, despite self-incompatibility.

The practical outcome is that the breeder has to accept and work with what he has got, to a great extent. Forced inbreeding of an outbreeder leads to severe loss of vigor (known as "inbreeding depression"), which must either be accepted or taken advantage of (see below). At all events, mating systems tend to dominate plant breeding plans.

Population Types in Plant Breeding

The result of considering both mating systems and propagation simultaneously is to define four populations, which the plant breeder must take into account. The first population includes inbred lines (IBL), in which prolonged selfing has reduced plants to a near-constant homozygosity, occasional recessives having been eliminated. Many annual cereals, such as wheat, barley, and rice, many annual legumes, such as chickpeas (*Cicer arietinum*) and broad beans (*Vicia faba*), and a few perennials, such as *Coffea arabica* come into this category. Second, some outbreeders can only be handled as though they remained outbreeders and a degree of genetic variability in the product must be accepted. The frequency of deleterious recessives can be reduced by selection but they cannot be eliminated. Many tropical corns, some legumes (pigeon pea, *Cajanus cajan*), brassicas, and many forages such as alfalfa (*Medicago sativa*) and grazing grasses come into this category. This category can conveniently be bred as open-pollinated populations (OPP). Third, some outbreeders can be forced to inbreed despite loss of vigor, whereupon crossing of lines restores vigor and sometimes high performance (beyond what can be expected of OPP). The excessive vigor of some hybrids is called hybrid vigor or "heterosis;" it has been recognized for decades, but a full genetic explanation is still elusive. Such varieties are called hybrids (HYB) and corn is the classic place for their exploitation. Several other crops have been bred as HYB, e.g., *Sorghum* spp., some cottons, and some tomatoes (*Lycopersicon esculentum*). The often formidable problems of HYB breeding usually relate to the difficulty of inventing an efficient emasculation system for the female parent to obtain sufficient quantities of commercial seed. Good genetic systems are preferred but Indian cottons are normally hand-emasculated. Finally, many plants, including many perennial outbreeders, can be propagated vegetatively, i.e., cloned, by means of tubers, cuttings, grafts, etc. Such populations can be called clones (CLO) and, in fact, they represent the most efficient use of genetic variability. Two outbred clones are crossed and selection is made from among the variable progeny. Extensive clonal trials are implied and necessary, but the system has great genetic attractions. Examples from this category include potato (*Solanum tuberosum*; from tubers), rubber (*Hevea brasiliensis*; from buddings), sugar cane (*Saccharum officinarum*; from cuttings), and innumerable fruits and nuts.

The above four populations that the plant breeder must consider are determined by the biology of the crop, but with some socioeconomic considerations. A few crops can reasonably be handled by two or more methods, depending on circumstances. Thus, a sorghum or a cotton may be a HYB or an IBL and a rubber CLO may occasionally be replaced by seedlings, i.e., by an OPP.

Response to Selection

Let us designate response to selection by R, meaning the advance in the character X (yield, quality, disease resistance, etc.) achieved in a single cycle of selection. Because the breeder is handling large numbers of specimens, he nearly always has to select visually in the first instance, though the products will be subjected to trials testing later. Progress of R in one generation is then given by $R = i.h^2.\ V_P$; for most purposes, this can be written as $R = i.\ \sqrt{h^2}\sigma_G$. Progress is directly dependent on i, a statistical feature representing the "intensity of selection" (which represents the proportion of the selected population carried over to the next generation), h^2 is heritability, the additive genetic share of variability, and σ_G is the additive genetic standard deviation. Progress depends on variability, heritability, and hard work. **Figure 2** gives an example of how heritability affects advance under a standard selection pressure; in this example, heritabilities range from low to medium (0.5) to large. It must be remembered that this is a very simple example, though surprisingly often applicable. In practice, values can often be attributed to other forms of variation besides the simply additive, and endless complications are possible in calculating heritabilities. The response equation can be rewritten as $h^2 = R/Vi$, which is "realized selection." In effect, heritability is estimated *a posteriori* from response. The equation obviously has no predictive value.

One other point must be made here. Heritabilities are biometrical devices based upon genetic assumptions. However, it is possible to treat many

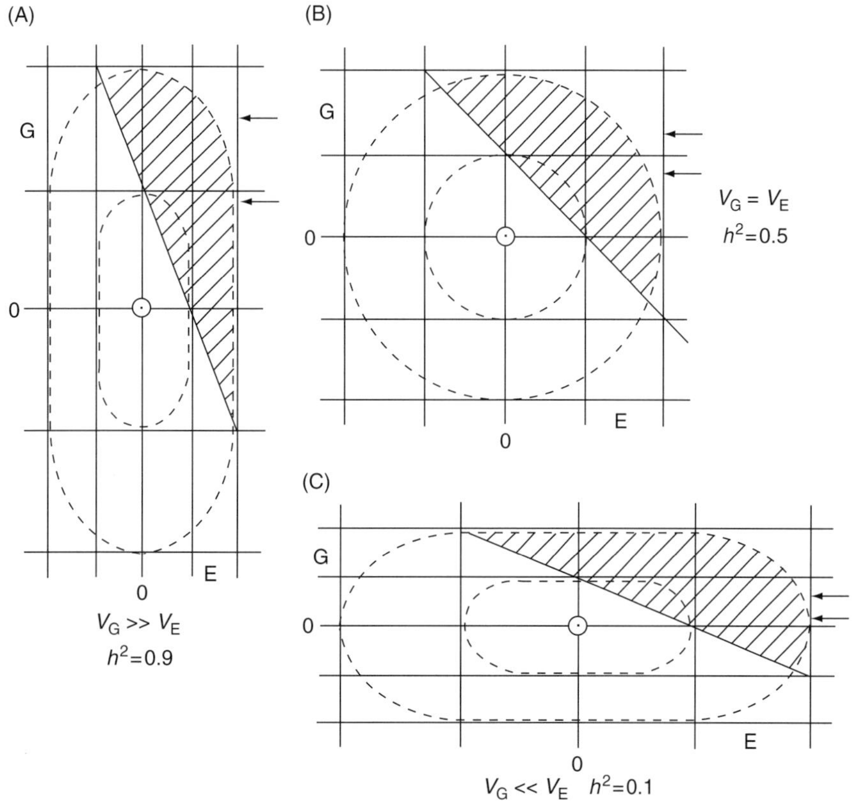

Figure 2 Genotype, phenotype, environment, and heritability. Genotypic values (G) plotted against environmental values (E) in the form of bivariate frequencies bounded by ellipses. Phenotypic values (P) are defined by diagonals. Scales are standardized about means of zero. The three cases are markedly contrasted, ranging from (A) high genetic variance and high heritability to (B) equal environmental and genetic variance to (C) low genetic variance and low heritability. The diagonal lines show the sectors cut off by phenotypic selection and the bold arrows mark the genetic advances, *R*, achieved by selecting at two levels. The higher the heritability, the greater the advance; also, and obvious to inspection, the more intense the selection, the greater the advance. Reproduced with permission of Blackwell Science from Simmonds NW (1998) Tropical crops and their improvement. In: Webster CC and Wilson PN (eds) *Agriculture in the Tropics*, pp. 257–293. Oxford: Blackwell Science.

data sets as though there were no assumptions, by using the idea of elements of variance from an analysis of variance. Additive constants (main effects) can thus be fitted and remainders are residual interactions

This treatment is called "combining ability analysis" and the two components are referred to as "general combining ability" (GCA) and "specific combining ability" (SCA). Heritabilities are not calculable, but valuable indications of breeding value and interactions between parents may be revealed. Useful predictions of the performance of a particular cross not yet made can thus be achieved.

Trials

Virtually all breeding program involve large numbers of entries and are thereby committed to predominantly visual selection, which is universally recognized to be inefficient in the selection of quantitative characters. Proper trials are essential and modern designs need to become ever more efficient; they depend heavily upon computers both in terms of design and analysis, because they generally lie beyond the limits of hand calculation. Such designs are usually nonorthogonal and this is the main source of complication. The computers are also useful for the increasing tendency to store all breeding data in data banks and to generate working and crossing plans from them. The hazards of misuse of computers have been widely recognized but not always observed. At all events, competent modern breeding programs produce varieties that have been well tested from an early stage.

Special Techniques

Polypoidy

Many plants have two sets of chromosomes and are called diploid. Others have four or more sets; these are called polyploids and are permanent hybrids with respect to homoeologous chromosomes. The most obvious kind of polyploid, an autopolypoid, which has replicated identical sets of chromosomes, is much less common because of sterility caused by pairing irregularities at meiosis. Many cultivated potatoes are probably near to this condition, but other examples of autopolyploidy are virtually all experimental. Some of the wheats, rutabaga (*Brassica napobrassica;* swede), tobacco (*Nicotiana tabacum*), and, *C. arabica* are plainly allopolyploids and, though derived from outbred diploids, are highly tolerant of inbreeding. Thus, the principal merit of allopolyploidy is probably the maintenance of hybridity. It is no accident that the few examples of attempts to produce new species experimentally (e.g., *Raphanobrassica, Triticosecale*) are allopolyploids, while the very numerous autopolyploids that have been studied, though often lush and vigorous, have been agriculturally useless; they sometimes have large grains, but only a few of them (e.g., in barley). There was a considerable vogue for autopolyploids in the 1940s, but the fashion soon died out.

So far we have considered increased chromosome numbers. However, lower numbers are sometimes also possible: thus, haploids (also called monoploids) are possible and have their uses, principally as a quick source of homozygotes after chromosome doubling to produce doubled haploids. The equivalent forms of autopolyploids are polyhaploids, e.g., dihaploids (i.e., diploids) out of tetraploid potatoes; these, too, have their uses, principally in formal genetic analysis. Making haploids and polyhaploids usually depends upon some genetic tricks, such as specific foreign pollination, often followed by *in vitro* culture of very young plants. The route to homozygosity may be greatly shortened through the production of doubled haploids if fluent haploid culture is possible; some good corn, barley, and oilseed rape (*Brassica napus*) inbred lines have been made in this way.

In vitro Culture

It has been known for decades that some sterile pieces of higher plants can be cultivated *in vitro*, given the correct medium. Thus, a juvenile mitotic phase can sometimes be restored to differentiated tissue. Recently, the techniques necessary to separate parenchymatous cells *in vitro* have been identified and the critical further step of inducing those cells to form growing embryos has also been achieved in many plants. This opens the way to biotechnology, transgenosis, transformation, or genetic engineering. The essence of the matter is the introduction of foreign DNA by any one of several quite diverse methods, details of which are irrelevant here. It will be obvious that biotechnology is dependent upon traditional botany for this critical phase. Another use for isolated cells lies in the production of "cybrids," which are somatic hybrids of cells grown *in vitro* and stripped of their cell walls. This is a potential source of allopolyploids, which may be very hard to make by sexual crossing. There were high hopes for this technique a few decades ago, but its success was limited and it has been overtaken by the much more fashionable biotechnology.

Embryo Culture/Embryo Rescue

Young plant embryos commonly grow in endosperm so why not grow them *in vitro* instead? The very youngest embryos cannot be treated in this way, but part-grown ones can often be successfully cultured. The problem was successfully tackled decades ago, with positive results. Thus, the way was open to raising seedlings from some difficult crosses and to raising embryos from cells transformed by biotechnological means.

Manipulations of chromosome numbers and *in vitro* operations rarely constitute more than a very small fraction of the plant breeder's effort. Indeed, many excellent programs have no need for them at all, but they can be valuable when applied locally for special purposes.

Genetic Resource Conservation

The response to selection equation discussed above shows us that genetic variability, intense selection, and a degree of heritability are all required for crop improvement. The very process of response ensures that variability is constantly lost as adaptation improves; old crops, competently bred for a long time, are therefore likely to be running short of genetic variability. What is to be done about it? We are all aware that natural variability in plants and animals in the wild is being rapidly depleted as a result of human activities. Crops are by no means immune from this loss. Setting aside the problems posed by the invention or introduction of new crops, there remains the leading question of action required for established crops that are running into problems associated with narrowing of the genetic base, that is, loss of the variability needed for further progress. This condition is probably a good deal more common than we yet know or recognize. What sources and methods can we look to?

First, we can reuse older stocks in the expectation that they have not been totally exploited. Many breeders practise this, though not explicitly as an attack on the genetic base. It is sometimes useful but clearly cannot go far. Second, breeders can look to collections as sources of variability and these are fundamental. Collections may be maintained *in situ* (in actual cultivation), which is generally difficult and uncertain. Or they may be taken away from the place of collection and maintained *ex situ*, generally in an experiment station and/or genebank. Methods must be adapted to the biology of the crop. Thus, some seeds can be stored cold and dry for very long periods of time (probably decades), while other seeds can be stored only for short periods and then must be regrown. Clones nearly always pose special problems, though techniques involving the frozen storage of meristems are improving apace. Seed storage, even if clonal identity is not preserved, is sometimes feasible, e.g., in potato and sugar cane (*S. officinarum*). All crops pose particular problems, but all these problems can be more or less overcome.

Given access to a good collection, the breeder can use it in one or other of two broad ways: he can seek specific characters (often disease resistance) for transfer by some sort of backcrossing technique, including marker-assisted backcrossing (this may be called "introgression"); or he can aim at massive improvement of large sectors of the collection to suit his circumstances (this may be called "incorporation"). The latter involves a large scale of operation and, probably, rather slow progression on a wide genetic base. Gentle mass selection with some element of natural selection is indicated. The technique has worked well in potato and sugar cane and could certainly be adapted to many other crops. It would also be the ideal way in which to try to adapt foreign crops to new environments (as in forest trees, for example). Nowadays, there is increasing interest in the subject of the genetic base, arising from the realization that more crops suffer from genetic shortage than we yet understand. A further consideration in recent years is that of evaluation and utilization of the conserved material. Collection and conservation play essential roles in maintaining a broad-based gene pool, but unless this "preserved" material can be made accessible to breeders, it remains just that – preserved. Considerable attention is now being paid to evaluating collections and making descriptive information available, including where and when the material was collected and its relationship to other collected material (such as species and subspecies identification, common and cultivar names). This is collectively known as "passport data," and includes information on plant descriptions and performance, uniformity within subsets, etc. Increasingly, molecular techniques that use molecular markers as indicators of uniformity/heterogeneity (such as molecular finger printing) are employed to facilitate this process.

Agencies and Plant Breeding

Major collections are usually accumulated and held by Government agencies, notably those in Russia, China, and the USA. In more recent years, one of the Consultative Group on International Agricultural Research (CGIAR) Institutes has been assigned a leading role in the tropics, though many institutes hold good collections. This is the International Plant Genetic Resources Institute (IPGRI), which promotes and sustains collections, though it is not itself a user of them.

Several different types of agency carry out plant breeding activities, including the following. (1) Commercial plant breeding is carried out for profit to provide farmers with satisfactory seed or clones; commercial breeding is a worldwide activity. (2) Government-supported experiment stations are widely responsible for much excellent breeding and are prominent throughout the world. (3) University departments of agriculture, especially in the USA, have been very important sources of new varieties. (4) Research institutes run by commercial industries themselves commonly have plant breeding departments, some of which have carried out excellent work. Work on sugar cane is outstanding and rubber, tea (*Camellia sinensis*), coffee, cocoa bean, and a few other crops are also competently handled.

See also: **Crop Improvement**: Chromosome Engineering; Comparative Genetics; Marker Assisted Selection; Mutation Techniques; Plant Breeding, Practice. **Diseases**: Breeding for Disease Resistance. **Genetic Modification**: Gene Cloning, General Principles; Transformation, General Principles. **Tissue Culture**: Clonal Propagation, *In Vitro*; General Principles. **Tissue Culture and Plant Breeding**: Somatic Hybridization; Somaclonal Variation.

Further Reading

Hill J, Becker HC, and Tigerstedt PMA (1998) *Quantitative and Ecological Aspects of Plant Breeding*, 1st edn. London: Chapman and Hall.

Simmonds NW (1962) Variability in crop plants. Its use and conservation. *Biological Review.* 37: 442–465.

Simmonds NW (1991) Genetics of horizontal resistance to diseases of crops. *Biological Review.* 66: 189–241.

Simmonds NW (1993) Introgression and incorporation. Strategies for the use of crop genetic resources. *Biological Review.* 68: 539–562.

Simmonds NW and Smartt J (1999) *Principles of Crop Improvement*, 2nd edn. Oxford: Blackwell Science. (the major source of this article)

Plant Breeding, Practice

P D S Caligari, Universidad de Talca, Talca, Chile
J Brown, University of Idaho, Moscow, ID, USA

Introduction

Plant breeding has been practiced since humans first began to cultivate crops. Over the past 100 years the intensity of plant breeding has increased, and is now recognized as an intricate integration of science (or sciences) and practicality. Over recent times the increasing need to feed the world's population, as well as an ever-greater demand for a balanced and healthy diet, has meant that there has been a continuing pressure to produce improved new crop cultivars. The strategies used to produce these are increasingly based on our knowledge of relevant science, particularly genetics, but involves a multidisciplinary understanding that optimizes the approaches taken.

A first requirement of any breeding program is to produce genetic variation in the characters that are to be improved. Once genetic variation is produced, it is necessary to select the desired types, which have a better expression of particular characters or combination of characters. Once identified the selected types need to be stabilized and propagated/multiplied for commercial use.

Plant breeding therefore appears to be a relatively simple process, and in many ways it is true that ideas of crop improvement are simple. However, the reality is more complex. It is possible to consider the three elements of the plant breeding processes (noted above as: to produce genetic variation, to select, and to stabilize and multiply for commercial use) in order to understand modern plant breeding. With this standpoint it is possible to realize what is being done and what alternative techniques might play a role in future cultivar development. However, each of these elements is tailored to be appropriate to the particular type of crop, or species, being improved.

Production of Genetic Variation

Conventionally, genetic variability is exposed, naturally, through sexual crossing to exploit the very principles that Mendel explained. Two parents, that between them have the expression of desirable characters, are intercrossed and the subsequent segregating generations examined for plants with the desired characters in new combinations (recombinants). This process therefore relies on the segregation of alleles at all the relevant genetic loci, during the normal process of meiosis. At fertilization there is a random fusion of gametes (pollen from the one plant and egg from the other) to give the embryo, which develops into the seed. So by the natural process of sexual reproduction, but between plants that the breeder has deliberately chosen, we get offspring, which contain novel combinations of the alleles that were originally dispersed between the two parents. It should therefore be clear that choice of parents is critical.

Breeders generally use the process noted above to take advantage of natural variation that already exists within crop species. Genetic variation can be observed for virtually all characters of interest in plant breeding. This naturally occurring genetic, therefore heritable, variation accounts for most of the responses that have been made in plant breeding. However, reliance on this one source of variation does limit the potential for long-term progress particularly in relation to improving specific characters. In these instances genetic variation can be produced by:

1. Wide intercrossing to wild, ancestral relatives of the crop itself, which may still be able to cross sexually (albeit sometimes only with the intervention of tissue culture techniques) with the crop species and which may be indigenous in another country.
2. Induced mutagenesis. Genetic variation in all plants arises initially from rare mistakes (mutations) that occur in the replication of DNA. Variation for all characters in today's crops is the accumulation of advantageous mutations over time. The frequency with which mutations occur can be increased using specific "mutagens," and the subsequent variants exploited.
3. Developments in the areas of molecular biology and biotechnology. These have extended the possibilities for introducing additional variation in the breeding process by transferring genes from different species, which would not be possible by more traditional means.

Selection amongst the Variation

The first difficulty is to decide which characters to select for. This may seem straightforward but in practice means trying to put in order of priority what characteristics are needed in future cultivars, and what level of expression will be necessary for each trait in say 10 to 12 years' time. Breeders also face several practical difficulties. First, it is impossible to measure every relevant character because there are simply too many for this to be practical. Second, evaluation of some characters takes excessive time and effort, and demands more resources than are available. Breeding programs need to test large numbers of genotypes but have available only small quantities of planting material. In addition, which characters can be effectively evaluated based on small plots (often a single plant)? How does the breeder grow plants such that they display characters under conditions that resemble those that they will experience in the hands of the grower?

Some selectable characters show variation that it is easy to classify into discrete classes (i.e., controlled by major genes) such as Mendel investigated, and can usually be selected efficiently. However, most important breeding traits are controlled by multiple genes and so show continuous variation. With these the genotypic performance can be extensively modified by the environment in which they are grown. For example, yield is a character of interest to all breeders, but is controlled by many genes and affected by environmental factors such as fertilizer levels, husbandry, and weather.

The general breeding objectives of all crops are to increase usable yield, improve quality/nutritive value, and increase stability of these traits over different environments and years.

Usable Yield

Crude yield is not important and it is the usable fraction that can actually be eaten, processed, etc. that is of major interest. This therefore brings in factors such as storage life, waste produced, and consumer acceptance. Therefore knowledge of the crop utilization (i.e., for direct human consumption, animal foodstuff, processing, etc.) must be gathered at the outset of the breeding program.

Quality of the Product

This includes, for food crops, nutritional quality, taste, and consumer preference. Although this can be related to usable yield, quality is also concerned with the nutritive value, calorific value, protein content, fat level, vitamin concentration, oil composition, etc.

Stability of Yield and Quality

Some cultivars do very well in some years or under some particular conditions but their performance is not consistent. This can lead to disaster when they fail because of changes in the growing conditions, a poor year for rain, no fertilizer available, too wet a period at harvest, etc. Thus breeding for resistance/tolerance to all biotic and abiotic stresses is a major aim. The need to evaluate breeding lines over several years to test for stability is one reason that it takes a number of years from starting to breed a cultivar until its release to the grower (often 10 years or more). This means that breeders require an ability to forecast the future. So for example a breeder might need to assess:

- What characteristics will growers be requiring in the future?
- What will happen in terms of the emphasis for growers, e.g., what subsidies will there be and what will the political situation be in the future?
- How will climate change have affected growing patterns?
- How will farming systems have changed?
- What will be the spectrum of diseases and pests?
- What will the end-users require in the future?

Stabilizing and Multiplying New Cultivars

One of the most important determinants in setting the breeding strategy is the natural breeding system of the crop species. The main natural breeding systems can roughly be classified into inbreeding, cross-pollinated (outbreeding or outcrossing), and vegetative reproduction (i.e., clonally propagated). These differences in reproduction lead to the main differences in classical breeding programs, and are briefly reviewed here.

Schemes for Inbreeding Crops

Examples of inbreeding crops include: wheat (*Triticum* spp.), barley (*Hordeum vulgare*), rice (*Oryza sativa*), soybean (*Glyzine max*), pea (*Pisum sativum*), tobacco (*Nicotiana tabacum*), tomato (*Lycopersicon esculentum*), millet (*Panicum miliaceum*), lentil (*Lens culinaris*), flax (*Linum usitatissimum*), and chickpea (*Cicer arietinum*). Inbreeding species naturally self-pollinate, and in commercial practice are grown as true-breeding, homozygous lines. So all the individuals of a particular cultivar are genetically identical.

Each generation is produced by allowing the plants to self-pollinate in each cycle of the breeding program so that while the testing and selection process is proceeding the plants are at the same time becoming more inbred. When finished cultivars are identified they will breed true from seed (homozygous). So inbred cultivars can be multiplied simply by letting them set seed in isolation from any other cultivars of the same species, so as to minimize the possibility that some pollen could pass between them causing contamination.

There are two main methods by which selection is achieved during this inbreeding process, bulk method and pedigree method, although many breeding programs employ refinements of those described or indeed a combination of both methods.

Bulk method Bulk methods start with the creation of genetic variation, usually by hybridization between two parents. The F_1 and several subsequent generations, often up to and including the F_6 generations, are grown as field populations, and the seed harvested and resown as a bulk. No conscious selection is imposed on these generations and it is assumed that the genotypes most suited to the environment in which they are grown will produce more offspring, by selfing, and hence predominate in future generations. It is therefore critical that these populations, or "bulks," are grown in an environment that will be similar to that expected for the developed cultivars. At the second stage, individual plants showing desirable characteristics are selected. From each selected plant, a plot is grown and the produce from the best plots is selected, bulk harvested, for initial yield trials, and for seed multiplication.

This method is one of the simplest and least expensive inbreeding methods. It also avoids conscious selection amongst segregating genotypes. The disadvantage is the length of time from initial crossing until yield trials are grown. In addition, it has often been found that the natural selection acting in the early bulked generations is not always that which favors characters desirable for agriculture.

Pedigree method In a pedigree breeding scheme single plant selection is carried out at the F_2 down to the F_6 generations. Again, the scheme begins by hybridization between chosen homozygous parental lines; F_2 populations are obtained by selfing the heterozygous F_1s. Single plants are selected by the breeder from amongst the segregating F_2 population. The progenies from these selected plants are grown in plots at the F_3 generation. The most desirable single plants are selected from the "better" F_3 plots, and these are grown in plant plots again at the F_4 stage. This process is repeated until plants are *near* homozygous (i.e., F_6). At this stage the most productive rows are bulk harvested and used as the seed source for initial yield trials and multiplication at F_7.

In addition to being laborious (as a considerable amount of record-keeping is required) and relatively expensive, individual plant selection is often ineffective and leads to the loss of valuable genotypes before they are fully tested. However, this is a preferred method in many crops.

Schemes for "Cross-Pollinated Crops"

Some examples of cross-pollinated crops include: alfalfa (*Medicago sativa*; lucerne), rye (*Secale cereale*), herbage grasses, forage legumes, red clover (*Trifolium pratense*), some corns (*Zea mays*; maize), perennial ryegrass (*Lolium perenne*), sugar beet (*Beta vulgaris*), and oil palm (*Elaeis guineensis*).

The selection of new cultivars of cross-pollinated crop species is a process that changes the gene frequency of desirable alleles within a population of mixed genotypes while trying to retain a high degree of heterozygosity. So it is the properties of the population that are vital, not individual genotypes (as in self-pollinating crops). Instead of resulting in a cultivar for release that is a uniform genotype, the population will be a complex mixture of genotypes, which together give the desired performance.

Selection of desirable characters is usually carried out by mass selection, recurrent phenotypic selection, or selection with progeny testing, or a combination of all three. The maintenance of these cultivars is through uncontrolled (random) mating.

Mass selection This is a simple selection scheme that uses natural environmental conditions to alter the genotypic frequency of an open-pollinating population. Initial seed that results from a set of controlled crosses is grown under field conditions over a number of seasons. It is assumed that mating will occur at random and so result in a population quickly moving towards equilibrium that can be maintained, as a population, for exploitation. Mass selection plots need to be isolated from other crops of this species to avoid any unwanted cross-pollination.

Recurrent phenotypic selection In general, this type of scheme tends to be more effective than mass selection. A population is created by cross-pollination between two (or more) populations to create a base population. A large number of plants is grown from the base population and a subsample of the most desirable phenotypes are selected and harvested as individual plants. These plants are then randomly mated to produce a new improved population. This process is repeated a number of times – it is a recurrent process. Populations that can be exploited as cultivars can be extracted at any stage, tested and distributed to growers.

Developing Hybrid Cultivars

High productivity in most outpollinating crops is often thought to be related to high heterozygosity. Plant breeders have utilized this feature by developing hybrid cultivars, which can be highly heterozygous and hence express hybrid vigor, or heterosis. In theory any species might be used in hybrid production but commonly it is outbreeding species, which are somewhat tolerant to inbreeding, such as Brussels sprouts (*Brassica oleracea* var. *gemmifera*), kale (*Brassica oleracea* var. *acephala*), corn, onions (*Allium cepa*), rape (*Brassica napus*), *Sorghum*, and tomato.

At the beginning of the twentieth century there was a general awareness, especially in the United States, that the means being used to develop new corn (maize) cultivars (mass selection and ear-row selection) were less effective than expected in producing cultivars with increased yield. Another approach was suggested from the knowledge that hybrids produced by cultivar × cultivar crosses often showed heterosis (i.e., produced yields greater than the better parent). It was then proposed that this could be exploited by manually detasseling one corn line (designated as the female parents, i.e., removing the male flowers) in plots also containing the second line, so that seeds produced on the line designated as female must have been pollinated by the pollen from the flowers of the male line.

There are hardly any agricultural crops where hybrid production has not been considered, although hybrids are exploited in relatively few crop species. The reasons behind this are, first, that not all crops show the same degree of heterosis, and, second, that it is not possible to find a commercial seed production system that is economically viable. Indeed if corn had not had separate male and female reproductive organs, thus allowing easy manual emasculation, hybrid cultivar development might never have developed, or would have been delayed at least 20 years, until cytoplasmic male sterile systems were available.

If hybrid cultivars are to be developed from a crop, then the species must: (1) show a high degree of heterosis; (2) be capable of producing inexpensive hybrid seed; (3) not easily be produced uniformly by other means; (4) and have a high premium for crop uniformity.

Hybrid cultivars have been developed in:

- sorghum, onions, and other vegetables using a cytoplasmic male sterile (CMS) seed production system;
- in sugar beet and some *Brassica* crops (mainly Brussels sprouts, kale, and rapeseed) using CMS and self-incompatibility to produce hybrid seed;
- in tomato and potato using hand emasculation and pollination.

There are two major steps in producing hybrid cultivars: (1) develop inbred lines to be used as parents; and (2) intercross these lines to identify parent combinations that give the best progeny. Inbred parental lines are developed in the same way as described above for inbreeding species, although pedigree methods are most common. After near-homozygous parents have been developed, their performance is first evaluated by test crosses, where each potential new parent is crossed to a parent of known worth and the progeny evaluated. After further parental selection in this manner, selected parents are intercrossed in different combination and the progeny evaluated in yield trials to identify the most productive hybrid cross combination.

It is believed by many breeders (and geneticists) that the magnitude of heterosis is directly related to the degree of genetic diversity between the two parents. For this reason, it is common in most hybrid breeding programs to maintain two, or more, distinct germplasm sources (heterotic groups). Breeding and development is carried out within each source and the different genetic sources are only combined in the actual production of new hybrid cultivars. For example, corn breeders in the United States found that they observed significant heterosis by crossing Iowa Stiff Stalk breeding lines with Lancaster germplasm. Since this discovery these two different germplasm sources have not been intercrossed to develop new parental lines but, rather, have been kept genetically separated.

Schemes for Clonal Cultivars

Examples of clonal cultivars are: bananas (*Musa* spp.), cassava (*Manihot esculenta*), *Citrus* spp., potatoes (*Solanum tuberosum*), rubber trees (*Ficus elastica*), soft fruit (raspberry (*Rubus strigosus*), blackberry (*Rubus fruticosus*), strawberry (*Fragaria*)), sugarcane (*Saccharum officinarum*), sweet potatoes (*Ipomoea batatas*), top fruit (apples (*Malus*), pears (*Pyrus*), plums (*Prunus*), etc.). Clonal crops are often perennial, although several crop species, particularly those where the actual unit of clonal reproduction is the part of the plant that is exploited (e.g., tubers of potato and sweet potato) are treated in agriculture as annual crops. Clonal crops also include many long-lived tree crops (e.g., apple, cherry, rubber, and mango (*Mangifera indica*)), which can be productive crops for many decades after being established.

Methods of propagation are various. Rosaceous top fruits, citrus, avocado (*Persea americana*), and grape (*Vitis vinifera*) involve budding and grafting onto various rootstocks. Leafy cuttings are used for pineapple (*Ananas comosus*), sweet potato, and strawberry. Leafless stem cuttings are used in sugarcane and lateral shoots are used for banana and palms. There is also, for a number of species, the potential for clonal reproduction via tubers (swollen stems, e.g., potatoes) bulbs, corms, etc.

In general, clonal crop species are often outbreeders that are basically intolerant to inbreeding. Individual clones are genetically heterozygous and so it is easy to exploit the presence of any heterosis.

The process of developing a clonal cultivar is, in principle, very simple. Breeders generate segregating progenies of seedlings, select the most productive genotypic combination, and simply multiply this asexually; thus there is no need for extra procedures to stabilize the genetic makeup (i.e., it relies on asexual reproduction, thus avoiding problems relating to genetic segregation arising from meiosis). Despite the apparent simplicity of clonal breeding it should be noted that while clonal breeders have shared in some outstanding successes, it has rarely been due to such a simple process.

In the case of potato, the length of the breeding process is partly related to a slow multiplication rate. In addition, seed tubers are bulky and require large amounts of storage space. To accommodate planting material for 1 acre of potatoes will require approximately 2000 lb of seed tubers. With many other clonal species there may be some considerable time between crossing to first selection. In apple breeding, for example, if a breeder is successful with the very first cross combination made, then it is still unlikely that a cultivar will be released (from that cross) by the time the breeder retires! Finally, many diseases and viruses are transmitted vegetatively and great care needs to be taken to ensure that breeding lines do not become infected.

New Genetic Approaches

Tissue Culture (*in Vitro*) Techniques

A variety of techniques (micropropagation, haploid production, protoplasts, embryo culture, apical culture, somatic embryogenesis, etc.) have been developed under the title of tissue culture and so just two particular examples are noted here to give an idea of the possible applications.

Haploid production Producing inbred (true-breeding) homozygous lines is an essential part of developing new cultivars in many crop species. These homozygous lines can be used directly as cultivars (i.e., for inbreeding crop species) or as parents to produce hybrids which are then exploited. Plant breeders in the past have used the process of selfing to achieve homozygosity, which is a time-consuming process. Therefore the possibility of producing plants from gametic, haploid cells has been an ambition as a way to produce "instant" inbred lines (after chromosomes of the haploids are doubled).

Haploid gametes are produced during normal meiosis and so are obvious targets for exploitation in obtaining homozygous lines. If such gametic, haploid cells can be stimulated to develop into plantlets, a haploid plant can develop, which can then be encouraged to double its chromosomes complement, to produce a completely homozygous line (a doubled haploid).

Although haploidy is a very attractive technique the natural rate of occurrence of haploid plants is low. However, it is possible to produce plants from gametic cells at a higher frequency by culturing plant tissue *in vitro*.

Although male and female sex cells might be used to give rise to haploid plants, it is the male cells (microspores or pollen) that have been exploited most successfully for the regeneration of large numbers of haploid and doubled-haploid lines. This is in part because of the ease with which pollen, as opposed to eggs, can be collected, and in part because pollen grains are produced in much greater numbers than eggs.

Other approaches to deriving haploids are also being used and one that has been successful in some specific cases is interspecific or intergeneric crossing. With the right cross, rather than giving rise to a hybrid, such crosses give haploid plants and examples of successful crosses are: *S. tuberosum* (cultivated European potato) crossed with a wild relative *Solanum phureja*; *H. vulgare* (cultivated barley) crossed with *Hordeum bulbosum*; and *Triticum aestivum* (bread wheat) crossed with *Zea mays* (corn). The results of these crosses are now leading to cultivars being widely released.

***In vitro* multiplication** *In vitro* multiplication of breeding lines has two major advantages in plant breeding programs: (1) plants propagated *in vitro* can generally be initiated and maintained in a disease-free state, and so can be used to maintain stocks of breeding lines, facilitate long-term germ plasm storage, and smooth the progress of international germ plasm exchange; (2) short generation times and fast growth mean that rapid increases in plant number can be achieved.

Both the above are important with clonal crops in which there is a tendency for relatively low multiplication rates (a property of their vegetative propagation) and which are particularly susceptible to viral and bacterial diseases (which tend to be multiplied and transmitted through each clonal generation). Good examples of maintaining high disease status and offering rapid plant regeneration potential include potato and strawberry.

Plant Transformation

Using plant transformation techniques makes possible the transfer of lengths of DNA (single genes, i.e., simply inherited traits) into plants, to have such "transgenes" expressed, and to have them function successfully in their new genome. In theory, any gene can be transferred from any source into a developed cultivar or advanced breeding line, and this can be achieved in a single step. So, recombinant DNA techniques allow breeders to transfer genes between completely unrelated organisms. For example, bacterial genes can be transferred and expressed in plants.

Thus plant transformation has added to the tools available to the breeder for genetic manipulation, but it does, as with all techniques, have its own limitations. Some of the limitations will diminish as the protocols are further developed; others are inherent to the basic approach. At present, recombinant DNA techniques can generally only transfer single genes. This means that they are very effective where the trait can be affected significantly by one gene, or a few genes, of large effect. This means that such genes need to have been identified and cloned. The number of such identified, desirable genes is still finite, although increasing rapidly.

Already there is a growing list of crop species that have proved successful hosts for transformation including: alfalfa, apple, barley, carrot (*Daucus carota*), cauliflower (*Brassica oleracea* var. *botrytis*), celery (*Apium graveolens* var. *dulce*), cotton (*Gossypium*), cucumber (*Cucumis sativus*), flax, horseradish (*Armoracia lapathifolia*), lettuce (*Lactuca sativa*), corn, peas, potato, oilseed rape, rice, rye, sugar beet, soybean, sunflower (*Helianthus*), tomato, tobacco, walnut (*Juglans*), and wheat.

The first transgenic crops were ones changed in terms of modifying or enhancing traits that related directly to traditional farming activities. These included the control of insects, weeds, and plant diseases. More recently work has focused on more consumer-orientated goals such as altering end-use quality (including oil composition, starch, vitamin level, and even the inclusion of vaccines).

There has been considerable public debate concerning the application of plant transformation technology, particularly as new transgenic crops have been released into commercial cultivation and as the products have entered the human food chain. Plant breeders must be aware of the concerns as well as the regulations that apply to plants derived using recombinant DNA. This means that alongside the general social and environmental concerns the breeder must be convinced that the techniques being used are the most effective for what is to be achieved and not simply assume that "high-tech" means most efficient!

Molecular Markers in Plant Breeding

As noted initially, plant breeders have practiced their 'art' for many centuries, but genetics, as a subject, was only recognized in the twentieth century, with the rediscovery of Mendel's work. From that point, research in genetics has developed rapidly but we still have little, or no, information in most crops about: (1) the locations of many of the loci of interest, or even which chromosome they are on; (2) the number of loci affecting any trait we are interested in; and (3) the effect of an individual allele at a locus on the observed phenotype, except where there is an obvious major effect (e.g., height and dwarfing genes).

One idea for helping increase our knowledge was to try to associate easily visualized markers with loci that were affecting variation in traits of interest, as proposed by Sax as early as 1923. The idea is that often characters of interest to plant breeders are difficult to evaluate, because they show continuous variation, or assessment is detailed and time-consuming, or the trait is only expressed after several years of growth. If the genes controlling these characters are closely linked to loci controlling a second, easily assessable trait, then it would be possible to select for the difficult characters by assessing the easily identifiable one.

The characteristics of a good marker system are:

- the markers are easy, quick, and inexpensive to score
- the markers have no deleterious effects on fitness
- they have no effects on other traits
- they show a high level of variation
- they are stably expressed in different environments
- they can be assessed in early growth phases of the plant (seedling level), and/or in tissue culture
- the scoring of the markers can be achieved by nondestructive means
- heterozygotes can be differentiated from either homozygous genotype (show codominance).

The types of markers that have been used in plant breeding include:

1. Morphological markers, ones whose variation can be seen by simply looking at a plant's phenotype, such as pigmentation, dwarfism, leaf shape, absence of petals, etc.
2. Biochemical markers, such as variant forms of an enzyme, which are functionally identical but can be distinguished by electrophoresis (isozymes). Different forms of the enzyme will migrate to different points depending on their charge, size, and shape.
3. Molecular markers, which represent the variation present at the DNA level. So molecular markers are simply differences in the DNA itself between individuals, groups, species, taxa, etc. The main characteristics of molecular markers are that they are found in virtually all organisms; are not affected by the environment; show high levels of polymorphism; have no effects on the phenotype; and can be detected using small pieces of tissue.

Given the above characteristics and their relatively unlimited numbers, it is no surprise that the advent of molecular markers in the 1990s was seen as a major step forward.

Future Potential

Plant breeding will continue to rely on classical techniques of crossing, trialing, and selection but will be helped by the application of the new approaches that are developing so quickly. These new techniques will be used, but in conjunction with the more established ones. Plant breeding will therefore continue to change to take into account new possibilities and new techniques but with a firm base in traditional approaches.

List of Technical Nomenclature

Allele	One of the alternative forms of a gene at a particular location on a chromosome (i.e., at a locus).
Haploidy	The condition of a cell or nucleus containing only a single set of unpaired chromosomes.
Heterosis	The superiority of the heterozygous progeny over its inbred parents.
Heterozygosity	The situation in which the alleles at a locus are identical by descent.
Homozygosity	The situation in which the alleles at a locus are not identical by descent.
Locus	The region of DNA in a chromosome that is occupied by a particular gene determining a biochemical function.
Meiosis	The process of cell division which produces daughter nuclei which have half the parental chromosome number.
Mutation	A change in the DNA of either sequence, structure, or amount.

See also: **Crop Improvement**: Marker Assisted Selection; Molecular Markers; Mutation Techniques. **Postharvest Physiology**: Genetic Engineering for Postharvest Quality. **Tissue Culture and Plant Breeding**: Somatic Hybridization.

Further Reading

Caligari PDS (2000) Plant breeding and crop improvement. In *Encyclopedia of Life Sciences*. London: Nature Publishing Group http://www.els.net/.

Hayward MD, Bosemark NO, and Romagosa I (1993) *Plant Breeding: Principles and Prospects*: London: Chapman & Hall.

Hill J, Becker HC, and Tigerstedt PMA (1998) *Quantitative and Ecological Aspects of Plant Breeding*: London: Chapman & Hall.

Poehlman JM and Sleper DA (1995) *Breeding Field Crops*. Ames: Iowa State University Press.
Simmonds NW and Smartt J (1999) *Principles of Crop Improvement*. Oxford: Blackwell Science.

Hybridization and Plant Breeding

B G Murray, University of Auckland, Auckland, New Zealand

Introduction

Hybridization remains an important component of many plant breeding programs despite its long history and the development of transgenic techniques for genetic engineering of plants. The first record of artificial hybrids was a cross between *Dianthus caryophyllus* and *D. barbatus*, produced in 1718 by Fairchild, and since then thousands of new hybrid combinations have been made. Hybridization can involve crosses between different species (interspecific hybridization), or crosses between genetically different individuals (selections, breeding lines, or cultivars) within a species (intraspecific hybridization). Hybridization is usually important for two main reasons: to transfer genes and, therefore, the characters they control, from one plant to another; or to exploit the vigor that is often observed when genetically different plants are crossed. This is often called heterosis or hybrid vigor. There is also an important commercial benefit in that F_1 hybrid varieties derived from inbred lines, though highly uniform, are not true breeding, so the grower needs to buy fresh seed each season.

Interspecific Hybridization

Interspecific hybridization, often called wide hybridization, is usually used when a specific character or group of characters is missing from a cultivated species. This may have come about as a consequence of the reduction in variation observed in most cultivated species due to the sieve of selection, or the crop many never have had the character. A search of related wild species is then required to identify which of them may be useful as potential gene donors.

Not all species are able to hybridize, and there can be considerable variation in the ease of hybridization, even between closely related species. This is not entirely unexpected since a restriction of gene flow between species is an important aspect of the speciation process itself, but clearly any restriction in gene flow can be accomplished in a variety of ways. Intrinsic barriers to hybridization are more likely to exist where related species grow in a sympatric or parapatric situation than in an allopatric one where they are physically remote from one another. There are also genotypic differences within species that influence the success or failure of interspecific hybridization. Amongst the best known are the crossability genes (*Kr*1 and *Kr*2) that influence the success of crosses between wheat (*Triticum aestivum*) and rye (*Secale cereale*). Cultivars of wheat that are recessive at both loci will cross readily with rye, but when dominant alleles are present, pollen tubes are inhibited at the base of the style. However, in most species, little is known about the genetic control of crossability between species, though there are many reports of intraspecific variation for this character.

Barriers to hybridization or isolating mechanisms are usually put into two categories: prefertilization barriers that act to prevent the formation of a hybrid embryo; and postfertilization barriers that reduce the vigor of any hybrid embryos that do form. In **Figure 1**,

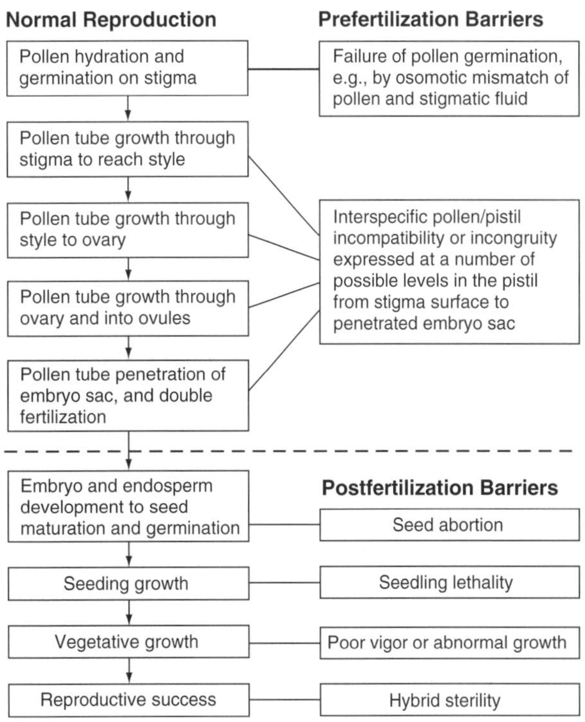

Figure 1 Schematic representation of stages in normal sexual reproduction (left) and related barriers to interspecific hybridization (right). Reproduced with permission from Williams EG (1987) Interspecific hybridization in pasture legumes. *Plant Breeding Reviews* 5: 237–305.

the normal processes of sexual reproduction are aligned with the related hybridization barrier, and it can be seen that these operate at a variety of stages during fertilization and subsequent growth and development. An understanding of these isolating mechanisms is important if problems are encountered with the production of interspecific hybrids and attempts are to be made to overcome the barriers.

The failure of pollen to germinate on an alien stigma is often very difficult to overcome, but in *Populus* spp., increased frequencies of hybrids have been obtained using mentor pollen. These and many other plants have recognition molecules contained within their pollen exine, and mentor pollination involves pollination with a mixture of compatible (self) pollen that has been killed by irradiation or heat/cold treatment and incompatible (alien) pollen. The recognition molecules from the compatible pollen block the recognition of the alien pollen with the result that fertilization takes place. Failure of pollen germination can also be associated with differences in the osmotic balance between pollen and stigma. This can be related to differences in the ploidy level of the species to be hybridized, as in crosses between diploids and tetraploids. The artificial manipulation of chromosome number, usually doubling of the diploid by colchicine treatment, may be effective in overcoming the barrier in these cases.

Prefertilization barriers can also be found after pollen germination, where alien pollen tubes may be slow-growing in the style, or may be inhibited completely at various places in the style in a similar manner to that seen in gametophytic incompatibility (**Figure 2**). Differences in style length between parental species have also been implicated in prefertilization failure. Crosses between corn (*Zea mays*; maize) and *Tripsacum* spp. can be made by shortening the corn style so that it is approximately the same length as the *Tripsacum* one, rather than approximately 15 times longer. Barriers within the style have been overcome using stylar amputation, and the pollination of the cut surface of the style. In crosses between *Lathyrus* species, the pollination of the cut surface of the style removed the barrier to pollen tube growth and allowed the production of new hybrid combinations. Plant growth regulators have also been used successfully where pollen tube growth rates are slow. In both orchard and field crops, successful hybrids have been produced, and early examples include the treatment of pear (*Pyrus communis*) stigmas with NAA prior to pollination with apple (*Malus pumila*) pollen, and the use of gibberellic acid (GA_3) and IAA in many cross combinations of the Triticeae. These treatments may increase both the rate of pollen tube growth and prevent abscission of the pollinated flowers, and together may allow fertilization to occur. *In vitro* fertilization can also be used where barriers in both stigma and style prevent pollen tube growth. Excised embryos are cultured on artificial media and then pollinated and allowed to develop either directly into an embryo or, in some cases, into a seed. When all attempts to overcome these species isolating mechanisms fail, protoplast fusion can be used (*see* **Tissue Culture and Plant Breeding**: Somatic Hybridization).

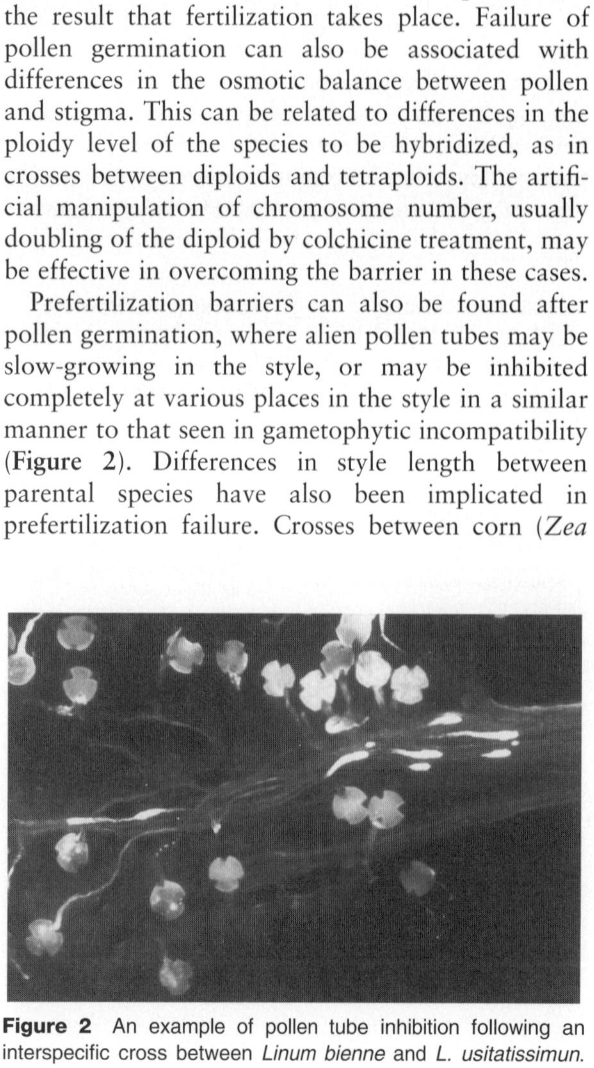

Figure 2 An example of pollen tube inhibition following an interspecific cross between *Linum bienne* and *L. usitatissimun.* Large amounts of callose have accumulated at the tip of the pollen tube, and these fluoresce when stained with aniline blue and observed under ultraviolet light.

Postfertilization barriers also operate in a variety of ways. The most common of these involves the abortion of seeds due to the failure of the endosperm to develop. The endosperm is usually triploid and contains two maternal genomes and a single paternal one; it provides an important source of nutrients for the developing embryo. There is probably no single reason why endosperm development fails in some hybrids but not in others, but differences in DNA quantity (which correlate with differences in cell cycle time) between the parent species, differing amounts of late-replicating heterochromatin, and the presence of genes that regulate endosperm and embryo development (endosperm balance number, EBN) have all been implicated. A commonly used technique to circumvent the problem of poor endosperm development involves embryo culture using *in vitro* techniques, a procedure also known as embryo rescue. This is most easily accomplished when the hybrid embryo has developed to the heart-shaped stage or beyond; very young embryos at the globular stage have proved much more difficult to rescue. The developing embryo is removed from the seed and placed on a usually solid medium with a high osmotic concentration and a variety of additional components, including coconut (*Cocos nucifera*) water, various amino acids, casein hydrolysate, and plant growth regulators. Although some embryos will grow on fairly simple media, in other cases, it is necessary to develop specific media on an empirical basis, as what works to stimulate growth in one hybrid combination may be inhibitory in another. In legumes, very young hybrid embryos have been cultured successfully by transferring them into normal endosperm of one or other parental species whose normal embryo has been removed (**Figure 3**).

Seedling lethality in specific crosses cannot be overcome, but the use of a wide variety of genotypes may result in specific gene combinations that are viable. These seedlings may not be very vigorous, but careful nurturing in the greenhouse can result in the plants reaching maturity, and eventually flowering. These plants may be used as parents in backcross breeding programs, or may produce more vigorous segregants in subsequent generations. Even when F_1 hybrids are vegetatively vigorous they may show hybrid sterility, either due to the failure of their chromosomes to synapse and segregate in a normal manner, or to genetic imbalance following the segregation of chromosomes in the production of haploid gametes. In both cases, a small number of fertile gametes is usually produced. The fertility of hybrids with irregular chromosome behavior can often be increased by doubling the chromosome number of the more or less sterile hybrid plants to produce fertile allopolyploids.

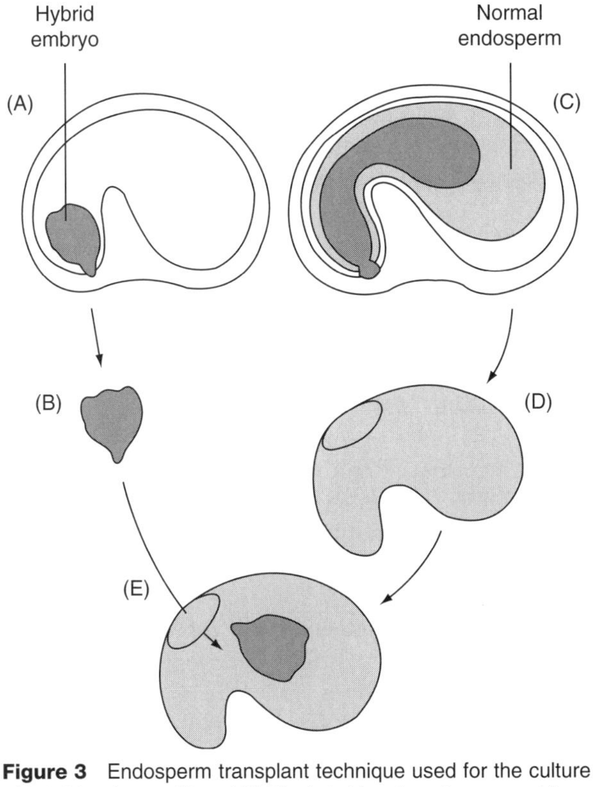

Figure 3 Endosperm transplant technique used for the culture of hybrid embryos. (A and B) The hybrid embryo is removed from its ovule in which endosperm development has failed. (C) To provide a nurse endosperm, a normally developing ovule is dissected to remove the sac of developing endosperm, and (D) the normal embryo is pressed out of the endosperm leaving an exit hole through which the hybrid embryo can be inserted (E). Reproduced with permission from Williams EG (1987) Interspecific hybridization in pasture legumes. *Plant Breeding Reviews* 5: 237–305.

Successful Examples of Interspecific Hybrids in Agriculture

The products of interspecific hybridization are seldom initially successful, as the desirable characters present in the wild species are often linked to deleterious ones, and the breaking of these linkage groups may prove difficult. Nevertheless, there have been many successes, and one outstanding example of an agronomically successful interspecific hybrid must be triticale, a hybrid between wheat and rye that combines the yield and nutritional quality of wheat with the hardiness of rye. Triticales can be made at various ploidy levels depending on whether tetraploid or hexaploid wheat is used as the parent, but, in general, the hexaploid triticales outperform the octoploid ones. Triticale is now widely grown over approximately 2.4 million hectares in Europe, Australia, South Africa, and North America with an annual yield of 6 million tonnes. Several other hybrids, such as those between *Raphanus* and *Brassica*, *Festuca* and *Lolium* and amongst the species of *Rubus*, have also been developed successfully. Many important crop plants have evolved from interspecific hybrids, sometimes involving several different species (**Table 1**).

In addition to combining whole genomes as outlined above, interspecific hybridization has also been used to transfer characters from wild relatives into cultivated species. Wheat again provides many examples of this with the transfer of important disease resistance genes for mildew and various rusts from species of *Aegilops*, *Agropyron*, and *Secale*.

Intraspecific Hybridization

There are usually no barriers to hybridization within plant species, and hybridization has been widely exploited in commercial plant breeding. It is a way of combining desirable characters from two or more different plants, and producing widely segregating populations that can then be subjected to selection. The considerable vigor for both vegetative and reproductive characters shown by F_1 hybrid cultivars has also been widely exploited in the development of both field and greenhouse crops. The genetic basis for

Table 1 The hybrid origin of a selection of common cultivated plants

Crop species	*Ploidy level*	*Proposed wild ancestors*
Bread wheat (*Triticum aestivum*)	6x	*Triticum monococcum*, *Aegilops squarrosa*, *Aegilops speltoides* (or close relative)
Cotton (*Gossypium hirsutum*)	4x	*Gossypium raimondii*, *G. arboreum* or *G. herbaceum*
Banana (*Musa acuminata*)	3x	*Musa acuminata*, *M. balbisiana*
Canola, oilseed rape (*Brassica napus*)	4x	*Brassica oleracea*, *B. rapa*
Strawberry (*Fragaria ananassa*)	8x	*Fragaria chiloensis*, *F. virginiana*
Tobacco (*Nicotiana tabacum*)	4x	*Nicotiana sylvestris*, *N. otophora*, or *N. tomentosiformis*
Dahlia (*Dahlia variabilis*)	8x	*Dahlia coccinea*, *D. pinnata*
Garden roses (*Rosa* spp.)	Various	*Rosa* spp., including *R. chinensis*, *R. damascena*, *R. gallica*, *R. foetida*

this hybrid vigor, or heterosis, remains to be finally resolved, but two main hypotheses have been presented. The first is based on the observed superiority of some heterozygotes over the corresponding homozygotes of particular genes, the overdominance hypothesis. Thus, the individuals that make up the F_1 hybrids will have the greatest number of heterozygous allele combinations, and will be more vigorous compared to either parent. The alternative dominance hypothesis proposes that the increased vigor of F_1 hybrids is a result of the combination of favorable dominant genes derived from the two different parents in the same individual. The dominance hypothesis assumes that beneficial alleles that increase vigor are dominant and harmful ones are recessive. This implies that heterosis could be fixed in a variety by combining all dominant alleles in the homozygous state, and these homozygotes would be as vigorous as the F_1 hybrids. It is widely accepted that inbreeding depression, the converse of heterosis, is a consequence of the accumulation of deleterious recessives that are expressed as homozygosity increases with each generation of inbreeding.

Hybrid Varieties

The development of hybrid varieties is intimately associated with the development of corn in North America in the twentieth century. At the beginning of that century, it appeared that little improvement in yield was coming from selection within corn varieties, but that crosses between varieties showed heterosis. The first attempts to develop hybrids were made in 1909 by G.H. Schull, who showed that crosses between pairs of exceptional inbred lines could outperform the source population, and had the added advantage of producing highly uniform progeny. A major problem with this approach was that the inbreds were poor pollen and seed producers, so that the yield of F_1 hybrid seed was low, and, together with the labor costs associated with detasselling or emasculating one of the lines, too expensive. Schull's single-cross hybrids were soon replaced by double-cross hybrids that used four inbred lines to produce two F_1s, which were then crossed with each other (**Figure 4**). This had the advantage that the first generation of F_1s was vigorous and highly fertile, so, consequently, in the second round of hybrid seed production, large numbers of hybrid seed were produced. The use of cytoplasmic male sterility (see below) significantly improved the economics of hybrid corn production, and by the late 1960s, detasselling had become a thing of the past. In more recent years, with the selection of better, more vigorous inbred lines, single-cross hybrids using cytoplasmic male sterility have replaced double-cross hybrids.

Several ancillary techniques are important in the production of hybrid varieties, in particular, haploid production and pollen or male sterility.

Haploids

Haploid plants contain the gametic (n) chromosome number, but it is necessary to differentiate between haploids produced from diploid species, where each gene will be represented only once, and haploids produced from polyploid plants (polyhaploids), where the number of gene copies will depend on the type of polyploid parent from which they are derived. Haploids arise spontaneously at low frequency in natural populations, usually following the failure of fertilization of embryo sacs and their subsequent development into haploid plants, a process known as parthenogenesis. The first haploids to be described, in *Datura stramonium*, were of this type. Apart from their value for basic genetic studies, haploids are useful in plant breeding as a way of producing completely homozygous or isogenic plants, as doubling the chromosome number of haploid plants will give rise to dihaploid plants with two identical chromosome sets. This can be a more rapid way of obtaining isogenic lines, and is particularly valuable

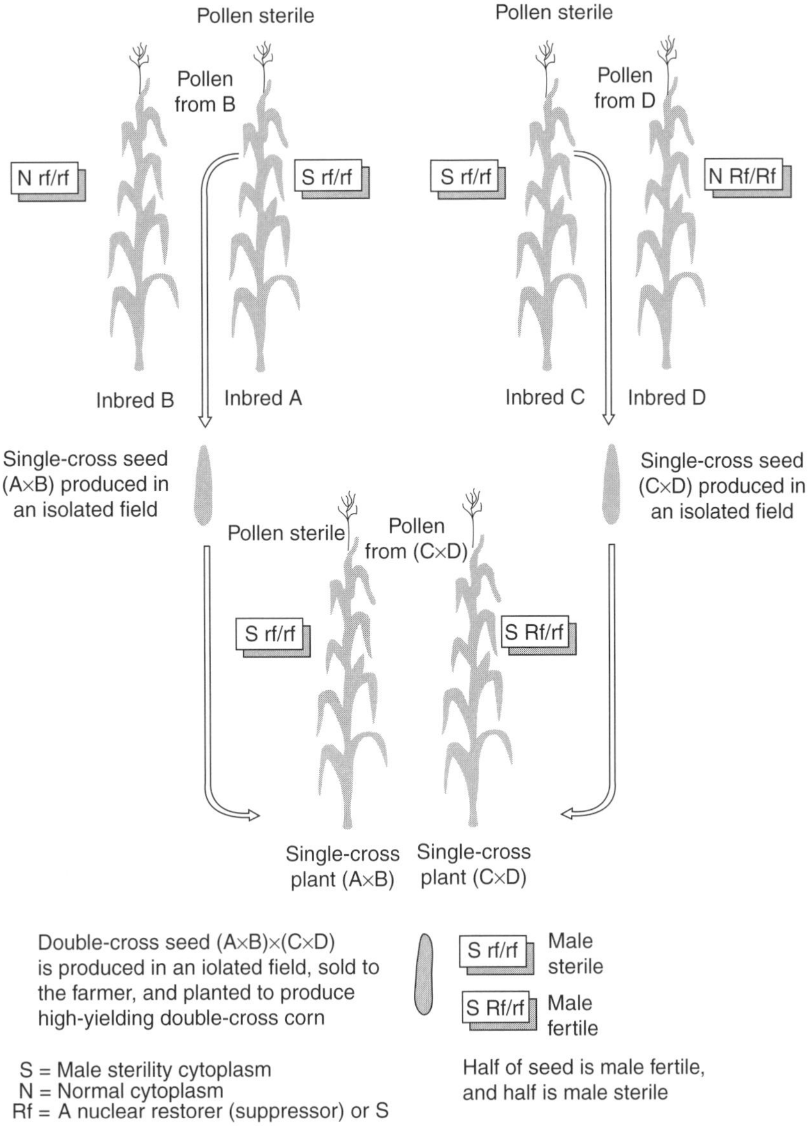

Figure 4 The use of cytoplasmic male sterility to facilitate the production of hybrid corn. The hybrid seed is produced from four inbred lines: A, B, C, and D. At each step, appropriate combinations of cytoplasmic and nuclear restorer genes ensure that the female parents will not self-pollinate and that the male parents have fertile pollen. Reproduced with permission from Griffiths AJF, Miller JH, Suzuki DT, *et al.* (2000) *An Introduction to Genetic Analysis*. New York: Freeman.

in plants with strong self-incompatibility systems, where inbreeding via self-fertilization is not possible.

Haploids can be produced artificially by three main techniques: female parthenogenesis, androgenesis/anther culture, and chromosome elimination. To be of value in plant breeding, it is necessary to be able to produce haploids from a wide variety of genotypes or parents. No one method for producing haploids is available across a variety of species, and for each crop, optimum conditions must be established by experimentation.

Parthenogenesis can be induced in many species by *in vitro* culture of unpollinated ovaries and ovules. There are often strong genotypic differences that influence parthenogenesis, but this has been successfully exploited in major crops such as rice (*Oryza sativa*), sugar beet (*Beta vulgaris*), and potatoes (*Solanum tuberosum*). More widely used is androgenesis, the development of haploid plants from microspores that involves the switch in development of the microspore from its normal pathway to an abnormal one, resulting in the production of new haploid plants via an embryogenic pathway. Immature anthers are harvested and grown on sterile medium where they may either behave like a zygote, and develop directly into an embryo, or undergo indirect androgenesis by first forming a callus of undifferentiated cells, which then differentiates into shoots and roots. A variety of conditions affect the efficiency of induction of androgenesis, and these include the developmental stage of the anther, the age and condition of the source plants, the composition of the culture medium, the genotype of the source plant, and whether the anthers have been subjected to a thermal shock prior to culture.

Chromosome elimination was first observed in crosses between barley (*Hordeum vulgare*) and its wild relative *H. bulbosum*. Fertilization occurred in these crosses, but the embryos died during the early stages of development. When these embryos were rescued by embryo culture, viable plants were produced, and these were found to be haploid, containing only chromosomes derived from the barley parent. Subsequent studies on the development of such plants showed that there was a progressive elimination of the *H. bulbosum* chromosomes in both embryo and endosperm during the initial stages of seed formation. Haploids have also been produced in wheat after crossing with *H. bulbosum*, corn, and *Sorghum* spp. Initially, the rate of recovery of haploids from these combinations was low, but selection of appropriate genotypes and improved culture conditions have greatly increased the efficiency of the process.

Male Sterility

Populations of wild plants all show a few individuals that fail to produce viable pollen and are therefore male sterile. These individuals can be very useful in plant breeding programs, as they eliminate the need for manual emasculation of flowers, and, consequently, result in a huge saving in labor costs in the production of hybrid varieties. Male sterility may have either a genetic or a cytoplasmic basis, and there are also "restorer" genes that can restore pollen function in male sterile plants. Genetic male sterility (GMS) is usually controlled by recessive alleles of genes that can have a range of effects on vital pollen functions. These genes may also cause abnormalities in chromosome behavior and cytokinesis, but may also cause male sterility due to the excessive development of the pollen exine, or the failure of the formation of a germinal aperture as in mutants of barley. Genetic male sterility can be readily exploited by producing lines homozygous for the recessive alleles, which can then be used in hybridization programs. However, GMS has one major drawback, the male sterile lines cannot be maintained alone and need to be regenerated by crossing with heterozygous maintainer lines. The progeny of these crosses will segregate for male sterility, and male sterile individuals need to be selected either manually, or by using tightly linked genes with clear phenotypes.

Cytoplasmic male sterility (CMS) is much less common in plant populations than the genetic variety, but is by far the most widely used form of male sterility for the production of hybrid varieties. As the name implies, this type of male sterility is not under nuclear control, but results from mitochondrial mutations that are highly organ-specific and cause abnormal development of pollen, but leave female function unimpaired. Since the majority of angiosperms show maternal inheritance of mitochondria, CMS is also maternally inherited. Species that show cms usually contain plants that carry genes that can restore pollen fertility in plants with the sterile cytoplasm, and these two features have been exploited for the production of hybrid varieties in crops such as corn (**Figure 4**).

List of Technical Nomenclature

Allopatric	Populations or species that occupy separate geographical locations.
Androgenesis	The production of plants from microspores or pollen.
GA_3	Gibberellic acid.
Heterosis	Superiority of an F_1 hybrid compared to its parents.
IAA	Indole-3-acetic acid.
NAA	1-Napthaleneacetic acid.
Overdominance	When the phenotypic expression of the heterozygote exceeds that of either homozygote.
Parapatric	Populations or species that have contiguous but nonoverlapping geographical distributions.
Parthenogenesis	The production of plants from unfertilized embryo sacs.
Sympatric	Populations or species that occupy the same geographical location.

See also: **Flowering and Reproduction**: Fertilization; Gametophytic Self-Incompatibility. **Tissue Culture and Plant Breeding**: Somatic Hybridization.

Further Reading

Bajaj YPS (1990) *Biotechnology in Agriculture and Forestry 12: Haploids in Crop Improvement I*. Berlin and Heidelberg: Springer-Verlag.

Hayward MD, Bosemark NO, and Romagosa I (1993) *Plant Breeding, Principles and Prospects*. London: Chapman and Hall.

Kalloo G and Chowdhury JB (1992) *Distant Hybridization of Crop Plants*. Berlin and Heidelberg: Springer-Verlag.
Knox RB, Willing RR, and Ashford AE (1972) Role of pollen-wall proteins as recognition substances in interspecific incompatibility in poplars. *Nature* 237: 381–383.
Raina SK (1997) Doubled haploid breeding in cereals. *Plant Breeding Reviews* 15: 141–186.
Sharma DR, Kaur R, and Kumar K (1996) Embryo rescue in plants – a review. *Euphytica* 89: 325–337.
Williams EG (1987) Interspecific hybridization in pasture legumes. *Plant Breeding Reviews* 5: 237–305.
Williams EG, Maheswaran G, and Hutchinson JF (1987) Embryo and ovule culture in crop improvement. *Plant Breeding Reviews* 5: 181–236.

Comparative Genetics

P J Sharp, Plant Breeding Institute, University of Sydney, Australia
M D Gale, John Innes Centre, Norwich, UK

Definitions

"Comparative genetics" is derived from "comparative mapping," which began in plants in the late 1980s. Comparative mapping, initially using common DNA probes to map related but distinct species, demonstrated that gene order and content was much more conserved than previously believed. The presence and location of a gene on the chromosomes of one species could be used to predict the location on the chromosomes of other species. Comparative genetics involves more than map location, and includes predictions of the mode of gene action, and the physiological and biochemical pathways controlling traits in one species from knowledge derived in another. Comparative genomics extends this by using the massively parallel tools of genomics such as DNA chips from a model species, e.g., *Arabidopsis* or rice, for the analysis of gene expression of related crop plant species.

Beginnings and Significance

Soon after the rediscovery of Mendel's work early in the twentieth century, effects of large numbers of single gene differences in plants were documented. The basic rules and variations of transmission genetics – dominance, linkage, and epistasis, were established in many species of both plants and animals. Among these early publications, the famous Russian geneticist Nicholai I Vavilov (1887–1943) foretold comparative genetics with his book *The Law of Homologous Series in Variation* (1920), using examples from crop plants and their wild relatives. In this work Vavilov documented that there were similar changes, or parallel mutations, in closely related but distinct species. He showed that closely related species and genera are so regularly characterized by similar series of mutations that, knowing the series of mutations in one species or genus, one can predict the existence of parallel mutations in other species and genera. Whole plant families are generally characterized by a pattern of variation through all genera and species in a family. For example, cereals, such as rye (*Secale cereale*), wheat (*Triticum aestivum*), millet (*Panicum miliaceum*), and corn (*Zea mays*; maize) have varieties with and without husks surrounding the grains, with nonshattering versus brittle ears, and possessing or lacking ligules on their leaves. Each difference is controlled by a single gene in each species, so that it is clear that there is a great deal of similarity between the genes in the different species, and these are related to each other. Vavilov's theory also predicts that other cereals and grasses would have a gene controlling each of these three traits.

This theory has now been greatly extended. The two main model plant species used, a small eudicotyledonous weed (*Arabidopsis thaliana*, Brassicaceae), and the monocot, cultivated rice (*Oryza sativa*, Poaceae), are intensely studied and both now have very detailed genetic maps, and "essentially complete" genome DNA sequences available. The impetus underlying the large effort that resulted in these two major scientific resources is the idea, descended from Vavilov and spurred on by the early comparative mapping results, that the genes and general principles of plant growth and development found in the model species will be generally applicable to all plants.

Studying Comparative Genetics

Vavilov could only analyze visible characters, using the similarity of both the phenotype, and the genetic control of the phenotypic variation, to infer that the same gene was present in the different species. The same principles are still used, except that the phenotype and genotype are now studied with the tools of molecular genetics. This move to the molecular genotype in comparative genetics has helped overcome, but not eliminate, a perennial problem of comparative genetics – distinguishing between orthologs and paralogs. Orthologs (also known as homoeologs) are those genes that are the equivalent ones in the two species being compared, therefore giving correct comparative genetic information. In contrast, analyzing paralogs across species while assuming the genes are orthologs, will give erroneous comparative genetic information. Paralogous genes arise from duplication within a genome after which the different copies may have evolved different gene functions while retaining some DNA sequence similarity.

Comparative Genetic Mapping

Comparative mapping followed on from the early attempts to construct maps of plant genomes using molecular markers. These first maps, which predated the polymerase chain reaction (PCR) technique, were constructed with restriction fragment length polymorphism (RFLP) loci, often using anonymous or random cDNAs as the probes. cDNA clones represent active genes rather than random fragments of the nonfunctional DNA that make up much of plant genomes. It was soon found that cDNA RFLP markers could often be used successfully across species, i.e., in a comparative genetics approach. If a particular cDNA clone from one species detects a locus by RFLP mapping in both its source species and another species, it can be assumed with a reasonable level of certainty that these two loci represent orthologous genes in the two species. The comparisons were occasionally confused by detection of paralogs, but the method was biased towards the detection of orthologs by the use of stringent hybridization methods.

The first report of the use of RFLP-based maps in comparative genetics was from the laboratory of Tanksley, then in New Mexico, in 1988. His group used cDNA probes in genetic mapping in both potato (*Solanum tuberosum*) and tomato (*Lycopersicon esculentum*). The work showed that cDNAs from one species worked well as RFLP probes in the other, and that there were extensive similarities between the chromosomes of the two species. In fact, only five chromosomal inversions differentiated the 12 chromosomes of potato and tomato. Although the maps were of fairly low resolution, the implication of these results was that whole chromosomes and large chromosomal regions had remained intact during the evolutionary divergence of the potato (*Solanum* genus) and tomato (*Lycopersicon* genus) lineages. It also implied that between the marker loci specifically used in the comparative mapping experiment, there were large numbers of undiscovered orthologous genes in a colinear arrangement in each of the two species.

Grass Comparative Mapping

It was in grasses that early research suggested that chromosomes could be conserved across plant species. This started with the work of Sears in the 1940s, in wheat and its relatives. The gross visible phenotypes and viability of wheat aneuploid lines suggested that bread wheat, an allohexaploid, had seven groups of three chromosomes, from each of the ancestral species, and that each chromosome of a group of three had similar gene contents – they contained the same set of orthologous (homoeologous) genes. Further, similar phenotypic observation of lines of bread wheat with chromosomes added or substituted from relatives such as rye also suggested that the presence of largely orthologous (homoeologous) chromosomes also applied to other cereals related to wheat.

The use of RFLP mapping with cDNA probes, first carried out at the Plant Breeding Institute in Cambridge, UK, largely confirmed this in the Triticeae, the wheat tribe, at about the same time as the work in potato and tomato mentioned above. The three genomes in bread wheat, and the chromosomes of barley (*Hordeum vulgare*), were shown to have a colinear arrangement of marker genes, with only a few major evolutionary translocation differences between a small number of chromosomes. The larger of these translocations had previously been predicted from cytogenetic analysis. This and similar work in other grasses enabled a set of cDNA clones to be identified that gave good hybridization in RFLP studies across a wide range of grasses, representing an "anchor" probe set. These probes were then used in the genetic mapping of species from the main grass families; rice representing the Bambusoideae, finger millet (*Eleusine coracana*) representing the Chloridoideae, foxtail millet (*Setaria italica*), sorghum (*Sorghum bicolor*), pearl millet (*Pennisetum glaucum*), corn and the highly polyploid sugar cane (*Saccharum officinarum*) representing the Panicoideae, the wheat group (wheat, barley, and close relatives), the forage grasses (*Lolium* and *Festuca*) and oats (*Avena* spp.) representing the Pooideae. The resulting genetic maps of each species were then aligned to detect regions of colinear linked markers across the species.

The key finding was that all the grasses studied had large syntenous linkage blocks, up to the size of complete chromosomes, within which gene order and content is highly conserved. This is despite a range of haploid chromosome numbers from 5 to 12, the nuclear DNA content varying over 40-fold between these species, and the grasses being the result of evolutionary diversification over more than 50 million years.

The chromosome maps could all be related to the rice chromosome maps. A way of representing this was by a series of circles (**Figure 1**). Each circle represents the chromosomes of a species, broken and transposed where required to align with the rice chromosomes. The order of chromosomes or segments in a species circle is determined by the syntenic relationships of these to the rice chromosomes in the center. The rice genome is presently known to consist of something like 25 linkage blocks that form the basis, in various combinations, of the chromosomes of the other grass species. This relationship is not restricted to markers defined by cDNA probes. Major genes, and also quantitative trait loci for important traits, such as vernalization, flowering time, plant height, and shattering also showed orthologous, colinear relationships across barley, wheat, corn, and rice chromosomes. The grasses can therefore be treated as a single genetic system.

Another important result of this comparative mapping in the grasses was in corn. The comparative results showed that this chromosomally diploid species was an ancient tetraploid – there were basically two copies of each locus and chromosome. Corn is therefore represented in **Figure 1** by two circles of chromosomes. This result confirmed earlier suggestions that corn was an ancient tetraploid that was based on the genetics of a small number of genes controlling morphology and isozymes.

Evidence from rice genomic sequencing indicates that up to 70% of the genome is duplicated. Most of these duplications occurred before the grass species divergence over the last 60 million years and therefore do not impact the syntenic relationships between grasses greatly. However, the last major duplication involving the duplication of the short arms of rice chromosomes 11 and 12 may have occurred as recently as 25 million years ago. Syntenic predictions in these areas are difficult between the Panicoideae, which display the duplication, and the Pooideae, where an equivalent duplication event is not known.

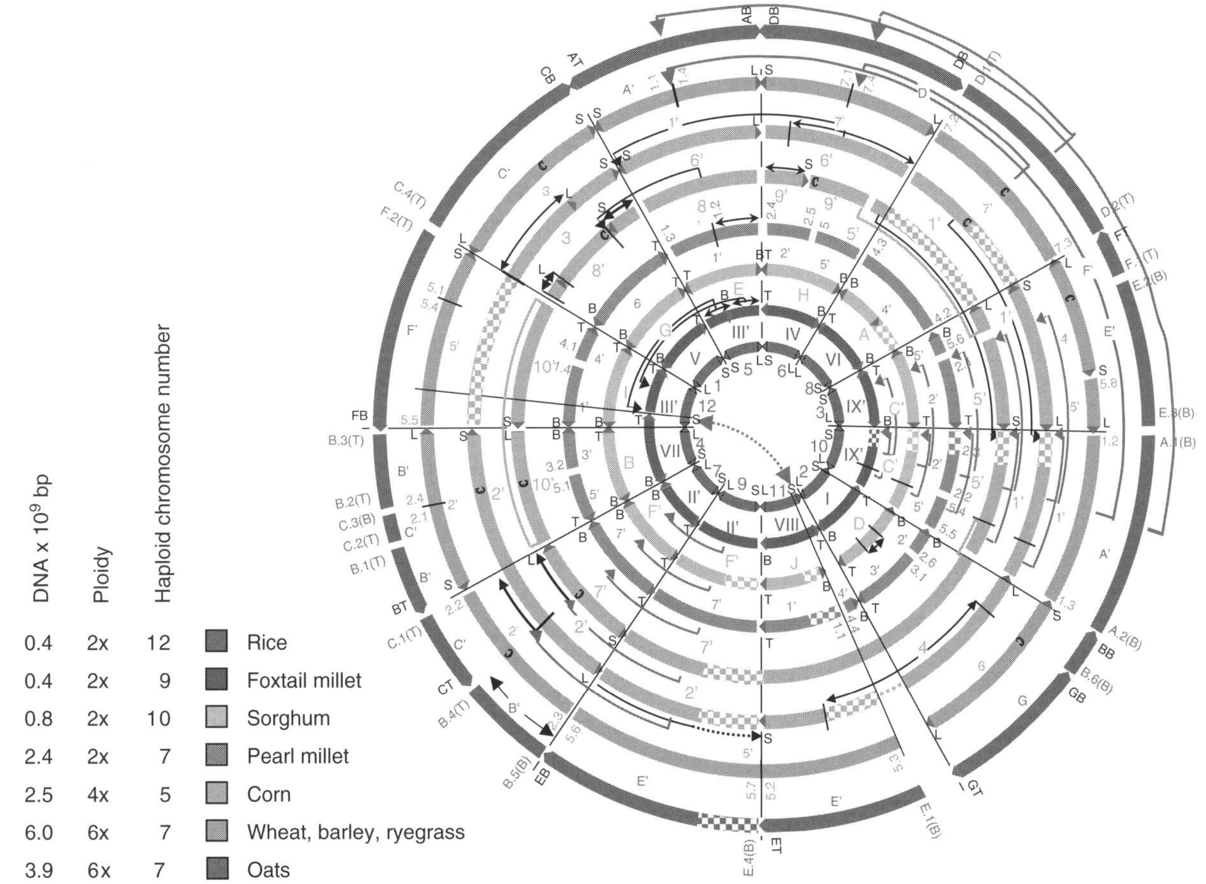

Figure 1 (see color plate 4) Genome relationships among the major cereals. The chromosomes of the several economic grasses are aligned against the 12 rice chromosomes that make up the innermost circle. The locations of orthologous genes, QTLs or DNA sequences in the different genomes can be predicted to lie on radii in the diagram. The major insertions, inversions and translocations which distinguish the other genome from rice are indicated by the arrows. The hatched areas represent chromosomal regions where the relationships are uncertain.

The overall trend is toward great synteny in the grasses, but some classes of genes, especially those that are likely to evolve particularly rapidly, are prone to break the rules of colinearity. For example, genes related to disease resistance, the resistance gene analogs (RGAs), have been found in many examples to be absent in a species when their presence would be predicted from another. It is possible that the high selection pressures applied by rapidly changing and multiple pathogen populations could cause individual genes to rapidly become redundant and be eliminated from the genome.

Comparative Mapping in Crucifers

The Brassicaceae family is also well studied by comparative mapping. The family encompasses the *Arabidopsis* model species, for which the entire genomic DNA sequence is known, and a considerable number of economically important crops of the genus *Brassica*; oilseed rape (canola), the mustards, and the cabbage/broccoli group of vegetables.

Comparative mapping within the *Brassica* genus has revealed, as in the Triticeae wheat group, there is extensive colinearity between the diploids *Brassica nigra*, *B. oleracea*, and *B. rapa*, but that the three species are differentiated by many chromosomal rearrangements. Most syntenic segments are therefore only relatively small chromosomal regions. The amphidiploid species *B. napus* (canola) and *B. juncea* each have chromosomes that are basically unchanged from their respective diploid progenitors.

Mapping has also shown that most chromosome segments in these chromosomal diploid *Brassica* species are triplicated, implying that the species have an ancient and common hexaploid ancestor. The ancient hexaploid nature of the present diploid *Brassica* species was borne out by comparison with *Arabidopsis* and *Capsella rubella*, species with much smaller genome sizes. These two species have essentially colinear chromosomes, disrupted by a few breaks to give their different chromosome number, and a few inversions. Most segments of *Arabidopsis* chromosomes are present three times in the diploid *Brassica* species, but associated with dozens of chromosomal rearrangements resulting in the conservation of only short colinear chromosome segments.

Microcolinearity

Microcolinearity studies are usually performed by comparing large-insert (bacterial artificial chromosomes (BAC) or yeast artificial chromosomes (YAC)) clones from different species. The clones are selected on the basis of having a gene of interest in common. The comparisons can be made by cross-hybridization of small segments between the two species, but they are now usually carried out through the complete DNA sequencing of the two regions in clones which are hundreds of kilobases long.

Comparisons of the regions around the *Sh2/a1* loci in rice, corn, and sorghum revealed good conservation of gene content and order across the three species though the distance between the genes varied between the species, but demonstrated clearly that synteny extended only to the genes themselves. The predicted protein sequences of the gene products were highly conserved, as were the intron–exon structures of the genes. However, there is little conservation of sequence in the intergenic regions, and large variation in the location and nature of interspersed transposable elements that make up the bulk of the DNA in these regions. Similar studies around the *Adh1* locus in corn and sorghum indicate that the nine genes in corn are present in sorghum in a colinear arrangement. However, sorghum has three further genes present. Two of these are present on the other corresponding syntenous corn chromosomal region, and can be assumed to have been redundant and lost from the *Adh1* region. The third can only be assumed to have been imported from elsewhere in the genome, which indicates that unpredictable nonsyntenous events can be expected to be widespread. Comparison of a region defined by a barley disease resistance gene of the barley and rice genomes has also indicated a lack of complete microcolinearity, with a 10–15-kb region having moved to a nonsyntenic location. Microcolinearity studies in crucifers reveal similar situations.

The general result is that in small regions of chromosomes that show overall synteny based on mapping studies, closely related species have good microcolinearity of the genes themselves, but there is divergence of the intergenic regions that are largely made of various classes of transposable elements. With more distantly related species the same regions can be expected to display considerable disruptions to microcolinearity, due to gene deletion, translocation, and inversion events. This is particularly the case where major duplication events (including polyploidization events) have occurred subsequent to separation by speciation. This is often the case in the eudicots where evolutionary times are relatively long, e.g., 150 million years between *Arabidopsis* and tomato, and of course the 200 million year monocot–eudicot divide. These disruptions mean that the use of synteny between model and crop plants for the isolation of genes of agronomic importance from the crop species should be approached with caution. The approach may require comparisons of the crop species with close wild relatives, as well as the model species in order that the exceptions to microcolinearity of pairwise comparisons can be overcome to enable routine gene isolation by syntenic comparison.

Examples of the Use of Comparative Genetics

Comparative genetic information has resulted in significant advances in knowledge in plant science. At the very simple level, the knowledge of the broad syntenic relationships between species allows genetic mapping research to use targeted groups of probes from one species to help saturate the molecular marker map of another species. Successful work in this area has used the RFLP and expressed sequence tag (EST) resources of a model species, rice, as a source of probes for a more complex crop plants species such as sugar cane, barley, and wheat. This increase in marker map density can allow the development of targeted research resources in the complex crop species that allow isolation of genes of economic and scientific interest.

The more "classical" level of comparative genetics, the similarity of mutant phenotypes between the orthologs, coupled with genomic resources of model species, has also been a successful strategy for gene isolation across a wide taxonomic divide, that between dicots and monocots. One of the "green revolution" semidwarfing genes in wheat (*Rht-1*) gives a gibberellic acid-insensitive dwarf phenotype. Chromosome walking and other techniques enabled cloning of a gene controlling a similar gibberellic acid-insensitive dwarfed phenotype in *Arabidopsis*, and database searches revealed that this gene had strong sequence similarity with certain rice ESTs. These rice ESTs were then used as probes to isolate wheat cDNA clones, as potential clones of the *Rht-1* gene of wheat. Genetic mapping using these as RFLP probes then enabled wheat cDNA clones that cosegregated with the *Rht-1*, and also with a gibberellic acid-insensitive dwarf mutant (*d8*) of corn. Comparison of the DNA sequence of the *Rht-1* "tall" and "dwarf" alleles then enabled the precise mutant change in the "dwarf" allele to be identified. This work is now enabling the use of this dwarfing gene in other crops.

Many genes of economic significance in crop plants are currently known as quantitative trait loci (QTL), that control proportions of the genotypic variation in particular quantitative traits. Except for some disease resistances, and some simple morphological and quality traits with simple qualitative genetic control, QTL are the main loci that plant breeders are attempting to manipulate in their programs. QTL are identified and mapped on chromosome maps using the combined tools of quantitative genetics and linkage analysis. Because of their importance in plant breeding there is considerable interest in further characterizing QTL of crops. Comparative genetics has allowed this to occur, especially in the crucifers. The control of flowering time in *Brassica* was characterized using knowledge from studies in *Arabidopsis*. Using the cloned *CO* gene of *Arabidopsis* (*CO* influences flowering time) as probes in genetic mapping revealed that *CO*-related sequences were located at QTL identified in *Brassica nigra*, and these regions are colinear with three of the six flowering time QTL identified in *B. rapa*. Similar cosegregation studies have allowed the characterization of the economically important loci controlling erucic acid content of oilseed rape (canola). Orthologs of the cloned *Arabidopsis* fatty acid elongase (*FAE1*) gene colocate with the two loci determining oilseed rape erucic acid content. This knowledge allows molecular markers derived from the *FAE1* clone to be used as perfect markers in marker assisted selection for the two loci in oilseed rape breeding programs.

List of Technical Nomenclature

Colinearity Conservation of gene order within a chromosomal segment between different species.

Homoeologs The same as orthologs (see below), but usually applied to orthologous genes in quite closely related species, such as within the wheat tribe, the Triticeae, where the term was first used. Also spelt "homoeologues" (cf. homologs, describing the members of a pair of chromosomes in a diploid).

Microcolinearity The conservation of gene order in small chromosomal segments as observed at the DNA sequence level, usually studied by DNA sequencing large-insert clones such as bacterial artificial chromosomes and yeast artificial chromosomes.

Orthologs Genes that arose from a common ancestor and are conserved in various species. The term orthologous can also be used to describe a region of the chromosomes in common between species (cf. paralogs).

Paralogs Genes that have arisen by duplication within a species and which may have now diverged from each other in structure, function and/or regulation, and gene location.

Synteny	The original use of the word was to describe genes located on the same chromosome by physical mapping within a species. However, the term has now become widely used to indicate the conservation of the order of orthologous genes between different species, i.e., the same as "colinear."

See also: **Crop Improvement**: Genetic Maps; Molecular Markers; Marker Assisted Selection; Plant Genomes; Chromosome Engineering.

Further Reading

Bancroft I (2001) Duplicate and diverge: the evolution of plant genome microstructure. *Trends in Genetics* 17: 89–93.

Feuillet C and Keller B (2002) Comparative genomics in the grass family: molecular characterization of grass genome structure and evolution. *Annals of Botany* 89: 3–10.

Gale M and Devos K (1998) Plant comparative genetics after 10 years. *Science* 282: 656–659.

Keller B and Feuillet C (2000) Colinearity and gene density in grass genomes. *Trends in Plant Science* 5: 246–251.

Moore G, Devos KM, Wang Z, and Gale MD (1995) Grasses, line up and form a circle. *Current Biology* 5: 737–739.

Schmidt R (2000) Synteny: recent advances and future prospects. *Current Opinion in Plant Biology* 3: 97–102.

Schmidt R, Acarkan A, and Boivin K (2001) Comparative structural genomics in the Brassicaceae family. *Plant Physiology and Biochemistry* 39: 253–262.

Shimamoto K and Kyozuka J (2002) Rice as a model for comparative genomics of plants. *Annual Review of Plant Biology* 53: 399–419.

Plant Genomes

J S Heslop-Harrison, University of Leicester, Leicester, UK

Introduction

The plant nuclear genome is contained within chromosomes and consists of all the DNA, including the complete set of genes. The chromosomal DNA is composed of the four chemical bases adenine, thymine, guanine, and cytosine, in a linear sequence unique to each genotype of a species. Since the early 1990s, extensive and detailed knowledge about the DNA sequence of fragments of the genome, many individual chromosomal regions, and larger stretches of many plant genomes, has become available and this DNA sequence is published in extensive nucleotide databases. The impact of genomics – the science of studying the genome – is now such that a complete study of genetics, physiology, or development usually includes DNA sequence data. Mitochondria, chloroplasts, and viruses have genomes (DNA, or RNA in the case of some classes of viruses), and there are extensive studies of their genomes and interactions with the nucleus.

The Sizes of Genomes

Nuclear genome size may be measured as the number of base pairs (in the two strands of the double helix) of DNA present in the nucleus, or as the mass of DNA present in a nucleus (as picograms, pg, 10^{-12} g, which can be converted to base pairs by dividing by the average mass of a nucleotide, so 1 pg is equivalent to 1000 million base pairs (Mb) within the limits of DNA amount measurement). As nuclei move through the cell cycle and development, their duplication and replication stages alter. The amount of DNA in an unreplicated haploid (n, as found in a gametophyte) nucleus is known as the 1C value, and this value is normally quoted in measurements of numbers of base pairs. The 2C value is twice the 1C value and is the genome size of a replicated haploid (n) or unreplicated diploid ($2n$, as found in the sporophyte) nucleus, while the 4C value is found in a replicated diploid nucleus (such as a prophase in the sporophyte). As with the values of the chromosome number, where $2n$ is the number of chromosomes in the nucleus of the sporophyte, genome size measurements are the actual size of the genome so the ploidy level of the organism (see below) is not taken into account. A database of plant genome sizes (currently some 4000 taxa) is maintained at http://www.rbgkew.org.uk/cval/homepage.html.

Plant nuclear genomes have a huge range in size, from some 125 Mb of DNA in a range of species including crucifers, roses, some trees, pteridophytes, and bryophytes, up to more than 70 000 Mb found in other pteridophytes and angiosperms (one lily has a genome of more than 120 000 Mb). The DNA in the genome is divided among discrete chromosomes, the number of which varies from a diploid number of $2n = 4$ (e.g., in the genera *Brachycome, Zingeria,* and *Haplopappus*, where gametes have only two chromosomes) to over 150 in some angiosperms and pteridophytes. Polyploidy has played an important part in the evolution of many plant species, so a plant may have inherited complete genomes from several ancestors to make up its own genome: bread wheat (*Triticum aestivum*), with 42 chromosomes, is described as $2n = 6x = 42$, and is a hexaploid species with complete diploid genomes inherited from its three ancestors, each described as $2n = 2x = 14$.

DNA Sequences in Genomes

The genome of plants, as in other species, includes not only the genes and their regulatory DNA, but motifs of DNA sequences ranging in size from two nucleotides (or even just one nucleotide) to 10 000 or more base pairs, which are repeated tens, thousands, or millions of times in the genome (**Figure 1**). This repetitive DNA may comprise more than 90% of all the DNA in the genome. The major component of the repetitive DNA in most species is retrotransposons – DNA elements typically 4000–10 000 bp long which amplify through transcription to RNA and reverse transcription back to DNA, before the DNA inserts into the genome. The elements themselves encode reverse transcriptase and integrase, and used the cellular RNA polymerase for transcription. As expected from their mode of amplification, these sequences are dispersed among other sequences within the genome. Retroelements and their recognizable derivatives represent between 30% and 50% of all the DNA in plant genomes that have been closely analyzed, although many of the sequences are incomplete or have mutated open reading frames which cannot encode functional enzymes.

DNA sequences can occur as tandem repeats, where a motif is present as an array of multiple copies within discrete blocks of DNA. Tandem repeats are abundant in all plant genomes which have been analyzed, typically representing 5–20% of all the DNA. Characteristic tandem arrays are often present in subtelomeric and centromeric regions of chromosomes, and these motifs often have a length of 170–190 bp, corresponding to the length of DNA which wraps around a single nucleosome. Function and structure is now being linked through genome analysis. Centromeric and telomeric tandem arrays may have functions in chromosome stability, recombination, and segregation. The ribosomal genes and intergenic spacer regions are also present as tandem arrays in plant genomes. The 45S rDNA unit, which includes the 18S, 5.8S, and 28S rRNA genes, is a unit some 10 000 bp long. In *Arabidopsis thaliana*, the 45S rDNA is present on two pairs of chromosomes, and the 600 copies represent some 5% of all the DNA.

Microsatellites, or simple sequence repeats, are motifs of typically one to five base pairs that are in

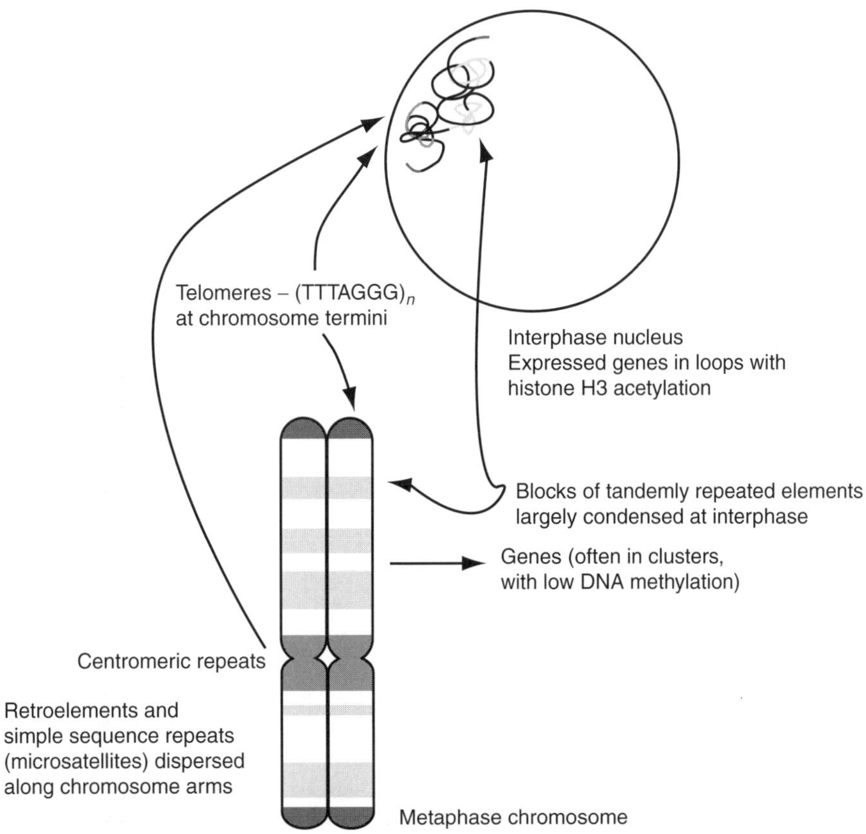

Figure 1 Key features of plant genomes and their organization in a metaphase and interphase chromosome.

tandem repeats of 10 to 50 copies. These motifs are often flanked by single copy DNA, dispersed throughout the genome, and very variable in copy number between different accessions of a plant species, so are widely used as molecular markers for DNA diversity or genotyping.

Transposable elements, and particularly the class of MITEs (miniature inverted tandem elements) also make up an abundant class of repetitive elements in the genome. Additional motifs including tandem and dispersed repeats with no known function and relationship to retroelements are also known. Beyond sequence information, much exciting information is appearing about epigenetics and the modulation of genomes involving DNA and histone methylation and histone acetylation (**Figure 1**).

Genome Sequencing Projects

The genome sequence of the flowering plant *Arabidopsis thaliana* was published in 2000, and the species was the second multicellular eukaryote to be sequenced after the nematode worm *Caenorhabditis elegans*. *Arabidopsis* is a dicotyledonous plant (eudicotyledon) in the Brassicaceae (formerly known as the Cruciferae). Two chromosomes of a monocotyledonous grass, rice (*Oryza stativa*) were sequenced in 2002, and the genome will be completed in 2003. Genome sequencing initiatives are underway for the legume alfalfa (*Medicago sativa*, lucerne), corn (*Zea mays*, maize) and perhaps banana (*Musa* spp.) so substantially more sequence data are likely to be available in the next few years.

Arabidopsis was chosen for analysis because of its advantages for genetics and genomics: it has a relatively small genome size, small plant size, and rapid generation time. The genome sequence is some 125 Mb long and covers all the chromosomes with the exception of the termini, some regions consisting largely (although perhaps not entirely) repetitive DNA including the rDNA and repetitive regions at the centromere. The sequences have been analyzed to find all domains that correspond to genes, and some 26 000 genes have been identified, a number that was below most expectations (but agreeing with data from human sequencing, where only 35 000 genes were revealed), and accounting for about half of the genomic DNA sequenced. Of the remaining DNA in *Arabidopsis*, about 10% is retrotransposons, 5% is the rDNA, and some 5–10% is represented by blocks including tandemly repeated sequences around the centromeres, not all of which have been sequenced.

Results from the rice genome sequencing program allow valuable comparisons with *Arabidopsis*. Rice has a genome size of about 430 Mb, three times the size of that of *Arabidopsis*. Although the average size of exons of genes is similar to *Arabidopsis*, the introns are substantially larger: an apparently larger number of genes, and an increased proportion of retrotransposons, MITEs and other repetitive elements cause much of the increase in DNA amount across the genome. More than half of the genes identified from the rice sequence show no homology to *Arabidopsis*, indicating that there are substantial differences in the genomes of the two species, and if the results from chromosomes 1 and 4 remain true for other chromosomes, there may be some 60 000 genes in rice. Knowledge of the sequences of genes and analysis of their expression and proteins is allowing major progress in understanding the function of plant genes.

For examination of structures in plant genomes and subsequent analysis, large blocks of DNA, typically 70–200 kb long, are usually cloned, using yeast (YAC) or bacterial artificial chromosomes (BAC) as vectors. The use of these large-insert clones is a prerequisite for sequencing, and also a valuable approach to physical mapping. In mapping the end sequences of the artificial chromosomes are matched with others to give a contiguous tiling-path (known as BAC or YAC contigs) through the genome, where every piece is covered by a clone. Genes or expressed sequence tags (ESTs) (see below) can be located by hybridization onto these contigs and hence their physical location determined. In genetic terms, the chromosomes correspond to linkage groups (the genes of which show independent segregation at meiosis) while genes on each chromosome or linkage group, as a whole group, do not show independent segregation (although those which are far apart on the chromosome may not be linked genetically). By analysis of the segregation frequency of molecular markers in hybrid mapping populations, genetic distances between markers can be obtained and correlated with physical distances from contig locations.

ESTs and Gene Sequences

Another way to look at the gene composition of plant genomes is the sequencing of mRNA molecules, with the aim of sequencing all the possible genes that are present. Often, only one end of the clones of mRNA is sequenced, giving ESTs, which allow classification and comparison of all the genes in a species. The approach avoids sequencing large amounts of repetitive DNA without genes, and is particularly important in species such as barley (*Hordeum vulgare*) (with a genome size 12 times larger than rice, and impractical to sequence fully with current technology), wheat (three times larger than barley), or pines (*Pinus* spp.) with genomes typically larger than 20 000 Mb. The method is also important for comparative analysis of genes between species, and large numbers of ESTs have been analyzed from many legume species, and also from crops and wild relatives, such as wild wheat species or *Lycopersicon hirsutum* and *L. pennellii* for comparison with tomato (*L. esculentum*).

Applications of Genomics and Comparative Analysis

Many of the DNA sequences obtained from plant genome studies are proving valuable as genetic markers. The segregation of characters is hard to study as most do not have a clear phenotype or suitable variation, but the use of DNA probes gives an essentially unlimited number of markers across the genome.

There have been major developments in the comparative analysis of plant genomes in the last 5 years. Major crop species have extensive genomic research. Particular targets include the cereals (barley, wheat, corn (maize), oat (*Avena sativa*), rice, *Sorghum*), the solanaceous crops (potato, tomato, and tobacco (*Nicotiana tabacum*)), the legumes (soybean (*Glycine max*), alfalfa (lucerne), lotus (*Nelumbo nucifera*), and pea (*Pisum sativum*)) and crucifers (oilseed rape (*Brassica napus*; canola)). The community of researchers within individual crops has been vital to the progress, combined with large genetic and mutant stock collections, and open exchange of raw data and resources: sequence data from international programs are routinely made available soon after production, and genetic stocks, large insert libraries, and other resources are widely available. The ways that we can apply and exploit data by making comparisons of the different species and using publicly available resources will make discoveries critical to both fundamental plant science and its application in plant breeding and environmental biology.

See also: **Crop Improvement**: Comparative Genetics; Genetic Maps; Molecular Markers.

Further Reading

Feng Q, Zhang Y, Hao P, *et al.* (2002) Sequence and analysis of rice chromosome 4. *Nature* 420: 316–320.

Fransz P, de Jong JH, Lysak M, Castiglione MR, and Schubert I (2002) Interphase chromosomes in *Arabidopsis* are organized as well defined chromocenters from which euchromatin loops emanate. *Proceedings of the National Academy of Sciences of the USA* 99: 14584–14589.

Gindullis F, Desel C, Galasso I, and Schmidt T (2001) The large-scale organization of the centromeric region in *Beta* species. *Genome Research* 11(2): 253–265.

Heslop-Harrison JS (2000) Comparative genome organization in plants: from sequence and markers to chromatin and chromosomes. *Plant Cell* 12: 617–635.

Sasaki T, Matsumoto T, Yamamoto K, *et al.* (2002) The genome sequence and structure of rice chromosome 1. *Nature* 420: 312–316.

Schmidt T and Heslop-Harrison JS (1998) Genomes, genes and junk: the large scale organization of plant chromosomes. *Trends in Plant Science* 3: 195–199.

The Arabidopsis Genome Initiative (2000) Analysis of the genome sequence of the flowering plant *Arabidopsis thaliana*. *Nature* 408: 796–815.

Genetic Maps

R M D Koebner, John Innes Centre, Norwich, UK

Introduction

Genetic maps display the arrangement of genetic loci (genes and genetic markers) linearly along the chromosomes that make up the genome of an organism. Distances between points on the map represent an estimate of the frequency of meiotic recombination between them and these largely, but imperfectly, also reflect their physical separation in base pairs. These estimates are affected by a number of not readily controllable factors; therefore, map distances must be viewed as estimates, rather than as absolute measures. Genetic map distances fail to correspond to physical separation in genomic regions, both where recombination is significantly suppressed or enhanced and where unlinked loci cosegregate because of genetic interactions. Maps are used for the genetic analysis of complex characters, especially those that are quantitatively assessed; for deriving marker-assisted selection assays; and for the isolation of genes by positional cloning. The overwhelming majority of landmarks in genetic maps are anonymous fragments of DNA, usually referred to as molecular markers. Polymorphisms in these markers are assayed either by DNA:DNA hybridization, or by the polymerase chain reaction (PCR); a short description is given of the important variations in these assays.

What is a Genetic Map?

A genetic map is a representation of the linear order of, and the relative separation between, genes or anonymous sequences (together referred to as "loci") along a defined stretch of DNA. In higher organisms, maps can be constructed for either the nuclear or the nonnuclear genomes (mitochondria, chloroplasts, etc.), the latter being universally much smaller than the former, and consisting of only a single DNA molecule. Generally speaking, genetic maps relate to the nuclear genome, which is physically separated into the haploid number of chromosomes. Thus, where the haploid chromosome number is greater than one, the nuclear genetic map will consist of the same number of definable groups of loci or "linkage groups" as there are chromosomes. The lowest known plant haploid chromosome number is two (as in the grass *Zingeria biebersteiniana* and four other angiosperms), while among the common crop species, the number ranges from seven (e.g., in barley, *Hordeum vulgare*) to about 60 (in sugar cane, *Saccharum officinarum*). When a genome has been incompletely mapped, the consequence of a gap between two groups of linked loci which are, in reality, physically contiguous on a single chromosome is that this chromosome will be resolved into two separate linkage groups. In such a situation, the number of linkage groups will thus exceed the haploid chromosome number. Therefore, the completeness of a genetic map can be initially assessed by the degree of correspondence between the number of linkage groups and the haploid chromosome number, the latter being directly observable by cytological means.

Some of the points or "landmarks" along the map will correspond to known genes, but more frequently they represent defined, but anonymous, sequences of DNA, which are assayed by a variety of methods (*see* below and **Crop Improvement**: Molecular Markers). This is because in most organisms, the number of cloned genes and/or the number of direct gene assays is rather low; in contrast, there is effectively no limitation on the number of assays based on anonymous sequences. In a complete genetic map, covering the entire genome, both ends of each chromosome will be marked by the telomere, a structure that in most plant species is composed of tandem arrays of a well-conserved DNA sequence. Despite this, it has not always proved to be a straightforward matter to confirm the position of the telomere on a genetic map. In its ultimate form, a complete genome sequence defines the physical location of every genetic component in the organism. Genome sequences have recently been completed for the model plants *Arabidopsis thaliana* and rice, *Oryza sativa*.

How is a Genetic Map Constructed?

Meiotic recombination provides the underlying basis for genetic mapping. Recombination occurs as a result of the formation of a chiasma between two nonsister chromatids of the tetrad chromatid structure present in the first meiotic division cell. Thus, when a chiasma forms between two loci, only half of the meiotic products will be of crossover (recombinant) type, the other half remaining as parental type. The map distance separating a pair of loci is defined as the average number of crossovers per chromatid, and an average of one crossover per chromatid is defined as a distance of one morgan (abbreviated M). Map distance equates to one half of the chiasma frequency. The assumption is that the probability that a chiasma is formed (and therefore recombination occurs) between two loci increases with the extent of their physical separation (but see "What is the Difference Between a Genetic and a Physical Map?" below). When this distance is sufficiently great, the likelihood that the loci will be separated by recombination rises to 1 and the loci are said to be unlinked. In this situation, 50% of the gametes will be expected to be of parental type and 50% of recombinant type.

Where two loci are so close to one another that recombination never occurs (100% parental type gametes, 0% recombinant) there is complete linkage, with intermediate levels (partial linkage) arising when the loci are physically sufficiently closely positioned with respect to one another so that they are only sometimes separated by recombination. As map distance increases, the chance of a double crossover rises. Where these occur, the linear relationship between map distance and recombination frequency breaks down. To re-establish the property of additivity of distances over long stretches of chromosome, recombination percentages are converted to map units, and a generally accepted choice is the Kosambi mapping function, which calculates map units (in centimorgans, cM, where $1\,M = 100\,cM$) according to the formula: $25 \log_e(1 + (0.02 \times \text{recombination } \%)/(1 - (0.02 \times \text{recombination } \%)$. The Kosambi map unit lies, in theory, between zero and infinity, although in practice the range is from 0 to 100, as the cM equivalent of a recombination percentage of 49% is only slightly more than 100. When the recombination percentage is small, its value approximates to the cM map distance (1% is equivalent to 1.00 cM; 5% to 5.02 cM; 20% to 21.2 cM), but as recombination frequency increases, the two measures diverge (30% is equivalent to 34.7 cM; 40% to 54.9 cM, etc.).

Genetic maps are assembled by analyzing patterns of cosegregation between multiple loci in a segregating population. At its simplest, two-point linkage estimates the recombination frequency between two loci *A* and *B*, at which the two parents carry, respectively, the alleles *A1 B1* and *A2 B2*. It represents the proportion of the total number of gametes that are recombinant, i.e., (number of *A1 B2* gametes + number of *A2 B1* gametes)/total number of gametes. For three loci *A*, *B* and *C*, three separate distances can be estimated: *AB*, *BC* and *AC*. With a linear order *ABC*, the distance *AC* should approximate to the sum of *AB* and *BC*, unless a significant number of double recombination events disturbs this additivity. Three point linkage analysis, therefore, allows an estimate of the frequency of double crossovers, and thereby corrects for the distance *AC*. As the number of loci in a linkage group increases, the complexity of these corrections requires computer-aided algorithms, a number of which have been developed. These programs differ from one another by the choice of criteria for assessing the most probable locus order; overall they generate similar maps, showing variation mostly in regions having a number of closely linked loci.

The choice of parents of mapping populations dictates the genomic coverage of the map, since only loci at which the parents carry contrasting alleles can be mapped. Thus, closely related parents will produce populations in which only a few loci can be mapped, while wide crosses maximize the number of mappable loci. In wide crosses, too much allelic variation can be a disadvantage when the target(s) of the mapping are quantitative traits. Since these are often genetically determined by the action and interaction of a number of genes, it becomes difficult to determine the contribution of any one genetic component to the phenotype of individual segregants unless some of the components are nonvariable.

Choices also need to be made with respect to the type of mapping population generated. In self-pollinating species, the possibilities include F_2, doubled haploid, recombinant inbred and BC_1; in out-crossing species, the first three of these are often not obtainable, as they require one or more generations of self-pollination. In some species, this is prevented by self-incompatibility mechanisms while, in others, significant levels of homozygosity induce inbreeding depression and thus poor plants. As a common alternative, top-cross ((a × b) × c) populations are employed. In certain special cases, mapping can be carried out directly in the gametophyte itself, thereby avoiding any need for a sporophytic generation.

In the simplest of populations, the F_2 population, each individual is the product of two independent gametes, both of which can be recognized provided that the alleles being assayed are all codominant (**Table 1**). Where dominance is a factor, F_2 populations are inefficient; since *Aa* and *AA* genotypes are indistinguishable, the only gametes that can be unequivocably genotyped are *aa*, representing only 25% of the population (**Table 2**). In this case, BC_1 (segregating *Aa* or *aa*), doubled haploid (*AA*, *aa*), or recombinant inbred (*AA*, *aa*) populations are more suitable. The type of population used will have an effect on the mapping algorithm, for example, in recombinant inbred populations, each successive selfing generation gives an opportunity for recombination, whereas in an F_2, back-cross or doubled haploid population, recombination will have occurred only once.

What is the Difference Between a Genetic and a Physical Map?

Whereas a genetic map records the recombinational separation between loci, a physical map indicates the

Table 1 F_2 cosegregation from a cross between *A1A1B1B1* × *A2A2B2B2*, where *A* and *B* are two loci with codominant alleles

Male/female gamete	*A1B1* p	*A1B2* r	*A2B1* r	*A2B2* p
A1B1 p	A1B1 p+p	A1B1B2 p+r	A1A2B1 p+r	A1A2B1B2 p+p
A1B2 r	A1B1B2 p+r	(A1B2 r+r)	(A1A2B1B2 r+r)	A1A2B2 p+r
A2B1 r	A1A2B1 p+r	(A1A2B1B2 r+r)	(A2B1 r+r)	A2B1B2 p+r
A2B2 p	A1A2B1B2 p+p	A1A2B2 p+r	A2B1B2 p+r	A2B2 p+p

p, parental; r, recombinant gamete.
Phenotype A1A2B1B2 results from the fusion of two parental (p+p) or two recombinant (r+r). gametes. Otherwise all gametes are fully scorable, since r+r types will be rare if *A* and *B* are linked. Rare combinations are indicated in parentheses.

Table 2 F_2 cosegregation of two loci from a cross *AABB* × *aabb*, with dominant alleles

Male/female gamete	*AB* p	*Ab* r	*aB* r	*ab* p
AB p	AB p+p	AB p+r	AB p+r	AB p+p
Ab r	AB p+r	(Ab r+r)	(AB r+r)	Ab p+r
aB r	AB p+r	(AB r+r)	(aB r+r)	aB p+r
ab p	AB p+p	Ab p+r	aB p+r	ab p+p

Phenotype AB can be p+p or p+r. Rare combinations (where *A* and *B* are linked) are indicated in parentheses.

length of DNA between them. Thus, while genetic distances are measured in cM, physical distances are given in the form of base pairs (bp). A critical relationship defines the equivalence of cM and bp. The global ratio of these measures (covering the whole genome) has turned out to be misleading, as the ratio varies by orders of magnitude across the genome. Most genomes consist of a proportion – in some cases a substantial proportion – of repetitive DNA, assumed to be noncoding. The repetitive DNA is not evenly dispersed throughout the genome and is concentrated in particular regions, such as around the chromosome centromere, leaving regions toward the tips of the chromosomes relatively gene rich. Because chiasmata tend not to be formed in regions with a high repetitive DNA content, cM/bp ratios in these regions are generally relatively small, i.e., recombination is more likely to occur in gene-rich rather than in gene-poor regions. The consequence of this unequal distribution of recombination is that a small cM distance in the centromeric region of a plant chromosome can represent a substantial proportion of the physical map of the chromosome; and as a corollary, in gene-rich regions, a small physical distance can be equivalent to a large genetic one. Since genome size is largely dictated by the relative content of repetitive DNA, and because chiasma formation (and thereby recombination) is so heavily suppressed in highly repetitive regions, distortions in the cM/bp ratio are most exaggerated in large genome species. Further noncongruency between the physical and the genetic maps can be brought about in the region surrounding a recombinational hotspot (loci where recombination appears to be enhanced), and also in certain special cases where unlinked loci cosegregate as a result of genetic interactions.

What Factors Affect Genetic Distance Estimates?

Since recombination is the product of crossing-over, genetic distances are inevitably affected by any disturbance to chiasma formation. A number of such factors have been identified, both directly from cytological observations, and indirectly by comparisons of genetic distance estimates.

Sex

For back-cross or top-cross populations, the direction of the cross dictates whether the map has been assembled from male or from female meiosis. For doubled haploid populations, it is possible in some situations to target one gametic type: thus for male gametes, haploid plants may be regenerable from microspores, while female gametes can be obtained by exploiting the phenomenon of chromosome elimination that occurs in some interspecific crosses (such as wheat × maize). Where F_2 and recombinant inbred populations are used, male and female meioses are integrated. Experiments have repeatedly shown that recombination frequency, and therefore map distances (but not marker order) and overall map length, are often sex-dependent. In the human and the mouse, female maps are in general longer than male ones, although this tendency is reversed at the tips of the chromosomes. In plants, the pattern has been found to be less predictable, but differences are frequently observed.

Environment

The major environmental factor known to affect meiosis is temperature. At its extreme, low temperature at a critical phase in meiosis can lead to complete asynapsis (pairing failure, and therefore absence of recombination). As for many biological phenomena, increases in temperature generally promote activity, in this case chiasma formation and, as a result, map distance. Although there has been little direct experimental proof, much of the variation observed between estimates of map distance within a given genetic background is likely to be due to uncontrolled temperatures at meiosis. There is circumstantial evidence that other stresses (drought, salinity) can also influence chiasma frequency, but these are less likely to be pertinent to the experimental estimation of map distances.

Genetic Background

Estimates of map distance along a given chromosomal segment frequently vary from population to population, although this variation is often confounded with environmental effects where experiments are not conducted simultaneously. Part of this variation merely reflects sampling error, since the occurrence of recombination in a particular segment is a probabilistic event. The standard error associated with an estimate of map distance reflects the size of the mapping population. For practical reasons, this size is typically of the order of 100; in such a population, a map distance of 5 cM is associated with a standard error of about 2 cM. However, the genetic background of the population can itself affect estimates of map distances, either locally or globally. Chromosomal events, in particular translocations, deletions/insertions and inversions have a dramatic effect on maps. The effect of a reciprocal translocation difference between two genotypes will be to generate a single linkage group joining the two chromosomes involved. Where mapping parents differ by a deletion/insertion, the differential region will appear as a single locus, even though it may encompass a substantial chromosome segment. Similarly, the effect of an inversion is to suppress recombination, thereby telescoping what may be a significantly sized physical segment into a rather small genetic one. Localized distortion in segregation can also be caused by certain well-characterized "segregation distorter" genes, which act to bias the gametic output in the direction of one of the two parental alleles. This inevitably affects genes in their neighborhood to a decreasing degree as their linkage distance from the distorter locus increases. Genetic variation has also been invoked to explain global, rather than localized differences in map distances, although the nature of these genes, if they exist, is obscure.

What can Crop Genetic Maps be Used for?

Quantitative Trait Locus (QTL) Analysis

A major use of genetic maps is to identify the number and location of genes determining traits of agricultural interest, and particularly quantitative ones such as yield, adaptation, and quality. The principle of marker-mediated location of genes is to derive an association between the segregation of known marker alleles with differences in the phenotypic expression of the trait, a procedure involving: (1) the establishment of an appropriate recombinant population between the genotypes chosen as differing in phenotype for the character(s) of interest; (2) the characterization of the individual lines in this population for marker loci dispersed throughout the genome (ideally, the target is to generate markers spaced about 20–30 cM apart on each chromosome arm, spanning the centromere); (3) the evaluation of the recombinant lines in appropriate replicated and randomized experiments in appropriate environments, chosen to maximize the expression of genotypic variation whilst minimizing random environmental variation; and (4) the use of statistical procedures to partition the genotypic variation into components attributable to variation at the individual marker loci, either singly or in linked combinations. The simplest approach, and the preliminary to more detailed analysis, is to use analysis of variance techniques to partition the variation within and between genotypes for each marker in turn. Further analysis can define the exact location of the quantitative trait locus (QTL) on the assumption that only one QTL is in the vicinity of the marker.

Marker-Assisted Selection (MAS)

Having located a gene and particular alleles of value to a breeding program, the information on map location can be turned into a selection tool – marker-assisted selection (MAS). The principle of MAS is that indirect selection for a trait can be performed by selecting for a marker closely linked to the trait, rather than directly for the trait itself. This is of particular relevance to traits that are oligogenically determined, since monogenic traits can usually be directly selected, unless the expression of the gene is heavily masked by variation due to environmental heterogeneity. The potential advantages of MAS are considerable. Selection can be made at very early stages in the pedigree, and often at a juvenile plant age. Using the PCR, with its requirement for very little DNA, individual seeds can be screened, and only positive selections need ever to be grown. This has large resource implications for breeding programs, as only relatively small populations of segregating material need to be grown to maturity, since undesirable allelic variation at a number of important genes has been greatly reduced or even eliminated by MAS. Furthermore, both the environment, and genotype × environment interaction effects, which can be large in conventional breeding programs, where trait measurement is the basis of selection, are irrelevant for those traits selected for by MAS. This means, in some cases, that generations can be advanced in off-season conditions or in locations where a particular environmental limitation (disease or abiotic stress) is absent.

Map-Based Gene Isolation

A major route to gene isolation relies on fine-scale genetic mapping, and this approach is particularly suited to the situation where the phenotype generated by the target gene is defined, but the biochemical mechanism of its action is not understood. The strategy relies on being able to identify genetic markers very closely linked to and, if possible, flanking the target gene, and this close linkage can only be achieved by mapping in large populations. These populations are typically an order of magnitude larger than a normal mapping population, but the number of loci included in these fine-scale maps is necessarily very small, since only those linked to the target are included. Commonly, a bulk segregant strategy is applied to identify critical markers; for this purpose, DNA from individuals with each of the two alternative alleles is pooled, and the two pools are then fingerprinted. A marker(s) which differentiates one pool from the other is likely to be linked to the target, since the "background" of the two pools is identical. Closely linked flanking markers represent the starting points of a chromosome walk, which is effected by assembling a contig of large insert clones spanning the region containing the target.

What Forms can Genetic Markers Take?

Markers fall broadly into three classes: morphological, based on monogenically determined plant traits; biochemical, based on gene products; and molecular, based on DNA sequence. The first of these (prominent examples being race-specific seedling disease resistances and dwarfness) are the simplest of all markers to assay, but cannot be widely applied as they are limited to the small number of monogenic phenotypic traits. The second class exploits electrophoretic or chromatographic differences between allelic products. A considerable advantage of biochemical markers in breeding and selection is that they generally have a neutral effect on phenotype, unlike the morphological markers, which directly reflect plant form. However, they also suffer from relatively small number, as well as from the complexity and variety of the assays. Molecular markers are assessed by a variety of DNA profiling techniques.

Hybridization-Based Molecular Markers

The dominant technology in this class is restriction fragment length polymorphism (RFLP). The basis of RFLP is the hybridization of a cloned piece of DNA ("probe") to one or more restriction fragments, generated by the cleavage of genomic DNA by endonucleases. Allelic fragments vary in length either as a result of point mutations at restriction recognition sites or via insertion/deletion events in the genomic DNA sequences flanking the probe. The restriction fragments are separated from one another using gel electrophoresis, immobilized on a nylon membrane, denatured and interrogated by a probe derived from either a library of genomic plant DNA (gDNAs) or of reverse transcriptase products of messenger RNA (complementary DNAs, cDNAs). Where binding occurs as a result of sequence homology, the detection signal is concentrated in a band(s), which is(are) visualized by the appropriate detection system.

PCR-Based Molecular Markers

The common components of polymerase chain reaction (PCR) are a thermostable DNA polymerase, the four nucleotides, template DNA from the species being analyzed and one or more (usually two) short synthetic oligonucleotides (primers). The specificity of the reaction to a particular locus or loci depends only on the choice of primers. Amplification is achieved by cycling the reaction temperature a preset number of times, between the melting temperature of DNA and the annealing temperature of the primer(s) to the template, which depends on both primer length and nucleotide sequence. The reaction products are commonly separated by size (although occasionally also by conformation) on a gel. Polymorphism can be either of the on/off type (due to sequence variation at one or both priming sites); or as length variants (where either part of the amplified sequence is deleted, or an extra sequence has been inserted). Where no length variants are detected, internal sequence variation can sometimes be exposed by digesting the PCR product with a range of restriction enzymes before electrophoresis. Where this fails, amplification products from different genotypes are themselves sequenced in a search for internal sequence variation. Any base-pair differences revealed can then be exploited to design primers, which are locus- or allele-specific.

The Future of Genetic Mapping

Since the first description of DNA profiling some 20 years ago, genetic maps have been developed in most of the significant agricultural and horticultural crops. Comparisons between these maps have allowed substantial insights into the evolution of crop species. A prominent finding has been that the small grain cereals – rice (*O. sativa*), corn (*Zea mays*; maize), sorghum, barley (*H. vulgare*), oats (*Avena sativa*), rye (*Secale cereale*) and wheat (*Triticum aestivum*) – share substantial levels of similarity with respect to gene order, and relative (but rarely absolute) map distance, despite large disparities in chromosome number and genome size. These comparisons also extend to more distant monocotyledenous species, but break down across the monocotyledon/dicotyledon divide. Similar comparisons have also been demonstrated within certain dicot groups. Apart from the evolutionary interest that cross-species gene order conservation evokes, there has come the important realization that genetic data can be transferred from one crop to another, so that crop-specific data bases can begin to be merged.

The variety and power of DNA genotyping assays is constantly increasing, but each major technical advance is tending to be associated with a concomitant increase in the sophistication of the infrastructure required. The cost/benefit balance of

MAS for plant breeding programs is worsening, generating a pressure toward centralization of plant breeding effort. However, for other applications, the increasing coverage and resolution of genetic maps should continue to simplify and facilitate their use in genetic analysis and finally in the isolation of genes, which can be manipulated in the wide context of crop improvement.

List of Technical Nomenclature

AFLP	Amplified fragment length polymorphism.
Back cross (BC)	The (repeated) crossing of an F_1 individual with one of its two parental genotypes.
Bulk segregant analysis	A strategy to derive markers for a gene(s) determining a selectable phenotype. DNAs from segregants are pooled according to phenotype and pools are interrogated for informative markers.
cDNA	Complementary DNA. Synthesized from messenger RNA via reverse transcription.
Chiasma (pl. chiasmata)	The cytological structure representing the crossing-over of homologous chromatids during meiosis.
Chromosome elimination	The loss of chromosomes of the pollen parent in the first few mitoses following fertilization in some interspecific crosses.
cM	Centimorgan.
Contig	A physical length of DNA defined by a set of overlapping clones.
Deletion/insertion	The loss/gain of a segment of DNA.
Doubled haploid	A plant derived by the induction of chromosome doubling from a haploid individual. Doubled haploid individuals are necessarily homozygous at all loci.
Gametophyte	A haploid plant developed from mitotic division of the gamete (pollen or egg).
gDNA	Genomic DNA.
Genome	The entirety of DNA of an organism. Split into "nuclear" (DNA in the nucleus) and "cytoplasmic" (in the plastids or mitochondria).
Haploid chromosome number	Half the number of chromosomes present in the normal somatic cell; the number of chromosomes present in the standard gamete.
Inbreeding depression	Intolerance of homozygosity, expressed as increasingly weak and inviable plants as the level of homozygosity rises.
Inversion	Reversed orientation of a segment of DNA.
Linkage group	A set of genetic markers/genes that are linked by mapping.
Locus (pl. loci)	The physical position of a genetic marker/gene on a chromosome.
MAS	Marker-assisted selection.
Microspore	An immature pollen grain.
Monogenic	Controlled by a single (Mendelian) gene.
Oligogenic	Controlled by a small number of genes.
PCR	Polymerase chain reaction.
Polymorphism	Difference.
QTL	Quantitative trait locus.
Recombinant inbred	A population derived by successive self-fertilization over a number (>6) of generations. The lines are genetically distinct from one another and largely, but not completely, homozygous at all loci.
RFLP	Restriction fragment length polymorphism.
Self-incompatibility	Genetically determined barrier to fertilization by pollen from the same genotype as the egg.
Segregation distortion	Departure from Mendelian segregation.
SNP	Single nucleotide polymorphism.
Sporophyte	Diploid plant developed from mitotic division of the zygote.
SSR	Simple sequence repeat (microsatellite).
Telomere	The physical structure capping each end of a chromosome. Normally composed of tandemly repeated DNA motifs.
Top cross	The crossing of an F_1 individual with a third genotype.
Translocation	A novel chromosome formed by the fusion between either part of, or an entire chromosome with a chromosomal segment derived from an unrelated chromosome.

See also: **Crop Improvement**: Doubled Haploid Production; Marker Assisted Selection; Molecular Markers; Plant Genomes.

Further Reading

Bonnema G, Schipper D, van Heusden S, Zabel P, and Lindhout P (1997) Tomato chromosome 1: High resolution genetic and physical mapping of the short arm in an interspecific *Lycopersicon esculentum* × *L. peruvianum* cross. *Molecular and General Genetics* 253: 455–462.

Gale MD and Devos KM (1998) Plant comparative genetics after 10 years. *Science* 282: 656–659.

Gardiner JM, Coe EH, Meliahancock S, Hoisington DA, and Chao S (1993) Development of a core RFLP map in maize using an immortalized-F2 population. *Genetics* 134: 917–930.

Harushima Y, Yano M, Shomura P, *et al.* (1998) A high-density rice genetic linkage map with 2275 markers using a single F-2 population. *Genetics* 148: 479–494.

Koebner RMD, Powell W, and Donini P (2001) Contributions of dna molecular marker technologies to the genetics and breeding of wheat and barley. *Plant Breeding Reviews* 21: 181–220.

Mohan M, Nair S, Bhagwat A, *et al.* (1997) Genome mapping, molecular markers and marker-assisted selection in crop plants. *Molecular Breeding* 3: 87–103.

Song WY, Wang GL, Chen LL, *et al.* (1995) A receptor kinase-like protein encoded by the rice disease resistance gene, *Xa21*. *Science* 270: 1804–1806.

Yano M and Sasaki T (1997) Genetic and molecular dissection of quantitative traits in rice. *Plant Molecular Biology* 35: 145–153.

Molecular Markers

R M D Koebner, John Innes Centre, Norwich, UK

Introduction

Molecular markers are diagnostic assays for genotype, and are specifically based on variation at the DNA level. Each marker defines a particular DNA segment within a genome. The enabling technology was initiated ca. 1980, and has developed at a rapid rate, such that the number of molecular markers potentially available in key species no longer presents any technical limitation. Current methods rely on three core techniques: DNA/DNA hybridization, the polymerase chain reaction (PCR), and DNA sequencing. This article describes the principles underlying each of the major types of molecular marker, and summarizes their suitability for a range of applications.

What is a Molecular Marker?

Molecular markers represent a subset of genotype diagnostics (or "genetic markers"), specifically those that rely on a direct DNA assay, rather than, for example, on an inherited phenotypic trait (e.g., flower color in pea, *Pisum sativum*, or semi-dwarfness in cereals, referred to as "morphological markers") or a protein-based assay of a gene product (e.g., amylase or seed storage proteins – "biochemical markers"). Strictly defined, a molecular marker identifies a segment of genomic DNA, within which allelic variation in sequence has allowed its position to be genetically mapped, and where the map location thereby derived is sufficiently close to that of a locus of interest that the two are in linkage disequilibrium (i.e., they are most frequently inherited together). Its less narrowly defined, but much more widely used, meaning extends to a DNA-based assay of a locus, whether or not it is genetically linked to any particular gene of interest.

The attractiveness of genetic markers in genetic research and breeding has long been recognized. The ability to follow specific chromosome regions through pedigrees is critical to the uncovering of the inheritance of complex traits, whether this complexity arises because the trait is determined by more than a single gene and/or because variation in the environment affects the expression of the trait significantly enough to mask the genetic contribution. It is also critical to the isolation of genes based on their location in the genome ("map-based cloning"), since a basic requirement of this technique is the identification of molecular markers tightly linked to the gene to be isolated. Until the development of DNA assays, the number of available markers was limited to a relatively restricted collection of single gene morphological variants and, more recently, a small number of biochemical assays. With the advent of molecular markers, however, this constraint has been largely removed and, as a result, a wide range of applications in both breeding and fundamental genetic research has become available.

The technology related to the detection and analysis of variation at the DNA sequence level is changing fast. The driver for these developments is in part directed at making the technology more efficient and in part aimed at accessing to date untapped sources of genetic variation via novel analytical means. This article outlines the principles behind the major types of molecular markers and briefly addresses their suitability (or otherwise) for applications ranging from marker assisted selection to gene isolation.

The Major Categories of Molecular Marker

The assay of DNA variation is based, singly or in combination, around three basic technologies:

DNA/DNA (or RNA/DNA) hybridization, the polymerase chain reaction (PCR) and DNA sequencing. Within these broad categories a profusion of techniques has evolved, each with its own merits and demerits, particularly with respect to practicality, unit cost and resolution.

Hybridization-Based Technologies

The recognition of complementarity between pairs of purine and pyrimidine molecules, i.e., the affinity of A (adenine) for T (thymine) in DNA and U (uracil) in RNA, and G (guanine) for C (cytosine), along two single strands of nucleic acid is the basis of hybridization; the degree of mismatching that occurs, *in vitro* at least, is governed by the stringency of the hybridization reaction conditions. Stringency is increased both as the temperature rises and as the salt concentration of the hybridization medium falls, while the time allowed for the hybridization reaction also has an effect on the amount of mismatch pairing that is retained. In overly stringent conditions, even correctly matching base pairs fail to bind, while overly lax reaction conditions result in an ever greater extent of nonspecific (incorrect or mismatch) base pairing.

Restriction Fragment Length Polymorphism

The most widespread and flexible system within this category is the restriction fragment length polymorphism (RFLP) technique. For this assay, genomic DNA is digested with one or more of a wide range of (largely bacterial origin) restriction endonucleases, which cut double-stranded DNA wherever an enzyme-specific sequence of bases (most commonly 4, 6 or 8 bp long) occurs, thereby generating a mixture of discrete fragments. In general, the larger the number of bases in the recognition sequence, the less frequently the DNA is cut, and thus the greater is the mean length of the fragments generated. On the simplistic assumption of random sequence in a sample of DNA, an enzyme with a 6-bp recognition sequence will cut, on average, every 4096 (4^6) bases, giving rise, in a large genome species such as barley (*Hordeum vulgare*), to approximately 10^6 fragments; in practice, the fragment size range resulting from cleavage with a typical 6-bp cutter is a few hundred to a few tens of thousands of bp long.

The DNA fragments are separated one from another by electrophoresis through agarose or polyacrylamide gels, then are denatured (made single-stranded) and fixed, by a passive or an electrophoretic transfer, to a membrane (nylon or nitrocellulose) or "filter." Filters are "hybridized" with a single-stranded "probe," a DNA molecule obtained from a variety of sources, but typically derived either by the cloning or polymerase chain reaction (PCR) amplification of genomic DNA (gDNA probes), or by *in vitro* reverse transcription of mRNA (cDNA, complementary DNA, probes). The assay relies on the detection of the double-stranded structure that results from the binding of the probe to a restriction fragment(s) with which it shares sufficient sequence homology. In order to visualize these, probes are labeled either isotopically, or by coupling with a fluorescent dye. The latter approach has now largely replaced the former one. The double-stranded structure can thereafter be readily denatured, so that a single filter can be reprobed a number of times. Probes are either homologous (originating from the species under investigation) or heterologous (originating from a related species). Some sequences (particularly cDNAs) are so well conserved across genera, families and even phyla, that they can serve as universal probes, but in general, heterologous probes require lower stringency conditions in the hybridization reaction.

Allelic variation between two genotypes at an RFLP locus occurs where the hybridizing targets are of unequal length, as this can be recognized by a difference in the migration of the fragments through the gel. Such a difference can arise as a result of either a point (single base) or micro (a few bases) mutation, which abolishes one or both of the two flanking restriction enzyme recognition sequences in either allele present, or as a result of structural polymorphism, such as the insertion/deletion of a sequence within the restriction fragment. The two possibilities are easily distinguished as, in the former case, RFLP will only be uncovered by a specific restriction enzyme while, in the latter case, RFLP will be recognized following digestion with many different restriction enzymes. In most crop species, insertion/deletion events are the commonest source of naturally occurring RFLP.

A given probe detects either a single- or multi-copy locus. Single-copy RFLPs have the advantage that their scoring, as a single band, is unambiguous and that allelism (and therefore codominance) can be immediately recognized. In contrast, probes that give a multi-banded profile allow an assay that simultaneously targets more than one locus. In general, most cDNA probes, but only a proportion of gDNA probes (particularly those generated from digests with methylation-sensitive enzymes, such as the commonly used *Pst*I), produce single or low copy hybridization profiles. Some gDNA probes – particularly those derived from libraries made from DNA cut with a methylation-insensitive enzyme – detect repetitive sequences that generate multi-banded patterns. These can be useful as a DNA fingerprinting technique, but are not usually suited to genetic analysis or marker-assisted selection. The same is true of multi-gene families assayed by a cDNA probe, where the individual members are dispersed throughout the genome.

A special case of RFLP is represented by the variable number of tandem repeats (VNTR) technique, which takes advantage of the presence in eukaryotic genomes of dispersed repeated short sequences of 1–20 bp in length. Allelic variation is generated by the number of repeat units present at a locus, as this determines the length of the target. The repeat sequence itself is used as a probe so that, unlike standard RFLP probes, VNTR probes consist of synthesized oligomers of commonly occurring repeat units (e.g., $[GATA]_4$). Some repeat units are common enough to target small restriction fragments (less than 1 kbp in length) and, in this case, separation is achieved in polyacrylamide gels and the hybridization can be performed in the gel itself, avoiding the need for transfer to a membrane.

The Dot Blot Assay

Where a probe sequence is present in one individual, but absent in another, the gel separation part of the assay can be omitted. Undigested genomic DNA is applied directly to the membrane, which is then hybridized with a probe in the normal way. This represents a drastic simplification, as neither restriction digestion, gel separation nor transfer are necessary. Its disadvantage is that very few alleles are truly "null" (i.e., complete absence of the hybridizing sequence), which is necessary for this assay within a species. The technique finds particular application where contrasts between species are relevant. In these cases, advantage can be taken of species-specific repetitive DNA, which is not usually difficult to identify. Since the target consists of many copies, this allows the assay to be carried out on impure extracts, and exposure times to film (for autoradiography and some fluoresceins) or suitable screens (for many fluoresceins) can be very much shortened over the standard single copy RFLP protocol.

A miniaturized version of the dot blot is possible in a DNA microchip approach, which allows large numbers of DNA preparations to be displayed over a small area; this is achieved by the use of roboticized microprinting technology. The hybridization of a chip in this format is no different from that used for membrane-immobilized DNA; the only difference being in the scale of throughput achievable. Probes for DNA chip analysis are invariably labeled with fluorescent, not radioactive, tracers.

PCR-Based Technologies

The PCR assay requires at least one to two orders of magnitude less DNA ("template") than does RFLP; it is also generally less demanding in its requirement for sample purity and it usually allows for a quicker turn-around than do hybridization assays. There is not necessarily a need for labeling, because the PCR is designed to generate many copies of a single (or a few) specific molecule, sufficient for sensitivity levels which can be achieved by a variety of methods, the most simple of which is the staining of bands in gels with ethidium bromide. For all these reasons, PCR presents an attractive alternative to hybridization approaches. All PCR assays rely for their specificity on the complementarity of the DNA sequence of the one or more oligomers ("primers") included in the reaction to a portion of the sequence at the target locus/loci; the function of the primers is to allow the formation of a localized double-stranded DNA structure, as this is a requirement for DNA polymerase to initiate the synthesis of a new strand of DNA. Where primers are designed to amplify a defined segment of DNA, the assay is generally referred to as a "sequence tagged site" (STS). An important class of PCR markers uses one or more primers that are partially or completely random in sequence, and these assays therefore target loci that are individually *a priori* undefined.

STS Assays

Typically, the STS assay targets a single copy sequence either within, or close to, a defined locus, utilizing a primer located at each end of the sequence. In principle, any DNA segment whose DNA sequence is known can be amplified in this way (within the normal technical limits of amplifiable length, i.e. $\leqslant$ 2000 bp). However, in practice, some sequences do not readily amplify, for reasons probably related to the secondary structure of the template. Commonly, there is little or no recognizable length polymorphism between the STS product amplified from a pair of related individuals, but this problem can often be side-stepped by detecting sequence polymorphism within the amplicon, especially if the variation creates (or destroys) an endonuclease restriction sequence. Where this occurs, the amplified products are simply digested by the relevant restriction enzyme(s), and the resulting fragments separated electrophoretically, usually on an agarose gel; if this allows alleles to be distinguished, the assay is referred to as a CAPS (cleaved amplified polymorphic sequence) marker assay. Where no such restriction site can be uncovered, it may still be possible to detect polymorphism electrophoretically using the SSCP (single strand conformation polymorphism) technique, an electrophoretic method that discriminates between DNA fragments more on the basis of their sequence than on their size. In the SSCP assay, double-stranded DNA is denatured and then separated on a polyacrylamide gel in which renaturing is favored. Because the homologous strands have been separated, renaturing tends to occur within a strand, leading to distinctive secondary structures of the single-stranded molecules, and these structures, which are sequence-determined, can migrate differently from one another through the gel. If this approach also fails, one (or both) of the PCR primers can be redesigned to directly target the variant sequence site(s), thereby ensuring that only one of the templates is amplified by each primer pair. In this situation, the primers become allele-, as opposed to merely locus- (or gene-) specific.

Sequence Tagged Microsatellite Site

The sequence tagged microsatellite site (STMS) assay – alternatively and frequently referred to as "simple sequence repeats"(SSR) – is based on variation between individuals in the number of mono-, di-, or trinucleotide repeats present at a given locus in a single-copy sequence context. These repeats (e.g., ATATAT ... or GATAGATA ...) are highly represented in most genomes, and the number of repeats present can be highly variable. When the repetitive region is amplified by priming from the flanking sequences, any variation in the number of repeats within the locus and/or any deletion/insertion in the nonrepetitive region flanking the repeat, but internal to the priming site, results in amplicons that vary in length. These can be recognized electrophoretically, generally using denaturing polyacrylamide gels. Thus, the assay, in practice, consists of a standard reaction, followed by visualization on a gel using one of a number of alternative conventional DNA staining methods. STMS markers are codominant, and the loci have proven to be well dispersed throughout the genome. Thanks to appreciable levels of polymorphism in a growing range of both crop and noncrop species, they have become the standard marker type for most genotyping applications, especially those concerned with genetic mapping and gene tagging.

Inter-Simple Sequence Repeat, Inter-Retrotransposon Amplified Polymorphism and Retrotransposon-Microsatellite Amplified Polymorphism

Inter-simple sequence repeat (ISSR), inter-retrotransposon amplified polymorphism (IRAP) and retrotransposon-microsatellite amplified polymorphism (REMAP) constitute a broad class of marker that exploits the existence of large numbers of copies of highly conserved repetitive sequences throughout the genomes of many plants. In some crop plants, notably the small grained cereals – wheat (*Triticum aestivum*), barley, rye (*Secale cereale*), and oat (*Avena sativa*) – this repetitive fraction of the genome can be as high as 90%. "Simple sequence repeats" and "microsatellites" are synonyms, referring to short (1–5 bp) sequences (see VNTR, above), while retrotransposons represent much larger (3–12 kbp) virus-like mobile genetic elements, which, because they are copied by reverse transcription, generate daughter sequences at dispersed locations. Each copy is flanked by a characteristic repeat sequence (the "long terminal repeat"), which tends to be highly conserved, and this sequence provides the basis for the design of the primers. In IRAP, the primers amplify the DNA between two adjacent retrotransposons; in REMAP, between a microsatellite and a retrotransposon; and in ISSR between two adjacent microsatellites. The amplified DNA is separated electrophoretically in the normal way, and any polymorphism in size of these PCR products can potentially be used as a molecular marker.

Single Nucleotide Polymorphism

At its simplest, DNA sequence variation consists of a single base difference in a defined segment of DNA present in two individuals. The single nucleotide polymorphism (SNP) marker identifies such a difference. In any but the most closely related individuals, the potential number of such markers is enormous; while human STMS loci have been estimated to occur on average every few tens of kbp, SNP frequency is an order of magnitude higher. The overwhelming attraction of SNPs lies both in a high density of markers and in the potential for non-gel-based assays, since the processing of samples by electrophoresis has become the most rate-limiting step in marker analysis, and is the step least easy to automate. A number of alternative SNP assays has been described in recent years, but as yet no "industry standard" has emerged.

All the major SNP detection techniques rely on an initial PCR amplification of the target DNA segment.

The SNP is targeted either internally within the amplicon or at/immediately downstream of the 3′ end of one of the amplification primers. For the former case, the presence of an SNP may be detected by a method analogous to a CAPS marker assay – if the SNP destroys (or creates) a restriction enzyme recognition sequence, then post-PCR digestion with the relevant enzyme will reveal whether the variant allele is present. Alternatively, non-gel-based assays can be assembled by the inclusion, along with the standard PCR primers, of a fluorescently labeled oligonucleotide probe whose sequence incorporates that surrounding the SNP site. In this case, one primer is designed to recognize the variant sequence, while a second one recognizes the wild-type. If these two primers are labeled with different fluorochromes, then the amplicons derived from the alternative templates can be distinguished by the color of their fluorescence.

Non-STS Assays

In non-STS assays, the amplicon is not defined *a priori*, because the primers are not directed at a predetermined sequence of DNA. Instead, the primer sequences are chosen on a random or semirandom basis, and the content of the amplicon, which is usually heterogeneous, is then determined *post facto*. The main advantage of this approach is that no (or little) prior knowledge of genomic sequence is necessary, but the major disadvantage is that the genetic basis of the amplicon is unknown. Furthermore, the successful amplification of a similar sized product from two unrelated templates does not guarantee that the same locus has been amplified from each. This latter consideration undermines the transferability of this class of marker between populations.

Multiple Arbitrary Amplicon Profiling: DNA Amplification Fingerprint, Random Amplified DNA Polymorphism, and Arbitrarily Primed PCR

The simplest form of random priming relies on the simultaneous occurrence of the primer sequence on one strand of the genomic DNA template and its presence within ca. 2000 bp in inverted orientation on the second strand. In this situation, the intervening segment of genomic DNA will be amplified by a single primer complementary to this sequence. Because the probability of this alignment of sequence falls as the length of the primer increases, the complexity of the amplicon decreases as the length of the primer increases. DNA amplification fingerprint (DAF) uses primers as short as 5 bp, random amplified DNA polymorphism (RAPD) primers are commonly 10 bp, while arbitrarily primed PCR (AP-PCR) primers can be as long as 20 bp. Thus DAF profiles consist of a large number of products, which need polyacrylamide gels to be resolved from one another; both RAPD and AP-PCR products are generally separated on agarose gels. It may be desirable on efficiency grounds to obtain a large number of products from a single reaction, but as the primer is shortened, so the optimal temperature at which it anneals to the target sequence in the template is reduced. A disadvantageous consequence of low annealing temperatures is that the stringency of annealing, and therefore the specificity of the reaction, is also reduced. The 10mer primers used in RAPD represent a compromise between the number of products amplified per amplicon and the specificity of the reaction. Despite this, nonreproducibility of the assay remains a persistent problem, and transferability of markers between populations is often problematical.

Amplified Fragment Length Polymorphism

Amplified fragment length polymorphism (AFLP) was designed to deliver complex banding patterns, using a high level of stringency, and so has found a particular niche as an efficient means of scanning the genome for markers and for DNA fingerprinting. The technique is a hybrid between PCR and RFLP, and involves generating a large number of amplifiable restriction fragments, and then scanning for length differences within these. Unlike the majority of PCR assays, the genomic DNA first needs to be digested, normally with two restriction enzymes, one cutting frequently and the other rarely. This requires a level of purity for the template, which is disadvantageous in terms of practicality. The digestion results in three classes of restriction fragment: most commonly, those cut by the frequent cutter at each end; infrequently, those cut at one end by the frequent cutter and at the other end by the rare cutter; and very rarely, those cut at both ends by the rare cutter. The restriction fragments are made amplifiable by ligating a known short synthetic sequence to each cut end. This sequence, plus the remaining bases marking the cut restriction site, allow primers to be designed that, in principle, stringently amplify all the restriction fragments present. In practice, the majority of the amplified fragments are formed by those having dissimilar ends (the second class described above), and not all such molecules are equally efficiently amplified. In order to reduce the number of distinct amplified fragments to a number that can be resolved electrophoretically, a second (or "selective") PCR is usually then carried out, in which a variable number of randomly chosen bases is added to the 3′ end of either or both the primer(s), and one of the two primers is labeled either radioactively or fluorescently. In this way, each primer pair amplifies a distinct subset of all the restriction fragments present in the initial amplicon. The constituent fragments in the selective PCR amplicon are separated by denaturing polyacrylamide gel electrophoresis and the labeled products are visualized by the appropriate means.

Typical AFLP fingerprints consist of 50–100 bands, but only a small proportion of these is polymorphic in typical varietal comparisons. Competition between template molecules during the preselective and selective amplification stages is a particular problem in large genome species, and can compromise AFLP comparisons, because not all fragments that are theoretically amplifiable are in practice visualized. A further problem is that AFLP markers often have a tendency to map in clusters. AFLP markers usually behave as dominant markers, and they appear to be largely based on repetitive DNA, at least when nonmethylation-sensitive restriction enzymes are used to generate the template for the selective amplification. When methylation-sensitive restriction enzymes are used to generate the initial restriction fragments, the template is enriched for genic regions, since these sequences tend to be undermethylated compared with the nongenic fraction.

Hybrid Systems

A number of hybrid systems have been developed in which a specific class of repetitive DNA or a well-distributed gene family is targeted in an AFLP approach. The aim of these approaches is to derive DNA fingerprints where the individual components are likely to be members of the class

Table 1 Molecular marker types and their suitability for a range of applications

	Genetic mapping	*Genetic fingerprinting*	*Marker-assisted selection*	*Comparative mapping*	*Map-based cloning*
RFLP	Reliable. Unsuited to high throughput	Inefficient, slow	Unsuited to high throughput	Well suited	Unsuited to high throughput
Dot blot	Few suitable probes	Few suitable probes	Good where suitable probe available	Not suited. Probes are usually species-specific	Good where suitable probe available
STS	Not generally polymorphic, except as CAPS	Not generally polymorphic, except as CAPS	Well suited	Well suited	Well suited
MSTS	Well suited	Need many simultaneous assays	Well suited	Unsuitable. Assay usually species-specific	Well suited
ISSR REMAP IRAP	Unsuitable, as may be cross-specific	Well suited	Well suited	Unsuitable. Assay usually species-specific	Well suited
SNP	Well suited	Need many simultaneous assays	Well suited	Probably unsuitable as assay likely to be species-specific	Well suited
MAAP	Well suited	Well suited	Unsuitable, as often cross-specific	Unsuitable. Assay usually species-specific	Well suited
AFLP	Well suited	Well suited	Inefficient at high throughput	Unsuitable. Assay usually species-specific	Well suited
S-SAP SAMPL	Well suited	Well suited	Inefficient at high throughput	Unsuitable. Assay usually species-specific	Well suited

of sequence being targeted. Examples of this include sequence-specific amplification polymorphisms (S-SAP), which is directed to retrotransposon ends, selective amplification of microsatellite polymorphic loci (SAMPL), which targets a microsatellite repeat, and RGA profiling, in which the consensus sequences of a large class of plant disease resistance genes determines the specificity of the assay.

Molecular Marker Applications

The major applications of molecular markers are covered in detail elsewhere in this encyclopedia. **Table 1** addresses the suitability of marker types to a range of applications.

List of Technical Nomenclature

AFLP	Amplified fragment length polymorphisms.
Amplicon	The amplified DNA molecule(s) generated by the PCR.
AP PCR	Arbitrarily primed PCR.
CAPS	Cleaved amplified polymorphic sequence.
cDNA	Complementary DNA; derived via reverse transcription from mRNA.
DAF	DNA amplification fingerprint.
gDNA	Genomic DNA.
IRAP	Inter-retrotransposon amplification polymorphism.
ISSR	Inter-simple sequence repeat.
MAAP	Multiple arbitrary amplification profile.
mRNA	Messenger RNA.
PCR	Polymerase chain reaction.
RAPD	Random amplified polymorphic DNA.
REMAP	Retrotransposon-microsatellite amplification polymorphism.
RFLP	Restriction fragment length polymorphism.
RGA	*R* (resistance) gene analog.
SAMPL	Selective amplification of microsatellite polymorphic loci.
S-SAP	Sequence-specific amplification polymorphism.
SNP	Single nucleotide polymorphism.
SSCP	Single strand conformation polymorphism.
SSR	Simple sequence repeat (synonymous with STMS).
STMS	Sequence tagged microsatellite site.
STS	Sequence tagged site.
VNTR	Variable number of tandem repeats.

See also: **Crop Improvement**: Genetic Maps; Marker Assisted Selection. **Genetic Modification**: Gene Cloning, General Principles.

Further Reading

Botstein D, White RL, Skolnick M, and Davis RW (1980) Construction of a genetic linkage map in man using restriction fragment length polymorphisms. *American Journal of Human Genetics* 32: 314–331.

Kalendar R, Grob T, Suoniemi A, and Schulman AH (1999) IRAP and REMAP: Two new retrotransposon-based DNA fingerprinting techniques. *Theoretical and Applied Genetics* 98: 704–711.

Koebner RMD, Powell W, and Donini P (2001) The contribution of current and forthcoming DNA molecular marker technologies to wheat and barley genetics and breeding. *Plant Breeding Reviews* 21: 181–220.

Powell W, Machray GC, and Provan J (1996) Polymorphism revealed by simple sequence repeats. *Trends in Plant Science* 1: 215–222.

Vos P, Hogers R, Bleeker M, *et al.* (1995) AFLP: a new technique for DNA fingerprinting. *Nucleic Acids Research* 23: 4407–4414.

Welsh J and McClelland M (1990) Fingerprinting genomes using PCR with arbitrary primers. *Nucleic Acids Research* 18: 7213–7218.

Marker Assisted Selection

P Langridge, A R Barr and K J Chalmers,
University of Adelaide, Glen Osmond, SA, Australia

Definition and Background

Marker assisted selection (MAS) involves the use of genetic markers to follow regions of the genome that encode specific characteristics of a plant. For example, a marker genetically linked to a disease resistance locus can be used to predict the presence of the resistant or the susceptible allele. The reliability of the prediction will depend upon the closeness of the genetic linkage. Markers that cosegregate with the target trait are absolutely reliable and can be regarded as diagnostic.

To be effective, the markers must detect a polymorphism between the plants being analyzed. The nature of the polymorphism will vary depending on the marker system being used. Initially, markers based on protein differences were widely used. These were based on variation in protein size detected by size fractionation electrophoresis or differences in protein charge or isoelectric point. Specific proteins or enzymes could be detected by staining for total proteins or using the enzyme's activity to produce or remove a colored substrate. Isoelectric variants are referred to as isozymes and were, for many years, extremely important markers for specific chromosomes and chromosome regions. They suffer from two major weaknesses:

1. They are limited in number and are often difficult to detect or assay. The total number of isozyme loci that can be assayed is generally fewer than 100 in a well-characterized species.
2. They detect only low levels of polymorphism. The assays are dependent on revealing a shift in protein mobility either through an altered size or altered isoelectric point, involving a change in at least one amino acid in the protein. Therefore, only changes in the coding sequence of a limited number of genes can be detected, and changes that lead to an amino acid substitution or deletion. In addition, many such changes are not selection neutral but may involve a change in enzyme function that can be detrimental to other characteristics of the plant.

Detection of sequence variation at the DNA level offers several important advantages over protein-based markers. There are essentially an unlimited number of such DNA markers, since variations in sequence, in the form of single base changes, insertions and deletions, or large sequence differences, are abundant. Indeed, in many systems where detailed analysis has been carried out, a sequence difference occurs between two sexually compatible individuals at a frequency of greater than one in every 300 bases. Most of these changes will have no detectable effect on plant performance, and are selection neutral. In addition, a large range of techniques are now available for detecting sequence variation. The available techniques differ in their ability to detect variation, the ease of assay, and the number of loci that can be assayed simultaneously. While early work focused on the use of restriction fragment length polymorphisms (RFLP) detected via southern hybridization, polymerase chain reaction (PCR)-based marker systems are now more widely used, in particular, microsatellite or simple sequence repeat (SSR) and amplified fragment length polymorphisms (AFLP). New techniques not requiring gel electrophoresis are gaining in importance. As sequence information becomes more widely available for the major crop species, we can expect to see alternatives to gel-based detection systems applied to identify single nucleotide polymorphisms (SNPs).

Steps in Marker Development – from Discovery to Application

Marker development and implementation can be divided into the following steps:

- identify parents differing in the trait of interest;
- develop a population of plants segregating for the trait of interest (using doubled haploids, single seed descent or F_2 populations);
- screen the population for the traits of interest;
- construct linkage maps of the cross, use bulked segregant analysis, or genomic fingerprinting;
- identify molecular markers that cosegregate with the trait of interest;
- test the applicability and reliability of the markers in predicting the trait in related families (also referred to as marker validation);
- produce clear and simple protocols for assaying the markers;
- modify breeding strategy to optimize use of MAS relative to alternative selection techniques; and
- implement into the breeding programs.

Finding Markers Linked to Traits of Interest

The technique used to identify a molecular marker linked to a trait of interest will depend upon the type of trait and the resources and marker systems available. There are several features of the trait that will be important in devising the most efficient marker development strategy. These include:

- the mode of inheritance; simple or polygenic;
- the heritability of the trait and the reliability of the alternative (bioassay, biochemical analysis, etc); and
- the ease of bioassay (this may restrict the size of the population that can be assayed).

The process used to decide on the best approach to marker development is summarized in **Figure 1**. There are five key questions to be asked:

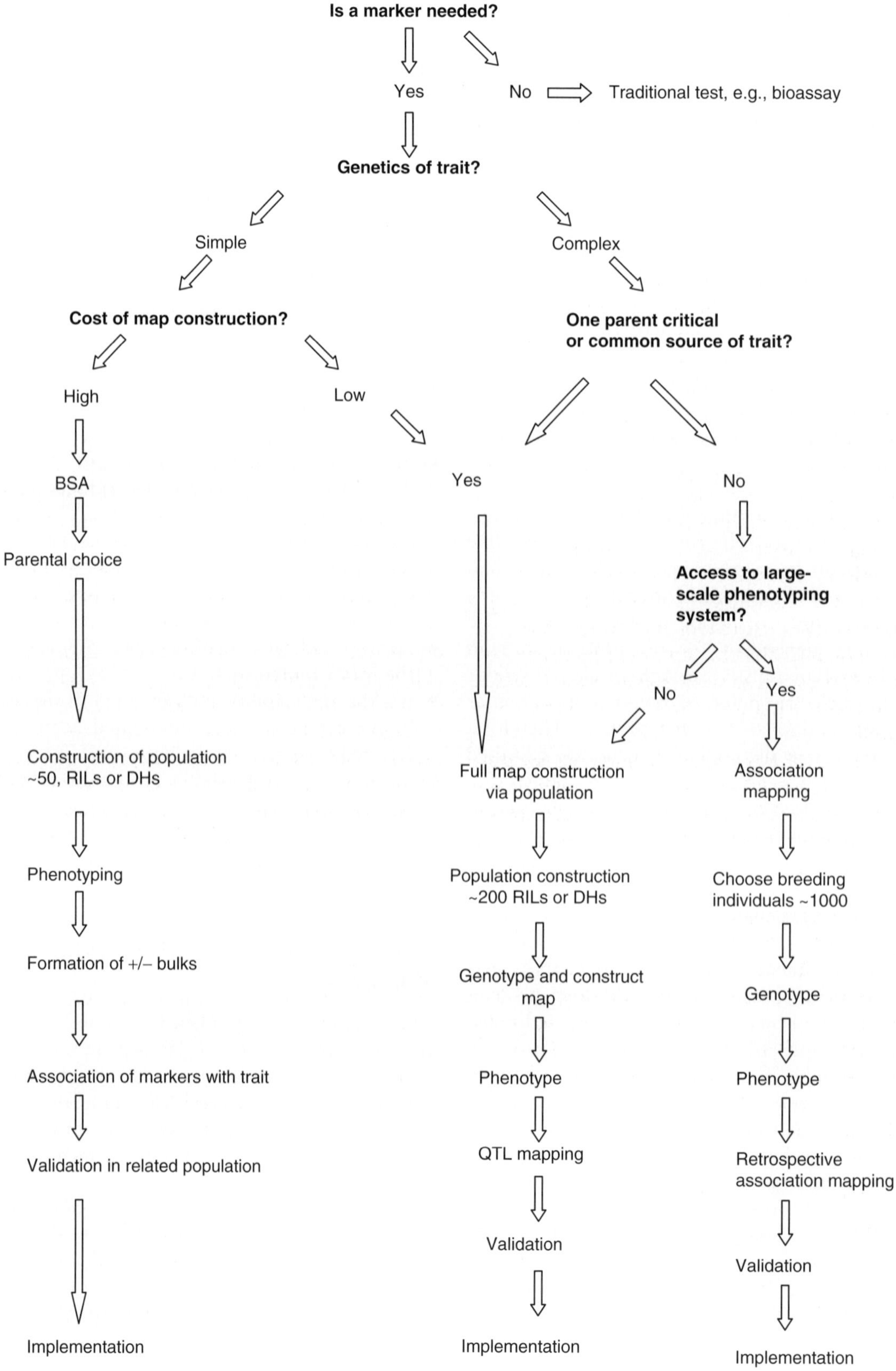

Figure 1 Marker development schemes.

1. Is a marker needed? Marker development is slow and expensive so it must be clear that there will be a real benefit to the breeding program through access to a molecular marker. This will involve comparing the cost of marker development and use, to more traditional screening methods.
2. Is the trait simply inherited or is it controlled by many genes? If the inheritance is simple, then bulked segregant analysis would usually be the preferred method for marker identification. However, if inheritance is controlled by several genes or is strongly influenced by the environment (low heritability), full map construction or linkage disequilibrium mapping would be the best options.
3. Cost of map construction? The shift to automation and high throughput marker analysis is dramatically reducing the cost and time involved in marker development. For many species, map construction will be sufficiently cheap to replace bulked segregant analysis (BSA).
4. Is one parent critical or a common source of trait? The question relates to the diversity of germplasm available for screening. If there are multiple sources of the desirable trait, linkage disequilibrium mapping offers the option of localizing loci from multiple sources, while full map construction will only allow the genes from one source to be localized.
5. Is there access to large-scale phenotyping systems? Linkage disequilibrium mapping is dependent upon the analysis of a large pool of germplasm and will require substantially more marker assays than full map construction.

The marker systems that are currently available, and their relative advantages and disadvantages are summarized in **Table 1**, where the data represent averages for these marker systems. Although the data in **Table 1** suggest that randomly amplified polymorphic DNA (RAPD) markers are the most efficient, the poor reliability of these markers and the inability to transfer them between crosses has greatly limited their use.

Essentially, three basic systems have been used to develop markers linked to traits of interest. The relative merits of each and their applicability are outlined below.

Map Construction

Complete linkage maps generated from screening the progeny of a cross have provided the basis for most early marker development work. In this process, a population is produced from parents differing in the trait of interest. Molecular markers, able to detect polymorphisms between the two parents, are screened against the population so that each line will have been assayed for the target trait or traits and a large number of markers. Linkage between the markers is assessed based on the segregation pattern, and several public and commercial software packages are available to assist in the map construction. When this kind of analysis has been used to investigate the linkage relationships of a large number of markers in a cross, the markers have always been found to occur in a distinct set of linkage groups, with each linkage group corresponding to one of the pairs of homologous chromosomes.

It is often difficult to assign marker based linkage groups to know chromosomes. For most major crop species, good-quality linkage maps have been published. Markers from these maps can be used as reference points to determine chromosome designations. For some species, such as wheat (*Triticum* spp.), chromosome substitution, translocation, or deletion lines can be used to determine the chromosome designation for specific probes and hence linkage groups. Markers that fall into a linkage group are then ordered through two and three point analysis. With a good-quality linkage map, the target trait can be mapped relative to the molecular markers by measuring the frequency of trait cosegregation with markers.

Full map construction is difficult and labor intensive, particularly in species with a large number of linkage groups, where linkage maps must be

Table 1 Comparison of major marker systems

Marker system	*Usual loci/assay*	*DNA amount*	*Time/assay*	*Comments*
Restriction fragment length polymorphisms (RFLPs)	1	5.0 μg	5 days	Codominant, reliable, often low level polymorphism
Simple sequence repeats (SSRs)	1	0.2 μg	5 h	Codominant, reliable, large number of alleles
Amplified fragment length polymorphisms (AFLPs)	50	0.2 μg	1 day	Mostly dominant, reliable, low level of polymorphism, but high multiplexing capacity
Randomly amplified polymorphic DNAs (RAPDs)	10	0.2 μg	5 h	Dominant, unreliable in most situations

constructed for each chromosome. Usually, 10–20 markers are desirable for each chromosome to give reasonable genome coverage (about 1 every 10–20 cM). The work involved in screening this number of markers is considerable if markers such as RFLPs are used. For some cultivated species, the germplasm base may be small and the level of polymorphism between varieties or cultivars low. In order to reveal sufficient polymorphisms for full genome coverage, it may be necessary to screen several thousand probes. Combining RFLP with various other marker systems offers an alternative approach. In many cases, SSR markers can substitute for the use of RFLPs, and AFLP markers are particularly useful due to the high multiplexing ratio. Even though the level of polymorphism for AFLP bands is usually lower than for RFLPs, the large number of bands that can be scored in a single assay makes this a valuable marker system for enhancing linkage maps.

It is also often important to consider how many traits of significance may be segregating in the population used for marker development. If several important traits are segregating, the benefits derived from constructing a full linkage map may outweigh the time and costs associated with full map construction relative to BSA.

Linkage maps of most major crop species have now been constructed. These established maps are a valuable resource and a good starting point in identifying the most appropriate markers to use to obtain good genome coverage.

Bulked Segregant Analysis

Bulked segregant analysis (BSA) is a technique that allows rapid, cheap development of markers for simply inherited traits. The first step involves pooling individuals from the two phenotypic extremes of a segregating F_2, doubled haploid, or similar population. DNA isolated from the two gene pools are then screened with DNA markers, usually RAPD or AFLP, and polymorphic bands identified. Clear polymorphisms seen between the two pools will be derived from regions of the genome that are common between the individuals that made up the pools. The remainder of the genome will be randomly contributed by the parents and should show no polymorphisms between the pools. The principle of this technique is illustrated in **Figure 2**.

This technique has clear advantages for marker identification relative to construction of complete linkage maps. Only a few weeks are required to screen the pools with the AFLP or RAPD primers, and the cost will be between 20 and 30% of that for full map construction. For traits controlled by a single gene, such as many disease resistances, only 5–10 plants are required for each bulk, meaning that only small populations are required. However, there are also the following disadvantages to this technique:

- The method is not effective (although not impossible) for complex traits controlled by several unlinked loci. The technique can be used for traits controlled by fewer than three major loci.

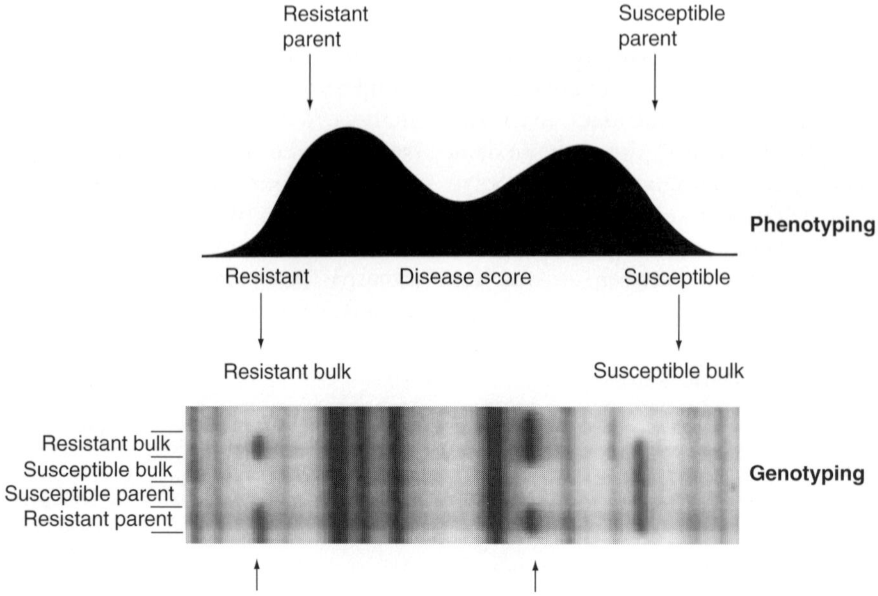

Figure 2 Bulked segregant analysis in the identification of AFLP markers linked to a disease resistance locus. Parents, differing in their resistance phenotype, are used to produce a population segregating for resistance. This population is phenotyped for resistance and two bulks are generated; resistance and susceptible. The parental DNA and the DNA pooled from the bulks are analyzed with a range of AFLP markers to identify bands present in the resistant parent and the resistant bulk (arrows).

- The actual location of the gene of interest is not revealed, as only linked markers are detected.
- The method can only be efficiently used with PCR-based marker systems such as SSR, RAPD, and AFLP markers. For AFLP and RAPD markers, it is usually necessary to convert the marker to a sequenced tagged site (STS) marker, and this is not a simple task. This problem is largely overcome through the use of SSR markers, or where very good-quality maps are available of AFLP markers. The effort involved in the initial screening of the pools may be greater for SSR markers, and a large number will be required, but the benefits in having markers that are immediately applicable in selection programs may overcome this drawback. Good collections of SSRs are available for most major crop species, but are limited for minor species.

Linkage Disequilibrium Mapping

The principle techniques used to identify marker trait associations in crop species have been based around the construction of linkage maps or the use of BSA. Both techniques are based around the production and phenotyping of special populations. The populations are generally developed from two varieties that show a major difference in the traits targeted for mapping. This leads to several problems:

- There is usually a high cost associated with phenotyping, particularly for traits requiring extensive field trials or complex analysis, such as many aspects of processing quality and yield. Consequently, the number of replicates and sites is often limited, reducing the sensitivity of some of the analyzes.
- Many marker development projects for annual crops are using populations that were established 5 or more years before the marker development work. This reduces the value of the information gathered and the scope of its implementation.
- The structure of the populations limits the types of traits that can be mapped and many of the subtleties of adaptation cannot be analyzed.
- Mapping is frequently restricted to known traits for which a well-defined bioassay is available.

These limitations in existing mapping strategies can be addressed through association or linkage disequilibrium (LD) mapping. LD mapping is based on seeking associations between phenotype and allele frequencies.

Linkage disequilibrium is based around the association of marker loci with traits at the population level. Equilibrium is seen in large populations over many generations where there is no selective advantage, or disadvantage, associated with a particular allelic combination in a region of the genome. Disequilibrium appears where selective pressure increases, or decreases, the frequency of particular alleles or allelic combinations. It can be measured through an estimation of changes in allele frequencies as a result of the selective pressure. In this way, particular regions of the genome can be associated with particular traits.

There are four potential advantages to this approach in mapping genes in crop species:

- It provides a new perspective for trait mapping, because it uses population structures (based largely around pedigree) and phenotypic data that differ from those used for full map construction of BSA. Consequently, we can expect to see new marker trait associations and targets for more detailed analysis.
- LD mapping also provides detailed fingerprinting information on a large number of lines and varieties and this information will be valuable in several of the breeding strategies outlined below.
- The LD method uses real breeding populations, the material is diverse and relevant, and the most important genes (for adaptation, etc.) should be segregating in such populations. The breeder is also involved in the process, and this may lead to improved rate and efficiency of validation and adoption. Plant breeders are often reluctant to grow and assess a huge number of lines with little or no potential for direct commercial outcome, such as required for complete map construction. The advantage of LD mapping to the breeder is that mapping and commercial variety development is conducted simultaneously.
- Pattern analysis of marker data has the potential to detect complex combinations (even epistatic interactions) between alleles at several loci that underlie the superior individuals in a breeding population. This might prove difficult to isolate and validate using the full mapping approach.

Taking a Marker through to Application in a Breeding Program

There are three important requirements for the effective implementation of molecular markers:

- Cost-effective markers. Markers need to be easy and cheap to use. The most widely used marker systems have been based on RFLPs; these are expensive and technically demanding to apply to many species, particularly for polyploid species or those with a large genome size, such as wheat or onion (*Allium cepa*). Emphasis has recently shifted to microsatellite markers and AFLPs; both are PCR-based assay systems, which are cheaper and more easily used than RFLPs.

- Marker validation and implementation strategies. These concern the availability of reliable and efficient strategies for marker use in the breeding programs, and involve the analysis of marker behavior in different genetic backgrounds and the analysis of polymorphism. Importantly, the marker must be able to detect polymorphisms in between the lines, or varieties used in the breeding and selection program. Breeders need to evaluate carefully their breeding methodology to ensure that MAS is used effectively, both in terms of selection and cost.
- Implementation. This involves establishing a skill and equipment base to ensure ready access by breeders to the markers.

The identification of markers associated with a trait of importance represents a first step in marker application. Several further steps are needed before a marker can be used in a practical breeding program (outlined in **Figure 3**). The key element is marker

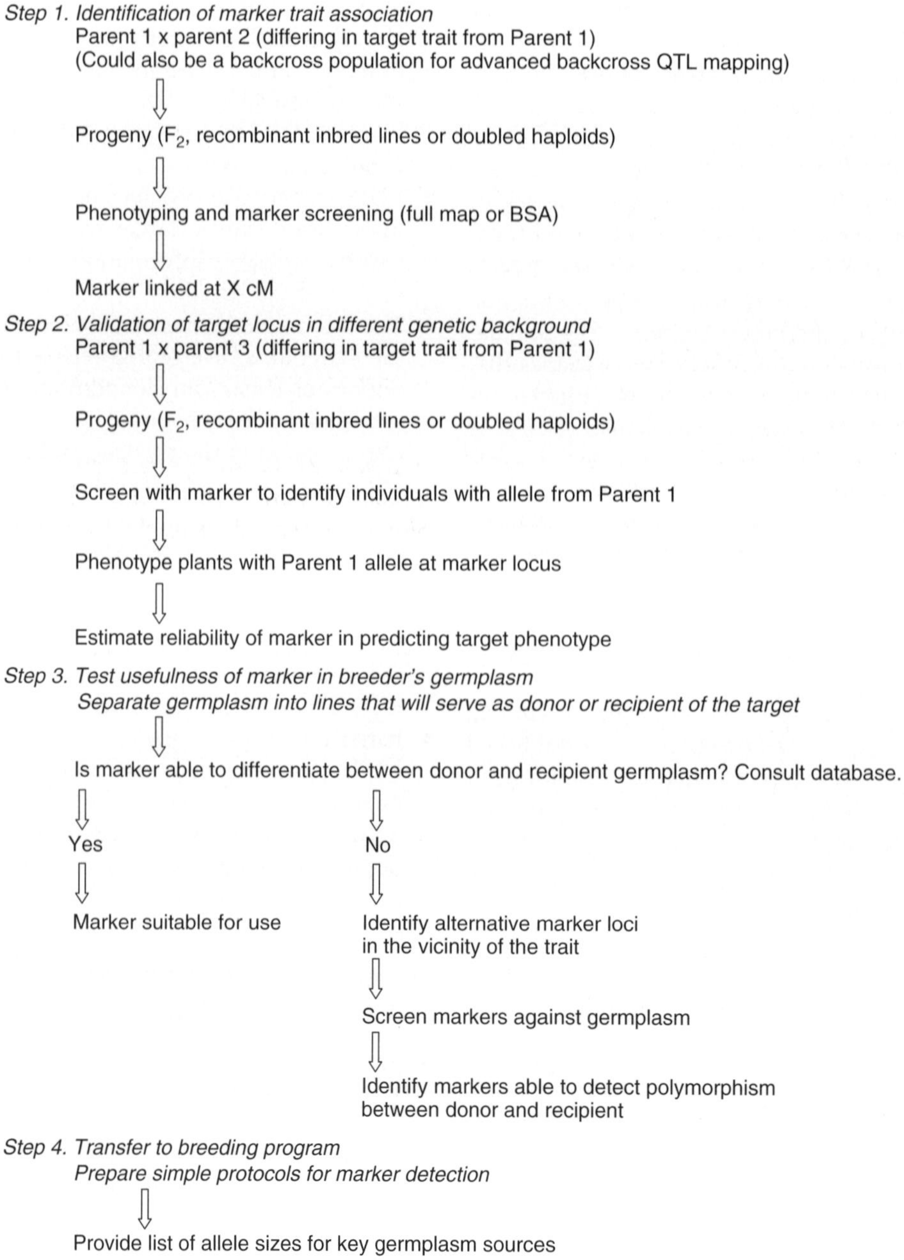

Figure 3 Development validation and implementation of markers for MAS schemes.

validation. Irrespective of the technique used to identify a marker linked to a target trait, the association has been found in a particular set of circumstances; usually in the progeny of a specific cross. The reliability of the marker can be estimated from the closeness of linkage. As the next step, the ability of the marker to predict the target phenotype must be tested in further populations. Usually one would aim to keep one parent in common in the first test. Therefore, the predictive value of the marker in identifying the phenotype of the plant can be estimated and compared with the original mapping result.

The usefulness of the marker in screening and assaying germplasm in use in the breeding program is assessed as the next step. The parents likely to be used in the marker screening program are screened for polymorphisms between those that have and do not have the target trait. Given the narrow germplasm base of many crop plants, a high proportion of markers will fail this test of applicability. It is then necessary to see if other markers can be found in the vicinity of the target locus that could be used in screening. Ideally, one would aim to have about 10 marker loci within 10 cM of the target locus to have a reasonable chance of being able to track the locus in the germplasm used in a breeding program. At this point, markers such as SSRs are particularly valuable, since they tend to detect a greater level of diversity than RFLP-based marker systems.

Finally, a clear protocol for use of the markers must be provided. This will include the following information:

- Which molecular marker to use for a given germplasm combination.
- The protocol for the marker assay and the expected size of the alleles.
- The reliability of the marker in terms of the actual success rate of the assay. For example, SSR markers can vary in their ease of assay, some working almost 100% of the time, and others showing less than 80% success rate. This is often due to the sequence of the primers.
- The closeness of linkage. This provides a measure of the success of the marker in predicting the target locus or allele.

Practical Application of Marker-Assisted Selection in Breeding Programs

Transfer of a single dominant gene This is perhaps the simplest MAS application attempted by most breeders. In 1986, Beckman and Soller showed the frequency of the favorable allele was substantially increased where MAS was applied during backcrossing.

*Example 1: transfer of resistance to cereal cyst nematode (*Heterodera avenae *Woll) into the malting cultivar Sloop* In the case of the South Australian Barley Improvement Program (SABIP), the transfer of the gene *Ha2* by Kretschmer and coworkers for resistance to cereal cyst nematode (CCN) from Chebec to the malting quality cultivar Sloop of barley (*Hordeum vulgare*) was initiated in 1994. MAS was applied using the RFLP marker *Xawbma21* for three cycles of backcrossing. Doubled haploid plants were produced from the BC_3F_1, and 66% of the regenerants were classed as CCN resistant. This exceeds the expected proportion (48%, for a marker 3cM distal from *Ha2*), as it seems that a region associated with improved regeneration from tissue culture is linked in coupling to the *Ha2* gene from Chebec. It is estimated that the use of MAS saved at least 2 years compared with phenotypic selection. The best of these CCN resistant BC_3-derived Sloop types were released in 2002 and registered under "Plant Breeders Rights" in Australia as "Sloop SA". In 2002, all 200 tonnes of breeder's seed was sold to farmers eagerly awaiting the new malting variety with the resistance required to manage cereal cyst nematode infestations in their cropping rotations.

Example 2: transfer of resistance to BYDV (Barley Yellow Dwarf Virus) into the malting cultivar Sloop A second example from the SABIP is the successful transfer of BYDV tolerance conditioned by the gene *Yd2* from Franklin into Sloop using MAS with the PCR marker YLM by Paltridge and coworkers. MAS was employed through two backcross cycles, and the BC_2F_2-derived lines were field-tested in plots inoculated with viruliferous aphids. *Yd2* was successfully transferred to the BC_2 lines, and losses due to BYDV were dramatically reduced (**Table 2**).

Transfer of a single recessive gene Provided a codominant marker (such as a RFLP or SSR) is available, it is possible to save the selfing step required between cycles of phenotypic selection in a conventional backcrossing program. This then makes the transfer of a recessive gene, conceptually and practically, little different to the above example of a major gene. In barley, an example of a recessive trait with linked markers is *mlo* (marker derived from the cloned gene).

Selection of several genes simultaneously One of the most exciting applications of MAS in backcrossing is the potential to transfer multiple traits. It is difficult to conceive and manage practical

Table 2 Effect of *YLM* variant on grain yield and grain plumpness (% >2.5 mm) in F_2-derived BC_2 lines infected with BYDV at Strathalbyn, South Australia, 1998

	YLM variant	*– Aphids*	*+ Aphids*	*% of – Aphids*	*Sig. of diff. (+/–) (P)*
Grain yield	Sloop	1023a	316a	29	<0.0001
(g/plot)	Franklin	1257b	1096b	87	<0.05
Plump grain	Sloop	63a	29a	34	<0.0001
(%>2.5 mm)	Franklin	72a	69b	97	NS

Significance of difference (*P*) between plots inoculated with viruliferous aphids (+Aphids) and those not inoculated (–Aphids) is provided. a and b, significant difference of pair-wise comparison between YLM variant classes at $P<0.001$. NS, not significant.

backcrossing strategies using phenotypic selection to simultaneously transfer two traits, let alone more. In contrast, the application of MAS is only limited by handling an appropriate BC_xF_1 population to ensure recovery of the required number of individuals heterozygous for all target loci. In a defect elimination strategy with the Sloop, the South Australian Barley Improvement Program (SABIP) introduced genes for resistance to CCN (either *Ha2* or *Ha4*), tolerance to BYDV (*Yd2*), resistance to spot form of net blotch (*Rpt4*), and manganese efficiency (*Mel1*) into Sloop in parallel backcross streams. Beginning in 1996, these independent streams have been progressively merged. Sloop types with combinations of these traits entered field trials in 2000.

The following difficulties have been encountered with this approach:

- There have been problems with the availability of polymorphic markers as the number of donor parents increases, although this difficulty could be reduced by fixing genes in each backcross stream, by selfing or doubled haploid production, prior to intercrossing lines from each stream.
- There have been difficulties encountered in managing populations of an appropriate size.

Example 3: backcross conversions – transfer of resistance to spot form net blotch and CCN to Gairdner malting barley from Keel feed barley Gairdner is a widely adapted malting variety bred by Agriculture Western Australia. It has a good disease resistance profile with resistance to leaf scald, BYDV tolerance (*Yd2 locus*), powdery mildew (presumably Franklin's two genes), and net form net blotch (unknown), but it is susceptible to leaf rust, spot form net blotch, and cereal cyst nematode. The high-yielding, feed variety Keel has resistance to spot form net blotch (*Rpt4* plus another locus (or loci) conditioning adult plant resistance), leaf scald (unknown), and cereal cyst nematode (*Ha4*).

We designed a "fast-track" strategy using single seed descent, which was initiated in 1998, to rapidly introduce resistance to spot form net blotch (SFNB) and cereal cyst nematode from Keel into Gairdner. We used MAS and bioassays in large populations of the BC_1F_1 and BC_1F_2 to select for *Rpt4* – the gene conferring seedling resistance to spot form net blotch. BC_1F_4 individuals were multiplied over summer and placed into stage 1 field trials in 2000. Following phenotypic selection for SFNB and CCN resistance and "Gairdner" plant type, the remaining 43 individuals were tested for *Yd2*, *Rpt4*, *Bmy1* (Gairdner carries the SD1 allele for β-amylase), and three malt extract quantitative trait loci (QTLs; 1H, 2HL, 5H). The malt extract QTLs are important to maintain the good malting quality of Gairdner in the face of introgression of potentially undesirable alleles from the feed variety Keel. Four lines were promoted to stage 4 trials in 2001, based on the marker profile and their field performance in 2000. In this cross, the Gairdner allele for malt extract on 1H was the most significant, providing 2% higher extract than the Keel allele. This strategy would have been greatly assisted by markers for CCN resistance (we did not know Keel carried *Ha 4* until 2000) and the "second" adult plant gene in SFNB resistance. Further, it would have been useful to have access to a wider range of SSR markers and the ability to deploy them cost effectively. Whilst this project started with a BC_1F_1 population of over 150 individuals per generation, there were still too few selected lines by the BC_1F_7. Nevertheless, promising lines completed stage 4 trials in 2001 with excellent yield, malt quality, and disease resistance. 500–1000 kg of seed will have been produced from the best three lines by May 2002.

The next phase in the Gairdner "defect elimination" is to introduce genes for boron tolerance, leaf rust resistance, leaf scald resistance, and the SD3 allele for thermostable β-amylase. The crossing strategy involves merging five streams (**Table 3**). By the time the seed of this complex cross is available, it is hoped that markers for the "missing" genes will be available and that it will be practically possible to select for the Gairdner background genotype using SSR markers. We have avoided using AFLPs for selecting recurrent parent genotype because of the cost of implementing a commercial licence.

Table 3 Trait donors for five streams of the Gairdner defect elimination program

CCN+SFNB stream	*Boron stream*	*SD3 stream*	*Leaf scald stream*	Leaf rust stream
Gairdner/Keel// Gairdner	Gairdner/DH115// Gairdner	Gairdner/SD3// Gairdner	Novel alleles from *Hordeum spontaneum*	Complex cross of Gairdner and Fanfare

Transfer of QTL There are now several examples of backcrossing to transfer QTLs in barley. These include, quantitatively inherited barley stripe rust (*Puccinia striiformis* f. sp. *hordei*) resistance, distilling quality, malting quality, and boron toxicity tolerance. The Australian National Barley Molecular Marker Program has progressed with the analysis of the genetic control of malt quality traits to the point where MAS is possible for malt extract, diastatic power, and wort viscosity.

In its simplest form, there is little difference in the backcross strategy required to transfer a QTL compared to a major gene. However, there are important differences. Firstly, there is ambiguity about the location of the QTL, often making the use of flanking markers necessary. This usually increases the size of the introgressed segment, unless the region of interest is finely mapped. Secondly, the most important issue becomes "validation" of the QTL. This process involves testing the phenotypic effect of the particular chromosomal region derived from a mapping population in a range of, first, related, then progressively less related, germplasm. Collins and Coventry and their teams have undertaken this task for malt extract and diastatic power, respectively, and have recommended QTLs that are amenable to marker assisted introgression. The opportunity to manipulate complex quality traits via MAS/backcrossing is now available. This was not previously possible without time-consuming progeny testing cycles in a conventional backcrossing system using phenotypic selection. The other issue important to QTL transfer is accuracy of selection. Two separate studies by Lande and Thompson and Zhang and Smith showed that MAS significantly improved the accuracy of selection for traits of low heritability.

Example 4: "Feed barley conversion" – Transfer of malt quality alleles from Alexis, AC Metcalfe, and Haruna nijo to the disease-resistant feed barley Keel Typically, when breeding malt quality barley cultivars, we attempt to limit the percentage of "feed" germplasm in a pedigree to a maximum of 25%, as the genetic control of malting quality is complex, and top quality is difficult to retain if we introgress too much genetic material from inferior lines. However, resistance to five leaf diseases plus cereal cyst nematode plus adaptation to our tough Australian growing conditions is also a complex genetic assignment. Recent advances in the genetics of malt quality and in marker assisted selection for malt quality have provided the possibility of "converting" feed quality lines into malting varieties – a strategy opposite to the conventional wisdom.

In our program, progress in breeding for yield and disease resistance has been more rapid in feed barley than in malting barley for a number of reasons:

- The number of breeding priorities set for malting barley (>12) compared to feed barley (6–7) diminishes the rate of progress.
- Physical quality specifications have been much easier to achieve in feed barley.
- A wider range of germplasm is of direct use in the improvement of feed barley than in malting barley.

Consequently, there is at least 10% difference in yield between malt and feed varieties, as well as significantly poorer disease resistance in the malt varieties.

A current project aims to introgress the key genetic loci influencing malt extract, diastatic power, and fermentability from Canadian, European, and Japanese material into elite Australian feed lines. The initial focus is on the feed variety Keel (released in 1999), the second phase will focus on WI3385 (expected commercial release 2003), and new recurrent parents will be added as elite lines are identified in the feed barley breeding program.

A breeding strategy to convert elite feed varieties into malting quality lines has been developed based on transferring the important genes for malting quality from Alexis, Haruna nijo, and AC Metcalfe, into feed varieties. The strategy (outlined in **Figure 3**) is based on backcrossing with extensive use of molecular markers for the most rapid and efficient development of new lines. In the first year of the project, this approach has successfully developed Keel backcross lines (BC_1 and BC_2) that carry the key quality genes from the three malting barley parents, including malt extract on chromosomes 1H, 2HS, 2HL, and 5H, and the β-amylase gene on 4H, which influences diastatic power and fermentability. These lines will enter field trials in 2002 for selection based on agronomic performance, prior to yield trials and malt quality evaluation in 2003.

The key malting quality genes from Alexis, Haruna nijo, and AC Metcalfe are also being combined by merging the three Keel backcrossing programs. BC_3 generations will also be developed for each of the Keel backcrossing programs. Molecular markers will be used to identify elite individuals from these intercrosses and the BC_3F_1 generations. The elite lines will also be used to develop doubled haploid populations by the end of 2003.

Future Directions

For much of this article, we have assumed that there are sound reasons for the development and implementation of a MAS selection scheme. Having developed and implemented the tools for MAS, it is time to re-evaluate this assumption and realistically assess the value of MAS compared to traditional selection methods.

We should answer the following questions:

1. How did the cost compare per individual?
2. Did it improve flexibility, for example, was a field test constrained to one generation per year, whereas the MAS allowed 2–3 generations per year?
3. What was the accuracy of traditional selection (error rate, heritability, etc.) versus MAS (recombination between the marker and locus, error rate, heritability, etc.)?
4. Was it an advantage to characterize the genotype – is it better for systematically accumulating QTL to exploit additive genetic variation or to build a specific disease resistance pyramid?
5. Was it a disadvantage to know the genotype? For example, MAS only works for well-characterized loci, it can't detect an unknown locus for, say, disease resistance, which would be identified in a traditional test based on phenotype.
6. Did the MAS scheme expand or contract the germplasm pool used in the breeding program? There are examples of both scenarios. Expansion may be achieved by implementation of the "advanced backcross QTL" approach to introgress novel alleles from wild relatives, whereas contraction may occur if parental combinations lacking polymorphisms for the available markers (but not for the actual trait) are ignored.

In the early stages of MAS development, when markers, resources, and throughput are limited, breeders opt to use MAS to tackle bottlenecks, or attack key leverage points in their program, such as:

- backcrossing for defect elimination;
- backcrossing for a single recessive trait;
- introduction of a quarantine trait, where traditional selection methods are not possible;
- building a crucial parent; and
- enriching F_1s from complex crosses.

As confidence, skill, and resource base increases, breeders may then tackle:

- high throughput early generation screening; and
- whole genome approaches, such as selection for recurrent parent in backcrossing or analysis of linkage blocks.

What is now clear about those programs that have explored MAS is that these programs are irreversibly changed. In most cases, improvements in the heritability, ability, speed, and flexibility of selection have been measured, resulting in more cost-effective breeding programs that are better able to respond to the changing demands of the clients. At worst, breeders are much the wiser about the genetics of the traits being manipulated, and of the characteristics of the germplasm pool at their disposal.

List of Technical Nomenclature

AFLP Amplified fragment length polymorphism.

Backcross A population created by crossing an individual back to one of its parents.

BSA Bulked segregant analysis; marker analysis applied to two bulked DNA samples prepared from individuals selected, in a segregating population, to be contrasting with respect to a trait of interest.

cM Centimorgan; a unit of measurement of recombination frequency. One centimorgan is equal to a 1% chance that a marker at one genetic locus will be separated from a marker at a second locus due to cross over in a single generation.

Codominant marker Two alleles of a marker that result in distinctly different patterns, but when they are both inherited together in an individual, the individual has both of the marker patterns.

DH Doubled haploid; homozygous plant produced by chromosome doubling from a haploid gamete.

Dominant One allele is said to be dominant with respect to an alternative allele if a heterozygous individual bearing the two alleles is indistinguishable from an individual homozygous for the dominant allele.

F_1 First filial generation; the offspring resulting from a cross.

F_2	Second filial generation; the offspring resulting from a cross between members of the F_1 generation.
Genotype	The genetic constitution of an organism.
Heritability	Relative importance of hereditary and environmental influences on the expression of a specific trait.
Isozyme	Any of a group of related enzymes that catalyze the same reaction, but have different structural, chemical, or immunological characteristics.
LD	Linkage disequilibrium; population-based form of genetic analysis based on associations of phenotypic with genotypic information.
Locus	The position on a chromosome of a gene or other chromosome marker; also, the DNA at that position. The use of locus is sometimes restricted to mean regions of DNA that are expressed.
PCR	Polymerase chain reaction; a process that permits making, *in vitro*, unlimited numbers of copies of a DNA sequence.
Phenotype	A visible physical characteristic or behavior that is a result of combined environmental/ecological conditions and genetic expression.
Polymorphism	The difference in DNA sequence among individuals.
Probe	A biomolecule that is labeled with radioactive isotopes or with a fluorescent marker that selectively binds to a specific gene so it can be isolated or identified.
Quantitative trait	A trait that takes a quantitative rather than a qualitative value.
QTL	Quantitative trait locus; a locus contributing to the expression of a quantitative trait.
RAPD	Randomly amplified polymorphic DNA; DNA polymorphisms amplified by arbitrary primers using PCR.
RFLP	Restriction fragment length polymorphism; variation between individuals in DNA fragment sizes cut by specific restriction enzymes. Polymorphic sequences that result in RFLPs are used as markers on genetic linkage maps.
RIL	Recombinant inbred lines; type of population formed from an initial outcross between two inbred strains followed by several generations of inbreeding of the individuals generated from the F_1.
SNP	Single nucleotide polymorphism.
Southern hybridization	A technique used for searching for a specific DNA fragment.
SSD	Single seed descent; referring to the advancement of generations by propagating one seed per plant from the previous generation. Used to advance rapidly heterozygous segregating individual generations to fixation.
SSR	Simple sequence repeat (microsatellite); a short sequence of repeated nucleotides in a genome.
Validation	The process by which a marker trait association is verified, usually in a different family to that used for the marker development, prior to its use for marker assisted selection.

See also: **Crop Improvement**: Comparative Genetics; Double Haploid Production; Genetic Maps; Molecular Markers; Mutation Techniques; Plant Breeding, Practice; Plant Breeding, Principles; Plant Genomes. **Diseases**: Breeding for Disease Resistance. **Genetic Modification**: Transgene Stability and Inheritance.

Further Reading

Beckmann JS and Soller M (1986) Restriction fragment length polymorphisms and genetic improvement of agricultural species. *Euphytica* 35: 111–124.

Buschges R, Hollricher K, Panstruga R, *et al.* (1997) The barley Mlo gene: A novel control element of plant pathogen resistance. *Cell* 88: 695–705.

Collins HM, Logue SJ, Jefferies SP, Stuart IM, and Barr AR (1999) *A Study of the Physical, Biochemical and Genetic Factors Influencing Malt Extract*. Proceedings of the 9th Australian Barley Technical Symposium.

Coventry S, Barr AR, Jefferies SP, *et al.* (1999) *Validating Quantitative Trait Loci for Diastatic Power in Populations of Malting Barley (Hordeum vulgare* L.). Proceedings of the 9th Australian Barley Technical Symposium.

Cushman JC and Bohnert HJ (2000) Genomic approaches to plant stress tolerance. *Current Opinion in Plant Biology* 3: 117–124.

Han F, Ullrich SE, Kleinhofs A, *et al.* (1997) Fine structure mapping of the barley chromosome-1 centromere region containing malting-quality QTLs. *Theoretical and Applied Genetics* 95: 903–910.

Jefferies SP, Barr AR, Karakousis A, *et al.* (1999) Mapping of chromosome regions conferring boron toxicity tolerance in barley (*Hordeum vulgare* L.). *Theoretical and Applied Genetics* 98: 1293–1303.

Jefferies SP, Pallotta MA, Paull JG, *et al.* (2000) Mapping and validation of chromosome regions conferring boron toxicity tolerance in wheat (*Triticum aestivum*). *Theoretical and Applied Genetics* 101: 767–777.

Kearsey MJ and Farqhar AGL (1998) QTL analysis in plants; where are we now?. *Heredity* 80: 137–142.
Kretschmer JM, Chalmers KJ, Manning S, *et al.* (1997) RFLP mapping of the *Ha2* cereal cyst nematode resistance gene in barley. *Theoretical and Applied Genetics* 94: 1060–1064.
Kumar LS (1999) DNA markers in plant improvement: An overview. *Biotechnology Advances* 17: 143–182.
Lande R and Thompson R (1990) Efficiency of marker-assisted selection in the improvement of quantitative traits. *Genetics* 124: 743–756.
Liu BH (1998) *Statistical Genomics: Linkage, Mapping and QTL Analysis*. Baco Raton: CRC Press.
Lynch M and Walsh B (1998) *Genetics and Analysis of Quantitative Traits*. Sunderland, USA: Sinauer Associates Inc.
Paltridge NG, Collins NC, Bendahmane A, and Symons RH (1998) Development of YLM, a codominant PCR marker closely linked to the *Yd2* gene for resistance to barley yellow dwarf disease. *Theoretical and Applied Genetics* 96: 1170–1177.
Plant Genome Information Centre, USDA-ARS National Agricultural Library. http://www.nal.usda.gov/pgdic/
Rommens CM and Kishore GM (2000) Exploiting the full potential of disease-resistance genes for agricultural use. *Current Opinion in Biotechnology* 11: 120–125.
Thomas WTB, Powell W, Swanston JS, *et al.* (1996) Quantitative trait loci for germination and malting quality characters in a spring barley cross. *Crop Science* 36: 265–273.
Toojinda T, Baird E, Booth A, *et al.* (1998) Introgression of quantitative trait loci (QTLs) determining stripe rust resistance in barley: An example of marker-assisted line development. *Theoretical and Applied Genetics* 96: 123–131.
Young ND (1999) A cautiously optimistic vision for marker-assisted breeding. *Molecular Breeding* 5: 505–510.
Zhang W and Smith C (1992) Computer simulation of marker-assisted selection utilizing linkage disequilibrium. *Theoretical and Applied Genetics* 83: 813–820.

Alien Cytoplasms

K Tsunewaki, Fukui Prefectural University, Fukui, Japan

Introduction

In plants, three organelles carry genetic information (**Figure 1**): the nucleus, chloroplast, and mitochondria, which are specialized to function in inheritance, photosynthesis, and respiration, respectively. The entire genetic information present in the nucleus of a gamete, chloroplast, and mitochondria is called the genome (sometimes, nuclear genome), chloroplast genome (or plastome), and mitochondrial genome (or chondriome), respectively. The cytoplasmic genome is the sum of the chloroplast and mitochondrial genomes. Here, the terms nucleus and cytoplasm are used synonymously with the terms nuclear and cytoplasmic genome, respectively, according to their customary use in genetics.

Alien cytoplasm is the cytoplasm of a species, subspecies, or variety that has been introduced from another species, subspecies, or variety. In this article, only the alien cytoplasms introduced into crop species are dealt with. Cytoplasmic variation found within crop species, such as male-sterile cytoplasms of onion (*Allium cepa*), sugar beet (*Beta vulgaris*), and corn (*Zea mays*; maize), or that found among wild species but not introduced into crop species, is not included here.

This article consists of three main sections. The first section deals with the transfer of alien cytoplasms into crop species, including the method of the transfer, source of alien cytoplasms, genetic constancy of the transferred alien cytoplasm, and examples of alloplasmic lines produced. The second section is concerned with the genetic effect of alien cytoplasms, including their importance relative to the genotypic effect, and the effect on individual characters. The last section handles cytoplasmic male sterility, which is a major concern in the practical use of alien cytoplasms. In this section, sources of alien male-sterile cytoplasms, fertility restoration in the presence of a male-sterile cytoplasm, and use of alien male-sterile cytoplasm in hybrid breeding are described.

Alien Cytoplasm Transfer

Method

Only one method of transferring alien cytoplasm is currently available: the repeated backcross method (**Figure 2A**). Nucleus transplantation (**Figure 2B**) has not been successful in higher plants as yet. Asymmetrical somatic fusion has been used in some crops to transfer alien male-sterile plasmagene; however, it is not certain whether the entire alien cytoplasm is transferred in such studies.

Principle The repeated backcross method originates from different modes of transmission of the nuclear and cytoplasmic genes, namely, biparental transmission of the former and maternal transmission of the latter (**Figure 2A**). To transfer the cytoplasm of an alien species (cytoplasm donor) to a crop species

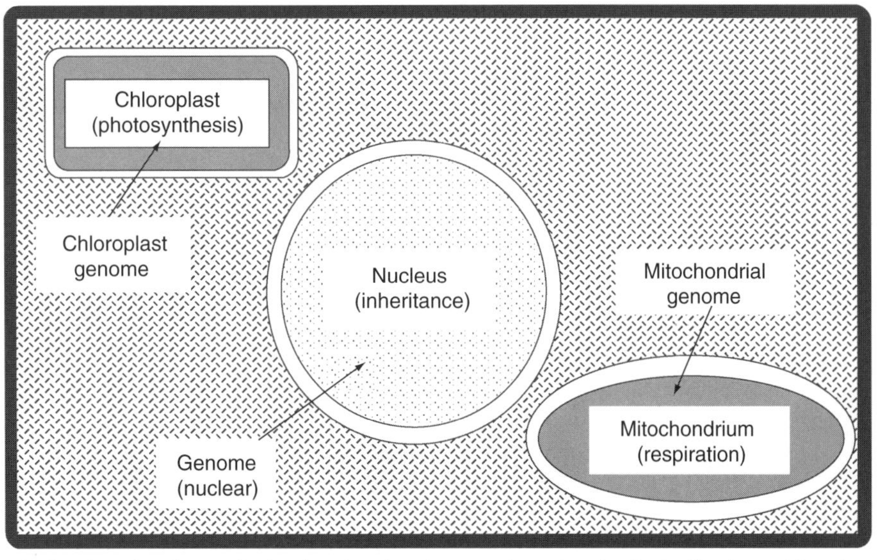

Figure 1 Schematic drawing of a plant cell, showing the tripartite genetic system in plants. Nucleus, chloroplast, and mitochondria are specialized to function in inheritance, photosynthesis and respiration, respectively. They have their own genetic information, called the nuclear, chloroplast, and mitochondrial genomes. The nuclear genome (or, simply, genome) is biparentally transmitted to the offspring, whereas the chloroplast and mitochondrial genomes are usually transmitted uniparentally only through the female gamete. The number of genes in chloroplast and mitochondrial genomes in vascular plants is about 100 and 50, respectively.

(nucleus donor), the F_1 hybrid between the alien species as female and the crop species as pollen parent is backcrossed with the pollen of the same crop species. Then, the B_1 hybrid obtained (B_1 refers to the first backcross generation) and its subsequent offspring are backcrossed repeatedly with the pollen of the crop species. The hybrid ultimately obtained retains the cytoplasm of the alien species and restores the nucleus of the crop species, and is called an alloplasmic line of the latter.

Limitations in application This method cannot be applied to crops in which the chloroplast or mitochondrial genome or both are transmitted through the pollen, even in small quantities. This method is also applicable only when viable hybrids are obtained between the cytoplasm and nucleus donors. Alloplasmic lines often show severe weakness and their maintenance may fail within a few generations of the backcrosses, usually as a result of incompatibility between nucleus and alloplasm. Lower ploidy is more sensitive to nucleus–cytoplasm incompatibility than higher ploidy. The cytoplasm donor often carries cytoplasm-specific viability gene(s) on its specific chromosomes. Cotransfer of such chromosomes with its cytoplasm gives normal vigor and fertility to the alloplasmic lines, facilitating the production of otherwise inviable alloplasmic lines.

Sources of Alien Cytoplasm

The sources of alien cytoplasm incorporated into main crops are shown in **Table 1**; sources are limited in diploid crops compared to polyploids. In barley (*Hordeum vulgare*), its two closest relatives are the source of alloplasm and in rice (*Oryza sativa*), alloplasms are transferred only from some A-genome diploids. In contrast to this, a wide range of species has provided alloplasm to polyploid crops, e.g., tetraploid tobacco (*Nicotiana tabacum*) and hexaploid common wheat (*Triticum aestivum* strain Salmon). More than 15 *Nicotiana* species provided alloplasm to tobacco, though only one species represents each section in **Table 1**. For common wheat, about 50 sources of alloplasm are enumerated from four genera. Again, one species is only indicated for each ploidy level of the respective wheat (*Triticum*) group or goatgrass (*Aegilops*) section in **Table 1**.

Genetic Constancy of the Alien Cytoplasm

Genetic constancy of the alloplasm during repeated backcrosses has been proved in an alloplasmic line of common wheat. An F_1 hybrid between *Aegilops caudata* as female and a bread wheat (*Triticum aestivum*) as pollen parent was produced in 1951, and its subsequent offspring have been repeatedly backcrossed with the pollen of the same bread wheat for 50 generations until 2001. This alloplasmic line showed complete pollen sterility first in the B_3 generation, which is an expression of the genetic incompatibility between the cytoplasm of *Ae. caudata* and the nucleus of bread wheat. This alloplasmic line has never recovered pollen fertility at any

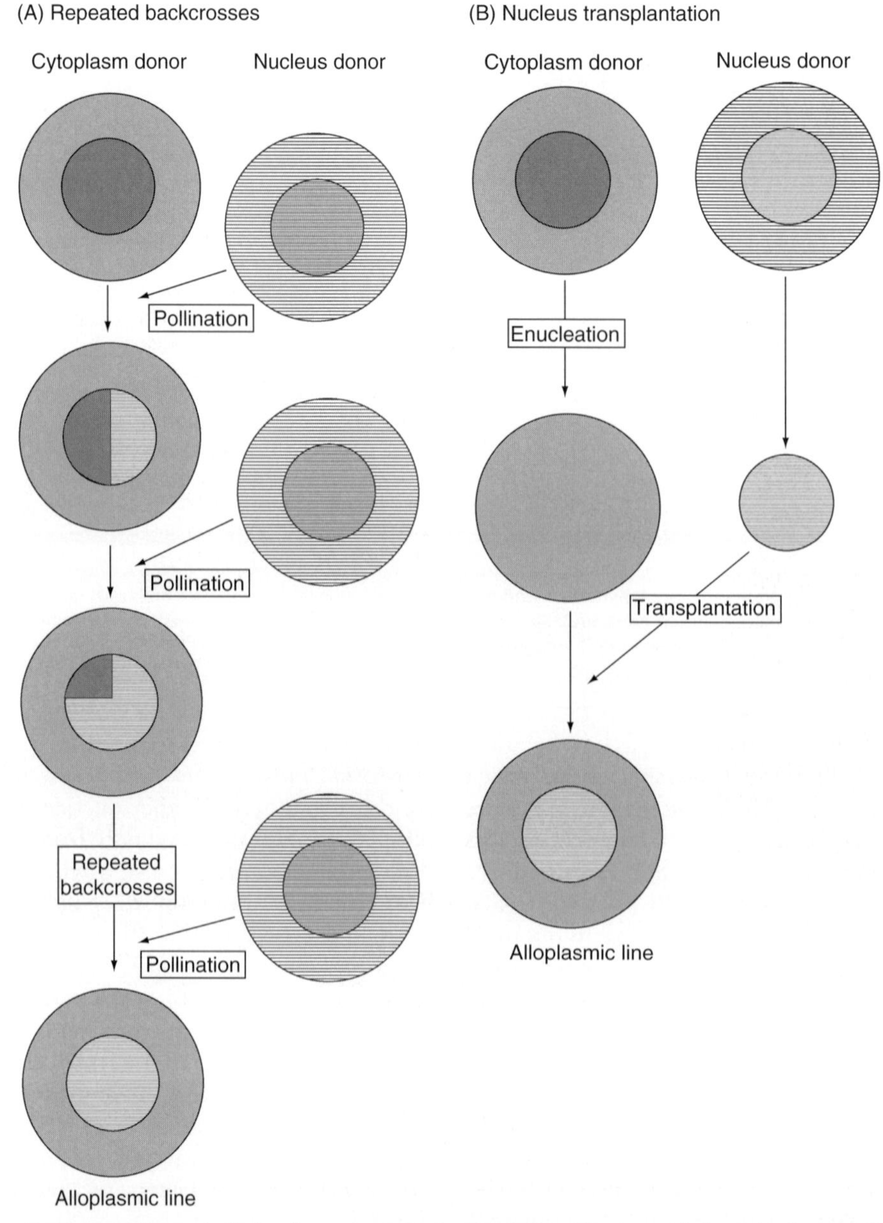

Figure 2 Two methods of nucleus substitution to produce an alloplasmic line. The inner circle and outer zone indicate the nucleus and cytoplasm, respectively. (A) Repeated backcross method. An F_1 hybrid between the cytoplasm donor as female parent and the nucleus donor, and its subsequent offspring are repeatedly backcrossed with the pollen of the nucleus donor. When the backcross is repeated 10 times, the probability of a nuclear gene of the cytoplasm donor being retained will be less than 1×10^{-3}, and the plants of the B_{10} or later generation can be regarded as alloplasmic. (B) Nucleus transplantation. The nucleus of the cytoplasm donor is enucleated using a micropipette and the nucleus of the nucleus donor is transplanted.

extent even after repeated pollination with the normal wheat pollen for a further 47 generations. Its chloroplast and mitochondrial DNAs are the same as those of *Ae. caudata* so far as their restriction fragment patterns are concerned. These facts prove that the cytoplasm of *Ae. caudata* has not changed its hereditary characteristics even after 50 years of coexistence with the genome of bread wheat.

Production of Alloplasmic Lines

In rice, male-sterile alloplasmic lines of many *indica* cultivars are produced by transfer of the cytoplasm of a wild rice, *Oryza rufipogon*. Similarly, the cytoplasm of an *indica* cultivar, Chinsurah Boro II, is introduced into *japonica* cultivars for the same purpose. In sunflower, cytoplasm of a wild relative, *Helianthus petiolaris*, is transferred to convert its

Table 1 Sources of alien cytoplasm for some crop species

Recipient of alien cytoplasm			*Donor of alien cytoplasm*		
Common name	*Scientific name*	*Ploidy*	*Common name*	*Scientific name*	*Ploidy*
Barley	*Hordeum vulgare*	2x	Wild barley	*Ho. vulgare* subsp. *spontaneum*	2x
Barley	*Hordeum vulgare*	2x	—	*Ho. bulbosum*	4x
Pearl millet	*Pennisetum glaucum*	2x	—	*Pen. glaucum* subsp. *monodii*	2x
Rice	*Oryza sativa*	2x	Rice	*O. sativa* spp.	2x
Rice	*Oryza sativa*	2x	Wild rice	*O. rufipogon*	2x
Rice	*Oryza sativa*	2x	African rice	*O. glaberrima*	2x
Sugar beet	*Beta vulgaris*	2x	Sea beet	*Be. maritima*	2x
Sugar beet	*Beta vulgaris*	2x	Spinach beet	*Be. vulgaris* var. *cicla*	2x
Sunflower	*Helianthus annuus*	2x	Prairie sunflower	*He. petiolaris*	2x
Petunia	*Petunia hybrida*	4x	—	*Pet. axillaris*	2x
Oilseed rape	*Brassica napus*	4x	Oilseed rape	*Br. napus* cultivars	4x
Oilseed rape	*Brassica napus*	4x	Indian mustard	*Br. juncea*	4x
Oilseed rape	*Brassica napus*	4x	Radish	*Raphanus sativus*	2x
Sorghum	*Sorghum bicolorrace kafir*	4x	Sorghum	*So. bicolorrace milo*	4x
Tobacco	*Nicotiana tabacum*	4x	Sect. Alatae[a]	*N. plumbaginifolia*	2x
Tobacco	*Nicotiana tabacum*	4x	Sect. Bigeloviana[a]	*N. bigelovii*	4x
Tobacco	*Nicotiana tabacum*	4x	Sect. Paniculatae[a]	*N. glauca*	2x
Tobacco	*Nicotiana tabacum*	4x	Sect. Repandae[a]	*N. repanda*	4x
Tobacco	*Nicotiana tabacum*	4x	Sect. Suaveolentes[a]	*N. debneyi*	4x
Tobacco	*Nicotiana tabacum*	4x	Sect. Undulatae[a]	*N. undulata*	2x
Upland cotton	*Gossypium hirsutum*	4x	—	*G. harknessii*	2x
Bread wheat	*Triticum aestivum*	6x	Einkorn wheat	*T. monococcum*	2x
Bread wheat	*Triticum aestivum*	6x	Emmer wheat	*T. turgidum*	4x
Bread wheat	*Triticum aestivum*	6x	Timopheevi wheat	*T. timopheevi*	4x
Bread wheat	*Triticum aestivum*	6x	Sect. Amblyopyrum[b]	*Aegilops mutica*	2x
Bread wheat	*Triticum aestivum*	6x	Sect. Comopyrum[b]	*Ae. comosa*	2x
Bread wheat	*Triticum aestivum*	6x	Sect. Cylindropyrum[b]	*Ae. Caudata*	2x
Bread wheat	*Triticum aestivum*	6x	Sect. Cylindropyrum[b]	*Ae. Cylindrica*	4x
Bread wheat	*Triticum aestivum*	6x	Sect. Polyeides[b]	*Ae. Umbellulata*	2x
Bread wheat	*Triticum aestivum*	6x	Sect. Polyeides[b]	*Ae. Triuncialis*	4x
Bread wheat	*Triticum aestivum*	6x	Sect. Polyeides[b]	*Ae. Triaristata*	6x
Bread wheat	*Triticum aestivum*	6x	Sect. Sitopsis[b]	*Ae. Speltoides*	2x
Bread wheat	*Triticum aestivum*	6x	Sect. Vertebrata[b]	*Ae. Squarrosa*	2x
Bread wheat	*Triticum aestivum*	6x	Sect. Vertebrata[b]	*Ae. Ventricosa*	4x
Bread wheat	*Triticum aestivum*	6x	Sect. Vertebrata[b]	*Ae. Juvenalis*	6x
Bread wheat	*Triticum aestivum*	6x	Rye	*Secale cereale*	2x
Bread wheat	*Triticum aestivum*	6x	Wheatgrass	*Agropyron ciliare*	4x
Bread wheat	*Triticum aestivum*	6x	Wheatgrass	*Ag. Trachycaulum*	4x
Bread wheat	*Triticum aestivum*	6x	Wheatgrass	*Ag. Glaucum*	6x
Bread wheat	*Triticum aestivum*	6x	Wheatgrass	*Ag. Trichophorum*	6x

[a] Genus *Nicotiana*.
[b] Genus *Aegilops*.

cultivars to the male sterile. In tobacco, alien cytoplasms are transferred to cultivars from more than 14 wild species, belonging to the sections Alatae, Bigeloviana Paniculatae, Repandae, Suaveolentes, and Undulatae. Most of them cause male sterility in association with morphological anomalies of corolla, petals, and/or stamens. In common wheat, cytoplasms of seven species of the wheat genus *Triticum* and 21 species of the goatgrass genus *Aegilops* are transferred to 12 divergent genotypes of common wheat. Some of those cytoplasms are also introduced into durum wheat (*Triticum durum*). In addition, cytoplasms of rye (*Secale cereale*) and wheatgrasses (*Agropyron glauca* and *Ag. trichophorum*) are introduced into several common wheat cultivars. Most of them induce male sterility in one or other genotypes of common and emmer wheat.

Effects of Alien Cytoplasm on the Phenotype of the Crop

Relative Genetic Effects of the Nucleus, Cytoplasm, and Their Interactions

Investigations in wheat, using 552 alloplasmic lines of 12 common wheat genotypes combined with 46 alloplasms and 12 euplasmic lines as controls, have revealed that the genetic effects of all the nuclear genotypes, cytoplasms, and their interactions are significantly high for every character studied, compared to the effect of the error component. **Figure 3** shows the relative effects of the three genetic components on 12 representative characters. The genotypic effect on most vegetative characters is within the range of between 30% and 50% of the total genetic variation. The length of the top internode, heading date, and ear length show the least cytoplasmic effects, including the interaction. The proportion of genotypic effects on these traits is as high as 60–80% of the total genetic variation. These characters are highly homeostatic against genetic disharmony caused by alloplasms. In contrast, selfed seed fertility that is mainly determined by pollen fertility is extremely sensitive to alloplasms, the proportion of the genotypic effect only being 2.5%. This coincides with the fact that the most frequently encountered alloplasmic effect is male sterility.

Different genotypes have different sensitivities to the same cytoplasm (genotypic effect). Similarly, different cytoplasms exert different effects on the same character of a given genotype (cytoplasmic effect). In addition, certain combinations of the genotype and cytoplasm exhibit more serious effects than some other combinations on the same character (the genotype–cytoplasm interaction). Typical examples of these effects on the seed fertility of common wheat are shown in **Figure 4**.

Characters Affected by Alien Cytoplasms

Although all characters are affected by alloplasms, a few examples of characters observed in common wheat are briefly described below.

Growth vigor Dry weight is one of the indices of growth vigor. Of the 46 alloplasms studied, one reduced dry weight to about 10% of the normal, another to 30%, 13 to 50–75%, 29 to 75–95%, and the remaining three did not show any effect.

Growth period Thirteen alloplasms delayed heading for 1–3 days, seven for 4 days to 1 week, eight for 1–2 weeks, and two for 3–5 weeks. No alloplasms accelerated heading.

Pistillody of stamens Cytoplasms of *Ae. caudata*, *Ae. crassa*, *T. timopheevi*, and their relatives induced pistillody of stamens in certain genotypes of common wheat. The best studied is the pistillody expressed in an alloplasmic line of the bread wheat cultivar Norin 26, having the *Ae. crassa* cytoplasm. This particular alloplasmic line shows pistillody of all three stamens in each floret, when day length is longer than 15 h

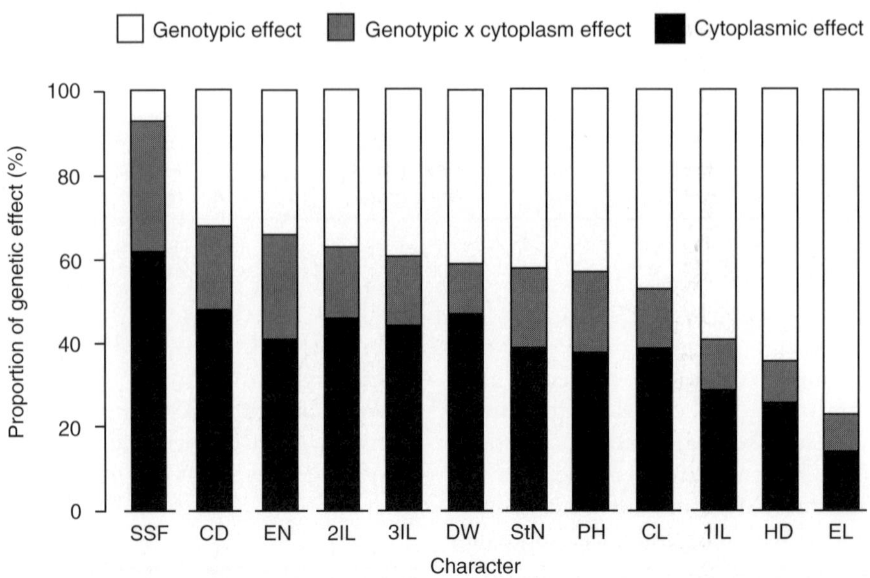

Figure 3 Relative proportion of the genetic effects of the nuclear genotype, cytoplasm, and their interaction on 12 characters of common wheat, in terms of the genetic variance. The amount of genetic variance attributable to each of the genotype, cytoplasm, and their interaction has been estimated by analysis of variance of the original data. The relative proportion of the genotypic effect is obtained by (variance for the genotype) × ((variance for the genotype) + (variance for the cytoplasm) + (variance for the interaction))$^{-1}$. Relative proportions for two other components are obtained in a similar way. Abbreviations: SSF, selfed seed fertility; CD, culm diameter; EN, ear number per plant; 1IL–3IL, length of the first to third internode from the top; DW, dry matter weight; StN, spikelet number per ear; PH, plant height; CL, culm length: HD, heading date, EL, ear length.

during the floral differentiation stage, resulting in complete male sterility. This type of male sterility is called photosensitive cytoplasmic male sterility.

Male sterility The main cause of selfed seed sterility is pollen sterility (**Figure 4**). There are no cytoplasms that cause male sterility in all genotypes. This is evident from the fact that the donor of the male-sterile cytoplasm has normal male fertility. Because of this, it is not appropriate to generalize that a certain cytoplasm is male-sterile or fertile.

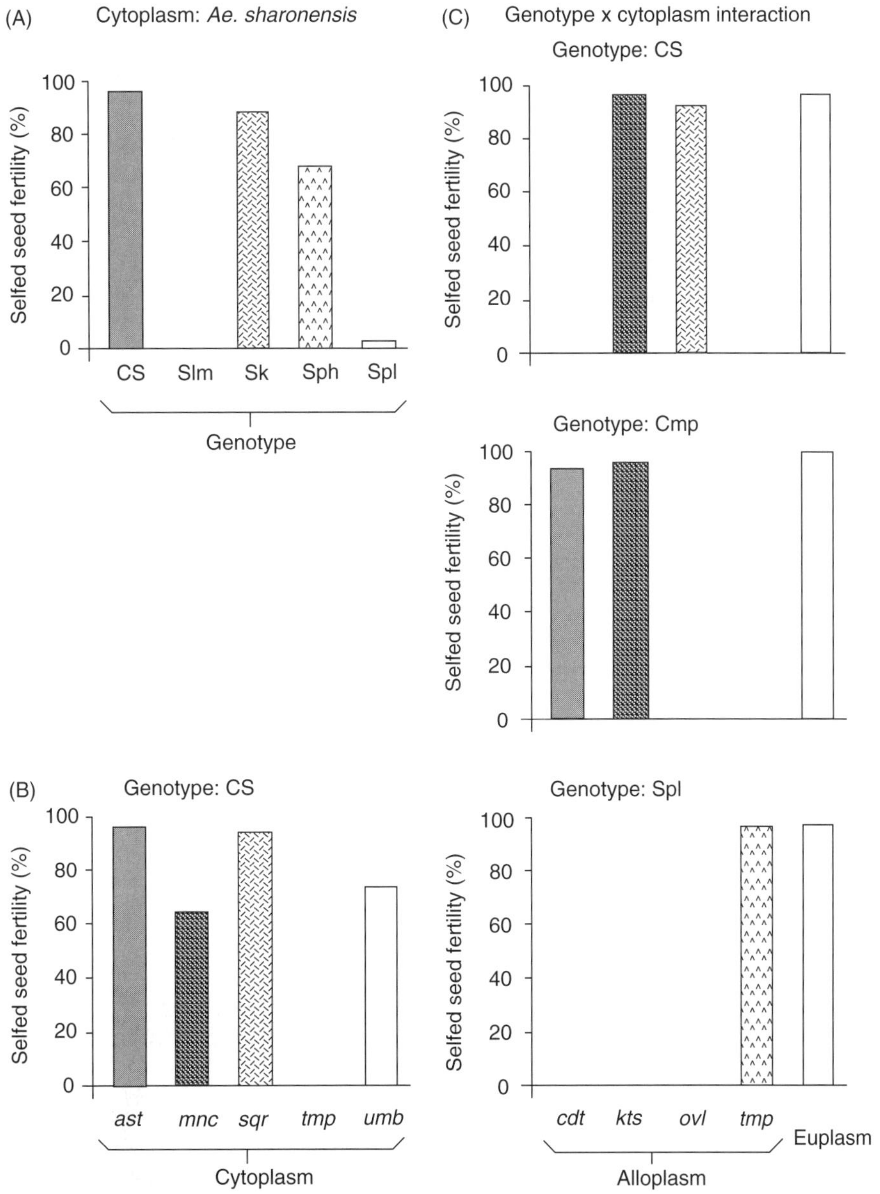

Figure 4 Examples showing differences in the effect of the genotypes (A), cytoplasms (B), and their interactions (C) on the selfed seed fertility of common wheat. Each line is indicated by its genotype and cytoplasm. Euplasm is own cytoplasm of the respective genotype. (A) *Ae. sharonensis* cytoplasm causes different levels of sterility in five common wheats, namely, complete sterility in Slm, almost complete sterility in Spl, and partial sterility in Sph, but almost no sterility in CS and Sk. (B) Common wheat cultivar, Chinese Spring, responds differently with four alien cytoplasms, namely, complete sterility with the *tmp* cytoplasm, partial sterility with the *mnc* and *umb* cytoplasms, and no sterility with the *sqr* cytoplasm. (C) Complete sterility is induced by specific genotype–cytoplasm combinations, namely, in CS with the *cdt* and *tmp* cytoplasms, in Cmp with the *ovt* and *tmp* cytoplasms, and in Spl with the *cdt, kts*, and *ovt* cytoplasms. Abbreviation of genotypes: CS, Slm, and Sk refer to *Triticum aestivum* cv. Chinese Spring, Salmon, and Selkirk, respectively; Cmp, *T. compactum* cv. No. 44; Sph, *T. sphaerococcum*; Spl, *T. spelta*. Abbreviation of cytoplasms: *ast, mnc*, and *tmp* refer to *Triticum aestivum*, *T. monococcum*, and *T. timopheevi* cytoplasms, respectively; *cdt, kts, ovt, sqr*, and *umb* refer to *Aegilops caudata*, *Ae. kotschyi*, *Ae. ovata*, *Ae. squarrosa*, and *Ae. umbellulata* cytoplasms, respectively.

No alloplasms seriously disturb meiosis of pollen mother cells. Male-sterile alloplasmic lines usually produce normal pollen tetrads. However, in these lines, cytoplasms affect the process of pollen maturation, including pollen mitoses. In some genotype–cytoplasm combinations, pollen development is arrested at the uninucleate stage before pollen mitoses, at the binucleate stage after the first pollen mitosis, or at the trinucleate stage after the second pollen mitosis. In the last case, pollen development is arrested before or after two male nuclei undergo morphological change from the spherical to wedge shape.

Although concrete evidence still is lacking, many reports suggest that one or other mitochondrial genes involved in the generation of ATP are responsible for pollen abortion (**Table 2**). The genes *atpA*, *apt6*, and *atp9* encode the corresponding subunits of ATP synthetase, *coxI* and *coxII* code for those of cytochrome oxidase, and the gene *cob* encodes apocytochrome *b* protein. The function of other possible genes (*orf* or *urf*) is unknown.

Parthenogenesis In the common wheat strain, Salmon, certain alloplasms induce parthenogenetic development of the unfertilized egg cell. When flowers that contain the parthenogenetic embryo are pollinated, seeds that give rise to haploid or twin seedlings are produced. The twins are derived from parthenogenesis of the egg cell and fertilization of either synergid in the same embryo sac.

Premature sprouting of the seed Cytoplasms of *T. timopheevi*, including its relatives, and *Ae. ovata* induce premature sprouting of seeds in certain common wheat genotypes. This is associated with precocious production of some isozyme forms of α-amylase in developing seeds, which normally appear after seed germination. Premature sprouting causes serious reduction in seed germination rate.

Enlarging the Cytoplasmic Variability of the Crop

One of the most striking disasters in modern crop breeding was the destructive epidemic of hybrid corn with T-type cytoplasm caused by a new race, T, of Southern corn blight (*Bipolaris maydis*). The use of alloplasms with no harmful effects is one strategy that will help to protect crops from such cytoplasm-specific diseases.

The cytoplasms of some *Aegilops* species do not impair both growth and fertility of common wheat. These cytoplasms have been employed experimentally for widening the cytoplasmic variability of wheat. The alloplasmic lines of a wheat cultivar, Norin 26, possessing the cytoplasms of *Ae. squarrosa*, *Ae. kotschyi*, and *Ae. speltoides*, have been crossed with Norin 61, the outstanding cultivar in Japan. With selection through the F_2 to F_4

Table 2 Mitochondrial genes and their products which are reported to be responsible for male sterility caused by respective cytoplasms

Crop species		*Male-sterile cytoplasm*		*Mitochondrial gene*	*Responsible product*
Common name	*Scientific name*	*Species*	*Type*		
Sugar beet	*Beta vulgaris*	*Be. vulgaris*	S type (Owen)	*atpA*	Abnormal transcript
Oilseed rape	*Brassica napus*	*Br. napus*	*pol* type	*atp6*	Abnormal transcript
Cotton	*Gossypium hirsutum*	*G. harknessii*	D2-2	?	?
Sunflower	*Helianthus annuus*	*He. petiolaris*	?	*atpA*	Abnormal transcript
Sunflower	*Helianthus annuus*	*He. petiolaris*	?	*orfH522*	16 kDa Protein
Perennial ryegrass	*Lolium perenne*	*L. perenne*	S type	*coxI*	Abnormal transcript
Tobacco	*Nicotiana tabacum*	*N. bigelovii*	?	*atpA*	Abnormal transcript
Rice	*Oryza sativa* subsp. *indica*	*O. rufipogon*	?	?	?
Rice	*Oryza sativa* subsp. *japonica*	*O. sativa* subsp. *indica*	Boro II	*orf79*	?
Pearl millet	*Pennisetum glaucum*	*P. glaucum* subsp. *monodii*	S-A1 type	*coxI/rrn18-rrn5*	?
Petunia	*Petunia hybrida*	*Pet. axillaris*	?	*pcf*[a]	25 kDa Protein
Rye	*Secale cereale*	*Se. cereale*	Pampa type	*cob*	Abnormal transcript
Sorghum	*Sorghum bicolorrace kafir*	*So. bicolorrace milo*	A3 type	*orf107*[b]	11.85 kDa Protein
Bread wheat	*Triticum aestivum*	*Aegilops kotschyi*	S^v type	?	?
Bread wheat	*Triticum aestivum*	*T. timopheevi*	G type	*atp6, orf25*	Abnormal transcript

?, Unknown
[a] A fused gene of *atp9*, *coxII*, and *urf3*.
[b] Includes a part of *atp9*.

generations, alloplasmic lines that are equally productive to the normal Norin 61 line have been obtained with all alloplasms, proving that widening of cytoplasmic variability in common wheat using certain *Aegilops* cytoplasms is an easy task.

Cytoplasmic Male Sterility

Source of Male-Sterile Cytoplasm

The alien cytoplasms that have practical value are limited so far to those that induce male sterility, because they are useful in breeding hybrid cultivars. Representative male-sterile alloplasms for different crops are given in **Table 2**. In the case of outcrossing crops, a male-sterile cytoplasm is often found among individuals of the same cultivar or among different cultivars of the same crop species. On the other hand, in self-pollinating crops, they are found only in different subspecies, species, or genera.

Fertility Restoration

In breeding hybrid cultivars of vegetable or ornamental crops, restoration of male fertility is not important. However, it is essential for hybrid cultivars of seed crops. Furthermore, the gene (or genes) for fertility restoration, called a fertility-restoring gene (*Rf*), should be dominant in order to

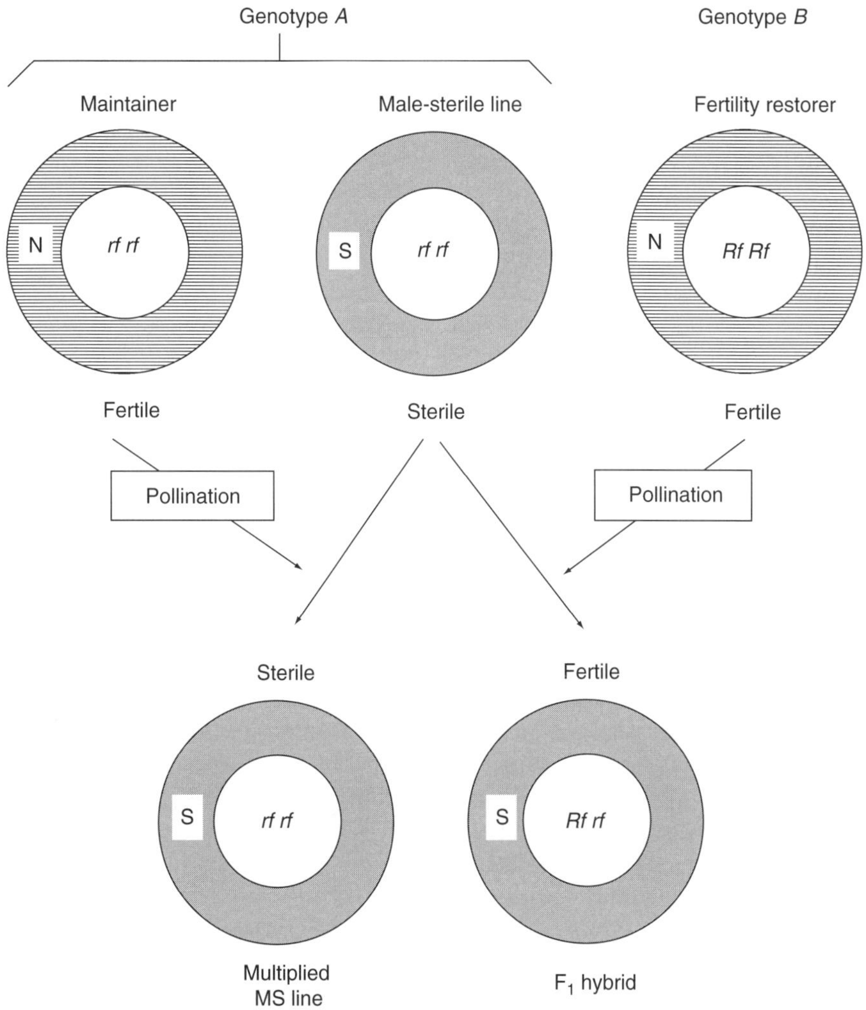

Figure 5 Method of breeding an F_1 hybrid cultivar of a seed crop, using cytoplasmic male sterility. The inner circle indicates the genotype for fertility restoration and the outer zone indicates the cytoplasm. *Rf* is a dominant allele for restoring male fertility and *rf* is its recessive allele inducing male sterility, in the presence of a male-sterile cytoplasm, S. N refers to normal cytoplasm that gives male-fertile plants in the presence of either the *Rf* or *rf* allele in the homozygous state. Three lines, namely, male-sterile and maintainer lines of a genotype *A*, and a fertility-restoring line of another genotype *B* are necessary to produce the F_1 hybrid. The first two lines carry the *rf* allele, whereas the third line should carry the *Rf* allele. Genotypes *A* and *B* are different true-bred cultivars or lines that produce a commercially acceptable F_1 hybrid upon hybridization. Multiplication of the male-sterile line is carried out by open pollination of this line with the pollen of the maintainer line. Open pollination of the male-sterile line of the genotype *A* with the pollen of the fertility-restorer line of the genotype *B* produces seeds, from which the F_1 hybrids between the two genotypes with restored pollen fertility are grown.

restore fertility in the hybrid. A reliable source of the *Rf* gene to a given male-sterile alloplasm is the donor of this cytoplasm or its close relatives. *Helianthus petiolaris* and *T. timopheevi* carry *Rf* genes for male sterility of sunflower and common wheat, respectively, that is induced by their cytoplasms. When the *Rf* gene of the cytoplasm donor is located on a chromosome nonhomologous with any chromosome of the crop, transfer of this gene into the genome of the crop is difficult, although addition of the entire chromosome or its arm carrying the *Rf* gene allows fertility restoration. Incorporation of such *Rf* genes into the crop genome requires special chromosome engineering techniques. Large-scale screening of the cultivars of crop species is sometimes useful for obtaining *Rf* genes, and such screening has been successful in both rice and common wheat.

Hybrid Breeding

Three types of parental lines are needed for breeding a hybrid cultivar of a seed crop (**Figure 5**): the cytoplasmically male-sterile (CMS) line, maintainer line, and fertility-restorer line. The first two lines should be of the same genotype, should not have any *Rf* genes, and are used for multiplication of the CMS line. Hybrid seeds are produced between the CMS line and a fertility restorer line of different genotypes by alternating their rows or blocks of rows in the nursery. Seeds set on the CMS line are hybrid between the two parental lines. They have the male-sterile cytoplasm transmitted from the female parent but give rise to pollen-fertile plants because they receive the dominant *Rf* gene from the pollen parent.

The method used for breeding a hybrid cultivar of a vegetable or ornamental crop is essentially the same as that for a seed crop, the only difference being that the pollinator for the CMS line does not necessarily need to have any *Rf* genes, being either *Rf* or *rf* homozygote. The seeds harvested from the CMS line are hybrids between the CMS and pollinator lines.

Breeding hybrid cultivars by the use of male-sterile alloplasms has been successful in bread wheat, cotton (*Gossypium* spp.), *indica* rice, pearl millet (*Pennisetum glaucum*), sorghum (*Sorghum* spp.), sunflower, tobacco, and some other crops.

List of Technical Nomenclature

Alloplasm	Alien cytoplasm.
Asymmetrical somatic fusion	Fusion of two kinds of protoplast, one of which is heavily treated with irradiation.
ATP	Adenosine triphosphate.
Backcross	Crossing of an F_1 hybrid to one of its parents.
Bread wheat	*Triticum aestivum*, a species of common wheat.
CMS	Cytoplasmically determined male sterility.
Common wheat	Hexaploid wheat, having AABBDD genomes.
Cultivar	A cultivated variety.
Cytoplasm	Used in a genetic sense, meaning all genetic information in the cytoplasm; synonymous to cytoplasmic genome or plasmon.
Euplasmon	Own cytoplasm.
Fertility restoration	Restoration of pollen fertility by a nuclear gene(s) in a cytoplasmically male-sterile line.
Genome	The nuclear genome.
Genotype	Synonymously used for a cultivar, line, or strain.
Nucleus	Used in a genetic sense, meaning all genetic information present in the nucleus.
Nucleus substitution	Replacement of the nucleus of a species (or genotype) with that of another species.
Organelle	Used in a genetic sense, i.e., organelle containing some genetic information.
Pistillody	A phenomenon in which other organs such as stamens are transformed into pistils, functional or nonfunctional.
***Rf* gene**	A fertility-restoring, nuclear gene.
Variety	The taxonomic variety.

See also: **Crop Improvement**: Chromosome Engineering; Hybridization and Plant Breeding; Plant Breeding, Practice. **Flowering and Reproduction**: Fertilization. **Seed Development**: Seed Quality. **Seed Dormancy**: Preharvest Sprouting. **Tissue Culture and Plant Breeding**: Somatic Hybridization.

Further Reading

Bajaj YPS (1990) *Biotechnology in Agriculture and Forestry 13. Wheat*. Berlin: Springer-Verlag.

Brennicke A and Kuck U (1993) *Plant Mitochondria: With Emphasis on RNA Editing and Cytoplasmic Male Sterility*. New York: Winheim.

Coors JG (1999) *The Genetics and Exploitation of Heterosis in Crops*. Madison: Crop Science Society of America.

Edwardson JR (1970) Cytoplasmic male sterility. *Botanical Review* 36: 341–420.
Frankel R (1983) *Heterosis: Monograph, Theoretical and Applied Genetics* 6. Berlin: Springer-Verlag.
Grun P (1976) *Cytoplasmic Genetics and Evolution*. New York: Columbia University Press.
Hancock JF (1992) *Plant Evolution and the Origin of Crop Species*. Englewood Cliffs, NJ: Prentice Hall.
He S, Abad AR, Gelvin SB, and Mackenzie SA (1996) A cytoplasmic male sterility-associated mitochondrial protein causes pollen disruption in transgenic tobacco. *Proceedings of the National Academy of Sciences, USA* 93: 11763–11768.
Michaelis P (1954) Cytoplasmic inheritance in *Epilobium* and its theoretical significance. *Advances in Genetics* 6: 288–401.
Reid RA and Leech RM (1980) *Biochemistry and Structure of Cell Organelles*. London: Blackie.
Tsunewaki K (1996) Plasmon analysis as the counterpart of genome analysis. In: Jauhar PP (ed.) *Methods of Genome Analysis in Plants*, pp. 271–299. New York: CRC Press.

Chromosome Engineering

P P Jauhar, United States Department of Agriculture – Agricultural Research Service[1], Fargo, ND, USA

Introduction

Conventional plant breeding, coupled with intuitive selection procedures, practiced over the centuries transformed the relatively useless weedy species into the useful crop plants that sustain us today. Sustained genetic improvement of crops has mainly been achieved through hybridization with land races and related species, resulting in high-yielding superior cultivars of food, fiber, oilseed, and other crops. Wild relatives of crop species are rich reservoirs of genes that have been and are being used for genetic enrichment of various crops, especially cereals. Thus, adoption of the tools of cytogenetics since the 1930s has facilitated wide hybridization of wheat (*Triticum* spp.) with its annual and perennial relatives, and has resulted in wheat cultivars with alien genes. The pioneering work of Sears (1956), in which he transferred a chromosome segment carrying a leaf rust resistance gene from a wild grass, *Aegilops umbellulata*, into common wheat (*Triticum turgidum*), heralded an era of chromosome engineering, facilitating the utilization of wild resources for wheat improvement. Because of limitations on space, this article will cover only the main facets of chromosome engineering as applied to genetic improvement of cereal crops, particularly wheat. This is not only because of the vital importance of wheat for humankind, but also because a great deal of work on chromosome manipulation has been done on this cereal. Much of this work and the technologies applied are of relevance and can be applied to other crops.

[1]Mention of a trademark or proprietary product does not constitute a guarantee or warranty of the product by the USDA or imply approval to the exclusion of other products that also may be suitable.

Cytogenetic Architecture of Wheat: The Concept of Chromosome Engineering

Because a preponderance of the following discussion will center on wheat, it is important to understand its cytogenetic makeup. Both bread wheat (*T. aestivum*) and durum wheat (*T. turgidum*) are true breeding natural hybrids with the genomic constitution of AABBDD and AABB, respectively. Durum wheat is the forerunner of bread wheat. Its two genomes, A and B, were derived from *T. urartu* and *Aegilops speltoides*, respectively. The two diploid progenitors, which are native to the Middle East, hybridized in nature, with *Ae. speltoides* as female parent, some half a million years ago and produced tetraploid wild emmer wheat (*T. dicoccoides*) in one step as a result of functioning of unreduced gametes in both parents. Then wild emmer through domestication gave rise to cultivated durum wheat.

Because homoeologous (genetically related) chromosomes such as 1A (i.e., chromosome 1 of the A genome) and 1B (i.e., chromosome 1 of the B genome) are closely related and hence capable of pairing with one another, some sort of control of pairing would be necessary. Therefore, at the time of origin of tetraploid emmer, a spontaneous mutation gave rise to the homoeologous pairing suppressor gene, *Ph1*. The presence of a *Ph1*-like gene in one of the diploid parents cannot be ruled out, although such a gene would have no regulatory function in a diploid species. *Ph1*, located in the long arm of chromosome 5B, permitted pairing only among homologous chromosomes ensuring diploid-like pairing and disomic inheritance, helping to maintain precisely the meiotic integrity of the A and B genomes. But for the rigid regulation of chromosome pairing, these two genomes might have converged during the period of half a million years they have been together. Clearly, *Ph1* conferred meiotic regularity and reproductive stability to tetraploid wheat, and subsequently to hexaploid bread wheat (AABBDD), that evolved about 8000 years ago as a result of natural hybridization of tetraploid emmer with a third diploid progenitor, *Aegilops tauschii*, that contributed the D genome.

The origin of a rigid regulatory mechanism in the form of *Ph1* was essential for reproductive stability and survival of polyploid wheats. Because this regulatory mechanism restricts pairing to precisely homologous partners it does not permit "adultery" among related (homoeologous) chromosomes. However, in as much as plant breeding depends mainly on "adultery" among related chromosomes, *Ph1* offered an obstacle to the incorporation of alien genes into wheat. Elegant means of incorporating such variation into wheat were discovered and applied by Sears and other wheat researchers. The cytogenetic manipulations, including those based on suppression of the *Ph1* regulation, for engineering desirable alien chromatin into wheat were termed "chromosome engineering." Although other methods of transferring alien chromatin into the wheat genome are known, e.g., through irradiation, the occurrence or promotion of homoeologous pairing offers a better means of "chromosome surgery," and a precise and more desirable means of alien gene transfer.

Genetic Control of Chromosome Pairing in Other Polyploid Crops

Although allopolyploidy has provided the basis for the evolution of some of our most important grain, forage, and fiber crops, a genetic control of chromosome pairing is essential for the reproductive stability of these polyploid crop species. As in cultivated wheats, the diploid-like chromosome pairing in hexaploid oats, *Avena sativa* (AABBCCDD genomes), is under genetic control. Both these regulatory mechanisms are essentially similar and can be suppressed by appropriate genotypes of their wild relatives, leading to homoeologous pairing and thereby gene transfers.

Tall fescue (*Festuca arundinacea*), an important forage crop, is also an allohexaploid like bread wheat and oats. However, the genetic control of chromosome pairing in tall fescue is hemizygous ineffective or haplo-insufficient, i.e., one dose of the pairing control genes is not sufficient to suppress homoeologous pairing. Such a genetic control in tall fescue (and other polyploid fescues) permits in their hybrids with species of the companion genus *Lolium*, for example, extensive pairing, facilitating gene flow and hence reshuffling of characters across species. The ease with which the genetic control breaks down could prove an asset to the breeder in conducting chromosome engineering in the *Festuca-Lolium* group of grasses.

Genomic Relationships and Chromosome Pairing

Genome analysis helps in deciphering the cytogenetic architecture of polyploid species and in elucidating phylogenetic relationship among species and taxa. The rationale of genomic analysis is that the chromosomes that pair are closely related, and those that do not pair are not related. Conversely, the knowledge of genome relationships provides useful information on chromosome pairing relationship, and would help in planning strategies on transferring desirable traits from one species to another. Breeders have long recognized the potential of combining desirable attributes of related species into new cultivars. The successful assembly of the units of recombination, which may be individual genes, chromosome segments, whole chromosomes, or genomes, depends largely on the chromosome pairing relationships of the species involved. Thus, chromosome engineering depends largely on chromosome pairing.

Genomic Reconstruction and Crop Improvement

The occurrence or induction (by methods described below) of pairing between chromosomes of a crop species and an alien species in their hybrids leads to transfer of alien genes or gene clusters into the crop genome. This type of chromosome engineering leads to genomic reconstruction resulting from fruitful recombination of parts of genomes. Thus, plant breeding essentially involves genomic rebuilding or genomic reconstitution. The *Ph1*-regulated pairing can be manipulated to reconstruct chromosomes and genomes to bring in new desirable traits or remedy existing defects.

Hybridization and Synthesis of Hybrids: A Prerequisite for Genomic Reconstruction

The synthesis of hybrids between a crop species and an alien species is the single most important step for alien gene transfer into a crop plant. This step must therefore be planned very carefully using appropriate, crossable genotypes. A crop species is generally used as the female parent. The cytogenetic manipulation to be adopted will depend on the longevity of the synthetic hybrid, which in turn depends upon whether an annual or perennial donor is involved. A wheat cultivar can, for example, be crossed with its annual or perennial relative depending on the alien trait to be transferred.

Annual vs. Perennial Nature of Hybrids: Their Cytogenetic Manipulation

F_1 hybrids of wheat with another annual species, such as barley (*Hordeum vulgare*) or rye (*Secale cereale*), are annual and hence short lived. Such hybrids generally have some female fertility. Therefore, they (as a female parent) should be backcrossed immediately to the wheat parent to produce hybrid derivatives. However, if an F_1 hybrid is totally sterile, it may be treated with colchicine to make it an amphidiploid to restore fertility. Chromosome elimination and low seed set in such wide hybrids is a major problem, particularly in the wheat–barley hybrids, for example. Production of wheat–barley hybrids using highly crossable barley genotypes may help overcome the problem of barley chromosome elimination and low seed set at the first backcrossing.

On the other hand, hybridization between wheat and perennial grasses mostly yields perennial F_1 hybrids. These hybrids generally live for several years and are therefore amenable to cytogenetic manipulation over an extended period of time. The F_1 hybrids are male-sterile but generally have some female fertility, which permits their backcrossing to the wheat parent. Wheat–perennial grass hybridizations coupled with manipulation of chromosome pairing have resulted in introgression of desirable alien genes into bread wheat.

Some workers in the field find it easier to use amphidiploids for transferring desired alien genes into wheat. In addition to their use in alien gene transfers, amphidiploids help produce alien addition and substitution lines of wheat.

Major Considerations in a Wheat Hybridization Program

A hybridization program should be planned carefully. For greater success, the following points must be considered.

A gene pool as source of a desirable trait The ease with which a desirable gene can be transferred from a related species into a crop plant depends largely on the evolutionary distance between the species involved. Therefore, a desired trait should first be sought in such donors as a closely related species or infraspecific categories or landraces, which form a "primary gene pool" with the crop plant in question. Species or infraspecific categories belonging to the primary gene pool of bread and durum wheats, for example, share one or more closely related (homologous) genomes. This donor group includes landraces of common wheat, wild emmer wheat (*T. turgidum* var. *dicoccoides*), and donors of the A, B, and D genomes. These donors cross relatively easily and form at least partially fertile hybrids with wheat. The hybrids show high chromosome pairing and recombinations, facilitating gene transfer into wheat. Several genes conferring resistance to diseases and pests have been successfully transferred from these donors into wheat (see **Table 1**).

The species in the "secondary gene pool" of common wheat include the tetraploid species, *T. timopheevii*, polyploid species of *Aegilops*, and diploid S-genome species of the section Sitopsis. These species can be crossed with wheat but with some difficulty, and the hybrids have high sterility and poor chromosome pairing. The transfer of desirable genes from these species into wheat cultivars is nevertheless possible.

The "tertiary gene pool" species are more distantly related with a higher degree of reproductive isolation with bread wheat and durum wheat. Hybridization with the wheats is possible, but with greater difficulty, using specialized techniques. Moreover, the resultant hybrids are highly sterile with minimal pairing, if any, between parental chromosomes. Radiation treatments may help induce chromosomal interchanges and transfer of traits across species.

Based on the above considerations, the main sources for the genetic improvement of bread wheat and durum wheat will, in order of preference, be: (1) intraspecific diversity of *T. aestivum* (for bread wheat) and that of *T. turgidum* (for durum wheat); (2) other *Triticum* spp., e.g., the donor of the A genome; (3) other relatives of wheat, i.e., various genome donors (e.g., *Aegilops* spp.); (4) other genera of the Triticeae; and (5) more distantly related grasses. It is always preferable, where possible, to use donor material in the primary gene pool.

Crossability of the recipient crop cultivar An important factor in the success of a hybridization program is the crossability of the crop cultivar used. Considerable genetic variation exists, for example, in crossability of wheat cultivars with rye and with wild relatives. A number of crossability genes, designated *kr*, have been identified. Crossability of wheat and rye, for example, is controlled by two recessive genes, *kr1* and *kr2*, located on chromosomes 5B and 5A, respectively. Using telocentrics, the *Kr1* was mapped to 5BL (the long arm of chromosome 5B), and *Kr2* was located on 5AL. Chinese Spring is a crossable cultivar because it has *kr1* and *kr2*, and it has been extensively used in wide hybridization programs. Dominant alleles of the crossability genes, i.e., *Kr1* and *Kr2*, not only reduce crossability of wheat with rye, but also with other species of the Triticeae. A third crossability allele *kr3*, located on

Table 1 Some recent examples of chromosome engineering and transfer of desirable alien genes into wheat

Trait	*Donor species*	*Analysis*	*Reference*
Bread wheat (2*n* = 6*x* = 42; AABBDD)			
Leaf rust resistance	*Aegilops ventricosa*	RFLP	Bonhomme *et al.* (1995)
Leaf rust resistance genes, *Lr9, Lr19, Lr24, Lr32*	*Aegilops sp.* and *Lophopyrum sp.*	RFLP	Autrique *et al.* (1995)
Leaf and stem rust	*Triticum cylindricum*	RFLP	Bai *et al.* (1995)
BYDV (Barley yellow dwarf virus) resistance	*Thinopyrum intermedium*	RFLP, isozymes	Banks *et al.* (1995)
			Larkin *et al.* (1995)
	Th. intermedium	C-banding, GISH, RFLP	Hohmann *et al.* (1996)
	Th. intermedium, *Th. ponticum*	Tissue block immunoassay, GISH	Fedak *et al.* (2001)
	Lophopyrum ponticum (= *Th. ponticum*)	GISH	Chen *et al.* (1998)
Powdery mildew resistance	*Ae. longissima*	RFLP, meiotic pairing	Donini *et al.* (1995)
	Ae. speltoides	Meiosis, isozymes, RAPD	Eser (1998)
	Triticum dicoccoides (= *T. turgidum* var. *dicoccoides*)	Genetic analysis, RAPD, and screening	Rong *et al.* (2000), Liu *et al.* (2002)
	Povos wheat–*Ae. ovata* addition VI	Monosomic analysis, AFLP, and RFLP	Zeller *et al.* (2002)
Stripe rust resistance	Synthetic wheat (durum x *Ae. tauschii*)	Testing with pathotypes	McIntosh and Lagudah (2000)
Cephalosporium stripe resistance	*A. elongatum*	C-banding, GISH	Cai *et al.* (1996)
Eyespot resistance	*Haynaldia villosa*	RFLP	Yildirim *et al.* (1998)
Wheat curl mite resistance	*L. ponticum*	GISH	Thomas *et al.* (1998)
Cold tolerance	*Agropyron. cristatum*	Northern blot	Limin *et al.* (1995)
Durum wheat (2*n* = 4*x* = 28; AABB)			
Pm13, Lr19, storage proteins	*A. elongatum* (= *L. elongatum*)	GISH, SDS-PAGE	Ceoloni *et al.* (1996)
Grain protein content	*T. turgidum* var. *dicoccoides*	GPC, RFLP	Joppa *et al.* (1997)
Stripe (yellow) rust resistance	*T. turgidum* var. *dicoccoides*	RAPD	Sun *et al.* (1997)

Scab resistance	*L. elongatum* and *Th. junceiforme*	GISH	Jauhar and Peterson (2000, 2001)
Kna1 (K^+/Na^+ selectivity gene)	*T. aestivum*	RFLP	Luo *et al.* (1996)
Hessian fly resistance genes, *H21* and *H25*	*T. aestivum* (originally from rye)	C-banding, GISH	Friebe *et al.* (1999)

Autrique E, Singh RP, Tanksley SD, and Sorrells ME (1995) Molecular markers for four leaf rust resistance genes introgressed into wheat from wild relatives. *Genome* 38: 75–83.

Bai D, Scoles GJ, and Knott DR (1995) Rust resistance in *Triticum cylindricum* Ces. (4x, CCDD) and its transfer into durum and hexaploid wheats. *Genome* 38: 8–16.

Banks PM, Larkin PJ, Bariana HS, *et al.* (1995) The use of cell culture for subchromosomal introgressions of barley yellow dwarf virus resistance from *Thinopyrum intermedium* to wheat. *Genome* 38: 395–405.

Bonhomme A, Gale MD, Koebner RMD, *et al.* (1995) RFLP analysis of an *Aegilops ventricosa* chromosome that carries a gene confering resistance to leaf rust (*Puccinia recondita*) when transferred to hexaploid wheat. *Theoretical and Applied Genetics* 90: 1042–1048.

Cai X, Jones SS, and Murray TD (1996) Characterization of an *Agropyron elongatum* chromosome conferring resistance to cephalosporium stripe in common wheat. *Genome* 39: 56–62.

Ceoloni C, Biagetti M, Ciafi M, Forte P, and Pasquini M (1996) Wheat chromosome engineering at the 4x level: The potential of different alien gene transfers into durum wheat. *Euphytica* 89: 87–97.

Chen Q, Ahmad F, Collin J, *et al.* (1998) Genomic constitution of a partial amphiploid OK7211542 used as a source of immunity to barley yellow dwarf virus for bread wheat. *Plant Breed* 117: 1–6.

Donini P, Koebner RMD, and Ceoloni C (1995) Cytogenetic and molecular mapping of the wheat–*Aegilops longissima* chromatin breakpoints in powdery mildew-resistant introgression lines. *Theoretical and Applied Genetics* 91: 738–743.

Eser V (1998) Characterisation of powdery mildew resistant lines derived from crosses between *Triticum aestivum* and *Aegilops speltoides* and *Ae. mutica*. *Euphytica* 100: 269–272.

Fedak G, Chen Q, Conner RL, *et al.* (2001) Characterization of wheat–*Thinopyrum* partial amphiploids for resistance to barley yellow dwarf virus. *Euphytica* 120: 373–378.

Friebe B, Kynast RG, Hatchett JH, *et al.* (1999) Transfer of wheat-rye translocation chromosomes conferring resistance to Hessian fly from bread wheat into durum wheat. *Crop Science* 39: 1692–1696.

Hohmann U, Badaeva ED, Busch W, Friebe B, and Gill BS (1996) Molecular cytogenetic analysis of *Agropyron* chromatin specifying resistance to barley yellow dwarf virus in wheat. *Genome* 39: 336–347.

Jauhar PP and Peterson TS (2000) *Progress in Producing Scab-Resistant Germplasm of Durum Wheat*. Proceedings of the International Symposium on Wheat Improvement for Scab Resistance, Nanjing Agricultural University, Nanjing, China, pp. 77–81.

Jauhar PP and Peterson TS (2001) Hybrids between durum wheat and *Thinopyrum junceiforme*: Prospects for breeding for scab resistance. *Euphytica* 118: 127–136.

Joppa LR, Du C, Hart GE, and Hareland GA (1997) Mapping a gene(s) for grain protein in tetraploid wheat (*Triticum turgidum* L.) using a population of recombinant inbred chromosome lines. *Crop Science* 37:1586–1589.

Larkin PJ, Banks PM, Lagudah ES, *et al.* (1995) Disomic *Thinopyrum intermedium* addition lines in wheat with barley yellow dwarf virus resistance and rust resistances. *Genome* 38: 385–394.

Limin AE, Houde M, Chauvin LP, Fowler DB, and Sarhan F (1995) Expression of the cold-induced wheat gene *Wcs120* and its homologs in related species and interspecific combinations. *Genome* 38: 1023–1031.

Liu Z, Sun Q, Ni Z, Nevo E, and Yang T (2002) Molecular characterization of a novel powdery mildew resistance gene *Pm30* in wheat originating from wild emmer. *Euphytica* 123: 21–29.

Luo M-C, Dubcovsky J, Goyal S, and Dvořák J (1996) Engineering of interstitial foreign chromosome segments containing the K^+/Na^+ selectivity gene *Kna1* by sequential homeologous recombination in durum wheat. *Theoretical and Applied Genetics* 93:1180–1184.

McIntosh RA and Lagudah ES (2000) Cytogenetical studies in wheat. XVIII. Gene *Yr24* for resistance to stripe rust. *Plant Breed* 119: 81–83.

Rong JK, Millet E, Manisterski J, and Feldman M (2000) A new powdery mildew resistance gene: Introgression from wild emmer into common wheat and RFLP-based mapping. *Euphytica* 115: 121–126.

Sun GL, Fahima T, Korol AB, *et al.* (1997) Identification of molecular markers linked to the *Yr15* stripe rust resistance gene of wheat originated in wild emmer wheat, *Triticum dicoccoides*. *Theoretical and Applied Genetics* 95: 622–628.

Thomas J, Chen Q, and Talbert L (1998) Genetic segregation and the detection of spontaneous wheat–alien translocations. *Euphytica* 100: 261–267.

Yildirim A, Jones SS, and Murray TD (1998) Mapping a gene conferring resistance to *Pseudocercosporella herpotrichoides* on chromosome 4V of *Dasypyrum villosum* in a wheat background. *Genome* 41:1–6.

Zeller FJ, Kong L, Hartl L, Mohler V, and Hsam SLK (2002) Chromosomal location of genes for resistance to powdery mildew in common wheat (*Triticum aestivum* L. em Thell.) 7. Gene *Pm29* in line Pova. *Euphytica* 123: 187–194.

chromosome 5D, and a fourth allele *kr4*, located on 1A, of wheat also influence its crossability with rye. There is a strong correlation between crossability with rye and with other members of the Triticeae.

Crossability of wheat is a quantitative trait and a higher number of *kr* alleles seems to confer higher crossability on wheat. Most European wheat varieties carry the dominant *Kr* alleles and hence have low crossability with rye. On the other hand, wheat landraces from China, Japan, and Iran are rich in *kr* alleles. From a cross between two wheat cultivars, it was reported that a major QTL involved in crossability was located on 5BS, i.e., the short arm of chromosome 5B.

A wheat cultivar's crossability is, of course, important. However, a crossable cultivar with desirable agronomic traits would be preferable.

Genomic relationships As stated above, genome analysis provides useful information on chromosome pairing relationships and, hence, on prospects of transferring desirable traits across various species. Material in the primary gene pool is genomically closer to wheat, and therefore more suitable for use in gene introgressions.

A remarkable recent development is the availability of comparative molecular maps for closely and even remotely related species. Comparative genome mapping provides an understanding of the spread of genetic information among divergent species. The synteny and colinearity among cereal genomes, for example, facilitates isolation of agronomically superior genes from complex cereal genomes. Considerable structural similarities among the wheat, rice, and corn (*Zea mays*; maize) genomes can be recognized in the form of clusters of genes or syntenic units.

Crossability of the donor parent Genetic variation for crossability has been observed in several species, which are potential donors of desirable traits to wheat. Rye has a single dominant gene for crossability with wheat, whereas wild barley (*Hordeum spontaneum*) carries a crossable factor(s) presumably similar to *Kr* alleles in wheat. Genotypic differences in crossability also exist in several other grasses, e.g., *Thinopyrum* spp., *Elymus scabrous*, and *Psathyrostachys juncea*.

Crossability of wheat with rye can be suppressed by chromosome 4E of *Lophopyrum elongatum*. Thus, all the disomic substitutions, i.e., 4E (4A), 4E (4B), and 4E (4D), and disomic addition involving the addition of 4E, showed significantly lower crossability than control Chinese Spring. Using 14 D-genome chromosome substitution lines of Langdon (which had an A- or a B-genome chromosome replaced with a homoeologous D-genome chromosome), it was observed that chromosome 5D was responsible for the crossability of durum wheat with *Aegilops squarrosa*.

Crossability with distantly related genera Identification of genotypes with enhanced crossability in wheat as well as in barley has enabled the production of wheat × barley hybrids. Wheat has not only been crossed with barley, but also with corn, sorghum, and pearl millet (*Pennisetum glaucum*). This kind of wide hybridization often leads to production of haploids of both bread wheat and durum wheat via chromosome elimination. In addition to their use in the production of haploids, wide crosses could help produce "partial" hybrids of wheat because of incomplete elimination of chromosomes of the pollinating parent. From oat × corn crosses, oat plants with 21 oat chromosomes plus 1–4 corn chromosomes have been recovered. This provides a mechanism for the production of oat–maize chromosome addition lines. Such addition lines help introgress desirable genes by transferring chromosome segments from corn into oat. Moreover, alien chromosome additions provide invaluable material for isolation of corn chromosome-specific sequences for use in gene mapping and gene cloning.

It is known that chromosome elimination in wide hybrids of the type described above is under genetic control. Genetic variation for chromosome elimination could, therefore, be exploited to retain some of the alien chromosomes in the wheat genome. The use of highly crossable barley genotypes will not only help to produce wheat–barley hybrids efficiently, but also reduce the problems of barley chromosome elimination and low seed set at the first backcrossing. Crosses between wheat and some *H. bulbosum* genotypes are also known to result in partial or complete wheat–*H. bulbosum* hybrids. Corn chromosomes have been observed in plants recovered from wheat × corn crosses. From durum wheat × corn crosses, we have produced a large number of durum haploids ($2n = 2x = 14$). We also observed some durum plants with 15 chromosomes, although the identity of the extra chromosome retained could not be established.

Environmental effects on hybridization Success in hybridization depends not only on the genotypes of the parental species but also on environmental factors. Environmental effects on interspecific hybridization are known, for example, in *Pennisetum*. The crossability of both tetraploid and hexaploid wheats with rye is also influenced by environmental factors at the location of crossing.

Possible effects of cytoplasm on crossability There are some indications of cytoplasmic effects on crossability. A significant improvement in crossing ($P = 0.001$) was noted by Simonenko and co-workers when they pollinated wheat by the (*Ae. squarrosa* × *S. cereale*) amphidiploids, compared to the reciprocal cross, who concluded that *Ae. squarrosa* cytoplasm decreased crossability in wide crosses.

Chromosome Pairing: The Essence of Gene Transfer and Crop Improvement

The synthesis of an F_1 hybrid between a crop species and an alien species is necessary, but is only the first step toward transferring alien genes into the crop plant. Pairing between the crop chromosomes and alien chromosomes is essential for successful gene transfer. However, as noted above, chromosome pairing is a function of genomic relatedness and may be minimal or absent in some hybrids. When pairing is absent or low, methods of promoting pairing must be adopted. Wheat researchers, mainly Sears (see "Further Reading"), introduced tools for promoting chromosome pairing that ushered in the era of chromosome engineering in wheat, making it possible to introduce alien genetic variation into this uniquely important cereal. Chromosome engineering methodologies, based on manipulation of the pairing control mechanism and induced translocations, have been applied successfully to transfer into wheat cultivars specific disease and pest resistance genes of alien origin.

Control of homoeologous chromosome pairing in wheat: its manipulation for inducing gene transfers As stated above, the *Ph1* gene in the long arm of chromosome 5B may suppress homoeologous pairing between wheat chromosomes and alien chromosomes, and consequently severely restrict alien gene transfer into wheat. In such cases, therefore, methods of promoting pairing between wheat and alien chromosomes must be used. In wheat–alien species hybrid derivatives, too much alien chromatin brings with it some undesirable characters of the wild parent. The most efficient method of shortening the alien segment is by induced homoeologous recombination.

Use of ph *mutants* A series of mutants that permit homoeologous pairing to various degrees have been produced in both bread wheat, and durum wheat, and may be employed for alien gene transfers into wheat. A high-pairing mutation involving a small, intercalary deficiency for *Ph1* was produced by Sears in the wheat cultivar Chinese Spring in 1977 and was designated *ph1b*. The level of homoeologous pairing obtained in the absence of *Ph1* is the highest obtainable, and a wheat cultivar carrying this mutation may be used for making hybrids with donor species. It has been found that homoeologous pairing and recombination occur between individual chromosomes of wheat and rye in their hybrids carrying the *ph1b* mutation, and that chromosomes of rye pair with chromosomes of the B genome of wheat more frequently than with those of the A or D genomes.

The centric wheat–rye translocations 1RS · 1BL (involving the short arm of rye chromosome 1R and the long arm of wheat chromosome 1B) have become very valuable in numerous commercial wheat cultivars worldwide, because the translocations confer resistance to several pests and diseases and have a positive effect on yield. Further reconstruction work has been done and is being done on this hybrid chromosome. Thus, to remedy the breadmaking quality defect brought in by this translocation in wheat, its rye arm 1RS has been recombined with wheat arm 1BS using the *ph1b* mutation.

A mutant, called *ph1c*, induced in durum wheat can be used in promoting homoeologous pairing in intergeneric hybrids involving durum. The mutants *ph1b*/*ph2b* and *ph1c* have been employed to bring about alien gene transfers into durum wheat (see "Further Reading").

Use of 5B-deficient stocks in durum wheat A relatively easy method of promoting intergenomic chromosome pairing, and thereby intergeneric gene transfers, in durum wheat is to use, as the female parent, a durum substitution line in which chromosome 5B is substituted by chromosome 5D. In such a 5D(5B) substitution line, the *Ph1* gene is absent, and hence the resulting hybrids with alien species also lack this gene. High intergenomic chromosome pairing in these hybrids facilitates alien gene transfer into durum wheat. To promote pairing between durum chromosomes and those of the wild grass, *Thinopyrum bessarabicum*, we made hybrids between these species using a 5D(5B) durum substitution line. Chromosome pairing in these hybrids increased more than fourfold and chiasma frequency per cell went up fivefold, compared with their counterpart with normal 5B. The same strategy was employed to introgress genes from *Th. curvifolium* into durum wheat.

Chromosome 5H of barley is known to have genes for early heading. To incorporate these genes into wheat, wheat–barley 5H recombinant lines were produced using nullisomy for chromosome 5B of wheat.

Use of appropriate genotypes of alien species Certain genotypes of alien species, e.g., *Agropyron cristatum* and *Elymus sibiricus,* inactivate the homoeologous-pairing suppressor *Ph1*. In fact, the ability to suppress the activity of *Ph1* or *Ph1*-like genes seems to be widespread in perennial grasses. An excellent method of inducing intergenomic pairing in hybrids is by crossing wheat with these genotypes, thereby promoting homoeologous pairing and hence alien gene transfer into wheat. *Aegilops speltoides*, for example, is known to suppress the activity of *Ph1* in its hybrids with wheat. The high pairing-inducing strains of *Ae. speltoides* have been used to transfer its genetic material into wheat, facilitating the incorporation of stripe rust resistance.

In an elegant approach to alien gene transfer, Bikram Gill's group at Kansas State University in the US transferred the *Ph1*-inhibiting gene Ph^I from *Ae. speltoides* into wheat, making it possible to manipulate pairing between wheat chromosomes and alien chromosomes. Because the Ph^I gene is dominant, wheat–alien exchanges can be induced by the presence of a single dose of Ph^I without the need to remove *Ph1*. However, the degree of homoeologous pairing induced by the Ph^I gene is not as high as that obtained by eliminating chromosome 5B or using the *ph1b* mutant.

Possible cytoplasmic effects Cytoplasmic effects on chromosome pairing have been observed, albeit rarely. The *Aegilops crassa* cytoplasm, for example, was found to promote homoeologous pairing in its hybrid with bread wheat. The cytoplasm of the Israeli ecotype of hexaploid tall fescue promoted multivalent formation (homoeologous pairing) in its hybrids with Spanish ecotypes.

Genetic compensation of translocations between homoeologous and alien chromosomes The success of gene transfers depends largely on the degree of chromosome homoeology that exists between the donor and recipient species. Homoeologous recombination allows unfavorable traits carried by the alien chromosome to be separated from the desired traits. On the other hand, noncompensating translocations cause undesirable duplications and deficiencies. For any wheat–alien exchange to be usable, the following criteria should be met: (1) the exchanged segment should be small; (2) the alien genetic material should compensate for the genetic material of the wheat parent that it replaces; and (3) the introgressed genetic material must be transmissible to the progenies, preferably in a Mendelian fashion. Such compensating translocations between homoeologous wheat and alien chromosomes, whether chromosome arms or chromosome segments, are agronomically desirable. DNA markers may be used to assess homoeology of alien chromosomes added into wheat, as has been done in wheat × *Leymus racemosus* crosses; the homoeology of the added *L. racemosus* chromosomes with wheat chromosomes was shown using DNA markers.

Absence of Chromosome Pairing: Gene Transfer by Irradiation

In wheat–alien hybrids with very low or no chromosome pairing, X-ray treatment may be employed to induce translocations between wheat and alien chromosomes. This process is random, but some translocations may help recover wheat chromosomes with a segment replaced by an alien segment carrying the desired gene. As early as 1956, using ionizing radiation, Sears was able to transfer a leaf rust resistance gene from *Ae. umbellulata* to chromosome 6B of bread wheat. Similar transfers have been made from rye and *Agropyron* to wheat. However, this procedure is laborious, and most translocations thus induced are deleterious because they involve nonhomoeologous chromosomes. With the availability of methods of inducing homoeologous pairing, it is advisable to adopt this approach to transfer alien genes.

Radiation-induced transfers have also been made across remotely related species. The irradiation of an oat line containing a single corn chromosome resulted in the transfer of corn chromosome segments into the oat genome.

Spontaneous Crop–Alien Chromosome Translocations

Chromosomes occurring as univalents at meiosis occasionally undergo misdivision due to breakage at the centromere, resulting in two telocentrics. These telocentrics may reunite, and if more than one chromosome has misdivided, the union may occur between arms of different chromosomes. This provides an excellent mechanism for substituting one arm of an alien chromosome for the corresponding arm of a wheat homoeologue. To pursue this possibility, a genetic stock is created that has an alien chromosome and its wheat homoeologue simultaneously in the monosomic condition and hence in the univalent state at meiosis. Then, spontaneous whole arm Robertsonian translocation may occur between them. The "hybrid" chromosome so formed may have the desired alien chromosome arm combined with a homoeologous wheat arm. If the added and deleted chromosome arms are homoeologous, Robertsonian exchanges can be genetically well compensated. Moreover, such substitutions are less likely to be deleterious than substitutions of the entire alien chromosome, although they may still have more alien chromatin than desired.

There are several instances of genetically compensated Robertsonian translocations in wheat. Thus, hexaploid wheats possessing a homoeologous exchange involving replacement of 1BS with 1RS from rye have been shown to have wider adaptability, resistance to powdery mildew, stripe rust, leaf rust, and stem rust, and yield performance equal or superior to that of comparable conventional wheats. The 1BL · 1RS translocations have been transferred from hexaploid bread wheat to durum wheat.

Spontaneous wheat–alien chromosome translocations are not uncommon. Wheat–rye translocations – 1BL · 1RS and 1AL · 1RS – that arose from centromeric breakage and reunion are perhaps the most successful wheat–alien translocations used for wheat improvement. Rye chromosome 2RL in a 2BS/2RL translocation has resistance to Hessian fly, and in a 2A/2RL translocation has resistance to leaf rust. Some mildew resistant wheat–rye translocations have also been isolated in the last few years.

Recently, two Robertsonian translocations, both presumably involving the reunion of the short arm of a group 6 chromosome of *Lophopyrum ponticum* and the long arm of chromosome 6D of wheat, have been isolated. These lines showed resistance to wheat curl mite (*Eriophyses tulipae*), a vector of wheat streak mosaic virus. Non-Robertsonian wheat–rye translocations have also been recorded.

In the last few years, wheat–barley addition lines with resistance to septoria leaf blotch (SLB) and common bunt were identified. Resistance to SLB was located on barley chromosomes 7 and 4, whereas almost complete resistance to common bunt was conferred by chromosome 6.

Other Methods of Inducing Wheat–Alien Translocations

Some novel techniques for inducing wheat–alien translocations have been tried successfully, and have, in some cases, led to the development of new cultivars.

In vitro Culture: Crop–Alien Chromosome Translocations

Wheat–alien chromosome translocations may be induced during *in vitro* culture of hybrids involving wheat. Such interchanges have been observed in wheat–rye hybrids regenerated from tissue culture, and in long-term callus cultures derived from embryos and inflorescences of wheat–*Th. intermedium* hybrids. The cell culture-induced recombinations of wheat and grass chromosomes in the latter hybrids conferred BYDV resistance to the material derived.

Gametocidal Chromosomes and Induction of Crop–Alien Chromosome Translocations

Another novel technique involves the use of the chromosome-breaking action of certain gametocidal chromosomes as described by Endo and coworkers in Japan. When introduced into common wheat, certain *Aegilops* chromosomes become gametocidal and induce chromosome breakage. By introducing gametocidal genes into a wheat–alien addition or substitution line, random wheat–alien chromosome translocations were recovered in the selfed progenies. Structural changes in barley chromosomes added to wheat have been induced by a gametocidal chromosome derived from *Aegilops cylindrica*. Most recently, in 2002, an international team, including Endo of Japan and McIntosh of Australia, transferred rye chromosome segments into wheat by employing a gametocidal system.

Crop–Alien Chromosome Translocations: Their Characterization

Using chromosome engineering, numerous chromosome-mediated alien gene transfers have been accomplished in cereal crops, including bread wheat and durum wheat. Some of the recent gene introgressions in wheat are listed in **Table 1**. Since chromosome identification is useful for gene manipulation in breeding programs, localizing alien chromatin in the wheat genome would be desirable. Several cytological, biochemical, and molecular techniques of characterizing alien chromatin have been used by different workers (reviewed by Jauhar and Chibbar). Giemsa C-banding provides an efficient tool to characterize alien chromatin integrated into the wheat genome.

A combination of techniques has been used effectively. Meiotic pairing combined with RAPD analysis, for example, helped to characterize powdery mildew-resistant lines derived from crosses between bread wheat and *Ae. speltoides* and *Ae. mutica*. Fluorescent genomic *in situ* hybridization (fl-GISH), a technique combining molecular biology with cytology, has proven particularly useful for detecting alien chromatin incorporated in the wheat genome. Because total genomic DNA is used as a probe to chromosome spreads, the genomic origin of alien chromatin can be determined (**Figure 1**). Combining GISH with C-banding provides a direct comparison of the molecular and cytological landmarks and facilitates identification of individual alien chromosomes or chromosome segments. Using this

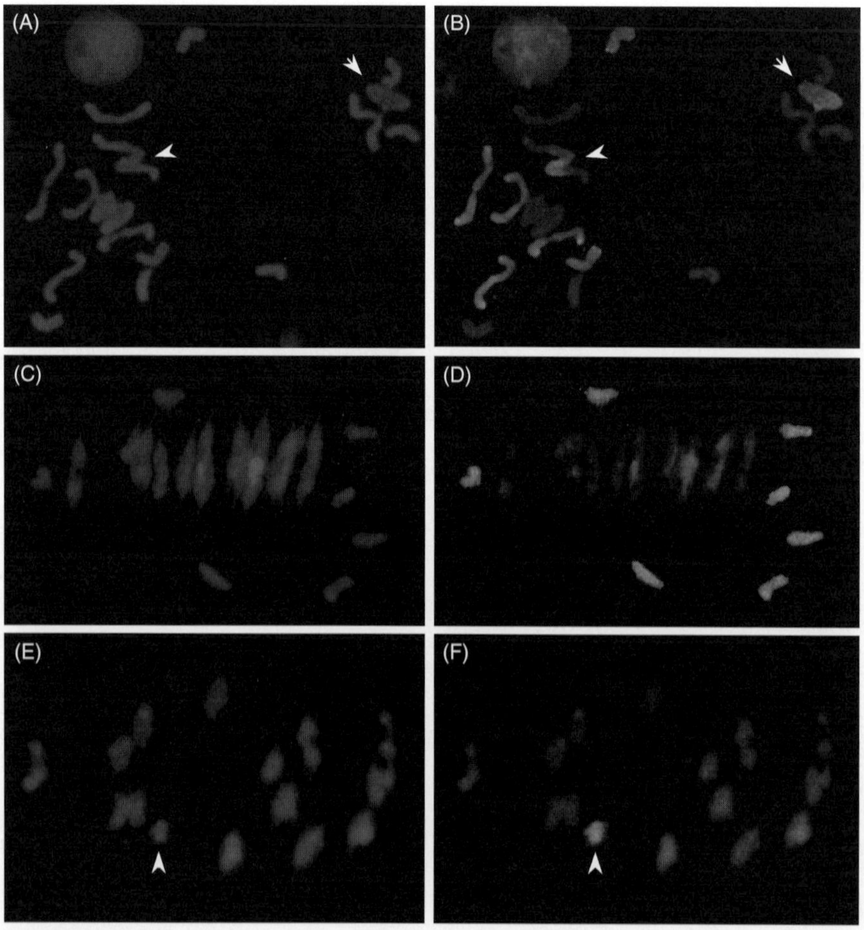

Figure 1 (see color plate 5) Incorporation of alien chromosomes and chromosome segments into the durum wheat genome. (A) Metaphase I chromosomes of durum wheat–*Thinopyrum junceiforme* hybrid, counterstained with propidium iodide (PI). Note a quadrivalent (arrow). (B) Same cell as in (A) after fluorescent GISH, showing grass chromatin integrated with durum chromosomes (arrow). (C) A pollen mother cell (PMC) of a durum–*Lophopyrum elongatum* hybrid, counterstained with PI, showing full chromosome complement of durum and seven grass chromosomes. (D) The same cell as in (C) after fluorescent GISH, showing seven fluorescing grass chromosomes. (E) A PMC (counterstained with PI) showing one grass chromosome (arrow) added to durum chromosome complement. (F) The same cell as in (E) after fluorescent GISH, showing a fluorescing grass chromosome (arrow). This grass chromosome confers a high degree of scab resistance to durum.

combination, *Haynaldia villosa* chromosomes added to wheat were identified.

Molecular markers also provide powerful tools for detecting alien chromosomes or chromosome segments. RAPDs and RFLPs, for example, are useful in diagnosing and screening large populations for traits such as disease resistance. The identification of markers tightly linked to the target gene would allow marker-assisted selection for introgression of this gene into wheat. Thus, RAPD markers were used to detect a powdery mildew resistance gene, *Pm21*, of *H. villosa* introduced into wheat (see **Table 1**).

Alien Chromatin in Durum Wheat Confers Scab Resistance

Fusarium head blight (FHB) or scab, caused by the fungus *Fusarium graminearum*, is a ravaging disease of wheat. Two wild relatives of wheat – tetraploid wheatgrass *Thinopyrum junceiforme* ($2n = 4x = 28$; $J_1J_1J_2J_2$) and diploid wheatgrass *Lophopyrum elongatum* ($2n = 2x = 14$; EE) – are excellent sources of resistance to scab. We have been able to transfer chromosome segments from these grasses into durum wheat, and have confirmed it by fluorescent GISH analysis. The alien chromatin conferred a high degree of FHB resistance to durum wheat.

Crop Improvement by Direct Gene Transfer

Alien gene transfers brought about by hybridization of wheat with its wild relative, coupled with manipulation of chromosome pairing, has resulted in the genetic enrichment of both bread wheat and durum wheat (**Table 1**). However, these chromosome-mediated gene transfers involve sexual hybridization, which also results in transmission of undesirable alien chromosomes. Consequently, adverse genetic interactions may lead to sterility. Recently developed tools of direct gene transfer are helping to engineer into crop plants new genes that are otherwise very difficult to introduce by conventional techniques.

Genetic engineering involves the insertion of a characterized gene(s) derived from an unrelated organism into plant cells, and the subsequent recovery of fertile plants with the inserted gene(s) integrated into their genome. Since this novel technique eliminates the constraint of sexual compatibility, it provides access to an unlimited gene pool and has added new dimensions to plant improvement programs. Using various tools of direct gene transfer, several crop cultivars with resistances to diseases and pests and improved nutrition have been produced, e.g., by Jauhar and Khush in 2002.

In the 1990s to the present day, the development of novel gene transfer techniques that allow direct introduction of DNA into regenerable calli has led to genetic transformation of bread wheat in laboratories in the US, Canada, and Germany. After establishing a suitable regeneration protocol for durum wheat, Jauhar's laboratory produced the first transgenic durum wheat in 1997. Since they showed that durum is amenable to genetic transformation, additional reports of the production of transgenic durum from laboratories in the UK and the international center CIMMYT in Mexico have appeared. This new technology of genetic engineering, combined with traditional chromosome engineering techniques, could help accelerate the genetic improvement of crop plants.

Interestingly, it has now become possible to use transgenic approaches to improve abiotic stress tolerance in important crops. By demonstrating the regulated overexpression of *Escherichia coli* trehalose biosynthetic genes (*otsA* and *otsB*) as a fusion gene, in 2002, Garg and coworkers showed the feasibility of engineering rice for increased tolerance of abiotic stress, resulting in an elevated capacity for photosynthesis under drought and low-temperature stress conditions. The current plant transformation techniques allow random integration of DNA in a crop genome, which may lead to changes in transgene expression due to position effect. With the availability of the recombinase mediated approach, suggested by David Ow and his colleagues at Berkeley, California, site-specific integration of DNA has become possible in rice, for example.

The combined application of chromosome engineering and genetic engineering could lead to accelerated genetic amelioration of crop plants. These techniques could very well supplement the traditional tools of crop improvement. Hopefully, the enormous potential of genetic engineering will be harnessed to the best advantage of the human race.

List of Technical Nomenclature

Allopolyploid A hybrid-derived plant containing two or more different sets of chromosomes from two or more diverse species; durum and bread wheat are the best examples.

Amphidiploid A plant of hybrid origin that contains the full chromosome complements of two species; also referred to as an allotetraploid.

Chiasma The site where two homologous (or sometimes homoeologous) chromosomes exchange genetic material during the first division of meiosis, resulting in a crossover.

Fl-GISH A technique that uses total genomic DNA along with fluorescent material as a probe in hybridization to distinguish chromosomes of a particular genome.

Genome A basic (haploid) set of chromosomes that contains all the genetic information needed to produce an organism.

Genome analysis Refers to the techniques of ascertaining genome relationships. Chiasmate chromosome pairing is a good indicator of genomic affinities.

Haploid A haploid plant is a sporophyte with half (gametic) chromosome number.

Homoeologous chromosomes Genetically and evolutionarily related chromosomes from related species.

Ph1 Pairing homoeologous gene, located in the long arm of chromosome 5B of wheat, that suppresses pairing among homoeologous chromosomes.

Robertsonian translocation A translocation involving the fusion of the arms of two chromosomes (or fusion of two acrocentric chromosomes) at the centromere, i.e., by centric fusion. Observed in hybrid derivatives.

Transgenic plant A plant created by introducing new gene sequences from alien sources by biotechnological means such as microprojectile bombardment.

See also: **Crop Improvement**: Alien Cytoplasms; Double Haploid Production; Marker Assisted Selection; Plant Breeding, Practice; Plant Breeding, Principles. **Genetic Modification, Applications**: Molecular Farming. **Genetic Modification**: Transformation, General Principles; Transgene Stability and Inheritance. **Tissue Culture and Plant Breeding**: Somaclonal Variation.

Further Reading

Ananiev EV, Riera-Lizarazu O, Rines HW, and Phillips RL (1997) Oat–maize chromosome addition lines: a new system for mapping the maize genome. *Proceedings of the National Academy of Sciences, USA* 94: 3524–3529.

Aghaee-Sarbarzeh M, Ferrahi M, Singh S, *et al.* (2002) Ph^{I}-induced transfer of leaf and stripe rust-resistant genes from *Aegilops triuncialis* and *Ae. geniculata* to bread wheat. *Euphytica* 127: 377–382.

Chen ZJ, Phillips RL, and Rines HW (1998) Maize DNA enrichment by representational difference analysis in a maize chromosome addition line of oat. *Theoretical and Applied Genetics* 97: 337–344.

Ceoloni C, Biagetti M, Ciafi M, Forte P, and Pasquini M (1996) Wheat chromosome engineering at the 4x level: The potential of different alien gene transfers into durum wheat. *Euphytica* 89: 87–97.

Ceoloni C, Forte P, Ciaffi M, *et al.* (2000) Chromosomally engineered durum wheat: the potential of alien gene introgressions affecting disease resistance and quality. *Options Méditerranéennes* A-40: 363–371.

Endo TR (1995) Chromosomes of *Aegilops* species inducing chromosomal structural changes in common wheat. In: Raupp WJ and Gill BS (eds) *Proceedings of US–Japan Symposium on Classical and Molecular Cytogenetic Analysis*, pp. 123–128. Manhattan, Kansas: Kansas Agricultural Experimental Station.

Friebe B, Jiang J, Raupp WJ, McIntosh RA, and Gill BS (1996) Characterization of wheat–alien translocations conferring resistance to diseases and pests: current status. *Euphytica* 91: 59–87.

Garg AK, Ju-Kon K, Owens TG, *et al.* (2002) Trehalose accumulation in rice plants confers high tolerance levels to different abiotic stresses. *Proceedings of the National Academy of Sciences, USA* 99: 15898–15903.

Jauhar PP (1975) Genetic control of diploid-like meiosis in hexaploid tall fescue. *Nature (London)* 254: 595–597.

Jauhar PP (1975) Genetic regulation of diploid-like chromosome pairing in the hexaploid species, *Festuca arundinacea* Schreb. and *F. rubra* L. (Gramineae). *Chromosoma* 52: 363–382.

Jauhar PP (1977) Genetic regulation of diploid-like chromosome pairing in *Avena*. *Theoretical and Applied Genetics* 49: 287–295.

Jauhar PP (1993) *Cytogenetics of the Festuca–Lolium Complex: Relevance to Breeding*. Berlin, Heidelberg, New York, Paris, Tokyo: Springer-Verlag.

Jauhar PP and Chibbar RN (1999) Chromosome-mediated and direct gene transfers in wheat. *Genome* 42: 570–583.

Jauhar PP and Joppa LR (1996) Chromosome pairing as a tool in genome analysis: merits and limitations. In: Jauhar PP (ed.) *Methods of Genome Analysis in Plants*, pp. 9–37. Boca Raton, London, Tokyo: CRC Press.

Jiang J, Friebe B, and Gill BS (1994) Recent advances in alien gene transfer in wheat. *Euphytica* 73: 199–212.

Liu B, Luan Y, Han F, Ji W, and He M (2001) Cell culture induced introgression of *Thinopyrum intermedium* chromatin into common wheat. *Plant Cell, Tissue and Organ Culture* 65: 9–13.

Lukaszewski AJ (1997) Further manipulation of centric misdivision of the 1RS · 1BL translocation in wheat. *Euphytica* 94: 257–261.

Lukaszewski AJ (2001) Breeding behavior of the cytogenetically engineered wheat–rye translocation chromosomes 1RS · 1BL. *Crop Science* 41: 1062–1065.

Masoudi-Nejad A, Nsauda S, McIntosh RA, and Endo TR (2002) Transfer of rye chromosome segments to wheat by a gametocidal system. *Chromosome Research* 10: 349–357.

Ow DW (2002) Recombinase-directed plant transformation for the pot-genomic era. *Plant Molecular Biology* 48: 183–200.

Riera-Lizarazu O, Rines HW, and Phillips RL (1996) Cytological and molecular characterization of oat × maize partial hybrids. *Theoretical and Applied Genetics* 93: 123–135.

Riera-Lizarazu O, Vales MI, Ananiev VE, Rines HW, and Phillips RL (2000) Production and characterization of maize chromosome 9 radiation hybrids derived from an oat–maize addition line. *Genetics* 156: 327–339.

Repellin A, Båga M, Jauhar PP, and Chibbar RN (2001) Genetic enrichment of cereal crops by alien gene transfers: new challenges. *Annual Review of Plant Biotechnology and Applied Genetics*, pp. 159–183 Dordrecht: Kluwer Academic Publishers.

Reynolds MP, Calderini DF, Condon AG, and Rajaram S (2001) Physiological basis of yield gains in wheat associated with the *LR19* translocation from *Agropyron elongatum*. *Euphytica* 119: 137–141.

Sears ER (1972) Chromosome engineering in wheat. *Stadler Genetics Symposium* 4: 23–38.

Sears ER (1981) Transfer of alien genetic material to wheat. In: Evans LT and Peacock WJ (eds) *Wheat Science – Today and Tomorrow*, pp. 75–89. Cambridge: Cambridge University Press.

Sears ER (1984) Mutations in wheat that raise the level of meiotic chromosome pairing. *Stadler Genetics Symposium* 16: 295–300.

Sears ER (1993) Use of radiation to transfer alien segments to wheat. *Crop Science* 33: 897–901.

Shi F and Endo TR (1999) Genetic induction of structural changes in barley chromosomes added to common wheat by a gametocidal chromosome derived from *Aegilops cylindrica*. *Genes and Genetic Systems* 74: 49–54.

Simonenko VK, Motsny II, Sechnyak AL, and Kulbida MP (1998) The use of wheat–alien and *Aegilops*–rye amphiploids for introgression of genetic material to wheat. *Euphytica* 100: 313–321.

Srivastava V and Ow DW (2001) Biolistic mediated site-specific integration in rice. *Molecular Breeding* 8: 345–350.

Taketa S, Takahashi H, and Takeda K (1998) Genetic variation in barley of crossability with wheat and its quantitative trait loci analysis. *Euphytica* 103: 187–193.

Tanksley SD, Ganal MW, and Martin GB (1995) Chromosome landing: a paradigm for map-based gene cloning in plants with large genomes. *Trends in Genetics* 11: 63–68.

Villareal RL, Bañuelos O, Mujeeb-Kazi A, and Rajaram S (1998) Agronomic performance of chromosomes 1B and T1BL·1RS near-isolines in the spring bread wheat Seri M82. *Euphytica* 103: 195–202.

Vasil V, Srivastava V, Castillo AM, Fromm ME, and Vasil IK (1993) Rapid production of transgenic wheat plants by direct bombardment of cultured immature embryos. *Bio/Technology* 11: 1553–1558.

Doubled Haploid Production

K J Kasha, University of Guelph, Guelph, ON, Canada

Definitions

"Haploid" is a general term used to refer to, in this case, a higher-order plant (sporophyte) containing the gametic chromosome number (n). The definition may be refined into categories with more specific terminology as illustrated in **Figure 1**. Euhaploids contain one or more complete genomes. A genome is a basic set (x) of chromosomes for a species which contain essentially a complete complement of the genes for that species. They may be derived from a diploid ($2n=2x$) species or from polyploid ($2n>2x$) species which may have two or more different genomes (alloploids) or more than two sets of one genome (autoploids). Alternatively, the haploids may be aneuhaploid, which have a chromosome number that does not contain a balanced number of chromosomes. At the 1st International Symposium on Haploids in Higher Plants (1974) a panel of researchers convened to set up and adopt common terminology to be used in haploid research (**Figure 1**).

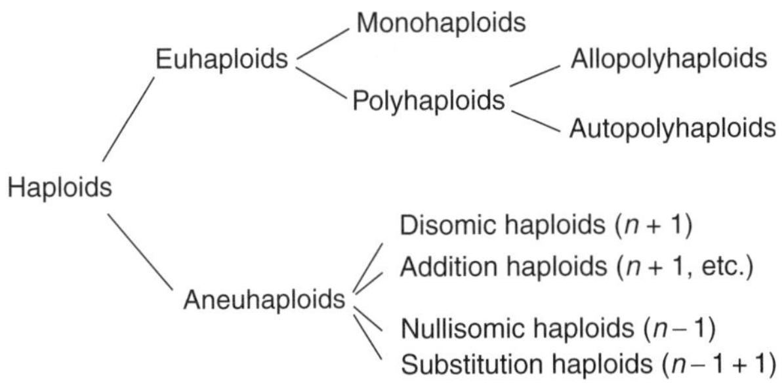

Figure 1 Classification of haploid terminology. Modified from de Fossard RA (1974) Terminology in "Haploid" work. In: Kasha KJ (ed.) (1974) *Haploids in Higher Plants, Advances and Potential.* Proceedings of the 1st International Symposium. Guelph, Ontario, Canada: University of Guelph.

"Doubled haploids" plants have originated from a haploid cell or plant but have had their somatic chromosome number ($2n$) doubled, either spontaneously or by induction. Since haploid plants from most species (whether from diploids or polyploids) are highly sterile, chromosome doubling frequency is an important aspect of haploid production methods and for utilization.

"Dihaploid" is one term that is frequently misused as doubled haploids of monohaploids. Dihaploid was originally used for the haploid plants from an autotetraploid which have two sets of chromosomes that are not genetically identical. Therefore, a dihaploid is genetically different from the genetically homozygous doubled haploid.

Androgenetic haploids are those derived from the male gamete (usually the microspore) while gynogenetic haploids refer to those derived from the female gamete.

Discovery of Haploid Plants

Bergner in 1921 was the first to recognize a haploid plant when he was working with jimson weed (*Datura stramonium*). Belling and Blaklee in 1924 first suggested that chromosomally doubled haploids could be valuable in crop improvement and much interest in identifying haploids and developing methods for their production ensued.

Methods of Doubled Haploid Production

To be useful in crop improvement the method of haploid production should attempt to meet three requirements:

1. The method should produce large numbers of haploids from most or all genotypes of the species.
2. These haploids should represent a random sample of the gametes produced at meiosis.
3. The haploids and subsequent doubled haploids should be genetically stable.

A list of potential methods of haploid production along with examples of some major crops where they can be utilized in breeding or crop improvement is summarized in **Table 1**. The first four methods tend to relate to seed production and provide a continuum from unfertilized parthenogenesis and its stimulation through to partial (semigamy) and complete fertilization followed by chromosome elimination. The fifth method, *in vitro* culture, is quite distinct and bypasses seed production. The origins of haploids relative to the life cycle in higher plants is illustrated in **Figure 2**. Each method will be described briefly. Specific genes or genetic stocks tend to influence the success of all methods including *in vitro* cultures and will be mentioned under each method. However, under parthenogenesis, a subsection on gene control provides examples of strong gene influences. *In vitro* culture systems tend to show strong genotype differences in most crops but anther and microspore culture are considered to have the greatest potential for haploid production based on the extremely large numbers that might be induced into embryogenesis and plants.

Parthenogenesis

Selection systems The first haploids observed arose spontaneously from parthenogenesis, i.e., by the development of an egg cell into an embryo without fertilization. These were rare and frequencies varied from genotype to genotype. Thus, systems were developed which utilized genotypes that produced higher frequencies along with marker genes that would enable one to recognize and select these haploids. There is a line in corn (*Zea mays*; maize) that gave 3.2% haploids (monoploids). The expected frequency of such gynogenetic haploids is 1 in 1000 seeds while the frequency of androgenetic origin haploids from seeds is 1 in 80 000. By using a dominant anthocyanin marker in the pollen parent to mark diploid seed, by 1949 haploid-derived inbreds were available in commercial corn hybrids.

Researchers working with various species tried techniques to stimulate parthenogenesis, such as delayed or sparse pollination, irradiation of pollen, or using pollen from a different species. The latter two were major efforts in forest tree species where haploidy could be very important in improvement due to their long life cycle. Moreover, trees are not readily amenable to inbreeding as they tend to be outcrossed and thus heterozygous for numbers of deleterious genes.

Table 1 Methods of haploid production with examples of major crops where they are used

	Method	Crop examples
1	Parthenogenesis	
	Selection	Corn, flax
	Gene control	Corn
2	Wide hybridization	Potato
3	Semigamy	Cotton
4	Chromosome elimination	Barley, wheat
5	*In vitro* culture	
	Anther/microspore	Canola, rice, barley, wheat, tobacco, asparagus
	Ovule	Sugar beet

Gene-controlled parthenogenesis There are some examples where the induction of haploids is strongly determined by a gene or genes. Examples are the *ig* gene (indeterminate gametophyte) in corn and the haploid inducer gene in barley (*Hordeum vulgare*). Similarly, semigamy in cotton (*Gossypium* spp.) could also be considered as genetically induced or controlled but will be covered separately. Alien cytoplasm may also induce parthenogenesis.

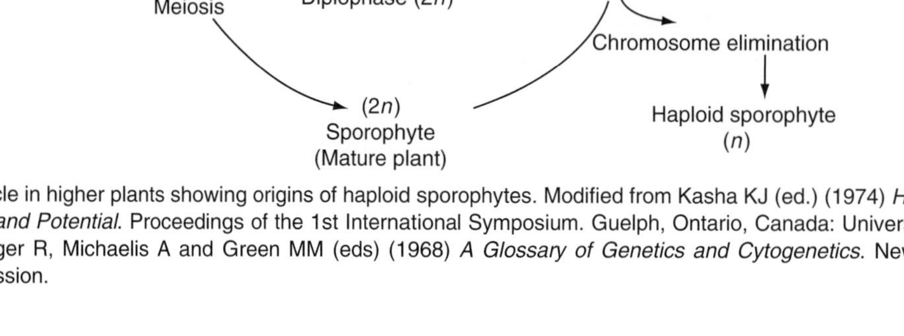

Figure 2 Life cycle in higher plants showing origins of haploid sporophytes. Modified from Kasha KJ (ed.) (1974) *Haploids in Higher Plants, Advances and Potential.* Proceedings of the 1st International Symposium. Guelph, Ontario, Canada: University of Guelph as modified from Rieger R, Michaelis A and Green MM (eds) (1968) *A Glossary of Genetics and Cytogenetics.* New York: Springer Verlag, with permission.

Indeterminate gametophyte in corn leads to much higher than normal frequencies of parthenogenesis. In this case the gametophyte does not mature but continues nuclear division and multiple cells can develop. Furthermore, haploids from both the male and female gametes can be formed, in a ratio of about one maternal to two paternal haploids. The frequency of androgenetic haploids ranges from 0.5% to 2%, much higher than the 1 in 80 000 that occur spontaneously. About 10% of the androgenetic-derived haploids are diploid and have been determined to be homozygous doubled haploids.

A haploid initiator gene occurs in barley; in the presence of this dominant gene, haploid progeny are produced in frequencies ranging from 30% to 40% under self-pollination of a plant homozygous for the gene. When crossed as female to a plant without the initiator gene, about 3–5% of F_1 will be haploid. One advantage is that the seeds germinate so embryo culture is not required. The difficulty with this gene is that it is present in doubled haploids so that subsequent progeny will contain sterile haploid plants, detracting from their suitability for new cultivars. Some diploid progeny from plants with this gene are doubled haploids.

Wide Hybridization

The hybridization of a crop species with other species from the same or different genera can lead to the production of haploids. This system has worked well in potato (*Solanum tuberosum*), an autotetraploid that can be reduced to the diploid (dihaploid) level for breeding. Up to 30% of fruit set from crosses of *S. tuberosum* (4*x*) by *Solanum phureja* (2*x*) may contain at least one seed that is dihaploid, which are considered to have arisen from parthenogenesis, the rest of the seeds being tetraploid or triploid hybrids. On occasion, it has been feasible to obtain a monohaploid by crossing the dihaploids with *S. phureja*. Other diploid species of *Solanum* can also be used as pollinators to produce potato dihaploids but the success rate is lower.

In essence, wide hybridization is an example of induced parthenogenesis where, due to the large numbers of seeds per fruit in potato, fairly large numbers of haploids can be achieved. It is similar to the use of foreign pollen used to produce low parthenogenetic frequencies in forest or fruit trees.

Another example of haploid production from wide hybridization is the induction resulting from alien cytoplasm in wheat. It has been observed that placing the hexaploid wheat (*Triticum aestivum*) genome in the cytoplasm of some *Aegilops* species resulted in wheat haploid plants by parthenogenesis. However, sterility caused by the alien cytoplasm has limited the application of this method.

Semigamy

Semigamy represents an incomplete fertilization process where the sperm nucleus enters the egg cell but does not fuse with the egg cell nucleus. The cell is thus stimulated to divide and grow. Each nucleus can form their own cell, resulting in an embryo and plant that is chimeric. This chimeric plant is usually partially haploid for the male gamete and partially haploid for the female gamete, but sometimes one type will become dominant. By using a visual marker in one parent, such as a gene for anthocyanin production, the sectors can be distinguished. After chromosome doubling, the seed from each sector can be harvested separately to maintain the two doubled haploids. It has been postulated that semigamy is controlled by a genetic factor or factors, but most likely a singular dominant factor. It has been used in cotton species *Gossypium hirsutum* and *G. barbadense* to produce doubled haploids. It has also been observed in other species that are not major crops.

Chromosome Elimination

During attempts to produce interspecific hybrids between cultivated barley and a related wild species *Hordeum bulbosum* a quite high frequency of haploid plants was found. The fertilization of barley ovules occurred using the *H. bulbosum* pollen and stimulated embryo development. However, during early development, the chromosomes from *H. bulbosum* were preferentially eliminated during cell division, leaving most embryos with only a single set of barley chromosomes. Elimination also occurred during endosperm development so that by 2 weeks after pollination, the embryos had to be rescued and grown into small plants on culture media. The keys to initially understanding that this was elimination and not parthenogenesis were the high frequencies of haploids, differences in response depending upon the ploidy level of the two parents, and similar results from reciprocal crosses (**Table 2**).

When crosses were made at the same ploidy level, e.g., 2*x* by 2*x* or 4*x* by 4*x*, elimination occurred to produce haploid barley plants. When 2*x* barley was crossed with 4*x* *H. bulbosum*, 3*x* triploid species hybrids were produced. Thus, chromosome balance was thought to be important in stability or elimination and it was later shown that this was most likely a balance of genes on specific chromosomes numbered 2 and 3. The exciting aspect of this system of haploid production is that nearly all barley genotypes would hybridize with selected diploid *H. bulbosum*

Table 2 Genome balance and elimination in progeny obtained from crosses between barley (*Hordeum vulgare*), designated V and *H. bulbosum*, designated B, at various ploidy levels and reciprocal crosses

Female parent	*Male parent*	*Constitution of progeny plants*
VV	BB	V, VB[a]
BB	VV	V
VV	BBBB	VBB
BBBB	VV	VBB
VVVV	BBBB	VV, VVBB[a]
BBBB	VVVV	VV

[a] A low frequency of hybrids in addition to barley haploids (V or VV) were also produced in these crosses.

plants which could be multiplied vegetatively and distributed. Thus, it would be suitable for barley breeding or improvement programs and allowed production of fairly large numbers of haploid plants. The efficiency of this method in barley has been greatly improved by selection of better *H. bulbosum* pollinators and improved crossing procedures.

Since the haploid plants were sterile, chromosome doubling procedures, usually colchicine treatment, were required as spontaneous doubling frequency was very low. Efficient chromosome doubling methods were developed so that 70–80% of the haploids were partially doubled and produced genetically homozygous seed.

Subsequent attempts to pollinate hexaploid bread wheats with *H. bulbosum* pollen showed that chromosome elimination occurs, resulting in rescued embryos developing into wheat haploids. However, most wheat genotypes could not be efficiently fertilized with *H. bulbosum* pollen due to the presence of KR (rye crossability) genes. Thus, while useful in producing some doubled haploids, the system was not suitable for breeding programs.

Chromosome elimination has been shown to be quite common among interspecific hybrids. Bread wheats pollinated with corn pollen and in combination with auxin treatment, induced good seed set. After embryo rescue, this system also produced wheat haploids across most genotypes. Thus, corn pollination of wheat is now quite commonly used for haploid production in wheat breeding and research programs. While not as successful, pollen from *Sorghum*, teosinte (*Zea mexicana*), pearl millet (*Pennisetum glaucum*), and Italian ryegrass (*Lolium multiflorum*) will also produce haploids in wheat. Corn pollination on oats (*Avena sativa*) results in oat haploids as well as some corn addition chromosomes in oats when elimination is not complete.

Chromosome elimination has been observed among many interspecific crosses in *Hordeum* as well as in some other genera and is an indication that chromosome elimination may occur quite often following wide hybridization. The rescue of haploid embryos is a key to detection while a good source of pollen is essential for developing a haploid production system.

In Vitro Culture Methods

In vitro cultures for haploid production may involve organs containing gametes, such as ovules or anthers, or the isolated immature male gametes (microspores). *In vitro* androgenetic procedures have much greater potential because there are thousands of microspores per anther versus a single ovule per floret for culture in monocots. This same advantage exists over methods such as wide hybridization, chromosome elimination, or parthenogenesis. Anther or isolated microspore culture has been achieved for hundreds of species, including most crop species, and is the method of choice in many.

Anther and microspore culture Years of research in laboratories around the world have gone into these methods in efforts to overcome the problems of: low response; strong genotype differences in response; a tendency for nonrandom samples of gametes; and genetic stability of the plants derived. The solution to most of these problems would be to obtain large numbers developed through a direct embryo induction procedure.

However, there are many factors which influence response and therefore could be a source of genotype response differences. These include growth conditions and health of donor plants, pretreatments for induction of callus or embryo development, isolation procedures, culture media, and regeneration media. In monocotyledonous species, albino plants may be produced and their frequencies are strongly genotype-influenced. In species where the procedures have become highly successful, the genotype factor is reduced and procedures, particularly isolated microspore culture, are used successfully in large breeding programs. These include canola (*Brassica napus*), barley, and tobacco (*Nicotiana tabacum*). One big advantage of the system is that in some crops such as cereals, a high proportion of the plants obtained are fertile doubled haploids, eliminating the need for an additional chromosome doubling step as well as producing much more seed from the doubled haploid plant.

Ovule culture Ovule culture has been achieved in a few species and is used in breeding programs in sugar beet (*Beta vulgaris*), a crop where it is currently the most efficient haploid method. Haploid plants are vigorous with no albinism problems but chromosome doubling procedures are required. Haploids from ovule culture have also been produced in crops such as barley, rice (*Oryza sativa*), and wheat but other haploid procedures are much more efficient in these crops.

Applications of Haploidy

The applications of haploid producing systems in crops will vary depending upon whether the crop is cross-pollinated or self-pollinated, has disomic or polysomic inheritance, and reproduces vegetatively or by seed. In self-pollinated crops, the usual breeding objective is to obtain homozygous true-breeding cultivars. Thus, doubled haploids provide potential homozygous cultivars in one step from F_1 or highly heterozygous materials and save generations of time otherwise required for their production by inbreeding. In cross-pollinated crops, the breeding objective is usually to produce hybrids or composite crosses. In such crops, haploid methods would rapidly produce the inbred lines for making hybrids or could improve progress in recurrent selection.

In disomic diploid species, the haploids have a single genome (monoploid) so that the doubled haploids are completely homozygous. In polyploids, the predominantly self-pollinating crops are usually disomic allopolyploids and cultivars produced would also be highly inbred and genetically uniform. Thus, potential cultivars can also be produced in one generation as is the case with diploids. The predominantly cross-pollinated polyploid crops are usually polysomic autopolyploids with multiple copies of one or more genomes. Thus they are highly heterozygous and heterogeneous in nature, traits which are maintained by their outcrossing nature. The polysomic breeding behavior combined with outcrossing allows deleterious or lethal genes to be maintained in the populations. Thus when they are inbred a rapid loss of vigor occurs due both to the approach to homozygosity (inbreeding depression) and to the exposure of deleterious genes. In crops, such as alfalfa (*Medicago sativa*; lucerne) and potato which have undergone breeding for some time, many of the deleterious genes have been eliminated from the breeding populations so that many haploids can be produced. Both alfalfa and potato are autotetraploid and so the haploid is a dihaploid with two copies of the genome with different alleles of genes at many loci. Such plants are highly variable in vigor and often show some gamete sterility. When chromosomally doubled back to the tetraploid level, this tetraploid is thought to be the equivalent of at least three generations of inbreeding and plants often lack commercial yield levels.

Many cross-pollinated species have evolved mechanisms to help maintain this system, such as self-incompatibility mechanisms or dioecious plants. Haploidy provides a quick method for producing lines homozygous for different self-incompatible alleles.

The discussion of applications so far has been about producing seed in crops. Vegetatively reproducing species usually have a sexual cycle for seed production which may be utilized for haploid production. In such cases, the doubled haploids could be used to produce potential lines to be used commercially via vegetative reproduction.

Self-Pollinating Crops

In self-pollinating species, it usually takes five to six generations of selfing after making a cross to reach a level of homozygosity and uniformity sufficient to be able to evaluate quantitative traits such as yield and quality. Selection efficiency of such traits in earlier generations is low due to the presence of dominance and heterozygosity. However, doubled haploids can achieve complete uniformity in a single generation, allowing for the evaluation of both simple and quantitatively inherited traits (**Table 3**). Furthermore, when grown in isolation, rouging in subsequent generations is not required, compared to lines produced by self-pollination.

Doubled haploids also improve the selection of fairly simply inherited traits earlier in the breeding program. Looking at rows of doubled haploid lines in the field provides confidence in selection for agronomic plant type, disease resistance, and physiological traits, as compared to selection on single plants in early generations of heterozygous segregating populations using other breeding methods. If the doubled haploid falls short of being a potential cultivar, it can be used as a parent in crosses aimed at combining the traits desired. Thus, the time for a breeding cycle can be shortened.

One question often debated is the size of doubled haploid population to provide sufficient variability for selecting an improved cultivar. Since doubled haploid breeding is essentially gamete selection, the frequency of the best gamete should be the square root of the frequency of having two superior gametes meet when crossing or self-pollinating the F_1 of a cross. Thus, 100 doubled haploid plants from a cross should provide the same odds of a superior genotype as 10 000 F_2 plants from self-pollination. This is the extreme situation as some of the F_2 plants may recombine to provide superior combinations upon inbreeding.

The theoretical aspects of the best generation for doubled haploid production following a cross,

Table 3 Relative breeding schedule in spring barley using doubled haploid and pedigree methods

Time	*Doubled haploid method*	*Time*	*Pedigree method*
Year 1	Grow parents, grow F_1 and produce doubled haploid	Year 1	Grow parents Grow F_1
Year 2	Grow doubled haploid rows Select best lines	Year 2	Grow F_2 Discard poor plants
Years 3–5	Evaluate in plots – yields, etc.	Years 3–6	Grow F_3, F_4 and F_5 rows Select best lines
Years 6–8	Coop trails for licensing	Years 7–9	Evaluate in plots – yields, etc.
Year 9	License and release	Years 10–12	Coop trails for licensing
		Year 13	License and release

Modified from Choo TM, Reinbergs E, and Kasha KJ (1985) Use of haploids in barley breeding. *Plant Breeding Review* 3: 219–252.

relative to the extent of recombination and linkage present, have been investigated. If traits were linked, production of doubled haploid lines from the F_2 and F_3 would provide more opportunity for recombination. If the desired recombinations have not been recovered by the F_3 it would be more efficient to select the best line and make additional crosses among plants with the desired traits. With a doubled haploid system, this process of selection is simplified with higher probabilities of success.

The doubled haploid system for breeding requires the ability to produce large numbers of doubled haploid plants from any genotype of this crop. Furthermore, they should represent a random sample of gametes and the resulting lines should be genetically stable for use as cultivars. The chromosome elimination (*H. bulbosum*) method of doubled haploid production in barley was the first to provide sufficient doubled haploid material to evaluate these desired traits. The *H. bulbosum* doubled haploid method has been compared to the pedigree, single seed descent, and the bulk methods of breeding barley. The superior lines from the doubled haploid method were found to be equal to the superior lines derived by the other methods. However, the doubled haploid lines were obtained in many fewer generations. In studies of adaptability the doubled haploid lines performed just as well across environments and were as genetically stable as lines produced by other methods.

Purity of self-pollinated crop cultivars often has very stringent requirements for licencing and therefore doubled haploid production of potential cultivars produced by inbreeding may be conducted at the F_6 or F_7 generation to ensure homozygosity and homogeneity.

In selection for quantitative traits, both recessive and dominant genes can be selected in homozygous doubled haploid plants. Due to the lack of heterozygosity, only the additive genetic variation will be expressed in the doubled haploid progeny of a cross. Thus, where additive genetic variance is important, such as in self-pollinating crops, selection for improved quantitative traits is enhanced using doubled haploid methods.

It has also been proposed to use doubled haploid methods in recurrent selection schemes with successive cycles of doubled haploid production and selection. This would be one way to overcome the concern about less opportunity for recombination when producing haploids from the F_1 plants.

While doubled haploid plants with good vigor and genetic stability were observed in barley and cotton where the female gamete was the source of haploids, a tendency towards reduced vigor and more genetic and morphological variation than expected has been reported in allotetraploid tobacco using anther culture. It has been postulated that the origin of such differences might be: limited numbers of cytoplasmic organelles or their degradation in microspores; the disruption of nuclear–cytoplasm relationships; residual heterozygosity in the polyploid parents; or the induction of mutations during haploid production or chromosome doubling procedures.

Many new cultivars of self-pollinated crops have been produced by haploid methods. A worldwide survey in barley reported that by 2002 there were at least 98 cultivars produced by haploid methods, 37 by anther culture, three by isolated microspore culture, and 58 by chromosome elimination following hybridization with *H. bulbosum*. Large numbers of new cultivars from anther culture have been produced in rice, wheat, and tobacco, with some in many other crops.

Haploids have many applications in crops besides breeding. Androgenetic methods of doubled haploid production from microspores offer advantages for use in mutation studies. Very large populations of haploid microspores can be subjected to mutation and selection pressure for some traits can be applied to the microspore cultures. In monoploids both dominant and recessive mutations would be expressed for easy selection. Haploid systems have also been used to recover stable lines from highly sterile plants produced by wide hybridization. Androgenetic haploids using the *ig* gene in corn provide a method of transferring a genotype into a new cytoplasm, such as cytoplasmic male sterility. Doubled haploid populations are very valuable for molecular mapping of crops and have been widely used in barley and wheat as well as in some other crops.

Cross-Pollinated Crops

Breeding of cross-pollinated crops is usually conducted using random mating populations or synthetic populations, or by producing single- or double-cross hybrids. Doubled haploid systems have been proposed for all these procedures.

In recurrent selection, the doubled haploids could be produced from a random sample of the population; these are then assessed and selected doubled haploid lines are crossed to provide a new random mating population. The cycle can then be repeated. This system would select more for additive genetic variance than dominance genetic variances of quantitative traits. It has been found that incremental changes in the population mean would be greater using doubled haploid populations, as compared to random mating populations. This advantage is maintained regardless of population size and is due to emphasis on selecting for additive genetic variance.

Debate in disomic corn has been focused on whether there is a need for producing doubled haploid inbred lines for evaluation as parents in hybrids. One argument is that via haploidy one would need fewer inbreds to find superior hybrids and that the hybrids would be more uniform. Anther culture derived inbred lines in corn have been evaluated and found to be stable over generations, enhancing their use as parents for hybrids.

Inbred lines are more difficult to develop by inbreeding in many cross-pollinating species due to the presence of self-incompatibility systems. However, production of doubled haploids circumvents the self-incompatibility system, providing a rapid procedure for the production of homozygous lines with a single self-incompatibility allele. These doubled haploid lines can be used directly in developing hybrids or synthetic cultivars.

In polysomic crops species, most evaluations of haploidy have been conducted in potato or alfalfa where the crop is essentially autotetraploid and cross-pollinated. Using wide hybridization, dihaploids can be produced, and on rare occasions, a monoploid from a dihaploid. Research with dihaploids and chromosome doubling has demonstrated that multiple alleles (more than two) per locus is important at the tetraploid level in order to obtain the greatest yields and seed fertility. To realize this advantage, the "analytical" breeding method has been developed. This involves producing dihaploids from two genetically diverse genotypes, crossing each type of dihaploid with a distinct diploid genotype, and then crossing the two resulting hybrids to produce a tetraploid via unreduced ($2n$) gametes. Unreduced gametes are found quite frequently in potato and alfalfa and are influenced by specific genes. Thus, production of tetraploids with alleles from four divergent genotypes is feasible.

The value of producing dihaploids in such crops is that breeding can be conducted at the diploid level where inheritance and selection are simplified and smaller populations are required to obtain recombinants. In addition, new germplasm from diploid species may be introduced through crossing them with dihaploids. The problem is that chromosomally doubled dihaploids show loss of vigor and productivity at the tetraploid level, due to the equivalence of three generations of inbreeding. Thus, there arises the need for an analytical breeding system or the production of hybrids or synthetics after chromosome doubling. An alternative proposal has been to use protoplast fusion at the diploid level to restore multiple allelic tetraploids.

In rare dioecious crop species haploidy can lead to the production of homozygous males (supermales YY) for use in the production of hybrids. Asparagus (*Asparagus officinalis*) has such an XY sex determination system and the most productive plants are the XY males. While the XX females can be obtained in progeny from crosses with XY males, the supermales (YY) are rarely found. By haploid procedures, both YY supermales and XX females can be obtained. Hybridization of supermales onto females produces all male (XY) progeny that are superior for yield in asparagus.

See also: **Crop Improvement**: Mutation Techniques; Plant Breeding, Practice; Plant Breeding, Principles. **Genetic Modification**: Transformation in Dicotyledons; Transformation in Monocotyledons. **Tissue Culture**: Somatic Embryogenesis. **Tissue Culture and Plant Breeding**: Somatic Hybridization.

Further Reading

Bajaj YPS (ed.) (1990) *Biotechnology in Agriculture and Forestry*, vol. 12, *Haploids in Crop Improvement*. Berlin: Springer-Verlag.
Chase SS (1969) Monoploids and monoploid derivatives in maize (*Zea mays* L.). *Botanical Reviews* 35: 117–167.
Choo TM, Reinbergs E, and Kasha KJ (1985) Use of haploids in barley breeding. *Plant Breeding Review* 3: 219–252.
Davies DR and Hopewood DA (eds) (1980) *The Plant Genome*. 2nd International Haploidy Conference. Norwich, UK: The John Innes Charity.
Jain SM, Sopory SK, and Veilleux RE (eds) (1996) *In Vitro Haploid Production in Higher Plants*, vols. 1–5. London: Kluwer Academic Publ.
Kasha KJ (ed.) (1974) *Haploids in Higher Plants, Advances and Potential*. Proceedings of the 1st International Symposium. Guelph, Ontario, Canada: University of Guelph.
Nitzsche W and Wenzel G (1977) *Haploids in Plant Breeding*. Berlin: Verlag Paul Parey.
Snape JW (1989) Doubled haploid breeding, theoretical basis and practical applications. In: Mujeeb-Kaz A and Sitch LA (eds) *Recent Advances in Plant Biotechnology, 1985–88*, pp. 19–30. Manila, Philippines: IRRI and Mexico: CIMMYT.

Mutation Techniques

M Maluszynski, IAEA, Vienna, Austria
I Szarejko and J Maluszynska, University of Silesia, Katowice, Poland

Definitions

Mutation, according to the classic definition, is a heritable change in the DNA structure not caused by the genetic recombination process. An organism carrying a mutated gene or genes is called a mutant. Mutation may change expression of a gene, leading to a different phenotype of the mutant in comparison to the form from which it arose. Mutations are classified as spontaneous or induced. Spontaneous mutations, a major initial source of natural variation conditioning all evolutionary changes, arise as a consequence of errors in DNA metabolism (e.g., replication, recombination, repair) or genome instability. Mutagenic agents, also called mutagens, are used to induce mutations. Any factor that increases the frequency of mutations above spontaneous levels is described as a mutagen. The process generated by a mutagen and leading to mutations is called mutagenesis. There are no qualitative differences between spontaneous and induced mutations.

Spontaneous Mutations

Spontaneous mutations have been observed in natural populations for a long time. These spontaneous changes in phenotype were observed by Darwin, who named them "sports." For many years, spontaneous mutants have been selected by farmers and breeders, propagated and grown as new cultivars. A rice spontaneous mutant "Dee-geo-woo-gen" (DGWG) observed at the end of the 1950s in a local Chinese variety of rice in Taiwan has been most widely used in crosses as a genetic source of semidwarfness, and has contributed greatly to the development of high-yielding varieties, generating (together with wheat semidwarfs) a green revolution in Asia. The DGWG mutation (*sd1* allele) causes significant changes in the phenotype rice producing plants with a short stature, and stiff straw, and that are fertilizer-responsive and photoperiod-insensitive. It was recently discovered that a spontaneous deletion in the gibberellin 20-oxidase gene is responsible for these phenotypic changes. Similarly, wheat "green revolution" genes, *Rht-B1b* and *Rht-D1b*, are spontaneous mutations. They are both nucleotide substitutions, which create stop codons that alter the N-terminal region of a protein that resembles a nuclear transcription factor involved in gibberellin signaling. In vegetatively propagated plant species, spontaneous mutations appear as sectors with different phenotypes. These sectors can easily be vegetatively propagated with conventional methods or with the use of tissue culture. New seedless orange varieties with improved taste and intensive juice color have been developed using this method and there are many other examples where this technology has been applied. It is estimated that spontaneous mutation rates in higher eukaryotes range from 0.1 to 100 per genome per sexual generation.

Similar or even identical mutations to those appearing spontaneously have been obtained as induced mutations. L.J. Stadler in the 1920s initiated the first experiments with the use of radiation to generate genetic changes in plants. The high frequency of mutants observed in the progenies of mutagen-treated populations generated hope of fast breeding success, and this breeding strategy became known as "mutation breeding." Currently, induced mutations are considered as one of the most efficient methods of germplasm enhancement for breeding and basic research, especially in functional genomics and plant physiology. The techniques and procedures leading to induction and selection of desired mutants are called mutation techniques.

Molecular Basis of Mutations

Mutations result from changes in nucleotides or from insertions, deletions, or rearrangements of DNA sequences. These mutations are classified as chromosomal or gene mutations. Those gene mutations that affect a single base or a few base pairs of DNA are called point mutations. The replacement of one nucleotide pair with another is called nucleotide or base substitution. When a base is replaced by another base of the same chemical category, e.g., a purine by another purine or a pyrimidine by another pyrimidine, the base substitution is termed transition. In contrast, a transversion mutation involves a change from a purine to a pyrimidine, or vice versa. Altogether, there are four possible transitions and eight possible transversions (**Table 1**). Among spontaneous base substitutions, the ratio of transitions to transversions is about 2:1. Another category of point mutations are those that involve the addition (insertion) or deletion of a small number of nucleotides.

Mutations that occur in protein-coding sequences can also be categorized according to their effect on the amino acid sequences in proteins. Missense mutations are base substitutions that change a codon for one amino acid to that of another. A base substitution that creates a stop codon resulting in a premature termination of protein synthesis is called a nonsense mutation. When a base substitution creates a synonymous codon (encoding the same amino acid), the mutation is called silent. A neutral mutation is a base pair substitution that produces a different codon and, in consequence, a different amino acid in the protein, but this replacement does not change the biological function of the protein. Mutations resulting from the insertion or deletion of one or a few nucleotides (but not multiples of three) cause a shift in the translational reading frame of the mRNA and are termed frameshift mutations. In the majority of cases, a frameshift mutation results in a nonfunctional protein. A mutation that causes a complete loss of function of a genetic locus is termed a null mutation. Mutations in noncoding sequences of DNA can also have pronounced phenotypical effects if they occur in promoters, enhancers, splice junctions or transcription termination signals. On the other hand, mutational changes outside of coding or regulating DNA regions may not have a directly detectable effect on phenotype.

Table 1 Types of point mutations

Name	*Mutation*
Transition	A→G, G→A, C→T, T→C
Transversion	A→T, A→C, G→T, G→C, C→A, C→G, T→A, T→G
Insertion/deletion	$\pm$ ($3n\pm1$ (or a few) bp), when *n* is an integer

Physical Mutagens

Ionizing radiation (including X-rays and gamma rays, alpha and beta particles, protons and neutrons) and UV light have been utilized for mutation induction in plants. All of these are electromagnetic radiations emitted as quanta, but they differ in wavelength and particular characteristics. The electromagnetic spectrum varies in wavelength: 0.001 nm is characteristic for gamma rays, 10 nm for X-rays, and 100–300 nm for UV-C radiation.

Ionizing radiation can randomly cause damage to all cellular components and induce a variety of DNA lesions. So-called direct effects of radiation result from the direct interaction of the radiation energy with the DNA. Indirect effects are caused by the interaction of DNA with highly reactive species such as ion radicals, free radicals, and excited molecules. In a cell, deposition of energy from ionizing radiation results in formation of a variety of excited and ionized molecules. Radiolysis of water is the major source of reactive species, such as H_2O^+, $\cdot OH$, $H\cdot$, H_2, O_2^-, H_2O_2. It has been estimated that about 65% of damage to DNA caused by ionizing radiation is due to hydroxyl radical attack, as compared to 35% damage resulting from direct ionization. The $\cdot OH$ radicals typically attack the C-5 = C-6 double bond of DNA bases creating a variety of ring-saturated derivatives with altered pairing characteristics. Damage to sugar residues in DNA results in single and double strand breaks. In summary, ionizing radiation can cause a wide range of DNA lesions that appear phenotypically as gene mutations and chromosomal aberrations.

UV light is a nonionizing radiation that creates mutations in DNA by photochemical changes. Absorption of UV light at a wavelength of about 260 nm can cause formation of abnormal chemical bonds between two adjacent pyrimidine bases in the same DNA strand or between pyrimidines on opposite strands of the double helix. Most frequently, two pyrimidines on the same strand are covalently linked by formation of C-5 and C-6 double bonds. The majority of pyrimidine dimers are rapidly repaired by a process called photoreactivation. However, if they are not repaired before replication, the presence of pyrimidine dimers in the DNA affects base-pairing and produces a bulge in the DNA strands. The DNA polymerase cannot traverse the dimer during replication, leaving a gap in the newly synthesized strand opposite the dimer. Point mutations are created when the gap is repaired by incorporation of random bases into a new strand.

Chemical Mutagens

Alkylating agents are the most commonly used chemical compounds in plant mutagensis. They belong to the class of base-modifying agents that act by directly modifying the chemical structure and properties of the bases. The sites of reaction for many alkylating agents have been identified in all four DNA bases, although not all of them are equally reactive. In general, the ring nitrogens of the bases are more reactive than the oxygens, with the N^7 position of the guanine and N^3 position of the adenine showing the highest nucleophilic potential. However, some alkylating agents tend to react most often with the other groups in the DNA, such as O^6 position of the guanine or phosphodiester groups of the DNA backbone. The alkylation leads to the direct mispairing of nucleotides, e.g., O^6 alkylguanidine pairs with thymine instead of cytosine, and results in G–C→A–T transition. Alkylating agents can be either monofunctional or bifunctional. The former interact covalently with single sites in DNA; the latter have two reactive groups, potentially capable of reaction with two different sites in DNA bases, which can result in inter- or intra-strand crosslinks. In general, alkylating agents cause every type of point mutations: transitions, transversions, deletions and frameshifts, as well as chromosome breakages. Detailed studies revealed differences in mutation spectrum caused by particular alkylating agents. For instance, *N*–methyl-*N*-nitrosourea (MNU) induces entirely G–C→A–T transitions in *Escherichia coli*, while *N*-ethyl-*N*-nitrosourea (ENU) causes both transitions (G–C→A–T and A–T→G–C), as well as transversions (G–C→C–G and A–T→C–G), with G–C→A–T transitions being in the majority. Ethyl methanesulfonate (EMS) an alkylating agent used very often in *A. thaliana* mutagenesis, generates mostly C–G→T–A transitions. The alkylating agents applied most frequently in plant mutagenesis are presented in **Table 2**.

Table 2 Alkylating agents most commonly used in plant mutagenesis

Name	*Abbreviation*
Ethyleneimine	EI
Diethyl sulfate	dES
Ethyl methanesulfonate	EMS
N-Ethyl-*N*-nitrosourea	ENU (ENH)
N-Methyl-*N*-nitrosourea	MNU (MNH)
N-Methyl-N^1-nitro-*N*-nitrosoguanidine	MNNG

Sodium azide is a chemical agent that becomes mutagenic only in some organisms, e.g., *E. coli*, yeast (*Saccharomyces cerevisiae*) barley (*Hordeum vulgare*), rice (*Oryza sativa*) and several other higher plants, but it fails to increase mutation frequencies in *A. thaliana, Triticum aestivum*, or *Drosophila melanogaster.* In barley, sodium azide becomes mutagenic only in acidic pH and its action causes mostly A–T→G–C transitions. On the other hand, sodium azide induces few, if any, chromosomal aberrations.

Base analogs (e.g., (*H. vulgare*) 5-bromo uracil, 2-amino purine, 8-ethoxy-caffeine, maleic hydrazide) and intercalating agents (e.g., (*Oryza sativa*) acridines) are also chemical mutagens. However, they are, only occasionally used to induce mutation in plants.

Mutated Generations

Plant material (e.g., seeds, vegetative organs, tissue cultures) before mutagenic treatment is referred to as M_0, and after treatment is referred to as the M_1 generations (**Figure 1**). If homozygous cells are treated with a mutagen, some genes mutate to new alleles, and most cells become heterozygous. As different genes mutate in each cell of treated tissues, the developing M_1 plant becomes chimeric (chimerisms of M_1 plants) and heterozygous. Seeds developed on M_1 plants are already the M_2 generation. The M_2 generation obtained from M_1 plants is segregating but following patterns of chimerism. A plant of the M_2 generation is not chimeric and its progeny, M_3, is homozygous for observed mutations or segregates following Mendelian laws.

Somatic Effect of Mutagenic Treatment

Mutagenesis can be divided into two phases: DNA primary damage and DNA repair. The final frequency of induced mutations on a cellular level mainly depends on the effectiveness of both processes. The level of the DNA primary damage depends on the nature of the mutagenic action of a particular mutagen and, especially, on its ability to penetrate organs, tissues, cytoplasm, and nucleus. For example, the penetration ability of the physical mutagen UV light is limited to one or a maximum of a few plant cell layers, while gamma rays can penetrate plant material almost without limitation. The DNA repair process depends mainly on the efficiency of enzymatic systems, which are themselves the subject of mutations, which affect their efficiency.

Figure 1 Descriptions of mutated generations in induced mutations programs of seed-propagated crops. sd, semidwarf.

This is one of the major reasons for variable radiosensitivity in different varieties from the same species. The same rule applies to the mutagenic effect of chemical mutagens. As mutagenesis is a random process, each cell in the treated plant material will have different levels of DNA damage in the nucleus, cytoplasmic organelles, and all components of the cytosol. This will, of course, influence the metabolism of the cell. The level of damage in cells of various tissues and organs influences the growth and development of M_1 plants. All abnormalities in the plant growth and development can be used for characterization and measurement of the somatic effect of mutagenic action.

M_1 Plant Growth and Development

Mutagens, in a range of doses that are used for induction of mutagenesis (**Table 3**), can generate chromosomal aberrations in M_1, disturbances of the cell cycle, delay in seed germination, reduction of plant emergence, growth reduction, appearance of chlorophyll defects, and reduction of fertility and plant survival. The term "reduction" indicates the change in expression of particular characters in relation to the control, which has, for example, 100% growth and 0% growth reduction.

The reduction in growth, following mutagenic treatment of seeds is expressed as early as the seedling stage. It can be measured using different tests depending on the plant species used. For cereals, the "growing-rack" test developed by C.F. Konzak is often used; however a pot test is more informative as it provides data, not only on seedling growth reduction, but also on emergence reduction. Seedling growth reduction is often used to compare the somatic effect of various mutagens and their doses of treatment. In other words, it indicates differences in sensitivity among various species and varieties of the same species (or radiosensitivity in cases using radiation) to a particular mutagen. Generally, higher doses of mutagen generate higher levels of somatic effects. The difference in sensitivity among varieties can often be significant. It is possible to observe stimulation of growth or emergence, instead of reduction, after mutagenic treatment with low doses; however, the same doses can induce significant growth reduction in another variety (**Figure 2**). Greater somatic effects of the mutagen reduce the size of the M_2 population due to high sterility and low survival rate.

Chromosomal Aberrations in M_1

Chromosomal aberrations in the M_1 generation are the first manifestation of the genetic effect of

Table 3 Doses of physical and chemical mutagens most often used for seed treatment

Crop species	*Mutagen*	*Range of doses*
Rice (*Oryza sativa*)	Fast neutrons	3–8 Gy
	X-rays	95–250 Gy
	Gamma rays	100–350 Gy
	MNH (MNU)	(0.7–1.5 mmol l^{-1}) × (3–5 h)
	ENH (ENU)	(1.7–2.5 mmol l^{-1}) × (3–5 h)
	EMS	(0.2–0.5%) × (8–20 h)
	NaN_3	(0.5–2 mmol l^{-1}) × (3–5 h)
Wheat (*Triticum aestivum*)	Fast neutrons	2–6 Gy
	Thermal neutrons	1×10^{11} N cm^{-2} – 8×10^{12} N cm^{-2}
	X-rays	150–250 Gy
	Gamma rays	50–350 Gy
	MNH (MNU)	(0.75–1.5 mmol l^{-1}) × 5 h
	EMS (ENU)	(0.01–0.04%) × (10–30 h)
	NaN_3	(0.5–2.0 mmol l^{-1}) × 5 h
Barley (*Hordeum vulgare*)	Fast neutrons	2–5 Gy
	Thermal neutrons	4.0×10^{7} N cm^{-2} – 6.5×10^{7} N cm^{-2}
	X-rays	60–200 Gy
	Gamma rays	150–400 Gy
	MNH (MNU)	(0.5–1.0 mmol l^{-1}) × 5 h
	ENH (ENU)	(1.0–2.5 mmol l^{-1}) × 5 h
	EMS	(0.02–2.5%) × (8–20 h)
	NaN_3	(0.5–1.5) × 5 h
Rape-seed (*Brassica napus*)	X-rays	350–450 Gy
	Gamma rays	350–1400 Gy
Soybean (*Glycine max*)	fN	3–6 Gy
	Gamma rays	120–300 Gy
	X-rays	100 Gy
	ENH (ENU)	1–3 mmol l^{-1} × 5 h
	EMS	(0.8–1.5%) × (6–10 h)

fN, fast neutrons.

mutagenic treatment. Analysis of the first mitotic cycle after mutagenic treatment and determination of the frequency of chromosomal mutations is a quick test for the evaluation of mutagenic effects. The cytogenetic test can be used to evaluate the genotoxicity of different agents and, compare the effect of different types of mutagenic treatment or different doses of the same mutagen.

The most frequent chromosomal rearrangements appearing after mutagenic treatment are acentric fragments (**Figure** 3) and dicentric chromosomes, seen as bridges at anaphase mitosis (**Figures** 3 and 4). Fragments and whole chromosomes lagging in anaphase can form micronulei in the daughter cells (**Figures 3E** and **4E**). The frequency of chromosomal aberrations, as with other effects, depends on the type of mutagen and its dose, the plant species, the physiological stage of the treated plants (**Figure** 5) and the treatment conditions. Mutagens may influence the cell cycle, delay cell divisions and reduce the mitotic index. This should be considered when seed samples are estimated for mutagenic treatment to have sufficient numbers of shoots or roots for fixation and chromosome preparation. The detection of chromosomal aberrations, such as translocations, deletions, inversions, or duplications, especially when small chromosome segments are involved, is difficult at mitosis, but they can be identified at meiosis in M_1 plants.

Chimerism of the M_1 Plants

The multicellular embryo is the major target of mutagenic agents in seed-propagated plants. It is expected that almost every single cell of an embryo will have mutations if a mutagenic agent is used in doses used for germplasm enhancement. Depending on the position of cells in an embryo, the chimeric sectors of various sizes are developed, following anticlinal or periclinal cell divisions. As a result of these processes, various tissues and organs building the somatic body of the M_1 plant have a different genetic constitution (**Figure 6A**). Visible effects of the chimeric structure of M_1 plants are chlorophyll sectors on leaves, tillers, and inflorescences. All these genetic changes in somatic tissues will not be inherited in the next generation if sexual reproduction of the M_1 plant is considered. The genetic structure of the generation following meiosis (M_2) depends on the number of initial cells on the

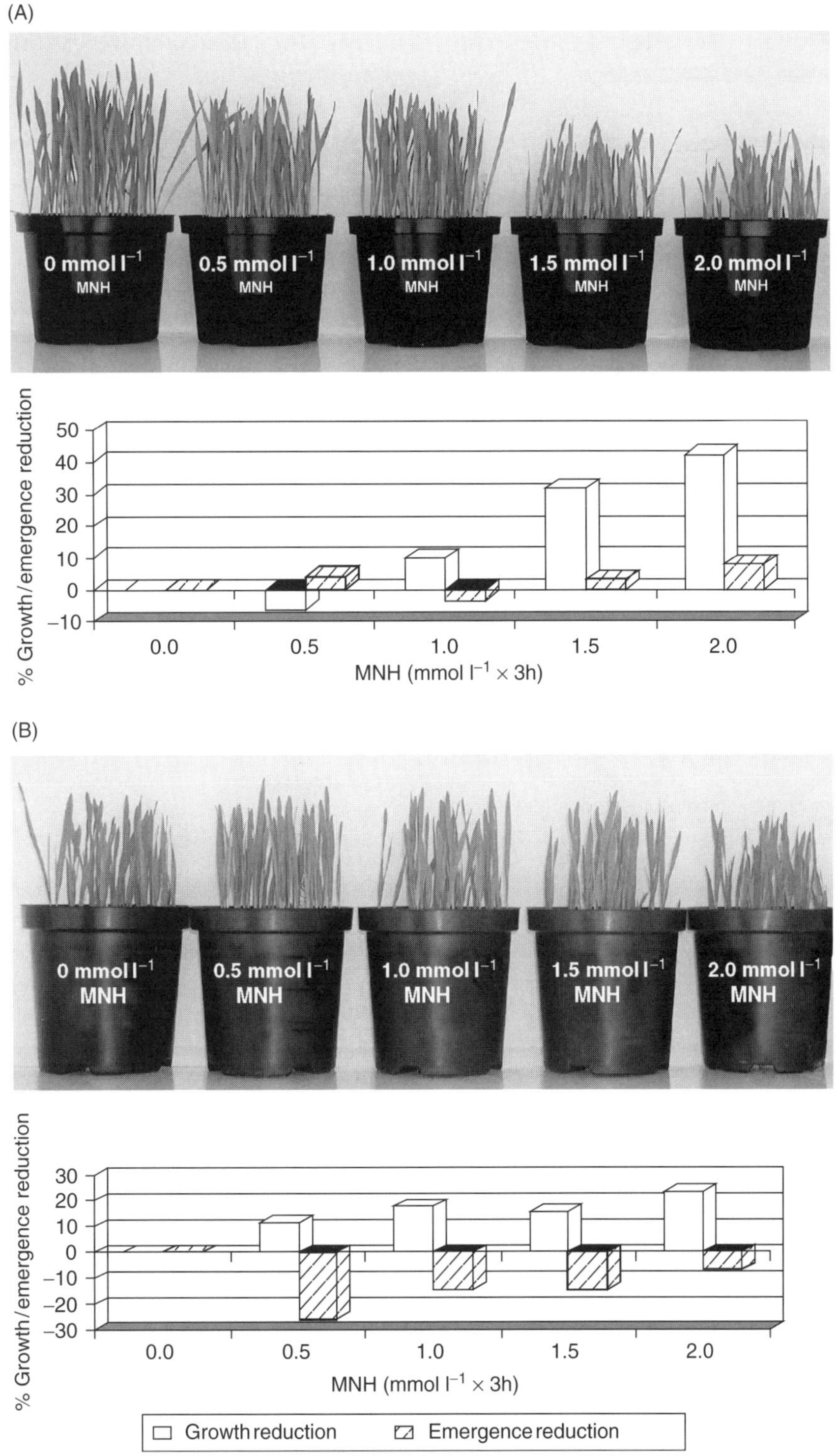

Figure 2 The reaction of two spring barley varieties "Acsad-176" (A) and "Start" (B) to the mutagenic action of MNH.

generative pathway present in the embryo during treatment and the DNA mutation rate induced in them. The location and a number of these initial cells are typical for a particular species, or even varieties, and they are responsible for the chimeric structure of inflorescences. Two to three initial cells contribute to the genetic structure of the M_2 generation in *A. thaliana* and six to twelve in barley (*H. vulgare*). Mutated initial cells contribute, on different levels, to the size of sector in generative tissue (inflorescent) depending on their location in the embryo. The chimeric structure of generative organs can be one

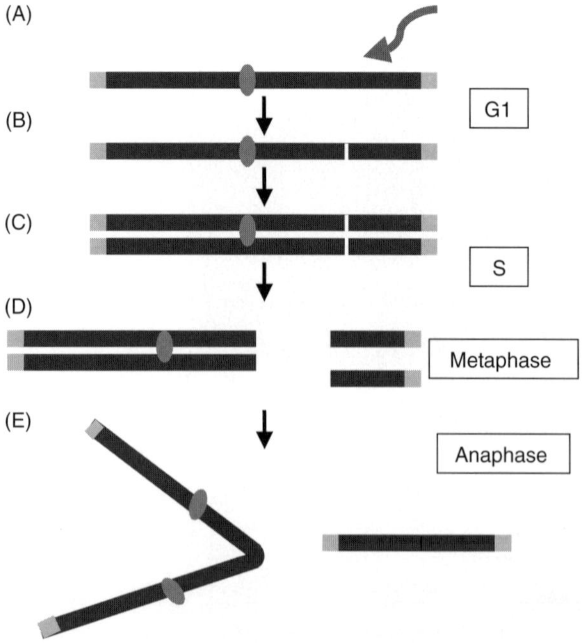

Figure 3 The origin of a dicentric chromosome and acentric fragment as a result of chromosome breakage induced by mutagenic treatment. (A) Chromosome in G1phase treated with mutagen; (B) single break in one arm of the chromosome; (C) replication of the chromosome with unrepaired break; (D) lack of telomeres at the end of the broken chromosome; (E) fusion of "sticky" ends and formation of chromosome bridge and acentric chromosome fragment in anaphase.

of the reasons for disturbed segregation in the following M_2 generation.

In vegetatively propagated species, multicellular organs such as cuttings, root stocks, bulbs, or tissues in *in vitro* culture are mutagenized. As a result of mutagenic treatment followed by cell divisions, a chimeric M_1 plant arises. The dissection of chimeric structure can be achieved by successive vegetative propagation. This leads to the reduction of the number of mutated sectors and to an increase of sector size. In mutated tissue cultures, the method of plant regeneration through somatic embryogenesis from single cell or organogenesis from numerous cells decides on the further genetic constitution of mutated plants.

Genetic Effect of Mutagenic Treatment

Point (Gene) Mutations (Molecular Basis of Mutation)

Many genes in each cell are usually changed as a result of the mutagenic action of higher doses of mutagens. As a simultaneous identical mutation in both alleles of a locus has infinitely low probability, a mutated allele changes from the homozygous (*AA*) to the heterozygous (*Aa*) status in the locus. The Mendelian segregation of 3:1 should be expected in the M_2 after meiosis. However, if the mutated sector in an inflorescence of a M_1 plant is very small, the deficit of recessive phenotypes is often observed in the M_2 progeny. High sterility of M_1 plants can distort segregation also. Some dominant mutations can already be observed in the M_1 generation if the mutated sector was large enough to phenotypically express a mutated allele in the tissue or organ.

A phenotypic mutant carrying a homozygous recessive mutation could be visible in the M_2 generation. Segregating mutants are easy to select in the M_2 generation if mutated characters drastically differ from the parental phenotype. This is the case for mutants with changes in flower or fruit color, leaf or fruit shape, plant height, seed size, shape and color, or plant architecture, but also some changes in disease and insect resistance. Identification of mutants with changes in physiological traits or quality characters require application of related analytical methods.

Chromosome Mutations

Chromosome mutations are variations from the parental type in chromosome number and/or chromosome structure.

Structural chromosome mutations DNA molecules in chromosomes are permanent targets of chemical or physical mutagens. DNA double-strand breaks (DSB) are the principal lesions in the process of chromosomal aberration formation. DBS can arise spontaneously with quite significant frequency through a variety of cellular processes such as DNA replication, transposition, and mitotic recombination, or they can be induced by chemical or radiation treatment. Most DNA breaks are repaired, but if left unrepaired or repaired improperly, may result in chromosomal rearrangements. UV radiation and the majority of chemical mutagens are not able to induce DSB directly, but induce other damage to DNA, which may result in DSB during repair or DNA replication, followed by chromosomal aberration formation.

Chromosome mutations can be detected either by genetic analysis or cytologically during mitosis or meiosis. The ability to detect different kinds of chromosomal aberrations depends on the size, morphology, and the number of chromosomes in the karyotype of the investigated species. Most frequently observed aberrations are acentric fragments and dicentric bridges. Differential chromosome staining (chromosome banding) and molecular cytogenetic techniques (fluorescence *in situ* hybridization, FISH, and genomic *in situ* hybridization, GISH) can

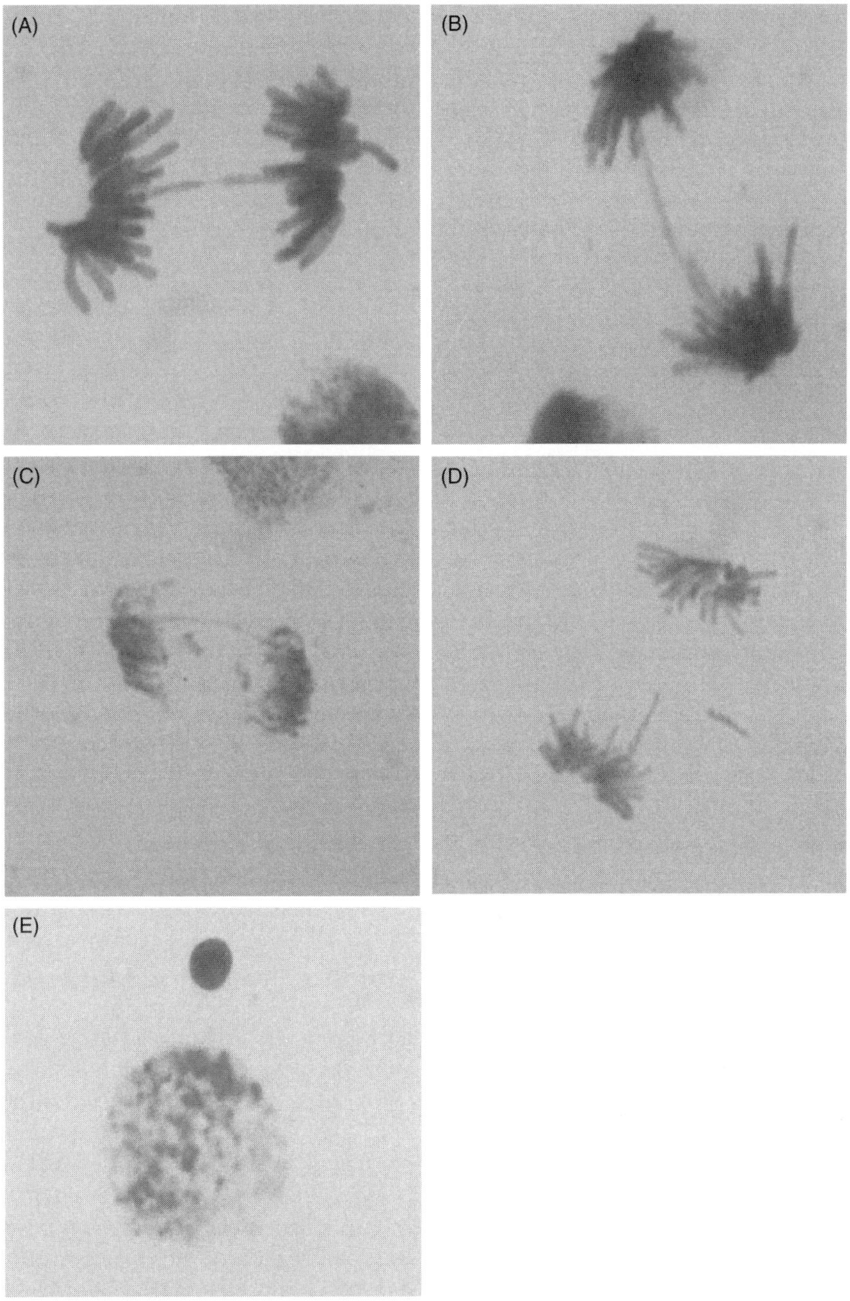

Figure 4 Chromosomal aberrations in M_1 induced by treatment with maleic hydrazide. (A, B) anaphase bridge; (C) bridge and acentric fragments; (D) acentric fragment; (E) micronucleus. (A,B) *Allium cepa,* (C,D) *Hordeum vulgare, and* (E) *Crepis capillaris* cells.

help in the identification of chromosomes and the detection of even small rearrangements (**Figure 7**). Chromosomal aberrations are initiated by one or more breaks in the chromosome. Broken chromosome ends without telomeres (specific DNA sequences at the ends of chromosomes) become "sticky", which means that they may fuse to the other broken chromosome end (**Figure 3**).

There are four main types of structural chromosome rearrangements: deletions, inversions, duplications and translocations (**Figures 8** and **9**). Deletion is the loss of a segment of a chromosome. It can be divided into terminal (**Figure 8B**) and intercalary (**Figure 8C**). Deletion results in an unpaired loop when two homologous chromosomes pair at meiosis, which may be detected cytologically. Inversion is the result of two breaks in one chromosome and reintegration of the chromosome segment in an orientation 180° from the original orientation. Inversion can be divided into: paracentric, where an inverted segment occurs in one arm and does not include the centromere (**Figure 8D**); and pericentric,

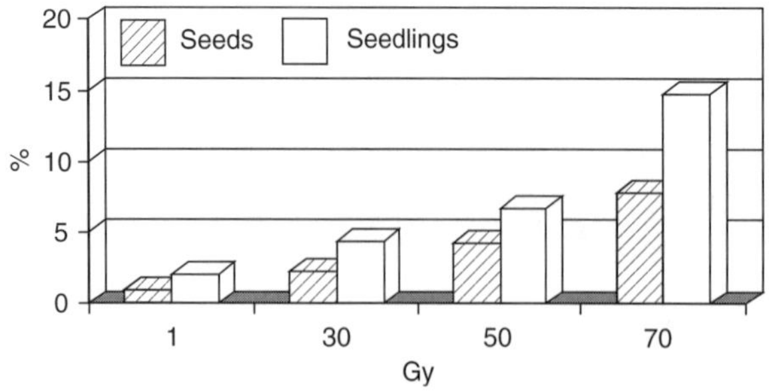

Figure 5 Frequency of chromosomal aberrations in root cells after treatment of dry seeds and seedlings of *Crepis capillaris* with X-rays. Seedlings are more sensitive and display a higher frequency of cells with aberrations.

which involves both arms of a chromosome and the inverted segment includes the centromere (**Figure 8E**). The inverted segment on one homologous chromosome forms an inversion loop during chromosome pairing at meiosis.

Duplication is the doubling of a segment of a chromosome. The size and location of the duplicated segment may vary. When a chromosome segment is transferred between homologous chromosomes, it can occur:

- at a different location (**Figure 8F**);
- adjacent to each other – tandem repeats (**Figure 8G**);
- at opposite orientation to the original location – inverse repeats (**Figure 8H**).

A chromosome segment can also be transferred between nonhomologous chromosomes. Duplications can result in unpaired loops at meiosis, similar to those described for deletions.

A translocation is a change in the position of a chromosomal segment. It can be divided into:

- intrachromosomal – translocation within a chromosome (**Figure 9B**)
- interchromosomal – translocation of a chromosome segment from one chromosome to a nonhomologous chromosome (**Figure 9C, D**) can be reciprocal when an exchange of segments between two chromosomes occurs (**Figure 9D**) and nonreciprocal when translocation occurs in one direction only (**Figure 9B, C**).

Reciprocal translocations can be recognized by a cross-like configuration in prophase I and as rings or chains in metaphase I of meiosis. Such chromosome configurations may affect the products of meiosis leading to a reduction of plant fertility.

Numerical chromosome mutations The basic number of chromosomes characteristic for a genome is described by x while n is the gametic number of chromosomes and $2n$ is the zygotic (somatic) chromosome number (in diploids $2n=2x$). Numerical chromosome variation is divided into two categories: euploidy and aneuploidy. Euploidy is further subdivided into autopolyploidy and allopolyploidy. A euploid is a cell or an individual that carries an exact multiple of the basic chromosome number x. The triploid ($3x$), tetraploid ($4x$), and pentaploid ($5x$) are autopolyploids because one basic genome is multiplied. Autopolyploids can be obtained after treatment with chemicals such as colchicine or oryzalin, which affects spindle formation and induces chromosome nondisjunction during the mitosis. The formation of unreduced gametes may also be a source of autopolyploid plants. Allopolyploid forms contain two or more genomes derived from different species and they arise as a result of interspecific hybridization and the formation of unreduced gametes.

An aneuploid is a cell or an individual with one or more whole chromosomes added or missing in the diploid or polyploid set (e.g., $2x+1$; $4x+1$; $2x-1$). They can appear as a result of irregular chromosome segregation at meiosis or as the offspring of translocation and inversion heterozygotes. Aneuploids are usually unbalanced and have reduced fertility.

Mutation Induction Methods

Irradiation of plant material by gamma rays can be achieved using a gamma irradiator, a gamma chamber, a gamma greenhouse and a gamma field. A gamma irradiator is a simple facility, which is easy to install. Radioisotopes of cobalt-60 (^{60}Co) or caesium-137 (^{137}Cs) are the most common sources of radiation. The gamma irradiator is mainly used for acute irradiation of seeds, cuttings, inflorescence and tissue cultures. The FAO/IAEA Agriculture and Biotechnology Laboratory in Seibersdorf, Austria offers a free-of-charge service for acute irradiation. A gamma chamber can be used for acute irradiation of seeds or whole plants in pots. Most national centers of nuclear technique applications have a gamma irradiation facility and an X-ray machine; however, hospital X-ray machines can also be used for irradiation of plant material. Gamma greenhouse and gamma field methods are used for chronic irradiation of plants. Such a facility is successfully operating in Ohmiya-machi, Japan (**Figure 10**).

The mutagenic treatment procedure for seeds is very simple. Usually, healthy dormant seeds are presoaked in beakers with an abundance of distilled

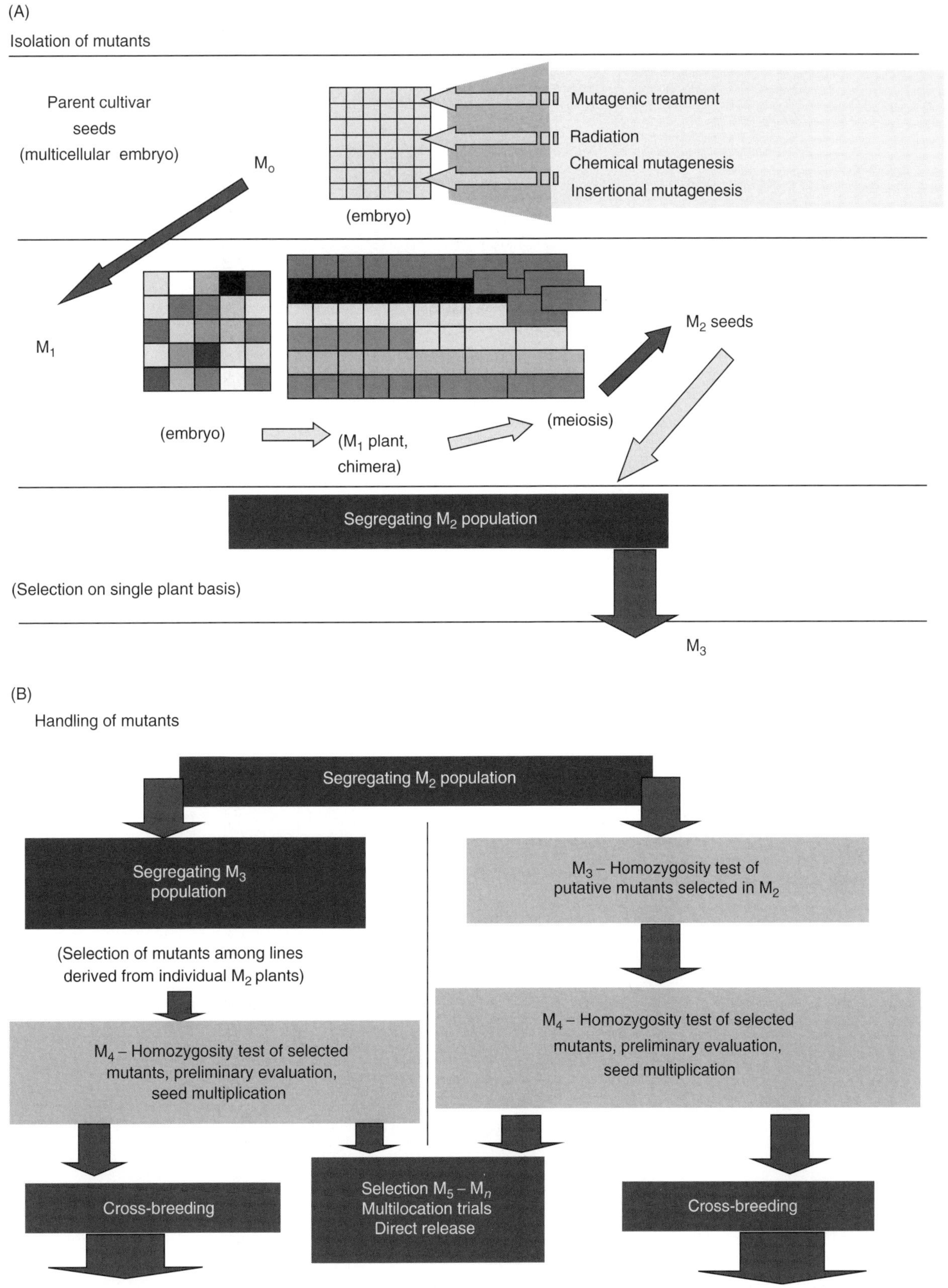

Figure 6 General procedure for the development of mutants in seed-propagated crops. (A) Chimerism of the M_1 plant; (B) Handling of mutated generations.

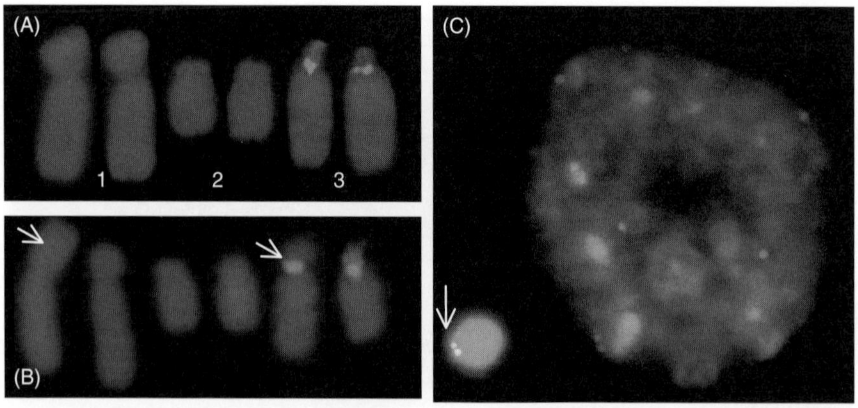

Figure 7 (see color plate 6) Detection of chromosomal aberrations using ribosomal DNA (rDNA) and telomeric sequences as probes for FISH of *Crepis capillaris* chromosomes and nuclei. (A,B) 5S rDNA – red, 25S rDNA – green; (C) telomeric sequences – red; (A,B,C) DAPI staining – blue. (A) Karyotype of *C. capillaris* ($2n = 6$) – control; (B) karyotype with translocated chromosomes (arrows). Chromosome no. 1 – longer short arm; chromosome no. 3 – lack of 25S rDNA signal and longer short arm in one homolog. (C) Micronucleus with double signal after FISH with telomeric DNA indicates that the micronucleus contains one whole chromosome, while 10 signals observed in the nucleus indicates that one chromosome is missing.

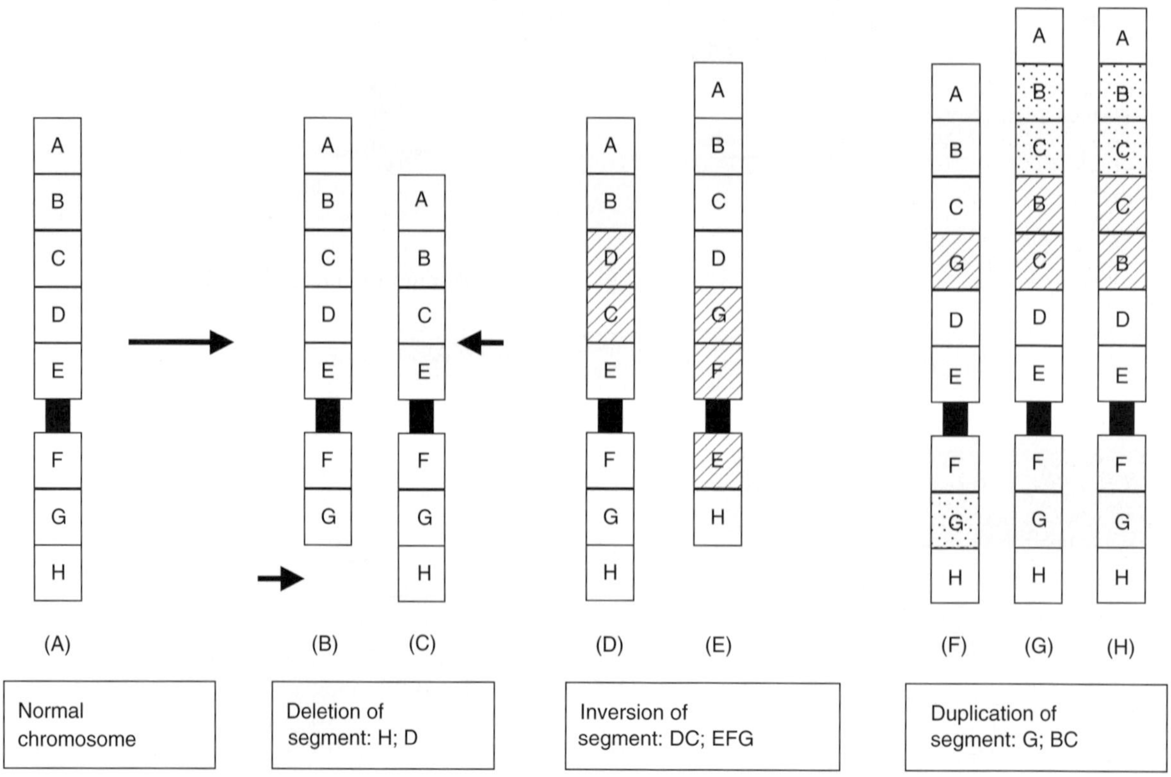

Figure 8 Structural chromosomal mutations: (A) normal chromosome; (B,C) Deletion of chromosome segment marked by arrows; (D,E) inversion – inverted segments appear shaded ▨; (F,G,H) duplications – duplicated segments are stippled ▩.

water at room temperature (5–10 h depending on the plant species) and treated in the same beaker with a mutagen solution for a few hours. The treatment dose is calculated on the basis of a concentration of mutagen and duration of mutagenic treatment (e.g., $1\,\mathrm{mol\,l^{-1}} \times 5\,\mathrm{h}$). The temperature of the treatment solution is also important and should be controlled. Seeds are rinsed several times in tap water after treatment. Caution: a simple biological laboratory, equipped with a fume hood, is sufficient for the induction of mutations with chemical agents. However, all operations with mutagen solutions should be carried out in plastic glows under a fume hood, as most chemical mutagens are strong carcinogens.

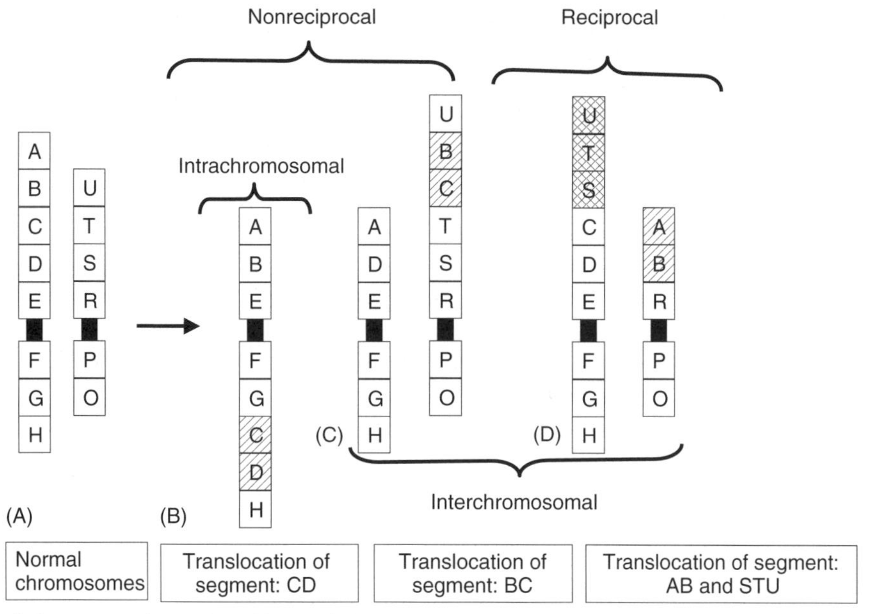

Figure 9 Structural chromosomal mutations. Types of chromosomal translocations: (A) normal chromosome; (B) intrachromosomal translocation; (C,D) interchromosomal translocations. Shaded boxes represent translocated segments of chromosomes.

Critical Dose of Mutagen

The choice of mutagen dose before initiation of a large-scale field experiment is extremely important for any induced mutations project. This could be done in a laboratory. Seedling growth reduction tests are most often used to find the "critical dose" of mutagenic treatment (**Figure 2A, B**). The critical dose, generating between 50% and 70% growth reduction at the seedling stage, results in very high sterility and a reduction in the survival rate observed during harvest, resulting in reduced numbers of seeds for the M_2 generation (**Figure 11**). For plant breeding projects aimed at improving one or two characters in a well-adapted variety, doses of mutagen generating below 30% growth reduction should be applied. Higher doses can be used in germplasm enhancement projects for genomics, where crop adaptability and productivity are not important.

Evaluation of the Frequency of Point Mutations

M_2 generation seeds are used to evaluate the frequency of point mutations generated by an applied mutagenic treatment. In cereals, the appearance of chlorophyll mutants in M_2 is used for this purpose. In the early 1940s, A. Gustafsson (Sweden) described and classified chlorophyll mutations in barley. This classification included the following color types of seedlings: *albina* (white), *xantha* (yellow), *viridis* (pale green), *alboviridis* (partially white then green), and *tigrina, striata, maculata* (striped or irregular chlorophyll spots). The frequency of mutants is defined as a ratio of mutated seedlings to total number investigated. For dicotyledonous *A. thaliana*, A. Müller (Germany), in the late 1960s, classified embryonic lethal mutations in immature siliques. Using these classifications, it is possible to compare not only the frequency of point mutations induced by different mutagens at various doses, but also the spectra of mutants induced by them.

General Procedure for Development of Mutants

The M_1 generation should be grown under good conditions. This is extremely important as plants developed from treated seeds carry a somatic effect of the mutagenic action. The size of the M_1 generation should be calculated to guarantee enough seeds for the M_2 generation. It is strongly recommended that large M_2 populations (about 30 000–50 000 plants) are developed to obtain a large mutation spectrum. In rice, it is necessary to grow 10 000–15 000 M_2 plants to select, for example, semidwarf mutants. Selection for less phenotypically visible characters and for characters strongly modified by the environment should be postponed to the M_3 generation. Homozygous M_2 mutants will give rise to genetically stable M_3 lines, which can be evaluated and characterized easily. Heterozygous M_2 plants with normal phenotype usually give a clear Mendelian segregation in the M_3 generation. Seeds of homozygous M_3 mutants are used for growing mutant lines of the M_4 generation (**Figure 6**). Frequently, homozygous M_4 mutant lines are used for allelism tests and for crosses with other genotypes to transfer the mutated character.

Figure 10 A gamma field (A) and a gamma greenhouse (B) in Ohmiya-machi, Japan. Arrows indicate the source of gamma rays.

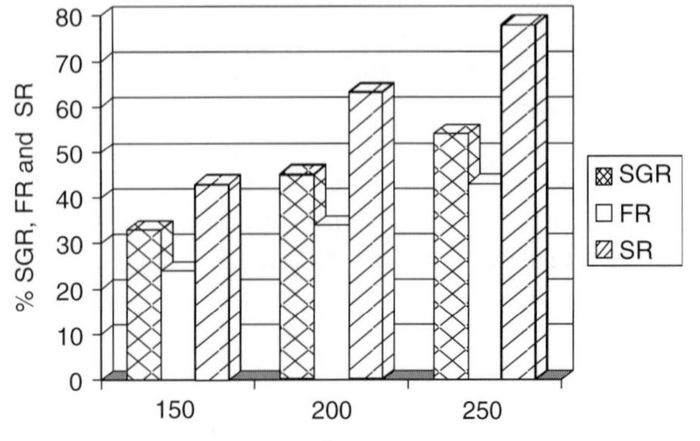

Figure 11 Seedling growth reduction (SGR), plant fertility reduction (FR), and survival reduction (SR) in spring barley after irradiation with different doses of gamma rays.

In vegetatively propagated species, when the *in vitro* culture phase is applied, the selection is postponed at least until the fourth vegetative generation of M_1 (M_1V_4). These four cycles of micropropagation reduce chimerism and significantly increase homogeneity of regenerated plants. If cuttings are mutated, similar results can be achieved through the grafting of mutated cuttings onto root stocks and the propagation of mutated material by successive development of new stems and their grafting onto root stocks.

Induced Mutations in Crop Improvement and Genomics

During the past 70 years, more than 2250 varieties have been released worldwide that have been derived

Figure 12 Induced mutants and mutant varieties in crops. (A) Early and short mutant of local Mexican variety of sorghum; (B) semidwarf mutant from the tall, high-grain-quality rice variety "Kashmir Basmati," USA; (C) well-adapted, high-yielding barley mutant variety "Una LaMolina 1995" (right) and parent variety (left) in Peruvian highland; (D) high-yielding, early cotton mutant variety NIAB78 in Pakistan; (E) drought-tolerant wheat mutant (M_3 generation) released later as mutant variety "Njoro-BW 1" in Kenya; (F–H) root hairs of spring barley, (F) parent variety, (G) root hairless mutant, and (H) short root hair mutant (H), SEM, Poland; (I) apple parent variety and its mutant "Golden Haidegg" (J) released in Austria.

Table 4 Economic impact of induced mutant varieties

Crop	*Country*	*Mutant variety and year of release*	*Mutagen or cross with mutant*	*Basis of value assessment*	*Value or area*
Rice (*Oryza sativa*)	Thailand	RD6 (1977) and RD15 (1978)	Gamma rays	Total crop value at farm gate for the period 1989–98	US$16.9 billion
	China	Zhefu 802 (1981)	Gamma rays	Cumulative planted area between 1986 and 1994	10.6 million ha
	Australia	Amaroo (1987)	Cross	Current annual planted area	60–70% rice growing area in Australia
	Myanmar	Shwewartun (1975)	Gamma rays	Total planted area in 1993	800000 ha
Bread wheat	Pakistan	Jauhar 78 (1979), Soghat 90 (1991), Kiran 95 (1996)	Fast neutrons Sodium azide Cross	Additional income to farmers during 1991–1999	US$87.1 million
Durum wheat	Italy	Creso (1974)	Cross	Additional income to farmers during 1983–1993	US$1.8 billion
Barley (*Hordeum vulgare*)	UK – Scotland Numerous European countries	Golden Promise (1967)	Gamma rays	Crop value during 1977–2001	US$417 million
		Diamant (1965) and derived varieties	X-rays	Total area planted in 1972	2.86 million ha
Chickpea (*Cicer arietinum*)	Pakistan	CM 88 (1994), CM 98 (1998)	Gamma rays Gamma rays	Additional annual income to the growers	US$9.6 million
Blackgram (*Vigna mungo*; urdbean)	India, Maharashtra State	TAU-1 (1985)	Cross	Value of increased production in the season 1998–1999	US$64.7 million
Cotton (*Gossypium* spp.)	Pakistan	NIAB-78 (1983)	Gamma rays	Total value of crop from 1983 to 1993	US$3 billion
				Additional income to growers from 1983 onward	US$486 million
Japanese pear (*Pyrus pyrifolia*)	Japan	Gold Nijisseiki (1991)	Gamma rays (gamma field)	Additional annual income to growers	US$30 million
Grapefruit (*Citrus paradisi*)	USA, Texas	Rio Star® Star Ruby (1971) Rio Red (1985)	Thermal neutrons Thermal neutrons	Currently grown area (year 2000)	7300 ha (75% of total area)

either as direct mutants or from their progenies. In reality, the number of released varieties is much higher than this, as many mutated genes have been used in cross breeding programs without any indication of the nature of the transferred gene. Induction of mutations with radiation has been the most frequently used method for directly developed mutant varieties.

Mutation techniques have mainly been applied to upgrade local, well-adapted plant varieties by altering one or two major traits that limit their productivity or to enhance their quality value. Earliness, semidwarfness, dwarfness, tallness, seedlessness, seed/grain quality (including cooking and processing quality), viral, bacterial and fungal disease resistance, and drought and salinity tolerance are the most frequently induced characters in crop breeding programs (**Figure 12**). The FAO/IAEA Mutant Varieties Database (MVD) is available with detailed information.

Crop mutant varieties have been officially released in 59 countries, mainly in Asia (1142 mutant varieties), Europe (847), and North America (160). The majority of mutant varieties were released in cereals (1072), legumes (311), and industrial (81), vegetable (66) and oil (59) crops. Mutant varieties have contributed billions of dollars to the economies of many countries, including China, India, Italy, Japan, Pakistan, Thailand, and the USA (**Table 4**).

Recently, many germplasm enhancement programs have been initiated to organize large-scale mutant collections for genomics projects. The assignment of biological function to genes lags behind our ability to obtain sequence data. Mutational analysis of plant traits appears to be one of the simplest current approaches to generate polymorphism and tag the function with the sequence of differentially expressed genes. Gene discovery by forward genetics and map based cloning is being enhanced by reverse genetics approaches. Insertional mutagenesis, in which each mutation is molecularly tagged, has been developed and widely applied in the functional genomics of corn, rice (*O. sativa*) and *Arabidopsis*; however, it is very labor and cost intensive, whereas radiation and chemical mutagenesis are more efficient but anonymous. The use of mutagenized populations in combination with microarrays and other high-throughput methods, such as TILLING (Targeting Induced Local Lesions IN Genomes), offers a potentially powerful way to detect gene-specific mutations and move toward the closure of the "phenotype gap," which is still evident for most genes and traits in most plant species.

List of Technical Nomenclature

Chimerism	The presence of two or more genetically different types of cells in an individual.
Chromosome mutation	Any change in chromosome structure or number.
Mutagen	An agent that leads to an increase in the frequency of mutations above the spontaneous mutation level.
Mutagenesis	The process by which mutations are produced.
Mutant	An organism that carries one or more mutations in its genome.
Mutant frequency	The proportion of mutants in a population.
Mutant variety (cultivar)	A crop variety developed through induced mutations and officially released as a direct mutant or as a recombinant from a cross with a mutant.
Mutation	A heritable change in the DNA structure of an organism's genome.
Mutation rate	The number of mutations occurring per gene per unit of time.
Point mutation	A mutation that results from the substitution of one base pair for another or from the addition or deletion of a small number of nucleotides.

See also: **Crop Improvement and Biotechnology**: Insertional and Transposon Mutagenesis. **Genetics of Crop Improvement**: Germplasm Conservation.

Further Reading

Broertjes C and van Harten AM (1988) *Applied Mutation Breeding for Vegetatively Propagated Crops*. p. 345. Amsterdam: Elsevier Science.

Friedberg EC, Walker GC and Siede W (1995) DNA Repair and Mutagenesis, p. 698. Washington, DC: ASM Press.

FAO/IAEA Mutant Varieties Database Website; http://www-mvd.iaea.org (2003).

IAEA (1977) Manual on Mutation Breeding: Technical Reports Series No. 119, p. 288. Vienna: IAEA.

Maluszynski M, Nichterlein K, van Zanten L, and Ahloowalia BS (2000) Officially released mutant varieties – the FAO/IAEA Database. *Mutation Breeding Review* 12: 1–84.

Maluszynski M, Szarejko I, and Maluszynska J (2001) Induced mutations in wheat. In: Bonjean AP and Angus WJ (eds) *The World Wheat Book*, pp. 939–977. Londres: Lavoisier Publishing.

Natarajan AT (2002) Chromosome aberrations: past, present and future. *Mutation Research* 504: 3–16.

D

DISEASES

Contents

Plant Pathology, Principles

G N Agrios, University of Florida, Gainesville, FL, USA

Introduction

The principles of plant pathology are statements that hold true for a large number and variety of plant diseases that share some basic common characteristics, for example, they are all caused by microorganisms. Regardless of their specific cause, such diseases share the same principles, as opposed, for example, to the abiotic (environmental) diseases that are caused by physical or chemical nonliving factors. The principles, however, often need to be modified somewhat to accommodate changes caused by the extreme differences among pathogens, such as those between fungi and viruses. The principles of plant pathology also multiply as we consider the various stages (infection, defense, genetics, epidemiology, control) of disease initiation, development, and control. Nevertheless, the principles of plant pathology provide some true statements that unify the extraordinary variety of events accompanying plant diseases.

The Nature and Effects of Plant Pests and Diseases

Plant pests and diseases interfere with the growth and cause damage to cultivated and naturally growing plants. The interference and damage often result in failure of plants to grow and produce or in destruction of parts or of total plants. Interference of pests and diseases with the growth of and production by plants, and their damage and destruction of plants removes plants and their products from availability for food by humans and animals. It is estimated that, as a minimum, pests and diseases destroy approximately 30–40% of the produce of cultivated plants grown annually by humans for food or feed. Losses are greater in less developed countries where knowledge about and materials for managing or control of pests and diseases are lacking or limiting. In addition, pests and diseases cause humans, in their efforts to manage or control pests and diseases, to spend a great deal of time and money building spraying and refrigeration equipment and vehicles, and to apply millions of kilograms of toxic pesticides on plants and the soil.

Plant pathology is the science that studies the causes of plant diseases, the mechanisms by which diseases develop in individual plants and in plant populations, and the ways and means by which plant diseases can be managed or controlled. Plants, like all living organisms, become diseased as a result of infection by living microbes and a few other organisms, and by exposure to adverse environmental factors. Plants have internal mechanisms of defense that protect them against diseases but, when defenses are insufficient, they become diseased. The study of plant diseases and how they spread enables humans to develop ways and means by which they can protect plants from becoming diseased and from allowing the disease to spread rapidly into individual plants or into plant populations, thereby significantly reducing losses of plant produce from diseases.

The Principles of Plant Health

A plant is healthy as long as it has its complete genetic make-up and is free of deleterious mutations; as long as it is growing under environmental conditions such as temperature, soil moisture, etc., that remain within the normal range of such conditions for the particular species and variety of plant; and as long as it is free of microorganisms and other organisms that injure or otherwise interfere with the normal functions of a healthy plant. The normal functions of healthy plants include: absorption of water and nutrients through the roots; translocation of water and nutrients from the roots through the vascular tissue (xylem) of the stem and shoots to the leaves and fruit; photosynthesis; translocation of photosynthates from the leaves to the fruits, roots, and other organs and tissues of the plant; transpiration; assimilation and/or storage of nutrients; respiration; and reproduction (**Figure 1**). As long as these functions can be carried out within the normal rates prescribed by the genetic material of the particular plant, the plant is considered healthy and grows normally, and produces at the expected

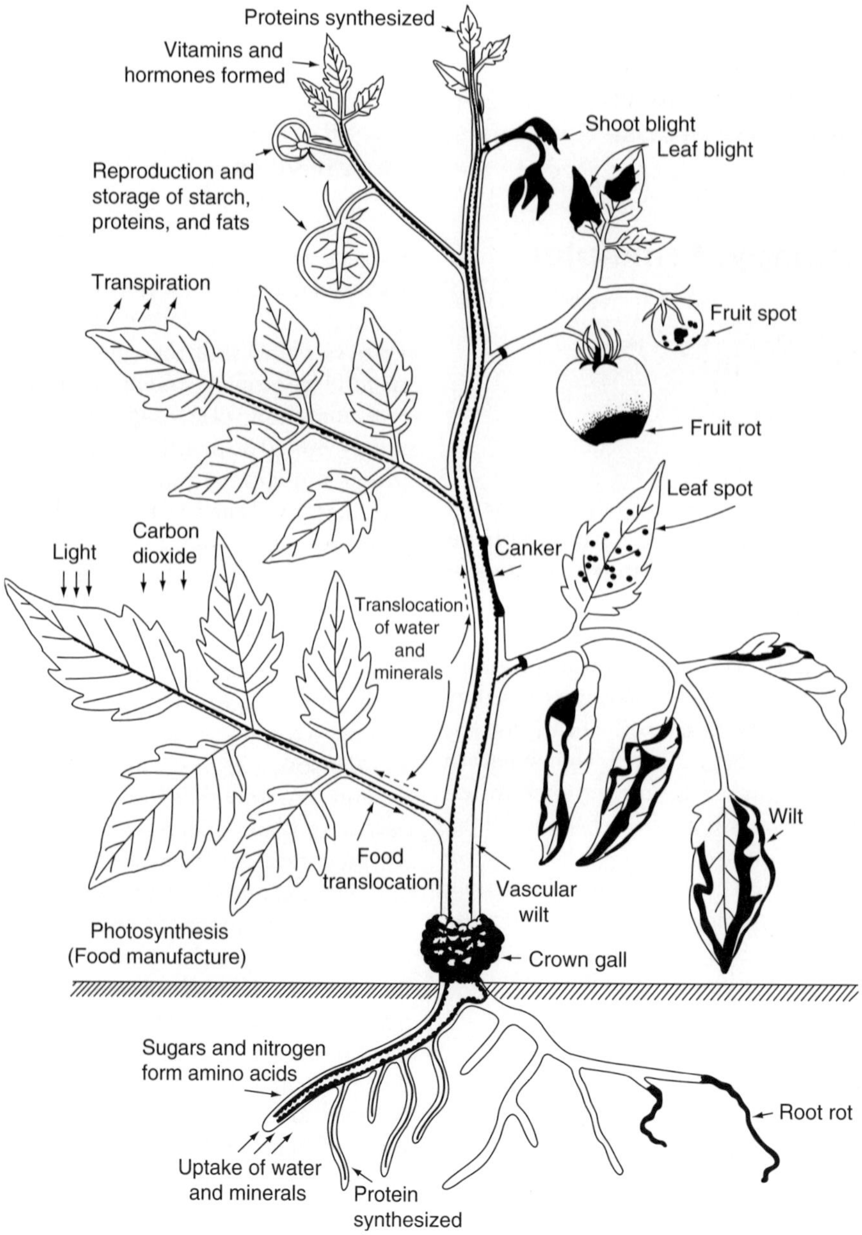

Figure 1 Schematic representation of the basic functions in a plant (left) and the interference with and some of the symptoms caused by several common types of plant diseases. Reproduced with permission from Agrios GN (1997) *Plant Pathology*, 4th edn. San Diego: Academic Press.

rate whatever it is the produce for which it is cultivated. On the contrary, interference with the carrying out of any of these necessary normal functions by either an abiotic (environmental) or a biotic factor, that is, another organism microscopic or not, the plant becomes diseased. In a diseased condition, the plant cannot grow normally, cannot grow as much or as well, it produces less yield or of inferior quality because parts of it may show different types of damage (symptoms), and parts of the plant or the entire plant may die.

The Principles of Plant Pathology

Plant pathology considers as disease the reduced growth or the damage or injury caused to plants by the interference with their normal functions of the plant by environmental (abiotic) factors and by microorganisms and certain other organisms. Plant pathology does not deal with injury or damage caused to plants by insects and mites and by other herbivore animals, nor by humans cutting or otherwise utilizing plants.

Principles of Abiotic Diseases

The recognized principles governing abiotic diseases are:

- abiotic diseases are caused by nonliving factors
- they do not spread from infected to healthy plants
- they may be caused by a single or by a combination of several environmental factors
- abiotic diseases usually appear suddenly, may cause minor to extremely severe losses, and may cover and affect at once from a few plants to all the plants over huge areas of millions of square miles
- control of abiotic diseases may involve minimal efforts or it may be impossible.

Causes of Abiotic Diseases

The main causes of abiotic diseases responsible for the vast majority of crop losses caused by abiotic factors are shown in **Figure 2**. The main principles of some of the major causes of abiotic diseases are as follows.

Extremes in temperature Extremes in low temperature slow down or completely disrupt normal cellular functions, while extremes in high temperature accelerate and finally destroy normal cellular functions. The normal functions of plants cells can be carried out by them only within a range of temperatures determined by the genetic make-up of the plant. Exposure of a plant to temperatures below that plant's low limit causes the plant cell contents to freeze and to slow down or to develop frost damage (death) in some or all tissues and organs. Even at temperatures slightly above the lower limit, the rate of the cellular functions is decreased and the plant may develop symptoms of disease. Similarly, in plants exposed to temperatures higher than the upper limit of their normal range, plant enzymes and hormones are hyperactivated, and the cellular functions are at first accelerated abnormally and later are disrupted and become uncontrolled, leading to malfunction and death of affected cells, organs, or entire plants.

Extremes in soil moisture Excessively low moisture slows down and finally makes impossible normal biochemical reactions in cells, while prolonged excessively high soil moisture excludes oxygen from reaching the roots, which asphyxiate and die. All normal functions of cells and plants are carried out in the presence of water. Different types of plants have different ways to obtain and to conserve water. Reduction in the amount of water below a certain level (drought) makes the plant wilt and die. On the other hand, since all cells require oxygen to carry out respiration and produce energy, water-saturated soil (flooded) drastically affects plant roots which suffocate and die. Such plants remain small and discolored, and eventually die.

Extremes in light Extremely low light does not allow plants to carry on photosynthesis and to produce appropriate hormones and so the plants starve and malfunction. Extremely high light is uncommon in nature but, under some circumstances, plants are injured by the excessive ultraviolet light in it. Plants require sunlight or appropriate artificial light in order to carry on photosynthesis. Photosynthesis is the only chemical function on earth that generates energy that is used by plants to combine water from the soil and carbon dioxide from the air into sugars. The sugars produced by plants are used to produce all the other organic compounds used by all living organisms. When plants are grown under reduced light, for example in shaded areas or during prolonged cloudy weather, their normal cell functions are affected in ways that result in shorter or taller spindly (etiolated) plants that are easily distinguished from plants receiving normal amounts of light.

Air and soil pollutants Plants need oxygen and carbon dioxide from the air and oxygen from the soil but they are sensitive to and become diseased when toxic gases or solutes reach their cells. Plants obtain oxygen and carbon dioxide from the air but they cannot exclude other airborne compounds some of

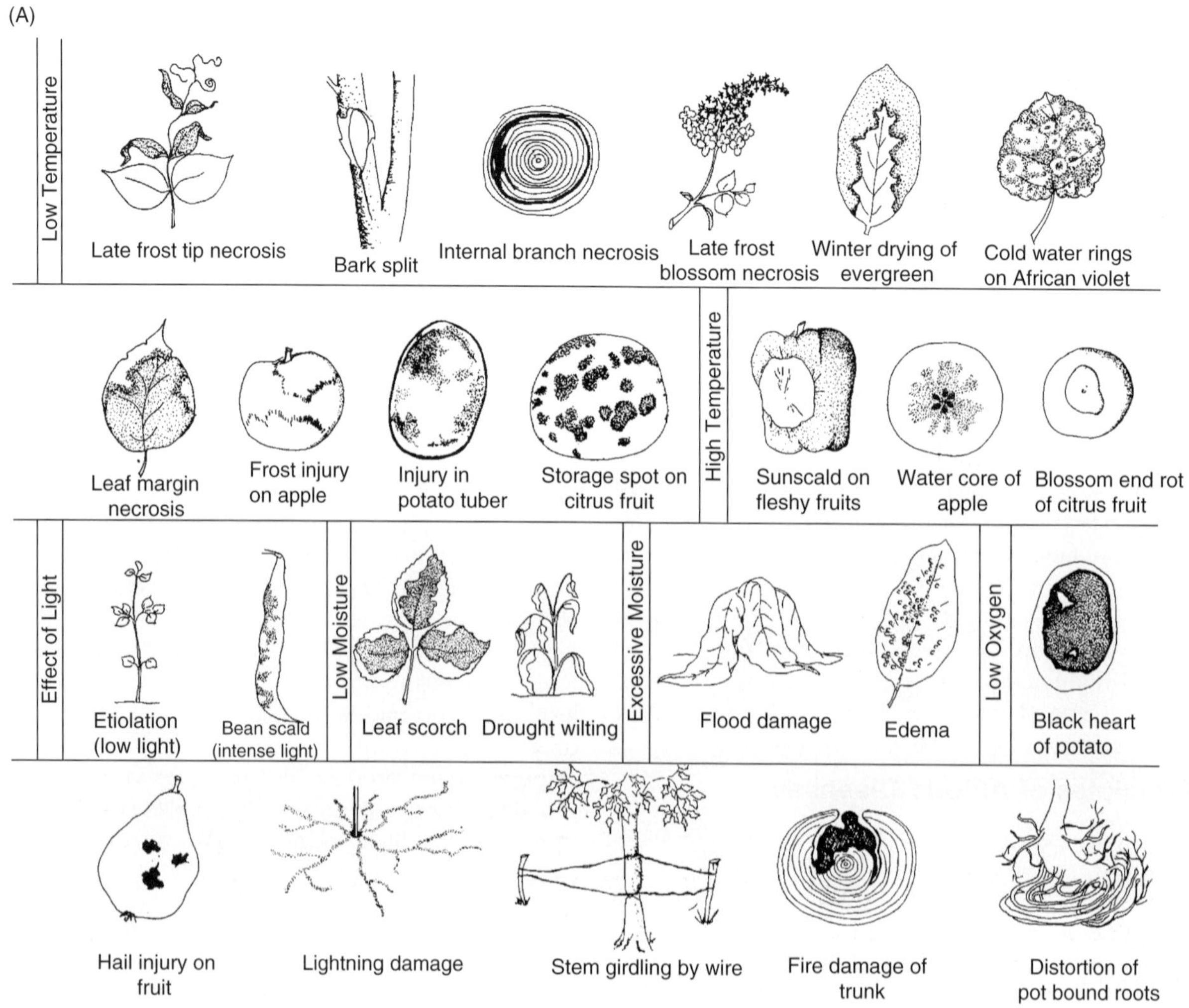

Figure 2 Schematic presentation of various types of symptoms caused by different abiotic (environmental) factors. Reproduced with permission from Agrios GN (1997) *Plant Pathology*, 4th edn. San Diego: Academic Press.

which may be toxic to plant cells. A number of such toxic air pollutants, including ozone, sulfur dioxide, nitrogen oxide, incompletely burned compounds resulting from automobile exhausts and from other sources, chlorine, fluoride, ammonia, etc., are often present in high concentration in the air surrounding the plants and upon absorption by the plant cause numerous small or large necrotic lesions on their leaves. Such plants are obviously diseased and grow and produce only a portion of the growth and yield attained by healthy plants. Similar are the effects of toxic compounds or toxic elements present in the soil and absorbed by the roots of the plant. Such soil-borne compounds exert their toxic effects on root, leaf or other cells many of which cause the entire plant to be affected, grow poorly, and, often, to die.

Nutritional deficiencies or toxicities Plants require varying amounts of a number of inorganic elements for their nutrition and well-being: shortages or excesses in the available amounts of essential elements lead to disease. Shortage in even one element below a required minimum affects the performance of the particular normal function or functions for which the element is required and this results in a disease expressed as some kind of foliar symptoms and reduced growth or death of the plant. On the other hand, excess amounts of one or more of even the needed elements present in the soil results in toxicity symptoms developing in affected plants with accompanying reduced growth or death of the plant.

Inappropriate cultural practices Adverse or untimely actions by humans often damage plants. Unlike native plants, that is, weeds, cultivated plants need and/or are subjected to cultural practices to improve their chances for better growth and yield. Humans often carry out cultural practices of plants that are inappropriate for the particular plant, are carried out using inappropriate materials or inappropriate

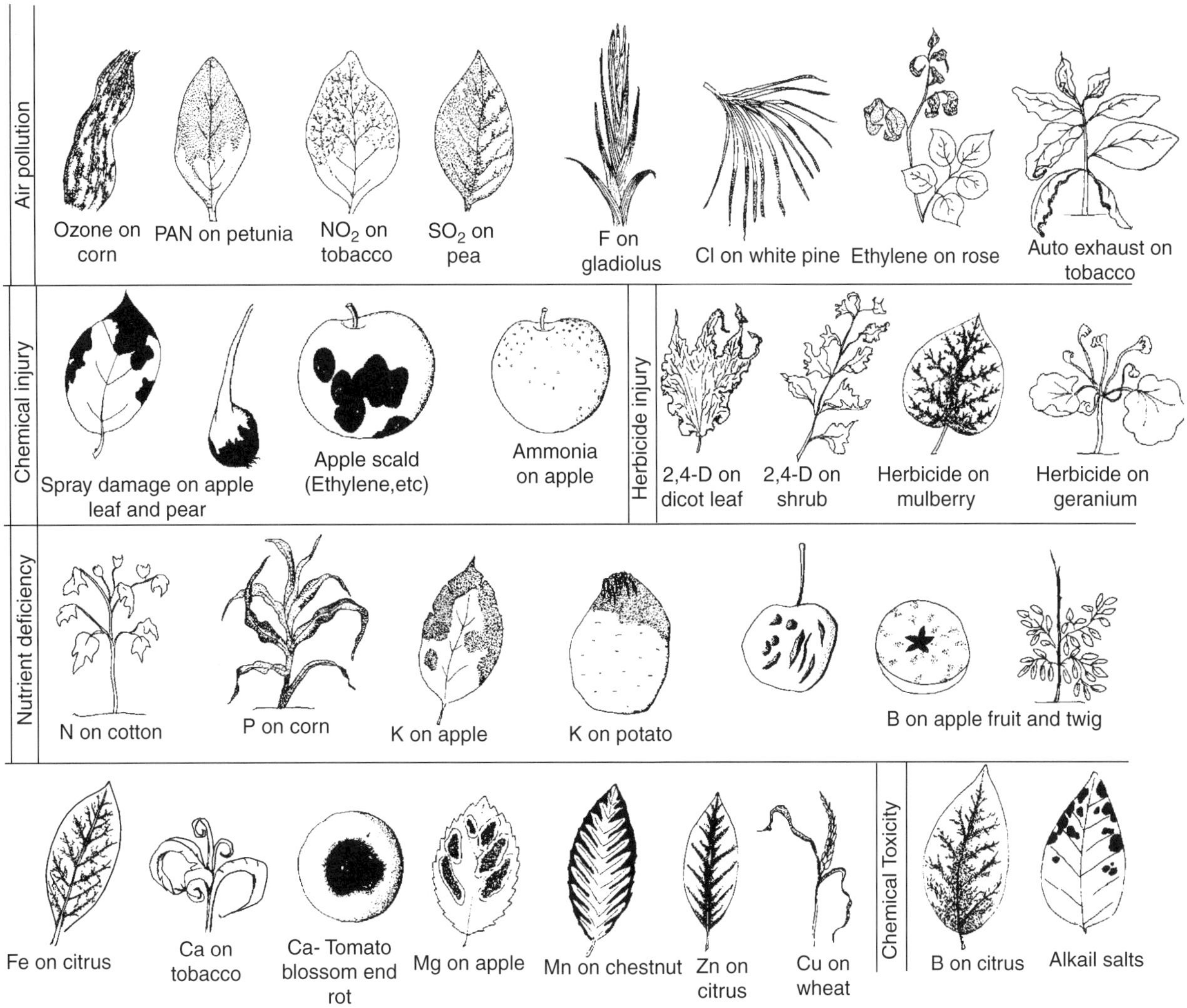

Figure 2 Continued.

dosages of materials, or are done in the wrong way or at the wrong time. Such practices may involve application of pesticides or fertilizers, hoeing, pruning, etc. Injured plants may develop symptoms that may vary in severity up to and including death of the plant.

Lightning, etc. Excessive physical energy (e.g., lightning) or force (e.g., wind, hurricane) can damage or kill plants. In some areas, lightning damage of plants is rather common. Both trees and field plants can be affected, the plants being seriously injured or killed by lightning.

Diagnosis and Control of Abiotic Diseases

Diagnosis of abiotic diseases is difficult because it must take place after the disease has occurred by which time the cause may no longer be present. Since in most abiotic diseases the cause of the disease may or may not be present in the field by the time disease symptoms are produced (for example in a flood, a frost, pesticide injury, etc.), one must examine the history of recent meteorological events and of cultural practices that may explain the observed symptoms. If the suspected cause of the disease can be replicated in an experiment, for example, a pesticide application, it should be tried to verify the diagnosis of the disease.

Control of abiotic diseases depends primarily on preparing for future occurrences. Once the cause of the disease has been determined, appropriate controls may be too late for the present condition but all possible measures should be taken to prevent the same situation from recurring in the future.

Principles of Biotic Plant Diseases

The recognized principles governing the causes, development, and control of biotic diseases are as follows: Biotic, infectious, or contagious diseases of plants are caused by individuals of the following types of microorganisms: fungi, bacteria, viruses, viroids, nematodes, and protozoa (**Figure 3**). These are the same types of microorganisms that cause diseases in humans and animals but the individuals are different. The relative shapes and sizes of plant pathogens in relation to a plant cell are shown in **Figure 4**. A number of plant diseases are also caused by parasitic higher plants and parasitic green algae. All organisms that can cause disease in plants are called plant pathogens.

Plant pathogenic organisms can spread from infeced plants to healthy ones where they initiate new diseases. Although the pathogens are the only things that move and spread, the diseases they cause are also said to "spread" from plant to plant. Biotic plant diseases are also known as "infectious diseases" or "contagious diseases." Plant pathogens multiply at various rates, some of them completing a life cycle (a new generation) in a few minutes while others require several days and some require several years to complete a life cycle.

Plants are attacked by plant pathogens because plants contain nutrients that the pathogens need as their food. Some pathogens, the viruses and viroids, do not consume nutrients of the plant as food. They utilize living plant cells, however, for their machinery and their compounds which are needed to reproduce more viruses and viroids.

Some pathogens complete one or many life cycles on the same plant but others, such as the apple scab fungus, pass part of their life cycle in infected plant debris on or in the soil. Some pathogens, such as some fungi causing the rust diseases, pass their life cycle on two alternate host plants very different from each other (for example, wheat (*Triticum aestivum*) and barberry (*Berberis* spp.)).

Some pathogens can infect, grow, and multiply only in living cells and tissues and are known as obligate parasites. Such are the fungi causing powdery mildews, downy mildews, and rusts; the xylem- and phloem-inhabiting bacteria and phytoplasmas, and all the other phloem-inhabiting plant pathogens. The remaining plant pathogenic fungi and bacteria can grow and multiply in living as well as dead plant tissues and are called facultative parasites or facultative saprophytes.

Principles of Pathogen Attack on Plants

The main principles governing the mechanisms by which pathogens attack plants are:

- each kind (race or species) of a pathogen attacks only certain kinds (varieties, species) of plants which are its "host plants"

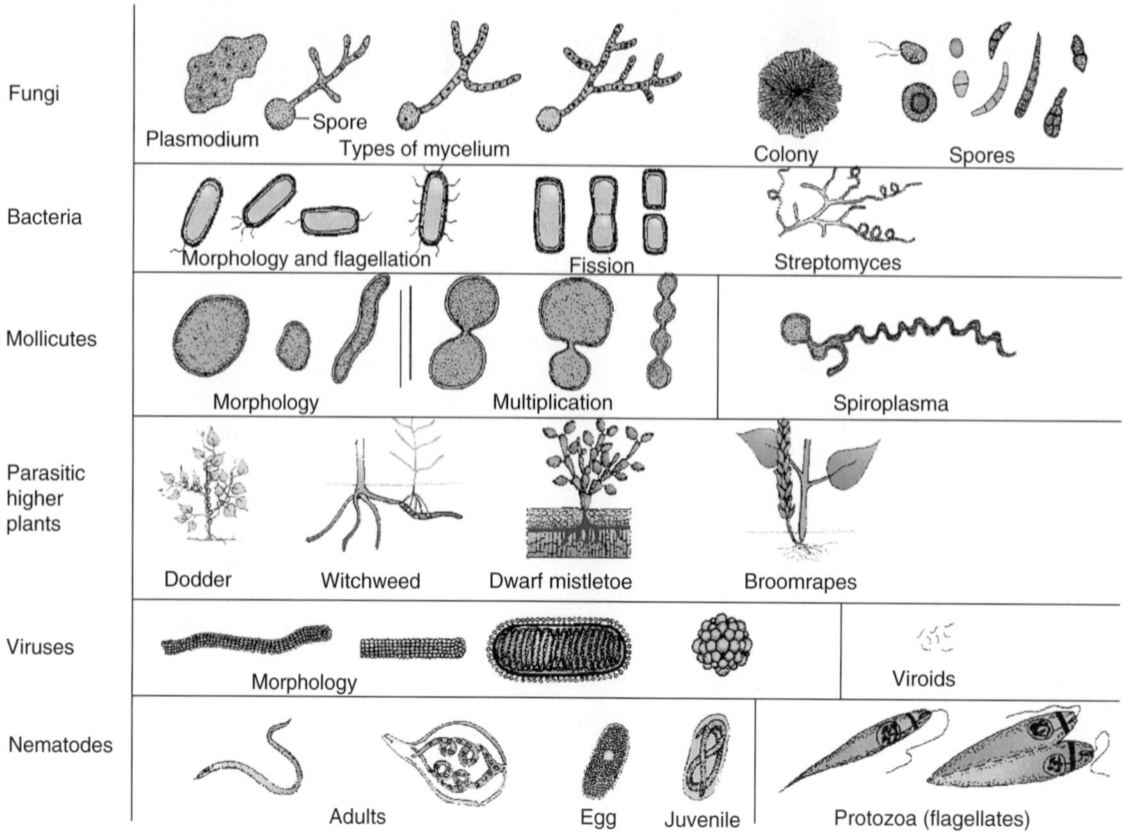

Figure 3 Morphology and multiplication of some of the groups of plant pathogens. Reproduced with permission from Agrios GN (1997) *Plant Pathology*, 4th edn. San Diego: Academic Press.

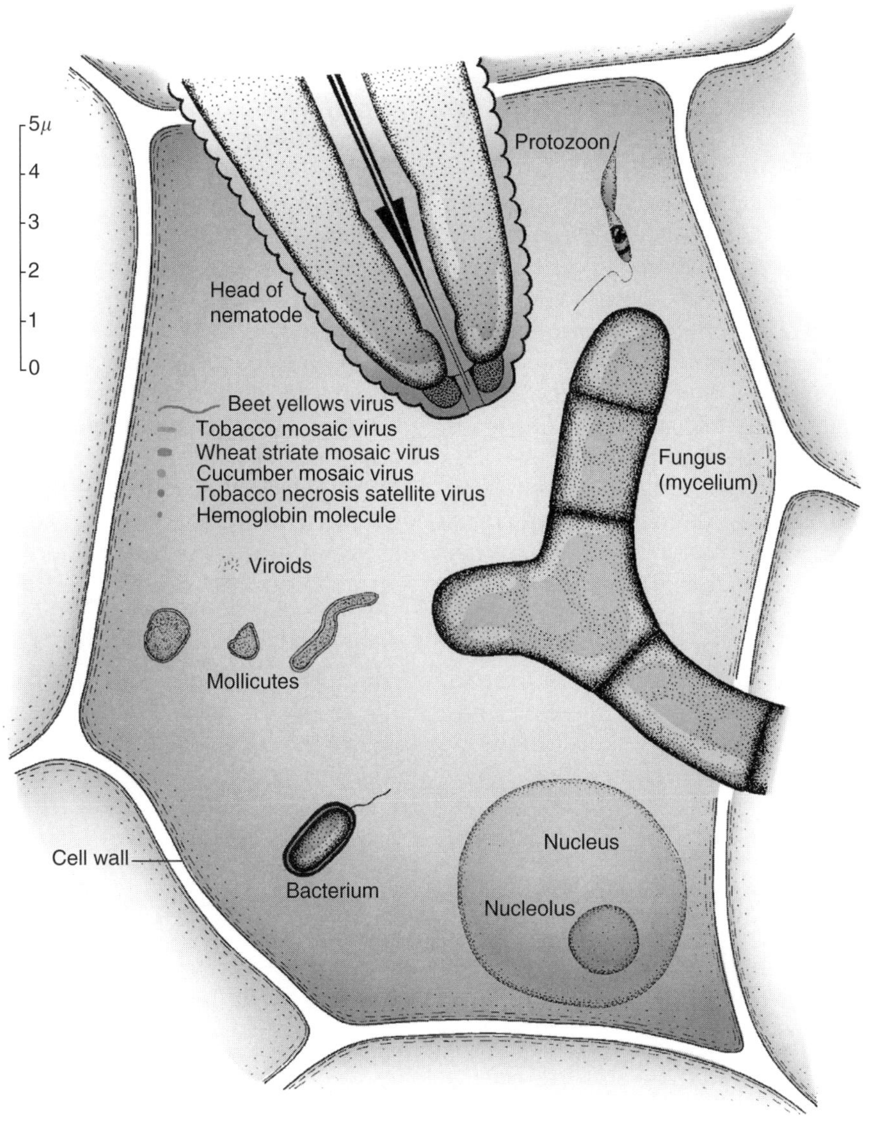

Figure 4 Schematic diagram of the shapes and sizes of certain plant pathogens in relation to a plant cell. Note: Plant pathogenic bacteria, mollicutes, and protozoa exist outside nucleated living plant cells. Reproduced with permission from Agrios GN (1997) *Plant Pathology*, 4th edn. San Diego: Academic Press.

- plants are either immune to a pathogen or they show various degrees of susceptibility or resistance to the pathogen.

In order to cause disease, pathogens must penetrate and enter the plant, and must be able to obtain nutrients and multiply in or on the plant. Some pathogens (bacteria, some fungi) enter plants through natural openings such as stomata and lenticels, others (some bacteria, fungi, viruses) must penetrate through fresh wounds, and others (some fungi, all nematodes, parasitic higher plants) penetrate host plants directly. Once inside the plant tissues, pathogens utilize the plant cell contents as food for their survival and grow and/or move through the plant intercellularly (between the cells) or intracellularly (through the plant cells). Some pathogens (some fungi, nematodes, parasitic higher plants) may use physical force to enter the plant and to penetrate through its tissues; others (bacteria, many fungi) use only chemical means (enzymes) by which they break down cellular barriers and then move; still others (viruses, viroids) move from living cell to living cell through part or all of the plant under forces not yet clear.

Pathogens secrete biologically active compounds (enzymes, toxins, growth regulators, etc.) into their host plants and through them disturb the structural integrity of the plant, injure or kill plant cells, and alter the ability of the plant cells to divide and enlarge appropriately. Infected plants can tolerate a certain amount of injury caused by pathogens without showing any symptoms but beyond that amount of injury plants show internal and external symptoms of disease that may range from minor to death of the plant. The symptoms caused by a disease depend on the types of plant cells or organs affected and the main normal functions of such cells that are inhibited. Many types of infections (leaf, fruit, or stem spots or lesions) by most pathogens may cause disease and may affect the growth and productivity of the plant but they seldom kill plants. On the other hand, infection by certain fungi or bacteria of the xylem vessels of plant roots and stems almost always leads to the death of the infected plant.

Some pathogens, although living and multiplying inside the plant, produce reproductive structures (fungi produce spores, bacteria release bacteria, nematodes eggs, parasitic higher plants seeds) outside the plant from where they spread to other plants. Many pathogens (xylem-inhabiting fungi and bacteria, phloem-inhabiting bacteria, and all viruses and protozoa) do not exit the plant ever and are disseminated to healthy plants by specific vectors, primarily specific insects, mites, nematodes, fungi, etc.

Principles of Plant Defense against Pathogens

Plants defend themselves against attacks by pathogens through physical and chemical means. The principles that govern these defense measures are as follows. All plants have always present on their periphery certain preexisting defensive structures (leaf hairs, small stomata, thick cuticle and cell wall, etc.) that stop some pathogens from penetrating. Also, all plants have in their epidermal and other cells certain preexisting chemical compounds (phenolics, etc.) that are toxic to most pathogens and prevent them from penetrating and becoming established in the plant. Some plants (for example, onion (*Allium cepa*)) even release in their surroundings certain toxic compounds (for example, protocatehuic acid) that kill spores of certain pathogens (onion smudge fungus) before the spores can attack the plant.

Plants often defend themselves against attack by pathogens by producing, after the initiation of attack, defensive cellular and histological structures (cork layers, tyloses in xylem vessels, abscission layers, gums, etc.) that often stop the further spread of a pathogen within the plant. Most frequently, plants defend themselves against attacks by pathogens by producing, after initiation of attack by a pathogen, a variety of defensive compounds that, acting alone or more commonly together, are toxic to the pathogen or otherwise interfere with the survival of the pathogen. In many cases, plants defend themselves against attacking pathogens through the hypersensitive response (RH). This occurs in the attacked and the surrounding cells in which there is a sudden burst of generation of active oxygen radicals followed by disruption of cell membranes, oxidation of phenolic compounds into more toxic quinones, and death of affected cells along with the attacking pathogen.

In response to infection, plants produce several pathogenesis-related proteins that show strong antimicrobial activity or are associated with strengthening the cell wall of plant cells against pathogens. Also in response to infection, there is increased production of phenolic compounds and activation of polyphenoloxidase enzymes that oxidize the phenolic compounds into quinones and lignins both of which are antimicrobial and serve to strenghthen the cell walls of surrounding cells. In many infections, surrounding plant cells produce antimicrobial compounds called phytoalexins. A phytoalexin compound is usually specific for the host that produces it and for the pathogen against which it is produced.

In most cases, plants receiving balanced and sufficient fertilizer and water can resist certain pathogens more effectively than under stressful conditions. In some infections, the toxin produced by a pathogen can be detoxified by the host plant and that stops further invasion by the pathogen.

A small area around the lesion produced by an attacking pathogen develops localized induced resistance and few or no lesions can be caused by the pathogen in that area. Following infection in part of a plant, systemic induced resistance develops in distant parts of the plant that slows and reduces infection of the plant by the same and by other pathogens. From the point of infection, systemic induced resistance in plants is transferred to increasingly distal parts by signal transduction systems and lasts for several weeks. Salicylic acid, a precursor of aspirin (methylsalicylic acid), and certain other substances are involved in the development of systemic acquired resistance. Systemic acquired resistance can also be induced in plants against pathogens by inoculating plants with avirulent and hypovirulent microorganisms, treating them with lesion-inducing compounds, and by treating them with "plant defense activator" compounds that increase the acquired systemic resistance without causing necrotic lesions on the plant and without being toxic to the plant or to the pathogen.

The resistance of plants to a disease can be increased for subsequent generations of plants by crossing the susceptible plants with resistant ones and selecting the resistant progeny. Plants can be made more resistant to a disease through genetic engineering techniques such as introducing into plants genes that code for enzymes or toxins detrimental to the pathogen, introducing resistance genes from resistant to susceptible plants; by introducing into the plant segments of genetic material of the pathogen that interfere with the growth and infectivity of the pathogen; and, in some cases, by introducing into the plant genes engineered to produce antibodies against the pathogen.

Principles of the Genetics of Plant Disease Resistance/Susceptibility and of Pathogen Virulence/Avirulence

The ability of a microorganism to cause disease is governed by its genetic material (presence of pathogenicity genes). The ability of a pathogenic microorganism to infect a specific variety or species of a host plant is governed by its gene(s) for virulence. The inability of a pathogenic microorganism to cause disease in a species or variety of plants is due to avirulence genes present in that microorganism for that variety or species. The ability of a plant variety or species to resist infection is governed by its gene(s) for resistance against that pathogen. In some plant/pathogen systems there is a gene for resistance in the plant for each gene for virulence present in the pathogen.

Genes for virulence in pathogens enable them to produce enzymes, toxins, etc. that enable the pathogen to attack and neutralize the integrity and defenses of the plant and subsequently to cause infection and disease. Genes for avirulence in pathogens produce molecules (elicitors) recognizable by the host plant which then expresses its resistance genes and stops the avirulence gene-carrying pathogen from causing infection. Plants carrying genes for susceptibility towards a pathogen have no defense against it and become infected and diseased. Genes for resistance can be strong (major genes) and effectively defend the plant against a race or species of a pathogen. Other resistance genes can be of lesser strength (minor genes) and can provide only partial protection against the pathogen but, usually, a plant has many minor genes. Major genes are very good but last only a few months or years because the pathogen produces new individuals some of which are mutants that can attack the variety or species which was resistant until now.

Many pathogen genes for virulence/avirulence, and many plant genes for resistance/susceptibility have been isolated and the proteins they code for have been identified. Several plant genes for resistance have been inserted by genetic engineering techniques into the genome of other plants which thereby behaved as resistant.

Principles of Development of Plant Disease Epidemics

An epidemic is any increase in the population of individuals that develop a particular disease. Only diseases caused by biotic agents can spread and develop into epidemics. The larger the number of individuals that become diseased and the wider the area in which the disease can be found, the larger and more severe the epidemic. The larger the amount of the initial (overwintering or oversummering) inoculum near the crop, the more severe the epidemic can be expected to be. The more susceptible the cultivated crop, the larger the epidemic can be expected to be. The more genetically uniform and the larger the area it covers, the more severe the epidemic can be expected to be. The more favorable the environmental conditions (temperature and moisture) are for disease development, the more severe will be the epidemic. The more favorable the weather conditions are for the multiplication and spread of pathogen propagules or its vectors, the more severe will be the epidemic. Different pathogens multiply at different rates. The rate of development of epidemics usually parallels the rate of multiplication of the pathogen. Depending on the rate of development of the pathogen, epidemics may be monocyclic (having one disease cycle per growth season), polycyclic (having more than one disease cycle per growth season), or polyetic (requiring more than one growth season to complete one disease cycle).

Principles of Plant Disease Management and Control

All plant diseases can be managed/controlled by a variety of appropriate measures. The best way to prevent losses by a disease is by keeping the pathogen out of the area where the crop is grown (exclusion). All propagating stock (seed, seedlings, grafts, bulbs, etc.) must be free of the pathogen. The soil where the crop will be grown must be free of the pathogen or it must be treated in ways (fallow, crop rotation, etc.) that reduce to a minimum the pathogen population in it.

The soil should be prepared (drainage, fertilization, etc.) so that the crop plants can germinate and grow unimpeded. The crop plants should be planted when temperature and soil moisture favor quick plant growth. Growing plants, if needed, should be provided with irrigation and additional fertilizer so they can continue growing without being unduly stressed. Barriers, trap crops, etc., should be used to keep pathogens and their vectors from moving into the crop early and in large numbers. The upper layer of the soil, where most plant roots are, can be freed from most of the pathogen the previous year by planting with nonhost plants, leaving fallow, by solarization, by flooding for a few months, and by treating with pesticides (fumigants, broadcast, etc.). Only resistant plants, or the most resistant varieties available, should be planted if it is known that the pathogen is going to be present anyway.

Plant diseases can be managed/controlled by different pesticides. These include fungicides, bactericides, nematicides, which may be inorganic or organic chemical compounds, localized in action or systemic. Pesticides can be used in a variety of ways: as soil fumigant, as soil drench, as seed treatment, or as a spray or a dust applied on the plants. Most pesticides control/manage pathogens by inhibiting germination of spores, inhibiting pathogen growth and division, or by directly killing the pathogen. Some pesticides are antibiotic compounds produced by microorganisms for their own defenses but used by humans against any pathogens susceptible to them.

Some plant diseases are now commercially controlled by treatment with nontoxic plant defense-activating compounds that induce development of systemic acquired resistance in the treated plants.

In some cases, plant diseases can be controlled or managed through biological control, that is by treating the plants with antagonistic microorganisms or with microorganisms that infect, parasitize and sometimes kill the plant pathogen. Similarly, some plant diseases can be controlled/managed by reducing or eliminating the vector of the pathogen.

Through genetic engineering it is now possible to control or manage plant diseases by:

1. Early detection and elimination of infected seed, seedlings, etc.
2. Inserting into plants of genes for resistance against certain pathogens.
3. Inserting into plants genes that code for proteins toxic to the pathogen.
4. Introducing into plants animal genes that code for antibodies aimed against a particular plant pathogen.
5. Introducing into plants genetic material from a virus that makes the plant resistant to that and to other viruses.

List of Technical Nomenclature

Abiotic	Without life; nonliving, environmental factor.
Biotic	Living, as in living microorganisms; living pathogens.
Life cycle	The time and stages required by an organism to complete a generation.
Pathogen	A microorganism that can cause disease.
Phloem	Conductive tissue of plants that carries photosynthates downward.
Plant diseases	Plant diseases caused by pathogenic microorganisms and plants.
Plant pests	Generally, insects and weeds that feed on or compete with plants.
Toxin	A compound that can kill cells and whole organisms.
Vector	An organism, e.g., an insect or fungus, that can carry a pathogen to a plant.
Xylem	Conductive tissue of plants that carries water and minerals upward.

See also: **Diseases**: Bacterial Diseases; Fungal Diseases; Viral Diseases. **Integrated Pest Management**: Disease Prediction Models.

Further Reading

Agrios GN (1997) *Plant Pathology*, 4th edn. San Diego: Academic Press.

Alexopoulos CJ, Mims CW, and Blackwell M (1996) *Introductory Mycology*, 4th edn. New York: John Wiley.

Carefoot GL and Sprott ER (1967) *Famine in the Wind*. Chicago: Rand McNally.

Horsfall JG and Cowling EB (1977–1980) Plant Disease, vols. 1–5. New York: Academic Press.

Hull R (2002) *Matthews' Plant Virology*, 4th edn. San Diego: Academic Press.

Oerke E-C, Dehne H-W, Schonbeck F, and Weber A (1994) *Crop Production and Crop Protection: Estimated Losses in Major Food and Cash Crops*. Amsterdam: Elsevier.

Sing US, Mukhopadhyay AN, Kumer J, and Chaube HS (1992) *Plant Diseases of International Importance*, vols. 1–4. Englewood Cliffs: Prentice Hall.

Singh US, Kohmoto K, and Singh RP (1995) *Pathogenesis and Plant Specificity in Plant Diseases: Histopathological, Biochemical, Genetics, and Molecular Basis*, vols. 1–3. Amsterdam: Elsevier.

Bacterial Diseases

M Dow, National University of Ireland, Cork, Ireland
H Slater, Lancaster University, Lancaster, UK

Introduction

Plants are subjected to attack by a number of biological agents able to incite disease; these include fungi, viruses, nematodes, and bacteria. Although least important amongst these biological agents in terms of the number of diseases incited, bacteria cause some of the most serious diseases of crop plants. Bacterial diseases occur in all parts of the world but are particularly important in tropical and subtropical regions where conditions of humidity and warmth promote pathogen growth and dispersal. In these regions the effects of disease can be widespread and sometimes catastrophic, leading to substantial yield losses. Bacterial blight of rice (*Oryza sativa*), caused by *Xanthomonas oryzae* pathovar *oryzae*, affects many Asian countries including China, Japan, the Philippines, Indonesia, and India where millions of hectares are severely infected. Bacterial wilt caused by *Ralstonia* (formerly *Pseudomonas*) *solanacearum* affects over several hundred species of plant in 44 families. Epidemics of bacterial wilt disease in plantains in Central America in the 1960s caused the loss of millions of plants. Outbreaks of citrus canker disease in Florida, caused by *Xanthomonas axonopodis* pathovar *citri*, required the destruction of millions of trees in the 1930s as an eradication measure. Concern that the disease had re-emerged in nurseries in the 1980s prompted the destruction of 20 million young trees. Economically important bacterial diseases also occur in temperate zones. Fireblight disease, caused by *Erwinia amylovora*, affects rosaceous plants and has been particularly deleterious to the apple and pear industries of a number of countries. Soft rot disease of potatoes (*Solanum tuberosum*), caused by *Erwinia carotovora*, is a major cause of losses affecting both tubers in the field and in postharvest storage. In this article the major groups of bacterial plant pathogens are described, with their classification and the types of disease that they elicit, with an overview of those bacterial factors that contribute to the ability to cause disease and of the mechanisms underlying pathogenesis. Prospects are assessed for disease control based upon an understanding of the molecular biology of plant disease and disease resistance.

Groups of Plant Pathogenic Bacteria and the Diseases They Incite

Bacteria that are able to invade and cause disease in otherwise healthy plants belong to a small number of genera (**Table 1**). The list includes Gram-positive bacteria (*Clavibacter, Curtobacterium*, *Rhodococcus*, and *Streptomyces*), Gram-negative bacteria and mollicutes (wall-free bacteria). Studies of pathogenicity in Gram-negative bacteria are considerably advanced over the other groups. Many Gram-negative pathogens can be cultured easily in the laboratory and can be readily inoculated into plants. In addition a range of tools for molecular genetic analysis developed in other Gram-negative bacteria can be deployed in plant pathogens. In contrast the paucity of information on Gram-positive and mollicute pathogens reflects both technical difficulties of growing them in the laboratory and the slower development of appropriate molecular genetic methods.

Table 1 Major genera of plant pathogenic bacteria

Genus	*Types of disease caused*	*Other particular features*
Gram-negative		
Agrobacterium	Tumors, root proliferation	Disease uniquely involves transformation of plant host by incorporation of bacterial DNA
Erwinia	Soft rot, wilt, necrosis	Soft rot erwinias such *as E. chrysanthemi* and *E. carotovora* produce copious plant tissue-macerating enzymes
Pseudomonas	Leaf spot, wilt, canker, galls	Several species. One species *Pseudomonas syringae* is subdivided into over 40 pathovars
Ralstonia	Wilt	*Ralstonia solanacearum* (formerly *Pseudomonas solanacearum*) is an extremely important plant pathogen
Xanthomonas	Wilt, leaf spot, rot, canker	In earlier classifications, *Xanthomonas campestris* was subdivided into over 120 pathovars. Currently classified as several species, some with pathovar subdivisions
Xylella	Wilt, chlorosis	Only one species *Xylella fastidiosa* is so far defined
Gram-positive		
Clavibacter	Wilt, canker, leaf spot, stunt	*Clavibacter, Curtobacterium*, and *Rhodococcus* were formerly classified as *Corynebacterium*
Curtobacterium	Wilt, leaf spot, stem spot	
Rhodococcus	Galls, fasciation	Bacteria produce plant growth hormones
Streptomyces	Rot, common scab	Actinomycetes do not constitute a major group of plant pathogens
Mollicutes		
Spiroplasma	Stunt, wilt	No cell wall, spiral in form, can be cultured
Phytoplasma	Phyllody, stunt	No cell wall, unculturable

Table 2 Types of disease incited by plant pathogenic bacteria

Disease type	*Typical symptoms*	*Typical causal organisms*	*Modes of entry into plant*
Leaf spot	Small localized lesions which may be chlorotic, necrotic, or water-soaked	Many *Pseudomonas syringae* pathovars and *Xanthomonas* spp.	Usually through stomata, sometimes wounds
Wilt	Vascular pathogens interfere with plant water flow, inducing wilting and necrosis	*Ralstonia solanacearum*, *Xanthomonas campestris* pv *campestris*, *Clavibacter* and *Curtobacterium* spp., *Xylella fastidiosa*	Usually through wounds. *Xanthomonas campestris* pv *campestris* can enter via hydathodes at leaf margins. *Xylella fastidiosa*, which inhabits xylem, is transmitted by insects
Soft rot	Extensive maceration of plant tissue	*Erwinia chrysanthemi*, *Erwinia carotovora*, some *Pseudomonas* spp.	Lenticels and wounds
Cankers and dieback	Tissue necrosis extending from initial infection site. Gives rise to lengthy canker lesions or dieback from the ends of branches or shoots	*Erwinia amylovora*, *Xanthomonas axonopodis* pv *citri*, *Pseudomonas syringae* pv *syringae*	Wounds and stomata (e.g. for *Pseudomonas syringae*), nectaries (for *Erwinia amylovora*)
Tumors, galls, and other abnormal plant development	Abnormal plant cell division and cell size (hypertrophy). Phyllody (replacement of floral parts by leaflike structures) and adventitious root formation	*Agrobacterium tumefaciens*, *Pseudomonas savastanoi*, and *Rhodococcus fascians* cause galls and tumors. Many phytoplasmas cause phyllody. *Agrobacterium rhizogenes* causes hairy roots	Mostly through wounds. Phytoplasmas, which inhabit phloem, are transmitted via insects

The diseases caused by plant pathogenic bacteria can be classified into a limited number of types that are detailed in **Table 2**. Some examples of the phenotypes of the different diseases are given in **Figure 1**. It should be noted that different species of bacteria from within the same genus are able to produce different disease types. For example, although many *Pseudomonas* pathogens cause leaf spots, others cause soft rot and dieback diseases.

Modes of Entry and Dissemination

In order to cause disease, bacteria have to gain entry into the internal tissues of a plant. This is achieved either through natural openings or wounds because, unlike some fungal plant pathogens, bacteria are not able to gain entry through the intact plant epidermis.

Natural openings such as stomata and lenticels form a portal of entry for a wide range of bacterial pathogens. In contrast a restricted number of pathogens are able to infect plants through hydathodes and floral parts. Different pathogens use different modes of entry, depending on the nature of the disease. Stomata are considered to be an important portal of entry for bacteria causing leaf spots. In contrast, hydathodes, which are structures located at the leaf margin at the vein ends, are the portal of entry for some vascular pathogens (**Table 2**). Lenticels are derived from stomata during the development of periderm, and are a major site of entry into potato tubers for *E. carotovora,* the soft rot pathogen. Infection of flowers is important in a few diseases including fireblight caused by *E. amylovora*. This bacterium can grow epiphytically on anthers and stigmata although the major site of entry is through nectaries.

A wide range of bacteria can enter plants through wounds. Wounds can be caused by insects, environmental factors such as hail, frost, and wind damage, agricultural practices such as pruning and as a consequence of plant development. The small breaks in the epidermis of roots, which occur as a result of normal root growth and branching, can act as portals of entry of *Ralstonia solanacearum*, the

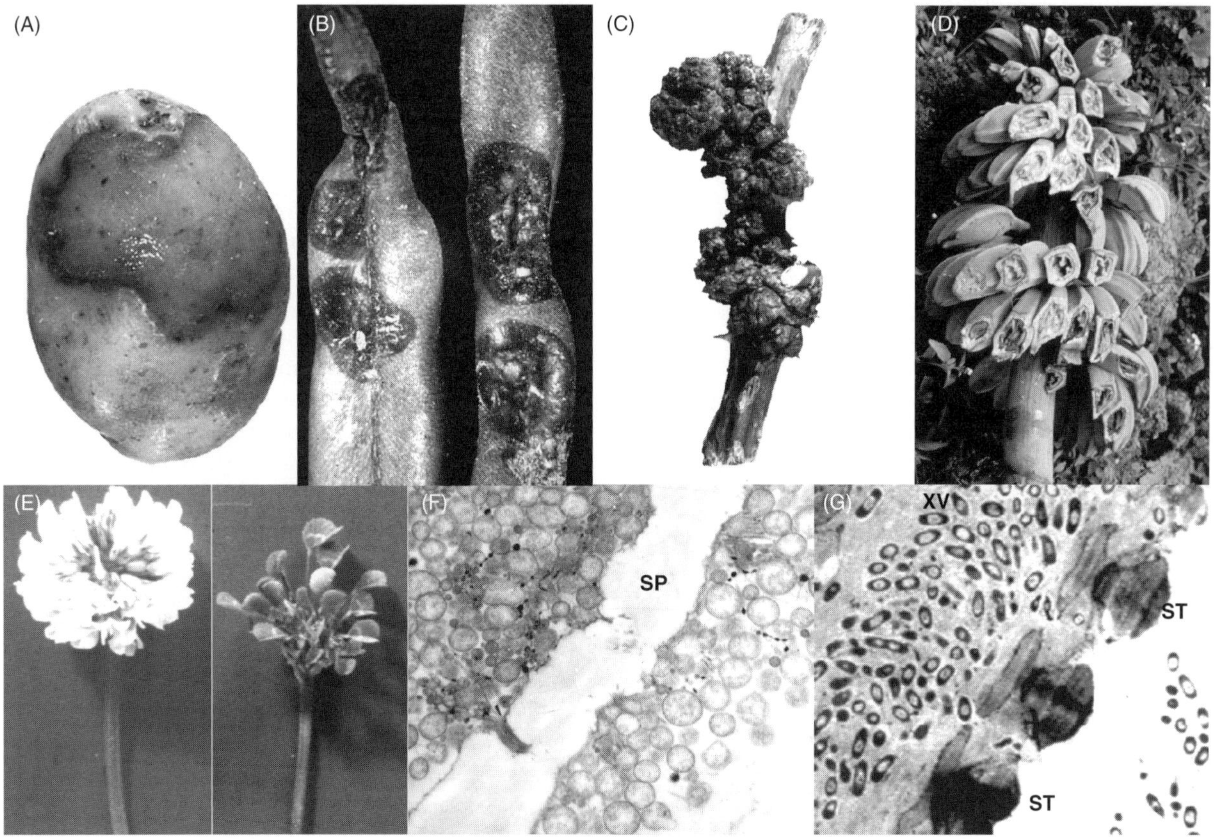

Figure 1 Bacterial diseases of plants and their causal agents. (**A**) Soft rot of potato caused by *Erwinia carotovora*. In the infected part of the tuber the tissue has been sufficiently macerated through the action of enzymes released by the bacteria so that it could be easily scraped out with a spatula. Figure provided by George Salmond. (**B**) Water-soaking lesions on bean pods caused by *Pseudomonas syringae* pathovar *phaseolicola*. (**C**) Tumors produced by *Agrobacterium tumefaciens* on loganberry. The formation of tumors by *Agrobacterium tumefaciens* depends on genetic transformation of the plant cells with bacterial DNA. (**D**) Banana blood disease caused by *Ralstonia solanacearum*. The bacteria infect the vascular system of the plant in a systemic fashion leading to the maceration of the bananas which is seen after they are cut. Figure provided by Sue Seal. (**E**) Phyllody in clover caused by Phytoplasma. The floral parts of the clover (left-hand section) have been replaced with leaflike structures in the infected plant (right-hand section). (**F**) Electron micrograph of phytoplasmas in the phloem elements of infected clover next to a sieve plate (SP). These wall-less bacteria have a rounded cross-section of approximately 0.5 μm in diameter. Figure provided by Mike Daniels. (**G**) Electron micrograph of xylem vessels (XV) of a cabbage plant infected with *Xanthomonas campestris* pathovar *campestris*, the black rot pathogen. Note that the bacteria appear to be suspended in less electron-dense material, perhaps bacterial extracellular polysaccharide. Bacteria are colonizing a newly infected xylem element by passage through the pit membranes between the spirals of secondary thickening (ST). This process may require degradation of the pit membranes by cell wall degrading enzymes secreted by the bacteria. Reproduced with permission from Wallis *et al.* (1973) *Physiological Plant Pathology* 3: 371–378.

vascular wilt pathogen. Wounds in other parts of plants can act as sites of initiation of infections by *Agrobacterium tumefaciens*. These bacteria may locate the wound site through chemotaxis to phenolic compounds produced by the wounded tissue. Insect vectors are important in the entry and spread of a number of bacterial pathogens. They are particularly crucial in the infection and dissemination of mollicutes, xylem-limited fastidious bacteria such as *Xylella fastidiosa*, and phloem-limited bacteria. In the case of these pathogens, the insect vectors are probably the only sites of survival of the organism other than in the diseased plant. The insect vectors of *X. fastidiosa* are sharpshooter leafhoppers that feed specifically on xylem sap. Bacteria in infected xylem sap colonize the foregut of the insect and can then be transferred to the xylem of an uninfected plant upon insect feeding. Other bacterial pathogens such as *E. amylovora* can be transmitted by insects, but only in the aforementioned cases does entry and dissemination appear to entirely depend on an insect vector.

The spread of many bacterial pathogens can depend upon environmental factors such as rain and wind, upon insects and upon agricultural practices such as pruning, bud grafting, lack of crop rotation, and the use of contaminated seeds. Rainwater from infected leaves may carry large numbers of bacteria to uninfected parts of the plants. The formation of small droplets of water (aerosols) carrying bacteria would aid dispersal by wind. Bacteria in the soil can also form the source of inoculum for future crops. Pathogens such as *Xanthomonas campestris* pathovar *campestris* can survive in soil in association with dead plant material whereas others such as *R. solanacearum* and some soft rot *Erwinia* spp. occur throughout the soil-water environment.

Host Range and the Concepts of Pathovar and Race

Although bacteria classified as belonging to individual species are often responsible for diseases in a wide range of plant hosts, in many cases particular strains within that species have only limited host range. In order to reflect this variation, some species have been further divided taxonomically into "pathovars." *P. syringae*, for example, includes at least 40 pathovars and in earlier classifications *X. campestris* contained over 140. Pathovars are usually named after the plant from which the bacterial strain was first (or is most commonly) isolated. The relationship between different pathovars is however unclear since the full host range of each of the strains has usually not been established. New taxonomic groupings within the genus *Xanthomonas* have recently been established based on molecular criteria. Although the new classification redefines the species structure within *Xanthomonas*, it retains the pathovar subgrouping for a number of species, in particular *X. axonopodis*. In some cases particular strains within a pathovar are able to cause disease on only certain cultivars of the host plant. These different strains are then considered to be different races of the pathogen. Strains of *R. solanacearum*, which affects a large number of plants, have also been subdivided into races, based on host range. However the term is used in a less specific fashion than in race–cultivar interactions seen with pathovars, since each race can still infect a wide variety of plants. Other classifications of pathogens are based on biochemical characters (biovars), immunological characteristics (serotypes), and sensitivity to bacteriophages (lysotypes).

So far we have discussed cases in which bacteria cause disease, i.e., compatible plant–pathogen interactions where the bacteria are said to be virulent and the host susceptible. What happens if bacteria are inoculated into plants or plant cultivars on which they do not normally cause disease? These are incompatible interactions where the bacteria are said to be avirulent and the host resistant. In many cases, inoculation of bacteria above a threshhold level leads to a rapid confluent collapse and necrosis of the inoculated tissue. This is termed the hypersensitive response or hypersensitive reaction (HR). HR is not a passive response of plants to a large bacterial inoculum. On the contrary it is considered to be a programmed cell death response which requires specific alterations in plant gene expression, protein synthesis and protein turnover. Relatively high numbers of bacteria (above 10^7 colony forming units per ml) are required to establish confluent HR, which usually occurs within 24 h and can occur as rapidly as 8–12 h after inoculation. The HR phenotype is often used to establish incompatibility in race–cultivar interactions. HR is also a feature of some, but not all, nonhost interactions, which operate at the pathovar or species level. These include the classical response of tobacco (*Nicotiana tabacum*) to a wide range of phytopathogenic *Pseudomonas* spp. Confluent HR, with the associated death and desiccation of the collapsed tissue, may serve to restrict bacterial growth and spread. HR is however only one of a number of defense responses that are activated in incompatible host–pathogen interactions.

Genetic Determinants of Host Specificity

The work of Flor on the interaction of flax rust fungus with different flax lines led to the development of the gene-for-gene hypothesis to explain the specificity of the interaction between races of pathogens and different cultivars of host plants. According to this hypothesis, incompatibility is specifically determined by the presence of genes in both the host and the pathogen (**Figure 2**). Accordingly the presence of an avirulence gene in the pathogen and a corresponding resistance gene in the plant leads to incompatibility. Both avirulence and resistance genes are dominant. Plant–pathogen combinations with recessive alleles of the avirulence gene (or where it is absent) or with recessive plant

	Plant resistance genotype	
Bacterial avirulence genotype	$R_A R_A$ or $R_A r_A$	$r_A r_A$
avrA	Incompatible	Compatible
No *avrA* (or recessive allele)	Compatible	Compatible

Figure 2 A hypothetical gene-for-gene interaction between different strains of a bacterial pathogen and different cultivars of a host plant. Incompatibility (resistance) is specifically determined by the presence of a dominant avirulence gene (*avrA*) in the pathogen and a cognate, dominant resistance gene (R_A) in the plant. All other combinations give rise to compatibility. Some avirulence genes in bacteria are dispensable. r_A is a recessive allele of R_A.

resistance genes are compatible. Gene-for-gene systems also appear to determine the outcome of interactions of plants with nematodes and insects, in addition to fungi and bacteria. The first experimental support for this hypothesis came through the cloning in 1984 of an avirulence (*avr*) gene from *P. syringae* pathovar *glycinea* that determined race-specific incompatibility on soybean (*Glycine max*). Subsequently over 30 avirulence genes have been cloned from pathovars of *P. syringae* and *Xanthomonas* spp. but not from other plant pathogens. A number of the matching plant disease resistance genes have also been cloned and again these are dominant as predicted by the gene-for-gene hypothesis. In addition to gene-for-gene interactions, avirulence genes may also determine incompatibility in some nonhost interactions.

A question that has vexed many plant pathologists is why avirulence genes should persist in bacterial populations when they apparently only serve the negative function to restrict host range. One answer is that these genes may have functions in virulence. For a limited number of avirulence genes it has been shown that mutation leads to a loss of fitness on normally susceptible hosts. In addition some avirulence genes have been shown to contribute to the production of specific symptoms in host plants. It is plausible that many avirulence genes have roles in aspects of the disease cycle such as epiphytic survival or competition with saprophytes that are not readily testable. In addition their functions in virulence may be redundant so that differences in behavior in plants as a result of mutation of individual genes are not observed. This could also explain why some avirulence genes are apparently dispensable or are only found in a limited number of bacterial strains.

The mechanistic basis for gene-for-gene determined resistance to bacteria has been the subject of considerable study. It has been suggested that direct interaction between the protein products of cognate resistance and avirulence genes triggers HR and other plant defense responses. Intriguingly, for those bacterial avirulence gene products tested, it is evident that recognition must occur inside the plant cell and not at the plant cell surface as envisaged in earlier models. Recognition (through binding) may occur in the cytoplasm in some cases. Some avirulence gene products however carry amino acid sequence motifs that would target them to the plant cell nucleus, although whether they directly interact with the cognate resistance gene product at that location is not known. The mechanism by which these avirulence gene products may be exported from the bacteria and delivered to sites within the plant cell will be considered later.

Molecular Biology of Bacterial Disease

Phytopathogenic bacteria produce a range of factors that potentially contribute to their ability to incite disease on susceptible plants. These factors include toxins, plant growth factors, extracellular polysaccharides, and extracellular enzymes capable of degrading plant components. Symptoms associated with the different types of plant disease suggest that these factors may have different relative importance in those diseases. For example the extensive plant tissue maceration associated with soft rot diseases suggests an involvement of plant tissue degrading enzymes, whereas the alterations in plant development associated with tumor formation by *Agrobacterium* suggest a role for plant growth factors of bacterial origin. Many leaf spot diseases involve chlorosis rather than extensive tissue maceration. In some cases toxins capable of eliciting the symptoms of disease have been isolated from the growth medium of the causal bacteria. The development of molecular genetic tools for many Gram-negative pathogens has enabled critical tests of the role of a number of suspected factors in the disease process. This has been achieved through the construction of specific mutants unable to make the factor(s) followed by testing of the behavior of these mutant strains in plants. In addition, random mutagenesis of bacterial genomes followed by plant testing has allowed the identification of further determinants of disease. Those factors whose loss leads to complete inability to cause disease on any plant are termed pathogenicity determinants, whereas those factors whose loss leads to a reduction in the aggressiveness of a strain are called virulence determinants. Conventionally loss of virulence or pathogenicity determinants should not affect the growth of bacteria in nutrient medium; those factors with established roles in virulence or pathogenicity are reviewed below.

The *hrp* Genes and Basic Pathogenicity

The presence of *hrp* genes is characteristic of phytopathogenic *Erwinia*, *Xanthomonas*, *Pseudomonas*, and *Ralstonia* spp. These genes are so named because mutations in them cause a loss of both HR in resistant plants and pathogenicity in susceptible plants. Most *hrp* genes occur in clusters spanning 20 to 40 kb of DNA. Some of the *hrp* genes encode components of a particular class of protein secretion system designated Type III (**Table 3**). Similar protein secretion systems are found in bacterial pathogens of animals including *Yersinia*, *Shigella*, and *Salmonella* spp., where they contribute to virulence. The conservation of components in Type III secretion

Table 3 Secretion systems for macromolecules in Gram-negative plant pathogenic bacteria

System	*Number of accessory proteins*	*Specific characteristics*	*Requirement for signal sequence dependent transfer across cytoplasmic membrane*	*Examples*
Type I	Usually 3 or 4	Secreted protein moves in a channel directly from the cytoplasm to the cell surface	No	Protease secretion in *Erwinia chrysanthemi*
Type II	Up to 14	A terminal branch of the general secretion pathway which moves proteins across the bacterial outer membrane	Yes	Secretion of pectinases and cellulases from *Erwinia* and *Xanthomonas* spp.
Type III	15 to 20	Moves proteins directly from cytoplasm to bacterial cell surface or to interior of plant cell	No	Hrp systems exporting avirulence proteins and harpins from *Xanthomonas, Pseudomonas*, and *Erwinia*
Type IV	At least 10	Moves T-DNA/protein complex from bacterial cytoplasm into plant cell	No	T-DNA transfer from *Agrobacterium* into plant cells
Type V	None	Autotransporter mechanism. The transported protein has a signal sequence and a domain which forms a pore in the outer membrane through which the "passenger" domain is exported to the cell surface	Yes	Putative mechanism for export of adhesins from *Xanthomonas* and *Xylella*

systems between these bacterial genera has led to the renaming of some *hrp* genes as *hrc* (for hypersensitive response and conserved). Some *hrp* genes encode protein products called harpins that are exported out of the bacterial cell via the Type III secretion system. Other genes within *hrp* clusters encode regulators that are thought to control *hrp* gene expression in response to those environmental signals encountered in plants. In general *hrp* genes are not expressed in rich nutrient medium, but expression is induced upon inoculation into plants.

Type III secretion systems in bacterial pathogens of animals are known to deliver bacterial proteins to sites within the eukaryotic host, where they promote disease by affecting specific host processes. This delivery of proteins depends upon close contact between the pathogen and the host cell. Although similar direct transfer has not yet been detected between plant pathogens and their hosts, there is considerable circumstantial evidence that this must occur. As outlined above, in gene-for-gene interactions it is evident that for HR to be triggered, the avirulence gene product has to be recognized within the plant cell. The available evidence is consistent with a role for the *hrp*-encoded Type III secretion system in the delivery process. A simple model depicting the role of the *hrp* genes in both pathogenesis and disease resistance is shown in **Figure 3**.

Extracellular Enzymes

A number of phytopathogenic bacteria produce extracellular enzymes capable of degrading components of the plant cell wall including pectic polymers, hemicelluloses such as xylan, cellulose, proteins, and glycoproteins (**Table 4**). Extracellular enzymes have a number of potential roles in plant disease, including overcoming host defense responses, mobilization of plant cell walls for nutritional purposes, facilitation of movement of bacteria into and between vascular elements, and promotion of bacterial survival on plant material in the soil.

The production of a range of enzymes able to degrade pectin is a particular feature of the "brute force" soft rot pathogens, *E. carotovora* and *E. chrysanthemi*. In isolation these pectic enzymes can cause the tissue maceration and cell killing, which are the primary symptoms of soft rot disease. Extracellular cell-wall degrading enzymes are also produced by pathogens inciting other types of disease, including the wilt-inducing *R. solanacearum* and *X. campestris* pathovar *campestris*, the black rot pathogen. *P. syringae* pathovars responsible for leaf spotting diseases do not appear to produce extracellular enzymes however, implying a "stealth" mode of pathogenesis. In addition although many *Xanthomonas* spp. responsible for leaf spots produce protease activity, the production of pectic enzymes is limited to certain strains.

Many bacteria produce several different proteins with the same enzymatic activity; these are called isoforms. This is particularly evident for the pectate lyase enzymes of the soft rot *Erwinia* spp. In *E. chrysanthemi* the different isoforms of pectate lyase may collectively contribute to the ability of this bacterium to cause disease on a wide range of hosts, with different isoforms varying in importance in different hosts. Detailed differences in the substrate preference, regulation, and pattern of products produced also suggest that all isoforms in *Erwinia* do not perform the same function.

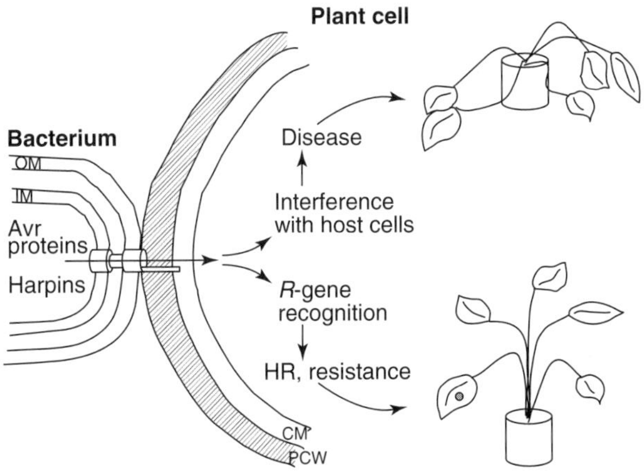

Figure 3 A simple model for the roles of *hrp* genes in pathogenesis and disease resistance. Some genes within the *hrp* gene cluster encode proteins that are components of a Type III secretion system that delivers bacterial effector proteins to the plant cell surface and to the plant cell interior. These transported proteins may be encoded by other genes within the *hrp* cluster or by unlinked genes. The effector proteins interfere with specific plant processes, thereby promoting disease. In this case they can be considered as virulence factors. In some plants or plant varieties, one or more of the effector proteins can be recognized by the plant cell so that resistance reactions are triggered. In this case the effector protein that is recognized is considered to be an avirulence determinant. Plant components responsible for recognition of avirulence determinants have been defined; the targets for the action of virulence determinants are as yet unknown. OM, outer membrane; IM, inner membrane; PCW, plant cell wall; CM, plant cytoplasmic membrane.

In several cases, mutations in genes encoding specific enzymes lead to a reduction in virulence of bacteria, establishing a role for those enzymes in disease. Examples would include polygalacturonase and cellulase in *R. solanacearum* and some of the pectic enzymes in *Erwinia* spp. In other cases however, loss of specific enzymes has little apparent effect. This may be because of redundant functions amongst enzyme isoforms, although it is also possible that the particular enzyme plays a role in an aspect of the disease not addressed by the plant tests used.

After synthesis within the cell, nearly all of the extracellular enzymes in plant pathogenic bacteria are exported out of the cell by Type II secretion systems (**Table 3**), a notable exception being the proteases of *Erwinia* spp., whose export is mediated by a Type I system. Mutation of genes encoding the Type II secretion system in *Erwinia* spp. leads to a complete loss of the ability to macerate plant tissue. In *R. solanacearum*, *X. campestris* pathovar *campestris*, and *Xanthomonas oryzae* pathovar *oryzae*, mutations affecting Type II secretion lead to a loss or reduction of virulence. These findings offer only indirect support for the role of extracellular enzymes in disease

Table 4 Extracellular plant cell-wall degrading enzymes produced by bacterial phytopathogens

Substrate	*Enzyme*	*Activity of enzyme*	*Representative producers*
Pectic polysaccharides	Pectin methyl esterase	Hydrolysis of methyl esters of pectin	*Erwinia chrysanthemi*
	Pectin acetyl esterase	Hydrolytic removal of acetyl groups	*Erwinia chrysanthemi*
	Pectate lyase	Transeliminative cleavage of pectin polymer	*Erwinia chrysanthemi*, *Erwinia carotovora*, *Xanthomonas campestris* pv *campestris*
	Polygalacturonase	Hydrolytic cleavage of pectin polymer	*Erwinia chrysanthemi*, *Erwinia carotovora*, *Ralstonia solanacearum*
Cellulose and hemicelluloses	Cellulase (endo-β(1,4)-glucanase)	Hydrolysis of cellulose or β(1,4)-linked glucans	*Erwinia chrysanthemi*, *Erwinia carotovora*, *Ralstonia solanacearum*, *Xanthomonas campestris* pv *campestris*, *Clavibacter michiganensis*
	Xylanase	Hydrolysis of xylan	*Erwinia chrysanthemi*, *Xanthomonas oryzae* pv *oryzae*
Proteins	Protease	Hydrolysis of internal peptide bonds in proteins	Many *Erwinia*, *Xanthomonas* spp., *Xylella*
Hydroxyproline-rich glycoproteins	(Glyco)protease	Hydrolysis of internal peptide bonds in these structural glycoproteins	*Erwinia carotovora*, *Xanthomonas campestris* pv *campestris*

however, since Type II secretion systems may export proteins other than plant cell wall degrading enzymes, with a role in disease.

Toxins and Extracellular Polysaccharides

Toxins are small, diffusible molecules that are able to incite many or all of the symptoms of a disease in the absence of the causal organism. Toxins are produced by a limited number of bacterial plant pathogens, primarily pathovars of *P. syringae* (**Table 5**). Different toxins are produced by different *P. syringae* pathovars. The cellular targets of many of the toxins have been established. The purified toxins, unlike the bacteria that synthesize them, show no host specificity, causing chlorotic or necrotic symptoms in a wide variety of plants. In general, bacterial toxins have a role in virulence but are not absolutely required for pathogenesis. In addition to development of disease symptoms, they can contribute to lesion size, systemic spread of bacteria, and bacterial multiplication in the host. Since many toxins have antimicrobial activity, it is plausible that they primarily function to eliminate competitors during epiphytic colonization of leaf surfaces.

The production of extracellular polysaccharides (EPSs) is a feature of most bacteria including many plant pathogens. Different phytopathogenic bacteria produce polysaccharides with different monosaccharide composition. A role for the EPS in the virulence of *R. solanacearum*, *E. amylovora*, *E.* (now *Pantoea*) *stewartii*, *X. campestris* pathovar *campestris*, and *P. syringae* pathovar *syringae* has been established by studying strains with mutations in the genes encoding the biosynthetic enzymes for these polymers. The precise role that these EPSs play in plant disease is not known and may be different in different diseases. EPSs may contribute to disease symptoms such as wilt, by blocking the flow of water in vascular elements and to water-soaking lesions by promoting the retention of water in intercellular spaces. Roles of benefit to the pathogen may include protection of bacteria against environmental stresses including the action of antimicrobial factors that may be generated in the plant, promoting pathogen multiplication by maintenance of water soaking and masking of bacterial determinants that may be recognized by plants to trigger resistance responses. EPSs may also contribute to epiphytic survival of many pathogens.

Agrobacterium spp. and the Transfer of Bacterial Genes to Plants

The induction of plant tumors by *Agrobacterium tumefaciens* and plant root hairs by *Agrobacterium rhizogenes* involves a permanent genetic change of the plant host cell in which a segment of bacterial DNA is transferred and integrated into the plant genetic material. This segment of DNA (the transferred or T-DNA) is part of a large plasmid in these bacteria, called Ti (for tumor-inducing) or Ri (for root-inducing). The T-DNA carries genes involved in the production of plant growth hormones and opines, compounds which are derivatives of amino acids or sugars and which are specific nutrients for *Agrobacterium* spp. The abnormal production of plant growth hormones causes the formation of tumors and hyperproduction of roots. These growths produce large amounts of opines which can be specifically catabolized by the infecting bacteria. Particular phenolic compounds released by wounded plant tissue act as a chemoattractant for the bacteria at the early stages of the infection process and also activate the expression of a number of genes (*vir* genes) which are located on the plasmid. The products of these genes are responsible for the excision of the T-DNA, its transfer into the plant cell, translocation into the nucleus, and integration into the plant genome (**Figure 4**). The ability to induce tumors also depends upon the

Table 5 Toxins produced by plant pathogenic bacteria

Producing organism	*Toxin*	*Target or function*
Pseudomonas andropogonis	Rhizobitoxine	β-Cystathionase, inactivating homocysteine synthesis
Pseudomonas corrugata	Corpeptin	Membrane active?
Pseudomonas fuscovaginae	Fuscopeptin	Membrane active?
Pseudomonas marginalis	Viscosin	Membrane active?
Pseudomonas syringae		
several pathovars	Coronatine	Molecular mimic of jasmonic acid, a plant signal molecule
pv *phaseolicola*	Phaseolotoxin	Inhibitor of ornithine carbamoyl transferase, inhibiting amino acid synthesis
pv *syringae*	Syringomycin	Increases host membrane permeability
	Syringopeptin	Increases host membrane permeability
pv *tabaci* and others	Tabtoxin	Tabtoxin derivative inhibits glutamine synthetase
pv *tagetis*	Tagetitoxin	Inhibitor of chloroplast RNA polymerase
Xanthomonas albilineans	Albicidin	Inhibition of plastid DNA replication

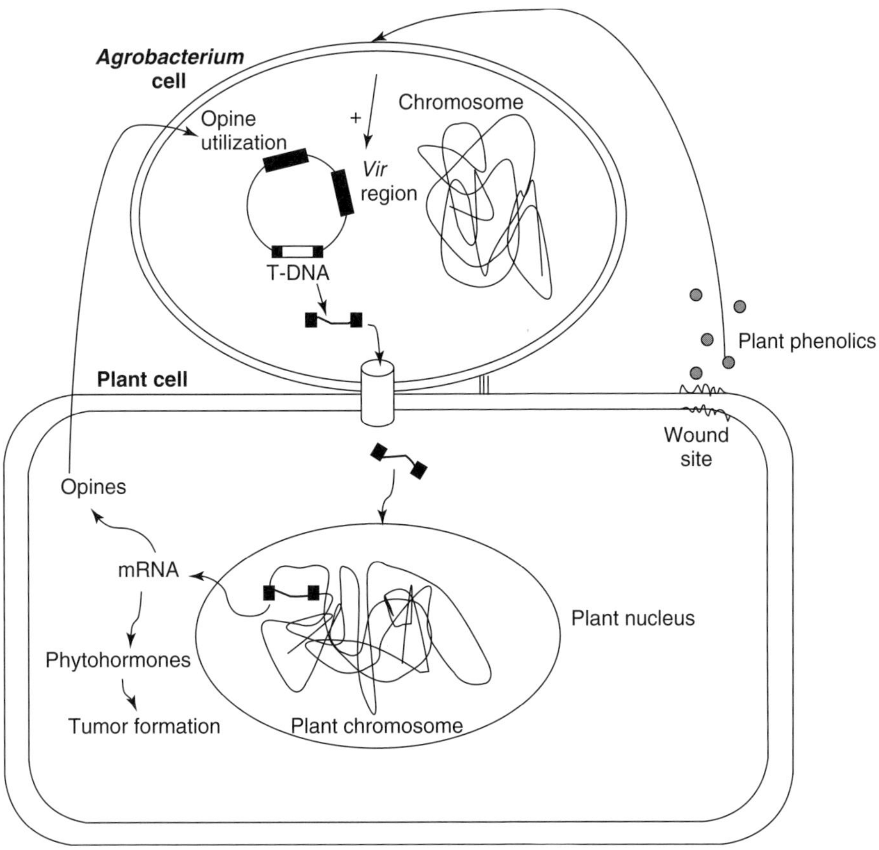

Figure 4 Infection by *Agrobacterium* spp. involves transfer of part of the bacterial DNA to the plant cell nucleus. The segment of DNA transferred is called the T-DNA and is part of a large plasmid within the bacteria. The recognition of plant wound signals by the bacteria leads to the activation of the bacterial *vir* genes. The *vir* gene products are responsible for the excision of the T-DNA, transfer into the plant cell via a Type IV secretion system, translocation into the nucleus and integration into the plant genome. The T-DNA carries genes involved in the production of plant growth hormones, whose actions give rise to tumor formation and of opines which are catabolized by the bacteria within the tumor.

function of genes that are located in the bacterial chromosome. Infection by *Agrobacterium* is now used routinely to introduce foreign genes, engineered into the T-DNA, into plants. The *Agrobacterium* bacteria used are "disarmed" by removal of the genes responsible for tumor formation and opine production.

Coordinate Behavior in Bacterial Pathogens: Regulation of Production of Virulence Factors

Although they are single-celled organisms, it is increasingly evident that, in many environments, bacteria behave in a colonial or group fashion. Molecular signals between the bacteria within a colony probably serve to coordinate changes in gene expression in response to the changing environment and growing cell population. These signals are known to have a role in regulating virulence to plants. In most cases the signaling molecules, which may be of different chemical nature in different bacteria, regulate the production of a number of factors (**Table 6**). The exact roles that cell–cell signaling plays in bacterial diseases of plants are unclear and may be different for different bacteria. The signals may serve as "quorum sensing" agents, ensuring that the synthesis of virulence factors occurs only when sufficient bacteria are present to produce an effect to overwhelm host defenses. They may serve as confinement sensors, whose external concentration (and hence effect) is influenced not only by bacterial numbers but also the volume of the space (such as a xylem element) in which the bacteria are growing. What is clear is that the synthesis of virulence determinants is strictly controlled during plant disease and occurs in response to environmental conditions, which may include the bacterial population size and density.

Concluding Remarks

An understanding of the molecular biology of plant pathogenic bacteria and the molecular mechanisms underlying plant disease and disease resistance forms the basis for development of strategies to improve

Table 6 Cell–cell signaling in phytopathogenic bacteria

Species	*Signaling molecule*[a]	*Processes regulated*
Agrobacterium tumefaciens	OOHL	Conjugal transfer of T-DNA
Erwinia carotovora	OHHL	Extracellular enzyme production; virulence
Erwinia chrysanthemi	OHHL, HHL	Extracellular enzyme production; virulence
Erwinia (Pantoea) stewartii	OHHL	Extracellular polysaccharide; virulence
Pseudomonas syringae pathovar *tabaci*	OHHL	Unknown
Ralstonia solanacearum	3-Hydroxypalmitic acid methyl ester	Production of extracellular polysaccharide, extracellular enzymes; virulence
	OHHL, HHL	Unknown
Xanthomonas campestris pathovar *campestris*	DSF, possible butyrolactone	Production of extracellular enzymes and extracellular polysaccharide; virulence
	DF, butyrolactone	Production of pigment and extracellular polysaccharide; epiphytic survival

[a]OOHL, *N*-(3-oxooctanoyl)-L-homoserine lactone; HHL, *N*-hexanoyl-L-homoserine lactone; OHL, *N*-(3-oxohexanoyl)-L-homoserine lactone; DSF, diffusible signal factor; DF, diffusible factor.

plant health. Improvements in agronomic practice have already resulted from rapid, sensitive, and specific detection of pathogens in infected seeds or seedlings. The identification of bacterial processes that are essential to the development of disease can define targets for specific interference. Mutational analysis suggests that, although many bacterial factors contribute to virulence, interference with processes that coordinately affect the synthesis or secretion of a number of factors profoundly affects the ability to cause disease. Can protein inhibitors of Types II and III secretion systems be developed and expressed into plants? Can we interfere with those bacterial processes that signal the production of virulence factors, for example by engineering degradation of intercellular signals? The identification of Type III-secreted virulence determinants that are widely distributed may facilitate the search for new sources of genetic resistance. For example there may be plants, perhaps noncrop species, which can recognize these virulence determinants to trigger resistance responses. The cognate resistance genes could perhaps be usefully deployed in agriculture. The work on bacterial pathogenesis of plants, in particular with relation to Type III secretion, has been heavily informed by the work on bacterial pathogenesis of animals. Some common strategies of pathogenesis involving plants and animals have already emerged, in addition to many obvious differences. It is to be hoped that continued work on plant–pathogen interactions in the postgenomic era will contribute to an understanding of bacterial disease in its broadest context.

List of Technical Nomenclature

Actinomycetes Gram-positive bacteria with filamentous growth habits.

Adhesin Protein component responsible for bacterial attachment to surfaces, including those in eukaryotic hosts.

Allele Variant of a gene which has altered function as a result of altered DNA sequence.

Chemotaxis Movement of bacteria in response to a concentration gradient of a chemical; can be positive (bacteria move towards higher concentration) or negative.

Chlorosis Yellowing of green plant tissues; loss of chlorophyll.

Epiphytic colonization Growth of bacteria on plant surfaces such as leaves, roots, and floral parts.

Gram-negative bacteria Type of bacteria whose cell envelope comprises cytoplasmic and outer membranes that bind the periplasmic space and peptidoglycan layer. Outer membrane carries lipopolysaccharide.

Gram-positive bacteria Type of bacteria whose cell envelope comprises single (cytoplasmic) membrane and cell wall.

Mollicutes Wall-less bacteria; at 0.2–0.5 μm they are smaller than walled bacteria (0.5–2 μm).

Necrosis Death of plant tissues, often associated with browning and drying of the tissue.

Quorum sensing Mechanisms by which bacteria regulate the expression of particular genes such that they are activated only at higher bacterial cell densities.

Signal sequence Amino acid sequence which directs the transfer of a protein across a biological membrane; in bacteria, specifically the cytoplasmic membrane.

See also: **Cell Walls and Fibers**: Cell Walls. **Diseases**: Breeding for Disease Resistance; Fungal Diseases; Plant Pathology, Principles. **Growth and Development**: Cells. **Water Relations of Plants**: Basic Water Relations.

Further Reading

Alfano JR and Collmer A (1996) Bacterial pathogens in plants: life up against the wall. *Plant Cell* 8: 1683–1698.

Cornelis GR and van Gijsegem F (2000) Assembly and function of Type III secretory systems. *Annual Review of Microbiology* 54: 735–774.

Dangl JL (ed.) (1994) *Bacterial Pathogenesis of Plants and Animals: Molecular and Cellular Mechanisms*. New York: Springer-Verlag.

Denny TP (1995) Involvement of bacterial polysaccharides in plant pathogens. *Annual Review of Phytopathology* 33: 173–197.

Goodman RN and Novacky AJ (1994) *The Hypersensitive Reaction in Plants to Pathogens: A Resistance Phenomenon*. St Paul: APS Press.

He SY (1998) Type III protein secretion systems in plant and animal pathogenic bacteria. *Annual Review of Phytopathology* 36: 363–392.

Lee I-M, Davis RE, and Gundersen-Rindal DE (2000) Phytoplasma: phytopathogenic mollicutes. *Annual Review of Microbiology* 54: 221–255.

Lindgren PB (1997) The role of *hrp* genes during plant–bacterial interactions. *Annual Review of Phytopathology* 35: 129–152.

Russel M (1998) Macromolecular assembly and secretion across the bacterial cell envelope: Type II protein secretion systems. *Journal of Molecular Biology* 279: 485–499.

Schumann GL (1991) *Plant Diseases: Their Biology and Social Impact*. St Paul: APS Press.

Sigee DC (1993) *Bacterial Plant Pathology: Cell and Molecular Aspects*. Cambridge: Cambridge University Press.

Strange RN (1993) *Plant Disease Control: Towards Environmentally Acceptable Methods*. London: Chapman & Hall.

Tzfira T, Rhee Y, Chen M-H, Kunik T, and Citovsky V (2000) Nucleic acid transport in plant–microbe interactions: the molecules that walk through walls. *Annual Review of Microbiology* 54: 187–219.

Williams P, Ketley J, and Salmond GPC (eds) (1998) *Bacterial Pathogenesis*. San Diego: Academic Press.

Fungal Diseases

K M Plummer, University of Auckland, Auckland, New Zealand

Introduction

Fungi are generally microscopic, eukaryotic organisms. They are placed in their own kingdom, Fungi, distinct from plants and animals. Fungi, like animals, are heterotrophs: they are entirely dependent on other organisms to supply them with food, as they are unable to manufacture their own source of organic carbon via photosynthesis. Fungi are highly adapted to degrade a wide variety of complex compounds, including many plant products that animals are unable to digest. For this reason fungi are crucial players in nutrient cycling. Most fungi live on dead organic matter and are known as saprobes. Some fungi are not exclusively saprobic and can attack living plants. Fungi are responsible for the majority of plant diseases and consequently have a major economic impact on crop production. This article discusses several aspects of fungal diseases of crops:

- economic and social impacts
- characteristics of plant pathogenic (disease-causing) fungi
- basic factors that influence fungal diseases
- identification and diagnosis
- current and possible future control strategies.

Economic and Social Impacts of Fungal Diseases

All crop plants, irrespective of their end use (food, fiber, ornamentals, wood products) are vulnerable to fungal attack. The impact of fungal diseases on crops ranges from minor yield reduction in individual plants, to total yield loss that can occur with devastating epidemics. Fungi can also affect the quality of a product. Cosmetic blemishes and reduced shelf-life due to postharvest spoilage lower crop value (**Figure 1**). This is particularly important for fresh fruit and vegetables and is an aspect of fungal diseases familiar to most people as consumers.

International transportation of plant products dramatically increased with the advent of refrigerated shipping containers and airfreight. Consumers enjoy seasonal produce throughout the year; however, this transcontinental trade in plant products also allows the spread of fungal pathogens beyond their original, natural range. Global travel also aids the spread of pathogens as microscopic fungal spores hitch a ride on unwitting human passengers and their luggage. Most countries have quarantine restrictions to guard against contaminated products in an attempt to control the entry of unwanted pathogens. The area of fungal detection and identification has thus become an extremely important issue, as restriction or denial of access to markets can result in massive income disruption.

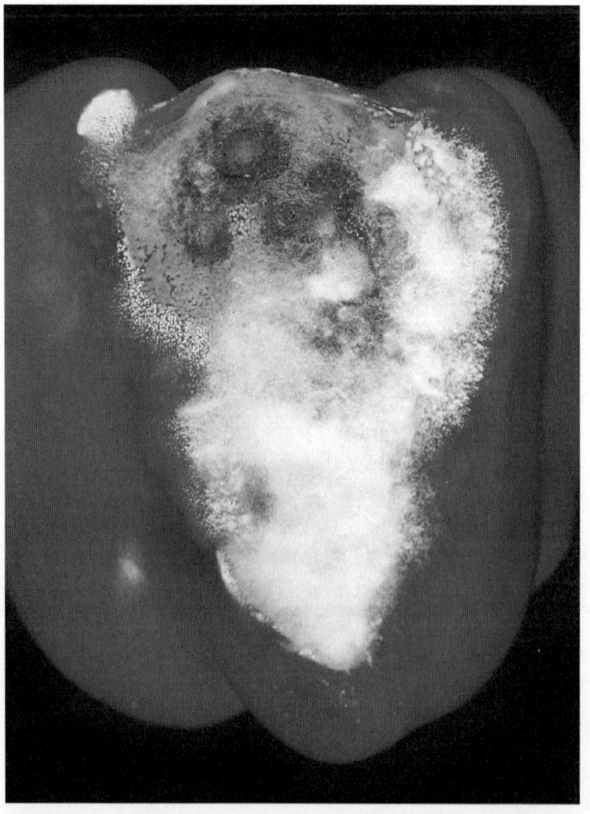

Figure 1 (see color plate 7) Rots such as this infection on *Capsicum* caused by *Glomerella cingulata* are a major cause of losses postharvest.

Fungal contamination of food can also have serious health consequences. Diseases caused by *Fusarium* on corn (*Zea mays*; maize) and *Aspergillus flavus* on nuts and stored grains can have a devastating impact as they produce contaminating toxins, known as mycotoxins, which are poisonous to both humans and livestock. These toxins can cause diseases including cancers and can even lead to death.

The presence of toxic fungal compounds can also be associated with fungi that produce few disease symptoms in their plant host. A group of fungi, known as endophytes, grow in a benign way within plants, commonly in grasses, and have a minor impact on yield. Endophytes often protect the plant from attack by herbivores or other microbial pathogens. The impact of endophytes is their ability to produce chemicals (e.g., alkaloids), which are highly toxic to animals, including cattle and sheep, which graze on infected grass. The symptoms in affected animals include lethargy, nervous system damage leading to staggers in grazing animals, gangrene, and eventually death.

Ergot of rye (*Secale cereale*), caused by the fungus *Claviceps purpurea*, is an example of such an endophytic association that has caused major outbreaks of disease (ergotism) in humans in the past. Ergots, also known as sclerotia, are tough structures made up of compact fungal threads, which are produced in place of some of the seeds in infected plants. Several different alkaloid compounds are found in ergots including lysergic acid diethylamide (the hallucinogen LSD). This explains some of the symptoms that occur in humans when they eat milled grain contaminated with ergots. Ergotism symptoms include hallucinations, burning and crawling sensation of the skin, convulsions, abortions, gangrene, and eventually death. In the Middle Ages the disease was commonly known as "Holy Fire" or "St Anthony's Fire" and one particular epidemic in Europe in the year AD 994 was responsible for more than 40 000 deaths. It is also thought that ergotism symptoms were interpreted as evidence during widespread accusations of witchcraft in Europe and the United States, most notably in Salem. As recently as 1951, in a village in France, five people died and 30 became temporarily insane due to ergotism. The problem is controlled today by modern management practices and monitoring ergot contamination. Alkaloid compounds from ergots are now widely used as pharmaceuticals, e.g., ergotamine, for the treatment of migraine.

Fungal disease epidemics throughout history have had major impacts resulting in famine. In India, the rice (*Oryza sativa*) disease known as brown spot, caused by *Bipolaris oryzae*, led to the Bengal Famine in 1943, resulting in the death of approximately 4 million people. Cereal rust and smut diseases are thought to reduce the world grain crop by approximately 10% per year. In 1970, a single epidemic of Southern corn leaf blight in the United States resulted in losses valued at approximately US$1 billion.

The late blight epidemic of potato (*Solanum tuberosum*), which led to the Irish potato famine during the 1840s, is perhaps the most famous historical event in plant pathology. It is estimated that up to 1.5 million people died of starvation and a similar number emigrated, mainly to North America, as a consequence of the famine. The impact of the epidemic was compounded by the Irish population's dependence on potato as their major food source and the requirement to send cereal crop harvest to English landlords as payment for rent.

Late blight of potato is caused by *Phytophthora infestans*. The name *Phytophthora* is derived from the Greek for plant ("phyto") destroyer ("phthora"), which is an accurate description of the effects this organism has on its host. *Phytophthora* is not a true fungus, but a member of the group variously known as water molds, lower fungi, or Oomycetes. Pathogens that cause downy mildews and white rusts are other examples of Oomycetes. The Oomycetes were removed from the kingdom of Fungi after research, using genetic and physiological analyses, showed that although these organisms look similar to true fungi, they are more closely related to brown algae. They are often referred to as "fungi-like" and they are classified within the diverse grouping of organisms known as protists, within the separate kingdom of Stramenopila.

Late blight of potato and several other diseases, including rice blast and rust diseases of coffee (*Coffea* spp.), sugar cane (*Saccharum officinarum*), and soybean (*Glycine max*), are predicted to have a major social impact in the future. Pathogens causing soybean rust and downy mildew of corn have recently spread to new continents. The importance of these diseases is due to the fact that they affect crops that are staple foods for humans or livestock, e.g., rice, wheat (*Triticum aestivum*), corn, *Sorghum*, soybean; or they are the major commodity crop for a region, e.g., sugar cane, coffee, cocoa (*Theobroma cacao*).

Characteristics of Plant Pathogenic Fungi

Fungi usually grow as microscopic filaments (approximately 1–2 μm in diameter and up to several centimeters long) known as hyphae. Masses of hyphae (known as mycelia) grow on or within a plant or other food source, absorbing nutrients over the massive surface area of their threadlike hyphae or through specialized absorption structures known as haustoria. Fungi have cell walls that are strong outer coverings similar to those of plant cells, but with a different composition. Fungal cell walls are made up of long chainlike molecules of repeated units of chitin or glucan. Insect outer coverings (exoskeletons) are also made of chitin. For many years fungi were considered to be microscopic plants; however, genetic analyses have shown that fungi are more closely related to animals.

Fungi are generally haploid (having only one set of chromosomes) throughout most of their life cycle, which simplifies the study of genetic inheritance in these organisms. Fungi reproduce, both asexually and sexually, by the production of spores, which can be single or multicelled and are highly variable in form and arrangement (**Figure 2**). Spores are often borne simply on hyphae or can be held in more elaborate fruiting bodies (**Figure 3**). The most familiar fruiting body is a mushroom or toadstool, but the forms of fruiting bodies are extremely variable. The various forms of fungal structures are utilized by plant pathologists and mycologists for the identification and classification of fungi and the diseases they cause (see below).

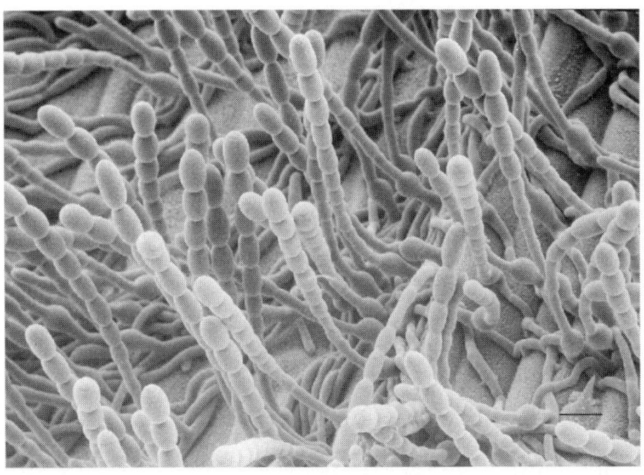

Figure 2 Scanning electron micrograph showing asexual spores and growth habit of the powdery mildew fungus (*Blumeria graminis*), on a barley leaf. Scale = 40 μm. Photograph courtesy of M Cromey, Crop and Food, New Zealand and P Sutherland, HortResearch, New Zealand.

Figure 3 Sexual fruiting bodies (dark spheres) of the blackleg fungus (*Leptosphaeria maculans*) on canola stubble (decaying plant matter) are a source of inoculum (infectious components) for new infections upon planting of a new oilseed crop. Scale = 0.5 mm.

Basic Factors that Influence Fungal Diseases

Fungal plant diseases are a result of the combined effects of the environment, the host, and the fungal pathogen. The progress of each disease is influenced by all three components. The fungal pathogen itself is not the disease. The optimal growth conditions for fungi vary widely, but moist, temperate climates are particularly favorable to many fungal pathogens. Most environments have a range of fungi and plant hosts that have adapted to the specific conditions.

Host Plant Defenses

Just as pathogens that attack humans are usually limited to humans, most fungi are limited in the range of plants that they can attack. Conversely it holds that most plants are resistant to most pathogens. In fact very few pathogens are able to attack a particular plant species and those that can have usually evolved specific mechanisms to do so.

Plants cannot run from an enemy or easily choose a new location to find a favorable environment. To cope with this they have evolved a range of defenses to protect themselves from the elements and from pest and pathogen attack. These defenses include primary physical and chemical barriers which are in place all the time. Physical barriers include plant cell walls and the thick waxy cuticle covering aerial plant surfaces. Chemical barriers include antimicrobial chemicals, that are toxic to prospective invaders, e.g., phenol or cyanide compounds.

Plants are also able to activate additional defenses after it is detected that the primary defenses have been breached by accidental wounding or by a fungal invader. These secondary defenses sometimes involve the sacrifice (death or walling off) of a small area of tissue to localize the pathogen to the site of invasion. The whole plant can also be alerted to the attack and can mount an overall defense response, including the production of additional antifungal compounds.

From this we can see that plants are far from being a passive meal for the invading fungus. Fungi that are pathogenic on a particular host must overcome the host's battery of physical and chemical defenses as well as avoid detection by surveillance systems that trigger the induced defense responses.

Host Range

The range of plant species that a particular fungus can infect is usually referred to as its host range and this is typical for a particular pathogen. This aids in the identification of plant disease, as relatively few fungi will be likely suspects for a particular symptom. In some cases there is variation within a fungal species as to the host plants that can be attacked. It is common to encounter additional subspecific designations for fungal pathogens, e.g., "pathovar," "special form," or *forma specialis* (f. sp.). These subclasses of pathogens are differentiated from each other not by differences in morphology (shape, structure, and size), but by differences in the host range that they can attack. *Puccinia graminis* is the species name for stem rusts of many cereals, but the subclass of stem rust that specifically attacks wheat (*Triticum* spp.), is referred to as *Puccinia graminis* f. sp. *tritici* (**Figure 4**). Fungal subspecies are often distinguished by differences in their genetic makeup, detectable by DNA sequencing.

Figure 4 Stem rust of wheat, caused by *Puccinia graminis* f. sp. *tritici.* (A) Rust symptoms on wheat stems; (B) closeup of infected wheat stem; (C) barberry leaves with fruiting bodies of the rust fungus. Reproduced with permission from Agrios GN (1997) *Plant Pathology*, 4th edn. New York: Academic Press.

Plants can also vary in their susceptibility to pathogens depending on the growth stage and the general health of the plant. Different plant parts may vary in their susceptibility to pathogens, and pathogens often specialize in the plant organ that they attack. Root pathogens rarely attack aerial parts such as fruit, leaves, or shoots and the converse is true.

Fungal Pathogens

Pathogenic fungi have an arsenal of biochemical weapons that enable them to penetrate host barriers and colonize plant tissues. The weapons that are used vary between fungal pathogens and are adapted to particular host plants. Enzymes are secreted to degrade particular plant components, e.g., cellulases attack cellulose of plant cell walls. Some fungi have developed a specialized structure, the appressorium, to penetrate the host. The appressorium is a swollen hyphal tip which adheres to the plant surface and then, with the assistance of concentrated pressure, a fungal thread, or peg, is physically forced through the plant cuticle and cell wall. This process is likely to be assisted by the secretion of various enzymes.

Some fungi produce enzymes that can detoxify antifungal compounds produced by a plant, e.g., cyanide. Cyanide is produced in clover (*Trifolium* spp.), soybean, and sorghum when the plants are under attack by a fungus. Fungi that are pathogens on these hosts must be able to cope with the cyanide. *Gloeocercospora sorghi*, a pathogen of sorghum, produces an enzyme, cyanide hydratase, which reduces cyanide to formamide, which is less harmful to the invader.

Once the host is penetrated the fungus will then use a wide variety of strategies to obtain nutrients from the host. The strategy depends very much on the lifestyle or physiology of the fungus. Some fungi secrete an array of toxins, which enable a fungus to quickly kill the host before it can mount a defense response. In addition, enzymes are secreted to degrade the host tissue in advance of the fungal hyphal growth. The fungus then grows through its host absorbing the degraded tissues. These thuggish fungi are known as necrotrophs, because they kill their host and feed ("troph") on the dead ("necro") remains. Necrotrophs often have a relatively wide host range and are able to survive as saprobes in the absence of a living host.

Many fungi, such as rust and mildew fungi, are obligate parasites and are referred to as biotrophs. Biotrophs require a living ("bio") host to supply them with food. Biotrophs utilize an arsenal of hormones to stimulate host plant growth, while avoiding recognition to prevent the plant from mounting a defense. Specialized extensions of hyphal tips, haustoria, are produced by some biotrophs, such as the rust and mildew fungi, to tap into living plant cells and drain the plant of nutrients. In the case of barley (*Hordeum vulgare*) mildew, caused by *Blumeria graminis*, the majority of the fungal body is on top of the barley leaf with only a few of the host cells in the leaf surface providing nutrients to the fungus via haustoria (**Figure 2**). The infected host remains alive for a significant time during the infection. Biotrophs are highly adapted to manipulate specific host plants into surrendering their nutrients. Eventually the host nutrients will be exhausted. By this time the pathogen will have produced spores to locate another host on which to prey, or structures to survive in the absence of a host.

Fungal Reproduction, Dispersal, and Survival

Fungi reproduce both sexually and asexually. Reproduction serves several purposes:

- propagation
- dissemination
- survival or hibernation
- genetic recombination, providing offspring with new mechanisms for survival or adaptation.

The relationship with the host plant is crucial in terms of strategies for the fungal pathogen. A fungus may need to survive the absence of a host. The host may be an annual species that dies off at the end of one season, or a deciduous perennial species that loses its leaves every autumn.

Fungi vary in their strategies for survival and reproduction. Some survive the absence of a living plant as saprophytes living on decaying matter, often on the fallen leaves or debris from a deciduous or annual host (**Figure 3**). Obligate parasites, such as rusts, which require a living host to grow on, can sometimes alternate between two distinct seasonal hosts. In cooler climates of the northern hemisphere, *Puccinia graminis* f. sp. *tritici*, the cause of stem rust of wheat (**Figure 4**), survives in the absence of the annual wheat crop by growing and reproducing on an alternate host, barberry (*Berberis* spp.), during spring. Wheat and barberry are extremely different types of plants and yet the fungus has adapted perfectly to these species to enable it to survive year round. Barberry was used as a common hedge plant in many wheat-growing regions until the connection between the fungi growing on these two hosts was recognized. Eradication of barberry from wheat-growing regions reduced the variation generated in the rust by removing the sexual stage; however, this rust is able to persist on its wheat host.

Another strategy for various pathogens is to go into hibernation and survive as resistant resting structures either within dormant host tissues, such as in bark or seeds, or in soil in the absence of the host. Sclerotia, the tough, dark and compact masses of hyphae produced by *Sclerotium* species and ergots produced by *Claviceps purpurea*, are examples of such survival structures. Sclerotia are resistant to both unfavorable environmental conditions (e.g., dehydration) and to degradation by other micro-organisms. They can persist in soil for up to 20 years in the absence of a host and for this reason they can be a source of infection that is extremely difficult to eradicate. The dormancy of these structures is often broken by the presence of specific chemicals exuded by the newly growing host early in the next season.

Figure 5 (see color plate 8) Apple scab disease, caused by *Venturia inaequalis*, leads to major apple crop loss worldwide.

Identification and Diagnosis of Fungal Plant Diseases

The diagnosis of a particular plant disease and identification of a causal organism is essential for the treatment of a crop and subsequent planning for crop protection. This is simplified by the fact that only relatively few fungal species are able to attack each plant species. Additionally, symptoms for a particular disease are usually characteristic of a specific fungus. Many diseases can be diagnosed by simple observations with the naked eye or with a light microscope.

Examples of Symptoms Caused by Fungi

Fungi cause leaf spots, wilts (usually the result of blocking the vascular system of the plant), cankers, scabs (**Figure 5**), blights, mildews (**Figure 2**), rusts (**Figure 4**), fruit rots (**Figure 1**), stem and root rots, and galls, to name a few. The symptom of a disease is the manifestation of the interaction between the environment, the fungus, and its host, with both organisms producing compounds that alter the physiology of the other. The resulting appearance of the diseased tissue will depend on the organ affected and the type of interaction. Necrotrophic rot fungi induce very different symptoms to the biotrophic rusts and mildews that maintain an intimate relationship with the living host during most of the infection period.

Diagnosis of a Causal Organism of a Fungal Disease

Diagnosis can be aided by the presence of obvious fungal structures. Infected tissues may be cultured (grown) on specific media to promote growth, to induce reproduction, and to isolate the infecting fungal pathogen. In most cropping situations the diagnosis may be complicated by the presence of more than one microorganism in the infected tissues. This is due to opportunistic infections by secondary parasites that take advantage of the reduced defenses of the host. Determining which organism is the primary cause of a particular disease symptom requires more investigation. Usually, organisms are isolated from the host to aid in the identification and subsequent verification of a diagnosis. This is quite straightforward if the disease is already well characterized and if the causal organism can be cultured. For plants with novel disease symptoms or hosts that have not been widely studied this process is more complicated. In the 1870s a set of rules known as Koch's postulates were devised that guide the determination of the cause of a disease. These postulates, whilst having significant utility for many diseases, are dependent on the ability of the pathogen to be cultured, which is not possible for many key diseases. A compromise of guilt by association must be struck under these circumstances.

Fungal reproductive structures (spores and other fruiting bodies) are generally the characters used for identification and classification of fungi. Fungi produce different spore types depending on the stage of their life cycle. One extreme example of this is wheat stem rust, where five distinct spore types are produced on two vastly different host plant species. Additionally, sexual and asexual reproductive structures are very different in appearance and often vary in their timing and location during the infection of a plant. In apple scab disease the asexual stage occurs on the living leaves and fruit of apple trees (*Malus pumila*) during the growing season (**Figure 5**). The sexual reproductive structures are produced during winter in the dead leaves at a time when the apple trees are dormant. This is a common strategy for survival and reproduction of plant pathogens. It is also the reason that the apple scab pathogen has two names, one named on the basis of the asexual stage, *Spilocaea pomi,* and one on the basis of the sexual reproductive structures, *Venturia inaequalis*. This is true of many pathogens where a sexual and an asexual phase are present. Often a name based on the asexual (anamorph) stage was assigned before a sexual phase (teliomorph) for that fungus was identified. Usually when a sexual phase is identified for a fungus, the sexual, or "perfect name," is the one that should be adopted, but the imperfect or asexual name is often maintained for organisms that rarely undergo sexual reproduction. The cause of grey mold in grape (*Vitis vinifera*) and many other fruits, *Botrytis cinerea*, for example, is rarely referred to by the perfect name, *Botryotinia fuckeliana*, as this stage is seldom observed in the field.

Fungi that have no known sexual reproduction are grouped together in a loose association known as the Fungi Imperfecti, which are classified solely on the basis of their asexual reproductive structures. Improved methods of identification, such as DNA sequencing and fingerprinting, have aided the clarification of relationships between sexual and asexually reproducing fungi.

Control of Fungal Diseases

In most cropping situations fungal pathogens are primarily controlled with chemicals. Most agrochemical usage for disease control is for fungi. The impacts of these control measures on human health and the environment are difficult to estimate. Concerns over these impacts have led to restrictions in the types and quantities of chemicals that can be used. Broad-spectrum chemicals that are toxic to nontarget organisms (humans, other animals, plants, and nonpathogenic microorganisms) are being phased out. Chemicals that are active against specific pathogens and degrade rapidly in the environment are a safer option, however fungi are extremely adept at developing resistance to these.

Disease forecasting is employed for some crops to minimize fungicide usage by targeting applications to the times when the conditions of the host, the environment and the pathogen are conducive to infection. Crop management strategies that decrease the incidence of fungal infections are also crucial to reducing chemical usage. The use of healthy, pathogen-free, and disease-resistant planting material is adopted where possible. Maintenance of overall plant health is also important as stressed plants, as with animals, are usually more susceptible to disease. Planting in soil that is free of pathogenic fungi is also desirable, although not always practical.

Annual crops, like wheat, are planted in rotation with unrelated crop species like canola (*Brassica napus*), to lessen their exposure to pathogens that remain on plant debris from a previous season. Good farm hygiene is important both in the growing season, where humans and farm equipment can easily spread infectious materials, and at postharvest where a clean, pathogen-free environment is essential for disease control.

The interactions that occur between all organisms in a crop are vital for fungal disease development. Maintaining and encouraging the presence of beneficial, nonpathogenic organisms is fundamental to disease control, especially in organic and integrated pest management systems. The application of biological control agents (microbial antagonists and competitors of pathogens) is an area of active research. Development of these agents is complicated by the fact that they are often delivered as a living organism, which will be influenced by and influence conditions in the cropping environment. Few biological control agents are currently used commercially although many are under development and show promise for the future. Many fungi in the genus *Trichoderma* are common soil saprobes, which have been shown to have utility as biological control agents.

Molecular technologies are also being utilized for the development of long-term solutions for fungal disease control. Traditional plant breeding programs are now using molecular markers (or tags) to assist selection of resistant varieties. In addition, these markers can aid in the identification of specific genes that confer natural plant disease resistance. These genes can then be transferred, using genetic engineering, from a resistant plant cultivar with poor agronomic qualities, to a susceptible cultivar with excellent qualities. A substantial effort to identify these genes is currently in progress, aided by recent developments in rapid genome sequencing.

The key to identifying solutions for disease control lies in understanding the complex interactions that result in disease. The challenge for future crop producers will be to develop and adopt alternatives that are sustainable for managing plant diseases, whilst ensuring delivery of products that are free of harmful chemicals and fungal contaminants.

List of Technical Nomenclature

Anamorph	Asexual stage of a fungus.
Biotroph	An organism that can live and multiply only on another living organism (synonym: obligate parasite).
Cell wall	Protective, resistant, but permeable structure secreted externally to the cell membrane in plants, bacteria, fungi, and certain other organisms.
Endophyte	Any fungus that grows in a benign way within the plant tissues, often producing compounds, e.g., alkaloids, that are toxic to animals.
Eukaryote	An organism containing a membrane-bound nucleus and other organelles, i.e., all higher plants, animals, fungi, and protists.
Host plant	Living plant attacked by or harboring a parasite or pathogen and from which the invader obtains part or all of its nourishment.
Host range	The range of plants on which an organism, particularly a pathogen, feeds.
Fruiting body	Any of various complex, spore-bearing fungal structures.
Fungicide	Any chemical that kills or inhibits the growth of fungi.
Haploid	Having a single complete set of chromosomes.
Haustorium	Specialized terminal branch of hyphae formed inside host cells to absorb nutrients.
Heterotroph	An organism that obtains nourishment from outside sources and must obtain its carbon from organic carbon compounds.
Hypha	A single, tubular filament of a fungal thallus or mycelium; the basic structural unit of a fungus.
Mycelium	The mass of hyphae that constitute the body of a fungus.
Necrotroph	A parasite that typically kills and obtains its energy from dead host cells.
Saprophyte	An organism that obtains nourishment from nonliving organic matter.
Spore	The reproductive structure of fungi and some other organisms, containing one or more cells.
Teliomorph	A sexual phase of a fungus.

See also: **Crop Improvement**: Plant Breeding, Practice; Plant Breeding, Principles. **Diseases**: Breeding for Disease Resistance; Genetic Modification of Disease Resistance, Fungal Pathogens; Plant Pathology, Principles. **Integrated Pest Management**: Disease Prediction Models; Practice; Principles. **Production Systems and Agronomy**: Organic Farming.

Further Reading

Agrios GN (1997) *Plant Pathology*, 4th edn. New York: Academic Press.

Alexopoulis CJ, Mims CW, and Blackwell M (1996) *Introductory Mycology*, 4th edn. New York: John Wiley.

American Phytopathological Society (2003) APS*net Education Centre – The Plant Health Instructor.* http://www.apsnet.org/education/top.html

Carefoot GL and Sprott ER (1967) *Famine on the Wind.* Chicago: Rand McNally.

D'Arcy CJ, Eastburn DM, and Schumann GL (2001) APS*net Education Centre – The Plant Health Instructor, Illustrated Glossary of Plant Pathology.* http://www.apsnet.org/education/IllustratedGlossary/top.htm

Hawksworth DL, Kirk PM, Sutton BC, and Pegler DN (1995) *Ainsworth and Bisby's Dictionary of the Fungi*, 8th edn. Kew: Commonwealth Mycological Institute.

Horsfall JG and Dimond AE (1959) *Plant Pathology.* New York: Academic Press.

Lucas JA (1998) *Plant Pathology and Plant Pathogens*, 3rd edn. Oxford: Blackwell Science.

Schumann GL (1991) *Plant Diseases: Their Biology and Social Impact.* St Paul: American Phytopathological Society Press.

Ulloa M and Hanlin RT (2000) *Illustrated Dictionary of Mycology.* St Paul: American Phytopathological Society Press.

Van der Plank JE (1963) *Plant Diseases: Epidemics and Control.* New York: Academic Press.

Viral Diseases

R Hull, John Innes Centre, Norwich, UK

Introduction

There are records of what we now know to be plant viral diseases from as long ago as 752 AD when a yellowing disease of *Eupatorium lindleyanum* was described in a poem by the Japanese Empress Koken. In the early seventeenth century, many paintings and drawings depicted tulips with flowers showing symptoms of virus disease (*Tulip breaking virus*). During that period, such flowers were much prized leading to "tulipomania" in the Netherlands. Trade in infected tulip bulbs resulted in hyperinflation and eventually the Netherlands government had to intervene.

There are records from the eighteenth and early nineteenth centuries of viral symptoms being transmitted by grafting, but the real significance of the etiological agent of viral diseases came from a series of experiments conducted in the 1880s and 1890s. Mayer, Iwanowski, and Beijerink showed that a disease of tobacco (*Tobacco mosaic virus*, TMV) could be transmitted by inoculation of sap extracts from infected plants and that the infective agent passed through filters that retained bacteria.

The study of virology, led in several respects by plant virology, developed through the twentieth century. The early stages were the demonstration of insect transmission of viruses and the recognition that many plant diseases are caused by viruses. The chemical nature of viruses, being protein and nucleic acid, was determined by Stanley and Bawden in the 1930s and the infectious nature of the nucleic acid was recognized in the 1950s. At that time, the shapes and structures of virus particles were determined by the new techniques of electron microscopy and X-ray crystallography. Detailed information on the composition of viral nucleic acids and proteins emerged slowly until the advent of molecular technology, which speeded up the process markedly. Now the genome sequence has been determined for one or more representative of each plant genus except for one. Other major impacts of molecular technology and the ability to transform plants with "foreign" DNA have seen a great increase in the understanding of how plant viruses replicate and how they interact with their hosts to induce disease.

There have been many attempts to define a virus. The most recent definition is: "A virus is a set of one or more of a single type of nucleic acid template molecule (RNA or DNA), normally encased in a protective coat or coats of protein or lipoprotein, that is able to organize its own replication only within suitable host cells. Within such cells, virus replication is (i) dependent on the host's protein-synthesizing machinery, (ii) organized from pools of the required materials rather than by binary fission, (iii) located at sites that are not separated from the host cell contents by a lipoprotein bilayer membrane, and (iv) continually giving rise to variants through various kinds of change in the viral nucleic acid."

Classification of Viruses

Over the years, there have been several systems for naming plant viruses, especially in the 1920s and 1930s when many new viruses were being recognized and described. The general system that is now being used is to name the virus after the host in which it was first found and from the characteristic symptoms in that host. Some exceptions to this have arisen from common usage. To make communication easier, viruses are also given acronyms, the rules for which are determined from guidelines issued by the International Committee on the Taxonomy of Viruses (ICTV).

The classification system for all viruses, be they from animals, plants, insects, or bacteria, is under the ICTV. The classification is dynamic with the rapid increase in information on viruses being used to adjust the existing taxa and to create new ones. The basic taxon is the virus species, of which there are more than 970 that infect plants. Species that share characters such as particle shape and genome organization are grouped into genera; there are 71 genera of plant viruses (**Table 1**). Some of the genera with common features are grouped together as families, of which there are 14 for plant viruses; it is likely that many of the unassigned genera will be grouped into families in the near future. Hence, the listing of the genera in **Table 1** brings together the genera with most similarities.

Most of the virus families are unique to plants, in some cases, having features in common with animal viruses. However, three of the families listed in **Table 1**, the *Reoviridae*, the *Rhabdoviridae*, and the *Bunyaviridae*, have species that infect higher animals and invertebrates. It is thought that these are invertebrate viruses that have become adapted to plants.

The rules for virus nomenclature and orthography of virus names are set by the ICTV. The last word of a species is "virus" and the suffix for a genus is "*....virus*," for a subfamily "*....virinae*" and for a family is "*....viridae*." In formal taxonomic usage, the name of the taxon should precede the name being used, the name being printed in italics with the first letter being capitalized; other words in a species name are not capitalized unless they are proper nouns or parts of proper nouns. Thus, formal usage should be "the family *Potyviridae*," "the genus *Potyvirus*," and "the species *Tobacco etch virus*." In informal use, the family, subfamily, genus, and species names are written in lower case Roman script, the taxon does not include the formal suffix, and the taxonomic unit follows the name being used. In the widely used less formal naming, the taxonomic unit is omitted and the name for higher taxa can be in the plural, e.g. "potyviruses" and "tobacco etch virus."

A common problem in determining whether a new virus is a distinct species is the variation of all viruses. A virus species is not a uniform population

Table 1 Features of plant viruses

Genome[a]	Particle shape[b]	No. of genome segments	No. of components	Family	Genus	Type species	No. spp.	Vector[c]
ds DNA (RT)	I	1	1	*Caulimoviridae*	*Caulimovirus*	*Cauliflower mosaic virus*	9 + 4t	as
	I	1	1		*Soymovirus*	*Soybean chlorotic mottle virus*	2	u
	I				*Cavemovirus*	*Cassava vein mottle virus*	1	u
	I	1	1		*Petuvirus*	*Petunia vein clearing virus*	1	u
	B	1	1		*Badnavirus*	*Commelina yellow mottle virus*	12 + 4t	m
	B	1	1		*Tungrovirus*	*Rice tungro bacilliform virus*	1	Lh
ss DNA	G	1	1	*Geminiviridae*	*Mastrevirus*	*Maize streak virus*	12 + 2t	Lc
	G	1	1		*Curtovirus*	*Beet curly top virus*	3 + 1t	Lc
	G	2	2		*Begomovirus*	*Bean golden mosaic virus*	76 + 8t	W
	G	1	1		*Topocuvirus*	*Tomato pseudocurly top virus*	1	Th
	I	6–11	6–11	*Circoviridae*	*Nanovirus*	*Subterranean clover stunt virus*	4 + 1t	Ac
ds RNA	IC	10	1	*Reoviridae*	*Fijivirus*	*Fiji disease virus*	8	Pp
	IC	10	1		*Oryzavirus*	*Rice ragged stunt virus*	2	Pp
	IC	12	1		*Phytoreovirus*	*Wound tumor virus*	3 + 1t	Lp
	I	2	1	*Partitiviridae*	*Alphacryptovirus*	*White clover cryptic virus 1*	16 + 10t	S
	I	2	1		*Betacryptovirus*	*White clover cryptic virus 2*	4 + 1t	S
	RR	2	2	No family	*Varicosavirus*	*Lettuce big-vein virus*	1 + 3t	F
ss RNA (−)	CB	1	1	*Rhabdoviridae*	*Cytorhabdovirus*	*Lettuce necrotic yellows virus*	8	lp, ap
	CB	1	1		*Nucleorhabdovirus*	*Potato yellow dwarf virus*	7	lp, ap
	CB	1	1		Unassigned		58	lp, ap, mtp
	CI	1	1	*Bunyaviridae*	*Tospovirus*	*Tomato spotted wilt virus*	8 + 5	Tp
	F	4–6	4–6	No family	*Tenuivirus*	*Rice stripe virus*	6 + 5	Pp
	F	3	3		*Ophiovirus*	*Citrus psorosis virus*	3	U
ss RNA (+)	I	3	3	*Bromoviridae*	*Bromovirus*	*Brome mosaic virus*	6	U
	B	3	3		*Alfamovirus*	*Alfalfa mosaic virus*	1	An
	I	3	3		*Cucumovirus*	*Cucumber mosaic virus*	3	An
	I - B	3	3		*Ilarvirus*	*Tobacco streak virus*	17	S
	B	3	3		*Oleavirus*	*Olive latent virus 2*	1	U
	I	2	2	*Comoviridae*	*Comovirus*	*Cowpea mosaic virus*	15	B

I	2	2		*Fabavirus*	*Broadbean wilt virus*	4	An
I	2	2		*Nepovirus*	*Tobacco ringspot virus*	31 + 9t	N
FR	1	1	*Potyviridae*	*Potyvirus*	*Potato virus Y*	91 + 88t	an, s
FR	1	1		*Ipomovirus*	*Sweet potato mild mottle virus*	1 + 1	W
FR	1	1		*Macluravirus*	*Maclura mosaic virus*	2	An
FR	1	1		*Rymovirus*	*Ryegrass mosaic virus*	4 + 1	Mt
FR	1	1		*Tritimovirus*	*Wheat streak mosaic virus*	2	Mt
FR	2	2		*Bymovirus*	*Barley yellow mosaic virus*	6	F
I	1	1	*Tombusviridae*	*Tombusvirus*	*Tomato bushy stunt virus*	13	u, s, f
I	1	1		*Avenavirus*	*Oat chlorotic stunt virus*	1	U
I	1	1		*Aureusvirus*	*Pothos latent virus*	1	U
I	1	1		*Carmovirus*	*Carnation mottle virus*	12 + 6t	u, s, f
I	1	1		*Machlomovirus*	*Maize chlorotic mottle virus*	1	S
I	1	1		*Necrovirus*	*Tobacco necrosis virus A*	5 + 2t	F
I	1	1		*Panicovirus*	*Panicum mosaic virus*	1 + 1t	U
I	2	2		*Dianthovirus*	*Carnation ringspot virus*	4 + 1t	U
I	1	1	*Sequiviridae*	*Sequivirus*	*Parsnip yellow fleck virus*	2	As
I	1	1		*Waikavirus*	*Rice tungro spherical virus*	3	Ls
FR	1	1	*Closteroviridae*	*Closterovirus*	*Beet yellows virus*	11 + 16	as, w, m
FR	2	2		*Crinivirus*	*Lettuce infectious yellows*	1	W
I	1	1	*Luteoviridae*	*Luteovirus*	*Barley yellow dwarf virus (MAV strain)*	2	Ac
I	1	1		*Polerovirus*	*Potato leaf roll virus*	5	Ac
I	2	2		*Enamovirus*	*Pea enation mosaic virus-1*	1	Ac
I	1	1		Unassigned		11	Ac
RR	1	1	No family	*Tobamovirus*	*Tobacco mosaic virus*	16 + 1t	C
RR	2	2		*Tobravirus*	*Tobacco rattle virus*	3	N
FR	1	1		*Potexvirus*	*Potato virus X*	26 + 19t	C
FR	1	1		*Carlavirus*	*Carnation latent virus*	31 + 29t	as, w, s, c
FR	1	1		*Allexivirus*	*Shallot virus X*	7 + 3t	Mt
FR	1	1		*Capillovirus*	*Apple stem grooving virus*	3 + 1t	S
FR	1	1		*Foveavirus*	*Apple stem pitting virus*	2 + 1t	U
FR	1	1		*Trichovirus*	*Apple chlorotic leafspot virus*	3	S
FR	1	1		*Vitivirus*	*Grapevine virus A*	4 + 1t	as, h, m, sc
RR	2	2		*Furovirus*	*Soil-borne wheat mosaic virus*	1 + 4t	F
RR	2	2		*Pecluvirus*	*Peanut clump virus*	2	f, s
RR	3	3		*Pomovirus*	*Potato mop top virus*	4	F
RR	4	4		*Benyvirus*	*Beet necrotic yellow vein virus*	2	F

Continued

Table 1 Continued

Genome[a]	Particle shape[b]	No. of genome segments	No. of components	Family	Genus	Type species	No. spp.	Vector[c]
	RR	3	3		*Hordeivirus*	*Barley stripe mosaic virus*	4	c, s
	I	1	1		*Marafivirus*	*Maize rayado fino virus*	3	Lp
	I	1	1		*Sobemovirus*	*Southern bean mosaic virus*	11 + 3t	b, s
	I	1	1		*Tymovirus*	*Turnip yellow mosaic virus*	21 + 2t	B
	I	2	2		*Idaeovirus*	*Raspberry bushy dwarf virus*	1	S
	B	3	3		*Ourmiavirus*	*Ourmia melon virus*	3	U
	N	1	N		*Umbravirus*	*Carrot mottle virus*	7 + 4t	H

Data from van Regenmortel MHV, Fauquet CM, Bishop DHL (eds) (2000) *Virus Taxonomy: Classification and Nomenclature of Viruses. Seventh Report of the International Committee on Taxonomy of Viruses*. San Diego: Academic Press.

[a]Genome type: ds DNA (RT) = double-stranded DNA (replicating by reverse transcription); ss DNA + single-stranded DNA; ds RNA + double-stranded RNA; ss RNA (−) = (−)-sense RNA; ss RNA (+) = (+)-sense RNA (mRNA).

[b]Particle shapes: B, bacilliform; CB, complex bacilliform; F, filamentous; FR, flexuous rod; G, geminate; I, isometric; IC, isometric complex; N, no (capsid); RR, rigid rod.

[c]Vector: ac, aphid, circulative; an, aphid, nonpersistent; ap, aphid, propagative; as, aphid, semipersistent; b, beetle; c, contact (mechanical); f, fungus; h, requires a helper virus, usually a luteovirus; lc, leafhopper circulative; lh, leafhopper but requiring helper virus; lp, leafhopper propagative; ls, leafhopper, semi-persistent; m, mealybug; mt, mite; mtp, mite, propagative; n, nematode; pp, planthopper propagative; s, seed; sc, scale insect; th, treehopper; tp, thrip, propagative; u, unknown; w, whitefly; b, bacilliform; bc, complex bacilliform; f, filamentous rod; fl*, flexuous particle but may be a complex particle; g, geminate; i, isometric; (i), possibly isometric; ic, complex isometric; ih, encapsidated by helper virus giving isometric particle; r, rigid rod.

but is heterogeneous due to errors in replication. Thus, a virus is a quasi-species, which is a type of population structure in which collections of closely related genomes are subjected to a continuous process of genetic variation, competition, and selection. Usually the distribution of variants is centered on a master sequence. However, if an advantageous mutant appears in the distribution, the equilibrium may collapse and a new population structure arise.

Virus Particles

The majority of plant viruses are composed of a nucleic acid genome that carries the information for infection surrounded by a protein coat (the capsid) made up of one or more virus-encoded protein species. The capsid protects the genomic nucleic acid from nuclease damage and also acts as the interface between the virus and the external environment. It can be involved in a variety of interactions, including the movement of the virus between plants by a specific type of vector and the movement of the virus within the plant from cell to cell and systemically through the plant.

Most plant virus particles have a simple structure, being either rod-shaped or based on isometric principles (see **Table 1**). Some viruses, notably those of the *Reoviridae*, *Rhabdoviridae*, and *Bunyaviridae*, which also have species that infect animals and invertebrates, have a complex structure. However, all the structures are based on strict principles relating the coat protein subunits to one another in the most efficient manner.

Isometric Virus Particles

The structure of isometric virus particles is based on several considerations, including crystallographic criteria and the use of identical subunits, which minimize the genetic information required for capsid formation, as does the size of the protein subunits. Virus particles have an icosahedral structure with 5-fold, 3-fold, and 2-fold axes of symmetry (**Figure 1A and B**). The basic icosahedral structure comprises 12 vertices with 5-fold rotational symmetry and 20 triangular faces with 3-fold symmetry. Thus, they are made up of 60 identical subunits (12×5 or 20×3). However, with constraints on the size of protein subunits, a particle made up of 60 subunits of a protein of about 30 kDa would not be able to contain a normal-sized genome. Thus, the basic icosahedron is modified by dividing the 20 faces into further triangles. Such modified icosahedra are described by the triangulation number (T), the basic one being T = 1, and the modified ones belonging to a permitted series of T = 3, T = 4, T = 7, etc. The number of subunits in these icosahedra is $20 \times 3 \times T$. Thus, T = 1 particles comprise 60 subunits, T = 3 particles 180 subunits and T = 4 particles 240 subunits. However, the insertion of extra subunits into the basic T = 1 icosahedron results in the loss of true symmetrical relationship between the subunits and in a quasiequivalent arrangement of the subunits. In the "quasi" symmetry, the subunits are identical but accommodate inconsistencies in symmetry by slight changes in tertiary structure.

The structure of many plant virus isometric particles, those with diameters of about 25–35 nm, is that of T = 3. Technically, comoviruses, which have particles in this range of diameters, have T = 1 particles but each "subunit" comprises two protein species, one of which has two structural domains; thus, the particles have 60 subunits but 180 structural domains. Some viruses with diameters in the range 15–20 nm have true T = 1 particles (e.g., *Satellite tobacco necrosis virus*) and caulimoviruses have T = 7 symmetry (420 subunits). Geminivirus particles (**Figure 1C**) comprise two T = 1 icosahedra fused together at 5-fold axes to give geminate particles with 110 subunits.

Isometric particles are stabilized by interactions that range from electrostatic protein:RNA bonds (e.g., *Cucumber mosaic virus*, CMV) to strong protein:protein bonds (e.g., comoviruses and nepoviruses). In the latter case, virus preparations have a proportion of "empty" particles, which do not contain nucleic acid. In some viruses (e.g., *Southern bean mosaic virus*), particle stabilization involves divalent cations such as Ca^{2+} as well as protein:RNA and unusual carboxyl: carboxylate bonds similar to those found in TMV.

The bacilliform particle structure found in several viruses (e.g., *Alfalfa mosaic virus* and badnaviruses) is considered a modification of the isometric structure. The curved ends of the particles are half icosahedra with the tubular part being made up of hexagonally arranged subunits. The diameter of the particle is determined by the icosahedral criteria described above and the length by the number of rings of hexagons inserted between the two half icosahedra. Thus, the bacilliform structure avoids the strict structural requirements of icosahedral symmetry and enables encapsidation of varying sizes of nucleic acids. Particles of ilarviruses are thought to have a loose quasi-icosahedral structure, which also overcomes the constraints of icosahedral symmetry.

Rod-Shaped Viruses

The shapes of rod-shaped particles of plant viruses range from short, rigid rods (e.g., TMV,

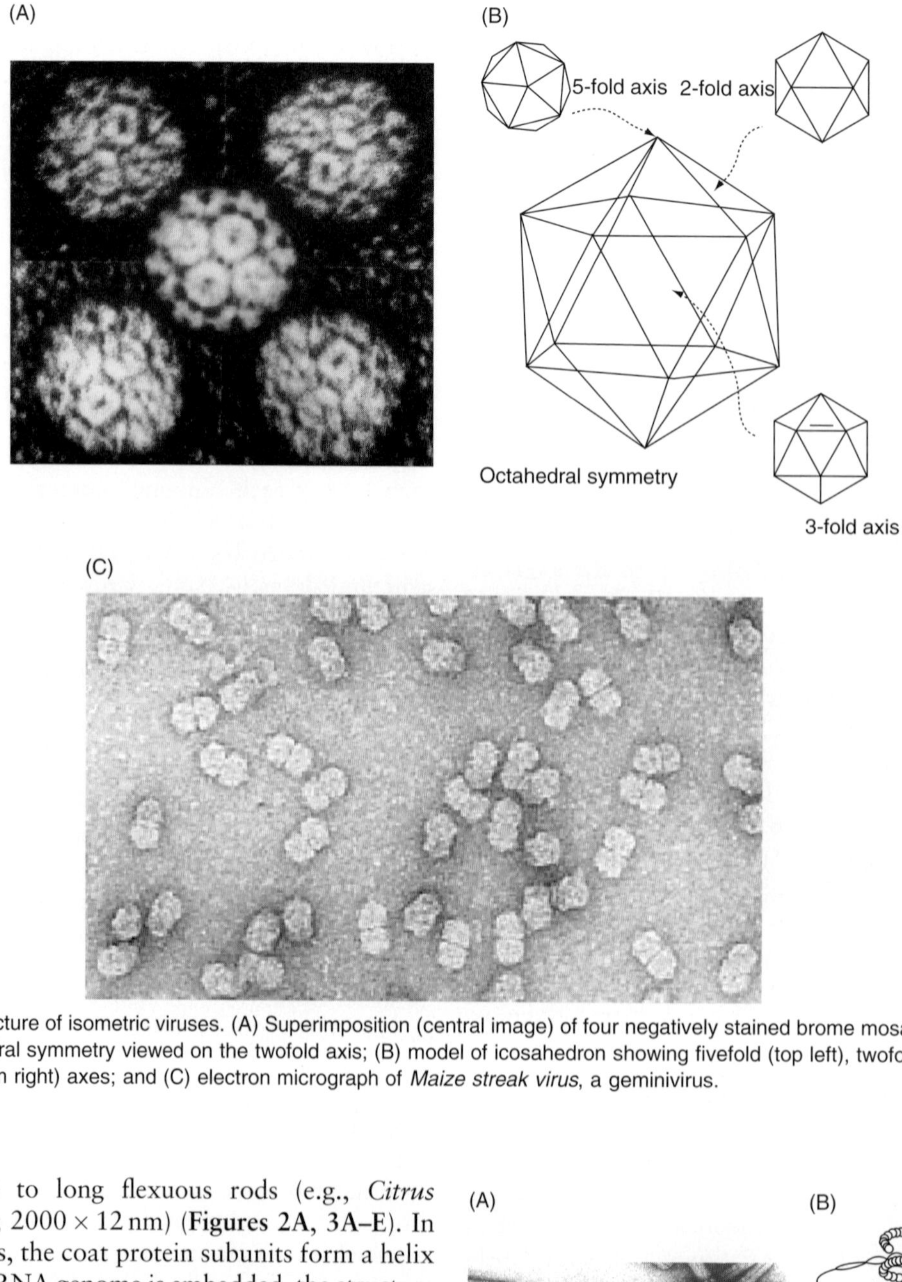

Figure 1 Structure of isometric viruses. (A) Superimposition (central image) of four negatively stained brome mosaic virus particles shows icosahedral symmetry viewed on the twofold axis; (B) model of icosahedron showing fivefold (top left), twofold (top right) and threefold (bottom right) axes; and (C) electron micrograph of *Maize streak virus*, a geminivirus.

300 × 18 nm) to long flexuous rods (e.g., *Citrus tristeza virus*; 2000 × 12 nm) (**Figures 2A, 3A–E**). In these particles, the coat protein subunits form a helix in which the RNA genome is embedded, the structure being stabilized by both protein:protein (including in TMV unusual carboxyl:carboxylate bonds) and protein:RNA links. The structure and assembly of TMV particles has been studied in detail (**Figure 2B**). These particles comprise about 2100 subunits of a single coat protein species arranged in a right-hand helix with a pitch of 2.3 nm and $16\frac{1}{3}$ subunits/turn; the RNA is set in a groove at about 4 nm radius. The particles have a central canal of radius 2 nm, which is involved in particle assembly.

To initiate particle assembly, a double-disk of 17 coat protein subunits per disk binds to a sequence-specific stem loop structure in the RNA (the origin of assembly) about 1100–900 nucleotides from the

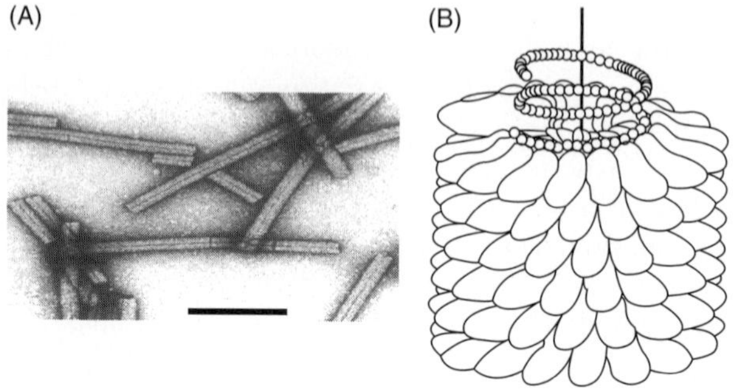

Figure 2 Structure of tobacco mosaic virus. (A) Electron micrograph of negatively stained virus particles showing rigid rods (300 × 18 nm); and (B) drawing to show relationship of the RNA and protein subunits. Scale bar = 100 nm. The RNA shown free of protein could not maintain this configuration in the absence of the protein subunits. Reproduced with permission from Klug A and Caspar DLD (1960) The structure of small viruses. *Advances in Virus Research* 7: 225–325.

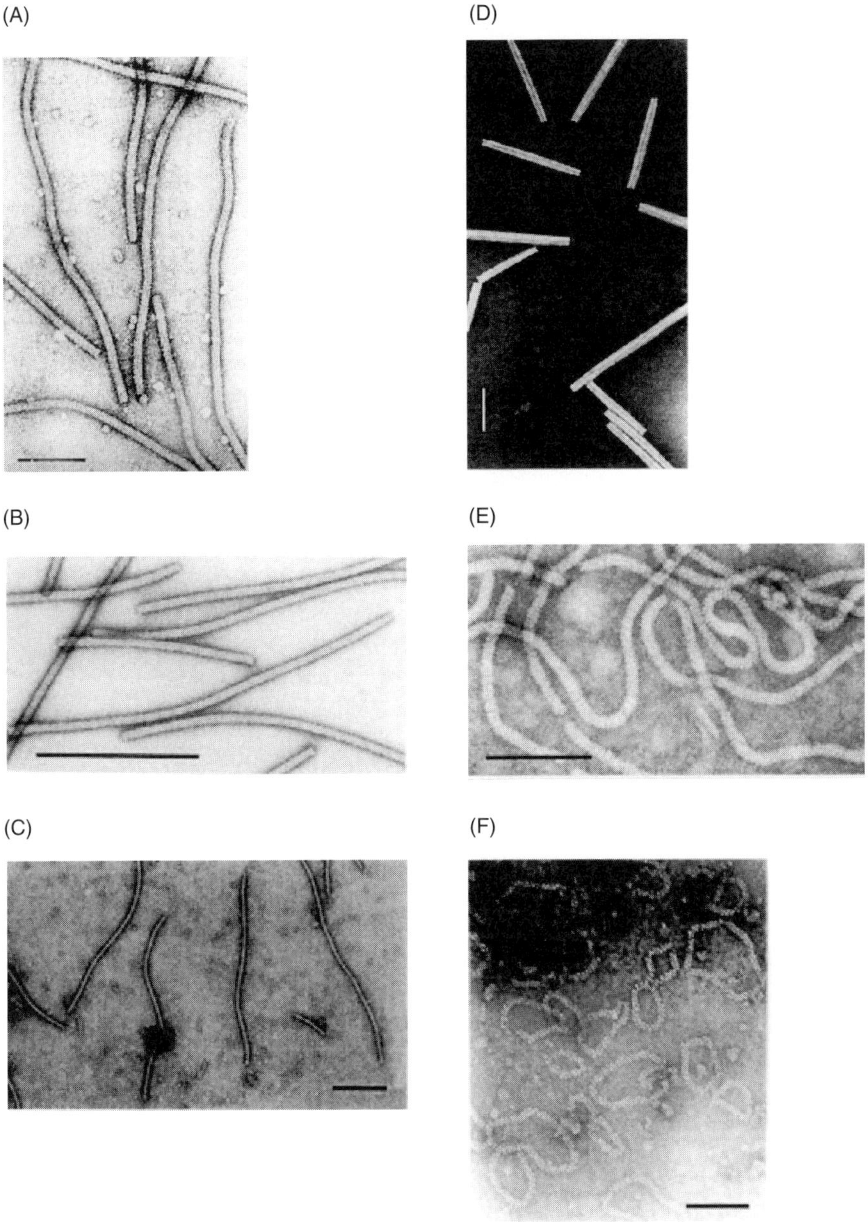

Figure 3 Electron micrographs of negatively stained virus particles as representatives of virus genera. (A) *Potato virus X*, *potexvirus*; bar = 100 nm (reproduced from Brunt AA, Foster GD, Martelli GP, and Zavriev SK (2000a) Genus Potexvirus. In: Regenmortel MHV, Fauquet CM, Bishop DHL, Carstens EB, Estes MK, Lemon SM, Maniloff J, Mayo MA, McGeoch DJ, Pringle CR and Wickner RB (eds) *Virus Taxonomy. Seventh Report of the International Committee on Taxonomy of Viruses*, pp. 975–981. San Diego: Academic Press); (B) *Carnation latent virus*, *carlavirus*; bar = 250 nm (reproduced from Brunt AA, Foster GD, Morozov S. Yu, and Zavriev SK (2000b) Genus Carlavirus. In: van Regenmortel MHV, Fauquet CM, Bishop DHL, Carstens EB, Estes MK, Lemon SM, Maniloff J, Mayo MA, McGeoch DJ, Pringle CR and Wickner RB (eds) *Virus Taxonomy. Seventh Report of the International Committee on Taxonomy of Viruses*, pp. 969–975. San Diego: Academic Press); (C) *Plum pox virus*, *potyvirus*; bar = 200 nm; (D) *Barley stripe mosaic virus*, *hordeivirus*; bar = 50 nm (reproduced from Lawrence DM, Solovyev AG, Morozov S. Yu, Atabekov JG and Jackson AO (2000) Genus Hordeivirus. In: van Regenmortel MHV, Fauquet CM, Bishop DHL, Carstens EB, Estes MK, Lemon SM, Maniloff J, Mayo MA, McGeoch DJ, Pringle CR, and Wickner RB (eds) *Virus Taxonomy. Seventh Report of the International Committee on Taxonomy of Viruses*, pp. 899–904. San Diego: Academic Press); (E) *Citrus tristeza virus*, *closterovirus*; bar = 100 nm (reproduced from Martelli GP, Agranovsky AA, Bar-Joseph M, Boscia D, Candresse T, Coutts RHA, Dolja VV, Duffus JE, Falk BW, Gonsalves D, Jelkmann W, Karasev AV, Minafra AV, Murant A, Namba S, Niblett CL, Vetten HJ, and Yoshikawa N (2000) Family Closteroviridae. In: van Regenmortel MHV, Fauquet CM, Bishop DHL, Carstens EB, Estes MK, Lemon SM, Maniloff J, Mayo MA, McGeoch DJ, Pringle CR and Wickner RB (eds) *Virus Taxonomy. Seventh Report of the International Committee on Taxonomy of Viruses*, pp. 943–952. San Diego: Academic Press); (F) *Rice hoja blanca virus*, *tenuivirus*; bar = 100 nm. (Reproduced from Mayo MA, de Miranda JR, Falk BW, Goldbach R, Haenni A-L and Toriyama S (2000) Genus Tenuivirus. In: van Regenmortel MHV, Fauquet CM, Bishop DHL, Carstens EB, Estes MK, Lemon SM, Maniloff J, Mayo MA, McGeoch DJ, Pringle CR, and Wickner RB (eds) *Virus Taxonomy. Seventh Report of the International Committee on Taxonomy of Viruses*, pp. 622–627. San Diego: Academic Press). All images reproduced with permission from Academic Press.

3′ end of the RNA. The double-disk changes to a helical (lock washer) form with both ends of the RNA protruding from the same side of the disk, the 5′ end passing through the central canal. The helical rod grows in the 5′ direction by the addition of further double-disks pulling the RNA through the axial hole until all the RNA 5′ of the origin of assembly is encapsidated. Assembly in the 3′ direction is much slower than assembly in the 5′ direction and is by addition of small aggregates of coat protein to the helical rod.

Most of the other viruses with short, rigid rod-shaped particles have a helical repeat similar to that of TMV, but those with longer flexuous particles have a more relaxed structure with helical repeat at about 3.5 nm. Some closteroviruses have a short section of a second virus-encoded protein, similar to the main coat protein, at one end of their particles; it is thought that this protein might be associated with insect transmission of these viruses.

The more complex structures of some viruses are also based on the isometric principles described above. For instance, the particles of members of the *Reoviridae* comprise two shells with icosahedral symmetry; other proteins give surface projections or spikes. Rhabdovirus particles are made up of a nucleoprotein structure (comprising the viral RNA associated with virus-encoded proteins) contained within a bacilliform structure, which is encased in a membrane through which virus-encoded glycoprotein spikes protrude. Tospoviruses have the nucleoprotein structure in a quasi-isometric particle, again surrounded by a glycoprotein-containing membrane. The glycoprotein-containing membrane and the surface spikes are involved in cell-to-cell spread in the animal (insect) vector in which these viruses also multiply, but are probably not required for infection of plants.

The nucleoprotein filamentous rods of tenuiviruses (**Figure 3F**) are usually found in plants without any surrounding membrane, but it is thought that more organized membranous structures might be found in their insect vectors.

Viral Genome Structure, Expression, and Replication

Five genome types are found in plant viruses (**Table 1**). About 80% of plant viruses have RNA genomes, with nearly 65% having a genome of single-stranded (+)-sense RNA. The double-stranded DNA viruses are pararetroviruses which replicate by reverse transcription but there are no plant viruses with double-stranded DNA genomes replicated by direct DNA > DNA mechanisms.

The genome organizations of representatives of most of the genera listed in **Table 1** have been determined by sequencing. Basically, each virus encodes products that facilitate its replication, all except the umbraviruses encode one or more coat protein species, and many encode proteins for movement within and between plants. Some also encode products such as proteases and/or translational control elements, which are involved in the full expression of the genome.

On entry into an uninfected cell, plant viruses, as with all viruses, go through four stages. Initially the particle is uncoated or the coat structure is relaxed and if the virus does not have a (+)-strand RNA genome, mRNA is synthesized. The mRNA undergoes translation, which can often be divided into two phases: early, which produces proteins involved in replication, and late, producing proteins for virus movement and encapsidation. Usually, between the two translation phases, the genome is replicated. Details of these stages differ between the types of viral genomes. Replication leads to the accumulation of virus particles ranging from less than 10^5 per cell to more than 10^{12} per cell, depending on the virus.

Various strategies are found that maximize the capacity of the viral genome to express a number of products. There are some examples of bicistronic translation in which both the 5′ and 3′ cistrons are translated (e.g., *Beet yellows virus*, BYV); in these cases, there are few, if any, extra translation initiation codons (AUG) in the 5′ cistron. Controlled expression of two adjacent open reading frames from the same mRNA can be achieved either by the ribosome changing reading frame between two overlapping open reading frames (frameshift) (e.g., the 5′ two cistrons of luteoviruses) or reading through of a weak stop codon (readthrough) (e.g., the 5′ two cistrons of TMV). Both these mechanisms yield two proteins, the larger of which has the same N-terminal sequence as the smaller together with a unique C-terminal sequence. Frameshift and readthrough occur at a low frequency, which controls the relative production of the two proteins. Other mechanisms of maximizing expression capacity include splicing (found in caulimoviruses, *Rice tungro bacilliform virus*, and some geminiviruses) and complex ribosome shunts and translational transactivation found in many of the double-stranded DNA viruses.

Double-Stranded DNA Viruses

Plant double-stranded DNA viruses have circular genomes of about 7.2–8.0 kb with one discontinuity in one strand and one or more in the other strand;

these discontinuities represent the priming sites for DNA synthesis. Members of the *Caulimoviridae* have between three and six open reading frames, depending upon genus, mostly encoding single proteins. However, one open reading frame of members of the *Badnavirus* and 'rice tungro bacilliform virus-like' genera encodes a polyprotein, which is processed by a virus-encoded protease to give the functional proteins. Viral DNAs are transcribed asymmetrically by the host DNA-dependent RNA polymerase II to give a more-than-genome length RNA with terminal redundancy (often termed the 35S RNA), which is the template for the reverse transcription phase of replication and is the mRNA for several of the gene products. Other mRNAs may be produced by splicing of the 35S RNA or by separate transcription, also asymmetric.

The replication of these viral genomes is in two phases: transcription in the nucleus and reverse transcription of the 35S RNA to give viral DNA in the cytoplasm using virus-coded reverse transcriptase and RNaseH.

Single-Stranded DNA Viruses

Geminiviruses have circular single-stranded DNA genomes of 2.7–3.0 kb. The genomes of Mastreviruses and Curtoviruses are monopartite, whereas those of many members of the Begomovirus genus are divided between two molecules. In all cases, the genome is transcribed bilaterally from a characteristic region (termed the common region) with the transcripts terminating on the opposite side of the genome, one transcript being from the virus-sense DNA and the other from the complementary DNA. Each of these transcripts expresses one or more proteins using mechanisms such as splicing.

The genomes of the *Nanovirus* genus comprise several (between 6 and 11) small circular single-stranded DNA molecules of about 0.8–1 kb. Each DNA species is thought to be monocistronic.

Viruses with single-stranded DNA genomes replicate by a rolling circle mechanism of DNA>DNA synthesis using both viral gene products and host proteins. Replication takes place in differentiated plant cells where host DNA replication has ceased. Viral gene products change the host cell cycle from the G phase to the S phase and DNA replication is effected by host enzymes.

Single-Stranded (+)-Strand RNA Viruses

The genome required for infection by (+)-strand RNA viruses can be either as one molecule (monopartite) or divided into two, three, or four molecules (multicomponent). Multicomponent genomes are often encapsidated into several different particles (components), e.g., comoviruses into two isometric components, hordeiviruses into three rod-shaped components. The genome sizes of (+)-strand RNA viruses range from about 4 kb in many single-component isometric viruses to more than 20 kb in some of the closteroviruses.

The RNA of these viruses acts directly as an mRNA on entry into the cell. As noted above, plant viruses express several genes from their genomic nucleic acid, which in many cases is polycistronic. As the expression is in eukaryotic cells, viruses have to overcome the problem of eukaryotic ribosomes only translating the 5′ cistron on a mRNA; usually open reading frames downstream of the 5′ one are effectively translationally closed. There are two basic mechanisms by which plant (and other eukaryotic) viruses overcome this problem: by dividing the RNA into several monocistronic units or translating the genetic information into a single protein (a polyprotein), which is then cleaved into functional units.

The mechanism of division of the RNA genome into several mRNAs is found in various forms in many plant viruses. Viruses such as TMV and *Potato virus X* (PVX) have a single genomic RNA from which 3′ co-terminal subgenomic mRNAs, which encode the downstream genes at their 5′ ends, are produced during virus replication. Thus, TMV expresses the 5′ replicase genes from the genomic RNA and the downstream movement and coat proteins from subgenomic mRNAs. In the multicomponent viruses, most, if not all, of the genome segments are effectively monocistronic. Each of the genome segments of members of the *Reoviridae* are monocistronic. The larger two RNAs of members of the *Bromoviridae* are monocistronic; the smallest RNA is bicistronic, the 3′ gene being expressed from a subgenomic RNA produced during replication.

The polyprotein mechanism is found in viruses such as potyviruses, comoviruses, and sequiviruses. In these viruses, the polyprotein is processed to the functional units by virus-encoded proteinases through often complex pathways.

Many of the (+)-strand RNA viruses are uncoated by a process termed cotranslational disassembly, which involves the particle structure becoming relaxed, say by chelation of stabilizing divalent cations or by protonation of the unusual carboxyl:carboxylate bonds, and ribosomes locating the 5′ end of the viral genome. Translocation of the ribosomes along the RNA translates the 5′ cistron and, at the same time, displaces viral coat protein subunits. For some viruses (e.g., sobemoviruses, AMV), the whole genome is probably uncoated by this process. The genome of TMV is uncoated from the 5′ end to the origin of assembly by ribosomes and then is thought to be uncoated from the 3′ end to the origin of assembly by the viral replicase, which is translated from the 5′ end of the RNA.

All RNA viruses encode an RNA-dependent RNA polymerase (RdRP) that synthesizes both (−)- and (+)-strands. The enzyme has several functional domains, including the catalytic, capping, methylase, and, often, helicase domains, and the replication complex also contains various host factors. Replication usually takes place on membranes where the complex is assembled and starts with priming at the 3′ end of the input RNA proceeding toward the 5′ end and giving a (−)-strand. This (−)-strand is then the template for synthesis of genomic (+)-strands and, if required, subgenomic RNAs. RdRPs do not have proofreading ability and thus there is a high error rate in newly synthesized RNA.

Double-Stranded RNA Viruses

Phytoreoviruses have similar genome organizations to those of animal-infecting reoviruses with the genome divided between 10 or 12 monocistronic segments (2.5–0.8 kb), depending on genus. These RNAs encode various proteins that make up the virion structure and nonstructural proteins, including the RdRp. The cryptic viruses (*Partitiviridae*) have genomes comprising two monocistronic segments, one encoding the RdRp and the other the coat protein.

The basics of replication of plant viruses with double-stranded RNA and (−)-strand RNA genomes are similar to those of (+)-strand viruses, except in the early stages. As the encapsidated genome cannot be translated directly, uncoating is not by cotranslational disassembly; the mechanism of uncoating is unknown. Virus particles contain RdRp, which effects the transcription of (+)-strands from which the rest of the replication cycle can then proceed.

Single-Stranded (−)-Sense RNA Viruses

As noted above, two of the three families that have (−)-sense RNA genomes (*Rhabdoviridae* and *Bunyaviridae*) have many genome organizational features in common with viruses that infect animals; they also replicate in their insect vectors. However, they also contain gene(s) for adaptation to plants. For instance, *Sonchus yellow net virus* has a genome organization identical to those of many animal rhabdoviruses except in having an extra gene product most likely involved in cell-to-cell movement in plants.

Tospoviruses have three genome segments, the largest being (−)-sense monocistronic encoding the RdRp and the other two each encoding two proteins, one from the virion-sense RNA and the other from the complementary RNA; this is termed ambisense expression.

The genomes of tenuiviruses comprise four RNA segments. The largest is (−)-sense encoding the RdRp and the other three have ambisense arrangement.

Virus–Plant Interactions

Host Range and Symptoms

There are different levels of virus interactions with plants ranging from complete immunity through replication in the initially infected cell but not spreading to adjacent cells (termed subliminal infection) to full systemic infection. Some viruses are restricted, usually by host defences, to the initially inoculated leaf, giving what is termed local infection. This variety of interactions causes problems in defining host range but the term is usually used for interactions that give overt symptoms.

Viruses differ in the number of plant species that they can infect either locally or systemically. Although obviously not in all plant species, many host–virus combinations have been tested, with most viruses reported to infect a few to a few tens of species. Some viruses have very wide host ranges; for example, CMV can infect more than 1000 species in more than 85 botanical families. Others have very restricted host ranges, infecting one or a few species (e.g., *Apple stem grooving virus*). Despite spreading systematically, some viruses are restricted to certain organs within the plant. For instance, luteoviruses are restricted to the vascular tissue and necroviruses are usually restricted to roots.

The breadth of the natural host range of a virus is usually narrower than that of the experimental host range. Among the factors that control this difference is the specificity of the biological vector in determining which plant species a virus can be naturally transmitted to.

Virus infection of plants results in a wide range of symptoms from no or very mild symptoms (latent or cryptic infection) to death of the plant. As it is usually to the "advantage" of the virus to maintain its host, it is generally considered that a virus causing minimal effect is more adapted to that plant species than one that kills the host.

The basic symptom type is often specific to the virus but can vary in detail with different environmental conditions, e.g., temperature or light, with virus strain, or with host genotype. In some virus–host combinations, a local reaction around the point of virus entry into the leaf gives chlorotic or necrotic lesions. This symptom type is used for virus quantitation, as the number of lesions can be related to virus concentration. Certain plant species, e.g., *Chenopodium* spp., are local lesion hosts for a number of viruses. Necrotic, and sometimes chlorotic, local lesions often limit the virus to around the point of entry. In other virus–host combinations, local infection can be expressed as ringspots with a necrotic (or chlorotic) ring surrounding a green or chlorotic area. The rings frequently spread and coalesce to give what is termed an "oak-leaf pattern."

The most common systemic symptom is a mottle or mosaic pattern of light and dark green and sometimes yellow areas on the leaf. The distribution of the different colors is usually delimited by the leaf veins giving an irregular pattern in dicotyledonous species and striping in monocotyledons. Mosaic-like patterns can be found in other parts of the plant usually giving striping on petioles and stems and striping or color-break in flowers. Other systemic color symptoms include general chlorosis or reddening of leaves, usually associated with phloem-limited viruses such as BYV, chlorotic bands alongside the veins (chlorotic vein bending), vein clearing (often an early symptom), chlorosis of the veins themselves, and chlorotic ringspots, which can resemble those found on inoculated leaves and which can become necrotic.

Viruses can cause growth abnormalities in plants. The most common of these is stunting of growth. Other less common growth abnormalities include reduction of the leaf lamina leading to strap-like leaves (e.g., infection of tomato, *Lycopersicon esculentum*, with CMV and/or *Tomato mosaic virus*), enations or outgrowths of the veins on the underside of the leaf (e.g., *Pea enation mosaic virus* in peas, *Pisum sativum*), swelling of the stem (e.g., *Cocoa swollen shoot virus* in cacao, *Theobroma cacao*), and tumor-like growths at wound sites (e.g., *Wound tumor virus*).

The cytological effects of virus infection on plants are twofold: effects on the plant cellular structure and organelles and manifestations of the virus itself. The most obvious effects are found on chloroplasts, which reflect the different color regions of mosaic or chlorotic symptoms. Some effects are also found on mitochondria and nuclei. Viral replication usually involves membrane surfaces and organellar membranes are implicated in some cases. For instance, tymovirus infection is associated with "flask-like" structures on the chloroplast membranes, tobravirus infections appear to associate with mitochondrial membranes, and nucleorhabdoviruses appear to involve the nuclear membrane.

Characteristic inclusion bodies are formed within cells from the accumulated virus particles or gene products of members of several virus groups and these are sometimes used for virus identification. For instance, crystalline inclusions are formed from the rod-shaped particles of viruses such as TMV, BYV, and PVX, and the isometric particles of viruses such as *Broad bean wilt virus* and *Turnip yellow mosaic virus*. Gene products can form inclusion bodies such as the characteristic nuclear and cytoplasmic inclusions of potyviruses, the X bodies of TMV, and the caulimovirus inclusions.

Molecular Interactions in Virus Infection

Viruses usually initially infect one or a few cells and spread to the rest of the plant by movement through other cells within the leaf (cell-to-cell movement) and then to distant parts of the plant through the vascular tissue (long-distance spread). For cell-to-cell spread, the virus crosses the cellulose cell wall bounding each cell through cytoplasmic connections termed plasmodesmata. The effective diameter of plasmodesmata (the size exclusion limit, about 1000 Da) is too small to enable virus particles to pass directly through, but viruses have found ways of overcoming this problem. Two basic mechanisms are recognized at present. In the first, exemplified by TMV, the virus encodes a protein (P30), termed the movement protein, which is thought to temporally increase the exclusion limit of the plasmodesmata to about 10 kDa. P30 also binds to the TMV RNA forming a long narrow nucleoprotein structure, which can pass through plasmodesmata. Details of the mechanisms are still being determined and recent observations indicate that plant cytoskeleton structures, microtubules and actin filaments, are involved. After it has been used for cell-to-cell movement, P30 is phosphorylated and "stored" near plasmodesmata where it can be observed by immunoelectron microscopy.

In the second mechanism, which is employed by *Cowpea mosaic virus* (CPMV) and *Cauliflower mosaic virus* (CaMV), one or more virus gene products (P58/48 of CPMV; gene 1 product of CaMV) are involved in forming tubules containing virus particles, which pass from cell to cell, most likely through plasmodesmata. The processes so far determined allow the movement of the virus through mesophyll tissues to the vascular bundle sheath. The indications are that in many plant species different mechanism(s) of cell-to-cell movement operate within the bundle sheath.

Long-distance movement of viruses is usually via the phloem sieve elements, though the sobemoviruses also use xylem elements. Virus movement in the phloem follows the source-sink movement of photoassimilates. Little is known of how viruses enter and leave the phloem. Persistently transmitted viruses, such as luteoviruses, are probably introduced directly into the phloem by their vectors and thus must be able to exit as complete virus particles. For many other viruses, the viral coat protein is required for long-distance movement.

A better understanding of the molecular interactions is emerging from the application of molecular techniques to isolate and modify host genes and to manipulate and mutate cloned infectious viral genomes. At least two forms of host response have now been recognized. These can be termed a "specific resistance" response and a "generalized resistance" response. The specific resistance response is directed by one or more host genes, the products of which interact with certain viral determinants and limit the spread of the virus from the site of initial infection. The limitation is usually by a hypersensitive reaction giving a necrotic or chlorotic local lesion. In at least one case, that of the N gene which controls necrotic local lesion production of TMV in *Nicotiana* spp., the gene has been isolated and shown to have a similar structure to genes giving hypersensitive reactions to fungi and bacteria in plants.

The generalized resistance is that mounted by the plant against "foreign" nucleic acids, in the case of viruses usually by posttranscriptional gene silencing. In this response, the recognition by the plant that the viral nucleic acid is "foreign" initiates a pathway that leads to the degradation of the nucleic acid. Successful viruses have one or more mechanisms to overcome this host defense response. One mechanism is the production of a viral gene product that suppresses the response; another mechanism is the rapid multiplication and movement of the virus to keep ahead of the response. The host defense response operates not only at the plant level but also at the cell and domain level. The green areas of a mosaic symptom do not contain much if any virus and it is considered likely that virus production in these areas is inhibited by the host response.

The actual details of the processes involved in the induction of disease symptoms are not well understood. Many of the biochemical changes involved may not be directly connected with virus replication. Stunting probably involves changes in the balance of growth hormones. The formation of mosaic patterns in virus-infected leaves involves events that occur in the early stages of leaf ontogeny. Even small changes in the viral sequence can give different symptoms. Thus, single nucleotide changes in the RNA genome of TMV or in the DNA genome of *Maize streak virus* results in different symptom expression. Similarly, the application of techniques such as microscopy coupled with serology and hybridization is opening the understanding of the changes occurring at the cellular level. For instance, in some potyviral infections, host gene expression is altered, either inhibited or enhanced, in the cells in which the virus is replicating.

Virus Transmission

To infect a plant, a virus particle must cross several barriers: the cuticle, the cellulose cell wall, and the plasma membrane. Mechanical damage is required for crossing of at least the first two barriers and all plant-to-plant virus spread involves mechanisms that inflict such damage. Many viruses can be mechanically transmitted experimentally by rubbing sap from the infected plant or purified virus, together with an abrasive, onto a leaf of the healthy plant. This method is used in many cases for host range studies and for local lesion assays.

Natural transmission of viruses is by a wide range of damage-inducing mechanisms (**Table 1**) which, in most cases, is virus-specific. Some viruses that occur at high concentration (e.g., TMV and PVX) are naturally mechanically transmitted through, say, broken hairs of infected and healthy plants rubbing together or through root contact. However, the vast majority of viruses require a biological vector. These include arthropods, which feed by piercing and sucking (e.g., aphids, leafhoppers, whiteflies), nematodes, fungi (*Plasmodiophoromycetes* or *Chytridiomycetes*) and pollen; some viruses are transmitted vertically by seed. The interactions between the virus and its vector are highly specific. Thus, if a virus is transmitted by an insect it will not be transmitted by nematodes or fungi. Furthermore, it will only be transmitted by one type of insect, say aphid and not leafhopper, and in many cases by only one or a few aphid species or even aphid biotypes. However, there may be transmission by different vector types within a viral genus (**Table 1**).

There are two basic forms of virus–vector interactions. In the externally borne interaction, the virus does not enter inside the vector but associates with the mouthparts. This interaction usually results in rapid transmission and short persistence of the virus association with its vector. Based on the length of time that the insect is able to transmit the virus, the interaction is termed nonpersistent or semipersistent (**Table 1**). Externally borne viruses interact with specific region(s) of the vector's anterior food canal: the nonpersistent with the ends of the stylets and the semipersistent with regions further down the food canal but anterior to the first valve. The interaction is either directly between the viral coat protein and a specific site in the vector's mouthparts (e.g., CMV) or with a virus-encoded nonstructural protein linking the particle to the specific site in the vector. This protein is termed a helper or transmission protein (component) and is found in, for example, potyviruses and caulimoviruses.

The other basic type of virus–vector interaction is termed internally borne (or persistent or circulative), as the virus particles enter the vector's body and circulate therein. In some cases, they multiply in the insect and thus are called propagative. Internally borne viruses require longer feeds (hours to days) than do externally borne viruses for acquisition; there is then a latent period before the vector becomes infective, but the vector remains able to transmit for long periods, even the whole of its life in propagative viruses. All circulative viruses have to cross several barriers, usually the gut wall to enter the hemocoel and the salivary gland basal lamina to gain entry to the salivary glands, to be transmitted. Luteoviruses cross these barriers in their aphid vectors by endocytosis and movement through various vesicles. Little is known about the detailed interactions between the virus and receptor sites at these barriers; it is these interactions that presumably control vector specificity.

Some viruses require the presence of other viruses to effect their transmission. For instance, the genomes of umbraviruses are encapsidated in the coat protein of their helper virus, usually a luteovirus, and are transmitted in the same manner as their helper virus. The leafhopper transmission of *Rice tungro bacilliform virus* is thought to be dependent on a transmission protein provided by its helper virus, *Rice tungro spherical virus*.

Control of Plant Viruses

There are three basic approaches to controlling plant viruses: avoidance of sources of infection, control of vectors, and protection of the plant against systemic infection.

Avoidance of Sources of Infection

The use of clean planting material is essential since viruses will pass to vegetatively propagated progeny and some viruses are seed transmitted. It is possible to free certain plant species from certain viruses by techniques such as tissue culture and heat treatment, but this is only economically viable for high-return crops such as ornamentals and fruit. For long-term perennial plantation crops, such as cacao and citrus, roguing of infected plants is used but this requires considerable labor in virus diagnostics and can lead to loss of crop production while the trees are being replaced. Cultural techniques such as removal of infected material from previous crops, planting away from sources of infection, and modifying planting and harvesting procedures so that there is no overlap can be very effective.

Control or Avoidance of Vectors

A common approach to controlling viruses of annual (and some perennial) crops is to control the vector. In the case of insects, the application of insecticides can be effective unless resistance to the chemical arises. The efficiency of control also depends on the virus–vector interaction and is less for those externally borne viruses that are acquired and transmitted rapidly. Various other cultural techniques such as planting to avoid the vector and removal of sources of infection can be effective.

Protecting the Plant from Systemic Infection

Plants can often be protected from exhibiting virus symptoms by inoculating them with a mild strain of the virus. Mild strain protection is only really effective in high-value crops such as tomatoes or citrus. In most cases, chemicals do not give plants protection against, or cure them from, virus infection. Much effort has gone into breeding for resistance to the virus. However, there are relatively few resistance genes available in sexually compatible species for the most prevalent viruses of major crops. Another approach to protecting plants is by transformation with transgenes that confer resistance. Many of these transgenes are derived from viral sequences, which will interfere with or block the normal infection cycle of the virus. There are two basic approaches to transgenic resistance. The initial approach was to express viral gene products, primarily the coat protein, which interfered with an early stage of virus replication. More recently, it has been shown that the expression of viral sequences from transgenes can elicit the cosuppression or gene silencing phenomenon described above. The use of transgenes is very promising but their durability has yet to be fully determined.

List of Technical Nomenclature

Bicistronic	A messenger RNA that expresses two adjacent cistrons.
Capsid	The closed shell or tube of a virus, usually composed of protein.
Cistron	The basic unit of genetic function. It usually refers to a gene or coding region for a protein.
Icosahedral symmetry	A type of cubic symmetry found in isometric viruses. The basic form is 12 pentameric subunits forming the vertices

	of the particle and giving twofold, threefold, and fivefold rotational symmetry axes. For larger icosahedra, the number of subunits is increased in a regular manner by the addition of hexameric subunits in defined numbers and places.
Local lesion	A symptom sometimes found in a plant leaf inoculated with a virus caused by the death of, or changes in, cells around the original point of entry.
Monocistronic	A viral genomic or subgenomic RNA that contains one cistron.
Open reading frame	A set of codons for amino acids uninterrupted by stop codons.
Posttranscriptional gene silencing	A plant defense system that targets "foreign" RNA, which it directs into a degradation pathway.
Subliminal infection	Infection in which the virus multiplies in the originally infected cell but is unable to spread to adjacent cells.
Triangulation number	A description of the triangular face of an icosahedron indicating the number of triangles into which the face is divided.

See also: **Diseases**: Plant Pathology, Principles; Genetic Modification of Disease Resistance, Viral Pathogens.

Further Reading

Bisaro DM (1996) In: DePamphilis ML (ed.) *DNA Replication in Eukaryotic Cells*, pp. 833–854. New York: Cold Spring Harbor.

Brunt AA, Foster GD, Martelli GP, and Zavriev SK (2000a) Genus Potexvirus. In: van Regenmortel MHV, Fauquet CM, Bishop DHL, Carstens EB, Estes MK, Lemon SM, Maniloff J, Mayo MA, McGeoch DJ, Pringle CR, and Wickner RB (eds) *Virus Taxonomy. Seventh Report of the International Committee on Taxonomy of Viruses*, pp. 975–981. San Diego: Academic Press.

Brunt AA, Foster GD, Morozov SYu, and Zavriev SK (2000b) Genus Carlavirus. In: van Regenmortel MHV, Fauquet CM, Bishop DHL, Carstens EB, Estes MK, Lemon SM, Maniloff J, Mayo MA, McGeoch DJ, Pringle CR, and Wickner RB (eds) *Virus Taxonomy. Seventh Report of the International Committee on Taxonomy of Viruses*, pp. 969–975. San Diego: Academic Press.

Bos L (1963) *Symptoms of Virus Diseases of Plants*. Wageningen: Centre for Agricultural Publications and Documentation.

Caspar DLD and Klug A (1962) Physical principles in the construction of regular viruses. *Cold Spring Harbor Symposium on Quantitative Biology* 27: 1–24.

Covey SN (2000) Silencing gene silencing genes. *Trends in Plant Science* 5: 405–406.

Granoff A and Webster RG (1999) *Encyclopedia of Virology*, 2nd edn. San Diego: Academic Press.

Hull R (2001) *Matthews' Plant Virology*. London: Academic Press. (The standard text on plant virology.)

Lawrence DM, Solovyev AG, Morozov SYu, Atabekov JG, and Jackson AO (2000) Genus Hordeivirus. In: van Regenmortel MHV, Fauquet CM, Bishop DHL, Carstens EB, Estes MK, Lemon SM, Maniloff J, Mayo MA, McGeoch DJ, Pringle CR, and Wickner RB (eds) *Virus Taxonomy. Seventh Report of the International Committee on Taxonomy of Viruses*, pp. 899–904. San Diego: Academic Press.

Martelli GP, Agranovsky AA, Bar-Joseph M, Boscia D, Candresse T, Coutts RHA, Dolja VV, Duffus JE, Falk BW, Gonsalves D, Jelkmann W, Karasev AV, Minafra AV, Murant A, Namba S, Niblett CL, Vetten HJ, and Yoshikawa N (2000) Family Closteroviridae. In: van Regenmortel MHV, Fauquet CM, Bishop DHL, Carstens EB, Estes MK, Lemon SM, Maniloff J, Mayo MA, McGeoch DJ, Pringle CR, and Wickner RB (eds) *Virus Taxonomy. Seventh Report of the International Committee on Taxonomy of Viruses*, pp. 943–952. San Diego: Academic Press.

Mayo MA, de Miranda JR, Falk BW, Goldbach R, Haenni A-L, and Toriyama S (2000) Genus Tenuivirus. In: van Regenmortel MHV, Fauquet CM, Bishop DHL, Carstens EB, Estes MK, Lemon SM, Maniloff J, Mayo MA, McGeoch DJ, Pringle CR, and Wickner RB (eds) *Virus Taxonomy. Seventh Report of the International Committee on Taxonomy of Viruses*, pp. 622–627. San Diego: Academic Press.

Pavord A (1999) *The Tulip*. London: Bloomsbury.

Rothnie HM, Chapdelaine Y, and Hohn T (1994) Pararetroviruses and retroviruses: a comparative review of viral structure and gene expression strategies. *Advances in Virus Research* 44: 1–67.

Scholthof K-BG, Shaw JG, and Zaitlin M (eds) (1999) *Tobacco Mosaic Virus: One Hundred Years of Contributions to Virology*. St Paul, MN: American Phytopathology Society Press.

van Regenmortel MHV, Fauquet CM, Bishop DHL, *et al.* (eds) (2000) *Virus Taxonomy: Classification and Nomenclature of Viruses. Seventh Report of the International Committee on Taxonomy of Viruses*. San Diego: Academic Press.

Breeding for Disease Resistance

H S Bariana, University of Sydney, Camden, NSW, Australia

Introduction

Breeding for disease resistance requires the long-term commitment of skilled human resources and testing facilities. It is only feasible in host–pathogen systems of significant economic importance. The initiation of a resistance breeding program depends on production statistics of the target crop, crop losses due to the particular disease, an understanding of the extent of variability of the pathogen, cost–benefit aspects of other control measures, and a reliable means of assay. The success of a program also depends on effective communication among scientists from key discipline areas. **Figure 1** depicts the information flow strategy of the Australian Cereal Rust Control Program (ACRCP) centered at the University of Sydney Plant Breeding Institute, Cobbitty. The information generated in the host and pathogen components underpins the germplasm screening and enhancement activities of the program. In turn, germplasm screening identifies genetic aspects to be researched in detail. The following critical factors are discussed in detail:

- economic importance of the target disease
- types of resistance and their genetic control
- knowledge of genetic variation for resistance
- knowledge of pathogen variation
- cost-effective and high-throughput screening tools
- breeding methodology.

Economic Importance of Disease

The importance of a disease can vary between countries and also between environments within a country. For example, rust diseases of wheat (*Triticum aestivum*) are of global importance and uredospores can be wind-dispersed over long distances irrespective of national and geographical boundaries. By contrast, the fungi causing necrotophic diseases such as yellow spot, fusarium head blight, and septorias are of more localized significance. Rust epidemics in various parts of the world have caused losses in the magnitude of millions of dollars. In Australia, breeding for resistance to rust diseases of wheat is estimated to save the industry in excess of Aus$300m annually. Disease epidemics in a range of crops of regional importance have affected agricultural productivity throughout history. Late blight of potato (*Solanum tuberosum*) and brown spot of rice (*Oryza sativa*) caused famines in Ireland (1845–6) and Bengal (1942), respectively, and rice blast epidemics have brought famine to parts of Japan. Southern corn (*Zea mays*; maize) blight devastated the corn industry in the United States in 1970.

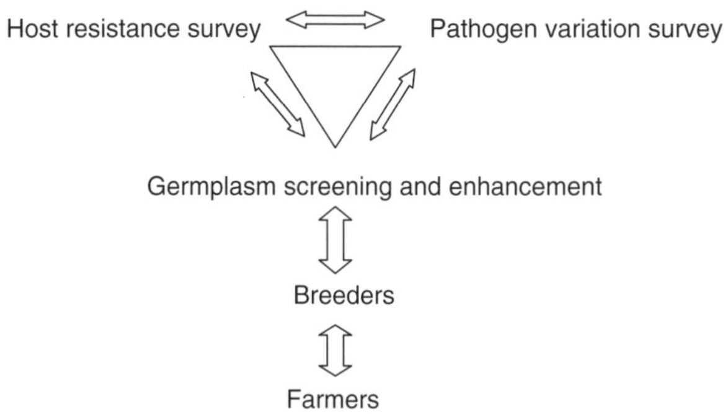

Figure 1 Information flow strategy of the Australian Cereal Rust Control Program based at the University of Sydney Plant Breeding Institute, Cobbitty.

Types of Resistance and their Genetic Control

Types of Resistance

Various contrasting terms have been used to describe alternate types of resistance (**Table 1**). Differences are based on the mode of inheritance and/or how resistance is expressed. One type is often simply inherited and controlled by single or few genes of large effects, whereas the alternate type is often controlled by minor effect additive resistance genes.

Simply inherited resistances are those that are either overcome by virulent pathotypes or are likely to be overcome by new pathotypes. The expression of resistance is often characterized by hypersensitivity. However, some simply inherited resistances have remained effective after long term deployment; for example, wheat stem rust resistance genes *Sr2* and *Sr26* and the gene *mlo* for resistance to powdery mildew in barley (*Hordeum vulgare*). This type of resistance is likely to prove pathotype specific. The gene-for-gene hypothesis of Flor states that "for each gene conditioning rust reaction in the host there is a specific gene conditioning pathogenicity in the parasite." The gene for gene system has been reported to operate for many diseases, for example; wheat rusts, wheat powdery mildew, bean anthracnose, corn rusts, late blight of potato, rice blast, and bacterial leaf blight of rice. Another system that

Table 1 Terms used for describing disease resistance in plants

Simple	Complex
Qualitative	Quantitative
Seedling	Adult plant
Race-specific	Race-nonspecific
Specific	General
Overall	Mature plant
Vertical	Horizontal
Oligogenic	Polygenic
Complete	Partial
Major	Minor
Durable	Nondurable, transient

mirrors the gene-for-gene system for obligate parasitic system is host–toxin interaction. In this case toxin produced by the pathogen is lethal to host cells in genotypes that specifically recognize it. In the gene-for-gene system, a genotype is resistant when a gene for resistance (producing a receptor) is matched by a gene for avirulence (producing an elicitor) in the pathogen, whereas in the host–toxin interaction the resistance phenotype is a result of insensitivity (no receptor or ability to neutralize the toxin) in the host to the toxin. The inheritance of resistance in this case is recessive. The latter interaction is reported to operate in oat (*Avena sativa*) Victorian blight, corn southern leaf blight, and wheat yellow spot toxin systems.

The alternate resistance type that may show complex inheritance is often described as race-nonspecific. Other terms used are minor gene resistance, polygenic resistance, horizontal resistance, and rate-reducing resistance. This type of resistance produces a susceptible phenotype with varying levels of disease severity and reduced rate of disease development. Individually, these genes do not provide adequate resistance and genotypes possessing them singly are often hard to distinguish from susceptible genotypes under high disease pressure. Clearly this makes it hard to detect pathotypic variation for such genes even if variation does occur. For example, wheat genotypes carrying the linked rust resistance genes *Lr46* and *Yr29* are hard to differentiate from susceptible genotypes at the terminal stages of disease development; however, periodic scores on disease development provide differentiation. One school of thought described this type of resistance as race-nonspecific, whereas the other considered that only those resistances that are shown to be race-specific could be described in a definitive way. This author is in agreement with the latter viewpoint and believes that many examples of such resistance if studied in detail under appropriate experimental conditions, could be proved race-specific. The slow mildewing resistance of wheat cultivar Genesee reported to be under complex genetic control was shown to carry a single gene when studies were conducted under controlled conditions.

The literature on simply inherited resistances clearly distinguishes between the contrasting disease response phenotypes, whereas in the case of resistances controlled by additive interaction of minor effect genes, phenotypic contrasts are often not very well presented in the published literature. This necessitates a need for genetic definition of susceptibility. In a genetic analysis the susceptible parent must show a clearly susceptible response and susceptible segregates in a genetic analysis must be equivalent in disease response to the susceptible parent. Certain examples of resistance to stripe rust and septoria diseases in wheat and resistance to blast in rice are reported to be controlled by combinations of minor effect resistance genes. Often such resistances are heritable enough to be selected in breeding programs.

As an alternative to the above terminologies, Johnson and Law used the term "durable resistance" to describe resistance that remained effective when a cultivar is grown over a large area for a long period of time. They used French wheat cultivar Hybride-de-Bersee (Bersee) as an example of a wheat that had remained resistant for over 20 years of wide cultivation in the United Kingdom and France. Johnson stated:

> unlike other terms the term durable resistance requires no knowledge of the genetic control of resistance, its mechanism, its degree of expression or any relevant aspects of race specificity and its phenotype might be of an intermediate non-hypersensitive type.

However, this author believes that a detailed knowledge of the genetics of durable resistance is crucial to formulate strategies for its use in breeding programs. Current knowledge of the genetics of durable resistance to stripe rust clearly demonstrates the involvement of more than two genes in many sources of durable resistance including French wheat cultivar Bersee.

Genetic Control

A thorough understanding of the inheritance of resistance will enhance the use of resistance in a breeding program. Various methods used for genetic analyses in self-pollinated plants include tests on F_2 populations and F_3 families, backcross/test cross (BC/TC) F_1 and F_2 families, doubled haploid (DH) or single-seed descent (SSD) populations and combinations of more than one of these approaches. In cross-pollinated crops genetic analysis based on tests on F_2 populations and F_3 families, tests on backcross derivatives of inbred lines, and tests on recombinant inbred lines (RILs) are commonly used. Methods used for genetic analyses of disease resistance affect interpretation of data. Different methods used for genetic analyses and the expected genetic ratios (for simply inherited traits) are listed in **Table 2**. For many host–pathogen systems tests on F_3 lines or BC F_2 populations are preferred because they permit the allocation of genotype to the F_2 or BC F_1 parental plants, provide a source of repeatability of individual plant results, and permit the derivation of single gene lines in instances of genotypes carrying combinations of resistance genes. The SSD (where individual F_2

Table 2 Comparison of genetic analysis method and expected segregation ratios assuming dominance for different numbers of independent heterozygous loci

Method	*Number of heterozygous loci/segregation ratios*			
	1	*2*	*3*	*4*
F_2 population[a,e]	3:1[b]	15:1	63:1	255:1
Single-seed descent families	1:1[c]	3:1	7:1	15:1
Doubled haploid population	1:1[c]	3:1	7:1	15:1
Testcross F_1[e]	1:1[d]	3:1	7:1	15:1

[a] Segregation is based on single plants.
[b] Resistant: susceptible.
[c] Homozygous resistant: homozygous susceptible.
[d] Segregating: homozygous susceptible.
[e] F_2 and testcross ratios will vary if recessive genes are involved. Further categories are possible if individual genes confer distinctive symptoms.

plants are self-pollinated for several generations prior to bulking as near-homozygous lines) and DH methods are important for genetic analysis of minor gene controlled resistances because of the more stable expression of resistance in homozygotes and availability of unrestricted replication.

The advantage of the BC, DH, or SSD method over the F_3 analysis is the smaller number of genotypes required to establish genetic ratios when three or more genes are involved (**Table 2**). The advantage of BC/TC and F_3 methods is that the single gene segregation can be recognized and those genes can be isolated during the process of screening. In the case of DH and SSD populations further crossing may be necessary to confirm the identities of single gene lines.

Knowledge of Genetic Variation for Resistance in Target Host Species

Since durability of resistance cannot be assured, then genetic variability must be deployed to buffer against widespread losses in the event of a pathotypic change. Thus a resistance breeding program relies on genetic diversity for resistance. In order to exploit genetic variation it is important firstly to search continually for diverse sources of resistance, secondly to perform genetic characterization of different sources of resistance, and thirdly to ensure their transfer to desired agronomic backgrounds. Cultivars from other geographic regions are the most attractive sources of resistance. The Brazilian wheat cultivar Frontana has provided good levels of resistance to rust diseases and moderate resistance to fusarium head blight (FHB) for many breeding programs. Similarly, Chinese wheats have displayed resistance to FHB. In Australia, Japanese spring cultivars and French winter cultivars have been used as sources of black leg resistance in *Brassica napus* breeding programs. Japanese germplasm contributed towards resistance to rice blast. Sometimes pathotypic variation in different regions is unique and sources of resistances that are no longer effective in one region may be effective in another. For example, the wheat stripe rust resistance gene *Yr27* is ineffective in Mexico, whereas it is effective against Australian *Puccinia striiformis* f. sp. *tritici* pathotypes.

Table 3 Examples of introgression of disease resistance from undomesticated relatives

Donor species	*Resistance to pathogens*
Tomato	
Lycopersicon hirsutum	*Septoria lycopersici*, TMV
L. chilense, L peruvian	*Heterodera maioni*
Potato	
Solanum demissum	*Phytophthora infestans*
S. stoloniferum	*Phytophthora infestans*, PVA and PVY
S. acaule	Potato virus X
Common wheat	
Triticum tauschii	*Puccinia* spp.
T. speltoides	*Puccinia* spp.
T. timopheevi	*Puccinia* spp. and *Erysiphe graminis*
T. ventricosum	*Puccinia* spp. and cereal cyst nematode
Thinopyrum intermedium	Barley yellow dwarf virus and *Puccinia triticina*
Rice	
Oryza rufipogon	*Sclerotium oryzae*
O. nivara	Grassy stunt
Corn	
Tripsacum floridanum	*Helminthosporium turcicum*

Both cultivated and wild relatives of crop plants have been a great source of resistance to diseases. Many resistance genes not only of wheat but also of many other crop species have been transferred from undomesticated related species (**Table 3**). Mutation has been used to create variation in disease responses in a number of crops. One successful example of exploitation of such a mutation has been the widespread use of the gene *mlo* for barley mildew resistance. Cloned resistance genes offer the potential to increase diversity through manipulation of the structure of resistance genes.

Knowledge of Pathotypic Variation in the Pathogen

Pathogens, especially obligate pathogens, causing diseases of economic importance are known to show pathotypic variation. In some systems pathogenic variation with respect to host resistance is so great that the use of resistance seems almost a futile exercise. In other systems, resistance once deployed has been extremely stable. It is common for scientists working with many disease systems, such as the airborne cereal rusts, to undertake annual surveys that enable deductions on the nature of pathogenic variability, and epidemiology. These surveys also provide key updated pathotypes for the screening of breeding populations.

Gene frequencies in pathogens are affected by the usual evolutionary process of migration, mutation, selection and chance. The introduction of *P. striiformis* f. sp. *tritici* to Australia in 1979 and to South Africa in 1996 are significant examples of the introduction and establishment of a pathogen in new geographical areas. Clearly, subsequent introductions of the same species are not so spectacular but may have even more significance. Following its introduction to Australia, the wheat stripe rust pathogen was windborne to New Zealand in 1980, but has not been reported in the wheat areas of Western Australia until 2002 despite the cultivation of susceptible cultivars. This illustrates the effects of the predominating weather patterns. Subsequent to introduction to these countries the pathogen has undergone individual mutational changes leading to arrays of many pathotypes. In crop season 2002 another introduction of *P. striiformis* f. sp. *tritici* took place in Western Australia. This pathotype differed from existing Australian *P. striiformis* f. sp. *tritici* pathotypes with respect to combined virulence for *Yr8* and *Yr9*, and avirulence for stripe rust resistance genes carried by European cultivars.

Amongst pathogens and pests sexual recombination may be important in generating new pathotypes (genotypes) permitting the expression of recessive genes. Although many of the cereal rust pathogens have a sexual phase, this source of variation is often not important. The organisms, however, are known to undergo somatic recombination at least to a limited extent and evidence for such events has been documented in a number of populations.

Once resistance genes are used, selection has an important bearing on the pathotype structure of the pathogen population, however, chance effects are also important in small populations. With many pathogens there are huge changes in a population size with epidemic cycles and from year to year as the pathogen must survive on living hosts during the noncropping phase of the year. The cereal rust pathogens are highly polymorphic and unrecorded virulences may be present in a population that have never been exposed to resistance genes. For example, virulence for leaf rust resistance gene *Lr28* derived from *T. speltoides* is almost universal in North American *P. triticina* populations, whereas avirulence for this gene occurs in Australia and Indian subcontinent. When the *T. tauschii*-derived gene *Lr39* was introduced in Australia, virulence was already present in some pathotypes. In 2000, virulence in *P. triticina* for wheat leaf rust resistance gene *Lr24* was identified after almost 20 years of its exploitation in Australia. At the time of its downfall *Lr24* was present in wheats occupying about 25% of the total wheat area. By contrast when used in North America and South Africa *Lr24* remained effective for no more than a few years.

Despite a reputation for durability, some sources of adult plant resistance have proved highly race-specific and have been no more useful than those sources where resistance genes can be detected in seedling tests. English workers reported a pathotype virulent on cultivar Joss Cambier, which carries adult plant resistance to stripe rust, indicating that pathotypic evolution could take place to acquire virulence for nonhypersensitive type of adult plant resistance.

Effective Screening Tools

A key criterion of a resistance breeding program is a robust and cost-effective screening system. The main selection tools used for breeding for disease resistance are:

- seedling tests in the greenhouse
- field screening
- enzyme linked immunosorbent assay (ELISA)
- disease response linked markers
- combinations of the above.

Seedling Tests

Where a resistance of interest to a breeder is controlled by seedling-effective genes, a cost-effective screening test can often be developed. However, the pathogen isolate used for the assay must be relevant for the purpose. Selection based on seedling tests is often used by cereal rust workers. In a few instances breeders have considered the option of selecting against seedling effective resistances, preferring to concentrate on adult plant resistance sources, particularly those with a reputation of durability.

Various greenhouse assays have been developed for a range of cereal disease/parasites such as cereal cyst nematode (CCN) and root lesion nematode (RLN), crown rot, and yellow or tan spot. Even in instances where evidence for pathogenic variability is lacking it is worthwhile using or adding cultures recently isolated from current cultivars and/or the regions where new cultivars are likely to be grown. The responses of materials selected under greenhouse or special environment test conditions should be confirmed in field tests.

Field Screening

Breeders must have effective field testing facilities. These can be disease-prone agricultural areas or special nurseries where conditions for effective testing have been developed by creation of special environments and/or encouragement of high disease pressure through artificial inoculations. In the case of highly variable airborne pathogens, dependence on natural inoculum in disease-prone areas might not be as effective as the use of controlled epidemics with known pathotypes. There are many opinions and methods of effective resistance breeding and they change with the particular disease, importance of resistance as an objective, the resources and facilities available, knowledge of host–pathogen system, and simplicity of the test.

ELISA

Immunological tests such as ELISA have been widely used in the selection of resistance to viral diseases in a range of crops. They have also been used in a limited way to select for resistance against other diseases.

Disease Response-Linked Markers

Markers linked with low disease responses include morphological markers, protein markers and DNA markers.

Morphological markers The genetic association of phenotypic markers with low disease response allows indirect selection of the target resistance. Linkages of morphological markers with low disease responses have been used to select for resistance to rust diseases of wheat. For example, in appropriate crosses brown chaff color is linked with stripe rust resistance gene *Yr10*, pseudo black chaff and seedling chlorosis are linked with *Sr2*, and leaf tip necrosis is associated with genes for resistance to leaf rust (*Lr34*), stripe rust (*Yr18*), and barley yellow dwarf virus (*Bdv1*).

Protein markers Linkages of gliadin alleles with *Yr10* and fusarium head blight resistance have been reported. The eyespot resistance gene *Pch1* in wheat has been shown to be linked with endopeptidase *Ep-D1b*. These genetic associations could be used to select for resistance genes in breeding populations.

Linked disease resistance genes Instances of linked inheritance of resistance to more than one disease have been reported (**Table 4**). Australian wheat cultivar Sunbri selected for stem rust resistance gene *Sr38* also carried linked leaf rust resistance gene *Lr37*, stripe rust resistance gene *Yr17* and cereal cyst nematode resistance gene *Cre5* and rice lines bred for rice stripe virus also exhibited field resistance to rice blast.

DNA markers DNA markers have been developed for many disease resistance genes. **Table 5** lists certain examples of markers linked to simply

Table 4 Examples of genetic association of disease resistance genes

Diseases	*Resistance genes*
Wheat stem rust, leaf rust, stripe rust, cereal cyst nematode	*Sr38/Lr37/Yr17* and *Cre5*
Wheat stem rust and leaf rust	*Sr24/L24*, *Sr25* and *Lr19*, *Sr39* and *Lr35*
Wheat stem rust and stripe rust	*Sr9g* and *Yr7*
Wheat leaf rust, stripe rust, barley yellow dwarf virus	*Lr34/Yr18/Bdv1*
Wheat stem rust, leaf rust, powdery mildew, root lesion nematode	*Sr15/Lr20/Pm1* and *Rlnn1*
Potato cyst nematode and potato virus X	*Gpa2/Rx1*

Table 5 Examples of identification of DNA markers linked with low disease response phenotypes

Disease	*Gene(s)*	*Type of marker*[a]
Wheat stem rust	*Sr36*, *Sr38*, *Sr39*	PCR
Wheat stripe rust	*Yr10*, *Yr17*, *Yr26*	PCR, SSR
Wheat leaf rust	*Lr19*, *Lr35*	PCR
Bacterial blight of rice	*Xa5*, *Xa21*	PCR
Cereal cyst nematode (wheat)	*Cre1*, *Cre3*, *Cre5*	PCR
Root lesion nematode (wheat)	*Rlnn1*	RFLP
Barley yellow dwarf virus	*Yd2*, *Bdv2*	PCR
Barley leaf rust	*Rph7*	PCR
Barley stem rust	*Rpg1*	PCR
Barley scald	*Rh2*	RFLP
Soybean cyst nematode	*Rhg4*	RFLP

[a] PCR, Polymerase chain reaction; RFLP, restriction fragment length polymorphism; SSR, simple sequence repeats.

inherited traits. Most of the tagged traits in wheat are alien and thus are easy targets for the detection of polymorphism. Quantitative trait loci (QTL) explaining phenotypic variation for disease responses have been identified for many crops (**Table 6**). The effective implementation of disease response-linked DNA markers relies on conversion of identified markers into simple user-friendly protocols followed by validation across different genetic backgrounds. **Figure 2** illustrates validation of *Sr36*-linked marker stm773 in different genetic backgrounds.

Table 6 Examples of low disease response-linked quantitative trait loci (QTL)

Disease	*Number of QTL identified*
Common bean bacterial blight	7
Corn gray leaf spot	>10
Mung bean powdery mildew	3
Ascochyta blight of pea	3
Potato cyst nematode	2
Late blight of potato	11
Rice blast	10
Bacterial wilt of tomato	3
Yellow leaf curl virus of tomato	3
Fusarium head blight of wheat	>5
Capsicum CMV	4
Soybean cyst nematode	8
Barley (against rice blast)[a]	4
Barley mildew	2
Barley stripe rust	>6
Wheat stripe rust	>6

[a]Nonhost QTL for resistance identified using rice blast on barley.

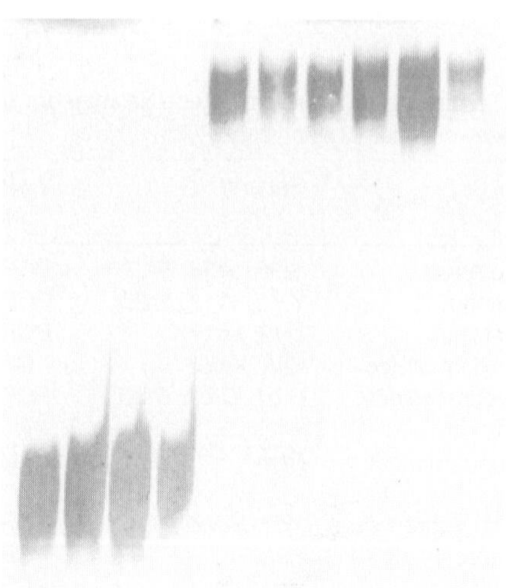

Figure 2 Validation of stem rust resistance gene *Sr36* linked microsatellite marker *stm773* across different genetic backgrounds. First four genotypes carry *Sr36* and last six genotypes lack *Sr36*.

Breeding Methodology

Following identification of diverse sources of resistance, breeding for disease resistance involves two phases:

1. Prebreeding.
2. Incorporation of resistance in the breeding populations.

Prebreeding

Backcrossing and/or convergent backcrossing have been used for transferring resistance to desirable agronomic backgrounds. The number of backcrosses needed to achieve agronomically similar derivatives depends on the donor genotype. Backcrossing method is also used for resistance gene pyramiding. Epistatic interaction of genes and linked markers are used to confirm combinations. **Figure 3** illustrates epistatic effect of stem rust resistance genes *Sr26* and *Sr38*. Stem rust resistance gene *Sr26* and *Sr38* produce distinct low infection types individually and they produce even lower infection type when combined in a single genotype.

The backcross method is also being used for producing disease-resistant inbred lines to be used as parents in hybrid breeding programs for various crops. Marker-assisted selection is beginning to be used in breeding programs. Barley workers in the United States pyramided QTL controlling stripe rust resistance and demonstrated that resistant derivatives carried combinations of QTL. Cereal programs in

1 2 3 4

Figure 3 (see color plate 9) Epistatic interaction of stem rust resistance genes *Sr26* and *Sr38*. 1, Sr26 + Sr38; 2, Sr26; 3, Sr38; 4, susceptible control.

Australia are using markers linked to rust resistance, cereal cyst nematode resistance, root lesion nematode resistance (neglectus), and barley yellow dwarf virus.

Incorporation of Resistance in Breeding Populations

The modified pedigree method has been used for improving disease resistance in various crops. The modification mainly included bulking of certain generations followed by selection. **Figure 4** illustrates the modified pedigree method being followed at the University of Adelaide, Australia. At each step of the program opportunistic selection for all simply inherited traits is made. For septoria tritici, genotypes with more than acceptable levels of infection are discarded. Some F_2 populations are screened under artificially created epidemics by the ACRCP staff and resistant F_2 head bulks are supplied to the breeders for incorporation in the breeding program. ACRCP staff also produce backcross derivatives with added rust resistance in advanced breeding material and commercial cultivars as an insurance against any important shift in pathotypic evolution. Such material is either used as donor source in the breeding programs or undergo agronomic evaluation for commercial release.

For parasites like yellow spot, crown rot, or root lesion nematode, the backcrossing approach coupled with pedigree program has been used in Australia. Doubled haploid technology is currently being used in many breeding programs. **Figure 5** illustrates breeding strategy combining backcross method and double haploid technology to transfer additive minor effect resistance genes. This method could also be used to achieve combinations of simply inherited resistance genes. The major advantage of the DH technology is the attainment of homozygosity. This technology is useful in combining and early testing of quantitatively inherited traits. It has been shown to save at least 4 years in breeding disease resistant wheat. Selection for and against many disease resistance traits that express in seedlings and where the markers are available can take place even at the haploid stage.

Double and four-way crosses coupled with multilocation testing have been exploited by the

1. Crosses made
2. F_2 populations are space planted and selection for highly heritable traits and plant type including diseases is performed
3. Individual plants harvested
4. Plant to row check for segregation performed
5. F_4 selections grown at three sites as single replication yield trials and opportunistic selection performed against prevailing diseases
6. About 10000 F_5 lines used for yield trial
7. 3000 F_6 lines selected based on preliminary yield and quality and disease nurseries
8. 3000 F_6 lines sent for rust testing by the ACRCP staff
9. Replicated yield trials and quality analyses performed for three years
10. 40–50 lines sent for detailed rust tests by the ACRCP staff
11. 10–12 lines sent for rigorous rust testing by the ACRCP staff
12. Cultivar registered and commercial partners sought
13. Cultivar seed multiplied and made available to growers

In the case of selection for rust resistance at F_2 level and or rust resistant backcrossed derivatives, material is incorporated in the program at step 4.

Figure 4 Detailed presentation of modified pedigree program used by the Roseworthy Wheat Breeding Unit of the University of Adelaide, Australia.

Parent A (susceptible) × Parent B (Carrying four minor effect resistance genes)

↓

F_1 × Parent A

↓

Generate 200–500 BC_1 F_1s

(To provide 50–150 individuals with at least three resistance genes)

(Number of individuals carrying at least three resistance genes will increase (150–350) if two genes each are contributed by each parent, and F_1 was backcrossed to agronomically superior parent. This scenario will also include 50–125 individuals carrying four genes.)

↓

Select plants with low response using appropriate selection technology (Marker-assisted selection can be performed, if markers are available)

↓

Progeny test and develop doubled haploid or SSD lines from superior plants

↓

Agronomic evaluation and release of cultivars

Figure 5 Diagrammatic presentation of backcross and doubled haploid/SSD based strategy to transfer minor effect resistance genes in self-pollinated winter cereals.

International Maize and Wheat Improvement Center in Mexico to accumulate several adult plant resistance genes that conferred moderate levels of resistance to leaf rust and stripe rust. Adult plant stripe rust and leaf rust resistances were likewise transferred to highly susceptible soft club wheats in Australia. It is important to keep in mind that material characterized in one location may behave differently at another location. For example, lines produced in Mexico to carry high levels of adult plant resistance to stripe rust and leaf rust are often found to carry effective simply inherited specific resistance genes when tested against Australian pathotypes of the respective pathogen. Breeders at the International Rice Research Institute used a combination of recurrent selection and backcrossing to achieve high levels of additive resistance to rice diseases such as sheath blight and rice blast.

Transformation capacity has been developed for many plant species at the experimental level. Transgenic expression of viral coat protein genes, pathogenicity related protein, bacterial toxins, and antisense technology has been reported. Its routine use is limited to only few crops and in few countries due to consumer acceptance issues and lack of robustness and diminishing resources in public plant breeding programs. A transgenic cultivar of papaya (*Carica papaya*) with resistance to ring spot virus was released recently in Hawaii. This event demonstrated the potential of transformation for improvement of diseases resistance in perennial plants where conventional technologies have proved laborious, time-consuming, and expensive.

Summary

Genetic resistance has commonly been used to control plant diseases. Success of a breeding program aiming at incorporating disease resistance depends on the availability of diverse sources of resistance and is greatly enhanced by knowledge of genetic control, evolutionary potential of the pathogen, appropriate choice of breeding methods, and, more importantly, robust selection tools. Germplasm development programs in place around the globe are continuously searching for new sources of resistance to important pathogens. Some international programs utilize multinational testing opportunities to ensure effectiveness of sources of resistance across wide geographical areas. Regional pathogen survey programs and international initiatives on global monitoring of pathogens are crucial for development of breeding strategies to achieve more durability of resistance. Breeding programs covering a range of crops use a modified pedigree method coupled with limited backcrossing as an overall disease resistance incorporation strategy. In the case of cross-pollinated crops, recurrent selection is often practiced, and parent building for hybrids also includes backcrossing or convergent backcrossing as means of transferring disease resistance to superior inbreds. Many breeding programs now use doubled haploid technology. The capacity of transformation to create genetic variability has been demonstrated in many crops but its routine use is limited to only a few crops and is currently restricted in many countries by consumer resistance. The ongoing development of molecular markers linked to effective disease resistance genes and the identification of genomic regions controlling quantitatively inherited minor effect resistance loci should enhance selection efficiency in the absence of appropriate pathogen isolates.

See also: **Crop Improvement**: Marker Assisted Selection; Plant Breeding, Practice; Plant Breeding, Principles. **Diseases**: Bacterial Diseases; Fungal Diseases.

Further Reading

Bariana HS and McIntosh RA (1995) Genetic of adult plant stripe rust resistance in four Australian wheat cultivars and French cultivar Hybride-de-Bersee. *Plant Breeding* 114: 485–491.

Bonman JM, Khush GS, and Nelson RJ (1992) Breeding rice for resistance to pests. *Annual Review of Phytopathology* 30: 507–528.

Dubin HJ and Rajaram S (1996) Breeding disease resistant wheats for tropical highlands and lowlands. *Annual Review of Phytopathology* 34: 503–526.

Ellingboe AH (1981) Changing concept of host–pathogen genetics. *Annual Review of Phytopathology* 19: 125–143.

Hallauer AR, Russell WA, Lamkey KR (1988) Corn breeding. In: Sprague GF, Fuccillo DA, Perelman LS, and Stelly M (eds) *Corn and Corn Improvement*, Agronomy Monograph no. 18, 3rd edn, pp. 463–564. Madison, WI: American Society of Agronomy.

Johnson R (1981) Durable resistance: definition, genetic control and attainment in plant breeding. *Phytopathology* 71: 567–568.

Johnson R and Lupton FGH (1987) Breeding for disease resistance. In: Lupton FGH (ed.) *Wheat Breeding and its Scientific Basis*, pp. 369–424. London: Chapman & Hall.

Jones SS, Murray TD, and Allan RE (1995) Use of alien genes for the development of disease resistance in wheat. *Annual Review of Phytopathology* 33: 429–443.

McIntosh RA (1995) Breeding wheat for resistance to biotic stresses. *Euphytica* 100: 19–34.

Meiners JP (1981) Genetics of disease resistance in edible legumes. *Annual Review of Phytopathology* 21: 189–209.
Salisbury PA, Ballinger DJ, Wratten N, Plummer KM, and Howlett BJ (1995) Blackleg disease on oilseed *Brassica* in Australia: a review. *Australian Journal of Experimental Agriculture* 35: 665–672.
Young ND (1996) QTL mapping and quantitative disease resistance in plants. *Annual Review of Phytopathology* 34: 479–501.

Genetic Modification of Disease Resistance, Fungal Pathogens

K M Plummer, University of Auckland, Auckland, New Zealand

Introduction

Genetic modification (GM) of crops has been shown to be a useful means of generating novel resistance for fungal diseases. The development of these GM crops has lagged behind GM for insect and viral resistance. This lag is partly due to the relatively simple nature of insect and viral resistance that can be achieved with monogenic insertions, e.g., *Bacillus thuringiensis* (*Bt*) and viral gene-antisense constructs. Also, many different insects and viruses have been able to be controlled using similar approaches. Fungal resistance, on the other hand, is usually more complex, involving multiple genes that are more difficult to manipulate.

Fungi interact with plants, causing disease in many different ways. This complicates the design of GM plants for fungal resistance. Genetic modifications that confer resistance to one fungus may also leave the plant more vulnerable to a different fungus (especially in the absence of chemical interventions). Our lack of knowledge of the genetic and molecular basis of fungal disease and resistance in major crops has also compounded the difficulties with developing GM crops with fungal resistance. The recent advent of rapid genome sequencing is likely to improve our knowledge, at least for major crop diseases that are currently the focus of significant research efforts.

The availability of economical alternatives (e.g., fungicides, cultivation practices, and disease-resistant plant cultivars) for protection against a wide range of fungi has also impacted on the commercialization of GM crops with fungal resistance. The long-term future of current fungal disease control approaches may be limited, however, by the abilities of fungal pathogens to develop resistance to chemicals and to overcome disease resistance in plant cultivars. Public concern over impacts of chemical use on the environment and health is also influencing the choice of controls available to producers. Similar concerns have impacted on the release of GM crops.

Decisions regarding crop production models are made based on the relative risk or benefit of the various options. Systems for sustainable crop management should utilize the best technologies and knowledge to gain the most from the available resources for current and future crop production. It is likely that GM crops, designed to reduce the need for fungicide applications, will play a significant role in this.

In this article the term "genetically modified" or "GM" organisms are strictly defined as those organisms that have had a specific gene (known as a transgene), or genes, transferred into them from other organisms using genetic engineering, i.e., using enzymes and laboratory techniques, such as *Agrobacterium*-mediated gene transfer or biolistics, rather than biological hybridization. Plants that have had genes transferred from one cultivar to another of the same plant species are also classified as GM, because genetic engineering technology has been used to create these plants. A discussion of the standard techniques used for genetic modification, also known as genetic transformation or gene transfer, is beyond the scope of this article.

The science and technology associated with GM of crops is also developing rapidly. Most developments in fungal disease control are reported in the scientific literature, which can be quite difficult for the nonscientist to interpret. A few general plant biotechnology texts and several excellent web sites have been recommended as additional sources of relevant information. By the time of publication of this volume it is likely that more advances will be made and indeed commercial releases of fungal disease-resistant GM crops may have occurred. For this reason it is recommended that the reader also seek out the most recent information.

Examples of GM Plants with Resistance to Fungal Diseases

The first commercially released GM crop that conferred resistance to fungal disease occurred in the mid 1990s; however, the resistance was a side benefit of controlling insect damage. GM corn (*Zea mays*, maize), that was modified to increase resistance to insect damage using the *Bt* gene from the bacterium *B. thuringiensis*, is also more resistant to ear and stalk diseases caused by a number of different fungal pathogens, e.g., *Fusarium* and *Aspergillus* species. The control is due to the reduction of feeding damage by insects, which limits infection by fungi that normally colonize the corn via wounds made by feeding caterpillars. A major benefit of the disease control was the reduction of contamination with compounds (fumonisins and aflatoxins) produced by these fungi, which are extremely toxic to mammals, including livestock and humans.

At the time of writing this article there were no commercial releases of GM crops specifically modified to confer fungal disease resistance. Many researchers have reported success in obtaining fungal disease resistance using GM in a variety of plants, both in experimental field and laboratory trials. Various transgenes have had utility in providing fungal disease resistance in these crops. There are several likely reasons for the lack of commercial release. Genetic engineering is a relatively new and complex technology and there is significant public confusion and concern over the safety (health and environmental) of GM crops, especially foods. The perceived risks of these crops must be considered against the genuine risks. The choice of production systems in a spectrum from GM, to conventional, to "organic," will be made considering the economic, social, and environmental advantages pertinent to the situation (i.e. crop, location, production scale). Cultivated crops of the future may well integrate aspects from all of these production models.

Various transgenes have been incorporated into crop species to achieve resistance against fungal pathogens in field and laboratory trials. Most transgenes are from other plants, but fungal and bacterial genes have also been used successfully. The genes that have been used for GM crops encode proteins that fall into two major categories:

- proteins with a direct activity against the fungal invader (antifungal compounds)
- proteins that increase the plant's own natural defenses.

Antifungal Compounds

Plants and antagonistic microbes are known to produce a range of antifungal compounds. Low levels of antifungal compounds are present in healthy plants and are often upregulated (production is increased) throughout the plant upon detection of an invading pathogen.

Pathogenesis-related (PR) proteins One group of antifungal proteins is known as the pathogenesis-related (PR) proteins. These proteins include enzymes, e.g. chitinases and glucanases, that attack the major components of fungal cell walls. Other PR proteins, e.g., PR1, have no known function.

Many experimental GM crops have additional PR proteins, especially chitinases that have originated from other plants or from fungi that are mycoparasites (fungi that are parasitic on other fungi, e.g., *Trichoderma* species). For example, expression of a chitinase from pea (*Pisum sativum*) is able to confer resistance in GM oil-seed rape (*Brassica napus*, carola) to the fungus *Rhizoctonia solani*. The hypha of the invading fungus is attacked by the additional chitinase and is unable to grow in the GM plant. Often the addition of two or more transgenes, e.g., chitinase and glucanase together, can act synergistically to achieve a higher degree of fungal disease control. Chitinases have been used to create a variety of disease resistant GM crop species, including rice (*Oryza sativa*), apple (*Malus pumila*), tobacco (*Nicotiana tabacum*), oilseed rape/canola, tomato (*Lycopersicon esculentum*), peanut (*Arachis hypogaea*); however, none is commercially available.

It is important to note that a major group of pathogens, known as Oomycetes, which are not true fungi, do not have chitin in their cell walls. For this reason diseases caused by these organisms, e.g., potato blight, grape downy mildew, and white rust of *Brassica* species, cannot be controlled with GM crops expressing additional chitinases.

Phytoalexins Many plants produce a range of natural antimicrobial toxins known as phytoalexins. Phytoalexins are not proteins and require a series of enzymes for their production. Phytoalexins are also part of the induced defense response of plants and their production is increased upon pathogen challenge. GM plants with transgenes for enzymes that increase the quantity and type of phytoalexin have been produced that have increased plant resistance to fungal infection.

The stilbene compound resveratrol is a phytoalexin that is produced by grape (*Vitis vinifera*). Most plants produce the precursors of this compound but require the addition of the grape enzyme stilbene synthase to produce the active, antimicrobial phytoalexin. Addition of the grape enzyme has been shown to increase disease resistance by production of resveratrol in GM tobacco and tomato.

Active oxygen species (AOS) Enzymes that generate active oxygen species, such as hydrogen peroxide (H_2O_2), are known to aid in fungal disease resistance. The expression of a transgenic enzyme, e.g., glucose oxidase, which produces H_2O_2 confers resistance in potato to several plant pathogens, including the

Oomycetes, *Phytophthora infestans* (the potato blight pathogen), and the *Verticillium* wilt pathogen.

Ribosome-inactivating proteins (RIP) Ribosome-inactivating proteins (RIP), produced by several plants, can deactivate the protein manufacturing mechanism of ribosomes in fungi. The action of these transgenic proteins is enhanced by the addition of transgenic chitinases, which are thought to increase the access of the RIP to the fungal ribosomes by degrading the fungal cell wall.

The pokeweed antiviral protein (PAP) is a type of RIP that is produced by American pokeweed (*Phytolacca americana*) and various other plant species. Studies have demonstrated that transgenic tobacco expressing PAP are also somewhat resistant to fungal pathogens, e.g., *R. solani*. The resistance in this case is probably due to the induction of natural plant defense PR proteins rather than a direct effect, as the PAP protein is not inhibitory to the fungus grown in culture.

Small peptides – defensins Another group of antifungal compounds include a family of small antifungal peptides, known as defensins. Thionins, lectins, and lipid transfer proteins are examples of these small antifungal peptides. Defensins are found in plant seeds and other tissues and transgenic expression can confer resistance against fungal pathogens.

Activation of Plant Disease Resistance Mechanisms

Disease resistance (*R*) genes Plant disease resistance (*R*) genes occur in naturally disease-resistant plants. *R* genes encode proteins that confer the ability of the plant to defend itself against the fungal invader. The first of these resistance genes to be cloned and sequenced was the *Hm1* gene from corn. This gene encodes a protein that detoxifies the HC-toxin, produced by the leaf spot fungus (*Cochliobolus carbonum*). Cultivars of corn that have been bred by traditional methods to contain the *Hm1* gene are resistant to infection by *C. carbonum*.

Most *R* genes mediate recognition and signaling to the plant that it is under attack from a pathogen. A series of defenses are then activated to prevent the pathogen from invading, for example:

- the hypersensitive response (HR), which involves rapid plant cell death
- the accumulation of antimicrobial toxins, such as phytoalexins
- cell wall modifications
- production of PR proteins.

Specific *R* genes from one plant cultivar or species can be transferred to another cultivar or species within the same genus. An example is tomato disease resistance genes, which can be transferred from a resistant cultivar (with poor agronomic characteristics) to a susceptible cultivar (with excellent agronomic characteristics). This process can convert a plant from susceptible to resistant to a particular disease with the insertion of a single gene. Undesirable qualities are not transferred, which often occurs in the conventional breeding process. The recognition and resistance is specific to a particular pathogen and is therefore, by definition, not broad spectrum. Single transgene insertions of specific *R* genes are likely to be as vulnerable as conventional crops with single major genes for resistance, which have resulted in boom-and-bust scenarios of crop protection and subsequent failures. For this reason gene pyramiding, or stacking, of resistance traits (both *R* genes and other) for a particular pathogen is more desirable. Indeed the main utility of GM is the ability to create plants with multiple disease resistances. These plants are likely to be much more difficult for pathogens to attack and will hopefully provide more durable disease resistance when released into field production. The major impediment to this approach is the fact that relatively few *R* genes have been identified. These genes are quite variable in their sequence and often exist in clusters in the plant genome, making them quite difficult to identify. Genome sequencing projects, currently under way for several major pathogens, are likely to ease the process of *R* gene identification, increasing the likelihood of their deployment in GM crops in the near future.

Why Use GM for Fungal Disease Resistance?

Sources of natural plant disease resistance are often wild relatives of the cultivated crop species. These wild species often have undesirable characters that prevent them from being commercially viable for production. Crops in cultivation have been selected over many years by plant breeders for qualities that are required by the producer, consumer or market, e.g., for qualities such as flavor, ability to be stored, size and shape of product, and yield. The existence of effective means for disease control, such as chemical fungicides, has enabled the plant breeders to focus on qualities other than disease resistance.

Breeding for durable disease resistance is very labor intensive and time consuming. Single gene resistance, which is the most straightforward type of quality to select, can take decades to incorporate into a marketable plant product. A cultivar, when released widely for crop production, will be subjected to a broad range of pathogens that within a few years may be able to overcome the plant disease resistance. Resistance conferred by a single gene is usually relatively easy for pathogens to overcome. For this reason plant breeders usually try to combine several different types of disease resistances so that the pathogen has to overcome more than one factor to enable it to grow on the plant. To generate a plant cultivar with multiple disease resistance traits is even more difficult and time consuming. Transferring multiple genes into a cultivar using genetic engineering is likely to aid this process.

GM is currently a major research tool for dissecting the plant disease process. The mechanisms by which microbes can attack certain plant species but not others are being investigated with the use of transgenic plants and fungi. These transgenic plants or fungi enable the effects of single genes to be determined. Strategies for control of the pathogens based on the knowledge of these mechanisms can then be devised.

Conclusions

Most plants are naturally resistant to attack by most pathogens. Fungi that are able to cause disease on a plant have adapted specifically to enable them to overcome the plant's defenses. GM crops are usually adapted to enhance the plant's ability to defend itself against pathogen invasion, either by producing substances that attack the fungus itself, or by increasing the plant's ability to recognize that it is under attack and then mount a counterdefense. As we learn more about how pathogens attack and plants defend themselves, we are likely to be able to devise better strategies to counter plant disease. It is likely that GM crops will become vital weapons in this battle, either as tools for research or for integration into commercial production systems.

List of Technical Nomenclature

Genetically modified (GM)	Description of an organism that has been modified to include specific genes transferred from other organisms using genetic engineering, i.e., enzymes and laboratory techniques, rather than biological hybridization (synonymous with transgenic).
Monogenic	Caused by a single gene.
Pathogenesis-related (PR) proteins	Plant proteins (e.g., chitinases, glucanases) that are antimicrobial in function and are synthesized as part of an induced defense mechanism upon pathogenic invasion of plants.
Phytoalexin	Any of the natural antimicrobial toxins (not proteins) produced by plants.
Resistance (*R*) genes	Specific plant disease-resistance genes that are involved in recognition of specific factors (of pathogen or plant origin) that signal to the plant that it is under attack from a pathogen. A series of defenses are then activated to prevent the pathogen from invading.
Transgenes	Genes inserted into one species that have originated from a different organism (usually, but not always, from a different species).

See also: **Crop Improvement**: Plant Breeding, Practice; Plant Breeding, Principles. **Diseases**: Breeding for Disease Resistance; Fungal Diseases; Genetic Modification of Disease Resistance, Viral Pathogens; Plant Pathology, Principles. **Pests**: Genetic Modification of Pest Resistance, Insect Pests.

Further Reading

Agrios GN (1997) *Plant Pathology*, 4th edn. New York: Academic Press.

Baron C and Zambryski PC (1995) Notes from the underground: highlights from plant–microbe interactions. *Trends in Biotechnology* 13(9): 356–361.

Birch RG (1997) Plant transformation: problems and strategies for practical application. *Annual Review of Plant Physiology and Plant Molecular Biology* 48: 297–326.

Chrispeels MJ and Sadava DE (2002) *Plants, Genes and Crop Biotechnology*, 2nd edn. Boston: Jones & Bartlett.

Galun E and Breiman A (1997) *Transgenic Plants*. London: Imperial College Press.

Hammond J (1999) Overview: the many uses and applications of transgenic plants. In: Hammond J, McGarvey P, and Yusibov V (eds) *Plant Biotechnology: New Products and Applications*, pp. 1–19. Berlin: Springer-Verlag.

Hammond-Kosack KE and Jones JDG (1997) Plant disease resistance genes. *Annual Review of Plant Physiology and Plant Molecular Biology* 48: 575–607.

Keen NT (2000) A century of plant pathology: a retrospective view on understanding host–parasite interactions. *Annual Review of Phytopathology* 38: 31–48.

Michelmore R (1995) Molecular approaches to manipulation of disease resistance genes. *Annual Review of Phytopathology* 33: 393–427.

Nielsen K and Boston RS (2001) Ribosome-inactivating proteins: a plant perspective. *Annual Review of Plant Physiology and Plant Molecular Biology* 52: 785–816.

Shah DM, Rommens CM, and Beachy RN (1995) Resistance to diseases and insects in transgenic plants: progress and applications to agricultures. *Trends in Biotechnology* 13(9): 362–368.

Genetic Modification of Disease Resistance, Viral Pathogens

J I Cooper, NERC Centre for Ecology and Hydrology, Oxford, UK

J A Walsh, Horticulture Research International, Wellesbourne, UK

Background and Discovery

Hundreds of viruses that infect green vascular plants have been assigned to dozens of genera and many of these have been targets for transgenic resistance/tolerance. In addition, diverse replicating nucleic acids (notably satellites) that depend for their multiplication and/or transmission on coinfection with another ("helper") virus are associated with and sometimes ameliorate the pathogenicity of helper viruses in plants. Satellites are a very heterogeneous collection of subviral agents including nucleic acid molecules that contain sequences of bases distinct from those of the genomes of both their helper viruses and their hosts. Within this definition it is appropriate to note that chimeras between satellites and viral genomes, plant chloroplast nucleic acids and viroids have been recognized. Possibly as a consequence, plants expressing RNA sequences from a viral satellite reportedly ameliorate damage from potato spindle tuber viroid. Satellites have not been associated with all known viruses but have been recognized in context with more than 30 of the viruses that infect angiosperms. In the People's Republic of China, the United States, and also in Europe, satellites were in the vanguard when experiments were done to find transgenes for virus tolerance; they have been evaluated both as transgenic sequences in plants and also as components of virus inocula causing acceptably mild disease. In the United Kingdom, the unpredictable interactions between satellites and naturally occurring viruses were judged unacceptable by regulators and the initial promising laboratory/glasshouse experiences were not followed up with field trials. Interestingly, one current US field trial with a satellite-derived transgene has been authorized.

In the space available, it is not possible to describe the properties of all the differing virus-like parasites and we therefore refer readers to reports from the International Committee on Taxonomy of Viruses where names are compiled and properties contrasted. Although viruses with genomes comprising double-stranded (ds) RNA, single-stranded (ss) DNA, or dsDNA infect plants, more than 90% of the known plant-infecting viruses have ssRNA genomes that are copied via double-stranded intermediates in processes catalyzed by virus-encoded RNA-dependent RNA polymerase enzymes (=replicases). This key element in viral life cycles is now suspected to provide the tool that plants naturally use in their defense. That view has been arrived at as a result of many attempts to explain inconsistencies experienced by scientists attempting to obtain transgene-mediated virus tolerance or resistance.

Terminology

The behavior of a virus in a host is often confused with the responses of the host to virus infection. Because viruses have properties that are not shared by other plant parasites it is necessary here to define some of the terms (including that of "host"). When viruses are introduced into a plant, there are two possible extreme consequences: either infection occurs, or it does not. Virus particles can enter cells in which they do not replicate and their constituents may move separately or collectively within cells entering nuclei, chloroplasts, etc. However, these acts do not constitute infection. Infection is a term properly restricted to the act of nucleic acid replication following nucleic acid or nucleoprotein entry into cells. There is no unanimously agreed term to describe either the absolute state of exemption from infection or the antithesis of this. The term "infectible" distinguishes hosts from nonhosts unambiguously but is not universally used. The terms "resistant" and "susceptible" denote points on a scale covering the effects on an infectible individual following virus invasion, multiplication, and infection. The terms "tolerant" and "sensitive" denote points on a scale covering the reaction of infectible individuals to virus infection and establishment. Resistance to infection generally occurs when an infectible individual is not very readily infected; another type of resistance reflects impediments to invasion mediated by a range of virus-coded "movement" proteins.

The GM Approaches Used to Obtain Virus Resistance/Tolerance

Laboratory methods now enable the production of green vascular plants with a diverse range of novel properties. "Genetic modification" (GM) is the most widely used colloquial term applied to the process but "transformation" is the more generally applicable "scientific" word. In the last 60 years, genetic transformation of flowering plants has been achieved in a variety of different ways but the methods of first choice are based on the integration into a plant's nuclear genome of a plasmid (from the bacteria *Agrobacterium tumefaciens* or *A. rhizogenes*). These *Agrobacterium* species facilitate the more or less stable incorporation of one or a small number of copies of defined DNA segments. Traits deriving from such 'transgenes' are inherited in the ways that have long been studied by "classical" plant breeders. Whereas *Agrobacterium*-mediated gene transfer has been applied in a diverse range of broadleaved plant genera and species, the process is not routinely successful for all plant genotypes (cultivars) and monocotyledonous plants are notably recalcitrant. For these, and for some other economically important plants (e.g., tree species), alternative (usually less efficient and more labor-intensive) "direct/physical" means of DNA transfer have been developed (e.g., electroporation, ballistic injection, vacuum infiltration).

The core requirements for transgene-mediated resistance are (1) reliable cultural characteristics in the target plant (ideally, chemically defined media on which tissues differentiate rapidly and predictably) and (2) DNA sequences conferring the appropriate traits. Until comparatively recently, plant-derived virus resistance genes were not available for use as transgenic genes. Now, several have been identified and one such sequence has been incorporated in a crop that is currently authorized for field assessment in the United States. The characterization of "natural" virus-resistance genes in plants required protracted effort and a number of alternative routes towards transgenic virus resistance were explored more expeditiously (as spinoffs from studies of virus characteristics). At present, the complex interactions between viruses and their plant hosts are not sufficiently well understood in molecular detail to allow many targeted attacks on key elements in these processes. Nevertheless, great ingenuity has been applied and, to indicate one successful approach towards virus resistance, ribosome-inactivating glycosidases that stop protein synthesis/elongation have been linked to virus-derived promoters so that specific virus infection triggers cell death and restricts invasion to the first cells inoculated.

Why Use "Alien" Sources of Virus Tolerance/Resistance?

Genetic modification of plants has the capacity to augment existing "natural" genetic variation or resistance (where they occur, they may be linked to undesirable agronomic characteristics; in many cases they are not available). Importantly, the exploitation of exotic resistance sources, provides extra breathing space in the plant breeder's race with pathogen evolution. This is particularly necessary because viruses are fecund and somewhat more intrinsically variable than other types of plant pathogen.

Parasite-Derived Systems for Virus Tolerance/Resistance

Although others had considered the possibility, Sandford and Johnston deserve credit for generalizing a principle whereby an infectible organism could be transformed with genes from one of its parasites to induce resistance to that agent. That concept, formulated in 1985, of "parasite-derived resistance" was rapidly proved applicable to plants. Since that time, a diverse range of examples of the phenomenon (in annual and perennial food plants, trees and industrial crops) have been described (**Table 1** lists some examples).

Parasite-derived resistance protocols have been successful against economically significant viruses with positive sense ssRNA genomes (i.e., the RNA is infectious and can directly facilitate synthesis of polypeptides). Numerous viruses have virions that contain nucleic acids lacking the property of direct infectiousness; their ssRNA genome message requires copying before translation into polypeptides and is described as negative sense. An intermediate (ambisense) is characteristic of tomato spotted wilt virus. A small number of plant pathogenic viruses have ssDNA genomes (e.g., vigna mungo yellow mosaic virus) or dsDNA genomes (e.g., cauliflower mosaic virus). Despite repeated attempts, viruses with dsDNA genomes have not yielded to any variant of the transgenic approach.

Tobamoviruses were among the first to be characterized in molecular detail and there was a parallel development in knowledge of tobacco genetics, tissue culture, and differentiation. Thus, when Sandford and Johnston published their theory, the transcriptional/translational sophistication of genome expression in tobacco mosaic virus was known and the "gene" coding for the principal structural protein (the capsid (or coat) protein) was available in precise molecular form. The first studies in the context of parasite-derived resistance concerned transgenic

Table 1 Three transgenic approaches effective against a range of viruses

Capsid mediated	*Replicase mediated*	*RNA mediated*
Arabis mosaic virus	African cassava mosaic virus	Papaya ringspot virus
Alfalfa mosaic virus	Cucumber mosaic virus	Pea early browning virus
Beet necrotic yellow vein virus	Cowpea mosaic virus	Plum pox virus
Cucumber mosaic virus	Cymbidium ringspot virus	Potato virus A
Cymbidium ringspot virus	Pea early browning virus	Potato virus X
Lettuce mosaic virus	Pea seedborne mosaic virus	Potato virus Y
Papaya ringspot virus	Pepper mild mottle virus	Tobacco etch virus
Poplar mosaic virus	Potato virus X	Tobacco mosaic virus
Potato aucuba mosaic virus	Potato virus Y	Tomato spotted wilt virus
Potato leafroll virus	Plum pox virus	Tomato yellow leaf curl virus
Potato virus S	Tobacco mosaic virus	Turnip mosaic virus
Potato virus X		
Rice stripe virus		
Strawberry latent ringspot virus		
Soybean mosaic virus		
Tobacco mosaic virus		
Tobacco rattle virus		
Tobacco streak virus		
Tomato spotted wilt virus		
Tomato yellow leaf curl virus		
Turnip mosaic virus		
Zucchini yellow mosaic virus		

tobacco plants and showed that seedlings from the primary transformant lines expressing the virus-derived capsid protein were infectible but disease was delayed, or indeed did not develop following challenge with the virus. In other words, the transgenic tobacco plants were tolerant of infection but also somewhat resistant to invasion. Given that promising experience of what came to be called "capsid-mediated" resistance, attempts were made to repeat the success with a diverse range of morphologically similar ssRNA viruses (with tubular virions) and subsequently viruses with ssRNA genomes and an icosohedral particle morphology.

How Do These Transgenic Approaches Work?

Notwithstanding an extensive literature, the mechanisms underlying parasite-derived transgenes for virus tolerance are not fully understood and the durability of benefit remains largely unproven because few of the transgenic crops have been challenged in the field for more than a few years. In some instances, it was observed that the more capsid protein expressed, the greater the efficacy. Therefore, efforts were aimed at maximizing the transcription and translation efficiency and consequently the accumulation of transgenic capsid protein in plants. For protein production, transgenes were used in a sense orientation but, to provide control plants, the same transgene was often inserted in an antisense orientation. Unexpectedly, even plants in which no virus-coded protein was detected, were protected against virus challenge.

Despite observations that protein expression was not always necessary for parasite-derived transgenic resistance, capsid-mediated resistance held the scientific spotlight and, in a climate of euphoria and hope that a new savior had been found for crops threatened by hitherto uncontrolled scourges, it was not widely acknowledged that there are substantial limitations to the capsid-mediated approach. Importantly, although there are a few exceptions, it is noteworthy that the capsid-mediated transgenic approaches usually only work well when the protecting and the target viruses share substantial numbers of amino acids in the capsid-coding regions. Furthermore, high concentrations of virus inoculum overcome capsid-mediated resistance. Importantly, even when capsid-mediated resistance/tolerances are observed in controlled glasshouse conditions (with manual inoculation of the viruses concerned), they are not, in all instances, apparent when natural vectors deliver the virus challenge. For example, although capsid proteins from diverse nepoviruses have been expressed in transgenic plants, only that from strawberry latent ringspot virus is currently known to provide protection to plant roots inoculated by soil-inhabiting trichodorid nematode vectors.

Coding Sequences for Viral Nonstructural Proteins as Transgenes

Viruses routinely encode genome copying (replicase) enzymes. In addition, many viruses encode specific proteases (which permit assembly of useful components from the primary polyproteins translated from genome transcripts). Furthermore, virus genomes encode proteins with a "movement" function that facilitate cell-to-cell movement within hosts, and while studies of capsid-mediated transgenic resistances were approaching their zenith, sequences coding for several of these nonstructural virus-coded proteins were shown to act as protective genes in transgenic plants. Replicase-mediated resistance was found to offer potential advantages over capsid-mediated systems, notably in being effective against greater amounts of challenge inoculum concentrations than capsid-mediated systems. Replicase-mediated transgenic approaches have been shown to be efficacious against a diverse range of different viral taxa including, in a tobacco model system, the ssDNA-containing geminivirus associated with African cassava mosaic disease. Deliberately defective replicase constructs have been reported to be only transiently effective as transgenes conferring virus tolerance/resistance but, when the enzyme is complete, replicase-mediated protection is characterized by robustness coupled with an extreme (but practically often undesirable) degree of specificity. Because of intrinsic variability, natural virus challenge is very rarely, if ever, attributable to narrowly defined virus genotypes. Whereas replicase-mediated GM virus resistances are characterized by a narrow spectrum of activity, the expression in GM plants of movement proteins rendered "nonfunctional" as a result of deletion has provided broad-spectrum and, it is hoped, durable resistance to challenge.

Satellite RNA (of cucumber mosaic virus; CMV) is not known to encode proteins and when expressed in transgenic plants, was shown to lessen disease attributable to CMV. Early studies showed that amounts of RNA that accumulated from a virus-derived transgene and amounts of transgene product did not always correlate with the degree of resistance induced. Thus, it was shown that parasite-derived transgenic resistance does not need a proteinaceous product and this was subsequently reaffirmed when untranslatable virus-derived sequences (used as transgenes) also induced resistance.

No single hypothesis adequately explains the molecular basis of replicase-mediated transgenic approaches. With the defective "movement" proteins, it is very likely that they compete for plasmodesmatal binding sites used by invading virions (or RNA) and thereby impede invasion or bind to viral RNA. Nucleic acid-mediated resistance is usually attributed to RNA silencing. The key trigger in that process is dsRNA that may either be expressed from a transgene or made while viruses replicate. The dsRNA is subsequently cleaved to small interfering RNAs (siRNA), 21–25 nucleotides in length, which are suspected to guide endonuclease complexes sequence specifically to target RNA which is then cut.

The realization that nucleic acid-mediated resistance occurs led to the demonstration of a link between transgenic virus resistance and posttranslational suppression (i.e., silencing) of transgene expression. Gene silencing is now a very active area of research and many examples of RNA-mediated resistance to viruses have been described (**Table 1** provides some examples). Importantly, chimeric transgenes derived from two distinct viruses have been shown to confer multivirus resistance through gene silencing. Furthermore, preliminary tests suggest that this approach offers promise of a cure for a virus with a ssDNA genome (vigna mungo yellow mosaic virus).

Although RNA-mediated resistance does not suffer from some of the perceived risks (e.g., transencapsidation) attributed to capsid protein-mediated resistances, RNA-mediated resistances are dependent on nucleotide sequence homology with target genomes and are not thought likely to provide such broad-spectrum benefit as capsid protein-mediated resistance. Additionally, and importantly, many viruses are now known to be capable of suppressing gene silencing associated with RNA-mediated resistance. This phenomenon contributes to the inconsistencies (and consequential commercial uncertainties) when challenge tests are done in the field. Silencing has been shown to be one of the mechanisms by which plants naturally defend themselves against viruses and hence, in evolutionary/ecological terms, viruses capable of suppressing the silencing may have a selective advantage in appropriate circumstances.

Other Transgenic Approaches

The expression in transgenic plants of modified antibodies resulting from hybridoma technology and targeting viral epitopes in capsid proteins (with a potential to limit vector-mediated transmission) offer an approach to the management of viruses that is not subject to the same concerns about environmental harm that pertain to other approaches. As with the expression of interferon, the claims of success against viral targets are few. Nevertheless, the claims have not been pursued, partly because sophisticated blood chemistry and hybridoma technology are not established in many plant virology laboratories.

Table 2 gives a flavor of the current approaches under trial in the United States but it is appropriate to record that there have been many others and that some of these were initially promising in small-scale

Table 2 Some authorizations in the United States of field trials with virus-resistant/tolerant transgenic plants containing non-structural protein genes

Source virus (and suspected function of gene product)	*Plant*	*Target*	*Number of trials*
Tomato yellow leaf curl virus (TYLCV) (replicase)	Tomato	TYLCV	2
Potato leafroll virus (PLRV) (replicase)	Potato	PLRV	89
Papaya ringspot virus (PRSV) (replicase)	Papaya	PRSV	2
Cucumber mosaic virus (CMV) (replicase)	Tomato	CMV	1
Potato virus Y virus (PVY) (replicase)	Potato	PVY	1
Barley yellow dwarf virus (BYDV) (replicase)	Oat	BYDV	1
Potato virus Y virus (PVY) (protease)	Potato	PVY	2
Raspberry bushy dwarf virus (RBDV) (movement)	Raspberry	RBDV	2
Tobacco mosaic virus (TMV) (movement)	Tomato	TMV	1
Cymbidium mosaic virus (CymMV) (movement)	Orchids	CymMV	1
Potato virus Y virus (PVY) (NIa)[a]	Potato	PVY	4
Papaya ringspot virus (PRSV) (NIa)[a]	Squash	PRSV	3
Papaya ringspot virus (PRSV) (NIa)[a]	Melon	PRSV	3
Papaya ringspot virus (PRSV) (NIa)[a]	Cucumber	PRSV	4
Papaya ringspot virus (PRSV) (NIb)[b]	Squash	PRSV	3
Wheat streak mosaic virus (WSMV) (NIb)[b]	Wheat	WSMV	1
Potato virus Y (PVY) (NIb)[b]	Potato	PVY	7
Papaya ringspot virus (PRSV) (NIb)[b]	Melon	PRSV	3
Papaya ringspot virus (PRSV) (NIb)[b]	Cucumber	PRSV	4
Potato virus Y (virus-coded genome-linked)	Potato	PVY	5
Tobacco rattle virus (TRV) (60k)	Potato	TRV	3
Cucumber mosaic virus (satellite RNA)	Tomato	CMV	1
Tobacco veinal mottle virus (HC pro)[c]	Potato	PVY?	2
Tobacco veinal mottle virus (HC pro)[c]	Tobacco	TVMV	1
Tobacco veinal mottle virus (protease)	Potato	PVY?	2
Tobacco veinal mottle virus (TVMV) (cylindrical inclusion – ATPase and helicase)	Tobacco	TVMV	1

Source: http://www.aphis.usda.gov
[a] Polyprotein including protease.
[b] Amino acid motifs suggesting replicase equals RNA-directed RNA polymerase.
[c] Helper of vector transmission with associated protease.

contained studies. Some also worked in the field. For example, when expressed in transgenic potato (*Solanum tuberosum*), a rat-derived oligoadenylate synthetase (that is suspected to play a role in "scavenging" dsRNA) protected the plants from challenge with potato virus X – even under field conditions.

Commercial Use of Parasite-Derived Virus Tolerance

Trials of GM crops with obtained virus resistance/tolerance have been held in the People's Republic of China, Australia, and, uncommonly, in states of the European Union, but the range of such transgenic crops authorized for field release in North America is greater and includes both herbaceous annuals and trees. Most of the crops known to be under investigation express virus-derived capsid-coding sequences (**Table 1**) but transgenic wheat (*Triticum aestivum*), potato, and pea (*Pisum sativum*) expressing genes from yeast (notably dsRNases) are also currently undergoing field assessment in the context of "obtained" virus resistance in the United States. Unregulated (i.e., commercial) use, in the United States, of squash (*Cucurbita pepo* subsp. *ovifera*) containing two potyvirus-derived capsid coding sequences was authorized in 1994. By 2001, replicase-mediated potato leafroll virus resistant potatoes had been added to the short list of crops authorized for commercial release and it is noteworthy that 89 trials of potatoes containing such transgenes are currently authorized for field trials, an indication of perceived commercial potential for this approach.

Papaya ringspot virus (PRSV) which is ubiquitous in plantations and is responsible for 70–90% yield (largely market quality) reduction in papaya and also cucurbit crops throughout tropical and subtropical regions of the world, provide current success stories for transgenic resistance. Using microprojectile bombardment, papaya (*Carica papaya*) was transformed with the capsid coding sequence from PRSV and selections from those transgenic events collectively showed a range of degrees of resistance to the virus – including immunity. Transgenic virus resistant papaya was the second crop with obtained virus resistance to be commercialized in the United States (principally for use in the state of Hawaii). If favorable economic and social benefits are sustained and if appropriate technology transfer agreements can be agreed, it is anticipated that derivatives of these transgenic lines will form a basis for revivified commercial papaya production outside the United States. Many less-developed countries currently suffer from the effects of a diverse range of variants of the two major types of PRSV. For many of these regions, the current technology is likely to need site-specific tailoring.

Market share and potential for payback on investment are key elements in each crop/pathogen commercial decision and, in the current climate of public uncertainty and nonacceptance, very few transgenic crops justify the risk and the investment in regulatory approvals. An important reason for the relatively slow uptake of virus-resistant crops obtained by genetic engineering is the perception (particularly in Europe) that too few data are available to form a reliable scientific basis for the assessment of risk to the environment. This, with concerns for public health, is the key issue that must be satisfied under regulations that respond to EU Directives.

Before crops expressing virus-derived sequences were close to market there had been few studies of virus impacts outwith crops. Consequently, there was essentially no knowledge about whether or how much "ecological release" might occur with the nullification of pathogens that hitherto constrained genetic diversity or kept specific wild species in check. Driven by public sensitivity to possible biodiversity change following pollen (seed)-mediated transgene flow, some research has been done and there are now data on the natural prevalence rates and also the diversity and potential fitness impacts of a handful of viruses that infect crop relatives. However, there is still too little general knowledge to justify broad generalizations about potential "ecological/biodiversity" impacts. Hence, risk assessments continue on a case-by-case basis.

One perceived risk of possible harm from transgenic virus-tolerant crops was recognized early and derived from concerns about the possible role that the transgenic plants might play in the enhancement of virus diversity as a result of expressed virus capsid protein (transencapsidation/heterologous encapsidation). The concern was attributed to a lack of baseline knowledge of frequency and significance for vector relations providing scope for consequential (transient) extension of a virus. Whenever viral sequences are expressed in plants there is the possibility of new genome combinations resulting as a result of recombination between the transgene transcripts and infecting viral RNA. However, concerns about recombination are hard to discount because realistic rates of recombination are not easy to measure and knowledge of virus genome diversity is fragmentary.

See also: **Diseases**: Viral Diseases.

Further Reading

Fitch MMM, Manshardt RM, Gonsalves D, Slightom JL, and Sandford JC (1992) Virus resistant papaya plants derived from tissues bombarded with the coat protein gene of papaya ringspot virus. *Bio/Technology* 10: 1466–1472.

Gonsalves D (1998) Control of papaya ringspot virus in papaya: a case study. *Annual Review of Phytopathology* 36: 415–437.

Hackland AF, Rybicki EP, and Thompson JA (1994) Coat protein-mediated resistance in transgenic plants. *Archives of Virology* 139: 1–22.

International Committee on Taxonomy of Viruses (2003) http://www.ncbi.nlm.nih.gov/ICTV/

Pooggin M, Shrivaprasad PV, Veluthambi K, and Hohn T (2003) RNAi targeting of DNA virus in plants. *Nature Biotechnology* 21: 131–132.

Sandford JC and Johnston SA (1985) The concept of parasite-derived resistance-deriving genes from the parasite's own genome. *Journal of Theoretical Biology* 113: 395–405.

Slot KAE van't and Knogge W (2002) A dual role for microbial pathogen-derived effector proteins in plant disease and resistance. *Critical Reviews in Plant Sciences* 21: 229–271.

Tepfer M (2002) Risk assessment of virus-resistant transgenic plants. *Annual Review of Phytopathology* 40: 467–491.

Tien P and Wu GS (1991) Satellite RNA for the biocontrol of plant disease. *Advances in Virus Research* 39: 321–339.

Vance V and Vaucheret H (2001) RNA silencing in plants: defense and counterdefense. *Science* 292: 2277–2280.

Voinnet O, Pinto YM, and Baulcome DC (1999) Suppression of gene silencing: a general strategy used by diverse DNA and RNA viruses of plants. *Proceedings of the National Academy of Sciences of the USA* 96: 14147–14152.

Walsh JA (2000) Transgenic approaches to disease resistant plants as exemplified by viruses. In: Dickinson M and Beynon J (eds) *Molecular Plant Pathology Annual Plant Reviews*, vol. 4, pp. 218–252. Sheffield: Sheffield Academic Press.

Dormancy *see* **Seed Dormancy**: Development of Dormancy; Genetics of Dormancy. **Regulators of Growth**: Photoperiodism.

E

ENERGY CROPS

Contents

Biofuels from Crop Plants

D J Murphy, University of Glamorgan, Cardiff, UK

Introduction

Plants are the oldest sources of fuels and, even today, over 30% of total wood production is used as a combustion fuel. Over the past two centuries, fossil-derived materials, such as coal, mineral oil, and natural gas, have gradually come to replace plant biomass as more readily transported and calorie-rich fuels for both industrial and commercial use. More recently, however, there has been a resurgence of interest in alternative fuels that could be produced locally and are carbon-neutral. These have been termed "biofuels" and include the following three main crop categories:

1. carbohydrate crops for conversion into bioalcohol (sometimes called fuel ethanol or gasohol);
2. oil crops for conversion into biodiesel; and
3. wood coppice for direct combustion in power stations.

This article will focus on the first two categories of biofuels, namely bioalcohol and biodiesel, and especially on their use as alternative fuels for transportation. The market for biofuels is driven by the perceived need for security of fuel supply (i.e., local production to replace imported fuels) and the recognition that greenhouse gas emissions are contributing to global climate change (hence the desire for renewable, carbon-neutral fuels). In the USA, the transportation sector is responsible for more than 70% of the petroleum consumed and over 30% of the carbon dioxide emissions. Statistics are similar in Europe, where a Commission to the European Parliament put out a proposal to promote the use of biofuels for transport in November 2001 that introduced the objective of 20% substitution of alternative fuels in the road transport sector by the year 2020. The Commission issued the proposed directives in response to Kyoto Protocol emission reductions goals and to gain energy security for the members of the European Union.

Bioalcohol

Bioalcohol or fuel ethanol is the most widely used biofuel for transportation applications, especially in the Western Hemisphere, where it far outstrips biodiesel in importance. The major global producer of bioalcohol is Brazil, which alone produces 50% of the world's fuel ethanol. The source of all the Brazilian bioalcohol is sugar cane juice and molasses, for which it produces over 300 million tonnes of raw sugar cane a year. This represents about 25% of all sugar cane production worldwide. The much higher yield of ethanol precursors in tropical crops like sugar cane is the reason for the concentration of bioalcohol production in countries like Brazil. Little bioalcohol production occurs in Europe and only a small, but growing amount in North America. In the USA, 95% of the fuel ethanol produced comes from corn (*Zea mays*; maize) while more temperate countries like Canada produce ethanol from other less efficient starchy crops like wheat (*Triticum* spp.), corn, and barley (*Hordeum vulgare*).

As the world's largest bioalcohol producer, Brazil also uses nearly 15 billion liters year^{-1} of ethanol fuel. Approximately 40% of cars in Brazil run on 100% ethanol fuel and the other 60% use a 22% ethanol–78% gasoline blend. Well over a million Brazilians, many of them rural workers, depend directly on the bioalcohol market for their livelihoods. In recent years, the demand for Brazilian bioalcohol has declined as the government reduced its subsidy versus the still-cheaper gasoline from mineral oil. Since 1995, the Brazilian bioalcohol fleet has been reduced to 4 million vehicles, so by 2000 it comprised less than 40% of the national fleet. Only 1% of new vehicles are designed to use alcohol, down from 80% of new production in the 1980s. However, there are indications that production of bioalcohol vehicles is now rising again. An important factor in encouraging the uptake of bioalcohol has been the establishment of "green fleets" by local governments, as well as the tests with the alcohol–diesel mix for municipal buses.

Apart from Brazil, the USA is emerging as a major producer and user of bioalcohol. The USA uses about 12% of ethanol in fuel, typically in a 10% ethanol–90% gasoline blend. A major driving force to encourage bioalcohol production in the USA is the perceived economic and security advantages. For example, in late 2001, it was projected that implementing a renewable fuels standard in the USA would create 300 000 additional jobs by 2016; use of renewable fuels in motor fuel would increase net farm income by $6.6 billion over the next 15 years; the deficit would reduce by $63.4 billion by 2016 by replacing crude oil imports and renewable fuel use would replace nearly 2.9 billion barrels of crude oil between 2002 and 2016. Also in the USA, the Renewable Fuel Association said that a tripling of domestic ethanol consumption could replace Iraqi oil imports, for example, and make the USA less dependent on foreign oil imports. The USA consumed about 1 million barrels/day of Iraqi oil in January 2002, and the USA produced 1.77 billion litres of ethanol in 2001. The ethanol industry expects the demand for clean energy and energy security to enhance continued future growth for ethanol demand and supply. The US General Accounting Office indicated that prices could remain stable and production would meet demand until 2005 based on the ethanol industry's projections, which include planned new ethanol capacity, assuming new capacity comes online as planned. Bioalcohol is now the third largest market for US corn, after animal feed and corn destined for export. More than 16.5 million tonnes of corn are converted to ethanol annually, which provided as much as $3 billion in additional income to farmers in 2000.

The future of bioalcohol fuels is uncertain and relies to a great extent on the willingness of governments to favor these biofuels over fossil-derived fuels. Ethanol still costs more to produce than gasoline, but the cost to produce it has dropped substantially over the past 15 years. In August 2002, a USDA report found that increased corn yields and technological efficiencies in how corn is converted into ethanol in recent years has made ethanol production much more efficient than in the past. The report states that ethanol production now yields 34% more energy than is used in growing and processing corn. It is believed that biotechnology can be used to dramatically reduce ethanol production costs even more by using low-cost biomass feedstocks such as agricultural and forest residues, prairie grass, kudzu, waste wood, used paper products, and even urban waste.

In addition, researchers are working to develop a collection of new biocatalysts, i.e., genetically engineered enzymes, yeasts, and bacteria, that may make it possible to produce biomass ethanol from just about any plant, tree, or agricultural waste. This would dramatically reduce the cost to produce ethanol because the industry would no longer have to rely solely on corn as a feedstock.

In summary, a comparison can be made between today's ethanol industry and the petroleum industry of a century ago when it first "cracked" the petroleum molecule. That breakthrough allowed more petroleum to be used to produce high-performance gasoline, thereby reducing waste and the cost. Likewise, new biocatalysts and processing techniques are being developed that will allow ever more diverse plant products, such as lawn clippings or wood waste from sawmills, to be used to make renewable fuels. Fast-growing switchgrass, which was once common on the American plains, is also being considered as a bioethanol feedstock.

Biodiesel

Vegetable oils have been used as liquid fuels since the earliest days of the internal combustion engine. Indeed, only 5 years after Rudolph Diesel patented his engine, it was demonstrated at the 1900 world exhibition in Paris using peanut (*Arachis hypogaea*) oil as a fuel. Some of the properties of vegetable oils, which make them attractive as environmentally friendly alternative lubricants and fuels, include high biodegradability, low toxicity, low evaporation (reducing risk of inhalation), high flashpoint (reducing risk of fire), and reduced emissions (e.g., of CO_2, sulfur dioxides, soot, or aromatic compounds). It is obvious that renewable oils from oilseed crops can never replace the volume of fuel that is currently derived from mineral oil and is the basis of much of our electricity generation and transport fuels. Indeed, it has been calculated that even if we utilized the entire cultivable landmass of the earth for oilseed crops, the resulting biofuel could only supply 10–20% of our current energy requirements. Nevertheless, there is probably scope for the use of significant quantities of biodiesel fuels for transportation in certain environmentally sensitive areas, such as city centers and inland waterways. The technical specifications of fossil-derived diesel and biodiesel fuels are compared in **Table 1**.

Vegetable oils cannot be used directly in the long-term running of diesel engines. The presence of natural impurities and a high viscosity eventually results in clogging of the injectors and a reduction in engine efficiency. Conversion of the oils to their methyl ester derivatives produces a fuel, which can be mixed with either mineral diesel or diluted with unmodified vegetable oil, which is suitable for running normal diesel engines. Biodiesel, either in pure form, or mixed with mineral diesel, is now available commercially in most of western Europe, where current production capacity is in excess of 1.1 million tonnes per year.

Like bioalcohol production, the economics of biodiesel production are less favorable than those of fossil fuels. Therefore, current European and US production of biodiesel relies on high levels of subsidy driven agriculture and the availability of special government incentives to encourage its use. It is generally admitted that biodiesel prices will not be competitive with mineral diesel on a simple cost basis for the foreseeable future. Nevertheless, it has recently been estimated that a biodiesel plant capable of producing 100 000 tonnes per year could do so at a cost of about US 31 cents per liter. This assumes that the glycerol byproduct of methyl ester production can be sold for about US$1700 per tonne. There has been much scepticism about whether the large quantities of extra glycerol produced as a result of biodiesel manufacture could be absorbed by world markets. However, it now appears that, given the relatively buoyant state of glycerol and glycerine markets, such concerns may not be as serious as first thought.

Table 1 Comparison of fossil fuel and biodiesel fuel specifications[a]

Property	*Diesel*	*Biodiesel*
Carbon chain length (no. of C atoms)	11–20	16–18
Density at 15°C (g cm^{-3})	0.82–0.86	0.875–0.885
Viscosity at 20°C (mm^2 s^{-1})	3.0–8.0	6.0–9.0
Flashpoint (°C)	> 55	> 55
CFPP[b] summer (°C)	> +5	> −5
CFPP[b] winter (°C)	−20	> −15
Calorific value (MJ kg^{-1})	42.9	37.2
(MJ dm^{-3})	35.6	32.9
Efficiency at 1200 r.p.m.[c]	38.2	40.7

[a]Compiled from Körbitz W (1995) Utilisation of oil as a biodiesel fuel. In: Kimber D and Magregor I (eds) *Brassica Oilseeds, Production and Utilization*, pp. 353–372. Boca Raton: CAB International.

[b]CFPP, cold-filter plugging point.

[c]Efficiency expressed as the ratio of energy input (fuel) to energy output (engine performance).

It is probable that, in North America and Europe, biodiesel can be produced from conventional oilseed crops such as soybean (*Glycine max*), oilseed rape (*Brassica napus*), and sunflower (*Helianthus* spp.), at a cost of US 30–40 cents per liter (US$1.40–1.80 per gallon pretax). Although this cannot compete with mineral diesel, the environmental advantages of biodiesel may outweigh its higher cost where factors such as biodegradability are important. These factors may become more important if diesel users are charged a price based upon a complete environmental audit of the consequences of the use of such fuels. For example, the costs of environmental remediation in inland waterways, which is currently borne by local authorities, could be passed on to diesel users, possibly in the form of a fossil fuel tax. Such measures are now under active consideration in European legislatures. The resulting increases in the price of mineral diesel fuels may well render biodiesel economical for such niche markets.

Another area of interest for biodiesel is within urban areas, where it has been claimed that it gives rise to significantly decreased particulate emissions compared to mineral diesel fuels. Numerous bus companies in Europe are now conducting trials with biodiesel. The biodiesel industry in Europe has developed out of subsidy driven oilseed production and surplus land capacity. This, together with government support in several countries such as France and Austria, has resulted in the setting up of a considerable biodiesel production capacity. This capacity may be expanded in the coming years if demand for biodiesel rises due to fossil fuel taxes based on environmental audit data. Many of these factors are also relevant in oilseed producing areas of North America, although the high capital cost of converting existing plant or building new plant for biodiesel production may hinder its development in the absence of active government support. Nevertheless, by 2002, over 100 national vehicle fleets in the USA had committed themselves to using biodiesel as part of a government (Environmental Protection Agency) act that gives them special fuel credits to offset the higher cost of biodiesel fuel. The importance of proactive government policies to promote biodiesel can be seen by comparing the uptake of biodiesel in different European countries. In countries with more interventionist governments, e.g., France and Austria, biodiesel production is strongest. In contrast, in countries with more *laissez faire* policies, such as the United Kingdom, relatively little biodiesel is produced or consumed.

Although there is undoubtedly a niche market for biodiesel, the long-term future of the product must remain uncertain in view of the ever-changing politico-economic situation regarding oil crops in developed countries. The levels and even the existence of crop subsidies, set-aside payments, and fossil fuel taxes are all in doubt. At the same time, genetic engineering coupled with advanced breeding methods are now beginning to allow us to produce novel and potentially valuable semirefined industrial oils from crops. These products will compete effectively with biodiesel once their industrial markets are fully established. Looking even further ahead to the eventual exhaustion of fossil derived hydrocarbons, it is more likely that we will tend to husband our valuable renewable oils for use as lubricants and organic synthols, rather than burn them as fuels (**Table 1**).

List of Technical Nomenclature

Biodiesel A hydrocarbon fuel, normally methyl esters, derived from oil crops that is used in diesel engines.

Bioethanol An ethanol fuel derived from starchy crops, such as sugar cane, that is normally used in vehicle engines as a blend with gasoline.

See also: **Energy Crops**: Biomass Production. **Primary Products**: Oils.

Further Reading

Croy C and Dotson K (1995) Glycerine. *Inform* 6: 1104–1118.

DiPardo J (2000) *Outlook for Biomass Ethanol Production and Demand*, Energy Information Administration, Washington, DC, last updated April 26, 2000. http://www.eia.doe.gov/oiaf/analysispaper/biomass.html

EU proposals to encourage alternative fuels (Nov 2001) http://europa.eu.int/comm/taxation_customs/publications/official_doc/IP/ipfuel.pdf

Körbitz W (1995) Utilisation of oil as a biodiesel fuel. In: Kimber D and Magregor I (eds) *Brassica Oilseeds, Production and Utilization*, pp. 353–372. Boca Raton: CAB International.

Lugar RG and Woolsey J (1999) The new petroleum. *Foreign Affairs* 78: 88–102.

Murphy D (1994) Future perspectives for oil crops. In: Murphy DJ (ed.) *Designer Oil Crops*, pp. 297–310. Weinheim: VCH Press.

National Renewable Energy Laboratory (2002) *Fuelling the Environment: Switchgrass Review.* Biofuels News 1: 3–4. http://www.ott.doe.gov/biofuels/pdfs/fall98.pdf

Sanders GS and Turner K (1995) Is biodiesel production economically viable in the UK? *Lipid Technology* March: 29–33.

The National Corn Growers' Association (2002) *US Corn Usage Statistics.* http://www.ncga.com

Urbanchuk JM (2001) *An Economic Analysis of Legislation for a Renewable Fuels Requirement for Highway Motor Fuels*, AUS Consultants. http://www.ncga.com/ethanol/pdfs/Urbanchuk_Final_Report.pdf

USDA (2002) *USDA Report Finds Ethanol is Energy Efficient*, United States Department of Agriculture press release, Aug 1, 2002. http://www.usda.gov/news/releases/2002/08/0322.htm

Biomass Production

E O' Connor, Bio-Renewables Ltd, Ely, UK

Introduction

Plant biomass is the weight of plant dry matter production dried to a constant moisture. It is usually measured on a plant or unit of land basis. Total biomass of a plant is closely related to the capacity for carbon assimilation of the plant. About 90% of biomass is derived from the incorporation of carbon into organic compounds via photosynthesis. It is estimated that 200 billion tonnes of CO_2 is converted to biomass annually; 40% by marine plankton and 60% by higher plants. Photosynthesis occurs in the leaves of plants, and any other green tissues such as stems. The final biomass of a plant is an interaction of gene expression and environmental conditions that occur throughout the growing season. Plants rarely reach their full biomass potential. The environmental limits to biomass accumulation come into effect before genetic limits by affecting photosynthesis. This article reviews the environmental regulation of biomass accumulation. First, the major environmental influences, radiation (sunlight), temperature and water, are discussed. Then, the effect of these factors on biomass accumulation throughout the life cycle of plant growth is reviewed. Lastly, crops grown specifically for biomass are considered.

The Main Environmental Factors Governing Biomass Accumulation

The Köppen Climate Classification System divides the world into five climates based on annual and monthly temperature and precipitation patterns, e.g., tropical moist climates and moist mid-latitude climates with cold winters. These five climates generally coincide with world patterns of vegetation and soil types. Thus, we have regions such as tundra, steppe, savannah, and rainforest. A climate region provides a certain growing season characterized by incident radiation, rainfall, and temperature. Each region has a distinct profile of flora associated with it. How do these plants match their pattern of growth and reproduction to the changing season to ensure survival? They use cues from the changing environment as signals to regulate growth and development, i.e., they are adapted to a particular region. All plants have a restricted range of adaptation, although this range may be large.

Solar Radiation

Carbon is assimilated using solar radiation energy (sunlight). The amount of incident solar radiation varies from place to place and is mainly influenced by latitude and season. While tropical regions, close to the equator, receive radiation in a relatively consistent manner throughout the year, temperate regions, in the higher latitudes, experience changes in intensity and daily duration of solar radiation throughout the year. Temperate countries receive maximum solar radiation intensity and duration in May/June (about $1000\,\mathrm{W\,m^{-2}}$ and up to 18 h of daylength) and minimum in December/November (as low as $200\,\mathrm{W\,m^{-2}}$ and about 7 h daylength). More regional influences, such as altitude and increased cloud cover in maritime regions, also modify the amount of solar radiation incident on an area. Plants do not exploit the full radiant energy received. Plants absorb energy at certain wavelengths, and only 40% of solar radiation corresponds to the absorption spectrum of chlorophyll (400–700 nm). Therefore, only 40% of solar radiation is photosynthetically active radiation (PAR).

Temperature

Temperature affects biomass productivity by controlling the initiation and rate of many growth and development processes. Mean annual and mean monthly temperature values delimit the range of a species. The length of the growing season for a crop is defined as the number of days that mean daily temperatures remain above a species dependent minimum level for growth. This may be most of the year in tropical regions. Plants also respond to daily fluctuations in temperature. Photosynthesis is one of the most temperature sensitive aspects of growth, and both low and high temperatures can limit plant growth. In temperate zones, the maximal rates of photosynthesis occur between 20 and 30°C. Species from warmer climates tend to have higher temperature optima for maximum photosynthetic rate. Reduced temperature, below the optimum range for a particular species, is often accompanied by a reduction in leaf expansion rates.

Species in temperate regions have adapted to seasonal changes in temperature and respond to temperature variation to ensure that growth and reproduction occurs at an appropriate time to ensure survival. Often, thermal time is important, rather than chronological time or temperature separately, and determines the physiological age of a plant. For plants that respond to thermal time as a growth signal there is a base temperature above which physiological ageing occurs for that species. Thermal time is measured in accumulated day-degrees, calculated by using the following formula:

$$\text{No. of days with mean temperature above base (species dependent)} \times (\text{mean temperature} - \text{base temperature}) = \text{day-degrees}$$

As an example, if a mean daily temperature of 10°C occurs for 50 days and the temperature above which physiological ageing occurs for a particular species has been determined as 4°C, then the accumulated day-degrees during the 50-day period is 300 day-degrees ($50 \times (10 - 4)$). Temperate plants, growing where there are large variations in summer and winter temperatures, are more adapted to seasonal temperature changes than tropical plants, which are more adapted to seasonal rainfall changes. For many species, thermal time governs the length of the growing season by influencing germination or dormancy break, the initiation of new leaves, and controls the change from the vegetative to the reproductive stage of growth.

Water Availability

Actively growing nonwoody species consist of 80–95% water (woody species, which are comprised of mainly dead cells, contain less water). There is a constant turnover of water through a plant, crop, or ecosystem as water is taken up by roots and transpired through leaves. Water availability is determined by the water content of the soil, which in turn is determined by the level of rainfall/precipitation of the region. Flooding, natural or by design, or irrigation can be used to exploit precipitation occurring at different times or in other regions. The productivity of different ecosystems is tightly linked to annual precipitation levels. As the major cytoplasmic constituent, water is the medium for all cellular reactions and is involved in reactions requiring the acceptance or donation of hydrogen. The transpiration of water from pores on the leaves (stomata) dissipates excess radiation energy and prevents the plant from overheating. Plants possess a number of strategies to cope with limited water supply that allow survival but interfere with biomass productivity. Under water limited conditions, plants can reduce transpiration from the leaves by reducing the pore area of the leaf stomata (stomatal control). However, reducing the stomatal aperture also restricts the diffusion of CO_2 into the leaf and, therefore, limits photosynthesis. Plants constantly adjust stomatal aperture to maximize photosynthesis and minimize transpiration. Water use efficiency (WUE) is a measure of the amount of water transpired (liters) for weight of biomass accumulated (kg).

While radiation, temperature, and water availability are the main environmental determinants, many other environmental factors influence biomass accumulation. Low nutrient availability, particularly low nitrogen availability, due to soil type or fertilization regime, retards growth. Some elements, such as copper and zinc, are essential for normal growth, but become toxic at high levels. High saline content in the soil can also limit growth.

Environmental Regulation of Biomass Accumulation throughout the Plant Life Cycle

In the 1940s, D. J. Watson determined that, in temperate regions, variation in solar interception accounts for most of the differences in crop yields that occur between sites and seasons. Since then, increasing knowledge of the effect of environmental parameters on yield, and the advance of computer technology, has resulted in the development of productivity models for many crop species and ecosystems. These models are used to predict yield for a particular area, and to determine the circumstances under which intervention, such as irrigation, would be cost effective. These models are becoming increasingly refined. Yet, fundamentally, biomass accumulation is governed by components that affect the amount of solar radiation intercepted, and the efficiency at which intercepted radiation is used to incorporate carbon:

$$\begin{aligned}\text{Biomass} = {} & \text{incident solar radiation} \\ & \times \text{proportion of the solar radiation intercepted} \\ & \times \text{net assimilation rate}\end{aligned}$$

Modeling productivity is dependent on the input of data that affect these three components. Incident solar radiation is dependent on location. The interception of solar radiation refers to accumulated interception throughout the growing season. This component can be altered as a result of an increase in the growing season, leaf area, leaf area duration, or daily duration of photosynthesis. The net assimilation rate is a measure of the efficiency with which PAR is used to increase dry matter accumulation. While beyond the remit of this article, readers should note that three types of photosynthetic pathways exist, C_3, C_4, and CAM, and the three pathways vary greatly in net assimilation rates in different environments. Generally, only about 5% of the energy reaching the surface of a leaf is available for biomass accumulation, 60% of the radiation energy occurs outside the photosynthetically active wavelength range, about 16% is lost due to reflection and heat dissipation, and about 19% of the energy captured is used by the plant in current metabolism, including respiration. Typical assimilation efficiencies (UK) for cereals and potatoes are 2.0% and 2.4%, respectively.

In a living system (plant, crop, ecosystem), radiation, temperature, and water interact to control growth and development. To provide a better understanding of the influence of these factors on biomass, at the different plant growth and developmental stages, the next section discusses the influence of these factors on biomass accumulation throughout the life cycle of plants. Particular reference is made to agricultural crops.

Germination

Seeds of most species, especially in temperate regions, undergo a period of dormancy, during which they can withstand extremes of desiccation, temperature, and light without losing viability. The timing of germination determines the beginning of the growth season. In temperate regions, germination is usually closely regulated by environmental factors to ensure that germination occurs at a time when survival of the seedling is likely. Water availability is crucial to germination of all seeds as they imbibe a lot of water, relative to their size, during the initial stages of germination, which restores the water content from about 10% in the dormant seed to the 80–95% level found in actively respiring (energy using) tissue. Temperature can effect germination in two ways: first, to break seed dormancy, and then to control initiation and rate of ensuing germination. A day-degree requirement to signal that spring has arrived involves a temperature and time interaction. Warm autumn temperatures could result in germination in the same year as seed production. To prevent germination as winter approaches, seeds of some species may need to experience exposure to an extensive period of low temperatures, stratification, before germination will commence. Subsequent to the breaking of dormancy, a base temperature controls initiation and the rate of germination (a linear response between the minimum and optimum temperature). Light is used less frequently to control germination, but, when used, can be either inhibitory or inducing. When inhibitory, germination occurs when the seeds are buried, ensuring good contact with soil. When light dependent, the germination of one generation of seed is spread out over a long period, as buried seeds can remain dormant until exposed to light.

Commercial plant production often employs techniques that mimic the environmental controls of germination to break dormancy, speed germination, and provide uniform emergence. Earlier germination using seed priming, or breeding to modify the germination response to environmental factors, provides a route to lengthen the growing season and increase biomass yield. However, this route is limited by exposure of the crop to adverse temperatures early in the season, which could damage long-term productivity or even kill the crop. Earlier germination must be matched with altered physiology that can cope with adverse climatic circumstances or with protection of the seedlings, under glass, for example. Some crops are amenable to early germination in protected environments followed by modular planting. Seed trays with individual sections for each seed are used. The resulting "plugs," containing seedlings with two or more leaves, are then mechanically planted in the field when climatic conditions for continued growth have been reached. Environmental factors also affect the initiation of spring growth in biennial and perennial crops.

Vegetative Growth

In the vegetative growth stage, above-ground biomass accumulation is due to leaf and stem production. The formation of new leaves and their longevity is a major factor in the fraction of solar radiation intercepted. The leaf area index (LAI) is the leaf area (one side) per unit of ground area, $m^2\ m^{-2}$ or dimensionless. The critical leaf area of a crop (LAIc) occurs when the crop reaches a LAI that intercepts all the available radiation. Therefore, reaching LAIc, or canopy closure, as quickly as possible increases the integrated radiation interception for the season. To maximize biomass yield, seasonal LAI is matched with solar radiation variation.

Temperature and time interact to control the phyllochron interval, i.e., the interval between successive leaf initiation. Once initiated, leaf growth occurs due to cell division and cell expansion. Leaf elongation rate is dependent on temperature. Each species has a minimum temperature below which growth does not occur. Generally, this minimum temperature for growth is lower for temperate species (generally 4–6°C) than for tropical species (about 10°C). Early season biomass accumulation is limited primarily by temperature and, often, in the spring and early summer, interception opportunity is lost as the preceding weeks were not warm enough to enable canopy development. Leaf growth is extremely sensitive to water stress. Carbon assimilation is decreased when stomatal area is reduced in response to water deficits. Even mild water stress early in the season can delay canopy closure.

Plants grow in continuously changing environments and function within certain environmental ranges by adjusting their metabolism with no long-term damage. These ranges are particular for each species and reflect their place of origin. Extreme environmental events, outside the normal coping range, induce stress. The effect of environmental stress on a plant is dependent on its stage of growth and the nature of the stress. The consequences for biomass accumulation can include severely reduced productivity during the stress, damage to the plant so that accumulation rates remain low after environmental conditions have returned to within the normal range, and may even lead to plant death. Temperatures above the normal range can cause excess water loss as transpiration is used to stabilize leaf temperature. Freezing temperatures result in cell death as water in the plant expands as it freezes. High irradiences, at intensities above those that can be used in photosynthesis, can damage the photosynthetic system, especially when combined with other stresses, such as below normal temperatures.

Flowering

A major change in plant development is the switch from leaf production to the formation of floral organs. Some stimulus induces the vegetative meristem(s) to become a reproductive meristem. Plants use environmental signals, such as temperature and light, to ensure that flowering occurs at a time that allows seed development at an appropriate time in the season, before the onset of adverse conditions such as frost or drought. Biomass accumulation responses to environmental factors, such as temperature, are different in the vegetative growth phase and reproductive growth phase, so that two separate functions are needed to model accurately total biomass accumulation in the pre- and postflowering stages. Flowering also affects biomass accumulation in regard to the length of the growing season as, in many species, the onset of senescence follows seed production.

Many species use photoperiod or daylength as the signal to induce flowering, as changing daylength throughout the year is directly related to season and, unlike temperature, is not susceptible to unseasonable aberrations. Plants with a vernalization requirement will remain in the vegetative state if not exposed to low temperatures for a minimum period during the vegetative growth stage. Flowering response to photoperiodism and temperature is an adaptation to a specific location, and, thus, displays intraspecies variation as well as interspecies variation in the photoperiods, temperatures, and time involved. During domestication, temperature and photoperiod responses in crop species have been modified to allow them to be grown over a wider geographical range, but maintained to match growth patterns to seasonal variables, and to ensure that the crop flowers and matures in a uniform manner. Not all agricultural crops flower or are allowed to flower, e.g., sugar beet (*Beta vulgaris*). Sugar beet is a biennial species. In the vegetative phase, large reserves are stored in the modified roots to support flowering and seed growth in the next season; the economic harvest of sugar beet is comprised of these reserves. Flowering would occur in the following spring after exposure to cool winter weather that had induced a change to the reproductive phase. Therefore, the crop is allowed just one season before it is harvested in the vegetative stage.

Senescence

Plant senescence occurs at the end of the growing season and, in relation to solar interception, it is the reverse of crop canopy closure discussed above as it involves the loss of photosynthetic capacity of the leaves. During senescence, photosynthetic capacity is progressively lost until all green area is gone and at this stage the growing season is over. Senescence also involves the redistribution of biomass. Senescence results in death for annual plants, and biomass reserves are mobilized to the seeds for use by the next generation. For perennials and biennials, biomass reserves are redistributed to the overwintering structures for reuse the following spring. Leaf senescence is a complex developmental process that is under hormonal control, but is also influenced by nutritional status and environmental factors such as temperature and daylength. Delaying leaf senescence to increase the duration of photosynthesis offers an opportunity of increasing the total amount of carbon accumulated by a crop. Green area duration can be increased by delaying the onset of senescence or modifying the rate at which senescence occurs. Delaying the senescence of the assimilatory system by only a few days can have a big effect on yield. However, in a multicropping system, there is little chance of drastically extending the growing season as adverse environmental factors occur on either side of the season, and a delayed harvest date will impact on subsequent crops.

Biomass Crops

Carbon dioxide (CO_2) is a greenhouse gas that promotes global warming; rising atmospheric levels are mainly due to the release of carbon into the atmosphere when fossil fuels are used as an energy source. Because of the close association between plant biomass accumulation, and carbon assimilation from the atmosphere, there has been increasing emphasis on exploiting plant biomass to control atmospheric CO_2.

Plant biomass can be used as a carbon sink. This involves afforestation of vast tracts of land not needed for food production. Carbon assimilated during the accumulation of biomass would be locked into the standing crop. Forests are favored for this purpose as, due to their large biomass, they can lock up large amounts of carbon, about 600 tonnes of carbon per hectare in a mature forest. While certainly having an immediate effect, this approach is limited in its long-term effect as the biomass, and hence carbon status, of a mature forest community reaches a plateau when death and regeneration of the system reaches equilibrium.

The stored carbon in plant biomass represents an energy source, and this provides another route for using plants to control atmospheric CO_2; by replacing the use of fossil fuels with crops for the production of energy. During photosynthesis, both energy and carbon are stored in carbon rich compounds such as sugars, oils, or ligno-cellulose (cell wall material). Combustion, thermochemical (e.g., gasification), biochemical (e.g., fermentation), or chemical (e.g., catalyst conversion) processes are used to release the stored energy in a usable form such as heat, gas, electricity, or transport fuel.

Biomass accounts for the largest supply of nonfossil fuel energy providing 13% of global primary energy demand; mostly for cooking and heat in developing countries. Biomass energy can be derived from agricultural and forestry residues or from crops grown specifically for this purpose, i.e., dedicated energy crops. The carbon released when a biomass crop is used equates with the carbon assimilated throughout the growing season, and will be matched by carbon uptake in the subsequent season. Therefore, in this respect, energy production from biomass cropping is considered carbon neutral and, unlike fossil fuels, is a renewable source of energy. However, the energy ratio (the energy output of the crop/energy input into establishing, growing, harvesting, and transport of the crop) must be kept positive to maximize the environmental benefits. For example, the application of nitrogen to a crop uses energy in the production of the fertilizer and in the application of the fertilizer. It also results in emissions of NO_2, another greenhouse gas, into the atmosphere.

The profile of a suitable crop for biomass production has been identified as: perennial growth, high net assimilation rate, high dry matter yields, efficient use of water and nutrients, and good pest and disease resistance. In addition, the harvested material must have good fuel characteristics such as high energy content and low moisture content. The increased urgency for biomass energy crops to check rising atmospheric CO_2 levels has resulted in the investigation of different species for their ability to meet these criteria. Globally, a diverse array of crop species is grown for biomass energy generation. Examples of species used include fast-growing trees, such as poplar (*Populus* spp.), willow (*Salix* spp.) and *Eucalyptus* spp., and perennial grass species, such as sugar cane (*Saccharum officinarum*) and *Miscanthus*. The grass species are harvested annually. Trees that have the ability to grow back after being cut close to the ground may be harvested every 3–8 years, i.e., coppice. Biomass crops may consist of native species traditionally used for energy production, e.g., willow coppicing in Europe. Biomass crop development may involve a new end-use for established crops, e.g., the development of bioethanol production from sugar cane in Brazil, or may involve the introduction of a species to a new region, e.g., *Miscanthus* from Southeast Asia grown in Europe.

The selection of an appropriate biomass crop for a particular region consists of identifying species that obtain the highest harvest of energy (fixed carbon) with the lowest possible energy inputs. Energy contents of dry plant biomass generally range between 17 and 19 GJ tonne^{-1} compared to about 24 GJ tonne^{-1} for coal and 45 GJ tonne^{-1} for crude oil. High biomass yields are achieved by using fast-growing species and matching crop characteristics

Figure 1 *Miscanthus*, a C_4 perennial grass native to Southeast Asia, growing as a biomass crop in the UK. Courtesy of Bio-Renewables Ltd.

with environmental factors governing biomass accumulation to maximize annual accumulation (see above). Both grasses given as examples have the efficient C_4 carbon assimilation pathway. Low energy input into a biomass cropping system is achieved by selecting plants that manage resources efficiently. The perennial growth habit has a number of advantages over annual establishment for positive energy ratios:

- It removes the need for annual establishment inputs.
- Reduced fertilizer input is achieved by translocation of nutrients from the harvested material to the roots or woody tissues during senescence for recycling in the following spring.
- Nutrient recycling also produces a clean fuel, which is low in elements that produce acidifying gases (causing acid rain), e.g., chlorine, or greenhouse gases, e.g., nitrogen, when processed.
- A perennial growth habit also allows deep rooting systems to be developed for efficient water and nutrient uptake.
- The no-till management and deep rooting system results in increased carbon sequestration in the soil.

Disease and pest resistance reduces the need for pesticide inputs. Other characteristics that will bestow an advantage and make one species more suitable than another are rapid growth in spring, which increases canopy duration and out-competes weeds (reduced herbicide input), and natural drying in the field.

Miscanthus, a C_4 perennial grass native to Southeast Asia, growing as a biomass crop in the UK is illustrated in **Figure 1**.

Summary

Owing to their stationary nature, growing plants have evolved strategies for coping with a constantly changing environment by adjusting their metabolism. Within given parameters, specific for each species, plants have adapted to their environment to ensure survival, not for maximum biomass accumulation. In this article, the three main factors of solar radiation, temperature, and water relations are discussed. These three factors interact and alter the photosynthesis, respiration, and transpiration rates, which control the energy balance of the plant and determine the prevailing net carbon accumulation rate. Within crop species, current crop varieties have increased annual intercepted solar radiation (achieved via matching LAI with seasonal parameters or extending green area duration) when compared to older varieties. Modification of net assimilation rate per unit leaf area appears harder to achieve. This may, in part, be due to the fact that the net assimilation rate is the conclusion of the very complex metabolic pathway between light capture, carbon assimilation, and carbon accumulation. It is very difficult to assess in field situations due to the dynamic response to the environment and compensatory processes that occur between the different components of the system.

Biomass crops, along with other renewable sources of energy such as wind, offer the potential to slow down rising atmospheric CO_2 levels by replacing the use of fossil fuel. Biomass crops also offer some protection against fluctuating world oil prices, and decrease reliance and expenditure on imported energy sources, particularly for countries that do not have a fossil fuel resource. Currently, these issues are the main driving force in the development of biomass renewable energy production as many technical, economic, and environmental constraints apply. Understanding the physiological basis of plant biomass accumulation allows one to select an appropriate biomass crop for a given region, and indicates what direction crop improvements should take to improve energy ratios and profitability.

List of Technical Nomenclature

Leaf area index (LAI)	The leaf area (one side) per unit of ground area, $m^2\ m^{-2}$ or dimensionless.
Photoperiod	Day length duration.
Plant biomass	The weight of plant dry matter production dried to a constant moisture.
Solar radiation	Photon energy from the sun.
Thermal time	Interaction of temperature and time to measure physiological ageing in plants.
Water use efficiency (WUE)	A measure of the amount of water transpired (l) for weight of biomass accumulated (kg).

See also: **Abiotic Stresses**: Cold Stress. **Growth and Development**: Growth Analysis, Crops. **Photosynthesis and Partitioning**: C_3 Plants; C_4 Plants; CAM Plants; Photoinhibition. **Postharvest Physiology**: Senescence, Leaves. **Regulators of Growth**: Photomorphogenesis; Vernalization. **Seed Development**: Germination. **Water Relations of Plants**: Drought Stress; Uptake, Loss and Control.

Further Reading

Alexander AG (1985) *The Energy Cane Alternative*. Oxford: Elsevier.

Jones HG (1992) *Plants and Microclimate: A Quantitative Approach to Environmental Plant Physiology.* Cambridge: Cambridge University Press.
Monteith JL (1977) Climate and efficiency of crop production in Britain. *Philosophical Transactions of the Royal Society of London* 281: 277–294.
Speller CS (1993) The potential of growing biomass crops for fuels on surplus land in the UK. *Outlook on Agriculture* 22: 23–29.
Steinmans SJ, Prather TS, and Holt JS (2000) Estimation of base temperatures for nine weed species. *Journal of Experimental Botany* 51: 275–286.
Taiz L and Zeiger E (1991) *Plant Physiology*, 2nd edn. Sunderland, MA: Sinauer Associates/Benjamin.
Watson DJ (1952) The physiological basis of variation in yield. *Advances in Agronomy* 4: 101–145.
Zelitch I (1982) The close relationship between net photosynthesis and crop yield. *BioScience* 32(10): 796–802.

ETHICS AND BIOSAFETY

Contents

Development and Commercialization of Genetically Modified Plants

P W B Phillips, University of Saskatchewan, Saskatoon, SK, Canada

Introduction

The advent of modern biotechnology has engendered a wide ranging and rambunctious debate about the sources, uses, and purposes of the food we eat. Food plays a unique role in society. While it has become a global commodity, it is a core part of local human societies and how they portray themselves. As such, recent efforts to change what foods we produce, how we produce them, and where we eat them challenge many social norms. Hence, while many scientists view genetic modification – the transfer of genes between species through genetic engineering – as a relatively useful and benign technology, many in society see it as a threat.

Ultimately the debate about biotechnology and its uses in the global agrifood world is more than about whether it is useful or not. Rather, it is a debate about the values we want to underpin our economy and society. Ethicist Paul Thompson has posited that ultimately

> ethical judgments are not stand-alone judgments; rather they are all-things considered judgments that take into account a multitude of values.

In determining the ethics of biotechnology, one must consider: the distribution of winners and losers, who owns and controls the technology; food safety and the consumer acceptance; and the impact on the environment. Ultimately, individuals and the state must debate and balance these considerations to find an ethical position.

What Is Biotechnology?

Three different definitions of biotechnology, as applied to plants, are used, depending on who is involved in the debate. At its broadest, plant biotechnology includes all efforts over the past 6000 years to select seeds in order to improve yield or quality or to reduce susceptibility to diseases or environmental risks. Through traditional selection and breeding, plant scientists have created highly efficient and effective plant products that form the basis for the world's food supply. Traditional breeding, however, is limited because it ultimately depends on selecting for traits at the level of the whole organism level and attempting natural reproduction, which can occur only within or between close species. The fundamental difference between traditional and modern biotechnology is that the new technologies allow breeders to work at the molecular level. Many date the beginning of modern biotechnology from 1970, when for the first time DNA, often called the "blueprint of life," was moved between unrelated organisms. A large number of the techniques since developed – such as maps of the genome and marker-assisted breeding – are used by conventional plant breeders to selectively enhance their breeding efforts. Transgenic biotechnology, the third and most controversial approach, involves technologies that enable scientists to isolate, multiply, insert, and activate genes from unrelated species, thereby changing the molecular makeup of the host organism. This last is what is commonly understood by the term genetically modified (GM) as applied to plants.

How Is Biotechnology Being Used?

Although all the foods we eat have been modified from their initial state, people are especially interested in those that have been transgenically modified to incorporate one or more genes from another species. Between 1986 and 1997 biotechnology was used to genetically modify 60 crops for 10 different classes of traits. During that period, more than 25 000 field trials were conducted in over 45 countries. As of 2002, 15 crops modified for one or more of 47 phenotypic traits were commercialized, most with attributes related to input and yield performance. Insect resistance (e.g., *Bacillus thurengensis* in cotton (*Gossypium hirsutum*) and corn (*Zea mays*, maize)) and herbicide tolerance (e.g., Roundup Ready^{TM} soybeans (*Glycine max*), corn, and canola (*Brassica napus*)), as single or stacked constructs, accounted for 99% of the GM world acreage in 2001. Plants have also been engineered to resist viruses (e.g., NewLeaf^{TM} viral resistant potato (*Solanum tuberosum*)) and to be sterile (e.g., the Invigor^{TM} hybrid system for canola). Several output trait crops have also been developed and commercialized, including delayed ripening tomatoes (*Lycopersicon esculentum*) (i.e., FlavrSavr^{TM}), new or modified oils (e.g., Laurical^{TM} canola), and blue carnations (*Dianthus caryophyllus*).

Over the first 7 years of commercial cultivation, those 15 crops were planted on >180 million ha. Four main crops – soybeans, corn, cotton, and canola – made up 99% of the total acreage planted to GM crops in 2001. Overall, the market value of GM seeds has risen from US$1 million in 1995 to an estimated US$3 billion in 2001. This production has been highly concentrated. Monsanto is estimated to have a market share of between 88% and 91%, with Bayer and Novartis holding the remainder of the market. Although not as dramatic, production is also geographically concentrated. While 16 countries have produced one or more transgenic crops, the United States, Argentina, Canada, and China accounted for 99% of the acreage over the first 7 years of production. In 2001, 46% of the global soybean acreage used GM seeds, as did 20% of global cotton, 11% of global canola, and 7% of global corn production. By 2001, there was at least one commercial introduction on each continent and in all of the major agrifood producing and/or exporting countries in the world.

Many question how the industry will evolve. Industry optimists see great potential for the science and the market. Corporate and public researchers are already working on crops to improve tolerance to abiotic stresses (e.g., heat), to enhance nutritional or functional attributes (e.g., GoldenRice^{TM}), to express nutraceutical or pharmaceutical proteins and enzymes, and to use in the phytoremediation of contaminated soils. If the science and markets advance as hoped, market watchers have speculated that the value of GM seed sales could double by 2005 and reach US$25 billion by 2010. The optimists note that their forecasts of sustained growth were made more likely with approvals in 2001–2 of GM cotton in China, India, and Indonesia and GM corn and soybeans in South Africa.

The Ethical Dimension

This new technology raises a host of ethical considerations. Individuals and groups expecting to gain from the technology are often strong proponents, while those expecting to lose demand social justice. While this cost–benefit approach offers some insights, it ignores issues that have nothing to do with material return. Many of the concerns being raised involve values that cannot be handled in the context of tradeoffs. Some citizens are simply angry that they do not have a choice about whether or not to consume GM food products because of the absence of effective labeling systems while others are concerned about the private ownership of these technologies and of "life" itself through gene patents and whole organism patents. Still others worry that the increasingly global agrifood industry will accelerate globalization and further erode national sovereignty, limiting domestic choices for social and economic development. While most of these concerns are not unique to this area, the rapid development and commercialization of biotechnology has added urgency to these debates.

At its root, the debate about biotechnology and its role in society touches on and challenges our concept of the role of humankind in the world. Many environmental, consumer, and political groups

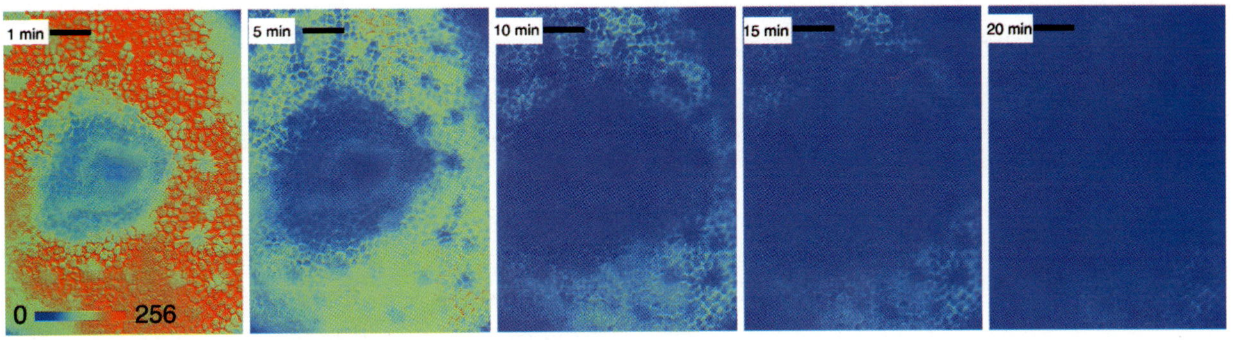

Plate 1 Abiotic Stresses Oxidative burst in the epidermal cells of the leaves of *Pseudowintera colorata* at various times after puncturing with a needle. Hydrogen peroxide (red color), produced immediately after wounding, is gradually scavenged by antioxidants (increasing blue coloration). Scale bar = 200 μm. Reproduced with permission from Gould KS, McKelvie J, and Markham KR (2002) Do anthocyanins function as antioxidants in leaves? Imaging of H_2O_2 in red and green leaves after mechanical injury. *Plant, Cell & Environment* 25: 1261-1269.

Plate 2 Cell Walls and Fibers Cross-section of mature flax stem, stained with Cellofluor under UV.

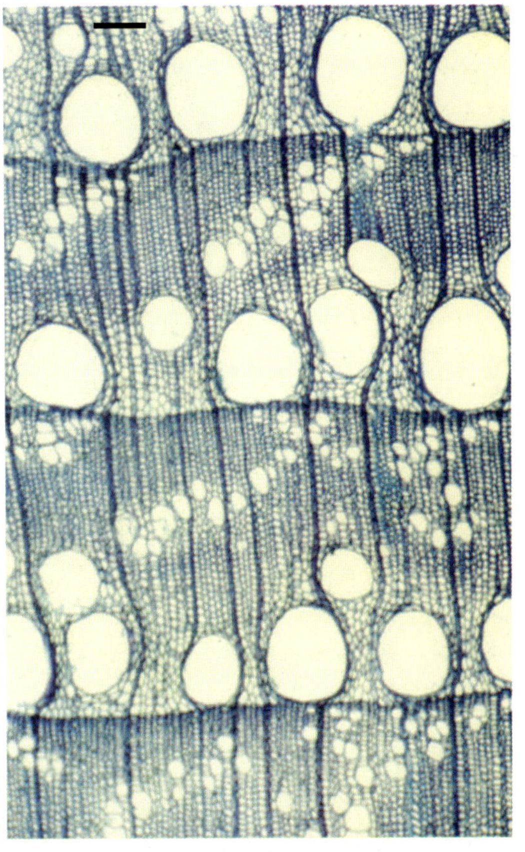

Plate 3 Cell Walls and Fibers Water conduction and the multilayered nature of wood at the microscopic level. In temperate zones with their successions of warm and cold seasons, there is generally a dormancy period in winter when primary and secondary growth are stopped. During spring, there is a very active period of growth (new leaves, flowers, primary growth) during which water conduction from roots to leaves is of primary importance. "Conductive" cells (i.e., vessels or thin-walled tracheids) are produced in greater numbers than supporting cells (i.e., fibers or thick-walled tracheids). The reverse is true in summer, so each annual ring is composed of a very porous "spring wood" layer and a denser "summer wood" layer. For many woods, the mechanical properties of summer wood is two to four times higher than spring wood. In this cross-section of chestnut (*Castanea* spp.) wood, the spring wood has very large vessels, typical of ring porous species. Scale bar, 0.5mm= 500 μm.

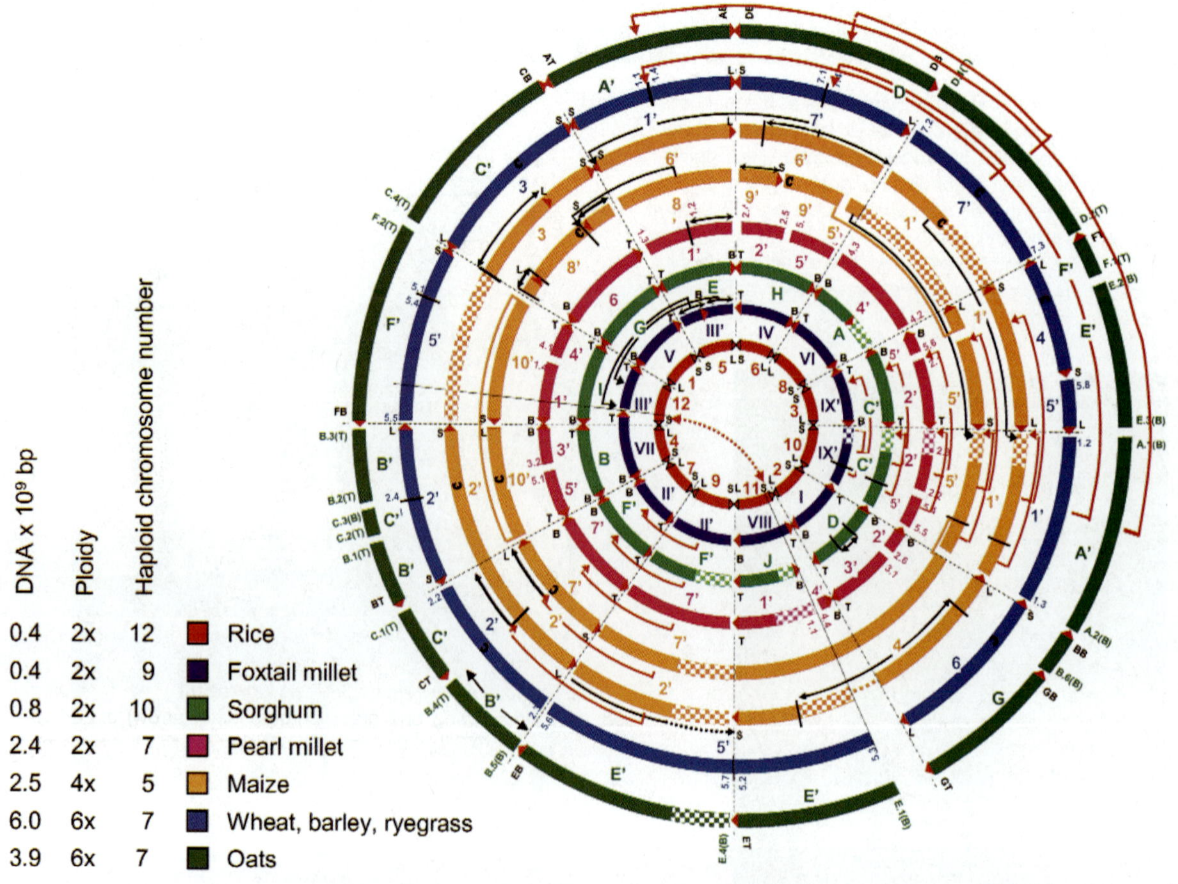

Plate 4 Crop Improvement Genome relationships among the major cereals. The chromosomes of the several economic grasses are aligned against the 12 rice chromosomes that make up the innermost circle. The locations of orthologous genes, QTLs or DNA sequences in the different genomes can be predicted to lie on radii in the diagram. The major insertions, inversions and translocations which distinguish the other genome from rice are indicated by the arrows. The hatched areas represent chromosomal regions where the relationships are uncertain.

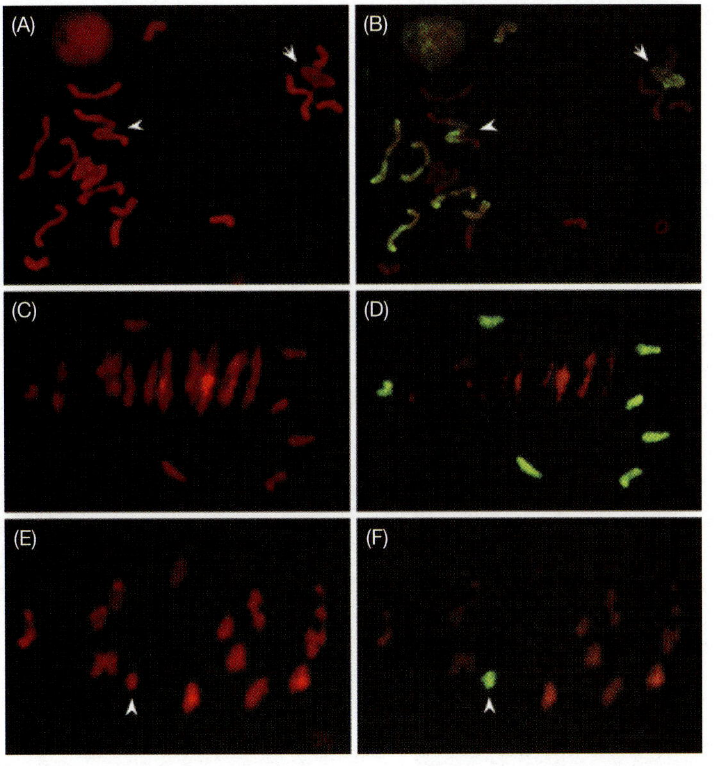

Plate 5 Crop Improvement Incorporation of alien chromosomes and chromosome segments into the durum wheat genome. (A) Metaphase I chromosomes of durum wheat–*Thinopyrum junceiforme* hybrid, counterstained with propidium iodide (PI). Note a quadrivalent (arrow). (B) Same cell as in (A) after fluorescent GISH, showing grass chromatin integrated with durum chromosomes (arrow). (C) A pollen mother cell (PMC) of a durum–*Lophopyrum elongatum* hybrid, counterstained with PI, showing full chromosome complement of durum and seven grass chromosomes. (D) The same cell as in (C) after fluorescent GISH, showing seven fluorescing grass chromosomes. (E) A PMC (counterstained with PI) showing one grass chromosome (arrow) added to durum chromosome complement. (F) The same cell as in (E) after fluorescent GISH, showing a fluorescing grass chromosome (arrow). This grass chromosome confers a high degree of scab resistance to durum.

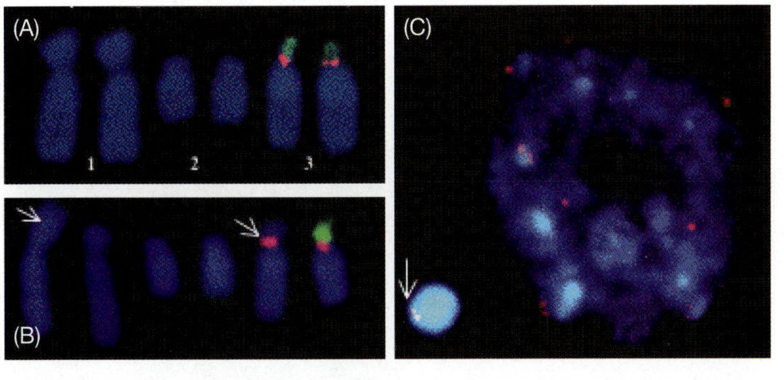

Plate 6 Crop Improvement Detection of chromosomal aberrations using ribosomal DNA (rDNA) and telomeric sequences as probes for FISH of *Crepis capillaries* chromosomes and nuclei. (A,B) 5S rDNA – red, 25S rDNA – green; (C) telomeric sequences – red; (A,B,C) DAPI staining – blue. (A) Karyotype of *C. capillaris* (2*n*=6) – control; (B) karyotype with translocated chromosomes (arrows). Chromosome no. 1 – longer short arm; chromosome no. 3 – lack of 25S rDNA signal and longer short arm in one homolog. (C) Micronucleus with double signal after FISH with telomeric DNA indicates that the micronucleus contains one whole chromosome, while 10 signals observed in the nucleus indicates that one chromosome is missing.

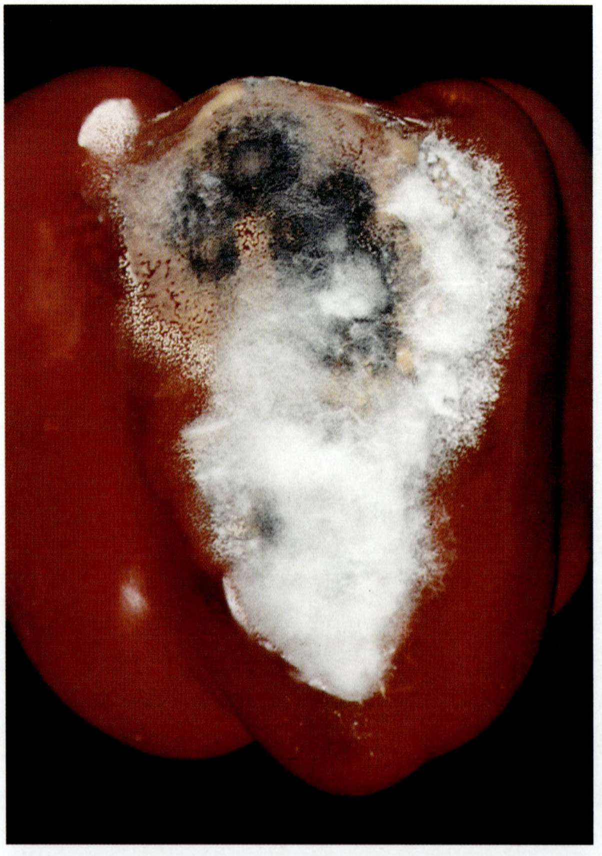

Plate 7 Diseases Rots such as this infection on *Capsicum* caused by *Glomerella cingulata* are a major cause of losses postharvest.

Plate 8 Diseases Apple scab disease, caused by *Venturia inaequalis*, leads to major apple crop loss worldwide.

Plate 9 Diseases Epistatic interaction of stem rust resistance genes *Sr26* and *Sr38*. 1, Sr26 + Sr38; 2, Sr26; 3, Sr38; 4, susceptible control.

Plate 10 Flowering and Reproduction Pollen behavior in RNase-based GSI. In squashes of *Solanum chacoense* styles (genotype S_1S_2) stained with aniline blue and observed by epifluorescence 48 h after pollination, pollen tubes fluoresce bright green. In incompatible crosses (A), the pollen tubes exhibit arrested growth at mid style, whereas in compatible crosses (C) they enter the ovary. (B) Schematic representation. Note the compatibility of diploid S-heteroallelic pollen (S_1S_2).

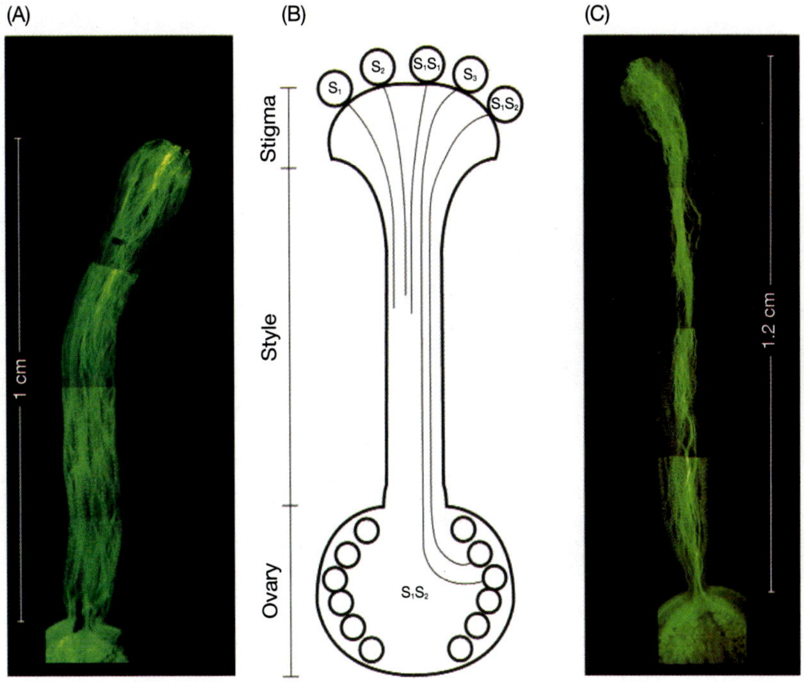

Plate 11 Flowering and Reproduction Typical features of S-RNases. (A) Hydrophobicity profile of the *S. chacoense* S_{11}-RNase sequence. (B) A three-dimensional model of the S_{11}-RNase produced using SWISS-MODEL to thread the S_{11}-RNase sequence through the structure of RNase Rh from *R. niveus*. Note the superficial location of the HV regions, well separated from the inner core of the C2 and C3 regions forming the active site (the two conserved histidines are black). (C) The sequence of the HV regions of the S_{11}-RNase, the phenotypically distinct S_{13}-RNase and various mutated S-RNases. The amino acids in bold are those that differ between the S_{11}- and the S_{13}-RNases. The HVapb RNase displayed a dual-specificity $S_{11/13}$ phenotype.

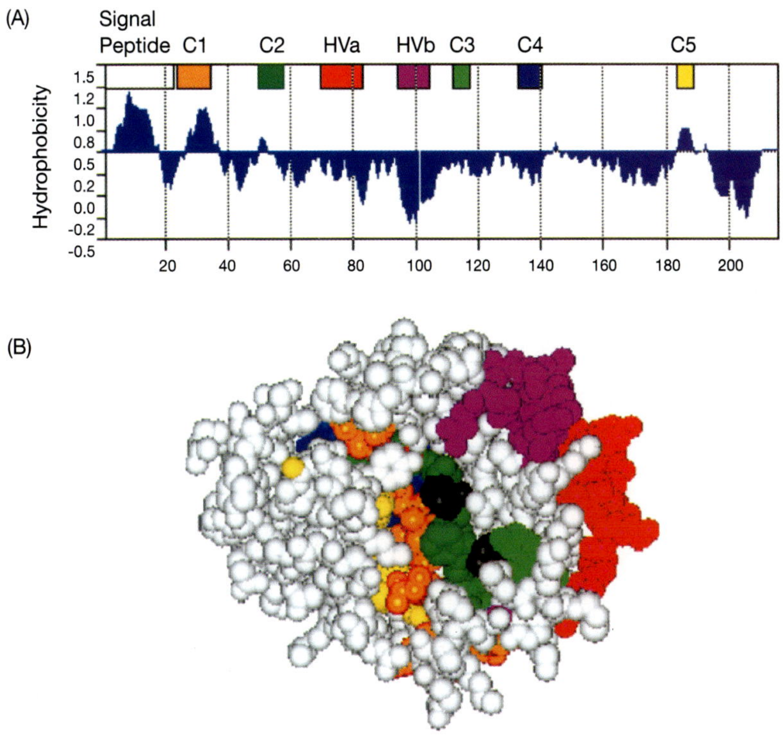

(C)

S-RNase	HVa	HVb
S_{11}	KPKL**TY**N**Y**F SDKML.....	IDQASA**R**KDQP
HVapb	KPKL**N**YN**F**F SDKML.....	IDQASA**L**KDQP
HVa	KPKL**N**Y**KF**F SDKML.....	IDQASARKDQP
HVab	KPKL**N**Y**KF**F SDKML.....	IDQASA**L**KDQP
S_{13}	KPKL**N**Y**KF**F SDKML.....	IDQASA**L**KDQP

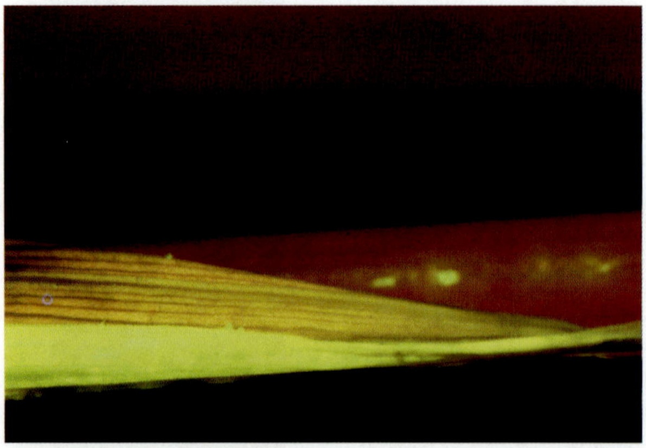

Plate 12 Genetic Modification Green fluorescent protein used as an excision marker in rice. Rice plants were transformed with a construct in which the *Ac* element is interposed between the *gfp* gene and its promoter. The observed mosaic pattern reflects late excision events affecting restricted clones of cells.

Plate 13 Genetic Modification Selection of transplastomic clones by spectinomycin resistance. (A) Spectinomycin inhibits callus formation, greening, and shoot regeneration from tobacco leaf segments on shoot regeneration medium. Transplastomic clones are resistant to spectinomycin and are identified as green shoots (or calli). (B) The shoots are chimeric, visualized by accumulation of green fluorescent protein in transplastomic sectors. Shoots regenerated from the transformed sector are homoplastomic. Reproduced with permission from Maliga P (2003) Progress towards commercialization of plastid transformation technology. *Trends in Biotechnology* 21: 20-28.

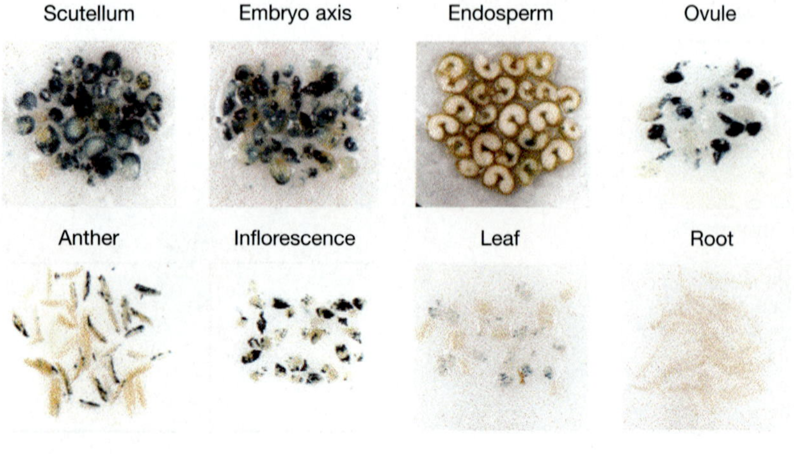

Plate 14 Genetic Modification Expression of the GUS (beta-glucuronidase) marker gene in transgenic wheat tissues. Blue coloration indicates expression of the GUS transgene.

(A)

(B)

(C)

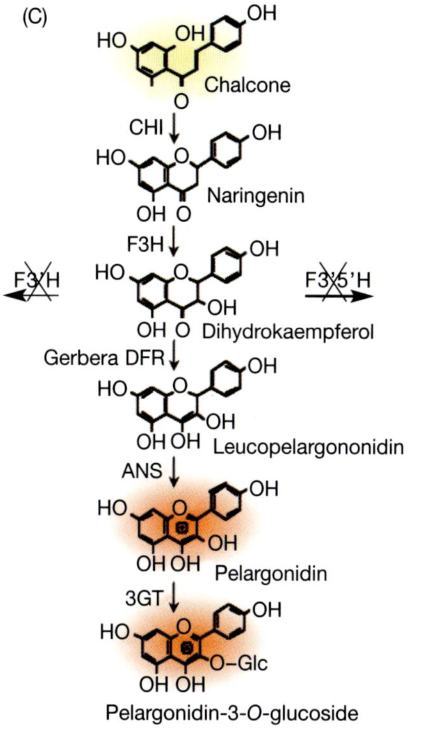

Plate 15 Genetic Modification, Applications A gerbera gene in petunia opens pathway to pelargonidin glycosides. (A) The petunia line RL01 has mutations in genes coding for F3'H and F3'5'H activities and accumulates dihydrokaempferol that is not a substrate for petunia dihydroflavonol 4-reductase (DFR). (B) Transformation of RL01 petunia with a gerbera gene for DFR results in conversion of dihydrokaempferol into leucopelargonidin, which is further converted by petunia enzymes to the brick-red/orange colored pelargonidin glycosides. (C) The pathway from chalcone to pelargonidin glycosides in transgenic petunia. The genetic blocks of the RL01 mutant in flavonoid 3' and 5' hydroxylation are shown by crosses.

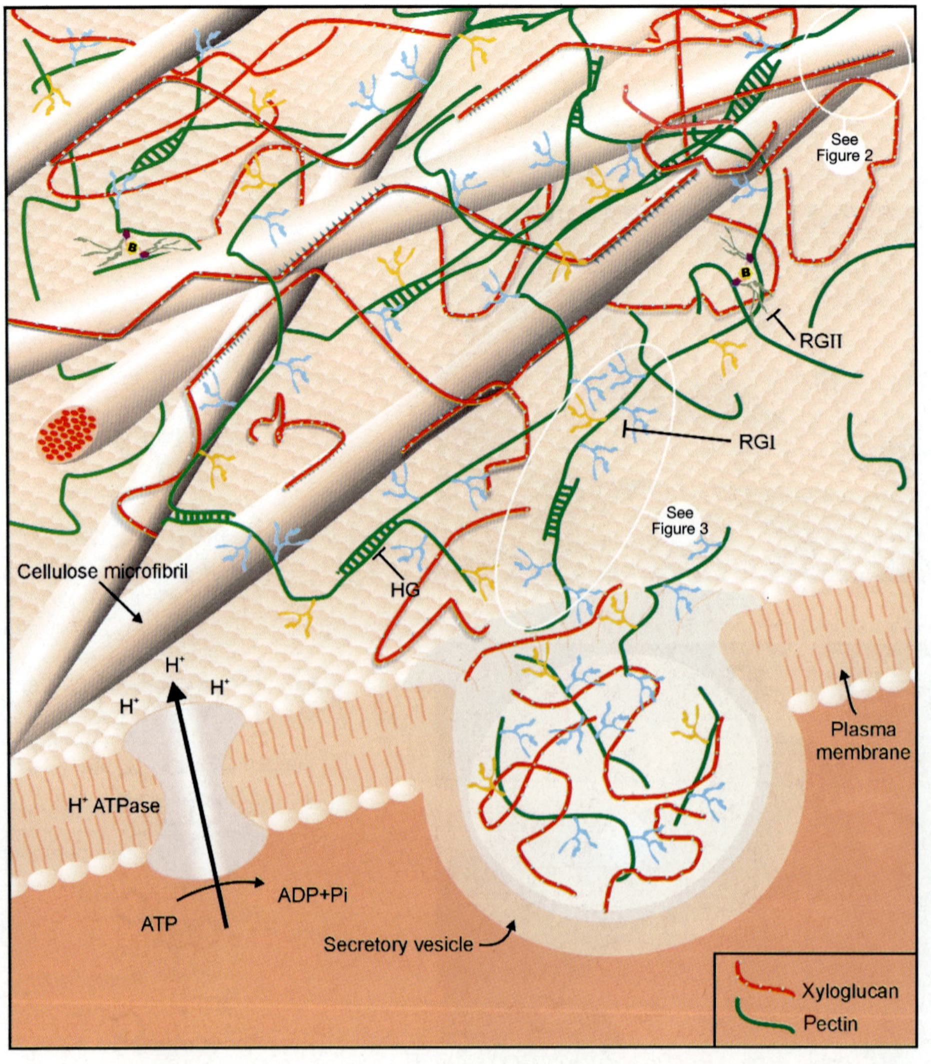

Plate 16 Growth and Development A generalized model of the molecular structure of type I primary cell walls. Cellulose and xyloglucans form the main framework of the wall (see **Figures 1** and **2**). Xyloglucan polymers hydrogen bond to the surface of microfibrils and may span between microfibrils serving to anchor them to one another. A network of pectin polymers is coextensive with the cellulose/xyloglucan network throughout the wall. Pectins have regions of homogalacturonan (HG). HG chains may be crosslinked to one another through electrostatic interactions between HG carboxyls and calcium ions. Pectins are also associated through the formation of dimmers of rhamnogalacturonan II (RGII), through covalent linkages between apiose residues of neighboring chains and boron ions (see **Figure 3**). Domains of rhamnogalacturonan I (RGI) contain extensive galactan and arabinan side chains that may provide steric hindrance to pectic associations and thus limit crosslinking of the network. Cellulose microfibrils are produced at the plasma membrane (see **Figure 1**), whilst all other polysaccharides are synthesized in the Golgi system and released into the wall from the fusion of secretory vesicles with the plasma membrane.

(A)

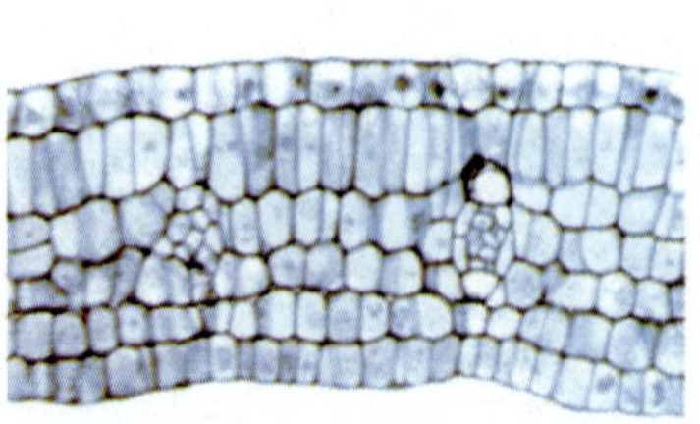

(B)

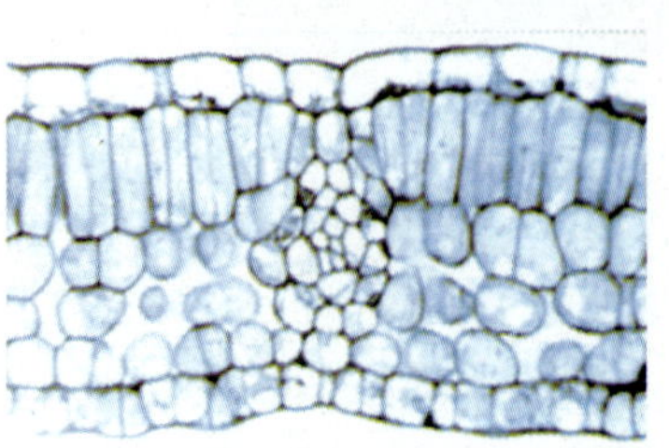

(C)

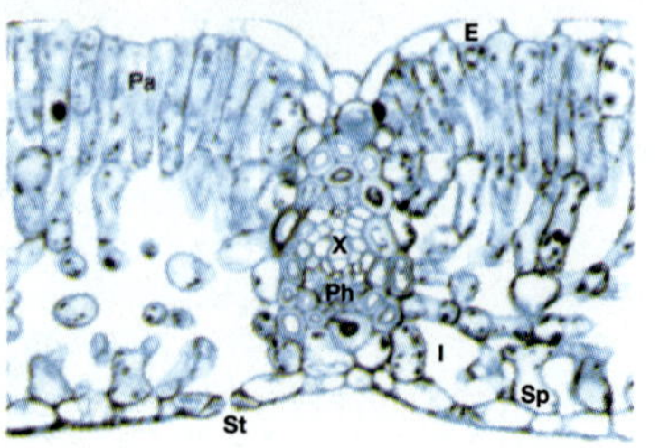

Plate 17 Growth and Development Histogenesis as leaves expand from overwintering bud of American beech (*Fagus grandifolia*). (A) Beginning of bud swelling; (B) seven days after bud break; (C) 28 days after bud break. E, epidermis; I, Intercellular airspace; Pa, palisade mesophyll; Sp, spongy mesophyll; St, stomata; X, xylem; P, phloem. Scale = 100 μm. Reproduced with permission from Dengler NG, Mackay LB, and Gregory LM (1975) Cell enlargement and tissue differentiation during leaf expansion in beech, *Fagus grandiflora. Canadian Journal of Botany* 53: 2846-2865.

express a real sense of outrage at the introduction of this technology, citing concerns about the newness of the technology and the potential for undetected risks, or disgust at the profit-seeking activities of the owners of the technology. Some object on the basis that changing DNA is "God's work." Many concerns relate less to the products or processes of food and agriculture biotechnology than to the social institutions that develop, promote, and regulate them. It has been suggested that one reason for this unease is that most of the social institutions related to biotechnology suffer from a deficit of public trust, which is partly due to opaque processes and limited opportunity for public involvement in decision-making.

This debate has also opened a rift between wealthy industrial consumers, who can afford to forgo the technology, and some consumers and governments from less-developed countries that see the application of biotechnology as one of their best opportunities to increase food supplies and reduce malnutrition. While there is potential for biotechnology to assist poorer farmers in many developing areas, thereby increasing food availability in food-insecure parts of the world, others express concern that this technology may exacerbate many of the trends started by the Green Revolution – i.e., destabilizing societies by the infusion of technology and capital into traditional agricultural systems in peasant economies.

It has been suggested that from the standpoint of technology management, ethics weighs in on whether people must be informed and their consent obtained before they can become bearers of risk, and how tradeoffs between risk and benefit are to be evaluated. The rest of this article examines a number of those tradeoffs and how society is attempting to respond to them.

The Distribution of Winners and Losers

Much of the biotechnology debate is about who wins and who loses. The conventional economic view is that yield-enhancing research (e.g., herbicide-tolerant plants) ultimately benefits consumers more than producers because the expansion of supply lowers prices, which transfers many of the gains to consumers. Quality-enhancing innovations (such as GoldenRice™), in contrast, tend to provide greater benefits to farmers because prices usually rise due to a tradeoff between yield and quality and because consumers are often willing to pay more than before.

The extension and widespread private use of intellectual property rights somewhat changes the above analysis because the innovator also seeks a share of the value generated by the technology. It can be argued theoretically that with private use of intellectual property rights and drastic inventions – i.e., one that is absolutely superior to and ultimately replaces all existing technologies – most of the benefits go to consumers or other producers, as discussed above. Thus the innovator will need to price the technology lower to get producers to adopt the technology, which producers then do, which causes the product output supply to expand and prices in output markets to fall, ultimately shifting much of the benefit to consumers. But for nondrastic inventions – i.e., where the invention is better than most but not absolutely superior to all existing technologies – the main beneficiary is the inventor or holder of the intellectual property rights. In this circumstance, the owner of the technology will price their invention where users are almost indifferent between the existing and new technologies, so that output stays relatively constant and product prices remain steady. As a result, the innovator captures most of the benefits.

While economists accept these conclusions generally, governments and consumers want to see evidence of the distribution of benefits and costs. So far, most of the analyses concerning GM crops have been partial. The applied studies of yield-enhancing inventions suggest that a significant portion of the benefits appear to go to consumers. Most analyses are similar, estimating that consumers gain from 10% to 75% of the benefits of new yield gains. The difficulty is that the gains often are in almost invisible increments, often amounting to a minor fraction of any purchase. As such, while the overall consumer share of the benefit is large, it is usually overlooked or discounted in most advanced industrial economies. Of course, this benefit is shared wherever the good is sold and consumed, which means that for many of these products, a large share of the benefit is moved to importing countries around the world. In developing countries, however, small changes in prices of some important foods (especially proteins and edible oils) could have a major impact on quality of life.

At the product level, the evidence of returns to the producer is mixed. Recent studies on *Bacillus thurengensis* corn, *Bt* cotton, Roundup Ready™ soybeans, and herbicide-tolerant canola show that gross returns can be variable, depending critically on weed or insect infestation rates (if the infestation rates are low, returns will be low). While some productivity gains have been observed, producers also lose due to lower market prices caused by increased productivity and supply. These estimates may underestimate farmer returns as many of the reported benefits (e.g., ease of use and reduced risk) have yet to be quantified. Regardless of the size of the benefits, the distribution among producers will not be equal, because farmers with different locations, farm structures, and management capacity will gain differently. In the first instance, those producers who adopt the new technologies have a greater opportunity to gain. In general, these tend to be better-capitalized and more innovative farmers, especially in North America. Studies have shown that early-adopting producers have tended to gain up to one-third to one-half of the benefits. Late adopters often neither gain nor lose, as any productivity gains are offset by lower prices, caused by greater supply. Nonadopters unambiguously lose, as they face lower prices without any compensating rise in productivity. Many of these nonadopters are smaller, less well-capitalized farmers, often in depressed regions in developed or developing nations, which often raises concerns about the plight of farmers and rural agrarian regions of the world.

Innovating companies are in many cases the big winners. As both the major investors in the technology and the major beneficiaries of the returns, they have a significant influence on the distribution of benefits. It is important to include innovators' monopolistic or oligopolistic profits in the total calculation of the returns on the technology. A number of studies have estimated that innovators capture between 37% and 60% of the gross benefits generated by new nondrastic, yield-enhancing technologies. Although gross annual returns of biotechnology applications appear large, it is necessary to consider the costs of developing the innovations and any externalities of adopting the technology. A longitudinal assessment of the adoption of herbicide-tolerant canola concluded that although the gross annual returns were large, the gross investment by the industry was only recovered in the seventh year of adoption and, even then, industry had not recouped the value of pre-existing chemical markets that were cannibalized by the new technology. There are also possible externalities of introducing new GM varieties into a market. Losses caused by comingled supplies or, alternatively, the investments made to segregate supplies, must be included in any cost–benefit analysis of new products. Because few of those costs are borne by innovators, there will be a continued public debate about how new GM products can be made to benefit society as a whole, and not simply corporate innovators.

One final issue that generates significant debate is the geographic distribution of benefits and costs. Given the unevenness of many research programs and the potential for increasing returns to scale from research and development clustered in a few locales, those countries that are able to attract corporate research programs often can gain some long-term economic development benefits. In the first instance, they can capture a large share of the investment and jobs in development of a new technology or variety. More importantly, perhaps, they can also capture early-adopter benefits if the research results are commercialized locally. One must keep in mind, however, that once a product is extensively adopted, innovator returns are often reassigned to the home countries or regions of the innovators and their shareholders, which are located almost exclusively in the United States or the European Union.

Biotechnology and Industrial Change

One overriding concern with biotechnology is its rapidly changing economic structure. In the past 20 years governments around the world have extended private property rights for new biotechnologies, for genes and gene constructs and (in some countries) for the entire plants transformed by new genes. As a result, private investment has flourished. Historically the majority of the research into new crop varieties – especially those without hybrids or other reproductive controls – was funded by and undertaken in public institutions and universities. Today more than half of the total global agrifood research activity is privately funded and directed. Furthermore, even research that is publicly managed is now treated in much the same way as private research, and results are usually protected by property rights and exploited commercially.

As a result, the entire agrifood sector has begun to change. Research capacity and technologies were increasingly bought during the late 1990s by a handful of large agrochemical and pharmaceutical companies – e.g., Monsanto, Bayer, Advanta, Syngenta, Dupont, and Dow – which strategically aligned themselves with marketing, processing, and retail companies. These global networks both integrated markets through trade and redistributed the benefits by concentrating their research and early commercialization efforts in a few advanced developed economies (e.g., corn, cotton, and soybeans in the United States and canola in Canada). While pharmaceutical enterprises moved after 2000 to divest their agricultural operations, significant consolidation remains.

Corporate concentration poses some potential problems. Many development agencies and nongovernmental organizations are concerned that these large consortia command significant market shares and can extract significant profits at the expense of farmers and consumers. Furthermore, few of these companies have any interest in subsistence or staple crops of critical importance to farmers and consumers in developing countries. More disturbing to the development community is that these companies charge for access to their technologies, which may limit their use for small and subsistence crops. The absence of new market entrants and the low incidence of legal battles over intellectual property suggests that there may be socially undesirable corporate concentration. These may be symptoms of a highly concentrated industry that is extracting excessive monopoly profits. An alternate view is that the unexpectedly low and uncertain net returns realized by most investors in the agricultural biotechnology has discouraged new entrants. In short, the mere absence of competition does not mean the industry is extracting inequitable profits. If market prospects improve in the agricultural biotechnology sector, we may find that the sector becomes more visibly competitive.

GM Foods and Consumer Acceptance

Consumers are often confused about GM food. They express great concern about GM ingredients, yet are for the most part incapable of describing in even the simplest terms what they dislike. Generally, the response among consumers in developed countries around the world is that they want to be able to choose whether to eat GM foods or not.

The top public concern about GM foods is whether they are safe. While the scientific community has concluded that there are no new measurable or anticipated risks to human or animal health from the crops currently grown, consumers in wealthy countries are not convinced. In absence of any perceptible consumer benefits, consumers focus on the uncertainties of the new technology in the food system. A variety of surveys in the past few years that directly ask consumers about biotechnology have concluded that a large majority of consumers are aware of GM elements in their food and that most, if given a choice, would avoid buying them. In most cases, only a minority of consumers indicated that they would be willing to pay a premium for GM labeling. Meanwhile, when consumers are asked more generally about their concerns about the food they eat, most place the GM issue well down on their list of priorities. They are more concerned about food cost, quality, presentation, and a variety of conventional food-safety risks (e.g., chemical residues).

The ultimate challenge is to provide a credible, transparent, and accountable system that provides effective choice to consumers between GM and non-GM food. While the principle is simple, and most governments agree with it, there are many difficulties in actually delivering such a system. Part of the problem is that consumers are presented with few of the GM crops in their raw form. The three top GM food crops – soybeans, corn, and canola – are mostly processed into oil and meal components, which are then incorporated into up to 60% of the processed foods on grocery store shelves. Most of those ingredients have been so highly processed, however, that there are no detectable GM traits left in the foods when they are presented to consumers. Given the credence nature of these markets – i.e., claims are not testable by consumers directly – most of private industry has been unwilling to voluntarily label any products.

Thus, the debate has moved to the regulatory systems around the world. As of 2002 more than 26 countries had either adopted provisions or announced plans for rules to assist the market to develop and deliver labeled products. At one extreme, the regulators in the United States, Canada, Argentina, and Hong Kong have concluded that mandatory labels will only be required to signal known public health and safety aspects, such as nutritional and compositional changes or new allergens. Food processors may use voluntary labels in those markets to signal their product either includes GM traits or that it is free of GM traits, provided the claims are credible and verifiable. At the other extreme, 22 countries, plus the countries of the European Union, have adopted or announced plans to implement mandatory labeling systems, with a range of thresholds (e.g., 1%, 3%, 4%, or 5% by volume of ingredient).

While mandatory labels seem to address consumers' concerns, they are far from a perfect solution. Some studies suggest that labeling systems could impose a one-time set-up cost of 1% to 6% of annual food expenditures and an ongoing cost of 0.5% to 3%. Given that there is no identifiable public good from this investment, the cost–benefit is high and for many unacceptable. Furthermore, given that only a minority of consumers state or show in experiments that they would pay for GM-free or proactively labeled GM products, there is some question about whether mandatory labels are efficient. Economic theory suggests that voluntary labeling would be more economically efficient, but there remains a crucial question of who has the right to choose.

Consumers and governments from developing countries have a substantially different perspective. With an estimated 800 million malnourished citizens, many in those countries are vigorous supporters of the new technology. They look to its potential to increase domestic food production and, with new products such as β-carotene-enhanced rice, to substantially reduce malnutrition and disease. Even so, there has been a debate in the aid community about the appropriateness of donating GM grains and oilseeds as food aid. This culminated in 2002 in a food aid and trade dispute when the Zambian government rejected an aid shipment of corn from the United States, on the grounds that it likely incorporated GM varieties that had not been approved in Zambia. While food safety concerns were raised, it would appear that the main concern was that in the long run the presence of GM seeds in Zambia could jeopardize future trade with the European Union.

GM Foods and the Environment

Regulators, scientists, and public interest groups have raised a number of concerns about the environmental impact of this new technology, which has been released on all continents. Some believe that the transgenic traits in these new crops and animals are superior to natural traits and will make the host crops "super organisms" that could dominate and destabilize other ecosystems. Others are concerned that outcrosses between GM organisms and other domesticated or wild organisms will create long-range environmental problems, possibly destroying landraces in developing countries that adopt these varieties. There is also some concern that nontarget species could be affected by some of the technologies and that pollen flow within and between species could jeopardize organic and other speciality crop production. In contrast, some argue that the increased specificity of transgenic modification offers lower risk than traditional plant breeding techniques because only selected genes rather than the entire plant are involved. In addition, many believe that the traits involved so far (herbicide tolerance, insect and viral resistance, delayed ripening, and modified nutrition) are not the type that would make weeds "super." Against the risks, proponents point to the demonstrated benefits of reduced chemical usage. Most of the targeted crops have for generations been intensively cropped, requiring significant amounts of fertilizers, herbicides, and pesticides. A 2002 review of 40 major studies concluded that the adoption of eight biotechnology cultivars in the United States increased crop yields by 2 million tonnes, lowered annual growers production costs by US$1.2 billion, and reduced pesticide use by 20 000 tonnes.

Lower chemical usage reduces the risks to farmers, farming families, and the broader population caused by direct exposure to the chemical or by ingestion through residues on the food or in our drinking water. Furthermore, there is some evidence that regions adopting *Bt* crops are seeing regeneration in populations of beneficial insects that previously were killed by sprayed pesticides. Ultimately, increased productivity from traditional farms lessens the pressure to cultivate lands in marginal areas.

Our Social Response

In the first instance, citizens and consumers turn to governments to address their concerns about new technologies. All of the GM foods on the market today have been subjected to extensive regulatory review in one or more countries, often involving between 3 and 7 years of testing and evaluation.

New GM foods, in particular, have led to different responses in different markets. Those markets lacking domestic regulators that command the confidence of consumers have tended to act in a "precautionary" way, either reviewing the products more slowly or imposing temporary bans on the introduction of the new products. This is a sharp break from the international food safety system that evolved over the past 100 years, where importers tended to accept the food and environmental safety judgments of regulators from the developed and exporting countries. One result of this "renationalization" of agrifood safety regulation is that national systems have tended to diverge. The United States, Canada, Mexico, and Japan, among others, generally make consistent safety rulings based on science and have approved most of the new GM products for production and consumption. Regulators in the European Union, Australia, and New Zealand, in contrast, have expressed concerns about uncertainties in the science, and have postponed most approvals in recent years. Another 20 or so countries have developed domestic regulatory systems consistent with one of the other approaches but most of the rest of the countries in the world do not have any domestic regulatory capacity and are seeking guidance and help from international institutions.

One concern about the diverging domestic regulatory systems is that production and trade is shifting. The new technology is for the most part pro-trade, with development and adoption of new GM varieties concentrating in the main producing and exporting countries. Up to 88% of trade in some products with GM varieties comes from the four lead GM-adopting countries. Meanwhile, the products that have been modified have been for the most part marketed as commodities, often with significant blending and mixing, ultimately flowing into the international marketplace to as many as 177 countries around the globe. Diverging domestic trade rules are causing the four lead GM adopters, especially the United States and Canada, to abandon or lose markets and divert trade to new markets. US exports of corn to the European Union fell by 70% between 1996 and 2002, US exports of soybeans to the European Union dropped by 48%, and Canadian exports of canola to the European Union dropped 96%. Meanwhile, the EU developed new GM-free sources of soybeans from Brazil and canola from Australia, both markets that had not approved GM varieties for those crops. So far these changing trade flows have not significantly affected producer returns as trade has simply been reallocated between adopting and nonadopting countries, but over time such policies have the potential to seriously distort trade and offset many of the benefits of freer international trade in these products.

There currently are nine international bodies vying to coordinate and regulate different aspects of food safety. Conceptually they represent a progression: five largely science-based (the International Plant Protection Convention, International Epizootics Organization, *Codex Alimentarius*, Food and Agricultural Organization, and the World Health Organization), one trade-based (World Trade Organization), and three others that have broader objectives such as environmental protection and other social or political goals (the Organization for Economic Cooperation and Development, various regional initiatives, and the BioSafety Protocol). Together they seek to develop standards for health, safety, and labeling for GM foods, establish testing procedures to ensure the standards are met, provide rules for allowable policies and create systems to manage disputes. While there is substantial effort being undertaken, the difficulty is that there is no common view on the goal of international regulation. While most agree that safety is the bottom line, few can agree on what that means, whose opinions should hold most weight (scientists or citizens), or how to handle issues other than safety, such as social, economic, or ethical concerns.

See also: **Ethics and Biosafety**: Ethics of Genetically Modified Crops; Ownership of Plant Genetic Resources.

Further Reading

Alston J, Norton G, and Pardey P (1995) *Science under Scarcity: Principles and Practice of Agricultural Research Evaluation and Priority Setting*. Ithaca: Cornell University Press.

Buckingham D and Phillips P (2001) Issues and options for the international regulation of GM foods. *Estey Journal for Trade, Law and Economics* 2: 178–189. http://128.233.58.173/estey/index.htm.

Carpenter J and Gianessi L (2001) *Agricultural Biotechnology: Updated Benefit Estimates*. Washington: National Center for Food and Agricultural Policy. http://www.ncfap.org/reports/biotech/updatedbenefits.pdf.

Gianessi L, Silvers C, Sankula S, and Carpenter J (2002) *Plant Biotechnology: Current and Potential Impact for Improving Pest Management in US Agriculture – An Analysis of 40 Case Studies*. Washington, DC: National Center for Food and Agricultural Policy. <http://www.ncfap.org/pubs.htm#Biotechnology>.

Glaeser B (ed.) (1987) *The Green Revolution Revisited: Critique and Alternatives*. London: Allen & Unwin.

James C (2002) *Global Status of Transgenic Crops, ISAAA Briefs 1, 5, 8, 12, 21, 23, 24*. http://isaaa.org/publications/briefs/isaaa_briefs.htm.

Kalaitzandonakes N (ed.) (2003) *The Economics of Transgenic Crops*. Dordrecht: Kluwer Academic Publications.

Marra M (2002) Agricultural biotechnology: a critical review of the impact evidence to date. In: Pardey P (ed.) *The Future of Food: Biotechnology Markets and Policies in an International Setting*, pp. 155–184. Baltimore: Johns Hopkins University Press.

Moschini G and Lapan H (1997) Intellectual property rights and the welfare effects of agricultural R&D. *American Journal of Agricultural Economics* 79: 1229–1242.

Moschini G, Lapan H, and Sobolevsky A (2000) Roundup Ready soybeans and welfare effects in the soybean complex. *Agribusiness* 16(1): 33–55.

OECD (2000) *Overview and Compendium of International Organizations with Food Safety Activities*. SG/ADHOC/FS(2000)4/FINAL, May 11. Paris: Organization for Economic Cooperation and Development.

Phillips P and Khachatourians GG (2001) *The Biotechnology Revolution in Global Agriculture: Invention, Innovation and Investment in the Canola Sector.* Wallingford: CAB International.

Thomson P (2000) *Food and Agriculture Biotechnology: Incorporating Ethical Consideration*. http://www.cbac-cccb.ca/documents/en/FoodAgric_Thompson.pdf.

Ethics of Genetically Modified Crops

R Straughan, University of Reading, Reading, UK

Introduction

Genetically modified (GM) crops have become the subject of heated controversy and media attention. This is partly because they raise not only scientific and technological issues but also what are often referred to as moral and ethical ones, about which it seems difficult to reach any substantial degree of consensus. These concerns, to take a few common examples, may arise from unease that the technology interferes with the workings of Nature and Creation, that it involves irresponsible risk-taking for commercial profit, or that it exploits and harms vulnerable individuals and communities. Such concerns are of importance for a variety of reasons, not least because many studies have shown them to be significant in shaping public perceptions of the technology and attitudes towards it. The exact nature of these concerns, however, is often far from clear, and the purpose of this article is to outline the role of ethics in discussions of GM crops and to examine the main issues at stake.

The Function of Ethics

Ethics is an ambiguous term. In discussions about GM crops we tend to find the words "moral" and "ethical" used interchangeably, but this can be confusing and some writers have found it useful to try to distinguish between them along the following lines.

We all can be said to have moral views, beliefs and concerns, to the effect that certain things are right or wrong and that certain actions ought or ought not to be performed. Such "moral concerns" may result from a lot of deliberation and reflection or from very little; they may be firmly grounded in a consistent set of carefully considered rational principles or they may not; their justification may have been consciously analyzed or it may not.

Ethics can be a narrower concept than morality. It can refer to a particular branch of philosophy which tries to analyze and clarify the arguments that are used when moral questions are discussed and to probe the justifications that are offered for moral claims. So ethics in this sense puts our moral beliefs under the spotlight for scrutiny.

To call something a moral concern, then, does not necessarily mean that it is of much ethical significance. A number of surveys have shown that, if asked, people will express moral concern about genetic engineering in general and GM crops in particular, but this does not tell us whether they have done any ethical thinking about the issues.

The function of ethics is not merely to provide factual information about the issue in question. Most philosophers agree that conflicting moral views cannot be resolved by a simple appeal to "the facts," for sets of facts, whether about GM crops or anything else, can never prove something to be morally right or wrong. However, this does not mean that facts are unimportant or irrelevant when considering ethical questions, particularly when we are trying to predict the likely or possible consequences of whatever is under ethical examination.

Surveys have also shown that moral concerns are of considerable practical importance in influencing public perceptions of and attitudes towards GM crops. Our attitudes are not of course wholly determined by our moral views alone. Nevertheless, our moral views about any subject will exert a powerful effect upon our perceptions of that subject and upon our attitudes towards it, which in turn will help to shape our choices and behavior.

Worries are frequently expressed by those eager to promote the development of the new technology that the alleged potential benefits of GM crops may be lost if the new processes and products fail to gain consumer acceptance because of moral concerns. Ethics, then, can be a useful tool in examining the basis of such concerns, not in order to try paternalistically to persuade people that GM crops are really a good thing, but to raise the level of debate and to encourage judgments to be made on a more rational and considered basis.

Moral Concerns about GM Crops

The moral concerns expressed about GM crops cover a wide range, but may for convenience be divided into two basic categories, which have been labeled "intrinsic" and "extrinsic." Genetic modification in general and GM crops in particular may for a variety of reasons be thought to be either intrinsically wrong in themselves or extrinsically wrong because of their consequences.

Intrinsic arguments cut deeper than extrinsic ones. If GM crops are thought to be intrinsically wrong, no further considerations are accepted as morally relevant because nothing can reverse that intrinsic wrongness; the possible consequences of this innovation do not have to be taken into account. Intrinsic arguments about the wrongness of something also focus attention upon its precise nature and its distinguishing characteristics, e.g., what precisely is morally objectionable about GM crops, and what distinguishes them in this respect from crops produced by more traditional selective breeding?

Intrinsic Concerns

1. It is possible to hold religious views to the effect that any form of genetic modification is blasphemous. These views may rest upon the belief that God has created a perfect, natural order; for people to attempt to "improve" that order by manipulating DNA, the basic ingredient of all life, thereby crossing species boundaries instituted by

God, is thought to be not merely presumptuous but sinful.

The essence of this concern, then, is that modern biotechnology is trying to "play God" – a view given support by the Prince of Wales in the recent debate in the United Kingdom. Such concerns rest upon theological considerations and as such will carry moral weight only for those holding certain religious beliefs based upon certain religious assumptions. However, by no means would all religious believers share these concerns. Even among Christians there is no unanimous condemnation of genetic modification *per se*. There is, for example, scriptural support for the view that humanity has been given by God an approved, privileged position of "dominion" over Nature. Some modern theologians even see biotechnology as a challenging, positive opportunity for us to work with God as "co-creators."

In GM crops genes may be moved from one species to another, but religious believers do not necessarily hold that the boundaries between species are sacred and immutable, nor indeed that they are so regarded by God. For all except the "creationists," who believe in changes only within fixed limits of originally created kinds of plants and animals, evolutionary theory is more likely to suggest to religious believers a view of species as provisional and fluid, with each one playing its part in a developing process initiated by God.

2. A belief that GM crops are intrinsically wrong need not rest upon a religious basis. Agnostics and atheists would be unmoved by arguments about blasphemy, but might still share what seems to be a widely felt concern that biotechnology is in some sense "unnatural" and therefore wrong. Several studies have suggested that it is this concern which lies at the root of much of the public unease about GM crops. The first task of ethics here is to clarify what exactly is being objected to and why.

   However, it is no easy task to identify and agree about what is to count as "natural" and "unnatural." Depending on the context in which it is used, the word "natural" may mean "usual," "normal," "right," "fitting," "appropriate," "uncultivated," "innate," "spontaneous," and no doubt many other things as well. Perhaps most commonly, "natural" is contrasted with "artificial" or "man-made," but on the basis of that distinction practically every element of our modern Western lifestyle is "unnatural." Nor can more traditional products and processes avoid such a charge of "unnaturalness," for the progress of civilization has been largely dependent upon humanity's "interference with Nature." Yet if every domestic or farm animal, every garden plant or agricultural crop, every item of food or clothing is thought of as the result of "unnatural interference," then the concept of "unnaturalness" becomes so broad as to be meaningless.

   The more specific and serious charge of "unnaturalness" that has been leveled against biotechnological products such as GM crops, however, is that they breach natural species boundaries and violate the natural integrity of species, but one problem with this argument is that biologists are unsure about the extent and even the definition of "natural species boundaries."

   Another basic ethical question to be asked about these intrinsic concerns is "What is good about being 'natural'?" Philosophers have pointed out that a "natural" event, product, process, or tendency (however defined) is not automatically good or desirable. Many "natural" substances are harmful; many of our "natural" tendencies and reactions, such as jealousy and aggression, are not normally thought morally praiseworthy; many "natural" events, such as earthquakes and hurricanes, create destruction and suffering, and are indeed usually labeled "natural" disasters; many "natural" organisms cause pain, disease, and death. We cannot simply deduce what is morally desirable from certain facts about the world and about Nature, and to assume that we can is to commit what is usually referred to as the "naturalistic fallacy." Simply because something happens in Nature does not mean that we are morally unjustified in interfering with it. So even if natural species barriers can be identified (which may be difficult), their mere existence provides no clear ethical directives about what ought to be done about them.

Extrinsic Concerns

These represent a less uncompromising position than the intrinsic variety, because they invite discussion about what the most likely consequences really are and suggestions about how they might be avoided or modified. Extrinsic concerns about GM crops focus on two main areas – safety and socioeconomic effects – and both of these involve making predictions about future events.

Safety and risk Much concern has been expressed about the possible risky consequences of GM crops, but riskiness is not in itself a moral or ethical matter. Some activities are inevitably more risky than others, though none can be totally risk-free, and it does not follow that low-risk activities are morally superior to high-risk ones. Risk and safety become matters of moral concern only when they raise further questions about responsibility and justifiability. Moral concern is appropriate when irresponsible and unjustifiable risks are thought to be taken, which may result in harm to innocent parties.

What kinds of risk are claimed to attach to GM crops, and are they sufficiently "irresponsible" and "unjustifiable" to merit moral concern? The main fears center upon the release of genetically manipulated organisms into the environment and the possible harmful consequences of such procedures. Such consequences might include, it has been suggested, "genetic pollution" of the countryside resulting from the spread of indestructible weeds, resistant to pests and herbicides. Further fears have been voiced that GM crops might lead to a loss of genetic diversity and a loss of wildlife habitats for various creatures.

The risks envisaged here are of such a nature that no one would feel justified in turning a blind eye to them. However, it is impossible to prove that a particular event will or will not happen in the future. No activity or process can ever be guaranteed to present no risk whatever and to be 100% "safe," and the development of GM crops is no exception to this logical rule. Any activity could conceivably lead to catastrophic consequences.

Critics argue that the risks involved here are of such a level as to make the further development and application of the technology irresponsible; it is the particular and peculiar risks associated with these techniques, they claim, that make them morally unjustifiable. The issue here is partly one of technical assessment, but a more general counterargument has also been proposed to the effect that excessive caution does not necessarily remove the risk of future catastrophes. By banning research and development in any new technology that is thought to involve risks, we may run the greater risk of failing to produce an innovation which will be desperately needed in some future, unforeseen crisis. The history of science has proved to be highly unpredictable, and there can be no guarantee that "playing safe" by abandoning research and development in GM crops will not deny us a technique or product which may prevent an environmental disaster in 50 years' time.

Nevertheless, as it is logically impossible to guarantee the total safety of GM crops, a value judgment has to be made about the acceptability and justifiability of the possible risks, and that judgment will involve the weighing of potential risks against potential benefits. The ethical basis of this approach is provided by utilitarianism, the theory that actions are right or wrong in proportion to the total amount of pleasure or pain that they produce; this makes "the greatest good of the greatest number" (or in more modern terms, the lowest cost–benefit ratio) the ultimate ethical criterion. There are, however, a number of problems with this approach, the most practical one being how exactly these calculations of harm and benefit are to be made. In the case of GM crops, for example, it is not obvious which option should be identified as the potentially harmful one, as harm may result from developing the technology or from not developing it. Moreover, different groups will be involved in the likely distribution of harm and benefit, and this is what lies at the center of the second area of extrinsic concern.

Socioeconomic effects Concerns have been raised about the economic vulnerability of poorer farmers and poorer countries to some of the possible effects of the new technology, and about the economic disadvantage they are thought likely to suffer. The technology will be expensive, it is claimed, and will favor large-scale, capital-intensive styles of agriculture; poor farmers will not be able to afford the new products and will lose out in the increased competition that will result. On a broader scale whole economies as well as individual farmers could be threatened, and the economic gap between the "developing" and the "developed" world could widen yet further.

More specific charges have been made about the degree of profit-motivated "control" exercised by agrochemical companies over small farmers, particularly again in poorer countries. Another problem is that the technology may lead to products at present obtainable only from developing countries being made in Western laboratories.

A further emotive and complex issue is that of patenting. The most general moral concern expressed about patenting is that it is wrong to think of "life forms" as objects that can be "invented" and "owned." More specifically, however, it is claimed that biotechnological patenting violates the ethical principle of fairness or justice and will produce various "unfair" consequences, including:

1. Greater control by large corporations over small farmers through the payment of royalties, which could reduce their independence and even force them out of business.
2. Misappropriation of Third World genetic resources by First World countries.
3. Restriction of the free exchange of genetic material and the flow of information about it.

On the other hand it can be argued that there is an obvious ethical justification for patenting in terms of fairness and justice. If a revolutionary new product is developed, perhaps at great expense, which may benefit millions of people, is it fair or just to deny the inventors recompense for their work and protection against others who might otherwise capitalize upon their efforts? Also the charge of "patenting and owning life" loses much of its force when it is realized that the ownership involved is of the invention itself and not of organic matter. In the case of GM crops, patenting implies the ownership not of a seed or plant but of the invention of the "genetic kit" which produces a particular attribute of that crop.

Critics have argued, however, that if the possibility of undesirable socioeconomic effects is added to the possibility of risk, these factors together constitute a strong moral argument against GM crops. Value judgments about the acceptability and justifiability of the possible costs of GM crops, however, need to take into account another set of factors concerning the potential benefits, which have somehow to be weighed against the potential costs.

Various ethical arguments have been developed in support of policies and practices which involve the consideration of others' interests, particularly where those others are in a disadvantaged position. Such other-regarding principles are in line with the precepts of Christianity and other religions, but ethical theorists have also laid weight from a nonreligious viewpoint upon the need to make moral judgments by "standing in the other person's shoes," and to approach questions of justice from the perspective of those who are the most disadvantaged. Any ethical appraisal of GM crops, then, must take into account what the possible benefits of such crops are and who is most likely to reap those benefits.

The potential benefits of GM crops are claimed by many to be considerable. Farmers are finding it increasingly difficult to grow enough food for an increasing world population, and environmental conditions in many traditionally productive agricultural areas are deteriorating. New techniques and products, it is claimed, are needed to feed a hungry world and to start to redress the balance between the food supplies available to rich and poor countries. GM crops may provide a possible, partial solution in the form of new, more productive strains, which will be more able to resist diseases and pests and to withstand hostile environmental conditions. Opponents maintain, however, that most research into GM crops is not aimed at reducing hunger, as it is heavily concentrated in a number of large corporations whose main interest is in the demands coming from their major markets, leading to the development of herbicide-resistant crops, for example, rather than high-yielding crops.

Other crops, it is argued, may offer specific and valuable benefits, such as the so-called "golden rice," which could increase the dietary intake of vitamin A. Its proponents maintain that this could save the sight of half a million children a year, but these claims have been hotly disputed and provide a good example of the difficulty both of establishing the precise consequences of any new development and of ethically assessing these possible consequences.

There is no simple, straightforward way of weighing the potential costs and benefits to produce a clear ethical decision. Priorities have to be assigned to the competing costs and benefits, and those priorities will in turn reflect more fundamental value judgments, which can be rationally defended but can never be conclusively proved right or wrong. Ethics can offer no easy or obvious answers here, but it can help us to clarify the issues at stake and to examine the principles and value judgments underlying the debate about GM crops.

Conclusion

The examples of moral and ethical issues mentioned above are intended to illustrate the main areas of current concern and to demonstrate the role of ethics in examining them. One of the challenges presented by our ever-developing range of knowledge and technologies is that it continues to throw up novel and more complex questions about what it is right and wrong to do. These questions cannot be answered by referring back to some previously agreed moral rule-book, partly because no such rule-book exists and partly because, even if it did, it would be inadequate to deal with fresh and often unforeseen developments. Science and ethics, then, have to proceed hand in hand in exploring new and sensitive areas such as GM crops, if rational and responsible judgments are to be made about them.

See also: **Ethics and Biosafety**: Development and Commercialization of Genetically Modified Plants; Intellectual Property Rights; Ownership of Plant Genetic Resources; Plant Genetic Engineering, Ecological Issues; Plant Genetic Engineering, Food Safety Issues.

Further Reading

Chadwick R (1997) *Encyclopedia of Applied Ethics*. London: Academic Press.

Committee on the Ethics of Genetic Modification and Food Use (1993) *Report on the Ethics of Genetic Modification and Food Use*. London: HMSO.

European Federation on Biotechnology Task Group on Public Perceptions (1999) *Ethical Aspects of Agricultural Biotechnology*. The Hague: Cambridge Biomedical Consultants/EFB Task Group.

Macer D (1997) *Food, Plant Biotechnology and Ethics*. Paris: UNESCO.

Mepham TB, Tucker GA, Wiseman J (eds) (1996) *Issues in Agricultural Bioethics*. Nottingham: Nottingham University Press.

Nuffield Council on Bioethics (1999) *GM Crops: Ethical and Social Issues*. Oxford: Nuffield Council.

Reiss M and Straughan R (1996) *Improving Nature? The Science and Ethics of Genetic Engineering*. Cambridge: Cambridge University Press.

Select Committee on the European Community (1998–9) *Regulation of GM in Agriculture*. House of Lords Report. London: HMSO.

Straughan R and Reiss M (1996) *Ethics, Morality and Crop Biotechnology*. Swindon: BBSRC.

Thompson PB (2000) *Food Biotechnology in Ethical Perspective*. London: Blackie.

von Wartburg WP and Liew J (1999) *Gene Technology and Social Acceptance*. Lanham: University Press of America.

Wheale P, McNally R (eds) (1990) *The Bio-Revolution: Cornucopia or Pandora's Box?* London: Pluto Press.

Intellectual Property Rights

G Dutfield, Queen Mary College, London, UK

Introduction

Intellectual property rights are legal and institutional devices to protect creations of the mind such as inventions, works of art and literature, and designs. They also include marks on products to indicate their difference from similar ones sold by competitors. Over the years, the rather elastic intellectual property concept has been stretched to include not only patents, copyrights, trademarks and industrial designs, but also trade secrets, plant breeders' rights, geographical indications, and rights to layout designs of integrated circuits, among others.

To proponents, intellectual property rights contribute to the enrichment of society through (1) the widest possible availability of new and useful goods, services and technical information that derive from innovative activity, and (2) the highest possible level of economic activity based on the production, circulation, and further development of such goods, services, and information. These objectives are supposed to be achieved because owners can seek to exploit their legal rights by turning them into commercial advantages. The possibility of attaining such advantages, it is believed, encourages innovation. But after a certain period of time, these legal rights are extinguished and the now-unprotected inventions and works can be freely used by others.

Intellectual property rights have never been more economically and politically important than they are in today's so called "knowledge-based society." Neither have they ever been so controversial. Intellectual property rights are frequently mentioned in discussions and debates on such diverse topics as human rights, agriculture, health, education, trade, industrial policy, public health, biotechnology, information technology, the entertainment and media industries, and the widening gap between the income levels of the developed countries and the developing countries.

Such importance and controversy arise from the fact that today's international intellectual property rules require both developed and developing countries to provide unprecedentedly high standards of protection. The 1994 Agreement on Trade-Related Aspects of Intellectual Property Rights (TRIPS), one of the main outcomes of the Uruguay Round of the General Agreement on Tariffs and Trade (GATT), which is administered by the Geneva-based World Trade Organization (WTO), is of special importance in that it establishes enforceable global minimum (and high) standards of protection and enforcement for virtually all the most important intellectual property rights in one single agreement.

However, there is much more to today's global intellectual property regime than TRIPS. The architecture of the global intellectual property regulatory system includes an increasing diversity of multilateral agreements, international organizations, regional conventions and instruments, and bilateral arrangements. In sum, the international law of intellectual property in its present form consists of three types of agreement. These are multilateral treaties, regional and supranational treaties or instruments, and bilateral agreements. Of these, the agreements that affect the greatest number of countries are TRIPS and some of the multilateral treaties administered by the World Intellectual Property Organization (WIPO), a specialized United Nations agency that is also located in Geneva. Prominent examples of the latter include the 1883 Paris Convention for the Protection of Industrial Property and the 1886 Berne Convention for the Protection of Literary and Artistic Works. Most multilateral intellectual property rights agreements are administered by WIPO, but there are some important non-WIPO treaties such as the 1961 International Convention for the Protection of New Varieties of Plants, which is administered by the

Union Internationale pour la Protection des Obtentions Végétales (UPOV), and of course TRIPS. Important regional agreements and instruments include the 1973 European Patent Convention and the 1998 European Union Directive on the Legal Protection of Biotechnological Inventions.

Intellectual Property and the Plant Sciences

Patents

In the plant sciences, the two most significant intellectual property rights are patents and plant breeders' rights. Patents provide inventors with legal rights to prevent others from using, selling, or importing their inventions for a fixed period, nowadays normally 20 years. Applicants for a patent must satisfy a national patent-issuing authority that the invention described in the application is new, susceptible of industrial application (or merely 'useful' in the United States), and that its creation involved an inventive step or would be unobvious to a skilled practitioner.

It is commonly assumed that patents are for mechanical devices. Mousetraps are frequently mentioned as typical examples. In fact, this has never been accurate. To illustrate this, in 1724, Thomas Greening was granted an English patent for "grafting or budding the English elm upon the stock of the Dutch elm." In 1785, Philip Le Brocq acquired a patent for "rearing, cultivating, training, and bringing to perfection, all kinds of fruit trees, shrubs, and plants; protecting their leaves, blossoms, flowers and fruits."

Nonetheless, the patenting of plants, seeds, and plant parts does not appear to go much further back than the 1930s. The United States Plant Patent Act of 1930 provided a type of patent to asexually reproduced plants. In Europe, patents were occasionally granted at around this time. In 1936, for example, the German Appeal Board accepted an application for a patent that claimed seed material of certain varieties of lupin. During the 1940s and early 1950s, countries adopted a range of approaches, from denying all intellectual property protection (for example, Britain and Denmark), to allowing patents (Italy from 1948, France from 1949, and Belgium from 1950), and creating specific intellectual property systems for plant varieties (for example, The Netherlands in 1941, Austria in 1946, and Germany in 1953). South Africa followed the United States example by introducing a modified patent system for plants in 1952.

In 1985, the United States led the way in extending the patent system more clearly to plants and plant material when the Patent and Trademark Office's Board of Patent Appeals and Interferences reversed the PTO's earlier rejection of a patent claiming corn plants and seeds as well as plant tissue cultures. The plants were produced through conventional cross-breeding, but relied on new techniques such as cell culture and genetic analysis. By 1988 over 40 patents on crop plants had been issued. To date there are more than 1800 US patents with claims to plants, seeds, or plant parts or tissues.

Nonetheless, the 1985 decision did not permanently settle the question of whether or not plants are patentable. In December 2001, the US Supreme Court finally confirmed the legality of patents on plants. The opportunity to do so arose because lawyers representing a company called J.E.M. Ag Supply, that was being sued by Pioneer Hi-Bred for patent infringement, requested the court to determine whether plant-related patents are invalid because of the existence of two intellectual property laws designed specifically to protect plants: the 1930 Plant Patent Act and the 1970 Plant Variety Protection Act.

The situation in Europe with respect to the patenting of plant-related inventions has been plagued by legal uncertainties. The 1973 European Patent Convention (EPC) states that patents shall not be granted in respect of "plant or animal varieties or essentially biological processes for the production of plants or animals."

This did not settle matters completely. Although from 1988, the European Patent Office (EPO) began to grant patents on plants, in 1995, the EPO Technical Board of Appeal in *Greenpeace* v. *Plant Genetic Systems* ruled on an appeal against the upholding of a plant-related patent and determined that a claim for plant cells contained in a plant is unpatentable since it does not exclude plant varieties from its scope. This implied that transgenic plants *per se* were unpatentable because of the plant variety exclusion. Consequently, for the next 4 years, the EPO stopped accepting claims on plants *per se*. However, in December 1999, the EPO Enlarged Board of Appeal decided that, while genetically modified plant varieties are unpatentable;

> a claim wherein specific plant varieties are not individually claimed is not excluded from patentability under Article 53(b), EPC even though it may embrace plant varieties.

In 1989, the European Commission, concerned about the legal uncertainties which, it was felt, could be prejudicial to the future of biotechnology in

Europe, and fearing that some European countries might respond to mounting controversy by banning patents on living organisms and genes, drafted a Directive on the Legal Protection of Biotechnological Inventions. The aim was to harmonize patent law relating to biotechnology around high minimum standards, while preventing member states from "backsliding." In 1998, the Directive was finally adopted.

According to Article 3(2) of the Directive,

> biological material which is isolated from its natural environment or produced by means of a technical process may be the subject of an invention even if it previously occurred in nature.

As with the Convention, animal and plant varieties and essentially biological processes for the production of plants and animals are excepted. Article 2(2) clarifies that

> a process for the production of plants or animals is essentially biological if it consists entirely of natural phenomena such as crossing or selection.

This definition has been accepted by the EPO.

Outside Europe and the United States, the legal situation is even more uncertain and has not yet been settled by the international rules. The TRIPS Agreement echoes the EPC by allowing countries to exclude from patentability

> plants and animals other than micro-organisms, and essentially biological processes for the production of plants or animals other than non-biological and microbiological processes.

It also states that countries "shall provide for the protection of plant varieties either by patents or by an effective *sui generis* system or by any combination thereof." Yet many developing countries have yet to fulfil their obligations in this respect, and it would be fair to say that many of them are unclear on how they should best do so. The majority do not allow plants to be patented. However, many of those governments expressing a preference for alternative intellectual property systems to protect plant varieties are generally reluctant to extend patent protection to biotechnological inventions, and have criticized the distinction between microorganisms and other types of organisms, and between microbiological processes and other biological ones, as being artificial and contrary to the basic tenets of patent law.

Since the 1980s, sequences of isolated and purified DNA from plants and other organisms are increasingly being claimed in patent applications. This has been criticized by those who wonder how, if patents apply to inventions but not discoveries, naturally occurring genes isolated in a laboratory can be patented.

Companies and researchers defending such patenting argue that locating, isolating, and describing molecular biological matter requires considerable ingenuity. If in doing so they have revealed to the world something that was previously unknown, and are able to explain its function and its possible industrial application, then they have an invention that should be patentable. Moreover, many claims are not actually for DNA in its raw state but for complementary DNA (cDNA). Therefore, defenders argue, as with any other synthetic chemicals, cDNA sequences should be patentable provided that they fulfil the normal criteria of inventiveness and that the application discloses a credible function. This view is consistent with European practice. In 1995, the Opposition Division of the EPO declared DNA to be "not 'life', but a chemical substance which carries genetic information," and therefore patentable just as any other chemicals are.

Many critics are sceptical that the deletion of "junk DNA" is inventive enough to deserve the reward of a patent on the grounds that a claimed cDNA molecule is likely to be obvious to somebody "skilled in the art" who knew the sequence of its naturally occurring equivalent. Furthermore, as some critics point out, gene function is so complex that treating genes as patentable inventions on the basis of a single discovered function is more a reflection of ignorance than of insight. Another argument is that the patenting of genes and gene fragments used in basic research may be placing undesirable restrictions on the ability of other scientists to use these in their own research.

Plant Breeders' Rights

Plant breeders' right (PBR) certificates provide legal rights for breeders over the varieties they produce which fulfil the criteria of novelty, distinctness, uniformity, and stability. Virtually all national PBR systems are based to a greater or lesser degree on the UPOV Convention.

The existence of the UPOV Convention can be attributed largely to two organizations, the International Association for the Protection of Industrial Property (AIPPI), and the International Association of Plant Breeders (ASSINSEL). At the 1952 AIPPI Congress, the delegates, partly at the urging of ASSINSEL's representatives, discussed the issue of plant varieties. There was general agreement that plant varieties should be legally protected in some way. The most concrete ideas came from the German AIPPI group, which submitted a detailed technical report. As the report noted, it is a normal requirement of patentability that other people skilled in the art should be able to reproduce the invention described in the specification. That is to say, following the instructions provided in the specification should result in the invention as claimed. However, as the authors explained, when it comes to plant breeding, being able reliably to reproduce the new variety from the beginning is difficult because it depends on natural processes over which breeders do not have total control, and which are to some extent random. But repeating the whole breeding process is not necessarily important or even necessary. What really matters is that the new plant that has been brought into existence can be directly propagated. For sexually reproducing plants, this means that they must breed true so that the offspring are identical to their parents.

As the authors saw it, patents in their present form would accommodate a certain amount of innovation in plant breeding and should be made available to breeders. But, for many new varieties, workable intellectual property protection would require a relaxation of the novelty and inventive step requirements so that varieties reflecting incremental improvements on existing ones and that were already known about could nonetheless be protected.

The 1952 Congress could not reach a consensus on the means of protection, as AIPPI failed also to do at its next Congress two years later. One of the main reasons was that some of the patent lawyer members of AIPPI opposed the patenting of plant varieties on the grounds that doing so would stretch basic patent law concepts like inventiveness to the point of undermining the credibility of the patent system. ASSINSEL's members decided at their own Congress in 1956 to abandon the patent route and to call for an international conference to consider the possibility of developing a new international instrument for protecting plant varieties. ASSINSEL requested the French government to organize what became the International Conference for the Protection of New Varieties of Plants. The Conference, which convened in May 1957 in Paris, established the basic principles of plant breeders' rights that were later incorporated into the UPOV Convention. At the second meeting of the Conference in November 1961, the International Convention for the Protection of New Varieties of Plants (the UPOV Convention) was adopted. It was revised in 1972, 1978 and 1991. The 1978 Act entered into force in 1981, and the 1991 Act in 1998. The Convention established the International Union for the Protection of New Varieties of Plants (UPOV), which is based in Geneva and has a close association with WIPO.

To be eligible for protection under the UPOV system, plant varieties must be novel, distinct, stable, and uniform (in UPOV 1991) or homogeneous (in UPOV 1978). To be novel, the variety must not have been offered for sale or marketed, with the agreement of the breeder or his successor in title, in the source country, or for longer than a limited number of years in any other country. To be distinct, the variety must be distinguishable by one or more characteristics from any other variety whose existence is a matter of common knowledge. To be considered as stable, the variety must remain true to its description after repeated reproduction or propagation. Unlike patents there is no disclosure requirement. Instead, applicants are required to submit the plant material for which protection is sought to the responsible governmental authority for testing to ensure that the above eligibility requirements have been met. This material may then be used by a government institution to demonstrate stability and homogeneity conclusively through propagation trials.

UPOV 1978, to which several countries are still contracting parties, defines the scope of protection as the breeder's right to authorize the following acts:

> the production for purposes of commercial marketing; the offering for sale; and the marketing of the reproductive or vegetative propagating material, as such, of the variety.

The 1991 version extends the scope of the breeders' rights in two ways. First, it increases the number of acts for which prior authorization of the breeder is required. These include:

> production or reproduction; conditioning for the purpose of propagation; offering for sale; selling or other marketing; exporting; importing; stocking for the above purposes.

Second, such acts are not just in respect of the reproductive or vegetative propagating material, but also encompass harvested material obtained through the use of propagating material, and so-called 'essentially derived' varieties.

However, the right of breeders both to use protected varieties as an initial source of variation for the creation of new varieties and to market these varieties without authorization from the original breeder (the "breeders' exemption") is upheld in both versions. One difference is that UPOV 1991 extends rights to varieties which are essentially derived from the protected variety. The idea here is that breeders should not be able to acquire protection too easily for minor modifications of extant varieties. This provision is also intended to ensure that patent rights and plant breeders' rights operate in a harmonious fashion.

There is no reference in the 1978 version to the right of farmers to resow seed harvested from protected varieties for their own use (often referred to as "farmers' privilege"). The Convention establishes minimum standards such that the breeder's prior authorization is required for at least the three acts mentioned above. Thus, countries that are members of the 1978 Convention are free to uphold farmers' privilege or eliminate it.

The 1991 version is more specific about this. Whereas the scope of the breeder's right includes production or reproduction and conditioning for the purpose of propagation (Article 14), governments can use their discretion in deciding whether to uphold the farmers' privilege. According to Article 15, the breeder's right in relation to a variety may be restricted

> in order to permit farmers to use for propagating purposes, on their own holdings, the product of the harvest which they have obtained by planting ... the protected variety.

Although the seed industry generally dislikes the farmers' privilege, until recently most countries upheld it, either explicitly or by default. However, since 1994, European Community plant breeders' rights restrict farmers' privilege to certain crops, and breeders must be remunerated through the payment of royalties unless they are small farmers, in which case they are exempted. The United States is less strict in this regard, but seed saving must be restricted to the amount necessary for on-farm replanting.

UPOV 1991 extends protection from at least 15 years to a minimum of 20 years. This later version is silent on the matter of double (that is, both patent and PBR) protection whereas the earlier version stated that members

> may recognise the right of the breeder provided for in this Convention by the grant either of a special title of protection or of a patent.

Removing the bar on double protection was intended to ensure that the United States remained compliant with UPOV.

Until recently, the overwhelming majority of UPOV members were developed countries, reflecting the fact that in many developing countries, especially in Africa, private sector involvement in plant breeding and seed supply is quite limited. Moreover, in many of these countries small-scale farming communities are responsible for much of the plant breeding and seed distribution, as they have been for centuries. Consequently, until recently there would have been few domestic beneficiaries of a PBR system, especially if state involvement in breeding was also quite limited.

However, many developing countries are now joining UPOV. In many if not most cases, this is not because of any strong domestic demand for PBRs, but because of their obligations under Article 27.3(b) of TRIPS. The UPOV system is the only *sui generis* system for plant varieties that exists in international law. This is not to say that other models have not been developed, but that these have not yet been implemented by any country and remain to be tested. Moreover, the UPOV system is being actively promoted world wide by the organization itself as well as by the United States and the European Community.

Despite the increasing membership of UPOV, one may reasonably wonder why breeders still tend to prefer PBRs to patents, and also whether this particular intellectual property right has a future. After all, patents provide much stronger legal protection. Moreover, breeders nowadays tend to work not for small independent seed firms, but for large life-science corporations that invest huge sums in biotechnological research and hold massive patent portfolios. Probably the main reasons for this preference for PBRs is the breeders' exemption, which allows them such broad access to breeding material, and their concerns that the patenting of biotechnological research tools may jeopardize this access. This suggests that as long as there is profit-motivated plant breeding, PBRs will continue to exist alongside patents, and sometimes in tension with them.

See also: **Crop Improvement**: Plant Genomes. **Ethics and Biosafety**: Ownership of Plant Genetic Resources.

Further Reading

Barton JH (1998) The impact of contemporary patent law on plant biotechnology research. In: *Intellectual Property Rights*, Vol. 3, *Global Genetic Resources: Access and Property Rights*, pp. 85–97, Madison: Crop Science Society of America and American Society of Agronomy.

Bent SA, Schwaab RL, Conlin DG, and Jeffery DD (1987) *Intellectual Property Rights in Biotechnology Worldwide*. Basingstoke: Macmillan.

Bently L and Sherman B (2001) *Intellectual Property Law*. Oxford: Oxford University Press.

Bugos GE and Kevles DJ (1992) Plants as intellectual property: American practice, law, and policy in a world context. *Osiris* 7: 75–104.

Drahos P and Blakeney M (eds) (2001) *Intellectual Property in Biodiversity and Agriculture: Regulating the Biosphere*. London: Sweet & Maxwell.

Dutfield G (2000) *Intellectual Property Rights, Trade and Biodiversity: Seeds and Plant Varieties*. London: Earthscan Books.
Dutfield G (2003) *Intellectual Property Rights and the Life-Science Industries: A Twentieth Century History.* Aldershot: Ashgate.
Fowler C (1994) *Unnatural Selection: Technology, Politics, and Plant Evolution*. Yverdon: Gordon & Breach.
Funder J (1999) Rethinking patents for plant innovation. *European Intellectual Property Review* 21: 551–577.
Heitz A (1987) The history of plant variety protection. In: *The First Twenty-Five Years of the International Convention for the Protection of New Varieties of Plants*, pp. 53–96. Geneva: International Union for the Protection of New Varieties of Plants.
Kloppenburg J Jr. (1988) *First the Seed: The Political Economy of Plant Biotechnology.* Cambridge: Cambridge University Press.
Llewelyn M (2000) The patentability of biological material: continuing contradiction and confusion. *European Intellectual Property Review* 22: 191–197.
Overwalle G van (1999) Patent protection for plants: a comparison of American and European approaches. *IDEA – The Journal of Law and Technology* 39: 143–193.
Tripp R (1997) *New Seed and Old Laws: Regulatory Reform and the Diversification of National Seed Systems*. London: Intermediate Technology Publications.
Wuesthoff F (1957) Patenting of plants. *Industrial Property Quarterly* 2: 12–30.

Ownership of Plant Genetic Resources

P W B Phillips, University of Saskatchewan, Saskatoon, SK, Canada

Introduction

The international development community has been engaged in recent years in discussions and negotiations about how to encourage invention and innovation in the global agrifood sector, while respecting and preserving much of the traditional knowledge and culture that is linked to agronomic practices in many communities around the world. This is not a new problem. By some reports, the majority of the indigenous landraces that have developed since man began to cultivate the environment have been lost due to disuse or neglect.

During the past century, various national governments and international agricultural research centers developed a range of programs and institutes to identify and preserve indigenous plant genetic resources either *in situ* (i.e., in use by local farmers) or *ex situ* (i.e., in discrete gene bank centers). While these collections are now quite large, with more than 6 million discrete accessions globally, it is unclear whether the agronomic vitality or traditional knowledge surrounding the germplasm is being preserved. Furthermore, there is significant debate about who owns the rights to determine access and use of the materials and who should benefit from any discoveries or inventions using those materials.

This debate has become much more urgent in recent decades due to two complementary developments. The development of new biotechnology tools has lowered costs and widened the possibilities of finding economically valuable traits in indigenous plants, animals, or microbes and then successfully developing them as products for commercial markets. Meanwhile, countries around the world have extended rights to intellectual property, including plant varieties, which has created new incentives to undertake research related to plant genetic materials. The combination of means and motive has combined to create a gold-rush mentality among many bioprospectors, who often ignore social norms, cultural artefacts, and pre-existing economic interests.

The issue of how to protect and manage plant genetic resources is highly complex, involving economic, social, cultural, and ethical dimensions. Any resolution of the issues will by necessity be complex and multifaceted.

What are Plant Genetic Resources?

Plant genetic resources (PGRs) have meaning at a number of levels, which vary depending on whether one is interested in the practical economic application in agriculture or more interested in their symbolic or cultural value to society.

The Food and Agriculture Organization's (FAO) *Glossary of Biotechnology for Food and Agriculture* defines plant genetic resources as the reproductive or vegetative propagating material of: cultivated varieties (cultivars) in current use and newly developed varieties; obsolete cultivars; primitive cultivars (landraces); wild and weed species; near relatives of cultivated varieties; and special genetic stocks (including elite and current breeder's lines and mutants). At that level, then, plant genetic resources are based on the original natural distribution of genetic materials before the intervention of man. Once man began to select specific foods and cultivate them proactively, key crops and animals became domesticated. Scientists and historians have identified that between 5000 and 10 000 years ago, people in "centers of origin" created the germplasm base for our main sources of plant protein today, including wheat (*Triticum* spp.) in south-western Asia, rice (*Oryza sativa*) in China, corn (*Zea mays*, maize) in Mesoamerica, potato (*Solanum tuberosum*) in the Andes and Amazonia, sorghum (*Sorghum* spp.) in the Sahel, oil palm (*Elaeis guineensis*) in tropical west Africa, and sugar cane (*Saccharum officinarum*) and banana (*Musa sapientum*) in New Guinea. Over the intervening millennia, farmers, and more recently organized public and private plant breeding programs, have dramatically altered the phenotypes of these plants, to the point where they would be unrecognizable to the original developers. Along the way, these plants were introduced and adapted to new regions, so that most of the world's production of these crops is now in areas far removed from the centers of origin. The accumulated genetic resources of the collective effort of mankind are the subject of debate. The FAO estimates that there are about 6 million accessions being stored world-wide in a total of 1308 genebanks, in 77 countries.

Many people and societies vest much more in their plants than the simple genetic and agronomic vitality of the seeds. Many plants play a major role in the economic, social, cultural, and historical life of traditional communities, and thus citizens are concerned about how the plants they have developed and nurtured are viewed and used by others. The 1993 Crucible Group directly considered the link between plant genetic resources and cultural vitality. They concluded that local and indigenous peoples who speak ancestral languages are severely threatened by loss of sovereignty over land, resources, and cultural traditions, which tends to be linked to control over their indigenous PGRs. Furthermore, as communities become increasingly marginalized, local people lose local scientific knowledge, innovative capacity, and wisdom about species and ecosystem management. Ultimately, there is concern that the reductionist approach of modern science, and biotechnology in particular, does not respect the vast body of traditional knowledge about plants and their place in the ecosystem built up over the millennia.

Intellectual Property Rights and PGRs

The debate about plant genetic resources has to a large degree been precipitated by the extension of private rights to intellectual property (IP). While there is a long history of countries trying to protect their indigenous plants – India in the 1700s, for instance, had a penalty of death for anyone illegally exporting rice for propagation – it was really only in the twentieth century that these rules were adopted to encourage investment in invention and innovation.

By the 1930s, the US patent system granted rights for asexually reproduced plants and in the 1940s a number of European nations granted plant breeders' rights (PBR) for new varieties developed domestically. PBRs were extended internationally in the 1970s through negotiation of the International Union for the Protection of New Varieties of Plants (UPOV). During the 1980s, rights were revised and extended as the new biotechnologies enabled new inventions. By 1985, the US Patent and Trademark Office had granted its first patent on a whole plant. This effort was internationalized in the 1990s, with the negotiation of the World Trade Organization Agreement in 1995 and the Trade-Related Aspects of Intellectual Property (TRIPS) Agreement, which requires all member states (144 in 2002) to either grant full patent protection to plants or to provide a "sui generis" (i.e., purpose built) system of protection for plant inventions. All developed countries were required to comply with TRIPS by 2001, while developing countries have until 2006 to comply.

The extension of private rights to plant genetic resources has worked to divide the research and farming community, clearly creating owner and user classes. Formal intellectual property right (IPR) structures – including patents, plant breeders' rights and trademarks – are all designed to grant a limited monopoly (by time, application, or market) for the "fruit of one's labors" in exchange for full disclosure of how the invention works. The various mechanisms are designed to provide a right to exclude for a finite, limited term others from the use of an invention, in order to allow the inventor or creator to exploit and benefit from the effort. All of these mechanisms require a clearly identified inventor or creator, a clearly defined inventive step (that is new and not obvious), and a proscribed use. In that sense, they are not designed to protect or support collective or public efforts that happen over a long period of time.

Private IPRs have been deliberately adopted in order to encourage greater private, for-profit research and to encourage more rapid commercialization of inventions. While many microeconomic studies indicate that in discrete areas the extension of property rights has accelerated investment, invention, and commercialization, the profusion of investment "races" leads to significant duplication of effort. Furthermore, many scientists are concerned that private rights to fundamental processes, specific genes, or whole organisms could block other users. In some cases, farmers and communities are concerned that they may lose legal access to plant genetic resources that they themselves developed and conserved if it becomes included in a patented variety. Researchers are concerned that subsequent invention will be reduced as licences to use IP are either strategically withheld to limit competition or the cost of negotiating and enforcing arrangements exceeds the capacity of programs or is greater than the value that would be generated by access. Thus, there is some debate about whether private IPRs are the only or the best method of supporting and encouraging research into plants. Other strategies, which might be more compatible with traditional PGRs, are more public investment though consortia, public research programs, or subsidies to private companies to undertake directed research.

Ethical Issues Related to PGRs

There are five specific ethical issues that the advent of private IPRs has raised around plant genetic resources. First, there is concern that many of the "inventions" being granted licences are simply rediscoveries or thefts of traditional knowledge. Second, there is concern that the rapid privatization and commercialization of the seed trade will cause accelerated loss of biodiversity. Third, there is real concern that as PGRs are increasingly privatized, farmers may be excluded from using their own traditional knowledge. Fourth, communities that have nurtured landraces with attractive commercial traits are concerned about how they will be able to share in any benefits resulting from greater use. Finally, the advancement of science, specifically genomics, is challenging the definition and role of PGR.

Invention, Discovery, and Biopiracy

One of the main concerns about the introduction of private investment in the development of plant genetic resources is that there is the potential for people to either inadvertently "rediscover" traditional knowledge or to knowingly misappropriate the communal property of others. Either way, there is some evidence that some individuals and groups have acquired patents or other property rights to traditional varieties. Many call this biopiracy.

In theory, patents or plant breeders' rights should not be available for pre-existing varieties, as the rules clearly state that disclosure and use anywhere in the world before filing for property rights invalidates a claim. Patents and PBRs are solely to reward invention and not discovery. The problem is that it is very difficult for patent examiners or variety reviewers in North America or Europe to know of every single variety in existence around the world. There has been a tendency in many countries to grant applications of claim and then to let the courts or marketplace determine whether the claims are valid. While that works in some instances, it creates potential for mistakes.

The Erosion, Technology, and Concentration (ETC) Group (formerly the Rural Advancement Foundation International) reports that they found 147 suspected cases of inappropriate PBRs in Australia, equal to 6% of all the applications at that point. Others argue that the problem is global in nature. There are a number of high profile cases that illustrate the concern. One company acquired a patent in both the US and EU on Basmati rice, a traditional variety in India. Once this became known, public interest groups in Europe and the US launched challenges and the patents were revoked. Meanwhile, a US-based seed company acquired a US patent to a yellow variety of bean, which subsequently turned out to be an established variety known as the Mexican golden bean; that patent was also challenged.

Remedies exist in all national IPR systems, but they are not fully accessible by everyone. They are often time consuming and expensive, especially if the patent is granted in multiple jurisdictions. For example, there was significant concern recently in Canada about a US patent issued in the 1990s to yellow-seeded canola, which was developed decades ago in Canada but was not commercialized extensively. When the Canadian government and canola research community investigated an appeal through the US Patent and Trademark Office, they learned that it could cost them up to US$1 million to prosecute the case. Given that the variety had limited commercial application, the industry has so far decided to let the patent stand unchallenged.

PGRs and Biological Diversity

In 1996, the FAO issued its first State of the World's Plant Genetic Resources, warning of a large scale loss of plant genetic resources. The report argued that "the spread of modern, commercial agriculture and the introduction of new varieties of crops has been the main cause of the loss of genetic diversity." The erosion of biodiversity and the loss of genes in those species was viewed as critical for agriculture and food security. The report argued that currently there are 800 million people in developing countries who do not have adequate food, of which 200 million are chronically undernourished children. The loss of biodiversity threatens food security, especially for the poor, who rely on biological products for 85–90% of their livelihood needs (i.e., food, medicine, fuel, fiber, clothing, shelter, energy, transportation). While loss of biodiversity predates the advent of biotechnology and private IPRs, both developments have the potential to accelerate the problem.

No one knows absolutely how much diversity once existed in domesticated species, so it is impossible to say exactly how much has been lost. One study estimates that 34 000 species of plants – estimated to be about 12.5% of the world's current flora – are facing extinction. The FAO report on the State of the World's Plant Genetic Resources issued in 1996 cites some examples of the loss of biodiversity. In China, of the nearly 10 000 wheat varieties in use in 1949, only 1000 remained in the 1970s, while in the USA, 95% of the cabbage (*Brassica oleracea* var. *capitata*), 91% of the field corn, 94% of the pea (*Pisum sativum*), and 81% of the tomato (*Lycopersicon esculentum*) varieties cultivated in the nineteenth century have been lost. Many other species are suffering a similar fate in Europe, Africa, Asia, and South America.

Loss of biodiversity poses two main concerns. First, the concomitant increase in uniformity of plant genetic material can lead to greater risk and uncertainty. Considerable genetic uniformity now exists in crops such as hybrids of rice and sunflowers (*Helianthus* spp.). The FAO notes that uniformity in the rootstock of California wine grapes (*Vitis vinifera*) and the resulting uniform susceptibility to a virulent disease has caused wine producers to dig up and replace their vines at the cost of hundreds of millions of dollars. Second, it has been argued that for every plant that becomes extinct, 30 other species go with it, many of them microorganisms. This is viewed with concern by some biologists because many plant pathogens (including fungi, viruses, and bacteria) are believed to play a vital role in the functioning of ecosystems. While commercial agricultural practices are not the only reason for loss of biodiversity – degradation and destruction of forests and bush lands, overgrazing, wars, and civil strife also threaten ecosystems – the international agricultural research community believes this is one area that they have the potential to influence.

One solution has been to develop *ex situ* genebanks to house extensive collections of landraces and other varieties. While there are more than 6 million accessions now in place, the infrastructure supporting those collections is at risk. The FAO reports that equipment problems (particularly cooling and humidity control units), insecure electrical supply, difficulties in seed drying (especially in humid regions), and inadequate investment in regeneration is endangering seed quality. The FAO has further stressed that seed conservation in genebanks (*ex situ*) and in ecosystems and natural habitats (*in situ*) are complementary, and both should be strengthened.

Access to PGRs

There is significant concern that the increased private involvement in plant breeding will put plant genetic resources out of the reach of all but the wealthiest farmers and societies. Concerns arose in the 1970s that developed nations and private companies were using developing country PGRs to breed new plant varieties, which were protected by private property rights and sold back to developing country farmers on a for-profit basis. More recently, there has been concern that the extension of patents to genes and whole plants raises the possibility that not only farmers but researchers may be unable to use proprietary germplasm for breeding. Hence, they will have lost not only the benefits of their past investments, but access to future improvements.

The differential operation of the patent and plant breeders' rights systems is at the root of this concern. Plant breeders rights laws in all countries provide two key exemptions: farmers are allowed to save seed from a crop for subsequent planting, without further payment of royalties; and researchers are allowed full and free access to germplasm protected by PBRs for follow-on research and breeding. Patents, however, have neither of these exemptions. Thus, farmers are legally prohibited from saving patented seed for replanting and researchers must negotiate licences to commercialize new varieties that are based on patented germplasm. This "pull-through" nature of the patent system has raised concern throughout the public and private plant breeding community world-wide.

One solution offered was the International Undertaking on Plant Genetic Resources, adopted by the FAO in 1983, which deemed that all plant germplasm in international genebanks would be "the common heritage of mankind." This has been hard to put into operation. Many of the countries where the accessions were gathered have claimed that the genetic material was taken without authority or compensation. As a result, it is not clear that the Undertaking covers all of the accessions in the international genebanks. In the 1990s, the Consultative Group on International Agricultural Research, which operates the 13 largest international genebanks, agreed with the FAO that all duly deposited accessions in their facilities would be covered by the undertaking. In practice, this means that breeders are given free access to any germplasm but in exchange they may not apply private property protection on any resulting varieties.

At the product level, there have been a number of cases that have raised practical concerns about how developing nations and their farmers can gain access to improvements that are based on their traditional varieties. The most celebrated case was GoldenRice^{TM}, a beta-carotene-enhanced variety bred by a Swiss research group using mostly public funds. Once the inventors had produced a variety they thought had potential, they discovered that they had accessed more than 70 different proprietary technologies, owned by more than 40 different entities. In this case, they made an arrangement with Astra-Zeneca, holder of many of the key IP rights. Astra-Zeneca agreed to acquire freedom to operate for the variety and licence it without royalty to ten developing countries with vitamin A deficiencies. In exchange, the company received the world-wide rights to exploit the variety for profit. Some economists have suggested that this case could provide a model for dual-marketing of inventions, differentiating between for-profit exploitation in developed nations and not-for-profit dissemination in developing regions. One of the complicating factors in the above case was that there were different arrays of rights in different countries, and searching for rights holders and negotiating licences in each market was deemed to be very expensive. One solution to this might be for developing regions to pursue regional systems of intellectual property rights, whereby inventors would register in one system that represents multiple countries. This would enable researchers seeking owners to negotiate single licences for a region.

Other models are being tried or considered. The International Service for the Acquisition of Agri-biotech Applications (ISAAA) has undertaken a number of international collaborations with national research centers and multinational companies to mediate the transfer of technology from developed countries and multinational corporations to developing countries. Meanwhile, some private companies have simply donated some of their IP to developing countries and plant breeders. The most notable example was in 2001 when Monsanto announced that it would make its working draft of the rice genome freely available to any researcher.

Benefits Sharing

One of the main concerns about the widespread use of traditional PGR is that the traditional developers and conservators of that material often do not benefit in any real way from follow-on development. There is great suspicion that few enhancements add much value; rather, most of the value is actually a return to past investments. Many of the plants, animals, and microbes nurtured and conserved by indigenous peoples have value but they are often not in the optimal form for effect. Research often simply distils and refines the active ingredient and then transforms it into a more commercially valuable formulation. In these cases, there is a basis for claiming that much of the value in the process already existed, and often there are demands for benefits sharing. The challenge is that the focus of the modern intellectual property system has been almost exclusively on creating incentives for new invention, and less on compensating those from the distant past. Given that most PGRs in genebanks and *in situ* were developed long ago and often through collective action, there is no basis for a claim through the various national patent or PBR systems for benefits sharing. Nevertheless, there appears to be both moral and practical benefits in negotiating some form of benefit sharing.

At the moral level, there is a case to be made that many communities have been vital to the development and conservation of unique traits in our plants that otherwise might have been lost to mankind. Especially in the past century, with rapid loss of biodiversity in many areas, the role of conservators has been vital. The argument often made is that the investment by these peoples should be respected and rewarded. Vandana Shiva of India, the Crucible Group and ETC Group, among others, argue that this is the right thing to do. This principle is a prominent feature of the Convention on Biological Diversity (CBD) and of Farmers' Rights as discussed and supported in the FAO Commission on Genetic Resources for Food and Agriculture.

Even if the moral obligation is not convincing, there are strong practical reasons for benefit sharing. There is some evidence that without access to the traditional knowledge surrounding a species, its role in an ecosystem, and its potential food or therapeutic uses, much bioprospecting for new traits may be uneconomic. Much of the biological surveying done in the tropical rainforests, in arid and desert regions, and in richly endowed river ecosystems, such as the Amazon, would be daunting without the guides and interpreters often provided by local communities. The traditional lore surrounding many foods and plants often has a grounding in the genetics or ecosystem context for the plants. Being able to access that advice in an efficacious manner can sharply lower the cost of bioprospecting.

While there are both moral and practical bases for benefits sharing, there are also some considerations that limit its effective application. In the first instance, it is often difficult, if not impossible, to determine the ownership of many PGRs. There are multiple ownership claims to most PGRs, reflecting the fact that few PGRs are unique to a particular country or community. Second, the sheer size of the value generated by an invention imposes a constraint on benefits sharing. While many point to the handful of lucrative pharmaceutical compounds and food varieties that have emerged from bioprospecting, they ignore the fact that the entire bioprospecting industry is not overly profitable. Some suggest that only about one in a million candidate organisms offer adequate commercial returns to justify the investment. So far, only about 30 pharmaceutical organic compounds bioprospected have been successfully commercialized.

Even if ownership could be delineated appropriately and adequate returns existed to justify the effort, the diffusion of potential owners and lack of effective institutions would make many negotiations on rights highly complex, costly, and uncertain. The International Indigenous Forum on Biodiversity argued that the presence of multiple sellers of PGRs will inevitably lead to competition and low prices for PGRs. The forum argued that if some pooling mechanism could be developed that would hold monopoly rights to PGR, then they would be able to extract a greater share of their value. The difficulty is that no such monopoly presently exists and there is no obvious candidate to undertake the task. This would require countries and communities to subordinate their interests, which does not appear to be popular.

Nevertheless, there are a number of examples from both the developed and developing world where private research efforts have engaged in some form of benefits sharing. The more successful arrangements have usually been negotiated before any significant investment in bioprospecting. These arrangements often involved jobs in the bioprospecting stage (including service and supply contracts) and clearly agreed upon provisions for ownership of any discoveries and resulting inventions, including provisions for royalties. There are documented agreements with the Costa Rican government, with the Indians in the Brazilian rainforest and with First Nations' people in the forest area in Western Canada.

Genomics and PGR

While the FAO's definition of PGRs focuses on whole, reproducing plants, the advent of new biotechnology tools that allow for identification, isolation, and use of genes that express for specific traits has widened the debate about what could or should be interpreted as PGRs. Many interest groups, communities, and some countries have argued that the definition of plant genetic resources should also include the respective genomes. While this appears reasonable to many, it creates a technical challenge.

The ability to sequence parts of entire genomes and then to isolate, extract, and move selected genes of potential commercial value has sharpened plant breeders' focus from plant phenotypes to the genetic material itself. In the process of sequencing a wide range of plant, animal, and microbial genomes, scientists have discovered that most organisms share a large portion of their genetic material with other species, often from other kingdoms. For example, we now know that humans and mice share about 90% of the same genes, while humans even have many common genes with plants. Thus biotechnologists are no longer limited to examining the stock of plant genetic material in a species or its close relatives to find useful traits. Breeders using biotechnology can now look in a wide range of organisms, and can often find identical or similar genes in many of them. As a result, it is going to be harder for discrete communities to claim unique rights to any individual gene. Rather, their claim to uniqueness may be forced back to the FAO definition, where it is the unique combination of genes in a specific species rather than any individual trait.

See also: **Crop Improvement**: Plant Breeding, Principles. **Ethics and Biosafety**: Development and Commercialization of Genetically Modified Plants; Intellectual Property Rights. **Genetic Modification**: Transgene Stability and Inheritance.

Further Reading

Crucible Group (2000) *Seeding Solutions: Policy Options for Genetic Resources*, Vol. 1. Convention on Biological Diversity. Available at: http://www.biodiv.org/convention/articles.asp

ETC Group (Undated) Available at: http://www.rafi.org/main.asp

FAO Commission on Genetic Resources for Food and Agriculture (1983) *International Undertaking on Plant Genetic Resources.* Available at: http://www.fao.org/ag/cgrfa/IU.htm

Food and Agriculture Organization (FAO) (1998) *State of the World's Plant Genetic Resources for Food and Agriculture.* Rome: FAO.

Kryder RD, Kowalski SP and Krattiger AF (2000) *The Intellectual and Technical Property Components of Pro-Vitamin A Rice (GoldenRice™): A Preliminary Freedom-To-Operate Review.* ISAAA Briefs No. 20. Ithaca, NY: ISAAA.

Shiva V (1997) *Biopiracy: The Plunder of Nature and Knowledge.* Toronto: Between the Lines.

World Trade Organization (1994) *The Results of the Uruguay Round of Multilateral Trade Negotiations: The Legal Text*. Geneva: GATT. Available at: http://www.wto.org/wto/english/docs_e/legal_e/legal_e.htm#TRIPs

Plant Genetic Engineering, Ecological Issues

J M Dunwell, University of Reading, Reading, UK

Introduction

Transgenic crops are now grown on more than 40 million ha in 13 countries throughout the world. This article will consider various ecological aspects related particularly to the cultivation of transgenic herbicide and insect-resistant crops. This will include the significance of gene transfer to related plants or to microbes, the effect of insect-resistant plants on nontarget organisms, and the possibility of other inadvertent effects of the transgenic plant on the environment. It will concentrate on recently published data and reviews, and the reader is referred to "Further Reading" for further information.

Background

Approximately 600 species of insects, 1800 species of plants, and numerous species of nematodes and microbes are considered to be serious pests in agriculture and if left uncontrolled they would significantly reduce yields. For thousands of years since the start of agriculture, these pests were controlled by a variety of cultural and biological means. For example, prior to the development of synthetic pesticides after the Second World War, weeds were controlled by tillage, mowing, site selection, crop rotation, hoeing, or pulling by hand. The first means of chemical control of pests included such toxic chemicals as mercury, arsenic, copper compounds, and sulfur that were used as insecticides and fungicides. Following the development of synthetic compounds in the middle of the last century, farmers rapidly adopted these products such that between 1950 and 1980, herbicide use on corn reached 100% of the acreage. By 1997, the total amount spent on agricultural pesticides in the USA was US$8.8 billion, with approximately two-thirds on herbicides and one-fifth on insecticides.

Environmental concern about pesticide use has developed for a number of reasons, although many of these concerns are based on theoretical considerations rather than real experimental data. For example, most insecticides used in agriculture are toxic to honeybees and wild bees and there is concern about the effect of these compounds on both wild and cultivated plants; in the USA, approximately one-third of annual agricultural production comes from insect-pollinated crops, and flowering plants in natural ecosystems may not thrive because of fewer pollinators. However, it is generally considered that intensification and habitat loss are the most frequent cause of pollinator impoverishment reducing crop yield; there is no clear example of low crop yield resulting from the effect of pesticide (or transgenic plants – see below) on pollinators. In addition, many pests become resistant to pesticides after repeated exposure. It is reported by the US congress that in the USA, over 180 insect and arachnid pests are resistant to one or more insecticides, and 18 weed species are resistant to herbicides. Such concerns, along with changes in agricultural policy, meant that annual pesticide use as measured by weight of active ingredient peaked in the USA at a level of about 270 million kg in 1982, at a time when land used for agriculture was at a record high. Since that time, the use of most pesticides, particularly herbicides, has fallen. This trend is now being affected by the introduction of genetically modified (GM) or transgenic crops. These products, which depend upon the insertion of a specific gene or genes into an existing variety, can be produced more rapidly and at a lower cost than an equivalent new chemical product. It is estimated that production of a GM variety may take 6 years at a cost of US$10 million compared with 11 years at a cost of US$50–70 million for a chemical pesticide.

The rapid uptake of the first generation of these GM crops, those with herbicide (principally glyphosate or glufosinate) and/or insect tolerance (these plants express an insecticidal protein from the soil bacterium *Bacillus thuringiensis* or Bt), can be seen from the latest data, which show the estimated global area of transgenic crops for the year 2000 to be 44.2 million ha or 109.2 million acres. Putting this global area of transgenic crops into context, 44.2 million ha is equivalent to almost twice the area of the UK. During the 5-year period 1996 to 2000, the global area of transgenic crops has increased by more than 25-fold, from a base of 1.7 million ha. This high rate of adoption reflects the growing acceptance of transgenic crops by farmers using the technology in both industrial and developing countries, the present total being 13. Considering the type of transgenic crop, herbicide tolerance has consistently been the dominant trait with insect resistance (Bt) being second. In 2000, herbicide tolerance, deployed in soybean (*Glycine max*), corn (*Zea mays*; maize), and cotton (*Gossypium* spp.), occupied 74% of the total acreage, with 8.3 million ha planted to Bt crops, equivalent to 19%, and stacked genes for herbicide tolerance and insect resistance deployed in both cotton and corn occupying 7% of the global transgenic area. The trend for stacked genes to gain an increasing share of the global transgenic crop market is expected to continue.

Adoption of herbicide-tolerant crops has been particularly rapid. Herbicide-tolerant soybeans became available to farmers for the first time in limited quantities in 1996, but use expanded to about 17% of soybean acreage in the major States surveyed in 1997, and to more than 50% in 2000. Herbicide-tolerant cotton expanded from 10% of surveyed acreage in 1997 to 26% in 1998, and reached 46% in 2000.

The overall impact of these changes of agricultural practice on the environment may be assessed in several ways. For example, the adoption of herbicide-tolerant soybeans has led to the substitution of glyphosate herbicides for those used previously. The most recent calculations suggest that an estimated 2.5 million kg of glyphosate replaces 3.3 million kg of other synthetic herbicides, such as imazethapyr, pendimethalin, and trifluralin. Glyphosate has a half-life in the environment of 47 days, compared with 60–90 days for the herbicides it commonly replaces. These latter herbicides are also 3.4–16.8 times more toxic than glyphosate, according to a chronic risk indicator based on the Environmental Protection Agency (EPA) reference dose for humans. Thus, the substitution due to herbicide tolerance in soybeans results in glyphosate replacing other synthetic herbicides that are at least three times as toxic and that persist in the environment nearly twice as long. Similarly, it has been reported that since the introduction of Bt cotton varieties in the USA, the amounts of pesticides used on this crop has fallen by ~ 1.2 million kg year^{-1}; about 40% of all the insecticides in the world are applied to cotton.

However, there are several other ecological aspects to the growth of transgenic crops that need to be considered and some of these are addressed below.

Gene Flow

The possibility, and consequences, of gene transfer from a transgenic crop to another organism may be divided into two aspects. First, there is the possibility of pollen transfer to the same species or a closely related species and, second, of gene transfer from the plant to another nonrelated species, particularly a microbe growing on or close to the plant. This latter process is usually known as horizontal gene transfer.

Pollen Transfer

It is estimated that of the world's 13 most important crop plants, 12 are known to hybridize with wild relatives in some part of their agricultural distribution. For this reason, the possibility of pollen transfer has been amongst the first considerations of the regulatory authorities responsible for the licensing of any transgenic crop. For gene flow to occur via normal sexual transmission, certain conditions must exist: the two parents must be sexually compatible; their period of flowering must coincide; a suitable pollen vector must be present and capable of transferring pollen between the two parents; and any resulting progeny must be fertile and ecologically fit for the environment in which they are situated. Unless these conditions are met there will be no progeny to be assessed.

Amongst the most obvious consequences of pollen transfer is the possible transfer of the herbicide-tolerant trait from a cultivated species to a related weed. The concept of thus generating a "superweed" that could not be killed by conventional herbicides has been one of the most frequent fears raised by critics of transgenic technology. The two most commonly investigated species in studies of pollen transfer are *Brassica napus* or oilseed rape (also known as canola in North America, or colza in France) and sugar beet (*Beta vulgaris* subsp. *vulgaris*). Results from recent investigations are summarized below.

There have been several reports on pollen transfer using *B. napus* as a model, both with non-GM and GM material. Possibly the largest scale study was that conducted using remote sensing from satellites to identify possible sites of sympatry between conventional varieties of *B. napus* and its progenitor species over 15 000 km^2 of southern England. Two sympatric populations with *B. rapa* and one with *B. oleracea* were identified over the entire survey area, and within these populations only one hybrid was found in the 505 *B. rapa* plants screened and none of the nine *B. oleracea* plants were hybrid. These important conclusions provide the context in which the consequences of any pollen transfer from transgenic varieties must be considered.

In a related experimental study designed to examine the possible genetic costs associated with the acquisition of a *B. napus* transgene by a weedy relative *B. rapa*, it was found by analysis of 457 plants from the third backcross generation that there were no significant differences between transgenic and nontransgenic plants in survival or number of seeds per plant. It was concluded that the costs associated with the transgene were probably negligible.

The other main crop used in this type of studies is sugar beet (*B. vulgaris* subsp. *vulgaris*), a species that is sexually compatible with the wild sea beet species *B. vulgaris* subsp. *maritima*. Since cultivated beet seed production areas are sometimes adjacent to sea beet populations, gene flow from cultivated beets has the potential to alter the genetic composition of the nearby wild populations. In one study, isozyme allele frequencies of 11 polymorphic loci were measured in 26 accessions of cultivated beet, 20 sea beet accessions growing near a cultivated seed production area, and 19 sea beet accessions growing far from such areas. Evidence was found for two alleles that provided evidence for gene transfer to the sea beet grown close to commercial beet seed. Similar studies on 54 French populations showed that weed beets in northern France are intermediates between sugar beet and inland wild beets in southwestern France. There have been various assessments of the frequency of pollen transfer in beet, including one that recommended the use of cytoplasmic male sterile plants as bait plants. Of course, the use of such plants will significantly overestimate the frequency of gene transfer under natural conditions. A related study on the performance of hybrids between Swiss chard (*B. vulgaris* subsp. *cicla*) and rhizomania-resistant transgenic beets expressing a beet necrotic yellow vein virus (BNYVV) coat protein suggested that ecological implications due to the introduction and spread of virus-resistant hybrids will only be observed in those feral chard populations in which fitness is significantly influenced by high infestations of BNYVV.

In one of a number of similar studies comparing the frequencies of transfer of native and introduced genes, it was found that the morphological marker cotyledon color in melon (*Cucurbita melo*) was transferred at the same frequency as the transgene *npt II*. There is no evidence from any study that transgenes are transferred at a frequency different than that for native genes.

There has been some concern expressed that the commercial development of crop varieties, each with resistance to a different herbicide, may lead to the inadvertent production of plants with multiple resistances to several herbicides and thus to a problem in controlling "volunteer" plants remaining from a previous season's crop. Indeed, examples of such multiple resistances in oilseed rape have been reported in Canada; volunteer plants were tested for resistance to glyphosate, imidazolinone, and glufosinate and several were found to contain double resistance with two seedlings from 924 tested having resistance to all three compounds. However, such multiply resistant plants are still susceptible to a range of other herbicides. It should be noted that there is some evidence that relict plants of feral oilseed rape can persist in a seminatural habitat (road verges) for at least 8 years.

It has been suggested that much of the potential pollen transfer from GM crops could be reduced by extensive isolation distances or the use of border rows of nontransgenic plants. Various experimental studies of transgenic *B. napus* have confirmed data obtained many years previously with non-GM material, namely that outcrossing rates decline exponentially from 0.7% at 0 m to 0.02% at 30 m with more than 80% of the outcrossing events detected in the first 10 m of the border area. Current Canadian Government regulations for field testing of experimental GM material require either isolation zones of 200 m or 10-m border rows of synchronously flowering non-GM plants. Other experimental studies have shown that the effectiveness of borders is greatly influenced by both the relative numbers of donor plants and by environmental variables.

Horizontal Gene Transfer

Horizontal gene transfer (HGT) has been defined as the nonsexual exchange of genetic material between organisms belonging to the same or different species. In the present context, this refers to the transfer of a transgene from a plant, either from a live specimen or its residue, to a microbe on the surface of the plant or in the vicinity of the root. Although there is no doubt that there has been extensive genetic exchange between microbes and higher eukaryotes over the course of evolutionary time, the evidence for actual gene transfer under natural conditions is much less certain, principally because such transfer would suggest the automatic compromising of species identity.

For HGT to occur in the field a large number of conditions must be fulfilled. These range from the presence of plant DNA to the induction of "competence" in the recipient microbe. The first of these conditions is the most likely to be met, since DNA from transgenic sugar beet plants has been shown to persist in the soil for up to 2 years, despite the conventional wisdom that high molecular weight DNA is rapidly degraded in the soil environment. It is thought that the DNA may be protected from degradation by its adsorption onto soil constituents such as clay, feldspar, and heavy minerals, as well as humic acids. Estimates of the actual frequency of any transfer are very difficult to make because of the normally very infrequent occurrence. It has been shown experimentally that the *npt II* gene present in transgenic potato plants transforms naturally competent cells of the soil bacteria *Pseudomonas stutzeri* and *Acinetobacter* BD413 harboring a plasmid with a defective *npt II* gene at a relatively high frequency. However, the transformation frequency is reduced by 10^8–10^9-fold for recipient cells that lack the homologous sequences. This suggests a very low probability of nonhomologous DNA fragments are integrated by illegitimate recombination events during transformation. The best overall estimates of the occurrence of HGT under natural conditions is 10^{-17}. It should also be noted that this phenomenon only has an ecological consequence if the presence, and expression, of the transgene affects the behavior of the recipient microbe in such a manner that it has a consequential effect on its ecosystem. It is important to note that the introduced genes that determine both herbicide and insect resistance in the crops commercialized to date have a microbial origin and are present in many soil microbes.

Among the various studies related to HGT have been investigations of the microbial community (rhizosphere) around the roots of transgenic plants. The composition and diversity of this community is influenced by many factors including root density and presence of root exudates. In a study of three oilseed rape cultivars, it was found that the transgenic variety had a lower diversity in the endophytic flora than the two non-GM, though not isogenic, varieties. The importance of testing individual transgenes has been confirmed by results that show that roots of an experimental potato (*Solanum tuberosum*) expressing a T4 lysozyme show a 1.5–3.5-fold higher killing *of Bacillus subtilis* in the rhizosphere than do roots of non-GM plants.

Nontarget Organisms

Probably the majority of ecological investigations of transgenic plants have involved an examination of the possible impact of these plants on organisms other than those that were the target of the deliberate gene introduction. In particular, there have been numerous studies of the effects of insect-resistant Bt plants on a variety of beneficial, nontarget insects, such as bees and butterflies. Some of these are summarized below.

Most studies with bees have been conducted on oilseed rape, with the particular research method determined by the type of transgene involved. For example, the possible direct effects are examined by an assessment of the transgene product in the nectar, pollen, and resin, observation of the foraging behavior, and by measuring the lifespan of worker bees exposed to the product. The indirect effects are studied by assessment of pollen and nectar production, flower size and density, and by observation of the ability of bees to cross-visit transgenic and conventional plants, or cultivated oilseed rape and wild relatives. Results obtained so far suggest that any impact on pollinators will depend on a case by case analysis of the gene concerned and on its expression in the parts of the plant ingested by bees.

In a similar type of study on worms, it has been observed that there were no significant differences in the per cent mortality and weight of earthworms after 40 days in soil plants with Bt or non-Bt corn, although the Bt protein was present in the guts and casts of worms exposed to soil with Bt plants. This finding confirmed that the Bt protein released in root exudates or transgenic biomass was bound to surface-active particles in the soil, and thus partially protected from biodegradation.

Most studies on the ecological effects of Bt crops have focused on the possible toxicity to nontarget moths and butterflies, particularly the monarch butterfly (*Danaus plexippus*). This followed a preliminary report of the toxic effect of Bt corn pollen applied to milkweed (*Asclepias* spp.) leaves in a laboratory. Subsequent large-scale studies conducted under field conditions have not confirmed this finding.

A related study was conducted on the black swallowtail (*Papilio polyxenes*), whose host plants in the midwestern US are located primarily in narrow strips between roads and field. Using host plants placed at intervals from the edge of a field of Bt corn, there was found to be no relationship between larval mortality and proximity to the field or deposition of pollen on the host plants. Moreover, Bt pollen failed to cause mortality in the laboratory even at the highest level tested (10 000 grains cm^{-2}), a level that greatly exceeded the highest level observed in the field (200 grains cm^{-2}).

Other results from a series of more recent studies, conducted in the USA and Canada, may be summarized as follows:

1. Expression of Cry1b proteins, the active Bt protein produced in transgenic corn plants, varies considerably in different corn hybrids, particularly so in the pollen.
2. A comparison of three Bt "events" (different transgene introductions) showed that monarch larvae first instars exposed for 5 days to low doses (22 grains cm^{-2}) of event-176 pollen gained 18% less weight than those exposed to Bt11 or Mon810 pollen. In addition, larvae exposed in an event-176 field had 60% lower

survival and 42% less weight gain than larvae from outside the field. However, this particular strain of Bt corn is at present grown on <2% of the corn acreage and is not being reregistered. The other Bt events are likely to have no acute effects on monarch butterfly larvae under field conditions.

3. Pollen density was highest (maximum: 1400 grains cm^{-2}; average: 171) inside the cornfield and fell to a value of 14 grains cm^{-2} 2 m outside the edge of the field. There was found to be no differences between Bt and non-Bt plants or pollen in terms of the oviposition behavior of adult butterflies.
4. Bt pollen drifting onto adjacent rows of non-Bt corn has no effect on the incidence of corn borer (*Ostrinia nubilalis*) and therefore should not increase the risks related to resistance management.
5. In a similar study, it was found that Bt potato plants resistant to the Colorado potato beetle (*Leptinotarsa decemlineata*) had no effect on the relative abundance of naturally occurring generalist predators. In contrast, potato plants sprayed with biweekly applications of the broad spectrum pyrethroid permethrin as a means of beetle control had significantly reduced levels of several generalist predators such as spiders (Araneae), big-eyed bugs (*Geocorus* spp.), damsel bugs (*Nabid* spp.), and minute pirate bugs (*Orius* spp.), along with significantly increased numbers of green peach aphid (*Myzus persicae*), a vector of virus disease. It was concluded that such transgenic potatoes were quite compatible with the development of integrated pest management (IPM) regimes.

In addition to studies of the direct effects of Bt corn on nontarget pests, there have been a series of studies of tritrophic interactions with this material. The conclusions from these studies were that there was no effect on the relative abundance of the herbivorous prey *Anaphothrips obscurus* (a thysanopteran pest of corn) and the predator *Orius majusculus*. In a similar study using the two prey species *Rhopalosiphum padi* and *Spodoptera littoralis*, it was shown that larvae of the prey *Chrysoperla carnea* could not discriminate between *R. padi* fed on Bt or non-Bt corn, whereas they showed a significant preference for *S. littoralis* fed on the non-Bt plants. Similarly, Bt corn was found not to have a significant effect on *Orius insidiosus*, a natural predator of the European corn borer, the main target in such corn.

Phytoremediation

Amongst the most interesting potential ecological benefits of transgenic plants is their use in phytoremediation. Such experimental plants have already been shown to be capable of reducing the contamination of soils with heavy metals (e.g., mercury), explosives (e.g., trinitrotoluene, TNT), chlorinated solvents (e.g., trichlorethylene), and hydrocarbons (oil and petroleum residues). A similar benefit is provided by the use of transgenic plants as biomonitors of toxic chemical and radiological compounds. For example, GM plants have been produced that are capable of detecting the increased radioactive pollution resulting from the Chernobyl accident.

Other Indirect Ecological Effects

In a recent theoretical modeling study of the effect of GM herbicide-resistant crops on weed populations, it has been suggested that weed populations might be reduced to low levels or practically eliminated, thus having an indirect, but dramatic, effect on bird populations. In contrast, a detailed ecological study of glufosinate-resistant winter oilseed rape grown in central Germany found that weed control could be provided by a single application of the relevant herbicide. Under these conditions, a high diversity of epigeal arthropod species (about 120), including many in the red-data books, was observed in both years, and there was no evidence of any reduction in diversity in the transgenic crop.

Another recently reported aspect of Bt corn relates to the physiological characteristics of the transgenic plants. Ten Bt corn hybrids, representing three different transformation events, were examined by fluorescence microscopy and staining with toluidine blue, and the results from all hybrids indicated a higher content of lignin in the vascular bundle sheaths and in the sclerenchyma cells surrounding the vascular bundle than in their respective non-Bt isolines. Chemical analysis confirmed that the lignin content of all the Bt hybrids, whether grown in a plant growth room or in the field, was significantly higher (33–97%) than that of their respective non-Bt isolines. It was suggested that as lignin is a major structural component of plant cells, modifications in lignin content may have ecological implications.

Ecosystem Approach

The alternative to simply considering the characteristics of an individual transgenic crop is to adopt the more wide-ranging ecosystem approach to the assessment of transgenic plants, an approach endorsed by the Conference of Parties to the

Convention on Biological Diversity. The term "Ecosystem" is defined as a dynamic complex of plant, animal, and microorganism communities and their nonliving environment interacting as a functional unit. This policy is exemplified by decisions such as Directive 85/337/EEC, which pays particular attention to the environmental sensitivity of geographical areas, for example, the special protection areas of the Natura 2000 network. It also forms the basis for the European Union (EU) biodiversity action plan for agriculture. According to directive 2001/18/EC, the information requirements for field trials of GM higher plants must include:

1. location and size of the release site;
2. release site ecosystem, including climate, flora, and fauna;
3. presence of sexually compatible wild relatives or cultivated plant species;
4. proximity to officially recognized biotopes or protected areas that could be affected.

This ecosystem, or rather agroecosystem approach, has been particularly recommended in countries such as Austria, a region that harbors high habitat diversity, extending from the Alps to the Pannonian continental lowlands. In these Alpine regions, transgenic grasses might eventually cause profound changes in community structure, whereas the latter flora includes potential crossbreeding partners of several crops, which do not occur in western Europe.

Conclusion

It is recognized that despite the use of 2.5 million tons of pesticides worldwide more than 40% of all potential food production is lost to various pests prior to harvest and an additional 20% is lost after harvest. The beneficial aspects of pesticides in protecting food production are offset by the associated costs of pesticide usage worldwide; these include an estimated 26 million poisonings and 220 000 fatalities, and an environmental and public health cost in the US estimated to be $9 billion per year. There is, therefore, a continuing need for alternative pest control strategies, which might include a role for transgenic crops. However, the various types of study summarized above have demonstrated some of the complexities associated with the first generation of transgenic crops and the future of such crops remains uncertain.

Whilst some campaigning groups insist that transgenic crops are of no value, and indeed are dangerous additions to agriculture, the opinion of other organizations and governments is generally more positive, though still cautious. For example, it was concluded in a recent report issued by the Royal Society of London and the Academies of Science from the USA, Brazil, China, India, and Mexico that: "Historically, both poverty and structural change in rural areas have resulted in severe environmental deterioration. The adoption of modern biotechnology should not accelerate this deterioration. It should instead be used in a way that reduces poverty and its deleterious effects on the environment".

According to a recent document from the UK government's Department of Environment, Food, and Rural Affairs (DEFRA): "Genetic modification of crop plants provides a powerful tool in research and in the breeding of new crop varieties that have the potential to dramatically reduce the environmental impact of crop production and improve crop quality". It is hoped that this potential will be fulfilled in the next few years.

List of Technical Nomenclature

Competence	A physiological condition in bacteria, required for the uptake of exogenous DNA.
Endophyte	A microbe present in the interior of a plant.
Nontarget species	Species other than the intended target of a pesticide, or of a plant carrying a gene with pesticide properties.
Phyllosphere	The microbial community on the surface of a leaf.
Rhizosphere	The microbial community on the surface of a root.
Sympatric	Species that grow in the same location.
Tritrophic	A three-tier ecological system comprising a plant, an herbivorous insect, and a prey species.

See also: **Pests**: Genetic Modification of Pest Resistance, Insect Pests. **Weeds**: Herbicide Resistance.

Further Reading

Altieri MA (2000) The ecological impacts of transgenic crops on agroecosystem health. *Ecosystem Health* 6: 13–23.

Ammann K, Jacot Y, Simonsen V, and Kjellsson G (eds) (1999) *Ecological risks and prospects of transgenic plants, where do we go from here? Methods for Risk Assessment of Transgenic Plants*, vol. 3. Basel: Birkhäuser.

Barton JE and Dracup M (2000) Genetically modified crops and the environment. *Agronomy Journal* 92: 797–803.

Bertolla F and Simonet P (1999) Horizontal gene transfers in the environment: natural transformation as a putative process for gene transfers between transgenic plants and microorganisms. *Research Microbiology* 150: 375–384.
Conference of Parties to the Convention on Biological Diversity. http://www.biodiv.org/
Gould F (1998) Sustainability of transgenic insecticidal cultivars: integrating pest genetics and ecology. *Annual Review of Entomology* 43: 701–726.
Hails RS (2000) Genetically modified plants – the debate continues. *Trends in Ecology and Evolution* 15: 14–18.
Jank B and Gaugitsch H (2001) Assessing the environmental impacts of transgenic plants. *Trends in Biotechnology* 19: 371–372.
Kjellsson G, Simonsen V, and Ammann K (eds) (1997) *Methods for Risk Assessment of Transgenic Plants, Vol. 2: Pollination, Gene-Transfer, and Population Impacts*. Basel: Birkhäuser.
Lutman PJW (ed.) (1999) *Gene Flow and Agriculture: Relevance for Transgenic Crops*. Farnham: British Crop Protection Council.
Marvier M (2001) Ecology of transgenic crops. *American Scientist* 89: 160–167.
Paoletti MG and Pimentel D (2000) Environmental risks of pesticides versus genetic engineering for agricultural pest control. *Journal of Agricultural and Environmental Ethics* 12: 279–303.
Saxena D and Stotzky G (2001) *Bacillus thuringiensis* (Bt) toxin released from root exudates and biomass of Bt corn has no apparent effect on earthworms, nematodes, protozoa, bacteria, and fungi in soil. *Soil Biology and Biochemistry* 33: 1225–1230.
Schuler TH, Poppy GM, Kerry BR, and Denholm I (1999) Potential side effects of insect-resistant transgenic plants on arthropod natural enemies. *Trends in Biotechnology* 17: 210–216.
Sears MK, Hellmich RL, Stanley-Horn DE, *et al.* (2001) Impact of Bt corn pollen on monarch butterfly populations: a risk assessment. *Proceedings of the National Academy of Sciences, USA* 98: 11937–11942.
Wolfenbarger LL and Phifer PR (2000) Biotechnology and ecology – the ecological risks and benefits of genetically engineered plants. *Science* 290: 2088–2093.

Plant Genetic Engineering, Food Safety Issues

M J Reiss, Institute of Education, University of London, UK

Introduction

An enormous amount has been written both about the ethics of plant genetic engineering and about the safety of foods resulting from plant genetic engineering. This article aims to clarify what is meant by food safety, to discuss how it might be established whether or nor foods made from genetically engineered plants are safe, and to examine the ethical issues raised by these issues. Too much of the debate about plant genetic engineering consists of opposing camps shouting across each other. My hope is that this article might establish some areas of agreement and indicate where more work needs to be done.

What Do We Mean by Safety?

Dictionary definitions talk about safety as meaning freedom from danger or risks. So a safe food might simply be taken as one that helps the person who eats it avoid such dangers and risks as malnutrition and food poisoning. Immediately, though, we note that food safety is not a property only of a food. It is also a property of the person who consumes the food. Consider certain unpasteurized cheeses. These can present a risk to the unborn children of pregnant women yet be safe for healthy adults. Similarly, many foods, notoriously peanuts, are safe for over 99% of all people yet cause life-threatening allergies in others. In this sense safety, unlike mass or temperature, is therefore not an absolute property of a food; food safety is rather the result of an interaction between a particular food and the person who consumes it.

But what are the risks or harms that might attend the eating of an unsafe food? These range from death (e.g., from anaphylactic shock) through permanent illness or disability (e.g., eating of foods without enough vitamin A can lead to permanent eye damage) to temporary illness (e.g., mild food poisoning, most cases of which are never even reported to doctors). Of these, loss of human life is obviously the most severe harm. However, identifying loss of human life as the most severe harm that can occur in relation to food intake leads to the suggestion that it may be unsafe (in the sense of risking human life) to grow crop varieties with less than the maximum likely yields in situations where there is serious food shortage. Such suboptimal varieties may be grown for a variety of reasons: because higher-yielding ones are not available, because alternative crops are preferred (e.g., because they yield more cash), because there are principled objections to the higher yielding varieties, and so on.

Is Safety Always a Good?

Thus far I have written as if safety, however understood, is the ultimate good. But it isn't. If it was we would not only ban such dangerous pursuits as skiing, we would forbid certain foods from being eaten (notably fugu, human deaths from which still reach double figures each year in Japan). Indeed, it can be quite wrong to act so as to maximize one's own safety. It is safer to watch someone drown than to attempt to rescue them. More mundanely, it may be safer to watch someone hitting a child in a shop than it is to remonstrate with them. Safety is a good but it isn't the ultimate good. For virtue ethicists this means that safety isn't a virtue – since you can't have too much of a virtue (such as honesty or trustworthiness).

How Can We Decide What Actions Are Good?

Whether there is a single ultimate good is a fundamental question in ethics but one that still lacks an agreed answer. The simplest approach to deciding whether an action would be right or wrong is to look at what its consequences would be. No one supposes that we can ignore the consequences of an action before deciding whether or not it is right. This is obvious when we try to consider, for example, whether imprisonment is the appropriate punishment for certain offences, e.g., robbery. We need to look at the consequences of imprisonment, as opposed to alternative courses of action such as imposing a fine or requiring community service. Even when complete agreement exists about a moral question, consequences will still have been taken into account.

The deeper question is not whether we need to take consequences into account when making ethical decisions but whether that is all that we need to do. Are there certain actions that are morally required – such as telling the truth – whatever their consequences? Are there other actions – such as betraying confidences – that are wrong whatever their consequences?

Consequentialism and Utilitarianism

Those who believe that consequences alone are sufficient to let one decide the rightness or otherwise of a course of action are called consequentialists. The most widespread form of consequentialism is known as utilitarianism. Utilitarianism begins with the assumption that most actions lead to pleasure (typically understood, at least for humans, as happiness) and/or displeasure. In a situation in which there are alternative courses of action, the desirable (i.e., right) action is the one which leads to the greatest net increase in pleasure (i.e., excess of pleasure over displeasure, where displeasure means the opposite of pleasure, that is, harm).

Utilitarianism now exists in various forms. For example, some utilitarians – preference utilitarians – argue for a subjective understanding of pleasure in terms of an individual's own conception of his/her well-being. What all utilitarians hold in common is the rejection of the view that certain things are right or wrong in themselves, irrespective of their consequences.

Consider the question as to whether or not we should label genetically modified (GM) foods or even always tell the truth. A utilitarian would hesitate to provide an unqualified "yes" as universal answers. Utilitarians have no moral absolutes beyond the maximization of pleasure principle. Instead, it would typically be necessary for a utilitarian to look in some detail at particular cases and see in each of them whether telling the truth would indeed lead to the greatest net increase in pleasure.

There are two great strengths of utilitarianism. First, it provides a single ethical framework in which, in principle, any moral question may be answered. It doesn't matter whether we are talking about the genetic modification of crops, the legalization of cannabis, the age of consent, or the patenting of DNA; a utilitarian perspective exists. Second, utilitarianism takes pleasure and happiness seriously. The general public may sometimes suspect that ethics is all about telling people what not to do. Utilitarians proclaim the positive message that people should simply do what maximizes the total amount of pleasure in the world.

However, there are difficulties with utilitarianism as the sole arbiter in ethical decision-making. For one thing, an extreme form of utilitarianism in which every possible course of action would have consciously to be analyzed in terms of its countless consequences would quickly bring practically all human activity to a stop. Then there is the question as to how pleasure can be measured. For a start, is pleasure to be equated with well-being, happiness, or the fulfilment of choice? And, anyway, what are its units? How can we compare different types of pleasure, for example sexual and aesthetic? Then, is it always the case that two units of pleasure should outweigh one unit of displeasure? Suppose two people each need a single kidney. Should one person (with two kidneys) be killed so that two may live (each with one kidney)?

Utilitarians claim to provide answers to all such objections. For example, rule-based utilitarianism accepts that the best course of action is often served by following certain rules – such as "Tell the truth," for example. Then, a deeper analysis of the kidney example suggests that if society really did allow one person to be killed so that two others could live, many of us might spend so much of our time going around fearful that the sum total of human happiness would be less than if we outlawed such practices.

Alternatives to Utilitarianism

The major alternative to utilitarianism is a form of ethical thinking in which certain actions are considered right and others wrong in themselves, i.e., intrinsically, regardless of the consequences. So, for example, while utilitarian analyses of GM crops have looked at all the likely pros and cons of growing them, some have argued that GM crops are intrinsically unacceptable.

There are a number of possible intrinsic ethical principles and because these are normally concerned with rights and obligations of various kinds, this approach to ethics is often labeled deontological (i.e., "rights discourse"). Perhaps the most important such principles are those of autonomy and justice.

People act autonomously if they are able to make their own informed decisions and then put them into practice. At a common-sense level, the principle of autonomy is why people need to have access to relevant information, for example farmers before deciding whether or not to plant a new crop variety.

Autonomy is concerned with an individual's rights. Justice is construed more broadly. Essentially, justice is about fair treatment and the fair distribution of resources or opportunities. Considerable disagreement exists about what precisely counts as fair treatment and a fair distribution of resources. For example, some people accept that an unequal distribution of certain resources (e.g., educational opportunities) can be fair provided certain other criteria are satisfied (e.g., the educational opportunities are purchased with money earned or inherited).

If it is the case that arguments about ethics should be conducted solely within a consequentialist framework, then the issues are considerably simplified. Deciding whether anything is right or wrong now reduces to a series of detailed, in-depth studies of particular cases. As far as food safety issues are concerned, ethicists still have a role to play but of greater importance are scientists, doctors, and others who know about risks and safety, while sociologists, psychologists, policy-makers, and politicians who know about people's reactions and public opinions also have significant roles.

Much energy can be wasted when utilitarians and deontologists argue. The safest conclusion is that it is best to look both at the consequences of any proposed course of action and at any relevant intrinsic considerations before reaching an ethical conclusion.

How Can We Establish That a Food Is Safe?

Suppose someone argues that apples aren't safe to eat. I might first check if they mean that they aren't safe if large amounts of pesticide residues are ingested or that they aren't safe to those who have allergies to apples. But suppose that the person means that they aren't safe for anyone to eat. I would dismiss the argument, probably on three grounds. First, that I have eaten plenty of apples myself and believe I have come to no harm; second, that many different people – female and male, young and old, pregnant and otherwise – in many countries of the world have eaten apples without any harm coming to them; and third, that apples have been eaten safely for hundreds of years.

None of these answers is available to argue for the safety of a novel food. The very fact that it is novel, whether genetically engineered or not, means first that I can't vouch for its safety from personal experience; second that I can't cite others who have safely eaten it; and third that I cannot appeal to the lessons of time.

There is nothing particular to novel food in this. The same objections obtain to the safety of any new product or existing product made in a new way. It might logically be that mobile phones prove dangerous to the grandchildren of heavy users. We cannot disprove this possibility with certainty. Indeed, this point is not raised merely in a spirit of logical disputation. We now realize that X-rays and the radiation from nuclear explosions are dangerous in ways that were not initially realized. On the other hand, certain technologies, such as bicycles which were initially considered by some to be harmful to women, are not now held to be so dangerous. It takes history to demonstrate conclusively that something is safe – and a history is the one thing that a new technology lacks.

So how can one establish (or at least investigate) the safety of a novel food? In 1993 the Organization for Economic Co-operation and Development (OECD) introduced the principle of "substantial equivalence." Substantial equivalence is based on the principle that if a novel (including GM) food is shown to have no material (i.e., relevant or potentially relevant) differences in composition from an existing food, then it can be considered as safe as the existing food. The principle has been widely endorsed, though it remains rejected by some, and now lies at the heart of EU and some other international food safety regulatory processes.

So what happens when a novel food is not sufficiently substantially equivalent to an existing safe food? Some sorts of tests are required. For some, the gold standard of safety testing is found in the way many countries test for the safety of new medicines, pesticides, and certain other products. A key component of this involves feeding or otherwise introducing (e.g., by inhalation) the products in question to laboratory animals (often one rodent and one nonrodent mammalian species) at a range of doses. Some of these dose levels are intentionally set far in excess of those at which the product would be expected to be administered to or encountered by humans. This allows a considerable margin of safety. The problem, though, is that most foods (aside from components such as additives and vitamins that are normally taken in small amounts) cannot be ingested at levels far in excess of normal.

In practice a wide range of approaches is used to investigate the food safety of GM and other novel foods:

- the principle of substantial equivalence
- biochemical studies to investigate the likely significance of any nonsubstantial equivalence
- animal testing, sometimes over more than one generation
- human volunteer studies (including tests for allergenicity)
- postmarket monitoring.

It is often pointed out that if the same rigor had been used in the past as is now applied in Europe to novel food safety we would be most unlikely to be eating such familiar foods as potatoes. On the other hand, it has been argued that most of the tests undertaken to investigate the safety of novel, including GM, foods are carried out by the companies developing the foods and are rarely published in peer-reviewed journals.

What Particular Issues Are Raised by GM Plants?

Much, though not all, plant genetic engineering entails introducing one or more genes into a crop from other sources. This raises the possibility of causing a crop to become allergenic. Around 1–2% of adults and 6–8% of children have food allergies. "Decision trees" for assessing the allergenic risks of GM foods have been developed by the Food and Agriculture Organization, World Health Organization and others and are required by some regulatory bodies.

Various plant viral sequences are commonly used in the genetic engineering of crops and concern has been expressed, though by only a very small number of scientists, that these might have adverse health consequences, for example by producing new viruses through recombination or by activating transposable elements already present in the human genome.

A further fear is that genes introduced into a crop might become incorporated into the genetic makeup of a person ingesting the food. Although most ingested DNA is broken down in the gut, some uptake of gene-sized DNA into cells lining the gut does occur. It is difficult to see how this is likely to have major adverse health consequences. For some time it has been argued that the use of antibiotic-resistance genes as markers might make some intestinal bacteria resistant to antibiotics. While this effect is likely to be small compared to the misdiagnosis and use of medical and veterinary antibiotics, many have argued that such markers should no longer be used in the human or farm animal food chain.

A final issue raised by GM plants is that the time is at hand when GM foods will be produced in an attempt to enhance health. Such foods might, for example, contain nutrients otherwise present in a population in inadequate amounts (e.g., vitamin A and iron) or be designed so as not to contain certain potentially harmful chemicals (e.g., genes for allergenic proteins).

How Much of a Solution is Food Labeling?

Given the objections that a large number of people have to foods made from genetically engineered organisms, it can be argued that labeling is the way forward on the grounds that labeling provides information, that information allows choice, and that choice is a good thing. However, it may be a mistake to presume that choice requires mandatory labeling, though this is what is now being introduced throughout the EU in the case of foods and animal feeds containing or made from GM organisms. It can be argued that it is better not to require such labeling but to permit retailers and restaurants that provide foods to decide, in a free market environment, whether or not to label. In that way the principle of choice holds both at the level of retailers and restaurants and at the level of individual consumers. If consumers really want to know whether the food they are buying or eating is from GM organisms or from conventional organisms they will seek out those retailers and restaurants who label.

The issue can be examined more generally by asking the question "Why do we need regulations (e.g., in regard to the labeling of food) rather than simply allowing people harmed by actions to take those responsible for the harm to court?" One answer is that people often want regulation to prevent such harms from happening. (I may be able to sue you for harming me because you have failed to show a duty of care, for example by driving too fast, but I may prefer that the harm had never happened, for example by my government having passed laws about speeding limits.) Another answer is that legal recourse is a very imperfect way of attempting to redress harms. In particular, those with little money, power, or persistence stand only a small chance of successfully taking anyone, certainly a large company, to court.

These considerations suggest that the particular duties of a regulatory system are (1) to prevent certain harms, and (2) to provide especial protection for those unable to take legal actions against those responsible for harms. This view largely derives from John Stuart Mill's arguments about liberty. Mill held that whereas it can be appropriate for the law to intervene to prevent harm to others, it is never appropriate for it to intervene to prevent harm to oneself. In the case of GM crops, I would therefore expect regulators to pay particular attention (a) to those especially likely to be harmed, (b) to those who lack agency. Some actors, for example farm animals and young children, potentially fall into both categories. Other actors, for example adults with certain food allergies, fall into just one. Indeed, it is precisely when it can be argued that significant harm would (or might frequently enough) be prevented by labeling that there is a powerful argument for mandatory labeling. For this reason few are likely to object to the mandatory labeling of foods that contain important allergens. (Importance here combines aspects both of the severity of the allergy and of the number of people with it.) Whether it can validly be argued that mandatory labeling of all GM foods is required, given the enormous amount of effort made in the United States, Europe, and a number of other regions to ensure that those GM foods that can be bought are at least as safe as conventional foods, is more open to debate.

List of Technical Nomenclature

Consequentialism — Ethical framework in which actions are judged solely by their results (i.e., consequences).

Deontology — Ethical framework in which actions are judged in terms of rights and moral obligations.

Substantial equivalence — Principle used for judging the safety of a novel food by comparing it with the nearest established food.

Utilitarianism — Ethical framework in which the right action is that which maximizes the amount of overall pleasure in the world.

See also: **Ethics and Biosafety**: Ethics of Genetically Modified Crops; Plant Genetic Engineering, Ecological Issues.

Further Reading

Adams J (1995) *Risk*. London: UCL Press.

Advisory Committee on Novel Foods and Processes (2002) *Annual Report 2001*. London: Food Standards Agency.

Diamand E (2001) *The Great Food Gamble: An Assessment of Genetically Modified Food Safety*. London: Friends of the Earth.

FAO/WHO (2000) *Safety Aspects of Genetically Modified Foods of Plant Origin*. Report of a Joint Food and Agriculture Organization/World Health Organization Consultation. Rome: Food and Agriculture Organization/World Health Organization.

Gasson M and Burke D (2001) Scientific perspectives on regulating the safety of genetically modified foods. *Nature Reviews Genetics* 2: 217–222.

ILSI International Food Biotechnology Committee (2002) *International Documents and Scientific Publications on Plant Biotechnology and the Safety Assessment of Food Products Derived from Plant Biotechnology*. Washington, DC: ILSI International Food Biotechnology Committee.

Menlowe MA (1998) Safety laws. In: Chadwick R (ed.) *Encyclopedia of Applied Ethics*, Vol. 4, pp. 1–7. San Diego: Academic Press.

Mill JS (1861/1991) *On Liberty and Other Essays*, edited by J Gray. Oxford: Oxford University Press.

Reiss MJ and Straughan R (1996) *Improving Nature? The Science and Ethics of Genetic Engineering*. Cambridge: Cambridge University Press.

Royal Society (2002) *Genetically Modified Plants for Food Use and Human Health: An Update, Policy Document 4/02*. London: Royal Society. Available at www.royalsoc.ac.uk

Royal Society of Canada (2001) *Elements of Precaution: Recommendations of the Regulation of Food Biotechnology in Canada*. Ottawa: Royal Society of Canada.

F

FLOWERING AND REPRODUCTION

Contents

Flower Development

R M Twyman, University of York, York, UK

Introduction

Reproductive development in higher plants involves the production of flowers, structures that contain male and/or female reproductive organs. The switch from vegetative to reproductive development is, in most plants, controlled by an autonomous pathway, which is governed by developmental age, and thus prevents flowering in juvenile plants that have yet to build up a sufficient nutrient supply. However, flowering is also influenced by a number of environmental signals, such as temperature, day/night length, and light quality. This ensures maximum reproductive success by inhibiting flower development when the environment is unsuitable, and by helping to synchronize flowering time in outcrossing species. Genetic analysis in *Arabidopsis thaliana* has uncovered many of the genes involved in the regulation of flowering time, and recent experiments have revealed how these signals are integrated. Although different species show diverse reproductive strategies, many of them possess orthologs of the *Arabidopsis* flowering time genes. Therefore, the control of flowering time may involve a conserved set of genes, the regulation of which varies among different species.

The initiation of flower development involves the activation of floral meristem identity genes, which inhibit further shoot growth and make the meristem competent to produce floral organs instead of leaves. Flower patterning is achieved by the spatially restricted expression of floral organ identity genes within the floral meristem. Although the morphology of flowers varies widely, all are based on the general theme of four concentric whorls with sepals on the outside, then petals, stamens, and inner carpels. The establishment and organization of the floral organs represents a paradigm in pattern formation, and many genes involved in this process have been characterized in *Arabidopsis*, *Antirrhinum majus*, and other species. Notably, *Arabidopsis* has flowers that are more or less radially symmetrical while other species, such as *Antirrhinum majus*, have flowers with a distinct dorsoventral polarity.

This article provides a brief summary of the molecular basis of the major events in flower development, focusing on *Arabidopsis thaliana*, beginning with the transition to flowering and ending with the control of floral patterning.

Control of Flowering Time

Overview

The initiation of flowering in *Arabidopsis* is controlled by both autonomous and exogenous pathways, which converge to regulate three genes responsible for floral initiation: *LEAFY* (*LFY*), *SUPPRESSOR OF OVEREXPRESSION OF CO 1* (*SOC1*, also known as *AGL20*), and *FLOWERING LOCUS T* (*FT*). These are known as floral integrator genes (**Figure 1**). The autonomous pathway promotes flowering irrespective of environmental signals, while separate pathways mediate responses to day/night length, temperature, and stress. Flowering in *Arabidopsis* is promoted by long days and vernalization

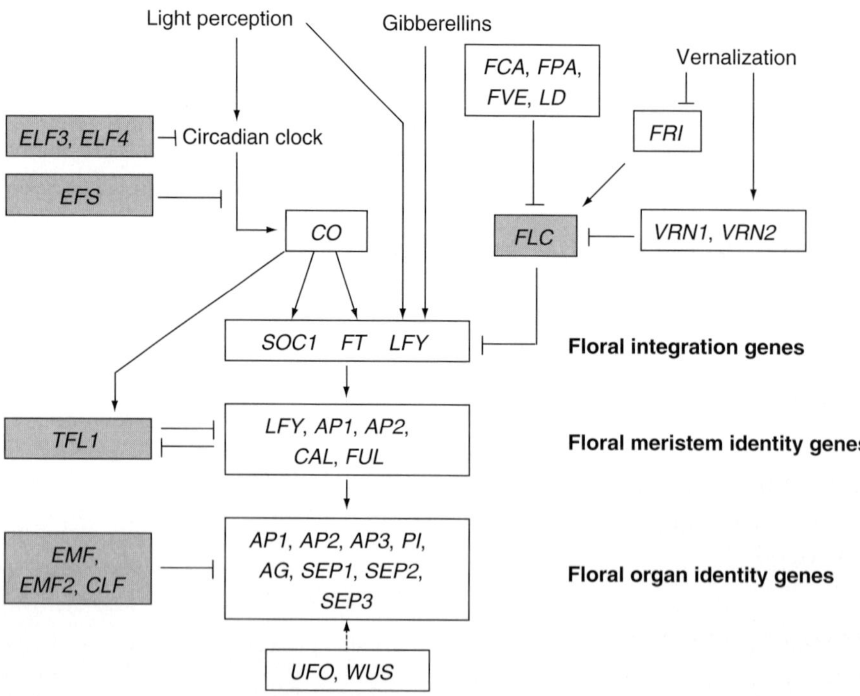

Figure 1 An overview of the genes controlling flowering time, the transition to floral development, and floral patterning in *Arabidopsis*. Genes shown in shaded boxes are floral repressors.

(prolonged cold temperatures), both of which are indicative of seasonal change, but it will eventually flower under prolonged short day conditions. In contrast, flowering in other species may be promoted by different cues. In rice (*Oryza sativa*), for example, flowering is promoted by short days.

The Autonomous Pathway

An autonomous pathway is required to prevent flowering before the plant has undergone sufficient vegetative growth to enjoy reproductive success. Without an autonomous pathway, shoots emerging from the soil under convivial environmental conditions would flower immediately. Mutations that show late flowering phenotypes under both long and short day conditions have enabled the identification of proponents of the autonomous pathway. These include genes such as *FLOWERING LOCUS CA* (*FCA*), *FPA*, *FVE*, and *LUMINIDEPENDENS* (*LD*). A common molecular feature of the mutants is the accumulation of the mRNA for another gene, *FLOWERING LOCUS C* (*FLC*), indicating that their normal role is to repress *FLC* expression. *FLC* encodes a MADS box transcription factor that inhibits (probably indirectly) the three floral integration genes listed above. While LD is a homeodomain containing transcription factor, FCA and FPA are RNA binding proteins. These products may therefore downregulate *FLC* gene expression at both the transcriptional (LD) and posttranscriptional (FCA, FPA) levels. While *FCA*, *FPA*, *FVE*, and *LD* promote flowering, the *EARLY FLOWERING IN SHORT DAYS* (*EFS*) gene encodes an inhibitor of flowering in the autonomous pathway, since the early flowering mutant phenotype is more noticeable in short days.

The Vernalization Pathway

A long period of cold temperature (below 10°C) promotes flowering in many species including *Arabidopsis*, and is a useful reproductive strategy because it delays flowering until the spring. This vernalization response is epistatic to the autonomous pathway, i.e., vernalization overcomes the late-flowering phenotypes of *fca*, *fpa*, and *ld* mutants.

The molecular basis of this observation is the intersection of the vernalization and autonomous pathways at *FLC*. The expression of *FLC* is downregulated at low temperatures, and although the direct regulatory factors have not been identified, an important component of this pathway is the product of the *FRIGIDA* (*FRI*) gene. In naturally occurring late-flowering ecotypes, the *FRI* gene promotes the accumulation of *FLC* mRNA, whereas vernalization (and the autonomous pathway) act to reduce *FLC* expression. How is the repression of *FLC* maintained when warm temperatures return? This is revealed by the analysis of mutants such as *vrn2* in which the vernalization response in a constitutive late-flowering genetic background is abolished. The *VERNALIZATION RESPONSE 2* (*VRN2*) gene encodes a nuclear protein of the Polycomb family, which maintains the structure of chromatin domains through successive cell divisions. This indicates that *FLC* repression is established in a manner that is dependent on the loss of FRI activity, but maintained epigenetically. The epigenetic control of flowering time is confirmed by the observation that *Arabidopsis* mutants deficient in DNA methylation also show flowering time phenotypes.

The Photoperiod Pathway

Flowering time in *Arabidopsis* is promoted under long-day conditions. The perception of long versus short days requires both a mechanism for perceiving light and a stick by which to measure it. Therefore, mutants that affect the photoperiod dependent regulation of flowering time reveal genes involved in light signal transduction and the circadian oscillator. Information from both these systems converges at the *CONSTANS* gene (*CO*), which encodes a transcription factor with two zinc fingers. The net result is that the CO protein is active only under long-day conditions, where it promotes flowering by activating the floral integration genes (direct upregulation of *FT* and *SOC1* has been demonstrated, and there may be an indirect mechanism for activating *LFY*). In *co* mutants, flowering is delayed under long-day conditions, but occurs normally under short-day conditions, showing that the autonomous pathway is not affected. Not all light perception signals are channeled through the circadian oscillator and CO. Several of the components of the light perception pathway bypass CO and regulate the floral integration genes directly. For example, the product of the *PHYB* gene inhibits the expression of *LFY* in response to red light.

Other genes have been identified that act as repressors in the light response pathway, and, in contrast to the *EFS* gene discussed above, the mutant phenotypes are characterized by early flowering under both long- and short-day conditions. These genes, which include *EARLY FLOWERING 3* (*ELF3*) and *ELF4*, are often involved in circadian clock function.

Other Pathways

Fewer details are known about other environmental pathways that affect flowering time, such as heat shock, drought, and nutrient depletion. Overcrowding stress induces early flowering, and this may involve the reflection of red light from nearby vegetation, which is perceived as a change in light quality. Many types of stress induce the production of plant hormones, and these are also known to affect flowering times. For example, *Arabidopsis* mutants that have impaired gibberellin biosynthesis or signal perception have delayed flowering phenotypes and may never flower in short days, while those with constitutively active gibberellin signaling pathways show precocious flowering. It should be emphasized, however, that signaling pathways used in one species may have no effect on flowering time in others.

Floral Initiation

Floral Meristem Identity Genes

Endogenous and exogenous signals controlling flowering time in *Arabidopsis* converge on the three floral integration genes: *SOC1*, *FT*, and *LFY*. The expression of all three genes is dependent on the downregulation of *FLC*, which in turn relies on autonomous developmental signals or vernalization. *SOC1* and *FT* are positively regulated by CO, while *LFY* is regulated by the photoperiod pathway independently of CO and by further pathways such as gibberellin signaling. The control circuits that regulate *SOC1*, *FT*, and *LFY* expression are likely to be much more complex than set out above, and variations or even completely distinct pathways may operate in other plants. It is known that the molecular signal or signals that result in the switch from vegetative to reproductive development originate outside the meristem, probably in the leaves. Grafting experiments have shown that the signal travels at a defined rate along the vasculature, but years of investigation have thus far failed to isolate a specific molecule, which has been termed the florigen. What is clear, however, is that once the signal has been perceived in the meristem, each of the floral integration genes encodes a product that can help to regulate genes that act downstream in the flower development pathway. *SOC1* and *LFY* encode transcription factors, while FT appears to be a protein kinase inhibitor. The immediate targets of these regulators are the floral meristem identity genes.

Several floral meristem identity genes have been identified on the basis of mutations that cause flower-to-inflorescence or inflorescence-to-leaf transformation phenotypes. *LFY* is one of these genes, and thus has a dual role as a floral integrator and floral meristem identifier. Other meristem identity genes include *APETALA1* (*AP1*), *AP2*, *CAULIFLOWER* (*CAL*), and *FRUITFUL* (*FUL*). The floral meristem identity genes have two major functions. First, they confer upon the shoot meristem the ability to produce flowers instead of leaves on its flanks and prevent indeterminate growth of lateral organs. Second, they confer upon the meristem the ability to produce correctly patterned flowers. In this respect, it is interesting to note that two of the meristem identity genes, *AP1* and *AP2*, also function at the flower patterning stage. There is a degree of cooperation both among the meristem identity genes to maintain their expression, and between these and other genes that are expressed in the shoot meristem to determine the expression domains of the flower patterning genes. Such interactions are discussed in more detail below.

Floral Repressors

Mutations that result in early or inappropriate flowering have revealed the existence of a whole range of floral repressors, indicating that vegetative growth is promoted by (and dependent on) floral repression. These repressors function at all levels of floral initiation. We have already discussed genes that inhibit the signaling pathways that activate the floral integration genes (e.g., *ELF3*, *ELF4*, and *EFS*) and those that inhibit the activity of the floral integration genes themselves (e.g., *FLC*). Other inhibitors act later. For example, the *TERMINAL FLOWER1* gene (*TFL1*) encodes a transcriptional repressor that binds to the upstream promoters of *LFY* and *AP1*, and inhibits their expression. Interestingly, the transcription factors encoded by *LFY* and *AP1* can repress the *TFL1* gene, indicating that a balance between these gene products is important in flower initiation. In the leaf primordium, the balance is tipped in favor of *TFL1* and flowering is repressed, while in the flower primordium the situation is reversed. Ectopic expression of *TFL1* delays flowering, whereas in *tfl* mutants there is early flowering characterized by a single terminal flower. The *EMBRYONIC FLOWER1* gene (*EMF1*) encodes a transcription factor that is thought to repress the flower patterning genes. *emf1* mutants flower without first passing through a rosette development phase, suggesting that the *EMF1* gene product is absolutely required for vegetative development.

The products of many floral inhibitor genes have been identified as proteins that modify chromatin structure, and mutations are often pleiotropic. This observation suggests that epigenetic mechanisms may be important in the regulation of flowering, but that this may be part of a more general strategy of developmental regulation in plants. Examples of such genes include *VRN2* (discussed above), *EMBRYONIC FLOWER 2* (*EMF2*), *EARLY BOLTING IN SHORT DAYS* (*EBS*), and *CURLY LEAF* (*CLF*).

Floral Patterning

The Classical ABC Model

The typical flower of a dicotyledonous plant comprises four concentric whorls of floral organs. From the outside inwards these are sepals (protective), petals (attractive), stamens (male reproductive organs), and carpels (female reproductive organs) (**Figure 2**). Despite their very different structures and functions, all these organs begin as undifferentiated cell clumps on the surface of the floral meristem. Therefore, an important aspect of flower development is pattern formation, i.e., the regional specification of cells in the meristem so that each forms the correct structure in the correct place.

The genes that confer positional information in the flower are known as floral patterning genes or floral organ identity genes. They are also called homeotic genes because when their functions are abolished (through mutation or epigenetic effects) then the floral organs undergo homeotic transformation, i.e., one whorl of floral organs develops with the likeness of another. This is analogous to the situation in *Drosophila*, where certain mutations cause body segments to take on incorrect regional identities, resulting in mutants with misplaced appendages (e.g., legs growing out of the head instead of eyes). The similarities go further in that in both *Drosophila* and plants, the genes responsible for homeotic mutations encode transcription factors that appear to provide each cell with positional information to confer its regional identity.

The "ABC model" of flower development is based on the mutant phenotypes and expression patterns of the first floral organ identity genes to be discovered (**Table 1**). The model maintains that each whorl of floral organs is defined by a combination of three

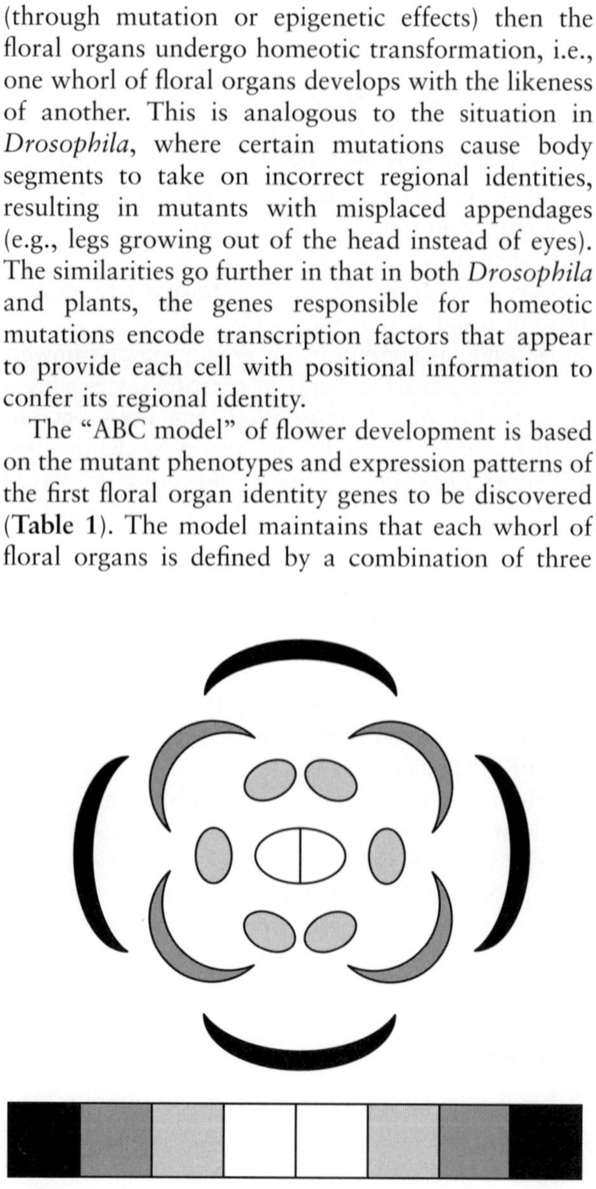

Figure 2 Transverse section of a wild-type *Arabidopsis* flower, showing the organization of the four whorls. 1 = sepals (4), 2 = petals (4), 3 = stamens (6), 4 = carpels (2).

classes of genes: A, B, and C. Expression of class A genes alone specifies sepals (whorl 1), class A and B genes in combination specify petals (whorl 2), class B and C genes in combination specify stamens (whorl 3), and class C genes alone specify carpels (whorl 4). Class A genes are therefore expressed in whorls 1 and 2, class B genes in whorls 2 and 3, and class C genes in whorls 3 and 4. The last component of the model is that the class A and C genes negatively regulate each other so that their expression patterns never overlap. Therefore, in class A gene mutants, the domain of class C gene expression expands to cover all four whorls, and the converse is true in class C gene mutants. Examples of all three classes of genes have been identified in *Arabidopsis* and *Antirrhinum* (**Table 1**). Flower development in wild-type plants and in all the homeotic mutants can be explained in terms of this model (**Table 2**).

Table 1 Floral organ identity genes grouped by class in *Arabidopsis thaliana* and *Antirrhinum majus*

Class (and expression)	Arabidopsis *genes*	Antirrhinum *genes*
Class A (whorls 1 and 2)	*APETALA1 (AP1)* *APETALA2 (AP2)*	*SQUAMOSA (SQA)*
Class B (whorls 2 and 3)	*APETALA3 (AP3)* *PISTILLATA (PI)*	*DEFICIENS (DEF)* *GLOBOSA (GLO)*
Class C (whorls 3 and 4)	*AGAMOUS (AG)*	*PLENA (PLE)*
Class X (whorls 2, 3, and 4)	*SEPALATA1 (SP1)* *SEPALATA2 (SP2)* *SEPALATA3 (SP3)*	

Control of ABC Gene Expression

Several of the *Arabidopsis* floral organ identity genes are known to be direct targets of LFY, which is expressed from the very earliest stages of flower development. For example, LFY has been shown to bind to the promoters of the class C gene *AG* and the class A gene *AP1*, and therefore activate them directly. While this explains the order of gene activation in flower development, it does not explain how the floral organ identity genes are expressed in precise and overlapping patterns. Furthermore, *AP1* and *AP2* are also floral meristem identity genes with initially broad expression patterns that resolve to the outer two whorls. Recent experiments have shown that LFY can both activate these downstream genes and define their restricted expression patterns by cooperating with other, regionally specific regulators (**Figure 3**).

One of the earliest targets for LFY is *AG*, which is expressed in the inner two whorls of the developing flower. This restricted expression pattern arises because LFY cooperates with the product of another gene expressed in the shoot meristem, *WUSCHEL* (*WUS*). The WUS protein is restricted to the central, proliferative region of the shoot meristem. The activation of *AG* initiates a negative feedback loop in which *WUS* gene expression is repressed. Therefore, the cooperation between LFY and WUS not only localizes the expression domain of *AG*, but also converts the indeterminate shoot apical meristem into a determinate floral meristem.

Table 2 Explanation of *Arabidopsis* floral patterning mutant phenotypes in terms of the extended ABC model of floral organ specification

Gene class and genes	*Normal whorl expression domains* 1	2	3	4
A: *AP1, AP2*				
B: *AP3, PI*				
C: *AG*				
X: *SEP1, SEP2, SEP3*				

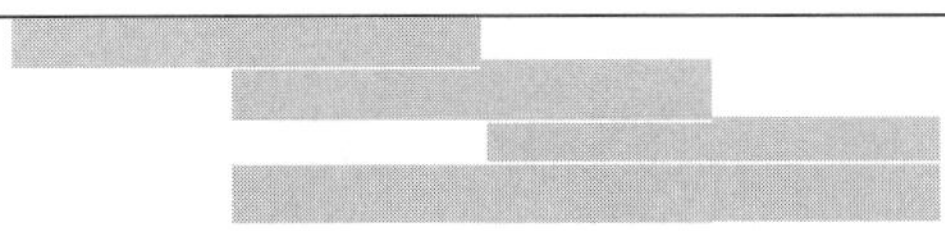

Genotype	*Whorl phenotype* 1	2	3	4	*Whorl expression domain pattern* 1	2	3	4
Wild-type	Sepal	Petal	Stamen	Carpel	A	A + B + X	B + C + X	C + X
A mutant (*ap1, ap2*)	Carpel	Stamen	Stamen	Carpel	C	B + C + X	B + C + X	C + X
B mutant (*ap3, pi*)	Sepal	Sepal	Carpel	Carpel	A	A + X	C + X	C + X
C mutant (*ag*)	Sepal	Petal	Petal	Sepal	A	A + B + X	A + B + X	A + X
AB mutant, e.g., *ap1 + ap3*	Carpel	Carpel	Carpel	Carpel	C	C + X	C + X	C + X
AC mutant, e.g., *ap1 + ag*	Sepal/petal intermediate structure					B + X	B + X	X
BC mutant, e.g., *ap3 + ag*	Sepal	Sepal	Sepal	Sepal	A	A + X	A + X	A + X
X mutant, *sep1 + sep2 + sep3*	Sepal	Sepal	Sepal	Sepal	A	A + B	B + C	C
ABC mutant, e.g., *ap1 + ap3 + ag*	Leaf	Leaf	Leaf	Leaf	0	0	0	0

Upper part of table shows expression patterns of homeotic genes of classes A, B, C, and X in wild-type plants. Lower part of table shows expression patterns and resulting phenotypes in wild-type and mutant flowers.

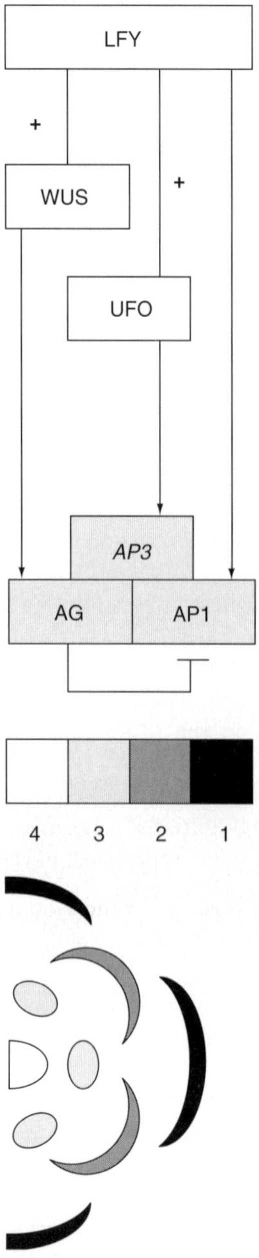

Figure 3 A model for the establishment of floral organ identity gene domains of expression in the floral meristem. LFY protein is expressed throughout the meristem and activates *AP1* throughout. However, it must cooperate with WUS to activate *AG* specifically in whorls 3 and 4, and the AG protein represses *AP1*, restricting its expression to whorls 1 and 2. In order to activate *AP3*, LFY must cooperate with UFO. Bar shows expression domains with respect to the four whorls in a half flower, as shown in the lower panel.

LFY also activates *AP1*, which is initially expressed in a broad domain in a similar way to *LFY* itself. However, as the centrally restricted *AG* gene is expressed and the protein begins to accumulate, the expression of *AP1* is restricted to the outer two whorls because the AG protein is a negative regulator of *AP1*. Similarly, *AP2* is restricted to these whorls, and together the AP1 and AP2 proteins reinforce the central restriction of *AG*.

LFY also activates *AP3*, although the mechanism involved in its localization is more complex than discussed for *AG*, *AP1*, and *AP2* above. LFY may cooperate with the product of another gene, *UN USUAL FLORAL ORGANS* (*UFO*), for the activation of *AP3*. Unlike most of the genes discussed in this section, *UFO* does not encode a transcription factor, but instead a protein that may be involved in protein turnover. When *AP3* expression commences, *UFO* is already restricted to the second and third whorls, and this is mirrored by *AP3*. *AP3* expression is strongly reduced in either *lfy* or *ufo* mutants, and a current model suggests that LFY is a transcriptional activator of *AP3* throughout the meristem whereas UFO helps to break down a transcriptional repressor of *AP3*, a repressor that is also expressed throughout the meristem.

Extension of the ABC Model

Studies of *Petunia* spp. have shown that certain genes, when overexpressed, promote the formation of ectopic ovules. Although such genes have not been functionally characterized in other plants (including *Arabidopsis*), their existence in *Petunia* has led to the development of an ABCD model in which the innermost whorl of the flower can be divided into two distinct subdomains. Carpels would be specified in one subdomain by the expression of class C genes alone, while ovules would be specified in the other subdomain, either by the expression of class D genes alone or by a combination of class C and D genes. *Petunia* class D genes include *FLORAL BINDING PROTEIN7* (*FBP7*) and *FBP11*.

Another class of homeotic genes has been identified in *Arabidopsis* that adds a further layer of complexity to the ABC model. These genes encode AG-like proteins that are expressed in the three inner whorls of the developing flower. Individual mutations have subtle phenotypes, but triple mutants generate flowers in which all four whorls develop as sepals, and in which the meristem remains indeterminate. The genes have been named *SEPALLATA1* (*SEP1*), *SEP2*, and *SEP3*. The triple mutant phenotype is very similar to double mutants for class B and class C genes (**Table 2**), but the expression of class B and C genes remains undisturbed in the *sep1* ; *sep2* ; *sep3* triple mutants. Furthermore, the expression of *SEP1*, *SEP2*, and *SEP3* is undisturbed in the class B and C double mutants. Insights into the molecular basis of these observations come from studies of how the products of the floral patterning genes regulate each other.

Interactions among the Floral Patterning Gene Products

All of the floral patterning genes in *Arabidopsis* encode transcription factors containing a MADS box, with the exception of *AP2*. *In vitro* experiments have shown that the four proteins (AG, AP1, AP3, and PI) can physically interact, and it has been suggested that they combine to form dimers in the regulation of gene expression. *In vivo*, it appears that only class-specific interactions result in gene activation. That is, genes can be activated by AG homodimers, AP1 homodimers, and AP3 and PI can either form homodimers or heterodimers with each other. More recently, it has been shown that the SEP proteins can also form complexes with the products of the classical ABC genes, and there is evidence for the formation of larger complexes.

From these observations, the "quartet model" has been derived (**Figure 4**). This model suggests that floral organ identity is defined by four different combinations of a quartet of transcription factors. In this model, sepals would be specified by an AP1 dimer in combination with an unknown component (perhaps another AP1 dimer). Petals would be specified by AP1, AP3, PI, and SEP (SEP represents any of the three *SEP* gene products, which can be considered more or less redundant for the purposes of the model). Stamens would be specified by AP3, PI, AG, and SEP. Finally, carpels would be specified by an AG dimer and a SEP dimer. This model explains all the homeotic mutant phenotypes, including those caused by the triple SEP mutant.

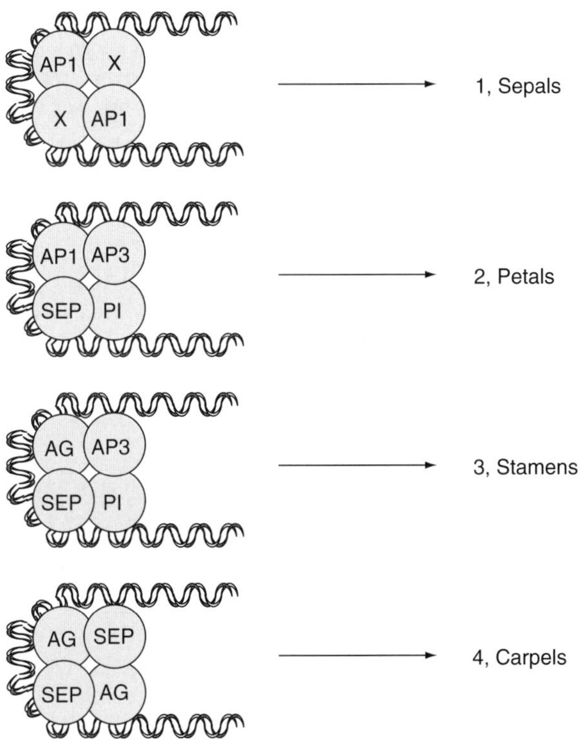

Figure 4 Quartet model of floral organ identity. This model suggests that each floral organ is specified by a particular combination of four homeotic proteins. In each of whorls 2, 3, and 4, organ identity is specified by a combination of Class A, B, and C proteins plus one of the three SEPALATA proteins, generically represented by SEP. In whorl 1, where the SEP proteins are not expressed, two AP1 dimers might form the quartet, or an AP1 dimer might cooperate with an as yet unknown dimer (represented by X).

Other Influences on Floral Patterning

As well as floral organ identities, the floral developmental program must include systems to regulate floral organ number, size, and shape. In *Arabidopsis*, there are four sepals, four petals, six stamens, and two partially fused carpels. The organization of the outer two whorls is radially symmetrical while the inner whorls show dorsoventral polarity. Several mutants have been identified that alter the number and distribution of these organs and reveal underlying molecular mechanisms. For example, the *perianthia* mutant has five equally spaced organs in whorls 1, 2, and 3. As in the wild-type plant, these are arranged in an alternating manner, with petals arising at the position of the junction of two sepals and stamens at the junction of two petals. These results strongly suggest that the position of individual floral organs is controlled by lateral inhibition from other organs both within the same whorl and in adjacent whorls. Other genes may influence the arrangement of floral organs by controlling cell division. The *FLORAL ORGAN NUMBER1* (*FON1*) gene is thought to limit the proliferation of cells in the central whorls. In *fon1* mutants, the size of the fourth whorl is larger than normal and there are extra carpels. The other whorls appear normal. The role of the *SUPERMAN* gene may be to limit cell proliferation at the boundary of whorls 3 and 4. In *superman* mutants, the cells of whorl 3 overproliferate and effectively annex whorl 4, resulting in flowers with extra stamens, but no carpels.

While much of what we know about the molecular basis of flower development has been learned from the analysis of *Arabidopsis*, *A. majus* is a valuable additional model because the flowers show pronounced dorsoventral polarity. *Antirrhinum* flowers comprise four concentric whorls, with five sepals, five petals, four stamens, and two carpels. Dorsoventral polarization is most apparent in the second and third whorls. There are three distinct petal identities: two dorsal, two lateral, and one ventral (**Figure 5**). The ventral petal is bisected by the axis of symmetry. As in *Arabidopsis*, the stamens alternate with the petals. Therefore, there is a single dorsal stamen (this does

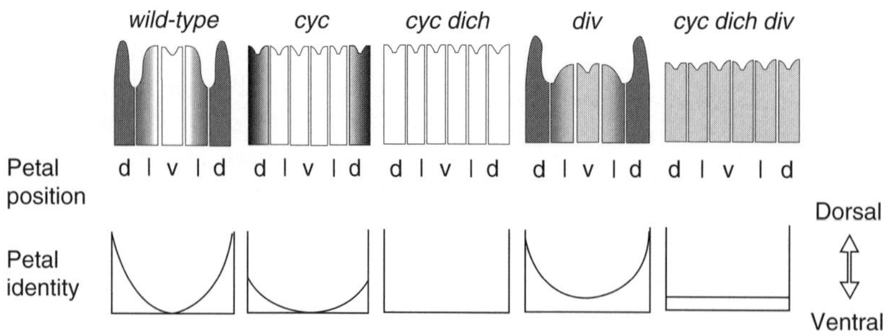

Figure 5 Genetic control of floral organ shape in *Antirrhinum*. The petals are shown arranged schematically with the ventral petal in the center, the dorsal petals on the outside and the lateral petals in between. Positional identities are identified by shading. Modified with permission from Weigel D (1998) From floral induction to floral shape. In: *Current Opinion in Plant Biology* 1: 55–59. Oxford: Elsevier Ltd.

not develop fully, and is termed the staminode), two ventral stamens, and two lateral ones.

A number of mutants have been identified in which the dorsoventral polarity of the flower is reduced. Mutations in the *CYCLOIDIA* (*CYC*), *DICHOTOMA* (*DICH*), and *RADIALIS* (*RAD*) genes cause ventralization. The *cyc* mutants have six organs in each of the three outer whorls, with the lateral petals and stamens becoming more ventral in character and the dorsal petals and stamens becoming more lateral in character. The *cyc* phenotype is much stronger than the *dich* phenotype, in which there is partial transformation of the dorsal petals toward lateral petals. However, the double mutant *cyc* ; *dich* is strongly ventralized and almost perfectly radially symmetrical. The *CYC* mRNA has been shown to be restricted to the dorsal floral meristem, but its expression domain is significantly smaller than the effect of the mutation would suggest, indicating that it acts in a cell-nonautonomous fashion. Mutations in the *DIVARICATA* (*DIV*) gene are dorsalized, with the ventral petal taking on more lateral characteristics. Interestingly, in *cyc* ; *dich* ; *div* triple mutants, the radially symmetrical flowers have lateral rather than ventral petals, indicating that *DIV* is epistatic to *CYC* and *DICH*, and its activity may be repressed by *CYC* and *DICH* in the dorsal meristem (**Figure 5**).

List of Technical Nomenclature

ABC model — A model of pattern formation in the flower where the four floral organs are specified by the overlapping patterns of three classes of genes.

Autonomous pathway — A pathway that controls flowering time according to the developmental stage of the plant, and which is not influenced by the environment.

Circadian oscillator — The molecular basis of circadian rhythms, i.e., biological processes that cycle with a periodicity close to 24 h under constant environmental conditions. Based on oscillating gene expression with feedback regulation and synchronized by the light–dark cycle.

Determinate — Of meristems, able to form a fixed number of structures, such as leaves, before flowering.

Dorsoventral polarity — Of a flower, having an axis of bilateral symmetry.

Epigenetic — Control of gene expression not mediated by *cis*-acting elements surrounding the gene. Includes mechanisms such as chromatin remodeling, DNA methylation, and RNA interference.

Floral integrator genes — Genes regulated by the various pathways controlling flowering time.

Floral meristem identity genes — Genes required to convert the shoot apical meristem into an inflorescence meristem.

Floral organ identity genes — Genes whose combined expression patterns define the four different floral organs.

Floral repressors — Genes (and their proteins) that inhibit flowering. Required for vegetative growth.

Flower — Reproductive organ of a flowering plant.

Homeotic genes — Genes that control the regional identity of particular organs or structures.

Homeotic transformation — A situation in which one organ or structure develops with the likeness of another usually found in a different position.

Indeterminate — Of meristems, able to form a variable number of structures, such as leaves, before flowering.

Long-day plants — Plants that flower when days are long.

MADS box	A conserved DNA binding domain found in most of the transcription factors that control floral organ identity.
Photoperiod	Length of the daily light period.
Quartet model	A model of pattern formation in the flower where the four floral organs are specified by particular combinations of four transcription factors in a tetrameric complex.
Vernalization	A prolonged period of low but not freezing temperatures.

See also: **Flowering and Reproduction**: Flower Development. **Growth and Development**: Molecular Biology of Development. **Regulators of Growth**: Circadian Rhythms; Vernalization.

Further Reading

Battey NH and Tooke F (2002) Molecular control and variation in the floral transition. *Current Opinion in Plant Biology* 5: 62–68.

Blazquez MA (2000) Flower development pathways. *Journal of Cell Science* 113: 3547–3548.

Mouradov A, Cremer F, and Coupland G (2002) Control of flowering time: interacting pathways as a basis for diversity. *Plant Cell* 14: S111–S130.

Ng M and Yanofsky MF (2000) Three ways to learn the ABCs. *Current Opinion in Plant Biology* 3: 47–52.

Simpson GG and Dean C (2002) *Arabidopsis*, the Rosetta Stone of flowering time? *Science* 296: 285–289.

Sung ZR, Chen L, Moon Y-H, and Lertpiriyapong K (2003) Mechanisms of floral repression in *Arabidopsis*. *Current Opinion in Plant Biology* 6: 1–7.

Theissen G (2001) Development of floral organ identity: stories from the MADS house. *Current Opinion in Plant Biology* 4: 75–85.

Pollination

J L Osborne, Rothamsted Research, Harpenden, UK
J B Free, University of Wales, Cardiff, UK

The Pollination Process

Pollination, in angiosperms, is the transfer of pollen (male gametophytes) from the male part of a flower (anthers) to the female parts (stigma) of the same flower, or another flower of the same plant species. Pollination comprises three phases: pollen release, pollen transfer from source to recipient, and pollen deposition (**Figure 1**). If pollination is successful, the pollen will germinate (**Figure 2**) and fertilize the ovule, and a seed will subsequently develop.

Self-pollination is the transfer of pollen from an anther to a stigma of a flower of the same plant. Self-compatible plants are those capable of producing viable seed when self-pollinated. Sometimes the flower parts are so arranged that pollen falls from the anthers of a flower onto its own stigma so that pollination occurs automatically without the intervention of an external agent; this is autopollination and results in autogamy if the plant is self-compatible (**Figure 1**). Usually, autopollination is prevented or limited and pollen has to be transferred by an external agent. Geitonogamy is fertilization resulting from pollination between two flowers on the same plant (**Figure 1**), so the plant has some degree of self-compatibility, but an external agent is required to transfer the pollen. Cross-pollination is the transfer of pollen from an anther to a stigma of a flower on a different plant of the same species, resulting in xenogamy if the fertilization is successful (**Figure 1**). Self-incompatible plants are those incapable of producing viable seed unless cross-pollinated. Full self-compatibility and self-incompatibility are at opposite ends of a continuum. Even if self-pollination leads to successful fertilization, cross-pollination has the advantage of mixing genes to give the population an opportunity to adapt to changing environments. Hybrid vigor may lead to increased quality and/or quantity of seeds and/or fruit. Repeated self-pollination can lead to inbreeding depression, but self-pollination does have advantages. It does not require the presence of external agents to transfer the pollen and the plant does not need to expend energy attracting pollinators.

Plants have complex systems promoting cross-pollination, including the spatial and temporal separation of male and female sexual organs. Floral morphology and phenology help to attract animals to transfer pollen, or to launch pollen into the air or water, increasing the chances of cross-pollination. Floral mechanisms for attracting pollinators will favor both male and female reproductive success. When pollination does not limit female success (i.e., when pollen receipt no longer limits seed production), floral attraction of pollinators can still benefit male success by increasing pollen dispersal.

Pollinators

A pollinator is an agent (abiotic or biotic) which transfers pollen from anthers to stigma either within one flower or between flowers. Wind and insects are

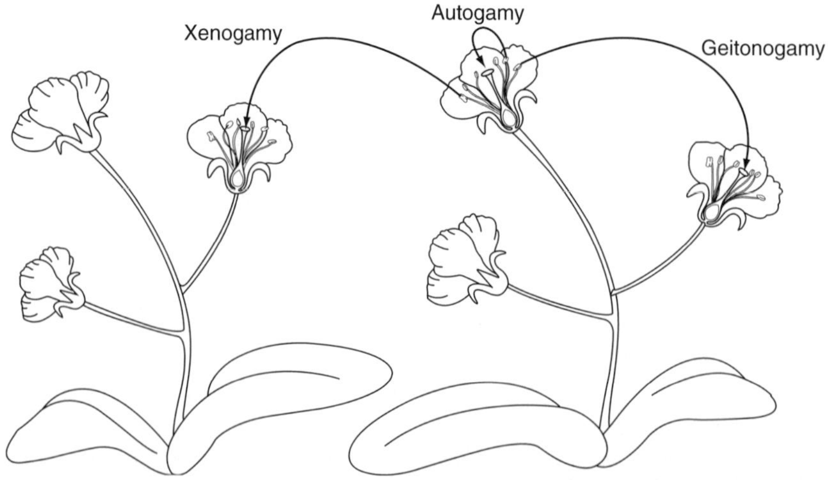

Figure 1 Diagrammatic representation of the pollination process, distinguishing between self-pollination (within a plant) and cross-pollination (between plants).

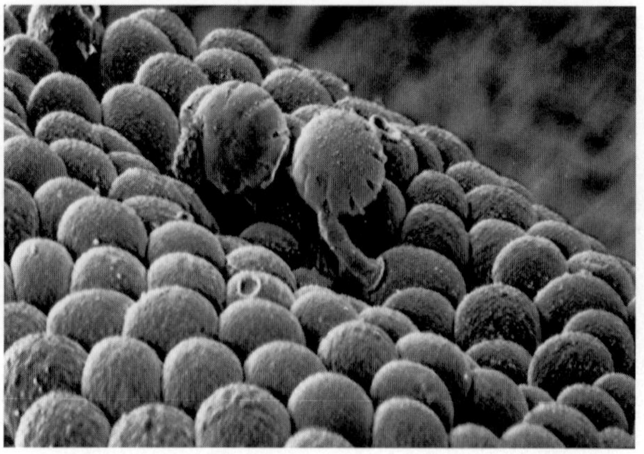

Figure 2 Electron micrograph of a germinating pollen grain penetrating the stigma of a borage flower (*Borago officinalis*). Scale: height of figure equivalent to 1.2 mm.

Figure 3 Honey bee (*Apis mellifera*) visiting canola or oilseed rape (*Brassica napus*). Photograph courtesy of J. B. Free.

the main external agents of pollination but water, birds, and mammals (primarily bats) are also pollinators. Of the insects, bees are the most important pollinators (**Figure 3**), but beetles, flies, butterflies, and moths also pollinate plants.

Suites of floral characteristics or pollination syndromes can be associated with particular pollination systems. However these characteristics should not be considered prescriptive, and detailed, objective experiments are required to compare the effectiveness of different potential pollen vectors. A plant may be visited by a range of animals, but only certain species may be successful pollinators. A visitor does not equate to a pollinator. It has generally been assumed that most plants are specialized to be pollinated by one or a few species, but evidence is growing that many plants have generalized pollination systems, taking advantage of pollinator communities which fluctuate in space and time.

Nevertheless, it is useful to list the characteristics usually associated with wind- and insect-pollinated plants, although some plants may be pollinated by a combination of wind and insects. Wind-pollinated plants usually have inconspicuous flowers with prominent stigmas and large exposed anthers producing copious dry, small pollen grains. It is common to find the male and female function either in separate flowers, or separated in time in these species, which assists wind-mediated cross-pollination. Insect-pollinated plants have large flowers attracting insects often by a combination of visual and olfactory cues (color and scent) combined with a reward for the pollinator such as pollen, nectar, or oil. The pollen is often large, nutritious, and sticky to adhere to the insect's body.

Wind Pollination

Almost all the grasses, sedges, and rushes are wind-pollinated, as are the majority of forest trees in temperate climates, including conifers. Airborne pollen cannot be targeted to stigmas of compatible plants as specifically as insect-borne pollen. The success of wind pollination therefore relies on huge quantities of light, dry pollen being produced and transported into air currents so that at least a small proportion lands on the exposed stigmas of other plants. Millions of grains of pollen fill the air when trees or grasses are in flower, and it is this pollen which is largely responsible for hay fever, an allergic reaction to a variety of pollens. Wind pollination has the advantage that it does not rely on the presence and seasonality of insects.

The rate of fall of pollen grains in calm air varies considerably from about 2 cm s^{-1} for small buoyant pollen grains (e.g., hazel (*Corylus avellana*), birch (*Betula* spp.)) to 50 cm s^{-1} for heavy windborne pollen (e.g., corn (*Zea mays*; maize)). The distribution and dynamics of pollen flow away from the source will depend not only on the plant and atmospheric conditions, including turbulence, but also on the vegetation and surrounding landscape. In temperate deciduous trees, the flowers are borne and pollination is effected before the leaves unfurl and obstruct air movement around the trees. Airborne pollen dispersal often shows a leptokurtic distribution (**Figure 4**). This means that pollen deposition close to the source is greater than expected from a normal distribution. There is often a long thin tail to

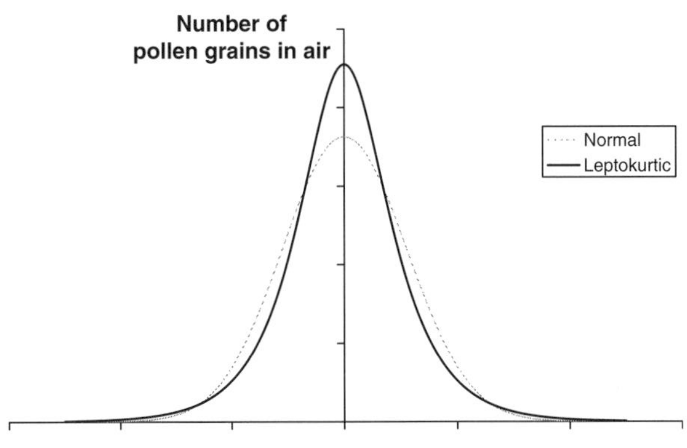

Figure 4 Simulation of a leptokurtic distribution of airborne pollen grains at different distances from a wind-pollinated plant source. This is the typical form of the dispersal curve when pollen is sampled using traps. Dotted line shows a normal distribution for comparison.

the distribution: tiny quantities of pollen are distributed very long distances. Since wind pollination is most effective at short distances, it is most often found in plant species growing at high density, for example grasses and temperate forest trees.

Insect Pollination

Although thousands of animals, mainly insects, visit flowers for nectar or pollen or both, relatively few are consistently good pollinators. To be an efficient pollinator an insect must:

- visit several flowers of the same species in succession
- move frequently from one flower to another
- carry pollen on its body
- brush against the stigmas of flowers, and so transfer the pollen.

Many insects visiting flowers fail to pollinate them because of their behavior, small size, and smooth body surfaces. Bees (wild and domesticated) are extremely effective pollinators. Pollen and nectar are the sole source of protein and carbohydrate for bees and their brood, and all bees are covered with branched hairs, specially adapted for collecting pollen. Bees learn to manipulate flowers effectively and demonstrate flower constancy, moving systematically from flower to flower of the same species in succession. Hoverflies are good pollinators because their hairy bodies carry as much pollen as bees, but they do not work the flowers as consistently.

The effectiveness of different insects as pollinators can vary in different conditions and will depend on the pollinator assemblage. For example, a bee which removes large quantities of pollen, but deposits very little, may be useful when no other pollinators are present but, if a more effective pollinator is present which deposits a larger proportion of removed pollen, then the first pollinator may act to decrease the pollen available. These interactions and balances can be extremely complex and demonstrate the importance of considering the mechanisms of pollen transfer from anther to stigma, in an empirical fashion.

The Study of Pollination

The relevance of pollination to the reproductive success of plants was discovered in the late seventeenth century, and the first systematic study of the relationship between flowers and insects was published by Christian Konrad Sprengel in 1793. There was great interest in the natural history of animal–flower relationships throughout the nineteenth century, encouraged by Charles Darwin's observations and writings. After a period of reduced activity in the early twentieth century, increased interest in genetics and interactions between animals and plants in the 1950s and 1960s led to renewed enthusiasm for the study of pollination ecology. This resulted in the publication of several comprehensive textbooks, and research in this field is currently flourishing.

Pollination ecology is multidisciplinary. It is studied by botanists, zoologists, ecologists (theoretical, behavioral, community, or evolutionary), plant breeders, and agronomists. This inevitably leads to a variety of experimental approaches. A behavioral ecologist may want to find out how pollinators move between flowers and plants to optimize energy returns. An evolutionary ecologist may be more interested in how pollination systems can shed light on the evolution of plant species. A community ecologist may be interested in whether plants are competing for pollinators, or pollinators for plants, whilst a plant breeder or agronomist will be keen to find out how best to optimize yield or purity of seed. Experimental approaches range from theoretical, through mechanistic and functional, to applied (e.g., predicting how many bee colonies are required on a field). Each has benefits, but all would be enhanced by further integration of approaches for a more complete understanding of ecosystem function.

Techniques in the Study of Pollination

Pollination is studied at a variety of spatial scales, from gene flow within and between plant populations, to aspects of the transfer of individual grains between anther and stigma. Mechanisms of pollen transfer cannot be determined simply by examining floral morphology and phenology. Systematic experiments are required to determine (1) whether a plant needs an external agent to transfer pollen, (2) whether the most effective agent is wind (or water) or a flower-visiting animal, and (3) if it requires flower visitors then whether those present are adequate. **Figure 5** shows a simplified scheme for investigating these alternatives. Seed set is compared in caged and open-pollinated plants. Although seed set does not strictly represent pollination (the transfer of compatible pollen grains from anther to stigma), it is often used as a simple and useful index of pollination success.

If a plant is primarily wind-pollinated, then the pollen dispersal curve is normally investigated. Airborne pollen grains (before impact with a stigma) are counted from a variety of traps, including sticky traps, "rotorods," and suction traps, or successful fertilizations are quantified using genetic markers. Neither technique measures the actual number of pollen grains reaching compatible stigmas, which would be a more accurate estimate of wind pollination.

If a plant requires visitors to transfer pollen, then we must establish which flower visitor is most effective. The simplest way of comparing different species of flower visitor is to introduce them into separate cages of plants, and compare seed set in these cages with that achieved from open pollination or from plants caged without flower

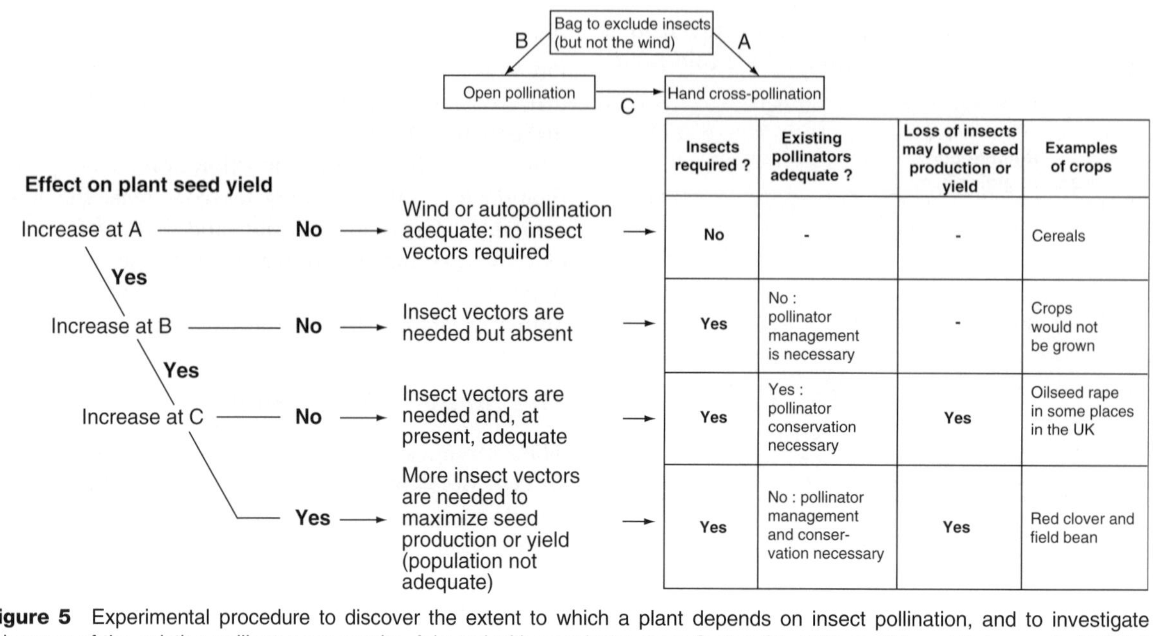

Figure 5 Experimental procedure to discover the extent to which a plant depends on insect pollination, and to investigate the adequacy of the existing pollinator community. Adapted with permission from Corbet SA, Williams IH, and Osborne JL (1991) Bees and the pollination of crops and wild flowers in the European Community. *Bee World* 72: 47–59.

Figure 6 Pollination cages, for examining pollinator effectiveness, on a field of linseed (*Linum usitatissimum*). Photograph courtesy of I. H. Williams.

visitors (**Figure 6**). A more comprehensive picture is obtained if pollinator effectiveness is studied in greater detail. The effectiveness of different flower visitors as pollinators will depend on the quality and quantity of their visits. The quality of a pollinator's visits can be measured by counting the number of pollen grains removed during a visit, the number transferred to other flowers of the same species, the number of grains deposited on other stigmas, and the viability of that pollen. Sometimes, instead of counting pollen grains, the success of an individual visit is measured by scoring germinating pollen tubes, counting fertilized ovules, or scoring seed set. Genetic or phenotypic markers are often used in this context to study pollen flow between plants. These fine-scale studies originated on wild plants but are now being applied to economically important plants, where knowledge of the most effective pollinators can help a farmer to maximize yield.

The Importance of Pollination

Pollination is a keystone process in natural and managed ecosystems. The reproductive success of plants, and consequently their survival and evolution depend on it. A large range of insects depend on the pollen and nectar provided by flowers, and many animals are dependent on the seeds and fruits produced as a result of pollination.

Although the pollination systems of many plant genera have been studied in detail, the pollination requirements of the majority of wild plants are suprisingly unknown. There are lists of flower visitors and inferences are made regarding their importance as pollinators but the actual effectiveness of these visitors has not been studied in detail for most plant species.

In agriculture and plant breeding, most crops require wind or insect pollination, either to set any seed at all, or to improve the quality, size, or quantity of seed and fruit. Pollination is often, therefore, an essential component of yield, and the majority of insect-pollinated crops need bees.

Estimates of the economic value of pollination to crop production are extremely variable, partly because of the great variety of crops and the degree to which they depend on pollination, and partly because of the variety of environmental conditions in which the crops are grown. Because of the preponderance of honey bees as managed pollinators, attempts have been made to evaluate their economic importance on commercial crops. Early attempts relate the value of bee pollination to the entire value of the crops that the bees visit. Recently, only the estimated proportion of pollination performed by the honey bee itself has been considered. An annual sum of US$14.6 billion has been proposed as the contribution honey bees made to pollination of 57 crops in the United States from 1996 to 1999. Other estimates range from US$1.6 billion to US$5.7 billion.

Pollination Requirements of Crops

Many crops have a variety of pollination mechanisms depending on environmental conditions and variety; and unfortunately we do not know the detailed requirements of many species, particularly those grown in the tropics.

Some crops are mainly self-pollinated, for example wheat (*Triticum aestivum*), rice (*Oryza sativa*), barley (*Hordeum vulgare*), peas (*Pisum sativum*), and bananas (*Musa sapientum*), whilst wind is the principal pollinating agent of many cereals, other grasses, nut trees, and a few other crops. Self and wind pollination are therefore responsible for seed yield in the most widely grown, and important, global food crops: wheat, rice, maize, and sorghum (*Sorghum bicolor*). Pollination does not normally limit seed set or yield in these crops. However, insect pollination improves seed or fruit production in a much larger number of crop species. Of the crops harvested for seed or fruits, around 20 are primarily self- or wind-pollinated, while over 80 crops rely on insect pollination to some extent, including the majority of fruit, pulse, and oilseed crops. Some crop species (e.g., oilseed rape (*Brassica napus*)) are pollinated by wind and insects although their relative importance is unclear and will depend on the abundance of insects, the strength of the wind, and other climatic factors. Some crops give a moderately increased yield when pollinated by insects, others give a greatly increased yield, and yet others require insects to set any fruit or seed at all. There may be other advantages to insect pollination, such as larger or better-formed berries or fruits, or a greater proportion of early flowers setting seed, giving an earlier and more uniform harvest. Bees are by far the most important pollinators of insect-pollinated crops.

Growers can improve levels of pollination in insect-pollinated crops by encouraging and enhancing the natural pollination service provided by wild insects, usually bees. The following measures can be taken:

- grow crops appropriate to the local pollinator community
- conserve and augment areas with nesting sites and wild forage to sustain pollinators when crops are not in flower
- ensure that the crop is in flower when the most important wild pollinators are at their seasonal peak
- employ a pollinator-friendly spray regime; for example, avoid spraying insecticides while the crop is in flower, use those that are least toxic to pollinators, and avoid spraying field margins and hedgerows.

Wild pollinators are subject to large fluctuations, and populations can be small. Therefore natural pollination services often need to be supplemented with introduced, "managed" pollinators.

Introducing Insect Pollinators to Crops

Honey bees (*Apis mellifera*) are the most well-known, commonly managed pollinators, and were the only managed pollinator available until the 1960s. Most recommendations for boosting crop pollination are therefore based on introducing honey bee colonies. Honey bees are good generalist pollinators, but are not always the most effective, and there is a variety of managed pollinators now available (**Table 1**). If possible, growers should therefore choose the correct insect for the job. For example, bumble bee colonies, primarily *Bombus terrestris*, are reared and transported to pollinate glasshouse crops, whilst leaf-cutter bees (*Megachile rotundata*) pollinate alfalfa (*Medicago sativa*, lucerne).

There are many other species of bumble bee and solitary bee that could potentially be managed for various crops (**Table 2**). However, the cost of developing new species as managed pollinators may be prohibitive if their advantages over currently managed species are not verified.

Honey Bees

Although most honey bee colonies are kept for honey production their value as pollinators far exceeds that of the honey and wax they produce. Introduced colonies are vital components to the yield of many agricultural and horticultural crops. Many growers have an agreement with a local beekeeper to provide colonies when their crops flower. In North America, migratory beekeepers travel around the country placing hundreds of colonies on flowering crops throughout the year (**Figure 7**). Growers pay the beekeepers for this pollination service; providing sufficient colonies for adequate pollination is seldom compatible with honey production.

The number of honey bee colonies placed on a crop will be a balance between the cost of the

Table 1 Bee species managed as pollinators of various crops around the world

Bee species	*Common name*	*Crops and country*	*Managed throughout life cycle*	*Attracted to artificial nest sites*
Apis mellifera	Honey bee	Many	*	
Bombus spp.	Bumble bee			
B. terrestris		Tomatoes	*	
B. hortorum		Red clover, New Zealand		*
B. ruderatus		Red clover, New Zealand		*
Megachile rotundata	Leaf-cutter bee	Alfalfa	*	
Nannotrigona testaceicornis	A stingless bee	Strawberries, Japan	*	
Nomia melanderi	Alkali bee	Alfalfa	*	
Osmia spp.	Orchard bee			
O. cornifrons		Orchards, Japan		*
O. lignaria propinqua		Almonds, USA	*	
O. cornuta		Orchards, USA, EU	*	

Data summarized from Crane (1990), Kevan *et al.* (1990), Torchio (1991), and Free (1993).

Table 2 Some bee species suggested as potentially suitable for management as crop pollinators

Bee species	*Common name*	*Crops and country*
Osmia rufa	Orchard bee	Greenhouse fruits, apples
O. ribifloris	Orchard bee	Highbush blueberry, USA
Megachile species (other than *M. rotundata*)	Leaf-cutter bee	Alfalfa, USA
Peponapis pruinosa	Squash bee	Cucurbits, India
Xylocopa spp.	Carpenter bee	Vine fruits, tropics
Rhophitoides canus		Alfalfa, Russia

Data summarized from Kevan PG, Clark EA, and Thomas VG (1990) Insect pollinators and sustainable agriculture. *American Journal of Alternative Agriculture* 5: 13–22, Torchio PF (1991) Bees as crop pollinators and the role of solitary species in changing environments. *Acta Horticulturae* 288: 49–60, Free JB (1993) *Insect Pollination of Crops*, 2nd edn. London: Academic Press, and Williams IH (1996) Aspects of bee diversity and crop pollination in the European Union. In: Matheson A, Buchmann SL, O'Toole C, Westrich P and Williams IH (eds) *The Conservation of Bees*, pp. 63–80. Academic Press, London.

colonies, and the benefits in terms of increased yield of the crop. Recommendations of the number of colonies required are based on:

- pollination requirements of the crop
- size of the field and density of flowers
- nectar and pollen available, and attractiveness of the flowers to bees
- number of pollinating insects already present
- competition from other forage
- environmental conditions.

Bumble Bees

Bumble bees are extremely effective pollinators of tomatoes (*Lycopersicon esculentum*), more effective than either honey bees or the conventional technique of vibrating flowers individually by hand. Tomato flowers do not contain nectar; the only reward is pollen. Honey bees do not forage consistently on the flowers and, if put into a glasshouse of tomato plants, they will recruit their nest mates to more profitable forage sources outside. Bumble bees gain plentiful rewards of pollen from tomato flowers because they can vibrate their wings at a very high frequency, or "buzz" the flowers, to release the pollen. They do not communicate and recruit in the same way as honey bees, so do not leave the glasshouse en masse to forage elsewhere.

The effectiveness of bumble bees as pollinators of tomato crops led, in the 1980s, to a new industry in rearing bumble bee colonies. The key to successful domestication of the bumble bee lay in finding a way to break the overwintering period so that new queens would begin a new nest without having to hibernate for 6 months. Once a protocol was established, boxed *B. terrestris* colonies began to be sold to tomato-growers in 1987. Hundreds of thousands of colonies are now sold worldwide (**Figure 8**). Besides tomatoes, the colonies are used on a variety of other high-value horticultural crops in glasshouses, such as sweet peppers (*Capsicum frutescens*) and eggplants (*Solanum melongena*). This highlights the profits that can be achieved by choosing the most effective pollinator for the crop in question.

Commercial bumble bee colonies are not used on field crops in Europe. They are too small and expensive. But in New Zealand, colonies of bumble bees are placed in red clover (*Trifolium pratense*) fields. Red clover has a very long corolla and only long-tongued bees can reach the nectar effectively. Bumble bees are thus the most effective at pollen transfer. Other field crops benefiting from bumble bee pollination include raspberries (*Rubus* spp.) and field beans (*Visia faba*).

Figure 7 Honey bee hives being moved onto a field of canola (*Brassica napus*) in Canada. Photograph courtesy of J. L. Osborne.

Figure 8 Commercial bumble bee colonies (*Bombus terrestris*) in a tomato glasshouse. Photograph courtesy of J. L. Osborne.

Solitary Bees

The United States and Canada developed commercial leaf-cutter bee (*M. rotundata*) rearing when they found that these bees were more effective than honey bees in pollinating alfalfa. They forage more consistently on the crop and they can be stored overwinter as pupae, unlike honey bees. The leaf-cutter bee shelters contain boards with semicircular grooves which form tunnels in which the leaf-cutter bees build cells and lay eggs. When the alfalfa field begins to flower, pupae from the previous year are placed in the field. The adults emerge and lay eggs in the shelters provided. They provision the brood with pollen from the alfalfa, pollinating the crop at the same time. At the end of the season, the cells in the shelter are removed and stored for use the following year.

Another species of solitary bee, the alkali bee (*Nomia melanderi*) is also used on alfalfa crops. Orchard bees (*Osmia* spp.) have been successfully managed on orchard crops where they have been found to be more effective than honey bees.

Flies

Blowflies are used for crop pollination in glasshouses or small cages. They only pollinate open or relatively shallow flowers such as parsnip (*Pastinaca sativa*), celery (*Apium graveolens dulce*), onion (*Allium cepa*), and *Brassica* spp. with easily accessible nectaries. A handful of blowfly pupae, from which adults are about to emerge, are put on the floor of the glasshouse or cage as flowering begins. More pupae are added as necessary during flowering.

Other Challenges in Crop Pollination

Certain types of crop production require special attention to be paid to pollination. In the commercial production of hybrid seed crops, insect pollinators (normally bees) are required to transfer pollen between rows of male plants and rows of female (male sterile) plants. Often, bees will concentrate on the rows of male plants producing pollen. The bees need to be encouraged to visit both male and female plants to ensure effective pollination. This may be effected by careful arrangement of the plants.

A similar situation occurs in fruit orchards needing bee-mediated cross-pollination. Bees must transfer pollen between the pollinizer trees and the main variety trees. Individual foragers are usually constant to a relatively small area of forage during a single foraging trip, and in an orchard this may only embrace one or two large fruit trees. Planning of fruit orchards therefore needs to take this into account. Other means of spreading compatible pollen to the main variety trees include devices fitted to hive entrances, which either dispense pollen onto outgoing foragers, or increase the mixing of pollen between bees (removing pollen from incoming bees and depositing it on outgoing bees).

Obtaining Pure Seed

The relatively small foraging areas of individual bees are advantageous when growing pure, uncontaminated seed of a particular cultivar. Effective isolation distances depend on the crop species, whether it is primarily wind- or insect-pollinated, the pollinator community, and the environmental conditions. For many insect-pollinated crops, little contamination occurs when large plots of cultivars are isolated by 100 m or more. The degree of contamination that can be tolerated depends on what the seed is used for. Whether wind- or insect-pollinated, a small amount (1–2%) of contamination of a seed crop often persists far from the undesirable pollen source. Pollen can be transferred between bees as they brush against each other within the hive between foraging trips, increasing distances of pollen movement.

Genetically Modified Crops

The release of genetically modified (GM) crops into the environment has raised concerns about the movement of transgenes, via cross-pollination, into other crops or wild relatives in the vicinity, producing offspring containing the genetic modification. This highlights the need to further quantify mechanisms of pollen flow if appropriate isolation distances are to be recommended.

If GM crops are male fertile, it is not possible to ensure that no pollen will travel from these crops to others, whatever the isolation distances. It is more important to understand whether the incorporation of the genes into conventional crops or wild relatives presents any environmental or health risk, and whether the GM crops themselves have the potential to harm pollinators. The evidence so far suggests that, even for GM crops with insecticidal properties, the risks to bees are extremely low.

Assuming genetic incorporation into other plants via pollination is "safe" at an acceptable level, then a more productive approach to planting GM crops would be to accept a certain limit on contamination from GM crops, based on extensive knowledge and guidelines gained from plant breeding and the production of high-purity seed, and recommend isolation distances for organic crops accordingly.

A Pollination Crisis?

Diverse pollinator communities are essential to maintain the diversity and integrity of wild and managed ecosystems. In many parts of the world, there is increasing evidence that populations of managed bees, wild bees, and other insects are becoming inadequate to pollinate crops, and probably also wild plants. The numbers of wild bees and other pollinating insects are thought to have diminished for some or all of the following reasons. Intensive cultivation and land management, particularly in industrialized countries, have eliminated large areas of seminatural vegetation and

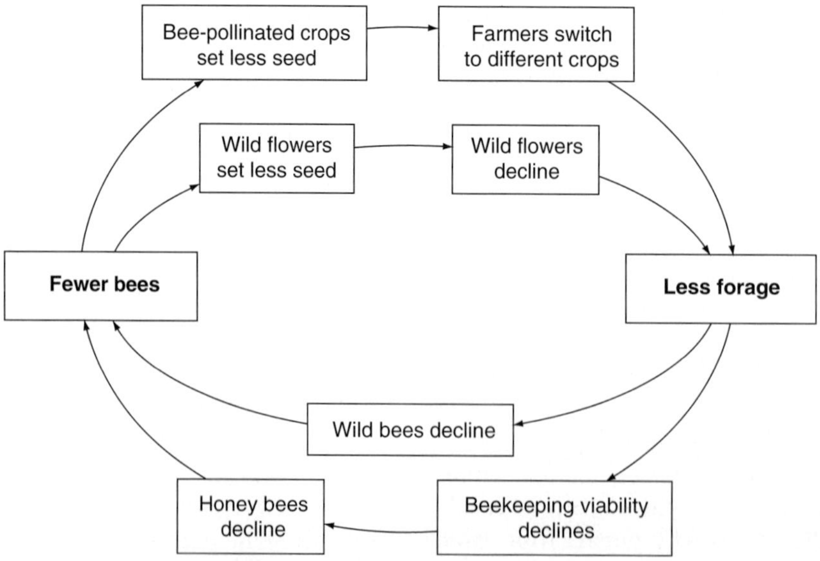

Figure 9 A vicious circle of pollinator and plant declines, mediated through reduced crop and wild flower pollination.

consequently reduced forage sources and the number of nesting and hibernation sites available. Although some crops provide copious nectar and pollen, they are likely to flower over a short period and cannot sustain pollinator populations in the area over the whole season. Insecticides can kill the insects directly and herbicides reduce the availability of nectar and pollen sources.

The decline in numbers of pollinating insects may be leading to a pollination crisis, illustrated by the vicious circle in **Figure 9**. Positive actions are required to retain and enhance pollinator communities. Every effort should be made to encourage the use of spare land, for growing flowering plants that provide food for bees and other pollinating insects, whilst sustained research into the functioning of pollination processes within the ecosystem will enhance our ability to manage pollination systems of wild and cultivated plant species.

List of Technical Nomenclature

Anther	Floral organ in which pollen development takes place.
Autogamy	Self-fertilization within one flower.
Geitonogamy	Fertilization between flowers on the same plant.
Leptokurtic	A statistical distribution that is more heavily concentrated about the mean than the normal distribution (i.e., with a tall narrow peak), and thus has a large kurtosis.
Pollination	Transfer of pollen (male gametophytes) to the stigma.
Pollinator	A vector, aiding transfer of pollen between anther and stigma.
Stigma	Floral organ, at distal end of style, where pollen is deposited.
Xenogamy	Cross-fertilization, between flowers on different plants (or different genets).

See also: **Flowering and Reproduction**: Fertilization; Reproductive Strategies. **Seed Development**: Seed Production.

Further Reading

Allen-Wardell G, Bernhardt P, Bitner R, *et al.* (1998) The potential consequences of pollinator declines on the conservation of biodiversity and stability of food crop yields. *Conservation Biology* 12: 8–17.

Corbet SA, Williams IH, and Osborne JL (1991) Bees and the pollination of crops and wild flowers in the European Community. *Bee World* 72: 47–59.

Crane E and Walker P (1984) *Pollination Directory for World Crops*. London: International Bee Research Association.

Delaplane KS and Mayer DF (2000) *Crop Pollination by Bees*. Wallingford: CAB International.

Free JB (1993) *Insect Pollination of Crops*, 2nd edn. London: Academic Press.

Jay SC (1986) Spatial management of honey bees on crops. *Annual Review of Entomology* 31: 49–65.

Kearns CA and Inouye DW (1993) *Techniques for Pollination Biologists*. Boulder: University Press of Colorado.

Kearns CA, Inouye DW, and Waser NM (1998) Endangered mutualisms: the conservation of plant–pollinator interactions. *Annual Review of Ecology and Systematics* 29: 83–112.
Kevan PG and Baker HG (1983) Insects as flower visitors and pollinators. *Annual Review of Entomology* 28: 407–453.
Lloyd DG and Barrett SCH (eds) (1996) *Floral Biology: Studies on Floral Evolution in Animal-Pollinated Plants.* New York: Chapman & Hall.
Matheson A (ed.) (1996) *Bumble Bees for Pleasure and Profit.* Cardiff: International Bee Research Association.
McGregor SE (1976) *Insect Pollination of Cultivated Crop Plants.* ARS-USDA, Washington, DC: Agricultural Research Service, US Department of Agriculture. (Also available at http://gears.tucson.ars.ag.gov/book/index.html
Morse RA and Calderone NW (2000) The value of honey bees as pollinators of US crops in 2000. *Bee Culture*, March 2000.
Proctor M, Yeo P, and Lack A (1996) *The Natural History of Pollination.* London: Harper Collins.
Roubik DW (ed.) (1995) *Pollination of Cultivated Plants in the Tropics.* Rome: Food and Agriculture Organization.

Fertilization

J S Carmichael, University of North Dakota, Grand Forks, ND, USA

Introduction

The processes of fertilization and embryogenesis are two of the most fundamental phenomena studied by both plant and animal biologists. The recognition of sperm and egg, and subsequent fusion of those gametes, is of paramount importance in sexual reproduction in plants and most other eukaryotes. Although we have known about the basic underlying principles of fertilization in higher plants for nearly 200 years, it is only in the past few decades that we have begun to understand the mechanisms responsible for successful fertilization. This article will focus on the events that take place during fertilization in flowering plants (angiosperms), recent advances in *in vitro* fertilization techniques, and the relationship between the cell cycle and gamete development.

History of Fertilization Studies

Giovanni Amici is credited as the first person to observe and report the germination of pollen grains and subsequent growth of pollen tubes back in 1824. Amici also recognized the presence of "germination vesicles" within ovules prior to the arrival of a pollen tube, which we now recognize as true egg cells. Amici's reports were later confirmed by Wilhelm Hofmeister, a botanist who revolutionized our understanding of plant biology with his studies on embryogenesis and alternation of generations. After studying many groups of plants, Hofmeister concluded that all land plants alternate between two phases during their life cycle: a diploid spore-bearing phase (sporophyte) and a haploid gamete-bearing phase (gametophyte) (**Figure 1**). Among flowering plants, the sporophyte is the larger, dominant phase and consists of the major organs including roots, stems, leaves, and parts of flowers. The gametophytes are much smaller than sporophytes and are produced within flowers. Pollen grains are haploid, sperm-producing structures and therefore represent the male gametophyte (or microgametophyte). Embryo sacs are haploid, egg-producing structures which constitute the female gametophyte (or megagametophyte).

In most groups of plants, fertilization occurs only between male and female gametes and yields zygotes. However, flowering plants are characterized by a

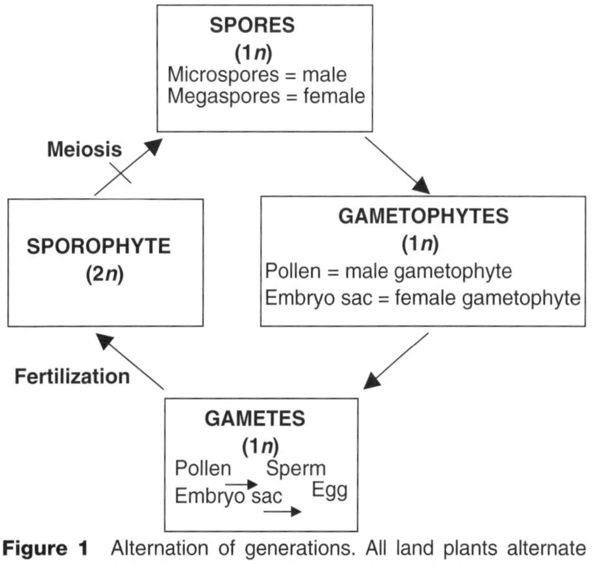

Figure 1 Alternation of generations. All land plants alternate between a diploid sporophyte phase and a haploid gametophyte phase as part of their life cycle. In angiosperms, the sporophyte includes all the tissues associated with roots, stems, leaves, and some flower tissues. Special cells within flowers undergo meiosis to form separate male and female haploid spores (microspores and megaspores, respectively). The spores then develop into haploid gametophytes. Within anthers, microspores develop into pollen (male gametophyte). Within ovules, megaspores develop into embryo sacs (female gametophyte). After pollination and fertilization, the gametes unite and give rise to an embryo contained within a seed. If the seed germinates, the embryo will grow into a mature sporophyte.

process of double fertilization whereby a single pollen tube delivers two sperm to an embryo sac. One of the sperm fertilizes the egg cell to form a zygote while the second sperm fertilizes the central cell, which ultimately develops into endosperm, an embryo-nourishing tissue found exclusively in flowering plants. This process of double fertilization was observed and reported independently by two investigators well over a century ago. Sergius Nawaschin and Leon Guignard both reported this phenomenon in 1899 with their studies on *Lilium martagon*. Remarkably, Nawaschin and Guignard, as well as Amici and Hofmeister, were able to produce accurate and detailed studies despite the relatively primitive types of microscopy available at the time. Although fertilization has been studied in virtually all groups of land plants, most of our understanding comes from studies on angiosperms since the majority of economically important plants are flowering plants.

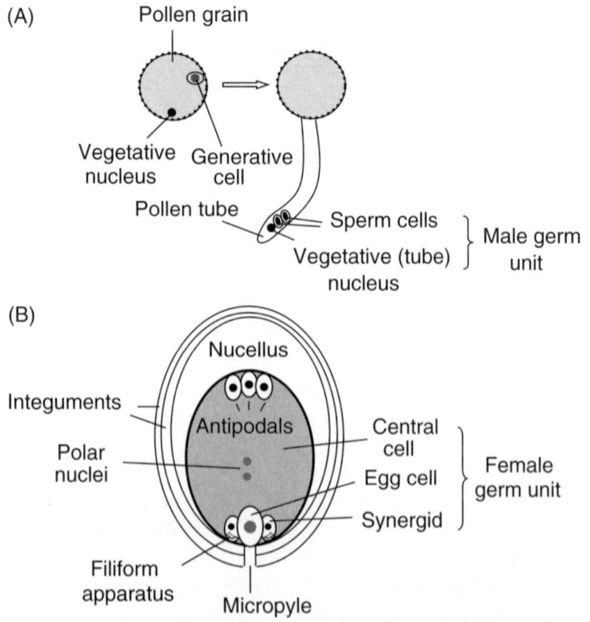

Figure 2 Male and female gametophytes of angiosperms. (A) Male gametophyte. Pollen grains consist of two or three cells at the time they are released from the plant. This figure shows a two-celled pollen grain consisting of a tube cell and a generative cell. The generative cell ultimately divides to yield two sperm cells well after pollination takes place. In plants with three-celled pollen, the generative cell divides before the pollen is released. Thus, pollen in these plants includes a tube cell and two sperm. Pollen tubes grow through the stigma and style carrying sperm to the embryo sac. (B) Embryo sac (female gametophyte). Although there is significant variation among taxa, the embryo sac of most angiosperms consists of an egg cell, two synergids, three antipodals, and a binucleate central cell. The embryo sac is embedded within the nucellus and the ovule is surrounded by integuments (precursors to seed coats). A filiform apparatus is commonly found within many synergids. The filiform apparatus appears to play a role in pollen tube guidance and the synergids are the cells that receive the pollen tube for fertilization.

Male Gametophyte Development

Development of pollen grains (male gametophytes) takes place in the anther. Pollen development begins when specialized cells (microsporocytes) differentiate in young anthers. There are generally hundreds or thousands of microsporocytes per anther and each passes through meiosis to produce four haploid microspores. Each microspore becomes highly vacuolate and the haploid nucleus moves toward the periphery of the cell. The microspore then undergoes a highly unequal cell division producing a larger vegetative cell and a smaller generative cell. Microtubules and endoplasmic reticulum have been shown to facilitate this asymmetric division by maintaining the peripheral position of the nucleus prior to mitosis and keeping plastids away from the generative cell as it is walled off from the rest of the vegetative cell. Thus, the vegetative cell typically contains most of the plastids of the male gametophyte. The vegetative cell undergoes high levels of transcription, completely surrounds the smaller generative cell, and ultimately develops into a pollen tube once successful pollination has taken place. The generative cell, which has a wall rich in cellulose and callose, will ultimately divide to yield two sperm. This mitotic division occurs either before or after pollination, depending on the species. Each sperm is membrane-bound and thus represents a true cell. Once the sperm cells are formed, they remain closely appressed to the vegetative nucleus collectively forming the male germ unit (**Figure 2**). The male germ unit is carried within pollen tubes directly toward the embryo sac.

Female Gametophyte Development

Development of the embryo sac (female gametophyte) takes place in tissues embedded within the ovule. Although several variations exist, the majority of flowering plants follow the monosporic pattern of embryo sac development. As part of this pattern, a single megasporocyte differentiates within an ovule. That cell passes through meiosis to yield four haploid megaspores. Typically, three of the megaspores degenerate while one survives and develops into a mature embryo sac. The death of the three megaspores appears to be a genetically programmed phenomenon. Callose accumulates around the three abortive megaspores and, as a result, plasmodesmata are not established around these cells and they subsequently undergo apoptosis.

The single surviving megaspore undergoes three mitotic divisions in the absence of cytokinesis. As a result of these free-nuclear divisions, a total of eight nuclei are found within the confines of the former megaspore. The nuclei then become walled off and differentiate into distinct cells of the embryo sac, which include three antipodal cells formed in the chalazal region, two synergids and one egg cell in the micropylar region, and a large, binucleate central cell containing the two polar nuclei (**Figure 2**). Collectively, the egg, two synergids, and large central cell constitute the female germ unit.

The cellularization of the embryo sac occurs without the formation of preprophase bands of microtubules. The nuclei themselves may serve as microtubule-organizing centers and may determine the deposition of cell wall material. In addition, the position of nuclei, even during the four nucleate stage, is an important component of embryo sac maturation. At the four nucleate stage, two nuclei are positioned at the micropylar region while two are positioned at the chalazal region. Of the two micropylar nuclei, one divides to form the two synergids while the other divides to form the egg cell and one of the two polar nuclei. Of the two chalazal nuclei, one divides to form two antipodals while the other divides to form one antipodal and the other polar nucleus.

At the completion of cellularization, the egg and synergids are small, wedge-shaped cells, which then expand, their walls becoming discontinuous and bead-like. Initially, the egg and synergids are similar in structure and appearance. However, the synergids soon develop characteristic wall ingrowths similar to those found in transfer cells. These convoluted walls form the filiform apparatus that is characteristic of the synergids of most angiosperms. The synergids play an important role in fertilization by attracting pollen tubes toward the embryo sac and facilitating the fertilization process (see below). It is not entirely clear what role, if any, the antipodals play during sexual reproduction in plants. One thought is that the antipodals may facilitate the movement of nutrients into the developing embryo sac. In some plants, e.g., wheat, corn, and other grasses, the antipodals may proliferate into a small mass of cells. Nevertheless, antipodals do not take part directly in fertilization processes.

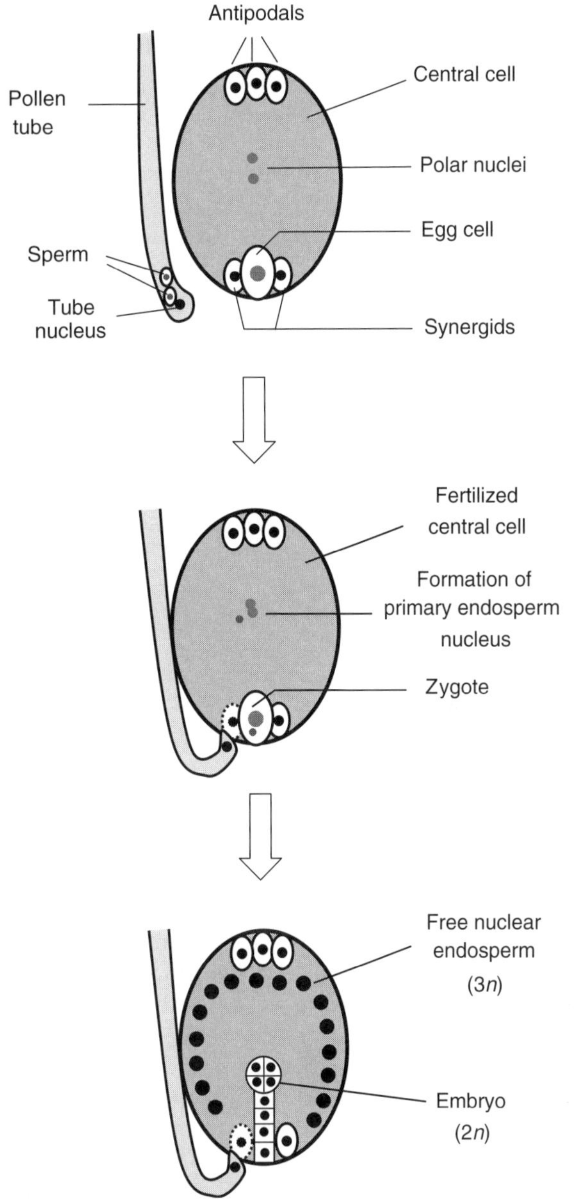

Figure 3 Double fertilization. A pollen tube carries two sperm toward the micropylar region of the embryo sac. Typically, one of the two synergids degenerates and the pollen tube enters the embryo sac through this degenerate cell. The pollen tube releases its gametes. One of the sperm fertilizes the egg cell to form a zygote. The zygote will then develop into an embryo within the seed. The second sperm fertilizes the binucleate central cell. The product of this fertilization event will develop into endosperm.

Events Leading to Fertilization

After a pollen grain lands on the stigma of a flower of the same species, it hydrates, germinates, and the pollen tube carries male gametes toward the embryo sac for fertilization (**Figure 3**). The following sequence of events leading to fertilization in flowering plants based on available evidence as well as hypothetical scenarios was published in 1981: (1) pollen lands on the stigma of a flower; (2) stigma, style, and ovary respond via recognition reactions, nutrient mobilization, and hormone synthesis; (3) hormones reach the synergids; (4) hormones activate a synthetic mechanism in at least one synergid; (5) synergids synthesize autolytic and chemotropic compounds; (6) one synergid begins to degenerate by autolysis; (7) chemotropic and autolytic compounds are secreted into the micropyle via the filiform apparatus; (8) pollen tubes reach the ovary and are attracted to the micropyle by a chemotropic agent; (9) a pollen tube comes into contact with the autolytic compound and its cytoplasm begins to degenerate; (10) a pollen tube penetrates into the cytoplasm of the degenerated synergid and discharges its contents; (11) sperm migrate to their respective female nuclei; and (12) syngamy and double fertilization take place.

Growth of Pollen Tube Toward the Ovule

Pollen growth begins at the stigmatic surface. Once pollen adheres to a stigmatic cell, the extracellular pollen coat actually migrates toward the site of contact, a phenomenon referred to as coat inversion. This process is immediately followed by hydration and commencement of metabolic activities. Lipids have been found to play an important role in allowing water transport from stigma to pollen. The pollen then begins tube growth in the proper orientation and in many species the pollen tube grows through whichever pore is adjacent to the stigmatic surface.

After germination, a pollen tube must navigate through several different types of tissues as it makes its way to an embryo sac and multiple signals guide the pollen tube along the way. Some of the likely mechanisms of pollen tube guidance include chemotaxis, mechanical control, electrotaxis, and the active promotion of growth by transmitting tissues in the stylar matrix. Both sporophytic and gametophytic tissues are ultimately responsible for guiding a pollen tube into the embryo sac.

Initially, pollen tubes grow through the stigmatic cells on their way to other female tissues. In *Arabidopsis*, the pollen tube actually burrows into the wall of a stigma cell and grows within the wall until it reaches the transmitting tract tissues within the style. Cells lining the transmitting tract are coated with a nutrient-rich extracellular matrix that completely fills the intercellular spaces in species with closed styles. Nutrients and signal molecules aid pollen tube growth while in the transmitting tract.

Pollen tubes grow through the style and into the ovary but eventually exit the transmitting tract at specific locations as they grow toward an ovule. In fact, in plants with multiple ovules, a pollen tube will emerge from the transmitting tract at a region where the first available (unfertilized) ovule is located. Plants with mutations that prevent embryo sac formation do not show this normal pattern of pollen tube guidance indicating that separate signals control pollen tube movement through, and exit from, the transmitting tract.

After exiting the transmitting tract, a pollen tube may grow along the inner surface of the ovary and along the funiculus toward an ovule. In most flowering plants, the pollen tube enters the ovule through the micropyle, a condition known as porogamy. However, there are some plants that display chalazogamy, whereby the pollen tube enters through the chalazal region of the ovule, and also mesogamy, whereby the pollen tube enters through the side of the ovule. Whatever the case, the tip of the pollen tube always seems to enter the micropylar region of the embryo sac (**Figure 3**).

Receipt of the Pollen Tube

Synergids have been found to play an important role in the short-range guidance of pollen tubes to the embryo sac. The two synergids that are positioned adjacent to the egg have been found in every family of angiosperms studied to date. At the microscopic level, the two synergids appear structurally similar and display similar cytoplasmic organelles and ultrastructure. The filiform apparatus, found within each synergid, is thought to secrete compounds into the micropylar region of the ovule which serve to attract pollen tubes toward the egg cell. Studies show that if cells of the embryo sac other than the synergids are ablated then pollen tubes will still grow toward the micropyle. However, if synergids are ablated then the pollen tube is not attracted toward the micropyle. A variety of substances have been documented in or near the micropyle including carbohydrates, proteins, and even calcium in the synergid vacuoles of cotton and some grasses. Some angiosperms, such as *Ceratostigma*, *Dyerophytum*, *Plumbago*, and *Plumbagella*, lack synergids. In these plants, a filiform apparatus may be present in the egg cell itself suggesting its possible role in guiding pollen tubes toward the embryo sac.

There are a variety of ways in which pollen tubes can enter the embryo sac via the synergids, including: (1) between two intact synergids; (2) between the egg cell and synergid; (3) between the embryo sac wall and one or both synergids; (4) penetrating the synergid; (5) destroying the synergid it encounters; and (6) directly entering the cytoplasm of the degenerated synergid. In general, pollen tubes seem to enter degenerate synergids. Typically, only one of the two synergids degenerates. In many plants, the synergid degenerates only after the pollen tube reaches the ovule; however, synergids have been found to degenerate even before pollination in *Brassica* (mustards), *Triticum* (wheat), and *Oryza* (rice). In those angiosperms that lack synergids, there is no degeneration of the egg cell or the embryo sac itself.

Double Fertilization

Once a pollen tube enters a synergid, it ruptures near its apex and discharges its contents. It is not clear what, exactly, causes the pollen tube to rupture once it enters the synergid. However, rupture is probably triggered by differences in osmotic potential between the pollen tube and synergid or by the effect of calcium ions on the pollen tube. Sperm are reported to share the following characteristics at the time of fertilization: (1) male gametes are contained in a male germ unit (sperm closely associated with a tube nucleus) which is severed during pollen tube discharge; (2) the additional plasma membrane surrounding both of the sperm cells (i.e., the inner pollen plasma membrane) is removed during pollen tube discharge, directly exposing the sperm plasma membrane; (3) progressive loss of cell wall material during development of the pollen tube until there is no cell wall present at the time of gamete delivery; and (4) male gametes are true cells containing a cytoplasm with a diminished number of organelles.

After discharge, male gametes usually contact both the egg cell and central cell simultaneously. The membrane of one of the sperm cells fuses with that of the large central cell. By that time, the two polar nuclei typically have already fused forming a diploid secondary nucleus. The sperm nucleus then migrates toward the secondary nucleus and karyogamy ensues forming a triploid primary endosperm nucleus within the central cell (**Figure 3**). Fusion also occurs between the second sperm and the egg cell to form a zygote, which undergoes repeated rounds of mitosis and cytokinesis as it develops into an embryo. Typically, two sperm are discharged from just one pollen tube. However, in some plants, several pollen tubes arrive at the embryo sac and the sperm of one may fertilize the egg cell while the sperm of a separate pollen tube may fertilize the central cell. This phenomenon is referred to as heterofertilization. In most flowering plants, the endosperm tends to develop in advance of the embryo and is characterized by numerous rounds of free nuclear divisions prior to cellularizing as endosperm tissue. However, some plants lack the free nuclear phase and display cellular endosperm development *ab initio*.

After fertilization, the pollen tube typically becomes sealed by wall formation in the synergid. A callose plug may also form in the pollen tube, but this may be a general wound response and may not be caused by the fertilization process directly. Pollen tubes typically form many callose plugs during growth and development that form a barrier between the apex of the tube and the rest of the cell.

Preferential Fertilization

Although two sperm are found within each pollen tube, they are often dimorphic and one of the sperm may be more tightly associated with the tube nucleus. Therefore, there is polarity in the male germ unit, which may extend to cytoplasmic and potentially genetic differences in some instances. In fact, cytoplasmic differences between two sperm within a pollen tube are common among angiosperms. Usually, the sperm associated with the tube nucleus contains more mitochondria than the sperm not associated with the tube nucleus. In cases of sperm polarity, one of the sperm typically has a greater chance of fusing with the egg than the other. In terms of genetic differences, corn (*Zea mays*; maize) contains B chromosomes that do not separate during sperm formation resulting in one sperm with extra B chromosomes. The sperm with the extra B chromosomes appears to preferentially fertilize the egg cell over the central cell. However, this may be an unusual situation among flowering plants.

Although preferential fertilization occurs in some taxa, it is not clear whether this represents a widespread phenomenon among angiosperms. Two alternative hypotheses have been proposed for double fertilization: (1) the chance hypothesis; and (2) the specific receptor hypothesis. The chance hypothesis suggests that both of the sperm in a pollen tube have equal chances at fertilizing the egg. The specific receptor hypothesis indicates that fusion of one of the sperm with the egg is predetermined by cell surface recognition molecules. In other words, only one of the two sperm is destined to fertilize the egg while the other sperm is destined to fertilize the central cell. Considering the sperm dimorphism found in certain plants and the B chromosome phenomenon in corn, the specific receptor hypothesis will probably continue to receive support as more taxa are studied. In fact, we are now able to study gene expression in single isolated sperm and egg cells using reverse transcriptase polymerase chain reaction. Techniques such as these will continue to expand our knowledge of the molecular interactions between male and female gametes.

Cytoplasmic Inheritance

After fertilization, the nuclear material of the sperm represents half of the genetic material in the zygote, but cytoplasmic organelles (e.g., plastids and mitochondria) vary between sperm and egg cells. The vast majority of angiosperms display uniparental, maternal cytoplasmic inheritance of plastids and mitochondria, consequently, all of the plastids and mitochondria found in the zygote originate from the egg cell. In some flowering plants, plastids have been detected in the sperm or generative cells (recall that the generative cell divides mitotically to yield the two sperm), representing the possibility of biparental cytoplasmic inheritance. Uniparental, paternal cytoplasmic inheritance, i.e., where the zygote contains mitochondria and plastids derived exclusively from the sperm cell, is rare among angiosperms, but common in many gymnosperms (e.g., Araucariaceae, Cephalotaxaceae, Cupressaceae, and Taxodiaceae).

Nuclear Fusion

Karyogamy is initiated by the fusion of the outer envelope of the male and female nuclei; eventually, the inner envelopes fuse also. There are several variations observed with respect to the timing of karyogamy and the onset of zygotic mitosis. In premitotic karyogamy, the male and female nuclei merge prior to zygotic mitosis. This is the most common condition in flowering plants. Some plants display postmitotic karyogamy where the male and female nuclei approach each other, remain close, but do not fuse until after mitosis is initiated. There is also an intermediate form of karyogamy whereby the male nucleus enters mitosis precociously, the egg then enters mitosis and, during metaphase, a common metaphase plane is formed between the two sets of dividing nuclei.

Relationship Between the Cell Cycle and Gamete Development

With the advent of *in vitro* fertilization techniques and regeneration of genetically engineered zygotes and embryos there is renewed interest in the relationship between the cell cycle and sexual reproduction in higher plants. Unfortunately, reproductive cell cycles have not been adequately investigated in most groups of seed plants and, with a few exceptions, relationships between cell cycle patterns and sexual reproduction have been derived exclusively from a small number of studies on angiosperms.

There have been few studies of seed plants that have documented nuclear DNA levels of gametes as well as zygotes. One of the first, and perhaps most compelling, studies on the relationship between the cell cycle and sexual reproduction revealed that egg cells of *Tradescantia paludosa* contain the 1C quantity of DNA through fertilization and newly formed zygotes contain the 2C amount. Sperm DNA levels were not measurable but were inferred to contain the 1C quantity of DNA. Thus, in *T. paludosa*, gamete nuclei contain the 1C level of DNA and remain positioned within the G_1 phase of the cell cycle throughout karyogamy, and zygotes pass through the S phase of the cell cycle only after male and female gamete nuclei have completely fused. This mode of fertilization, in which gamete nuclei contain the 1C quantity of DNA throughout karyogamy, has been reported in several other flowering plants, including barley, maize, and *Pennisetum ciliare*, and has been defined as "G_1 karyogamy" (**Figure 4**). Recently, a second and distinct fertilization pattern has been documented in *Ephedra trifurca*, a nonflowering seed plant (gymnosperm). In *Ephedra*, male and female gamete nuclei each contain the 1C quantity of DNA at the onset of

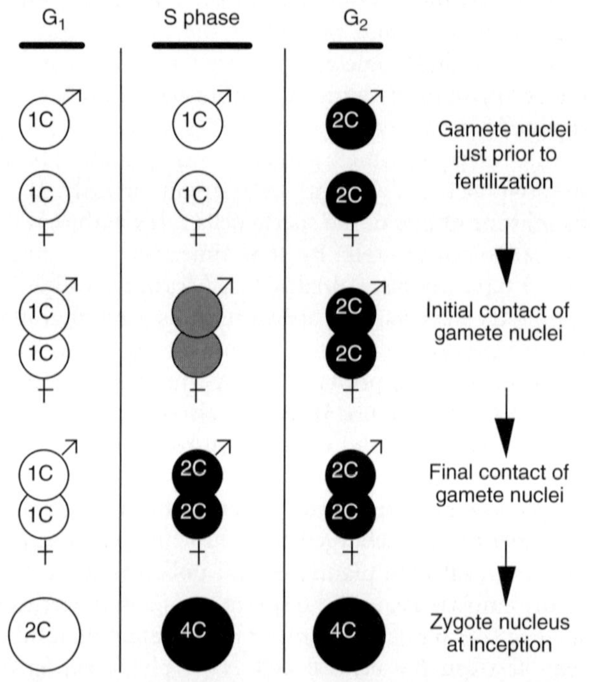

Figure 4 Relationships between the cell cycle and gamete karyogamy. In G_1 karyogamy, gamete nuclei remain in the G_1 phase of the cell cycle and contain the 1C amount of DNA throughout fusion. The resulting zygote contains the 2C quantity of DNA at inception and subsequently passes through S phase before zygotic mitosis. In S-phase karyogamy, gamete nuclei contain the 1C quantity of DNA at the onset of fusion but enter the S phase of the cell cycle after the initial contact between two nuclei is made. Nuclear fusion is not completed until the gamete nuclei pass through S phase. In G_2 karyogamy, gamete nuclei pass through the S phase of the cell cycle and increase their DNA content from 1C to 2C prior to fusion. ○, nuclei in the G_1 phase of the cell cycle; ●, nuclei in S phase of the cell cycle; ●, nuclei in G_2 phase of the cell cycle. Reproduced with permission from Carmichael JS and Friedman WE (1995) Double fertilization in *Gnetum gnemon*: the relationship between the cell cycle and sexual reproduction. *The Plant Cell* 7: 1975–1988. This material is copyrighted by the American Society of Plant Biologists.

fertilization. However, once sperm nuclei are deposited in an egg cell, male and female gamete nuclei remain in close contact for an extended period of time while they pass through the S phase of the cell cycle synchronously. Only after DNA synthesis is complete do gamete nuclei completely fuse to form a diploid zygote nucleus with the 4C compliment of DNA that is competent to enter mitosis. This cell cycle pattern in which gamete nuclei synchronously initiate DNA replication at the onset of fertilization is defined as "S-phase karyogamy" (**Figure 4**). Another recent study on the nonflowering seed plant *Gnetum gnemon*, demonstrates a third and distinct mode of fertilization in which male and female gamete nuclei pass through the S phase of the cell cycle and double their DNA content prior to the onset of fertilization. This reproductive cell cycle pattern can be defined as 'G_2 karyogamy' (**Figure 4**). *Arabidopsis* has also been shown to exhibit this cell cycle pattern.

By the time a pollen tube enters the female gametophyte in *Gnetum*, each of the two sperm nuclei contains the 2C quantity of DNA. Thus, male gametes complete DNA replication prior to the onset of fertilization. The tube (vegetative) nucleus does not enter the S phase of the cell cycle and does not exceed the 1C quantity of DNA. Since the tube nucleus and sperm nuclei are genetically identical, remain appressed to each other within a pollen tube and exhibit disparate cell cycles, it is likely that either the biochemical or molecular cues that trigger DNA synthesis are found within the sperm cell itself, or that the tube nucleus is not competent to enter the S phase of the cell cycle.

Except in apomictic species, an unfertilized egg cell will typically not divide. So, what is the role of the sperm in cell division? The expression of cell cycle genes in isolated gametes and zygotes has been studied in corn. Sperm cells have been found to express several cell cycle-specific genes (e.g., *cdc2ZmA/B* and Zea-ma;*CycA1;1*). However, other mitotic cyclins such as Zeama;*CycB1;2* and Zeama;*CycB2;1* are not zex-pressed in sperm cells, but are expressed during the first embryonic cell division cycle. Therefore, the first embryonic cell cycle in corn appears to be regulated zygotically rather than maternally as in many animal zygotes. Further studies are needed in diverse taxa of angiosperms to determine the extent of zygotic and maternal regulation of embryonic cell cycles.

In vitro Fertilization

Although it has been well over 100 years since the discovery of double fertilization in flowering plants, only recently have advances been made on *in vitro* fertilization techniques in angiosperms. Most of the *in vitro* studies have concentrated on corn, but other major crop plants continue to be studied. *In vitro* fertilization studies have been difficult in flowering plants primarily because of the technical challenges associated with isolation of male and female gametes. Unlike most of the "lower" plants, such as mosses and ferns which produce free-swimming sperm and readily accessible egg cells, the gametes of angiosperms are concealed in pollen tubes or are deeply embedded within surrounding nucellar tissue. Thus, they must be isolated and must remain fertile if they are to be useful in *in vitro* fertilization studies. Fortunately, researchers are now able to produce zygotes successfully and those zygotes are able to develop into embryos and even fertile plants with the help of cell culture techniques. Furthermore, fertilized central cells are able to develop into endosperm tissue with the use of cell culture techniques. Indeed, embryos and endosperm can develop to maturity *in vitro* without the surrounding parental tissue normally found in plants and even when kept separate from each other. This growth pattern indicates that these fertilization products do indeed possess their own developmental growth regulators. *In vitro* fertilization techniques are especially useful for producing various crop plant hybrids. However, they do not eliminate potential problems associated with zygotic and postzygotic incompatibility.

Intact sperm cells are typically isolated by grinding pollen or bursting pollen tubes by osmotic shock. Embryo sacs and embryos have been isolated mainly from monocots, such as corn, barley (*Hordeum vulgare*), wheat (*Triticum aestivum*), and rye (*Secale cereale*) but have also been successfully isolated from a few dicots, including rape-seed (*Brassica napus*) and tobacco (*Nicotiana tabacum*). In corn, the microdissection of egg cells and zygotes is generally facilitated by treating nucellar tissue with enzymes such as cellulase, hemicellulase, and pectinase. Collectively, these enzymes help to degrade cell wall material and allow individual cells to be easily isolated. Laser microbeam dissection may also be used to isolate egg cells. A proficient worker is able to isolate 20 to 50 corn egg cells per day.

Gametes are basically protoplasts once they are isolated from pollen tubes or embryo sacs. Egg and central cell protoplasts are spherical and contain clusters of organelles at their periphery. They also display cytoplasmic polarity in their isolated state. These protoplasts can be maintained in nutritive medium prior to fertilization. Some of the fusion techniques commonly employed in *in vitro* fertilization include electrofusion, treatment with polyethylene glycol or calcium, or even laser-mediated fusion.

Micromanipulation is very important to achieve successful fertilization as it is necessary to bring male and female gametes in close contact with one other.

Isolated sperm, egg cells, and central cells can be aligned using a glass microneedle or an AC electrical current. Electrofusion is a very proficient technique for fusing sperm and egg cells and in corn has a success rate of 85%. For some reason, the fusion of sperm and central cell using this technique results in a lower success rate of about 44%. Electrofusion is accomplished by applying short pulses of DC electricity. This may potentially mimic conditions found in the plant as electrical currents have been documented in plant cells. Fusion occurs rather quickly and is usually completed in less than 1 s. The spindle-shaped nature of the sperm of some plants is maintained by the cytoskeleton. Often, the pointed tip region of the sperm aligns with the egg cell membrane. This type of directed fusion was observed well over 100 years ago.

Under *in vitro* conditions, sperm are able to fuse not only with egg and central cells, but also with mesophyll and other nonreproductive cells. Thus, the conditions used to accomplish fertilization *in vitro* may not necessarily represent those found in the plant itself.

Corn has been the most widely studied plant for *in vitro* fertilization. *In vitro* fertilization has resulted in fertilized egg cells of corn developing into zygotes, embryos, and even mature fertile plants. Fertilized central cells are also able to develop into mature endosperm tissue. However, the generation of fertile plants from *in vitro* zygotes has been quite difficult to achieve in other plants. For example, *in vitro* zygotes of wheat have only developed into multicellular structures. *In vivo* zygotes have been a little more successful as fertile plants have been derived from barley, wheat, and corn. Thus, although fusion of gametes is relatively easy to accomplish, the exact conditions required for proper development continue to be optimized.

Researchers studying zygote development have focused on zygotes formed by two different methods. A true *in vitro* zygote is formed after fusion of isolated sperm and egg cells. An *in vivo* zygote is one that is isolated after pollination and fertilization in the plant itself. With *in vitro* zygotes obtained from corn, new cell wall material was observed around the newly formed zygote within 30 s of inception. This rapid formation of cell wall material may represent a block to polyspermy. After initial fusion of the cells, karyogamy typically occurs within 35–45 min in the zygote and about 1 h in the fertilized central cell.

In both *in vitro* and *in vivo* zygotes of corn, polarity is evident in the form of uneven distribution of cytoplasm within the egg itself and an asymmetry is evident in the plane of the first cell division. These phenomena are thought to be regulated by mechanisms found within the egg cell. For example, corn eggs fused with sperm from barley or wheat result in an asymmetric first cell division, much like that which occurs when corn sperm and egg are fused. However, when an egg cell from wheat and a sperm cell from corn are fused, this results in a symmetrical division plane similar to that observed when wheat sperm and egg cells are fused. The molecular mechanisms involved in the regulation of zygote formation and maturation will continue to represent active areas of research in the near future.

List of Technical Nomenclature

Double fertilization The process in flowering plants whereby a single pollen tube delivers two sperm cells, one of which fertilizes the egg to produce a zygote and the other fertilizes the central cell to yield endosperm.

Embryo sac (female gametophyte) A multicellular structure found within the ovule that forms the egg cell. In most flowering plants, the embryo sac consists of one egg cell, two synergids, three antipodals, and a binucleate central cell.

Female gametophyte The embryo sac, bearing the female gametes.

Filiform apparatus A subcellular structure typically found in synergids. This structure is thought to play a role in guiding pollen tubes toward the embryo sac.

Generative cell The cell found either in the pollen grain or pollen tube that divides to form two sperm cells.

***In vitro* fertilization** The technique of fusing sperm isolated from pollen tubes with eggs and central cells isolated from embryo sacs.

Male gametophyte The pollen grain and its associated pollen tube, bearing the male gametes.

Micropyle The opening near the apex of an ovule through which pollen tubes grow to reach the embryo sac.

Synergid One of the embryo sac cells found adjacent to the egg cell. Typically, these are the cells that actually receive pollen tubes during fertilization.

See also: **Flowering and Reproduction**: Reproductive Strategies. **Growth and Development**: Cells. **Seed**

Development: Embryogenesis; Endosperm Differentiation; Seed Production.

Further Reading

Friedman WE (1990) Double fertilization in *Ephedra*, a nonflowering seed plant: its bearing on the origin of angiosperms. *Science* 247: 951–954.

Heslop-Harrison J, Heslop-Harrison JS, and Heslop-Harrison Y (1999) The structure and prophylactic role of the angiosperm embryo sac and its associated tissues: *Zea mays* as a model. *Protoplasma* 209: 256–272.

Huang BQ and Russell SD (1992) Female germ unit: organization, isolation, and function. *International Review of Cytology* 140: 233–293.

Jensen WA (1973) Fertilization in flowering plants. *BioScience* 23: 21–27.

Johnson MA and Preuss D (2002) Plotting a course: multiple signals guide pollen tubes to their targets. *Developmental Cell* 2: 273–281. (This work was a significant source of information used in this article.)

Knox RB and Singh MB (1987) New perspectives in pollen biology and fertilization. *Annals of Botany* 60: 15–37.

Knox RB, Zee SY, Blomstedt C, and Singh MB (1993) Tansley Review No. 61: Male gametes and fertilization in angiosperms. *New Phytologist* 125: 679–694.

Kranz E and Kumlehn J (1999) Angiosperm fertilisation, embryo and endosperm development *in vitro*. *Plant Science* 142: 183–197. (This work was a significant source of information used in this article.)

Mogensen HL (1996) The hows and whys of cytoplasmic inheritance in seed plants. *American Journal of Botany* 83: 383–404.

Russell SD (1992) Double fertilization. *International Review of Cytology* 140: 357–388. (This work was a significant source of information used in this article.)

Russell SD (1996) Attraction and transport of male gametes for fertilization. *Sexual Plant Reproduction* 9: 337–342.

Gametophytic Self-Incompatibility

M Cappadocia, Université de Montréal, Montréal, Canada

Introduction

Self-incompatibility (SI), a classic example of cell–cell recognition, represents the most widespread of the genetic mechanisms exhibited by angiosperms to prevent inbreeding and promote outcrossing. Estimated to be present in more than 50% of all flowering plant species, SI is the inherited ability of a fertile hermaphrodite seed plant to prevent the production of zygotes after self-pollination. The phenomenon involves interactions between gene products expressed in the pollen, and those expressed in specialized cells of the pistil. As a result, self-incompatible plants reject pollen that expresses an allele also present in their pistils, whereas they accept any pollen with different alleles (**Figure 1**). In SI, pollen either does not germinate or germinates to produce a pollen tube whose growth is arrested either at the surface of the stigma or in the style. The term SI was introduced by Stout in 1917 to replace "self-sterility," a phenomenon first discovered in *Verbascum phœniceum* by Kölreuter in 1764. One century later, Darwin in his books *Effects of Cross- and Self-Fertilization in the Vegetable Kingdoms* and *Different Forms of Flowers on Plants of the Same Species* not only reviewed the literature on the subject, but precisely described heteromorphic SI (see below) as well as the effects of the environment on SI. Since then, a considerable amount of information has become available on the distribution, physiology, and genetics of SI. Some of the biochemical and molecular processes involved in the recognition and rejection of self-pollen, however, have become known only in the past 20 years, thanks to the development of molecular techniques, which have provided an answer to some of the many questions concerning SI. In this article, the current state of knowledge on gametophytic self-incompatibility is summarized and crucial experiments performed in the Solanaceae, the most studied family, are highlighted.

Self-Incompatibility Systems

Two SI types are recognized on the basis of floral morphology: heteromorphic and homomorphic. In heteromorphic SI (present in at least 28 families) there are four classes of stylar polymorphism (distyly, tristyly, stigma height dimorphism, enantiostyly) that differ in their relative positions within a flower and for the number of floral morphs that occur within populations. Crosses succeed only when they involve flowers with different morphology. By contrast, in the homomorphic type (present in about 90 families), all the flowers of the interbreeding plants have the same morphology. In the vast majority of species characterized by homomorphic SI, the interactions between pollen and pistil are under the control of elements of one of the most polymorphic loci in plants, the S locus.

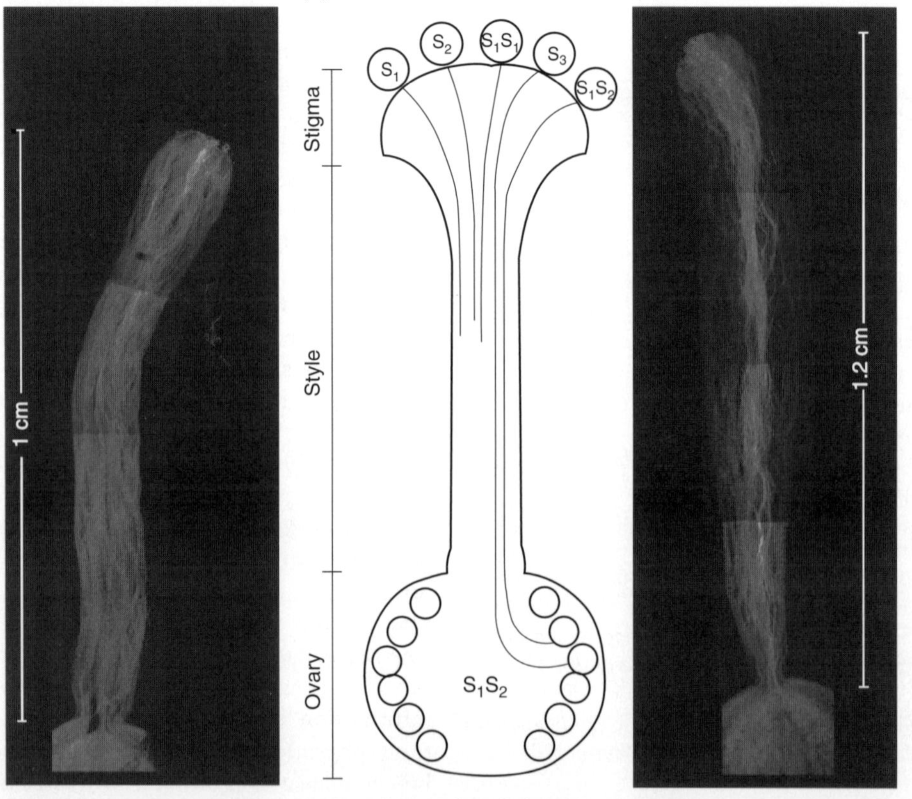

Figure 1 (see color plate 10) Pollen behavior in RNase-based GSI. In squashes of *Solanum chacoense* styles (genotype S_1S_2) stained with aniline blue and observed by epifluorescence 48 h after pollination, pollen tubes fluoresce bright green. In incompatible crosses (A), the pollen tubes exhibit arrested growth at mid style, whereas in compatible crosses (C) they enter the ovary. (B) Schematic representation. Note the compatibility of diploid S-heteroallelic pollen (S_1S_2).

The homomorphic type is further subdivided into two systems: the sporophytic and gametophytic. This subdivision is based on the genetic control of pollen–pistil interactions following pollination. The sporophytic system (SSI) exists in only six families, is characterized by stigmatic rejection and is usually found in species with dry stigma and trinucleate pollen. In this system, the breeding behavior of the pollen, whether compatible or incompatible, is determined by the allelic constitution at the S locus of the pollen-producing plant. The far more widespread gametophytic system (GSI) is generally characterized by stylar rejection and is typically found in species with wet stigmas and binucleate pollen. Here, the breeding behavior of the pollen is determined by the S allele constitution in each individual pollen grain. Pollen rejection occurs when the S haplotype of the pollen grain matches any of the S alleles (usually two) expressed in the pistillar tissues (**Figure 1A,B**). Species belonging to Brassicaceae, Asteraceae and Convolvulaceae display SSI, whereas typical examples of GSI are found among Solanaceae and Rosaceae.

GSI is Not Monophyletic

The SI reaction involves components from both pollen and pistil, but so far only the female components have been identified in five families with GSI. In the Solanaceae, Rosaceae, Schrophulariaceae, and Campanulaceae, the product of the pistil *S* gene has been identified as a special type of RNase, termed S-RNase (see below). In the Papaveraceae, however, small stigmatic S glycoproteins trigger a Ca^{2+}-based signal transduction cascade leading to S-specific inhibition of incompatible pollen. These glycoproteins neither display RNase activity nor share sequence similarity with the S-RNases. These findings strongly suggest that GSI is not monophyletic and, together with the molecular data available for the SSI system in Brassica, point to several independent origins of SI in angiosperms. Since the bulk of the current knowledge on the molecular mechanisms involved in GSI comes from results obtained on Solanaceae, in the following section the most significant studies performed on three model systems, i.e., *Nicotiana*, *Petunia*, and *Solanum*, will be emphasized. The case of the Papaveraceae will be treated separately.

General Features of GSI Based on S-RNases

Products of the *S* Gene in the Pistil

The first cloning and sequencing of a cDNA encoding an S protein that cosegregates with an *S* allele was obtained in 1986 by Clarke and coworkers in *Nicotiana alata*. S proteins are glycoproteins with molecular masses ranging between 20 and 40 kDa and pI values between 6.2 and 9.5. They are relatively abundant (micrograms per pistil), are secreted by the cells of the transmitting tissue of the style, and have ribonuclease activity (hence their identification as S-RNases), although it is unclear which type of pollen RNA they degrade. Sequence identity among S-RNases ranges from 35 to 95%, with S-RNases of different species sometimes being more closely related than those of the same species. Hence, S-RNase polymorphism in the Solanaceae may predate species divergence, a suggestion confirmed by molecular phylogenetic studies of S-RNase sequences from *N. alata* and *Petunia inflata*, species that diverged some 27 million years ago. Similar data and conclusions have been obtained in the Rosaceae.

The sequence analyses also showed a number of conserved features, including: (1) an N-terminal signal peptide, 22 amino acids long (not present in the mature protein), that allows the S-RNases produced by the cells of the transmitting tissue of the style to enter into the secretory pathway; (2) conserved cysteines (usually eight, although their number can vary from seven to ten), presumably involved in disulfide bridges; (3) Asn-X-Ser/Thr motifs (usually one, although up to five have been found) for the attachment of N-linked glycosyl chains; (4) hydrophilic and hydrophobic regions; and (5) a distinct pattern of short conserved and longer variable regions. Five of the conserved regions, designated C1 through C5, are very highly conserved among the S-RNases (**Figure 2A**). In particular, the sequence of amino acids in the C2 and C3 regions is very similar to the catalytic domains of the T2 and Rh ribonucleases from the fungi *Aspergillus orizae* and *Rhizopus niveus*, respectively. This initially helped Clarke's group to identify the S proteins as S-RNases. The two conserved histidines in C2 and C3 are in the active site and are essential for RNase activity. The conserved regions C1 and C5 contain mostly hydrophobic residues and are thought to be involved in stabilizing the core of the protein. The function of C4 (a region absent in the Rosaceae) is not known, yet its hydrophilicity and predicted localization at the surface of the protein suggest it can interact with other molecules.

In addition to the conserved regions, multiple alignment of S-RNases has revealed two regions with the highest degree of diversity among the sequences analyzed. These two highly hydrophilic, hypervariable regions (termed HVa and HVb), were predicted to form the regions that interact with the pollen S-gene product, and thus carry the determinants for the S-allele specificity (**Figure 2A,B**), a hypothesis confirmed by transformation experiments.

S-RNases in the Pistil are Responsible for SI Reaction

In 1994, Kao's and McClure's groups showed independently that S proteins are responsible for recognition and rejection of self pollen, using the plant transformation approach in self-incompatible *Petunia inflata* and *Nicotiana alata*, respectively. In petunia, both loss-of-function and gain-of-function were attempted. To show loss-of-function, an S_3-RNase cDNA, in an antisense orientation and driven by the S_3-gene promoter, was used to transform plants of the S_2S_3 genotype. Six transgenic plants were fully self-compatible and failed to reject S_3 pollen. Of these, three did not produce detectable levels of S_3-RNase, as assessed by cation exchange chromatography. Three others produced very low levels of both S_3- and S_2-RNases, a result interpreted as cosuppression due to sequence similarity between the two S alleles. Interestingly, when S_2-RNase was also low, S_2 pollen was accepted. To demonstrate gain-of-function, Kao's group transformed an S_1S_2 genotype with a construct containing the S_3-RNase cDNA in a sense orientation. Four transgenic plants had acquired the S_3 specificity, since they could recognize and reject S_3 pollen. Taken together, these results proved that the stylar S proteins were necessary and sufficient for the recognition and rejection of self-pollen. In addition, since the pollen behavior of the transgenic plants was unaltered, these studies also supported the hypothesis that the S locus contains at least two distinct genes, one (the S-RNase) controlling the pistil function and the other (still unidentified) controlling the pollen function.

Additional evidence that S-RNases alone are sufficient to control SI in the pistil was provided by McClure's group, working on an interspecific ornamental tobacco hybrid (*N. langsdorffii* × SC *N. alata*) and using the gain-of-function approach. Two of the transgenic plants expressing high levels of the S_{A2}-RNase from SI *N. alata* rejected S_{A2} pollen,

(A)

(B)

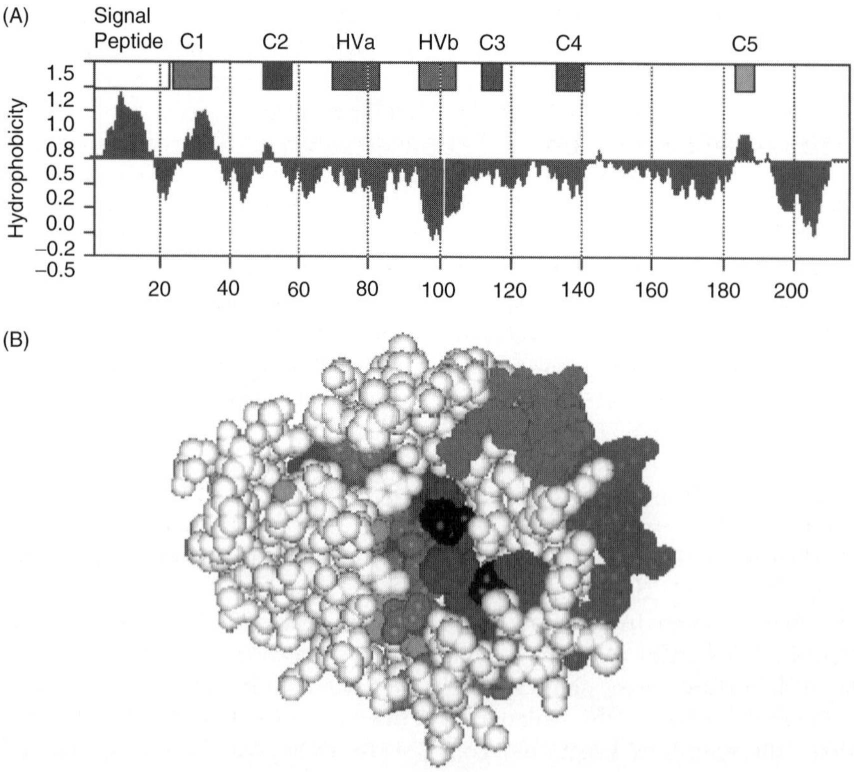

(C)

S-RNase	HVa	HVb
S_{11}	KPKL**T**Y**NY**F SDKML	IDQASA**R**KDQP
HVapb	KPKL**N**YN**F**F SDKML	IDQASA**L**KDQP
HVa	KPKL**N**Y**KF**F SDKML	IDQASARKDQP
HVab	KPKL**N**Y**KF**F SDKML	IDQASA**L**KDQP
S_{13}	KPKL**N**Y**KF**F SDKML	IDQASA**L**KDQP

Figure 2 (see color plate 11) Typical features of S-RNases. (A) Hydrophobicity profile of the *S. chacoense* S_{11}-RNase sequence. (B) A three-dimensional model of the S_{11}-RNase produced using SWISS-MODEL to thread the S_{11}-RNase sequence through the structure of RNase Rh from *R. niveus.* Note the superficial location of the HV regions, well separated from the inner core of the C2 and C3 regions forming the active site (the two conserved histidines are black). (C) The sequence of the HV regions of the S_{11}-RNase, the phenotypically distinct S_{13}-RNase and various mutated S-RNases. The amino acids in bold are those that differ between the S_{11}- and the S_{13}-RNases. The HVapb RNase displayed a dual-specificity $S_{11/13}$ phenotype.

while accepting unrelated S_{C10} pollen. Conversely, expression of the S_{C10}-RNase in the same hybrid resulted in rejection of the S_{C10} pollen and acceptance of the S_{A2} pollen. The promoter used was obtained from the stylar-expressed chitinase gene (*ChiP;1*) from tomato, and was quite efficient in expressing high levels of the stylar expressed S transgene.

RNase Activity is Required for Pollen Rejection

In 1990, Clarke's group reported that in *N. alata* S-allele-specific pollen, RNA degradation occurred *in vivo* after self-pollinations, but not after compatible pollinations. The demonstration that RNase activity is essential for pollen rejection, however, was obtained in 1994 by Kao's group using *P. inflata* transformed with an S_3 cDNA in which the conserved histidine in the C3 region (one of the two histidines present in the active site) was replaced with an asparagine. The mutated S_3 allele introduced into an S_1S_2 genotype resulted in two transgenic plants that produced normal amounts of the mutant S_3-RNase, but were unable to reject S_3 pollen. In the same year, a naturally occurring S-RNase in *Lycopersicon peruvianum* that co-segregated with a mutant SC allele and lacked the conserved histidine in C2, was also described. Taken together, these observations proved that RNase activity of the stylar S-glycoproteins is required for the rejection of self-pollen.

Determinants for the S-Allele Specificity in S-RNases

Recognition between a stylar S-RNase and its pollen counterpart probably involves interactions at the surface of both molecules. All solanaceous S-RNases contain S-allele-specific hydrophilic HV regions exposed at the surface of the molecule (**Figure 2A,B**). In addition, all are glycosylated in an S-allele-specific manner, with highly variable carbohydrate chains that constitute a major surface characteristic of the molecule. Thus, the S-allele-specificity determinants could lie either in the AA sequence at the two HV regions, or in the carbohydrate moiety, or both. A partial answer came from the use of transgenics in *P. inflata*. Here, the codon for the Asn-29 of the S_3 allele, involved in the single N-glycosylation site present in the S_3-RNase, was replaced by site-directed mutagenesis with that of the Asp. The mutated gene was then introduced into an S_1S_2 genotype. Interestingly, six plants expressing wild-type levels of the nonglycosylated S_3-RNase could reject S_3 pollen. This demonstrated that the carbohydrate moiety is not involved in pollen recognition and rejection.

The exclusion of the glycans as determinants for the S-allele specificity in S-RNases left little doubt that recognition must lie in the primary sequence of the protein. However, whether the specificity lay at the HV regions alone or at the HV regions, as well as somewhere else in the molecule, required clarification. Domain swap experiments involving interchange of large regions of two phenotypically distinct S-RNases were attempted in both petunia and nicotiana to address this issue. In petunia, the sequences encoding either both HVa and HVb regions, or only the HVb region of the S_1 allele, were substituted with the corresponding sequences of the S_3 allele and the construct introduced into an S_2S_2 genotype. Allelic recognition was disrupted in both chimeric S-RNases, since the transformants produced S proteins with RNase activity, which did not reject either S_1 or S_3 pollen. All plants maintained the ability to reject S_2 pollen. The conclusion drawn was that HV regions are necessary but not sufficient for S-allele specificity. In the case of *Nicotiana*, nine chimeric S-RNases were constructed, combining the variable sequences of the S_{A2} allele with those of the S_{C10} allele, and introduced into a *N. langsdorffii* × SC *N. alata* hybrid. Once again, none of the transformants harboring any of the chimeric genes could recognize and reject either S_{A2} or S_{C10} pollen. These results suggested that S-allele specificity is not localized to a specific region of the S-RNase.

Direct evidence that HV regions are both necessary and sufficient to determine allelic specificity came in 1997 from studies on *Solanum chacoense*. Three years earlier, two remarkably similar S-RNases, termed S_{11} and S_{13}, which share 95% identity at the protein sequence level, had been identified. Only 10 AA changes differentiate the mature proteins from each other, four of which are found in the HV regions (**Figure 2C**). The wild-type S_{11} under control of the style-expressed tomato chitinase promoter gene introduced into an $S_{12}S_{14}$ genotype resulted in transgenic plants that rejected S_{11} pollen. In another experiment, a chimeric gene construct termed HVab-RNase (**Figure 2C**), in which the four AA in the HV regions of the S_{11}-RNase were replaced with the corresponding ones of the S_{13}-RNase, was introduced into the same host. The transformants accepted S_{11} pollen and, more importantly, could reject S_{13} pollen. This switch from an S_{11} to an S_{13} phenotype by changing only the primary sequence of the HV regions confirmed the key role they play in determining allelic specificity. This also supports the hypothesis that point mutations are the mechanism for the generation of new S alleles.

Generation of New S-Alleles

For nearly 50 years, one of the most intriguing questions about GSI has been the mechanism(s) whereby new S alleles are generated. There is a startling contrast between the amazingly large number of S alleles present in natural populations and the failure of numerous attempts to generate new S alleles. Classic mutagenic treatments always resulted in the inactivation of the existing S alleles. Thus, in addition to point mutations, several mechanisms such as gene conversion, intragenic recombination and other extensive sequence rearrangements have been proposed to operate in nature. If one accepts that SI interactions are controlled by separate genes operating in pollen and style, the major difficulty with these hypotheses is the inevitable breakdown of the SI system should a new S-allele sequence arise in either the pollen or the style alone. The availability of S-RNases with almost identical sequences, such as S_{11} and S_{13}-RNases of *S. chacoense* described above, and the possibility to transform one into the other by site-directed mutagenesis, has allowed the issue of new S-allele generation to be addressed. This was done by investigating the phenotype of intermediate forms (**Figure 2C**).

In the most interesting of the intermediate forms, two of the amino acids of the HVa and the single amino acid of the HVb region of the S_{11}-RNase were substituted with those of the S_{13}-RNase to produce a chimeric S-RNase (**Figure 2C**, HVapb-RNase). The HVapb-RNase has a dual S-allele specificity ($S_{11/13}$), as the plants expressing wild-type levels of it acquired the unprecedented property of rejecting both S_{11} and S_{13} pollen. A careful examination of **Figure 2** reveals that only one amino acid differs between the dual-specific HVapb-RNase and the monospecific HVab-RNase, displaying a dual-specific $S_{11/13}$ and a monospecific S_{13} phenotype, respectively. Hence, a single amino acid substitution can confer an altered recognition phenotype to an S-RNase. However, it is important to note that not all random changes of three amino acids produce a new S-allele specificity. In another experiment, only the amino acids of the HVa region of the S_{11}-RNase were substituted with the corresponding amino acids of the S_{13}-RNase (**Figure 2C**, HVa-RNase). Notably, the plants expressing the transgene at wild-type levels did not reject either S_{11} or S_{13} pollen.

The production of a dual-specific S-RNase suggests a previously unsuspected mechanism for the generation and evolution of new S alleles. In this mechanism, point mutations accumulating in either the pistil or pollen SI genes can generate a dual-specific allele, which allows the cognate SI component to evolve to a new specificity without loss of the SI phenotype. However, as yet, naturally occurring S alleles with dual specificity have not been described.

Breakdown of SI: Stylar Mutations

Although SI species usually display a stable SI phenotype, self-compatible mutants have been described in many species. Since SC mutations at the S locus can affect the pollen or the style independently, Lewis proposed that the S locus is tripartite, i.e., it is composed of three closely linked parts: an S-specificity (which determines the S haplotype in both pollen and style); an activity part that allows expression of the S-specificity in the pollen; and a stylar part that allows expression of the S-specificity in the style. The detailed knowledge of the S locus required to verify this hypothesis is still unavailable. However, we know mainly from molecular analyses that stylar mutants either no longer possess the S-RNase gene (as found in Japanese pear (*Pyrus serotina*), for example), or have an S-RNase lacking RNase activity (described in *L. peruvianum* or artificially produced in *P. inflata*, for example). In other cases, stylar SC has been attributed to the presence of modifier loci epistatic to the S locus (as found in *Petunia axillaris, N. alata,* or *S. chacoense*). An exception is self-compatible *Nicotiana sylvestris*, where blockage of the recognition site of the S-RNase is due to a mutation in its HVa region, which generates an additional glycosylation site that appears to block interaction with the pollen component to SI.

Products of the S-Gene in the Pollen

The physical size of the S locus in GSI plants is not known, but it may be considerably larger than the SSI S locus in *Brassica*. In addition, apart from the S-RNase expressed in the mature style, the genetic content of the S locus is also largely unknown. However, using RNA differential display in *P. inflata*, Kao's group recently identified 10 pollen-expressed genes very closely linked to the S-RNase gene. Unfortunately, none turned out to be the pollen-expressed S gene, whose identity remains unknown.

The early hypotheses that the same gene is responsible for both pistil and pollen components to GSI are neither supported by transformation experiments in *Lycopersicon peruvianum*, where an S-RNase has been expressed in pollen without affecting the SI phenotype, nor by analyses of a self-compatible cultivar of *Pyrus serotina*, where a deletion affected the pistil SI function, yet the function of the pollen S component was unaffected. Thus, it seems unlikely that the S-RNase will also represent the pollen component of SI. The consensus view regarding the pollen component to SI assumes that it is highly polymorphic, is encoded by another gene at the S-locus, and possesses structural features that permit allele-specific interactions with the S-RNase.

Models for Allele-Specific Rejection of Pollen

Clearly, all S-RNases contain an RNase activity domain essential for pollen rejection and a recognition domain, constituted by the two hypervariable regions, involved in allele-specific pollen recognition. Hence, much thought has been devoted to possible mechanisms that result in allele-specific activation of an S-RNase in pollen. Two fundamentally different models have been proposed, which are often referred to as the gatekeeper and the inhibitor models. In the gatekeeper model, recognition between S-RNase and pollen occurs at the surface of the growing pollen tube. This model proposes that the product of the pollen S gene is a membrane- or wall-bound receptor that recognizes and translocates its cognate S-RNase inside the pollen tube, where pollen RNA degradation follows (**Figure 3A**). Selective activation thus occurs by compartmentalization. In the inhibitor model, the recognition phase occurs in the cytoplasm of the pollen tube. Here, the product of the pollen S gene is a protein that, like the S-RNase, contains two

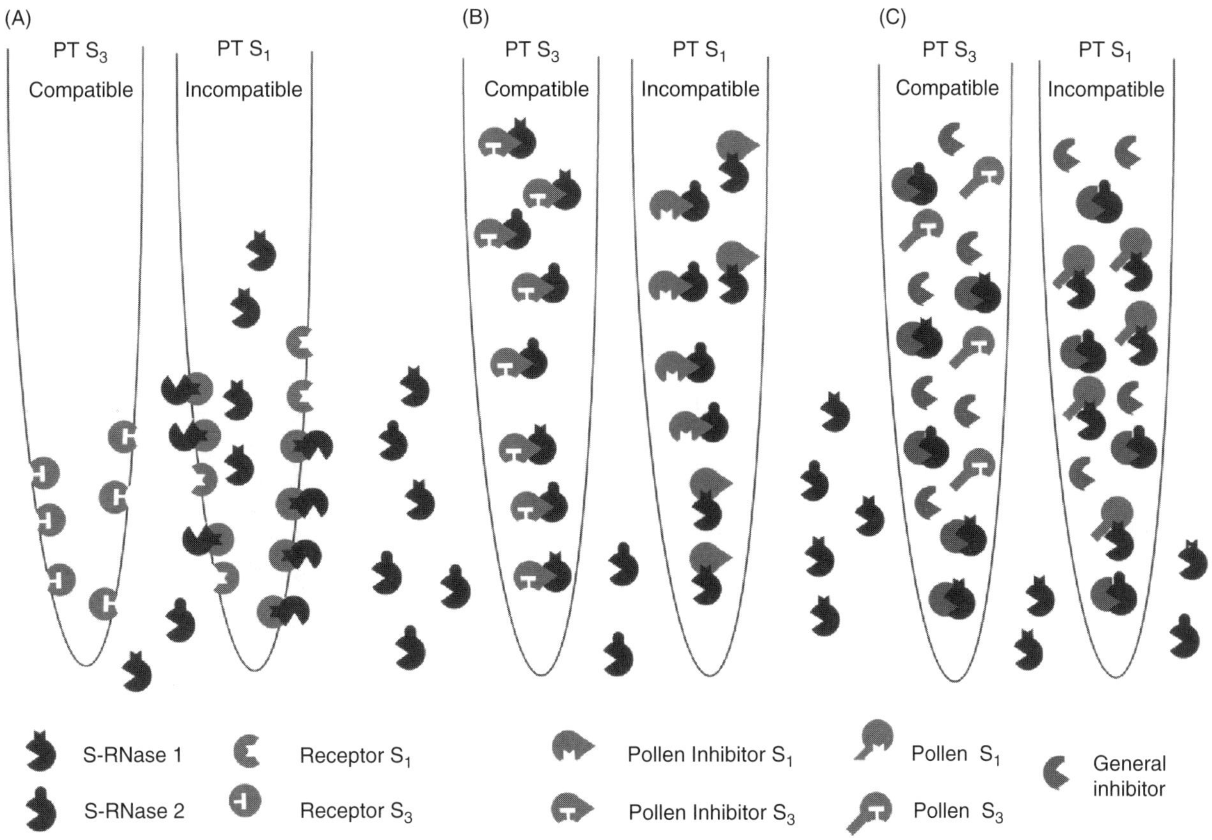

Figure 3 Models for S-RNase-mediated pollen rejection. The crosses represented in A, B and C involve pistils of S_1S_2 haplotype pollinated with pollen from an S_1S_3 genotype. The behavior of individual pollen tubes (PT) is either compatible (left; no active RNase inside pollen tubes) or incompatible (right; active RNase inside pollen tubes). (A) Receptor model. A pollen-S protein recognizes and allows entry only of its cognate S-RNase into pollen tubes. (B) One component inhibitor model. Pollen-S proteins have two binding domains, one for the allelic specificity determinants on the S-RNases and one for the RNase activity domain. Binding to the two domains is assumed to be mutually exclusive, and although a pollen S could bind to either domain of its cognate S-RNase, binding to the allelic specificity determinants is assumed to be thermodynamically favored (S-RNase 1 at right). (C) Two-component inhibitor model. Pollen-S, binding to the allelic specificity determinants of the S-RNase, is a different protein from a general inhibitor binding to the RNase activity domain. Once again, binding to the two domains is assumed to be mutually exclusive and binding to the allelic specificity determinants is assumed to be thermodynamically favored.

domains. One of the domains binds in an allele-specific way to the recognition domain of its cognate S-RNase, whereas the other domain binds, and thus inhibits, the RNase activity domain of the noncognate S-RNases (**Figure 3B**). This model was considered the less likely of the two because in order for it to work it is also necessary to propose that binding to the two domains is mutually exclusive, and that binding to a cognate recognition domain is thermodynamically favored over binding to the RNase activity domain. Thus, while all S-RNases enter any pollen tube, noncognate S-RNases are detoxified by the product of the pollen S gene, an S-RNase inhibitor that blocks the RNase domain, and only the S-RNase recognized by its cognate pollen S component maintains a free RNase domain that can degrade pollen RNA. Nonetheless, recent results in *S. chacoense* have provided strong support for the inhibitor model. Immunocytochemical labeling of pollen tubes growing in compatible and incompatible styles indicates that the stylar S_{11}-RNase accumulates in the cytoplasm of all pollen tube haplotypes. It is possible that RNase inhibition and allele-specific recognition functions are carried out by two separate proteins, a general RNase inhibitor present in all pollen tubes that would bind and inactivate any S-RNase and the product of the pollen S-gene (**Figure 3C**) that would bind to the recognition domain of its cognate S-RNase, leaving the RNase activity domain free to destroy the pollen tube RNA.

Breakdown of SI: Pollen Mutations

In addition to stylar mutations, occurrence of permanent SC in otherwise SI species can also occur through mutations affecting only the pollen. In most cases, the SC phenotype is associated with an additional copy of the S locus. Thus, this appears similar to the SC found in tetraploids generated from SI diploids, a phenomenon referred to as the competition effect. In this case, diploid pollen containing two different S loci (termed S-heteroallelic pollen) displays compatibility in any cross, including self-pollinations or crosses to the diploid SI parental plant (**Figure 1B**). In contrast, diploid pollen carrying two identical S loci follows the same rules as haploid pollen and remains incompatible on pistils displaying the same S haplotype (**Figure 1B**). In other cases, SC could not be attributed to the competition effect, and thus pollen compatibility was attributed to deletion or inactivation of the pollen S gene. This has been difficult to explain using the inhibitor model, since deletions of a pollen S gene able to inhibit S-RNases should be lethal. This objection, however, could be bypassed if the RNase inhibition and allele-specific recognition were carried by two distinct proteins. In this version of the inhibitor model, pollen contains a general RNase inhibitor that can inactivate any S-RNase that has entered the pollen tube. It also contains a polymorphic pollen S gene product that, by binding preferentially to the recognition domain of its cognate S-RNase, would free the RNase domain. This activation of pollen RNA degradation would lead to the SI reaction.

GSI in Papaveraceae

GSI in Papaveraceae is quite unique and has been studied almost exclusively in the field poppy, *Papaver rhoeas*. The system is controlled by a single highly polymorphic S locus (with over 65 S alleles) and is genetically identical to that displayed by the Solanaceae. The site of the SI reaction, however, is unusual as it occurs at the stigmatic surface (instead of the stylar region). Furthermore, the reaction involves rejection of binucleate pollen (instead of trinucleate pollen, typical of stigmatic SI). Interestingly, the events leading to pollen rejection may be reproduced in an *in vitro* system. The bioassay, devised by the group of Franklin and Franklin-Tong at the end of the 1980s, is based on the development of a culture medium suitable for pollen germination. The incorporation of purified stigmatic extracts into the culture medium permits discrimination of compatible from incompatible pollinations, since, in the latter, pollen tube growth is inhibited in an S-allele-specific manner. Such a bioassay has been very useful for the identification of stigmatic proteins (S proteins) cosegregating with the S alleles, as well as for the analysis of events that occur in incompatible pollen during the SI response.

Characteristics of the S Proteins

The stigmatic S proteins of *P. rhoeas* are small (~17 kDa) secreted glycoproteins present at very low levels (nanograms per pistil). They have no RNase activity and do not share sequence similarities with either S-RNases or S-gene products from other SI systems, although they share substantial similarity with members of a predicted gene family called SPHs (S-protein homologs) found in *Arabidopsis*. Like S-RNases, the stigmatic S proteins of *P. rhoeas* display a high degree of polymorphism, sharing between 51 and 65% sequence identity. Their predicted secondary structure, however, is believed to be identical, and consists of six β-strands followed by two α-helices, all linked by seven hydrophilic loops. Four cysteine residues occupy conserved positions and are possibly involved in the configuration of the tertiary structure. Site-directed mutagenesis performed on the S_1 protein has revealed that several amino acids in loop 6 participate in the recognition and rejection of self-pollen. Changes in the other six loops had little or no effect on the stigmatic S_1-protein activity and they did not affect allele-specific inhibition of incompatible S_1 pollen.

Products of the S-Gene in the Pollen

In vitro bioassay has revealed that the stigmatic S protein of *P. rhoeas* presumably acts as a signal molecule that, as a result of incompatible pollinations, triggers a complex cascade of signal transduction events inside the incompatible pollen, which results in inhibition of pollen tube growth and ultimately in pollen death. To initiate such events, the S proteins are believed to interact with the product of the pollen S gene. This product has not been identified as yet, but it is thought to be a transmembrane kinase receptor. A plasma membrane 70–120-kDa glycoprotein called SBP (S-binding protein) has indeed been identified that is involved in pollen rejection. Since SBP binds to the stigmatic S proteins in a non-allele-specific way, it may not represent the product of the pollen S gene itself. However, it could form a recognition complex with stigmatic S proteins that, in incompatible pollinations, could bind to the pollen S gene product and thus trigger the cascade of signal transduction events.

Model for GSI in Papaveraceae

In vitro bioassay techniques have also permitted some of the events involved in the SI response to be followed that take place after pollen is exposed to stigmatic S proteins of the same S haplotype. These events start with an increase in cytosolic calcium levels inside the incompatible pollen tubes induced by the inositol-1,4,5,triphosphate (IP3) protein. Free calcium appears to act as a secondary messenger and calcium-dependent protein kinases induce phosphorylation of a pollen 26.1 kDa pyrophosphatase and other proteins. This results in changes in pollen gene expression, leading to pollen tube inhibition and programmed cell death.

Conclusions and Perspectives

Understanding the various and complex events that characterize pollen recognition and its possible rejection, opens the possibility of using SI in practical applications in agriculture. In self-incompatible species, especially fruit trees, the removal of SI would result in improved efficiency of land management and more reliable fruit formation, whereas in self-compatible crop species, the introduction of SI would facilitate the production of hybrid seeds with higher vigor and therefore higher yield. In this regard, the successful production of self-incompatible *Arabidopsis thaliana* plants by transfer of two S-locus genes (the stigma receptor kinase SRK and its cognate pollen ligand SCR) from the related SI species *A. lyrata*, appears particularly encouraging.

In conclusion, considerable progress in our understanding of GSI has been made in the past 20 years. Disappointingly, however, a key element is still missing, namely the identity of the pollen S gene in both RNase-based and poppy-type GSI. Also unknown are the physical size of the S-locus and its genetic elements. Nevertheless, we do have many helpful clues concerning the mechanism of pollen rejection and it is hoped that these clues will shed more light on the events that characterize GSI.

Recent analyses of the genomic structure of the S-locus region in Rosaceae and Scrophulariaceae, suggest that in these families the physical size of the S-locus could be about 70 kb. Among the identified genes, a polymorphic pollen-expressed gene linked to the S-RNase gene has been found. The corresponding product, an F-box protein potentially involved in uniquitin/26S proteasome proteolytic pathway, could thus represent the pollen-S gene product. In this case, it would probably target non-self S-RNases for degradation, leaving intact (active) self S-RNases.

List of Technical Nomenclature

Anther	The pollen-bearing portion of the stamen.
Glycoprotein	Polypeptide modified by the addition of specific carbohydrate groups.
GSI	Gametophytic self-incompatibility.
Inbreeding	Mating between genetically closely related plants, usually by self-pollination.
Outbreeding	Mating between genetically different plants.
Pistil	The seed-bearing portion of the flower, composed of stigma, style and ovary.
Pollen	The male gametophyte.
Pollination	The transfer of pollen from an anther to a stigma.
Polyploid	An individual with more than two genomes in its somatic cells.
S allele	A specific variant of any polymorphic gene within the S locus.
SC	Self-compatibility.
S haplotype	The specific genotypic constitution of the S locus.
SI	Self-incompatibility.
SSI	Sporophytic self-incompatibility.
S-RNase	S-ribonuclease.

See also: **Flowering and Reproduction**: Fertilization; Reproductive Strategies; Sporophytic Self-Incompatibility. **Genetic Modification**: Transformation in Dicotyledons; Transformation, General Principles; Transgene Stability and Inheritance.

Further Reading

Anderson M, Cornish E, Mau SL, *et al.* (1986) Cloning of cDNA for a stylar glycoprotein associated with expression of self-incompatibility in *Nicotiana alata*. *Nature* 321: 38–44.

Charlesworth D (1985) Distribution of dioecy and self-incompatibility in angiosperms. In: Greenwood P, Harvey P, and Slatkin M (eds) *Evolution*, pp. 237–268. Cambridge: Cambridge University Press.

de Nettancourt D (1977) *Incompatibility in Angiosperms.* New York: Springer-Verlag.

de Nettancourt D (1997) Incompatibility in angiosperms. *Sexual Plant Reproduction* 10: 185–199.

de Nettancourt D (2001) *Incompatibility and Incongruity in Wild and Cultivated Plants*, 2nd edn. Berlin: Springer-Verlag.

Franklin F, Lawrence M, and Franklin-Tong V (1995) Cell and molecular biology of self-incompatibility in flowering plants. *International Review of Cytology* 158: 1–64.

Kao T-H and McCubbin A (1996) How flowering plants discriminate between self and non-self pollen to prevent inbreeding. *Proceedings of the National Academy of Science, USA* 93: 12059–12065.

Kao T-H and McCubbin A (1997) Molecular and biochemical bases of gametophytic self-incompatibility in Solanaceae. *Plant Physiology and Biochemistry* 35: 171–176.
Karunanandaa B, Huang S, and Kao T-H (1994) Carbohydrate moiety of the *Petunia inflata* S3 protein is not required for self-incompatibility interactions between pollen and pistil. *Plant Cell* 6: 1933–1940.
Lee H-S, Huang S, and Kao T-H (1994) S proteins control rejection of incompatible pollen in *Petunia inflata*. *Nature* 367: 560–563.
Luu D, Qin X, Morse D, and Cappadocia M (2000) S-RNase uptake by compatible pollen tubes in gametophytic self-incompatibility. *Nature* 407: 649–651.
Matton DP, Maes O, Laublin G, *et al.* (1997) Hypervariable domains of self-incompatibility RNases mediate allele-specific pollen recognition. *Plant Cell* 9: 1757–1766.
Matton DP, Luu DT, Xike Q, *et al.* (1999) The production of an S-RNase with dual specificity suggests a novel hypothesis for the generation of new S-alleles. *Plant Cell* 11: 2087–2097.
McClure B, Haring V, Ebert P, *et al.* (1989) Style self-incompatibility products of *Nicotiana alata* are ribonucleases. *Nature* 342: 955–957.
McCormick S (1998) Self-incompatibility and other pollen–pistil interactions. *Current Opinion in Plant Biology* 1: 18–25.
McCubbin A and Kao T (2000) Molecular recognition and response in pollen and pistil interactions. *Annual Review of Cell and Developmental Biology* 16: 333–364.
Nasrallah ME, Liu P, and Nasrallah JB (2002) Generation of self-incompatible *Arabydopsis thaliana* by transfer of two S-locus genes from *A. lyrata*. *Science* 297: 247–249.
Zurek D, Mou B, Beecher B, and McClure B (1997) Exchanging sequence domains between S-RNases from *Nicotiana alata* disrupts pollen recognition. *Plant Journal* 11: 797–808.

Sporophytic Self-Incompatibility

D A Tabah and S J Hiscock, University of Bristol, Bristol, UK

Definition

Self-incompatibility (SI) can be defined as "the inability of a plant producing functional gametes to set seed upon selfing." In most cases, SI is under the control of a single multiallelic locus (*S*). *S* loci are predicted to consist of a minimum of two closely linked genes encoding, respectively, the male and female determinants of SI. Recombination between these genes is believed not to occur so the genes behave as a single locus. Because *S* specificity is determined by at least two different genes, allelic forms of the *S* locus are now more accurately referred to as *S* haplotypes. When pollen and pistil share the same *S* haplotype, the SI response is activated, so preventing fertilization by self-pollen or pollen from other individuals sharing the same *S* haplotype.

Two main types of SI are recognized according to how the *S* incompatibility phenotype of the pollen is determined. In gametophytic SI (GSI) systems, expression of pollen *S* occurs in the haploid pollen, so that each pollen grain carries a single *S* allele specificity. In sporophytic SI (SSI) systems expression of pollen *S* occurs in the diploid sporophyte and therefore pollen grains carry two allelic male SI specificities and behave as though diploid. This sporophytic determination of SSI allows for dominance interactions to evolve between *S* alleles in pollen and stigma.

The Genetic Basis of SSI

SSI systems were first described in the Asteraceae and soon after in the Brassicaceae. SSI also occurs in the Convolvulaceae, Geraniaceae, Betulaceae, Polemoniaceae, Caryophyllaceae, and possibly in the Apocynaceae and Malvaceae. In the Asteraceae, Brassicaceae, and Convolvulaceae, recognition and rejection of incompatible pollen occurs at the stigma surface. Five characteristics are true for all SSI systems.

Crosses Are Always Either Fully Compatible or Fully Incompatible

Each pollen grain carries one *S* haplotype in its haploid genome. In GSI systems each pollen grain expresses the single male determinant encoded by its haploid genome, but in SSI, expression of the male determinant(s) occurs in the diploid sporophytic tissues of the anther. Each pollen grain will therefore carry male products of both *S* haplotypes. Expression of the male determinant distinguishes SSI from GSI. In GSI systems, when two plants sharing only one *S* haplotype are crossed, half of the pollen grains will be compatible. These crosses are said to be half-compatible and are diagnostic of GSI systems. For the equivalent cross in SSI systems, if both *S* haplotypes are codominant, all pollen grains will be incompatible (**Figure 1**).

The *S* Locus Consists of at Least Two Closely Linked Genes

Early work on self-incompatibility revealed that the *S* locus in GSI systems consists of at least two closely

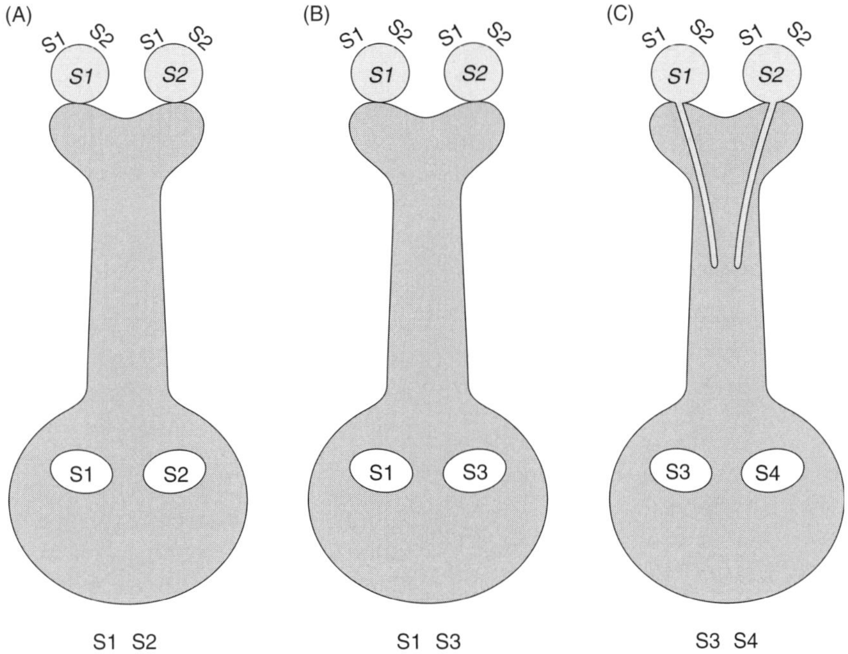

Figure 1 Pollen grains are haploid and therefore carry only one *S* haplotype in their genome. However, in sporophytic self-incompatibility, expression of the male determinants is controlled by the parent sporophyte, and therefore pollen grains will carry both parental *S* specifities (assuming codominance). In most cases, for a compatible reaction to occur, pollen cannot share an *S* specificity with the stigma. Note that in gametophytic self-incompatibility (GSI), (B) would lead to a half-compatible reaction.

linked genes, one encoding the male determinant and the other encoding the female determinant. This work was carried out on pollen part mutants, which had a fully functional female determinant of SI but no functional pollen determinant. In SSI systems, such mutants occur but are extremely difficult to detect. Molecular studies of SSI in *Brassica* later confirmed this predicted structure of the *S* locus by showing that it consists of separate male and female determinants. Dominance relationships between *S* haplotypes often differ in the pollen and stigma, further suggesting that factors determining dominance relationships between *S* haplotypes are acting on two separate male and female determinants.

The *S* Locus Is Highly Polymorphic

Natural populations usually maintain large numbers (30 to 40) of *S* haplotypes. Such high levels of polymorphism at the *S* locus arise as a consequence of negative frequency-dependent selection. The rarer an *S* haplotype is in a population, the more likely it is to encounter compatible *S* haplotypes and lead to fertilization. Conversely, very common *S* haplotypes will be more likely to be involved in incompatible crosses. Novel functional *S* haplotypes will therefore be favored in any SI system. Nevertheless, the emergence of new *S* haplotypes seems to be a very rare event because *S* haplotype divergence predates speciation events in the Solanaceae, Rosaceae, and Brassicaceae. This is thought to be the case because the evolution of new *S* haplotypes involves complementary mutations in both the female and male determinants.

The negative frequency-dependent selection acting on *S* haplotypes presents a dilemma regarding the stability of SI. A new self-compatible *S* haplotype would spread rapidly through the population but would not be subject to negative frequency-dependent selection, as the self-compatible haplotype will not be involved in incompatible crosses. Selective pressure must act against such self-compatible *S* haplotypes in SI species. The high cost of inbreeding depression is the selective force acting to avoid the invasion of self-compatible haplotypes. SI is therefore an adaptation to avoid inbreeding depression.

Plants Are Normally Heterozygous for *S*

As a rule, plants homozygous at the *S* locus are rare in natural populations. This is because the crosses leading to such a homozygote are usually incompatible, because both parental plants share at least one *S* haplotype. Nevertheless homozygotes for *S* do occur in SSI systems when *S* haplotypes exhibit recessive relationships with other *S* haplotypes. Plant breeders can generate *S* locus homozygotes through the application of a saline solution to stigmas prior to self-pollination. This leads to a breakdown of SI and concomitant self-fertilization, but the mechanism is unknown. Certain environmental factors such as carbon dioxide, temperature, and humidity can also lead to a breakdown of SSI and plant breeders have successfully exploited these factors to generate *S* homozygotes for breeding purposes.

Dominance Relationships Occur Between S Alleles

In SSI, *S* haplotypes can exhibit dominance relationships in either "pollen" (anther) or in the stigma. Such interactions are not seen in GSI where all female *S* haplotypes must be codominant. Dominance interactions act to increase the number of compatible crosses compared with a fully codominant system. When an *S* haplotype is recessive, it will be hidden and fail to initiate an incompatibility response, thus allowing plants that share only one recessive *S* haplotype to set seed when crossed. In this way *S* homozygotes can arise naturally in SSI species. It has been proposed that the relative frequencies of recessive and dominant *S* haplotypes reflect a dynamic balance between reproductive assurance (favored by recessive haplotypes) and the benefits of outbreeding (favored by dominant haplotypes).

Dominance relationships between *S* alleles can differ in anther and stigma such that an *S* haplotype may be dominant in the stigma but recessive in the anther of the same plant. This is possible if dominance affects the female determinant in the stigma and the male determinant in the pollen independently of each other.

SSI in the Brassicaceae

Most efforts to characterize the molecular mechanism of SSI have focused on the Brassicaceae, for which the male and female determinants are now well characterized. The female determinant of SI is an *S*-receptor kinase (SRK). This protein consists of a conserved kinase domain, a transmembrane domain, and a variable extracellular receptor domain with 12 conserved cysteine residues. SRK is located specifically in the plasma membrane of specialized epidermal cells of the stigma (papillae), and only limited expression has been detected in pollen.

The male determinant of SSI is the small *S*-locus cysteine-rich protein (SCR or SP11, depending on preference). SCR/SP11 is a member of a large family of pollen coat proteins (PCPs) located within the coating of the pollen wall (exine). The pollen coating is mostly composed of secretions from tapetum cells, deposited on the pollen grains following degradation of the tapetum. Some PCPs are produced by the pollen grains themselves, thus arising from gametophytic expression. It is now known that *SCR/SP11* is expressed by the pollen and the tapetum cells in the Brassicaceae, with recessive alleles being expressed by tapetum cells and immature pollen, and dominant alleles being expressed by tapetum cells and microspores as well as by mature pollen.

A third well-characterized gene at the *S* locus encodes the *S*-locus glycoprotein (SLG). SLG is a secreted form of SRK lacking the kinase and transmembrane domains and is thought to aid SRK in the recognition of SCR/SP11. Nevertheless *SLG* is not essential for fully functional SSI. For certain haplotypes, two *SLG* genes are found at the *S* locus. These genes are described as *SLGA* and *SLGB*. *SLGA* has two alternative products, one secreted and the other bound to the plasma membrane. SLGB is found only as a secreted glycoprotein. *Arabidopsis lyrata*, a self-incompatible member of the Brassicaceae, lacks *SLG*.

The incompatibility reaction involves the haplotype-specific interaction of SCR/SP11 and SRK (**Figure 2**). SLG may also be involved in this process in certain cases, perhaps facilitating the binding of SCR/SP11 to SRK. SRK in its inactive state is associated with a thioredoxin and dephosphorylated. Binding of SCR/SP11 causes SRK to dimerize and become autophosphorylated. Binding of SCR/SP11 also causes thioredoxin to become dissociated from the kinase domain of SRK. ARC1 (an arm-repeat protein) then associates with SRK and becomes phosphorylated, possibly triggering a downstream signal transduction cascade ultimately leading to the inhibition of pollen germination. Arm-repeat proteins are known to participate in interactions with the cytoskeleton, but the exact components of the signal transduction cascade downstream of ARC1 have yet to be determined.

Molecular Determinants of Dominance

A key factor in the genetics of SSI is the fact that dominance interactions can occur between *S* haplotypes. These dominance interactions can be very complex and have been studied using classical genetics in only a small number of families. Recent key molecular studies of dominance relationships between *S* alleles have centered around the SSI system of the Brassicaceae.

Until recently dominance was thought to occur in a linear fashion, with two main groups of *SRK* alleles, Class I and Class II. Class I alleles tend to be dominant over Class II alleles in the pollen and within each type, alleles follow a dominance hierarchy or are codominant. The sequences of

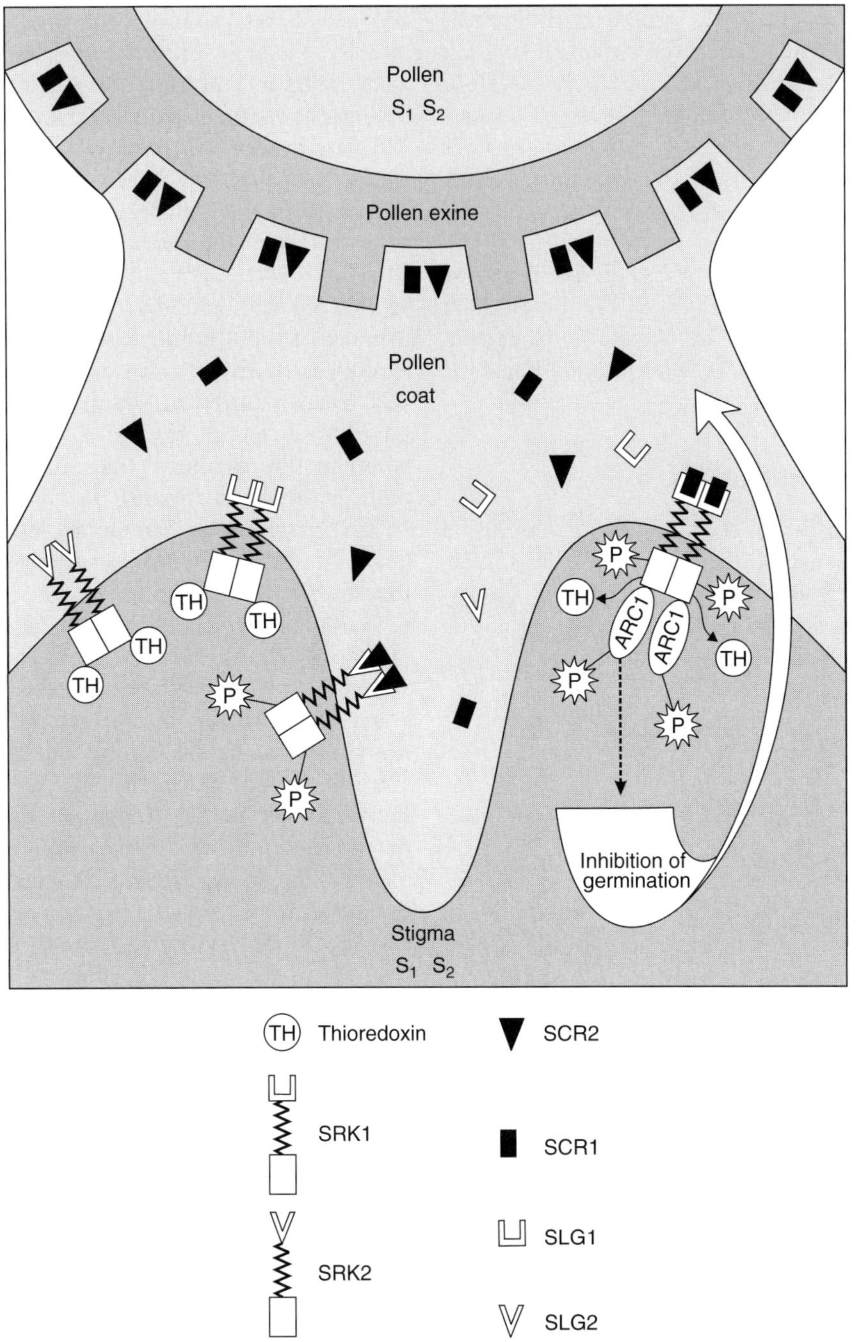

Figure 2 Model of the molecular interactions that occur during pollination in an incompatible cross. See text for details.

SRK alleles have highly conserved regions and hypervariable regions. The hypervariable regions are thought to be responsible for the *S* specificity of SRK, as well as the dominance relationships between different *SRK* alleles. It has been shown that *SRK* alone is sufficient to elicit the dominance relationship between two *SRK* alleles. However, this is not necessarily the case for all *SRK* alleles and there may be modifying elements that affect the dominance relationships of other *SRK* alleles.

Dominance relationships between *SCR/SP11* alleles have also been investigated, although studies are in their infancy due to the recent discovery of SCR/SP11. Because dominance relationships between *S* alleles differ in pollen and stigma, the factors determining dominance relationships between *SRK* alleles and between *SCR/SP11* alleles may be different. Studies have shown a high sequence divergence between dominant (Class I) and recessive (Class II) *SCR/SP11* alleles. This indicates that, as with SRK, structural differences between SCR/SP11 alleles may be sufficient to elicit the dominance relationships observed between *SCR/SP11* alleles. A further difference between Class I and Class II *SCR/SP11* alleles is that the latter are expressed in tapetal cells and immature pollen, whilst Class I alleles are expressed by tapetal cells and immature pollen as well as by mature pollen. Dominance appears to act by repression of the expression of the recessive *SCR/SP11* alleles, such that no mRNA transcripts are detectable in heterozygotes consisting of a Class I and Class II haplotype.

SSI in Other Families

SSI has been described in six families other than the Brassicaceae: the Asteraceae, Convolvulaceae, Geraniaceae, Betulaceae, Polemoniaceae, and Caryophyllaceae and may also occur in the Apocynaceae and Malvaceae. These families are not closely related, so it is likely that SSI, as with GSI, has arisen on multiple occasions during the diversification of the angiosperms. Due to a lack of molecular data from these other families with SSI, it is not known whether SSI in species from these families operates through the same SCR/SP11 and SRK system as in the Brassicaceae.

Whilst the Brassicaceae is the only SSI system for which the determinants of SI are known, SSI is also being studied at a molecular level in the Convolvulaceae and the Asteraceae. *SRK* homologs are known to occur in *Ipomoea* stigmas (Convolvulaceae) and *Senecio* stigmas (Asteraceae), but current evidence suggests that these SRK-like proteins are not the female determinants of SSI in these families.

Thus preliminary data from the Convolvulaceae and Asteraceae suggest that molecular mechanisms of SSI in these two families are different from that of the Brassicaceae. It has recently been confirmed that the *S*-RNase-based GSI systems of the Rosaceae, Solanaceae, and Scrophulariaceae share a common origin. It is therefore of interest to determine whether the SSI mechanisms also share a common origin. This is particularly relevant for the Convolvulaceae and the Asteraceae, which are more closely related to each other than either is to the Brassicaceae.

Evolution of SSI

In 1950, Whitehouse proposed that the evolution of SI coincided with the evolution of the angiosperm carpel and was a key factor contributing to the rapid diversification of the first flowering plants. This view however has not been supported by molecular data because SI systems across the angiosperms appear to operate through different molecular mechanisms. This suggests multiple origins for SI after the divergence of the angiosperms. Nevertheless, the *S*-RNase-based GSI systems evolved in a common ancestor of the Rosaceae, Solanaceae, and Scrophulariaceae. In this conclusion, *S*-RNase based GSI is ancestral to all the families within the Asteridae and Rosidae subclasses and these clades include all known SSI families. Self-compatibility is also present in most of these families, so the most parsimonious evolutionary explanation for the occurrence of SSI and self-compatibility within this clade is that these are derived characters. The question then arises as to whether SSI evolved once or (more likely) on multiple occasions (**Figure 3**).

The Brassicaceae, Asteraceae, and Convolvulaceae (all SSI families) are not closely related. In fact, the Brassicaceae is more closely related to the Rosaceae than to the Convolvulaceae, which is the sister family of the Solanaceae. Both the Rosaceae and Solanaceae exhibit very well-described *S*-RNase-based GSI systems. It is therefore expected that SSI is a derived character that evolved independently in these three families. Indeed, research showed that homologs of *Brassica SRK* do occur in the Convolvulaceae and Asteraceae, but that these are unlikely to be involved in the SSI response. These findings suggest that the Brassicaceae SSI system probably evolved independently. The SSI systems of the Convolvulaceae and Asteraceae have not yet been fully characterized, and therefore homology between them is as yet unknown.

In terms of its potential for tighter regulation of biparental inbreeding, SSI is considered a more effective outcrossing mechanism than GSI. If *S* haplotypes are codominant, SSI systems can cause rejection of all pollen from closely related plants that carry a shared *S* haplotype. In the equivalent cross under a GSI system the outcome would be half-compatible. For this reason, GSI has been traditionally seen as less efficient at enforcing outcrossing and therefore as a likely ancestral condition. Recent data initially appear to be consistent with this view. However, the question of how SSI systems evolve and replace GSI systems has not been fully addressed. There are three possible evolutionary pathways from GSI to SSI systems. The first is via self-compatibility followed by SSI. The second is that GSI is altered in some way so as to lead to the evolution of a SSI system, that utilizes the same molecules as the ancestral GSI system. The third pathway involves the evolution of an SSI system without the loss of the ancestral GSI system – the SSI system thus evolves on top of the GSI system.

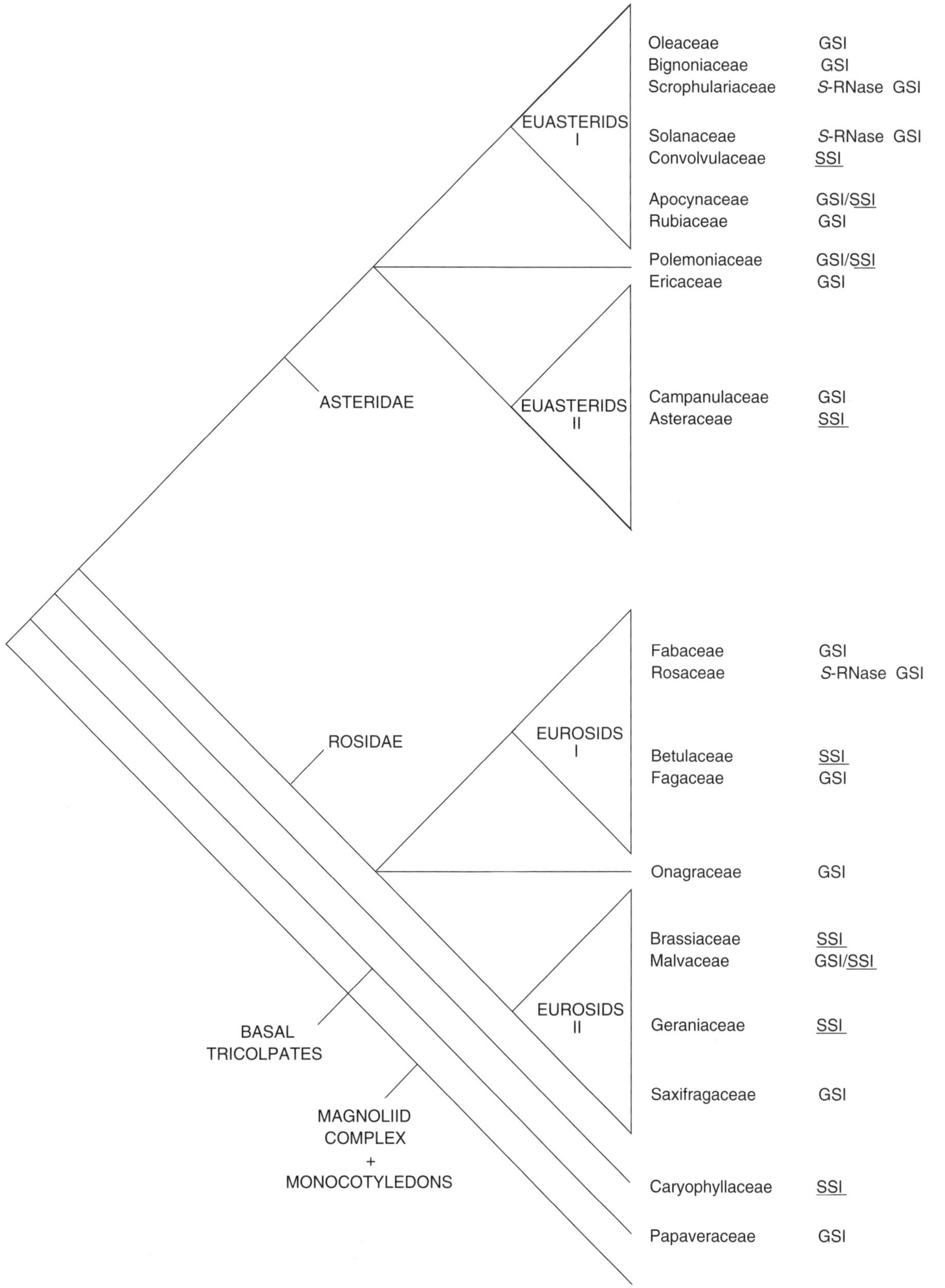

Figure 3 Distribution of sporophytic self-incompatibility (SSI) and gametophytic self-incompatibility (GSI) in the angiosperms. Note that the Poaceae (monocotyledons) also exhibit GSI, but this system is sufficiently different from SI mechanisms discussed here. Diagram adapted from Igic B and Kohn JR (2001) Evolutionary relationships among self-incompatibility RNases. *Proceedings of the National Academy of Sciences of the USA* 98: 13167–13171.

Evolution of SSI through Self-Compatibility as an Intermediate

This is perhaps the simplest view of SI evolution. If *S*-RNase-based GSI is ancestral in the Rosidae and Asteridae, self-compatibility has clearly arisen on multiple occasions, as many families within the derived clade are self-compatible. The evolution of SSI systems is assumed to occur within self-compatible families, perhaps due to changes in the plant's environment resulting in selective pressure for outcrossing. Self-compatibility has arisen even within predominantly SI families. This is known to occur in cases of polyploid speciation, where new species arise due to hybridization and a change in the number of chromosomes. The exact reasons why polyploid speciation has led to self-compatibility has not been fully explained. Polyploid speciation is not the only way in which self-compatibility can arise. Plants such as *Arabidopsis thaliana* in the Brassicaceae and *Senecio flavus* in the Asteraceae are examples of self-compatible species in predominantly SSI families that are not polyploid hybrids and have lost SI through mutation. In the case of *A. thaliana*, the *S* locus has been deleted from its genome. These are examples of self-compatible plants within SSI families but it is not unreasonable to postulate that the changes in selective pressure that allowed these two species to become self-compatible may occur within GSI families. There is no clear evidence as to how a new SSI mechanism would then evolve.

Evolution of SSI from Established GSI Systems

Whilst many different SI systems have been described in the angiosperms, normally for a given family, there is only one type of SI system. Intriguingly however, both GSI (*Phlox drummondii*) and SSI (*Linanthus parviflorus*) have been described in the Polemoniaceae, but there is some controversy as to the validity of this claim. *S* haplotypes in the SSI system of the Brassicaceae are known to have diverged millions of years ago, before speciation events within the family. Theoretically, this may be widespread amongst SI systems. If this is also the case for the Polemoniaceae, it is more parsimonious to assume that the SI mechanism has switched from gametophytic to sporophytic control. In such a scenario, the molecular mechanisms behind GSI and SSI would be identical. The Polemoniaceae therefore may provide evidence that GSI and SSI systems can evolve from one another.

Three possible explanations for switches from GSI to SSI have been proposed. The first hypothesis arose from classical genetic studies on the Caryophyllaceae. It was found that some *S* haplotypes seemed to behave as if expressed sporophytically (quick) and others seemed to behave as if gametophytically (slow). This suggested that the timing of expression of the male determinant of SI could lead to gametophytic or sporophytic *S* alleles. Quick *S* alleles are expressed before meiosis (or in the tapetum) and therefore the pollen grains carry the products of expression of a diploid genome. Slow *S* alleles are expressed after meiosis in pollen and therefore pollen grains carry the product of a haploid genome.

The second hypothesis is that expression of *S* can switch between tapetum cells (sporophytic) and pollen cells (gametophytic), as is suggested by work on *SCR/SP11* alleles in the Brassicaceae. The timing of expression of *SCR/SP11* alleles of *Brassica* can vary. Recessive *SCR/SP11* alleles are expressed by tapetum cells and immature pollen and dominant alleles are expressed by tapetum cells and immature pollen and by mature pollen grains. Such observations suggest that expression of *S* by the gametophyte alone is a possibility. A male determinant allele that was expressed solely by the pollen would behave as an *S* haplotype from a GSI system.

The third hypothesis is that the SSI system has a male determinant that is expressed entirely gametophytically. The male determinants are then secreted onto neighboring pollen grains. In this way all pollen grains will carry both parental male determinants as for an SSI system, even though these are the products of gametophytic expression.

Evolution of SSI without the Loss of the Ancestral GSI System

There is evidence from work on the Brassicaceae and the Asteraceae that a GSI system may be working alongside SSI. These GSI systems may be relics of an ancestral GSI system that has remained active in certain families, but is often difficult to detect due to action of the SSI system.

It has been shown that *Brassica* and *Raphanus* (Brassicaceae) present a gametophytically expressed factor encoded by the *G* locus that overrides the SSI system in certain crosses. For certain *S* haplotype combinations, crosses expected to be incompatible are instead half-compatible or compatible. Crosses are half-compatible when plants share only one *G* allele or compatible if the plants do not share *G* alleles. A similar phenomenon has been invoked for the Asteraceae.

Such a gametophytic element acting on SSI in the Brassicaceae and possibly the Asteraceae could be a relic of an ancestral GSI system. If the *G* locus does encode an *S*-RNase, it would be compelling evidence in support of the hypothesis that SSI evolved within GSI clades.

Future Research

The field of SI research is advancing rapidly, particularly for SSI in the Brassicaceae where the male and female determinants of SSI have now been identified. Research into the downstream signaling pathway leading from the recognition response to pollen growth inhibition is already the matter of intensive research.

An important landmark in SSI research has been the recent success in transforming self-compatible *A. thaliana* (Brassicaceae) with the determinants of SSI from the close relative *A. lyrata*. The genome of self-compatible *A. thaliana* contains no homologous regions to the *S* locus, indicating that the components of the downstream signaling pathway involved in the SSI response have been conserved since these two species diverged approximately 5 million years ago, in spite of the breakdown of SSI. Hence the signaling pathway that leads to pollen rejection during the SI response may be involved in other processes that have been conserved in self-compatible *A. thaliana*. Data from the *A. thaliana* genome project will no doubt provide important insights into the components of this signaling pathway in the near future.

Further research will continue into the SSI mechanisms of families other than the Brassicaceae. Once the determinants of SI in these families have been identified, hypotheses about the evolution of SSI can be based on more stringent evidence. This will generate important information on the evolution of SI, and the origin of SSI.

List of Technical Nomenclature

Clade	Phlylogenetic lineage of related taxa originating from a common ancestral taxon.
Exine	Component layer of the pollen grain wall.
Haplotype	A single genetic locus consisting of two or more nonrecombining genes.
PCP	Pollen coat protein.
SCR, *SCR*	*S*-locus cysteine-rich protein.
SLG, *SLG*	*S*-locus glycoprotein.
SRK, *SRK*	*S*-receptor kinase.

See also: **Flowering and Reproduction**: Gametophytic Self-Incompatibility; Reproductive Strategies. **Seed Development**: Seed Production.

Further Reading

De Nettancourt D (2002) *Incompatibility in Angiosperms*. Berlin: Springer-Verlag.

Goodwillie C (1997) Self-incompatibility in *Linanthus parviflorus* (Polemoniaceae). *Heredity* 79: 424–432.

Hatakeyama K, Takasaki T, Suzuki G, *et al.* (2001) The *S*-receptor kinase gene determines dominance relationships in stigma expression of self-incompatibility in *Brassica*. *Plant Journal* 26: 69–76.

Hiscock SJ and Kues U (1999) Cellular and molecular mechanisms of sexual incompatibility in plants and fungi. *International Review of Cytology* 193: 165–295.

Igic B and Kohn JR (2001) Evolutionary relationships among self-incompatibility RNases. *Proceedings of the National Academy of Sciences of the USA* 98: 13167–13171.

Kusaba M, Tung C, Nasrallah ME, and Nasrallah JB (2002) Monoallelic expression and dominance interactions in anthers of self-incompatible *Arabidopsis lyrata*. *Plant Physiology* 128: 17–20.

Levin DA (1993) *S*-gene polymorphism in *Phlox drummondii*. *Heredity* 71: 193–198.

Lewis D (1994) Gametophytic–sporophytic incompatibility. In: Williams EG, Clarke AE, and Knox RB (eds) *Genetic Control of Self-Incompatibility and Reproductive Development in Flowering Plants*, pp. 88–101. Dordrecht: Kluwer.

Lundqvist A (1994) "Slow" and "quick" *S*-alleles without dominance interaction in the sporophytic one-locus self-incompatibility system of *Stellaria holostea* (Caryophyllaceae). *Hereditas* 120: 191–202.

McCubbin AG and Kao T (2000) Molecular recognition and response in pollen and pistil interactions. *Annual Review of Cell and Developmental Biology* 16: 333–364.

Ockendon DJ (1974) Distribution of self-incompatibility alleles and breeding structure of open-pollinated cultivars of Brussels sprouts. *Heredity* 33: 159–171.

Richards AJ (1997) *Plant Breeding Systems*, 2nd edn. London: Chapman & Hall.

Whitehouse HLK (1950) Multiple-allelomorph incompatibility of pollen and style in the evolution of the angiosperms. *Annals of Botany* 14: 199–216.

Reproductive Strategies

G A Slafer and D J Miralles, University of Buenos Aires, Buenos Aires, Argentina
J L Araus, University of Barcelona, Barcelona, Spain

Introduction

Reproduction is the process by which plants and animals are able to produce offspring allowing the propagation of a species in natural systems. In agricultural systems, it leads to the propagule for the crop to be sown or planted, and in most major crops (grain crops) and orchards (fruit trees) it results in a harvest. Thus, understanding the reproductive mechanism of a plant species is critical for plant ecology in general and for agronomy in particular.

During reproduction, all or some (depending on the type of reproduction) of the parental attributes are passed to the descendants. The reproductive behavior of plants is variable depending on attributes of the particular species such as size, longevity, growth habit, and ecological niche. For instance, the reproduction of an invader plant may require it to be largely independent of the presence of other individuals of the same species. However, this may reduce the chances of producing genetic variation to improve fitness but may maximize the chance of producing offspring that will colonize new habitats.

Depending on the species, plants may reproduce sexually or asexually, although in many cases both forms coexist. For instance, some of the world's most important weeds, e.g., johnsongrass (*Sorghum halepense*), or crops, e.g., potato (*Solanum tuberosum*), have evolved both sexual and asexual reproductive mechanisms. Sexual reproduction requires the fusion of gametes, produced by a haploid generation, termed the gametophyte, which develops within the dominant sporophyte generation. Thus, plants exhibit an alternation by haploid and diploid generations. When both generations coexist in an individual (sexual reproduction) the gametophytic (haploid) generation is part of the sporophytic structure (e.g., wheat (*Triticum aestivum*) as shown in **Figure 1**). Within the sporophytic generation, flowers are specialized organs that have evolved to provide structures for the gametophytic generation to maximize the likelihood of sexual reproduction. The gametophytic cycles produce the female and male gametes (within the ovary and the anthers, respectively) and result in the process of fertilization (*see* **Flowering and Reproduction**: Fertilization).

Flowers are classified as perfect or imperfect. The former are hermaphroditic with male and female organs, while imperfect flowers are those that only possess either male or female organs. Plants with imperfect flowers may still be hermaphroditic if they have both male and female flowers in the same individual (monoecious plants). Alternatively, each type of flowers may be present in different plants with male and female individuals (dioecious plants). The vast majority of higher plants are hermaphroditic with only about 5% being dioecious.

Sex or No Sex: That Is the Question

A first major division of strategies for reproductive development of plants distinguishes between producing offspring with or without the intervention of sex. The distinction brings about important issues on

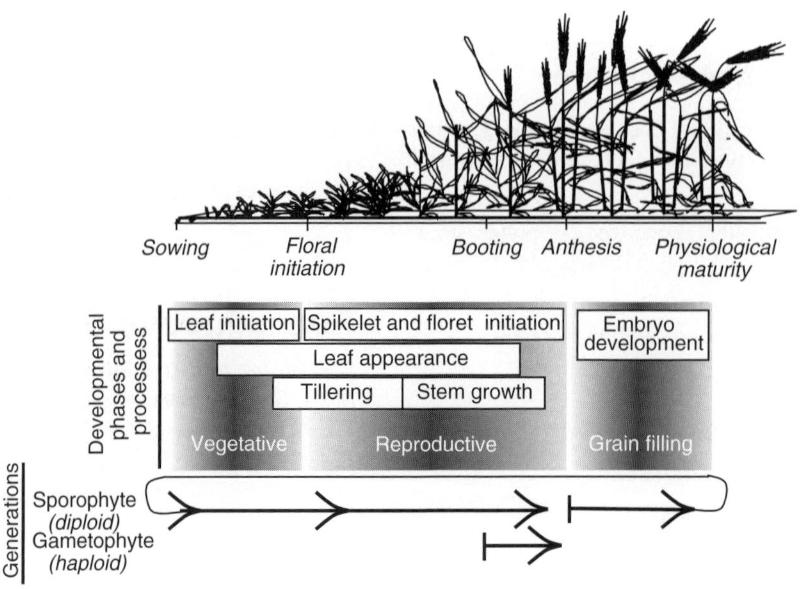

Figure 1 Life cycle of a wheat crop in which the sporophytic and gametophytic generations are shown. The relative duration of the different phases shown in the illustration are not fixed, strongly depending on the climate. Adapted from Miralles DJ and Slafer GA (1999) Wheat development. In: Satorre EH and Slafer EA (eds) *Wheat: Ecology and Physiology of Yield Determination*, pp. 13–43. New York: Food Product Press.

the resource economy of plants (whatever is invested in sexual organs is diverted from what otherwise would be more vegetative growth) and has ecological and agronomic implications.

To reproduce sexually, plants must spend a variable amount of resources on flowering and seed growth, and natural selection would not have selected for sexual reproduction unless it implied ecological benefits for the species in excess of the costs involved for flowering and seed production. The fact that a plant possesses sexual or asexual reproduction indicates whether the investments in producing the organs for sexual reproduction are smaller or greater, respectively, than the benefits in adaptive terms.

Adaptive Implications

The main difference between plants reproducing asexually or sexually is based on the genetic constitution of the offspring. In the case of asexual reproduction, the descendants are genetically identical to the parent plant: they are clones. The only chance for genetic variation in these offspring is the occurrence of mutation. Sexual reproduction, on the other hand, determines that the genetic constitution of the descendants is a mixture of the genes from each of the female and male parents. That is, the most relevant characteristic of sexual over asexual reproduction is the creation of genetic variability, potentially improving the fitness of the species.

The two relevant steps, which are responsible of the creation through sexual reproduction of genetic variability that can be observed in the offspring, are:

1. *Recombination–segregation*, taking place during the sporophytic generation in the ovary and anthers, a process through which somatic chromosome number ($2n$) is halved during meiosis creating gametes with a single set of chromosomes ($1n$). During recombination a random selection of one of the two alleles located at each unlinked locus takes place, rearranging the allelic constitution. In other words, alleles of genes are exchanged between homologous chromosomes during meiosis, this exchange resulting in the generation of new combinations of alleles on segments of chromosomes. Then a random incorporation of the chromatids of each chromosome is included in each gamete. Thus the possible number of gamete genotypes from a single parent is 2^x for a diploid species, x being the number of chromosomes in the haploid form.
2. The formation of the diploid zygote from the fusion of the haploid gametes enlarges the possible variability. Chromosome pairs form the genome of the descendants, each individual chromosome of the pair coming from the male or the female gametes. These gametes may originate in the same hermaphrodite, perfect flower, or in different flowers; which in turn may be on same or different individuals. In any case the resultant size of the expected variability is $(2^x)^2$.

Depending on the alternatives for reproduction (defined below), the expected degree of genetic variation may be increased or decreased (**Figure 2**).

At least intuitively, we should assume that there must be clear advantages of sexual over asexual reproduction, as most higher plants exhibit sexual reproductive strategies (some species possess both systems; the offspring produced sexually are called genets, while the asexual descendants are called ramets). Although it is true that all the descendants of an asexually reproduced plant will possess the genotype proven to be well adapted to the actual biotic and abiotic environmental conditions in which the parent has grown and developed, (1) conditions are not stable, at least in the long term in which natural selection acts, and (2) recombination and sexual reproduction may also produce some genotypes that are even better adapted to the prevailing conditions than the progenitor(s). Therefore, even when as a result of sexual reproduction the successful genotype of the parent(s) is likely to be lost, chances are that within the variable progeny, some descendants can be fitter than their parent(s).

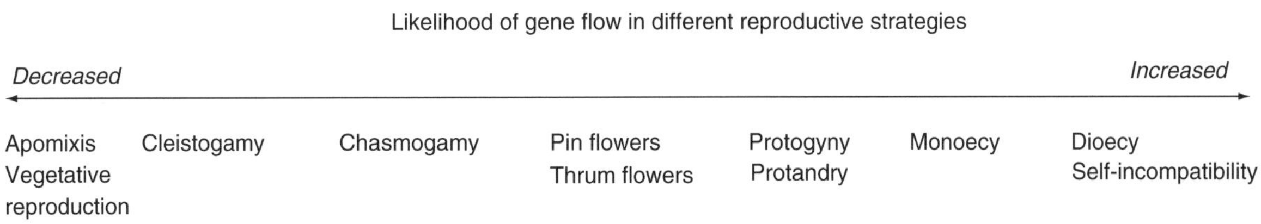

Figure 2 Schematic diagram showing several of the different possibilities for reproductive strategies and their influence in the likelihood of gene flow. More than one strategy may be found in a single species, or even in a single plant.

Resource Balance

There are several resources, carbohydrates and proteins, invested in sexual reproduction. Some of these resources are simply invested in producing the gametes as well as the structures where they are generated and contained. That is, in the case of the male the cost required is the investment for producing the stamen, while the female costs include not only those related to the pistil (ovary, style, and stigma) but also the investments in seed growth and dispersal structures.

There are, in addition, many other resource investments related directly to sexual reproduction but not necessarily associated only with male or female functioning in perfect flowers. The most important ones are the investments in floral structures related to physically support sexual organs (e.g., peduncles) or to protect them (e.g., sepals and bracts), and those associated with the attraction (e.g., petals with particular patterns of colors) and the reward offered to pollinators (e.g., nectar production). Although some authors consider that the investments in structures and products to attract and reward pollinators in hermaphrodite flowers should be ascribed to male costs, it is clear in many cases that female functioning also benefits from these structures and products. Clearly the investments are not necessarily the same for any reproductive system, and in general it is higher in cross-pollinated than in self-pollinated species. For instance, self-pollinated plants do not need to invest in attractants and rewards which are essential in plants that depend on animal pollinators nor do they need to produce a great amount of pollen to maximize the number of descendants which is essential in wind-pollinated plants.

In general, the female-related investments of resources are far greater than male investments. This particularly relates to the large costs involved in seed growth, particularly the accumulation of resources in endosperm or cotyledons, and where applicable in the formation of dispersal structures in the seeds. However, the analysis considers investments largely in terms of biomass, which is likely underestimating the male cost, as the energetic cost of producing pollen may be higher than that of producing female structures and gametes. These costs may be relevant in agronomy. For instance in a monoecious crop such as corn (*Zea mays*; maize) in which yield is produced in the axillary female inflorescences. It has been frequently demonstrated that plants acting as females, with a reduced or impaired male structure (this is done either mechanically, genetically, or by chemical agents) exhibit a higher growth of the ears and therefore express higher yields than the wild-types.

Asexual Reproduction

Asexual reproduction is based on mitosis, producing offspring (ramets) genetically identical to the parent, maintaining genotypes adapted to a particular set of conditions (again something quite important in many agronomic applications, particularly in horticulture and fruit tree culture). Although variability cannot be created by asexual reproduction, levels of genetic variation in the populations can be maintained.

There are two major modes of asexual reproduction, vegetative reproduction and apomixis. Techniques for inducing asexual reproduction (e.g., cuttings and tissue culture) are valuable in agronomy and breeding.

Vegetative Reproduction

Most of the vegetative reproductive systems found under natural conditions are based on modified stems, growing above- or belowground, and reserve organs that may become a new but identical individual. They include mechanisms based on runners or other vegetative modified organs. The dynamics of formation of these reproductive structures is frequently regulated by environmental stimuli, such as photoperiod and temperature, which also regulate floral development in sexual reproduction. Plants store reserves in these specialized bud-bearing organs, which remain dormant in the soil after the shoot dies and sprout new shoots at the onset of the next growing season. When the new shoots are formed (either regenerating the natural population or as a consequence of planting vegetative propagules), the reserves of these organs are consumed. Vegetative reproduction provides a source of clonal material for a wide variety of crops such as sugar cane (*Saccharum officinarum*), potato, and many ornamental flower crops.

Runners, the specialized stems responsible for vegetative reproduction, are generally developed from meristems in the leaf axil or crown that grow horizontally with respect to the ground. They are divided into stolons that are aboveground and rhizomes that are belowground runners. In both cases, the horizontal stems may be able to produce a new plant at the nodes, when they are physiologically mature and are freed from intraplant competitive restrictions. Strawberry (*Fragaria x ananassa*) is an example of a common agricultural plant that reproduce by stolons and many grasses reproduce by rhizomes. Doubtless the existence of this reproductive system (either stolons or rhizomes) in some weeds is particularly complex for applying efficient agronomic measures. *Cynodon dactylon* and *Sorghum halepense* are examples of stoloniferous and rhizomatous weeds of worldwide importance.

Within the group of storage organs that become vegetative reproductive structures, the two most common are tubers, e.g., potato, and bulbs, e.g., onions (*Allium cepa*). Tubers are modified stems, while bulbs are specialized organs derived from buds and leaves. Tubers have several "eyes," which are meristematic buds, representing in the modified stem what would have been nodes in the actual stem. In fact, tubers exhibit the apical dominance typical of the stems, for which the apical "eye" dominates the sprouting and, agronomically, a single tuber may be divided to multiply the propagules.

Apomixis

Apomixis is asexual reproduction by seed (agamospermy). Thus, apomictic plants have the advantage of seed reproduction (mainly dispersal), but conserve the identical (successful) genotype of the parent. Also, it is a way to assure reproduction even in the absence of pollinators (for an otherwise cross-pollinated species), unless apomictic seed formation requires the stimulus of pollination (see below).

An asexual seed requires the production of the zygote without the fusion of the genome of the male and female gametes. The mechanisms that lead to apomictic seed production are diverse but in all of them the succession of steps during sexual reproduction leading to haploid gametes is avoided so that diploid, somatic reproductive cell gives rise to the embryo. The simplest of these mechanisms is where a diploid embryo sac develops directly from a somatic cell of the nucellus, a layer of diploid cells surrounding the embryo sac, by mitosis. This mechanism is known as sporophytic apomixis, as the somatic cell gives rise directly to the embryo, as well as adventitious embryony, as the embryo is produced at an unusual site. If the unreduced cells initially develop as a megagametophyte, and then the apomictic process takes place to produce the diploid embryo sac, the mechanism is termed gametophytic apomixis. There are two main types of gametophytic apomixis depending on the origin of the embryo sac. If it is derived from the megaspore mother cell it is called diplospory, while if it originates from nucellar cells it is called apospory; the latter is the most common of the apomictic mechanisms in higher plants. In some apomictic species the stimulus of pollination is required to trigger embryo development or the production of endosperm; this is called pseudogamy.

Apomixis is widespread and present in diverse plant families. Among crops, apomixis is present in species such as citrus and numerous perennial forage grasses. It is also found in one of the most important agronomic weeds in the world, dandelion (*Taraxacum officinale*). There has been a growing interest in apomixis as a tool in plant breeding. If a trait with an apparently simple genetic control can be introgressed into the genetic backgrounds of superior hybrids it may become a tool for permanently fixing hybrid vigor (see below); a superior F_1 genotype could be clonally propagated.

Techniques for Inducing Asexual Reproduction

Several techniques have been developed to maximize the advantages of sexual reproduction of plants. The most common ones include vegetative propagation from fragments of plant organs, known as cuttings, and the tissue culture of some cells that, under appropriate conditions, can regenerate adult plants.

Cuttings are essentially a fragment of a plant that is induced to produce a ramet. Cuttings are commonly taken from stems (including modified stems, such as rhizomes, tubers, and stolons), though in several plants cuttings can be also made from leaves or roots. Cuttings have the advantage of being easy, cheap, and fast, and simply involve placing the stem fragment into a medium with growth regulators to stimulate rooting and then left in a wet medium (likely soil) to root and propagate the whole plant. In many fruit crops grafting of shoot cuttings (scions) onto different rootstocks is common practice. This is often used when the rootstock is resistant to biotic constraints such as pest and diseases, e.g., phylloxera in grapes (*Vitis vinifera*), or may improve productivity in the shoot by reducing vegetative vigor of the plant, resulting in dwarf cultivars e.g., apples (*Malus pumila*) and citrus. In the grafting process the rootstock and the cutting tissues are united when their cambial regions make contact and subsequently grow as a single plant.

Tissue culture is a micropropagation (*in vitro*) technique where a piece of excised tissue (explant), from leaves, buds, stems or roots regenerate new plants that are ramets of the plant whose cells were originally cultured. The genetic potential of these cells to reproduce the entire organism is called totipotency. The success of a particular case of micropropagation depends on the genotype (there is a large degree of inter- and intraspecific variation in the ability of a whole plants to regenerate from an explant), the environment (both biotic, largely due to internal and external diseases, and abiotic, mainly the culture media and factors such as temperature and photoperiod), and their interaction (genotypes vary in their response to different media).

Sexual Reproduction

Sexual reproduction has the advantage of recombining parental genotypes, but the magnitude of the variability that can be created depends on the breeding system of the plant.

Inbreeding and Outbreeding

The main characteristic defining to what degree new variability is created by sexual reproduction is whether the ovules are fertilized by pollen from individuals which are more (inbreeding) or less (outbreeding) closely related to the particular female.

Inbreeding reduces heterozygosity in plant populations and can result in populations of pure lines, due to the persistent trend to lose allelic diversity. This produces the so-called inbreeding "depression," a decrease in fitness of offspring with respect of that of the parents, as a consequence of the breeding between close relatives. Conversely, outbreeding creates variability and increases the likelihood of heterozygosity (the presence of different allelic forms). This is called heterosis or hybrid vigor, a relevant characteristic in agriculture, when the expression of the vigor in terms of harvestable productivity justifies the extra cost associated with its usage. Corn is a species with a clear expression of hybrid vigor, with the hybrids being mostly far higher-yielding than either of the parental inbred lines; the yields of the parental lines are not a good predictor of the yield expected in the hybrid, as the expression of vigor is a consequence of the interaction between different allelic forms rather than additive effects (**Figure 3**).

In some species, to secure the availability of viable pollen for fertilization can be advantageous despite the expected inbreeding depression, whereas in other species the opposite is true. The former species have evolved mating systems emphasizing inbreeding, while the latter have evolved systems preventing inbreeding. Many species have both systems and are mostly inbreeding (but a burst of new variability is created by some occasional episodes of outbreeding) or mostly outbreeding (with the capacity for inbreeding if there is a lack of pollen from unrelated individuals).

Mating Systems

The extreme case of inbreeding is selfing or self-fertilization, where the pollen from the same genet fertilizes the ovule. This can be done from different

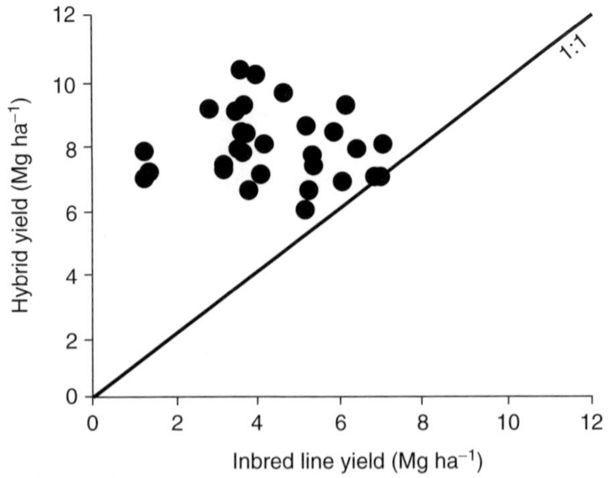

Figure 3 Yield of experimental corn single hybrids plotted against the yield of the lines that produced the hybrids. The straight line represents the 1:1 ratio. Data from Pico P and Slafer GA, Morgan Seeds–Argentina and University of Buenos Aires, unpublished data.

individuals of the same genet, i.e., different ramets providing the pollen and ovule, or within the same ramet, for which the species must be hermaphrodite. Within hermaphrodites the selfing may be based on the fertilization of the ovule of a flower with pollen from a different flower from the same plant (in monoecious, self-compatible plants), or from the same perfect flower.

Species in which ovules are (predominantly) fertilized by pollen from the same flower are known as autogamous, whilst allogamous species are those in which the ovule of a flower is fertilized by pollen from another flower, frequently of another plant. Autogamy is a system maximizing the likelihood of inbreeding. Within autogamous species, cleistogamous species are those in which pollination occurs before the opening of the flowers while chasmogamous species are those where the pollen is shed after the flower opens.

Obviously, allogamy is required for outcrossing but does not ensure that it occurs. For instance, selfing occurs in allogamous plants in which pollen from one flower fertilizes the ovule of another flower on the same plant, a mating system known as geitonogamy. An allogamous plant will still be self-fertilized if both gametes forming the zygotes are from flowers in different plants provided they are ramets asexually reproduced from the same genet. Geitonogamy occurs when self-compatible species either have a monoecious habit or when male and female functions in perfect flowers are separated in time or in space.

True outcrossing (xenogamy) requires that pollen from the stamen of a particular plant be carried to a flower on a genetically different plant where it fertilizes the ovule producing a heterotic zygote. An extreme case to secure xenogamy is the evolution of dioecious species. Plants have also developed several different strategies to maximize the chances of xenogamy, while still permitting geitonogamy or autogamy, mainly by separating in time or space pollen shedding and stigma receptivity.

Strategies to Increase Outcrossing

A very common strategy to increase the likelihood of outcrossing involves the separation of male and female functions, even within the same hermaphrodite flower, so that self-pollination is reduced. The two main separations that are possible are those in time (dichogamy) and those in space (herkogamy) (**Figure 4**).

In dichogamous species the receptivity of the stigmas and shedding of the pollen occur at different times within the same flower (or plant in monoecious species). There may be protandry, with pollen shedding before the stigmas are receptive, or protogyny, where stigma receptivity precedes pollen shedding.

In hermaphrodite flowers the existence of heterostyly, a polymorphism in the length of stamen and pistil that changes between different plants of the same species, may prevent the fertilization of receptive stigmas by viable pollen of the same flower. There are two forms of heterostyly, distyly (with two coexistent morphs) and tristyly (with three morphs). In the former, one morph produces pistils with long styles and stamens with short filaments (pin flowers), while the opposite is true for the other morph (thrum flowers). Tristyly occurs when individuals present either long, intermediate, or short styled pistils, with anthers being at the other two positions by means of the length of the filament.

Finally outcrossing is also promoted by the development of self-incompatibility, a system by which pollen that reaches a stigma of the same genet (or sufficiently closely related) does not produce viable seed. There are several mechanisms that may result in self-incompatibility; some simply restrict fertilization from pollen of the same genet while others impede the production of viable zygotes from self-fertilization. The different systems are broadly divided into sporophytic or gametophytic self-incompatibility systems (*see* **Flowering and Reproduction:** Gametophytic Self-Incompatibility; Sporophytic Self-Incompatibility). In addition, there are also male sterility systems governed by alleles that cause a normally hermaphrodite plant to behave as a female, if the allelic constitution prevents the normal development of male gametes. The system is relevant in agriculture as male sterility is introduced in lines that are designed to be female for hybrid production in certain crops. It is possible to manipulate the system relatively simply, because there are cytoplasmic genes that control male sterility, and there are other independent nuclear genes that may be used to restore male fertility (*see* **Crop Improvement:** Hybridization and Plant Breeding).

Some Agronomic Implications

With the demonstration of sex in plants, humans realized that plants could be improved by crossing superior parents, followed by selection within the

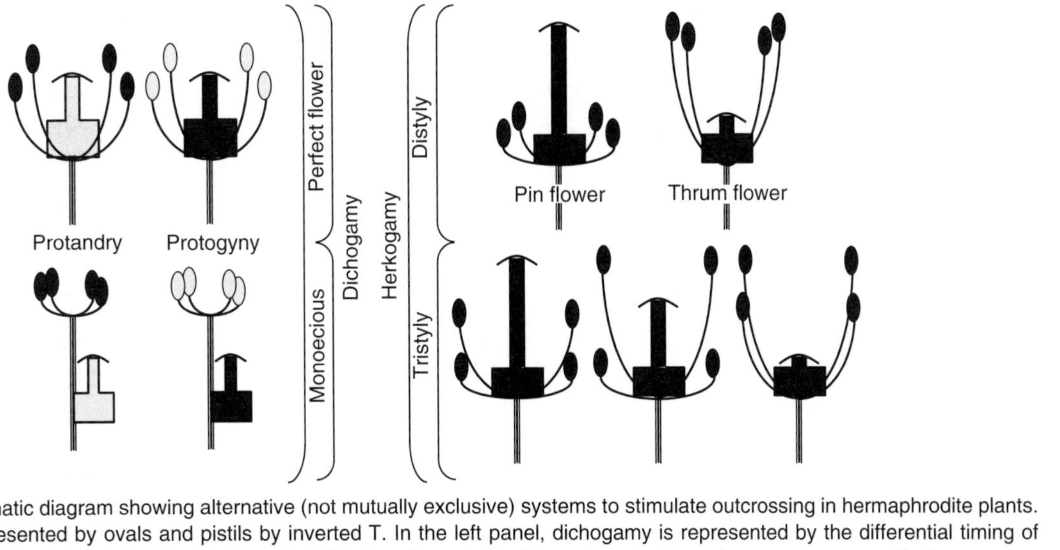

Figure 4 Schematic diagram showing alternative (not mutually exclusive) systems to stimulate outcrossing in hermaphrodite plants. Stamens are represented by ovals and pistils by inverted T. In the left panel, dichogamy is represented by the differential timing of maturity of stamen and pistil, being the dark and light filled surfaces indicating mature and immature organs, respectively.

created plant population with an increased likelihood of producing individuals of superior performance in the descendant population. Thus, modern cultivars of major crops represent combinations of favorable genes. Therefore, the choice of parents to be crossed is one of the most critical decisions made by a plant breeder. There are many examples of manipulations of reproductive systems by different methods to developing new genotypes for commercial seed production. Two major aspects are the maintenance of prospective parental genotypes in a breeding program and the development of breeding products e.g., commercial cultivars.

Once prospective parental material is collected, it must be maintained ready to be used when required. The type of reproduction of each species determines the way in which parent material should be maintained. In the case of self-pollinating species, such as wheat, barley (*Hordeum vulgare*), and rice (*Oryza sativa*), the maintenance of parent material as pure lines is relatively simple. Additionally, off-type plants are easily detected in a uniform plant population. Parent materials of cross-pollinated crops have to be maintained as inbred lines. This can be done easily by selfing, if the species is self-compatible with reasonable plant vigor (e.g., corn). In some species vigor declines dramatically after two or more generations of selfing e.g., alfalfa (*Medicago sativa*, lucerne), and material is more easily maintained as heterogeneous populations. In the case of maintaining inbred lines that are used to produce hybrids with cytoplasmic male sterility, the maintenance requires the multiplication of three different lines of the same (for all other traits than male fertility) genotype: the 'wild' inbred line (i.e., male fertile line with normal cytoplasm and nucleus), the male sterile line with cytoplasmic genes controlling pollen sterility and no nuclear genes for restoring male fertility, and the restorer line, using a male parent with the nuclear genes for male fertility restoration. The male-sterile line is normally developed by crossing a maintainer line as male to a male-sterile female with sterile cytoplasm and non-restoring genes; after that it is necessary to make repeated backcrosses to the maintainer line. In crops cultivated for vegetative organs (e.g., potato, sugar cane, and several pasture grasses) parent material is maintained in the vegetative form as tubers or cuttings. Maintaining the material through tissue culture is also possible.

Parental material can be used to develop different types of products as pure-line cultivars, hybrids, synthetics, or cultivars propagated by apomixis or vegetatively. Generally self-pollinated species are commercially propagated by pure-line cultivars. The two main steps for developing new cultivars consists of (1) generating a breeding population by hybridization of the parental material selected, and (2) allowing the self-pollination/fertilization of the progeny to develop pure lines, applying selection criteria defined by the aim of the program. The resulting pure lines can be, and frequently are, used as parental lines for further crosses. Breeding hybrids to take advantage of heterosis is common when this plays an important role in increasing yield performance. In general, open-pollinated species have greater hybrid vigor than self-pollinated plants, in which additive gene action predominates. Commercial hybrids are extensively used in corn, *Sorghum*, sunflower (*Helianthus annuus*), and sugar beet (*Beta vulgaris*), among other crops. Normally the use of hybrids for commercial crops is limited to those species in which the increased cost of hybrid seed production is counterbalanced by higher yield associated with hybrid vigor. Self-incompatibility is common in those species propagated by open-pollinated cultivars as legumes, rye, and grasses. In these species, the open-pollinated cultivar obtained for commercial purposes can be a synthetic or an improved population. Cultivars propagated by apomixis include perennial forages, and breeding this sort of crops requires the production of variability from which one can select for improved performance. Most apomictic species have close relatives, which allows crossing with an asexual plant with obligate apomixis and subsequent selection of the apomitic F_1 hybrids. Once the apomictic barrier is broken there is generally an important rise in variability resulting from new genotypes. Further understanding of apomitic mechanisms will allow the control and manipulation of apomixis as a breeding tool, solving the reproductive problems associated with this particular nonsexual reproduction method.

List of Technical Nomenclature

Allogamy — Sexual reproduction in which the ovule of a flower is fertilized by pollen from another flower.

Apomixis — Asexual reproduction by seed: plants have seed reproduction conserving the identical genotype of the parent. Also termed agamospermy.

Asexual reproduction — Reproduction system producing offspring genetically identical to the parent (clone). This may be through rizomes, stolons, or other vegetative organs (vegetative reproduction) or seeds (apomixis).

Autogamy Sexual reproduction in which the ovule is fertilized by pollen from the same flower, an extreme case of inbreeding (which is the reproduction between related individuals).

Chasmogamy Pollination characterized by the shedding of pollen after the flower opens.

Cleistogamy Pollination characterized by the shedding of pollen before the flower opens.

Dioecy The condition in which male and female flowers occur on separate plants (with male and female individuals, as in most higher animals); obligatorily outcrossing plants.

Gametophytic apomixis Type of apomixis where a megagametophyte (female gametophyte) is developed first, followed by the apomictic process which produces the diploid embryo sac.

Hermaphrodite An individual with female and male organs, having either perfect flowers (that is flowers with both male and female parts) or female and male flowers separately (monoecious plants).

Hybrid vigor Increased performance of hybrid individuals compared to that of the parents; the opposite of inbreeding depression.

Monoecy The condition in which a hermaphroditic plant bears separated female and male flowers. These are necessarily allogamous plants (though inbreeding and even selfing may occur if the female flower is fertilized by pollen of the same individual).

Pin flower Short stamens and long styles within a perfect flower.

Protandry Strategy to reduce autogamy, in which the plant sheds pollen before the stigma are receptive. Then, when the female organs are receptive, they may be only or predominantly fertilized by pollen from other plants.

Protogyny Strategy to reduce autogamy in which the female organs of the flowers (or female flowers in monoecious plants) mature and become receptive prior to male organs, so that pollen shed by a plant may only fertilize stigmas that are receptive in other plants.

Pseudogamy A type of apomixis in which pollination stimulates embryo development with no fusion of the egg nucleus with that of the sperm.

Self-incompatibility Inability of a viable pollen grain to fertilize an ovule of the same flower.

Sexual reproduction Reproduction system producing a mixture of attributes from female and male gametes from either a unique parent (necessarily hermaphrodite) or two different parents.

Sporophytic apomixis A type of apomixis where the somatic cells (nucellus) give rise directly to the embryo; also known as adventitious embryony.

Thrum flower Long stamens and short styles within a perfect flower.

See also: **Crop Improvement**: Hybridization and Plant Breeding; Plant Breeding, Principles. **Flowering and Reproduction**: Fertilization; Flower Development; Gametophytic Self-Incompatibility; Sporophytic Self-Incompatibility.

Further Reading

Barret SCH and Harder LD (1996) Ecology and evolution of plant mating. *Trends in Ecology and Evolution* 11: 73–79.

Crawley MJ (1997) Sex. In: Crawley MJ (ed.) *Plant Ecology, 2nd edn*, pp. 156–213. Oxford: Blackwell Science.

Fehr WR and Hadley HH (1980) *Hybridization of Crop Plants*. Madison: Crop Science Society of America.

Hartmann HT, Kester DE, Davies FT, and Geneve RL (2002) *Hartmann and Kester's Plant Propagation: Principles and Practices, 7th edn*. Upper Saddle River: Prentice Hall.

Hayward MD, Bosemark NO, and Romagosa I (1993) *Plant Breeding: Principles and Prospects*. London: Chapman & Hall.

Jarne P and Charlesworth D (1993) The evolution of the selfing rates in functionally hermaphrodite plants and animals. *Annual Review of Ecology and Systematics* 24: 441–466.

Kaufman PB, Carlson TF, Dayanandan P, *et al.* (1989) *Plants: Their Biology and Importance*. New York: Harper & Row.

Kay QON (1987) The comparative ecology of flowering. In: Rorison IH, Grime JP, Hunt R, Hendry GAF, and Lewis DH (eds) *Frontiers of Comparative Plant Ecology*, pp. 265–281. London: Academic Press.

Lovett Doust J and Lovett Doust L (1988) *Plant Reproductive Ecology: Patterns and Strategies*. Oxford: Oxford University Press.

Otegui ME and Slafer GA (2000) *Physiological Bases for Maize Improvement*. New York: Food Product Press.

Richards AJ (1997) *Plant Breeding Systems*, 2nd edn. London: Chapman & Hall.

Stalker HT and Murphy JP (1992) *Plant Breeding in the 1990s*. Wallingford: CAB International.

Forestry *see* **Production Systems and Agronomy**: Plantation Crops and Plantations. **Tissue Culture and Plant Breeding**: Clonal Propagation, Forest Trees; Clonal Forestry.

G

GENETIC MODIFICATION

Contents

Gene Cloning, General Principles

R M Twyman, University of York, York, UK

Introduction

The analysis of plant genes requires a method for cloning, i.e., the preparation of millions of copies of the same DNA sequence so that enough DNA is present for laboratory scale experiments. Cloning methods can be divided into two broad categories:

- Cell based cloning, in which the plant gene of interest is joined to a replicon (such as a plasmid or bacteriophage vector) and amplified by repeated replication cycles inside a particular host cell (usually the bacterium *Escherichia coli*).
- *In vitro* cloning, by the polymerase chain reaction (PCR), where DNA amplification is achieved using purified enzymes.

The two techniques are compared in **Figure 1**.

These routine procedures can be applied to any DNA sequence of interest. Therefore, the bottleneck in cloning is not the amplification process itself, but the isolation of the target gene from its source. Under most circumstances, the source is very complex, and the desired sequence is diluted in a background of many millions of nontarget sequences. There are two typical sources of DNA for cloning:

- Genomic DNA. Whole genomic DNA represents the entire plant genome (usually the nuclear genome – plastid genomes are considered separately). Genomic DNA isolated from any cell type or organ of an individual plant is generally the same. Each gene is equally represented. Regulatory sequences and introns are present. The gene density in some species is very low, and in such cases much of the genome may be taken up by intergenic spacer regions, transposable elements, and tandem repeat sequences. In corn (*Zea mays*; maize), for example, only 10% of the genome is represented by genes.
- Complementary DNA (cDNA). Complementary DNA is reverse transcribed from mRNA. It does not represent the entire genome, only the genes expressed in the cell type or organ from which the mRNA was isolated. This means that cDNA prepared from different tissues and organs contains different sequences, and different genes are not represented equally – strongly expressed genes will produce more transcripts and give rise to more cDNA copies. Regulatory elements and introns are not present. Splice variants are represented by different cDNAs.

Both cell based cloning and the PCR can be used to isolate specific genes from complex sources. PCR based approaches can be applied directly to the source DNA, while in the case of cell based cloning, the typical approach is to create a clone library. For genomic DNA, this first involves breaking the source DNA into manageable fragments. The fragmented genomic DNA or whole cDNA is then cloned nonselectively into vectors, with the aim of generating a collection of clones representing the entire source population of DNA molecules. While each plant gene

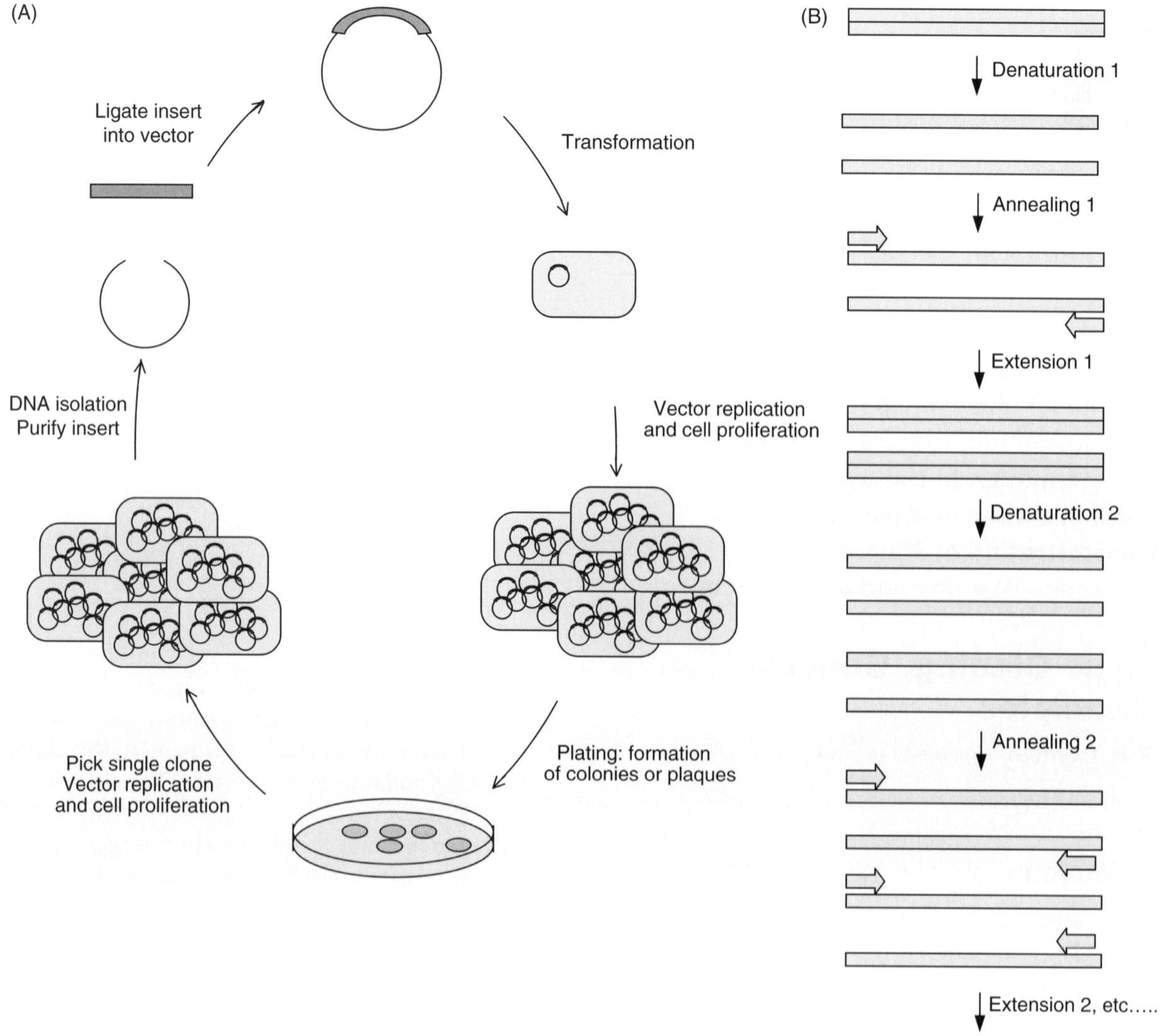

Figure 1 Principles of (A) cell based cloning and (B) the polymerase chain reaction.

is represented equally in genomic DNA, the random sampling involved in nonselective cloning means that genomic libraries are not equally representative of all genes. The DNA must be broken up randomly (to ensure no systematic exclusion of particular sequences), and enough clones must be generated to cover the genome several times (to avoid exclusion due to sampling errors). For example, to ensure a 95% probability of including every gene, there must be approximately fivefold coverage of the genome. The precise number of clones required depends on the genome size and the average insert size, which depends on the type of cloning vector used (**Table 1**).

The clone library must then be screened to identify the gene of interest. For genomic and cDNA libraries, a standard approach is to use hybridization or PCR assays to identify clones containing particular genes. Some knowledge of the target gene's sequence is generally required, as is the case for direct PCR amplification from the source DNA. In contrast, since cDNAs are discrete functional sequences, they can be expressed by including appropriate regulatory elements in the cloning vector. Expression libraries can then be produced, and these can be screened by exploiting properties of the gene product, rather than the gene itself.

Sequence-Dependent Cloning Strategies

Many gene isolation strategies are based on the identification of specific sequences by hybridization or PCR. In each case, the limitation is that there must already be some knowledge of the desired sequence before the screening process begins. Thus, sequence-dependent screening is like the chicken and the egg: without a clone, you cannot produce a probe (or primer set), and without a probe (or primer set) you cannot isolate a clone.

Table 1 Different types of vectors used in the cloning of plant genes

Type of vector	*Main uses*	*Maximum insert size*
E. coli plasmid	Subcloning, cDNA cloning, direct gene transfer to plants	10–15 kb
Binary vector	*Agrobacterium* mediated gene transfer to plants	20–40 kb
λ insertion	cDNA cloning	7–8 kb
M13, phagemids	Sequencing	5 kb
λ replacement	cDNA cloning, genomic cloning	18–20 kb
Cosmid, fosmid	Genomic cloning	40 kb
P1	Genomic cloning	125 kb
BAC	Genomic cloning, genome projects	300 kb
PAC	Genomic cloning, genome projects	300 kb
BIBAC/TAC	Transfer of large inserts to plants	300 kb
YAC	Genomic cloning, genome projects	500–1500 kb

Abbreviations: BAC, bacterial artificial chromosome; PAC, P1 artificial chromosome; BIBAC, binary BAC; TAC, transformation-competent artificial chromosome; YAC, yeast artificial chromosome.

Fortunately, there are several ways to generate probes/primers without knowing the full and precise sequence of the target gene:

- In some cases, it may be possible to obtain a partial sequence to use as a probe or for primer design. For example, expressed sequence tags (ESTs) generated by random sequencing of cDNA libraries can be used to isolate homologous full length cDNAs or genomic clones.
- Gene or cDNA sequences isolated from one species can often be used as probes to identify homologous genes in other species. Depending on the evolutionary conservation of the target gene, it may even be possible to use sequences from bacteria, yeast, or animals. For example, plant homeobox genes have been isolated using animal sequences as the basis for probe/primer design.
- Gene or cDNA sequences can be used to isolate homologous genes (paralogs) or partly homologous genes (domain sharing genes) from the same species.
- Sequences corresponding to a highly conserved domain can be used to isolate entire gene families. A good example in this case is the use of the first *Arabidopsis* MADS box genes to isolate related genes in *Arabidopsis* and other species.
- Probes or primer pairs can be designed and synthesized *in vitro* based on known amino acid sequences (**Figure 2**).

In the first strategy listed above, the partial sequence is identical to part of the target gene, and a suitable screening strategy would be to use high stringency hybridization or highly specific PCR primers. In all the others, however, there is likely to be a degree of disparity between the probe and its target(s). In such cases, hybridization assays are carried out with reduced stringency, which means a degree of mismatch can be tolerated between the probe and its complementary target. Similarly, PCR primers can be designed to match the most highly conserved regions within a family of sequences, and/or a degree of degeneracy can be built in to primer design. Degeneracy is always required where probes or primers are designed on the basis of amino acid sequences, since more than one codon is used to specify most amino acids. Protein sequences that contain tryptophan and methionine are useful for probe or primer design because these amino acids are each specified by a single codon, thus reducing the total amount of degeneracy required (**Figure 2**).

```
PHE  ASN  LYS  MET  HIS  TRP

  C    C    A         C
UU   AA   AA   AUG  CA   UGG
  U    U    G         U

UUY  AAY  AAR  AUG  CAY  UGG
```

Figure 2 Designing a probe/primer based on a known amino acid sequence. The amino acid sequence shown on the top line can be encoded by any of 16 different nucleotide sequences as represented by the middle line, and is therefore said to show 16-fold degeneracy. The bottom line shows the nucleotide sequence with the standard degenerate representations (Y is any pyrimidine, R is any purine). Although primer combinations have been used successfully with up to 128-fold degeneracy, the total amount should be limited as much as possible. This is helped by choosing sequences containing the nondegenerate residues methionine and tryptophan.

Map Based Cloning Strategies

Map based cloning is another term for positional cloning, a strategy developed for the isolation of human disease genes in cases where no sequence or biochemical information is available. The strategy uses linkage analysis to define a candidate region of the genome flanked by the two closest genetic markers. A bridge of physical clones is then built across the interval using the genetic markers as a starting point (**Figure 3**). This technique, known as chromosome walking, is laborious if large numbers of clones are involved. However, in plants it is often possible to identify markers that lie within a few hundred kilobases of the target gene, and with the availability of large insert vectors such as YACs and BACs (**Table 1**) the chromosome walks can be very short indeed. It is sometimes possible to identify a single clone spanning the candidate interval, whereupon chromosome walking becomes chromosome landing.

The isolation of the rice (*Oryza sativa*) *Hd1* gene provides a useful example of map-based cloning. Genetic mapping initially established a candidate region of 100 kb containing a quantitative trait locus (QTL) affecting sensitivity to day length. Physical mapping, using the flanking genetic markers as probes, identified a single BAC clone spanning the entire candidate interval. This BAC clone was used to establish a very fine genetic map based on CAPS markers (cleaved amplified polymorphic sequences). Further mapping then refined the interval to 12 kb, which contained two genes. One of these genes, *Hd1*, happened to be a rice homolog of the *Arabidopsis* gene *CONSTANS*, which regulates flowering time in response to day length. Prior to this map based cloning strategy, investigators had absolutely no idea of the sequence of the gene or its biochemical function – only its approximate location and its physiological effect.

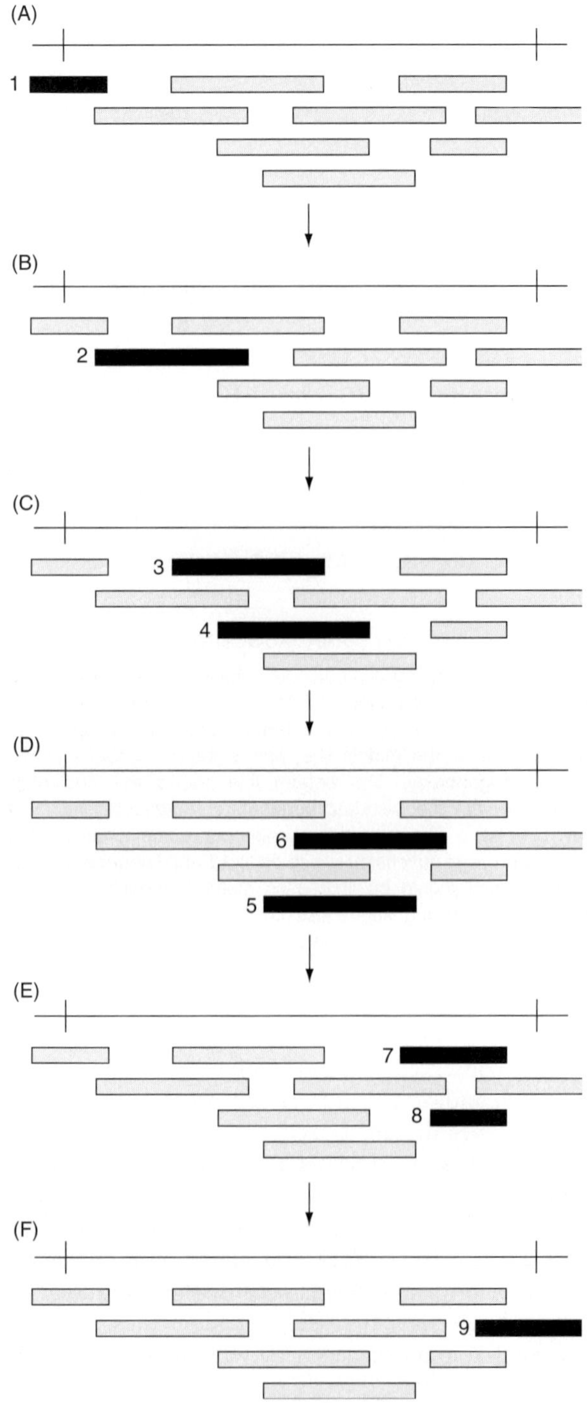

Figure 3 The principle of map-based (positional) cloning. The thin line represents a candidate interval of the genome defined by two genetic markers. (A) If the left flanking marker is used as a probe, genomic clone 1 is identified. (B) If clone 1 is used as a probe, overlapping clone 2 is identified. (C) Clone 2 identifies clones 3 and 4, as well as clone 1, which can be ignored. (D) Similarly, clone 4 identifies clones 5 and 6. (E) Clone 6 identifies clones 7 and 8, but note that clone 5 would identify clone 7 only, because it does not overlap with clone 8. (F) Clone 8 identifies clone 9, which also hybridizes to the right flanking marker. The genetic interval is now spanned by a bridge of physical clones.

Shortcuts to Map-Based Cloning

The genetic mapping of plant genes often requires laborious crosses between lines with alternative (mutant or wild-type) alleles at a given locus. Positional cloning on this basis is a labor-intensive and, sometimes, disappointing venture, which involves systematic library screening to identify contigs of overlapping clones spanning a given candidate interval. A useful short cut is to generate mutants by insertional mutagenesis, for this leaves a DNA tag within the interrupted gene. Several cloning strategies have been devised to isolate such genes, including the following:

- Genomic libraries can be produced from the tagged plant line and then screened with a probe (or PCR primers) corresponding to the insert. Genomic clones thus isolated generally contain flanking DNA sequences that can be used in wild-type libraries to isolate corresponding intact genes or cDNAs. It is important to make sure the insertion sequence is unique in the genome, or irrelevant flanking sequences could be obtained.
- Plasmid rescue. In this technique, the insertion element contains the origin of replication and

antibiotic resistance marker of a bacterial plasmid. Whole genomic DNA from the insertion line is simply digested with a restriction enzyme that does not cut within the insert, diluted, and circularized with DNA ligase. If this mixture of genomic DNA circles is used to transform *E. coli*, only bacteria containing the insert and its immediate flanking sequences will survive under antibiotic selection (**Figure 4**).

- Inverse PCR. In this technique, genomic DNA from the insertion line is digested with a restriction enzyme that does not cut within the insert, diluted, circularized with DNA ligase, and then amplified using PCR primers that anneal to the insert but face outward. This amplifies not the insert but the flanking sequences. These can be used as probes to isolate intact clones (**Figure 4**).
- Other PCR methods. A number of more sophisticated PCR approaches are available that do not require circularization of the genomic DNA prior to amplification. These include transposon display and amplification of insertion mutagenized sites (AIMS), both of which involve the digestion of genomic DNA with restriction enzymes, the addition of adaptor oligonucleotides, and several rounds of nested PCR. In transposon display, amplification is exponential and requires a transposon specific primer and an adaptor specific primer. In AIMS, only the transposon specific primer is used and amplification is linear. Another nested PCR technique is thermal asymmetric interlaced (TAIL) PCR, which involves the use of three nested transposon specific primers and a small arbitrary primer. As an alternative to

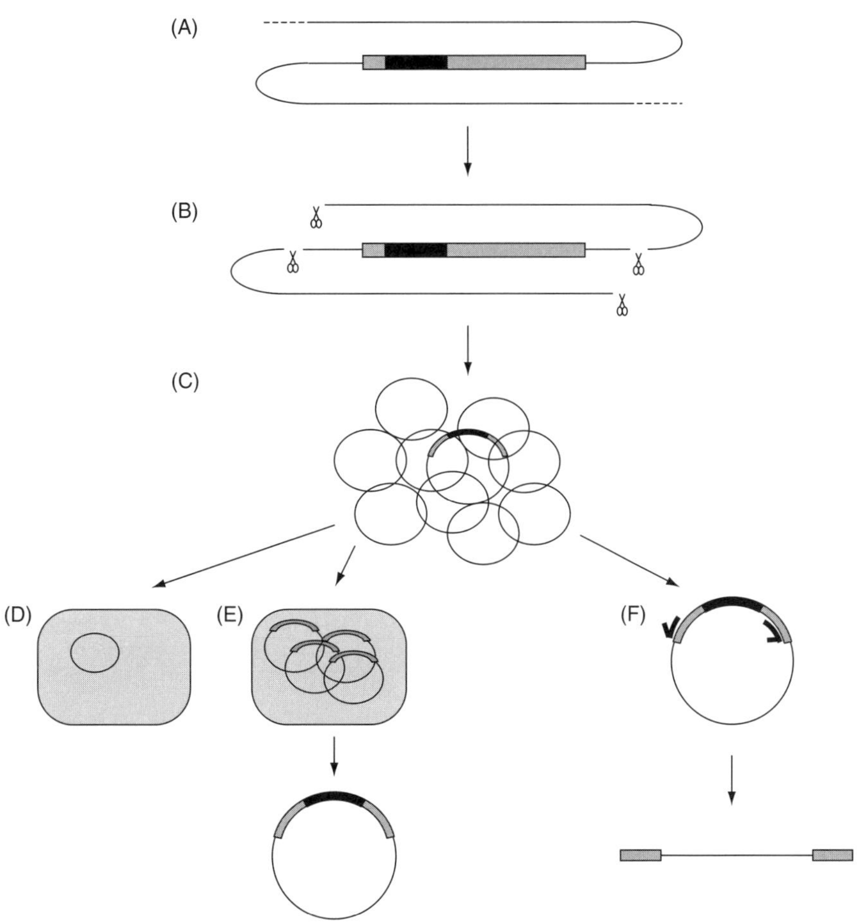

Figure 4 Cloning the flanking sequences adjacent to a transposon or T-DNA insertion element. (A) The element integrates into genomic DNA. (B) The genomic DNA is isolated and cleaved with restriction enzymes that do not cut within the element so that flanking sequences remain associated with it. The genomic DNA can then be used to construct a clone library, and the interrupted gene can be isolated by hybridization with a probe corresponding to the insertion element (not shown). (C) If the insert contains a plasmid origin of replication and selectable marker gene (black bar), recircularization of the genomic fragments with DNA ligase followed by random transformation of bacteria will produce (D) some bacterial cells containing random genomic fragments and (E) some carrying the insertion element and flanking sequences as a functional plasmid. Only bacteria containing the rescued plasmid will proliferate under selection, allowing the flanking sequences to be recovered. (F) Alternatively, the random circles of DNA can be amplified by PCR using primers that anneal to the insertion element but face outward, producing the flanking sequences as an amplification product.

genomic DNA, insertion events can be detected at the mRNA level by rapid amplification of cDNA ends (RACE).

Expression Cloning and Functional Cloning

Any cloning strategy that relies on the expression of the target gene is an expression cloning strategy. Included within this category is functional cloning, which refers to cloning strategies in which the biochemical function of the gene is exploited to identify it in a background of nontarget sequences. The most common expression cloning strategy is the immunological screening of cDNA libraries. This involves the use of labeled antibodies (or more commonly, antibody sandwich reactions) to isolate expression clones based on the structure of the corresponding protein. An antibody must be available, which generally means that the corresponding protein must already have been purified to some extent. Sometimes this approach is used because the alternative technique of degenerate probe design has failed, and it is often favored where an antibody is already available. In any case, a further requirement is that the protein folds properly when it is expressed, so that the epitope recognized by the antibody probe is retained.

For certain classes of protein, alternative probes may be used. For example, DNA-binding proteins can be isolated using labeled DNA probes corresponding to the protein's recognition site, a technique called southwestern screening due to its similarities to Southern blot and western blot methods. It is necessary to ensure that nonspecific DNA binding is blocked since many proteins bind to DNA, but have no particular sequence specificity. Similarly, northwestern screening involves the use of RNA probes to identify RNA binding proteins. Occasionally, protein ligands have been used to isolate interacting proteins from expression libraries. However, it is now much more common practise to use high-throughput screening techniques to identify cDNA clones corresponding to interacting proteins (see below).

Functional cloning strategies have been widely used in animals, but much more rarely in plants. Some functional cloning approaches include:

- Screening for the biochemical activity of a protein, e.g., enzyme activity, effect on plant growth.
- Functional complementation of mutant phenotypes in transformed plant cells or transgenic plants (**Figure 5A**). The development of high-capacity plant transformation vectors (binary bacterial artificial chromosome (BIBAC) and transformation-competent artificial chromosome

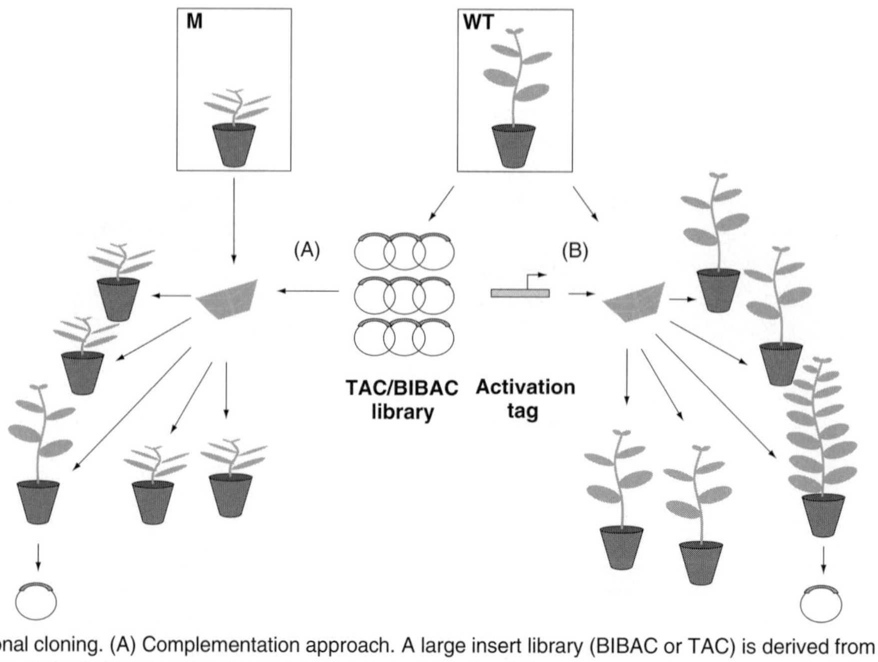

Figure 5 Functional cloning. (A) Complementation approach. A large insert library (BIBAC or TAC) is derived from a wild-type plant. The clones are systematically introduced into explants of a mutant plant and the transgenic clones regenerated. Any clone capable of rescuing the mutant phenotype of the plant must carry the corresponding gene. (B) Gain-of-function approach. Wild-type plants are transformed with an activation trap construct, essentially a T-DNA or transposon carrying a strong, outward-facing promoter. This integrates into the genome randomly and may activate adjacent genes. In some plants, a strong gain-of-function phenotype may be recognized. The activated gene can be cloned by the methods shown in **Figure 4**. M, mutant; WT, wild-type.

(TAC) vectors) has greatly facilitated the isolation of plant genes by functional complementation in transgenic plants.

- Analysis of gain-of-function phenotypes in transformed plant cells or transgenic plants. This can be achieved by overexpression or ectopic expression. An interesting variation is the use of activation tag constructs, insertional vectors based on transposons or T-DNA that contain a strong, outward facing promoter, which boosts the expression of any adjacent gene (**Figure 5B**). Such constructs have been used, for example, to identify genes encoding transcription factors that coordinately regulate several genes of a metabolic pathway.

Difference Cloning

It is often desirable to identify cell-specific genes or genes that are expressed in particular states or in response to particular stimuli. Difference cloning describes any cloning strategy that identifies sequences represented in one source of DNA, but not another, or sequences that differ significantly in abundance between sources. In most cases, this means differentially expressed cDNAs, representing genes that are expressed in one sample, but inactive in another (e.g., two different tissues, two different developmental stages, stimulated vs. nonstimulated, healthy vs. diseased). There are various cell based and PCR based difference cloning approaches, but all of them follow one of two principles: enrichment or display.

Enrichment techniques work by removing sequences that are shared between two DNA sources, therefore leaving behind those sequences that are unique. One common approach is subtraction cloning (**Figure 6**), in which mRNA from one source (known as the driver) is used to subtract homologous cDNA sequences from a second source (known as the tester). This is often achieved by labeling the driver mRNA sample universally with biotin, and then adding it in great excess to the tester sample, which has been used for first strand cDNA synthesis. The single strands are allowed to anneal. Since the biotinylated mRNA is present in such great excess, most of the hybridized cDNA strands that represent shared genes will form duplexes with biotinylated mRNA, allowing them to be selectively removed by affinity to streptavidin. Several rounds of subtraction can be carried out to generate a first strand cDNA population that is highly enriched for clones unique to the tester. This can be used to generate a subtracted cDNA library. A reciprocal library can be generated by using the second sample as the driver

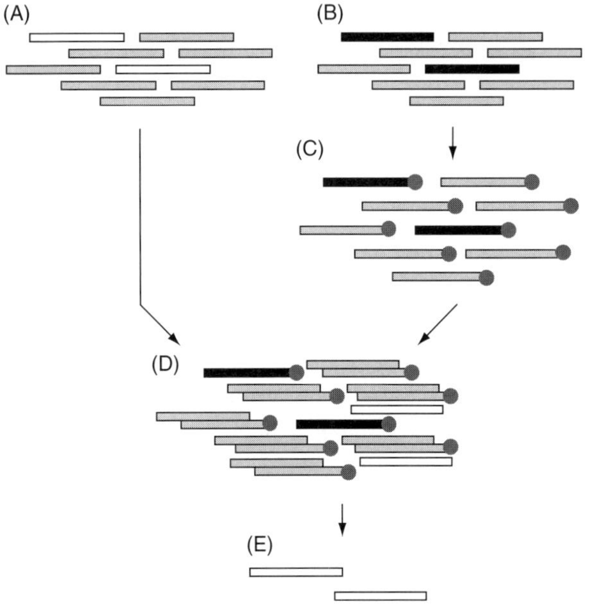

Figure 6 The principle of subtraction cloning. Two different mRNA samples are shown, representing the tester (A) and representing the driver (B) Transcripts unique to the tester population are white, while those unique to the driver population are black. Gray transcripts are common to both populations. (C) The mRNA from the driver population is universally labeled with biotin. (D) The mRNA from the tester population (A) is converted into cDNA and mixed with a large excess of labeled driver mRNA. The two populations hybridize. (E) The mixture is passed through a column of streptavidin beads. Most of the tester cDNAs that are common to both populations are sequestered into biotinylated duplexes and are retained on the column. cDNAs unique to the tester population are eluted and can be used to make a subtracted library.

and the first as the tester. A similar, PCR based approach is representational difference analysis, in which linkers are added to the tester population and mixed with a large excess of the driver population, which lacks the linkers. Most of the shared products form heteroduplexes in which only one strand has linkers, and therefore cannot be amplified using primers that anneal to the linkers. However, products unique to the tester population can only form homoduplexes in which both strands have linkers, and these can be successfully amplified.

Display methods work on the principle that all clones (or a representative subset thereof) are visually displayed, allowing differentially expressed sequences to be identified by eye. The most popular display method is differential display PCR, in which arbitrary PCR reactions are used to amplify pools of cDNA fragments equivalent to ESTs, and display them side-by-side on a sequencing gel. Bands that are present in one sample, but not the other, are selected for more detailed characterization, as these potentially correspond to differentially expressed genes.

Cloning Strategies: From Genes to Genomes

The cloning strategies discussed above have been developed to identify individual genes of interest with the assumption that the rest of the genome is an unknown void. However, genome projects are ongoing for a number of important plant species and two genome sequences, those of *Arabidopsis thaliana* and rice, have already been completed. With the sequence in hand, the next tasks are gene finding (structural annotation) and the elucidation of gene functions (functional annotation). Due to the large amounts of data arising from the genome projects, bioinformatics methods are employed for the heavy work of front-end annotation. The integration of database search methods and *ab initio* gene prediction can build up a virtual gene landscape, which can be refined by the assignment of known genes and ESTs onto the physical clone map. The increasing use of databases to share information between laboratories means that quite a lot can be learned about the function of a gene without doing any experiments – functional annotations based on sequence or structural homology, conserved expression patterns, mutant phenotypes, and protein interactions are available across many different species. Therefore, when map-based cloning reveals a number of possible genes in the region defined by linkage analysis, it is often possible to home in on candidate genes, which are the most likely to be involved in the process of interest. In species with well developed physical maps, it may even be possible to obtain the corresponding DNA clone from a central resource.

The isolation of the corn *Ivr2* gene provides an example of this candidate gene approach. Genetic mapping initially established a very large region containing a QTL affecting the response of corn plants to water stress. However, this interval was far too large to consider a search for individual genes. In a separate series of experiments, corn plants were subjected to water stress and cDNAs were isolated corresponding to genes expressed specifically under stress conditions. These cDNAs were assigned to the physical map and one of them mapped to the same locus as the previously identified QTL. In this case, the correct gene was identified from the many alternatives in the candidate interval because of its relevant expression profile.

The reductionist strategy of studying individual genes is being supplanted progressively by a new holistic approach that involves the study of whole genomes. At the beginning of this article, it was stated that the bottleneck in cloning is not amplifying the DNA sequence of interest, but isolating that sequence from the nonspecific background. We are rapidly approaching the day when this bottleneck will also be removed and all genes will potentially be available "off the shelf" from central genome project resources, at least for a range of model species and the commercially important crops. The new bottleneck will therefore not be a cloning strategy, but a strategy to link genes with their corresponding phenotypes. Genomics has provided us with a whole spectrum of high throughput functional analysis tools to make this possible, and this article closes with some examples:

- DNA microarrays. These are small devices with thousands of different DNA sequences immobilized in a grid. Genome-wide arrays are available for several microbes, and it is likely that an *Arabidopsis* genomic array will be available in the near future. They can be interrogated with complex hybridization probes, and differentially labeled cDNAs from two samples can be hybridized simultaneously to identify genes that are expressed more strongly in one sample than in another. Microarrays have been used to identify genes induced by drought stress, disease, oxidative stress, wounding, insect feeding, and many other processes in several species, but particularly in *Arabidopsis* and rice. In *Arabidopsis*, microarrays have been used for a methodical analysis of the molecular basis of systemic acquired resistance.
- Large scale EST sampling. This is an alternative to microarrays and involves the statistical analysis of EST data from different cDNA libraries to identify differentially expressed genes. It has been used to identify genes induced by salt stress in *Arabidopsis* and in several salt tolerant plants.
- Proteomics. The proteome is the entire collection of proteins in the cell. A high-resolution separation method, such as two-dimensional electrophoresis or high performance liquid chromatography, is used to resolve the thousands of proteins in a plant cell into individual fractions, which are identified by mass spectrometry. This approach has been used to characterize proteins involved in the response to a number of biotic and abiotic stresses, and to identify polymorphisms that can be used as genetic markers.
- Large-scale transposon and T-DNA mutagenesis programs have been established in rice, corn, and *Arabidopsis*. In the case of corn and *Arabidopsis*, databases of insertion flanking sequences have been established. Therefore, an investigator studying any particular gene can search the database for insertions and order seeds for the corresponding mutant plant line. Viral based gene silencing and RNA interference represent newer technologies

for systematic gene inactivation, which are more rapid than the production of mutants.
- Large-scale studies of protein interactions are possible using mass spectrometry to analyze protein complexes or library-based techniques such as phage display and the yeast two-hybrid system to probe interactions systematically. Interactions among the MADS box transcription factors that act combinatorially in the specification of cell fates within the developing floral organs have been studied in this way.

Summary

There are two established techniques for cloning DNA sequences – cell based molecular cloning and *in vitro* amplification by PCR. Cell based cloning procedures are particularly versatile because they can be used to identify genes based on the properties of the gene product, as well as the sequence itself. The current bottleneck in plant gene cloning is the isolation of specific genes from the large background of irrelevant sequences in genomic DNA or cDNA. Sequence based, map based, immunological, and functional approaches have been described, as well as strategies that help to enrich for differentially expressed genes. As more genomes are sequenced and annotated, this bottleneck will be removed and genes will be available to order. A whole-genome sequencing strategy has been used for the relatively small rice and *Arabidopsis* genomes, whereas for the larger genomes of crops such as wheat (*Triticum* spp.) and corn, the large excess of repetitive DNA may make a gene-focused strategy more applicable. In either case, the challenge ahead will not be cloning, but identifying associations between genes and their functions.

List of Technical Nomenclature

Ab initio gene prediction	Identification of genes in genomic DNA by looking for gene-like features (promoters, splice sites) rather than relying on homology to known sequences.
Activation tag	An insertion element that carries a strong, outward facing promoter, which can activate nearby genes.
Adaptor, linker	A double-stranded oligonucleotide that fits over the sticky ends left by a restriction digest (adaptor) or blunt ends left by a restriction digest or PCR (linker), and provides a primer annealing site.
AIMS	Amplification of insertion mutagenized sites. Nested PCR using a transposon specific primer for linear amplification.
Candidate gene approach	Map-based cloning strategy in which the choice of genes within the candidate interval is assisted by functional information.
CAPS markers	Polymorphic markers identified by PCR followed by the digestion of the PCR products with restriction enzymes.
Chromosome landing	Situation where a single genomic clone contains both the target gene and the nearest genetic marker, avoiding the necessity for chromosome walking.
Chromosome walking	Assembly of a physical map by identifying a series of overlapping clones.
Clone library	A collection of DNA clones in vectors.
Codon	The sequence of three nucleotides in a mRNA that specify an amino acid.
Degeneracy	The situation in which an amino acid can be specified by two or more codons.
Differential display PCR	Side-by-side display of equivalent RT-PCR products from two sources, showing differentially expressed genes.
Domain sharing genes	Genes that can be aligned over part of their lengths because they share a particular domain, but may be completely unrelated elsewhere in the sequence. Such genes have arisen by domain shuffling during evolution.
Driver	The source of DNA that is used to drive hybridization and therefore remove common sequences during subtractive cloning.
Ectopic	Outside the normal spatial or temporal domain.
Epitope	The specific region on a protein recognized by an antibody.
Expressed sequence tags (ESTs)	Short cDNA fragments obtained by random sequencing of clones from cDNA libraries.
Expression qcloning	Any cloning strategy that requires the expression of the target gene.
Expression library	A clone library in which each cDNA insert is expressed.
Functional cloning	An expression cloning strategy in which the biochemical, cellular, or biological function of the gene is exploited.
Functional complementation	Restoration of the wild-type phenotype by supplying a mutant cell or organism with the missing gene product.
Gene families	Homologous genes, often used to describe complete sets of paralogous genes in one genome.

Genetic mapping, linkage analysis	Mapping based on recombination frequencies (frequency of meiotic crossovers) between genes and/or markers.
Genetic marker	A polymorphic DNA sequence that can be used in genetic mapping.
Homeobox	DNA sequence encoding the homeodomain, a DNA binding protein domain found in many transcription factors.
Homologous	Sequences related by divergence from a common ancestor.
Host	The cell used for cell based cloning.
Immunological screening	Library screening with an antibody probe.
Insertional mutagenesis	Mutation produced by the insertion of a transposon or a T-DNA sequence.
Inverse PCR	PCR on a circularized template in which the primers face outward from an insertion sequence and amplify flanking DNA.
MADS box	A DNA binding protein domain found in many transcription factors, including those controlling floral organ specification.
Map based cloning, positional cloning	Gene cloning strategy dependent entirely on map position, with no biochemical, sequence, or functional information.
Mass spectrometry	Technique for determining accurate molecular masses, which can be applied to the identification of proteins.
Molecular cloning, DNA amplification	Production of many copies of the same DNA sequence.
Nested PCR	Sequential PCRs in which one or both primers are represented by nested sets placed progressively further inward along the template to increase specificity.
Northwestern screening	Detection of RNA binding proteins using an RNA probe.
Paralogous	Homologous sequences in the same genome that have arisen by gene duplication and divergence.
Phage display	Detection of interacting proteins by testing bait proteins with libraries of phage carrying different proteins on their surfaces, achieved by nonselective cloning into the phage coat protein gene.
Physical mapping	Any mapping technique in which the distances between genes and other markers are measured in real physical distances (kilobases, megabases).
Plasmid rescue	The inclusion of a plasmid origin of replication and antibiotic resistance marker in a transposon so that the flanking genomic region can be isolated and cloned as a bacterial plasmid.
Polymerase chain reaction	Exponential *in vitro* cloning method.
Quantitative trait locus (QTL)	Locus that contributes to a trait showing continuous variation.
RACE	Rapid amplification of cDNA ends. A PCR technique for the cloning of full-length cDNA sequences.
Replicon	A DNA element that can replicate autonomously.
RNA interference	Posttranscriptional silencing induced by double-stranded RNA. Probably the underlying mechanism of viral gene silencing.
Southwestern screening	Detection of DNA binding proteins using a DNA probe.
Splice variant	Alternative transcript/cDNA produced by differential splicing of the original gene.
Stringency	Conditions that determine how perfectly matched two complementary sequences need to be in order to anneal.
Subtraction cloning	Construction of a library enriched for differentially expressed genes by removing cDNA clones common to two samples.
TAIL PCR	Thermal asymmetric interlaced PCR. Similar to AIMS, but using a short arbitrary primer rather than an adaptor-specific primer.
Tester	The source of DNA that has common sequences subtracted by the driver.
Transposon display	Nested PCR using a transposon-specific primer and a primer annealing to an oligonucleotide adaptor.
Two-dimensional gel electrophoresis	Protein separation technique in which proteins are fractionated according to their charge in the first dimension and then by their mass in the second dimension.
Vector	A replicon used for cloning.
Viral gene silencing	Deliberate posttranscriptional silencing of plant genes using viral vectors carrying plant transgenes.
Yeast two-hybrid system	A high-throughput method for detecting protein interactions in which interacting fusion proteins assemble a functional transcription factor.

See also: **Crop Improvement**: Chromosome Engineering; Genetic Maps; Molecular Markers; Mutation Techniques; Plant Genomes. **Genetic Modification**: Transformation, General Principles.

Further Reading

Brown TA (2001) *Gene Cloning and DNA Analysis: An Introduction*, 4th edn. Oxford: Blackwell Science.

Jannink J-L, Bink MCAM, and Jansen RC (2001) Using complex plant pedigrees to map valuable genes. *Trends in Plant Science* 6: 337–342.

Maes T, De Keukeleire P, and Gerats T (1999) Plant tagnology. *Trends in Plant Science* 4: 90–96.

Primrose SB and Twyman RM (2002) *Principles of Genome Analysis and Genomics*, 3rd edn. Oxford: Blackwell Science.

Primrose SB, Twyman RM, and Old RW (2001) *Principles of Gene Manipulation*, 6th edn. Oxford: Blackwell Science.

Rabinowicz PD, McCombie WR, and Martinessen RA (2003) Gene enrichment in plant genomic shotgun libraries. *Current Opinion in Plant Biology* 6: 1–7.

Shibata D and Liu YG (2000) *Agrobacterium*-mediated plant transformation with large DNA fragments. *Trends in Plant Science* 5: 354–357.

Waugh R, Dear PH, Powell W, and Machray GC (2002) Physical education – new technologies for mapping plant genomes. *Trends in Plant Science* 7: 521–523.

Insertional and Transposon Mutagenesis

R M Twyman, University of York, York, UK
A Kohli, John Innes Centre, Norwich, UK

Introduction

Insertional Mutagenesis in Forward Genetics

The most direct way to establish the function of a plant gene is by mutation. Traditionally, gene–function relationships have been studied in a phenotype driven approach, sometimes described as forward genetics. Mutagenesis has been achieved by irradiating seeds or treating them with chemical mutagens, and then searching for interesting mutant phenotypes in subsequent generations. If a large enough population of seeds is mutagenized, it should be possible to recover mutants affecting any gene in the genome. Dominant mutations are often revealed in the first generation of plants, but recessive mutations require breeding to homozygosity. If such mutations are lethal, they can be maintained in heterozygous plants.

The disadvantage of radiation and chemical mutagens is that they tend to generate point mutations, which are difficult to map and make the process of gene isolation laborious and time consuming. Insertional mutagenesis is the use of DNA sequences as mutagens. Insertional mutagenesis is advantageous because genes tend to be completely disrupted, which favors the recovery of null mutants. Furthermore, the interrupted gene becomes tagged with the insertion, which can be used to clone flanking sequences, either by traditional means or strategies based on the polymerase chain reaction (PCR). The approach is generally known as gene tagging or signature tagged mutagenesis (STM) and is discussed in greater detail below.

Insertional Mutagenesis in Reverse Genetics

Reverse genetics is a gene-driven approach, in which a sequence with no known function is deliberately mutated allowing the resulting mutant phenotype to be studied. In certain lower plants, such as the moss *Physcomitrella patens*, it is possible to introduce exact changes into preselected genes by a process called gene targeting. This relies on homologous recombination between the target gene (in the genome) and a homologous sequence that has been mutated *in vitro* and introduced into the cell. Unfortunately, homologous recombination occurs at a very low frequency in higher plant cells. Therefore, although gene targeting has been achieved in *Arabidopsis thaliana* and in tobacco (*Nicotiana tabacum*) the process is so inefficient that it is unsuitable for routine application.

Insertional mutagenesis has recently become established as an alternative reverse genetics strategy in plants. Although it is not possible to systematically target specific genes for disruption, it is possible to saturate the genome with mutations and maintain a resource of seeds representing interruptions in every single gene. A researcher discovering a particular gene and wanting to know its function can then search for insertional mutants affecting the gene of interest, either by screening the mutant seed bank or by searching databases of flanking sequences. We consider the practicalities of this approach below.

Insertional Mutagenesis Systems

A number of different insertional mutagenesis systems have been developed for plants, but all are based either on transposons or the integration of T-DNA. Transposons are mobile genetic elements that jump from site to site in the genome when supplied with the enzyme transposase. A number of transposons have been used for mutagenesis programs, including *Activator* (*Ac*), *Suppressor-mutator* (*Spm*) and *Mutator* (*Mu*) from corn (*Zea mays*; maize), *Tam3* from *Antirrhinum majus*, and *Tos17* from rice (*Oryza sativa*). Some of these transposons function only in their natural host plant, while others are promiscuous and can be used in a variety of heterologous species. The corn *Ac* element, for example, has been shown to function in *Arabidopsis*, tobacco, rice, potato (*Solanum tuberosum*), petunia (*Petunia hybrida*), tomato (*Lycopersicon esculentum*), and many other plants. T-DNA is not a transposon but is a short segment of DNA that is transferred from a bacterial plasmid to the plant genome when plants are infected by the soil pathogen *Agrobacterium tumefaciens*. T-DNA transfer by *A. tumefaciens* has been used extensively to generate transgenic plants, and, more recently, for insertional mutagenesis.

Properties of Transposons and T-DNA

Transposons and T-DNA each have their advantages and disadvantages as insertional mutagens (**Table 1**). T-DNA is a bacterial sequence, and therefore always provides a unique signature when introduced into the plant genome. However, not all plants are susceptible to *A. tumefaciens*, so transposons remain the only choice for insertional mutagens in many species. All plants contain some form of transposable element population, but this means that several copies of any given transposon may already be present in the genome making it difficult to identify novel insertion events. This problem can be overcome by introducing unique foreign sequences into a transposon that is going to be used as an insertional mutagen in its normal host species. Alternatively, a transposon from one species may be introduced into a heterologous plant. In this case, either the *Agrobacterium* T-DNA system or a direct transfer approach such as particle bombardment must be used to achieve initial gene transfer.

T-DNA and transposons also differ in their behavior. Once integrated, T-DNA is unable to jump ever again because it lacks the ability to self-mobilize. Therefore, insertions generated by T-DNA are usually stable and permanent. In contrast, if a transposase is present, transposons may continue to jump in subsequent generations of plants. The advantage of this continued activity is that large numbers of insertion events can be generated with a small number of starter lines. However, the disadvantage is that the first generation of insertion events may be unstable, since the transposons used as mutagens in plants tend to move by a cut-and-paste mechanism, which involves excision from the original insertion site and integration elsewhere. There is a number of strategies available to generate stable transposon insertions. The most common is to use a two-component transposon system, which comprises an autonomous (self-mobilizing) element and a non-autonomous derivative. For example, the corn transposon *Activator* (*Ac*) is autonomous because it encodes its own transposase, but shorter derivatives of *Ac* called *Dissociation* (*Ds*) lack the transposase gene. *Ds* elements can transpose if transposase is provided by a copy of *Ac* in the same genome, but not when *Ac* is absent. Therefore, if *Ac* and *Ds* are introduced into the same genome, both elements can be mobile, but if *Ac* is removed by crossing, the *Ds* insertions will be stable. Another strategy is to use a retrotransposon system, such as *Tos17* in rice. Retrotransposons mobilize at the RNA level and move by a replicative mechanism involving transcription of an integrated element and the insertion of a cDNA copy of the transcript at a different location. The original insertion remains intact during this process.

Both transposons and T-DNA appear to favor integration into gene-rich DNA, which may reflect some preference for active chromatin structure. However, while the integration of T-DNA is essentially random within the gene space of the plant genome, transposons all appear to show greater or lesser degrees of target site preference, often defined

Table 1 Comparison of transposon and T-DNA derived insertional mutagens

Transposons	*T-DNA*
Endogenous in all plants	Bacterial in origin but can be introduced into many plants
Ideally, transposon based insertion vectors need to be modified or introduced into heterologous species to distinguish them from endogenous elements	Bacterial sequence; provides unique sequence signature in all plants
Insertions may be unstable	Insertions generally stable
Small number of founder lines can generate multiple insertions	Large number of founder lines required
Integrates preferentially into gene rich DNA	Integrates preferentially into gene rich DNA
Shows target site preference and, in some cases, local reinsertion	Random integration within gene space
Generally clean insertions	Often generates complex insertions

by a weak consensus sequence at the precise integration site. Transposons tend to generate clean insertions that facilitate the isolation of flanking sequences. In contrast, T-DNA insertions are often messy. There may be two or more copies arranged as tandem or inverted repeats, and genomic DNA at the integration site is more likely to be rearranged. The isolation of flanking sequences can be difficult at such complex insertion sites.

Many transposons, including *Ac*, also demonstrate a phenomenon known as local reinsertion, where transposition preferentially occurs to linked sites. This can be useful if the aim is to saturate a particular region of the genome with mutations, but it can make genome-wide saturation mutagenesis difficult. One way to approach this problem is to generate founder lines with regularly spaced *Ac* insertions. Another way to avoid local reinsertion is to use one of several available counterselection strategies, i.e., strategies that select against local reinsertion, to avoid transposition to linked sites. Alternatively, a different transposon system should be used. The corn transposon *Mu*, for example, does not demonstrate preferential local reinsertion. Local reinsertion may also depend to a certain extent on the genetic background. *Ac*, for example, transposes preferentially to linked sites in corn and *Arabidopsis*, but not in rice.

Identifying Interrupted Genes

Forward Genetics – Isolation of Flanking Sequences

Any phenotype generated by a basic transposon or T-DNA insertion is likely to occur because the insertion element has disrupted the corresponding gene, which means that the gene's sequence will flank the insertion site. Insertional mutagenesis, therefore, provides a short cut to isolating the gene in question without recourse to linkage mapping and the assembly of large clone contigs. Conceptually, the simplest method of gene isolation is to generate a genomic library from the mutated plant, and isolate clones containing the insertion element (and several kilobases of flanking DNA) by hybridization with the cloned transposon or T-DNA. However, there is a number of even more convenient methods, which allow the flanking sequences to be isolated in a matter of hours without the need for library preparation.

Plasmid rescue In the simple plasmid rescue technique, the T-DNA or transposon is modified to contain the origin of replication from a bacterial plasmid and an antibiotic resistance marker functional in *Escherichia coli*. Genomic DNA is isolated from a particular mutant plant of interest and cleaved with a restriction enzyme that does not cut in the insertion element or cuts only once, but leaves the origin and selectable marker gene intact. The DNA is then diluted and DNA ligase is added, promoting self-circularization. This mixture of random genomic circles is then used to transform *E. coli*, and bacteria are cultured under antibiotic selection. Only the genomic fragment containing the plasmid sequences will be maintained as a plasmid in the bacterial culture (**Figure 1**).

PCR based approaches The first PCR based technique developed for flanking sequence isolation was inverse PCR. Genomic DNA from a particular mutant plant line is isolated, cleaved with a restriction enzyme as discussed above for plasmid rescue, and self-circularized. PCR amplification is then carried out using primers that specifically anneal to the insertion element (transposon or T-DNA) but point outward rather than inward. In a circular DNA molecule, this will result in the amplification of sequences flanking the insert (**Figure 2**).

A number of more sophisticated PCR approaches is available that do not require circularization of the genomic DNA prior to amplification. These include transposon display and amplification of insertion mutagenized sites (AIMS), both of which involve the digestion of genomic DNA with restriction enzymes, the addition of adaptor oligonucleotides, and several rounds of nested PCR. In transposon display, amplification is exponential and requires a transposon specific primer and an adaptor specific primer. In AIMS, only the transposon specific primer is used and amplification is linear. Another nested PCR technique is thermal asymmetric interlaced (TAIL) PCR, which involves the use of three nested transposon specific primers and a small arbitrary primer. As an alternative to genomic DNA, insertion events can be detected at the mRNA level by rapid amplification of cDNA ends (RACE).

Reverse Genetics – Identifying Insertions in Specific Genes

Two general strategies are available for the identification of inserts in specific genes, one based on high-throughput sequencing and one based on gene-by-gene PCR analysis. In both strategies, the sequences or PCR data must be correlated with a seed bank or collection of plant lines.

The first strategy is the systematic sequencing of all flanking sequences, and their deposition into a

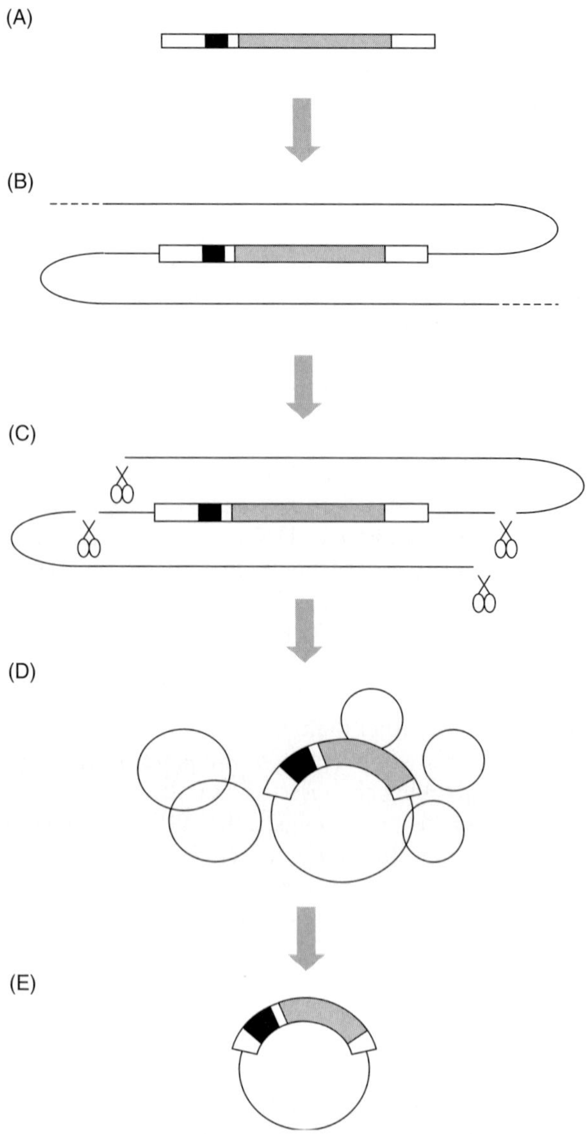

Figure 1 Principle of plasmid rescue technique. (A) Insertion construct contains plasmid origin of replication (black) and bacterial antibiotic resistance marker (gray). (B) The construct inserts into genomic DNA. (C) Genomic DNA is cleaved with restriction enzymes that do not cut in the construct. The construct is liberated with flanking genomic sequences. (D) The cleaved DNA is diluted and ligated to form circles. One of the circles forms a functional plasmid. (E) Transformation of *E. coli* followed by antibiotic selection allows the plasmid, and the genomic sequences it contains, to be cloned.

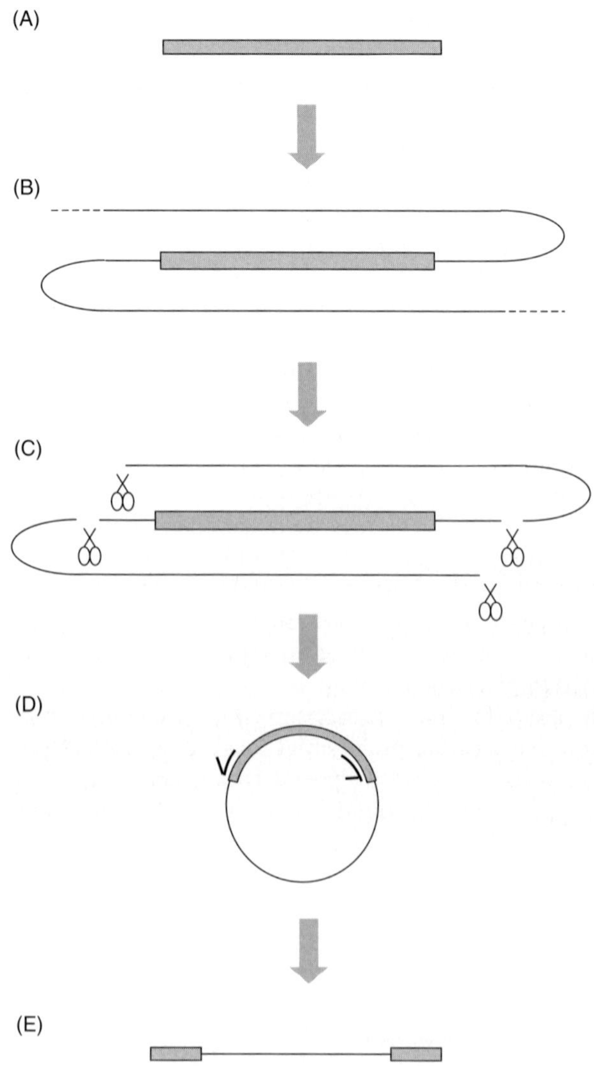

Figure 2 Principle of inverse PCR technique. (A) The insertion construct is shown as a gray box. (B) The construct inserts into genomic DNA. (C) Genomic DNA is cleaved with restriction enzymes that do not cut in the construct. The construct is liberated with flanking genomic sequences. (D) The cleaved DNA is diluted and ligated to form circles. The circle containing the insertion construct is recognized by insertion-specific primers. (E) Amplification using this primer pair allows the flanking sequences to be cloned.

database that can be searched by researchers working on particular genes. Ideally, it should be possible to search the database over the Internet using bioinformatic tools such as BLAST or FASTA. For example, several genome-wide transposon mutagenesis projects have been carried out in *Arabidopsis*, and many of the resulting mutant lines are maintained at the Nottingham *Arabidopsis* Stock Centre in the UK and/or the US *Arabidopsis* Biological Resource Center in Columbus, Ohio. The flanking sequences have been annotated, and the results are available on searchable databases allowing individual researchers to identify inserts in specific genes and obtain the corresponding plants. Another 130 000 plant lines, with T-DNA inserts and associated flanking sequence data, are maintained at the University of Wisconsin *Arabidopsis* Knockout Facility.

The alternative strategy is the maintenance of a genomic DNA bank that can be screened by PCR. Screening is achieved simply by providing one PCR primer representing the gene of interest and one primer representing the insertion element. Typically, DNA samples from seeds are pooled in a hierarchical manner, so that individual PCRs do not have to be carried out on thousands of DNA samples. Such screening facilities are available, for example, at the Pioneer Hi-Bred Trait Utility System for Corn (TUSC) project, and the Maize Targeted Mutagenesis project (MTM). In each case, PCR primers facing away from the transposon are used in combination with a gene specific primer to identify insertions into specific genes (**Figure 3**). The Maize Gene Discovery Project uses a similar approach on a library of plasmids obtained by plasmid rescue with a modified *Mu* transposon called *RescueMu*.

Specialized Insertion Constructs

Another advantage of insertional mutagenesis over traditional random mutagenesis is that the insertion construct can be modified to provide information about the interrupted gene, or even to modify genome structure and activity. Several different types of specialized insertion vector are discussed below.

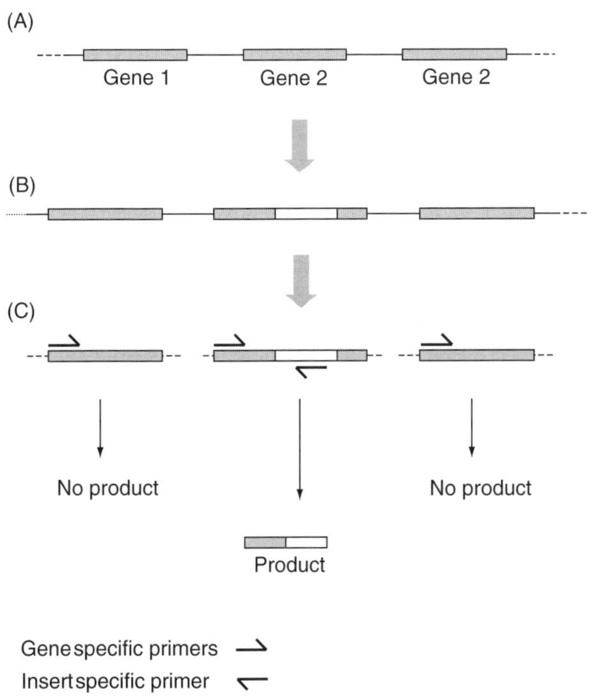

Figure 3 Identification of insertions in specific genes by PCR. (A) A section of genomic DNA is shown containing three genes. (B) An insertion construct (white box) interrupts gene 2. (C) Individual PCRs carried out using one gene specific primer and one insert specific primer generate a product only if the plant contains an insert in the gene being tested.

Enhancer Traps

An enhancer trap is an insertional construct that contains a weak promoter upstream of a reporter gene, such as *gusA* (**Figure 4A**). The promoter is too weak to generate anything other than minimal expression on its own, but if the construct inserts in the vicinity of an endogenous enhancer, this may activate the gene and generate a strong and specific expression pattern. Enhancer traps based on T-DNA vectors have been used in plants, but their usefulness for gene isolation is limited since enhancers may act over distances of hundreds of kilobases.

Gene Traps

A gene trap is an insertional construct containing a reporter gene or even a selectable marker gene that is activated only when the construct integrates within a

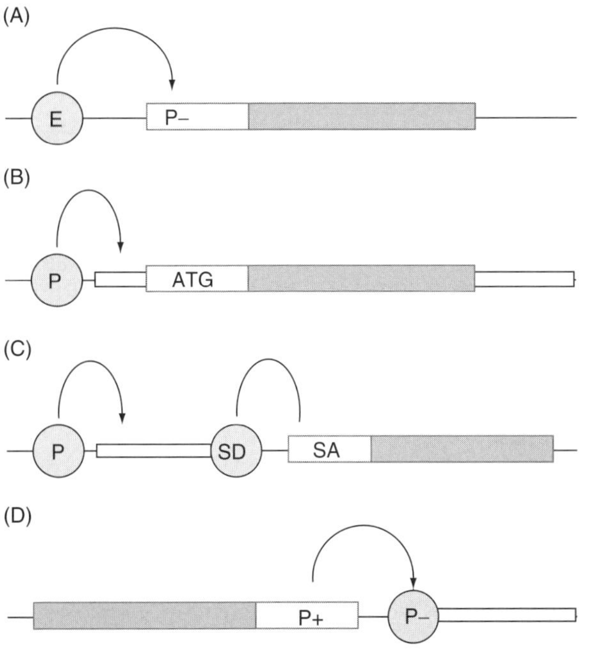

Figure 4 Specialized insertion elements. In each case the insertion element is shown as a wide box, and plant sequences that respond to the insertion are shown as circles. (A) Enhancer trap. The reporter gene (gray box) is driven by a weak promoter (P –). If it inserts in the vicinity of a local enhancer (E) the reporter gene is activated. (B) Promoter trap. The reporter gene (gray box) has no promoter, just an initiation codon (ATG). The reporter can only be activated if it inserts within a plant gene (narrow white box) with an active promoter (P). (C) Splice trap. The reporter gene (gray box) is downstream of a splice acceptor site (SA). This can only be activated if it inserts within a gene and downstream of a splice donor site (SD), enabling it to be expressed as a transcriptional fusion. The gene must also have an active promoter (P). (D) Activation tag. This construct has an outward facing strong promoter (P +). If it inserts adjacent to a gene with a weaker or more restricted promoter (P –) the gene may be upregulated or expressed ectopically.

gene. There are several different types, but all work on the basis that marker gene expression is dependent somehow on the gene's potential or actual activity. The simplest constructs are known as promoter traps and comprise a completely promoterless marker gene, which is activated only if the construct inserts downstream of an active plant promoter (**Figure 4B**).

More sophisticated splice traps (**Figure 4C**) contain an intron and splice acceptor site before the marker gene. In this case, marker gene expression is dependent not only on an active promoter, but also on intron splicing and the inclusion of the marker gene as a transcriptional fusion. Gene traps of this nature have been widely used in plants, and several functional genomics programs have been initiated. For example, a T-DNA gene trap vector incorporating the screenable marker *gusA* has been used to produce 22 000 rice mutants. Over 5000 tagged lines were analyzed for reporter gene expression in leaves and roots, 7000 lines for expression in flowers, and 2000 for expression in seeds. Overall, about 2% of the lines showed marker gene activity, in many cases restricted to highly specific cell types or tissues.

A disadvantage of the current gene trap systems used in plants is that they are dependent on gene activity. Integration into a gene that is not expressed when the plants are analyzed will not yield a reporter phenotype. In the future, this problem could be addressed by the use of more sophisticated vectors in which expression of the marker is driven by its own promoter, but is dependent on the gene for polyadenylation.

Activation Tag Constructs

An activation tag carries a strong, outward facing promoter (**Figure 4D**). Unlike other insertion constructs, which tend to generate loss-of-function phenotypes by gene disruption, the principle of the activation tag is to generate gain-of-function phenotypes by promoting overexpression or ectopic expression of adjacent genes. This strategy has been used, for example, to identify genes regulating alkaloid biosynthesis in the medicinal plant *Catharanthus roseus*. A T-DNA construct containing the strong and constitutive CaMV35S promoter adjacent to the right border was introduced into the plant, and the resulting population of transformants selected for increases in tryptophan decarboxylase activity. This enzyme is a key early component of the alkaloid biosynthesis pathway, and its increased expression is necessary for any increase in downstream alkaloid production. Using this method, a novel transcription factor called ORCA3 was identified.

Monitoring Excision Activity

A transposon construct, in which the transposon is sited between a promoter and a reporter gene, allows excision events to be monitored directly. Where transposition has not occurred, the reporter gene remains silent. However, upon excision of the transposon, the promoter and reporter gene are brought into juxtaposition, and the reporter is activated. This direct screening method can be used in conjunction with PCR based tests to characterize excision activity. An example of green fluorescent protein used as an excision marker for *Ac* in rice is shown in **Figure 5**.

Chromosome Engineering

Transposon mutagenesis can be used to introduce more striking changes to the plant genome. Large chromosomal deletions have been introduced into the *Arabidopsis* genome by introducing a DNA construct containing a *Ds* transposon and two *loxP* sites, which are recognized by the Cre site specific recombinase. The construct also contains a reporter gene and promoter arranged to monitor for excision as discussed above. The design of the construct is shown in **Figure 6**. The critical feature is that one of the *loxP* sites lies within the *Ds* element and one lies without. Integration is random, but excision of the *Ds* element results in preferential reinsertion at a linked site, and carries the *loxP* site with it. In this manner, a large section of genomic DNA is delimited by two *loxP* sites. If Cre recombinase is then supplied, the two *loxP* sites recombine resulting in the deletion of the intervening genomic DNA.

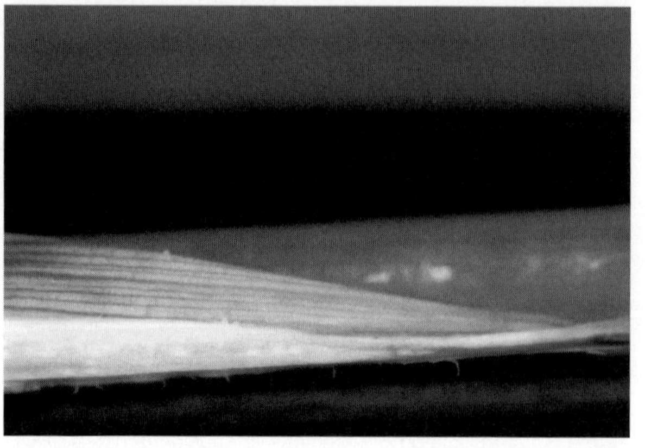

Figure 5 (see color plate 12) Green fluorescent protein used as an excision marker in rice. Rice plants were transformed with a construct in which the *Ac* element is interposed between the *gfp* gene and its promoter. The observed mosaic pattern reflects late excision events affecting restricted clones of cells.

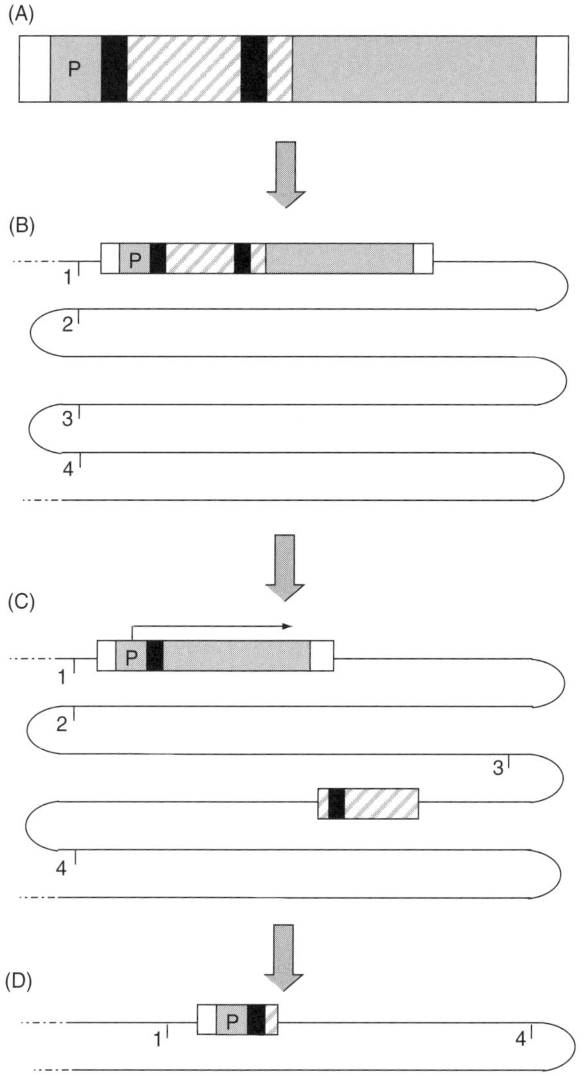

Figure 6 Chromosome engineering with *Ds*/Cre-*loxP* vectors. (A) Construct design. The gray box represents a reporter gene (P is its promoter). Between the promoter and reporter gene is a *Ds* element (hatched box) with one 5′ *loxP* site and one internal *loxP* site (black boxes). (B) The construct inserts into genomic DNA. The ticks and numbers represent physical markers along the chromosome. (C) The *Ds* element excises from the construct and reinserts 10–50 kb away, carrying the *loxP* site with it. The reporter gene is activated (arrow). (D) If Cre recombinase is made available, site specific recombination occurs between the *loxP* sites and deletes a large segment of genomic DNA, as shown by the loss of markers 2 and 3.

Conclusions

Insertional mutagenesis using either T-DNA or transposons is being developed into an increasingly sophisticated palette of tools for functional genomics, providing the means to generate genome-wide libraries of mutants that allow the interrupted genes to be identified rapidly. In combination with other genomic resources, such as genome sequences and EST collections, this provides an efficient route to high-throughput functional annotation. While basic insertion constructs generate loss-of-function phenotypes, additional refinements show how the interrupted genes are expressed, and may even produce gain-of-function phenotypes resulting from overexpression and ectopic expression. The use of such transposon or T-DNA libraries in concert with DNA arrays, expression proteomics, protein structural analysis, and protein interaction mapping provides an unprecedented integrated approach to the functional analysis of plants.

List of Technical Nomenclature

Activation tag	An insertional DNA element containing a strong, outward facing promoter. The function of such an element is to activate neighboring genes, causing dominant overexpression or ectopic expression phenotypes.
Adaptor oligonucleotide	A double-stranded oligonucleotide that fits over the sticky ends left by a restriction digest and provides a primer annealing site.
AIMS	Amplification of insertion mutagenized sites. Nested PCR using a transposon-specific primer for linear amplification.
Autonomous transposon	A transposon encoding its own transposase and able to self-mobilize.
Clean insertion	Insertion of one copy of a DNA element with minimal disturbance to surrounding DNA.
Complex insertion	Insertion of two or more copies of a DNA element and concomitant rearrangement of the surrounding DNA.
Cre recombinase	An enzyme from bacteriophage P1 that catalyzes recombination between short DNA sequences known as *loxP*.
Cut-and-paste	The movement of transposons by excision from the original site and insertion at a new site. Does not increase transposon number unless transposition occurs from replicated to nonreplicated DNA.
Enhancer trap	An insertional element containing a reporter gene downstream of a weak promoter, such as a naïve TATA box. When this inserts in the vicinity of an enhancer, the reporter gene is expressed with the same pattern as the gene normally activated by that enhancer.
Gene space	Gene rich regions of the plant genome.

Gene tagging Tagging an interrupted gene with the insertional DNA element so that it can be identified and cloned.

Gene targeting Introducing a specific change into a preselected gene by homologous recombination.

Gene trap An insertional element containing a reporter gene whose activity is dependent on insertion within a gene.

Homologous recombination Recombination occurring between two similar DNA sequences carried out by enzymes provided by the host plant.

Insertional mutagenesis Mutagenesis by gene disruption with a DNA sequence.

Inverse PCR PCR on a circularized template in which the primers face outward from an insertion sequence and amplify flanking DNA.

Local reinsertion The tendency of active transposons to transpose to sites linked to their origin.

Nested PCR Sequential PCRs in which one or both primers are represented by nested sets placed progressively further inward along the template to increase specificity.

Non-autonomous transposon A derivative of a full-length transposon lacking transposase, but able to mobilize in *trans* if transposase is supplied by an autonomous element in the same genome.

Plasmid rescue The inclusion of a plasmid origin or replication and antibiotic resistance marker in a transposon, so that the flanking genomic region can be isolated and cloned as a bacterial plasmid.

Promoter trap A gene trap activated by insertion downstream of an active promoter.

RACE Rapid amplification of cDNA ends. A PCR technique for the cloning of full length cDNA sequences.

Replicative transposition Transposition in which a copy of the element is left behind at the original site. Leads to increases in transposon number.

Retrotransposon A DNA element that mobilizes using an RNA intermediate.

Signature tagged mutagenesis The same as gene tagging. The insertional DNA element is the "signature" that tags the interrupted gene.

Splice trap A gene trap containing a splice acceptor site activated by incorporation into a transcriptional fusion product.

Starter line The first generation of plants carrying a newly introduced transposon or T-DNA sequence.

TAIL PCR Thermal asymmetric interlaced PCR. Similar to AIMS but using a short arbitrary primer rather than an adaptor specific primer.

Target site preference The tendency for transposons to preferentially integrate at certain target sites, often defined by a weak consensus sequence.

Transgenic plant A plant carrying extra (usually foreign) DNA in every cell.

T-DNA Transferred DNA, the segment of DNA on the Ti-plasmid of *A. tumefaciens*, which is transferred to the plant genome.

Transposase The DNA cutting and joining enzyme that allows transposons to move.

Transposon A DNA sequence that can jump from site to site in the genome.

Transposon display Nested PCR using a transposon-specific primer and a primer annealing to an oligonucleotide adaptor.

Two-component transposon system A system comprising an autonomous transposon and a defective transposon.

See also: **Crop Improvement**: Mutation Techniques. **Genetic Modification**: Gene Cloning, General Principles; Transformation in Dicotyledons; Transformation in Monocotyledons; Transformation in Plastids; Transformation, General Principles. **Growth and Development**: Control of Gene Expression, Post Transcriptional Regulation; Control of Gene Expression, Regulation of Transcription.

Further Reading

Azpiroz-Leehan R and Feldmann KA (1997) T-DNA insertion mutagenesis in *A. thaliana*: going back and forth. *Trends in Genetics* 13: 146–152.

Coelho PSR, Kumar A, and Snyder M (2000) Genome-wide mutant collections: toolboxes for functional genomics. *Current Opinion in Microbiology* 3: 309–315.

Hamer L, DeZwaan TM, Montenegro-Chamorro MV, Frank SA, and Hamer JE (2001) Recent advances in large-scale transposon mutagenesis. *Current Opinion in Chemical Biology* 5: 67–73.

Jeon J-S, Lee S, Jung K-H, *et al.* (2000) T-DNA insertional mutagenesis for functional genomics in rice. *Plant Journal* 22: 561–570.

Kempin SA, Liljegre SJ, Block LM, *et al.* (1997) Targeted disruption in *Arabidopsis*. *Nature* 389: 802–803.

Krysan PJ, Young JC, and Sussman MR (1999) T-DNA as an insertional mutagen in *Arabidopsis*. *Plant Cell* 11: 2283–2290.
Maes T, De Keukeleire P, and Gerats T (1999) Plant tagnology. *Trends in Plant Science* 4: 90–96.
Maize Targeted Mutagenesis Project (MTM). http://mtm.cshl.org.
Nottingham *Aribidopsis* Stock Centre. http://nasc.nott.ac.uk.
Osbourne BI and Baker B (1995) Movers and shakers: maize transposons as tools for analyzing other plant genomes. *Current Opinion in Cell Biology* 7: 406–413.
Parinov S and Sundaresan V (2001) Functional genomics in *Arabidopsis*: large-scale insertional mutagenesis complements the genome-sequencing project. *Current Opinion in Biotechnology* 11: 157–161.
Parinov S, Sevugan M, Ye D, *et al.* (1999) Analysis of flanking sequences from *Dissociation* insertion lines: a database for reverse genetics in *Arabidopsis*. *Plant Cell* 11: 2263–2270.
Ramachandran S and Sundaresan V (2001) Transposons as tools for functional genomics. *Plant Physiology and Biochemistry* 39: 243–252.
The Maize Gene Discovery Project. http://www.zmdb.iastate.edu/zmdb/education/overview.php.
Tissier A, Marillonnet S, Klimyuk V, *et al.* (1999) Multiple independent defective *Suppressor-mutator* transposon insertions in *Arabidopsis*: a tool for functional genomics. *Plant Cell* 11: 1841–1852.
US *Arabidopsis* Biological Resource Center, Columbus, Ohio. http://www.biosci.ohio-state.edu/~plantbio/Facilities/abrc/abrchome.htm.
Walbot V (2000) Saturation mutagenesis using maize transposons. *Current Opinion in Plant Biology* 3: 103–107.
Weigel D, Ahn JH, Blazquez MA, *et al.* (2000) Activation tagging in *Arabidopsis*. *Plant Physiology* 122: 1003–1013.

Transformation, General Principles

H D Jones, Rothamsted Research, Harpenden, UK

Introduction and Historical Perspective

Over a period of two decades, research in plant genetic transformation has progressed from proof-of-concept to a mature, platform technology in plant science. With the increasing availability of tissue-specific, developmentally regulated and constitutive promoters, reporter genes and techniques such as insertional mutagenesis and gene tagging, it offers many opportunities for both basic research in model species and for the modification of economically important traits in crops.

Genetic transformation involves the insertion of nucleic acid molecules, usually in the form of an expression cassette, into plant cells. These molecules are integrated, via illegitimate recombination by the cell's repair machinery, into the nuclear genome and can be either maintained in culture as a cell line, or regenerated into adult plants and transmitted via gametes into subsequent generations. Alternatively, the introduced DNA may be intended for short-term "transient" expression studies. Specialized transformation procedures involving homologous recombination can be used for targeting genes into the genome of plastids.

In the 1970s, progress in plant transformation lagged behind the successes in bacterial, yeast and animal systems. The early attempts to transfer DNA to plant cells targeted protoplasts with *Agrobacterium*, polyethylene glycol (PEG) or electroporation delivery methods. The first reports that foreign genes could be inserted and expressed in cultured plant cells, principally tobacco and petunia, came in the early 1980s. The development of protoplast-to-plant regeneration systems allowed the production of adult transformed plants that transmitted the inserted DNA into the next generation. By the mid-1980s, strategies for *Agrobacterium*-mediated T-DNA delivery and direct gene transfer for several species were available. Since then, many of the major model plant species, as well as human food, animal feed, fiber and ornamental crops, have been transformed by one route or another. This section describes the general principles of genetic transformation and should be read in conjunction with the other sections in this text that focus on specific applications, ethics, etc.

DNA Delivery and Integration

There are two fundamental ways to deliver DNA into the nucleus of target plant cells, either by "direct gene transfer" or by utilizing the capacity of the soil bacterium *Agrobacterium tumifaciens*, the causative agent of crown gall disease. Direct methods utilize a physical or chemical stimulus to deliver a naked, double-stranded DNA molecule into the target cell nucleus. In contrast, *Agrobacterium*-mediated gene transfer involves the transportation of single-stranded T-DNA, with associated virulence (vir) proteins, from bacterial to plant cell via specialized intercellular channels. Once in the plant cell nucleus, the T-DNA is made double-stranded after the displacement of bound vir proteins. The principles underlying these methods are discussed below. Regardless of the method used to deliver DNA into the nucleus, the process of transgene integration into the plant genome appears to be similar. Molecular analysis of integration sites suggests that transgenes insert via double-stranded illegitimate recombination, utilizing the plant's own DNA repair machinery, at a single (or sometimes more than one) locus. The locus may contain one or multiple, concatameric copies of the transgene, which may have undergone rearrangements and/or may have generated short lengths of "filler" DNA homologous to flanking plant genomic DNA. *In situ* hybridization data suggest that physical transgene integration occurs at random within and between plant chromosomes; however, other analyses demonstrate a preference for actively transcribed regions in the plant genome with the possibility that native plant genes are disrupted in the process.

Direct Gene Transfer

A range of direct gene transfer methods have been used to generate transgenic plants, including microinjection, vortexing with silicon carbide whiskers, electroporation, particle bombardment, laser-mediated uptake, PEG-mediated uptake, etc. Each method has been used with particular species, explant sources and regeneration protocols, with varying degrees of efficiency and reproducibility. The most widely used and versatile direct delivery method is particle bombardment (or biolistics), particularly in cereal crops where *Agrobacterium* delivery has proved difficult. This procedure involves accelerating submicron gold or tungsten particles (microcarriers) coated with DNA into regenerable plant tissues. Parameters considered to have a marked effect on successful biolistic-mediated DNA delivery include (i) preparation of the microcarrier suspension, (ii) preparation and precipitation of DNA onto microcarriers, (iii) speed and spread of particles when they hit the explant, (iv) the choice of callus or regenerable plant tissue to be bombarded, and (v) the use of preculture or osmotic shock treatments. Hand-held and bench-top biolistic devices are used extensively for both stable transformation and for studies involving transient gene expression.

Agrobacterium-Mediated DNA Delivery

It was first suspected that *Agrobacterium tumifaciens* was the causative agent for crown gall disease in the early 1900s; however, it was not until the 1970s that the large (Ti) plasmid actually responsible for tumor induction was identified. Another key discovery 10 years later, that only the T-DNA portion of the Ti plasmid actually integrates into the plant genome, led to the development of disarmed (i.e. nononcogenic) *Agrobacterium* vector systems for transformation. Nowadays, the most widely used vectors are based on binary systems in which an expression cassette containing the gene of interest is cloned between the left and right border sequences of an artificial T-DNA. A plasmid containing this T-DNA is then transformed into competent cells of a suitable *Agrobacterium* strain already containing a disarmed "helper" Ti plasmid supplying the *vir* genes. Finally, *Agrobacterium* cells are co-incubated with regenerable plant tissues such as leaf disks, cotyledons, stem segments or callus suspension cultures to allow the bacterial cells to attach and form a transporter complex which mediates T-DNA import. This step can be aided by physical stimuli such as wounding, application of vacuum, or the addition of phenolic inducers such as acetosyringone.

In Planta Transformation versus Regeneration via Tissue Culture

Assuming DNA has been successfully delivered and integrated into the genome of a target plant cell, there remains a significant challenge to recover whole, fertile, nonchimeric plants. Most transformation systems require a tissue culture stage, which is labour- and materials-intensive and increases the chance of genetic instability due to unpredictable somaclonal variation. There has been considerable interest in various *in planta* transformation methods because of their simplicity and lack of any tissue-culture requirement. The basic principles of the two approaches are described below.

In Planta Transformation

This concept depends on the ability to introduce naked or T-DNA into gametes, pregametic tissue or zygotes around the time of fertilization. Attempts at injecting, electroporating or bombarding DNA into pollen, floral tissues or germline meristems of various species have proved to be generally unreproducible. However, one notable exception is the success of *Agrobacterium*-mediated germline transformation of *Arabidopsis*. This is now the universal method of choice for *Arabidopsis* transformation, and involves applying *Agrobacterium* directly to the mature flowers with the aid of a surfactant or vacuum. Seeds from the treated plants are germinated on selective media, and although efficiencies are low (less than 1%), the large number of seeds obtained, and ease of screening, make this a reproducible, facile method.

Regeneration via Tissue Culture

For species other than *Arabidopsis*, successful transformation depends on the remarkable plastidity and totipotency of plant cells to recover adult plants from transformed somatic cells. There are two principal routes by which plants can be recovered via tissue culture: somatic embryogenesis and organogenesis. Some species, such as sugar beet (*Beta vulgaris*), can be regenerated via either route, but with most species one prevails as the method of choice. Somatic embryogenesis is a nonsexual propagation process where somatic cells differentiate into embryo-like structures with shoot and root meristems. With appropriate phytohormones and other culture medium additions, somatic embryos can be "germinated" and give rise to viable adult plants. In most species, somatic embryos are derived indirectly from callus, cells, but in a few cases, somatic embryos can be produced directly with no callus phase, for example, from citrus (*Citrus* spp.) ovaries or cassava (*Manihot* spp.) leaves. Organogenesis refers to the ability of some plant tissues (e.g., hypocotyls, cotyledons, leaf bases or callus derived from them) to reorganize into shoot meristems, which can subsequently be rooted and generate complete plants. If successful rooting is not possible, then grafting onto a root-stock is possible. Selection pressure is applied at appropriate stages to inhibit the growth of nontransformed tissue. **Figure 1** summarizes the main tissue culture routes through which plants can be recovered, and the stages suitable for DNA delivery.

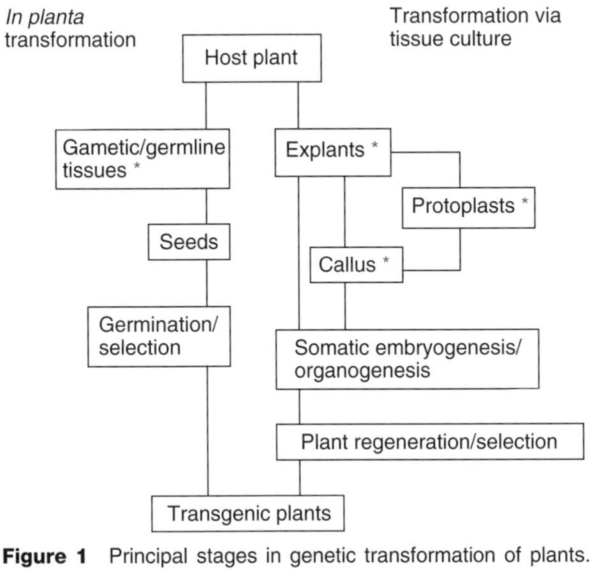

Figure 1 Principal stages in genetic transformation of plants. Asterisks denote tissues suitable for DNA delivery. The left-hand route denotes *in-planta* transformation, and the right-hand route, transformation via tissue culture.

Occasionally, genetic instability may result in somatic cells as a consequence of cell/tissue culture, referred to as "somaclonal variation." This has been identified in many early tissue culture protocols, and has been widely exploited (e.g. in potato (*Solanum tuberosum*)) as a source of variation for breeding material. Factors that can promote genetic instability in micropropagated plant cells, such as high levels of 2,4-D and long culture periods, are now actively minimized in transformation methods.

Selection

Selectable marker genes are a vital part of most transformation protocols. They are delivered alongside the gene of interest, either on the same plasmid or on a separate plasmid. A wide range of selectable marker regimes is available and is particularly important in species where transformation efficiencies are low. Selectable marker genes can be categorized into those based on resistance genes that confer the ability to grow in the presence of toxic compounds such as antibiotics or herbicides which kill or otherwise compromise untransformed tissue (negative selection). Alternatively, a range of positive selection systems are available which provide transformed tissues with an enhanced ability to utilize, for example, an unusual carbohydrate or amino acid supply and thus enrich the culture for transformed tissue expressing the marker gene. Concerns about the environmental spread of selectable markers, particularly antibiotic resistance genes, from field-planted GM plants prompted the development of approaches to eliminate selectable markers. One concept is to co-transform with a gene of interest and a marker gene at nonlinked loci, thus the primary transformant contains both genes, but genetic segregation in the progeny plants allows identification of individuals lacking the marker but possessing the gene of interest. This concept can be augmented to include co-integrated marker and trait genes by incorporating a transposition system in the vector design. Thus, the marker gene can be separated from the gene of interest by transposition in later generations and then removed by segregation. Yet other approaches, such as *cre/lox* or *flp/frt*, use excision/recombination to physically remove selected sequences after transformation.

Predictability of Transgene Inheritance and Expression

In most cases, the transgene insertion is inherited by subsequent generations as a dominant trait in predictable Mendelian ratios, depending on locus number and linkage (3:1 for a single locus, 15:1 for two unlinked loci, etc.). However, distorted inheritance ratios have also been observed, which may persist or revert to Mendelian in subsequent generations. Possible causes of distorted segregation patterns include chimerism in the primary transformant, counterselection of the chromosome containing the transgene, or disruption of plant genes involved in DNA replication, gametogenesis, fertilization, etc. In addition, variation in levels of gene expression can occur between individuals transformed with a given expression cassette. Factors that can influence this include copy number, integration pattern, and interactions between foreign and native DNA that vary depending upon the precise point of insertion (so-called "position effect"). The traditional strategy to overcome nonMendelian inheritance ratios and unpredictable levels of expression has been to create a large number of individual transgenic events, and then to select suitable individuals for further study. However, there is also considerable interest in modifications that increase the predictability of transgene expression. For instance, targeted homologous recombination in the nuclear genome has been demonstrated at low frequency in some species, and the use of matrix attachment regions to insulate the transgene may modulate variability of expression.

Applications of Transformation for Basic Research and Crop Biotechnology

The *Arabidopsis* genome has been completely sequenced, but analysis of the resulting DNA and protein sequences has failed to identify the function of most of the genes. The challenge of functional genomics is to ascribe functions to the vast amount of sequence data becoming available. Genetic transformation can be used for functional genomics research, both in forward genetics studies (to go from phenotype to sequence) and reverse genetics studies (from sequence to phenotype). Depending on the scientific rationale, a wide range of strategies involving transformation can be used, including (i) the overexpression or silencing of native or heterologous genes, (ii) the use of reporter genes under the control of specific promoters, either in stable or transient expression, and (iii) the generation of a randomly tagged population for gene or promoter identification. In addition to basic research, transformation is also a platform technology for the introduction of novel traits and improvement of commercial cultivars. As an adjunct to conventional plant breeding, transformation has the potential to increase crop yield, reduce chemical inputs, or provide the means by which pharmaceuticals or nutraceuticals can be made in "plant factories." The principle of these strategies is described below and summarized in **Figure 2**.

Overexpression/Gene Silencing

The most common application of genetic transformation is for the functional analysis of a cloned gene. This can be achieved by the overexpression of the sense strand to generate more protein, and/or downregulation via, for instance, antisense RNA expression, co-suppression or other gene silencing mechanisms. Both these approaches can be "constitutive," or targeted to specific tissues or developmental stages by the careful choice of promoter (see below). The choice of background species or wild-type/mutant genotype provides another layer of experimental choice.

Targeting Gene Expression

Many biological processes are controlled, spatially and temporally, at the level of transcription. The core promoter, along with enhancer elements and other *cis*-acting regulatory sequences, are nontranscribed DNA sequences that form part of the mechanisms that regulate this gene expression. The choice of promoter is critical for efficient transgene expression, and the scope of experimental design is improved by the availability of a well-characterized "catalog" of promoters that can be used to "drive" target gene expression in particular organs or developmental stages of the species in question. The availability of well-defined inducible, tissue-specific or developmentally regulated promoters has been a bottleneck in the design of transformation experiments in the past, but screens of promoter-tagged populations have begun to identify more. The use of promoter::reporter gene fusions in transgenic plants continues to be a powerful tool in the study of regulated gene expression.

Reporter Genes

Reporter genes encode products that can readily be detected in transformed tissue, and are a powerful tool for the investigation of *cis*- or *trans*-acting gene regulatory factors in both transient and stable transformation experiments. In an ideal reporter system the gene product should be nontoxic, and the assay should be facile, *in situ*, nondestructive and quantifiable. Unfortunately, no single gene meets all of these requirements, each having specific strengths and limitations. Examples of commonly used reporter systems include GUS, GFP, luciferase and anthocyanin, and depending on the purpose of the

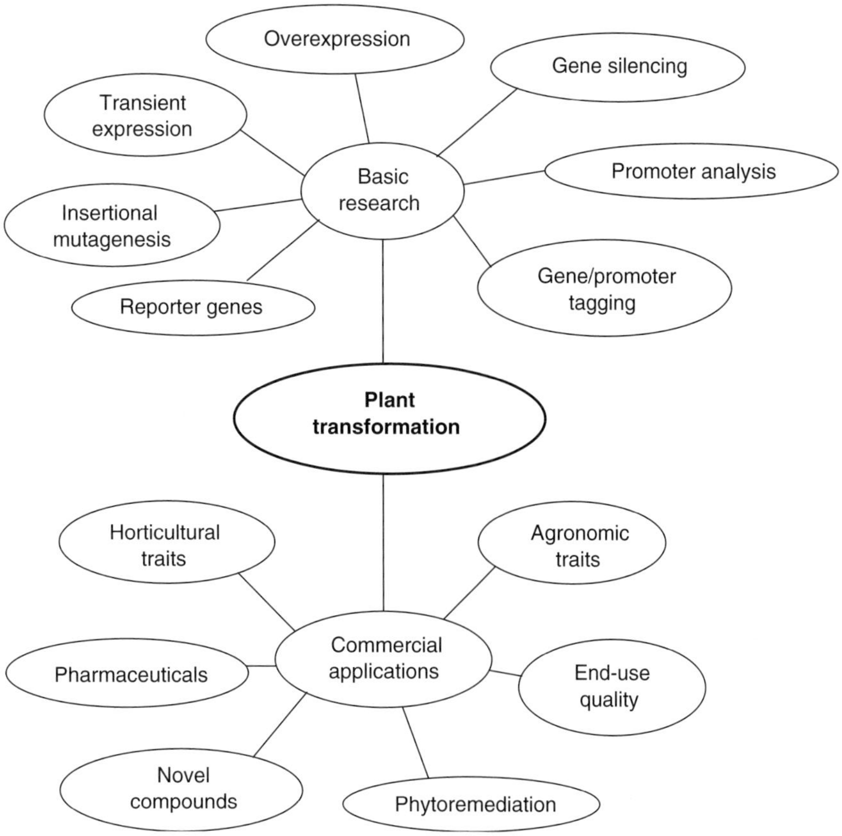

Figure 2 Applications of plant transformation: strategies for basic research and target traits for commercial applications.

experiment, these may be used as a transcription tag, a transcriptional fusion or a translational fusion. Reporter genes have been successfully used to characterize native and heterologous gene expression as well as protein trafficking.

Insertional Mutagenesis

The disruption of a random gene by the insertion of a known segment of DNA, and the subsequent observation of a mutant phenotype, has been used to great effect in several plant species. The inserted sequence not only knocks out or modifies the expression of the host gene, but also provides a known "tag" in the gene that facilitates the cloning and sequencing of flanking sequences, and therefore the identification of the disrupted gene. In species where the generation of large numbers of primary transformants is facile, particularly *Arabidopsis*, direct T-DNA insertions can be used. T-DNAs are thought to insert randomly within the genome, and it has been estimated that 100 000 independently transformed lines would be more than enough to saturate the *Arabidopsis* genome. Large T-DNA collections have been generated and systematic "reverse genetic" screens attempted. In other species where transformation is more difficult, transposons can be utilized. Transposons are short mobile genetic elements that can jump from one locus to another within the plant genome. The most commonly used transposons originally came from corn (*Zea mays*; maize), e.g. *Ac-Ds* (Activator-Dissociation), although other species, such as petunia (*Petunia* spp.) and snapdragon (*Antirrhinum* spp.), also possess several endogenous transposons. The excisable element can, for example, be engineered to flank a conventional expression cassette containing a scorable or screenable marker, and used to create a large number of tagged knockouts in a heterologous plant host by transformation and transposition.

Gene/Promoter Tagging

A major limitation of insertional mutagenesis is that it fails to generate a phenotype in the many genetically redundant (although functional) genes that exist in any organism. In addition, there is considerable interest in isolating regulatory sequences with specific expression patterns for use in studies involving heterologous tissue-specific or developmentally regulated overexpression. Various alternative approaches to highlight such genes and promoters are enhancer traps, promoter traps, gene traps and activation traps. These methods rely on either the activation of introduced marker genes by native *cis*-acting regulatory sequences, or the activation of native genes by introduced regulatory sequences. An elegant extension of the enhancer trap approach is a two-component system incorporating GAL4 fused to a minimal promoter which, in turn, drives GFP expression under the control of the GAL4 activating sequence. Work in *Arabidopsis* has shown that lines with highly specific patterns of GPF expression can be obtained, and that other genes fused to the GAL4 activating sequence show the same patterns when transformed into these lines.

Transient Expression

The production of stably transformed plants is not a trivial process and some biological questions can be usefully addressed by using short-lived "transient expression" from noninherited, extrachromosomal foreign DNA sequences. Transient expression typically lasts for 1–7 days and can be induced from conventional transformation vectors or from viral vectors that are capable of systemically spreading through the plant after mechanical inoculation. It can be used as a rapid screen for assessing promoter activity, or comparing the effects of induced point mutations in a gene, etc. Facile cloning methods to generate viral vectors make large-scale screening of whole cDNA libraries possible.

See also: **Ethics and Biosafety**: Development and Commercialization of Genetically Modified Plants; Ethics of Genetically Modified Crops. **Genetic Modification**: Transformation in Plastids. **Tissue Culture**: General Principles. **Tissue Culture and Plant Breeding**: Somaclonal Variation.

Further Reading

Barcelo P, Rasco-Gaunt S, Thorpe, C, and Lazzeri PA (2001) Transformation and gene expression. *Advances in Botanical Research Incorporating Advances in Plant Pathology* 34: 59–126.

Galun E and Breiman A (1997) *Transgenic Plants*. London: Imperial College Press London.

Gelvin SB and Schilperoort RA (eds) (2000) *Plant Molecular Biology Manual*. Dordrecht: Kluwer Academic Publishers.

Hall RD (ed.) (1999) *Plant Cell Culture Protocols*. Totowa, NJ: Humana Press.

Hansen G and Wright MS (1999) Recent advances in the transformation of plants. *Trends in Plant Science* 4: 226–231.

Lindsey K (ed.) (1998) *Transgenic Plant Research*. New York: Harwood Academic.

Potrykus I and Spangenberg G (eds) (1995) *Gene Transfer to Plants*. Berlin: Springer.

Reynolds PHS (ed.) (1999) *Inducible Gene Expression in Plants*. Wallingford: CABI Publishing.

Springer PS (2000) Gene traps: tools for plant development and genomics. *Plant Cell* 12: 1007–1020.

Zupan J, Muth TR, Draper O, and Zambryski P (2000) The transfer of DNA from *Agrobacterium tumifaciens* into plants: a feast of fundamental insights. *Plant Journal* 23: 11–28.

Transformation in Dicotyledons

D J Murphy, University of Glamorgan, Cardiff, UK

The production of transgenic crops relies on three distinct technologies:

1. The isolation and manipulation of small sequences of DNA (often called gene cloning).
2. The insertion of such exogenous DNA sequences into plant tissues (i.e., transformation).
3. The selection of transformed cells or tissues and their regeneration into adult plants capable of further propagation.

Early attempts to transform plants with exogenous DNA focused on the integration of naked DNA, normally in the form of plasmids, into isolated plant cells. Typically, the cells were pretreated with cell wall digesting enzymes to create protoplasts before being shocked, e.g., by electrical treatments or reagents like EDTA, in order to allow the DNA to enter transient pores in their plasma membranes. However, these methods proved to be relatively inefficient and unreliable. The real breakthrough in plant transformation technology came with the introduction, in the early 1980s, of *Agrobacterium* vector systems. The great advantage of *Agrobacterium* vectors was that they not only delivered DNA into plant cells at reasonably high efficiency, but they also facilitated the incorporation of the exogenous DNA into the genome of the host plant cell.

For most of the first decade or so of plant transformation research and development, applications tended to be limited to dicotyledenous crops. There were two main reasons for this; firstly, the natural host range of the common *Agrobacterium* vectors was restricted to certain dicotyledons; and, secondly, the easiest plants to regenerate from transformed cells of tissue slices were also dicotyledons. Hence, much of the early work on plant transformation was concentrated on Solanaceous crops such as tobacco (*Nicotiana tabacum*), tomato (*Lycopersicum esculentum*), and potato (*Solanum tuberosum*). Later, in the 1980s and early 1990s, other important dicotyledonous crops, especially many of the brassicas, were successfully transformed, although their transformation is often highly genotype dependent. In practice, this means that one particular variety of a brassica species may be used for the initial transformation – in the case of oilseed rape (*Brassica napus*; canola), this is often the variety "Westar" – but the transformed plants then have to be backcrossed into more agronomically suitable elite breeding lines for commercial production. Hence, the process can be slower and more costly than if the elite lines themselves could be used for direct transformation.

Nowadays, the use of improved strains of *Agrobacterium* and the refinement of direct DNA delivery via projectile mediated techniques (biolistics) means that virtually all dicotyledonous species can be transformed with reasonable efficiency. Nevertheless, significant challenges remain before the transformation of dicotyledons is optimized, with regard to ease, efficiency and predictability. Transformation of plants is still fairly crude and inefficient compared with, for example, the transformation of microbes or animals. Transgenes in plants insert more or less randomly into the genome, the number of copies inserted can range from one to a dozen or more, and the introduced DNA sequences are frequently modified and/or fragmented in the process. For example, pieces of DNA may lose part of one or both of their end regions, or they may be cleaved at several points and the resulting fragments inserted into different parts of the genome. Both the position of insertion in the genome and the copy number of a transgene can significantly affect its expression in the resulting plant. The presence of multiple copies of a transgene can result in instability of its expression. Finally, transgenes are normally inserted into the recipient genome as part of a multigene construct that also contains regulatory elements and a selectable marker, often an antibiotic or herbicide resistance gene. There is already public concern about the use of antibiotic resistance markers, and many researchers and breeders are also concerned that, as transgenic crops become more widespread, they will inevitably be subject to many further rounds of transformation as additional genes are inserted in order to keep improving the crop. The accumulation of many different selectable marker genes would soon present problems as breeders ran out of new benign markers for further rounds of transformation. The continued use of the same marker can lead to cosuppression and the loss of expression of the transgene itself, and also of genes with related sequences.

Although there are important challenges in optimizing transformation technologies, there has been much progress in addressing these challenges in research laboratories around the world. These include the development of excisable markers, the use of plastid transformation to achieve precise transgene insertion, and the use of novel selection systems that do not involve antibiotic resistance genes. In the next decade, dicotyledon transformation should become much more reliable, as well as cheaper and easier. This may facilitate its uptake in developing countries where, due to its cost and complexity, the current transformation technology remains largely out of reach of their scientists.

See also: **Genetic Modification**: Gene Cloning, General Principles; Transformation in Monocotyledons; Transformation in Plastids; Transformation, General Principles.

Further Reading

Bent AF (2000) *Arabidopsis* in planta transformation. Uses, mechanisms, and prospects for transformation of other species. *Plant Physiology* 124: 1540–1547.

Birch RG (1997) Plant transformation: problems and strategies for practical application. *Annual Review of Plant Physiology and Plant Molecular Biology* 48: 297–326.

Crispeels MJ and Sadava DE (2003) *Plants, Genes and Crop Biotechnology*. Sudbury, MA: Jones & Bartlett.

Daniell H, Muthukumar B, and Lee SB (2001) Marker free transgenic plants: engineering the chloroplast genome without the use of antibiotic selection. *Current Genetics* 39: 109–116.

De la Riva GA, González-Cabrera J, and Vázquez-Padrón R (1998) *Agrobacterium tumefaciens*: a natural tool for plant transformation. *Electronic Journal of Biotechnology* 1: no. 3: http://www.ejb.org/content/vol1/issue3/full/1/#51

Newell CA (2000) Plant transformation technology: Developments and applications. *Molecular Biotechnology* 16: 3–65.

Transformation in Monocotyledons

P A Lazzeri and P Barcelo, DuPont Wheat Transformation Laboratory, Barcelona, Spain

Introduction

Many of the world's most important crop plants are monocotyledonous. These include temperate and tropical cereals, which are the single crop group most exploited for human and livestock nutrition, and other major groups of crops, such as forage and turf grasses, palms, and vegetable and ornamental bulb species. Thus, there are many opportunities for applying transformation technology in monocotyledon crop improvement, both as a basic research technique and as a tool in monocotyledon crop breeding.

The first demonstrations of plant transformation were made in the early to mid 1980s using dicotyledon species and soon after this attention turned to the major cereal crops such as corn (*Zea mays*; maize), wheat (*Triticum* spp.), and rice (*Oryza sativa*). However, it quickly became apparent that the early transformation techniques, which were based either on *Agrobacterium* as a vector or on DNA uptake into protoplasts, worked well in dicotyledon species such as tobacco (*Nicotiana tabacum*) and *Petunia*, but did not function in monocotyledon species. Monocotyledons, and the cereals in particular, were much more difficult to culture and regenerate *in vitro* and were not naturally susceptible to infection by *Agrobacterium*. Thus began more than a decade of research aimed at the development of efficient transformation technology applicable in monocotyledon crop species. This work, which focused on overcoming the recalcitrance of cereal species to transformation, resulted in the invention of a range of novel tissue culture and gene transfer methodology.

Owing to the great economic importance of the major monocotyledon crops, much of the research on transformation technology has been done in industry, although academic scientists have also made major contributions. The field has progressed very quickly, so that the time span between the development of new transformation methods and their application to produce genetically modified (GM) commercial varieties has only been a few years. Today, routine transformation methods are available for all the major cereals and grasses and most of the other important monocotyledon crops can be transformed, although bulb species such as onion (*Allium cepa*) and garlic (*Allium sativum*) are still recalcitrant and further development in the technology is required.

In this article, the transformation procedures commonly used for monocotyledons are introduced; the target tissues, culture and selection systems, and the gene transfer technologies available are considered. The vectors used as vehicles for DNA delivery and the way in which transgenes are integrated into the host genome, are expressed, and inherited are examined. Finally, the status of application of transformation in monocotyledon crops and the prospects for future developments in this field are discussed.

Transformation Target Tissues and Culture Systems

The basis of all transformation systems is the integration of a new gene into the genome of a recipient cell, followed by the recovery of a whole plant from the transformed cell. Under normal circumstances, there are a limited number of cells in a plant that can give rise to new plants. These are the germline cells in meristems, which form floral organs, or shoot meristems, which may be induced to root and thus produce new plantlets. Efforts have been made to develop transformation systems that deliver genes into germline cells in whole plants, but until the very recent development of inflorescence infiltration techniques there had been little success (see section below). Therefore, the majority of established transformation techniques target cells in cultured tissues, which are more accessible to gene delivery.

There are clear differences between the response to tissue culture of monocotyledon and dicotyledon plants. In dicotyledon species, it is generally possible to take explants from a range of different organs, such as leaves, stems, or floral parts, and to put these into culture and regenerate plants. In contrast, in monocotyledons, a much more limited range of tissues may be used to establish cultures and regenerate new plants. With few exceptions, these explants are taken from immature plant parts whose cells are either still dividing or have recently ceased division. The explants that are most frequently used for monocotyledon transformation systems are immature embryos and inflorescences, immature leaf tissue, isolated shoot meristems and cultured anthers or microspores (see **Table 1**). The use of immature tissues as transformation targets has the disadvantage that a constant supply of donor plants must be grown in order to supply explants. In addition, the physiological status of the donor plants has a major effect on the response of explants in culture and to achieve efficient and reliable transformation it is necessary to grow these plants under controlled environment conditions and free from pathogens. This, together with laborious manual isolation of explants, makes the production of target tissue costly. To avoid these constraints, transformation techniques targeting cultures derived from mature seeds have recently been developed for rice and there is interest in applying this simple approach in other monocotyledon species.

Table 1 Target tissues and culture systems used for monocotyledon transformation

Target tissue	*Advantages*	*Disadvantages*
Microspore	May target haploid cells, many target cells available	Inefficient transformation and regeneration
Immature embryo	Good target tissue, efficient plant regeneration	Need constant supply of donor plants
Immature inflorescence	Good source of target tissue	Regeneration less efficient than from embryos
Seedling leaf base	Good source of target tissue from young seedlings	Low efficiency transformation and regeneration
Shoot meristem	Good source of target tissue from young seedlings	Low efficiency transformation and regeneration
Embryogenic callus	Good target tissue, callus lines may be maintained	Difficult to establish lines with good regeneration capacity
Cell suspension	Renewable tissue source, good target for transformation	Regeneration efficiency often low, abnormal regenerants
Protoplast	High efficiency transformation possible	Regeneration often very difficult, abnormal regenerants common

A further aspect of the choice of target tissue for transformation is that it is possible to deliver the transforming DNA into either primary or secondary cultures. Primary cultures are the original explants isolated from the donor plant, which may be targeted immediately after isolation, or cultured for one to several days before transformation. Secondary cultures are tissues derived from proliferation and growth of the original explant, which have been subcultured one or more times. Examples of secondary cultures are callus lines, propagated meristem cultures, cell suspension cultures, and suspension cell protoplasts. Secondary cultures have the advantages that they can provide large numbers of target cells from the proliferation of the original explant tissue and it may be possible to visually select tissue types or cell lines that are amenable to transformation and regeneration. However, in many monocotyledons it is difficult to establish secondary cultures that retain a good capacity for plant regeneration. This problem is most acute in long-term secondary cultures, which typically accumulate genetic aberrations such as chromosome number changes, so that plants regenerated from them are abnormal.

Gene Delivery Methods

There are two systems employed for transformation of plants: the use of *Agrobacterium* as a biological gene transfer vector, or the use of physical or chemical methods to effect DNA delivery, termed direct gene transfer (DGT) techniques. Most monocotyledon species are naturally recalcitrant to infection by *Agrobacterium* and early research attempting to transform monocotyledons using techniques developed in dicotyledon species showed little success. As a result, emphasis was given to using DGT techniques for these species.

The transformation techniques applied in monocotyledons are evaluated in **Table 2**. As the table shows, a number of DGT techniques have been used to achieve transient gene expression or occasional stable transformation in monocotyledons, but only four techniques are used routinely for the production of transgenic plants. These techniques are protoplast transformation, tissue electroporation, silicon carbide fiber vortexing, and particle bombardment.

Protoplast Transformation

This was the method used to generate the first transgenic monocotyledon plants. It is based on the use of chemical or electrical stimuli to cause the uptake of DNA into protoplasts derived from suspension cultures. While the technique can give high transformation efficiencies, it has the disadvantages that suspension cultures are difficult to produce in most monocotyledons, regeneration from protoplast cultures is poor and regenerants may be variant, and protoplast transformants often have complex integration patterns, leading to unreliable transgene expression (see section below).

Tissue Electroporation

This transformation method uses high-voltage electrical pulses to create temporary pores in the cell membranes of tissue pieces suspended in a buffer containing DNA. This allows DNA uptake into the cells, which are subsequently cultured and regenerated into plants. The method has been shown to function in several monocotyledon species, but has only been used for reproducible production of transgenic plants in rice and corn.

Silicon Carbide Fiber Vortexing

In this technique, plant tissue and microscopic silicon carbide fibers are vortexed together in a buffer

Table 2 Gene transfer techniques and evaluation of their use in monocotyledons

Technique	*Evaluation*
Imbibition of tissues in DNA solutions, injection of DNA into plant organs	Occasional transient gene expression, no confirmed stable transformation
Ultrasound treatment of tissues in DNA-containing buffer solution	Transient gene expression, stable transformation not confirmed
Laser-mediated puncture of plants cells in DNA-containing buffer solution	Transient gene expression, stably transformed callus but no transgenic plant
Microinjection of cells with DNA solution	Transient gene expression, transgenic plants produced in limited species, but technique technically very difficult
Vortexing of tissues with silicon carbide fibers in DNA-containing buffer	Transgenic plants produced from suspension cells in a few species, but not suited to multicellular explants
DNA uptake into protoplasts	Transgenic plants produced in cereals, forage grasses, and other selected monocotyledon species. Technically demanding, plants recovered may be variant
Electroporation of cells or tissues	Transgenic plants produced in several species, but with limited transformation efficiency. Applicable to particular cell/tissue types only
Particle bombardment	DNA delivery to all cell types in all plants. The standard transformation technique for most monocotyledon species; widely applied in commercial GM crop breeding
Agrobacterium transformation	Applied in the major cereal species; use being extended to other monocotyledons, but requires specific culture types. Can be highly efficient in amenable varieties. Used in commercial GM crop breeding

containing DNA. The cells are punctured by the fibers allowing DNA uptake and transformation. The technique is most efficient when applied to suspension culture cells, which have thin walls and present a high surface area for targeting, and is less suitable for transforming multicellular explants. It has been applied in monocotyledons, such as corn and rice in which regenerable suspensions are available.

Particle Bombardment

The development of particle bombardment technology first made the routine transformation of monocotyledon species possible. This is still the method most widely used for monocotyledon GM, although in cereals such as rice, corn, and barley (*Hordeum vulgare*), *Agrobacterium* techniques are being used increasingly. In particle bombardment, dense microscopic particles, usually of gold, have DNA precipitated onto them and are then accelerated into plant cells. DNA is subsequently released from the particles, allowing transformation to take place. A number of different particle acceleration systems have been developed, using explosive discharge, air, gas, or electrostatic propulsion, but the standard device used in most laboratories is the PDS-1000He particle gun, which uses helium gas as propellant (see **Figure 1**). A major advantage of particle bombardment for gene delivery is that it is essentially species and tissue independent, although the recovery of

Figure 1 The PDS 1000 helium-powered particle gun, the instrument most commonly used for DGT transformation of monocotyledons. Photo courtesy of Huw Jones.

transgenic plants obviously requires the ability to culture and regenerate the targeted cells.

Agrobacterium Transformation

After more than a decade of research, the breakthrough in the development of viable *Agrobacterium*-mediated transformation in monocotyledons came when it was shown in 1994 by Hiei and colleagues that japonica rice callus cultures could be transformed at high efficiency using bacterial strains containing highly active virulence genes. This research, and following studies, demonstrated that although monocotyledon cells do not exhibit the wound response processes associated with the *Agrobacterium* interaction in dicotyledons, cultured monocotyledon cells are susceptible to infection and the gene transfer mechanism functions as in normal host plants. Following the *Agrobacterium*-mediated transformation of rice, the technique has been applied in several other cereals, including corn, barley, wheat, and sorghum (*Sorghum* spp.) It seems that when the correct type of culture system can be developed, which is typically an actively developing embryogenic callus culture, and these cultures are exposed to *Agrobacteria* containing appropriate virulence functions, transformation is possible in a number of monocotyledon species. Under optimal conditions, *Agrobacterium* transformation in monocotyledons can be highly effective and up to 25–50% of treated calluses may yield transgenic plants in amenable varieties of rice and corn. Although the mechanisms of gene delivery by *Agrobacterium* and in particle bombardment are clearly quite different, it is notable that the same monocotyledon species and varieties are either amenable or recalcitrant to transformation by either method; for example, japonica rice varieties are good subjects, while bulbaceous monocotyledons, such as onion or tulip (*Tulipa* spp.), are difficult targets. This suggests that DNA delivery is not the only limiting factor and that the capability for manipulation and regeneration *in vitro* is of primary importance. In this context, the recent demonstration that rice may be transformed by *Agrobacterium* infiltration into developing inflorescences (panicles) of intact plants, a technique first developed in *Arabidopsis*, is a significant advance, as it avoids the need for tissue culture.

A general feature of *Agrobacterium* transformation, in comparison with DGT techniques, is that transgene insertions tend to be simple, with low transgene copy numbers. This is a positive characteristic for the reliability of transgene expression and for handling transgenic lines in breeding programs. In consequence, there is general interest in moving toward *Agrobacterium* techniques for commercial transformation of monocotyledon crop species.

Transformation Vectors for Gene Delivery

The primary function of a plant transformation vector is to deliver genes into a recipient genome. The "active components" of a transformation vector are similar, regardless of the transformation method (i.e., *Agrobacterium* or DGT), and are promoter, structural gene, and terminator sequences. The promoter is responsible for controlling expression of the gene, the structural gene codes for the new protein/enzyme, and the terminator sequence terminates transcription of the transgene unit. The promoter–gene–terminator unit is cloned in this order into a DGT or *Agrobacterium* plasmid DNA molecule, whose features and functions are detailed below (**Figure 2** shows the structures of typical transformation vectors).

Vectors for DGT transformation are based on circular DNA plasmids and have three major features. The first feature is a sequence that allows the insertion (cloning) of the promoter–gene–terminator sequences described above. This feature is termed a multiple cloning site, which includes target sites for several "rare-cutter" restriction enzymes ideal for cutting and pasting desired DNA sequences into the site. The second and third features of the vector allow it to be propagated (amplified) to provide the relatively large quantities of DNA needed for DGT transformation procedures (for example, a single particle gun discharge may use 0.3 μg of DNA and an experiment may comprise 20 "shots"). For plasmid amplification, an origin of replication of the plasmid in the host bacterial cells is required. The third feature is a bacterial selectable marker gene, to ensure only bacteria containing the transformation vector are proliferated. The most common marker genes confer resistance to antibiotics such as ampicillin. There is concern about the incorporation of antibiotic marker genes in DGT vectors as they will subsequently be present in the genome of transgenic plants. It has been hypothesized (but not demonstrated) that horizontal gene transfer might occur, leading to their appearance in microorganisms in the wider environment. This has motivated changes in the design of transformation vectors. One strategy substitutes antibiotic resistance genes for other markers such as genes complementing a metabolic mutation in the bacterial genome, so that only bacterial cells containing the complementary gene are able to grow on minimal culture medium. A second strategy uses restriction enzymes to isolate the functional promoter–gene–terminator fragment from

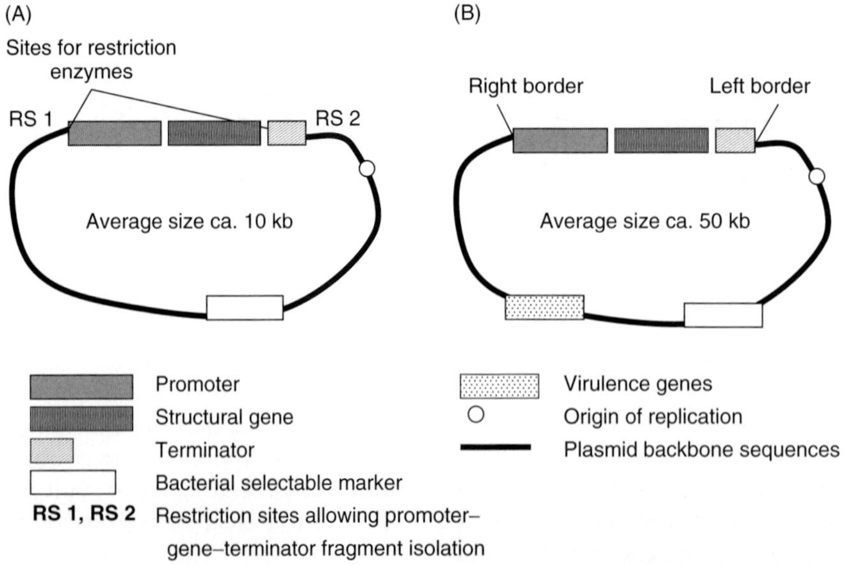

Figure 2 Structure of typical vectors used for (A) direct gene transfer (DGT) or (B) *Agrobacterium* transformation of monocotyledons.

isolated transformation plasmid DNA. The isolated fragment preparation is then used for transformation. This strategy has the advantage that strictly defined sequences are integrated into the target genome, avoiding all nonessential DNA.

Vectors for *Agrobacterium* transformation have a more complex structure than DGT plasmids, as apart from the basic transgene they also contain sequences involved in the mechanism of transfer from the bacterial cell to the plant cell and in integration into the plant genome. One feature of *Agrobacterium* transformation vectors (T-DNA vectors) is a pair of short palindromic sequences ($\sim$25 bp), the so-called "right" and "left" borders, between which the promoter–gene–terminator sequences described above can be cloned. The vector then operates by a set of *Agrobacterium* genes called virulence genes, recognizing the T-DNA borders, cutting the sequence at these points and packing, transferring, and integrating the DNA sequences cloned between them into the recipient plant genome. The virulence genes can be located either on the same T-DNA plasmid or on another plasmid molecule present in the bacterium. Monocotyledon species have been considered recalcitrant to *Agrobacterium* transformation and this only became routine when "supervirulent" plasmids containing extra copies of the virulence genes (e.g., *vir*G, *vir*C and *vir*B) were used for transformation. A last important difference between DGT and *Agrobacterium* transformation vectors is that the latter have a much larger "payload" of DNA, which means that they are more suitable for the transfer of multiple genes in a single cassette.

Selection Systems

Most plant transformation systems target cells randomly and result in the production of a few transgenic cells in a population of nontransformed cells. In order to identify these cells and recover transgenic plants from them, there is the need for a selection system. This can function at the level of visual selection, in which a gene whose expression gives a visual phenotype, may be used to track transformed cells as they divide and regenerate (**Figure 3** illustrates the use of the beta-glucuronidase (GUS) marker gene to visualize transgene expression). However, in most cases, it is necessary for the transformed cells to show preferential growth over nontransformed cells, as it is otherwise impractical to recover transformants from an overwhelming excess of nontransgenic cells. This is particularly important in monocotyledon transformation systems, which tend to show relatively low transformation efficiencies.

The most widespread selection strategy is the use of a resistance (marker) gene, which gives transformed cells the ability to grow in the presence of a toxic selection agent, while wild-type cells are inhibited or killed. The first selection genes applied in monocotyledon transformation were antibiotic resistance genes, used in combination with antibiotics such as kanamycin or hygromycin (see **Table 3** for examples of selection systems used in monocotyledon transformation). However, these genes were not fully satisfactory, as monocotyledons were found to be relatively resistant to most antibiotic selection agents and there were concerns over the release of antibiotic resistance genes from GM cereal

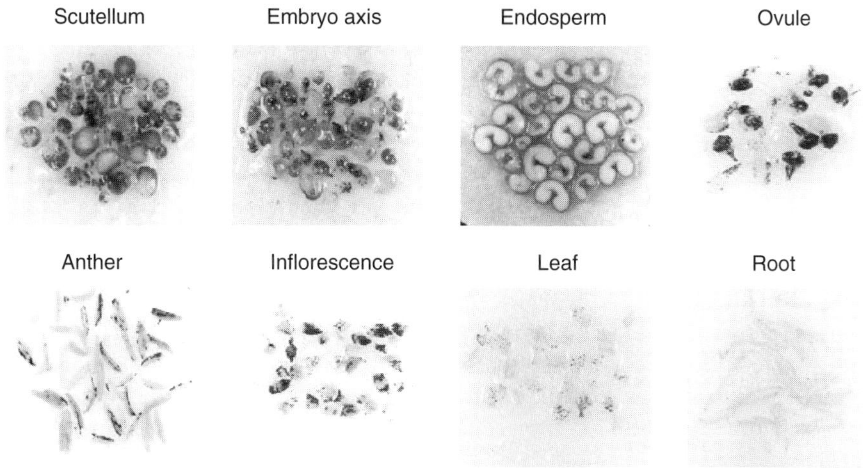

Figure 3 (see color plate 14) Expression of the GUS (beta-glucuronidase) marker gene in transgenic wheat tissues. Blue coloration indicates expression of the GUS transgene.

Table 3 Examples of selection systems used in monocotyledon transformation

System type	*Marker gene*	*Selection agent*
Antibiotic	*neo*/*nptII* (neomycin phosphotransferase)	Kanamycin, geneticin, paromomycin
Cytotoxin	*hpt* (hygromycin phosphotransferase)	Hygromycin
Cytotoxin	Mutant *dhfr* (dihydrofoliate reductase)	Methotrexate
Antibiotic	*aadA* (spectinomycin resistance)	Spectinomycin
Herbicide	*bar*/*PAT* (phosphinothricin acetyl transferase)	Phosphinothricin (PPT), glufosinate (Basta™/ Liberty™), bialaphos
Herbicide	*EPSPS* (enolpyruvate shikimate phosphate synthase) + *GOX* (glyphosate oxidoreductase)	Glyphosate (Roundup)
Herbicide	*ALS* (acetolactate synthase)	Sulfonylurea herbicides (e.g., Rimsulfuron, Nicosulfuron)
Positive selection	*manA* (phosphomannose isomerase)	Mannose
Positive selection	*xylA* (xylose isomerase)	Xylose

crops. In consequence, attention moved to herbicide resistance genes, in particular, the *bar* or *PAT* gene, encoding the enzyme phosphinothricin acetyl transferase. This gives resistance to phosphinothricin (PPT) and related compounds, the active ingredient in herbicides such as Basta™ (Liberty™) and bialaphos. Other herbicide marker genes that have been used with success in monocotyledons are *EPSPS* + *GOX* and *ALS*, giving resistance to glyphosate (Roundup) and sulfonylurea herbicides, respectively (see **Table 3**).

A more recent development in selection technology is the use of positive marker genes, which give transformed cells a metabolic and hence growth advantage over nontransgenic cells. A system that has been used with success in several monocotyledon species, including wheat and corn, is the phosphomannose isomerase gene, which allows transgenic cells to use mannose as carbon source. This, and similar positive selection genes (see **Table 3**) have the advantage for practical application in GM agriculture that they may be viewed as being more acceptable in terms of potential health and environmental impact than antibiotic or herbicide resistance markers.

Transgene Integration, Expression and Stability

Following transgene delivery, two further steps are required: first, the foreign DNA has to be introduced into the plant cell nucleus; and, second, the DNA must be integrated into the plant genome. The major difference between the two transformation methods lies in the way in which foreign DNA is prepared and introduced into the nucleus. In the case of DGT methods, foreign DNA is introduced into the nucleus as a naked DNA molecule. In contrast, in *Agrobacterium*-mediated gene transfer, a T-complex, consisting of a T-DNA covered with virulence proteins, is produced, which is transported from the bacterium into the plant cell through a channel assembled by other virulence proteins. The T-complex is then targeted to the nucleus.

Independent of the DNA delivery method, transgenes are usually integrated into the plant genome by illegitimate recombination at one or two independent locations (loci) in the genome. At the transgenic locus, transgene units (promoter–gene–terminator) can integrate completely and properly with no truncations or can be integrated as rearranged/truncated insertions. Low number and simple transgene insertions are most desirable for subsequently predictable transgene expression, as complex patterns of integration tend to lead to transgene instability and silencing.

The number of insertions of the transgene and the complexity of these insertions can be highly variable between transformation methods, experiments, and among plant species. In general, protoplast-based methods tend to produce transformants with larger numbers of insertions than other DGT methods, such as particle bombardment and also *Agrobacterium*-mediated transformation, which tends to produce the lowest insertion numbers. Considering transgene expression, once transgenes are integrated into the genome, the single most important factor controlling and regulating transgene expression is the promoter sequence controlling the transgene. These DNA sequences are responsible for the level as well as for the spatial and temporal expression of a given gene and are crucial for expressing genes in the desired manner.

There are three major classes of promoters currently used in monocotyledon transformation; those able to drive expression constitutively, those driving expression specifically in certain tissues and at certain developmental stages, and those only able to drive expression under inductive conditions. Constitutive promoters are mainly used to drive expression of selectable marker genes for the identification of transgenic tissues *in vitro*; this is generally achieved by using promoters with strong and ubiquitous activity. The most important promoter for monocotyledon transformation has been the corn ubiquitin promoter, which functions well across a range of species. This promoter, in common with several other highly active promoters used in monocotyledons, contains a 5′ intron sequence. This sequence is an integral part of the promoter unit, since its removal greatly reduces the activity of the promoter.

Tissue-specific and developmentally regulated promoters allow the expression of transgenes only in specific tissues or under certain developmental conditions, leaving the rest of the plant unmodified by transgene expression. For engineering monocotyledon crops, there are many circumstances in which such type of regulation is required; e.g., genes involved in grain quality may only need to be expressed in endosperm cells. In cereals, in particular, promoters have been characterized for a number of tissues, such as endosperm, embryo, anther tapetum, meristem, mesophyll, and phloem.

A third class of promoters is those that are only expressed under specific induction conditions. To date, relatively few promoters in this category have been characterized in monocotyledons, but sequences regulated by the action of stresses, such as pathogens or wounding, or by the action of chemicals, e.g., ethanol- and safener-induced promoters, are available.

The stability with which transgenes are inherited and expressed over generations is of central importance in plant transformation. While transgene instability has been described in many plants, including monocotyledon species, a general rule is that lines containing simple, low copy number transgene insertions show relatively stable and predictable transgene expression. Poor or unstable expression is, in many cases, related to plants that do not contain an intact functional copy of the transgene, or which contain multiple and complex insertion patterns. While the underlying mechanisms are still not fully understood, inactivation of the transgenes often involves methylation or the inactivation of plant defense mechanisms, which degrade DNA sequences identified as "foreign." In this context, heterologous sequences from distant species may be more likely not to function properly in the recipient genome than sequences from related (i.e., other monocotyledon) plants.

One aspect of transgene stability in transgenic crops concerns the potential for unwanted transfer of transgene sequences from one crop variety into another variety or related species via cross-pollination, or into another plant or microorganism by a horizontal gene transfer mechanism. Discussion has focused on the problems that could arise from the spread of antibiotic or herbicide resistance genes, and a number of approaches, which have been termed "clean transformation" methods, have been proposed to address the problem. One technique is the use of recombinase systems such as Cre-Lox or Flp-Frt, incorporated into the transformation vector. On activation, the recombinases allow unwanted sequences, such as marker genes, to be excised from the transgenic plant. An alternative approach involves the targeting of the plant transformation vector to the chloroplast, where it is integrated at a specific site by a homologous recombination mechanism. In this case, pollen transfer of transgene sequences is greatly reduced, as plastid transmission to pollen is minimal in most plants. These techniques have been shown to work at reasonable efficiency in several dicotyledon species and have also been demonstrated in monocotyledons. Unfortunately, the present levels of efficiency achieved in monocotyledons would make their routine use a major hindrance to applied transformation programs. Currently, the more pragmatic solution is the use of marker genes, which are judged to have neutral environmental or health impact (see "Selection Systems" above), delivered in transformation vectors containing only essential sequences. In the future, however, as gene excision systems become more sophisticated and efficient, it is likely that these will become standard tools in monocotyledon transformation to ensure that transgenic crops contain only those sequences needed for expression of the desired new trait.

Applications

Considering that the transformation of monocotyledons is a relatively young technology, dating from the end of the 1980s, it has already had considerable impact in both basic and applied research and in commercial breeding of monocotyledon crops. In the field of basic research, transformation technology allows the investigation of the function of endogenous (native) genes by either up- or downregulating their expression and then examining the effect on the plant phenotype. The extreme level of downregulation, when a gene is silenced completely, is the transgenic equivalent of a classical loss-of-function mutation, such as those produced artificially by mutagenic chemicals or irradiation. In the transgenic case, however, the mutation can be directed at a specific target gene and there is the possibility of controlling the level of reduction in activity. In contrast, other mutagenic treatments are essentially random in terms of target and severity. In monocotyledons, transgenic approaches to gene overexpression and downregulation are increasingly used to understand metabolic processes. They can be applied in any case in which the gene for a target enzyme has been isolated. The approach has been used with success to study carbohydrate (starch) and amino acid biosynthesis in species such as corn and rice, by manipulating the expression of a series of synthetic enzymes. The power of this approach is such that in those species that are readily transformable, it is becoming standard practice to consider a knockout experiment to determine the function of any new gene that is isolated. A second basic research application of monocotyledon transformation is the testing of genes from other plant species to determine their function in monocotyledons. This approach is today providing important information on plant evolution and speciation and shows how essentially the same sets of genes are able in different plants to produce an enormous variety of differing structures and growth habits.

The main application of monocotyledon transformation is, however, its use in applied research and direct crop improvement by GM. In this application, transformation is used to create novel genetic variation, as the result of transgene expression, which can then be used to breed improved GM varieties. As in basic research applications, the targets for manipulation may be either endogenous or heterologous genes. Useful mutants can be produced by downregulation approaches, to reduce levels of unwanted molecules such as allergens or toxins, or to remove the activity of particular enzymes. Alternatively, the activity of desirable enzymes and the levels of valuable metabolites may be increased by overexpression. Transformation may also be used to express heterologous genes, resulting in the production of novel molecules in the monocotyledon crop. The first generation of GM crop varieties was produced using the latter approach, by expressing genes conferring resistance to normally non-selective herbicides (e.g., RoundupReady crops resistant to glyphosate or LibertyLink crops resistant to glufosinate) and creating insect-resistant varieties by expression of the *Bacillus thiuringiensis* (Bt) toxin gene.

These first commercial products have demonstrated the enormous potential of transformation technology to deliver monocotyledon crops with valuable new traits. As gene discovery and transformation research progresses, we can expect to see GM monocotyledons with a much broader range of modifications, including pest and disease resistance, increased tolerance to abiotic stresses, such as drought and salt, and with modified architecture, flowering, and productivity.

List of Technical Nomenclature

Direct gene transfer	The transformation method in which gene transfer is effected by a non-biological stimulus.
Explant	The plant tissue isolated or excised from the parent organ and used to initiate a tissue culture.
Protoplast	The plant cell after the removal of the cell wall (normally by enzymatic digestion).

Regeneration	The formation of a new plant by cultured plant tissue.
Stable transformation	The transformation in which the transferred transgene is stably integrated into the new host genome.
Suspension culture	Plant cells grown in shaken liquid culture.
Transgene	A gene artificially inserted (transformed) into a host genome.
Transient gene expression	The temporary expression of a transgene in a cell without stable integration into the host genome.

See also: **Crop Improvement**: Marker Assisted Selection; Molecular Markers. **Genetic Modification**: Gene Cloning, General Principles; Insertional and Transposon Mutagenesis; Transgene Stability and Inheritance. **Growth and Development**: Control of Gene Expression, Regulation of Transcription. **Tissue Culture**: Clonal Propagation, *In Vitro*.

Further Reading

Abelson PH and Hines PJ (1999) The plant revolution. *Science* 285: 367–389.

Barcelo P, Rasco-Gaunt S, Thorpe C, and Lazzeri PA (2001) Biotechnology of cereals: transformation and gene expression. *Advances in Botanical Research* 34: 59–128.

Birch RG (1997) Plant transformation: problems and strategies for practical application. *Annual Reviews of Plant Physiology and Molecular Biology* 48: 297–326.

Finer JJ, Finer KR, and Ponappa T (1999) Particle bombardment mediated transformation. *Current Topics in Microbiology and Immunology: Plant Biotechnology* 240: 60–80.

Hansen G and Chilton MD (1999) Lessons in gene transfer to plants by a gifted microbe. *Current Topics in Microbiology and Immunology: Plant Biotechnology* 240: 21–57.

Hansen G and Wright MS (1999) Recent advances in the transformation of plants. *Trends in Plant Science* 4: 226–231.

Komari T, Hiei Y, Ishida Y, Kumashiro T, and Kubo T (1998) Advances in cereal gene transfer. *Current Opinion in Plant Biology* 1: 161–165.

Lazzeri PA and Shewry PR (1993) Biotechnology of cereals. *Biotechnology and Genetic Engineering Reviews* 11: 79–146.

Lindsey K (ed.) (1998) *Transgenic Plant Research*. Amsterdam: Harwood Academic.

Wilmink A and Dons JJM (1993) Selective agents and marker genes for use in transformation of monocotyledonous plants. *Plant Molecular Biology Reporter* 11: 165–185.

Transformation in Plastids

P Maliga, Rutgers University, Piscataway, NJ, USA

Introduction

DNA in a higher plant cell is found in the nucleus, plastids, and mitochondria. Most efforts to improve crops focus on engineering the nuclear genome which encodes >25 000 genes. The plastid genome (ptDNA) of higher plants encodes only ~120 genes most of which are photosynthetic genes and genes required for plastid maintenance. In addition, >10% of nuclear genes target proteins to plastids. These proteins complement plastid-encoded genes required for plastid maintenance and photosynthesis, and are involved in starch metabolism, lipid and amino acid biosynthesis, and nitrogen assimilation. Plastid transformation in a higher plant, tobacco (*Nicotiana tabacum*), was reported in 1990. Research in tobacco was inspired by the pioneering work on plastid transformation in *Chlamydomonas reinhardtii*, a unicellular alga, described in 1988. Transformation of the plastid genome was 7 years delayed relative to nuclear gene transformation (1983). Transformation of the mitochondrial genome in higher plants has yet to be accomplished.

More than 10 years after the initial breakthrough plastid transformation in higher plants is still routinely obtained only in tobacco. However, efforts are under way to extend plastid transformation to all major agronomic crops. Readily obtainable high protein levels, feasibility of expressing several proteins from polycistronic mRNAs, and lack of pollen transmission make plastids an attractive alternative to transgene expression in the nucleus. Progress has been made to engineer plastid-based herbicide resistance and insect resistance traits and to express recombinant proteins for molecular farming applications. The plastid engineering toolkit is complete with an efficient marker gene excision system to obtain marker-free transplastomic plants.

The Plastid Genetic System

There are several plastid types in a plant cell. Small rapidly dividing proplastids (0.5 μm to 1 μm) are found in meristematic cells; green photosynthetic chloroplasts (~10 μm) are present in leaves; yellow (lemon) or red (tomato) chromoplasts are characteristic of fruits; starch-containing amyloplasts are found in tubers and oil-containing elaioplasts in seed storage tissue. All plastid types have the same genome that is represented as a double-stranded circular DNA

120–180 kb in size (**Figure 1**). Staining of ptDNA revealed that the genome copies are clustered in plastid nucleoids, a location where proteins anchor ptDNA to membranes. Proplastids contain only one to two nucleoids. Each nucleoid contains ~10 ptDNA. Since meristematic cells contain 10–14 proplastids, the total number of ptDNA copies in a meristematic cell is 100 to 200. Leaf cells contain ~100 chloroplasts; each chloroplast contains ~10 (7–14) nucleoids, therefore the total number of ptDNA copies is ~10 000 per leaf cell. Given the large number of copies it is not surprising that ptDNA in a leaf cell can make up 10% of total cellular DNA. The challenge of plastid transformation is to uniformly alter all genome copies, the condition of obtaining a genetically stable plant.

Approaches to Stable Transformation of the Plastid Genome

Stable transformation of the plastid genome in higher plants is obtained by incorporation of the transforming DNA in the plastid genome where it is replicated with the rest of the genome. Plastid transformation vectors are *Escherichia coli* plasmid derivatives, which carry sequences (1–2 kb) to direct integration at specific sites in the plastid genome. The plastid vector is propagated in *E. coli*, and then introduced into plastids where the marker gene and the gene of interest integrate by two homologous recombination events in the targeted region of the plastid genome (**Figure 2**). The *E. coli* vector part is not inserted in the plastid genome. Since it lacks a plastid replication origin, it is subsequently lost. Plastid targeting sequences do not have special properties other than being homologous and may come from any part of the plastid genome. By now insertion of foreign DNA has been accomplished in at least 20 locations in the ptDNA (**Figure 1**). Stable integration of foreign DNA by homologous targeting is the general approach used for plastid transformation in higher plants.

An alternative to stable integration of transforming DNA in the plastid genome has been incorporation of plastid replication origins in plastid

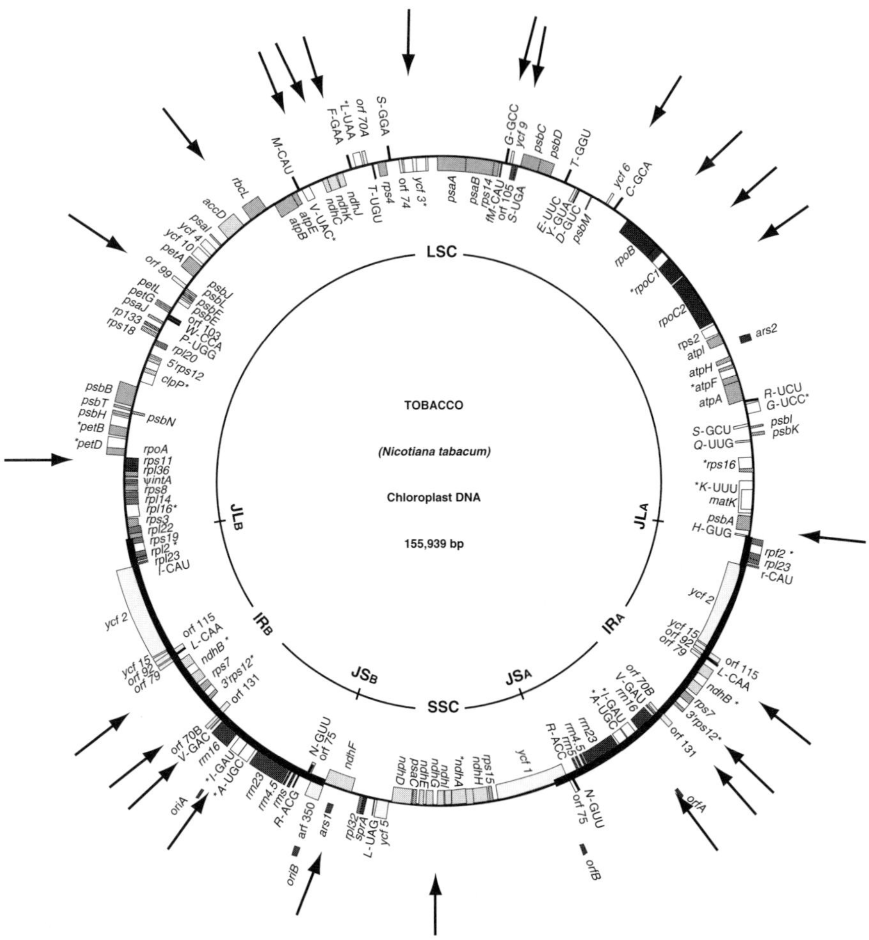

Figure 1 The map of the tobacco plastid genome. Arrows point at sites where foreign DNA has been inserted in the plastid genome. Modified from unpublished tobacco plastid genome map, kindly provided by M. Sugiura, with permission.

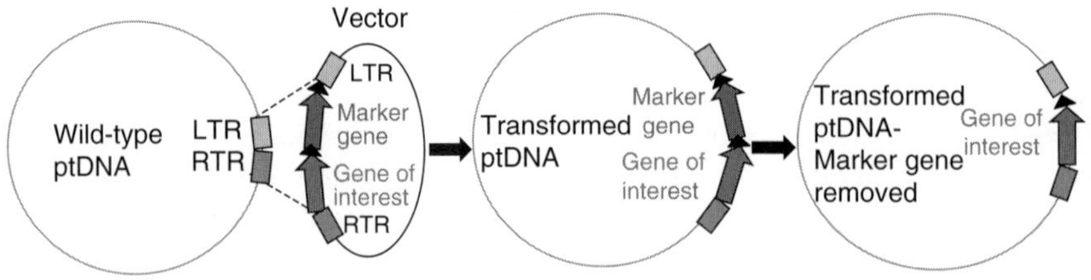

Figure 2 Plastid transformation involves incorporation of foreign DNA by two homologous recombination events via the left (LTR) and right (RTR) targeting sequence and removal of the marker gene using the CRE-*lox* site-specific recombination system. Reproduced with permission from Maliga P (2003) Progress towards commercialization of plastid transformation technology. *Trends in Biotechnology* 21: 20–28.

transformation vectors, which therefore can be independently maintained as episomal elements. Such plasmids are termed shuttle vectors as they are maintained in *E. coli* as plasmids by replication using the ColE1 replication origin and in plastids using a plastid *ori* sequence. These first-generation shuttle vectors are not practical because they are rapidly lost in the absence of selection. However, stable forms may be obtained in the future by improving vector design.

Methods for Introduction of Transforming DNA

The most commonly used approach for plastid transformation is biolistic DNA delivery. The transforming DNA is precipitated on the surface of microscopic (~1 μm) tungsten or gold particles. The ~100 000 particles delivered in a charge are then accelerated in a suitable biolistic device to a speed such that the heavy DNA-coated metal particles penetrate the cell wall and settle in the plant tissue. In case of a tobacco leaf, the penetrated layer is at least four to five cells deep. The transforming DNA dissolves inside the chloroplasts from particles lodged inside the chloroplast or comes in contact with ptDNA by entering the chloroplast through holes punched by the metal particles. Each particle carries about 20 to 50 copies of the plastid vector DNA. Targets for biolistic transformation can be chloroplasts in intact tobacco leaf tissue or proplastids in cultured rice (*Oryza sativa*) embryogenic cells.

An alternative approach for DNA delivery utilizes polyethylene glycol (PEG) treatment to facilitate introduction of the transforming DNA into protoplasts. PEG treatment was originally developed for the transformation of the nucleus. For plastid transformation, protoplasts are obtained by enzymatic removal of the cell wall, and are then suspended in a DNA solution and treated with PEG. PEG treatment triggers shrinking of the protoplasts accompanied by engulfing some of the cell membrane and results in eventual disintegration of protoplasts. The PEG treatment, however, is terminated before the membrane is irreversibly damaged and the protoplasts are nursed back to health. The transforming DNA apparently finds its way into chloroplasts in a process not fully understood yielding transplastomic clones. Isolation and handling of protoplasts requires advanced cell-culture skills, which is the reason why the technically less demanding biolistic delivery to plastids in intact tissues is the commonly used method.

Another approach for the delivery of DNA into plastids is microinjection. Successful DNA delivery by microinjection was shown by reporter gene expression. However, utility of microinjection remains to be shown for stable transformation of the plastid genome.

Genetic Markers for Identification of Transplastomic Clones

Plastid transformation in higher plants was made feasible by marker genes that allowed selective amplification of transformed plastid genomes. First transplastomic tobacco plants were obtained by selection for spectinomycin resistance. Utility of spectinomycin resistance as a plastid marker has been known through the study of plastid-encoded spectinomycin resistance mutants. Spectinomycin blocks shoot regeneration from tobacco leaf sections and formation of green callus due to inhibition of plastid protein synthesis. Spectinomycin (and streptomycin) resistant mutants were identified by their ability to regenerate shoots and form green callus on selective spectinomcyin medium (**Figure 3**). Spectinomycin resistance is due to mutations in the plastid 16S rRNA that prevents spectinomycin binding. Initial plastid transformation vectors utilized plastid targeting sequences that contained mutant plastid *rrn16* genes as markers. The mutant *rrn16* genes were rapidly abandoned in favor of the 100 times more efficient *aadA* marker gene, a bacterial gene that also confers spectinomycin/streptomycin

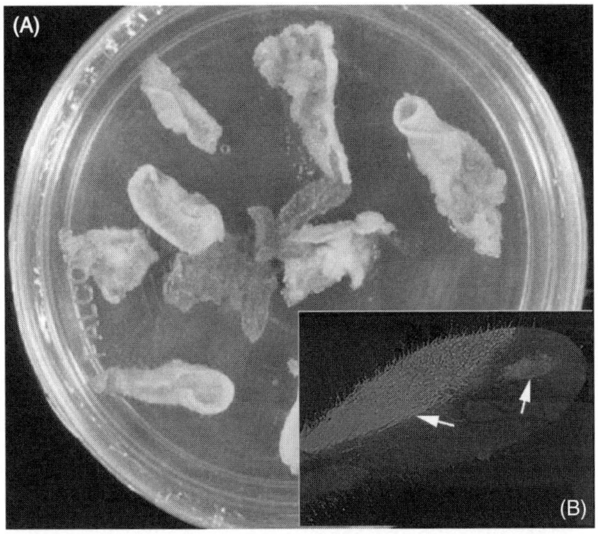

Figure 3 (see color plate 13) Selection of transplastomic clones by spectinomycin resistance. (A) Spectinomycin inhibits callus formation, greening, and shoot regeneration from tobacco leaf segments on shoot regeneration medium. Transplastomic clones are resistant to spectinomycin and are identified as green shoots (or calli). (B) The shoots are chimeric, visualized by accumulation of green fluorescent protein in transplastomic sectors. Shoots regenerated from the transformed sector are homoplastomic. Reproduced with permission from Maliga P (2003) Progress towards commercialization of plastid transformation technology. *Trends in Biotechnology* 21: 20–28.

resistance. Spectinomycin resistant clones transformed with *aadA* are selected by the same procedure as the lines transformed with mutant *rrn16* genes (**Figure 3**). An alternative marker for plastid transformation is kanamycin resistance conferred by the *neo* or *aphA-6* genes, which encode different neomycin phosphotransferases. The enzyme encoded by *neo* has narrow substrate specificity and confers resistance only to kanamycin. The *aphA-6* gene product confers resistance to multiple drugs.

A potentially useful, new positive selection scheme involves identification of transplastomic tobacco lines by resistance to betaine aldehyde (BA). Resistance to BA is conferred by expression of betaine aldehyde dehydrogenase (BADH) in plastids. BADH is the product of a plant nuclear gene. Identification of transplastomic clones apparently relies on a phenotype unique to plastid expression. The applicability of the selection protocol is yet to be confirmed.

Not every marker gene is suitable for selective amplification of the rare, initial transplastome although resistance may be obtained when most plastid genomes carry the transgene. The *bar* gene is an excellent selectable marker for nuclear gene transformation conferring resistance to phosphinothricin (PPT) herbicides. Expression of the *bar* gene in plastids confers PPT resistance when it is introduced by selection for a linked *aadA* (spectinomycin resistance) gene. However, *bar* was not suitable for direct selection of transplastomic lines even if expressed at a high level (>7% TSP (total soluble cellular protein)). Thus, it appears that subcellular localization of the gene encoding the detoxifying enzyme is critical when directly selecting for herbicide resistance and initial, low-level expression from a few transformed plastid genome copies is insufficient to protect the cell from PPT.

A negative plastid selection scheme has been developed based on expressing the bacterial enzyme cytosine deaminase. Cytosine deaminase catalyzes deamination of cytosine to uracil. 5-fluorocytosine (5FC) is toxic for cytosine deaminase-expressing cells, since the enzyme converts 5-fluorocytosine to the toxic 5-fluorouracil. Plant cells lack cytosine deaminase but become sensitive to 5FC when the enzyme is expressed in plastids. Resistance to 5FC was utilized as a marker to identify cells in which the gene encoding cytosine deaminase was removed by the CRE-*lox* site-specific recombinase.

In addition to the positive and negative selection schemes, visual markers are also available in plastids. β-Glucuronidase (GUS) enzymatic activity can be visualized by histochemical staining and the green fluorescent protein (GFP) can be visualized in UV light. A gene encoding GFP was fused with *aadA* and the fusion gene was used as visual and selective (spectinomycin resistance) bifunctional marker.

Plastid Transformation Vectors

There are at least as many plastid transformation vectors (>20) as insertion sites targeting insertions at unique locations in the plastid genome (**Figure 1**). However, there are only two vector families that have been refined to provide cloning convenience by combining a plastid targeting sequence, a selective marker, and a multiple cloning site. Both of these utilize *aadA* as selective marker (**Figure 4**). The *trnV/rps12* intergenic region as insertion site has the advantage that there is no read-through transcription from the direction of the *rrn* operon. The *trnV/rps12* intergenic region is in the inverted repeat region of the plastid genome. Genes at this site are present in two copies per genome.

Vector development is critical for extending the technology to new crops, and for new applications. In the near future new vector families are expected to appear that will utilize kanamycin resistance as a marker gene and/or will have recognition sequences for site-specific recombinases to allow efficient removal of marker genes.

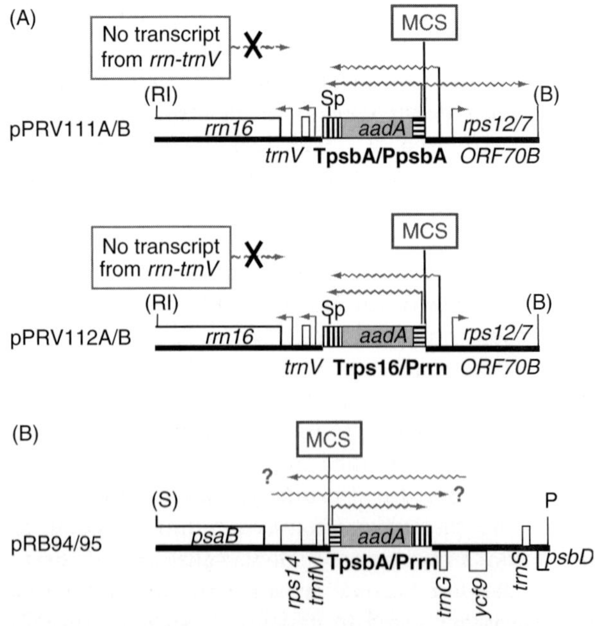

Figure 4 Plastid transformation vectors. (A) The plastid repeat vector (pPRV) family (GenBank accession numbers U12809-U12815) (Zoubenko (1994) #900). Cloning convenience is ensured by alternative expression signals for *aadA* in the pPRV111 and pPRV112 series, and the pUC multiple cloning sites (MCS) available in both orientations (A or B). There is no read-through transcription of transgenes in MCS from the *rrn* operon. (B) The pRB94/pRB95 vectors (EMBL Accession numbers AJ312392, AJ312393). Vectors differ with respect to the orientation of the Bluescript MCS. No information is published on read-through transcription at the MCS. Reproduced with permission from Maliga P (2003) Progress towards commercialization of plastid transformation technology. *Trends in Biotechnology* 21: 20–28.

Status of Implementing Plastid Transformation in Crops

Plastid transformation has been most successful in the solanaceous crops tobacco, potato (*Solanum tuberosum*), and tomato (*Lycopersicon esculentum*). Plastid transformation in potato has opened up the possibility to engineer tuber-specific plastid gene expression. Transformation of plastids in tomato is the starting-point for attempts for fruit-specific expression of recombinant proteins. Tomato fruits are good candidates for delivery of oral vaccines. Plastid transformation was also feasible but inefficient in species that belong to the mustard (Brassicaceae) family, including *Arabidopsis thaliana*, *Brassica napus*, and *Lesquerella fendleri*. Out of the cereals, heteroplastomic plants have been obtained in rice. Plastid transformation in each of these species has some unique features. Plastid transformation of tobacco is discussed below as it is the best characterized system.

Sterile leaves of culture-grown tobacco plants are bombarded with vector DNA-coated tungsten particles. The leaves are dissected into small squares ($1\,cm^2$) and cultured on nonselective shoot induction medium to induce cell division. After 2 days, the leaves are transferred onto spectinomycin-containing ($500\,mg\ l^{-1}$) shoot induction medium. Nontransformed cells proliferate slowly and form white callus on the selective medium. Cells carrying transformed ptDNA copies regenerate green shoots in 3–12 weeks. These shoots are chimeric and consist of sectors with transformed and nontransformed plastids (**Figure 3B**). Shoots regenerated from the transformed sectors yield genetically stable, homoplastomic plants. Starting with vector DNA, it takes about 3–5 months to obtain genetically stable transplastomic plants. Removal of the marker gene may extend this time period by about one month (see below).

Plastid transformation efficiency is measured as transplastomic clone per bombarded sample. In tobacco, one leaf (5–7 cm) constitutes one sample. Plastid transformation efficiencies are difficult to reproduce: the efficiency is typically in the range between 0.5 and 5.0 transplastomic clones per bombarded sample.

Mechanisms Yielding Genetically Stable Transplastomic Plants

Plastid transformation involves transformation of one or a few ptDNA copies, then gradual elimination of the nontransformed copies by dilution on a selective medium. Achieving the homoplastomic state in cultured tobacco cells takes ~20 cell divisions. Homoplastomic cells derive from heteroplastomic cells because somatic cell division does not involve exact duplication of the cytoplasm and because genomes of the organelles that carry the selective marker are preferentially replicated on a selective medium in tissue culture. Formation of homoplastomic cells is accelerated by the ptDNA being clustered in nucleoids and that the nucleoids themselves are being localized in independently segregating plastids. Shoot regeneration from cultured tobacco leaf segments involves formation of meristematic cells with concomitant reduction in plastid number from ~100 chloroplasts in leaves to ~12 proplastids in meristematic cells and concomitant reduction in nucleoid number from 7 to 14 (in chloroplasts) to 1 to 2 (in proplastids), then rebuilding the organelle and genome copy number in leaf cells on a selective medium. Plastid division during the chloroplast to proplastids transition, in the absence of an increase in nucleoid number, rapidly yields plastids which carry the nucleoid that came in contact with the transforming DNA (**Figure 5A**; nucleoid #1). Replication and segregation of transformed genome copies eventually yield proplastids that are homoplastomic for the transgenic ptDNA (**Figure 5A**; event #1a). Reduction of plastid number in meristematic cells, coupled with random segregation of plastids (only some of the progeny cells carry transformed plastids), accelerates elimination of plastids which do not carry transforming DNA (**Figure 5B**; progeny cell #1). DNA replication and segregation at the nucleoid and organellar levels yield the homoplastomic sectors that are visible in the leaves regenerated on spectinomycin medium (**Figure 3B**). Shoots regenerated from these sectors are typically homoplastomic and carry only transformed genome copies.

The observed reduction of ptDNA copy and organelle number in tissue culture cells is not universally applicable to all species. It is known, for example, that embryogenic rice tissue culture cells contain as many plastid genome copies as leaf cells. Thus, obtaining the homoplastomic state after bombardment of rice embryogenic cells is likely to take more cell generations than in tobacco.

Engineering of Plastid Genes for High-Level Expression

To facilitate expression of foreign genes in chloroplasts, modular cassettes have been designed with suitable restriction sites at the boundaries of the cassettes (**Figure 6**). The 5′ PL cassette (PL, promoter and leader) includes the promoter and translation control sequences. The translation control sequences can be the mRNA 5′-UTR (untranslated region) (**Figure 6A**), or the 5′-TCR (translation control region) that includes the 5′-UTR and an N-terminal segment of the coding region (**Figure 6B**). The mRNA 5′-UTR includes a stem–loop structure required for mRNA stability and sequences that facilitate loading of mRNAs onto ribosomes. The 3′ T-cassette encodes the mRNA 3′-UTR, which also includes a stem–loop structure. The 3′-UTR functions as an inefficient terminator of transcription and is required for mRNA stability.

High-level expression of foreign genes is dependent on transcription from a strong promoter. In most cases the strongest plastid promoter, driving expression of the plastid rRNA operon, is used. Targets for engineering are the 5′-UTR and coding region N-terminus because dramatic (10 000-fold) changes in protein accumulation have been obtained by choosing alternative 5′-UTRs. Success of improving protein expression by engineering the 5′-UTR is underlining the importance of posttranscriptional regulation in plastid gene expression.

The need for synthetic genes is a concern when expressing a protein in a heterologous host. Infrequent use of specific codons in the host organism normally indicates limiting supply of cognate tRNAs. Although the tobacco plastid genome is relatively AT-rich, there is no example for extremely low codon usage frequencies. Indeed, codon optimization in tobacco plastids yielded only two- to threefold increases in protein levels, much less increase than the 10 000-fold change obtained by engineering translation control signals.

Premature transcription termination and degradation of mRNAs was a problem when expressing high-AT bacterial genes in the yeast and plant nuclei. Examples are the *Bacillus thuringiensis* insecticidal protein gene and the gene encoding the tetanus toxin C-terminal fragment. The same coding sequences yielded stable mRNAs and high levels of protein expression in plastids indicating that plastids are a versatile system for expression of foreign genes of diverse origin.

Potential and Limits of Transplastomic Technology

It may be more attractive to incorporate transgenes in the plastid genome than in the nucleus. Given the size variability of plastid genomes it is assumed, that the plastid genome can be enlarged by about 50 kb. Dependent on the size of the genes, 50 kb of foreign DNA may encode 20 to 30 genes. The largest inserts published thus far are about 5 kb in size, encoding a *Bacillus thuringiensis* insecticidal protein gene and a selectable marker gene. The genes have been inserted in the inverted repeat region, therefore the plastid genome expressing the two genes is increased by a total of ~10 kb. Some of the plastid genes are expressed in operons therefore expression of enzymes in the same metabolic pathways may be practical from polycistronic mRNAs. The largest protein expressed, Cry1A(c), is 133 kDa in size. Protein levels may be readily obtained in the 5–25% range.

Localization of a transgene in the plastid genome may not always be advantageous. No plastid-encoded protein is known to be exported, a limitation of plastid expression when subcellular localization other than the plastid is desired. Plastid proteins may be modified posttranslationally. Glycosylation, however, is a protein modification that is known not to occur in plastids. Also, little is known about tissue-specific expression of plastid genes other than

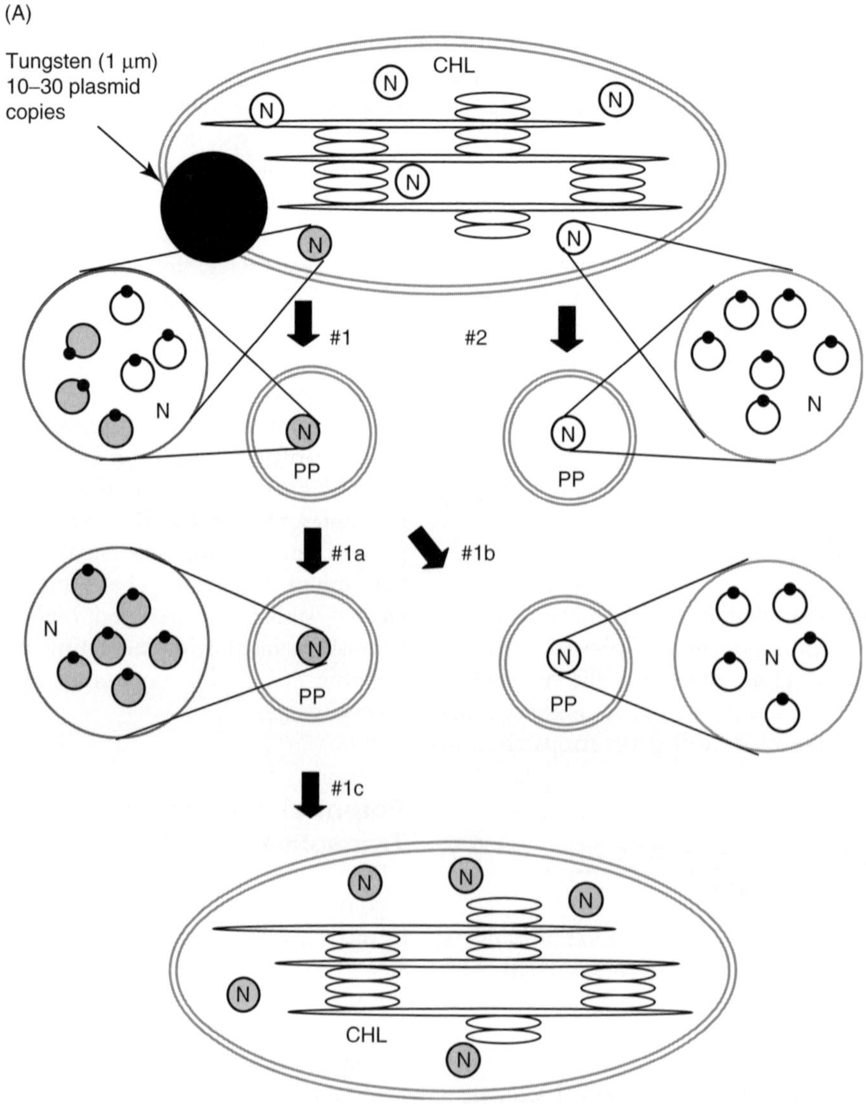

Figure 5 Chloroplasts (CHL) to proplastid (PP) dedifferentiation accelerates formation of homoplastomic cells during shoot regeneration. (A) Nucleoids are the units of segregation during the chloroplast to proplastids transition. Transformed and nontransformed nucleoids (N) are marked as gray and open circles, respectively. In enlarged nucleoids, transformed and nontransformed ptDNA is depicted as gray and open circles, respectively; ptDNA are anchored to membranes by proteins (black dots). Nucleoid #1 is heteroplastomic, and is the progenitor of homoplastomic transgenic (#1a) and wild-type (#1b) proplastids. Homoplastomic proplastid differentiates into chloroplast (#1c; bottom). Wild-type proplastid (#2) is antibiotic sensitive and divides slowly. (B) Reduction in plastid number during chloroplasts (~100; elongated forms) to proplastid (~12; circles) transition and lack of exact duplication of the cytoplasm accelerates formation of homoplastomic cells. On top is shown leaf mesophyll cell with one transformed chloroplast (unshaded) and nucleus (Nu). Meristematic cell #1 is heteroplastomic. Cleavage of the cytoplasm yields one meristematic cell with transformed chloroplasts only (#1A) and one with wild-type plastids (#1b). Meristematic cell with transformed plastids is the progenitor of homoplastomic mesophyll cells (1c). Meristematic cells with wild-type plastids divide slower and are eventually diluted out.

expression in chloroplasts. Information on plastid gene expression in nongreen tissue is needed to aid the design of tissue-specific expression in fruits, tubers, epidermis cells, etc. Expression tools borrowed from prokaryotic organisms may fill the need. Nuclear control of plastid gene expression was reported by expressing a phage RNA polymerase in the plant nucleus that was driving transgene expression from a phage promoter in plastids.

Marker Gene Elimination Systems

Marker genes are essential to obtain transplastomic plants. However, as soon as plastid transformation is accomplished, removal of the marker genes is highly desirable. There is only a small number of efficient plastid marker genes; removal of the marker allows a second round of engineering using the same marker gene. Also, release of crops with selectable marker

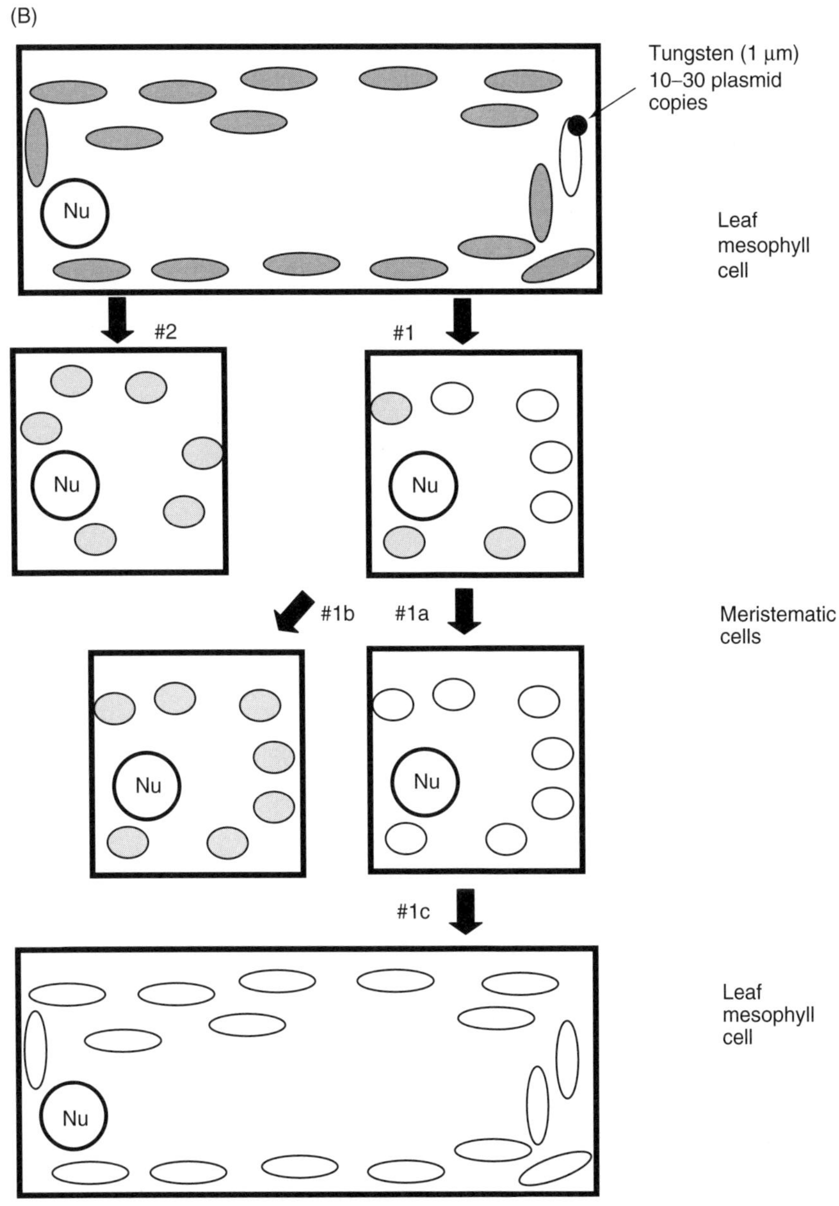

Figure 5 Continued.

genes is undesirable due to regulatory concerns. Furthermore, the marker genes may be expressed at a high level, imposing an unnecessary metabolic burden on the crop.

Multiple approaches have been developed for the removal of marker genes from the plastid genome. The most efficient system uses the P1 phage CRE-*loxP* site-specific recombination system (**Figure** 7). According to the CRE-*loxP* scheme, the marker gene and the gene of interest are introduced into the plastid genome in the absence of CRE activity. The marker gene is flanked by two directly oriented *lox* sites. When elimination of the marker gene is required, a gene encoding a plastid-targeted CRE site-specific recombinase is introduced into the nucleus which, subsequent to its import in plastids, excises sequences between the *loxP* sites. In tobacco, removal of the marker gene extends the time needed to obtain marker-free plants only by 1 month.

Applications of Plastid Transformation

Applications in Basic Science

Plastid transformation in basic science is widely utilized to understand plastid gene function and gene regulation. By now knockout mutants have been described for more than 17 plastid genes. Study of these mutants helped to identify the genes' function. For example, study of *ndh* knockout plants

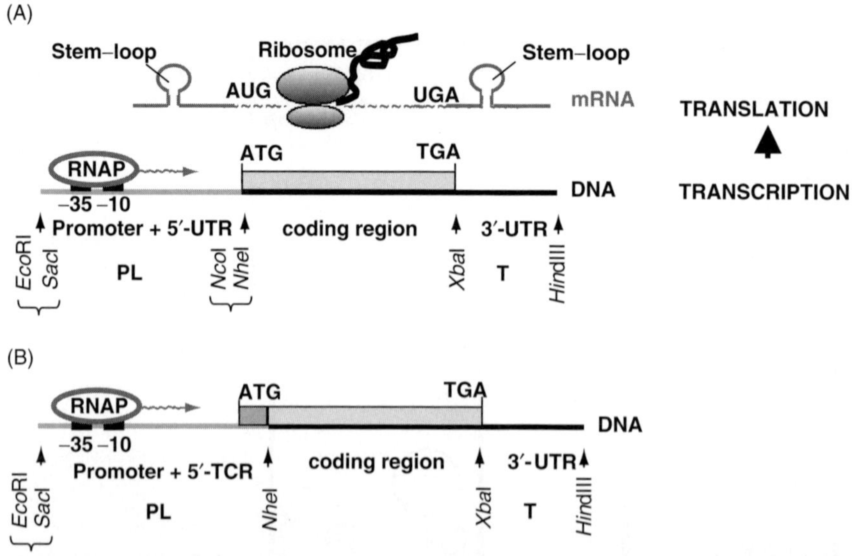

Figure 6 Modular design of plastid promoter-leader (PL) and terminator (T) cassettes. PL cassettes encode a promoter and (A) 5′-untranslated region (5′-UTR) or (B) a 5′-translation control region (5′-TCR, 5′-UTR plus a segment of the coding region). Proteins are expressed from *NheI/XbaI* or *NcoI/XbaI* fragments. Stem–loop structures formed by nucleotide pairing in the 5′-UTR and 3′-UTR and the AUG translation initiation and UGA stop codons are marked. Reproduced with permission from Maliga P (2002) Engineering the plastid genome of higher plants. *Current Opinion in Plant Biology* 5: 164–172.

yielded the first insights into the function of the "chlororespiratory pathway." Deletion of the plastid-encoded RNA polymerase subunits revealed promoters for a nuclear-encoded, plastid targeted second plastid RNA polymerase and shed light on the division of labor between the two transcription machineries. Plastid transgenes were utilized to identify plastid promoter elements and sequences regulating mRNA translation. Study of transgenes incorporating segments of edited genes yielded new information about the functional significance of posttranscriptional C-to-U conversion that affects only 20 to 31 specific C nucleotides in 19 transcripts. Plastid transformation is also utilized to explore the feasibility of improving photosynthetic efficiency.

Applications in Agriculture and Molecular Farming

Currently, no transplastomic plant is approved for field release. This will change as soon as plastid transformation is implemented in major agronomic crops as groundwork has been laid to engineer commercially important traits. Progress has been made to understand engineering resistance to herbicides with glyphosate and phosphinothricin (Bialaphos, Liberty) as active compounds. Also, *B. thuringiensis* insecticidal protein genes were expressed at a high level in chloroplasts, allowing implementation of a high-dose strategy. A second area with good potential for applications of plastid transformation is expression of recombinant proteins for molecular farming. The examples include expression of antibodies, vaccines, and industrial enzymes.

Biosafety and Transplastomic Plants

In general, transplastomic plants are considered to be the safest form of GM crops for transgene containment as in most crops plastids are not transmitted by pollen. In contrast, if a transgene is incorporated in the nucleus, each pollen carries the transgene. Examples for strict maternal inheritance of plastids are corn (*Zea mays*, maize), wheat (*Triticum aestivum*), rice, cotton (*Gossypium* spp.), soybean (*Glycine max*), tobacco, tomato, oilseed rape (*Brassica napus*), and sunflower (*Helianthus annuus*). Examples for biparental inheritance of plastids are potato and alfalfa (*Medicago sativa*, lucerne).

Gene flow should be considered within a species and between species, by hybridization with weedy relatives. Plastid localization eliminates outgoing transgene flow via pollen. However, opportunities still exist for the much less frequent hybridization at the field's edge, or by hybridization due to fertilization with pollen from outside the field. Thus, the possibility for hybridization with weedy relatives should be considered. Danger is presented only if the hybrids of the transplastomic crop and the weedy relative are viable. An alternative source of plastid genome transfer to weeds could be extremely rare, paternal transmission of plastids detected under laboratory conditions in plants with an alien cytoplasm or in interspecific crosses. Such plastids,

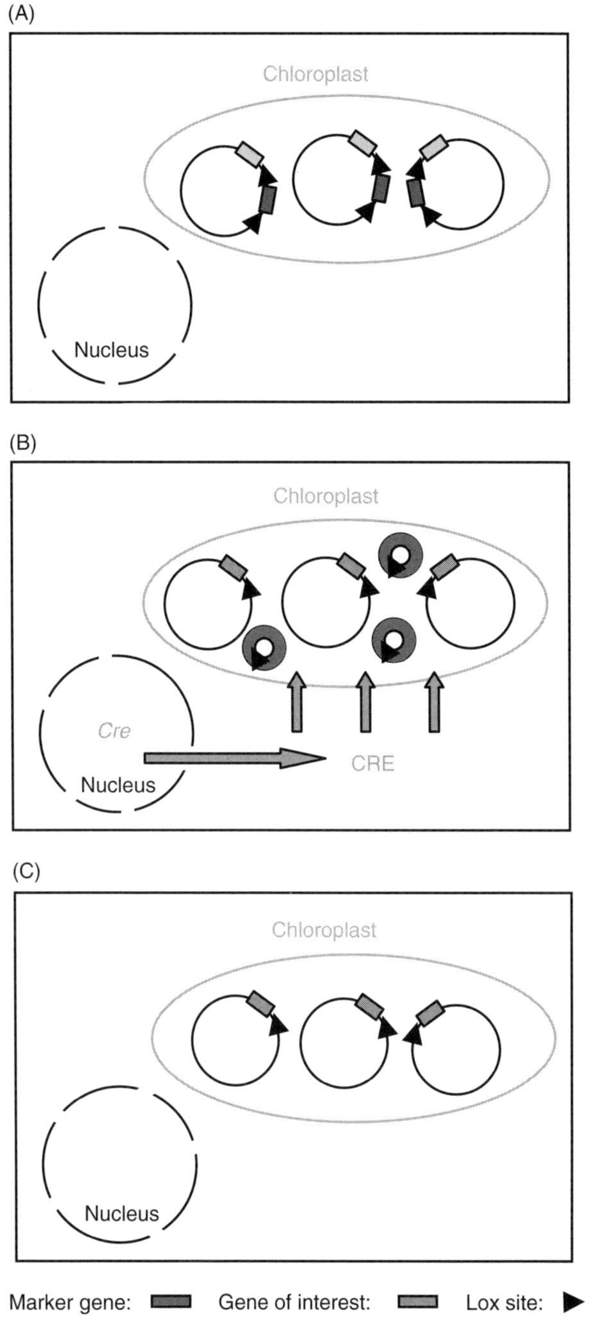

Figure 7 Removal of marker genes by the CRE-*lox* site-specific recombination system. (A) Plant cell with transformed plastid genome. Marker gene flanked by *lox* sites. The gene of interest is represented by light gray box. (B) The *Cre* gene is introduced into the nucleus. The plastid-targeted CRE enters the organelle and simultaneously excises the marker genes from all genome copies. Note both products, the circularized marker gene, and modified plastid genome. (C) The marker gene is rapidly lost in somatic cells. The nuclear *Cre* gene segregates out in the seed progeny. Only one *lox* site and the gene of interest are present in the plastid genome.

even if transfer occurs, are very unlikely to get established in the recipient weed in the light of difficulties to obtain homoplastomic plants. Nevertheless, the possibility of plastid genome transfer from future transplastomic crops to weedy relatives and their potential impact should be evaluated.

List of Technical Nomenclature

Heteroplasmic	Refers to the presence of different cytoplasmic genetic determinants determined by segregation in the seed progeny. Term was commonly used in the past when plastid or mitochondrial localization of a maternally inherited trait was not known.
Heteroplastomic	Refers to the presence of different plastid genomes.
Homoplasmic	Refers to lack of segregation of a maternally inherited trait. Term was commonly used in the past when plastid or mitochondrial localization of a maternally inherited trait was not known.
Homoplastomic	Refers to a uniform population of plastid genomes.
Integrating plastid vector	Integrates in plastid genome via targeting sequences. Episomal element (plasmid) in *E. coli*.
Plastid shuttle vector	Episomal element in plastids and *E. coli*.
Plastome	The genome of plastids.
Transformation of plastids	Stable transformation of plastids that yields nonsegregating seed progeny.
Transient gene expression in plastids	Expression of genes in plastids from introduced DNA that is not stably maintained.
Transplastome	Transformed plastid genome.
Transplastomic	Pertaining to the transformed plastid genome.

See also: **Crop Improvement**: Alien Cytoplasms. **Ethics and Biosafety**: Development and Commercialization of Genetically Modified Plants. **Genetic Modification, Applications**: Cell Factories; Molecular Farming; Plantibodies. **Genetic Modification**: Transformation in Dicotyledons; Transformation in Monocotyledons. **Growth and Development**: Control of Gene Expression, Regulation of Transcription; Field Crops. **Pests**: Genetic Modification of Pest Resistance, Insect Pests. **Weeds**: Herbicide Resistance.

Further Reading

Bendich AJ (1987) Why do chloroplasts and mitochondria contain so many copies of their genome. *Bioessays* 6: 279–282.

Bock R (2001) Transgenic plastids in basic research and plant biotechnology. *Journal of Molecular Biology* 312: 425–438.
Heifetz PB (2000) Genetic engineering of the chloroplast. *Biochimie* 82: 655–666.
Koop HU, Steinmüller K, Wagner H, *et al.* (1996) Integration of foreign sequences into the tobacco plastome via peg-mediated protoplast tranformation. *Planta* 199: 193–201.
Kuriowa T (1991) The replication, differentiation, and inheritance of plastids with emphasis on the concept of organelle nuclei. *International Review of Cytology* 128: 1–62.
Maliga P (2002) Engineering the plastid genome of higher plants. *Current Opinion in Plant Biology* 5: 164–172.
Maliga P (2003) Progress towards commercialization of plastid transformation technology. *Trends in Biotechnology* 21: 20–28.
Pyke KA (1999) Plastid division and development. *Plant Cell* 11: 549–556.
Staub JM (2002) Expression of recombinant proteins via the plastid genome. In: Parekh SR and Vinci VA (eds) *Handbook of Industrial Cell Culture: Mammalian, Microbial and Plant Cells*, pp. 261–280. Totowa: Humana Press Inc.
Sugiura M (1992) The chloroplast genome. *Plant Molecular Biology* 19: 149–168.
Thomas MR and Rose RJ (1983) Plastid number and plastid ultrastructural changes associated with tobacco mesophyll protoplast culture and plant regeneration. *Planta* 158: 329–338.
Whitney SM and Andrews TJ (2001) Plastome-encoded bacterial ribulose-1,5-bisphosphate carboxylase/oxygenase (RubisCO) supports photsynthesis and growth of tobacco. *Proceedings of the National Academy of Sciences of the USA* 98: 14738–14743.

Transgene Stability and Inheritance

H D Jones, Rothamsted Research, Harpenden, UK
M Cannell, Syngenta, Jealotts Hill International Research Centre, Bracknell, UK

Introduction

The improvement of crop performance by genetic engineering requires that inserted genes are expressed at the desired level, in the intended tissue, at the required stage of development, and/or following exposure to a specific stimulus. The ideal outcome of transgenic overexpression is that, for a given promoter/gene cassette, the number of transgene copies would correlate with the level of protein expression and this would be identical between independent lines. In addition, the transgenes would be faithfully inherited accordingly to Mendelian principles, giving rise to the desired phenotype in the descendants and derivatives of the primary transformant. However, since the first reports of plant transformation in the early 1980s, it has become obvious that not every transformation event with a particular gene construct expresses or transmits its transgene(s) to its progeny in the intended (or even a predictable) way. Different expression levels are often found between different events with the same transgene. Some of the possible causes of this variation are copy number, ploidy, and so-called "position effects." In addition, various forms of gene silencing, unintended insertional mutagenesis, somaclonal variation, and/or unexpected transgene inheritance patterns can be found. The molecular basis for much of this variation is poorly understood and, as a consequence, crop genetic engineers generally have to produce large numbers of transformants in order to identify the individuals with the appropriate number of transgene copies that exhibit and predictably inherit the required phenotype. This article will cover some aspects of current thinking on transgene stability and inheritance, the factors influencing them, and strategies employed to avoid instability.

Transgene Integration and Inheritance

The method of transgene delivery (most commonly *Agrobacterium*-mediated or by direct gene transfer such as particle bombardment) can impact on the complexity of transgene integration and inheritance. There are reports indicating that *Agrobacterium*-mediated transformation results in fewer transgene copies with a more simple integration pattern than direct methods. Both methods generally result in plants in which integration occurred at a single genetic locus and although single or multiple copies of the transgene may be present within a single locus, this locus segregates and is inherited as a single dominant allele, at a Mendelian ratio of 3:1. Inheritance studies indicate that the primary transformed plant possesses the transgene in the hemizygous state (the term hemizygous is used here in preference to heterozygous because there is no corresponding allele present on the homolog of the transformed chromosome (see **Figure 1**)).

Occasionally, plants might be recovered in which integration occurred at more than one genetic locus,

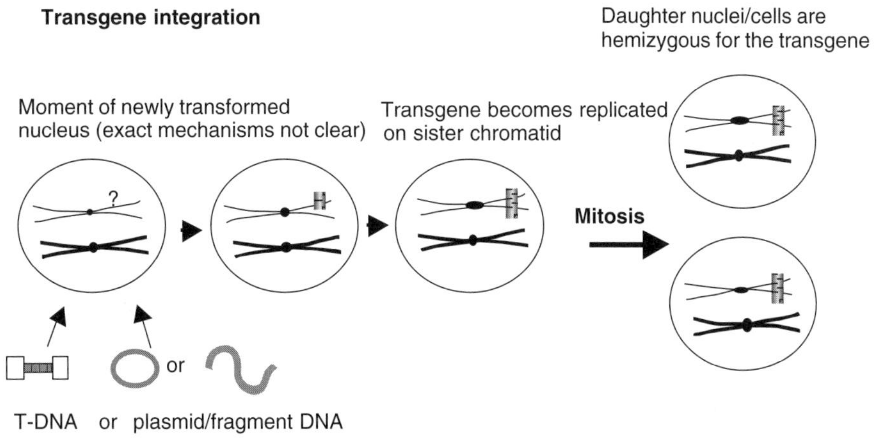

Figure 1 Transforming DNA is integrated at one or more loci and becomes replicated by mitosis resulting in daughter cells that are hemizygous for the transgene.

leading to independent segregation of two or more traits. This situation can sometimes be used to facilitate the removal-by-segregation of unwanted selectable marker genes. The relative spatial organization of the selectable and trait genes on the DNA vector is an important consideration. For example, different transgene sequences may be delivered simultaneously, either on whole plasmids or as linear fragments of DNA, and the selectable marker and trait genes may reside either on the same or on separate plasmid/fragment molecules. If they are on the same molecule prior to delivery (this approach is sometimes called cointegration), clearly it is unlikely that they will separate by segregation. However, if they are delivered on different molecules (called cotransformation), a proportion of transformants with the marker and trait genes at separate loci may be recovered. Traditionally *Agrobacterium* delivers a single transfer (T-DNA) fragment, flanked by left and right border sequences. Such an arrangement leads to the insertion of only one type of T-DNA-bounded transgene, resulting in transgenes being inseparable by segregation. Relatively recently, however, vector systems have been developed in which two different T-DNA fragments may be inserted during the same transformation experiment. As with cotransformation in direct gene delivery methods, these may generate transformants in which the marker and trait genes segregate independently.

Integration at a Molecular Level

The exact order and nature of the molecular events leading to the stable integration of a transgene into the host plant chromosome are not clear. However, it is likely that with the exception of a few specific applications and regardless of the method of DNA-delivery, genetic transformation involves double-stranded illegitimate recombination at one or multiple loci, utilizing the cell's nuclear repair machinery. The locus may contain one, or multiple, copies of the transgene, which may have undergone rearrangements and/or may have generated short lengths of "filler" DNA homologous to flanking plant genomic DNA. The moment of integration most likely coincides with a DNA metabolic event such as replication or transcription. Some *in situ* hybridization data suggest that physical transgene integration occurs at random within and between plant chromosomes; however, other analyses demonstrate a preference for distal chromosomal locations and actively transcribed regions in the plant genome with the possibility that native plant genes are disrupted in the process.

Variability in Transgene Expression

It is common to find differences in transgene expression between independent transgenic events with the same construct. There has been much discussion as to the basis of this variation. It is likely that the major contribution is from line to line variation in gene silencing modulated by factors such as copy number, transgene rearrangements, and position effect. The physical location of a transgene within the genome may modulate an increase in expression when inserted near to native enhancer elements or decrease it when in regions of condensed chromatin. Other factors known to have a direct effect on transgene expression are gene dosage via homo/heterozygosity and ploidy.

Gene Silencing

Gene silencing is a mechanism that downregulates or degrades specific genes and can propagate this effect to homologous sequences. It was once thought to be a quirk of the transformation process but it is now recognized as part of the plant's innate gene regulatory mechanisms. Transgene silencing probably reflects a manifestation of a natural "plant defense system" against invasive nucleic acids and also an inherent genome maintenance and regulatory system. The evolutionary drivers for its occurrence were most likely pathogenic attack (exogenous) and disruptive mobile DNA elements such as transposons and retrotransposons (endogenous). The combination of this surveillance system for nonself DNA and the fact that the most commonly used sequences in transgene cassettes are widely heterologous with respect to the host must certainly contribute to transgene silencing as a widespread phenomenon. It can occur naturally or be triggered by transgenes or viruses and involves some degree of DNA or RNA sequence homology. Gene silencing can operate at transcriptional and posttranscriptional levels and our understanding of the precise molecular components involved has increased enormously over the last 2 or 3 years. Silencing effects can be triggered by transgenes and by viral infection. Various approaches have been used in attempts to intentionally induce selective gene silencing in transgenic organisms, including antisense RNAs, chimeric RNA/DNA oligonucleotides, double-stranded RNAs, cosuppression, various virus-induced silencing methods, etc., but precise downregulation of targeted genes can be unpredictable and unintentional silencing of native or introduced genes in transformed plants may be observed. Additionally, transgene-mediated silencing does not necessarily occur in the primary transformant but can be triggered during subsequent clonal propagation, self-fertilization, or hybridization.

Mechanisms of Gene Silencing

One of the main mediators of transcriptional transgene silencing in response to genome perturbation is DNA methylation. Although the precise signals and order of events leading to *de novo* methylation of transgenes are not fully understood, it is clear that DNA–DNA or RNA–DNA interactions are crucially involved and that shared homology within integrated copies of transgenes can induce methylation. In some cases, it is directed specifically at the promoter region, and affects transcription probably by attracting factors that are able to form chromatin structures that are transcription repressive. These effects are most commonly seen within an integration event, where end-to-end insertional repeats or rearrangements may lead to the formation of secondary structures via DNA–DNA pairing and induce nucleotide methylation or chromatin condensation. Transcriptional gene silencing has also been observed in *trans*, where an insertion at one locus leads to silencing of other transgene insertions or native genes at distant genetic locations. Posttranscriptional gene silencing involves the formation of double-stranded RNA, which leads to sequence-specific RNA degradation. This type of gene silencing is also termed cosuppression because it has been observed in plants where sense-overexpression sometimes leads to homology-dependent downregulation of native and transgenes. However, these are rare events with the vast majority of sense transformation events giving rise to conventional overexpression phenotypes.

Chimerism

With the exception of germline transformation, transgenes are delivered into cultured protoplasts or explants of dividing plant tissue and it is assumed that different cells become transformed independently from one another. Many of the individual transformed cells (and nontransformed cells) have the potential to divide, somatically differentiate, and regenerate into a whole plant, and tissue culture conditions are manipulated to encourage this process. To allow transformed cells to grow preferentially, a selectable marker gene is delivered along with the trait gene, and the appropriate positive or negative selection pressure is applied during the plant culture phase. Tissue culture and selection procedures are generally intended to allow only whole plants that are derived from one original transformed cell to regenerate. However, some poorly optimized systems can give rise to plants that survive selection but are derived from a single cell that was not transformed. In some cases, these may be nontransgenic plants that simply escaped selection. In other cases, the transformation may have occurred in a cell or cells that were derived from the original progenitor cell and from which subsequent divisions lead to an unequal distribution of transformed and nontransformed tissue within the same whole plant. Transgenic sectors develop that may or may not incorporate gametic tissue. Such plants are chimeras and their existence is sometimes an explanation of poor or zero transgene inheritance.

Strategies to Avoid Instability

With our current state of knowledge it is impossible to predict the expression profile and inheritance pattern of every *de novo* transgenic event. However, strategies are emerging that are likely to increase the proportion of those displaying stable, expected phenotypes. These include low number of transgene copies with no rearrangements, low sequence homology between different introduced sequences and to native sequences, no viral/pathogenic sequences in the introduced DNA, short tissue-culture phase with low auxin concentrations to reduce somaclonal variation, and selection regimes that reduce the likelihood of chimeras. In addition, there is evidence that the inclusion of intronic or matrix attachment region (MAR) sequences into vector designs influences expression stability.

Introns

The inclusion of introns between promoter and transgene coding region has been shown to positively influence expression stability (intron-mediated enhancement, IME). For example, the most commonly used promoter for high-level constitutive transgene expression in dicotyledons is the 35S promoter of the cauliflower mosaic virus (CaMV35S). However, this promoter works poorly in monocotyledons unless linked to an intron such as the first one from the corn (*Zea mays*; maize) *adh* gene (*adh*1 intron). Similarly, the widely used monocotyledon promoter of the corn ubiquitin gene drives only weak expression unless coupled to the first intron of the gene. IME is often dependent on which promoter and/or gene is used and the precise mechanism of action is not fully understood. Many introns have been shown to increase expression levels and, interestingly, IME appears to occur at higher levels in monocotyledons than dicotyledons. From this knowledge, one would intuitively expect the presence of introns within transgene coding sequences to also have a positive effect on expression stability. However, a systematic comparison of expression levels from cDNAs vs genomic transgenes has yet to be carried out. The reason for the widespread use of cDNA sequences as transgenes rather than genomic clones is twofold. Foremost is the greater availability of cDNA sequences from ESTs (expressed sequence tags) and gene cloning projects and, secondly, it is generally easier to subclone them into cassettes containing different promoters and other DNA elements for expression analysis. Thus, although cDNAs have greater flexibility in everyday use than their genomic counterparts, it is possible that their widespread use has contributed to some of the transgene expression problems observed to date.

Matrix Attachment Regions

Matrix attachment regions (MARs) are A/T-rich DNA sequences, which bind to the protein nuclear matrix. They are frequently located in gene-rich areas associated with transcribed units and there is evidence that they can reduce or eliminate some forms of gene silencing. Genes flanked by MAR sequences have been shown to exhibit less variation in expression among independent transgenic plants or cell culture lines. In particular, MARs appear to prevent silencing caused by multiple transgene copies *in cis*. There is conjecture as to the precise role of MAR sequences in transgene cassettes, but current theories suggest that they may act as boundary elements facilitating the formation of decondensed chromatin loops that in turn enhance transgene transcription and reduce some forms of gene silencing.

Conclusions

The long-term expression profile of any particular transgene is influenced by several factors and our current understanding of these various factors is incomplete. There are, however, strategies that can be adopted to increase the chances of predictable long-term transgene expression and these will be refined with improved understanding of gene silencing, in particular.

See also: **Crop Improvement**: Mutation Techniques. **Tissue Culture and Plant Breeding**: Somatic Hybridization.

Further Reading

Grierson D, Lycett GW, and Tucker GA (1996) *Mechanism and Applications of Gene Silencing*. Nottingham: Nottingham University Press.

Hannon GJ (2002) RNA interference. *Nature* 418: 244–251.

Lindsey K (ed.) (1998) *Transgenic Plant Research*. Amsterdam: Harwood Academic.

Matzke MA and Matzke AJM (2000) Plant gene silencing (special issue). *Plant Molecular Biology* 43: 1–298.

Reznikoff W and Gold L (1986) *Maximising gene expression*. Burlington, MA: Butterworths.

Vance V and Vaucheret H (2001) RNA silencing in plants – defence and counter-defence. *Science* 292: 2277–2280.

Vaucheret H and Fagard M (2001) Transcriptional gene silencing in plants: targets, inducers and regulators. *Trends in Genetics* 17: 29–35.

GENETIC MODIFICATION, APPLICATIONS

Contents

Flower Color

T H Teeri and P Elomaa, University of Helsinki, Helsinki, Finland

Introduction

Biosynthetic routes to colored secondary metabolites, especially anthocyanins, have been under genetic and biochemical studies for decades before the advent of gene technology. Being dispensable, easy to evaluate and classify by eye, and harboring plenty of natural genetic variation, they have been the ideal targets for biochemical genetics. It is thus no surprise, that anthocyanins, or more generally phenylpropanoids, are maybe the best-known branch of secondary metabolism in plants, with nearly every step of the skeletal pathway characterized both at the enzyme and the gene level. Genetic modification of the anthocyanin pathway has been shown to be possible with a variety of ornamental plant species. However, anthocyanin-based flower color has more aspects than the composition of anthocyanins in the vacuole. Knowledge of regulation of the amount of pigments, of copigments, and of vacuolar pH has been accumulating more slowly, and many challenges in engineering flower color still exist.

Although anthocyanins are best known for their biosynthesis, they are not the only pigments found in flowers. Some plant species accumulate betalains, a second class of water-soluble pigments, instead of anthocyanins. All higher plants are able to synthesize carotenoids, which have essential roles in photosynthetic tissues, but which also are the basis of yellow to red coloration in flowers. Biosynthesis of carotenoids is not as well known as that of anthocyanins, but very important progress has taken place during recent years.

Biochemical Pathway to Anthocyanins

Anthocyanins are water-soluble pigments, which reside in vacuoles of plant cells as glycosides or acylglycosides. They belong to flavonoids, which is an important group of the phenylpropanoid secondary metabolites of plants. The first enzyme of the phenylpropanoid pathway, which links these metabolites to primary metabolites, is phenylalanine ammonia lyase (PAL). PAL catalyzes deamination of phenylalanine to *trans*-cinnamate, which then becomes hydroxylated and activated by linking it with coenzyme A through a thioester bond (**Figure 1**). The (hydroxy)cinnamic acid CoA thioesters are substrates for chalcone synthase (CHS), the first dedicated enzyme of the flavonoid pathway. CHS is a polyketide synthase, which adds two-carbon units from malonyl-CoA to a growing polyketide chain, which finally folds into a new aromatic ring structure. After stereospecific isomerization and hydroxylation in the heterocyclic ring, a crucial step of differential hydroxylation of the cinnamate derived aromatic ring occurs. After a step of a reductase and an oxygenase reaction, this leads to the three major classes of anthocyanins. Pelargonidins have a single hydroxyl group at the 4′ position, cyanidins two hydroxyl groups (at 3′ and 4′ positions) and delphinidins three (at 3′, 4′, and 5′ positions). Although the final color of anthocyanins is a result of several factors (see below), pelargonidins are generally of orange to scarlet hues, cyanidins of crimson to magenta, and delphinidins of mauve to blue hues.

The aglycones of anthocyanins are not stable in the physiological conditions of the vacuole where they are deposited. The simplest stable anthocyanins are 3-*O*-glucosylated (in the heterocyclic ring), but glycosylation may occur in other positions as well. Glycosylation increases also water solubility of anthocyanins and is a prerequisite for acylation of the carbohydrate moiety by carboxylic acids. Aromatic acylation changes the hue of anthocyanins towards blue. In addition to glycosylation and acylation, methylation of the hydroxyl groups of anthocyanidins is known to occur. Altogether, the variety of possible modifications is the basis of the

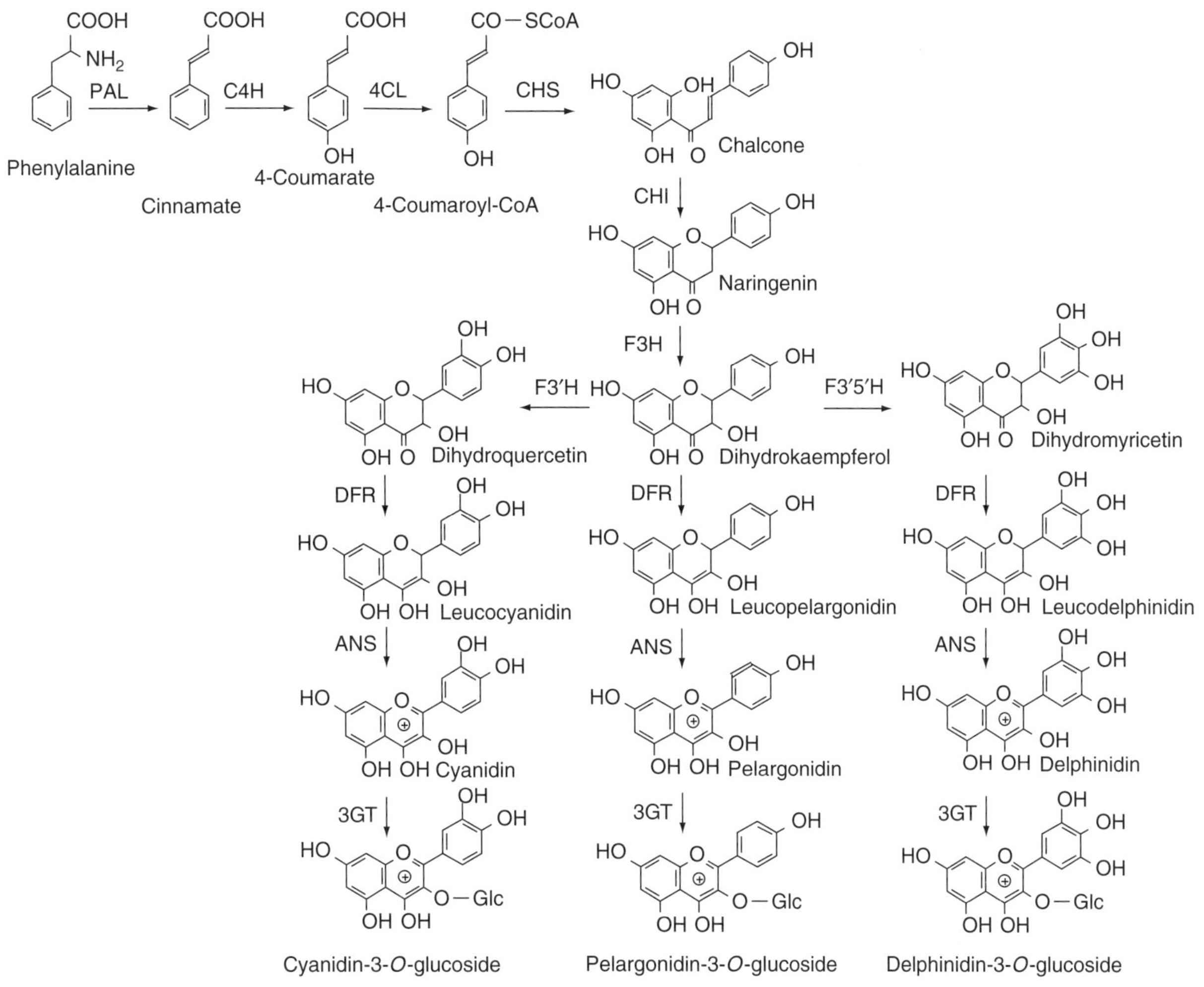

Figure 1 Biosynthetic pathway to anthocyanins. Phenylalanine, a primary metabolite, is converted in a series of enzymatic reactions to hydroxycinnamates, flavonoids, and finally to colored anthocyanins. Chalcone may be responsible for yellow flower color; cyanidins are generally colored crimson to magenta; perlagonidins range from orange to scarlet in color; and delphinidins are colored mauve to blue. Plants rarely express all three main branches of anthocyanin biosynthesis, e.g., often the blue delphinidins are missing because of no F3′5′H activity. Petunia lacks pelargonidins because the DFR enzyme does not accept dihydrokaempferol as substrate. PAL, phenylalanine ammonia lyase; C4H, cinnamate 4-hydroxylase; 4CL, 4-coumaroyl-CoA ligase; CHS, chalcone synthase; CHI, chalcone-flavanone isomerase; F3H, flavanone 3-hydroxylase; F3′H, flavonoid 3′-hydroxylase; F3′5′H, flavonoid 3′,5′-hydroxylase; DFR, dihydroflavonol 4-reductase; ANS, anthocyanidin synthase; 3GT, UDPG-flavonoid-3-*O*-glucosyltransferase.

fact that about 300 different anthocyanins have been isolated from nature.

Copigments, Metal Ions, and Vacuolar pH

Paradoxically, anthocyanins are colorless in the physiological pH range of the vacuole (pH 3–7). The strongly colored flavylium cation exists only at pH below 3, and the blue anions only at alkaline pH. Anthocyanins are nearly planar molecules with highly delocalized electrons, which allow the molecules to easily make noncovalent molecular complexes. This molecular stacking can take place between the anthocyanin molecules themselves (self-association) or with related or unrelated colorless compounds. The latter are called copigments, as they are able to intensify anthocyanin color in the physiological pH range of the vacuole without being colored themselves. In the molecular stacks, access of water molecules to the chromophore is limited, and the change in chemical surrounding leads to the variety of flower color observed in nature. Chelation of metal ions has also been shown to be important in intensifying anthocyanin color and in shifting the color towards blue.

Although copigmentation allows anthocyanins to display spectral properties that they do not have in the pH range of vacuoles, the vacuolar pH itself does play an important role in flower color. Several mutant lines of petunia are known where the vacuolar pH is increased, with a coincident bluing of the anthocyanin pigments. However, analysis of these genes has, as yet, not been informative of the cellular mechanisms responsible for regulation of vacuolar pH.

Biosynthesis of Carotenoids

Carotenoids are important compounds for all photosynthetic tissues, where they have roles in light harvesting and protection from photooxidative damage. However, often the intense yellow, orange, and sometimes red pigments in flowers are due to carotenoids accumulating in the petals. Carotenoids are water insoluble, and they reside in specialized plastids, chromoplasts, as protein complexes. They belong to the category of isoprenoid metabolites, and it has recently become clear that they are synthesized inside the plastids by a pathway that is independent of the pathway to cytoplasmic isoprenoids. Most enzymatic steps of carotenoid biosynthesis and their corresponding genes have been defined in plants. However, as carotenoids are synthesized also by many microbes, prokaryotic genes have been isolated as well, and have been important in genetic engineering of carotenoid biosynthesis.

The 40-carbon (C_{40}) skeleton of all carotenoids, phytoene, is synthesized from a C_{20} precursor, which itself is composed of four C_5 isoprene units. The colorless phytoene is desaturated in a series of reactions into pink lycopene which, with its highly conjugated double-bond system, can be considered to be the chromophore of carotenoid pigments (**Figure 2**). Two cyclases act on the ends of the molecules to produce the bicyclic α- and β-carotenes, and hydroxylases, ketolases, and epoxidases act on the ring structures to produce oxygenated carotenoids, or xanthophylls.

Carotenoid biosynthesis forms a complex and reticulated system of enzymatic reactions, and by combining enzymes from different organisms it is becoming evident that an enormous and unexplored potential exists for engineering of these compounds. Utilization of genetic engineering of carotenoids as flower pigments has only started, with rerouting of carotenoid biosynthesis in tobacco (*Nicotiana tabacum*) floral parts from the yellow β-carotene and violaxanthin to intensely red astaxanthin using a gene for a microbial β-carotene ketolase.

Transgenic Downregulation of Pigment Biosynthesis

Anthocyanin biosynthesis can easily be blocked at various stages of the pathway using antisense or sense transgenes. The first examples of these approaches were published as early as 1988 and 1990, and demonstrated suppression of petunia flower pigmentation using chalcone synthase transgenes in antisense and sense orientation, respectively. Ever since, as a result of development of routine transformation protocols for many ornamentals, the number of new target species has increased. Still, the resulting transgenic phenotypes seem to follow very similar rules in all species studied so far.

In the antisense approach the transgene is cloned in reverse orientation under a promoter element. This kind of gene construct produces RNA complementary to the RNA transcribed from the endogenous gene. Suppression of the target gene expression,

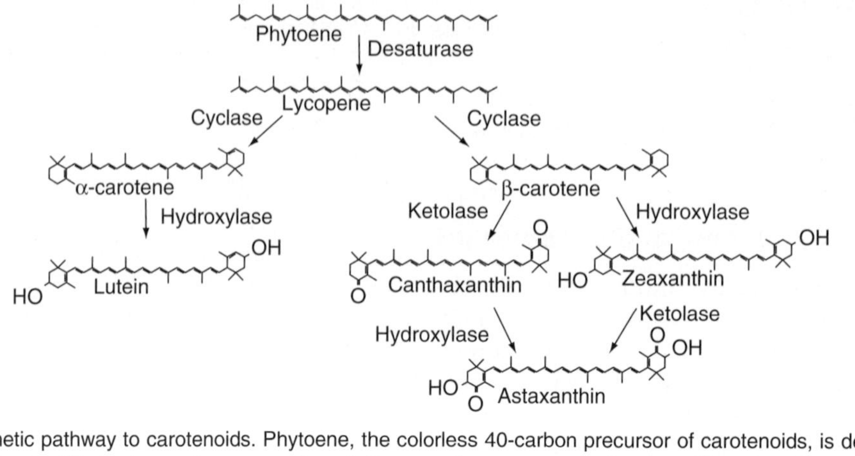

Figure 2 Biosynthetic pathway to carotenoids. Phytoene, the colorless 40-carbon precursor of carotenoids, is desaturated into the pink lycopene. In this simplified scheme, the enzymatic activities are given as generic, i.e., two different cyclases are responsible for formation of the two types of ring structures at the ends of α- and β-carotenes. The pathway is reticulated, i.e., the hydroxylation and oxidation reactions that convert β-carotene to astaxanthin may take place independently at the two ends of the molecule and in either order, resulting in a large number of intermediates not shown here. Lycopene, canthaxanthin, and astaxanthin are pink to red in color; α-carotene and β-carotene are orange; lutien and zeaxanthin are yellow in color.

which eventually prevents pigment biosynthesis, results from sequence specific, posttranscriptional RNA degradation, the details of which are only gradually unfolding. Typically, in individual transgenic lines, reduction of pigmentation to different degree is achieved. In many cases, transformation of a sense gene also leads to suppression of target gene expression, and it seems probable that at least in some cases these two approaches share common mechanisms. In addition to petunia, pale or fully white-flowering transgenic crops have been produced in chrysanthemum (*Dendranthema grandiflora*), gerbera (*Gerbera hybrida*), rose (*Rosa hybrida*), carnation (*Dianthus caryophyllus*), lisianthus (*Eustoma grandiflorum*), and torenia (*Torenia hybrida*) by blocking anthocyanin biosynthesis at the chalcone synthase and/or dihydroflavonol-4-reductase step of the pathway. Variegated pigmentation patterns such as sectors, stripes, and star patterns have been observed in petunia and lisianthus while in torenia more simple patterns were obtained. Especially in petunia, transformation of the sense chalcone synthase gene led to complex color patterns and was correlated with complex transgene integration patterns as well as with environmental effects. Pattern formation seems to be associated with species like petunia and lisianthus, which also show white-sectored floral forms in nature. Still, the primary cause for the phenomenon is unclear and actually not much studied. If stabilized, these novel patterned transgenic varieties may possess of more commercial value in future, while in general the suppressed pigmentation phenotypes benefit the studies on gene silencing mechanisms in plants more generally.

Opening New Pathways

Orange Pelargonidins in Petunia

No ornamental crop provides us colors at full spectrum. The very first example of flower color modification in plants, over a decade ago, introduced synthesis of orange pelargonidin pigments in petunia, which normally shows only red, white, and purple colors in nature. This is due to the limited substrate specificity of the petunia dihydroflavonol-4-reductase which cannot convert dihydrokaempferol into pelargonidin (**Figure 1**). Genetic modification of petunia took advantage of profound knowledge of anthocyanin biosynthesis in this species. Pelargonidin producing petunias were obtained by transformation of the corn (*Zea mays*; maize) *A1 (dfr)* gene into petunia mutant RL01 that accumulates dihydrokaempferol. The corn enzyme has the capability of using this compound as a substrate and thereby the transformation opened up a completely new biosynthetic route to pelargonidins in petunia. The resulting brick-red transgenic lines, however, showed significant instability in pigmentation due to methylation of the CaMV35S promoter used to control the transgene's transcription. Traditional breeding was applied to stabilize pelargonidin production and the novel color was introduced into commercially important elite varieties. Transformation of the dicot *dfr* gene isolated from *G. hybrida* led to similar brick-red phenotypes in RL01. However, in this case, no epigenetic instability was observed suggesting that the gerbera gene, with fewer putative methylation sites, probably fits better the genomic organization in petunia and is prevented from transgene inactivation (**Figure 3**).

Yellow Deoxychalcones in Petunia

In many ornamental plants yellow flower colors are due to carotenoids, but chalcones, important intermediates of the anthocyanin pathway, can also be responsible for yellow flower color. In petunia, suppression of chalcone isomerase activity, which would lead to accumulation of yellow hydroxychalcones, has been unsuccessful. Even if it would be possible, chalcones with 2′- and 6′-hydroxyl groups are probably spontaneously isomerized to flavanones and further synthesized to anthocyanins. In yellow *chi* mutants of carnation, chalcone is stabilized via glycosylation of the 2′ position that prevents its isomerization to naringenin. In 6′-deoxychalcones, which are known to contribute to yellow colors in, for example, *Dahlia* and *Cosmos* species, the isomerization rate is even lower and further the endogenous chalcone-flavanone isomerase (CHI) cannot act on them. In petunia, production of 6′-deoxychalcones was achieved by transformation with a chalcone reductase (CHR) cDNA from alfalfa (*Medicago sativa*, lucerne). CHR together with the endogenous petunia CHS produces 6′-deoxychalcones and in a white-flowered background this led to formation of pale yellow phenotypes. In a cyanic background, reduced flower color was obtained due to rerouting of the substrates into chalcones. This provides a second example of opening completely new pathways and engineering of novel flower colors.

Rerouting Pigments

Along the anthocyanin pathway there are branch points where common substrates are shared by several enzymes. The ability to control expression levels of these genes has allowed rerouting of the

Figure 3 (see color plate 15) A gerbera gene in petunia opens pathway to pelargonidin glycosides. (A) The petunia line RL01 has mutations in genes coding for F3′H and F3′5′H activities and accumulates dihydrokaempferol that is not a substrate for petunia dihydroflavonol 4-reductase (DFR). (B) Transformation of RL01 petunia with a gerbera gene for DFR results in conversion of dihydrokaempferol into leucopelargonidin, which is further converted by petunia enzymes to the brick-red/orange colored pelargonidin glycosides. (C) The pathway from chalcone to pelargonidin glycosides in transgenic petunia. The genetic blocks of the RL01 mutant in flavonoid 3′ and 5′ hydroxylation are shown by crosses.

pathway towards different end products and thereby modifications of flower color. For example, flavonol synthase (FLS), needed for flavonol (copigment) production, and dihydroflavonol-4-reductase (DFR) both act on dihydroflavonol substrates. In petunia, suppression of FLS activity using antisense transformation led to formation of redder flowers due to decreased flavonol production. Also in tobacco, expression of the petunia anti-FLS construct caused reduction of flavonol production and even more pronounced increase of anthocyanin production. Similarly, dihydrokaempferol serves as the substrate for four different enzymes (DFR, FLS, F3′H, F3′5′H (see below)). In torenia, cosuppression of F3′5′H turns the pathway from delphinidin to mainly cyanidin (and slightly to pelargonidin) and changes flower color from blue to pink. Also enzymes not involved in flavonoid biosynthesis have been shown to affect metabolite flow to anthocyanins. Stilbene synthases (STS) use the same substrates as CHS (phenylpropanoid-CoA and malonyl-CoA esters), but synthesize colorless stilbenes instead of precursors to anthocyanins. High activity of grapevine (*Vitis vinifera*) STS in tobacco, introduced using gene transfer, exhausts the pool of 4-coumaroyl-CoA, resulting in almost white flowers.

Genetic Modification of the Regulation of Pigment Biosynthesis

Transcriptional regulators of anthocyanin biosynthetic genes have been characterized in an increasing number of species, most importantly in corn, petunia, and snapdragon (*Antirrhinum majus*). These structurally conserved proteins are encoded by two gene families, *myb* and *bHLH*. In corn, a member of the *myb* gene family (*C1, Pl*) and a member of the *bHLH* family (*R, Sn, Lc, B*) are both needed for tissue specific activation of all the biosynthetic genes. In petunia, the loci *an1*, *an2*, and *an11* control floral pigmentation. Cloning of *an1* and *an2* showed that also these genes code for bHLH- and MYB-type proteins, respectively, which activate the late biosynthetic genes (starting from the *dfr*) of the anthocyanin pathway. The *an11* locus encodes a novel cytosolic protein that is most likely involved in posttranslational modification of AN2. Anthocyanin regulators in combination with the constitutive CaMV35S promoter or tissue specific promoters can be used to create especially vegetative pigmentation patterns of potential ornamental value. Furthermore, production of anthocyanins at very high concentrations would perhaps lead to appearance of "black" flower color in the target species.

There are some examples where regulatory genes have been used for color modification, but none of these has been directly for breeding purposes. Corn activators have been shown to be able to induce anthocyanin pigmentation in various dicot species. In *Arabidopsis* and tobacco (*Nicotiana tabacum*), ectopic expression of the corn *R* enhanced anthocyanin pigmentation in tissues that in wild-type plants were faintly pigmented. C1 alone had no effect. However, in *Arabidopsis*, crossing of the *R* and *C1* expressing lines led to production of anthocyanins in roots, petals, and stamens, i.e., in tissues that normally are acyanic. Another *bHLH* gene of corn, *Lc*, enhanced pigmentation in transgenic petunia both in vegetative and floral tissues. Interestingly, in addition to the strong activation of the late biosynthetic genes, *Lc* also weakly activated some of the early biosynthetic genes (*chs, chi,* and *f3h*). Furthermore, anthocyanins accumulated to the subepidermal layers of the flower limb while they normally are located in epidermal cells. The *Antirrhinum* gene *delila* (*bHLH*) has been shown to induce pigmentation in tobacco and in tomato (*Lycopersicon esculentum*). In tobacco, enhanced pigmentation was restricted to flowers while in tomato, vegetative pigmentation was strongly increased. Similarly, the *delila* cDNA under CaMV35S promoter strongly induced vegetative anthocyanin pigmentation in gerbera. These examples prove that the whole anthocyanin pathway can be modified by transformation of single or few regulatory genes. However, we still need more functional knowledge of the various regulators in order to be able to create specific and controlled pigmentation patterns in ornamental plants.

The Challenge of Blue Color

The production of a blue rose via genetic modification of the anthocyanin pathway is still a dream, which has, however, come closer to reality. The final blue hue in flowers is affected by multiple factors. Prerequisite for blue is the presence of 3′,4′,5′-hydroxylated delphinidin pigments, which are synthesized by flavonoid-3′5′-hydroxylase (F3′5′H), a cytochrome P-450 enzyme, from dihydrokaempferol (**Figure 1**). Species like rose and carnation lack F3′5′H activity and are unable to produce blue flower colors. Genes encoding F3′5′H have been isolated from many different plant species and have already been applied to flower color modification. The world's first commercialized transgenic flowers are the violet and mauve carnations (Moondust^{TM} and Moonshadow^{TM}), which were created by expression of the petunia *f3′5′h* and *dfr* genes in a white variety of carnation.

Recently, a gene coding for cytochrome b_5 protein in petunia has been shown to affect the activity of F3′5′H. The gene was identified among cDNA clones that were downregulated in an *an1* regulatory mutant. Targeted inactivation of this gene resulted in reduction of F3′5′H activity and in a change of flower color. The exact mechanism of how cytochrome b_5 affects the hydroxylase activity is still unclear, but the finding may have useful applications in enhancing the production of blue pigments in transgenic ornamentals.

The structural modification by aromatic acylation has been shown to stabilize anthocyanins and to intensify blue colors due to intra- and/or intermolecular copigmentation. The effect of bluing and stabilization increases with increasing number of aromatic moieties. Anthocyanin 5-aromatic acyltransferase (5AT) has been purified from gentian (*Gentiana*) petals which contain a novel anthocyanin, gentiodelphin, acylated with caffeic acid at 5- and 3′-glucose of the delphinidin 3,5,3′-triglucoside. The corresponding cDNA has been also isolated and the expression of the gene has been shown to be restricted to petals in coordination with other biosynthetic genes (*f3′5′h*, *dfr*, and *3gt*) in this species. Genes encoding aromatic acyltransferases may provide another important group of genes for applied purposes.

In addition to accumulation of properly modified delphinidin pigments, sufficiently high vacuolar pH and copigment levels are required for obtaining blue flowers. In petunia, *an1*, *an2*, and *an11* regulatory genes are known to modify the pH of petal extracts. Furthermore, seven loci (*ph1–ph7*), when mutated, increase the pH of petal extracts and cause bluing of the flower without affecting the anthocyanin composition. The cloning of the corresponding genes and analysis of their functional roles will provide us much knowledge of regulation of various aspects of flower color, other than simply the biosynthesis of the chromophores.

Conclusion

Many molecular tools for genetic modification of flower color are today available, e.g., the isolation of genes responsible for anthocyanin biosynthesis in any given application plant is relatively easy using the wealth of sequence information of the corresponding genes in other plant species. Also, producing flowers with white or pale color by blocking anthocyanin biosynthesis using these genes is straightforward once a technique of gene transfer is worked out. However, controlled rerouting of metabolites of the anthocyanin pathway is more difficult, and requires profound knowledge of the genetic background of the target species as well as kinetic properties of the corresponding enzymes. The many examples including petunia are a reflection of the very good classical genetics of anthocyanin pathway in this species, especially availability of well-characterized mutants.

The slow progress in producing blue flower color in species where blue does not occur naturally demonstrates that there is more than the particular chemical species of anthocyanins involved. Apart from the biochemical and chemical aspects addressed here, also morphological features come to play. Some flower "color" mutants in *Antirrhinum* show that the conical shape of the epidermal cells has a profound effect on how light is reflected on the surface of the petal. Flat epidermal cells have altered optical properties and appear lightly pigmented although their anthocyanin content was not changed.

Transgenic modification of flower color is still far from engineering in the sense that a given plan could be fulfilled in a given ornamental. Much interesting progress is still in sight.

List of Technical Nomenclature

Antisense — RNA synthesized from the opposite strand of DNA than mRNA; pairs with mRNA.

Promoter — Genetic element regulating expression of a gene.

Secondary metabolite — Natural product with no immediately obvious function for cell growth.

Silencing — Genetic phenomenon where a transferred gene causes downregulation of the endogenous gene (and of itself).

See also: **Genetic Modification**: Transgene Stability and Inheritance. **Flowering and Reproduction**: Flower Development.

Further Reading

Cunningham FX Jr. and Gantt E (1998) Genes and enzymes of carotenoid biosynthesis in plants. *Annual Review of Plant Physiology and Plant Molecular Biology* 49: 557–583.

Davies KM, Bloor SJ, Spiller GB, and Delores SC (1998) Production of yellow colour in flowers: redirection of flavonoid biosynthesis in *Petunia*. *Plant Journal* 13: 259–266.

De Vetten N, Ter Horst J, Van Schaik H-P, *et al.* (1999) A cytochrome b_5 is required for full activity of flavonoid 3′5′-hydroxylase, a cytochrome P-450 involved in the formation of blue flower colors. *Proceedings of the National Academy of Sciences of the USA* 96: 778–783.

Harborne JB (1994) *The Flavonoids: Advances in Research Since 1986*. London: Chapman & Hall.

Holton TA, Brugliera F, and Tanaka Y (1993) Cloning and expression of flavonol synthase from *Petunia hybrida*. *Plant Journal* 4: 1003–1010.

Lloyd AM, Walbot V, and Davis RW (1992) *Arabidopsis* and *Nicotiana* anthocyanin production activated by maize regulators *R* and *C1*. *Science* 258: 1773–1775.

Mann V, Harker M, Pecker I, and Hirschberg J (2000) Metabolic engineering of astaxanthin production in tobacco flowers. *Nature Biotechnology* 18: 888–892.

Meyer P, Heidmann I, Forkmann G, and Saedler H (1987) A new petunia flower colour generated by transformation of a mutant with a maize gene. *Nature* 330: 677–678.

Mol J, Grotewold E, and Koes R (1998) How genes paint flowers and seeds. *Trends in Plant Science* 3: 212–217.

Mol J, Cornish E, Mason J, and Koes R (1999) Novel coloured flowers. *Current Opinion in Biotechnology* 10: 198–201.

Napoli C, Lemieux C, and Jorgensen R (1990) Introduction of a chimeric chalcone synthase gene into petunia results in reversible co-suppression of homologous genes in *trans*. *Plant Cell* 2: 279–289.
Tanaka Y, Tsuda S, and Kusumi T (1998) Metabolic engineering to modify flower color. *Plant Cell Physiology* 39: 1119–1126.
Van Der Krol AR, Lenting PE, and Veenstra J (1988) An antisense chalcone synthase gene in transgenic plants inhibits flower pigmentation. *Nature* 333: 866–869.
Van Der Krol AR, Mur LA, Beld M, Mol JNM, and Stuitje AR (1990) Flavonoid genes in petunia: addition of a limited number of gene copies may lead to a suppression of gene expression. *Plant Cell* 2: 291–299.

Nitrogen Fixation

L Skøt, Institute of Grassland and Environmental Research, Aberystwyth, UK

Introduction

Nitrogen is the most important plant nutrient, and is often limiting for growth and survival, despite its abundance as N_2 gas in the atmosphere. The reason is that nitrogen in this form is unusable by plants. The triple bond in the N_2 molecule gives it very high thermal stability, and it requires large amounts of energy and highly reducing conditions to reduce N_2 to NH_3 (NH_4^+), the form in which nitrogen can be assimilated by plants. The majority of plants depends on reduced nitrogen in the soil in the form of nitrate (NO_3^-), most of which is formed from microbial mineralization of organically bound nitrogen. However, global agricultural crop production is increasingly dependent upon input of industrially fixed nitrogen fertilizer to maintain yields and feed an increasing world population. A lot of energy (usually fossil fuels) is required for industrial production of NH_3 from N_2. According to FAO, 82 million tonnes of fertilizer nitrogen was used in 1998. Apart from the high costs involved in producing fertilizer nitrogen, both economically and in terms of energy, contamination of water supplies and evaporation of nitrous oxide gases are other unfortunate consequences of their use.

A number of prokaryotic organisms can, however, reduce N_2 to NH_4^+, using an enzyme, nitrogenase, thereby supplying themselves with nitrogen for survival and growth. Globally, biological N_2 fixation is estimated to contribute approximately 140 million tonnes of N on an annual basis. Some of these diazotrophs (see **Table 1**) are free-living, while others form associations or symbioses with higher organisms, including plants, most notably of the legume family, providing the host organism with reduced nitrogen in return for a carbon source. These symbioses have been the subject of intense research, particularly over the last 25–30 years. One of the aims is to gain sufficient understanding of the phenomenon to enable transfer of this trait to some of the world's most important crop plants such as rice (*Oryza sativa*), corn (*Zea mays*, maize), and wheat (*Triticum aestivum*). This would have a major impact on agricultural production, reducing the need for fertilizer nitrogen, and is perceived to be environmentally more benign and sustainable.

In this article some important properties of nitrogenase, the key enzyme in biological N_2 fixation, will be described, with an account of how they have influenced the development of both symbiotic and nonsymbiotic N_2 fixing organisms. The development and functioning of the agriculturally important symbioses between legumes and rhizobia is described in some detail to provide a context for the subsequent discussion of strategies for genetic manipulation of N_2 fixation.

Nitrogenase Is Key

The nitrogenase enzyme complex responsible for biological N_2 fixation has been studied in greatest

Table 1 Examples of diazotrophic organisms

	Diazotrophic organisms			
	Autotrophic organisms		*Heterotrophic organisms*	
Cyanobacteria	Heterocystous filamentous forms	*Anabaena* *Nostoc*	Aerobic	*Azotobacter vinelandii* *Beijerinckia* *Derxia*
	Nonheterocystous filamentous forms	*Lyngbya* *Trichodesmium*	Microaerobic	*Azospirillum* *Rhizobium*
	Unicellular forms	*Gloeothece* *Synechococcus*	Facultative anaerobic	*Klebsiella pneumoniae* *Enterobacter*
Other phototrophic bacteria		*Rhodospirillum* *Rhodobacter*	Anaerobic	*Clostridium*

detail in free-living diazotrophs such as *Klebsiella pneumoniae*. Nitrogenase consists of two protein components, the homodimeric Fe-protein, the structural subunits of which are encoded by the *nifH* gene, and the heterotetrameric MoFe-protein encoded by the structural genes *nifD* and *nifK*. Metalloclusters are associated with both components. The FeMo-cofactor is believed to be the substrate-binding site. For every electron transferred from the Fe-protein to the FeMo-protein two MgATP molecules are hydrolyzed to MgADP + P_i. The translation products of the structural *nifHDK* genes require processing by a suite of other *nif* gene products before becoming active. A total of 16 *nif* genes are necessary to synthesize an active nitrogenase enzyme complex in *K. pneumoniae*.

It appears that no N_2-fixing eukaryotic organisms have evolved. The reason is not clear, but may be important for evaluating the prospects of transferring N_2 fixation to crop plants. A key property of nearly all nitrogenase enzyme systems studied so far is their sensitivity to oxygen. The enzyme is inactivated within seconds of exposure to air. The importance of this feature is emphasized by the varied strategies diazotrophs have developed to avoid exposure of nitrogenase to O_2. *Klebsiella pneumoniae* is a facultative anaerobe, and the *nif* genes are only expressed under anaerobic conditions. In some photosynthetic filamentous cyanobacteria (blue–green algae), N_2 fixation takes place in specialized cells, the heterocysts. There, the O_2 developing step of the photosynthesis, photosystem II, is absent, thus avoiding exposure of nitrogenase to oxygen. In the filamentous, nonheterocystous, and the unicellular cyanobacteria, as well as the phototrophic bacteria, nitrogenase is either expressed at night, avoiding photosynthetically developed O_2, or the organisms have other means of protecting the enzyme from oxygen. Protection of nitrogenase from O_2 has thus played a major role in the evolution of N_2-fixing organisms. Even so, N_2 fixation in most species is restricted to anaerobic or microaerobic conditions. However, there are exceptions. *Azotobacter vinelandii* is an obligate aerobic diazotroph, but the nitrogenase complex is protected from O_2 by conformational changes and respiratory protection. Furthermore, the recent discovery of an oxygen-tolerant nitrogenase from *Streptomyces thermoautotrophicus* demonstrates that nitrogenase activity is possible in the presence of O_2. In this species reduction of N_2 is coupled to the oxidation of CO to CO_2. The electrons are transferred to O_2, thereby producing a superoxide anion radical. This is reoxidized to O_2, and the electrons are transferred to the dinitrogenase enzyme for reduction of N_2. However, the structure of this nitrogenase complex appears to be quite different from other known nitrogenases.

Symbiotic Nitrogen Fixation

Some cyanobacteria form symbiotic associations with various eukaryotic organisms, including lichens, ferns, and angiosperms. The actinomycete *Frankia* fixes N_2 in symbioses with woody plant species of eight different nonleguminous families by forming root nodules. Endophytic diazotrophs such as *Acetobacter diazotrophicus* and *Herbaspirillum* spp. fix N_2 under microaerobic conditions. They have been found in large numbers in both root and aerial tissue of sugar cane (*Saccharum officinarum*), rice, and other graminaceous plants. However, these microorganisms do not induce formation of separate organs in the plant.

By far the most important symbiotic diazotrophs in agriculture are the soil bacteria, broadly termed rhizobia. They form symbiotic associations by inducing the formation of nodules on the root system of their legume host plants, providing optimal conditions for N_2 fixation. Soybean (*Glycine max*), pea (*Pisum sativum*), and peanut (*Arachis hypogaea*) are examples of some of the world's most important legume crops, providing a vital source of protein. The following section will deal with the biology of symbiotic associations in legume nodules, to provide a context for discussing possible ways of manipulating legume symbioses or transferring this ability to other species.

The Legume–*Rhizobium* Symbiosis

Phylogenetic studies based on ribosomal RNA sequences have demonstrated great genetic diversity among rhizobia, and they are now separated in five different genera: *Azorhizobium, Bradyrhizobium, Mesorhizobium, Rhizobium,* and *Sinorhizobium.* Host specificity is a prominent feature with, for example *Sinorhizobium meliloti* nodulating alfalfa (*Medicago sativa,* lucerne) while *Bradyrhizobium japonicum* nodulates soybean exclusively. **Table 2** lists different rhizobia and their host plants.

The formation of legume nodules is the culmination of a complex interaction between rhizobium bacteria and the host plant. In the best-studied symbioses it is initiated by the rhizobium-induced curling of root hairs, followed by localized hydrolysis of the cell wall in the curls. Concomitantly, rhizobium triggers the activation of mitotic activity in certain cortical root cells, initiating the formation

Table 2 Rhizobium species and their host plant specificities

Rhizobium species	*Host plant*
Azorhizobium caulinodans	*Sesbania*
Bradyrhizobium elkani	Soybean (*Glycine max*), cowpea (*Vigna unguiculata*)
Bradyrhizobium japonicum	Soybean (*Glycine max*)
Mesorhizobium loti	Bird's-foot trefoil (*Lotus* sp.)
Rhizobium leguminosarum	
bv *phaseoli*	Bean (*Phaseolus vulgaris*)
bv *trifolii*	Clover (*Trifolium* spp.)
bv *viciae*	Pea (*Pisum sativum*), field bean (*Vicia faba*)
Rhizobium etli	Bean (*Phaseolus* spp.), cowpea (*Vigna unguiculata*)
Rhizobium tropici	Beans (*Phaseolus* spp.), *Leucaena*
Sinorhizobium fredii	Soybean (*Glycine* sp.), bean (*Phaseolus* spp.), cowpea (*Vigna unguiculata*)
Sinorhizobium meliloti	Alfalfa (*Medicago sativa*), sweet clover (*Melilotus* spp.)
Sinorhizobium NGR234	Broad host range

of a nodule primordium. The plant cell membrane invaginates in the curls, and new cell wall material is deposited. The resulting tubelike structure, the infection thread, containing the invading rhizobia, continues to grow and branch through the outer cortical cell layers until it reaches the developing nodule primordium in the inner cortex. In an endocytosis-like process the rhizobia are released into the cytoplasm of some of the primordial cells, and become surrounded by a membrane derived from the infection thread, in a structure called the symbiosome, which divides and eventually fills the infected cells. The rhizobia differentiate to N_2-fixing bacteroids in the symbiosome. The infected tissue is surrounded by a peripheral nodule cortex with vascular bundles. Fixed nitrogen, assimilated into amino acids or ureides is exported through the xylem to the plant, while sucrose from the plant enters the root nodule through the phloem (**Figure 1**).

Other forms of invasion by rhizobia exist, most notably in peanut, and other tropical legumes. In these, rhizobia invade directly by "crack entry" in spaces between epidermal cells at junctions with lateral roots. Infection threads are not formed, and the rhizobia enter the host cells directly by endocytosis. *Azorhizobium caulinodans* forms both stem and root nodules on its host *Sesbania rostrata*. The stem nodules are formed by crack entry through dormant root primordia along the stem.

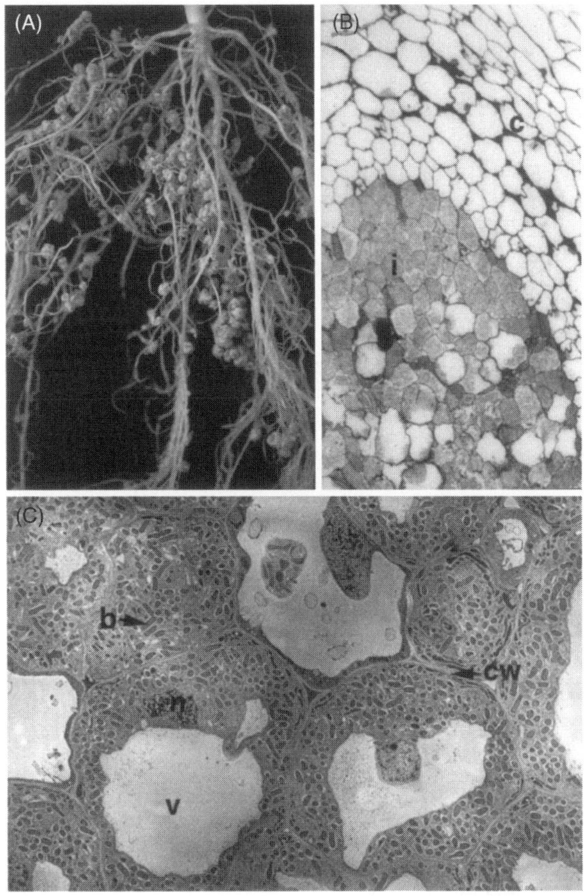

Figure 1 Legume root nodule structure. (A) Nodulated root of soybean plant. (B) Light micrographic section of soybean root nodule with cortex and central infected tissue. (C) Electron micrograph (magnification × 2000) of white clover nodule cells infected with rhizobium bacteroids. c, cortex; i, infected tissue; v, vacuole; n, nucleus; b, bacteroids; cw, cell wall.

Signaling in the Legume-*Rhizobium* Symbiosis

Legume nodules form as a result of a complex series of signal exchanges and gene activation sequences. In symbioses such as *Medicago–Sinorhizobium*, *Glycine–Bradyrhizobium*, and *Lotus–Mesorhizobium*, the product of the regulatory *nodD* gene from rhizobium interacts with flavonoids or betaines secreted from the roots of the host plant, and activates transcription of a whole suite of rhizobial nodulation genes (*nod*, *nol*, and *noe*). Activation of the *nod–nol–noe* gene regulon in turn leads to the biosynthesis of the Nod factor (**Figure 2**). When the host plant perceives this signal molecule, it triggers root hair deformations, preinfection thread formation, and initiation of cortical cell divisions, the first steps in nodule morphogenesis, as well as induction of expression of certain plant genes.

Rhizobial Nod factors all have the same basic molecular structure (**Figure 3**). They consist of a

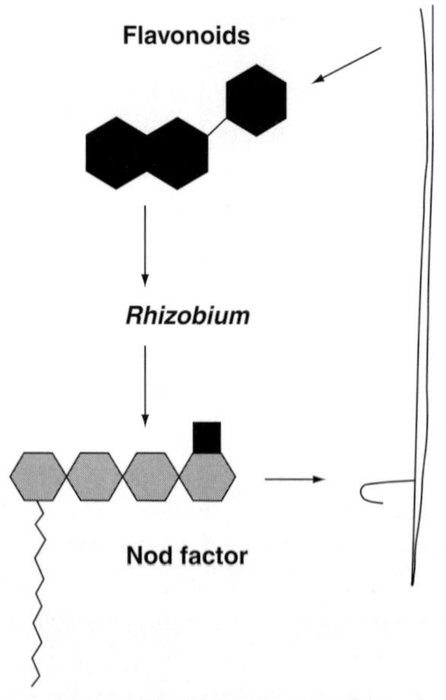

Figure 2 Schematic illustration of molecular signaling events during legume root nodule initiation.

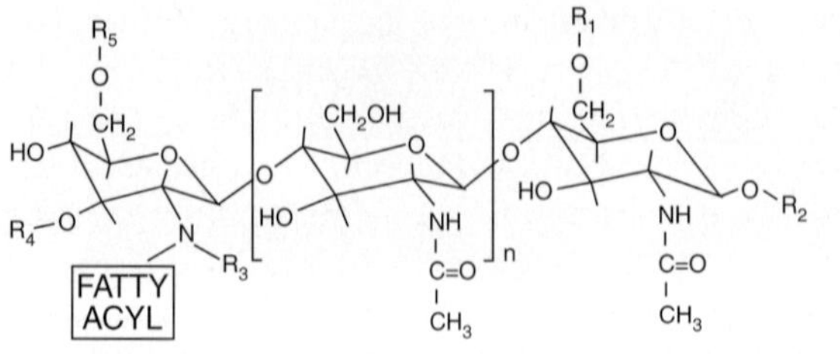

Figure 3 Chemical structure of Nod factors in rhizobium. R_1, hydrogen, acetyl, sulfate, fucose, methylfucose, sulfo-methylfucose or acetyl-methylfucose. R_2, ethyl, glycerol, or mannosyl. R_3, hydrogen or methyl. R_4, hydrogen or carbamoyl. R_5, hydrogen, acetyl, or carbamoyl. n, number of *N*-acetyl-glucosamine residues.

backbone of three to six β-1,4-linked *N*-acetyl-D-glucosamine residues. The nitrogen of the nonreducing residue has a fatty-acyl group attached, in which the length and degree of saturation varies between rhizobial species. The terminal reducing sugar moiety carries other substitutions such as acetyl, sulfuryl, fucosyl, acetyl-fucosyl, mannosyl, and arabinosyl units. Substitutions occur in other places as well. These strain- or species-specific "decorations" on the *N*-acetyl-D-glucosamine backbone are responsible for some of the host specificities of rhizobia. For example, the sulfuryl substitution at the reducing moiety, found in *S. meliloti,* confers host-specific nodulation of *M. sativa*. Strains with broad host ranges have various fucosyl substituents on the reducing sugar moiety. However, not all host specificities are explained by variations of the Nod factor. Identical Nod factors have been found in both *R. etli* and *M. loti*, but these species have very different host plants (*Phaseolus* spp. and *Lotus* spp., respectively), while *R. tropici* and *R. etli* produce different Nod factors, but nodulate the same host species (*Phaseolus* spp.). Thus, it would seem that Nod factors are just one, albeit important, determinant of host specificity. The mechanism of perception of the Nod factor signal by the host plant is not yet known, but is the subject of intense research.

Rhizobia also produce extracellular polysaccharides (EPS), lipopolysaccharides (LPS), and other complex carbohydrates. They play important roles in the infection process, possibly by suppressing host plant defense responses.

Regulation of N_2 Fixation in Legume Nodules

The high requirement for MgATP and reductant and the sensitivity of nitrogenase to O_2 are major factors determining how the N_2 fixation process is regulated, and the architecture of legume root nodules. Maintaining microaerobic conditions in the infected nodule cells whilst supplying sufficient O_2 for bacterial respiration, which is essential to produce enough ATP to support N_2 fixation, has been achieved in two ways. Firstly, a variable gaseous diffusion barrier is located in the inner nodule cortex surrounding the infected tissue. The intercellular air spaces in these cell layers vary, thereby controlling gaseous diffusion into the N_2-fixing tissue. This does not appear to present problems for diffusion of N_2 gas into the nodule, but O_2 becomes limiting for fixation. Numerous physiological and biochemical studies have shown that the barrier varies in response to requirements for energy to support fixation. However, the mechanism of operation of the variable barrier is still unknown. Secondly, the infected cells contain an O_2-carrying plant-encoded hemoprotein, leghemoglobin, which facilitates delivery of oxygen for bacterial respiration, while maintaining the concentration of free O_2 sufficiently low in the infected cells to avoid inactivation of nitrogenase.

A complex sensory and regulatory circuitry in rhizobia, responds to low free O_2 concentration to control transcriptional activation of the *nif* gene regulon. Conversely, the presence of fixed nitrogen (NH_4^+) inhibits expression of nitrogenase by negative regulatory mechanisms at the transcriptional level.

Host Plant Genes

The rhizobial Nod factor induces the expression of certain host plant genes, which presumably are involved in the early phases of nodule initiation and morphogenesis. Plant genes with enhanced expression in root tissue and nodules in response to rhizobial inoculation have been termed nodulin genes, and their gene products, nodulins. Nodulin genes, expressed prior to the onset of nitrogenase activity, are called early nodulin genes. Examples of early nodulins include ENOD12 and ENOD5. The first is assumed to be a cell wall protein. It is expressed in root hairs and epidermal cells, as well as in the nodule primordium and invasion zone of nodules. The latter is expressed in the nodule primordium and the infected zone of nodules. Another one is a gene from the forage legume *Lotus japonicus*, encoding a calcium-binding protein, the expression of which is enhanced in response to inoculation with *M. loti*. These early nodulins are probably involved in signal transduction and nodule morphogenesis processes, while late nodulins are expressed after completion of nodule morphogenesis. They are involved in housekeeping processes in the nodule. Glutamine synthetase catalyzes the formation of glutamine from glutamate and NH_4^+, the first step in the assimilation of fixed nitrogen by the plant. Other late nodulins, such as sucrose synthase, are involved in the catabolism of sucrose and starch, derived from host plant photosynthesis, leading to the formation of dicarboxylic acids, the carbon source exported to the bacteroids. The oxygen carrier leghemoglobin is another example of a late nodulin.

Numerous other nodulins have been identified and characterized to various degrees, and homologs of nodulins have also been found in many nonsymbiotic nonlegumes. Nodulin genes may have been recruited by rhizobia to achieve the formation and function of nodules. Genomic sequencing and identification of expressed sequence tags (ESTs) in plant species will no doubt lead to the identification of which plant genes are contributing to establishing and maintaining the symbiosis.

Manipulation of Nitrogen Fixation

If the premise is accepted that crop production should rely less on industrially produced fertilizer nitrogen, and more on biologically fixed nitrogen, then how is this best achieved? Four strategies will be considered here.

1. Improve N uptake and utilization by N_2 fixation in existing symbioses either by manipulating the host or rhizobium.
2. Improve the N_2-fixing associations between endophytic bacteria and nonlegume plants, such as rice, wheat, corn, and sugar cane.
3. Formation of nodule-like symbioses between nonfixing crop plants and rhizobium.
4. Direct transfer of N_2 fixation genes to nonfixing crop plants.

Improving Existing Symbioses by GM

On the plant side, differential gene expression studies, *in situ* hybridization, reverse genetics with transposon or T-DNA tagging approaches, transformation with reporter genes or antisense constructs, chemical mutagenesis, and map-based cloning have, together with numerous physiological, biochemical, and agronomic studies, contributed to the understanding we have of the legume nodule symbiosis. Despite this, no transgenic legumes with improved N_2 fixation or nitrogen assimilation properties have been considered for release. Perhaps one of the more promising long-term approaches involves the use of *in situ* hybridization to establish spatial and temporal expression patterns of key genes involved in nodule carbon and nitrogen metabolism, with a view to designing strategies for manipulating gene expression to achieve improved nitrogen and carbon utilization efficiency.

Improving the legume–rhizobium symbiosis by inoculation with superior rhizobium strains has been used for a long time in agriculture. In terms of increased N_2 fixation and plant performance, this practice has been most successful in soils with a low number of indigenous rhizobia. In soils with a large indigenous rhizobium population, the inoculant strain is often outcompeted by the indigenous, but less efficient rhizobia, resulting in no improvement in plant performance. A genetic engineering approach to improve the competitiveness of inoculant rhizobia makes use of genes encoding the naturally occurring trifolitoxin (TFX) antibiotic. This peptide antibiotic was first found in a nonfixing *Rhizobium leguminosarum* bv *trifolii* strain. It is not only toxic to rhizobia that do not produce it, but also to other soil microorganisms. A cluster of genes coding for the biosynthesis and resistance to TFX has been characterized, and transferred to other rhizobium strains. The competitiveness and symbiotic performance of such transgenic strains are being evaluated in the field in the United States. The creation of a "biased rhizosphere," i.e., one that favors the inoculant strain, is another approach being explored. It involves the exploitation of "rhizopine" biosynthesis and catabolism genes found in certain *R. leguminosarum* and *Sinorhizobium meliloti* strains. Rhizopines are unusual metabolites, providing the inoculant rhizobia possessing the genes for their biosynthesis and catabolism, with an exclusive carbon and nitrogen source for its survival and growth. This may give such strains a competitive advantage in the rhizosphere.

Rhizobia engineered for improved symbiotic performance have also been made. *S. meliloti* strains with extra copies of *nifA/dct* genes improve symbiotic performance of their host alfalfa. The *nifA* gene is a key regulatory gene, the product of which activates transcription of the structural nitrogenase genes, and *dct* genes are involved in the transfer of dicarboxylate from the host plant to the bacteroids, where it is used as a carbon source. Such strains have been approved for commercial release as inoculants in the United States.

Expression of uptake hydrogenase genes by some rhizobia can improve the efficiency of the symbiosis. An intrinsic property of nitrogenase is that for every mol N_2 reduced, at least one mol H_2 is evolved leading to wastage of ATP and reductant. In the 1970s it was found that some root nodules of soybean and pea evolved little or no H_2. This was due to rhizobium bacteroids expressing uptake hydrogenase, thus recycling H_2. The hydrogenase complex also contains genes allowing for the coupling of H_2 oxidation to oxidative phosphorylation, thus saving ATP. Soybean nodulated by *hup*$^+$ rhizobia is being evaluated in field experiments.

Rhizobia have also been engineered to contain insecticidal toxin genes cloned from *Bacillus thuringiensis* with a view to conferring resistance to insect larvae that specifically eat root nodules. Growth room experiments suggested that nodules formed by these rhizobia were more resistant to feeding by *Sitona* weevil larvae, an important pest of clover (*Trifolium* spp.) and pea.

Endophyte Associations

One of the best-studied N_2-fixing endophytic associations is that between sugar cane and microaerobic diazotrophs, such as *Acetobacter diazotrophicus* and *Herbaspirillum seropedica*. They colonize both root and stem tissue, and ^{15}N isotope studies have suggested that their N_2 fixation can contribute significantly to the N supply of sugar cane. Diazotrophic endophytes have also been found in rice plants, although there is as yet no evidence that they contribute significant amounts of fixed N to the host plant. Research efforts aim to select rice–endophyte associations in which endophyte N_2 fixation contributes at least half of the N required by the plant. Creation of biased rhizospheres, analogous to the rhizopines in legumes described above, has been considered for improving colonization by the desired endophyte. This would involve engineering the host plant to produce and secrete a "rhizopine" serving as an exclusive carbon and nitrogen source for a compatible endophyte strain engineered to contain "rhizopine" catabolism genes. Other approaches involve interfering with endophytic NH_3 assimilation by inhibiting glutamine synthetase, the first step in this process, or by manipulating regulatory genes to minimize inhibition of nitrogenase by NH_3.

Rhizobium Symbioses with Nonfixing Plants

In recent years, many investigations have reported the induction of nodule-like structures on roots of rice, wheat, and corn by rhizobia. Initially, this was achieved by applying cell-wall degrading enzymes or plant hormones. However, the induction of nitrogenase activity and N_2 fixation has not been demonstrated. This is perhaps not surprising since these are simple structures compared with legume nodules with a cortical oxygen diffusion barrier, leghemoglobin, and symbiosome, to reconcile the requirement for aerobic respiration with the need to protect nitrogenase from oxygen. Although homologs of many nodulin genes, including leghemoglobin, have been found in nonsymbiotic plants, we are still some way from knowing which ones will be required to achieve conditions favorable for N_2 fixation. Microarray expression studies should help to reduce these gaps in our understanding.

In this context it is worth noting that legumes and the actinorhizal symbioses of nonlegume families all belong to one clade, the Rosidae, suggesting that the predisposition to form nodular symbioses only evolved once. The implications of this, in terms of the prospects for developing nodular symbioses on presently nonsymbiotic plant species outside this clade, are difficult to assess.

Direct Transfer of *nif* Genes

Transformation of virtually every major crop species is now possible, so this is no longer a limitation for transfer of *nif* genes. When considering the direct transfer to nonnodulating plants such as cereal crops, the crucial question is how to achieve sufficient expression of nitrogenase without exposure to damaging levels of oxygen. Secondly, a sufficiently large supply of ATP and reducing power must be available, in addition to a mechanism for assimilation of the product, NH_3. At first glance, plant chloroplasts may seem an odd choice as target for *nif* gene expression, because of their photosynthetic oxygen production. However, the *nif* genes could be engineered for nocturnal expression, as is the case in many diazotrophic cyanobacteria. It is widely accepted that plant chloroplasts are derived from endosymbiotic cyanobacteria, possibly even diazotrophs. Many free-living cyanobacterial species have developed mechanisms to reconcile photosynthesis with N_2 fixation. Furthermore, chloroplastic photosynthetic products provide a potential energy source for nitrogenase, and enzymes for ammonium assimilation, such as glutamine synthetase, are present in chloroplasts. Tobacco (*Nicotiana tabacum*) plants have been transformed with *nifH* (the Fe-protein of nitrogenase) and *nifM* (essential for activation of the Fe-protein). They expressed both genes in the chloroplast, but no nitrogenase enzyme activity was detected, probably due to poor translation or degradation of the protein.

Recently, three chloroplast genes encoding enzymes involved in light-independent chlorophyll biosynthesis were discovered. They have a high degree of similarity to the *nifHDK* genes in diazotrophs, and catalyze analogous chemical reactions. This supports the theory that chloroplasts are derived from diazotrophic cyanobacteria. One of these genes, *chlL* was replaced by its "homolog" *nifH* from *Klebsiella pneumoniae* by direct chloroplast transformation of the unicellular green alga *Chlamydomonas reinhardtii*. The *nifH* gene product was able to partially complement the *chlL*$^-$ phenotype, indicating that the oxygen-sensitive *nifH* gene product can function in chloroplasts. Despite these interesting results, there are still formidable obstacles to overcome, including the introduction of the remaining necessary *nif* genes and regulatory sequences. It is also not certain that sufficient levels of ATP and reductants or sufficient respiratory capacity to remove oxygen are available to support N_2 fixation in the chloroplast.

The recent discovery of an oxygen-tolerant nitrogenase from *Streptomyces thermoautotrophicus* raise the possibility of introducing this nitrogenase complex to chloroplasts or mitochondria to ensure sufficient levels of reducing power and ATP to support N_2 fixation. However, the plant has to be able to tolerate the superoxide radical required by this enzyme. This nitrogenase may only be expressed in plants exposed to photooxidative stress, unless an alternative electron donor can be found.

Conclusions and Future Prospects

Perhaps it is not surprising that genetic manipulation of nitrogen fixation in plants has progressed furthest in the rhizobial symbiont, in terms of producing inoculants of potential use in agriculture. It has been the easiest partner in the symbiosis to manipulate. The development of nodular symbioses in cereals appears to be some way off, due to our still incomplete understanding of complexity of the legume–rhizobium interaction. However, forward and reverse genetic approaches to identify and study "symbiosis" genes combined with genomic and EST sequencing efforts and other functional genomic approaches will help close these gaps. Manipulating cereals for further exploitation of endophytic diazotrophs to improve the contribution of N_2 fixation to their nitrogen supply, would appear to have good prospects for success in the shorter term. Obtaining N_2-fixing cereals by direct transfer of *nif* genes is still a distant prospect. However, the recent discovery of "*nif*-like" genes in chloroplast genomes and microbes with oxygen-tolerant nitrogenase enzymes is fascinating, and may offer new opportunities for achieving this challenging goal.

See also: **Nutrition**: Nitrogen Fixation. **Root Development**: Mycorrhizae; The Rhizosphere and its Microorganisms.

Further Reading

Dixon R, Cheng Q, Dowson-Day M, and Day A (2000) Prospects for engineering nitrogen-fixing photosynthetic eukaryotes. In: Pedrosa F, Hungria M, Yates G, and Newton W (eds) *Nitrogen Fixation: From Molecules to Crop Productivity*, pp. 635–639. Dordrecht: Kluwer Academic.

Doyle JJ (1998) Phylogenetic perspectives on nodulation: evolving views of plants and symbiotic bacteria. *Trends in Plant Science* 3: 473–478.

Gualtieri G and Bisseling T (2000) The evolution of nodulation. *Plant Molecular Biology* 42: 181–194.

Hirsch AM and LaRue TA (1997) Is the legume nodule a modified root or stem or an organ *sui generis? Critical Reviews in Plant Sciences* 16: 361–392.

Ladha JK, de Bruijn FJ, and Malik KA (eds) (1997) Opportunities for biological nitrogen fixation in rice and other non-legumes. *Plant and Soil* 194: 1–215.

Maier RJ and Triplett EW (1996) Toward more productive, efficient, and competitive nitrogen-fixing symbiotic bacteria. *Critical Reviews in Plant Sciences* 15: 191–234.

Mytton L and Skøt L (1993) Breeding for improved symbiotic nitrogen fixation. In: Hayward MD, Bosemark NO, and Romagosa I (eds) *Plant Breeding, Principles and Prospects*, pp. 451–472. London: Chapman & Hall.

Spaink HP (2000) Root nodulation and infection factors produced by rhizobial bacteria. *Annual Review of Microbiology* 54: 257–288.

Oxidative Stress

C H Foyer and K Groten, Rothamsted Research, Harpenden, UK
K Kunert, University of Pretoria, Pretoria, South Africa

Introduction

Environmental stresses constitute the most significant factors leading to substantial and unpredictable decreases in crop yield in agriculture today. Plants constantly monitor their surroundings and make appropriate metabolic, structural, and physiological adjustments to accommodate environmental changes. Within the framework of genetic background, plant productivity is dependent on this constant adjustment of gene expression in response to environmental cues. The genome–environment interaction is an essential focus for the elucidation of the nature of the phenotypic variation leading to successful stress tolerance responses. Moreover, this interaction is also a key determinant to plant tissue composition related to crop quality factors, as well as plant anatomy, morphology, and development.

Plants integrate a diverse range of environmental and metabolic signals via a network of interacting signal transduction pathways that together regulate gene expression during stress. Plants make use of an interacting network of common pathways and components to optimize the stress tolerance responses. This phenomenon, which is known as cross-tolerance, allows plants to acclimate to a range of different stresses after exposure to one specific stress. A common signaling system involving hormones, oxidants, and antioxidants has evolved to provide adequate defense and protection against hazardous environments.

Redox Chemistry and Oxidative Stress in Plants

Plants are autotrophic organisms powered by photosynthesis, a process dominated by light-driven redox chemistry. It is therefore no surprise that redox signals from the light reactions of photosynthesis initiate profound changes in gene function. Through the necessity of harnessing light energy to drive photosynthetic metabolism, plants have optimized strategies for redox control. It is now widely accepted that redox signals exert control on nearly every aspect of plant biology from chemistry to development, growth, and eventual death. It is most probable that "redox signaling" was the first type of sensory regulation that evolved in nature, since it prevented uncontrolled changes in energy production, utilization, and exchange. There followed further pathways of redox control through regulation of gene expression at many levels. It is also likely that intermediates and other systems of signal transduction arose from this central core. The systems in the chloroplast that sense redox changes and control redox homeostasis have been intensively studied; these include posttranslational modification of proteins by phosphorylation, redox modulation of assimilatory reactions, and control of gene transcription and translation. Moreover, redox controls and signals also ensure that the expression of plastid-encoded photosynthetic proteins is precisely coordinated with that of nuclear encoded chloroplast components. Photosynthetic control of the expression of genes located in chloroplasts occurs at both transcriptional and posttranscriptional levels. Redox signals arising in the chloroplast also regulate the transcription of nuclear genes.

The basic photoelectrochemistry of photosynthesis first evolved in an anaerobic world. However, about 2 billion years ago, molecular oxygen became intimately involved with the essential energy exchange reactions on which life is based (photosynthesis and respiration), allowing use of the very high electrochemical potential ($Em_7 = +815\,mV$) of the O_2/H_2O redox couple. Cyanobacteria, the prokaryotic relations of higher plant chloroplasts, created the oxygen-rich atmosphere of the Earth through oxygenic photosynthesis. Oxygenic photosynthesis and aerobic respiration are the only processes that are able to undertake the concerted, four-electron exchange between water and oxygen, without release of reactive, partially reduced intermediates. However, many processes in plants catalyze only partial reduction of oxygen, and so generate active oxygen species (AOS) such as superoxide ($O_2^{\bullet -}$) and hydrogen peroxide (H_2O_2). Photosynthesis has a very high capacity for the production of these reduced intermediates and also for the formation of singlet oxygen. Singlet oxygen is an extremely reactive molecule, which can cause extensive damage to proteins and membranes. Similarly, while H_2O_2 participates widely in plant metabolism it can be converted via Fenton-type reactions to the hydroxyl radical ($OH^{\bullet}$). This extremely reactive radical can initiate lipid peroxidation, cause protein degradation or modification, and mediate DNA damage. The accumulation of AOS increases the probability of hydroxyl radical formation. Failure to control redox reactions leading to the formation of singlet oxygen and hydroxyl radical leads to the phenomenon known as oxidative stress.

In addition to formation through processes intrinsic to photosynthesis, superoxide and/or H_2O_2 are generated at significant rates by photorespiration, oxidative phosphorylation, fatty acid β-oxidation, and also by many types of oxidase activity. The intracellular levels of these oxidants are tightly controlled by antioxidant system, comprising a network of enzymatic and nonenzymatic components. In most cases such defence reactions are linked to the two major redox buffers of plant cells, ascorbate and glutathione.

In animals, mitochondria are the major source of AOS, and are involved in aging, degenerative diseases, cancer, and programmed cell death (PCD). Very little is known about the role of mitochondria in related processes in plants because of the overall higher fluxes of electron transport processes in chloroplasts and peroxisomes. Mitochondria, like chloroplasts, originated as bacterial endosymbionts, and retain their own genomes. Like photosynthetic electron transport, mitochondrial electron transport has the capacity to generate superoxide. The main sites of mitochondrial AOS formation in plants lie in respiratory complexes I and III. Despite the fact that the generation of superoxide and H_2O_2 by the plant mitochondrial electron transport chain is well known, the production of these AOS by leaf mitochondria had, until recently, been considered to be of only secondary importance. However, several recent reports have suggested that, in plants as in animals, mitochondria are involved in the tolerance to oxidative stress. In particular, mitochondria may have a key role in the hypersensitive response (HR), which is a form of programmed cell death induced in response to plant pathogens.

H_2O_2 and Signal Transduction: What Does not Kill Me Makes Me Stronger!

Superoxide and H_2O_2 are produced as primary signals in the stress signal cascade. However, they are also produced as second messengers during growth, tropic, and movement responses as a result of hormone action. Rapid compartment-specific differences in redox state can be achieved either by modifying the rates of superoxide and H_2O_2 production or by repression or activation of the antioxidant defenses or both. For example, H_2O_2 accumulation in the apoplast, which is important in defense and growth responses, is controlled by regulation of the activity of enzymes that generate H_2O_2 particularly NADPH oxidases. This allows very precise control of the generation of H_2O_2 either in localized microenvironments or in more global bursts. H_2O_2 accumulation can either lead to enhanced expression of antioxidant enzymes and increased expression of other defence proteins, such as glutathione s-transferases, or it can initiate PCD (particularly when nitric oxide (NO) is also produced) depending on the intensity of the oxidative signal or oxidative load on the tissues. A large increase in oxidative load triggers PCD while smaller localized bursts of H_2O_2 production (as occur for example during abscisic acid-induced stomatal closure) strengthen stress resistance. For example, local and systemic signaling via H_2O_2 during occurs during the adaptation of leaves to high light.

Much remains to be resolved concerning the components of the H_2O_2-induced signaling cascade and the mechanism(s) by which information on redox status is used to modify gene expression. H_2O_2-responsive elements are present in the promoters of a number of plant genes. One of the most important developments in the next few years will be the elucidation of transcription factors that are involved in the H_2O_2 signaling cascade. This is important as a single transcription factor can orchestrate the expression of many genes to improve oxidative stress tolerance. Multiple protein–protein and/or protein–DNA interactions frequently dictate the rate of transcription by activation/repression of a given promoter under given environmental conditions. The *Arabidopsis thaliana* glutathione-*S*-transferase-6 promoter, for example, contains Dof-binding sites closely linked to a 20 bp octopine synthase (ocs) element. This element, which responds to H_2O_2 is the binding site for bZIP proteins. Downstrean signaling is modulated by other factors including calcium mobilization and protein phosphorylation. Mitogen-activated protein kinases (MAPKs) have been found to modulate protein phosphorylation as a result of H_2O_2 signaling. H_2O_2 activates MAPKs, which repress auxin-inducible promoters. Such interactions demonstrate the presence of intensive metabolite and hormone cross-talk between oxidative stress tolerance and plant growth responses.

One of the most important biotic stress responses in plants is HR, which is a defense-associated process characterized by rapid PCD at the site of pathogen attack. Recent studies have suggested that PCD associated with HR has many similarities with animal apoptosis, inferring that plant mitochondria play a crucial role. One of the earliest events in HR is an oxidative burst that involves dramatic localized increases in superoxide and H_2O_2 generation. This occurs on the apoplastic face of the plasma membrane. In comparison to the cytoplasm, the apoplastic space and cell wall have relatively little antioxidant defence. Hence, a strong oxidative signal can persist for much longer in the apoplast than in the cytoplasm. The redox state of the apoplast can thus be maintained at a very different level from that of the cytoplasm and a redox gradient can exist between the two compartments.

A second important feature of HR is systemic acquired resistance (SAR) that occurs in tissues not only around the areas undergoing PCD but also in tissues and organs far removed from the site of attack. SAR involves the preemptive deployment of gene expression to modify cell metabolism to cope with future attacks. Increases in salicylic acid (SA) are observed at the site of pathogen attack, and to a lesser extent at remote sites.

Many studies of the type described above confirm that H_2O_2 is a signaling molecule that mediates plant responses to a range of biotic and abiotic stresses. It is therefore not surprising that a wide range of different stresses can induce similar patterns of gene transcription. Increases in antioxidant gene expression are frequently observed following exposure to abiotic and biotic threats. Similarly, pathogenesis-related (PR) proteins are induced by stresses as diverse as pathogen attack and exposure to ozone. Decreases in the ability to scavenge H_2O_2 induce similar changes in gene expression.

Overall, plants are much more tolerant of H_2O_2 than animals, and their antioxidant systems appear to have been designed to ensure control of the cellular redox state rather than to facilitate the complete elimination of H_2O_2. Together with a number of antioxidant enzymes, two hydrophilic low molecular weight antioxidants, ascorbate and glutathione, determine the lifetime of H_2O_2 in plant cells. These antioxidants ensure that the cytoplasm is strongly buffered to prevent frequent oscillations in redox state and signaling.

Antioxidant Defenses

An appropriate intracellular balance between AOS generation and scavenging therefore exists in all cells. This "redox homeostasis" requires the efficient coordination of reactions in different cell compartments and is governed by a complex network of pro- and antioxidant systems. The latter include nonenzymatic scavengers such as ascorbate, glutathione, and hydrophobic molecules (tocopherols, carotenoids, and xanthophylls) in addition to enzymes that operate in the different cellular organelles. These enzymes include superoxide dismutases which catalyze the dismutation of $O_2^{\bullet -}$ to H_2O_2 and O_2. Three classes of superoxide dismutases (SOD) have been identified in plants on the basis of their metal cofactor content. FeSODs are found only in chloroplasts, MnSODs are found mainly in mitochondria, while Cu/ZnSODs are located in the chloroplasts, cytosol, apoplast and peroxisomes. The main enzymatic H_2O_2 scavengers of photosynthetic cells are catalases (CAT), which convert H_2O_2 to H_2O and O_2, and ascorbate peroxidases, which use ascorbate as the electron donor for H_2O_2 reduction. In peroxisomes and glyoxysomes, catalase predominates. Catalase isoforms are distinguished on the basis of organ specificity and responses to environmental stress. Chloroplasts contain stromal, thylakoid-bound, and lumenal ascorbate peroxidase (APX) forms. Ascorbate peroxidases use reduced ascorbate as an electron donor. At the thylakoid membrane, the oxidized form monodehydroascorbate is reduced back to ascorbate via reduced ferredoxin, generated as a result of photosynthetic electron flow. Together, these reactions comprise the Mehler–peroxidase cycle in which electrons are transferred from water to O_2 and back to water, consuming reducing power while producing ATP. In addition to the Mehler–peroxidase cycle, chloroplasts contain a second mechanism involving ascorbate, that removes H_2O_2 and consumes reducing power while producing ATP; this is the ascorbate–glutathione cycle. In the ascorbate–glutathione cycle, monodehydroascorbate generated via the ascorbate peroxidase reaction is converted back to ascorbate via monodehydroascorbate reductase. Monodehydroascorbate that escapes this route of recycling rapidly disproportionates to dehydroascorbate and ascorbate. Dehydroascorbate is converted back to ascorbate by the action of dehydroascorbate reductases, which utilize reduced glutathione (GSH) as reductant. This is regenerated from its oxidized form, glutathione disulfide, by glutathione reductase. The ascorbate–glutathione cycle is not only an important H_2O_2-detoxifying system in the chloroplast but it also serves a similar function in the cytosol, peroxisomes, and mitochondria. There is a vast array of other antioxidant components that function in synchrony with the Mehler–peroxidase and ascorbate–glutathione cycles to maintain low levels of H_2O_2 and other oxidants in cells. Of particular note are the peroxiredoxins, glutathione peroxidases and glutathione *S*-transferases (GST) that fulfil essential roles in the elimination of lipid peroxides and other lipid-derived oxidants in cellular membranes. Moreover, many compounds such as lipoic acid, flavonoids, and polyamines have auxiliary antioxidant functions. Together, these antioxidants form a formidable garrison of defenses that prevent accumulation of superoxide and H_2O_2 in order to minimize the possibility of hydroxyl radical formation and consequent oxidative stress.

The Role of Antioxidants in Signal Transduction

The amounts and redox states of the ascorbate and glutathione pools in the different compartments of plant cells have considerable significance for redox signaling. These antioxidants determine cellular H_2O_2 concentrations, and therefore influence the known roles of this oxidant in signal transduction. Changes in antioxidant redox state and concentration can be much more pronounced than changes in leaf H_2O_2 contents. Gene transcription is differentially regulated by changes in the absolute amounts of ascorbate and glutathione and by changes in the ratios of ascorbate to dehydroascorbate and GSH to glutathione disulfide. Thiols such as thioredoxin, dithiothreitol, cysteine, and GSH, enhance NF-κB DNA binding activity in animal systems. GSH-responsive elements are present in the promoters of *GST* genes and in genes involved in the synthesis of phytoalexins. Although glutathione has long been considered as a transcriptional regulator, ascorbate-mediated regulation of gene expression has only recently been demonstrated. Ascorbate deficiency in the *A. thaliana* mutant *vtc1* causes slow growth and late flowering. This is not due to changes in photosynthesis or increased oxidative stress. Ascorbate deficiency has its own unique molecular signature, inducing transcript changes that can be reversed by adding ascorbate. Transcript changes in *vtc1* indicate that growth and development are constrained by modulation of the balance between abscisic acid and gibberellic acid signaling.

GM Strategies for the Amelioration of Defense against Oxidative Stress

In many areas of the world stress tolerance is a vital trait. Stresses such as drought and extremes (high and low) of temperature are major limitations on crop productivity particularly in developing countries. These stresses are complex environmental phenomena and plant responses are hence inherently multigenic in nature. While the vast majority of these genes remain to be identified, some transcription factors and regulatory sequences in plant promoters that are involved in oxidative stress tolerance have been described. In a wide range of plant species, *cis*-elements and their corresponding binding proteins are involved in stress tolerance. A number of these *cis*-elements and transcription factors have been implicated in tolerance to several types of stress. Such observations may be explained by common requirements such as protein stability during dehydration, a component that is inherent to drought, salt stress, and high and low temperature stresses as well as other environmental conditions.

Our increasing knowledge of the genetic basis for stress tolerance coupled to the use and application of transformation technologies has allowed the development of increasingly more sophisticated strategies for crop improvement. An array of genetically modified (GM) crops has been produced in order to explore the effects of specific transgenes. The use of GM plants constitutes an important mechanism whereby many of the current limitations on conventional breeding can essentially be bypassed. However, it remains essential to identify stress-regulated transcription factors and to characterize the proteins and signaling mechanisms that control their function. The *cis*-responsive elements in the promoters of genes involved in stress signaling may hold the key with which to unravel the underlying mechanisms conferring stress tolerance. However, to date relatively few studies with GM plants have been undertaken in this regard. Rather, limitation of oxidative damage has formed the basis for many strategies designed to improve crop performance upon exposure to stress. A large number of GM studies have been based on the enhancement or modulation of individual components of H_2O_2-detoxification systems (**Table 1**). While such strategies are able to prevent oxidative damage to cell proteins and membranes, they also have the potential to shorten the lifetime of the signal and hence prevent induction of the endogenous defence systems. This represents something of a paradox and may form the basis for some of the disparities in the results obtained in studies on overexpression of antioxidant enzymes shown in **Table 1** and discussed below.

In the last 15 years a large number of reports have appeared in the literature concerning the effects of overexpression of antioxidant enzymes on stress tolerance. These studies were based on the following observations:

1. Antioxidant enzyme activities and/or transcripts increase following exposure to oxidative stress.
2. There is good correlation between enhanced antioxidant capacity and improved stress tolerance.
3. Exposure to oxidative stress gives cross-tolerance to a variety of other stresses.

Table 1 provides information on the results of studies conducted in the last 5 years with regard to antioxidant enzymes while **Table 2** provides a summary of literature information on studies involving modified expression of the biosynthetic enzymes involved in the synthesis of ascorbate and glutathione. Taken together, the results of these studies clearly show that improved stress tolerance can be achieved under laboratory conditions by amelioration of the antioxidant defenses. However, this approach is not always successful. Moreover, the outcome of such experiments depends on number of

Table 1 Summary of recent results on the manipulation of antioxidant enzymes for engineering stress tolerance

Plant species	*Enzyme*	*Expression*	*Stress tolerance*	*Reference*
Tobacco (*Nicotiana tabacum*)	Glutathione *S*-transferase with glutathione peroxidase activity	Overexpression	Reduced oxidative damage under thermal or salt stress Increased growth under chilling or salt stress	Roxas *et al.* (2000) *Plant Cell Physiol.* 41: 1229–1234 Roxas *et al.* (1997) *Nat. Biotechnol.* 15: 988–991
Poplar (*Populus tremula* × *P. alba*)	Glutathione reductase	Overexpression	No effect on acute ozone stress	Strohm *et al.* (1999) *J. Exp. Bot.* 50: 365–374
Cotton (*Gossypium hirsutum*)			Enhanced protection against moderate chilling at high light intensity	Payton *et al.* (2001) *J. Exp. Bot.* 52: 2345–2354
Tobacco (*Nicotiana tabacum*)			Depending on the transformed cultivar protection of PSII from photoinhibitory damage, but no enhancement of the repairing mechanism	Tyystjaervi *et al.* (1999) *Physiol. Plant.* 105: 409–416
	Ascorbate peroxidase	Suppression (single antisense)	Hyperresponsive to pathogen attack	Mittler *et al.* (1999) *Proc. Natl Acad. Sci. USA* 96: 14165–14170
			Increased susceptibility to ozone injury	Oervar and Ellis (1997) *Plant J.* 11: 1297–1305
	Ascorbate peroxidase and catalase	Suppression (double antisense)	Reduced susceptibility to oxidative stress compared with single antisense plants	Rizhsky *et al.* (2002) *Plant J.* 32: 329–342
Arabidopsis thaliana	Ascorbate peroxidase	Overexpression	Enhanced tolerance to heat stress	Shi *et al.* (2001) *Gene* 273: 23–27
			Increased protection against oxidative stress caused by aminotriazole	Wang *et al.* (1999) *Plant Cell Physiol.* 40: 725–732
Cotton (*Gossypium hirsutum*)			Enhanced protection against moderate chilling at high light intensity	Payton *et al.* (2001) *J. Exp. Bot.* 52: 2345–2354
Tobacco (*Nicotiana tabacum*)	Ascorbate oxidase	Overexpression	No stress effect reported	Kato and Esaka (2000) *Planta* 210(6): 1018–1022
Poplar (*Populus tremula* × *P. alba*)	Fe-Superoxide dismutase	Overexpression	No protection against photoinhibition of PSII	Tyystjaervi *et al.* (1999) *Physiol. Plant.* 105: 409–416
			Protection of PSII against overreduction when CO_2 is depleted, less sensitivity to methyl viologen	Arisi *et al.* (1998) *Plant Physiol.* 117: 565–574
Alfalfa (*Medicago sativa*)			Increased winter survival without detectable improved oxidative stress (methyl viologen) tolerance	McKersie *et al.* (2000) *Plant Physiol.* 122: 1427–1437
			Stress effects not tested	Rubio *et al.* (2001) *Mol. Plant Microbe Inter.* 14: 1178–1188
Corn (*Zea mays*; maize)			Enhanced tolerance towards methyl viologen	Van Breusegem *et al.* (1999) *Plant Cell Physiol.* 40: 515–523
	Mn-Superoxide dismutase	Overexpression	Stress effects not tested	Kingston-Smith and Foyer (2000) *J. Exp. Bot.* 51: 1867–1877
			Increased antioxidant capacity of the leaves	Van Breusegem *et al.* (1999) *J. Exp. Bot.* 50: 71–78
Cotton (*Gossypium hirsutum*)			No enhanced protection against moderate chilling at high light intensity	Payton *et al.* (2001) *J. Exp. Bot.* 52: 2345–2354

Table 1 Continued

Plant species	*Enzyme*	*Expression*	*Stress tolerance*	*Reference*
Alfalfa *(Medicago sativa)*			Stress effect not tested	Rubio *et al.* (2001) *Mol. Plant Microbe Inter.* 14: 1178–1188
			Enhanced winter survival rate without effects on primary freezing injury	McKersie *et al.* (1999) *Plant Physiol.* 119: 839–847
		Pyramided overexpression	No stress effect tested, but joint expression resulted in lower shoot and storage organ biomass compared with plants having only one transgene	Samis *et al.* (2002) *J. Exp. Bot.* 53: 1343–1350
Tobacco (*Nicotiana tabacum*)	Mn-Superoxide dismutase and ascorbate peroxidase	Pyramided overexpression	High level of protection against methyl viologen oxidative stress. Single SOD expression only with slightly enhanced protection	Kwon *et al.* (2002) *Plant Cell Environ.* 25: 873–882
	CuZn-Superoxide dismutase and ascorbate peroxidase			
	Catalase	Suppression	Enhanced resistance to tobacco mosaic virus	Takahashi *et al.* (1997) *Plant J.* 11: 993–1005
			Enhanced resistance against *Pseudomonas syringae*	Chamnongpol *et al.* (1998) *Proc. Natl Acad. Sci. USA* 95: 5818–5823
			Increase in CO_2 compensation point at high temperatures	Brisson *et al.* (1998) *Plant Physiol.* 116: 259–269
		Overexpression	Impaired defense responses to tobacco mosaic virus	Talarczyk *et al.* (2002) *Plant Physiol.* 129: 1032–1044
			Better resistance to pathogen attack under high light conditions, one line less sensitive to methyl viologen	Polidoros *et al.* (2001) *Transgenic Res.* 10: 555–569
			Increased tolerance to methyl viologen treatment	Miyagawa *et al.* (2000) *Plant Cell Physiol.* 41: 311–320
			Increased tolerance to drought stress at high light intensity	Shikanai *et al.* (1998) *FEBS Lett.* 428: 47–54
			Reduced CO_2 compensation point in a temperature-dependent manner	Brisson *et al.* (1998) *Plant Physiol.* 116: 259–2699
Potato (*Solanum tuberosum)*			Enhanced resistance to *Phytophthora infestans*	Yu *et al.* (1999) *Plant Mol. Biol.* 39: 477–488
Arabidopsis thaliana	2-Cysteine peroxiredoxin	Suppression	Enhanced activities and expression of enzymes associated with ascorbate metabolism	Baier *et al.* (2000) *Plant Physiol.* 124: 823–832
			Impaired photosynthesis in young plants	Baier and Dietz (1999) *Plant Physiol.* 119: 1407–1414
Potato *(Solanum tuberosum)*			Increased damage of the photosynthetic apparatus under oxidative stress conditions	Broin *et al.* (2002) *Plant Cell* 14: 1417–1432
Tobacco (*Nicotiana tabacum)*	Anionic peroxidase	Suppression	Some reduction in wound-induced deposition of lignin-like polymers	Lagrimini *et al.* (1997) *Plant Physiol.* 114: 1187–1196
		Overexpression	Diminished root mass, causes wilting	Lagrimini (1997) *Plant Mol. Biol.* 33: 887–895
Tomato *(Lycopersicon esculentum)*			Enhanced insect resistance	Behle *et al.* (2002) *J. Econ. Entomol.* 95: 81–88

continued

Table 1 Continued

Plant species	*Enzyme*	*Expression*	*Stress tolerance*	*Reference*
Sweetgum (*Liquidambar styraciflua*)			Higher resistance to feeding by caterpillars and beetles, more susceptible to feeding by the corn earworm	Dowd *et al.* (1998) *Cell Mol. Life Sci.* 54: 712–720
Tobacco (*Nicotiana tabacum*)	Cationic peroxidase	Overexpression	Increased susceptibility to *P. parasitica nicotinoae,* but increased suppression of *E. carotovora*	Elfstrand *et al.* (2001) *Planta* 425: 681–716
	Mn-Peroxidase	Overexpression	Detoxification of pentachlorophenol for phytoremediation	Iimura *et al.* (2002) *Appl. Microbiol. Biotechnol.* 59: 246–251
	Cell wall peroxidase	Overexpression	Increased germination rate under high salt or osmotic stress	Amaya *et al.* (1999) *FEBS Lett.* 457: 80–84

Table 2 Studies on manipulation of the reported enzymes for antioxidant synthesis reported in the last 5 years

Plant species	*Enzyme*	*Expression*	*Stress tolerance*	*Reference*
Arabidopsis thaliana	γ-Glutamyl-cysteine synthase	Suppression	Enhanced sensitivity towards cadmium	Xiang *et al.* (2001) *Plant Physiol.* 126: 564–574
Poplar (*Populus tremula* × *P. alba*)		Overexpression	Enhanced tolerance towards chloroacetanilide herbicides	Gullner *et al.* (2001) *J. Exp. Bot.* 52: 971–979
Tobacco (*Nicotiana tabacum*)			Increased oxidative damage (light intensity-dependent chlorosis or necrosis)	Creissen *et al.* (1999) *Plant Cell* 11: 1277–1291
Arabidopsis thaliana			No significant increase in cadmium tolerance	Xiang *et al.* (2001) *Plant Physiol.* 126: 564–574
Indian mustard (*Brassica juncea*)			Enhanced cadmium tolerance and accumulation	Zhu *et al.* (1999) *Plant Physiol.* 121: 1169–1177
	Glutathione synthase	Overexpression	Enhanced cadmium tolerance and accumulation	Zhu *et al.* (1999) *Plant Physiol.* 119: 73–79
Poplar (*Populus tremula x P. alba*)			No effect on acute ozone stress	Strohm *et al.* (1999) *J. Exp. Bot.* 50: 365–374
Tobacco (*Nicotiana tabacum*)	Cysteine synthase	Overexpression	Enhanced tolerance to sulfur-containing environmental pollutants	Noji *et al.* (2001) *Plant Physiol.* 126: 973–980
	O-Acetyl-serine(thiol) lyase	Overexpression	Enhanced tolerance to oxidative stress (methyl viologen, acute SO_2)	Youssefian *et al.* (2001) *Plant Physiol.* 126: 1001–1011
Tobacco (*Nicotiana tabacum*)	L-Galactono-1,4-lactone dehydrogenase	Suppression	Stress effects not tested	Tabata *et al.* (2001) *Plant J.* 27: 139–148
Tobacco (*Nicotiana tabacum*) Corn (*Zea mays*; maize)	Dehydroascorbate reductase	Overexpression	Stress effects not tested	Chen *et al.* (2003) *PNAS* 100: 3525–3530

crucial factors, including the type of enzyme used for overexpression analysis, the cellular compartment to which the enzyme is targeted, and the type and duration of stress used to test tolerance. As shown in **Tables 1** and **2** there is a general lack of consistency between studies. Clearly, enhanced protection is not always guaranteed even when significant increases in the activities of enzymes such as Cu/ZnSOD and glutathione reductase (GR) have been achieved. This may be due in part to effects of constitutive enhancement of antioxidant enzymes on endogenous signal transduction cascades. Increased GSH in the chloroplast, for example, has not always been found to have a positive effect.

A wide range and number of other proteins and factors implicated in the web of plant defence reactions have been expressed in GM plants with a view to improving stress tolerance (**Figure 1**). In some cases, improved stress tolerance attained in this way has been attributed to increased H_2O_2 production. For example, overexpression of choline oxidase, an enzyme that converts choline to glycinebetaine, in *A. thaliana* enhances tolerance to high light and other stresses. Since choline oxidase produces H_2O_2 as well as glycinebetaine, it is probable that much of the enhanced protection is due to stimulation of antioxidant defenses by H_2O_2. Moreover, elevated production of H_2O_2 in such transgenic plants activates many defense-related genes in addition to antioxidants that make a major contribution to overall disease resistance as well as tolerance to abiotic stress.

Other types of enzymes/cofactors used in GM studies to achieve better protection against oxidative stress include serine acetyltransferases, aldose/aldehyde reductases, various transcription factors, and vacuolar Na^+/H^+ antiporter proteins. Expression of the *Escherichia coli* serine acetyltransferase in GM tobacco (*Nicotiana tabacum*) caused a two- to threefold increase in cysteine and glutathione levels. The most striking physiological consequence of this modification in sulfur metabolism, however, was a large increase in resistance to the oxidative stress generated by exogenous hydrogen peroxide. Engineered high synthesis of an aldose/aldehyde reductase in GM tobacco yielded considerable increases in tolerance to the herbicide paraquat and to exposure to heavy metals. Overexpression of a glutathione *S*-transferase with glutathione peroxidase activity (GST/GPX) in tobacco enhanced seedling growth under high temperature stress or upon exposure to salt stress (**Table 1**). It was considered that the increased capacity for glutathione-dependent H_2O_2-scavenging prevented the inhibition of growth found in the wild-type seedlings exposed to these stresses. Enhanced protection against salt stress has been achieved by overexpression of a vacuolar Na^+/H^+ antiporter.

The constitutive expression of transcription factor proteins has been found to be associated with simultaneous alteration of several target genes in GM plants. For example, the transcription factor Alfin1 was found to improve salt tolerance in alfalfa (*Medicago sativa*, lucerne) by enhancing root growth under normal and saline conditions. Alfin1 binds to promoter fragments of the NaCl-inducible *MsPRP2* gene regulating its expression in a tissue-specific manner. In many cases, plants survive stress by the arrest of metabolism, growth, and development. These dramatic changes may be enhanced as a result of activation or deregulated expression of such transcription factors in transformed plants. Unpredictable performance is an important outcome of such manipulations and is a crucial limiting factor in terms of crop quality and yield.

Agriculturally important parameters, such as yield and biomass, are critical points in any consideration of such manipulations in crop plants. It is therefore crucial to have information concerning how such GM plants perform under field conditions. In the field, plants face a variety of stresses either individually or in combination. To date, only a very limited number of field experiments have been carried out. GM alfalfa expressing SOD had greater field survival rates under freezing temperatures after two winters than non-GM plants despite the absence of convincing enhanced stress tolerance traits in the GM plants examined under laboratory and greenhouse conditions. Improved survival was also found in GM

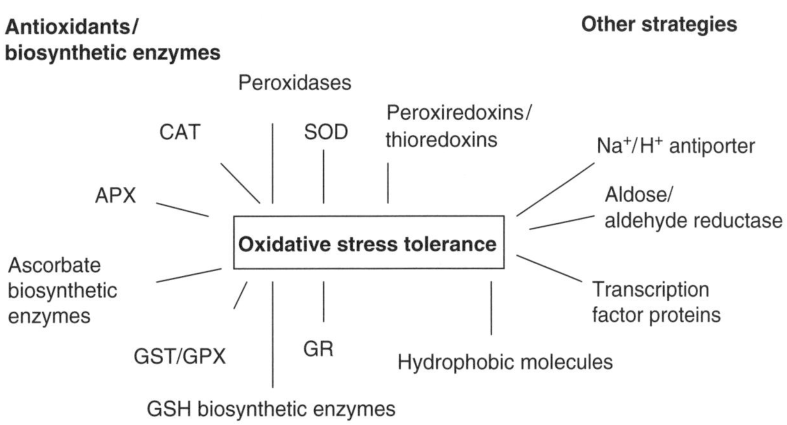

Figure 1 Enzymes/proteins overexpressed in GM plants to obtain improved stress tolerance. APX, ascorbate peroxidase; CAT, catalase; GPX, glutathione peroxidase; GR, glutathione reductase; GSH, glutathione; GST, glutathione *S*-transferase; SOD, superoxide dismutase.

alfalfa plants where SOD had been directed to the mitochondria or chloroplasts. In these experiments neither subcellular targeting nor the type of SOD selected for expression seemed to effect primary freezing injury. In this study improved stress tolerance was considered to be a consequence of better protection against secondary injury symptoms. Results of such studies indicate that SOD overexpression has a beneficial effect. This could be due either to better superoxide scavenging during freezing or to increased H_2O_2 production leading to redox signaling as discussed above.

Perspectives and Possibilities for Future Development

In the last 10 years our perception of the function of AOS in plants has dramatically changed. The viewpoint has shifted from one where oxidants such as H_2O_2, were considered to be intrinsically harmful (and therefore should be removed at all costs) to one where they are regarded as essential and beneficial components of the plant stress response process. This shift has meant that we have had to rethink GM strategies for improved stress tolerance. Strategies for engineering enhanced stress tolerance via amelioration of antioxidant defense capacity are therefore less logical than they appeared 10 years ago. Other strategies that involve enhanced H_2O_2 formation such as occurs in glycinebetaine production in plants overexpressing choline oxidase now appear much more attractive than previously. GM approaches have been invaluable in advancing current concepts of the role of oxidants and antioxidants in whole plant stress tolerance. The laboratory/greenhouse studies carried out over the last 10 years form the basis for some cautious optimism for the potential of GM applications. However, because of a lack of information regarding the complexities of the oxidant/antioxidant interaction in plant stress tolerance responses, the strategies applied so far to GM plants are less straightforward than was originally thought and they are often not particularly convincing.

An accumulating body of evidence suggests that overexpression of single selected genes might be too simplistic and ultimately ineffective in engineering stress tolerance. Often such GM studies are very focused on stress tolerance; this means that other important potential effects of high constitutive expression of targeted enzymes and proteins on crop yield and plant development are not examined in detail, especially under suboptimal growth conditions. This might be especially relevant since such GM crops may be forced to invest large amounts of nitrogen into the products of the transgenes under stress conditions. It is important also to determine whether stress-related traits depend on the interaction of a number of different genes. A balanced interaction involving the simultaneous increase in several protective enzymes using gene pyramiding might be required to obtain substantial and more reliable increases in stress tolerance. In two pioneering studies, the potential of gene pyramiding has already been demonstrated. GM tobacco with enhanced cytosolic activities of both GR and SOD was found to have higher tolerance to paraquat when compared with GM plants overproducing either one of these enzymes. Recently, higher tolerance against paraquat was also found in GM tobacco when both SOD and APX were expressed in chloroplasts. On the other hand, in a recent study the simultaneous expression of two SOD genes, Mit-MnSOD and Chl-MnSOD, in alfalfa resulted in lower shoot and storage organ biomass compared to plants having only one or the other transgene (see **Table 1**). It was postulated that tissue specific or a less strongly expressed promoter might be necessary if closely related transgenes are to be pyramided in the same plant.

In addition to the above concerns it is necessary to note that many of the major world food crops such as wheat (*Triticum aestivum*), rice (*Oryza sativa*), and corn (*Zea mays*; maize) are monocotyledonous plants whereas most transformation studies for stress tolerance are conducted on dicotyledonous species. Caution must be exercised in the use of information on stress tolerance mechanisms obtained in one plant species for application in another. Indeed, the extent to which information derived from GM studies on dicotyledonous models can be applied to the development of GM strategies for improvement of stress responses in cereals, such as wheat, is not known. Large differences in stress tolerance exist between the major cereals and this limits the suitability of direct comparisons. For example, wheat shows the acclimatory property known as cold hardening that enables survival of freezing, whereas 18°C is a suboptimal growth temperature for corn. Agroindustries demand that genetic improvement studies for traits, such as stress tolerance are specifically conducted on the major crops particularly wheat. Commercially grown wheat varieties are hexaploid with a large genome compared to commonly studied model species such as *A. thaliana*, and even with other monocot models, such as rice and corn. The extensive synteny among cereals and the rapid development of functional genomics in major crop species, particularly rice whose genome will be soon fully sequenced, provide a wealth of vital information with applications in crop improvement. Routine robust wheat transformation systems with mean transformation frequencies of

5–6% for elite varieties are now available. This allows the testing and application of GM strategies for improvement of stress tolerance directly in wheat.

Reverse genetic approaches utilizing large-scale insertion mutagenesis and identification of insertion sites is one of the best methods for unraveling the function of specific genes. Transposable elements have been successfully used as mutagens and have proved to be effective tools for the cloning of tagged genes. We are currently using the corn Activator/Dissociation (Ac/Ds) system to create a population of wheat plants tagged with a promoterless *uidA* gene generated by transposon insertion. The approach consists essentially of crossing two populations of GM wheat, one containing the *Ac* transposase gene and the other a promoterless reporter gene (*uidA*) flanked by two *Ds* elements (*Ds* lines). The crossing promotes transactivation and random insertion of *Ds-uidA* elements throughout the wheat genome. Stable tagged *Ds* lines are selected after removal of the *Ac* transposase gene by segregation and used for screening of the reporter gene *uidA*. We have already confirmed the ability of the corn *Ac* transposase to promote Ds transposition in wheat. Tagged wheat populations can be screened for modified tolerance or sensitivity to oxidative stress and/or biotic and abiotic stresses. This provides an additional objective GM approach towards improving stress tolerance and moreover, it allows the identification of *cis*-responsive elements and interactive *trans*-acting factors in cereals such as wheat.

Ultimately, key questions that concern all GM crops including the commercialization and consumer acceptance need to be addressed. Without doubt, any agronomically important crop expressing enhanced oxidative stress tolerance would be of great commercial value, because many limiting factors for plant performance involve oxidative stress. Such traits might confer additional benefits; for example, they could be used effectively in conjunction with other strategies to produce high value proteins, such as vaccines, in plants. When coexpressed, proteins that enhance stress tolerance could effectively ensure better production of more value-added proteins for pharmaceutical purposes since they would confer improved protection against pathogens and pests. Other benefits can accrue; for example constitutive enhancement of antioxidants, such as ascorbate and GSH in plant organs via improved biosynthesis for example, has the potential to increase the nutritional value of food crops and also enhance the viability and the quality of seeds, vegetables and other organs during storage.

As with traditional breeding programs, commercial success for GM crops depends on many factors particularly the consistent expression of the desired trait in the progeny at a required level in a wide range of natural stressful environments. Such GM crops must give consistently higher yields or improved quality traits under field conditions than non-GM controls. In this regard, relatively few data are available to date on the long-term stability of introduced transgenes or on the effects of silencing phenomena in different environments. Tolerance against abiotic stress is unlike tolerance against biotic stress where tolerance often breaks down with the evolution of the pathogen. Therefore environmental concerns with regard to biodiversity must also be considered. Most importantly, the success of any GM crop depends on the ability to overcome the current trend for negative public perception. This represents the most important barrier to be crossed, even in countries where crop losses due to stress are most severe and which would therefore benefit most from the successful development of GM crops with enhanced stress tolerance.

See also: **Abiotic Stresses**: Free Radicals, Oxidative Stress and Antioxidants. **Genetic Modification of Primary Metabolism**: Photosynthesis.

Further Reading

Allen JF (1993) Control of gene expression by redox potential and the requirement for chloroplast and mitochondrial genomes. *Journal of Theoretical Biology* 165: 609–631.

Allen RD, Webb RP, and Schake SA (1997) Use of transgenic plants to study antioxidant defenses. *Free Radical Biology and Medicine* 23: 473–479.

Dunwell JM (2000) Transgenic approaches to crop improvement. *Journal of Experimental Botany* 51: 487–496.

Foyer CH, Descourvieres P, and Kunert KJ (1994) Protection against oxygen radicals: an important defence mechanism studied in transgenic plants. *Plant Cell and Environment* 17: 507–523.

Foyer CH and Noctor G (1999) Leaves in the dark see light. *Science* 284: 599–601.

Foyer CH and Noctor G (2000) Oxygen processing in photosynthesis: regulation and signalling. *New Phytologist* 146: 359–388.

Grover A, Pareek A, Singla SL, *et al.* (1998) Engineering crops for tolerance against abiotic stresses through gene manipulation. *Current Science* 75: 689–696.

Karpinski S, Reynolds H, Karpinska B, *et al.* (1999) Systemic signaling and acclimation in response to excess excitation energy in *Arabidopsis*. *Science* 284: 654–657.

Kuzniak E (2002) Transgenic plants: an insight into oxidative stress tolerance mechanisms. *Acta Physiologiae Plantarum* 24: 97–113.

McKersie BD, Bowley SR, and Jones KS (1999) Winter survival of transgenic alfalfa overexpressing superoxide dismustase. *Plant Physiology* 119: 839–847.

McKersie BD, Murnaghan J, Jones KS, and Bowley SR (2000) Iron-superoxide dismutase expression in transgenic alfalfa increases winter survival without a detectable increase in photosynthetic oxidative stress tolerance. *Plant Physiology* 122: 1427–1437.

Neill S, Desikan R, and Hancock J (2002) Hydrogen peroxide signalling. *Current Opinion in Plant Biology* 5: 388–395.

Noctor G and Foyer CH (1998) Ascorbate and glutathione: keeping active oxygen under control. *Annual Review of Plant Physiology and Plant Molecular Biology* 49: 249–279.

Noctor G, Veljovic-Jovanovic S, and Foyer CH (2000) Peroxide processing in photosynthesis: antioxidant coupling and redox signalling. *Philosophical Transactions of the Royal Society B* 355: 1465–1475.

Pastori GM and Foyer CH (2002) Common components, networks and pathways of cross tolerance to stress: the central role of "Redox" and ABA-mediated controls. *Plant Physiology* 129: 460–468.

Pfannschmidt T, Nilsson A, and Allen JF (1999) Photosynthetic control of chloroplast gene expression. *Nature* 397: 625–628.

Smirnoff N (1998) Plant resistance to environmental stress. *Current Opinion in Biotechnology* 9: 214–219.

Cell Factories

R M Twyman, University of York, York, UK
E Stoger, RWTH Aachen, Aachen, Germany
P Christou, Fraunhofer Institute of Molecular Biology and Applied Ecology, Schmallenberg, Germany

Plant Cells as Factories

The plant cell can be viewed as a minute biomolecular factory, taking in simple molecules and assimilating them into complex ones. Human beings attempt to replicate this process by chemically synthesizing a variety of useful molecules, particularly to use as drugs, but we can achieve nothing like the specificity or selectivity of a living cell, especially when dealing with large and complex molecules. The alternative is therefore to harness the intricacy of the cell and exploit it for the production of specific desirable or useful molecules. There are three major ways in which the plant cell factory is exploited. The first is to introduce new genetic information into the cell enabling it to synthesize an entirely new protein which is itself of value. This approach is called molecular farming. The second is to introduce new genetic information which in some way alters the metabolism of the cell, resulting in the production of a valuable product. This could be a small metabolite, or even a large polymer. This approach is called metabolic engineering. The third is to introduce new genetic information into the cell which modifies the properties of the plant in a new or desirable way, e.g., improving its nutritional qualities, changing the flower color, or prolonging the storage lifetime of its fruit. This approach can be called trait modification.

Engineering the Plant Cell Factory

Strategies for Plant Transformation

Manipulation of the plant cell factory requires a stable and permanent genetic change in the host plant. Stable DNA transfer to plants can be achieved by two approaches. The first is dependent on the soil pathogen *Agrobacterium tumefaciens*, which is known to induce tumors (crown galls) in certain plant species. The tumor results from the transfer of a small segment of DNA (the transferred DNA, T-DNA) from a resident tumor-inducing plasmid (Ti-plasmid) to the plant genome. The genes that cause tumor growth are contained in the T-DNA but they are not required for gene transfer. They can therefore be deleted and replaced with foreign DNA, leading to the production of transgenic plants carrying any foreign gene of interest. The second approach is called direct DNA transfer and encompasses several diverse techniques linked only by their ability to physically introduce DNA into plant cells. Such techniques include particle bombardment (where DNA-coated microprojectiles are accelerated into the cell), protoplast transformation mediated by polyethylene glycol, electroporation (where DNA is taken up through transient pores generated by a brief electric pulse), and whisker transformation (where silicon carbide whiskers perforate the cell wall allowing DNA to enter). Once inside the cell, some of the DNA finds its way to the nucleus where it may be expressed episomally. In rare cases, the DNA integrates into the genome, and these stably transformed cells can be selectively propagated if an appropriate marker gene is introduced into the cell along with any transgenes of interest.

Production of Novel Proteins

The simplest strategy for manipulating the cell factory is to express a new protein in the plant. This is always the case in molecular farming applications, where the protein itself is the desired molecule. For metabolic engineering, it is often necessary to express one or more new enzymes in order to divert metabolic flux in a particular direction and achieve the synthesis of a desired product. Trait modification may involve the production of new proteins (e.g., insecticidal toxins) and/or incorporate aspects of metabolic engineering (e.g., the synthesis of β-carotene in rice (*Oryza sativa*) grains). In each case, the plant undergoes a "gain of function" when the introduced transgene is expressed. The success of this strategy relies principally on the design of the expression construct, which should normally include a strong promoter, elements to optimize the efficiency of protein synthesis, and if necessary a modified coding region to match the codon preference of the host plant. For molecular farming applications it may be necessary to target the protein to a particular tissue, especially if the protein is toxic to the plant. Seed-specific promoters are often used for this purpose so that the recombinant protein does not accumulate in vegetative tissues and affect plant growth. Subcellular targeting may also be an important consideration, especially for the expression of complex proteins and glycoproteins. Transgene expression, and hence the accumulation of recombinant proteins, may also be influenced by epigenetic phenomena and it may be necessary to screen a moderately sized population of plants to identify individual lines showing strong and stable transgene expression.

Interfering with Endogenous Gene Expression

Metabolic engineering and trait modification sometimes require that the expression of certain endogenous genes is either reduced or increased. First we consider ways in which endogenous genes can be inactivated, resulting in a loss of function in the host plant. Conceptually the simplest way to achieve this is to mutate the gene, which can be carried out by homologous recombination with a suitable targeting construct (a procedure called gene knockout). Unfortunately, while it is an efficient process in some plant systems (particularly the moss *Physcomitrella patens* and in chloroplast genomes) homologous recombination occurs with such low efficiency in the nuclear genomes of higher plants that it cannot be used as a routine approach for manipulating the cell factory. The alternative approaches each work by reducing gene expression, either at the RNA or protein levels. The methods include blocking transcription and/or protein synthesis through the expression of antisense RNA, destroying messenger RNA (mRNA) through the expression of antisense RNA joined to ribozymes, overproduction of sense transcripts to provoke cosuppression of endogenous gene expression, the use of double-stranded RNA or small interfering RNA to induce RNA interference, and the expression of antibodies or dominant negative alleles to interfere with the activity of endogenous proteins.

When it comes to increasing levels of an endogenous gene product, one simple strategy is to introduce extra copies of the gene by transformation, usually under the control of a strong constitutive promoter. However, as discussed above, the overproduction of sense transcripts can trigger cosuppression, resulting in the opposite effect to that desired. Another way in which increased levels of gene expression have been achieved is activation tagging, where a T-DNA carrying a strong, outward-facing promoter integrates adjacent to an endogenous gene and enhances its expression level. Other strategies that have been explored include the introduction of a transgene encoding a transcription factor that positively regulates the target gene. This is a particularly useful approach in metabolic engineering because genes acting in a single metabolic pathway are often coregulated by the same transcription factors, so this is a convenient method of controlling the activity of several relevant genes in one experiment.

Cell Factories: Applications

Molecular Farming

Molecular farming involves the manipulation of the cell factory to produce a valuable protein, often with therapeutic potential in humans. Human proteins are popular targets because the use of proteins isolated from natural sources risks contamination with viruses or prions that are wholly absent from plants. Plants have similar protein synthesis and modification pathways to mammals, so recombinant proteins expressed in plants are generally soluble and functional, which is a significant advantage over microbial expression systems. However, plants can also be used to produce industrial enzymes, technical proteins used in research, food and feed additives, and biopolymers. A selection of proteins that have been expressed in plants is listed in **Table 1**.

The first human protein to be expressed in plants was growth hormone, and this was produced in transgenic tobacco (*Nicotiana tabacum*) and sunflower (*Helianthus annuus*) as a fusion with the *Agrobacterium* nopaline synthase enzyme. Tobacco plants have been widely used for molecular farming because the yield of recombinant proteins is high, and there is an established infrastructure for agriculture and downstream processing. However, alternative expression hosts such as cereals, legumes, and plants with edible fruit are becoming more popular

Table 1 A selection of recombinant proteins that have been produced by molecular farming in plants, including pharmaceutical proteins, industrial/processing enzymes, food additives (nutriceuticals), technical proteins, and biopolymers

Species	*Recombinant protein*[1]
Pharmaceutical proteins: human blood products, enzymes, hormones, and growth factors	
Tobacco (leaves, chloroplasts, seeds), sunflower	Growth hormone
Tobacco, potato (*Solanum tuberosum*)	Serum albumin
Tobacco	Epidermal growth factor
Rice	α-Interferon
Rice (cell suspension cultures)	α1-Antitrypsin
Tobacco	Erythropoietin
Pharmaceutical proteins: recombinant antibodies	
Tobacco	VH domain, substance P
Tobacco	IgG, scFv, human creatine kinase
Tobacco (cell suspension culture)	ScFv-immunotoxin, CD-40
Soybean (*Glycine max*)	Humanized IgG, herpes simplex virus
Tobacco, rice, wheat (*Triticum aestivum*), pea (*Pisum sativum*) (whole plants and culture systems)	ScFv, carcinoembryonic antigen
Tobacco	IgG, IgA, *Streptococcus mutans* adhesin
Tobacco (virus-infected plants)	ScFv, 38C13 murine B-cell lymphoma
Pharmaceutical/veterinary proteins: vaccines	
Tobacco	Hepatitis B virus surface antigen
Tomato (*Lycopersicon esculentum*)	Rabies virus glycoprotein
Potato, tobacco	Cholera toxin B subunit
Alfalfa (*Medicago sativa*), *Arabidopsis thaliana*	Foot-and-mouth disease virus VP1
Tobacco, potato	Norwalk virus capsid protein
Industrial/processing enzymes	
Alfalfa, barley (*Hordeum vulgare*), potato, tobacco	1,4-β-D-endoglucanase
Barley, canola (*Brassica napus*), tobacco	Xylanase
Canola, rice, tobacco, wheat	Phytase
Tobacco, bean (*Phaseolus vulgaris*), pea	α-Amylase
Food additives (nutriceuticals)	
Potato	β-Casein
Potato	Lactoferrin
Technical proteins	
Maize (*Zea mays*)	Avidin
Maize	Aprotinin
Maize	β-Glucuronidase
Biopolymers	
Tobacco	Synthetic elastin
Tobacco	Human collagen
Tobacco, potato	Synthetic spider silk

[1]The table lists selected recombinant proteins that have been expressed in plants. Under recombinant antibodies, the description indicates the type of antibody/antibody derivative and the corresponding antigen. Under vaccines, the description indicates the pathogen for which the vaccine was developed.

because recombinant proteins produced in such crops can be consumed directly without processing. This is particularly relevant in the case of vaccines and antibodies designed for passive immunotherapy. Furthermore, different plants have advantages for particular applications. Recombinant proteins produced in cereal grains, for example, are stable at ambient temperatures for years without loss of stability or functionality.

The production of antibodies in transgenic plants is a particularly useful example of molecular farming, since antibodies are multisubunit glycoproteins that are used diagnostically and therapeutically in humans. Such proteins thus test molecular farming to its limit; they must not only be expressed at high enough levels to be economically purified from the host plant, but they must also be correctly folded (to preserve their antigen-binding capability), correctly glycosylated (to avoid inducing an immune response when administered to human subjects), and correctly assembled from individual chains. A typical monoclonal immunoglobulin G (IgG) comprises two heavy chains and two light chains. The corresponding transgenes must be introduced into the plant either at the same time (by cotransformation) or into separate plant lines followed by crossing to generate hybrids. Both these strategies have been used successfully, but an alternative is to express single-chain Fv antibodies, which contain the antigen-binding regions of the heavy and light chains joined by a flexible peptide linker. A number of principles have been learned from the expression of antibodies in transgenic plants. These include the importance of subcellular targeting to the secretory pathway, which ensures correct glycosylation and improves stability compared to antibodies accumulating in the cytosol. Furthermore, the inclusion of a retrieval signal that returns the antibody to the lumen of the endoplasmic reticulum significantly increases the yield of recombinant protein. Remarkably, it has proven possible to express secretory immunoglobulin A (IgA) antibodies in transgenic plants. These have four separate components which must assemble correctly in a single plant cell. Even in mammals, two different cell types are required to assemble these molecules. A plant-derived recombinant sIgA against the oral pathogen *Streptococcus mutans* has reached phase II clinical trials as the drug CaroRx^{TM}, which should help to reduce the incidence of dental caries.

Metabolic Engineering

Metabolic engineering involves the use of gene transfer technology to modify or extend metabolic pathways in the plant cell. In some cases, plants that produce a particular metabolite in small quantities are modified so that flux along that particular pathway is increased resulting in the accumulation of larger amounts of the product. In other cases, existing pathways are extended or entirely new pathways are imported so that the plant produces one or more novel metabolites.

First, we consider how to increase the production of a particular metabolite already synthesized in the plant cell. Plant secondary metabolism produces an extremely diverse array of small molecules, many of which are therapeutically relevant in humans. For example, alkaloids are isolated from plants and used as drugs even though they are produced in only minute quantities. Isolation is technically demanding and very expensive, but because the molecules are too complex to produce by direct chemical synthesis, it is currently the only feasible approach. Metabolic engineering offers the prospect of producing designer plants or plant cell cultures that produce specific alkaloids in large amounts. This has been achieved in a few cases, the most obvious examples being the drugs paclitaxel and shikonin. Unfortunately, in many cases, secondary metabolic pathways are long and complex. They may have cell type-specific components, and they may be regulated by numerous overlapping feedback mechanisms. To produce more of a particular metabolite, a simple strategy would be to identify the rate-limiting step in the metabolic pathway and introduce a transgene encoding that enzyme, thereby removing the bottleneck and allowing increased flux down the pathway. However, where this has been tried a common result is that another form of feedback regulation then limits the activity of the pathway, leaving the investigator no better off. Even where flux is increased through the bottleneck, this approach generally only serves to reveal the next bottleneck, frequently at the next step in the pathway!

As an example, let us consider production of the valuable antineoplastic alkaloids vinblastine and vincristine in the Madagascar periwinkle (*Catharanthus roseus*). These molecules are produced in very small amounts and are isolated at great expense, accounting for their current market value of over US $1 million per gram. Two pathways converge to make these alkaloids, the terpenoid pathway and the shikimate pathway (**Figure 1**). At the end of the shikimate pathway, tryptophan must be decarboxylated to generate tryptamine, and this is catalyzed by the enzyme tryptophan decarboxylase (TDC). Tryptamine is then condensed with the terpenoid molecule secologanin to produce strictosidine, the precursor of all terpenoid indole alkaloids. The enzyme strictosidine synthase (STR) is responsible for this reaction. TDC is a rate-limiting step in alkaloid biosynthesis, so the *C. roseus tdc* gene was overexpressed under the control of a strong and constitutive promoter. The resulting transgenic cell cultures produced high levels of tryptamine, but no alkaloids, because strictosidine synthase then became the rate-limiting step in the pathway.

Multistep regulation is a possible solution to this problem. It has been shown that several of the biosynthetic genes in this pathway are coinduced by externally applied agents such as methyl jasmonate, including tryptophan decarboxylase (*tdc*), strictosidine synthase (*str*), and at least six others. Transcription factors have been identified that bind to jasmonate response elements in the promoters of these genes and have been shown to upregulate the genes coordinately when overexpressed. A more thorough understanding of the regulatory mechanisms controlling this pathway may eventually allow precise control over the levels of valuable metabolites such as vinblastine and vincristine produced in this plant.

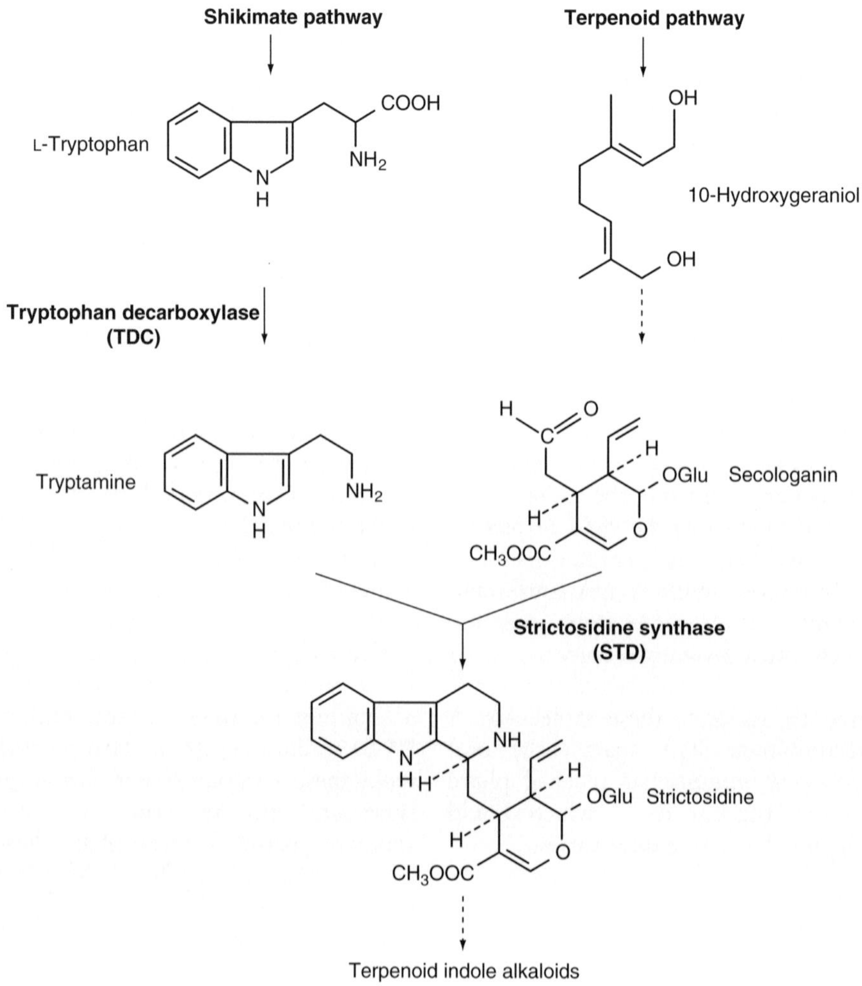

Figure 1 Convergence of the shikimate and terpenoid pathways for the biosynthesis of terpenoid indole alkaloids.

Now we consider how plants can be persuaded to manufacture entirely novel compounds. One example is the production of polyhydroxyalkanoates in plants by diverting metabolic flux from the fatty acid biosynthesis pathway. Such products are not known to be made naturally in any plant species, and the relevant transgenes were imported from bacteria. The polyhydroxyalkanoates can be used to make biodegradable thermoplastics. Another example is the production of scopolamine, an antihypertensive alkaloid drug, in *Atropa belladonna*. Scopalomine is synthesized from hyoscyamine in a two-step reaction carried out by the enzyme hyoscyamine-6-hydroxylase (H6H). This enzyme is present in the plant *Hyoscyamus niger*, which produces scopalomine, but not in *A. belladonna*, which accumulates hyoscyamine. However, when a *H. niger* cDNA encoding H6H was expressed in *A. belladonna*, the hyoscyamine pool was used to produce scopalomine, i.e., the endogenous metabolic pathway was extended beyond its natural endpoint in this species.

Trait Modification

At the simplest level, any manipulation of the plant genome which results in a new phenotype can be considered a trait modification. However, while gene transfer can be used to engineer disease resistance, herbicide resistance, pest resistance, or tolerance of abiotic stresses such as drought, cold, and waterlogging, the result is still a plant which looks, and where appropriate tastes, the same as normal. Other types of trait modification make more of an impact on the quality of the plant.

In other cases, trait In some cases, trait modification overlaps with metabolic engineering. For example, modifying the flower color of petunia generates a striking phenotypic change, but the underlying theme is still one of increasing or decreasing the levels of various enzymes in the developing plant, in this case those involved in pigment biosynthesis. Nutritional improvement is another example. Cereal grains lack β-carotene (provitamin A), which is required in the diet for vitamin A biosynthesis. A lack of vitamin A in childhood leads to blindness, and since almost two-thirds of the population of the world eat rice as their staple diet, improving the vitamin A content of this cereal would make a real impact on world health. Initially, rice plants were transformed with the phytoene synthase gene from the daffodil (*Narcissus pseudonarcissus*) and accumulated phytoene, an intermediate in the vitamin A biosynthesis pathway. More recently, rice plants were generated expressing two daffodil enzymes and one bacterial enzyme, recapitulating the entire β-carotene pathway and producing grains that were golden in color. In this example, an entire metabolic pathway was assembled and imported into rice.

modification overlaps with molecular farming. The expression of a single heterologous protein, the iron-storage protein ferritin, can increase the levels of bioavailable iron in mineral-depleted foods, such as rice. In a slightly different approach, rice was transformed with a fungal gene encoding the enzyme phytase. This removes phytic acid from cereal grains, a molecule which inhibits iron absorption from the gut.

Future Directions

In other cases, trait The scope for manipulating the cell factory is limited only by the imagination of scientists and the availability of genes allowing desirable products to be synthesized. Molecular farming is the simplest application and improvements in this area will come mainly from the development of better expression constructs, particularly in terms of helping recombinant proteins to accumulate to high levels in transgenic plants. Metabolic engineering and trait modification often require more complex manipulations, which may involve the transfer of multiple genes into the same plant, or the modification of endogenous gene expression levels. Improvements in this area will come largely from a more thorough understanding of plant metabolism, particularly how the various steps in complex pathways are regulated. There is no doubt that the completion of the genome projects for several crop plants plus the increasing amounts of available expressed sequence tags (ESTs) and proteomic data will provide the basis for such research, and that new functional genomics tools such as high-throughput transcriptional profiling, systematic mutagenesis, genome-wide protein interaction screening, and rapid determination of protein structures will be critical in the accumulation of knowledge about metabolic pathways.

List of Technical Nomenclature

Activation tagging The isolation of genes through the integration of a DNA element containing a strong promoter that can activate genes adjacent to the insertion site.

Alkaloids Complex nitrogen-containing heterocyclic organic compounds, mainly of plant origin, with potent pharmacological properties in humans.

Antisense RNA RNA that is complementary to a given mRNA. Expression of antisense RNA results in stoichiometric binding to the target message, blocking transcription and protein synthesis and leading to degradation.

Codon preference The frequency with which particular codons are used to specify the same amino acid in different species.

Constitutive promoter A promoter that drives transcription continuously and in all cell types.

Cotransformation Simultaneous transformation with multiple genes.

Epigenetic Refers to phenomena that affect the phenotype of a plant without altering the nucleotide sequence of the genome. Includes factors such as DNA methylation, repression of gene expression or protein function, and environmental effects.

Gain of function A new function conferred on a plant by gene transfer or mutation.

Gene knockout A form of directed homologous recombination in which a functional allele is replaced by a nonfunctional (null) allele.

Homologous recombination A ubiquitous recombination process occurring between two DNA sequences with long regions of homology but no particular sequence specificity.

Homology-dependent gene silencing Transcriptional or posttranscriptional epigenetic effect in which gene expression is reduced or eliminated, brought about by the presence of multiple copies of the same DNA sequence.

Metabolic engineering The deliberate alteration of metabolism by gene manipulation, in order to modify the levels of a particular metabolite.

(Methyl) jasmonate An intermediate in elicitor-stimulated signaling which can also induce defense responses when applied to plant cells.

Molecular farming The production, in transgenic plants or transformed plant cells, of valuable recombinant proteins.

Monoclonal	Describing an antibody produced by a clonal hybridoma cell line, which recognizes a single antigenic determinant.
Nutritional improvement	Genetic engineering to improve the nutritional content or quality of a particular crop plant.
Position effects	The effects of surrounding chromatin or genomic sequence context on the level of transgene expression. Repressive position effects lead to silencing even if the transgene is present as a single copy.
Recombinant protein	A protein encoded by a transgene and produced by a transformed cell or transgenic plant.
Ribozyme	A catalytic RNA molecule that cleaves other RNAs. Can be targeted to specific mRNAs if linked to an appropriate antisense RNA molecule.
RNA interference	The ability of double-stranded RNA to induce potent and specific silencing of a homologous gene by targeting the mRNA for degradation via the RNA induced silencing complex.
Secondary metabolism	The sum of metabolic pathways producing molecules with specialized functions, not involved in general homeostasis.
Secretory IgA	An antibody with four component polypeptides, which is generally found in mucosal secretions.
Single chain Fv	A derivatized antibody molecule in which the antigen-binding regions of the heavy and light chains are expressed as a single polypeptide.
Targeting construct	A transformation vector containing a homology region matching an endogenous gene, designed to stimulate homologous recombination and replace the endogenous gene with the sequence carried in the vector.
Trait modification	The use of gene transfer technology to express one or more new proteins or alter endogenous metabolism in order to improve a certain trait in a crop plant.

See also: **Genetic Modification**: Gene Cloning, General Principles; Transformation, General Principles. **Genetic Modification, Applications**: Flower Color; Molecular Farming; Plantibodies. **Genetic Modification of Secondary Metabolism**: Alkaloids; Terpenoids. **Tissue Culture**: Secondary Metabolism in Plant Cell Cultures.

Further Reading

Fischer R and Emans N (2000) Molecular farming for pharmaceutical proteins. *Transgenic Research* 9: 279–299.

Fray RG and Grierson D (1993) Molecular genetics of tomato fruit ripening. *Trends in Genetics* 9: 438–443.

Hiatt A (1990) Antibodies produced in plants. *Nature* 344: 469–470.

Lucca P, Hurrell R, and Potrykus I (2001) Genetic engineering approaches to improve the bioavailability and the level of iron in rice grains. *Theoretical and Applied Genetics* 102: 392–397.

Mason HS and Arntzen CJ (1995) Transgenic plants as vaccine production systems. *Trends in Biotechnology* 13: 388–392.

Murphy DJ (1999) Production of novel oils in plants. *Current Opinion in Biotechnology* 10: 175–180.

Primrose SB, Twyman RM, and Old RW (2001) *Principles of Gene Manipulation*, 6th edn. Oxford: Blackwell Science.

Verpoorte R, van der Heijden R, ten Hoopen HJG, and Memelink J (1999) Metabolic engineering of plant secondary metabolic pathways for the production of fine chemicals. *Biotechnology Letters* 21: 467–479.

Verpoorte R, van der Heijden R, and Memelink J (2000) Engineering the plant cell factory for secondary metabolite production. *Transgenic Research* 9: 323–343.

Walmsey AM and Arntzen CJ (2000) Plants for delivery of edible vaccines. *Current Opinion in Biotechnology* 11: 126–129.

Ye XD, Al-Babili S, Kloti A, *et al.* (2000) Engineering the provitamin A (beta-carotene) biosynthetic pathway into (carotenoid-free) rice endosperm. *Science* 287: 303–305.

Molecular Farming

R M Twyman, University of York, York, UK
E Stoger, RWTH Aachen, Aachen, Germany
P Christou, Fraunhofer Institute for Molecular Biology and Applied Ecology, Schmallenberg, Germany

Introduction

Molecular farming refers to the production, in transformed cells or transgenic organisms, of recombinant proteins with potential therapeutic or commercial value. Molecular farming began in the 1980s when the first recombinant human proteins – insulin and growth hormone – were synthesized in the bacterium *Escherichia coli*. Since that time, a wide variety of alternative expression systems has been explored, including other species of bacteria, yeasts and other fungi, animal and plant cells in culture, and transgenic animals and plants.

Bacterial systems have the advantage of being both inexpensive and relatively high in yield, but bacteria lack eukaryotic chaperones and fail to carry out most of the posttranslational chemical modifications, such as glycosylation, that are characteristic of eukaryotic cells. The lack of chaperones means that many recombinant animal proteins do not fold properly in bacteria, and accumulate as insoluble inclusion bodies, which must be extracted and solubilized *in vitro*. Even when an expressed protein is soluble, the absence of glycosylation and other forms of modification means that the protein may not function correctly, or, in the case of recombinant human proteins for therapeutic use, may elicit an immune response. Yeast and insect cells do modify expressed proteins but the patterns of glycosylation often differ considerably from those found in mammals. Mammalian cell cultures and transgenic animals are thus optimal expression systems in terms of achieving correct folding and protein modification, but these are the most expensive to initiate and scale up.

Advantages of Molecular Farming in Plants

Plants have certain advantages over all other expression systems for the synthesis of recombinant proteins, but it is the combined benefits of plants that make them such an attractive system for molecular farming. Unlike microbial expression systems, plants possess a very similar protein synthesis and modification pathway to animals. Only minor differences are found in glycosylation patterns, and thus far these have been shown neither to influence the functionality of recombinant animal proteins nor make them immunogenic when administered to humans. The expression levels of recombinant proteins produced in plants compare favorably with those produced in other systems, but unlike animal cells and transgenic animals, the costs associated with production are not high. For example, scaling up recombinant protein production in plants simply involves growing more plants, whereas scaling up animal cell expression systems requires expensive media and fermentors, and scaling up transgenic animal systems involves increased costs of housing, feeding, and veterinary care. Perhaps the most important advantage of plants over animals is that plants do not carry potentially dangerous contaminants, such as viruses and prions, which could cause disease in humans. Therefore, even for therapeutic proteins, the costs of downstream processing are considerably lower in plant than in animal systems. Indeed, in some cases, there is no need for any downstream processing at all. For example, protein subunit vaccines can be produced in fruits or seeds, which are then eaten as a form of direct oral administration.

Practical Considerations

Transformation Strategies

Methods of gene transfer Recombinant protein production in plants requires the introduction of foreign DNA into the plant cell. Gene transfer can be achieved essentially by three routes. The first involves the soil pathogen *Agrobacterium tumefaciens*, which can transfer a small segment of DNA from a resident plasmid (Ti plasmid) into the plant genome. This natural system can be exploited by removing the bacterial genes from the transferred DNA segment (T-DNA) and replacing them with foreign DNA. The second route is direct DNA transfer, which encompasses a range of unrelated techniques that force DNA across the cell wall and/or membrane. Such techniques include particle bombardment, electroporation, transformation of protoplasts mediated by polyethylene glycol, and transformation of intact cells following perforation of the cell wall with silicon carbide whiskers. The third route exploits the ability of plant viruses to penetrate cells naturally and carry nucleic acids into the interior. The genomes of plant viruses can be modified to carry foreign genes, such that viral infection results in recombinant protein expression as well as the production of viral gene products.

Transient expression and stable transformation *Agrobacterium*-mediated transformation and direct DNA transfer can both result in stable transformation, which means that foreign DNA integrates into the genome and is maintained as a new genetic locus. Plants regenerated from stably transformed cells or explants are described as transgenic because they carry the foreign DNA in all cells including those destined to form gametes. Thus, the transgene can be passed to progeny resulting in a transgenic plant line. However, only a small proportion of cells that receive foreign DNA undergo stable transformation, and these must be identified by prolonged selection for properties conferred by the foreign DNA. A selectable marker gene conferring resistance to antibiotics or herbicides, or conferring a novel metabolic property on the plant tissue, is co-introduced with the transgene of interest to facilitate this.

During transformation, a large number of cells receive foreign DNA but do not undergo transgene integration. These cells are capable of transient expression before the foreign DNA is diluted and degraded, and this provides a convenient assay to test for successful expression of the foreign gene construct. The key point here is that transient expression assays are rapid and inexpensive, whereas the production and verification of transgenic plant lines is a lot more time consuming and labor intensive. A technique often used to test expression constructs is agroinfiltration, where *A. tumefaciens* is infiltrated into the plant tissue and transgene expression is monitored in the absence of selection. A key advantage of this method is that multiple transgene constructs can be tested simultaneously allowing the verification of recombinant protein production and, if necessary, the assembly of multiple proteins into an oligomeric complex. To test this process using transgenic plants, it would be necessary either to transform plants simultaneously with multiple transgenes or generate separate lines expressing individual genes and interbreed them.

Expression Systems

Transgenic plants Once the function of a transgene construct has been verified by transient expression, transgenic plants offer the most attractive expression system for long-term production. This is because, once a useful transgenic line has been produced, it represents a continuous and stable source of the recombinant protein over many generations, and scaling production is simply a matter of increasing the number of hectares of land used for a particular crop. The disadvantage of transgenic plants lies in the investment and time-scale required to produce the primary transformants and verify that the transgene is stably transmitted and expressed over several generations. Once this has been achieved, normal agricultural practices can be used to sow and harvest the crops.

Virus-infected plants Plant viruses do not integrate into the genome, so infecting plants with recombinant viruses does not lead to the production of transgenic crops. However, virus-infected plants represent a useful expression system because the onset of transgene expression is rapid following infection, and many plant viruses are systemic, so infection (and transgene expression) spreads throughout the plant. Thus, high yields of recombinant protein are obtained in a much shorter time than is possible with transgenic plants, plus there are only limited set-up costs. Although not a permanent resource like transgenic lines, mechanical inoculation of plants is relatively straightforward and viruses can also be transmitted by natural infection routes or by grafting infected scions onto virus-free rootstock.

Choice of species Different plant species have different production advantages and these must be taken into account when considering an expression host for molecular farming. Important factors include efficiency of recombinant protein production, total biomass yield, storage properties, edibility, land area required for growth, labor required for harvesting the crop, existing infrastructure for distribution and processing, and downstream processing costs. These must be evaluated for each recombinant protein on a case-by-case basis. For example, cereal seeds are particularly useful expression vehicles where transport or long-term storage is available only at ambient temperatures, since recombinant proteins stored in cereal endosperm have been shown to remain stable and functional over several months under these conditions. This would not be the case for proteins synthesized, for example, in tobacco (*Nicotiana tabacum*) leaves or tomato (*Lycopersicon esculentum*) fruits; these would have to be dried or stored in chilled containers, respectively. However, tobacco is one of the highest yielding plants in terms of recombinant protein production per hectare, so is optimal for recombinant proteins with a low to moderate market value. An advantage of tomatoes is that they are grown in greenhouses and are therefore beneficial where containment of the transgenic crop is a particular concern. Where recombinant proteins are going to be orally administered, e.g., vaccines or antibodies for passive immunity, cereals and fruits are more appropriate expression vehicles than tobacco, since the former can be eaten with minimal processing, while the latter is not edible and contains toxic compounds such as alkaloids.

Cell suspension cultures As an alternative to transgenic plants, plant cells can be grown in suspension culture, either in flasks or fermentors, and transformed cell lines can be produced relatively easily either by *Agrobacterium*-mediated transformation or a range of direct DNA transfer methods. Compared to transgenic plants, the set-up costs for recombinant protein production are low, and transformed tobacco and rice (*Oryza sativa*) cell lines can be obtained and verified in a relatively short time. However, scaling up production in cell suspension cultures is expensive, because large fermentors and skilled labor are required. This system is therefore advantageous for the production of high-value recombinant proteins intended for clinical use, since the culture conditions are sterile and precisely controlled. A further advantage is the ease of protein purification. Recombinant proteins can either be isolated from the cells themselves or, if suitable expression constructs are used, the protein can be secreted and isolated from the culture medium.

Optimizing Production

Optimization of transcription To maximize recombinant protein expression in plants, it is often beneficial to use a strong and constitutive promoter. In tobacco and other dicotyledons, the cauliflower mosaic virus 35S RNA promoter (CaMV 35S) is the most widely used, and derivatives with duplicated enhancer regions and other modifications show even greater activity. Alternatives include the promoters of the *Agrobacterium* nopaline synthase (*nos*) and octopine synthase (*ocs*) genes. The CaMV 35S promoter shows much lower activity in monocotyledons, so different promoters tend to be used where high-level expression is required in cereals, e.g., the rice actin-1 and corn (*Zea mays*; maize) ubiquitin-1 promoters. In all cases, the presence of an intron has been shown to markedly increase expression levels. Constitutive promoters are not always beneficial, since the recombinant proteins may be toxic to the host plant. One alternative strategy is to use a tissue-specific promoter, such that the protein accumulates in a particular organ. Seed-specific promoters, such as those from the corn zein, wheat (*Triticum aestivum*) glutenin, and rice glutelin genes have been used for this purpose in cereals. Promoters that are inducible by tetracycline and other externally applied agents can also be used.

Optimization of protein synthesis The transgene construct can also be modified to increase the efficiency of protein synthesis. For example, the 5′ leader sequence of the tobacco mosaic virus RNA (the omega sequence) and a similar region from the petunia (*Petunia* spp.) chalcone synthase gene have been shown to act as translational enhancers, increasing the level of expression over 50-fold. Further considerations include making the translational start site match the optimal "Kozak consensus" for plants and modifying the coding region of the transgene to match the codon usage preferences of the expression host. Different organisms prefer to use different codons for the same amino acid, so the preferred codon for valine in humans is not necessarily the one preferred in rice or tobacco. When the ribosome encounters a "rare" codon, it can pause, skip over, or introduce a frameshift, thus reducing the efficiency of protein synthesis considerably.

Subcellular protein targeting The inclusion of subcellular targeting information in the expression cassette is an important consideration when developing a strategy for molecular farming. Targeting to the secretory pathway may be necessary for some recombinant proteins, particularly if they are normally glycosylated, since this process takes place in the endoplasmic reticulum (ER) and Golgi apparatus. The environment of the ER may also be required for correct folding, and this is demonstrated by the differing stability of some recombinant proteins when they are targeted to the secretory pathway or to the cytosol. Targeting to the secretory pathway is achieved by incorporating an N-terminal signal sequence into the expression construct, such as that from the mouse immunoglobulin heavy chain or the pea legumin protein. In plant cells, proteins are secreted to the apoplastic space, and this may be beneficial for downstream protein purification. However, much higher production levels have been achieved by storing recombinant proteins in the ER through the use of a retrieval signal (the amino acid sequence H/KDEL) at the C-terminal end of the recombinant protein. Targeting to other compartments, such as the chloroplast or vacuole, may also be useful under certain circumstances. For example, plants expressing high levels of recombinant avidin in the cytosol are not viable because high levels of avidin interfere with endogenous biotin metabolism, but large amounts of avidin can be stored in the vacuole without affecting the host plant.

Pharmaceutical Proteins Expressed in Plants

Blood Products, Hormones, Growth Factors and Cytokines

Some of the most widely used therapeutic human proteins include blood products (e.g., human serum albumin, factor VIII), hormones (e.g., insulin, growth hormone), growth factors (e.g., platelet-derived growth factor, nerve growth factor), and cytokines (e.g., interleukins, interferons). The first human protein to be expressed in plants was growth hormone, and this was achieved in 1986. Although the protein was expressed as a fusion to the *Agrobacterium* nopaline synthase enzyme and was therefore unsuitable for administration to humans, this was perhaps the first example of foreign gene expression in transgenic plants that could properly be considered as "molecular farming". In 1990, recombinant human serum albumin was produced in plants for the first time. The global demand for human serum albumin is high (over 500 tonnes per year) and the protein is normally isolated from blood.

Although careful screening is carried out on all blood products, there are still risks associated with this procedure, so there is considerable interest in safe alternative sources for this and other human proteins. For this reason, many of the early examples of molecular farming in plants concerned the synthesis of human blood products, hormones, growth factors, and cytokines, and these continue to be popular products today. A selected list of human proteins that have been expressed in plants is provided in **Table 1**.

Antibodies (Plantibodies)

Unlike the proteins discussed above, the demand for antibodies is relatively low (1–50 kg per year, depending on the particular antibody). However, antibodies are used for both therapeutic and diagnostic procedures and are traditionally produced in hybridoma cell lines, which are expensive to scale up. The expression of human antibodies in plants is of particular interest because these proteins are oligomeric complexes. This means they must not only be correctly expressed and glycosylated, but also assembled in the plant cell. In the case of secretory IgA antibodies (sIgA), which have four separate components, assembly must occur in a single plant cell, even though two different cell types are required for this process in humans. Antibody expression in plants was first demonstrated in 1989, with the production of full-size immunoglobulins. The strategy used at this time was to generate transgenic tobacco plants, individually expressing the heavy and light chains, and then breed them to produce hybrids expressing both molecules. Since then, functional antibodies have been produced in many different plant species and systems, including transgenic tobacco and cereals, virus-infected tobacco, and rice cell cultures. Various classes of full-size antibodies have been produced, including secretory IgA, as well as a range of derivatives for specialized applications; a selected list is provided in **Table 2**. Transgenic plants producing a recombinant sIgA active against the oral pathogen *Streptococcus mutans* have been produced and this molecule is now commercially available as the drug CaroRx^{TM}, marketed by Planet Biotechnology Inc. *S. mutans* is the causative agent of dental caries (tooth decay) and the secretory antibody has been shown to confer passive immunity to the pathogen.

Vaccines

One aspect of molecular farming in plants that warrants special consideration is the development of

Table 1 A selection of recombinant human proteins that have been produced by molecular farming in plants

Year	*Species*	*System*	*Recombinant human product*
1986	Tobacco, sunflower (*Helianthus annuus*)	Plants	Growth hormone[1]
1990	Tobacco, potato (*Solanum tuberosum*)	Plants	Serum albumin
1993	Tobacco	Plants	Epidermal growth factor
1994	Rice	Plants	α-Interferon
1995	Tobacco	Cells	Erythropoietin
1996	Tobacco	Plants	Protein C serum protease
1997	Tobacco	Plants	Muscarinic cholinergic receptor
1998	Tobacco	Cells	Interleukin-2
1999	Rice	Cells	α1-Antitrypsin
2000	Tobacco	Chloroplasts	Growth hormone

[1] Growth hormone was expressed as a fusion protein with the *Agrobacterium* nopaline synthase enzyme.

Table 2 The production of different classes of antibody and derivatized antibody molecule in plants (the first known report for each type of antibody is listed)

Year	*Species*	*Type of antibody*[1]	*Antigen*
1989	Tobacco	IgG	Phosphonate ester
1990	Tobacco	IgM	Neuropeptide hapten
1991	*Nicotiana benthamiana*	V_H	Substance P
1992	Tobacco	ScFv	Phytochrome
1993	Tobacco	Fab	Creatine kinase
1995	Tobacco	IgA/G	*S. mutans* adhesin
1997	Tobacco (culture)	scFV-toxin	CD-40
1998	Soybean (*Glycine max*)	Humanized IgG	HSV-2
1999	Tobacco (cells)	bi-scFv	Tobacco mosaic virus

[1] V_H, variable domain of the heavy chain; scFv, single chain Fv region; Fab, fragment antigen binding; HSV, herpes simplex virus.

vaccines. As discussed above, plants are an inexpensive, efficient, and, most importantly, safe expression system for recombinant proteins, which means that recombinant subunit vaccines produced in plants can be orally administered by the consumption of unprocessed or partially processed plant material. The first example of vaccine expression in plants again involved the oral pathogen *S. mutans*. A surface antigen from this organism, which can induce a mucosal immune response in the oral cavity and prevent colonization by the bacteria, was expressed in tobacco. A number of edible transgenic plants have been generated expressing antigens derived from animal viruses such as rabies, hepatitis B, and cholera (**Table 3**). A related application is the infection of plants with recombinant cowpea mosaic viruses (CMV) expressing various antigen epitopes on their surfaces. Recent successes using this system include vaccination trials with recombinant CMV vectors expressing epitopes of HIV gp41 and canine parvovirus. The first clinical trials using a plant-derived vaccine were carried out in 1997, using transgenic potatoes expressing the B subunit of the *E. coli* heat labile toxin, a causative agent of diarrhea.

Future Perspectives

Plant transformation is now a routine procedure and the focus of research in this area has shifted away from transformation itself to the control of transgene expression. Molecular farming in plants benefits from any new developments in plant expression technology, and this is where most improvements in the field are likely to originate in the near future. In particular, the field is likely to benefit from the discovery and development of stronger promoters and translation regulatory elements, more robust subcellular targeting signals, and facile methods to control epigenetic gene silencing phenomena. One recent development that deserves a special mention is the possibility of high-level protein production following the transformation of chloroplasts. Plastid transformation is advantageous because the transgene copy number is amplified by the large number of chloroplasts in the cell. However, the chloroplast genome does not seem to be affected by the position effects and silencing phenomena that often afflict nuclear transgenes. Transplastomic tobacco plants containing a human growth hormone gene have been shown to express the recombinant protein at a level equivalent to 7% total cellular protein. Hopefully, this type of achievement can be repeated with other human proteins.

Table 3 Selected vaccines against human and animal pathogens that have been produced in plants

Year	*Species*	*Antigen*
1992	Tobacco	Hepatitis virus B surface antigen
1995	Tomato	Rabies glycoprotein
1996	Tobacco, potato	Norwark virus coat protein
1997	Produced in a recombinant plant virus	Mink enteritis virus
1998	Potato	Cholera toxin B subunit
1999	Alfalfa (*Medicago sativa*)	Foot and mouth disease VP1 protein

List of Technical Nomenclature

Agroinfiltration Infiltrating *Agrobacterium* into plant tissue under vacuum in order to test for transient expression.

Chaperone A protein whose function is to assist with the correct folding of other proteins during or after translation.

Codon usage The frequency with which particular codons are used to specify the same amino acid in different species.

Constitutive promoter A promoter that drives transcription continuously and in all cell types.

Electroporation Transformation mediated by a brief electric pulse, which causes transient pores to open in the cell membrane allowing the uptake of DNA.

Epitope A short peptide sequence that has the ability to stimulate the immune response when presented as part of a larger molecule. A single antigenic determinant.

Inclusion body A protein aggregate that forms in bacterial cells when they overexpress a recombinant protein that does not fold properly and thus remains insoluble.

Kozak consensus The sequence surrounding and including the start codon, which is optimal for translational efficiency.

Molecular farming The production, in transgenic plants or transformed plant cells, of valuable recombinant proteins.

Particle bombardment Transformation mediated by the acceleration of DNA-coated metal particles into the plant cell.

Passive immunity Immunity conferred not by stimulating the immune system to produce antibodies, but by providing those antibodies directly.

Position effects The effects of surrounding chromatin or genomic sequence context on the level of transgene expression. Repressive position effects lead to transgene silencing.

Protoplast A plant cell deprived of the cell wall.

Retrieval signal A short consensus sequence H/KDEL found at the C-terminal end of proteins that function in the endoplasmic reticulum (ER) lumen, allowing it to be retrieved from further down the secretory pathway and retained in the ER.

Selectable marker gene A gene that confers a selectable property on plant cells, such as resistance to herbicides or antibiotics, allowing stably transformed cells to be selectively propagated.

Signal sequence A short amino acid sequence at the N-terminal end of the protein which targets it to the endoplasmic reticulum during translation.

Stable transformation Permanent transformation that occurs when DNA integrates into the plant genome. The DNA becomes a new genetic locus, which is transmitted to progeny.

T-DNA Transferred DNA, the segment of DNA on the Ti-plasmid which is transferred to the plant genome.

Ti-plasmid A tumor-inducing plasmid found in virulent strains of *A. tumefaciens*. The T-DNA carries hormone biosynthesis genes that cause unregulated cell proliferation (hence tumor growth) when the plant is transformed.

Transgene silencing Loss of transgene expression caused by epigenetic processes rather than mutation. Includes position effects and homology-dependent silencing.

Transplastomic A plant containing transgenes in the chloroplast genome.

See also: **Genetic Modification**: Gene Cloning, General Principles; Transformation in Monocotyledons; Transformation in Dicotyledons; Transformation in Plastids. **Genetic Modification, Applications**: Cell Factories; Plantibodies; Transformation, General Principles.

Further Reading

Barta A, Sommergruber K, Thompson D, *et al.* (1986) The expression of a nopaline synthase-human growth hormone chimaeric gene in transformed tobacco and sunflower callus tissue. *Plant Molecular Biology* 6: 347–357.

Dalsgaard K, Uttenthal A, Jones TD, *et al.* (1997) Plant-derived vaccine protects target animals against a virus disease. *Nature Biotechnology* 15: 248–252.

Daniell H, Streatfield SJ, and Wycoff K (2001) Medical molecular farming: production of antibodies, biopharmaceuticals and edible vaccines in plants. *Trends in Plant Science* 6: 219–226.

Doran PM (2002) Foreign protein production in plant tissue cultures. *Current Opinion in Biotechnology* 11: 199–204.

Fischer R and Emans N (2000) Molecular farming for pharmaceutical proteins. *Transgenic Research* 9: 279–299.

Fischer R, Twyman RM, and Schillberg S (2003) Production of antibodies in plants and their use for global health. *Vaccine* 21: 820–825.

Giddings G, Allison G, Brooks D, and Carter A (2000) Transgenic plants as factories for biopharmaceuticals. *Nature Biotechnology* 18: 1151–1155.

Hiatt A (1990) Antibodies produced in plants. *Nature* 344: 469–470.

Hood EE (2002) From green plants to industrial enzymes. *Enzyme & Microbial Technology* 30: 279–283.

Ma JK, Hikmat BY, Wycoff K, *et al.* (1998) Characterization of a recombinant plant monoclonal secretory antibody and preventive immunotherapy in humans. *Nature Medicine* 4: 601–606.

Mason HS, Warzecha H, Mor T, and Arntzen CJ (2002) Edible plant vaccines: applications for prophylactic and therapeutic molecular medicine. *Trends in Molecular Medicine* 8: 324–328.

Primrose SB, Twyman RM, and Old RW (2001) *Principles of Gene Manipulation*, 6th edn: Oxford: Blackwell Science.

Sijmons PC, Dekker BMM, Schrammeijer B, *et al.* (1990) Production of correctly processed human serum albumin in transgenic plants. *Biotechnology* 8: 217–221.

Staub J, Garcia B, Graves J, *et al.* (2000) High-yield production of a human therapeutic protein in tobacco chloroplasts. *Nature Biotechnology* 18: 333–338.

Stoger E, Sack M, Perrin Y, *et al.* (2002) Practical considerations for pharmaceutical antibody production in different crop systems. *Molecular Breeding* 9: 149–158.

Plantibodies

P Langridge and J Langridge[†], Adelaide University, Glen Osmond, SA, Australia

Definition and Background

Various heterologous systems have been used for the production of antibodies. Antibodies produced in plants or plant cells are known as "plantibodies" or "phytoantibodies." The prerequisites for plantibody production are cloned genes encoding specific antibodies, effective regulatory signals for gene

[†]Deceased.

expression in the target plant, and transformation and selection procedures for the plant.

Antibodies are proteins that show specific binding properties for a wide range of compounds including other proteins. The specificity and sensitivity of the interaction between an antibody and its target molecule has led to a wide range of medical and veterinary applications for diagnosis and therapy. Antibodies are also used in purification processes and in the detection and removal of specific compounds. Antibodies are widely used in many aspects of biological research.

Transgenic plants and plant cells can be used to produce a wide range of proteins for medicinal and other purposes, including biopharmaceuticals, vaccines, and antibodies. Plants offer several advantages over other systems for the production of antibodies.

- Protein production in plants is cheaper than using fermentation systems.
- Large-scale production of plants and plant parts, such as seeds, is well established.
- Plant storage organs, such as seeds and tubers, can serve as storage systems for the antibodies.
- Processing of plant material for extraction of food components is widely used.
- Purification of the antibodies is simplified where the plant is a normal human food, since the likelihood of hazardous contaminants is reduced.
- Plants can produce antibodies in large amounts and in stable form through targeting expression to specific compartments in the cell, such as chloroplasts or endoplasmic reticulum.

Antibodies, or immunoglobulins, have a common structure of two heavy (H) chain and two light (L) chain polypeptides, which are held together by disulfide bridges and noncovalent bonds. This structure allows for the production of several modified forms and components that remain as active antibodies and opens the opportunity for the engineering of novel antibodies. **Figure 1** shows the broad structure of antibodies and the structures of modified forms that have been produced in plants. The two key options available are the production of full-length antibodies or a variety of fragments that retain the antigen-binding specificity of antibodies.

Genes encoding antibodies or antibody fragments can be identified or produced by a range of techniques. Monoclonal antibodies are produced from cells resulting from the fusion of lymphocytes with myeloma. The cells can be grown in culture and produce antibody with only one specificity. The genes encoding the antibody can be isolated using standard gene cloning procedures. More recently, phage display libraries have provided a valuable technique for cloning specific antibody fragments.

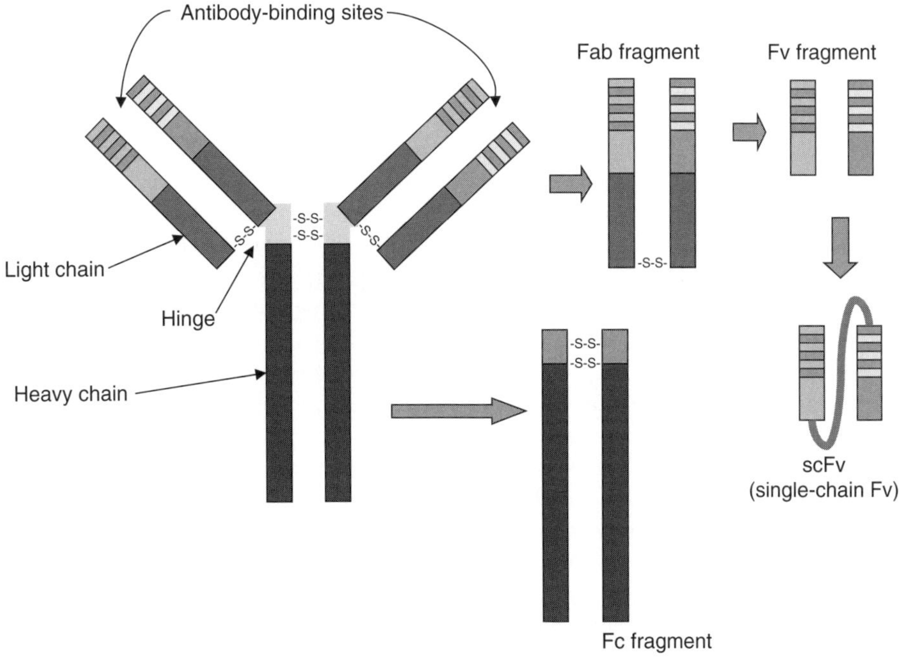

Figure 1 Structure of an antibody and the various active fragments that can be generated from an antibody. Fab and Fv fragments retain the antigen-binding specificity of the parent antibody. The single-chain Fv fragment (scFv) has the antigen-binding specificity of the Fv fragment and the parental antibody, but allows the production of an active antibody fragment as a single polypeptide. Fc has no antigen-binding specificity, but carries the regions required for complement binding.

Full Length Antibodies

An important advantage of plants over other recombinant expression systems is the ability to assemble full-length heavy chains with light chains to form full-length antibodies efficiently. Bacterial expression systems are not usually able to assemble full-length antibodies. Indeed, bivalent antibody molecules can only be produced in *Escherichia coli* by complex molecular engineering. This becomes important, since smaller antibody fragments, such as Fab or single-chain molecules, may be insufficient to show the desired characteristics. For example, even though the constant region of the antibody has no role in antigen recognition or binding, it has marked effector functions and contains several important signals involved in glycosylation, complement activation, phagocyte binding, and the hinge region. The constant region is also involved in association with a joining (J) chain and secretory components. However, one of the most important aspects of the constant region is that it allows bivalent antigen binding. Bivalent binding uses both antigen binding sites and the full flexibility of the antibody molecules at the hinge region, which is important in aggregation and is a major protective mechanism. In addition, binding antigen bivalently considerably affects the strength of the antigen/antibody interaction due to the increase in affinity.

Antigen recognition and binding is a critical and highly sensitive test for correct assembly. Individual light or heavy chains or misfolded antibody molecules are usually not functional. In plants, passage through the endoplasmic reticulum (ER) is required and assembly mechanisms broadly similar to those in mammals are probably present. In the absence of a signal peptide related to either the light or the heavy chain gene, assembly of the antibody does not take place. However, signal sequences from both plant and nonplant sources may be used. Within the plant ER, proteolytic processing of the signal sequence occurs. Plant chaperones occur within the ER and aid in the folding and assembly of foreign proteins to form assembled antibodies. This confirms that the native conformation of antibody molecules is important for stability.

As in mammalian cells, antibodies are secreted by plant cells following assembly and posttranslational processing. Accumulation of secreted antibody into the apoplasm could be advantageous as this is thought to be a stable environment with minimal hydrolytic processing. Also, extraction of antibodies from apoplastic fluid may be simpler and might be achieved under milder conditions than those required for proteins located elsewhere.

The concentration of full-length and Fab antibodies in plants varies greatly among antibody genes, plant species, and plant tissue. For example, an antibody that accumulated to only 0.4% of total soluble protein in tobacco (*Nicotiana tabacum*) leaves, accumulated to 1–3% in *Arabidopsis* leaves.

The ability of plants to produce full-length antibodies and to complement mammalian protein assembly can be used to produce antibody molecules with altered properties. The light and heavy chain genes have been cloned together on a single T-DNA (the region of the Ti plasmid of *Agrobacterium tumefaciens* that is transferred to the plant genome during transformation) in order to introduce both the heavy and light chain genes into the same molecule simultaneously. Some particularly complex antibodies have also been produced in plants. For example, SIgA is the predominant form of immunoglobulin found in mucosal secretions, such as the gastrointestinal tract. This molecule is complex, consisting of two basic Ig monomeric units (heavy and light chains) that are dimerized by a J chain and further associated with a fourth polypeptide, secretory component (SC). These modifications are believed to enhance the activity of SIgA in the mucosal environment. Dimerization by the J chain increases the affinity of binding of the antibody and enhances the potential for bacterial aggregation, while the secretory component confers a degree of resistance against proteolysis, an important property for antibodies within the harsh environment of the gastrointestinal tract.

Four transgenic *N. tabacum* plants were generated to express either a murine monoclonal antibody kappa chain, the hybrid IgA-G antibody heavy chain, the murine J chain, or the rabbit SC. A series of sexual crosses were made between these plants and filial recombinants to generate plants in which all four protein chains were expressed simultaneously. In the final quadruple transgenic plant, three forms of antibody were detectable. The assembly was very efficient, with greater than 50% of the SC being associated with dimeric IgA-G, and the SIgA-G yield from fully expanded leaves was 200–500 $\mu g\,g^{-1}$ fresh weight material. Functional studies confirmed that the SIgA-G molecule bound to its native antigen.

Antibody Fragments

The various fragments of antibodies are illustrated in **Figure 1**. The requirements for assembly of functional, recombinant antibody fragments, such as Fab, Fv, single-chain (scFv) and single-domain (sdFv) antibodies, are usually less demanding than for full-length antibodies. These can be produced in *E. coli* and other heterologous expression systems, and are small in size, which may be an advantage in some applications. However, they have reduced binding affinity and cannot elicit secondary effector mechanisms. As processing through the ER is not essential, the antibody fragments can be accumulated intracellularly if required, or a signal sequence can be added to direct secretion into the extracellular space. Most of the antibody fragments described in *E. coli* have also been produced in plants. These include a single-domain antibody (dAb) in tobacco, single-chain Fv molecules (scFv), and Fab production in tobacco and *Arabidopsis*.

Glycosylation

Mammalian antibodies and immunoglobulin-related proteins are normally glycosylated and this modification also occurs in plant antibodies. Complex plant glycans may be heterogeneous, but they tend to be smaller than mammalian complex glycans and also differ in terminal sugar residues. These differences in glycosylation of plant antibodies have no effect on antigen binding or specificity, but they may affect the antigenicity of antibodies and this may be a problem for pharmaceuticals. Two options have been used to deal with potential problems associated with different glycosylation patterns between mammals and plants. First, glycosylation recognition sequences can be removed from the coding sequence or the antibody can be retained in the ER. Both procedures lead to the formation of unglycosylated proteins. Second, transgenic plants expressing human glycosyl transferases such as human β-1,4-galactosyltransferase can be used. Combined glycosyltransferase-expressing plants can then be crossed with antibody-producing plants to result in the formation of antibodies with human glycoslyation profiles.

Transformation of Plants and Expression of Antibody Genes

The improvement in transformation methods and the availability of a wide range of tissue-specific promoters allows targeting of antibody production to specific tissues of the plant. This can offer major advantages for antibody purification or delivery. Seeds have been particularly important targets for four reasons:

1. Seeds can be readily harvested in large quantities using conventional farm machinery.
2. Protein expression in seeds has been extensively studied and good selection of seed-specific promoters are available for use.
3. Seeds on major crop species are desiccated and dormant, and antibodies remain stable in these tissues over an extended period of time. This means there is additional time available to deliver the seed to purification or processing sites, relative to antibodies that might be produced in leaf tissues. Stability of an scFv antibody for over 1 year has been reported.
4. The breeding and selection process has led to the concentration of nutrients and resources in seeds relative to other parts of the plant; therefore, there is minimal wastage of resources in plant residues.

An alternative tissue for expression of antibodies has been in the fruit of some species, particularly banana (*Musa* spp.). The key advantage here is in delivery of pharmaceutical antibodies to patients. However, issues of stability and uptake by the patient have limited the application of this method and it is largely used to produce pharmaceutical antigens rather than antibodies.

Stable transformation of the target tissue may not always be required. Where only low to intermediate amounts of antibody are required, transient expression after microprojectile bombardment or the use of viral vectors may allow sufficient antibodies to be produced. In these cases, the target tissue will usually produce the antibody for 1–2 weeks. The great advantage of transient transformation is that it allows the rapid production of antibodies or other proteins in plant tissues.

Issues of stability of the transgene over time and multiple generations are similar to the production of other heterologous proteins in plants. The same sort of precautions are needed to ensure consistent levels and stability of antibody expression.

Whole Plants or Plant Cell and Tissue Cultures

In most cases, antibody production is within a whole plant or plant organ, such as the seed or fruit. However, it is also feasible to produce antibodies in plant cell or organ cultures. This approach is particularly useful where the antibody needs to be produced under contained or highly controlled conditions. Tissue cultures may also be useful where the antibody can be secreted from the plant cells into the growth medium, and can then be purified from the medium. Several culture systems have been used for antibody production, the most common being cell suspension cultures or hairy root cultures.

Obtaining High Expression Levels

The level of expression has a large impact on the commercial production of pharmaceutical plantibodies. There are three key criteria in optimizing expression level:

1. The promoter used to drive expression.
2. The codon usage should be optimized for the target plant used. Codon usage in plants differs significantly from animals and there are also major differences between monocotyledons and dicotyledons. Tables of codon usage for different species can be readily generated from the public sequence databases.
3. Targeting expression to specific tissues or cell compartments for stable accumulation of antibodies will help obtain high levels of protein. Antibodies can also be accumulated to higher levels if the protein is targeted to the apoplast of ER rather than to the cytoplasm. ER retention can usually be achieved by adding Lys-Asp-Glu-Leu (KDEL) to the C-terminus of the antibody.

The stability of the antibody has been shown to vary depending on the intrinsic properties of the protein. The evaluation of a variety of candidate fragments may be important in obtaining high levels of expression.

Recent advances in the transformation of the chloroplast genome has offered a new site for accumulation of large quantities of antibodies in plant cells. The chloroplast allows not only high-level accumulation, but also isolation of the protein away from cytoplasmic factors. It appears that this may be particularly important for the production of multimeric antibodies. The site of accumulation of the antibody may be related to the function. Where the antibodies are being used to confer resistance or tolerance to viral infection, the intracellular accumulation of the antibody must be the same as the site where the viral target accumulates.

Applications and Requirements

There are two key applications of plantibodies: first, to modify the characteristics of the plant behavior, such as engineering novel disease resistance; and, second, to use the plant as a bioreactor to produce a commercially or scientifically valuable product, the antibody. The antibodies within the plant can be designed to interact with native plant proteins or compounds, or proteins produced by a plant pathogen or symbiont. In the case of antibodies targeted to a native plant protein, the objective may be to inactivate a catalytic activity of an enzyme, disrupt a normal interaction of a protein with another protein or compound, or create an artificial sink that removes a compound, such as a hormone. Antibodies against a wide range of plant proteins and metabolites have been produced. In this way, a metabolic or developmental pathway may be altered. In the case of antibodies targeted to a pathogen or pest, the aim is to disrupt the normal infection process. The most common example has been with antibodies targeted to viral proteins, particular enzymes involved in viral replication. However, antibodies have also been produced that have the potential to disrupt the feeding of sap-sucking insects and fungal infections.

There are now many examples of the production of antibodies with pharmaceutical potential in plants, some of which are shown in **Table 1**. The focus of work to date has largely been on high-value antibodies or complex antibodies that cannot be effectively produced in other heterologous expression systems, such as IgA being produced against dental caries. Several such plantibodies are now used in clinical trials. It is also possible to use antibodies produced in plants for direct delivery to humans or farm or companion animals, such as cats and dogs, by eating the plant tissue containing the antibodies. There are no examples of this application to date, since it requires antibody activity and uptake after passage through the digestive system or the antibody must be targeted to an antigen (for example, a bacterium) in the mouth or throat. Furthermore, in this type of application, control of dosage can present problems.

Issues

Although plants offer a cost-effective approach to large-scale antibody production, they also present some special problems, which are of particular importance where the antibodies are for human use. In deciding on the best approach to antibody production in plants, a series of issues need to be addressed:

1. The level of expression and suitability of a particular regime for large-scale production needs to be evaluated. This will include decisions about the most appropriate species to use and the range of issues that relate to the level of production (outlined above). However, in some cases, large-scale production may not be required. A careful evaluation of the size and value of the market will give an idea of the production scale needed.

Table 1 Some examples of plantibodies that have been produced

Plant	*Antibody type*	*Specificity*	*Application*	*Reference*
Tobacco	IgA	Streptococcus antigen I or II	Dental caries	Ma *et al.* (1998)
Alfalfa (*Medicago sativa*)	IgG	Antihuman IgG	Diagnostic	Khoudi *et al.* (1999)
Wheat (*Triticum* spp.), rice (*Oryza sativa*) and tobacco Potato (*Solanum tuberosum*)	scFv	Carcinoembryonic protein	Cancer treatment	Stoger *et al.* (2000)
Tobacco	scFv	Ideotype vaccine	B-cell lymphoma treatment	McCormick *et al.* (1999)
Tobacco	IgG	Surface antigen	Colon cancer	Verch *et al.* (1998)
Rice	IgG	Herpes simplex virus 2	Antiviral	Briggs *et al.* (2001)

Modified from: Ma JKC, Hikmat BY, Wycoff K *et al.* (1998) Characterisation of a recombinant plant monoclonal secretory antibody and preventative immunotherapy in humans. *Nature Medicine* 4: 601–606; Khoudi H, Laberge S, Ferullo J-M, *et al.* (1999) Production of a diagnostic monoclonal antibody in perennial alfalfa plants. *Biotechnology and Bioengineering* 64: 135–143; Stoger E, Vaquero C, Torres E, *et al.* (2000) Cereal crops as viable production and storage systems for pharmaceutical scFv antibodies. *Plant Molecular Biology* 42: 583–590; McCormick AA, Kumagai MH, Hanley K, *et al.* (1999) Rapid production of specific vaccine for lymphoma by expression of the tumour-derived single chain Fv epitopes in tobacco plants. *Proceedings of the National Academy of Sciences, USA* 96: 703–708; Verch T, Yusibov V, and Kopranski H (1998) Expression and assembly of a full length monoclonal antibody in plants using a plant virus vector. *Journal of Immunological Methods* 220: 69–75; Briggs K, Zeitlin L, Wang F, *et al.* (2001) An anti-HSV antibody produced in rice plants prevents vaginal HSV-2 infection in mice. *AIDS* 15: S19–S20.

2. The issues of storage and processing of the plant tissues containing the plantibody need to be considered.
3. The physical production of the plant or plant parts must be considered. Here, the proximity of the production site to the processing facility will be important. For highly stable plantibodies, this may not be a problem, since the seed, for example, can be transported to the processing site. However, if the protein is to be purified from leaves, the processing site will probably have to be on farm or facilities for fixing and stabilizing the plantibody in the plant may be needed. Related to this may also be issues of containment. It may, for example, be desirable to grow the crop well away from similar crops in order to minimize issues related to pollen flow.
4. The antibodies must be fully functional and free of contaminants. This will require an efficient purification process and must result in a product that is homogeneous and stable. Protein degradation can be a serious problem, since it may result in non- or poorly functional antibodies or antibodies with altered allergenic properties. Degradation during protein deposition and accumulation in the plant or during storage and transport of the plant parts may require changes in the species, target tissue, or other aspects, as discussed above. Degradation during extraction and processing can usually be minimized through the use of proteinase inhibitors.
5. Purification is a major cost for the commercial production of pharmaceuticals. However, for plantibodies it may be possible to use standard food processing techniques and facilities for the early-stage purification and stabilization of the antibody.

Regulatory Issues

The regulatory framework for the production and use of plant-produced pharmaceuticals is not yet fully defined. However, plants do offer some regulatory advantages over mammalian-based antibody production systems. Contamination risks from viral pathogens or prions are clearly very low for plantibodies. However, there may be issues related to residues of pesticides or herbicides that may come through the antibody purification process. Plantibodies that are produced for pharmaceutical use are also likely to require the full testing and regulatory regime used for other pharmaceutical products. These tests and approvals can be both time-consuming and costly.

In many countries, the large-scale production of transgenic plants remains a problem and regulatory requirements make release of the plants difficult. In many countries, there are also specific requirements for the testing of transgenic crops and there may also be restrictions on the type of transformation vector and selectable markers that can be used. Issues of transgene "escape" through pollen flow may also be of concern. The use of self-pollinated or male sterile lines may help overcome this restriction. For particularly high-value and sensitive plantibodies, plant cell cultures may provide a viable alternative to the use and release of whole plants.

List of Technical Nomenclature

Antigen A substance that is capable of inducing a specific immune response and reacting with the antibodies or sensitized T-cells produced through that response.

Apoplasm The region outside the plasma membrane of plant cells. It includes cell walls and xylem vessels.

Chaperones Cytoplasmic proteins that bind to unfolded polypeptides and ensure their correct folding and transport.

Complement A functionally related series of distinct serum proteins that act as the effector in the immune response.

Endoplasmic reticulum Membrane system that runs through the cytoplasm and contains an internal space, which forms a separate compartment in the cytoplasm.

Epitope The site of an antigen to which an antibody is produced and to which the antibody will bind.

Fab A fragment of IgG containing the light chains plus the variable region of the heavy chain with the antigen-combining sites. Two Fabs can be produced from one IgG (see **Figure 1**).

Fc The constant sequence region of the heavy chains with cell and complement binding sites. One Fc fragment can be produced from an IgG (see **Figure 1**).

Glycosylation The addition of sugar units to proteins.

Golgi Stacks of membrane-bound vesicles in the cytoplasm where protein glycosylation occurs and where proteins are prepared for secretion.

H chain The heavy polypeptide chains found in immunoglobulins.

IgA A major class of immunoglobulin in external secretions of mammals. They protect mucosal surfaces from infection.

IgG A classical class of immunoglobulin. They act on pathogens by agglutinating them, activate complement-mediated reactions, and neutralize toxins. They are composed of two Fabs and an Fc fragment.

J chain A polypeptide found in IgA and IgM that joins the heavy chains to form dimers of IgA and pentamers of IgM.

Kappa chain The light polypeptide chains found in immunoglobulins.

Lymphocyte A type of white blood cell, grouped into two types: T lymphocytes, which are responsible for cell-mediated immunity and B-cell stimulation; and B lymphocytes, which are responsible for antibody production.

Monoclonal antibody An antibody produced in the laboratory from a single clone of cells. It is used for the production of antibodies with specific affinities.

Phage display library Bacteriophage libraries where the protein products of genes inserted into the phage genome are displayed on the surface of the bacteriophage.

Phagocyte A cell that is capable of phagocytosis.

scFv Single-chain Fv antibodies (see **Figure 1**).

See also: **Photosynthesis and Partitioning**: Primary Products of Photosynthesis, Sucrose and other Soluble Carbohydrates. **Primary Products**: Proteins. **Secondary Products**: Pharmaceuticals, Plant Drugs.

Further Reading

Conrad U and Manteuffel R (2001) Immunomodulation of phytohormones and functional proteins in plant cells. *Trends in Plant Science* 6: 399–402.

Daniell H, Streatfield SJ, and Wycoff K (2001) Medical molecular farming: production of antibodies, biopharmaceuticals and edible vaccines in plants. *Trends in Plant Science* 6: 219–226.

Hiatt A (1990) Antibodies produced in plants. *Nature* 344: 469–470.

Larrick JW, Yu L, Chen J, Jaiswal S, and Wycoff K (1998) Production of antibodies in transgenic plants. *Research in Immunology* 149: 603–608.

Larrick JW, Yu L, Naftzger C, Jaiswal S, and Wycoff K (2001) Production of secretory IgA antibodies in plants. *Biomolecular Engineering* 18: 87–94.

Ma JKC, Hiatt A, Hein M, *et al.* (1995) Generation and assembly of secretory antibodies in plants. *Science* 268: 716–719.

Moffat AS (1995) Exploring transgenic plants as a new vaccine source. *Science* 268: 658–660.

Peters K, De Wilde C, De Jaeger G, Angenon G, and Depicker A (2001) Production of antibodies and antibody fragments in plants. *Vaccine* 19: 2756–2761.

Smith MD (1996) Antibody production in plants. *Biotechnology Advances* 14: 267–281.

Stoger E, Sack M, Fischer R, and Christou P (2002) Plantibodies: applications, advantages and bottlenecks. *Current Opinion in Biotechnology* 13: 161–166.

Stoger E, Sack M, Perrin Y, *et al.* (2002) Practical considerations for pharmaceutical antibody production in different crop systems. *Molecular Breeding* 9: 149–158.

Walmsley AM and Arntzen CJ (2000) Plants for delivery of edible vaccines. *Current Opinion in Biotechnology* 11: 126–129.

Wynn RL, Meiller TF, and Crossley HL (1999) Tobacco "plantibodies" for caries prevention. *General Dentistry* 47: 450–454.

GENETIC MODIFICATION OF PRIMARY METABOLISM

Contents

Carbohydrates

R N Chibbar, National Research Council Canada, Saskatoon, SK, Canada
M Båga, University of Saskatchewan, Saskatoon, SK, Canada

Introduction

Plant carbohydrates are produced in photosynthetic cells to provide energy and substrates for the growing plant and the early stages of the next generation. Between 50% and 80% of plant dry matter is composed of carbohydrates, with the percentage being not only species and tissue specific, but also varying with the plant's developmental stage. Most of the plant carbohydrates are complex molecules formed by polymerization of monosaccharides or simple sugars. A wide variety of carbohydrate structures are produced through structural variations of the basic carbohydrate unit, and subsequent linkage of similar or different units to form complex carbohydrate molecules of various sizes (**Figure 1**). The different forms of structural and storage carbohydrates differ in tissue and subcellular localization and play different roles in plant growth and development. Cell wall carbohydrates, such as cellulose, hemicellulose, and callose, provide much of the structural integrity to the plant, whereas storage tissues contain energy rich sugars, gums, arabinoxylans, and starches. In this article, the structure, occurrence, and uses of some of the common storage and structural carbohydrates in plants are described. The biosynthesis of these carbohydrates will be briefly described to identify some of the targets for genetic modification by mutation breeding and/or genetic transformation technologies. A few examples of genetically modified carbohydrates in plants will be presented.

Biosynthesis of Carbohydrates

Green plants have the unique capability of harvesting light energy to fix reduced carbon dioxide and water into a simple carbohydrate backbone, CH_2O, through the process of photosynthesis (*see* **Genetic Modification of Primary Metabolism**: Photosynthesis). The reactions take place in chloroplasts, in which CO_2 entering the reductive pentose phosphate cycle is incorporated into 3-phosphoglycerate (3-PGA) molecules, which are subsequently converted to triose-phosphates (**Figure 2**). In addition, the photosynthetic reaction results in the production of reducing equivalents and ATP from the captured light energy. The triose-phosphates are either transported by triose-phosphate transporters to the cytosol, or converted to phosphorylated compounds, including fructose-6-phosphate in the plastid. The plastidial fructose-6-phosphate is used both for regeneration of ribulose-1,5-bisphosphate and production of glucose-1-phosphate through glucose-6-phosphate. The enzyme ADP-glucose pyrophosphorylase (AGPase) converts glucose-1-phosphate and ATP to ADP-glucose, which initiates starch biosynthesis as described below. Starch synthesized in the chloroplasts during the day is known as transitory starch, as it is degraded primarily to maltose, glucose, hexose phosphates, and triose phosphates at night. A fraction of the starch degradation products are transported to the cytosol and converted to the major transport sugar, sucrose. Through symplastic or apoplastic transport, sucrose enters the phloem and is further mobilized to storage (sink) tissues, where symplastic or apoplastic uptake into sink cells occurs. Sucrose accumulating in the sink cells may constitute the major storage carbohydrate, like in sugar beet (*Beta vulgaris*), or be converted to substrates for, for example, synthesis of structural carbohydrates of the cell wall or storage carbohydrates like starch and fructan (**Figure 2**). Depending on the final form of stored sucrose derived products, different uses have been found for crop plants.

(A)

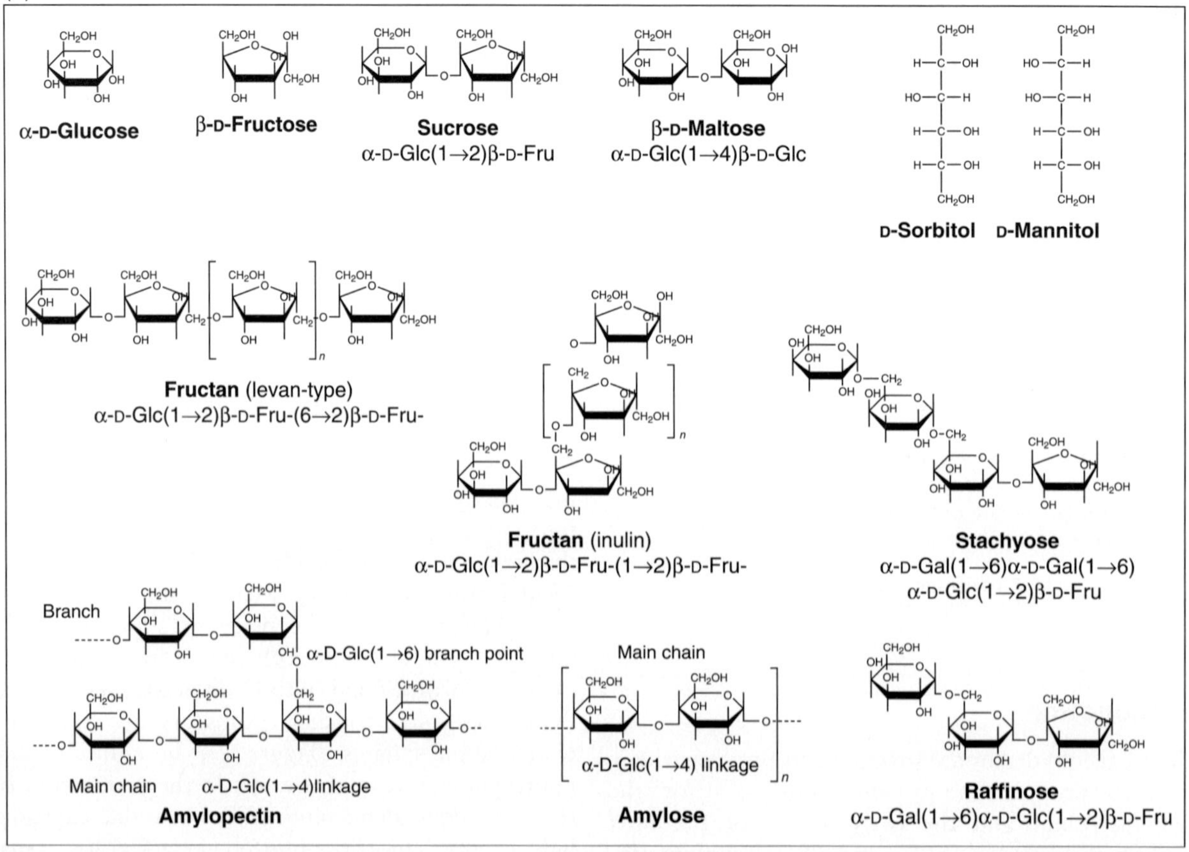

Figure 1 Diversity of carbohydrates found in the plant kingdom. Representative examples of storage (A) and structural (B) carbohydrates found in plants. The storage carbohydrates generally have glucan molecules linked by α(1→4) linkages as compared to structural carbohydrates, which have β(1→4) linkages, except for callose which has β(1→3) linkages, and β-glucan, which contains both β(1→3) and β(1→4) linkages.

Genetic Modification

Carbohydrates are one of the most abundant storage compounds in plants and are an important source of food and fiber for humans and feed for animals. The large diversity of carbohydrate structures results in a multitude of physical properties, which is reflected in the numerous uses of carbohydrates for human needs. Postharvest changes to the carbohydrate structure through chemical or physical processes have until recently been the only method to increase the diversity and utility of many plant derived carbohydrates. The increased knowledge of biosynthetic pathways for major carbohydrate compounds obtained from studies of natural mutants, identification and characterization of substrates and genes encoding key enzymes, and development of genetic transformation techniques for plants has opened up possibilities to modify carbohydrates *in planta*. Recently, transgene technology has been successfully used to introduce modified genes encoding carbohydrate biosynthetic enzymes to provide novel and modified carbohydrates in plants, as discussed below.

Targets for Genetic Modification

Sucrose

Sucrose, a nonreducing sugar, accounts for most of the CO_2 fixed during photosynthesis and constitutes the major form of transported carbon in plants. The translocation of sucrose from source to sink tissues is very active during seed maturation when storage carbohydrates are synthesized and during seed germination when stored carbohydrates are degraded to provide energy for the germinating embryo. In some plants, such as sugar beet and sugar cane (*Saccharum officinarum*), sucrose is the major storage carbohydrate. Sucrose is primarily synthesized in the cytosol of photosynthetic cells from triose phosphates by the action of fructose-1,6-biphosphatase (EC 3.1.3.11; FBPase) and the associated control by fructose-2,6-biphosphate and pyrophosphate:fructose-6-phosphate-1-phosphotransferase (EC 2.7.1.90; PFP). Four enzymes, UDP-glucose pyrophosphorylase (EC 2.7.7.9), sucrose phosphate synthase (EC 2.4.1.14; SPS), sucrose phosphatase (EC 3.1.3.24), and sucrose synthase (EC 2.4.1.13;

(B)

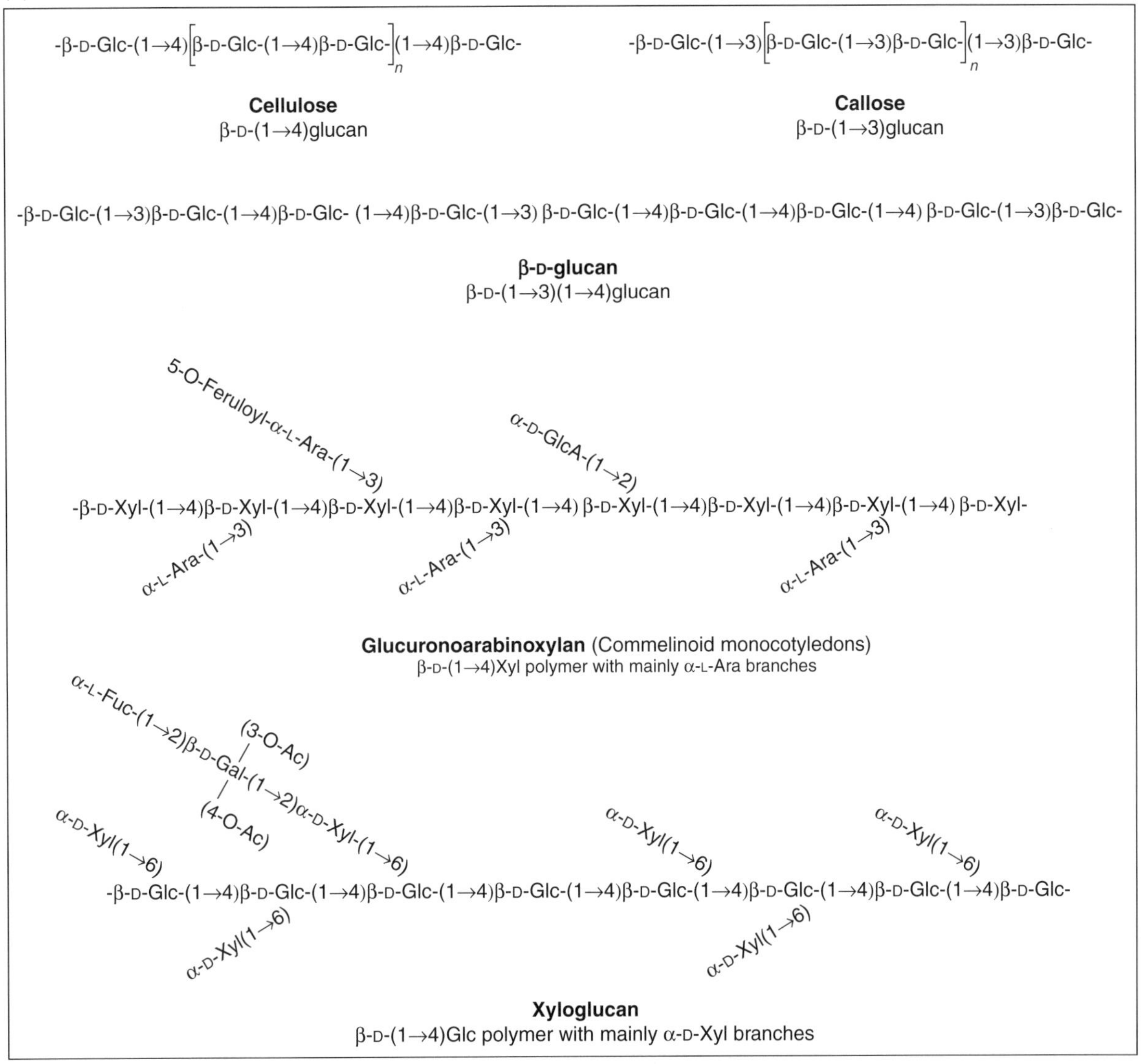

Figure 1 Continued.

SS), participate in sucrose synthesis. Sucrose synthase catalyzes a reversible reaction and is known to participate both in sucrose synthesis and degradation. In the green photosynthetic tissues, FBPase and SPS are the key enzymes regulating sucrose synthesis, whereas in the sink tissues, SPS regulates sucrose synthesis and is related to the hexose-phosphate pool in the cells.

To increase the sucrose concentration, genetic transformation technology has been used to elevate the concentration of FBPase and SPS, the two key regulatory enzymes in sucrose biosynthesis. Transgenic expression of the spinach (*Spinacia oleracea*) FBPase gene in tobacco (*Nicotiana tabacum*) plants increased the FBPase activity four- to sixfold. However, the transgenic plants did not show any significant change in growth and development, but the mature leaves showed a slightly lower level of starch concentration as compared to nontransgenic controls. Using another approach, antisense reduction of the PFP activity in potato (*Solanum tuberosum*), the transgenic potato tubers showed reduced starch content, but only a slightly higher sucrose concentration. Sucrose concentration could be increased up to twofold by overexpressing SPS in tomato (*Lycopersicon esculentum*), but at the expense of a 25–50% reduction in starch concentration. Although the overall growth of the sucrose overproducing transgenic plants was normal, there was a tendency toward an increase of shoot dry matter and reduced root growth. On the other hand, antisense reduction of SPS in potato leaves resulted in reduced sucrose concentration and increased starch concentration. These results show the possibility of

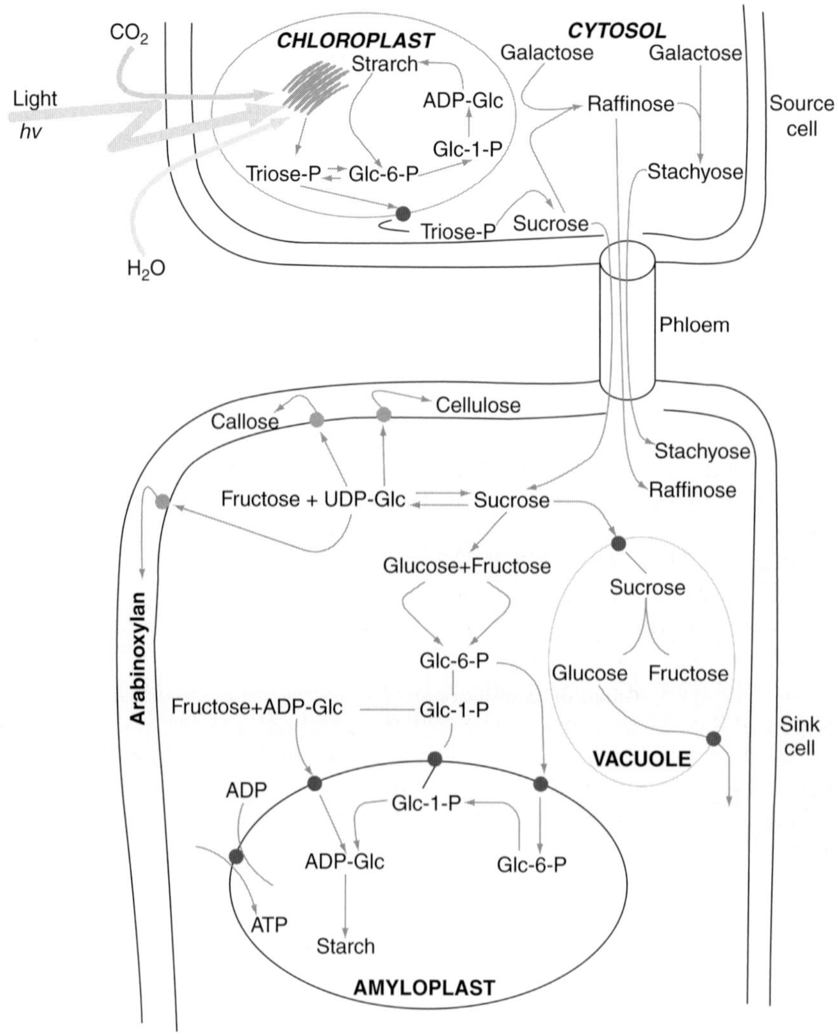

Figure 2 Schematic representation of the metabolic pathways showing fates of photosynthetically fixed carbon into different storage and structural carbohydrates. See text for description of the pathways and abbreviations of enzymes and substrates.

altering sucrose concentration in plants by regulating the SPS activity. Nevertheless, it must be kept in mind that overexpression of the enzyme should be in concert with appropriate posttranslational regulation of the enzyme.

Sucrose Oligosaccharides

Raffinose oligosaccharides are a family of soluble sucrose derivatives that constitute an important form of transported carbon in some plants (e.g., pumpkin). These oligosaccharides are also stored in seeds, roots, and tubers of many plant species. Besides being a storage and transport form of carbohydrate, the raffinose members play a role in abiotic stress tolerance. The major raffinose compounds are raffinose (trisaccharide) and stachyose (tetrasaccharide), but higher oligomers can also be found in some plants. In Lamiaceae, Cucurbitaceae and Oleaceae, raffinose is the predominant carbohydrate translocated in the phloem. Raffinose oligosaccharides have generally been considered to have low nutritional value, due to their indigestibility and promotion of flatulence in humans and animals. However, recently it has been found that raffinose oligosaccharides have a beneficial effect on the gut microflora, and therefore are recommended in human diets to prevent cancer in the digestive tract.

Raffinose oligosaccharides are synthesized in specialized companion cells of the phloem, in which galactose units provided by galactinol are added stepwise to sucrose (**Figure 2**). The synthesized trisaccharide, raffinose, serves as an acceptor for another galactosyl residue from galactinol, giving rise to the tetrasaccharide, stachyose. These two reactions are catalyzed by raffinose synthase (EC 2.4.1.82) and stachyose synthase (EC 2.4.1.67), respectively. A specific galactosyltransferase (EC

2.4.1.123) utilizes UDP-galactose and myo-inositol to produce galactinol for raffinose oligosaccharides. The complementary DNA (cDNA) encoding the raffinose biosynthetic enzymes have been recently isolated and characterized from plants. However, there are no reports on genetic modification of raffinose oligosaccharides in plants.

Fructans

Fructose polymers (fructans) serve as soluble storage carbohydrates in about 12–15% of all flowering plants including cereals (e.g., wheat (*Triticum* spp.), barley (*Hordeum vulgare*), oat (*Avena sativa*)), forage grasses (e.g., *Lolium*, *Festuca*), vegetables (e.g., chicory (*Chichorium intybus*), lettuce (*Lactuca sativa*), onion (*Allium cepa*)), and ornamentals (e.g., dahlia, tulips). In cereals, fructans can be found in young internodes, from which fructan derived carbohydrates are mobilized to kernels during grain filling. Besides being an energy reserve in plants, fructans have a role in regulation of osmotic pressure, sink strength, and resistance to cold and drought. The sweet taste makes fructans useful as low calorie sweeteners and they have also found some use as fat replacers in foods. Fructans with a low degree of polymerization are promoted as a soluble dietary fiber in the human diet as they are poorly digested in the small intestine, but stimulate growth of beneficial microbes in the large intestine.

Sucrose is the primary substrate for fructan biosynthesis, which takes place in the vacuoles by the action of two or more fructosyltransferases. Sucrose:sucrose 1-fructosyltransferase (EC 2.4.1.99; 1-SST) catalyzes the addition of a fructosyl residue derived from one sucrose molecule to another sucrose to form the trisaccharide 1-kestose. 1-Kestose can be further elongated by fructan:fructan fructosyltransferase (EC 2.4.1.100; 1-FFT), which reversibly adds a fructosyl residue from one fructan (DP$\geqslant$3) to another fructan or sucrose molecule. The concerted action of 1-SST and 1-FFT results in a mixture of fructans with 10–20 fructose residues. Plant fructans show a variety of branching patterns. For example, barley produces bifurcose, which is composed of both inulin (β-2,1-linked)-type and levan (β-2,6-linked)-type fructans. The enzyme Suc:-fructan 6-fructosyltransferase (EC 2.4.1.10; 6-SFT), which only uses sucrose as fructosyl donor, catalyzes the synthesis of bifurcose. The liliaceous species, such as onion (*Allium* spp.) and asparagus (*Asparagus* spp.), have a fructosyltransferase, 6G-FFT (EC unassigned), which catalyzes the formation of the trisaccharide neokestose produced by the transfer of a fructose residue from 1-ketose to the C6 of the glucose moiety of sucrose. The 1-FFT catalyzes extension of the fructan chain at either end of the glucose molecule. In the last few years, fructan degrading exohydrolytic enzymes have been purified from many plant species. Two exohydrolytic enzymes, one with a β-(2-6)-linkage-specific fructan-β-fructosidase and another with β-(2-1)-linkage-specific activity, from *Lolium perenne* and Jerusalem artichoke (*Helianthus tuberosus*), respectively, have been isolated.

The cDNA encoding 1-SST, 1-FFT, 6-SFT, and 6G-FFT in various plants have been isolated and expressed in heterologous plants. For example, it has been possible to alter sugar beet to primarily store low molecular weight fructans instead of sucrose by expression of 1-SST cDNA isolated from *H. tuberosus*. Similarly, large amounts of short chain fructan molecules can be produced in transgenic potato by expressing a 1-SST cDNA isolated from artichoke blossom disks. Other transgenic experiments include expression of an onion 6G-FFT in chicory, which alters the inulin profile in this fructan-accumulating crop. In potato, a full spectrum of fructan molecules is produced in lines constitutively expressing globe artichoke 1-SST and 1-FFT. In summary, genetic transformation technology has been successful in fructan production in nonfructan producing plants, as well as altering the fructan structures in fructan accumulating plants. It is also expected that increased fructan synthesis may improve stress tolerance in some plants.

Sugar Alcohols

Sugar alcohols are primary photosynthetic products that are accumulated temporarily in leaves during light and are translocated to other plant organs during dark. As compared to the corresponding sugar, sugar alcohols have an additional hydroxyl group (**Figure 1**), and, therefore, are also designated as polyols, polyalcohols, or polyhydric alcohols. Mannitol, sorbitol, galactitol, and gluctinol are the main sugar alcohols that have been studied in plants. These compounds are rare in monocotyledons but contribute significantly to transported and stored carbon in some horticultural plants, such as members of the Rosaceae, Rubiaceae, and Plantaginaceae families. Sugar alcohols have been implicated in abiotic stress tolerance. As compared to sucrose, sugar alcohols are more metabolically sequestered, and this has important implications in their physiological role in translocation and storage of carbohydrates.

Sugar alcohols are synthesized from hexoses or hexose-phosphates, through the consecutive action of reductases and phosphatases. In green celery (*Apium graveolens*) tissues, three cytosolic enzymes, mannose-6-phosphate isomerase (EC 5.3.1.8), NADPH-dependent mannose-6-phosphate reductase (EC 1.1.1.224), and mannitol-1-phosphate phosphatase (EC 3.1.3.22) convert fructose-6-phosphate to mannose-6-phosphate, mannitol-1-phosphate, and mannitol, respectively. Glucitol is similarly synthesized from glucose-6-phosphate by the action of a glucose-6-phosphate reductase (EC 1.1.1.200) and a sorbitol (glucitol)-6-phosphate phosphatase (EC 3.1.3.50). The genes for some of the sugar alcohol biosynthetic enzymes have been isolated from bacterial species and plants like celery and apple (*Malus* spp.), and used to genetically engineer elevated concentrations of sugar alcohols in tobacco, apple, persimmon (*Diospyros* spp.), and *Arabidopsis*. Some of the transgenic lines producing elevated levels of sugar alcohols show enhanced tolerance to drought, salinity, or oxidative stress.

Starch

Starch distribution and structure Starch is the most common storage carbohydrate in plants. Besides being an important energy source in the human diet and animal feed, starch is also utilized in many nonfood applications, e.g., paper, pulp, and textile production, because of its excellent gelling, pasting, and adhesive properties. Since starch is a renewable and biodegradable resource, it has received increased interest for the production of environmentally friendly biomaterials.

Plant starches are composed of two large glucan polymers, amylose (10^5–10^6 Da) and amylopectin (10^7–10^9 Da), in which the glucose residues are mainly linked by $\alpha(1\rightarrow4)$ bonds to form a chain. The glucan chain is branched by $\alpha(1\rightarrow6)$ linkages, which are frequently found on the heavily branched amylopectin molecule, but relatively few on the essentially linear amylose molecule. Both polymers are packaged into granules to form a three-dimensional, semicrystalline structure. Amylose and amylopectin occupy about one-quarter and three-quarters, respectively, of the granule, which also contains small amounts of lipids, proteins, and phosphate. Both the proportion of amylose to amylopectin and structure of amylose and amylopectin molecules varies with plant species, tissue, and sometimes with developmental stage. In addition, starch granules from different sources have different sizes and shapes, thus increasing the natural diversity of starches (**Table 1**). Most of the gelling and pasting behavior of extracted starch is determined by the amylose/amylopectin ratio, amylopectin structure, and starch granule size distribution.

Studies of starch mutants have so far generated a general consensus that ADP-glucose pyrophosphorylase (EC 2.7.7.27; AGPase), soluble starch

Table 1 Characteristic features of starch granules isolated from some common crops

Plant species	*Storage organ*	*Shape*	*Average size (nm)*	*Average amylose (%)*	*Average amylopectin (%)*
Corn	Endosperm	Polygonal, Round	5–30	25	75
Wheat	Endosperm	A-type[a] (lenticular)	15–35	35	65
		B-type[a] (round)	3–15	25	75
Barley	Endosperm	A-type[a] (lenticular)	15–25	22	78
		B-type[a] (round)	4–15	19	81
Rye	Endosperm	A-type[a] Lenticular	22–36	25	75
		B-type[a]		22	78
Sorghum	Endosperm	Polygonal, Round	4–25	27	73
Oats	Endosperm	Polyhedral	3–10	26	74
Rice	Endosperm	Polygonal	3–8	19	81
Pea	Embryo	Oval	30	30	70
Potato	Tuber	Oval, Spherical	5–100	21	79
Cassava (*Manihot* spp.; produces tapioca)	Root	Truncated, Oval	4–35	17	83

[a] Wheat, barley, rye, and triticale produce two distinct types of starch granules with distinct chemical and physical properties.

synthases (EC 2.4.1.21; SS) and granule bound starch synthases (GBSSI), starch branching enzymes (EC 2.4.1.18; SBE), starch debranching enzymes (DBE; EC 3.2.1.41, pullulanase; EC 3.2.1.68, isoamylase) and possibly also disproportionating enzymes (EC 2.4.1.25; D-enzyme) catalyze the final steps leading to amylose and amylopectin (**Figure 3**). Several of the enzymes exist in different isoforms, some of which vary in their subcellular distribution, substrate specificity, temporal activity, and interaction with other enzymes, thus adding to the complexity of the pathway. The plant AGPase is composed of two large and two small subunits, and catalyzes production of building blocks, ADP-Glu, derived from Glu-1-P and ATP. This reaction is considered to be the major rate limiting step in starch biosynthesis in plants like potato and corn, due to strong allosteric control of AGPase by inorganic phosphate (Pi) and 3-phosphoglycerate (3-PGA). In addition, AGPase has been found to undergo redox modification in potato. The priming mechanism for starch glucan polymerization in plants is unclear, but it probably involves short-chain maltodextrins that are extended and branched by the action of GBSSI, SBEI, and SSI to form a molecule with amylopectin-like structure. The primary glucan polymer is further polymerized through cycles of chain extensions and branching catalyzed mainly by SSI and SBEII. Two models have recently been proposed to explain the involvement of starch debranching enzymes in starch synthesis. In the glucan trimming model, debranching enzymes are required to debranch the ends of water soluble pre-amylopectin molecules to produce amylopectin that can be crystallized and packaged into granules. The glucan recycling model suggests that the role of debranching enzymes is to degrade water soluble glucans synthesized by starch synthases and starch branching enzymes to prevent accumulation of a water soluble, highly branched glucan polymer (phytoglycogen). Observations made in *Chlamydomonas* suggest that the D-enzyme is also involved in starch synthesis and granule formation in addition to starch degradation. Polymerization of amylose is presumed to be initiated from soluble malto-oligosaccharides that serve as primers for GBSSI, which is the only enzyme known to be involved in amylose biosynthesis.

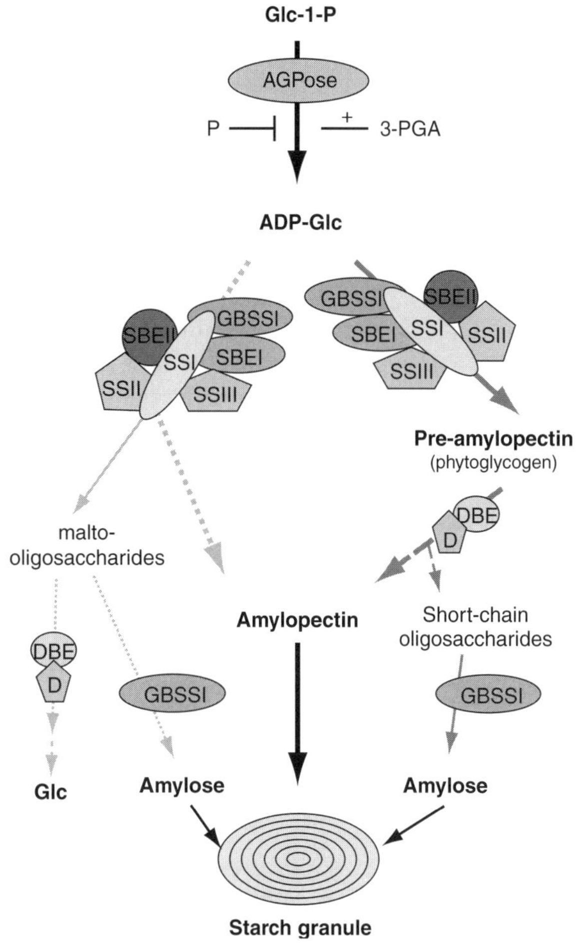

Figure 3 Schematic illustration showing two models for starch biosynthesis in plants. Solid thick arrows represent steps in common for both models, dotted arrows (left route) illustrate steps according to the glucan recycling model, and thin arrows (right route) represent the pre-amylopectin trimming model. See text for description of proposed pathways and abbreviations of enzymes and substrates.

Modification of starch quantity Since starch accounts for two-thirds to three-quarters of dry weight in cereals and potato tubers, it is an important contributor to yield in these crops. Therefore, efforts to increase crop production have included strategies to modulate the highly regulated AGPase, which synthesizes the monomer precursors for starch synthesis. This has been successful in some transgenic potato lines, in which expression of an unregulated *E. coli* AGPase has increased starch content by 30%. Similar experiments conducted in corn using a Pi insensitive AGPase large subunit gene has so far not led to a higher percentage of starch in the endosperm, although the seed weight is increased by 11–18%. An alternative approach to increased starch production in potato tubers has been to increase the ATP pool in the amyloplasts. A 60% increase in starch production and 84% increase in tuber yield was achieved by antisense inhibition of a plastidial adenylate kinase gene. Increased import of ATP into plastids through overexpression of an amyloplastidial ATP/ADP translocator also resulted in higher starch levels in the tubers. Other successful experiments in potato include expression of an *E. coli* inorganic pyrophosphatase in the cytosol, which improved starch production by 20–30%.

Modification of starch properties Natural mutants lacking GBSSI activity produce starch granules containing only the amylopectin glucan polymer (**Table 2**). These so-called *waxy* mutants exist, for example, in corn, rice, and barley, but have not been identified in crops with multiple genomes, such as wheat. However, through breeding it has been possible to recombine the three null GBSSI present in different wheat germplasms to develop lines with <1% amylose. "Partially waxy" starch, i.e., starch with amylose levels between normal (25%) and *waxy* starch, is desirable for the production of certain foods, e.g., Japanese noodles. Partially waxy wheat starch has been obtained through breeding of germplasm carrying one or two GBSSI null alleles. The partially waxy wheat starches contain about 12% amylose and have higher swelling power; they show crystallization patterns and gelatinization properties distinct from both normal and waxy starch.

An increase in the amylose/amylopectin ratio and higher average amylopectin branch length in starch is found in corn, rice, and pea (*Pisum sativum*) mutants, which lack expression of one of the two SBEII isoforms. The starch in these *amylose extender* mutants contains significant amounts of "intermediate material," a glucan polymer with properties distinct from both amylose and amylopectin. In contrast, the effects of SBEI mutations on starch biosynthesis are subtler than those observed in *amylose extender* mutants. Reduction of SBEI activity to 5% of wild-type levels in transgenic potato tubers does not affect the proportions of amylose and amylopectin or the amylopectin branch length distribution. However, there are small changes in average granule size and the physicochemical properties of the transgenic potato starch. In corn, a mutation in *Sbe1* does not cause any apparent change in the leaf or endosperm starch structure.

Studies of various mutants and transgenic plants carrying antisense SS constructs have demonstrated that this class of enzymes has a role in determining the fine structure of amylopectin. For example, structural analysis of amylopectin produced in pea and wheat SSII mutants and transgenic antisense SSII potato show an increased number of short chains (degree of polymerization, dp<10) and a reduced number of intermediate-sized chains (dp 15–25). The glucan modifications alter the physicochemical properties of starch and cause deformation of the starch granules. The starch content in the SSII mutant lines is also lower as compared to wild-type lines. Changes to the amylopectin branch length distribution affecting starch content, granule morphology, and starch physicochemical properties has also been observed in a corn SSIII mutant, *dull1*, and transgenic potato carrying an antisense SSIII construct. To date, no SSI mutants have been found in plants.

The seeds of corn and rice lines lacking DBE activity produce a highly branched and water soluble polysaccharide called phytoglycogen in addition to starch. In these mutants, the phytoglycogen content is approximately 25–30% of total carbohydrates, as compared to nondetectable concentrations in control

Table 2 Mutations affecting storage starch quantity and quality in crop species

Crop	*Mutation*	*Affected enzyme*	*Characteristics*
Corn	*waxy*		
Wheat	*waxy*		
Barley	*low amylose*	Reduced activity or lack of GBSSI	Low or complete lack of amylose
Rice	*waxy or glutinous*		
Pea	*low amylose (lam)*		
Corn	*sugary-2*	Lack SSIIa	Increased amylose/amylopectin ratio, lower starch content; abnormally branched amylopectin; abnormal starch granules
Pea	*rugosus 5*	Lack SSII	
Wheat	Null *Sgp-A1, -B1, -D1*	Lack SSII-A, SSII-B, SSII-D	
Corn	*dull-1*	Lack SSIII (DU-1), lower SBEIIa activity	Produce, in addition to amylose and amylopectin, polymers of intermediate size. Abnormally branched amylopectin
Corn	*amylose extender*	Lack SBEIIb	Increased amylose/amylopectin ratio; lower starch content; abnormal starch granules
Pea	*rugosus*	Lack SBEI[a]	
Rice	*amylose extender*	Lack SBEIII	
Corn	*sugary-1*	Lack isoamylase, lower limit dextrinase activity	Accumulation of phytoglycogen-like polymers. Elevated levels of source and reducing sugars. Altered amylopectin structure
Rice	*sugary-1*		
Barley	Riso 17, Notch-2	Lack isoamylase	Reduced starch content; accumulation of phytoglycogen-like polymers; abnormal granule size distribution
Corn	*brittle-1*	Lack adenylate translocator	Lower starch content; increased ADG-Glc concentration
Corn	*shrunken-2*	Lack large subunit of AGPase	Lower starch content; increased levels of source and reducing sugars
	brittle-2	Lack small subunit of AGPase	
Pea	*rb*	Low AGPase activity	

[a]Homologous to SBEII in other plant species.

lines. An increase of phytoglycogen content above 30% is expected to reduce the viscosity, gel formation, and retrogradation rate and increase the water holding capacity of storage starch. It is further speculated that such starch would have increased digestibility, which could have an economic impact on production of corn for livestock feed.

Potato starch differs from cereal starches in that a higher amount of phosphate is covalently bound to the glucose monomers, leading to better functional qualities over cereal starches for many applications. Phosphorylation of tuber starch is catalyzed by an α-glucan water dikinase (R1 protein) and down-regulation of R1 gene expression consequently reduces starch phosphate content leading to negative effects on starch viscosity. Thus, it is anticipated that development of cereal starches with increased phosphate content would enhance the utility of cereal starches for many food and industrial purposes.

Starch granule size is an important factor for mouth-feel and overall texture of food products, smoothness of cosmetic products, coating of films, paper and carbonless copy paper, and extractability of cereal grain during the malting process. Cereals like wheat, barley, and rye (*Secale cereale*) differ from other crop plants, in that both large and small granules are produced in the endosperm (**Table 1**). However, very little is known about the factors determining starch granule size in plants. Genetic mapping studies in barley have indicated that several loci regulate starch granule size and granule size distribution in this crop. The involvement of several factors is further strengthened by studies of barley mutants with altered starch granule size distribution in which the activity of several starch biosynthetic enzymes is affected. Studies in potato have indicated that SBEI may have a role, as an increased starch granule size is seen in plants expressing an antisense *Sbe1* construct. It has also been suggested from studies in wheat that an SBEI isoform, SBEIc, may be involved in determining bimodal starch granule size distribution in this crop.

Cellulose

Cellulose is the major component of cell walls and accounts for 28–30% of dry matter in forage grasses and 42–45% of wood. It is the most abundant renewable biomass synthesized on earth, and since the early days of human civilization has been used for fuel, timber, fiber, and forage. Today, cellulose has additional uses as a raw material for the production of paper, pulp, and many chemicals. Cellulose is composed of several dozen glucan chains that are arranged parallel to each other and hydrogen bonded to form a microfibril with an average thickness of 36 chains. The chains contain a few thousand glucose units giving an individual length of 2–3 μm, but the chains start and end at different points on the microfibril, thus extending the length of the microfibril to hundreds of micrometers.

In higher plants, cellulose synthesis occurs in a plasma membrane bound complex known as "rosette TC." The structure contains six subunits, each composed of six glucan (cellulose) synthase molecules that are symmetrically arranged into a rosette. The subunits of the complex synthesize simultaneously 36 (1→4)β-D-glucan chains, which are bundled into a microfibril emerging from the rosette TC. The basic biochemical understanding of cellulose synthase has mainly been obtained from studies of the bacterium, *Acetobacter xylinum*. Cellulose synthase (EC 2.4.1.12) consists of at least two subunits, one with catalytic and the other with a regulatory role. The catalytic subunit of cellulose synthase (Ces A) is a progressive glycosyltransferase, which binds the substrate, UDP-glucose, derived from sucrose by the action of a membrane bound sucrose synthase (**Figure 2**). At the initiation of polymerization, two UDP-glucose molecules are present in the substrate binding site and as the chain elongates, glucose is added to the nonreducing end of the chain. The first plant cellulase synthase genes were characterized in cotton and have been followed by cellulase synthase (*CesA*) and cellulase synthase-like (*Csl*) genes from *Arabidopsis thaliana*, rice, and poplar (*Populus* spp.). In addition to the cellulose synthase, it has been reported that membrane anchored endo-1,4-β-glucanases (EC 3.2.1.4) may also play a role in cellulose biosynthesis, based on studies with the *Arabidopsis KORRIGAN* mutant and the cellulose producing bacterium *Agrobacterium tumefaciens*.

Several cellulose deficient mutants have been reported in *A. thaliana*, where the lesion has been shown to be in the cellulose synthase (*CesA*) genes or in a reduced number of rosette TC. In barley, a reduced cellulose concentration is found in the *brittle culm* mutant. As the genes encoding cellulose biosynthesis are characterized in other plants of economic value and our understanding of cellulose biosynthesis in plants improves, attempts will be made to alter this economically important carbohydrate in plants.

Other Cell Wall Carbohydrates

Callose is a cell wall component that is produced by a few cell types, such as growing pollen tubes, cell plates of dividing cells, or cells affected by wounding or fungal infection. Callose consists of (1→3) β-D-glucan chains, which can form helical duplexes and triplexes. Genes encoding callose synthase (EC

2.4.1.34) have been isolated from a few plant species, which opens up the possibility to genetically modify callose production.

Other important polysaccharide components of the cell wall include cross-linking glycans, which are hydrogen bonded to cellulose microfibrils. These glycans coat the cellulose microfibrils and link two or more cellulose microfibrils to form a network in the primary cell walls of all plants. The two major classes of the cross-linking glycans are xyloglucans (XyGs) and glucuronoarabinoxylans (GAXs), of which XyGs are present in cell walls of most dicotyledonous and almost half of all monocotyledonous species. In palms, gingers, cypresses, and grasses, GAXs are the major cross-linking polysaccharide. The XyGs are made up of linear chains of $(1\rightarrow4)\beta$-D-glucan with numerous α-D-Xyl units linked to the O-6 position of the glucan units. In some cases, the xylosyl units are further substituted by arabinose or galactosyl units. The GAXs in commelinoid monocotyledons contain arabinose units in the O-3 position instead of the O-2 position in plants with xyloglucans as the major cross-linking glycan. In addition to the two xyloglucans, some members of the Poaceae family, which includes the majority of cereal crops, such as wheat and barley, contain a third cross-linking glycan, β-glucan. This is a linear polymer that consists of 3-D-glucopyranosyl monomers joined by both $(1\rightarrow3)$ and $(1\rightarrow4)$ linkages, normally in a ratio of 2:1. The molecular weight of β-glucan ranges from 1.1×10^6 to 3×10^6, depending upon the plant species.

The GAXs and β-glucans are present in substantial quantities in the cell walls of the starchy cereal endosperm, and have a significant impact on the utilization of the grains for both food and animal feed. These cell wall carbohydrates are not completely digested in the stomach or small intestine of humans or other monogastric animals, but are digested by bacteria in the large intestine. They also appear to reduce the absorption of mono- and disaccharides in the gut by raising the viscosity of the gut contents, thus decreasing the levels of blood glucose and insulin. β-Glucans have been shown to significantly reduce blood cholesterol levels, when present in the human diet at high levels. In animal feed, both GAXs and β-glucans have antinutritional effects, as they may decrease the absorption of nutrients due to their high viscosity and indigestibility. As the cross-linking glycans differ in value depending on the end use, genetic modification of these cell wall components will depend on the end use of the grain.

The cell wall polysaccharides are synthesized from sucrose, which through the action of sucrose synthase (EC 2.4.1.13) provides UDP-glucose, which subsequently through the action of UDPG dehydrogenase (EC 1.1.1.22), UDP glucuronate decarboxylase (EC 4.1.1.35), xylan synthase (EC 2.4.2.24), and arabinosyl transferase (EC 2.4.2.34) sequentially give rise to UDP-glucuronate, UDP-xylose, 1,4β-xylan, and arabinoxylan, respectively. There are also three epimerases, which can convert UDP-glucose to UDP-galactose, UDP-glucuronate to UDP-galactouronate, and UDP-xylose to UDP-arabinose, respectively. All of these enzymes have been purified and characterized from *Phaseolus*, and the UDP-glucose dehydrogenase gene has been isolated and characterized from soybeans (*Glycine max*). Therefore, it is likely that in the near future, attempts will be made to alter the structure and content of these important cell wall carbohydrates.

Future Perspectives

Plant carbohydrates, which range in structure from simple sugars to complex macromolecules, are important constituents of plants. They provide the mechanical basis for strength, help plants overcome adverse environmental conditions, act as signaling molecules, and provide energy and substrates needed for growth and development. For humans, it is the storage function of plant carbohydrates that has been the main reason for domestication of plants, which forms the basis of modern day agriculture. Genetic modification of carbohydrates in plants by breeding, or recently by transgene technologies, has diversified the use of plant carbohydrates beyond food, feed, and fiber. However, there are many uncharacterized carbohydrates in plants, some of which may be of value as raw materials for industrial applications or food ingredients. High throughput genomics technologies will significantly help in the isolation and characterization of genes involved in key steps in the biosynthesis of carbohydrates. Such discoveries will identify novel targets for genetically modifying carbohydrates in plants. As demonstrated above, genetic and transgenic technologies will result in the production of both modified and novel carbohydrates in plants.

Acknowledgments

NRCC# 45252. National Research Council Canada, Natural Sciences and Engineering Research Council, Canada Saskatchewan Agri-Food Innovation Fund, Saskatchewan Agricultural Development Fund, Brewing and Malting Barley Research Institute, and Genome Canada/Genome Prairie are gratefully acknowledged for supporting research in our laboratories.

See also: **Genetic Modification of Primary Metabolism**: Photosynthesis. **Photosynthesis and Partitioning**: Primary Products of Photosynthesis, Sucrose and other Soluble Carbohydrates. **Primary Products**: Starch.

Further Reading

Ball SG, van de Wal HBJ, and Visser RGF (1998) Progress in understanding the biosynthesis of amylose. *Trends in Plant Science* 3: 462–467.

Boyer C (1996) Biochemical genetics of carbohydrate metabolism in source and sink tissue. In: Zamski E and Schaffer AA (eds) *Photoassimilate Distribution in Plants and Crops – Source–Sink Relationships*, pp. 341–367. New York: Marcel Dekker.

Buchanan BB, Gruissem W, and Jones RL (2000) *Biochemistry and Molecular Biology of Plants*. Rockville, MD: American Society of Plant Physiologists.

Delmer DP (1999) Cellulose biosynthesis: exciting times for a difficult field of study. *Annual Review of Plant Physiology and Plant Molecular Biology* 50: 245–276.

Doblin MS, Kurek I, Jacob-Wilk D, and Delmer DP (2002) Cellulose biosynthesis in plants: from genes to rosettes. *Plant Cell Physiology* 43: 1407–1420.

Gregory ACE, O'Connell AP, and Bolwell GP (1998) Xylans. *Biotechnology and Genetic Engineering Reviews* 15: 439–455.

Kossmann J and Lloyd J (2000) Understanding and influencing starch biochemistry. *Critical Reviews in Biochemistry and Molecular Biology* 35: 141–196.

Loescher WH and Everard JD (2000) Regulation of sugar alcohol biosynthesis. In: Leegood RC, Sharkey TD, and von Caemmerer S (eds) *Photosynthesis: Physiology and Metabolism*, pp. 275–299. Dordrecht: Kluwer Academic Publishers.

Myers AM, Morell MK, James MG, and Ball SG (2000) Recent progress toward understanding biosynthesis of the amylopectin crystal. *Plant Physiology* 122: 989–997.

Smith AM (1999) Making starch. *Current Opinion in Plant Biology* 2: 223–229.

Vijn I and Smeekens S (1999) Fructan: More than a reserve carbohydrate. *Plant Physiology* 120: 351–359.

Winter H and Huber SC (2000) Regulation of sucrose metabolism in higher plants: localization and regulation of activity of key enzymes. *Critical Reviews in Plant Sciences* 19: 31–67.

Proteins

N D Hagan and T J V Higgins, CSIRO Plant Industry, Canberra, Australia

Introduction

Animals are unable to synthesize 10 of the 20 amino acids needed for protein production and must obtain these essential amino acids from their diet. Although cereals and legumes are major sources of protein for humans and livestock, individually these crops cannot supply the full complement of essential amino acids; cereal grains are deficient in lysine, whilst legume grains are deficient in the sulfur containing amino acids cysteine and methionine. For many years, traditional breeding has attempted to improve the balance of essential amino acids in seed proteins. Although good progress has been made in breeding high-lysine corn, breeding of high-methionine legumes has met with only limited success. More recently, gene technology has been used to modify proteins in cereals, legumes, and other plants, and has shown potential in addressing the nutritional deficiencies in these crops.

One GM approach to improve nutritional quality has been to alter the amino acid composition of abundant endogenous storage proteins. In addition, several attempts have been made to address deficiencies in crop plants by expressing genes for lysine or methionine rich proteins from other plants. For example, the brazil nut (*Bertholletia excelsa*) 2S and sunflower (*Helianthus annuus*) seed albumins are methionine rich proteins that have been used to increase levels of this amino acid in a variety of plants. Finally, entirely synthetic proteins containing high levels of lysine and methionine have been produced in GM plants.

Despite rapid advances in this area over the past 10 years, a number of issues remain outstanding. In some cases, the rate of synthesis of the target amino acid limits its accumulation in the new protein. This may require manipulation of key points in the production pathway to achieve the required amino acid content. In other cases, the introduction of a new protein has altered the expression levels of endogenous proteins, potentially affecting desirable traits. In addition, some of the most beneficial proteins in terms of amino acid composition have been identified as allergens.

Deficiencies in the Protein Nutritive Value of Cereal and Legume Grains

Cereal and legume grains provide a concentrated source of digestible protein for humans and form the basis of many livestock feeds. Cereal grains contain 7–14% protein by weight, and are also a good source of carbohydrates. Legume grains contain 25–40% protein, and provide the majority of the protein in mixed stockfeed formulations. However, these two feed sources cannot supply an adequate quantity of the essential amino acids individually.

The animal growth requirements for methionine and lysine are approximately 1.6 g and 7 g per 100 g of total protein, respectively. Animals require an additional 1.9 g of cysteine per 100 g protein but this can be met by either of the sulfur containing amino acids as animals can convert methionine to cysteine. Most cereal seeds contain sufficient methionine to meet the minimum growth requirement, but fall well short of the requirement for lysine. Conversely, legume seeds contain enough lysine, but insufficient methionine. For example, 100 g of pea (*Pisum sativum*) seed protein contains 7 g lysine but only 0.8 g methionine (**Table 1**). Consequently, a diet based on a single cereal or legume results in amino acid deficiencies. Part of these deficiencies can be overcome by mixing seed meals from legumes and cereals, but this alone is not sufficient for maximum animal growth. As a result, livestock feeds are often supplemented with purified methionine and lysine, which adds considerably to the cost of production.

Like other animals, humans require a minimum quantity of the essential amino acids for normal physical and mental development. The major source of protein in the human diet comes from seeds, especially from the cereal grains rice (*Oryza sativa*), wheat (*Triticum* spp.), and corn (*Zea mays*; maize). However, eating a vegetarian diet that relies wholly on these grains can result in a lysine deficiency and affect normal growth and development.

Alteration of Endogenous Proteins

The problem of protein quality in seeds arises from the amino acid composition of the most abundant proteins accumulated in the seed. Among legumes, the major storage proteins are the vicilin-like and legumin-like globulins, which account for 70–80% of the seed protein. As both globulins have few sulfur containing amino acids, they are primarily responsible for the nutritional deficiencies of legume grains. One approach to increasing the nutritional value of legume grains has been to alter these and other endogenous storage proteins to contain higher levels of cysteine and methionine.

Table 1 Lysine and methionine contents of common cereal and legume seeds

Seed protein	*Lysine content ($g\,100\,g^{-1}$ protein)*[a]	*Methionine content ($g\,100\,g^{-1}$ protein)*[a]
Corn	2.9	1.9
Wheat	2	1.6
Rice	3.4	1.8
Pea	7	0.8
Lupin	7	0.6
Soybean	6.9	1.2

[a] The animal growth requirement for lysine and methionine is approximately 7 g and 1.6 g $100\,g^{-1}$ protein, respectively.

This approach was used to attempt to increase the number of methionine residues in the vicilin-like and legumin-like globulins of faba bean (*Vicia faba*). A gene for a methionine free legumin subunit was altered by frameshift and site-directed mutagenesis to introduce four methionine residues in the new protein. The modified legumin gene was introduced into tobacco (*Nicotiana tabacum*), but it was shown that the amino acid changes interfered with correct folding and the protein was rapidly degraded. When eight methionine codons were introduced into the *Vicia faba* vicilin gene, the vicilin variant was stable in transgenic tobacco seed, but the protein was produced at levels too low to change seed protein amino acid composition.

Storage proteins from the common bean (*Phaseolus vulgaris*) have also been targeted for modification. β-Phaseolin, the vicilin-like globulin from *P. vulgaris*, was modified by the addition of a methionine rich region from the corn 15-kDa zein storage protein. The added peptide increased the number of methionine residues from three to nine in the new protein. The corn-derived peptide was predicted to form an α-helical structure and was inserted into an α-helical region of β-phaseolin. The level of mRNA transcript was shown to be high, but only a small amount of protein accumulated. It was concluded that despite the conserved α-helical insertion, the high methionine β-phaseolin was unstable in the developing seed. Phytohemagglutenin, an abundant lectin from *P. vulgaris* seeds, contains no cysteine or methionine residues and has also been targeted for modification. Four methionine residues were introduced by site directed mutagenesis on the basis of the location of methionine residues in homologous lectin proteins. The engineered gene was introduced into tobacco, where it was found that the protein underwent correct modification and accumulated in the vacuoles of the cotyledons.

It appears that overexpression, folding, and stability of the modified protein is a major limitation to this approach. In addition, all modified proteins produced so far contain relatively few methionine residues. To raise the methionine levels significantly in legume grains, the modified proteins would have to be expressed at unrealistically high levels. However, research so far suggests that this approach may be feasible, particularly if a high methionine protein is engineered using a fuller knowledge of the structural requirements of storage proteins than is currently available.

Expressing Genes for Proteins from Other Plants

The most promising GM approach for addressing amino acid deficiencies has been to express genes for proteins that are rich in the required amino acid and that have been isolated from other plants. This approach has the advantage that the introduced proteins are likely to be as stable in the new plant as they are in their plant of origin, particularly if they are deposited in the same organ and tissue type.

Although lysine rich proteins have been identified, they are not normally found at high levels in seeds. An example is a chymotrypsin inhibitor from barley (barley chymotrypsin inhibitor-2; CI-2) that contains 11.5% lysine. The effect of expressing a protein in seeds with such a biological function has yet to be determined, but could present some difficulties. Recent efforts have focused on engineering additional lysine residues into CI-2, based on the tertiary structure of the native protein.

A number of seed storage proteins have been identified that are rich in methionine. The most extensively studied of these have been the brazil nut 2S albumin (BNA) containing 18% methionine and 8% cysteine residues, and the sunflower seed albumin protein (SSA) containing 16% methionine and 8% cysteine residues (**Table 2**).

The Brazil Nut Albumin (BNA) Protein

Chimeric genes encoding the BNA protein have been introduced into a variety of plants, resulting in seed specific expression. An early study expressed BNA in tobacco seeds under the control of the β-phaseolin promoter. The BNA protein accumulated to a level of 8% of total seed protein, resulting in a 30% increase in methionine. More recently, chimeric BNA genes have been introduced into large seed legumes, such as the narbon bean (*Vicia narbonensis*). Transgenic narbon bean lines that expressed BNA under control of the faba bean legumin promoter showed accumulation of up to 5% of total seed protein. The highest BNA expressing line showed a doubling of total methionine, putting it in the range required for optimal animal growth. BNA was also expressed in transgenic soybean (*Glycine max*) using the β-phaseolin promoter. Expression levels were shown to be as high as 10% of total seed protein with a 50% increase in seed methionine content. An increase of this magnitude also places transgenic soybean in the methionine range required for optimal animal growth. In an approach that combined the use of BNA with protein modification, variants containing up to seven additional methionine residues were produced. The altered genes were expressed in potato (*Solanum tuberosum*) tubers, but only low levels of expression were achieved.

BNA is one of the major allergens in Brazil nuts. Allergenicity was maintained in BNA expressed in the seeds of transgenic soybean. Thus, any food produced from these soybeans would have to be labeled.

The Sunflower Seed Albumin (SSA) Protein

The SSA protein is another well studied sulfur rich protein that has been introduced in a variety of legumes and cereals, including subterranean clover (*Trifolium* spp.), alfalfa (*Medicago sativa*; lucerne), narrow leaf lupin (*Lupinus angustifolius*), pea, chickpea (*Cicer arietinum*), rice, and wheat. The SSA gene was placed under the control of the pea vicilin promoter and introduced into narrow leaf lupin, a legume commonly fed to stock because of its high level of protein and fiber. SSA accounted for 5% of total seed protein in the transgenic lupin, and resulted in a doubling of seed methionine and an overall increase in sulfur containing amino acids of 20%. In feeding trials with sheep, the transgenic seeds gave an 8% increase in live weight gain and an 8% increase in wool growth, compared to non-transgenic seeds. Grain containing the SSA protein has the potential to be very useful as feed for ruminant animals, as SSA has been demonstrated to be resistant to degradation in the rumen. This allows

Table 2 Proteins with the potential to address amino acid deficiencies in plants

Protein	*Feature*	*Plant/s tested*
Sunflower seed albumin (SSA)	Rich in methionine and cysteine	Lupin, chickpea, alfalfa, subterranean clover, rice, wheat
Brazil nut 2S albumin (BNA)	Rich in methionine and cysteine	*Arabidopsis*, tobacco, narbon bean, soybean, canola
Corn 15-kDa zein	Rich in methionine	Tobacco
Amaranth seed albumin (AmA1)	Rich in essential amino acids	Potato
Barley chymotrypsin inhibitor-2 (CI-2)	Rich in lysine	N/A
Synthetic CP3-5	Rich in lysine and methionine	Tobacco, soybean, corn

more sulfur containing amino acids to be delivered to the small intestine of the animal, bypassing incorporation into microbial protein. Transgenic subterranean clover (*Trifolium subterraneum*) and alfalfa expressing SSA in the vegetative tissues have also been produced for evaluation as pasture for stock animals.

Although cereal grains are not deficient in the sulfur containing amino acids, SSA was introduced in rice to gauge its effects on the levels of methionine, cysteine, and the endogenous sulfur containing storage proteins. Transgenic rice expressing SSA at a level of approximately 7% of total protein did not show any increase in cysteine, methionine, or total protein. However, significant changes in the protein profile of the seeds were observed. Endogenous sulfur rich storage proteins were downregulated in the SSA transgenic, presumably because SSA had sequestered most of the available sulfur. In addition, sulfur poor storage proteins were upregulated, allowing the seed to maintain a constant level of total protein. In SSA transgenic peas and chickpeas, SSA accumulation was associated with significant increases in seed methionine. Surprisingly, the seeds also had increased amounts of total protein, and so the seed protein was not enriched with respect to the sulfur containing amino acids.

Other Methionine Rich Proteins

Of the other methionine rich proteins, the 15-kDa zein gene from corn has been introduced into tobacco using viral regulatory regions to drive expression. The protein accumulated to 1.6% of total protein in the seeds, but was not broken down upon germination, potentially leading to problems with seed viability.

Creating Synthetic Proteins

Synthetic genes encoding proteins with high levels of lysine, methionine, tryptophan, threonine, and isoleucine have been formed by random ligation of oligonucleotides containing a high proportion of codons for essential amino acids. Limited expression of these proteins was demonstrated in potato tubers, but the protein failed to accumulate or significantly increase the levels of essential amino acids. This may have been due to inefficient expression or protein instability. It is not known whether the proteins produced would be likely to fold into stable structures.

To increase lysine and methionine content in seeds, a synthetic protein was produced containing 31% lysine and 20% methionine. The novel protein was designed to have an α-helical coiled coil structure in order to promote the formation of stable multimers and overcome stabilization problems encountered in previous research. Expression was controlled in tobacco seeds by the soybean β-conglycinin promoter, and resulted in a 30% increase in methionine and a 47% increase in lysine. If expressed at a similar level in grains of cereals or legumes, such a protein could significantly increase their nutritive value.

Manipulating Amino Acid Biosynthesis

Accumulation of foreign lysine or methionine rich proteins can be limited by the rate of biosynthesis of these amino acids in the plant. In order to increase the availability of limiting amino acids, one approach has been to deregulate the biosynthetic pathway involved. Recently, GM technology has been used to increase flux through amino acid biosynthetic pathways by introducing feedback insensitive versions of the key enzymes involved.

Transgenic canola (*Brassica napus*) and soybean seeds were produced with increased lysine by circumventing the normal feedback regulation of two enzymes in the lysine biosynthetic pathway, aspartokinase (AK) and dihydrodipicolinic acid synthase (DHPS). Plant DHPS is feedback inhibited by lysine, whilst AK is feedback inhibited by either lysine or threonine. Transgenic canola plants, containing a feedback insensitive bacterial homolog of DHPS, showed a doubling of total seed lysine. Transgenic soybean plants with both DHPS and AK showed an even greater increase in total seed lysine.

In the sulfur accumulation pathway, key control points are catalyzed by serine acetyl transferase (SAT) and O-acetylserine (thiol) lyase (OASTL). SAT converts serine to O-acetylserine and is an important enzyme in regulating the input of nitrogen into the sulfur assimilation pathway. OASTL is a multisubunit enzyme that catalyzes the formation of cysteine from O-acetylserine and sulfide (**Figure 1**). Feedback insensitive versions of these enzymes have now been isolated and characterized. Currently, research is being conducted that combines expression of sulfur rich proteins with biosynthetic pathway manipulation. For example, subterranean clover has been produced that expresses both SSA and feedback insensitive OASTL, in an attempt to further increase the sulfur containing amino acid content of the leaves.

Current Limitations and Future Prospects

Despite significant progress in recent years, the use of gene technology in this area has encountered some

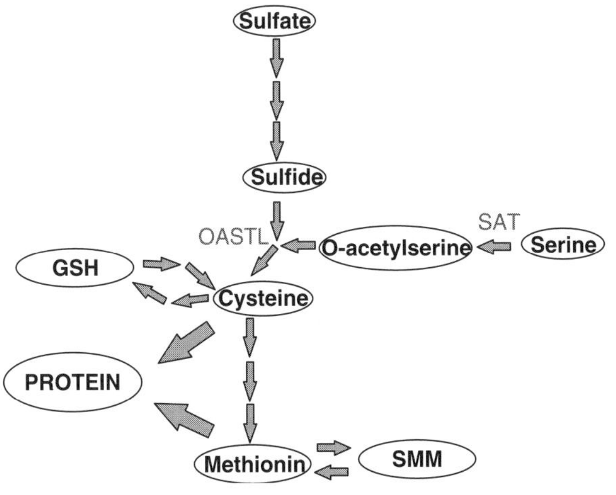

Figure 1 The sulfur assimilation pathway in plants. Inorganic sulfur in the form of sulfur is transported into cells and converted to sulfide. Nitrogen is introduced through *O*-acetylserine, which is catalyzed from serine by serine acetyl transferase (SAT). Sulfide and *O*-acetylserine are combined to form cysteine by the action of *O*-acetylserine (thiol) lyase (OASTL). Cysteine can then be incorporated into protein or converted to methionine. Glutathione (GSH) and *S*-methyl methionine (SMM) represent other major pools of organic sulfur in the cell.

problems. Some novel proteins, produced by either *de novo* synthesis or modification of existing proteins, have had problems with folding and stability. As more is learnt about protein engineering and the structure of storage proteins, these novel proteins will undoubtedly be refined.

One additional problem has been the low expression of the genes encoding proteins with high sulfur amino acid contents. More efficient promoters have been identified in recent years, which has helped increase the levels of these proteins to the extent that they can make a significant difference to the amino acid content of the plant. However, there are a variety of responses between species, which means that even a high level of expression does not necessarily result in a large increase in the desired amino acid. Good examples of this can be seen in transgenic lupin, chickpea, and rice expressing the gene for the SSA protein. Lupins expressing SSA at 5% of total protein show a greater total increase in methionine levels than chickpeas expressing SSA at 16%. Rice, expressing SSA at 7% of total protein, does not show any increase in seed methionine. High levels of transgene expression may also lead to altered levels of endogenous proteins, which may or may not affect the sulfur amino acid content of the seed protein.

At present, no high methionine grain legume has been made available for commercial production due to the problems listed above and the strict regulatory schemes associated with the release of GM plants. However, a high methionine lupin is at an advanced stage of evaluation. Commercially, the production of high lysine rice, wheat, and barley would be of even greater importance. High lysine corn is already available through plant breeding. Gene technology could be used in addressing amino acid deficiencies in these other important cereals. In particular, the introduction of nutritious, nontoxic proteins combined with the approach of using biosynthetic pathway manipulation shows great promise in creating food that is able to supply a balanced level of all the essential amino acids.

List of Technical Nomenclature

Albumin	A water soluble protein.
Allergen	An environmental antigen that promotes a hypersensitive response by the body's immune system.
Cotyledon	The one (monocotyledon) or two (dicotyledon) seed leaves of an angiosperm embryo.
Globulin	A salt soluble/water insoluble protein.
Lectin	A carbohydrate binding protein.
Mutagenesis	The creation of mutations.
Ruminant	An animal, such as a cow or sheep, with a multicompartmentalized stomach, specialized for a herbivorous diet.

See also: **Primary Products**: Amino Acids; Proteins.

Further Reading

Altenbach SB and Simpson RB (1990) Manipulation of methionine-rich protein genes in plant seeds. *TIBTECH* 8: 156–160.

Altenbach SB and Townsend JA (1995) Transgenic plants with improved protein quality. In: Wallace RJ and Chesson A (eds) *Biotechnology in Animal Feeds and Animal Feeding*, pp. 71–92. Weinheim: VCH.

Duranti M and Scarafoni A (1999) Modification of storage protein content and quality in legume seeds. *Journal of New Seeds* 1: 17–35.

Falco SC, Guida T, Locke M, Mauvais J, Sanders C, Ward RT, and Webber P (1995) Transgenic canola and soybean seeds with increased lysine. *Biotechnology* 13: 577–582.

Galili G and Hofgen R (2002) Metabolic engineering of amino acids and storage proteins in plants. *Metabolic Engineering* 4: 3–11.

Habben JE and Larkins BA (1995) Genetic modification of seed proteins. *Current Opinion in Biotechnology* 6: 171–174.

Hagan ND, Upadhyaya N, Tabe LM, and Higgins TJ (2003) The redistribution of protein sulfur in transgenic rice expressing a gene for a foreign, sulfur-rich protein. *Plant Journal* 34: 1–11.

Herbers K and Sonnewald U (1999) Production of new/modified proteins in transgenic plants. *Current Opinion in Biotechnology* 10: 163–168.

Lai J and Messing J (2002) Increasing maize seed methionine by mRNA stability. *Plant Journal* 30: 395–402.

Molvig L, Tabe L, Eggum BO, *et al.* (1997) Enhanced methionine levels and increased nutritive value of seeds of transgenic lupins (*Lupinus angustifolius* L.) expressing a sunflower seed albumin gene. *Proceedings of the National Academy of Sciences, USA* 94: 8393–8398.

Müntz K, Christov V, Jung R, *et al.* (1997) Genetic engineering of high methionine proteins in grain legumes. In: Cram WJ, De Kok LJ, Stulen I, Brunold C, and Rennenberg H (eds) *Sulfur Metabolism in Higher Plants*, pp. 71–86. Leiden: Backhuys Publishers.

Newman CW, Newman RK, and Lockerman RH (1988) Utilization of food legumes in human nutrition. In: Summerfield RJ (ed.) *World Crops: Cool Season Food Legumes*, pp. 405–411. Dordrecht: Kluwer Academic Publishers.

Tabe L and Higgins TJV (1998) Engineering plant protein composition for improved nutrition. *Trends in Plant Science* 3: 282–286.

Tabe L, Higgins CM, McNabb WC, and Higgins TJV (1993) Genetic engineering of grain and pasture legumes for improved nutritive value. *Genetica* 90: 181–200.

Acyl Lipids

D F Hildebrand and K Yu, University of Kentucky, Lexington, KY, USA

Introduction

Over the past several decades, tremendous knowledge has been obtained concerning oil biosynthetic pathways and their regulation in plants. Advances in modern genetics have led to successful cloning of many key enzymes responsible for the biosynthesis of oil. This has made it possible to improve the fatty acid compositions of seed oils or to lead to the accumulation of unusual fatty acids that have special value. Such genetic changes in oilseed lipid composition have long been pursued by traditional breeders with some notable successes. The possibility of transferring a relatively small number of specific genes that impact lipid metabolism from other living organisms, mainly from other plants, afforded by modern molecular genetics has greatly shortened the time needed to improve major oilseed crops.

This article reviews the progress that has been made on genetic improvement of oilseeds for edible and industrial uses. Current knowledge regarding plant lipids is given, and their relationship with the human world discussed. Plant lipid metabolism, including the diversity of fatty acids accumulated in seed oils and the diverted fatty acid synthesis pathways, and the major successful changes and progress in genetic engineering on fatty acid compositions of seed oils are reviewed. The difficulties and future prospects for the genetic engineering of oilseeds are described.

Plant Lipids and their Relationship with the Human World

Lipids are compounds found in nature that are insoluble in water and soluble in nonpolar organic solvents. In biological systems, major lipids include carotenoids, sterols and other isoprenoids, chlorophyll, and vitamins such as tocopherols. The most abundant group of lipids is the fatty acids and their derivatives or acyl lipids (**Figures 1** and **2**). Fatty acids are nonpolar long carbon chains (mostly with even carbon atoms and at least six total carbon atoms) with a carboxylic acid group at one end. Plant lipids are mainly classified as fats and oils, phospholipids, galactolipids, waxes, and other lipids, such as sterols. Fats (solid at room temperature) and oils (liquid at room temperature) are glycerol esters of fatty acids, usually triacylglycerol. Phospholipids are composed of glycerol esters of two fatty acids and one phosphoric acid that is usually esterified to other groups. The major galactolipids in plants are composed of glycerol esters of two fatty acids and

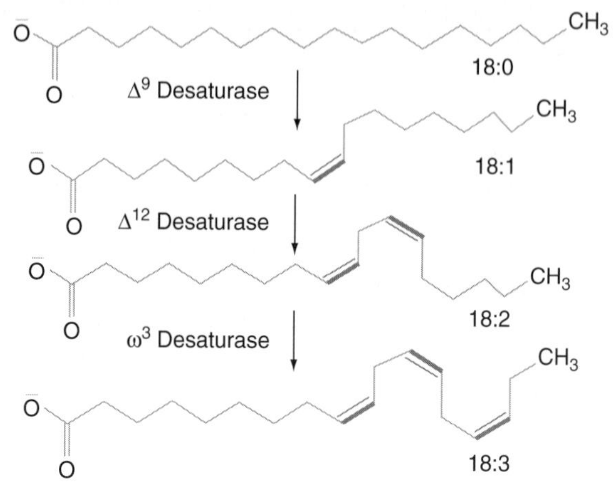

Figure 1 Structures of four of the five main plant fatty acids. Palmitic acid (16:0) has the same structure as stearic acid (18:0) but is two carbons shorter.

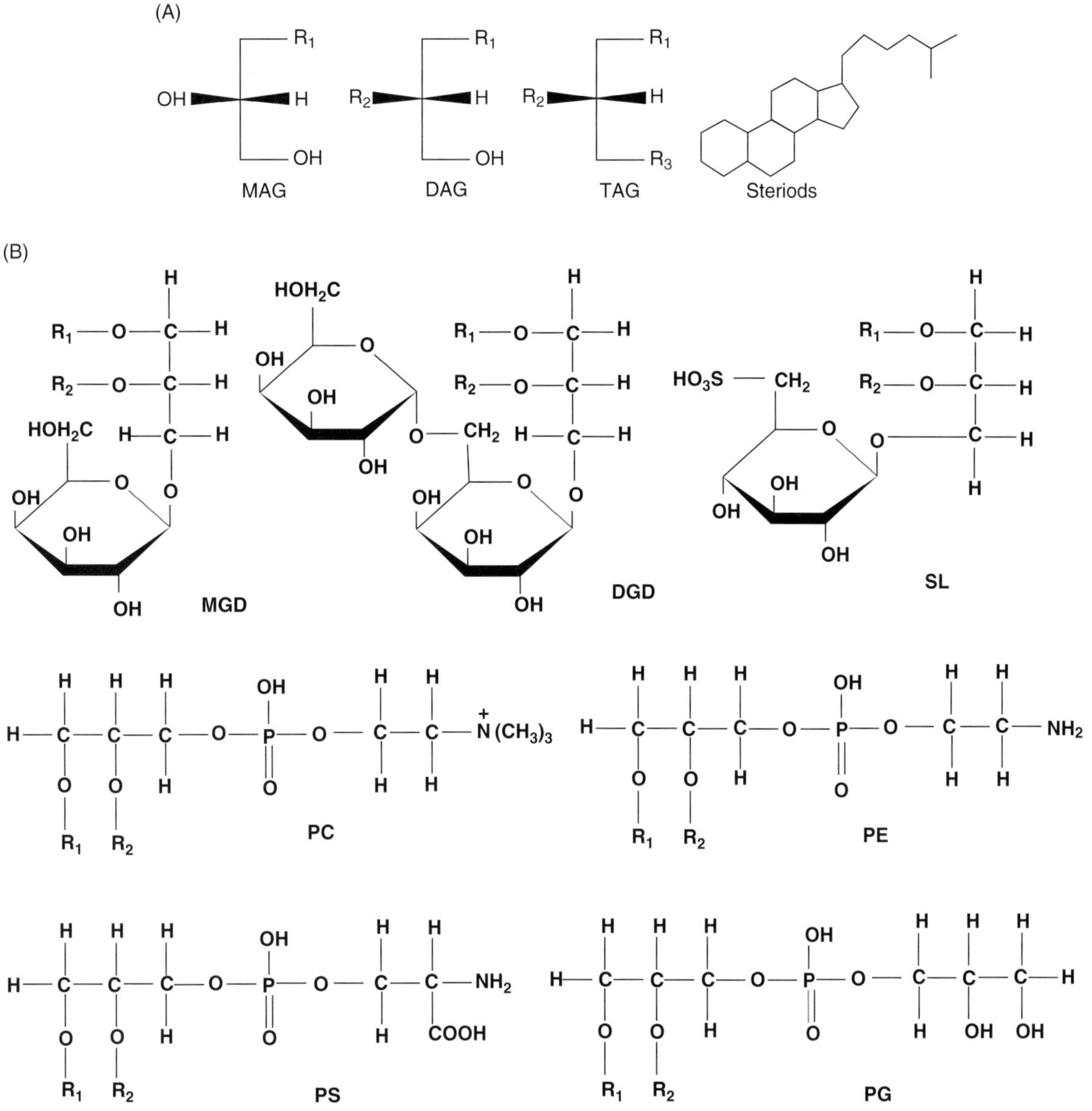

Figure 2 Structures of major plant lipids. (A) TAG, waxes, and steroids are the major lipids in plants. MAG and DAG are low abundance intermediates in lipid metabolism (see **Figure 4**); (B) major membrane phospholipids and glycolipids; and (C) other common but less abundant membrane lipid structures.

one or more (up to four) galactose residues. Waxes are composed of a fatty acid esterified with a long-chain alcohol. Steroids are sterols or their esters. Sterols share a four fused ring structure with variations on groups around the edges (**Figure 2**). In this article, only the acyl lipids will be covered.

Lipids have many functions in plants. Phospholipids are the essential components of cellular and subcellular membranes, which not only serve as a barrier for the cells and subcellular structures, but also control the passing of materials. Membranes also form subcellular compartments and allow many biochemical reactions to occur. Membrane fluidity can be controlled by adjusting the degree of fatty acid unsaturation as the environmental temperature fluctuates. As the temperature rises, the degree of desaturation decreases and vice versa. The ability of temperate plants to adjust membrane lipid desaturation allows survival in different climatic conditions. Fats and oils (mainly triacylglycerides) accumulated in seeds are a major energy reserve for seed germination and early seedling growth. Waxes physically protect the plant from water loss and attack by many pathogens. Five fatty acids, palmitic (16:0), stearic (18:0), oleic (18:1), linoleic (18:2) and linolenic (18:3) acids, are the main fatty acids of membrane lipids and most storage oils (**Figure 1**). Common polyunsaturated fatty acids are either ω^6 (18:2) or ω^3 (18:3) fatty

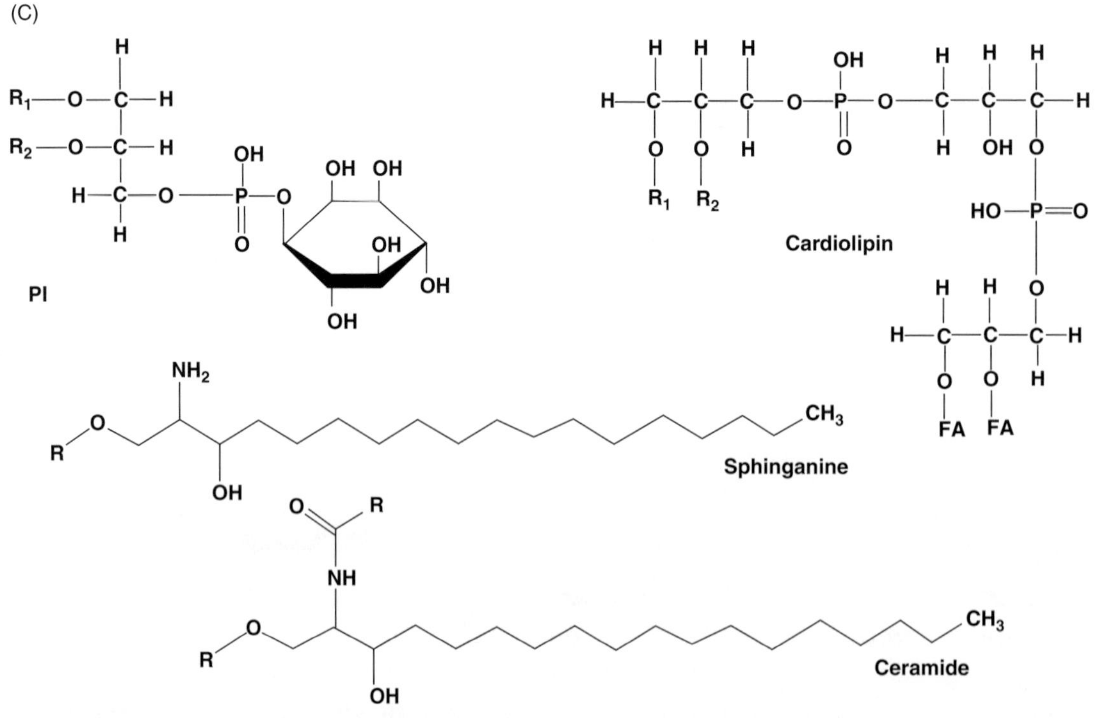

Figure 2 Continued.

acids. The double bonds of these fatty acids are also methylene (–C=C–C–C=C–) interrupted. Unlike membrane lipids, storage lipids have a wide diversity in their fatty acid species. To date, hundreds of unusual fatty acid species have been found to accumulate in storage lipids, mainly in seeds.

Currently, the major oilseed crops supply close to 100 million metric tonnes of oil worldwide (**Table 1**). About two-thirds of the oil is from soybean (*Glycine max*), palm, and canola (genetically improved oilseed rape; *Brassica napus*). The major fatty acids present in the world's oil supply are palmitic, linoleic, and oleic acids (**Table 2**). Many unusual fatty acids can accumulate in seed oils as their major components, sometimes in substantial amounts (**Table 3**). Unusual fatty acids include variations in carbon chain length and degree of unsaturation. Most fatty acids have double bonds in the *cis* configuration; however, *trans* double bonds are also found. There are also different positional isomers, conjugated unsaturated, acetylenates, hydroxy, epoxy, and keto fatty acids. Cyclopropenoids, cyclopentenoids, and even fluoro fatty acids add more diversity to the fatty acid species.

Fats and oils have many uses, and the type of use is mainly determined by the fatty acid composition. As energy sources, fats and oils are isotonic and can release about $9\,cal\,g^{-1}$ after complete catabolism, roughly twice the caloric value of proteins and carbohydrates. Fats and oils also facilitate the absorption of fat soluble vitamins. Some polyunsaturated fatty acids are nutritionally essential since humans cannot synthesize them. Long chain saturated fatty acids (C12–C16) increase serum cholesterol by reducing the activity of receptors for low-density lipoproteins (LDL), thus decreasing clearance of LDL particles from plasma. Oleic acid does not raise or lower serum cholesterol levels. Polyunsaturated fatty acids (PUFAs) can lower serum cholesterol. However, PUFA exceeding 10% of total calorie uptake may promote tumor growth. Linoleic and linolenic acids are oxidatively unstable because of the high degree of unsaturation. Thus, fats or oils containing high levels of linoleic and linolenic acids cannot be stored in air for a long period of time. Many fats and oils also have many nonfood uses, such as in soaps and other detergents and as a substitute for petroleum derivatives in manufacturing plastics. The chain length of the fatty acid influences the property of the soap made from it. Many of the unusual fatty acids currently have a number of uses, and others may potentially be useful in the future. For example, fatty acids with conjugated double bonds may have a drying property that is suitable for paints and inks, and epoxy fatty acids with just one epoxy group can prevent cross-linking and other reactions with the polymer chains and thus are ideal for many polyvinylchloride (PVC) plasticizer applications.

With the desire to maximally utilize the existing fatty acids that can be produced in nature, scientists have been trying to domesticate many unusual fatty

Table 1 Production of major world oil crops

Oilseed/oil	*Planted area*[a] *(Mha)*	*Production*[a] *(MMT)*	*Oil production*[a] *(MMT)*	*Oil %*[b]
Soybean	79.17	179.98	25.79	22
Palm	10.78	135.72	25.08	50
Canola	22.86	32.55	12.45	41
Sunflower	19.57	22.74	8.18	50
Peanut (*Arachis hypogaea*; in shell)	25.86	34.50	5.35	50
Cotton (*Gossypium hirsutum*) seed	30.00	33.07	4.00	26
Coconut	10.79	49.32	3.56	35
Palm kernel	10.78	7.04	2.89	51
Olive (*Olea eucopaea*; with residues)	8.34	15.07	2.66	31
Corn (*Zea mays*; maize)[c]	138.9	602.03	2.02	4
Sesame (*Sesamum indicum*)	7.33	2.89	0.75	52
Rice, paddy (*Oryza sativa*)[c]	146.03	579.48	0.87	19
Linseed (*Linum* spp.)	2.95	2.28	0.68	41
Castor (*Ricinus communis*)	1.14	1.05	0.52	53
Safflower (*Carthamus tinctorius*)	0.83	0.61	0.17	40
Vegetable tallow (*Stillingia sebifera*)	N/A	N/A	0.13	26
Stillingia oil (*Stillingia sebifera*)	N/A	N/A	0.13	20
Tung	0.18	0.56	0.09	60

N/A, not available; Mha, million hectares; MMT, million metric tonnes.
[a] Data are of 2002 from Statistical Databases of Food and Agriculture Organization of the United Nations, except cotton seed area and production, which are from Foreign Agricultural Service of Production Estimates and Crop Assessment Division of the United States Department of Agriculture.
[b] Oil percentage are on a dry weight basis, except palm and coconut, which are based on mesocarp and flesh, respectively.
[c] Note: corn and rice are mainly for non-oil uses.

Table 2 Typical fatty acid profiles of major plant oils (%)

Oil/Crop	8:0	10:0	2,4-10:2	12:0	14:0	16:0	16:1	18:0	18:1	18:2	18:3	20:0	20:1	22:0	22:1	24:0
Soybeans						10.8		3.6	22.3	54.1	8.2		0.9			
Palm					1.1	44.2		4.8	38.9	10.9						
Canola						4.2		1.0	62.8	21.1	9.8				1.0	
Sunflower						6.5		4.1	23.0	65.0	0.7			0.7		
Sunflower (high oleic)						3.7		5.4	81.8	9.1						
Peanut						11.6		3.6	43.1	34.6		1.5	1.4	3.0		1.2
Cotton seed					1.0	28.4	1.0	3.5	21.1	44.9						
Coconut	7.8	7.2		48.2	16.3	8.5	1.0	2.7	6.0	2.3						
Palm kernel	2.2	3.5		48.2	18.3	8.9		2.5	14.9	1.6						
Olive						11.8	0.9	3.6	73.0	9.4	0.8	0.5				
Corn						13.8	0.6	2.7	29.3	52.4	1.2	0.6				
Sesame						10.0		5.9	43.5	40.6						
Rice bran						14.8		1.1	42.1	39.7	1.3	0.5	0.6			
Linseed						5.8		3.7	25.8	14.1	50.7					
Safflower						6.8		1.0	17.8	73.3		0.5	0.5			
Safflower (high oleic)						5.0		1.0	80.5	12.2				1.2		
Vegetable tallow						63.5		2.9	31.8	1.8						
Stillingia sebifera			4.5			5.2		2.0	11.0	38.8	38.5					

acid producing oilseed crops. Unfortunately, the domestication of many unusual oilseed crops has a diversity of problems such as climatic requirements, low yield, harvesting, and management, etc. On the other hand, it is desirable to change many seed oil fatty acid compositions to meet our nutritional and functional requirements. Hydrogenation of oils was once a popular way to produce margarines and shortenings, and more stable oils. However, the *trans* fatty acids produced by this process have been found to increase serum LDL cholesterol and reduce HDL cholesterol, and thus increase the risk of cardiovascular diseases. Genetic engineering can result in fatty acid compositions with the same functional properties as oils modified by chemical hydrogenation without the *trans* fatty acids, and it eliminates the need and costs of the chemical processing. Traditional breeding techniques have been very successful

Table 3 A list of high unusual fatty acid accumulators

Accumulator	*Unusual fatty acid*	*Oil %*	*Unusual fatty acid %*
Cuphea pulcherrima	8:0 (caprylic)	N/R	94.4–95.6
Cuphea avigera	8:0 (caprylic)	N/R	94.0
Cuphea painteri	8:0 (caprylic)	30–36	73.0
Cuphea koehneana	10:0 (capric)	N/R	91.6–95
Cuphea procumbens	10:0 (capric)	16.7–28.7	87.7–94.6
Cuphea schumannii	10:0 (capric)	N/R	93.8
Cuphea ignea	10:0 (capric)	30–36	87.0
Cuphea inflata	10:0 (capric)	N/R	86.4
Cuphea lanceolata	10:0 (capric)	20–25.6	80–86.5
Cuphea llavea	10:0 (capric)	30–36	83–86
Hydnocarpus anthelmintica	11-(2-cyclopenten-1-yl), 11:0 (hydnocarpic)	16.3–18.4	60–67.8
Lindackeria dentata	11-(2-cyclopenten-1-yl), 11:0 (hydnocarpic)	24	50.8
Hydnocarpus wightiana	11-(2-cyclopenten-1-yl), 11:0 (hydnocarpic)	32.4–43	43–65
Litsea stocksii	12:0 (lauric)	N/R	94.0
Actinodaphne hookeri	12:0 (lauric)	48–48.9	93.9
Cuphea melanium	12:0 (lauric)	N/R	85.9
Cuphea hyssopifolia	12:0 (lauric)	N/R	78.5
Cuphea urens	12:0 (lauric)	N/R	75.6
Cuphea confertiflora	12:0 (lauric)	N/R	73.3
Litsaea sebifera	12:0 (lauric)	46.6	71.0
Cuphea acinifolia	12:0 (lauric)	N/R	65.1
Cuphea cetosa	12:0 (lauric)	N/R	62.2
Cuphea carthagenensis	12:0 (lauric)	30–36	57.0
Hydnocarpus alcalae	13-(2-cyclopentene-1-yl),13:0 (chaulmoogric)	65	90.0
Lindackeria dentata	13-(2-cyclopentene-1-yl),13:0 (chaulmoogric)	24	51.0
Caloncoba glauca	13-(2-cyclopentene-1-yl),13:0 (chaulmoogric)	38	31.2–45.1
Hydnocarpus wightiana	13-(2-cyclopentene-1-yl),13:0 (chaulmoogric)	32.4–43	35.0
Caloncoba welwitschii	13-(2-cyclopentene-1-yl),13:0 (chaulmoogric)	46	30.7–31
Taraktogenos kurzii	13-(2-cyclopentene-1-yl),13:0 (chaulmoogric)	23.1	29.6
Taraktogenos kurzii	13-(2-cyclopenten-1-yl), 6-13:1 (gorlic)	23.1	25.1
Caloncoba glauca	13-(2-cyclopenten-1-yl), 6-13:1 (gorlic)	38	7.3–20
Knema globularia	14:0 (myristic)	N/R	92.0
Myristica fragrans	14:0 (myristic)	24–30	76.6
Virola sebifera (ucuuba)	14:0 (myristic)	49.2–62.3	73.0
Cuphea salvadorensis	14:0 (myristic)	N/R	64.5
Cuphea palustris	14:0 (myristic)	N/R	63.7
Myrica cerifera	16:0 (palmitic)	N/R	92.0
Thunbergia alata (black-eyed susan vine)	6-16:1	20	82.2
Doxantha unguis-cati (cat's claw)	6-16:1	29	64.0
Securidaca longipedunculata	13h,9,11t-18:2	50.6	15.0
Heritiera littoralis	8me-17:1 (malvalic)	8.3	53.7
Pterospermum acerifolitum	8me-17:1 (malvalic)	21.5	32.2
Nesogordonia routak	8me-17:1 (malvalic)	1.2	29.9
Nesogordonia macrophylla	8me-17:1 (malvalic)	14.7	19.3
Nesogordonia thouarsii	8me-17:1 (malvalic)	22.4	18.8
Jodina rhombifolia	8a,10t,16-17:3 (pyrulic)	31.9	20.1
Garcinia cornea	18:0 (stearic)	N/R	65.0
Litchi chinensis	9m-18:0	<5	35–48
Tragopogon porrifolius	9e-18:0 (epoxystearic)	22	43.0
Dioscoreophyllum cumminsii	5-18:1	26	84.5
Hedera helix (ivy)	6-18:1 (petroselinic)	29.54	62–86
Petroselinum sativum	6-18:1 (petroselinic)	18–20	75.0
Petroselinum hortense	6-18:1 (petroselinic)	19.25	76–79.3
Daucus carota (carrot)	6-18:1 (petroselinic)	10.57–16	54.47–76.25
Heracleum candicans	6-18:1 (petroselinic)	9	66.5
Coriandrum sativum (coriander)	6-18:1 (petroselinic)	17.1–28.4	65.7
Apium graveolens	6-18:1 (petroselinic)	14	51–58
Euphorbia lathyris (caper spurge)	9-18:1 (oleic)	48–50	84–85
Euodia fargesii	10-18:1	N/R	31.7
Euodia meliaefolia	10-18:1	N/R	28.5
Entandrophragma clindricum	11-18:1 (vaccenic)	45.2	49.7

Table 3 Continued

Accumulator	*Unusual fatty acid*	*Oil %*	*Unusual fatty acid %*
Entandrophragma angolense	11-18:1 (vaccenic)	62.4	39.4
Entandrophragma utile	11-18:1 (vaccenic)	53.7	30.8
Ricinus communis (castor)	12h, 9-18:1 (ricinoleic)	43–53.7	84–91
Hiptage benghalensis	12h, 9-18:1 (ricinoleic)	40.2	84.3
Hiptage madablota	12h, 9-18:1 (ricinoleic)	67	70.0
Agonandra brasiliensis	12h, 9-18:1 (ricinoleic)	35	44.85–65
Wrightia coccinea	9h, 12-18:1 (strophanthus)	23	76.1
Wrightia tinctoria	9h, 12-18:1 (strophanthus)	30	69.0
Holarrhena antidysenterica	9h, 12-18:1 (strophanthus)	33.7	73.4
Chamaepeuce afra	9h,10h,18h,12-18:1	24.4	14.0
Bernardia pulchella	12e, 9-18:1 (vernolic)	26.4	91.0
Vernonia galamensis	12e, 9-18:1 (vernolic)	38.6 (30.2–42)	76(68.9–81)
Stokesia laevis	12e, 9-18:1 (vernolic)	27–44	65–79
Vernonia anthelmintica	12e, 9-18:1 (vernolic)	20–27.9	68–75
Crepis biennis	12e, 9-18:1 (vernolic)	34	68.0
Vernonia volkameriaefolia	12e, 9-18:1 (vernolic)	20	63.5
Cephalocroton cordofanus	12e, 9-18:1 (vernolic)	42	62.0
Euphorbia lagascae	12e, 9-18:1 (vernolic)	42–50	57–62
Erlangea tomentosa	12e, 9-18:1 (vernolic)	23.7	52.0
Lactuca sativa	9e, 12-18:1 (coronaric)	31	16.9
Siegesbeckia orientalis	9e, 12-18:1 (coronaric)	18.2	16.0
Diospyros melanoxylon	9o,13-18:1	6	29.0
Sterculia foetida	9me-18:1 (sterculic)	51–55	45–65.1
Sterculia tragacantha	9me-18:1 (sterculic)	23.4	30.2
Tarrietia utilis	9me-18:1 (sterculic)	29.7	20.2
Pachira aquatic	2h, 9me-18:1 (α-hydroxysterculic)	55.5	12.8
Picramnia tariri	6a-18:1 (tariric)	65–70	95
Picramnia sow	6a-18:1 (tariric)	58.6	89.8
Alvaradoa amorphoides	6a-18:1 (tariric)	58.8	58.0
Persea americana (butter fruit)	9o, 12-18:1	6.4	25.6
Pyrularia pubera	9a-18:1 (stearolic)	N/R	19.0
Taxus chinensis	5,9-18:2 (taxoleic)	N/R	16.4
Leucas cephalotes	5,6-18:2 (laballenic)	28.5	28
Dimorphotheca aurantiaca	9h, 10t,12t-18:2 (dimorphecolic)	38	70.0
Dimorphotheca pluvialis (rain daisy)	9h, 10t,12t-18:2 (dimorphecolic)	15.8–36	53.6–62
Monnina emarginata	13h,9,11t-18:2 (coriolic)	20	30.0
Securidaca longipedunculata	13h,9,11t-18:2 (coriolic)	50.6	27.0
Securidaca longipedunculata	11h,7,9t-16:2 (coriolic)	50.6	15.0
Crepis alpina	12a, 9-18:2 (crepenynic)	12	73.7(70–80)
Crepis rubra	12a, 9-18:2 (crepenynic)	21	54.7
Crepis foetida	12a, 9-18:2 (crepenynic)	24	36–60
Ixora chinensis	12a, 9-18:2 (crepenynic)	23.2	30.5
Santalum yasi	9a,11t-18:2 (ximenynic)	30.4	83.0
Santalum album	9a,11t-18:2 (ximenynic)	26.8–50	50–75
Santalum obtusifolium	9a,11t-18:2 (ximenynic)	17.8	71.5
Osyris alba	9a,11t-18:2 (ximenynic)	26.8–36	57–64
Exocarpus cupressiformis	9a,11t-18:2 (ximenynic)	32.9	60
Jodina rhombifolia	9a,11t-18:2 (ximenynic)	31.9	55
Calea urticaefolia	3t,9,12-18:3	18.3	31.2
Lamium purpureum	5,6t,16-18:3 (lamenallenic)	40.5	16
Larix decidua	5,9,12-18:3 (pinolenic)	N/R	28.5
Aquilegia vulgaris	5t,9,12-18:3 (columbinic)	N/R	61.2
Borago officinalis (borage)	6,9,12-18:3 (γ-linolenic)	24.6–35.1	22.8
Calendula officinalis (pot marigold)	8t,10t,12-18:3 (calendic)	42	47–63.6
Osteospermum spinescens	8t,10t,12-18:3 (calendic)	43	30.0
Euphorbia parryi	9,12,15-18:3 (α-linolenic)	44	76.0
Euphorbia tinctoria	9,12,15-18:3 (α-linolenic)	24	72.0
Euphorbia myrsinites	9,12,15-18:3 (α-linolenic)	32	72.0
Euphorbia anacampseros	9,12,15-18:3 (α-linolenic)	30	69.0
Mercurialis annua	9,12,15-18:3 (α-linolenic)	37–48	68–70

continued

Table 3 Continued

Accumulator	*Unusual fatty acid*	*Oil %*	*Unusual fatty acid %*
Euphorbia cybirensis	9,12,15-18:3 (α-linolenic)	42	66.0
Euphorbia medicaginea	9,12,15-18:3 (α-linolenic)	41	66.0
Euphorbia thamnoides	9,12,15-18:3 (α-linolenic)	34	66.0
Sebastiana commersoniana	9,12,15-18:3 (α-linolenic)	19.7	65.0
Phyllanthus maderaspatensis	9,12,15-18:3 (α-linolenic)	16.3	64.4
Perilla frutescens	9,12,15-18:3 (α-linolenic)	30–51	56.8 (50–70)
Nepeta cataria var. citriodora	9,12,15-18:3 (α-linolenic)	21.96	62.8
Dracocephalum moldavica	9,12,15-18:3 (α-linolenic)	22.8	59.4–63
Croton sparciflorus	9,12,15-18:3 (α-linolenic)	20.4	56.6
Garcia nutans	9,11t,13t-18:3 (α-eleostearic)	41.3–55	85–95
Aleurites fordii (tung)	9,11t,13t-18:3 (α-eleostearic)	60	82 (77–86)
Momordica cochinchinensis	9,11t,13t-18:3 (α-eleostearic)	47.7	62.2
Momordica bowmanii	9,11t,13t-18:3 (α-eleostearic)	38.5	60.9
Momordica dioica	9,11t,13t-18:3 (α-eleostearic)	33.5	54.9
Centranthus macrosiphon	9,11t,13t-18:3 (α-eleostearic)	26.2	50.4
Punica granatum (pomegranate)	9,11t,13-18:3 (punicic)	14–20	70–86.1
Trichosanthes nervifolia	9,11t,13-18:3 (punicic)	27.9	51.7
Trichosanthes bracteata	9,11t,13-18:3 (punicic)	31.6	41.8
Mallotus discolor	18h,9,11t,13t-18:3 (kamolenic)	50	65–70
Mallotus claoxyloides	18h,9,11t,13t-18:3 (kamolenic)	50	65–70
Mallotus philippinensis (kamala)	18h,9,11t,13t-18:3 (kamolenic)	35–50	46.5–68.8
Licania rigida	4o,9,11,13-18:3 (licanic)	45–65	30–75.4
Tanacetum corymbosum	12a,8t,10t-18:3	16.7	17.0
Heisteria silvanii	9a,7,11t-18:3 (heisteric)	46.6	22.6
Exocarpus cupressiformis	9a,11a, 13t-18:3 (exocarpic)	1.3 (root)	59
Buckleya lanceolata	9a,11a, 13t-18:3 (exocarpic)	37.2	40
Buckleya distichophylla	9a,11a, 13t-18:3 (exocarpic)	17.8	29
Ongokea gore	9a,11a,17-18:3 (isanic)	60	32
Ongokoa gore	8h,9a,11a,17-18:3 (isanolic)	60	15
Ixora chinensis	8,10,12,14t-18:4	23.2	38.0
Parinarium laurinum	9,11t,13t,15-18:4 (α-parinaric)	17–35	53.5
Impatiens capensis	9,11t,13t,15-18:4 (α-parinaric)	52	41.0
Sebastiana brasiliensis	9,11t,13t,15-18:4 (α-parinaric)	23.4–26.3	25.4–39.3
Impatiens balsamina	9,11t,13t,15-18:4 (α-parinaric)	22–32	34.0
Impatiens pallida	9,11t,13t,15-18:4 (α-parinaric)	43	22.0
Nephelium lappaceum	20:0 (arachidic)	N/R	33.0
Limnanthes douglasii (meadowfoam)	5-20:1	22–33	65.0
Limnanthes alba(meadowfoam)	5-20:1	25–31	60–61
Simmondsia chinensis (jojoba)	11-20:1	43.57–53.4	66–71
Koelreuteria	11-20:1	49.4 (kernel)	60.0
Teesdalia nudicaulis	11-20:1	29–32	56.0
Cardiospermum grandiflorum	11-20:1	35.2–37.5	48–49
Koelreuteria paniculata	11-20:1	22.2	46.0
Cardiospermun canescens	11-20:1	34.6	44.0
Paullinia meliaefolia	13-20:1 (paullinic)	38.4	47.1
Lesquerella fendleri	14h, 11-20:1 (lesquerolic)	19.8–28.1	45–51.4
Alchornea cordifolia	14e,11-20:1 (alchornoic)	46	51.2
Podocarpus nagi	5,11,14-20:3 (sciadonic)	30.8	26.39
Juniperus communis	5,11,14,17-20:4	20.13	19.8
Brassica tournefortii	22:0 (behenic)	N/R	48
Tropaeolum majus	13-22:1 (erucic)	6–15	62–82
Crambe abysinnica (crambe)	13-22:1 (erucic)	35.6 (23.3–44)	55–62.5
Eruca sativa	13-22:1 (erucic)	32–41	26.7–52.4
Alliaria petiolata	13-22:1 (erucic)	22–30	47.0
Brassica napus (high erucic acid oilseed rape)	13-22:1 (erucic)	41.5–47.8	45.8
Eranthis hyemalis	13,16-22:2	31.5	56.4
Adenanthera pavonina	24:0 (lignoceric)	14	18.25
Lunaria biennis	15-24:1 (nervonic)	48	21
Lunaria annua	15-24:1 (nervonic)	26.6	29

Note: t, trans; h, hydroxy; e, epoxy; a, acetylene; o, oxo; me, methylene with double bond; m, methylene.

in genetically improving many major oilseed crops, but modern molecular genetics has made much faster progress in this area.

Lipid Metabolism in Plants

Fatty Acid Synthesis

In plants, fatty acids are made in the stroma of plastids, and are mostly modified in the endoplasmic reticulum (ER) (**Figure 3**). Plant fatty acid biosynthesis starts with acetyl CoA carboxylase (ACCase) catalyzed condensation of acetyl CoA with CO_2 to form malonyl CoA. Subsequently, a transacylase catalyzes the reaction of acyl carrier protein (ACP) and malonyl CoA forming malonyl ACP. Then, a stepwise two-carbon condensation cycle starts with a β-keto acyl synthase III (KAS III) catalyzed irreversible reaction of malonyl ACP and acetyl CoA to form 3-ketobutyryl ACP and release of CO_2. The 3-ketobutyryl ACP is then reduced, dehydrated, and reduced again to produce butyryl ACP. This cycle is repeated seven or eight times producing palmitoyl ACP or stearoyl ACP using different enzymes KAS I and KAS II. The reaction can be terminated early by special thioesterases (FatB; **Figure 3**) allowing accumulation of shorter chain fatty acids.

The α-ketoacid elongation (α-KAE) pathway is an unusual pathway to synthesize saturated fatty acids that adds one carbon per cycle to the carbon chain. This pathway is mainly involved in formation of the alkyl components of trichome exudate (gum) sugar esters in *Nicotiana* and petunia.

Glycerolipid Synthesis

A soluble Δ^9 desaturase in the plastid catalyzes the desaturation of stearoyl ACP to form oleoyl ACP. Some acyl ACPs are used by stromal acyltransferases for plastidial membrane synthesis. However, most of the fatty acid synthesized in the plastid is hydrolyzed by acyl ACP thioesterases encoded by fatty acid thioesterase genes *FatA* and *FatB*, specific for oleoyl ACP and saturated acyl ACPs, respectively, to form free fatty acids terminating fatty acid synthesis in the plastid. Other than the small portion used for plastid membrane lipid synthesis, the free fatty acids are mostly transported out of plastids into the cytoplasm with esterification of coenzyme A (CoA) in plastid envelopes.

In developing seeds, the site of glycerolipid synthesis is the ER and begins with a transfer of an acyl group from fatty acid CoA to the *sn*-1 position of glycerol-3-phosphate (G3P) by glycerol-3-phosphate acyltransferase (GPAT) to produce lysophosphatidic acid (LPA) (**Figure 4**). A second acyl group

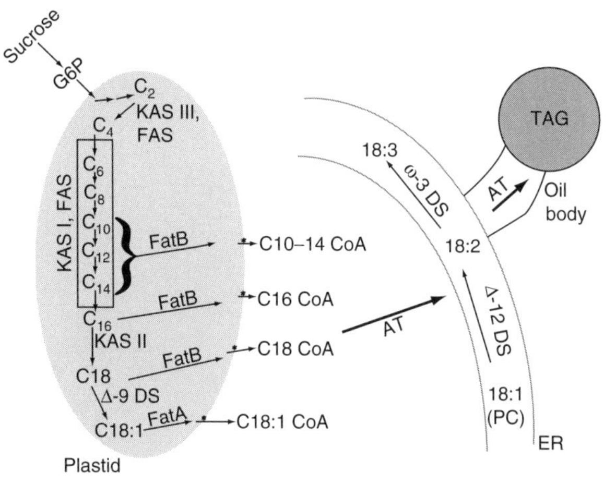

Figure 3 Overview of fatty acid and triacylglycerol biosynthesis. Reductant and chemical energy for oil synthesis in developing seeds starts with the arrival of sucrose from leaves via the phloem. Glucose-6-phosphate appears to be the main form of carbon uptake by developing seed plastids. After metabolism to acetate (C2), fatty acids are built two carbons per cycle of the fatty acid synthase complex (FAS). A FAS component, β-keto acyl synthase (KAS), is in three different forms depending on the starting carbon length. The FAS elongates fatty acids esterified to an acyl carrier protein (ACP) to 18 carbon lengths unless the process is terminated by a thioesterase encoded by *FatA* or *FatB* for monounsaturated or saturated acyl groups. DS, desaturase; *, acyl CoA synthetase; AT, acyltransferase.

is transferred to the *sn*-2 position by lysophosphatidic acid acyltransferase (LPAT) to form phosphatidic acid (PA). PA can be converted to diacylglycerol (DAG) by PA phosphatase. Diacylglycerol acyltransferase (DGAT) transfers a third acyl group from acyl CoA to *sn*-3 of DAG to form triacylglyceride (TAG). Another enzyme, phospholipid diacylglycerol acyltransferase (PDAT) can transfer the *sn*-2 chain from phosphatidylcholine (PC) to DAG forming lyso-PC and TAG. A third enzyme, a diacylglycerol transacylase (DGTA) can transfer an acyl group from one DAG to another forming a TAG and monoacylglyerol (MAG) (**Figure 4**). The TAG accumulates in oil bodies surrounded by a phospholipid membrane monolayer that contains oleosins as the major proteins.

For membrane lipid synthesis, the PA reacts with cytidine 5′-triphosphate (CTP) to form CDP-DAG, which reacts with myoinositol, serine, and G3P to form phosphatidyl inositol (PI), phosphatidyl serine (PS), and phosphatidylglycerol phosphate, which is the precursor of phosphatidylglycerol (PG). Alternatively, an activated head group such as CDP-choline, CDP-ethanolamine, uridine 5′-diphosphate (UDP)-galactose, and UDP-sulfoquinovose can react with DAG to form PC, phosphatidyl ethanolamine (PE),

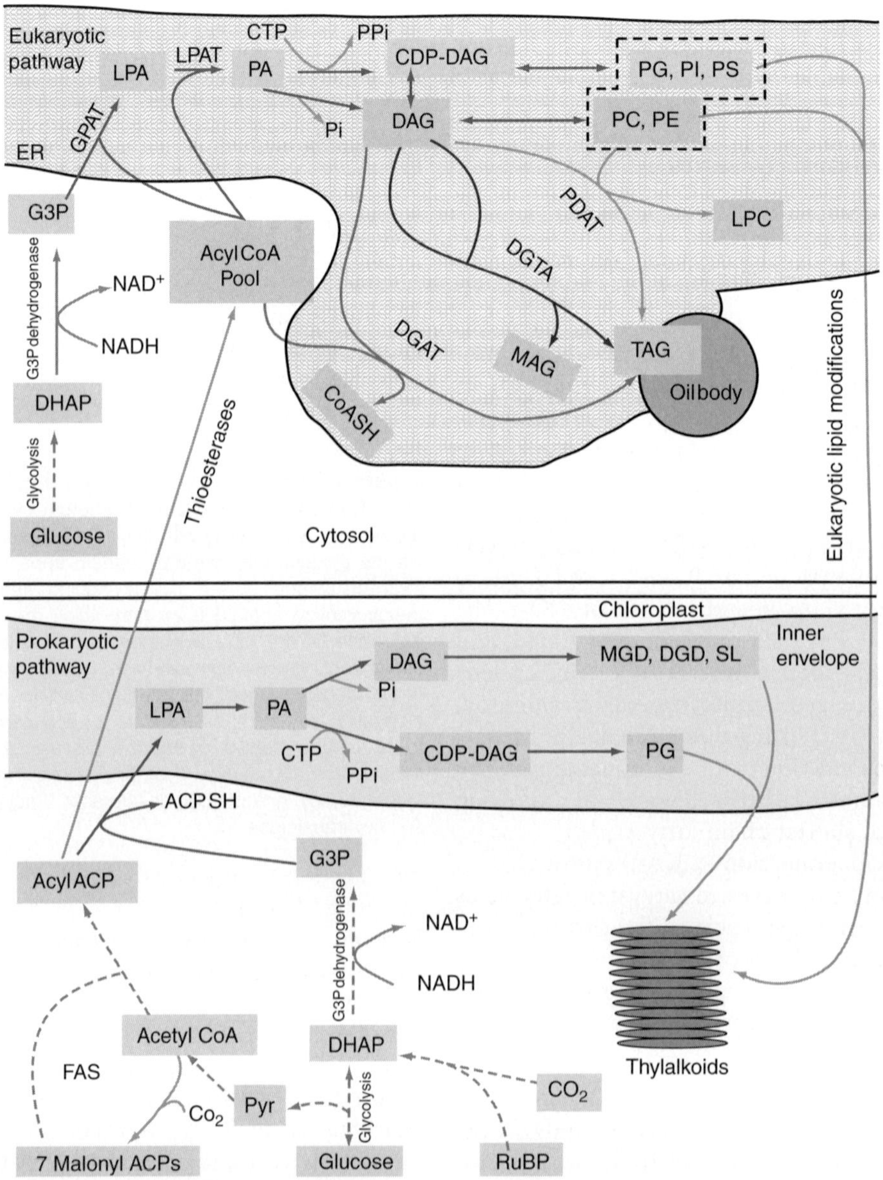

Figure 4 Overall flow of carbon to triacylglycerol (TAG) of oil bodies illustrating the three possible routes for TAG synthesis catalyzed by diacylglycerol acyltransferase (DGAT), phospholipid diacylglycerol acyltransferase (PDAT), and diacylglycerol transacylase (DGTA).

monogalactosyl diacylglycerol (MGD), and sulfoquinovosyl diacylglycerol (SL). Digalactosyl diacylglycerol (DGD) is synthesized from MGD.

Variations in Lipid Biosynthetic Pathways

Oilseeds that accumulate short and medium chain fatty acids (ranging from C8 to C14) have specialized thioesterases encoded by the *FatB* gene family to terminate their fatty acid elongation at shorter length. Specialized fatty acid synthase and LPAT have also been evolved for the synthesis and accumulation of saturated short chain fatty acids.

Fatty acid desaturases are nonheme di-iron oxidases that transfer four electrons to O_2, two of which are from the fatty acid substrate and two from a reductant. The plastid desaturases utilize reduced ferredoxin as the reductant. The soluble ACP desaturase contains two proximal iron atoms per enzyme monomer. The ER desaturases require reduced nicotinamide adenine dinucleotide (NADH), cytochrome b_5, and cytochrome b_5 reductase.

Monounsaturated fatty acids can have unusual double bond positions, for example, petroselinic acid (Δ^6 18:1), which is the result of a plastidial Δ^4 desaturase that converts palmitoyl ACP to Δ^4

hexadecanoyl ACP, which is then elongated to petroselinoyl ACP. This desaturase belongs to the soluble acyl ACP desaturase family (AAD1 gene family), which also includes a Δ^6 16:0-ACP desaturase from *Thunbergia alata* and a Δ^9 16:0-ACP desaturase from *Asclepias syriaca*. These genes share over 70% amino acid sequence similarity with standard Δ^9 18:0 desaturase, and substitution of as few as five amino acids can convert a Δ^6 16:0-ACP desaturase into a Δ^9 18:0 desaturase. However, a Δ^5 desaturase acting on eicosanoyl CoA from *Limnanthes* is membrane associated.

Plant sphingolipids have a *trans* Δ^8 double bond resulting from the action of a cytochrome b_5-fusion desaturase. γ-Linolenic acid (18:3 $\Delta^{6,9,12}$) (GLA) is the result of a cytochrome b_5-fusion Δ^6 desaturase closely related to the sphingolipid desaturases. Another member of this cytochrome b_5-fusion superfamily from *Ceratodon purprueus* has a bifunctional Δ^6 acetylenase and desaturase.

Chloroplasts contain four membrane bound desaturases. The *Fad4* gene product inserts a *trans* double bond at the Δ^3 position only into palmitic acid in the *sn*-2 position of PG. The *Fad5* product places a *cis* double bond at the Δ^7 position of palmitic acid mainly at the *sn*-2 position of MGD and possibly DGD. The enzyme encoded by *Fad6* places a second *cis* double bond three carbons after the first double bond toward the ω end of hexadecanoic acid or oleic acid at either position of MGD, DGD, or SL. Two desaturases, encoded by *Fad7* and *Fad8*, insert a *cis* double bond at the ω^3 position of 7,10-hexadecadienoic acid and linoleic acid.

In the ER, oleic acid is first incorporated into membrane PC and desaturated to linoleic acid by a membrane-associated Δ^{12} desaturase encoded by *Fad2*, and linoleoyl PC can be further desaturated to linolenoyl PC by another membrane associated ω^3 desaturase encoded by *Fad3* (**Figure 1**).

Plants can evolve new functions from existing enzymes to diversify fatty acid synthesis. For example, ω^6 hydroxylases from *Ricinus communis* and *Lesquerella fendleri*, ω^6 epoxygenases from *Vernonia galamensis* and *Crepis palaestina*, and ω^6 acetylenase from *Crepis alpina* are all encoded by diverged *Fad2* genes. *Fad2* related conjugases from *Mormordica charantia*, *Impatiens balsamica*, and *Calendula officinalis* resulting in α-eleostearic (18:3 $\Delta^{9,\ 11t,\ 13t}$), α-parinaric (18:4 $\Delta^{9,\ 11t,\ 13t,\ 15}$), and calendic (18:3 $\Delta^{8t,\ 10t,\ 12}$) acids have also been identified.

Fatty acid CoAs can be further elongated to very long chain fatty acids (VLCFAs) by membrane bound fatty acid elongases. The VLCFAs can be found in plant membrane sphingolipids and can occasionally be incorporated into TAG. Erucic acid (22:1 Δ^{13}) from oilseed rape is the result of a membrane bound fatty acid elongase adding two carbons in two cycles to oleoyl CoA.

Wax Synthesis

The VLCFAs are mainly used in the synthesis of wax, cutin, or suberin. In wax synthesis, some of the VLCFA CoAs are decarboxylated into odd chain pure hydrocarbons, some of which are hydroxylated in the middle to alcohols that may be further oxidized to ketones. Other VLCFA CoAs have the carboxylic acid reduced to aldehydes that may be further reduced to fatty alcohols. The fatty alcohols can be esterified to VLCFA CoA forming esters catalyzed by an acyl CoA fatty alcohol acyltransferase. All of these fatty acid derivatives are components of wax. Jojoba (*Simmondsia chinensis*), a desert shrub native to the southwestern United States is the only plant known to accumulate liquid waxes in its seeds for energy storage. Seeds of these plants can be up to 60% monounsaturated C20, C22, and C24 fatty acids and alcohols.

Fatty Acid Catabolism

Fatty acids are mainly catabolized through β-oxidation in plants, which occurs in the glyoxysomes or peroxisomes. Fatty acids are first released by specific lipases and form an acyl CoA that is degraded by two carbons each cycle, generating acetyl CoA, NADH, and reduced flavin adenine dinucleotide ($FADH_2$). In each cycle, acyl CoA is first oxidized to form enoyl CoA catalyzed by acyl CoA dehydrogenase. Then enoyl CoA reacts with water to form *L*-3-hydroxyacyl CoA catalyzed by enoyl CoA hydrase. The *L*-3-hydroxyacyl CoA is further oxidized to form keto acyl CoA catalyzed by *L*-3-hydroxyacyl CoA dehydrogenase. Finally, the keto acyl CoA is thiolyzed to form acetyl CoA and a two-carbon shorter acyl CoA catalyzed by β-keto-thiolase.

A less common α-oxidation degrades fatty acids one carbon atom at a time. An α-dioxygenase, which is inducible by pathogens, catalyzes the formation of long-chain aldehydes and hydroxy fatty acids, as well as odd-chain fatty acid catabolites.

Fatty acids can also be converted by lipoxygenase into hydroperoxy derivatives. The primary products of lipoxygenase, the fatty acid hydroperoxides can be rapidly converted into additional products. Hydroperoxide lyase can cleave fatty acid hydroperoxides forming six- and twelve-carbon aldehydes. The six-carbon aldehydes may be converted into alcohols. Allene oxide synthase converts the fatty acid hydroperoxides into an unstable epoxy intermediate that can be converted into a cyclic ketone by a cyclase.

A series of subsequent reactions can result in the formation of jasmonic acid, which can be converted into methyl jasmonate. Peroxygenases convert fatty acid hydroperoxides into epoxides, which can be converted to hydroxides by hydroxylases.

Lipoxygenase is a nonheme single iron containing dioxygenase that catalyzes the peroxidation of *cis,cis*-1,4 pentadiene structures such as in linoleoyl and linolenoyl groups. Some of the compounds formed from fatty acid oxidation are biologically active and important in signal transduction in response to biotic and abiotic stresses, such as pest attack, wounding, and other stresses. The induced release of pheromone-like compounds attracts predators and parasites to plants being fed upon by herbivorous insects.

Progress on Genetic Engineering of Oilseed Crops

Increasing oil yield is an obvious goal for genetic engineering of oilseed crops. However, regulation of the oil synthesis pathway is not completely understood. ACCase is found to be a limiting step for fatty acid synthesis. Two independent ACCase activities, homomeric and heteromeric, exist in plants. Overexpressing homomeric ACCase has been found to increase oil content by about 5%. Expressing a yeast LPAT mutant gene *SLC1-1* in *B. napus* increased seed oil content by 8–48%.

Oil produced from original varieties of oilseed rape known as high erucic acid rapeseed (HEAR) is inedible because it contains 25–50% erucic acid, which can cause myocardial lipidosis and necrosis. In the 1960s, low erucic acid rapeseed (LEAR) was selectively bred in Canada producing seed oils with less than 5% erucic acid. In 1974, cultivars yielding seed oil with less than 2% erucic acid and low glucosinolates, an antinutritional and anticarcinogenic compound, were developed by breeding plants for edible oil production. These varieties are known as canola in North America. This oil does not cause health problems and has good nutritional quality because of its relatively low saturated fatty acid content and high oleate content, and it is moderately high in ω^3 fatty acids. Today, canola oil is one of the major edible oils in the world.

When *Cuphea hookeriana* thioesterase was expressed in canola, caprylate accumulated to about 12% in the seed oil. Canola expressing California bay (*Umbellularia californica*) thioesterase under a napin promoter produced 48% laurate in the seed oil. However, the laurate was not incorporated in the *sn*-2 position of the TAG due to the LPAT discrimination. Hybridization of canola plants expressing the California bay thioesterase with plants expressing a coconut (*Cocos nucifera*) LPAT yielded seed oil with up to 67% laurate distributed in all three positions of the TAG.

High stearate oil has a high melting point and is suitable as a margarine or for shortening feedstock. Antisense expression of stearoyl ACP desaturase in *B. napus* resulted in a 10- to 20-fold increase in stearate production. A *FatA* thioesterase that prefers stearoyl ACP from mangosteen (*Garcinia mangostana*) was expressed in *B. napus* resulting in a 10-fold higher level of stearate in the seed oil. Use of a hairpin RNA mediated gene silencing technique on stearoyl ACP Δ^9 desaturase and *Fad2* genes resulted in up to 40% stearate and 77% oleate in the resulting cottonseed oil.

Hybrids of normal sunflower (*Helianthus* spp.) varieties yielding oils with greater than 80% oleate have been obtained by chemical mutagenesis. Mid-oleate (~60% 18:1) soybean oil has been produced by mutagenesis and breeding, and a soybean oil with ~85% oleate has been produced by antisence silencing of the seed specific *Fad2-1*. By silencing of the *Fad2* gene in canola, oleate level in the seed oil increased to 83% and the saturated fatty acids decreased at the same time. Similarly, by silencing of the *Fad3* gene, linolenate decreased from 9% to 1.4% and linoleate increased from 55% to 65% in the soybean oil.

Unusual monoenoic fatty acids have special properties including variations on melting points suitable for some commercial uses. One of them, petroselinic acid can be used to make adipic acid, a petroleum derived feedstock for nylon manufacturing. The responsible genes including Δ^4 palmitate desaturases from coriander (*Coriandrum sativum*) and *Asclepias syriaca*, Δ^6 16:0-ACP desaturase from *Thunbergia alata*, Δ^9 14:0-ACP desaturase from *Pelargoniun xhortorum*, Δ^9 16:0-ACP desaturases from *Asclepias syriaca*, and *Doxantha unguis-cati* (cat's claw) have been cloned. However, only less than 10% accumulation of monoenes from transgenic plants expressing these genes was found.

After random chemical mutagenesis and selection, a new linseed oil, marketed as linola, with high linoleic acid and almost zero α-linolenic acid was obtained via null mutations in two ω^3 desaturase genes.

GLA synthesized by a Δ^6 desaturase is a valuable pharmaceutically, nutraceutically, and nutritionally important fatty acid in mammals. Canola transformed by a Δ^6 desaturase from *Mortierella alpina* resulted in 13% GLA accumulation. However, when canola was cotransformed by Δ^6 and Δ^{12} desaturases from *M. alpina*, the accumulated GLA was over 40%. A complementary DNA (cDNA) from a fungus, *Pythium irregulare*, expressed in *Brassica juncea* under the control of the napin promoter resulted in up to 40% GLA in the seed oil. Borage Δ^6 desaturase was identified and was found to have a 5′ cytochrome b_5 domain. Two desaturases, Δ^6 and Δ^5, and a specific elongase were identified from a nematode *Caenorhabditis elegans* and found to be less than 2 kb apart. The elongase converts GLA to di-homo-γ-linolenic acid (DHGLA) while the Δ^5 desaturase converts DHGLA to arachidonic acid. A Δ^4 desaturase responsible for the conversion of docosapentaenoic acid (22:5) to docosahexaenoic acid (DHA; 22:6) was identified from a marine heterotroph *Thraustochytrium*. These genes may be used to transform major oilseeds species for very long chain PUFAs accumulation.

A castor oleate Δ^{12} hydroxylase gene was identified and functionally expressed in tobacco and *Arabidopsis*. A *Lesquerella fendleri* elongase gene (LfKCS3) was also identified and functionally tested in *Arabidopsis* by crossing with transgenic plants expressing the castor oleate Δ^{12} hydroxylase gene.

Epoxy fatty acids may be used as biodegradable replacements for petroleum derivatives such as PVC plasticizers, nylon monomers, heat stable lubricants, adhesives, epoxy coatings, paints, resins, and the formulation of carriers for slow-release pesticides and herbicides. Also, epoxy fatty acids can readily and economically be converted to similarly useful hydroxy fatty acids. *Vernonia galamensis* epoxygenase belonging to the ω^6-diiron desaturase gene family was cloned, and transgenic soybean lines expressing this gene can produce as much as 8% vernolic acid. Interestingly, vernolic acid is synthesized by a cytochrome P-450 monooxygenase in *Euphorbia lagascae*.

Conjugated fatty acids such as α-eleostearic acid have industrially important drying properties and are used in coating materials. A bifunctional Δ^6 desaturase/acetylenase gene was identified from the moss *Ceratodon purpureus*. The gene is grouped with Δ^{12} fatty acid desaturases and hydroxylases rather than conjugases. The gene encoding the desaturase responsible for introducing conjugated Δ^{11t}, Δ^{13t} double bonds forming α-eleostearic acid was identified from developing seeds of *Aleurites fordii*, which produces tung oil. Conjugases from *Mormordica charantia* and *Impatiens balsamica* were expressed in soybean forming α-eleostearic acid and parinaric acid, respectively. A divergent *Fad2* from *Calendula officinalis* catalyzing the formation of Δ^{8t}, Δ^{10t} double bonds forming calendic acid was identified and functionally expressed in somatic soybean embryos resulting in 22% calendic acid in the total fatty acids in the embryo oil.

An unusual Δ^5 20:1 fatty acid that accumulates in the seed oil of several *Limnanthes* species can be used as a precursor for the industrial compounds δ-lactones, and for cosmetics, surfactants, and lubricants. *Limnanthes douglasii* Δ^5 desaturase was cloned and when expressed in soybean somatic embryos, Δ^5 16:1 accumulated to 2–3% in embryo oil. When soybean was transformed by Δ^5 desaturase and fatty acid elongase 1 (FAE1) from *Limnanthes*, up to 12% of Δ^5 20:1 and Δ^5 22:1 was accumulated in the embryo oil.

Although HEAR oil is not edible, erucic acid can be used as a renewable and environmentally friendly source for plastic manufacturing and for lubricants and emollients. The LPAT of rapeseed does not incorporate erucic acid into TAG, thus limiting the erucic acid proportion in the seed oil to a theoretical 66% by traditional breeding. *B. napus* expressing a yeast LPAT mutant gene SLC1-1 under the tandem cauliflower mosaic virus 35S (CaMV35S) promoter increased the erucic acid proportion in the seed oil by up to 7%. HEAR expressing the *A. thaliana FAE1* gene under the control of the seed specific promoter napin also resulted in up to a 7% increase of erucic acid in the seed oil.

Long chain wax esters can be used as industrial lubricants and transmission fluids. Coexpression of *Lunaria annua* long-chain acyl CoA elongase, jojoba reductase, and acyl transferase was tested in *Arabidopsis*, which accumulated up to 70% of wax in the resulting seeds.

Summary and Future Developments

The improvement of oil quality by conventional breeding and genetics has had good success for some traits, and will continue to play a role in the improvement of oil quality. These developments, however, could be greatly accelerated by the use of genetic engineering and molecular genetic approaches, which are the only known means for development of oilseeds producing novel fatty acids.

Although many key genes for oil synthesis have been identified for the production of unusual fatty acids, only limited success has been achieved in the accumulation of high levels of products of these genes in major oilseed crops such as soybean and oilseed rape. High concentrations of these fatty acids are required in order for the products to be commercially viable. In order to determine the factors that limit the accumulation of unusual fatty acids in transgenic major oilseed crops, a far more complete understanding of the related cellular biochemistry is required. For example, the specificities of the acyltransferases in the natural high accumulators of unusual fatty acids might be one of the key factors for their high accumulation.

The opportunities to extract value from the diverse plant fatty acids are enormous. Many additional genes are expected to be discovered with the help of the advancing molecular genetic tools such as automated DNA sequencing, DNA microarrays, and bioinformatics. New enzymes may be designed by site-specific mutagenesis adding more opportunities for oilseed improvement.

In the future, oils from transgenic oilseeds may be tailored for optimum human nutrition, and used as tremendous renewable sources for various industrial applications replacing specialty petroleum derivatives.

List of Technical Nomenclature

AAD	Acyl ACP desaturase.
ACCase	Acetyl CoA carboxylase.
ACP	Acyl carrier protein.
AT	Acyltransferase.
CaMV 35S	Cauliflower mosaic virus 35S.
CDP	Cytidine 5′-diphosphate.
CoA	Coenzyme A.
CTP	Cytidine 5′-triphosphate.
DAG	Diacylglycerol.
DGAT	Diacylglycerol acyltransferase.
DGD	Digalactosyl diacylglycerol.
DGTA	Diacylglycerol transacylase.
DHA	Docosahexaenoic acid.
DHAP	Dihydroxy acetone phosphate.
DHGLA	Di-homo-γ-linolenic acid.
DS	Desaturase.
ER	Endoplasmic reticulum.
Fad	Fatty acid desaturase.
FADH2	Reduced riboflavin dinucleotide.
FAE1	Fatty acid elongase 1.
FatA	Fatty acid thioesterase A.
FatB	Fatty acid thioesterase B.
FAS	Fatty acid synthase.
G3P	Glycerol-3-phosphate.
G6P	Glycerol-6-phosphate.
GLA	γ-Linolenic acid.
GPAT	Glycerol-3-phosphate acyltransferase.
HDL	High-density lipoprotein.
HEAR	High erucic acid oilseed rape.
KASI	β-Keto acyl synthase I.
KASII	β-Keto acyl synthase II.
KASIII	β-Keto acyl synthase III.
LDL	Low-density lipoprotein.
LEAR	Low erucic acid oilseed rape.
LfKCS3	*Lesquerella fendleri* elongase.
LPA	Lysophosphatidic acid.
LPAT	Lysophosphatidic acid acyltransferase.
MAG	Monoacylglycerol.
MGD	Monogalactosyl diacylglycerol.
Mha	Million hectares.
MMT	Million metric tonnes.
NADH	Reduced nicotinamide adenine dinucleotide.
PA	Phosphatidic acid.
PC	Phosphatidylcholine.
PDAT	Phospholipid diacylglycerol acyltransferase.
PE	Phosphatidyl ethanolamine.
Pi	Phosphate.
PI	Phosphatidyl inositol.
Ppi	Pyrophosphate.
PS	Phosphatidyl serine.
PUFA	Polyunsaturated fatty acid.
PVC	Polyvinylchloride.
Pyr	Pyruvate.
RuBP	Ribulose-1,5-biphosphate.
SL	Sulfoquinovosyl diacylglycerol.
SLC1	Yeast lysophosphatidic acid acyltransferase.
TAG	Triacylglycerol.
UDP	Uridine 5′-diphosphate.
VLCFA	Very long-chain fatty acid.
α–KAE	α-Ketoacid elongase.

See also: **Abiotic Stresses**: Cold Stress. **Crop Improvement**: Mutation Techniques; Plant Breeding, Practice; Plant Breeding, Principles. **Ethics and Biosafety**: Plant Genetic Engineering, Food Safety Issues. **Primary Products**: Oils. **Production Systems and Agronomy**: Oil Palm and Coconut. **Secondary**

Products: Epicuticular Waxes; Wax Pathways. **Seed Development**: Germination. **Water Relations of Plants**: Uptake, Loss and Control.

Further Reading

Ackman RG (1990) Canola fatty acids – An ideal mixture for health, nutrition, and food use. In: Shahidi F (ed.) *Canola and Rapeseed – Production, Chemistry, Nutrition, and Processing Technology*, pp. 81–98. New York: Van Nostrand Reinhold.

Gunstone FD (2001) Oilseed crops with modified fatty acid composition. *Journal of Oleo Science* 50: 3–13.

Hildebrand DF, Rao SS, and Hatanaka T (2002) Redirecting lipid metabolism in plants. In: Kuo TM and Gardner HW (eds) *Lipid Biotechnology*, pp. 57–84. New York: Marcel Dekker.

Kinney AJ (2002) Perspectives on the production of industrial oils in genetically engineered oilseeds. In: Kuo TM and Gardner HW (eds) *Lipid Biotechnology*, pp. 85–93. New York: Marcel Dekker.

Miege C and Marechal E (1999) 1,2-*sn*-Diacylglycerol in plant cells: Product, substrate and regulator. *Plant Physiology and Biochemistry* 37: 795–808.

Murphy DJ (2002) Biotechnology and the improvement of oil crops – genes, dreams and realities. *Phytochemistry Reviews* 1: 67–77.

Schultz DJ and Ohlrogge JB (2002) Metabolic engineering of fatty acid biosynthesis. In: Kuo TM and Gardner HW (eds) *Lipid Biotechnology*, pp. 1–25. New York: Marcel Dekker.

Slabas AR, Hanley Z, Schierer T, *et al.* (2001) Acyltransferases and their role in the biosynthesis of lipids – opportunities for new oils. *Journal of Plant Physiology* 158: 505–513.

Thelen JJ and Ohlrogge JB (2002) Metabolic engineering of fatty acid biosynthesis in plants. *Metabolic Engineering* 4: 12–21.

Topfer R, Martini N, and Schell J (1995) Modification of plant lipid synthesis. *Science* 268: 681–686.

Voelker T and Kinney AJ (2001) Variations in the biosynthesis of seed-storage lipids. *Annual Review of Plant Physiology and Plant Molecular Biology* 52: 335–361.

Biopolymers

Y Poirier, University of Lausanne, Lausanne, Switzerland

Introduction

Plants naturally synthesize a variety of polymers that have been used by mankind as a source of useful biomaterials. For example, cellulose, the main constituent of plant cell wall and the most abundant polymer on earth, has been used for several thousand years as a source of fibers for various fabrics. Similarly, rubber extracted from the bark of the tree *Hevea brasiliensis*, has been a major source of elastomers until the development of similar synthetic polymers. In the last century, the usefulness of plant polymers as biomaterials has been expanded through the chemical modification of the natural polymers. For example, a number of plastics have been made by substituting the hydroxyl groups present on the glucose moiety of cellulose with larger groups, such as nitrate or acetate, giving rise to materials such as cellulose acetate, a clear plastic used in consumer products such as toothbrush handles and combs. Similarly, starch has been used in the manufacture of plastics by either using it in blends with synthetic polymers or as the main constituent in biodegradable plastics. The advent of transformation and expression of foreign genes in plants has created the possibility of expanding the usefulness of plants to include the synthesis of a range of biomolecules. In view of the capacity of certain crops to produce a large quantity of organic raw material at low cost, such as oils and starch, it is of interest to explore the possibility of using transgenic plants as efficient vectors for the synthesis of biopolymers. Such plant based biopolymers could replace, in part, the synthetic plastics and elastomers produced from petroleum, offering the advantage of renewability and sustainability. Furthermore, being natural products, biopolymers are usually biodegradable and can thus contribute to alleviate problems associated with the management of plastic waste.

In this article, the emphasis will be on the use of transgenic plants for the synthesis of two novel classes of industrially useful polymers, namely protein based polymers made from natural or artificial genes, and polyhydroxyalkanoates, a family of bacterial polyesters having the properties of biodegradable plastics and elastomers.

Protein Based Polymers

A number of proteins with interesting plastic or adhesive properties are synthesized in nature. For example, marine mussels (*Mytilus* spp.) can produce strong protein based adhesives that are effective in aqueous environments, allowing these organisms to stick to many surfaces. Several proteins composed of blocks of amino acids repeated extensively form tough and flexible fibers that could find applications in many areas of material science, including in the medical field. Such proteins include silk, collagen, and elastin. Plant cell walls also contain proteins that are partially composed of repeated blocks of amino acids, such as extensin, the glycine rich proteins, and the proline rich proteins. The mechanical properties of the pure form of these natural plant proteins have not been well studied.

The ability to synthesize protein polymers from genes allows the precise determination of both molecular mass and amino acid sequence, making it possible, in principle, to achieve a degree of control over physical properties and functionality that is outside the scope of chemical polymerization technologies. The development and use of these protein based biomaterials are, in many cases, limited by the difficulties in producing sufficient quantities of the material to establish the structure/function relationships and design applications, as well as difficulties in the economical large-scale production of polymers for commercialization. It is in this respect that plants have been explored as a potential production vector for protein based biopolymers.

Synthesis of Silk Proteins

Silks represent a broad class of polymers that can be loosely defined as externally spun fibrous protein secretions. Commercial silk used in the textile industry is produced from the domesticated worm *Bombyx mori*. Spiders also produce a variety of silks that are used in the construction of webs or as draglines. The different types of spider silks produced have impressive properties, ranging from lycra-like elastic fibers to kevlar-like superfibers. The exceptional material properties of spider silk have made it an early target for its synthesis in larger quantities in a more amenable production system. Cloning of the spidoin proteins forming the dragline of the spider *Nephila clavipes* enabled the exploration of their synthesis in *Escherichia coli*. However, expression of a synthetic gene coding for a protein containing highly repeated blocks of amino acids found in spidoins led only to low production of a mixture of proteins of variable sizes. This heterogeneity in the proteins synthesized was due to premature termination of translation. In contrast, expression of a similar gene construct in the yeast *Pichia pastoris*, as well as in tobacco (*Nicotiana tabacum*) and potato (*Solanum tuberosum*), resulted in the synthesis of a higher amount (between 2 and 10% of total proteins) of a full length protein, indicating that, at least in these eukaryotes, the repetitive nature of the gene construct was not an obstacle to the synthesis of these unusual proteins. Several synthetic genes containing various combinations of the repetitive peptides found in the *N. clavipes* spidoin protein have been expressed in tobacco, allowing an assessment of the link between protein structure and physical properties.

Synthesis of Artificial Protein Polymers

Elastin is a very strong natural fiber found in ligaments and arterial walls. Analysis of the primary structure of all mammalian elastins has shown the presence of a repeated block of amino acids with the sequence GVGVP. Synthetic proteins made from multiple repeats of this sequence display elastic properties. These protein polymers are also biodegradable as well as biocompatible, being nontoxic and naturally resorbed by animal tissues. Polymers based on the GVGVP sequence have been shown to prevent postsurgical adhesions and scars in rats. Other medical applications of similar bioelastics include tissue reconstruction, wound covering, and programed drug delivery. Expression in *E. coli* of a synthetic protein made of 121 repeats of the GVGVP peptide was shown to result in accumulation of polymer inclusion bodies that occupied up to 90% of the cell volume. However, expression of the same synthetic gene in the fungus *Aspergillus nidulans*, as well as in tobacco, only led to low polymer accumulation despite adequate accumulation of mRNA. It is thought that efficient expression in eukaryotes of such proteins made of repeated blocks of peptides is limited by the availability of certain amino acids or tRNAs. Thus, accumulation of high levels of artificial protein polymers in plants may be dependent on additional genetic engineering, in particular, modifications in the amino acid biosynthetic pathways and/or tRNA pools.

Polyhydroxyalkanoates

Polyhydroxyalkanoates (PHAs) are polyesters of hydroxyacids synthesized by a wide variety of bacteria (**Figure 1**). The best-characterized PHA is polyhydroxybutyrate (PHB), a homopolymer of 3-hydroxybutyrate. Although initially described by M. Lemoigne in 1926, it was not until 1962 that the value of PHB as a thermoplastic was recognized in a US patent, and not until 1982 when the first PHA produced by bacterial fermentation was

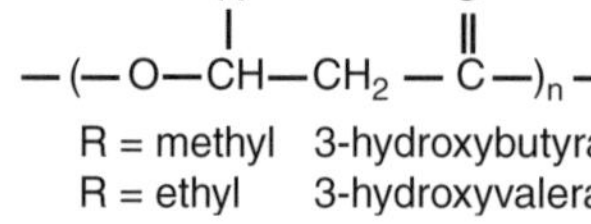

R = methyl 3-hydroxybutyrate
R = ethyl 3-hydroxyvalerate

Figure 1 Chemical structure of 3-hydroxyacid monomers constituting PHA. The monomer can range from 3 to 16 carbons in length, depending on the size of the pendant R group.

commercialized. Since then, there has been a growing interest in producing and commercializing PHA as a family of biodegradable thermoplastics and elastomers.

Synthesis of PHA in Bacteria

PHAs represent a large group of bacterial polyesters, which can include over 150 different hydroxyacids. Although the majority of PHAs are composed of R-3-hydroxyacids ranging from 3 to 16 carbons in length, some polymers also contain monomers of 4-hydroxy and 5-hydroxyacids. Furthermore, a diversity of functional groups have been found on the monomers forming PHAs, including unsaturated groups, as well as methyl, halogenated, cyano, and phenoxy groups, just to name a few. The majority of bacteria synthesizing PHAs can be broadly subdivided into two groups. One group produces short-chain-length PHAs (SCL-PHAs) with monomers ranging from 3 to 5 carbons in length, while a distinct group synthesizes medium-chain-length PHAs (MCL-PHAs) with monomers from 6 to 16 carbons. This classification is, however, not strict as a few bacteria have been shown to produce "hybrid" PHAs with monomers ranging from 4 to 10 carbons or higher. PHAs accumulate in bacteria as a high-molecular-weight polymer forming intracellular granules of 0.2–0.3 μm in diameter. Typically, PHAs accumulate to a significant proportion of the cell dry weight when bacteria are grown in a media that is limited in a nutrient essential for growth (typically nitrogen or phosphorus), but with an abundant supply of carbon (e.g., glucose). Under these conditions, bacteria convert the extracellular carbon into an intracellular storage form, namely PHA. When the limiting nutrient is resupplied, intracellular PHA is degraded and the resulting carbon is used for growth. PHAs have been shown to occur in nearly 100 genera of bacteria, encompassing Gram-positive and Gram-negative species, as well as in some archea.

The large diversity of monomers found in PHA translates into a wide spectrum of physical properties. The homopolymer PHB is a relatively stiff and brittle plastic with limited applications. In contrast, PHA copolymers composed primarily of 3-hydroxybutyrate with a fraction of longer chain monomers, such as 3-hydroxypentanoate, 3-hydroxyhexanoate, or 3-hydroxyoctanoate, are more flexible and tougher plastics. These PHAs can be used in a wide variety of consumer products, such as containers and bottles, as well as water resistant films. PHAs made of longer monomers, such as MCL-PHAs, are typically elastomers and sticky materials.

Numerous bacteria and some fungi have been found to degrade extracellular PHA and metabolize the breakdown products to CO_2 and water. Degradation of extracellular PHA is made possible by the secretion of PHA depolymerases, enzymes catalyzing the hydrolysis of PHAs to monomers and oligomers. PHAs are thus biodegradable polymers, and products made from them can be degraded completely in a variety of environments. Degradation is particularly rapid in the microbe rich environment of a compost (typically 3–9 months for a bottle made of PHB). PHAs can also be degraded, albeit slowly, in animal tissues. Since the breakdown products are 3-hydroxyacids, which are naturally found in animals, PHAs are biocompatible, making them useful for some medical applications, such as implants, gauzes, and resorbable suture filaments.

The main limitation in using bacterial PHAs as a source of biodegradable polymers is their production cost. Synthesis of PHAs in native or recombinant bacteria using cheap carbon sources, such as glucose or sucrose, has been estimated to cost a minimum of 2–4 US$/kg, making them 5–10 times more expensive than petroleum based polymers, such as polypropylene. Thus, although bacterial PHAs could be used in high-value applications where the cost of the material is small compared to the value of the final product, such as in medical products, bacterial PHA cannot compete in price with synthetic plastics for high-volume, low-value consumer products, such as packaging material and disposable items. It is in this context that agriculture has been regarded as a promising alternative for the production of PHA on a large scale and at low cost. In comparison to bacterial or yeast fermentation, crop plants are capable of producing very large amounts of a number of useful chemicals at low cost. For example, starch and lipids are two of the most industrially useful and versatile products harvested from crop plants. Starch is used extensively in the food industry and as a source of fermentation substrate for the production of a wide variety of chemicals, including polylactic acid. Several million tonnes of cornstarch is produced annually in the USA, with a market value of approximately US$0.25/kg, making it one of the cheapest commodity products. Plant lipids are also utilized by the food and nonfood industries. Nonfood industrial uses of lipids include the manufacture of soaps and detergents, paints, varnishes, lubricants, adhesives, and plastics, as well as biodiesel. The USA produces approximately 10 million tonnes of vegetable oil annually, with a market value of approximately US$0.50/kg. In view of the high productivity of crop plants, it was reasoned that if PHA could be synthesized in plants at a level comparable to reserve lipids in oilseed plants, i.e., approximately 10–40% dry weight, several million tonnes of PHA could potentially be produced at a cost comparable to vegetable oil, making it competitive with petroleum derived plastics.

Synthesis of PHB in Plants

The first PHA produced in plants was the homopolymer PHB. In bacteria, PHB is synthesized from the condensation of two acetyl CoA by a 3-ketothiolase to form acetoacetyl CoA, which is subsequently reduced to R-3-hydroxybutyryl CoA by an acetoacetyl CoA reductase (**Figure 2A**). The R-3-hydroxybutyryl CoA is finally polymerized by the PHA synthase to form PHB. Expression of the PHB biosynthetic pathway in plants was first demonstrated in 1992 in the plant *Arabidopsis thaliana*. Constitutive expression of the acetoacetyl CoA reductase and PHA synthase from the bacterium *Ralstonia eutropha* into the cytoplasm of transgenic *A. thaliana* plant cells led to the production of PHB up to 0.1% of the shoot dry weight (dry wt). Growth of these transgenic plants was however compromised, being severely stunted, although seeds could still be produced. It is thought that synthesis of PHB from acetyl CoA present in the cytoplasm leads to a limitation in the accumulation of isoprenoids and flavonoids, which are also synthesized from cytoplasmic acetyl CoA, resulting in a growth defect. PHB in transgenic plants accumulated in the form of intracellular granules with a size and appearance identical to bacterial PHA granules. Analysis of the chemical structure and physical properties of the polymer confirmed that PHA synthesized in plants was similar to the bacterial polymer. Similar low amounts of PHB synthesized in the cytoplasm of plants have also been demonstrated in tobacco, oilseed rape (*Brassica napus*), and cotton (*Gossypium* spp.) fiber cells (**Table 1**).

In plants, biosynthesis of fatty acids from acetyl CoA occurs in the plastid. The plastid is therefore a site of higher flux of carbon through acetyl CoA compared to the cytoplasm. This flux is particularly

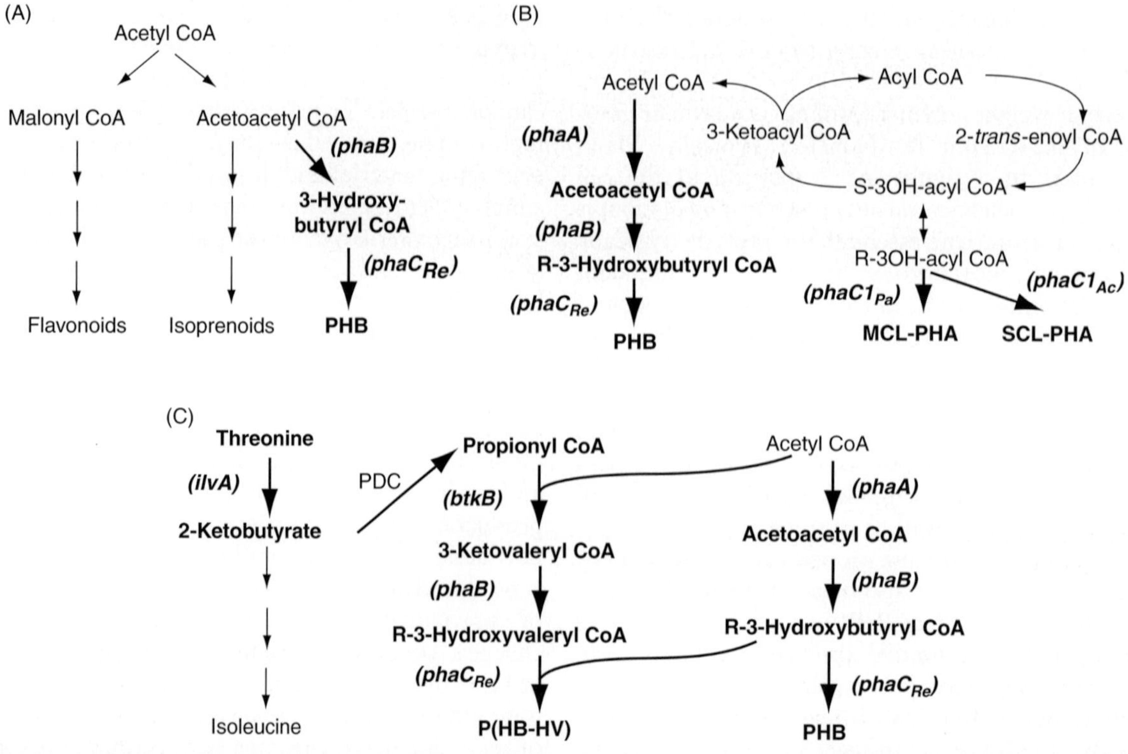

Figure 2 Modification of plant metabolic pathways for the synthesis of PHAs. (A) Synthesis of PHB from acetyl CoA in the cytoplasm; (B) synthesis of PHB, SCL-PHA, and MCL-PHAs in the peroxisome; and (C) synthesis of PHB and P(HB-HV) copolymer in the plastid. The pathways created or enhanced by the expression of transgenes are highlighted in bold, while endogenous plant pathways are in normal type. The various transgenes expressed in plants are indicated in parentheses. The *phaA*, *phaB*, and $phaC_{Re}$ genes encode a 3-ketothiolase, an acetoacetyl CoA reductase, and a PHA synthase from *R. eutropha*, respectively. The *btkB* gene encodes a second 3-ketothiolase isolated from *R. eutropha*, which shows high affinity for both propionyl CoA and acetyl CoA. The *ilvA* gene encodes a threonine deaminase from *E. coli*. The $phaC1_{Pa}$ and $phaC_{Ac}$ genes encode PHA synthases from *P. aeruginosa* and *A. caviae*, respectively. While the PHA synthase from *R. eutropha* can polymerize 3-hydroxyacyl CoAs ranging from 3 to 5 carbons, the PHA synthases from *A. caviae* and *P. aeruginosa* can polymerize 3-hydroxyacyl CoAs ranging from 3 to 7 carbons and 6 to 16 carbons, respectively. PDC refers to the endogenous plant pyruvate dehydrogenase complex.

Table 1 Summary of transgenic plants producing PHAs

Subcellular compartment	*Species*	*Tissue*	*PHA type*	*PHA quantity (%dry wt)*
Cytoplasm	*A. thaliana*	Shoot	PHB	0.1
	Oilseed rape	Shoot	PHB	0.1
	Tobacco	Shoot	PHB	0.01
	Cotton	Fiber	PHB	0.3
Plastid	*A. thaliana*	Shoot	PHB	40
	Oilseed rape	Seed	PHB	8
	Cotton	Fiber	PHB	0.05
	A. thaliana	Shoot	P(HB-HV)	0.8
	Oilseed rape	Seed	P(HB-HV)	2.3
Peroxisome	*A. thaliana*	Whole plants	MCL-PHA	0.6

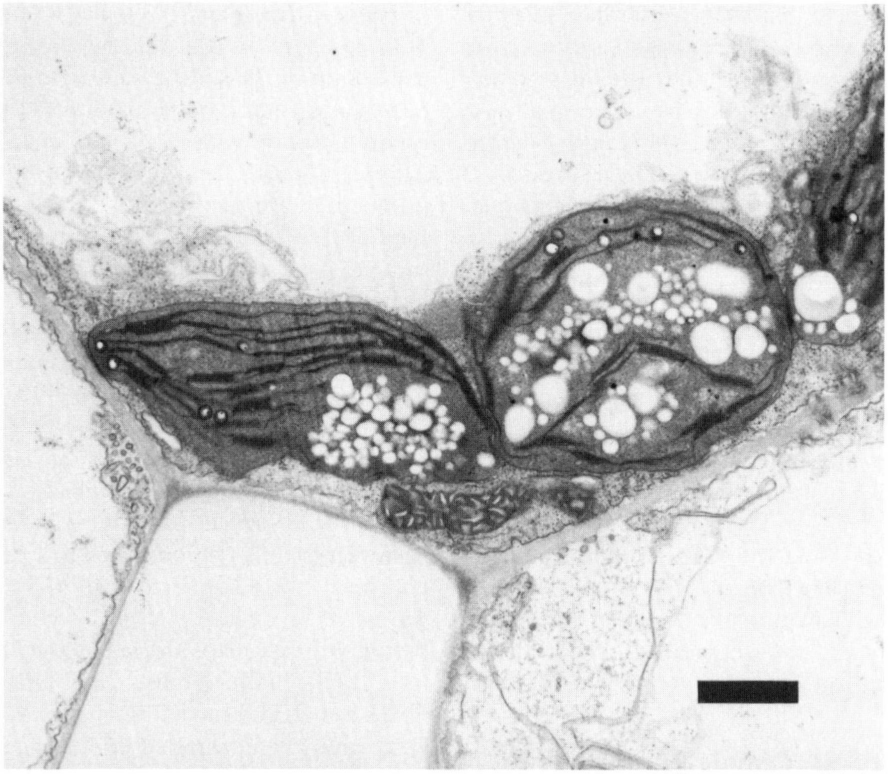

Figure 3 Accumulation of PHB granules in chloroplasts of transgenic *A. thaliana*. Granules are seen as agglomerations of electron-lucent granules in the stroma of the chloroplast. Scale bar = 1 μm.

enhanced in the seed of oil-accumulating plants, such as *A. thaliana*, where up to 40% of the seed dry weight is triacylglycerides. It was therefore hypothesized that the larger flux of acetyl CoA in the plastid may lead to a significant increase in polymer production in transgenic plants expressing the PHB pathway in chloroplasts. Constitutive expression of the PHB biosynthetic pathway in the chloroplasts of leaves of *A. thaliana* led to polymer accumulation up to 10–40% of the leaf dry weight. These experiments were done by modifying the *R. eutropha* enzymes 3-ketothiolase, acetoacetyl CoA reductase, and PHB synthase for plastid targeting by fusing the transit peptide of the small subunit of ribulose bisphosphate carboxylase to the N-terminal end of the proteins (**Figure 2C**). Transmission electron micrographs revealed that the plastids contained numerous inclusions typical of PHB granules (**Figure 3**). The accumulation of PHB in leaves was found to increase over the life span of the plant, with fully expanded presenescing leaves typically showing ten times more PHB than younger expanding leaves of the same plant. Plants producing approximately 5–10% dry wt PHB showed little signs of reduced growth, although chlorosis of leaves was observed, indicating that some aspect of chloroplast function was affected. In plants accumulating higher levels of PHB, a defect in growth was observed, being severe at 40% dry wt and accompanied by sterility. These experiments indicated that although the plastid was a much better site for the synthesis of PHB than the cytoplasm, accumulation of high amounts of PHB in such an important organelle as the chloroplast has its limits.

In contrast to the chloroplast, the leukoplast of developing seeds is not involved in photosynthesis and its metabolism is simpler, being devoted mainly to the synthesis of reserve lipids and protein in the case of oilseed crops. Restriction of the expression of the PHB biosynthetic pathway to the leukoplast of developing seeds of oilseed rape resulted in accumulation of PHB up to 8% of the mature seed dry weight. These transgenic seeds had a normal size, appearance, and germination frequency. Furthermore, since the transgenes were not expressed in the vegetative tissues, growth of the plants was normal.

Acetyl CoA, the building block of PHB, is found not only in the cytoplasm and plastid, but also in the mitochondria and peroxisomes, being implicated in these organelles in the tricarboxylic acid and β-oxidation cycles, respectively. Although no conclusive demonstration of PHB in plant mitochondria has been reported, synthesis of up 2% dry wt PHB was reported in transgenic black Mexican sweetcorn (*Zea mays* L. cv. Black Mexican Sweet) suspension cell cultures expressing the PHB biosynthetic pathway in the peroxisomes (**Figure 2B**). Further experiments are needed to explore the potential advantages and disadvantages of synthesizing PHB in the peroxisomes of plants growing in the field.

Synthesis of PHA Copolymers in Plants

Since PHB is a polymer with relatively poor physical properties, being too stiff and brittle for use in most commodity products, successful production of PHA in crops depends on the synthesis of polymers having better physical properties. In the late 1970s, the copolymer poly(hydroxybutyrate-hydroxyvalerate) (P(HB-HV)) was identified as a valuable polymer with sufficient flexibility and impact resistance for its use as a commodity plastic. This co-polymer, marketed under the trade name Biopol^{TM}, was first synthesized by bacterial fermentation using *R. eutropha* growing in media supplemented with propionate in order to create an intracellular pool of propionyl CoA that could be condensed to acetyl CoA to form 3-ketovaleryl CoA.

In order to synthesize P(HB-HV) in the plastids of plants, propionyl CoA was generated by modifying the branched amino acid biosynthetic pathway (**Figure 2C**). An *E. coli* threonine deaminase modified for plastid targeting was overexpressed in plants, leading to the accumulation of 2-ketobutyrate, which can be converted, albeit at low efficiency, to propionyl CoA by the plant pyruvate dehydrogenase complex. A novel 3-ketothiolase from *R. eutropha* showing high affinity for both propionyl CoA and acetyl CoA was also targeted to the plastids along with an acetoacetyl CoA reductase and a PHA synthase. Expression of all four genes in *A. thaliana* under the control of the constitutive CaMV 35S promoter resulted in the accumulation of P(HB-HV) of between 0.2% and 0.8% dry wt in shoots, with a 4–17 mol% HV content in the polymer. Seed specific expression of the same pathway in the leukoplast of oilseed rape embryos resulted in PHA accumulation up to 2.3% dry wt in seeds, with up to 6.5 mol% HV. Although these levels of P(HB-HV) copolymer production are lower than for PHB, synthesis of Biopol^{TM} in plants is an important milestone on the route to commercialization of PHA-producing crops.

Synthesis of the copolymer MCL-PHAs in several pseudomonads was shown to be mediated by two distinct pathways. In one pathway, the intermediate 3-hydroxyacyl-acyl carrier protein (ACP) of the fatty acid biosynthetic pathway is first converted to 3-hydroxyacyl CoAs by a transacylase before being polymerized by the PHA synthase. The creation of this pathway in the plastids of plant cells has not been successfully demonstrated to date. In the second pathway, the 3-hydroxyacyl CoA intermediates generated by degradation of fatty acids through the β-oxidation cycle are utilized by the PHA synthase to form MCL-PHAs (**Figure 2B**). Synthesis of MCL-PHAs using intermediates of β-oxidation has been successfully demonstrated in *A. thaliana* expressing a PHA synthase from *Pseudomonas aeruginosa* modified at the C-terminal end by the addition of a tripeptide allowing targeting of the protein to the peroxisomes. MCL-PHA production in peroxisomes was, however, low, reaching a maximum of 0.4% dry wt in 7-day-old germinating seedlings. The monomer composition of the polymer was complex, including saturated and unsaturated monomers ranging from 6 to 16 carbons. Growth of these transgenic plants in liquid media supplemented with various fatty acids resulted in a significant increase in MCL-PHA containing monomers derived from the β-oxidation of the external fatty acids. For example, addition of either tridecanoic acid, tridecenoic acid (C13:1, Δ12), or 8-methyl-nonanoic acid resulted in the production of PHA containing mainly saturated odd-chain, unsaturated odd-chain, or branched-chain 3-hydroxyacid monomers, respectively. These studies demonstrated that the plant β-oxidation pathway was capable of generating a large spectrum of monomers from fatty acids, which can be included in MCL-PHAs.

In a further extension of the study of MCL-PHA synthesis in plant peroxisomes, it has been shown that an endogenous control over the flux of substrates toward β-oxidation and PHA synthesis could be achieved through the combined expression of a peroxisomal PHA synthase with an acyl ACP thioesterase. Expression of a plastidial caproyl ACP thioesterase from *Cuphea lanceolata* in *A. thaliana* was used to channel decanoic acids toward peroxisomal β-oxidation, resulting in a 8-fold increase in synthesis of MCL-PHA in mature leaves, with the polymer being composed of approximately 40 mol% 3-hydroxydecanoic acid, 32 mol% 3-hydroxyoctanoic acid, and 4 mol% 3-hydroxyhexanoic acid. This strategy initially applied to the synthesis of MCL-PHAs in vegetative green tissues was further extended to the synthesis of PHA in developing seeds of *A. thaliana*. In dry seeds, a maximal amount of 0.1% dry wt was detected. From these studies, a working model was established whereby enzymes and genes involved in the synthesis of unusual fatty acids in plants can be used to modulate the quantity and quality of substrates channeled toward MCL-PHAs. The absolute amount of MCL-PHA accumulating in mature shoots or seeds of transgenic plants expressing both thioesterase and PHA synthase remains relatively low ($\leqslant$0.6% dry wt) compared to PHB synthesis in plastids (**Table 1**). The reasons for this difference could be multiple, including differential activity of bacterial enzymes in various plant subcellular compartments, and the relative ability of the created and endogenous metabolic pathways to compete for the same substrates (acetyl CoA for PHB and 3-hydroxyacyl CoAs for MCL-PHAs).

Synthesis of a small amount (0.04% dry wt) of SCL-PHA containing 3-hydroxyhexanoic, 3-hydroxyvaleric, and 3-hydroxybutyric acids has also been reported in *A. thaliana* expressing the PHA synthase from the bacterium *A. caviae* in the peroxisomes. Whereas the 3-hydroxyhexanoic acid and 3-hydroxybutyric acid monomers are derived from the β-oxidation of fatty acids, the source of the 3-hydroxyvaleric acid is unknown.

PHA to Modify the Properties of Plant Fibers

Although the synthesis of PHA in plants is mainly focused on the extraction and utilization of the polymer as a thermoplastic or elastomer, a novel inventive use of PHA in plants was demonstrated by the expression of the PHB biosynthetic pathway in the cytoplasm of cotton fiber cells. In this system, the polymer is produced in plants only to change the physical properties of the fiber. Accumulation of 0.3% dry wt PHB in the cytoplasm of the fiber was shown to be sufficient to significantly decrease the rate of heat uptake and cooling of the fiber, resulting in a higher heat capacity and an improvement in the fiber insulating properties. Interestingly, in contrast to the large increase in accumulation of PHB in leaves of *A. thaliana* expressing the pathway in the chloroplast, expression of the PHB pathway in the plastid of the cotton fiber resulted in a lower amount of polymer, indicating a difference either in the number or metabolic activity of the plastids in fiber cells compared to leaves. It is also tempting to speculate whether the properties of wood, rubber, or starch could also be positively modified through the coaccumulation of PHAs.

Conclusions

A spectrum of PHAs ranging from the stiff PHB to the more flexible P(HB-HV) plastics and MCL-PHA elastomers have now been successfully synthesized in plants. The challenge for the future is to succeed in high-level production ($\geqslant$15% dry wt) of a limited number of useful PHAs without a decrease in crop yield. Although quite complex, this task appears to be feasible. Initial experiments with expression of protein based polymers in plants indicate that, like PHAs, engineering of various metabolic pathways is also likely to be necessary before high-level accumulation of polymer is accomplished. For both protein based polymers and PHAs, synthesis of polymers for the modification of the properties of plant products, such as fibers, are attractive alternatives to the extraction of polymers for industrial uses. PHAs and protein based polymers are only two out of a large group of polymers produced in nature that have interesting properties. In the future, novel industrially useful polymers could perhaps also be developed through the *in situ* modification of natural plant compounds.

See also: **Genetic Modification of Primary Metabolism**: Proteins. **Primary Products**: Proteins.

Further Reading

Daniell H and Guda C (1997) Biopolymer production in microorganisms and plants. *Chemistry and Industry* 14: 555–558.

Hinman MB, Jones JA, and Lewis RV (2000) Synthetic spider silk: a modular fiber. *Trends in Biotechnology* 18: 374–379.

Mittendorf V, Robertson EJ, Leech RM, *et al.* (1998) Synthesis of medium-chain-length polyhydroxyalkanoates in *Arabidopsis thaliana* using intermediates of peroxisomal fatty acid β-oxidation. *Proceedings of the National Academy of Sciences USA* 95: 13397–13402.

Mittendorf V, Bongcam V, Allenbach L, *et al.* (1999) Polyhydroxyalkanoate synthesis in transgenic plants as a new tool to study carbon flow through β-oxidation. *Plant Journal* 20: 45–55.

Nawrath C, Poirier Y, and Somerville C (1995) Plant polymer for biodegradable plastics: cellulose, starch and polyhydroxyalkanoates. *Molecular Breeding* 1: 105–122.

Poirier Y (2002) Polyhydroxyalkanoate synthesis in plants as a tool for biotechnology and basic studies of lipid metabolism. *Progress in Lipid Research* 41: 131–155.

Poirier Y, Nawrath C, and Somerville C (1995) Production of polyhydroxyalkanoates, a family of biodegradable plastics and elastomers, in bacteria and plants. *Biotechnology* 13: 142–150.

Steinbüchel A (1991) Polyhydroxyalkanoic acids. In: Byrom D (ed.) *Novel Biomaterials from Biological Sources*, pp. 123–213. New York: MacMillan.

Steinbüchel A and Füchtenbusch B (1998) Bacterial and other biological systems for polyester production. *Trends in Biotechnology* 16: 419–427.

Urry DW, McPherson DT, Xu J, *et al.* (1996) Protein-based polymeric materials: synthesis and properties. In: Salamone JC (ed.) *The Polymeric Materials Encyclopedia: Synthesis, Properties and Applications*, pp. 7263–7279. Florida: CRC Press.

Yamamoto H (1995) Marine adhesive proteins and some biotechnological applications. *Biotechnology and Genetic Engineering Reviews* 13: 133–165.

Photosynthesis

M J Paul and D W Lawlor, Rothamsted Research, Harpenden, UK

Introduction

Photosynthetic organisms use the physical energy of light to form reduced compounds ("reductants" ferredoxin and NADPH), and the energy source adenosine triphosphate (ATP): these are required for the synthesis of organic products, primarily sugars and amino acids, from inorganic starting materials, carbon dioxide (CO_2), and nitrate (NO_3^-) and sulfate (SO_4^{2-}) ions. This article considers the general mechanisms of photosynthetic CO_2 assimilation for plants with the C_3 type of assimilation (**Figure 1**), with brief comments on the C_4 type, and the feasibility of altering them by genetic modification to increase the photosynthetic rate (*A*). This depends on environmental conditions, particularly light, CO_2 concentration and temperature, and on many genetically determined subprocesses. These may be separated into: (1) light reactions, including light capture, electron transport (et), and synthesis of reduced pyridine nucleotide (NADPH) and adenosine triphosphate (ATP), and (2) CO_2 assimilation by the enzyme ribulose bisphosphate (RuBP) carboxylase/oxygenase (rubisco) in which RuBP, synthesized by the photosynthetic carbon reduction (PCR) cycle, is carboxylated (CO_2 added) forming sugar phosphates. These are metabolized to sucrose. In addition, reaction of RuBP with oxygen (O_2), a process called oxygenation, and also catalyzed by rubisco, results in photorespiration (PR), a major inefficiency in C_3 photosynthesis. Regulation of these subprocesses is complex, with extensive interactions which determine the response to environmental conditions, and also to use of assimilates by the plant for growth and respiration. The effects on photosynthetic rate of genetically modifying amounts, activities and characteristics of enzymes and other components involved in carbon assimilation are outlined, and the interaction with environmental conditions discussed.

The Need for Improved Photosynthesis

Photosynthesis is fundamental to all biological activity on the planet and underpins all agricultural production of crops, and thereby also animals, for the world's human population of over 6 billion. Selection breeding has increased yield potential of many crops but has not increased the potential rate of CO_2 assimilation per unit leaf area (A_{max}) nor potential biomass production, rather it has increased harvest index (yield of required product/total biomass ratio). Improved agronomic practices, such as increasing fertilization and irrigation and pest and disease control, have maximized the actual rates of CO_2 assimilation per unit leaf area (*A*) and the leaf area index (leaf area per unit area of ground surface) to the point where biomass and yields in intensive agriculture are approaching their maximum. To increase yields further, A_{max} and *A* must be increased. A major aim of current studies of photosynthesis is to understand the mechanisms and to manipulate them and so to increase A_{max}; another aim is to control the types and amounts of products formed.

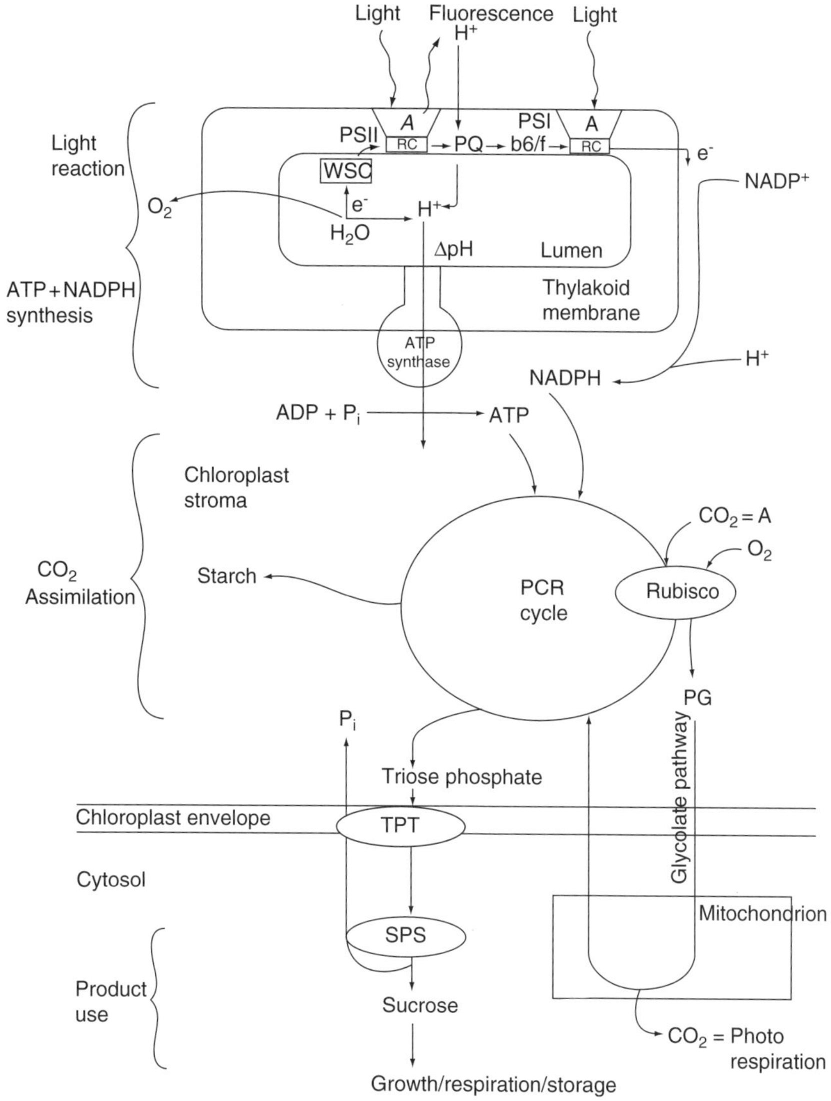

Figure 1 Schematic of photosynthetic metabolism in a C_3 plant, indicating some of the principal components. Genetic modifications required to increase photosynthetic CO_2 assimilation (*A*) depend on the environment, principally light and CO_2 supply. In weak light, synthesis of ATP and NADPH is limiting so the capacity to increase light capture etc. is essential to increase *A*. In strong light, CO_2 and the activity of rubisco are limiting, so requiring increase. If the light reactions and CO_2 assimilating systems were increased then the capacity to use assimilates would have to be increased. TPT, triosephospate–phosphate transporter; WSC, water splitting complex; *A*, photosynthetic CO_2 assimilation; PCR, photosynthetic carbon reduction.

Role of Genetic Modification

Genetic modification (GM), using molecular biological techniques, is widely regarded as having the potential to increase *A*. This is a considerable challenge, however. Basic photosynthetic mechanisms are very conservative, and the subprocesses are very tightly interconnected and highly regulated within cells, but there is large quantitative variation between species, suggesting that the processes may be manipulated. Only in the last decade has GM been attempted to alter *A*: largely *A* has been decreased with few reports of increases. Great advances in understanding of regulation of *A* have already accrued from specifically perturbing amounts and activities of enzymes combined with biochemical and physiological analysis of GM plants. However, further understanding of basic mechanisms, particularly their regulation, is needed before effective increase in *A* can be achieved.

Genetic Modification Approaches

To increase A_{max}, it will be necessary to increase the amounts and (or) activities, of components (e.g., enzymes, electron transport), and probably also to improve their characteristics by altering the genome. The genome sets the potential capacity of *A*. Current

GM techniques (e.g., antisense methods), allow the total amount of components to be changed, to decrease the amount of specific proteins. Also, enzyme characteristics may be altered, although this generally decreases activity and efficiency and rarely increases them. Probably, as is discussed, many components must be altered to improve *A*. It is also possible to introduce new genes, from other organisms: incorporation of enzymes of C_4 photosynthesis into C_3 plants may be a way of decreasing the loss of CO_2 and of improving *A*, especially under adverse conditions. Such changes are not possible by selection breeding, but the difficulties experienced in experimental studies suggest that the complexities of cell development and metabolism have been underestimated. Often, increased number of gene copies does not result in more active protein because of regulatory processes in gene expression and protein synthesis. There may be difficulties of integrating novel proteins (for which the required chaperone systems are lacking) into the functional cell. In addition, there is substantial impact of the environment – light, CO_2 and water supply, temperature – on gene expression. This genotype × environment interaction (G × E) is very important for optimization of processes to particular environments and must be carefully considered as an essential component of genetic modification.

Mechanisms of Photosynthesis

Figure 1 summarizes the mechanism and interactions between subprocesses.

Light Reactions, Electron Transport, NADPH and ATP Synthesis

Photons of suitable wavelength interact with electrons in chlorophylls of the light-harvesting antenna complexes in thylakoid membranes of chloroplasts (**Figure 1**). The resultant excitation energy is transferred to the reaction centers (RC) of photosystems (PSII) and excites electrons (e^-) there. The e^- passes to an acceptor, and then down an electron transport (et) chain to another type of photosystem (PSI), where it is again energized. Eventually e^- and protons (H^+) pass to $NADP^+$, reducing it to NADPH. The empty ("open") PSII RC is filled by e^- from water, which is oxidized, releasing O_2. Protons accumulate in the thylakoid lumen, and are also "pumped" in by et, forming an H^+ concentration gradient (ΔpH) across the thylakoid membrane. This ΔpH provides the energy for ATP synthesis, by the enzyme complex ATP synthase or coupling factor. Thus, light drives the synthesis of NADPH and ATP, which are essential for CO_2 assimilation.

Light responses of photosynthesis The response of *A* to light may be generalized for C_3 plants (see below) as in **Figure 2A**: In darkness there is no photosynthesis, and respiration produces CO_2, which is emitted. With increasing flux of photons of correct wavelength (400–700 nm: PPF = photosynthetic photon flux), *A* increases until it equals respiration (which continues in the light). With further increase in PPF, *A* increases linearly. At yet larger PPF, *A* slows until a plateau is reached (light saturation, A_{max}). The magnitude of *A* depends on other conditions: when $[CO_2]$ in the atmosphere (C_a) is large ($> 1000\,\mu l\,l^{-1}$) and O_2 small ($0.21\,l^{-1}$), then *A* and A_{max} are larger than in the current atmosphere ($360\,\mu l\,l^{-1}\,CO_2$ and $0.021\,l^{-1}\,O_2$) because of the characteristics of rubisco, as outlined later.

The rate of et depends not only on the PPF, but also on the rate at which ATP and NADPH are consumed and ADP and $NADP^+$ are recycled, so using ΔpH and e^-. Recycling depends on *A*, which is

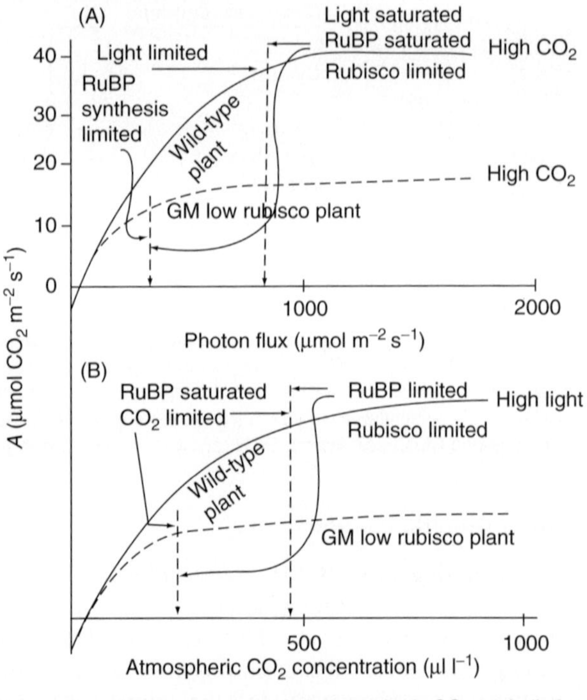

Figure 2 (A) Light response of photosynthetic CO_2 assimilation related to light (photon flux) at large $[CO_2]$ for a wild-type and genetically modified (decreased rubisco) plant, showing the effects of environmental conditions on the response and the effects of reducing the amount of rubisco. (B) Response of A to $[CO_2]$ in high light for wild-type and decreased-rubisco plants. In both, limitations caused by light (and thus RuBP supply) and by CO_2, and the consequences of decreasing rubisco capacity are shown.

the primary use of e^- and energy, although other processes, predominantly PR, also use e^- which is transferred to O_2. To achieve very high *A* requires that the light reactions capture sufficient photons to generate large amounts of NADPH and ATP, and that the CO_2 fixation mechanisms have large capacity. The balance between photon capture, et, and reductant and ATP synthesis is very variable between plant species. Some species are genetically adapted to large PPF (obligate sun plants), with relatively small capacity for photon capture but large capacity for et and synthesis of ATP and NADPH, plus also for CO_2 assimilation. Such plants are inefficient at low PPF. Others are adapted to small PPF (obligate shade plants), with large capacity for photon capture but less for et and *A*; they are inefficient in strong sunlight. As they capture but cannot use the energy in bright light, they may be damaged by photoinhibition. Thus, different plant species exhibit optimization of components and capacities for capturing and using light in different conditions. C_4 plants generally have a large capacity for et, generation of ATP and NADPH and for CO_2 assimilation and so can better exploit bright light.

When energy capture increases the potential e^- flux but *A* does not increase (this is particularly the case in C_3 plants due to CO_2 limitation), consumption of e^- decreases, and the excess captured photon energy is dissipated as heat by the xanthophyll cycle. This process is called nonphotochemical quenching (qN) and is measured via chlorophyll a fluorescence: qN decreases photoinhibition. Thus, for very efficient use of light by leaves over a long period, not only must the light reactions, et and ATP synthesis be very rapid, but CO_2 assimilation and energy dissipation also. Different steps in light capture, et, and NADPH and ATP synthesis must be regulated so that the amounts and activities of components of energy capture are related to the energy consumption in *A* (which depends on the capacity of the enzyme systems, and on use of products by metabolism). Energy dissipation then serves as a "safety valve" to achieve a functional balance.

How *A* depends on the capacity of et, and the considerable interaction between processes, has been shown by genetic modification. By decreasing the Rieske FeS component of the cytochrome b_6/f complex, part of the et chain, NADPH synthesis is slowed, decreasing *A* (**Figure 2A**). The blockage to et comes after the plastoquinone pool, so the pool becomes fully reduced, whilst synthesis of ATP is decreased. Lack of NADPH and ATP slows *A* and decreases qN quenching, so captured light-energy is not dissipated effectively, leading to photoinhibition. Changes at other points have different consequences. Alterations in the amount of ATP synthase (by antisense reduction of the δ-subunit) and its activity decreases ATP synthesis, thus slowing the efficiency of PSII and et but, unlike the loss of the cytochrome b_6/f function, it does not decrease but increases ΔpH so that qN increases. However, the result of all these changes is decreased *A*. Effects are particularly pronounced at large PPF because e^- fluxes are much greater than at small PPF. The extent of the genetic changes is often not proportional to the response of the whole system. For example, when the protein of the Rieske iron–sulfur (Fe–S) component of et was decreased (by up to 90%) the amount synthesized was unaltered and there was no effect on et or *A*. Yet other studies found that when the protein was decreased by 60–86% there were large effects on et. The effect depends on the PPF during growth: large PPF stimulates production of the protein via the redox state of electron transport components. Often such transformations are unstable, changing with age.

Carbon Dioxide Assimilation

The Calvin or photosynthetic carbon reduction (PCR) cycle is the fundamental, autocatalytic, CO_2 assimilatory process in all photosynthetic organisms. In the basic form (**Figure 3** and **Table 1**), called C_3 photosynthesis, a three-carbon compound, 3-phosphoglycerate (3-PGA) is the first stable product. C_3 photosynthesis occurs in many crop plants, e.g., wheat (*Triticum aestivum*), rice (*Oryza sativa*), potatoes (*Solanum tuberosum*), yams (*Dioscorea* spp.), and sugar beet (*Beta vulgaris*). Two other types of photosynthesis, C_4 and crassulacean acid metabolism (CAM), occur. In both, the CO_2 concentration at which the PCR cycle operates is increased by CO_2 concentrating mechanisms of different types based upon phosphoenolpyruvate carboxylase (PEPc), which has high affinity for CO_2. These concentration mechanisms are associated with anatomical, biochemical, and physiological differences of considerable magnitude, particularly a "bundle sheath" in which rubisco is located (see below). C_4 and CAM do not replace the PCR cycle, but rather add to it. In C_4 photosynthesis the first compound formed is a four-carbon organic acid, hence the name; C_4 crops include corn (*Zea mays*; maize), *Sorghum*, and sugar cane (*Saccharum officinarum*), which are very productive. CAM occurs in few crops (e.g., pineapple (*Ananas comosus*)). C_4 photosynthesis is very characteristic of grasses of a hot, dry areas, and CAM plants are frequently succulents of a hot, dry environment: both types use water more efficiently than C_3 plants.

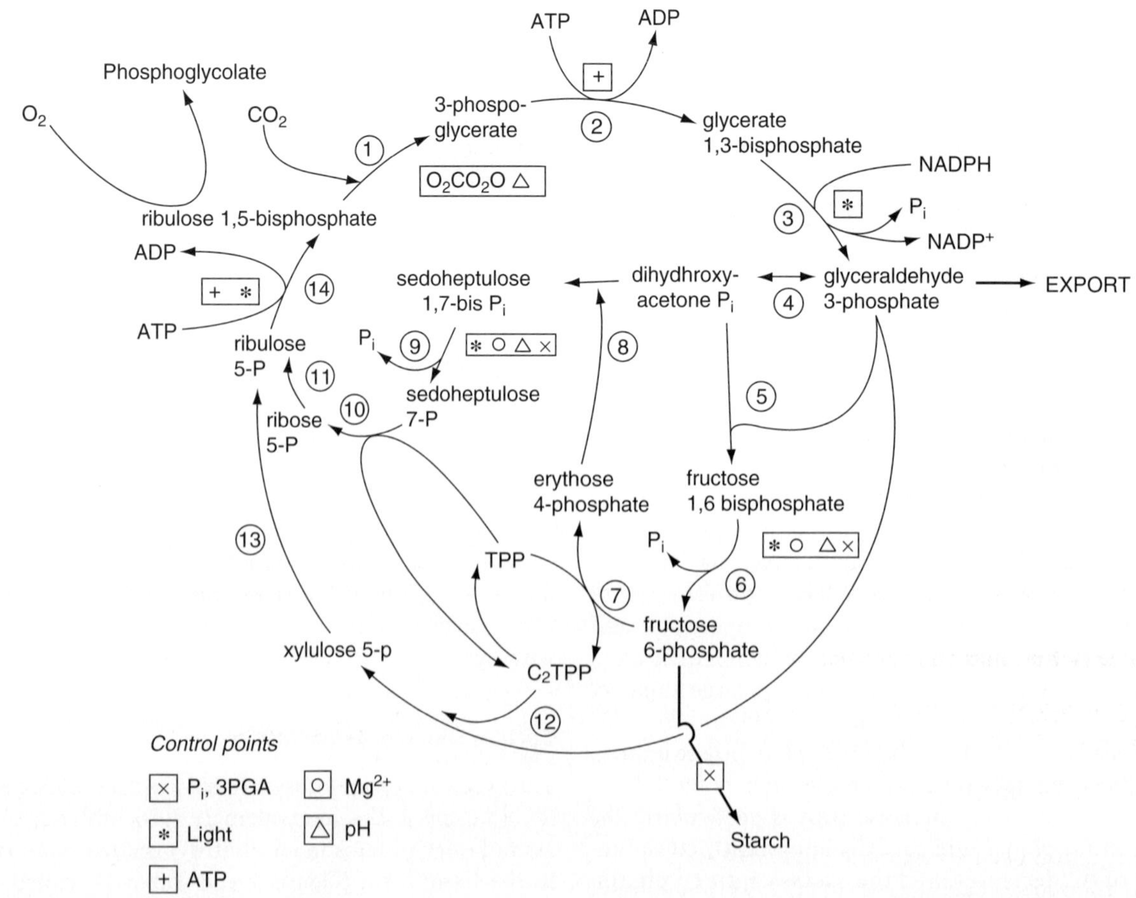

Figure 3 The photosynthetic carbon reduction cycle, with reactions numbered as in **Table 1**, where the enzymes are shown. Carboxylation and oxygenation of RuBP are shown in the upper part of the figure, RuBP regeneration in the lower. After Lawler DW (2001) *Photosynthesis*, 3rd edn. Oxford: BIOS Scientific Publishers, with permission.

Response of A to CO_2

The response of C_3 leaves to C_a is shown in **Figure 2B** under saturating light. Without CO_2, leaves emit CO_2 from dark respiration and photorespiration. As C_a increases so does A: at the CO_2 compensation point, A equals respiration. After this A rises linearly before slowing with further increase in CO_2, reaching a plateau as the full capacity of the biochemical reactions is reached (A_{max}). If the oxygen concentration ($[O_2]$) is decreased A increases because PR (see below) is inhibited. To increase A requires an increase in carboxylation efficiency or in A_{max} or both. With the current C_a, A_{max} is limited by inadequate CO_2 in the chloroplast ($[C_c]$) due to both the low C_a and stomatal restriction on CO_2 diffusion. So the progressive increase in C_a which is occurring due to burning of fossil fuels will increase A. In the long term, at very large C_a, accumulation of carbohydrates in the plant caused by large A relative to demand may result in decreased capacity for photosynthesis (so-called down regulation).

The PCR Cycle

The chemistry of CO_2 assimilation is described by:

$$3CO_2 + 9ATP + 6NADPH + 5H^+ = C_3H_5O_3P + 9ADP + 8P_i + 6NADP^+ + 3H_2O + 468\ \text{kJ mol}^{-1} \quad [1]$$

The outline of this autocatalytic process, with 11 enzymes in a 13-step carboxylation cycle is given in **Figure 3** (the component enzymes are listed in **Table 1**). The enzymes and metabolites are in the chloroplast stroma, where ATP and NADPH from the light reactions, together with CO_2, are the substrates. The function of the PCR cycle is regulated by the characteristics of the individual enzymes, by changes in concentrations of substrates and products and those of other reactions (metabolic effectors).

Table 1 Calvin cycle enzyme properties

Reaction number	*Enzyme*	*Free energy change* (ΔG^s)	*Increased by*	*Decreased by*
1	Ribulose bisphosphate carboxylase/oxygenase	−41	High pH, CO_2, Mg^{2+}, FBP	Gluconate, SBP?
2	Phosphoglycerate kinase	+16	3PGA, ATP, Mg^{2+}	DGPA, ADP
3	Triosephosphate dehydrogenase (NADP glyceraldehyde P dehydrogenase)	−6.7	Light, NADPH, ATP	P_i? DHAP
4	Triosephosphate isomerase	−7.5	Alkaline pH	PEP, RuBP, FBP glycolate P, PGA
5,8	Aldolase	−1.6	High GAP/DHAP ratio, pH 8	RuBP, ADP, PGA
6	Fructose bisphosphatase	−27	High pH, Mg^{2+}, ATP, reductant, light	P_i
7,10,12	Transketolase	−5.9	Mg^{2+}, high pH	
9	Sedoheptulose bisphosphatase	−29.7	Light, pH, Mg^{2+}	P_i
11	Ribose-5-phosphate isomerase	−0.5	Freely reversible, pH 8	
13	Ribulose-5-phosphate 3 epimerase	−0.6	Freely reversible	
14	Ribulose-5-phosphate kinase	−15.9	Reductant, ATP, energy charge?, pH>7	ADP

Modified from Lawler DW (2001) Photosynthesis, 3rd edn. Oxford: BIOS Scientific Publishers, with permission.

Carboxylation and Oxygenation by Rubisco

The primary step is the reaction of CO_2 with RuBP, catalyzed by rubisco, producing 3-PGA;

$$\text{RuBP} + CO_2 = 2 \times \text{3-PGA} \qquad [2]$$

The reaction occurs in all photosynthetic organisms, with rubisco comprising up to 50% of the soluble protein (6 mg mg^{-1} chlorophyll or 15 g m^{-2} leaf) in leaves of C_3 plants.

Rubisco also catalyzes oxygenation of RuBP by O_2 giving 3-PGA and phosphoglycolate (PG):

$$\text{RuBP} + O_2 = \text{PG} + \text{3-PGA} \qquad [3]$$

The reaction is very important, using RuBP to make PG, which is metabolized outside the chloroplast by the glycolate pathway (**Figure 1**), with release of CO_2 in PR, which substantially decreases *A*.

Oxygenation occurs at the same enzyme active site as carboxylation. Carboxylation requires large [C_a] (>800 μl l^{-1}) for saturation and thus, in the current atmosphere of 360 μl l^{-1} CO_2, it is not saturated, especially as [C_i] is only 50% of C_a due to stomatal and mesophyll resistances to CO_2 transport. The oxygenase requires large [O_2] for saturation. If [C_a] is high and [O_2] low, carboxylation is much greater than the oxygenation, but as O_2 increases so oxygenation, and hence PR, become more important. The characteristics of rubisco differ between higher plants and cyanobacteria and algae, but the causes are not understood. They probably relate to very small differences in quaternary structure. Differences between rubisco of higher plant are small. Eliminating (or decreasing) oxygenation is desirable but not achieved during evolution and probably impossible as the catalytic site is the same. Changing amino acids in rubisco has generally decreased carboxylation relative to oxygenation.

Rubisco is a large, protein complex with eight reaction sites per complex, and a very slow reaction rate; an active site turns over about every 2.5 s at 25°C. Thus A_{max} of 40 μmol CO_2 m^{-2} s^{-1} in saturating light and CO_2 (achieved by many C_3 plants) requires a large number of active sites per unit leaf area, hence a large number of rubisco molecules and large mass of rubisco. This demands much nitrogen (N) for protein synthesis and is a major reason for the large N requirement of crops. Because C_4 plants have a mechanism which gives large [C_c] around rubisco in the leaf bundle sheath, oxygenation and PR are prevented, and high rates of *A* occur. This reduces the amount of rubisco required and the demand for N, compared to C_3 plants.

Rubisco is the main limitation to *A* in the current atmosphere. If PR could be prevented, then C_3 plants would increase *A* and require less N. Raising C_i and C_c, as occurs in C_4 plants, overcomes the problem and the future increase in C_a due to burning of fossil fuels etc. will stimulate *A* and production of C_3 plants.

Alteration of Rubisco Amounts and Characteristics

Amounts of rubisco have been decreased by antisense methods, and have been very important in showing the complexity of photosynthetic regulation and the regulation of *A*. When generation of ATP and NADPH is limited by light, a large proportion (60–80%) of rubisco activity may be removed without affecting *A*. However, with a large PPF, *A* is much smaller in leaves with decreased rubisco (**Figure 2B**), because there are too few enzyme sites to use all the RuBP synthesized. Thus, there is a strong interaction of enzyme activity and environment. Effects of increasing Rubisco have not been tested experimentally, but is unlikely to be effective unless RuBP supply is also increased. However, GM tobacco with altered trehalose phosphate synthase has more Rubisco and larger A without increased capacity of RuBP generating enzymes, suggesting that alterations are possible. The problem of low CO_2 and high O_2 would remain.

Rubisco Activation

Rubisco requires activation by Mg^{2+} and CO_2, and the presence of another protein called rubisco activase increases the rate of catalysis. Activase uses ATP to structure the active site and remove inhibitors which accumulate there; when it is decreased rubisco is very slow or inhibited. When activase is decreased (compared to normal wild-type plants) by antisense methods, *A* is much decreased, especially under varying and high light. Probably there is an optimum between activase and rubisco amounts and activities, which depends on CO_2 and PPF, and is sensitive to conditions (e.g., ATP concentration) in the plant.

Prevention of Photorespiration

PR can be prevented by large $[CO_2]$ and low $[O_2]$ but cannot be stopped by inhibiting the glycolate pathway (e.g., by GM of glycine decarboxylase) as PG accumulates and recycling of carbon is prevented so RuBP is not made and *A* is inhibited. Alternative use of PG might decrease PR, but such GM approaches are inadequately tested. As the changes in metabolism required are substantial they may be unsuccessful. Increasing $[C_i]$ in C_3 plants by introducing a "CO_2 pump" as in C_4 plants, by inclusion of PEPc, or by incorporating specialized CO_2 concentrating structures, called carboxysomes, from algae, is also a potential way of reducing PR. Alternatively, refixation of CO_2 by positioning the glycolate pathway reactions close to chloroplasts, as in C_3–C_4 intermediate plants, may be advantageous. C_3–C_4 intermediates are more efficient at reassimilating CO_2, especially under hot, dry and bright conditions. These methods of decreasing PR will require substantial alterations in the genome, gene expression, etc. to give the required anatomical changes, and are being assessed.

As PR uses energy (as reductant), it is actually an essential process for C_3 plants with current C_a under high light, particularly when stomatal conductance is small and C_i low (as under water stress conditions), when it helps to dissipate excess reductant. Thus, mechanisms that decrease PR may lead to energy imbalance and damage.

Regeneration of RuBP

The 3-PGA from the rubisco reaction is used to regenerate RuBP: the excess is metabolized to form a temporary store of starch in the chloroplast, or is used for sucrose synthesis in the cytosol. The rate of cycle turnover depends on rates of NADPH and ATP synthesis, on enzyme activities, and on the export and use of carbon, as well as on CO_2 supply. There is great complexity in regulation, with strong interactions between fluxes of carbon, the size of the pools of intermediates and the characteristics, amounts and activation states of the enzymes. Such complexity stabilizes the system (dynamic homeostasis), so that *A* and growth can function over a wide range of PPF, CO_2, and temperature. An important regulatory mechanism of the PCR cycle is control by PPF. Several enzymes of the PCR cycle, fructose bisphosphatase (FBPase), sedoheptulose bisphosphate (SBPase), $NADP^+$glyceraldehyde phosphate dehydrogenase (GAPDH), phosphoribulokinase (PRK), and ATP synthase, are activated via the redox state of proteins called thioredoxins which are reduced by et, allowing the PCR cycle to start as PPF increases.

There are several enzymes, in addition to rubisco, which may have some control over *A* depending on condition. Amounts of phosphoglycerate kinase (PGK), GAPDH, FBPase, SBPase, and PRK have been decreased by GM and decrease *A*. Greater catalytic activity per unit of protein and increased concentrations of substrates, etc. often compensate for loss of protein so that substantial decreases may be needed before *A* decreases. GM of enzymes not considered as regulating the PCR cycle has shown that the system is not fully understood, e.g., decreasing activity of aldolase has substantial effects. Although genetic modification to decrease any enzyme will eventually stop *A*, the idea that particular enzymes exert particular control over the rate of cycle turnover is too simplistic. Rather, the reactions in the cycle regulate in a more dynamic way, depending on the amount and activity of the enzymes, and on substrate and effector concentrations, which depend on environmental conditions. Hence, a number of the fully reversible PCR cycle enzymes, once regarded as having no regulatory functions, can exert control under particular conditions.

Transport of Assimilates out of the Chloroplast

PCR cycle products are transported from the chloroplast stroma (although transitory starch may accumulate there before use when assimilation slows or stops) across the envelope, to the cytosol. Transport is by specific translocators which regulate many aspects of cell metabolism by affecting concentrations of substances in organelles. The triose phosphate/inorganic phosphate translocator (TPT) controls the counterexchange of triose phosphates with P_i from the cytosol. A 30% decrease in TPT, by antisense, increased 3-PGA concentration and decreased P_i in the chloroplast, stimulating starch synthesis there. In darkness, starch was remobilized to glucose and then transported to the cytosol, fully compensating for the much smaller sucrose concentration, so that plants with low TPT activity grew at the same rate as wild-type. A similar effect resulted from decreasing the starch synthesizing enzyme ADP glucose pyrophosphorylase. However, the mechanism is quite different: less starch was made and more triose phosphate exported to make sucrose, so growth did not suffer. However, if both the TPT and ADP glucose pyrophosphorylase are inhibited, then growth is decreased as neither starch nor sucrose is available at any time.

Sucrose Synthesis

Triosephosphate is converted to sucrose phosphate by sucrose phosphate synthase (SPS) in the cytosol. SPS has complex characteristics and is regulated by P_i and G6P, and also via phosphorylation/dephosphorylation (addition and removal of P) by a protein kinase, which uses ATP. SPS is at the interface between sink and source in the plant and may regulate sucrose supply to the sink; probably its complex regulation reflects the role of sucrose in many processes (e.g., osmotic regulation) as well as being a carbon source. Accumulation of sucrose in cells balances supply and demand between source and sink, but if persistently large then *A* falls due to feedback mechanisms, including reduction in synthesis of photosynthetic enzymes. If SPS activity is limited, then increasing it by GM should stimulate sucrose supply to sinks, thus increasing the growth; such effects have been observed but not consistently. Probably, stimulation will not occur when demand (growth, respiration) is already saturated or when the supply of other assimilates (e.g. amino acids) is limiting and increasing or decreasing SPS would then unbalance the system.

Regulation of CO_2 Assimilation and Metabolism

Transgenic plants have enabled quantitative analyses of how the physiological state of the plant, interacting with the environment, affects *A*. Free energies of enzyme reactions were thought to show sites of regulation in the PCR cycle: reversible reactions, near equilibrium, with very small free energy change and with enzyme capacity (activity per unit of protein × amount of protein) exceeding the flux, were considered not rate limiting. Reactions far from equilibrium, with large negative free energy change and total activity close to the rate of the cycle (particularly if regulated by effectors or "signals" of some type, such as those triggered by light) were considered limiting. Analysis of GM plants has shown that this does not accurately describe regulation of the cycle and associated processes. Decreasing amounts of different enzymes had unexpected results: some not considered important decreased *A* substantially (e.g., aldolase), but others known to be highly regulated and thought to be important could be altered greatly with little effect (PRK). Hence, simply increasing the amount of one enzyme, even one as important as rubisco, may not increase *A*, although evidence from plants genetically modified with trehalose phosphate synthase suggests that this conclusion is premature.

Rate limitation Essentially *A* depends on the slowest or rate-limiting step, as suggested by the Blackman "law of limiting factors," but with great complexity, so that one or other subsystem may regulate and – within each – control may be determined by one or several components depending on conditions. When PPF is limiting, energy limits synthesis of NADPH and ATP (light activation of enzymes requires only low PPF) and so controls RuBP synthesis and *A*. At high PPF, NADPH and ATP are abundant and C_i limits (this depends on C_a and stomatal conductance). If C_i increases then activity of rubisco limits. Under some conditions (e.g., slow growth) sink demand may limit *A*. Some models incorporate RuBP regeneration and rubisco kinetics, and successfully simulate photosynthesis.

The concept of "limit of compensatory ability" (LCA) for an enzyme (**Figure 4**) states that if activity is above a particular value then the flux will be insensitive to decreasing activity but decrease when activity falls below it. The LCA will depend on the conditions and total flux. The response of *A* to changing activity of a particular enzyme depends on the total activity and how much control it exerts.

Another approach to understanding regulation and control of *A* is metabolic flux control analysis (MCA). This assesses, quantitatively, the contribution of reactions to flux through the PCR cycle. It does not assume that *A* is determined by a limiting factor on grounds of enzyme characteristics, but evaluates the contribution of all, including the

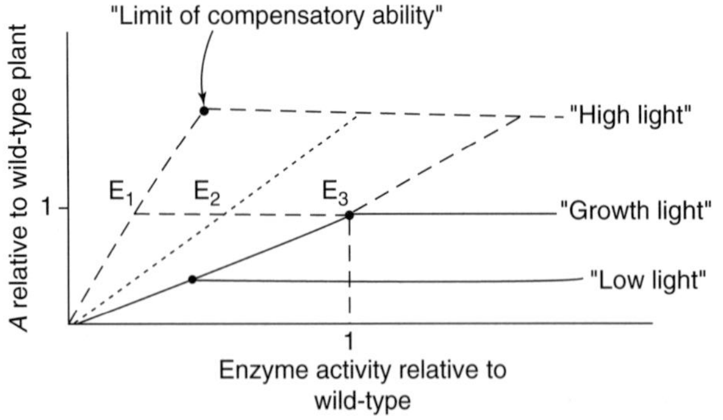

Figure 4 Illustration of the relation between activity of three enzymes, E1, E2, and E3, relative to the wild-type plant grown under three different light conditions. For E1, at which the plant is grown, substantial decrease in activity is required before *A* is decreased, even greater decrease is needed at low light but much less at high light. A small change in E3, in contrast, affects *A*, even in growth light. Increasing the enzymes would not increase *A* in low light but E3 would have to be increased in high light, although E2 would just be sufficient. This simple model does not include any nonlinear responses or regulatory features which would alter the responses. Modified from Lawler DW (2001) Photosynthesis 3rd edn. Oxford: BIOS Scientific Publishers, with permission.

influence of changing substrates and products, etc., and quantifies the metabolic flux for the whole system and the consequences of changing enzyme activity and metabolites. MCA shows that photosynthetic metabolism exhibits "distributed control," which changes with conditions, or with genetic modification. Application of MCA to analyze the limitations in the PCR cycle by rubisco activity at low CO_2 illustrates the approach. When PPF is small, et limits RuBP regeneration and rubisco and C_i are not limiting (low flux control coefficient, CC). With increasing PPF, *A* rises, until CO_2 or rubisco limits (large CC). In an environment to which the plant is adapted, rubisco has a low CC, which is balanced with that of RuBP regeneration. Substantial decrease in rubisco activity of transgenic tobacco (*Nicotiana tabacum*) was required before it became limiting (large CC) in part because RuBP increased and compensated. However, rubisco was limiting (high CC) at smaller PPF than in the wild-type.

From these analyses a consensus emerges about control of the PCR cycle. Different combinations of enzymes are important under particular conditions: identifying a "limiting enzyme" from reaction free energy is not valid and the control changes with conditions. Probably, particular combinations of enzyme characteristics are important under specific conditions, and the characteristics must change to allow the flexibility in metabolism required to cope with fluctuating conditions and varying demand for assimilates from growth, etc.

Potential for Genetic Modification of Photosynthesis

Photosynthetic systems are approximately optimized in composition and function to deal with a range of environmental conditions and also (often disregarded) plant requirements. Absence of a single controlling or limiting enzyme has implications for genetic engineering to increase *A*, as changing a particular step is unlikely to achieve the goal, as overall control (and "limitation") will then reside in other parts of the system. Altering activity is not itself easy as it depends on increasing gene expression (largely by increasing copy number rather than stimulating rate of expression of existing genes) and processes such as cosuppression may prevent this. Also, regulation of protein turnover may influence the amount of activity. Components of the system will not interact linearly: change in one part will not cause a proportional increase in another. For example, changing the amount of enzyme may not result in a linear change in the activity or of fluxes; activation state of enzymes may change as concentrations of effectors, substrates and products vary.

Just as decreasing activity is problematic, so is increasing activity, but is not well studied. Theoretically, not only are increased capacities of several PCR enzymes required but also of light harvesting and et to increase ATP and NADPH supply. Additionally, appropriate adjustments to assimilate transporting and consuming systems would be needed. The magnitude of adjustments would depend on the magnitude of changes and on environmental conditions. One approach to increasing *A* is to stimulate the activity of the whole system by increasing the capacity of all components, by introducing cassettes of genes, to increase copy number. Modifications by introduction of novel genes may have unexpected consequences. It has recently been shown that the expression of trehalose phosphate synthase in tobacco substantially increased *A* by increasing rubisco, and by affecting carbohydrate utilization and sugar signaling mechanisms. Thus, the potential exists for modifying photosynthesis and there are grounds for pursuing it: modification of such feedback mechanisms may be the best way forward.

Carbon assimilation is frequently and substantially limited by deficiency of nutrients and water, adverse temperatures, and by pests and diseases. Amelioration (using well-established technology even with limited resources) could produce substantial gains in photosynthesis. Using GM to increase *A* to boost photosynthesis and production of current staple crops, under very limiting environmental conditions and supply of resources, will require deeper investigation particularly of how photosynthesis integrates with other whole-plant processes. Potential exists for modifying photosynthesis: the task is considerable as extensive modifications (and hence large input of still uncertain technology, resources and expertise) may be required, but the gains may be large.

See also: **Energy Crops**: Biofuels from Crop Plants. **Photosynthesis and Partitioning**: C_3 Plants; C_4 Plants; CAM Plants; Photoinhibition; Photorespiration; Sources and Sinks. **Plants and the Environment**: Phytoremediation.

Further Reading

Atwell BJ, Kriedemann PE, and Turnbull CGN (1999) *Plants in Action: Adaptation in Nature, Performance in Cultivation*. South Yarra, Australia: Macmillan Education.

Barber J (ed.) (1994) *Advances in Molecular and Cell Biology, vol. 10, Molecular Processes of Photosynthesis*. Greenwich: Jai Press.

Caemmerer S von (2000) *Techniques in Plant Science, vol. 2, Biochemical Models of Leaf Photosynthesis*. Collingwood: Australia, CSIRO Publishing.

Flügge U-I (1999) Phosphate translocators in plastids. *Annual Review of Plant Physiology and Plant Molecular Biology* 50: 27–45.

Fridlyand LE, Backhausen JE, and Scheibe R (1999) Homeostatic regulation of enzyme activities in the Calvin cycle as an example for general mechanisms of flux control: what can we expect from transgenic plants? *Photosynthetic Research* 61: 227–239.

Huber SC and Huber JL (1996) Role and regulation of sucrose-phosphate synthase in higher plants. *Annual Review of Plant Physiology and Plant Molecular Biology* 47: 431–444.

Lawlor DW (2001) *Photosynthesis*, 3rd edn. Oxford: BIOS Scientific Publishers.

Lawlor DW and Paul MJ (2000) Genetic modification of photosynthesis. *Journal of Experimental Botany* 51 (GMP Special Issue).

Leegood RC, Sharkey TD, and von Caemmerer S (eds) (2000) *Photosynthesis: Physiology and Metabolism*. Dordrecht: Kluwer Academic.

Parry MAJ, Loveland JE, and Andralojc PJ (1999) Regulation of rubisco. In: Bryant JA, Burrell MM, and Kruger NJ (eds) *Plant Carbohydrate Biochemistry*, pp. 10–18. Oxford: BIOS Scientific Publishers.

Paul MJ, Pellny T, and Goddijn O (2001) Enhancing photosynthesis with sugar signals. *Trends in Plant Science* 6: 197–200.

Poolman MG, Fell DA, and Thomas S (2000) Modelling photosynthesis and its control. *Journal of Experimental Botany* 51: 319–328.

Reynolds MP, van Ginkel M, and Ribaut J-M (2000) Avenues for genetic modification of radiation use efficiency in wheat. *Journal of Experimental Botany* 51: 459–473.

Ruelland E and Miginiac-Maslow M (1999) Regulation of chloroplast enzyme activities by thioredoxin: activation or relief from inhibition? *Trends in Plant Science* 4: 136–141.

Schulze E-D and Caldwell MM (eds) (1994) *Ecophysiology of Photosynthesis*. Berlin: Springer-Verlag.

GENETIC MODIFICATION OF SECONDARY METABOLISM

Contents

Alkaloids

R M Twyman, University of York, York, UK
R Verpoorte, Leiden University, Leiden, The Netherlands
J Memelink, Leiden University, Leiden, The Netherlands
P Christou, Fraunhofer Institute for Molecular Biology and Applied Ecology, Schmallenberg, Germany

Alkaloids

Alkaloids are complex organic molecules containing a heterocyclic nitrogen ring, which have been widely exploited for their diverse pharmacological properties. Such compounds are produced by many different organisms, including animals and microbes, but a particularly diverse array of alkaloids is produced by plants. Alkaloids are often isolated from plants and used as pure compounds, although they may also be administered as crude extracts. Approximately 10% of all plant species are thought to produce alkaloids as secondary metabolites, where they function predominantly in defense against herbivores and pathogens. Although such ecochemical functions are important in their own right, most of the interest in alkaloids stems from their potent and specific pharmacological effects in humans, and this is also the reason why metabolic engineering of alkaloid biosynthesis is an important area of research. The structures of over 16 000 different alkaloids have been elucidated, and some important examples are listed in **Table 1**. The chemical structures of selected alkaloids are illustrated in **Figure 1**.

Reasons for Engineering Alkaloid Biosynthesis

Due mainly to their useful pharmacological properties, alkaloids are often isolated from plants and produced commercially as fine chemicals. The complexity of the molecules means that they are in most cases impossible to produce by total chemical synthesis, so extraction from the source plant remains the most economically viable strategy. However, plants often produce complex mixtures of alkaloids at low levels, with the result that commercially produced specific alkaloids are very expensive. As the genetic manipulation of plants becomes more sophisticated, research has focused on the engineering of alkaloid biosynthesis to generate transgenic plants or cell lines that overproduce specific alkaloids. This can be achieved by increasing the synthesis of a particular alkaloid and/or inhibiting the synthesis of related compounds to increase the ease of purification.

Complexities of Alkaloid Biosynthesis

Before considering engineering strategies to exploit alkaloid biosynthesis it is necessary to have a thorough understanding of the endogenous biosynthetic pathways. Despite the interest in engineering alkaloid biosynthesis, progress in this area has been slow primarily due to the lack of knowledge of such pathways and their regulation. Predominantly, this is because alkaloid biosynthesis involves very complex multistep pathways, which have been difficult to dissect. Traditional pathway analysis has involved feeding plants with radiolabeled metabolic intermediates and determining whether the label was incorporated into particular end products. This type of analysis is frustrated by the difficulties involved in synthesizing some intermediates and the fact that, in many cases, there is a significant degree of cross-talk between pathways, such that the label can be incorporated into various products via different routes. A major step towards elucidating a metabolic pathway is to define the enzymes involved. Again, this can be a complex task due to the fact that some enzymes cannot be isolated in an active form or that intermediates cannot be synthesized, so no convenient assay for enzyme activity can be employed. Furthermore, it is possible that some biosynthetic steps occur as spontaneous chemical reactions without the use of enzymes at all, e.g., conversion of the intermediate neopine into codeinone in the morphine biosynthetic pathway. Also, some enzymes may

Table 1 Sources and pharmacological uses of selected plant-derived alkaloids

Alkaloid	*Source*	*Properties*
Ajmaline	*Rauvolfia serpentina*	Antiarrhythmic, antihypertensive
Caffeine	*Coffea arabica*	Stimulant, insecticide
Camptothecin	*Camptotheca acuminata*	Antineoplastic
Cocaine	*Erythroxylon coca*	Analgesic, narcotic, local anesthetic
Codeine	*Papaver somniferum*	Analgesic, antitussive
Emetine	*Uragoga ipecacuanha*	Antiamoebic, expectorant, emetic
Hyoscyamine	*Atropa belladonna* and others	Anticholinergic
Morphine	*Papaver somniferum*	Analgesic, narcotic
Nicotine	*Nicotiana tabacum*	Stimulant
Pilocarpine	*Pilocarpus jaborandi*	Cholinergic
Quinidine	*Cinchona* spp.	Antiarrhythmic
Quinine	*Cinchona* spp.	Antimalarial
Reserpine	*Rauvolfia serpentina*	Tranquilizer
Scopolamine	*Hyoscyamus niger* and others	Sedative, anticholinergic
Strychnine	*Strychnos nux-vomica*	Stimulant, poison
Taxol	*Taxus brevifolia*	Antineoplastic
Vinblastine and vincristine	*Catharanthus roseus*	Antineoplastic

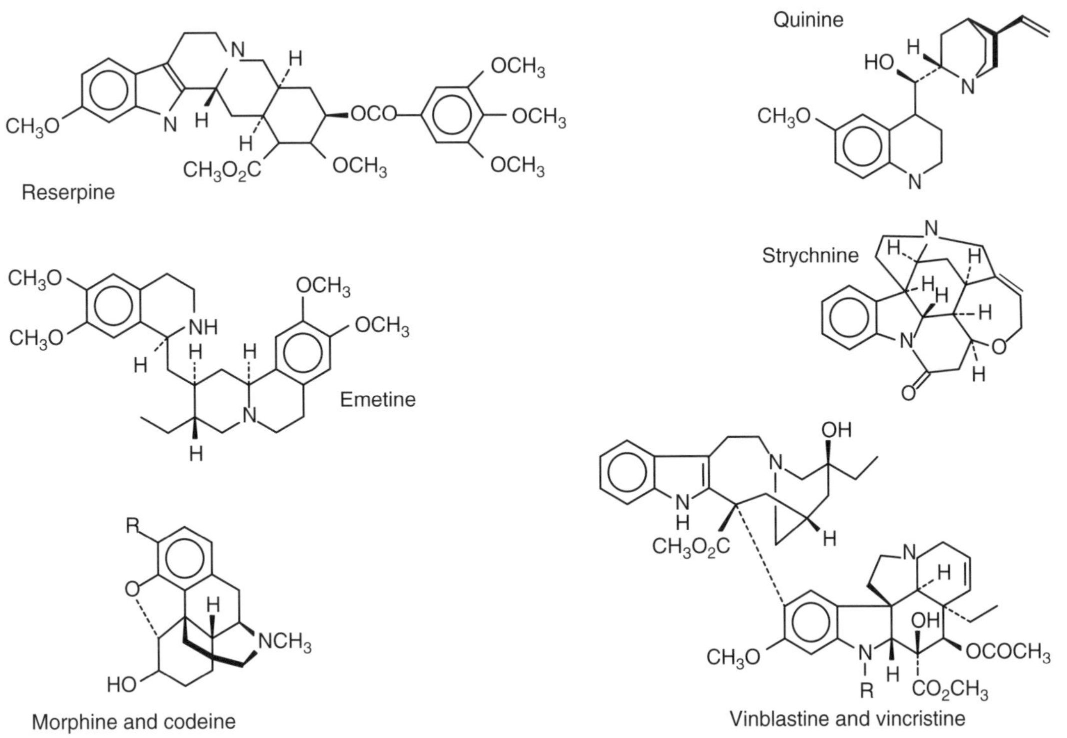

Figure 1 Structures of selected alkaloids. Note that the structures of morphine and codeine are based on the same skeleton, but are decorated with different functional groups in the position represented by 'R'. In morphine, this group is –OH, while in codeine it is CH_2O. Similarly, vinblastine and vincristine are based on the same skeleton, but differ in the nature of the R-group, which for vinblastine is $–CH_3$ and for vincristine is –CHO.

catalyze two or more separate reactions, e.g., hyoscyamine 6-hydroxylase, which carries out two consecutive steps in the scopolamine biosynthetic pathway.

Another complication in alkaloid biosynthesis is compartmentalization. The biosynthesis of the indole alkaloids vinblastine and vincristine in *Catharanthus roseus* provides an example where different enzymatic steps are carried out in different cellular compartments (**Figure 2**). The final steps in the pathway are also carried out in a different cell type to the early steps, which requires the intercellular transport of metabolic intermediates. Similarly, scopolamine biosynthesis requires two different cell types.

Strategies for Engineering Alkaloid Biosynthesis

There are three basic goals of metabolic engineering: (1) producing more of a specific compound, (2) producing less of a specific compound, and (3) producing a novel compound (this may be a known compound that is heterologous in the expression system being used or a completely novel compound). Strategies for achieving these goals can involve the engineering of single steps in a pathway to increase or decrease metabolic flux to target compounds, or block competitive pathways, and the simultaneous modulation of multiple enzymatic steps. An alternative is to stimulate or inhibit the catabolism of the target compound, although to date this has not been used with respect to alkaloids.

Single-Step Engineering

The simplest strategy to increase the levels of a specific target alkaloid is to increase the metabolic flux towards that compound. As discussed above, this requires a thorough understanding of the endogenous pathway because it is necessary to identify the rate-limiting step in the pathway and to use this as the primary target for engineering. For example, consider the biosynthesis of terpenoid indole alkaloids such as vinblastine and vincristine in *C. roseus* (**Figure 2**). Although synthesized in whole plants, these valuable alkaloids are not synthesized in cell culture, probably due to the tissue-specific compartmentalization in whole plants discussed above. The universal terpene indole alkaloid precursor strictosidine is produced in culture, and is metabolized to form other alkaloids such as ajmalicine and catharanthine. For indole alkaloid biosynthesis, a rate-limiting step is the conversion of the amino acid tryptophan to tryptamine, catalyzed by the enzyme tryptophan decarboxylase (TDC).

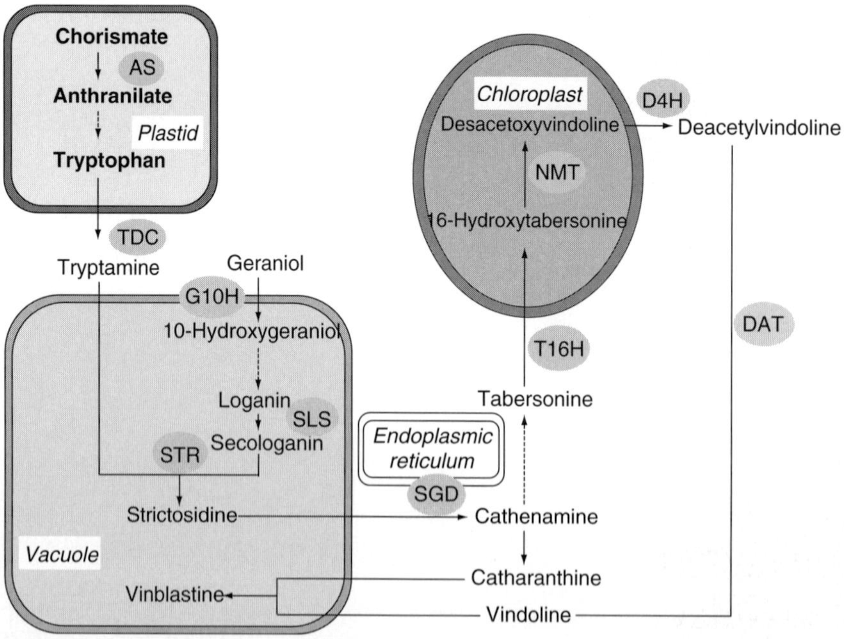

Figure 2 Very simplified scheme showing the compartmentalization of terpenoid indole alkaloid biosynthesis in *Catharanthus roseus*. En route to vinblastine synthesis, different enzymes (shown as disks) are located in distinct compartments, including the plastid, vacuole, endoplasmic reticulum membrane, mature chloroplast, and cytosol (the location of some enzymes is not precisely known). Alkaloid biosynthesis therefore involves extensive trafficking of metabolic intermediates. Although not shown in this figure, certain reactions are also cell-type specific, so intermediates must also be transported between cells. Broken arrows represent multiple steps, while intact arrows represent single reactions. Abbreviations: AS, anthranilate synthase; TDC, tryptophan decarboxylase; G10H, geraniol-10-hydroxylase; SLS, secologanin synthase; STR, strictosidine synthase; SGD, strictosidine β-D-glucosidase; T16H, tabersonine-16-hydroxylase; NMT, *S*-adenosyl-L-methionine: 16-methoxy-2,3-dihydro-3-hydroxy-tabersonine-*N*-methyltransferase; D4H, desacetoxyvindoline 4-hydroxylase; DAT, deacetylvindoline 4-O-acetyltransferase.

Tryptamine is then condensed with the terpenoid molecule secologanin to produce strictosidine. To alleviate the regulation of TDC, the *C. roseus tdc* gene was overexpressed under the control of a strong and constitutive promoter. The resulting transgenic cultures produced high levels of tryptamine, but not further-downstream alkaloids. Thus, overcoming the rate-limiting step merely revealed the next rate-limiting step, at the level of the enzyme strictosidine synthase (STR). This is likely to reflect a limitation in the supply of the terpenoid precursor secologanin.

Single-step engineering can also be used to extend a metabolic pathway in a heterologous plant. For example, the alkaloid scopolamine is produced in *Hyoscyamus niger* but not in *Atropa belladonna*, where the tropane alkaloid pathway stops at L-hyoscyamine. However, introduction of a cDNA from *H. niger* encoding hyoscyamine-6-hydroxylase into *A. belladonna* resulted in the production of scopolamine in this plant (**Figure 3**).

Multiple-Step Engineering

Over the last few years it has become increasingly common for multiple genes to be introduced into plants either stepwise (by crossing independent single gene transgenics) or by simultaneous multiple gene transfer. For the most part, such experiments have involved marker genes rather than genes of agronomic interest, but examples of the latter are beginning to be reported, particularly involving the simultaneous transfer of multiple genes conferring resistance to different pests and pathogens. In terms of metabolic engineering, one way to overcome the limitations of the single-gene approach above would be to simultaneously transform plants with genes encoding enzymes that act at different steps of a biosynthetic pathway. For example, transgenic tobacco (*Nicotiana tabacum*) plants and *C. roseus* cell cultures have been generated simultaneously overexpressing both *tdc* and *str* transgenes, the latter showing high levels of strictosidine accumulation when fed with loganin. Particle bombardment has been used to transform plants simultaneously with 13 different genes, so it is theoretically possible to transfer genes encoding the enzymes of an entire pathway into a suitable expression host.

Engineering Regulatory Genes

In most organisms, transcription factors act as master regulators of complex pathways of

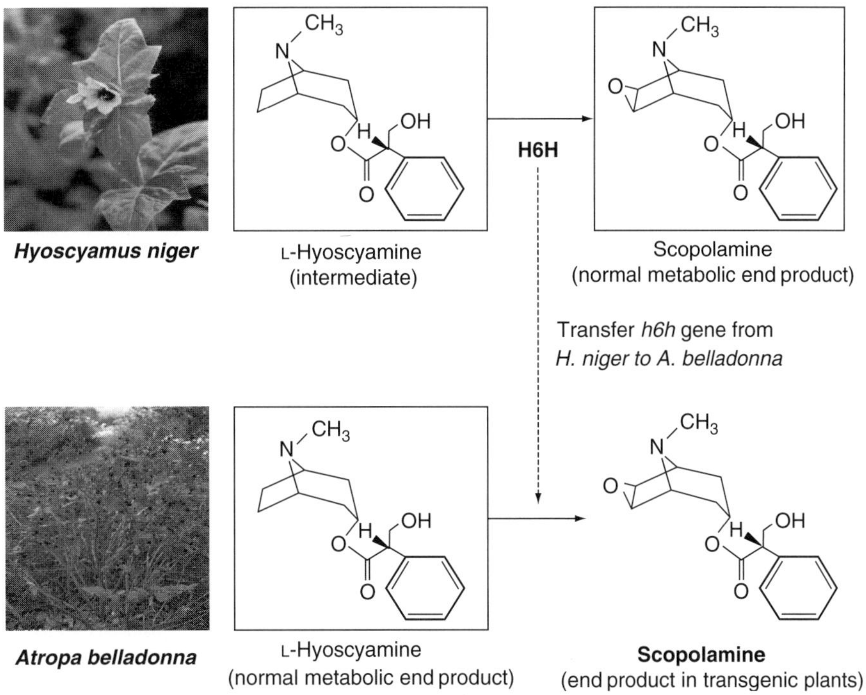

Figure 3 An example of single-gene metabolic engineering. The enzyme hyoscyamine-6-hydroxylase (H6H) is responsible for two reactions that convert L-hyoscyamine into scopolamine. *Atropa belladonna* accumulates L-hyoscyamine because the enzyme H6H is not produced in this species. However, by transferring the enzyme from another plant such as *Hyoscyamus niger*, which does produce scopolamine, the metabolic pathway in *A. belladonna* can be extended.

downstream genes, with important examples recognized in development, signal transduction and metabolism. The important feature of such regulatory systems is that a number of downstream genes are coordinately regulated, either in response to the developmental program of the organism or in response to external inductive signals. As well as responding to the developmental program in plants, a number of external stimuli are known to induce alkaloid biosynthesis, including UV light, fungal elicitors, auxin starvation, and the signaling molecule jasmonic acid. Cell cultures are valuable for testing such responses because signals can be applied to the cultured cells and the expression of different biosynthetic genes can be assayed directly by northern blot or reverse transcriptase polymerase chain reaction (RT-PCR), or using reporter genes linked to the endogenous gene promoters. In this way, it has been shown that the *C. roseus* genes for tryptophan decarboxylase (*tdc*), strictosidine synthase (*str*), geraniol-10-hydroxylase (*g10h*), NADPH:cytochrome P450 reductase (*cpr*), strictosidine β-D-glucosidase (*sgd*), deacetylvindoline 4-O-acetyltransferase (*dat*), and desacetoxyvindoline 4-hydroxylase (*d4h*) are all inducible by methyl jasmonate and/or fungal elicitor in cell culture. In addition, *tdc* and *str* are known to be UV-B-inducible in leaves.

The common regulation of many of the genes involved in terpenoid indole alkaloid biosynthesis in *C. roseus* suggests that a useful strategy for increasing the levels of alkaloids in this system would be to identify and manipulate transcription factors that control the expression of these genes. Such transcription factors have been sought in a number of ways. The yeast one-hybrid system (**Figure 4A**) has been very successful, yielding two AP2/ERF-type transcription factors called ORCA1 and ORCA2 (octadecanoid-responsive *Catharanthus* AP2) that bind to the jasmonate/elicitor response element (JERE) in the *str* gene promoter. While ORCA1 appears not to be involved in jasmonate/elicitor induction, ORCA2 has been directly shown to activate the *str* gene by binding to the JERE, and the *orca2* gene is itself induced by jasmonate and elicitor stimulation. A further transcription factor called ORCA3 has been identified using an alternative insertional mutagenesis strategy called activation tagging (**Figure 4B**). Briefly, *Agrobacterium tumefaciens* was used to transform *C. roseus* with a T-DNA construct containing the strong and constitutive CaMV35S promoter adjacent to the right border. The rationale behind this strategy was that the promoter would activate genes adjacent to the site of integration. Occasionally, the construct might integrate near a gene regulating alkaloid biosynthesis, resulting in the

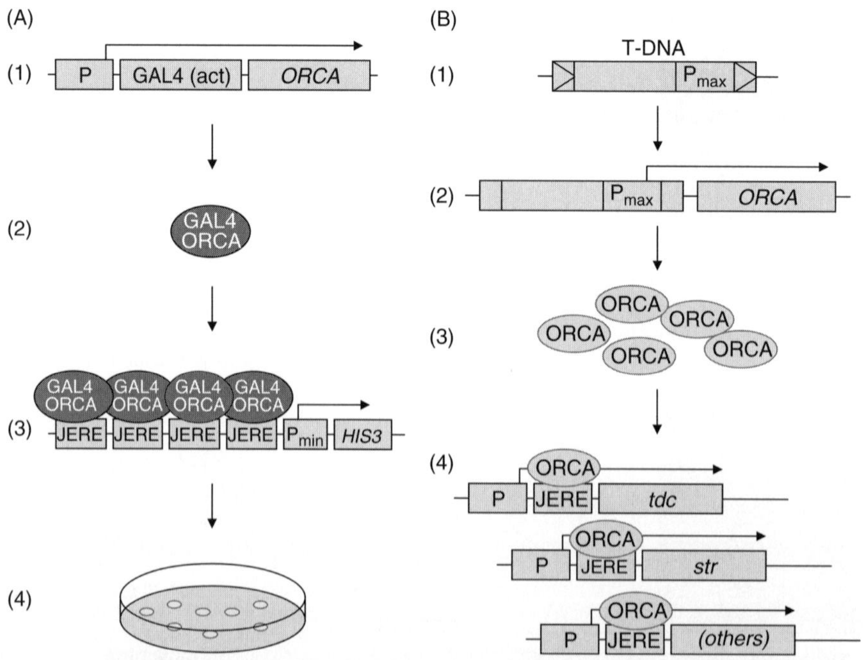

Figure 4 Approaches that have been used to isolate regulatory genes controlling alkaloid biosynthesis in *Catharanthus roseus*. (A) The yeast one-hybrid system. A *C. roseus* cDNA expression library is generated in yeast. Each cDNA is expressed as a fusion protein with the activation domain of the yeast transcription factor GAL4 as shown in (1). If the *C. roseus* cDNA encodes a transcription factor such as ORCA2, which regulates alkaloid biosynthesis, the fusion protein (2) will activate a *HIS3* reporter gene via a minimal promoter (P_{min}) linked to four jasmonate/elicitor response elements (JERE) as shown in (3). Yeast cells in which the construct is active are able to grow on minimal medium lacking histidine (4). These cells are likely to contain fusion proteins in which the *C. roseus* component is a transcription factor that binds to the JERE. (B) Activation tagging. An *Agrobacterium* T-DNA construct contains a strong promoter (P_{max}) adjacent to the right border as shown in (1). If this integrates adjacent to a gene that encodes a relevant transcription factor such as ORCA3 (2), its expression will be stimulated (3) resulting in the upregulation of genes such as *tdc* and *str*. Cells with elevated TDC activity can be selected using 4-methyltryptophan, and the expression levels of *tdc*, *str* and other genes can be analyzed.

upregulation of one or more structural genes in the pathway. Populations of *C. roseus* cells transformed in this manner were selected in a medium containing the tryptophan analog 4-methyltryptophan (4-mT). The levels of 4-mT used in the experiment would be toxic to normal cells, but not those in which the activity of the *tdc* gene was stimulated, since TDC converts 4-mT into a harmless derivative, 4-methyltryptamine. Of course this selection strategy identifies not only global regulators of alkaloid biosynthesis, but also the *tdc* gene itself. To specifically identify regulatory genes, surviving cell lines with elevated levels of *tdc* mRNA were screened for overexpression of other biosynthetic genes, such as *str*, and this resulted in the isolation of ORCA3. The ORCA2 and ORCA3 transcription factors are both induced by jasmonate/elicitor with similar kinetics and they both bind to the *tdc* and *str* promoters. The two proteins are highly conserved in the AP2/ERF DNA-binding domain but are otherwise dissimilar in sequence.

Cells that overexpress ORCA3 show increased expression of *tdc*, *str*, *sgd*, *cpr*, and *d4h*, although other genes are not induced, even though they are normally responsive to jasmonate/elicitor stimulation. This suggests that although the structural genes of the pathway may be coordinately induced, this may be under combinatorial control by ORCA2, ORCA3 and perhaps other regulators. ORCA3-overexpressing cells accumulate tryptamine but no downstream alkaloids, suggesting the incoming terpenoid branch of the pathway is regulated independently of ORCA3.

Future Perspectives

The valuable pharmacological properties of alkaloids together with their low level of production by natural sources means that active research into novel strategies for metabolic engineering will continue to be of great importance in the future. The approaches discussed above – single-step engineering, multiple-step engineering, and the engineering of regulatory genes – are likely to be used in combination with new high-throughput strategies for defining cellular biochemistry at a global level. This "metabolomics" approach will become more applicable as the number of plants with completed genome sequences increases, and it becomes possible to analyze the expression of all genes simultaneously using microarrays and oligonucleotide chips. In the future, it may be possible to identify the most suitable route to metabolic engineering simply by carrying out a small number of array-based hybridization experiments to identify genes and external conditions that upregulate appropriate sets of biosynthetic genes correlating with increased metabolite biosynthesis.

List of Technical Nomenclature

Activation tagging	The isolation of genes through the integration of a DNA element containing a strong promoter that can activate genes adjacent to the insertion site.
Alkaloid	Any of the complex nitrogen-containing heterocyclic organic compounds, mainly of plant origin, with potent pharmacological properties in humans.
Compartmentalization	The situation in which different steps in a metabolic pathway are carried out by enzymes localized in different compartments within the cell.
Coordinate regulation	The situation in which one transcription factor controls several genes in the same manner, so that they are activated as a unit.
Elicitor	Molecules produced by microorganisms, which are recognized by receptors on the plant cell and induce a defense response in the plant.
Marker gene	A gene whose product confers an easily recognizable property on a cell, such as the ability to grow in the presence of antibiotics or the production of a visible product.
Metabolic engineering	The deliberate alteration of metabolism by gene manipulation, in order to modify the levels of a particular metabolite.
Metabolic pathway	A series of enzymatic steps that converts a particular precursor into a given end product through a number of intermediates.
Metabolomics	The global study of metabolism. The simultaneous analysis of metabolites produced by a cell or organism.
(Methyl)jasmonic acid	An intermediate in elicitor-stimulated signaling which can also induce defense responses when applied to plant cells.
Microarray	A collection of cDNA sequences arranged on a miniature solid support allowing the simultaneous analysis of many gene expression profiles.
Oligonucleotide chip	A collection of oligonucleotides printed onto a miniature solid support allowing the simultaneous analysis of many gene expression profiles.
One-hybrid system	A yeast-based expression cloning system in which DNA-binding proteins are preferentially isolated.
Particle bombardment	Gene transfer achieved by firing DNA-coated microprojectiles into a plant cell.
Rate-limiting step	A bottleneck in a metabolic pathway, where conversion from one intermediate to another is slowest.
Response element	A DNA sequence in a gene's promoter that responds to external stimuli by binding transcription factors whose activity is dependent on signal transduction pathways stimulated by external molecules binding to receptors.
Secondary metabolism	The sum of metabolic pathways producing molecules with specialized functions, not involved in general homeostasis.
Single-step engineering	Metabolic engineering through the modification of one step in a metabolic pathway.
Transcription factor	A protein that controls gene expression by binding to regulatory elements and controlling transcription.

See also: **Genetic Modification, Applications**: Cell Factories. **Genetic Modification of Secondary Metabolism**: Terpenoids. **Tissue Culture**: Secondary Metabolism in Plant Cell Cultures.

Further Reading

De Luca V and St Pierre B (2000) The cell and developmental biology of alkaloid biosynthesis. *Trends in Plant Science* 5: 168–173.

Facchini PJ, Huber-Allanach KL, and Tari LW (2000) Plant aromatic L-amino acid decarboxylases: evolution, biochemistry, regulation, and metabolic engineering applications. *Phytochemistry* 54: 121–138.

Kutchan TM (1995) Alkaloid biosynthesis: the basis of metabolic engineering of medicinal plants. *Plant Cell* 7: 1059–1070.

Kutchan TM (1998) Molecular genetics of plant alkaloid biosynthesis. In: Cordell GA (ed.) *The Alkaloids*, vol. 50, pp. 257–316. San Diego: Academic Press.

Memelink J, Verpoorte R, and Kijne JW (2001) ORCanization of jasmonate-responsive gene expression in alkaloid metabolism. *Trends in Plant Science* 6: 212–219.

Primrose SB, Twyman RM, and Old RW (2001) *Principles of Gene Manipulation*, 6th edn. Oxford: Blackwell Science.

Verpoorte R and Alfermann AW (eds) (2000) *Metabolic Engineering of Plant Secondary Metabolism*. Dordrecht: Kluwer Academic Publishers.

Verpoorte R, van der Heijden R, and Memelink J (1998) Plant biotechnology and the production of alkaloids: prospects of metabolic engineering. In: Cordell GA (ed.) *The Alkaloids*, vol. 50, pp. 453–508. San Diego: Academic Press.

Verpoorte R, van der Heijden R, ten Hoopen HJG, and Memelink J (1999) Metabolic engineering of plant secondary metabolic pathways for the production of fine chemicals. *Biotechnology Letters* 21: 467–479.

Verpoorte R, van der Heijden R, and Memelink J (2000) Engineering the plant cell factory for secondary metabolite production. *Transgenic Research* 9: 323–343.

Terpenoids

J Degenhardt and J Gershenzon, Max Planck Institute for Chemical Ecology, Jena, Germany

Introduction

Terpenes, also known as terpenoids or isoprenoids, form the largest group of plant natural products, comprising about 30 000 substances, with new compounds being discovered at the rate of about 1000 per year. The large structural diversity of terpenes in plants is mirrored by their extremely high functional diversity. They play a myriad of roles in both primary metabolism, as pigments, membrane components and hormones, and in secondary metabolism, as defenses against herbivores and pathogens, and attractants for pollinators and herbivore enemies.

In human society also, terpenes play many important roles. They are indispensable in nutrition (vitamins A, D, and E) and in providing certain industrial raw materials (rubber, resins, and turpentine). They also impart flavors and fragrances to foods, beverages, perfumes, soaps, toothpaste, tobacco products, and many other items of daily commerce. Because the majority of customers usually prefer "natural" odors and tastes, rather than synthetic or semisynthetic substitutes, the cultivation of crops for flavor and fragrance production is an economically important sector of the agricultural industry, serving a world market of several billion dollars annually. In medicine, terpenes are generally used as antiseptics, antioxidants, or anti-inflammatory agents, as well as for treatment of diseases, such as cancer and malaria. For example, a high demand exists for paclitaxel (**Figure 1**), a highly substituted diterpene with potent anticancer activity. Since paclitaxel can only be isolated from plants in low yields, it is currently produced by semisynthesis from a more widely available plant product of related structure.

After decades of conventional breeding, researchers have recently begun to employ genetic engineering methods for the manipulation of terpene metabolism, in order to improve the overall yield and composition of terpenes from plants with high economic importance. These efforts have often been hampered by the complexity of terpene biosynthetic pathways and their regulation, but a few notable successes have already been achieved. This article will discuss the different phases of terpene biosynthesis and the strategies that have been pursued for altering each phase to modify the quantity and quality of terpenes in plants.

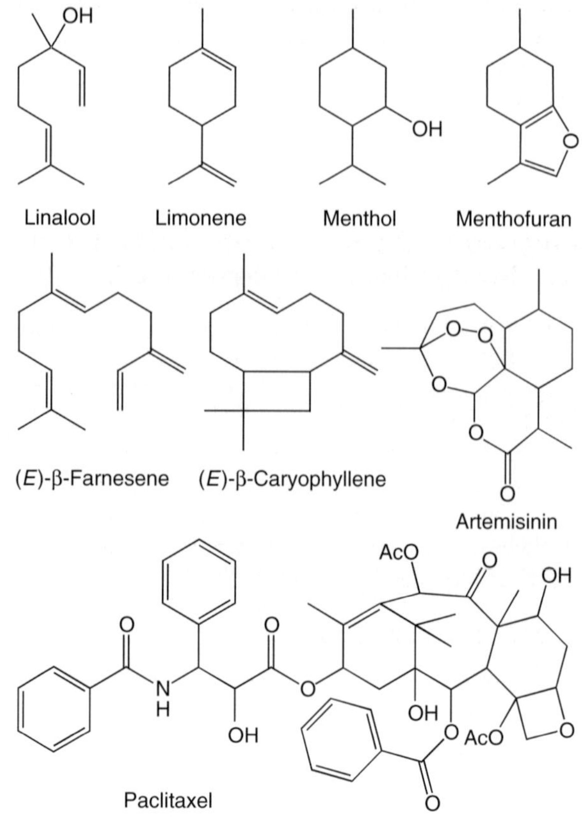

Figure 1 Examples of terpenes of economic importance.

Outline of Terpene Biosynthesis

In the last few years, our understanding of plant terpene biosynthesis has been greatly clarified and expanded by the discovery of the methylerythritol phosphate pathway, a second route for the formation of the basic five-carbon units of terpenes, in addition to the mevalonate pathway. The biosynthesis of terpenes can be conveniently divided into four phases (**Figure 2**):

1. The basic five-carbon units, isopentenyl diphosphate (IPP) and dimethylallyl diphosphate (DMAPP), are formed from primary metabolic intermediates (acetyl-CoA, pyruvic acid, glyceraldehyde-3-phosphate) via both the mevalonate pathway and the methylerythritol phosphate pathway.
2. The five-carbon units are condensed by prenyltransferases to form prenyl diphosphates of different length, which are the precursors for the major classes of terpenes (**Table 1**).
3. Terpene synthases convert the prenyl diphosphates to a wide variety of terpene skeletons, which usually have one or more rings.
4. Some terpenes are further modified by oxidation, reduction, and conjugation.

All four phases of terpene biosynthesis have been targeted by genetic engineering with the aim of altering plant terpene formation. Some of this work is outlined below, and the potential of metabolic manipulation for producing economically important terpenes is discussed.

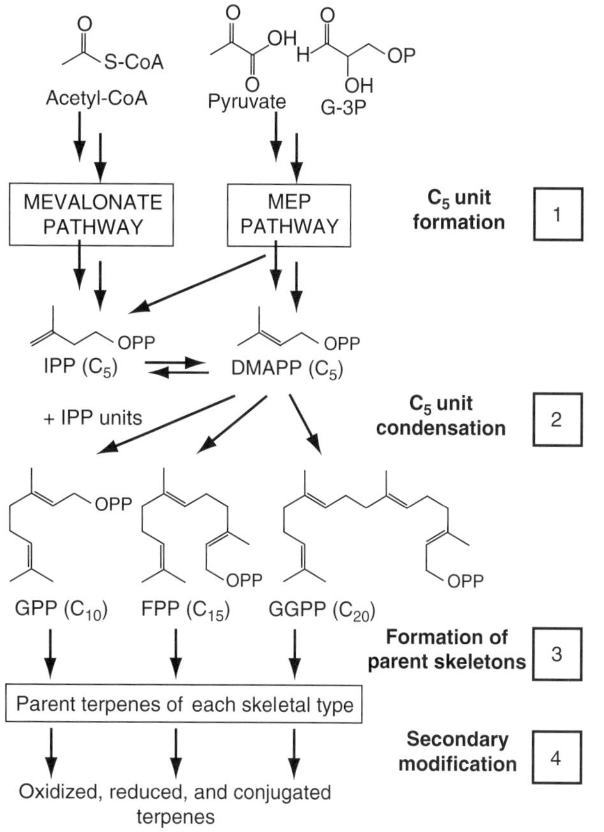

Figure 2 The four major phases of terpene biosynthesis.

Table 1 Terpenes are classified by the number of five-carbon units they contain

Terpene class	*C_5 units*	*C atoms*	*Examples*
Hemiterpenes	1	5	Isoprene
Monoterpenes	2	10	Menthol, linalool
Sesquiterpenes	3	15	Farnesene, artemisinin
Diterpenes	4	20	Abietic acid, paclitaxel
Triterpenes	6	30	Digitoxin, cholesterol
Tetraterpenes	8	40	β-Carotene, lycopene
Polyterpenes	9–30 000	>40	Rubber

Engineering the Early Steps Affects the Flux through Terpene Biosynthesis

Terpene biosynthesis in higher plants is compartmentalized, with the synthesis of monoterpenes, diterpenes, and tetraterpenes largely confined to the plastids, and the formation of sesquiterpenes and triterpenes occurring in the cytosol. Since each compartment primarily utilizes a different pathway to synthesize the basic five-carbon units, it is theoretically possible to manipulate both these groups of terpenes independently.

The Mevalonate Pathway

Sesquiterpenes, sterols, and triterpenes are synthesized in the cytosol from IPP and DMAPP produced by the mevalonate pathway. Beginning from acetyl-CoA, the six steps of the mevalonate pathway are regulated by the activity of the third step, 3-hydroxy-3-methylglutaryl CoA reductase (HMGR), which reduces HMG-CoA to mevalonate. An attempt to increase HMGR activity by constitutive overexpression of a heterologous HMGR gene in tobacco had no effect on the production of sesquiterpene phytoalexins, but did lead to a 3- to 10-fold increase in total sterols, suggesting that crude attempts to modify HMGR activity may not necessarily affect all terpenes formed by the mevalonate pathway. Overexpression of an endogenous HMGR gene in *Arabidopsis* downregulated HMGR activity, and so did not promote terpene biosynthesis.

The Methylerythritol Phosphate Pathway

Manipulation of the methylerythritol phosphate pathway has been more successful in promoting increased terpenoid formation in plants. This recently discovered, plastid-localized pathway supplies IPP and DMAPP for monoterpene, diterpene, and tetraterpene (carotenoid) formation. Overexpression of the first step of the pathway, the condensation of pyruvate and glyceraldehyde-3-phosphate to form 1-deoxyxylulose-5-phosphate (DXP), catalyzed by DXP synthase, resulted in elevated levels of carotenoids in *Arabidopsis thaliana*, as well as greater amounts of chlorophylls and tocopherols, both of which contain terpene moieties. In contrast, the disruption of DXP synthase in *A. thaliana* resulted in albino phenotypes with reduced chlorophyll and arrested chloroplast development. It was recently discovered that plants have two distinct types of DXP synthase genes that share only 70% similarity in their corresponding amino acid sequences. Since one type seems to specifically regulate the formation of terpenoid secondary metabolites, it might be a good candidate for future efforts to engineer increased production of any terpenes used as flavors, fragrances, pharmaceuticals, and industrial materials, all of which can be classified as secondary metabolites.

When the second step of the methylerythritol phosphate pathway, the conversion of DXP to methylerythritol-4-phosphate, was overexpressed in peppermint (*Mentha piperita*), the result was an increase of about 50% in the monoterpenes that constitute the essential oil of this species. Cosuppression of this same enzyme in peppermint resulted in deficiencies in chlorophyll and decreases in other terpenes. Thus, the flux through the methylerythritol phosphate pathway in plastids can be altered by manipulating either of its first two steps.

Modification of the two basic terpenoid pathways must be performed cautiously, however, since elevating the levels of many terpenoids indiscriminately could have unforeseen negative consequences for the plant. Some terpenoids have critical regulatory roles as hormones (gibberellins, abscisic acid, brassinosteroids), or as components of signal transduction pathways (prenylated proteins). When the desired terpenes are known to be made in specific cell types, it might be wise to attempt to elevate flux through the basic terpenoid pathway in those cells only. For instance, monoterpene formation in peppermint is localized in glandular trichomes found on the leaf surface. Hence, the use of trichome-specific promoters might be judicious for engineering monoterpene metabolism in this species.

The Condensation of Five-Carbon Units by Prenyltransferases

In the second phase of terpene biosynthesis, IPP and DMAPP are condensed to form geranyl diphosphate (GPP), farnesyl diphosphate (FPP), and geranylgeranyl diphosphate (GGPP), which are the precursors of monoterpenes, sesquiterpenes, and diterpenes, respectively. Manipulation of the prenyltransferases that catalyze these condensations could allow one to raise or reduce selectively the levels of each of these terpene classes. Although prenyl transferases forming GPP, FPP, and GGPP have been identified, there have been only a few attempts to employ them to modify terpene biosynthesis. Overexpression of a cotton FPP synthase in *Artemisia annua* led to a more than twofold increase in the concentration of artemisinin (**Figure 1**), an antimalarial sesquiterpene of this plant.

Manipulating the Later Steps of Terpene Biosynthesis Controls the Formation of Specific End Products

Terpene Synthases Form the Basic Terpene Skeletons

The first committed step in the biosynthesis of individual terpenes is catalyzed by terpene synthases, which convert GPP, FPP, or GGPP into an abundance of different carbon skeletons. A wide variety of terpene synthase genes has been isolated and characterized from plants providing a choice of compounds for introduction or elimination. Extensive investigation has been carried out with linalool synthase from *Clarkia breweri*, which converts GPP to (*S*)-linalool (**Figure 1**), a major constituent of many flower fragrances. The linalool synthase gene was transformed into tomato (*Lycopersicon esculentum*) under control of a promoter conferring gene expression during fruit maturation with the aim of enhancing aroma and flavor in fruit. Fruits from transgenic plants contained significant concentrations of *S*-linalool and 8-hydroxylinalool, a product of allylic hydroxylation of linalool, demonstrating that the addition of a heterologous terpene synthase gene can result in production of a novel monoterpene. The same strategy was pursued to add a linalool fragrance to odorless varieties of the ornamentals, petunia (*Petunia* spp.) and carnation (*Dianthus caryophyllus*). Unfortunately, in petunia, the linalool produced was efficiently converted to a nonvolatile, nonodorous glucoside derivative by the action of an endogenous glucosyltransferase. In carnation, a large fraction of the *S*-linalool product was oxidized to *trans*-linalool oxide. Both studies demonstrate that endogenous enzymatic activities need to be considered when plants are subjected to metabolic engineering with the aim of introducing a novel terpene.

Interestingly, the use of endogenous GPP by transgenic monoterpene synthases did not measurably affect the biosynthesis of carotenoids in these studies, although both of these branches of terpenoid biosynthesis are supplied by the methylerythritol phosphate pathway. However, when a direct attempt was made to increase carotenoid biosynthesis in tomato by expression of phytoene synthase (which converts two molecules of GGPP to the carotenoid precursor phytoene), the transformants showed severely reduced growth. The dwarf phenotype was caused by reduced formation of gibberellins, a group of diterpene hormones, due to the redirection of the common pathway intermediate GGPP toward carotenoid biosynthesis. In contrast, efforts to engineer the β-carotene biosynthetic pathway into rice (*Oryza sativa*) by introducing phytoene synthase and one or more desaturating enzymes of carotenoid synthesis did not negatively affect plant growth and development. Carotenoid levels were increased significantly in these transformed lines creating a new rice variety with greater nutritional value. This "golden" rice could help prevent vitamin A deficiencies in many areas of the world.

Bioengineering of terpene metabolism can be used not only to introduce new compounds, but also to alter the existing terpene composition of a plant. The introduction of 4*S*-limonene synthase from spearmint (*Mentha spicata*) into the closely related species peppermint and field mint (*Mentha arvensis*) resulted in transformants with alterations in essential oil composition, which were mostly consistent with overexpression of the transgene or cosuppression of the endogenous limonene synthase. In addition to cosuppression, elimination of a specific enzyme activity can also be achieved by transformation with antisense constructs or transposon insertion into the corresponding gene. The use of transposon insertion mutants is restricted to corn (*Zea mays*; maize) or other plants harboring well-characterized transposable elements, but these mutants have the advantage of not being transgenic, and thus may gain wider acceptance in the current political and regulatory climate. Cosuppression, antisense DNA, and transposon insertion can be utilized to decrease or eliminate the biosynthesis of specific terpenes, which might be desirable if they impart noxious odors or flavors, or attract herbivores to crop plants.

The majority of terpene synthases, including 4*S*-limonene synthase from spearmint, do not form a single terpene product, but rather a complex mixture of several terpene structures. Therefore, alteration of terpene synthase activity may affect the amounts of several terpene end products, some of which may have critical roles in plant defense or other processes. However, ongoing studies to elucidate the reaction mechanisms of terpene synthases should eventually facilitate the engineering of specific, single-product terpene synthases, which will greatly simplify efforts to modify terpene composition.

Modification of Terpene Hydrocarbons

Some of the initially formed terpene synthase products are further modified by oxidation, reduction, and conjugation. The few attempts to alter these modifications with transgenic methods have focused on reducing enzymatic activities by cosuppression and transformation with antisense constructs. For example, the concentration of the economically undesirable (+)-menthofuran (**Figure 1**) in the essential oil of peppermint was significantly reduced by transformation with an antisense construct of the cDNA for menthofuran synthase, a cytochrome P-450 oxidase. Similarly, the suppression of a tobacco (*Nicotiana tabacum*) P-450 hydroxylase reduced the hydroxylation of a diterpenoid cembratriene-ol to a cembratrien-diol, resulting in enhanced resistance to aphids. Both studies are very promising in that these modifying enzymes seem to have a very specific activity that can be directly targeted by transgenic approaches.

Conclusion

The manipulation of plant terpene metabolism has long been a focus of interest in plant biotechnology because many terpenes have high economic importance as pharmaceuticals, insecticides, flavors, and fragrances. Unfortunately, progress in this area has been hampered by lack of basic tools and knowledge. For many branches of terpenoid biosynthesis, the genes encoding the critical biosynthetic steps are not yet available, and even the reactions themselves are not known with confidence. However, progress in the understanding of plant terpene biosynthesis and the development of transformation techniques applicable to terpene-containing crops should continue to broaden the possibilities of engineering terpene metabolism in the coming years. Bioengineering of terpene crops may then replace conventional breeding efforts, since it is a much more precise tool for manipulating terpene metabolism.

List of Technical Nomenclature

DMAPP	Dimethylallyl diphosphate.
DXP	1-Deoxyxylulose-5-phosphate.
FPP	Farnesyl diphosphate.

GGPP	Geranylgeranyl diphosphate.
GPP	Geranyldiphosphate.
HMGR	3-Hydroxy-3-methylglutaryl-CoA reductase.
IPP	Isopentenyl diphosphate.

See also: **Secondary Products**: Antifeedant Substances in Plants.

Further Reading

Barkovich R and Liao JC (2001) Metabolic engineering of isoprenoids. *Metabolic Engineering* 3: 27–39.

Gershenzon J and Kreis W (1999) Biosynthesis of monoterpenes, sesquiterpenes, diterpenes, sterols, cardiac glycosides and steroid saponins. In: Wink M (ed.) *Biochemistry of Plant Secondary Metabolism: Annual Plant Reviews*, vol. 2, pp. 222–229. Sheffield: Sheffield Academic Press.

Grierson D (1998) Manipulation of fruit ripening by genetic modification. In: Lindsey (ed.) *Transgenic Plant Research*, pp. 109–124. Amsterdam: Harwood Academic Publishers.

Mahmoud SS and Croteau R (2001) Strategies for transgenic manipulation of monoterpene biosynthesis in plants. *Trends in Plant Science* 7: 266–373.

McCaskill D and Croteau R (1999) Strategies for bioengineering the development and metabolism of glandular tissues in plants. *Nature Biotechnology* 17: 31–36.

Rodriguez-Conception M and Boronat A (2002) Elucidation of the methylerythritol phosphate pathway for isoprenoid biosynthesis in bacteria and plastids. A metabolic milestone achieved through genomics. *Plant Physiology* 130: 1079–1089.

Wood Quality

A-M Boudet, Université Paul Sabatier, Toulouse, France

Introduction

Wood is a composite material mainly composed of three different polymers: cellulose, hemicelluloses, and lignins (**Figure 1**). These macromolecules are deposited in the cell walls which delimit each plant cell either alive or not. Cellulose is the most abundant biopolymer on earth, and it is of crucial importance for the feeding of herbivores and for industrial purposes including the pulp industry which represents a huge economic activity worldwide (**Figure 2**). For example, it has been estimated that the production value of the paper industry in the United States is higher than the value of crude agricultural crops. Lignins, the biopolymers second in abundance after cellulose, are complex three-dimensional phenolic polymers deposited in the cell walls of specialized plant cells. Lignification was undoubtedly a crucial stage in the development and evolution of land plants. The primary function of lignins is to reinforce cell walls, rendering them rigid, hydrophobic, and impermeable to water. Lignins impart resistance to compressive stresses imposed by water transport and by the mass of the plant. However, despite their importance in the adaptive strategies of tracheophytes (pteridophytes, gymnosperms, and angiosperms), the presence of lignins in plants dramatically affects their usefulness for agroindustrial purposes. The digestibility and dietary conversion of grasses are altered by the presence of lignins and in the pulp industry lignins are undesirable components: complex, expensive, and polluting processes are needed to extract them in order to recover cellulose, the main component of pulp.

Over the last 35 years, the production of paper has more than tripled. To remain competitive, to reduce the uncontrolled exploitation of natural forests, and to limit energy and environmental problems, the forest products industry must develop genetically improved woody materials. Until now, genetic improvement of trees was achieved using conventional breeding methods, which are particularly difficult to exploit for woody species due to long generation times and the need to evaluate specific characters at the adult stage. Among the various facets of plant biotechnology, genetic engineering offers new opportunities and attractive perspectives for improving woody species.

Genetic engineering is a powerful tool used to modify plants that are of economic interest. This is accomplished by introducing genes that confer desirable traits such as pathogen resistance, herbicide tolerance, and improvement of chemical composition. The introduction of foreign genes (transgenes) is routine for many annual crop species and many examples are undergoing field trials and commercialization. However, similar transgene research on forest trees is less advanced.

The first transgenic tree was obtained in 1987 and since that time the development of genetic modification (GM) techniques for several tree species has led to an exploitation of genetic engineering bringing improvements in forest products.

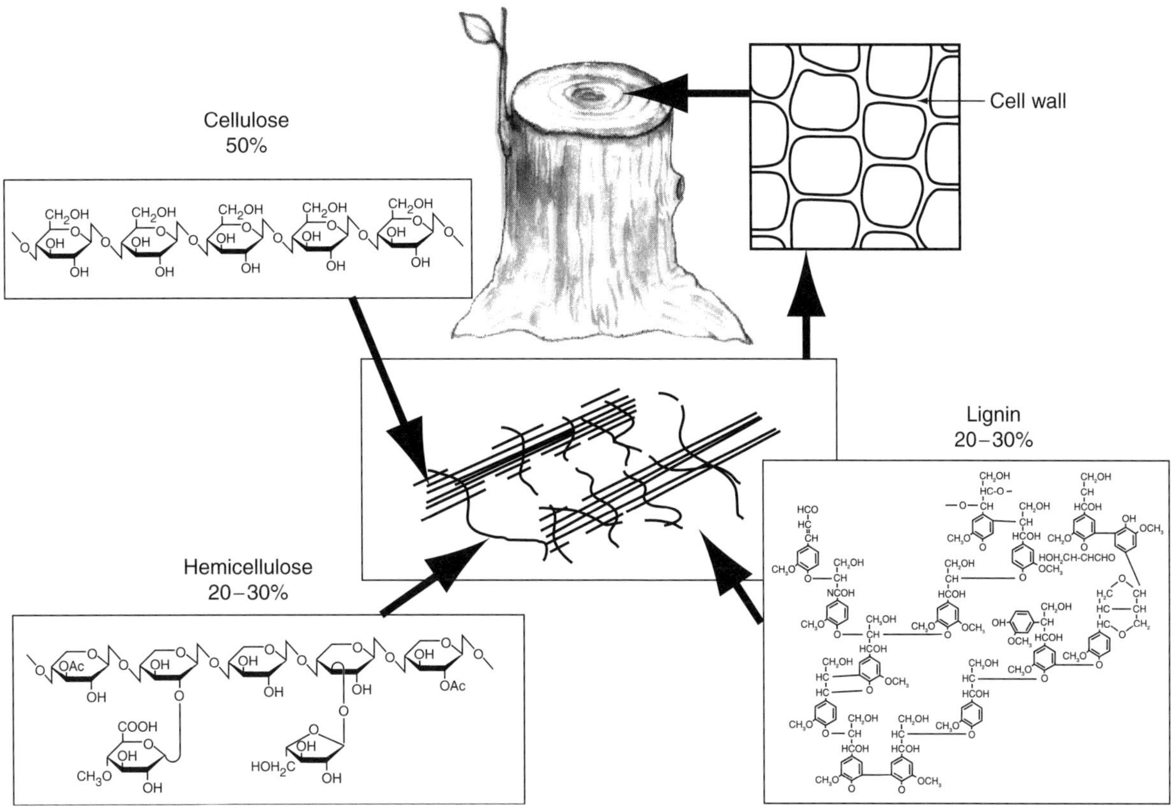

Figure 1 Major components of wood.

- **World consolidated sales of pulp and paper US $735 000 million**
- **Estimated value of pulp US $120 000 million**
- **The share of the world market splits roughly:**

  - North America 44%
  - Europe 29%
  - Asia 15%
  - Latin America 7%
  - Africa 3%
  - Australia 6%

Latin America, Asia, and Africa are dominated by tropical hardwoods (notably *Eucalyptus*) while Europe and North America are dominated by pines.

Figure 2 Pulp and paper market at the world scale.

As the composition of wood is very important for the pulp industry and other economic purposes, lignin genetic engineering is a very active area of research which has been stimulated within recent years by the characterization of important genes controlling lignification. A significant number of GM plants exhibiting qualitative and quantitative changes in their lignins has already been obtained and, in a limited number of cases, preliminary data have demonstrated the industrial usefulness of the resulting products. It is clear that GM trees which produce modified lignins (in lower amounts or that can be more easily processed) would lead to substantial industrial and environmental benefits. In the long term the successful extension of this strategy to forestry of cloned trees has at least four prerequisites: (1) convenient germplasm, (2) large-scale vegetative propagation, (3) convenient transformation techniques, and (4) long-term field trials to confirm the effects observed. Investigations are being actively conducted along these different lines and, in parallel, investigations are conducted on easy-to-transform model plants in order to define the best ways to engineer new cell wall compositions.

Lignin Structure and Major Objectives of Lignin Genetic Engineering

Lignin is a phenolic heteropolymer resulting from the apparent random polymerization of three monomers or monolignols (**Figure 3**). More precisely, this polymerization concerns the phenoxy radicals derived from the oxidation of the monolignols through the potential intervention of various oxidases including peroxidases or laccases. These radicals are associated by different chemical linkages and the resulting polymer has great chemical complexity. When these monomers are integrated into the polymer they give rise by chemical degradation to three different units called: H units, G units, and S units (**Figure 3**). The relative proportion of these units greatly varies according to the plant species and the developmental stage but in any case the final polymer is hydrophobic and highly resistant to biological degradation. It forms a cement in the cell wall which creates difficulties for the industrial extraction of cellulose or its assimilation by cattle.

One of the first objectives of breeders and research workers involved in lignin GM is to obtain plant species with a lower lignin content or with a lignin that is easier to extract. Indeed, it is clear, in the case of the pulp industry, that there is an inverse correlation between the lignin content of different woody species of the same genus and the pulp yield. In addition, the lignin from gymnosperms which do not contain S units are more resistant to the chemical degradation during the Kraft pulping process than angiosperm lignins which contain both G and S units in high proportion. The reason for this difference is that there is a more limited amount, in the angiosperm lignins, of very resistant carbon–carbon bonds, since the highly substituted sinapyl alcohol monomers present in angiosperms cannot contract this kind of linkage. Consequently lignins enriched in S units are more convenient for the pulp industry. An increase in free phenolic units making the lignins more water soluble would also be an advantage for the pulp industry.

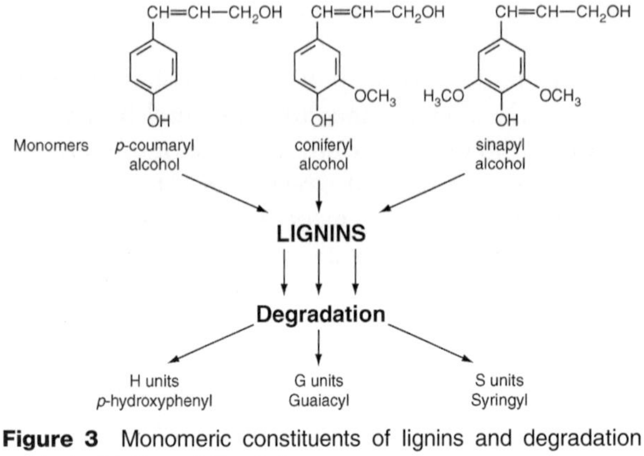

Figure 3 Monomeric constituents of lignins and degradation products (H, G, and S units).

Recent Achievements in GM Lignin

Among the polymers of the plant cell wall, little has been done with genetic manipulation of cellulose and hemicelluloses due to the lack of characterization of convenient genes. In contrast, over the last decade, it has become clear that genetic engineering can tackle lignin composition and content. This is due to a reasonable knowledge of the biosynthetic pathways.

The synthesis of lignins from phenylalanine proceeds through three different sets of reactions (**Figure 4**):

1. The common phenylpropanoid pathway which provides general intermediates (cinnamoyl-CoAs) for different phenolic syntheses.
2. The monolignol specific pathway which gives rise to the three building units of the lignin polymer – the coumaryl, coniferyl, and sinapyl alcohols.
3. The polymerization of monolignols for which peroxidases and laccases are candidates.

Most of the genes encoding the enzymes along these pathways have now been characterized. These genes have been exploited through antisense or cosuppression strategies in order to reduce the activity of the corresponding enzymes and alter the functioning of the lignin pathway. In a limited number of cases the overexpression of these genes has also been achieved.

Changes in Lignin Composition

Caffeic acid 5-hydroxyferulic acid *O*-methyltransferase (COMT) downregulated plants have been obtained by several different groups. The resulting transgenic lines all exhibit a strong alteration of lignin structure with a higher content of condensed bonds, a reduction of the proportion of syringyl units, and an incorporation of 5-hydroxyguaiacyl units into the polymer. Recent results obtained by Jouanin's group showed that a strong downregulation of COMT activity in poplar (*Populus*) induced a significant reduction in lignin level (up to a 17% decrease). Despite the reduction of lignin content which may positively affect the pulp yield the higher degree of condensation of these transgenic lignins makes them more resistant to chemical degradation and not suitable for the pulp industry. The alternative methylation enzyme caffeoyl-CoA 3-*O*-methyltransferase (CCoAOMT) which seems to control the

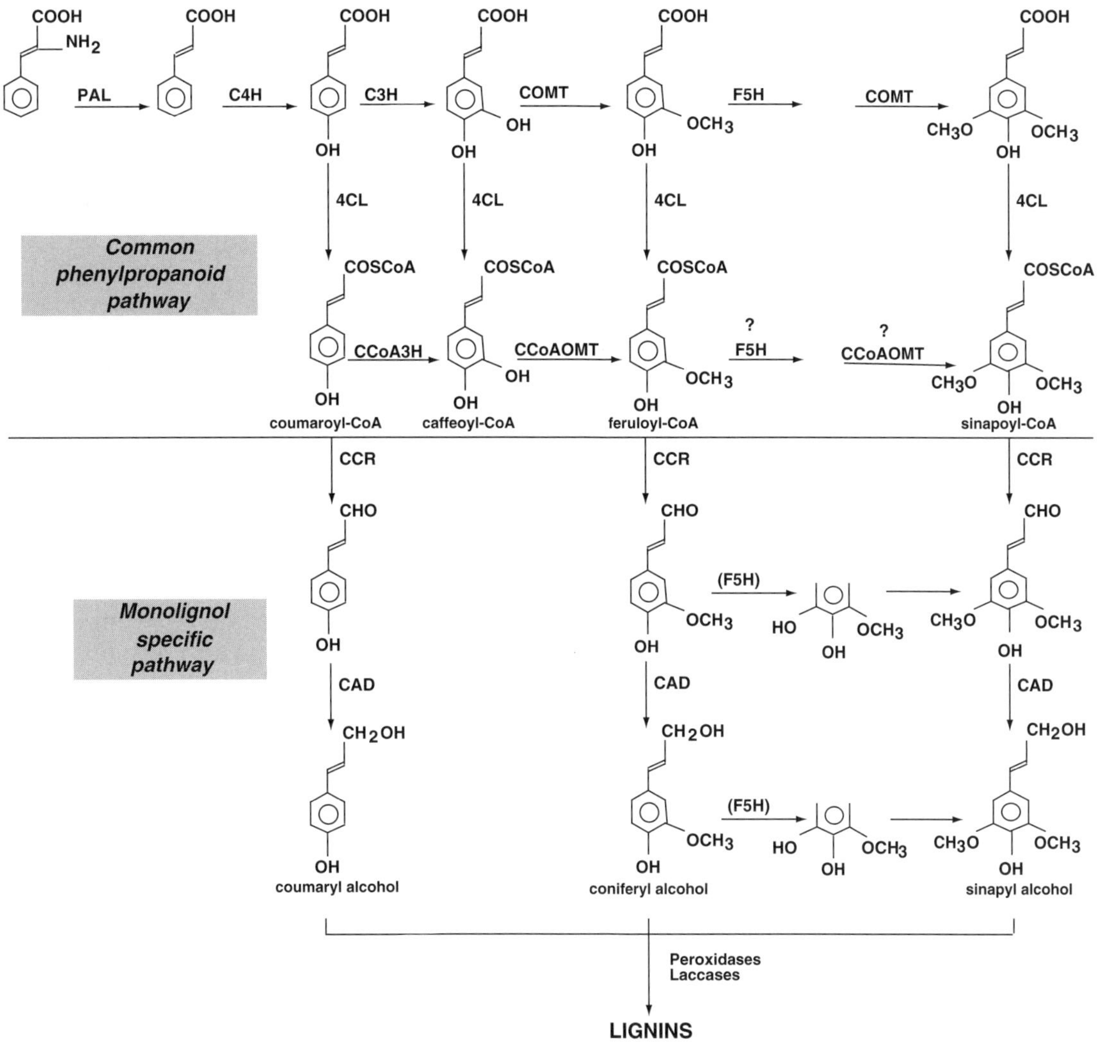

Figure 4 The lignin biosynthetic pathway. PAL, phenylalanine ammonia-lyase; C4H, cinnamate-4-hydroxylase; C3H, 4-coumarate-3-hydroxylase; COMT, caffeic acid 5-*O*-methyltransferase; CCoAOMT, caffeoyl-CoA 3-*O*-methyltransferase; F5H, ferulate-5-hydroxylase; 4CL, hydroxycinnamate-CoA-ligase; CCR, cinnamoyl-CoA-reductase; CAD, cinnamyl alcohol dehydrogenase. F5H may hydroxylate coniferyl aldehyde and coniferyl alcohol with more efficiency than ferulic acid.

synthesis of G units has been shown to be an interesting target for decreasing lignin content, but additional studies are still necessary to confirm the interest of this strategy, particularly in woody species.

Interesting results have been obtained through the control of ferulate-5-hydroxylase (F5H/CAld 5H) level by the group of Chapple. In the model plant *Arabidopsis thaliana*, the downregulation or the overexpression of the corresponding gene completely changes the composition of lignins which are respectively depleted or strongly enriched in S units. Recently, overexpression of this gene in poplar has been shown to induce a significant increase in syringyl monomer content. The expression of this enzyme in gymnosperms should allow the conversion of softwood lignins lacking S units into hardwood lignins that are easier to process during the pulping process.

Cinnamyl alcohol dehydrogenase (CAD) down-regulated transgenic lines have also been obtained. Even though the lignin content is not affected the transgenic lignins are enriched in hydroxycinnam-aldehydes and these structural changes induce both a pink/red coloration of the xylem and an increased extractability of these lignins reflected by a decrease in kappa number (a measurement of the amount of chemicals necessary to remove residual lignins from the pulp during the bleaching process).

Results obtained within the TIMBER consortium funded by the EC have shown that in a poplar clone of economic interest (OGY), for a strong reduction of CAD activity, the lignin content is slightly reduced, and that the higher percentage of free phenolic groups makes these lignins particularly soluble in mild alkaline conditions. These characteristics are very positive for the pulp industry and will generate chemical and energy savings during the pulping process.

More recent results have extended these data and have demonstrated that these advantages do not induce significant changes in the development and agronomic characteristics of the tree, nor cause negative environmental impacts.

Interestingly, field trials have demonstrated the stability of transgene expression and of changes in lignin profiles for 4-year-old CAD downregulated poplar trees. These data show that the strategy of CAD downregulation may be of immediate utility and can be potentially extended to a large number of woody species.

Changes in Lignin Content

The downregulation of target genes of the common phenylpropanoid pathway 4-hydroxycinnamate CoA ligase (4CL) or of the monolignol specific pathway cinnamoyl coenzyme A reductase (CCR) has demonstrated that the lignin content of the resulting transgenic plants can be strongly decreased (by up to 50%). Surprisingly, this reduction in lignin content and (or) some associated effects induce opposite effects on the development of the plants. Aspen (*Populus tremula*) trees that are 4CL downregulated exhibit an increased size whereas CCR downregulated tobacco plants (*Nicotiana tabacum*) display a severe reduction in growth and other alterations in shape and morphology. The explanation for these discrepancies is not clear at the moment. However, recent data suggest that it is not the reduction of the amount of lignins *per se* which is responsible for the altered development of the tobacco plants. Indeed double transformants (downregulated both for CCR and CAD) display a decrease in global lignin content similar to the CCR downregulated line (**Figure 5**) but with a normal development. In fact these two kinds of transformants differ by the spatial deposition of lignins. The CCR downregulated line exhibits a reduction in lignin deposition in the walls of the different xylem cell types whereas this reduction was selectively targeted to the fibers in the double transformant. Thus it appears that different events of transformation which lead in the same species to a similar reduction of lignin content may induce different patterns of lignin deposition at the cell and subcellular levels.

These results demonstrate that lignification is tightly regulated in individual cell types and that it should be possible to target accurately lignin modifications to specific cells in the future. Transgenic lines with reduced lignin content and normal (CAD × CCR) or improved (4CL) development, which have until now been obtained in growth chambers or greenhouses, could be of great interest if it can be demonstrated that they behave in the same way as control plants in field conditions. These transgenic lines exhibit in both cases a significant increase in their cellulose:lignin ratio with corresponding expected benefits.

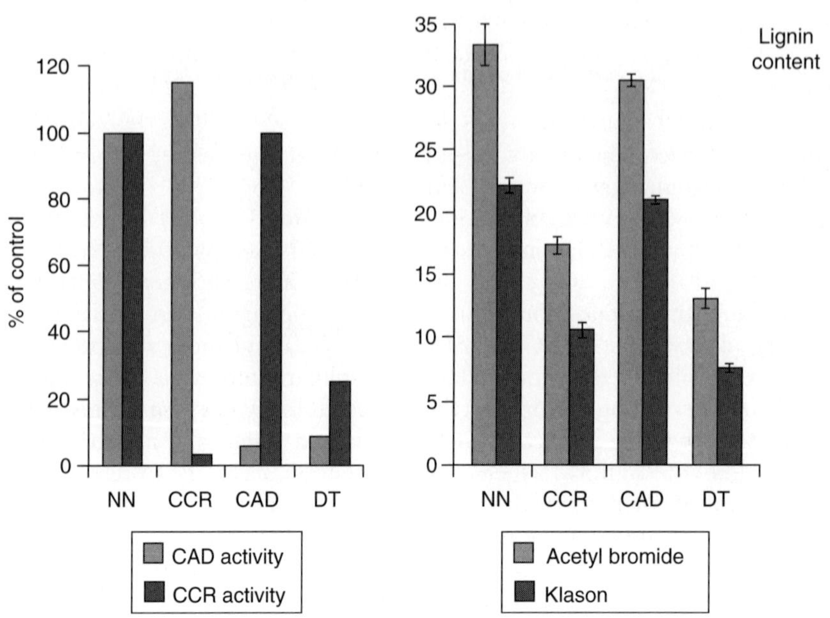

Figure 5 Enzyme activities and lignin content of wild-type tobacco plants (NN) and different trangenic lines downregulated for cinnamoyl-CoA-reductase (CCR), cinnamyl alcohol dehydrogenase (CAD) and for both activities (DT).

Potential Side Effects and Compensation Mechanisms Associated with Lignin Modifications

One potential drawback related to lignin manipulation in plants would be the attenuation of defense mechanisms against pathogens. Indeed lignification is considered as one of the adaptive responses of plants to pathogen attack. Downregulating lignin synthesis could then reduce the ability of plants to protect themselves against pests and diseases. Such a hypothesis has not been substantiated using the transgenic plants already obtained, but the use of chemical inhibitors of lignification resulted in decreasing the hypersensitive response in wheat (*Triticum aestivum*) plants challenged by the fungus *Puccinia graminis*. In order to minimize any reduction in these defense mechanisms, which are usually most important at the surface of the tissues, the use of specific promoters to drive transgene expression has been envisaged. Ideally, such a promoter should be xylem-specific, avoiding a decrease in stress lignins produced on the plant surfaces.

Another important aspect of lignin manipulation lies in the potential disturbance of plant development. Reduction or modification of the lignin content of plant cell walls may theoretically exert pleiotropic effects on plant functions through changes in the strength of plant organs, sap conduction through the xylem, or permeability of cell wall barriers (caspary band). Changes in plant development have been observed for strongly depressed CCR tobacco lines (or for some CCoAOMT downregulated tobacco lines) even though it has not been clearly demonstrated whether the resulting phenotype (reduced growth) is due to changes in cell wall properties or to changes of soluble phenolic profiles.

It is clear that for commercial purposes the GM plants should behave like wild-type plants in terms of biomass production and agronomic performance. At the moment this has been verified for 4-year-old CAD downregulated poplars, which appear to be promising candidates for immediate application. The same evaluation procedures at the field level must be performed over long timescales for each new transformed line.

Plants with altered lignin synthesis seem to adopt compensatory strategies for maintaining the integrity of their cell walls. This is clearly illustrated in the case of CAD downregulated plants which utilize cinnamaldehydes instead of cinnamyl alcohols as building-blocks of their transgenic lignins. CCR downregulated tobacco plants partially compensate the lack of lignin in their cell walls by a striking increase in tyramine ferulate. Additional reinforcement of the cell wall may also be provided in these transgenic lines by a very significant increase in wall associated phenolic compounds.

The question is still open as to whether these new biochemical phenolic decorations of the cell wall or of the cytosolic fraction have a positive or a negative impact on the pulping process. A negative example is provided by *Eucalyptus camaldulensis* CAD downregulated plants which overproduce tannins interfering with the bleaching step during the pulping process. In contrast, if the compensation strategy could target useful components of the cell wall the benefits would be obvious. This situation is illustrated by the exciting results obtained by Chiang's group for 4CL downregulated aspen for which the downregulation of lignin was accompanied by an increase in cellulose leading to a nearly doubled cellulose: lignin ratio. If the required defense, water transfer, and other properties survive, as they appear to in these particularly vigorous plants, the potential for improving plant utilization is enormous (**Table 1**).

Perspectives and Future Trends

At the moment, the number of significant modifications of lignin profiles obtained through genetic engineering is already impressive. In addition, new target genes or combinations of genes should be exploited in the future. Among these are the genes encoding the oxidases responsible for the formation of phenoxy radicals or the dirigent proteins potentially controlling lignin monomer assembly. Recent results have indicated that genetic engineering of transcription factors involved in lignin biosynthesis can also be an efficient strategy for producing plants with a reduced lignin content.

Thus the main problems we are faced with in the area of lignin genetic engineering are not related to a lack of potential target genes but: (1) to an extensive evaluation of the effects of the individual transformations on various aspects of plant metabolism and development; and (2) to the transfer of the best technologies to crops and woody species of major interest.

Lignin Patterns

Despite important progress, our knowledge of lignin chemistry is still incomplete, even though the analysis of GM plants has confirmed and extended our views concerning chemical variability and flexibility of lignins. Future studies should clearly identify the structural changes of the polymer induced by the transformations. Interestingly, the chemical changes in lignin induced by ectopic modifications of lignification genes could generate new lignins or new byproducts with direct practical applications.

Table 1 Strategies for modifying wood quality by transgenic means

Tree species and transferred genes	*Observed phenotype*	*Reference*
Altered expression of the anionic peroxidase (*prxA1*) in poplar (*Populus* sp.)	Resulted in increased total peroxidase activity but lignin characteristics and structures were not evaluated	Kajita S, Osakabe K, Katayama Y, *et al.* (1994) *Agrobacterium*-mediated transformation of poplar using a disarmed binary vector and the overexpression of a specific member of a family of poplar peroxidase genes in transgenic poplar cell. *Plant Science* 231–239
Antisense expression of an *O*-methyltransferase (OMT) in poplar hybrids	No reduction of total lignin but reduced syringyl : guaiacyl ratio (beneficial for Kraft pulping), novel lignin component identified (5-hydroxyguaiacyl residue)	Van Doorsselaere J, Baucher M, Chognot E, *et al.* (1995) A novel lignin in poplar trees with a reduced caffeic acid/5-hydroxyferulic acid O-methyltransferase activity. *Plant Journal* 8: 855–864
Downregulated cinnamyl alcohol dehydrogenase (CAD) in poplar	A 70% reduced CAD activity that was associated with a red coloration of the xylem tissue. Modified lignin was observed (higher lignin extractability was observed) but the lignin quantity was identical	Baucher M, Chabbert G, Pilate J, *et al.* (1996) Red xylem and higher lignin extractability by down-regulating a cinnamyl alcohol dehydrogenase in poplar. *Plant Physiology* 112: 1479–1490
Suppression of the caffeic acid *O*-methyltransferase (COMT) gene in the xylem of quaking aspen (*Populus tremula*)	Incorporation of a higher amount of coniferyl aldehyde residues in the lignin. Characterization of the lignins in those trees showed that a high amount of coniferyl aldehyde could be the origin of the specific coloration observed	Tsai C-J, Popko JL, Mielke MR, *et al.* (1998) Suppression of O-methyltransferase gene by homologous sense transgene in quaking aspen causes red-brown wood phenotypes. *Plant Physiology* 117: 101–112
Altered expression of the caffeic acid/5-hydroxyferulic acid *O*-methyltransferase (COMT) gene in two 2-year-old poplar trees	A 90% lower COMT activity did not change lignin content but considerably increased the frequency of guaiacyl units and resistant biphenyl linkages in lignin. This alteration severely lowered the efficiency of Kraft pulping	Lapierre C, Pollet B, Petit-Conil M, *et al.* (1999) Structural alterations of lignins in transgenic poplars with depressed cinnamyl alcohol dehydrogenase or caffeic acid O-methyltransferase activity have an opposite impact on the efficiency of industrial kraft pulping. *Plant Physiology* 119: 153–164
Altered expression of cinnamyl alcohol dehydrogenase (CAD) gene in two 2-year-old poplar trees	Xylem displayed a red coloration mainly in the outer layers. No change in the lignin content but a greater proportion of syringaldehyde and diarylpropane structures were observed. An increased content in free phenolic units results in an easier lignin solubilization and fragmentation during Kraft pulping	Lapierre C, Pollet B, Petit-Conil M, *et al.* (1999) Structural alterations of lignins in transgenic poplars with depressed cinnamyl alcohol dehydrogenase or caffeic acid O-methyltransferase activity have an opposite impact on the efficiency of industrial kraft pulping. *Plant Physiology* 119: 153–164
Antisense inhibition of the 4CL gene encoding 4-coumarate:coenzyme A ligase in aspen	A 45% reduction of lignin and a 15% increase in cellulose. The total lignin–cellulose mass remained essentially unchanged. Leaf, root, and stem growth were improved	Hu WJ, Harding SA, Lung J, *et al.* (1999) Repression of lignin biosynthesis promotes cellulose accumulation and growth in transgenic trees. *Nature Biotechnology* 17: 808–812
Overexpression of the GA 20-oxidase gene from *Arabidopsis* in hybrid aspen	Faster growth in height and diameter, larger leaves, more numerous and longer xylem fibers were observed. Also, overexpression of the transgene resulted in an increased biomass that could also affect wood quality	Erikkson ME, Isrealsson M, Olsson O, and Moritz T (2000) Increased gibberellin biosynthesis in transgenic trees promotes growth, biomass production and xylem fiber length. *Nature Biotechnology* 18: 784–788
Gene silencing for caffeic acid *O*-methyltransferase (COMT) in poplar	Reduced lignin level in 6-month-old trees showing a structure with a higher content in condensed bonds (lack of syringyl units) and an important incorporation of 5-hydroxyguaiacyl units. A higher cellulose content and condensation degree of the lignin was observed	Jouanin L, Goujon T, de Nadaï V, *et al.* (2000) Lignification in transgenic poplars with extremely reduced caffeic acid O-methyltransferase activity. *Plant Physiology* 123: 1363–1374
Field trial of transgenic poplar with modified CAD and COMT gene expression	The growth indicators and performance were normal. The improved Kraft pulping characteristics were maintained during the 4 years' trial	Pilate G, Guiney E, Petit-Conil M, *et al.* (2002) Field and pulping performances of transgenic trees with altered lignification. *Nature Biotechnology* 20: 607–612

Metabolic Profiling of Transformed Plants

The metabolic side effects of lignin manipulation are not well known at present but since lignins are the major carbon sink in phenolic metabolism, a decrease in lignin content should make upstream precursors available for the synthesis of other phenolics. Indeed dramatic quantitative and qualitative changes in soluble phenolic profiles have been observed as a result of 4CL, CCR, and CCoAOMT downregulation. The impact of these changes on the industrial or nutritional properties of the transformed lines should be carefully evaluated in the future as well as their potential effects on plant–environment interactions.

In addition to phenolic compounds, potential changes in the proportion of cell wall polysaccharides, which have already been demonstrated for 4CL and CCR and CAD × CCR downregulated plants, should also be investigated in the future. It is still a matter of debate whether the carbon that is not used for lignin synthesis can be, at least in part, redeployed to the synthesis of polysaccharides in lignin-depleted plants or if the ratio of polysaccharides to lignins is simply increased by the reduction of lignin content. Whatever the mechanisms involved, plants with an increased polysaccharides : lignins ratio are interesting for the pulp industry.

Finally, apart from soluble phenolics and cell wall polysaccharides, other plant metabolites could be altered as a result of changes in lignification gene expression. Indeed, cross-talk between genes which have already been identified in some transgenic lines are likely common phenomena. Microarrays for probing gene expression and metabolic profiling of different classes of biochemical compounds will reveal the extent of such secondary effects which have to be clearly identified for the acceptance of the transgenic products.

Additional Issues

Due to a higher C : OH ratio, lignins have a higher calorific value than cellulose and their increase may be beneficial if the conversion of biomass to energy is envisaged. Very few studies have been initiated in this area. Despite the overexpression of different individual enzymes of the lignification pathway no reports of a significant increase in lignin content have been published to date. It is clear that, as already shown for flavonoid synthesis, the overexpression of transcription factors controlling several genes of the pathway could be an efficient way to enhance lignin synthesis in plants. As already discussed for the reduction of lignin quantity, the impact of such modifications on plant development should be carefully assessed.

Future work is also necessary to demonstrate clearly the advantages of the defined technologies when applied to species that are still difficult to transform genetically. As an example, we have recently successfully introduced CCR and CAD antisense constructs in *Eucalyptus globulus*, *E. camaldulensis*, and the *E. grandis* × *E. urophylla* hybrid. The target enzymes have a reduced activity level for some transgenic lines and the analysis of lignins is under way.

Finally, one of the main concerns about GM trees is their acceptance by public opinion. The fact that these transformed plants are nonfood products designed for industrial purposes should facilitate public acceptability. However, a careful examination of these new transgenic products is then absolutely necessary in order to evaluate the potential impacts, if any, which could outweigh the already demonstrated advantages for competitiveness of the pulp industry. The results of these studies should be clearly communicated to the public and if the lack of risks can be clearly demonstrated it will be necessary to highlight the advantages of the GM plants in terms of environmental benefits, energy savings, and economic competitivity.

In the event of continuing strong opposition at the public and political levels to the release of transgenic tree species in the environment, the data resulting from lignin GM experiments could be exploited in an alternative way. It is still possible, in most cases, to go back to natural populations in order to find genetic variability at the gene locus or find a mutant for this gene that exhibits a similar phenotype to the original transgenic species.

List of Technical Nomenclature

Latex — A viscous mixture produced in the bark of some tropical trees from which natural rubber is derived.

See also: **Cell Walls and Fibers**: Wood Growth and Development. **Secondary Products**: Lignin.

Further Reading

Baucher M, Monties B, Van Montagu M, and Boerjan W (1998) Biosynthesis and genetic engineering of lignin. *Critical Reviews in Plant Science* 17: 125–197.

Boudet A-M, Lapierre C, and Grima-Pettenati J (1995) Biochemistry and molecular biology of lignification, Tansley Review no. 80. *New Phytologist* 129: 203–236.

Boudet A-M, Goffner D, Marque C, Teulières C, and Grima-Pettenati J (1998) Genetic manipulation of lignin profiles a realistic challenge towards the qualitative improvement of plant biomass. *Agricultural Biotechnology News and Information* 10(9): 295N–304N.

Grima-Pettenati J and Goffner D (1999) Lignin genetic engineering revisited. *Plant Science* 145: 51–65.

Hu WJ, Harding SA, Lung J, *et al.* (1999) Repression of lignin biosynthesis promotes cellulose accumulation and growth in transgenic trees. *Nature Biotechnology* 17: 808–812.

Pilate G, Guiney E, Petit-Conil M, *et al.* (2002) Field and pulping performances of transgenic trees with altered lignification. *Nature Biotechnology* 20: 607–612.

GROWTH AND DEVELOPMENT

Contents

Cells

S C Fry, The University of Edinburgh, Edinburgh, UK

The Multicellular Plant

All embryophytes ("land plants;" a term which includes mosses, liverworts, ferns, and all seed-bearing plants) have bodies that are partitioned into numerous cells, each cell being bounded by a cellulose-rich cell wall (*see* **Cell Walls and Fibers**: Cell Walls). The only unicellular green "plants" are certain algae, e.g., *Chlorella*, but even most algal species are multicellular.

The cellular structure of a plant tissue was first observed in 1665 by Robert Hooke using the newly invented light microscope (LM). The tissue first described by Hooke was cork, in which the cell walls are very thick and therefore easy to see, but in which the cell contents are dead at maturity. Many mature plant cells, in contrast to this, have living contents.

The cell is the smallest unit of a plant that can remain viable in isolation. The cell, then, is the elementary unit of plant life. Indeed, isolated plant cells (even some that are highly differentiated) can be kept alive indefinitely when incubated on suitable artificial media. They will continue (or resume) cell division and even go on to develop into whole new plants: the single cell evidently contains all the genetic "know-how" necessary for production of an entire plant. This biotechnological feat of plant regeneration is much more commonplace than the recent successes in mammalian regeneration exemplified by "Dolly" the cloned sheep.

In a few plant tissues, the protoplasm is not partitioned into cells; instead, numerous nuclei are bathed in a common cytoplasm. An example of such a multinucleate structure (coenocyte) is the endosperm in its early stages of development.

Why have cells? There are several answers to this question, including: (1) the network of cell walls, which ramifies through the plant's body, confers strength; (2) in the event of accidental or microbial damage, a small number of cells can die without the demise of the whole organ ("bleeding to death"); (3) division of labor – the different cells within an organism can take on different specialized roles, the whole being more successful than the sum of its parts; and (4) the system of air spaces between neighboring cells facilitates gas exchange between the atmosphere and the internal cells that are buried far deeper in a tissue than would possibly permit gas exchange by diffusion of dissolved gases through cytoplasm alone.

The cell normally consists of a protoplast and a cell wall. The protoplast, whose outer layer is called the plasma membrane, is sometimes regarded as the "living part" of the cell, and the wall as a nonliving product secreted by the protoplast. However, the distinction is not clearcut, and metabolism – one of the phenomena that defines "living" – takes place in the wall as well as in the protoplasm. In addition, not all parts of the protoplast are "living:" for example, tannin bodies and starch grains, within the protoplast, would normally be regarded as nonliving ergastic substances. In fact, no individual part of the cell can really be regarded as "living:" the mitochondrion, for example, is merely a power-station (making ATP, etc.) and the nucleus is merely an encyclopedia (storing DNA), neither in itself viable but nevertheless necessary for the life of the cell. In the light of the above, the wall is also regarded as part of the living cell, albeit external to the protoplast.

The protoplast can most easily be seen by light microscopy after plasmolysis: if the cell is placed in a concentrated, nontoxic solution, e.g., of a sugar or salt, the protoplast loses water by osmosis (through the selectively permeable cell membrane) and thus shrinks. The cell wall does not normally shrink to the same extent, so the protoplast becomes detached from the wall over much of its area, and can be seen as a separate, often almost spherical, entity. This is especially clear if the cell has previously been loaded with a nontoxic, colored solute such as methyl red. Susceptibility to plasmolysis is a simple, rapid test of whether a cell is alive since on death a cell quickly loses its ability to plasmolyze because the cell membrane ceases to be selectively permeable.

A related test of whether a cell is alive involves application of a solution of fluorescein diacetate (FDA), which is hydrophobic enough to pass through the cell membrane. Inside the protoplast, esterases hydrolyze the FDA to fluorescein, which, unlike FDA, is too hydrophilic to exit from the protoplast and is highly fluorescent under ultraviolet light. An FDA-treated cell thus only becomes fluorescent if it (1) contains active esterases and (2) possesses an intact plasma membrane – i.e., if it is, by these criteria, "alive."

Protoplasts can be isolated, i.e., freed from their cell walls, by digestion with a mixture of suitable enzymes (cellulase, pectinase, etc.). Isolated protoplasts are spherical, indicating that the original shape of the cell had been dictated by the wall. In addition, the diameter of the protoplast is governed by the osmotic pressure of the bathing solution, indicating that the original cell volume had been dictated by the wall. When placed in a dilute solution or pure water, the isolated protoplast swells so much that it bursts. On a suitable medium, with an adequate osmotic pressure to prevent bursting, the isolated protoplast can regenerate a cell wall; cell division and growth can then resume.

The smallest plant cells, found in embryos and apical meristems, are somewhat under 10 μm in all dimensions (i.e., they are isodiametric) and the nucleus occupies most of the cell's volume. Such cells tend to be closely packed together with no air spaces between them. The absence of air spaces, and thus reliance on the diffusion of dissolved (not gaseous) O_2 to enable respiration in the center of the meristem, limits the maximum possible diameter of an apical meristem to roughly 3 mm. In fact, the diameter of the shoot apical meristem varies from 50–100 μm (e.g., in *Arabidopsis*) to roughly 3 mm (e.g., in *Chrysanthemum*).

Mature plant cells, in contrast, are usually much larger than 10 μm and are often elongated. Mature cells of the pith and cortex are often 30–100 μm in diameter and may be 500–1000 μm or more in length. Therefore, a 1000-fold increase in cell volume during maturation is commonplace in plants. Such extreme cell expansion is rare in animals – a mature tree can contain fewer cells than a single human liver, such is the difference in mature cell volume between animals and plants. The vast majority of the volume within a mature plant cell is vacuole.

Once a cell has divided to form a pair of daughter cells, these normally remain attached to each other throughout the development of the plant. This contrasts with animal development, where sister cells can migrate to different destinations, e.g., within an embryo. Plant cells cannot slither in this way because they are joined to each other via the middle lamella. Indeed, so permanent is the linkage between neighboring cells that, at certain points, tiny pores in the cell wall are precisely and permanently lined up with similar pores in the wall of the neighboring cell, forming intercellular channels of protoplasmic continuity called plasmodesmata.

Parts of the "Typical" Plant Cell

The cell is typically composed of a cell wall and a protoplast (see above). The protoplast usually consists of the cytoplasm plus one nucleus. (The term "protoplast" is sometimes used as a shorthand for "isolated protoplast," i.e., a cell with its wall removed; however, this usage is not standard. Normal protoplasts do have walls.) The cytoplasm consists of cytosol plus several distinct organelles, described later. Thus, chloroplasts, mitochondria, Golgi bodies, ribosomes, cytosol, etc. are all part of the cytoplasm.

Subcellular components larger than about 0.3 μm can be seen (i.e., their shape and size can be perceived) by LM; smaller ones only by electron microscopy (EM). However, structures smaller than 0.3 μm can still be detected by LM for example if they mark a boundary between one phase and another; for example, the presence of a primary cell wall (commonly only ~80 nm thick) can be detected by LM. Indeed, many objects much smaller than 0.3 μm can be detected (even if not strictly "seen") by LM if they are first made fluorescent by application of a suitable reagent or fluorescently labeled antibody.

There is no such thing as "the typical" plant cell: each cell type has a unique set of structural and functional features. Nevertheless, there are certain structural components, described below, that are common to many plant cells. Many of these components are membrane-bounded organelles, so it is relevant first to consider membranes in general.

Membranes

A membrane is a very thin, lipid-rich sheet. The thickness, as seen by transmission electron microscopy (TEM) is just less than 10 nm; the membrane therefore contrasts with the cell wall, which usually contains little lipid and is >80 nm thick. Membranes are based on a bilayer of polar lipids, usually mainly phosphatidyl choline, phosphatidyl ethanolamine, and related saponifiable phospholipids, and usually with a high proportion of unsaturated fatty acids. The lipids are arranged with their polar groups (phosphate, choline, etc.) facing outwards and their fatty acid chains meeting in the center of the membrane. A bilayer is a stable arrangement for a flat membrane in a biological milieu because both sides of the membrane then have a high affinity for the major surrounding solvent, water, whereas the interior of the membrane is a self-coherent nonpolar ("oily") structure held together by hydrophobic interactions. Other, less polar, lipids are also present, e.g., sterols.

Embedded within the membrane, and protruding from it on one or both sides, are numerous proteins (enzymes, channel proteins, permeases, receptors, etc.), making up roughly half the dry weight of the membrane. Some of the proteins are covalently anchored to specific minor lipids, such as glycosylated phosphatidylinositols (GPIs). Often, some of the proteins in a membrane are glycosylated, usually far more on the side facing away from the cytosol.

Most biological membranes are selectively permeable, i.e., they permit water and certain other small molecules to pass through, whereas they prevent the passage of most hydrophilic molecules (sugars, inorganic ions, amino acids, organic acids, phosphorylated compounds, etc.). Exceptions include the outer membranes of plastidial and mitochondrial envelopes, which are unusual in allowing most small, hydrophilic molecules to pass through. Other membranes can also permit the passage of specific solutes if there are suitable proteins (carriers, pumps, or channels) in the membrane. Hydrophobic solutes, e.g., cinnamic acid, readily pass through all plant cell membranes; to prevent this, the cell usually attaches hydrophilic groups (e.g., sugar residues) to such molecules.

The Three Genomic Organelles

The plant has three genomes, located in the nucleus, plastids, and mitochondria respectively. Each of these three organelles is bounded by an envelope comprising a double membrane (i.e., two bilayers). Nuclei, plastids, and mitochondria can be partially separated from each other (after homogenization of the tissue) by differential centrifugation: the rate of sedimentation is usually nuclei > chloroplasts > mitochondria, i.e., in proportion to the mass of the organelle.

Nucleus Most of the plant's genes (10^8–10^{11} base pairs of DNA) are in the nucleus, a relatively large organelle, often roughly oval in shape, and first noted by Robert Brown in 1833. In an apical meristem cell, the nucleus is often roughly spherical and may occupy as much as 70% of the cell's volume. In vacuolated cells, the nucleus is often squashed into an oval shape. Almost all living plant cells (except mature sieve tube members) possess nuclei, usually one per cell.

Scanning electron microscopy (SEM) shows the nuclear envelope to be characterized by the presence of structurally complex nuclear pores that penetrate both the membranes of the nuclear envelope and thus connect the contents of the nucleus (nucleoplasm) with the cytosol. These pores recognize specific polymer molecules and permit these to enter and leave the nucleus. The outer membrane of the nuclear envelope may have continuities with the endoplasmic reticulum (see below).

The DNA of the nucleus is associated with specific proteins (basic histones and numerous relatively acidic proteins), in a macromolecular complex called chromatin. During interphase (the period when the nucleus is not undergoing division (mitosis)), most of the chromatin is relatively dispersed (euchromatin), although ~10% remains highly condensed (heterochromatin). The genes that are acting as templates for RNA synthesis (transcription) are located within the euchromatin. During mitosis, the chromatin condenses very tightly such that the individual chromosomes can be clearly distinguished by LM.

The nucleolus is a membraneless substructure visible within the nucleus by both LM and TEM; it is the site at which rRNA and ribosomal subunits are synthesized.

Plastids and mitochondria A small proportion of the cell's genes are housed in plastids and mitochondria, a group of semiautonomous organelles that probably descended from endosymbiotic bacteria and that contain a small circular nucleoid ("chromosome") made up of ~145 000 and ~200 000 base pairs of DNA in plastids and mitochondria respectively. This DNA is not complexed with histones and does not form chromosomes that are visible by LM. Although the organelles' genomes are small, there are multiple copies per cell; for example, spinach (*Spinacia oleracea*) leaves contain ~5500 copies of the plastid genome per cell, accounting for ~23% of the cell's total DNA. Both plastids and mitochondria have an envelope consisting of two very different membranes: an outer one which is unusual in allowing almost any small molecule to pass through it, and an inner one that allows only a restricted subset of small molecules to enter and leave. Most of the volume inside the organelle is a "bag of enzymes" called the stroma (in plastids) or matrix (in mitochondria), which also contains DNA, RNA, ribosomes, intermediary metabolites, and (in plastids) starch. The plastids and mitochondria in an individual plant are maternal (derived only from the egg cell, not from the pollen). They increase in number during plant growth only by division of existing plastids and mitochondria.

The best-known plastids are chloroplasts, which contain the pigments (chlorophylls and carotenoids) responsible for photosynthesis. Typically there may be ~50 chloroplasts per mesophyll cell, each shaped somewhat like a convex lens roughly 5 μm in diameter and 2–3 μm thick. The stroma is the location of the dark reactions of photosynthesis (Calvin cycle). Chlorophylls and carotenoids are hydrophobic substances located in internal plastid membranes called thylakoids. Thylakoid membranes form flattened sacs (lamellae), some of which are stacked within the matrix to form dark green structures (grana) that are just visible by LM; others (stromal lamellae) are suspended more loosely between adjacent grana. The light reactions of photosynthesis occur on and in the thylakoid membranes – electron transport (from H_2O to $NADP^+$), leading to NADPH and O_2 production and setting up a pH gradient between the stroma (alkaline) and the thylakoid lumen (acidic). The energy represented by this gradient is then conserved in ATP synthesis (photophosphorylation), catalyzed by ATP synthase, which is also located in the thylakoids.

Nongreen plastids are named mainly according to characteristic substances that they contain, for example, amyloplasts (storing starch grains, e.g., in tubers, rhizomes, seeds, etc.), chromoplasts (rich in carotenoids but not chlorophylls, e.g., in some petals and fruits), and elaioplasts (oils). All plastids are derived ultimately from very small, colorless "proplastids," which occur in the zygote, the embryo, and other meristematic tissues. Proplastids increase in number by fission and they then differentiate, depending on which cells they find themselves in, into chloroplasts, chromoplasts, amyloplasts, elaioplasts, or etioplasts (small, colorless plastids found in etiolated shoots). In addition, mature plastids can be converted from one differentiated form to another, e.g., chloroplasts → chromoplasts in ripening fruit (*see* **Postharvest Physiology**: Ripening).

Mitochondria are smaller (e.g., 1 × 0.5 μm; just visible by LM) but more numerous (e.g., 500 per cell) than plastids. Mitochondria constantly change in shape – often tubular, sometimes lobed, but also sometimes spherical. The matrix is the location of the Krebs cycle. The inner membrane of the mitochondrial envelope is characterized by extensive infoldings, forming cristae. Membranes of the cristae are the site of the respiratory electron transport chain (from NADH to O_2), whose operation sets up a pH gradient between the matrix (alkaline) and the intermembrane space (acidic). The energy invested in this gradient is then conserved in ATP synthesis from ADP (oxidative phosphorylation), catalyzed by ATP synthase, which is also located in the cristae.

The brief summary given above indicates that there are remarkably fundamental similarities – metabolic, structural, and genetic – between chloroplasts and mitochondria.

Nongenomic Organelles with Single Membranes

The cell contains several types of organelle that differ from nuclei, plastids, and mitochondria in lacking DNA and in being bounded by a single membrane instead of a double one. Although the various single-membraned organelles are distinguishable from each other, they may be related to one another in the sense that pieces of membrane can move from one to another. Some of these organelles can be separated, after homogenization of the cell, by isopycnic centrifugation in a preformed density gradient of sucrose or Ficoll (a synthetic polymer of sucrose).

Typically their buoyant densities are in the order endoplasmic reticulum < Golgi body < plasma membrane.

Vacuoles Plant cells often undergo enormous expansion as they mature from their meristematic origins. At maturity, the vast majority of the cell's volume may be occupied by a sap-filled vacuole, the other organelles such as plastids, mitochondria, etc. being confined to a very thin layer of cytosol squashed between the vacuole and the plasma membrane. Meristematic cells have numerous, much smaller vacuoles.

Vacuolar sap is often acidic and contains a high concentration of small molecules, such as citrate, malate, sucrose, and inorganic salts, which are actively transported there from the cytosol. As a result, the vacuolar sap has a high osmotic pressure, so there is a constant tendency for water to enter, ensuring that the cell maintains a high hydrostatic (turgor) pressure. The vacuole also contains numerous enzymes, especially hydrolases (such as α-mannosidase and proteases), which, when released by programmed or accidental cell death, may digest other cellular components so as to recycle their building-blocks. Vacuolar sap also contains secondary metabolites, e.g., phenolic compounds and the red pigment (betanin) of beetroots (*Beta vulgaris*). Some of these compounds (e.g., dopamine, present in vacuoles of banana (*Musa sapientum*) skins) are highly reactive and, when released from the vacuole by cellular damage, come into contact with oxidases elsewhere in the cytoplasm, resulting in a rapid browning reaction. The vacuole may also contain visible particles, e.g., crystals (see below).

The membrane surrounding the vacuole is called the tonoplast. Among the specific proteins that it contains are some that can utilize ATP to "pump" solutes from the cytosol into the vacuole. Sometimes the vacuole is traversed by strands of cytoplasm (still separated from the vacuolar sap by tonoplast), and sometimes these strands may contain other visible organelles such as plastids or the nucleus.

Endoplasmic reticulum The endoplasmic reticulum (ER) is composed of membrane-bounded, flattened sacs (cisternae), each enclosing a narrow lumen. There are connections between the ER and the outer membrane of the nuclear envelope.

Rough ER (RER) has its surface covered with active ribosomes and is the site of synthesis of those proteins that are destined for any of several other membrane-bounded organelles (e.g., Golgi bodies, plasma membrane, tonoplast, and the ER itself), or the apoplast. Such proteins have a hydrophobic "signal peptide" at the N-terminus, which, even before translation is complete, docks with a signal recognition particle, enabling binding to the outside of the ER. The nascent protein, led by its signal peptide, then threads through the membrane into the ER lumen, where the signal peptide is cleaved by a specific proteinase. Once inside the ER, these proteins are often glycosylated on the side chain N atom of certain asparagine residues.

Smooth ER is free of ribosomes and is involved in lipid synthesis; its cisternae tend to be more tubular than flattened. Intermediate forms of ER (with few ribosomes) also occur.

Golgi body and its vesicles Most plant cells contain several (sometimes more than 100) Golgi bodies (dictyosomes), each of which consists of a stack of 3–10 membrane-bounded, flattened sacs (cisternae). Vesicles bud off the ER and carry newly synthesized proteins to one side (the *cis*-cisternae) of a Golgi body. Further vesicles then bud off this, progressively carrying membrane and lumen contents along the endomembrane traffic system: *cis*-cisternae → *medial* cisternae → *trans*-cisternae → *trans*-Golgi network → secretory vesicles → plasma membrane. When the secretory vesicles fuse with the plasma membrane, their contents are exuded onto the inner face of the cell wall, a process called exocytosis. The approximate transit time from Golgi to plasma membrane is 15–30 min. During this journey, proteins are further modified, e.g., by addition and removal of sugar residues. In addition, most of the noncellulosic polysaccharides destined for the cell wall (pectins and most hemicelluloses except callose) are synthesized in the various Golgi cisternae.

Plasma membrane The whole protoplast is surrounded by a single membrane — the plasma membrane (PM) or plasmalemma. Topologically, the outside of the PM is equivalent to the inner face of the ER and Golgi cisternae; this face is particularly rich in glycosylated proteins. The PM normally adheres to the cell wall, but can be pulled away from it, e.g., during plasmolysis. The PM is the location of numerous proteins that enable the uptake and export of specific, small, hydrophilic solutes, e.g., certain sugars, amino acids, and inorganic ions. There are also proteins that act as water channels (aquaporins), although it seems clear that enough water can pass through most cells' PMs to satisfy requirements without the aid of aquaporins; those cells (especially the root endodermis) whose PMs stand in the way of the transpiration stream may be exceptions. The PM is also the site at which cellulose is synthesized – on 'terminal rosettes', observed by

Table 1 Selected stains and related methods used for the microscopic localization and partial characterization of plant cell wall components

Stain	*Mode of action*	*Wall components typically detected*	*Types and parts of wall typically stained*
Ruthenium red	Stain has positive charge, therefore binds to polyanions	Acidic polymers, e.g., pectins, arabinogalactan-proteins, mucilages	Middle lamella, primary walls, root cap
Calcofluor white	Intercalates with β-(1→4)-linked polysaccharides, and fluoresces under UV	Cellulose	Virtually all walls of higher plants
Phloroglucinol–HCl	Reacts chemically with phenolic aldehydes to form a red product	Lignin	Xylem vessels and tracheids, some fibers, endodermis of mature root
Sudan III	Partitions into hydrophobic phases of specimen, imparting red color to them	Cutin, suberin	Epidermis, cork, Casparian strip
Toluidine blue O	Metachromatic. (a) Stain has a positive charge, so binds to polyanions (→ purple); (b) forms molecular stacks with aromatic components (→ green). A combination of (a) and (b) gives blue cell walls	(a) Pectin, etc.; (b) lignin, etc.	Most plant cell walls: tissues distinguished by color reactions
None (autofluorescence)	Aromatic wall components themselves fluoresce under UV	Feruloyl groups	Cuticle of most plants; primary walls of parenchyma in grasses and Chenopodiaceae
None (UV absorbance)	Aromatic wall components themselves absorb UV giving dark zone on UV "photograph"	Lignin	As phloroglucinol–HCl
Fluorescently[a] labeled *Ulex* lectin	Lectin recognizes and binds to α-L-fucose residues	Xyloglucan (and some glycoproteins?)	Primary walls of dicotyledons
Fluorescently[a] labeled antibody	Antibody, raised against a specific wall component ("antigen"), recognizes and binds to that component	Many diverse examples, e.g., β-(1→4)-galactans, arabinogalactan-proteins (AGPs), enzymes, lignin (when present in trace amounts)	Diverse. For instance, meristematic cells with specific destinies (e.g., future xylem, future endodermis, etc.) may have specific AGPs in their walls

[a] Or gold labeled, for use in electron microscopy.

EM at the growing end of each microfibril (*see* **Primary Products**: Cellulose).

Microbodies Microbodies (microsomes) are small, almost spherical organelles (diameter 1–2 μm) bounded by a single membrane. There are two main types: peroxisomes and glyoxysomes. Peroxisomes occur in green cells and contain oxidases, especially glycollate oxidase, which catalyzes a reaction important in photorespiration

$$\text{glycollate} + O_2 \rightarrow \text{glyoxylate} + H_2O_2,$$

and catalase, which rapidly destroys the potentially dangerous H_2O_2 thus formed. Glyoxysomes occur in tissues that are actively converting lipid to sugar (e.g., cotyledons of oily seeds after germination), and contain the enzymes isocitrate lyase and malate synthase necessary for this pathway.

Oleosomes (lipid bodies; spherosomes) In oily seeds, the oils are stored in approximately spherical organelles that are bounded by a "half-membrane" – a phospholipid monolayer (*see* **Primary Products**: Oils). The outer face of this oleosome membrane is relatively normal (phospholipids, with their polar head-groups facing the cytosol, and including membrane proteins), but the inner face consists of the fatty acid residues of the phospholipids in contact with (and effectively dissolved in) the stored oil droplet. Oleosomes are formed by budding off from the smooth ER. The latex-storing organelles of laticifers (*see* **Secondary Products**: Rubber Production) are also bounded by similar "half-membranes."

Cell Components Lacking Membranes

Cytosol The soluble phase of the cytoplasm, not contained within any other organelle, is termed the cytosol. Besides water, it contains soluble proteins (a very high concentration, e.g., $300\,mg\,ml^{-1}$), mRNA, tRNA, and intermediary metabolites (e.g., those of glycolysis). The cytosol (plus many of the

Table 2 Distinguishing features of diverse plant cells

Cell type	*Cell location; function*	*Cell shape and size*	*Any special features of protoplast*	*Cell division*	*Special chemical components or metabolic pathways*	*Wall thickness*	*Wall chemistry (in dicotyledons)*	*Air spaces*
Apical meristems	Shoot apex, leaf primordium, root apex. Divide to form new cells that grow into stem, leaf, root	Isodiametric. The plant's smallest cells (e.g., 5–10 μm in each dimension).	Nucleus occupies over half cell volume. Many very small vacuoles. Proplastids. Numerous ribosomes.	Rapid: e.g., 1–2 divisions per day.	DNA, RNA, protein, and phospholipid synthesis.	Very thin (~80 nm). Only 1°.	Cellulose ≈ xyloglucan ≈ pectins > xylans, etc. No lignin, cutin, suberin.	None.
Vascular cambium	Thin cylinder of cells between 2° xylem and 2° phloem. Divide to give new cells that form new xylem, phloem and rays	Fusiform initials: slablike cells (e.g., 150 × 20 × 5 μm). Ray initials: isodiametric (~7–20 μm).	Large vacuole. Numerous ribosomes, Golgi bodies, much endoplasmic reticulum.	±Rapid: e.g., 50 per year.	DNA, RNA, protein, and phospholipid synthesis.	Very thin (~80 nm). Only 1°.	As above.	None.
Cork cambium (phellogen)	Inner surface of cork in bark of tree. Divide to give new cells that form cork.	Slablike cells: tangential wall, e.g., 20–50 μm; radial wall, e.g., ~10 μm.	Large vacuole.	E.g. 5 per year.	Presumably as above.	Presumably as above.	Presumably as above.	None.
Parenchyma	Cortex and pith of stem and root; pericarp of fruit; also similar cells within other tissues (e.g., rays, phloem parenchyma). Sometimes called "packing cells."	Varied. Often ± isodiametric, or somewhat elongated along axis of organ. Size: moderate to very large, e.g., 25–250 μm.	Extremely large vacuole.	Very rare at maturity.	Often starch; sometimes oils, tannins ("tanniniferous cells": Figure 1) or crystals.	Usually thin, 1° wall.	Usually as above.	Large.
Chlorenchyma (including mesophyll)	(Special case of parenchyma.) Mesophyll, outer layers of cortex in stem.	Spongy mesophyll: ± isodiametric (e.g., 30–80 μm), with short "arms" pushing out between neighboring cells. Palisade mesophyll: ±cylindrical; e.g., 15–25 μm diameter, ~100 μm long in anticlinal direction.	Numerous chloroplasts. Large vacuole.	Very rare at maturity.	Photosynthesis. Starch made by day, degraded at night. Synthesis and export of sucrose. Chlorophyll. Sulphated galactolipids.	Usually thin, 1° wall.	Presumably as above.	Large.

Aerenchyma	(Special case of parenchyma.) In pith and cortex of anaerobic organs; e.g., roots in waterlogged soil. Facilitate air transport.	±Isodiametric but with long "arms" (e.g., 25–50 μm) between neighboring cells.	Large vacuole	Very rare at maturity.		Usually thin, 1° wall.	Presumably as above.	Extremely large.
Collenchyma	Within cortex of stem and petiole, either just beneath the epidermis or a few cells away from it. Often mainly beneath "ribs" on the stem. Protective and strengthening, but resilient.	Often square or pentagonal in TS (e.g., 30–100 μm diameter), elongated (e.g., 1000 μm) along long axis of stem. Walls thickened in angular, lacunar, or lamellar pattern.	Alive at maturity, despite the thickened walls. May be some chloroplasts.	Very rare at maturity.	Abundant pectins.	Irregularly thickened 1° wall, often especially near cell corners.	As above, or richer in pectins.	Sometimes; if so, with wall thickenings facing air spaces.
Sclerenchyma (fibers)	Roots, stems, leaves, fruits. Phloem fibers: exterior of vascular bundles (as bundle cap). Xylem fibers: mixed with xylem vessels and tracheids. In monocotyledon leaves, fibers may form a sheath round the vascular bundle, extending to both epidermises. Protective, strengthening; less resilient than collenchyma.	Often ± hexagonal in TS (e.g., 15–25 μm diameter); long or extremely long cells in LS (1–50 mm; ramie fibers up to 0.5 m). Elongate (by tip growth) during stem elongation; may continue to elongate thereafter.	Dead after lignification, but can stay unlignified (alive) for long periods.	Very rare.	Wall polysaccharides and lignin synthesis.	Very thick (e.g., 10–15 μm), especially 2° wall.	2° wall: cellulose, xylans, mannans (no xyloglucan or pectin); often rich in lignin at maturity.	Usually none.
Sclerenchyma (stone cells or sclereids)	Small clusters within pith and cortex of stems and petioles. Complete layer in seed coat. Protective. Feeding deterrent?	Very diverse shapes/sizes. Brachysclereids ±isodiametric (e.g., 50 μm); other sclereids up to 10-fold elongated, sometimes H- or star-shaped (e.g., 500-μm arms).	Often dead at maturity (after lignification).	Very rare.	Wall polysaccharides and lignin synthesis.	Thick, especially 2° wall. Often with prominent pits.	Usually lignified.	Usually none.

continued

Table 2 Continued

Cell type	*Cell location; function*	*Cell shape and size*	*Any special features of protoplast*	*Cell division*	*Special chemical components or metabolic pathways*	*Wall thickness*	*Wall chemistry (in dicotyledons)*	*Air spaces*
Epidermis	Outer surface of whole (young) plant. Usually one cell thick.	Brick-shaped, or with wavy side walls. Variable size: e.g., $25 \times 50 \times 200\,\mu m$ in stem, $20 \times 30 \times 30$ in leaf.	Usually no chloroplasts. Large vacuole. Large Golgi bodies.	Continues during differentiation, becoming rarer with maturity.	Synthesis of cutin and waxes. Flavonoids in vacuole (shoot epidermis).	Outer wall much thicker than inner wall.	Covered by cutin (much thicker in shoot than root).	None.
Hairs	Outgrowths of epidermal cells, forming trichomes (sometimes multicellular; in shoots) or root hairs (usually unicellular; in roots).	Long cylindrical outgrowth from epidermal cell (trichoblast). Very long in the case of cotton ovule hairs ("cotton fibers").	Usually no chloroplasts. Large vacuole.	Rare.	Variable.	Variable.	May be silicified. Cotton ovule hair walls ± pure cellulose.	None.
Guard cells	Modified epidermal cells (in shoot, not root).	In dicotyledons: a pair of kidney-shaped cells, each; e.g., $15 \times 30\,\mu m$. In grasses: a pair of dumbell-shaped cells.	With chloroplasts (unlike other epidermal cells) and vacuole.	No.	Periodic K^+ influx/efflux.	Thick on side facing stoma.		Stoma (pore) between pair of guard cells.
Cork (phellem)	Near exterior of plant. Often originating in cortex, later in phloem. Also formed on wounded surfaces. Protective, indigestible, waterproof cover.	Often ± cubic; length, e.g., $30\,\mu m$.	Dead at maturity.	No.	Suberin synthesis.	Moderate.	Impregnated with suberin.	No (except at special "vents" called lenticels).

Xylem	In stele / vascular bundles, usually towards center of stem (facing pith) or of root; towards adaxial surface of leaf. 1° xylem formed from procambium, 2° xylem from cambium. Transport of water and salt; strength.	Tracheids (in all vascular plants): long (e.g., 300 μm), fairly narrow (e.g., 30 μm), ±pointed at both ends. Vessel elements (in angiosperms): fairly long (e.g., 150–500 μm), and wide (e.g., 30–100 μm); arranged end-to-end with perforation plates (holes in end walls) aligned to transport water.	Much endoplasmic reticulum and many Golgi bodies in immature cell. Dead at maturity (after lignification).	No.	Lignin	2° wall thickened in characteristic patterns: e.g., annular, helical, reticulate, or pitted.	Impregnated with lignin.	No.
Phloem	In association with xylem, but more external in stems and in mature roots, and abaxial in leaves. Translocation of sugars, etc. from source to sink tissues.	Sieve elements: narrow to moderate diameter (e.g., 5–40 μm), long (e.g., 200–500 μm), arranged end-to-end with enlarged plasmodesmata in sieve plates (end walls) aligned to conduct sap. Companion cell: similar length but narrower (e.g., 3–10 μm), lying along side of sieve element.	Sieve tube members: lose nucleus but remain alive at maturity. Plasma membrane intact. Companion cell: retains nucleus, densely cytoplasmic, alive at maturity.	No.	Protein fibrils within sieve tube cytoplasm.	Side walls fairly thin (e.g., 200 nm); sieve plate thicker (e.g., 400 nm).	Callose in sieve plates.	No.
Endodermis	Innermost layer of cortex. Control of salt uptake by root; graviperception in shoots.	Brick-like cells: e.g., in TS 30 × 50 μm in shoot; e.g., 15 × 20 μm in root.	In shoot, rich in amyloplasts (which fall under gravity, i.e., act as stotoliths).	Rare at maturity.	Starch, suberin synthesis.	Often fairly thick.	In roots: suberin in radial walls, i.e., the Casparian band; often lignified at maturity.	No.

TS, transverse section; LS, longitudinal section; 1°, primary; 2°, secondary; ±, more or less.

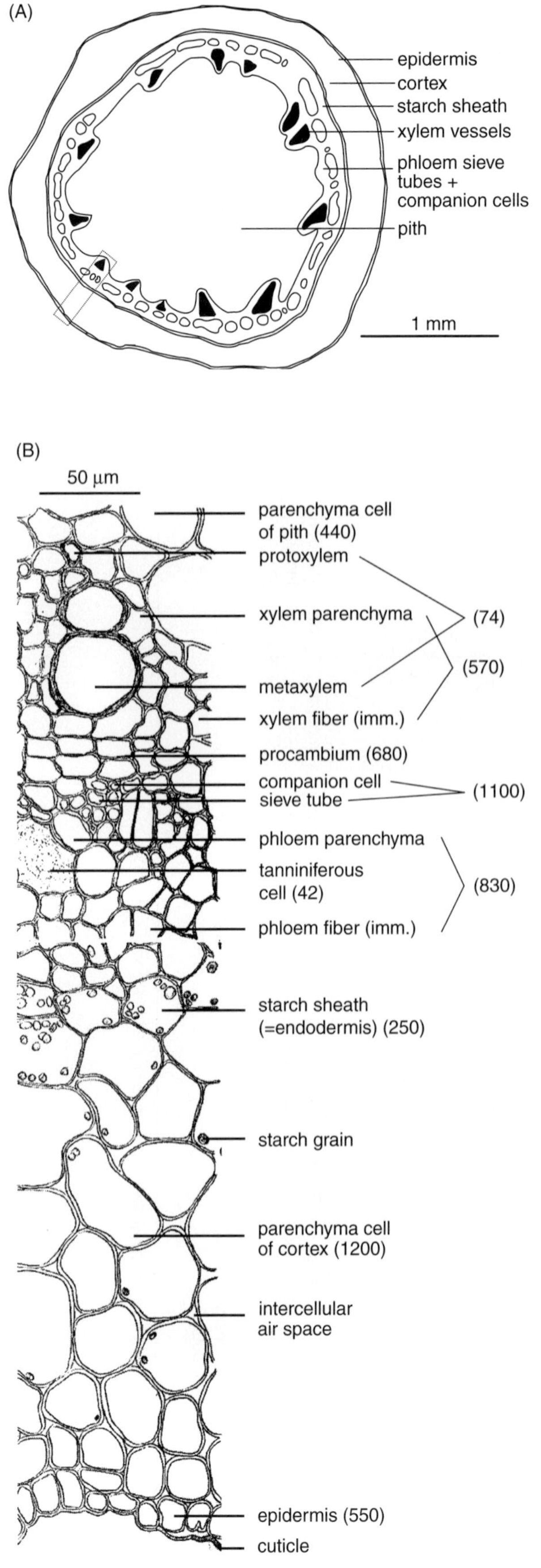

cytoplasmic organelles suspended in it) constantly moves around a healthy plant cell, a phenomenon called cytoplasmic streaming, readily visible by LM. Other organelles, e.g., the nucleus, are relatively immobile.

Ribosomes Ribosomes are very small macromolecular complexes, composed of rRNA and proteins, and are responsible for protein synthesis (translation) on mRNA templates. Each ribosome is composed of a large and a small subunit. Ribosomal subunits are made in the nucleolus and then exported into the cytoplasm, where they pair off to form whole ribosomes. Often several ribosomes are translating a given mRNA molecule simultaneously; such a structure is called a polysome. Plastids and mitochondria contain ribosomes that are slightly smaller (similar to those of bacteria) than those in the cytosol.

The cytoskeleton A three-dimensional scaffolding runs throughout the cytosol. One component of this scaffolding is the microtubule, a hollow cylinder 25 nm in diameter and many μm long. Microtubules are built up of globular proteins called tubulins. Also present are microfilaments, which are long chains (7 nm in diameter; not hollow) of a globular protein called G-actin. There are also "intermediate filaments," which are ~10 nm in diameter and composed of various rod-shaped proteins. Microtubules and microfilaments can be constantly lengthened (by addition of new protein molecules, mainly at one specific end called the + end) and shortened.

Microtubules are concerned with the movement of chromosomes during mitosis. Initially, microtubules form a preprophase band around the cell (just inside the PM), encircling the nucleus and marking the position at which the new cell wall will form. Later, microtubules form a mitotic "spindle" and associated structures, which guide chromatids to opposite poles of the dividing cell and direct ER and vesicles to the equator between the two daughter nuclei that form in telophase, where they are organized into a disk called the phragmoplast. Within the phragmoplast a cell plate and new cell wall then form. The movement of chromatids, vesicles, etc. is achieved by

Figure 1 Differentiated plant cells. (A) Cross-section of a beanstalk. Tissue distribution diagram of a transverse section of the hypocotyl of an 11-day-old etiolated seedling of *Phaseolus vulgaris* (cv Chevrier Vert). The section was taken just below the cotyledons. (B) More detailed histological drawing of the boxed part of (A). The numbers indicate the number of such cells visible in the whole area of (A). imm., immature.

certain motor proteins, e.g., kinesins and dyneins, which associate with microtubules. Between successive mitoses (i.e., during interphase), microtubules are found just below the PM, where they often appear to orientate the deposition of new cellulose microfibrils in the growing cell wall.

Microfilaments are responsible for cytoplasmic streaming. Motor proteins called myosins bind to microfilaments and can then hydrolyze ATP, the resulting free energy being conserved as motion. Microfilaments may also guide Golgi vesicles to appropriate areas of the PM prior to exocytosis.

Crystals Diverse crystals are often located within the vacuole(s). Most of the crystals found in plants are composed of calcium oxalate. They vary in shape: for example, they may either be simple prism-shaped crystals, or numerous prisms may be clustered into a single spiky body called a druse. There are also long, needle-shaped crystals (raphides), pointed at both ends, found for example in asparagus (*Asparagus officinalis*) foliage. Other crystals, known as cystoliths, are composed of $CaCO_3$. Certain cells also contain visible particles of phenolic material, referred to as tannin bodies. Crystals of protein, or other particles of protein called aleurone grains, are found in many seeds, e.g., in the cereal endosperm. Protein crystals may be storage products; calcium oxalate crystals may be waste products and/or herbivore deterrents.

Most crystals are found in parenchyma cells, e.g., of the cortex, pith, and secondary phloem. Sometimes the crystals are confined to specific cells within the parenchymatous tissue; such cells are called idioblasts and are often much larger than their neighbors.

Cell Walls

The cell wall is a highly characteristic component of plant cells and is mainly composed of polysaccharides (cellulose, pectins, and hemicelluloses), with smaller amounts of glycoproteins and, in specialized cells, other components. Such components help to characterize different cell types (see below). Although plant cell walls are often described as "rigid," this is not a valid generalization; "strong" and "inextensible" would be more helpful descriptions. Primary walls are usually very flexible (the opposite of rigid) (*see* **Cell Walls and Fibers:** Cell Walls).

The walls of neighboring cells are glued together by a pectin-rich middle lamella. A selection of cytological stains which enable one to see the various chemical components of the cell wall is listed in **Table 1**.

Different Cells

So far we have discussed the "generalized" cell – which, however, is fictitious. Real cells within the plant differ widely in shape, size, chemistry, and physiological role. **Table 2** lists the main cell types found in higher plants, and some of their distinguishing features. It should be noted that the data listed here are themselves rather gross generalizations, especially the attempt to provide approximate sizes for the various cell types. Examples of diverse cell types are illustrated in **Figure 1**, which gives an idea of the approximate abundance of each of the main cell types.

Acknowledgments

I am very grateful to the late Dr Elisabeth Wangermann, University of Leicester, for introducing me to plant cells. I also thank Dr Zoë Popper, The University of Edinburgh, for critical reading of the manuscript.

See also: **Secondary Products**: Rubber Production. **Cell Walls and Fibers**: Cell Walls. **Postharvest Physiology**: Ripening. **Primary Products**: Cellulose; Oils.

Further Reading

Buchanan BB *et al.* (ed.) (2000) *Biochemistry and Molecular Biology of Plants*. Rockville: American Society of Plant Physiologists.

Cutter EG (1969) *Plant Anatomy: Experiment and Interpretation, Part 1, Cells and Tissues*. London: Edward Arnold.

Kaufman PB (1989) *Plants: Their Biology and Importance*. New York: Harper & Row.

Mauseth JD (2003) *Botany: An Introduction to Plant Biology*, 3rd edn. Sudbury: Jones & Bartlett.

Oparka KJ (undated) *The Living Plant Cell*. (Video and accompanying booklet) Invergowrie, Dundee: Scottish Crop Research Institute.

Stern KR (2003) *Introductory Plant Biology*, 9th edn. New York: McGraw-Hill.

Cell Growth

S McQueen-Mason, C Darley, P Roberts and L Jones, University of York, York, UK

Introduction

Growth can be defined as an irreversible increase in size or volume. In plants, growth is brought about by a combination of cell division and cell enlargement. Cell division alone does not constitute growth but rather increases the potential for growth by increasing the number of cells that can subsequently undergo expansion. Growth itself relies on the process of cell expansion and, in plants, this mechanism is highly dependent on the mechanical properties of the cell wall. These properties are, in turn, governed by the molecular components of the cell wall. The cell wall is key to much that exemplifies higher plants, providing both a skeletal role as well as determining many developmental processes that give rise to specialized plant structures. This article covers the molecular architecture of plant cell walls, especially as it relates to the process of cell expansion. With reference to this molecular architecture, an overview of the processes governing wall expansion and biosynthesis during cell growth is given below.

Almost every cell produced during the life of a plant is encased in a tough polymeric extracellular matrix, or cell wall. The presence of the wall allows plant cells to exist in osmotic disequilibrium with their extracellular fluids. Typically, the concentration of osmotica within the cell is much higher than that in the surrounding "apoplastic fluids", providing a driving force for water to enter the cell. In a nongrowing cell, this driving force is counterbalanced by the development of hydrostatic (or turgor) pressure within the protoplasts. Turgor pressure in plant cells can be very high with typical values around 0.5 Mpa and reaching as high as 5.0 Mpa in specialized cell types such as stomatal guard cells. This pressure is maintained by the strength of the cell wall, which effectively prevents water uptake and cell expansion. The amount of turgor generated in growing and nongrowing cells is about the same. However, the ability of growing cells to expand is determined by the ability of the wall to extend in response to this pressure, whilst the walls of nongrowing cells are effectively rigid. Growing plant cells may increase their volumes at rates in excess of 10% per hour, representing a very substantial increase in surface area. It is now known that the extensibility of the wall determines the rate of cell growth and that localized differences in extensibility of the walls of a single cell determine isotropic expansion giving rise to the diversity in shapes of plant cells. One of the central mysteries of plant life, which is now beginning to be unraveled, is that of the molecular basis enabling such a composite material (the cell wall) to combine remarkable strength with dynamic flexibility.

The aim of this article is to provide a general understanding of how the molecular components of the plant cell wall associate with one another to produce an outstanding composite material capable of fulfilling the demanding performance requirements described above. The components of the primary plant cell wall will be considered first.

Composition of the Primary Wall

The properties and components of the plant cell wall reflect the developmental status of the cell. The majority of plant cells possess a relatively thin primary cell wall, whereas certain specialized cells (such as those in vascular tissues) have a modified, thickened secondary cell wall that is laid down subsequent to cell expansion. Because the extensibility required for growth is a unique feature of the primary wall, this structure will form the focus of this article. The composition and structure of primary walls varies according to the age and physiology of the cells they surround. Although the precise components of the primary wall can vary throughout the plant kingdom and even between cells of different tissues in an individual plant, the basic composition is essentially the same. Plants have developed a high-performance composite material based predominantly on polysaccharides (polymers with a relatively low metabolic cost). The wall is effectively a fiber composite based around cellulose microfibrils in which matrix polysaccharides play a role loosely analogous to the resin in man-made fiber-composite materials. The extraordinary aspect of this composite material lies not only in its strength, but also in the dynamic manner in which its mechanical properties are controlled. A general model of primary cell wall structure and detail of the major polysaccharide components are presented in **Figures 1–4**.

Cellulose Microfibrils

Cellulose microfibrils form the major structural elements in the wall and, as such, cellulose is one of the most abundant components accounting for between 15% and 30% of the dry mass of primary walls. Cellulose microfibrils are composed of crystalline linear assemblies of (1-4)-β-D-glucan chains hydrogen bonded to one another. Each glucan chain may be 1–2 μm long, but because they start and end in different positions within the wall, a microfibril can reach several hundred micrometers in length. The degree of crystallinity within fibrils reflects the degree of hydrogen bonding between the glucan chains. Highly organized three-dimensional lattices will form if all available sites for hydrogen bonding

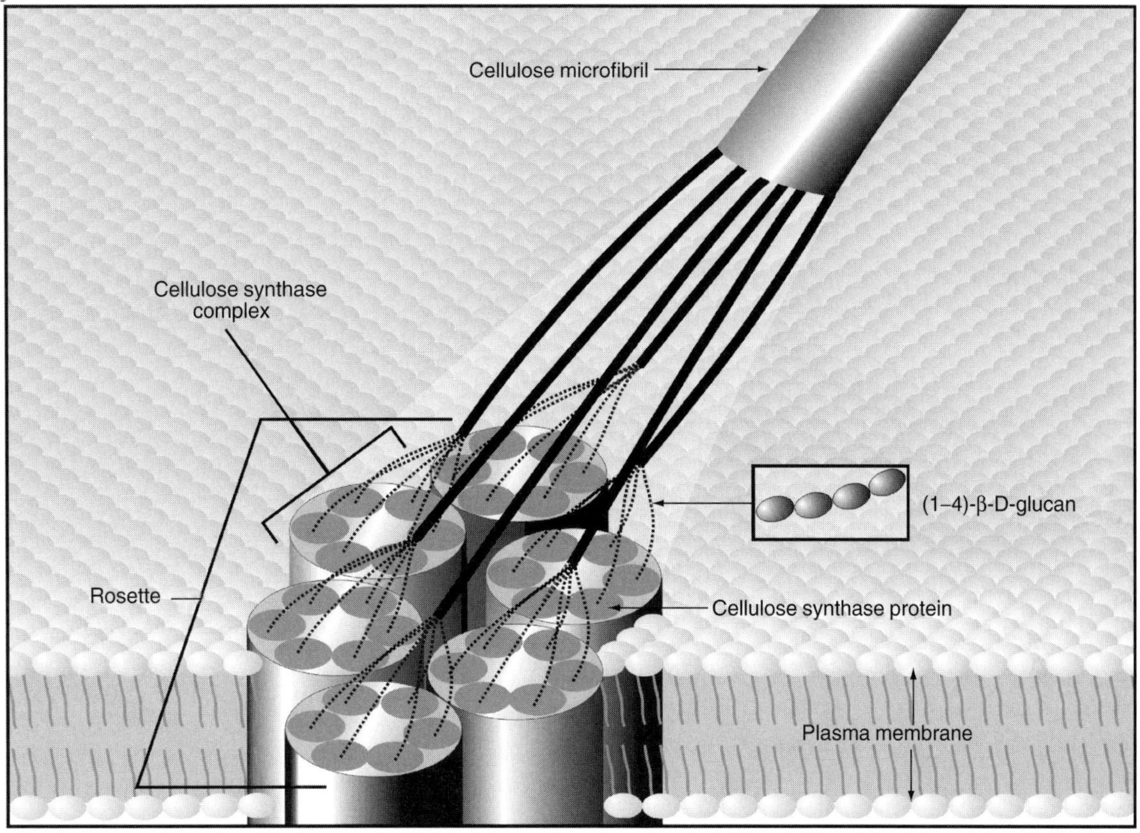

Figure 1 Cellulose microfibrils are synthesized by large protein complexes called rosettes, located in the plasma membrane. Each rosette contains six large protein complexes. Each protein complex in turn contains six cellulose synthase catalytic units. Each cellulose synthase produces a single chain of (1–4)-β-D-glucan. The six glucan chains from each complex may coalesce as they are produced to form subfibrils. The subfibrils produced by each of the complexes in the rosette come together to form a cellulose microfibril.

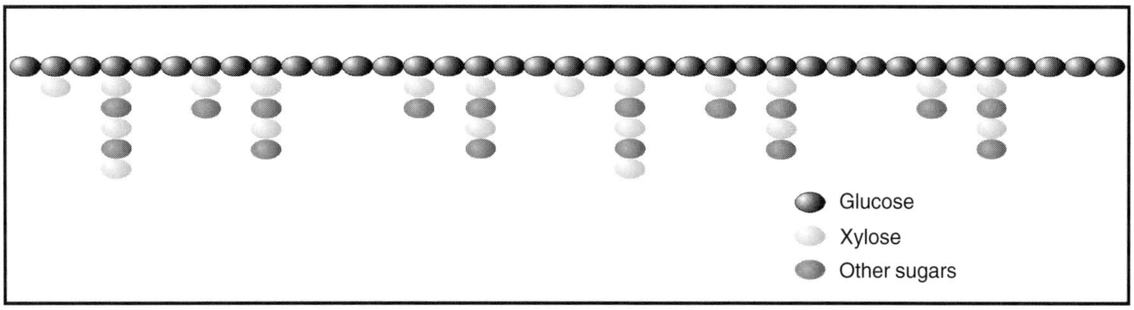

Figure 2 Detail of xyloglucan structure. Xyloglucans have a backbone of (1–4)-β-D-glucan, identical to that of the polymers that form cellulose microfibrils. In contrast to those glucans that form microfibrils, xyloglucans have extensive side chains that prevent them from forming crystalline associations.

between the chains are occupied. In less well-organized sections, the glucan chains are arranged in a paracrystalline assembly and this allows cross-linking glycans (discussed below) to hydrogen bond to the surface of the microfibril.

Cellulose microfibrils are examples of a highly organized biochemical material. A typical plant microfibril contains 36 glucan chains in cross-section. The secret of how these chains come together to form a crystalline microfibril lies in the nature of their biosynthesis. Each glucan chain is synthesized from UDP-glucose monomers by a protein called cellulose synthase. Cellulose synthase subunits appear to be organized into a

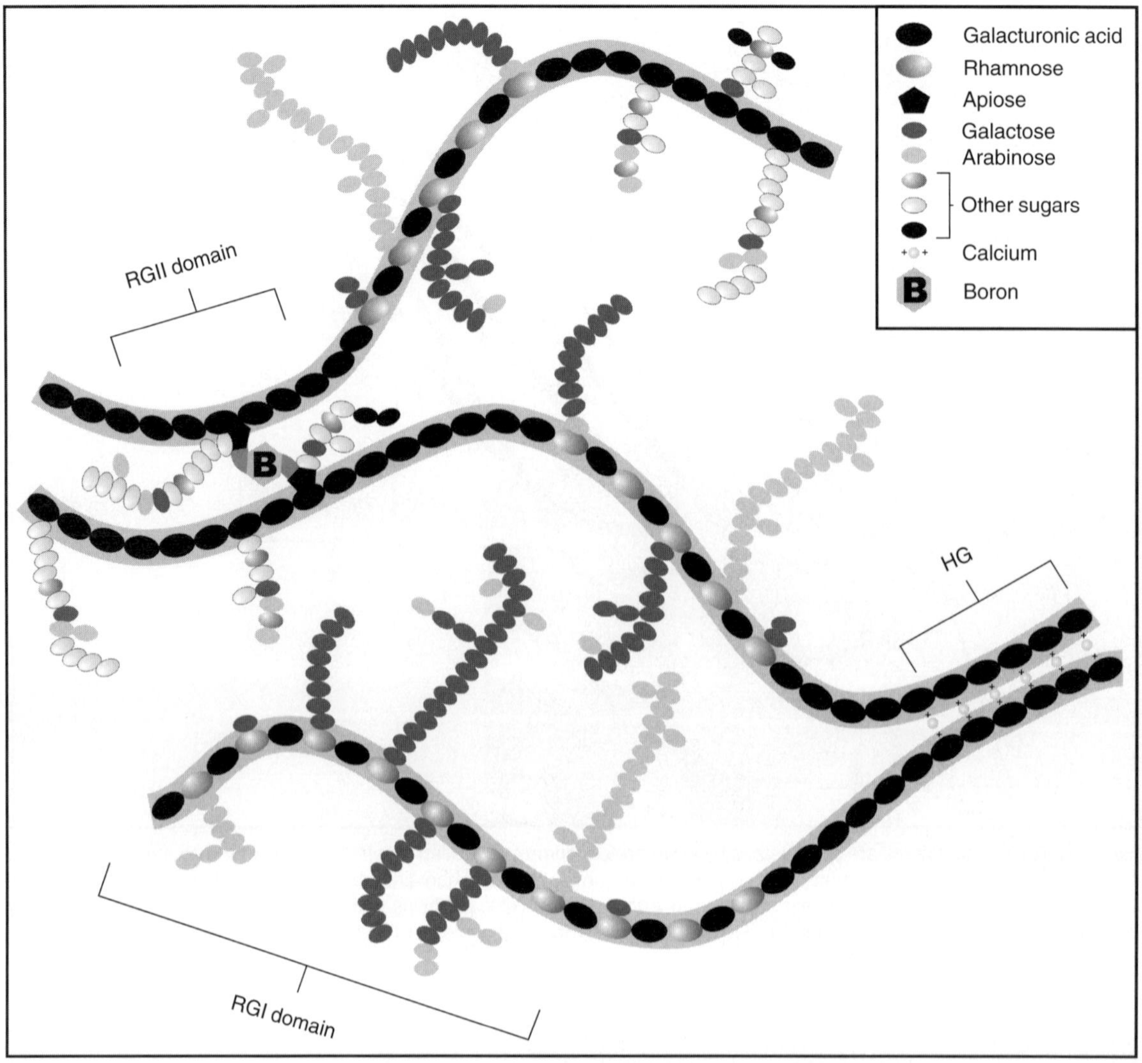

Figure 3 Detail of pectin structure. The unifying feature of pectins is the presence of galacturonic acid residues in the backbone. Unsubstituted chains of homogalacturonan (HG) are able to form stable calcium crosslinked associations through their carboxyl groups. Rhamnogalacturonan II (RGII) has a homogalacturonan backbone but is extensively decorated with complex side chains containing a wide range of sugars. RGII units may be covalently linked through boron ions to apiose residues in the side chains to form stable dimers. Rhamnogalacturonan I (RGI) has a backbone of alternating galacturonic acid and rhamnose residues. The rhamnosyl residues may be decorated with extensive side chains of arabinosyl and galactosyl residues.

membrane-bound multienzyme complex that involves six catalytic units (see **Figure 1**). These complexes are, in turn, organized to form a macromolecular unit known as a rosette, containing six complexes. These rosettes can be visualized by freeze-fracture electron microscopic studies of the plasma membrane. It is the close association of these biosynthetic units that allows the 36 strands of glucan chains to coalesce and assemble into microfibrils at the cell surface. The rosettes are mobile in the fluid plasma membrane and travel around the cell as the microfibrils are deposited. Since cellulose microfibrils are so strong, their orientation in the wall can lead to anisotropy in wall mechanical properties, serving to reinforce the wall in a given plane, and the orientation of microfibril deposition is believed to be a determining factor in the direction of cell expansion. Anisotropically expanding cells have randomly oriented microfibrils, whereas elongating cells have microfibrils aligned transversely to the axis of elongation.

Cross-linking Glycans

Cross-linking glycans form hydrogen bonds with the paracrystalline, amorphous regions of cellulose microfibrils to form a strong network of interlinking fibers within the wall matrix, which, together with

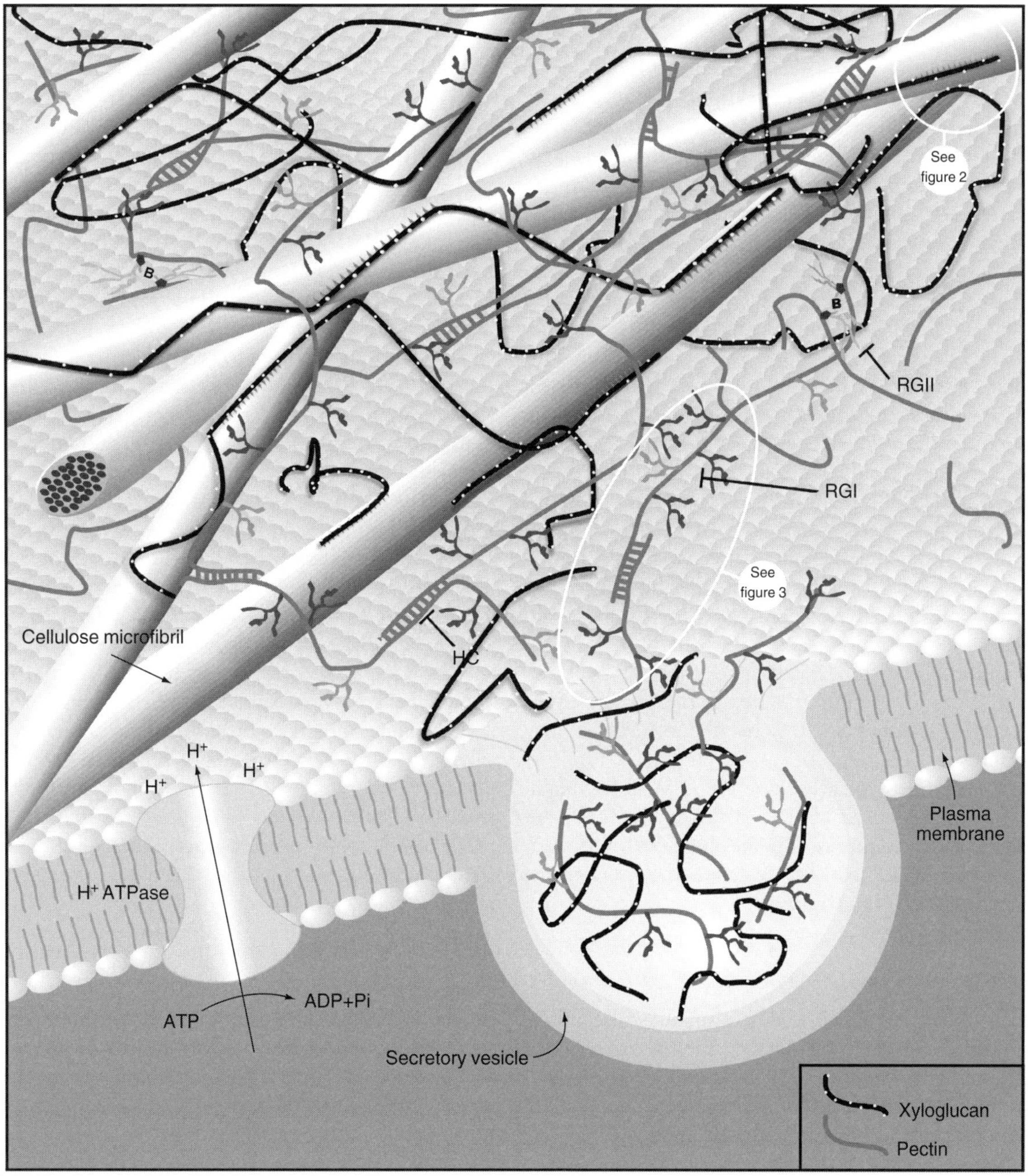

Figure 4 (see color plate 16) A generalized model of the molecular structure of type I primary cell walls. Cellulose and xyloglucans form the main framework of the wall (see **Figures 1** and **2**). Xyloglucan polymers hydrogen bond to the surface of microfibrils and may span between microfibrils serving to anchor them to one another. A network of pectin polymers is coextensive with the cellulose/xyloglucan network throughout the wall. Pectins have regions of homogalacturonan (HG). HG chains may be crosslinked to one another through electrostatic interactions between HG carboxyls and calcium ions. Pectins are also associated through the formation of dimers of rhamnogalacturonan II (RGII), through covalent linkages between apiose residues of neighboring chains and boron ions (see **Figure 3**). Domains of rhamnogalacturonan I (RGI) contain extensive galactan and arabinan side chains that may provide steric hindrance to pectic associations and thus limit crosslinking of the network. Cellulose microfibrils are produced at the plasma membrane (see **Figure 1**), whilst all other polysaccharides are synthesized in the Golgi system and released into the wall from the fusion of secretory vesicles with the plasma membrane.

cellulose, provide the main load-bearing components of the cell wall. The two main crosslinking glycans in angiosperms are xyloglucans (XG) and glucuronoarabinoxylans (GAX); a third group, found exclusively in the cell walls of grasses, are the mixed linkage (1–3), (1–4)-β-D-glucans (MLGs). Crosslinking glycans are highly abundant in the wall accounting for roughly 30% of its dry weight. A common feature of these polymers is that they all have a backbone that is capable of forming extensive hydrogen-bonding interactions with cellulose microfibrils. In the case of XGs, for example, the backbone is identical to that of cellulose, but the presence of substantial side chains on the backbone of crosslinking glycans prevents them from forming crystalline structures. Thus, the crosslinking glycans essentially form "hairy coatings" on the outside of the cellulose microfibrils. These crosslinking glycans are long enough that individual chains may associate at either end with separate microfibrils effectively tethering them to one another. The glucan backbone of XGs is decorated at regular intervals at the O-6 position with various sugar residues (**Figure 1**). These side chains are predominantly xylosyl but, depending on the plant species, xylosyl residues are in turn substituted with fucosyl-, glucosyl-, galactosyl-, arabinosyl-, or mannosyl residues. GAX forms the main crosslinking glycan in commelinoid plants (such as grasses) but is also found in small quantities in other angiosperms. This crosslinking glycan consists of a backbone of (1–4)-β-D-xylose units decorated at the O-3 position with arabinosyl residues, interspaced with occasional glucuronic acid residues attached to the O-2 position of xylosyl residues in the backbone. MLGs are unbranched polymers of (1–3) and (1–4)-linked glucan residues. Typically these are (1–4)-β-D-glucan polymers interspersed with (1–3)-linkages every 4–6 residues. This mixture of linkages produces a kinked structure that is unable to form the crystalline associations seen in cellulose.

Pectins

Pectins encompass a diverse group of acidic polysaccharides characterized by the presence of D-galacturonic acid residues. It is generally considered that pectins form an independent polymeric network that is coextensive with that formed by cellulose and crosslinking glycans. The degree of this independence is unclear as there are reports of covalent associations between the two networks and, without doubt, they have profound effects on one another through a variety of noncovalent interactions. The presence of the acidic residues gives pectins quite different properties to other glycans, allowing them to interact electrostatically and to form highly hydrated complexes. In particular, pectins interact with divalent metal ions, especially calcium. These cations associate with pectic carboxyl groups, linking them together in paired associations allowing extensive junctions to form between neighboring pectic chains and producing a cohesive network (see **Figure 3**).

Three types (or domains) of pectin (homogalacturonan, rhamnogalacturonan-I, and rhamnogalacturonan-II) can be found in plant cell walls. Homogalacturonan (HGA) is a linear polymer of (1–4)-α-linked-D-galacturonic acid. Rhamnogalacturonan-I (RG-I) has a backbone of repeating galacturonyl and rhamnosyl sugars and is extensively decorated with oligosaccharide side chains. Rhamnosyl residues in the RG-I backbone may be decorated with polymeric (1–4)-β-D-galactan and (1–5)-α-L-arabinan chains. The structure of RG-I appears to be quite varied and the backbone is thought by some authors to be covalently attached to that of HGA. RG-I may represent a highly branched domain within pectin, preventing chains of HGA from forming tight associations and thus increasing pectin fluidity.

RG-II is composed of a homogalacturonan backbone and a complex and highly conserved branching structure. The backbone of RG-II is coextensive with HGA and may be thought of as a branched domain within the general HGA backbone. These branches contain 11 different sugar groups. RG-II molecules form dimers in the presence of boron ions that form stable diesters with apiose residues in RG-II side chains. These linkages probably help associate the polymers of the pectin network and their formation has been shown to be essential for normal cell growth.

Together, these pectin macromolecules serve a variety of roles within the cell wall matrix; they influence wall porosity, serve as pH and ionic buffers, act as sites for cell-to-cell adhesion and provide a binding site for some cell wall enzymes. The charge state of pectins appears critical to their function in the wall. HGA is synthesized with a large percentage of the carboxyl groups masked by methyl or acetyl esters. Enzymes in the wall remove these ester groups during maturation increasing the overall charge of pectins and increasing their ability to form ionically crosslinked structures. Highly charged (de-esterified) HGA is largely found in the outer regions of the wall, where it appears to stabilize cell-to-cell adhesion, thereby effectively serving as a glue to bind adjoining cells and maintain tissue integrity.

Structural Proteins

Structural proteins are relatively minor but influential components of the primary cell wall. These proteins fall into three discrete groups but share some features. Generally, they are characterized by a long repeat structure in their peptide sequence and often have characteristic posttranslational modifications such as amino-acyl hydroxylations and glycosylations. The groups are named according to their major repeating amino acid, thus the three main groups are: hydroxyproline-rich glycoproteins (HRGPs), proline-rich proteins (PRPs), and glycine-rich proteins (GRPs). Many of these structural proteins are glycosylated, and most are capable of crosslinking to one another through oxidatively produced dityrosyl groups. It is generally believed that these proteins form covalently linked polymers in the wall, serving to reinforce the polysaccharide matrix. These proteins are particularly abundant in cells under mechanical stress and their presence is believed to strengthen the wall. The polymerization of this network typically occurs once cell expansion has ceased, leading to a rigid crosslinked structure.

There is an additional group of nonenzymatic glycoproteins called arabinogalactan proteins (AGPs), about which much is known of their structure but little regarding function. These proteins are extensively decorated with arabinogalactan side chains and the sugar residues can make up to 95% of the molecular mass of an AGP. AGPs are often anchored in the membrane through glycosylphosphatidylinositol anchors. Some evidence suggests a role for these proteins in cell-to-cell signaling, as well as a role in nucleating wall synthesis and in the cohesion of the primary wall.

Wall Architecture

Although components of the cell wall can be analyzed independently, in order to study their role in cell wall extension we must examine how these networks interact and how they assemble *in vivo*. As mentioned above, plant cell walls have been divided into two classes based on their composition. Type 1 walls are found in the majority of dicotyledonous species. These walls are built around cellulose microfibrils with xyloglucan acting as the major crosslinking glycan. Type II walls are found in Commelinoid plants, which include the grasses. Type II walls are essentially similar to type I walls except that GAX and MLG make up the major crosslinking glycans and they contain generally less pectin. A generalized model based on type I walls is presented in **Figure 4**.

The major crosslinking glycan is xyloglucan, which binds to cellulose microfibrils via hydrogen bonds, tethering them to other microfibrils or, by entanglement, to other XG polymers. Much of the surface of a microfibril may be coated with XGs and the extensive tethering of microfibrils to one another produces the major mechanical load-bearing network in the wall. Coextensive with this is the pectin network, which is held together through the calcium crosslinking between HGA chains, as well as by boron diesters of RG-II regions. It can be speculated that the RG-I regions serve to hinder backbone associations between regions in the pectins and hence maintain polymer fluidity. Thus, removal of RG-I side chains and HGA esters can potentially both serve to increase pectin crosslinking and rigidify the wall. Structural proteins are believed to form another interlocking network in the wall once growth has ceased, but these have been left out of this model for the sake of simplicity.

The Growing Primary Wall

A key feature of plant cell walls undergoing extension is that it is a composite material in which the individual polymers are held to one another mostly through noncovalent interactions. Because the polymers of the wall are long, these noncovalent interactions are sufficiently strong, *en masse*, to maintain wall strength and cohesion. The advantage of this is that noncovalent bonds typically require little energy to disrupt and can spontaneously reform. This is important because the cell wall is devoid of biochemical potential energy in the form of ATP or reducing equivalents, and this imposes limits on the types of restructuring that can occur within the wall.

As mentioned earlier, wall extensibility can be altered over very short time scales. Such modulations in extensibility appear to be mediated through the action of enzymes or other chemical agents in the wall. A number of proteins are found in growing walls that directly modify the properties of the cellulose/crosslinking glycan network. β-glucanases can hydrolyze the glucan matrix and thus weaken interactions holding microfibrils to one another. Endotransglycosylases are functionally related to glycanases, but instead of hydrolyzing glycans, a glycosidic link is broken within a polymer and then remade between one end of the original donor and another polymer. Best known are xyloglucan endotransglycosylases (XETs) that break bonds in one XG molecule and then remake the bond, effectively stitching two polymers together. The assumed advantage of this type of action is that breakage in a load-bearing polymer will relieve stress in the wall and allow polymers to slip. The bond is then remade in a new position maintaining overall wall integrity. XETs are thought to play a role both in loosening wall structure during growth as well as in the integration of new XG polymers into the wall as they are secreted from the cell.

A third group of wall-active proteins, called expansins, catalyze so-called acid-induced extension in primary cell walls. It has been known for many years that growing cell walls can extend rapidly in response to low pH but are more rigid at neutral pH. This acid-induced extension was shown to be lost following proteolytic treatment of walls. The proteins responsible for this effect were subsequently isolated and named expansins. Expansins are unusual proteins in that they have dramatic direct effects on wall extension but do not alter the covalent structure of the wall. Indeed, their mechanism of action appears to involve the disruption of hydrogen bonds between XG molecules and cellulose microfibrils. Expansins are very effective at low concentrations and it seems likely that they work by progressively "unzipping" load-bearing tethers as they go. Expansins appear to represent one of the primary mechanisms of wall loosening in plant cell walls. The oxidative scission of wall polymers by reactive oxygen species has recently been shown to be important during wall extension and may represent a key nonenzymatic mechanism of wall extension.

As well as the potential wall-loosening mechanisms just described, there are a number of activities in the wall that can lead to increased wall rigidification and serve to slow down growth. These mechanisms essentially involve the formation of additional crosslinks between polymers, thus decreasing the fluidity of the wall. Several mechanisms for wall rigidification have been proposed. One of these involves the removal of ester groups from the pectin backbone by pectin esterases. This increases the potential for calcium-bridged crosslinks within the pectin network, effectively increasing its rigidity. Indeed, it has been shown that the degree of pectin methylation decreases in cell walls in a manner commensurate with the cessation of cell growth. A second mechanism involves the crosslinking of wall polymers through phenolic esters. A number of wall polymers are secreted with phenolic esters attached to their side chains. These esters are particularly abundant in type II walls and are commonly found on arabinan side chains of GAX and RG-I, as well as galactan chains on RG-I. Pairs of these phenolic esters can be oxidatively dimerized to form stable crosslinks. This process may be catalyzed by wall peroxidases. A third wall-stiffening process involves the polymerization of structural proteins, again through the action of peroxidases. It has also been proposed that removal of side chains from wall polysaccharides may be involved in rigidification. For example, removal of side chains from crosslinking glycans may increase their ability to hydrogen bond with one another and with cellulose, thus stiffening the wall. A generalized concept would see that wall material is deposited in a relatively fluid but cohesive form, allowing extension whilst maintaining integrity. This material is subsequently "stitched" together with increasing crosslinks as it matures and growth comes to an end.

Wall Biosynthesis

The wall is an extracellular structure made from components synthesized in the protoplast. The majority of wall components are secreted via two different routes (outlined in **Figure 4**). As already described, cellulose microfibrils are produced by large complexes in the plasma membrane that take UDP-glucose from the cytosol and polymerize the glucose units to form glucan chains that are extruded into the extracellular space to form microfibrils. Most of the other major wall components are produced via the secretory pathway. All the proteins of the wall (both structural and enzymatic) are produced on the rough endoplasmic reticulum (ER) and then processed through the ER where any appropriate posttranslational hydroxylations and glycosylations occur. These then progress through the Golgi system, where protein O-glycosylation is potentially carried out. The Golgi is also the site of synthesis of all of the matrix polysaccharides of the wall and thus contains a large complement of the various glycosyltransferases and other enzymes involved in their synthesis. Matrix polysaccharides are then deposited, along with wall proteins, into the wall space through the fusion of secretory vesicles with the plasma membrane. The deposited matrix components and cellulose microfibrils then need to assemble and integrate with existing wall components to produce a cohesive network. Much of this process may be attributed to spontaneous assembly, but is likely also to be assisted by the presence of matrix-modifying enzymes, such as glycanases, endotransferases, and expansins.

Obviously, it is essential that wall extension and biosynthesis are closely coordinated to ensure that wall integrity is maintained. This requires both temporal and spatial coordination. Clearly, in a cell expanding in a purely isodiametric fashion, wall extension should occur evenly over the whole surface of the cell and it would be simply a matter of ensuring even wall deposition to keep pace with expansion. However, cells often expand in a highly polarized manner. For example, in an elongating cell, the sides of the long axis are expanding rapidly, whilst the end walls barely expand at all. This requires the deposition of material to specific regions of the cell. The most extreme form of this is seen in tip-growing cells such as pollen tubes and root hairs. Here, growth occurs only at the very tip of the expanding cell giving rise to long, thin, tubular forms. In tip-growing cells, almost all new wall material is directed to this one highly active area. The direction of wall deposition is largely controlled by the cytoskeleton. Secretory vesicles from the Golgi system, for example, travel along actin filaments in the cytoplasm and the pathway to the appropriate part of the cell is determined by these dynamic structures. Similarly, it has been suggested that the patterning of cellulose microfibril deposition may be determined by the arrangement of cortical microtubules that run around the periphery of the cell.

Our understanding of the regulation of wall synthesis is extremely poor. However, it is becoming increasingly clear that there are a number of potential signaling pathways between the wall and the cytoplasm that could be involved in coordinating wall extension and synthesis. It is apparent that there are proteins that link the wall and the plasma membrane physically to one another. Clearly, as the wall extends, it is essential that the membrane also extends to keep pace with it, otherwise it may rupture at the site of growth. One suggestion is that when a membrane is under tension, calcium is able to pass into the cell through stretch-activated channels setting up local gradients that serve to stimulate vesicle direction and fusion. Similarly, a class of wall-associated protein kinases have now been identified. These are transmembrane proteins that span the plasma membrane and have a cytoplasmic protein kinase domain and a wall-associated domain on the other side of the membrane. The wall-associated domains of these proteins exist in a variety of forms: some as potential protein–protein interaction domains, some as structural protein-like domains, and others as domains that might interact with wall glycans. The exact roles of these signaling proteins are only beginning to emerge, but it seems likely that some at least may be involved in coordinating events either side of the membrane.

The Control of Wall Extension

As mentioned above, wall extensibility represents a key control point for growth in plant cells. As such, extensibility itself is regulated by a wide range of factors, including hormones and environmental factors such as light, gravity, and various stresses. The exact means by which these factors act on extensibility are still unclear, but there are a number of general mechanisms by which this is likely to occur. Effects on extensibility may be either long term or short term or involve a combination of the two. Long-term effects are likely to involve significant changes in wall composition and thus be affected through changes in gene expression and wall deposition. In the shorter term, a relationship between wall pH and growth has long been described and is generally referred to as the acid growth hypothesis. Wall extensibility is generally much greater at pH values around 4.5 and much lower at values above 5.5. As mentioned earlier, expansins are believed to be responsible for much of this effect as their activity is optimal around pH 4.5. A number of hormones and stimuli that affect growth rates appear to do this by short-term changes in wall pH and it is likely that these are transduced into effects on extensibility by the activity of expansins in the wall. Wall pH itself is directly regulated by the activity of the plasma membrane proton pump. This membrane protein excretes protons using energy from the hydrolysis of ATP, and its action sets up the main electrochemical gradient across the plant plasma membrane used in driving many transmembrane processes. Many factors that exert rapid effects on growth rate, for example, some hormones and light treatments, do so through altering the activity of the proton pump. This type of mechanism allows for very close and rapidly responsive modulations of wall extensibility at the cellular level and even within particular regions of the wall of an individual cell. Thus, the cell is able to control wall extensibility over long and short timescales in order to adjust wall extensibility and produce an appropriate rate of cell expansion.

See also: **Cell Walls and Fibers**: Cell Walls. **Growth and Development**: Cell Division and Differentiation; Cells.

Further Reading

Anderson CM, Wagner TA, Perret M, *et al.* (2001) WAKs: cell wall-associated kinases linking the cytoplasm to the extracellular matrix. *Plant Molecular Biology* 47: 197–206.

Carpita N and McCann M (2000) The cell wall. In: Buchanan BB, Gruissem W, and Jones RL (eds) *Biochemistry and Molecular Biology of Plants*, pp. 52–110. Rockville. MD: American Society of Plant Physiologists.

Cosgrove DJ (2000) Expansive growth of plant cell walls. *Plant Physiology and Biochemistry* 38: 109–124.

Darley CP, Forrester AM, and McQueen-Mason SJ (2001) The molecular basis of plant cell wall extension. *Plant Molecular Biology* 47: 179–195.

Doblin MS, Kurek I, Jacob-Wilk D, and Delmer DP (2002) Cellulose biosynthesis in plants: from genes to rosettes. *Plant Cell Physiology* 43: 1407–1420.

Willats WGT, McCartney L, Mackie W, and Knox JP (2001) Pectin: cell biology and prospects for functional analysis. *Plant Molecular Biology* 47: 9–27.

Cell Division and Differentiation

B Veit, AgResearch, Palmerston North, New Zealand

Evolutionary Considerations

Analysis of fossil material and DNA sequence comparisons among living organisms suggest that plant and animal lineages diverged roughly 1 billion years bp, prior to the evolution of multicellularity. With this perspective, it is reasonable to expect that many aspects of cell division and differentiation might differ between plants and animals. Indeed, when their growth and development are compared, a number of key differences do emerge. For example, whereas embryogenesis in animals involves migration of cell layers to establish the fundamental endoderm and mesoderm tissues, no cell migration occurs during plant embryogenesis. Instead, the formation of fundamental vascular, cortical and epidermal tissues is achieved through localized patterns of cell division and is guided by what appear to be highly position-dependent cues. Moreover, while the fully developed animal embryo shows a highly complex organization comparable to that of the adult, the plant embryo has a rudimentary organization, consisting of a vascularized bipolar axis which is terminated by the shoot and root apical meristems (SAM and RAM).

Programs of Cell Division and Differentiation

Embryogenesis

Embryogenesis in *Arabidopsis* provides a clear example of the linkage between morphogenesis and specific patterns of cell division and differentiation. In *Arabidopsis*, an asymmetric division of the zygote produces a large cell destined to form the suspensor while the smaller, more distal cell undergoes a stereotyped series of divisions to form the globular stage embryo. As morphological differentiation of the embryo proceeds, division patterns become less predictable leading to the formation of the root and hypocotyl from the basal half of the embryo proper while the cotyledons and shoot apex derive from the upper half. In many other plant species, patterns of embryonic cell division are more variable, but still yield embryos having a consistent form and organization.

Apical Meristems

Apical meristems provide further examples of regulated patterns of cell division and differentiation. Located at the tips of growing shoots and roots, these structures serve both to maintain a reservoir of undetermined, pluripotent cells, as well as to coordinate the differentiation of the mitotic derivatives of these cells into the various tissue and organ systems that make up the plant. Polarized patterns of cell division lead to the formation of clonally distinct layers. For example, cells in superficial layers of the SAM divide predominantly in an anticlinal plane (new cell walls are oriented perpendicular to the tissue surface) to produce a so-called tunica layer (**Figure 1**). More internally positioned cells, which form the corpus, divide in a variety of planes to increase the diameter of the shoot apex. Analogous, though not identical, patterns of cell division can also be described for the RAM.

The frequency of cell division within the SAM and RAM also varies according to position. The central zone of the SAM and the quiescent center of the RAM both are distinguished by the relatively slow division rates. In both cases, these regions contain what are termed apical initials. As a consequence of their distal position and their persistent undifferentiated state, apical initials generate large clonally related lineages of cells. For example, in the shoot, lineage analysis has shown that these clones of cells are arranged in two or more concentric layers. The most superficial of these corresponds to the so-called L1, while more internal layers are designated L2, L3, etc. Each of these layers can be traced back to two to three apical initials located within the central zone (**Figure 1**).

Histogenesis

The primary plant body can be viewed as containing three basic tissue systems, usually referred to as the epidermal, vascular, and fundamental tissues. An epidermal layer is first established in the embryo through anticlinal divisions, which set off a superficial cell layer, while vascular progenitors are recruited from more internal layers. During later stages of development, including vegetative and reproductive growth, additional cells that are recruited into these three fundamental tissues originate from the root and shoot apical meristems.

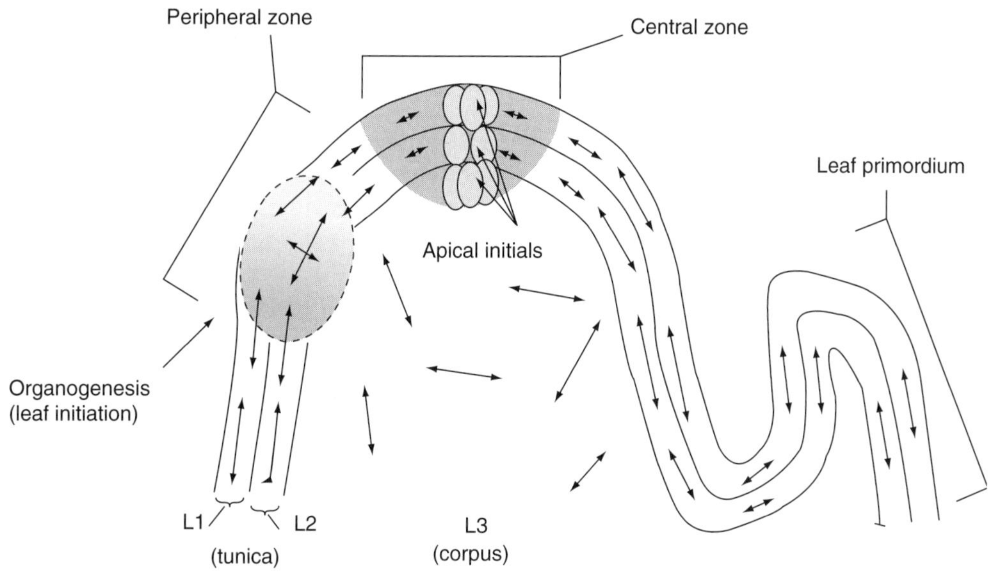

Figure 1 Schematic view of cellular behaviors in the shoot apical meristem. Polarized patterns of anticlinal divisions in the L1 and L2 maintain a tunica layer, which is characterized by its uniform thickness and superficial position. Within the L3, divisions in various planes maintain the internally positioned corpus. Each of the L1, L2, and L3 layers can be traced back to separate sets of apical initials. Distally positioned cells in all three layers lie within the central zone, a region characterized by relatively slowly dividing, undifferentiated cells. As cells enter the peripheral zone, they divide more rapidly and become organized into structures such as leaf primordia.

The epidermis of both the shoot and root can be traced to clones of cells that arise from superficial layers of the SAM and RAM. As these clones of cells are displaced from the apical meristems, they gradually acquire characteristic traits, which may include specialized structures such as trichomes and stomata, as well as the capacity for specialized metabolism for the production of epidermal specific compounds, including waxes, oils, and other secondary metabolites. Though the epidermis typically exhibits a relatively uniform thickness of one cell layer, cells making up this layer exhibit a wide variety of shapes that characterize their position within the plant. For example, aerial portions of the plant typically exhibit trichome and guard cells types, whereas root epidermal cells may be differentiated according to the presence or absence of root hairs.

Vascular tissues of the stem and root generally differentiate in a progressive manner as more undifferentiated tissue is laid down by the SAM and RAM. Within leaves, vascular differentiation often appears discontinuous, with both acropetal and basipetal extension of isolated provascular elements that ultimately join up with more proximal preexisting vascular elements. Cells destined to become vascular tissue first differentiate as provascular tissue, which can be recognized by the elongated shape of cells, before terminal differentiation into functional phloem and xylem elements.

Cortical tissues encompass a diverse collection of cell types which serve various functions including photosynthesis and storage. Despite their specialized functions, parenchyma-derived cells show considerable developmental plasticity, and may further differentiate under appropriate conditions, such as in tissue culture, or *in planta* through wound induction or adventitious programs of development.

Organogenesis

Changes in the polarity of cell division and growth provide the first overt signs of organ initiation. For example, periclinal divisions in subepidermal layers on the flank of the shoot apex signal the location of a new growth axis that will ultimately develop into a leaf. Similarly, foci of cell divisions in the root pericycle layer mark the position of incipient lateral roots. In the case of leaves, polarized divisions also accompany the anisotropic growth responsible for the production of the flattened leaf blade. The adaxial and abaxial surfaces of the leaf blade typically show a pronounced asymmetry which usually involve the distribution of specialized cell types such as trichomes and stomata which differentiate as the leaf grows. The distribution of spongy mesophyll, palisade tissues, and vasculature provide examples of internal asymmetries within the leaf.

Phase Transitions

Global changes in patterns of cell division and differentiation can be linked to specific stages of a plant's development. For example, during the transition to the reproductive phase, division events that were previously concentrated in the organogenic peripheral zone of the vegetative SAM become more uniformly distributed. Subsequently, a characteristic series of floral organs is initiated whose form and position differs markedly from their vegetative counterparts. Several lines of evidence, beginning with morphological comparisons by Goethe and followed more recently by molecular and genetic analyses, argue for the existence of basic developmental programs that are common to vegetative and floral organogenesis. Current evidence suggests that upon floral induction, these basic programs are modified to yield specific floral organs that have a characteristic shape and cellular makeup.

Regulation of Cellular Behavior

Control of Cell Division

Recent sequencing of the *Arabidopsis* and rice (*Oryza sativa*) genomes has revealed plant counterparts of many animal and fungal genes known to regulate cell division, and in many cases, the function of these genes appears conserved. As in animals, cyclin-dependent kinases (CDKs) are thought to regulate progression through the cell cycle through phosphorylation of key regulatory proteins. As implied by their name, the activity of CDKs depends on their association with specific classes of proteins known as cyclins. Several classes of cyclins can be differentiated both by their association with specific CDKs, and by their pattern of synthesis and turnover during the cell cycle. Major transitions in the cell cycle can be traced to changes in the levels of these cyclins (**Figure 2**). The transition from G_1 to S, for example, is promoted by D class cyclins binding CDK-a. This activated form of CDK-a phosphorylates the retinoblastoma protein (Rb), causing it to dissociate from E2F transcriptional factors. In this dissociated form, E2Fs promote transcription from genes required during S phase, such as those involved in DNA synthesis. The activation of CDKs by cyclin B appears to trigger the G_2-to-M transition, again through the phosphorylation of stage-specific control proteins. The activities of CDKs are also positively and negatively regulated by other classes of kinases as well as phosphatases. Cyclin dependent kinase inhibitors (CKIs) inhibit CDK activity through the addition of inhibitory

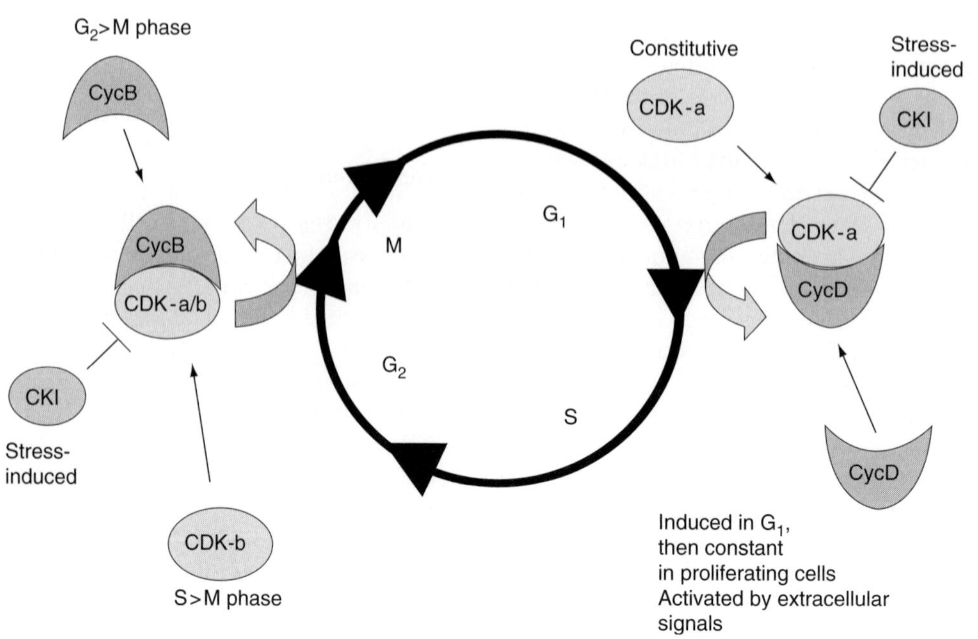

Figure 2 Basic mechanisms controlling cell cycle progression in plants. The cell cycle is depicted as a circle broken up into four successive phases (G_1, S, G_2, and M). Progression between phases is promoted by the periodic synthesis and turnover of cyclins which bind to and activate cyclin-dependent kinases (CDKs). Once activated, CDKs phosphorylate specific classes of target proteins that mediate phase specific processes (e.g., DNA replication during S phase, or spindle assembly and chromosome separation during M phase). Cyc, cyclins; CKI, cyclin-dependent kinase inhibitors.

phosphates at specific sites. Alternatively, CDK activity may be promoted by removal of inhibitory phosphates via specific phosphatase activities, or through addition of phosphates at specific sites by kinases.

Cellular Aspects of Differentiation

All cell types in plants manifest some degree of polar differentiation in terms of their shape, subcellular asymmetry, or division behavior. The root and shoot apical meristems as well as developing organ primordia exhibit strongly polarized patterns of cell division. Once these cells are laid down, they may then manifest polar patterns of elongation and terminal differentiation. Both division and elongation polarity can be attributed to actin- and microtubule-based cytoskeletal elements. The contributions of microtubules and actin to the cytoskeleton have been distinguished through their differential sensitivity to specific drugs, while genetic and biochemical approaches have provided further insight into the dynamic behavior of the cytoskeleton and its composition. Microtubules appear to play a major role in establishing cellulose microfibril orientation, apparently by acting as a structural template upon which synthesis occurs. So-called diffuse patterns of cellular expansion are largely limited to planes that are transverse to the orientation of these microfibrils. The role of microtubules in maintaining cell shape is illustrated by temperature sensitive mutations to *MOR1*, a gene encoding a microtubule associated stabilizing protein, which lead to a loss of cellular polarity throughout the plant upon shifts to the restrictive temperature. Recent studies suggest that proteins belonging to the kinesin family are also involved in establishing and maintaining microtubule networks.

Actin microfilaments, by contrast, appear to play essential roles in localized growth of cells, such as observed in root hairs and trichomes. Actin-based cystoskeletal elements mediate this type of growth through the localized deposition of wall and membrane precursors. Actin cytoskeletal elements appear to play important roles in establishing the polarity of cell division through influencing the dynamics of cell plate formation.

Localized differences in the mechanical properties of cell walls may also lead to polarized patterns of cell growth. Proteins such as expansins change wall extensibility by disrupting crosslinks between adjacent microfibrils, while extensin class proteins influence cell extensibility through changes in wall synthesis.

Experimental Analysis of Cell Division and Differentiation

Microscopy

Beginning with van Leeuwenhoek's observations in the seventeenth century, microscopy has provided increasingly detailed insights into the composition and behavior of plant cells. Superficial aspects of cellular architecture, as well as division frequency and polarities, are readily described by light microscopy. More recently, finer aspects of cellular and subcellular organization have been revealed by electron microscopy while microsurgical approaches on living plants have enjoyed a renaissance through the recent development of low-vacuum scanning electron microscopy (SEM). The power of microscopy based techniques has been further extended by the development of *in situ* hybridization and immunohistochemistry techniques by which the distribution of specific RNA and protein species can be described. The use of β-glucoronidase (GUS) reporter constructs whose expression is driven by tissue specific promoters has revealed additional examples of tissue specific transcription patterns. With the development of vital reporters such as green fluorescent protein (GFP), these same approaches can be used to monitor changes in gene expression in living tissue.

Analysis of Cell Lineages

Many aspects of cell division patterns in plants that were initially suggested by microscopic observations have been further explored by analysis of cell lineage relationships. By genetically marking a cell or subset of cells in a manner that produces a visible trait, it becomes possible to follow the contribution of that cell to the plant body. Early examples involved the use of colchicine to induce the formation of polyploid cells in the SAM, leading to an easily observed increase in nuclear volume. By tracing the distribution of these polyploid cells, Satina and coworkers were able to show that shoot system of *Datura* contains three fundamental layers, L1, L2, and L3, each of which can be traced back to a distinct set of apical initials. Plants such as these, in which one of the three layers is genetically distinct, are termed periclinal chimeras. Common examples of periclinal chimeras are seen in variegated plants, in which cells within one fundamental layer differ in their capacity to form pigment. Careful examination of these plants has shown that each of the fundamental tissue layers within the shoot apex contributes to the shoot in a predictable though not completely invariant manner. Thus, a typical dicotyledonous periclinal chimera that has an albino L2 will produce variegated leaves with green L1-derived margins surrounding albino L2-derived tissues.

Additional insight into patterns of cell division and differentiation has been gained through the use of mutagens to create chimeras. Poethig and coworkers exploited this approach to mark individual cells in plants at various stages of development to show that in most cases, the developmental fate of cells is more strongly influenced by their position than by their lineage history. Thus, when clones of L2-derived cells are displaced into an epidermal position, they invariably adopt an epidermal identity, despite their clonal relationship to internal tissues. Even very small clones displaced late in development show this position-dependent identity, arguing that plant cells do not become determined in a position-dependent manner until very late in their development. Lineage analysis has also been used to describe the nature or programs regulating organogenesis. The extent and shape of clonal sectors within leaves has been used to estimate the number of founder cells, suggesting that in tobacco (*Nicotiana tabacum*) and corn (*Zea mays*, maize), leaves arise from approximately 100–200 cells contained within several layers on the flank of the shoot apex.

More recently, transgenic tools have been exploited to examine cell lineage relationships. Sporadic excision of an artificially introduced transposable element in *Arabidopsis* embryos was used to generate marked clones of cells during early stages of development. The boundaries of large sectors, although nonrandom, were clearly not invariant, suggesting that positional cues must operate to maintain consistent patterns of growth and differentiation. Analogous methods were used to demonstrate that multicellular trichome complexes have a polyclonal origin rather than arising from a single determined cell.

Surgical Studies

Determination of cells to a specific differentiated fate has been approached by a variety of surgically based methods. Grafting studies have provided a number of examples of determination. For example, extended cold treatment, which may be required to promote flowering upon subsequent inductive conditions, has been shown by grafting to act primarily on the shoot apex rather than other parts of the plant. Similar grafting approaches have been used to establish when shoot apices become committed to flowering as a result of signals originating elsewhere in the plant as well as indicating leaves are the source of these signals. In a related approach, explants to defined synthetic media have been used to assess the determination state of lateral organ primordia. For example, lateral primordia from the fern *Osmunda* developed as shoot-like organs if transplanted soon after initiation, or leaf-like primordia, if harvested later, suggesting that the leaf-like identity of the primordia becomes established gradually as they age.

Surgical methods have also been applied to examine determination events in the shoot apex. When large fractions of the shoot apex were removed, regeneration of a functional apex occurred, suggesting that cells may adopt the identity of apical initials based on their position. Surgery has also addressed questions concerning phyllotaxy. Classic studies in which incisions to the shoot apices altered the positions at which leaves formed suggested that the position at which leaf primordia arise becomes determined shortly before they begin overt development, and is influenced by preexisting primordia, which appear to inhibit the initiation of new primordia. Incisions between the developing leaf and shoot apex led to the development of radially symmetrical organs, suggesting that development of adaxial/abaxial tissues requires communication between these two regions.

Ablation

Although transplantation of small amounts of tissues is a commonly used method in animal systems to assess whether a tissue is committed to a particular developmental fate, wound responses make the use of this technique in plants problematic. In recent years, however, novel ablation techniques have been developed to address how cells and tissues become committed to a particular developmental fate. By exploiting the transparent nature of *Arabidopsis* roots, Scheres and colleagues used highly focused lasers to ablate specific cells to demonstrate inductive interactions between cells. Ablation of specific cell types was shown to alter the developmental fate of neighboring cells. It was further shown that more differentiated proximal regions of the root appear to provide signals that guide the differentiation of less differentiated cells.

Ablation of specific groups of cells has also been achieved by transgenic means by fusing a gene encoding diphtheria toxin to a tissue specific promoter. With this technique, Irish and colleagues selectively removed precursors to the petal and stamen whorls of the developing *Arabidopsis* flower. The development of adjacent stamen and carpel whorls was largely unaffected, arguing against models in which normal whorl development depends on inductive interactions with adjacent whorls.

Physiological and Biochemical Analyses

The highly evolved capacity of plants to adapt to their environment through regulated changes in development has afforded numerous opportunities to investigate differentiation processes through manipulation of physiological parameters. The response of plants to hormones, such as auxin and cytokinins, provide several examples of this approach, including the response of plants grown in cell culture. Classic work by Skoog and coworkers demonstrated how relative levels of plant hormones determine specific determination pathways. High levels of auxin relative to cytokinins promote differentiation of root from callus culture, while higher relative levels of cytokinins promote shoot growth. Auxin has also been shown to play a key role in vascular development. Work of Sachs and others has demonstrated that auxin can induce the formation of vascular elements from parenchymatous tissue, both *in planta* and in cultured explants. These experiments have been cited in support of models for auxin dependent vascular development in which the basipetal movement of auxin from sites of synthesis towards sinks in the root becomes canalized along certain paths, resulting in the induction of vascular elements. Biochemical changes associated with such differentiation have been examined in *Xinnia* cell culture in which leaf parenchyma tissue is induced by auxin to differentiate into xylem tissue. As well as playing a key role in vascularization, auxin has also been implicated in establishing polarity in the embryo as well as influencing sites of leaf initiation.

Genetic Approaches

The development of molecular genetic tools in model species has quickly generated a vast literature on the role of specific genes in regulating cell division and differentiation. In a typical example of so-called forward genetics, mutations that affect some aspect of division or differentiation are isolated in phenotypic screens for mutant plants with altered patterns of development (e.g., the absence of trichomes). The relationship of genes defined in such screens may be explored through the analysis of double-mutant phenotypes, in which the traits of one mutant may be modified by the activity of other genes. Insights into the mechanism by which the gene acts may be gained by creating genetic mosaics in which genetically mutant and wild-type tissues are juxtaposed. Nonautonomy, in which the phenotype of genetically mutant cells is rescued by adjacent normal tissue, would suggest the wild-type gene may act via a diffusible signal. Additional insight into the mechanism of gene function may be gained by cloning the gene through tagging the gene with a known insertional element, or map-based cloning.

Genetic analyses of epidermal differentiation pathways have been notably successful. In the case of root hairs and trichomes, a number of genes have been identified that appear to specifically regulate the development of both structures, but do so in an opposing manner. For example, several genes have been described that are required for normal trichome development, but are simultaneously required to suppress ectopic trichome development in the root. Highly stereotyped patterns of cell division and differentiation associated with stomatal development are also becoming understood at the genetic level.

Genetic studies have also provided evidence for molecules that appear to act over a relatively short distance in both the shoot and root. For example the *CLAVATA 1* gene of *Arabidopsis*, which encodes a leucine rich receptor kinase, limits the proliferation of uncommitted cells in the SAM. *CLAVATA 1* activity depends on it binding a small peptide encoded by a second gene, *CLAVATA 3*. Functional genomic analyses of structurally related genes reveal a large family that may play a prominent role in controlling other cellular interactions.

Emerging Themes in the Control of Cell Division and Differentiation in Plants

Long-standing debates continue as to whether development in plants is driven by cellular behaviors or whether these behaviors are a passive consequence of developmental programs that operate at a supracellular level. Detailed analysis of plants in which specific aspects of division are perturbed has been instructive. Artificial inhibition of cell division through lowered CDK activity leads to plants with reduced stature and cell number, but which have a relatively normal shape. Similarly, mutant plants in which characteristic patterns of polarized cell division are disrupted often have normal shapes. These observations argue in favor of morphogenetic programs whose execution does not rely on fixed patterns of cell division. On the other hand, regulated changes in the levels of certain D cyclins seem related to the elaboration of shape. Differential regulation of this class of cyclins is closely correlated with specific patterns of cell division observed during floral organogenesis in *Antirrhinum*. In transgenic systems, their overexpression can lead to increased rates of division and the formation of misshapen leaves containing abnormal undifferentiated cells. Thus, it would appear that while programs that drive development can accommodate some variability in cell division, at some levels, control of cell division appears more crucial.

Several genes have been described that appear to integrate cellular behaviors into larger organogenic programs. *Arabidopsis* plants mutant for the *ROTUNDIFOLIA* gene show decreased elongation at both the cellular and organ level along the proximal/distal axis of the leaf, while plants mutant for the *ANGUSTIFOLIA* gene show analogous changes, but which are directed along the lateral axis. The activity of these genes would suggest that cellular polarity is in part established through axial patterning cues that operate throughout the plant. The involvement of the *ROTUNDIFOLIA* gene as well as others thought to mediate brassinosteroid activity suggests this class of hormone may play a key role regulating cell expansion behavior. Insight into the relationship between organ size and cell division has been provided by analysis of the *AINTEGUMENTA* gene. Expression of this gene promotes increases in organ size by acting to prolong the period of cell division that occurs during early stages of organ development.

Despite the compartmentalization of the plant by cell walls, cytoplasmic continuity between cells is provided by plasmodesmata which have been shown to mediate the intercellular transport of both protein and RNA. Indeed, recent evidence has demonstrated the biological relevance of such movement in both the root and shoot. The behavior of artificially injected macromolecules suggest that specific regions of the plant exhibit a high level of cytoplasmic continuity, and therefore may function as a syncytium in which molecules move relatively freely. As previously discussed, the active transport of plant hormones, such as auxin, is also likely to play a crucial role in coordinating cellular behaviors.

Summary

An understanding of basic programs that control the division and differentiation behavior of plant cells is beginning to emerge through a combination of molecular and genetic analyses. Many aspects of cell division and differentiation appear conserved with other eukaryotes. Less clear is how these cellular behaviors are integrated at an organismal level. Molecules such as auxin, abscisic acid, brassinosteroids, and cytokinins may play key roles. Signals that operate over a shorter range are less well understood, and are likely to involve complex receptor–ligand interactions. Analyses of these and other genetically defined developmental cues promise new insights into how cell division and differentiation are integrated into larger programs of development.

List of Technical Nomenclature

Acropetal Describing the movement, growth or differentiation from the base towards the tip.

Anisotropic Not uniform, as in growth or expansion along a particular axis.

Basipetal Describing the movement, growth, or differentiation from the tip towards the base.

Cyclins A group of proteins conserved among eukaryotes which bind to and activate cyclin dependent kinases (CDKs). The periodic synthesis and destruction of cyclins appears to be a key mechanism in promoting the progression of cells through the cell cycle.

G_1 An interval in the cell division cycle which follows mitosis and precedes chromosomal DNA synthesis (S phase).

G_2 A stage in the cell division cycle which follows chromosomal DNA synthesis, but precedes mitosis (M phase).

Green fluorescent protein (GFP) A commonly used protein derived from a gene from jellyfish that is used to monitor gene expression *in vivo*, either through gene promoter–GFP fusions or through protein–GFP fusions.

β-glucoronidase (GUS) A commonly used protein derived from a gene from bacteria that is used to monitor gene expression in fixed tissue, either through gene promoter–GUS fusions or through protein–GUS fusions.

Hypocotyl The region of the seedling stem between the radicle and cotyledons.

M phase An interval of the cell cycle during which mitosis occurs.

Palisade A form of mesophyll consisting of elongated cells that in many plants is normally distributed along the adaxial interior half of the leaf.

Pericycle The outermost layer of the stele, located adjacent and interior to the endodermis.

Phyllotaxy The arrangement of leaves on the stem.

S phase An interval of the cell cycle during which chromosomal DNA synthesis occurs.

Spongy mesophyll Leaf tissue composed of variously shaped photosynthetic cells which are interspersed with large air spaces.

Suspensor A stalk-shaped zygotically derived structure to which the developing plant embryo is attached.

Syncytium A single compartment in which macromolecules such as RNAs and proteins freely diffuse.

See also: **Cell Walls and Fibers**: Cell Walls. **Flowering and Reproduction**: Flower Development. **Growth and Development**: Cells; Cell Growth; Leaf Development; Molecular Biology of Development; **Root Development**: Genetics of Primary Root Development; Lateral Root Initiation.

Further Reading

Hulskamp M, Schnittger A, and Folkers U (1999) Pattern formation and cell differentiation: trichomes as a genetic model system. *International Review of Cytology* 186: 147–178.

Jackson D (2001) The long and the short of it: signaling development through plasmodesmata. *Plant Cell* 13: 2569–2572.

Lyndon RF (1998) *The Shoot Apical Meristem*. Cambridge: Cambridge University Press.

Meijer M and Murray J (2001) Cell cycle controls and the development of plant form. *Current Opinion in Plant Biology* 4: 44–49.

Nakajima K and Benfey PN (2002) Signaling in and out: control of cell division and differentiation in the shoot and root. *Plant Cell* 14: 265–276.

Poethig RS (1987) Clonal analysis of cell lineage patterns in plant development. *American Journal of Botany* 74: 581–594.

Sachs T (1991) *Pattern Formation in Plant Tissues*. Cambridge: Cambridge University Press.

Scheres B (2001) Plant cell identity: the role of position and lineage. *Plant Physiology* 125: 112–114.

Steeves TA and Sussex IM (1989) *Patterns in Plant Development*. Cambridge: Cambridge University Press.

Traas J and Doonan JH (2001) Cellular basis of shoot apical meristem development. *International Review of Cytology* 208: 161–206.

Tsukaya H (2002) Interpretation of mutants in leaf morphology: genetic evidence for a compensatory system in leaf morphogenesis that provides a new link between cell and organismal theories. *International Review of Cytology* 217: 1–39.

Vandepoele K, Raes J, De Veylder L, *et al.* (2002) Genome-wide analysis of core cell cycle genes in *Arabidopsis*. *Plant Cell* 14: 903–916.

von Groll U and Altmann T (2001) Stomatal cell biology. *Current Opinion in Plant Biology* 4: 555–560.

Molecular Biology of Development

R M Twyman, University of York, York, UK

Introduction

Development is a complex process during which a multicellular organism with organized tissues and organs arises from (in many cases) a single cell. Plant development can be divided into two distinct phases. The first is embryonic development, in which the single cell is converted into a miniature body plan of the seedling. The second, postembryonic development phase involves the proliferation of meristems laid down in the embryo to form the shoot and root systems of the mature plant.

Development is progressive. An initially simple structure with crude organization and few cell types is gradually refined into a complex organism with many cell types organized in specific patterns. Five major processes are involved: growth, cell division, differentiation, pattern formation, and morphogenesis. Growth and cell division go hand in hand, and are required to increase the biomass and cell number of the plant. Differentiation is the formation of cells with specialized structures and functions, such as xylem, cortex, and epidermis. The number of different cell types always increases during development. Pattern formation involves the organization of cells, for example, the formation of concentric circles of different cell types in the shoot and root. Morphogenesis is the formation of shapes and structures, and this depends on aspects of cell behavior such as differential cell growth and proliferation, and the plane of cell division. All these processes are influenced by the environment, but are controlled by genes. The most important genes in plant development encode transcription factors, components of signal transduction pathways, and regulators of the cell cycle.

Major Differences between Plant and Animal Development

General Biological Aspects

Plants are sessile organisms and use development as a strategy to adapt to their environment. In contrast, animals are motile and can exploit their environment by moving around it. Therefore, animals have a rather rigid developmental program, but plant development is much more flexible in order to take environmental variation into account.

This is shown by the different relationship between the embryo and adult in animals and plants. The animal embryo is a miniature representation of the adult body plan. Postembryonic development in animals generally involves growth and refinement, but no major departures from the embryonic body plan are expected. In contrast, the plant embryo provides a basic body plan for the seedling, but postembryonic growth generates an entirely new adult body plan based on the continuous proliferation of shoot and root meristem populations. The contrasting lifestyles of animals and plants are also reflected in their reproductive strategies. Most animals set aside a germ line early in embryonic development, and this is the sole source of gametes. In contrast, there is probably no germ line in the plant embryo. The somatic cells of plants can switch from vegetative to reproductive growth during postembryonic development.

Cell Fate and Potency

The plasticity of plant development is most clearly demonstrated by the ability of differentiated plant cells to regenerate into whole plants when isolated and placed in the appropriate culture medium. This shows that many plant cells remain totipotent, i.e., able to recapitulate the entire developmental pathway and produce all the normal cell types of a complete plant. Both embryonic and postembryonic development can be recapitulated (somatic embryogenesis, organogenesis) although the capacity for these processes differs from species to species and shows strong genotype dependence. In contrast, the potency of animal cells is progressively restricted during development. No matter how many embryonic signaling proteins and regulatory factors are added to the culture medium, human liver cells will never give rise to a human embryo.

The molecular basis of this distinction between animals and plants reflects the mechanisms used to establish cell fate. The fate of a cell is the sum of cell types that cell and its progeny will produce during normal development (development unperturbed by experimental manipulation). In both animals and plants, cell fate is restricted throughout development by a series of hierarchical decisions. In plants, however, these decisions are reversible because they depend on the position of the cell and signals from surrounding cells. Epidermal cells normally arise from the L1 layer of the embryo, and normally remain as epidermal cells throughout the lifetime of the plant. However, an unusual periclinal cell division that causes an epidermal cell to invade the underlying cortex results in the respecification of the epidermal cell as a cortical cell. The fact that this cell comes from a different lineage is not important; it is the new position that counts. In animals, position is also important in development, but there comes a time when each cell becomes determined in its developmental pathway, which means it becomes irreversibly committed to its fate. Under these circumstances, moving the cell to a new site has no effect on that cell – it will carry on in its determined path. The fate of the cell then depends on its lineage (i.e., developmental decisions that were made in its past) rather than on its interactions with its neighbors. Determination and differentiation each result from changing patterns of gene expression. Put simply, these patterns of gene expression become "locked" in animal cells, through a variety of genetic and epigenetic mechanisms, while in most plant cells they are labile.

Pattern Formation and Morphogenesis

Patterns of cell organization in plants and animals appear to be generated using a similar set of molecular mechanisms. These include:

- The use of gradients of signaling molecules to set up axes (e.g., polarized auxin transport specifies the apical–basal axis of the embryo).
- Signaling between adjacent cells or groups of cells (e.g., in the control of stereotyped cell divisions in the root).
- The use of transcription factors with overlapping and combinatorial expression domains to confer positional information (e.g., MADS box transcription factors in the specification of floral organs).
- The use of lateral inhibition to achieve optimal spacing between structures (e.g., the initiation of lateral organ primordia by the shoot apical meristem, and the spacing of floral organs).
- The use of developmental compartments, within which cells are constrained by clonal boundaries (e.g., organ primordia in the flower).
- A limited capacity for spontaneous pattern formation (e.g., the ability of disrupted meristems to form lateral organs with correct organization).

In contrast to pattern formation, there are some major differences between plants and animals in terms of morphogenesis. In animals, the formation of cell and tissue structures is strongly dependent on cell–cell interactions, particularly the gain and loss of cell adhesion. This allows cells to move relative to each other either as whole tissues or as individual migrating cells. One of the landmark events in animal development – gastrulation – is driven almost entirely by the relative movement of cells. In contrast, the rigid plant cell wall and the way cells are tightly cemented together means that such mechanisms cannot be used in plants. There is no equivalent to gastrulation in plant development. Changes of shape and structure in the developing plant are therefore mediated by alternative mechanisms, such as controlling the plane and symmetry of cell division, and by increasing and decreasing the size of cells by varying the amount of water in the vacuole.

The structure and organization of plant cells also means that cell–cell communication in development is different to that seen in animals. Animal cells lack a cellulose wall, and bristle with receptors for all manner of molecules. Furthermore, certain signaling molecules can diffuse through the plasma membrane and interact directly with intracellular receptors. The diversity of signaling pathways is somewhat more limited in plants, and cell–cell communication often occurs through plasmodesmata, the cytoplasmic threads that connect plant cells into a continuous symplasm.

With the constraints on space in this encyclopedia, it would be impossible to discuss every aspect of plant development comprehensively. Therefore, the aim of this article is to illustrate various molecular principles of development with case studies from individual systems in developing plants. The following examples have been chosen:

- the use of signal gradients for axial patterning in the plant embryo;
- balancing cell division and differentiation in the shoot apical meristem (SAM);
- lineage versus position in the specification of cell fates in the shoot;
- lineage versus position in the specification of cell fates in the root;
- lateral inhibition for the initiation of lateral organ primordia and the spacing of trichomes on the surface of the leaf;
- floral integration as an example of how multiple environmental and endogenous signals can be channeled into a single developmental pathway; and
- positional information in the flower as a paradigm of pattern formation.

Case Studies

The Use of Signal Gradients for Axial Patterning in the Plant Embryo

The two major developmental processes occurring in plant embryonic development are the specification of regions along the apical-basal axis (axial patterning) and the specification of the three radial cell layers (radial patterning). Before discussing these processes in detail, however, a brief overview of embryonic development in flowering plants is necessary. Fertilization is followed by elongation of the zygote along the future apical–basal axis of the embryo, which corresponds to the shoot–root axis of the seedling. The elongated zygote divides asymmetrically to produce a small apical cell (which gives rise to most of the embryo) and a larger basal cell (which forms part of the root as well as the suspensor that attaches the embryo to the ovule). In order to form the suspensor, the basal cell divides many times in the same plane, while the apical cell undergoes a stereotyped sequence of horizontal and vertical divisions to generate a ball of cells called the proembryo. As cell divisions continue, the embryo is organized into regions representing the layout of the seedling – the shoot apical meristem (SAM), the cotyledon(s), the epicotyl, hypocotyl, radicle (embryonic root), and root apical meristem (RAM). The embryo also becomes radially organized into three fundamental layers that pervade the adult plant – L1, L2, and L3. The identification of mutations in *Arabidopsis thaliana* and other plants that perturb these processes has shown that they are largely independent.

In animals, axial patterning of the embryo is often dependent on graded signals. Mutations in the genes that provide these signals or respond to them generate patterning defects in which, characteristically, parts of the axis are missing. Very similar mechanisms appear to be involved in the patterning of the plant embryo. Polarized auxin transport has been shown to play an important role in establishing the apical–basal axis, and mutations in the genes that are required either to set up this gradient or respond to it generate phenotypes in which the apical, central, or basal regions of the embryo are affected. For example, the *MONOPTEROUS (MP)* gene is usually expressed in the bottom tier of cells of the proembryo, which gives rise to the hypocotyl, radicle, and initial cells of the root apical meristem (RAM). In *mp* mutants, these structures are absent. The product of the *MP* gene is a transcription factor that binds to auxin-responsive elements, so the *mp* mutant is likely to be insensitive to the basal-to-apical gradient of auxin. The product of the *GNOM (GN)* gene is expressed just after fertilization, and appears to be important for the localization of auxin efflux pumps, such as PIN1, specifically in the basal membranes. In *gn* mutants, the zygote divides symmetrically rather than forming apical and basal cells of different sizes. There is no progressive localization of the PIN1 auxin transporter to basal membranes and no accumulation of auxin in the basal region of the embryo. At the heart stage, *gn* mutants are more rounded than wild-type embryos, and the RAM is not specified. Note that not all axial patterning mutants affect auxin transport or perception. For example, the defect in the *gurke* (*gk*) mutant, in which the apical region of the embryo is absent, appears to reflect abnormal control of cell division.

Balancing Proliferation and Differentiation in the Shoot Apical Meristem

Meristems are small populations of rapidly proliferating cells that produce all the adult organs of a flowering plant. Two meristem populations are established in the embryo, the SAM (which gives rise to the aerial parts of the plant) and the RAM (which gives rise to the root system). The SAM comprises a central zone of slowly dividing mother cells surrounded by a peripheral zone of rapidly dividing apical initial cells. As cells leave the meristem proper, they enter the rib meristem, where files of cells are laid down to extend the shoot. Adjacent to the rib meristem are the lateral morphogenesis zones, where organ primordia begin to differentiate. The spatial arrangement and timing of lateral organ initiation is discussed later.

The patterning mutants discussed above, such as *mp* and *gk*, produce plants in which either the SAM or the RAM is never established, and there is consequently a complete failure of postembryonic development. In other mutants, the meristems are established correctly, but there is a defect that prevents cell division. This occurs in the *shoot meristemless* (*stm*) mutant of *Arabidopsis* and the *knotted-1* (*kn1*) mutant of corn (*Zea mays*; maize). In *Arabidopsis*, *STM* gene expression begins in the embryo at the early globular stage. Null mutations cause the arrest of development after rudimentary cotyledons have formed, indicating that the meristem cells have differentiated prematurely. In postembryonic development, *STM* is expressed strongly in the apical initial cells but is downregulated in the rib meristem and morphogenesis zones, where differentiation takes place. Weaker *stm* mutants produce shoots that give rise to a few leaves and flowers with reduced numbers of floral organs. This indicates that the meristem is "consumed" during the formation of leaves so that a smaller number of cells is available for floral organ formation. Other mutations cause the opposite phenotype, i.e., a larger meristem. For example, the *PRIMORDIA TIMING* (*PT*) gene influences meristem size in the embryo, and strong mutants show progressively increasing meristem size from the heart stage until germination. Loss-of-function mutants for the *CLAVATA1* (*CLV1*) gene show increased meristem size during postembryonic development. It appears the function of this gene is to recruit cells leaving the meristem into organogenesis. *CLV1* encodes a receptor kinase and functions in a signaling pathway that also includes the genes *CLV2* and *CLV3*. Mutations in two other genes, *MGOUN 1* (*MGO1*) and *MGO2* cause cells leaving the central zone of the SAM to accumulate in the peripheral zone of initial cells rather than forming organ primordia. All these data suggest a model in which *STM* maintains the meristem by blocking differentiation, which is facilitated by the *CLAVATA* and *MGOUN* genes acting sequentially (**Figure 1**). The *WUSCHEL* gene, which is expressed as early as the 16-cell stage, may be required to maintain meristem identity in the postembryonic meristem since in *wus* mutants the presumptive meristem cells differentiate, but are not recruited into lateral organs. The function of *WUS* in the embryo prior to the formation of the SAM is unknown. Another gene, *ZLL*, is required to maintain the meristem population in the embryo, and probably acts upstream of *STM* because in *zll* mutants, *STM* expression is not maintained.

Lineage versus Position in the Specification of Cell Fates in the Shoot

The specification of cell fates in the SAM has been investigated using cell chimeras, i.e., plants in which the meristem contains two genetically and phenotypically distinct cell populations. In periclinal chimeras, the distinct cells are found in different radial layers. Most of the cells in the outer two layers (L1 and L2) undergo anticlinal divisions, and remain within the same layer, but there are occasional periclinal divisions, which allow cells from one layer to invade the other. Under these circumstances, the fate of the invading cell is respecified, i.e., it begins to behave in the same manner as its neighbors rather than its forebears. As discussed earlier, this indicates

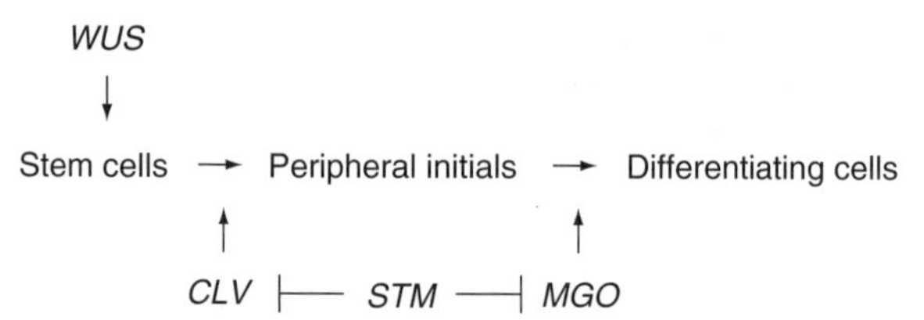

Figure 1 Genes involved in the maintenance of the SAM. *WUS* is required to maintain the population of stem cells. *STM* inhibits the differentiation of stem cells, and its activity is opposed by genes of the *CLAVATA* (*CLV*) and *MGOUN* (*MGO*) families.

that the radial fate of cells in the stem is maintained by cell–cell contacts rather than being dependent on lineage.

In mericlinal chimeras, segments of the meristem (comprising all three radial layers) are genetically marked, and these can be used to investigate the specification and determination of organ primordia. Although the details differ between different species, a common factor is that the fate map of the meristem is organized roughly as a series of concentric circles, with the outer circles representing nodes near the base of the stem and the inner circles representing nodes nearer the growing tip. There are always more cells around the edge of a circle than in its center, so marked cell clones of equivalent size at the periphery and in the center of the meristem fate map behave differently. At any stage of development, a small clone of marked cells from the periphery of the meristem will generate a marked sector of tissue in a basal leaf or stem, whereas a similar clone of marked cells from the center will probably give rise to several apical leaves derived entirely from marked cells (**Figure 2**).

Meristem fate maps have been produced in corn and *Arabidopsis*. The growth of corn plants is determinate, i.e., there is a predetermined number of nodes (18 plus a terminal spike and tassel, representing the male reproductive organs). Conversely, the growth of *Arabidopsis* is indeterminate, i.e.,

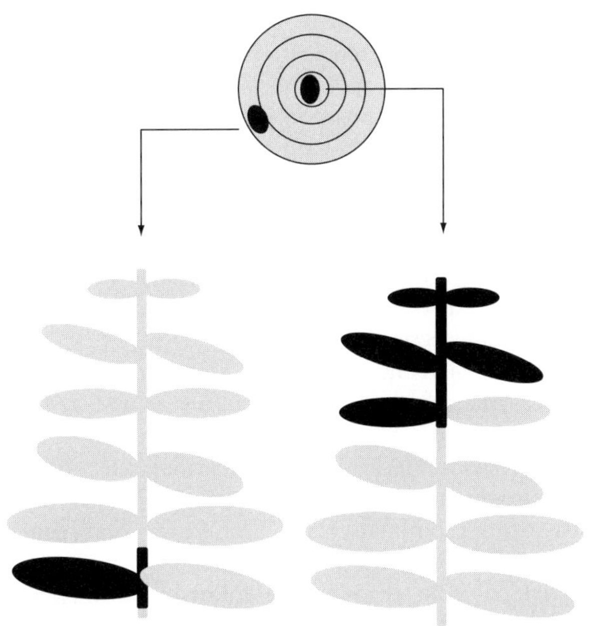

Figure 2 Cell fate in the SAM. Marked cell clones from the periphery of the SAM fate map give rise to older, lower nodes, while those from the center give rise to younger, more apical nodes. Peripheral cells may contribute to one leaf or a sector within one leaf, whereas central clones may cover several phytomers.

the number of leaves is not fixed. However, the determinacy of the corn SAM is reversible. The SAM removed from a corn plant after the formation of six or seven leaves can be deprogramed in culture, and will regenerate into an entire plant, not just the terminal portion of the plant from which it was removed. Therefore, it is not only radial cell types that are maintained by signaling, but also the positional identity of cell types within the meristem itself.

Lineage versus Position in the Specification of Cell Fates in the Root

The root system of the plant derives from the RAM, which is established at the basal pole of the embryo, and derives partly from the embryo proper and partly from the terminal cell of the suspensor (known as the hypophysis). While the SAM and RAM have similar functions, there are also some important differences between them. The SAM not only elongates the stem, but also gives rise to primordia that form branches and lateral organs (leaves, thorns, etc.). Conversely, lateral roots arise from the differentiated primary root well behind the growing tip. Also, the RAM gives rise not only to the differentiated cells of the root (which lie behind the meristem), but also to a distal root cap that is continually replenished as the cells are sloughed off by abrasion.

The RAM is organized in a similar way to the SAM with a quiescent center surrounded by proliferative initial cells that give rise to the file meristem and the distal root cap. Together, the quiescent center and the initial cells comprise the promeristem. The *Arabidopsis* promeristem has three layers. The lower layer comprises 12 central cells that produce the root cap and 16 surrounding cells that generate the lateral root cap and the epidermis. The middle layer comprises the four cells of the quiescent center and eight surrounding cells that give rise to the endodermis and cortex. Finally, the upper layer of stele initials gives rise to the pericycle and vascular bundles (**Figure 3**). Unlike the situation in the shoot, the division and differentiation of cells in the root meristem is invariant. For example, the cortical/endodermal initial cells divide to produce one initial cell and one daughter cell. The daughter cell then goes on to divide to produce one cortical cell and one endodermal cell, which enter the file meristem. The process is repeated throughout root growth (**Figure 4A**).

Ablation experiments have shown that although the cell divisions are stereotyped, cell fates are still dependent predominantly on position rather than

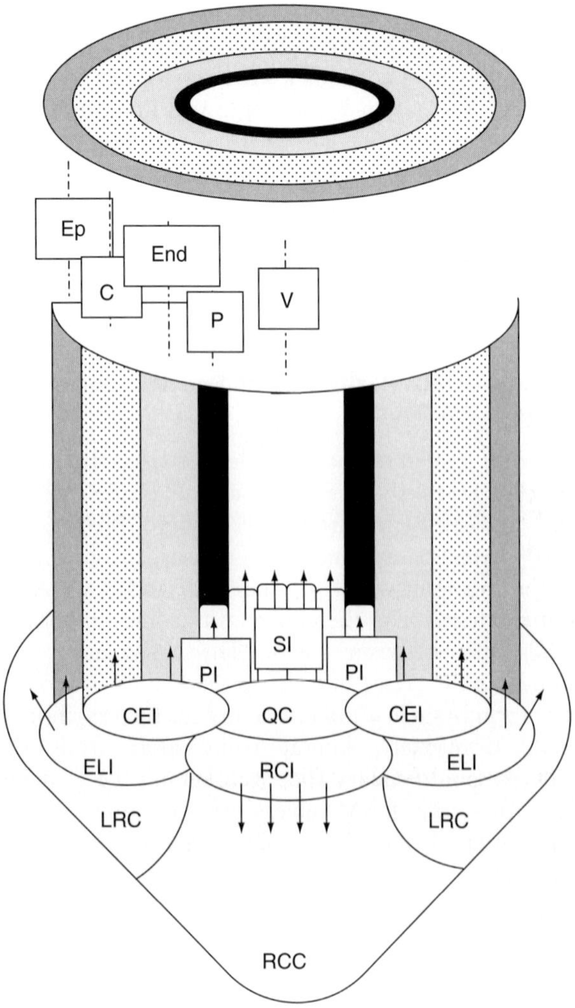

Figure 3 Organization and activity of the root apical meristem. The different cell types in the file meristem and mature root are shown in different shades and are labeled as follows: Ep, epidermis; C, cortex; End, endodermis; P, pericycle; V, vasculature. The cells of the root cap are labeled as follows: LRC, lateral root cap; RCC, root cap columella. They are derived from the initial cells of the promeristem, which are identified as follows: SI, stele initials; PI, pericycle initials; CEI, cortical/endodermal initials; ELI, epidermal/lateral root cap initials; QC, quiescent center; RCI, root cap columella initials.

lineage. For example, the ablation of one cortical/endodermal (CE) initial cell is followed by the unexpected division of an adjacent pericycle initial cell, so that the CE initial is replaced by a pericycle daughter cell (**Figure 4B**). The replacement cell then behaves in the same manner as a CE initial. If root development were lineage dependent, the pericycle initial would define a separate lineage (i.e., it would only be capable of producing pericycle cells), and such a replacement would not occur. If the daughter cell of a CE initial is ablated, the underlying CE initial cell divides again to replace it (**Figure 4C**). However, if three or more daughter cells lying above the CE initial are ablated, the CE initial becomes unable to divide asymmetrically. This shows that cell behavior is determined by signals from adjacent cells in the same file. Recently, it has been shown that these signals may be constrained within layers of cells by channeling through plasmodesmata.

Lateral Inhibition to Produce Regular Spacing Patterns

Lateral inhibition is a situation where the differentiation of one cell or group of cells is prevented by signals emanating from another. In plant development, this developmental strategy is used in two ways:

- to delay the initiation of successive lateral organs; and
- to generate regular spacing patterns in sheets of cells.

The periodic initiation of lateral organs by the SAM defines a developmental unit known as a plastochron, which gives the shoot time to extend before the next organ is initiated. Shoots are thus composed of nodes and internodes, the nodes bearing the lateral organs (usually leaves). A simple model for the regular initiation of lateral organs is that the formation of one primordium inhibits the formation of the next until a certain time has passed. This can be demonstrated by removing organ primordia as soon as they have formed. For example, if a new leaf primordium is ablated in the lupin (*Lupinus albus*) shoot apex, the next primordium emerges as expected, but the position of the one after that is altered. It is shifted toward the site of excision. This strongly suggests that lateral inhibition is used to delay the emergence of organ primordia to ensure the correct spacing of leaves. This also confirms that cell fates in the SAM are labile until just before the emergence of each primordium, and are maintained by signaling between cells.

Regular spacing patterns are common in development, and in plants one example is the regular spacing of trichomes on the dorsal surface of the leaf. Trichomes are derived from epidermal cells, and all epidermal cells have the capacity to form these structures. In animals, it has been shown that regular spacing patterns can arise spontaneously in fields of equivalent cells through the following mechanism. Initially, all the cells are undifferentiated and send out signals to their neighbors to suppress differentiation. However, if one cell begins to differentiate, its inhibitory signals become stronger and its ability to respond to inhibitory signals from adjacent cells begins to decline. The system becomes unstable due

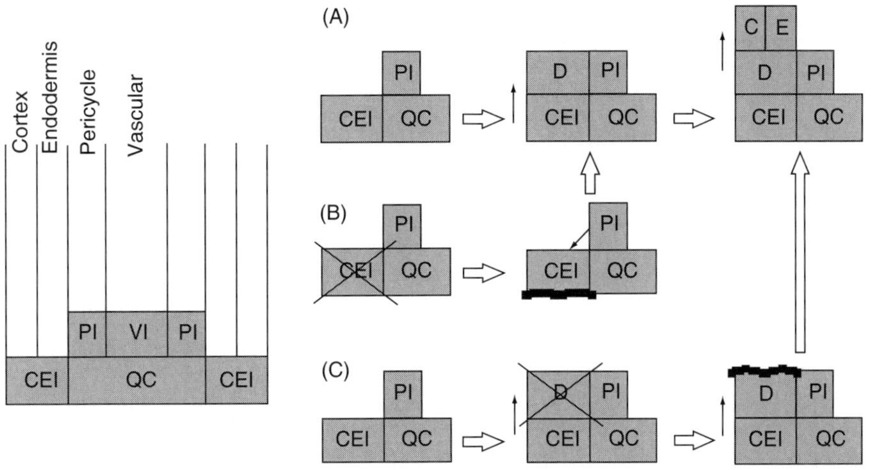

Figure 4 Laser ablation studies in root development. Left panel shows schematic representation of the promeristem (CEI, cortical/endodermal initials; QC, quiescent center; PI, pericycle initial; VI, stele (vascular) initials). Cell layers in the file meristem are shown above the corresponding initial cells and are identified at the top of the figure. (A) In normal development, each CEI gives rise to a daughter cell (D), which divides to form cortical and endodermis cells (C, E) adding to existing files. (B) If the CEI cell is ablated, an unexpected division of the PI cell replaces it. (C) If the daughter cell is ablated, the CEI undergoes an extra division to replace it. Wavy lines indicate squashed remains of ablated cells. Reproduced with permission from Twyman RM (2000) *Instant Notes in Developmental Biology*. Oxford: BIOS Scientific Publishers.

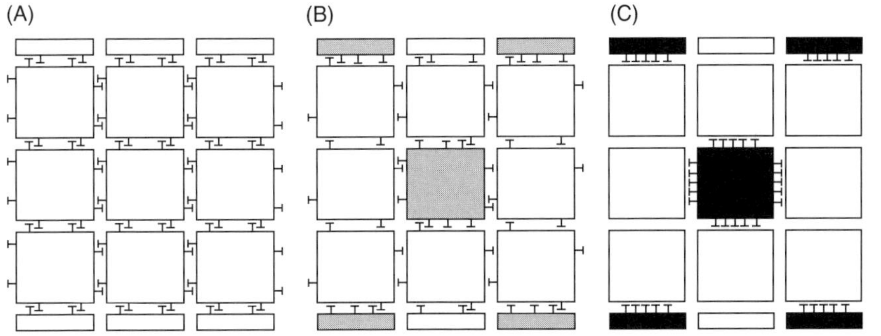

Figure 5 How spacing patterns are produced by lateral inhibition. (A) Initially all cells are undifferentiated (white) and inhibit each other's differentiation equally (T-bars). (B) Random fluctuations in the level of inhibition then cause some cells to assert dominance, and cause the surrounding cells to lose their inhibitory activity. (C) Eventually, the dominant cells differentiate and completely inhibit the surrounding cells. Reproduced with permission from Twyman RM (2000) *Instant Notes in Developmental Biology*. Oxford: BIOS Scientific Publishers.

to random fluctuations that transiently increase or reduce the inhibitory signals produced from each cell. Eventually, individual cells become dominant over their neighbors, and the spacing of these cells depends on the range of the signal (**Figure 5**).

The *Arabidopsis* leaf is a useful model to study the genetic basis of trichome spacing because the trichomes are very distinct, and they are dispensable when plants are grown in the laboratory. Genes have been identified that regulate trichome differentiation at various stages, from initiation to final morphogenesis. Those acting early in the pathway are the most interesting from a spacing perspective, and they include *TRYPTICHON* (*TRY*), *GLABROUS1* (*GL1*), *GL2, GL3*, and *TRANSPARENT TESTA GLABRA* (*TTG*). Mutations in *TRY* cause the formation of clusters in which four or five adjacent epidermal cells can form trichomes. *TRY* therefore appears to be involved in the lateral inhibition process that normally causes the trichomes to be interspersed by four or so epidermal cells. Mutations in *GL1, GL2, GL3*, and *TTG* reduce the number of trichomes, suggesting that these genes promote trichome development. However, weak *ttg* and *gl1* alleles also produce trichome clusters, so *TTG* and *GL1* also appear to play a role in trichome spacing. A recent model suggests that GL1 and TTG form a trimeric complex that contains either GL3 or TRY. The balance of the availability of GL3 and TRY would therefore determine whether the complex promoted or inhibited trichome development through the production of a diffusible inhibitory signal (**Figure 6**). The model is attractive in its simplicity, but is complicated by the existence of a number of other regulators, such as the MYB-like protein AtMYB23 that functionally overlaps with GL1 and the putative negative regulators CAPRICE (CPC) and COTYLEDON TRICHOME1 (COT1).

Integration of Multiple Environmental Signals into one Developmental Pathway

The transition from vegetative growth to flowering has to be timed correctly in order to maximize reproductive success, especially in out-crossing species where flowering must be synchronized. The initiation of flowering is controlled by both an autonomous pathway that acts as a developmental clock and exogenous pathways that respond to different cues in the environment. Many of the components of these pathways have been identified in *Arabidopsis*, a facultative long day plant (i.e., a plant where flowering can occur at any time, but is promoted on long days). The transition to flowering represents a paradigm in signal integration. It appears that all the signals, internal and external, converge on three so called floral integrator genes that act as master switches in reproductive development. These are *LEAFY* (*LFY*), *SUPPRESSOR OF OVEREXPRESSION OF CO 1* (*SOC1*) and *FLOWERING LOCUS T* (*FT*).

An autonomous pathway is required to prevent flowering before the plant has undergone sufficient vegetative growth to enjoy reproductive success. Mutations that show late-flowering phenotypes under both long- and short-day conditions identify proponents of the autonomous pathway. These include genes such as *FLOWERING LOCUS CA* (*FCA*), *FLOWERING LOCUS PA* (*FPA*), and *LUMINIDEPENDENS* (*LD*). The common molecular feature of such mutants is the accumulation of the mRNA for another gene, *FLOWERING LOCUS C* (*FLC*), whose product is a MADS box transcription factor that inhibits (probably indirectly) the three floral integration genes listed above. The inhibition of *FLC* is therefore a central step in the transition to flowering.

A long period of cold temperature (below 10°C) promotes flowering in many species including *Arabidopsis*, and is a useful reproductive strategy because it delays flowering until the spring. This vernalization response is epistatic to the autonomous pathway, and

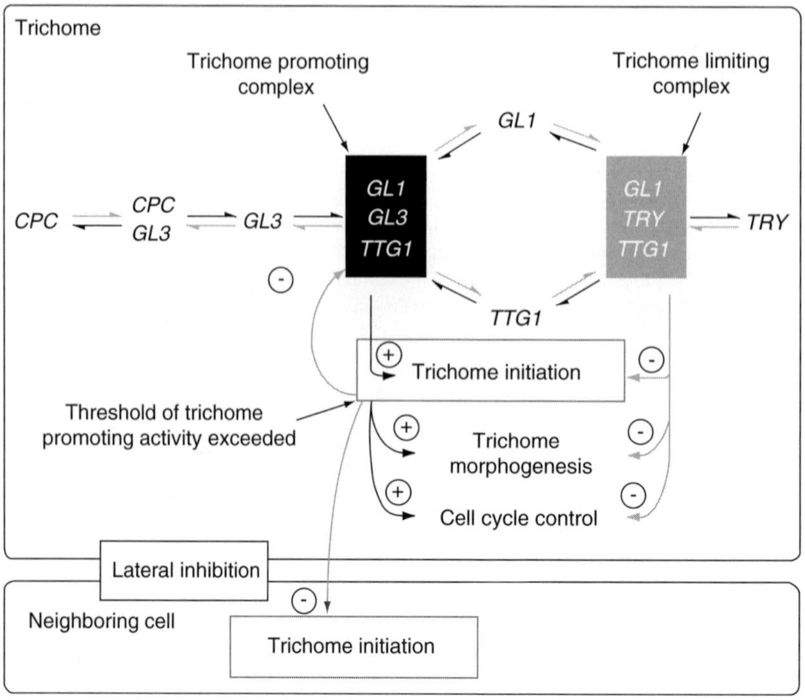

Figure 6 A model for the equilibrium reactions that establish trichome fate in the developing leaf. The upper box symbolizes a cell that has exceeded a threshold of trichome promoting activity and is inhibiting an adjacent, lower cell symbolized by the lower box. Arrows are shown between components that can disturb the equilibrium, either favoring trichome formation or inhibiting it. The major complexes are boxed and both contain the GL1 and TTG gene products. The trichome promoting complex also contains GL3 while this is replaced by TRY in the trichome inhibiting complex. Reproduced from Szymanski DB, Lloyd AM, and Marks MD (2000) Progress in the molecular genetic analysis of trichome initiation and morphogenesis in *Arabidopsis*. *Trends in Plant Science* 5: 214–219, with permission of Elsevier Ltd.

this reflects the fact that vernalization response genes such as *FRIGIDA* (*FRI*) also regulate the expression of *FLC*. The expression of *FLC* is downregulated (indirectly) by FRI at low temperatures, and then maintained in a repressed state when the weather gets warmer (and when FRI activity is reduced) by epigenetic mechanisms dependent on genes such as *VERNALIZATION RESPONSE 2* (*VRN2*), which encode chromatin remodeling proteins.

Flowering time in *Arabidopsis* is promoted under long-day conditions, which means that genes for light perception and circadian rhythm must be involved. Information converges on the *CONSTANS* gene (*CO*), whose product is active only under long-day conditions. The CO protein is a transcription factor that directly regulates two of the floral integrator genes (*FT* and *SOC1*). In *co* mutants, flowering is delayed under long-day conditions, but occurs normally under short-day conditions, showing that the autonomous pathway is not affected. Other genes have been identified that act as repressors in the light response pathway, e.g., *EARLY FLOWERING 3* (*ELF3*), *ELF4*, and *EARLY FLOWERING IN SHORT DAYS* (*EFS*). Other environmental signals also affect flowering, including overcrowding, heat, drought, nutrient depletion, and changes in light quality. It is unclear how many of these signals are integrated, although some may result in the increased synthesis of plant hormones. All plant hormones that have been identified thus far have been shown to influence flowering time, although only the gibberellin pathway, which promotes flowering through gibberellin response elements in the *LFY* promoter, has been investigated in any detail.

Positional Information in the Flower as a Paradigm of Pattern Formation

Floral integration, described above, results in the conversion of the SAM into a floral meristem. The latter is a determinate structure that no longer gives rise to periodic lateral organs, but to concentric whorls of floral organs (sepals, petals, stamens, and carpels) that are required for reproduction. Despite their very different structures and functions, all these organs begin as undifferentiated cell clumps on the surface of the floral meristem. Therefore, an important aspect of flower development is pattern formation, i.e., the regional specification of cells in the meristem so that each forms the correct structure in the correct place.

Many mutants that show floral patterning defects have been identified, but the most interesting class is the homeotic mutants, in which one whorl of floral organs develops with the likeness of another. This is analogous to the situation in *Drosophila*, where certain mutations cause body segments to take on incorrect regional identities, resulting in mutants with misplaced appendages (e.g., legs growing out of the head instead of eyes). The similarities go further in that in both *Drosophila* and plants, the genes responsible for homeotic mutations encode transcription factors that appear to provide each cell with a "positional code" to confer its regional identity. In plants, these are known as floral organ identity genes and many (but not all) of them encode transcription factors of the MADS box family.

The control of floral organ identity is explained by the ABC model, in which each whorl of floral organs is defined by a combination of three classes of genes: A, B, and C. Expression of class A genes alone specifies sepals (whorl 1), class A and B genes in combination specify petals (whorl 2), class B and C genes in combination specify stamens (whorl 3), and class C genes alone specify carpels (whorl 4). Class A genes are therefore expressed in whorls 1 and 2, class B genes in whorls 2 and 3, and class C genes in whorls 3 and 4 (*see* **Flowering and Reproduction**: Flower Development). The last component of the model is that the class A and C genes negatively regulate each other so that their expression patterns never overlap. Examples of all three classes of genes have been identified in *Arabidopsis* and all the homeotic mutant phenotypes can be explained in terms of this model. Additional complexities are generated by so-called class D genes, which are required for ovule development, and genes of the *SEPALATA* class, which are expressed in whorls 2, 3, and 4.

Conclusions

Seven case studies have been presented to illustrate the variety of developmental mechanisms that operate in plants. As can be seen from the foregoing discussion, the genetic basis of many of these processes is now becoming clear and many of the relevant genes have been identified. For applied plant science, the next step is to find ways of exploiting these genes to modify and improve plants. For example, genes controlling stem or root growth could be used to alter plant architecture; those controlling leaf development could be used to modify leaf number, shape or distribution; those controlling trichome development could be used to increase the number of root hairs and facilitate nutrient uptake; and those controlling flower development could be used to regulate flowering time externally and perhaps even increase the number of flowering cycles (and therefore harvests) per year. Relationships among the many plant genes controlling development are likely to become more clear as functional genomics is used to elucidate functional pathways, hierarchies and networks.

List of Technical Nomenclature

ABC model	A model of flower development in which the identity of floral organs is specified by the overlapping expression patterns of three classes of gene.
Anticlinal division	A cell division within the plane of a cell sheet, which expands the sheet but does not increase its thickness.
Apical initial cells	Rapidly proliferating cells of the shoot apical meristem that give rise to stem files and lateral organs.
Axial patterning	The organization of cells into particular functional regions along the axis of a structure.
Body plan	The basic outline of the body with all the organs in their correct places.
Cell chimeras	Plants containing two populations of genetically and phenotypically distinct cells, such as normal cells and cells that do not produce chlorophyll. The abnormal cells are said to be marked.
Cell cycle	The series of events between successive cell divisions.
Cell division	Increase in cell number.
Cell fate	The full spectrum of cell types a given cell can give rise to in normal development.
Central zone	The source of stem cells in the shoot apical meristem.
Determinate	Able to form a fixed number of structures, such as leaves, before development ceases.
Determination	The situation where a cell becomes irreversibly committed to its fate.
Developmental compartment	A structure within which a clone of cells is constrained.
Differentiation	Specialization of cells to carry out different functions in the body.
Embryonic development	Development of the plant embryo within the seed.
Epistatic	An effect where a mutation in one gene cancels out the effect of a mutation in another, often because both genes function in the same developmental pathway, but the first gene acts at a later stage.
Fate map	A map of an embryo, or meristem, showing the fate of each cell.
Floral integrator gene	A gene that acts as a master switch for flowering and responds to multiple internal and environmental signals.
Floral organ identity gene	A gene that helps to establish the identity of a floral organ by conferring positional information.
Gastrulation	A landmark event in animal development when the primary germ layers are formed.
Germ line	A group of cells that gives rise to gametes.
Growth	Increase in biomass and size.
Homeotic mutant	A mutant in which one whorl of floral organs develops with the likeness of another because it has received the wrong positional information.
Hypophysis	The terminal cell of the suspensor, which gives rise to part of the root apical meristem.
Indeterminate	Able to go on forming structures, such as leaves, indefinitely.
Lateral inhibition	The ability of one cell to suppress the differentiation of neighboring cells.
Lineage	The line of descent of a cell or organism.
Long-day plant	A plant in which flowering is promoted by long days.
Mericlinal chimera	Chimeras in which a sector of marked cells covers all radial cell layers.
Morphogenesis	The formation of shapes and structures.
Morphogenesis zone	The regions adjacent to the shoot apical meristem where lateral organs develop.
Motile	Free living; able to move around.
Normal development	Development unperturbed by experimental manipulation or mutation.
Organogenesis	Shoot and root development from cultured cells or explants.
Pattern formation	Organization of cells into specific patterns.
Periclinal chimera	Chimeras in which the marked and unmarked cells exist in different radial layers.
Periclinal division	A cell division parallel to the surface of a cell layer, which increases the thickness of that layer or ejects one of the daughter cells.
Phytomer	A node and internode.
Plasticity	Flexibility in developmental behavior.

Plastochron	A unit of developmental time between the initiation of successive lateral organs by the shoot apical meristem.
Positional information	Information in the form of the concentrations of molecules inside or outside a cell that tells that cell where it is in relation to other cells in the organism.
Postembryonic development	Development of the adult plant from shoot and root meristems.
Potency	The full spectrum of cell types a given cell can give rise to in any environment.
Quiescent center	The source of stem cells in the root apical meristem.
Radial patterning	The organization of cells, particularly different cell types, in concentric layers.
Regeneration	The formation of a complete plant from a particular cell or explant.
Regional specification	A process that gives a cell positional information.
Sessile	Attached to a substrate; stationary.
Signal gradient	A signal diffuses from a particular source and establishes a gradient over a field of cells, such that cells near the source are exposed to a higher concentration of the signal than those further away. If this causes the cells to behave differently, the signal is called a morphogen.
Somatic embryogenesis	Development of plant embryos from cultured cells or explants.
Specification	The situation where a cell is committed to a certain fate, but that commitment can be reversed by a change in the environment.
Spontaneous pattern formation	The ability of disorganized tissues to produce organized structures.
Suspensor	The structure that connects the plant embryo to the ovule.
Totipotent	Able to form all the cells of the adult organism.
Transcription factor	A protein that regulates gene expression at the level of transcription.
Trichome	Specialized epidermal cell on the surface of a leaf.
Vernalization	The promotion of flowering by prolonged cold temperature.

See also: **Flowering and Reproduction**: Flower Development. **Growth and Development**: Cell Division and Differentiation; Cell Growth; Leaf Development. **Root Development**: Genetics of Primary Root Development.

Further Reading

Battey NH and Tooke F (2002) Molecular control and variation in the floral transition. *Current Opinion in Plant Biology* 5: 62–68.

Blazquez MA (2000) Flower development pathways. *Journal of Cell Science* 113: 3547–3548.

Costa S and Dolan L (2000) Development of the root pole and cell patterning in *Arabidopsis* roots. *Current Opinion in Genetics and Development* 10: 405–409.

Fletcher JC and Meyerowitz EM (2000) Cell signaling within the shoot meristem. *Current Opinion in Plant Biology* 3: 23–30.

Harada JJ (1999) Signaling in plant embryogenesis. *Current Opinion in Plant Biology* 2: 23–27.

Howell SH (1998) *Molecular Genetics of Plant Development*. Cambridge: Cambridge University Press.

Lenhard M and Lauz T (1999) Shoot meristem formation and maintenance. *Current Opinion in Plant Biology* 2: 44–50.

Mouradov A, Cremer F, and Coupland G (2002) Control of flowering time: interacting pathways as a basis for diversity. *Plant Cell* 14: S111–S130.

Ng M and Yanofsky MF (2000) Three ways to learn the ABCs. *Current Opinion in Plant Biology* 3: 47–52.

Paquette AJ and Benfey PN (2001) Axis formation and polarity in plants. *Current Opinion in Genetics and Development* 11: 405–409.

Scheres B and Berleth T (1998) Root development: new meanings for root canals. *Current Opinion in Plant Biology* 1: 32–36.

Simpson GG and Dean C (2002) *Arabidopsis*, the Rosetta Stone of flowering time? *Science* 296: 285–289.

Szymanski DB, Lloyd AM, and Marks MD (2000) Progress in the molecular genetic analysis of trichome initiation and morphogenesis in *Arabidopsis*. *Trends in Plant Science* 5: 214–219.

Theissen G (2001) Development of floral organ identity: stories from the MADS house. *Current Opinion in Plant Biology* 4: 75–85.

Twyman RM (2000) *Instant Notes in Developmental Biology*. Oxford: BIOS Scientific Publishers.

Control of Gene Expression, Posttranscriptional Regulation

R M Twyman, University of York, York, UK

Introduction

The biological information that specifies the phenotype of a plant is provided by genes. Crop improvement by genetic modification introduces new genes, and therefore new biological information, into the plant genome. This results in a change of phenotype that is hopefully beneficial. However, the presence of new genes is not in itself sufficient to cause a change in phenotype. In order for this to occur, genes must also be expressed, first to produce RNA transcripts and subsequently proteins. The control of gene expression is therefore of paramount importance in the development of useful transgenic plants. Transgene expression is regulated by sequences present in the expression construct and factors intrinsic to the host plant.

The first level of gene expression is transcription (*see* **Growth and Development**: Control of Gene Expression, Regulation of Transcription). While transcription is the predominant level of gene regulation, at least for nuclear genes, research over the last decade has revealed that the abundance and distribution of a gene's final product also depends on posttranscriptional regulation, which occurs at many different levels and provides immense scope for increasing the diversity of gene products. After transcription, the primary transcript must be processed in a variety of ways to yield a mature transcript. Most nuclear genes of higher plants contain introns that must be removed, and variations in the pattern of splicing can produce different messenger RNAs (mRNAs) called splice variants that encode distinct proteins. An important and often-overlooked level of regulation is the control of mRNA stability, since the steady-state level of a transcript depends not only on its rate of synthesis but also its rate of degradation. Most transgenes encode proteins, so protein synthesis is another potential opportunity for gene regulation. In the plastid genome, regulation at the level of mRNA stability and translation is thought to be more important than transcriptional regulation. Once the protein has been synthesized it folds into its functional conformation and is often chemically modified, e.g., by phosphorylation or glycosylation. Folding and modification are important determinants of protein stability and depend to a large extent on where the protein is localized in the cell. The abundance of each protein depends not only on its rate of synthesis, but also on its rate of degradation. This may be rapid for some proteins (e.g., the D1 protein in photosystem II) but extremely slow for others (e.g., seed storage proteins).

The posttranscriptional regulation of gene expression is controlled by regulatory elements present in the primary transcript and mature transcript, and by peptide signal sequences in the protein. All of these elements are provided by the gene and to a certain extent, they can be recombined in expression constructs to control the posttranscriptional regulation of transgene expression precisely. In addition to these genetic mechanisms, however, nuclear genes are also subject to host-dependent epigenetic regulation, which can reduce the steady-state level of mRNA in the cell by increasing the rate of mRNA degradation. This is known as posttranscriptional silencing or RNA silencing and appears to be triggered by double-stranded RNA (dsRNA). The silencing mechanism is thought to have evolved to protect plants against RNA viruses, which utilize a dsRNA replication intermediate. Integrated transgenes can produce dsRNA if they are arranged as inverted repeats or if normal transcripts undergo self-priming. If the integrated transgenes are homologous to endogenous genes, the latter can also be silenced at the posttranscriptional level, a phenomenon known as cosuppression. There appears to be some cross-talk between RNA silencing and transcriptional silencing mechanisms because there is evidence of chromatin remodeling and DNA methylation in some transgenes and endogenous genes that are silenced at the RNA level (*see* **Growth and Development**: Control of Gene Expression, Regulation of Transcription).

RNA Processing

Splicing

Higher plant genes usually contain introns and these must be spliced out after transcription. Intron splicing in plants bears all the characteristic hallmarks of other eukaryotic systems, although plant introns are on average much smaller than those of vertebrates and have a much higher proportion of uridine residues. Like those of vertebrates, most plant introns follow the so-called GU–AG rule, with AG↓GUAAGU and UGCAG↓G consensus sequences representing the 5′ and 3′ splice sites respectively. Splicing involves two sequential transesterification reactions, the first joining the 5′ splice site to an internal branch point in the intron, creating a lariat intermediate, and the second joining the 5′ and 3′ exons and eliminating the intron. This process is preceded by the assembly of a ribonucleoprotein complex called the spliceosome from small nuclear ribonucleoprotein particles (snRNPs) and associated proteins. As is the case for vertebrates, a minority of plant genes contain AU–AC introns that are processed by an alternative spliceosome assembly.

One of the major differences between plant and vertebrate genes is the presence of AU-rich sequences in plant introns. These are required for efficient splicing, and may either reduce secondary structure or bind plant-specific components of the splicing apparatus. In support of the latter hypothesis, a protein called UBP-1 isolated from tobacco (*Nicotiana tabacum*) nuclei has been shown to recognize oligouridylate sequences *in vitro* and enhance splicing *in vivo*. Transgene expression in plants is occasionally disrupted by the presence of cryptic AU-rich sequences that are recognized by the splicing apparatus, as occurs in unmodified versions of *Bacillus thuringiensis* (*Bt*) toxin genes. The other major difference between plant and vertebrate genes is the difference in intron size, which may reflect the mechanism of intron recognition. In vertebrates, which typically have small exons and very large introns, the introns are thought to be delimited by protein complexes that span the exon (exon-definition). In plants, with smaller introns, both exon-definition and intron-definition mechanisms may operate.

Alternative splicing, resulting in multiple mRNAs from the same primary transcript, is a relatively common phenomenon in vertebrates. The prevalence of alternative splicing in plants is unknown, and where alternative splicing has been documented its biological significance is often unclear. A recent report with relevance to crop improvement is the demonstration of alternative exon inclusion in the tobacco *N* gene for resistance to tobacco mosaic virus (TMV). The *N* gene produces two mRNAs, N_S and N_L, of which *N*s is more abundant in non-infected plants. *N*1 becomes prevalent after infection and is required to confer complete resistance to TMV. As in animals, SR proteins (serine/arginine-rich proteins) play an important role in regulating splice-site selection by binding to splicing enhancers or silencers. Varying levels of specific SR proteins in different tissues might be responsible for cell-specific or developmentally regulated differences in mRNA splicing patterns.

One major practical application of splicing in transgenic plants is the improvement of transgene expression levels by intron-mediated enhancement (IME). In the best cases, the inclusion of an intron enhances transgene expression several hundred-fold, although this depends on the intron, the flanking exon sequences, the promoter, the cell type and the plant species. IME is generally more effective in monocotyledons than dicotyledons, although 30-fold enhancement has been achieved in *Arabidopsis*. The molecular basis of intron-enhanced transgene expression has yet to be investigated in detail.

Messenger RNA Stability

The steady-state level of a given mRNA depends on its rate of transcription and its rate of turnover. In both animals and plants, the half-life of endogenous mRNAs varies from minutes to days depending on the transcript in question. This intrinsic stability is determined by sequence and/or structural determinants on the transcript, which influence its interaction with the RNA degradation machinery. For some mRNAs, the turnover rate is fixed, while for others it is regulated by internal or external signals. This indicates that *trans*-acting factors must interact with the transcript to modulate its stability. An example is the pea (*Pisum sativum*) ferrodoxin (*Fed-1*) mRNA, whose stability is regulated by light.

Sequences that confer mRNA stability have been identified in vertebrates but not thus far in plants. Such elements may eventually be found in highly stable transcripts such as those of the cereal storage protein genes. The motif AUUUA has been shown to confer instability upon a number of naturally occurring mammalian mRNAs including those encoding short-lived products such as growth factors and cytokines. The same sequence is able to confer instability on recombinant plant mRNAs, although endogenous genes containing this sequence remain to be identified. Another motif, known as the downstream element (DST), has been shown to confer instability on a number of plant genes. The DST motif was originally identified in the *small auxin-up RNA* transcript (*SAUR*) and increases the turnover rate of reporter gene transcripts in transgenic plants. Such sequences should be eliminated from transgenes to help achieve maximum expression levels. For example, the presence of AU-rich instability sequences in unmodified *Bt* toxin transgenes is another reason why such genes are expressed at very low levels in transgenic plants.

Endogenous mRNAs are broken down by the general RNA degradation machinery. Several pathways have been identified in yeast, the most important of which is the deadenylation-dependent decapping decay pathway. Components of this pathway have been identified in plants but it is not clear if the mechanism is completely conserved. A minor pathway in yeast, the deadenylation-independent decapping decay pathway, is thought to mediate the degradation of mRNAs with incorrectly-placed nonsense codons. This process is sometimes termed nonsense-mediated decay and is a form of mRNA surveillance, avoiding the accumulation of nonfunctional RNA molecules. Components of this pathway have been identified in plants and the accelerated decay of transcripts with 5′ nonsense codons has been documented.

As discussed above, a separate RNA degradation pathway appears to have evolved specifically to protect plants from RNA viruses. The substrate for this pathway is dsRNA, an essential intermediate in most RNA virus replication cycles. The same pathway can be activated by transgene expression if dsRNAs are adventitiously synthesized, as often occurs if transgenes are integrated as inverted repeats. The impact of RNA-mediated silencing on transgene expression is considered in more detail below.

Protein Synthesis

Initiation of Translation

The initiation of protein synthesis in eukaryotes is usually dependent on the 5′ 7meG cap. This is recognized by a cap-binding complex comprising various initiation factors as well as the polyadenylate-binding protein (PABP). The assembly of these proteins brings the 5′ and 3′ ends of the transcript together in a circular RNP complex, which is required for initiation and may facilitate reinitiation with the same ribosome (**Figure 1**).

Global regulation of protein synthesis in animals is accomplished by reversible phosphorylation of the initiation factors that are required to assemble the ribosome on the mRNA. In plants, the global regulation of translational initiation is complicated by the presence of two distinct cap-binding complexes, one comprising the initiation factors eIF4E and eIF4G, and the other comprising the initiation factors eIFiso4E and eIFiso4G. The remaining components of the initiation complex are the same in each case: the RNA helicase eIF4A, the RNA-binding protein eIF4B, and polyadenylate-binding protein (PABP) (**Figure 1**). The abundance and activity of these alternative initiation factors differs according to cell type, developmental stage, and external signals. The genes encoding these proteins are expressed at different levels and the proteins themselves are reversibly phosphorylated under various forms of stress.

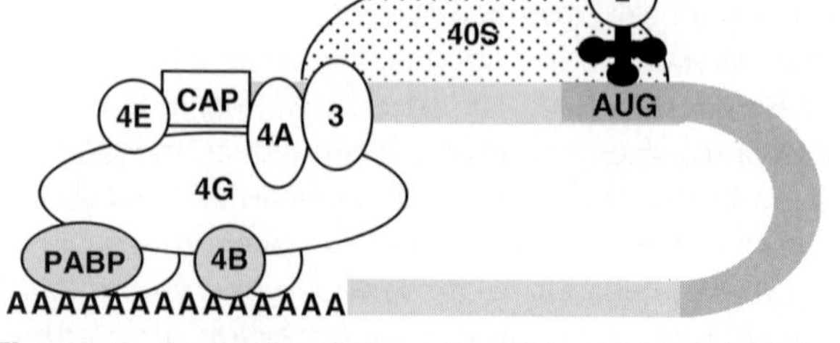

Figure 1 Assembly of the circular translational initiation complex. Each initiation factor is represented by a number, e.g., 2 = eIF2 and 4A = eIF4A. Initiation factor eIF4E (4E) or eIFiso4E binds to eIF4G (4G) or eIFiso4G, and the 5′-7mGpppN cap of the mRNA (shown as CAP). Interactions between eIF4 subunits and polyA-binding protein (PABP) cause the mRNA to circularize. Initiation factor eIF4B may also interact with the 5′ untranslated region (UTR) of the mRNA. Components of the initiation complex that can be phosphorylated, thus affecting their interactions with each other and the mRNA, are shaded.

Once assembled, the cap-binding complex then interacts with eIF3 to position the preinitiation complex (ribosomal small subunit and eIF2α-GTP-$tRNA_{met}$ ternary complex) near the 5′ end of the mRNA, allowing it to scan along the 5′ leader sequence towards the initiation codon. The correct initiation codon is embedded within a short sequence known as the Kozak consensus. Once this has been identified, eIF2α-GDP is released from the initiation complex and the large subunit of the ribosome is recruited, processes that require the activities of eIF5 and eIF5B. The efficiency of translational initiation can also be regulated by sequences that influence the efficiency of ribosomal scanning. In plants, as in other eukaryotes, a common regulatory mechanism is the presence of short open reading frames upstream of the genuine initiation codon. In many cases these interfere with translational initiation, but they can also confer a translational advantage, as is the case for the polycistronic CaMV 35S RNA.

Transcript-specific regulation can be superimposed over global regulation and can be used to protect certain genes from global repression. For example, the 5′ leader sequences of the corn (*Zea mays*; maize) *HSP70* mRNA and the TMV genomic RNA confer a translational advantage in heat-shocked cells because they recruit proteins that maintain the circular organization of the transcript, while in other transcripts the interactions are disrupted due to the phosphorylation of initiation factors.

In transgenic plants, optimal translational initiation can be achieved by removing negative regulatory elements from the 5′ and 3′ untranslated sequences and replacing them with translational enhancers such as the TMV 5′ leader, also known as the omega sequence. The translational start site should contain a single AUG codon within an optimized Kozak consensus sequence. If the transgenic plant is intended to tolerate of a particular form of biotic or abiotic stress, it may be desirable to model the untranslated regions of the transgene on endogenous transcripts that are known to escape global repression under those stressful conditions.

Polypeptide Elongation

Translational elongation factors are also subject to global regulation but there is little that can be done in terms of expression construct design to exploit this. However, one feature of construct design that can have a significant effect on polypeptide elongation in transgenic plants is the codon bias in the transgene's coding region in comparison to the codon preferences of the host. Different species have very different codon preferences when specifying degenerate amino acids. Taking the amino acid arginine as an example, the codon CGU is preferred in alfalfa (*Medicago sativa*, lucerne), and is 50 times more likely to occur than the rarest codon, CGG. In contrast, both of these codons are equally prevalent in corn, but the preferred choice is CGC. In yeast, AGA is by far the most popular codon for arginine, and this is also the case in potato (*Solanum tuberosum*), soybean (*Glycine max*), tobacco, and tomato (*Lycopersicon esculentum*). The expression of foreign transgenes in transgenic plants can therefore be very inefficient if infrequently used codons are abundant. The ribosome pauses at such codons and this generally reduces the rate at which the recombinant protein accumulates. However, more serious problems can also occur such as premature termination, misincorporation, frameshifting, and skipping. This is yet another reason why native *Bt* genes are not expressed efficiently in transgenic plants. Codon optimization is the process of modifying a foreign transgene by introducing silent mutations that bring codon preference in line with that of the host plant.

Protein Stability

Protein stability is an often-overlooked component of transgene expression, but in many cases this is the most crucial factor in the development of useful transgenic plants. Proteins emerging from the ribosome must fold to achieve their native conformation, they are often chemically modified, and some must then assemble into multiprotein complexes. These posttranslational processes are interdependent and often depend critically on the site within the cell where the protein accumulates. Therefore a failure of any one process can lead to the accumulation of a misfolded and nonfunctional polypeptide that is rapidly degraded.

Many simple proteins fold spontaneously to form their native conformation. In the case of soluble globular proteins, this process is driven by hydrophobic collapse, i.e., the shielding of hydrophobic groups in the core of the protein while hydrophilic residues form the external loops. Local secondary structures form within a few milliseconds and then the protein gradually adopts its final tertiary conformation over the next few seconds or minutes. In many cases, this folding process can be recapitulated if proteins are denatured and then allowed to refold *in vitro*.

More complex proteins are unable to fold or refold spontaneously. This suggests that multiple folding pathways are available and that it is possible for proteins to get "stuck" in an unproductive pathway. To prevent this outcome, the cell produces proteins known as molecular chaperones whose function is to help other proteins fold to achieve their correct conformations. In plants, molecular chaperones are not ubiquitously available. Many are synthesized specifically in the secretory pathway and this is also the compartment wherein glycosylation and other forms of posttranslational modification take place. The chemical environment of the endoplasmic reticulum is also the most suitable for the formation of disulfide bonds (which are required both to fold individual polypeptides and to assemble them into complexes).

The control of protein targeting in the cell is therefore an important component of the design of plant expression constructs. There is always a trade-off between the efficiency of protein accumulation, the functionally most suitable site of accumulation, and the health of the plant. For example, full-size recombinant antibodies produced in plants accumulate to very low levels if they are targeted to the cytosol. The reasons for this include the absence of molecular chaperones (resulting in inefficient folding), the unsuitable environment for disulfide bond formation, and the lack of glycosylation. Much better levels are obtained if antibodies accumulate in the secretory pathway, which can be achieved by the inclusion of an N-terminal signal peptide. Both plant and animal signal peptides appear to work equally well. The default site of accumulation for proteins thus targeted is the apoplast, the space between the plasma membrane and the cell wall. However, the inclusion of a KDEL signal at the C-terminus of the recombinant protein causes it to be retrieved from the Golgi apparatus and returned to the lumen of the endoplasmic reticulum in the manner of a resident endoplasmic reticulum protein. This strategy is appropriate if the goal is maximal accumulation, but if the antibody is designed to protect the plant from viral pathogens then the cytoplasm is a functionally more suitable accumulation site because this intersects with the viral infection cycle. Thus, a cytosolic antibody has been found to confer better protection than an equivalent antibody in the secretory pathway even though the stability of the former was so low that the secretory pathway antibody was 40 000 times more abundant. Molecules that interfere with the normal physiology of the plant might be better targeted to other compartments, such as the chloroplast or the vacuole. For example, recombinant avidin expressed in plants accumulated in the vacuole to much higher levels than possible in other compartments because of toxicity effects resulting from the interference with endogenous biotin metabolism.

Posttranscriptional Gene Silencing

Transgene silencing and cosuppression occur at both the transcriptional and posttranscriptional levels. Transcriptional silencing often occurs due to the integration of the transgene at a genomic position that is already repressed (position-dependent silencing) (*see* **Growth and Development:** Control of Gene Expression, Regulation of Transcription). However, even transgenes in active regions of the genome may be silenced if they are recognized as foreign invaders. This context-dependent silencing, which may detect unusual DNA structure or base composition, is a defense mechanism against invasive DNA, i.e., transposons and other mobile elements. Similarly, transcriptional silencing can occur in complex transgene loci containing tandem or inverted repeats, again because the cell may recognize certain features shared with active transposons.

In transcriptional silencing, the silenced locus is generally hypermethylated and sequestered into inactive chromatin, and no mRNA is produced. In the case of posttranscriptional silencing, mRNA is produced (indeed transcription is necessary for silencing) but it is rapidly degraded. This is confirmed by nuclear run-on assays, which measure the amount of pre-mRNA in the nucleus. Like transcriptional silencing, posttranscriptional silencing appears to have evolved as a defense against invasive nucleic acids. In this case the target is viruses and the trigger is dsRNA. Similar defense pathways have been identified in many animals and are starting to be exploited to deliberately interfere with endogenous gene expression (RNA interference). These posttranscriptional mechanisms can be grouped under the collective term "RNA silencing."

RNA silencing occurs in transgenic plants if the transgene produces dsRNA. At multicopy loci, this can occur by the adventitious production of hairpin RNAs, either through transcriptional read-through from adjacent transgene copies arranged as inverted repeats, or through the transcription of inverted partial transgene copies. The resulting dsRNA is a substrate for the ribonuclease Dicer, which cleaves it into short fragments 21-23 bp in length. Dicer endonucleases are multidomain RNase III enzymes that have been identified in many eukaryotes. In *Arabidopsis*, there is evidence for at least four Dicer homologs, one of which is encoded by the gene *CARPEL FACTORY* (*CAF*) also known as *SHORT INTEGUMENT 1* (*SIG1*). The activity of Dicer alone would be insufficient to destroy invading RNA viruses, since chopping up the replicative intermediate would inhibit replication but would not prevent transcription of single-stranded copies of the genome. In a second phase of the process, however, the short interfering RNA (siRNA) fragments produced by Dicer then associate with other proteins to assemble a RNA-induced silencing complex (RISC) which recognizes homologous single-stranded RNAs and degrades them (**Figure 2**). This is a very efficient process which can eradicate even quite abundant viral RNAs. If dsRNA corresponding to a transgene or an endogenous gene is produced in the cell, the same potent defense mechanism eradicates the corresponding mRNA as soon as it is synthesized. The silencing mechanism is systemic, reflecting the ability of siRNAs to spread throughout the plant.

It would appear that RNA-mediated transgene silencing could be avoided by making sure transgenic loci were organized in such a fashion that hairpin RNA could not form. However, posttranscriptional silencing of actively transcribed single-copy loci has also been documented, so another mechanism of dsRNA formation must be possible. It has been suggested that there is a threshold level of mRNA above which the Dicer enzyme is activated, to prevent runaway expression of viral genes. One possible mechanism is saturation of the polyadenylation machinery. This would result in the accumulation of unadenylated transcripts which might be capable of self-priming and the formation of hairpins.

Avoiding Transgene Silencing in Transgenic Plants

From the discussion above, several conclusions can be drawn about transgene silencing and several general strategies can be employed to help prevent it:

- the transgenic locus should be protected from position effects
- the transgene should have a similar base composition to the host plant
- superfluous sequences, particularly the plasmid backbone, should be avoided
- the transgene copy number should be kept as low as possible
- the locus structure should be simple, preferably a single intact copy of the transgene
- inverted repeat structures should be avoided wherever possible
- maximized transcription may be counterproductive.

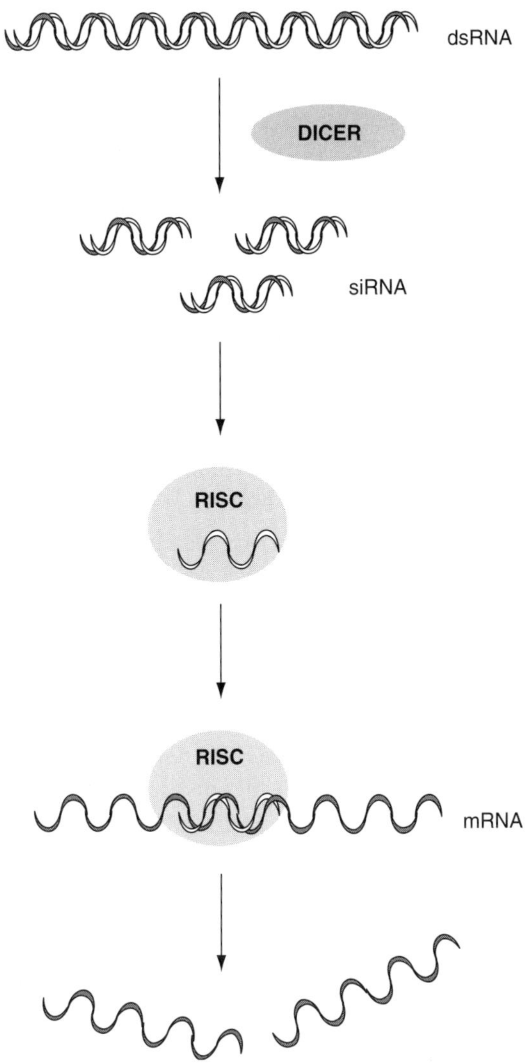

Figure 2 Mechanism of epigenetic RNA silencing. Double-stranded RNA (dsRNA), the trigger for RNA silencing, is the substrate for the enzyme Dicer, which in *Arabidopsis* may be encoded by the *CARPEL FACTORY/SHORT INTEGUMENT1* (*CAF/SIN1*) gene. The enzyme cleaves the dsRNA into short interfering RNAs (siRNAs) that are 21–25 bp in length with 2-nt overhangs at each end. The siRNAs are incorporated into the RNA-induced silencing complex (RISC) which has a conserved domain with Dicer that may facilitate their interaction. The RISC then interacts with and cleaves homologous mRNAs using the siRNA to achieve sequence specificity.

Chloroplast Gene Expression

The chloroplast genome differs in many respects from the nuclear genome and most of these differences reflect the prokaryotic origin of plastids. One distinction with particular relevance to this article is the relative importance of transcriptional and posttranscriptional gene regulation. While chloroplast genes are regulated at the transcriptional level, posttranscriptional gene regulation is predominant. As might be expected, light plays an important role in the regulation of many chloroplast genes. Notably, neither transcriptional nor posttranscriptional silencing has been demonstrated for chloroplast genes which means that very high levels of transgene expression can be achieved in plants with genetically transformed chloroplasts.

Posttranscriptional Regulation of Chloroplast Gene Expression

RNA processing, degradation, and translation are the predominant levels of gene regulation in the chloroplast genome. Several distinct forms of RNA processing occur in chloroplasts. Like bacterial transcripts, chloroplast transcripts are not capped and generally not polyadenylated. Introns occur in some chloroplast genes but these are generally of the self-splicing variety. An interesting feature of chloroplast gene regulation that is rare in the nuclear plant genome is RNA editing, the replacement of individual cytosine residues with uracil after transcription. Translation of chloroplast mRNAs has many features in common with bacterial translation, including the use of polycistronic transcripts, the requirement for similar initiation factors, the assembly of a eubacterial-like ribosome, and the formylation of the initiator amino acyl-tRNA. Chloroplast ribosomes exist either free in the stroma or attached to the thylakoid membranes, the latter allowing the cotranslational insertion of nascent polypeptides into the membrane. The initiation codon in chloroplast genes is almost always AUG, but occasionally GUG may be used as is the case in eubacteria. In higher plants, the ribosome often binds to a sequence reminiscent of the eubacterial Shine–Dalgarno (SD) motif, which is complementary to the 16S rRNA and helps position the ribosome over the initiation codon. In some cases, the mechanism appears similar to that in bacteria (i.e., the SD-sequence places the 30S ribosomal subunit at the correct position on the transcript for translational initiation; an example of this is the barley (*Hordeum vulgare*) *rbcL* gene). In others the ribosomal subunit may bind to the SD-sequence, but must then scan along the transcript to find the initiation codon (e.g., the barley *psbA* gene). In algae, about half of the chloroplast genes lack an SD-sequence and initiation must occur through a different mechanism involving the binding of specific translational initiation factors to sequences in the 5′ untranslated region. Where SD sequences do exist, mutagenesis experiments suggest that the secondary structure formed by such elements is more important than the sequence itself.

The stability of some chloroplast RNAs is regulated by RNA-binding proteins that respond to light. For example, nuclear-encoded protein factors have been identified in *Chlamydomonas reinhardtii* that bind to the 5′ untranslated region (UTR) of the *psbB* and *psbD* transcripts and increase their stability. Similarly, light-dependent proteins are also known to increase the efficiency of translation. The well-studied example of *C. reinhardtii psbA* mRNA is particularly informative since the components of the translational regulatory complex have been identified. Four nuclear-encoded proteins assemble on the 5′ UTR, one of which is a polyadenylate-binding protein. Although chloroplast transcripts are generally not polyadenylated, this protein recognizes a similar adenine-rich element in the *psbA* 5′ UTR. Another component of the complex, RB60, is structurally modified by light-dependent signals. For example, it is phosphorylated by an ADP-dependent kinase, which accumulates in the dark, therefore leading to the dissociation of the complex in the absence of sunlight. The redox state of the chloroplast, which is influenced by photosynthetic activity, also changes the structure of RB60 by reversibly oxidizing and reducing critical disulfide bonds. Certain chloroplast genes may also be regulated at the translational level by attenuation, i.e., the binding of the gene product to the mRNA. An example is the *C. reinhardtii* cytochrome *f* protein, which inhibits its own transcript (*petA*).

Conclusions

Posttranscriptional gene regulation is a complex, multistage phenomenon that helps to fine tune the levels of active proteins in the cell. Transcription is followed by RNA processing to produce a mature transcript for translation. The stability of this transcript and the efficiency of translational initiation are important regulatory steps, particularly in the chloroplast genome. After protein synthesis, the protein must be folded and chemically modified, processes that depend largely on the site of accumulation. In transgenic plants, all stages of transgene expression can be controlled by expression construct design, but other factors are involved that are influenced by the integration of the transgene and the structure of the transgenic locus. Successful transgene expression depends to a large extent on the transgene being accepted as a genuine locus and not branded as a foreign sequence. Of particular importance, the production of dsRNA can trigger a potent and long-lasting posttranscriptional silencing mechanism that can affect both transgenes and homologous endogenous genes.

List of Technical Nomenclature

Cap-binding complex	A protein complex that assembles on the 5′ end of the mRNA and helps to recruit the small ribosomal subunit.
Codon bias	The species-specific preference for certain codons to represent particular degenerate amino acids.
Codon optimization	The modification of the coding region of a transgene to match the codon bias of the expression host.
Context-dependent silencing	The recognition of a transgene as foreign DNA due to dissimilarities in base composition compared to surrounding genomic DNA.
Cosuppression	The epigenetic silencing of an endogenous gene that is homologous to an integrated transgene.
Dicer	An endonuclease that recognizes and cleaves double-stranded RNA molecules.
Exon	The segments of the primary transcript that are retained in the mature transcript.
Hairpin RNA	An RNA molecule that is self-complementary and folds back on itself.
Homology-dependent silencing	A broad term that includes any form of gene silencing (transcriptional and post-transcriptional) provoked by the presence of multiple copies of the same sequence in the genome.
Initiation codon	The first codon in an open reading frame. Usually AUG but occasionally others.
Initiation complex	The complex of proteins required to initiate protein synthesis. Required to help the ribosome interact with the mRNA.
Intron	An intervening sequence in the transcribed part of a gene that is removed from the mature transcript.
Intron-mediated enhancement	The ability of a heterologous intron to increase the transcriptional activity of a promoter in transgenic plants.
KDEL	A C-terminal tetrapeptide that causes proteins to be recycled from the Golgi apparatus to the endoplasmic reticulum.
Kozak consensus	The sequence surrounding the initiation codon, which is required for efficient assembly of the initiation complex.
Mature transcript	The final (processed) product of transcription. In the case of protein-

encoding genes, this is also known as a messenger RNA (mRNA).

Molecular chaperone A protein whose function is to help fold or refold another protein.

Nonsense-mediated decay An alternative mRNA degradation pathway that specifically acts on mRNAs with misplaced termination codons.

Posttranscriptional silencing Gene silencing that requires transcription and results from the rapid degradation of RNA (also known as RNA silencing). When applied deliberately to inactivate particular genes, it is known as RNA interference.

Primary transcript The initial (unprocessed) product of transcription, typically containing introns that need to be spliced out. In the case of protein-encoding genes, this is also known as pre-mRNA.

Regulatory element A *cis*-acting element that regulates gene expression.

RNA editing A special form of RNA processing in which the information content of the coding region is changed after transcription by the substitution, insertion, or deletion of one or more nucleotides.

Shine–Dalgarno sequence The sequence in bacterial genes upstream of the initiation codon, which is complementary to the 16S rRNA and helps to position the ribosome correctly. Similar sequences are found in some plastid genes but not others.

Signal peptide A short peptide sequence present at the extreme N-terminus of a protein that directs the ribosome to receptors on the endoplasmic reticulum and inserts the protein into the lumen, the first compartment of the secretory pathway.

Small interfering RNA (siRNA) Short segments of double-stranded RNA, the products of the Dicer endonuclease, which interact with homologous RNA molecules and result in their rapid degradation.

Spliceosome The complex of RNAs and proteins that carries out splicing.

Splicing Removal of introns from the primary transcript.

Steady-state level The amount of a specific RNA or protein present in the cell at any one time, a function of the rate of synthesis and the rate of degradation.

Transcriptional silencing Gene silencing involving the loss of transcriptional activity. It results from the sequestration of DNA into inactive and often hypermethylated chromatin.

Transgene A foreign gene introduced into a plant, such plants being described as transgenic if the transgene is nuclear or transplastomic if the transgene is integrated in the plastid genome.

Transgene silencing Epigenetic loss of transgene expression. Can occur at either the transcriptional or posttranscriptional levels.

See also: **Genetic Modification**: Gene Cloning, General Principles; Transformation in Monocotyledons; Transformation in Plastids; Transformation, General Principles. **Growth and Development**: Control of Gene Expression, Regulation of Transcription.

Further Reading

Ambrose V (2001) Dicing up RNAs. *Science* 293: 811–813.

Brown JWS and Simpson CG (1998) Splice site selection in plant pre-mRNA splicing. *Annual Review of Plant Physiology and Plant Molecular Biology* 49: 77–95.

Browning KS (1996) The plant translational apparatus. *Plant Molecular Biology* 32: 107–144.

Bruick RK and Mayfield SP (1999) Light-activated translation of chloroplast mRNAs. *Trends in Plant Science* 4: 190–195.

Gallie DR (1993) Posttranscriptional regulation of gene expression in plants. *Annual Review of Plant Physiology and Plant Molecular Biology* 44: 77–105.

Gallie DR (1998) Controlling gene expression in transgenics. *Current Opinion in Plant Biology* 1: 166–172.

Grant SR (1999) Dissecting the mechanisms of posttranscriptional gene silencing: divide and conquer. *Cell* 96: 303–306.

Gutiérrez RA, MacIntosh GC, and Green PJ (1999) Current perspectives on mRNA stability in plants: multiple levels and mechanisms of control. *Trends in Plant Science* 4: 429–438.

Hammond SM, Caudy AA, and Hannon GJ (2001) Posttranscriptional gene silencing by double-stranded RNA. *Nature Review Genetics* 2: 110–119.

Kawaguchi R and Bailey-Serres J (2002) Regulation of translational initiation in plants. *Current Opinion in Plant Biology* 5: 460–465.

Lorkovic ZL, Wieczorek Kirk DA, Lambermon MHL, and Filipowicz W (2000) Pre-mRNA splicing in higher plants. *Trends in Plant Science* 5: 160–167.

Twyman RM, Stoger E, Kohli A, and Christou P (2002) Foreign DNA: integration and expression in transgenic plants. In: Setlow JK (ed.) *Genetic Engineering: Principles and Practice*, vol. 24, pp. 107–136. New York: Plenum Press.

Voinnet O (2002) RNA silencing: small RNAs as ubiquitous regulators of gene expression. *Current Opinion in Plant Biology* 5: 444–451.

Control of Gene Expression, Regulation of Transcription

R M Twyman, University of York, York, UK

Introduction

The phenotype of a plant depends on gene expression. Therefore, where one aims to improve the phenotype of a plant by introducing new genes, the control of gene expression is of paramount importance. Over the last decade, much has been learned about the complexities of gene expression, and the simplistic view that "DNA makes RNA makes protein" has been superseded. It is now realized that the abundance and distribution of a gene's final product is regulated at many different levels, and that each level provides scope for increasing the diversity of gene products. The first stage of gene expression is transcription, and in the nuclear genome this is the primary level of gene regulation. Complex and overlapping controls govern where and when genes are transcribed, and such control mechanisms are discussed below. The primary transcript must then be processed in a variety of ways to yield a mature transcript, which is translated into protein. The nascent protein folds into its functional conformation and is often chemically modified, e.g., by phosphorylation or glycosylation. These posttranscriptional events can also be regulated (*see* **Growth and Development**: Control of Gene Expression, Posttranscriptional Regulation).

Transcription is controlled by regulatory elements present in and around each gene. To a certain extent, such elements can be recombined in expression constructs to regulate transgene expression precisely. In addition to these genetic mechanisms, however, nuclear genes are also subject to epigenetic regulation mechanisms, which include DNA methylation and the modification of chromatin structure. These processes provide an additional and global level of control over endogenous gene transcription, but they have unpredictable effects on transgenes and can result in a phenomenon known as transgene silencing. There are several triggers for transgene silencing. In some cases, the position of integration is to blame. In other cases, it is the presence of repeats or the recognition of the transgene as "foreign" due to its unusual base composition. Both these transcriptional silencing mechanisms have evolved to protect plants against the destructive effects of transposable elements. Silencing can also occur at the posttranscriptional level, e.g., the transgene is transcribed but the RNA fails to accumulate because it is rapidly degraded (*see* **Growth and Development**: Control of Gene Expression, Posttranscriptional Regulation). It is worth noting that there may be a degree of cross-talk between transcriptional and posttranscriptional silencing mechanisms since RNA mediated silencing often induces hypermethylation of the corresponding transgene. If a transgene is similar in sequence to an endogenous gene, that gene can also be silenced, a phenomenon known as cosuppression.

The first part of this article provides a basic overview of nuclear gene transcription, emphasizing the features that are unique to plants. We discuss the regulatory elements that control gene expression, and how these have been adapted to be used in plant expression constructs. The second part focuses on epigenetic gene regulation, and explores strategies to prevent transgene silencing caused by position effects and locus organization. Finally, we consider the transcription of genes in the chloroplast genome, although it is evident that posttranscriptional regulation plays a critical role in the regulation of chloroplast genes (*see* **Growth and Development**: Control of Gene Expression, Posttranscriptional Regulation).

Genetic Control of Transcription

Overview of Transcriptional Initiation in Plants

The general machinery of transcriptional initiation appears to be highly conserved between animals and plants. The core promoters of protein encoding genes in plants, as in animals, often contain a TATA box and/or an initiator element just upstream of the transcriptional initiation site, and these help to position the general transcription factor TFIID, which contains the TATA binding protein (TBP) and various TBP associated factors (TAFs). Other components of the transcriptional initiation complex then assemble: TFIIA, TFIIB, RNA polymerase II (recruited to the complex by TFIIF), TFIIE, and TFIIH (**Figure 1**). These proteins have been isolated and functionally characterized in a number of plants and appear closely related to their animal counterparts.

The stability of the initiation complex is generally very low unless further transcriptional activators are bound to elements upstream of the core promoter, and at more distant sites called enhancers. Some of these activators are constitutive while others may be cell specific, stage specific, or inducible by external signals. Although activation and the cessation of activation would be sufficient to control gene expression, there are also transcriptional repressors

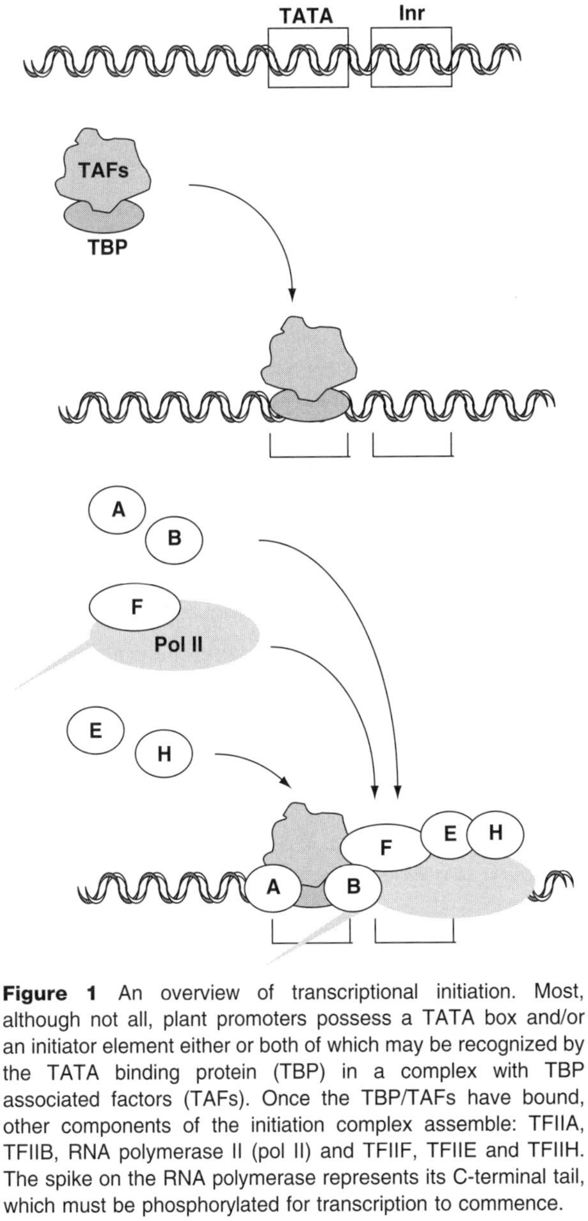

Figure 1 An overview of transcriptional initiation. Most, although not all, plant promoters possess a TATA box and/or an initiator element either or both of which may be recognized by the TATA binding protein (TBP) in a complex with TBP associated factors (TAFs). Once the TBP/TAFs have bound, other components of the initiation complex assemble: TFIIA, TFIIB, RNA polymerase II (pol II) and TFIIF, TFIIE and TFIIH. The spike on the RNA polymerase represents its C-terminal tail, which must be phosphorylated for transcription to commence.

that bind to the upstream promoter and to more distant elements known as silencers. Transcriptional activators and repressors function in various ways to influence the stability of the initiation complex. Some are DNA binding proteins, while others interact at the protein level. Some may form direct contacts with the initiation complex, or indirect contacts through a bridging complex called mediator, while others modify chromatin structure (see below) or introduce bends or kinks into the promoter to facilitate other interactions. Some regulatory proteins can act as either activators or repressors depending on promoter context and the other proteins that are present.

Control of Transcriptional Initiation in Transgenic Plants

Constitutive promoters Transcriptional initiation is the primary level of nuclear gene regulation, so the choice of promoter is one of the most important considerations when thinking of gene expression in transgenic plants. In the simplest case, it may be desirable to maximize transcription, and a constitutive promoter would be the most suitable. When the term "constitutive" is used to describe a promoter, it is usually taken to mean that the promoter is active in all cells all of the time. In reality, constitutive promoters are widely but not ubiquitously active, and the transcriptional activity of any gene regulated by that promoter will vary in different cell types.

The most popular constitutive promoters for transgenic plants are renowned for their strong activity. For example, the cauliflower mosaic virus 35S promoter (CaMV 35S) is very active in many dicotyledonous plants, and is often used to drive transgene expression. Alternatives include the octopine synthase (*ocs*), nopaline synthase (*nos*), and mannopine synthase (*mas*) promoters from *Agrobacterium tumefaciens*. Note that all these promoters are derived from plant pathogens, and have evolved to promote maximum activity of pathogen genes in a wide range of host species.

None of the above promoters is particularly active in monocotyledons, but, in some cases, it has been possible to increase expression levels by artificial modification. The activity of the CaMV 35S promoter, for example, has been increased by duplicating the upstream enhancer region and adding a heterologous intron to the expression construct (intron mediated enhancement). Another way to achieve high-level constitutive expression in monocotyledons is to use monocotyledon-derived promoters. Those from the rice (*Oryza sativa*) actin-1 and corn (*Zea mays*; maize) ubiquitin-1 genes are widely utilized.

Cell specific and stage specific promoters Constitutive expression is not suitable for all transgenes. For example, there are many benefits associated with the restriction of transgene expression to fruits or seeds. In molecular farming applications, seed specific expression is desirable because high levels of recombinant protein may be toxic to the vegetative organs of the plant. Also, proteins accumulating in seeds remain more stable than those expressed in green tissue, and the absence of recombinant proteins in the vegetative parts of the plant helps to prevent the product leaching into the environment. In the case of plants modified to increase their resistance to pests or diseases, the site of transgene expression may be restricted in order to kill the pest without affecting beneficial organisms. In the case of plants with enhanced uptake of nutrients or toxic metal ions from the environment, root specific transgene expression might be appropriate.

Seed specific promoters have been characterized in both dicotyledons and monocotyledons, and appear to share a number of common motifs. For example, a G-box (CACGTG) or E-box (CACCTG) is found in many dicotyledonous seed specific promoters, while the promoters of monocotyledonous seed storage protein genes often contain elements such as the prolamin box (TGNAAAK), GCN4 motif (TGAGTCA), and AACA motif (AACAAACTCAATC). Several of these promoters have been exploited in transgenic plants, including those from rice glutelin, corn zein, and wheat (*Triticum* spp.) glutenin genes.

Several other classes of cell- or stage-specific promoter are often exploited in transgenic plants:

- pollen or tapetum specific promoters, which are useful for introducing male sterility; for example, the rice *G6B* promoter;
- phloem specific promoters, which are useful for the expression of toxins directed against sap-sucking pests; for example, the rice sucrose synthase (*SS1*) promoter;
- promoters from genes expressed specifically in green tissues, which are useful for the analysis of photosynthesis; for example, the *rbcS* promoter;
- root specific promoters, useful for modulating nutrient uptake, bioremediation strategies, and transgenic strategies allowing the use of marginal soils; for example, the *Agrobacterium rhizogenes rolC* promoter; and
- meristem specific promoters, which restrict transgene expression to dividing cells; for example, promoters from S-phase specific genes such as those encoding histones and components of DNA polymerase.

Inducible promoters Inducible promoters are used to control transgene expression externally, which can have many benefits. In molecular farming applications, for example, inducible transgene control allows the desired recombinant protein to be expressed just before or after harvest, therefore avoiding toxicity to the host plant and possible leaching of the product into the environment. Many disease-resistant transgenic plants have resistance transgenes placed under the control of a promoter that is induced by pathogen attack. In basic research, inducible promoters allow conditional effects of transgene overexpression to be investigated.

Some endogenous plant promoters are inducible by physical stimuli (e.g., many plant genes are activated by light). Others are responsive to chemicals, and these may be native plant promoters or heterologous/artificial systems introduced as multiple transgenes. The advantage of chemically induced promoters is that they can be controlled by the application of agrochemicals in the field. In the native category, the *PR-1a* gene promoter is responsive to benzothiadiazole. Heterologous and artificial inducible promoters include those responsive to tetracycline, copper, insect or mammalian steroid hormones, and alcohol (a tetracycline inducible expression system is shown in **Figure 2**). The ideal inducible promoters have minimal background activity (in the absence of inducer) and a high induction ratio (elevation of expression over background levels in the presence of inducer). These properties are essential if the transgene encodes a toxic product. For example, barnase is a cytotoxic ribonuclease that can be used as a suicide gene in transgenic plants. It has been expressed under the control of a pathogen induced promoter so that infected plant cells are killed, thus halting the infection. However, leaky background expression of the transgene would lead to widespread cell death in the uninfected plant. To avoid this, the barnase inhibitor barstar is simultaneously expressed using a strong constitutive

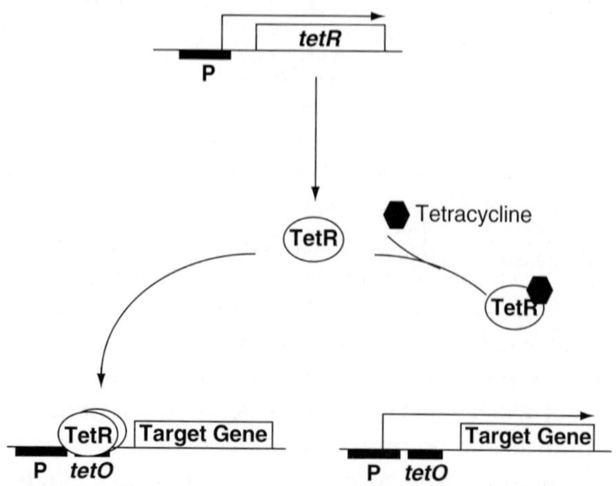

Figure 2 A tetracycline inducible expression system for use in transgenic plants. The components of the system are derived from the *Escherichia coli tet* operon, which contains genes for the catabolism of tetracycline. The operon is usually switched off by a repressor protein (TetR) that binds to operator sequences (*tetO*) in the promoter (P). However, in the presence of tetracycline or a similar inducer, the repressor is prevented from binding and the operon becomes active. If the TetR repressor is expressed constitutively in transgenic plants then any transgene with a *tet* operator sequence in the promoter will be switched off. However, if tetracycline or a similar molecule is administered to the plant, the transgene will be expressed.

promoter, thus abolishing barnase activity in the absence of the pathogen.

Transcriptional Elongation and Termination

While transcriptional initiation is the most highly regulated stage of transcription for the majority of nuclear genes, it is also necessary to consider the events following promoter clearance by the transcription complex. Successful escape from the promoter is dependent on capping of the transcript, which involves the addition of a 7-methylguanosine (7-meG) residue by way of an unusual $5' \rightarrow 5'$ phosphodiester bond. Continued elongation depends on the activity of elongation factors, which suppress pausing at regions rich in secondary structure and prevent arrest and disassembly of the transcription complex prior to termination.

The termination of transcription is associated with polyadenylation, the addition of approximately 250 adenosine residues to the 3′ end of the transcript. This occurs at an AU rich sequence called the polyadenylation site, which recruits the components of a multisubunit polyadenylation complex, and results in cleavage of the nascent transcript. In the absence of a polyadenylation site, the transcript is unstable and rapidly broken down. Therefore, a polyadenylation site is always included in plant expression constructs. Different polyadenylation sites show a 100-fold variation in efficiency. The strongest motifs are preferred for use in transgenic plants, and include those from the CaMV 35S sequence and the *A. tumefaciens nos* gene. An equally important consideration is ensuring the absence of "cryptic" polyadenylation sites in foreign transgenes, which could lead to truncation and loss of expression. For example, the presence of AU rich sequences resembling polyadenylation sites in the native toxin genes from *Bacillus thuringiensis* (Bt) is one of several reasons for the low-level accumulation of such toxins in transgenic plants. Other reasons for these poor expression levels are discussed elsewhere (*see* **Growth and Development**: Control of Gene Expression, Posttranscriptional Regulation).

Transgenes lacking a functional polyadenylation site are not expressed but this can be exploited in the development of gene traps that are not constrained by the expression pattern of the gene they interrupt. Most gene traps exploit the promoter of the gene they interrupt, either directly (promoter trap; insert consists of a reporter gene with a naïve open reading frame and relies on the activity of an adjacent promoter) or indirectly (splice trap; insert consists of a reporter gene with an upstream splice acceptor and relies on forming a transcriptional fusion with upstream exons of the interrupted gene). Therefore, if the interrupted gene is not expressed, the gene trap is not activated. A more refined gene trap has the reporter gene driven by a constitutive promoter and has a downstream splice donor site but no polyadenylation site. Therefore, transcriptional initiation occurs independently, but polyadenylation is dependent on the formation of a transcriptional fusion to downstream exons and the use of the endogenous gene's polyadenylation site (**Figure 3**).

Epigenetic Control of Transcription

Regulation of Endogenous Genes

We turn now from genetic to epigenetic mechanisms of transcriptional regulation, which are based on DNA methylation and the modification of chromatin structure. The basic properties of chromatin are thought to be similar in all eukaryotes. There are, however, features of chromatin-mediated gene regulation that are unique to plants, probably reflecting their distinct growth and development strategies. DNA methylation is used to regulate gene expression in vertebrates and higher plants, but not other eukaryotes. However, there appear to be many similarities between vertebrates and plants in the way that DNA methylation is achieved and the way in which is it linked to chromatin structure. Both

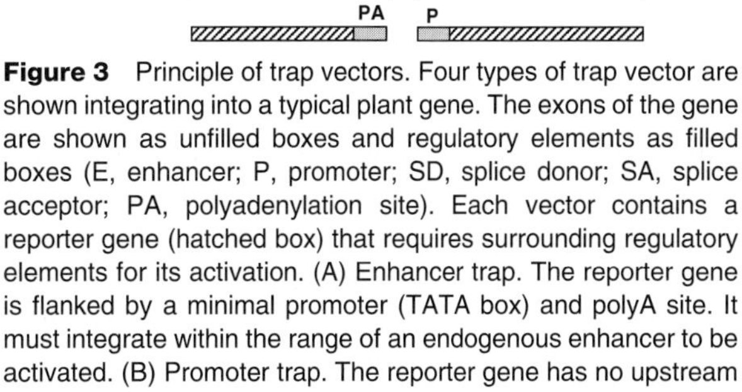

Figure 3 Principle of trap vectors. Four types of trap vector are shown integrating into a typical plant gene. The exons of the gene are shown as unfilled boxes and regulatory elements as filled boxes (E, enhancer; P, promoter; SD, splice donor; SA, splice acceptor; PA, polyadenylation site). Each vector contains a reporter gene (hatched box) that requires surrounding regulatory elements for its activation. (A) Enhancer trap. The reporter gene is flanked by a minimal promoter (TATA box) and polyA site. It must integrate within the range of an endogenous enhancer to be activated. (B) Promoter trap. The reporter gene has no upstream regulatory elements and must integrate downstream of an endogenous promoter to be activated. (C) Typical gene trap. The reporter gene lies downstream of a splice acceptor site and becomes fused to upstream exons of the interrupted gene if it inserts within the transcribed region. (D) Expression-independent gene trap. The reporter gene is driven by its own constitutive promoter but relies on the interrupted gene to provide a polyadenylation site.

vertebrates and plants go through global changes in DNA methylation during development, and genes can be methylated in cell specific patterns to reinforce developmental decisions.

The fundamental unit of chromatin is the nucleosome, which comprises a histone octamer (containing two copies each of histones H2A, H2B, H3, and H4) encircled by two loops of DNA. Additional linker histones are associated with the nucleosome, and serial nucleosomes are thought to be organized into a complex solenoid configuration to generate a 30-nm fiber, the typical structure of open chromatin in the interphase nucleus. This fiber is thought to be arranged as loops attached to the nuclear matrix, each loop perhaps corresponding to a functional chromatin domain.

The activity of chromatin in animals and fungi is determined to a large degree by histone modification. The acetylation of lysine residues on the core histones may reduce the affinity of histones for DNA and result in a loose form of association that is compatible with transcription. Alternatively, or additionally, it is possible that the function of histone acetylation is to regulate histone interactions with other proteins in the transcriptional assembly. There is a strong correlation between histone acetylation patterns and gene activity. Heterochromatin is rich in nonacetylated histones, and this may facilitate the sequestration of DNA into a more condensed and inaccessible packaging state. In plants, recent experiments have shown that histone acetylation patterns change during the cell cycle and correlate with replication rather than transcription, suggesting the relationship between chromatin structure and gene expression may be more complex.

Enzymes called histone acetyltransferases (HATs) transfer acetyl groups onto histones and help to activate gene expression, while histone deacetylases (HDACs) catalyze the reverse reaction and function as silencing proteins. HATs and HDACs with different specificities have been identified in many species, and appear to function in all cases as components of large, multiprotein complexes. Generally, both HATs and HDACs are highly conserved in plants, animals, and yeast, but plants possess two families of HDACs not represented in other taxa, which may provide further evidence of the more complex role of chromatin in plants. HATs and HDACs are often associated with the transcriptional initiation complex, indicating that chromatin structure plays an important role in endogenous gene regulation. HDACs are often associated with methyl-DNA binding proteins, providing a link between DNA methylation and gene silencing. DNA methylation is perpetuated through rounds of DNA replication by enzymes called maintenance methylases, and this may act as a "molecular lock" to maintain domains of inactive chromatin. DNA methylation may also suppress gene expression directly, by interfering with DNA–protein interactions.

Other forms of histone modification, including methylation and phosphorylation, are also involved in the regulation of chromatin structure, and may help to establish a "histone code" to divide the genome into functional domains. The prevalence of different forms of modification varies in plants, animals, and fungi. For example, histone H4 is the primary target for acetylation/deacetylation in animals and fungi, while histone H3 is the primary target in higher plants.

Transcriptional Gene Silencing

Integrated transgenes in plants are often subject to epigenetic gene silencing at the level of transcription. This involves hypermethylation of the transgenic locus and its sequestration into heterochromatin. Several different triggers for this type of silencing are apparent.

Position dependent silencing Different transgenic plant lines carrying the same transgene construct integrated at different positions in the genome often show variable expression levels, a phenomenon known as the position effect. This reflects the influence of surrounding chromatin on the transgenic locus. In some cases, the integrated transgene is silenced. If the transgene integrates into heterochromatin, the repressive chromatin structure and DNA methylation spreads into the transgene, preventing access by transcription factors. In the absence of any other triggers for silencing, the position effect can be avoided in one of two ways:

- Protect the transgene from the position effect by establishing an open chromatin domain. As discussed above, chromatin is thought to be divided into functionally independent domains, perhaps because it is organized as loops attached to the nuclear matrix. Two types of DNA sequence are involved in this process: boundary elements (insulators) and matrix attachment regions (MARs). A boundary element can insulate a test gene from an enhancer if interposed between them, while a transgene flanked by boundary elements is protected from position effects in transgenic plants. MARs interact with the nuclear matrix and may organize chromatin into loops, but the relationship between boundary elements and MARs is complex. Sometimes two MARs flanking a transgene provide insulator activity, but other MARs appear to facilitate gene expression rather than establishing boundaries, since one element is sufficient for the protective effect. Similarly, not all boundary elements associate with the nuclear matrix, so the insulator function may not require physical division of the chromatin into loops.

- Integrate the transgene specifically at a site that is permissive for transcription. This can be achieved using the Cre-*loxP* or FLP-*FRP* site-specific recombination systems. If a *loxP* or *FRP* site is integrated at a permissive site through standard transformation procedures, any transgene flanked by such elements can then be introduced at that site by recombination, if Cre or FLP activity is supplied. The recombination site must be mutated to avoid re-excision of the transgene.

Context dependent silencing Transgene silencing can occur even following integration at a permissive site if the plant recognizes the transgene sequence as "foreign." The trigger in this case is unknown. Many single-copy transgenes have been inactivated in this way and sequestered into heterochromatin following *de novo* DNA methylation. It is possible that plants possess some form of genome scanning apparatus that looks for sequences with unusual base composition compared to the surrounding genomic context. The mechanism has probably evolved to prevent the proliferation of transposable elements in regions of active chromatin. Prokaryotic DNA appears to be a strong trigger for *de novo* silencing, so the co-integration of vector backbone sequences (which are generally of bacterial origin) should be avoided. Unfortunately, vector sequences are cotransferred in as many as 75% of *Agrobacterium* mediated transformation experiments, while direct transformation methods are typically carried out with whole, supercoiled plasmids. The use of minimal linear cassettes instead of whole plasmids has been shown to significantly increase the frequency of stable transgene expression.

Homology dependent silencing Transgenes may be silenced at the transcriptional level if there are multiple copies. For transcriptional silencing, homology in the promoter region is important and may reflect some form of DNA–DNA interaction. Inverted repeats are particularly strong triggers for this type of silencing, and such structures are often found in transgenic loci. Interactions between transgenes at different sites, or between transgenes and homologous endogenous genes, may also occur. This form of transgene silencing can be avoided by choosing transformation strategies that favor the creation of simple and preferably single copy transgenic loci. In the case of *Agrobacterium* mediated transformation, different bacterial strains and different types of vector produce distinct integration patterns. A novel strategy is to interrupt each transgene with a *loxP* site and then express Cre recombinase in the transgenic plant. At multicopy loci, recombination between direct repeats of *loxP* results in the elimination of all but one copy of the transgene, and has been shown to significantly increase transgene expression frequencies.

Transgene silencing and cosuppression also occur at the posttranscriptional level. This is confirmed by nuclear run-on assays, which show that the transgene is transcribed but the mRNA is rapidly degraded. Many of the factors that induce transcriptional silencing may also induce posttranscriptional silencing, particularly the presence of DNA repeats, and, as stated above, there is a degree of cross-talk between the pathways, which may result in DNA hypermethylation at posttranscriptionally silenced transgenes and endogenous genes. However, the major trigger for posttranscriptional silencing is thought to be double-stranded RNA, which can be produced from complex transgene loci (especially transgenes arranged as inverted repeats) and from highly expressed sense transgenes (*see* **Growth and Development**: Control of Gene Expression, Posttranscriptional Regulation).

Avoiding Transgene Silencing in Transgenic Plants

From the discussion above, several conclusions can be drawn about transgene silencing, and several general strategies can be employed to help prevent it:

- the transgenic locus should be protected from position effects;
- the transgene should have a similar base composition to the host plant;
- superfluous sequences, particularly the plasmid backbone, should be avoided;
- the transgene copy number should be kept as low as possible;
- the locus structure should be simple, preferably a single intact copy of the transgene;
- inverted repeat structures should be avoided wherever possible; and
- maximized transcription may be counterproductive.

Transcription of Chloroplast Genes

Overview

The chloroplast genome differs in many respects from the nuclear genome and most of these differences reflect the prokaryotic origin of plastids. An important distinction is that chloroplast DNA is neither methylated nor associated with histones, so the epigenetic mechanisms responsible for transcriptional silencing in nuclear transgenic plants do not exist. This means that very high levels of transgene expression can be achieved in plants with genetically transformed chloroplasts.

Chloroplast transformation is achieved by introducing expression cassettes in which the foreign transgene is embedded within a chloroplast homology region. Integration occurs by homologous recombination and is therefore site specific. There are no position effects in transplastomic plants. Since *Agrobacterium* mediated transformation involves nuclear targeting of the T-DNA complex, this method is unsuitable for plastid transformation, and direct DNA transfer strategies such as particle bombardment or protoplast transformation must be used. Another important distinction is that chloroplast genes, like bacterial genes, are typically organized into operons. Transgenes can therefore be targeted to insert into intergenic spacer regions within an existing operon, such that the transgene comes under the control of endogenous regulatory elements. A third important feature of chloroplast gene expression is that, while transcriptional regulation is important, the predominant level of gene regulation is posttranscriptional (*see* **Growth and Development:** Control of Gene Expression, Posttranscriptional Regulation). As might be expected, light plays an important role in the regulation of many chloroplast genes.

Pros and Cons of Chloroplast Transgenes

The special properties of the plastid genome provide several advantages for transgene expression. Owing to the absence of silencing and the large plastid copy number, very high levels of transgene expression are possible. Indeed, in the best case reported thus far, a recombinant protein has been expressed at levels approaching 50% total soluble protein. The polycistronic organization of the genome means that multiple transgenes can be expressed as part of the same transcript and thus coregulated. The accumulation of proteins within the chloroplast is advantageous if the protein is toxic at high levels. Finally, in most species, chloroplasts are not present in the pollen, providing a natural form of transgene containment. Recombinant proteins expressed in chloroplasts have been shown to fold properly and form disulfide bonds. However, other forms of modification such as glycosylation do not occur, so it is unclear whether chloroplasts can be used for the expression of complex glycoproteins.

Transcriptional Regulation of Chloroplast Gene Expression

The transcription of chloroplast genes is carried out by at least two different RNA polymerases. One of these is a multisubunit complex similar to eubacterial RNA polymerase. The subunits are encoded by chloroplast genes, and the enzyme is therefore known as plastid encoded polymerase (PEP). The promoters of PEP transcribed genes are similar in their architecture to eubacterial promoters, containing the characteristic −10 and −35 boxes. Genes encoding components of the photosynthetic apparatus are regulated by PEP promoters. The other polymerase is a single polypeptide related to RNA polymerase from bacteriophage T7. This enzyme is encoded by a nuclear gene and is imported into the chloroplast; it is therefore known as nuclear encoded polymerase (NEP). A few genes encoding housekeeping functions are controlled solely by NEP promoters, while most housekeeping genes have hybrid promoters that can be activated by PEP and NEP. NEP promoters, often contain an AT rich element surrounding the transcriptional start site, containing the conserved motif YRT. In some cases, a further motif with the core consensus GAA is found about 15 nt upstream. Other NEP promoters lack both these features and have an alternative consensus sequence around the start site.

Genes controlled by NEP promoters are generally activated early in plastid development, while those solely under the control of PEP promoters are inactive. Genes controlled by PEP promoters are activated in response to illumination, although the rate of transcription declines in most cases over a period of days. Exceptionally, the transcription of *psbA* and *psbD* is maintained at a high rate under constant light, since these genes encode labile components of photosystem II. NEP-transcribed genes include those encoding the components of PEP. It therefore appears that NEP activity corresponds to chloroplast differentiation. The light dependent activity of PEP may reflect the light dependent dephosphorylation of a sigma-like factor that helps the plastid polymerase recognize PEP promoters.

Conclusions

Gene expression is a complex process occurring in several phases. Before transcription can occur in the nuclear genome, chromatin remodelling must take place to make the appropriate regulatory elements accessible to transcription factors. These transcription factors are synthesized or activated in complex overlapping patterns, dependent on cell type, developmental stage, and a host of external signals. In transgenic plants, all stages of transgene expression can be controlled by expression construct design, but other factors are involved that are influenced by the integration of the transgene and the structure of the transgenic locus. Successful transgene expression depends to a large extent on the transgene being accepted as a genuine locus and not branded as a foreign sequence. Strangers in the plant genome are treated as enemies and are silenced at either the transcriptional or posttranscriptional levels. While often an annoyance in terms of producing plants with stable and appropriate levels of transgene expression, these defense strategies can also be exploited to deliberately shut down endogenous plant genes.

List of Technical Nomenclature

30-nm fiber	The typical organization of euchromatin in the interphase nucleus. Thought to be accessible by transcription factors and potentially transcriptionally active.
Boundary element	A *cis*-acting element that is able to block other *cis*-effects, such as the activity of enhancers and the effect of surrounding chromatin on integrated transgenes.
Chromatin	A complex of DNA and protein, the natural state of nuclear DNA.
Context dependent silencing	The recognition of a transgene as foreign DNA due to dissimilarities in base composition compared to surrounding genomic DNA.
Constitutive promoter	A promoter that is active in many cell types and under most circumstances.
Cryptic	A DNA, RNA, or protein sequence that resembles a genuine signal (e.g., a promoter, splice site, polyadenylation site, etc.) and is used by the cellular machinery in error resulting in an aberrant product.
Cosuppression	The epigenetic silencing of an endogenous gene that is homologous to an integrated transgene.
***De novo* methylase**	Enzymes that transfer methyl groups onto cytosine residues on fully non-methylated DNA molecules. Involved in gene regulation and genome defense.
DNA methylation	In vertebrate and plant genomes, conversion of cytosine residues in the sequence 5′CG3′ or 5′CNG3′ to 5-methylcytosine. The presence of many methylated cytosine residues is hypermethylation, the opposite phenomenon is hypomethylation. The presence of methylated residues on one DNA strand only is hemimethylation.
Enhancer, silencer	*Cis*-acting elements that can act over great distances to activate or repress gene expression in cooperation with the basal promoter. Enhancers cannot initiate transcription in the absence of a promoter.
Epigenetic	Factors that affect gene expression but are independent of the nucleotide sequence, e.g., DNA methylation, chromatin structure, and posttranscriptional silencing.
Genetic	Pertaining to the nucleotide sequence.
Gene trap	An insertional mutagen containing a defective reporter gene that is activated only when the construct inserts within an endogenous gene.
Heterochromatin	Densely packaged chromatin which is inaccessible and transcriptionally inert.
Histone	A basic protein that interacts with DNA and packages it into chromatin.
Histone acetyltransferase (HAT)	An enzyme that acetylates histones. Associated with transcriptional activation.
Histone deacetylase (HDAC)	An enzyme that deacetylates histones. Associated with transcriptional silencing.
Histone modification	The chemical modification of histone proteins (acetylation, phosphorylation, methylation) which alters their DNA binding properties and hence the accessibility of chromatin.
Homology dependent silencing	A broad term that includes any form of gene silencing (transcriptional and post-transcriptional) provoked by the presence of multiple copies of the same sequence in the genome.
Inducible promoter	A promoter that is active in response to a particular environmental stimulus, either physical or chemical. This usually reflects the presence of a transcription factor that is dependent on that stimulus for activity.
Initiation complex	The complex of proteins required to initiate transcription. Required to help

RNA polymerase interact with the DNA template in a stable fashion.

Intron mediated enhancement The ability of a heterologous intron to increase the transcriptional activity of a promoter in transgenic plants.

Maintenance methylase Enzymes that transfer methyl groups onto cytosine residues on one strand at the sequence 5′CG3′ or 5′CNG3′ if the cytosine residue on the other strand is already methylated. A mechanism for the perpetuation of methylation patterns through several rounds of replication.

Matrix associated region (MAR) A DNA sequence that interacts with the nuclear matrix and may help organize chromatin into topologically isolated loops.

Mature transcript The final (processed) product of transcription. In the case of protein encoding genes, this is also known as a messenger RNA (mRNA).

Mediator A ubiquitous protein complex that is thought to help transcriptional activators and repressors interact with the transcriptional initiation complex.

Methyl-DNA binding proteins Proteins that bind specifically to 5-methylcytosine residues in DNA.

NEP Nuclear encoded polymerase. A single-subunit RNA polymerase encoded by the nuclear genome and imported into the chloroplast. Responsible for the transcription of housekeeping genes, such as those with roles in transcription and translation.

Operon A form of gene organization found in bacteria, plastids, and occasionally in eukaryotes, in which several genes are coordinately transcribed as part of the same polycistronic mRNA. Individual regulation occurs at the posttranscriptional level.

PEP Plastid encoded polymerase. A eubacterial-type RNA polymerase encoded by the chloroplast genome, which is responsible for the transcription of photosynthetic genes and some housekeeping genes.

Phenotype The biological characteristics of an organism.

Primary transcript The initial (unprocessed) product of transcription, typically containing introns that need to be spliced out. In the case of protein encoding genes, this is also known as pre-mRNA.

Polyadenylation The addition of 2–300 adenosine residues to the end of a transcript. This process is associated with transcriptional termination and appears necessary for efficient protein synthesis.

Position effect The influence of chromatin structure at the site of transgene integration on transgene expression.

Promoter The core promoter is a *cis*-acting element immediately adjacent to the 5′ end of the gene, which is recognized by the components of the basal pre-initiation complex. It often contains an initiator element and/or a TATA box. Upstream promoter elements bind other transcriptional activators and repressors and dictate the pattern of gene expression.

Regulatory element A *cis*-acting element that regulates gene expression.

Site-specific recombination Recombination catalyzed by specific proteins and occurring at short sequences. Contrasts with homologous recombination, which involves ubiquitous proteins and requires long regions of homology.

Transgene A foreign gene introduced into a plant, such plants being described as transgenic if the transgene is nuclear or transplastomic if the transgene is integrated in the plastid genome.

Transgene silencing Epigenetic loss of transgene expression. Can occur at either the transcriptional or posttranscriptional levels.

Transcription The synthesis of RNA using a DNA template.

See also: **Genetic Modification**: Gene Cloning, General Principles; Transformation in Monocotyledons; Transformation in Plastids; Transformation, General Principles. **Growth and Development**: Control of Gene Expression, Post Transcriptional Regulation.

Further Reading

Gallie DR (1998) Controlling gene expression in transgenics. *Current Opinion in Plant Biology* 1: 166–172.

Guilfoyle TJ (1997) The structure of plant gene promoters. In: Setlow JK (ed.) *Genetic Engineering: Principles and Methods*, vol. 19, pp. 15–47. New York: Plenum Press.

Kohli A, Twyman RM, Abranches A, *et al.* (in press) Transgene integration, organization and interaction in plants. *Plant Molecular Biology.*

Kumpatla SP, Chandrasekharah MB, Iyer LM, Li G, and Hall TC (1998) Genome intruder scanning and

modulation systems and transgene silencing. *Trends in Plant Science* 3: 97–104.
Lusser A, Kölle D, and Loidl P (2001) Histone acetylation: lessons from the plant kingdom. *Trends in Plant Science* 6: 59–65.
Matzke AJM and Matzke MA (1998) Position effects and epigenetic silencing of plant transgenes. *Current Opinion in Plant Biology* 1: 142–148.
Plasterk RHA and Ketting RF (2000) The silence of the genes. *Current Opinion in Genetics and Development* 10: 562–567.
Spiker S and Thompson WF (1996) Nuclear matrix attachment regions and transgene expression in plants. *Plant Physiology* 110: 15–21.
Stern DB, Higgs DC, and Yang J (1997) Transcription and translation in chloroplasts. *Trends in Plant Science* 2: 308–315.
Twyman RM, Stoger E, Kohli A, and Christou P (2002) Foreign DNA: integration and expression in transgenic plants. In: Setlow JK (ed.) *Genetic Engineering: Principles and Practice*, vol. 24, pp. 107–136, New York: Plenum Press.
Tyagi AK (2001) Plant genes and their expression. *Current Science* 80: 161–169.

Leaf Development

N Dengler and J Kang, University of Toronto, Toronto, ON, Canada

Introduction

Leaves are the most conspicuous part of the shoot system of most vascular plants. They range in size and shape from the giant leaves of the raphia palm (*Raphia ruffia*) to the tiny floating "fronds" of aquatic duckweeds (*Lemna* spp.). In addition to differences among species, leaves can vary in size and shape within the same individual. Seedlings usually produce successively larger and more complex leaves. Also, environmental factors such as shading can influence mature leaf size and anatomy. Despite this diversity, leaves share certain features of construction and carry out the same primary function: photosynthesis. In addition, all leaves undergo the same four basic stages of leaf development: (1) initiation, (2) morphogenesis, (3) expansion, and (4) histogenesis. Differences in leaf shape between different groups of plants arise early in leaf development, during the initiation and morphogenesis stages. These shapes are either maintained or minimally altered during the expansion and histogenesis stages. In contrast, environmental factors such as shading or water stress have the greatest impact on the expansion and histogenesis phases, adjusting leaf development so that it is in line with available resources or so that leaves can function better in an altered environment. Genes that regulate several important aspects of leaf development have been recently identified, making it possible to genetically manipulate leaf characters that contribute to photosynthetic efficiency.

Leaf Function and Design

Leaf Morphology

Leaves are designed to capture light and carbon dioxide for photosynthesis, while at the same time reducing water loss and maintaining tissue temperatures within an optimum range. In dicotyledonous plants, leaf blades are typically broad and thin and are held in a horizontal plane by the leaf petiole. The flattened shape of the blade facilitates the penetration of light to the photosynthetic tissues within, while the petiole positions the blade so that it is shaded as little as possible by neighboring leaves. In monocotyledonous plants such as the grasses, leaf blades are held in a vertical position that reduces self-shading.

Leaf Anatomy

All leaves share a similar organization of internal leaf tissues (**Figure 1**). Leaf veins function in the import and distribution of water (xylem tissue), as well as accumulating the products of photosynthesis and exporting them to the rest of the plant (phloem tissue). In addition, the larger leaf veins serve to support the leaf against the forces of gravity and

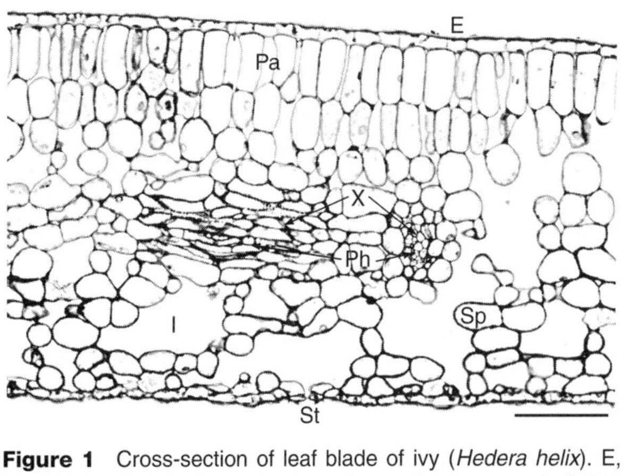

Figure 1 Cross-section of leaf blade of ivy (*Hedera helix*). E, epidermis; I, intercellular airspace; Pa, palisade mesophyll; Ph, phloem tissue of vein; Sp, spongy mesophyll; St, stomate; X, xylem tissue of vein. Scale = 100 μm. Photograph courtesy of H. Chan and A. Cornell.

wind. Photosynthetic mesophyll tissue consists of both palisade and spongy parenchyma cells. The extensive intercellular airspace between the cells permits the free diffusion of gases. These internal tissues are completely enclosed in a layer of epidermal tissue that functions to reduce water loss, while permitting light and carbon dioxide to enter the leaf. Epidermal cells secrete a cuticle, a hydrophobic layer on the surface of the leaf that reduces evaporation from leaf tissues. The epidermis contains stomates, each a small pore surrounded by two specialized guard cells. Stomatal pores are open during the day to allow diffusion of carbon dioxide into the leaf, but close at night to prevent the evaporative loss of water from internal tissues.

Variation among Species

Most of the familiar diversity in leaf form is found among dicotyledonous plants (**Figure 2**). Different plant species, genera, or families tend to be characterized by distinctive leaf shapes. Dicotyledon leaves typically consist of a blade, petiole, and leaf base region. The leaf base often bears two appendages called stipules (**Figure 2C**). Stipules may be conspicuous in mature leaves (as in peas (*Pisum* spp.) and roses (*Rosa* spp.)) and function in photosynthesis. In most species, however, stipules function only to provide protection for young leaves in the bud and then become inconspicuous as the leaf matures. Leaves are classified as simple when there is only one blade (as in spinach (*Spinacea oleracea*), lettuce (*Lactuca sativa*), and tobacco (*Nicotiana tabacum*)) (**Figure 2A,B**) and compound when the blade is divided into two or more leaflets (as in peas, beans (*Phaseolus* spp.), and tomato (*Lycopersicon esculentum*)). Simple leaves may still have relatively complex shapes: the single blade may form lobes (as in ivy (*Hedera helix*), oaks (*Quercus* spp.), and maples (*Acer* spp.)) (**Figure 2B**), and the blade edge or margin may bear teeth (as in elm (*Ulmus* spp.)). Compound leaves may be either pinnately compound (as in roses, legumes and tomato) (**Figure 2C**) or palmately compound (as in horse chestnuts (*Aesculus hippocastanum*)) (**Figure 2D**).

Monocotyledonous plants, such as the grasses, sedges, irises, and lilies, tend to have simple strap-shaped leaves (**Figure 2E,F**). Typical monocotyledon leaves have a conspicuous leaf base that wraps around the stem, but they lack a petiole and special appendages such as stipules. Grass leaves such as those of wheat (*Triticum aestivum*), barley (*Hordeum vulgare*), and corn (*Zea mays*; maize) have a specialized leaf sheath at the base of the blade (**Figure 2E**). The leaf sheath serves to protect the delicate growing tissues of the stem and younger leaves that

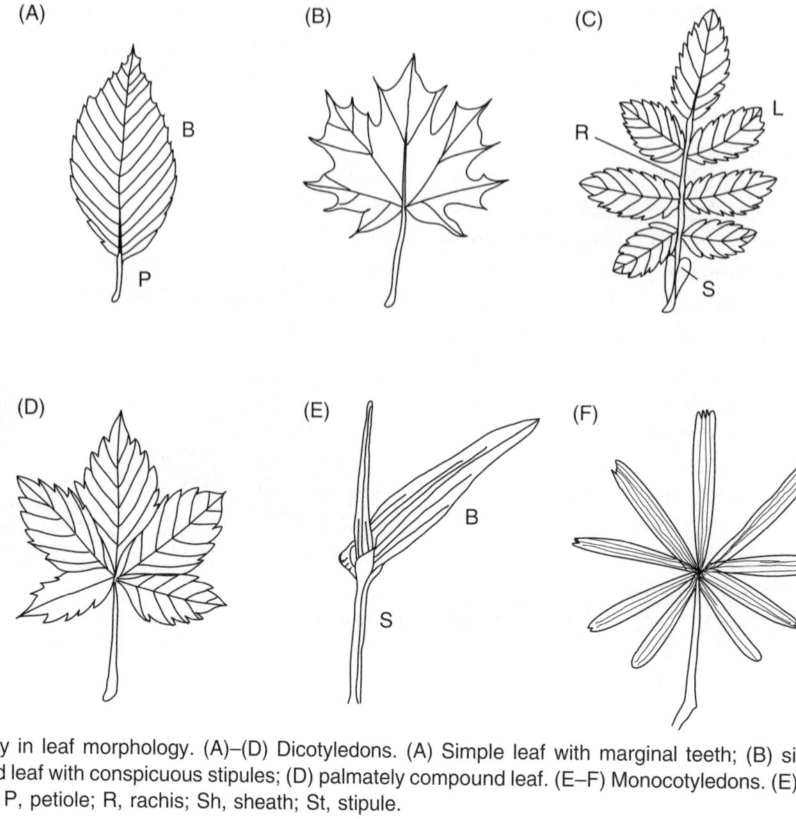

Figure 2 Diversity in leaf morphology. (A)–(D) Dicotyledons. (A) Simple leaf with marginal teeth; (B) simple leaf with lobes; (C) pinnately compound leaf with conspicuous stipules; (D) palmately compound leaf. (E–F) Monocotyledons. (E) Grass leaf; (F) palm leaf. B, blade; L, leaflet; P, petiole; R, rachis; Sh, sheath; St, stipule.

are enclosed within it. Certain other groups of monocotyledons have much more complex shapes (**Figure 2F**). For instance, members of the palm family (Arecaceae) have either pinnately compound or palmately compound leaves, with a conspicuous petiole that develops between the sheathing leaf base and the blade. Members of the aroid family often have highly lobed leaf blades, and unusual species such as the 'Swiss cheese plant' (*Monstera deliciosa*) have a series of holes that develop within the leaf tissue.

Why there is such a great diversity of leaf shapes is unknown. One contributing factor is the need for plants in warm environments to dissipate heat (which is absorbed along with light energy) so that internal tissues can carry out photosynthesis within an optimal temperature range. The lobes and marginal teeth of leaf blades contribute to heat loss by serving as vanes that radiate heat through conduction and convection. A high ratio of perimeter to area increases the rate of heat loss, so breaking up a single leaf blade into a series of separate leaflets may represent one way of maintaining large photosynthetic surface, but preventing heat buildup. The many kinds of compound leaves may simply represent different means of achieving the same end.

Heteroblasty

In almost all plant species, some variation in leaf shape and anatomy occurs within a single individual. The most conspicuous form of this intraindividual variation is called heteroblasty, the change in leaf form that occurs as a shoot becomes older. Typically, seedlings produce small simple leaves at their first-formed nodes (the juvenile phase) and then larger and more complex leaves at later-formed nodes (the "adult" phase). Such changes usually occur gradually, with intermediate leaves forming a transition between juvenile and adult leaves. For example, in *Arabidopsis thaliana*, juvenile leaves are round in shape, thin, and bear trichomes only on the adaxial surface, while adult leaves are elongate in shape, thicker, and bear trichomes on both surfaces (**Figure 3**).

The functional significance of heteroblasty is unknown in most cases. Since shoot apical meristem size is correlated with leaf size and complexity in most heteroblastic series, it is thought that heteroblasty might simply arise from a constraint on the raw material required to make a leaf. The small shoot apices of new seedlings form small leaf primordia, while the larger shoot apices of older plants can form larger leaf primordia with the potential to form larger leaves. In a few examples, heteroblasty has been shown to be an adaptive trait. In certain Australian acacias (*Acacia* spp.), the juvenile leaves produced near the ground surface are exposed to intense heat. However, the small leaflet size of the compound juvenile leaves can dissipate heat efficiently. Adult leaves are produced when the shoot system has grown about a meter above the ground surface. Here, air temperatures do not reach the same highs, and adult leaves have a simple shape.

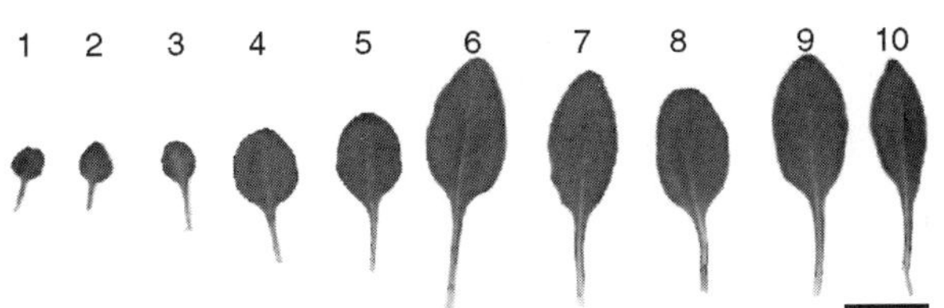

Figure 3 Heteroblasty in *Arabidopsis thaliana*. Leaves are numbered in order formed on seedling. Note changes in leaf size and shape with position. Scale = 1 cm.

Environmentally Induced Heterophylly

Heterophylly is an inclusive term that refers to any variation in leaf characteristics within an individual plant. In addition to heteroblasty, leaf shape variation can occur among leaves of the same position, e.g., among adult leaves. This variation is termed environmentally induced heterophylly, that is, differences in leaf shape induced by environmental factors. The most dramatic examples occur in certain aquatic species such as the amphibious buttercup (*Ranunculus flabellaris*) (**Figure 4**). These plants are perennials that grow in ephemeral pools. Early in the growing season, they produce underwater shoots with leaves that are divided into many filament-like lobes. Later in the growing season, the pools dry, providing a signal for a different pattern of leaf development. Now, the same shoots produce terrestrial leaves, with fewer, broader lobes. Both kinds of leaves allow the plant to function more efficiently in its environment.

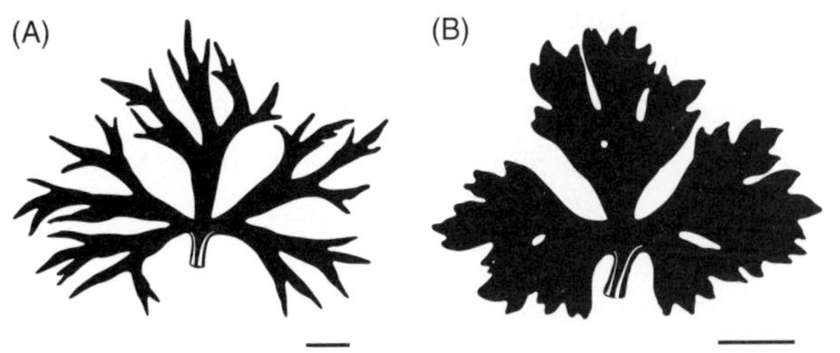

Figure 4 Environmentally induced heterophylly in the aquatic buttercup (*Ranunculus flabellaris*). Note differences in shape of leaf lobes expanding under water (A) or under terrestrial conditions (B). Scale = 1 cm. Reproduced with permission of University of Chicago Press from Young JP, Dickinson A, and Dengler NG (1995) A morphometric analysis of heterophyllous leaf development in *Ranunculus flabellaris*. *International Journal of Plant Sciences* 156: 590–602.

Underwater leaves lack stomates, but have a high surface-to-volume ratio, facilitating the penetration of light and diffusion of carbon dioxide to the photosynthetic cells. Terrestrial leaves possess stomates, but a lower surface-to-volume ratio; this combination reduces water loss by transpiration, while permitting the entry of carbon dioxide for photosynthesis.

Four Stages of Leaf Development

Despite the great range of variation in leaf size and shape seen among different plant species, and even within a single individual, leaves are all formed by similar developmental processes. Leaves develop from minute primordia that consist of uniform embryonic cells. Each leaf primordium gives rise to a mature leaf that is typically several thousand times larger than it is, has a distinctive shape, and is composed of many different specialized cell and tissue types. This overall process of leaf development involves four overlapping stages: (1) leaf initiation, (2) morphogenesis, (3) expansion, and (4) histogenesis.

Leaf Initiation

Leaves are formed on the flanks of the shoot apical meristem through an alteration in growth direction that results in a short finger-shaped protrusion, the leaf primordium (**Figure 5**). Formation of a leaf primordium usually involves a change in both the plane and rate of cell division, reflecting the new axis of growth. In most dicotyledons, the zone of initiation usually extends only a short distance from the original locus of leaf formation, resulting in a narrow leaf base at maturity. In contrast, the zone of leaf initiation in most monocotyledons extends circumferentially around the shoot apex, encircling the stem and giving rise to a sheathing leaf base. In monocotyledons such as onion (*Allium cepa*), the sheathing leaf base elongates, forming a closed tube. In other monocotyledons such as corn, the zone of leaf initiation does not extend all the way around the apex, leaving an open leaf base with overlapping edges.

Morphogenesis

The term morphogenesis is used to describe the developmental processes that establish the basic form of the leaf. This phase occurs early in leaf development, usually when the leaf is less than 1 mm in length and is well protected by older leaves in the apical bud. Primary morphogenesis involves local regions of growth enhancement and suppression, so

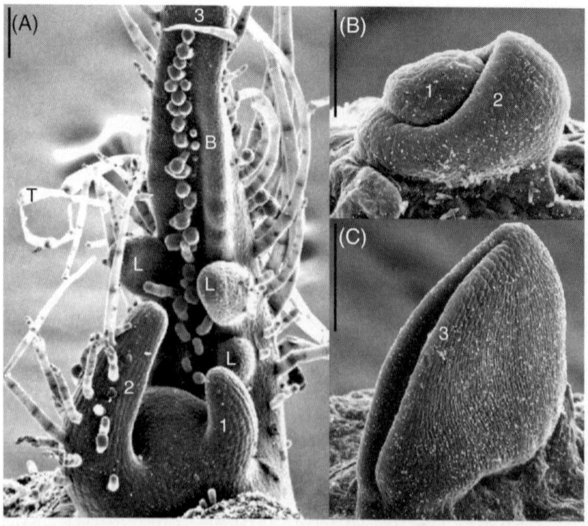

Figure 5 Leaf primordia of (A) a dicotyledon, tomato (*Lycopersicon esculentum*) and (B,C) a monocot, St Augustine's grass (*Stenotaphrum secundatum*). Leaves are numbered in order of increasing age (1, 2, 3). B, blade forming on terminal leaflet; L, lateral leaflets; T, trichomes. Scale = 100 μm. (A) Reproduced with permission of University of Chicago Press from Dengler NG (1984) Comparison of leaf development in normal (+/+), entire (e.e) and lanceolate (LA/+) plants of tomato, *Lycopersicon esculentum* "Ailsa Craig". *Botanical Gazzette* 145: 66–77. (B,C) Reproduced with permission from Sud R and Dengler NG (2000) Cell lineage of vein formation in variegated leaves of the C_4 grass, Stenotaphrum secundatum. *Annals of Botany* 86: 99–112.

that a leaf primordium with an initially simple shape becomes more complex. In typical dicotyledons, the first indication of primary morphogenesis is the formation of the leaf blade (**Figure 6**). This involves an alteration of growth direction so that the leaf begins to grow in the transverse plane as well as along its longitudinal axis. Usually, blade formation is restricted to the distal portion of the primordium, delimiting the more proximal petiole and leaf base regions.

Growth is uniformly distributed throughout a newly formed leaf blade, except in leaf blades that are lobed or have a serrate margin. Lobes and teeth are formed early in blade development and arise from regions of localized growth enhancement that are separated by regions of growth suppression. Compound leaves are formed when growth is completely suppressed in portions of the primordium margin, leading to the formation of separate leaflets. In species with pinnately compound leaves, the region between the leaflets (called a rachis) elongates, separating the leaflets as the leaf expands. In contrast, palmately compound leaves are formed when little or no rachis extension occurs (**Figure 6**).

Most monocotyledons follow a quite different pattern of primary morphogenesis. After leaf

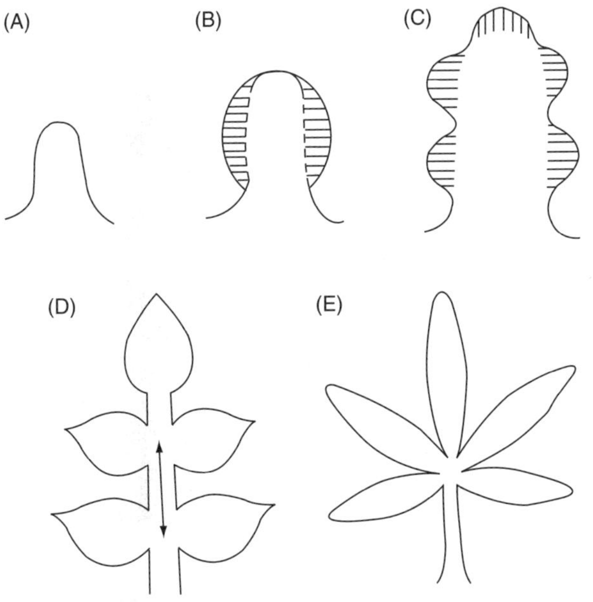

Figure 6 Leaf morphogenesis. (A) Primordium stage; (B) blade formation; (C) lobe formation in simple leaf blade; (D) leaflet formation in pinnately compound leaf; (E) leaflet formation in palmately compound leaf. Shaded areas represent enhanced growth. Arrow indicates main axis of growth.

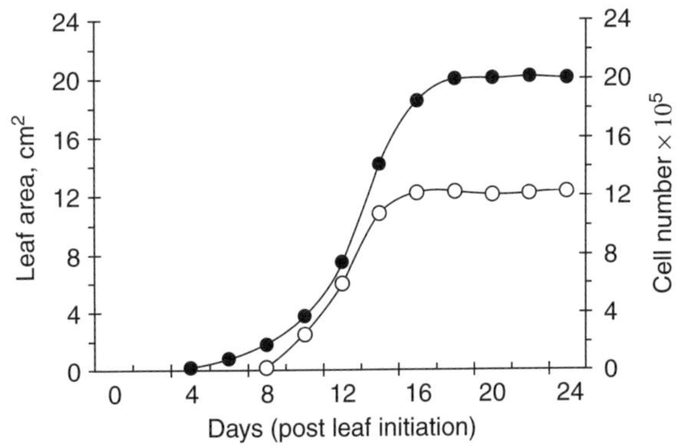

Figure 7 Leaf expansion (solid circles) and increase in epidermal cell number (open circles) in sunflower (*Helianthus annuus*).

initiation, growth is largely vertical, resulting in a short ensheathing leaf base and a strap-shaped distal region. In the grasses, the leaf becomes differentiated into a distal leaf blade and an elongated leaf sheath; the boundary between these regions is marked by a flap of tissue called a ligule.

Leaf Expansion

The third phase of leaf development, expansion, encompasses a much longer time period and represents an increase in surface area and volume of several thousand-fold. Measurement of leaf growth rates usually reveals a typical growth curve: a short lag phase, followed by a period of exponential growth, and finally a slowing of growth until mature size is reached (**Figure 7**). All parts of the leaf expand at similar rates, but regions near the tip of the leaf cease expanding first and those near the base last. As many as 95% of all leaf cells are formed during leaf expansion, and cell divisions may occur until a leaf has reached 75% of its final area. The predominant plane of cell division is anticlinal (perpendicular to the surface), a pattern that perpetuates the layered arrangement of leaf tissues. As with overall expansion, cells stop dividing in a wave, starting near the leaf tip and proceeding toward the leaf base (basipetal direction). This longitudinal gradient is strongly accentuated in grass leaves. Growth and accompanying cell divisions occur for a long period near the base of a grass leaf and form a region called an intercalary meristem. Growth from the intercalary meristem is stimulated when the grass is grazed by a herbivore or when a lawn is mowed.

The expansion phase can also contribute to overall leaf shape. Surface growth may be isometric (equal in all directions). Isometric growth produces a mature leaf with the same shape as that acquired by the primordium during the morphogenesis stage. More commonly, surface growth along one axis exceeds another and, thus, growth is allometric. Allometric growth can change leaf shape during expansion, resulting in leaves that are proportionately longer or wider at the end of expansion. Localized differences in expansion rate may also change leaf shapes. For instance, cultivars of the squash (*Cucurbita argyrosperma*) have a similar leaf shapes at the end of primary morphogenesis, but different rates of expansion in the lobes lead to differences in leaf lobe length to width ratio among cultivars. Leaves of the common nasturtium (*Tropaeolum majus*) are lobed at the end of primary morphogenesis, but accelerated expansion in the sinus region between the lobes leads to a rounded mature leaf shape. The long strap-shape of many monocotyledon leaves is enhanced by allometric expansion: leaves expand primarily along their longitudinal axis, but very little in width.

Histogenesis

Histogenesis involves the differentiation of tissues and their component cell types. All leaf tissues are derived from one of three so-called primarily meristematic tissues: protoderm, ground meristem, and procambium. Precursors of the protoderm and ground meristem are present from initiation of the leaf primordium from the shoot apical meristem. Protoderm is derived from the surface layer of the meristem and ground meristem from subsurface layers. In contrast, procambium, the precursor of the vascular tissues, extends acropetally (from base to tip) from the stem vascular tissues and ramifies within the leaf during the primary, morphogenesis and expansion stages of leaf development. The three primary meristematic tissues become distinct through different patterns of cell division in relation to overall tissue expansion. For instance, the ground meristem cells that give rise to palisade parenchyma continue to divide long after divisions in adjacent protoderm cells have ceased. Since both tissues are growing at the same rate, palisade parenchyma cell size is small in comparison to epidermal tissue, reflecting the greater number of times that each cell is subdivided. Procambial cells continue to divide longer than other cells in the leaf, resulting in narrow, elongate xylem and phloem cells.

As tissue precursors stop dividing, they begin the final stages of cell differentiation (**Figure 8**). Distinctive patterns of specialized cell types appear within each tissue. In the epidermis, specialized guard cells and trichomes differentiate in a more or less regular spacing pattern from undifferentiated protoderm cells. Within the mesophyll, palisade parenchyma and spongy parenchyma cells develop their distinctive cell shapes from similar-appearing precursors in the ground meristem. An important aspect of mesophyll differentiation is the splitting of regions of the middle lamella layer between adjacent cells, forming the extensive intercellular airspace which facilitates diffusion of carbon dioxide to the photosynthetic mesophyll cells. Within the procambial strands, specialized conducting cells of the xylem (vessel elements) and of the phloem (sieve tube elements) also differentiate. Specialized parenchyma cells adjacent to the vascular bundles differentiate as bundle sheath cells.

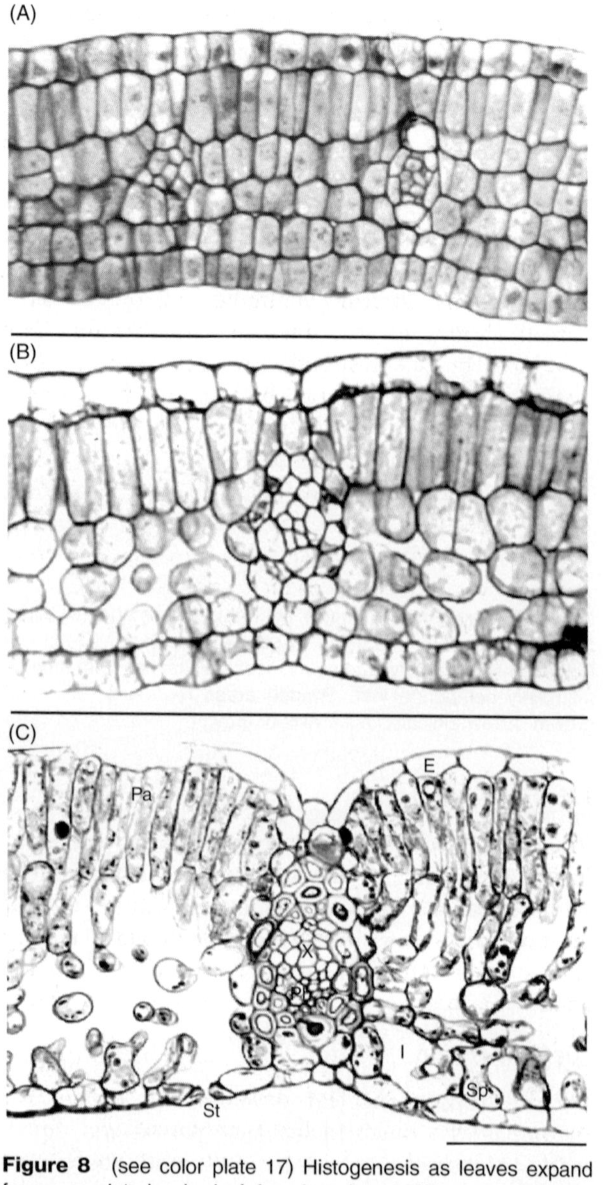

Figure 8 (see color plate 17) Histogenesis as leaves expand from overwintering bud of American beech (*Fagus grandifolia*). (A) Beginning of bud swelling; (B) seven days after bud break; (C) 28 days after bud break. E, epidermis; I, Intercellular airspace; Pa, palisade mesophyll; Sp, spongy mesophyll; St, stomate; X, xylem; P, phloem. Scale = 100 μm. Reproduced with permission from Dengler NG, Mackay LB, and Gregory LM (1975) Cell enlargement and tissue differentiation during leaf expansion in beech, *Fagus grandiflora. Canadian Journal of Botany* 53: 2846–2865.

Influence of the Environment on Leaf Development

Generally, leaves are buffered from environmental perturbations during the early stages of development–leaf initiation and morphogenesis. Leaves at these very young stages are well protected within the bud, and the distribution or resources within the plant, such as water and carbohydrates, are channeled to the young leaves, even during times of stress. As leaves undergo the expansion and histogenesis stages, they are much more vulnerable to environmental stresses, particularly water availability, temperature, and light. Adjustments are often made during these late stages of leaf development, allowing the mature leaf to function more efficiently under new conditions. Environmentally induced heterophylly in the aquatic buttercup is a good example of this. Leaf primordia of underwater and terrestrial leaves are very similar in size and shape and no differences occur until leaves have reached about 25% of mature size. After this stage, expansion of the underwater leaves is highly allometric, resulting in long filamentous leaf lobes. Histogenesis is also affected, and differentiation of certain specialized cell types is suppressed in underwater leaves.

In *Ranunculus* and in other plants, changes in the abiotic environment must be perceived by the developing leaves and translated into an alteration of developmental processes. In some cases, physical factors have a direct effect on leaf developmental processes. For instance, the rate of leaf development is positively correlated with temperature. Leaves are initiated and expand more quickly under higher temperatures, but otherwise, leaf size, morphology, and anatomy appear unaltered. In other cases, growth hormones are involved in changing leaf development in response to an environmental change. In the aquatic buttercup, the growth hormone abscisic acid is produced in response to low water availability. Abscisic acid is thought to act as an intermediate signal that induces development of the specific shape and anatomy of terrestrial leaves, at least in this species. Support for this comes from experiments in which abscisic acid was provided to submerged plants, resulting in the development of leaves with all the features of terrestrial plants.

Effects of Light

The natural light environment of plants varies in at least two important ways which affect leaf development. First, the quantity of light striking the leaf surface varies from almost 2000 $\mu mol\,m^{-2}\,s^{-1}$ in full sunlight to less than 40 $\mu mol\,m^{-2}\,s^{-1}$ in the shade of a forest or crop canopy. Second, the quality of light, usually expressed as the ratio of red to far-red wavelengths (R:FR ratio), varies depending on the wavelengths that are absorbed and transmitted by other objects. Green leaves preferentially absorb in the red part of the light spectrum, so that the dense shade cast by a deciduous forest canopy reduces the high R:FR ratio of full sunlight (1.2) to values as low as 0.1. Shade-intolerant plants generally respond to the lower quantity of light and the altered R:FR ratio of shade by reducing the rate of leaf expansion, resulting in smaller leaves. This response is characteristic of most crop plants which require full sunlight for normal growth. Shade-tolerant plants, on the other hand, have a different response to shading. They increase the rate of leaf expansion, resulting in leaves with a larger surface area in the shade. The relationship between cell division and leaf expansion remains the same, so that cell sizes are generally similar between sun and shade leaves. Other aspects of shade leaf histogenesis are affected, however, and leaves are usually thinner, with more intercellular airspace and less palisade parenchyma. These features allow shade leaves to capture more light energy and photosynthesize more effectively under low light conditions.

Effects of Water Availability

Leaf expansion is very sensitive to water availability. Placing plants under water stress at any time during the period of leaf expansion results in a smaller leaf area. Plant tissues expand because they take up water. Expansion also requires an increase in the extensibility of plant cell walls above a baseline level called the yield threshold. Extensibility is controlled by the cell through the release of protons (H^+) into the cell wall, lowering its pH, and activating cell wall proteins called expansins. Expansins release the hydrogen bonds between cell wall cellulose and hemicellulose molecules, allowing these fibrillar components to slip past each other and resulting in wall expansion. A local control of cell wall extensibility is required for leaf morphogenesis where minute regions of the leaf primordium expand more than others. Increased wall extensibility throughout the whole leaf is required for the expansion phase of leaf development. Since both water and increased wall extensibility are essential for growth, leaves developing under conditions of water stress are small. Surprisingly, leaf area is strongly correlated with cell number, indicating that the process of cell division somehow monitors overall tissue expansion and that cells only divide when they reach a critical size. Thus, cell sizes and other anatomical characters required for leaf function are preserved, even though overall leaf size is reduced.

Genes that Regulate Leaf Development

Genes regulating each of the four phases of leaf development have been identified, but most is known about those affecting: (1) morphogenesis, (2) expansion, and (3) perception of environmental factors. Genes that regulate leaf morphogenesis have been identified through studying mutants that have abnormal leaf morphology (**Figure 9**). For example, normal wild-type tomato plants have compound leaves in which some of the leaflets are themselves compound. Mutations in the *SOLANIFOLIA* gene suppress the formation of secondary leaflets, while mutations in the *LANCEOLATE* gene completely suppress leaflet formation. Genes that regulate allometric leaf expansion have been identified from studying *Arabidopsis thaliana* mutants. These include the *ROTUNDIFOLIA* gene that regulates growth in leaf length and the *ANGUSTIFOLIA* gene that regulates growth in leaf width. The third group of genes that are important for leaf

Figure 9 Leaf shape mutants in tomato (*Lycopersicon esculentum*). (A) Wild-type; (B) *solanifolia* mutant; (C) *lanceolate* mutant. Scale = 8 cm. Reproduced with permission from Dengler NG (1984) Comparison of leaf development in normal (+/+), entire (e.e) and lanceolate (LA/+) plants of tomato, *Lycopersicon esculentum* "Ailsa Craig". *Botanical Gazzette* 145: 66–77.

development and for understanding how environmental variables affect leaf development are those involved in sensing the physical environment, such as shading. Changes in the R : FR ratio are sensed by the plant pigment phytochrome. Genes such as the *CONSTITUITIVELY PHOTOMORPHOGENETIC* and *DE-ETIOLATED* genes have been identified by studying the responses of mutant *Arabidopsis* seedlings to different light wavelengths. These genes have been shown to be involved in translating the phytochrome signal into the tissue response, so that leaves expand more when shaded. Identification of these genes and understanding their mode of action at a molecular level makes it possible to genetically manipulate the size and shape of leaves. Since these characteristics contribute to photosynthetic efficiency and ability to withstand water stress, knowledge of these genes may have an impact on crop improvement.

List of Technical Nomenclature

Acropetal	Process occurring from base toward the apex.
Allometric	Expansion along one axis differs from expansion along another.
Basipetal	Process occurring from apex toward the base.
Dicotyledon	One of two major groups of flowering plants with two cotyledons in the embryo, netlike leaf venation, flower parts in fours and fives.
Ground meristem	Undifferentiated precursor tissue to mesophyll.
Heteroblasty	Variation in leaf size and shape among positions within an individual plant.
Heterophylly	Variation in leaf size and shape within an individual plant, usually induced by the abiotic environment.
Isometric	Expansion along all axes is equivalent.
Monocotyledon	One of two major groups of flowering plants with one cotyledon in the embryo, striate leaf venation, flower parts in threes.
Morphogenesis	Development of shape.
Primordium	A rudimentary structure, the product of leaf initiation.
Procambium	Undifferentiated precursor tissue to vascular tissue.
Protoderm	Undifferentiated precursor tissue to epidermis.
Shoot apical meristem	Actively dividing population of cells at tip of stem; produces leaf primordia.

See also: **Cell Walls and Fibers**: Cell Walls. **Genetic Modification**: Gene Cloning, General Principles. **Growth and Development**: Canopy Architecture; Cell Division and Differentiation; Cell Growth; Cells; Growth Analysis, Individual Plants. **Regulators of Growth**: Abscisic Acid; Photomorphogenesis; Phytochromes and other Photoreceptors. **Water Relations of Plants**: Basic Water Relations; Drought Stress.

Further Reading

Bruni NC, Young JP, and Dengler NG (1996) Leaf developmental plasticity in *Ranunculus flabellaris* in response to aerial and aquatic environments. *Canadian Journal of Botany* 74: 823–837.

Dale JE (1992) How do leaves grow? *BioScience* 42: 423–432.

Donnelly PM, Bonetta D, Tsukaya H, Dengler RE, and Dengler NG (1999) Cell cycling and cell enlargement in developing leaves of *Arabidopsis. Developmental Biology* 215: 407–419.

Granier C, Turc O, and Tradieu F (2000) Co-ordination of cell division and tissue expansion in sunflower, tobacco and pea leaves: dependence or independence of both processes? *Journal of Plant Growth Regulation* 19: 45–54.

Maksymowych R (1990) *Analysis of Growth and Development of* Xanthium. Cambridge: Cambridge University Press.

Raven PH, Evert RF, and Eichhorn SE (1999) *Biology of Plants*. New York: WH Freeman.

Sinha N (1999) Leaf development in angiosperms. *Annual Review of Plant Physiology and Plant Molecular Biology* 50: 419–446.

Steeves TA and Sussex IM (1989) *Patterns in Plant Development*. Cambridge: Cambridge University Press.

Van Volkenburgh E (1999) Leaf expansion: an integrating plant behavior. *Plant Cell Environment* 22: 1463–1473.

ISBN 0-12-227050-9

9 780122 270505

ENCYCLOPEDIA
OF
APPLIED PLANT SCIENCES

ENCYCLOPEDIA
OF
APPLIED PLANT SCIENCES

Editor-in-Chief

BRIAN THOMAS

Editors

DENIS J MURPHY

BRIAN G MURRAY

ELSEVIER
ACADEMIC
PRESS

Amsterdam Boston Heidelberg London New York Oxford
Paris San Diego San Francisco Singapore Sydney Tokyo

Elsevier Ltd., The Boulevard, Langford Lane, Kidlington, Oxford, OX5 1GB, UK

Elsevier Inc., 525 B Street, Suite 1900, San Diego, CA 92101-4495, USA

First edition 2003

Library of Congress Control Number: 2003109025

A catalogue record for this book is available from the British Library

ISBN: 0-12-227050-9 (set)

This book is printed on acid-free paper

Printed in the United Kingdom

EDITORIAL ADVISORY BOARD

FOREWORD

I often tell visitors and policy makers who want to learn about current research in the plant sciences and agriculture that 'This is not your father's plant science'. As amply evidenced in these three volumes of the Encyclopedia of Applied Sciences, plant biology is undergoing a significant change in emphasis. The change engages far greater numbers of plant scientists in genetics, molecular biology, and biotechnology than ever before. This is not surprising given the growing amount of information on genome structure and functional genomics and the tools of genetic transformation of plants. These changes makes it possible for laboratory scientists to have a far greater impact on agriculture production and products than in any other period of time in the long history of applied plant sciences. The future promises to bring even more changes that bring sustainability to production agriculture and greater variability to its products.

These volumes contain chapters from world authorities in a range of topics and carry information that will inform the advanced reader, as well as the newcomer to the field, about the latest advances in the expanding fields of plant science. The reader will learn about the use of transgenic plants with altered metabolism that enhances accumulation of proteins and carbohydrates, alkaloids and terpenoids, and other components. Other chapters report the state of the science and recent breakthroughs in genetics and biochemistry of photosynthesis, plant growth and development, and plant nutrition. There is also a much needed update on the many experimental approaches to develop transgenic plants with novel characteristics, including enhanced resistance to pests and pathogens; biofortification and enhanced food safety; altering seed biology; reducing ripening times and post-harvest losses; bioremediation, and so on. The advances that have been made in the field are truly remarkable and are a testament to the many opportunities that modern genetics and biotechnology have brought to plant sciences.

However, while the advances have been great the informed reader will realize that very few of the discoveries reported in these chapters have yet found their way into commercialization. Furthermore, there has been a disturbing trend in recent years to slow the delivery of new biotechnology products to the marketplace. Governmental regulatory agencies in the US and the EU, as well as countries around the world, have shown reluctance to grant approval of new products, or to approve for sale products that have gained regulatory approval in another country.

At this juncture the products that make it to the marketplace are those for which significant profit can be accurately predicted: to date, the direct beneficiaries of the few approved products are the farmers. It remains to be seen whether subsequent products will be limited to plants whose traits benefit the farmer/producer, or whether products that have more obvious 'consumer benefit' will be approved for release. And, given the troubling trend of governments to expand regulatory oversight, and therefore cost of bringing a product to market, it also remains to be seen whether public sector institutions will or will not be a major player in developing genetically enhanced crops, or whether this opportunity will be left to major private sector companies.

Nevertheless, the quality of plant science research continues to improve, enhancing our knowledge of plant biology, and creating more and more opportunities to make contributions to sustainably produce sufficient amounts of nutritious and safe foods to feed an ever-growing population. And, to bring plant made

pharmaceuticals and other specialty products to market. The concerns aside, I remain optimistic and hope that that you share this optimism after reading these volumes.

Roger N. Beachy, Ph.D.
President
Donald Danforth Plant Science Center
St. Louis, Missouri, USA

INTRODUCTION

Plants have a huge impact upon human existence. All of our food and much of our other energy needs are ultimately derived from the products of plant photosynthesis. Plants also have their impact through the social and economic dimensions of human existence. All agriculture, which provides us with our food, is ultimately based on plant production, be it directly through arable crops such as cereals, horticultural crops including fruits and vegetables or indirectly through pasture, forage and plant derived animal feedstuffs. Industrial crops such as fiber crops or woodlands and forests are a vital source of renewable materials for clothing, building and recreation, that we take for granted in everyday life. Flowers and ornamental plants delight us with their beauty, fragrance and their seemingly infinite variety.

Societies and economies all over the world have been able to gain increasing benefits from plants as our scientific understanding of their properties and potential has increased rapidly over recent decades. The range and complexity of this new information about plants is now potentially bewildering. A real challenge is how can such accumulated knowledge be made available to those taking an interest in the area for the first time. Equally how can a specialist in one aspect of plant applications gain access to information from wider or complementary areas of the subject? This is where we hope that this work will be of the most value.

The Encyclopedia of Applied Plant Sciences provides a friendly and comprehensive portal to the huge amount of knowledge that is available on plants and how we make use of them. The Editors, assisted by a distinguished Editorial Board have commissioned articles from an unparalleled group of experts across the plant sciences to provide introductory overviews and summaries of the state-of-the-art for their areas. Each article is sufficient for the reader to gain an understanding of the essentials of the topic covered and also provides an entry to the substantial amount of literature that lies beneath the articles. Rigorous editing and comprehensive indexing by the publishers ensures that the articles are understandable and that information can be located as needed. Articles covering the breadth of plant science are grouped under clearly described entries that will help the reader in finding not only the article they require but also articles on closely related topics.

Wherever possible authors have tried to show how the underlying science and information links through to applications in practical situations. Plant science undoubtedly has an exciting future and in this Encyclopedia we signpost this through a range of entries on the potential of the new genetic technologies, both in addressing traditional challenges and in making novel contributions to agriculture and other applications of plant science.

Putting the encyclopedia together has been a long and involved project. We are grateful for the help and support from the staff at Elsevier, the expertise and time given by the editorial board and the huge amount of work by the authors of the articles. We believe that the end result is a unique assemblage of the key information in this important field and are privileged to have been involved in the project. The value in the work is down to the quality of the contributors. Omissions or lack of balance are entirely the responsibility of the Editors. We hope that the work proves to be not only of practical value but can act as a catalyst for readers to gain entry to the exciting world of plant science.

Brian Thomas
Denis Murphy
Brian Murray

GUIDE TO USE OF THE ENCYCLOPEDIA

Structure of the Encyclopedia

The material in the Encyclopedia is arranged as a series of entries in alphabetical order. Some entries comprise a single article, whilst the majority of entries are on more diverse subjects and consist of several articles that deal with various aspects of the topic. In the latter case the articles are arranged in a logical sequence within an entry.

To help you realize the full potential of the material in the Encyclopedia we have provided three features to help you find the topic of your choice.

1. Contents List

Your first point of reference will probably be the contents list. The complete contents list appearing in each volume will provide you with both the volume number and the page number of the entry. On the opening page of an entry a contents list is provided so that the full details of the articles within the entry are immediately available.

Alternatively you may choose to browse through a volume using the alphabetical order of the entries as your guide. To assist you in identifying your location within the Encyclopedia a running headline indicates the current entry and the current article within that entry.

You will find "dummy entries" where obvious synonyms exist for entries or where we have grouped together related topics. Dummy entries appear in both the contents list and the body of the text. For example, a dummy entry appears for Arable Crops which directs you to Production Systems and Agronomy: Agricultural Crops where the material is located.

Example
If you were attempting to locate material on Biotechnology via the contents list.

BIOTECHNOLOGY *see* DISEASES: Genetic Modification of Disease Resistance, Fungal Pathogens; Genetic Modification of Disease Resistance, Viral Pathogens. GENETIC MODIFICATION: Gene Cloning, General Principles; Insertional and Transposon Mutagenesis; Transformation, General Principles; Transformation in Dicotyledons; Transformation in Monocotyledons; Transformation in Plastids; Transgene Stability and Inheritance. GENETIC MODIFICATION, APPLICATIONS: Flower Color; Nitrogen Fixation; Oxidative Stress; Cell Factories; Molecular Farming; Plantibodies. GROWTH AND DEVELOPMENT: Control of Gene Expression, Posttranscriptional Regulation.

At the appropriate location in the contents list, the page numbers for these articles are given.

If you were trying to locate the material by browsing through the text and you looked up Biotechnology then the following information would be provided.

Biotechnology *see* **Diseases**: Genetic Modification of Disease Resistance, Fungal Pathogens; Genetic Modification of Disease Resistance, Viral Pathogens. **Genetic Modification**: Gene Cloning, General Principles; Insertional and Transposon Mutagenesis; Transformation, General Principles; Transformation in Dicotyledons; Transformation in Monocotyledons; Transformation in Plastids; Transgene Stability and Inheritance. **Genetic Modification, Applications**: Flower Color; Nitrogen Fixation; Oxidative Stress; Cell Factories; Molecular Farming; Plantibodies. **Growth and Development**: Control of Gene Expression, Posttranscriptional Regulation.

Alternatively if you were looking up Genetic Modification the following information would be provided.

GENETIC MODIFICATION

Contents

2. Cross References

All of the articles in the Encyclopedia have been extensively cross referenced. The cross references, which appear at the end of an article, have been provided at three levels:

i. To indicate if a topic is discussed in greater detail elsewhere

ROOT DEVELOPMENT/Root Growth and Development.
See also: **Growth and Development**: Cell Division and Differentiation Cell Growth. **Nutrition**: Ion Transport; Mineral Uptake; Nitrogen Fixation. **Root Development**: Genetics of Primary Root Development; Lateral Root Initiation; The Rhizosphere and its Microorganisms. **Water Relations of Plants**: Phloem; Xylem.

ii. To draw the reader's attention to parallel discussions in other articles

ROOT DEVELOPMENT/Root Growth and Development.
See also: **Growth and Development**: Cell Division and Differentiation; Cell Growth. **Nutrition**: Ion Transport; Mineral Uptake; Nitrogen Fixation. **Root Development**: Genetics of Primary Root Development; Lateral Root Initiation; The Rhizosphere and its Microorganisms. **Water Relations of Plants**: Phloem; Xylem.

iii. To indicate material that broadens the discussion

ROOT DEVELOPMENT/Root Growth and Development.
See also: **Growth and Development**: Cell Division and Differentiation; Cell Growth. **Nutrition**: Ion Transport; Mineral Uptake; Nitrogen Fixation. **Root Development**: Genetics of Primary Root Development; Lateral Root Initiation; The Rhizosphere and its Microorganisms. **Water Relations of Plants**: Phloem; Xylem.

3. Index

The index will provide you with the volume number and page number of where the material is to be located, and the index entries differentiate between material that is a whole article, is part of an article or is data presented in a table. Detailed notes are provided on the opening page of the index.

4. Color Plates

The color figures for each volume have been grouped together in a plate section. The location of this section is cited in the contents list. Color versions of black and white figures are cited in figure captions within individual articles.

5. Contributors

A full list of contributors appears at the beginning of each volume.

CONTRIBUTORS

W W Adams III
University of Colorado
Boulder, CO, USA

S R Adams
Horticulture Research International
Wellesbourne
UK

G N Agrios
University of Florida
Gainesville
FL, USA

B S Ahloowalia
International Atomic Energy Agency
Vienna, Austria

G Alibert
INP-ENSAT
Auzeville, France

T Altabella
Universitat de Barcelona
Barcelona, Spain

A Altman
The Hebrew University of Jerusalem
Rhovot, Israel

J L Araus
University of Barcelona
Barcelona, Spain

S Arcioni
Institute of Plant Genetics
Perugia
Italy

A E Arthur
John Innes Centre
Norwich, UK

S N Azam-Ali
University of Nottingham
Nottingham, UK

R A Azevedo
University of São Paulo
Piracicaba, Brazil

M Bacon
University of Lancaster
Lancaster, UK

M Båga
University of Saskatchewan
Saskatoon, SK, Canada

P Barcelo
DuPont Wheat Transformation Laboratory
Barcelona, Spain

H S Bariana
University of Sydney
Camden, NSW, Australia

A R Barr
University of Adelaide
Glen Osmond, SA, Australia

L Bastiaans
Wageningen University
Wageningen, The Netherlands

M H Behboudian
Massey University
Palmerston North, New Zealand

J N B Bell
Imperial College London
Ascot, UK

G D Bending
Horticulture Research International
Wellesbourne, UK

R L Benech-Arnold
IFEVA
Buenos Aires, Argentina

L R Benjamin
Rothamsted Research
Harpenden, UK

P Berjak
University of Natal
Durban, South Africa

A Bernadac
UMR 990 INRA-ENSAT
Toulouse, France

I J Bingham
Scottish Agricultural College
Aberdeen, UK

D Blakesley
Horticulture Research International
West Malling, UK

J M Bonga
Canadian Forest Service – Atlantic Forestry Center
Fredericton, NB, Canada

A-M Boudet
Université Paul Sabatier
Toulouse, France

M Bouzayen
UMR 990 INRA-ENSAT
Toulouse, France

J Brown
University of Idaho
Moscow, ID, USA

V Buchanan-Wollaston
Horticulture Research International
Wellesbourne, UK

M A S Burton
University of Hertfordshire
Hatfield, UK

P D S Caligari
Universidad de Talca
Talca, Chile

R W F Cameron
University of Reading
Reading, UK

S Cameron
Canadian Forestry Service
Fredericton, NB, Canada

M Cannell
Syngenta, Jealotts Hill International Research Centre
Bracknell, UK

M Cappadocia
Université de Montréal
Montréal, Canada

J S Carmichael
University of North Dakota
Grand Forks, ND, USA

A C Cassells
National University of Ireland
Cork, Ireland

K J Chalmers
University of Adelaide
Glen Osmond, SA, Australia

B Chanson
Université Montpellier II
Montpellier, France

D Chávez
University of Illinois
Chicago, IL, USA

S C Cheah
Malaysian Palm Oil Board
Kuala Lumpur, Malaysia

R N Chibbar
National Research Council Canada
Saskatoon, SK, Canada

V L Chopra
National Research Centre on Plant Biotechnology
New Delhi, India

L P Christensen
Danish Institute of Agricultural Sciences
Aarslev, Denmark

P Christou
Fraunhofer Institute of Molecular Biology and Applied Ecology
Schmallenberg, Germany

S D Clouse
North Carolina State University
Raleigh, NC, USA

K E Cockshull
Horticulture Research International
Wellesbourne, UK

J I Cooper
NERC Centre for Ecology and Hydrology
Oxford, UK

A Cordeiro
Universitat de Barcelona
Barcelona, Spain

C Darley
University of York
York, UK

E Dayan
Agricultural Research Organization
Hanegev, Israel

J Degenhardt
Max Planck Institute for Chemical Ecology
Jena, Germany

G-J de Klerk
Center for Plant Tissue Culture Research
Lisse, The Netherlands

J M DeLong
Agriculture and Agri-Food Canada
Kentville, Nova Scotia, Canada

B Demmig-Adams
University of Colorado
Boulder, CO, USA

N Dengler
University of Toronto
Toronto, ON, Canada

K Denyer
John Innes Centre
Norwich, UK

M Dow
National University of Ireland
Cork, Ireland

J G Dubrovsky
Universidad Nacional Autónoma de Mexico
Cuernavaca Morelos, Mexico

E M Duffy
National University of Ireland
Cork, Ireland

J M Dunwell
University of Reading
Reading, UK

G Dutfield
Queen Mary College
London, UK

P Elomaa
University of Helsinki
Helsinki, Finland

M R Emmett
University of Reading
Reading, UK

D E Evans
Oxford Brookes University
Oxford, UK

J D Everard
DuPont Experimental Station
Wilmington, DE, USA

W E Finch-Savage
Horticulture Research International
Wellesbourne, UK

K Fisher
Virginia Polytechnic Institute and State University
Blacksburg, VA, USA

M E Foley
USDA
Agricultural Research Service
Fargo, ND, USA

B V Ford-Lloyd
University of Birmingham
Birmingham, UK

C H Foyer
Rothamsted Research
Harpenden, UK

J B Free
University of Wales
Cardiff, UK

C Freeman
University of Wales
Bangor, UK

S C Fry
The University of Edinburgh
Edinburgh, UK

M D Gale
John Innes Centre
Norwich, UK

G Galiba
Agricultural Research Institute of the Hungarian Academy of Sciences
Martonvásár, Hungary

J Gershenzon
Max Planck Institute for Chemical Ecology
Jena, Germany

K G Gilbert
University of Bristol
Bristol, UK

T A Gorshkova
KIBB, RAS
Kazan, Russia

K S Gould
University of Auckland
Auckland
New Zealand

J Grace
University of Edinburgh
Edinburgh, UK

D J Gray
University of Florida
Apopka, FL, USA

J Gril
Université Montpellier II
Montpellier, France

K Groten
Rothamsted Research
Harpenden, UK

X Gu
North Dakota State University
Fargo, ND, USA

N D Hagan
CSIRO Plant Industry
Canberra, Australia

D L Hallahan
Central Research and Development
E I DuPont de Nemours & Company
Wilmington, DE, USA

J B Harborne[†]
University of Reading
Reading, UK

F Hay
Royal Botanic Gardens Kew
Ardingly, UK

P Hedden
Rothamsted Research
Harpenden, UK

J S Heslop-Harrison
University of Leicester
Leicester, UK

T J V Higgins
CSIRO Plant Industry
Canberra, Australia

D F Hildebrand
University of Kentucky
Lexington, KY, USA

H W M Hilhorst
Wageningen University
Wageningen
The Netherlands

S J Hiscock
University of Bristol
Bristol, UK

P J Holloway
The Long Ashton Research Station
Bristol, UK

R Hull
John Innes Centre
Norwich, UK

M O Humphreys
IGER
Aberystwyth, UK

R Hunt
University of Sheffield
Sheffield, UK

[†]Deceased

B S Jalani
Malaysian Palm Oil Board
Kuala Lumpur, Malaysia

P E Jameson
Massey University
Palmerston North, New Zealand

M F Jardinaud
Biotechnology and Plant Breeding, ENSAT
Auzeville, France

M C Jarvis
Glasgow University
Glasgow, UK

P P Jauhar
United States Department of Agriculture – Agricultural Research Service
Fargo, ND, USA

C E Jeffree
University of Edinburgh
Edinburgh, UK

R Johnson
Colby College
Waterville, ME, USA

D L Jones
University of Wales
Bangor, UK

H D Jones
Rothamsted Research
Harpenden, UK

L Jones
University of York
York, UK

V G Kakani
Mississippi State University
MS, USA

J Kallerhoff
INP-ENSAT
Auzeville, France

J Kang
University of Toronto
Toronto, ON, Canada

K J Kasha
University of Guelph
Guelph, ON, Canada

S P Kell
University of Birmingham
Birmingham, UK

R E Kendrick
Wageningen University
Wageningen, The Netherlands

A R Kermode
Simon Fraser University
Burnaby, British Columbia, Canada

A D Kinghorn
University of Illinois
Chicago, IL, USA

A J Kinney
DuPont Experimental Station
Wilmington, DE, USA

J D Klein
ARO-Volcani Center
Bet Dagan, Israel

R M D Koebner
John Innes Centre
Norwich, UK

A Kohli
John Innes Centre
Norwich, UK

M J Kropff
Wageningen University
Wageningen, The Netherlands

K Kunert
University of Pretoria
Pretoria, South Africa

A Kurian
Kerala Agricultural University
Vellanikkara, India

A Kushairi
Malaysian Palm Oil Board
Kuala Lumpur, Malaysia

M Labra
University of Milan
Milan, Italy

J Langridge[†]
Adelaide University
Glen Osmond, SA, Australia

P Langridge
Adelaide University
Glen Osmond, SA, Australia

E Larsen
Danish Institute of Agricultural Sciences
Aarslev, Denmark

A Latché
UMR 990 INRA-ENSAT
Toulouse, France

D W Lawlor
Rothamsted Research
Harpenden, UK

P A Lazzeri
DuPont Wheat Transformation Laboratory
Barcelona, Spain

P J Lea
Lancaster University
Lancaster, UK

R C Leegood
University of Sheffield
Sheffield, UK

R A Leigh
University of Cambridge
Cambridge, UK

Z Li
UMR 990 INRA-ENSAT
Toulouse, France

S E Lid
Agricultural University of Norway
Ås, Norway

A M Litterick
Crop and Soil Research Group, SAC
Aberdeen, UK

R E Litz
University of Florida
Homestead, FL, USA

[†]Deceased

G J Loake
University of Edinburgh
Edinburgh, UK

U Lüttge
Darmstadt University of Technology
Darmstadt, Germany

P Maliga
Rutgers University
Piscataway, NJ, USA

J Maluszynska
University of Silesia
Katowice, Poland

M Maluszynski
IAEA
Vienna, Austria

T Marks
Horticulture Research International
West Malling, UK

J M Martínez-Zapater
Centro Nacional de Biotechnología, CSIC
Madrid, Spain

G Matthews
Imperial College of Science, Technology and Medicine
Ascot, UK

N Maxted
University of Birmingham
Birmingham, UK

M R McAinsh
Lancaster University
Lancaster, UK

S McQueen-Mason
University of York
York, UK

J Memelink
Leiden University
Leiden, The Netherlands

M Mencuccini
University of Edinburgh
Edinburgh, UK

K Meyer
DuPont Central Research and Development
Wilmington, DE, USA

W B Miller
Cornell University
Ithaca, NY, USA

D J Miralles
University of Buenos Aires
Buenos Aires, Argentina

D J Murphy
University of Glamorgan
Cardiff, UK

B G Murray
University of Auckland
Auckland, New Zealand

D Nadella
North Dakota State University
Fargo, ND, USA

R M Napier
Horticulture Research International
Wellesbourne, UK

R E L Naylor
Trelareg Consultants
Banchory, Scotland, UK

P Newell
Formerly of The University of the South Pacific
Suva, Fiji Islands

W E Newton
Virginia Polytechnic Institute and State University
Blacksburg, VA, USA

P S Nobel
University of California
Los Angeles, CA, USA

P L Nurmberg
University of Edinburgh
Edinburgh, UK

A M Nuutila
VTT Biotechnology
Espoo, Finland

E O' Connor
Bio-Renewables Ltd
Ely, UK

C E Offler
University of Newcastle
New South Wales, Australia

K-M Oksman-Caldentey
VTT Biotechnology
Espoo, Finland

O-A Olsen
Pioneer Hi-Bred International
Johnston, IA, USA

J L Osborne
Rothamsted Research
Harpenden, UK

N W Pammenter
University of Natal
Durban, South Africa

Y S Park
Canadian Forest Service – Atlantic Forestry Center
Fredericton, NB, Canada

J W Patrick
The University of Newcastle
New South Wales, Australia

M J Paul
Rothamsted Research
Harpenden, UK

J-C Pech
UMR 990 INRA-ENSAT
Toulouse, France

K V Peter
Kerala Agricultural University
Vellanikkara, India

M Petitprez
Biotechnology and Plant Breeding, ENSAT
Auzeville, France

P W B Phillips
University of Saskatchewan
Saskatoon, SK, Canada

M Piñeiro
Centro Nacional de Biotechnología, CSIC
Madrid, Spain

A H Pickering
Massey University
Palmerston North, New Zealand

M Pletsch
Universidade Federal de Alagoas
Maceió, Brazil

K M Plummer
University of Auckland
Auckland, New Zealand

Y Poirier
University of Lausanne
Lausanne, Switzerland

S Pond
Canadian Forestry Service
Fredericton, NB, Canada

S A Power
Imperial College London
Ascot, UK

R K Prange
Agriculture and Agri-Food Canada
Kentville, Nova Scotia, Canada

P V V Prasad
University of Florida
Gainesville, FL, USA

F Pupilli
Institute of Plant Genetics
Perugia, Italy

C B Purrington
Swarthmore College
PA, USA

A S Raghavendra
University of Hyderabad
Hyderabad, India

O Ramli
Malaysian Rubber Board
Selangor, Malaysia

K R Reddy
Mississippi State University
MS, USA

M J Reiss
Institute of Education
University of London, UK

P Roberts
University of York
York, UK

D Robinson
University of Aberdeen
Aberdeen, UK

T L Rost
University of California
Davis, CA, USA

E C Rowe
Wageningen University
Wageningen, The Netherlands

G Russell
The University of Edinburgh
Edinburgh, UK

P R Ryan
CSIRO Plant Industry
Canberra, Australia

F Sala
University of Milan
Milan, Italy

R A Sánchez
IFEVA
Buenos Aires, Argentina

S Sanogo
New Mexico State University
Las Cruces, NM, USA

G E Schaller
University of New Hampshire
Durham, NH, USA

P J Sharp
Plant Breeding Institute
University of Sydney, Australia

T Shepherd
Scottish Crop Research Institute
Dundee, UK

P R Shewry
Rothamsted Research
Harpenden, UK

K Shinozaki
RIKEN Tsukuba Institute
Tsukuba, Japan

N W Simmonds[†]
University of Edinburgh
Edinburgh, UK

L Skøt
Institute of Grassland and Environmental Research
Aberystwyth, UK

G A Slafer
University of Buenos Aires, Buenos Aires
Argentina

H Slater
Lancaster University
Lancaster, UK

S M Smith
Institute of Cell and Molecular Biology
University of Edinburgh, UK

D Sparkes
University of Nottingham
Loughborough, UK

E Stoger
RWTH Aachen
Aachen, Germany

R Straughan
University of Reading
Reading, UK

J Sutka
Agricultural Research Institute of the Hungarian Academy of Sciences
Martonvásár, Hungary

I Szarejko
University of Silesia
Katowice, Poland

D A Tabah
University of Bristol
Bristol, UK

A G Taylor
Cornell University
Geneva, NY, USA

[†]Deceased

T H Teeri
University of Helsinki
Helsinki, Finland

B Thibaut
Université Montpellier II
Montpellier, France

B Thomas
Horticulture Research International
Wellesbourne, UK

J M G Thomas
University of Florida
Gainesville, FL, USA

G A Thompson
University of Arkansas
Little Rock, AR, USA

A F Tiburcio
Universitat de Barcelona
Barcelona, Spain

K Tsunewaki
Fukui Prefectural University
Fukui, Japan

R M Twyman
University of York
York, UK

A J E van Bel
Justus-Liebig-University
Giessen, Germany

J E G van Dam
WUR, Wageningen
The Netherlands

B Veit
AgResearch
Palmerston North, New Zealand

R Verpoorte
Leiden University
Leiden, The Netherlands

J A Walsh
Horticulture Research International
Wellesbourne, UK

H L Wang
University of Arkansas
Little Rock, AR, USA

C B Watkins
Cornell University
Ithaca, NY, USA

C A Watson
Crop and Soil Research Group, SAC
Aberdeen, UK

A D Webster
Maidstone, Kent, UK

J L Weller
Univeristy of Tasmania
Hobart, Australia

P J White
Horticulture Research International
Wellesbourne, UK

E J Woltering
Wageningen University
Wageningen, The Netherlands

X B Yang
Iowa State University
Ames, IA, USA

K Yu
University of Kentucky
Lexington, KY, USA

J A D Zeevaart
Michigan State University
East Lansing, MI
USA

R L Zimdahl
Colorado State University
Fort Collins, CO
USA

M Ziv
The Hebrew University of Jerusalem
Rohovot
Israel

CONTENTS

VOLUME 1

C

D

E

F

R

Volume 3

GROWTH AND DEVELOPMENT

Canopy Architecture

G Russell, The University of Edinburgh, Edinburgh, UK

Introduction

The concept of a plant canopy was first developed in tropical rainforests. It has since been applied to a wide range of situations ranging from short grass swards, through arable crops to forest plantations. A canopy is more than the foliage from a population of plants. It exhibits emergent properties, i.e., properties that are not directly deducible from the attributes of the constituent individuals, but which depend on their interaction. The growth of the above-ground parts of individual plants is modified by the proximity of neighboring plants through alteration of their aerial environment. As a stand of plants is rarely uniform, larger plants tend to grow at the expense of their neighbors. However, there can be a considerable change in the plants making up the canopy without any significant change in the functioning of the canopy as a whole. As with individual plants, the functioning of the canopy can be analyzed in terms of its integrated response to environmental factors. When, as is often the case, a canopy is not single-layered, the upper canopy has a strong impact on the one below. In some respects the leaf canopy above ground is functionally equivalent to the root system below ground, and indeed the two can be linked through the functional balance hypothesis (see below). Both the canopy and the root system capture the resources needed for plant growth. The canopy intercepts light and absorbs carbon dioxide while the roots capture water, nitrogen, and other nutrients needed for leaf growth.

Canopy Structure and Function

Canopies are usually described in terms of the attributes that affect light transmission and absorption. Where canopies are essentially horizontally homogeneous, leaf area index (LAI), the ratio of the area of leaf to the ground area, is an important measure of canopy size. If the light extinction coefficient is known then the attenuation of light down the canopy can be calculated using the Lambert-Beer extinction law:

$$I = I_0(1 - e^{-kL})$$

where I and I_0 are the irradiances below and above the canopy, L is the LAI, and k is the light extinction coefficient. Light transmission through many canopies appears to be adequately represented by Beer's law, even though the extinction coefficient may change down the canopy. The extinction coefficient can be calculated from the geometrical arrangement of the leaves and their optical properties. An increase in k has the same effect as an increase in LAI, so the same leaf area index can have very different effects on the quantity of light penetrating the canopy depending on the leaf angle distribution. Optical methods using the absorption of light are a well-studied alternative to more laborious methods of estimating the LAI of canopies involving harvesting representative samples of foliage. In heterogeneous canopies, such as those typical of broad-leaved woodland, the gap fraction, i.e., the proportion of the sky not obscured by foliage, can be assessed using hemispherical photography (**Figure 1**) and alternative models of light attenuation used to estimate LAI. However, in these canopies, the three-dimensional distribution of leaf area density, i.e., the leaf area divided by the volume of space occupied, can be more informative than a single value of LAI. Although averaging properties of the canopy is often useful, it

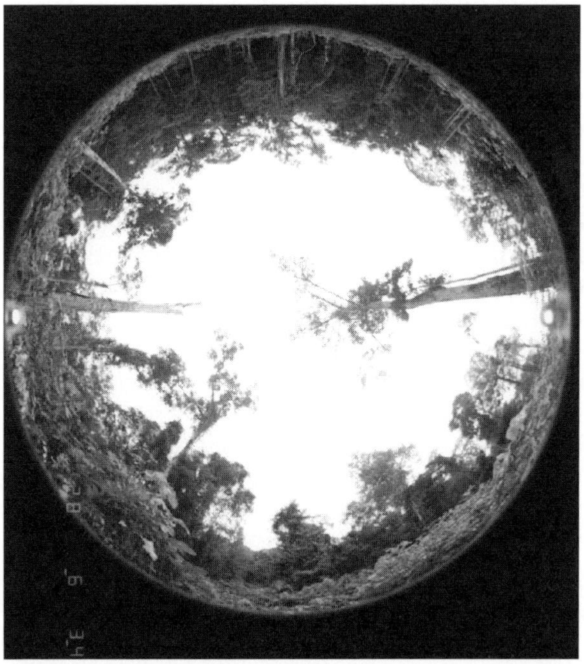

Figure 1 Hemispherical photograph taken from a gap in a forest canopy in Indonesia. Photograph kindly supplied by Dr P. van Gardingen with permission.

must always be done with caution and carried out in relation to the question being investigated. In a tropical rainforest, for example, the average irradiance at ground level may be too low to sustain plant life, although sunflecks passing through gaps in the canopy may allow individuals to survive.

Techniques such as eddy flux correlation can be used to measure the fluxes of gases such as CO_2 and water from extensive areas of canopy. This technique is now being widely applied in studies of carbon sequestration by extensive canopies. Previously, net carbon flux was estimated by scaling up measurements of photosynthesis made on individual leaves or branches. The physics of water loss from canopies through transpiration and evaporation of surface water is now well understood. Once the canopy is closed, the rate of evapotranspiration becomes essentially independent of LAI. However, if an increase in LAI is accompanied by a decrease in the surface roughness of the canopy, the evapotranspiration can even decline. This makes estimating canopy water loss by scaling up from measurements on a single plant rather unreliable.

Canopy properties can be measured from satellites for areas that may be many km^2 in size. The normalized difference vegetation index (NDVI), which is effectively a measure of the greenness of the canopy, can be used to study seasonal changes in canopy functioning caused by environmental factors such as drought superimposed on the normal phenological progression of leaf growth and death. Satellite imagery can also be used to estimate the water status of a canopy as water stress on a sunny day leads to the surface temperature exceeding air temperature. Portable instruments have been developed so that both of these methods can be used much closer to the canopy.

Canopies vary considerably in their complexity. At one end of the scale there are even-aged single species stands that can be remarkably homogeneous. In the early stages of annual crops and forest plantations, heterogeneity is imposed by the space between individuals. However, once the foliage of individual plants begins to overlap, the canopy becomes more homogeneous. Plant death caused by frost or waterlogging, or by attacks by pests or pathogens, can result in gaps in the canopy, although branching (tillering in the case of grasses) allows the surviving plants to colonize the space occupied by the dead plants. This is an example of the self-organizing abilities of canopies. At the other end of the spectrum are mixed species, uneven-aged stands. The original studies on canopies were in tropical rain forests that were multilayered as well as heterogeneous in the horizontal plane. It is important to recognize that the degree of heterogeneity depends on the scale of observation. Agroforestry systems, forests managed on the shelterwood system, and herb-rich grasslands are examples of commercially important systems characterized by complex canopy structures.

Plant Attributes Affecting Canopies

If all aspects of plant growth and development were purely genetically determined then canopy properties could be deduced by scaling up from individual plants, which is clearly not the case. However, although environmental factors are important, some genetic features are hardly affected by the proximity of other plants, and it could be argued that the response of the plant to an environmental factor is itself genetically determined. Basic plant architecture is genetically determined as can be seen from consistent differences between species across a wide range of environments. Studies of plant genomes are producing large amounts of information about the genetic composition of plant species. However, with the exception of single gene effects, there is a long way to go before we are able to predict the phenotype of a plant from its genotype and the environment in which it is growing. Leaf size, phenology, and leaf optical properties all have a strong genetic basis, although they are also influenced by environment.

Many wild species growing in stressful environments exhibit diurnal leaf movements (heliotropism), which either minimize or maximize the solar radiation loading. This strategy for tracking the sun is only effective where there is a high ratio of direct beam to diffuse radiation so that the incident solar radiation is strongly directional. Plant breeding programs have often removed this characteristic from crop plants. Some species with pinnate leaves show nyctinasty in which the leaves fold up at night thus reducing the effective canopy cover. Plant species from more than 20 families exhibit either heliotropism or nyctinasty. **Figure 2** shows an unusual feature characteristic of *Dryobalanops lanceolata*; the crowns of individual trees do not overlap resulting in a fractured canopy. The reasons for this are unclear.

Plant species often show plastic responses to changes in the environment that are mediated through changes in source and sink strength. The functional balance theory hypothesizes that there is a balance between the capacity of the root system to capture water and nutrients, and the size of the photosynthetic apparatus with feedbacks that lead the plant along a path toward optimization of resource use. Laboratory experiments in which the sizes of the root system and leaf canopy were

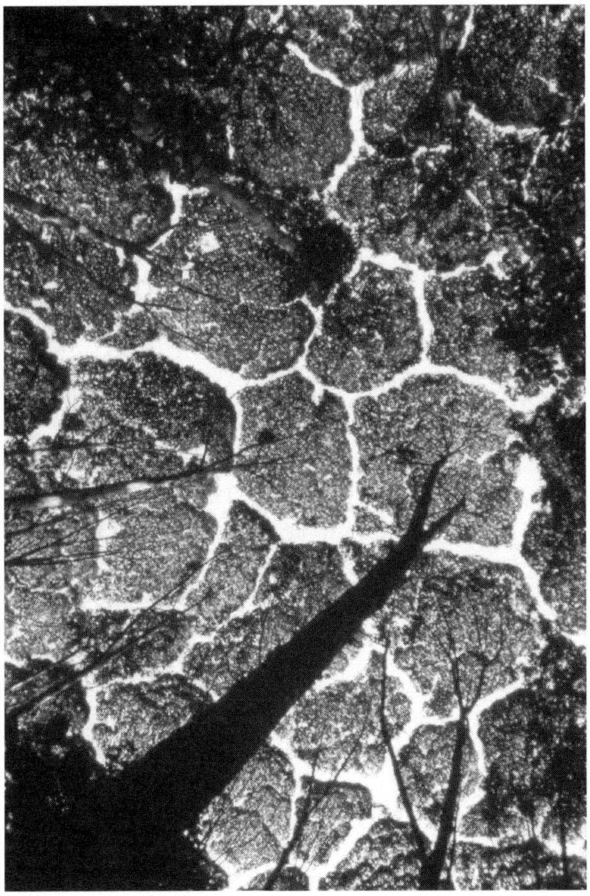

Figure 2 (see color plate 18) View of the unusual canopy of the dipterocarp *Dryobalanops lanceolata* characterized by gaps between the foliage of the individual trees. Photograph kindly supplied by Dr P. van Gardingen with permission.

manipulated have produced data that agree with this theory. However, the environment of field grown plants is much less predictable, and there would be an evolutionary advantage in plants evolving insurance strategies, such as the laying down of storage compounds rather than maximizing the use of the scarcest resource.

Environmental Factors Affecting Canopies

The availability of water can be a key factor affecting canopy size. Where the rate of water supply is less than the rate of potential transpiration, increasing water stress can lead to stomatal closure, a reduction or cessation of leaf growth, and even the loss of existing leaves, depending on the species. Abscisic acid produced in response to soil drying is implicated in these processes.

In many environments, canopy size is limited by the availability of nitrogen in the soil. As most of the nitrate is in the upper part of the soil profile, which dries first, what appears to be a water limitation may in fact turn out to be a nitrogen one. The concentration of nitrogen in leaves depends on its rate of uptake and has a consequent effect on leaf optical properties and photosynthetic rate. However, leaf survival depends on the nitrogen concentration remaining above a critical threshold. In environments characterized by low levels of available nitrogen, including those where the temperature is sufficient for leaf growth but not for organic matter mineralization, the production of new leaves depends on the recycling of nitrogen from the old leaves and their subsequent death. In old growth forests, much of the nitrogen is retained within the canopy in this way, and as the canopy is no longer increasing in size, the demand from the soil is low.

Considerable research has been carried out into the effect of light on canopies and vice versa. Leaves in parts of the canopy exposed to full light often differ in their morphology and biochemical functioning from those in shade. A canopy can thus be considered as two populations of leaves: one of sun leaves and another of shade leaves. As light penetrates the canopy it becomes weaker and its spectral quality changes. In dense canopies, or where daily solar irradiation is low, there will be a critical depth in the canopy at which the light level, when averaged over a period of a few days, becomes too low to sustain the leaf, which senesces and ultimately dies. It was formerly thought that leaves in deep shade were parasitic on the plant, but we now know that this is not the case and that their biochemistry changes to reduce respiratory losses. As green leaves preferentially absorb in the red and blue parts of the visible spectrum, the ratio of red to far red light decreases toward the ground surface. This change in red : far red ratio is thought to be an important mechanism by which plants sense the proximity of their neighbors. Weeds in agricultural crops also perceive this spectral shift, which results in responses such as an increase in internode length. The signal is perceived by the phytochrome system.

Other factors can also affect canopy development. For example, the soil pH affects the ability of plants to take up nutrients, and elements such as phosphorus may be in short supply. Plant species vary in their ability to take up water and nutrients through their leaves due to differences in leaf anatomy. Canopy architecture affects the number of leaves hit by a raindrop as it passes down through the canopy and thus the possibility of water and dissolved nutrients being captured and taken up by lower leaves. It is thought that acacias, corns and possibly even European beech possess canopies that facilitate this process. However, these factors are normally less significant than the ones mentioned above. Of more importance are biotic impacts, such as infestation by phytophagous insects and attack by pathogens, both of which can affect the survival of individual leaves or impair their functioning. Pathogens, in particular, can be influenced by temperature and humidity conditions within the foliage that are more favorable to their colonization and spread than those in the open. Dense canopies in which leaves of one plant frequently touch those of another also provide good conditions for the dispersal of insects and the spread of infection.

Canopy Dynamics

The dynamics of canopy growth, as represented by the changes in LAI with time, are due to a number of factors, some related to the plants themselves, and others to the environmental effects mentioned above. Canopies can be considered as a population of leaves with changes being explained in demographic terms as difference between the recruitment of new leaves and the death of old ones. Leaf recruitment is linked to the plants' phenological calendar. In annual crops, new leaves are generally continually produced up to the time of flowering or even beyond, while in perennials, such as forest trees, leaf initiation and appearance may be restricted to one or more short periods of the year. Given that there is a minimum irradiance at which leaves can survive, it follows that there is a maximum sustainable LAI, which will be highest toward the summer solstice, and beyond which recruitment of new leaves has to be balanced by the death of the old ones. Leaf death can be due simply to old age but can be hastened by adverse environmental conditions as explained above.

Canopy Management

Many agricultural and forestry operations are designed to manipulate the canopy to increase growth rate, decrease competition, or increase the efficiency with which an input such as nitrogen fertilizer is used.

To achieve the potential growth rate, the canopy needs to intercept as much of the incident photosynthetically active radiation as possible. A key goal in this respect is to reduce the loss of intercepted radiation early in the season when the canopy is incompletely formed. Early canopy development can be encouraged by ensuring adequate nitrogen is available and that there is an appropriate plant population density. The latter depends on the seed rate and the effectiveness of establishment. On the other hand, excessive nitrogen produces a dense canopy resulting in an unnecessarily large investment of photosynthate in the foliage at the expense of roots and stem, thin walled cells in the leaves, and a humid environment within the canopy. The latter features are particularly conducive to the spread of pathogens. The concept of canopy management has been developed for cereals as a way of optimizing the external inputs, such as fertilizer and pesticides. Target values of LAI are identified for particular stages of development, and situations and applications of nitrogenous fertilizer are increased or reduced depending on the actual LAI. In environments characterized by a shortage of water, particularly in the later stages of development of a crop as in a Mediterranean climate, the optimum canopy size for maximum productivity may be considerably less than in a humid environment. However, there have to be clear gaps in the canopy before the canopy water loss declines significantly.

In forest plantations, canopies have been modified for centuries by thinning, i.e., the selective removal of trees to increase the space for the others. Many decades of research have provided empirical justification for the appropriate thinning regimes in particular circumstances. Agricultural crops such as brassicas and beets used to be established by sowing an excessive number of seeds and then thinning to a predetermined optimum population. However, mechanization has led to the general cessation of this practice, at least in developed agriculture.

Canopies are also modified to increase their competitiveness against weeds. A careful choice of sowing date, seed rate, and method of establishment can reduce the risk of gaps in the canopy that allow weeds to become established and outcompete the crop. Competition for light may only be the proximate cause of weed success as effective uptake of nutrients by weeds may increase their ability to develop a canopy at the expense of the crop. Trials have also been carried out on crop cultivars that are better able to compete with weeds because their leaf angle distribution is more horizontal, thus casting more shade on the ground and germinating weeds.

Modeling Canopies

Simple models of light interception have been produced for more than 40 years. However, a major advance was made with the development of architectural models of individual plants by Lindenmeyer (L-system models). These models, which are based on the concepts of fractal geometry, consider plants as a series of modules from which new modules bud according to a set of simple rules. Subsequent developments have enabled populations of virtual plants to be grown, the light environment to be modeled, and the interactions between foliage elements and light to be characterized by another set of rules. Realistic representations of canopy development can be obtained in this way.

Research Areas

Research on canopies peaked some time ago and resulted in the development of methods and procedures for analyzing canopy structure and function in a wide variety of situations. On the other hand, interest in sustainable farming systems has led to research into canopy management as a means of optimizing input use. However, the exciting prospect of using information about a plant's genome to parameterize an architectural model of crop growth so that yield can be investigated under a wide range of climatic and management scenarios is still a long way off.

List of Technical Nomenclature

Diffuse radiation	The component of solar radiation that is scattered by molecules and particles in the air.
Direct beam	The component of solar radiation that comes directly from the sun.
Eddy flux correlation	A micrometerological method of estimating fluxes of gases from the three-dimensional movement of air parcels and their gaseous composition.
Gap fraction	The proportion of sky not obscured by foliage looking up from below a plant stand.
Heliotropism	The growth of a plant toward the sun, e.g., expressed as solar tracking by leaves of flowers.
Irradiance	The amount of solar energy received per unit time per unit area; expressed as $W m^{-2}$.
Leaf area density	The leaf area of a plant stand divided by the volume of space occupied.
Leaf area index	The ratio of the leaf surface area to the ground area occupied by a plant stand.
Light extinction coefficient	A constant that categorizes the exponential attenuation of light with depth in a homogeneous medium.
Normalized difference vegetation index	An index of ground surface greenness used in remote sensing that makes use of the relative brightness of the surface in the red and near infrared bands.
Phenology	The study of the timing of developmental stages, such as flowering, in organisms.
Sunfleck	A beam of light passing through a gap in the foliage to illuminate a small patch of ground; its position will change with the relative movement of earth and sun.
Sun leaves	Leaves that have developed in full sunlight and can be distinguished anatomically and physiologically from shade leaves, being thicker and with higher rates of photosynthesis at saturating irradiances.

See also: **Growth and Development**: Field Crops; Leaf Development. **Photosynthesis and Partitioning**: Sources and Sinks. **Production Systems and Agronomy**: Plantation Crops and Plantations. **Regulators of Growth**: Phytochromes and other Photoreceptors. **Root Development**: Root Growth and Development. **Weeds**: Weed Competition.

Further Reading

Hallé F, Oldemann RAA, and Tomlinson PB (1978) *Tropical Trees and Forests*. Berlin: Springer Verlag.

Lindenmeyer A and Prusinkiewicz (1988) Developmental models of multicellular organisms: a computer graphics perspective. In: Langton CG (ed.) *Artificial Life*, pp. 221–249. Redwood City: Addison Wesley.

Marshall B and Roberts JA (2000) *Leaf Development and Canopy Growth*. Sheffield: Sheffield Academic Press.

Moffett MW (2000) What's "up"? A critical look at the basic terms of canopy biology. *Biotropica* 32: 569–596.

Russell G, Marshall B, and Jarvis PG (1989) *Plant Canopies: Their Growth Form and Function*. Society for Experimental Biology Seminar Series, vol. 31. Cambridge: Cambridge University Press.

Growth Analysis, Individual Plants

R Hunt, University of Sheffield, Sheffield, UK

Introduction

The term plant growth analysis refers to a useful set of quantitative methods that describe and interpret the performance of whole plant systems grown under natural, seminatural, or controlled conditions. Plant growth analysis provides an explanatory, holistic, and integrative approach to interpreting plant form and function. It uses simple primary data such as weights, areas, volumes, and contents of plants or plant components to investigate processes within and involving the whole plant or crops. Originating at the end of the nineteenth century, plant growth analysis first illuminated plant physiology, then agronomy and, most recently, physiological and evolutionary plant ecology.

In plant growth analysis, there is an historical and practical distinction between the organism (and below) and the population (and above). These two areas demand different approaches, though each is similar in concept and consistent in application. The two are treated separately elsewhere (*see* **Growth and Development:** Growth Analysis, Crops), so this article deals only with the analysis of whole plants grown as spaced individuals.

"Growth" in the context of the individual plant means an irreversible change with time. Such changes are mainly in size (however measured), often in form, and occasionally in number. For example, classical experiments have shown that in both annual and perennial plants raised under productive conditions, growth initially follows the same typical course (**Figure 1**). Because the changes in total dry weight over these periods vary many-fold, very little of the earliest phases of development can be seen. When the data are transformed to logarithms, however, several benefits are gained (**Figure 2**). A visual benefit of the logarithmic transformation is that all the successive phases of growth series become equally clear; a statistical benefit is that variability between replicates is homogenized across the whole series, and a mathematical benefit is that a first step is taken toward calculations of rates of growth or efficiency of function, many of which require logarithmic transformation of primary data in order to be practical.

In the logarithmic plots, a zero or negative initial phase of growth is followed by a "grand period" of near-exponential (log-linear) increase and then by an approach to a plateau in the plant's maturity, with senescence and decay following. This pattern is quite general to annual plants grown in a productive environment, though there is great variation in the magnitude of the dry weights attained, in the symmetry of the curve, and in the time-scale covered. In perennial plants, the pattern is similar at first, but later, at least in a temperate climate, dry weight increase proceeds in a series of annual steps, which may be separated by periods of negative growth. Environmental conditions affect the magnitude of growth at all stages.

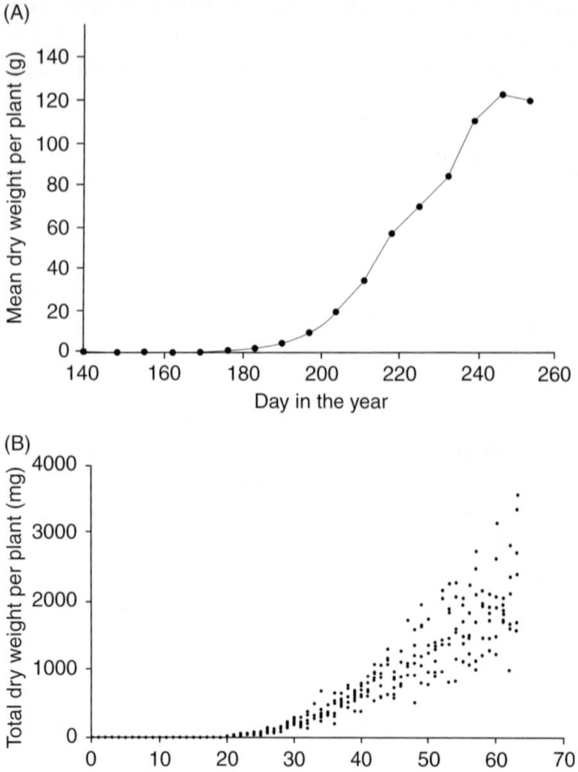

Figure 1 The accumulation with time of total dry weight per plant: (A) in corn, showing mean values each derived from 40 to 120 individuals harvested weekly from an outdoor environment. Reproduced from Kreusler U, Prehn A, and Hornberger R (1879) Beobachtungen über das Wachsthum der Maispflanze (Bericht über die Versuche vom Jahre 1878). *Landwirtschaftliche Jahrbücher* 8: 617. (B) in the grass *Holcus lanatus*, showing individual plants harvested daily from a productive, controlled environment.

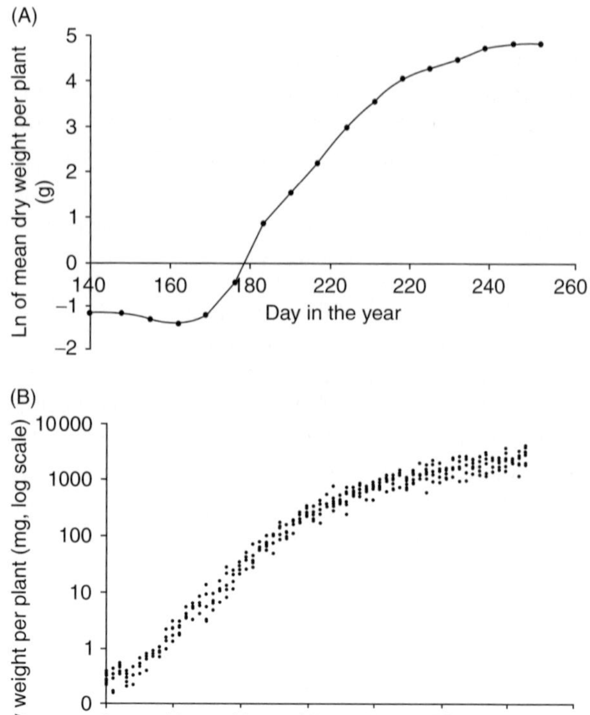

Figure 2 The same data as in **Figure 1**, but with weights plotted on a logarithmic scale.

Plant growth analysis addresses data such as these by means of powerful comparative tools. These aim to negate, as far as possible, the inherent differences in scale that can exist between different organisms or phases of growth. For example, rates of growth in the near-exponential phase can be below 10% per day in tree seedlings, even under very favorable conditions, but values rise much higher as we move in turn through herbaceous plants, algae, fungi, and microorganisms, to as much as 20 000% per day in an anti-*Escherichia coli* phage. The larger, older, or more complex the organism, the lower the maximum rate of dry weight increase that is normally possible. Plant growth analysis aims to compare such performance like for like.

Basic Technical Concepts

Classical and Functional Approaches

Two distinct approaches to the growth analysis of individual plants have evolved. In the so-called "classical" approach, events are followed through a series of relatively infrequent, large, often destructive, harvests (with much replication of measurements). In the so-called "functional" approach, the individual harvests supply data for curve-fitting, but they are each smaller (having less replication of measurements) and often much more frequent. The two approaches are not mutually exclusive if time and space are no object (harvests may be large and frequent), but in most cases, the experimenter is forced to choose one or other in advance, as this influences the design of the experiment.

Where to Start?

The stock in trade of plant growth analysis is a collection of simple primary data, the measured quantities upon which the subsequent analyses depend. These data may be determined either for the whole plant or for different components such as roots, stems, and leaves, as required. Naturally, all of the analytical and statistical techniques should be chosen before practical work is begun. In the growth analysis of individual plants, the primary data are used to calculate values of one or more of four distinct types of derivate.

Absolute Growth Rates

These are simple rates of change involving only one plant variate and time, examples being the whole plant's rate of dry weight increase, or the rate of increase in number of roots per plant.

Relative Growth Rates

These are more complex rates of change, but still involving only one plant variate and time, an example being the whole plant's rate of dry weight increase per unit of dry weight (for example, the percentage rate of growth mentioned above).

Simple Ratios

These involve ratios between two quantities, and may either be ratios (or more strictly quotients) between two like quantities, such as total leaf dry weight and whole plant dry weight, or ratios between two unlike quantities, such as total leaf area and whole plant dry weight.

Compounded Growth Rates

These are rates of change involving more than one plant variate, such as the whole plant's rate of dry weight increase per unit of its leaf area.

Basis of Calculation

It is usual to define all of the derived quantities and their interrelationships instantaneously: that is, as they stand at a single point in time. Though this provides a mathematically precise specification, it was for many years necessary to base the estimates upon mean values over a stated time interval, using the "classical" approach. (To appreciate the difference between the two, note that in the context of a car journey, the instantaneous and mean velocities are obtained from momentary speedometer readings, on the one hand, and from whole-journey calculations involving total times and distances, on the other.) Most of the instantaneous values can be determined only as mathematical derivatives from curves fitted in a "functional" approach, a method that was not generally practical until the 1960s.

Only instantaneous values may properly be represented as single points on a progression plotted against time; mean values should appear as a histogram with one column for each harvest interval. **Table 1** provides a synopsis of instantaneous and mean (harvest-interval) definitions of the four principal types of derivate.

Absolute Growth Rates

Absolute Growth Rate in Size

This is the simplest index of plant growth: a rate of change in size, which is an increment in size per unit of time. Most commonly if W is the total dry weight of the plant, then G is its absolute growth rate in total dry weight. The dimensions and units of G are

Table 1 Formulae defining instantaneous and mean values of each type of derivate (the formula for the mean value is approximate except in the case of absolute growth rate)

Type of derivate	*Instantaneous definition*	*Formula for mean value over the time interval t_1 to t_2*
Absolute growth rate	dY/dt	$(Y_2 - Y_1)/(t_2 - t_1)$
Relative growth rate	$(1/Y)(dY/dt)$	$(\ln Y_2 - \ln Y_1)/(t_2 - t_1)$
Simple ratio	Y/Z	$((Y_1/Z_1) + (Y_2/Z_2))/2$
Compounded growth rate	$(1/Z)(dY/dt)$	$((Y_2 - Y_1)/(t_2 - t_1)) \times ((\ln Z_2 - \ln Z_1)/(Z_2 - Z_1))$

mass per time, e.g., g day^{-1}. Instantaneously, $G = dW/dt$, and its mean value over the interval t_1 to t_2 is $(W_2 - W_1)/(t_2 - t_1)$. The instantaneous G can be derived from functions fitted to W versus t; thus if $W = f_W(t)$, then $G = f_W'(t)$. Mean values are obtained from the separate destructive estimates W_1 and W_2 made at times t_1 and t_2, respectively.

The mean values of absolute growth rate in the corn (*Zea mays*; maize) series (**Figure 1**) are shown in **Figure 3**; the greatest weekly increases occur relatively late in the series.

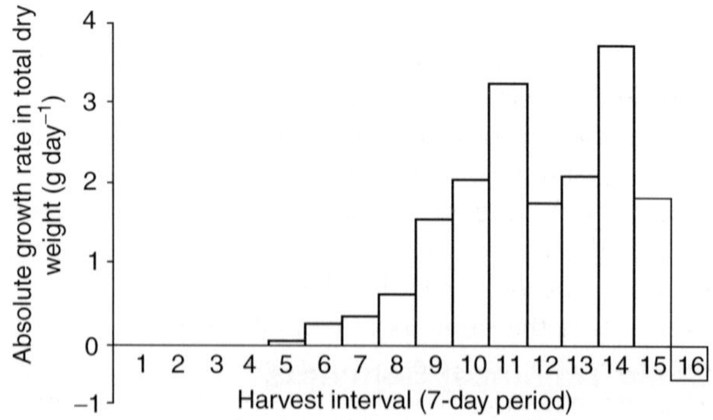

Figure 3 The progress of mean absolute growth rate in total dry weight in outdoor grown corn (see **Figures 1A** and **2A**).

Absolute Growth Rate in Number

This is the simplest index of growth in number; a rate of change in number, which is an increment in number per unit time. In individual plants, its use is confined to describing growth in numbers of discrete organs, such as leaves or roots. The definitions and calculations are the same as above, but the primary data differ, with the number of plant organs N being used in place of W.

Figure 4 illustrates growth in flower number in cotton (*Gossypium* spp.) crops; varietal and seasonal differences are discernible.

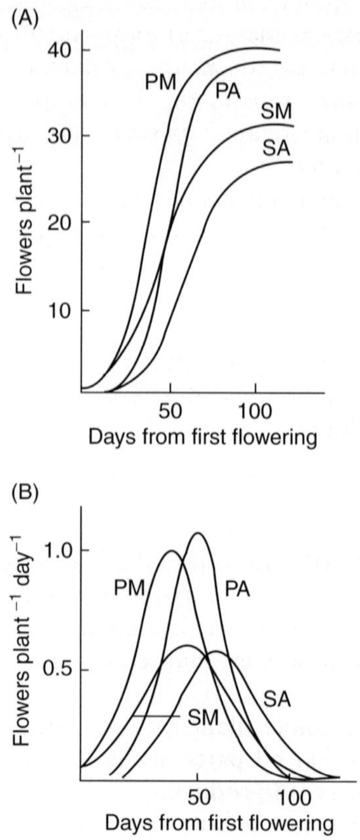

Figure 4 Flowering in Egyptian cotton: (A) logistic growth curves fitted to flower number per plant in varieties P (Pilion) and S (Sakellaridis) sown in M (March) or A (April); (B) instantaneous absolute growth rates in flower number, the slopes of the curves shown in (A). Reproduced with permission from Prescott JA (1922) The flowering curve of the Egyptian cotton plant. *Annals of Botany* 36: 121–130.

Relevance

The absolute growth rates are the simplest possible measures of plant growth rate. They can be valuable comparative tools when they are used within like bodies of data (as in **Figure 4**). When used to compare unlike systems, however, their usefulness declines. To compare the overall performance in such circumstances requires other approaches. For example, if two different species are grown for equal periods of time and both put on equal amounts of dry weight, they will both exhibit the same absolute growth rate even if the species differed in initial dry weight. Some measure of growth is needed that also takes into account this original difference in size. That measure is relative growth rate.

Relative Growth Rate

Explanations and Examples

This was originally termed an "efficiency index" because it expresses growth in terms of a rate of increase in size per unit of size. As such, it permits more equitable comparisons between organisms than does absolute growth rate. Normally, relative growth rate deals with total dry weight per plant, though other measures of size have also been used. In the financial world, relative growth rate is analogous to the rate of compound interest earned on capital. Negative relative growth rates are called relative decay rates.

The relative growth rate, R, is the rate of increase of total dry weight per plant, W, expressed per unit of W. Its dimensions are mass per mass per time, typically in units of $g\,g^{-1}\,day^{-1}$ or $g\,g^{-1}\,week^{-1}$ (if both weights appear in identical units they can be cancelled out); per cent per time may also be used. Instantaneously, $R=(1/W)(dW/dt)$. In calculus notation, this is the same as $d(\ln W)/dt$. The mean value over the interval t_1 to t_2 is $(\ln W_2-\ln W_1)/(t_2-t_1)$. The instantaneous R is derived from functions fitted to $\ln W$ versus t; thus, if $\ln W=f_W(t)$, then $R=f_W'(t)$. Mean values are obtained from the separate destructive estimates W_1 and W_2 made at times t_1 and t_2, respectively.

Figure 5 shows the drift in mean relative growth rate in a classical analysis of the corn series and, from the functional approach, the drift in instantaneous relative growth rate in the *Holcus* series. In each case, maximum relative growth rate occurs sooner rather than later.

Screening experiments have shown that maximum mean relative growth rate varies approximately tenfold between different temperate herbaceous species, but only twofold between different geographical populations of the same species, or between different genotypes within one population ("genets"), or between different clones within one genotype ("ramets").

Relevance

Relative growth rate is useful wherever current size realistically controls current increase in size. It provides a convenient integration of the many component processes that contribute to the perform-
form-
ance of the whole plant, but it depends upon the assumption that all parts of the relevant "size" are equally capable of producing further amounts of the same quantity (in the same way that invested capital grows through the accrual of compound interest). However, as most plants grow, the proportion of their mass that is largely supporting material (i.e., not directly productive) increases, for much the same reason that larger animals develop proportionally more bulky bones than smaller ones. So, relative growth rate soon declines with time (**Figure** 5), and the interest then passes to the components of relative growth rate in the hope of explaining how this decline comes about.

Simple Ratios

Leaf Area Ratio

This is a morphological index describing the leafiness of the plant. A measure of the "balance of payments" between income and expenditure, it deals with the potentially photosynthesizing and the potentially respiring components of the plant. It is the ratio, F, between total leaf area per plant, L_A, and total dry weight per plant, W. Its dimensions are area per mass; typical units are $mm^2\,mg^{-1}$ or $m^2\,g^{-1}$. Instantaneously, $F=L_A/W$ and its mean value over the interval t_1 to t_2 is $((L_{A1}/W_1)+(L_{A2}/W_2))/2$, which is, of course, $(F_1+F_2)/2$. The instantaneous values can be derived from functions fitted to $\ln L_A$ and $\ln W$ versus t; thus, if $\ln L_A=f_L(t)$ and $\ln W=f_W(t)$, then $F=L_A/W=\exp(f_L(t)-f_W(t))$. Unsmoothed instantaneous values are also available directly from the defining formula.

Mean leaf area ratio in Kreusler's corn series is shown in **Figure 6**. The relatively early position of the maximum in these values is common among herbaceous species.

Specific Leaf Area

This is an index of the "leafiness of the leaf": a measure of density or of relative thinness, which involves an assessment of the leaf's area in relation to its dry weight. No particular symbol is in general use. Leaf area ratio is the ratio between total leaf area per plant L_A and total leaf dry weight per plant L_W. (The term "specific" means divided by weight.) The dimensions are area per mass, typically $mm^2\,mg^{-1}$ or $m^2\,g^{-1}$. Instantaneously defined as L_A/L_W, the mean value over the interval t_1 to t_2 is $((L_{A1}/L_{W1})+(L_{A2}/L_{W2}))/2$. From functions fitted to $\ln L_A$ and $\ln L_W$ versus t, $f_A(t)$ and $f_W(t)$, the instantaneous L_A/L_W is derived as $\exp(f_A(t)-f_W(t))$. Unsmoothed instantaneous values are also available directly from the defining formula.

Specific leaf areas in three successive crops of glasshouse-grown winter lettuce (*Lactuca sativa*) are shown in **Figure** 7. The differences between the three curves are environmentally driven, principally by the level of photosynthetically active radiation.

Leaf Weight Ratio

This is an index of leafiness of the plant on a dry weight basis: a measure of the "productive

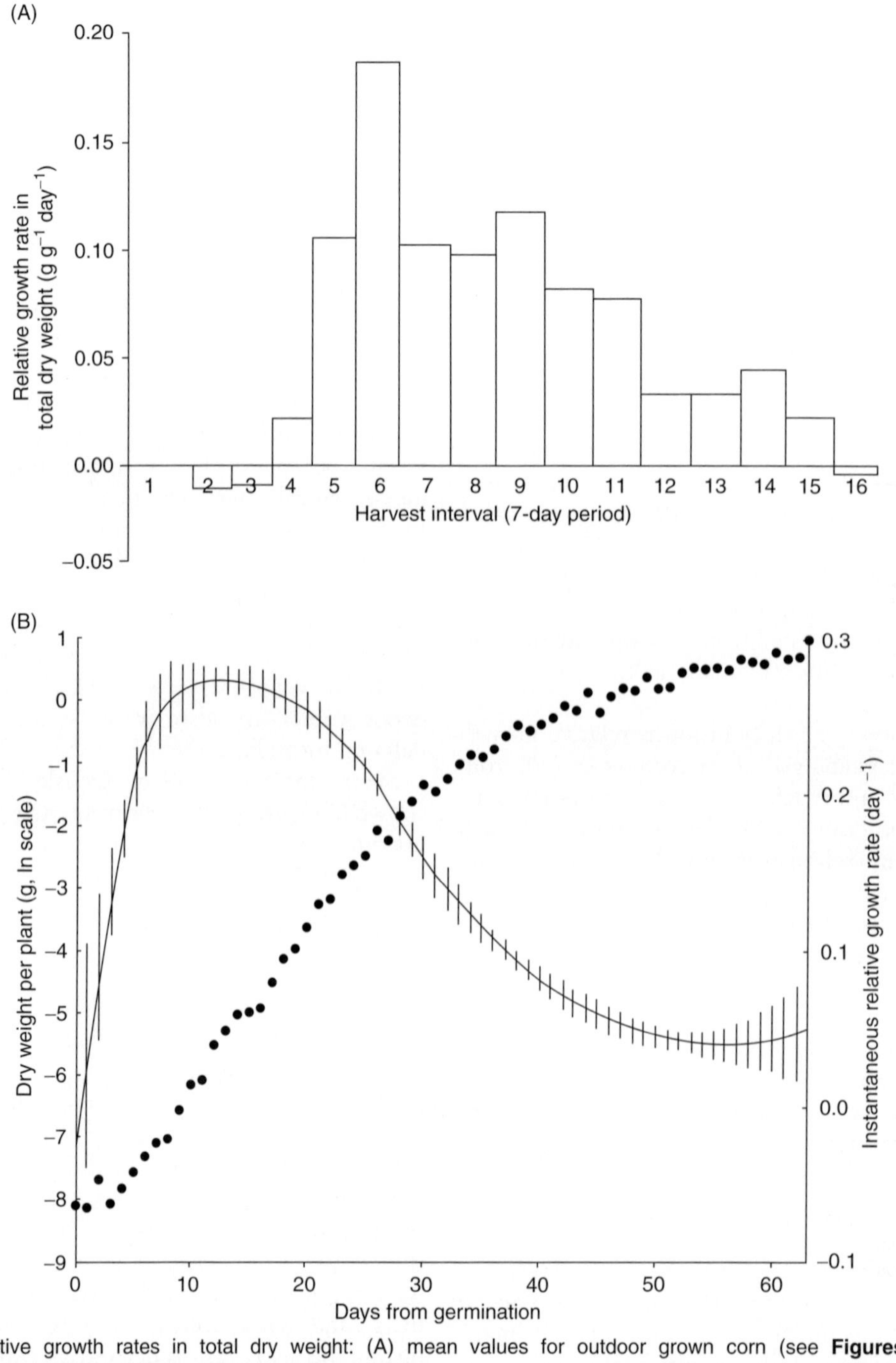

Figure 5 Relative growth rates in total dry weight: (A) mean values for outdoor grown corn (see **Figures 1A** and **2A**); (B) instantaneous values, with 95% confidence limits and harvest means of total dry weight, for *Holcus lanatus* in a productive, controlled environment (see **Figures 1B** and **2B**).

investment" of the plant, dealing with the relative expenditure on potentially photosynthesizing organs. No particular symbol is in general use. The leaf weight ratio is the ratio between total leaf dry weight per plant, L_W, and total dry weight per plant, W. It is more strictly a leaf weight fraction, but the term "ratio" is widely used. Its dimensions are mass per mass, which is effectively dimensionless. Numerically, the values are $0<x<1$. Instantaneously, L_W/W, the mean value over the interval t_1 to t_2 is given by $((L_{W1}/W_1)+(L_{W2}/W_2))/2$. The instantaneous values can be derived from functions fitted to $\ln L_W$ and $\ln W$, $f_L(t)$ and $f_W(t)$, so that $L_W/W = \exp(f_L(t) - f_W(t))$. Unsmoothed instantaneous values are also available directly from the defining formula.

Table 2 shows that species and varietal differences play an important part in determining leaf weight ratio.

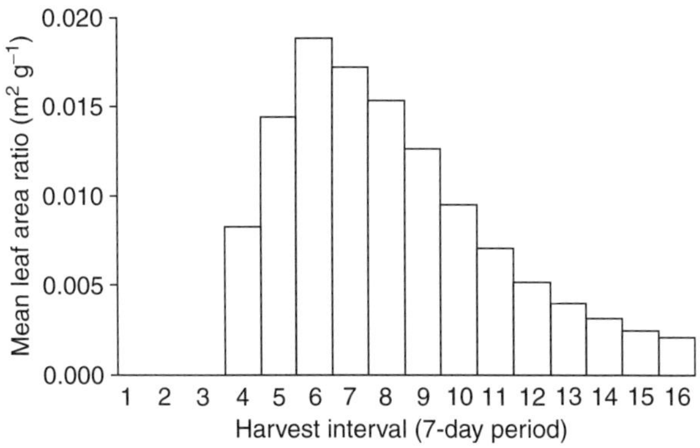

Figure 6 The progress of mean leaf area ratio in outdoor grown corn (see **Figures 1A** and **2A**).

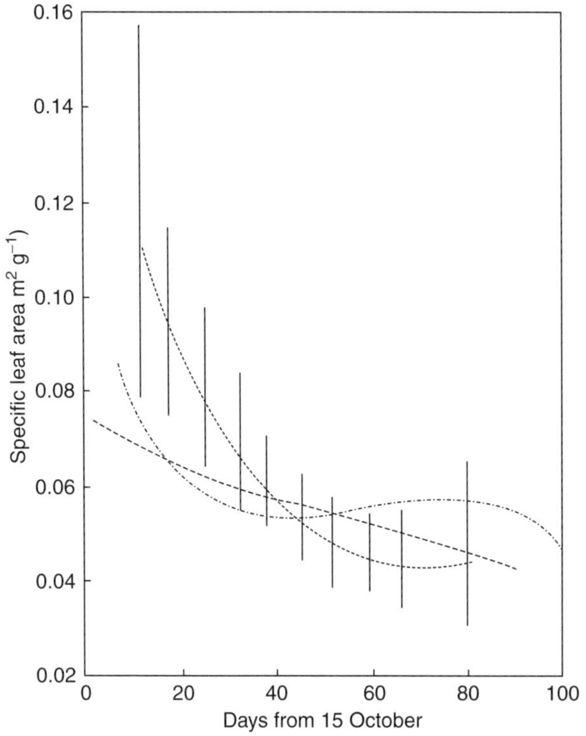

Figure 7 The progress of instantaneous specific leaf area in winter lettuce grown in a glasshouse in three successive seasons (95% confidence limits are added to the curve for the earliest year). Reproduced with permission from Hunt R, Warren Wilson J, Hand DW, and Sweeney DG (1984) Integrated analysis of growth and light interception in winter lettuce. I. Analytical methods and environmental influences. *Annals of Botany* 54: 743–757.

Root–Shoot Allometric Coefficient

This is an index of the balance of growth between root and shoot components of the plant integrated over a period of time. Effectively, a ratio between root and shoot mean relative growth rates (q.v.). The usual symbol is K; and the allometric relationship is defined as $R_W = bS_W{}^K$, where b is a constant. The allometric coefficient is dimensionless within the range $0 < K < \infty$. Equally balanced growth above and below ground gives a K value of unity. In "shooty" growth $K < 1$ and in "rooty" growth $K > 1$. Because $R_W = bS_W{}^K$, then $\log R_W = \log b + K \log S_W$ and $K = (\log R_W - \log b)/\log S_W$. Which base of logarithms is used is not important.

In practice, the coefficient K is derived from a series of paired measurements of R_W and S_W. The function $\log R_W = f(\log S_W)$ is fitted. Usually this is a linear regression of the form $\log R_W = \log b + K \log S_W$, but a bivariate principal axis should be used wherever possible because the x-variate (S_W) cannot be determined without error.

The allometric coefficient is highly sensitive to environmental conditions and generally exhibits a partitioning of material toward the more resource-deficient part of the plant (**Table 3**).

Table 2 Leaf weight ratio in ryegrass (*Lolium* spp.) and clover (*Trifolium* spp.): 16 cultivars examined over a 3-week period (95% confidence intervals in parentheses)

Species	*Cultivar*	*Leaf weight ratio*
Lolium perenne	S23	0.505 (0.023)
	S24	0.500 (0.017)
	Melle	0.485 (0.022)
	Pelo	0.512 (0.010)
	Standion	0.512 (0.017)
	Endura	0.497 (0.024)
Lolium multiflorum	Tiara	0.547 (0.014)
Lolium perenne × *Lolium multiflorum*	Sabrina	0.534 (0.029)
Trifolium repens	S100	0.452 (0.036)
	Hiua	0.458 (0.011)
	S184	0.483 (0.018)
	Blanca	0.471 (0.019)
	Kersey	0.444 (0.032)
	Pajbjerg	0.457 (0.039)
Trifolium pratense	Tetri	0.467 (0.016)
Trifolium hybridum	"British"	0.467 (0.015)

Source: Elias CO and Chadwick MJ (1979) Growth characteristics of grass and legume cultivars and their potential for land reclamation. *Journal of Applied Ecology* 16: 534–537.

Table 3 Root–shoot allometry in perennial ryegrass (*Lolium* spp.): values of *K*, the allometric coefficient, for the vegetative growth of perennial ryegrass under different conditions (95% confidence intervals in parentheses)

	Full light	*Shade*
Full nutrients	0.84 (0.04)	0.69 (0.10)
Low nitrogen	0.95 (0.04)	0.73 (0.10)

Source: Hunt R (1975) Further observations on root–shoot equilibria in perennial ryegrass. *Annals of Botany* 39: 744–757.

Relevance

Mathematically, these ratios are alone capable of instantaneous evaluation without recourse to fitted growth curves. However, if the functional approach is followed, the component ratio Z/Y should always be evaluated from the separate functions $\ln Z = f_Z(t)$ and $\ln Y = f_Y(t)$, rather than directly from $Z/Y = f_{Z/Y}(t)$, which can have difficult statistical properties. Where Z and Y are like quantities, their ratio is simply an index of allocation; where they are unlike quantities, the ratio is a "snapshot" of the functional balance between two related or antagonistic components. In the case of the allometric coefficient, there is the added advantage that it often remains approximately constant across substantial intervals of time, which increases its value as a comparative tool.

Compounded Growth Rates

Unit Leaf Rate

This is an index of the productive efficiency of plants calculated in relation to total leaf area. It is synonymous with the term net assimilation rate. The usual symbol is E: the rate of dry weight production expressed per unit of total leaf area, L_A. Its dimensions are mass per area per time, typically mg mm^{-2} day^{-1} or g m^{-2} day^{-1}. Instantaneously, $E = (1/L_A)(dW/dt)$. The mean value over the interval t_1 to t_2 is approximately $(W_2 - W_1)/(t_2 - t_1) \times (\ln L_{A2} - \ln L_{A1})/(L_{A2} - L_{A1})$. The unit leaf rate can be obtained instantaneously from functions fitted to $\ln W$ and $\ln L_A$ versus t, $f_W(t)$, and $f_L(t)$, such that $E = (1/L_A)(dW/dt) = f_W'(t) \times \exp(f_W(t) - f_L(t))$. Mean values are obtained from the formula given above, using the separate estimates (W_1, L_{A1}) and (W_2, L_{A2}) from harvests at t_1 and t_2, respectively.

Mean unit leaf rate in the corn series (**Figure 8A**) exhibits a relatively constant general level modified by fluctuations in the weather during the growing season. In the seedling phases of herbaceous and woody species grown in a productive, controlled environment (**Figure 8B**), there are two- to threefold variations between species within a functional group and also moderate overall differences between the groups themselves.

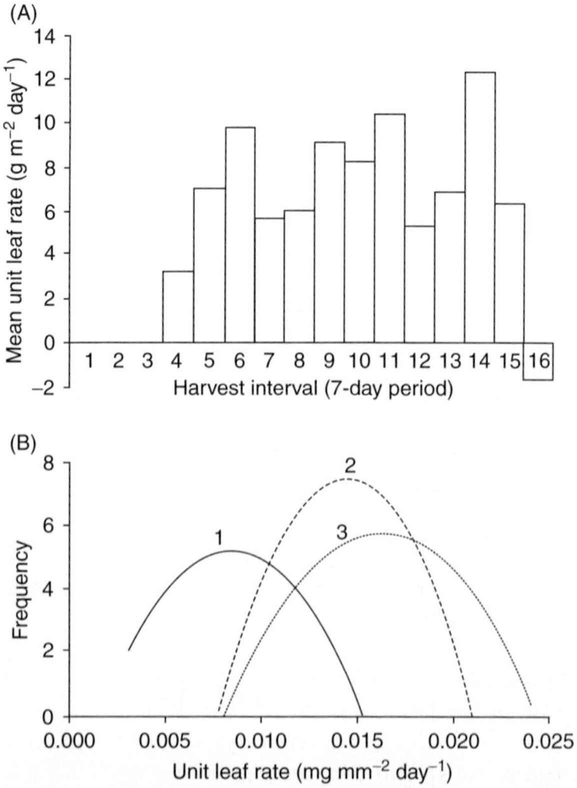

Figure 8 Unit leaf rates: (A) mean values for outdoor grown corn (see **Figures 1A** and **2A**); (B) smoothed frequency distributions for mean values obtained for (1) woody dicotyledons ($n = 16$), (2) herbaceous dicotyledons ($n = 22$), and (3) herbaceous monocotyledons ($n = 21$), all grown as seedlings in a productive, controlled environment. Reproduced with permission from Hunt R and Cornelissen JHC (1997) Components of relative growth rate and their interrelations in 59 temperate plant species. *New Phytologist* 135: 395–417.

Specific Absorption Rate

This is an index of the uptake efficiency of roots, calculated in relation to some measure of root size. The usual symbol is A, defined as the rate of uptake of mineral nutrient, M, expressed per unit of total root size, which may be root dry weight, R_W, or, alternatively, root length, area, volume, or number. The dimensions are mass per mass per time and typical units are μg mg^{-1} day^{-1} or mg g^{-1} day^{-1}. Instantaneously, $A = (1/R_X)(dM/dt)$, where R_X is some measure of root size. The mean value over the interval t_1 to t_2 is approximately $(M_2 - M_1)/(t_2 - t_1) \times (\ln R_{X2} - \ln R_{X1})/(R_{X2} - R_{X1})$. Term M may be the combined total content of more than one mineral nutrient element. Instantaneously, A is obtained from functions fitted to $\ln M$ and to $\ln R_X$ versus t; if $\ln M = f_M(t)$ and $\ln R_X = f_{RX}(t)$, then $A = (1/R_X)(dM/dt) = f_M'(t) \times \exp(f_M(t) - f_{RX}(t))$.

Mean values are obtained from the formula given above, using the separate estimates (M_1, R_{X1}) and (M_2, R_{X2}) from times t_1 and t_2, respectively.

Figure 9 shows curves for specific absorption rate in cultivated cranberry (*Vaccinium macrocarpon*); mycorrhizal infection increases nitrogen uptake but soil irradiation does not.

Relevance

These are rates of production of something per unit of something else. In plant growth analysis, provided that the "something" Y is of interest to the experimenter and that the "something else" Z may reasonably be held responsible for its production, then $(1/Z)(dY/dt)$ is an analytical tool of fundamental

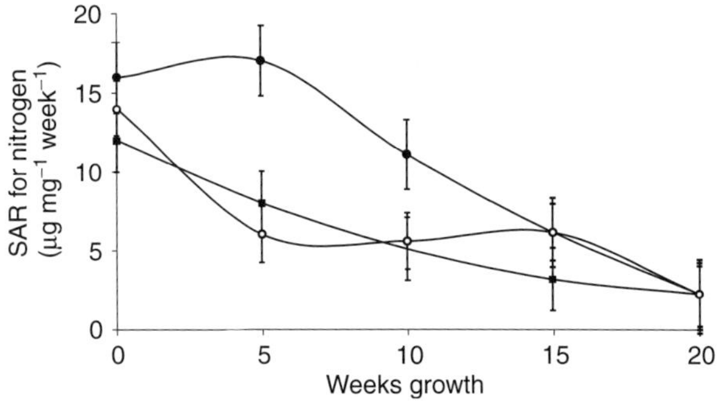

Figure 9 The progress of instantaneous specific absorption rate (SAR) for nitrogen in cultivated cranberry grown in a controlled environment: ■, mycorrhizal plants grown on poor soil; •, mycorrhizal plants grown on the irradiated soil; ○, nonmycorrhizal plants grown on the irradiated soil (95% confidence limits are added). Reproduced with permission from Stribley DP, Read OJ, and Hunt R (1975) The biology of mycorrhiza in the Ericaceae. V. The effects of mycorrhizal infection, soil type and partial soil-sterilization (by gamma-irradiation) on growth of cranberry (*Vaccinium macrocarpon* Ait.). *New Phytologist* 75: 119–130.

importance. As such, it has been used to describe plant processes from the molecular level, through organs, and whole herbaceous plants up to perennial woody crops. There is no reason why such applications should not flourish still further.

Interrelations

In General

Although the terms involved in plant growth analysis have the individual meanings described above, their strength as analytical tools owes much to their interrelations with one another, or the ways in which an individual term may be decomposed into others. Various types of interrelation can occur, but they all have the status of mathematical identities, not conditional equations. (A mathematical identity states a logical truth, not an hypothesis that is open to disproof.)

Simple

Where both parts of a simple ratio or fraction bear the same units, they provide an index of the importance of one component of the plant in relation to another. All such components can be linked into a single scheme. As a simple example, in young grasses if R_W/W is the root weight ratio (where R_W is the total root dry weight of the plant), S_W/W is the stem weight ratio (where S_W is the total stem dry weight of the plant), and L_W/W is the leaf weight ratio (where L_W is the total leaf dry weight of the plant), then the three are related by the expression $R_W/W + S_W/W + L_W/W = 1$.

Interrelations between more heterogeneous quantities can also occur. For example, a subdivision of leaf area ratio is $L_A/W = L_A/L_W \times L_W/W$, where L_A/L_W is the specific leaf area and L_W/W is the leaf weight ratio. By looking simultaneously at all three of these terms it is possible, for example, to establish that the much less leafy nature of Scots pine (*Pinus sylvestris*), in comparison with sunflower (*Helianthus* spp.), is due almost entirely to the relatively greater density of the pine needles and hardly at all to variation in leaf weight ratio (the productive investment of the plant), which, in fact, shows a small difference in favor of pine.

More Complex

It is often useful to subdivide an index of overall performance, such as relative growth rate, into indices that represent the individual performances of components of the system. In fact, unit leaf rate and leaf area ratio originally evolved as subdivisions of relative growth rate. So, it is by definition that $(1/W)(\mathrm{d}W/\mathrm{d}t) = (1/L_A)(\mathrm{d}W/\mathrm{d}t) \times L_A/W$. Simply expressed, the growth rate of the plant depends simultaneously upon the efficiency of its leaves as producers of new material and upon the leafiness of the plant itself. (Except in very special circumstances, this relation holds only approximately for the three mean values of these quantities; instantaneous values are needed for the interrelation to be exact.) Also, as $L_A/W = L_A/L_W \times L_W/W$, these subdivisions of leaf area ratio may be inserted into the equation for relative growth rate to give $(1/W)(\mathrm{d}W/\mathrm{d}t) = (1/L_A)(\mathrm{d}W/\mathrm{d}t) \times L_A/L_W \times L_W/W$ otherwise, relative growth rate expressed as the product of unit leaf rate, specific leaf area, and leaf weight ratio.

Tools for Performing the Calculations

Classical Approach

The instantaneous mathematical definitions of the various terms are usually not amenable to direct substitution of experimental data. That is why the harvest interval mean formulae were developed. When using these formulae, however, if there has been any replication of the measurements that require logarithmic transformation and it is wished to work with harvest mean values of these variables, it is important to calculate these as (for example) mean(ln W) and not ln(mean W).

The statistical properties of the harvest interval means are important and often neglected. It is incorrect simply to calculate, say, the variance of a group of several unit leaf rates spanning the same harvest interval. Not only does this involve difficult decisions as to how to "pair" the primary data across the harvest interval, but it also ignores the fact that the statistical properties of a derived term depend upon those of its primary data and not upon those of its sibling values. Spreadsheet tools are available within the growth analytical literature that will perform these calculations correctly.

Functional Approach

This involves curve fitting. The great advantages are that the instantaneously defined terms can be obtained directly from the fitted curves, seen in their exact interrelation to one another, and provided with statistics derived only from the primary data. The disadvantage is that an appropriate type of fitted curve has to be selected and applied, without under-fitting (forcing the data into too simple a strait-jacket) or over-fitting (chasing off after outliers, which should really be eliminated or smoothed). The growth analytical literature contains tools for fitting low-order polynomials, nonlinear asymptotic functions with up to four parameters, and splined curves (smoothly joined polynomials) which offer almost unlimited flexibility.

List of Technical Nomenclature

A	Specific absorption rate.
b	Intercept coefficient in allometry.
E	Unit leaf rate (net assimilation rate).
F	Leaf area ratio.
f	Suitable mathematical function.
f′	Slope of function *f*.
G	Absolute growth rate.
K	Allometric coefficient.
L_A	Total leaf area per plant.
L_W	Total leaf dry weight per plant.
M	Total mineral nutrient content per plant.
N	Integer number of individual organs.
R	Relative growth rate.
R_W	Total root dry weight per plant.
R_X	Root size, generalized.
S_W	Total shoot dry weight per plant or total stem dry weight per plant (according to context).
t	Time.
W	Total dry weight per plant.
x	Generalized real number.
Y	Generalized plant variate.
Z	Generalized plant variate.

See also: **Growth and Development**: Growth Analysis, Crops.

Further Reading

Causton DR (1983) *A Biologist's Basic Mathematics.* Edward Arnold: London.

Causton DR and Venus JC (1981) *The Biometry of Plant Growth.* London: Edward Arnold.

Evans GC (1972) *The Quantitative Analysis of Plant Growth.* Oxford: Blackwell Scientific Publications.

Hunt R (1978) *Plant Growth Analysis.* London: Edward Arnold.

Hunt R (1982) *Plant Growth Analysis: Statistical Checklist No 4.* Cambridge: Institute of Terrestrial Ecology.

Hunt R (1982) *Plant Growth Curves: the Functional Approach to Plant Growth Analysis.* London: Edward Arnold.

Hunt R (1990) *Basic Growth Analysis: Plant Growth Analysis for Beginners.* London: Unwin Hyman.

Hunt R and Cornelissen JHC (1997) Components of relative growth rate and their interrelations in 59 temperate plant species. *New Phytologist* 135: 395–417.

Hunt R and Evans GC (1980) Classical data on the growth of maize: curve-fitting with statistical analysis. *New Phytologist* 86: 155–180.

Hunt R and Lloyd PS (1987) Growth and Partitioning. In: Rorison IH, Grime JP, Hunt R, Hendry GAF, and Lewis DH (eds) *Frontiers of Comparative Plant Ecology,* pp. 235–249. London: Academic Press.

Hunt R, Causton DR, Shipley B, and Askew AP (2002) A modern tool for classical plant growth analysis. *Annals of Botany* 90: 485–488.

Mead R and Curnow RN (1983) *Statistical Methods in Agriculture and Experimental Biology.* London: Chapman and Hall.

Sattler R (1986) *Biophilosophy: Analytic and Holistic Perspectives.* Berlin: Springer-Verlag.

Growth Analysis, Crops

L R Benjamin, Rothamsted Research, Harpenden, UK

Introduction

The cardinal concepts of plant growth analysis (*see* **Growth and Development**: Growth Analysis, Individual Plants) are those of:

- rates of growth (such as absolute or relative growth rate);

- simple ratios (such as leaf weight to plant dry weight);
- compounded growth rates (such as increase in dry weight per unit leaf area);
- tools for performing growth analysis (such as fitting growth curves to weight data).

In this article, these growth analysis concepts and methods are applied to the performance of entire crops. In particular, growth analysis is applied to provide insights to the interactions between individuals that come to make up the crop community.

Plant Density

Plant density is a simple yet critically important variable that links the individuals to crops. Plant density is simply the number of individuals per unit ground area. In many circumstances, the identification of the individual is obvious. The unit of the population can clearly be identified as the individual plant in a crop of, for example, sunflowers (*Helianthus annuus*) or carrots (*Daucus carota*). All plant species display plasticity in their morphology and modular growth. Often this causes difficulties in identifying an appropriate unit of the population. For example, in moist conditions, many plants with a prostrate habit produce runners, which root at various locations. Each rooting position can be regarded as a new individual, or the whole assemblage derived from a single seed may be regarded as an individual. Tillering in grasses and cereals can also cause similar difficulties in deciding upon the unit of the population. Similarly, in potatoes (*Solanum tuberosum*), the number of stems arising from a seed tuber rather than the number of seed tubers sown can be a more practical measure of plant density.

There is no single clear-cut answer to what the unit of the population should be, but it is important to be aware that this difficulty can arise and there is a need for consistency in choice of the unit of the population.

Leaf Area Index

Leaf area is important in determining plant growth rate (*see* **Growth and Development**: Growth Analysis, Individual Plants). For crops, a new term is introduced called leaf area index (LAI). The leaf area index is the ratio of the leaf area of all the plants to the ground area they occupy. Hence, if L_A is the total leaf area in an area of ground, P, then:

$$\mathrm{LAI} = L_A/P \qquad [1]$$

Note also that if $\bar{L}_A$ is the mean leaf area per plant, and n is the plant density, then also

$$\mathrm{LAI} = \bar{L}_A \cdot n \qquad [2]$$

Leaf area index is an important physiological measure that determines light interception by the crop and the resulting crop growth rate. It has been proposed that there is an exponential decrease in light intensity with increasing distance from the top of the crop canopy, and that the slope of the decrease is dependent on the accumulative leaf area index from the top of the canopy. Hence, the amount of light intercepted, S, by the entire canopy is given by:

$$S = S_0 \cdot [1 - \exp(-k \cdot LAI)] \qquad [3]$$

where S_0 is the light intensity at the top of the canopy and k is the extinction coefficient of foliage for light. The value of k typically varies from 0.5 to 0.8, and is dependent upon the orientation of the leaves, and direction of the light. The interception of direct and diffuse light down the entire depth of the canopy and over an entire day is the basis of many complex process-based physiological models.

Crop Growth Rate

In growth analysis, leaf area index is also used as a component of crop growth rate, C. The unit leaf rate, E, is defined as the increase in plant weight per unit leaf area (*see* **Growth and Development**: Growth Analysis, Individual Plants). Hence

$$E = \frac{1}{L_A} \cdot \frac{\mathrm{d}W}{\mathrm{d}t} \qquad [4]$$

where W is total plant dry weight and t is time. If there is a single plant in an area of ground, P, it follows that

$$\mathrm{C} = \frac{1}{P} \cdot \frac{\mathrm{d}W}{\mathrm{d}t} = \frac{1}{L_A}\frac{\mathrm{d}W}{\mathrm{d}t} \cdot \frac{L_A}{P} \qquad [5]$$

Hence

$$\mathrm{C} = E \cdot \mathrm{LAI} \qquad [6]$$

The mathematical expansion of a term, as in eqn [5], is called decomposition and helps to clarify how a complex process can be broken down into less complex processes. This decomposition of crop growth rate into two components, E and LAI, can be interpreted as crop growth being the product of the efficiency of the leaves to intercept light and the leafiness of the crop. An analogous decomposition of relative growth into unit leaf rate and the leaf area ratio can be described (*see* **Growth and Development:**

Growth Analysis, Individual Plants). It should also be possible to predict the crop growth rate by measuring each of its components. In practice, this is cumbersome, and a more pragmatic approach is to measure directly the differences in sum of total plant weights per unit ground area, Y, between two dates, $t_2 - t_1$.

Hence, at time t, in a unit area of ground containing a density of n plants

$$Y_t = \sum_{i=1}^{i=n} W_{t.i} \qquad [7]$$

The $\sum$ sign in eqn [7] means sum all the values of W observed at time t from plant 1 to plant n. Mean crop growth rate, $\bar{C}$, can be calculated as

$$\bar{C} = \frac{Y_2 - Y_1}{t_2 - t_1} \qquad [8]$$

Methods for estimating growth rates based on the classical and/or functional approaches can be applied to observations of crop yield to estimate crop growth rates (*see* **Growth and Development:** Growth Analysis, Individual Plants). Such application, however, has to be done with caution. For individual plants, the weight versus time graph is usually sigmoidal, which can be described by high-order polynomials or by the Richards function. Change in crop yield with time, however, is often not sigmoidal, but consists of an initial exponential phase, which gradually changes to a linear phase of growth (**Figure 1**). At the very start of plant growth, zero or negative growth rates can occur. This can be followed by a short phase of increasing relative growth rates. Increasing relative growth rate with time is called hyperexponential growth. The periods of negative and then hyperexponential growth are not normally considered in crop growth analysis as they are of such small duration compared with the other periods of growth.

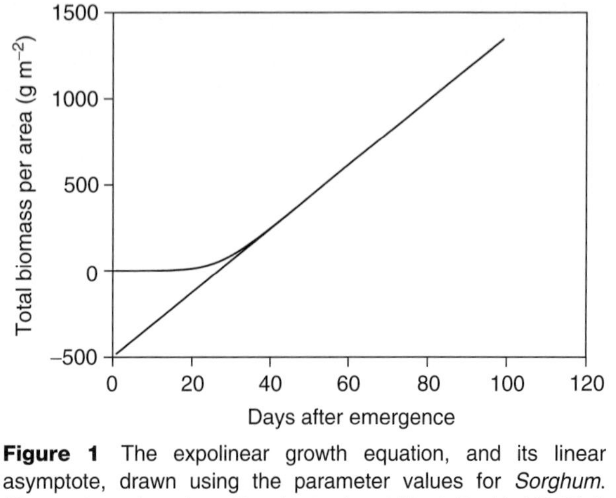

Figure 1 The expolinear growth equation, and its linear asymptote, drawn using the parameter values for *Sorghum*. (Parameter values from Goudriaan J and Monteith J L (1990) A mathematical function for crop growth based on light interception and leaf area expansion. *Annals of Botany* 66: 695–701.)

The feature of exponential growth slowly transforming to linear growth is further reinforced by the development of crops. Initially, crops consist of a collection of widely spaced seedlings. Each seedling efficiently converts any light it intercepts into dry matter, and its individual relative growth rate is high. Because the seedlings are inevitably small compared with their spacing, a large proportion of the light incident on the crop falls onto bare ground. Therefore, despite the high individual relative growth rates, the crop absolute growth rate is low. The seedlings rapidly fill the interplant gaps and form a closed canopy. Up to canopy closure, the crop growth rate increases approximately exponentially. After canopy closure, the lateral spread of individual plants is constrained by the presence of neighboring plants. During and after canopy closure, the interplant competition will cause dominant plants to continue growth rapidly at the expense of their suppressed neighbors. The overall effect for the crop is for nearly all the incident radiation each day to be converted to dry matter. Consequently, after canopy closure crop yield increases linearly.

An expolinear equation has been developed to describe the changes in crop yield, Y, with time, t. Hence

$$Y_t = \frac{C_m}{R_m} \cdot log_e\{1 + \exp[R_m(t - t_b)]\} \qquad [9]$$

where C_m, R_m, and t_b are constants. C_m is the maximum crop growth rate when the foliage intercepts all the light and in the UK, is typically 0.2 t dry matter ha^{-1}day^{-1}. R_m is the maximum relative growth rate of the crop and will be identical to the values of plant maximum relative growth rate. t_b is a form of shape parameter, that determines the position of the curve on the time axis. t_b is a measure of the time taken from sowing to canopy closure. In this period the crop fails to intercept a large proportion of the incident radiation and is influenced by the seedling size and density of the crop.

The expolinear equation has been developed further using a model which relates leaf area and crown zone area of individual plants to their weight. This model simulated the transition of growth from isolated plants to that of a crop by applying a set of rules for the contest for space by individual plants. Leaf area index of each plant plays an important role in determining its success in occupying space at the expense of its neighbors during the period of canopy closure.

Dynamics of Leaf Production

Clearly, light interception is important in determining crop growth rates, and the two key factors affecting light interception are unit leaf rate and leaf area index. Different crops have different patterns of leaf area index development with time of year. In **Figure 2**, the wheat (*Triticum aestivum*) crops have a marked peak in leaf area index, whereas the sugar beet (*Beta vulgaris*) has a much broader peak of leaf area index against day of the year.

The area under the curve relating leaf area index to day of the year is an important determinant of light interception during the entire growth of the crop; this is called integral leaf area duration, LAD. Leaf area duration will determine crop yield, providing that unit leaf rate is constant.

In addition to the leaf area duration, the timing of leaf area development should ideally be targeted to coincide with the annual periods of maximum solar radiation. The leaf production by wheat closely coincides with the annual variation in solar radiation, but by sowing in the autumn rather than the spring, the leaf area development is maximized to coincide with maximum solar radiation (**Figure 2**). The more effective deployment of leaf area in autumn-drilled wheat provides an explanation for the greater yield for autumn over spring-drilled wheat. Indeed, currently the greater part of the wheat crop in the UK is drilled in the autumn.

Sugar beet displays an interesting contrast to wheat. It is sown in the spring but its development of leaf area is much delayed compared with spring wheat. Sugar beet achieves a large LAI, but at about 100 days after the peak daily solar radiation values (**Figure 2**). Earlier sowing of sugar beet would synchronize radiation and leaf area deployment, but as the sowing date is brought forward, there is an increasing proportion of plants that bolt and flower. The relationship between crop growth rate, leaf area index, and unit leaf rate can be addressed by removing different fractions of the leaf area in crops of kale (*Brassica oleracea* var. acephala) and sugar beet. The leaf area index that gave the maximum crop growth rate was about 3 to 4 in kale and between 6 and 9 in sugar beet. Work that is more recent has altered LAI by manipulating fertilizer rates and has suggested an optimal LAI in the range 2.5 to 4.5.

The data in **Figure 2**, which are derived from growth analysis, provide explanations for the factors that determine crop yield, and allow strategies for crop improvement to be more readily identified.

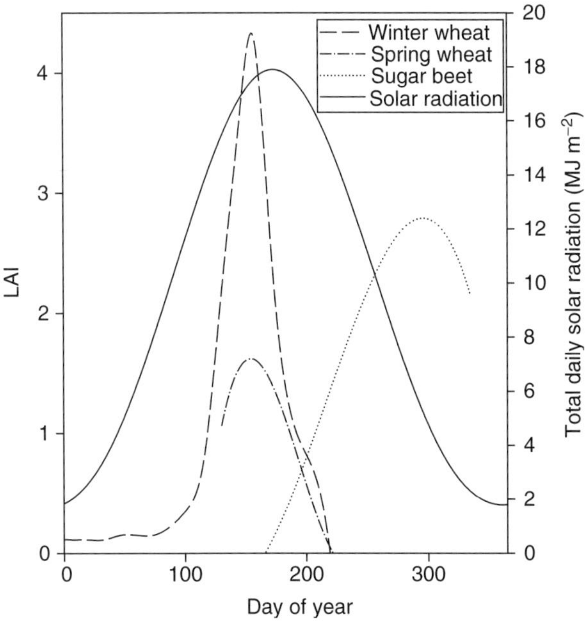

Figure 2 Changes in leaf area index (LAI) of spring and autumn sown wheat and sugar beet at 52° N in the UK and for the change in average daily solar radiation at 52° N with day of the year. LAI values redrawn from Watson DJ (1947) Comparative physiological studies on the growth of field crops. 1. Variation in net assimilation rate and leaf area between species and varieties and within and between years. *Annals of Botany* 11: 41–76. The solar radiation values are based on Hamer PJC (1999) An algorithm to provide UK global radiation for use with models. *Computers and Electronics in Agriculture* 22: 41–49.

Variability in Relative Growth Rate

Crops are an assemblage of individuals and there is much current interest in assessing the role of variation in relative growth rate in determining the performance of crops.

Seedling Relative Growth Rate Pre-Competition

There have been only few studies on relative growth rate of seedlings prior to emergence. Indeed, dry weight decreases from imbibition to emergence.

Detailed growth measurements have been made on carrot and sugar beet during and immediately after the period from germination to emergence. Seed bed factors that extend the period from germination to emergence reduce relative growth rate postemergence. The physiological basis of the reduction in relative growth rate with delayed emergence can be investigated by decomposition of the relative growth rate

$$R = \frac{1}{W} \cdot \frac{dW}{dt} = \frac{1}{L_A} \frac{dW}{dt} \cdot \frac{L_A}{W} \qquad [10]$$

that is

$$R = E \cdot F \qquad [11]$$

where E is unit leaf rate and F is leaf area ratio.

Leaf area ratio can be further decomposed to specific leaf area (SLA) and leaf weight ratio (LWR).

$$F = \frac{L_A}{W} = \frac{L_A}{L_w} \cdot \frac{L_w}{W} \qquad [12]$$

The reduction in postemergence relative growth rate with longer duration of preemergence growth is strongly associated with reductions in unit leaf rate (**Figure 3A**). The seedlings that had suffered a longer duration of preemergence growth invested only a low proportion of their dry matter in cotyledon weight (**Figure 3C**). This may be because most of the dry matter seed reserves had been consumed in the energy required for emergence. The plants attempted to compensate by producing higher specific leaf area (**Figure 3B**). Despite this compensation, the reductions in leaf weight ratio damaged the photosynthetic capacity of the cotyledons, expressing itself as a reduction in unit leaf rate. A linear tradeoff between leaf weight ratio and specific leaf area implies that leaf area ratio could not be increased to compensate for the reduction in unit leaf rate (**Figure 3D**).

Deep placement of the seeds in the soil and the development of soil surface crusts are typical factors that extend the period of heterotrophic growth in the dark. Hence, agronomic factors such as variability in drill performance and weather conditions can cause within-crop variation in seedling growth rates. This initial variation can be the basis for further variability in the growth of individual plants within the crop.

Plant Relative Growth Rate – The Phase of Competition

Following seedling emergence, further variation in relative growth rates is induced by competitive interactions between neighboring plants. Increasing plant density increases the intensity of the competitive interactions between plants. The role of competition in generating variability in relative growth rate within crops can be determined by measuring the change in coefficient of variation or Gini coefficient with increasing density. In the vast majority of experiments in which plant density had been varied and variability in plant weight recorded, the observation was that variation increased with density. This supports the view that competition magnifies initial variation in plant weight.

Experimental manipulation of the initial variability of populations produced evidence that variation in relative growth rate induced by competition may cause a doubling of the weight variability in crops. Carrot seedling populations were thinned to either

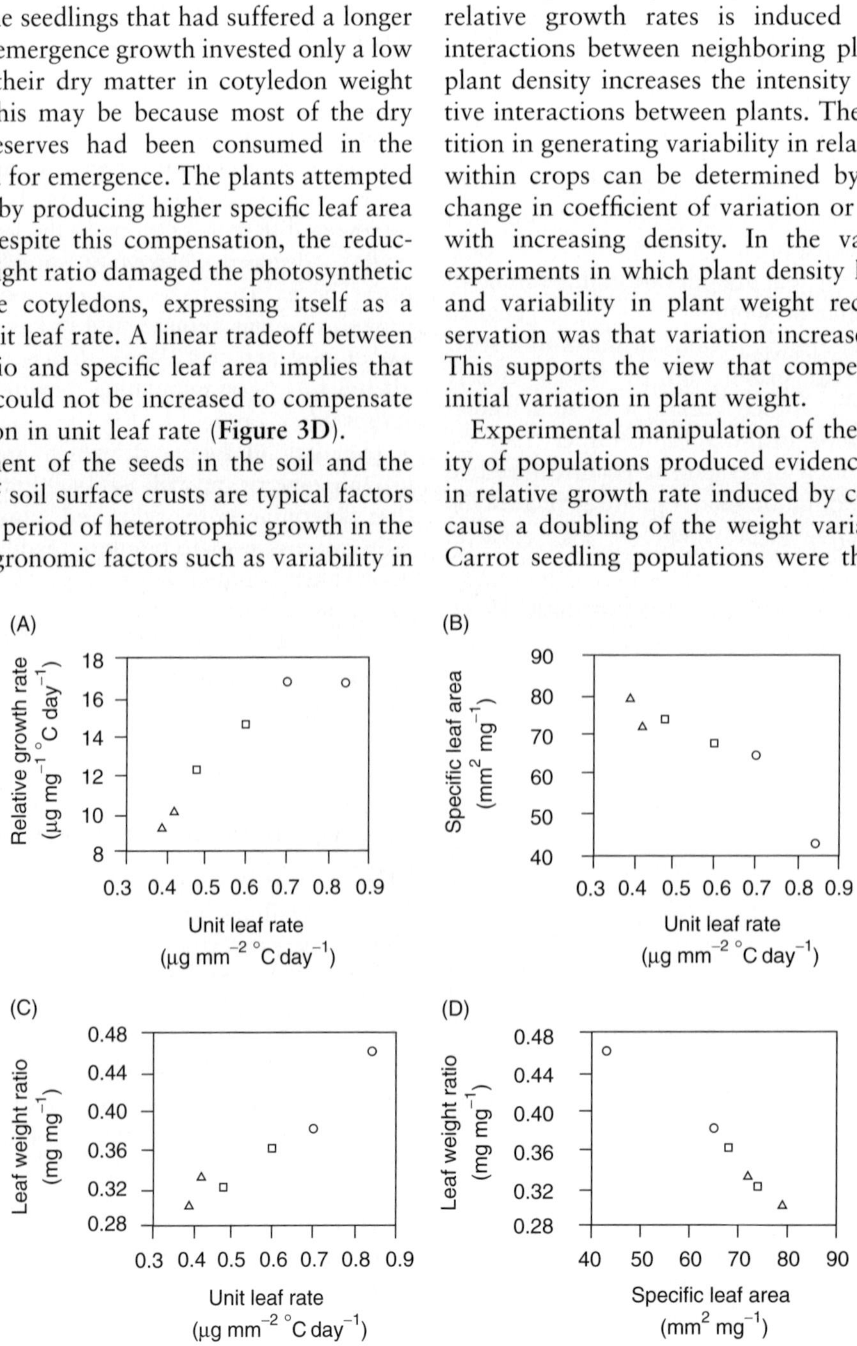

Figure 3 Relation between growth analysis parameters for carrot seedlings from emergence to first leaf production; (A) Relative growth rate and unit leaf rate; (B) Specific leaf area and unit leaf rate; (C) Leaf weight ratio and unit leaf rate; (D) Leaf weight ratio and specific leaf area. Time to emergence in day degrees above a base 3.5°C; (○) 65–70; (□) 100–110; (△) 130–150. Values from Tamet V, Boiffin J, Durr C, and Souty N (1996) Emergence and early growth of an epigeal seedling (*Daucus carota* L.): influence of soil temperature, sowing depth, soil crusting and seed weight. *Soil and Tillage Research* 40: 25–38.

25 or 400 plants m^{-2}, and at each density, populations were established which varied in the extent of seedling weight variation. This was achieved by retaining seedlings that had emerged synchronously or asynchronously. At 25 plants m^{-2}, the variation in carrot weight 4 months after sowing increased slightly with increase in seedling weight variation (**Figure 4**). In contrast, at 400 plants m^{-2} the variability of carrot root weight increased sharply with increase in variability in seedling weight. Competitive interactions had magnified the initial precompetitive differences in relative growth rate of the populations and these had marked impacts on the variability in size of plant through to a stage when they would be harvested commercially. In the UK, commercial densities of carrots are in the approximate range of 100 to 250 plants m^{-2}, but higher densities are common for the production of smaller roots. Hence, in most commercial production of this crop, variability in relative growth rates will play an important role in generating commercially undesirable variation in the harvested crop.

Competition for light is considered to increase the relative growth rate of taller (and probably heavier) individuals, whilst suppressing the relative growth rate of shorter plants. Such competition is called one-sided or asymmetric. In contrast, where competition is for belowground resources, it is not possible for roots of one plant to "over-top" those of a neighbor. Hence, competition for belowground resources reduces the mean relative growth rate but does not increase the variation in relative growth rates. This type of competition is called two-sided or symmetric.

This view has recently been challenged. The reasoning behind the challenge arises from decomposition of relative growth rate into three terms:

$$R = \frac{1}{W} \cdot \frac{dW}{dt} = \frac{1}{W_X}\frac{dU_X}{dt} \cdot \frac{W_X}{W} \cdot \left(\frac{dW}{dt} \bigg/ \frac{dU_X}{dt}\right) \qquad [13]$$

where W is total plant dry weight and W_X is the dry weight associated with the uptake of resource, X, and dU_X/dt is the rate of uptake of this resource. The first term, $(1/W_X \cdot dU_X/dt)$ is the relative rate of resource uptake, the second term (W_X/W) is a biomass partitioning factor, and the third term $(dW/dt/dU_X/dt)$ is the resource to biomass conversion efficiency. The first term can be further decomposed to

$$\frac{1}{W_X} \cdot \frac{dU_X}{dt} = \frac{1}{V_X}\frac{dI_X}{dt} \cdot \frac{V_X}{W_X} \cdot \left(\frac{dU_X}{dt} \bigg/ \frac{dI_X}{dt}\right) \qquad [14]$$

where V_X is the space occupied by W_X and dI_X/dt is the rate of resource supply. The first term, $(1/V_X \cdot dI_X/dt)$ is a resource availability factor, the second term (V_X/W_X) is an allometry (that is shape) factor, and the third term $(dU_X/dt/dI_X/dt)$ is the relative rate of resource uptake.

Equations [13] and [14] were applied to the observations of leaf area, ground cover, and dry weight of paired millet (*Panicum miliaceum*) plants at contrasting densities. Although the heavier of the plants in a pair intercepted a greater proportion of the incident light per unit ground area, they occupied less ground area relative to their size than smaller plants. Consequently, the heavier plants initially did not have a competitive advantage over lighter plants. Although competition for light was occurring, the competition was symmetric. Later in the crop's development, there was a greater difference in height between heavier and light plants, and at high density this height difference outweighed the taller plant's allometric disadvantage. Consequently, at high

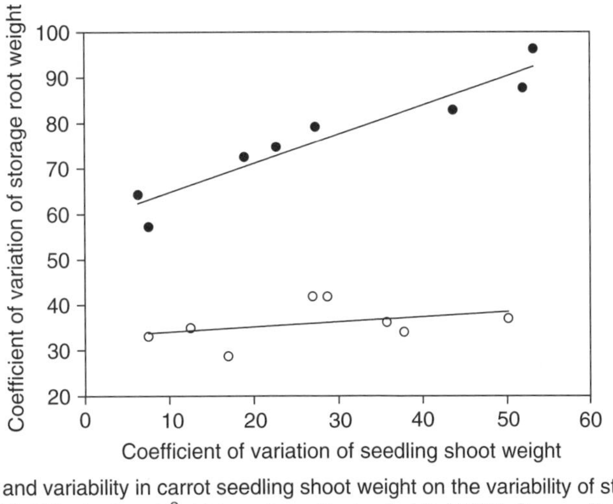

Figure 4 Effect of plant density and variability in carrot seedling shoot weight on the variability of storage root weight 4 months after sowing. (○) 25 plants m^{-2} and (•) 400 plants m^{-2}. Values from Benjamin LR (1982) Some effects of differing times of seedling emergence, population density and seed size on root-size variation in carrot populations. *Journal of Agricultural Science* 98: 537–545.

density later in the crop's development, the competition had become asymmetric, but at low density competition remained symmetric. Hence, competition for light does not always cause a divergence in relative growth rates between large and small plants.

It has also been argued that competition for belowground resources could be asymmetric. The soil medium is a source of several resources necessary for plant growth. Soils are also notoriously spatially heterogeneous in the availability of these resources. Over the range of an individual plant's root system there may be one or more patches enriched in one resource and impoverished in another. Several patches that differ widely in the ratios of resource availability may occur within the confines of the root system of a large plant. The large plant root system may be able to integrate this spatially heterogeneous resource availability, but this integration may be impossible for the root system of a small plant. There is no evidence yet collected to demonstrate divergence in growth rates due to competition for belowground resources, but it should not be excluded as a factor in determining the variability in plant size in crops.

By applying a rigorous growth analysis approach, it is possible to infer the mechanisms of competition between individuals with crops. These mechanisms of competition determine the range of relative growth rates within a community. A log normal distribution in weights will develop in a crop if a growth rate at each time step is subject to plant-to-plant variability and if many sources of variability are combined by multiplying a proportionate error. This log normal distribution will develop irrespective of the shape of the initial frequency distribution. An alternative approach has been taken by considering how changes in the shape of frequency distributions are generated by time-invariant responses of growth rate to starting weight. The relation between starting size and increase in size is called a "distribution modifying function." The distribution modifying function concept has been taken further by calculating it for any one time as the product of the mean growth rate, the variance in growth rate, and the mortality rate.

Variability in size at harvest, which results from variation in growth rates, has implications for the calculation of growth analysis statistics. Tools have been developed for the classical and functional approaches of growth analysis to handle correctly the statistical complications in deriving the growth rate and ratio statistics from variable data.

Variability in weight is also of great economic importance in most crops. Most horticultural crops and many arable crops have to meet strict size limits to be marketable. Increased variation in size causes yield loss and increases the costs of grading and packaging. Variation in size can also cause yield loss even in crops that do not have to meet marketable size criteria. This can occur if some threshold size has to be reached before a developmental process essential for marketable production (such as flowering or storage organ initiation) occurs. Excessive vegetative growth in cereals can result in production of a large proportion of tillers that do not form ears at the expense of smaller fertile neighboring plants.

Conclusion

The development of growth analysis was part of the transformation of plant sciences from a descriptive natural-history approach to a more precise understanding of underlying mechanisms. In chemistry, the weighing of the amounts of reactants and products in a chemical reaction led to rediscovery of the atomic theory. Similarly, in plant sciences, the quantification of the growth rates has been the catalyst for further experimentation, which has led to a deeper understanding of how environmental resources determine growth and how neighboring plants interact with one another in their competition for growth resources.

List of Technical Nomenclature

A	Area.
b	Base time required for canopy closure.
C	Crop growth rate.
E	Unit leaf rate, net assimilation rate (increase in weight per unit leaf area).
F	Leaf area ratio (ratio of plant leaf area to plant weight).
I	Supply of a resource of growth.
k	Light extinction coefficient.
L	Total leaf per plant.
LAD	Leaf area duration (integral of leaf area index over duration of the crop).
LAI	Leaf area index (ratio of leaf area to ground area)
LWR	Leaf weight ratio (ratio of leaf area to leaf weight).
m	Maximum.
n	Plant density (number per ground unit area).
R	Relative growth rate (increase in weight per unit weight).
S	Daily light integral.
SLA	Specific leaf area (ratio of leaf area to leaf weight).
U	Uptake of a resource for growth.
V	Space occupied by a plant or plant part.

W	Total dry weight per plant.
w	Weight.
x	General resource.
Y	Crop yield (total weight of all plants per unit ground area).
o	Initial value.

See also: **Growth and Development**: Growth Analysis, Individual Plants.

Further Reading

Benjamin LR and Hardwick RC (1986) Sources of variation and measures of variability in even-aged stands of plants. *Annals of Botany* 58: 757–778.

Evans GC (1972) *The Quantitative Analysis of Plant Growth*. Oxford: Blackwell Scientific Publications.

Hara T (1988) Dynamics of size structure in plant populations. *Trends in Ecology and Evolution* 3: 129–133.

Hunt R (1978) *Plant Growth Analysis*. London: Edward Arnold.

Hunt R (1982) *Plant Growth Analysis: Statistical Checklist No. 4*. Cambridge: Institute of Terrestrial Ecology.

Hunt R (1982) *Plant Growth Curves: The Functional Approach to Plant Growth Analysis*. London: Edward Arnold.

Hunt R (1990) *Basic Growth Analysis: Plant Growth Analysis for Beginners*. London: Unwin Hyman.

Hunt R, Causton DR, Shipley B, and Askew AP (2002) A modern tool for classical plant growth analysis. *Annals of Botany* 90: 485–488.

Schwinning S (1996) Decomposition analysis of competitive symmetry and size structure dynamics. *Annals of Botany* 77: 45–57.

Watkinson AR (1985) Plant responses to crowding. In: White J (ed.) *Studies on Plant Demography: A Festschrift for John L. Harper*, pp. 275–289. London: Academic Press.

Weiner J (1988) Variation in the performance of individuals in plant populations. In: Davy AJ, Hutchings MJ, and Watkinson AR (eds) *Plant Population Biology*, pp. 59–81. Oxford: Blackwell Scientific Publications.

Field Crops

D Sparkes, University of Nottingham, Loughborough, UK

Growth

The definition of growth is "quantitative and irreversible changes in the length, area, or weight of individual organs."

Plants are autotrophs: they obtain their energy from solar radiation. This is used to fix carbon dioxide and water into carbohydrates by the process of photosynthesis. By accumulating carbohydrate in different forms such as sugar and starch, the plant increases in size. Therefore, solar radiation drives plant growth. It has been shown that C_3 plants (whose first product of photosynthesis is the 3-carbon compound phosphoglycerate) convert solar radiation to dry matter with an efficiency of approximately 1.4 g MJ^{-1} of radiation intercepted. C_4 plants (whose first product of photosynthesis is the 4-carbon compound oxaloacetic acid) are more efficient, producing, on average, 2.0 g MJ^{-1} radiation intercepted. It is important to realize that these efficiencies can only be maintained when other factors such as water and nutrients are not limiting.

The direct relationship between radiation interception and crop growth means that to maximize yield, growers aim to produce a crop canopy that intercepts the majority of incident radiation. The proportion of incident radiation that is intercepted by a crop depends on the leaf area index (leaf area per unit ground area) and the extinction coefficient (k).

Leaf Area Index

The leaf area index (L) of a crop can be expressed as follows:

$$L = N_p \times N_l \times A_l \qquad [1]$$

where N_p is the number of plants per unit area, N_l is the number of leaves per plant, and A_l is the mean area per leaf. The number of plants per unit area depends on the number of seeds sown and the percentage establishment. Establishment will be influenced by a number of environmental (e.g., temperature, soil moisture content) and management (e.g., soil aggregate size, sowing date, residues from previous crops) factors.

The number of leaves at any given time will depend on the sowing date and the leaf emergence rate (see "Development"). The expansion of individual leaves is also a developmental event related to temperature, but it is influenced by nitrogen: leaves with a plentiful supply of nitrogen expand faster, and have a larger final size than those deficient in nitrogen.

Extinction Coefficient

The extinction coefficient (k) is defined as the inverse of the leaf area that produces 1/e of the incident light intensity. It describes the rate at which light is extinguished as it passes through a plant canopy. For the same leaf area index, a canopy with a large value of k will intercept a greater proportion of the incident light than a canopy with a small k value. Canopies of plants with large flat leaves such as rhubarb (*Rheum rhapoponticum*) have a k value of ~0.9; those with erect leaves such as wheat (*Triticum aestivum*) have a k value of ~0.4. Leaf parameters that affect the k value include angle, orientation, size, dispersion, shape, thickness, and surface properties.

Fractional interception of radiation (**Figure 1**) can be related to leaf area and extinction coefficient by the following equation, which is derived from Beer's law:

$$F = 1 - e^{-kL} \quad [2]$$

where F is the fractional interception (proportion of incident radiation intercepted by the crop canopy), k is the extinction coefficient, and L is the leaf area index.

This equation can be used to illustrate the impact of the extinction coefficient on the proportion of radiation intercepted by a canopy: take a canopy with a leaf area index (L) of 3, a crop with relatively prostrate leaves, and an extinction coefficient of 0.8:

$$F = 1 - e^{-0.8\times3} = 0.909$$

In other words, 91% of the incident radiation would be intercepted. Comparing this with a crop with the same leaf area index, but an extinction coefficient of 0.4:

$$F = 1 - e^{-0.4\times3} = 0.699$$

or 70% of the incident radiation would be intercepted. In fact, a crop with $k=0.4$ would require a leaf area index of 6 to intercept the same amount of radiation as a crop with a leaf area index of 3 and $k=0.8$.

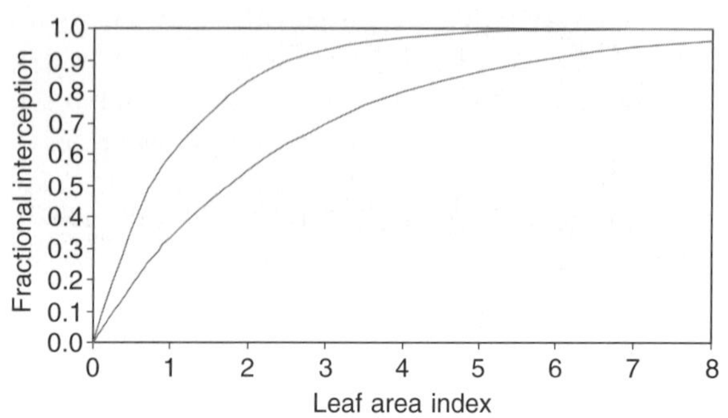

Figure 1 The relationship between leaf area and fractional interception for two canopies: $k=0.4$ (lower curve) and $k=0.8$ (upper curve), derived from $F=1-e^{-kL}$. (For further explanation see text.)

Does Temperature Affect Growth?

Plants have a base temperature, below which no growth occurs. This varies between species, but is usually between 0 and 5°C for temperate species, and above 9–13°C for tropical species. Plants also have a maximum temperature above which photosynthesis is inhibited.

Development

The definition of development is "progression through a series of discrete changes in the structure or number of individual organs." Plant development is driven by temperature and photoperiod.

To progress to the next developmental stage, the plant must experience one, or more, of the following: (1) a certain "thermal time" (measured in day degrees), (2) a certain daylength, and (3) vernalization.

Thermal Time

Thermal time is expressed by:

$$T' = (T_{max} + T_{min})/2 - T_{base}; \quad T' = 0 \text{ if } (T_{max} + T_{min})/2 \leq T_{base} \quad [3]$$

$$T_t = \sum_{i=1}^{i=t} T' \quad [4]$$

where T' is the daily temperature sum (°C), T_{max} is the daily maximum air temperature (°C), T_{min} is the daily minimum air temperature (°C), T_{base} is the base temperature (°C), and T_t is the accumulated thermal time (degree days) after t days.

For an example, see **Table 1**. Here the base temperature is 2°C. Therefore, this does not contribute to the thermal time. The accumulated thermal time for the 4 days is 31°C days. Note that if the mean temperature of a given day is less than the base temperature, then the accumulated thermal time for that day is 0.

Leaf emergence is an example of a developmental event controlled by temperature. The number of emerged leaves is linearly related to thermal time. The number of degree days between the emergence of

Table 1 Example of a calculation of thermal time

Day	*1*	*2*	*3*	*4*	*Total*
T_{base} (°C)	2	2	2	2	
T_{min} (°C)	5	6	4	4	
T_{max} (°C)	15	18	12	14	
T_{mean} (°C)	10	12	8	9	
T_t (°C day)	8	10	6	7	31

each leaf is known as the phyllochron. **Figure 2** shows the relationship between thermal time and leaf number per plant in the sugar beet (*Beta vulgaris*) crop.

Daylength

Many plants respond to changes in daylength; a response to the length of daylight is known as photoperiodism. One of the main responses of plants to daylength is flowering. Plants have been divided into three groups depending on their response to daylength:

1. Short-day plants: those that flower or whose flowering is accelerated by days shorter than a critical value.
2. Long-day plants: those that flower or whose flowering is accelerated when daylength exceeds a critical value.
3. Day-neutral plants: those that do not respond to changes in daylength.

Winter wheat is a long-day plant while soybean (*Glycine max*) is a short-day plant. This response can limit the growth of soybean in cool temperate climates because flowering may occur too late in the growing season for the crop to reach maturity.

Vernalization

Vernalization is the promotion of flowering by a period of low temperature. Plants must be exposed to a certain temperature for a given length of time to become and remain vernalized. If the plant is exposed to warm temperatures before the required time has passed then the process may be reversed; this is known as devernalization. Vernalization is perceived by the meristem of a plant.

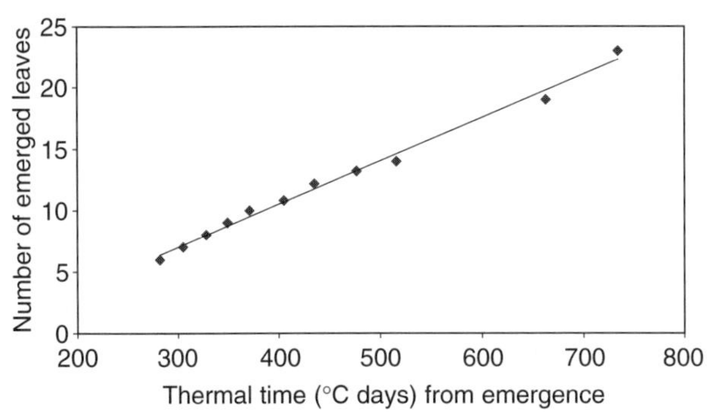

Figure 2 Relationship between temperature (expressed as thermal time from emergence) and number of emerged leaves in the sugar beet crop. Equation of the fitted line: $y = 0.0352 \times - 3.5613$, $R^2 = 0.99$.

Growth and Development Interactions

The temperature and/or photoperiod determine the duration of any developmental stage of a crop, while the amount of growth that occurs is determined by interception of solar radiation. It is important to understand how growth and development interact. For example, say that one developmental period of a plant has a thermal duration of 500°C. If the average temperatures are high, then the length of the developmental period (number of days) will be shorter than if the temperatures are cool. Regarding growth, there will be more growth on bright, sunny days, when incident radiation receipts are high, than cloudy days. Therefore, if the developmental period is warm and dull, less dry matter will be accumulated than if conditions are cool and bright (**Figure 3**).

This can be illustrated with an example. If a developmental period is 500 degree days:

- if mean temperature is 25°C (warm) then developmental period will last 20 days.
- if mean temperature is 15°C (cool) then developmental period will last 33 days.

Under bright, sunny conditions in summer in Britain, incident radiation will be approximately $20\,MJ\,m^{-2}\,day^{-1}$. If conditions are overcast, incoming radiation is likely to be about $8\,MJ\,m^{-2}\,day^{-1}$. Assuming that the crop intercepts 90% incident radiation:

1. If conditions are cool and bright: $33\,days \times 18\,MJ\,m^{-2}\,day^{-1} = 594\,MJ\,m^{-2}$ radiation will be intercepted during the developmental period. We know that C_3 plants convert radiation to dry matter with the approximate efficiency of $1.4\,g\,MJ^{-1}$, therefore this crop would increase in

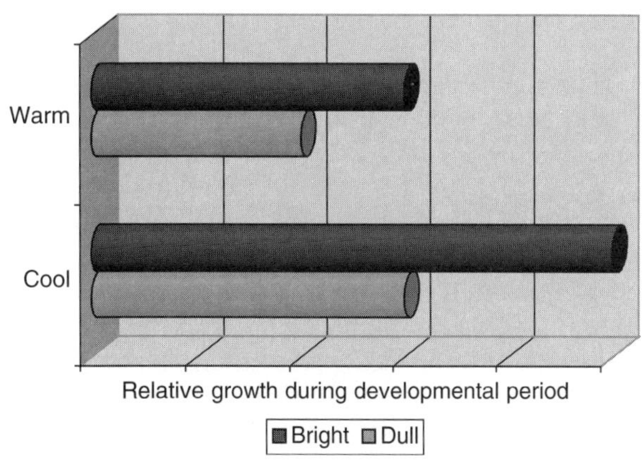

Figure 3 Schematic diagram of the influence of temperature and incident radiation on the amount of growth during a developmental period.

mass by 832 g m^{-2} during the 500 degree day developmental period.

2. If conditions are warm and dull: 20 days × 7.2 MJ m^{-2} day^{-1} = 144 MJ m^{-2} × 1.4 g MJ^{-1} = 202 g m^{-2}; then the crop would only increase in mass by 202 g m^{-2}.

Applications to Field Crops

Sugar Beet

Sugar beet is a biennial plant, naturally completing its life cycle in two years. In the first year leaves are produced, and the energy from photosynthesis is stored as sucrose in a swollen tap root. In the second year, this energy would be used to produce a tall seed head. To harvest sugar from the beet, the plants are grown for one year, and the sugar extracted from the tap root.

Vernalization In sugar beet, the vernalization requirement is obligate: the shoots of unvernalized sugar beet plants continue to produce new leaves, without producing flowers, for several years. The optimum temperature for vernalization is 5–10°C, and the plant needs to experience 40 days of cool temperatures to move into the reproductive phase of its life cycle. There is no juvenile phase and both seeds and seedlings can be vernalized. If sugar beet is sown too early in the year, the seedlings may experience enough cold weather to trigger the switch to the reproductive phase, leading to seed production in year 1. Plants that produce seed in year 1 are called "bolters." The sugar yield from bolters is very low; the sugar content being low and difficult to extract. Therefore, the sowing date of the sugar beet crop is chosen to avoid vernalization. In the United Kingdom, this means that the earliest safe sowing date across most of the country is mid-March. There is also a photoperiodic requirement after vernalization, as sugar beet is a long-day plant.

Radiation interception The amount of sugar stored in the root will depend on photosynthesis and therefore the amount of radiation intercepted. To intercept 95% of the incident radiation, the sugar beet crop needs to achieve a leaf area index of approximately 3. In the United Kingdom, maximal radiation receipts occur in June. However, sugar beet cannot be sown until mid-March at the earliest, due to the reasons described above, meaning that by the time maximum canopy size is achieved, incident radiation per day is declining. The leaf surface area needs to be as large as possible, as soon as possible, and to be maintained for as long as possible. The grower needs to plant the crop as soon as possible after the earliest safe sowing date so that the plants have time to expand their leaves, and intercept the maximum amount of radiation. There is a linear relationship between radiation and sugar yield.

Nitrogen application Sufficient nitrogen must be taken up by the crop to promote rapid expansion of the canopy (through its influence on leaf expansion), and to maintain a canopy that can intercept radiation and convert this to dry matter efficiently. However, if excessive nitrogen is available, particularly if this nitrogen is mineralized from organic matter or recently added manures, then a change in the partitioning of biomass means that much leaf material is produced, at the expense of root. Hence the harvest index (the proportion of the crop biomass which is harvested) is reduced. Excessive nitrogen uptake also results in nitrogenous compounds, notably α amino acids, accumulating in the root. Large concentrations of these compounds impair crystalization of sugar during processing.

Summary To maximize sugar yield from the sugar beet crop it is necessary to avoid vernalization at the seed or seedling stage; this would lead to premature seed production and poor sugar yield. As sugar yield is directly related to intercepted radiation, yield is maximized by promoting early canopy growth through early sowing and the supply of sufficient nitrogen to enhance leaf expansion and leaf duration. Excessive nitrogen application must be avoided as, in addition to detrimental environmental impact, too much nitrogen changes the partitioning within the plant. This leads to more assimilate being partitioned to aboveground parts of the plant and less to the roots, and can also cause problems with sugar extraction.

Winter Wheat

Wheat is an annual plant with a determinate growth habit – there is a marked transition from the vegetative to the reproductive phase. Development starts with germination and seedling growth. Leaves emerge on the seedling at intervals determined by the phyllochron. The phyllochron is determined partly by genotype and partly by sowing date, with later sown crops generally having a shorter phyllochron. Most winter wheat cultivars have a vernalization requirement for floral initiation, followed by exposure to long daylengths for full reproductive development. For winter wheat to be vernalized it needs to experience approximately 40 days within the temperature range of 3–10°C.

Tillers (side shoots) arise from buds in leaf axils where the leaf sheath joins the stem. Primary tillers are formed on the main stem, and these can bear secondary tillers. Tillering often ceases at around the onset of stem extension as the developing ear and growing stem compete with tillers for assimilate. However, in low plant populations, where there is less competition, tillering continues beyond this stage. Once tiller production stops, some tillers will die: the youngest tillers are the first to die in the competition for resources.

Leaves continue to emerge as the stem extends until flag leaf emergence. The ear then emerges, shortly followed by anthesis (flowering). The grain filling period then begins; the assimilate for grain filling mainly comes from current photosynthesis although there is some contribution from the redistribution of soluble carbohydrates which have been stored in the stem. When conditions during this period are unfavorable, perhaps due to drought or disease, soluble stem carbohydrates are important in buffering the final yield of the crop. Finally the crop senesces, and the grain matures and is harvested.

Wheat yield can be expressed as:

$$\text{number of ears m}^{-2} \times \text{number of grains per ear} \times \text{grain size.}$$

Number of ears m^{-2} is determined by the number of established plants × number of ears per plant. The number of ears per plant is determined by the number of tillers produced and the proportion of these which survive.

Shoot production and survival In nonlimiting conditions, tiller production follows the Fibonacci series, at intervals determined by the phyllochron. However, in stressed conditions, such as a shortage of water or nitrogen, some tillers may fail to develop. By the time wheat moves from the vegetative to the reproductive phase of growth, most plants have more tillers than will survive through to harvest. This is illustrated in **Figure 4** which shows the number of shoots (i.e., tillers plus the main stem) per plant from early vegetative growth through to maturity. Two contrasting crops are shown: when wheat was grown at a high plant population (230 plants m^{-2}), an average of nine shoots per plant were produced, but by harvest there were only three or four shoots per plant. When wheat was grown at a low plant population (60 plants m^{-2}), an average plant produced 15–16 shoots and a larger proportion of these shoots survived until harvest, finishing with an average of 10 shoots per plant. In this way, crops with very different plant populations produce similar numbers of ears per unit area. In this example, the crop with 230 plants m^{-2} produced 800 ears m^{-2} and the crop with 60 plants m^{-2} produced 600 ears m^{-2}. Competition for radiation and nutrients within large canopies (produced by early sowing, high plant populations or excessive nitrogen supply) leads to fewer shoots surviving to produce ears.

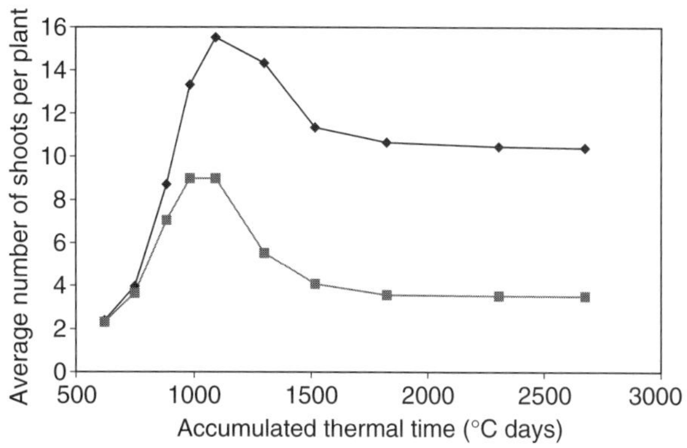

Figure 4 Shoot production and survival in two winter wheat crops: ■ 230 plants m^{-2}, ♦ 60 plants m^{-2}.

Number of grains per ear The inflorescence of a wheat plant consists of a rachis (stalk) to which spikelets (basic unit of a grass flower) are attached. Each spikelet contains between two and nine florets (individual flowers). The rate and duration of spikelet initiation and the survival of those spikelets and the florets within them determines the number of grains per ear.

Both delayed sowing and high plant populations accelerate the rate of development of wheat. This leads to a shortened phase of spikelet initiation in both cases but the impact on grains per ear is different. At high plant populations the rate of spikelet initiation is unaffected, hence the higher the plant population density, the fewer grains per ear. When sowing date is delayed, an increase in the rate of spikelet initiation compensates for the shorter duration of this phase of development, hence the effect on grain number per ear is negligible.

The amount of assimilate produced during the period of rapid ear growth (between flag leaf emergence and anthesis) has a strong influence on floret survival, and therefore grain number per ear. In a crop with a low plant population density, more radiation will be absorbed per shoot, floret survival will be greater, hence more grains will be produced per ear.

Grain size Assimilates for grain filling are derived from current photosynthesis and stored carbohydrates, the latter of which are particularly important when the crop experiences stress during the grain filling period. For instance, if the crop is subjected to drought during this period then yield may be "buffered" by stored carbohydrates. Grain filling has a thermal duration of approximately 700°C days. Therefore, if the grain filling period is particularly warm, the duration of this phase will be shorter and vice versa. However, because warm summer temperatures are usually associated with bright, sunny weather, the shorter duration will be, at least partially, compensated by the increased photosynthesis. Conversely, cooler temperatures are usually associated with dull conditions and hence reduced photosynthesis.

In practice, the main variable determining yield appears to be number of grains per unit area: grain size is relatively conservative. Potential grain size is limited by endosperm cell number, which is determined around anthesis.

Summary Wheat demonstrates a number of compensatory mechanisms which facilitate yield maintenance. There is an inverse relationship between the number of ears per square meter and the number of grains per ear, such that the number of grains per square meter is maintained across a wide range of plant populations, sowing dates and environmental conditions.

List of Technical Nomenclature

Anthesis	Flowering.
Extinction coefficient	The inverse of the leaf area that produces 1/e of the incident light intensity.
Floret	A small flower which is part of a composite flower.
Leaf area index	Leaf area per unit ground area.
Photoperiodism	Response to changes in daylength.
Phyllochron	The interval between emerging leaves.
Spikelet	The basic unit of a grass flower.
Vernalization	The promotion of flowering by a period of low temperature.

See also: **Growth and Development**: Canopy Architecture; Growth Analysis, Crops. **Photosynthesis and Partitioning**: C_3 Plants; C_4 Plants. **Production Systems and Agronomy**: Agricultural Crops. **Regulators of Growth**: Vernalization.

Further Reading

Azam-Ali SN and Squire GR (2001) *Principles of Tropical Agronomy*. Wallingford: CAB International.

Cooke DA and Scott RK (eds) (1993) *The Sugar Beet Crop: Science into Practice*. London: Chapman & Hall.

Hay RKM and Walker AJ (1989) *An Introduction to the Physiology of Crop Yield*. Harlow: Longman Scientific.

Squire GR (1990) *The Physiology of Tropical Crop Production*. Wallingford: CAB International.

Whaley JM, Sparkes DL, Foulkes MJ, *et al.* (2000) The physiological response of winter wheat to reductions in plant density. *Annals of Applied Biology* 137: 165–177.

H

Hormones *see* **Regulators of Growth**: Auxins; Abscisic Acid; Cytokinins; Gibberellins; Ethylene; Brassinosteroids; Jasmonates.

Horticulture *see* **Plants and the Environment**: Amenity Horticulture. **Production Systems and Agronomy**: Horticultural Crops, Protected; Orchard Crops; Commercial Flower Production Methodology; Nursery Stock and Houseplant Production.

I

INTEGRATED PEST MANAGEMENT

Contents

Principles

G Matthews, Imperial College of Science, Technology and Medicine, Ascot, UK

Introduction

Integrated pest management is defined as a pest management system that, in the context of the associated environment and population dynamics of the pest species, utilizes combinations of suitable techniques and methods in as compatible a manner as possible and maintains the pest population levels below those causing economic injury. The implication is that selected cultural, biological, and chemical controls will be used together with an appropriate crop variety. An organism that adversely affects the availability, quality, or value of a human resource is regarded as a pest.

Principles

The use of the term integrated pest management (IPM) became widespread in response to the growing concerns about the extent to which chemical pesticides were being used, especially in the USA in the 1950s. Inorganic chemicals, such as calcium arsenate, had been used to control insect pests prior to 1940, but the introduction of potent organic pesticides, such as DDT, 2,4-D, and others in the post-Second World War period, heralded a new era, initially seen as a panacea to end man's struggles to control pests. However, scientists soon became aware of the conflict of these chemicals with the natural environment and the potential for pests to become resistant to the new pesticides. Rachel Carson's book "Silent Spring" is often quoted as the catalyst for studying the effects of overuse of chemical pest control and, in particular, the persistent use of organochlorine insecticides. However, earlier studies in California had already led to the recognition that insecticides should be applied only when the pest population exceeded a threshold, above which there would be economic damage to the crop. The politics of the 1960–1970s period led to the promotion of IPM as the way forward in crop protection, and initially much of the early studies within the USA were led by entomologists, although the Food and Agriculture Organization (FAO) subsequently included a wider definition of pests. In addition to insect pests, plant pathogens, nematodes, mammalian and avian pests, as well as weeds, were included. IPM strategies vary among crops and different agroecological zones, and depend on several factors, including climatic conditions and farming practices.

Tactics That Can Be Used in an IPM Program

Host Plant Resistance

Since the earliest days of agriculture, farmers have selected seeds from their crops that have the ability to withstand pest invasions. However, the deliberate selection of wheat varieties (*Triticum* spp.), based on their ability to survive attack by diseases, began in the early nineteenth century, and initiated the development of plant breeding as a scientific means of increasing yields, utilizing Mendelian genetics (*see* **Diseases**: Breeding for Disease Resistance). The constant alteration in the virulence of different strains of plant pathogens has led to a continuing need to select crop cultivars that have some degree of resistance to the most important pathogens. Resistance to a narrow range of strains of a particular pathogen is often referred to as vertical resistance.

Ideally, the breeder selects for horizontal or durable resistance, but this is generally more difficult to achieve. In some cases, a mixture of varieties has been advocated to broaden the range of resistance or slow down the spread of the disease if some plants are susceptible. In selecting resistant cultivars, it is equally important to maintain yields and the quality of the crop, for example, a resistant cultivar may not be acceptable owing to poor taste of the harvested produce. Most of the major food crops have been selected to minimize any naturally occurring poisons, thus making them more palatable; unfortunately, these are often more susceptible to pest infestations. Plant breeders have concentrated on resistance to plant pathogens, including fungi and viruses. Nevertheless, some cultivars have been selected for resistance to insect pests. The approach may be biochemical, where plants are selected with increased levels of certain toxins, or resistance may be achieved by growing plants that are physically less attractive, e.g., with tough or pubescent leaves. When there is a choice of different cultivars, insects may choose to oviposit or feed on a preferred variety leaving other plots/varieties apparently unaffected. When there is no choice, the insects will infest an apparently more resistant variety.

A recent development has been the introduction of transgenic crops, in which it has been possible to insert specific genes to provide certain traits in the crop (*see* **Pests**: Genetic Modification of Pest Resistance, Insect Pests). In particular, selection of plants that can express a toxic gene from certain strains of *Bacillus thuringiensis* (Bt) are able to withstand infestations of particular insect pests, notably among the lepidoptera, but other Bt strains in potatoes (*Solanum tuberosum*) and corn (*Zea mays*; maize) provide resistance to coleopteran pests. However, such crops are still susceptible to certain other pests, such as aphids. Crops with a gene that allows them to break down certain herbicides, such as glyphosate, are now widely available, enabling the herbicide that would normally kill the crop to be applied. One advantage claimed is that weeds can be controlled later, allowing some weed growth in the inter-row to provide a habitat for natural enemies of insect pests. A potential disadvantage is that the herbicide-resistant crop could become a weed in a subsequent crop, although it would still be susceptible to herbicides with a different mode of action.

Biological Control

The presence of parasitic organisms and predators, including ants and spiders, generally provides a check on pest populations in the natural environment, but the action of farming, ploughing fields, and growing large areas of individual crops tends to disturb this balance. The balance is also adversely affected by the transport of organisms from their natural habitat to a new environment. Man has been guilty of moving plants between countries, often accompanied by a pest species, but not always with the pest's natural enemies.

Various techniques have been used to try and restore the balance between the pest population and its natural enemies. Classical biological control refers to searching for a suitable natural enemy in the original environment of a pest, and releasing it in the areas where the pest is unchecked. An example of this was the release in Africa of the parasitoid *Epidinocarsis lopezi* from South America to control the cassava mealy bug *Phenococcus manihoti*. The hope is that a single inoculation will allow the natural enemy population to increase and maintain pest populations below an economic threshold. Classical biological control has been most evident in perennial crops and in geographically isolated areas.

In some situations, the natural enemy may be present, but not in sufficient numbers to exert control, particularly at a critical stage of crop development. In this case, a periodic release may be required to augment the natural enemies that are present. Where several releases of a natural enemy are required, usually in large numbers, to obtain a rapid decline in a pest population, it is referred to as an inundative release. This has been used mostly for high-value crops in glasshouses; for example, *Encarsia formosa* is effective against whiteflies, and the predatory mite *Phytoseiulus persimilis* is released to control the two spotted spider mite *Tetranychus urticae*. In the former Soviet Union, the parasitoid *Trichogramma* has been released extensively on a number of crops, most notably on cotton (*Gossypium* spp.)

An important development of IPM has been to study the requirements of the natural enemies and endeavor to develop strategies that will conserve their populations throughout the year. This conservation may involve retaining other host plants in the vicinity, such as those in hedgerows, creating specific areas for natural enemies within a field, and avoiding the application of persistent pesticides. Maintaining wild flowers in a hedgerow provides a source of food for many natural enemies. One technique to conserve natural enemies in large cereal fields is to establish earth ridges sown with grass across the center of the field to provide a habitat, similar to that of a field boundary. This system favors overwintering populations of polyphagous predators. The extra cost of implementing the above strategies is offset by a reduction in the use of insecticides for aphid control.

Some biological agents are now applied as "biopesticides" in a similar manner to chemical pesticides. These biopesticides to control insects include baculoviruses, certain fungi (e.g., *Metarhizium anisopliae*), and entomopathogenic nematodes. Grasshoppers and locusts are controlled by the application of *Metarhizium anisopliae* var. *acridum*, as conidia in an oil suspension, and can even be applied in an arid environment. Although less rapid in action than a chemical insecticide, kills of 90% were achieved in 7–21 days without the decimation of natural enemies. Antagonistic fungi such as *Trichoderma* spp. are also used to control plant pathogens.

Cultural Controls

Cultivation techniques were initially developed to suppress weeds. Plowing was originally intended to bury weeds, but it will also bring buried weed seeds back to the surface, so after sowing farmers have traditionally also cultivated their fields, e.g., inter-row weeding, to prevent weeds from competing with their crops (*see* **Weeds**: Weed Technology and Control). The introduction of chemical herbicides has allowed the concept of minimum or zero tillage where the characteristics of the soil enable the establishment of sown crops, without plowing and preparing a seedbed. This reduces the adverse effects of plowing on soil organisms such as earthworms, but maintains weed management.

An increase in soil pests, especially plant parasitic nematodes, and certain persistent weeds forced farmers to introduce crop rotation, which, together with leguminous crops in the rotation, helped to improve the fertility of the soil. Crop rotation is also important where insect and other pests are specific to a particular crop, by reducing the pest survival between crops. This reduction of pest survival on an alternative crop has been extensively utilized by employing intercropping, that is the sowing of different crops, such as corn and cowpea (*Vigna sinensis*), in alternate rows. In some cases, one crop is sown under the first crop or later as a relay crop.

Certain plants are considered to repel or in some way reduce the incidence of a pest population by their odor or secretions from their roots. For example, Mexican marigold (*Tagetes* spp.) has been planted in strawberry beds to suppress plant parasitic nematodes. Conversely, some plants may be more attractive to an insect pest and may be grown as trap crops, but the sowing of trap crops is not popular unless it is also of value to the farmer. The trap crop may also be attractive to the pest for a limited period only, so synchronization of one or more sowings is required. This may be possible where the trap crop can be irrigated. Irrigation may also allow some flexibility in the sowing date of crops in an effort to avoid an insect infestation, but in rain-fed areas, there is little flexibility in sowing dates, which are dictated by the onset of sufficient soil moisture for germination.

Legislation has dictated the date of uprooting of some crops or the destruction of crop residues at the end of the season to minimize the survival of pests and pathogens from one season to the next. Farm animals can be used to consume certain crop residues, or crop residues can be mechanically shredded and plowed back in or used in compost to return some organic matter to the soil.

Interference Techniques

Insect pheromones affect intraspecific interactions. Sex pheromones that in most cases attract males to virgin females have been the most extensively used in IPM, especially with lepidopteran pests. These powerful sex pheromones can be used in place of light traps to selectively trap insects. Alarm and aggregation pheromones are also used in IPM. For certain pests, the decision to use an insecticide can be based on the number of insects trapped at critical periods of crop growth. Mass trapping in the absence of insecticide sprays has been less successful, as a large number of inexpensive traps is required. Mass trapping has been more successful, with aggregation pheromones as both male and female insects are trapped.

The use of sex pheromones as a control method has been more successful in bringing about mating disruption. This technique is based on the assumption that males will be disorientated if there is sufficient pheromone permeating in the air over and within a crop canopy, and consequently they will fail to mate. Slow-release formulations, either as a strip of an impregnated polymer that can be applied by hand or as microcapsules sprayed on the crop, allow the pheromone to remain effective over a period of several weeks. One example of this is the application of gossyplure to control *Pectinophora gossypiella*, the pink bollworm that attacks cotton. Alternatively, pheromones can be used to attract insects to a surface treated with insecticide – the "lure and kill" technique. Food attractants are sometimes used in place of sex pheromones, for example, the protein hydrolysate acts as a bait for fruit flies.

Interfering with the feeding of insects by applying chemicals known as antifeedants has seldom been very successful, as the pests are sufficiently mobile to find an area of foliage that has not been sprayed or where the surface deposit has been diluted by plant growth. However, interfering with the development of an insect by adversely affecting the molting between instars has been very successful when chitin inhibitors and other insect growth regulators have been used. They are stomach poisons and so tend to be more selective in controlling herbivorous insects, rather than natural enemies, and are well suited to IPM programs. Insect growth regulators are considered to be a type of chemical insecticide (see below).

For certain insects, it is possible to eradicate populations within discrete areas by the release of sterile males that successfully compete with the wild population. This technique is only suitable if there is a single mating. The technique, which was initially developed for controlling the screwworm (*Callitroya macellaria*), a pest of cattle, has also been used against fruit flies. The technique is usually integrated with chemical control, which reduces populations initially so that fewer sterile flies need to be released.

Chemical Control

An extensive range of pesticides is now available – insecticides, fungicides, herbicides, nematicides, acaricides, rodenticides – and within each group there are several types with different modes of action (**Table 1**). Some act on contact, while others must be absorbed into plants or ingested by the pest. Each pesticide must be registered in the country in which it is marketed, although there are efforts to harmonize the requirements for registration of active ingredients. When first registered, the pesticide's active ingredient is assigned a common name and when formulated into a commercial product, it is usually available under a specific trade name only from the company that developed it. Subsequently, it may be available from other companies under a range of trade names. The extent to which a particular pesticide or type with a specific mode of action can be applied may be restricted, where resistance management programs are being implemented. The restriction may be on the total amount or number of applications that may be applied to a crop within a year, or that individual applications should be alternated with a product with a different mode of action.

Pesticides vary in their toxicity to organisms so the registration process takes account not only of mammalian toxicity, but also a range of other organisms to assess the risk of its use in the environment. Many of the older pesticides are now banned due to their persistence and accumulation in food chains, or due to their extremely high toxicity. A good practice is to indicate the mammalian

Table 1 Examples of some pesticides

Type	*Group*	*Examples (common name)*
Insecticides	Organophosphates	Malathion, pirimiphos methyl
	Organochlorines	DDT, dieldrin
	Carbamates	Aldicarb, carbaryl, pirimicarb
	Benzoyl-ureas[a]	Diflubenzuron, teflubenzuron
	Juvenile hormone mimics[a]	Methoprene
	Ecdysone antagonists[a]	Tebufenozide
	Pyrethroids	Cypermethrin, lambda-cyhalothrin, deltamethrin
	Neonicotinoids	Imidacloprid
	Botanicals	Pyrethrins, rotenone, neem, nicotine
	Microbial	*Bacillus thuringiensis*
	Spinosyn	Spinosad (derived from *Saccharopolyspora sinosa*)
	Soaps	
	Desiccants	Abrasive dust
Fungicides	Azoles	Difenoconazole
	Benzimidazoles	Carbendazim
	Copper salts	Bordeaux mixture
	Dithiocarbamates	Mancozeb
	Strobilurins	Axoystrobilurin
	Sulfur	
	Antibiotics	Streptomycin

Herbicides (see section on Herbicides).
[a] Insert growth regulators.

toxicity of pesticides by the color of the label – purple, red, orange, and green being progressively less hazardous to the operator. Pictograms on the label simplify some key instructions. In integrated pest management programs, the choice of pesticide depends to a large extent on the pest and how the pesticide is formulated and applied, although preference is given to the more selective pesticides of low persistence in the environment. Botanical insecticides, which usually have a very low persistence, have been used on "organic" crops.

Many recommendations have been based on applying pesticides at a particular time of crop development and then repeating at set calendar intervals when the crop is susceptible to an infestation. A spray may have a devastating effect on the natural enemies, at a time when the crop can compensate for some damage. This may result in increased pest numbers (pest resurgence) in the absence of their natural enemies. In IPM, it is crucial to assess a pest infestation and only apply a pesticide if the pest population exceeds an economic or "action" threshold.

Various techniques are used to sample pest populations, including pheromone traps, but usually it is important to inspect the crop, a technique often referred to a "scouting" to assess pest numbers or damage in different areas of a field. Such techniques can, in some circumstances, confine a treatment to an infested "patch." Where farm machinery is equipped with global positioning instruments, geographical information systems can be linked to application equipment to facilitate spraying specific patches within a field. In many countries, there are national guidelines providing advice on choice of pesticide and recommended dosages, but ultimately the user has to follow the instructions provided by the manufacturer on the product label.

Most pesticides are formulated so that they can be diluted in water and applied with a sprayer using hydraulic nozzles. Initially, the most common formulation was an emulsifiable concentrate (EC) in which the active ingredient was dissolved in a solvent with an emulsifier so that it dispersed as an emulsion when mixed with water. Concern about the spraying of solvents has led to the use of a particulate suspension concentrate (SC) or a wettable granule (WG) formulation that readily disperses in water. Wettable powder formulations are also available, although these are ideally packed in water-soluble sachets, so that the user cannot inhale the small particles. Some pesticides, which are usually the most hazardous to apply, are formulated as dry granules and dispensed through special applicators. These can be selectively applied alongside seeds so that the pesticide concentrates in the root zone of young seedlings. Systemic pesticides absorbed through the roots can protect seedlings from aphids and similar early season pests for up to 6 weeks. Similarly, a pesticide may be applied selectively as a seed treatment. (*see* **Seed Development**: Seed Treatments). Where several cultivars with varying resistance to a pest are sown together, it may be advantageous to treat the seed of the susceptible cultivar with an appropriate pesticide.

Application equipment varies from small manually operated sprayers to large-scale motorized and aerial equipment, the choice of which depends on the size of the farm, the crop, and terrain. In IPM, the aim is to optimize application of the minimum dosage at the most critical time to prevent economic damage by the pest. Ideally, the spray is targeted selectively to where the pest is located. However, in most applications on field crops, the pesticide is inefficiently applied downward over the top of a crop, so coverage of secluded pests, such as whitefly nymphs on the underside of leaves, is very poor. Such pests are unaffected unless the insecticide has a translaminar effect. Improved penetration of crop canopies is achieved by using an air-assisted sprayer. Air-assisted equipment is also used to project spray upward into tree crops.

Integration

IPM programs need to vary according to the crop, production system, and local environment. Strategic models have been developed involving complex insect/plant interactions with vast inputs of data in an attempt to devise a tool to aid the implementation of an IPM program. Such models may be of interest to researchers and may indicate the impact of various factors in an ecosystem. Although an IPM program may ideally require more detailed ecological research, most existing programs have been developed pragmatically using existing information. It is important that the pest is correctly identified, and suitable assessments of pest populations can be made and developed into a monitoring tool. The aim of IPM is to utilize different control tactics so that they work together harmoniously (**Figure 1**). In some crops or areas, one particular tactic may be dominant. Important control tactics, such as crop rotation, may not be employed as much as they should, as the alternate crops are less suited to the local environment and provide a lower income. In some situations, monocultures can be easier to manage than highly diverse cropping systems, where relatively minor crops may be the key food source for a pest to survive between seasons.

Thus, there are often conflicts of interest and users have to make value judgements about which tactics can be implemented in a particular program or region. Farmers have readily adopted chemical control as it can be used quickly and can be independently managed on their farms. In contrast, many of the nonchemical methods depend to a large extent on universal adoption within an agroecological area. Legislation has enabled certain techniques to be adopted on a national scale; these include dates for sowing or uprooting specific crops and the use of specific varieties in certain areas. Area-wide adoption of pheromone mass disruption techniques has depended on the willingness of Government or farmer associations to distribute the pheromone collaboratively over an extensive area. With the selection of pests resistant to pesticides used extensively within certain areas, the need to collaborate on resistance management strategies has also become

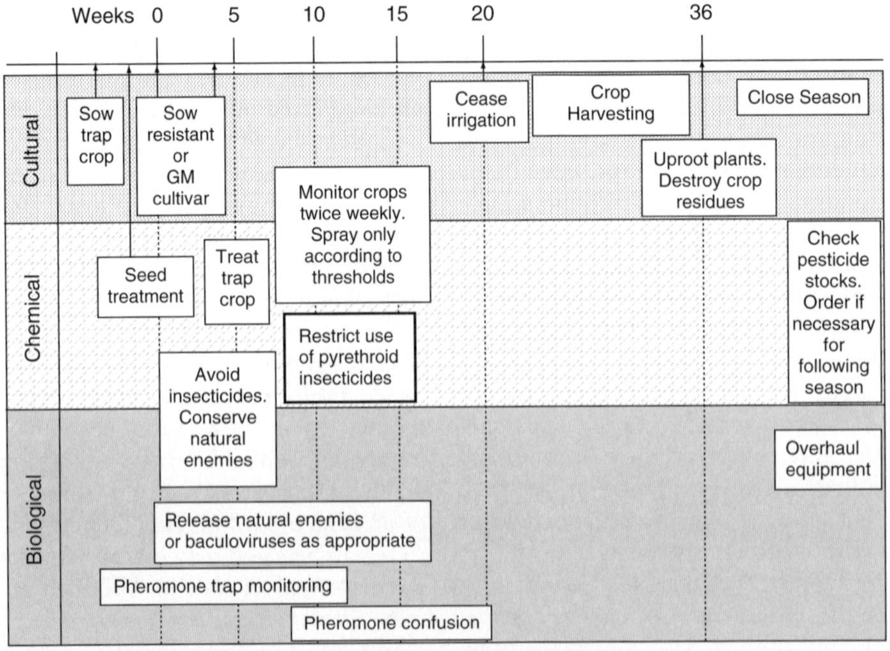

Figure 1 Different tactics integrated during the season into an IPM cotton program.

increasingly important to conserve the effectiveness of a limited range of effective pesticides.

An important advantage of growing resistant cultivars is that, even with partial resistance to some of the pests or pathogens, it may enable farmers to minimize the number of spray applications, or reduce the dosage needed. This has the added advantage that application of a lower dosage of pesticide reduces any adverse effects it might have on natural enemies, thus assisting integration of chemical and biological controls. Using a selective or nonpersistent insecticide, it may be possible to reduce a pest population, yet leave some survivors, on which its natural enemies can feed. Conversely, using a more selective pesticide can lead to a relatively minor pest increasing in importance as it is no longer controlled by the application of a broad spectrum chemical.

Much emphasis has been given to biological control, but unless there are very effective natural enemies that can be conserved in the crop area, alternative tactics, especially chemical control, may be needed if the pest population has to be reduced rapidly.

Implementation

Many have argued that pest management is a component of crop production and generally refer to integrated crop management (ICM). Although IPM requires harmonious integration of several control tactics, it is sometimes referred to as rational pest control. Many groups that are opposed to the use of pesticides promote "organic farming" methods and emphasize the importance of using biological and cultural controls.

Many aspects of implementation of IPM depend on Government regulations to provide a framework in which crop protection specialists can work. Where farmers have adopted computer technology, assistance has been provided by the development of "Expert Systems" or tactical models that help educate farmers, so that they can ultimately make their own decisions. This applies particularly to where treatments are needed on an area-wide basis to be effective. Those giving advice need to have sufficient training and to be aware of environmental considerations, as well as knowledge of the different pests and which control tactics are best suited for an IPM program. Such training requires the establishment of effective guidelines on IPM, which are appropriate to the crops and agroecosystem in a particular country.

See also: **Diseases**: Breeding for Disease Resistance. **Integrated Pest Management**: Practice. **Pests**: Genetic Modification of Pest Resistance, Insect Pests. **Production Systems and Agronomy**: Organic Farming. **Seed Development**: Seed Treatments. **Weeds**: Herbicide Resistance; Weed Technology and Control.

Further Reading

Carson R (1962) *Silent Spring*. Boston: Riverside Press.

Debach P and Rosen D (1991) *Biological Control by Natural Enemies*. New York: Cambridge University Press.

Dent D (2000) *Insect Pest Management*. Wallingford: CABI Publishing.

Flint ML and Gouveia P (2001) *IPM in Practice: Principles and Methods of Integrated Pest Management*. Oakland: University of California. University of California Publication 3418.
Norton GA and Mumford JD (eds) (1993) *Decision Tools for Pest Management*. Wallingford: CABI Publishing.
Stern VM, Smith RF, Van den Bosch R, and Hagen KS (1959) The integrated control concept. *Hilgardia* 29: 81–100.
Van Emden HF and Peakall DR (1996) *Beyond Silent Spring: Integrated Pest Management and Chemical Safety*. London: Chapman and Hall.

Practice

G Matthews, Imperial College of Science, Technology and Medicine, Ascot, UK

Introduction

Integrated pest management (IPM) in practice is in an intermediate position between the routine use of chemical control and the adoption of "organic" farming which avoids the use of pesticides or in some circumstances has allowed limited use of naturally occurring botanical insecticides, and the fungicides sulfur and copper. The extent to which IPM has been adopted varies considerably between countries. This is to some extent due to disputes about how to define IPM, with some preferring emphasis on biological control, while others accept rational use of pesticides. The latter approach tends to occur in areas of intensive farming with large fields where key crops are grown in monoculture. In areas with mixed farming and low inputs, IPM does rely more on cultural and biological controls. IPM is not necessarily the least expensive option, as it requires to a certain extent regular crop monitoring, in order to make appropriate management decisions throughout each cropping season. The aim is to have a more ecologically friendly and sustainable system. However, with world markets of some crops providing low income, there is often insufficient incentive to invest in more inputs, including the provision of trained managers.

Sampling Techniques

Crop Walking

The most important sampling systems involve visiting different areas within a field to determine what the pest is actually doing in the crop and how the crop is growing. It is the method most suitable for assessing weed populations. Knowledge of the pest will allow the time required for routine sampling to be minimized. Ideally a "scout" can concentrate on examining a defined number of leaves/plants, part of plant, or other sample that will show changes in pest numbers or if damage is increasing (**Figure 1**). For some pests, sampling can be simplified to record presence or absence of the pest on a plant (**Figure 2**). Searching for eggs or small insects depends on the observer having good eyesight and not being colorblind. Information needs to be recorded so that the trends can be assessed. It is important that samples are stratified across a field, as pests may be localized within a field. Sequential sampling programs have been devised so that once an action threshold has been reached, further sampling within a field is not required. Suitable action thresholds will depend on the crop variety and locality. Increasingly this information is provided by the nearest university or extension service on their web page on the internet.

Visual inspection of plants over a defined length of row or number of plants is usually adequate for readily seen and relatively slow-moving or sessile pests. For some pests it may be necessary to use a sweep net, or to tap the plants to dislodge and shake the insects on to a collecting tray.

Suction Traps

Various suction traps have been devised, but in all cases a motorized fan is used to suck a volume of air so that any airborne insects within it are collected on a mesh filter or deposited by a vortex into a container. Some of the larger suction traps are operated continuously in fixed positions to sample the air at 40 m and these have been used especially for aphids and other small wind-carried insects. These detect migrating populations. Small versions

Figure 1 Cotton being scouted for pests by an African farmer using a pegboard. Reproduced with permission from Matthews GA (2000) *Pesticide Application Methods*, 3rd edn., p. 13. Oxford: Blackwell Science.

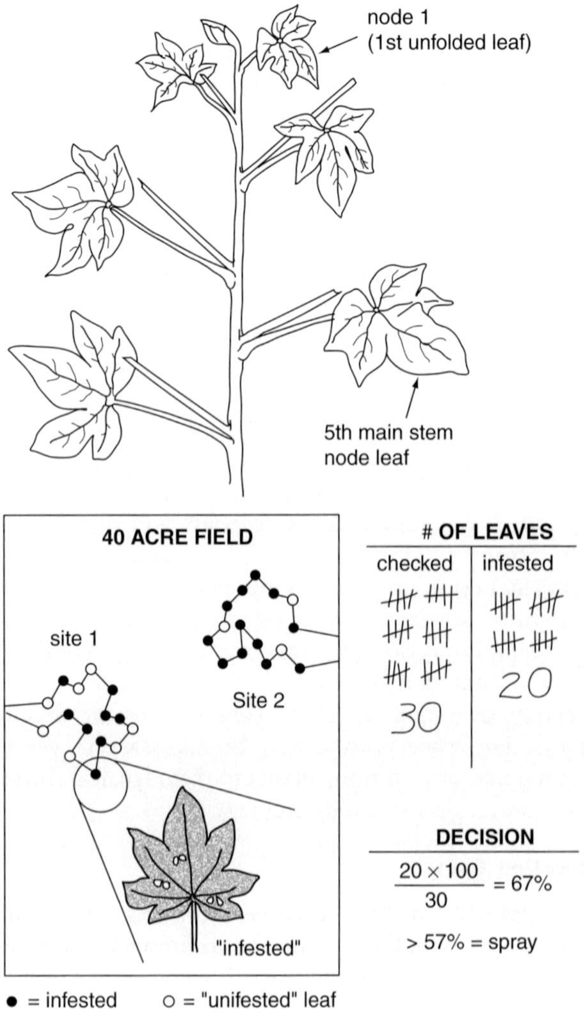

Figure 2 The action threshold for whiteflies in cotton in Arizona is 57% or more fifth main stem node leaves infested. Infestation levels are determined by sampling leaves on 30 plants in two areas of the field with presence/absence sampling. Reproduced with permission from Flint ML and Gouveia P (2001) IPM in Practise: Principles and Methods of Integrated Pest Management, University of California ANR Publication 3418, Oakland, CA.

can also be set in a field or are carried on a knapsack frame, so that different parts of a crop can be sampled. Care is needed to avoid damaging plants. Suction traps are used mainly to sample the smaller, more mobile insects, including natural enemies.

Sticky Traps

Some insects can be sampled by hanging colored cardboard or plastic strips, coated with a sticky substance, so that the attracted insects adhere to the trap. Yellow is the color most commonly used as it attracts many different insect species. Blue sticky traps will attract thrips. Unfortunately sticky traps will also attract some of the natural enemies. The sticky surface may need to be replaced if it gets coated with dust.

Light Traps

An ultraviolet fluorescent light positioned above a container with a lethal agent will attract a wide range of night-flying insects. These traps were widely used, but require an electricity supply so they may not be ideally situated to assess pest populations for a particular crop.

Pheromone Traps

These traps are of various designs, but all have a slow-release dispenser to emit a pheromone. Usually the male of a species is attracted when a sex pheromone is used, but some traps may be baited with an aggregation pheromone. The main advantage of pheromone traps is that they are much more selective than light traps, so the catch is easily counted, and they can be positioned in crops without needing a supply of electricity.

Other Sampling Systems

To sample larger crop canopies, especially trees, it is possible to spray a nonpersistent “knockdown” agent, such as a natural pyrethrin spray, into the foliage and sample the insects that fall out on a white tray or sheet.

Diagnostic Techniques

The presence of certain plant pathogens can be confirmed by using field test kits, using an enzyme-linked immunosorbent assay (ELISA). Early detection of a specific disease inoculum can allow the timing of a treatment to be optimized.

Meteorological Monitoring

Weather conditions have an important influence on the development of pest populations. In some situations it is economic to operate a small meteorological station close to or within a crop to ensure relevant weather information is immediately available to assist decision-making. Phenological models can be used to predict when an infestation may begin by accumulating day degrees above a defined threshold. Certain diseases develop when leaf surface is continuously wet for a defined period while above a minimum temperature.

Examples of IPM in Practice

Rice

Large areas of rice (*Oryza sativa*) are grown in the lowland irrigated areas of southeast Asia. In the 1970s, large areas were extensively treated prophylactically with insecticides as well as fertilizers, partly because of incentives to maximize yields of the new high-yielding varieties of the "Green Revolution." Later it was realized that in the attempt to grow several crops a year, a relatively minor pest, the brown planthopper, had become of major importance in the absence of its natural enemies destroyed by the sprays. Synchronized sowing, throughout a region, with gaps between successive crops was needed to interrupt the continuous food supply for the planthoppers. This was later combined with varieties selected for greater resistance to hopper damage. Unfortunately in many areas crop rotation is limited as few other crops, such as mung beans (*Phaseolus aureus*), can be grown in wet lowland areas.

Much emphasis is now given to farmer field schools to train farmers to recognize the many natural enemies that can suppress insect pest populations. Sprays of insecticide are now only recommended if a pest population exceeds an action threshold. These are avoided during the first 40 days after sowing, as plants can compensate for early damage by tillering.

Herbicide use is increasing. Farmers traditionally transplanted rice seedlings and controlled water levels as a means of weed management. Higher labor costs for transplanting is increasing the area of direct seeding of rice, so the farmer is increasingly dependent on chemical weed control. In some areas fungal diseases, such as sheath blight and rice blast, are sufficiently severe to warrant chemical control, although the use of disease-resistant varieties is preferred.

Corn

Corn (*Zea mays*, maize) is grown in many countries and may be affected by root pests (e.g., *Diabrotica virgifera*) and stem borers (e.g., *Ostrinia nubialis*, *Busseola fusca*). Ideally corn is grown in a rotation and in many tropical areas it is interplanted with other crops, but in areas of intensive monoculture, soil-applied insecticides have been extensively used. New transgenic corn cultivars may change this situation. In Africa, the first-generation stem-borer has been a major problem and in the absence of any suitably resistant varieties, many farmers apply insecticide granules into the leaf whorl. Later generations inflict less damage as individual plants are then sufficiently large to withstand feeding by a stem-borer larva. This pest overwinters in the base of the dry stem, so destruction of crop residues is an important part of the control program. The corn earworm (*Heliothis zea*) infestations are more serious on the sweeter varieties, which are normally harvested early. Where the sheath surrounding the cob is tighter, the larvae seldom inflict serious damage except at the extreme tip of the cob.

In parts of the tropics where more than one corn crop may be grown in a year it could be affected by other pests, including *Cryptophlebia leucotreta* and *Cicadulina mbila*, the latter being a vector of corn streak virus. This is most serious where corn is grown under irrigation during the dry season, when the wild grasses are not green and attractive for the leafhoppers.

In some areas of corn, sorghum and similar crops, where *Spodoptera exempta* armyworm outbreaks may occur, pheromone traps are used to detect concentrations of moths, so that farmers can be warned to inspect their fields and spray insecticide if needed. Surveillance schemes also operate for locust control in conjunction with forecasts issued internationally by the Food and Agricultural Organization (FAO). Efforts are made to control the hopper stage before plagues can build up and a mycoinsecticide is now available where locusts require control in ecologically sensitive areas.

Following harvest, a whole array of storage pests may infest the grain unless it is dried adequately and kept in cleaned stores. Recently *Prostephanus truncatus* has been of major importance in Africa, due to its spread from Central America. Apart from fumigation of infested grain, insecticides may have to be applied to the surfaces of stores or as an admixture with the grain, although in some countries controlled atmospheres are used to check postharvest insect pests.

Corn varieties are generally resistant to fungal diseases, although in some countries severe leaf disease such as grey leaf spot (*Cercospora zeae-maydis*) has been treated with fungicides. Southern leaf blight (*Cochliobolus heterostrophus* = *Helminthosporium maydis*) caused serious damage in the United States where large areas were sown with a susceptible variety. Seed is frequently treated with fungicides to facilitate crop establishment.

Corn yields will be seriously affected if weeds are not controlled during the initial growth stages. The main weed species are *Cyperus rotundus*, *Echinochloa colonum*, *Elusine indica*, and *Digitaria sanguinalis*, followed by *Cynodon dactylon*, *Sorghum halepense*, *Echinochloa crus-galli*, and *Portulaca oleracea*. In small tropical farms, traditionally weeding has been with hoes, but this is often slow and weed regrowth limits the area an individual farmer can cultivate. The parasitic weed *Striga* is also an early-season weed problem, although it is seldom noticed until much later. A suitable herbicide treatment is most effective, if applied early to prevent the weed becoming established. Herbicide-resistant corn varieties are now available and may form an important part of future corn IPM programs in Africa.

Cotton

Cotton (*Gossypium* spp.) has a reputation for receiving more chemical control than most other arable crops. However, it also provides a good example of diversity of IPM programs as it is grown in many different areas of the world. Cotton varieties with smooth leaves suffer extensively from jassid infestations (**Figure 3**), so early in the development of the largely rainfed cotton industry in sub-Saharan Africa, as in India it was recognized that hairy leaf was a host plant resistance (HPR) character that was essential to establish the crop. Later varieties with resistance to the disease *Xanthomonas axonopodis* pv *malvacearum* combined with jassid resistance were grown. Despite this HPR, yields of cotton remained low even with considerable research on different nonchemical control tactics, throughout the 1930s and 1940s. Ants and spiders were very effective predators but sufficient pests survived to cause economic damage to the cotton crops. A closed season with a specified uprooting date was introduced to minimize the population surviving from one season to the next, but this failed to affect insects that form diapause pupae in the soil. The arrival of synthetic insecticides allowed major increase in yield initially in the United States, but by 1960 it had been shown that yields even on small-scale farms in Africa could be doubled or trebled by applying insecticides such as carbaryl or dichlorodiphenyltrichloroethane (DDT).

In some countries these sprays were applied on a routine calendar schedule, but in Central Africa, a system of crop scouting was introduced to assess the number of bollworm eggs and other pests present in fields and determine when to spray and with which insecticide. Different insecticides were shown to be most effective against the two main bollworm pests, *Helicoverpa armigera* and *Diparopsis castanea.*

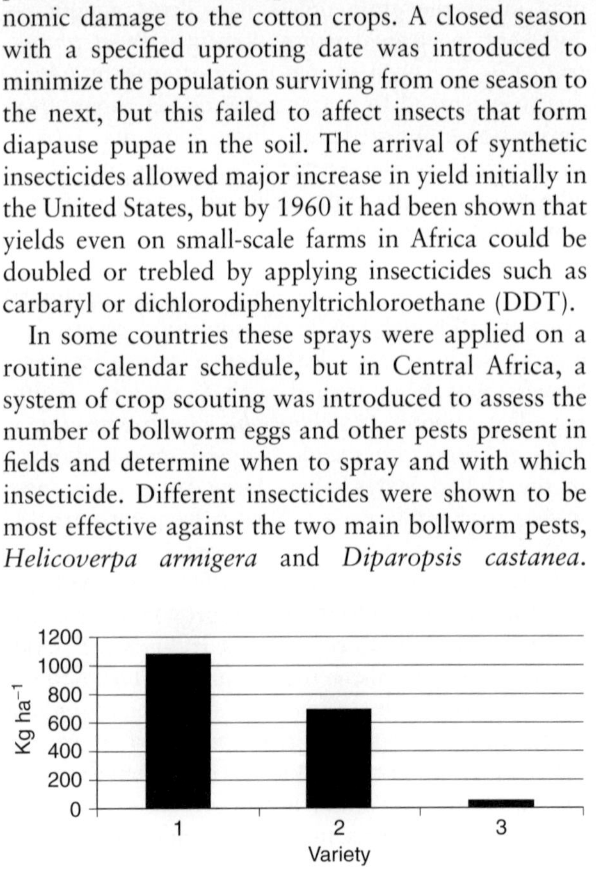

Figure 3 Yields of three varieties of cotton in Africa subject to jassid infestations, with different degrees of leaf hairiness: (1) albar cotton, (2) albacala, and (3) deltapine smooth leaf. Resistance to jassids reduces the period of chemical control to the time when bollworms are damaging the crops.

Applications were timed to control first instar bollworm larvae, as larger larvae were hidden within the bolls (fruit), so the dosage applied was kept low. This undoubtedly minimized any adverse effects on natural enemies.

Unfortunately the insecticides used to control bollworms reduced populations of predatory mites, so the red spider mite became a secondary pest. The organophosphate insecticide recommended for control of the mites was also used on irrigated vegetable crops throughout the year, so resistant mite populations were soon detected. To counteract this, Zimbabwe was one of the first countries to introduce a pragmatic rotation of different acaricides to control the tetranychid mites on cotton.

This contrasts with the situation in Central Asia and parts of China, where the crop is grown in very arid areas fed by irrigation water from the melting snows on the mountains. Here rotation of cotton with the fodder crop alfalfa (*Medicago sativa*, lucerne) has allowed natural enemies to suppress bollworm populations, which are reduced by extremely low winter temperatures. Cotton varieties were selected to provide resistance to the soilborne disease verticillium wilt. While many insecticides were used in these areas in the 1950s and 1960s, problems of ill health due to exposure to aerially applied insecticides led the Uzbek government to encourage the implementation of biological control. This led to the mass rearing of the parasitoids *Trichogramma* and *Bracon* and their release on a national scale on state farms. High yields of irrigated cotton were maintained, but it is not known whether effective insect control was due to greater survival of natural enemies following cessation of spray applications, or if the augmentation of natural enemies was the key factor.

In the United States, the boll weevil (*Anthonomus grandis*) which originated in Central America became the most important cotton pest. Insecticide sprays against bollweevil resulted in increased bollworm (*Helicoverpa zea* and *Heliothis virescens*) infestations (**Figure 2**). The crop was heavily sprayed, usually by aircraft, initially with cocktails such as DDT + methyl parathion + toxaphene and then with pyrethroid insecticides from the 1950s to the 1990s. Late-season sprays were emphasized against bollweevils to reduce the population entering diapause. In some areas, "weevil sticks" impregnated with its pheromone, grandlure, and an insecticide have been deployed as a "lure and kill" technique. This was aimed at the weevils entering cotton fields after their diapause. The situation has changed again more recently with the widespread adoption of transgenic Bt cotton. The number of sprays against bollworms has significantly declined, but treatments against a few relatively minor pests, notably certain sucking pests has increased in some areas.

In Australia, following failures to control *H. armigera* with pyrethroid insecticides there has been a resistance management strategy to minimize the use of any one type of insecticide. This remains in force even in areas where Bt cotton is grown as a proportion of the cotton area is sown with non-Bt transgenic cotton in an attempt to avoid the selection of insect pests resistant to the Bt toxin. At present, some new insecticides with a different mode of action are available to reduce reliance on the pyrethroid insecticides that have dominated cotton insect control. In Australia too, growing strips of alfalfa within cotton areas and spraying a food attractant on the cotton has been studied to conserve natural enemies and attract them into the cotton. Computer programs EntomoLogic, and previously SIRATAC, have been developed to assist farmers decide when to treat their cotton and with which insecticide.

In Egypt, a major pest of cotton is the pink bollworm, *Pectinophora gossypiella*. It survives from one season to the next as diapause larvae that remained in bolls attached to the dry stalks collected for fuel and stored on the roof of their houses. Insecticide sprays are not very effective against this pest as the larvae feed on seeds inside the bolls. Populations were however effectively suppressed by an area-wide deployment of its pheromone gossyplure as a mating disruption technique by attaching a slow release formulation to a sample of plants in each field. Use of the pheromone fitted in with biological control of the leafworm *Spodoptera littoralis* by the lacewing *Chrysoperla carnea* that overwinters on a fodder crop berseem. In Egypt, the bollworm *H. armigera* has not been a pest possibly due to the high gossypol content of the glands in long staple *Gossypium barbadense* cotton varieties but also due to the attractiveness of wide range of alternative crops, such as maize and tomatoes (*Lycopersicon esculentum*) grown in close proximity. In contrast in neighboring Sudan, fields are much larger and diversity of cropping is less so bollworm and whiteflies became major problems when aerial sprays were applied extensively at too low an action threshold.

In these examples of cotton IPM, the actual tactics depend very much on local circumstances so that the appropriate mix of tactics fits the farming system and climatic conditions. Where biological control is inadequate, any insecticide that is used must be carefully selected and applied at as low a dosage as possible to integrate with and conserve natural enemies.

Apples

IPM for apples (*Malus pumila*) is increasingly a part of integrated fruit production (IFP) to meet quality assurance schemes required by the retail market. Changes in cultivation techniques have led to a decrease in tree size, which facilitates harvesting and spray applications when needed. Windbreaks provide a habitat for some of the important natural enemies such as *Typhlodromus pyri*, while overwintering of earwigs (*Forficula auricularia*) which prey on certain pests such as summer fruit tortrix (*Adoxophyes orana*) is encouraged by provision of small boxes with straw.

Although there are many applications of fungicide in orchards to control scab and powdery mildew, crop monitoring has reduced the total amount of pesticide used and the broad-spectrum insecticides have been phased out. Encouragement of natural enemies and use of microbials and insect growth regulators has reduced insect pest pressure. The time at which an insecticide is needed, for example to control tortrix larvae, is determined increasingly by pheromone trap data. In some countries mating disruption has been used, although supplemental spraying is needed at the edge of the area-wide deployment of the pheromones.

To assist individual growers, computer-based decision support systems, such as GABY in The Netherlands, generate recommendations in accordance with local guidelines. Growers have to input monitoring records regularly for the recommendations to be relevant to their needs. The use of small meteorological stations has also assisted with the timing of fungicide applications in relation to factors such as leaf wetness over a period, when temperatures exceed a minimum value. Choice of fungicide is also important to avoid any adverse effects on natural enemies.

Vegetable Crops

In contrast to the large areas of crops referred to above, vegetable production tends to be on smaller areas with greater diversity of cropping, although to meet the demand of large supermarkets, some growers specialize in a limited number of horticultural crops. IPM in these crops is less advanced than in some other crops because retailers demand high-quality produce and do not tolerate any damage. Crops may reach maturity quite rapidly, so there is little time for biological controls to act on major pests. Crop rotation, seed treatment, including heat treatment, and the use of resistant cultivars, where available, are major components of IPM. However, pesticide sprays may still be required. The choice is limited as the agrochemical companies are reluctant to register products for these crops due to the relatively small areas of diverse crops, and the enormous costs of obtaining residue data when their potential income from pesticide sales is relatively low. Organic producers put considerable emphasis on botanical insecticides, but these are seldom sufficiently persistent or sufficiently effective.

Glasshouse/Protected Crops

In the confined environment of individual glasshouses, it is much easier to maintain control of the environment. Where spray had been applied too frequently to produce blemish-free produce, pests resistant to pesticides soon became apparent, and with crop failures, there was a more rapid acceptance of alternative control strategies. The distribution of *Phytoseiulus persimilis* was one of the first successful biological controls on cucumbers (*Cucumis sativus*). This was soon followed by release of *Encarsia formosa* for whitefly control. Various other natural enemies are now commercially available, including entomopathogenic nematodes. In some situations, a nonpersistent insecticide, such as bioresmethrin, may be needed to reduce a pest population prior to release of natural enemies, or applied very selectively if there is a localized infestation. Fine mesh is also used to reduce insect pests going into glasshouses from neighbouring weeds or crops.

Polyphagous Pests

One of the difficulties with certain pests, such as whiteflies, is their wide host range. Where horticultural crops are grown throughout the year or in glasshouses between outdoor field crops, the pest can easily migrate between hosts. Thus, whatever control measures or pesticides are used, resistance management strategies must be integrated over all crops within an area. When harvesting of a crop is completed and the plants are uprooted, an insect pest may then fly to another host, so an insecticide treatment may be required before destruction of the crop residues to stop this occurring. Nearby weeds, even outside a glasshouse, may have to be removed to reduce pest survival between crops.

In each of the crops illustrated above, there is a continuing need for research to provide updated recommendations to suit the dynamic changes in the pest situation, the availability of suitable cultivars and the influence of market forces on produce income. Careful recording of pest situations and the actions taken help to build up an important database to facilitate future decision-making.

See also: **Diseases**: Breeding for Disease Resistance. **Integrated Pest Management**: Principles. **Pests**: Genetic Modification of Pest Resistance, Insect Pests. **Seed Development**: Seed Treatments. **Weeds**: Weed Technology and Control; Herbicide Resistance.

Further Reading

Flint ML and Gonveia P (2001) *IPM in Practice – Principles and Methods of Integrated Pest Management*, 296 pp. Publication no. 3418. Oakland, California: University of California.

Heong KL and Escalada MM (1999) Quantifying rice farmers pest management decisions, beliefs and subjective norms in stem borer control. *Crop Protection* 18: 315–322.

Kenmore PE (1996) Integrated pest management in rice. In: Persley GJ (ed.) *Biotechnology and Integrated Pest Management*, pp. 76–97. Wallingford: CAB International.

Matthews GA (1997) Implementing cotton integrated pest management. *Experimental Agriculture* 33: 1–14.

Matthews GA and Tunstall JP (eds) (1994) *Insect Pests of Cotton*. Wallingford: CAB International.

Van Lenteren JC (1995) Integrated pest management in protected crops. In: Dent DR (ed.) *Integrated Pest Management*, pp. 311–343. London: Chapman & Hall.

Van Lenteren JC and Woets J (1988) Biological and integrated pest management in greenhouses. *Annual Review of Entomology* 33: 239–269.

Way MJ and van Emden HR (2000) Integrated pest management in practice: pathways towards successful application. *Crop Protection* 19: 81–103.

Disease Prediction Models

X B Yang, Iowa State University, Ames, IA, USA
S Sanogo, New Mexico State University, Las Cruces, NM, USA

Introduction

The development of a disease in a plant population is a dynamic process, having the nature of population growth of any biological organism. The population of a pathogen – which can be a fungus, a bacterium, or virus – develops when the host plant is susceptible and environmental conditions are suitable for the reproduction of the pathogen. Under normal production conditions, a population of a plant pathogen is at an equilibrium level and/or does not affect normal growth of plants or cause any economical losses.

Outbreaks of Disease

Outbreaks or epidemics of a plant disease occur when weather and host conditions are highly favorable to the growth of a plant pathogen. Disease outbreaks are the sudden increase of plant pathogens in a plant population within a short period of time, often a few days or weeks. Severe outbreaks of a disease in a crop can reduce yields to various levels, push individual producers out of business, or result in significant social consequence. For instance, the outbreaks of potato late blight in Ireland in the 1840s resulted in the Great Famine and the migration of about 1 million Irish people to North America. A recent example was the epidemic of southern maize blight in the 1970s in the US cornbelt, which caused farmers estimated losses valued over US$1 billion.

Prediction of the likelihood of a disease outbreak plays an important role in the prevention of disease epidemics. Timely and accurate predictions on the likelihood of disease outbreaks enable farmers to apply various preventative measures, such as spraying fungicides, to control the development of a pathogen population before it reaches damage levels, while reducing unnecessary application of fungicides. A reliable prediction that calculates the thresholds of disease outbreaks on a farm can be used by producers in making fungicide application decisions. Fungicides are applied only when necessary, reducing environmental damage. Disease prediction at a regional or national level provides useful information to assess potential risk of a disease, which is critical to making farming-related policies by governmental agencies.

Disease Prediction and Disease Management

A disease prediction model (also known as disease warning system) aims to estimate the probability of occurrence of and magnitude of a disease given a set of conditions of weather, host plants, and pathogens. A disease prediction model is made of a mathematical expression for the relationship between disease dynamics and these weather variables. It is the central tool for disease prediction. A disease prediction model often consists of mathematically structured parameters that capture the dynamic nature of a disease. Most disease prediction models use a set of weather variables and previous disease measurement as inputs to predict future development of a disease several days, weeks, or even months ahead.

Empirical and Mechanical Models

Botanical epidemiologists, who study plant disease prediction models, classify disease prediction models into empirical models and mechanical models. An empirical prediction model is derived from experimental data collected on a system or knowledge of a system accumulated through experience. Relationships are established among variables using regression models. Steps in the development of such models are as follows: (1) collection of quantitative or qualitative data on disease outbreaks for several years; (2) collection of weather data for each year of disease outbreak; and (3) development of regression models by relating data on disease outbreaks to data on weather. In these models, several variables (dependent variables) are expressed as a function of one or several variables (independent or predictor variables). For example, a regression model may be developed in which late blight severity (Y) is expressed as a function of temperature (T), relative humidity (RH), and rainfall (RF) as follows: $Y = b^0 + b^1T + b^2\mathrm{RH} + b^3\mathrm{RF}$. This expression signifies the importance of temperature, relative humidity, and rainfall in accounting for variation in late blight severity. However, this expression does not indicate any cause-and-effect relationship among variables, meaning that it cannot be inferred that changes in late blight severity are caused by temperature, relative humidity, and rainfall.

Mechanical models are based on information derived from a conceptual cause-and-effect relationship among variables. Such information is gathered from the literature or from formulated theories on interactions between weather, host plant, and pathogen. The models are tested by computer to assess how closely they represent (simulate) actual disease development. Improvements to the prediction may be made by altering components of weather, host plant, or pathogen.

Early Model Systems

Efforts to develop disease prediction models began in the first half of the twentieth century and since then have intensified. At first, these models were qualitative, that is they provided prediction merely in terms of discrete events such as disease occurrence versus no disease occurrence. Progressively, more models were formulated quantitatively, and provided prediction of disease occurrence with measurable levels of probability. Development of quantitative disease prediction models started in the 1950s and several disease prediction models using regression equations were developed. These systems provided disease warnings to guide fungicide applications. Early disease prediction models were less applicable because the amount of calculation needed was beyond the capability of farmers. Current disease prediction models are computer-based to handle large amounts of data collection, analysis, and prediction.

Computer-Based Disease Warning Systems

Computer-based disease warning systems have been developed since the 1970s. The first computerized disease prediction model – EPIDEM – was developed by Horsfall and Waggoner in 1969 on tomato (*Lycopersicon esculentum*) early blight (caused by *Alternaria solani*). To date disease prediction models have been developed for major horticulture and vegetable crops where fungicide applications are common. Most of these systems provide risk prediction based on environmental conditions favorable to disease development. The BLITECAST system was designed to provide risk prediction for late blight of potato (*Solanum tuberosum*) caused by *Phytophthora infestans*, based on weather variables such as temperature, relative humidity, and rainfall. The FAST and TOM-CAST systems were built to provide risk prediction for early blight and anthracnose in tomato. The WISDOM disease warning system was developed to predict the conditions which are favorable to early blight development in potato based on temperature and crop growth.

Many disease warning systems have been used in production. These prediction models are an important component in integrated pest management (IPM) for reducing fungicide usage. Commercial services for disease prediction are also available in California for controlling diseases in grape (*Vitis vinifera*) production. At the regional level, disease outbreaks have been correlated to extreme climatic events, such El Niño events. In the People's Republic of China, an annual prediction of the occurrence of wheat (*Triticum aestivum*) scab has been made using the Southern Oscillation Index, a measure of El Niño, and the results are published in national newspapers 3 months before the spring growing season.

See also: **Diseases**: Fungal Diseases; Plant Pathology, Principles. **Integrated Pest Management**: Principles; Practice.

Further Reading

Coakley SM (1988) Variation in climate and prediction of diseases in plants. *Annual Review of Phytopathology* 26: 163–181.

Gleason ML, MacNab AA, Pitblado RE, *et al.* (1995) Disease-warning systems for processing tomatoes in Eastern North America: Are we there yet. *Plant Disease* 79: 113–121.

Krause RA, Massie LB, and Hyre RA (1975) BLITECAST, a computerized forecast of potato late blight. *Plant Disease Reporter* 59: 95–98.

Madden L, Pennypacker SP, and MacNab AA (1978) FAST, a forecast system for *Alternaria solani* on tomato. *Phytopathology* 68: 1354–1358.

Zadoks JC (1984) A quarter century of disease warning. *Plant Disease* 68: 352–355.

N

NUTRITION

Contents

Mineral Uptake

R A Leigh, University of Cambridge, Cambridge, UK

Essential Nutrients and the Need for Selective Uptake

When plants are grown in disease free conditions with an adequate supply of light and water, their growth is limited by the availability of key mineral nutrients in the soil. These essential nutrients are needed by plants to complete their life cycle, and can be classified according to whether they are needed in large amounts (macronutrients) or only as trace quantities (micronutrients). A list of essential macro- and micronutrients is given in **Table 1** along with their approximate concentrations in plants. In addition to these elements, there are also beneficial nutrients. Plants can grow in the absence of beneficial nutrients, but the growth of some species is enhanced by their presence. Beneficial nutrients include sodium, silicon, and aluminum.

The concentrations of the different mineral elements in plants are not a simple reflection of those in the soil because root are able to absorb nutrients selectively. For instance, calcium (Ca) and magnesium (Mg) may be present in the soil solution at higher concentrations than potassium (K), but the reverse is often found in plants. Further, plants grow best when the concentrations of individual nutrients within their tissues fall within certain limits. When concentrations are too low, growth is limited by nutrient deficiencies; when they are too high, toxicities may occur. Therefore, roots must match the rate of nutrient uptake to growth to avoid dilution or over-concentration of nutrients within the plant's tissues. As any particular soil will not be able to supply all nutrients in optimal amounts, mechanisms have evolved that give plants the ability to enhance the supply of essential nutrients when they are poorly available. The operation of the uptake processes and their regulation are key to the performance of plants in agriculture, where nutrient supplies are artificially manipulated through fertilizer applications, and in natural ecosystems.

Nutrient concentrations in soil can vary over many orders of magnitude. For instance, there may be up to 10^{-2} mol l^{-1} K^+ in the soil solution of a well-fertilized soil, whereas Fe^{3+} can be as low as

Table 1 Essential nutrient elements with typical concentrations in tissues

Element	*Chemical symbol*	*Concentration in plant dry matter (mg kg^{-1})*
Macronutrients		
Nitrogen	N	15 000
Potassium	K	10 000
Calcium	Ca	5000
Magnesium	Mg	2000
Phosphorus	P	2000
Sulfur	S	1000
Micronutrients		
Chlorine	Cl	100
Iron	Fe	100
Manganese	Mn	50
Zinc	Zn	20
Boron	B	20
Copper	Cu	6
Nickel	Ni	0.1
Molybdenum	Mo	0.1

$10^{-15}\,mol\,l^{-1}$, especially in aerobic conditions and at alkaline pH. Nonetheless, the roots must ensure appropriate amounts are taken up at both ends of this range if growth, and ultimately reproduction, are not to be adversely affected. When roots fail to do this, plants show a range of visual deficiency symptoms, including poorer growth and discoloration.

Root Structure and Delivery of Nutrients to the Shoot

Roots form highly branched structures that provide a large surface area for nutrient and water absorption, and they ramify through the soil profile to tap new sources of nutrients. Further, their structure responds to local conditions, often proliferating in regions of high nutrient availability. The ability of roots to access and absorb nutrients is enhanced by the development of root hairs and by the formation of symbiotic relationships with mycorrhizal fungi, which grow in intimate contact with roots and extend their hyphae into the surrounding soil. Both root hairs and mycorrhizal fungi increase the volume of soil exploited and are particularly important in the acquisition of relatively soil immobile elements, like phosphorus (P) and zinc (Zn). In addition, secreted chemicals can modify the root environment to increase nutrient solubility, or promote the formation of beneficial associations with mycorrhizal fungi and, in legumes, with nitrogen fixing bacteria.

The structure of the root (see **Figure 1**) is key to understanding the transport processes that govern the movement of nutrients from the soil solution to the shoot. The epidermis and cortex are the sites of nutrient and water uptake, while the cell layers of the stele are responsible for loading of nutrients into the xylem for delivery to the shoot. The xylem is a dead tissue and essentially acts as a pipe connecting the roots with the aerial tissues. Transport up it is by mass flow of solution driven by the transpiration of water from the leaves. The stele is surrounded by the endodermis, which has a hydrophobic layer, the Casparian strip, in its cell wall, and this blocks the extracellular movement of ions and water at this point. Cells within the root are linked by plasmodesmata that cross the cell walls and link the cytoplasms of neighboring cells. The plasmodesmata have a complex internal structure, and are regulated transport pathways that permit water, ions, and small macromolecules to pass between cells.

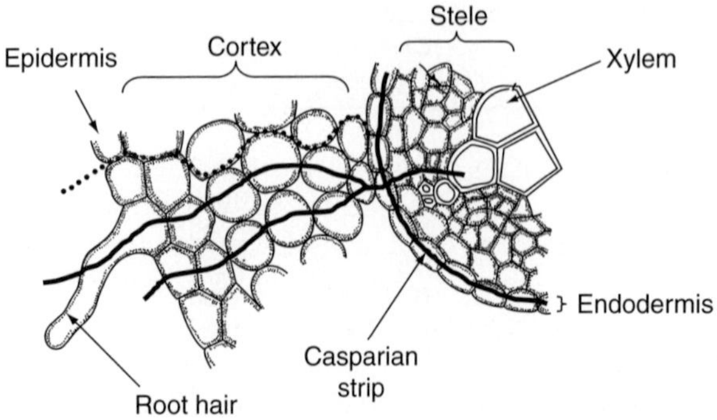

Figure 1 Spatial relationships between different cell types in a typical root. Ions may reach the stele by an apoplastic pathway (dotted line) or via the symplastic pathway (solid line). The Casparian strip blocks the apoplastic pathway at the endodermis. The Casparian strip does not traverse the cells of the endodermis in the way shown, but encircles them in the cell wall. Adapted from Sutcliffe JF (1962) *Mineral Salts Absorption in Plants.* London: Pergamon Press.

Nutrients moving from the soil solution to the root xylem can reach the endodermis by two different pathways. They can move through the cell walls (apoplast) of the epidermal and cortical cells without entering a cell. This route is not particularly ion selective because the processes governing movement through it are purely physical (e.g., by mass flow of solution driven by transpiration). However, the apoplast is blocked by the Casparian strip and to enter the stele, nutrients must be taken up across the plasma membrane of the endodermal cells. An exception to this occurs in locations where the Casparian strip is not present (e.g., root apices) or is disrupted (e.g., where lateral roots form). At these sites, the ions can reach the stele without leaving the apoplast. Alternatively, the nutrients may be taken up into epidermal or cortical cells, moving to the stele through the symplast, the system of plasmodesmatally linked cytoplasms. Thus, in the main part of the root with a functioning Casparian strip, any nutrient delivered to the stele has to be transported into a cell, either in the epidermis, the cortex, or the endodermis. Once in the stele, ions must be transported out of the stelar cells to reach the xylem. Hence, there is a minimum of two membrane transport processes, one an uptake step, the other a secretion, involved in delivering ions to the xylem. It is at these points that ion selectivity can be imposed because the suite of transport proteins present in the cell membranes will determine which ions can be taken up by the cells or secreted from them.

Uptake Properties of Roots

Roots are able to absorb nutrients over a wide range of external ion concentrations, and, provided the rate of delivery to the root surface is high enough, plants can grow at maximal rates in the presence of very low external nutrient concentrations. This is exemplified by experiments conducted by Asher and Ozanne in Australia in the 1960s. They grew plants in rapidly flowing nutrient solutions containing a range of potassium ion (K^+) concentrations. The speed with which the solution flowed past the roots' surface ensured that the rate of delivery of K^+ was high. Their results showed that individuals of all 14 species they grew could survive in 1 μmol l^{-1} K^+, a few grew maximally in 8 μmol l^{-1} K^+, while the majority needed only 24 μmol l^{-1} K^+ for maximal growth. These concentrations are very much lower than those present in soil solutions of fertile soils (1–10 mmol l^{-1} K^+). Asher and Ozanne also found that increases in the external K^+ concentration were not matched by proportionately large changes in shoot or root K^+ concentrations. Thus, a 1000-fold increase in the external K^+ concentration (from 1 to 1000 μmol l^{-1}) resulted in increases in plant weight of between twofold and 30-fold, depending on the species, and a fivefold to 15-fold change in the K^+ concentration in the shoots. These observations indicate that K^+ uptake by roots can operate at very low concentrations, and that the rate of uptake is matched to growth to regulate tissue K^+ concentrations. Other experiments with nitrate and phosphate have provided similar results, indicating that the same principles operate for other nutrients.

The concentration dependence of ion uptake by roots indicates two distinct kinetic mechanisms. One operates at low external concentrations (typically below 0.5–1 mmol l^{-1}), the other at higher concentrations (>1 mmol l^{-1}). This was intensively investigated for K^+ by Epstein and coworkers, who showed that, at external K^+ concentrations below approximately 1 mmol l^{-1}, uptake has a high affinity for K^+, is saturable, and is highly selective for K^+ (System 1); whereas at higher external K^+ concentrations, a low affinity mechanism (System 2) operates, which does not discriminate strongly between K^+ and Na^+. These properties suggest that two distinct transport processes, with quite different properties, are operating in K^+ uptake into roots. Similar low and high affinity systems also operate for other nutrients.

Active and Passive Uptake

Plant cells usually contain higher concentrations of nutrients than the solution surrounding them. This is most easily demonstrated with K, which is present both in the soil and the plant as K^+ and hence is not metabolized or incorporated into other molecules, so a direct comparison of internal and external concentrations is easy. In the experiments of Asher and Ozanne, the roots of barley (*Hordeum vulgare*) grown in 1 μmol l^{-1} K^+ contained 10 mmol l^{-1} K^+ (10 000-fold difference), while those grown in 1000 μmol l^{-1} K^+ contained 135 mmol l^{-1} (135-fold difference). Similarly, others have shown that roots of barley plants grown in 20 mmol l^{-1} NO_3^- contain about 80 mmol l^{-1} NO_3^-. Although these results indicate that uptake is against a concentration gradient, defining whether this requires the direct input of metabolic energy needs more detailed investigations to determine if the transport is thermodynamically "downhill" (passive) or "uphill" (active). Only the latter requires the direct input of energy.

To make an assessment of whether transport of a particular ion is active or passive requires measurement of the difference in its electrochemical potential on either side of the membrane. The electrochemical potential gradient is a measure of the difference in free energy between pools of ions in two locations, in this case outside and inside a cell. Ions move passively from regions where they have a high electrochemical potential to regions where they have a low one. Active transport is required if movement is in the opposite direction. If there is no difference in electrochemical potential then there is thermodynamic equilibrium. To measure the electrochemical potential difference in relation to ion uptake, the voltage difference (membrane potential) between the cell and the external solution and the concentrations of the ion on each side of the plasma membrane must be measured. Typically, plant cells have a membrane potential of -80 to -150 mV (negative inside the cell with respect to outside), although values much more negative than this have been measured. This inside-negative membrane potential provides a driving force for the passive uptake of positively charged cations such as K^+ and Ca^{2+}, and for the passive efflux of anions such as NO_3^- or SO_4^{2-}. At thermodynamic equilibrium, the membrane potential and concentrations of ions on each side of a membrane at 20°C are related by the following equation derived by Walter Nernst, a 1920 Nobel Prize winner:

$$E = \frac{58}{z} \log_{10} \frac{a_o}{a_i}$$

where E is the membrane potential, z is the valency of the ion, and a_o and a_i are its thermodynamic activities in the compartments on either side of the membrane. As thermodynamic activities are more difficult to measure than concentrations, most studies have used the latter (i.e., c_o and c_i), although this can lead to inaccuracies at high (>100 mmol l^{-1}) concentrations, where values of activities and concentrations diverge. For convenience, the rest of this discussion will refer to concentrations rather than activities, but the last caveat should be borne in mind.

What the Nernst equation shows is that, at equilibrium at 20°C, a membrane potential of -58 mV can be balanced by a 10-fold higher concentration of a monovalent cation (e.g., K^+) within the cell, or a 10-fold lower internal concentration of a monovalent anion (e.g., Cl^- or NO_3^-). For divalent ions (e.g., Ca^{2+} or SO_4^{2-}), a membrane potential of only −29 mV is needed to achieve the same differences in concentration. Hence, simply comparing concentrations inside and outside a cell without knowledge of the membrane potential does not allow any conclusions to be drawn about whether the concentration difference results from active or passive transport. However, if E, c_o and c_i (or better still a_o and a_i) are known, these can be inserted into the Nernst equation to see if there is thermodynamic equilibrium. If there is equilibrium, passive transport can be assumed; if there is not, the need for active transport must be invoked.

The external concentration (c_o) of a nutrient can be experimentally manipulated and so is known. As the solution bathing the roots is often dilute, c_o and a_o are usually similar in value. The membrane potential can be measured by inserting a fine electrode into the cell, while c_i (or a_i) can be accurately measured with ion selective microelectrodes containing a sensor that responds to changes in the concentration or activity of the ion in question. Such studies for K^+ have shown that uptake at low external concentrations (in the range of System 1) is active, while uptake at higher concentrations (System 2) is passive. This is also true for other monovalent cations such as NH_4^+, while divalent cations such as Ca^{2+} and Zn^{2+} can usually enter passively even at low external concentrations. For instance, at typical Ca^{2+} concentrations in the cytoplasm ($<1\,\mu mol\,l^{-1}$), and with a membrane potential of −116 mV, passive transport of Ca^{2+} can occur from an external solution containing only $10\,nmol\,l^{-1}$ Ca^{2+}. Active transport is generally required for uptake of anions such as NO_3^-, SO_4^{2-}, and $H_2PO_4^-$, unless the cells are very deficient, when passive transport can operate. For instance, NO_3^- concentrations are very low in the cytosol of NO_3^- starved root cells, and, with a sufficiently high external nitrate concentration, passive uptake is possible if the membrane potential is about −80 mV or more positive. However, after the cells have been bathed in $10\,mmol\,l^{-1}$ NO_3^-, the concentration in the cytoplasm rises to about $5\,mmol\,l^{-1}$, and this can only be maintained by active inward transport of nitrate. **Table 2** gives a summary of the need for active and passive transport in the uptake of a selection of nutrient ions at different external concentrations. For some ions, e.g., Na^+ and Ca^{2+}, their concentrations in plants are much lower than expected from the Nernst equation and active extrusion must be occurring.

Table 2 Role of active and passive uptake in the acquisition of a selection of essential macronutrient ions by plants

Ion	*External concentration ($mmol\,l^{-1}$)*	*Active or passive uptake*	*Probable mechanism*
K^+	<0.5	Active	H^+ Symport
K^+	>1	Passive	Ion channel
NH_4^+	<0.2	Active	H^+ Symport
NH_4^+	>1	Passive	Ion channel
NO_3^-	Most concentrations	Active	H^+ Symport
$H_2PO_4^-$	Most concentrations	Active	H^+ Symport
SO_4^{2-}	Most concentrations	Active	H^+ Symport

The transport of nutrients from the stelar cells into the xylem appears to be largely passive. Measurement of the electrochemical gradient for K^+ and Cl^- across files of cells between the epidermis and xylem has shown that the major difference occurs in the uptake step into the cells. The secretion of ions into the xylem is often energetically downhill for both ions. However, for divalent cations such as Ca^{2+} and Zn^{2+}, there must be active transport to the xylem because the thermodynamic gradients do not favor passive secretion out of the stelar cells. Relatively little is known about the transport processes operating in the stelar cells, although the development of methods to separate these cells is beginning to allow some of their properties to be investigated. Studies of isolated cortical and stelar cells have shown that both possess inwardly and outwardly directed K^+ transporters and that, in some species at least, the outward K^+ transport is higher in stelar cells, as expected if they play a role in the loading of K^+ into the xylem. Whether these differences between the two cell types are due to different suites of transporters, or differential regulation of similar transporters, is unclear, although stelar-specific K^+ transporters have been identified (see "Molecular Identities of Root Ion Transporters"). Studies of root cells from corn (*Zea mays*; maize) have shown that outwardly directed K^+ transport in stelar cells is downregulated in response to water stress and the hormone abscisic acid, but this does not occur in cortical cells.

Energization and Mechanisms of Transport

The mechanisms responsible for active and passive transport of ions into plant cells are now understood in general terms. Transport at the plasma membrane is energized by an ATP dependent H^+ pump (H^+-ATPase) that catalyzes the outward active transport of H^+. This creates a negative membrane potential and a pH gradient across the plasma membrane. These differences in voltage and pH can be used to

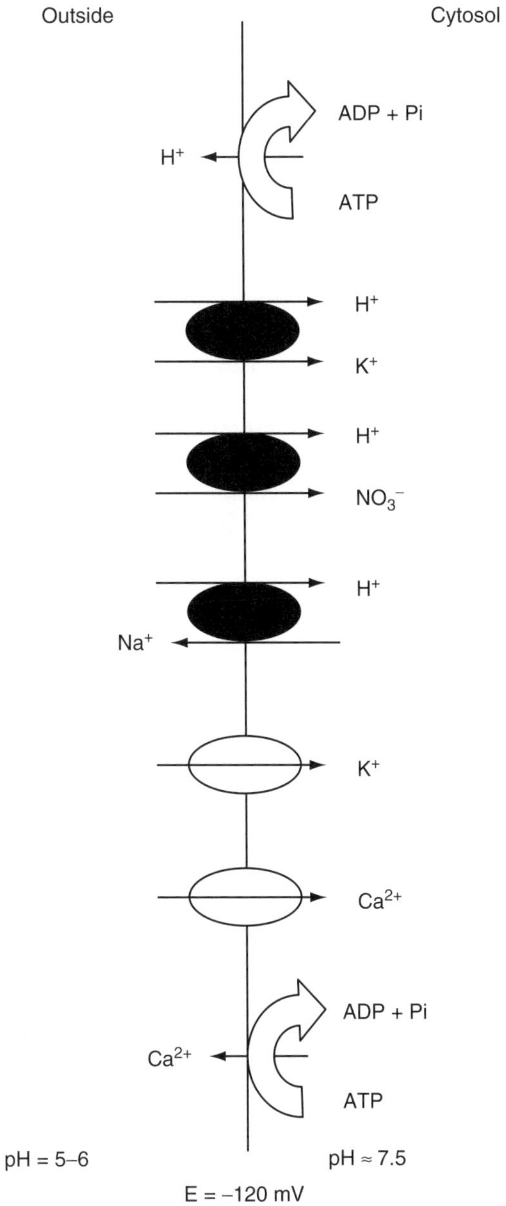

Figure 2 Role of ATP dependent pumps, H^+ linked cotransporters, and ion channels in the uptake of some macronutrients. The ATP dependent H^+ pump actively transports H^+ out of the cell and generates a pH gradient and membrane potential across the plasma membrane. These H^+ and electrical gradients are then used to drive the uptake of the other ions either actively via H^+ symports (e.g., for NO_3^-), or passively through ion channels (e.g., Ca^{2+}). Some ions, such as Ca^{2+} and Na^+, are at much lower internal concentrations than expected from the Nernst equation, and are actively extruded by H^+ antiports (Na^+) or ATP dependent pumps (Ca^{2+}). The values of pH and membrane potential are only indicative, and may vary considerably with time and environmental conditions.

drive the uptake of ions through ion channels and H^+ linked cotransporters (**Figure 2**).

In general, passive ion transport is mediated by ion channels. These are regulated protein pores which, when open, allow the very rapid transport of ions down their electrochemical potential gradient. Thus, depending on the direction of this gradient, the transport can be into or out of a cell. However, some channels will only open under conditions that permit transport in one direction, so restricting the opportunity for ion flow in the opposite direction (they are said to show rectification). In addition, channels can be very selective and permit only particular ions (e.g., K^+ or Ca^{2+}) to flow through them, whereas others are relatively unselective. However, all ion channels are characterized by very high rates of transport (10^6–10^8 ions s^{-1}), which is sufficient to flood or deplete a cell of ions in a very short time. Therefore, ion channels are highly regulated and open only at particular membrane potentials (voltage gated channels), or in the presence of certain chemical effectors such as Ca^{2+}, Ca^{2+}/calmodulin, ATP, cyclic nucleotides, or glutamate (ligand gated channels). Investigations in root cells have shown that uptake of K^+ within the range of System 2 is mediated by voltage gated, K^+ selective channels that permit inward, but not outward, transport of K^+. Such channels have been shown in epidermal and cortical cells of roots from a range of species. The activity of these channels has been found to increase following K^+ starvation, suggesting an important role for them in enhancing K^+ uptake in K deficient plants. Other K^+ permeable channels have also been described in roots, such as nonselective cation channels, but their significance in overall K^+ uptake is unknown. However, there is growing evidence that these nonselective channels may be involved in Na^+ influx in saline conditions, and so they are now being studied intensively because they may be important in understanding how to make crops more salt tolerant. Many other channel types have been detected in plants with probable involvement in the uptake of Ca^{2+}, Zn^{2+}, and Fe^{2+}, and anion influx (at low internal concentrations) or anion efflux (at high internal concentrations).

Active ion transport of the majority of nutrient ions is mediated by H^+linked cotransporters that use the energy stored in the pH gradient across the plasma membrane to drive the uphill transport of another ion. Thus influx of H^+ can be used to energize uptake or efflux of other ions. Active uptake processes at the plasma membrane (e.g., for K^+ and NO_3^-) are catalyzed by H^+ symports (H^+ and the ion both move into the cell), while active efflux (e.g., of Na^+) involves H^+ antiports (H^+ moves in as the ion moves out). Physiological experiments have provided evidence to support the involvement of H^+ linked cotransporters in the uptake of K^+, NO_3^-, $H_2PO_4^-$, and Cl^-, and the efflux of Na^+. For instance, the provision of NO_3^- to cells previously deprived of this ion causes transient changes in the membrane potential and external pH. The membrane potential initially becomes less negative before returning to its initial value, and the external pH becomes more alkaline, then also returning to close to its initial value. These effects are explained by the operation of a H^+ linked transporter in which two or more H^+ are cotransported with each NO_3^-. The stimulation of the transporter by the addition of its substrate, nitrate, causes a net influx of positively charged H^+, which results in the effects on membrane potential and external pH. These effects are transient because the activity of the H^+-ATPase is enhanced by these changes, and this increases H^+ efflux, which restores the starting values.

In addition to energizing H^+ pumps, ATP hydrolysis is also directly coupled to the outward transport of Ca^{2+}. The concentration of free Ca^{2+} is kept very low (below 1 $\mu mol\,l^{-1}$) in the cytosol of all eukaryotic cells because it is an important component of intracellular signaling pathways, with rises in its concentration triggering a range of responses. At typical membrane potentials and prevailing external Ca^{2+} concentrations, there is a massive driving force for the passive uptake of Ca^{2+} across the plasma membrane, and there is insufficient energy in the H^+ gradient to maintain the required low cytosolic concentration using only H^+ linked Ca^{2+} antiports. Therefore, ATP dependent Ca^{2+} pumps (Ca^{2+}-ATPases) operate to transport Ca^{2+} out of the cytoplasm into the external solution (and into the vacuole). Recently, metal ion dependent ATPases have been discovered in plants, and are thought to play a role in the transport of micronutrients such as Zn^{2+} and Cu^{2+}, but a role in uptake into roots has not yet been demonstrated.

Molecular Identities of Root Ion Transporters

The identification of the proteins responsible for active and passive uptake of ions has progressed rapidly in the last decade, mainly through the use of the yeast *Saccharomyces cerevisiae* as a system in which both to clone genes and to characterize their products, and more recently through the completion of the sequencing of the genome of the model plant *Arabidopsis thaliana*. A major outcome of these efforts has been the realization that plant ion transporters are encoded by large multigene families, and that for any single nutrient ion there may be a plethora of proteins involved in its transport. For instance, there are 12 plasma membrane H^+-ATPase genes in *Arabidopsis*, 14 Ca^{2+}-ATPase genes, and over 150 genes for cation transporters of which 35 are putative K^+ transporters (both channels and H^+ linked cotransporters). There are also multiple genes encoding NO_3^-, NH_4^+, and $H_2PO_4^-$ transporters. Having multiple proteins for the same task allows plants flexibility in their adaptations to environmental conditions, and is probably an evolutionary adaptation to their sessile lifestyle. However, this remains speculative because in only a few cases have the cellular location and roles of particular proteins been fully elucidated. Determining the role of each individual protein, and how it is expressed during development, or in response to changing environmental conditions, is now a major challenge. The scale of this challenge can be highlighted with reference to attempts to identify genes encoding proteins involved in K^+ uptake into roots.

A number of plant K^+ transporters have been identified by functional complementation of the K^+ transport-deficient *trk1 trk2* mutant of yeast. This mutant is unable to grow at low external K^+ concentrations, and plant genes encoding K^+ transporters have been identified by their ability to restore growth of the mutant on low K^+. Transporters that have been identified include channels responsible for inward and outward K^+ transport, as well as high-affinity active K^+ transporters. Among those that have been shown to be expressed in roots are AKT1 (an inward voltage gated K^+ channel), SKOR (a voltage gated K^+ channel), HKT1 (initially identified as a H^+ linked K^+ symport, but now thought to function as a $Na^+:K^+$ or $Na^+:Na^+$ symport), and members of the KUP/HAK family (dual affinity transporters related to fungal $H^+:K^+$ symporters).

AKT1 and SKOR are K^+ channels from *Arabidopsis* and are structurally very similar. They both have six membrane spanning domains (S1–S6), a preponderance of positive charges in S4 that probably represents the voltage sensing domain, and a hydrophilic domain (P domain) between S4 and S5 that forms part of the ion conducting pore, and is where K^+ selectivity is imposed on the transport process. Despite this high level of structural similarity between AKT1 and SKOR, they transport K^+ in opposite directions, with AKT1 catalyzing uptake, and SKOR catalyzing efflux. This emphasizes that physiological function cannot be inferred from structural similarities, and so every putative transporter for each individual ion must be studied to determine its physiological function and context. Localization studies for AKT1 and SKOR have shown that they are expressed in different parts of the roots. AKT1 is expressed mainly in root epidermal and cortical cells, while SKOR is localized in stelar cells. Thus, AKT1 provides a route for K^+ uptake, and SKOR a pathway for secretion into the xylem. These functions have been confirmed using *Arabidopsis* mutants in which the *AKT1* and *SKOR* genes have been disrupted. Mutants lacking AKT1 are compromised in K^+ uptake, while those lacking SKOR have diminished K^+ transport to the shoot.

HKT1 was cloned from wheat and is preferentially expressed in cortical cells of roots. Its gene product was initially characterized as a $H^+:K^+$ symport mediating high-affinity active transport of K^+. However, subsequent work showed that it probably functions as a $Na^+:K^+$ symport at low external Na^+ concentrations, and as a $Na^+:Na^+$ symport at higher external Na^+ concentrations. Physiological experiments have provided no support for Na^+-coupled K^+ transport in terrestrial plants, so the precise role of HKT1 remains unclear. It currently seems more likely that it is important in Na^+ uptake. Nonetheless, its involvement in high-affinity K^+ transport is supported by the observation that its expression is rapidly upregulated in response to removal of external K^+, and with a time course that matches changes in K^+ uptake.

Potassium transporters of the KUP/HAK subtypes have been isolated from a variety of plants, and their structural and sequence similarities indicate that they form a family of related transporters with 13 members in *Arabidopsis*. They have 12 membrane-spanning domains and a long cytosolic C-terminal domain. The proteins are similar to both bacterial and fungal high-affinity K^+ transporters. The plant proteins have been shown to mediate high-affinity K^+ transport when expressed in yeast, but it is unclear whether this transport is active or passive.

How K^+ uptake is partitioned between different transporters remains unclear. A simple explanation that assigns active transport by HKT1, HAK, and KUP transporters to System 1 and passive transport by AKT1 or other ion channels to System 2 is complicated by observations that suggest that all of these transporters may be capable of mediating transport at both $\mu mol\,l^{-1}$ and $mmol\,l^{-1}$ external K^+ concentrations. For instance, the *Arabidopsis* mutant lacking *AKT1* was found to be compromised in K^+ uptake at external K^+ concentrations as low as $10\,\mu mol\,l^{-1}$, well below the range in which AKT1 was presumed to be important in catalyzing influx. However, it was found that the root cells in the mutant had very negative membrane potentials, and this provided a sufficient driving force for passive K^+ uptake from this low external K^+ concentration. Thus, the relative contributions of active and passive transport to nutrient uptake may turn out to be more complicated than predicted.

Regulation of Nutrient Uptake

Nutrient acquisition and uptake are regulated to maximize the chances that nutrient concentrations in tissues will be above the minimum needed for growth and reproduction, and to ensure they do not become so high that toxicities occur. The highly branched root system ensures the soil is adequately explored and, under nutrient deficient conditions, the root : shoot ratio increases, indicating diversion of additional photosynthate to the roots to enhance nutrient capture. When a root system encounters a nutrient-rich patch, lateral roots often proliferate through branching and/or increased elongation to ensure efficient exploitation of the resource. This enhanced root growth is controlled by both local signals in the root environment and long-distance signals from the shoot. In *Arabidopsis*, the rate of lateral root elongation in response to localized nitrate is controlled by direct sensing of the nitrate, and involves a signaling pathway that depends on the *ANR1* and *AXR4* genes. Other experiments have shown that the nitrogen status of the shoot is also important in controlling root proliferation, although the long-distance signal involved has not been identified. Root hairs and mycorrhizal associations increase the volume of soil exploited around each root. In addition, roots secrete chemicals (e.g., organic acids, reductants, and chelators) that enhance the solubility of nutrients to increase their availability. Some species, e.g., lupin (*Lupinus albus*), form so-called “proteoid roots” when they are deficient in P and/or Fe. These are clusters of small rootlets that grow from the main root axis and secrete citric acid and other chemicals to lower the pH and increase the solubility of P and Fe.

Acquisition of Fe is a major problem for the majority of plants because it is present in the soil solution at very low concentrations, often below the level needed to support adequate growth. Therefore, plants have evolved special mechanisms for increasing the availability of this essential element. In most species, growth in Fe deficient conditions results in enhanced acidification of the soil, the secretion of reductants, and an increase in the activity of a NAD(P)H dependent Fe^{3+} reductase in the plasma membrane. The low pH and secreted chemicals increase the solubility of Fe^{3+} in the root environment, and the reductase reduces it to Fe^{2+}, which is then taken up through an Fe^{2+} permeable channel. In grasses, a different mechanism operates involving secretion of specific Fe^{3+} chelators called phytosiderophores. These are simple molecules that are able to solubilize and chelate Fe^{3+} from insoluble Fe^{3+} compounds, and deliver it to the root surface where the Fe^{3+}–phytosiderophore complex is taken up and the Fe released within the plant.

Induction of transport proteins is another common response to lack of nutrients. Thus, under K-, S-, and P deficient conditions, the rate of uptake of K^+, SO_4^{2-}, and $H_2PO_4^-$ is increased. This is achieved by the expression of genes encoding the appropriate transporters. These effects are specific, so lack of S only increases the expression of SO_4^{2-} transporters, but has no effect on those for other anions. This enhanced expression is reversible, so once the plant reaches sufficiency in a particular nutrient, the expression of the transporter genes is downregulated, so ensuring that the suite of transporters present in the plasma membrane is adapted to the needs of the plant. The internal signals involved in this up- and downregulation are not known with certainty, but may be the ion itself or one of its early metabolites. For instance, the expression of NH_4^+ and NO_3^- transporters is inversely related to the concentration of glutamine in roots, suggesting that this amino acid is a component of the signaling pathway that controls transcription of the genes. In addition, internal NH_4^+ and NO_3^- concentrations may also directly regulate the activities of transport proteins, inhibiting uptake when internal ion concentrations are high. These multi-level effects, which have been observed for many nutrient transporters, ensure that the roots respond both to long-term levels of supply and to short-term fluctuations in the immediate availability of an ion within the cell.

Concluding Remarks

The uptake of nutrients is fundamental to the growth of plants, and influences their ecological competitiveness and their agricultural productivity. In addition, plants are the major entry point for nutrients into food chains, so the uptake processes that occur in roots are of fundamental importance to the health of all animals, including humans. Despite their simple morphology, roots are highly sophisticated organs that can adapt their structure and biochemistry to regulate nutrient uptake, so ensuring that plants maximize their chances of acquiring essential nutrients in adequate amounts. In the future, the challenge is to modify these processes to enhance the efficiency of nutrient use in agriculture, and to ensure plants contain adequate essential nutrient elements to meet the dietary needs of societies where deficiency diseases could be overcome if crops were more nutritious.

List of Technical Nomenclature

Active transport Transport against the electrochemical potential gradient, requiring the input of energy.

Apoplast The extracellular space.

Beneficial nutrient A nutrient that is not essential for growth, but that enhances growth in some species.

Casparian strip The hydrophobic barrier in the cell walls of the endodermis.

Cortex The layers of cells between the root epidermis and the stele.

Electrochemical potential Measure of the free energy of an ion in a particular location.

Endodermis The single layer of cells surrounding the stele and separating it from the cortex.

Epidermis The outer cell layer of roots.

Essential nutrient A nutrient that is absolutely required for growth.

H^+ Antiport A cotransporter that couples the downhill movement of H^+ to the active transport of an ion or solute in the opposite direction across the membrane.

H^+ Linked cotransporters Transporters that link the active uptake of an ion or compound to the transport of H^+ ions down their electrochemical potential gradient.

H^+ Symport A cotransporter that couples the downhill movement of H^+ to the active transport of an ion or solute in the same direction across the membrane.

Ion channels Regulated protein pores that span membranes and allow the transport of ions down their electrochemical potential gradients.

Membrane potential Voltage difference across a membrane.

Mycorrhizae Specialized fungi that grow in symbiotic relationships with plant roots.

Passive transport Transport down the electrochemical potential gradient.

Phytosiderophores Fe^{3+} chelating molecules secreted by grasses to increase the availability and uptake of Fe.

Plasmodesmata Regulated junctions between cells.

Proteoid roots A cluster of rootlets formed on the main root axes of some species as an adaptation to P or Fe deficiencies.

Rectification Ability of an ion channel to permit transport of ions only in one direction.

Root hairs Tubular outgrowths of root epidermal cells.

Stele The central tissues of the root.

Symplast The linked cytoplasms of adjacent cells, connected by plasmodesmata.

Xylem The long-distance transport pathway responsible for delivering water and nutrients to the shoot.

See also: **Nutrition**: Deficiency Diseases, Principles; Ion Transport. **Root Development**: Mycorrhyzae; Root Growth and Development; The Rhizosphere and its Microorganisms. **Water Relations of Plants**: Salt Stress; Xylem.

Further Reading

Asher CJ and Ozanne PG (1967) Growth and potassium content of plants in solution cultures maintained at constant potassium concentrations. *Soil Science* 103: 155–161.

Barbier-Brygoo H, Vinauger M, Colcombet J, *et al.* (2000) Anion channels in higher plants: Functional characterization, molecular structure and physiological role. *Biochimica et Biophysica Acta* 1465: 199–218.

Epstein E (1972) *Mineral Nutrition of Plants: Principles and Perspectives*. New York: John Wiley.

Forde BG (2002) Local and long-range signaling pathways regulating plant responses to nitrate. *Annual Review of Plant Biology* 53: 203–224.

Marschner H (1995) *Mineral Nutrition of Higher Plants*, 2nd edn. London: Academic Press.

Mäser P, Thomine S, Schroeder JI, *et al.* (2001) Phylogenetic relationships within cation transporter families of *Arabidopsis*. *Plant Physiology* 126: 1646–1667.

McCully ME (1999) Roots in soil: unearthing the complexities of roots and their rhizospheres. *Annual Review of Plant Physiology and Plant Molecular Biology* 50: 695–718.

Nobel PS (1999) *Physicochemical and Environmental Plant Physiology*. San Diego: Academic Press.

Palmgren MG (2001) Plant plasma membrane H^+-ATPases: Powerhouses for nutrient uptake. *Annual Review of Plant Physiology and Plant Molecular Biology* 52: 817–845.

Raghothama KG (1999) Phosphate acquisition. *Annual Review of Plant Physiology and Plant Molecular Biology* 50: 665–693.

Taiz L and Zeiger E (2002) *Plant Physiology*, 3rd edn. Sunderland, Mass: Sinauer Associates.

Tester M and Leigh RA (2001) Partitioning of nutrient transport processes in roots. *Journal of Experimental Botany* 52: 445–457.

Williams LE and Miller AJ (2001) Transporters responsible for the uptake and partitioning of nitrogenous solutes. *Annual Review of Plant Physiology and Plant Molecular Biology* 52: 659–688.

Ion Transport

P J White, Horticulture Research International, Wellesbourne, UK

Introduction

Ions are defined as the positively (cations) and negatively (anions) charged particles into which the atoms or molecules of salts, acids, or bases are dissociated by solution in water. They include simple inorganic ions as well as complex organic compounds, such as amino acids, peptides, nucleotides, and organic acids. The electrical charge and hydrophilicity of ions inhibit their movement across lipid bilayers and transport proteins are required to accelerate their movement across biological membranes. This allows plants to control the movement of ions within both cells and tissues and thereby to regulate the ionic content of cells and cell compartments.

The ability to control the flux of ions across membranes allows the plant: (1) to maintain aqueous compartments of contrasting solute composition and to generate ionic and electrical gradients across cell membranes compatible with cell metabolism, (2) to regulate the transfer of ions involved in electrical (e.g., action potentials) and/or ionic (e.g., Ca^{2+}) signals, (3) to generate cell turgor (largely through the accumulation of ions within vacuoles), and (4) to distribute ions selectively within the plant. Thus, ion transport across membranes is fundamental to cytoplasmic homeostasis and cellular detoxification, energy generation in mitochondria and chloroplasts, the compartmentation of intermediary metabolism, cell signaling, cell volume regulation and extension growth, and nutrient uptake and distribution within the plant. This article focuses on the basic principles and mechanisms of ion transport across both cell membranes and over greater distances in the xylem and phloem.

The Movements of Ions: Basic Principles

To describe the movement of ions within the plant, it is necessary to introduce various kinetic and thermodynamic concepts. These allow ion fluxes, electrical potentials and the energy requirements of ion transport to be described quantitatively. In this section, we first consider the driving forces for "passive" diffusion in terms of concentration gradients and electrical fields and then the energetics of "active" transport processes that move ions across membranes against their electrochemical gradient.

Diffusion is a consequence of the random thermal movement of ions. It results in a redistribution of ions from a more concentrated to a more dilute region. It is considered as a passive process and rates of diffusion are directly proportional to the driving forces causing the movement. In the absence of electrical or gravitational fields, hydrostatic pressure or temperature gradients, diffusion can be described quantitatively by Fick's Laws (**Figure 1**). Fick's Second Law can be used to illustrate the distances over which diffusive fluxes can be effective. For

Fick's First Law states that the amount of species j passing a unit area per unit time (the flux density; J_j) is related to the change in its concentration (c_j) with distance (x) by the equation:

$$J_j = -D_j(\delta c_j/\delta x)$$

The negative sign indicates net diffusion away from the region of high concentration. The coefficient of proportionality, D_j, is known as the diffusion coefficient. It is not constant, but varies with concentration and temperature.

Fick's Second Law describes how the concentration of a species changes with position and time as a result of diffusion. For the special case in which D_j is constant and j is placed initially in a plane at the origin ($x = 0$ at $t = 0$) it holds that:

$$c_j = \{M_j/[2(\pi D_j t)^{0.5}]\}e^{-x2/4Djt}$$

where M_j is the total amount of j per unit area initially, and c_j is its concentration at position x at any later time t. Although this is only one solution to the second-order differential equation representing Fick's Second Law, its conditions can be met experimentally by inserting a radioactive tracer at $t = 0$. This equation can also be used to calculate the time taken for a proportion of j to diffuse a certain distance from the origin. For example, the time taken for 50% of an original population of molecules ($t_{0.5}$) to diffuse a given distance ($x_{0.5}$) is given by:

$$x_{0.5}^2 = 2.8\ D_j\ t_{0.5}$$

Figure 1 Fick's Laws.

The application of Fick's First Law to the diffusion of an ion (j) across a membrane requires an estimate of the concentration gradient of j within the membrane. This is approximated as the concentrations in the aqueous solutions on the outside (c_{jo}) and inside (c_{ji}) of the membrane multiplied by a partition coefficient (K_j), which expresses the ratio of the concentrations in the aqueous solution to those in the membrane, divided by the thickness (Δx) of the membrane:

$$J_j = D_j K_j (c_{ji} - c_{jo})/\Delta x$$

or $\quad J_j = P_j (c_{ji} - c_{jo})$

where $\quad P_j = D_j K_j/\Delta x$

where P_j is known as the permeability coefficient. Although the values D_j, K_j, and Δx are unknown, P_j is a measurable quantity and can, in theory, be determined empirically.

Figure 2 Diffusion of ions through a membrane.

example, it can be calculated that it takes 0.5 s for half a population of KCl ions, which have a diffusion coefficient (D_j) approximating $1.9 \times 10^{-9}\,m^2\,s^{-1}$ at 20°C, to traverse the distance equivalent to a cell (50 μm). This figure contrasts starkly with the estimated time taken for half a population of KCl ions to travel to the height of a small plant (0.5 m), which is over $1\frac{1}{2}$ years. Thus, although diffusion may be adequate to distribute ions within a cell (where movement may also be aided by cytoplasmic streaming), it will not suffice to distribute ions throughout a plant.

The diffusion of ions across membranes is orders of magnitude slower than that in aqueous solutions. Furthermore, there is a basic problem in applying Fick's Law to describe the diffusion of ions through a membrane, since their concentration in the membrane is unknown. Generally, the concentration gradient within the membrane is approximated from the concentrations in the aqueous solutions on the outside (c_{jo}) and inside (c_{ji}) of the membrane (**Figure 2**). However, it is difficult to estimate c_{ji} and c_{jo} in practice since a boundary layer exists in which the concentration of ions changes with distance from the membrane surface. This is a consequence of the solution close to the membrane being unstirred, and the concentrations of ions in this unstirred layer are influenced by their rate of transport across the membrane and by the charge associated with the membrane and, in intact plant cells, with the cell wall. Ions cross the unstirred layer by diffusion and, under some circumstances, this can become rate-limiting for those ions that are transported rapidly across cell membranes. Estimates of permeability coefficients range from about $10^{-6}\ m\ s^{-1}$ for ethanol to $10^{-9}\,m\,s^{-1}$ for a small ion such as K^+.

In addition to the concentration gradient, other factors influence the diffusion of ions. A more complete definition of the driving force for the movement of an ion (the negative gradient of its chemical potential, $-\delta\mu_j/\delta x$) combines the effects of temperature, ion concentration (or more strictly activity, a_j), hydrostatic pressure, electrical potential, and gravity (**Figure 3**). The parameter μ_j is a measure of the Gibbs free energy associated with a chemical species. The changes in μ_j between the initial and final states of a process define whether energy is required. The Gibbs free energy decreases for a spontaneous process at constant temperature and pressure and this decrease equals the maximum amount of energy available for work. Conversely, an increase in Gibbs free energy represents the minimum amount of work required to effect a transition between two states. This equation is important when considering the energetics of ion transport. For ion transport across membranes, under most biological conditions, differences in the concentration and electrical terms dominate this equation, and the driving force for ion movements is generally considered to be the electrochemical gradient.

The chemical potential (Gibbs free energy) of an ionic species (μ_j) is defined as:

$$\mu_j = \mu_j^* + RT \ln a_j + V_j P + z_j FE + m_j gh$$

The term μ_j^* is a reference chemical potential (defined at molar concentration, atmospheric pressure, no electrical field, and zero height). Other terms include the absolute temperature (T), ion activity (a_j), hydrostatic pressure (V_jP, where V_j is the partial molal volume of j and P is pressure in excess of atmospheric), electrical potential (z_jFE, where z_j is the valency of j, F is Faraday's constant, and E is voltage), and gravity (m_jgh, where m_j is mass per mole and h is the vertical position in the gravitational field).

Figure 3 Chemical potential of an ion.

When ions are transported across a membrane a charge separation occurs. This creates an electrical potential difference termed the membrane diffusion potential. The magnitude of the diffusion potential is related to the mobility of ions crossing the membrane. The diffusion potential of a plant cell is dominated by the fluxes of K^+, Na^+, and Cl^-. If it is assumed that ions move independently, and that the voltage difference decreases linearly across the membrane (the "constant field" assumption), then the measured membrane potential (E_m) is given by the Goldman–Hodgkin–Katz (GHK) voltage equation (**Figure 4**). Since the diffusion potential is frequently not the only component of the membrane potential (since energized H^+ and Ca^{2+} fluxes may also contribute significantly), the GHK voltage equation can be used to determine the "transport state" of the membrane (i.e., whether transport is dominated by diffusional or active transport processes). If ionic activities are known, and the zero-current (reversal) potential (E_{rev}) is determined, permeability ratios for ions (but not absolute permeabilities) can be calculated for diffusional processes using the GHK voltage equation. If there is only one permeant ion, E_{rev} becomes the Nernst potential (E_j) for that ion. The Nernst equation (**Figure 4**) allows the direction of the net diffusive ion fluxes to be determined at a given membrane potential. At $E_m < E_j$, cations can move into the cell, and anions out of the cell, by diffusion. At $E_m > E_j$ the opposite will occur. Under physiological conditions, it is observed that many cations are in flux equilibrium across the plasma membrane. For example a membrane potential of about -107 mV is required to maintain a difference in K^+ activities between the cytoplasm ($70\,\mathrm{mmol\,l^{-1}}$) and the external solution ($1\,\mathrm{mmol\,l^{-1}}$) at 20°C. These values are typical for many plant cells. However, anions cannot be concentrated in the cytoplasm by diffusion

The *Goldman-Hodgkin-Katz (GHK) voltage equation* relates the measured membrane potential (E_m) to the fluxes of ions across the membrane. For the monovalent ions generating the diffusion potential of a plant cell (K^+, Na^+, and Cl^-) the GHK voltage equation simplifies to:

$$E_m = RT \ln(P_K a_{Ko} + P_{Na} a_{Nao} + P_{Cl} a_{Cli}) / (P_K a_{Ki} + P_{Na} a_{Nai} + P_{Cl} a_{Clo})$$

Where P and a are the permeability coefficient and ionic activity for the subscripted ions, respectively. The GHK voltage equation becomes substantially more complex if divalent ions are also included. If there is only one permeant ion, the zero-current (reversal) potential (E_{rev}) becomes the *Nernst potential* (E_j) for that ion, and the *Nernst equation* defines the condition where there is no net flux of ions and the membrane potential balances the concentration gradient:

$$E_j = RT/z_jF \ln (a_{jo}/a_{ji})$$

Figure 4 The Goldman–Hodgkin–Katz voltage equation.

across the plasma membrane, and there is a considerable electrochemical gradient driving both Ca^{2+} influx from the external solution (millimolar Ca^{2+}) to the cytoplasm ($100\,\mathrm{nmol\,l^{-1}}$ Ca^{2+}) and Na^+ influx in saline environments.

The GHK current equation (**Figure 5**) allows the voltage dependence of diffusive ion fluxes to be calculated under specified ionic conditions, and also the absolute permeability of an ion to be calculated from a current measurement if ionic concentrations and membrane potentials are known. It also allows the independent influx (I_j^{in}) and efflux (I_j^{out}) of ions (as might be measured using tracer flux experiments) to be calculated. The ratio of unidirectional fluxes has been used by plant physiologists to test for the independent transport of ions by Ussing's criterion. In the absence of flux coupling, the ratio of unidirectional fluxes is directly proportional to the ratio of electrochemical activities of the permeant ion on either side of the membrane.

The GHK, Nernst, and Ussing equations can all be used to describe diffusion at the macroscopic (membrane) and microscopic (molecular) scales. At the molecular level, diffusion of ions across membranes is facilitated by ion channels. Here, there is a serious caveat. The assumptions embedded in the GHK equations are frequently inappropriate at the molecular level. Firstly, the permeation of different ions is not independent. Permeability coefficients vary with ionic concentration, fluxes saturate, and ratios of unidirectional fluxes do not conform to Ussing's criterion. Secondly, the electrical field does not drop linearly across the pore of an ion channel since both channel structure and permeating ions perturb the local electrical field. For these reasons the

The *GHK current equation* defines the relationship between ionic fluxes and voltage. It is often convenient to express the fluxes of ions in terms of their electrical current (I), which is expressed in amperes. One ampere is equivalent to a flux of a monovalent ion of 10 μmol s^{-1}. The net current carried by ionic species j is given by:

$$I_j = P_j z_j^2 EF^2/RT \{[a_{ji} - a_{jo} \exp(-v_j)]/[1 - \exp(-v_j)]\}$$

where: $v_j = z_j EF/RT$.

In addition, because this equation is derived assuming independent ionic movements, the independent influx (I_j^{in}) and efflux (I_j^{out}) of ions is given by:

$$I_j^{in} = P_j z_j F v_j\ a_{jo}/[1 - \exp[(v_j)]$$

$$I_j^{out} = P_j z_j F v_j\ a_{ji}/[1 - \exp(-v_j)]$$

The ratio of unidirectional fluxes has been used by plant physiologists as a test for the independent transport of ions. According to *Ussing's criterion*, in the absence of coupling between ion fluxes, the ratio of unidirectional fluxes is directly proportional to the ratio of electrochemical activites of the permeant ion in the solutions on either side of the membrane:

$$I_j^{out}/I_j^{in} = a_{ji}/a_{jo} \exp(z_j EF/RT)$$

Figure 5 The Goldman–Hodgkin–Katz current equation.

GHK equations should be regarded as (over)simplified descriptions of electrodiffusion through ion channels. A number of more complex models describing the permeation of ions through channels are available, which are based on energy profiles or protein structure.

In addition to transport processes that facilitate diffusion, ion transport across membranes can be catalyzed by active transport processes that move ions against their electrochemical gradient. This is achieved by coupling the transport of a particular ion to either the hydrolysis of an energy substrate (generally ATP or pyrophosphate), termed primary active transport, or to the movement of a second ion down its electrochemical gradient, termed secondary active transport. Since the electrochemical gradient of a second ion must first be generated by primary active transport, the different forms of active transport are consequently linked. In plant cells the primary active transport of H^+ against its electrochemical gradient is used to power secondary active transport of other ions.

In quantitative terms, the available energy in the cleavage of ATP is about 40 kJ mol^{-1} and 20 kJ mol^{-1} is available from the cleavage of pyrophosphate (PPi) under physiological conditions. The available energy in the proton electrochemical gradient is defined in terms of the proton electrochemical potential (μ_{H+}) (**Figure 3**), which can be simplified to $\Delta\mu_{H+} = EF - 2.3\ RT\ \Delta pH$. When the movement of one ion is coupled to the movement of another ion or to the hydrolysis of an energy substrate, it is necessary to consider the energy budget for the complete process to determine the direction of transport or whether a certain mechanism can occur. This can be calculated simply by adding the energy inherent in the electrochemical gradients of individual ions and the energy available from any hydrolytic reactions.

It should be noted that ionic fluxes catalyzed by both passive and active transport processes may saturate at high ion concentrations. This phenomenon results from the interactions (binding) between a transported ion and the transport protein. It has commonly been interpreted quantitatively in terms of the Michaelis–Menten equation, where K_m (Michaelis constant) and $I_j^{in}{}_{max}$ (conventionally referred to as V_{max}) represent the ion concentration at which flux is half-maximal (often referred to as the affinity of the transport process) and the maximal flux, respectively: $I_j^{in} = (I_j^{in}{}_{max}\ c_{jo})/(K_m + c_{jo})$.

Through the conventional relationships of inhibitor kinetics this relationship has also been used to quantify the interactions between ions competing for the same transporter and, thereby, to define the specificity of a transport process. However, the Michaelis–Menten equation presents a simplistic interpretation of transporter-catalyzed fluxes, and a single transporter protein may generate flux–concentration relationships that are far more complex than the Michaelis–Menten equation. Thus, care must be taken in the interpretation of the many flux versus concentration relationships observed under physiological conditions that are described by the summation of multiple Michaelis–Menten equations. Such data may not indicate that several mechanisms contribute to the transport of a particular ion, and further confirmatory evidence is required to support this supposition.

Biochemical Mechanisms for Ion Transport across Membranes

Transport proteins (transporters) are required to catalyze the movement of ions across biological membranes. They can be divided into several categories (**Figure 6**; **Table 1**). The activities of primary active transporters (pumps) are directly dependent upon the hydrolysis of energy substrates such as ATP and PPi. The secondary active transporters or "coupled transporters" harness the electrochemical gradient of a driving ion, which is generally the proton (H^+) in plants, to the movement of another ion in either the same (symport) or opposite (antiport) direction. Both pumps and

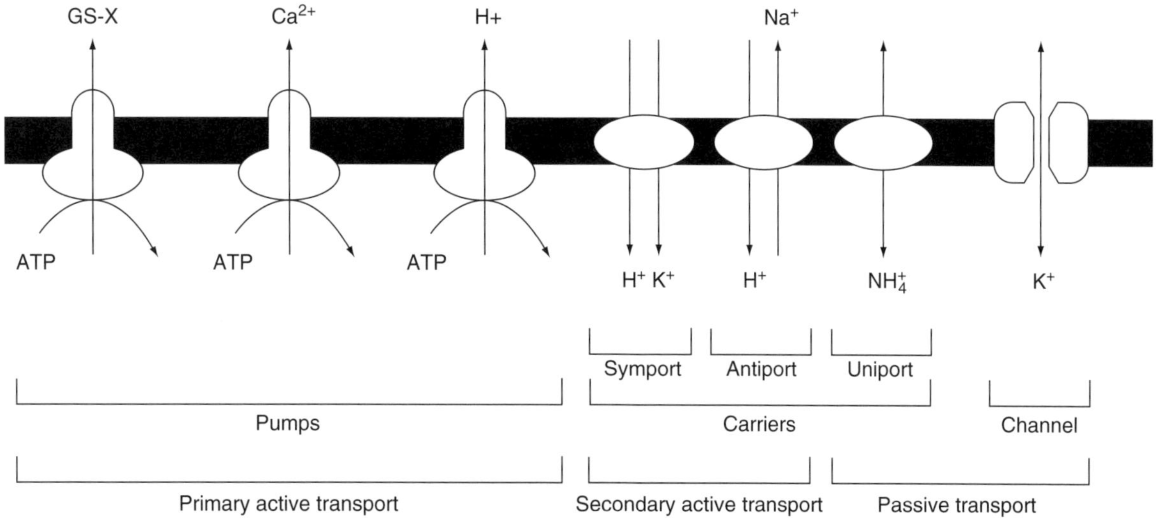

Figure 6 Schematic representation of primary active transport mechanisms, such as ABC transporters (e.g., glutathione conjugate pump), metal transporters (e.g., Ca^{2+}-ATPase) and H^+-ATPases, secondary active transport mechanisms, such as the K^+/H^+ symport or the Na^+/H^+ antiport, and passive transport mechanisms, such as the NH_4^+ carrier or a K^+ channel.

coupled transporters can transport ions against their electrochemical gradient. The passive transporters catalyze the movement of ions down their electrochemical gradient. This category includes a variety of ion carriers (uniporters) and channels.

Primary Active Transporters

Primary active transporters couple the transport of ions to the generation (in mitochondria or chloroplasts) or hydrolysis of an energy substrate such as ATP or PPi. They fall into several distinct classes. The F-type ATPases of mitochondria and chloroplasts produce the ATP for cellular processes. The P-type ATPases, so called because they form a phosphorylated intermediate during their reaction cycle, include: (1) the plasma membrane H^+-ATPases that acidify the cell wall and generate the extremely negative membrane potential and H^+ electrochemical gradient required for the uptake of solutes, (2) the Ca^{2+}-ATPases that maintain the low cytoplasmic Ca^{2+} concentration required for Ca^{2+} signaling and energy metabolism, (3) the heavy-metal pumps (CPx-ATPases) that detoxify the cytoplasm of Zn, Cu, Cd, and Ag, and (4) the phospholipid flippases that generate lipid asymmetries between the two faces of the bilayer. The V-type H^+-ATPases, which are composed of three major and several minor polypeptide subunits, and H^+-PPiases acidify the vacuole and the lumen of the endoplasmic reticulum/Golgi and contribute to both the membrane potential and H^+ electrochemical gradient required for the accumulation of solutes within these cellular compartments. The ABC transporters, so called because they contain an ATP-binding cassette motif, transport a variety of unusual compounds. They contribute to the detoxification of herbicides and other organic xenobiotics, to the removal of oxidation products conjugated to glutathione, to the sequestration of heavy metals, and the vacuolation of natural pigments and chlorophyll catabolites. There are many protein families within the ABC superfamily of transporters, and each family generally has its own specificity.

Carrier Proteins

Carrier proteins bind a solute on one side of the membrane and deliver it to the other side through a change in conformation. The transported solutes may be small organic molecules or inorganic ions. Their transport may be passive, down their electrochemical gradient, or actively coupled to the electrochemical gradient of another solute. Carrier proteins are often highly selective, and the complement of carrier proteins in a particular membrane is uniquely specialized to the requirements of the compartment that the membrane surrounds. In the plasma membrane there are carriers to import inorganic ions and nutrients such as sugars and amino acids. Since these must frequently be accumulated against their electrochemical gradient, their transport is generally coupled to the H^+ electrochemical gradient by a symport mechanism. In the tonoplast, carrier proteins serve to remove cytotoxic solutes, such as Na^+ and Ca^{2+}, from the cytoplasm. Again the transport of these ions is coupled to the H^+ electrochemical gradient, but by an antiport mechanism. A variety of carrier proteins are present

Table 1 Families of selected transport proteins in *Arabidopsis thaliana*, listed by proposed mechanism and including an estimate of the number of genes according to The Institute for Genomic Research (TIGR) *Arabidopsis* gene index (January 2002)

Mechanism	*Gene family names*	*Estimated number*
1. Primary active transporters		
H^+-ATPase (P-type)	AHA	12
Ca^{2+}-ATPase (P-type)	ACA, ECA	14
CPx-ATPase (P-type)	PAA, RAN, HMA	7
Phospholipid flippase (P-type)		12
H^+-ATPase (V-type)		19 (all subunits)
H^+-PPiase	AVP	3
ABC transporters	PGP, MRP, PDR, TAP	128
2. Secondary active transporters		
H^+/K^+ symport	KUP, HAK, AtKT	13
H^+/K^+ antiport	KEA	6
H^+/Na^+ antiport	NHX, NaD, CHX	27
H^+/Ca^{2+}, Cd^{2+} antiport	CAX	11
H^+/Fe^{2+}, Mn^{2+}, Zn^{2+}, Co^{2+}, Cd^{2+}, Cu^{2+}	Nramp	7
H^+/NO_3^- symporter	NRT, NTP	16
$H^+/H_2PO_4^-$	APT	10
H^+/SO_4^{2-}	AST	12
H^+/sugar	STP, SUC, SUT	30
H^+/amino acids	CAT, AAP, ProT, AUX, LAX, ANT	25
H^+/oligopeptide	PTR	5
3A. Passive transporters: carriers		
NH_4^+	AMT	5
Zn^{2+}, (Co^{2+}, Cd^{2+})	ZAT, MTP	8
Cu^{2+}	COPT	3
Zn^{2+}, Cd^{2+}	IRT, ZIP	15
Fe^{2+}, Zn^{2+}, Mn^{2+}, Cd^{2+}, Co^{2+}	MGT	10
3B. Passive transporters: channels		
Aquaporins	PIP, TIP, NLM	30
K^+ channels		
Inward rectifying	AKT, KAT, AtKC	7
Outward rectifying	GORK, SKOR, KCO	8
Beta subunit	KAB	1
Cl^- channels	CLC	4
Ca^{2+} channels (annexins)	Ann	1
Cation channels	CNGC	20
Cation channels	GLR	20

on the inner membranes of mitochondria and chloroplasts. These are required for the specific biochemistry of these organelles.

Ion Channels

Ion channels are integral membrane proteins. They span the phospholipid bilayer to form an aqueous pore through which ions cross the membrane. They can be distinguished from uniport carriers by their turnover rate. When a single ion channel is open it can transport up to 10 million ions per second. This is several orders of magnitude greater than the turnover rate of a carrier protein. Ion channels are selective for the ions they pass, this selectivity being generated by a structure termed the selectivity filter situated at the mouth of the pore. In addition to the pore-forming subunit, some ion channels also have beta (regulatory) subunits. The activity of ion channels is regulated by gating mechanisms. These are generally controlled by changes in membrane potential, the binding of ligands, covalent modification, or block by cytoplasmic solutes. Ion channels serve four basic functions: (1) to transport nutrients across membranes, (2) to accommodate osmotically significant fluxes over short periods, for example during the closing of stomata or the movements of touch-sensitive plants, (3) to propagate signals along or across membranes, which would include the propagation of electrical signals via action potentials or of chemical signals (such as elicitors) via ligand-gated channels, and (4) to maintain an optimal membrane potential or constant electrochemical gradient; this is important, for example, in controlling nutrient uptake in roots.

Genes for Transport Proteins

Analysis of the recently completed genome of *Arabidopsis thaliana* provides an insight into the diversity of genes encoding pumps, carriers, and channels in plants (**Table 1**). The *Arabidopsis* genome contains 25 500 genes. During its evolution it underwent a duplication of the entire genome, followed by gene loss and extensive local gene duplications. Hence, genes encoding many proteins belong to large families. The functions of 69% of the *Arabidopsis* genes have been classified based on sequence similarity to proteins of known biological function and over 600 genes have been identified that encode putative membrane transport systems.

The *Arabidopsis* genome contains an abundance of primary ATP-dependent transporters. There are large numbers of genes encoding the P-type ATPases involved in metal transport and a multitude of ABC transporters. By contrast, genes encoding the H^+-PPiase and subunits of the V-ATPase are present in lower numbers. Genes encoding a variety of H^+-coupled transporters are present. These include transporters for monovalent and divalent cations, and for inorganic anions such as phosphate, sulfate, and nitrate. Proton-coupled sugar and amino acid transporters are well represented, and a small family of genes encoding H^+-coupled oligopeptide transporters is present. However, few genes encoding transporters for carboxylates have been identified. There are small families of genes encoding putative carrier proteins for ammonium and various divalent cations and a significant proportion of the genes for transporters encode channel proteins. Genes encoding aquaporins that facilitate water movement across membranes account for a substantial number of the genes for channel proteins. Genes for Cl^- channels and channels that mediate K^+ fluxes both into (AKT, KAT, AtKC) and out of (GORK, SKOR, KCO) plant cells have been identified. Although the *Arabidopsis* genome lacks channels with significant homologies to animal Ca^{2+} channels, nonspecific cation channels, such as cyclic nucleotide gated channels (CNGC) and ionotropic glutamate receptor (GLR) channels, are well represented.

Transport of Ions between Cells and within Tissues

The cytoplasm of neighboring cells is connected by plasmodesmata. These structures allow the movement of solutes of low molecular mass, proteins, and nucleic acids between cells. They also provide a low-resistance pathway to the movement of electrical currents. The cytoplasmic continuum formed by plasmodesmatal connections is termed the symplast. Most cells have between 1 and 10 plasmodesmata per μm^2, which occupy 0.1–0.5% of the surface area of the cell. However, the density of plasmodesmata varies according to cell type and also spatially within the cell.

Plasmodesmata arise both during and subsequent to cell division. They have a diameter of about 60 nm. They are lined with a plasma membrane and also contain a central structure termed the desmotubule, which is thought to consist of a tightly rolled cylinder of membrane with no central lumen. The transport of solutes occurs in the space between the desmotubule and the plasma membrane lining, which is termed the cytoplasmic sleeve. The bore of the cytoplasmic sleeve is about 5.6 nm, but it is occluded by "particles" of 4.5–5 nm in diameter. Based on these observations, the cross-sectional area of spaces in which solutes can travel has been calculated to be between 2.6 and 3.6 nm^2. It has also been calculated that diffusion through the aqueous channels of plasmodesmata may increase the potential flux of ions between adjacent cells by several orders of magnitude.

Plasmodesmata are dynamic structures and can control the passage of solutes through the symplast. The size of macromolecules able to diffuse through plasmodesmata is increased substantially by viral movement proteins. This allows infective virus to spread throughout the plant. Endogenous plant proteins perform a similar function and it is thought that this is important for the movement of mRNA and other signals during cell differentiation. Plasmodesmata are shut by increasing cytoplasmic Ca^{2+} concentration. This may be related to a primitive role of cytoplasmic Ca^{2+} in cell signaling. It serves to isolate a cell experiencing an abiotic or biotic challenge, for example following pathogen attack. Plasmodesmata can also be blocked by the apposition of new wall material. This happens to the plasmodesmata formed between xylem parenchyma and immature xylem vessels as they become conductive.

Ions may also move in the aqueous solution outside the plasma membrane in the cell walls, intercellular spaces, and the lumina of tracheary elements. This region is called the apoplast. When ions diffuse through the apoplast they interact with the fixed (predominantly negative) charges in cell walls and cell membranes. Their movement is restricted by the suberized Casparian bands that develop in the anticlinal (radial and transverse longitudinal) walls of the endodermal and epidermal cells of most roots and by the periderm that develops from the cork cambium in the pericycle of some dicotyledonous plants during secondary thickening. The Casparian band is thought to ensure that solutes destined for the shoot first pass through the root symplast and, thereby, to effect selectivity in the long-distance transport of ions. Thus, the development of the Casparian band gives rise to contrasting longitudinal profiles in the delivery to the xylem of ions that traverse the root symplastically (K^+) or apoplastically (Ca^{2+}).

Long-Distance Transport of Ions in the Xylem and Phloem

Ions are circulated within the plant between root and shoot. Ions move from the root to the shoot in the xylem, and from the shoot to the root in the phloem. The principal role of ion transport in the xylem is to supply the shoot with mineral elements. The phloem is used to supply photosynthate to developing or nonphotosynthetic tissues, distributing assimilated nitrogen and sulfur compounds and redistributing nutrients during seed set, senescence, or periods of deficiency. In addition, since the phloem ionic composition often reflects the nutrient status of the shoot, it has been speculated that this might provide a signal to align ion uptake and translocation with shoot demand.

Transport in the Xylem

The xylem is composed of dead cells with thick, lignified secondary cell walls. These form a low resistance pathway for the movement of solution. It comprises both narrow tracheids and vessel elements. The hollow vessel elements are joined in files and their adjoining end walls are perforated (perforation plates). The diameter of vessel elements can vary considerably, between 8 and 500 μm. The vessel elements are closely associated with living xylem parenchyma cells, which store carbohydrates and may regulate the fluxes of solutes into and out of the vessel elements.

Xylem sap is quite dilute, containing only about $10\,\text{mmol}\,\text{l}^{-1}$ inorganic nutrients plus some organic nitrogen metabolites. The driving force for the movement of xylem sap is hydrostatic pressure and xylem sap flows according to Poiseuille's Law: $J_v = -r^2\Delta P / 8\eta l$, where J_v is the volume of solution with viscosity η flowing per unit area in a cylinder of radius r and with a difference in (hydrostatic) pressure ΔP over a distance l. Measurements of ΔP and estimates of the ΔP required to attain the solution flows observed both indicate that a ΔP of about $-0.03\,\text{MPa}\,\text{m}^{-1}$ exists in the xylem of rapidly transpiring plants. This can be compared with the pressure gradient required for an equivalent solution flow through the small interstices of cell walls, which approximates $-3\times10^5\,\text{MPa}\,\text{m}^{-1}$. This is the reason why the xylem is the preferred pathway for the transport of solution from the root to the shoot.

The rate and direction of transport of xylem sap is governed primarily by transpiration, which manifests itself in marked diurnal changes in ΔP in the xylem. Plants transpiring rapidly generate large tensions (negative hydrostatic pressures) in the xylem, which are maintained by the cohesion of water molecules resulting from hydrogen bonding. By contrast, the tension in the xylem diminishes, and hydrostatic pressures may become positive, under conditions of low transpiration. In some circumstances this draws water from the surrounding cells into the xylem and results in guttation.

Xylem sap is distributed within the shoot through the veins in the leaves. The subsequent movement of ions may be apoplastic, following the transpiration stream, or symplastic. Some ions are selectively accumulated by leaf cells and, as a consequence of evaporation, ions accompanying the transpiration stream may accumulate in the apoplast. These processes generate the characteristic ionic composition of the different cell types within the leaf. For example, P accumulates in bundle sheath and mesophyll cells, whereas Ca, Na, and Cl all accumulate in epidermal cells, especially those closest to the veins. Such differences between cell types in ion accumulation may be important in the adaptation of plants to saline or metalliferous soils.

Transport in the Phloem

The phloem is composed of living cells specialized for the long-distance transport of soluble organic compounds. In gymnosperms transport occurs through sieve cells, which are interconnected by structures resembling plasmodesmata. In angiosperms transport occurs through sieve tubes. Sieve tubes are composed of a file of cells called sieve elements (diameter about 24 μm) that are symplastically connected by sieve pores. Sieve pores are greater in diameter than plasmodesmata (1–5 μm) and do not contain a desmotubule. Sieve elements have neither nucleus, ribosomes, nor tonoplast, but do retain a plasma membrane, an endoplasmic reticulum, and mitochondria. However, they are connected to companion cells, which contain subcellular organelles and a nucleus, by plasmodesmata. It is these companion cells, and the analogous albuminous cells present in gymnosperms, that are presumed to control the metabolism, loading, and unloading of the phloem.

Phloem sap has a relatively high pH (7–8). It is viscous ($\eta = 1.3$ mPa) and contains high concentrations of sucrose (0.2 to 0.7 mol l^{-1}), organic acids, amino acids, and amides. The dominant inorganic ion is K^+. Although many ions are transported in the phloem, it is noteworthy that the divalent cations Ca^{2+}, Ba^{2+}, and Sr^{2+} are relatively immobile, and boron, copper, iron, manganese, molybdenum, and zinc are not transported readily. This explains the susceptibility of phloem-fed tissues to deficiencies in these elements. In contrast to xylem transport, which is unidirectional, phloem transport is bidirectional.

Ion fluxes in the phloem are too fast to occur by simple diffusion. The flow of phloem sap (about 170 $\mu m\,s^{-1}$) is driven by hydrostatic pressure and it has been calculated from Poiseuille's equation that a ΔP of about -0.04 MPa m^{-1} is required. This ΔP can be generated comfortably by the loading of solutes into the phloem in a "source" tissue and their unloading in a "sink" tissue. This hypothesis, originally proposed by Ernst Münch, neatly explains both the mechanism of transport and the direction of flow towards sink tissues. Ion transport can take place in opposite directions in a vascular bundle, but not within the same sieve tube. Energy is required indirectly, in the loading and unloading of solutes across the plasma membranes of the sieve tube/companion cell complexes.

Interestingly, there is evidence that solutes are transferred directly between the xylem and phloem within the vasculature. This may be of importance for mineral nutrition because xylem transport is directed to tissues with high transpiration, which are not usually those with the highest demand for nutrients.

List of Technical Nomenclature

Physical constants	Definitions	Unit
a_j	activity of j	mol m^{-3}
c_j	Concentration of j	mol m^{-3}
D_j	diffusion coefficient of j	$m^2\,s^{-1}$
E	voltage	V
E_j	Nernst potential for j	V
E_m	membrane potential	V
E_{rev}	zero-current (reversal) potential	V
F	Faraday's constant	9.65×10^4 J $mol^{-1}\,V^{-1}$
g	Gravitational acceleration	9.8 m s^{-2}
h	position in gravitational field	M
I_j	electrical current carried by j	A
I_j^{in}	inward current carried by j	A
$I_{j\max}^{in}$	maximal I_j^{in} (V_{max})	A
I_j^{out}	outward current carried by j	A
j	ionic species	
J_j	flux density of j	mol $m^{-2}\,s^{-1}$
J_v	volume flow	m s^{-1}
K_j	partition coefficient Dimensionless	
K_m	Michaelis constant	M
l	length of a cylinder	m
m_j	mass per mole	kg mol^{-1}
M_j	j per unit area	mol m^{-2}
P	pressure in excess of atmospheric	Pa = 1 J m^{-3}
P_j	Permeability coefficient	m s^{-1}
r	radius of a cylinder	m
R	gas constant	8.314 J $mol^{-1}\,K^{-1}$
t	time	s
T	absolute temperature	K
V_j	partial molal volume	$m^3\,mol^{-1}$
x	distance	m
z_j	valency of j	
η	viscosity	Pa
μ_j	Chemical potential	J mol^{-1}

See also: **Nutrition**: Deficiency Diseases, Principles; Mineral Uptake. **Plants and the Environment**: Land Reclamation and Remediation, Principles and Practice. **Photosynthesis and Partitioning**: Sources and Sinks. **Water Relations of Plants**: Basic Water Relations; Phloem; Xylem.

Further Reading

Aidley DJ and Stanfield PR (1996) *Ion Channels.* Cambridge: Cambridge University Press.

Alberts B, Bray D, Johnson A, *et al.* (1998) *Essential Cell Biology: An Introduction to the Molecular Biology of the Cell.* New York: Garland Publishing.

Flowers TJ and Yeo AR (1992) *Solute Transport In Plants*. London: Blackie Academic and Professional.
Hille B (2001) *Ionic Channels of Excitable Membranes*, 3rd edn. Sunderland: Sinauer Associates.
Karley AJ, Leigh RA, and Sanders D (2000) Where do all the ions go? The cellular basis of differential ion accumulation in leaf cells. *Trends in Plant Science* 5: 465–470.
Kragler F, Lucas WJ, and Monzer J (1998) Plasmodesmata: dynamics, domains and patterning. *Annals of Botany* 81: 1–10.
Marschner H (1995) *Mineral Nutrition of Higher Plants*, 2nd edn. London: Academic Press.
Nicholls DG and Ferguson SJ (1992) *Bioenergetics 2*. London: Academic Press.
Nobel PS (1999) *Physicochemical and Environmental Plant Physiology*, 2nd edn. London: Academic Press.
The Arabidopsis Genome Initiative (2000) Analysis of the genome sequence of the flowering plant *Arabidopsis thaliana*. *Nature* 408: 796–815.
White PJ, Biskup B, Elzenga JTM, *et al.* (1999) Advanced patch-clamp techniques and single-channel analysis. *Journal of Experimental Botany* 50: 1037–1054.
Williams LE (ed.) (2000) Special issue: Plant membrane transporters: structure, function and regulation. *Biochimica et Biophysica Acta (Biomembranes)* 1465: 1–361.

Nitrogen Fixation

K Fisher and W E Newton, Virginia Polytechnic Institute and State University, Blacksburg, VA, USA

Introduction

Nitrogen fixation research has intensified since the 1960s with the goal of improving the usable nitrogen (fixed nitrogen) supply to plants. Although commercial fertilizer is increasingly viewed as both a potential pollutant and health risk, it is widely accepted that there would be a major decline in crop productivity without it. The earth's atmosphere contains enormous amounts of nitrogen gas (N_2), but plants and animals cannot use N_2 as they do CO_2 and O_2. Instead, plants and animals must rely on bacteria to "fix" N_2 to form inorganic compounds that they can use. Biological nitrogen fixation by bacteria offers a means to reconcile acceptable crop productivity with safeguarding the environment.

The amount of N_2 being fixed at any given time by natural processes represents only a small addition to the pool of previously fixed nitrogen that cycles among the living and nonliving components of the earth's ecosystems. Most of that nitrogen is also unavailable to plants and animals because it is locked up in soil organic matter that must first be decomposed by soil microbes to release nitrogen as either ammonium or nitrate. The two major natural sources of new fixed nitrogen entering the nitrogen cycle are through N_2-fixing organisms and lightning. N_2-fixing organisms include a relatively small number of microorganisms. Many of them live free in the soil, but the most agriculturally important ones are bacteria that form close symbiotic relationships with higher plants. These bacteria all synthesize essentially the same enzyme, called nitrogenase, which enables them to convert N_2 directly into plant-usable forms. The structures of the two nitrogenase component proteins, their genetic basis, the mechanism of action, and the stringent means of regulation have all been determined to some degree. However, many challenges remain with this complex enzyme system, which has the unique ability to break one of the strongest covalent bonds occurring in nature.

Biological Nitrogen Cycle

The nitrogen cycle describes the major sinks, nitrogen-transfer pathways, and transformations involved in the global cycling of nitrogen (**Figure 1**). Atmospheric N_2 is 'fixed' by both biological and nonbiological processes and results in formation of either reduced or oxidized compounds of nitrogen. These nitrogen compounds are added to the soil and become available for assimilation by bacteria and plants. They return the fixed nitrogen to the soil when they die and decay. Other soil bacteria convert ammonia into nitrate and then nitrate back to N_2 to complete the cycling of nitrogen. These latter processes lead to a loss of usable fixed nitrogen from the soil. Overall, the input rate from biological, spontaneous (e.g., lightning), and industrial nitrogen fertilizer (by the Haber–Bosch process) production is slightly higher than the loss rate. Of these, the major input of fixed nitrogen is through biological nitrogen fixation by microorganisms known collectively as diazotrophs. The nitrogenase in these important organisms operates at ambient temperature and pressure in the soil with the sun as the ultimate energy source. N_2-fixing organisms have little in common with each other apart from being prokaryotic and diazotrophic. They are generally classified into two main groups; those that are free-living, and those that are symbiotic, i.e., that fix N_2 in association with a plant, usually with its root. Free-living bacteria may be either phototrophic or chemotrophic and either heterotrophic or autotrophic. Moreover, they may fix N_2 under anaerobic, microaerobic, or aerobic conditions.

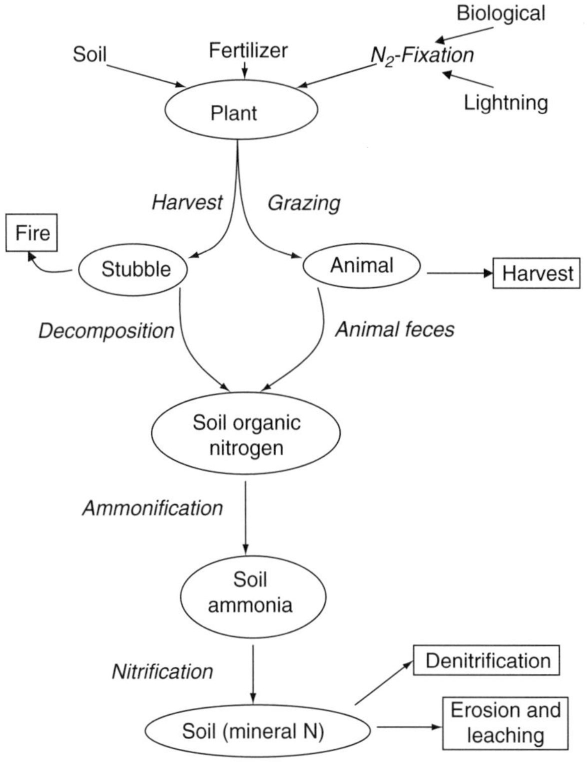

Figure 1 Nitrogen flow cycle. Fixed nitrogen enters the nitrogen cycle (top) through sources in the soil and through the biological, environmental, and industrial fixation of atmospheric N_2. Fixed nitrogen exits the system (boxed items) after nitrate uptake by plants through plant cropping, burning of field stubble, removal of livestock, soil erosion, weathering, and volatilization. Areas of temporary fixed-nitrogen storage are shown within ellipses. Fixed nitrogen moves through the cycle via the italicized processes.

Free-Living Microbes

Free-living organisms do not normally have a symbiotic partner, such as a plant, and are, therefore, rarely of great significance in the terrestrial nitrogen economy. Despite their agricultural insignificance, these bacteria are of great scientific importance because they are easy to culture and their nitrogenase is virtually identical to that from the agriculturally important symbiotic diazotrophs, called rhizobia. The only free-living organisms of any major importance are the photosynthetic cyanobacteria, which are self-sufficient microorganisms that are not limited by the carbon substrates in the soil.

The most-studied aerobes are members of the genus *Azotobacter*, all of which are obligate aerobes that efficiently fix N_2 in air. Most other aerobic bacteria are less well adapted to fix N_2 while growing in air. All of these organisms possess systems to protect nitrogenase from O_2 damage. Anaerobic bacteria are represented primarily by the N_2-fixing obligately anaerobic clostridia, which can be isolated from almost any soil or water sample. The clostridia metabolize carbon sources, such as glucose, to butyric acid, CO_2, and H_2. Although not universal, the ability to fix N_2 is not uncommon among clostridia. A second group of N_2-fixing anaerobes is the sulfate-reducing bacteria, e.g., *Desulfovibrio*. These organisms use the oxygen atoms of sulfate for respiration and so produce sulfide, which is responsible for the smell of polluted waters. *Desulfovibrio* has a special ecological role in the sea as the principal nonphotosynthetic N_2-fixing contributor to the formation of nitrogen-containing marine sediments. Another important group of anaerobic N_2-fixers is the methanogens, which are responsible for the methane produced by ruminant animals. Methanogens belong to the third domain of living things, the Archeae.

Facultative anaerobic bacteria are able to grow either with or without O_2 when provided with fixed nitrogen, but can only fix N_2 anaerobically. A member of this group, *Klebsiella,* is related to *Escherichia coli* and contains a number of species that fix N_2. They can be isolated from soils, water, and animal intestines. Although no diazotrophic strain of *E. coli* has been isolated naturally, strains have been genetically modified for N_2 fixation. Other genera, *Citrobacter*, *Enterobacter*, and *Bacillus*, include N_2-fixing strains. The bacilli, like clostrida, are characterized by their ability to form spores and, thus, survive desiccation and heating.

N_2 fixation by photosynthetic bacteria was first observed with the purple nonsulfur bacterium, *Rhodospirillum rubrum*. Because these bacteria lack photosystem II, photosynthesis does not produce O_2, which would compromise nitrogenase action. Photosynthetic bacteria are usually colored and include both sulfur and nonsulfur bacteria and, like cyanobacteria (see below), they can utilize CO_2 as sole carbon source. The second major group of phototrophs, the cyanobacteria, e.g., *Anabaena*, use a different O_2-protection strategy and locate nitrogenase in specialized cells, called heterocysts, which are incapable of photosynthesis.

Plant Associations

The best-known symbiotic systems, in which bacteria fix N_2 in association with the roots (usually) of various plant hosts, are the *Rhizobium*–legume associations. In addition, there are: actinorhizal symbioses on the roots of nonleguminous plants, such as alder (*Alnus* spp.), which harbor *Frankia* as the microsymbiont; cyanobacterial associations with fungi, ferns, and cycads; and the associative symbioses of various grasses with *Acetobacter*, *Azoarcus*, *Azotobacter*, and *Azospirillum*.

Rhizobium–Legume Associations

The Fabaceae is a family of flowering plants found in both temperate and tropical climates that ranges from clover to bushes and large trees, such as *Acacia*. The family comprises the Papilionoideae, the Mimosoideae, and the Caesalpinioideae. Of the species tested, about 90% of the Papilionoideae form root nodules compared with only about 25% of the Mimosoideae and relatively few of the Caesalpinioideae. All legume nodules result from root colonization by rhizobia, which were grouped into three genera; the fast-growing *Rhizobium*, which possesses plasmid-borne *nif* genes and a narrow range of host plants; the slow-growing *Bradyrhizobium* with chromosomal *nif* genes and often a broad range of hosts; and *Azorhizobium*, which colonizes the stem of *Sesbania*. More recently, the rhizobia family has expanded; the *Rhizobium* genus has been divided into *Mesorhizobium* and *Sinorhizobium* genera; *Photorhizobium* has been proposed but not yet accepted; and species of both *Burkholderia* and *Ralstonia* have been isolated from legume root nodules. There is also a single example of rhizobia nodulating a nonlegume, *Parasponia*, which is a woody member of the elm (*Ulmus*) family. The best-studied associations occur with important crops, such as peas (*Pisum*), beans (*Phaseolus*), clovers (*Trifolium*), and alfalfa (*Medicago sativa*, lucerne). Rhizobia often show specificity and only infect certain host plants. However, some legumes may be nodulated by several rhizobial species and a single rhizobial species may nodulate more than one legume host. Even so, rhizobia are usually divided into species and biovars based on their range of acceptable hosts.

The process of colonization starts with the secretion of flavonoids by the roots of the host plant, which attracts the appropriate rhizobial species to enter. The bacteria produce complex sugar derivatives, called lipochitooligosaccharides (Nod factors), which cause root hair deformation, branching, and/or curling. This communication results in the rhizobia entering the root by way of an infection thread. As the infection thread elongates, root inner cortical cells are induced to divide and these cells become the nodule primordium. The thread enters these cells and releases the rhizobia, which remain confined within vesicles bound by the plant-derived peribacteroid membrane. Within the infected plant cells, the bacteria cease multiplying and become misshapen and rich in nitrogenase. These nitrogen-fixing cells are called bacteroids.

The ability of rhizobia to invade plants and stimulate the host to develop nodules depends on the expression of a variety of genes, some of which are carried by the plant and others by the bacteria. Of particular interest are the nodule-specific plant genes (*nod* genes), which encode plant proteins called nodulins. Nodulins are expressed in a controlled sequence that influences the infection process, bacterial migration, nodule development, and bacteroid formation. A crucial rhizobial gene is *nodD,* which codes for a protein that recognizes the host plant flavonoid. It induces expression of the other *nod* genes and results in the synthesis of the bacterial signal molecules called Nod factors. These Nod factors consist of three to five β1-4-linked *N*-acetylglucosamine residues. The sugar at the nonreducing end is acylated with a variety of either saturated or unsaturated fatty acids and, in addition, may be either O-acetylated or carbamylated. Moreover, substitution of the reducing sugar with either sulfate in *Rhizobium meliloti* or 2-O-methylfucose in *Bradyrhizobium japonicum* is required for biological activity. The *nodABC* genes are required for the core oligosaccharide, whereas the other host-specific *nod* genes are responsible for the appended side chains, which likely bestow host specificity on the Nod factors.

Some rhizobia will fix N_2 *ex planta* but only when the microaerobic conditions, i.e., less than 0.5% O_2, existing in the nodule are simulated. The low O_2 requirement mimics in nodules the role of leghemoglobin, which is to supply a high flux of O_2 at a low concentration to the bacteroids for metabolism without causing nitrogenase damage. The genes for leghemoglobin are not restricted to the legume nodules, suggesting it may act generally to signal an O_2 deficit and, thus, the need for the plant to shift from oxidative to fermentative metabolic processing.

In some symbioses, nitrogenase-produced H_2 is evolved from the nodule, whereas others recapture and recycle this energy via an uptake hydrogenase. This recapture has been suggested as both an indicator of efficiency and a criterion for the selection of rhizobia for agricultural use, although any potential benefit is yet to be realized.

Some tropical legumes, like *Sesbania*, form stem nodules in association with *Azorhizobium caulinodans*. In contrast to root nodules, some stem nodules are photosynthetic and contain, in the case of *Aeschynomene indica*, rhizobia themselves capable of photosynthesis. This close relationship of photosynthesis to N_2 fixation could ease the high-energy demands of nodules.

Nodulated Nonleguminous Angiosperms (Actinorhizal Associations)

The broad variety of hosts nodulated by actinomycetes are unrelated taxonomically, but most are woody trees and shrubs. A well-studied example involves alder, which has root nodules (called actinorhizae) harboring actinomycetes of the genus *Frankia*. The plants are often the first to colonize either poor or devastated soils and, therefore, have an important ecological role. Although distinct *Frankia* species exist, there is no classification system, but some species have been named on the basis of their ability to infect the same group of plants. Unlike other N_2-fixing microsymbionts, *Frankia* is multicellular and differentiated. When *Frankia* colonizes a plant, it induces root hair deformation, and its hyphae-like filaments penetrate the plant tissue and end in club-shaped vesicles, which are the site of N_2 fixation. Cell-free preparations of nitrogenase isolated from these vesicles exhibit properties similar to those of the enzyme from the free-living bacteria. Most actinorhizal nodules have either little or no hemoglobin present and the only barrier to O_2 diffusion appears to be the vesicle envelope. Thus, most actinorhizae show maximum rates of N_2 fixation at atmospheric O_2 levels. However, some actinorhizae both produce a hemoglobin-like protein and have a physical barrier. Here, hemoglobin likely plays a similar role in O_2 diffusion as in the legume root nodule.

Cyanobacterial Associations

These symbiotic associations include lichens, liverworts, pteridophytes, gymnosperms, and a single angiosperm. Most of the cyanobacteria involved in associations are capable of growth and N_2 fixation independent of the host. In the symbiosis, the cyanobacterium's main function is to supply fixed nitrogen for both partners through specialized N_2-fixing cells, called heterocysts, which increase in number for symbiosis. No specialized structure is needed to accommodate the symbiosis. The microsymbiont usually invades normal host structures, like the leaf cavity in *Azolla*, although modifications of these structures may subsequently occur.

Lichens are exceptional in that the symbiotic state is classified as a separate "organism." All N_2-fixing lichens have a fungal and cyanobacterial symbiont (often either *Nostoc* or *Calothrix*), but some may also have a green alga as a partner. Each lichen genus accommodates only one cyanobacterial genus either in layers just below the surface or restricted to spherical bodies called cephalodia. In bryophytes (mosses and liverworts), the cyanobacteria (often *Nostoc*) are enclosed in cavities in the underside of the thallus.

The tiny water fern *Azolla* is the only known fern genus to associate symbiotically with a N_2-fixing cyanobacterium, *Anabaena azollae*. The microbe is rich in heterocysts, which protect its nitrogenase from the oxygen that is generated during its own and the plant's photosynthetic reactions. *Azolla* is an important green manure for rice (*Oryza sativa*) culture but is also known as a troublesome waterway-blocking weed. Either as a green manure or grown in dual culture, it can supply fixed nitrogen to rice in amounts comparable to those supplied by the rhizobial symbiont to the legume. The symbiotic *Anabaena*, which resides in a cavity in the dorsal leaf lobe of the fern, is difficult to grow away from its host plant. The fern can grow quite well without the cynaobacterium, if a fixed-nitrogen source is provided. Transmission from fern to fern is ensured by retention of some *Anabaena* filaments within the sexual reproductive organs so that, as the young fern develops through these filaments, its leaves become rapidly infected.

Within the gymnosperms, only the Cycadales form N_2-fixing associations. These plants produce morphologically distinct nodule-like structures, called coralloid roots, on the plant roots near the soil surface. These roots are well colonized by heterocystous cyanobacteria, such as *Nostoc*. The site of entry of the microsymbiont is unknown but it is likely through a crack in the dermal layer. *Gunnera* is unique among angiosperms in developing a symbiotic relationship with cyanobacteria. *Nostoc* invades through secretory glands just behind the shoot apex and, as a result, nitrogen-fixing nodules are formed at the bases of the leaves. The intracellular location of the bacteria and the well-developed vascular system surrounding the microsymbiont coupled with the high rates of fixation and efficient transport of fixed nitrogen ensure that all the needs of the host are met.

Associative Symbioses

Heterotrophic bacteria are often found associated with the roots of some grasses (family Poaceae), but no specialized structure is developed. The first-characterized examples are the close association of the bacterium *Azotobacter paspali* with the tropical grass *Paspalum notatum*, and that of *Azospirillum lipoferum* with the grass *Digitaria decumbens*. With *Paspalum*, a mucilaginous sheath forms around the roots within which the bacteria live and fix N_2. The bacteria do not invade the plant tissue. In the *Digitaria–Azospirillum* example, the spirilla are not in a sheath around the roots but instead invade the root tissue where they form a layer beneath the epidermis and stop growing but continue fixing N_2. The extent to which the plants benefit from the association is uncertain. More formalized endophytic associations involve both *Acetobacter diazotrophicus* and *Herbaspirillum* spp. with sugar cane (*Saccharum officinarum*) and *Azoarcus* spp. with Kallar grass (*Leptochloa fusca*) and rice. Some of these associations can supply up to 60% of the fixed nitrogen needed for the host plant's growth, indicating a significant agricultural and economic potential.

Casual Associations

It is often difficult to differentiate between N_2 fixation by free-living species on a plant's surface from that intimately associated with plant organs. Free-living diazotrophs are routinely found around the roots of dicotyledonous plants probably living on plant secretions and benefiting themselves rather than the plant. Some casual associations involve species of either *Pseudomonas* or *Azospirillum*, which are often found inside grass roots, but their function there is unclear. Certain tropical plants, e.g., *Ardesia* and *Psychotria*, have small nodules on their leaves that are filled by bacteria; the *Psychotria* nodules contain *Klebsiella*. Again, the benefit to the plant is unclear because the bacteria are not always diazotrophic and, even if they are, they cease fixing N_2 in the "leaf-nodule." Most likely, the plant benefits from bacterial hormone-like substances rather than from N_2 fixation. Mycorrhizae, a group of symbiotic fungi, were once thought to fix N_2 in association with the roots of the gymnosperm *Podocarpus*, but more recent evidence suggests that they simply create a favorable environment for casual fixation by soil bacteria.

Nitrogenase

There are four genetically distinct nitrogenase systems capable of catalyzing the reduction of N_2 to ammonia. Three are quite similar except that one is based on molybdenum (Mo-nitrogenase), the second on vanadium (V-nitrogenase), and the third which functions without molybdenum or vanadium (Fe-nitrogenase). The most commonly encountered nitrogenase is the Mo-nitrogenase. The component proteins of Mo-nitrogenase are generally known by their trivial names, the Fe protein and the MoFe protein, which reflect their metal contents. The physicochemical properties of all Mo-nitrogenases are remarkably similar, regardless of the N_2-fixing organism from which they are isolated. The fourth nitrogenase is from the thermophilic organism *Streptomyces thermoautotrophicus*, and is quite different to the conventional nitrogenases (see "Alternative Nitrogenases").

All nitrogenases comprise two separately purifiable component metalloproteins. All three conventional nitrogenase systems contain a specific Fe protein that functions best with its complementary MoFe protein or VFe protein or FeFe protein. The V- and Fe-nitrogenases are less characterized but their structure and mechanism of action appear very similar to those of Mo-nitrogenase. In some organisms, aspects of both the genetics and physiology of N_2 fixation are relatively well understood.

Reproducible N_2 fixation by cell-free extracts from *Clostridium pasteranium* was first demonstrated in 1960. Earlier, workers had achieved limited success and were unable to show consistent levels of activity *in vitro*. It was only after techniques for handling the extremely O_2-sensitive nitrogenase enzyme had been developed and the requirement for adenosine triphosphate (MgATP) had been recognized that sufficient highly purified material became available for investigation. Since then, Mo-nitrogenase has been isolated from at least 20 different species and has been extensively purified from five of them. In addition to a supply of MgATP, all three conventional nitrogenases require an anaerobic environment and a reductant of sufficiently low redox potential in order to function. The structures of the two component proteins of Mo-nitrogenase, both separately and as a complex, were solved by X-ray diffraction techniques and reported in the period of 1992–2002.

Mo-Nitrogenase Component Proteins

The Fe-protein component of the nitrogenase complex is a γ_2 homodimer with a M_r of 64 kDa. It contains one solvent-exposed Fe_4S_4 cluster that bridges a cleft between two identical subunits. Each subunit possesses both a Walker A- and Walker B-type nucleotide-binding motif that together form a MgATP-binding site. The Fe protein makes three major contributions to the mechanism of N_2 fixation. First, it contains the signal-transducing mechanism whereby MgATP hydrolysis is coupled to efficient unidirectional electron transfer. Second, it is the obligate electron donor to the MoFe protein. Third, it is required for both the biosynthesis of the FeMo-cofactor and its insertion into the immature MoFe protein.

The MoFe protein is an $\alpha_2\beta_2$ heterotetramer with a M_r of 230 kDa and a metal content of 2 Mo, 30 Fe, and 32 S^{2-} atoms per molecule. These metal atoms are organized into two pairs of unique metal clusters, the FeMo-cofactors and the P clusters. One FeMo-cofactor is located within each of the α-subunits, whereas each P cluster is situated at an $\alpha\beta$-subunit interface. The current supposition is that the P cluster, which has a Fe_8S_7 composition, is involved in the acceptance and storage of electrons from the Fe protein, prior to delivery to bound substrate at the FeMo-cofactor. The FeMo-cofactor, a $Mo_1Fe_7S_9$-homocitrate metallocluster and the site of substrate binding and reduction, is buried 10 Å below the surface of the α-subunit. A recent high-resolution (1.16 Å) crystallographic analysis reveals a light atom, most plausibly nitrogen, in the central cavity of this FeMo-cofactor cluster. This observation has important implications for the mechanism of N_2 reduction and suggests that this cluster may open upon N_2 binding and activation.

Component Protein Interactions

Complex formation between the MoFe and Fe proteins is a critical step in the overall mechanism of biological nitrogen fixation. Only in the two-protein complex does both MgATP hydrolysis and intermolecular electron transfer occur. X-ray crystallography shows that this nitrogenase complex from *Azotobacter vinelandii* is an $(\alpha\beta\gamma_2)_2$ octamer with one Fe protein γ_2-dimer per MoFe protein αβ-subunit pair, i.e., two Fe proteins are bound to each MoFe protein. **Figure 2** shows a representation of one half of the octamer $(\alpha\beta\gamma_2)$. When comparing the X-ray crystal structure of the complex with those of the individual proteins, most of the structural changes resulting from protein–protein interaction occur in the Fe protein. The major movement is a $\sim 13^\circ$ rotation of each γ-monomer around the subunit interface.

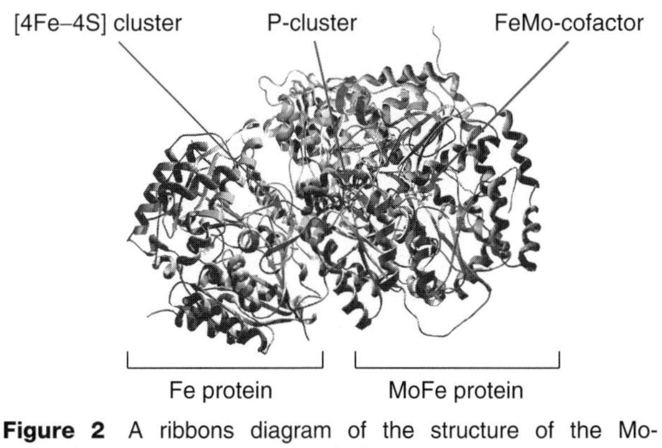

Figure 2 A ribbons diagram of the structure of the Mo-nitrogenase complex. Only one half of the nitrogenase complex is shown with the docking of one Fe protein to one αβ dimer of the MoFe protein. Also indicated are the locations of the Fe protein's (4Fe–4S) cluster and both the P cluster and the FeMo cofactor of the MoFe protein.

Mechanism of Mo-Nitrogenase Action

The minimum requirements for biological N_2 fixation are the MoFe and Fe proteins, the substrate N_2, a source of reductant (either flavodoxin or ferredoxin *in vivo*; sodium dithionite, *in vitro*), MgATP (plus an ATP-regenerating system), and an anaerobic environment. The ATP-regenerating system is needed because MgADP is an inhibitor of nitrogenase catalysis by competing for the MgATP-binding sites on the Fe protein. The optimal stoichiometry for N_2 reduction (see below) involves hydrolysis of four molecules of ATP for each pair of electrons incorporated into substrate:

$$N_2 + 8e^- + 8H^+ + 16MgATP \rightarrow 2NH_3 + H_2 + 16MgADP + 16P_i$$

The direction of electron flow is from reductant to Fe protein to MoFe protein and finally to bound substrate. A comprehensive kinetic model of nitrogenase action, the Thorneley–Lowe model, has described and quantified many of the individual steps of the nitrogenase catalytic reaction. This model comprises two interconnected cycles, the Fe-protein cycle and the MoFe-protein cycle. The Fe-protein cycle describes MoFe protein–Fe protein complex formation, electron transfer from the Fe protein to the MoFe protein with concomitant hydrolysis of MgATP, and the events necessary for re-reduction of the oxidized Fe protein in order to initiate another round of the cycle. The MoFe-protein cycle depicts the sequence of partial reactions that occur during N_2 reduction to ammonia (NH_3). Because N_2 reduction is accompanied by H_2 evolution, there are eight individual partial reactions where each progressive step represents one turn of the Fe-protein cycle.

Reducible Substrates

In addition to N_2 reduction, nitrogenase can reduce several other small molecules, most notably, the reduction of acetylene to ethylene. This reaction is often used as an indicator of nitrogenase in both cell lysates and purified preparations of suspected nitrogen-fixers. An abbreviated list of alternative substrates includes hydrogen cyanide, azide, hydrazine, nitrite, nitrous oxide, cyanamide, and cyclopropene. In the absence of all other substrates, nitrogenase will reduce protons to H_2. Carbon monoxide (CO) inhibits the reduction of all substrates except proton reduction to H_2. H_2 can serve as a reductant for nitrogenase via hydrogenase and it is a specific competitive inhibitor of N_2 reduction.

Alternative Nitrogenases

All conventional N_2-fixing species examined so far contain the Mo-nitrogenase. In some species (e.g., all rhizobia and *Klebsiella pneumoniae*), it occurs alone, whereas in others, it is found in all permutations with the V- and Fe-nitrogenases. The V-nitrogenase is present in several organisms, including *A. vinelandii*, *Azotobacter chroococcum*, *Anabaena variabilis*, and the archeon *Methanosarcina barkeri*. The VFe protein has α- and β-subunits that are highly similar in amino-acid sequence to those of the MoFe protein. The VFe protein, however, also possesses a small (M_r ~15 kDa) δ-subunit that is essential for activity. Thus, its overall subunit composition is an $\alpha_2\beta_2\delta_2$ hexamer, all of which are encoded by the *vnfDGK* genes. The Fe-nitrogenase has been cloned and sequenced from *A. vinelandii*, *Rhodobacter capsulatus*, and *R. rubrum*. Its FeFe protein also consists of three subunits encoded by the *anfDGK* genes.

The fourth nitrogenase system occurs in the thermophilic bacterium *S. thermoautotrophicus*. This microbe grows by reducing CO_2 to organic matter, but instead of using sunlight as reductant, like a plant, it uses CO. Unlike with the conventional nitrogenase systems, CO does not inhibit this enzyme, acetylene is not a substrate, and O_2 does not denature the enzyme. However, this unique nitrogenase does consist of two proteins. One is a heterotrimeric (αβγ; M_r 145 kDa) protein that contains Mo, Fe, and S and harbors the substrate-reduction site. The second protein is a homodimeric (α_2, M_r 48 kDa) protein that contains Mn and oxidizes superoxide radicals generated during CO metabolism to O_2. These electrons are transferred to the Mo protein and subsequently to substrate.

Genetics of Nitrogen Fixation

In the early 1980s, the suite of genes (*nif* genes) required for N_2 fixation by Mo-nitrogenase was established using the facultative species *K. pneumoniae*. The 20 genes are clustered in eight contiguous transcriptional units spanning about 23 kb on the chromosome. These transcriptional units are not all read in the same direction and so either single or groups of genes can be read independently of each other. Using the cloned *K. pneumoniae nifHDK* structural genes as a heterologous probe, many similar genes were found in other N_2-fixing organisms. However, not all *nif* genes are found in all organisms, e.g., there is no *nifJ* gene in *A. vinelandii*. Significant differences in organization also occur. The putative amino acid sequences of all *nif*-gene products are known and the majority of the products identified. The *nif* genes are listed in **Table 1** with the function of the gene product. The genes unique to the V-nitrogenase are designated by *vnf* and those for Fe-nitrogenase by *anf*. In all cases, the same following capital letter indicates a similar sequence for the gene and function for the gene product. However, the products of some *nif* genes are required for full activity of both the V- and Fe-nitrogenases.

Regulation of Nitrogen Fixation

The high-energy demand for N_2 fixation, the O_2 sensitivity of conventional nitrogenases, and the metal-ion requirement necessitate rigorous control of *nif* gene expression. Regulation is exerted at several different levels in different organisms. Three main environmental factors regulate Mo-nitrogenase synthesis. First, the fixed-nitrogen status is very important and bacteria will totally deplete any fixed-nitrogen source before expressing nitrogenase to fix N_2. This control mechanism is particularly significant for free-living N_2 fixers, like *A. vinelandii*, but less so for symbiotic rhizobia, which are adapted to export fixed nitrogen to their host. Photosynthetic

Table 1 The 20 *nif* genes involved in N_2 fixation by *Klebsiella pneumoniae* Mo-nitrogenase; their order on the chromosome, and product function.

nif *gene*	*Gene; product; known or proposed function*
J	Pyruvate oxidoreductase; generates electrons from pyruvate to reduce flavodoxin
H	Fe-protein subunit; required for active Fe protein and FeMo-cofactor biosynthesis
D	MoFe protein α-subunit; required for active MoFe protein
K	MoFe protein β-subunit; required for active MoFe protein
T,Y	Unknown
E,N	Two polypeptides that combine to form a template for FeMo-cofactor biosynthesis
X	A protein involved in FeMo-cofactor biosynthesis
U	A protein that likely supplies Fe for cluster biosynthesis
S	Cysteine desulfurase; supplies sulfur for cluster biosynthesis
V	Homocitrate synthase; makes homocitrate required for FeMo-cofactor
W,Z	Uncertain; but may be involved in FeMo-cofactor biosynthesis or insertion
M	Peptidyl prolyl isomerase; activates Fe protein
F	Flavodoxin; reduces the Fe protein
L	Negative regulatory protein; responds to O_2, fixed-N and energy status
A	Positive regulator protein; activates transcription at *nif* promoters
B	NifB-cofactor; a FeMo-cofactor precursor
Q	A protein involved in Mo uptake for FeMo-cofactor biosynthesis

bacteria use a reversible posttranslational modification to inactivate the Fe protein in response to NH_3. Second, O_2 tension is important because nitrogenase proteins are remarkably susceptible to oxidative damage. In strict aerobes, Mo-nitrogenase can undergo "conformational protection" against O_2 damage by forming a transient protein aggregation. In contrast, most filamentous cyanobacteria developmentally regulate *nif* gene expression so that nitrogenase is only synthesized in the specialized, nonphotosynthetic heterocysts. Third, because nitrogenase consists of metal-containing proteins, the availability of certain metal ions is regulatory. For example, the presence of molybdate in the growth medium represses the expression of both the V- and Fe-nitrogenases.

In many of the above examples, the effects are mediated through a common mechanism of transcriptional regulation of the *nif* genes by the products of specific regulatory genes, *nifLA*, which are in turn controlled by the *ntr* (nitrogen regulation) system. The *nifA*-gene product (NifA) is an enhancer-binding protein that activates transcription from all *nif* promoters, whereas NifL is a flavin-containing sensor protein that binds to and inactivates NifA in response to either high fixed-nitrogen or high O_2 concentrations.

Approaches, Limitations, and Future Outlook for Increasing Input of Fixed Nitrogen from Biological Nitrogen Fixation

Biological nitrogen fixation is both a direct and indirect contributor to crop productivity. It contributes directly to the grain harvested for human/animal consumption and indirectly by adding fixed nitrogen to the soil and so enhancing its fertility. Efforts to enhance the input of biologically fixed nitrogen to agricultural systems are concerned with: (1) farming practices, (2) environmental constraints, (3) the bacterial microsymbiont, usually *Rhizobium*, and (4) the host plant, usually a legume. The timescale for the predicted improvement from these efforts varies from the immediate to the long term. The most immediate gains would arise from the simplest approaches. With farming systems, the introduction of nodulated legumes in areas where they are not currently grown would improve soil fertility as well as providing a new and nutritious crop. Environmental constraints, such as soil acidity, phosphorus deficiency, or high soil nitrogen content, can be dealt with using the current technologies of liming the soil, fertilizer use, and crop rotations, respectively. Other immediate impacts would come from inoculating the soil with effective rhizobial strains in areas where none (or only poorly effective indigenous populations) are present.

Medium-term benefits of biological nitrogen fixation will accrue from the use of select rhizobial strains matched with efficiently nodulating legume plants. A major problem with this approach is "competition" among rhizobia and whether the select rhizobial strain can compete for nodule occupation in the field with the indigenous strains. So, the select strain must be both highly competitive and highly N_2-fixation effective. Moreover, if the resulting associations could be manipulated either to start N_2 fixation earlier or to continue it longer, a substantial yield, nutritional and economic benefit could result. Similarly, increased fixed-nitrogen input would accrue if rhizobia would continue to fix N_2 in the presence of fixed-nitrogen sources.

In the longer term, the ongoing discovery of new nonleguminous, associative symbioses indicates new avenues through which N_2 fixation and the delivery of fixed nitrogen to crop plants may be enhanced. Because these systems appear less formalized than the legume symbioses, they may provide insights into how to engineer new associations involving the principal food crops (corn (*Zea mays*; maize), wheat (*Triticum aestivum*), and rice) that presently have no N_2-fixing symbiotic partner. These associations could have dramatic effects on both fertilizer usage and food production. Probably, genetic manipulation is the ultimate solution for providing fixed nitrogen input to crop plants and so increasing global food supplies. Recently, research has shown that many plants already have most of the genes necessary to produce a nodule-like structure, which opens up the possibility of both enhancing current symbiotic associations and producing new ones. Furthermore, the success of transferring the *nif* genes from one bacterial genus to another has opened up the possibility of their transfer to crop plants. However, genetic transfer alone is insufficient because, in addition, any newly formed N_2-fixing plant would also have to provide, at a minimum, protection of nitrogenase from O_2 and a sufficient supply of energy. One possible location to supply these minimal needs would be the chloroplasts of leaves where, if properly protected from the O_2 evolved by photosynthesis, nitrogenase could take advantage of the directly available reducing equivalents produced from sunlight. However, no matter which of these efforts is successful, it is certain that humankind is destined to remain dependent on plants for a supply of fixed nitrogen for the foreseeable future.

List of Technical Nomenclature

Anaerobic	Having no O_2 in their immediate environment.
Bacteroid	Variously modified bacteria, such as those occurring in the root nodules of leguminous plants.
Diazotrophy	The ability to fix atmospheric N_2 from the air.
Heterocyst	A cell that is specialized for nitrogen fixation by some filamentous cyanobacteria.
Legume	The collective name for a large family of dicotyledonous plants that form nitrogen-fixing root (or infrequently stem) nodules in symbiosis with rhizobia.
Nitrogenase	A two component metalloenzyme system capable of reducing atmospheric N_2 to ammonia.
Nitrogen cycle	The continuous natural cycle by which nitrogen passes from the atmosphere to soil to organisms and back to the atmosphere.
Nitrogen fixation	The conversion of atmospheric N_2 to forms that are readily utilizable by living things through either natural or synthetic processes.
Nodulation	The process by which rhizobia induce the formation of nodules on the roots of their specific hosts.
Nodule	A swelling on a legume plant root induced by colonization by nitrogen-fixing bacteria.
Rhizobia	The name for several genera of bacteria that have the ability to infect the root of legumes and to produce root nodules.
Symbiosis	The living together in intimate association of two dissimilar organisms, so that the cohabitation is mutually beneficial.

See also: **Genetic Modification, Applications**: Nitrogen Fixation. **Nutrition**: Deficiency Diseases, Principles. **Root Development**: Mycorrhyzae; The Rhizosphere and its Microorganisms.

Further Reading

Burgess BK and Lowe DJ (1996) Mechanism of molybdenum nitrogenase. *Chemistry Reviews* 96: 2983–3011.

Christiansen J, Dean DR, and Seefeldt L (2001) Mechanistic features of the Mo-containing nitrogenase. *Annual Review of Plant Physiology and Plant Molecular Biology* 52: 269–295.

Eady RR (2003) Current status of structure function relationships of vanadium nitrogenase. *Coordination Chemistry Reviews* 237: 1–8.

Finan TM, O'Brian MR, Layzell DB, Vessey JK, and Newton WE (2002) *Nitrogen Fixation: Global Perspectives*. Wallingford: CAB International.

Howard JB and Rees DC (1996) Structural basis of biological nitrogen fixation. *Chemistry Reviews* 96: 2965–2982.

Ladha JK, de Bruijn FJ, and Malik KA (1997) Opportunities for biological nitrogen fixation in rice and other non-legumes. *Plant and Soil* 194: 1–216.

Leigh GJ (ed.) (2002) *Nitrogen Fixation at the Millennium*. Amsterdam: Elsevier Science.

Pedrosa FO, Hungria M, Yates MG, and Newton WE (2000) *Nitrogen Fixation: From Molecules to Crop Productivity*. Dordrecht: Kluwer Academic.

Peters JW, Fisher K, and Dean DR (1995) Nitrogenase structure and function: a biochemical–genetic perspective. *Annual Reviews of Microbiology* 49: 335–366.

Postgate JR (1998) *Nitrogen Fixation*. Cambridge: Cambridge University Press.

Deficiency Diseases, Principles

M H Behboudian and A H Pickering, Massey University, Palmerston North, New Zealand
E Dayan, Agricultural Research Organization, Hanegev, Israel

Introduction

The basic concepts of plant nutrition are discussed elsewhere in this volume. Here we note that so far 17 elements have been recognized as essential for normal growth and development of all higher plants. These elements are grouped into macronutrient elements or major elements, of which plants need a higher quantity, and micronutrient elements or minor elements which are used in lower quantities. The nine major elements and their concentrations (mmol kg^{-1}) that result in normal growth in plants are: carbon (C, 40 000), oxygen (O, 30 000), hydrogen (H, 60 000), nitrogen (N, 1000), phosphorous (P, 60), potassium (K, 250), calcium (Ca, 125), magnesium (Mg, 80), and sulfur (S, 30). The minor elements are: iron (Fe, 2.0), copper (Cu, 0.1), chlorine (Cl, 3.0), manganese (Mn, 1.0), nickel (Ni, 0.05), molybdenum (Mo, 0.001), boron (B, 2.0), and zinc (Zn, 0.3). Sodium (Na), cobalt (Co), and silicon (Si) are beneficial to some plants or might be considered as essential for a particular plant, but they are not considered essential to all higher plants and therefore are not covered in this chapter. Vanadium (V) and iodine (I) are considered to be beneficial to some plants especially to lower plants. More recently the essential elements have also been grouped into four functional groups: (1) those forming organic compounds in plants (C, H, O, N, S), (2) those important in energy storage or structural integrity (P, B), (3) those remaining in ionic form (K, Mg, Ca, Cl, Mn), and (4) those involved in electron transfer (Fe, Cu, Ni, Zn, Mo).

Here we describe the important mineral deficiency disorders in plants. Because disorders are more often related to function of elements in plants, a summary of functions will be given for each of the elements discussed. **Tables 1–3** list general guidelines for identifying nutrient deficiencies in plants and some examples of nutrient deficiency disorders are found in **Tables 4** and **5**. The abundance of C, O, and H in nature means that their deficiency in plants does not occur, and therefore our discussion covers the remaining 14 elements listed above.

Nitrogen

Nitrogen is a constituent of amides, hexosamines, amino acids, proteins, nucleotides, nucleic acids, enzymes, and coenzymes. It is thus implicated in enzymatic reactions and plays an active role in energy metabolism. Nitrogen is also contained in plant hormones. Many plants require large amounts of nitrogen which is often a limiting factor in agricultural systems. The most overt symptom of N deficiency is the reduced synthesis of proteins. Nitrogen deficiency is associated with low photosynthetic activity and slow vegetative growth. Plants lacking N remain stunted, to an extent depending on the severity of the deficiency. In early stages of growth the vegetative organs are affected, and if deficiency develops later, it affects mainly reproduction and growth-reserve proteins. Under a shortage of N, synthesis of indole acetic acid (IAA) and cytokinins slows down and that of abscisic acid (ABA) is encouraged.

Deficiency of N in conjunction with other essential nutrients can cause deposits of carbohydrates in vegetative cells. Therefore N deficiency often resembles xeromorphous features (the features characteristic of drought-resistant succulent plants) probably due to a relative surplus of carbohydrates. Some crops may become more susceptible to diseases or insects if deficient in N.

Phosphorus

Phosphorus is a structural component of sugar phosphates, nucleic acids, nucleotides, coenzymes, phosphoproteins, phospholipids, and phytic acid. It has a key role in reactions in which ATP is involved. Active plants, particularly at an early stage when growth is vigorous, need high quantities of P. An adequate supply of P early in the life of plants is important for the development of their reproductive structures. Large quantities are found in seeds and fruits, and are essential for seed formation. Plants need more P during formation of seed, fruit, and reproductive parts in general.

A good supply of P is necessary for root growth. Phosphorus deficiency inhibits protein synthesis and leads to poor lignification. Severe P deficiency inhibits sugar synthesis. It increases respiration so that growth is reduced. The deficiency retards overall growth depending on the severity. Phosphorus deficiency has a marked effect on reproductive organs. Reproductive development is retarded due

Table 1 A general guide for identifying essential element deficiencies whose symptoms first start on older leaves

Symptoms	*Element*
Symptoms on whole plant, reduced growth, rigid appearance	
Pale to yellow discoloration or chlorosis, purplish color on stem, petioles and undersides of leaves. Stems short and thin. Leaves are withered or dried. Flower formation restricted	Nitrogen
As above but necrotic symptoms appear quickly at leaf margins	Molybdenum
Intense green coloration of leaves. Sometimes stem, petioles, and veins develop red and purple color. Leaves senesce and die moving from yellowish/red to greenish/brown. Reproductive parts markedly reduced	Phosphorus
Symptoms localized	
Yellowing and drying of leaves starting at tips. Leaves drooped and plants wilted	
Dicotyledons: chlorosis of the leaf between veins with a broad green margin on either side of vein. Leaf becomes pale yellow and then brown and necrotic. Tips and margins turned or cupped upwards	Magnesium
Poaceae: necrotic spots within chlorotic strips. Mottled or chlorotic leaves	Magnesium
Old leaves dry up, yellow/brown color. Plant wilts, stems become short and weak. Roots long, yellow, slimy and few secondary rootlets	
Dicotyledons: chlorosis near dotted necrotic lesions first appearing at tips and along margins, later spreading to center	Potassium
Monocotyledons: scorched appearance on margins of leaves, no chlorosis	Potassium

Table 2 A general guide for identifying essential element deficiencies whose symptoms first start on younger leaves

Symptoms	*Element*
Mainly younger leaves affected	
Leaves not wilted	
Whitish brown necrotic lesions distributed over chlorotic patches, beginning in areas furthest from veins then move towards veins. Roots sparsely developed and flowering reduced	Manganese
Stalks slender, new leaves remain small, short brown roots with many rootlets and club shape tips. Poor fruit set and color. Chlorosis yellowish green forms in intercostal areas without necrosis	Iron
As above but also affects veins and leaves turn white	Iron
Leaves yellowish green to yellow. Chlorosis may involve veins which may be reddish color. Leaves thick and juicy, resembling succulents. Many ramified white roots. Reduced growth	Sulfur
Leaves wilted. Leaves withered and dead but terminal bud still alive. Nonspecific chlorosis with or without necrosis	
Poaceae: stems thin, ripening time delayed, more tillering than normal, ear formation inhibited	Copper
Dicotyledons: inflorescence misshapen, branch tips dying, root growth restricted, rootlets have long thin white appearance	Copper
Young/old leaves affected depending on time of appearance	
Marginal chlorosis of young leaves and marginal necrosis of older leaves. Generally affecting areas between veins but eventually affecting veins. Unevenly distributed clusters or rosettes of small stiff leaves formed at ends of young shoots. Bud formation and bud break reduced. Root growth restricted	Zinc
Necrotic spots on leaf tips	Nickel

Table 3 A general guide for identifying essential element deficiencies whose symptoms first start on terminal buds

Symptoms	*Element*
Terminal buds die	
Browning of veins and brown/violet chlorotic spots and blotches with necrotic lesions along margins beginning at tips and spreading to intercostal areas. Tissue twisted and deformed and growth inhibited. Hooked appearance of young leaves. Roots short, dark brown with bristly appearance. Fruits detrimentally affected, symptoms depending on species	Calcium
Abnormal or retarded growth of apical growing points. Young leaves misshapen, thicker and dark blue/green color. Chlorosis may be present. Terminal buds thick and stiff before dying. Roots sparse, brown and numerous, short rootlets with thickened tips	Boron

to delay in flowering and ripening. Inflorescences are small and misshapen, as exemplified by P deficiency in sunflower (*Helianthus*), and fruits fall prematurely. Germination of seeds and growth of seedlings is reduced and frost resistance of cereals, fruit trees, and many other plants over the winter is greatly impaired. The quality of fruits, forage vegetables, and grain crops will be lower. Phosphorus deficiency also lowers the tolerance of certain cereals to root diseases.

Potassium

Potassium is required as a cofactor for more than 40 enzymes. It is the major cation in establishing cell turgor and maintaining cell electroneutrality. Because it is mobile within the plant, K is important as the main intracellular agent influencing osmotic relations. Potassium-deficient plants have smaller carbohydrate reserves, thinner cell walls, poorly developed conducting tissues, and a lower structural strength. Potassium deficiency depresses total nitrate and protein content. Translocation of sugars from leaves is greatly reduced. The deficiency induces low starch content in cereals and potato (*Solanum tuberosum*) tubers. Adequate amounts of K are important for cell extension, growth, and maintenance of yield. With adequate K, cell walls are thicker and provide more tissue stability. Thicker cell walls improve resistance to lodging and to pests and diseases. The stalks and internal vascular system of K-deficient plants have unusually small cross-sections.

Tomato (*Lycopersicon esculentum*), cucumber (*Cucumis sativus*), and other thick-fleshed fruits are softer and more susceptible to bruising during transport if deficient in K. Fruits and vegetables grown with K deficiency seem to have a shorter shelf-life. Potassium deficiency reduces frost and infection resistance, and can increase plant sensitivity to stresses caused by temperature, water deficit, diseases, and wind. Other outcomes are small and broken chloroplasts and starch accumulation in cells because of slower transport of assimilates. If K is deficient, plants are less able to withstand water stress mainly because of their inability to make full

Table 4 Examples of mineral deficiency disorders for three macronutrient elements

Element	*Disorder*	*Plant*	*Description*
Ca	Brown heart	Leafy vegetables	Margin necrosis appears on all leaves in the course of their growth
Ca	Empty pods	Peanut	Kernels fail to develop
Ca	Blossom end rot	Tomato, watermelon	Rotting tissue near the blossom end of young fruit
Ca	Black heart	Celery	Leaf blades that have opened turn brown and the pinnules become twisted and blackened
Ca	Tip burn	Lettuce, cabbage, strawberries	Death and puckering along the edge of youngest leaves. Strawberries are hard and acid to taste
Ca	Cavity spot	Carrot, parsnip	Irregular shaped, sometimes scablike depressed blotches directly beneath the epidermis of the roots
Ca	Bitter pit	Apple	Indentations in the skin which are followed by brown corky spots in the flesh just below skin
Ca	Green blotch	Pears	Green blotches, spots or longish stripes, usually at stalk end of fruit. Depressed scabs on surface of fruit at flower end may also appear
Mg	Brush disease	Apple	Chlorosis appears between veins of leaves. Color turns reddish-brown and become necrotic. Leaves curl and then wither and fall. Only a few leaves arranged in a rosette at the tip remain
Mg	Stem dieback/stalk necrosis	Grapevine	Chlorosis between leaf veins occurs followed by brown necrosis of the leaves. Margins look dry
K	White spot	Barley	Caused by K deficiency and N excess. Firstly oval pale green blotches which look like waterlogged tissue appear. Then oval, round, or irregular shaped blotches along veins appear. Whole leaf may become white
K	Grey speck	Rice	Plants are stunted and leaves are small, narrow and bluish-green. Chlorotic spots, which become necrotic appear on older leaves, starting at tips. Leaves finally die
K	Brown center/hollow heart	Potato	Browning in middle of tuber which precedes hollow heart, where cavities are lined with brown necrotic tissue
K	Green back	Tomato	Hard green clearly defined irregular blotches around the end of the stalk. May also be helped with application of Mg
K	Green blotch/yellow blotch	Tomato	Green or yellow blotches on the skin. May also be helped with application of Mg

Table 5 Examples of mineral deficiency disorders for three micronutrient elements

Element	*Disorder*	*Plant*	*Description*
B	Crown and heart rot	Sugar beet	Main growing point dies, older leaves become brittle and chlorotic, rot and infection sets in
B	Internal cork	Apples and pears	Begins with clear water-soaked areas in flesh which later turn brown and dry out
B	Superficial or external cork	Apples and pears	Brown, rough, russetted skin on fruit with deep cracks extending into flesh. Fruit become misshapen
B	Pitting	Cherries	Depressed brown marking of fruit skin, usually on one side towards blossom end. Characteristic shape is star-shaped. Fruit become flattened and wedge-shaped
B	Blossom blast or blossom blight	Apples and pears	In mid-blossom, flowers wilt and turn black or brown, but remain on tree for some time (similar damage caused by frosts results in flowers dropping quickly)
Mo	Whiptail	Cauliflower	Margins of leaves curl upwards, forming cup-like shape. Youngest leaves produce only midribs without leaf blades
Mo	Yellow spot	Citrus	Glassy round or oval blotches appear between veins in spring, later turn golden yellow
Cu	Reclamation disease	Cereals	Young leaves show rolling or curling and wilt. Leaves turn ashen or grayish-green. Leaves then turn white and dry out

use of available water and their reduced ability to prevent excessive transpiration in hot dry weather. Therefore, plants lacking K tend to wilt more rapidly in dry conditions. Potassium deficiency first slows down growth and later growth stops completely. Due to growth inhibition, young leaves of K-deficient plants are smaller than those of normal plants. Leaf blades are smaller and plants have short internodes and a stocky appearance. Many dicotyledonous plants take on a rosette-like appearance, and potatoes and beans become bush-like. A lack of K when the plant is young cannot be compensated for later in development.

Calcium

Calcium is a constituent of the middle lamella of cell walls, is required as a cofactor by some enzymes involved in the hydrolysis of ATP and phospholipids and it acts as a second messenger in metabolic and genetic regulation. Fruits and leaves receive their Ca supply from xylem sap and hence are affected by factors limiting their transpiration rates. Problems in translocation mean that Ca levels are lower in fruit compared to leaves as exemplified by apple (*Malus*), tomato, and bean. Low Ca levels in fruit may develop because of a high concentration of K, high rate of assimilate transport into the fruit when xylem flow is reduced, by precipitation of calcium oxalates in fruit stalks, or by N excess.

Meristematic tissues are sensitive to Ca deficiency because of calcium's role in cell division, synthesis of cell walls, and the maintenance of membrane integrity. As Ca is essential for cell elongation and division, its deficiency results in failure in growth of shoot and root apices, and plant growth therefore ceases. Calcium deficiency increases the permeability of the cell membranes either causing ionic inundation of the vacuole or permitting ions to escape. It can lead to the breakdown of intercellular compartmentalization, and the escape of enzymes. Calcium deficiency reduces plant tolerance of salinity because more sodium will accumulate in the absence of competition from calcium. It affects germination of pollen and seeds and sprouting of potatoes. Plants low in Ca do not survive beyond the seedling stage. When growth is inhibited plants have a bushy appearance.

Flowers of field bean (*Phaseolus* spp.), clover (*Trifolium* spp.), and alfalfa (*Medicago sativa*; lucerne) drop prematurely if Ca is deficient. Ca is one of the most important factors influencing fruit quality and storage properties. Low concentrations of Ca in fruit accelerate ripening and reduce firmness and suitability for storage. A number of puzzling physiological disorders have been recognized as Ca deficiency and are attributed to the low mobility of Ca in the phloem.

Magnesium

Magnesium is required by many enzymes involved in phosphate metabolism and it is a constituent of chlorophyll. It can move fairly rapidly and if the supply from roots is inadequate, it is transferred from older leaves to young growing organs and especially to seeds and fruits. Thus the activities of enzymes influenced by Mg are initially reduced, but without a sudden collapse of metabolism. Damage to the blossom or fruit, if any, will be only slight at first. There are differences between varieties and species in the speed with which Mg is transported. Plants with enough Mg in their old leaves can recover from a temporary deficiency.

Magnesium deficiency leads to a rapid breakdown of chlorophyll because Mg incorporated into pectin does not act as a reserve as it cannot be remobilized but Mg from chlorophyll can. As this Mg is relatively mobile within the plant, chlorosis due to Mg deficiency is most pronounced in older leaves. Leaves damaged by Mg deficiency are usually shed prematurely.

Sulfur

Like nitrogen, S is involved in forming many organic compounds in plants such as the important amino acids methionine, cysteine, and cystine. These amino acids form a redox system, and activities of many biochemically active agents are governed by them. Owing to its roles in synthesis of proteins and important enzymes and in redox reactions, S is also important for chloroplast and chlorophyll synthesis. Sulfur is a constituent of lipoic acid, necessary for production of plant oils. Its deficiency causes absence of important amino acids and enzymatic systems. The deficiency disrupts nitrogen metabolism, reduces protein quality, and induces carbohydrate (starch) accumulation. Concentrations of various soluble amino acids are reduced and that of some other amino acids increased. With reduced protein synthesis, the ability to form stable chlorophyll protein complexes is impaired. Plants deficient in sulfur are smaller than normal, their leaves are smaller and narrower than usual and they have thin, short stems whose growth is affected more severely than is root growth. If S deficiency occurs at an early stage, plants are stunted and have a rigid erect appearance similar to that induced by a lack of N.

Iron

Iron is in the group of essential elements involved in electron transfer. It is a constituent of cytochromes and nonheme iron proteins involved in photosynthesis, nitrogen fixation, and respiration. A lack of iron consequently reduces the respiration rate, thus restricting the amount of energy available for synthetic activity and growth. Iron deficiency impairs nitrate uptake, inhibits protein synthesis from amino acids, and encourages protein hydrolysis. Its deficiency leads to a simultaneous loss of chlorophyll and degeneration of chloroplast structures. Cell division slows down or even ceases. New leaves often remain small. Fruit trees usually set less fruit, and the fruit they bear is often poorly colored. Severe deficiencies can cause serious damage and loss of yields.

The greatest practical problem associated with Fe deficiency is the chlorosis induced by lime or by high pH. The plants most frequently affected by this chlorosis are grape (*Vitis vinifera*), fruit trees, *Citrus* species, peanuts (*Arachis hypogaea*), soybeans (*Glycine max*), vegetables, *Sorghum*, and several ornamentals. Tolerant plants are alfalfa, barley (*Hordeum vulgare*), corn (*Zea mays*; maize), cotton (*Gossypium*), oats (*Avena sativa*), and rice (*Oryza sativa*).

Copper

Copper is involved in electron transport. It is a component of ascorbic acid oxidase, tyrosinase, monoamine oxidase, uricase, cytochrome oxidase, phenolase, laccase, and plastocyanin. In Cu-deficient plants lignification of cell walls and other structures is reduced and their elasticity and strength decreased and then totally inhibited. Severe deficiency is characterized by unlignified cell walls, abnormal vessel formations (in stem, stalk, and leaf) affecting both strength and ability to take up water and nutrients. Permanent wilting and limp leaves associated with Cu deficiency are probably caused by the disruption of lignification. Copper-deficient plants contain fewer chloroplasts in their cells, less copper in the chloroplasts, and less lipids than in normal plants. Concentrations of cytochrome are low. The membrane system of cells is abnormal too. Copper deficiency partly inhibits the differentiation of protoxylem into xylem vessels. Because of the importance of copper for protein metabolism and stabilization, if deficient, the protein fraction decreases and the soluble N fraction, organic acids, and amino acids increase. Because of its compensating role for negative effects of Mn excesses, Cu deficiency accelerates uptake of Mn to the extent of inducing toxicity. The two ions compete for binding sites across plasma membranes in uptake and transport. If Cu levels are low when applying heavy nitrogenous dressings, e.g., to cereals, both growth and yields will decrease and other elements like N, P, K, Mn, Ca, and Mo are taken up and accumulated.

The effects of Cu deficiency are diverse, from growth inhibition and yield loss without visible symptoms to severe damage and death, depending on the intensity of growth and development. Copper deficiency impairs the generative phase (and thus the seed production) of plants by reducing number of flowers and inducing the production of sterile pollen. The formation of flowers and fruits is more seriously affected than growth, at least in the initial stages of development, because the synthesis of compounds needed for generative organs is partly or completely inhibited at an early stage. The importance of Cu for respiration is shown by the lower germination capacity of seeds from Cu-deficient plants. Copper-deficient plants are less cold resistant.

Chlorine

Chlorine is required for the photosynthetic reactions involved in oxygen evolution. Because of its abundance in nature, deficiency of Cl is not common and therefore mainly occurs by withholding Cl from hydroponic systems. In these cases cell division in the leaves slows down, thus reducing growth and sizes of leaves and shoot. If deficiency is severe there is no fruit setting and yields can be reduced by up to 70%. Germination ability of seeds from plants lacking Cl is also poor. Chlorine-deficiency symptoms appear first as wilting tips then chlorosis and necrotic lesions appear. Chlorine ions are very mobile within the plant and are retranslocated, yet symptoms of deficiency usually appear in younger leaves first.

Manganese

Manganese is required for activity of some dehydrogenases, decarboxylases, kinases, oxidases, and peroxidases. It is involved in other cation-activated enzymes and photosynthetic oxygen evolution. Manganese deficiency depresses oxygen production and phosphorylation.

Because of its important role in respiration, its deficiency leads to the accumulation of certain acids, such as citric acid, and is accompanied by a reduction in sugar and cellulose content of plants. Plants lacking Mn accumulate considerable amounts of nitrate and nitrite, thus disrupting protein production. As Mn is implicated in the control of water status in plants, those with enough Mn need less water than those that are deficient. Manganese deficiency is associated with reduced flowering of plants, reduced resistance to cold, and a higher susceptibility to viruses.

Nickel

Nickel has been determined only recently as an essential element for plants and little is known about the exact processes that require it. Nickel is a constituent of urease and hydrogenases, and is found in nitrogen-fixing bacteria. It is particularly mobile in the phloem. High amounts of Ni are transported to seeds and fruit after uptake from the soil. For some plants, including nitrogen-fixing legumes, growth on urea but not ammonium is limited by Ni deficiency due to a Ni-dependent urease activity.

Molybdenum

Molybdenum is an essential component of nitrogenase and nitrate reductase which are two major enzymes in plants. Molybdenum deficiency in plants resembles that of nitrogen deficiency since the most important function of Mo in plant metabolism is in nitrate ion reduction. Molybdenum-deficiency symptoms appear in middle and old leaves first as chlorotic forms, but unlike nitrogen deficiency symptoms, necrotic symptoms appear very quickly at the leaf margins.

Leguminous plants in particular show symptoms similar to nitrogen deficiency as the Mo deficiency restricts N nutrition by affecting nitrate ion reduction and nitrogen fixation. The Brassicaceae also have a high demand for Mo, in particular cauliflower and cabbage. The deficiency is called "whiptail" due to the whiplike appearance of leaves in Mo-deficient plants. Molybdenum deficiency has also been observed in *Eucalyptus* seedlings grown in glasshouses in acidic soil with nitrate as the nitrogen source.

Boron

Boron is required for cell division, cell growth, and membrane function, though its role in cell metabolism is still being discovered. It is required for the synthesis of nitrogen bases like uracil, which is an essential component of RNA. Therefore RNA synthesis, ribose formation, and the synthesis of proteins all depend on B and all are important processes in meristemic growth. Like calcium, B is not retranslocated in the phloem with shoot and root tips being seriously impaired. Pollination and the development of viable seeds, and therefore fruit development, are also affected.

Boron deficiency results in a reduction in leaf and root expansion and deficiency symptoms resemble those of other nutrients such as calcium. Deficiency in B has also been found to reduce the translocation of other nutrients such as Ca and the translocation of assimilates due to the reduction of a coenzyme essential in the formation of sucrose. This causes poor movement of assimilates out of the leaves, where they are formed, to other parts of the plant. Enhanced callose production can also occur which creates callose plugs which in turn block sieve plate pores and detrimentally affect phloem transport.

Zinc

Zinc is among the nutrients involved in electron transport and therefore it is important in photosynthesis. It is a constituent of alcohol dehydrogenase, glutamic dehydrogenase, and carbonic anhydrase. It is closely involved in the N metabolism of the plant and Zn deficiency causes a decrease in the levels of RNA and the ribosome content of cells. It is also a requirement for the synthesis of auxin. Zinc deficiency inhibits the activity of enzymes involved in the detoxification of membrane damaging oxygen radicals and therefore increasing membrane leakiness. Symptoms appear in young expanding leaves as zinc is not readily distributed from old leaves.

List of Technical Nomenclature

Abscisic acid	(ABA) A hormone variously inducing abscission, dormancy, stomatal closure, growth inhibition, and other responses in plants.
Adenosine triphosphate	(ATP) A high-energy organic phosphate of great importance in energy transfer in cellular reactions.
Chlorosis	Failure of chlorophyll development because of a nutritional disturbance or because of an infection of virus, bacteria, or fungi.
Coenzyme	A substance, usually nonprotein and of low molecular weight, necessary for the action of some enzymes.
Cytochrome	A class of several electron transport proteins serving as carriers in mitochondrial oxidations and in photosynthetic electron transport.
Cytokinins	A group of plant hormones important in regulation of nucleic acid and protein metabolism, in cell division, delaying senescence, and organ initiation.

Indole-3-acetic acid (IAA) A plant hormone of the auxin group able to stimulate cell elongation and to influence other developmental processes.

Necrosis Death of a circumscribed piece of a tissue or of an organ.

Photophosphorylation A reaction in which light energy is converted into chemical energy in the form of ATP produced from ADP and inorganic phosphate.

Redox reaction A process in which electrons are transferred from an electron donor or reducing agent to an electron acceptor or oxidizing agent.

Ribonucleic acid (RNA) A substance found in all living cells and some viruses, the main function of which is the translation of the genetic code in protein synthesis.

See also: **Nutrition**: Ion Transport; Mineral Uptake; Nitrogen Fixation.

Further Reading

Atwell B, Kriedemann P, and Turnbull C (1999) *Plants in Action: Adaptation in Nature, Performance in Cultivation*. Melbourne: Macmillan Education Australia.

Bennett WF (1993) *Nutrient Deficiencies and Toxicities in Crop Plants*. St Paul: APS Press.

Bergmann W (1992) *Nutritional Disorders of Plants: Development, Visual and Analytical Diagnosis*. Jena: Gustav Fischer Verlag.

Bould C, Hewitt EJ, and Needham P (1983) *Diagnosis of Mineral Disorders in Plants*, vol. 1, *Principles*. London: HMSO.

Hopkins WG (1995) *Introduction to Plant Physiology*. New York: John Wiley.

Laegreid M, Bockman OC, and Kaarstad O (1999) *Agriculture Fertilizers and the Environment*. Wallingford: CAB International.

Marschner H (1995) *Mineral Nutrition of Higher Plants*, 2nd edn. London: Academic Press.

Mengel K and Kirkby EA (1987) *Principles of Plant Nutrition*, 4th edn. Worblaufen-Bern: International Potash Institute.

Raven PH, Evert RF, and Eichhorn SE (1999) *Biology of Plants*, 6th edn. New York: WH Freeman.

Salisbury FB and Ross CW (1992) *Plant Physiology*, 4th edn. Belmont: Wadsworth Publishing Co.

Scaife A and Turner M (1983) *Diagnosis of Mineral Disorders in Plants*, vol. 2, *Vegetables*. London: HMSO.

Taiz SL and Zeiger E (1998) *Plant Physiology*, 2nd edn. Sunderland: Sinauer Associates.

Wallace T (1961) *The Diagnosis of Mineral Deficiencies in Plants by Visual Symptoms*, 3rd edn. London: HMSO.

Weir RG and Cresswell GC (1993) *Plant Nutrient Disorders*, vol. 1, *Temperate and Subtropical Fruit and Nut Crops*. Melbourne: Inkata Press.

Winsor G and Adams P (1987) *Diagnosis of Mineral Disorders in Plants*, vol. 3, *Glasshouse Crops*. London: HMSO.

Iron Chlorosis

P V V Prasad, University of Florida, Gainesville, FL, USA

Introduction

Iron (Fe) is an essential micronutrient belonging to group VIII in the periodic table. Iron is the most abundant of all micronutrients in the earth's crust constituting about 5% of the lithosphere; it occurs in both ferrous (Fe^{2+}) and ferric (Fe^{3+}) forms. Most Fe in the earth's crust is in the form of ferromagnesium silicate. In soils, the predominant iron oxides are hematite, goethite, lepidocrocite, magnetite, maghaemite, and ferrihydrite. Iron is also a structural component of silicates and other primary or secondary minerals (e.g., pyroxenes, amphiboles, pyrite, and siderite). The concentration of Fe in agricultural soils ranges from 7000 to 500 000 parts per million (mean of 38 000 ppm or 3.8%). Although the concentration of Fe is high in soil, it is classified as a micronutrient because the requirement by plants is low. Iron deficiency generally occurs not because of low Fe in soil, but because of various soil, plant, and climatic factors that affect Fe availability, uptake, or metabolic use. Iron deficiency in plants is one of the major nutritional disorders prevalent on calcareous and sandy soils in arid and semi-arid regions of the world. It has been estimated that about one-third of the world's soils are calcareous and have a high potential for Fe deficiency. It is becoming a major problem worldwide, causing economic yield loss in numerous crops.

Location and Function

Iron was among the first micronutrient elements to be discovered and known to be essential for plant life. Essential nutrients are those without which plants cannot complete their life cycle; they are irreplaceable by other elements, and directly involved in plant metabolism. Almost all Fe in the plant system is located in the chloroplasts, the organelles that are responsible for photosynthesis. It is also distributed in the cytoplasm and other cell organelles, which contain additional heme and/or iron sulfur proteins.

In plants, Fe has a direct physiological role in photosynthesis, respiration, and nitrogen fixation, and is involved in enzyme activation and electron transfer. Iron acts as an enzyme activator or cofactor in chlorophyll synthesis, and activates several other enzymes including catalase, peroxidase, nitrate reductases, and nitrogenase. Iron plays an important role in protein synthesis and in a series of metabolic activities in respiratory enzymes and photosynthetic reactions. One of the most important Fe containing plant proteins is ferredoxin, which is involved in various metabolic processes. Iron also plays a primary role in legumes for nodulation and nitrogen fixation. It is the essential element required by legume host plants and rhizobia. Failure of infecting rhizobia to obtain adequate amounts of Fe from the plant results in arrested nodule development and failure of the host plant to fix nitrogen in adequate amounts.

Absorption and Translocation

Although Fe is one of the most abundant elements in the earth crust, its uptake and subsequent utilization depend on the efficiency of its uptake across the plasma membrane of root cells. Iron has low mobility in the plant because it is strongly bound within plant cells. Virtually no Fe is transferred from older to younger leaves; therefore, its deficiency first occurs in young leaves. After uptake into the plant, Fe is mainly translocated in chelated form, with aliphatic hydroxy acids (such as malic or citric acid), phenols, thiols, polysaccharides, and amino acids.

Strategies of Fe Uptake

Plant species and genotypes differ in their ability to absorb Fe from the soil. There are two specific Fe absorption mechanisms, strategy I and strategy II, for increasing the solubility of Fe in the rhizosphere and its rate of uptake. Both of these mechanisms are present in apical parts of the roots, and respond when Fe deficiency occurs.

Strategy I is mostly exhibited by dicotyledons and nongraminaceous monocotyledons. The main mechanisms are (1) enhanced reduction of Fe^{3+} to the soluble Fe^{2+} form by membrane-bound reductases; (2) increased H^+ efflux at the root to the lower pH of the rhizosphere, favoring formation of Fe^{2+}; (3) release of chemicals capable of reducing Fe^{3+} to Fe^{2+}; (4) increased production of organic acids, particularly citrate; and (5) induction of the Fe^{2+} transport system to absorb the reduced Fe. Examples of crops exhibiting strategy I are soybean (*Glycine max*), peanut (*Arachis hypogaea*), sunflower (*Helianthus* spp.), pea (*Pisum sativum*), tomato (*Lycopersicon esculentum*), and other dicotyledons.

Strategy II is mostly adopted by graminaceous monocotyledons and is characterized by the production and release of Fe solubilizing low-molecular-weight compounds termed "phytosiderophores," which are capable of forming complexes with sparingly soluble Fe^{3+}, and increase its availability for uptake by plants. The reason Fe–siderophore complexes are absorbed without breaking down is, possibly, due to a proton–anion cotransport system, which is produced in response to Fe deficiency. Examples of crops exhibiting Strategy II are rice (*Oryza sativa*), wheat (*Triticum* spp.), corn (*Zea mays*; maize), barley (*Hordeum vulgare*), oats (*Avena sativa*), and other grasses.

Diagnosis of Fe Deficiency

Diagnosis of nutrient deficiency is usually carried out using three methods: visual deficiency symptoms, soil analysis, and plant analysis. Integration of all three methods is essential for accurate diagnosis of Fe deficiency.

Visual Symptoms

In most of the plant species, Fe deficiency is characterized by interveinal chlorosis with fine reticulate patterns observed in newly formed leaves (Fe chlorosis). The dark green veins are clearly visible against a yellow background (**Figure 1**). In severe cases, the youngest leaves are completely white and devoid of chlorophyll. Iron deficiency can easily be mistaken for nitrogen or magnesium deficiency; however, Fe deficiency affects the young emerging

Figure 1 (see color plate 19) Typical progressive symptoms of Fe deficiency in common bean (*Phaseolus vulgaris*). Reproduced with permission of the American Phytopathological Society Press from *Nutrient Deficiencies and Toxicities of Plants* – digital image collection CD Rom (2000). APS Press: Minnesota.

leaves, whereas nitrogen deficiency affects old leaves. This is because Fe is relatively immobile in the phloem compared with nitrogen or magnesium.

Soil Analysis

Soil analysis is useful for determining whether a particular nutrient is limiting. Several methods have been devised to extract Fe from soil, the most commonly used being the diethylenetriaminepenta-acetic acid (DTPA) extraction method. The critical range for Fe for this test is 2.5–4.5 ppm. If the soil contains less than 2.5 ppm it is considered deficient in Fe. However, even if soils are sufficient in Fe, this does not mean that the plant will not suffer Fe chlorosis. The uptake of Fe by plants depends on many other factors besides extractable Fe in the soil. These factors are discussed below.

Plant Analysis

The prediction of micronutrient deficiencies based on tissue analysis has been successfully used for all the micronutrients except for Fe. Total Fe concentration in the plant does not correlate well with the occurrence of Fe chlorosis. This is mainly because only 10–20% of the total Fe in plants is "physiologically active" or "available Fe." For example, *o*-phenanthroline extractable Fe or HCl extractable Fe, which estimates Fe available to plants, is well correlated with the occurrence of Fe deficiency symptoms, and is a better method to diagnose Fe deficiency in plants. The other quantitative measure of diagnosing Fe deficiency is by estimating chlorophyll content, but this should be used with caution as chlorophyll deficiency can be caused by the deficiency of other nutrients (for example N, S, Mg, and Mn) as well.

Causes of Fe Deficiency

The various factors related to soil, climate, and plant that can contribute to Fe chlorosis are summarized below.

Soil Factors

The availability of Fe to plants largely depends on soil factors, including soil pH, lime, and bicarbonate content, soil organic matter, and the concentrations of other mineral nutrient ions.

Soil pH The availability and uptake of nutrients by plants is highly dependent on pH. The solubility of Fe is highly pH dependent, and the activities of Fe^{3+} and Fe^{2+} decrease by 1000-fold and 100-fold, respectively, for each unit increase in pH. Under alkaline conditions, Fe^{2+} is oxidized to Fe^{3+}, which is relatively unavailable to plants because it precipitates as ferric oxide, which has extremely low solubility.

Lime content Iron availability is low in calcareous soils, which contain a high concentration of lime (calcium carbonate). In the presence of water and carbon dioxide, lime decomposes, resulting in accumulation of bicarbonates and calcium, as shown in eqn [1], which in turn increases pH:

$$CaCO_3 + CO_2 + H_2O \rightarrow Ca^{2+} + 2HCO_3^- \qquad [1]$$

Iron chlorosis occurs severely in soils that have >20% total $CaCO_3$ and/or >10% active lime. Chlorosis caused by lime is often termed lime induced chlorosis. Studies have shown that leaves suffering from lime induced chlorosis often have a high concentration of total Fe similar to or even greater than that of green leaves, indicating a physiological inactivation of Fe in the leaf apoplast due to the alkalization process. Furthermore, it was observed that the severe symptoms of lime induced chlorosis are often correlated not only with higher levels of total Fe in leaves (on a dry weight basis), but also with severe inhibition of leaf growth and chloroplast development.

Bicarbonate content Bicarbonates (HCO_3^-) in soil and water are an important cause of Fe chlorosis. Bicarbonate ions can be formed in calcareous soils by the reaction of CO_2 and water on calcite. Poor soil moisture, accumulation of CO_2 produced by roots, and microbial respiration under high soil moisture conditions enhanced the accumulation of HCO_3^- in the rhizosphere to the extent of 400–500 ppm. High HCO_3^- concentration inhibits root growth, thus decreasing the rate of cytokinin export to the shoot, which is necessary for protein synthesis and chloroplast development, resulting in chlorosis.

Organic matter Available Fe in soils is primarily present as part of an organic complex. Organic matter in soils thus exerts a pronounced effect on Fe availability. The formation of soluble Fe complexes by naturally occurring chelates may enhance the solubility of Fe. However, heavy manuring in alkaline soils decreases Fe availability as it is strongly adsorbed on the surface of organic matter; however, on decomposing, it is slowly supplied to plants.

Nutrient interactions Iron deficiency can be induced by the interaction of Fe with various other nutrients, for example, N, P, K, Ca, Mg, and Zn. The form of nitrogen applied may affect the availability of Fe. Increased uptake of NO_3-N (nitrate nitrogen) may cause an imbalance in the cation/anion ratio, resulting in exudation of HCO_3^- into the rhizosphere with a subsequent decrease in Fe uptake. Nitrate uptake also leads to alkalization of the root zone, which can lower Fe solubility and availability. In contrast, NH_4-N (ammonical nitrogen) fertilizers produce acidity when plants utilize ammonical N.

High phosphorus (P) levels in soils are antagonistic to Fe and decrease its availability to plants due to formation of insoluble Fe phosphates. Similarly, at high P concentration, the plant tissue may also induce Fe chlorosis due to the immobilization of Fe in the veins of the leaves.

Studies have shown that certain Fe efficient plants are unable to respond to Fe deficiency stress in the absence of potassium (K). Potassium seems to play a very specific role in plants for maximum utilization of Fe. This is due mainly to its involvement in production of protons and phytosiderophores.

Environmental Factors

Climatic factors greatly influence the occurrence of Fe deficiency in plants under field conditions, the most important being temperature and moisture content.

Temperature Since Fe absorption and translocation from root to shoots is an active process, temperature influences the occurrence of Fe deficiency. Temperature could influence the severity of Fe deficiency in plants in the following ways: (1) low soil temperature reduces root growth and metabolic activity, thus reducing Fe uptake; (2) low soil temperature could reduce the production of phytosiderophores, and the resultant mobilization and uptake of soil Fe; (3) low soil temperature could increase HCO_3^- levels in the soil and severity of chlorosis by increasing the solubility of CO_2 in soils; (4) high soil temperature decreases Fe uptake by increasing microbial decomposition of phytosiderophores; and (5) high soil temperature could increase HCO_3^- levels and Fe chlorosis by stimulating microbial activity and CO_2 production.

Soil moisture A high soil moisture level has a strong effect on Fe chlorosis through its effect on plant metabolism. Many studies have indicated that excess irrigation or prolonged wet periods in calcareous soils result in Fe chlorosis because of HCO_3^- build up. Increased Fe chlorosis of plants subsequent to irrigation is sometimes due to high levels of HCO_3^- in irrigation water. In addition, high HCO_3^-, high pH, and low Fe content in poorly aerated soils caused by excess water destroy many of the smaller roots and reduce the absorptive capacity of the whole root system, which may induce Fe chlorosis.

However, in waterlogged soils (especially in paddy rice fields), Fe^{3+} compounds, particularly the amorphous forms, are reduced to Fe^{2+} by anaerobic bacteria that use Fe^{3+} as an electron acceptor during respiration (see eqn [2]). This results in a high solubility of Fe in flooded soils, leading to toxic levels of Fe ($>50\,mg\,kg^{-1}$).

$$Fe(OH)_3 + 3H^+ + e^- \rightarrow Fe^{2+} + 3H_2O \quad [2]$$

Plant Factors

Different species and even cultivars within the same species vary in their susceptibility to Fe deficiency. The presence or lack of ability to absorb and translocate Fe has been observed in many plant species. Some species or cultivars respond to Fe deficiency by inducing biochemical and physiological reactions that make Fe more available for absorption by the roots. These plants are classified as Fe efficient. Plants that are unable to develop such mechanisms are classified as Fe inefficient. Fe efficient species have a greater tendency to lower the pH of the rhizosphere due to their ability to accumulate phenols. These reactions enable Fe efficient species to absorb more Fe rapidly. Some species are also stimulated to form more root hairs and develop cells that increase the capacity to reduce Fe^{3+} in roots. Examples of the sensitivity of species to Fe deficiency are given in **Table 1**.

Management of Fe Deficiency

Some of the important practices adopted to alleviate Fe chlorosis are soil amendments, foliar application

Table 1 Relative sensitivity of various plants to Fe deficiency

Susceptible	*Moderately susceptible*	*Relatively tolerant*
Upland rice (*Oryza sativa*)	Corn (*Zea mays*; maize)	Wheat (*Triticum* spp.)
Sorghum spp.	Barley (*Hordeum vulgare*)	Rye (*Secale cereale*)
Dry (common) bean (*Phaseolus vulgaris*)	Oat (*Avena sativa*)	Sunflower (*Helianthus* spp.)
Soybean (*Glycine max*)	Alfalfa (*Medicago sativa*)	Cotton (*Gossypium* spp.)
Grape (*Vitis* spp.)	Peanut (*Arachis hypogaea*)	*Amaranthus* spp.
Maple (*Acer* spp.)	Potato (*Solanum tuberosum*)	Peppermint (*Mentha piperita*)

of Fe compounds, genotypic selection, and agronomic crop management practices.

Soil Additives

Soil additives for control of Fe chlorosis are mainly categorized into inorganic Fe salts, Fe chelates, and other organic and acidifying soil amendments. A list of some inorganic and organic Fe fertilizers is given in **Table 2**.

Inorganic Fe compounds The most common inorganic source of Fe is $FeSO_4$. Soil application of inorganic $FeSO_4$ can alleviate Fe chlorosis when soils are deficient in Fe. Soil applications of inorganic Fe sources are usually not very effective in supplying Fe for crops unless applied in high doses, which is not economical for field crops. Inorganic Fe sources, when applied to soil, are rapidly converted into forms of Fe that are not available to plants, especially in calcareous soils.

Iron chelates The term "chelate" refers to chemicals that surround certain micronutrients, protecting them from becoming unavailable by high content of Ca or other nutrient elements. Soil applications of chelate compounds are more effective than inorganic iron salts in correcting Fe chlorosis, but are too expensive for general use, except for commercial crops of high value.

Table 2 List of various sources of Fe fertilizers and their Fe content (%)

Fertilizer	*Formula*	*Fe (%)*
Inorganic		
Ferrous sulfate	$FeSO_4\ 7H_2O$	19
Ferric sulfate	$Fe_2(SO_4)_3\ 4H_2O$	23
Ferrous oxide	FeO	77
Ferric oxide	Fe_2O_3	69
Ferrous ammonium sulfate	$FeSO_4\ (NH_4)_2SO_4.6H_2O$	14
Ferrous ammonium phosphate	$Fe(NH_4)PO_4\ H_2O$	29
Iron ammonium polyphosphate	$Fe(NH_4)HP_2O_7$	22
Organic		
Na-Fe ethylene diamine tetraacetate (EDTA)		5–14
Na-Fe diethylene triamine pentaacetate (DTPA)		10
Na-Fe hydroxyl ethylene diamine tetraacetate (HEDA)		5–9
Na-Fe ethylene diamine di(*O*-hydroxyphenol acetate) (EDDHA)		6

Other soil amendments Organic materials such as plant residues, manures, sewage sludge, and peat are good carriers of Fe and are effective in alleviating Fe chlorosis. Similarly, the application of farmyard manure and green manure to soil has shown improvement of Fe deficiency.

One of the best ways in which to increase the availability of Fe in the soil is to reduce the pH of the soil. Soil amelioration to prevent Fe chlorosis by acidification of the entire root zone is not practical. Therefore, part of the soil near the root zone is acidified by the application of sulfur or other acidifying agents. The amount of acidifying materials required varies with the per cent $CaCO_3$ present in the soil.

Foliar Management

As soil applications of most Fe sources are generally less effective, foliar application is widely used for the correction of Fe chlorosis. Both inorganic and organic Fe sources are effective as foliar sprays. The spraying of $FeSO_4$ solution together with some surfactant is very effective in correcting Fe chlorosis in many crops. However, because of the poor translocation of applied Fe within the plant, the applied Fe does not readily move from sprayed parts to other parts of the plant. The rate of translocation varies according to the type of crop, but is always less than 50% of applied Fe to a given leaf or leaflet. Therefore, under field conditions, growers need to apply foliar sprays of Fe more than once to provide adequate amounts to the developing canopy.

Cultivar Selection

Certain cultivars are more efficient in utilizing Fe and are less affected by Fe deficiency. The main mechanisms involved in Fe efficient cultivars of dicotyledons include: (1) the ability to reduce pH due to H^+ efflux from roots; (2) enhanced root associated Fe^{3+} reduction; (3) secretion of citrate in the roots; (4) increased root hair development; and (5) greater ability to form chelates with Fe to improve absorption and translocation. In monocotyledons, it is the ability of the cultivar to increase production of phytosiderophores that makes them more efficient in utilizing Fe. Since Fe efficiency is genetically determined, the best long-term possible solution would be the development of cultivars that are tolerant to Fe deficiency; this could easily be adopted by farmers and would be the most efficient way to alleviate Fe deficiency in crops. Examples of Fe efficient cultivars of major food crops are given in **Table 3**.

Table 3 Examples of Fe efficient cultivars in important food crops

Crop	*Fe efficient cultivars*
Rice	ARC 10372, Cauvery, TR 23, TR 25
Wheat	N 85021
Corn (maize)	WF9
Oat	Cocker 227
Soybean	Bragg, Pioneer, A7, Dawson, lines A11 through 15
Peanut	ICGV 86031, 71–234, 71–58, 234, 238
Chickpea (*Cicer arietinum*)	K 850
Mungbean	UT1, CN 36, CN 60
Sunflower	RHA 273 and 274
Tomato	Roza

Effects of Fe Deficiency on Plants

Nodule Development

Severe Fe deficiency decreases root growth and elongation. In most legumes, early nodule development after nodule initiation was most sensitive to Fe deficiency. For example, in peanut, Fe deficiency decreased the number of excisable nodules, nodule mass, number of bacteroids and concentrations of leghemoglobin, nitrogenase activity, and nitrogen fixing ability (**Table 4**). Fe deficiency did not limit the growth and soil or rhizosphere populations of peanut *Bradyrhizobium*, and there was no effect on root infection processes or nodule initiation. This suggests that nodule development processes were more sensitive to Fe deficiency than nodule initiation processes.

Photosynthesis and Electron Transfer

Iron is an integral part of chlorophyll and is important for its synthesis. Fe chlorosis decreases the level of chlorophyll and other carotenoids in all plant species, e.g., in grape (*Vitis* spp.; **Table 5**). As chlorophyll is essential for photosynthesis, Fe deficiency leads to decreased photosynthesis. Among the various processes of photosynthesis, photosystem II is more sensitive to Fe chlorosis than photosystem I. Fe chlorosis decreases all components of the electron transport chain. Owing to the direct involvement of Fe in protein synthesis, Fe deficiency decreases total proteins including ribulose-1,5-bisphosphate (rubisco), which plays an important role in the carbon cycle of C_3 plants.

Growth and Yield

Fe deficiency can significantly decrease growth and yield of crops. For example, in peanuts grown in calcareous soils, Fe deficiency decreased leaf area and dry matter production of leaves, stems, and root, leading to yield losses of 20% (**Table 6**). Yield losses to the extent $\geqslant$25% due to Fe deficiency are reported for several other crops such as cereals (rice, wheat, corn, and sorghum), legumes (soybean, dry bean, mungbean (*Phaseolus aureus*), and lupin (*Lupinus albus*)), vegetables (tomato, capsicum (*Capsicum frutescens*), potato (*Solanum tuberosum*), and spinach (*Spinacia oleracea*)), and fruit crops (citrus, grape, pears (*Pyrus communis*), and apple (*Malus pumila*)). Similarly, the growth and value of commercial ornamental crops such as rose (*Rosa* spp.), chrysanthemum

Table 4 Effects of Fe deficiency on different aspects of nodulation in peanut

Trait	*−Fe*	*+Fe*
Number of nodule initials (plant^{-1})	55	52
Nodule numbers on tap root	11	32
Nodule mass (mg plant^{-1})	15	60
Leghemoglobin concentration (nmol g^{-1} fresh weight nodule)	6	86
Bacteroid content (10^6 plant^{-1})	2	430
Bacteroid concentration (10^6 g^{-1} fresh weight nodule)	170	7300
Acetylene reduction activity (pmol min^{-1} plant^{-1})	3	35

Treated plants received foliar application of 0.5% (w/v) $FeSO_4$ in 0.75% Tween 80 on day 10. Nodules were harvested from plants on day 15 for analyses. Adapted with permission from O'Hara GW, Dilworth MJ, Boonkerd N, and Parkpian P (1988) Iron deficiency specially limits nodule development in peanut inoculated with *Bradyrhizobium* sp. *New Phytologist* 108: 51–57.

Table 5 Effects of different stages of Fe deficiency on photosynthetic pigments, electron transport and photosystem activity, rubilose bisphosphate carboxylase (rubisco) activity, and protein content on per unit fresh weight basis in grape

Trait	*Control*	*Mild chlorosis*	*Severe chlorosis*
Chlorophyll (a + b)	2.25	1.30	0.36
Carotenoids	0.85	0.56	0.20
Electron transport chain	148.60	96.60	50.20
Photosystem I	368.50	351.80	324.70
Photosystem II	240.00	119.30	48.40
Rubisco activity	56.40	32.70	14.60
Proteins	48.60	36.90	23.30

Adapted with permission from Bertamini M, Nedunchezhian N, and Borghi B (2001) Effect of iron deficiency induced changes in photosynthetic pigments, ribulose-1,5-bisphosphate carboxylase, and photosystem activities in field grown grapevine (*Vitis vinifera* L. cv. Pinot noir) leaves. *Photosynthetica* 39: 59–65.

Table 6 Effects of Fe chlorosis on growth and yield of peanut

Trait	*−Fe*	*+Fe*
Leaf area (cm^2)	429.00	562.00
Leaf dry weight (g plant^{-1})	4.00	4.90
Stem dry weight (g plant^{-1})	4.20	4.90
Root dry weight (g plant^{-1})	0.41	0.48
Pod yield (kg ha^{-1})	1340.00	1660.00

Treated plants received foliar application of 0.5% (w/v) $FeSO_4$ at 40, 60, and 90 days after sowing.

(*Chrysanthemum* spp.), and tulips (*Tulipa* spp.) have been shown to decrease under Fe deficiency.

Engineering Fe Efficient Crops

In strategy I plants, reduction of Fe^{3+} to Fe^{2+} is dependent on the enzyme ferric reductase; therefore, Fe efficiency of plants can be improved by manipulating expression of this enzyme. It has been established that the *FRO2* gene in *Arabidopsis* encodes expression of ferric chelate reductase and is expressed under Fe deficiency. Similarly, the ferric reductase gene *FRE1* was identified in yeast. Genetically modified lines of tobacco (*Nicotiana tabacum*) with *FRE1* genes showed greater tolerance to Fe deficiency.

In strategy II plants, there is a strong correlation between the amount of phytosiderophores released and tolerance to Fe deficiency. Graminaceous plants secrete natural Fe chelator (mugineic acid phytosiderophores, MAs) from their roots to solubilize Fe in the soil. These MAs are synthesized from L-methionine through nicotinamine. Nicotinamine synthase and nicotinamine aminotransferase (NAAT) are the two critical enzymes in the biosynthesis of MAs, and in Fe tolerant species, activity of these two enzymes is increased under Fe deficiency. Among strategy II plants, rice is particularly sensitive to Fe deficiency when compared with barley. Recently, barley genomic DNA fragments containing two *NAAT* genes (*NAAT-A* and *NAAT-B*) that encode enzymes involved in the biosynthesis of phytosiderophores were successfully introduced in rice, using transformation techniques. The resultant transgenic rice showed greater tolerance to Fe deficiency and produced greater grain yields. These results suggest there are possibilities for genetically engineering Fe efficient plants.

Fe deficiency is also a major problem in humans, and is considered to be widespread across the globe, affecting more than 30% of the world's human population, especially in developing countries. Fe deficiency is mainly the result of plant based diets, which contain low levels of poorly bioavailable Fe. Research is currently underway to improve the uptake of Fe by plants from the soil, and to produce rice plants (and other food crops) that store higher concentrations of the Fe containing protein ferritin.

Acknowledgements

Approved for publication as FAES No. R-09448.

List of Technical Nomenclature

Adsorption	Accumulation of ions on a solid surface caused by ion exchange or other reactions.
C_3 plant	Plants in which the first product of carbon dioxide fixation is the three carbon compound, phosphoglyceric acid.
Chelate	Complex organic molecules that can combine with cations, do not ionize, and can supply nutrients to plants at slow and steady rates.
Chlorophyll	The green pigment in chloroplasts that absorbs radiant energy and transforms it into chemical energy.
Chloroplast	A membrane-bound cell organelle, which contains photosynthetic chemicals and structures.
Chlorosis	Yellowing of leaves due to lack of chlorophyll.
Dicotyledons	A class of angiosperms characterized by two cotyledons, and usually with net (reticulate) venation on the leaves.
DNA	Dioxyribose nucleic acid; carrier of genetic information.
Ferredoxin	An electron transferring, Fe containing protein that functions in photosynthesis.
Gene	A segment of the DNA double helix that acts as a template for the production of specific ribonucleic acid, and governs hereditary characteristics of an organism.
Monocotyledons	A class of angiosperms characterized by one cotyledon, and usually with parallel venation on leaves.
Nitrogen fixation	Process in which gaseous nitrogen is converted into nitrogenous compounds.
Nodule	Enlargement or swelling on root, which contains bacteria that fix nitrogen.
Ppm	Parts per million or mg kg^{-1}.
Rhizobia	Bacteria of the genus *Rhizobium*, which are associated with roots of leguminous plants and help in nitrogen fixation.
Soil pH	Negative logarithm of the hydrogen ion activity of a soil.
Translocation	Movement of dissolved materials in cell sap from one plant organ or tissue to another.

See also: **Plant Nutrition**: Deficiency Diseases, Principles; Ion Transport; Mineral Uptake; Nitrogen Fixation.

Further Reading

Abadia J, Morales F, and Abadia A (1999) Photosystem II efficiency in low chlorophyll, iron-deficient leaves. *Plant and Soil* 215: 183–192.

Bergmann W (1992) *Nutritional Disorders of Plants: Development, Visual and Analytical Diagnosis*. New York: Gustav Fischer Verlag Jena.

Bertamini M, Nedunchezhian N, and Borghi B (2001) Effect of iron deficiency induced changes in photosynthetic pigments, ribulose-1,5-bisphosphate carboxylase, and photosystem activities in field grown grapevine (*Vitis vinifera* L. cv. Pinot noir) leaves. *Photosynthetica* 39: 59–65.

Brady NC and Weil RR (2002) *The Nature and Properties of Soil*. New Jersey: Prentice Hall.

Cakmak I (2002) Plant nutrition research: priorities to meet human needs for food in sustainable ways. *Plant and Soil* 247: 3–24.

Jolley VD, Cook CA, Hansen NC, and Stevens WB (1996) Plant physiological responses for genotypic evaluation of iron efficiency in Strategy I and Strategy II plants: a review. *Journal of Plant Nutrition* 19: 1241–1255.

Marschner H (1995) *Mineral nutrition of higher plants*. London: Academic Press.

O'Hara GW, Dilworth MJ, Boonkerd N, and Parkpian P (1988) Iron deficiency specially limits nodule development in peanut inoculated with *Bradyrhizobium* sp. *New Phytologist* 108: 51–57.

Prasad PVV, Satyanarayana V, Potdar MV, and Craufurd PQ (2000) On-farm diagnosis and management of iron chlorosis in groundnut. *Journal of Plant Nutrition* 10: 1471–1483.

Aluminum Toxicity

D L Jones, University of Wales, Bangor, UK
P R Ryan, CSIRO Plant Industry, Canberra, Australia

Introduction

Aluminum (Al) is the third most abundant element in the earth's crust and is a primary constituent of many minerals in soil. Under most conditions, Al in soil remains in a solid form, either as Al-oxyhydroxides (e.g., $Al(OH)_3$) or as clay minerals. At the low pH values observed in acids soils (below pH 4.5), however, mineral dissolution increases and the most prevalent soluble Al species is the toxic Al^{3+} cation. The Al^{3+} cations effectively inhibit root growth in many species and this hinders plant development and reduces productivity. Understanding the basis of this Al induced rhizotoxicity has received continued attention for over a century and the cause of toxicity remains controversial to this day. The main reason for the lack of consensus on the cause of growth inhibition has been due to the continued search for a sole mechanism of toxicity. This article discusses the multiple sites of Al toxicity in plants and describes the mechanisms proposed to explain the greater tolerance to Al toxicity exhibited by some species.

Alongside salinity and water stress, Al toxicity is recognized as being one of the biggest limitations to crop production in the world and is characterized in all cases by the inhibition of root growth. It has been estimated that approximately 30% of the world's potential food-producing area is covered by acid soils. These soils are mainly associated with regions of high rainfall where base cations (Ca^{2+}, Mg^{2+}, K^+, Na^+, etc.) have been leached from the soil and replaced by Al^{3+} released from soil mineral weathering. In some continents, including North America and Europe, this soil acidity is counteracted by the application of lime ($CaCO_3$) in order to raise the pH to a level where Al^{3+} is no longer soluble (pH 6.0–7.0). In other regions of the world, however, where lime is unavailable as a resource to poor farmers, acid soils still provide a major constraint to food production (e.g., South America, Sub-Saharan Africa). This is particularly important considering that these areas of the world have the highest rates of population growth and the greatest demand for food production. To try and overcome these limitations, much recent attention has focused on attempting to understand the mechanisms responsible for Al toxicity and the genes that contribute to Al tolerance in crop plants. In the last two decades, crop plant varieties that possess resistance to Al have been identified alongside noncrop plants that show the ability to hyperaccumulate Al. The mechanisms for Al tolerance, as well as those of toxicity, are outlined below.

Understanding the causes of Al toxicity in plants is also of relevance in forest ecosystems and particularly forest decline. This mainly relates to determining the impact of increased anthropogenic emissions of SO_2 and NO_2 on forest health in industrial areas of the world (e.g., Europe). Current thought suggests that the increased deposition of anthropogenically derived HNO_3 and H_2SO_4 in forest ecosystems induces an acidification of the soil, thereby increasing free Al^{3+} in the soil solution, which consequently inhibits root growth. This has serious consequences, not only in reducing nutrient and water uptake, but also for biomass production and the increased potential for trees to be blown over.

Symptoms of Al Toxicity

In plants, Al appears only to be toxic to roots causing an inhibition of root growth, branching and root hair development. This poor root growth results in a reduced exploration of the soil for resources such as water and nutrients and also leads to poor formation of symbiotic relationships with mycorrhizae and N_2-fixing bacteria. Generally, shoot growth, even during the passage of the hypocotyl through the soil during germination, is relatively unaffected by the presence of Al.

Within the root, it appears that only root cells currently undergoing cell division and expansion are affected by Al (i.e., those in the apical, elongation, and root hair zones), while mature, nonexpanding, root cells appear to be unaffected. The rate and extent of Al-induced growth inhibition is critically dependent on both the concentration and duration of Al exposure. Generally, the impact of Al on root cells is very rapid with an initial inhibition of cell growth observed within 10–60 min of exposure in most crop plants (**Figure 1**). Although Al affects cell division through a disruption of mitosis, it is now clear that the inhibition of root growth is primarily due to a disruption of cellular expansion (i.e., the root elongation zone). Root growth inhibition of most crop plants is typically observed at Al^{3+} concentrations ranging from 1 to 50 μmol l^{-1} when grown in hydroponic culture and soil at a pH <4.5 (**Figure 2**). Above this pH, however, Al readily precipitates as $Al(OH)_3$, which is nontoxic to plants.

Al in solution can exist in a variety of ionic forms depending upon the pH of the solution. Owing to the observation that Al toxicity only becomes manifest at soil pH values less than 4.5, it has generally been assumed that the Al^{3+} species is responsible for toxicity. However, recently it has been shown that at low pH, polynuclear forms of Al, such as Al_{13} ($AlO_4Al_{12}(OH)_{24}(H_2O)_{12}^{7+}$), can also occur which are 50 times more toxic than monomeric Al^{3+}. Although the concentration of polynuclear Al in the soil solution remains unknown, from mass balance calculations and Al speciation experiments, it appears likely that the Al^{3+} is the main cause of toxicity. At low solution pH values, Al^{3+} has the tendency to form complexes with inorganic anions, including SO_4^{2-} and PO_4^{3-}, and organic anions (e.g., organic matter with carboxyl groups, organic acids). All of these Al complexes appear to be nontoxic to plants.

Sites of Al Toxicity in Plant Cells

Current evidence suggests that Al has multiple target sites within the plant. As Al^{3+} is initially present in the soil solution outside the root it must interact firstly with the cell wall and outer face of the plasma membrane before passing into the cytoplasm and interacting with intracellular processes.

Al Interactions with the Cell Wall and Plasma Membrane

One of the main causes of Al-induced root cell growth disruption is the result of interactions of Al with the cell wall and outer face of the plasma membrane. Both the cell wall and plasma membrane contain many negatively charged binding sites such as pectate and phospholipids with which Al^{3+} binds, often irreversibly. It has been determined that typically more than 95% of the Al associated with

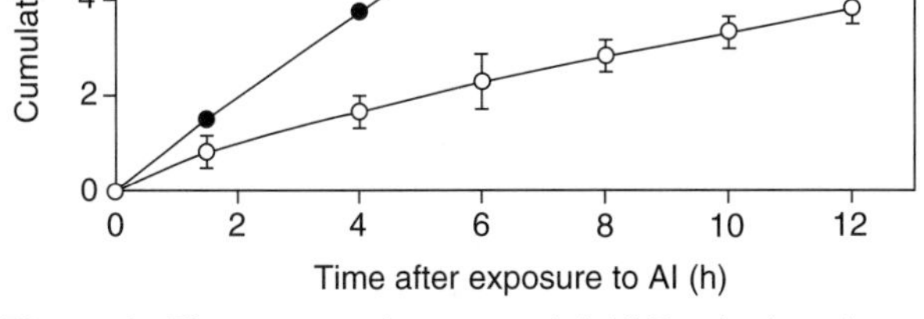

Figure 1 Time course of root growth inhibition in the primary root axis of an Al-sensitive cultivar of corn (*Zea mays*) grown hydroponically in minimal nutrient media (200 μmol l^{-1} $CaCl_2$). Roots were either grown in a solution containing Al (50 μmol l^{-1}; pH 4.40) or grown under control conditions (0 μmol l^{-1} Al; pH 4.40). Values represent means ±SEM ($n=4$).

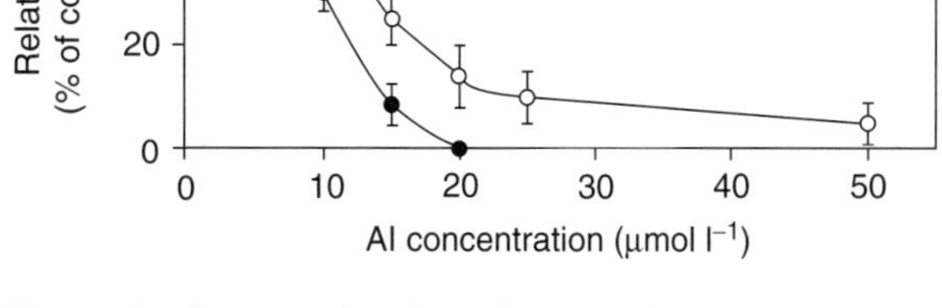

Figure 2 Concentration-dependent inhibition of growth by Al in the primary root axis of an Al-sensitive cultivar of wheat (*Triticum aestivum*) and in the root hairs of *Limnobium stoloniferum* (Amazon Frogbit) grown in hydroponic culture (solution pH of 4.40; 200 μmol l^{-1} $CaCl_2$). Values represent means ±SEM ($n=4$).

roots is bound to the cell wall. The probable site of this binding is through a replacement of Ca^{2+} in the galacturonic acid polymer pectate. It has been hypothesized that Al^{3+}, which is known to bind to pectin far more strongly than Ca^{2+} due to its higher charge and smaller ionic radius, causes the cell wall to become overrigidified, thus preventing loosening, which is a prerequisite for cellular expansion. In addition, it is likely that during this interaction with pectate, Al detrimentally alters the pH and free Ca^{2+} concentration in the apoplast that controls enzymes involved in wall loosening and cellular expansion.

It is probable that Al also affects the production and deposition of cellulose in growing cells by causing the plasma membrane to become more rigid so preventing free movement of membrane protein complexes associated with the extrusion process (e.g., cellulose synthase), or possibly by interacting with the microfibrils directly. Exposure of roots to Al induces production of the cellulose-related 1,3-β-glucan polymer callose, although this is thought to be a cellular defense response rather than a cause of toxicity.

Although no direct evidence exists for an interaction of Al with cellulose microfibrils, there is evidence to suggest that Al does cause the plasma membrane to become rigidified, which presumably affects protein movement in the bilayer. The mechanism by which this occurs is through the replacement of Ca^{2+} bridges between phospholipid head groups by Al^{3+} causing increased phospholipid packing (due again to its smaller size and higher charge in comparison to Ca^{2+}). One consequence of this increased rigidity of the plasma membrane will be to significantly affect metabolism both in the symplastic and apoplastic compartments. Al has also been shown to induce lipid peroxidation in plasma membranes as a direct result of the increased lipid packing and this is also believed to be a cause of toxicity owing to the consequent loss of membrane integrity. Of more importance, however, is the effect of Al-induced increased membrane rigidity on the fusion of membrane vesicles at the site of cell expansion. Studies *in vitro* have suggested that Al may significantly enhance (up to 10-fold) the rate of liposomal fusion in comparison to Ca^{2+}, but that the fusion events are not spatially coordinated leading to the improper construction of the plasma membrane. Al has also been shown to bind strongly to membrane-associated proteins; however, there is little evidence to suggest this binding has any detrimental effect on either protein conformation or activity.

Another potential site of Al toxicity in root cells is the blockage of Ca^{2+} uptake at high external Al concentrations. Ca^{2+} uptake is known to be integrally involved in cellular expansion, especially in root hairs, and Al appears to block the cation channels responsible for Ca^{2+} uptake. In contrast, at low external Al concentrations, root Ca^{2+} uptake appears to be unaffected; however, root growth is still inhibited suggesting that a disruption of Ca^{2+} uptake is not a primary Al target site.

Al Uptake and Concentrations in Root Cells

The evidence currently available suggests that the transport of Al into plant cells is a slow process and the exact mechanism of uptake remain to be identified. Using secondary ion mass spectrometry it has been shown that Al preferentially accumulates in the cytoplasm of the epidermis reaching concentrations of 10–70 $\mu mol\,l^{-1}$ within a few hours and 100–400 $\mu mol\,l^{-1}$ within 24 h.

Al Interactions Inside the Cell

Al can rapidly enter the cytoplasm of root cells and accumulate to significant levels (10–100 $\mu mol\,l^{-1}$). The question therefore arises as to what and where are the most likely target sites in the symplasm? On crossing the plasma membrane, Al is confronted by a marked change in its chemical environment, the most important probably being the pH shift from 4.5 outside to 7.1 within the cytoplasm. From chemical equilibria calculations it has been predicted that most of the Al in the cytoplasm will quickly (within seconds) become associated with ligands such as carboxylic acids (e.g., citrate) and compounds containing PO_4^{3-} groups (e.g., ATP, GTP) leaving only picomolar concentrations of Al^{3+} in the cytoplasm. As root respiration and general cell metabolism appears to be largely unaffected by the presence of Al, it is likely that the concentration of free Al^{3+} is kept low in the cytosol or that most of the ligand:Al complexes that do form are nontoxic. In contrast, the lipid phosphatidylinositol bisphosphate ($PtdInsP_2$), which is localized in the plasma membrane, appears to be a high-affinity binding site for Al. This is likely to be of importance as $PtdInsP_2$ is integrally involved in plant signal transduction, cytoskeletal modulation and root hair growth. It appears that the binding of Al to $PtdInsP_2$ prevents the production of the secondary messengers inositol(1,4,5)trisphosphate and diacylglycerol through an inhibition of the enzyme phospholipase C (PLC). Such an inhibition probably causes disruptions in signal transduction cascades within the plant.

It has been suggested that Al occupies Ca^{2+}-binding domains in enzymes, thus disrupting their activity/activation in the cell. However, recent evidence suggests that owing to the low concentrations of Al^{3+} in the cytoplasm and the inability of Al^{3+} to occupy the Ca^{2+}-binding domain of either calmodulin or a variety of other Ca^{2+}-dependent enzymes, this is not likely to be a major site of toxicity.

One of the visual symptoms of Al toxicity appears to be a nonsite-directed swelling of root cells (in contrast to their normal site-directed cylindrical expansion pattern) suggesting that cytoskeletal regulation may be one of the principal Al target sites in plants. It is known that in plants actin is regulated in part through the action of profilin, and that the concentration of profilin in plants is regulated by its binding to the lipid anchor $PtdInsP_2$. Whether the binding of Al to $PtdInsP_2$ is capable of disrupting actin cytoskeletal dynamics through the blocking of profilin binding is still unclear. Additionally, the actin cytoskeleton may be regulated through the interaction of Ca^{2+}-dependent phospholipid-binding annexins and profilin. The question therefore arises as to the possible effects of Al on the Ca^{2+}-activation, phospholipid-binding, and profilin-binding properties of annexins. Furthermore, annexins are thought to be involved in the regulation of vesicle fusion. Studies have indicated that Al causes the actin network to become rapidly and dramatically rigidified. Similar studies have indicated that Al significantly affects microtubule dynamics in plant cells affecting both the orientation and stability of the microtubule arrays. Generally, as occurs for actin, the microtubules become disorganized within the cell and become rigidified in comparison to untreated cells.

Another potential major target of Al may be associated with disruptions in cytoplasmic Ca^{2+} homeostasis, which is known to regulate many processes in plants and is also believed to be integrally involved in cell growth. A number of studies have demonstrated that Al can induce both positive and negative changes in cytoplasmic Ca^{2+} concentrations in plant cells.

Mechanisms and Genetics of Al Resistance

Plants vary widely in their ability to grow in acid soils where Al^{3+} toxicity is severe. Many species that are native to the naturally acid soils of tropical regions are very resistant to Al stress and some of these can accumulate high concentrations of Al in their leaves (>1% dry weight). Rice (*Oryza sativa*) and cereal rye (*Secale cereale*) are examples of another group of species that, while tolerant to Al^{3+} stress, do not accumulate high concentrations of Al internally. Some commercially important species such as alfalfa (*Medicago sativa*, lucerne), barley (*Hordeum vulgare*), tomato (*Lycopersicon esculentum*) and oilseed rape (*Brassica napus*) are sensitive to Al^{3+} and other species such as wheat (*Triticum aestivum*) and corn (*Zea mays*; maize) display a significant variation in Al^{3+} sensitivity among their genotypes.

Genetics of Al Resistance

The genetics of Al resistance has only been examined in detail in a small number of crop species. Although resistance appears to be a multigenic trait in the majority of plants examined, in some species most of the tolerance appears to be encoded by a single dominant locus. Wheat and barley are examples of this latter group and the major resistant loci have been mapped to the long arms of chromosome 4D and 4 H, respectively. However, even in these species, the variation in Al tolerance observed among cultivars indicates that allelic variability of the tolerance loci or the presence of additional tolerance genes could also be contributing to tolerance. However, none of the genes encoding any mechanism of Al resistance in plants have been isolated as yet.

Mechanisms of Al Resistance

A single mechanism is unlikely to be responsible for resistance to Al in all species. Different mechanisms are likely to operate in different species and since resistance is largely a multigenic trait, it is probable that several mechanisms act together in the same plant. Many hypotheses have been proposed to explain why some plants perform better than others on Al-toxic soils. These hypotheses can be divided into two general types of mechanisms: those that exclude Al from the root cells and those that enable the plant to tolerate Al^{3+} once it has entered the plant cells (**Figure 3**).

Exclusion mechanisms operate by reducing the potentially harmful interactions between Al^{3+} and the root cell wall and plasma membrane as well as minimizing the uptake of Al into the plant cells. Exclusion of Al might be achieved by: (1) releasing substances from the root that chelate and detoxify Al in the apoplasm (organic acids, phenolic compounds, proteins, mucilage); (2) raising the pH in the rhizosphere to reduce the activity of the Al^{3+} cations near the root surface; (3) immobilizing the Al^{3+} on the fixed negative charges in the cell wall; (4) maintaining a smaller net negative charge on the surface of the root cell membranes so that the electrostatic accumulation of the Al^{3+} cations toward the membrane surface is minimized; and (5) reducing the capacity of cells to transport Al into the cytosol or by actively transporting Al out of the cell to maintain a low internal concentration. Mechanisms of internal tolerance enable plants to cope with the Al once it enters the root cells. This could be achieved by: (1) detoxifying Al^{3+} in the cytosol by chelating it with organic anions, proteins, phenolic compounds, tannins, polyphosphates, etc.; (2) sequestration to metabolically less sensitive compartments, such as the vacuole; and (3) evolution of metabolic pathways that are less sensitive to interference by Al or that have greater capacity to overcome the effects of toxicity (e.g., oxidative stress). The amount of experimental support for each of these hypotheses varies widely but recent studies indicate the operation of several mechanisms of resistance. Two of these mechanisms rely on the ability of organic acid anions to complex Al^{3+} and either exclude Al from

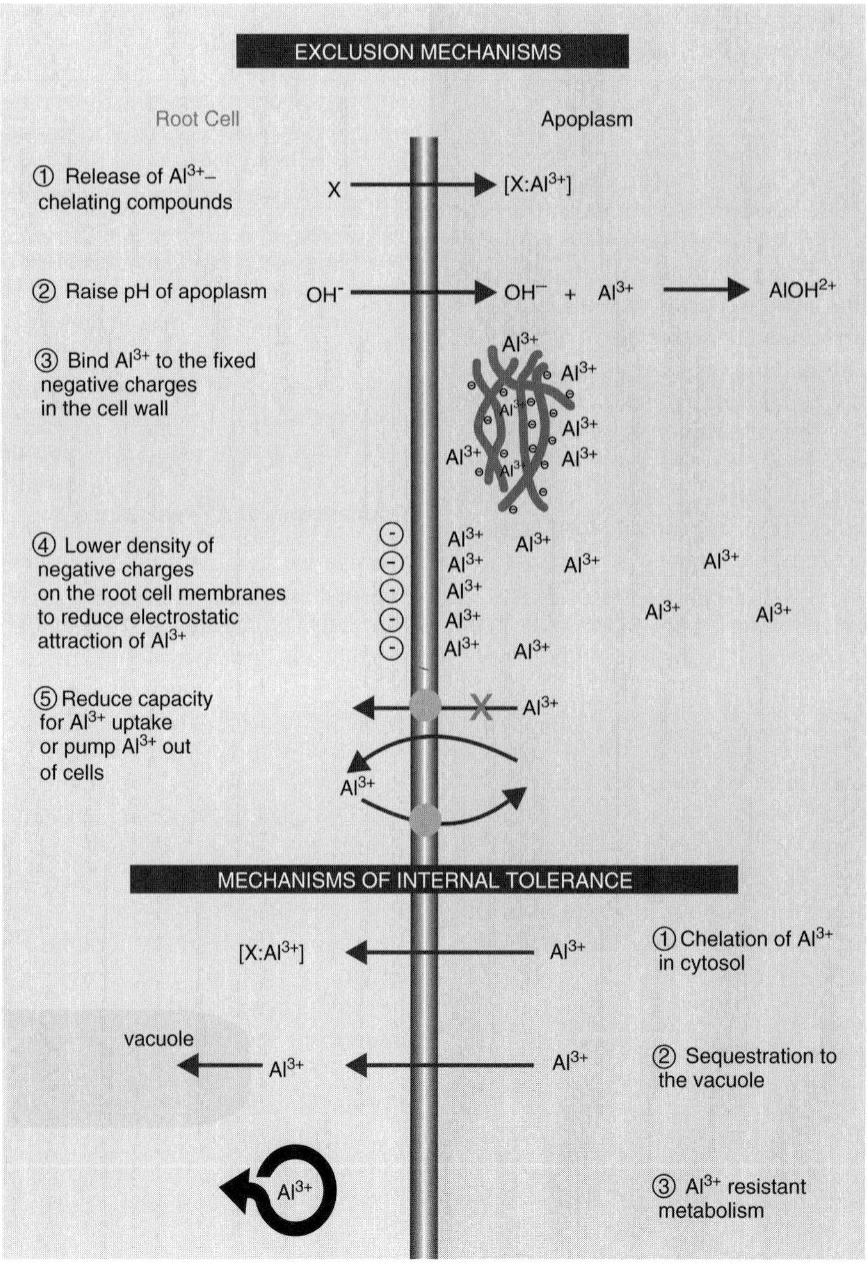

Figure 3 Cartoon illustrating some of the proposed mechanisms for aluminum resistance in plants. Aluminum could be excluded from the root tissue by: (1) the release of chelating substances; (2) raising the pH in the rhizosphere; (3) immobilizing the Al^{3+} in the cell wall; (4) minimizing the electrostatic accumulation of the Al^{3+} cations toward the membrane surface; and (5) reducing the influx or increasing the efflux of Al. Deleterious effects of Al could be minimized once it enters the root cytosol by: (1) chelating the Al ions with suitable compounds; (2) sequestering Al to the vacuole; and (3) metabolic pathways that are less sensitive to the harmful effects of Al.

the root cells or detoxify Al internally. These and other mechanisms are described further below.

Exclusion Mechanisms

Al can be excluded from plant tissues by the release of organic anions from roots. These anions complex with and detoxify the Al^{3+} cations at the root–soil interface and so prevent it from being taken up by the root cells. In some species, this release is localized to the root apex, a critical region for Al toxicity because it contains the zones of cell division and elongation. The specific organic anions released from the roots vary between species, but citrate, oxalate and malate are most commonly detected (**Table 1**). Malate and citrate are present in all cells because they are involved in the mitochondrial respiratory cycle. All three anions are able to form complexes with Al^{3+} with the following order of strength: citrate > oxalate > malate. Wheat has provided some of the most convincing evidence that organic anion release is a mechanism of Al resistance in plants. It shows a large natural variation in Al resistance and many of the key studies have been conducted on near-isogenic lines of wheat that differ in Al tolerance at a single dominant locus. In the F_2 progeny derived from a cross between these lines, tolerance to Al was shown to cosegregate with the capacity to release malate from the root apices during Al^{3+} treatment. Results collected from a set of ditelosomic wheat lines generated from the Chinese Spring cv. also support the involvement of malate efflux in Al resistance and confirm that the long arm of chromosome 4D contains a major tolerance locus in wheat.

A closer examination of the kinetics of organic anion release in several species reveals two main types of response (**Table 1; Figure 4**). In one response observed in wheat, buckwheat (*Fagopyrum esculentum*), and tobacco (*Nicotiana tabacum*), the activation of organic anion efflux by Al is rapid and occurs without any discernible delay after exposure to Al^{3+}. This rapid response suggests that all the necessary metabolic "machinery" is constitutively expressed in the root cells and that organic anion efflux is simply triggered by the Al^{3+} cations. In the second response observed in rye, *Cassia tora*, and Triticale, a lag of several hours is observed between the addition of Al and the start of organic anion release, with the rate of exudation then increasing gradually with time. This second response suggests that intermediate steps occur between the reception of the Al stimulus and organic anion release, which could involve signal-transduction pathways, gene transcription and *de novo* protein synthesis. Details of these signals or how organic anion release is regulated in either response is unknown. However, electrophysiological data collected with the patch-clamp technique has demonstrated that Al^{3+}-activated anion channels are present in protoplasts isolated from the root apices of wheat and corn. The genes encoding these anion

Table 1 A selection of species that show an Al^{3+}-dependent release of organic acid anions from their roots

Plant species	*Major organic anions released*	*Response to Al^{3+}*	*Efflux rate (nmol g^{-1} fresh wt h^{-1})*	*Tissue measured*
Wheat (*Triticum aestivum*)	Malate	Rapid	4000[a]	Root apices
Corn (*Zea mays*; maize)	Citrate	Rapid + slow	55[a]	Root apices
Tobacco (*Nicotiana tabacum*)	Citrate	Rapid	240[b]	Root apices
Cassia tora	Citrate	Slow	280[c]	Whole roots
Triticale (*Tritico secale*)	Citrate, malate	Slow	7, 9[d]	Whole roots
Rye (*Secale cereale*)	Citrate, malate	Slow	70, 35[d]	Whole roots
Buckwheat (*Fagopyrum esculentum*)	Oxalate	Rapid	70[d]	Whole roots
Soybean (*Glycine max*)	Citrate	?	55	Whole roots
Taro (*Colocasia esculenta*)	Oxalate	?	46	Whole roots
Oats (*Avena sativa*)	Citrate	?	6[d]	Whole roots
Radish (*Raphanus sativus*)	Citrate	?	23[d]	Whole roots
Oilseed rape (*Brassica napus*)	Citrate, malate	?	4, 4[d]	Whole roots
	Citrate		38[d]	Whole roots

The table shows the major organic anions released from each species, whether the response to Al^{3+} is known to be rapid or slow, and an estimate of the rate of exudation (root material used in these calculations is provided). Modified with permission from Ryan PR, Delhaize E, and Jones DL (2001) Function and mechanisms of organic anion exudation from plant roots. *Annual Review of Plant Physiology and Plant Molecular Biology* 52: 527–560.

Assumptions used in calculating efflux to common units:

[a] root apex = 2 mm × 0.5 mm diam.

[b] root apex = 6 mm × 1 mm diam.

[c] root apex = 6 mm × 0.4 mm diam.

[d] dry wt = 7% fresh wt. ? = not determined.

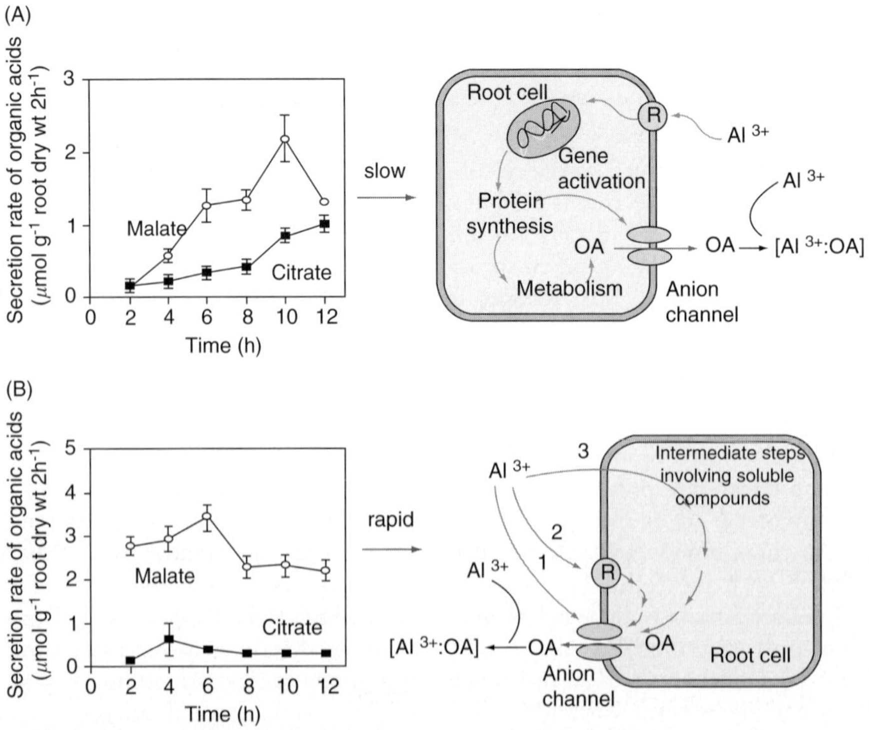

Figure 4 Aluminum-dependent release of organic acid anions (OA) from plant roots. In rye (A), the Al^{3+}-dependent release of OA increases with time indicating that the mechanism is induced by Al^{3+}, perhaps requiring gene activation and protein synthesis. In wheat (B), the activation of OA release is rapid and efflux is relatively constant suggesting the mechanism is expressed constitutively and simply triggered by Al. Possible pathways for this trigger are shown: (1) direct interaction with an anion channel; (2) activation via a membrane-bound receptor®; and (3) indirect activation via intermediate steps involving soluble compounds. Data is copyrighted by the American Society of Plant Physiologists and the data was adapted with permission from Li XF, Ma JF, and Matsumoto H (2000) Patterns of aluminum-induced secretion of organic acids differ between rye and wheat. *Plant Physiology* 123: 1537–1543.

channels have not been isolated and this remains an important challenge for the future.

There is evidence for other exclusion mechanisms in wheat and corn. For example, when the progeny from a cross between Al-resistant and Al-sensitive wheat genotypes were examined, it was found that resistance to Al cosegregated with the exudation of a 23-kDa polypeptide from the roots. Since this peptide was able to bind to Al^{3+}, its exudation could increase resistance to Al by chelating Al^{3+} in the rhizosphere, where it could detoxify the cations and prevent their uptake by root cells. Similarly, Al induces the exudation of certain phenolic compounds, such as catechin and quercetin, from the root tips of corn roots. Since Al can form stable complexes with these compounds, it is possible that they also contribute to tolerance but more work is required to confirm this hypothesis.

Internal Tolerance Mechanisms

Organic acid anions are also implicated in resistance mechanisms that rely on internal tolerance. Once the Al enters the cell, the concentration of free Al^{3+} cations in the cytosol will be very low (subnanomolar), owing to the high pH of the cytosol (~pH 7.0). But even at these concentrations Al^{3+} remains a hazard. The very high affinity of Al^{3+} for oxygen ligands allows it to compete with other ions for metabolically important sites despite a large disparity in their concentrations. For example, Al^{3+} binds to ATP almost 10^7 times more strongly than Mg^{2+}. However, the stability constant for the Al^{3+}:ATP complex is significantly smaller than for Al^{3+}:citrate or Al^{3+}:oxalate complexes, indicating that these organic anions could protect the plants by chelating Al^{3+} in the cytosol. The metal:anion complex could then be transported around the plant for storage. Indeed, the accumulation of Al in the tissues of some very acid-tolerant species (*Hydragenea*, buckwheat, and *Melastoma malabathricum*) occurs as complexes with citrate or oxalate. In buckwheat, the organic anions chelating Al are different in different tissues. For example, in the root cells and vacuoles of leaf cells, Al:oxalate (1:3) is detected, but translocation to leaves in the xylem stream occurs as an Al:citrate (1:1) complex.

Using Biotechnology to Develop Al^{3+}-Resistant Plants

Since Al stress is a major limitation to plant production on acid soils there is interest in developing new cultivars with a greater degree of Al resistance. Any natural variation of Al resistance within a species can be exploited by traditional plant breeding practices to improve the resistance of commercial lines. In species where the natural variation is small or absent, other approaches are necessary. Gene manipulation enables the expression of certain genes to be changed or for genes from one species to be expressed in different species. Several strategies have already been employed to increase the Al resistance of plants. Some success has been reported by artificially generating mutations in plants. For example, in *Arabidopsis*, barley, and wheat, irradiating seeds with gamma rays or treating them with chemical mutagens (*N*-methyl-*N*-nitroso urea or ethyl methane sulfonate) induced mutations that conferred greater resistance to Al toxicity. Examination of *Arabidopsis* mutants indicated that tolerance was probably caused by an increased organic anion exudation in one mutant and the maintenance of a higher rhizosphere pH in the other.

In a different approach, genes induced by Al treatment in several plant species were isolated and then transformed into *Arabidopsis*. Three Al-induced genes from tobacco, (glutathione-*S*-transferase, peroxidase, GDP-dissociation inhibitor) and one from *Arabidopsis* (blue-copper binding protein) conferred a small degree of Al^{3+} resistance to the transgenic plants. These studies demonstrated that some of the genes induced by Al^{3+} will help to ameliorate the damage caused by Al, such as oxidative stress. A third approach to increase the Al resistance of plants has relied on the important role of organic acids in the natural mechanisms of Al resistance. Genes involved in the synthesis of citrate and malate were overexpressed in plants in the hope that increasing their synthesis might also increase their release from the roots. This strategy has had mixed success. A carrot gene that encodes for the enzyme citrate synthase conferred a small but significant increase in Al resistance when expressed in *Arabidopsis*. In alfalfa, Al resistance was increased by transforming plants with an endogenous malate dehydrogenase gene isolated from root nodules. The resistant alfalfa plants contained higher internal concentrations of organic acids and showed enhanced efflux of several organic anions. By contrast, overexpression of an endogenous citrate synthase gene in tobacco plants caused no change in citrate release or Al resistance. In addition, conflicting reports have been published on the ability of a bacterial citrate synthase gene from *Pseudomonas aeruginosa* to increase citrate release and enhance Al resistance in transgenic tobacco plants. Attempts to alter the expression of other enzymes involved in organic acid metabolism are continuing and these studies offer opportunities for developing Al-resistant germplasm in the future.

List of Technical Nomenclature

Actin	A protein used to make filaments that compose part of the plant's skeleton.
$Al(OH)_3$	A nonplant toxic precipitated form of Al.
Anion channels	Membrane proteins that facilitate the energetically passive flow of anions through membranes.
Annexins	Calcium-binding proteins that interact with membrane phospholipids.
Apoplasm	The space between cells in the root.
ATP	Adenosine 5′ triphosphate; a plant energy source.
Calmodulin	A Ca^{2+}-binding protein that regulates cytoplasmic Ca^{2+} concentrations.
Complex	The chemical association of positively charged Al with a negatively charged compound.
Diacylglycerol	A secondary messenger in plants used for cell signaling.
Exclusion	The prevention of Al entering the root.
Forest decline	The damage of tree canopies by atmospheric acid deposition.
GTP	Guanosine 5′-triphosphate; required in protein synthesis, the assembly of microtubules, and for the activation of G-proteins.
Inositol(1,4,5)-trisphosphate	A secondary messenger in plants used for cell signaling.
Microtubule	Used to make up part of the plant's cytoskeleton.
Organic anions	Negatively charged compounds such as citrate, malate, and oxalate.
Pectate	A polysaccharide used to bind cellulose together in the cell wall.
Profilin	An actin-binding protein that forms a complex with G-actin.
Rhizotoxicity	The inhibition of root growth by a toxic metal.

See also: **Cell Walls and Fibers**: Cell Walls. **Ethics and Biosafety**: Development and Commercialization of Genetically Modified Plants. **Growth and Development**: Cell Growth. **Nutrition**: Aluminum Toxicity; Ion Transport; Mineral Uptake. **Plants and the Environment**: Land Reclamation and Remediation, Principles and Practice; Plants as Pollution Monitors. **Root Development**: Root Growth and Development; The Rhizosphere and its Microorganisms.

Further Reading

Andersson M (1988) Toxicity and tolerance of aluminum in vascular plants. A literature review. *Water, Air and Soil Pollution* 39: 439–462.

Barceló J and Poschenrieder C (2002) Fast root growth responses, root exudates, and internal detoxification as clues to the mechanisms of aluminium toxicity and resistance: a review. *Environmental and Experimental Botany* 48: 75–92.

Cronan CS and Grigal DF (1995) Use of calcium aluminum ratios as indicators of stress in forest ecosystems. *Journal of Environmental Quality* 24: 209–226.

Delhaize E and Ryan PR (1994) Aluminum toxicity and tolerance in plants. *Plant Physiology* 107: 315–321.

Godbold DL, Jentschke G, Winter S, and Marschner P (1998) Ectomycorrhizas and amelioration of metal stress in forest trees. *Chemosphere* 36: 757–762.

Horst WJ (1995) The role of the apoplasm in aluminum toxicity and resistance of higher plants: a review. *Zeitschrift für Planzenernährung und Bodenkunde* 158: 419–428.

Jones DL (1998) Organic acids in the rhizosphere – a critical review. *Plant and Soil* 205: 25–44.

Kinraide TB (1994) Use of Gouy–Chapman–Stern model for membrane-surface electrical potential to interpret some features of mineral toxicity. *Plant Physiology* 106: 1536–1592.

Kochian LV (1995) Cellular mechanisms of aluminum toxicity and resistance in plants. *Annual Review of Plant Physiology and Plant Molecular Biology* 46: 237–260.

Li XF, Ma JF, and Matsumoto H (2000) Patterns of aluminum-induced secretion of organic acids differ between rye and wheat. *Plant Physiology* 123: 1537–1543.

Ma JF, Ryan PR, and Delhaize E (2001) Aluminum tolerance in plants and the complexing role of organic acids. *Trends in Plant Science* 6: 273–278.

Marschner H (1995) *Mineral Nutrition of Higher Plants.* London: Academic Press.

Matsumoto H (2000) Cell biology of aluminum toxicity and tolerance in higher plants. *International Review of Cytology* 200: 1–46.

Rout GR, Samantaray S, and Das P (2001) Aluminum toxicity in plants: a review. *Agronomie* 21: 3–21.

Ryan PR, Delhaize E, and Jones DL (2001) Function and mechanisms of organic anion exudation from plant roots. *Annual Review of Plant Physiology and Plant Molecular Biology* 52: 527–560.

Taylor GJ (1995) Overcoming the barriers to understanding the cellular basis of aluminium resistance. *Plant and Soil* 171: 89–103.

P

PESTS

Contents

Nematodes and Mollusks

P Newell, formerly of The University of the South Pacific, Suva, Fiji Islands

Nematodes

Plant-feeding nematodes are small and inconspicuous animals. They are pseudocoelomate invertebrates with a fairly simple body plan. They are the second most important group of crop pests after insects and their damage can be either by direct attack on leaves, stems, flowers, or roots or indirect attack through transmission of viruses during feeding.

Nematodes are part of the normal soil fauna where they feed on fungi, bacteria, protozoa, and decaying organic matter. Some, however, are herbivores, and feed on living plants, either living externally or internally (**Table 1**). Because of their small size taxonomy has been a problem and the nematodes important to agriculture and horticulture have often been classified according to the damage that they cause and observed on susceptible crops; an "ecological classification."

Nematodes are important crop pests because agricultural methods favor the transmission of infected material from one part of the world, country, or farm to the other. In addition, if infected crops are grown on the same ground, year after year, soil populations reach levels such that complete crop failures can occur.

Vegetable production in the tropics would seem to be a clear example of a cropping system that is, or will be, catastrophically attacked by nematodes. In such systems, the same crops are grown in the same ground year after year from vegetative material, which has had no phytosanitary controls. On islands in the Pacific, with very restricted land areas, there are good reasons why fallow rotations are not practiced.

Nematodes, unlike most pulmonate mollusks, have separate sexes, mate, and produce eggs that hatch into juveniles, which after a series of molts become adult. Damage to crops occurs when the nematodes either feed on, or penetrate, the host plant where they feed and reproduce. They can feed on almost any part of the plant; roots are commonly attacked leading to a stunting of the root system (*Xiphinema* spp., *Trichodorus* spp., and *Tylenchorhynchus* spp.). The first symptoms noticed by the farmer may be a patchy crop appearance with stunted, poorly growing plants in the center of the patches; for example, Docking disorder of sugar beet (*Beta vulgaris*) in the United Kingdom caused by the feeding of needle and stubby root nematodes.

The life cycles of nematodes are quite varied, with different groups adopting different strategies for finding suitable host plants. Some include the formation of resistant egg cases (cysts) from the bodies of the females. These cysts protect the eggs until they hatch after they receive a suitable biochemical stimulus from compounds released by host plants. Research has been directed towards identifying these cues, so those manmade analogs might be tested on contaminated soils. In theory, if sufficient of the cyst nematodes could be stimulated to hatch without a suitable host plant being present, then populations would become depleted, effecting a control. So far this idea has eluded commercialization.

Other species, such as the stem and bulb nematode *Ditylenchus dipsaci*, are able to survive for long periods in a dormant phase (cryptobiosis). In this state they become desiccated and can adhere to seeds or bulbs and so are carried by farmers to new areas. Once conditions become favorable to plant growth, normally water and moderate temperatures, the nematodes become active and reinfect susceptible plants.

In all specialist works on plant-feeding nematodes there is discussion about physiological races or "feeding types." It is clear that certain races feed on

Table 1 Herbivorous forms of nematodes that are important agricultural pests

Migratory		*Nonmigratory*	
A. Forms living outside their food plants (ectoparasites)			
Tylenchorhynchus spp.	Stunt nematodes	*Hemicriconemoides* spp.	Sheath nematodes
Longidorus spp.	Needle nematodes	*Hemicycliophora* spp.	Sheath nematodes
Xiphinema spp.	Dagger nematodes	*Criconemoides* spp.	Ring nematodes
Trichodorus spp.	Stubby root nematodes	*Pratylenchus* spp.	Pin nematodes
B. Forms living partially outside their food plants (semiendoparasites)			
Rotylenchus spp.	Spiral nematodes	*Rotylenchulus reniformis*	Reniform nematode
Helicotylenchus spp.	Spiral nematodes	*Tylenchulus semipenetrans*	Citrus nematode (wide range of crops attacked in warm countries)
Scutellonema spp.	Spiral nematodes		
Hoplolaimus spp.	Lance nematodes		
Tylenchorhynchus spp.	Stunt nematodes		
C. Forms living within their food plants (endoparasites)			
Ditylenchus dipsaci	Stem and bulb nematode	*Heterodera* spp. and *Globodera* spp.	Cyst nematodes; potatoes (*Solanum tuberosum*) and sugar beet (*Beta vulgaris*) in Europe but some species attack grasses (Poaceae), tobacco (*Nicotiana tabacum*), rice (*Oryza sativa*), and sugar cane (*Saccharum officinarum*)
Ditylenchus destructor	Potato tuber rot nematode		
Ditylenchus angustus	"Ufra disease" of rice in East Pakistan		
Pratylenchus spp.	Lesion nematode		
Radopholus similis	Burrowing nematode		
Aphelenchoides spp.	Bud and leaf nematodes		
A. besseyi	Seed-borne in rice		
		Meloidogyne spp.	Root-knot nematodes
		Nacobbus spp.	False root-knot nematodes
		Anguina spp.	Flower and leaf gall nematodes
		A. tritici	Important in southern Europe and Asia in wheat crops

certain crops, or even varieties of crops. Certain varieties are resistant to attack by other races of the same species. However, whilst certain nematodes are very specific in their suitable food plants others show much more catholic tastes and so feed on a wide range of plants. These species with wide host ranges make control by crop rotation, or fallowing, a very difficult goal to achieve in practice. Resistance in crop plants to nematode attack is well known and so the transfer of the genes for resistance from one variety to another by suitable genetic manipulation is widely used by plant breeders. The genes for resistance to potato cyst eelworm (*Globodera rostochiensis*) were transferred from the wild potato (*Solanum tuberosum* subsp. *andigena*) into modern cultivars.

Phytosanitary regulations with regard to controlling the spread of nematodes are widespread and quarantine restrictions on the import of many different types of plants are found in many different countries. Seeds of crops such as alfalfa (*Medicago sativa*, lucerne), onions (*Allium cepa*), and clover can be treated with a methyl bromide fumigation to control *Ditylenchus dipsaci*. Vegetatively propagated crops, such as strawberries (*Fragaria x ananassa*), onions, and *Narcissus* bulbs can be hot-water treated to eliminate nematodes. Very precise control of temperatures and timings is necessary if the viability of the plants is to be maintained.

Table 2 Nematodes that are important on tropical and subtropical tree crops

Species	*Crop*	*Area*
Pratylenchus coffeae	Coffee (*Coffea* spp.), *Citrus* spp.	Indonesia, Caribbean, USA, India
Radopholus similis	*Citrus* spp. (spreading decline of citrus in Florida)	USA
	Banana (*Musa sapientum*) (black-head toppling disease)	Caribbean and other areas
Meloidogyne sp.	Coffee, tea (*Thea sinensis*) seedlings	Found in many growing areas

Nematodes Attacking Tropical and Subtropical Tree Crops

Crops that are harvested from woody perennial plants represent important areas of tropical and subtropical agricultural production (**Table 2**). Farmers often have to wait for several years before a marketable crop is produced and so the establishment of such orchards and plantations is a considerable investment.

Table 3 Viruses spread by certain migratory herbivores to host plants

Species	*Virus*	*Host plant*
Xiphinema diversicaudatum	Arabis mosaic	Cherry (*Prunus* spp.), grapes (*Vitis vinifera*), raspberry (*Rubus* spp.), strawberry (*Fragaria x ananassa*), and lettuce (*Lactuca sativa*)
	Brome grass mosaic	Cereals, grasses (Poaceae)
	Carnation ringspot	Carnations (*Dianthus caryophyllus*)
	Strawberry latent ring spot	Blackcurrant (*Ribes nigrum*), cherry (*Prunus* spp.), celery (*Apium graveolens dulce*), peach (*Prunus persica*), plum (*Prunus domestica*), raspberry (*Rubus* spp.), rose (*Rosa* spp.), strawberry (*Fragaria ananassa*)
	Cherry leaf roll	Cherry (*Prunus* spp.)
Trichodorus spp.	Pea early browning	Pea (*Pisum sativum*), alfalfa (*Medicago sativa*)
	Tobacco rattle	Potato (*Solanum tuberosum*), tobacco (*Nicotiana tabacum*), bulbs, sugar beet (*Beta vulgaris*), etc.

Specific areas developed for intensive vegetable production are often related to soil type or proximity to markets. Intensive production, such as the use of glasshouses or polythene tunnels can lead to the development of soil pests such as nematodes, and these can be very important, as control measures are not very efficient.

Virus Transmission

The fact that nematodes can transmit viruses adds a further dimension to their economic significance. Crops damaged include orchard crops, such as cherries (*Prunus* spp.), and long-lived plantation crops such as hops (*Humulus lupulus*) and grapes (*Vitis vinifera*). As one might expect the genera that have been implicated in virus transmission include the migratory herbivores (**Table 3**), in a situation analogous to the supreme position of aphids as insect vectors of virus diseases of crop plants.

Infected nematodes remain potent transmitters of viruses for long periods of time but the viruses are lost during molting. It is thought that this is because the virus particles are attached to cuticular structures, which are lost during the molting process. Many weeds are important reservoirs for viruses and so once a field is infected they can prove difficult to eradicate.

Control Measures

Legislation Many countries have adopted rigorous legislation to prevent the importation or spread of plant parasitic nematodes. These can be in the form of quarantine controls at the external borders of a country for both imports and exports, domestic quarantine controls and certification schemes aimed at assuring growers that, for example, the seed potatoes, or strawberry plants, that are being offered for sale are not nematode infested.

Chemical control Most plant parasitic nematodes spend part of their life cycle in the soil. Soil sterilization by heating (steam) or chemicals can then be an effective way of eliminating, or controlling, nematode populations. Chemical treatment of the soil is often very expensive but it can be considered as a possible solution to a pest buildup, particularly in confined areas or where high-value crops are being produced. Many of these materials are fumigants, which have to be applied when the soil conditions are exactly right, dry enough to allow penetration of the vapor yet wet enough to seal the surface after injection. Sometimes the soil surface is covered by polythene sheeting after applications of methyl bromide, to increase the length of time that toxic materials are retained within the soil.

Nonvolatile nematicides are applied to the soil surface and then mixed into the soil by rotary cultivation to crop root depth. More localized treatment can be achieved by application into the seedbed along the seed furrows. More precise targeting of these potent materials has been one of the main developments in nematode control. The use of drip feed irrigation systems in plantation crops has allowed nematicides to be introduced into the irrigation lines. Contamination of groundwater with nematicides has been observed in several areas and more precise targeting may be an answer to this contamination problem.

Nonlethal control techniques: resistant crops Breeding crops for resistance to nematode attack is probably the most elegant solution to this problem. The advantages of this approach include: (1) there are no toxic residues in the soil and there can be no contamination of the groundwater; (2) they do not require special application techniques or equipment and pose no safety problems to the farmers or their workers; (3) they can be absolutely effective in preventing nematode reproduction, unlike chemical controls; (4) rotations can be shortened and high-value crops grown making the best use of the land; and (5) they are cost free.

At the present time most crops do not have nematode resistant cultivars. Most crops have a variety of cultivars, to suit different markets and soils. To produce nematode resistance in this varietal range would be daunting. A further problem is that many nematodes exist as pathotypes, or races, and so resistance to one pathotype does not necessarily mean that this cultivar is resistant to all pathotypes.

Nonlethal control techniques: cultural practices
Crop rotations By planting nematode-susceptible crops only every 4 or 5 years, with nonsusceptible crops between, it is possible to reduce the populations of soil-borne nematodes. Such systems work well for potato cyst nematodes. However, some nematodes have very wide host ranges and so it is difficult to find a suitable rotation. Weeds can also act as suitable host plants, and so augment a population that should be being reduced by the rotation.

Soil conditions Many nematodes attack, and so damage, crop plant rooting systems. By providing crop plants with very good soil conditions root growth can be stimulated which may counteract some of the crop damage. Nitrogenous fertilizers applied to tea crops in Sri Lanka attacked by *Pratylenchus loosi* increased yields by improving root growth. Nematode populations were also increased. Improving soil conditions by the application of materials that improve soil structure and the nutrient supply can also influence nematode populations. Some of these materials improve the environment for nematode pathogens and so a measure of biological controls seems to be effected.

Heat treatment Heat treatment of bulbs and material for the vegetative propagation of certain plants, such as strawberries, are heat treated to kill nematodes as a part of normal commercial routine. The nematodes are killed by water temperatures that are sufficiently high to kill the parasites but just not high enough to kill the plants. Very precise thermal control has to be achieved if plant viability is to be maintained.

Steam sterilization of soils is an effective way to kill soil nematodes but is uneconomic except in small areas, such as glasshouses. Use of polythene sheets on field soils in summer months has been used in the northern parts of the United States to successfully control *G. rostochiensis*.

Similarly, some species of trichodorid nematode (e.g., *Paratrichodorus teres*) can be killed by double rotavation of coarse soils, presumably by a mixture of mechanical action and drying of the soil during the cultivation process. This technique is not widely used as it causes a loss of soil structure.

Biological control Many agents capable of killing nematodes within the soil have been identified. At present no biological control agents are commercially exploited. Natural biological control of *Heterodera avenae* occurs in fields that are continuously sown with cereals.

Mollusks

Only one Class of the Phylum Molluska is important as crop pests. These are members of the Class Gastropoda, which include the familiar slugs and snails. Many terrestrial and freshwater snails are pulmonates, as are all slugs, having lost their gills during their evolution and adaptation to life on land. However, in some tropical and subtropical areas, such as the Caribbean, most of the land snails are prosobranchs, and so share with their marine cousins the possession of a gill.

The basic anatomy of slugs and snails can be found in any textbook of invertebrate biology but the following features are worthy of particular attention. All snails have shells, some of which are heavy whereas others are thin, glasslike, and much lighter. They are made from calcium carbonate crystals laid within an organic matrix. Because they need access to calcium salts for shell-building they are normally restricted to base-rich soils with plentiful supplies of calcium. Where plants are particularly good at extracting calcium from acid, base-depleted, soils (such as tea (*Thea sinensis*) plants) which are common in the tropics, then snail populations can flourish by obtaining the necessary calcium from their grazing activities on the plant leaves. Slugs either have a greatly reduced shell, or no shell at all, which has allowed them to colonize a wider range of land sites, being freed from the constraint of shell-building.

The shell is an important front-line defense against predators, desiccation, and solar radiation. In the tropics many snail species have white shells which reflect the heat and so reduce thermal loading. Most have complex behavioral mechanisms, which restrict their activities to moist and shady areas, such as the soil surface or subcrop environments. Snails are able to aestivate and so survive in semiarid deserts, their water losses being minimized by fundamental changes to the mantle collar epithelium, which plugs the mouth of the shell when the animal retreats into its shell. These losses are reduced to the same level as insect cuticle, enabling the animals to survive for long periods, maybe years, without access to water. However, it must be emphasized that this is only possible if the animals remain inactive; the unique trick of insects to reduce water losses yet remain active is not paralleled in snails.

Slugs are unable to regulate water loss in this way and so the risk of desiccation is great for these animals. They also have complex behavioral mechanisms, including homing, which ensure that after feeding or mating on the soil surface they are able to find their way back to a suitably moist site. They are mostly nocturnal or crepuscular, and are rarely seen active during the day. Most species become active when there is a change of light intensity, which normally signals dusk or sundown. Sometimes the light intensity changes associated with thunderstorms and rain can induce surface activity.

Both slugs and snails feed by using a feeding apparatus unique to mollusks, the radula. This is a rasplike series of teeth held on a flexible ribbon of connective tissue which is found in the buccal cavity at the front end of the gut of many mollusks. This is drawn across the surface of food substances and rasps off fragments which are then ingested. Most are herbivores feeding on plant material, but carnivorous slugs and snails are known, and most have modified radulas, which equip them for their different diet. Most snails feed on aboveground plant structures, such as leaves, stems, and shoots. Some, such as *Helix aspersa*, will climb up into orange trees to feed on the newly pollinated oranges when only 2–3 mm in diameter. This specific damage to the future crop causes large losses, and where the fruitlets are not damaged enough to fall from the tree the mature oranges bear ugly brown skin blemishes. It seems that the snails are specifically attracted to the flowers by aromatic molecules produced by the scent glands. Again, homing in on food plants, using olfactory cues, and maybe crawling tens of meters to find suitable food are all features of molluskan feeding.

Rasping by the radula scrapes off small quantities of tissue and produces wounds to plant tissue, which can sometimes be confused with insect damage. However, slugs and snails often feed from the centers of the leaves, producing a skeletonization where the ribs and harder vascular tissue resists their feeding activities. In all cases mucus is left behind on the plant surface after they have crawled on it and it is this material that provides corroborative evidence for the source of the damage.

Attack by slugs can be on the underground parts of plants, such as potato (*Solanum tuberosum*) tubers. They can burrow through the soil and large populations can so damage a crop as to make it unsaleable. It is clear that certain varieties of potato are more susceptible to slug damage, varietal susceptibility seemingly following our own perceptions of excellence in flavor. At one stage it was thought that slugs could not make holes in tubers, and that they merely enlarged earlier holes made by insect larvae, such as wireworms (*Agriotes* spp.). This has clearly been shown to be incorrect, and slugs and snails are well equipped to deal with most living plant tissues. Many potato crops are grown to contract for a specific purpose, certain specialized purposes requiring certain specified varieties of potato. Crops for potato chip (crisp) manufacture often have maximum acceptable levels of slug damage, and if this figure is exceeded then the crop is unfit for chip manufacture, and so is rejected.

Slug and snail damage is not normally at such a level that a large proportion of the crop is consumed. However, their significance as agricultural and horticultural pests should not be underestimated. Often they feed on very specific parts of plants, which whilst not forming a large part of the overall plant biomass nevertheless produces a large economic loss. Some slugs, often the grey field slug (*Deroceras reticulatum*), feed in the autumn on the newly sown seeds of cereals. The animals eat the embryos out of the seeds making them unable to germinate, and so nonviable. One slug can consume several seed embryos per day and so autumn-sown winter wheat (*Triticum aestivum*) crops are found in the spring to have large patches of low plant densities, reducing the yields significantly.

As with so many insect pests, introduced species of slugs and snails can cause major pest problems in countries where they are novel species. Clearly *Helix aspersa* snail damage in citrus orchards in Florida, California, and South Africa fall into this category. However, a potentially very serious problem has emerged in the Philippines where *Pomacea* spp. snails have been introduced from South America via Japan and have become important pests of rice (*Oryza sativa*) crops.

Damage to flower crops will produce unsightly leaf or flower damage, reducing the value of the crops being grown. Salad crops, grown specifically for lush, undamaged leaves, are frequently attacked; again modern marketing methods tend to emphasize the production of crops with very low levels of foliar damage, crops with damaged leaves being rejected by wholesale buyers or supermarket quality-control officers. In parts of Africa and Southeast Asia giant land snails (*Achatina fulica* and *Archachatina marginata*) are important pests of vegetable crops. Hand collecting is often used for control. However, as some of the areas infested by these snails are also areas where local human populations suffer from dietary protein deficiencies, snail populations should not be regarded in a wholly unfavorable light. Many people eat snails and so they could be seen as converting plant tissue into a much needed protein-rich food resource.

Where crops are grown intensively high levels of organic fertilizers are often added to the soil, as well as regular irrigation. Field-scale vegetable production using organic fertilizers is often considered to produce the best yields, but all such systems run the risk of enhancing slug and snail populations, which are favored by moist, well-aerated, organic soils, which form ideal environments for egg hatching. Paradoxically, it is often farmers and growers who are farming in a responsible and sustainable manner, trying to maintain soil fertility and structure, that are most affected by these pests.

Control

Integrated pest management Slugs and snails are often cryptic animals and unlike insects, such as aphids, cannot be estimated by direct field measurements. They are nocturnal animals, but not all are active each night. However, surface activity includes both feeding and mating and so probably the majority of the sexually mature members of a population of the grey field slug will be active over a warm and damp period of 3–4 days. Populations reach pest proportions on wet heavy soils and often build up in some crops, such that the succeeding crop will have damage. This is also true for seasons; wet autumns or springs will allow populations to increase such that damage will be heavy later on in the year. Contact sprays cannot attack slugs and snails, as the active ingredient is not absorbed through the skin, which is protected by a covering of mucus. Irritant chemical sprays instantly provoke a production of mucus, which is then sloughed off, leaving the animal unharmed.

Slug and snail numbers can be estimated by a variety of methods. Soil sampling using large soil samples, which are then either washed through a sieve, or flooded to expel the animals, is probably the only way to obtain an absolute figure. However, it is very time consuming, generates huge amounts of wet soil as a waste, and is usually not considered necessary, as relative, and not absolute, figures are all that is normally required to make a decision about pest control.

Relative numbers can be estimated by using refuge traps, made from tiles or wet sacking. These act as surface slug shelters and the numbers aggregating under them give an estimate of the total numbers present. They have the advantage that they can be relatively easily monitored, but have the disadvantage that the numbers recorded under the trap are related to the weather conditions; and they are also inefficient in recording immature animals.

Estimating risk Crops that are planted after oilseed rape (*Brassica napus*) or lush clover (*Trifolium* spp.) leys, such as autumn-sown cereals, are often seen as high-risk crops. This is because slug populations are able to increase in numbers under the canopy of the preceding crop, survive in the soil after the harvest and cultivation process, and be waiting in the soil for the newly sown cereal crop. Likewise, incorporation of crop residues into the soil to improve structure and moisture-holding capacity is good farming practice but may exacerbate a slug pest problem. Often mild winters and wet summers provide ideal conditions for slugs, and so crops grown in these conditions are likely to be attacked.

Potato growers, for example, could have a management strategy that included an initial estimation of the slug densities; a decision would then be made as to whether the intended potato variety was vulnerable. If so, then perhaps a less susceptible variety might be planted. This is often a good management option, for often slug damage is restricted to a part of a field, and this could be planted with a different variety to the remainder. As slug damage often occurs late in the season, planting an early-maturing variety and lifting it before it is damaged would be another strategy to reduce damage. If a crop were at high risk then application of chemical controls would have to be considered.

The decision as to whether to apply control measures, or not, can be a difficult one for growers. Damage is unpredictable, often being related to the seasonal climatic conditions. For example, a very wet autumn in Europe may well delay harvesting of potato crops until they have accumulated maximal levels of starch and alkaloids, thus making them very attractive as food, and so susceptible to heavy slug damage. The control decision depends on balancing various "upside" and "downside" factors. The upside ones would include the value of the crop at harvest, the increase in yield, and the increase in market value. The downside would include the cost of the control chemical, cost of application, and the damage to the crop caused by the application. Some methods of application, such as use of helicopters, reduce the in-crop damage to insignificant levels.

Cultural control methods If one is trying to reduce damage to autumn-sown cereals one must bear in mind that the slugs have to be able to move through the soil in order to find and attack the newly sown grain. Production of a fine seedbed, which is compacted after sowing, reduces slug damage. However the extra costs involved in producing a fine seedbed can be prohibitive. Sowing the seeds deeper into the soil can also reduce damage. For potato crops early harvesting and growing resistant cultivars, particularly in wet areas of fields, are all excellent strategies. The potato cultivar Maris Piper is four times more susceptible to slug damage than Majestic. One of the most slug resistant is Storm Enterprise, which suffers only 23% of the damage recorded for Majestic. However, slug-susceptible cultivars such as Maris Piper are often grown for high premiums on contract for specialist processors such as potato chip manufacturers, and so such high-value crops are often worth protecting.

Chemical control methods All current methods of chemical control are based on pellets, which have to be eaten by the slugs. Modern formulations are all based on a mixture of active ingredient, normally about 5%, in a cereal bran carrier. These are treated with fungicides, waterproofing materials, and mammalian feeding repellents and blue coloring to make them less attractive to birds. The pellets are extruded in fine rods (2 mm) which are then cut into 2 mm lengths. The advantage of these "minipellets" is that when they are broadcast on the surface of the soil there are many more baiting points and so the chance for surface-active animals to encounter one is greatly increased.

There are just two current active ingredients used in slug pellets, methiocarb and metaldehyde. Both are poisonous to mammals but the mammalian toxicity of methiocarb is about 10 times that of metaldehyde. Both are effective as slug killers but recovery from sublethal ingestion of baits can occur when wet cloudy weather occurs after the application of baits. Slugs that ingest small amounts of poison are often paralyzed, and if the weather is dry and sunny then they are exposed on the soil surface and killed by the dry conditions. Conversely, wet conditions will minimize the risks of desiccation, and so aid recovery.

Carbamates such as methiocarb are broad-spectrum poisons and kill beetles, earthworms, and other beneficial components of the soil fauna. Higher animals such as birds and mammals have been poisoned by both types of pellets but modern formulations with feeding repellents and bright blue colors have minimized these hazards. In general, metaldehyde is regarded as a more selective molluskicide than methiocarb. Pellets remain active on the soil surface for about 2–3 weeks for metaldehyde, which then depolymerizes into carbon dioxide and water, leaving no potentially harmful residues in the soil. Methiocarb is much more persistent, pellets remaining active for perhaps months in certain field conditions.

Most pellets are applied to agricultural crops, which include potatoes, autumn-sown cereals, salad, and brassica crops.

Seed dressings of metaldehyde for autumn-sown cereals have been shown to be an effective way of protecting crops. However the search for natural compounds has shown that cartap, a natural toxin from marine polychaete worms, is also effective as a seed dressing, but its mammalian toxicity is too high for commercial use.

Biological control It is clear that slugs have a number of natural predators, and these would include the carabid beetles *Nevria brevicollis* and *Harpalus rufipes*. Thus, agricultural systems that encourage populations of beneficial predatory species are likely to suffer fewer slug attacks than those that do not. Treatment of fields with methiocarb may reduce populations of beneficial carabid beetles.

Slugs are susceptible to attack by parasitic nematodes ("nemaslug," *Phasmarhabditis hermaphrodita*) which invade the slug's body. The nematodes introduce a bacterium during invasion, which is the organism that causes mortality. Similar methods have been used against insects and the "nemaslug" resting larvae are produced in a similar way. Control can be effective but the costs are too high for field-scale agricultural use. These nematodes are not persistent in the soil, do not migrate, and only attack their target species. This approach may be useful for certain high-value horticultural crops grown in small areas.

List of Technical Nomenclature

Crepuscular	Appearing at or active at twilight.
Cryptobiosis	Ability to survive for long periods in a dormant state.
Ectoparasitic nematodes	Herbivorous nematodes that live outside their food plants.
Endoparasitic nematodes	Herbivorous nematodes that live within their food plants.
Nematicide	A substance or preparation used for killing nematodes parasitic to plants.
Radula	Rasplike series of teeth forming a feeding apparatus unique to mollusks.

See also: **Crop Improvement**: Plant Breeding, Practice. **Integrated Pest Management**: Practice.

Further Reading

Brusca RC and Brusca GJ (1990) *Invertebrates*. Sunderland: Sinauer Associates.

Evans K, Trudgill DL, Webster JM (eds) (1993) *Plant Parasitic Nematodes in Temperate Agriculture*. Wallingford: CAB International.

Godan D (1983) *Pest Slugs and Snails: Biology and Control*, transl. Sheila Gruber. Berlin: Springer-Verlag.

Henderson I (ed.) (1989) *Slugs and Snails in World Agriculture*. Thornton Heath: British Crop Protection Council.

Kerry BR (1987) Biological control. In: Brown RH and Kerry BR (eds) *Principles and Practice of Nematode Control in Crops*, pp. 233–263. Sydney: Academic Press.
Mead AR (1979) Economic malacology with particular reference to *Achatina fulica*. In: Fretter V and Peake J (eds) *Pulmonates*, vol. 2B, pp. 1–150. London: Academic Press.
Nickle WR (ed.) (1991) *Manual of Agricultural Nematology*. New York: Marcel Dekker.

Genetic Modification of Pest Resistance, Insect Pests

D J Murphy, University of Glamorgan, Cardiff, UK

The enhanced protection of crops against damage by insect pests is currently the second most important commercial application of GM technology in agriculture (herbicide tolerance is the most widespread GM trait). In 2002, there was a total of 14.5 Mha (million hectares) of GM crops with insect resistant traits, virtually all of which were expressing one of the insecticidal *Bacillus thuringiensis* (Bt) proteins. This total was made up of 10.1 Mha of crops expressing insect resistant traits alone plus 4.4 Mha of crops expressing insect resistant traits in combination with herbicide tolerance traits. The most common GM insect resistant crops were cotton (*Gossypium hirsutum*) and corn (*Zea mays*; maize), although numerous other crops are in the pipeline for future commercial release. Although other insecticidal agents, such as lectins and protease inhibitors from a variety of sources, are being developed for use in GM crops, the Bt toxins are by far the most important and this article will focus on these.

The reason for working on insect resistance as a desirable agronomic trait for modification is straightforward. Crop yields are regularly reduced by insect attack and even in intensively managed agronomic systems, losses of >30% from various pests and diseases can occur during bad seasons. Of course, in developing countries, crop losses from insect damage can be much higher, and control agents like pesticides are relatively expensive and potentially harmful to the operators. Insect resistance in transgenic corn, cotton, and potato (*Solanum tuberosum*) crops has now been conferred by insertion of a gene encoding a protein toxin from the Gram positive soil bacterium *B. thuringiensis* (Bt). The use of insect control sprays containing a live toxin producing *B. thuringiensis* suspension has been commonly used for over 30 years in organic farming, but the widespread use of Bt toxins in transgenic crops is much more recent. The Bt toxins are a family of so-called crystalline (cry) proteins that are converted into their active form during digestion in the gut of a range of insect larvae, resulting in a disruption of potassium ion transport that rapidly becomes lethal. Mammals do not convert the toxins into their active forms and are therefore unaffected by them. Early indications suggest that transgenic Bt crops are effective in controlling insects and improving yields (by 7%), while also reducing the need for spraying with more toxic and less desirable pesticides that often affect beneficial organisms such as insectivorous birds (saving growers some US$40 ha^{-1}).

The obvious danger in relying on a single class of control agents, like the Bt toxins, is that this strategy tends to establish a strong selection pressure favoring the survival of insects that are able to sequester the toxin or otherwise render it harmless. Before the mid-1990s, only a few thousand hectares of land on organic farms were sprayed annually with live Bt, but in the last few years, the cultivation of transgenic Bt crops has expanded to over 14 Mha, and this area is still increasing. The resultant chronic and widespread exposure of hundreds of insect species to the Bt toxin must increase the likelihood that resistance will eventually develop in some species. Indeed, as long ago as 1995, at least two insect species had already become resistant to Bt toxins in the field, with at least another 10 species showing the potential for the acquisition of resistance in laboratory studies.

Another risk with the expression of xenoproteins like Bt toxin in crops is that their accumulation may be curtailed if the plants are stressed. For example, in 1996 a combination of heat and drought caused a reduction in levels of Bt toxin in Monsanto's transgenic "Nu Cotn" variety of cotton. Although deleterious to the crop, these climatic conditions actually favored the development of the cotton bollworm (*Heliothis zea*), resulting in severe infestation in nearly 1 Mha of "resistant" cotton crops. The infestation was eventually controlled with conventional chemical pesticides, but the more serious long-term danger is that exposure of the bollworms to sublethal doses of the Bt toxin would be an ideal mechanism for the development of resistance by the insects.

To combat the development of resistance to Bt toxins in the field, growers are now advised to set aside refugia. The refugia are areas adjacent to the main transgenic crop that are sown with nontransgenic seeds of the same crop. In these areas, nontolerant insect populations can continue to thrive and hopefully will out-compete those of their conspecifics that develop Bt tolerance. This strategy relies upon the cooperation and enforcement of good management practices by growers, and can fail if not implemented rigorously on all farms. Another strategy to prevent or delay the acquisition of resistance by insects is the inclusion of several unrelated toxin genes in a transgenic crop, but this so-called "gene pyramiding" will be a much more expensive and much longer term option.

A potential problem with any pest control agent is that it might inadvertently affect harmless or even benign organisms. On this score, the Bt toxins appear to perform relatively well, and certainly better than many synthetic pesticides. In 1999, Losey and coworkers reported possible adverse effects of Bt-containing corn pollen on monarch butterfly (*Danaus plexippus*) larvae. Although this finding generated a great deal of controversy at the time, subsequent detailed studies have shown that there is no significant impact of Bt corn pollen on monarch butterfly larvae.

List of Technical Nomenclature

Refugia Refuges where pest populations are tolerated in order to reduce the possibility of them developing resistance to control agents, such as pesticides or pest resistant GM crops.

See also: **Diseases**: Breeding for Disease Resistance. **Integrated Pest Management**: Disease Prediction Models; Practice; Principles. **Weeds**: Herbicide Resistance.

Further Reading

Carozzi N and Koziel M (eds) (1997) *Advances in Insect Control: the role of transgenic plants*. London: Taylor & Francis.

de Maagd RA, Bravo A, and Crickmore N (2001) How *Bacillus thuringiensis* has evolved specific toxins to colonize the insect world. *Trends in Genetics* 17: 193–199.

Flexner JL and Belnavis DL (1999) Microbial insecticides. In: Rechcigl JE and Rechcigl NA (eds) *Biological and Biotechnological Control of Insect Pests*, pp. 35–62. Boca Raton, FL: CRC Press.

Gatehouse AM, Ferry N, and Raemaekers RJ (2002) The case of the monarch butterfly: a verdict is returned. *Trends in Genetics* 18: 249–251.

Kaiser J (1996) Pests overwhelm Bt cotton crop. *Science* 272: 423.

Losey JE, Rayor LS, and Carter ME (1999) Transgenic pollen harms monarch larvae. *Nature* 399: 214.

McGaughey WH, Gould F, and Gelernter W (1998) Bt resistance management. *Nature Biotechnology* 16: 144–146.

Shelton AM and Sears MK (2003) The monarch butterfly controversy: scientific interpretations of a phenomenon. *Plant Journal* 27: 483–488.

PHOTOSYNTHESIS AND PARTITIONING

Contents

C_3 Plants

A S Raghavendra, University of Hyderabad, Hyderabad, India

Summary

The basic route of carbon assimilation in plants is the Calvin cycle, where the first product of carboxylation is 3-phosphoglycerate (PGA), a 3-carbon compound. There are two variants of carbon assimilation, where plants use the C_4-pathway or CAM, as a CO_2 concentrating mechanism, to raise the level of CO_2 at the vicinity of rubisco in chloroplasts. Plants possessing only the Calvin cycle are called C_3 plants, and these constitute almost 90% of the plant kingdom. The Calvin cycle consists of three phases: (1) formation of PGA from ribulose-1,5-bisphosphate (RuBP) and CO_2 (carboxylation phase); (2) reduction of PGA to triose-P (reduction phase); and (3) regeneration of the CO_2 acceptor, RuBP, from triose-P (regeneration phase). The Calvin cycle is regulated in several ways: autocatalysis, flux control, light-activation of key enzymes, and through regulatory metabolites. Intense efforts are being made to genetically manipulate enzymes/proteins related to the Calvin cycle, such as rubisco, rubisco activase, fructose-1,6-bisphosphatase and phosphate translocator.

Introduction

Carbon assimilation by plants and other photosynthetic organisms is a very important event in the global carbon cycle. Plants fix carbon primarily into 3-phosphoglycerate (PGA, a 3-carbon compound) and hence the process is named the C_3 photosynthesis or C_3 pathway, or Calvin cycle (named after Nobel laureate, Melvin Calvin). The other two variants of photosynthetic carbon assimilation are C_4 photosynthesis (or C_4 pathway) and crassulacean acid metabolism (CAM). However, the carbon from C_4 acids formed initially during these two pathways has to be refixed ultimately through the C_3 or Calvin or Benson–Calvin cycle. Thus, the C_3 photosynthesis is the basic route of carbon assimilation while the C_4 pathway and CAM function as carbon accumulating mechanisms and form adjuncts of the Calvin cycle. Plants possessing only the Benson–Calvin cycle are called C_3 plants, while the other two categories include C_4 plants and CAM plants.

Biochemistry/Reactions

The path of carbon assimilation was mapped quite elegantly by Professor Melvin Calvin and his group during the early 1950s, by the combined use of two-dimensional paper chromatography and radioisotopic carbon (^{14}C); therefore, this pathway is called the Calvin cycle. It is also called the reductive pentose phosphate (RPP) pathway, or photosynthetic carbon reduction cycle (PCRC).

C_3 photosynthesis has three principal phases: carboxylation, reduction, and regeneration. During the first phase of carboxylation, carbon dioxide is accepted by ribulose-1,5-bisphosphate (RuBP) to give two molecules of PGA. During the next phase of reduction, PGA is reduced to triose phosphate (triose-P), by using assimilatory force or assimilatory power (generated in light reactions) of ATP and NADPH. The last phase is the regeneration of the primary acceptor of CO_2, RuBP, from triose phosphate through a series of reactions. For each molecule of CO_2 fixed, three ATP and two NADPH molecules are required.

The first step of C_3 photosynthesis is catalyzed by the enzyme ribulose 1,5-bisphosphate carboxylase/oxygenase (rubisco), which prefers CO_2 to the dissolved form of carbon (bicarbonate) as the substrate (**Figure 1**). Since the concentration of CO_2

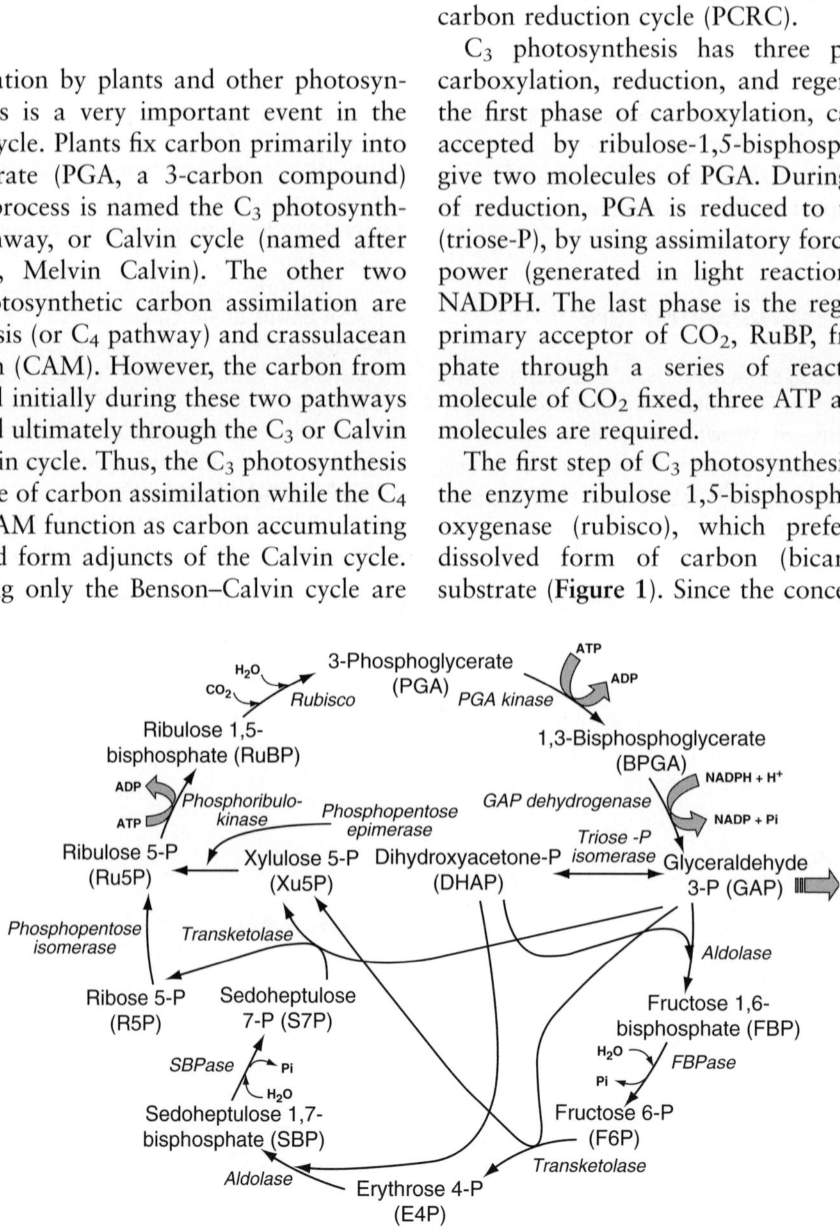

Figure 1 Schematic representation of the Calvin cycle or C_3 photosynthesis. The individual enzymes are as follows: rubisco (EC 4.1.1.39); PGA kinase (EC 2.7.2.3); GAP dehydrogenase (EC 1.2.1.13); triose-P isomerase (EC 5.3.1.1); aldolase (EC 4.1.2.13); FBPase (EC 3.1.3.11); SBPase (EC 3.1.3.37); transketolase (EC 2.2.1.1); phosphopentose epimerase (EC 5.1.3.1); phosphopentose isomerase (EC 5.3.1.6); and phosphoribulokinase (EC 2.7.1.19). The major product of the cycle is GAP, which is either exported out of the chloroplast to form sucrose or used within the chloroplast to form starch. A part of the GAP is used for the autocatalysis of the cycle.

within the plant cell is quite low (about 10 μmol l^{-1}), the activities of rubisco are rather low and are limited at the normal atmospheric CO_2. However, this phenomenon is compensated by the presence of large amounts of rubisco protein, which could be up to 50% of leaf protein. Apart from carboxylation, rubisco also catalyzes oxygenase activity, where O_2 reacts with RuBP to form phosphoglycolate, in addition to PGA. The affinity of rubisco for CO_2 and O_2, at typical *in vivo* concentrations, is quite similar, and as a result the carboxylase and oxygenase reactions are unavoidable in the present atmospheric levels of O_2 and CO_2.

The next phase of reduction of PGA to triose-P involves three steps: activation of PGA to 1,3-bisphosphoglycerate or BPGA (catalyzed by PGA kinase); (2) reduction of BPGA to glyceraldehyde 3-phosphate (GAP) with the help of NADPH (catalyzed by GAP dehydrogenase or GAPDH); and (3) interconversion of GAP and dihydroxyacetone phosphate (DHAP), by the enzyme triose-P isomerase (**Figure 1**). Both GAP and DHAP are termed triose-P.

During the third phase, RuBP (the CO_2 acceptor) is regenerated from triose-P through a series of reactions involving condensation and rearrangement. The triose phosphates, GAP and DHAP, are condensed by aldolase to form fructose 1,6-bisphosphate (FBP). This six-carbon sugar is then irreversibly hydrolyzed to fructose 6-phosphate (F6P) by fructose-1,6-bisphosphatase (FBPase). A two-carbon entity is transferred from F6P to GAP by the enzyme transketolase to form xylulose 5-phosphate (Xu5P) and erythrose-4-phosphate (E4P). Then, E4P is combined with DHAP to form sedoheptulose 1,7-bisphosphate (SBP), by the enzyme aldolase. This 7-carbon product, SBP, is hydrolyzed by sedoheptulose-1,7-bisphosphatase (SBPase) to yield sedoheptulose 7-phosphate (S7P). Two carbons from S7P are transferred to GAP by transketolase producing Xu5P and ribose-5-phosphate (R5P). The resulting R5P is converted to ribulose-5-P (Ru5P) by phosphopentose isomerase. The two molecules of Xu5P are converted into Ru5P by phosphopentose epimerase. The final step of the regeneration phase is the irreversible conversion (using ATP) of Ru5P to RuBP by phosphoribulokinase (PRK).

An important feature of the Calvin cycle is its autocatalysis, where plants are able to reach steady-state rates of photosynthetic conversion of CO_2 after a short delay, as the intermediates build up to the appropriate levels. Autocatalysis of the Calvin cycle is facilitated by the cyclic operation of the pathway as well as the controlled removal of the intermediates from the pathway for other purposes.

Products

The major product of the Calvin cycle is triose-P (GAP and DHAP) and further metabolism of triose-P is the branching point of the Calvin cycle. The available triose-P is used for conversion into two major products of photosynthesis: (1) starch (a glucose polymer), which accumulates during the day inside the chloroplast; and (2) sucrose, which is formed in the cytosol. Triose phosphate is exported from the chloroplast via the phosphate translocator (on the inner chloroplast envelope membrane) and is further metabolized to sucrose in the cytosol. Sucrose is largely exported from the leaf to different sink tissues, e.g., roots, developing leaves, reproductive organs, and other heterotrophic tissues. Sucrose may also accumulate in the vacuole during the day.

Besides the conversion into starch and sucrose, a significant portion of the triose-P is also converted within the chloroplasts to amino acids and fatty acids. Only 1/6 of triose phosphate is utilized/exported out for various biosynthetic processes, while the other 5/6 is needed to regenerate RuBP.

Rubisco

Rubisco is a unique and interesting enzyme, mediating the key reaction of photosynthetic CO_2 assimilation: conversion of one molecule of RuBP and one of CO_2 into two molecules of PGA (**Figure 2**). Besides the carboxylation reaction, Rubisco reacts with

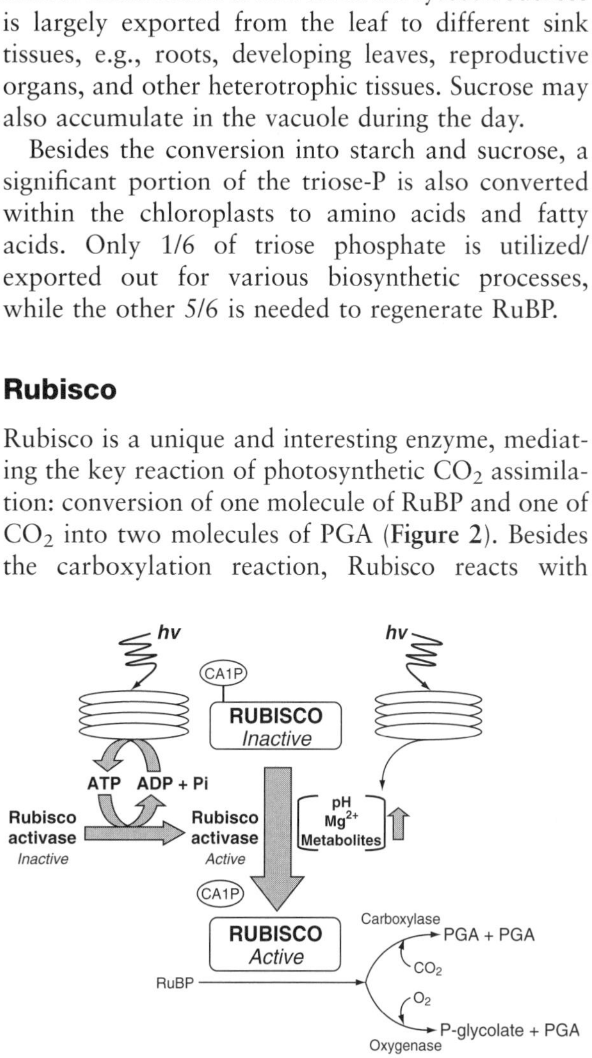

Figure 2 The reactions and regulation of rubisco. The enzyme functions as both a carboxylase and an oxygenase. The product of the carboxylase reaction, PGA, is metabolized in the Calvin cycle. One of the products of oxygenase, P-glycolate, is metabolized through photorespiration. Carboxyarabinitol 1-phosphate (CA1P) binds tightly to rubisco and makes it inactive (e.g., in darkness). Rubisco is activated in light in two ways: (1) by rubisco activase, which itself is activated in light through ATP; and (2) changes in the stromal microenvironment (increase in pH, Mg^{2+} levels, and Calvin cycle metabolites). Both the processes facilitate the release of CA1P and activation of rubisco.

oxygen to form one molecule of 2-phosphoglycolate and one of PGA. The plants try to retrieve part of the carbon diverted into phosphoglycolate through the process of photorespiration or C_2 cycle. Photorespiration also helps in the recycling of nitrogen and dissipation of excess energy and reductants (ATP and NADPH) in the chloroplasts. A major drawback of rubisco is its very slow catalytic capacity. The turnover number of the enzyme is extremely small, and such unfavorable kinetic behavior requires large amounts of enzyme (up to 50% of the total soluble protein) to ensure the carbon flux.

Rubisco of higher plants is a complex protein, with eight large subunits (mass of about 58 kDa) and eight small subunits (mass of about 18 kDa). The large subunits have the catalytic centers, while the small subunits seem to stabilize the rubisco complex. In cyanobacteria, rubisco is also made up of eight large and eight small subunits, while in purple bacteria, rubisco is a dimer of two large subunits only. The large subunits are encoded by the plastid genome and the small subunits in the nucleus. The regulation of rubisco is described below.

Regulation

C_3 photosynthesis is regulated in multiple ways, but is primarily initiated by illumination. One of the most important phenomena in the pathway is the light activation of key enzymes, through either dithiol-reduction of cysteines on protein or through changes in the microenvironment (e.g., magnesium or pH levels) of the stroma. Then, changes in substrate availability and metabolite flux rapidly set in motion an efficient autocatalysis of the Calvin cycle. In a long-term mode, the levels and turnover of rubisco protein also modulate the capacity of carbon fixation in the Calvin cycle. For example, limitation of nitrogen availability decreases the levels of rubisco protein and restricts the photosynthetic carbon assimilation in the leaves.

Regulation of Rubisco

The activity of rubisco, the first enzyme in the Calvin cycle, is highly regulated. Rubisco is inactive in the dark and is converted to an active form on illumination, which catalyzes fixation of CO_2. Activation of rubisco is the result of carbamylation, which involves the binding of CO_2 and Mg^{2+} to a lysine residue near the catalytic site.

Rubisco is active only when lysine-201 reacts with CO_2 near the catalytic site to form a carbamate and allows the binding of the Mg^{2+} ion. Carbamylation changes the conformation of the large subunit and activates the enzyme, while the active conformation is stabilized by the formation of a complex with Mg^{2+}. Carbamylation is essential for rubisco activation, as the noncarbamylated rubisco binds RuBP too tightly to allow catalysis.

Another protein, rubisco activase, is also involved in mediating the light activation of rubisco. On illumination, rubisco activase releases the inhibitor compounds, such as 2-carboxyarabinitol 1-phosphate or CA1P, which are bound to the active site of rubisco; otherwise, for example in darkness, these inhibitors prevent activation (carbamylation) of the enzyme. Rubisco activase itself is activated in light by utilizing ATP produced from photosynthetic electron transport. Rubisco activase and rubisco activation provide another mechanism of strong regulation by light of carbon assimilating reactions of photosynthesis.

CA1P, which occurs naturally in the leaves of several plants, is a strong inhibitor of rubisco. The affinity of rubisco for CA1P is much stronger than that for RuBP, the substrate. As a result, CAP, which accumulates in leaves during the night, inactivates rubisco by blocking the binding sites. During the day (or on illumination), the bound CA1P is released from rubisco, and this process is further accelerated by rubisco activase. However, the physiological role of CA1P is still debated, as it is not found in all plant species.

Light Activation of Enzymes

Light directly activates five enzymes of the Calvin cycle, namely rubisco, GAPDH, FBPase, SBPase, and PRK. Among these, rubisco is activated both indirectly (due to changes in the microenvironment of the chloroplast stroma) and directly (through rubisco activase). The other four enzymes are activated as a result of the reduction of dithiols on the enzyme. Interestingly, four of the five light-activated enzymes (rubisco, FBPase, SBPase, and PRK) catalyze irreversible reactions. Although the dark activities of these enzymes may vary, the Calvin cycle is essentially inactive until illumination takes place.

The Calvin cycle enzymes are activated as a result of the reduction of dithiols on the protein, mediated by a ferredoxin–thioredoxin system. On illumination, ferredoxin is reduced by the thylakoids, which in turn reduce thioredoxin. Reduced thioredoxin can reduce the dithiols at or near the active site of FBPase, SBPase, and PRK, thus activating the enzymes of the Calvin cycle. The thioredoxin–ferredoxin system (reduced in light) also activates two other chloroplastic enzymes, F_1-ATP synthase and NADP-malate dehydrogenase.

Light not only activates several enzymes, but also inactivates the key enzymes of the oxidative pentose phosphate pathway, again through thioredoxin mediated reduction of enzymes. Glucose-6-phosphate dehydrogenase is one such enzyme, which is inactivated on illumination to ensure that the oxidative pentose phosphate pathway does not operate simultaneously with the reductive pentose phosphate pathway, thus avoiding a futile cycle.

Light causes a marked change in the microenvironment of the chloroplast stroma. The photosynthetic electron transport chain in the thylakoid membrane decreases the proton concentration, raises the pH, and increases the Mg^{2+} concentration in the stroma. Rubisco, FBPase, and SBPase respond to such changes in pH and Mg^{2+} concentrations. These enzymes are almost inactive at pH 7.2 and low Mg^{2+}, and their activity increases with the rise in pH and Mg^{2+} concentration.

The enzymes are also under metabolite control. The stromal enzymes FBPase and SBPase are activated by their substrates and allosterically inhibited by their products. High levels of FBP or SBP activate these enzymes, while F6P or S7P inhibit the activity of the corresponding enzyme, to avoid excessive product accumulation and sequestration of phosphate. Similarly, PRK is inhibited by 3-PGA and ADP. As both these metabolites inhibit PRK, phophorylation by PGA kinase can proceed even under limiting ATP supply, thus preventing the accumulation of PGA.

Multiprotein Complexes

The occurrence and operation of multienzyme complexes involving Calvin cycle enzymes have been reported. Such complexes could be an important mode of regulation. For example, GAP and PRK interact with a small nuclear encoded chloroplast protein, CP12, and form a complex, which may be an additional mechanism for light regulation of PPK *in vivo*. Complexes involving different combinations of C_3-enzymes have been found, and these may facilitate rapid channeling of intermediates between enzymes, improving the efficiency of the cycle. Multienzyme complexes of phosphoribose isomerase, PRK, GAPDH, and SBPase have been located on the thylakoid membrane by immunoelectron microscopy. A complex of sucrose phosphate synthase and sucrose phosphate phosphatase has also been reported.

Flux Control

Flux control is an important concept, and the relative importance of individual enzymes in controlling the flux of carbon fixation through the Calvin cycle has been examined. Studies of transgenic plants with altered levels of individual enzymes in the C_3 cycle have revealed that no single enzyme has complete control of carbon fixation. Control is shared among a number of enzymes, with rubisco, SBPase, and aldolase having the most prominent roles. Enzymes with highly regulated activity, such as PRK, do not seem to have strong control on carbon fixation, whereas aldolase, which catalyzes a reversible reaction, strongly influences the rate of carbon fixation.

C_3 Plants: Limitations and Prospects for Improvement

C_3-photosynthesis is a feature of even the primitive lower groups and appears to have evolved much earlier than the CAM or C_4 pathway. Of about 300 000 plants known on earth, ~90% are C_3 plants, while the CAM and C_4-species constitute about 10% and 1%, respectively. Furthermore, most crops (particularly cereals, legumes, and oilseed crops) are of the C_3-type; therefore, C_3 plants have attracted the attention of several scientists. The performance and productivity of C_3 plants is restricted by at least three major factors: high photorespiration (a nonavoidable consequence of oxygenase activity of rubisco), a high water requirement, and a preference for temperate regions.

Attempts were made to evolve varieties with reduced photorespiration or high photosynthetic rates, but single features were not able to improve the productivity of plants. Another approach was to introduce a set of C_4 traits into C_3 plants, but hybridization of C_3 and C_4 species of *Atriplex* resulted in only C_3-type plants. Mutants deficient in one or more enzymes of photorespiration were unable to grow at atmospheric levels of CO_2, indicating that the process of photorespiration (as well as rubisco oxygenase activity) was an adaptation to the present atmospheric levels of CO_2/O_2. However, a very well known factor that could improve the photosynthetic performance and productivity of C_3 plants is elevated CO_2.

Molecular Biology and Biotechnology

The research before the 1950s involved studies mainly on the physiology (e.g., Blackman's law of limiting factors) and biochemistry (e.g., Calvin cycle) of C_3-photosynthesis. By the end of the twentieth century, focus had shifted onto the molecular biology of chloroplasts (and their components) and biotechnology to manipulate some of the proteins/enzymes.

Genes

In the past decade, enormous progress has been made in identifying the genes involved in not only photosynthesis but also other metabolic processes. The genomes of plants such as *Arabidopsis* and rice (*Oryza sativa*) have been completely sequenced. With reference to photosynthesis, the initial work was carried out with genes encoding the components of photosynthetic electron transport, but later covered most of the genes encoding the enzymes of the Calvin cycle and related processes.

The genes encoding the enzymes of the Calvin cycle, with the exception of the large subunit of rubisco, are located in the nuclear genome; the large subunit of rubisco is encoded by genes in the chloroplast. The preproteins are synthesized in the cytosol with N-terminal extensions called transit sequences, which direct them to the chloroplasts, where they are cleaved to form mature proteins. In recent years, cloning and analysis of the genes encoding the enzymes of the Calvin cycle have facilitated further studies on the control of their expression.

Regulation of Gene Expression

There is a perfect coordination of mRNA synthesis and protein accumulation during leaf development and chloroplast maturation. Much of the control has been found at the level of gene transcription. Light has a central role in regulating the biogenesis of the photosynthetic apparatus during chloroplast development, including switching on the expression of genes encoding enzymes of C_3 photosynthesis. These processes appear to be complex, involving many regulatory proteins and cross-talk between the different photoreceptor signaling pathways.

The expression of C_3 photosynthesis genes in leaves is also modulated by environmental and metabolic signals. High levels of glucose or sucrose reduce levels of a number of Calvin cycle mRNAs, including those encoding the small subunit of rubisco, SBPase, and FBPase. This feedback mechanism, acting at the level of transcription, might be important for source–sink regulation in the plant. Photosynthesis genes respond remarkably to nutrient status, particularly nitrogen and phosphorus levels, and this is influenced again by the carbohydrate status. Thus, an interaction between carbohydrate and nutrient status may be a long-term strategy to control carbon metabolism.

Genetic Manipulation

Several enzymes of the Calvin cycle have been genetically engineered in plants (**Table 1**). Attempts have been made to either overexpress the target enzyme or suppress the enzyme levels. The overexpression of these enzymes causes only a small improvement in the overall performance of the plant. However, the suppression of photosynthetic enzymes invariably led to a marked decrease in photosynthetic rate. A reduction in the levels of rubisco, rubisco activase, and NADP-GAPDH resulted in a significant

Table 1 Examples of genetic manipulation of enzymes or proteins involved in photosynthesis of C_3 plants; the effect (suppression or overexpression) of the target enzyme/protein, on the test plant and the consequence of such genetic manipulation are indicated

Target enzyme or protein/test plant	*Effect*	*Consequence*
Rubisco/tobacco	Suppression	Marginal effect on photosynthesis at high N and moderate light; significant limitation on photosynthesis at limiting N and high light
Rubisco activase/tobacco	Suppression	Reduction in activation/carbamylation of rubisco in light; photosynthesis decreased at both low and high CO_2
GAP dehydrogenase/tobacco	Suppression	No significant reduction in growth; increase in PGA indicating the export and reduction of PGA in cytosol
Aldolase/potato (*Solanum tuberosum*)	Suppression	Strong reduction in growth
Transketolase/tobacco	Suppression	Small decrease in transketolase led to marked decrease in growth, photosynthesis, and phenyl propanoid metabolism
Sucrose-P synthase/ *Arabidopsis*	Suppression	Decrease in sucrose synthesis as well as photosynthetic rate
Phosphate translocator/potato	Suppression	No effect on overall rate of carbon fixation, but increased starch accumulation in leaves in light
Single cyanobacterial FBP/ SBPase/tobacco	Overexpression	Enhanced photosynthesis and growth; increased allocation of carbon into sucrose
Mutated form of sucrose 6-P synthase (to avoid down-regulation)/tobacco	Overexpression	High sucrose-P synthase activity and stimulation of sucrose synthesis
Rat liver 6-phosphofructo-2-kinase/tobacco	Overexpression	Higher level of fructose-2,6-bisphosphate; sucrose synthesis reduced at the beginning of light period

decrease in photosynthesis, particularly at high light intensity. The manipulation of stromal FBPase or triose-P/Pi translocator affected the growth pattern only marginally, but significantly changed the pattern of assimilate translocation. The above genetically modified plants were obviously trying to compensate for the deficit of each enzyme/protein; such is the adaptive capacity of plants, due to their metabolic flexibility. In plants, FBPase and SBPase are separate enzymes, while in cyanobacteria, a single enzyme (FBP/SBPase) performs both of their functions. Overexpression of the single cyanobacterial enzyme from *Synechococcus* enhanced the photosynthesis and growth of tobacco (*Nicotiana tabacum*) plants.

Future Outlook

The performance of C_3 plants improves considerably when the atmospheric CO_2 level is elevated. Thus, under conditions of global warming with its elevated CO_2 levels, the use and cultivation of C_3 plants offer much greater potential than that of C_4 plants. However, the C_3 plants may require additional nutrients under high CO_2. This aspect needs to be studied further so that the C_3 plants can be exploited effectively under elevated CO_2.

Despite detailed studies, the functional mechanism of rubisco continues to be a marvel and a mystery. Further experiments are warranted to understand the conditions that would ensure the activated state of rubisco. Specific alterations in the amino acid residues with the help of site-directed mutagenesis may be of interest to achieve this target. An increase in the ratio of carboxylase to oxygenase of rubisco would be of immense potential to increase the photosynthetic efficiency. However, neither extensive screening of higher plant species and their cultivars, nor the use of different metabolites/inhibitors helped to alter the ratio of carboxylase/oxygenase activity. However, some of the plants from lower groups (particularly red algae) have been found to possess rubisco with a much higher carboxylase to oxygenase ratio than that of higher plants. These forms of rubisco are termed "super rubiscos," in view of their vast potential. Attempts are currently being made to incorporate "super rubisco" into higher plants.

Another possibility of improving the performance of C_3 plants is to supplement with CO_2 concentrating mechanisms. Nature has already achieved this in C_4 and CAM plants. An encouraging and important step in this direction is the success of expressing corn (*Zea mays*; maize) phosphoenolpyruvate carboxylase (PEPC) in rice leaves to achieve >100-fold increase in the activity of PEPC. There are plants, such as *Hydrilla verticillata*, *Egeria densa*, and *Borszczowia aralocaspica*, where CO_2 concentrating mechanisms or C_4 photosynthesis operates within a single cell. It would be quite exciting to analyze these plants and exploit the extracted information. With the introduction of an efficient CO_2 concentrating mechanism, the performance of C_3 plants is bound to improve.

Acknowledgments

The preparation of this manuscript and current work on photosynthesis in our laboratory is supported by grant No. SP/SO/A12/98 from the Department of Science and Technology, New Delhi.

List of Technical Nomenclature

Carbon assimilation	Conversion of simple inorganic carbon dioxide into complex organic carbon compounds, enriched with energy, such as sugars.
Chloroplast stroma	Soluble compartment of chloroplasts, enclosed by the envelope membranes, and contains several biosynthetic enzymes, metabolites, nucleic acids and lipid granules.
Gene transcription	The passing of genetic message contained in DNA to RNA, facilitating the translation of information into assembly of a specific chain of amino acids and biosynthesis of proteins.
Sink	The tissues which import and utilize the assimilated organic compounds, e.g. young developing parts, growing fruits, or storage organs, such as tubers.
Source	The organ which makes and supplies the energy-rich organic compounds, usually the leaf.

See also: **Genetic Modification of Primary Metabolism**: Photosynthesis. **Photosynthesis and Partitioning**: C_4 Plants; CAM Plants; Photoinhibition.

Further Reading

Blankenship RE (2002) *Molecular Mechanisms of Photosynthesis*. London: Blackwell Science.

Grossman AR (2000) *Chlamydomonas reinhardtii* and photosynthesis: Genetics to genomics. *Current Opinion in Plant Biology* 3: 132–137.

Gutteridge S and Jordan DB (2001) Dynamics of photosynthetic CO_2 fixation: Control, regulation and productivity. In: Aro E-M and Andersson B (eds) *Regulation of Photosynthesis*, pp. 297–312. Dordrecht: Kluwer Academic Publishers.

Heldt HW (1997) *Plant Biochemistry and Molecular Biology*. Oxford: Oxford University Press.

Lawlor DW (2001) *Photosynthesis: Molecular, Physiological and Environmental Processes*, 3rd edn. Berlin: Springer Verlag.

Leegood RC (1999) Photosynthesis in C_3 plants: The Benson–Calvin cycle and photorespiration. In: Lea PJ and Leegood RC (eds) *Plant Biochemistry and Molecular Biology*, 2nd edn, pp. 29–50. Chichester: John Wiley & Sons.

Malkin R and Niyogi K (2000) Photosynthesis. In: Buchanan BB, Gruissem W, and Jones RL (eds) *Biochemistry and Molecular Biology of Plants*, pp. 568–628. Rockville, MD: American Society of Plant Physiologists.

Martin W, Scheibe R, and Schnarrenberger C (2000) The Calvin cycle and its regulation. In: Leegood RC, Sharkey TD, and Von Caemmerer S (eds) *Photosynthesis: Physiology and Metabolism*, pp. 9–51. Dordrecht: Kluwer Academic Publishers.

Quick WP and Neuhaus HE (1997) The regulation and control of photosynthetic carbon assimilation. In: Foyer CH and Quick WP (eds) *A Molecular Approach to Primary Metabolism in Higher Plants*, pp. 41–62. London: Taylor and Francis.

Rao KK and Hall DO (1999) *Photosynthesis*, 6th edn. Cambridge: Cambridge University Press.

Sharkey TD (1998) Photosynthetic carbon reduction. In: Raghavendra AS (ed.) *Photosynthesis: A Comprehensive Treatise*, pp. 111–122. Cambridge: Cambridge University Press.

Spreitzer RJ and Salvucci ME (2002) Rubisco: Structure, regulatory interactions, and possibilities for a better enzyme. *Annual Review of Plant Physiology and Plant Molecular Biology* 53: 449–475.

Stitt M (1996) Metabolic regulation of photosynthesis. In: Baker NR (ed.) *Photosynthesis and the Environment*, pp. 151–190. Dordrecht: Kluwer Academic Publishers.

Von Caemmerer S and Quick WP (2000) Rubisco: physiology *in vivo*. In: Leegood RC, Sharkey TD, and Von Caemmerer S (eds) *Photosynthesis: Physiology and Metabolism*, pp. 85–113. Dordrecht: Kluwer Academic Publishers.

C_4 Plants

R C Leegood, University of Sheffield, Sheffield, UK

Introduction

Plants with the C_4 pathway of photosynthesis, typically grasses from warm habitats, are responsible for about a quarter of global terrestrial productivity. The C_4 mechanism involves a division of labor between two different photosynthetic cell types, the mesophyll and the bundle sheath. The CO_2-concentrating function of C_4 photosynthesis concentrates CO_2 at the site of Rubisco in the bundle sheath, providing a mechanism that overcomes the photorespiration incurred by the oxygenase activity of Rubisco, and which accounts for losses of around a quarter of the recently fixed carbon in C_3 plants, particularly at higher temperatures. The C_4 pathway also results in improved water and nitrogen-use efficiency compared with C_3 plants.

In a sense, C_4 plants were discovered, but not recognized as such, in the nineteenth century, when Haberlandt described their distinctive Kranz (German, "wreath") anatomy: a distinct layer of cells rich in chloroplasts surrounding the vascular bundle (**Figure 1**). The true nature of these plants did not become apparent until the 1960s, from work by a Russian, Karpilov, on corn (*Zea mays*, maize), and by Kortschak and colleagues on sugar cane (*Saccharum officinarum*) in Hawaii, who reported that the first products of photosynthesis were malate and aspartate, and not glycerate 3-P, as would be expected from CO_2 fixation via the C_3 Benson–Calvin cycle. In the mid-1960s, Hatch and Slack, in Australia, elucidated the path of carbon in C_4 photosynthesis.

Agricultural and Ecological Importance

Although the number of terrestrial plant species that are C_4 is quite modest (about 3%), about 50% of the grasses are C_4. Corn is one of the three principal cereal crops, together with wheat (*Triticum aestivum*) and rice (*Oryza sativa*) (both C_3 plants), each of which constitutes about 30% of total global cereal production (**Table 1**). Other important C_4 grass crops include millet (*Panicum miliaceum*), *Sorghum* spp., sugar cane, and pasture grasses (Poaceae). C_4 plants have two primary requirements for success: a warm growing season and high light, although aridity may also influence their distribution. Fourteen out of 18 of the world's worst weeds are C_4. C_4 grasses are the dominant (>90%) components of warm temperate and tropical grasslands, such as prairies and savannas. In prairies, C_3 grasses tend to be active during the spring, whereas C_4 grasses tend to be active during the summer, when temperatures are higher and soil water lower. Similarly, decreases in temperature with altitude tend to favor C_3 grasses. Other biomes with a high representation of C_4 plants, particularly at low (tropical) latitudes, include deserts and semideserts, arid steppes and salt deserts, beach dunes and saltmarshes, wetlands, and disturbed ground in arid regions.

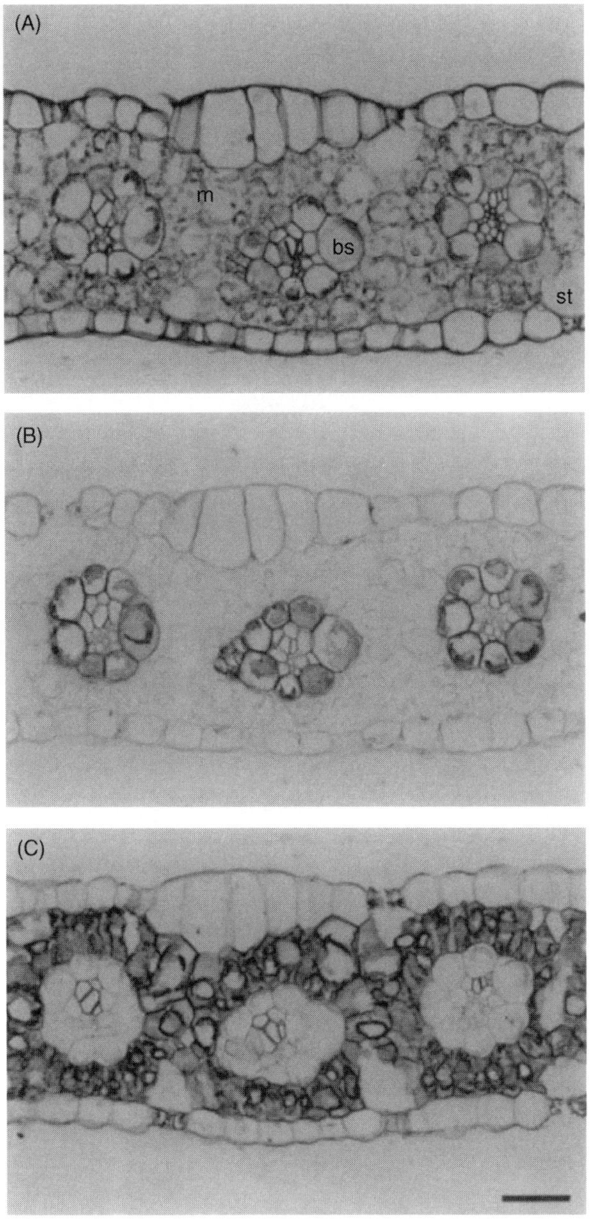

Figure 1 Consecutive transverse sections of a maize leaf. The structure (A) shows thin-walled mesophyll cells (m) and thick-walled bundle sheath cells (bs) surrounding the vascular bundle (v). A stomatal pore can also be seen (st). Immunolocalization indicates the presence of PEPC in the mesophyll (B) and Rubisco in the bundle sheath (C). Scale bar = 50 μm.

Table 1 Global crop production in the year 2002.

Crop	*Production (metric tonnes × 10^6)*
Total cereals	2032
Wheat (C_3)	568
Rice (C_3)	579
Barley (C_3)	132
Rye (C_3)	21
Oats (C_3)	28
Corn (C_4)	602
Sorghum (C_4)	55
Millet (C_4)	26
Other crops	
Potatoes (C_3)	308
Sugar beet (C_3)	252
Sugar cane (C_4)	1288

Source: United Nations Food and Agriculture Organization (http://apps.fao.org/)

Table 2 The essential characteristics of CO_2-concentrating mechanisms in plants, algae and cyanobacteria, as exemplified by C_4 plants.

Component of CO_2-concentrating mechanism	*Mechanism present in C_4 plants*
1. An active CO_2 capture system	PEPC in the cytosol of the specialized mesophyll cells
2. Photosynthetic energy supply	ATP provided by chloroplasts for PEP regeneration
3. An intermediate pool of captured CO_2	C_4 acids, malate and aspartate
4. A mechanism to release CO_2 from the intermediate CO_2 pool	Specific C_4 acid decarboxylating enzymes in the bundle sheath cells
5. A compartment to concentrate CO_2 around Rubisco	The bundle sheath compartment with specialized bundle cells and chloroplasts
6. A means to reduce CO_2 leakage from the site of CO_2 elevation	The bundle sheath structure, including suberized layers, as well as an absence of carbonic anhydrase in the bundle sheath
7. Modification of the kinetic properties of Rubisco	An intermediate affinity for CO_2, high specificity factor and increased V_{max}

The CO_2-Concentrating Mechanism of C_4 Photosynthesis

C_4 plants exemplify the key requirements for the operation of a CO_2-concentrating mechanism (**Table 2**), that are based on a sophisticated division of labor between the mesophyll and bundle sheath cells.

Anatomy

The photosynthetic mesophyll cells are thin-walled, and radiate from the bundle sheath into the intercellular spaces (**Figure 1**). The bundle sheath forms a compartment in which to concentrate CO_2 around Rubisco. It contains numerous chloroplasts, is not in contact with the intercellular spaces, and has thickened, often suberized, cell walls. Their conductance to CO_2 is low, and hence the leakage of CO_2 from the bundle sheath compartment is restricted. In grasses, about 10% of the CO_2 leaks

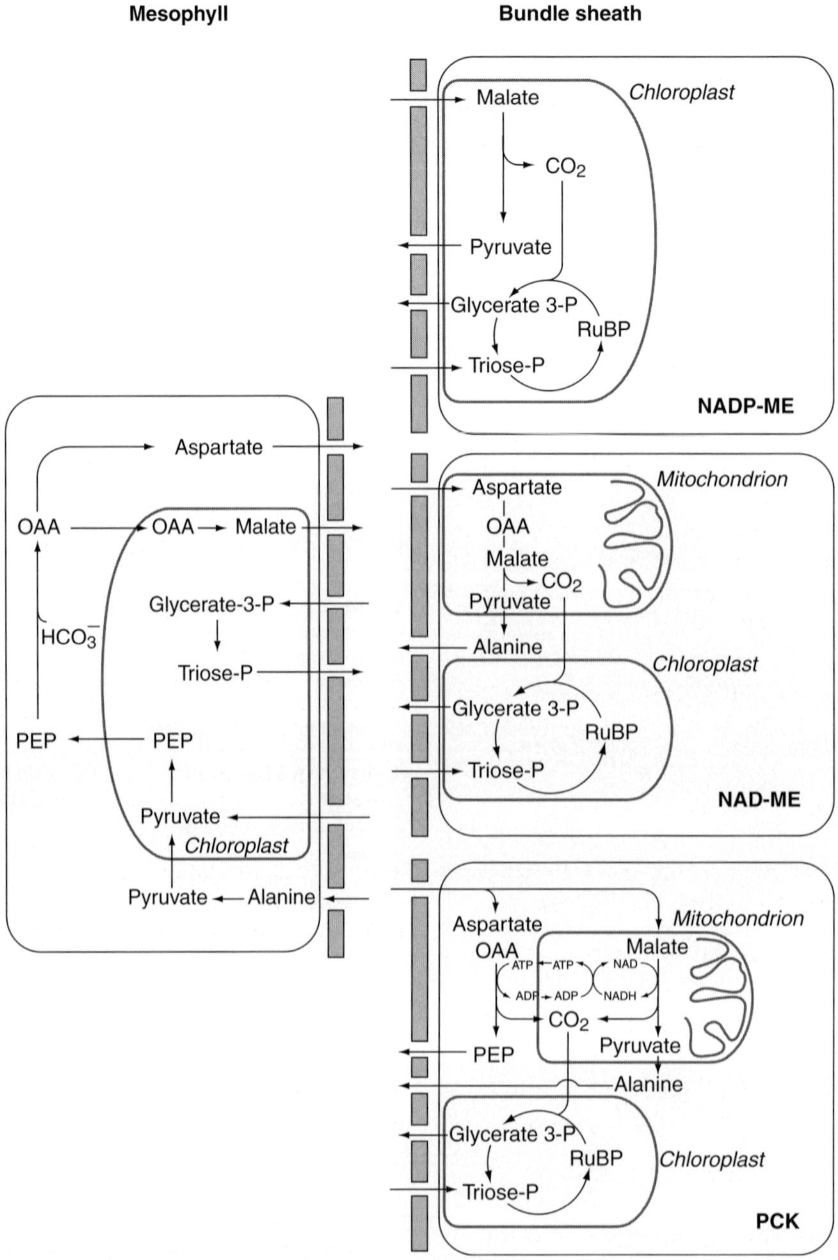

Figure 2 Pathways of photosynthetic carbon assimilation in the three different C_4 subgroups. The left-hand side shows the reactions in the mesophyll. Regeneration of phosphoenolpyruvate (PEP) from pyruvate, catalyzed by PPDK, and carboxylation of PEP, catalyzed by PEPC, to form oxaloacetate (OAA) that is then converted to four carbon acids (malate and aspartate), is common to all the C_4 subtypes. The right-hand side shows the decarboxylation reactions in the NADP-ME, NAD-ME, and PCK subtypes. Carboxylation of ribulose 1,5-bisphosphate (RuBP) via the Benson–Calvin cycle occurs in all of them, as does shuttling of glycerate-3-phosphate for reduction to triose-phosphate in the mesophyll. The shaded rectangles represent the mesophyll–bundle sheath cell wall, permeated by plasmodesmata to allow the exchange of metabolites. Most cofactors are omitted for the sake of clarity.

out of the bundle sheath. This loss is relatively small, although it increases the light energy required to operate the CO_2 pump, but it is inevitable in a bundle sheath that is connected to the mesophyll by numerous plasmodesmatal connections. Plasmodesmata are necessary to allow metabolite fluxes between the mesophyll and bundle sheath, which are driven by concentration gradients of the various metabolites (**Figure 2**). For this reason mesophyll cells are always closely associated with bundle sheath cells, hence the interveinal distance is lower than in the leaves of C_3 plants. The intracellular compartmentation of C_4 photosynthesis also requires rapid transport of metabolites between organelles, resulting in modification of existing (C_3-type) transmembrane metabolite transporters.

Until recently, it had been thought that all terrestrial plants with C_4 photosynthesis possessed Kranz anatomy. However, two chenopods have recently been shown to perform C_4 photosynthesis within single cells. These cells have very specialized anatomy that prevents diffusion of CO_2 away from the site of Rubisco.

Biochemistry

In contrast to many aquatic organisms that operate biophysical CO_2 pumps to raise the concentration of CO_2 at the site of carboxylation by Rubisco, C_4 plants operate a biochemical CO_2 pump – a cycle of reactions that increases the concentration of CO_2 from some 100 μbar in the mesophyll intercellular spaces to about 2000 μbar in the bundle sheath cells. This is achieved by a cycle of carboxylation in the mesophyll, leading to the formation of C_4 acids (malate and aspartate) and decarboxylation of those C_4 acids in the bundle sheath, yielding a C_3 compound that can be returned to the mesophyll.

In all C_4 plants, the carboxylation step of the C_4 cycle that occurs in the mesophyll is catalyzed by phosphoenolpyruvate (PEP) carboxylase (PEPC):

$$\text{PEP} + \text{HCO}_3^- \rightarrow \text{oxaloacetate} + \text{Pi}$$

This is preceded by the formation of bicarbonate from CO_2, catalyzed by carbonic anhydrase:

$$\text{CO}_2 + \text{H}_2\text{O} \leftrightarrow \text{HCO}_3^- + \text{H}^+$$

The mechanism of decarboxylation in the bundle sheath varies. Three C_4 subtypes have traditionally been recognized (**Figure 2**), those that decarboxylate via NADP-ME in the cytosol (e.g., sugar cane):

$$\text{malate} + \text{NADP} \leftrightarrow \text{pyruvate} + \text{NADPH} + \text{CO}_2$$

via NAD-ME in the mitochondria (*Amaranthus*):

$$\text{malate} + \text{NAD} \leftrightarrow \text{pyruvate} + \text{NADH} + \text{CO}_2$$

or via a combination of phosphoenolpyruvate carboxykinase (PCK) in the cytosol:

$$\text{oxaloacetate} + \text{ATP} \leftrightarrow \text{PEP} + \text{ADP} + \text{CO}_2$$

and NAD-ME in the mitochondria (*Panicum maximum*). These different mechanisms reflect the diversity of origin of C_4 photosynthesis, but the biochemical diversity is probably even greater, since some NADP-ME types may have PCK (corn) or NAD-ME (*Flaveria bidentis*) as auxiliary decarboxylases. Pyruvate or PEP then returns to the mesophyll to be regenerated to PEP by the action of pyruvate, Pi dikinase (PPDK):

$$\text{pyruvate} + \text{ATP} + \text{Pi} \leftrightarrow \text{PEP} + \text{AMP} + \text{PPi}$$

followed by hydrolysis of PPi and the phosphorylation of AMP by ATP, catalyzed by adenylate kinase:

$$\text{AMP} + \text{ATP} \leftrightarrow 2\,\text{ADP}$$

Hence 2 ATP molecules are used to regenerate PEP from pyruvate.

One of the essential features of the biochemistry is a marked compartmentation of photosynthetic enzymes. PEPC and PPDK are essentially confined to the mesophyll, while the decarboxylases and Rubisco are confined to the bundle sheath (**Figure 1**). In addition, the Benson–Calvin cycle is partially split between the two cell types, because all C_4 plants shuttle glycerate 3-phosphate from the bundle sheath to the mesophyll, where it is reduced to triose-phosphate (**Figure 2**). In NADP-ME types the only source of reductant in the bundle sheath is NADPH deriving from NADP-ME. This is because the bundle sheath chloroplasts have agranal chloroplasts that lack photosystem II and make ATP by photosystem I-dependent cyclic electron transport. In the other subtypes, export of glycerate 3-phosphate is probably both a mechanism of reducing photosynthetic O_2 evolution in the bundle sheath, so improving the $CO_2:O_2$ ratio and reducing photorespiration.

Since the carboxylation and decarboxylation phases of the C_4 cycle occur in two separate cells, regulation of enzymes of the cycle must take the form of metabolic or other signals passing between the mesophyll and bundle sheath. Regulation of the activity of the C_4 cycle occurs at a number of levels, all of which may interact in the regulation of a single enzyme:

1. Light-dependent changes in pH (alkalization) and increases in Mg^{2+} concentration in the chloroplast stroma. These appear to be particularly important in the regulation of the activity of NADP-ME in the bundle sheath chloroplast.
2. Changes in the concentrations of substrates, effectors, and cofactors, such as adenylates. For example, adenylates decrease the affinity for malate of NAD-ME from *Atriplex spongiosa* and accentuate the sigmoidicity of response to malate concentration. Hence if the mitochondrial malate concentration were low in darkness or in low light, malic enzyme activity would be greatly reduced. In some C_4 plants, such as *P. maximum*, dark phosphorylation of PCK amplifies its response to adenylates and decreases its activity at low substrate concentrations.

3. Posttranslational modification, either by phosphorylation (PEPC, PPDK, and PCK) or by the thioredoxin system (NADP-malate dehydrogenase).

PEPC is sensitive to both feedback inhibition, from malate and aspartate, and feedforward activation, from glucose 6-phosphate and triose-phosphate. Thus a slowing of decarboxylation of malate and aspartate in the bundle sheath will lead to their accumulation, and an inhibition of PEPC. Conversely, triose-phosphate accumulation will signal an increased rate of Benson–Calvin cycle turnover and greater demand for the decarboxylation of four-carbon acids. Phosphorylation of PEPC, which occurs in the light on an N-terminal serine residue, amplifies the response to these effectors, so that the phosphorylated enzyme is less sensitive to inhibition by malate and more sensitive to activation by glucose 6-phosphate (**Figure 3**). The low abundance kinase that phosphorylates PEPC is the smallest known protein kinase and its activity appears to be regulated by protein turnover.

PPDK is also regulated by phosphorylation, but this enzyme is unusual in that regulation is part of the catalytic mechanism of the enzyme. Modulation of the activity is achieved by a low-abundance regulatory protein that is a bifunctional enzyme, with two active sites, and which catalyzes the interconversion between active and inactive forms.

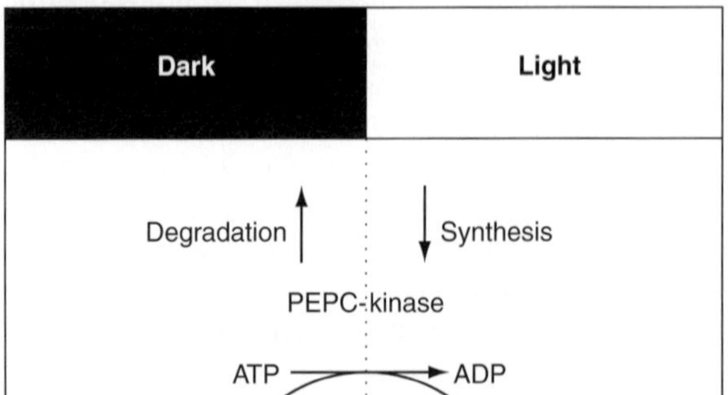

Figure 3 Light–dark regulation of PEP carboxylase in a C_4 plant. PEPC is phosphorylated on a serine residue by PEPC-kinase in the light, and dephosphorylated by a type 2A protein phosphatase (PP2A) in the dark. Phosphorylated PEPC is more active, less sensitive to feedback inhibition by malate and more sensitive to activation by glucose 6-phosphate (Glc 6-P). PEPC-kinase protein undergoes diurnal turnover.

Resource Use Efficiency

The possession of a CO_2-concentrating mechanism that overcomes photorespiration has impacts on the efficiency with which light, nitrogen, and water are used by plants and on the temperature optimum for photosynthesis. In C_4 plants, photosynthesis is saturated by CO_2 in air and rates of CO_2-saturated photosynthesis are often higher than in C_3 plants (**Figure 4**). The C_4 mechanism confers advantages in water-use efficiency to C_4 plants because photosynthesis can operate at low intercellular concentrations of CO_2 (about 100 μbar), and hence lower stomatal conductances, than in C_3 plants (**Figure 4**). Water-use efficiency in C_4 plants is double that of C_3 plants (e.g., in the field, wheat (C_3) and corn (C_4) use, respectively, about 600 g and 300 g H_2O per g dry matter produced). Although C_4 plants may use less water, they are not necessarily more tolerant of water stress.

The responses to temperature are different in C_3 and C_4 plants. Carbon dioxide assimilation shows an optimum of 20–30°C in C_3 leaves, and 30–40°C in C_4 leaves (**Figure 5**). The temperature dependency in C_3 leaves may be altered by changes in the rate of photorespiration, particularly the decline at higher temperatures. At high concentrations of CO_2 or low concentrations of O_2 (2%), which suppress photorespiration, the temperature response of a C_3 leaf becomes similar to that of a C_4 leaf.

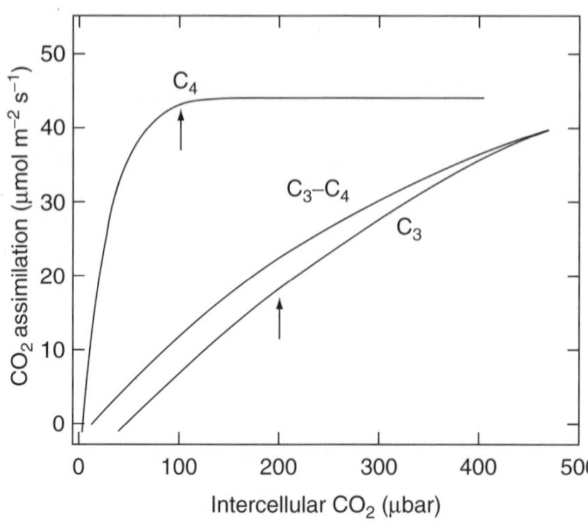

Figure 4 Changes in the rate of CO_2 assimilation in relation to the intercellular partial pressure of CO_2 in leaves of a C_4 plant (*Flaveria bidentis*), a C_3–C_4 intermediate (*Flaveria floridana*), and a C_3 plant (*Flaveria pringlei*), measured in high light (1600 μmol quanta m^{-2} s^{-1}). The arrows indicate the internal CO_2 partial pressure at ambient external CO_2 (350 μbar).

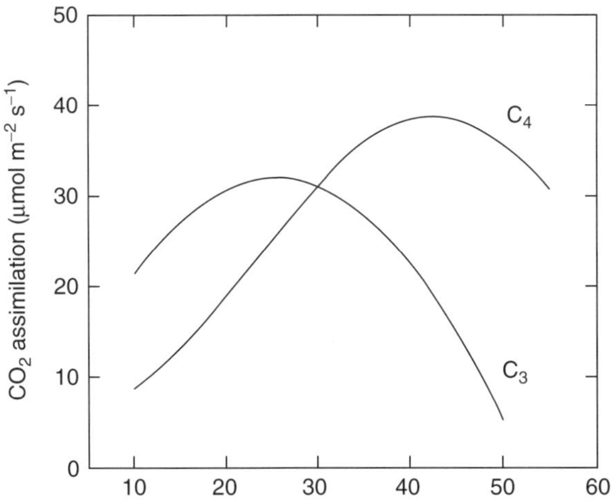

Figure 5 The temperature response of photosynthesis in two prairie grasses, *Agropyron smithii* (C_3) and *Bouteloua gracilis* (C_4). Reproduced with permission from Monson RK, Littlejohn RO, and Williams GJ (1983) Photosynthetic adaptation to temperature in four species from the Colorado shortgrass steppe – a physiological model for coexistence. *Oecologia* 58: 43–51.

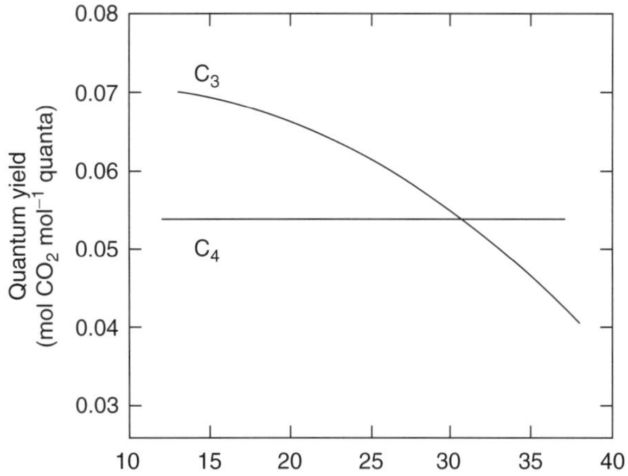

Figure 6 The quantum yield for net CO_2 uptake in air as a function of temperature for leaves of C_3 (*Encelia californica*) and C_4 (*Atriplex rosea*) plants. Reproduced with permission from Ehleringer J and Björkman O (1977) Quantum yields for CO_2 uptake in C_3 and C_4 plants. *Plant Physiology* 59: 86–90.

Rubisco is the major (up to 50%) soluble protein in the leaves of C_3 plants. Why? First, it is an inefficient enzyme with a low specific activity. Second, it catalyzes the wasteful oxygenation reaction and is, therefore, an N-containing resource that is inefficiently used. The CO_2-concentrating mechanism increases the efficiency with which this Rubisco is used, and Rubisco contents are consequently lower in the leaves of C_4 plants, and N-use efficiency higher.

Light-use efficiency is inherently lower in C_4 plants than in C_3 plants in the absence of photorespiration because the CO_2 pump requires an extra input of ATP (an extra 2 ATP per CO_2 fixed in an NADP-ME type C_4 plant, such as sugar cane). However, photorespiration also results in a large increase in the energy required for net CO_2 fixation in C_3 species. At rate-limiting photon flux densities, the amount of CO_2 taken up can be related to the amount of light absorbed by the quantum yield (mol CO_2 fixed mol^{-1} absorbed photons). With the proportion of oxygenation occurring in air at 25–30°C, the theoretical cost of net fixation of CO_2 in a C_3 plant is approximately equal to the theoretical cost of fixing CO_2 in C_4 photosynthesis (4.5 ATP and 3 NADPH in air in C_3 plants and 5 ATP and 2 NADPH in NADP-ME type C_4 plants) and the quantum yield for both C_3 and C_4 species is comparable. If C_3 plants have lower rates of photorespiration (at lower temperatures or at a higher intercellular concentration of CO_2) then this will require less photosynthetic energy and result in a higher quantum yield. If photorespiratory rates are high (at higher temperatures or at lower intercellular concentrations of CO_2), then more energy will be required. The quantum yield of leaves of C_3 plants thus changes with temperature, falling as the temperature is raised, but it remains constant in leaves of C_4 plants (**Figure 6**).

Carbon Isotope Discrimination

$^{13}CO_2$ is a naturally occurring stable isotopic form of CO_2 that forms about 1% of total atmospheric CO_2. Compared with a standard limestone, the present atmosphere over open oceans shows a $\delta^{13}C$ value of −7‰ (i.e., depleted in ^{13}C). Fractionation between the source (the atmosphere) and the product (e.g., starch or leaf material) occurs at many different stages. Fractionation by diffusion in air is +4‰ ($^{13}CO_2$ diffuses more slowly than $^{12}CO_2$); for the solution of CO_2 in water, catalyzed by carbonic anhydrase, it is −8‰. Fractionation during fixation of bicarbonate by PEPC is slight (+2‰) whereas fractionation against $^{13}CO_2$ by Rubisco is large (+34‰). The extent of discrimination is therefore an indication of the mechanism of carboxylation operating in photosynthesis. Although C_4 plants ultimately fix CO_2 via Rubisco, as in C_3 plants, the CO_2-concentrating mechanism means that Rubisco is effectively prevented from discriminating against $^{13}CO_2$ and that PEPC therefore determines the carbon isotope discrimination by photosynthesis. Thus in leaf material of C_4 (and CAM) plants, $\delta^{13}C$ values are −10‰ to −17‰, whereas in C_3 plants $\delta^{13}C$ values are typically in the region of −29‰.

Evolution

Phylogenetic evidence suggests that the first C_4 plants evolved at least 25 million years ago, and that C_4 photosynthesis originated independently at least 31 times. The oldest (12.5 million years) documented C_4 fossil is a grass, *Tomlinsinia*, from the Ricardo formation in California. It has Kranz anatomy, small interveinal distances, and a $\delta^{13}C$ of $-13.7‰$. Carbon isotope analysis of soil carbonates and of the fossil tooth enamel of mammalian grazers (which contains carbon from their diet) indicates a large expansion of C_4 biomass between 8 and 6 million years ago, possibly associated with a decline in atmospheric CO_2. There has been a general decline in atmospheric CO_2 concentrations over the last 100 million years, from 1000 µbar to about 200 µbar by 2 million years ago. This would have promoted photorespiration in C_3 plants, and favored C_4 plants. The current anthropogenic rise in the CO_2 concentration in the atmosphere will tip the balance back in favor of C_3 plants, though any associated global warming will favor C_4s (**Figure** 7).

About 30 species of plants exhibit photosynthetic characteristics that are intermediate between C_3 and C_4 plants in that they show reduced rates of photorespiration and CO_2 compensation points in the range 7–15 µbar compared with typical values in C_3 plants of 50 µbar and in C_4 plants of less than 5 µbar (**Figure 4**). Although all show Kranz anatomy, though often poorly developed, both mesophyll and bundle sheath cells contain Rubisco. Two main types of C_3–C_4 intermediate have been distinguished. In one type (e.g., *Moricandia arvensis*), there is no evidence for a functional C_4 acid cycle that donates CO_2 from C_4 acids to the Benson–Calvin cycle. However, glycine decarboxylase, the step that results in photorespiratory CO_2 release, is confined to the bundle sheath, as in all C_4 plants. It has been proposed that glycine is shuttled from the mesophyll to the bundle sheath and that refixation of photorespiratory CO_2 occurs in the bundle sheath cells, reducing the CO_2 compensation point. It has been proposed that these plants represent a first step in the evolution of true C_4 photosynthesis. In other C_3–C_4 intermediates (e.g., *Flaveria* spp.), the C_4 pathway enzymes, PEPC, PPDK, and NADP-ME, are present in appreciable amounts. These plants show varying capacities to fix ^{14}C into C_4 acids during short-term exposure to $^{14}CO_2$, and to transfer this to products of the Benson–Calvin cycle, indicating the operation of a partial C_4 cycle, in addition to the glycine shuttle.

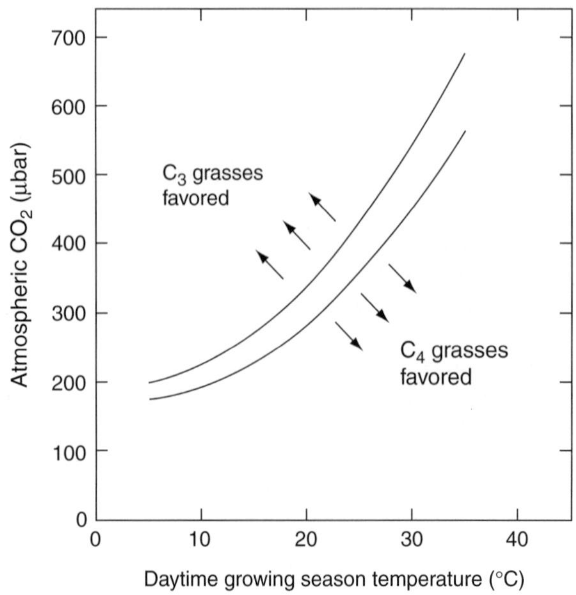

Figure 7 Model predicting C_3 or C_4 dominance of grasses related to temperature and partial pressure of CO_2, based on crossover temperatures (see **Figure 6**) of the quantum yield for CO_2 uptake. Reproduced with permission from Ehleringer JR, Cerling TE, and Helliker BR (1997) C_4 photosynthesis, atmospheric CO_2, and climate. *Oecologia* 112: 285–299.

Development

From a developmental point of view, the C_4 leaf requires three modifications of the basic C_3 pattern. First, the pattern must change to decrease vein spacing. Second, cell patterns must be modified (organization of the bundle sheath and mesophyll cells). Third, mechanisms must be present to allow differentiation of bundle sheath and mesophyll cells. All the enzymes present in the C_4 cycle have nonphotosynthetic isoforms that are widely expressed in C_3 plants, although usually in much lower amounts. Hence altered expression of existing genes occurs in C_4 plants. Genes coding for C_4 enzymes probably arose by gene duplication followed by the addition of new regulatory elements to confer high expression and cell specificity. The promoter for corn PEPC, for example, directs high mesophyll cell-specific expression in rice, while the promoter for the C_4 isoform of PEPC from *Flaveria trinervia* directs a high level of expression in the palisade mesophyll in tobacco (*Nicotiana tabacum*). A major change in the evolution of the C_4 PEPC genes was the acquisition of *cis*-acting DNA sequences in the 5′ promoter region. These appear to be recognized by *trans*-acting factors in unrelated C_3 plants to give a C_4-like pattern of expression.

Interestingly, there are examples of inducible C_4 systems in plants. The freshwater monocotyledon *Hydrilla verticillata*, when exposed to low CO_2 concentrations, develops high activities of C_4

enzymes (PEPC in the cytosol, NADP-ME and PPDK in the chloroplast) that form an intracellular, single-celled, C_4 cycle. Since *H. verticillata* lacks a compartment in which to concentrate CO_2, it is probable that diffusion of CO_2 away from Rubisco is restricted by the unstirred layer of water that surrounds the leaves. The amphibious leafless sedge *Eleocharis vivipara* develops C_4-like biochemical traits, as well as Kranz anatomy, under terrestrial conditions, but it develops C_3-like biochemical traits, without Kranz anatomy, under submerged conditions. When submerged plants are exposed to aerial conditions, they rapidly produce new photosynthetic tissues with C_4 traits. This change can be induced by abscisic acid, a plant growth regulator produced in response to water stress.

Biotechnology and C_4 Photosynthesis

The adverse effects of photorespiration on carbon gain and nitrogen and water-use efficiency and the demands of a steadily increasing population on the world's food supply are leading to attempts to introduce C_4 characteristics into C_3 plants, so as to improve their productivity and resource-use efficiency. This effort has largely concentrated on introduction of the single-celled C_4 cycle found in *H. verticillata*. The individual overexpression of PEPC, PPDK, and other C_4 enzymes in transgenic rice and other C_3 plants has led to little or no change in their photosynthetic characteristics. In principle, it seems unlikely that such attempts will be successful without also introducing the complete C_4 cycle and some of the structural characteristics of C_4 photosynthesis (**Table 2**). This includes a compartment in which CO_2 could be concentrated. If the anatomical characteristics of C_4 plants were to be incorporated, this means understanding the complex factors that regulate the development of the different cell types in C_4 plants, or else engineering pyrenoids or carboxysomes (intracellular compartments for the concentration of CO_2 that are found in algae and cyanobacteria) into C_3 plants. The alternative is to express, in C_3 plants, engineered forms of Rubisco in which the oxygenase reaction, and hence photorespiration, is suppressed.

List of Technical Nomenclature

Bundle sheath — The cells surrounding the vasculature in the leaves of C_4 plants have thick cell walls and abundant chloroplasts, unlike those of most C_3 plants.

C_3 plant — Any plant in which the first product of photosynthesis is a C_3 compound, glycerate 3-phosphate, generated by the carboxylation reaction of Rubisco.

C_4 plant — Any plant with a CO_2-concentrating mechanism in which C_4 acids (malate, aspartate) are the first product of photosynthetic CO_2 assimilation.

CO_2-concentrating mechanism — A biochemical (in terrestrial plants) or biophysical pump (in many aquatic photosynthetic organisms) which concentrates CO_2 around Rubisco so as to reduce photorespiration.

Photorespiration — The process by which glycollate 2-phosphate, which results from the oxygenation reaction of Rubisco, is recovered for the Benson–Calvin cycle. This results in the loss of a quarter of the carbon in glycollate 2-phosphate and the release of ammonia.

See also: **Genetic Modification of Primary Metabolism**: Photosynthesis. **Photosynthesis and Partitioning**: C_3 Plants; CAM Plants; Photorespiration.

Further Reading

Cerling TE, Ehleringer JR, and Harris JM (1998) Carbon dioxide starvation, the development of C_4 ecosystems, and mammalian evolution. *Philosophical Transaction of the Royal Society of London B* 353: 159–171.

Chollet R, Vidal J, and O'Leary MH (1996) Phospho*enol*pyruvate carboxylase: a ubiquitous, highly regulated enzyme in plants. *Annual Review of Plant Physiology and Plant Molecular Biology* 47: 273–298.

Collatz GJ, Berry JA, and Clark JS (1998) Effects of climate and atmospheric CO_2 partial pressure on the global distribution of C_4 grasses: present, past, and future. *Oecologia* 114: 441–454.

Edwards GE, Furbank RT, Hatch MD, and Osmond CB (2001) What does it take to be C_4? Lessons from the evolution of C_4 photosynthesis. *Plant Physiology* 125: 46–49.

Ehleringer JR and Monson RK (1993) Evolutionary and ecological aspects of photosynthetic pathway variation. *Annual Review of Ecology and Systematics* 24: 411–439.

Ehleringer JR, Cerling TE, and Helliker BR (1997) C_4 photosynthesis, atmospheric CO_2, and climate. *Oecologia* 112: 285–299.

Hatch MD and Osmond CB (1976) Compartmentation and transport in C_4 photosynthesis. In: Stocking CR and Heber U (eds) *Encyclopedia of Plant Physiology*, vol. 3, pp. 144–184. Berlin: Springer-Verlag.

Henderson S, Hattersley P, von Caemmerer S, and Osmond CB (1994) Are C_4 pathway plants threatened by global climatic change? In: Schulze ED and Caldwell MM (eds)

Ecophysiology of Photosynthesis, pp. 529–549. Berlin: Springer-Verlag.
Leegood RC (2002) C_4 photosynthesis: principles of CO_2 concentration and prospects for its introduction into C_3 plants. *Journal of Experimental Botany* 53: 581–590.
Leegood RC, Sharkey TD, and von Caemmerer S (eds) (2000) *Photosynthesis: physiology and metabolism*. Dordrecht: Kluwer Academic.
Osmond CB, Winter K, and Ziegler H (1980) Functional significance of different pathways of CO_2 fixation of photosynthesis. In: Lange OL, Nobel PS, Osmond CB, and Ziegler H (eds) *Encyclopedia of Plant Physiology*, vol. 12B, pp. 479–547. Berlin: Springer-Verlag.
Sage RF (2002) C_4 photosynthesis in terrestrial plants does not require Kranz anatomy. *Trends in Plant Science* 7: 283–285.
Sage RF and Monson RK (eds) (1999) *C_4 Plant Biology*. San Diego: Academic Press.
Sheehy JH, Mitchell PL, and Hardy B (eds) (2000) *Redesigning Rice Photosynthesis to Increase Yield*. Makati City: International Rice Research Institute.
von Caemmerer S and Furbank RT (eds) (1997) C_4 photosynthesis: 30 (or 40) years on. *Australian Journal of Plant Physiology* (Special Issue) 24: 409–554.

CAM Plants

U Lüttge, Darmstadt University of Technology, Darmstadt, Germany

Definition

Crassulacean acid metabolism (CAM) is a shunt of carbon flow before the actual photosynthetic assimilation of CO_2 (**Figure 1**). Carbon dioxide fixation first gives organic acid, which is subsequently decarboxylated. The CO_2 regained is reduced and assimilated to carbohydrate. In essence, this process is identical to the C_4 pathway in photosynthesis, the fundamental difference being the means by which the organic acid shunt is separated from light-dependent assimilation of the inorganic carbon. In the C_4 pathway, there is spatial separation by different types of cells, i.e., in the mesophyll (organic acid synthesis) and the bundle sheaths (assimilation); in CAM, there is temporal separation between the dark period (organic acid synthesis) and the light period (assimilation) (**Figure 1**). Thus, CAM is a day/night cycle occurring in single green cells. C_4 photosynthesis and CAM can be distinguished from C_3 photosynthesis, where the latter has no organic acid shunt and CO_2 is assimilated directly in the light.

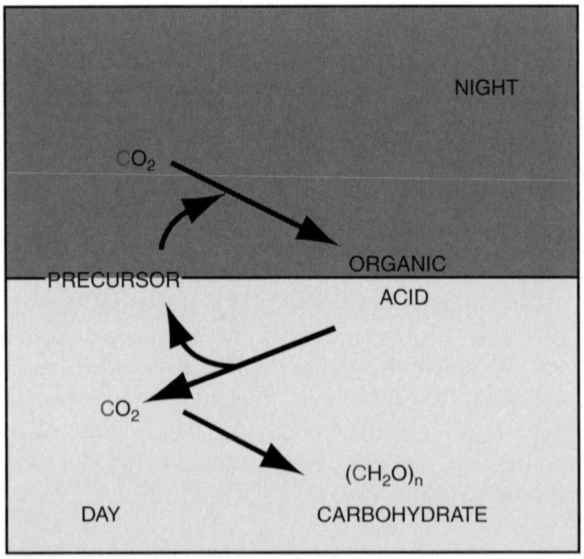

Figure 1 Basic day/night cycle of CAM.

Flow of Carbon in CAM

The major features of the flow of carbon in CAM are shown in **Figures 2** and **3** for the dark and light period, respectively. In the dark period, CO_2 is fixed via phosphoenolpyruvate carboxylase (PEPC), the precursor (PEP) being generated via glycolysis. The malic acid formed is stored in the central cell sap vacuole. Some CAM plants also form citric acid.

In the light period, organic acid is remobilized from the vacuole and decarboxylated. With respect to the mechanism of decarboxylation, two types of CAM plants are distinguished: NAD(P)-ME- and PEPCK-type species. In the former type, NAD-dependent mitochondrial or NADP-dependent cytosolic malic enzyme (ME) serves decarboxylation, while in the latter type, this is brought about by cytosolic PEP-carboxykinase (PEPCK). The CO_2 regenerated is refixed via ribulose-*bis*-phosphate carboxylase/oxygenase (Rubisco) and is assimilated in the Calvin cycle. Carbohydrate, both as a photosynthetic product for building up new biomass and as a source for PEP as a precursor for CO_2 fixation in the subsequent night, is formed in photosynthesis and via gluconeogenesis. Gluconeogenesis in malic enzyme-type CAM plants starts from pyruvate produced in the ME reactions generating PEP via pyruvate, Pi-dikinase (PPDK) in the chloroplasts. These plants have no PEPCK activity. In PEPCK-type CAM plants, PEP originates directly from malate decarboxylation, and these plants often have no PPDK activity. PEPCK-type CAM plants predominantly store free sugars in the vacuole as a reservoir for PEP formation in the dark period, while NAD(P)-type CAM plants store starch in the

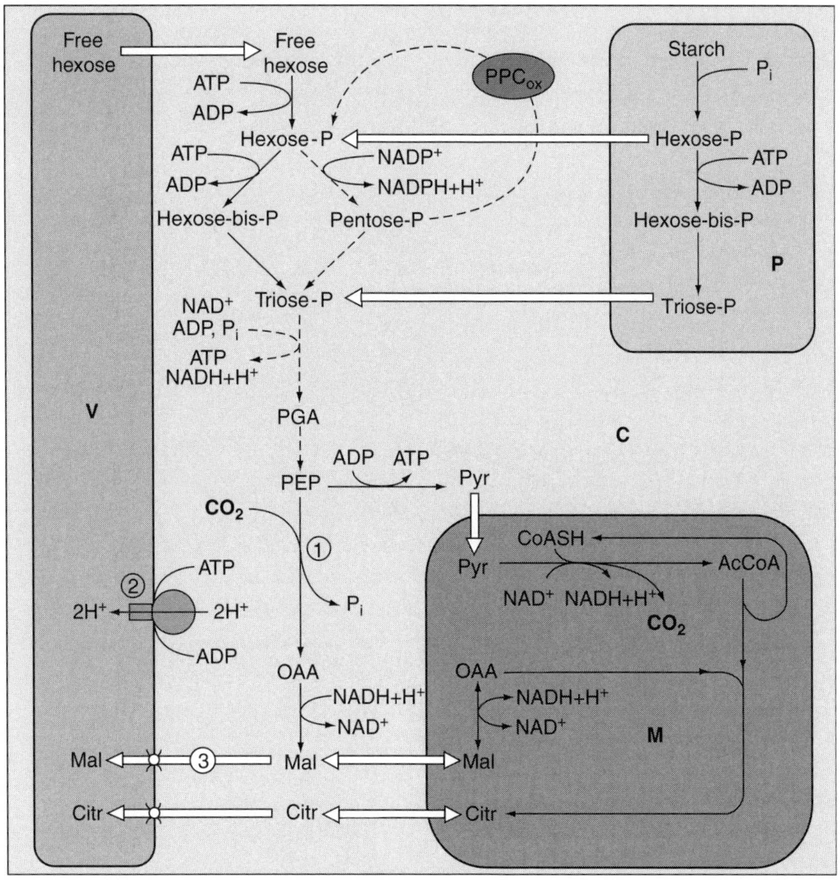

Figure 2 Nocturnal (dark period) flow of carbon in CAM. Simplified scheme of metabolic pathway and compartmentalization. Key elements of the CAM pathway are: (1) PEPC (phosphoenolpyruvate carboxylase for malate synthesis); (2) proton pumping at the tonoplast by the V-ATPase; and (3) malate transport into the vacuole via an organic acid anion channel. Citrate may be formed via mitochondrial reactions. The following abbreviations are used: AcCoA, acetyl-coenzyme A; Citr, citrate; CoASH, coenzyme A; Di-PGA, 2,3-diphosphoglyceric acid; Mal, malate; OAA, oxaloacetate; P_i, inorganic phosphate; PPC$_{ox}$, oxidative pentose phosphate cycle; PEP, phosphoenolpyruvate; Pyr, pyruvate; V, vacuole; C, cytosol; P, plastid; M, mitochondrium.

chloroplasts. However, this distinction is not strict, as many combinations are possible and both free sugars and starch may even play a role as nocturnal PEP precursors in the same species.

Nocturnal vacuolar acid accumulation is a secondary active process where the driving force is an electrochemical gradient of protons ($\Delta\tilde{\mu}_{H^+}$) built up by proton pumps at the vacuolar membrane, the tonoplast, i.e., mainly the H^+-transporting V-type ATPase and to some extent also a vacuolar H^+-transporting pyrophosphatase (V-PPase). Malic acid follows $\Delta\tilde{\mu}_{H^+}$ via an organic acid anion channel. Daytime acid efflux from the vacuole is a passive process.

Expression of CAM and the Quest for CAM-Specific Enzymes

Expression of CAM can be traced in phylogeny (evolution), ontogeny (developmental programs), and plasticity (environmental responses). In the entire complex flow of carbon in CAM depicted by **Figures 2** and **3**, there is no single type of enzyme that is unique to CAM and not found to also have housekeeping functions in non-CAM plants. However, while there does not appear to be a unique CAM enzyme, the quest for genes encoding CAM-specific forms of enzymes (isoenzymes) can be pursued following CAM expression in phylogeny, ontogeny, and plasticity.

Phylogeny

Since there are no CAM-unique enzymes it is no surprise that CAM originated many times, i.e. polyphyletically, on the phylogenetic tree of vascular plants (**Figure 4**). It is considered that CAM is a recent event in evolution and that C_3 photosynthesis is ancestral to C_4 photosynthesis and CAM. CAM does not occur among nonvascular plants; however,

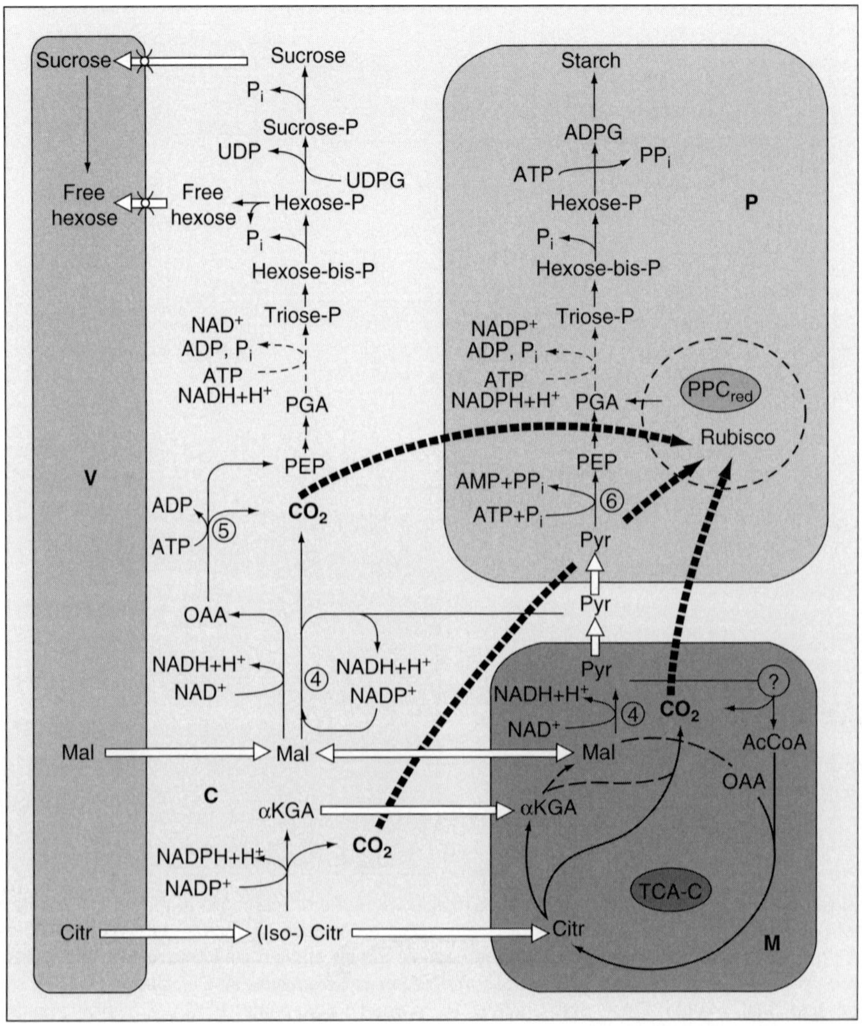

Figure 3 Daytime (light period) flow of carbon in CAM. Simplified scheme of metabolic pathway and compartmentalization. Key elements of the CAM pathway are: (4) NAD(P)-dependent malic enzymes; (5) PEP-carboxykinase (PEPCK); and (6) pyruvate, P_i-dikinase (PPDK). Citrate may be decarboxylated in the cytosol (isocitrate dehydrogenase) or in the mitochondria. Pyruvate is phosporylated by PPDK in the chloroplasts; its decarboxylation in the mitochrondria during the light period of CAM is not proven. In addition to those given in the legend to **Figure 1**, the following abbreviations are used: ADPG, adenosine-diphosphate-glucose; αKGA, α-ketoglutaric acid; PP_i, inorganic pyrophosphate; PPC_{red}, reductive pentose phosphate cycle (Calvin cycle); TCA-C, tricarboxylic acid cycle; UDP, uridine-diphosphate; UDPG, uridine-diphosphate-glucose.

in vascular plants, it is even present as basically as among the pteridophyta.

Molecular phylogenies of PEPC, a key enzyme in CAM, confirm the polyphyletic origin of CAM. There are different lineages of molecular modifications of PEPC in the evolution among families of different subclasses, i.e., Orchidaceae (in the Liliidae), Crassulaceae (in the Rosidae), and in families of the Caryophyllidae. However, two families of the latter subclass, i.e., Aizoaceae and Cactaceae, share the same lineage.

On the scale of plant families, in the Bromeliaceae, H. Griffiths and J.A.C. Smith present the case that CAM evolution is related to the evolution of epiphytism. Although CAM is a very effective adaptation to stresses associated with epiphytism (see below: **Table 2**), both CAM and epiphytism evolved independently several times in the three subfamilies of Bromeliaceae (**Figure 5**), the hypothetical evolution of which is based on comparative morphology of root and tank formation and solute-absorbing scales. In the genus *Kalanchoë* of the Crassulaceae, molecular PEPC evidence obtained by H. Gehrig and M. Kluge and the degree of CAM expression, with weak and strong CAM, respectively, as well as C_3/CAM intermediate behavior correlates with evolutionary advancement of subsections of the genus. Similar observations have been made for the *Sedum-Aeonium* group of the Crassulaceae.

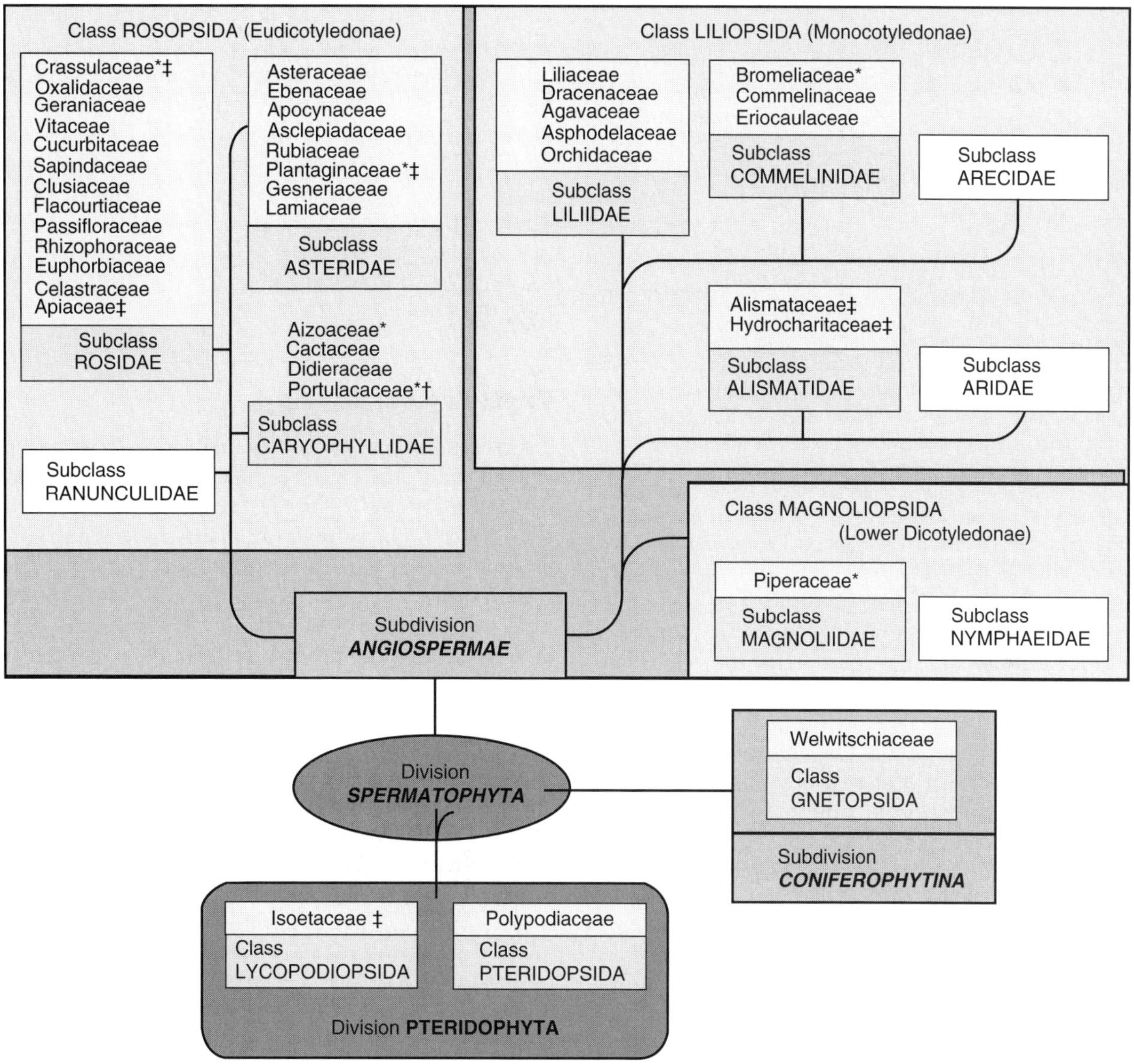

Figure 4 Phylogenetic distribution of plant families having CAM species among the major taxa of vascular plants: two families with CAM species in the division of Pteridophyta; among taxa of the division of Spermatophyta in the subdivision of Coniferophytina one possible (weak) CAM-species, i.e., *Welwitschia*, and in the subdivision of Angiospermae 10 families in three of the five subclasses of Liliopsida (Monocotyledonae), one family in one of the two subclasses of Magnoliopsida (lower Dicotyledonae), and many families in three of the four subclasses of Rosopsida (Eudicotyledonae). *, Known occurrence of C_3/CAM-intermediate species; †, C_4-CAM-intermediate species; and ‡, submerged aquatic CAM species.

Ontogeny

Many plant species in phyletically very different taxa are C_3/CAM intermediate, and there are also a few C_4/CAM intermediate species (**Figure 4**). Switches from C_3 photosynthesis to CAM frequently occur in ontogeny. In *Kalanchoë*, often the young developing leaves perform C_3 photosynthesis, i.e., even in the so-called obligate CAM species where the mature leaves only perform CAM. In *Kalanchoë blossfeldiana* cv. Tom Thumb, photoperiod determines the expression of C_3 photosynthesis (long days) and CAM (short days), respectively, a truly developmental cue in this short-day plant.

There are several other examples (see **Figure 4**) where developmental and environmental (see below) cues interact in determining C_3/CAM-intermediate behavior. However, the therophyte *Mesembryanthemum crystallinum* (Aizoaceae) emerged as a most vigorously studied model system. The switch from C_3 photosynthesis to CAM in this annual species, occurring at sites with Mediterranean-type climate with strong seasonality of wet and dry periods, in response to salinity, water stress, and osmotic stress, was first discovered by K. Winter and D. von Willert

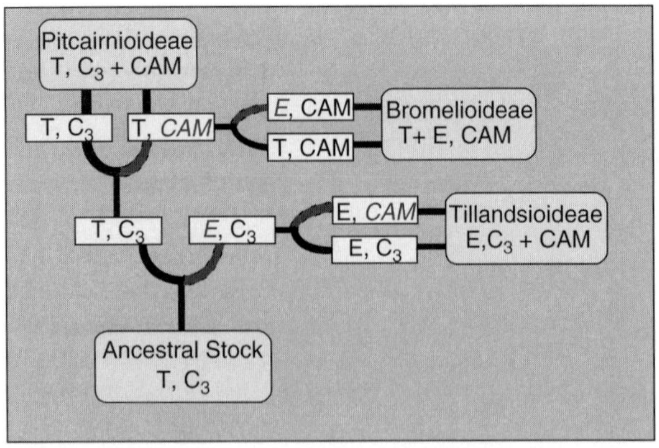

Figure 5 Evolution of CAM and epiphytism in the family Bromeliaceae. From an ancestral stock of terrestrial (T) C_3 plants, three subfamilies evolved, and CAM and epiphytism (E) evolved independently several times as indicated by light shading and italics. In some genera, there are both C_3 and CAM species, e.g., the terrestrial Pitcarnioideae *Puya* and the epiphytic Tillandsioideae *Tillandsia*. The epiphtic Tillandsioideae *Guzmania monostachia* is even C_3/CAM intermediate.

in the early 1970s; recently, the work of H. Bohnert and J. Cushman has shed light on the molecular basis of this switch. It is evident that C_3/CAM transition is inherent in the plant's developmental program, which is strongly amplified by environmental cues.

The amplification of activity and the induction of expression of enzymes during CAM induction allowed the identification of key enzymes involved in the carbon flow of CAM and the search for CAM-specific genes. For several of these enzymes (**Table 1**) there are multigene families with at least one CAM gene. Other enzymes involved in both housekeeping functions and CAM are encoded by single genes. Posttranslational regulation also occurs among these enzymes (**Table 1**).

The V-ATPase of the tonoplast driving nocturnal vacuolar acid accumulation is a complex multi-subunit enzyme having a V_1-domain with a head and a stalk and a membrane integral V_0-domain. In *M. crystallinum*, it covers 20–36% of the area and 30–35% of the total amount of protein of the tonoplast (**Figure 6**). The membrane integral subunit *c* forming the H^+-channel of the V-ATPase has a CAM-specific gene from a multi-gene family.

Plasticity

The C_3–CAM switch in the annual *M. crystallinum* is not strictly plastic, since it is underlain by a developmental program and stress responses are only reversible to the extent not yet given by the developmental stage (plant age). True plasticity with readily reversible C_3–CAM–C_3 shifts occurs in leaves of the perennial species of the mainly neotropical genus of woody plants, *Clusia* (Clusiaceae, Theales) (**Figure 7**).

Clusia has rapidly become an intensely studied model plant, along with *M. crystallinum*. However, in contrast to the latter species, studies of responses at the molecular level in the perennial C_3/CAM-intermediate species of *Clusia* are still in their infancy. Despite being hindered by its strong latex production, research on *Clusia* is beginning to advance rapidly, for example, by the PEPC studies of A. Borland.

Physiological Ecology

CAM is a mechanism for the acquisition of inorganic carbon (C_i) and also for elevating the internal concentration of C_i by daytime organic acid decarboxylation. In submerged plants, nocturnal CO_2 fixation avoids the competition for C_i with other photosynthesizing plants during the day. These plants only have CAM activity in submerged parts, and when amphibic they lose it as they emerge. In terrestrial plants, CAM supports water relations as stomata are mainly open in the dark period when evaporative demand is low and remain closed in the light period while internally regenerated CO_2 is assimilated. This much increased water use efficiency (WUE) of CAM in terrestrial species, where WUE = (CO_2 taken up) : (H_2O transpired), has led to the incorrect assumption that CAM species are mainly desert plants. While attractive succulent cacti, agaves, and euphorbias often characterize the physiognomy of dry habitats and deserts in the tropics and subtropics, the diversity of CAM species is much larger in other ecosystems, especially in rainforests with their epiphytic niches (**Table 2**).

The genus *Clusia*, which has about 300 species that all belong to one morphotype of woody plants with entire, somewhat succulent, leathery leaves, has an extraordinarily wide ecological amplitude in the neotropics, covering coastal sand dunes and rocks, savannas, gallery forests, inselbergs (rock outcrops), lowland rainforests, lower and higher montane rainforests, cloud forests, and elfin forests. This is supported by reversible C_3/CAM intermediate behavior, a plasticity that allows variable stress responses to a variety of factors ("stressors"). Ecophysiological CAM relations are given by the expression of CAM phases (**Figure 8**) and adaptations to various stressors (**Table 3**).

CAM Phases

CAM phases *sensu* C.B. Osmond are explained in **Figure 8**. The main phases are I and III corresponding

Table 1 Key enzymes of carbon flow in CAM; their genetic basis and regulation

Enzymes and their functions	*Multigene families with at least one CAM-specific gene*	*Single gene*	*Induction during CAM induction*	*Posttranslational regulation*	
				Phosphorylation/ dephosphorylation (P) thiol-reduction (SH)	*Feedback loops positive (+) negative (−)*
Nocturnal carboxylation:					
Phosphoenolpyruvate carboxylase (PEPC)	+		+	+(P)	Glucose-6-P (+); triose-P (+); malate (−)
PEPC-kinase			+		
Vacuolar malate accumulation:					
V-ATPase			+		
V-ATPase subunit *c*	+				
Malate-influx channel			+		
Daytime malate remobilization:					
Mitochondrial NAD-malic enzyme			+		
Cytosolic NADP-malic enzyme	+		+		
Phosphoenolpyruvate carboxy kinase (PEPCK)			+	+(P)	
Daytime oxidative stress:					
Superoxide dismutases			+		
Glycolysis/ Gluconeogenesis:					
NADP-glycerolaldehyde-phosphate dehydrogenase (GAPDH)		+		+(SH)	
Phosphoglyceromutase	+				
Enolase	+			+(P)	
Pyruvate-P_i-dikinase (PPDK)		+	+		
Malate-oxaloacetate conversions:					
NAD(P) malate dehydrogenase	+		+		

to the major processes determining the dark period and light period, respectively, of the CAM cycle (**Figure 1**). Phases II and IV are transition phases. Expression of CAM phases is influenced by environmental factors, such as availability of water, irradiance, and sun exposure (**Figure 8**). The manifestation of CAM phases may also relate to the performance of weak and strong CAM, respectively. Some plants show a C_3-photosynthesis-like day/night gas exchange with exclusively daytime CO_2 uptake and no night-time CO_2 uptake, which is nevertheless accompanied by nocturnal accumulation of organic acid due to internal recycling of respiratory CO_2. This is referred to as weak CAM or "CAM-cycling". Conversely, strong CAM is the expression of all four phases of CAM and the gradual suppression of the daytime phases of net CO_2 uptake (II and IV) as stress increases. Eventually, i.e., under severe water stress, phase I net CO_2 uptake is also gradually suppressed, while nocturnal acid accumulation may remain prominent due to recirculation of respiratory CO_2. In extreme cases, there is no net CO_2 exchange at all during day and night, and nocturnal organic acid accumulation is exclusively due to CO_2 recycling. This is called "CAM-idling." Thus, succulent CAM plants can overcome extended dry periods without gaining new carbon, but they benefit from losing only a little water via cuticular transpiration. They survive unless they lose more than 50% of the water reserves from their water-storing tissues. Thus, the expression of CAM phases with variation of daytime CO_2 uptake (phases II and IV) and nocturnal CO_2 recycling from 0 to 100% of malic acid accumulated (phase I)

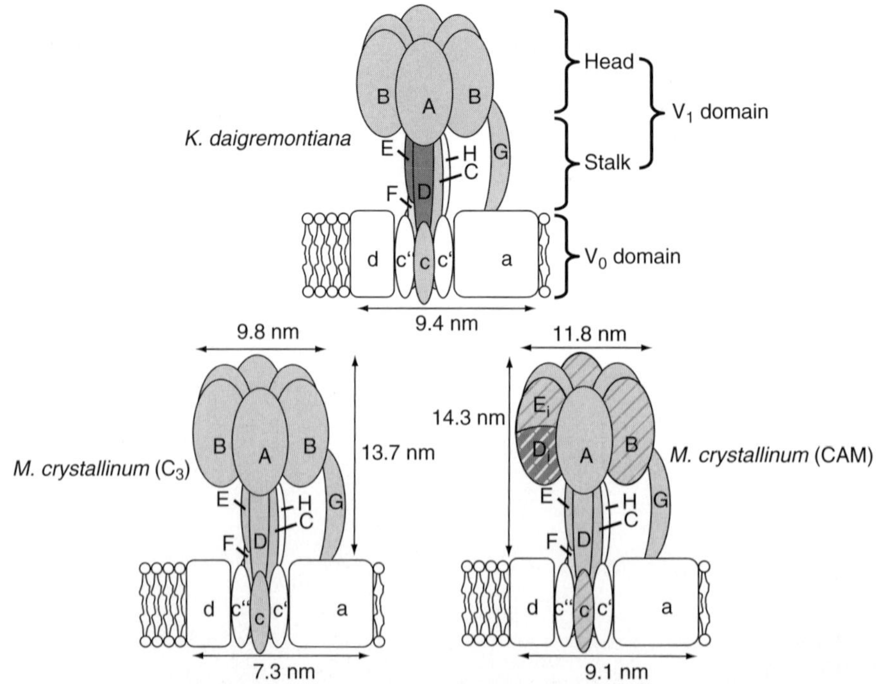

Figure 6 V-ATPase of the tonoplast of CAM plants, i.e., the obligate CAM species *K. daigremontiana* and the C_3/CAM intermediate species *M. crystallinum* as studied by U. Lüttge, R. Ratajczak, and associates. Subunits of the V_1 domain and the V_0 domain are marked by upper case and lower case letters, respectively. Occurrence of the subunits in the V-ATPases of the CAM plants is assumed from analogy to V-ATPases of other materials (no shading) or given by immunological (pale gray) and amino acid sequencing evidence and matrix-assisted laser-desorption ionization (MALDI) mass-spectrometry (dark gray). *A* is the catalytic and *B* a regulatory subunit of the V-ATPase head, respectively, where each of the two subunits is present in three copies. *c*-Subunits constitute the H^+ channel in the membrane. Upon a switch from C_3 photosynthesis to CAM, the dimensions and the subunit composition of the V-ATPase are changed; and specific proteolysis generates the new E_i and D_i subunits from a *B* subunit. Reproduced with permission from Drobny M, Fischer-Schliebs E, Lüttge U, and Ratajczak R (2003) Coordination of V-ATPase and V-PPase at the vacuolar membrane of plant cells. *Progress in Botany* 64: 171–216.

allows versatile stress responses even in obligate constitutive CAM plants without the option of C_3–CAM switches.

Environmental Factors

Water Water relations of CAM plants are determined by stomatal opening and closure in the CAM phases (**Figure** 8) and WUE of C_i acquisition. Morphological and anatomical structures, such as tanks (phytotelmata), and capillary water absorption structures, such as trichomes (leaves of bromeliads) or the dead cells of a velamen radicum (CAM-performing green aerial roots of orchids), are also involved. Most CAM plants are succulent. Among the ferns and orchids, K. Winter showed that CAM expression and leaf succulence were closely correlated. Cellular water relations are given by the simple relation of $\psi = P - \pi$, where ψ is the water potential, *P* the turgor potential, and π the osmotic potential, which are subject to night/day variations in CAM. Nocturnal vacuolar organic acid accumulation, i.e., of osmotically active solutes, strongly increases π with a linear relation between the two. The osmotic uptake of water associated with this can make up 2–11% of the cell volume. The diurnal dynamics of water relation parameters as studied by U. Lüttge, P.S. Nobel, and associates are depicted in **Figure** 9.

Light CAM plants are strongly affected by irradiance as seen, for example, in *Bromelia humilis* (see **Figure** 8C). The CO_2-concentrating effect of organic acid decarboxylation in phase III saturates Rubisco with its substrate and reduces its reaction with oxygen, and hence also photorespiration. This ensures optimal use of photosynthetically active photon flux density (PPFD) for the photochemical work of CO_2 reduction. Besides preventing loss of C_i, this is one important consequence of "CAM-idling."

Some CAM plants also accumulate citric acid in addition to malic acid. This is intriguing because nocturnal citric acid accumulation is not associated with a gain of C_i. This follows from the metabolic pathway (see **Figure 2**, where one CO_2 is fixed via PEPC and one CO_2 is lost in oxidative pyruvate

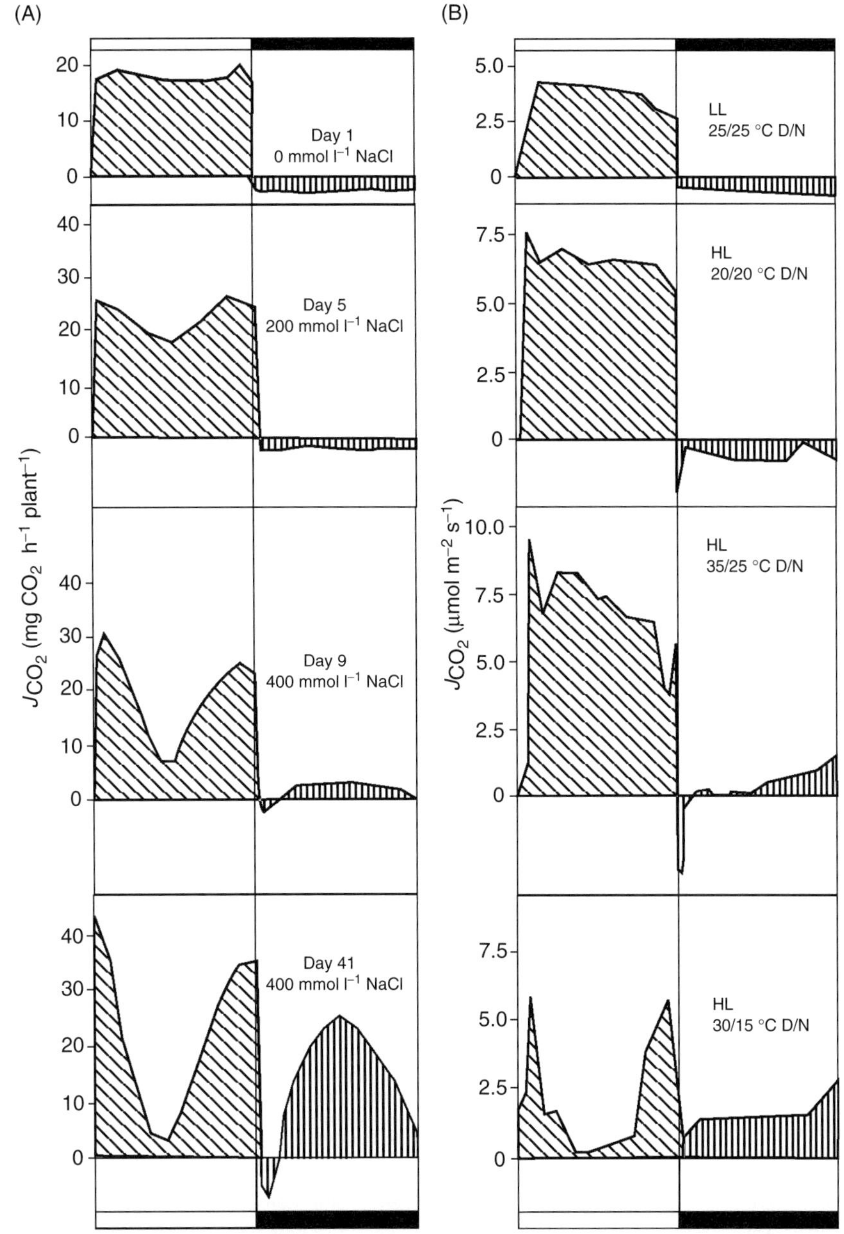

Figure 7 C_3/CAM-intermediate plants: (A) the annual species *M. crystallinum* and (B) the perennial species *C. minor*. In *M. crystallinum*, NaCl salinity elicits a shift from C_3 photosynthesis, with daytime net CO_2 uptake and nocturnal release of respiratory CO_2, to CAM with the four CAM phases (see **Figure 8**), i.e., night-time net CO_2 uptake (phase I), net CO_2 uptake in the early and late light period (phases II and IV, respectively), and stomatal closure during most of the day (phase III). In *C. minor*, similar gas exchange patterns are obtained under the influence of different combinations of external parameters such as light intensity (LL, low light intensity: 1.3–2.2 mol photons $m^{-2} day^{-1}$; HL, high light intensity: 11.2–13.0 mol photons $m^{-2} day^{-1}$) and day/night (D/N) temperature differences. Dark bars on the abscissa indicate dark periods and open bars indicate light periods.

decarboxylation as citrate is generated from malate/oxaloacetate and acetyl-CoA) or simply from the stoichiometry of six carbon atoms in both the hexose precursor and the product citrate.

The maximum levels of organic acid accumulated vary among CAM plants: in *M. crystallinum* this is 60–80 $mmol\,l^{-1}$; in *Kalanchoë* species up to 220 $mmol\,l^{-1}$; and in *Clusia* species, it may be as much as 350 $mmol\,l^{-1}$ malic acid plus 200 $mmol\,l^{-1}$ citric acid or a record of 1.3 $mol\,l^{-1}$ titratable protons. These variations appear to be due to vacuolar buffer capacity, since the vacuolar storage capacity for the organic acids seems to be mainly limited by the electrochemical proton gradient ($\Delta\bar{\mu}_{H^+}$) at the tonoplast against which the V-ATPase has to pump protons. A predominant term in $\Delta\bar{\mu}_{H^+}$ is

Table 2 Relevance of CAM among various life forms and habitats of vascular plants

Life forms or habitats	*Plant family*	*No. of species*	*No. of CAM species*	*No. of epiphytic species*
Desert succulents	Cactaceae	1500	1500 (100%)	150
	Agavaceae	300	300 (100%)	
			1800	
Rainforest species	Orchidaceae	19 000	9500 (50%)	14 000
	Bromeliaceae	2500	1250 (50%)	1500
			10 750	
Vascular epiphytes (~10% of vascular species)	85 Families	23 500	13 400 (57%)	

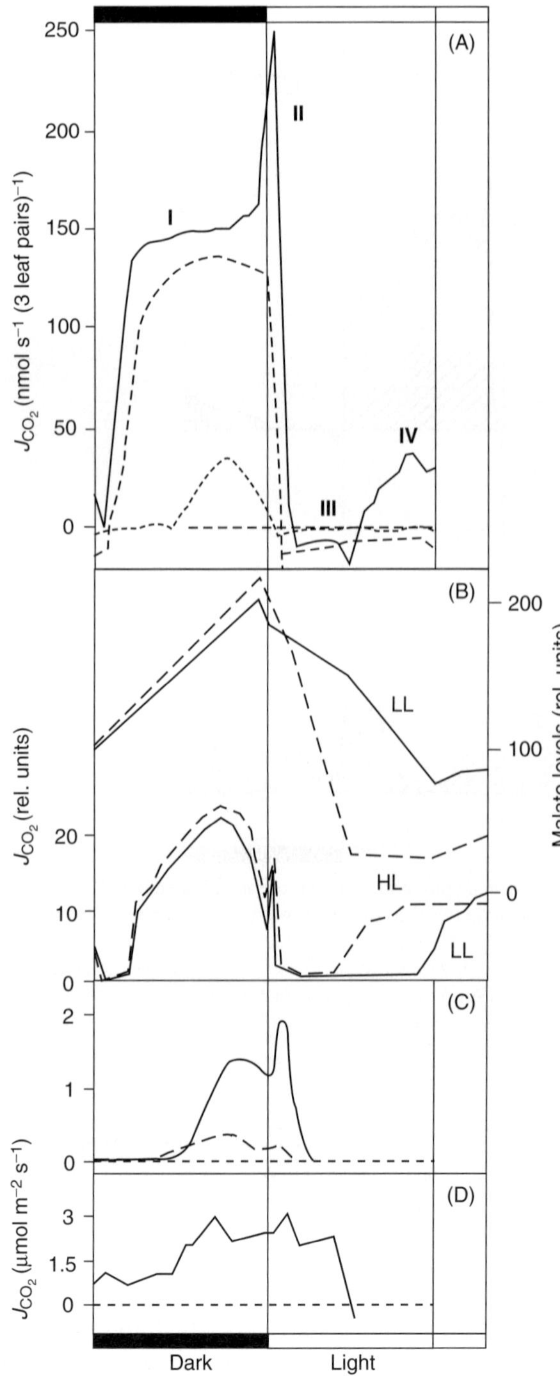

Figure 8 CAM phases and their modulation by environmental cues. (A) *Kalanchoë daigremontiana*. Well watered (——) with clear expression of the four CAM phases: I, nocturnal stomatal opening for CO_2 uptake and dark fixation of CO_2 via PEPC and vacuolar accumulation of malic acid; II, transition with a morning peak of net CO_2 uptake in the early light period, when both PEPC and Rubisco are active; III, daytime remobilization of malate, decarboxylation, refixation of the internally regained CO_2 via Rubisco, and assimilation in the Calvin cycle behind closed stomata; and IV, afternoon net CO_2 uptake via open stomata and assimilation in the Calvin cycle. With increased stress due to limited water supply first phases II and IV are suppressed and phase I CO_2 uptake is somewhat reduced (- - - - - -) and then the commencement of nocturnal CO_2 uptake is delayed in the dark period and increasingly strongly reduced (......). (B) *Kalanchoë tubiflora*. Shortening of phase III and earlier start of phase IV by high light intensity (HL, - - - - - -), which leads to more rapid consumption of nocturnally stored malate as compared to low light intensity (LL, ——) under well-watered conditions. Upper curves, malate levels; lower curves, J_{CO_2}. (C) *Bromelia humilis*. Shaded (——) and sun-exposed plants (- - - - - -) in the field, where nocturnal CO_2 uptake is much reduced in the exposed plants (7 mmol m^{-2} leaf area night^{-1}) as compared to shaded plants (22 mmol m^{-2} night^{-1}), but nocturnal vacuolar acid accumulation is similar (156 and 126 mol titratable H^+ per m^3 cell sap, respectively) so that internal CO_2 recycling was 87% and 56%, respectively. Note that 100% recycling would be "CAM idling." (D) *Clusia rosea* in the field with an extended phase II. Black and white bars on the abscissa indicate dark and light periods, respectively, i.e., 12 h each, except in (B) where the light period is extended. Reproduced with permission from Kluge M (1968) Untersuchungen über den Gaswechsel von Bryophyllum während der Lichtperiode. II. Beziehungen zwischen dem Malatgehalt des Blattgewebes und der CO_2-Aufnahme. *Planta* 80: 359–377.

the pH gradient between cytoplasm and vacuole (**Figure 10**). *Clusia* species are not only the strongest accumulators of titratable acidity, but also the strongest accumulators of citric acid among CAM plants. Citric acid is known as an effective buffer compound. Thus, citric acid cycling supports particularly high carbon cycling, where the extraordinarily high levels of organic acids (both malic and citric) available during the day for decarboxylation greatly support photochemical use of light excitation energy. This is also important in view of the relationship shown in **Figure 8B**, where high light intensity enhances organic acid breakdown and hence depletion of internal CO_2 sources, and the latter clearly is delayed when the original organic acid pool is larger.

However, CO_2 reduction at saturating substrate concentrations is, of course, also accompanied by photosynthetic O_2 evolution, which leads to much increased internal O_2 concentrations in the leaves

Table 3 Environmental factors (stressors) and adaptations of CAM plants

Environmental factor	*Adaptation*		*Specific role of CAM*
	Type	*Function*	
CO_2	Metabolic	Concentrating inorganic CO_2	Dark fixation of CO_2
H_2O	Morphological, anatomical	H_2O storage: tanks; hydrenchyma H_2O absorption: trichomes, velamen radicum	
	Physiological	Stomatal regulation, increasing water use efficiency	Dark fixation of CO_2, CO_2 recycling
	Biophysical	Osmotic activity	Nocturnal vacuolar acid accumulation
High light intensity	Metabolic	Photochemical activity with substrate saturated Rubisco, photorespiration; energy dissipation as heat (xanthophyll cycle); superoxide dismutases	Increased $p^i_{CO_2}$ and $p^i_{O_2}$ in phase III
Temperature (absolute)	Metabolic	Homoviscous adaptation of membranes	
Temperature (day/night regime)	Metabolic	C_3–CAM shifts	Temperature sensitivity of PEPC
Salinity	Physiological, morphological, metabolic	Avoidance by salt exclusion, decay of fine roots; C_3–CAM shift	Dark fixation of CO_2, CO_2 recycling
Nutrition (nitrogen)	Physiological, metabolic	Capacity of photosynthetic apparatus	

p^i, internal partial pressures of the gases.

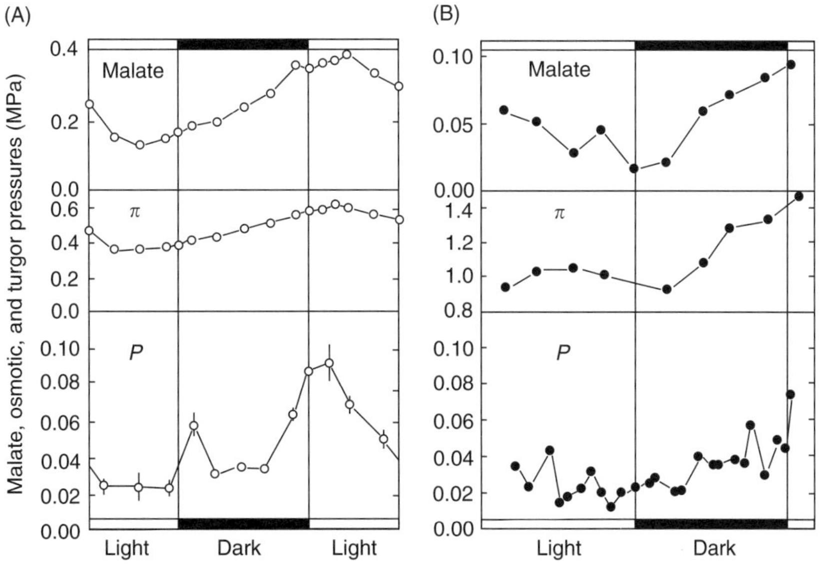

Figure 9 Correlations between day/night cycles of malate levels, cell sap osmotic pressure (π), and hydrostatic turgor pressure (P) measured by a cellular pressure probe in (A) the cactus *Cereus validus* (*C. forbesii* (syn.: *C. validus*)) and (B) the bromeliad *Tillandsia usneoides*. Malate levels were expressed as pressures (MPa) using the van't Hoff equation $\pi = c \cdot R \cdot T$, where c is concentration, R the gas constant, and T temperature in Kelvin. Dark bars on the abscissa indicate the dark period (10 h for *C. validus*, 12 h for *T. usneoides*). (A) reproduced with permission from Lüttge U, and Nobel PS (1984) Day night variations in malate concentration, osmotic pressure, and hydrostatic pressure in *Cereus validus*. *Plant Physiology* 75: 804–807. American Society of Plant Biologists. (B) reproduced with permission from Lüttge U (1987) Carbon dioxide and water demand: Crassulacean acid metabolism (CAM), a versatile ecological adaptation exemplifying the need for integration in ecophysiological work. *New Phytologist* 106: 593–629. Blackwell Publishing.

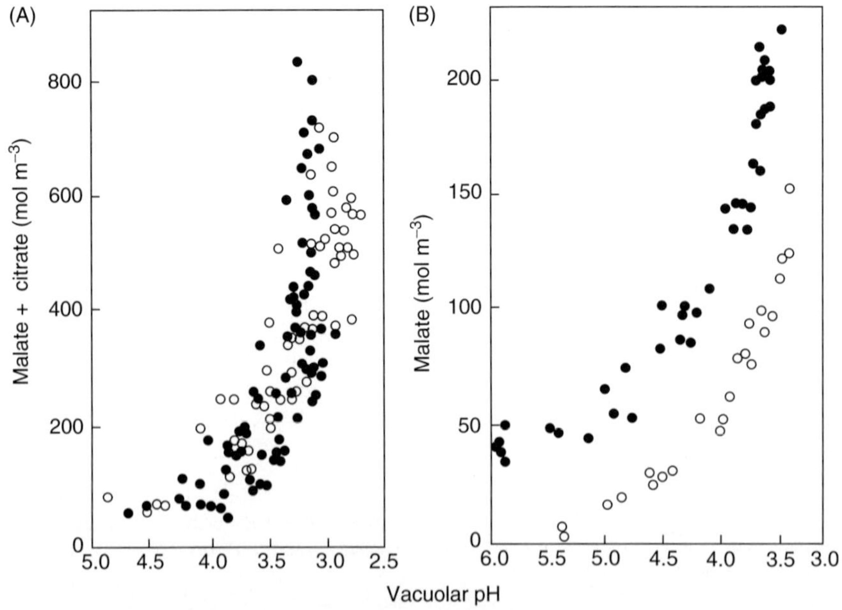

Figure 10 Levels of organic acids and pH in the vacuolar cell sap as they change in the day/night cycle of CAM. (A) *Clusia minor* (•) and *Clusia rosea* (○), which accumulate both malic acid (up to 350 mmol l^{-1}) and citric acid (up to 200 mmol l^{-1}) and where the accumulated titratable acid from the two carboxylic malic acid plus the tricarboxylic citric acid may amount to 1.3 mol l^{-1} H^+. (B) *Kalanchoë daigremontiana* (•) and *Kalanchoë tubiflora* (○), which accumulate mainly malic acid. In all cases, acid accumulation asymptotically approaches a vacuolar pH below which a $2H^+$/ATP pumping of protons by the V-ATPase would not be possible, when the electrochemical proton gradient at the tonoplast is $\Delta\tilde{\mu}_{H^+} = RT \cdot \ln(10^{-pH_{vac}}/10^{-pH_{cyt}}) + F\Delta E$, where R is the gas constant, T temperature in Kelvin, pH_{vac} vacuolar pH, pH_{cyt} cytoplasmic pH taken to be 7.5, F the Faraday, and ΔE transtonoplast electric potential difference taken to be a few tens of mV, i.e. $\Delta\tilde{\mu}_{H^+} \leq \Delta G'_0\text{ATP}/2$, where $\Delta G'_0$ ATP is the Gibb's free energy of ATP hydrolysis, i.e., about 55 kJ mol^{-1}. Reproduced with permission from Lüttge U, Smith JAC, Marigo G and, Osmond CB (1981) Energetics of malate accumulation in the vacuoles of Kalanchoë tubiflora cells. *FEBS Letters* 126: 81–84.

behind the closed stomata in phase III, as shown by Spalding and associates in the late 1970s. Thus, CAM plants in phase III are not exempt from oxidative stress. The enzymes of photorespiration are expressed in CAM plants and may not only be important in phase IV with its C_3-type photosynthesis. CAM plants are also subject to photoinhibition and this includes phase III. Elements of nonphotochemical dissipation of excitation energy are expressed, such as the xanthophyll cycle well studied by P. Horton and B. Demmig-Adams and associates, where energy is directly dissipated via zeaxanthin in the light-harvesting complexes or via the epoxidation/de-epoxidation cycle of zeaxanthin–antheraxanthin–violaxanthin. This also occurs in *Clusia*. Oxidative stress in CAM has been studied by A. Haag-Kerwer, M. Miszalski, R. Ratajczak, and associates. They found that elements of the antioxidative stress response system of CAM plants include superoxide dismutases (SODs) (**Table 1**), where CuZn SOD is increased on transcript and protein levels, as well as in activity, Mn SOD shows diurnal variations in activity, and Fe SOD is increased in activity during CAM induction.

Thus, in addition to internal CO_2 enrichment in phase III, CAM plants use the whole complement of protective mechanisms against overenergization by high light intensity. This makes them more versatile in their response to variable stress. A comparison between the C_3 species *Clusia multiflora* and the C_3/CAM-intermediate species *Clusia minor* illustrates this. When both species were grown at low light intensity (4 mol m^{-2} day^{-1}) and then suddenly transferred to high light levels (24.5 and 33.5 mol m^{-2} day^{-1}, respectively), leaves of the C_3 species showed a strong decline in photosynthesis after just 5 days at 33.5 mol m^{-2} day^{-1}. Although initially it could make use of the increased daily PPFD at 24 mol m^{-2} day^{-1} for photosynthesis at both daily light doses, leaves became necrotic and died. Conversely, the C_3/CAM-intermediate species switched to CAM and its leaves survived (**Figure 11**).

Temperature Day/night temperature differences have long been thought to be important for CAM due to a low temperature optimum of PEPC. In fact, in *C. minor*, large day/night temperature differences have a strong effect, inducing CAM. Obligate CAM plants, e.g., the Crassulaceae *Echeveria quitensis* (syn. *E. columbiana*) and various cacti, occur in the high Andes at elevations between 3000 and 4700 m a.s.l., where the typical day/night climatic changes of páramos in the tropics cause regular nocturnal frosts. In deserts in the southwestern USA and Mexico, CAM plants seasonally also may be subject to frost in the night. Not much is known of how Andean CAM plants cope with regular night frosts when strong metabolic activity is required for nocturnal synthesis of organic acid. Interestingly, the established obligate CAM plant *Kalanchoë tubiflora* has been shown by M. Kluge to switch to C_3 photosynthesis when the temperature is reduced to 7°C (**Figure 12**).

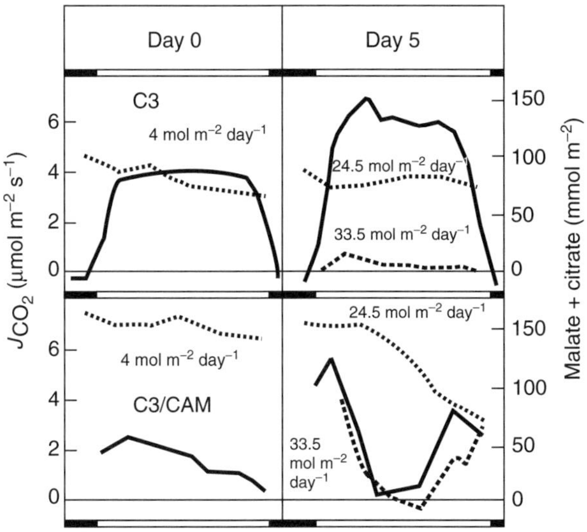

Figure 11 Light responses of the C_3 species *Clusia multiflora* and the C_3/CAM-intermediate species *Clusia minor*. Plants of *C. multiflora* and *C. minor* were grown at a daily photosynthetic photon flux density (PPFD) of 4 mol m^{-2} day^{-1} until day 0 when they were transferred to a PPFD of 24.5 and 33.5 mol m^{-2} day^{-1}, respectively. For day 0 all lines correspond to 4 mol m^{-2} day^{-1} (solid line, J_{CO_2}; dotted line, malate plus citrate). For day 5, solid line (J_{CO_2}) and dotted line (malate + citrate) refer to 24.5 mol m^{-2} day^{-1}, and dashed line (J_{CO_2}) refers to 33.5 mol m^{-2} day^{-1}. Dark bar and open bar (= 12 h) indicate dark and light period, respectively. Reproduced with permission from Herzog B, Hoffmann S, Hartung W, and Lüttge U (1999) Comparison of photosynthetic responses of the sympatric tropical C_3-species *Clusia multiflora* H.B.K. and the C_3-CAM intermediate species *Clusia minor* L. to irradiance and drought stress in a phytotron. *Plant Biology* 1: 460–470.

However, large day/night temperature differences are not a general prerequisite for CAM. In tropical rainforests with a high diversity of CAM plants (**Table 2**), day/night temperature differences are often only a few degrees at high temperatures of around 25–30°C. Moreover, as demonstrated in detail by M. Kluge and associates, *Kalanchoë* species are capable of "homoviscous adaptation" of the tonoplast membrane. Temperature during growth determines the lipids and proteins of the membrane and also affects the protein–lipid interactions. This leads to the formation of membranes of a higher state of order or lower fluidity and thus lower permeability at higher growth temperatures and lower order and higher permeability at lower growth temperatures (**Figure 13**). Since momentarily effective temperature also immediately influences membrane fluidity, this adaptation to the temperature experienced during growth is important because it ensures adjustment of permeability for the actually effective temperature and allows adequate balance of organic acid storage and remobilization in the CAM cycle.

Salinity As CAM is an effective adaptation to water stress, it may be surprising to note that there is little diversity of CAM species in relation to saline habitats. The mangrove tree *Rhizophora mangle* was suggested to have some CAM capacity under the influence of salt stress, as it may also be given in some of the Macronesian Sempervivioideae (Crassulaceae), e.g., *Aeonium sedifolium*. Conversely, many CAM plants of deserts are quite sensitive to salinity, e.g., *Agave deserti*. P.S. Nobel, U. Lüttge, and associates have found that desert cacti use an avoidance strategy. They are strong salt excluders at the root level (**Figure 14**), even sacrificing their

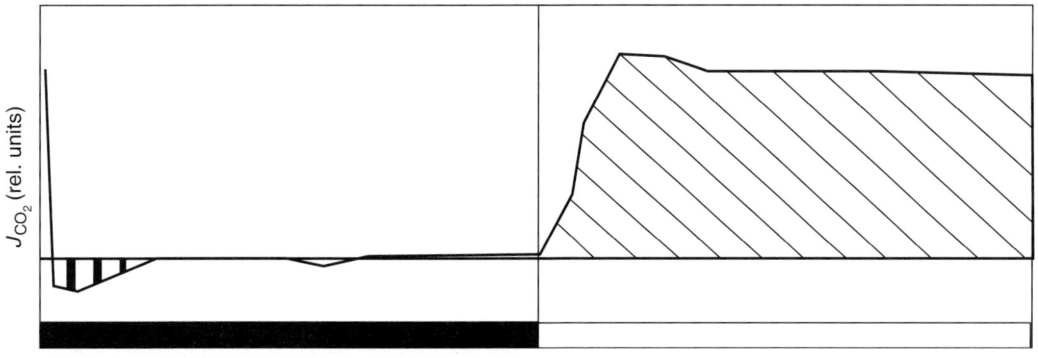

Figure 12 Net CO_2 exchange of the so-called obligate CAM plant *Kalanchoë tubiflora* at 7°C: the plant shows a C_3-type gas exchange pattern with net CO_2 uptake only in the light period. Dark bar and open bar indicate dark and light period, respectively, of 12 h each. Reproduced with permission from Kluge M (1969) Veränderliche Markierungsmuster bei 14CO2-Fütterung von Bryophyllum tubiflorum zu verschiedenen Zeitpunkten der Hell/Dunkelperiode. I. Die 14CO2-Fixierung unter Belichtung. *Planta* 88: 113–129.

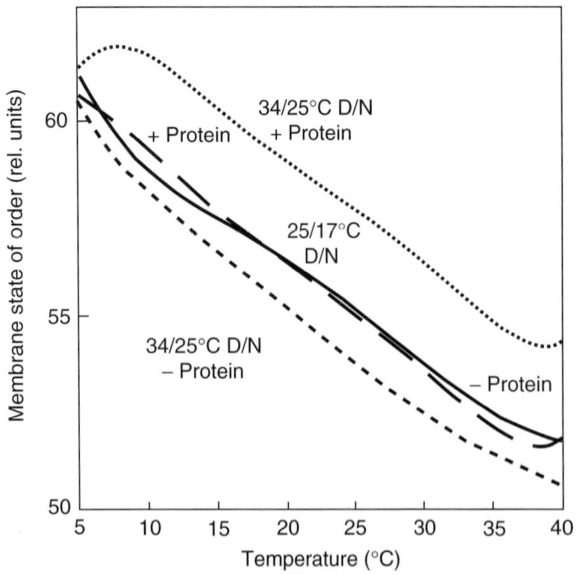

Figure 13 Homoviscous adaptation of the tonoplast membrane of *K. daigremontiana* as measured by M. Schomburg and M. Kluge. Growth temperatures of the plants were 34/25°C and 25/17°C day/night (D/N), respectively. The membrane state of order (ordinate) was measured at varied incubation temperature (abscissa) of native tonoplast vesicle preparations (+ protein) and of protein-deprived preparations (– protein). Reproduced with permission from Schomburg M, and Kluge M (1994) Phenotypic adaptation to elevated temperatures of tonoplast fluidity in the CAM plant *Kalanchoë daigremontiana* is caused by membrane proteins. *Botanica Acta* 107: 328–332. Nobel PS, Lüttge U, Heuer S, and Ball E (1984) Influence of applied NaCl on crassulacean acid metabolism and ionic levels in a cactus, *Cereus validus. Plant Physiology* 75: 799–803.

absorptive fine roots when under salinity stress during the dry seasons and rapidly regrowing new ones as water becomes available again.

The highly regulated salinity responses of the C_3/CAM intermediate plant *M. crystallinum* appear to be a very special case. There are at least three important overlapping processes: (1) the response to salinity and the metabolic requirements of coping with salinity; (2) the induction of CAM and its metabolic requirements; and (3) the developmental program of the annual plant. It is not always very clear to what extent induction of enzymes during CAM induction (**Table 1**) is really CAM-specific or whether it is a pure response to salinity. It seems that some of the changes in the V-ATPase found by U. Lüttge, R. Ratajczak, and associates (**Figure 6**) may be salinity effects. In *M. crystallinum*, the V-PPase is downregulated under salinity and as plants age, an effect that was not observed in other C_3/CAM intermediate species (e.g., *K. blossfeldiana* cv. Tom Thumb). It also appears that the upregulation of SODs as studied by R. Ratajczak, Z. Miszalski, and associates in *M. crystallinum* is usually more a response to salt stress than associated with CAM induction.

Endogenous Circadian Rhythmicity

The CAM cycle has several regulatory feedback loops (**Table 1**) and therefore is subject to nonlinear

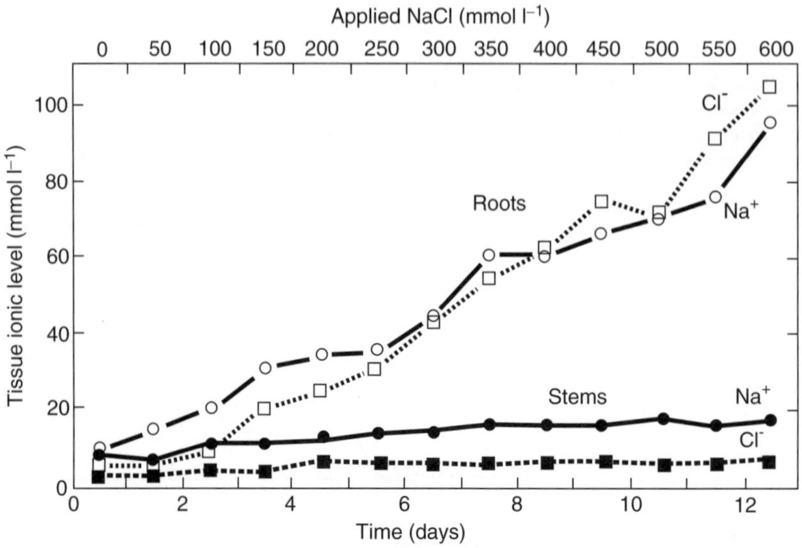

Figure 14 Salt exclusion by 10 cm tall seedlings of the cactus *Cereus validus* as measured by P.S. Nobel and U. Lüttge. As NaCl salinity in the root medium (upper abscissa) was increased over time (lower abscissa), Na^+ and Cl^- levels increased strongly in the root tissue but only very slightly in the stem tissue (green tissue as well as hydrenchyma). Reproduced with permission from Nobel PS, Lüttge U, Heuer S and Ball E (1984) Influence of applied NaCl on crassulacean acid metabolism and ionic levels in a cactus, *Cereus validus. Plant Physiology* 75: 799–803. American Society of Plant Biologists.

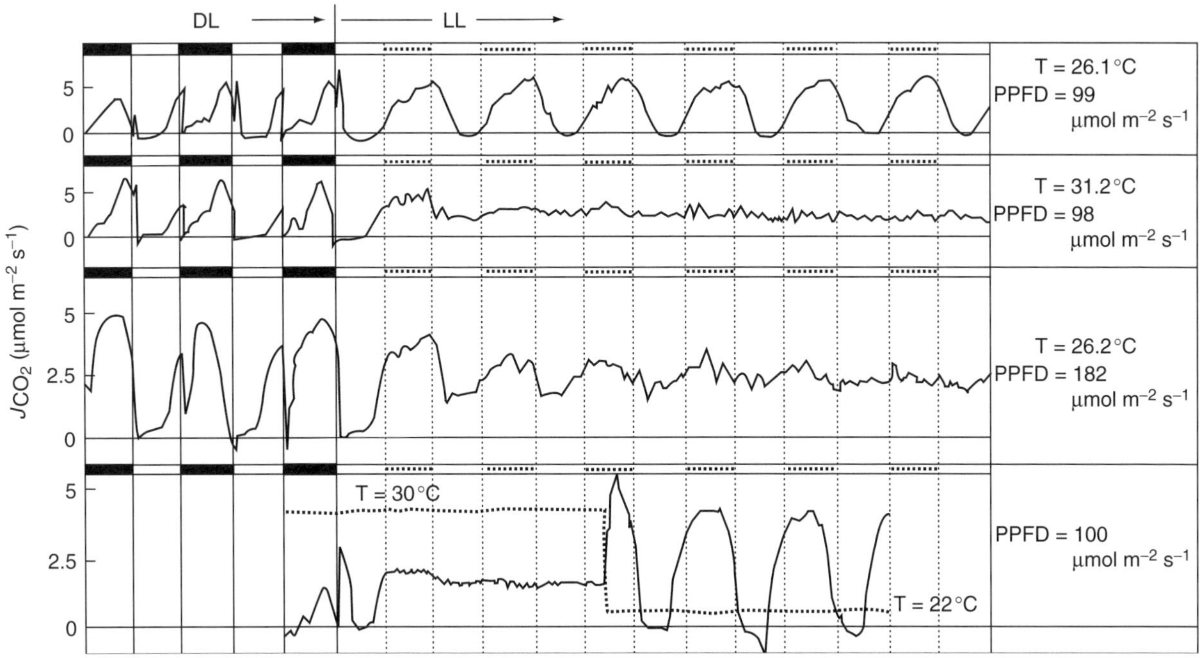

Figure 15 Rhythmic and arrhythmic behavior of gas exchange (net CO_2 exchange, J_{CO_2}) of *Kalanchoë daigremontiana* under the influence of the external control parameters photosynthetic photon flux density (PPFD) and temperature (T). DL: normal rhythm with 12-h dark and 12-h light periods; LL: continuous illumination with constant PPFD as indicated. Note that under the appropriate conditions there is long-lasting endogenous rhythmicity (upper time series); at too high a temperature or PPFD the time series becomes arrhythmic (center time series), which is reversible, e.g. when temperature is reduced from being too high into the appropriate temperature domain (lower time series).

dynamics. Such a situation frequently leads to oscillatory behavior. Indeed, CAM shows endogenous free-running rhythmicity under constant environmental conditions, as described earlier by M.B. Wilkins and is currently being studied as an interdisciplinary approach of biological experimentation, theoretical physics, modelling, and computer simulations by F. Beck, U. Lüttge, and associates. Rhythmicity of net CO_2 exchange (J_{CO_2}) under constant conditions in air and in continuous light occurs in long-lasting oscillations (**Figure 15**). At critical upper (**Figure 15**) and lower (not shown) threshold levels of PPFD and temperature this rhythmicity reversibly changes to an arrhythmic behavior, which is not stochastic but shows a distinct structure with the harmonic overtones and undertones of the basic circadian frequency as revealed in power spectra obtained by fast Fourrier transform analysis of time series of J_{CO_2}. Mathematical minimal models of CAM have been developed with a dramatic reduction of the complex reaction network of the flow of carbon in CAM (**Figures 2** and **3**) to only a very small number of metabolic pools connected by metabolite flows and feedback loops described by coupled nonlinear differential equations and influenced by external control parameters (**Figure 16**). These models simulate experimental time series of endogenous CAM oscillations astonishingly well. They are heuristic in that simulation can check whether a building block of the model really is essential or not.

There are two major points in CAM that may be basic hysteresis switch points in the oscillatory behavior, namely PEPC regulation by phosphorylation/dephosphorylation of the enzyme and vacuolar influx/efflux of malic acid. PEPC phosphorylation displays endogenous rhythmicity because transcription of the phosphorylating enzyme PEPC-kinase is under circadian control. However, both experiments in the laboratories of H. Nimmo and A. Borland and computer simulations by B. Blasius, F. Beck, U. Lüttge, and colleagues show that this is an epiphenomenon downstream of the basic hysteresis switch governing the endogenous oscillations. PEPC-kinase expression is governed by the malic acid status of the cytoplasm, i.e., by malate compartmentalization between cytoplasm and vacuole. The master switch must reside in the tonoplast determining phases of net malic acid influx and efflux, respectively. A theoretical thermodynamic membrane model in the phase space of membrane order, space available for individual lipid molecules, and temperature suggests that the tonoplast, in fact, can operate as such a hysteresis switch or beat oscillator

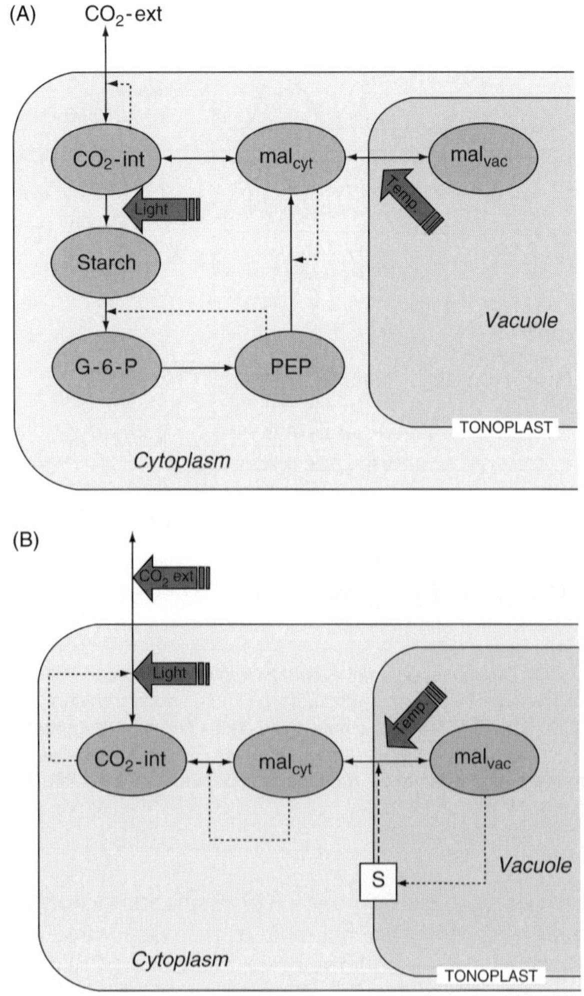

Figure 16 Heuristic minimal models of CAM. (A) The model retains six pools of metabolites, i.e., internal CO_2 (CO_2-int), starch, glucose-6-phosphate (G-6-P), phosphoenolpyruvate (PEP), and cytoplasmic and vacuolar malate, mal_{cyt} and mal_{vac}, respectively. Feedback loops (dashed lines) are the known feedback inhibitions of glycolysis by PEP and of PEP-carboxylase by malate (see **Table 1**), as well as effects of CO_2-int on CO_2 uptake from the outside (CO_2-ext). External control parameters are temperature and light. (B) The model has the number of metabolite pools still more reduced, i.e., to three, namely internal CO_2, cytoplasmic malate and vacuolar malate. The feedback loops of CO_2-int on CO_2 uptake and mal_{cyt} on PEP-carboxylase are retained; an additional important feedback loop is now vacuolar malate acting on vacuolar net influx and efflux of malate via the state of order (S) of the tonoplast membrane. External control parameters are temperature, light, and CO_2. Pools are connected by fluxes (solid lines). This is formulated by appropriate sets of nonlinear coupled differential equations allowing computer implementation and simulations. The simulations with both models reproduce experimentally observed behavior of gas exchange in the rhythmic and arrhythmic domains of CAM astonishingly well, which means that the absolutely required elements of CAM were extracted appropriately from the complex system of carbon flow (compare with **Figures 2** and **3**), which also shows the heuristic value of minimal models. Reproduced with permission from Lüttge U (2000) The tonoplast functioning as the master switch for circadian regulation of crassulacean acid metabolism. *Planta* 211: 761–769.

(**Figure 17A**). Thermodynamic feasibility is ensured by comparison with the actual changes of tonoplast surface resulting from osmotic adjustments in the natural day/night cycle of CAM (**Figure 9**). Moreover, experiments show that at high turgor

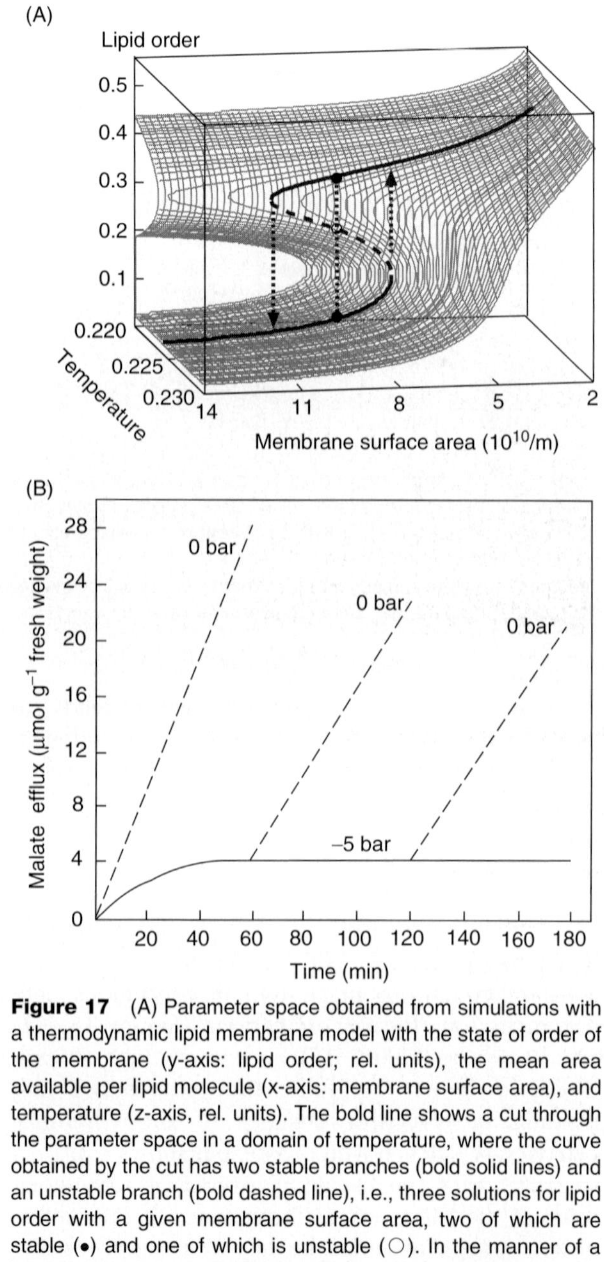

Figure 17 (A) Parameter space obtained from simulations with a thermodynamic lipid membrane model with the state of order of the membrane (y-axis: lipid order; rel. units), the mean area available per lipid molecule (x-axis: membrane surface area), and temperature (z-axis, rel. units). The bold line shows a cut through the parameter space in a domain of temperature, where the curve obtained by the cut has two stable branches (bold solid lines) and an unstable branch (bold dashed line), i.e., three solutions for lipid order with a given membrane surface area, two of which are stable (•) and one of which is unstable (○). In the manner of a hysteresis switch, the state of order can jump between the two stable branches (thin dashed lines with arrows). (B) Malate efflux from leaf slices of *Kalanchoë daigremontiana* submersed in solutions of 0 bar (high turgor) and −5 bar mannitol (low turgor). Reproduced with permission from Neff R, Blasius B, Beck F, and Lüttge U (1998) Thermodynamics and energetics of the tonoplast membrane operating as a hysteresis switch in an oscillatory model of crassulacean acid metabolism. *Journal of Membrane Biology* 165: 37–43. Lüttge U, Ball E, and Greenway H (1977) Effects of water and turgor potential on malate efflux from leaf slices of *Kalanchoë daigremontiana*. *Plant Physiology* 60: 521–523.

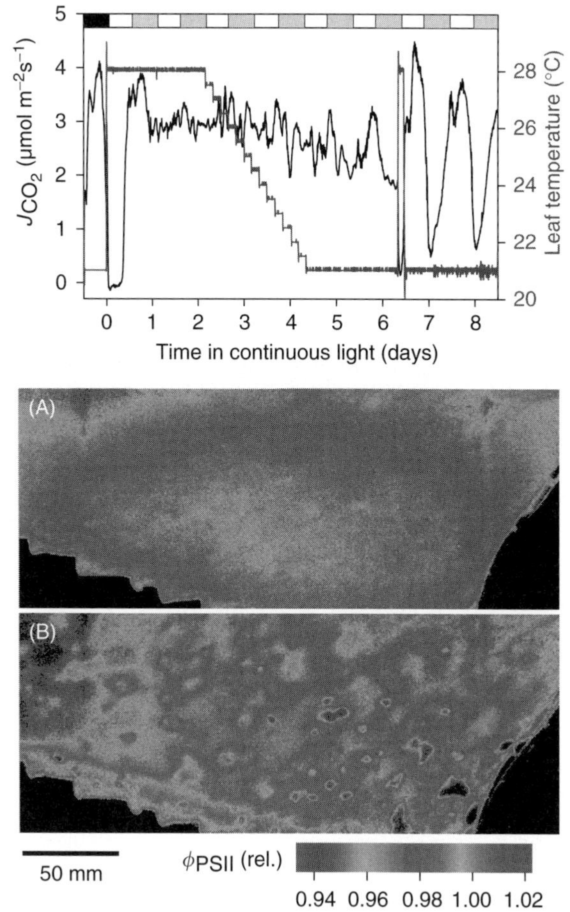

Figure 18 (see color plate 20) Synchronization and desynchronization of leaf patches of *Kalanchoë daigremontiana* functioning as separate oscillators in endogenous rhythmicity. The graph shows a normal dark/light cycle of net CO_2 exchange (J_{CO_2}, black line) with dark period (black bar) and light period (first white bar on the upper abscissa) followed by continuous light (with the gray bars on the upper abscissa indicating "subjective" or previous dark periods) in a temperature domain not allowing rhythmicity. When temperature (red line) is reduced gradually into a temperature domain normally allowing rhythmic oscillations, no rhythmicity of J_{CO_2} appears, but after a strong temperature signal of abrupt increase and decrease of temperature (at ca. 6.5 h) oscillations are regenerated immediately. The pictures of the leaf of *K. daigremontiana* show relative quantum efficiency of photosystem II (ϕ_{PSII} (rel.)) in false colors as indicated, with (A) homogeneity (synchronization) at the maximum of J_{CO_2} and (B) inhomogeneity (patchiness or desynchronization) during the transition from maximum to minimum J_{CO_2} in endogenous oscillations at a constant photosynthetic photon flux density (PPFD) of 190 μmol m^{-2}s^{-1} and temperature of 21°C. Reproduced with permission from Rascher U, Blasius B, Beck F, and Lüttge U (1998) Temperature profiles for the expression of endogenous rhythmicity and arrythmicity of CO_2 exchange in the CAM plant *Kalanchoë daigremontiana* can be shifted by slow temperature changes. *Planta* 207: 76–82.

Table 4 Crops and economic use of CAM plants

Plant family	*Plant species*	*Product*
Bromeliaceae	*Ananas comosus*	Fruit
Cactaceae	*Opuntia* spp. *Opuntia ficus indica* *Nopalea* spp.	
Orchidaceae	*Vanilla planifolia*	Spice from unripe seed coats
Cactaceae	*Opuntia* spp. *Nopalea* spp.	Fodder (spineless spp. and cultivars)
Euphorbiaceae	*Euphorbia antisphylitica*	Wax
Agavaceae	*Agave* spp. *Yucca* spp.	Fiber
Agavaceae	*Agave* spp. *Yucca* spp. *Agave tequilana*	Alcohol Tequila
Aizoaceae	*Mesembryanthemum crystallinum* *Carpobrotus edulis*	Salad, vegetable (occasionally)
Cactaceae	*Opuntia ficus indica*	Red pigment "cochineal" from shield louse *Dactylopius coccus* feeding on the cactus (cosmetics)
Many families, but mainly the Bromeliaceae, Cactaceae, and Orchidaceae		Ornamentals
Aizoaceae		Ground cover
Clusiaceae	*Clusia* spp.	Aforestation

pressure with more space available for membrane molecules (lipids), malic acid efflux is observed, which can be stalled reducing turgor (**Figure 17B**).

The results of modeling also suggested that synchronization/desynchronization of individual leaf patches is involved in endogenous CAM rhythmicity. Experiments by U. Rascher showed that a strong temperature signal was needed to regenerate rhythmicity out of the arrhythmic domain above the upper critical temperature threshold (**Figure 15**). Slow temperature reduction had no effect. Fast temperature reduction or a strong temperature signal was needed to regenerate rhythmicity by, most possibly, synchronizing individually oscillating patches. Such dynamical patchiness was demonstrated by U. Rascher using the noninvasive photographic imaging of chlorophyll fluorescence. This reflects photochemical activity, i.e., quantum efficiency of photosystem II, determined by internal CO_2 concentration as dependent on CO_2 fixation/regeneration or vacuolar organic acid accumulation/remobilization (**Figure 18**). Thus, CAM rhythmicity is recognized as a phenomenon realized not only in time but also in space. The experimental CAM plant *Kalanchoë daigremontiana* has developed into an important model system for the study of the spatiotemporal function of the biological clock in the circadian oscillations of a plant's metabolic cycle.

Table 5 Comparisons of productivity parameters of CAM, C_3 and C_4 plants: water use efficiency and net CO_2 uptake (for CAM separated for dark period (D), phase I, and light period (L), phases II and IV), daily productivity, and maximum annual productivity of most productive crops where CAM plants were grown under optimal spacing and watering

	CAM	C_3	C_4
Water use efficiency (WUE)	D: $(6–30) \times 10^{-3}$		
(mol CO_2 : mol H_2O)	L: $(1–4) \times 10^{-3}$	$(0.6–1.3) \times 10^{-3}$	$(1.7–2.4) \times 10^{-3}$
Net CO_2 uptake (μmol m^{-2} s^{-1})	D: 0.5–2.5		
	L: 7–8	10–25	25–50
Daily productivity (g dry weight m^{-2} day^{-1})	1.5–1.8	50–200	400–500
Crop productivity (Mg ha^{-1} $year^{-1}$)	40–50	35–45	50–90

Crops, Productivity and Global Change

Crops and other economic usages of CAM plants are listed in **Table 4**. Naturally, CAM is an adaptation for survival under various regimes of stress and not for high productivity. Daily productivity of CAM plants generally is much lower than that of C_3 and C_4 plants (**Table 5**). However, when studying the annual crop productivity of CAM plants, P.S. Nobel found that under optimal conditions of spacing and watering they may perform just as well as C_3 crops and reach the lower end of the range of C_4 crops (**Table 5**). This may be due to a significant contribution by the C_3-type CO_2 fixation of phase IV of CAM.

P.S. Nobel and his coworkers also studied the response of CAM plants to elevated atmospheric CO_2, $p^{a}_{CO_2}$. They found that on doubling $p^{a}_{CO_2}$, productivity could increase by 35%, on average, mostly as a result of increases in both night-time and daytime CO_2 uptake. These responses are similar to those predicted for C_3 plants, but much greater than those for C_4 plants. There appears to be no expectation of downward acclimation. Therefore, cultivation of CAM crops may be an interesting option in possible future global increases in $p^{a}_{CO_2}$.

List of Technical Nomenclature

Calvin cycle Flow of carbon for CO_2 assimilation after fixation to the acceptor ribulose-*bis*-phosphate and with regeneration of the acceptor; discovered by Melvin Calvin.

Chlorophyll fluorescence Energy dissipation by emission of light (fluorescence) after excitation (see "excitation") of chlorophyll in photosystem II.

C_3 Photosynthesis Direct CO_2 fixation via ribulose-*bis*-phosphate carboxylase/oxygenase; the first stable product is the C_3 acid 3′-phosphoglyceric acid.

C_4 Photosynthesis Direct CO_2 fixation via phosphoenolpyruvate carboxylase in mesophyll cells; the first stable product is the C_4 acid malic acid.

Electrochemical gradient Gradient – mostly across a membrane – determined by two components, i.e., distribution of electrical charge and chemical ion concentration.

Epiphytism Growth of plants (epiphytes) on other plants (phorophytes).

Evaporative demand Water vapor pressure deficit of the atmosphere driving water loss (evaporation) of wet surfaces including plant tissues.

Excitation (of chlorophyll and photosystems) Absorption of photosynthetically active photons leading to an energized state.

Fourier transformation Mathematical algorithm extracting frequencies of oscillations underlying complex time series (power spectrum).

Gluconeogenesis Pathway regenerating hexose from metabolic precursors such as pyruvate or phosphoenolpyruvate.

Glycolysis Pathway generating phosphoenolpyruvate and pyruvate from sugars (hexose).

Homoviscous adaptation (of membranes) Adaptation of membrane components (lipids, proteins) during growth for fluidity/rigidity homeostasis at varied temperatures.

House-keeping enzyme An enzyme required for basic functions.

Hydrenchyma The tissue of plants (often succulents) with large cell vacuoles serving water storage.

Ion channels Integral membrane proteins, usually highly regulated (gated, rectified), to allow passage of ions.

Isoenzymes The enzymes encoded by different genes, catalyzing the same reaction type but varying slightly in amino acid sequence and biophysical properties and often in compartmentalization within cells and whole organisms.

Morphotype Morphologically characterized type of – possibly – different species.

Ontogeny History of the development of an individual organism.

Photoinhibition Reduction of photosynthetic efficiency by protective mechanisms of energy dissipation (acute photoinhibition) or photodamage (chronic photoinhibition).

Photoperiod Length of the daily light period.

Photorespiration O_2 fixation by ribulose-*bis*-phosphate carboxylase/oxygenase, formation of glycolate, and subsequent flow of carbon (glycolate pathway).

Phylogeny The history of the evolution of an organism type.

Plasticity Versatile morphological and physiological responses (expression of phenotypes) in response to environmental cues.

Short day plant A plant expressing a certain trait (mostly flowering) only if a specific minimal photoperiod (see "photoperiod") is not surpassed.

Stomata Small openings in the epidermis of aerial parts of plants between and regulated by two cells, i.e., the stomatal guard cells.

Stressor External factor causing stress.

Tank A cistern or reservoir for water formed by tightly appressed leaf bases of rosette plants (mostly in the Bromeliaceae).

Therophyte Annual plant completing its entire life cycle within one growing season.

Vacuole The large central compartment of plant cells with aqueous solution surrounded by the membrane of the tonoplast.

See also: **Abiotic Stresses**: Cold Stress. **Energy Crops**: Biomass Production. **Genetic Modification, Applications**: Oxidative Stress; **Genetic Modification of Primary Metabolism**: Carbohydrates; Photosynthesis. **Growth and Development**: Leaf Development; Molecular Biology of Development. **Photosynthesis and Partitioning**: C_3 Plants; Photoinhibition. **Plants and the Environment**: Global Warming Effects. **Regulators of Growth**: Circadian Rhythms. **Water Relations of Plants**: Drought Stress; Salt Stress; Stomata.

Further Reading

Adams PA, Nelson DE, Yamada S, *et al.* (1998) Growth and development of *Mesembryanthemum crystallinum* (Aizoaceae). *New Phytologist* 138: 171–190.

Cushman JC and Bohnert HJ (1999) Crassulacean acid metabolism: molecular genetics. *Annual Review of Plant Physiology and Plant Molecular Biology* 50: 305–332.

Kluge M and Ting IP (1978) *Crassulacean Acid Metabolism: Analysis of an Ecological Adaptation. Ecological Studies*, vol. 30. Berlin, Heidelberg, New York: Springer-Verlag.

Lüttge U (1987) Carbon dioxide and water demand: Crassulacean acid metabolism (CAM), a versatile ecological adaptation exemplifying the need for integration in ecophysiological work. *New Phytologist* 106: 593–629.

Lüttge U (ed.) (1989) *Vascular Plants as Epiphytes: Evolution and Ecophysiology. Ecological Studies*, vol. 76. Berlin, Heidelberg, New York: Springer-Verlag.

Lüttge U (1993) The role of crassulacean acid metabolism (CAM) in the adaptation of plants to salinity. *New Phytologist* 125: 59–71.

Lüttge U (2000) The tonoplast functioning as master switch for circadian regulation of crassulacean acid metabolism. *Planta* 211: 761–769.

Osmond CB (1978) Crassulacean acid metabolism: A curiosity in context. *Annual Reviews of Plant Physiology* 29: 379–414.

Ting IP (1985) Crassulacean acid metabolism. *Annual Reviews of Plant Physiology* 36: 595–622.

Winter K and Smith JAC (1996) *Crassulacean Acid Metabolism: Biochemistry, Ecophysiology and Evolution. Ecological Studies*, vol. 114. Berlin, Heidelberg, New York: Springer-Verlag.

Photorespiration

B Thomas, Horticulture Research International, Wellesbourne, UK

Photosynthesis and Photorespiration

The primary biochemical reaction in higher plant photosynthesis is the light-dependent fixation of carbon dioxide by the enzyme ribulose bis-phosphate carboxylase (Rubisco). However, when leaves of C_3 plants are illuminated they also show an increase in the apparent rate of respiration and evolve carbon dioxide in an oxygen-dependent process. This light-stimulated form of respiration takes place in the absence of any increase in true (mitochondrial) respiration and is called "photorespiration."

The carbon dioxide released during photorespiration derives from the ability of Rubisco to function as an oxygenase, using oxygen as an alternative substrate to carbon dioxide. The product of the reaction between oxygen and ribulose bis-phosphate (RuBP) is glycollate 2-phosphate (**Figure 1**). Photorespiration results in a decrease in the efficiency of net carbon assimilation and is an energy-inefficient process.

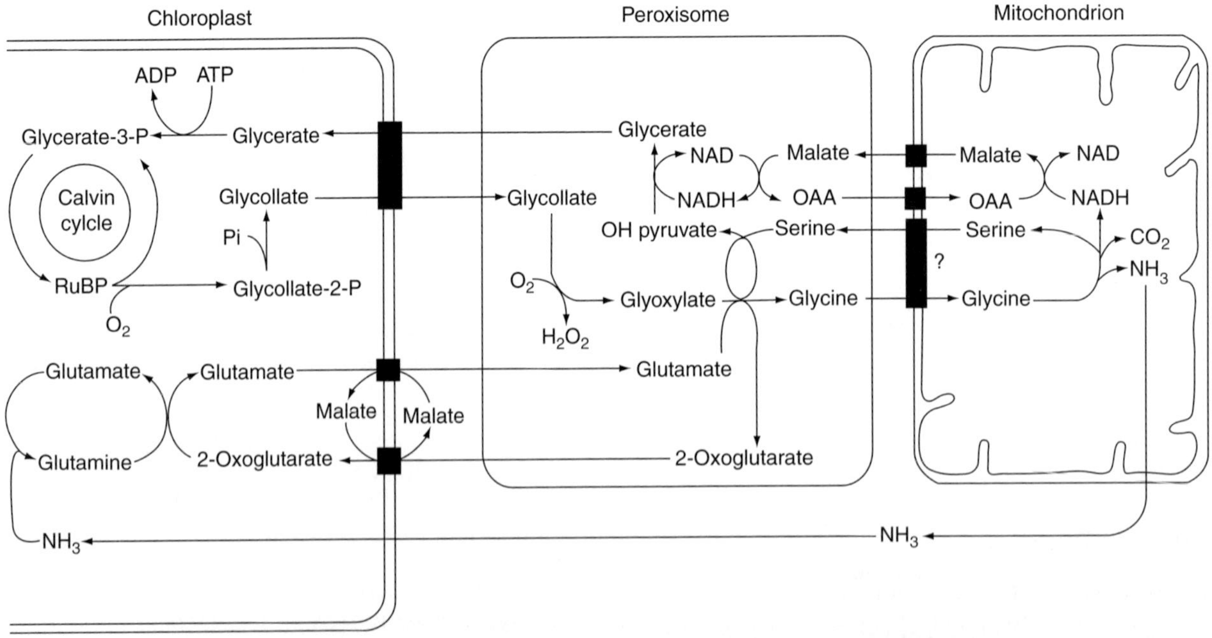

Figure 1 The photorespiratory pathway showing the salvage of carbon as glycerate coupled with the production of CO_2 and NH_3 and the localization of reactions within the chloroplast, peroxisome, and mitochondrion. Reproduced with permission from Leegood RC, Lea PJ, Adcock MD, and Häusler RE (1995) The regulation and control of photorespiration. *Journal of Experimental Botany (Special Edition)* 46, 1397–1414.

Loss of Carbon

The product of oxygenase action, glycollate 2-phosphate, cannot be used in the carbon assimilatory cycle (the Calvin cycle) or for other reactions within the chloroplast. Losses of carbon to the plant by this reaction are minimized through the involvement of a salvage pathway, which involves three plant organelles, the chloroplast, the peroxisome, and the mitochondrion (**Figure 1**). In a complex series of reactions, carbon skeletons are recovered as glycerate-3-phosphate and fed directly into the Calvin cycle. The photorespiratory pathway is not fully effective in carbon recovery and one-quarter of the carbon is lost as carbon dioxide. There is also a net release of NH_3, equal to the photorespiratory loss of carbon dioxide, although the leaf has mechanisms for the efficient reassimilation of NH_3 through the glutamine synthetase/glutamate oxyglutarate aminotransferase (GS/GOGAT) system.

The kinetics of oxygen and carbon dioxide use by Rubisco vary differentially with temperature so that at higher temperatures, oxygen competes with increasing effectiveness for the Rubisco binding site. This is because of two factors. Firstly, the solubility of oxygen decreases less with increasing temperature than that of carbon dioxide. Secondly, oxygenation has a higher activation energy than carboxylation and therefore increases more rapidly with temperature. Typically, a result of this is that net carbon assimilation may be depressed by up to 40% at temperatures above 30°C.

What Is Photorespiration For?

There has been considerable speculation as to why plants exhibit photorespiration. It does not seem to be an essential component of plant metabolism. Plants grow extremely well in atmospheres highly enriched with carbon dioxide where photorespiration is highly suppressed, indicating that the photorespiratory pathway is not supplying essential intermediates. Photorespiration appears to be an unavoidable consequence of the properties of Rubisco, but it may have important beneficial consequences under certain conditions. One of these may be the dissipation of light energy at low levels of carbon dioxide fixation, e.g., when stomata close in response to water stress, to avoid photoinhibition.

One can speculate that, particularly for plants growing at higher temperatures and in high light levels, there has been selective pressure to minimize photorespiration. Both C_4 and crassulacean acid metabolism (CAM) plants have developed mechanisms for concentrating carbon dioxide at the site of Rubisco action, thus minimizing oxygenase and hence photorespiratory activity.

List of Technical Nomenclature

Carbon assimilation	The process of converting carbon from its gaseous carbon dioxide form to organic constituents of plants.
Photoinhibition	Reduction of photosynthesis at high light levels resulting from photo-oxidative damage.
Respiration	Process in which organic compounds are broken down to produce energy usually through an oxygen-dependent process to produce carbon dioxide.

See also: **Photosynthesis and Partitioning**: C_3 Plants; C_4 Plants; CAM Plants.

Further Reading

Leegood RC (1999) Photosynthesis in C_3 plants: the benson–calvin cycle and photorespiration. In: Lea PJ and Leegood RC (eds) *Plant Biochemistry and Molecular Biology*, pp. 29–50. Chichester: John Wiley.

Leegood RC, Lea PJ, Adcock MD, and Häusler RE (1995) The regulation and control of photorespiration. *Journal of Experimental Botany (Special Edition)* 46: 1397–1414.

Ogren WL (1984) Photorespiration: pathways, regulation and modification. *Annual Review of Plant Physiology* 35: 415–442.

Portis AR (1992) Regulation of ribulose 1,5-bisphosphate carboxylase/oxygenase activity. *Annual Review of Plant Physiology and Plant Molecular Biology* 43: 415–437.

Photoinhibition

B Demmig-Adams and W W Adams III, University of Colorado, Boulder, CO, USA

Introduction and Definitions

Photosynthesis provides the energy and building-blocks for life on this planet as well as being a major carbon sink. Therefore changes in photosynthesis rates have the potential to affect global ecosystems and climate. In response to environmental stress, the efficiency with which solar energy is collected and utilized by plants can be diminished for prolonged periods of time.

What Is Photoinhibition?

Photoinhibition of photosynthesis is defined as a persistent decrease in the efficiency of solar energy conversion into photosynthesis in combination with a decreased overall capacity for photosynthesis. Environmental stresses that trigger photoinhibition include, for example, adverse temperatures, limited nutrient or water availability, and salinity. Evergreen coniferous forests in temperate climates, for example, can show a complete shutdown of photosynthesis during the winter, and the efficiency of solar energy conversion in photosystem II can drop to negligible levels in needles that remain green and continue to absorb considerable amounts of solar energy. Furthermore, high light stress can result from sudden increases in growth irradiance that may occur naturally when a canopy gap opens in a forest.

What is causing this loss in energy conversion efficiency and photosynthetic capacity under environmental stresses? And what does it mean for the productivity of plants? Does photoinhibition of photosynthesis limit the productivity of plants in environments with intermittent stress periods? Or does it reflect a photoprotective process that down-regulates photosynthesis during stressful times when the growth of plants is arrested? While many researchers in this field tacitly assume that photoinhibition of photosynthesis results from photodamage, others have suggested that this phenomenon is a regulatory and protective adjustment to environmental stress and change. While many assume that photoinhibition lowers plant productivity – and consequently expect that more photosynthesis and carbon uptake would occur in the absence of photoinhibition – others argue that photoinhibition does not lower the productivity of plants and that it only occurs when the opportunity for growth and the activity of the plant's sink tissues are low or absent.

It is quite surprising that a phenomenon as important as photoinhibition, which affects photosynthesis of many species in a profound way, is not better understood.

Features of Photoinhibition

What features of this phenomenon of photoinhibition are generally accepted? "Photoinhibition" of photosynthesis may include some or all of the following characteristics:

1. Lasting decreases in the efficiency with which absorbed solar radiation is converted into photosynthesis are detected as sustained decreases in the level of CO_2 fixed or O_2 evolved per unit of photons absorbed at light intensities limiting to the rate of photosynthesis. In particular, the level of photosystem II photochemistry per unit of photons absorbed in the photosystem II light-collecting

antennae is decreased persistently. Photosystem II photochemical efficiency is quantified from chlorophyll fluorescence signals as the ratio of variable to maximal fluorescence (F_v/F_m), with maximal values of 0.85 equaling an 85% efficiency of the conversion of absorbed light into photochemistry.

In sun-exposed leaves, solar energy conversion efficiency undergoes pronounced changes over the course of a clear day (**Figure 1**; summer day). This occurs under favorable environmental conditions when growth rates and carbon exchange rates are maximal. Even maximal rates of utilization of absorbed light for photosynthesis in the fastest-growing species do not consume all of the energy absorbed at peak irradiance, and the excess energy is dissipated as heat in the light-harvesting antennae of photosystem II before it reaches the reaction center. This phenomenon is sometimes described as "dynamic photoinhibition" (as opposed to "chronic photoinhibition" for sustained changes). However, there is no reason to address this phenomenon as photoinhibition, as long as these decreases in solar energy conversion efficiency relax quickly upon return to low light levels (cf. **Figure 1**, summer day).

2. Decreases in the maximal, light-saturated, and CO_2-saturated capacity of photosynthetic electron flow and photosynthesis are commonly observed features in response to longer exposure to conditions that induce photoinhibition.
3. Inactivation and degradation of the photosystem II reaction center in which high-energy electrons for electron transport are generated. The photosystem II reaction center core consists of two proteins, the D1 and D2 proteins. During photoinhibition, inactivation of the D1 protein is observed, followed by degradation of photosystem II reaction center cores (**Figure 2**). While short-term exposures can result mainly in an inactivation of D1 function, naturally growing plants exhibiting photoinhibition commonly show decreased levels of D1 and photosystem II cores during periods of stress. Much work has focused on the turnover of the D1 protein that is extremely rapid in plants growing under favorable conditions. D1 protein turnover is frequently addressed as the damage/repair cycle of the D1 protein.
4. Maximization of photoprotection through conversion of excess absorbed solar radiation into harmless thermal energy. This thermal dissipation of excess absorbed energy is catalyzed by the xanthophylls zeaxanthin and antheraxanthin of the xanthophyll cycle (**Figure 3**) and a specialized protein of the family of light-harvesting proteins. In nonphotoinhibited leaves, the rate of thermal energy dissipation increases and decreases rapidly in response to the level of excess light (**Figure 1**; summer day), and these rapid fluctuations are triggered by changes in the magnitude of the

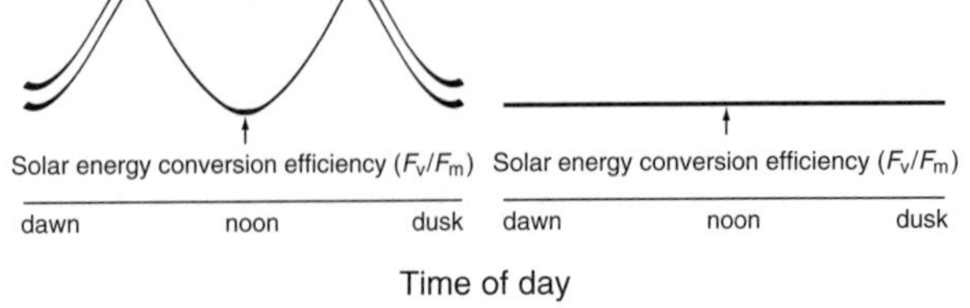

Figure 1 Diurnal patterns of formation of zeaxanthin + antheraxanthin in the xanthophyll cycle as well as changes in thermal energy dissipation rate and in solar energy conversion efficiency of photosystem II (F_v/F_m) under favorable conditions versus environmental stress.

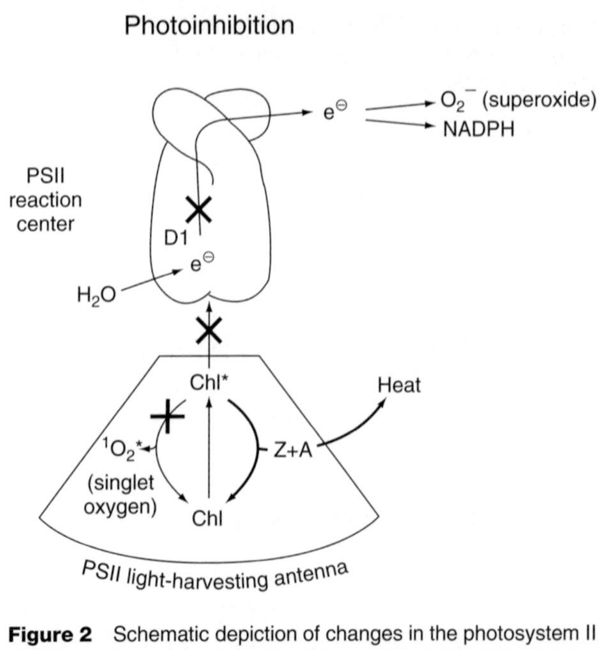

Figure 2 Schematic depiction of changes in the photosystem II (PSII) reaction center (with the D1 protein) and light-harvesting antenna in photoinhibited leaves. Photoinhibition causes an inactivation of D1 (or removal of PSII cores), and sustained (Z + A)-dependent dissipation of energy as heat. Z + A, zeaxanthin + antheraxanthin; Chl*, singlet excited state of chlorophyll. The cross marks stand for D1 protein inactivation/removal (leading to a suppression of the formation of high-energy electrons and of superoxide), a decrease in energy delivery from light-collecting antennae to PSII reaction centers, and a suppression of singlet oxygen formation by enhanced thermal dissipation.

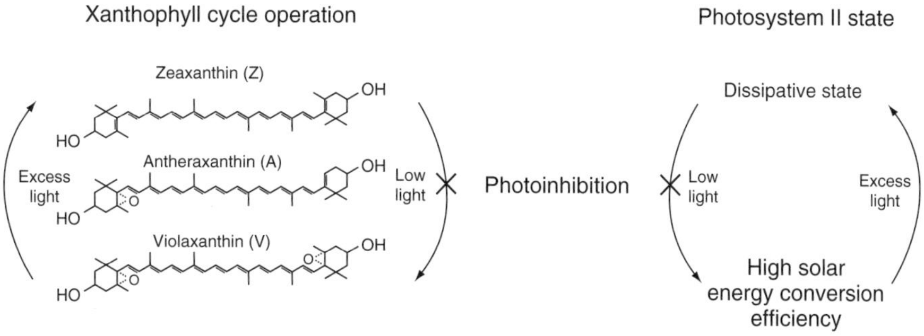

Figure 3 Changes in xanthophyll cycle operation and photosystem II state during photoinhibition. Photoinhibition inhibits (shown by the cross mark) the reconversion of zeaxanthin and antheraxanthin to violaxanthin as well as the return of photosystem II to a state of high solar energy conversion efficiency in low light.

proton gradient across the photosynthetic membrane (presumably sensed by the specialized protein). In photoinhibited leaves, these rapid fluctuations no longer occur and maximal thermal energy dissipation is "locked in" (**Figures 1** and **3**). Zeaxanthin and antheraxanthin no longer become removed at the end of the day and can be retained at high levels for the entire duration of the stress period. Likewise, photosystem II remains in a highly dissipative state (**Figures 1** and **3**).

Relationship between Inactivation of Photochemistry and Dissipation of Absorbed Energy as Heat

In theory, decreases in the efficiency of the conversion of absorbed energy into photochemistry in photoinhibited leaves can result either from a loss of photochemical competency or from an increase in thermal energy dissipation, or from a combination of both. Any decrease in photochemical competency will lower the energy conversion efficiency – and so will removal of excess absorbed energy in the chlorophyll pigment bed as heat before this energy reaches the photochemical reaction centers (**Figures 1** and **2**). Consequently, an observation of sustained decreases in photosynthetic energy conversion cannot tacitly be taken to indicate either the presence of photochemical disabling, photodamage, or the presence of sustained high levels of thermal energy dissipation. Further characteristics of these photoinhibited plants need to be established to make such an assessment. Recent studies do, however, indicate that these two phenomena co-occur under a range of environmental stresses, particularly in evergreen species. This may not be unexpected. Under environmental stress there is typically an imbalance between (greater) absorption of solar energy and (decreased) utilization of this energy in carbon fixation. Under these conditions of excess light, reactive oxygen species can be formed during light-harvesting and photochemistry (**Figure 2**). These reactive oxygen species have the potential to destroy cellular components directly and lead to cell death. They can also trigger signal transduction pathways leading to arrest of protein turnover, net degradation of proteins, and, ultimately, programmed cell death. Two main sites of reactive oxygen formation in chloroplasts are (1) the light-absorbing chlorophyll pigments that can catalyze singlet oxygen formation and (2) photochemical and electron transport reactions that can lead to superoxide formation (**Figure 2**). Dissipation of excess excitation energy in the chlorophyll pigment bed via the xanthophyll cycle counteracts singlet oxygen formation, and a lowering of the rate of photochemical charge separation (via D1 inactivation or removal) should decrease the transfer of electrons to oxygen to form superoxide (**Figure 2**). A lasting combination of the two processes may be a prerequisite for the maintenance of evergreen leaves through seasons with extreme environmental stress by some species, such as overwintering conifers or desert evergreens.

Photoinhibition: Friend or Foe?

Based on periods during which plants experience photoinhibition (frequently quantified as decreases in solar energy conversion from F_v/F_m), extrapolations have been made in the literature of how much more carbon uptake would be expected in the absence of these decreases in solar energy conversion into photochemistry. In such extrapolations, it is tacitly assumed that photochemistry is limiting carbon gain in these situations.

As will be discussed in the following section, the highest levels of photoinhibition are exhibited by evergreen species with a lifespan of many years. In climates with seasonal environmental stress, these species typically exhibit growth during the favorable season(s) and arrest growth during the stressful season(s). This growth arrest is associated with photosynthetic downregulation.

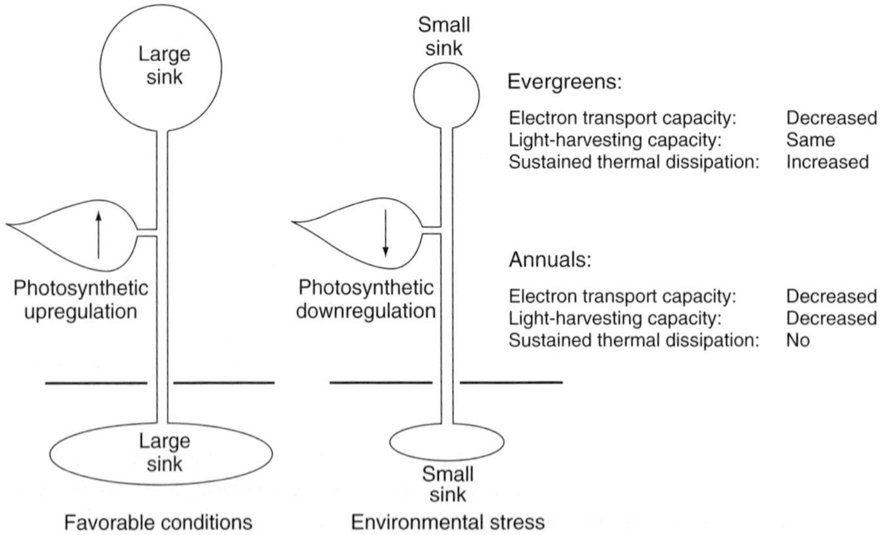

Figure 4 Schematic depiction of the regulation of photosynthesis by demand in different groups of plants.

The demand for products of photosynthesis at the whole-plant level regulates the level of photosynthetic proteins, resulting in increases in the capacity for photosynthesis when growth rates are high and decreases in photosynthetic capacities when growth rates are low (**Figure 4**). These changes in the capacity for photosynthesis are achieved via changes in the levels of photosynthetic proteins and result from modulation of gene expression at various levels. In evergreens it is during the times when downregulation of photosynthesis occurs that photoinhibition is observed. It thus appears that photoinhibition is unlikely to limit productivity in these species.

More studies are needed to establish a cause-and-effect relationship of photoinhibition and growth arrest in a large variety of plant species. For any decreases in solar energy conversion efficiency in response to limitations in the demand for photosynthate it would be highly inappropriate to extrapolate to "potential plant productivity in the absence of photoinhibition."

Does Photoinhibition Reflect Damage or Regulation?

Even if photoinhibition were not limiting to plant productivity, the argument has been made that it could still be the consequence of either photodamage or photoregulation.

When rapid D1 turnover was first discovered, it was assumed to reflect a regulatory process. However, soon thereafter the focus shifted to considering this process as a damage and repair cycle, and to *in vitro* characterization of sites of inactivation, the effect of blocking D1 synthesis, observing accumulation of inactivated D1 and PSII reaction centers, and altered structure and levels of the D1 protein.

Current research is uncovering a remarkable level of regulation of D1 synthesis and degradation. This involves environmental control of D1 synthesis at several levels as well as environmental control of the expression and activity of D1-degrading proteases. The D1 protein also possesses a motif characteristic of key regulatory proteins that turn over rapidly to allow rapid adjustments in metabolism. For such proteins, the investment in a high turnover rate would thereby allow a high level of metabolic control.

Key components in the regulation of D1 turnover are reactive oxygen species and other redox processes. For example, faster degradation of the D1 protein is seen in the presence of elevated levels of reactive oxygen species. This has been widely interpreted as support for the notion that the D1 protein becomes damaged and repaired. However, elevated levels of reactive oxygen species also arrest D1 synthesis at multiple levels.

An explosive development is taking place in the general field of oxidative stress. New insights are leading to a reinterpretation of many processes that used to be regarded as damage, and are now being recognized as components of redox-regulated signal tranduction pathways. Furthermore, the distinction between damage and regulation is beginning to blur, and these labels may be merged into a term such as regulation by oxidative modification. It is time that the process of photoinhibition of photosynthesis be subjected to the same sort of re-examination.

Reactive oxygen species oxidize proteins, among which the most sensitive to oxidation are signaling proteins whose oxidation results in a stress-related signal. Likewise, the most readily oxidized lipids are precursors of signaling molecules eliciting stress responses. In both animals and plants, polyunsaturated membrane lipids become oxidized to hormone-like stress messengers that orchestrate stress responses via the regulation of gene expression. It is attractive to speculate that downregulation of photosynthesis and development of photoinhibition under environmental stress may be triggered by a combination of hormonal signals from sink tissues (**Figure 4**) with hormonelike and other signals originating in the chloroplast under excess light stress. Candidates for such stress-triggered signals originating in the chloroplast include reactive oxygen species themselves, other redox signals, and messengers formed as a result of lipid peroxidation. A more meaningful way of phrasing the question of damage versus regulation is whether plants would be better or worse off without photoinhibition. It may be such that evergreens, which experience regular seasonal stresses and downregulate photosynthesis periodically, would not be able to preserve their chlorophyll without photoinhibition in the form of a lowered energy conversion efficiency. But virtually no data are available to address this issue directly. There has been but one study with transgenic algae maintaining a high rate of photosystem II photochemistry under environmental stress, which turned out to die quickly, while the wild-type exhibiting photoinhibition of photosystem II survived. Studies with mutant or transgenic higher plants are needed to ascertain whether or not plants that naturally undergo D1 inactivation and degradation under photoinhibitory conditions may suffer when forced to maintain high levels of photochemically active D1 under light stress. The only clear conclusion is that, based on the evidence available to date, the tacit assumption that photoinhibition must entail damage should be abandoned.

Examples for Photoinhibition: Response to Environmental Stress as Dependent on Plant Species

Decreases in photosynthetic electron transport capacity can occur with or without lasting decreases in solar energy conversion efficiency (**Figure 4**). The type of response displayed varies among plant species. Evergreens maintain green leaves, and thus a high light-harvesting capacity, and exhibit sustained thermal energy dissipation when electron transport capacities are decreased. Annuals often lower light-harvesting capacity together with overall photosynthetic capacity and do not exhibit sustained thermal energy dissipation.

These two phenomena represent the opposite extremes of a continuum. Preservation of a high chlorophyll content, i.e., preservation of a high capacity to collect light during times of diminished utilization of solar energy, necessitates a high capacity for harmless removal of excess absorbed light (**Figures 2** and **4**). On the other hand, a pronounced degradation of chlorophyll lowers the capacity for light collection and prevents the absorption of large amounts of excess light. This latter strategy puts relatively less emphasis on the thermal dissipation of excess absorbed light during times when overall photosynthetic capacity is strongly decreased. These two strategies may be viewed as different ways of achieving the same effect, i.e., lowering the levels of excess excitation energy.

In a strict sense, only the combination of decreased electron transport capacities with lasting decreases in solar energy conversion efficiency should be termed photoinhibition. A concomitant decrease of light-harvesting capacity with overall photosynthetic capacity may be viewed as part of the regulation of photosynthesis by demand via concomitant regulation of the expression of chlorophyll *a*/*b*-binding and CO_2-fixing enzymes.

Why don't all species simply lower the capacity for light collection? Short-lived species seem to do this most consistently, whereas many perennial evergreens take the approach of maintaining dark green leaves and a high light-collecting capacity (**Figures 2** and **4**). The reason for this may lie in their different lifespans and growth habits. Evergreen coniferous forests instantly resume maximal photosynthesis rates after spending the winter in a strongly photoinhibited state. This rapid response is facilitated by the fact that the complement of light-harvesting complexes does not have to be resynthesized. In contrast, annual species have evolved accelerated metabolic responses to complete their life cycle before the onset of a season with severe stress. Leaves of these species typically yellow and senesce when environmental conditions no longer permit high growth rates.

Photoinhibition Involving Chlorophyll Preservation and Strong, Lasting Decreases in Solar Energy Conversion Efficiency

Photoinhibition can occur as a consequence of high temperatures, high levels of ultraviolet irradiation, or drought. Here we will focus on two additional conditions, i.e., cold temperatures and increases in growth light intensity.

Conifers and other winter-hardy evergreens and winter stress The case of photoinhibition in overwintering conifers has been studied rather extensively. While maintaining their proverbial green needles, many conifers do arrest growth during the winter season and downregulate photosynthetic capacity. This involves pronounced decreases in the level of the D1 protein of the photosystem II reaction center but maintenance of considerable levels of light-harvesting proteins (**Figure 5**; cf. **Figure 2**). Furthermore, this preservation of the light-harvesting system is accompanied by retention of high levels of zeaxanthin and antheraxanthin at all times as well as lasting decreases in the efficiency of energy conversion into photochemistry in photosystem II (from F_v/F_m) (**Figure 5**; cf. **Figure 1**). Once again, maintenance of a high light-harvesting capacity in the absence of a sink for excitation energy in the form of photochemistry calls for a drastic lowering of the efficiency of solar energy conversion in the light-collection system, which presumably occurs via sustained high levels of xanthophyll-dependent thermal energy dissipation (**Figures 1** and **2**).

The response of chilling-sensitive species to cold stress will be addressed in a separate section.

Transfer of evergreen shade-grown plants grown in deep shade to high light Tropical evergreen tough-leafed species, like *Schefflera arboricola* and *Monstera deliciosa*, have been examined extensively in this respect. Both species occur naturally over a wide range of irradiances and are tolerant of deep shade as well as full sunlight. When deep shade-grown plants of these species are suddenly transferred to high light levels, an enormous excess of light is absorbed, owing to their dark green leaves with a high light-collecting capacity and their low maximal capacity for photosynthesis. The result of such a transfer is strong photoinhibition, with leaves remaining dark green for many days and exhibiting a strong decrease in the efficiency of energy conversion into photochemistry in photosystem II. This photoinhibition involves a locking-in of maximal dissipation of absorbed energy, with maintenance of the xanthophyll cycle in a state of maximal conversion to the photoprotective pigments zeaxanthin and antheraxanthin (cf. **Figure 3**). As is the case in overwintering evergreens, these shade-grown leaves strongly degrade the D1 protein of photosystem II while maintaining the light-harvesting proteins. In addition, we have observed a continuing accumulation of carbohydrates in these photoinhibited leaves over many days and conclude that carbon-export capacity of deep shade-grown leaves of evergreen *M. deliciosa* is permanently low. Yet, when these plants are grown for extended periods of time in high light, new leaves develop with higher capacities for photosynthesis and no signs of photoinhibition. In shade-grown herbaceous species with soft leaves, growth of new leaves with higher photosynthetic capacities and no photoinhibition occurs much more rapidly, i.e., within a few days upon an increase in growth irradiance, whereas this takes months in the case of the evergreens. Furthermore, in herbaceous species even the shade-grown leaves themselves show some capacity for high light acclimation (photosynthetic upregulation), whereas this capacity is very limited in tough-leafed evergreens.

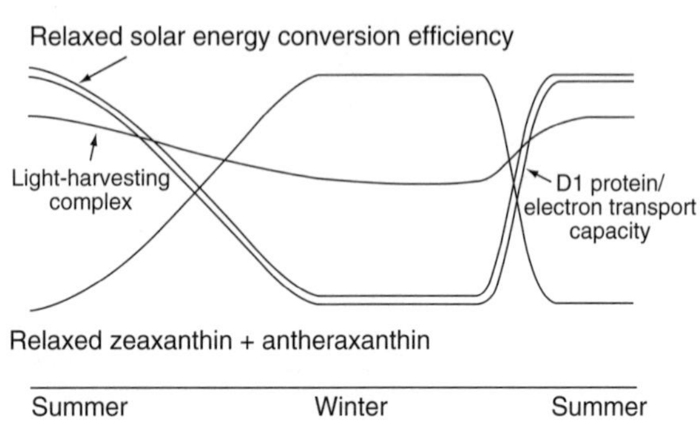

Figure 5 Seasonal changes in the utilization of solar energy in photosynthesis in conifers that exhibit strong photoinhibition during the winter season. Relaxed solar energy conversion efficiency (F_v/F_m) and relaxed zeaxanthin + antheraxanthin levels are representative of needles collected predawn and kept at warm temperatures for several hours.

Availability of soil nitrogen can have a strong impact on the responses during transfer from low to high irradiance. Transfer to high irradiance at low soil nitrogen levels can result in photoinhibition, while high soil nitrogen levels can prevent photoinhibition. In the absence of adequate levels of soil nitrogen, plants are presumably unable to grow sufficiently rapidly – or upregulate their photosynthetic capacity sufficiently – to utilize the much increased level of available light energy. In evergreen perennials this tends to induce sustained decreases in solar energy conversion efficiency, whereas in annual species a lowering of light-harvesting capacity appears to be the predominant response, as is described below.

Stress-Induced Degradation of Chlorophyll and Decreases in Photosynthetic Capacity on a Leaf Area Basis, but No or Little Decrease in Energy Conversion Efficiency

Herbaceous species under nitrogen limitation In herbaceous species (like spinach (*Spinacea oleracea*), tobacco (*Nicotiana tabacum*), corn (*Zea mays*;

maize), or wheat (*Triticum* spp.)) nitrogen limitation leads to decreased photosynthetic capacities with decreased levels of chlorophyll and CO_2-fixing enzymes. These features are seen only in sun-exposed but not in shaded plants. Does that mean that photodamage is involved? The most severe limitation under limiting soil nitrogen is experienced in the growing points of the plant, particularly in potentially fast-growing sun-exposed individuals. Soil nitrogen limitation affects primarily the growing points of the plant and severely limits sink activity (**Figure 4**). Carbohydrates build up in the leaves as a consequence, and the rate of synthesis of chlorophyll-binding and CO_2-fixing proteins decreases. This leads to a downregulation of photosynthetic capacity in response to the imbalance between a high availability and a low utilization of carbohydrates (**Figure 4**), at the same time remobilizing some of the nitrogen from photosynthetic enzymes to the growing points of the plant. In the chloroplast, an increase in the capacity for rapidly reversible xanthophyll cycle-dependent thermal energy dissipation is observed, but no sustained increase in thermal dissipation or photoinhibition (as a sustained decrease in energy conversion efficiency). This is likely related to the strong decrease in chlorophyll content, and thus light-collection capacity.

Species of tropical origin growing in temperate climates When plants are grown outside the range of environmental conditions (with respect to temperatures, water availability, salinity, etc.) to which they are well adapted, yellowing of leaves is frequently observed together with decreasing photosynthetic capacities, sometimes over prolonged periods of time. This applies not only to chilling-sensitive annual crops in areas with, for example, early-season cold spells, but also to tropical evergreen species in cold or dry climates. Photoinhibition is typically observed in sun-exposed locations and leaves. However, that also does not allow the conclusion that photoinhibition is caused by photodamage since sun exposure has dramatic effects on parameters such as leaf and plant temperature, soil and plant water deficit, and water transport throughout the plant.

One likely scenario in the case of water stress is a disruption of water transport throughout the plant in species adapted to warm, humid climates by embolisms as a result of soil water deficits or frequent freeze–thaw cycles in cold climates. Either of these conditions is likely to disrupt water delivery to the shoot of plants, causing stomatal closure and downregulation of photosynthetic capacity via suppression of the synthesis of chlorophyll-binding proteins as well as other photosynthetic proteins. If this were the sequence of events, any efforts to increase plant productivity and survival would have to target the limiting steps of inherent drought or temperature tolerance. An interesting study for the case of temperature intolerance is available from cyanobacteria. Engineering cyanobacteria to contain a membrane lipid desaturase (that increases membrane fluidity under moderately low temperatures) dramatically increased survival of the bacteria at these temperatures. In addition, this change dramatically decreased D1 degradation and photoinhibition in these bacteria at the low temperatures. High levels of D1 degradation under low-temperature stress in the chilling-intolerant wild-type of these cyanobacteria have been interpreted as a sign of damage, but might also be viewed as a protective response to avoid massive reactive oxygen formation. Removing the apparent source of the chilling intolerance by ameliorating membrane fluidity either prevented the damage to D1 or made the inactivation of photochemistry unnecessary.

Maintaining Electron Transport Capacity during Environmental Stresses

The herbaceous weed *Malva neglecta* is a sun-loving species with high growth rates and high photosynthetic capacities that are maintained even during the harsh winter conditions of a montane climate. Never throughout its lifecycle does *M. neglecta* show either photoinhibition or extremely high capacities for photoprotection (in the form of very high capacities for thermal energy dissipation). Whereas *M. neglecta* does not show appreciable levels of photoinhibition, this species does maintain photosystem II in a dissipative state with high levels of retained zeaxanthin and antheraxanthin as long as leaf temperatures remain substantially below freezing in the field. This is probably caused by maintenance of a proton gradient across the photosynthetic membrane that keeps thermal energy dissipation engaged at all times. Since this cold-sustained energy dissipation disappears promptly upon warming of the leaves, it should not be considered photoinhibition. This response offers protection during cold periods in the winter. During intermittent milder periods, photosystem II quickly returns to a photochemically highly competent state and photosynthesis resumes. Whereas biennial and annual herbaceous species, such as *M. neglecta*, spinach, winter rye (*Secale cereale*), mullein (*Verbascum thapsus*), and *Arabidopsis thaliana*, persist throughout a single winter season and maintain growth and high photosynthetic capacities throughout the challenging season, they are unable to survive the more extreme conditions that some perennial evergreens can tolerate.

Engineering Plants for Enhanced Stress Tolerance: Where Does Photoinhibition Fit In?

Correlations exist between high rates of D1 turnover and a high resistance to photoinhibition among various plant species and varieties. Based on this observation, two contrasting hypotheses have been formulated. One is that the rate of D1 synthesis or D1 turnover is limiting stress resistance in some species or varieties. In this case, plants engineered to possess higher rates of D1 synthesis and turnover would be expected to exhibit greater stress tolerance. In the alternative hypothesis, the rate of D1 synthesis and turnover is controlled by signals reflecting the opportunity for utilizing solar energy for plant growth. In this case, manipulations of growth rate would automatically modulate D1 synthesis and turnover, but increase of D1 synthesis rate *per se*, without addressing growth rates, may actually lead to greater rates of membrane lipid peroxidation. Studies with transgenic plants are needed to determine which hypothesis is correct.

Another area of interest is antioxidant systems. Conditions triggering photoinhibition commonly involve an imbalance between light absorption and light product utilization in the chloroplast. Such an imbalance has the potential to increase the levels of various messengers that trigger a net degradation of photosynthetic proteins. Future research may clarify whether these events involve any damage or solely regulation. However, the consequences may be rather similar, in terms of efforts to manipulate plant stress tolerance. Antioxidation and other processes that counterbalance the oxidative processes may be expected either to prevent damage or to counteract the generation of signals that trigger degradation. Future research will have to show whether overexpression of the capacity for thermal energy dissipation or for antioxidation in the chloroplast will result in less photoinhibition and/or chlorophyll degradation in any plant species. It has to be noted that all species examined to date appear to possess a remarkable plasticity in their ability to increase the capacity of photoprotective processes. At the same time, the response of different species to conditions that limit growth and sink activity varies widely as outlined above. While many crops respond with massive chlorophyll degradation and premature senescence, evergreens maintain green, albeit photosynthetically inactive leaves. Might it be possible to engineer crops able to maintain dark green and fully protected leaves throughout intermittent periods with harsh environmental conditions that prevent continued sink activity? Future research needs to address species-dependent differences in the signaling processes that result in contrasting responses.

See also: **Abiotic Stresses**: Cold Stress; Free Radicals, Oxidative Stress and Antioxidants. **Genetic Modification, Applications**: Oxidative Stress. **Genetic Modification of Primary Metabolism**: Photosynthesis. **Genetic Modification of Secondary Metabolism**: Terpenoids. **Photosynthesis and Partitioning**: C_3 Plants; Sources and Sinks. **Primary Products**: Oils. **Regulators of Growth**: Jasmonates. **Water Relations of Plants**: Drought Stress.

Further Reading

Adams WW III, Rosenstiel TN, Demmig-Adams B, Ebbert V, and Brightwell AK (2002) Photosynthesis and photoprotection in overwintering plants. *Plant Biology* 4: 545–557.

Critchley C (1999) Molecular adaptation to irradiance: the dual functionality of photosystem II. In: Singhal GS, Renger G, Sopory SK, Irrgang K-D, and Govindjee (eds) *Concepts in Photobiology: Photosynthesis and Photomorphogenesis*, pp. 571–587. New Dehli: Narosa Publishing House.

Dalton TD, Shertzer HG, and Puga A (1999) Regulation of gene expression by reactive oxygen. *Annual Review of Pharmacology and Toxicology* 39: 67–101.

Demmig-Adams B and Adams WW III (1996) The role of xanthophyll cycle carotenoids in the protection of photosynthesis. *Trends in Plant Science* 1: 21–26.

Koch KE (1996) Carbohydrate-modulated gene expression in plants. *Annual Reviews of Plant Physiology and Plant Molecular Biology* 47: 509–540.

Melis A (1999) Photosystem-II damage and repair cycle in chloroplasts: what modulates the rate of photodamage *in vivo*? *Trends in Plant Science* 4: 130–135.

Osmond B, Badger M, Maxwell K, Björkman O, and Leegood R (1997) Too many photons: photorespiration, photoinhibition and photooxidation. *Trends in Plant Science* 2: 119–121.

Primary Products of Photosynthesis, Sucrose and other Soluble Carbohydrates

J D Everard, DuPont Experimental Station, Wilmington, DE, USA

Introduction

Plants are autotrophic organisms and thus have the capacity to create all the components required for their existence by harnessing light energy to drive the biochemical synthesis of complex components from simple molecules, principally CO_2, H_2O, nitrogen, and mineral nutrients. Performance and ultimately survival of a plant requires that the photoassimilation of carbon is coordinated and balanced with the many competing demands for reduced carbon and the ability of the plant to acquire the nitrogen, H_2O, and other resources needed for the synthesis of proteins, lipids, nucleic acids, and structural and other components. This coordination has to be maintained not only on a biochemical level at the sites of carbon assimilation (the source tissues), but must also be integrated with the demands of spatially separated tissues, particularly the heterotrophic sinks, which are completely dependent on the source tissues for their metabolic needs. From an anthropic perspective, a plant's ability to allocate resources to the components used for human consumption have been the basis of selective plant breeding since such endeavors became common practice. For this reason, resource allocation and partitioning in plants have been, and continue to be, the subject of intense research effort. From this research comes the realization that not only are the control mechanisms involved very complex and multilayered, but also that carbohydrates and their metabolism play central roles.

A ubiquitous player in plant carbohydrate metabolism is sucrose. As a generalization, when mature leaves are pulse labeled with radioactive $^{14}CO_2$ in the light, after an extended period of darkness, most of the label in the soluble carbohydrate fraction will be recovered as sucrose. Such experiments demonstrate that sucrose is often the principal stable product of photosynthetic carbon assimilation in source tissues. They also show that once synthesized, the carbon in the sucrose pool is somewhat sequestered from general metabolism in source tissues, and that it can therefore act as a temporary storage form for newly assimilated carbon. Sucrose is also the form that most plant species use to transport assimilated carbon from the sites of synthesis to the sites of utilization or storage. This article reviews the characteristics that make sucrose suitable for the roles it plays in plants and a few exceptions where it does not perform these roles exclusively. The biochemistry and regulation of sucrose synthesis and degradation is covered along with a brief consideration of the potential for sucrose and its metabolites to act as signaling molecules involved in the coordination of whole plant carbon partitioning and resource allocation.

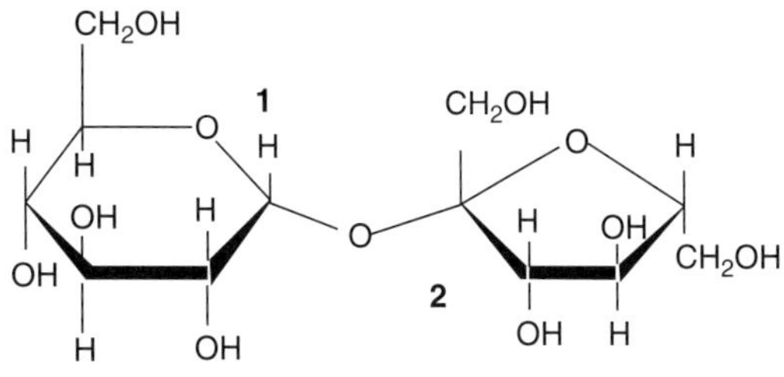

Figure 1 Sucrose (α-D-glucopyranoside (1-2) β-D-fructofuranoside).

Sucrose and its Properties

Sucrose (α-D-glucopyranoside (1-2) β-D-fructofuranoside) is a disaccharide consisting of α-D-glucose glycosidically linked to β-D-fructose. The glycosidic bond occurs between the anomeric carbons of the glucose (C-1) and fructose (C-2) units (**Figure 1**). This arrangement has the effect of obscuring the functional groups of the glucose (aldehyde) and fructose (ketone) moieties. Sucrose, therefore, is a nonreducing sugar and is chemically relatively inert. The glycosidic bond has a high free energy of hydrolysis, and a number of biochemical transformations involving sucrose led to the preservation of some of this bond energy, or use it to drive the catalytic process. Sucrose is extremely soluble and, at least at concentrations of relevance to plants, has little effect on solution viscosity; phloem sieve element concentrations as high as 1.8 $mol\,l^{-1}$ have been reported and cytosolic concentrations of 0.2 $mol\,l^{-1}$ are not uncommon. The carbon and energy contained in sucrose can be returned directly into metabolism simply by hydrolyzing the glycosidic bond to release the component hexoses (glucose and fructose). Hexoses or, more accurately, their phosphorylated and nucleotide forms are "ideal" metabolites as they lie at the junction between many different metabolic pathways (**Figure 2**). They also play an important role in sugar signaling, and may be part of a mechanism that plants use to monitor metabolic activity.

Other Carbohydrates Associated with Primary Production

Few other carbohydrates have the properties of sucrose, and the vast majority of plants use it almost exclusively as their primary photoassimilated and translocated carbohydrate. The only other major class of soluble carbohydrates that have been found to play a significant role in primary carbohydrate assimilation are the acyclic sugar alcohols (polyols). Mannitol (**Figure 3**) is by far the most widely distributed polyol, being found in over 50 higher

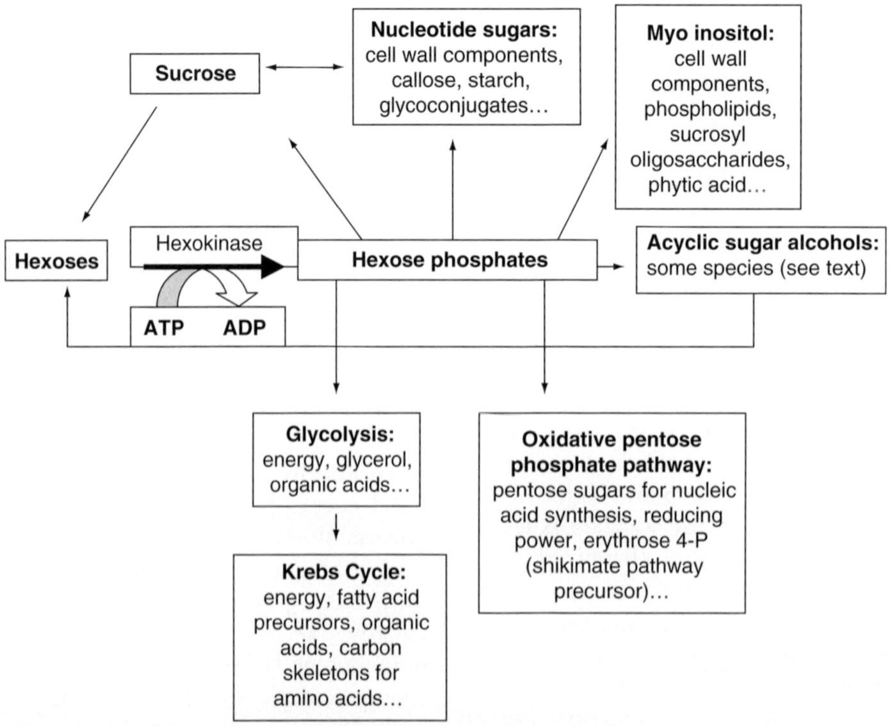

Figure 2 An illustration of the central role that hexose sugars play in plant metabolism.

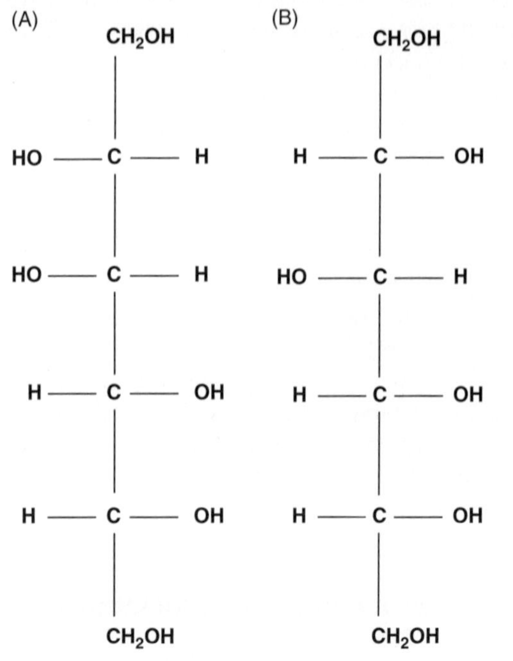

Figure 3 The acyclic sugar alcohols D-mannitol (A) and D-sorbitol (B).

plant families including many of commercial importance in the Rubiaceae (coffee), Oleaceae (olive), and Apiaceae (celery). Mannitol is also a primary photosynthetic product in many marine algae, especially members of the Phaeophyta (brown algae and kelps). Sorbitol (**Figure 3**) is a common primary product in many of the woody Rosaceae, including apple, peach, plum, and cherry. Labeling studies indicate that polyols are synthesized via the hexose phosphate pool, and that fixed carbon often appears in the polyol and sucrose pools concomitantly. Like sucrose, polyols are effectively sequestered from metabolism in source tissues. Phloem saps from species that synthesize polyols contain high levels (up to 90% of the total carbohydrate), indicating roles in carbohydrate translocation. In sink tissues, oxidation of the polyols to their respective aldose or ketose sugars brings the carbon contained within them back into the hexose pool, in a manner analogous to that illustrated for sucrose in **Figure 2**. To date, all vascular plants that have been found to synthesize polyols as primary photoassimilates also synthesize (often at higher rates) and translocate sucrose, indicating that polyols cannot fully substitute for sucrose; this observation lacks a full physiological explanation. Little more will be said about polyols here except to point out that an estimated 30% of annual global primary production occurs through them. The interested reader is referred to reviews listed at the end of the article.

Free glucose and fructose are ubiquitous in most plant tissues, but usually represent less than 20% of the total soluble carbohydrate pool in source leaves. Woody perennials may be exceptions and glucose may account for upward of 50% of the soluble carbohydrate pool in source leaves. The physiological significance of this large glucose pool has not been determined, but may play an osmoprotective role during periods of stress. Labeling studies, however, indicate that the glucose is likely to be a secondary rather than a primary product as label enters the pool very slowly, usually through sucrose.

Another group of carbohydrates that are produced as secondary products of sucrose synthesis in the source leaves of many species are the sucrosyl oligosaccharides raffinose, stachyose, and verbascose. These compounds may account for between 20 and 80% of the carbohydrate in phloem saps. In most cases, synthesis of these compounds occurs in intermediary (companion) cells within the vascular tissues of species that use a symplastic strategy for phloem loading, i.e., their site of synthesis is spatially removed from the sites of carbon assimilation.

Carbon Assimilation and the Steps Leading to Sucrose Synthesis

De novo carbon assimilation takes place in the soluble (stromal) compartment of photosynthetic plastids. Numerous variations in the way plants initially capture CO_2 have evolved to enhance plant fitness within certain environments (e.g., C4 and CAM species), but in all photosynthetic organisms the biochemical centerpiece of carbon assimilation is the photosynthetic carbon reduction cycle (PCR cycle; also known as the Calvin cycle, reductive pentose phosphate pathway, or dark reactions of photosynthesis). Thirteen biochemical steps make up the PCR cycle, but only those leading to the triose phosphates (glyceraldehyde 3-phosphate, GAP, and its isomer, dihydroxyacetone phosphate, DHAP), will be considered here. The triose phosphates are the substrates through which newly assimilated carbon enters primary metabolism, either within the plastid or in the cytosol (**Figure 4**). The amount of triose phosphate (principally DHAP) exported from chloroplasts or used for primary product synthesis within the plastid represents only a small fraction of that synthesized. This is because the biochemical priority of the PCR cycle is the regeneration of the carboxylation substrate ribulose 1,5-bisphosphate (RUBP). If too much carbon is withdrawn from the cycle, further carbon fixation rapidly ceases as the RUBP pool is depleted. In theory, one out of every six triose phosphate molecules is in excess of the demand for RUBP regeneration. In practice, other metabolic demands within the plastid and CO_2 losses due to photorespiration translate into about one in every eight molecules of triose phosphate being available for the

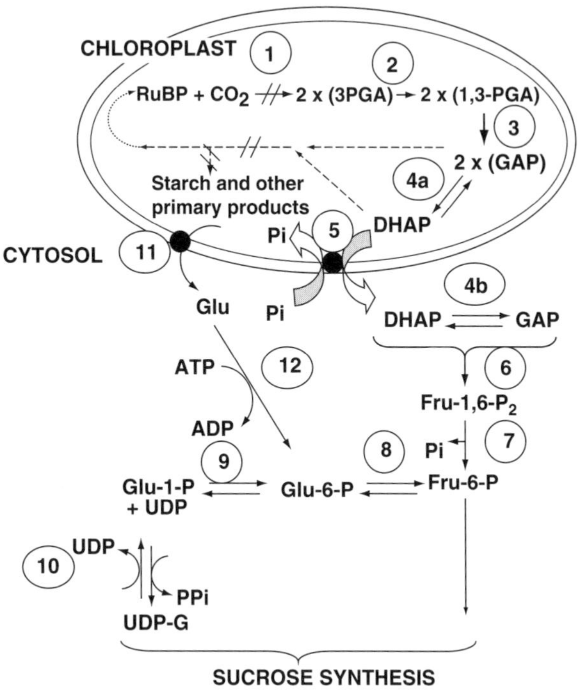

Figure 4 Carbon fixation is initiated by the ribulose bisphosphate carboxylase/oxygenase (Rubisco; EC 4.1.1.39; **1**) mediated carboxylation of ribulose 1,5-bisphosphate (RuBP). An unstable six-carbon intermediate is hydrolyzed to yield two molecules of 3-phosphoglycerate (3PGA). Phosphorylation (by 3-phosphoglycerate kinase; EC 2.7.2.3; **2**) results in two molecules of 1,3-diphosphoglycerate (1,3-PGA), which are reduced (by NADP: glyceraldehyde 3-phosphate dehydrogenase (phosphorylating); EC 1.2.1.13; **3**) to two molecules of glyceraldehyde 3-phosphate (GAP). Isomerization of GAP, by plastidic triose phosphate isomerase (EC 5.3.1.1; **4a**), produces dihydroxyacetone phosphate (DHAP), the principal triose phosphate transported from illuminated plastids. Transport of DHAP into the cytosol is mediated by the triose phosphate translocator (TPT; **5**), located in the inner plastid membrane, and occurs in a strict counter exchange for inorganic phosphate (Pi) from the cytosol. Pathways for the regeneration of RuBP from triose phosphate retained within the plastid are represented by the broken arrows. Under certain conditions (see text), a portion of the retained triose phosphate will be converted to starch within the plastid. DHAP in the cytosol and its isomer GAP (equilibrium concentrations are maintained by a cytosolic triose phosphate isomerase; **4b**) are condensed, by fructose-bisphosphate aldolase (EC 4.1.2.13; **6**), to form fructose-1,6-bisphosphate (Fru-1,6-P_2). The phosphate on carbon 1 is released by fructose-1,6-bisphosphatase (FBPase; EC 3.1.3.11; **7**) to form fructose-6-phosphate (Fru-6-P). Hexose phosphate transformations can now take place through the actions of isomerases (e.g., glucose 6-phosphate isomerase; EC 5.3.1.9), which catalyzes the interconversion of Fru-6-P to glucose-6-phosphate, Glu-6-P; **8**) and mutases (e.g., phosphoglucomutase; EC 5.4.2.2), which catalyzes the interconversion of Glu-6-P to glucose-1-P, Glu-1-P; **9**). The final step in the generation of the precursors for sucrose synthesis (Fru-6-P and uridine diphosphate-glucose, UDP-G) is the transformation of Glu-1-P to UDP-G, a reaction mediated by UDP-glucose pyrophosphophosphorylase (UDPGase; EC 2.7.7.9; **10**). Amylolytic starch breakdown within the chloroplasts at night leads to the release of glucose and maltooligosaccharides, which are exported to the cytosol by a glucose translocator in the inner plastid membrane (**11**). Cytosolic hexokinase (**12**) brings the glucose into the hexose phosphate pool.

synthesis of primary carbohydrate products. Transport of DHAP into the cytosol is mediated by the triose phosphate translocator (TPT), located in the inner plastid membrane, and occurs in a strict counter exchange for inorganic phosphate (Pi) from the cytosol (**Figure 4**).

Once DHAP has been transported through the plastid membrane, the next steps in the pathway to sucrose synthesis are common to the gluconeogenic pathway, and involve the generation of hexose phosphates and interconversions between hexose phosphates and nucleotide sugar phosphates (**Figure 4**) to generate fru-6-P and UDP-G. Before moving on to cover the enzymatic steps unique to sucrose metabolism, it is worth considering the control mechanisms that are represented in the processes and enzymatic reactions described above and in **Figure 4**, as they play a crucial role in the regulation of carbon partitioning. Given the demands for triose phosphates in the plastid and cytosol, the TPT plays a primary role in regulating carbon allocation between these two compartments. The TPT is not regulated by anything other than the plastidic triose phosphate pool size and the availability of Pi in the cytosol, both of which are under extremely tight biochemical regulation. If the cytosolic pool of Pi decreases it is an indication that plastidic triose phosphate synthesis and transport into the cytosol are exceeding the cytosolic demand (closely linked to the capacity to synthesize sucrose in illuminated tissues) because the Pi that should be released during sucrose synthesis would remain "locked up" in the form of phosphorylated intermediates. As the Pi is not available in the cytosol the TPT has nothing to exchange for plastidic triose phosphate, thus transport out of the plastids is curtailed. The consequence of this is that stromal triose phosphate levels increase and Pi levels decrease, a combination that leads to the activation of ADP-glucose pyrophosphorylase (AGPase; EC 2.7.7.27) and starch synthesis within the plastids (**Figure 4**). Regulation of triose phosphate consumption in the cytoplasm is believed to be primarily mediated by cytosolic fructose-1,6-bisphosphatase (FBPase; EC 3.1.3.11). The allosteric regulators AMP, Mg^{2+}, Fru-6-P (the enzyme's reaction product), and a potent inhibitor, Fru-2-6-bisphosphate, all interact with FBPase and effectively "signal" to the plastid, by way of Pi availability, the current cytosolic demand for triose phosphate. Cytosolic triose phosphates are not only utilized for sucrose synthesis, but can also feed into the glycolytic or oxidative pentose phosphate pathways, and demand through these pathways is also essentially "monitored" by FBPase.

The biochemistry of the pathways leading to carbon allocation between the plastids and the cytosol has been very elegantly described, but work with genetic mutants and transgenic plants, in which the activities of key players have been substantially attenuated, do not lead to the devastating metabolic consequences that might be predicted from biochemical principles. This is not an indication that the biochemical predictions are invalid, but rather illustrates the immense metabolic flexibility of plants. This is demonstrated by several experiments in which mutation or transgenic methods have been used to reduce the levels of enzymes involved in cytosolic triose phosphate metabolism, and one where the TPT was attenuated. As might be expected photosynthetic carbon assimilation and growth were reduced, but for the most part plants were able to compensate for the various metabolic lesions. Compensation often resulted in the accumulation of unusually large quantities of starch within the plastids during the light period. This starch was mobilized (to glucose) and transported out of the plastid during the dark period via a plastidic glucose transporter, and entered the cytosolic sucrose biosynthetic pathway at the level of hexose, thereby bypassing the TPT and any limitations for triose phosphate metabolism in the cytosol (**Figure 4**).

Enzymes Unique to Sucrose Biosynthesis

The synthesis of sucrose in source tissues occurs through the concerted action of two cytosolic enzymes, i.e., sucrose phosphate synthase (SPS; EC 2.4.1.14):

$$\text{UDP-G} + \text{Fru-6-P} \rightleftharpoons \text{sucrose-6-phosphate} + \text{UDP} + \text{H}^+$$

and sucrose-6-phosphate phosphatase (SPP; EC 3.1.3.24):

$$\text{sucrose-6-phosphate} + \text{H}_2\text{O} \rightarrow \text{sucrose} + \text{Pi}$$

The reaction catalyzed by SPS has been drawn as being reversible, and under *in vitro* conditions the reverse reaction can take place. However, a large amount of free energy is lost from sucrose-6-phosphate when the phospho–ester bond is hydrolyzed by SPP. Consequently, the action of SPP effectively drives the combined reaction series in the direction of sucrose synthesis. There is evidence that SPS and SPP are physically associated *in vivo*, and may form a metabolic complex in which the product of one reaction feeds directly into the reaction site of the second enzyme, a mechanism that would further favor sucrose synthesis.

SPS is subject to transcriptional and posttranslational control mechanisms. During development, SPS is regulated at the transcriptional level and there is a close relationship between SPS gene expression, SPS enzyme activity, and the development of source status. Transcription of SPS is also sensitive to light.

At the protein level, SPS is subject to allosteric regulation. It is repressed by Pi and activated by G6P. Its sensitivity to these allosteric effectors is greatly influenced by posttranslational modification. Numerous phosphorylation sites have been identified, but, to date, only two of these, serine-158 and serine-424, have been associated with changes in SPS's catalytic activity. A third site, serine-229, is believed to represent a target for 14-3-3 protein binding. Although the physiological significance of this latter interaction has not yet been determined for SPS, the observation is very evocative because 14-3-3 protein interactions have been associated with processes as diverse as metabolic coordination, regulation of gene expression, metabolite transport processes, targeting of protein to various cellular compartments, "tagging" proteins for degradation, and providing "scaffolding" to facilitate protein–protein interactions.

Reversible phosphorylation of SPS at serine-158 (in spinach (*Spinacia oleracea*); serine-162 in corn (*Zea mays*; maize)) has been shown to correlate with the activation state of the enzyme. In some species, changes in the phosphorylation status of this highly conserved serine residue have been associated with light/dark activation/inactivation, i.e., SPS is phosphorylated (inactivated) in the dark and dephosphorylated (activated) in the light. The phosphorylation status of SPS has also been correlated with the enzyme's activation state in heterotrophic sinks and in the leaves of species that do not display obvious changes in maximal enzyme activities when illuminated. Therefore, it is important to realize that, in most species, dephosphorylation/phosphorylation does not affect the maximal activity of SPS (V_{max}), i.e., the enzyme should not be envisioned as being turned "on and off" as a result of its phosphorylation state. Rather, the enzyme's phosphorylation state changes its sensitivity to allosteric regulators and its substrate affinity. This is an important point as the need for sucrose synthesis in source tissues does not cease at the onset of darkness, and sucrose translocation to sinks continues, essentially unabated, throughout the dark period. Some of the sucrose translocated at night will have been synthesized during the light period and stored in the vacuole, but much of it will have been synthesized *de novo* from precursors exported from the plastids, the products of starch reserve breakdown. The latest evidence suggests that the mobilized starch is not metabolized by the plastidic glycolytic pathway to triose-phosphates prior to export through the TPT, as newly assimilated carbon is during the light period, but is exported in the form of glucose (the product of amylolytic starch degradation). Starch breakdown products therefore enter the cytosolic hexose phosphate pool (**Figure 4**), and that not metabolized by competing pathways (**Figure 2**) is converted to sucrose via SPS/SPP and exported from the leaf.

Sucrose is not only a central player in the assimilation, storage, and distribution of photoassimilates in plants, but it has also been proposed to be an osmoprotectant accumulated during periods of abiotic stress (e.g., drought, cold, osmotic). For example, osmotically stressed spinach leaves accumulate sucrose, apparently through activation of SPS. This has recently been shown to be associated with phosphorylation at serine-424 and is therefore distinct from, and appears to override, the activation state changes associated with serine-158 phosphorylation described above. Sucrose accumulation has been reported in response to cold adaptation in *Arabidopsis*, drought stress in peach (*Prunus persica*), and osmotic stress in cultured sweet potato (*Ipomea batatas*) cells to name but a few cases. In some experiments, increases in SPS transcript levels have been reported, but this is not universal. It will be interesting to see whether phosphorylation of serine-424 (or its equivalent) is a general mechanism leading to adaptive sucrose accumulation in stressed tissues.

Other Roles for SPS

SPS is found in all tissues in which *de novo* sucrose synthesis occurs, and it is not confined to sites of primary photosynthetic assimilation. For example, in germinating oil seeds storage lipids are mobilized and converted into sucrose by the combined actions of the glyoxylate cycle, gluconeogenesis, and SPS/SPP. SPS has also been shown to be expressed and active in sinks such as developing corn kernels, fruit tissues, and potatoes (*Solanum tuberosum*). Its role in these tissues is unknown, it may be involved in the resynthesis of sucrose from sucrose hydrolytic products generated during phloem unloading, and it has been proposed to take part in futile cycles in which sucrose is hydrolyzed and resynthesized, a process suggested to play a role in metabolic control.

Regulatory Kinases and Phosphatases

As discussed above, SPS is the substrate for numerous regulatory phosphatases and kinases. Characterization of these regulatory proteins is an area of active research, and a number have been purified and cloned. One finding of particular interest is that the kinase that phosphorylates spinach SPS at serine-158 is a close relative of the SNF1 (sucrose nonfermenting 1) kinase family. As a group, SNF1 kinases are members of kinase cascades important in the regulation and coordination of numerous metabolic pathways in yeast and mammalian systems. They were first described in yeast as global repressors of gene expression during periods of glucose sufficiency. Whether they play such global roles in coordinating metabolism in relation to carbohydrate availability in plants has yet to be determined, but some intriguing results are beginning to emerge. To date, SNF1 related kinases have been implicated not only in the modulation of SPS activity in plants, but have also been shown to regulate nitrate reductase (the primary step in nitrate assimilation) and HMG CoA reductase (a key step in isoprenoid and sterol synthesis) by protein phosphorylation. SNF1 related kinases have also been implicated in regulating the expression levels of sucrose synthase (see below). These and other data are shedding light on the possible mechanisms involved in regulating and coordinating primary metabolism in plants.

Transgenic Manipulation of SPS Levels in Plants

From the above it will be apparent that carbon flux through SPS/SPP represents the sole pathway for sucrose synthesis in photosynthetic source tissues, and that together they may represent a rate limiting step. If this were the case, overexpression of SPS in plant tissues might be expected to enhance sucrose synthesis and increase yields. To date, at least four independent laboratories have reported on transgenic monocotyledonous and dicotyledonous species in which either corn or spinach SPS was expressed under the control of an array of promoters. In all cases, increased SPS levels were detected (up to fivefold increases above controls), and some spectacular increases in yield and changes in carbohydrate profiles have been reported. However, the yield increases have proved to be highly inconsistent and greatly dependent on developmental and environmental factors. These results should not be taken as an indication that the SPS/SPP pathway is unimportant in sucrose synthesis and carbon partitioning, but are rather an indication of the ability of plants to maintain metabolic control, even when the potential for greater metabolic flux through a pathway exists. How this is achieved is not clear, but much of the heterologous SPS in the transgenic plants appeared to have been downregulated through phosphorylation. However, point mutations to remove the serine phosphorylation site did not lead to further yield increases and greater partitioning to sucrose, indicating that other levels of regulation were involved.

Temporary Storage

During the light period, the production of soluble photoassimilated carbohydrate in source leaves may exceed the capacity of the translocation system to distribute it to sinks. Under such circumstances, sucrose (and polyols) are accumulated in vacuoles, which act as temporary storage buffers. What makes this process significant is that the majority of a photosynthetic cell's volume ($>70\%$) is vacuole and so represents a substantial reservoir for sucrose accumulation. The mechanisms involved in the translocation of sucrose into the vacuole remain obscure, and, to date, no tonoplastic sucrose transporter has been characterized at the molecular level, although there is substantial physiological evidence for the presence of tonoplastic sucrose transporters in sink tissues.

Some species, particularly the temperate grasses, use sucrose in the vacuoles of source tissues for the synthesis of fructose polymers (fructans). The enzymes involved use sucrose to prime the reaction (each polymer chain starts with sucrose), and then extend the primer by fructose unit addition, releasing free glucose. The polymers can be quite elaborate in their branching patterns, and can vary in degree of polymerization (dp) from 3 (1 sucrose + 1 fructose) to dp$<$200, depending on the species. Energy released during the hydrolysis of the sucrose glycosydic bond helps drive the polymerization reaction. Fructan accumulation in the leaves of temperate grasses has been associated with periods when sucrose supply exceeds demand and during cold acclimation. For this latter reason, fructans have been associated with cold tolerance in species that accumulate them.

Long-Distance Transport

Sucrose synthesized in source tissues is translocated to heterotrophic tissues in the phloem. Phloem transport occurs by mass flow, a process that is dependent on a pressure gradient between the source, where osmotically active compounds are accumulated, and the sink, where osmotically active compounds are "removed" (usually by sequestration into large molecules, with a low osmolality, e.g., starch). For such a mechanism to function, active loading of assimilates, against a concentration gradient, has to take place at the source. Transport of sucrose from the sites of synthesis to the phloem sieve element/companion cell complex (SE/CC) can occur via the symplast (cellular route via plasmadesmatal connections), a characteristic of species with very abundant plasmodesmatal connections between the SE/CC and adjacent cells. A second phloem loading pathway occurs by an apoplastic route, a characteristic of species that effectively have no such plasmodesmatal connections. In these latter species, sucrose is believed to move to the vicinity of the vasculature, via the symplast, but to gain access to the SE/CC complex, assimilates are released into the extracellular space (apoplast) and actively taken up into the phloem complex via specific carriers. Yeast complementation screens have helped identify a plethora of plant plasma membrane transporters, a subgroup of which have been shown to be sucrose/proton symporters. To date, six putative sucrose transporter genes have been identified in *Arabidopsis*, and, in a recent study, one of the putative transporters associated with phloem tissues was inactivated by a disruptive insertional mutation. The mutant plants grew very slowly and accumulated large amounts of starch in their source leaves. They also failed to translocate radioactive sucrose (fed to source leaves) to the roots, indicating that the transporter played an important role in sucrose phloem loading and translocation. Some phloem associated sucrose transporter homologs appear to have no capacity to transport sucrose (despite structural similarity to bona fide transporters), and it is proposed that they may act as transmembrane sucrose sensors.

Phloem unloading mechanisms in sinks are thought to involve symplastic transport out of the SE/CC complex. Seeds lack vascular connections between the maternal, and filial tissues and sucrose is released into the apoplast at the interface to be taken up by the filial tissues by energy dependent plasma membrane sucrose/proton symporters. There is considerable evidence, from work with grain legumes, that the efflux of sucrose from maternal tissues at the maternal/filial interface is sensitive to turgor pressure, which in turn must be governed by the ability of the filial tissues to remove the sucrose and other carbohydrates from the sites of phloem unloading, and to sequester them into less osmotically active storage macromolecules.

Sucrose Hydrolysis in Heterotrophic Sink Tissues

In any cell that cannot meet its carbohydrate requirements for maintenance and growth by on-site *de novo* carbon assimilation, the capacity to hydrolyze sucrose (or a related carbohydrate) to a metabolizable form (the hexoses) is a prerequisite. Sucrose delivered to these tissues is returned to the metabolic pool by either the invertase pathway or the sucrose synthase pathway:

$$\text{Sucrose} + \text{UDP} \rightleftharpoons \text{UDP-G} + \text{fructose.}$$

Each pathway is characterized by numerous isoforms, the expression of which is subject to developmental (sink/source), temporal, and environmental factors, as well as the tissue's metabolic status. As a general rule, invertase and SS expression is reciprocally related and the "preferred" pathway is loosely associated with the nature of the sink. In young tissues undergoing rapid expansion that have a high metabolic demand, the hydrolysis of sucrose to glucose and fructose by invertases provides both osmotica, to drive turgor dependent cell expansion and hexoses for metabolism. This pathway is also likely to be important in bulb, fruit, petiole, and taproot tissues that store glucose and fructose. In developing sinks that are laying down starch reserves, degradation via SS is thought to be favored as its product, UDP-G, can feed directly or after transformation into ADP-G, into the synthesis of structural and nonstructural carbohydrate macromolecules.

In monocotyledonous species, SS is encoded by at least two independent loci (sh1 and sus1 in corn) that display differential expression patterns; dicotyledonous species also appear to have at least two independent copies. Corn mutants lacking sh1 have a compromised ability to accumulate starch in the endosperm of kernels, but appear unaffected in any other aspect of growth and development. Expression of sh1 and sus1 genes is regulated by tissue carbohydrate status, but their expression is reciprocally related. In root tips, maximal expression of sh1 occurred under conditions of sugar depletion, whereas the sus1 gene product was more highly expressed under sugar repletion, and vice versa. It is now recognized that not only are genes involved in sucrose metabolism influenced by tissue levels of sucrose, glucose, and fructose, but that whole suites of genes appear to be influenced in a coordinated manner. The terms feast and famine have been coined to describe these effects. In general, high levels of carbohydrates repress the expression of genes involved in carbon acquisition and upregulate those involved with resource allocation and sequestration (the feast genes). Carbohydrate depletion reverses these effects and the so-called famine genes are activated; carbon assimilation and reserve mobilization are upregulated.

As with SPS and SPP, SS is a cytosolic enzyme. However, SS can also be associated with the plasma membrane and forms associations with the cytoskeleton under certain conditions. Membrane association of SS was first reported in developing cotton fibers, and it was proposed that plasma membrane associated SS might be involved in channeling sucrose (via UDP-G) directly into cellulose and callose synthase complexes. Subsequently, it was shown that SS can be phosphorylated, at serine-15, and that the phosphorylated form was soluble. Whether the SS implicated in cellulose synthesis in cotton fibers represents the dephosphorylated form has not been definitively shown, but it seems likely as serine-15 is highly conserved among the SS genes so far sequenced. The significance of SS association with the cytoskeleton has yet to be determined as are the precise conditions that cause this partitioning, although preliminary evidence suggests that it may be related to sucrose concentrations. How compartmentation of SS affects its kinetic properties and activity is not clear. SS activity is freely reversible *in vitro*, and there is evidence from *in vivo* labeling studies that it can synthesize sucrose under certain conditions. How important a role SS plays in sucrose synthesis in sink tissues that also have SPS and SPP activity is not known, but will obviously be dependent on the concentrations of substrates and allosteric regulators of SPS and inhibitors of SS, which is strongly inhibited by hexoses, particularly fructose.

The term invertase is a generalized one and in plants there are numerous isoforms that are characterized by their pH optima (acid, basic, and neutral) and solubility (soluble or insoluble). They all catalyze the irreversible hydrolysis of sucrose to glucose and fructose:

$$\text{Sucrose} + H_2O \rightarrow \text{glucose} + \text{fructose}$$

Acid invertases are located in the vacuole (soluble) and extracellular (soluble and insoluble) compartments in plants where they are believed to modify soluble sugar profiles, although recent transgenic experiments have led to the realization that they play far more fundamental roles in carbon partitioning and plant development. Very briefly, antisense repression of cell wall and vacuolar invertase in carrots (*Daucus carota*) caused severe stunting and deformation during early development, and "normal" seedlings could only be recovered if they were cultured on sucrose containing media supplemented with glucose and fructose. At later stages of development, the transgenic plants were capable of autotrophic growth in soil, but the root to shoot ratio was massively modified, with the severity of the symptoms dependent on which invertase isoform had been repressed. Plants with lowered cell wall invertase were essentially unable to develop tap roots but shoot growth was greatly enhanced. Tap roots did develop on plants deficient in vacuolar invertase, but they were much smaller than those on the controls. Carbohydrate contents and compositions in the tissues of transgenic plants were also considerably altered with the shoots of invertase deficient plants having much higher sucrose and starch levels than the controls. These and other experiments, in which ectopic expression of yeast invertases in cytosolic and vacuolar compartments of potato have been investigated, illustrate the importance of invertases in carbohydrate partitioning and plant development. They also give indirect evidence that the metabolism of sucrose also plays a vital role.

As with SS, the expression of acid invertase genes is influenced by spatial, developmental, temporal, and environmental factors, as well as tissue metabolic status. The soluble acid invertases ivr1 and ivr2 have been classified as famine and feast genes, respectively, because their expression is sensitive to tissue carbohydrate status in corn root tips. Aside from modulation of gene expression, little is known about the regulation of invertase enzyme activities; although an invertase inhibitor protein has been identified in the cell walls of tobacco (*Nicotiana tabacum*), its physiological significance has yet to be determined.

Basic or neutral invertases are soluble and located in the cytosol where they are thought to cleave sucrose in order to provide hexoses for maintenance metabolism. This is somewhat supported by the observation that expression of a yeast invertase in the cytosol of potato led to enhanced rates of respiration at the expense of starch synthesis. The first neutral invertase from plants has recently been cloned, and this will undoubtedly lead to a greater insight into their roles in sucrose metabolism. As with the acid invertases, little is known about the regulation of neutral invertase enzyme activity *in vivo*.

Sugar Signaling and Final Comments

It is clear that sucrose plays a central role in the metabolism of higher plants and that its synthesis, distribution, and catabolism are subject to a whole array of control mechanisms. Not only are these control mechanisms important in maintaining sucrose metabolism and partitioning at levels sufficient to support growth and development, but they must also be sensitive to the metabolic status of the plant as a whole. The evidence that carbohydrate status appears to play a role in the coordination of whole suites of enzymes, by regulating their expression levels, is intriguing as it is consistent with what is known about carbohydrate modulation of metabolism in yeast and mammalian systems. Carbohydrate metabolism in photosynthetic higher plants is undoubtedly unique, and drawing direct analogies from nonplant systems may be misleading, but there seems little doubt that plants, like other organisms, can sense their own carbohydrate status and modulate metabolism appropriately. What is unclear at present is to what extent the signals for metabolic regulation involve direct detection of the carbohydrates themselves as opposed to more indirect mechanisms that meter their rate of metabolism, either directly or indirectly, through effects on cellular energy status. There is evidence for each. Sucrose responsive elements have been identified in the 5′ promoter regions of some, but not all, genes shown to be responsive to sucrose levels, and evidence for plasma membrane sucrose sensors is beginning to emerge. There is also a developing body of literature showing that the rate of glucose phosphorylation, by specialized isoforms of hexokinase, may be used to monitor tissue carbohydrate status. Other mechanisms are also likely to come into play and carbohydrates other than sucrose and its direct metabolites may also be involved in sugar sensing and the regulation of metabolic status. For example, the disaccharide trehalose (α-D-glucopyranosyl 1-α-D-glucopyranoside) and its metabolites have recently been implicated as having roles in sugar sensing in plants. More detailed information on this and its implications for plant productivity can be obtained from references given in the Further Reading section.

With respect to the mechanisms by which the perceived signals are transduced into metabolic responses, the discovery of the involvement of 14-3-3 proteins and the SNF1-related kinases in sucrose metabolism is highly suggestive, given their known roles in metabolic regulation in other systems. Early evidence in plants shows that they may provide mechanisms by which biochemical cross-talk can occur in order to keep essentially competing pathways (e.g., sucrose synthesis, glycolysis, nitrogen assimilation, lipid and secondary product synthesis) coordinated and in sync with the available resources. *Arabidopsis* mutant screens have led to the identification of numerous genes involved in the sensing of carbohydrates, and one fascinating aspect of this work is that a number of the genes identified have also been shown to be involved in hormone responses. As more of these mutants and mechanisms are characterized, a clearer picture of the role of sucrose in the global regulation of plant growth and development will begin to emerge.

List of Technical Nomenclature

ADP	Adenosine diphosphate.
AMP	Adenosine monophosphate.
ATP	Adenosine triphosphate.
DHAP	Dihydroxyacetone phosphate.
FBPase	Fructose-1,6-bisphosphatase.
Fru	Fructose.
Fru-1,6-P_2	Fructose-1,6-bisphosphate.
Fru-6-P	Fructose-6-phosphate.
GAP	Glyceraldehyde 3-phosphate.
Glu	Glucose.
Glu-6-P	Glucose-6-phosphate.
Glu-1-P	Glucose-1-P.
ivr 1 and ivr 2	Transcripts for soluble acid invertases.
Pi	Inorganic phosphate.
3PGA	3-Phosphoglycerate.
1,3-PGA	1,3-Diphosphoglycerate.
Rubisco	Ribulose bisphosphate carboxylase/ oxygenase.
RuBP	Ribulose 1,5-bisphosphate.
sh1	Shrunken 1 locus of corn sucrose synthase.
SNF1	Sucrose nonfermenting 1 kinase.
SPP	Sucrose phosphate phosphatase.
SPS	Sucrose phosphate synthase.
SS (Susy)	Sucrose synthase.
Sus1	Corn sucrose synthase locus 1.
TPT	Triose phosphate translocator.
UDP	Uridine diphosphate.
UDP-G	Uridine diphosphate-glucose.
UDPGase	UDP-Glucose pyrophosphosphorylase.
V_{max}	Maximal catalytic velocity.

See also: **Genetic Modification of Primary Metabolism**: Carbohydrates. **Photosynthesis and Partitioning**: C_3 Plants; Sources and Sinks. **Primary Products**: Starch. **Seed Development**: Nutrient Loading of Seeds. **Water Relations of Plants**: Phloem.

Further Reading

Avigad G and Dey PM (1997) Carbohydrate metabolism: storage carbohydrates. In: Dey PM and Harborne JB (eds) *Plant Biochemistry*, pp. 143–204. London: Academic Press.

Bowyer JR and Leegood RC (1997) Photosynthesis. In: Dey PM and Harborne JB (eds) *Plant Biochemistry*, pp. 50–110. London: Academic Press.

Brownleader MD, Harborne JB, and Dey PM (1997) Carbohydrate metabolism: primary metabolism of monosaccharides. In: Dey PM and Harborne JB (eds) *Plant Biochemistry*, pp. 111–141. London: Academic Press.

Dennis DT and Blakeley SD (2000) Carbohydrate metabolism. In: Buchanan BB, Gruissem W, and Jones RL (eds) *Biochemistry and Molecular Biology of Plants*, pp. 630–675. Rockville MD: American Society of Plant Physiologists.

Farrar J, Pollock C, and Gallagher J (2000) Sucrose and the integration of metabolism in vascular plants. *Plant Science* 154: 1–11.

Geiger DR and Servaites JC (1994) Diurnal regulation of photosynthetic carbon metabolism in C_3 plants. *Annual Review of Plant Physiology and Plant Molecular Biology* 45: 235–256.

Gibson SI (2000) Plant sugar-response pathways. Part of a complex regulatory web. *Plant Physiology* 124: 1532–1539.

Halford NG and Hardie DG (1998) SNF1-related protein kinases: Global regulators of carbon metabolism in plants? *Plant Molecular Biology* 37: 735–748.

Koch KE (1996) Carbohydrate-modulated gene expression in plants. *Annual Review of Plant Physiology and Plant Molecular Biology* 47: 509–540.

Lalonde S, Tegeder M, Throne-Holst M, Frommer WB, and Patrick JW (2003) Phloem loading and unloading of sugars and amino acids. *Plant, Cell and Environment* 26: 37–56.

Loescher WH and Everard JD (1996) Sugar alcohol metabolism in sinks and sources. In: Zamski E and Schaffer AA (eds) *Photoassimilate Distribution in Plants and Crops: Source-Sink Relationships*, pp. 185–207. New York: Marcel Dekker.

Malkin R and Niyogi K (2000) Photosynthesis. In: Buchanan BB, Gruissem W, and Jones RL (eds) *Biochemistry and Molecular Biology of Plants*, pp. 568–628. Rockville MD: American Society of Plant Physiologists.

Patrick JW (1997) Phloem unloading: Sieve element unloading and post-sieve element transport. *Annual Review of Plant Physiology and Plant Molecular Biology* 48: 191–222.

Paul M, Pellny T, and Goddijn O (2001) Enhancing photosynthesis with sugar signals. *Trends in Plant Science* 6(5): 197–200.

Roberts MR (2000) Regulatory 14-3-3 protein–protein interactions in plant cells. *Current Opinion in Plant Biology* 3: 400–405.

Smeekens S (2000) Sugar-induced signal transduction in plants. *Annual Review of Plant Physiology and Plant Molecular Biology* 51: 49–81.

Stitt M (1999) The first will be last and the last will be first: non-regulated enzymes call the tune? In: Bryant JA, Burrell MM, and Kruger NJ (eds) *Plant Carbohydrate Biochemistry*, pp. 1–16. Oxford: BIOS.

Stitt M and Sonnewald U (1995) Regulation of metabolism in transgenic plants. *Annual Review of Plant Physiology and Plant Molecular Biology* 46: 341–368.

Van Bel AJE (1993) Strategies of phloem loading. *Annual Review of Plant Physiology and Plant Molecular Biology* 44: 253–281.

Winter H and Huber SC (2000) Regulation of sucrose metabolism in higher plants: location and regulation of activity of key enzymes. *Critical Reviews in Biochemistry and Molecular Biology* 35(4): 253–289.

Sources and Sinks

A J E van Bel, Justus-Liebig-University, Giessen, Germany
C E Offler and J W Patrick, University of Newcastle, New South Wales, Australia

Introduction

Sources and Sinks

Biomass production of higher plants essentially depends on the photosynthetic activities of mature green leaves, since some 90% of plant biomass consists of carbon, hydrogen, and oxygen. Photosynthetically reduced carbon products (photoassimilates) are transported from sources, sites of production, to sinks, sites of growth and storage. There, photoassimilates support maintenance and growth of tissues with a negative carbon balance (such as terminal sinks: root and shoot apices, flower initials, etc.). Photoassimilates in excess of growth and maintenance requirements are accumulated in parenchyma storage tissues of organs such as stems, roots, fruits, seeds, bulbs, and tubers.

Functional Phloem Zones

Photoassimilates are transported throughout plants in sieve tubes of the phloem. In angiosperms, phloem transport involves several successive steps which occur in three functional zones of the phloem, each of which has a specific task (**Figure 1**). Following production in the leaf mesophyll, photoassimilates are loaded into sieve element/companion cell complexes (SE/CC complexes) in the collection of phloem. From there, photoassimilates are translocated by the transport phloem, located in leaf major veins, petioles, stems, and roots, to be distributed

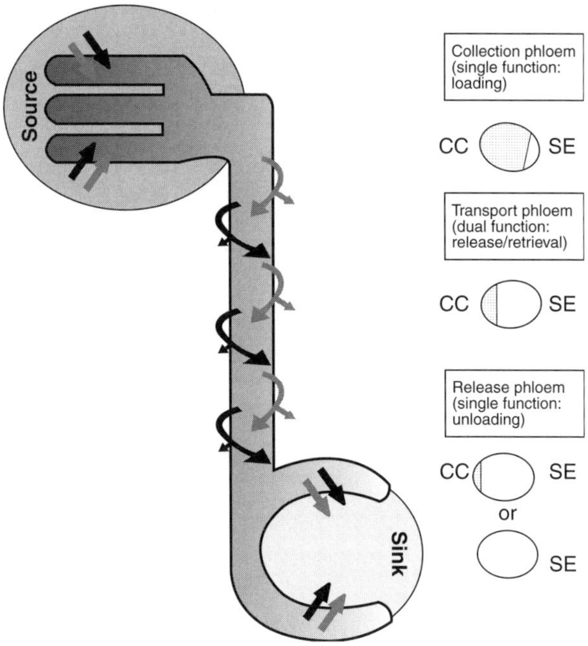

Figure 1 A dynamic version of the Münch pressure flow model, the local fluxes of photoassimilates (stippled arrows) and water (black arrows) and the relative proportion of sieve elements (SE) and companion cells (CC) in the respective phloem zones. Photoassimilates are translocated via the phloem through essentially leaky instead of hermetically sealed pipes (sieve tubes). The solute content, and implicitly the turgor, are controlled by release/retrieval systems located on the plasma membrane of the sieve element/companion cell complexes (SE/CC complexes). The retrieval mechanisms are energized by the proton-motive force. Differential release/retrieval balances along the pathway control the influx/efflux of sugars and water in the respective phloem zones. In collection phloem (where phloem loading takes place), the retrieval or uptake dominates, in the release phloem (where phloem unloading takes place) the release dominates. In transport phloem having a dual task (nourishment of axial sinks along the pathway and terminal sinks), the balance between release and retrieval varies with the requirements of the plant. The gradual loss of solute and the commensurate amounts of water has been ascribed to a slightly decreasing proton-motive gradient along the phloem pathway. Alternatively, the relative size reduction of companion cells along the source-to-sink path may explain a decreasing retrieval capacity of the SE/CC complexes in the direction of the sink. The massive photoassimilate delivery in the sinks is assigned to symplasmic and/or apoplasmic loss of photoassimilates from the SEs driven by the high consumption and/or storage rates in the sink tissues. Modified, with permission from van Bel AJE (2003) The phloem, a miracle of ingenuity. *Plant, Cell and Environment* 6: 125–149. Oxford: Blackwell Publishing.

between sink organs. In sinks, photoassimilates are unloaded from the sieve elements of release phloem into growing or storage cells, where they are metabolized or sequestered, respectively.

Phloem Loading and Unloading

Phloem loading includes the transport of photoassimilates through the pre SE-pathway and their collection into the SEs. These events take place in the phloem loading zone (**Figure 2A**). Phloem unloading encompasses release from SEs and the subsequent transport through the post-SE pathway and takes place in the phloem unloading zone (**Figure 2B**). It is likely, that the differential local functions of the SE/CC complexes, reflected by decreasing volume ratios between CC and SEs towards sinks, strongly draw on their physiology (**Figure 1**). Moreover, it should be noted that collection phloem being located in a thin layer of tissue is more directly exposed to environmental variations (e.g., in temperature or drought) than transport phloem which is embedded in several buffer layers. As for release phloem, the direct environmental impact is dependent on the nature and location of the sink.

The Mass Flow Concept

Photoassimilates are concurrently the cargo and fuel of phloem transport (**Figure 3**). Phloem loading and unloading of photoassimilates drive mass flow through the sieve tubes. Loading of photoassimilates boosts turgor at the source ends while unloading of photoassimilates reduces turgor at the sink ends of the sieve tube system. The turgor pressure difference drives a bulk flow (J_v) through the sieve tubes from source to sink according to the Hagen–Poiseuille equation:

$$J_v = (\Delta P \times \pi r^4)/(8\eta \times \Delta L) \qquad [1]$$

in which ΔP is the difference in turgor pressure between source ($\Psi_p$so) and sink ($\psi_p$si) ends of the sieve tubes, r the radius of the channel (i.e., the pores of the sieve plates), η is the viscosity of the sieve tube sap, and ΔL the length of the sieve tube channel. The turgor pressure difference (ΔP) equals $\psi_p\text{so} - \psi_p\text{si}$ or $(\psi_{\pi o}\text{so} - \psi_{\pi i}\text{so}) - (\psi_{\pi o}\text{si} - \psi_{\pi i}\text{si})$, in which $\psi_{\pi o}$ is the water potential of the local apoplasm and $\psi_{\pi i}$ is the osmotic potential in the corresponding sieve tube section (**Figure 3**). In conclusion, turgor pressure difference throughout the sieve tubes essentially results from source supply and sink consumption of photoassimilates.

Under conditions limiting photosynthesis, sink demand exceeds production of photoassimilates and sinks have spare capacity for growth and storage (source-limited growth). Mostly, when sink demand limits amounts of photoassimilates transported, sources have a spare capacity for photosynthesis (sink-limited growth). Sinks control photoassimilate production by setting the size to which the leaf surface expands during the sink stage and, once fully expanded, by feedback control of leaf photosynthesis by sink demand for photoassimilates.

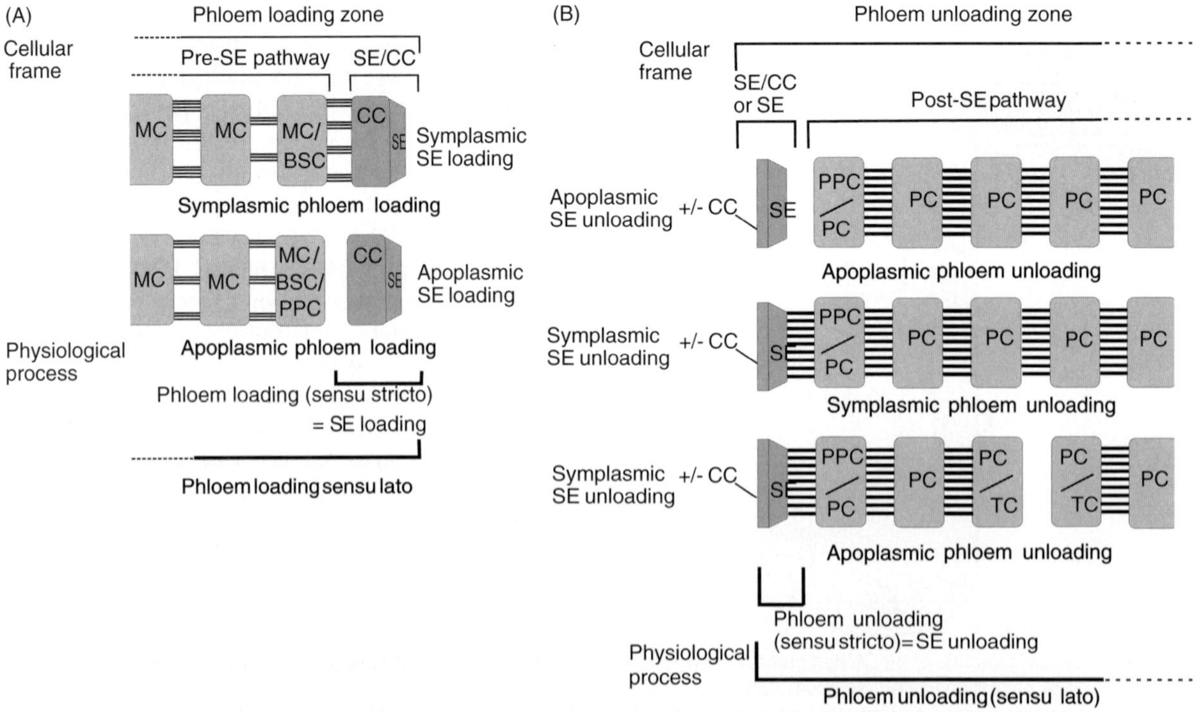

Figure 2 Survey of the terminology regarding phloem loading/unloading processes. (A) Phloem loading *sensu lato* is a series of transport events taking place in an array of cells (MC, mesophyll cell; BSC, bundle sheath cell; PPC, phloem parenchyma cell; CC, companion cell; SE, sieve element) designated the phloem loading zone. It is difficult to define the length of the array as indicated by dashing. Phloem loading into the sieve element/companion cell complex (SE/CC), the transport unit of the plants, is called SE loading or phloem loading *sensu stricto*. SE loading is preceded by photoassimilate transport through the preSE pathway. Phloem loading *sensu stricto* and *sensu lato* are often confused in the literature. If the SE/CC complex is symplasmically detached from the surrounding parenchyma cells, SE loading is defined as being apoplasmic. In the case where SE/CC complexes are symplasmically coupled to the bordering parenchyma cells and phloem loading is not reduced by membrane transport inhibitors, SE loading is defined as being symplasmic. In the latter instances, phloem loading (*sensu lato*) is defined as being apoplasmic or symplasmic, respectively. (B) Phloem unloading *sensu lato* is a series of transport events taking place in an array of cells (SE with or without CC; PC, parenchyma cell; TC, transfer cell) designated as the phloem unloading zone. In principle, definitions for phloem loading and unloading are made on the same basis (see above) with one additional definition. Phloem unloading *sensu lato* is defined as apoplasmic, if there is an apoplasmic step present at some distance removed from the SE/CC (lower row).

Feedforward control by environmental effects (e.g., photon flux density, elevated carbon dioxide, mineral nutrition) on photosynthesis determines photoassimilate supply to the collection phloem. In contrast, temperature and water stress have a greater impact on sink activity, as cell division and expansion are more sensitive to these factors than photosynthesis. Feedback control to source leaves is through leaf sugar levels regulating expression of genes encoding sugar transporters as well as sugar metabolizing and photosynthetic enzymes.

Demarcation of the Focus Area

Both sources and sinks contain separate compartments for metabolic processing and transport, being located at either end of the transport phloem. Major environmental impact on overall photoassimilate flow could be on any of these compartments. Environmental effects on photosynthesis and respiration are not discussed, although they are integral components of source and sink function. The focus is on how environmental factors – in particular, drought and temperature – affect development and functioning of source and sink organs. In conjunction with environmental impacts on sources and sinks, environmental effects on development and functioning of the phloem system is also overviewed.

Effects on Sources

Leaf Development

Leaf primordia originate from a foliar buttress transforming into an axillar phyllopodium which, in turn, extends laterally by a complex interaction of several meristems. During these and subsequent developmental stages, when the major veins differentiate, the leaf is a net importer of reduced carbon and functions as a sink. Further expansion of the leaf

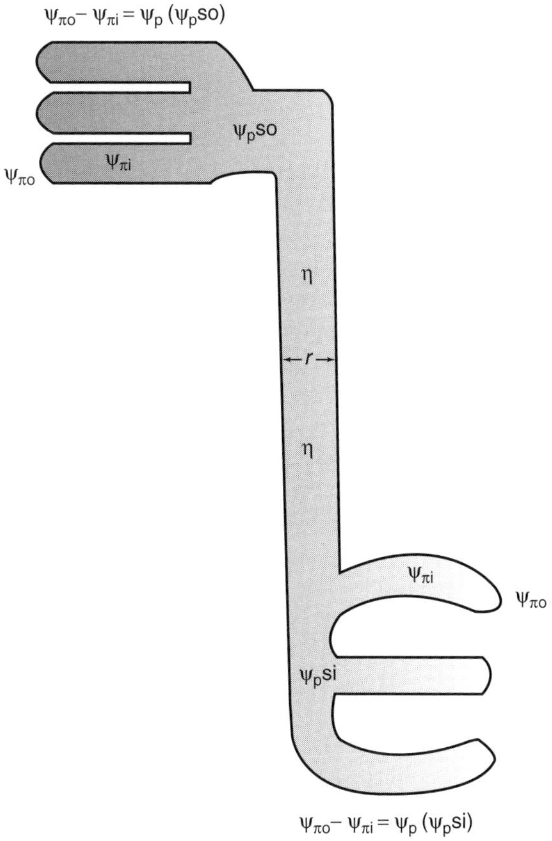

Figure 3 The driving force for pressure flow depends on the turgor pressure difference between the sieve tube ends in source and sinks. The local turgor pressure (ψ_p) in source ($\psi_p$so) and sink ($\psi_p$si) results from the difference in water potential between the local apoplasm ($\psi_{\pi o}$) and the adjacent sieve element cytoplasm ($\psi_{\pi i}$). Mass flow is further determined by the radius (*r*) of the tube (in this instance the functional radii of the sieve pores) and the viscosity (η) of the solute (mainly the viscosity of the sugars translocated) (see eqn [1]).

coincides with differentiation of stomata and gas spaces and maturation of the minor veins collectively transforms the developing leaf into a photoassimilate source.

In dicotyledons, the vein system is mostly reticulate. From a mid-vein (lowest-order or first-order vein), vein orders of reducing complexity diverge. Leaves are compartmentalized into areoles, mesophyll areas enclosed by higher-order veins in which the highest-order vein ends blindly. The venal network develops initially from the leaf base. During this stage, the lower-order veins differentiate acropetally. After completion of the major vein system, a second basipetal wave of vein differentiation begins. Structural and functional vein maturation proceeds in a narrow distinctly demarcated band.

In monocotyledons, the vein system extends from the leaf base and the longitudinal veins are mostly parallel to one another through the leaf blade. The vein system consists of small, intermediate and large longitudinal veins interconnected by transverse veins (also termed lateral veins or cross-veins), the smallest components of the venal network. Structural and physiological evidence suggests that the small and intermediate longitudinal veins are responsible for phloem loading. Strong support in favor of the smaller longitudinal veins as the sites of phloem loading is that aphids exclusively feed on small and intermediate veins of barley (*Hordeum vulgare*). Transverse veins may be engaged in lateral transport of photoassimilates between the longitudinal veins and in temporary storage rather than in phloem loading. As in dicotyledons, vein differentiation in monocotyledons takes place in two waves: an acropetal differentiation of the major longitudinal veins and subsequent basipetal differentiation of the smaller longitudinal and transverse veins.

Environmental Effects on Leaf Development

Environmental parameters mainly affect leaf development by impacting on expansion of the leaf surface (leaf size) and hence photosynthetic machinery to generate photoassimilates for phloem loading. Leaf size is a function of the mitotic activity in the leaf primordia, the rate of cell expansion, and the plastochron index.

Effects of temperature on leaf growth are strongly related to a species-specific temperature-dependence of cell division. Arctic and mountain species and plants flowering in spring start growing at temperatures just above zero. In temperate-zone, subtropical, and tropical plants, the onset of growth by division and cell expansion is observed at 10 °C, 12 °C, and 15 °C, respectively. The optimal temperature for leaf growth is 30–40 °C for tropical plants, and 15–30 °C for other plants.

Salt stress reduces leaf size and sometimes induces leaf succulence. In general, the number of mesophyll cells is reduced in contrast to the vascular tissues. Most responses to salt stress are similar to those to drought stress, as a salt stress imposed by lack of water is what is sensed by the plant.

Shading triggers a range of morphological effects which are consistent with plants striving for sufficient illumination. This photomorphogenic shading syndrome includes elongation of the organs (internodes, petioles, grass leaves) involved in lifting the photosynthetic apparatus and a coincident lesser investment in photosynthetic activities. Furthermore, leaf thickness and leaf surface area are reduced as well as chlorophyll synthesis and plastid development.

The first wave of vein development – the emergence of the major vein pattern – seems to depend primarily on genetic programs for differentiation and spacing. Minor vein patterns are more variable between individual leaves and between areoles within one leaf. The variations may be brought about by local physiological triggers and environmental factors experienced by the differentiating cells. Factors that increase drought stress such as a low soil water availability, low air humidity, and high wind speed increase leaf vein density in herbs. The consequences of a more negative leaf water potential seem to be compensated for by a higher vein density. However, the relationship is ambiguous. Many tropical rainforest species, which generally experience less negative water potentials than plants in other habitats, exhibit relatively high vein densities.

Physiological Events in the Phloem Loading Zone

Photoassimilate pool size available for phloem loading A number of cytosolic enzymes are involved in making photoassimilates available for phloem translocation in the form of sucrose. Some key enzymes in carbohydrate processing, notably sucrose phosphate synthase (SPS), respond to environmental influences (**Figure 4**). Several domains reactive to environmental conditions have been

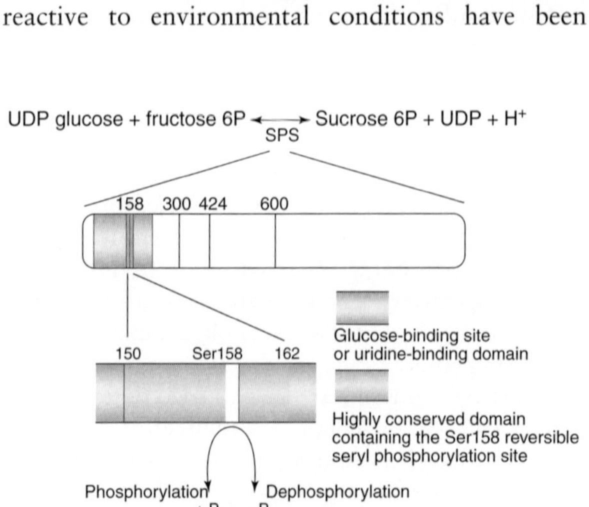

Figure 4 Sucrose phosphate synthase (SPS), an enzyme responsible for sucrose synthesis has domains which are sensitive to environmental factors. For instance, light activates SPS by dephosphorylation of the Ser158 amino acid residue. The light effect is probably indirect as inorganic phosphate (P_i) is withdrawn from the mesophyll cytosol by chloroplast reactions associated with photosynthesis. Likewise, environmental conditions that cause osmotic stress activate SPS by a kinase-mediated reaction taking place at the Ser424 site. By consequence, sucrose synthesis and phloem loading are controlled by environmental factors.

found in SPS. For instance, a Ser158 site has been identified responsible for the phosphorylation of SPS. This light/darkness modulation of SPS activity is a key event in the partitioning of photoassimilates between sucrose and starch and hence in making sucrose available for export. Furthermore, activation of SPS is controlled by drought stress, N nutrition and osmotic stress.

The activation/inactivation of SPS exemplifies the way in which numerous metabolic pathways are controlled by environmental conditions. The responses of SPS show that the relationship between enzyme activity and an environmental agent may often be indirect. For instance, light-induced SPS activation by dephosphorylation of the Ser158 site (**Figure 4**) probably results from the withdrawal of inorganic phosphorus from the mesophyll cytosol by chloroplasts to act in photosynthesis-associated metabolic reactions. This example moreover demonstrates how complex the relationships are: SPS activation at the Ser158 and Ser424 residues by light and osmotic stress, respectively, is mediated by different SPS kinases.

Intercellular photoassimilate transport in the mesophyll In the pre-SE pathway, photoassimilates successively move from the mesophyll cells (MCs) through one or occasionally two or three layers of bundle sheath cells (BSCs) to the vascular parenchyma cells (VPCs) or directly to the SE/CC complexes (**Figure 2A**). As symplasmic movement is generally accepted as the transport mechanism throughout this trajectory, plasmodesmata seem to be the corridor for photoassimilate transport. Documentation of plasmodesmal frequencies (PFs, total number of plasmodesmata per cellular interface) and plasmodesmal densities (PDs, number of plasmodesmata per square unit of cell interface) suggests a preferential radial transfer in the pre-SE pathway.

In dicotyledons, MCs of a poorly defined layer present in the plane of the minor veins in many leaves, possess significantly higher PDs than the other MCs suggesting a function in photosynthate collection similar to that of the paravenal mesophyll in legumes. In monocotyledons, the few records available show a relatively high plasmodesmal connectivity between the nonvascular cells contrasting with a low, in many instances nonexistent, symplasmic coupling between the vascular cells. These observations point to a symplasmic supply route of photoassimilates in the mesophyll followed by apoplasmic phloem loading (**Figure 2A**).

Downregulation of the diameter of plasmodesmata between the mesophyll cells is expected to decrease phloem loading, upregulation will do the opposite. Environmental regulation of the permeability of plasmodesmata (plasmodesmal gating) has hardly been investigated. Temperature may exert an effect on plasmodesmal gating. For instance in cereals, virus spread, which depends on plasmodesmal permeability, is negatively influenced by low temperature. Drought may also affect the loading pathway, as plasmodesmal gating may be sensitive to cell turgor. A difference in turgor between adjacent cells in leaf hairs of tobacco (*Nicotiana tabacum*) was shown to cause closure of plasmodesmata at their interface.

Modes of phloem loading In dicotyledons, the degree of symplasmic connectivity between the SE/CC complex and adjacent parenchyma cells (PPs) corresponds roughly with the ultrastructure of the CCs (**Figure 5**). Three types of CCs are distinguished: intermediary cells (ICs) with numerous vesicles of an unknown nature, transfer cells (TCs) with numerous cell wall invaginations, and ordinary or smooth-walled CCs (SCs) without

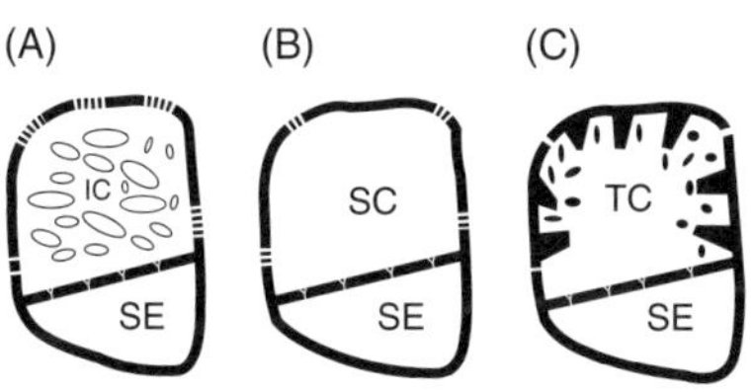

Number of plasmodesmata	Large >10 μm^{-2}	Intermediate 10–0.1 μm^{-2}	Small <0.1 μm^{-2}
Uphill sugar gradient	Gentle	Absent	Steep
Predominant transport sugars	Sucrose, galactosyl-oligo-saccharides	Mainly sucrose, often sugar alcohols	Mainly sucrose

Figure 5 Diagrams of the sieve element (SE)/companion cell complexes in the minor veins of angiosperm leaves. Three types of companion cells are distinguished on the basis of the subcellular structure and of the plasmodesmal frequency at the interface with the adjacent parenchyma cells. (A) IC, intermediary cell, contains fragmented vacuoles or dilated endoplasmic reticulum. (B) SC, smooth-walled cell or ordinary companion cell, only shows the basic features of a companion cell. (C) TC, transfer cell, possesses conspicuous cell wall ingrowths. Associated features are summarized in the table attached (the uphill sugar gradient expresses the concentration gradient between mesophyll and SE/CC complex, the number of plasmodesmata refers to the plasmodesmal density at the interface between CCs and adjacent mesophyll cells, transport sugars are the sugars in the sieve tubes). Modified with permission from van Bel AJE (2003) The phloem, a miracle of ingenuity. *Plant, Cell and Environment* 6: 125–149. Oxford: Blackwell Publishing.

special features. At the IC/PP interface, the PD varies between 10 and 60 plasmodesmata μm^{-2}, at the SC/PP interface between 10 and 0.1 plasmodesmata μm^{-2} and, at the TC/PP interface, PD is mostly lower than 0.1 plasmodesmata μm^{-2}. The mode of phloem loading has been demonstrated to correlate with the type of SE/CC. ICs are involved in symplasmic phloem loading and TCs in apoplasmic phloem loading, while the mode of phloem loading executed by SCs remains uncertain. In apoplasmically phloem-loading species, the major transport sugar is sucrose; in symplasmically phloem-loading species it is sucrose with varying amounts of raffinose-related sugars.

In monocotyledons, two types of SEs occur within the metaphloem of the smaller longitudinal veins of grasses. The thin-walled SEs develop before the thick-walled SEs which lie closest to the xylem vessels. The thin-walled SEs fit the classical description of phloem sieve tube members, in that they are associated with CCs. The thick-walled SEs are considered to be virtually isolated from the adjoining parenchyma. In small longitudinal veins, the ratio of thin-walled to thick-walled SEs is much lower (between 1 and 3) than that in large longitudinal bundles (between 3 and 10). Vein ultrastructure in grasses, the group most studied, suggests apoplasmic phloem loading, the transport sugar being mainly sucrose.

Effects on Environmental Phloem Loading

Temperature was postulated to exert a disparate effect on symplasmic and apoplasmic phloem loading. On the basis of electron microscope studies, phloem loading in symplasmically loading species may be more suppressed by low temperatures through its effect on opening of plasmodesmata. Entry of assimilates into the ICs was therefore postulated to be impeded by cytoskeleton-mediated closure of the plasmodesmal corridor under low temperatures. However, using radiolabeled photoassimilates, phloem loading was demonstrated to be equally affected in symplasmically and apoplasmically loading species under lowered temperature (from 20 °C to 10 °C). Yet, temperature may have species-specific effects on phloem loading. For instance, in some temperate-zone evergreens with a symplasmic vein configuration, phloem loading operates best at temperatures around 10 °C and phloem loading continues at near-zero temperatures.

Propulsion of water is the driving force for mass flow in the sieve tubes; water must be withdrawn from tissues adjacent to the collection phloem. Thus, water status of the mesophyll is important for phloem loading and transport. Osmotic adjustment of the sieve tube content by upregulation of photoassimilate accumulation reduces the negative effects of drought stress on phloem loading. Under extreme drought conditions, however, lack of water in the leaf decreases phloem loading. Apart from the direct poisonous effects on cell physiology, salt stress is often sensed by the plant as drought stress and the effects on phloem loading may be similar. A slight increase in TC wall ingrowths was observed under salt stress, consistent with an increased capacity for salt export.

The response of export rates to elevated CO_2 varies between plant species. For instance, export of ^{14}C-labeled photoassimilates is stimulated in wheat (*Triticum aestivum*), *Sorghum*, catnip (*Nepeta cataria*), and rose (*Rosa* spp.) and suppressed in oat (*Avena sativa*), celery (*Apium graveolens*), cucumber (*Cucumis sativus*), pepper (*Capsicum frutescens*), and sunflower (*Helianthus annuus*). Thus, there is no clear-cut correlation between the export response to elevated CO_2 and the mode of phloem loading.

Diurnal rhythms can indirectly or directly effect phloem loading. Indirect effects are exerted through impacting on amounts of photoassimilates available for phloem loading by altering partitioning between carbohydrates in leaf mesophyll. A direct effect of diurnal rhythms on phloem loading is mediated through diurnal changes in expression of the SE membrane-bound sucrose transporter SUT1 in potato (*Solanum tuberosum*). Species-dependent diurnal rhythms in phloem loading have been observed: in some species, phloem export is higher at night, in others during the day. Further, floral induction in some day length-sensitive species is preceded by an elevated rate of phloem loading. It is unclear whether phytochrome directly regulates phloem loading by altering the membrane potential of SE/CC complexes to drive sucrose/proton symport or whether phytochrome-related sink factors regulating turgor pressure are involved.

Effects on the Transport Pathway

Development of the Vascular System along the Translocation Pathway

Differentiation of vascular tissue is influenced by seasonal rhythms in which light may play an important part. Vascular tissue formed during winter has a composition different from that formed during summer. In tomato (*Lycopersicon esculentum*) plants grown in winter, for instance, the conducting xylem elements are more lignified and the proportion of fibers in the xylem is larger than in summer. It thus appears that differentiation of cambial-derived cells is much influenced by seasonal effects, most of which are unknown.

Environmental Effects on Phloem Transport

After decades of controversy, mass flow has generally been accepted as the long-distance phloem transport mechanism. Velocities of sap flow measured by heat propagation, fluorochrome movement, and nuclear magnetic resonance matched those of mass flow rates calculated by theoretical models. Proteinaceous material deposited onto the sieve plates does not appear to entirely block the sieve pores *in vivo*, but leaves free corridors for mass flow. Environmental factors affecting the turgor difference between source and sink are mainly responsible for the mass flow rate, but mass flow is also dependent on physical resistance within the system (**Figure 3**). First of all, temperature may impact on the viscosity of the phloem sap. With sucrose, sieve tube sap is calculated to be less viscous at low temperatures than with raffinose-related sugars. It is not excluded that several rare sugars and sugar alcohols in the sieve tube sap function as fluidizers or as a protection against increased viscosity at low temperatures.

The size of the transport corridors within sieve pores, the bottlenecks in the transport pipes (**Figure 3**), may be regulated by the degree of protein deposition which may in turn depend on environmental factors. The Hagen–Poiseuille law predicts that hydraulic conductivity of sieve tubes is a function of the radii of the sieve pores and viscosity of the transported sap. Sieve pores interconnecting adjacent sieve elements determine the limiting radius of sieve tubes. Sieve pore radius is reversibly influenced by callose formation and protein deposition, both of which are responsive to temperature. Inefficiency of phloem transport at low temperatures may reflect clogging of sieve pores with phloem-specific proteins.

At high temperatures, large deposits of callose on sieve plate pores have been correlated with slowing of phloem translocation through localized regions of heat-treated stems. In addition, seasonal variation in callose deposition in deciduous perennials is correlated with phloem transport rates. Collectively these temperature impacts on hydraulic conductance of sieve tubes partially account for differences in chilling sensitivity of phloem transport between species. There is, however, a physiological component manifested in the continuous reloading of photoassimilates leaked to the phloem apoplasm (**Figure 1**). Indeed, recovery of transport rates through a stem section exposed to low temperatures is attributed to upstream reloading enhancing the pressure difference across the chilled stem portion to restore transport through the region of lowered hydraulic conductance.

Balance between leakage from and recovery into sieve tubes along the transport phloem determines whether there is a net unloading or loading of this pathway. These processes are influenced by prevailing environmental conditions. Net unloading and subsequent storage occurs under sink-limited conditions that can be imposed by environmental impacts such as water stress leading to downregulation of sink demand. A direct effect on unloading from the transport phloem is observed in stems colonized by holoparasites and bacteria that induce galls. In both cases, enhanced unloading is linked to a pronounced induction of a symplasmic pathway of unloading from the sieve tubes of the host plant. Conversely, environmental reduction in photoassimilate export from source leaves can lead to the onset of remobilization of stored reserves loading the transport phloem and subsequent delivery to nearby sinks. Thus, unloading from, and reloading into, transport phloem functions as a buffer to balance environmental impacts on source and sink processes.

Effects on Sinks

Definition of Sink Types and Sink Development

Growth sinks (cell division and expansion) may generate vegetative or reproductive organs with growth centers located in terminal (root/shoot apices, inflorescences) or intercalary (vascular cambium, stem meristems in monocotyledons) positions. Meristem initials, once committed to cell expansion, are capable of storing imported assimilates. A proportion of stored assimilates functions as osmotica to maintain favorable water potential gradients for water uptake as the cell expands.

Overall, sink types can be broadly categorized as growth (vegetative and reproductive) or storage (temporary or committed storage of solutes or polymers) sinks. The duration of storage varies from minutes to years and can be classified broadly into temporary and committed storage functions. Vegetative (e.g., tubers) and reproductive (fruits, seeds) propagules are committed to long-term storage of assimilates. Temporary storage sinks are characteristically located adjacent to axial transport pathways in stems and roots. Assimilates are temporarily stored when source outputs exceed photoassimilate demand by growth and committed storage sinks. Conversely, photoassimilates are mobilized from temporary storage when source output falls below assimilate demand.

Environmental Effects on Sink Development

In general, sink organs are not exposed to high evaporative loads and consequently exhibit low rates of transpiration. Therefore, most assimilates and water are imported through the phloem with water being used in cell expansion. In storage sinks, phloem import continues as cell expansion slows and excess water is dispersed by sink transpiration and by recirculation of water via the xylem to the plant body.

Sink growth in general is strongly dependent on temperature. For example, as for roots, the lower temperature limit for growth of temperate-zone species is about 5 °C, while the roots of tropical and subtropical plants need higher temperatures. Incidentally, soil temperature, rather than air temperature, may present a major factor in containment of subtropical and tropical plants to their present habitats.

Flower bud initiation and flower development often require disparate species-specific temperature thresholds. For instance, low temperatures and short-day conditions are necessary for the induction of flower primordia which develop and unfold only under subsequent long-day conditions in most grasses and many perennial herbs of the temperate zone. Temperatures required for ripening of fruits and seeds are mostly higher than those necessary for the development of the vegetative parts. The order of different temperature regimes is crucial for the development of sinks. Hence, it appears that an optimal temperature course is mostly more critical for development of sinks than an average optimal temperature for sink development.

Physiological Events in the Phloem Unloading Zone

General remarks Environmental impacts on assimilate import into sinks may result from effects on assimilate supply determined by source or path factors or from direct effects on sinks that determine their capacity to import assimilates for growth or storage. It is imperative that an analysis of environmental impacts on assimilate import by sinks distinguishes environmental effects on assimilate supply from direct environmental effects that influence the capacity of a sink for assimilate import. In addition, responses of assimilate transfer and transport effects in sinks may result from instantaneous changes to their thermodynamic properties or from induction of gene expression leading to more metabolic machinery and capacity for assimilate import.

Pathways of unloading Once unloaded from the importing SEs, assimilates move through postSE pathways (**Figure 2B**) to reach the cellular sites of growth, maintenance respiration, and/or storage. Most assimilates move along symplasmic routes from importing SEs. In some sinks, such as developing seeds and biotrophic relationships with other organisms, the two genomes are separated sympasmically. This separation necessitates the addition of membrane transport steps to and from the sink apoplasm. Symplasmic movement of assimilates through postSE pathways may be a combination of bulk flow and diffusion. The relative contribution of each transport mechanism is determined by hydrostatic pressure and concentration differences between importing SEs and recipient sink cells. Assimilate metabolism and compartmentation within sink cells influence the magnitude of these differences. In addition, membrane transporter activity can impact on differences in hydrostatic pressure and assimilate concentration, where a membrane transport step intervenes in the postSE pathway. For a set driving force, conductivities of interconnecting plasmodesmata modulate realized rates of photoassimilate transport.

Thus how, and the extent to which, environmental perturbations affect assimilate import by sinks will be influenced by sink function and the cellular pathway of the postSE transport route.

Environmental Effects on Phloem Unloading

Effects on assimilate supply to growth sinks Growth rates of vegetative and reproductive growth sinks are source-limited. This characteristic is demonstrated by their growth responses to environmental effects that alter photosynthetic rates of whole plants. Limiting photoassimilate supplies specifically impact on rates of cell division, while cell expansion is less compromised.

Meristematic sinks import less than 10% of the total photoassimilate pool available for growth and storage. Thus, it is puzzling why rates of cell division in these sinks are limited by supply of photoassimilates. A possible explanation is found in the relatively low conductance of the postSE pathway linking them with the most distal SE. This bottleneck imposes a marked influence on the forces required to drive assimilate transport through the postSE pathway. Furthermore, sucrose imported into meristematic sinks fulfils a dual function as a substrate to fuel, and as a signal to regulate cell division.

Impacts of drought stress, salinity, temperature, and light on assimilate import by terminal meristems may be exerted through altering assimilate supply or by acting directly on sink processes governing import. For instance, severe water shortage and heat stress as well as low light levels cause abortion of flowers, fruits, and seeds. Fruits and seeds that remain attached to the parent plant under these conditions invariably exhibit reduced cell numbers. This outcome diminishes their subsequent capacity for assimilate storage. In the case of drought and heat stress, organ abortion and reduction in cell numbers can be prevented by sucrose supplied as infusions to stems supporting the reproductive structures. Young sink leaves are more tolerant than reproductive structures to drought and heat stress, and this causes a shift in assimilate partitioning patterns to favor vegetative growth. Light effects are more difficult to interpret since selective shading of reproductive organs also results in fruit and seed abortion.

Direct effects on growth sink activity The relationship between cell turgor (P) and cell wall properties (m, extensibility; Y, yield threshold) in determining rates of volume change ($\Delta V/\Delta t$) of an expanding cell is encapsulated in the Lockhart equation:

$$\Delta V/\Delta t = m(P - Y) \qquad [2]$$

As cells expand, solutes must be produced or imported to maintain cell osmotic concentrations that otherwise are diluted by water uptake. Solute and water delivery to shoot meristems is exclusively through the phloem. In contrast, root apices also have access to solute and water supplies from the soil solution. On these grounds alone, it is not surprising that leaf compared to root elongation is more sensitive to drought stress and this behaviour accounts for decreases in shoot/root ratios by water stress.

Water stress exerts an immediate block on cell expansion if cell turgor falls below the yield threshold (eqn [2]) and this impact is offset by enhancing the import of photoassimilates. For example, transients of slowed root extension recovering to pretreatment rates on exposure to high osmotic solutions are interpreted in terms of commensurate transients in elevated rates of solute import through the postsieve element pathway. The elevated rates of solute import presumably raise the intracellular osmotic potential of the expanding cells and restore cell turgor above the yield threshold so that elongation resumes (eqn [2]).

The accelerated import is undoubtedly driven by steepened pressure differences across the post SE pathway. However, it could also be facilitated by increases in conductance of the post SE pathway as suggested by observed osmotic-induced pore enlargement of interconnecting plasmodesmata. Ultimately, a new equilibrium is reached between hydrostatic pressure differences to drive water and solute import and expanding cell turgors to adjust above the yield threshold. These short-term responses of assimilate import are replaced by more long-term inductive responses, mediated by plant hormones, that lead to adjustments in biochemical and biophysical properties of cell walls that collectively determine cell expansion rates and hence steady-state rates of assimilate import.

It is difficult to fully understand temperature effects, because most processes in a plant are sensitive to temperature and many interactions exist. However, some distinctions are possible. For instance, for roots it has been found that metabolism responds more rapidly to temperature than assimilate import. If transport through the symplasm is limited by plasmodesmal transit, then temperature effects across the physiological range are more likely to be mediated through alterations in driving force across them than conductance of the postSE pathway. Thus, temperature-enhanced metabolism will probably lower cytosolic pool sizes which in turn increase concentration differences driving diffusion through the postSE pathway. Over the longer term, these thermodynamic impacts on sucrose import will cause alterations in sucrose-regulated gene expression of sugar-metabolizing enzymes to match the new sugar fluxes.

Light quality and duration sensed by phytochrome causes profound alterations in the distribution of growth and dry matter in plants. The question of whether the changes in enzyme activity and carbohydrate levels are the cause or the consequence of altered growth patterns has only been addressed in a few systems. For example, light stimulates young tomato fruit to take up 30% more sucrose and accumulate almost twice as much starch and hexoses as dark-grown fruits. The increased levels of starch accumulated are accounted for by light-activation of ADP glucose phosphorylase. The resulting depression in sugar pools of fruit storage parenchyma cells could act to increase diffusion through the postSE pathway by increasing sucrose concentration to the importing SEs.

Effects on photoassimilate transport to and within storage sinks Under conditions of an adequate supply of water and minerals and favorable growth temperatures, accumulation of assimilates into storage pools of tubers, fruits, and seeds is limited by the storage capacity of these organs. This state is readily demonstrated by independence of rates of biomass gain from variations in leaf photosynthesis manipulated by light flux densities or carbon dioxide enrichment. In addition, substantive reductions in leaf photosynthesis have disproportionately small effects on biomass accumulation by storage sinks. This outcome results from photoassimilate supplies being buffered by remobilization of temporary storage pools located along the axial pathway. Only in those circumstances where the pool of storage reserves do not meet assimilate demand is biomass gain by storage organs compromised.

During development, seed and fruit become hydraulically isolated from the parent plant through imposition of xylem discontinuities which ensure that storage capacities of these sinks have a reduced sensitivity to drought stress. In this context, cell turgors of developing fruits and seeds are surprisingly low and stable. However, severe reductions in water potential impact on assimilate transport by dissipating favorable pressure gradients and downregulating key sugar-metabolizing enzymes.

Interestingly, membrane transporters, responsible for loading sugars into the embryo or endosperm from the seed apoplasm, appear to be more resilient to shortage of water. Indeed, sucrose transporter activity has been shown to increase in both monocotyledonous and dicotyledonous seeds subjected to drought stress. A possible regulator is the protoplasmic level of sucrose which has been shown to be a negative regulator of sucrose transporter activity. A balance with metabolic demand is achieved by sucrose acting as a positive regulator for gene expression of sugar metabolic enzymes involved in storage product accumulation.

During storage in seeds, temperature effects are mediated on the seed sink as illustrated by comparable temperature responses of *in vivo* and *in vitro* seed growth rates. Under high temperature conditions, seed growth rate usually increases, but the growing period decreases more with increasing temperature as ontogeny is accelerated. Heat stress causes development of "pinched" seeds by perturbing their capacity to store assimilates. Heat-stressed seeds exhibit a comparable resilience of sucrose uptake capacity to those subjected to water stress. However, storage function of heat-stressed seeds appears to be compromised by the high-temperature vulnerability of certain key starch-metabolizing enzymes such as starch synthase.

Conclusions

On the global scale, biomass production is most affected by drought and temperature. Thus environmental effects on source and sink ontogeny and on the functioning of sources and sinks are easy to observe, but difficult to understand. The major reason is that environmental influences are multifactorial and act simultaneously on development and physiology at several, mostly overlapping, levels.

Direct and indirect effects can hardly be disentangled. Information about the environmental impact on photoassimilate transport or distribution in the source-to-sink trajectory is scarce and fragmentary. To date, studies of the relationship between the environment and development of sources and sinks are mainly of a descriptive nature. Studies on the interaction between gene expression and environmental factors such as drought, temperature, salinity, and ultraviolet light – one of the key issues in developmental plant biology – are being carried out, but even the beginning of a coherent picture is not available yet.

List of Technical Nomenclature

Apoplasm	Extracellular spaces accessible to solutes, largely comprising fluid-filled spaces in cell walls and surface films of intercellular spaces.
Hydraulic conductivity	Capacity of a structure to support mass flow of water per unit cross-sectional area for a given hydrostatic pressure difference between the ends of the transport pathway.
Osmotic pressure	Potential pressure that can be developed as a result of separating pure water from a solution containing dissolved solutes; at low solute concentrations, osmotic pressure is a direct function of the combined solute concentration.
Plasmodesmata	Plasma membrane-lined micropores linking cytoplasm of adjacent cells.
Sieve plate	Region in lateral and transverse cell walls of sieve elements consisting of pores that allow phloem sap flow between or through sieve tubes.
Sink	An organ or tissue that is a net importer of a specified nutrient.
Source	An organ or tissue that is a net exporter of a specified nutrient.
Symplasm	Cytoplasmic continuity of adjoining cells interconnected by plasmodesmata.
Turgor pressure	Positive hydrostatic pressure in plant cells resulting from the cell walls resisting volume change brought about by water uptake.
Water potential	Free energy state of water in a system relative to pure water at standard temperature and pressure measured in pressure units derived from the chemical potential of water divided by the molal volume of water.

See also: **Energy Crops**: Biomass Production. **Flowering and Reproduction**: Flower Development. **Growth and Development**: Cell Growth; Cell Division and Differentiation; Leaf Development. **Seed Development**: Nutrient Loading of Seeds; Seed Production. **Water Relations of Plants**: Basic Water Relations; Drought Stress; Phloem; Salt Stress.

Further Reading

Behnke H-D and Sjolund RD (1990) *Sieve Elements: Comparative Structure, Induction and Development*: New York: Springer-Verlag.

Corbesier L, Lejeune P, and Bernier G (1998) The role of carbohydrates in the induction of flowering in *Arabidopsis thaliana*: comparison between the wild type and a starchless mutant. *Planta* 206: 131–137.

Gamalei YV, van Bel AJE, Pakhomova MV, and Sjutkina AV (1994) Effects of temperature on the conformation of the endoplasmic reticulum and on starch accumulation in leaves with the symplasmic minor-vein configuration. *Planta* 194: 443–453.

Grodzinski B, Jiao J, and Leonardos ED (1998) Estimating photosynthesis and concurrent export rates in C3 and C4 species at ambient and elevated CO2. *Plant Physiology* 117: 207–215.

Huber SC and Huber JL (1996) Role and regulation of sucrose-phosphate synthase in higher plants. *Annual Review of Plant Physiology and Plant Molecular Biology* 47: 431–444.

Körner C, Pelaez-Riedl S, and van Bel AJE (1995) CO2 responsiveness of plants: a possible link to phloem loading. *Plant, Cell and Environment* 18: 595–600.

Larcher W (2002) *Physiological Plant Ecology*: New York: Springer-Verlag.

Nelson T and Dengler N (1997) Leaf vascular pattern formation. *Plant Cell* 9: 1121–1135.

Patrick JW and Offler CE (2001) Compartmentation of transport and transfer events in developing seeds. *Journal of Experimental Botany* 52: 551–564.

Patrick JW, Zhang W, Tyerman SD, Offler CE, and Walker NA (2001) Role of membrane transport in phloem translocation of assimilates and water. *Australian Journal of Plant Physiology* 28: 695–707.

Roth-Nebelsieck A, Uhl D, Mosbrugger V, and Kerp H (2001) Evolution and function of leaf venation architecture: a review. *Annals of Botany* 87: 553–566.

Turgeon R (1996) The selection of raffinose family oligosaccharides as translocates in higher plants. In: Madore MA and Lucas WJ (eds) *Compartmentation and Source–Sink Interactions in Plants*, pp. 195–203: Rockville: American Society of Plant Physiology.

van Bel AJE (2003) The phloem, a miracle of ingenuity. *Plant, Cell and Environment* 6: 125–149.

van Bel AJE, Ehlers K, and Knoblauch M (2002) Sieve elements caught in the act. *Trends in Plant Science* 7: 126–132.

PLANTS AND THE ENVIRONMENT

Contents

Amenity Horticulture

R W F Cameron, The University of Reading, Reading, UK

Introduction

Amenity horticulture is the cultivation of plants in the landscape for the primary purpose of aesthetics. It embraces aspects of both science and art and combines elements of design and visual impact with effective management of the landscape based on scientific principles and practice. It covers a broad range of disciplines, including arboriculture, amenity grass management, the management of public gardens, municipal parks and open spaces, street planting and floral displays, interior design, and landscaping. Notably, it also comprises home gardening and even in this aspect alone amenity horticulture carries a high financial and sociological profile. In the UK, for example, the "home garden" industry is valued at £3200 million per annum (direct plant sales contribute approximately £1000 million) and amenity horticulture is estimated to employ approximately 90000 personnel.

The relationship between carefully designed and constructed landscapes and our psychological well-being has only recently become the focus for careful scientific examination; nevertheless, the long history of managed landscapes and gardens would clearly indicate that they have an important role in the development of society. In particular, the advent of urbanization has corresponded with an increased requirement and appreciation for "open green spaces" as a forum for leisure and recreation. Gardeners and landscape designers often select plants to provide a specific ambience or function within the landscape. Common attributes and criteria that can determine selection are outlined in **Table 1**.

Table 1 Attributes and other criteria used in the selection of amenity plants

Final size – scale and proportion in the landscape
Growth rate, e.g., ability to screen an unsightly object quickly
Form, e.g., pendulous or fastigiate
Line – the ability to form a visual line, e.g., hedges or avenues
Texture – roughness or smoothness of leaves or bark
Foliage color
Flower color
Flowering time and duration
Berrying or fruiting characteristics
Fragrance
Habitat for wildlife and conservation value
Absorb or deflect noise, e.g., traffic or industrial units
Reduce the impact due to particulate matter and aerial pollutants
Modify the environment – reduce wind speed, provide shade, or increase relative humidity
Security – sharp thorns or the formation of impenetrable barriers
Provide a horizontal, malleable surface for sport or recreation, i.e., grass lawns

Plants in amenity situations may either be grown *in situ* (e.g., sown grass swards, annual flowering plants) or, more likely, grown under intensive cultivation then moved and planted in the landscape. This transition from carefully managed intensive production systems in nurseries to planting in the wider landscape, where there may be little or no provision to provide additional nutrients or water, is one of the main challenges facing the landscape manager.

Sources of Ornamental Plants and their Commercial Production

Unlike other areas of agriculture and horticulture, which tend to rely on a limited range of plant families and species, amenity horticulture utilizes a wide variety of plant classes and genera. A typical garden will comprise annual flowering plants, herbaceous perennials, flowering bulbs, tubers and corms, ornamental grasses, and conifers, as well as deciduous and evergreen dicotyledonous trees and shrubs. The range of plants available has been widened even further by the selection of natural mutations and selective breeding to enhance the ornamental appeal of any given species. Therefore, in a temperate climate, such as the UK, there may be in the region of 70000 different plants and cultivars available for use in ornamental settings.

Flowering annual species and their cultivated forms tend to be commercially propagated by seed. The majority of other plant types, however, are propagated vegetatively by cuttings, division, or grafting. This is to ensure that clones remain "true to type" and that relatively large plants can be produced quickly. Many plants are grown-on in nurseries before reaching marketable size. Most plants are grown in containers using a peat- or bark-based growing media, although some (large trees especially) are field-grown and lifted "bare-root" in the dormant period. Container-grown material has the advantage of being suitable for retail throughout the year and does not require immediate planting. The size and age of plants sold at the nursery stage can vary tremendously, for example, trees may be sold as small "transplants" (30 cm tall) or "extra heavy standard" trees (>5 m tall).

Establishment of Ornamental Plants in the Landscape

Choice of Species and Size of Plant

Of primary importance is the choice of species that will tolerate the climate and prevailing soil conditions. Even within relatively local areas, soil chemistry, extent of soil compaction, moisture availability, solar radiation, and degree of exposure to wind or frost can vary considerably. Many urban situations may prove particularly problematic. Buildings re-radiate thermal energy and can increase local wind speed, thereby promoting hot, dry atmospheric conditions. Similarly, large areas of concrete or pavement allow little penetration of rainwater into the soil reducing the availability of soil moisture to roots. The soils in urban situations may contain large amounts of building rubble or other spoil material and tend to be deficient in organic matter. High levels of soil compaction (due to machinery and pedestrian traffic) and low organic matter content lead to poor soil structure. Impaired drainage and aeration will inhibit root development and accentuate stress due to drought or waterlogging. If planting occurs on old industrial sites, then soils may be contaminated with heavy metals and other phytotoxic elements and a combination of both soil amelioration and the use of tolerant species will be required. Similarly, in roadside locations, the species selected need to be salt tolerant to withstand the effects of de-icing salts and able to tolerate aerial pollutants such as nitric/nitrous oxides and ozone derived from traffic fumes. Appropriate choice of species is therefore critical.

The size of plant that is used will be determined by the requirements of the site or client (e.g., large trees for "instant" effect), cost, and the finances and other resources available for the subsequent management of the site. Where costs allow, it would be logical to assume that larger trees ("standards") would be the preferred option. It has often been assumed that standard specimens are able to compete more effectively against weeds, are less likely to be trampled on or damaged accidentally, and, because of their relatively large biomass, should be able to establish readily. In reality, the results from research suggest that the use of smaller trees may be more beneficial, especially when an instant effect is not required. This is due to the fact that standard trees may lose a proportionally greater amount of their root system at transplanting compared to that of small trees and they will also encounter high evaporative demands due to their relatively large total leaf surface area. As such, large standard trees may be more prone to water stress after planting, which can affect establishment and subsequent growth rates (**Table 2**). Secondly, even large trees will face significant competition from weed growth if it is not controlled effectively. Thirdly, although standards are unlikely to be trampled on by pedestrians and livestock, they may still require some protection to avoid damage from vandalism or wind-throw. Even when standard trees are chosen and the postplanting management is effective, shoot growth rates may be slow for a number of years after planting ("transplant shock") as the shoot to root ratio re-establishes an equilibrium. In contrast, smaller specimens may establish very effectively after transplanting and even outgrow larger specimens planted at the same time.

Minimizing Stress and Competition

Regardless of initial tree size, careful management is required during the transplanting process. Plants, and root systems in particular, should not become desiccated after removal from the nursery. Bare-root material should be kept damp and enclosed in damp sacking or polythene to minimize water loss by evapotranspiration. (**Table 3**). It is equally important that the appropriate irrigation management is imposed for the following weeks, months, or even years until the plants become fully established.

Water stress is considered to be the main cause of death for newly planted trees and shrubs in the

Table 2 The effect of tree size of *Acer rubrum* cv October Glory on water relations and growth responses in the year of planting (year 1) and 1 year later (year 2)

	Year 1		*Year 2*	
	Small	*Large*	*Small*	*Large*
Mean daily leaf water potential (MPa)	−5.2	−6.6[a]	−2.3	−2.4
Predawn leaf water potential (MPa)	−2.0	−3.2[a]	−1.2	−1.7[a]
Net photosynthesis (μmole m^{-2} s^{-1})	9.0	5.1[a]	12.8	8.7[a]
Height increase (mm)	73	53	343	57[a]
Stem girth increase (mm)	4.1	2.7	14.1	7.1[a]

Original stem girths were approximately 38 mm for small trees and 76 mm for large trees.
[a]Denotes significant differences by LSD (least significant difference) at $P = 0.05$ for comparisons between small and large trees at any given time. Reproduced with permission from Lauderdale DM, Gilliam CH, Eakes DJ, Keever GJ, and Chappelka AH (1995) Tree transplant size influences post-transplant growth, gas exchange and leaf water potential of 'October Glory' red maple. *Journal of Environmental Horticulture* 13: 178–181

Table 3 The effects of storage treatment on the survival of young trees of *Betula pendula* and *Sorbus aucuparia*

Storage treatment	*Survival (%)*	
	B. pendula	*S. aucuparia*
Lifted and immediately replanted	77	83
Lifted and stored bare-root for 1 day	42	82
Lifted and stored bare-root for 7 days	9	65
Lifted and stored with roots in polythene bags for 7 days	79	81

Modified, with permission from Bradshaw A, Hunt B, and Walmsley T (1995) *Trees in the Urban Landscape – Principles and Practice.* London: E &FN Spon, Thompson Professional.

landscape. In the UK, 15% of tree failure is associated directly with water stress, although the figure could be considerably larger if the indirect effects on water availability through weed competition are taken into account. Even when the drought period is not prolonged enough to kill the young plant, transient periods of water stress or limited supplies of soil moisture can compromise the early growth of a transplanted specimen. Root growth, in particular, may be inhibited in some species during periods of drought, thus limiting the plant's ability to explore new soil and seek new sources of water and nutrients. In addition, prolonged periods of stomatal closure will reduce photosynthetic capacity and carbohydrate accumulation, thus risking the long-term viability of the plant. How much water a plant requires varies considerably on a daily basis and depends largely on the evapotranspiration demand on the plant. This is mainly determined by environmental factors such as temperature, relative humidity, and wind speed. Plant size and relative leaf area strongly influence the water requirements; indeed, total leaf area appears to be a more important factor than any other effects associated with different plant species (**Figure 1**).

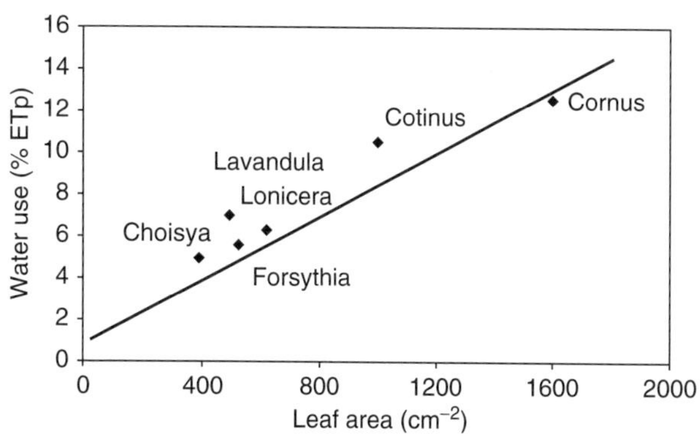

Figure 1 The relationship between leaf canopy area and water use (potential evapotranspiration, ETp) across a range of amenity shrub species. Evapotranspiration is largely determined by leaf area, with relatively little variation due to differences in leaf morphology among species. Reproduced with permission from Haerisan-Murray *et al.* (2002) Improving the control and efficiency of water use in container-grown hardy ornamental nursery stock. DEFRA Hortlink Report HL0132LHN.

There are three main techniques for improving water uptake in newly planted trees and shrubs: (1) establishing a larger root system in the transplanted specimen, as the volume of water available in the soil correlates with the diameter of the root system; (2) ameliorating the soil to improve its water-holding capacity; and (3) providing regular irrigation, either by hand, sprinkler, or drip systems. Developing a larger root system not only relies on ensuring as much of the original roots are recovered at the time of lifting, but also that planting time ensures the development of new roots prior to the onset of drought conditions in summer (i.e., autumn or early winter planting). Soil amelioration consists of incorporating organic materials into the soil. These include peat, coir, spent mushroom compost, and sewage sludge, as well as a range of "water-holding" polyacrylamide gels, although results on water-holding capacity and establishment have not always been consistent. The provision of permanent irrigation systems to new planting sites is not necessarily cost effective or feasible; nevertheless, the use of temporary irrigation, even for just a few weeks, via a water bowser and/or hose pipes, is important to ensure new root growth and the establishment of the plant.

Spacing and Weed Control

Plant establishment is strongly influenced by the competition for resources, particularly light, water, and nutrients, that occurs between neighboring plants. Correct plant spacing is important to ensure that individual ornamental plants have sufficient space to grow and develop. Over time, the formation of a foliar canopy will eliminate the competition from weeds. With mixed plantings, which utilize a range of different ornamental species, careful selection is required to ensure that the species compliment each other and that excessive competition is avoided. To this effect, good ornamental planting schemes often mimic nature, with different plant forms being used to exploit a range of "ecological niches." Even in simple planting schemes, such as trees grown in grass, the degree of competition can be high. Grass is a very effective competitor for water, in particular, and many new tree plantings fail because the presence of grass exacerbates the water stress imposed on the transplanted trees (**Figure 2**). Paradoxically, supplying nutrients to the planting site can increase the competition if remedial actions are not implemented (**Figure 3**).

The most effective and economical method of controlling weeds is through herbicides, although care is required to ensure desirable plants are not damaged by spray drift. Because of environmental concerns, there is a move away from a reliance on soil-acting residual herbicides in amenity horticulture to translocated types that break down readily after use. In complex ornamental plantings with large numbers of different plant species, weed eradication prior to planting following by regular handweeding may be the only effective means of control. Once perennial and persistent weeds are removed at the outset, then mulches can be used to suppress any further weed germination and development. Common mulching materials include chopped bark and wood chippings, which not only arrest weed development but conserve moisture and can be aesthetically pleasing in their own right.

Maintenance Procedures

Once plants have become established, the degree of maintenance required will vary with the type of plant and its role in the landscape. Areas of amenity grass, for example, may need regular mowing. To provide the smooth "billiard board" image of a fine-texture lawn, the grass may need to be mown twice a week and irrigated daily during dry periods. In contrast, mature trees in the landscape may only require an occasional inspection to ensure the structure of the tree remains sound and that there is little risk to the public or properties from diseased or damaged branches. For most woody plant types, regular

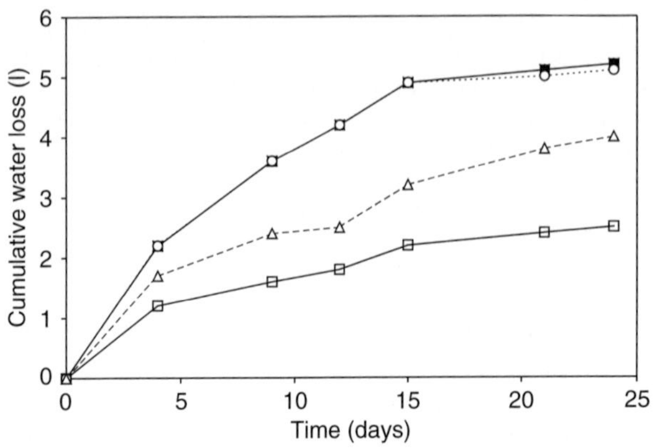

Figure 2 Water utilization from a limited volume of soil over time. Water use in soil alone (□), trees alone (Δ), grass alone (○), and trees and grass together (■). Reproduced with permission from Bradshaw A, Hunt B, and Walmsley T (1995) *Trees in the Urban Landscape – Principles and Practice*. London: E &FN Spon, Thompson Professional.

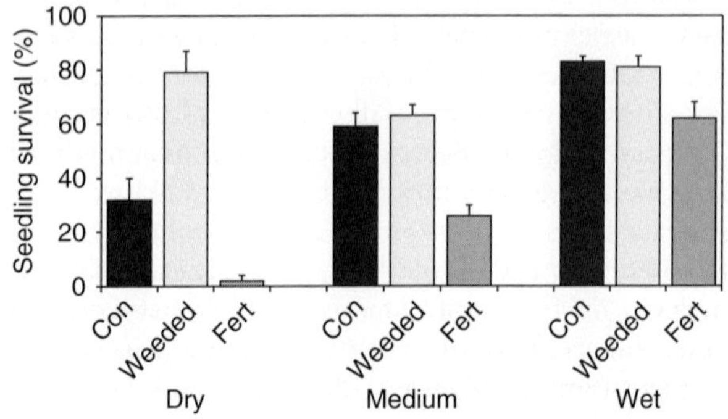

Figure 3 Seedling survival of *Quercus ellipsoidalis* grown in soils with different moisture contents (dry, medium, and wet) and herbaceous vegetation present (con) or removed (weeded) or the herbaceous vegetation present but with high nitrogen enrichment (fert). Reproduced with permission from Davis MA, Wrage KJ, Reich PB, Tjoelker MG, Schaeffer T, and Muermann C (1999) Survival, growth, and photosynthesis of tree seedlings competing with herbaceous vegetation along a water-light-nitrogen gradient. *Plant Ecology* 145: 341–350.

maintenance procedures could include pruning, the addition of nutrients, and protection from pests and diseases. In reality, the extent to which such practices are implemented will depend strongly on the location and purpose of the plants.

Pruning

The pruning of ornamental plants is an effective method of controlling growth and enhancing their performance and function in the landscape. The type of pruning and the desired effect will vary depending on plant form and habit. Trees are formatively pruned to produce a single stem (trunk) form with secondary branches forming at a set height above ground level (the crown), and are generally designed to provide access for pedestrians or vehicles beneath the tree canopy. Shrubs, on the other hand, are pruned to promote a multistemmed form with branches developing from near or at ground level. This facilitates the display of their flowers and foliage at an appropriate height for viewing, or provides ground cover; it can even obstruct access completely through the formation of hedges. Pruning alone can determine whether an individual species (e.g., beech, *Fagus sylvatica*) forms a standard landscape tree 40 m high or is a component of a tightly branched small boundary hedge only 0.5 m high. Pruning has a number of functions: training young plants, compensating for root loss after transplanting, maintaining or altering appearance, removing dead or diseased branches, controlling plant size, re-invigorating old plants, and promoting flower or fruit formation.

Pruning at the formative nursery stage is key to developing a well-shaped quality tree or shrub. Poorly formed nursery trees will rarely respond to later remedial treatment and, at best, are likely to be misshaped specimens and, at worse, become a risk in later years due to an asymmetrical center of gravity. The aim of nursery pruning is to produce stems that are structurally sound and induce the development of strong-growing, evenly spaced branches. Weakly growing lateral shoots or shoots that are growing at a very acute angle should be removed at this stage. At angles $<20^\circ$ a weak crotch develops between the trunk and a lateral branch, which can cause the tree to rupture through the middle heartwood, once it matures. To ensure a clean, straight trunk form, lateral shoots initiated below crown height are regularly removed. In shrubs, nursery pruning is generally aimed at providing an open "bowl"- or "vase"-shaped habit to the plant. To minimize production time and costs, pruning regimes are designed to maximize the number of axillary lateral buds that develop into new shoots after each pruning event.

Pruning is also frequently employed just after a tree has been planted. This is usually to reduce the resistance of the tree to wind to stop the young tree from being blown over (wind-throw) or to avoid movement of the stem loosening the contact between root ball and soil. As mentioned above, newly transplanted trees are likely to have lost a proportion of their roots and may experience water stress until fully established. To help counteract this, shoots are removed to reduce the future leaf area after budbreak (and hence potential transpiration) and to promote a more effective root to shoot ratio. This is a controversial area, however, as there is a counterclaim that shoot removal reduces the number of active apical buds on the plant. These buds are thought to be important in activating new root growth early on in the growing season, through the transport of auxin from the buds to the root zone. On unpruned trees, expanding buds and new leaves may stimulate enough additional root growth to more than compensate for any increased transpiration due to a larger leaf surface.

Mature trees and shrubs will often benefit from crown thinning to increase light transmission and improve aeration. Thinning helps to reduce wind resistance in the canopy and improved air movement will minimize the risk of fungal pathogens spreading, as well as allowing better penetration of chemical sprays into the canopy, should they be required. A number of shrub species tend to form flower buds on the current season's growth (new or 1-year-old wood) and, as pruning promotes the development of new shoots at the expense of old, a considerably more impressive display of flowers can be achieved by appropriate and regular pruning.

Nutrition

In reasonably fertile agricultural or woodland soils, the majority of large trees and shrubs rarely require supplementary nutrition as they are effective at exploiting the reserves of nutrients within the soil. Where planting occurs in urban soils with limited organic matter, however, additional sources of nitrogen may be necessary. The use of bulky organic manures, slow-release fertilizers, or even nitrogen-fixing complementary plants may be the best way of providing nitrogen in these soils over prolonged periods. Where plant parts are being continually removed, such as in pruning (e.g., roses; *Rosa* spp.) or mowing (amenity grass), then more attention needs to be paid to providing supplementary minerals to maintain vigorous growth. In ornamental lawns, it is desirable to keep the sward height low and this requires regular mowing and removal of the

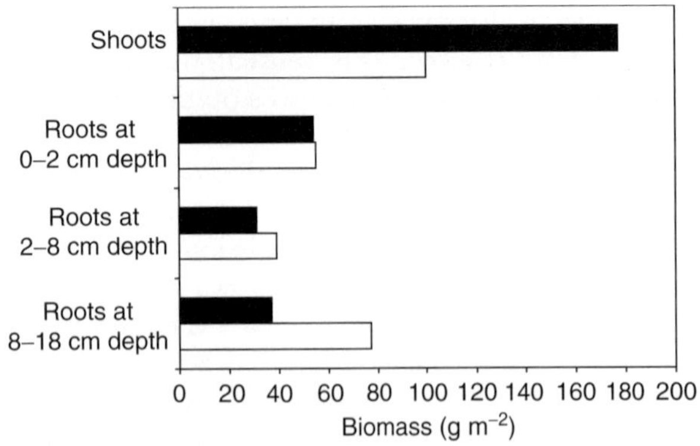

Figure 4 Shoot and root biomass of turf grass as affected by excessive nitrogen supply (10 mmol l^{-1} N, ■) compared to control levels (1 mmol l^{-1} N, □). Modified with permission from Adams WA and Gibbs RJ (1994) *Natural Turf for Sport and Amenity – Science and Practice.* Wallingford: UK CAB International.

clippings. Continual removal of clippings from the sward quickly reduces the mineral availability within the soil and fertilizer input is essential to maintain strong growth. Providing the right balance of nutrients is important. Excessive applications of nitrogen, for example, will reduce root growth at the expense of shoot growth, thereby inhibiting root penetration down a soil profile (**Figure 4**). A shallow root system increases the potential for damage should the turf be exposed to drought stress. Excessive applications of nutrients may also impact on soil and groundwater pollution. In a study in Germany, the use of nitrogen-based fertilizers in home gardens accounted for 27% of the total amount of leached nitrate in ground water, even though gardens only covered 3.5% of the total area in the survey.

Pests and Diseases

The vast range of plants used in the landscape and the limited situations where plants of any given species are grouped in large numbers (as occurs in agricultural monocultures) means that specific, economically important pest and disease problems are generally uncommon in amenity situations. There are a number of notable exceptions, however, where either a single species or groups of plants are highly susceptible to damage from an individual pathogen/pest or where a pest or pathogen is sufficiently nonspecific to be detrimental to a wide range of plants. The former category would include Dutch elm disease (*Ceratocystis ulmi*), rose blackspot (*Diplocarpon rosae*), fireblight (*Erwinia amylovora*), cherry blackfly (*Myzus cerasi*), red thread in grass (*Corticum fuciforme*) and the latter vine weevil (*Otiorhynchus sulcatus*), grey mold (*Botrytis cinerea*), and the field slug (*Deroceras reticulatum*). At a small scale, such as a garden, chemical application may be feasible as a control mechanism, but in the broader context of the landscape more strategic mechanisms may be required. Expanding the gene pool in *Rosa* through breeding to introduce increased tolerance in new cultivars or altering management practices to minimize the risk of damage or infection in the first place, such as removing dead or decaying organic matter to reduce slug populations or encouraging natural predators, are examples of such mechanisms.

List of Technical Nomenclature

Amenity	Relating to ornamental or recreation purposes.
Auxin	A plant growth substance involved in cell development.
Budbreak	The time of bud unfolding and leaf emergence.
Coir	A by-product of coconut husks.
Crown	The aerial part of a tree where structural branches have formed.
Evapotranspiration	Water lost through evaporation and transpiration.
Phytotoxic	Toxic to plants.
Soil amelioration	Improving the structure or chemical composition of soil.
Standard tree	A young tree with distinct trunk and crown sections.
Stomata	The pores in the underside of leaves that allow movement of gases and water into and out of the leaf.
Transpiration	Movement of water through the plant from the roots to the atmosphere via the leaves.
Transplanting	Moving a plant from one location to another.
Transplant shock	An inability for a plant to grow after transplanting.
Translocated	Moved from one organ of a plant to another.
Vegetative	Nonsexual.
Wind-throw	The blowing over of trees by wind force.
Water stress	Drought.

See also: **Growth and Development**: Leaf Development. **Production Systems and Agronomy**: Grassland; Nursery Stock and Houseplant Production. **Water Relations of Plants**: Drought Stress. **Weeds**: Weed Competition.

Further Reading

Acquaah G (ed.) (2002) Growing Plants Outdoors: ornamentals. In: *Horticulture Principles and Practices*, 2nd edn, pp. 503–659. Upper Saddle River, NJ: Prentice Hall.

Adams WA and Gibbs RJ (1994) *Natural Turf for Sport and Amenity – Science and Practice*. Wallingford: UK CAB International.

Bradshaw A, Hunt B, and Walmsley T (1995) *Trees in the Urban Landscape – Principles and Practice*. London: E &FN Spon, Thompson Professional.

Feucht JR and Butler JD (1988) *Landscape Management – Planting and Maintenance of Trees, Shrubs and Turfgrasses*. New York: Van Nostrand Reinhold.

Patch D (1985) *Advances in Practical Arboriculture – Forestry Commission Bulletin 65*. London: Her Majesty's Stationary Office.

Thoday P and Wilson J (1995) *Landscape Plants*. Proceedings of the Institute of Horticulture Conference on "Plants for Landscape Sites", Cheltenham & Gloucester College of Higher Education, Cheltenham, Gloucestershire, UK.

Land Reclamation and Remediation, Principles and Practice

D L Jones, University of Wales, Bangor, UK
E C Rowe, Wageningen University, Wageningen, The Netherlands

Introduction

This article describes the major problems affecting degraded land and some of the remediation options that may be employed to overcome these. Causes and types of agricultural and industrial dereliction are considered. Methods for overcoming contamination and for bringing soil conditions back into the range suitable for plant growth are discussed, with emphasis on practical considerations. Requirements for different end-uses, such as agriculture or nature conservation, are described.

Principles

Degradation and Reclamation

Degradation implies a reduction in the utility of land, whether this be in terms of its agricultural or forest productivity or its ability to support other beneficial processes, such as filtering pollutants or use for recreation purposes. Other ecosystem processes that are of less direct benefit to humanity, such as carbon sequestration or maintaining and promoting biodiversity, may also be viable aims for land reclamation schemes. The methods used for reclaiming land depend on the type of degradation that the land has undergone and, fundamentally, on the desired end-use. End-uses include agriculture, forestry, nature conservation, industrial development, and amenity/recreation. Therefore, the term "reclamation," which originally referred to creating productive agricultural land from wetlands or other inhospitable terrain, has come to encompass a variety of activities, including, for example, the conversion of arable land back to wetlands for nature conservation. The term "remediation" is most often associated with the removal and cleaning up of industrially contaminated sites that contain heavy metals, radionuclides, and/or organic pollutants. Within the text, the terms "remediation" and "reclamation" will be used synonymously to describe the process by which land is brought back into useful function.

In most reclamation schemes, end-users require the growth of various types of vegetation, which also implies that a biologically functioning and stable soil environment is required. Plant growth may be constrained by a variety of soil physical, chemical or biological factors as well as by wider environmental variables such as climate (**Table 1**). Physical or chemical stresses decrease productivity and extreme stress can prevent the growth of plants entirely. However, what is a stress for one species may benefit another. In particular, species that are adapted for unusual or extreme environments are easily outcompeted if stress is reduced. Even small changes in environmental stress are likely to affect the relative success of different species and thus change the composition of the species assemblage. The aim when reclaiming land for agriculture and forestry is to normalize soil conditions and reduce plant stress, but if the end-use is nature conservation it may be appropriate to maintain certain stresses to prevent the dominance of common species (e.g., grasses during heathland restoration).

Degrees of Degradation

The degree of land degradation can vary from slight disturbance or perturbation of a plant species assemblage to extreme toxic contamination. Between these two extremes, the terms "degraded" and "derelict" are used, particularly for agricultural land and postindustrial sites, respectively.

Causes and Types of Agricultural Degradation

In many areas of the world, inappropriate agricultural practices have stripped the available resources from

Table 1 Typical factors constraining plant growth on degraded land

Category (constraint)	*Examples*	*Effect*
Chemical		
Deficiency of plant macronutrients	N, P, K, Mg, Ca	Plant growth reduced
Deficiency of plant micronutrients	Cu, Zn, Mn, Fe	Plant growth reduced
Toxicity of metals	Pb, Cu, Zn, Al, Se, Hg	Root growth reduced or prevented
Salinity	Na	Root growth reduced or prevented
Toxicity of hydrocarbons	Diesel, crude oil	Root growth reduced or prevented
Toxicity of other organic compounds	Chlorinated solvents, pesticides	Root growth reduced or prevented, bioaccumulation
Exposure to extreme pH	$<$pH 4 or $>$pH 8	Growth of nonadapted plants prevented
Exposure to radionuclides	^{3}H, ^{14}C, ^{36}Cl, ^{90}Sr, ^{129}I, ^{137}Cs, ^{239}Pu	Mutagenesis
Physical		
Instability and disturbance	Coal waste, new spoil tips	Root establishment prevented, erosion
Soil compaction	Industrial sites, mining waste	Root exploration prevented, erosion
Excessive drainage	Hard rock waste	Water stress
Insufficient drainage	Mining waste	Waterlogging and anoxia
Exposure	Mountain or coastal sites	Wind damage, water stress
Biological		
Lack of propagules	New or isolated sites	Colonization prevented
Lack of symbionts	N_2-fixing bacteria, mycorrhizae	Plant growth reduced
Competition from other plants	Grasses	Growth of desired species suppressed
Herbivory from large animals	Rabbits, sheep, deer	Plant growth reduced
Herbivory from small animals	Caterpillars, snails	Plant growth reduced
Plant disease	Fungal and bacterial pathogens	Plant growth reduced

the land leaving the soil in a very degraded state and practically useless for agricultural production. Agriculture, and in particular arable cultivation, tends to accelerate the loss of soil and soil nutrients through erosion and leaching. The cultivation of soil, while often resulting in a release of nutrients in the short term, is also typically associated with a loss in soil organic matter. This leads to a loss of soil structure, reduction in soil microbial activity and diversity, and reduced chemical buffering capacity, all of which are crucial factors controlling plant growth. Loss of soil structure tends to reduce the rate of water infiltration during rainfall events, which consequently increases runoff and thus enhances soil erosion, placing agroecosystems into a downward degradation spiral. In addition, the continued removal of nutrient cations (e.g., NH_4^+, K^+, Ca^{2+}, and Mg^{2+}) from the soil by crop offtake and leaching leads to their replacement on soil cation exchange sites by protons (H^+) and phytotoxic aluminum (Al^{3+}), making the soil increasingly acidic and accelerating the spiral of degradation. Unless the nutrients are replaced (e.g., by fertilizers) and the pH rectified (e.g., by lime application), yields gradually decline over time, and eventually crops fail because of nutrient deficiencies or the toxic effects of the acidification of soil.

Overgrazing of pastures also leads to soil degradation, by removing the protective cover of vegetation and increasing surface compaction, thus increasing runoff and accelerating soil erosion by wind and water. In addition, large areas of irrigated land have become degraded because of excessive salt accumulation in surface soils causing phytotoxicity. Soil salinization is typically due to the high rates of water evaporation/evapotranspiration that occurs in irrigated areas, inducing the surface accumulation of salts previously present in the irrigation water or inducing the rise of saline groundwater to the surface. Localized sources of environmental damage may also result from overuse or spills of fertilizers, manures, and pesticides. In many regions of the world, large areas of agricultural land are eventually abandoned due to invasion by weeds, such as *Striga* spp. and *Imperata cylindrica*. It is not easy to assess the total amount of land that has been severely degraded by agriculture, but one estimate suggests the global figure is 430 million ha.

Causes and Types of Industrial Dereliction and Land Contamination

Human industrial activity has been driven by the possession of natural resources (e.g., coal, oil) and by global market forces and regional economics. Changing resource bases and increasingly dynamic markets have led to a continual movement of industry around the world and thus the continual production of postindustrial sites. Current estimates suggest that there are several million such sites around the world in need of restoration. In many cases, these sites are not heavily polluted, but redundant infrastructure remaining from the previous industrial activity requires removal before a desired end-use can be achieved (e.g., mineral waste tips, concrete buildings,

Table 2 Classes of contaminants found at postindustrial sites

Category	*Examples*
Organic contaminants	
Fuel hydrocarbons	Benzene, toluene, xylene, MTBE
Polynuclear aromatic hydrocarbons (PAHs)	Creosote, pyrene, anthracene
Polychlorinated biphenyls (PCBs)	Arochlor 1221, arochlor 1254
Chlorinated solvents	Trichloroethylene, vinyl chloride
Phenolics	Pentachlorophenol
Pesticides	Atrazine, DDT, lindane
Explosives	Trinitrotoluene (TNT)
Detergents	Nonylphenol ethoxylates
Phthalates	*Bis*-2-ethylhexylphthalate
Ketones	Acetone
Complexing agents	EDTA, nitriloacetic acid
Organic acids	Citric acid, benzoic acid
Alkyl phosphates	Tributyl phosphate
Inorganic contaminants	
Metals	Hg, Cr, Pb, Zn, Cu, Cd, As, Ni, Se
Radionuclides	^{3}H, ^{14}C, ^{36}Cl, ^{90}Sr, ^{129}I, ^{137}Cs, ^{239}Pu
Salts	NaCl, KCl
Acids	HCl, H_2SO_4
Alkalis	NaOH, Na_2CO_3
Biological contaminants	
Bacterial, fungal and viral pathogens	*Escherichia coli* O157, anthrax
Prion agents	BSE

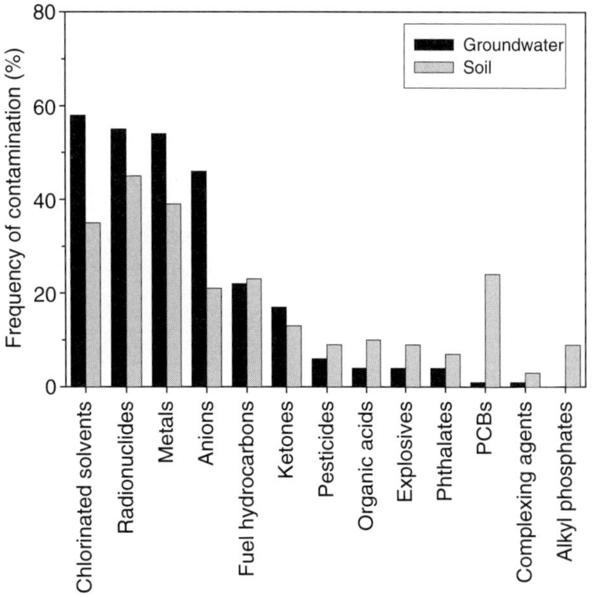

Figure 1 Frequency of occurrence of contaminant classes in either the groundwater or soil of 91 United States Department of Energy hazardous waste sites. Data from Lenhard RJ, Skeen RS, and Brouns TM (1995) Contaminants at U.S. DOE sites and their susceptibility to bioremediation. In: Skipper HD and Turco RF (eds) *Bioremediation: Science and Applications; SSSA Special Publication 43*, pp. 157–172. Madison, WI: Soil Science Society of America.

etc.). Sometimes, however, human activity has caused the contamination of land with organic and inorganic compounds, which pose threats to freshwater, marine, and terrestrial ecosystems, and to human health. The range of pollutants that may be found at these contaminated sites is shown in **Table 2** and **Figure 1**. The contamination of many sites is caused by human error during handling, allowing spills and leaks to occur. This can result in either a localized contamination of the environment (point source; e.g., from a single underground gasoline storage tank) or in the widespread contamination of the environment (diffuse source; e.g., benzene emissions from vehicles). In many cases, the contamination of industrial sites and the surrounding environment occurred decades ago when the environmental and human impact of chemical release was not fully appreciated (e.g., radionuclides released during weapons testing; chemical release into ground and surface waters from leach and disperse municipal solid waste sites).

Parallels between Land Reclamation and Natural Succession

Since the early days of plant ecology, many studies have looked at the changes in the plant species assemblage (i.e., the set of species growing in a locality) that occur following disturbance. The colonization of naturally denuded surfaces, such as on new volcanic islands or under retreating glaciers, is referred to as primary succession. Subsequent changes in plant species assemblages, and the colonization of disturbed surfaces that retain soil and plant propagules, are referred to as secondary succession. In the classic model of plant succession, one cohort of species replaces another until a climax assemblage is reached, which is predictable from soil and climatic conditions. Lichens and lower plants can colonize bare rock surfaces and ameliorate conditions sufficiently for "pioneer" species of herbs and woody plants to grow. These facilitate the development of soil and further ameliorate conditions, allowing the colonization of climax species. Where there is continuing stress, for example from herbivory, the succession may be maintained at an intermediate stable state known as a plagioclimax. This model has, however, been criticized. Some species are early colonists but also appear in climax assemblages; and stable climaxes may be rare because of variation in climate, or the effects of disease. The composition of an assemblage may also be affected by stochastic events such as seed dispersal.

Theories of plant succession have been of use in guiding reclamation of degraded land. Alleviating stresses may not be enough to encourage the growth of appropriate plants; propagules and symbionts may need to be introduced. Local species that act as pioneers on denuded substrates are obvious choices when selecting species to introduce. Intermediate stages may be necessary, in which "nurse species" chosen for their ameliorative qualities, such as nitrogen-fixing *Alnus* species, are used to improve conditions so that the desired plant species can then establish.

Practice

Site Assessment

The starting point for any reclamation project is an assessment of the problems and potentials of the land. The age of the site and amount of vegetation cover already achieved indicate the severity of constraints to plant growth. Surveys and analyses of vegetation and soil are useful. For nature conservation, any small areas of remaining habitat should be carefully retained, to act as sources of propagules. Surveys may also reveal rare species that are adapted to the degraded conditions, leading to a redefinition of the restoration objectives. It is also important to examine the history of the site in order to determine the cause of the degradation. This preliminary site assessment is vital before undertaking a survey and monitoring of pollutant levels, since cost and time constraints rarely make it possible to screen for all potential contaminants.

Surveys should aim to identify the factors that currently limit development of the target vegetation. These may not be the same as the initial cause of degradation; for example, grazing animals or competition from weeds may be preventing vegetation regeneration. Any continuing causes of stress must be brought under control. Laboratory experiments or field trials in which individual limitations are removed may be necessary to identify the main constraints. However, different limitations interact, and some may be easier to overcome than others. On loose mining waste, for example, drought is often the cause of plant death, but increasing the nutrient supply can allow plants to develop deeper root systems and increase their drought tolerance – applying fertilizer is considerably easier than applying soil. In the case of contaminated sites, the site survey must explore the type, concentrations, and spatial distribution of the contaminant. At most contaminated sites, it is common to find mixtures of pollutants present at different levels and of varying toxicity. This is often due to the primary contaminant being broken down *in situ* by soil microorganisms to produce secondary contaminants (e.g., dichloroethane, vinyl chloride from trichloroethylene). The presence of these pollutant cocktails may cause some of the chemicals to behave differently than if they were present on their own (e.g., altered water solubility, bioavailability, etc.). Further, pollutants are rarely distributed uniformly either vertically or horizontally throughout the site, but are usually concentrated in "hot spots" of contamination.

Required Soil Conditions

Soil characteristics must be brought into a range suitable for plant growth. **Table 3** shows the range of soil conditions necessary for the growth of most plants, though certain species may grow optimally under more extreme conditions. Nitrogen content is worth particular note as plants possess a high demand for nitrogen and require a continued supply (e.g., 100 kg N ha^{-1} year^{-1}), but available nitrogen (NH_4^+, NO_3^- and dissolved organic nitrogen compounds) is easily leached from soils. In established ecosystems, nitrogen is made continually available through the mineralization of soil organic matter. The aim for sustainable reclamation should therefore be to ensure a sustained supply of nitrogen and other nutrients by recapitalizing the soil with organic matter. Usually the best and cheapest way to achieve this is to incorporate nitrogen-fixing plant species into the reclamation scheme. Suitable trees, shrubs, and herbs are to be found in the Fabaceae ("legumes") as well as in certain other groups such as alders (*Alnus* spp.) and casuarinas (*Casuarina* spp.). The symbiotic microorganisms (e.g., *Rhizobium*, *Frankia*) that allow these plants to fix nitrogen

Table 3 Normal range of soil characteristics required for successful plant growth

Parameter	*Target range*
PH	5–7.5
Cation exchange capacity (meq kg^{-1})	100–300
Electrical conductivity (dS m^{-1})	0–4
Bulk density (g cm^{-3})	0.8–1.4
Volumetric water content (cm^3 cm^{-3})	0.15–0.35
Temperature (°C)	10–25
Redox potential (mV)	>500
Nutrients (μg g^{-1})	
K	100–300
Ca	500–2000
Mg	50–300
Fe	5–200
P (available)	5–20
NH_4^+-N	2–20
NO_3^-N	2–20
Mineralizable N	50–200
Total N	1000–10000

may also need to be introduced, for instance, by watering beds of seedlings with a suspension of homogenized nodules. It is also good practice to apply an inoculum of soil from established target vegetation, to introduce beneficial decomposers and mycorrhizal symbionts. In the case of contaminated land, the concentration of pollutants must be low enough to allow some degree of plant growth. It must be noted that some contaminants may be more phytotoxic in combination with other chemicals (i.e., in cocktails) than on their own (e.g., toluene in the presence of methyl tertiary butyl ether), while in other cases one pollutant may reduce the toxicity of another (e.g., heavy metals may be complexed by organic contaminants). Most heavy metals become increasingly soluble and therefore more toxic at low pH.

Strategies for Reclaiming Land Degraded by Agriculture

Reclamation of degraded agricultural land is commonly aimed at returning the land to productivity. Degradation is associated with loss of soil organic matter and nutrients, loss of topsoil through erosion, accumulation of salts, and/or weed infestation. Where nutrient deficiencies constrain production, these may be overcome by adding fertilizer or organic wastes. Increasing soil organic matter content is likely to be of benefit, providing slow-release pools of nutrients and increasing the protection of nutrients by increasing soil structure and ionic exchange capacity. Soil organic matter content may be increased by fallowing, though an addition of nutrients may still be necessary to ensure good growth of the fallow. Improved fallows and green manures have been successful in some situations, using introduced species, particularly nitrogen fixers, to increase soil fertility. Physical erosion of soil is worsened by a lack of vegetation, and erosion control often starts with overcoming nutrient constraints or controlling grazing to maintain cover. Erosion control measures should be carefully located in relation to water flows, for example, grass strips along contours. The reclamation of salinized soils is made difficult because high salt concentrations destroy soil structure and so make it difficult to wash out the salts. Adequate drainage is the first priority, providing an exit route for salts. Liming is another key process in reclaiming salinized soils, since the calcium ions from the lime are able to replace sodium ions on soil exchange sites, allowing clay particles to clump together and improving structure and drainage. Strategies for reclaiming weed-infested land include cultivation, herbicide, and the use of improved fallows, cover crops, and tree planting to shade out weeds. Adding fertilizer may help when the weed is adapted for infertile conditions.

While the biophysical factors described above are the most immediately obvious causes of agricultural degradation and indicate where interventions are likely to help restore productivity, many other factors profoundly affect agricultural livelihoods. Reversing land degradation thus also depends on examining the factors (e.g., land tenure, access to markets, availability of labor) influencing the investment of time and resources that people are able to make into land restoration.

Phytoremediation Strategies for Postindustrial and Contaminated Land

The first part of any remediation strategy is the creation of site conditions that will be conducive to plant growth. Typically, this will involve the removal of compacted layers and industrial structures, the control of water flow on the site, the addition of substrate amendments and the landscaping of the site. The substrate amendments may include protective covers to enhance substrate stability and prevent soil erosion (e.g., geosynthetic textile matting, plastic netting) alongside addition of fertilizers and organic materials to promote soil biological activity, long-term nutrient supply, pollutant buffering, and soil structural stability. Organic amendments commonly used at contaminated sites include by-products from other industrial processes, including paper processing waste, composted waste, water treatment aluminum and iron hydroxide biosolids, sewage sludge, etc. However, careful attention must be paid to the chemistry of these amendments so as not to cause further contamination (e.g., by heavy metals in sewage sludge) or nutrient imbalances in the soil (e.g., nitrogen immobilization and deficiency by the addition of paper waste with a high carbon:nitrogen ratio; phosphorus deficiency by the addition of water treatment aluminum and iron hydroxide biosolids). If topsoil is being brought back to the site after stockpiling elsewhere during the previous industrial operations, it is important that the prior storage does not promote soil degradation. Often, workers unfamiliar with the living nature of topsoil will store it in one big pile, causing excessive compaction and anaerobic conditions.

After the first stage, plants can be introduced onto the site. The type of plants employed depends on both the end-use of the land and the type of contamination. In some cases, these aims can be contradictory. One example of this can be seen during the restoration of municipal solid waste (landfill) sites. The planting of trees can significantly enhance the amenity value of these sites, but may increase the risk of environmental contamination by promoting drying and rupturing of the protective clay cap, releasing noxious and potentially explosive gasses in an uncontrolled manner. This can be avoided by careful construction of landfill caps with a layer of highly compacted soil through which tree roots cannot penetrate. The actual planting can also be problematic, expensive and time-consuming in some environments (e.g., remote locations with uneven and unstable topography). In addition, environmental constraints need to be overcome that may require significant technological investment (e.g., irrigation schemes, etc.). Plant propagation and planting techniques developed for silviculture, agriculture, and horticulture are appropriate. However, there are techniques specific to land reclamation. Hydroseeding is a method of delivering seeds over large or inaccessible areas using a low pressure hose to deliver a water-based slurry of organic material and seeds. This technique aids both in the adhesion of seeds to the ground and also enhances soil quality through the addition of nutrients, microbial propagules, and organic matter.

A wide variety of phytoremediation strategies can be employed to reclaim polluted sites and the major features of these are summarized in **Table 4**. In many of these remediation schemes, the contaminants in the soil are not actually removed, but their movement and bioavailability are controlled at least in the short term. Where actual decontamination of the pollutants occurs, this normally only takes place in the rooting zone, often leaving large amounts of contaminants remaining at depth. These pollutants must be treated by other mechanisms (e.g., pump and treat, air sparging, bioremediation using added bacteria, chemical remediation, natural attenuation using naturally present microorganisms). One major drawback with phytoremediation schemes is the uncertainties and risks associated with them (**Table 5**). In many instances, this has prevented

Table 4 Summary of phytoremediation strategies for cleaning up and restoring contaminated and postindustrial sites

Pollutant type	*Mode of operation*	*Examples of plants*
Inorganic pollutants		
Phytoextraction	High-biomass-yielding plants that take up and accumulate heavy metals (e.g., As, Cd, Zn), excess cations (e.g., Na) or nutrients (e.g., PO_4^{3-}, NO_3^-, NH_4^+) in the shoots. These are then harvested and disposed of safely. Metal complexing agents (e.g., EDTA) can be added to the soil to aid metal uptake by the plant	Specific *Brassica* spp., trees and ferns for heavy metals. Halophytic plants for Na. Most plants for NO_3^-, NH_4^+ and PO_4^{3-}
Rhizofiltration	Plants are grown on hydroponic rafts allowing the roots to grow into contaminated water. The roots take up or immobilize toxic metals (e.g., U, NO_3^-, PO_4^{3-})	Sunflowers, water hyacinth
Phytostabilization	Plants stabilize the soil and consequently the pollutants contained within them (e.g., Cu, Zn). Prevents erosion and reduces pollutant leaching. Normally the plants are nonaccumulators (excluders)	Many grasses and trees
Phytovolatilization	Plants take up the inorganic pollutants and release them from the leaves to the atmosphere in a gaseous (volatile) form (e.g., Se and Hg).	Willows (often genetically engineered to enhance volatalization potential)
Organic pollutants		
Phytodegradation	Pollutants are taken up into the plants and metabolized in the roots and leaves into nontoxic byproducts	Many grasses and trees
Rhizodegradation	Stimulation by plants (via root exudation and root turnover) of soil microorganisms living in the rhizosphere which biodegrade the pollutants	Many grasses and trees
Hydraulic pumping	Plant roots take up water lowering the water table preventing the sideways migration of pollutants and enhancing the loss of volatile organic compounds from the soil surface	Deep-rooted trees
Phytovolatilization	Plants take up the pollutants from the soil, transport them up to the shoots and then release them from the leaves into the atmosphere	Many grasses and trees
Phytosorption	The absorption of pollutants by plant leaves and roots preventing them from migrating through the soil	Many grasses and trees
Phytocapping	Plant interception of rainfall reducing the amount of leaching and runoff and consequently pollutant migration through the soil	Many grasses and trees

Table 5 Advantages and disadvantages of phytoremediation strategies for cleaning up and restoring contaminated and postindustrial sites in comparison to other remediation strategies (e.g., excavation and landfilling)

Positive points	*Negative points*
Cheap	Slower than other methods, with longer timescale of intervention and greater uncertainties
Likely to cause less pollution	May mobilize toxins into biological systems
Visually appealing	Expensive if it goes wrong
Faster than natural attenuation	Few plants to choose from for some restoration scenarios
Generates less secondary waste	Effectiveness is very dependent on the type and level of contaminants
Soils still useable at the end	Labor intensive
Requires little equipment	Mainly applicable to shallow contaminated sites
Plants are capable of being genetically engineering to enhance remediation	The use of genetically engineered plants is not allowed in many areas of the world

adoption of the technology by industrial companies who are looking for a rapid, economically viable, legally secure, and guaranteed way of decontaminating land. In many situations, phytoremediation of degraded sites is a long-term strategy in which complete pollutant removal or revegetation cannot be guaranteed; however, certainly these strategies are better than taking no action at all (i.e., relying on natural attenuation by leaching or by indigenous plants and microorganisms). Uncertainties mainly center on the lack of control over the environment (e.g., freak weather events, attack by pathogens or herbivores) and the inability to predict the long-term management of phytoremediation schemes.

Land Reclamation and Restoration of Biodiversity

Any of a range of habitats could be created on a degraded site, depending upon the resources available. Where biodiversity conservation is the main objective of land reclamation, some comparison of the conservation value of the possible habitats is essential. Costs must also be considered. While habitats of conservation value are likely to be cheap to establish and maintain compared with agricultural or amenity end-uses, this is not always the case. All habitats require monitoring and management, and operations such as maintaining grazing or removing weeds can be expensive. Before beginning reclamation works, a target habitat must be selected after weighing up conservation benefits and costs. The criteria used in selecting statutory nature conservation sites are useful in comparing the conservation value of potential targets. These include naturalness, diversity, typicalness, size, rarity, and fragility.

Local conservation experts will be able to advise on the habitats and species that are priorities for conservation locally. These may include habitats that are already present on the site, such as remnants of the original vegetation, or new habitats that have developed as a result of disturbance. Creating several habitats is likely to increase the total species diversity within a site, and favors species that require several habitats, or edge habitats. However, certain species require large contiguous areas of a single habitat.

Reclamation for nature conservation generally does not require all stresses to be removed. Soils that present no nutrient or water availability restrictions are rarely valuable for nature conservation, since they become dominated by common species. Postindustrial sites often support unusual species because their soils present restrictions to the growth of these common species. The diversity of soils on many postindustrial sites encourages biodiversity and should be retained. Areas that support little plant growth reduce the visual appeal of a site to most people, but not necessarily its conservation value. This depends on the relative value of skeletal habitats (e.g., scree, bare rock, sparse grassland) and more vegetated habitats (e.g., closed swards, heath, scrub, woodland), which will vary with local conservation priorities. Successional habitats have generally been considered a low priority for nature conservation, but perceptions of their value are changing, in part because of the obvious conservation value of many old quarries and other postindustrial sites. Shifting the emphasis from conservation of climax to dynamic habitats could, however, allow any impoverished habitat to be justified in nature conservation terms as a successional stage.

It is essential to establish criteria for assessing whether habitat targets have been achieved and to plan for a program of monitoring. Occurrence of priority species can be used as an indicator of the success of habitat creation. Alternatively, the target can be expressed as a threshold abundance or cover of species that are viewed as essential to the structure of the habitat. Measurements of biodiversity, whether simple species counts or more sophisticated biodiversity indices, are useful indicators of restoration success, but do depend on the identification of all species. This can be a very time-consuming process and accurate identification requires expert knowledge. Since demand for taxonomic expertise usually exceeds the amount of time available, it may be sensible to base a monitoring program on morphotypes that can be easily distinguished by nonexperts. Species assemblages can be compared with target assemblages by ordinating species abundance data and measuring the degree of similarity, for example, as distance on an ordination biplot.

When reclaiming land for nature conservation, particular attention must be paid to the provenance of seed and planting stock, and in most cases using stock of local provenance will be advantageous. As well as preserving local genetic diversity of the planting stock, this is likely to benefit invertebrates and other species that are adapted for the local genotypes. Careful attention to provenance can reveal genotypes particularly well suited to growing on denuded or extreme substrates. Mixed materials such as hay, brash, leaf litter, or soil can often be obtained from a similar ecosystem to the target and used to introduce propagules of a range of species.

Conclusions

The restoration of land to useful function will be a critical human activity so long as land is used in ways that are not sustainable, or desired end-uses are subject to change. The key components in any restoration strategy are a clear definition of the desired end-use, the identification of limiting factors, the design and management of measures to remove these constraints, and monitoring so that success can be evaluated and any necessary modifications made to the restoration scheme. Land restoration can be very expensive and its costs should be taken into account when considering the ecomomics of processes that result in the degradation of land.

List of Technical Nomenclature

Bioaccumulation	Becoming concentrated at higher trophic levels in a food web.
Fallow	Vegetation grown for its fertility benefit, without harvesting.
Morphotype	A group of similar, but not necessarily taxonomically identical, organisms.
Phytotoxic	Toxic to plants.
Sequestration	Storage for a long period.
Sparging	Injecting air into groundwater to volatilize contaminants and enhance biodegradation.
Symbionts	Organisms closely associated for mutual benefit.

See also: **Energy Crops**: Biomass Production. **Nutrition**: Aluminum Toxicity; Deficiency Diseases, Principles; Mineral Uptake. **Plants and the Environment**: Plants as Pollution Monitors; Waste Water Treatment. **Root Development**: The Rhizosphere and its Microorganisms.

Further Reading

Barrow CJ (1994) *Land Degradation: Development and Breakdown of Terrestrial Environments*. Cambridge: Cambridge University Press.

Bradshaw AD and Chadwick MJ (1980) *The Restoration of Land: the Ecology and Reclamation of Derelict and Degraded Land*. Oxford: Blackwell Scientific Press.

Cunningham SD, Anderson TA, Schwab AP, and Hsu FC (1996) Phytoremediation of soils contaminated with organic pollutants. *Advances in Agronomy* 56: 55–114.

Gilbert OL and Anderson P (1998) *Habitat Creation and Repair.* New York: Oxford University Press.

Harris JA, Birch P, and Palmer JP (1996) *Land Restoration and Reclamation. Principles and Practice.* Harlow: Addison Wesley Longman.

Lenhard RJ, Skeen RS, and Brouns TM (1995) Contaminants at U.S. DOE sites and their susceptibility to bioremediation. In: Skipper HD and Turco RF (eds) *Bioremediation: Science and Applications; SSSA Special Publication 43*, pp. 157–172. Madison, WI: Soil Science Society of America.

Newman LA (1999) Remediation of trichloroethylene in an artificial aquifer with trees: a controlled field study. *Environmental Science and Technology* 33: 2257–2265.

Rugh CL, Senecoff JF, Meagher RB, and Merkle SA (1998) Development of transgenic yellow poplar for mercury phytoremediation. *Nature Biotechnology* 16: 925–928.

Schnoor JL, Licht LA, McCutcheon SC, Wolfe NL, and Carreira LH (1995) Phytoremediation of organic and nutrient contaminants. *Environmental Science and Technology* 29: 318–323.

Schroll R, Bierling B, Cao G, and Dorfler U (1994) Uptake pathways of organic chemicals from soils by agricultural plants. *Chemosphere* 28: 297–303.

Skipper H and Turco R (1995) *Bioremediation: Science and Applications; SSSA Special Publication 43*. Madison, WI: Soil Science Society of America, American Society of Agronomy, and Crop Science Society of America.

Terry N, Zayed AM, de Souza MP, and Tarun AS (2000) Selenium in higher plants. *Annual Review of Plant Physiology and Plant Molecular Biology* 51: 401–432.

Ozone Depletion

P V V Prasad, University of Florida, Gainesville, FL, USA
V G Kakani and K R Reddy, Mississippi State University, MS, USA

Introduction

Oxygen (O_2), a gas that all life forms breathe, consists of two oxygen atoms. Ozone, a form of oxygen that has three atoms of oxygen (O_3), is formed in the atmosphere when an oxygen atom (O) combines with an oxygen molecule. Most of the atmospheric ozone is produced and is present in the upper layers of atmosphere called the stratosphere. The highest concentrations occur at an altitude of 19–30 km above the earth's surface, termed the "ozone layer." An estimated 3.5 billion tonnes of ozone exists in the stratosphere. The amount of atmospheric ozone varies both spatially (location to location) and temporally (season to season). Ozone absorbs and filters harmful ultraviolet (UV) radiation emitted from the sun, preventing it from reaching the earth's surface. UV radiation is harmful to cells of all life forms on earth including plants and animals. Life on earth would not have evolved without an atmospheric ozone layer.

Slight fluctuations in atmospheric ozone concentration have occurred over the scale of centuries, mainly in response to changes in solar energy output and volcanic eruptions. However, atmospheric ozone concentration has dramatically decreased since the 1950s. This was evident in 1985 with the discovery of an "ozone hole" over Antarctica. The ozone hole appears annually over Antarctica in September and October. Similar ozone depletion has also been recently observed over Arctic regions. The ozone hole is defined where stratospheric ozone levels fall below 200 Dobson Units (DU). Dobson Unit measures the total amount of ozone in the atmosphere, i.e., the ozone layer thickness when all the atmospheric ozone in a column of air is compressed into one layer at 0°C and with a pressure of one atmosphere above it. Every 0.01 mm thickness of the layer is equal to 1 Dobson Unit. The average amount of ozone in the stratosphere across the globe is 300 DU (thickness of 3 mm at 0°C and 1 atmospheric pressure).

Historical data have shown that atmospheric ozone concentrations were stable between the 1950s and late 1970s. Global ozone levels have declined by about 3% below the pre-1980 average values until 2001, and this decline was about three times faster than that recorded in the 1970s. A report released in 2002 showed that since the early 1990s, the monthly total atmospheric ozone concentrations in September and October have been about 40–50% below pre-ozone-hole levels, with up to a 70% decrease for a period of a week or so. In recent years, the ozone hole has persisted into early summer, increasing its impact on UV radiation. In addition to Antarctica and the Arctic, ozone depletion now occurs in several parts of the world and affects most of North America, Europe, Russia, Australia, New Zealand, and parts of South America.

In the early 1970s it was suggested that anthropogenic emissions of chlorofluorocarbons (CFCs) were likely the main source of stratospheric ozone depletion. CFCs contain chlorine, fluorine, and carbon atoms and are nontoxic. CFCs are widely used as coolants in air conditioners and refrigeration, as solvents in cleaners, as blowing agents in foam production, and as aerosol propellants. CFCs are extremely stable compounds in the stratosphere, having a lifetime of about 50–100 years. In the upper stratosphere, UV radiation breaks down these compounds and releases chlorine atoms (Cl), which are highly reactive with the ozone molecule, forming chloride (ClO) and oxygen. The chloride molecule then breaks down, liberating a chlorine atom and oxygen. The free chlorine atom again breaks down the ozone molecule to oxygen, leaving again chlorine atoms free to react with ozone. This process leads to stratospheric ozone depletion (**Figure 1**). It has been estimated that one molecule of CFC can cause the loss of 100 000 ozone molecules.

Emissions of CFCs account for about 80% of total stratospheric ozone depletion. In addition other halocarbons such as hydrofluorocarbons, fluorocarbons and bromocarbons also destroy ozone. After discovery of the ozone hole in 1985, the Vienna Convention was established for international cooperative research into the ozone layer and effects of ozone depleting compounds. Based on the Vienna Convention, the Montreal Protocol was drafted and signed by 24 countries and by the European Economic Community in 1987. This legislation was passed to stop production, imports, and exports of CFCs, halocarbons, and other ozone-depleting chemicals. It imposes strict penalties to violators. Many believe that if both developed and developing countries strictly follow the Montreal Protocol and subsequent amendments, it will require 50 years to repair the ozone layer and return atmospheric ozone concentrations to pre-1980 levels. Since CFCs

Approved for publication as FAES No R-09449.

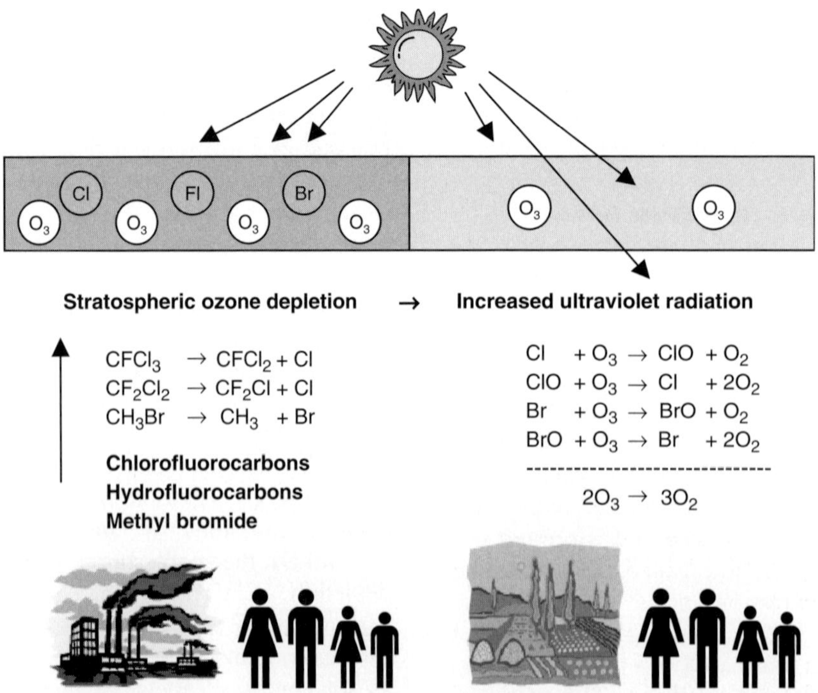

Figure 1 Process of stratospheric ozone depletion. Anthropogenic emissions of chlorofluorocarbons (CFCs), hydrofluorocarbons, and bromocarbons released in troposphere move up into stratosphere. Upon exposure to radiation, atomic chlorine or bromine are released which reacts with ozone and converts into oxygen, leading to depletion of ozone. Stratospheric ozone depletion leads to increase in ultraviolet radiation reaching the earth's surface.

remain in the atmosphere for long periods, it is probable that terrestrial plants will experience increased levels of UV radiation for many more years, even if there is full compliance with the Montreal Protocol.

Stratospheric Ozone Depletion and UV Radiation

The main consequence of ozone depletion is increased UV radiation reaching the earth's surface. Calculations based on relations with total ozone and total irradiance suggest that UV irradiance has increased since the early 1980s by 6–14% in middle and high latitudes of the northern and southern hemispheres. It is projected that every 1% decrease in ozone will increase UV exposure by 2–3% in the lower atmosphere. UV radiation is an electromagnetic form of energy that comes from the sun. The radiation energy coming from the sun is classified into several regions based on wavelength, which is measured in nanometers. The shorter the wavelength, the greater is the energy of the radiation. The main components of radiation in order of decreasing energy are gamma rays, X-rays, UV radiation, visible light, infrared radiation, microwaves, and radio waves. UV radiation is further divided into three categories based on wavelength: UV-A, between 320 and 400 nm; UV-B, between 280 and 320 nm; and UV-C, between 200 and 280 nm.

Shorter-wavelength radiation causes more damage to biological systems. UV-A is the least damaging component within the UV spectrum and reaches the earth's surface in large quantities. Both UV-B and UV-C are very harmful. Most of the UV-C radiation is absorbed by ozone, rarely reaching the stratosphere and never reaching the earth's surface. UV-B radiation is mostly likely to reach the earth's surface with increased ozone depletion. Factors such as altitude, latitude (distance from the equator), and time of day influence the amount of UV-B exposure. The current UV-B radiation level on the earth's surface ranges between 2 and $>8\,\mathrm{kJ\,m^{-2}\,day^{-1}}$depending on location. Small increases in the solar UV-B radiation can have substantial effects on plants at both the cellular and whole plant levels.

Methods in UV-B Studies

Several environments and different methods are used to impose UV-B radiation treatments. Most of the studies to illustrate UV-B effects are conducted in growth chambers and greenhouses with a few studies conducted under field conditions. An above-normal UV-B radiation dose is imposed by placing UV-B emitting fluorescent lamps at a height of 0.5 m above the canopy. In most studies, this process is compared with treatments in which UV-B is blocked by using a growth chamber or greenhouse casings (such as plexiglas, mylar plastic) that filter natural UV-B radiation. The UV-B emitting bulbs also emit a small amount of UV-C radiation that is filtered by wrapping the bulbs with cellulase diacetate film. In most studies the UV-B is delivered in a square wave fashion, i.e., a constant dose during a given time period. Recent studies are done using a modulated UV-B delivery method that mimics the natural diurnal trend.

It is important to consider the biological effectiveness of UV-B radiation while assessing the effects of ozone depletion on biological systems and processes. For this purpose, biological weighting functions are used which are derived from action spectra. The degree of absorption of radiation by important biological macromolecules is wavelength-dependent and as a result, the interfering effect of UV radiation is wavelength-specific. Consequently, it is necessary to weight the incident spectral irradiance with a function called the action spectrum ($S\lambda$), which expresses the relative effectiveness of UV radiation at various wavelengths for a particular biological process. For a selected biological process, the biologically effective UV irradiance (E_{eff}) is written as:

$$E_{\text{eff}} = \sum_{200}^{400} E_{\lambda}\, S_{\lambda}\, \Delta\lambda$$

where $E\lambda$ is the spectral irradiance in $\mathrm{W\,m^{-2}\,nm^{-1}}$ and $\Delta\lambda$ is the wavelength step in nm for the measurements of the spectral irradiance. The summation interval is the wavelength range from 200 nm to 400 nm. If the waveband consideration includes the vacuum UV region (<200 nm) then the lower limit can be extended. The biologically effective UV irradiance has units of irradiance.

Effects of UV-B Radiation on Plants

Plants are highly sensitive to UV-B radiation due to their sessile nature. In plants, UV-B radiation damages the cell membranes and all organelles within the cell, including the chloroplasts, mitochondria, and deoxyribose nucleic acid (DNA) within the nucleus. The damage to these cell organelles will directly or indirectly affect basic plant metabolic processes such as photosynthesis, respiration, growth, and reproduction. Consequently, UV-B damage harms yield and quality. A general overview of various processes affected by UV-B radiation at the cellular and plant levels is given in **Tables 1** and **2**. However, the effects of UV-B radiation will vary with intensity and duration of irradiation and stage of plant development. In addition, sensitivity to UV-B radiation varies widely among plant species and among cultivars of the same species.

Table 1 Effects of exposure to UV-B radiation on various physiological processes in plants

Trait	*Decreases*	*Increases*	*No effect*
DNA damage		✓	
Protein destruction		✓	
Fatty acid destruction		✓	
Photosynthesis	✓		
Photosystem I	✓		
Photosystem II	✓		
Rubisco	✓		
Stomata closure		✓	
Chlorophylls	✓		
Flavonoids		✓	
Waxes		✓	
Epidermal hairs		✓	
Cuticle thickness		✓	
Reproduction			
Pollen viability	✓		
Pollen tube growth	✓		
Fertilization	✓		
Cell division	✓		
Cell size			✓

Table 2 Effects of exposure to UV-B radiation on various growth and yield parameters in plants

Trait	*Decreases*	*Increases*	*No effect*
Photosynthesis	✓		
Stomatal conductance	✓		
Phenology			✓
Senescence		✓	
Plant height	✓		
Branching		✓	
Leaf area	✓		
Leaf growth and expansion	✓		
Leaf thickness		✓	
Specific leaf weight		✓	
Dry matter production	✓		
Flowering	✓	✓	
Fruit (grain) number	✓		
Fruit (grain) weight	✓		
Yield	✓		
Quality	✓		
Disease incidence			
Powdery mildew	✓		
Rust		✓	
Insect	✓	✓	

Genetic (DNA) Damage

The nucleus of each cell consists of genetic material in the form of DNA, which is highly sensitive to UV-B radiation. DNA is considered as the primary absorbing compound in the cell in the UV-B region of the spectrum. Exposure of DNA to UV-B radiation can result in (1) breakage of bonds in the DNA and DNA–protein crosslinks; (2) chromosomal breakage; (3) chromosomal aberrations; and (4) exchange and production of toxic and mutagenic photoproducts (for, e.g., cyclobutane pyrimidine dimmers, pyrimidine (6-4) pyrimidone products, thymine glycols, and pyrimidine hydrates). These changes in the DNA alter transcription, replication, and recombination of genes and cause significant changes in plant metabolic and genetic processes. Exposure to UV-B radiation can also cause damage to protein and lipid bodies in the cell. These damages can influence genetic makeup and affect protein synthesis, enzyme activities, and gene expression.

Ultrastructural Damage

Exposure to UV radiation can result in changes in ultrastructure of various cellular components in many plant species. The main ultrastructural changes include: damage and dilation of the nuclear membrane, rupture of the chloroplast wall, swelling of chloroplasts, dilation of thylakoids, swollen cisternae in the endoplasmic reticulum, damage to mitochondria and plastids, and vesiculation of plasmalemma and tonoplasts. In addition, UV-B radiation alters cell shape and structure. These changes in ultrastructure can potentially affect various physiological processes such as cell division, photosynthesis, respiration, and reproduction.

Photosynthesis

Photosynthesis is the process by which plants convert carbon dioxide and water into carbohydrate in the presence of sunlight. The photosynthetic apparatus is one of the important target sites of UV-B damage. The direct effects of enhanced UV-B radiation on photosynthesis include: (1) damage to ultrastructure of chloroplasts that are principal sites for photosynthesis; (2) impairment of light energy transfer, i.e., electron transport system of photosystem II (PSII) and to a lesser extent photosystem I (PSI); (3) decrease in activity of Rubisco (rubilose 1,5-bisphosphate carboxylase-oxygenase); (4) decreased carbon dioxide fixation and oxygen evolution; and (5) decreased starch and chlorophyll content. The components affected in PSII are the water-oxidizing system, light harvesting complex, and synthesis of chlorophyll a/b-binding proteins. Exposure to UV-B radiation decreases both activity and concentration of Rubisco. UV-B inactivation of Rubisco could be due mainly to modification of the peptide chain, degradation of the protein, and/or diminished transcription of the gene.

In addition to the direct effects of UV-B, photosynthesis may also be indirectly affected by: (1) induction of stomatal closure (stomata are small pores present on the leaf surface that allow gas exchange); (2) decreased individual leaf area and total canopy leaf area; (3) changes in thickness and anatomy; and (4) changes in canopy architecture and morphology. All these changes can potentially decrease light interception and gas exchange, which results in lower canopy photosynthesis. However, the stomatal closure mechanism reduces evapotranspiration water losses and increasing water use efficiency, which lead to increased plant growth and yield.

Plant Morphology and Architecture

Although UV-B radiation constitutes a small portion of the solar spectrum, it causes strong morphological changes in plants. In most plant species, leaves that are exposed to UV-B radiation initially develop irregular chlorotic patches (**Figure 2**). On continued exposure to UV-B, these chlorotic patches turn to brown necrotic spots and later die. Plant height is reduced under elevated UV-B radiation. A decrease of individual leaf size, tiller number, and branch lengths also occurs on exposure to UV-B radiation. These morphological changes result in a smaller canopy, which intercepts less UV-B radiation than a plant grown in a favorable environment. Plant architecture is modified by UV-B radiation. On exposure to UV-B radiation both attached and detached tendrils of pea (*Pisum sativum*) plants

Figure 2 (see color plate 21) Typical symptoms of exposure to UV-B radiation on leaves of cotton plants. (A) Control leaves showing no symptoms; and leaves exposed to UV-B radiation showing (B) initial chlorotic patches and (C) necrotic patches after prolonged exposure.

form spiral coils. Similarly, heliotropism of soybean (*Glycine max*) leaves differentiates between tolerant and sensitive cultivars. Soybean cultivars with leaves facing east or west are more sensitive to UV-B than cultivars with leaves facing south or north due to greater interception of light including UV-B radiation.

Phenology and Reproductive Processes

UV-B radiation does not influence early bud or flower development or the time to first flower. Any delay in time to first flower observed in crop species is due to shedding of early-formed floral buds. UV-B radiation affects flower size, anther number, pollen production, pollen germination, and pollen tube growth in many plant species. In general, reproductive organs of most plant species (pollen and ovules) are highly protected by sepals, petals, and ovary walls. In these plants, pollen is susceptible after it falls on the stigma. Exposure to UV-B radiation decreases pollen germination and rate of pollen tube growth by 10–25%. Consequently, it affects the fertilization process, which results in fewer seeds in sensitive plants. However, once the pollen tube penetrates the stigma surface, the walls of the style and ovary may provide some protection against UV-B radiation.

Growth and Dry Matter Production

Growth is the rate of increase in weight and size of plant organs such as leaf, stem, or root. Dry matter is the total weight of all plant organs. Exposure to UV-B radiation causes decreases in growth of leaves and stems in many plant species in both a controlled environment and field studies. The decrease in growth of leaves, main stem, and branches is due to reduced cell division rather than decreased cell size. Reduction in plant height on exposure to UV-B is due to decreased levels of growth hormone (indole acetic acid) levels in plants. The smaller and more compact canopy reduces the amount of UV-B intercepted by the plant but also reduces the potential or total photosynthetic area essential for growth. The combination of these various factors results in a decrease in total dry matter or biomass production. On exposure to UV-B radiation, the majority of crop species (60%) show a reduction in dry matter production, while a moderate 24% show no change; however, only 8% of crop species show an increase in dry matter production.

Yield and Yield Components

Yield is the economic product harvested from plants (e.g., grain from wheat (*Triticum aestivum*); seeds from pods of soybean; roots from carrot (*Daucus carota*); seed and lint from cotton (*Gossypium hirsutum*)). Similar to changes in dry matter production, change in yield of crop species on exposure to UV-B varies with species. Some species (e.g., pea, barley (*Hordeum vulgare*), and mustard (*Brassica nigra*)) show severe reduction, while others (e.g., cowpea (*Vigna unguiculata*), millet (*Setaria italica*), and tobacco (*Nicotiana tabacum*)) show less or no yield reduction. The main causes of yield loss are reduced fruit (grain) number due to failure in fertilization, abortion of fruiting structures, and decreased fruit size due to reduced supply of assimilates to the growing sink (fruits). UV-B radiation also affects the yield quality. For example, seed oil and protein content in soybean are reduced on exposure to UV-B radiation. Further research is required to evaluate and understand the impact of UV-B radiation on yield and yield quality of economically important crops.

Diseases and Insect Damage

Considerable progress has been made over the past few decades to understand the effects of UV-B on host plant species, whereas effects of UV-B on disease-causing pathogens (fungi and bacteria) and insect pests are limited. Research so far has shown both decreases and increases in disease and pest damage in response to increased UV-B radiation. The effects of UV-B on diseases and insects could be attributed to direct effects on their growth and indirect effects through changes in tissue characteristics and/or composition. A more definite understanding of the effects of UV-B on diseases and insects is needed to accurately predict consequences of ozone depletion on plant ecosystems.

Strategies for Protection against UV-B Radiation

Plants have evolved several mechanisms by which they try to protect themselves from the damaging effects of UV-B radiation (**Table 3**).

Repair Mechanisms

Plants have been constantly exposed to sunlight and UV radiation before the formation of the ozone layer. In the process of evolution, certain plants developed a tolerance to solar UV-B radiation that limits the amount of DNA damage they suffer. All cellular life forms possess DNA repair enzymes that recognize chemically modified bases, including those formed by UV radiation. Furthermore, cells have evolved a variety of biochemical mechanisms to restore the integrity of the genetic material after DNA damage and retain its stability. These processes are called "DNA repair mechanisms." These include photoreactivation, excision repair (nucleotide and base excision repair), and postreplication repair. Photoreactivation mainly involves photolyase, an enzyme responsible for the direct splitting of pyrimidine cyclobutane dimers. Excision repair is done by removing the damaged part of DNA, removing the bases in the damaged strand and synthesis of the gap. In postreplication repair, DNA damage is bypassed during DNA replication and the resulting gaps are filled in later using the information from the sister duplex. Such DNA repair mechanisms are observed in nuclear and chloroplast DNA. However, different plant species and varieties within a species vary in their ability to repair damage caused by UV-B.

Defense Mechanisms

When plants are exposed to UV-B radiation, it has to pass through epidermal layers to reach the sensitive sites. Therefore the surface structure, physiology, and composition of the epidermal layer can play an important role in protecting (shielding) cells from UV-B radiation. The important surface characteristics of the epidermal layer that can reduce penetration of UV-B radiation include protective structures such as hairs, trichomes, and wax coating. These structures have the capacity to attenuate, absorb, and reflect UV-B radiation due to their optimal structure and presence of chemical compounds.

Table 3 Protective mechanisms against damage by UV-B radiation in plants

Repair mechanisms

- DNA repair: photoreactivation enzymes (photolyase); excision repair by removing damaged part of DNA; bypass damaged DNA and fill gaps later from sister duplex

Defense mechanisms

- Increase reflectance to avoid entry of UV-B radiation through cuticle wax, leaf hairs, and trichomes
- Increase absorption of UV-B radiation at epidermal cells by production of pigments such as flavonoids, carotenoid, and anthocyanins
- Production of antioxidant enzymes (e.g., superoxide dismutase, ascorbate peroxidase, glutathione reductase) and compounds (ascorbates, α-tocopherol, and polyamines) that protect against oxidative stress caused by UV-B exposure

Studies have also shown that UV-B radiation leads to oxidative stress in plant systems as observed in many other abiotic (temperature and light) and biotic (insects and diseases) stresses. Therefore, as a result of UV-B exposure, plants increase the production of flavonoids and antioxidant enzymes (e.g., superoxide dismutase, ascorbate peroxidase, and glutathione reductase) which provide defense against UV-B radiation. Other chemicals such as α-tocopherol (vitamin E), peroxidases, ascorbates, β-carotene, and polyamines can provide protective function against UV-B damage.

Flavonoids are produced and mainly deposited in epidermal and mesophyll layers and leaf hairs. Presence and distribution of flavonoids at different locations provide an efficient screen to UV-B radiation. Flavonoids are very effective in screening (absorbing) UV-B radiation and reduce damage to sensitive cell organs (such as DNA, chloroplasts, and mitochondria). In addition to flavonoids, anthocyanins and carotenoids could have some potential to screen UV radiation, particularly in flowers, and provide protection to pollen grains. These compounds attenuate the damaging solar UV-B radiation but they transmit photosynthetically active radiation through the epidermis. Thus, these compounds do not directly influence photosynthesis and other physiological processes.

Selection or Genetic Improvement

There is a large variation in tolerance or susceptibility to UV-B radiation among plant species (**Table 4**) and cultivars within the same species (**Table 5**). Although the biological basis of this variation is not fully understood, factors such as variable stimulation

Table 4 Relative sensitivity of selected crops to increased UV-B radiation in controlled environments

Sensitive	*Moderately sensitive*	*Relatively tolerant*
Barley (*Hordeum vulgare*)	Common bean (*Phaseolus* spp.)	Corn (*Zea mays*)
Carrot (*Daucus carota*)	Lettuce (*Lactuca sativa*)	Cotton (*Gossypium hirsutum*)
Cucumber (*Cucumis sativus*)	Peanut (*Arachis hypogaea*)	Cowpea (*Vigna unguiculata*)
Mustard (*Brassica* spp.)	Pepper (*Piper nigrum*)	Clover (*Trifolium* spp.)
Oats (*Avena sativa*)	Petunia (*Petunia* spp.)	Millet (*Setaria italica*)
Pea (*Pisum sativum*)	Potato (*Solanum tuberosum*)	Radish (*Raphanus sativus*)
Soybean (*Glycine max*)	Rice (*Oryza sativa*)	Sunflower (*Helianthus annuus*)
Sweet corn (*Zea mays* var. *saccharata*)	Rye (*Secale cereale*)	Tobacco (*Nicotiana tabacum*)
Tomato (*Lycopersicon* spp.)	Sorghum (*Sorghum vulgare*)	Wheat (*Triticum aestivum*)

Table 5 Examples of differential response to UV-B radiation among cultivars in selected crops

Crop	Sensitive cultivars	Tolerant cultivars
Common bean (*Phaseolus* spp.)	BBL 290	Astro
Corn (*Zea mays*)	Braga, DK 498	Polo, REG.VR
Clover (*Trifolium* spp.)	Feathermark, Green, Lime	Luclair, Sustain, Tienshan
Cucumber (*Cucumis sativus*)	XPH 1484	Marketmore
Rice (*Oryza sativa*)	Carreon, IR 74, Norin 1	Kurkaruppan, IR 64, Sasanishiki
Soybean (*Glycine max*)	Essex, Shore, York, Yunuan	Forrest, Huanxianhuangdou, Williams
Wheat (*Triticum aestivum*)	Chuxiong 8807, Huining 18, Longchun 16	Dali 905, Liaochun 9, Mianyong 20

of protective mechanisms and repair mechanisms may be a potential cause. This variation can provide an opportunity for genetic improvement of plant species either through traditional plant breeding techniques (selection and crossing) or using modern molecular biology techniques such as plant transformation (genetic modification; GM). Screening wide germplasm from various locations and origin for UV-B tolerance including native plants and landraces may prove useful to identify tolerant species or cultivars.

List of Technical Nomenclature

Action spectrum The spectrum of light waves that elicits a particular response.

Anthocyanins Colored (red, purple, and blue) pigments found in plant cells.

Carotenoids Red to yellow pigments.

Chlorophyll Green pigment in chloroplasts that absorbs and transforms radiant energy to chemical energy.

Chloroplast Membrane-bound plant cell organelle in which photosynthetic chemicals and structures are contained.

Chromosome Structure in the nucleus that contains genetic information, consisting mainly of DNA and protein.

Ozone hole Region where stratospheric ozone levels drop below 200 Dobson Units.

Ozone layer Region in stratosphere where 90% of the earth's ozone presides.

Stratosphere Region between 15 and 50 km above the earth's surface.

Troposphere Region <15 km from the earth's surface, where most weather systems occur.

See also: **Crop Improvement**: Plant Breeding, Principles; Plant Breeding, Practice. **Diseases**: Bacterial Diseases; Viral Diseases. **Flowering and Reproduction**: Reproductive Strategies. **Growth and Development**: Growth Analysis, Crops. **Plants and the Environment**: Plants and Atmospheric Pollution. **Photosynthesis and Partitioning**: C_3 Plants; C_4 Plants; CAM Plants.

Further Reading

Anderson SO and Sarma KM (2002) *Protecting the Ozone Layer: The United Nations History.* Nairobi: United Nations Environment Programme.

Caldwell MM, Camp LB, Warner CW, and Fling SD (1986) Action spectra and their role in assessing biological consequences of solar UV-B radiation. In: Worrest RC and Caldwell MM (eds) *Stratospheric Ozone Reduction, Ultraviolet Radiation and Plant Life*, pp. 87–111. Berlin: Springer-Verlag.

Caldwell MM, Bjorn LO, Bornman JF, *et al.* (1998) Effects of increased solar ultraviolet radiation on terrestrial ecosystems. *Journal of Photochemistry and Photobiology B: Biology* 46: 40–52.

Cockell CS and Knowland J (1999) Ultraviolet radiation screening compounds. *Biological Reviews* 74: 311–345.

Hester RE and Harrison RM (2000) *Causes and Environmental Implications of Increased UV-B Radiation: Issues in Environmental Science and Technology.* Cambridge: Royal Society of Chemistry.

Hollosy F (2002) Effects of ultraviolet radiation in plant cells. *Micron* 33: 179–197.

Holmes MG and Keiller DR (2002) Effect of pubescence and waxes on the reflectance of leaves in the ultraviolet and photosynthetic wavebands: a comparison of a range of species. *Plant Cell and Environment* 25: 85–93.

Jansen MAK (2002) Ultraviolet-B radiation effects on plants: induction of morphogenic responses. *Physiologia Plantarum* 116: 423–429.

Jordan BR (2002) Molecular response of plants cells to UV-B stress. *Functional Plant Biology* 29: 909–916.

Krupa SV, Kickert RN, and Jager HJ (1998) *Elevated Ultraviolet (UV)-B Radiation and Agriculture.* Berlin: Springer-Verlag.

Lumsden P (1997) *Plant and UV-B: Responses to Environmental Change, Society for Experimental Biology Seminar Series No. 64.* Cambridge: Cambridge University Press.

Molina MJ and Rowland FS (1974) Stratospheric sink for chlorofluoromethanes: chlorine atom-catalysed destruction of ozone. *Nature* 249: 810–812.

Rozema J (1999) *Stratospheric Ozone Depletion: The Effects of Enhanced UV-B on Terrestrial Ecosystems.* Leiden: Backhuys.

United Nations Environment Programme (1998) *Environmental Effects of Ozone Depletion: 1998 Assessment.* New York: Elsevier Science.
United Nations Environment Programme (2002) *Scientific Assessment of Ozone Depletion: 2002. Executive Summary. Scientific Assessment Panel of the Montreal Protocol on Substances that Deplete the Ozone Layer.* Nairobi: UNEP.

Plants and Atmospheric Pollution

J N B Bell and S A Power, Imperial College London, Ascot, UK

Phytotoxic Air Pollutants

It has been known for at least 250 years that air pollution can have damaging effects on plants. Initially this was caused by coal-smoke arising from domestic and industrial sources. The main phytotoxin in the smoke was sulfur dioxide (SO_2), but particulates also played a role. Currently the main source of SO_2 in most countries is the combustion of coal and fuel oil, together with industries such as oil refineries and metal smelters. As a result of concerns over the adverse effects of SO_2, notably via the formation of "acid rain," stringent controls on emissions in developed countries are resulting in massive falls in ambient concentrations; in contrast, in many developing countries SO_2 levels are rising due to increased industrialization. A second major category of pollutant is the nitrogen oxides (NO_x), most of which is emitted as nitric oxide (NO), which is rapidly oxidized in the atmosphere to nitrogen dioxide (NO_2). In this case all high-temperature combustion processes produce NO, with motor vehicles being a very important source. Until recently NO_x levels were rising in both the developed and developing world, but are now falling in the former as emissions controls are tightened, particularly on motor vehicles. It is now recognized that in most parts of the world the most widespread and serious phytotoxic pollutant is ozone (O_3). This is produced as a result of photochemical reactions in the atmosphere involving NO_x and volatile organic compounds (VOCs) under conditions of high temperature, bright sunlight, and relatively calm air. The VOCs are also emitted in large quantities from motor vehicles, but there are numerous other sources, including industry, dry-cleaning, evaporating petrol, and vegetation. There is abundant evidence that worldwide the background levels of O_3 are rising. O_3 presents a particularly serious threat to vegetation in that it can cause injury not far above the maximum natural background concentration (*c.* 40 nl l^{-1}) and occurs at elevated levels across large areas of countryside, where it is invariably present at higher levels than in cities.

There are a number of other pollutants which can harm plants in the field, but these are usually restricted to fairly local sources. The most important of these is hydrogen fluoride (HF), which is emitted by specific industries, such as aluminum smelters, brickworks, and phosphate fertilizer factories. While injury can be caused by as little as 1–2 nl l^{-1} HF, probably of greater concern is that fluorine accumulated in forage can cause a bone and teeth disease, known as "fluorosis," in cattle and sheep. Another localized pollutant is the VOC, ethylene (C_2H_4), which is associated with vehicle emissions and the petrochemical industry. Also damage can occur on a localized basis as a result of ammonia (NH_3) emissions from intensive livestock production.

There are two classes of air pollutant, where the total deposition ("loads") to the earth's surface is the parameter of concern, rather than concentrations ("levels") in the atmosphere. One of these is acid deposition (primarily as acid rain), which is largely the result of atmospheric transformation of SO_2 and NO_x to sulfate and nitrate, respectively. The other is total nitrogen deposition in the form of both nitrate and ammonium, the latter arising from ammonia emissions which have risen rapidly in many countries in recent years. Both these forms of deposition are potentially implicated in widespread adverse impacts on certain sensitive ecosystems.

Types of Air Pollution Impacts on Plants

There are two main types of direct injury that air pollution can cause on plants: acute and chronic injury. Acute injury results from exposure to a high concentration of gas for a relatively short period and is manifested by clear visible symptoms on the foliage, often in the form of necrotic lesions. While this type of injury is very easy to detect (although not necessarily to diagnose), chronic injury is much more subtle: it results from prolonged exposure to lower gas concentrations and takes the form of growth and/or yield reductions, often with no clear visible symptoms. In the case of acute injury there are illustrated guidebooks which assist in diagnosis of the pollutants concerned and in separating out these effects from symptoms caused by other stress factors, such as frost or mineral deficiency.

In addition there are more subtle indirect effects that pollutants can have on plants. It is well known that a wide range of environmental factors, such as temperature, humidity, light intensity, drought, and soil nutrient status can markedly modify the response of plants to pollutants, in terms of both acute and chronic injury, in some cases via changes in stomatal aperture. Conversely air pollutants can change the response of plants to other environmental stresses, in most cases exacerbating injury. Such stresses can be abiotic, such as frost, drought, and salinity, or biotic, such as attack by fungal pathogens and herbivorous insects.

Most plant communities do not consist of monocultures, but rather of a mix of many different species. Because different species show differing responses to any given pollutant exposure, then air pollution can cause shifts in competitive advantage within a mixed sward, such that its species composition changes accordingly. It is well known that different cultivars of single species of crop plants can be differentially sensitive to pollutants. In the case of wild plants this genetic variation can result in selection for tolerance of an intraspecific nature in polluted locations.

At the physiological/biochemical level there is a wide range of effects that can be caused by pollutants, which may be manifested in the macroscopic effects already discussed. These include both increased opening or closure of stomata, depending on the gas and its concentration, the species, and environmental conditions. Such effects clearly have many implications for the plant, such as changes in pollutant uptake, photosynthesis, and transpiration. In addition, pollution can have direct inhibitory effects on photosynthesis. An almost universal impact of air pollution is damage to the integrity of membranes, so that they become leaky. This and other mechanisms are responsible for a very large number of enzymes being affected in their activity by pollutants. At very low concentrations of pollutants, stimulation of physiological processes can take place, which can result in concomitant small increases in growth. Current research is largely directed towards induction of active oxygen scavenging systems in the leaves, which play a major role in influencing the sensitivity of plants to O_3; this appears to be related to up- and downregulation of genes by O_3.

Techniques for Studying Effects

There is a wide range of experimental techniques for determining the impacts of air pollution, ranging from sophisticated highly controlled laboratory investigations to field surveys. In general the more carefully controlled and standardized forms of pollutant exposure, such as in controlled environment systems, then the more repeatable are the results, with the establishment of clear dose–response relationships. However, in the field the environment is not controlled but is fluctuating continuously, thereby modifying plant response to pollutants. Thus field studies give a good indication of the response of plants to the ambient situation, but each experiment is effectively unique and thus, theoretically, unrepeatable. In fact both types of study are required, with controlled studies being used to interpret field exercises.

Most of the early experiments involved fumigating plants with SO_2 in closed chambers located outside or in glasshouses. Subsequently these were largely replaced by open-top chambers (OTCs) employed for a range of pollutants. The latter have the advantage of microclimatic conditions which approximate more closely to the ambient, thus minimizing confounding effects of the chamber environment. They have been utilized in many large-scale studies in North America and Europe in establishing dose–response functions for crops, native vegetation, and trees. Both closed chambers and OTCs are also employed in filtration studies, in which plants are exposed to ambient air or the same air after it has been purified by passage through a charcoal filter. Such studies have provided the most unequivocal evidence for the adverse impacts of prevailing ambient air pollution. However, there remains the problem of some modification of microclimatic conditions, which also influence flux of the pollutant to the plant, and in identifying the causal pollutant(s) when a mixture of these is present.

In order to achieve greater realism in air pollution studies, systems of open-air fumigation have been developed. These involve fumigating a range of species, including trees, which are growing *in situ*, by a computer controlled system which releases pollutants from manifolds established around plots subjected to different concentrations. Thus the plants are exposed to pollutants without any modification of the microclimate and, also, any interactions with naturally occurring stresses, such as frost or insect pests can readily be detected. However, it has the disadvantage of confounding impacts of prevailing ambient air pollution. Less well developed, but of considerable value, are open-air filtration systems, in which ambient or filtered air is blown across low growing species in the field. In the case of mature trees, both filtration and fumigation experiments have been used with small chambers enclosing individual branches.

There are two further types of field study which have enormous potential. One of these is the transect approach, whereby plants are exposed as standard cultures along gradients of air pollution in the field. This has the advantage of a very high degree of realism, but the disadvantage of the potential confounding influence of differences in climatic conditions between sites, although these can be taken account of by appropriate multivariate analysis. The second technique is the application of a chemical protectant against a pollutant by application to soil or foliage. So far, this has only been utilized successfully for O_3, mainly by employing the antioxidant ethylenediurea (EDU). If complete protection against O_3 and no impact of EDU on its own can be demonstrated, this is probably the ideal method for field investigation, but it is not effective for some species.

Acute Injury

Plants have in-built mechanisms for detoxifying the breakdown products of most of the principal atmospheric pollutants. These include antioxidants and enzyme systems which convert toxic ions and free radicals into harmless products which can be dealt with through normal metabolic pathways. However, at high concentrations, these detoxification mechanisms cannot keep pace with the build-up of toxic products, frequently resulting in the development of visible signs of injury. The symptoms of acute injury vary according to both the type of pollutant and the plant species involved. Acute injury can take many different forms, varying from anthocyanin (red pigment) production and leaf chlorosis to leaf abscission and necrosis (cell death). These symptoms represent pigment breakdown, damage to cell membranes, and a disruption of normal physiological processes.

Classical symptoms of injury following acute exposure to SO_2 and NO_2 include leaf chlorosis and necrosis, especially in interveinal regions. In some species, injury is concentrated in the tips and margins of leaves. Tip necrosis and chlorosis are typical symptoms of acute fluoride injury, representing local accumulation following movement in the transpiration stream. Leaf age can be an important factor determining the extent of visible injury following a pollution episode, with, for example, older leaves typically exhibiting greater symptoms following exposure to high concentrations of NO_2. Ozone-induced injury is often characterized by the development of white flecks or spots, especially on the upper leaf surface, and represents the death of individual cells. In some species, anthocyanin production or generalized chlorosis is seen, indicating more diffuse visible symptoms of ozone injury.

Table 1 Species with particular sensitivity to the major air pollutants, as indicated by the extent of visible injury symptoms following acute exposure

Pollutant	*Plant species*
SO_2	Alfalfa (*Medicago sativa*), lichens, plantain (*Plantago major*)
NO_2	Wild celery (*Apium graveolens*), *Petunia*
O_3	Tobacco (*Nicotiana tabacum*) (Bel W3), clover (*Trifolium* spp.), radish (*Raphanus sativus*)
PAN	Annual meadow grass (*Poa annua*), small nettle (*Urtica urens*)
HF	Gladioli (*Gladiolus* spp.), *Freesia*, tulip (*Tulipa* spp.)
Ethylene	African marigold (*Tagetes erecta*), carnation (*Dianthus caryophyllous*)

Visible injury can also be seen in response to high concentrations of hydrogen ions in acidic (wet) deposition. The development of foliar lesions on sensitive species such as radish (*Raphanus sativus*) and kidney bean (*Phaseolus vulgaris*) depends on both the H^+ ion concentration and the duration of exposure. However, for most species, visible injury resulting from acidic deposition occurs only at concentrations typical of cloud water influenced by emissions from polluted industrial regions, rather than at more normal rainfall pHs.

The fact that different plant species are more or less sensitive to the various gaseous pollutants and produce defined visible responses, means that certain species can be used as indicators of a particular pollutant. **Table 1** summarizes those species which are particularly sensitive to the main atmospheric pollutants. The development of dose–dependent visible injury symptoms means that many of these species can be used as semiquantitative, biological indicators of air pollution, a subject which is covered in detail elsewhere (*see* **Plants and the Environment:** Plants as Pollution Monitors).

Chronic Injury and Effects on Biomass

Whilst exposure to high concentrations of gaseous pollutants often results in visible injury, pollutants can also affect plants in the absence of visible symptoms of damage. This is a phenomenon known as chronic or invisible injury and represents detrimental effects on plant physiology and growth. Adverse effects on photosynthesis and increased energetic costs of repairing cell damage frequently result in a reduction in plant growth. For many species, a linear reduction in photosynthetic rate has been shown with increasing pollutant concentration; sensitive species may exhibit significant decreases in the rate of photosynthesis even at ambient levels of pollutants such as O_3 and SO_2. Changes in stomatal resistance are also associated with pollutant exposure and a disruption in the plant's ability to regulate gas exchange and water loss through its leaves may also lead to reduced rates of photosynthesis and growth.

Over a whole growing season, a sustained reduction in photosynthesis will result in a decrease in plant growth, with reduced biomass being one of the most common responses to atmospheric pollution. Clearly the implications for yield reductions of agricultural crops and forestry are considerable, so much research effort has been directed at establishing dose–response relationships for economically important plant species and for key pollutants. One such program, the National Crop Loss Assessment Network (NCLAN), carried out in the United States in the 1980s, exposed a range of crop species to different concentrations of ozone, in OTCs. Linear reductions in yield with increasing O_3 concentrations were demonstrated for many species and it was estimated that a 40% reduction in ambient O_3 levels would result in an increase in crop yield equivalent to US$3 $\times 10^9$ per year; a substantial impact of just one ubiquitous air pollutant! Since plant response to pollution depends on many factors such as climate and plant genotype, a similar program (the European Open Top Chamber Network) was also established in Europe in the late 1980s. This involved the use of different crop cultivars grown under the diverse range of climatic conditions experienced in the nine participating countries. Similar dose–response relationships to the American study were found, establishing the generality of crop response to O_3 under a range of environmental conditions.

Although the effects of atmospheric pollution on crops and trees have been well documented, the chronic, growth responses of natural vegetation to realistic pollutant concentrations are still relatively poorly understood. Some plant groups are known to be particularly sensitive to certain pollutants, for example lichens and SO_2, but information on the growth responses of the diverse range of plant groups and different ecosystem types is relatively sparse. The UNECE International Cooperative Programme examining the effects of ambient O_3 on vegetation (ICP-Vegetation) has recently demonstrated a reduction in the yield of clover (*Trifolium* spp.) along a Europe-wide gradient of increasing O_3 exposure. (**Figure 1**). In addition there is evidence to suggest that the more sensitive species of natural and seminatural vegetation exhibit similar, if not bigger, growth reductions in response to pollutants such as O_3, compared with crop species.

Whilst exposure to even relatively low levels of some pollutants may result in reduced plant growth,

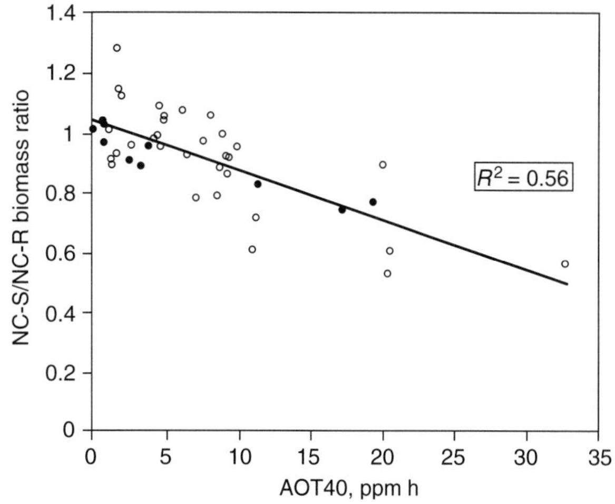

Figure 1 The relationship between ozone exposure (AOT40) and growth of white clover (*Trifolium repens*), along a natural pollution gradient from Scandinavia into southern Europe. Growth is expressed as the biomass ratio between an ozone sensitive (NC-S) and an ozone resistant (NC-R) clover clone, for three monthly harvests, to take into account effects of climatic and other environmental variation along a gradient of pollution into southern Europe. For explanation of AOT40, see section "Air Quality Guidelines." Figure courtesy of Dr Gina Mills, Chair, ICP-Vegetation programme.

low levels of others can, in fact, stimulate growth. This has been shown for SO_2, in soil-S deficient situations but is particularly the case with nitrogen-containing pollutants such as NO_2 and NH_3. In natural systems, low nitrogen availability typically limits plant growth. Therefore, levels of N-containing pollutants that are below those resulting in acute injury can provide an additional source of this nutrient for incorporation within the leaf or by root uptake following deposition to the soil.

Whilst a reduction in aboveground plant growth is the most commonly reported response to subacute levels of pollution, other, more subtle, chronic responses are also seen. For example, a reduction in biomass allocation belowground is often found following exposure to O_3, SO_2, NO_2, or NH_3. Other responses include effects on reproductive effort, timing, and success, changes in growth patterns, and increased sensitivity to cold temperature stress as a result of cell membrane damage. Clearly, plant response to air pollution is varied and can result in not only growth reductions, but also effects which will alter the way in which plants interact with their biotic and abiotic environment.

Community-Level Responses

Plant response to a given pollutant or combination of pollutants differs not only between species, but also between genotypes of the same species, with the potential for pollution-mediated changes in both inter- and intraspecific competition. Whilst this may have little impact amongst relatively homogeneous fields of a single crop species, the potential for effects on community composition in natural ecosystems is enormous. Differences in patterns of biomass allocation between neighboring plant species which result, for example, in a reduction in root : shoot ratio, will have implications for nutrient uptake and plant sensitivity to drought. Small advantages/disadvantages in a natural situation can lead to a change in the competitive balance between species and affect their ability to acquire limited resources. Furthermore, subtle effects on phenology or reproduction can, over realistic time scales, lead to the loss or increased dominance of a particular species in the field.

Experimental studies have found changes in species composition in response to both gaseous and wet-deposited pollutants. Both SO_2 and O_3 have been shown to increase the relative proportion of grass species in mixed grassland or shrubland communities, typically at the expense of herb and/or shrub species. It is inherently difficult to extrapolate community level responses of mature forest trees from chamber experiments using young tree seedlings. However, there is some evidence to suggest that air pollution, in particular ozone, may be responsible for changes in forest vitality and species composition. For example, in the San Bernardino Mountains downwind of Los Angeles, elevated ozone concentrations are believed to have been responsible for a shift in forest community composition, driven by an increased susceptibility of the formerly dominant tree species to drought and attack by bark beetles.

High deposition rates of nitrogen-containing pollutants (i.e., wet + dry deposition of NO_x and NH_3) have been linked with changes in the community composition of many natural and seminatural ecosystems in both Europe and the United States. Perhaps the best example is that of lowland heathlands, where increased sensitivity of heather to beetle attack and climatic stress have led to an increase in dominance of grasses in areas receiving high deposition inputs of nitrogen. Grasslands are also adversely affected by increased nitrogen availability, typically exhibiting an increase in one or more dominant grasses, and a reduction in overall species number. There is even evidence of an increase in more nitrogen-demanding species at a national level, following recent nation-wide vegetation surveys in the United Kingdom. Community-level responses to air pollution are generally associated with a decrease in species biodiversity and changes in the structure and function of ecosystems. However, it must be noted that most studies are of a relatively short duration, making it difficult to interpret the long-term effects of subtle changes in plant growth and phenology, and the possible evolution of tolerance over biologically meaningful timescales.

Evolution of Tolerance to Air Pollution

It has been long established that plants, mainly grasses, can evolve tolerance to heavy metals in areas with contaminated soils, via intraspecific competition. This phenomenon was identified at sites where extremely toxic levels were present. Thus it proved surprising that it has been shown unequivocally that evolution of tolerance to gaseous air pollutants appears to be widespread, even at widely prevailing ambient concentrations. The initial studies in this area represented an excellent example of parallel scientific developments, with demonstration of populations of a forb in Georgia and grasses in northern England being more tolerant to SO_2 than similar populations in "clean" areas. Furthermore, it appeared that this tolerance could develop over a timescale of a few years and with small changes in concentration. With falling SO_2 levels, tolerance disappeared, as sensitive genotypes were more fit under cleaner conditions (**Table 2**).

These earlier studies were carried out around point sources or in highly polluted areas. Thus it was not clear whether or not such evolution could develop for the widespread pollutant, O_3, where sharp gradients do not occur. Initial studies in the United States indicated that this occurred in poplar (*Populus*), a particularly sensitive species. Subsequently, a major program in the United Kingdom has demonstrated that the great plantain (*Plantago major*) shows a close positive correlation of degree of tolerance to O_3 with prevailing O_3 levels on both a temporal and spatial basis across Britain and Europe as a whole.

Interactions with the Abiotic Environment

The impact of air pollutants depends on a wide range of environmental factors, such as light, relative humidity, temperature, soil moisture deficit, and nutrient availability. These factors influence both the flux of gaseous pollutants into the leaf and the ability of plants to detoxify them once in the leaf environment. Low soil water availability and high vapor pressure deficits (VPD) both result in a reduction in stomatal conductance. This impedes leaf gas exchange, resulting in a lower pollutant flux for a given atmospheric concentration. The

Table 2 Changes in tolerance to SO_2 of grass species in relation to ambient pollution concentrations (% leaf area injured after an acute SO_2 fumigation)

Species	*Year of collection*					
	1976	*1978*	*1979*	*1980*	*1981*	*1982*
Perennial ryegrass (*Lolium perenne*)						
O.S.[a]	20.5	23.5	9.7	42.8	4.8	5.5
P.P.[b]	18.5	20.1	**4.3**	**33.0**	4.9	4.8
Smooth stalked meadow grass (*Poa pratensis*)						
O.S.[a]	6.1	9.4	3.4	44.5	12.7	–
P.P.[b]	5.6	8.8	4.4	**35.0**	10.7	–
Mean SO_2 concentration ($nl\,l^{-1}$)	45	46	36	26	25	28

Numerals in bold indicate statistically significant differences between O.S. and P.P.
[a]O.S., plants grown from original seed sown in the polluted Philips Park in Manchester in 1975.
[b]P.P., plants sampled at intervals from Philips Park from the plots sown in 1975.

magnitude of plant response to pollution is thus frequently lower when water is limiting. For example, yield reductions of clover following exposure to O_3 in OTCs were lower in droughted than in well-watered plants and lower concentrations of O_3 result in greater visible injury when VPD is low. However, other factors such as a pollutant-induced reduction in root biomass, or even the inhibition of stomatal closure in response to O_3 and SO_2 may occasionally result in an exacerbation of the effect of air pollution under conditions of water stress.

Light intensity can affect plant response to air pollution as a result of effects on photosynthesis, stomatal conductance, and plant development. Generally, the magnitude of plant response to pollutants is higher at lower light intensities, possibly as a result of reduced effectiveness of detoxification mechanisms when photosynthetic rates are low. Slow plant growth rates under conditions of low light intensity and low temperature have been shown to increase plant sensitivity to SO_2, with effects of a given concentration on both timothy grass (*Phleum pratense*) and winter barley (*Hordeum vulgare*) being more pronounced during winter months than in summer.

Changes in the structure and integrity of cell membranes following exposure to O_3 or SO_2 have been associated with reduced tolerance of low-temperature stress. For example, damage to heather shoots has been found following a period of very low temperatures during a long term O_3 fumigation study, compared to negligible damage at the same temperatures for plants grown in filtered air. Changes in plant phenology, particularly the onset of winter hardening and of bud-burst during spring are associated with high deposition rates of nitrogen-containing pollutants. Thus, increased sensitivity to low-temperature stress has been associated with high concentrations of NH_3 and also high deposition rates of nitrogen. The interaction between response to pollution and soil nutrient availability is another factor worth noting. Although this has not been so widely studied as most other environmental factors, plant response to O_3 in soils of differing fertility has been investigated for a limited number of species. Results suggest that in some situations, compensatory mechanisms reduce plant response to ozone in nutrient-poor soils, although certain aspects of plant performance (e.g., seed production) may be more sensitive to the pollutant under low nutrient conditions.

Since the atmosphere typically consists of mixtures of several air pollutants, rather than single ones on their own, it is important to consider the effects of combinations of pollutants. Effects of combinations of O_3, SO_2, NH_3, NO_2, and acidic deposition can be either additive, synergistic (greater than additive), or antagonistic (less than additive). For example, plant response to simultaneous exposure to both NO_2 and SO_2 is generally greater than response to the individual gases. The reason for this is that SO_2 appears to reduce the plant's ability to detoxify toxic nitrite ions associated with exposure to NO_2. However, examples of antagonistic and inconsistent responses are also seen for different pollutant combinations. Whilst it is difficult to make a general statement concerning the effects of pollutant mixtures on plant growth, it seems likely that the greater the pollutant stress on plants, the greater the long-term effects are likely to be at the community level.

Interactions with Biotic Stresses

Since the nineteenth century, there have been reports of changes in the incidence of attack by fungal pathogens and herbivorous insects on plants growing in polluted areas, compared with cleaner locations. In the case of insects most of these indicated an increase in populations in polluted locations, whereas for fungal pathogens both negative and positive relationships were observed. In the last 30 years the causality of these field observations has been examined for a limited number of pest–host and pathogen–host systems, mainly with respect to major agricultural crops.

In the case of insect pests, all the evidence points towards SO_2 and NO_2 stimulating the performance, in terms of relative growth rate and fecundity, of both sucking (e.g., aphids) and chewing (e.g., moth larvae) organisms. For O_3 the situation, albeit based on limited evidence, appears to be more complex, with indications that the direction of the response can be modified by factors such as ambient temperature and duration of exposure. A range of host plants of aphids have been fumigated with SO_2 or NO_2 and then removed to clean air. Aphids feeding on them subsequently have shown increases in performance compared with when they were feeding on plants previously grown in clean air. This demonstrates that the pollutants cause these stimulations via changes in the host plants rather than via a direct impact on the pest, as shown by a lack of effect when the aphids were fumigated directly when feeding on artificial diets (**Table 3**). Some of the studies on aphid–host–pollutant interactions have been conducted under field conditions. Thus exposure of plants along transects of air pollution gradients has shown that both sucking and chewing insects perform better when feeding on hosts from the more polluted sites. In the case of a SO_2 field fumigation it has been demonstrated that the natural build-up of aphids showed a positive relationship with pollutant concentration, with evidence that the pests were escaping control by their enemies, such as parasitoids. There is further evidence that O_3 fumigation can reduce the searching capability of predators of insect pests.

The evidence points towards chemical changes in the host plant rendering it a more favorable food source for the pest. This appears to be largely due to increases in nitrogen concentration and/or shifts in amino acid composition, this having been confirmed by studies of insect performance when feeding on artificial diets manipulated to mimic pollutant-induced chemical changes in the host.

As in the case of insects observations have been made for a long time that levels of fungal pathogen attack are changed in polluted areas. However, in this case both positive and negative relationships have been reported. As a general rule it appears that biotrophs are inhibited by pollutants, but with nonbiotrophs more mixed responses occur, albeit with a tendency towards stimulation of performance.

Table 3 Mean relative growth rate (d^{-1}) of black bean aphid (*Aphis fabae*) feeding on bean plants previously fumigated with SO_2 or NO_2, or feeding on artificial diet

	Pollutant	*Control*	*Significance*
SO_2 Plant	0.540	0.507	$p<0.01$
Artificial diet	0.395	0.394	n.s.
NO_2 Plant	0.575	0.530	$p<0.01$
Artificial diet	0.369	0.370	n.s.

These studies have been confirmed by controlled fumigation experiments, where a similar pattern has in general been observed for the two types of pathogen in response to SO_2 and O_3. However, it is not possible to be categorical, as there are a number of cases where stimulation of biotrophs has been demonstrated unequivocally. Conventional wisdom has hypothesized that in the case of biotrophs damage to the host's metabolism has an adverse impact on the pathogen with whose metabolism this is intimately linked. In the case of nonbiotrophs, damage to the leaf surface may provide courts for fungal infection. It has been shown that fungal performance can be changed in terms of spore germination and hyphal extension when the plant is fumigated before, during and after inoculation, thus indicating that effects on the pathogen can be mediated via the host. One of the best-known pathogens shown to be affected in the field by pollution is the blackspot of roses (*Rosa* spp.), *Diplocarpon rosae*. When SO_2 levels were high in British cities, urban gardeners enjoyed freedom from this serious disease, but now that air quality has improved it has invaded the areas where it was previously absent. A similar pattern has also been observed for the tarspot of sycamore (*Acer pseudoplatanus*), *Rhytisma acerinum* (**Figure 2**). Thus both these species have potential for use as biomonitors of changing SO_2 levels.

Effects on Lichens and Bryophytes

Lichens and bryophytes are well known for their extreme sensitivity to SO_2, but much less is known about their response to other pollutants. The main reason for such sensitivity is that they lack the protective cuticle found on higher plants, absorb the bulk of their nutrients directly from the atmosphere, and are only physiologically active (and hence susceptible to injury) when they are moist (thus absorbing large quantities of highly soluble SO_2). However, there is a wide range of sensitivities to SO_2

Figure 2 Tarspot of sycamore. Photograph courtesy of Dr Sally Power.

between different species. In general the shrubby lichens are the most sensitive, leafy lichens intermediate, and crustose lichens the least sensitive. Bryophytes have been studied to a much lesser extent and thus less is known about their relative sensitivity. It is well known that the more sensitive lichens were eliminated over large areas of many European countries, not only in cities but also in the countryside, after the Industrial Revolution. This differential sensitivity of different species has enabled lichens to be used as effective bioindicators of SO_2 based on zone scales which reflect different epiphytic communities that occur along a gradient of pollution. Now that SO_2 levels have fallen sharply in most of the developed world, there are signs of reinvasion of sensitive species of areas where they had previously been eliminated. Such reinvasion is relatively slow and thus lichen mapping to show SO_2 distribution is unsuitable in a situation of improving air quality. Much interest is now centered on developing lichens as bioindicators of current air pollution problems, notably NO_x and NH_3.

Probably the best example of a widespread impact of air pollution on bryophytes is the loss of *Sphagnum* moss from the blanket bogs of the southern Pennines in northern England following the Industrial Revolution. *Sphagnum* species are highly sensitive to SO_2 and examination of peat profiles of these bogs indicates their disappearance about 150 years ago, coinciding with the appearance of soot contamination. The loss of *Sphagnum* caused major changes in this upland ecosystem, leading to extensive erosion of the peat. Now that SO_2 concentrations have fallen, it is still not possible to reintroduce *Sphagnum* to the bogs, because it is sensitive to the high levels of nitrogen deposition.

Air Quality Guidelines

Most policies aimed at controlling pollutant emissions have been directed at the protection of human health. However, recognition of the magnitude and extent of pollutant impacts on vegetation and ecosystems has, more recently, led to the development of air quality guidelines and standards to protect agricultural productivity and the health and biodiversity of natural ecosystems. Developed countries all have national legislation in place to limit pollutant emissions, but the fact that air pollution does not respect national boundaries has led to the adoption of an international approach to emissions control in Europe, mediated through the United Nations Economic Commission for Europe (UNECE). Evidence supporting the role of transboundary air pollution in the acidification of freshwaters and soils, including a suspected link with forest decline, led to the introduction of policies aimed at reducing emissions of SO_2 and NO_x across Europe in the 1980s. This involved reduction targets of a fixed percentage of emissions, without particular consideration of ecosystem response to such reductions. More recently, however, an effects-based approach has been adopted, aimed at reducing emissions to levels which protect vegetation and ecosystems from harmful effects, based upon an understanding of plant and ecosystem responses to the different pollutants. These have drawn upon experimental evidence of exposure–response relationships, and, in some cases, on expert scientific judgment for those ecosystems where data are lacking and/or where a long-term perspective is considered important.

The World Health Organization has proposed air quality guidelines for the protection of sensitive vegetation and countries such as the United States and Canada also have national air quality standards and objectives which include effects on vegetation. An important distinction between guidelines or objectives and air quality standards is that the former are target concentrations or loads, and not backed by legislation, while the latter are legally enforceable. Within Europe, a critical-levels approach has been adopted to define threshold concentrations below which sensitive vegetation or ecosystem components do not experience significant harmful effects. These threshold (critical) levels apply to concentrations of a gaseous pollutant calculated over an appropriate timescale and vary according to vegetation type and averaging time (**Table 4**). For example, critical levels

Table 4 UNECE Critical Levels

Pollutant	*Receptor*	*Critical level (nl l^{-1}, i.e., ppb)*
SO_2	Agricultural crops	11 (annual or winter mean)
	Forests and natural vegetation	7 (annual or winter mean)
	Cyanobacterial lichens	4 (annual mean)
NO_2	All vegetation	16 (annual mean)
O_3	Crops	3000 ppb h^{-1} > 40 ppb for daylight hours over 3 months
	Forest trees	10 000 ppb h^{-1} > 40 ppb for daylight hours over 6 months

for O_3 are expressed as the cumulative exposure (parts per billion hours) above a threshold concentration of 40 ppb (nl l^{-1}), calculated over a 3-month growing season for crops and natural vegetation, or over a 6-month growth period for forests.

Critical loads represent threshold exposures of pollutants in the form of total deposition inputs (i.e., wet + dry deposition) and take into account the cumulative impact over long (biologically relevant) timescales. Ecosystem-specific critical loads have been established for acidity, sulfur, and nitrogen and are calculated using either mass balance or empirical approaches. Knowledge of soil and vegetation type in an area allows the construction of critical loads maps at local, regional, national, or international levels. Deposition maps can then be superimposed to determine the extent of critical load exceedance and thus identify those areas where vegetation is at risk. In this way, emissions reductions can potentially be targeted to minimize impacts on vegetation and maximize ecosystem protection in a cost-effective manner.

Future Directions

There has been a major shift in air pollution research over the past 30 years. Although high concentrations of SO_2 occur in less-developed regions of the world, falling levels in Western countries has meant that interest has focused more on NO_x, NH_3, and O_3 in recent years. Emissions of NO_x and VOCs from developed nations are also expected to decline in the future, largely as a result of emissions reductions policies in North America and Europe. However, increased emissions associated with industrialization and economic growth in developing countries means that globally, concentrations of O_3 and deposition rates of N are expected to continue to rise for some time. Further research effort is therefore required to establish the significance of changing regional patterns of air pollution in terms of agricultural productivity and their implications for the long-term sustainability of ecosystems and the valuable services they provide.

The adoption of a critical-levels approach has provided a useful tool to assess the potential impact of the current air pollution climate. However, it is recognized that atmospheric concentrations of a pollutant do not necessarily reflect accurately pollutant flux into a plant leaf. As indicated earlier, many factors can influence both the uptake of pollution and the magnitude of effect. A more sophisticated (level II) critical-levels approach is now being developed, to take into account the actual flux of a pollutant into a plant leaf. The role of soil moisture deficit, relative humidity, light, and temperature is under investigation, with the aim of providing refined, flux-based, critical levels in the near future. However, further work is needed in this area to consider the importance of other key factors, e.g., soil nutrient status and interactions with the biotic and abiotic environment in the context of critical levels of atmospheric pollution.

Other areas which merit further research in the near future include the establishment of plant response to: (1) the wide range of VOCs that are experienced in urban and industrialized areas; (2) ambient levels of nitric oxide, particularly since the discovery of its role in plant signaling and defense; (3) combinations of air pollutants, such as the complex mixtures of gaseous and particulate compounds associated with road transport in urban areas; and (4) interactions between global carbon and nitrogen cycles. Since most of the air pollution research carried out to date has focused on plants of agricultural (and to a lesser extent forestry) importance, relatively little is known about the impacts on natural vegetation. The complexity of natural ecosystems may mean that even relatively small effects can have important implications not only for individual plant species, but also for the functioning and services provided by natural ecosystems. Clearly, an understanding of plant and ecosystem level responses is essential in order to adopt appropriate strategies for the long-term protection of sensitive communities from the detrimental effects of atmospheric pollution.

List of Technical Nomenclature

Abiotic stress	A nonbiological stress, such as frost, drought or salinity.
Acute injury	Visible injury caused by a high pollutant concentration.

Air quality guideline/standard	Air pollutant concentration or deposition rate designed as a threshold for protection of a given receptor, e.g., crops or human health.
Biotic stress	A biological stress, such as herbivorous insects or plant pathogens.
Biotroph	Plant pathogen which requires a host plant for growth.
Brophyte	A moss or liverwort species.
Chronic injury	Growth effects caused by prolonged pollutant exposure.
Critical level	Threshold exposure in terms of atmospheric concentration to protect ecosystems.
Critical load	Threshold exposure in terms of deposition rate to protect ecosystems.
Ethylenediurea (EDU)	An antioxidant chemical used in studies on O_3 effects in the field.
Lichen	A species involving a symbiotic relationship between an alga and a fungus.
Nonbiotroph	Plant pathogen which can grow in the absence of a host plant.
OTC	Open-top chamber.
VOC	Volatile organic compound.

See also: **Abiotic Stresses**: Cold Stress. **Diseases**: Fungal Diseases. **Growth and Development**: Growth Analysis, Crops. **Pests**: Genetic Modification of Pest Resistance, Insect Pests. **Photosynthesis and Partitioning**: C_3 Plants; C_4 Plants. **Plants and the Environment**: Plants as Pollution Monitors; Ozone Depletion. **Water Relations of Plants**: Drought Stress; Stomata.

Further Reading

Agrawal SB and Agrawal M (eds) (1999) *Environmental Pollution and Plant Responses*. Boca Raton: Lewis Publishers.

Bell JNB and Treshow M (2002) *Air Pollution and Plant Life*, 2nd edn. Chichester: John Wiley.

Davison AW and Barnes JD (1998) Effects of ozone on wild plants. *New Phytologist* 139: 135–151.

DETR (2000) *The Air Quality Strategy for England, Scotland, Wales and Northern Ireland*. London: Department of the Environment, Transport and the Regions.

Flagler RB (1998) *Recognition of Air Pollution Injury to Vegetation: A Pictorial Atlas*, 2nd edn. Pittsburgh: Air and Waste Management Association.

Krupa SV (1997) *Air Pollution, People and Plants*. St Paul: American Phytopathological Society Press.

Lee JA and Caporn SJM (1998) Ecological effects of atmospheric reactive nitrogen deposition on semi-natural ecosystems. *New Phytologist* 139: 127–134.

Mansfield TA and Lucas PW (1996) Effects of gaseous pollutants on crops and trees. In: Harrison RM (ed.) *Pollution: Causes, Effects and Control*, pp. 266–292. Cambridge: Royal Society of Chemistry.

National Expert Group on Transboundary Air Pollution (2001) *Transboundary Air Pollution: Acidification, Eutrophication and Ground-Level Ozone in the UK*. London: Department for Environment, Food and Rural Affairs.

Wellburn A (1994) *Air Pollution and Climate Change*. Harlow: Longman.

WHO (2000) *Air Quality Guidelines for Europe*, 2nd edn. WHO Regional Publications, European Series no. 91. Copenhagen: World Health Organization.

Yunus M and Iqbal M (eds) (1996) *Plant Response to Air Pollution*. Chichester: John Wiley.

Plants as Pollution Monitors

M A S Burton, University of Hertfordshire, Hatfield, UK

History and Scope

Observations of plants showing a response to pollution in the environment in which they grow can be found in accounts dating back to the sixteenth century, when there were accounts of dying plants around smelting operations. In the nineteenth century, in industrialized towns and cities in Europe, paucity of epiphytic plants such as lichens and mosses was also considered to be due to air pollution, and gradually investigations allowed some attributions to be made as to which pollutants were causing such effects. For plants to be used as monitors they must show a measurable response to change in their environment.

Monitoring of pollutants is generally carried out using calibrated instruments. Such methods are needed to measure concentrations in air or water for comparison with standards and guidelines aimed at protecting health of humans and environmental components. Although legislation is formulated on the basis of observed and measured effects of specific pollutants on targets, it cannot encompass the range of organisms or the varied environmental conditions which interact with pollutants and may modify their effects. Biomonitoring methods attempt to redress this, by measuring accumulation of pollutants in and/or their effects on relevant organisms in different habitats. The results obtained are thus an integration of pollutant exposure and environmental factors and may be more useful for considering impacts on living organisms and ecosytems. Although chemical analysis is needed for measuring accumulation, the number of biomonitoring sites sampled can be much greater than for siting of instruments, on the basis of cost and feasibility.

Surveys during the twentieth century demonstrated that many different plants growing on land, in fresh water, and in the sea showed a response to pollution in their environment, ranging from visible injury to changes in their distribution. With the increasing availability of analytical methods for individual pollutants, it was possible to analyze plants and find out how much of one or more specific pollutants they had in their component parts and tissues. This was frequently carried out to show the relationship between concentrations in plants and distance from a source of emission or discharge. It was really only in the 1960s that results of some systematic surveys indicated that plants could be used to measure changes in pollutant concentrations in their habitat in a quantifiable way and thus be used as monitors. The surveys included plants in areas affected by heavy metals, radionuclides, hydrocarbons, pesticides, and, for example, sulfur and fluoride from gaseous emissions. Since then numerous publications have appeared from monitoring programs reporting concentrations of pollutants in plant tissues and measurements of some type of response or effect ranging from community level down to subcellular level. The protection of human health and of biodiversity is a primary concern in controlling pollution from industrial and agricultural sources and monitoring using relevant organisms can provide valuable information in this context.

Passive and Active Monitoring

The terms "passive" and "active" monitoring have been increasingly adopted to describe, respectively, the use of plants sampled in their natural habitat and those transferred/transplanted into new habitats for a known period of time. Both these approaches have value and can be used for most types of pollutants and most groups of plants both for measuring accumulation of pollutants and for measuring effects. A key aspect of all monitoring is the purpose of the program, which will determine the selection of the plants and the methodology. The development of standardized plant biomonitors for active monitoring illustrates the interest in this approach (**Table 1**).

Other important considerations are the availability of the species for the duration of the program in sufficient quantity and appropriate stage of development and, when collecting samples from natural habitats, conservation issues need to be taken into account. The resources available for a monitoring program often determine the methods that can be used. Whereas observation of visible injury may be cheap to carry out, sufficient numbers of trained observers are needed. Measurements of effects at a cellular level, either biochemical or structural, may provide evidence of stress at an earlier stage, but the preparation and measurements require access to more sophisticated instrumentation.

Table 1 Examples of plants used for active biomonitoring

Scientific name	*Common name*	*Pollutant(s)*	*Measurement*
Nicotiana tabacum Cv $BelW_3$	Tobacco	Ozone	Leaf injury
Trifolium repens Clone Regal	White clover	Ozone	Leaf injury
Populus nigra Clone Branderis	Poplar	Ozone	Leaf injury, leaf loss
Tradescantia Clone 4430	Spiderwort	Mutagenic substances	Formation of micronuclei
Lolium multiflorum subsp. *italicum* Cv Lema	Italian ryegrass	Heavy metals, trace elements, sulfur	Accumulation
Brassica oleracea acephala Cv Hammer/Grusa	Curly kale	Polycyclic aromatic hydrocarbons	Accumulation

Based on Klumpp A, Fomin A, Klumpp G, and Ansel W (2002) *Bioindication and Air Quality in European Cities.* Stuttgart: Gunter Heimbach and Zimmermann R-D, Wagner G, and Finck M (2000) *Guidelines for the Use of Biological Monitors in Air Pollution Control* (*Plants*), Part 1, *Methodological Guidance for the Drawing-up of Biomonitoring Guidelines* (*Plants*). Air Hygiene Report No. 12. Berlin: WHO Collaborating Centre for Air Quality Management and Air Pollution Control.

Selection of Plants: General Principles

It is important to appreciate that even closely related species may respond differently to pollutant exposure and, for interpretation of results, correct identification of the plants is essential. A knowledge of their biology is also needed to utilize samples most effectively. Some plants are able to bring protective mechanisms into play, thus not displaying injury which might otherwise manifest itself. These may be chemical complexes which interact with the pollutant within cell organelles; plant cell walls may bind metal pollutants within the root system in some higher plants on land and in water, affecting the distribution in aerial parts; in mosses, effective binding of metal cations in the cell walls is considered to be a major factor in their capacity to accumulate extremely high concentrations. There is also evidence of genetical adaptations of plants growing in chronically contaminated habitats. The conditions in which plants grow, including their nutritional status, the stage of their life cycle, seasonal influences, and the presence of other pollutants all affect the responses, so that plants should not be used for monitoring programs designed to influence decision-making until such aspects are understood.

For guidelines or standards for emissions and discharges of pollutants to be set, knowledge of concentrations at which harmful effects may occur is needed. Such studies are carried out under controlled conditions and with species whose biology is well understood in order to determine thresholds. Different organisms, even closely related species or plant varieties, show large differences in their responses, both in terms of their capacity to accumulate pollutants and their reaction to exposure. This, and the complexities of interactions between plants and their environment, which may increase or reduce their susceptibility to a specific pollutant, has meant a general reluctance to adopt their use as monitors for legislative purposes. There are, however, purposes for which data from their use are clearly of benefit (in addition to showing the state of contamination), for example, in predicting economic consequences in agriculture and forestry, for predicting human exposure via food chain components, and, in some cases, for predicting likely consequences for other ecosystem components.

Corresponding with the concept of using standard organisms to determine thresholds for response to pollutants, since the 1980s there have been proposals that standardized plant monitors, or bioindicators, should be used. These plants should be genetically identical, readily available in quantity and planted out for the monitoring period, thus reducing the interferences which may result from using less well-defined samples. For example, in Germany the Verein Deutscher Ingenieure (VDI) continues to promote the development of guidelines for the use of bioindicator plants for specific pollutants for air quality assessment in Europe. This "bioassay" approach can be very appropriate for certain types of monitoring, especially in urban areas and for managed environments where crops are grown for a season, but may be less appropriate for natural habitats where the diversity of interactions including age structure and seasonal changes may form an essential part of the responses to pollutants.

Approaches to Monitoring Using Plants

Sampling is a key factor in carrying out a monitoring program and will depend on the source of pollution and the type of environment into which it is dispersed. Quality control measures need to be in place and the possibility of contamination from other sources minimized.

Plants that are anchored to their substrate in terrestrial and aquatic habitats can be sampled consistently at the same locality at determined time intervals, ranging from hours to years, depending on the purpose of the monitoring, the type of pollutant, and its concentration. For monitoring, plants that are growing in the natural environment can be used (passive monitoring), but also it is also possible to transfer plants of known origin to sites identified for a monitoring program (active monitoring).

Retrospective data on pollutant concentrations can be obtained by chemical analysis of dated plant materials such as tree rings, peat cores, and herbarium specimens. In recent years there has been increased interest in setting up national environmental specimen banks, where representative samples are archived using standardized procedures and are available for examination or chemical analysis at a future time.

Chemical Analysis of Pollutants

Accumulation of pollutants in plant tissues can be measured by chemical analysis of plant parts of relevance to the monitoring program. There are many examples of measurements of metals and radionuclides in plants and also for some persistent organic pollutants, pesticides, and some elements originating in gaseous emissions. Whereas concentrations in the air or in water may be low in chronically polluted areas, the exposure of plants over periods of time may result in concentrations being very elevated in plant tissues. Monitoring of patterns of accumulation of pollutants from atmospheric deposition or around discharges in aquatic habitats generally involves collection of leaves, shoots, or fronds which have a large surface area and thus can represent the exposure.

Sampling strategy and quality assurance in sample analysis are essential parts of a monitoring program. Samples have to be prepared in a matrix suitable for the analytical method appropriate for the pollutant and may be in solid or, more frequently, in liquid form. Cleaning of plant samples depends on the purpose of the monitoring and whether the interest is in the total concentrations on plant surfaces as well as in the plant tissues, or whether only concentrations inside the tissues are of interest. For chemical analysis of plants access to suitable laboratory facilities for processing and analysis is needed although collection of samples may be relatively cheaply achieved. Where the purpose of a monitoring program is to examine the pattern of contamination from point sources, diffuse sources, or from agricultural amendments (such as fertilizers and pesticides), the prime consideration is that the plant accumulates the pollutant in relation to the extent of exposure. Thus, by implication, the plant must tolerate exposure to the range of concentrations in polluted areas. Measurement of concentrations does not in itself, however, provide information about whether damage is occurring to the plants.

Stress Responses to Pollutants

Measurable effects such as chlorotic flecking on leaves, leaf-tip necrosis, discoloration of lichen thalli, and changes in leaf morphology or growth rate are difficult to attribute solely to exposure to a specific pollutant, since many are indicators of general stress rather than stressor-specific. In some cases, this is the level of information needed, i.e., that the environmental quality is causing detectable adverse effects and thus further investigations are needed. Similar injury may, however, result from nutritional or climatic factors or from exposure to mixtures of pollutants. There are some plant species or varieties that show visible injury which relates to the dose or concentration of a specific pollutant and thus can be used for monitoring. In some plants ultrastructural changes and biochemical parameters such as changes in enzyme activity or pigment composition may be detected. These may be useful in detailed monitoring programs to detect effects at concentrations lower than those resulting in visible injury. Prior to field use, verified studies under experimental conditions are needed to ensure that the response is specific and can reflect the exposure concentration (or "zone" of concentrations) of a specific pollutant. Such monitoring is generally restricted to gaseous pollutants. The process of monitoring effects depends on the plants exhibiting sensitivity to the pollutant, as a measurable, generally adverse, response which reflects the exposure ranging from minor injury at low concentrations to severe injury, and, eventually, disappearance of the species from the most polluted sites.

Monitoring Plant Species Distribution

The monitoring of species occurrence in equivalent terrestrial or aquatic habitats exposed to different levels of pollution from industrial or urban sources may be used, with caution, to estimate the concentrations of pollutants in different zones. This can only be attempted once a relationship between a pollutant concentration threshold for species survival has been established and other factors, including habitat availability and climatic factors, are taken into account.

Terrestrial Plants

Most plant groups include species that can be used as monitors for at least some pollutants. Their relevance to the monitoring program is a major factor on which to base the selection of appropriate *in situ* species. Since individual species differ in their responses to pollutants, the same species should be used within a monitoring program, or, when this is impossible due to the range of sites investigated, species intercalibration may sometimes allow substitution of another related species.

Mosses have been used very extensively for monitoring concentrations of heavy metals, e.g., lead, zinc, copper, and cadmium and, to a lesser extent, for radionuclides and organic chemicals on land and in fresh water. Mosses and lichens are considered especially suited for monitoring atmospheric deposition since they do not have roots and vascular systems to translocate pollutants from their substrate and thus should reflect input from the atmosphere. Epiphytes such as the flowering plant genus *Tillandsia* share this property and are also used.

Conifer needles in particular and leaves from a variety of angiosperm trees, shrubs, and herbaceous plants have been used for monitoring the distribution of persistent pollutants deposited from atmospheric sources.

Plants may be used to monitor concentrations of pollutants in food components and the emphasis is then on the relevant dietary plants and plant parts within food chains. One of the best known examples relates to the monitoring of radionuclides in lichens to estimate transfer of radionuclides into human food chains at northern latitudes via reindeer/caribou which feed on lichens during times of food scarcity.

Similar studies on persistent chlorinated compounds such as polychlorinated biphenyls (PCBs) have shown that they can also be transferred to food chains in arctic regions, far from their source of origin.

Whereas leaves may be the most easily collected and prepared for analysis, they may not always be the most appropriate samples, and concentrations of pollutants in roots or fruit need to be determined when those are consumed.

Gaseous Pollutants

The best-known examples of plants for monitoring gaseous pollutants are concerned with visible injury to plant foliage and with monitoring the species distribution of pollution-sensitive plants. Methods are continually refined to improve the reliability of monitoring data.

Ozone monitoring Ozone monitoring using cultivars of tobacco (*Nicotiana tabacum*) was introduced in the early 1970s in the United States and in Europe, after the discovery by Feder and coworkers that there was a positive relationship between ozone concentrations in ambient air and the extent of leaf injury to cultivar Bel-W3. When grown under controlled conditions before transfer to sampling sites alongside a closely related cultivar such as Bel B which is not ozone-sensitive leaf injury is measured at regular intervals, generally each week in order to build up a picture of ambient ozone concentrations. In the past 30 years developments have included the use of seedlings which may respond more rapidly to fluctuations in ozone concentrations. Detailed evaluations of plant preparation and sampling design as well as training of personnel for recording damage have led to greater acceptance of this method which has been used to estimate predictions of crop losses. In addition, there are programs for biomonitoring using crop plants to gain information on appropriate varieties to grow in different regions and to design models to predict responses from combined data from meteorological and instrumental records of pollutant patterns. Monitoring data for the onset of injury and the yield measurements in successive years across a range of sites where ozone concentrations are also measured contribute valuable data of economic importance.

Sulfur dioxide Emissions of sulfur dioxide from fossil fuel combustion and from other industrial sources such as primary metal smelters affect the survival of sensitive plants such as lichens, which in clean air can colonize many types of surface, such as roofs, walls, and trees. Observations in the nineteenth century showed that although suitable substrates were present in urban areas, many species of lichens found in neighboring localities were absent. As a result of extensive studies in the United Kingdom during the 1960s, especially in the London area, the concept of lichen zones corresponding to zones of sulfur dioxide concentrations was proposed by Hawksworth and Rose. Subsequently, similar investigations in North America and in many countries in Europe showed that this was a general occurrence. Species recorded on trees varied in their distribution with distance from city centers, where sulfur dioxide from domestic coal-burning and industrial emissions were concentrated. On trees in the innermost zones, where mean winter sulfur dioxide concentrations exceeded 170 $\mu g\,m^{-3}$, lichens were absent. Such monitoring of species distribution clearly can be related only to long-term pollution monitoring, since it is based on the disappearance of species rather than monitoring injury before the stress was severe enough to eliminate the species. With the controls on coal-burning in many countries in urban areas, this approach to monitoring is no longer possible: improved air quality has resulted in recolonization, although rates for different species have been very varied. There may be a few individual species for which distribution may still be a valid way of monitoring the changes in air quality. The occurrence of species and the percentage cover by lichens on substrates can also be carried out in the vicinity of industrial emissions and, necessarily, has to take account of relevant environmental parameters affecting the sampling sites, which may modify the response to sulfur dioxide exposure. The acidity of the substrate is clearly a major factor, and the original zonation considered the species association on trees with acid bark and for neutral/alkaline bark. Contrasting with ozone, plants may accumulate sulfur which may provide additional evidence of the extent of pollution by sulfur dioxide. Samples of pine needles, for example, collected routinely as part of forest damage surveys may be analyzed for sulfur as well as for other elements, sometimes indicating the influence of more localized industrial emissions as well as contributing to a national or international picture of sulfur deposition patterns.

Hydrogen fluoride Hydrogen fluoride emissions from industries such as aluminum smelters and brickworks are perhaps a less widespread problem and thus efforts to identify a range of plants as monitors may have been fewer. The most well-known example is *Gladiolus hortulanus* cv White Friendship which, when used as a bioindicator transferred from standardized cultivation conditions to polluted sites, displays leaf tip injury which corresponds to the hydrogen fluoride in the ambient air. The distribution of hydrogen fluoride sources can be determined by chemical analysis of leaf samples, lichens, and mosses.

Metal Pollutants

The dispersal of metal pollutants in the environment is very widespread as a result of metal processing industries and the emissions resulting from the use and disposal of metal products. The persistence of metals and their presence in all environmental compartments has led to many programs to monitor them around point sources of emissions and discharges and in large-scale programs across Europe (e.g., coordinated by the United Nations Economic Commission (UNECE)), in North America, and a few examples from Asian and African countries. The greatest use of plants as monitors is to map distribution patterns of metal deposition on land by chemical analysis of samples. Mosses, lichens, and conifer needles are the main plant samples used.

By far the largest data sets are for mosses, which developed from studies in the late 1960s on roadside lead pollution in Sweden by Ruehling and Tyler. Harmonization of methods has been promoted, and the use of the same species (primarily *Pleurozium schreberi* and *Hylocomium splendens*) in many countries has produced evidence of deposition trends over time in different countries for a range of metals. Annual growth segments can be separated, providing a link with annual deposition rates of metals from instrumental measurements. The influence of environmental factors, some of which may be related to topography, needs more investigation, as do the possible effects of growth rates in polluted areas on metal concentrations in samples.

An additional application of monitoring data is to investigate sources of pollutant loadings, especially for metals, and involves the use of enrichment factors. Extensive data sets from monitoring metals in lichens are available and this source apportionment approach has been reported in North America and Europe and can be used to identify the major sources of pollutants on a regional basis.

Herbaceous plant leaves and tree foliage are widely sampled around localized sources of metal pollutants for monitoring the deposition in relation to emissions. Chemical analyses of conifer needles are included in national and international monitoring programs which specify, for example, sampling height and age of needles (first or second year). Many programs were set up to examine roadside pollution from lead added to gasoline as an antiknock agent. The decrease in lead contamination has been evident from continued monitoring of roadside plants after reduction and elimination of lead from gasoline in many countries.

Both mosses and lichens have been used for active monitoring of metal pollutants: samples from sites with low pollutant status are prepared in a consistent way before transfer to sites for exposure for specified time periods. This may be carried out using living material or, in an adaptation used since the early 1970s, as moss bags, where moss samples are cleaned, sometimes with acids, and exposed in mesh bags. Their cation binding capacity and particle retention may be similar to vegetation surfaces, but they are not really equivalent to living material. The concentrations of metals can be determined after known time intervals. The grass *Lolium multiflorum italicum* cv Lema has been suggested as a useful standard bioindicator for active monitoring of metal deposition.

Radionuclides

Although lichens were known to accumulate radionuclides effectively, following nuclear weapons testing in the 1960s, considerable attention focused on monitoring radionuclides in plants in many countries after the Chernobyl nuclear reactor explosion in 1986. Lichens and mosses can be used to estimate annual deposition rates, since growth increments can be separated in some species. Plants in the vicinity of nuclear power stations may be sampled as part of monitoring programs to examine deposition to grazing pastures and potential transfer to the human food chain via dairy milk. The emphasis is generally on food chain components, but a number of monitoring programs have continued in Europe to examine time trends in the concentrations of cesium and strontium in mosses and lichens and in conifer needles as indicators of ecosystem contamination.

Organic Chemicals

Despite some losses by volatilization, organic chemicals accumulated by plants can reflect the exposure in their environment. Chemical analysis of plant leaves and of mosses and lichens can indicate sources of contamination from pesticide use and from industrial and vehicular sources in urban areas. Monitoring concentrations in mosses and lichens has shown the distribution of organic pollutants such as

PCBs and other chlorinated hydrocarbons in arctic ecosystems. Relatively few data from monitoring programs are available, although the use of a standard bioindicator, curly kale (*Brassica oleracea* var. *acephala*), may be useful for active monitoring of polycyclic aromatic hydrocarbons.

Freshwater Plants

Plants in rivers and freshwater lakes may be exposed to contaminants from agricultural land, industrial discharges, and urban runoff, and also from atmospheric deposition. The same general points about species selection, sampling, and analytical procedures apply as for terrestrial species and their use in monitoring specific pollutants is primarily by measuring the accumulation in plants. Changes in species distribution may be monitored and related to organic pollution (eutrophication) to classify river water quality, but is not specifically related to exposure to concentrations of individual pollutants. Some monitoring programs have attempted to relate changes in pigment concentrations in aquatic mosses to metal pollution, but again this has not been specifically linked to a pollutant.

Monitoring the accumulation of metals, organic chemicals, and radionuclides in aquatic macrophytes can demonstrate the influence of industrial discharges and, if shoot tips of macroalgae or mosses are sampled, fluctuations in concentrations can be detected over short time periods. The use of shoot tips avoids the problem of epiphytes which may affect the metal concentrations in the macrophytes.

Active monitoring can be carried out using aquatic moss bags, containing shoots in mesh bags, or by transferring macrophytes attached to substrate rocks to monitoring sites for specified time periods and measuring the change in metal concentrations. Comparisons of metal concentrations in moss samples collected in a number of European rivers suggested that analysis of the mosses was correlated with the mean concentrations in the river water and were thus a valuable way of monitoring, provided standard procedures were followed for preparation and analysis.

Marine Plants

Widely distributed macroalgae such as *Fucus*, *Ascophyllum*, and *Enteromorpha* are effective accumulators of metals and radionuclides. The sampling of frond tips may provide a good indication of metal or radionuclide concentrations over short periods, but factors such as salinity and pH may affect uptake and need to be considered when concentrations are compared to monitor the distribution of pollutants in seawater. Monitoring using macroalgae has provided useful time series for concentrations in the vicinity of coastal nuclear power stations.

Methods of using plants as monitors of pollution are becoming increasingly standardized and accepted in some countries as providing a more direct link with environmental quality assessment than the reliance on instrumental methods alone. Careful experimental studies to identify thresholds for injury in species may increase the use of biomonitors for regulatory purposes, provided the design of the monitoring program and quality control procedures are addressed. Species used need to be selected to address the purpose of monitoring, but there is potential to extend the use of methods to other pollutants and groups of plants.

List of Technical Nomenclature

Bioassay	Use of standard organisms and procedures to measure quantities and/or effects of chemical substances in field or laboratory conditions over a specified time.
Biomonitor	An organism which shows response(s) to change in its environment measured as chemical composition or growth parameters (including metabolic factors).
Chlorotic flecking	Small areas on leaf surfaces lacking chlorophyll.
Ecosystem	Community of living organisms and its associated physical environment.
Epiphyte	Plant growing on other plant (usually a tree) which provides physical support.
Macrophytes	Plants, rooted or floating, growing in aquatic habitats and visible to the naked eye.

See also: **Biodiversity and Conservation**: Plant Diversity, Conservation and Use. **Plants and the Environment**: Plants and Atmospheric Pollution; Waste Water Treatment.

Further Reading

Burton MAS (1986) *Biological Monitoring of Environmental Contaminants (Plants)*. Monitoring and Assessment Research Centre, Report No. 32. London: King's College.

Klumpp A, Fomin A, Klumpp G, and Ansel W (2002) *Bioindication and Air Quality in European Cities*. Stuttgart: Gunter Heimbach.

Kovacs M (1992) *Biological Indicators in Environmental Protection*. London: Ellis Horwood.

Markert B (ed.) (1993) *Plants as Biomonitors*. Weinheim: VCH.
Mulgrew A and Williams P (2000) *Biomonitoring of Air Quality Using Plants*. Air Hygiene Report No. 10. Berlin: WHO Collaborating Centre for Air Quality Management and Air Pollution Control.
Pakeman RJ, Hankard PK, and Osborn D (1998) Plants as biomonitors of atmospheric pollution: their potential use in pollution regulation. *Review of Environmental Contamination and Toxicology* 157: 1–23.
Verein Deutscher Ingenieure (1999) *Biological Measuring Techniques for the Determination and Evaluation of Effects of Air Pollutants on Plants: Fundamentals and Aims*. Beuth, Berlin: VDI.
Wolterbeek B (2002) Biomonitoring of trace element air pollution: principles, possibilities and perspectives. *Environmental Pollution* 120: 11–21.
Zimmermann R-D, Wagner G, and Finck M (2000) *Guidelines for the Use of Biological Monitors in Air Pollution Control (Plants)*, Part 1, *Methodological Guidance for the Drawing-up of Biomonitoring Guidelines (Plants)*. Air Hygiene Report No. 12. Berlin: WHO Collaborating Centre for Air Quality Management and Air Pollution Control.

Waste Water Treatment

D L Jones and C Freeman, University of Wales, Bangor, UK

Introduction

The safe disposal of both solid and liquid waste is of ever increasing concern due to the growing scientific and public awareness about the impact of untreated wastes on both human and environmental health. Although the most successful way of preventing environmental pollution is to prevent the formation of waste in the first instance (e.g., by enhancing water use efficiency, reducing packaging, recycling, etc.), it is inevitable that there will be a continued need for society to dispose of waste in an ecologically friendly manner. The aim of this article is to demonstrate the effective use of different plant based strategies for the sustainable treatment of wastewater. However, some of the principles outlined below are also relevant to the treatment of other solid wastes. One example of this is the treatment of liquid effluent (leachate) produced as a consequence of the burial and subsequent decomposition of domestic waste in municipal solid waste (landfill) sites.

As a society, we produce many types of liquid waste, which can be broadly classified into three categories: agricultural, industrial, and domestic waste (**Table 1**). Although these can range from highly toxic wastewaters (e.g., landfill leachate) through to relatively nontoxic wastewaters (e.g., storm drain water from road/urban runoff), many features of their treatment remain independent of waste type (e.g., reduction in biological oxygen demand (BOD), chemical oxygen demand (COD), suspended solids and inorganic nutrients (NH_4^+, PO_4^{3-}, Na); (**Table 2**)). Without remedial treatment, the disposal of wastewater into freshwater and marine systems can have catastrophic consequences, having both short-term impacts (eutrophication, death of fish, human pathogens in recreation water, odor problems, etc.) and long-term effects (loss of wildlife, build up of toxic chemicals, etc.). Furthermore, improper disposal of wastewater onto land can result in the frequently irreversible loss of ecosystem productivity (e.g., due to salinization) and the pollution of groundwater resources. The treatment of wastewater by plant based systems therefore requires careful management. The two most widespread types of wastewater management strategies that employ plants are (1) wetland treatment systems, and (2) land based treatment systems.

Table 1 Types of wastewaters than can be treated by either wetland or land based plant based treatment systems

Domestic	*Industrial*	*Agricultural*
Untreated sewage	Acid minewater drainage	Dairy parlor washings
Partially treated sewage	Cooling water	Pig and cattle slurries
Road runoff	Metal processing	Sheep dips
	Pulp and paper waste	Vegetable washings
	Chemical production	
	Food processing	
	Abattoir waste	
	Landfill leachate	

Which Treatment System to Choose?

The design of a successful plant based wastewater treatment system is a complex process if the scheme is intended to operate over a long timescale. The main considerations when designing and implementing any waste treatment system are: (1) the quantity and quality of waste, and its temporal variability; (2) the amount and type of land available; (3) the economics of setting up and running the scheme relative to other waste disposal options; (4) the impact on the local environment; (5) public perception and social aspects; (6) legal considerations; (7) the fate of percolate water; (8) the types of

Table 2 Typical chemical composition (mg l^{-1}) of different types of wastewater commonly utilized in land based and wetland treatment systems

	Raw sewage	*Urban stormwater runoff*	*Paper processing effluent*	*Cooling tower water*	*Oil refinery wastewater*	*Cheese production wastewater*	*Acid mine drainage*	*River water*
pH	7.4	7.5	6.6	7.3	8.4	7.1	2.6	7.3
TSS	700	160	300	70	350	500	200	20
BOD_5	600	7	130	150	160	1200	0.5	2
COD	300	33	–	150	–	–	–	10
NO_3^--N	3.0	1	–	–	1	–	0.5	1.1
NH_4^+-N	51	0.2	–	50	87	38	0.1	0.1
Organic N	18	3	–	–	–	36	0.4	2
S	10	6		400	700	500	1100	30
P	7	5	–	6	49	–	–	0.2
Zn	0.4	0.05	–	–	–	–	–	0.001
Cu	1.6	0.03	–	2	–	–	–	0.005
Fe	0.8	0.6	–	1	–	–	130	0.4
Cl	20	–	–	162	310	–	12	50
Mg	10	–	–	–	–	–	90	12
Ca	9.8	–	–	–	–	–	140	23
Na	23	–	–	–	–	–	–	3
Fecal coliforms	10 000 000	100 000	–	–	–	<100	–	–
Fecal streptococci	2 000 000	16 000	–	–	–	<100	–	–

–, no data available, COD, chemical oxygen demand; BOD_5, biological oxygen demand with a 5-day assay time.

plants available; (9) the environmental conditions; and (10) the terminal land use upon site closure. In particular, understanding the long-term behavior of the soil and associated plants in response to the continued addition of wastewater is notoriously difficult to predict, even with the best models. In almost all situations, the design of plant based treatment systems requires a site specific recommendation due to the large number of factors to be considered.

Wetland Treatment Systems

Wetlands are often viewed as providing a cost effective and natural alternative treatment strategy for ameliorating pollutants during the transit of wastewater from terrestrial to aquatic systems. Natural wetlands have been used as convenient wastewater discharge sites for at least a century while artificially constructed wetlands are becoming increasingly important, as these can be bioengineered and carefully managed to cope with site specific waste problems. It is often assumed that, as in conventional treatment systems, physical and microbial processes drive the ameliorative mechanism. And yet, however, plants are known to improve the treatment efficiency of many pollutants in wetland systems. The precise role that the plants play in these systems remains controversial, however, the following discussion will highlight some of the main concepts.

Types of Wetland

There are three major types of wetland: (1) natural wetlands, (2) constructed surface flow wetlands, and (3) constructed subsurface flow wetlands (**Figure 1**). Surface flow constructed wetlands typically have a water depth of between 10 and 50 cm, and contain a dense array of different plants to which water is added at a rate of between 25 and 500 $m^3 ha^{-1} day^{-1}$. Subsurface flow wetlands contain plants growing in a saturated bed of gravel or soil to which wastewater is added. In these systems, wastewater flows horizontally through the wastewater saturated subsurface layer (ca. 0.4–0.6 m deep) due to gravity, with typical wastewater addition rates in the region of 100–2000 $m^3 ha^{-1} day^{-1}$. The plant species present in any given treatment wetland depends largely on the physical design of the system, due largely to the constraints imposed by water depth (**Figure 2**). Typically, concentrated wastewaters are not treated using wetlands until they have either been diluted or, more commonly, processed to a lower pollutant level (i.e. secondary or tertiary processed or "polished" water) so as not to adversely affect the functioning of the wetland environment.

The Role of Plants in Amelioration

The plants growing in both natural and constructed wetlands can include a wide variety of species such as mosses (e.g., *Fontinalis*, *Sphagnum*), ferns (e.g., *Salvinia*, *Azolla*), trees (e.g., *Salix*, *Alnus*, *Pinus*, *Taxodium*), large algae (*Cladophora*), and various

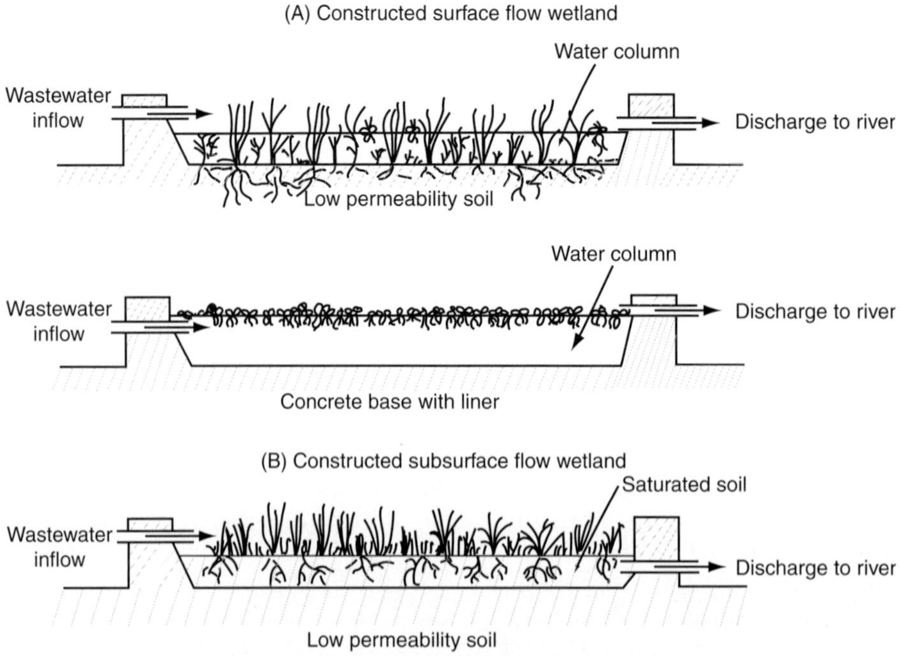

Figure 1 Schematic representation of (A) surface and (B) subsurface flow constructed wetlands.

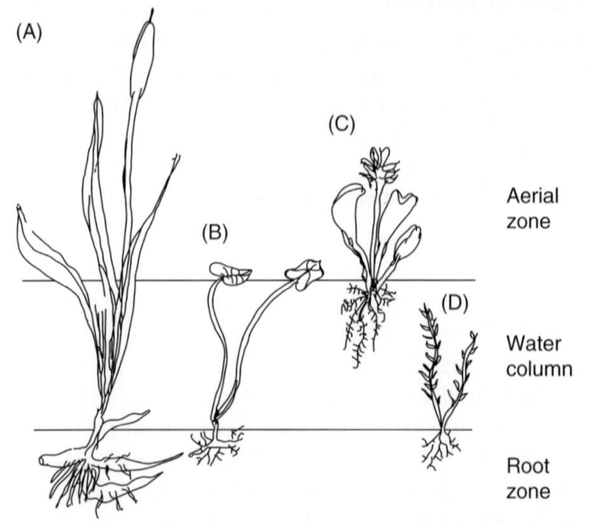

Figure 2 The four main classes of plants used in wetland construction. The type of wastewater treatment depends on the plant species and, in particular, whether plants are emergent (A) submerged with floating leaves (B), floating (C), or submerged (D).

monocotyledonous and dicotyledonous plants (e.g., *Phragmites*, *Juncus*, *Typha*). There are two ways in which plants improve the quality of wastewater passing through a wetland. The first involves a direct impact due to plant uptake and assimilation of organic and inorganic pollutants. The second is indirect, and involves modification of the wetland environment, in such a way that the efficiency of nonplant mediated processes are enhanced.

Direct Effect of Plants on Wastewater Quality in Wetlands

The uptake of nutrients from the water column and from the soil directly underneath is a feature common to the macrophytes of all aquatic systems. In treatment wetlands, this is of direct value as a mechanism for eliminating potentially polluting inorganic nutrients such as NO_3^- and PO_4^{3-} and the assimilation of organic pollutants (surfactants, phenols, oils, etc.). Wetland plants are among the most productive of any ecosystem and possess a substantial capacity for incorporating inorganic nutrients into their tissues. The plants require nutrients for growth and reproduction. Some nutrients can be assimilated through the stems and leaves immersed in the water, but in rooted plants, it is primarily the root systems that are responsible for nutrient uptake. The potential for uptake of nutrients varies according to the species dominating the system. The floating aquatic species, *Eichhornia crassipes* (water hyacinth) has one of the highest potentials, while plants with most of their biomass above the water surface (emergent plants) follow close behind (**Table 3**). Completely submerged plant species tend to have the lowest nutrient assimilation capacity. However, harvesting of the plant material at intervals can be critical in determining whether or not these potential uptake values are reached, because without harvesting of the above-ground biomass, most of the nutrients are returned to the

Table 3 Biomass production and nutrient uptake potential of typical plants used in constructed and natural wetlands

Category	*Biomass growth (kg dry weight ha^{-1} year^{-1})*	*N uptake rate (kg ha^{-1} year^{-1})*	*P uptake rate (kg ha^{-1} year^{-1})*
Floating aquatic species			
Eichhornia crassipes (water hyacinth)	50 000–100 000	1000–6000	250–1250
Pistia stratiotes (water lettuce)	50 000–75 000	1000–5000	100–750
Lemna minor (duckweed)	5000–25 000	250–2000	100–500
Herbaceous emergent plants			
Typha spp. (cattails)	5000–50 000	500–3000	50–500
Juncus spp. (rushes)	10 000–60 000	500–2000	50–400
Phragmites spp. (common reed)	10 000–60 000	250–1000	20–100
Swamp forest			
Taxodium/Nyssa spp. (cypress/tupelo forest)	5000–20 000	100–500	5–50

water course at the end of the growing season as the senescent plant tissues decompose. Often, the amounts of nutrients that can be harvested are extremely low compared to the loadings into the wetland. The only permanent mechanism for plant mediated long-term removal arises when the pollutants are incorporated into tissues, but then those tissues are poorly decomposed. This can arise either when conditions are unsuitable for effective decomposition, or when the nutrients are incorporated into recalcitrant compounds.

Indirect Impact of Plants on Wastewater Quality in Wetlands

The presence of plants in treatment wetlands has a number of effects that change the wetland environment in a manner that can influence wastewater treatment efficiency. These include physical effects, such as the trapping of suspended solids, modifying light intensity, reducing water flow velocity, lowering wastewater temperature, and reducing wind velocity. It also creates a surface for microbial colonization, and modifies the soil structure. Biological effects include the release by roots of oxygen and other organic exudates that modify conditions and stimulate the soil microbial community. The principle mechanism by which plants contribute to wastewater treatment largely depends on the design of the wetland, and, in particular, the constraints that the design imposes on the type of vegetation that can dominate the system, and whether those plants are, for example, emergent or floating types. The wastewater remediation processes operating in a wetland critically depends on whether the plant is in contact with the air, water, or soil in a wetland (**Figure 2**). In addition, operational factors such as inflow rate, inflow quality, water depth, climate/season, residence time, and length of the wetland are key determinants in the amelioration of wastewater by wetlands. Typical reduction profiles of wastewater components by constructed wetlands are shown in **Figure 3**.

The Aerial Wetland Zone

Plants producing biomass in the aerial zone can impact on the treatment process through five main mechanisms. First, with large amounts of tissue in the aerial zone, light reaching the soil surface may be reduced. This can reduce algal growth, which (although assimilating additional pollutants) creates the need for additional treatment in the form of filtration to remove them before discharge. Second, wind also creates movement in the plants, which has the benefit of keeping water column flow paths open and free of clogging. Third, the plant tissues also create a microclimate, which in winter creates an insulating layer that minimizes frost damage. Fourth, the aerial tissues are particularly amenable to harvesting, allowing the permanent removal of assimilated pollutants from the system, provided the harvesting is timed at an appropriate stage in the growth cycle. Finally, the presence of plant tissues in the aerial zone creates an aesthetically pleasing treatment system and provides ecologically significant habitats for wildlife (insects, birds, etc.). While this may not contribute to treatment efficiency, it certainly increases the likelihood of local acceptance of treatment systems in environmentally sensitive areas.

The Water Column

Submerged plant tissues lying in the water column impact upon treatment efficiency through several mechanisms. Perhaps the most important of these from the perspective of dissolved organic compounds and inorganic nutrients alike is that the tissues create a large surface area for biofilm colonization. Biofilms

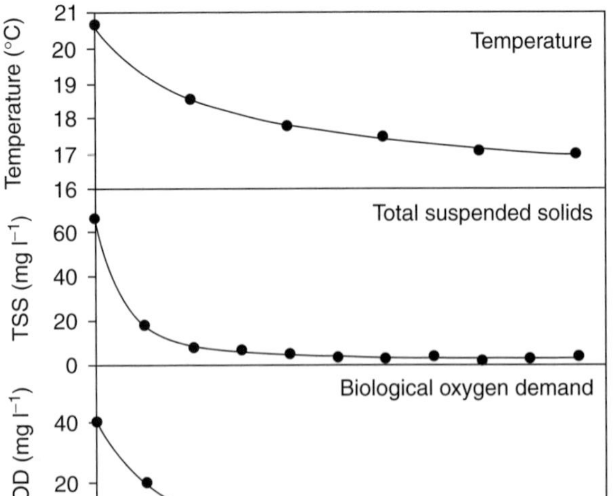

Figure 3 The decrease in concentration of various wastewater components with distance along a surface flow wetland. The wastewater inflow is located at 0 m. The pollutant reduction profiles are similar for subsurface flow wetlands, although generally the reductions occur over a shorter distance.

are mixed assemblages of algae, bacteria, fungi, and protozoa, embedded within a polysaccharide matrix. This matrix has been widely recognized to confer various functional benefits to the biofilm microflora in the ameliorative process. These include the ability of the matrix to act as: (1) a nutrient trapping system; (2) a medium for transport of compounds between cells, and a suitable environment for enzyme action maintaining biodegradative enzymes in close proximity to the microbial cells without washout into the overlying waters; and (3) as a buffer against changing organic substrate supplies. The fourth property is particularly of value, for it ensures that the biofilm can trap solutes from the water column using ion exchange mechanisms, even when conditions are suboptimal for microbial metabolism, and store those pollutants until conditions are more favorable for biodegradation.

For suspended solids, a further indirect benefit of the vegetation comes from the impact of the plant structures on flow velocities. The reduced flow of water creates conditions that favor sedimentation, while also minimizing erosion and resuspension. The lower flow rate also increases the water contact time with biofilm coated surfaces in the wetland. Localized release of photosynthetically produced oxygen can also aid aerobic microbial degradation processes. Plants in the water column can release as much as 0.25–9.6 g $O_2\,m^{-2}\,day^{-1}$.

Root Zone Impacts

Although direct uptake of wastewater pollutants by roots is a valuable component of the overall ameliorative success of a wetland, it could be argued that of even greater significance is the impact of plants in terms of the release of materials into the soil environment. The roots are a further location for biofilms, with all of their useful ameliorative properties as described above. There is much evidence to support the view that aquatic macrophytes contribute significantly to the oxygen content of wetland soils. This oxygen creates zones of aerobic conditions in what would otherwise be an anaerobic environment, and as such creates the potential for aerobic processes to occur, such as bacterial aerobic metabolism and nitrification (**Figure 4**). This oxygenation of the soil by roots also induces the formation of Fe(III)-hydroxide plaques on the surface of roots, which chemically immobilize PO_4^{3-}. Rates of oxygen release from lenticels and aerenchymous tissue can reach 0.5–10 g $O_2\,m^{-2}\,day^{-1}$. The complex mosaic of environments ranging from truly anaerobic to fully aerobic creates diversity of habitats for microorganisms, and the potential for treatment of a greater range of pollutants by a more diverse assemblage of microbial species.

A second and related impact of the plants comes from the exudation of organic compounds. Plants release a wide range of organic compounds from their roots, and it has been estimated that 2–15% of the photosynthetically assimilated carbon can be lost in this way. This mixture represents a valuable source of substrates for microorganisms living in the soil. This source of carbon has been proposed to drive various processes including selenium volatilization and denitrification induced nitrate removal. Enzymes play a pivotal role in amelioration of many pollutants, and the added carbon can be of value for promoting microbial enzyme synthesis for the

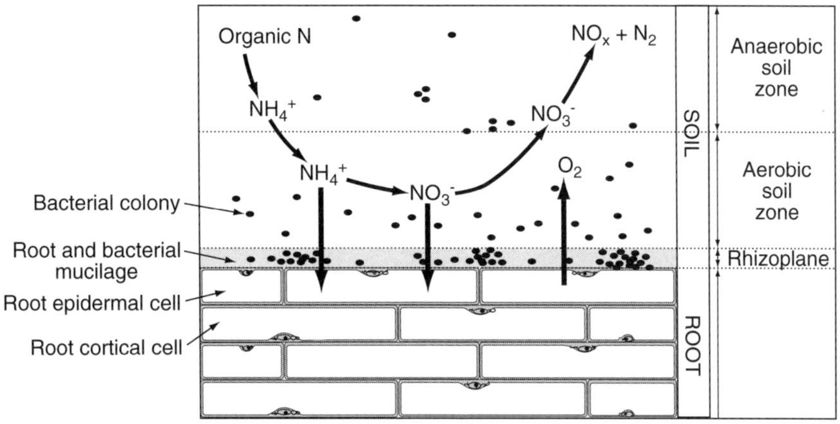

Figure 4 Schematic representation of the pathways of nitrogen transformation in the rhizosphere of a wetland plant root leading to the removal of N from wastewater. Organic N is broken down to NH_4^+ by microorganisms in the anaerobic (and aerobic) soil zone. The NH_4^+ then diffuses into the aerobic soil layer next to the root (due to O_2 release from the root) where it can be taken up by plant roots or nitrified by aerobic soil bacteria to NO_3^-. The NO_3^- can then be taken up by the root or it can diffuse back out into the anaerobic soil zone where it can be denitrified by anaerobic soil bacteria to N_2 and NO_x, which are eventually emitted into the atmosphere.

degradation of complex organic effluents. Other plant derived organic compounds also show biological activity. For example, phenolic compounds can have a substantial inhibitory impact on enzymic decomposition processes. Thus, these phenolics can benefit amelioration where plants play a primary role in the removal of inorganic pollutants from the effluent stream, as when enzymic decomposition is impaired, assimilated pollutants remain locked within plant detritus, rather than being rereleased into the environment at the end of the growing season. The exudates of some species are also considered to have antibiotic properties and are believed to be highly effective in the phytoremediation of bacterial pathogens present in wastewater.

The macrophytes also help to stabilize the soil, and minimize the formation of erosion channels. In vertical flow systems, plants have been shown to minimize clogging. There is also believed to be an impact on the hydraulic conductivity of horizontal flow systems, due to soils being loosened as roots and rhizomes grow, and then pores and channels being created as the structures eventually die and decay.

Problems with Wetlands

The most common difficulties experienced in wetlands, as for other land based wastewater treatments, occurs when an overloading of the system with wastewater takes place (volumes or pollutant load). In the case of wastewater with a high BOD, overloading can lead to highly reducing (anaerobic) conditions prevailing in the soil and water column, which causes plant stress, and will reduce the amount of nutrient removal and O_2 entering the subsurface zones. Water flow problems may also be encountered in wetlands due to violent hydrological events (e.g., storms) overloading the system, and conversely under low flow conditions due to the localized build up of suspended solids and plant material. Odor problems, floods, fires, and increases in local mosquito and pest populations have also been causes for public concern in some wastewater treatment schemes.

Land Based Treatment Systems

Land based wastewater treatment systems comprise a wide diversity of strategies for the removal of potential pollutants and pathogens. These techniques range from the poorly controlled spraying of wastewater onto land, through to carefully bioengineered and managed systems that allow complete control over the remediation process. Typical examples of wastes treated by land based disposal systems include partially treated domestic sewage, agricultural wastes (e.g., milking parlor washings, slurry, etc.), and low hazard industrial wastes (e.g., abattoir effluent, food processing waste, etc.). In many situations, the wastewater is primary or secondary treated by conventional means (e.g., sewage treatment works) before application of the "polished" wastewater to land. Typically, remediation of wastewater by these systems occurs due to interactions between plants, soil microorganisms, and the soil's solid phase. The main processes operating in these treatment systems include: (1) the breakdown of the organic compounds by the soil microbial community and plants; (2) the reduction of bacterial pathogen population numbers due to predation (e.g., protozoa) and UV exposure; (3) the sorption of cations and anions (e.g., heavy metals, excess nutrients) to the soil's solid phase; (4) the uptake of excess nutrients and metals by the plants; and (5) the removal of suspended solids. Although land based schemes without plants possess a significant ameliorative effect on wastewaters (e.g., in septic tank effluent treatment), typically, the addition of plants provides a more superior treatment system due to: (1) their capacity to absorb large quantities of nutrients compared to the soil microbial biomass; (2) their evapotranspiration capability, which removes excess water; (3) the release of root exudates and the death of roots, which stimulates soil microbial activity and, consequently, the breakdown of pollutants; (4) the ability of plants to stabilize soil structure preventing soil physical degradation during wastewater application; and (5) the capacity of plants to metabolize organic chemicals. The three main land based wastewater treatment strategies that are discussed below are "zero flow" treatment, "slow-flow" treatment, "high-flow" rapid infiltration, and "overland flow." If designed correctly, all industrial wastes, with the exception of radioactive wastes, can be treated with one of these techniques.

Zero Flow Treatment

This wastewater treatment option is normally associated with the subsurface injection of liquid waste material into soil with little movement of the liquid waste from the point of injection (**Figure 5A**). As this process is carried out using a tractor and attached waste tanker equipped with a multiple deep injector system, it is normally associated with the treatment of relatively small, but potentially hazardous, quantities of liquid waste (e.g., abattoir and food processing effluent). Normally, in all circumstances where this waste disposal option is practiced, plants form an integral part of the treatment for the reasons described above. Typical wastewater application rates range from 0.1 to 25 $m^3\ ha^{-1}\ day^{-1}$, with waste added to the land intermittently (ca. 1–6 times per year) to allow time for microbial assimilation of organic compounds, and to prevent excessive nitrate leaching and pathogen accumulation. These schemes are typically used for reducing the BOD, nutrient, and pathogen load of wastewaters.

Slow-Flow Treatment

This is the predominant land based treatment process and involves the spraying of wastewater (industrial and municipal) onto moderately flat agricultural land in order to meet the water demands of the crop (**Figure 5B**). Usually, the wastewater has been first treated by a conventional sewage treatment works; the typical plants used in these schemes include grasses, food crops (corn (*Zea mays*; maize), wheat (*Triticum* spp.), vegetables), and nonfood crops (trees, biofuels, fiber crops). This type of wastewater can also be used to irrigate recreational areas, e.g., golf courses. Depending upon the main purpose of the scheme, water application rates are designed either to achieve maximum agricultural production or maximum water treatment. Typical wastewater application rates range from 5×10^3 to $5 \times 10^4\ m^3\ ha^{-1}\ year^{-1}$ or 10 to 150 $m^3\ ha^{-1}\ day^{-1}$ with water added to the land intermittently (ca. 1–3 times per week) to prevent the formation of anaerobic conditions. These schemes are typically used for reducing the BOD, nutrient, and TSS load of wastewaters.

High-Flow Rapid Infiltration

This wastewater treatment strategy is similar to slow-flow treatment except this system often operates without plants, uses rapidly draining soils, and has very high wastewater loading rates (**Figure 5C**). Typical wastewater application rates range from 5×10^4 to $1 \times 10^6\ m^3\ ha^{-1}\ year^{-1}$ or 150 to 2000 $m^3\ ha^{-1}\ day^{-1}$ with most of the water percolating through the soil profile. These schemes are typically used for reducing the BOD and TSS load of wastewaters and are not commonly used.

Overland Flow

In these systems, pretreated wastewater is trickled down gently sloping vegetated land (1 to 7°; 20–50 m long) that possesses an intrinsically slow water infiltration capacity (**Figure 5D**). The most common plants used in these systems are pasture grasses. In this waste treatment option, the rates of percolation into the soil are low and evaporation rates are high, with a large proportion of the initially applied waster often collected as runoff at the base of the slope. Typical wastewater application rates range from 5×10^4 to $5 \times 10^5\ m^3\ ha^{-1}\ year^{-1}$ or 50 to 300 $m^3\ ha^{-1}\ day^{-1}$. These schemes are typically used for reducing the BOD, TSS, and pathogen load of wastewaters (**Figure 6**).

Problems with Land Based Treatment Systems

Land based treatment systems are not without their problems, particularly with long-term operations. In many cases, the plants present in the scheme are incapable of absorbing all the nutrients from the wastewater in equal quantities leading to excess leaching and groundwater contamination (e.g., NO_3^-) or the accumulation of metals to phytotoxic levels (e.g., NH_4^+, Na; **Figure 7**). Furthermore, effective nutrient removal only occurs during the main growing season, and therefore wastewater

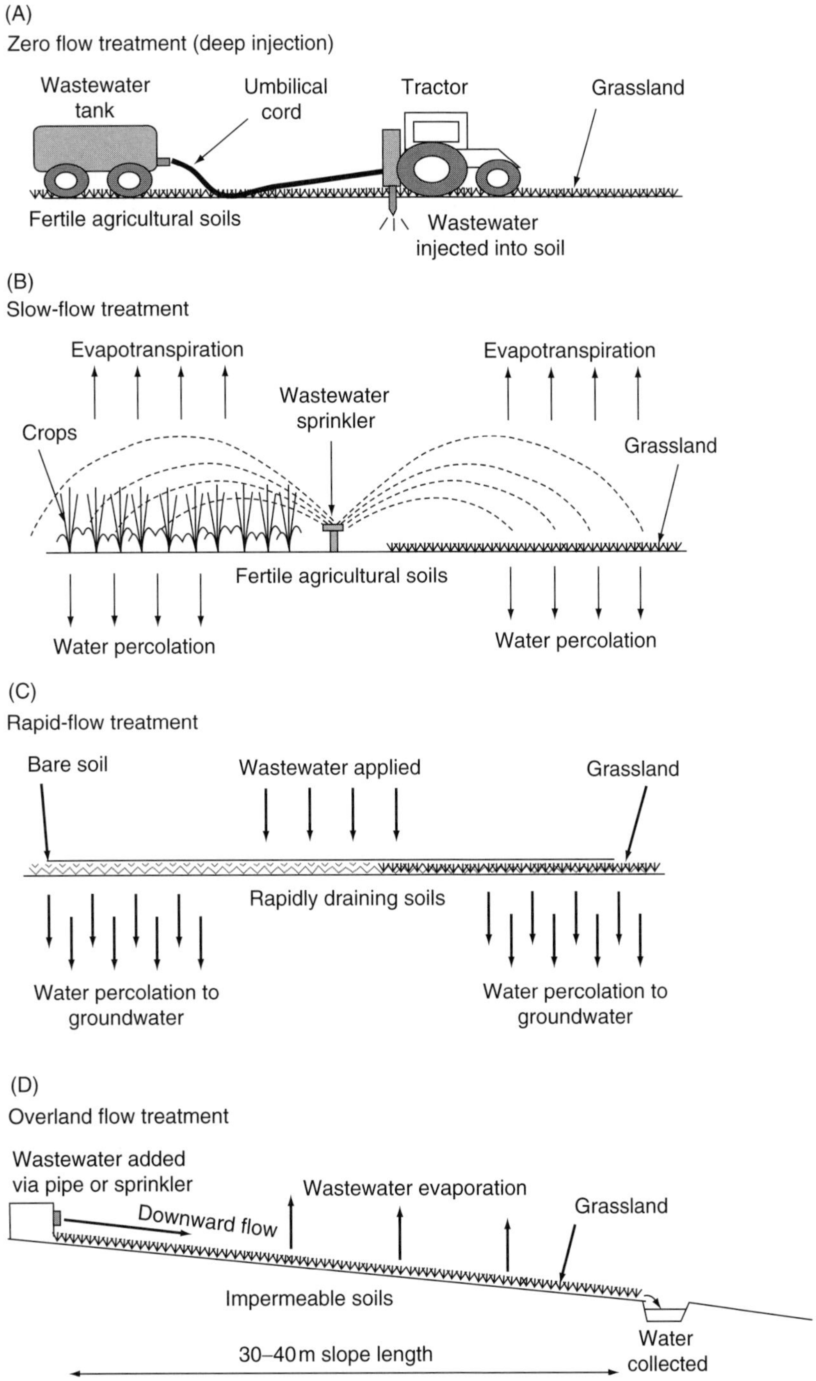

Figure 5 Schematic representation of four contrasting land based systems used for the treatment of wastewater.

holding tanks are required for wastewaters that are produced all year round in order to prevent pollution incidents. In addition, if the wastewater contains a cationic imbalance (e.g., too much Na^+), continued application of this imbalanced wastewater can lead to displacement of essential cations from the soil's exchange phase and excessive leaching of essential nutrients, leading to plant deficiencies (e.g., Na^+ displacing K^+). Further problems may include the capping or sealing of the soil during irrigation causing uncontrolled runoff and surface ponding. This occurs due to blockage of soil pores by added suspended solids and the production of polysaccharide slime by soil microorganisms feeding upon the carbon in the wastewater. Other problems may include an accumulation of pathogenic organisms on plant surfaces and in soil (e.g., human enteric bacteria) and chemicals (endocrine disrupting agents)

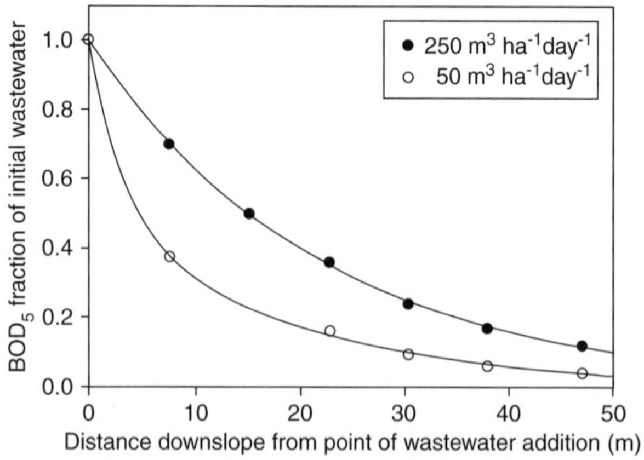

Figure 6 Reduction in biological oxygen demand (BOD_5) with distance along an overland flow wastewater treatment scheme. The results for two different wastewater addition rates are shown (50 and 250 $m^3\ ha^{-1}\ day^{-1}$).

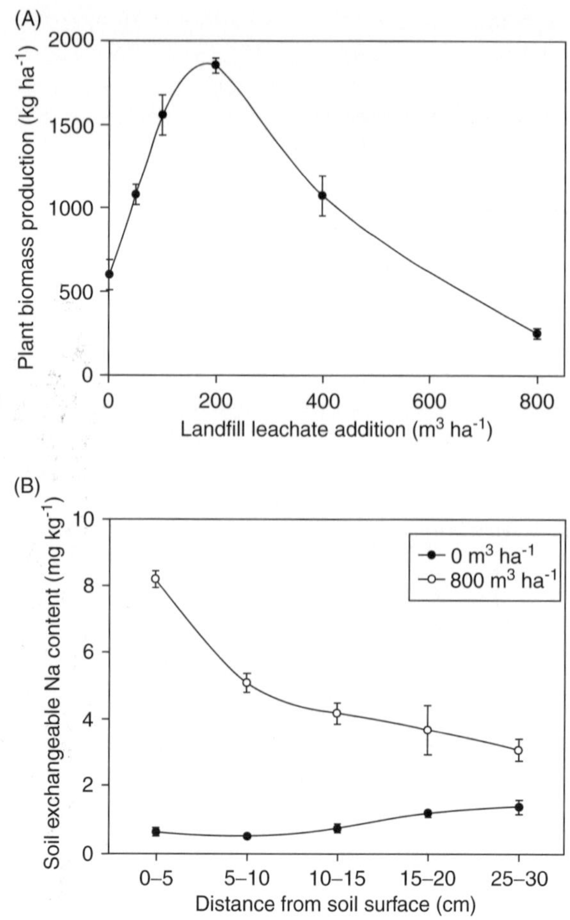

Figure 7 Impact of different repeated loadings of untreated municipal solid waste site wastewater (landfill leachate) on (A) plant biomass production (*Lolium perenne* grassland) and (B) soil Na content during a slow-flow wastewater treatment scheme. The main cause of toxicity in (A) at high leachate loadings is an excess accumulation of Na^+ and NH_4^+ in the soil.

that may subsequently enter the food chain. Odor problems and wind dispersal of potential pollutants may also be an issue, depending on the mode of wastewater application to the land (e.g., by spray irrigation).

List of Technical Nomenclature

Ammonification The conversion of organic nitrogen to NH_4^+ by soil microorganisms.

Biological oxygen demand The amount of dissolved oxygen consumed by organisms in wastewater during a 5-day incubation period.

Denitrification The conversion of NO_3^- to N_2O, NO, N_2 by soil microorganisms.

Macrophytes Large plants used in wetlands.

Nitrification The conversion of NH_4^+ to NO_3^- by soil microorganisms.

Polished water Water that has already been pretreated to remove most of the pollutant load.

Total suspended solids The amount of particulate matter in wastewater.

Wastewater Water that has previously been used for domestic, industrial, or agricultural purposes.

See also: **Energy Crops**: Biomass Production. **Nutrition**: Mineral Uptake. **Plants and the Environment**: Plants as Pollution Monitors; **Plants and the Environment**: Land Reclamation and Remediation, Principles and Practice. **Root Development**: The Rhizosphere and its Microorganisms. **Water Relations of Plants**: Salt Stress.

Further Reading

Cooper PF (1996) *Reedbeds and Constructed Wetlands for Wastewater Treatment*. Swindon: WRC.

Hammer DA (1989) *Constructed Wetlands for Wastewater Treatment: Municipal, Industrial and Agricultural*. London: Lewis Publishers.

Hammer DL and Knight RL (1994) Designing constructed wetlands for nitrogen removal. *Water Science and Technology* 29: 15–27.

Hegg RO and Turner AK (1983) Overland flow as a method of treatment for animal wastes – a review. *Agricultural Wastes* 8: 167–184.

Kivaisi AK (2001) The potential for constructed wetlands for wastewater treatment and reuse in developing countries: a review. *Ecological Engineering* 16: 545–560.

Knight R (2001) *Treatment Wetlands*. Boca Raton: CRC Press.

Knight RL, Payne VWE, Borer RE, Clark RA, and Pries JH (2000) Constructed wetlands for livestock wastewater management. *Ecological Engineering* 15: 41–55.

Metcalf and Eddy, Inc. (1991) *Wastewater Engineering – Treatment, Disposal and Reuse*. New York: McGraw-Hill.
Nerella S and Weaver RW (2000) Phytoremediation of domestic wastewater for reducing populations of *Escherichia coli* and MS-2 coliphage. *Environmental Technology* 21: 691–698.
Rogers F, Rogers K, and Buzer J (1997) *Wetlands for Wastewater Treatment*. Johannesburg: Withwatersrand University Press.
Sundaravadivel M and Vigneswaran S (2001) Constructed wetlands for wastewater treatment. *Critical Reviews in Environmental Science and Technology* 31: 351–409.
Younger PL (1997) *Minewater Treatment Using Wetlands*. London: Chartered Institution of Water and Environmental Management.

Phytoremediation

M Pletsch, Universidade Federal de Alagoas, Maceió, Brazil

Definitions and Brief Historical Retrospective

Phytoremediation (from Greek *phyto*, plant, and Latin *remedium*, cure or restoration) refers to the utilization of plants to clean contaminated environments. The use of this terminology is recent, even though its application is not. The exploitation of plants with the purpose of preventing leakage of hazardous compounds from landfills or of treating the movement of aqueous wastes has been carried out for decades in the form of revegetation or vegetative caps. Constructed wetlands (artificial ecosystems consisting of organic soils, aquatic plants, algae, and microfauna) which had been used in past centuries for the treatment of metal-polluted waste streams derived from coal mining, are still employed for treating municipal and industrial drainage. The first complete record concerning the unusual metal-accumulating ability of some plants dates from the mid nineteenth century and was published in the *Handbuch der Physiologischen Botanik* by J. Sachs. This compendium includes analytical data on the zinc content of four plants from the calamine area near Aachen, Germany, namely *Thlaspi alpestre* var. *calaminare*, *Viola calaminaria*, *Armeria vulgaris*, and *Silene inflata*. However, it was not until 1976 that T. Jaffré and others introduced the term "hyperaccumulator" in order to describe a plant, *Sebertia acuminata*, which contained highly elevated levels of nickel. According to Jaffré hyperaccumulators are plants capable of accumulating large amounts of metal, i.e., greater than 0.1–1.0% of the plant's dry weight, depending on the metal.

The term phytoremediation became popular in the beginning of 1990s, when plant researchers such as S. D. Cunningham and R. R. Brooks reviewed the progress achieved using plants for the remediation of contaminated soils, surface water, and groundwater. Phytoremediation technology is the result of intensive interdisciplinary research carried out in recent years in private companies, universities, and government research institutes, which has explored the abilities that plants have developed throughout millions of years of evolution to survive in inhospitable environments. The development of phytoremediation has created a new vocabulary which is summarized in **Table 1**.

Remediation of Inorganic Pollutants

Inorganic pollutants include common industrial contaminants such as the heavy metals (Pb, Zn, Cd, Cr, Co, Cu, Ni, and Hg), the semimetals (As), and the nonmetals (Se), as well as the radionuclides (^{3}H, ^{14}C, ^{36}Cl, ^{90}Sr, ^{129}I, ^{137}Cs, and ^{238}U) plus toxic cations and anions derived from mineral compounds (Na^{+}, NH_4^{+}, PO_3^{3-}, NO_3^{-} and ClO_4^{-}).

With the exception of radionuclides, these substances when present in trace amounts are not harmful, but when present at high concentrations they become toxic. Human activities, such as mining, smelting, dispersal of sewage sludge, and the use of phosphate fertilizers, increase the overall levels of inorganic contaminants in vegetation, including the levels in food crops, thus presenting a global environmental hazard. Inorganic contaminants are difficult to remove from the environment employing chemical or physical methods. Thus, the use of microorganisms and/or plants for the decontamination of soil and water is an attractive alternative in view of its low cost and high efficiency.

Phytoextraction and phytostabilization are generally accepted as cost-effective treatments for sites contaminated with heavy metals and radionuclides such as those which have been used by battery manufacturers, metal-finishing factories, mining companies, producers of solvents, coated glass, paints, leather, pottery, and chemicals, as well as those which are the responsibility of governmental agencies such as landfills, railroad yards, and nuclear power stations.

Phytomining (a form of phytoextraction) is an alternative to traditional, destructive mining practices. In this technique the harvestable parts of a

Table 1 Strategies used in phytoremediation

Strategy	*Objectives*
Phytoextraction	To absorb and accumulate contaminants into plant tissues
Phytostabilization	To prevent the movement of contaminants in the soil by incorporating them into either plant cell walls (lignification) or into the soil humus (humification), and sometimes even by precipitating them as insoluble forms
Phytomining	To take up precious metals from the soil and store them in the plant tissues from which they may be retrieved by further processing
Phytovolatilization	To take up the contaminants and release them into the atmosphere in a volatile form
Rhizofiltration	To absorb pollutants from aqueous environments and concentrate them into plant roots; this term also refers to the use of plants grown in hydroponic conditions
Blastofiltration	To remove pollutants from water by using seedlings rather than fully developed plants
Phytostimulation (rhizodegradation)	To promote the growth of microorganisms in the rhizosphere by providing physical support and nutritious exudates either to stimulate the uptake of contaminants by plants, or to enable their degradation outside of the plant
Phytotransformation (phytodegradation)	To absorb and convert toxic pollutants into innocuous compounds by activating plant metabolic systems
Hydraulic pumping	To lower the water table preventing the migration of pollutants and enhancing the loss of volatile organic compounds from the soil surface
Phytosorption	To absorb contaminants into either living or dead plant tissues

hyperaccumulating plant are reduced to ashes containing the valuable metals which can then be easily and safely separated and processed for further use. Phytomining can also be used to extract metals from low-grade ores that might otherwise not be cost-effective.

Phytovolatilization is feasible for Hg, As, and Se. The volatile forms of these elements (hydrides or methyl derivatives) can normally be liberated from the soil into the atmosphere as a result of the action of bacteria and fungi. There are no known examples of the application of higher plants in the volatilization of Hg except that involving transgenic *Arabidopsis* as described below. Regarding As, there are some aquatic plants such as *Ceratophyllum demersum* and *Egeria densa* that actively extract As from the water in which they grow. Recently, it was reported that the brake fern (*Pteris vittata*) is very efficient in extracting As from soil and accumulating it in its fronds. On the other hand the volatilization of Se by plants is well documented. For example, it has been shown that the emission rate of Se from the soil is greatly enhanced by the presence of the hyperaccumulator *Astragalus bisulcatus*. Furthermore, Se can also be released by nonaccumulator plants such as alfalfa (*Medicago sativa*, lucerne).

In the rhizofiltration approach, metals are retained within the roots or precipitated in the water by plants (large herbs or trees) grown hydroponically. This technique is particularly applicable to large volumes of water that contain low levels of contaminants. The plant biomass is harvested periodically, dried, and composted or burnt to ashes that can be buried or sold. Successful examples of plants used in rhizofiltration are *Helianthus annuus*, *Brassica juncea*, and *Populus* spp.

Table 2 Examples of known metal hyperaccumulators

Element	*Plant species*
Copper, cobalt	*Haumaniastrum robertii* (Lamiaceae) *H. katangense* (Lamiaceae)
Manganese	*Macadamia neurophylla* (Proteaceae) *Maytenus bureaviana* (Celastraceae)
Nickel	*Alyssum bertolonii* (Brassicaceae) *Phyllomelia coronata* (Rubiaceae)
Selenium	*Astragalus racemosus* (Fabaceae) *Lecythis ollaria* (Lecythidaceae)
Zinc, cadmium, lead	*Thlaspi caerulescens* (Brassicaceae) *T. rotundifolium* (Brassicaceae)

Blastofiltration is a novel technique whose full potential is yet to be evaluated. The possible advantage of blastofiltration over rhizofiltration is that some seedlings, whilst possessing the ability to absorb large quantities of metals, grow relatively independently of the environmental conditions (nutrition and light) during the first 10 days of their life cycle.

Plants often hyperaccumulate more than one metal and hence may be considered in three main groups, i.e., those which accumulate Cu and Co, those which accumulate Zn, Cd, and Pb, and those which only accumulate Ni. **Table 2** shows examples of metal hyperaccumulators and the families to which they belong. Natural hyperaccumulators not only grow slowly and have a small biomass, but also live in isolated habitats. For these reasons very little is known about the agronomic characteristics of hyperaccumulators such as their fertilizer requirements and their susceptibility to diseases.

The ultimate aim of inorganic phytoremediation is hyperaccumulation in defined tissues of plants. This form of remediation involves basically three strategies: (1) absorption of the elemental or ionic species from the soil, and sequestration in the roots thus preventing leakage from the contaminated site; (2) absorption and translocation of the toxic chemical form to the aerial parts which can then be harvested; and (3) absorption and biotransformation of the pollutant into less toxic forms within the roots or the aerial parts of the plant. Absorption, transportation, hyperaccumulation, and transformation are the main mechanisms that may be exploited in plants in order to eliminate inorganic pollution or to minimize its effects.

Metal ions are not only transported from the soil into the roots and distributed through the plant crossing membrane barriers but also enter the xylem from the apoplastic pathway across plant roots. Some toxic metals enter the plant through nutrient transport systems and others through specific transporters. Examples of such systems are the zinc transporter (ZIP) and iron transporter (ITR) protein families which are involved in Zn, Fe, and Cd uptake and transport. Translocation of metal and other ions in plants is also promoted by the release of metal chelators (organic acids) which liberate the metals from the soil and assist their transport through the plant tissues. Phytosiderophores, which have very strong affinities for Fe, are examples of this class of chelator.

Natural hyperaccumulators are able to concentrate metals at levels greater than 0.1–1% of the dry weight of the plant, depending upon the element. Two classes of low molecular weight proteins are involved in this hyperaccumulation: the metallothioneins (MTs) and the phytochelatins (PCs). Different types of plant MTs are responsible for binding specific metals to their cysteine residues forming metal–protein complexes. In normal metabolism, MTs function as an accessible pool of nutrient-essential metals (such as Fe, for example) and transport them to their final destination, possibly the prosthetic group of an enzyme. MTs may perform similar functions with toxic metals; however, their role as final ligands that protect plants from high concentrations of metals still has to be confirmed. On the other hand, the role of PCs in the detoxification of some heavy metals is quite clear, although the factors contributing to PC biosynthesis and function are not completely understood.

It is possible to select plants by classical breeding or to engineer plants genetically that encompass traits similar to those found in natural hyperaccumulators. **Table 3** summarizes some approaches available for the design of more efficient hyperaccumulators. One possible approach for enhancing metal tolerance and/or accumulation is to introduce genes of bacterial, plant, or animal origin which code for metal-binding proteins into plants, preferably those which are specific for toxic metals such as Cd, Hg, and Pb but not for essential elements such as Zn, for example. However, increasing metal-binding proteins may cause a physiological metal deficiency. Another method of overcoming metal toxicity is the transformation of toxic forms into harmless forms. Some elements (As, Hg, Fe, Se, and Cr) exist in a variety of states either as cationic and anionic species or organometallic forms. Mercury, for example, enters the environment in the volatile liquid Hg(0) form, Hg(II) ionic species and methylmercury, which is responsible for most of the poisoning accidents. It was demonstrated that transgenic *Arabidopsis thaliana* expressing the bacterial *mer* gene is able to convert ionic mercury Hg(II) into metallic mercury Hg(0), which can volatize from the plant tissues. Plants with this attribute can eliminate the ionic forms of mercury contamination from lakes, rivers, and wetlands, thereby preventing methylmercury formation by anaerobic bacteria.

Table 3 Approaches available for the development of inorganic phytoremediation

Approach	*Strategy to be improved*
1. Screening of natural variants that exhibit advantageous traits, such as tolerance towards high internal concentrations of metals, or ability to adjust soil acidity in order to control metal uptake	Phytoextraction
2. Engineering of root characteristics in order to improve root depth and density	Phytoextraction, rhizofiltration, phytostabilization
3. Breeding to increase aerial biomass	Phytoextraction
4. Manipulation of plant–microbe interactions	Phytostimulation
5. Genetic manipulation of inducible transporter proteins in order to increase the uptake and transport of specific metals to harvestable tissues or nonedible tissues of crop plants	Phytoextraction, phytovolatilization
6. Genetic manipulation of metal-binding proteins (metallothioneins and phytochelatins) in order to increase the accumulation of specific toxic metals	Phytoextraction

Remediation of Organic Pollutants

The organic pollutants that have been targeted for phytoremediation include petroleum hydrocarbons, nitro aromatic explosives, halogenated aliphatic and aromatic hydrocarbons, and halogenated phenols. **Table 4** lists some representative organic pollutants.

The ultimate goal of phytoremediation of organic xenobiotics is to degrade them chemically and possibly to metabolize them to carbon dioxide, nitrate, or ammonia. Degradation of contaminants can occur either in the plant body or in the soil; in the latter case, degradative enzymes may be derived both from the plant and from microorganisms present in the rhizosphere.

During evolution, plants have developed an arsenal of chemical weapons in order to defend themselves from attack by herbivores and from toxins released by viruses, bacteria, and fungi. A variety of compounds (for example, phenylpropanoids, terpenoids, and alkaloids) are synthesized by plants and used for defense purposes. The chemical structures and the mechanism of action of some of these natural products resemble those of major pollutants, hence it is not surprising that plants have enzymes capable of metabolizing these compounds. Some of the genes that code for degradative enzymes are expressed constitutively in plant tissues, but the expression of others is only induced by the presence of the pollutant. Examples of such enzymes are the nitroreductases (responsible for the degradation of nitro aromatics), dehalogenases (degradation of chlorinated solvents and pesticides), laccases (degradation of anilines), and peroxidases (polymerization of phenolics).

One example which proves that plant enzymes play a direct role in the degradative processes is the degradation of trichloroethylene (TCE) by axenically grown *Populus* sp. into trichloroethanol, chlorinated acetates, and carbon dioxide. The degradation of 2,4,6-trinitrotoluene (TNT), one of many explosives used in munitions, into triaminotoluene and subsequently to carbon dioxide and ammonium was demonstrated in axenically grown *Microphyllum aquaticum* plants and *Catharanthus roseus* hairy root cultures. Although this is an example of phytotransformation carried out *in vitro*, it demonstrates the potential contribution which these plants could give to the degradation of TNT in a contaminated site. **Table 5** gives other examples of plants which show the capability of biodegradation of organic pollutants.

Plants may also secrete degradative enzymes together with carbohydrates, amino acids, and other organic acids all of which serve to ease the penetration of roots into the soil and may provide nutrition for the indigenous microbes (bacteria, fungi, algae, and protozoa) present in the rhizosphere. Further, such microorganisms may solubilize compounds bound to the soil matrix and facilitate their uptake by plants. Of the bacteria, which are the most abundant of organisms found in surface soil, species of the genera *Arthrobacter*, *Pseudomonas*, and *Bacillus* are themselves able to degrade recalcitrant pollutants such as the polyaromatic hydrocarbons. Therefore, the association between plants and microbes is not only extremely beneficial for both partners but also environmentally important. The application of phytostimulation, which is based on such associations, is exemplified by the

Table 4 Some common organic pollutants: their sources and uses

Compound	*Source*	*Uses*
Hexane, cyclohexane, benzene, naphthalene	Petroleum	Gasoline constituents, percentage varies depending on the source of crude oil
Chloroform, trichloroethylene	Synthetically manufactured	Solvents, degreasing agents and in organic synthesis
Nonylphenol ethoxylates	Synthetically manufactured	Surfactants in cleaning products, paints, and emulsions
Phenol	Coal tar or synthetically manufactured	Disinfectant, manufacture of resins, dyes, and industrial chemicals
Pentachlorophenol (PCP)	Synthetically manufactured	Insecticide, defoliant, and wood preservative
Trinitrotoluene (TNT)	Synthetically manufactured	Organic explosive
Dioxins	Created during incineration processes, also contaminants associated with the manufacture of herbicides	
Atrazine	Synthetically manufactured	Selective herbicide
1,1,1-Trichloro-2,2-bis-(4-chlorophenyl)-ethane (DDT)	Synthetically manufactured	Contact insecticide

Table 5 Phytoremediation studies carried out with organic compounds

Compound tested	*Plant species used*	*Plant systems employed*
Pyrene	Rape (*Brassica napus*) Pepper(*Capsicum annuum*) Radish (*Raphanus sativus*)	Soil-grown plantlets
Atrazine	Poplars (*Populus* spp.)	Potted plants
Phenanthrene	Sunflower (*Helianthus annuus*) Pea (*Pisum sativum*)	Hydroponic cultures
Polychlorinated biphenyls	Black nightshade (*Solanum nigrum*)	Hairy root cultures
2,4,6-Trinitoluene (TNT)	*Datura innoxia*	Cell cultures
Phenol and chlorophenols	Carrot (*Daucus carota*) Duckweed (*Lemna gibba*)	Hairy root cultures Axenic cultures
Trichloroethylene	Poplars (*Populus* spp.)	Genetically modified cell cultures

biodegradation of fuel oil hydrocarbons in a soil microcosm. Here, the basic rate of degradation of contaminant by soil microbes was enhanced in the presence of rapidly growing corn (*Zea mays*; maize) plants: after about 4 months, when the corn plants had reached maturity, the rate of biodegradation returned to the original base level brought about by the microbe population.

Studies for improving the phytoremediation of organic materials are being carried out in the same manner as for inorganic pollutants. Some of the strategies which are being used for improving organic phytoremediation include: (1) study of biochemical and physiological processes related to adsorption, active uptake, translocation, accumulation, and inactivation of organic pollutants; (2) search and screening of plants which exhibit a high degree of tolerance towards organic pollutants; (3) engineering of root features such as root depth and root density; (4) creation of improved plants through classical breeding; (5) manipulation of plant–microbe interactions; and (6) genetic manipulation.

Main Advantages and Limitations of Phytoremediation

The major advantage of phytoremediation for decontamination purposes is its low cost (up to 40 times cheaper) compared to the alternative physical- and chemical-based technologies. The technology can be employed without disturbing the site or the neighborhood, and plants are aesthetically pleasing. In addition, for sites which are continuously being polluted, phytoremediation can be a permanent solution. In these aspects, the technology normally receives wide public acceptance.

On the other hand, many plants that would be suitable for remediation purposes are slow-growing and show seasonal dependence; thus plants possessing the correct characteristics in tropical and warm–temperate regions may not be appropriate for cool–temperate climates. The toxicity of the contaminant may also imply another disadvantage since plants, as all biological systems, have limited tolerance towards harsh conditions and their growth may be dramatically diminished by high levels of toxic materials. Furthermore, the technology is not 100% efficient (as is, for example, the total removal of surface soil) because biochemical reaction rates diminish as the concentration of contaminants decreases, thus such contaminants may not be completely removed.

Assessment of Phytoremediation

Although some phytoremediation approaches are already being commercialized (for example rhizofiltration), the technology is still in its infancy. Recent knowledge acquired in this area is based mainly on experiments performed in the laboratory through the use of plant tissue cultures and genetic modification techniques; these approaches are yet to be proven in the field. The full exploitation of phytoremediation depends on the continuity of intensive basic and applied research, together with the evidence derived from a string of irrefutable successes at the commercial level. Phytoremediation will be accepted as a remedial technology only when it proves to be effective and economical.

Furthermore, as is the case with other environmental technologies, phytoremediation activities are subject to governmental control and hence must obey the regulatory standards dictated by the relevant laws. The environmental consequences of the technology are not always totally clear. Whilst people prefer plants to heavy excavating machinery as a means of remediating a contaminated site, the possible risks to human and animal health must also be taken into serious account. Questions such as what happens to the contaminant within the plant after harvesting, what happens to birds or other animals that may feed on the (now contaminated) plants, whether pollen contains the contaminant, etc., must be addressed before implementing this innovative technology. In many cases, phytoremediation should not be considered as a unique solution but as part of an integrated strategy taken together with other conventional environmental methodologies as appropriate.

List of Technical Nomenclature

Hairy roots	Root tumors induced by infection of plant tissues with *Agrobacterium rhizogenes*.
Hydroponic cultures	Plants grown in liquid nutrient medium.
Hyperaccumulator	A plant that accumulates atypically high levels of a heavy metal.
Metallothioneins	Gene-encoded cysteine-rich proteins with high affinity for heavy metals.
Phytochelatins	Metal-binding peptides that are synthesized from the tripeptide glutathione by mechanisms not involving ribosomes.
Phytosiderophores	Nonprotein amino acids (mugineic acids) synthesized from methionine that efficiently chelate Fe(III).
Rhizosphere	Root zone.
Xenobiotic	Synthetic compounds not normally found in nature.

See also: **Nutrition**: Ion Transport; Mineral Uptake. **Plants and the Environment**: Land Reclamation and Remediation, Principles and Practice; Plants as Pollution Monitors; Waste Water Treatment.

Further Reading

Alleman BC and Leeson A (1999) *Phytoremediation and Innovative Strategies for Specialized Remedial Applications*. Columbus: Battelle Press.

Brooks RR (1998) *Plants that Hyperaccumulate Heavy Metals*. New York: CAB International.

Cunningham SD, Anderson TA, Schwab AP, and Hsu FC (1996) Phytoremediation of soils contaminated with organic pollutants. *Advances in Agronomy* 56: 55–114.

Kruger EL, Anderson TA, and Coats JR (1997) *Phytoremediation of Soil and Water Contaminants*. Washington, DC: American Chemical Society.

Meagher RB (2000) Phytoremediation of toxic elemental and organic pollutants. *Current Opinion in Plant Biology* 3: 153–162.

Raskin A, Raskin I, and Ensley BD (1999) *Phytoremediation of Toxic Metals: Using Plants to Clean Up the Environment*. New York: John Wiley.

Terry N and Banuelos G (1999) *Phytoremediation of Contaminated Soil and Water*. New York: Lewis.

Wickramanayake GB, Magar VS, Gavaskar AR, and Alleman BC (2000) *Bioremediation and Phytoremediation of Chlorinated and Recalcitrant Compounds*. Columbus: Battelle Press.

Global Warming Effects

J M G Thomas and P V V Prasad, University of Florida, Gainesville, FL, USA

Definition and Cause

"Global warming" refers to increases in the surface temperature of the earth's atmosphere, ocean, and landmass. Primary causes of global warming are the increases in concentrations of various atmospheric "greenhouse gases" which include water vapor, carbon dioxide, methane, nitrous oxide, and the chlorofluorocarbons (CFCs). Increasing concentrations of greenhouse gases will allow the atmosphere to trap higher than usual amounts of outgoing long wavelength (thermal) radiation, resulting in higher surface temperatures.

The earth receives energy from the sun as short wavelength radiation, about one-half of which is in the visible wavelengths, 400–700 nm. After reaching the earth's surface, solar irradiance is mostly absorbed and converted into heat. As the earth's surface warms up, it emits increasing amounts of long wavelength infrared radiation. The atmosphere absorbs a portion of this long wavelength radiation and reradiates it back to earth surface. The phenomenon is often called the "greenhouse effect" because the same effect occurs when solar radiation penetrates an enclosed greenhouse and the upwelling thermal radiation is contained, raising the greenhouse temperature. The higher the concentration of greenhouse gases in the earth's atmosphere, the more radiation is trapped, and the warmer the temperature.

The greenhouse effect is a natural phenomenon in the atmosphere and is essential to maintain the earth's temperature. In fact, without the greenhouse effect, the earth would be about 33°C colder than it is today, and would be uninhabitable for humans and most other life forms. However, increased concentrations of greenhouse gases will result in a continuous increase in earth's temperatures and this increase could significantly impact life on earth.

Most greenhouse gases occur naturally in the atmosphere. The most abundant greenhouse gas is water vapor, which reaches the atmosphere through evaporation from water surfaces and moist soil surfaces and through transpiration from plants. Since atmospheric water vapor concentration is extremely variable, the greenhouse effect attributable to water vapor is greatest in humid, hot zones, and least in dry, cold areas. The second most plentiful greenhouse gas is CO_2, the most important gas in the earth's biogeochemical cycles. Certain human activities such as burning fossil fuels and deforestation increase the concentration of CO_2 in the atmosphere. For this reason, increased emissions of greenhouse gases have been recorded over the last 50 years. For example, the CO_2 in pre-industrial times (before 1700) was approximately 280 parts per million, mole fraction basis (ppm or $\mu mol\ CO_2\ mol^{-1}$ air), while the current (2002) level of CO_2 is close to 370 ppm (or $\mu mol\ mol^{-1}$). At this point, emissions are expected to continue, and if the current rate of increase continues, the atmospheric CO_2 will double sometime during this century.

Atmospheric CO_2 would be the primary contributor to a temperature increase of anywhere from 1.6 to 5.8°C above current temperatures, depending on the actual rate of rise in the concentration of CO_2. Other greenhouse gases such as methane, nitrous oxide, and CFCs have also increased due to human activities and also contribute to global warming. Over the last 140 years, global average surface temperature has increased by 0.6 ± 0.2°C. The warmest decade in the instrumental record was the 1990s, with 1998 being the warmest year since 1861. Changes in precipitation are likely to accompany increases in temperature and drier inland areas could receive less water while coastal areas receive more.

Global Climate Models

Climate change is a result of interactions between changing global systems of air, earth, water, biosphere, and ever-incoming solar radiation. Over the last five decades, progress has been made in understanding the processes leading to past and present climate change and numerical models have been developed to simulate future changes in climate. The models work on the general principles of both fluid dynamics and thermodynamics and are commonly referred to as atmospheric/oceanic general circulation models (GCMs). These models have been developed to allow scientists to understand atmospheric processes and to predict the outcome of certain changes that could occur in the future. Inputs to the models can be used to describe scenarios that incorporate different variables such as concentration of greenhouse gases and amount of precipitation. Some of the models predict an increase of 1.6 up to 5.8°C in surface air temperature with selected rational scenarios of increases in greenhouse gases, especially CO_2.

Effects of Global Warming on Plants

Every plant has an optimum temperature and season for growth, development, and reproduction. Increased temperature will affect ecosystems and crops differently according to region. With warming, slight shifts in regional production of food crops will occur. For example, some plants require a chilling period to break dormancy and/or to produce flavorful fruit. Negative impacts are likely in areas where high temperature or less than adequate rain already limit productivity. Beneficial results are expected in areas where cold temperatures currently limit crop yield. Unfortunately, high temperature stress is among the least understood of all plant processes.

This discussion will focus on two aspects of global warming: (1) the effects of the projected temperature rise and (2) the effects of increased CO_2 on fundamental plant processes and on plant growth, development, and yield. Most of the examples used will be crop plants, because more research has been conducted on these plants essential to our food supply.

Physiological Processes

Photosynthesis Photosynthesis is the process by which plants convert atmospheric CO_2 into carbohydrates using the photochemical energy of light. Thus, the primary physiological process by which plants will respond to increasing CO_2 is photosynthesis. Carbon dioxide diffuses through small openings (stomata) on leaf surfaces to the interior of leaves, and then through the cells to the chloroplasts where photosynthesis occurs. Two biochemical pathways have been described and termed C_3 and C_4 based on the first carbon (C) products (either a 3-C or a 4-C) produced after the enzymatic fixation of CO_2 after it enters the leaf cells. A third process, called crassulacean acid metabolism (CAM), is found in many desert and epiphytic plants that fix CO_2 into a 4-C acid at night to avoid water loss during the day.

Most crop plants in temperate and tropical regions have the C_3 pathway and, depending on the plant species, an increase in daytime temperature, up to a maximum between 28 and 37°C, will increase the rate of photosynthesis. Photosynthetic rate decreases for many C_3 crop plants, like soybean (*Glycine max*) and rice (*Oryza sativa*), when air temperature is above 30°C, although photosynthesis in many cultivars of wheat (*Triticum aestivum*) decreases above 25°C. In contrast, most C_4 plants are adapted to hot dry climates and have a temperature optimum about 10°C higher than C_3 plants. Several important crop species, for example corn (*Zea mays*; maize), sugar cane (*Saccharum officinarum*), and sorghum (*Sorghum* spp.), are C_4 plants. However, many of the most troublesome weed species are also C_4 plants and for this reason, weed problems may become more severe due to global warming.

Elevated CO_2 is expected to enhance leaf level photosynthetic rates by 30–40% in plants with C_3 metabolism, like soybean, rice, and wheat. Much less effect on leaf photosynthesis will be observed in plants with C_4 metabolism because the C_4 enzyme system increases the concentration of CO_2 at the site of photosynthesis. Plants that utilize C_4 metabolism will be less responsive to increases in CO_2, because the enzyme system is nearly saturated at present-day atmospheric CO_2.

The photosynthetic process is directly affected by the interaction of elevated temperature and CO_2 and the biochemical aspects of photosynthesis have been utilized to create mechanistic plant growth models. The model equations have been repeatedly validated with field observations and then used to predict consequences of climate change on plants. Thus, the optimum temperature for C_3 photosynthesis is projected to increase with elevated CO_2 as shown in **Figure 1**. Notice how the assimilation rate will decrease once above the optimum temperature (T_{opt}).

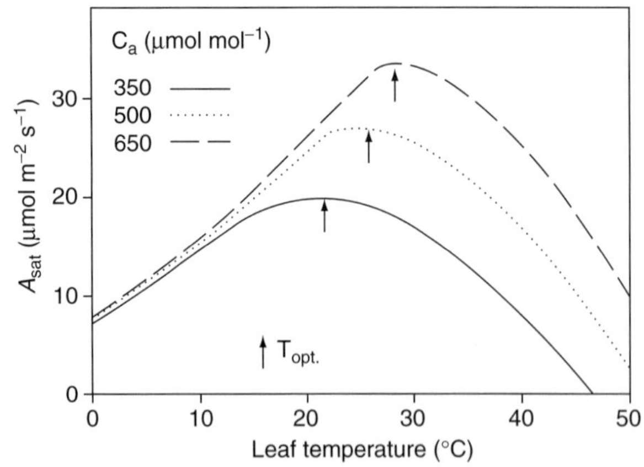

Figure 1 Predicted light-saturated rates of leaf CO_2 uptake (A_{sat}) with leaf temperature for three atmospheric CO_2 concentrations (C_a; μmol mol^{-1} of CO_2 in air). Values of A_{sat} were predicted following the procedure of Long (1991) using the improved temperature responses of the kinetic parameters described by Bernacchi CJ, Singsaas EL, Pimental C, Portis AR, and Long SP (2001) Improved temperature response functions for models of rubisco-limited photosynthesis. *Plant Cell and Environment* 24: 253–259 and Long SP (1991) Modifications of the response of photosynthetic productivity to rising temperature by atmospheric CO_2 concentration: has its importance been under-estimated? *Plant Cell and Environment* 14: 729–739; reprinted with permission.

Respiration Respiration is the process of breaking down complex carbohydrates all the way to CO_2 and water and in the process, generating chemical energy to drive metabolic processes. There are two types of respiration in plants: dark or mitochondrial respiration and light or photorespiration. Dark respiration is associated with biosynthesis of new tissue and maintenance of old tissue. As temperature increases, the rate of dark respiration increases, thereby using more carbohydrates for growth and maintenance. Elevated CO_2 has no effect on mitochondrial respiration.

Photorespiration is directly associated with the principle photosynthetic enzyme, rubisco (ribulose bis-phosphate carboxylase oxygenase). Rubisco functions in either of two ways: either it will fix CO_2 in photosynthesis (carboxylase activity) or it will fix oxygen (oxygenase activity), which leads to photorespiration. Elevated temperature will increase the extent of oxygenase activity because the solubility of O_2 increases in relation to CO_2 as temperature increases, as does the affinity of rubisco for O_2 relative to CO_2. Plants with C_3 metabolism produce glycolate as a product of oxygenation, providing a substrate for photorespiration. Elevated atmospheric CO_2 will decrease photorespiration at high temperatures by increasing the amount of substrate for carboxylation. This effect is illustrated in **Figure 1**. Mostly tropical C_4 plants have an additional enzyme (PEP carboxylase) which concentrates CO_2 at the site of photosynthesis, and avoid photorespiration because the concentrated CO_2 effectively competes against O_2 for rubisco-binding sites.

Stomatal conductance Stomata are tiny openings or valves distributed on leaf surfaces that allow for gas exchange. Many studies have shown a direct effect of CO_2 and temperature on stomatal conductance. As temperature increases, stomatal apertures open to allow evaporative cooling through the release of water vapor. When stomatal pores open, the supply of CO_2 available for the photosynthetic enzymes is increased.

In contrast, stomatal density and conductance decrease when plants are grown with elevated CO_2. For example, stomatal conductance decreases about 30–40% in many plant species when grown under a double concentration of CO_2. Since stomata provide the entry for CO_2 molecules into leaves, plants have developed the ability to sense higher CO_2 concentration and regulate the amount of CO_2 by closing the stomata. The degree of stomatal opening is decreased at elevated CO_2, thereby decreasing water loss through transpiration and also leading to a small increase in leaf temperature. Thus, just as elevated atmospheric CO_2 leads to increased atmospheric temperatures, the temperature of individual plant leaves and cells will increase. Since a steady increase in atmospheric CO_2 has been measured, the direct increase in leaf temperature due to increased CO_2 and the effect of increased temperature on physiological processes should be recognized as an important concern. Since leaf cooling through evapotranspiration is another function of these openings, decreased stomatal conductance due to elevated CO_2 may contribute to increased leaf temperature (**Figure 2**).

Transpiration, total water use, and water use efficiency The uptake of water and mineral nutrients by the roots is accomplished partially through the release of water vapor through leaf stomata, and the process is termed transpiration. Increased ambient temperature would increase transpiration because evaporation from the leaf increases with increased temperature; thus, higher temperature would lead to increased water use by increasing evapotranspiration. In contrast, water use efficiency (WUE), which is defined as the ratio of C assimilation to transpiration, will decrease at elevated temperatures due to increased transpiration. Decreased WUE will be a problem in areas that are already limited in water resources and will increase the amount of water required for irrigation. Crops grown under elevated CO_2 will increase WUE by 30–40%, because of an increase in C assimilation and a decrease in transpiration.

Several of the climate change models predict an overall decrease in precipitation in some areas. Water availability is an extremely important factor, and in crop models, the lowest yields are associated with the climate change scenarios that have the lowest rainfall and the highest temperature.

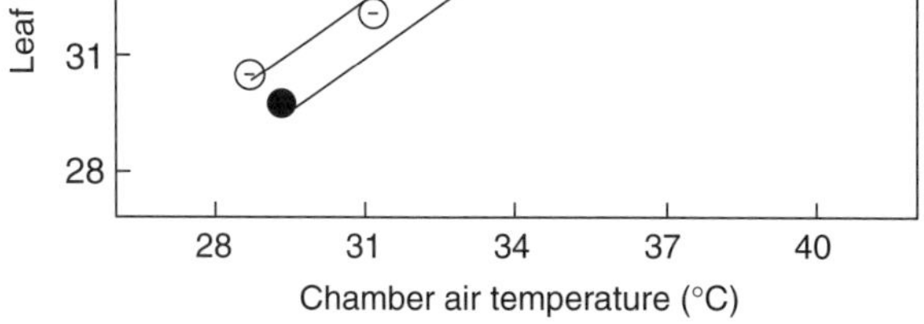

Figure 2 Relation between chamber air temperature and leaf temperature of kidney bean plants grown at different temperatures in either ambient (350 μmol mol^{-1}) or elevated (700 μmol mol^{-1}) CO_2 concentration. Reprinted with permission from Prasad PVV, Boote KJ, Allen LH Jr, and Thomas JMG (2002) Effect of elevated temperature and carbon dioxide on seed-set and yield of kidney bean (*Phaseolus vulgaris* L.). *Global Change Biology* 8: 710–721.

Flower initiation The switch from vegetative growth to flowering is one of the more complex processes in plant development. There is tremendous diversity among plant species in the initiation of flowering. On some annual plants, flowers develop after a certain number of leaves are produced. On some plants, flowering is triggered by precipitation, and in the tropics, flowering will often occur at the end of the dry season. Some plants will only flower after a period of low temperatures (vernalization) and with global warming, production of this type of crop may need to move further north (or south, in the southern hemisphere). A combination of shortened daylength (photoperiod) and cooler temperatures is necessary to induce flowering in many autumn season crops. High temperatures will often result in early flowering, and elevated CO_2 may also lead to early flowering.

Pollen production and viability Sexual reproduction of plants occurs within the floral organs. In a flower, male (pollen) and female (ovule) gametes function to recombine genetic information and produce seed for the next generation of plants. Temperatures above the optimum for each crop can have negative effects on plant reproductive processes. Pollen production and pollen viability are very sensitive to slight increases in temperature above optimum. Daytime temperature >34°C significantly reduces pollen production and viability of peanut (*Arachis hypogaea*), a warm-season crop. Cool-season plants, like wheat or some cultivars of dry bean (*Phaseolus* spp.), require cooler temperatures than peanut.

In rice, the 34°C critical temperature for spikelet sterility (as determined from the number of germinated pollen grains on stigma) was reduced by 1°C at elevated CO_2 (**Figure 3**). Similar observations on dry bean suggest that elevated CO_2 increases pollen susceptibility to high temperature by 1 to 2°C. Studies have shown that there is no benefit of elevated CO_2 as temperatures increase. For example, elevated temperature decreases pollen production and viability in pollen of dry bean at both ambient and elevated CO_2 (**Figure 4**).

Seed set Seed set is the process during which the fertilized ovule develops into a seed. Temperatures above the optimum decrease seed set in most seed-producing crops including rice, wheat, corn, peanut, cowpea (*Vigna unguiculata*), and dry bean. Since

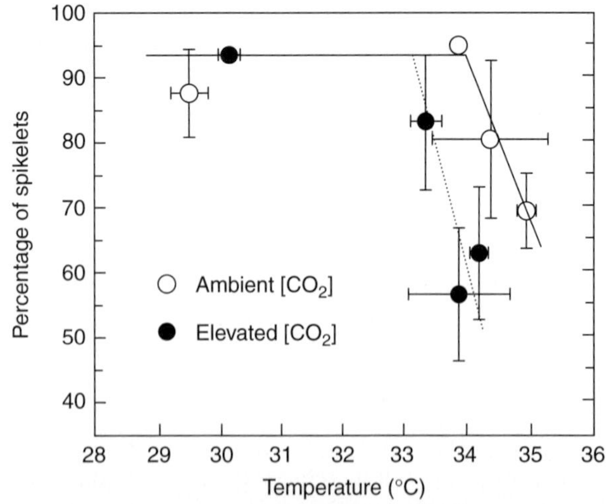

Figure 3 The correlation between the maximum temperature during heading and the percentage of rice spikelets having 10 or more germinating pollen grains. Reprinted with permission from Matsui T, Omasa T, and Horie T (1997) Effects of high temperature and CO_2 concentration on spikelet sterility in *indica* rice. *Field Crops Research* 51: 213–219.

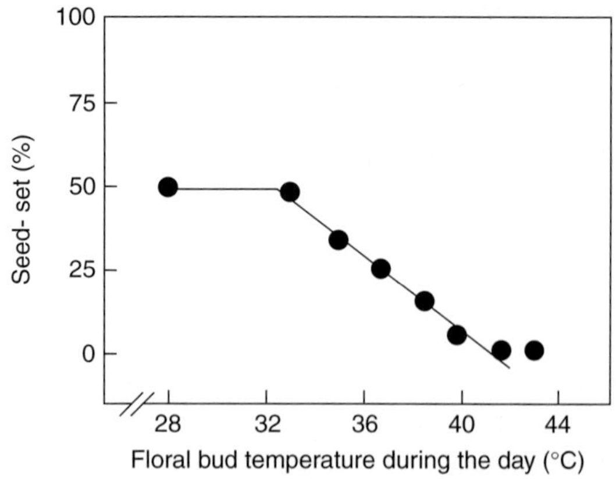

Figure 5 Relation between floral bud temperature and seed set when peanut flower buds were exposed to eight daytime floral bud temperatures from 28 to 44°C for 1 day during anthesis. Reprinted with permission from Prasad PVV, Craufurd PQ, Kakani VG, Wheeler TR, and Boote KJ (2001) Influence of high temperature during pre- and post-anthesis stages of floral development on fruit-set and pollen germination in peanut. *Australian Journal of Plant Physiology* 28: 233–240.

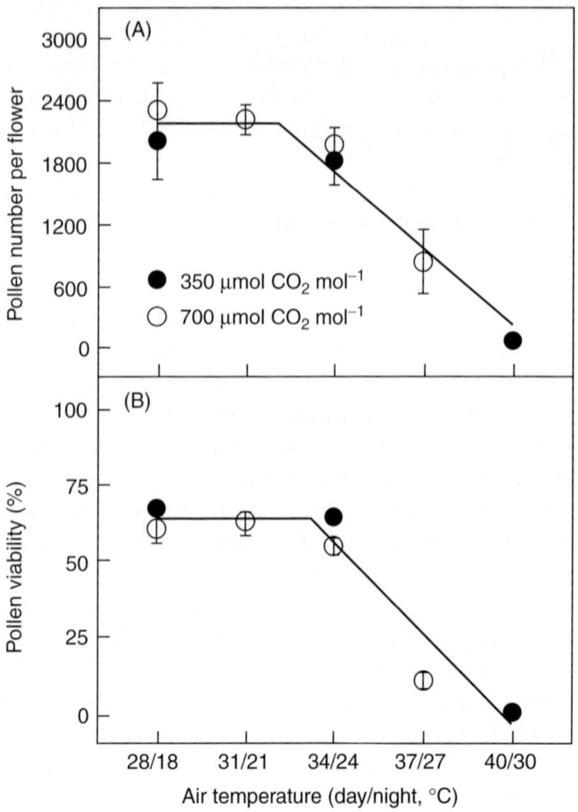

Figure 4 Relation of (A) pollen number per flower and (B) pollen viability to temperature of kidney bean grown at two concentrations of CO_2 and five daytime maximum/nighttime minimum temperature combinations. Reprinted with permission from Prasad PVV, Boote KJ, Allen LH Jr, and Thomas JMG (2002) Effect of elevated temperature and carbon dioxide on seed set and yield of kidney bean (*Phaseolus vulgaris* L.). *Global Change Biology* 8: 710–721.

seed provide approximately 90% of the world population's food supply, the importance of this process cannot be overemphasized. Exposure to daytime temperature >34°C, which is common on some hot summer days in temperate regions and during the growing season in tropical regions, is known to reduce seed set in peanut (**Figure** 5). A temperature greater than 35°C for as short a period as 1 h at flowering results in lower seed set in rice. Crops such as dry bean are even more sensitive.

The main reasons for decreased seed set (or fruit set) is decreased pollen production, lower pollen viability, reduced pollen germination, and less pollen tube growth. In general, male gametes (pollen) are more sensitive to elevated temperature than the female reproductive organ, the ovule. However, some studies have shown that ovule receptivity can be very sensitive to high temperature, which would also result in lower seed set.

Growth and Development

Plant development involves a progression of events in the plant life cycle, such as seed germination, emergence, vegetative growth, reproduction (flowering and pollination), and seed production. These processes are directly affected by environmental conditions like temperature and photoperiod. Temperature governs the rate of plant development; therefore, the thermal component (growing degree days) is extensively used for estimating growth rate and stage of development. Optimum temperatures for each crop's growth and development have been determined by breeders and agronomists in efforts to achieve the best crop yields under specified environmental conditions during the growing season.

Seed germination and emergence The optimum temperature for seed germination differs among plants. A typical temperature range for optimum germination of cool-season vegetable seed is from 3–17°C and for warm-season crops, 15–25°C. Temperatures above or below the optimum temperature inhibit the rate of seed germination.

In regions where cool temperatures limit crop seed germination; a slight increase in temperature would be beneficial for seedling establishment. However, warmer temperatures would also benefit weed seed germination. Therefore, as the climate warms, some of the more aggressive weeds may become a problem in northern areas, since cool temperatures now limit them. In regions where high temperature and lack of soil moisture currently limit seedling establishment, these effects will be even more severe.

Doubled concentration of CO_2 has no effect on seed germination. However, once the seedling has emerged from the soil and has grown roots, leaves will utilize light. Increased CO_2 will then be beneficial to seedling establishment by enhancing the rate of leaf photosynthesis.

Vegetative growth and development Vegetative development includes the initiation and expansion of roots, leaves, shoots, branches, and/or tillers, all processes that are strongly driven by temperature. In most of the determinate annual crop species, increases in temperature will not only hasten the initiation of organ growth, but will also shorten the duration of growth. This results in less time for photosynthetic carbon assimilation and biomass accumulation before the start of reproductive growth. Increasing temperature, up to about 25°C for wheat and 35°C for soybean, generally increases vegetative growth rates. Above a certain threshold temperature in each crop, the growth rate will decline, but within the range predicted for global warming (1.6–5.8°C above current temperatures), there should be profuse vegetative plant growth.

The effects of elevated CO_2 on crop development are smaller compared to those of temperature. However, elevated CO_2 enhances the total plant leaf area. In monocotyledons, increased plant leaf area is mainly because of increased tillering and not by increasing final individual leaf size. In dicotyledons, elevated CO_2 increases branching, leaf initiation rate, leaf expansion rate, and final leaf size. The greater leaf area or increase in leaf number at elevated CO_2 is due to increased photosynthesis. In addition, elevated CO_2 often increases leaf thickness by production of an extra layer of cells in dicotyledons. With elevated CO_2, increased structural cellulose, and increased starch and carbohydrates often occur. The combination of elevated temperature and CO_2 will increase the capacity for vegetative growth in most plants.

Reproductive development Reproductive growth is often depressed by the same conditions that enhance vegetative growth. Elevated temperature may increase the rate of reproductive development, depending on the plant; however, with slight increases in temperature, reproductive events are likely to fail. Flowers may abort before producing seed due to the sensitivity of pollen and fertilization to increased temperature. In addition, the effective filling period is often shortened at higher temperatures, leading to fewer and smaller seeds.

As most of the ontogenetic development and rate of initiation do not depend on carbohydrate supply, elevated CO_2 has little effect on timing and rates of reproductive development. Elevated CO_2 typically results in a greater number of seed produced, but has no effect on seed size.

Yield and Yield Components

Seed yield Seed yields of crop plants that are currently grown at optimum temperature may be decreased by further increases in temperatures. For example, above a mean temperature of 26°C, grain yields of rice declined by 10% per degree rise in temperature. Likewise, studies have shown that wheat yields decrease by 6–10% per every 1°C rise above the optimum temperature for each cultivar. Even for heat-tolerant crops such as sorghum, the higher temperature tolerance has been utilized and so it is being grown in warm areas already. Therefore any further increase in temperature is likely to reduce sorghum yield.

In contrast, elevated CO_2 can improve seed yields up to 30% because of increases in photosynthesis and growth. However, potential benefits of increased CO_2 on grain yield of wheat were offset by a temperature increase of only 1.0 to 1.8°C. A reduction in rice yield occurred because the percentage of spikelet fertility was lower at elevated CO_2 compared to ambient CO_2. The decrease in spikelet fertility is partially attributed to increased canopy temperatures at elevated CO_2. Similar observations were made in kidney bean (*Phaseolus vulgaris*) where the ceiling temperature for seed-set at elevated temperature is 1–2°C lower at elevated CO_2 compared to ambient CO_2.

Number of seeds and seed size Increased temperature may result in a lower number of seeds per plant. In contrast, elevated CO_2 increases the number of seeds due to increased tillering, branching, and growth. Increases in temperature generally decrease the seed size of many seed legume crops, such as soybean and dry bean. Plants growing at elevated temperatures often produce smaller seeds than those plants grown at moderate or cool temperatures. The decrease in seed size can be associated with lower seed quality in terms of percent emergence and uniformity of stand establishment. There is very little effect of elevated CO_2 on individual seed size.

Grain or seed filling period An increase in temperature will shorten the overall life cycle and grain filling period of many seed producing crops, leading to smaller seed. Studies have shown that an increase of 4–5°C above optimum temperature will decrease the duration of the grain filling period in wheat and corn resulting in lower yields, particularly in tropical regions. The duration of grain growth in wheat decreases by approximately 30 days when the temperature is increased from 15/10 to 30/25°C. Similar decreases in seed or grain filling at high temperatures were observed for other grain crops, including monocotyledons (for example rice and corn) and dicotyledons (for example soybean and kidney bean). In contrast, there is very little to no effect of elevated CO_2 on grain or seed filling period.

Harvest index Harvest index is a measure of the efficiency of plants in producing seed. It is defined as the ratio of grain yield to total aboveground biomass. Since elevated temperature decreases grain yield while either increasing or not affecting vegetative growth, an increase in temperature will result in a lower harvest index. Therefore, above a certain optimum temperature for each crop, increased temperature will decrease harvest index almost in direct proportion to the decrease in seed yield. For example in peanut, the rate of change of harvest index decreases at temperatures above a mean temperature of 26°C (**Figure 6**).

Effects of elevated CO_2 on harvest index may vary depending on the crop and cultivar. In most grain legumes, harvest index is decreased at elevated CO_2 because the increase in vegetative growth is greater than that of seed yields. In rice, there was no effect or increase in harvest index at elevated CO_2.

Seed composition Surprisingly few studies have addressed the effect of increased temperature and elevated CO_2 on seed composition. Warmer temperatures generally decrease mass per grain and starch concentration, while elevated CO_2 may increase the concentration of carbohydrates. Total oil concentration decreases in soybean when grown above optimum temperatures. Increased temperature affects oil composition because the composition of the oil itself can transmit temperature buffering capacity to plant membranes. A side benefit of this is that the number of double bonds in vegetable oil will naturally decrease to include a higher percentage of monounsaturated fatty acids like oleic acid, a more desirable effect for human nutrition.

Wheat composition was found to be much lower in protein concentration when grown in elevated CO_2 than that grown in ambient CO_2. Elevated CO_2 decreases the protein concentration because of the increased supply of carbohydrates from photosynthesis. This decrease in seed protein would affect the bread-making quality of the wheat grain. Protein concentration seems to be unaffected in grain legumes, probably because these plants have symbiotic nitrogen fixing bacteria growing in their roots that can supply protein in direct proportion to the needs of the plant.

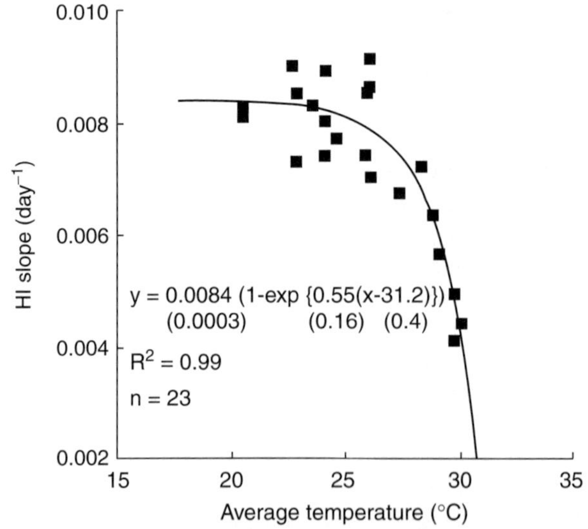

Figure 6 Slope of harvest index (HI) increase versus average daily temperature from sowing to end of leaf growth for numerous experiments on Early Bunch peanut. Reprinted with permission from Hammer G, Sinclair TR, Boote KJ, Wright GC, Meinke H, and Bell MJ (1995) A peanut simulation model. I. Model development and testing. *Agronomy Journal* 87: 1085–1093.

Strategies for Coping with Global Warming

Stress Avoidance

The most likely adoption strategy for coping with global warming would be to change suitable sowing dates to utilize a cooler season and to avoid high temperatures during the sensitive stages (flowering and grain filling) of crop growth. There is some evidence that certain rice cultivars avoid high temperature stress on reproduction by tending to flower and pollinate during earlier, cooler parts of the day. Similarly, use of soil mulches to reduce soil temperatures during seedling emergence for many summer crops may be beneficial. Management practices that alter crop microclimate, for example irrigation, would cool canopy air temperature and plant organ temperatures. Unfortunately, using irrigation to cope with global warming will increase the demand for crop irrigation water in regions where water supply is already limited.

Stress Tolerance

Selection of crops or crop genotypes that are tolerant to high temperature stress would be the best and easiest strategy for farmers to adopt. Very little work has been done to identify heat tolerance of cultivars of major food crops. However, crop species and genotypes within crops are known to vary in their responses to high temperature. For example, corn that originated in warm tropical regions is more heat tolerant than corn that originated in cooler highlands. Certain peanut cultivars are known to be able to set seed at elevated temperatures. Most of these lines were developed in sub-Saharan Africa and tropical Asia. Similar varietal differences are noticed in wheat, rice, and corn, and these differences appear to be related to place of genetic selection. For example, tropical lines of wheat are better adapted to high temperatures than European lines. Therefore, it is crucial to screen the available germplasm of crop species to identify heat tolerant cultivars and to use them in future breeding programs. Heat tolerant cultivars of major food crops will be essential to produce the necessary yields required by the world's population in the future.

Conclusions

Global warming will have a significant influence on the development, growth, and yield of all plants and particularly of seed-producing food crops. However, the effects of global warming will vary depending upon climatic conditions at various locations on earth. The effects of global warming on plants will be more detrimental in tropical regions where current temperatures are close to optimum, and any further increase in temperature will decrease yields. Regions where cold temperatures are currently limiting productivity may benefit by a slight increase in temperature. Certain regions of crop production will shift, which will change the importance of certain crop types.

Reproductive growth and development of seed producing crops is more sensitive to high temperature than is photosynthesis and vegetative growth. Global warming would not only decrease grain yields but may also decrease seed quality. Furthermore, this decrease in yield will occur despite the potential beneficial effects of elevated temperature and CO_2 on photosynthesis and vegetative growth. Future research needs to be focused on choosing the best crop management practices as far as planting date and irrigation. Cultivars with heat tolerance should be determined and used in plant breeding programs so that the negative effects of increased temperature on crop yields will be ameliorated under future climate change scenarios.

List of Technical Nomenclature

Atmosphere The gaseous envelope surrounding the earth.

Carbon dioxide (CO_2) A naturally occurring gas, also a by-product of burning fossil fuels and biomass, as well as land-use changes and other industrial processes; the principal anthropogenic greenhouse gas that affects the earth's radiative balance.

Global warming Increase in the surface temperature of the earth's atmosphere, ocean, and landmass.

Greenhouse effect Greenhouse gases effectively absorb infrared radiation emitted by the earth's surface, by the atmosphere itself, and by clouds; greenhouse gases trap heat within the surface–troposphere system.

Greenhouse gases Gaseous constituents of the atmosphere, both natural and anthropogenic, that absorb and emit radiation at specific wavelengths; natural greenhouse gases include: water vapor (H_2O), carbon dioxide (CO_2), nitrous oxide (N_2O), methane (CH_4), and ozone (O_3); human-made greenhouse gases include halocarbons and other chlorine and bromine containing substances, sulfur hexafluoride (SF_6), hydrofluorocarbons (HFCs), and perfluorocarbons (PFCs).

Harvest index The ratio of grain yield to total above-ground biomass.

Photosynthesis The process by which plants assimilate carbon from CO_2 to build carbohydrates, releasing oxygen in the process.

Respiration The process by which living organisms convert organic matter to CO_2, releasing energy and consuming oxygen.

Rubisco A short name for the principal photosynthetic enzyme of green plants: ribulose bisphosphate carboxylase-oxygenase.

Transpiration The release of water vapor through leaf stomata to accomplish uptake of water and mineral nutrients by plant roots.

See also: **Growth and Development**: Field Crops. **Photosynthesis and Partitioning**: Sources and Sinks; C_3 Plants; C_4 Plants; CAM Plants. **Seed Development**: Seed Quality. **Water Relations of Plants**: Stomata. **Weeds**: Weed Competition.

Further Reading

Allen LH Jr, Kirkham MB, Olszyk DM, and Whitman CE (eds) (1997) *Advances in Carbon Dioxide Effects Research*. ASA Special Publication No. 61. Madison: American Society of Agronomy.

Baker JT and Allen LH Jr. (1993) Contrasting crop species responses to CO_2 and temperature: rice, soybean and citrus. *Vegetatio* 104/105: 239–260.

Boote KJ, Bennett JM, Sinclair TR, and Paulsen GM (eds) (1994) *Physiology and Determination of Crop Yield*. Madison: American Society of Agronomy.

Bowes G (1993) Facing the inevitable: plants and increasing atmospheric CO_2. *Annual Review of Plant Physiology and Plant Molecular Biology* 44: 309–332.

Drake BG, Gonzalez-Meler MA, and Long SP (1997) More efficient plants: a consequence of rising atmospheric CO_2? *Annual Review of Plant Physiology and Plant Molecular Biology* 48: 609–639.

Intergovernmental Panel on Climate Change (IPCC) (2001) *Climate Change 2001: The Scientific Basis. Contribution of Working Group I to the Third Assessment Report of the Intergovernmental Panel on Climate Change*. Cambridge: Cambridge University Press.

Koch GW and Mooney HA (eds) (1996) *Carbon Dioxide and Terrestrial Ecosystems*. San Diego: Academic Press.

Reddy KR and Hodges HF (2000) *Climate Change and Global Crop Productivity*. Wallingford: CAB International.

Rosenzweig C and Hillel D (1998) *Climate Change and the Global Harvest*. New York: Oxford University Press.

POSTHARVEST PHYSIOLOGY

Contents

Ripening

S C Fry, The University of Edinburgh, Edinburgh, UK

Introduction

General Background

This article discusses various fruits (berries, drupes, and "false fruits") that soften as they ripen. Nonsoftening fruits (legumes, nuts, capsules, and others) are excluded.

A "true" fruit is what develops from the ovarywall and is therefore genetically maternal tissue. Usually it contains one or more seeds, developed from the fertilized ovule(s), although seedless fruits can arise parthenocarpically. In "false fruits" (e.g., strawberry (*Fragaria x ananassa*), apple (*Malus pumila*), pear (*Pyrus communis*), rosehip (*Rosa* spp.)), the "fruit" tissue is derived from the receptacle not the ovary wall.

Typically, a fruit grows to a maximum size, then ripens. Growth is not discussed here. "Ripening" cannot be precisely defined, being related to the (subjective) evaluation of when to harvest the crop ("ripe" and "reap" are etymologically related). Ripening is the sum of all the cytological and chemical changes that occur after a fruit has reached its full size and before it starts to deteriorate. "Deterioration" is also somewhat arbitrary – some individuals prefer bananas (*Musa paradisiaca*) black-flecked, others pale yellow, or (for cooking) green. "Overripening" mainly concerns excessive softening and is discussed briefly. Deterioration due to microbial infection is not discussed.

There is considerable commercial interest in manipulating ripening postharvest. For the mass market, fruit are best stored and transported before (full) softening. Knowledge of the processes that precede softening would facilitate the judicious timing of harvest in such cases. Postharvest, it would be valuable to interrupt ripening, resume it after transport, and stop it again before deterioration.

Some workers regard ripening as senescence; there are indeed similarities between the physiological changes (e.g., chlorophyll loss and membrane permeabilization) occurring in a ripening fruit and in a leaf or flower that is about to wither. However, fruit ripening involves additional processes, e.g., softening, sweetening, and the synthesis of carotenoids, flavors, and aromas that are not normally features of senescence in nonfruit tissues.

Climacteric and Nonclimacteric Fruits

Many, but not all, commercial fruits (**Table 1**) exhibit the "climacteric"–a sudden increase in ethylene and CO_2 output and O_2 consumption. This is a "point of no return" and many fruits have short shelf-lives after the climacteric. The classification of fruits as climacteric or nonclimacteric is not "all-or-none;" for instance, pineapples (*Ananas comosus*) show a small climacteric under certain conditions.

Many nonclimacteric fruits, e.g., *Citrus*, ripen slowly and have long shelf-lives. In the grape (*Vitis vinifera*), some features of ripening (sugar and acid metabolism) are slow and protracted; others (anthocyanin formation) are extremely rapid. Still other nonclimacteric species, e.g., strawberry, ripen precipitously and have very short shelf-lives.

Table 1 Some climacteric and nonclimacteric fruits

Climacteric species[a]	*Nonclimacteric species*
Apple	Bell pepper
Apricot	Blackcurrant
Avocado	Cherry
Banana	Cucumber
Cantaloupe melon	Grape
Kiwi fruit?	Grapefruit
Mango	Lemon
Olive	Melon
Papaya	Orange
Passion fruit	Pineapple?
Pawpaw	Strawberry
Peach	
Pear	
Persimmon	
Plum	
Raspberry?	
Tomato	

[a] In some cases, the climacteric response is slight or is only observed under certain circumstances, hence the question-marks.

Triggering and Delaying Ripening

Initial Endogenous Triggers

Fleshy fruits attract and are eaten by animals, promoting seed dispersal (but unfortunately also increasing the susceptibility to microbial decay). The development of fruits' attractive features usually starts when the seeds reach maturity. Logically, a reliable cue for ripening could be some factor emanating from (or failing to be absorbed by) mature seeds. The first signs of ripening are often in the locule (near the seeds), consistent with a seed-derived ripening trigger. However, any such trigger can obviously be bypassed since seedless varieties ripen.

Ethylene is an early trigger for ripening (see below), but may not be the primary one. The hypothetical initial ripening signals could include abscisic acid or oligosaccharins. Various oligosaccharins hasten ripening, including glycoprotein-derived oligosaccharides (in the presence of free galactose), pectic oligogalacturonides, and gentiobiose (**Figure 1**). Oligosaccharins may act by promoting ethylene synthesis. Gentiobiose is present in unripe tomato fruit (*Lycopersicon esculentum*) and increases towards ripening. Exogenous gentiobiose hastened ripening by around 2 days, exogenous ethylene by around 6 days. Such a difference is expected if the sequence of signals is oligosaccharin → ethylene → ripening. Further studies on the triggers of ethylene synthesis are required.

Ethylene

In climacteric fruits, an early ripening event is increased endogenous ethylene (ethene; C_2H_4) biosynthesis and a shift from system 1 ethylene synthesis (seen in most plant tissues) to system 2. System 2 is distinguished by being autocatalytic – small amounts of ethylene induce the synthesis of larger amounts. Increased ethylene biosynthesis occurs naturally in most climacteric fruit, but in banana it is usually triggered postharvest.

Many features of natural ripening (e.g., increased respiration, endo-polygalacturonase accumulation, tissue softening) are retarded by ethylene biosynthesis inhibitors (e.g., aminovinylglycine, aminooxyacetate) or ethylene action inhibitors (e.g., silver thiosulfate, 2,5-norbornadiene, and 1-methylcyclopropene). Similarly, ethylene absorbants (e.g., $Hg(ClO_4)_2$) delay the ripening of fruit placed in sealed bags. Furthermore, mutants and transgenic

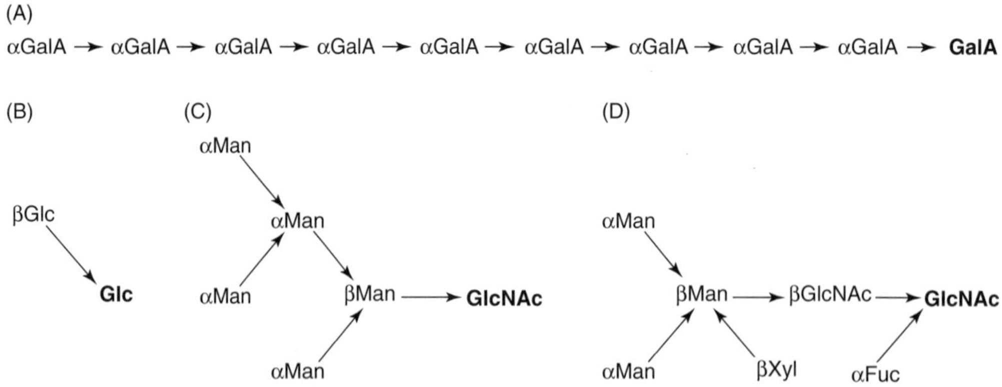

Figure 1 Four oligosaccharins reported to promote the ripening of tomato fruit. (A) A pectic oligogalacturonide; (B) gentiobiose; (C) and (D) Man/GlcNAc-rich oligosaccharides probably of glycoprotein origin. The arrows represent glycosidic bonds, and the direction of the arrow indicates the linkage between the neighboring sugar moieties: ↘, (1→6)-linkage; →, (1→4)-linkage; ↗, (1→3)-linkage; ↖, (1→2)-linkage. GalA, D-galacturonic acid; Glc, D-glucose; GlcNAc, *N*-acetyl-D-glucose; Xyl, D-xylose; Fuc, L-fucose. The sugar unit shown in bold is the reducing terminus.

plants defective in ethylene biosynthesis fail to ripen normally. All these observations point to a key signaling role for endogenous ethylene in ripening. However, certain ripening events, e.g., carotenoid biosynthesis, are ethylene-independent.

Silver thiosulfate interrupts ripening even after system 2 ethylene biosynthesis has been initiated; this suggests that continued ethylene action is required and ethylene does not simply act as a trigger.

Exogenous ethylene (administered either as the gas at $\sim$1–100 $\mu l\, l^{-1}$ or as a solution of 'Ethrel' (2-chloroethanephosphonic acid), which slowly releases ethylene) promotes ripening. In climacteric fruits, exogenous ethylene usually hastens the increase in respiration without changing the maximum respiration rate achieved; in nonclimacteric fruits, which do not naturally exhibit any rise in respiration rate, exogenous ethylene may induce one.

Ethylene is used commercially to trigger climacteric ripening in prematurely harvested fruits. In this capacity, ethylene is excellent since it is nontoxic and, being a gas, can be applied uniformly to a large batch of fruit and cleanly removed afterwards.

Nonclimacteric fruits do not increase in endogenous ethylene biosynthesis but may respond to exogenous ethylene. For example, in citrus fruits ethylene promotes a color change but does not initiate autocatalytic ethylene biosynthesis. In watermelons (*Citrullus lanatus*), ethylene induces undesirable effects, e.g., placental deterioration, water soaking, "off-odours", and excessive softening.

Other Regulators of Ripening

Exogenous gibberellins and auxins sometimes delay ripening, more in fruit slices than in whole fruit. However, such experiments can be complicated by the ability of auxins to induce ethylene biosynthesis. Gibberellins may delay plastid pigment changes (chlorophyll loss and carotenoid synthesis, e.g., in banana, citrus, tomato, and apricot (*Prunus armenaica*)) without retarding the respiratory climacteric.

Metabolic Changes during Ripening

Sugars and Starch

Sugars typically constitute 50–70% of a fruit's dry weight. The main sugars (**Figure 2**) are usually glucose ≈ fructose > sucrose, although sucrose varies from almost absent (grape, cherry (*Prunus avium*)) to predominant (peach (*Prunus persica*), pineapple). Many rosaceous fruits (pear, plum (*Prunus domestica*), cherry) contain a sugar alcohol, glucitol.

Usually sugar concentrations rise during ripening, whether pre- or postharvest. Simultaneously, starch concentrations fall, indicating starch → sugar conversion, apparently mainly by amylases rather than starch phosphorylase activity. However, even in ripe tomatoes, starch continues to be synthesized, and degraded to sugars, implying the operation of a "futile cycle." Sugar concentrations may fall again during prolonged fruit storage.

Often the sucrose : monosaccharide ratio falls during ripening; in tomato this appears to be due to both sucrose synthase (in the cytosol) and invertase (in the vacuole and apoplast) (**Figure 3**). In ripening rosaceous fruits, glucitol (**Figure 2**) may be oxidized to fructose during ripening.

Organic Acids

Most fruit juices are acidic owing to high vacuolar concentrations of (some or all of) citric, L-malic, oxalic, and L-tartaric acids (**Figure 4**). The vacuolar pH may be less than 2, e.g., in some *Citrus* species. Similar or higher concentrations of these compounds

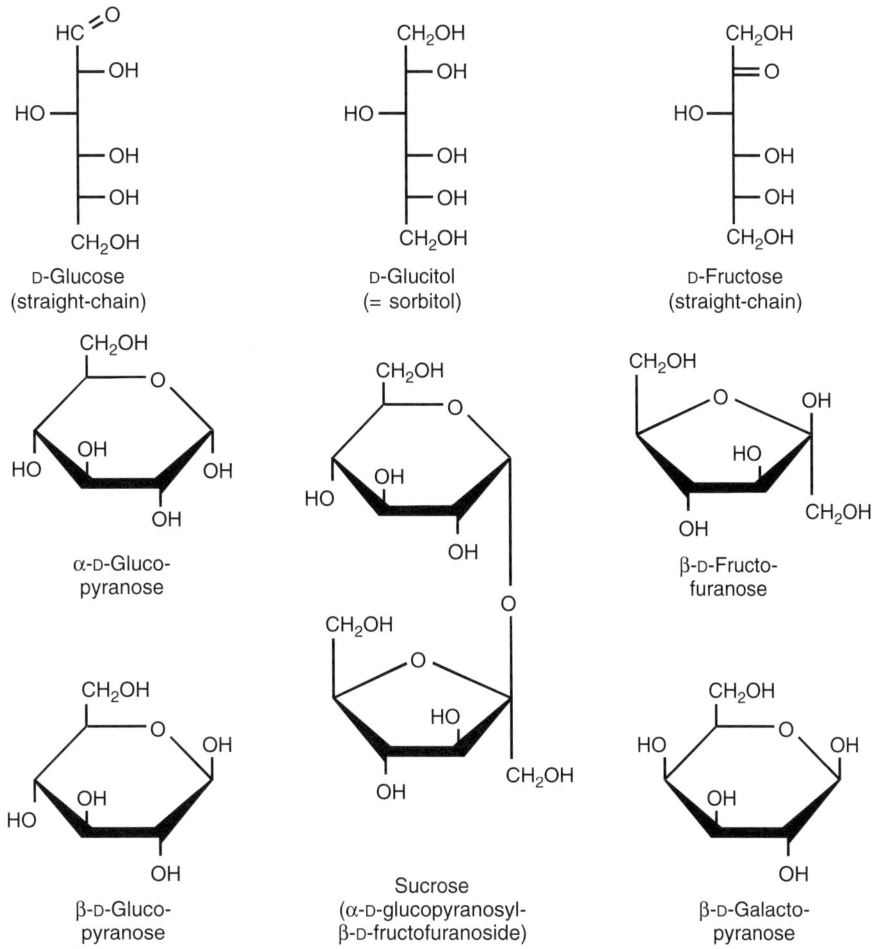

Figure 2 Major sugars (and a sugar alcohol) found in fruits. The top three structures are straight-chain forms, and are shown as Fischer projection formulae. Note that glucose and fructose possess an oxo (C = O) group and are therefore "reducing sugars" (they will reduce Fehling's solution). Glucitol is a sugar alcohol and can exist only as the straight-chain structure. The ring structures are shown as Haworth formulae. Glucose, fructose, and galactose can each form four different ring structures (α-pyranose, β-pyranose, α-furanose, β-furanose), some of which are shown here. In aqueous solution, all four ring-forms and the straight chain form are constantly interconverting. In the nonreducing disaccharide, sucrose, the glucose and fructose residues are each "locked" in one specific ring-form, such that no oxo group can form. In most fruits, glucose and fructose are the predominant sugars and traces of galactose are found. High glucitol concentrations are characteristic of some rosaceous fruits, e.g., pear.

are also present in leaves – but as salts, so leaf juice is less acidic. In most fruits, the acids decrease during ripening, enhancing palatability. The decrease is partly due to the acids' consumption by respiration; in addition, there is a decrease in the rate of organic acid synthesis from sugars.

One key enzyme, highly active in apples during the loss of acids is "malic enzyme," which catalyzes the reaction shown in **Figure 4**. The NADPH formed by this enzyme is important for the redox reactions occurring in ripening fruit; the pyruvate formed is consumed in the Krebs cycle, which operates rapidly during the respiratory climacteric.

Ascorbate (Vitamin C)

A major nutritional benefit of eating fruits is their L-ascorbate content (**Figure 4**). Concentrations vary from ~0.002% (some apples) to ~0.2% (blackcurrant (*Ribes nigrum*)) or ~0.5% (rosehip) on a fresh weight basis. The usual range is 0.01–0.1%.

Often, no dramatic changes in ascorbate concentration accompany ripening. For instance, in blackcurrant high concentrations accumulate during fruit growth and then remain relatively constant during ripening. However, in strawberry, the ascorbate concentration rises abruptly, coinciding with reddening.

Pigments

Ripening usually involves a change from green to yellow, orange, red, or purple. Chlorophyll breakdown occurs, initiated by chlorophyllase, and carotenoids (yellow–orange–red) and/or anthocyanins (red–pink–purple) are synthesized (**Figure 5**).

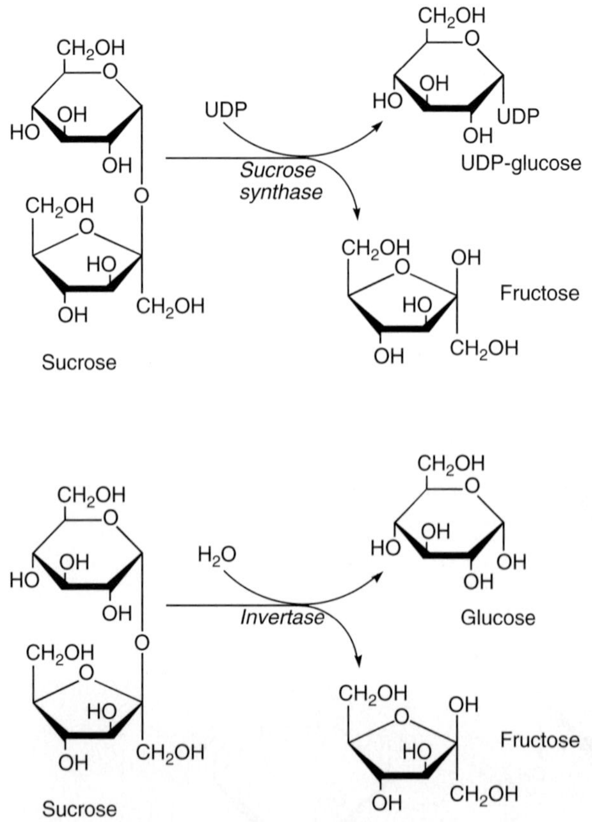

Figure 3 Two enzymic mechanisms for breaking down sucrose to yield monosaccharides. Sucrose synthase catalyzes a transglucosylation reaction with UDP to form UDP-glucose; this is a freely reversible reaction with a $\Delta G^{\circ\prime}$ close to zero. Invertase (β-fructofuranosidase) catalyzes a practically irreversible hydrolysis, in which the high energy content of the glycosidic bond of sucrose is "wasted."

Carotenoids include carotenes (C_{40} hydrocarbons with numerous, mainly *trans*, double bonds) and xanthophylls (similar but with oxygen(s)). Fruits with abundant carotenoids include tomato, orange (*Citrus sinensis*), rosehip and red pepper (*Capsicum frutescens*). In these species, the carotenoids greatly increase during ripening, whether on or off the plant. Fruit carotenoids differ from those of leaves: the main carotenoid in red pepper fruits is capsanthin, which is absent from leaves; and fruit xanthophylls, unlike leaf xanthophylls, are often esterified (e.g., with palmitate).

Chlorophylls are degraded, and carotenoids synthesized, in plastids, especially in the exocarp. Thus, chloroplast→chromoplast conversion occurs during ripening, entailing a loss of thylakoid membranes. However, fruits that were kept dark during their growth, and thus lack chloroplasts, still form normal chromoplasts in the dark; conversely, tomato mutants that fail to degrade chlorophyll still synthesize carotenoids normally. Thus, chlorophyll is not a required precursor for carotenoids.

Anthocyanins accumulate in the vacuole, not the plastids. Examples are cyanidin 3-glucoside (especially in blackberry and raspberry (*Rubus idaeus*)), cyanidin 3-galactoside (red apple skins and cranberry (*Vaccinium oxycoccus*)) and pelargonidin 3-glucoside (strawberry). Anthocyanin concentrations are usually highest in the epidermis, sometimes also in the underlying parenchyma. Concentrations rise dramatically during ripening.

Condensed Tannins (Proanthocyanidins)

Proanthocyanidins are colorless tannins found throughout most fruits, but usually most concentrated in outer tissues. They account for the astringency of many unripe fruits because they "tan" salivary proteins. Specific proanthocyanidins are probably responsible for the ability of American cranberry (*Vaccinium macrocarpon*) juice to combat urethral infections. Proanthocyanidins, unlike anthocyanins, often decrease in effective concentration during ripening; this may be because they polymerize to an M_r above the optimum for tanning.

Aromas

A bewildering variety of volatiles contribute to the, often attractive, aromas of fruits. Many are relatively simple chemically (**Figure 6**). Aroma compounds tend to be synthesized increasingly as ripening progresses. It is not always possible to achieve the full aroma when fruits are picked early and ripened postharvest.

Respiration

Increases in respiration rate occur in climacteric fruits in response to exogenous and/or endogenous ethylene, and in nonclimacteric fruits in response to exogenous ethylene. The increase in respiration can be about twofold in apple, threefold in tomato and banana, more in avocado (*Persea americana*). The respiratory climacteric is often an early event in ripening; however, in kiwi fruit (*Actinidia deliciosa*), the respiratory climacteric occurs after many other symptoms of ripening (e.g., softening) have occurred. Fruit mitochondria retain a high P : O ratio (i.e., oxidation of organic substrates remains tightly coupled to ATP synthesis); therefore the increased respiration boosts the ATP supply, which is necessary for transcription, translation, and other ripening-related biosyntheses. However, the respiratory climacteric appears greatly to exceed what is required to fuel such biosyntheses. The substrate for the respiratory climacteric differs between species: the respiratory quotient (RQ) decreases during ripening in bananas but increases in apples. High RQs are characteristic of malate or citrate oxidation.

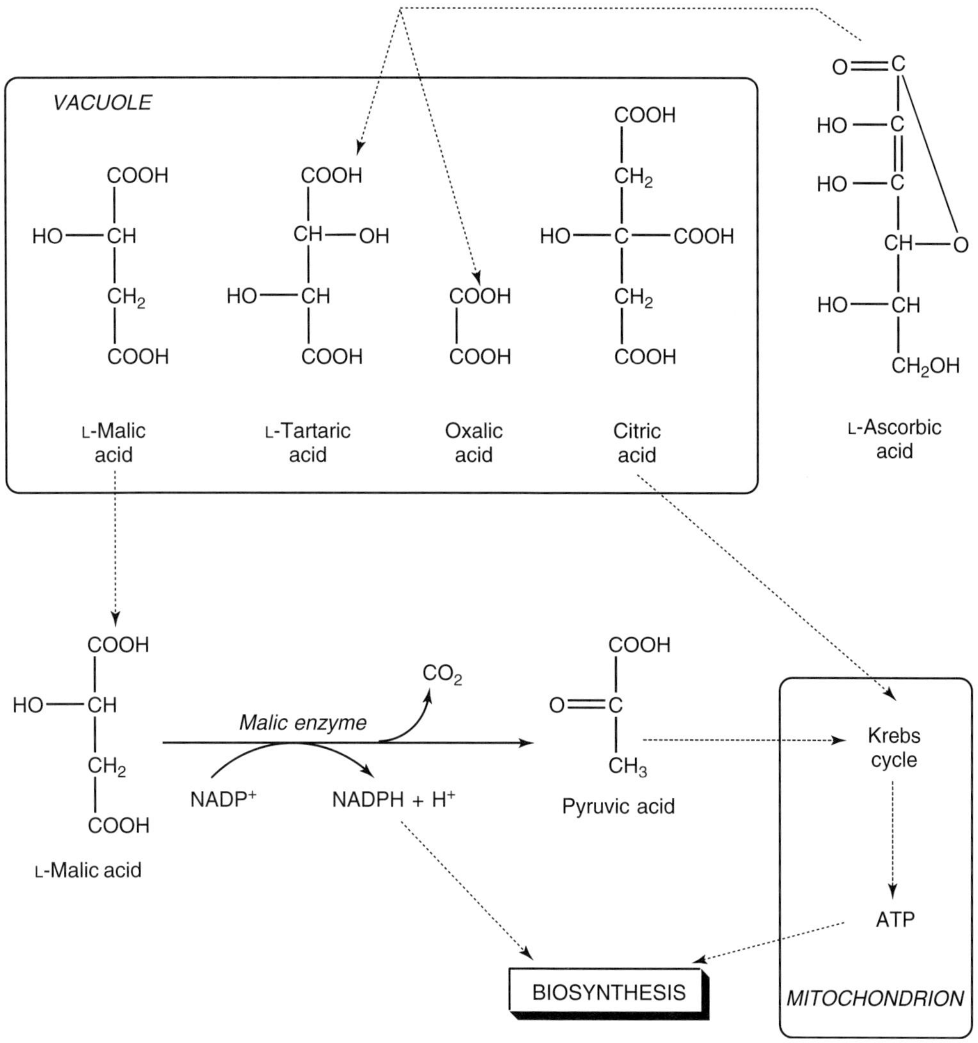

Figure 4 Five major organic acids found in fruits. Citric and oxalic acids are excellent calcium chelators. Oxalic and L-tartaric acids are degradation products of L-ascorbate. The majority of the citric, L-malic, L-tartaric, and oxalic acids are thought to be located in the vacuole – a relatively inert pool, metabolically speaking. However, after leaving the vacuole, L-malate can be oxidized by "malic enzyme" to form pyruvate and NADPH. Glucose and fructose can also be oxidized to pyruvate (glycolysis). Pyruvate does not normally accumulate to appreciable concentrations; instead it is rapidly oxidized via the Krebs cycle. Citric acid, after leaving the vacuole, can also be oxidized via the Krebs cycle. By any of these catabolic pathways, the cell can generate NADPH and ATP, which are required for the anabolic reactions that occur in ripening fruits.

RNA and Protein Synthesis

Total RNA and protein concentrations may or may not change appreciably during ripening, but there are always qualitative changes: some genes are upregulated, others downregulated. Blocking protein synthesis with cycloheximide severely impedes ripening. Changes in the concentrations of specific proteins and messenger RNAs (mRNAs) are discussed under relevant headings.

Polysaccharide Biosynthesis

Fruit cell walls consist of polysaccharides mainly synthesized during growth. However, some matrix polysaccharide biosynthesis continues throughout ripening (on or off the plant), as monitored by radiolabeling with $^{14}CO_2$ or [^{14}C]glucose; this is probably unrelated to the softening process.

Metallothioneins

Metallothioneins are small, cysteine-rich, metal-binding proteins, whose mRNAs often accumulate during fruit growth or ripening.

When zinc is the main bound metal, metallothioneins are potent scavengers of hydroxyl ($^{\bullet}$OH) radicals. However, it seems unlikely that metallothionein concentrations are sufficient to contribute appreciably to $^{\bullet}$OH scavenging *in vivo* since an

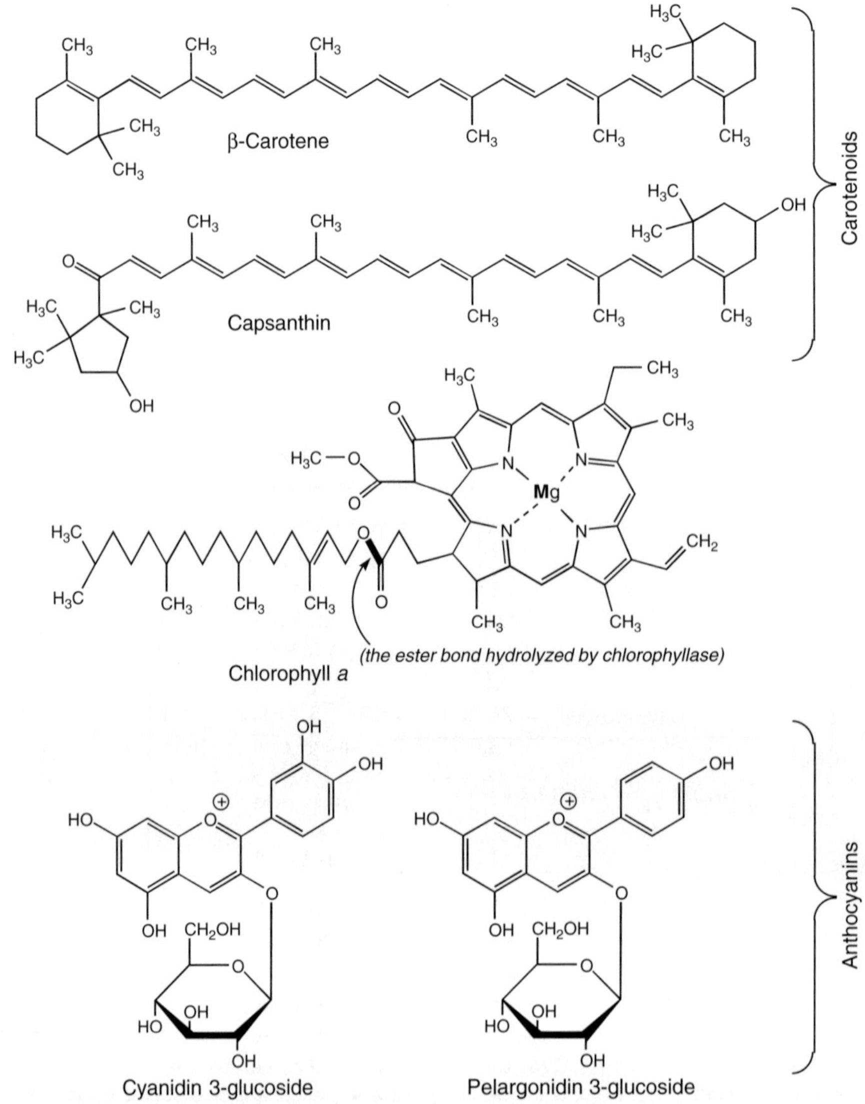

Figure 5 Pigments found in fruits. In general, chlorophylls are degraded and the non-green pigments (carotenoids and/or anthocyanins) are synthesized during ripening. Of the two carotenoids shown, capsanthin is a xanthophyll whereas β-carotene is a carotene – both are found in red pepper fruits. About 75% of the capsanthin molecules have one or both of their –OH groups esterified with fatty acid groups, e.g., palmitate. Chlorophyllase hydrolyzes the ester bond shown in bold, a first step in chlorophyll breakdown. Chlorophylls and carotenoids are found in plastids, anthocyanins in the vacuole.

enormous excess of other scavengers, e.g., glucose, is also present.

Conversely, when copper is the main bound metal, metallothioneins act as prooxidants to generate $^{\bullet}OH$ radicals, especially when H_2O_2 is also present. A possible role for $^{\bullet}OH$ in ripening is discussed later.

There is little information on the occurrence in fruits of metallothioneins themselves, only on their mRNAs; and metallothionein mRNA concentrations were not enhanced in bananas by ethylene. Nevertheless, these intriguing proteins deserve detailed investigation, especially their subcellular location.

Cellular "Degeneration"

Membrane Permeabilization

In 1928, Blackman and Parija suggested that ripening involved loss of "organizational resistance:" fruit cells lost the ability to maintain separate compartments owing to cellular degeneration. Later work, showing that ripening involves gene activation, led to the Blackman–Parija hypothesis losing favor. Instead, attention was focused on the synthesis of proteins, especially those involved in ethylene biosynthesis and polysaccharide hydrolysis. However, this evolution of ideas possibly went too far; a role for membrane permeabilization,

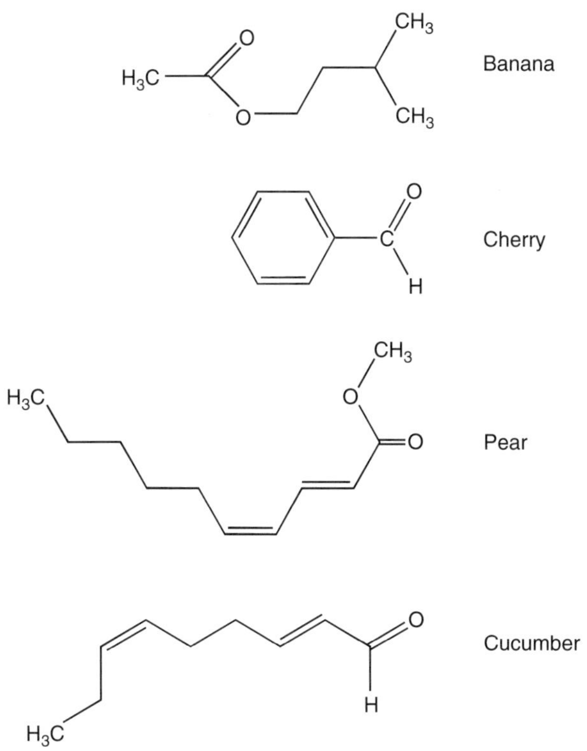

Figure 6 A selection of compounds that impart characteristic aromas to fruits.

itself perhaps triggered by gene expression, deserves re-examining.

Changes in membranes during ripening were demonstrated by electron microscopy and by chemical analyses, which showed net phospholipid losses, increased fatty acyl unsaturation, and increased lipid hydroperoxides.

Ripening-related membrane permeabilization is suggested by intracellular K^+ redistribution, amino acid efflux, and increased hydraulic conductivity. Such permeabilization is not total catastrophic loss of compartmentalization, but may promote the release of metabolites, e.g., ascorbate and organic acids, into the apoplast. Validly studying such changes is difficult because the manipulations required for efflux measurements (e.g., cutting tissue slices) may themselves do more damage to soft, ripe fruits than to crisp, unripe ones.

Pure apoplastic fluid, without protoplasmic contamination, is difficult to obtain owing to the fragility of highly vacuolated fruit cells. However, various techniques have been proposed, including blotting from intact surfaces onto absorbent paper, or gentle expression of fluid in a pressure bomb. The pH of apoplastic fluid decreases from 6.7 in fully sized green tomato fruit to 4.4 in ripe fruit. Such acidification may regulate apoplastic enzyme action and possibly contribute to softening. Also, apoplastic $[K^+]$ increased threefold during tomato ripening. As the concentration of osmotically active solutes in the apoplast rises to a value nearer that within the protoplast, cell turgor decreases.

Oxidative processes

There is steadily mounting evidence that fruit ripening involves oxidative as well as hydrolytic degradation. For example, in saskatoon fruit (*Amelanchier alnifolia*) and avocado peel, lipid peroxidation was the earliest symptom of ripening. Tomato fruit ripening (especially breaker stage, on-vine) was accompanied by increased H_2O_2 concentrations and the oxidation of lipids and proteins.

Lipoxygenases, which are often induced early during ripening, catalyze the reaction of O_2 with (*cis,cis-*) —CH=CH–CH$_2$–CH=CH— groups in lipids, such that hydroperoxide groups (>CH–OOH) are introduced. Active oxygen species may be generated during lipoxygenase action and as breakdown products of hydroperoxides. Thus, lipoxygenases can be prooxidants, and may initiate other oxidative reactions in fruits.

Several studies of oxidative damage to fruit lipids have focused on antioxidant enzymes (catalase, superoxide dismutase, etc.), on the assumption that lipid oxidation should be associated with an elevation in such enzymes, but no consistent trends have been found. Such trends may not be expected if oxidative membrane damage actually plays a positive role in ripening, e.g., increasing membrane leakiness (so that certain metabolites are released into the apoplast), or generating hydroxyl radicals that degrade wall polysaccharides and thus initiate fruit softening.

Fruit Softening

Structural Features

Neighboring cells adhere via the middle lamella. Loosening of the primary wall and/or middle lamella causes the attractive textural changes that characterize ripening. Unripe fruit is often crisp; ripe fruit should usually (though not in apple, etc.) be soft and juicy. “Juiciness” may be caused by release of vacuolar sap upon wall rupture (during biting) and/or by the presence of a hydrated layer on the cell surfaces. There is little difference in sap content between ripe and (fully sized) unripe fruit, but in ripe fruit the sap is more readily released because the wall is weaker. In addition, the wall polysaccharides of a ripening fruit become highly hydrated and the wall swells.

If excessive degradation of the middle lamella occurs and/or the wall fails to loosen, then cells tend to separate from one another when bitten, rather than breaking; such fruit has a "mealy," not juicy, texture.

Polysaccharide Breakdown

Pericarp cells have primary walls consisting of pectins (~35% of wall dry weight), hemicelluloses (~25%), and cellulose (~30%); small amounts of glycoproteins and phenolics (but rarely lignin) are also present. Epidermal walls are cutinized, and thicker than pericarp walls.

In softening fruits, the pectic polysaccharides become progressively less firmly wall-bound: they become extractable earlier in a conventional sequence of extractants such as:

- phenol/acetic acid/water (2:1:1; PAW)
- neutral 50 mmol l^{-1} CDTA
- 50 mmol l^{-1} Na_2CO_3
- 6 mol l^{-1} NaOH.

PAW (alternatively 15% formic acid) is a good first step because it denatures proteins and prevents digestion of polysaccharides by endogenous enzymes, and it is an excellent scavenger of •OH (which may be formed in crude homogenates by reactions of ascorbate and which can nonenzymically cleave polysaccharides).

In essentially all softening fruits examined, M_r reductions occur in the pectin fractions. This is partly due to cleavage of the pectic backbone, although it is difficult to determine whether such cleavage is enzymic hydrolysis (by endopolygalacturonase), enzymic elimination (by pectate lyase), or nonenzymic scission (by hydroxyl radicals). Also, neutral monosaccharides (especially galactose; sometimes also arabinose) are usually released from pectic side chains and then largely remetabolized. Such loss of pectic galactose probably does not in itself cause softening: it occurs in apples, which soften little, but not in plums, which soften greatly. Similarly, blocking pectic galactose loss by addition of 50 mmol l^{-1} free galactose to kiwi fruit disks did not prevent softening, showing that galactose loss was not essential for softening, although the possibility remains that the monosaccharide galactose itself (whether generated by polysaccharide hydrolysis or added exogenously) was essential.

Xyloglucan, the major hemicellulose, is partially depolymerized early in the softening of some fruits but not others. Xyloglucan chains probably "tether" adjacent microfibrils (*see* **Cell Walls and Fibers**: Cell Walls); partial depolymerization of xyloglucan could therefore loosen the wall.

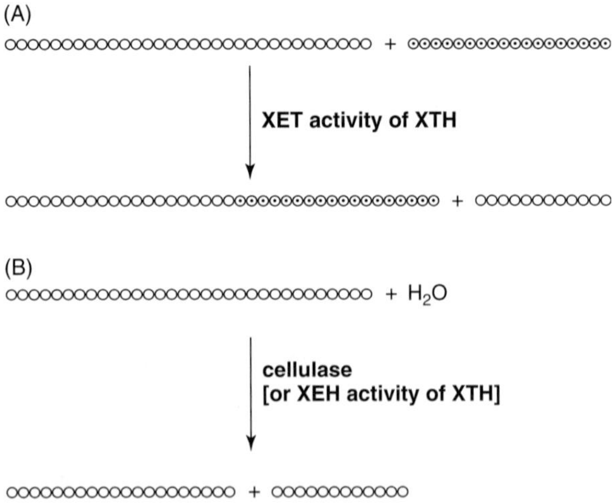

Figure 7 The action of two enzymes on xyloglucan, the major hemicellulose of fruit cell walls. (A) Xyloglucan endotransglucosylase (XET) activity catalyzes a "cutting-and-pasting" reaction, allowing the reorientation of polysaccharide chains with no net loss of integrity; (B) cellulase (endo-(1→4)-β-D-glucanase) activity catalyzes a simple "cutting" reaction. Each string of circles represents a xyloglucan chain; there need be no chemical difference between the two chains involved in the XET reaction. Some XTHs (see text) can catalyze both reactions (A) and (B).

Endoglycanases

There are two main endoglycanases of interest: cellulase (**Figure 7**) and endo-polygalacturonase (endo-PG) (**Figure 8**).

Extractable cellulase (endo-(1→4)-β-D-glucanase) activity often increases during ripening, e.g., in avocado, strawberry, and raspberry. In raspberry the increase may be associated with abscission of the compound fruit from the receptacle as well as with mesocarp softening. In avocado, the increase in activity involves accumulation of cellulase mRNA and of the protein (detected immunologically). Cellulase catalyzes the mid-chain hydrolysis of polysaccharides that possess (1→4)-β-D-glucan backbones, including the soluble "model" substrate, carboxymethylcellulose. Microfibrillar cellulose is a poor substrate owing to its crystallinity; the major natural substrate has therefore been suggested to be xyloglucan. However, at least some fruit cellulases (e.g., of avocado) are reported not to cleave xyloglucan, and their natural substrates are unknown. There is no clear evidence for *in vivo* cellulase action (as opposed to activity) during ripening. *In vivo*, it is impossible to distinguish cellulase action on xyloglucan from XTH action (see below). In transgenic tomatoes, blocking the increase in cellulase did not prevent softening.

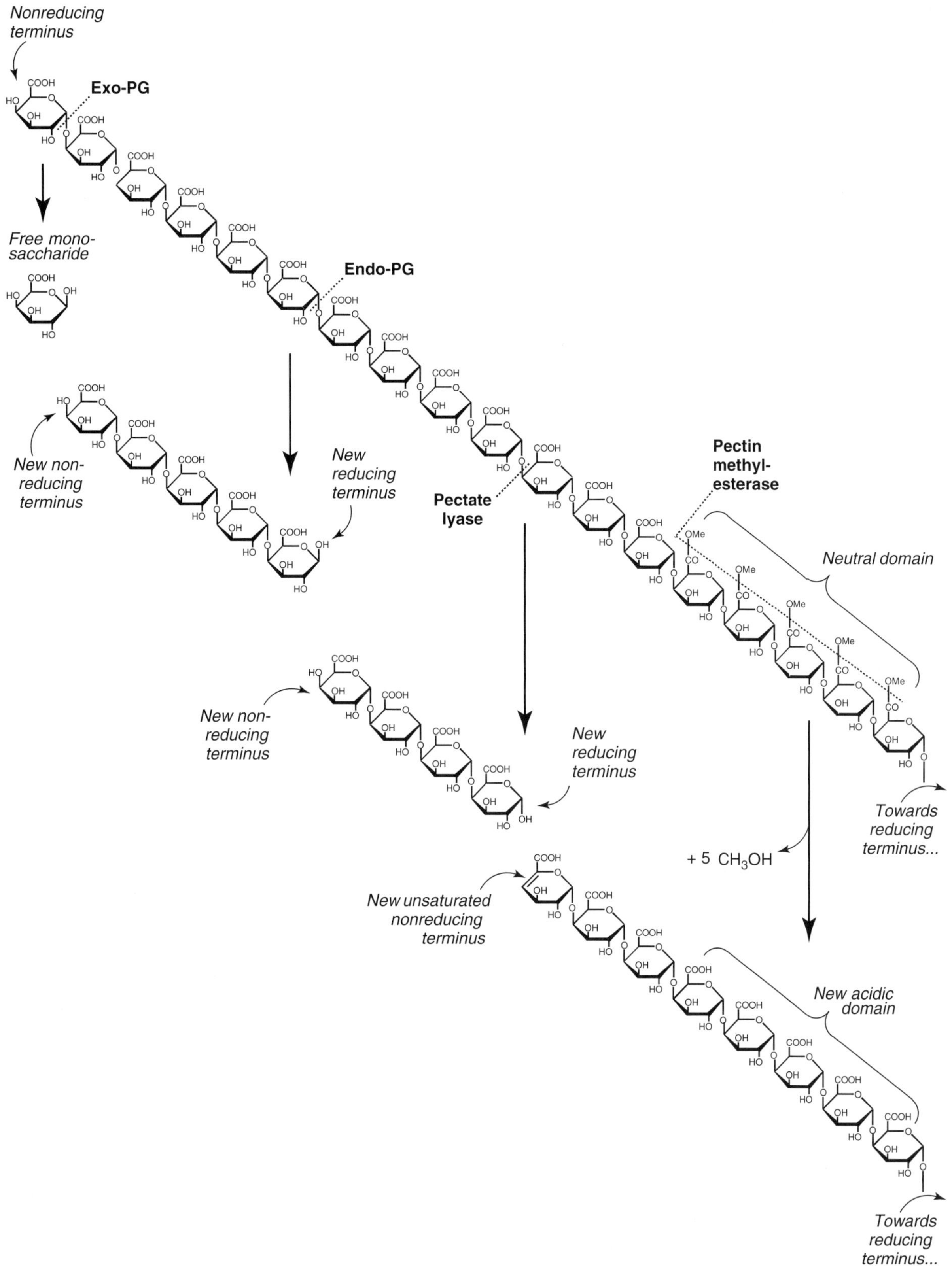

Figure 8 The action of four enzymes on partially methyl-esterified homogalacturonan (part of a representative pectic chain). The original chain is shown being subjected, by the four enzymes named, to exo-hydrolysis, endo-hydrolysis, endo-β-elimination and deesterification respectively; ······ represents bond(s) cleaved. The action of the esterase is block-wise, resulting in a "run" of contiguous acidic residues. The elimination reaction, catalyzed by pectate lyase, has not yet been shown to occur in healthy ripening fruit, but the mRNAs for putative pectate lyases are synthesized during ripening.

Endo-PG (=pectinase) catalyzes the mid-chain hydrolysis of acidic homogalacturonan, a pectic domain (**Figure 8**). Many fruits accumulate endo-PG mRNA and endo-PG activity during softening, suggesting a role in wall breakdown. In ripe tomato, all endo-PG is encoded by a single gene, a circumstance which has simplified experimentation. Neither the increase in endo-PG activity nor softening occurs in the tomato mutants *Rin* (*ripening inhibitor*) and *Nr* (*never ripe*), again compatible with a role for endo-PG in softening. However, restoration of endo-PG production in *Rin* fruit failed to restore softening despite successfully restoring apparently normal rates of pectin solubilization and depolymerization; this shows that endo-PG is not sufficient for softening. Furthermore, transgenic tomato fruit possessing only ~1% of normal endo-PG activity still softened normally. Therefore, >99% of the normal endo-PG is not necessary for softening. There are conflicting reports as to whether normal pectin solubilization occurs in low-endo-PG transgenic tomato fruit. The main practical consequences of an endo-PG deficit were a delay in the excessive softening associated with over-ripening, a greater resistance to damage during transport, and a greater viscosity of the manufactured tomato pastes.

Endo-PG is reported to be absent from several softening fruits, e.g., persimmon (*Diospyros kaki*) and cherry, and present only in traces in others, e.g., banana and strawberry – observations reinforcing the idea that endo-PG is not necessary for pectin solubilization and fruit softening. On the other hand, the *melting flesh* genetic locus of peach appears to be an endo-PG gene, supporting a role for endo-PG in the softening.

Exoglycanases

Exoglycanases hydrolyze off (usually single) sugar residues from the nonreducing termini of polysaccharides, glycoproteins, etc. For example, exo-polygalacturonase (exo-PG; α-D-galacturonidase) progressively removes single, nonesterified galacturonate residues from the nonreducing terminus of homogalacturonan (**Figure 8**). This "nibbling" action of exo-PG seems less likely than mid-chain cleavage by endo-PG to have any immediate, large impact on the properties of pectin and thus on fruit softening. Exo-PG activity peaks during fruit growth rather than softening.

β-D-Galactosidase and α-L-arabinofuranosidase are exoglycanases that appear to act *in vivo*, as evidenced by the loss of pectic galactose and arabinose residues during ripening. However, the ripening-associated "β-galactosidase II" of tomato appears somewhat after the initiation of fruit softening. There is evidence that reducing β-galactosidase II mRNA concentrations (by antisense technology) in tomato may retard normal fruit softening; however, these results are difficult to interpret because there was no difference from the wild-type in wall galactose residue content and concentrations of free galactose during ripening were normal.

Other exoglycanases without well-defined functions in fruits include α-D-galactosidase, α- and β-D-glucosidases, and β-D-xylosidase. The glucosidases, acting as exo-transglycosylases in the presence of the high concentrations of free glucose present in fruits, may synthesize disaccharides, e.g., nigerose and gentiobiose, some of which may have regulatory roles in ripening.

Pectin Methylesterase

Pectin methylesterase (PME) hydrolyzes several consecutive methyl ester groups (often in rapid succession) from homogalacturonans, releasing methanol and converting a neutral pectic domain into an acidic one capable of forming Ca^{2+} bridges (**Figure 8**). *Arabidopsis* has an astonishing 80 putative PME genes.

No clear role for deesterification *per se* in the mechanics of fruit softening has yet been established. Endo-PG can cleave pectic backbones only within acidic domains, so PME's role might be to prepare methylesterified domains for subsequent attack by endo-PG. However, transgenic tomato fruit with low PME activity (and measurably higher methylesterification) softened normally; they lost tissue integrity more rapidly than wild-types during fruit storage, supporting a role for Ca^{2+} bridges in resisting softening (see below).

Xyloglucan Endotransglucosylase/hydrolases (XTHs)

Many XTH isoenzymes exist in plants and some may participate in fruit softening. XTHs possess one or both of two enzyme activities: xyloglucan endotransglucosylase (XET) and xyloglucan endohydrolase (XEH) (**Figure 7**). Both activities first cleave a xyloglucan molecule in mid chain, forming a xyloglucan–XTH complex; the xyloglucan associated with this complex is then transferred either on to another xyloglucan chain (net result, transglycosylation), or to H_2O (net result, hydrolysis). Either transglycosylation or hydrolysis could potentially cause wall-loosening associated with fruit softening.

XET activity rises dramatically during fruit ripening in kiwi fruit and persimmon but not in apple (which softens less). However, no genetic work has yet supported the necessity of XTHs for softening.

Assays for pectic transglycosylase activities in fruits have so far not revealed the existence of such enzymes.

Pectate Lyases

Arabidopsis has many genes predicted to encode pectate lyases, enzymes that catalyze the mid-chain, eliminative (nonhydrolytic) cleavage of acidic homogalacturonans (**Figure 8**). Some putative pectate lyase genes are upregulated during ripening, e.g., in banana, which lacks detectable endo-PG. Pectate lyases could in principle fulfil the role exerted by endo-PGs in other fruits, but since there are no reports of pectate lyase activity in fruits, this idea remains conjectural. Note that most assays for "endo-PG" activity do not distinguish this from pectate lyase activity.

α-Expansins

α-Expansins are wall proteins that probably catalyze the breakage of hydrogen bonds between polysaccharides, e.g., between xyloglucan and cellulose, and might thereby promote softening. However, hydrogen-bond breakage is reversible, and the equilibrium lies strongly in favor of H-bond formation:

$$\text{xyloglucan} + \text{cellulose} \rightleftarrows \text{xyloglucan} \cdots \text{cellulose}$$

Catalysts do not affect the equilibrium position. Therefore, α-expansin could lead to net polysaccharide separation only if some force (e.g., cell turgor, or swelling of the wall's pectic gel) tended to move the detached polysaccharide domains apart, rendering their detachment permanent. There is no evidence that turgor is necessary for fruit softening, and pectic swelling occurs only after softening has begun, so an initiating role for α-expansins in softening is difficult to conceive.

One α-expansin, *Le-EXP1*, is strongly expressed in tomato fruits relatively late in ripening. Decreasing *EXP1* expression by antisense technology slightly inhibited fruit softening; overexpression promoted it. These observations support a role for α-expansin in the progression of softening, perhaps exposing matrix polysaccharides to further attack by hydrolases, XTHs, or hydroxyl radicals.

Hydroxyl Radical Action

Hydroxyl radicals (•OH) can nonenzymically cleave any polysaccharide chain. •OH may be generated in the cell wall *in vivo* by the nonenzymic action of ascorbate on O_2 in the presence of catalytic traces of Cu^{2+} or Fe^{3+}:

$$AH_2 + 2Cu^{2+} \rightarrow A + 2Cu^{+} + 2H^{+}$$

$$AH_2 + O_2 \rightarrow A + H_2O_2$$

$$Cu^{+} + H_2O_2 \rightarrow Cu^{+} + {}^{\bullet}OH + OH^{-}$$

(where AH_2 = ascorbate; A = dehydroascorbate)

The third of these reactions is a Fenton reaction. Ascorbate and Cu^{2+} in the fruit apoplast are thus of considerable interest.

•OH is very damaging to cells, but is so short-lived that, if generated in the cell wall, it would be most likely to react there, potentially cleaving a polysaccharide chain. Furthermore, in "throw-away" tissues such as fruit pericarp, some damage to DNA, proteins, and lipids may be tolerated. Indeed, some oxidative damage and increased membrane permeability are normal features of fruit ripening (see above). These considerations invite a reassessment of the prevailing view that proteins are the sole agents of fruit softening *in vivo*.

There is no proof that •OH contributes to fruit softening. However, in pear fruit, peroxides (and potentially also •OH generated from them) appear to promote softening, and the polysaccharides of softening pears display chemical "fingerprints" diagnostic of recent •OH attack. The ability of invading microbes to hasten ripening is also compatible with a role for active oxygen species, since these are often generated by plants as a defense response. Of two varieties of muskmelon (*Cucumis melo*), the one producing less •OH had the longer shelf-life. The evidence together suggests that fruit softening, often regarded as basically "hydrolytic," may also be "oxidative" – a suggestion raised by Brennan and Frenkel some time ago but largely ignored.

Chelators and Ca^{2+}

Ca^{2+} can ionically bridge adjacent pectins via their acidic homogalacturonan domains. Artificial chelating agents (e.g., EDTA) solubilize a proportion of the pectic polysaccharides and loosen fruit cell walls. Natural chelators, secreted into the fruit apoplast, might have a similar effect and contribute to softening. Natural Ca^{2+}-chelating agents abundant in fruits include citrate and sometimes oxalate; their location (protoplast versus apoplast) deserves rigorous study.

Recently, an attempt was made to analyze the apoplast of a wild tomato fruit (*Lycopersicon pimpinellifolium*) after isolation of fast-frozen wall particles in a water-immiscible organic solvent. A Ca^{2+} : organic acid ratio of 15 : 4 (charge basis) was found, suggesting that only enough organic acids were present in the apoplast to displace ~25% of pectin-bound Ca^{2+}. However, the apoplastic organic acid concentrations may have been underestimates because they had been "corrected" for protoplasmic contamination on the assumption that any phosphate found in the wall particles was indicative of protoplasmic contamination. In fact, phosphate is probably genuinely present in the fruit apoplast. Two additional observations – that added Ca^{2+} can delay fruit softening and that apple fruit firmness correlates positively with tissue [Ca^{2+}] – support the view that Ca^{2+}-bridging makes an adjustable contribution to fruit firmness.

"Necessity" and "Sufficiency"

Traditionally, the functional role of an agent, e.g., an enzyme, one assesses by attempting to demonstrate that it is both necessary and sufficient for the process under investigation, in this case fruit softening. So far, many agents have been shown to be unnecessary for softening, and no agent has been shown to be sufficient for it. This perhaps points to the conclusion that no single agent plays an overriding role in fruit softening. If any one agent is eliminated, softening as a whole continues almost unabated. Fruit softening is thus quite a robust process, and the fruit may continue to become soft and attractive to animals via any of several routes, including those discussed here. In the case of fruit ripening, it is an oversimplification to seek what is "necessary and sufficient;" future work will instead need to define which agents can and do contribute to softening.

List of Technical Nomenclature

α-, β-	Prefixes to designate the chirality at carbon atom number 1 of an aldose sugar (relative to that at its penultimate carbon atom). Starch is composed of α-linked glucose residues, cellulose of β-linked ones.
(1→3), (1→4), etc.	Locants, designating the points of linkage between two neighboring sugar residues within a polysaccharide.
Apoplast	The aqueous solution which permeates the cell wall (includes xylem sap).
Berry	A soft fruit in which the whole pericarp is soft, e.g., tomato, grape, orange, banana.
Cellulose	A polysaccharide composed entirely of β-(1→4)-linked glucose residues. The main component of microfibrils, the "skeleton" of the plant cell wall.
Compartment	A discrete subcellular volume, bounded by membrane(s).
Domain	A relatively large, distinct portion (normally >10 sugar residues) of a polysaccharide molecule, chemically distinguishable from other portions of the same molecule.
Drupe	A soft fruit in which the endocarp forms a hard casing around the seed ("stone"), e.g., cherry, plum, and each small edible sphere ("drupelet") of certain compound fruits such as raspberry and blackberry.
Endocarp	The inner layer of pericarp if particularly hard.
Endoglycanase	An enzyme that can split a polysaccharide molecule in mid-chain by hydrolyzing a glycosidic bond.
Exocarp	The outer layer of pericarp, if different (e.g., much less juicy) from the bulk of the pericarp.
Exoglycanase	An enzyme that can release sugars from a polysaccharide by hydrolyzing glycosidic bonds at a nonreducing terminus.
False fruit	A "fruit" in which the soft, edible portion is derived from the receptacle of the flower, not from the ovary wall, e.g., apple, pear, rosehip, strawberry.
Hemicellulose	Any of several nonpectic polysaccharides that (unlike cellulose) are extractable from the cell wall with alkali.
Homogalacturonan	A pectic domain whose only sugar residues are α-(1→4)-linked galacturonate (with or without methyl ester groups).
Hydroxyl radical	The highly unstable molecule, •OH, the most reactive known molecule.
Locule	The cavity, or extremely soft fruit tissue, within which the seeds are located.
Mesocarp	Juicy pericarp tissue between any exocarp and endocarp.
Middle lamella	The pectin-rich "glue" joining neighboring plant cells within a tissue.
Nonreducing terminus	A sugar residue that is glycosidically linked within a polysaccharide but has no other sugar residue glycosidically linked to it.
Oligosaccharide	As polysaccharide, but with fewer sugar residues (e.g., 3–30).

Oligosaccharin An oligosaccharide with biological signaling activity.

Organic acid A loosely defined term: a low-M_r compound with one or more carboxy groups, but usually taken to exclude amino acids, fatty acids, hormones, etc. Includes citric, malic, tartaric, oxalic, and pyruvic acids.

P:O ratio Ratio of ATP synthesized to every O atom of O_2 consumed during respiration.

Pectin Any of several galacturonate-rich polysaccharides of the plant cell wall.

Pericarp The major tissue of a fruit, excluding epidermis and vascular tissue.

Placenta The vascularized "core" of the fruit, to which the seeds are attached.

Polysaccharide A polymer composed of many sugar residues, linked to each other.

Primary cell wall A cell wall whose microfibrils were deposited while the cell was still (capable of) expanding.

Protoplast The "living" part of the cell, bounded by the plasma membrane, i.e., excluding the cell wall.

Respiratory quotient (RQ) Ratio of CO_2 evolved to O_2 consumed, during respiration. RQ values of <1, ~ 1, and >1 suggest the oxidation of fats, sugars, and organic acids, respectively.

Sugar residue A glycosidically linked sugar molecule.

Tannin A compound with several or many phenolic –OH groups, able to "tan" proteins (i.e., bind to them and precipitate them) and to give the sensation of astringency.

Transglycosylation The process of splitting a glycosidic linkage and reattaching the former sugar residue to a different molecule.

Xyloglucan A hemicellulose with a cellulose-like backbone plus side chains containing α-xylose residues; the major hemicellulose in dicot primary cell walls.

See also: **Cell Walls and Fibers**: Cell Walls. **Postharvest Physiology**: Genetic Engineering for Postharvest Quality; Postharvest Physiological Disorders of Fresh Crops; Storage. **Regulators of Growth**: Ethylene.

Further Reading

Almeida DPF and Huber DJ (1999) Apoplastic pH and inorganic ion levels in tomato fruit: a potential means for regulation of cell wall metabolism during ripening. *Physiologia Plantarum* 105: 506–512.

Brady CJ (1987) Fruit ripening. *Annual Review of Plant Physiology* 38: 155–178.

Brennan T and Frenkel C (1977) Involvement of hydrogen peroxide in the regulation of senescence in pear. *Plant Physiology* 59: 411–416.

Brownleader MD, Jackson P, Mobasheri A, *et al.* (1999) Molecular aspects of cell wall modifications during fruit ripening. *Critical Reviews in Food Science and Nutrition* 39: 149–164.

Brummell DA and Harpster MH (2001) Cell wall metabolism in fruit softening and quality and its manipulation in transgenic plants. *Plant Molecular Biology* 47: 311–340.

Dumville JC and Fry SC (2003) Gentiobiose: a novel oligosaccharin in ripening tomato fruit. *Planta* 216: 484–495.

Fry SC, Dumville JC, and Miller JG (2001) Fingerprinting of polysaccharides attacked by hydroxyl radicals *in vitro* and in the cell walls of ripening pear fruit. *Biochemical Journal* 357: 729–735.

Hulme AC (ed.) (1970) *The Biochemistry of Fruits and their Products*, vol. 1, London: Academic Press.

Hulme AC (ed.) (1971) *The Biochemistry of Fruits and their Products*, vol. 2, London: Academic Press.

Kays SJ (1997) *Postharvest Physiology of Perishable Plant Products*. Athens: Exon Press.

Macdougall AJ, Parker R, and Selvendran RR (1995) Non-aqueous fractionation to assess the ionic composition of the apoplast during fruit ripening. *Plant Physiology* 108: 1679–1689.

Priem B and Gross KC (1992) Mannosyl-containing and xylosyl-containing glycans promote tomato (*Lycopersicon esculentum* Mill.) fruit ripening. *Plant Physiology* 98: 399–401.

Redgwell RJ and Fischer M (2002) Fruit texture: cell wall metabolism and consumer perceptions. In: Knee M (ed.) *Fruit Quality and its Biological Basis*, pp. 46–87. Sheffield: Sheffield Academic Press.

Redgwell RJ and Harker R (1995) Softening of kiwifruit discs: effect of inhibition of galactose loss from cell-walls. *Phytochemistry* 39: 1319–1323.

Rose JKC and Bennett AB (1999) Co-operative disassembly of the cellulose–xyloglucan network of plant cell walls: parallels between cell expansion and fruit ripening. *Trends in Plant Science* 4: 176–183.

Rose JKC, Hadfield KA, Labavitch JM, and Bennett AB (1998) Temporal sequence of cell wall disassembly in rapidly ripening melon fruit. *Plant Physiology* 117: 345–361.

Seymour GB, Taylor JE, and Tucker GA (1993) *Biochemistry of Fruit Ripening*. London: Chapman & Hall.

Smith DL, Abbott JA, and Gross KC (2002) Down-regulation of tomato β-galactosidase 4 results in decreased fruit softening. *Plant Physiology* 129: 1755–1762.

Wakabayashi K (2000) Changes in cell wall polysaccharides during fruit ripening. *Journal of Plant Research* 113: 231–237.

Senescence, Leaves

V Buchanan-Wollaston, HRI Wellesbourne, Warwick, UK

What is Leaf Senescence?

Leaf senescence is clearly illustrated by one of the most dramatic changes that occurs annually in our environment. Every autumn, many trees in the temperate zones of the world lose their leaves to prepare for the winter season. Before the leaves are lost, the tree will reclaim much of the nutrient content of the leaf by the process of senescence. This process is made visible by the yellow, orange, and red coloration that results, and this is even visible from outer space as a sweeping wave of changing color. Another clear example of plant senescence is seen when fields containing crops such as wheat (*Triticum* spp.) or corn (*Zea mays*; maize) turn from green to golden as the grain fills and ripens before harvest.

Plants have evolved many complex mechanisms that allow them to survive in a hostile environment, and senescence is a key part of this adaptation. Unlike animals, plants cannot move when their immediate environment becomes unfavorable, and they have to cope with whatever conditions they are exposed to. They have developed a system of "selective death" whereby certain parts of the plant are sacrificed to increase the survival chances of the rest of the plant. Thus, leaves on a deciduous tree, which are not beneficial in the cold, dark winter months, are systematically dismantled and then dropped from the tree. Leaf nutrients are mobilized and stored in specialized cells within the tree trunk from which they can be distributed in the spring to support the growth of new leaves. Many plants only survive the cold of a winter as seeds. These annual plants, such as wheat, undergo complete senescence; mobilizable nutrients are transferred from the entire plant to the developing seeds to support the germination and growth of the next generation of plants in the spring.

Therefore, the senescence process that occurs as part of the development of the plant can be characterized primarily as a recycling process. The function of senescence is to mobilize the stored nutrients, mainly nitrogen, phosphorus, and metals, from the senescing tissue to another part of the plant, either for storage or for new growth. The final part of the senescence process is cell death, but this is actively delayed until maximum mobilization has occurred.

What Happens during Leaf Senescence?

Senescence takes place in a highly controlled manner. The first indication of the onset of senescence is a reduced efficiency of photosynthesis, which is paralleled by protein and chlorophyll degradation. By the time the leaf is visibly yellow, the bulk of the process has been completed. Up to this stage, senescence is reversible in many plants; a completely yellow leaf on a tobacco (*Nicotiana tabacum*) plant can be induced to regreen following the removal of growing parts of the plant and treatment with cytokinin. This shows that the organelle structures remain intact within the leaf. However, later in senescence, as the membranes surrounding the chloroplast and vacuole become disrupted, senescence enters the irreversible phase and cell death results.

The chloroplasts, which contain the majority of the mobilizable constituents of the leaf, undergo dramatic changes during the senescence process. **Figure 1** illustrates the changes that occur in *Brassica napus* chloroplasts as senescence progresses. Chloroplasts from mature green leaves contain substantial starch reserves laid down by their photosynthetic activity and the thylakoid membrane stacks are highly organized. In mid-senescence, approximately 50% of the chlorophyll and protein of the leaf has been degraded and the appearance of the chloroplast is very different. The membrane structures are disorganized, there is little stored starch, and there are large electron dense globules called plastoglobuli. However, at this stage the membranes surrounding the chloroplast and vacuole are still intact and there is some cytoplasm remaining. In late senescence, where there is very little protein or chlorophyll remaining in the leaf, the chloroplasts are completely dismantled, with only electron dense materials remaining. The chloroplast and vacuole membranes have ruptured, cellular contents have all been degraded, and cell death has probably occurred.

Other organelles, such as the mitochondria, which are required to provide energy for senescence, and the nucleus, which is necessary for continued gene transcription, remain intact until the last stages of senescence. The role of the peroxisomes changes in senescence. In green leaves, their main function is in photorespiration, while in senescing leaves they are converted to glyoxysomes, and contain enzymes that have a role in the remobilization of lipids via β-oxidation and the glyoxylate pathway.

Gene Expression

The dismantling of cellular constituents that occurs during senescence requires a high degree of control. Senescence is genetically regulated; this was shown

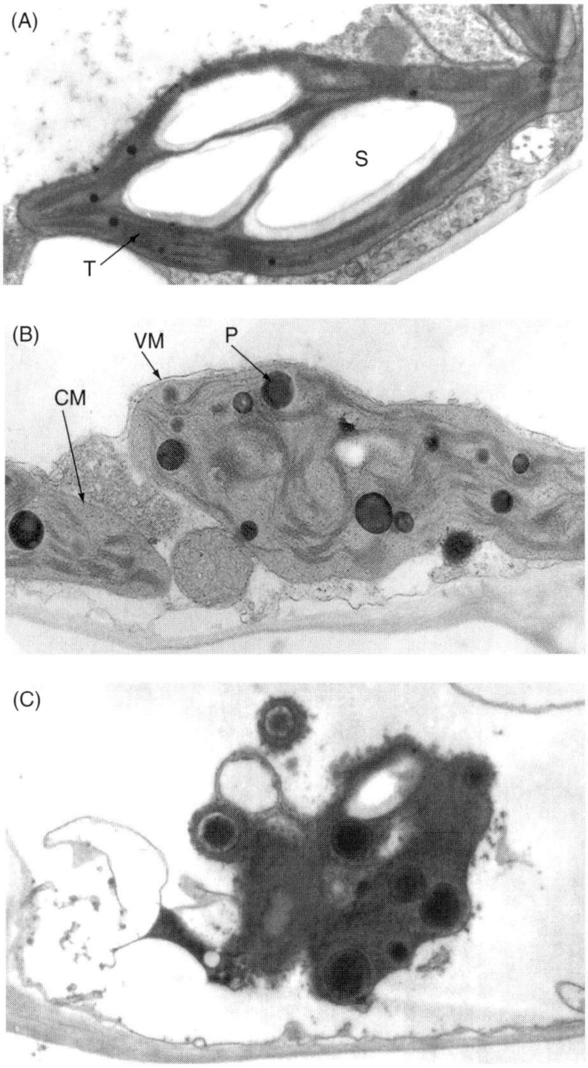

Figure 1 Electron micrographs showing a chloroplast in a *Brassica napus* leaf. (A) Mature green leaf, 100% chlorophyll and protein levels; (B) Early senescent leaf containing around 50% chlorophyll and protein; and (C) Late-senescent leaf containing around 20% chlorophyll and very little protein. S, starch granule; T, thylakoid membrane; P, plastiglobuli; VM, vacuole membrane; CM, chloroplast membrane.

some years ago by a number of different experiments. For example, it was found that the chloroplasts in enucleated protoplasts did not senesce, in contrast to those in protoplasts that still contained the nucleus. Also, leaves that had been treated with inhibitors of RNA or protein synthesis did not undergo senescence, indicating that expression of novel genes is required. Furthermore, a number of senescence mutants that show delayed or altered patterns of senescence in different plants have been identified. In the last few years, molecular techniques have been used to show that the patterns of gene expression change considerably as senescence progresses. The transcript levels of many genes, particularly those involved in photosynthesis, are reduced in abundance, while other genes show enhanced expression. A number of genes that show enhanced expression during leaf senescence have been cloned and characterized, and these provide further evidence that the senescence process is under genetic control.

Table 1 Potential functions of genes that show enhanced expression during leaf senescence

Potential function	*Examples of genes identified*
Protein degradation	Cysteine proteases, aspartic protease, vacuolar processing protease, polyubiquitin, Clp proteases
Nitrogen mobilization	Glutamine synthetase, glutamine dependent asparagine synthetase, branched chain ketoacid dehydrogenase
Carbon mobilization	Pyruvate orthophosphate dikinase, malate synthase, isocitrate lyase, sugar transporters
Nucleic acid degradation	RNase, bifunctional nuclease, acid phosphatase
Defense related	PR-1a, chitinase, osmotin, harpin induced gene (Hin1)
Antioxidants	Catalase, metallothionein, ferritin, glutathione peroxidase
Hormonal regulation	ACC oxidase (ethylene biosynthesis), 12-oxophytodienoate reductase (jasmonate biosynthesis)
Gene regulation	Transcription factors, serine threonine kinases, receptor kinases
Others	Cytochrome P-450s, glutathione-*S*-transferase, ABC transporters, many genes of unknown function

The DNA sequence analysis of senescence enhanced genes can, in some instances, give an indication of their potential function, and this information has allowed researchers further insight into the processes that are taking place during senescence. **Table 1** shows a list of the proteins encoded by some of these senescence enhanced genes, indicating their potential function in senescence. Thus, a number of events that may occur during senescence have been highlighted by the identification of genes and the implied role of their products in the senescence process.

Protein Degradation and Mobilization

Proteins, synthesized and stored in the leaf during development, represent a key resource for the plant, and efficient recycling of these is very important. The majority of the leaf protein (~70%) is stored in the chloroplast, and much of this is comprised of the soluble photosynthetic enzyme ribulose bisphosphate carboxylase (rubisco) and the chlorophyll binding, light harvesting complex proteins that are bound to the thylakoid membrane. The mechanisms by which plants degrade chloroplast proteins during senescence remains a mystery. Protein degradation starts to occur early in senescence, and structural changes are visible within the chloroplast well before the chloroplast membranes show any signs of rupture (**Figure 1**). Therefore, it is assumed that the initial protein degradation must occur within the chloroplast itself. Although a number of genes encoding different proteases have been shown to increase in expression during senescence (**Table 1**), most of the encoded enzymes have peptide sequences that target them to the endoplasmic reticulum or vacuole of the senescing cell rather than to the chloroplast. The exception to this are the genes that encode the subunits of the Clp protease (a homolog of a prokaryotic ATP dependent protease), some of which show senescence enhanced expression and may have a role in protein degradation in the chloroplast. However, a direct relationship between the activity of Clp protease and protein degradation in senescence has not yet been established. Therefore, the signal that initiates the onset of protein breakdown in senescence has not been identified.

Degradation of protein results in the release of amino acids, and there are two alternative destinies for these. Firstly, a large proportion of the nitrogen released from degraded protein will be transported from the leaf as free amino acids in the phloem. The levels of glutamine and asparagine in the phloem leaving the senescing leaf have been shown to be high; each of these amino acids contains two N moieties, in contrast to most other amino acids which contain a single N. Thus, the level of exported N is maximized. Increased expression of the gene encoding glutamine synthetase (GS) during senescence correlates with this observation; ammonia released from other amino acids by deamination reactions can be converted into glutamine by the action of GS. Secondly, the carbon skeletons from deaminated amino acids are likely to provide a source of energy, and the senescence enhanced expression of the genes encoding branch chain α-keto acid dehydrogenase indicates the importance of this pathway to generate energy. This enzyme is key in the catabolism of the products that result from deamination of the branch chain amino acids valine, isoleucine, and leucine.

Chlorophyll Degradation

Senescing leaves turn from green to yellow due to the degradation of their chlorophyll content. In general, the appearance of the yellow color is not due to the synthesis of novel pigments, but is due to the unmasking of carotenoids already present in the leaves. Increased synthesis of other pigments does occur in some plants, however, such as in leaves from deciduous trees that show vivid red and orange colors due to synthesis of novel anthocyanin pigments as the chlorophyll is removed. In mature green leaf cells, chlorophyll is bound to apoproteins in the thylakoid, which are responsible for light trapping during photosynthesis. During senescence, the N content of these apoproteins (which comprises around 30% of the total plastid nitrogen) must be remobilized, with the consequent release of chlorophyll. Since free chlorophyll is extremely toxic and would cause severe photodamage, it must be rapidly degraded. The mechanisms for this degradation have been difficult to elucidate but, in the last few years, much progress has been made in the identification of the degradation intermediates, and also in the isolation of some of the genes involved in the pathway.

The current picture of the pathway leading to chlorophyll degradation, which starts in the chloroplast and ends up in the vacuole, is shown in **Figure 2**. The first step in the pathway is carried out by the enzyme chlorophyllase (Chlase) which has recently been cloned from several plant species. The Chlase protein appears to be located in the plastid inner envelope, and the existence of a senescence related carrier protein (X), which transfers the chlorophyll from the chlorophyll–protein complex in the thylakoid to the Chlase in the envelope, has been hypothesized. Chlase starts the process by hydrolysis of the phytol residue, and then removal of the Mg atom by a Mg dechelatase is followed by cleavage of the tetrapyrrole ring by Pheophorbide *a* oxygenase to produce red chlorophyll catabolite (RCC). A stay green mutant of *Festuca pratensis* accumulates Phoephorbide *a* and is assumed to have a defect in the gene encoding Pheophorbide *a* oxygenase, but this gene has not, as yet, been cloned. The enzyme RCC reductase, which was recently cloned, then converts RCC to primary fluorescent chlorophyll catabolites (pFCCs) which, after modification, are transported from the chloroplast to the cytoplasm via an ATP dependent translocator. These undergo further modifications in the cytosol and are finally imported into the vacuole via a tonoplast located ABC transporter where they are nonenzymatically converted to nonfluorescent chlorophyll catabolites (NCCs), which remain in the vacuole and are lost when the leaf dies. Thus, the chlorophyll molecule is detoxified by a series of energy consuming steps, but very little of it is reused.

The signals that initiate chlorophyll degradation during senescence have not yet been found. The two genes so far identified that encode catabolic enzymes

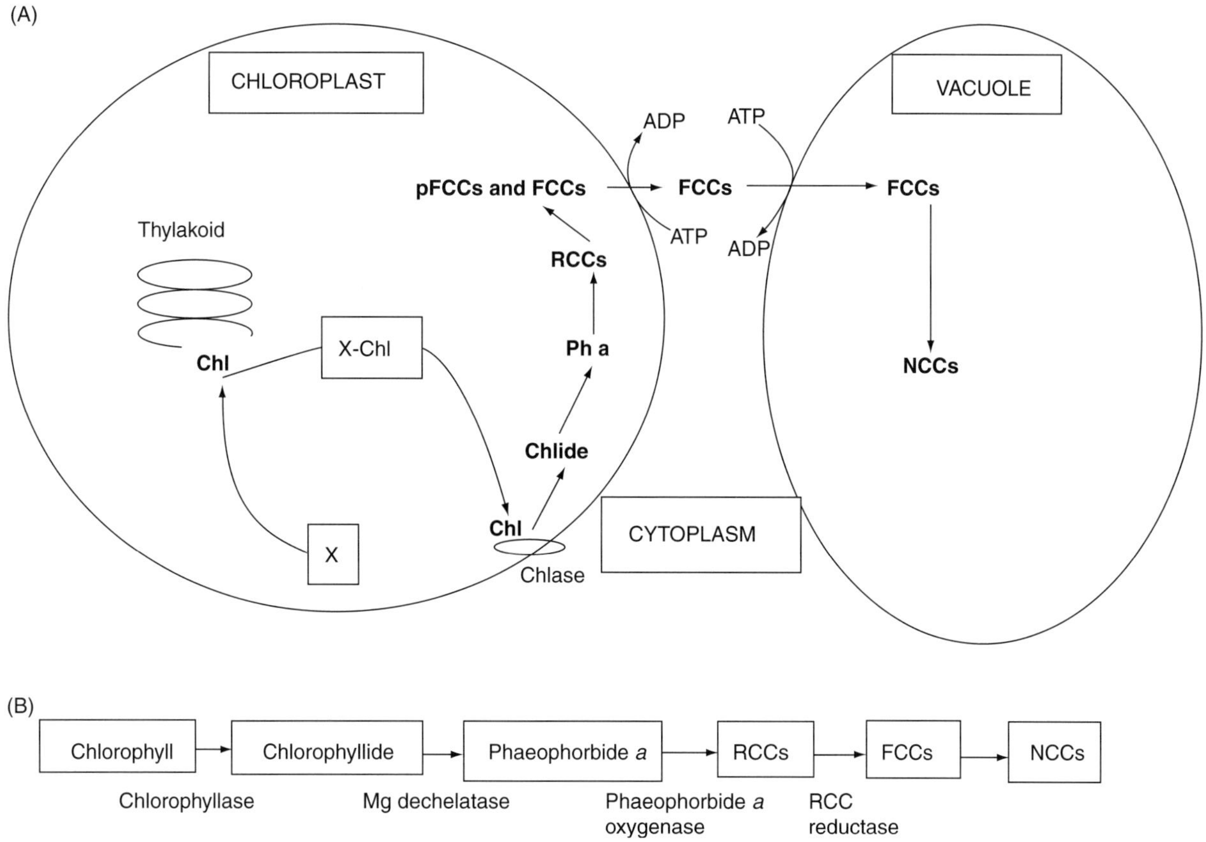

Figure 2 The pathway that is thought to lead to chlorophyll degradation in a senescing leaf is shown. (A) The degradation products and cellular location of the steps in the pathway; and (B) The degradation products and enzymes thought to be involved in the degradation. Unknown enzymes modify the FCC compounds and the NCCs are formed by nonenzymatic reactions in the vacuole. X, hypothetical transport protein that moves chlorophyll from its position in the chlorophyll/protein complex in the thylakoid to the chlorophyllase enzyme in the inner envelope of the chloroplast; FCCs, fluorescent chlorophyll catabolites; RCCs, red chlorophyll catabolites; NCCs, nonfluorescent chlorophyll catabolites; Ph a, Phaeophorbide *a*.

do not show increased expression at the onset of senescence, although one Chlase gene is induced by treatment with jasmonic acid, a compound that has been suggested to have a role in senescence. The activity of Pheophorbide *a* oxygenase increases dramatically during senescence, implicating this enzyme as a key and potentially controlling function in the process. This is also indicated by the phenotype of the stay green mutant that has no Pheophorbide *a* oxygenase activity, and which retains chlorophyll and much of the chlorophyll bound apoprotein in the thylakoid. Cloning and characterization of this gene would be a key step in the elucidation of the control mechanisms for chlorophyll degradation.

Lipid Mobilization

Photosynthetic activity declines during senescence, but the degradative and mobilization processes of senescence are highly energy requiring. The respiratory quotient of a senescing leaf is low, which indicates that some of the energy may be derived from lipid rather than carbohydrate degradation. The membranes of the thylakoid form a huge resource of mobilizable lipid, and this probably helps to fuel the senescence process. Conversion of lipids to sugars via the process of gluconeogenesis requires the activation of the glyoxylate cycle, which includes the enzymes isocitrate lyase and malate synthase, both of which have been shown to increase in activity during senescence. Also, increased expression of the gene encoding pyruvate orthophosphate dikinase during senescence indicates that the conversion of pyruvate to phosphoenolpyruvate can occur, which could lead to the synthesis of sucrose in the glycolysis pathway.

Nucleic Acid Degradation

Extensive nucleic acid reserves in leaves, mainly in the form of ribosomal RNA, provide a store of phosphate and also nitrogen. RNA levels fall during senescence, especially during the later stages, and enhanced expression of RNase genes has been found in a number of species. Inorganic phosphate released from the nucleic acid is a valuable resource that can be transported from the leaf. The sugar, pyrimidine, and purine components are probably recycled, and increased activity of the purine catabolizing enzymes, xanthine dehydrogenase and uricase, has been detected in the peroxisome during senescence. DNA in the nucleus appears to remain intact until very late senescence, presumably to maintain the genetic control of the process.

The Events Occurring during Leaf Senescence

The multiple pathways that take place during senescence are illustrated in **Figure 3**. Novel gene expression is required to start the process. Macromolecules such as protein, lipid, and nucleic acids are degraded into small mobilizable components, which are either transported out of the cell to other parts of the plant or are used to fuel the energy requiring steps that are taking place. Senescence involves many enzymatic functions, and all subcellular compartments have a role to play.

When is Senescence Induced?

A mature green leaf that is actively photosynthesizing is an asset to the plant, providing a steady supply of fixed carbon and amino acids to support further growth of the plant. However, at some stage during development, the nutrients stored within this leaf are required elsewhere in the plant and this is the signal for the onset of senescence. This may occur if the leaf becomes shaded by growth of younger leaves, or the development of seeds may initiate the senescence process.

Therefore, senescence occurs as the final stage of development of a leaf, when its contents become more important to the plant than its photosynthetic potential. The timing of this event depends on the plant species; leaves on rapidly growing annual plants such as *Arabidopsis* are fully expanded for only a few days before senescence is initiated. In contrast, needles on the long-lived bristle cone pine tree (*Pinus longaeva*) may last 30 years. However, senescence can be induced prematurely by a number of different stresses to which the plant may be exposed. Pathogen infection and oxidative stresses

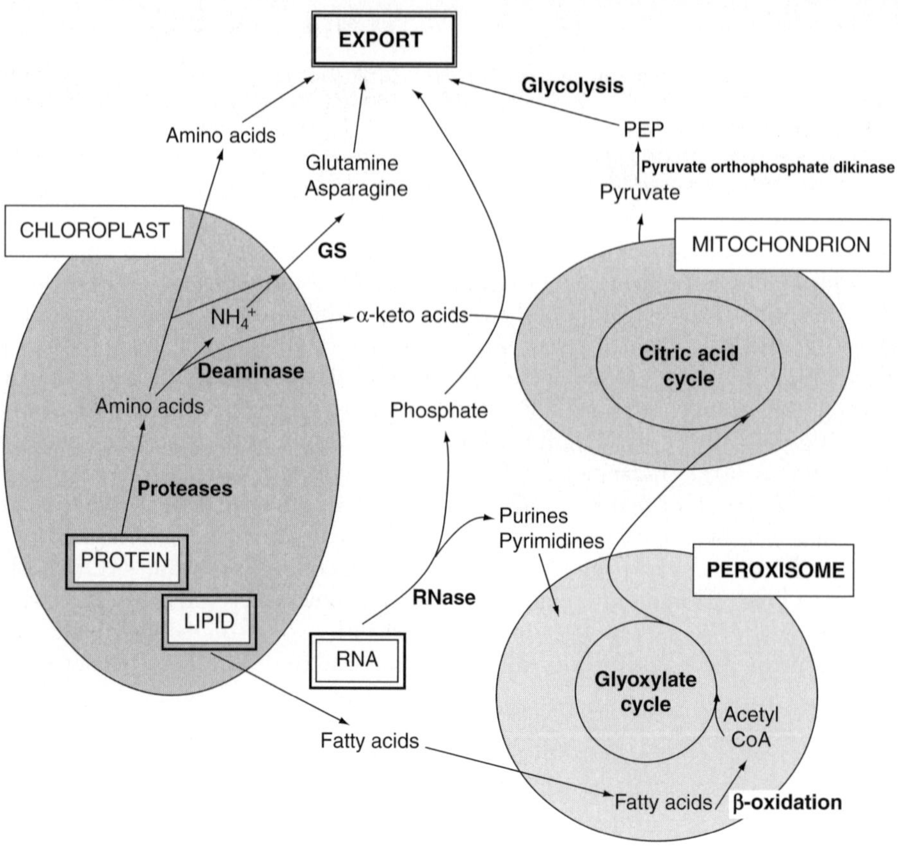

Figure 3 Cellular compartments and metabolic pathways involved in the degradation and mobilization of macromolecules during senescence. Macromolecules such as protein, lipid, and nucleic acids are degraded into smaller constituents, which are either used to fuel the senescence process or are exported out of the cell.

caused by ozone or UV-B can result in symptoms of premature senescence in the affected leaves and, in many cases, cell death will also occur. Nutrient stresses such as water or nitrogen deprivation can initiate premature senescence leading to advanced development and early seed formation. In addition, stresses imposed by harvesting leaves or other plant parts result in senescence-like symptoms in the harvested tissue. **Figure 4** illustrates a number of different types of plant senescence.

Gene expression studies have shown that many genes expressed in response to stress are also expressed during developmental senescence. Many pathogenesis related (PR) genes, which were originally

Figure 4 (see color plate 22) Different forms of plant senescence. (A) The onset of autumn with its shorter days and cooler temperatures induces the onset of senescence in a deciduous tree. The dramatic red and orange coloration is due to novel synthesis of anthocyanin pigments, which may act as antioxidants. (B) The yellowing of the leaves on this perennial shrub show that senescence starts at the edges of the leaves and progresses inwards. Often the areas closest to the leaf veins remain green the longest. This probably allows the final mobilization into the phloem cells to occur. (C) Leaves of increasing age from a *Brassica napus* plant. These leaves were all harvested off the same plant; yellow leaves were at the base of the plant with increasing greenness sequentially up the plant. The nutrient content of the lower leaves, which have become shaded by the leaves above, is needed for further development of the plant, and so senescence is initiated in these leaves. (D) Postharvest senescence in broccoli. Yellowing of the tissue is evident after several days storage. (E) Stress induced senescence in *Arabidopsis thaliana*. The lower leaves on this plant have been infected by the pathogen *Albugo candida*, and this infection results in yellowing of the leaf tissue.

identified as showing increased expression during pathogen infection, are also expressed during developmental senescence (**Table 1**). Senescing tissue may be more vulnerable to pathogen attack, and the protective PR genes may be switched on to prevent this. The degradative processes that occur during senescence are oxidative in nature, and the reactive oxygen species (ROS) levels in senescing cells increase. The plant responds to this by increased expression of certain antioxidant genes (**Table 1**) that presumably help to protect the senescing cell from severe oxidative stress, which would result in premature cell death by necrosis before the mobilization functions were complete.

Postharvest senescence is manifested by yellowing of the harvested tissue, which is accompanied by loss of protein and other nutrients. Recent work with postharvest broccoli (*Brassica oleracea* var. *italica*) has indicated that many genes expressed during developmental leaf senescence also show induced expression postharvest. At the biochemical level, a rapid loss of membrane fatty acids was detected after harvest. These fatty acids are probably used as an energy resource and broken down via the β-oxidation pathway, since increased levels of lipid peroxidation products were not detected until at least 50% of the fatty acid content of the cell was lost. This, together with an increased expression of protective antioxidant genes, indicated that, in the initial stages of postharvest senescence, an orderly dismantling of the cellular constituents occurs, using the available lipid as an energy source. Postharvest changes in broccoli florets, therefore, show many similarities to the processes of developmental leaf senescence.

How is Senescence Regulated?

Only a subset of senescence enhanced genes show increased expression in response to pathogens and other stresses as well as during developmental senescence. This subset includes LSC54 encoding a metallothionein protein, which appears to be expressed in response to increases in ROS and is induced in the hypersensitive pathogen response, during necrotic cell death, in response to wounding, UV-B and other oxidative treatments, as well as during leaf senescence. In contrast, other genes appear to be specific to senescence. For example, the SAG12 gene encoding a cysteine protease is only expressed in yellowing leaf tissue and senescing petals. Therefore, it is likely that multiple interlinking regulatory pathways control gene expression during senescence. Some of these pathways are common to other responses such as pathogen infection or oxidative stress, while other pathways are probably unique to senescence. (Different research groups have named cloned genes according to their expression in senescence, i.e., LSC (leaf senescence clone), SAG (senescence associated gene), SEN (senescence enhanced), etc.)

Hormonal Regulation of Senescence

Although many of the main plant hormones have been implicated in the control of senescence, the cases for ethylene and cytokinin are the most convincing. Ethylene is the ripening hormone and its presence is essential for the ripening of many fruit. Plants that are exposed to ethylene show premature senescence, and the older leaves on the plant are induced to go yellow. The younger leaves on the same plant remain green. The leaves of an ethylene insensitive mutant of *Arabidopsis* (*Etr1*) were delayed in their onset of senescence; similarly, an antisense tomato (*Lycopersicon esculentum*) plant that synthesized very low levels of ethylene showed delayed leaf senescence. However, senescence did occur normally once the process started. Hence, it was concluded that ethylene is a modulator of leaf senescence in these plants; its presence will speed up the senescence process, but is not essential for senescence to occur. Leaves have to be a certain age to be ready for the ethylene signal; young leaves treated with ethylene do not senesce.

It has been known for many years that cytokinin treatment could be used to inhibit leaf senescence. The importance of cytokinin in the regulation of senescence was confirmed in an elegant experiment by Gan and Amasino in which they fused the senescence enhanced SAG12 promoter to a cytokinin biosynthesis gene. Plants transformed with this fusion construct started to express the cytokinin biosynthetic gene as senescence commenced, when the gene promoter was induced. This resulted in cytokinin biosynthesis in the senescing leaf, which was sufficient to inhibit senescence and the gene expression was switched off. This autoregulation of cytokinin biosynthesis resulted in the leaves on these transgenic tobacco plants remaining green and nonsenescent. This experiment showed very clearly that loss of cytokinin is an early signal in the control of leaf senescence. However, the signal that causes the reduction of cytokinin in the leaf remains a mystery.

Other Senescence Inducing Signals

One of the first indications of senescence is a drop in the rate of photosynthesis and this would result in a reduced sugar level in the leaf. It has been proposed that this may be a signal for altered gene expression during senescence, and there is some preliminary evidence for this. Expression of some senescence enhanced genes can be inhibited by the application of sucrose and, also, transgenic plants with reduced internal levels of glucose and fructose due to overexpression of hexokinase showed accelerated leaf senescence. Therefore internal sugar levels may influence downstream processes of senescence.

Another potential signal for the onset or progression of senescence may be the redox status of the leaf cells, which may change due to increases in ROS levels. There are many sources of ROS, e.g., hydrogen peroxide (H_2O_2) and superoxide (O_2^-), produced by metabolic activities in the chloroplast and the peroxisome in both photosynthetic and senescent cells. The toxicity of these ROS is controlled by the antioxidant systems of the cell, particularly the ascorbate glutathione redox cycle. Increased age of the tissue may result in increased membrane damage and a decline in the antioxidant protection. The balance of antioxidant enzyme levels and ROS production changes during senescence, with consequent alteration in the redox state of the cell. This may be a signal for the expression of senescence enhanced genes, and it has been shown that the expression of a subset of senescence enhanced genes is induced in green leaves following treatments that induce oxidative stress. However, many senescence enhanced genes are not induced by these treatments, and hence their expression is not responding to redox changes.

The signaling molecules salicylic acid (SA), jasmonic acid (JA), and ethylene have been implicated in complex interconnecting pathways that control gene expression in plant pathogen responses and also in plant responses to oxidative stress such as ozone treatment. There is some evidence that similar pathways may have a role in gene expression during senescence. Recently, it was shown that expression of certain genes during leaf senescence depended on the presence of SA. Senescing leaves for a transgenic *Arabidopsis* plant carrying the NahG gene, which encodes a protein that degrades SA, showed no expression of certain senescence enhanced genes, including SAG12 and the PR genes. Other genes were reduced in levels in the senescing leaves from this transgenic plant indicating partial requirement for the SA pathway. Therefore, this pathway has a role in gene expression during leaf senescence.

Potential signals controlling gene expression during senescence are depicted in **Figure 5**. The onset of developmental senescence appears to be controlled by cytokinin levels and the removal of plant organs from the plant would also disrupt the cytokinin supply to the organ, which may be the signal that induces postharvest senescence. Reduced cytokinin may affect photosynthetic activity leading to reduced sugar levels. At the same time, an increase in ROS due to metabolic changes or deterioration of membranes due to ageing could induce gene expression and also activate pathways involving SA, JA, or ethylene. Further research is required to elucidate the complex signaling pathways that control gene expression during senescence.

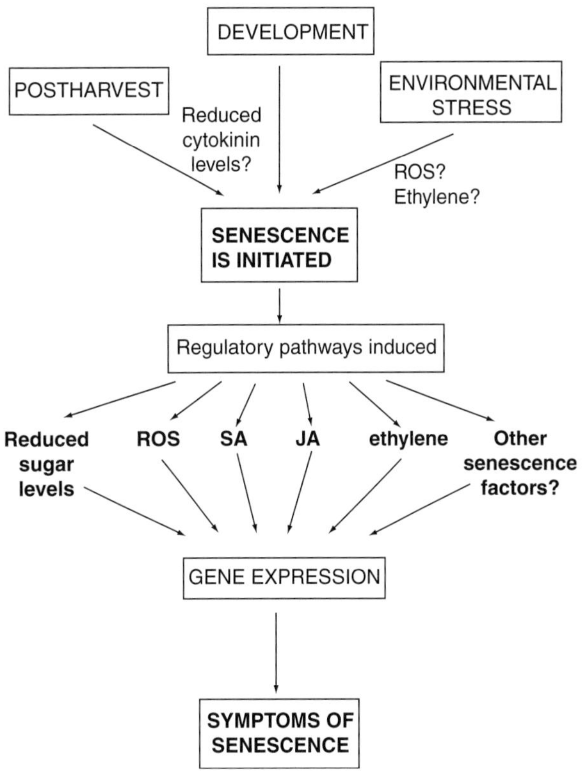

Figure 5 Signaling pathways that may be involved in the control of gene expression during senescence. A model. Senescence is initiated by different internal and external factors and the signals that initiate the process are not known. Reduced cytokinin levels during development or after harvest and increased ROS levels in response to stress may be early signals. The initiation signal induces the expression of regulatory genes, which activate signaling pathways such as the SA pathway, and these control the expression of the downstream genes that encode the enzymatic functions of senescence. SA, salicylic acid; JA, jasmonic acid; ROS, reactive oxygen species.

Future Applications of Senescence Research

Leaf senescence is a process that is unique to plants. Senescence in animals tends to be synonymous with ageing and results in death of the whole organism, while in plants, senescence is often necessary for the continued success of the plant. Senescence in plants is a form of programed cell death, in that the eventual outcome of the process is the death of the cell but, before this happens, the senescing cell is actively and carefully dismantled. Increased understanding of the genes and pathways by which plants control leaf senescence could be of key importance for future agronomic improvements. Premature senescence due to environmental stress can have severe consequences on the yield of a crop. The ability to delay senescence in a grain crop to extend the photosynthetic period could result in increased yields, and the stay green trait is actively pursued in crop breeding programs. Postharvest senescence affects the quality of the product both due to deterioration in the appearance and also in loss of health value due to degradation of nutrients. Research into the causes and consequences of senescence is a fascinating area of study that has important potential benefits for mankind.

See also: **Genetic Modification**: Gene Cloning, General Principles. **Growth and Development**: Control of Gene Expression, Post Transcriptional Regulation; Control of Gene Expression, Regulation of Transcription; Leaf Development. **Photosynthesis and Partitioning**: Sources and Sinks. **Regulators of Growth**: Cytokinins; Ethylene; Jasmonates.

Further Reading

Buchanan-Wollaston V (1997) The molecular biology of leaf senescence. *Journal of Experimental Botany* 48: 181–199.

Buchanan-Wollaston V, Earl S, Harrison E, *et al.* (2003) The molecular analysis of leaf senescence – a genomics approach. *Plant Biotechnology Journal* 1: 3–22.

Feller U and Fischer A (1994) Nitrogen metabolism in senescing leaves. *Critical Reviews of Plant Science* 13: 241–273.

Gan S and Amasino RM (1996) Cytokinins in plant senescence: from spray and pray to clone and play. *Bioessays* 18: 557–565.

Matile P, Hortensteiner S, and Thomas H (1999) Chlorophyll degradation. *Annual Review of Plant Physiology and Plant Molecular Biology* 50: 67–95.

Page T, Griffiths G, and Buchanan-Wollaston V (2001) Molecular and biochemical characterisation of post harvest senescence in broccoli. *Plant Physiology* 125: 718–727.

Quirino BF, Noh Y, Himelblau E, and Amasino RM (2000) Molecular aspects of leaf senescence. *Trends in Plant Science* 5: 278–282.

Smart CM (1994) Gene expression during leaf senescence. *New Phytologist* 126: 419–448.

Takamiya K, Tsuchiya T, and Hiroyuki O (2000) Degradation pathway(s) of chlorophyll: What has gene cloning revealed? *Trends in Plant Science* 5: 426–431.

Thompson JE, Froese CD, Madey E, Smith MD, and Hong Y (1998) Lipid metabolism during plant senescence. *Progress in Lipid Research* 37: 119–141.

Senescence, Flowers

E J Woltering, Wageningen University, Wageningen, The Netherlands

Flower Senescence, Wilting, and Abscission

The angiosperm flower is a complex terminally differentiated structure composed of several different organs, each of which has a variety of specific functions essential for successful reproduction. These include the production of pollen in the anthers, attraction of pollinators by the perianth, reception of pollen and guidance of pollen tubes by stigma and style, fusion of gametes in the ovules, and development of viable seeds in the ovary. Many flowers have complex corollas which serve to attract pollinators. Following pollination dramatic changes in appearance often occur. The maintenance of a showy corolla is costly in terms of required resources and it is believed that plants have evolved mechanisms to optimize floral longevity in the face of varying floral maintenance cost and varying contributions of flowers to fitness. Viewed in this way, flower senescence can be regarded as a process tightly controlled by developmental and environmental factors. The underlying biochemistry and molecular biology has only partly been elucidated.

Petal or corolla senescence is unwanted in the commercial flower trade. Therefore, senescence has been studied in species that are used as commercial cut flowers, potted plants or bedding plants such as orchids (Orchidaceae), carnation (*Dianthus caryophyllus*), iris (*Iris* spp.), day lily (*Hemerocallis* spp.), and *Petunia*. In addition to petal senescence, abscission of petals or whole corollas at the termination of flower life occurs in many species. Generally, flower parts abscise while still fully turgid and without any visible signs of deterioration. Although abscission of flower parts has been studied in great detail with respect to changes in the abscission zone, there are no reports on physiological or biochemical changes in the petal tissues during or after actual abscission.

Senescence is a highly regulated active process, which can be defined as the metabolic changes that are part of a genetically based program leading to death of the cells involved. Flower senescence often refers to deterioration of the perianth parts (petals, corolla, tepals) and not necessarily means that other flower parts such as the ovary or receptacle also senesce. In fact, in pollinated flowers, developmental

Plate 18 Growth Development View of the unusual canopy of the dipterocarp *Dryobalanops lanceolata* characterized by gaps between the foliage of the individual trees. Photograph kindly supplied by Dr P. van Gardingen with permission.

Plate 19 Nutrition Typical progressive symptoms of Fe deficiency in common bean *(Phaseolus vulgaris)*. Reproduced with permission of the American Phytopathological Society Press from *Nutrient Deficiencies and Toxicities of Plants* – digital image collection CD Rom (2000). APS Press: Minnesota.

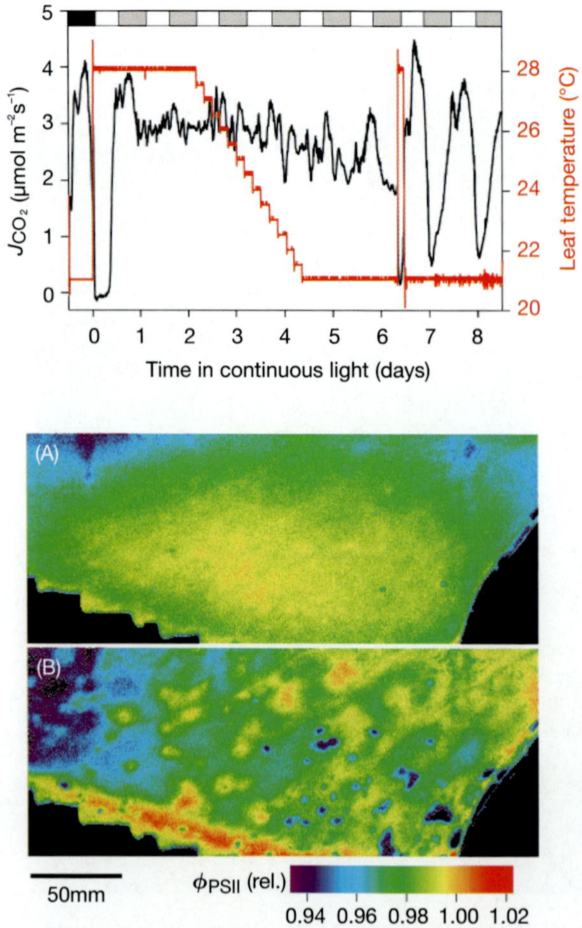

Plate 20 Photosynthesis and Partitioning Synchronization and desynchronization of leaf patches of *Kalanchoë daigremontiana* functioning as separate oscillators in endogenous rhythmicity. The graph shows a normal dark/light cycle of net CO_2 exchange (J_{CO_2}, black line) with dark period (black bar) and light period (first white bar on the upper abscissa) followed by continuous light (with the gray bars on the upper abscissa indicating "subjective" or previous dark periods) in a temperature domain not allowing rhythmicity. When temperature (red line) is reduced gradually into a temperature domain normally allowing rhythmic oscillations, no rhythmicity of J_{CO_2} appears, but after a strong temperature signal of abrupt increase and decrease of temperature (at ca. 6.5 h) oscillations are regenerated immediately. The pictures of the leaf of *K. daigremontiana* show relative quantum efficiency of photosystem II (ϕ_{PSII} (rel.)) in false colors as indicated, with (A) homogeneity (synchronization) at the maximum of J_{CO_2} and (B) inhomogeneity (patchiness or desynchronization) during the transition from maximum to minimum J_{CO_2} in endogenous oscillations at a constant photosynthetic photon flux density (PPFD) of 190 µmol m^{-2}s^{-1} and temperature of 21°C. Reproduced with permission from Rascher U, Blasius B, Beck F, and Lüttge U (1998) Temperature profiles for the expression of endogenous rhythmicity and arrythmicity of CO_2 exchange in the CAM plant *Kalanchoë daigremontiana* can be shifted by slow temperature changes. *Planta* 207: 76-82.

Plate 21 Plants and the Environment Typical symptoms of exposure to UV-B radiation on leaves of cotton plants. (A) Control leaves showing no symptoms; and leaves exposed to UV-B radiation showing (B) initial chlorotic patches and (C) necrotic patches after prolonged exposure.

Plate 22 Postharvest Physiology Different forms of plant senescence. (A) The onset of autumn with its shorter days and cooler temperatures induces the onset of senescence in a deciduous tree. The dramatic red and orange coloration is due to novel synthesis of anthocyanin pigments, which may act as antioxidants. (B) The yellowing of the leaves on this perennial shrub show that senescence starts at the edges of the leaves and progresses inwards. Often the areas closest to the leaf veins remain green the longest. This probably allows the final mobilization into the phloem cells to occur. (C) Leaves of increasing age from a *Brassica napus* plant. These leaves were all harvested off the same plant; yellow leaves were at the base of the plant with increasing greenness sequentially up the plant. The nutrient content of the lower leaves, which have become shaded by the leaves above, is needed for further development of the plant, and so senescence is initiated in these leaves. (D) Postharvest senescence in broccoli. Yellowing of the tissue is evident after several days storage. (E) Stress induced senescence in *Arabidopsis thaliana*. The lower leaves on this plant have been infected by the pathogen *Albugo candida*, and this infection results in yellowing of the leaf tissue.

Plate 23 Postharvest Physiology Some characteristics of ethylene-suppressed antisense ACC oxidase charentais cantaloupe melons (Antisense ACO, right) as compared to the nontransformed wild-type (WT, left). Note that ethylene-suppressed fruit do not develop the peduncular abscission zone (A); retain a green color of the rind on the vine (A) and after harvest (B); keep a firm flesh structure upon extended shelf storage (C), and do not develop chilling injury symptoms upon storage at low temperature (D). The color of the flesh is not affected by ethylene suppression (C).

Plate 24 Production Systems and Agronomy Field trials of oilseed rape cultivars near Cambridge, UK. Photograph courtesy of Dr M R Turner.

Plate 25 Production Systems and Agronomy Inside a modern "Venlo" glasshouse typical of those erected in northern Europe. Note the large glass panes, the lightweight structure, and the ridge mounted ventilators. Photograph courtesy of Dalsem Horticultural Projects B.V.

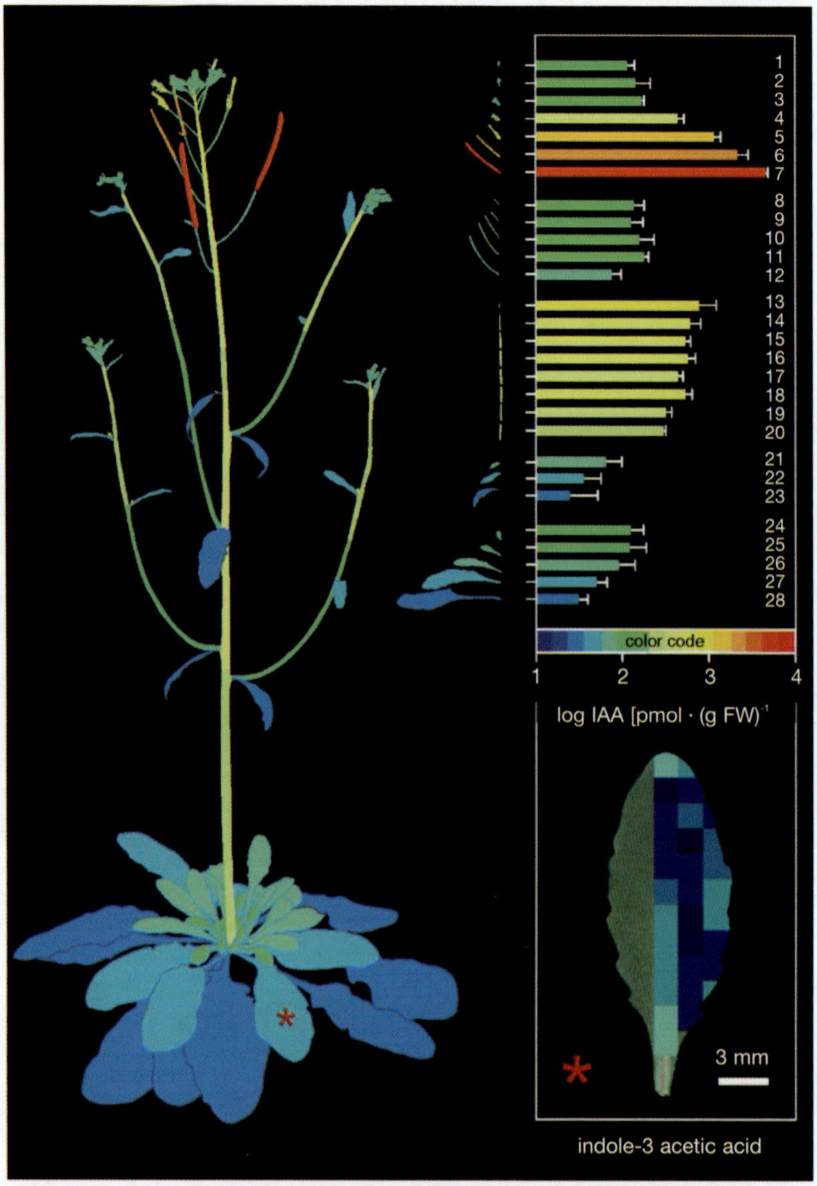

Plate 26 Regulators of Growth Auxin distribution in a mature *Arabidopsis* plant. The distribution of free IAA was measured using gas chromatography/mass spectrometry. Concentrations are illustrated by color-coding the organs, the scale is shown to the right. Reproduced with permission from Müller A, Düchting P, and Weiler EW (2002) A multiplex GC-MS/MS technique for the sensitive and quantitive single-run analysis of acidic phytohormones and related compounds, and its application to *Arabidopsis thaliana*. *Planta* 216: 44-56.

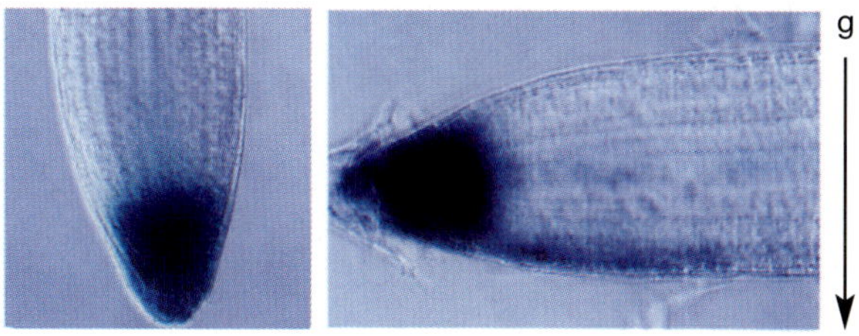

Plate 27 Regulators of Growth Auxin distribution after gravistimulation in roots. Expression of an auxin-responsive reporter gene (*DR5-GUS*) is observed by the accumulation of blueness. The vertical root tip is rich in auxin, but the cells behind the tip show little or no expression, indicating low auxin concentrations. However, if the root tip is turned through 90 degrees to lie horizontal, auxin is seen to accumulate in cells along the lower side in the root elongation zone. In roots, auxin inhibits cell elongation and so the cells on the upper side elongate to push the root tip back towards the vertical. Reformatted from the original; reproduced with permission from Rashotte AM, DeLong A, and Muday GK (2001) Genetic and chemical reductions in protein phosphatase activity alter auxin transport, gravity response, and lateral root growth. *Plant Cell* 13: 1683-1697. Copyrighted by the American Society of Plant Biologists.

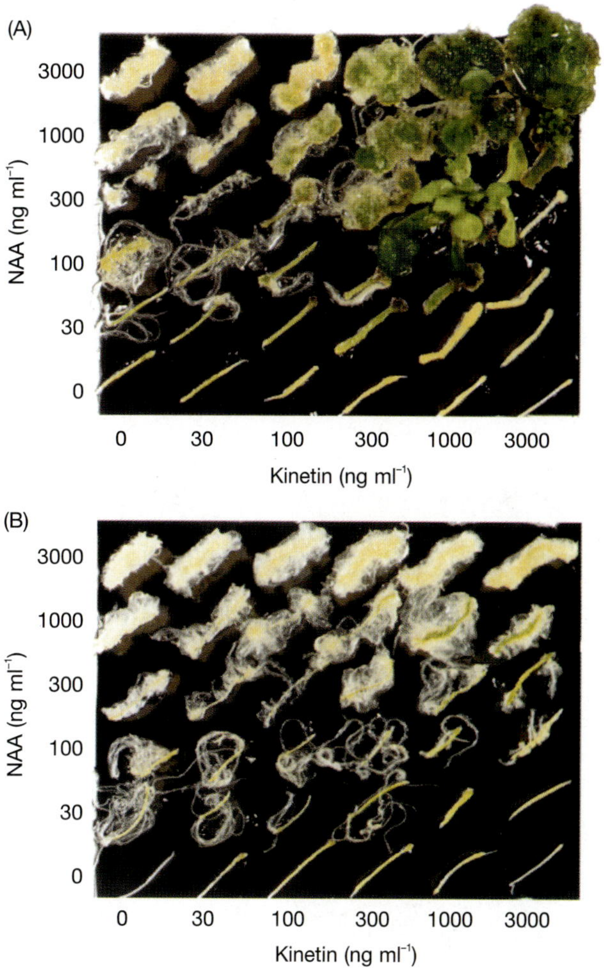

Plate 28 Regulators of Growth Callus growth, and shoot and root organogenesis of wild-type *Arabidopsis* and the cytokinin-resistant mutant *cre1-1* in media containing different cytokinin and auxin concentrations. Hypocotyl segments were excised and cultured on media containing different levels of kinetin and NAA. Wild-type explants (A) proliferated rapidly, turned green and produced shoots in the presence of high concentrations of cytokinins; the *cre1-1* explants (B) did not. Reproduced with permission from *Nature* from Inoue T, Higuchi M, Hashimoto Y, *et al.* (2001) Identification of *CRE1* as a cytokinin receptor from *Arabidopsis. Nature* 409: 1060-1063: ©2001 Macmillan Publishers Ltd.

Plate 29 Regulators of Growth Tobacco transformed with the bacterial *ipt* gene. Note release of axillary buds and inhibition of root growth in lines overexpressing the *ipt* gene. Reproduced with permission from McKenzie MJ, Mett V, Reynolds PHS, and Jameson PE (1998) Controlled cytokinin production in transgenic tobacco using a copper-inducible promoter. *Plant Physiology* 116: 969-977: © The American Society of Plant Biologists.

Plate 30 Regulators of Growth Phenotype of BR dwarf mutants. (A) Comparison of several 5-week-old BR-deficient or BR-insensitive (*bri1*) mutants of *Arabidopsis* with a wild-type *Arabidopsis* plant of the same age. See the text for a description of each mutant. (B) Close-up of an 8-week-old *bri1* mutant of *Arabidopsis* showing dark green, curled leaves and cabbage-like phenotype. (C) The *dpy* mutant of tomato (left) shows a dwarf phenotype with curled leaves and suppression of axillary shoots. When treated with 0.1 μM brassinolide twice weekly, the mutant is rescued to wild-type phenotype (right). (D) Phenotype of the *curl*-3 tomato BR-insensitive mutant. Note the similarity to *bri1* (B). Adapted with permission from Sakurai A, Yokota T, and Clouse SD (1999) *Brassinosteroids: Steroidal Plant Hormones.* Tokyo: Springer-Verlag and Koka CV, Cerny RE, Gardner RG, *et al*. (2000) A putative role for the tomato genes dumpy and curl-3 in brassinosteroid biosynthesis and response. *Plant Physiology* 122: 8-98.

processes in the ovary are often initiated as petals senesce. Developmental leaf senescence is regarded as a functional process in which nutrients are transported out of the leaf to sustain growth and development of other plant parts. Petal senescence may, however, merely function to affect flower attractiveness to pollinators and the senescence process may therefore differ from the process in leaves.

Here aspects of petal senescence are discussed as part of the floral developmental program. This process should not be confused with (acute) petal wilting induced by dehydration, e.g., in excised flowers with limited water uptake due to bacterial or embolism-induced vascular occlusion. The following discussion presents the present knowledge on the physiological, biochemical, and molecular events during senescence of showy flower parts, and will briefly introduce current techniques to delay postharvest flower senescence. For simplicity, the term petals will be used for different types of showy flower parts (separate petals, corolla, tepals).

Changes in Cell Membranes

One of the important events during senescence is loss of integrity of the cell membranes leading to leakage of the cell contents to the apoplast and intercellular spaces, increased water loss, and subsequent desiccation of the tissue. The resulting loss of turgidity (sometimes in specific cell layers) may give rise to typical senescence symptoms such as inrolling of the petals and closure of the flower. Membrane integrity changes are thought to result from changes in membrane lipid fluidity caused by lipid peroxidation and/or increased sterol:phospholipid ratio. Lipid peroxidation has been attributed to an increase in reactive oxygen species (e.g., O_2^-, $OH^\bullet$, and H_2O_2) resulting from, for example, increased lipoxygenase activity and decreased activities of protective enzymes such as catalase, ascorbate peroxidase, (cytosolic) superoxide dismutase, and glutathione-*S*-transferase or antioxidants such as ascorbate, glutathione, and α-tocopherol. The increased sterol: phospholipid ratio is thought to result from both an enhanced hydrolysis and decreased synthesis of phospholipids during senescence. Increased lipid peroxidation and sterol:phospholipid ratio in membrane are reflected in an increased leakage of ions from the petals when emerged in distilled water. The exact order in which these changes appear during senescence has not been accurately determined; also it is not clear whether the increases in ion leakage are due to an overall gradual disfunctioning of the cell membranes or due to disintegration of a limited number of cells. There is a general consensus that changes in membrane properties occur prior to increased ion leakage, subsequent fresh weight loss, and visual senescence.

Carbohydrate Metabolism during Senescence

Extensive breakdown of starch or of other carbohydrate storage forms in the petals often occurs during flower opening. The released sugars presumably function as osmotic compounds necessary for rapid petal expansion. During senescence, carbohydrate levels in the petals generally decline and the resulting "sugar starvation" may be one of the triggers of senescence. In support of this hypothesis is the observation that placing cut flowers in a sugar-containing solution delays senescence. However, even in petals from fully senesced flowers, substantial amounts of sugars may still be present, indicating that the subcellular localization of the remaining carbohydrates may limit availability. In flowering stems such as *Gladiolus* and lilies (*Lilium* spp.), redistribution of carbohydrates from open or senescing flowers to developing flower buds occurs. The presence of developing buds on the flowering stem shortens the life of the older flowers, indicating that the sugar status is an important factor in flower longevity.

In many cut flowers, respiration decreases during the postharvest life, which may be due to depletion of accessible respiratory substrate. In climacteric flowers (see below), there is often a small upsurge at the onset of senescence concomitant with the increase in ethylene production. During this respiratory upsurge in carnation flowers, the ratio of normal to cyanide (CN)-resistant respiration decreases suggesting that the toxic byproducts of CN-resistant respiration may contribute to senescence.

The remarkable changes in size and shape of petals during senescence implicate that considerable changes in cell wall properties must occur. Although there are some reports on decreases in cell wall constituents and changes in wall-based enzymes, their role in senescence is not clear and this is certainly an area where more research is needed.

Ethylene and Flower Senescence

Traditionally, flowers (like fruits), are categorized as being climacteric or nonclimacteric. In climacteric or ethylene-sensitive flowers such as carnations, *Gypsophila*, and orchids, senescence is accompanied by a sudden, transient increase in ethylene production and respiration while treatment of nonsenescent flowers with ethylene rapidly induces petal senescence. In nonclimacteric flowers such as *Gladiolus*, tulip (*Tulipa* spp.), and *Iris*, senescence may be as rapid as in climacteric flowers but generally, no increases in ethylene production and respiration are apparent. Although petal senescence in these flowers is insensitive to applied ethylene, other aspects of flower development such as flower opening, ovary development, or pedicel elongation may be responsive to ethylene. With respect to petal senescence, sensitivity to ethylene is roughly determined at the plant family level (**Table 1**). High sensitivity is found in Campanulaceae, Caryophyllaceae, Geraniaceae, Lamiaceae, Malvaceae, Orchidaceae, Primulaceae, Ranunculaceae, and Rosaceae species. Low sensitivity is found in Asteraceae and Iridaceae species and in most of the Amaryllidaceae and Liliaceae species. Sensitivity of species within one genus is generally comparable and the same is true for varieties within one species.

The sudden and transient increase in ethylene production observed in senescing (ethylene-sensitive) flowers, autocatalytic in nature, is caused by ethylene-induced expression of genes coding for the ethylene biosynthetic enzymes, 1-aminocyclopropane-1-carboxylic acid (ACC) synthase and ACC oxidase, thus leading initially to a pronounced increase in ethylene production (**Figure 1**). This positive feedback is mediated through a class of membrane-associated ethylene binding proteins and accompanying signal transduction components. Ethylene is known to affect a diversity of metabolic processes such as those involved in the loss of membrane integrity, cell wall changes, and cell death processes. It is generally believed that ethylene effects are primarily caused by ethylene-induced changes in expression of specific genes.

In ethylene-sensitive species, treatment with low concentrations of ethylene induces autocatalytic ethylene production and flower senescence whereas treatment with chemicals that block either ethylene biosynthesis or ethylene action greatly delay senescence. Such chemicals are widely used in commercial practice to prolong the life of flowers (**Figure 1**; see also below). Molecular genetic approaches have also demonstrated the pivotal role of ethylene in petal senescence in several flower species. Prevention of the accumulation of ACC (the direct precursor of ethylene) through downregulation of ACC synthase mRNA levels prolonged the vase life of carnation flowers. Similarly, overexpression of a bacterial ACC diaminase gene converting ACC into a metabolically inactive substance (α-ketobutyrate) greatly delayed petal senescence in petunia. Prevention of the

Table 1 Ethylene sensitivity of petal senescence and abscission

Plant family and species	*Ethylene sensitivity*[a]	*Symptoms*[b]
Acanthaceae	High	Abscission
Acanthus hungaricus		
Thunbergia alata		
Amaryllidaceae	Insensitive/low	Senescence
Alstroemeria pelegrina		
Narcissus pseudonarcissus		
Campanulaceae	Intermediate/high	Senescence
Campanula pyramidalis		
Trachelium caeruleum		
Caryophyllaceae	High	Senescence
Dianthus caryophyllus		
Gypsophila paniculata		
Asteraceae	Insensitive/low	Senescence
Chrysanthemum morifolium		
Liatris spicata		
Dipsacaceae	Intermediate/high	Senescence
Scabiosa caucasica		
Succisella inflexa		
Euphorbiaceae	Low	Senescence
Euphorbia fulgens		
Euphorbia pulcherrima		
Geraniaceae	High	Abscission
Geranium gracile		
Geranium sanguineum		
Gesneriaceae	High	Abscission
Saintpaulia ionantha		
Streptocarpus (hybrids)		
Iridaceae	Insensitive/low	Senescence
Gladiolus (hybrids)		
Iris (hybrids)		
Labiatae	High	Abscission
Physostegia virginiana		
Mentha sauveolens		
Liliaceae	Insensitive/low	Senescence
Lilium (hybrids)		
Tulipa (hybrids)		
Malvaceae	High	Senescence
Lavatera maritima		
Malva sylvestris		
Orchidaceae	Intermediate/high	Senescence
Cymbidium (hybrids)		
Phalaenopsis (hybrids)		
Primulaceae	Intermediate/high	Abscission
Lysimachia clethroides		
Primula denticulata		
Ranunculaceae	Intermediate/high	Abscission
Aconitum napellus		
Delphinium ajacis		
Rosaceae	Intermediate/high	Abscission
Potentilla grandiflora		
Rosa (hybrids)		
Scrophulariaceae	Intermediate	Abscission
Antirrhinum majus		
Veronica spicata		
Apiaceae	Insensitive	Senescence
Anethum graveolens		
Conium maculatum		

[a]Sensitivity to ethylene was determined following treatment of flowers with $3\,\mu l\,l^{-1}$ ethylene for 24 h at 20°C. High sensitivity, flower petal life reduced by >70%; intermediate sensitivity, flower petal life reduced by 30–70%; low sensitivity, flower petal life reduced by <30%; insensitive, virtually no response.
[b]Prevailing type of petal life termination is indicated. Senescence, petals senesce while attached to the flower; abscission, petals abscise before visual changes.

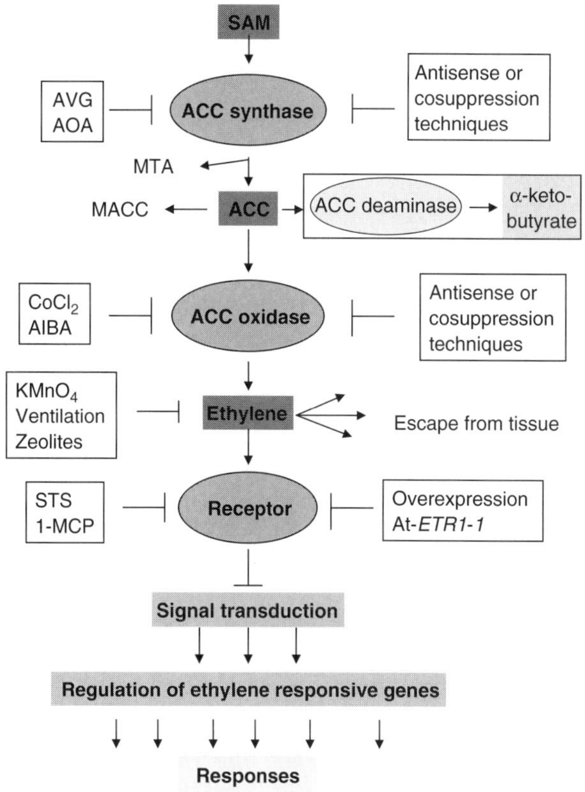

Figure 1 Strategies to inhibit ethylene responses by chemical (boxes left-hand side) and molecular genetic approaches (boxes right-hand side). In plants ethylene is synthesized from *S*-adenosylmethionine (SAM). The first enzyme, ACC synthase, converts SAM into 1-aminocyclopropane-1-carboxylic acid (ACC) and methylthioadenosine (MTA). The conversion of ACC into ethylene is catalyzed by ACC oxidase. Besides being converted into ethylene, ACC may also be converted into malonyl-ACC (MACC) which is believed to be an inactive end product. Ethylene presumably acts by binding, in a reversible way, to a membrane associated receptor protein. Ethylene binding blocks the signaling activity of upstream signal transduction components by which signaling activity of downstream components is activated leading to the enhanced transcription of specific genes and the synthesis and/or activation of enzymes and other proteins responsible for the physiological effects. The various steps in ethylene production and perception can be blocked by chemical inhibitors. Aminoethoxy-vinylglycine (AVG) and amino-oxyacetic acid (AOA) inhibit ACC synthase activity. Cobalt chloride and α-aminoisobutyric acid (AIBA) inhibit ACC oxidase activity. Silver thiosulfate (STS) and 1-methylcyclopropene gas (1-MCP) block the ethylene binding site. Most of these chemicals are used in floriculture for treatment of carnation and other ethylene-sensitive cut flowers. Ethylene, once produced, may diffuse out of the tissue and affect senescence of other commodities in the same storage room. Ethylene can be removed by ventilation with fresh air or by removal using, e.g., zeolites or other matrices impregnated with $KMnO_4$. Molecular genetic strategies to block ethylene production and perception have recently been developed. For both ethylene biosynthetic enzymes, activity has been blocked using antisense or cosuppression techniques. The resulting flowers produce less ethylene and have a longer vase life. Other successful approaches are overexpression of a bacterial gene to convert ACC into an inactive metabolite (α-ketobutyrate) and overexpression of the *Arabidopsis thaliana* mutated receptor gene (*ETR1-1*). The latter strategy leads to complete ethylene insensitivity and greatly prolongs flower life.

conversion of ACC to ethylene by downregulation of ACC oxidase gene expression was similarly effective in prolonging the life of carnation flowers. Blocking ethylene perception by overexpression of an *Arabidopsis thaliana* mutated receptor gene (ETR1-1) blocked function of the native receptor and effectively prolonged carnation vase life (**Figure 1**).

Apart from ethylene, other plant hormones may play regulatory roles in flower senescence. Cytokinins and gibberellins are considered antisenescent whereas abscisic acid is considered stimulatory. These effects may partly be related to the effects of such compounds on ethylene production and sensitivity; however, in ethylene-insensitive species also combinations of antisenescence hormones may delay petal senescence.

Programmed Cell Death (Apoptosis) in Petal Senescence

Programmed cell death (PCD) is a functional concept derived from the mammalian field, which refers to cell death that is part of the normal development of a multicellular organism; it involves controlled disassembly of the cell. PCD is employed during developmental events and in defensive mechanisms against infected or mutated cells. Deregulation of PCD is implicated in various human diseases ranging from cancer and autoimmune disorders to neurodegenerative diseases and AIDS.

In animal cells, PCD is often associated with the occurrence of a specific set of cellular morphological features such as condensation of the nucleus and the cytoplasm, fragmentation of DNA, and fragmentation of the cell into membrane confined DNA-containing vesicles (apoptotic bodies). The cell death process showing such morphological features is called apoptosis. Apoptosis is, in addition, associated with the degradation of chromosomal DNA into large (50–300 kb) and subsequently small (180 bp) fragments that can be visualized on agarose gels.

The core component of the apoptotic machinery is a proteolytic cascade involving a family of cysteine proteases named caspases. Caspase activation is controlled by interactions with both pro- and anti-apoptotic proteins and, once activated, caspases may process and activate downstream caspases. The activated executior caspases cleave numerous cellular proteins thereby activating degradative enzymes such as nucleases or deactivating protective enzymes such as DNA repair enzymes. Eventually this leads to dismantling of the cell and the apoptotic phenotype.

In some plant cell death processes, induced by a variety of biotic or abiotic factors, a striking similarity to animal apoptotic phenotype exists (e.g., nuclear condensation and fragmentation and formation of apoptotic bodies) and this process can be blocked by inhibitors specific to (animal) caspases. Furthermore, in some plant systems caspase-like activity has been shown during cell death. It is tempting to speculate that a process comparable to animal apoptosis may take place during petal senescence.

One of the hallmarks of apoptosis, fragmentation of DNA, can be detected *in situ* by the TUNEL method (see list of technical nomenclature) or by detection of DNA laddering patterns on an agarose gel. In petals of pea (*Pisum sativum*), *Gypsophila*, *Petunia*, and *Freesia*, TUNEL positive nuclei and/or DNA laddering were detected during senescence. The occurrence of these cell death markers was hastened by treatment with ethylene and delayed when senescence was delayed by treatment of the flowers with the ethylene perception blocker silver thiosulfate (STS). These observations indicate that during petal senescence, a cell death process exhibiting some similarities to animal apoptosis is induced and further elucidation of this pathway will provide information about the basic cell death mechanism.

During petal senescence decreases in RNA and protein content and increases in DNase, RNase, and protease activity are observed which fits in the idea of a programmed, ordered breakdown of the cells. The trigger(s) of this cell death process in petals have not yet been established. In suspension cultured plant cells, oxidative stress, in particular increased hydrogen peroxide, resulting from, for example, treatment with fungal elicitors can function as a signal to trigger apoptotic-like cell death. Ethylene was found to greatly potentiate the production of hydrogen peroxide during this process. Similarly, the increase in reactive oxygen species apparent during petal senescence may, in addition to their effect on lipid peroxidation, function as a trigger for ordered breakdown of the cells.

Pollination-Induced Flower Senescence and Interorgan Signaling

In many species, pollination (deposition of viable pollen onto the mature receptive stigma) induces rapid changes in the petals. This may include changes in pigmentation, senescence, or abscission. Pollination-induced changes in pigmentation have been reported in over 50 angiosperm families. Premature senescence of petals is another common response to pollination and has been studied in detail in flowers such as carnation, petunia, and several orchids. Pollination-induced petal abscission has been reported in foxglove (*Digitalis*), *Cyclamen*, and *Pelargonium*. Given the petal's role in attraction of pollinators, pollination-induced pigmentation changes, senescence, and abscission probably evolved to deter further visits by pollinators. Pollination also can induce developmental processes in the ovary that may affect senescence processes in the petals by changing source–sink relations. There are many species, however, in which pollination has no apparent effect on the rate of senescence and these are usually ethylene-insensitive types with a very short lifespan such as day-lily, iris, and many ephemeral flowers or Asteraceae species where floral attraction is expressed often through sterile ray florets. In long-lived species ethylene responsiveness of petal senescence may have evolved to optimize pollinator energy or flower life in face of the cost to maintain them.

In ethylene-sensitive species, pollination often leads to a pronounced increase in ethylene production eventually culminating in senescence or abscission of the petals. Despite extensive research, the pollen factor(s) responsible for the initial increased ethylene production and other postpollination responses, such as swelling and closing of the stigma in orchids and development of the ovary in many flower species, have not been determined with certainty. It has been established that neither the initial contact of the pollen with the stigmatic surface nor the wounding reactions associated with pollen tube penetration and subsequent growth through the stylar transmitting tissues elicit the early increase in ethylene production. Pollen of many species were found to contain high ACC concentrations. As stigmas and styles generally exhibit high ACC oxidase activity, pollen ACC was long thought to cause the early pollination-induced ethylene production. However, using ACC synthase inhibitors and transgenic pollen devoid of ACC it was clearly shown that pollination-induced ethylene is derived from newly synthesized ACC in the stigma/style region. Presumably, existing ACC synthase is activated or newly synthesized by pollination. Among different factors tested (auxin, methyl jasmonate, polygalacturonic oligomers, brassinolide, systemin, and different flavonols) only auxin (indole-3-acetic acid) was able to completely mimic the pollination response in orchid flowers. Auxin is also a prime candidate pollen factor responsible for ethylene production in carnation. Gene expression studies show that pollination rapidly induces a (auxin-responsive) style-specific ACC synthase responsible for a first relatively small upsurge in stylar ethylene production.

Following the initial increase in ethylene production in the stigma/style region, increases in ACC content and ethylene production are observed first in the ovary and then in petals. This sequence of events indicates that an ethylene-inducing factor moves through the flower and coordinates the senescence process. Gene expression studies in pollinated carnation, tobacco (*Nicotiana tabacum*), and orchid flowers have revealed the nature of the translocated factor. Pollination-induced ethylene in the stigmatic region triggers expression of additional (ethylene-responsive) ACC synthase and ACC oxidase genes and, as a consequence, ethylene production in the stigmatic region, and subsequently down the length of the style and in the ovary. Ovary-produced ethylene diffuses into the petal base where it may induce yet additional (petal-specific) ethylene dependent ACC synthase and ACC oxidase genes and ethylene production. Presumably through changes in gene expression and resulting biochemical events, ethylene eventually causes the observed senescence symptoms.

Diffusion of ethylene from one flower part to another is also thought to coordinate senescence in unpollinated flowers. Although in unpollinated flowers there is no direct reason for initiation of developmental processes associated with pollination and fertilization, ovary development is often observed during senescence and was found to be markedly stimulated by ethylene and inhibited by pretreatment of the flowers with the ethylene perception blocker STS. Removal of the ovary considerably increases carnation flower life. Changes in hormone homeostasis in the ovary may be the senescence initiating factor in unpollinated flowers.

Gene Expression during Petal Senescence

Petal senescence is accompanied by changes in the expression of numerous genes and, based on homology studies with other known proteins, putative roles of some of the proteins in petal senescence have emerged (**Table 2**). In ethylene-sensitive (climacteric) flowers, genes encoding the ethylene biosynthetic enzymes (ACC synthase and ACC oxidase) have been found to increase during petal senescence confirming the autostimulatory role of ethylene. On the contrary, genes encoding receptor proteins (two-component histidine kinases) have

Table 2 Genes and their putative functions whose expression is upregulated during petal senescence

Putative function	*Putative identity*	*Species*
(Membrane) lipid metabolism	Acyl lipid desaturase	Rose
	Lipase	Carnation
	Acyl CoA oxidase	*Phalaenopsis*
	In-chain fatty acid hydroxylase	Day lily
	Allene oxide synthase	Day lily
	Lypoxygenase	Rose
	Carboxy PEP mutase	Carnation
Cell wall metabolism	β-Galactosidase	Carnation
Protein degradation	Cysteine proteases	Carnation, day lily
	Cysteine protease inhibitor[a]	Carnation
	Aspartic protease	Day lily
Nucleic acid degradation	S1-type nuclease	Day lily
Oxidative events	Lipoxygenase	Rose
	Glutathione-*S*-transferase	Carnation
Nutrient redistribution	Asparagin synthetase	*Sandersonia aurantiaca*
Ethylene biosynthesis	ACC synthase	E.g., carnation, petunia, orchids
	ACC oxidase	E.g., carnation, petunia, orchids
Ethylene perception	Two component histidine kinase (ETR, ERS)[b]	Carnation, pea, rose
Others	Fatty acid elongase	Day lily
	β-Glucosidase	Carnation

[a]Cysteine protease inhibitor is downregulated which would imply increased protease activity.
[b]ETR and ERS are negative regulators. Gene expression may be upregulated or downregulated during senescence. Effect on ethylene sensitivity is not known yet.

been found to decrease in some flowers. As the ethylene receptor is thought to be a negative regulator, a decrease in number or activity would imply increased ethylene sensitivity. Genes that may be related to cell wall metabolism (β-galactosidase), detoxifying mechanisms (glutathione-*S*-transferase), membrane lipid metabolism (carboxyphosphonoenolpyruvate mutase, acyl CoA oxidase, lipase), and protein degradation (cysteine and aspartic proteases) have been found to increase during petal senescence in both ethylene-sensitive and ethylene-insensitive flowers. Together these functionalities support the results derived from biochemical studies. Although many of these gene products may play a role in the senescence syndrome, an overall picture of the sequence of events and their interactions is still lacking. Molecular genetic strategies to prolong flower life have, till now, been focused on manipulation of ethylene production or perception in ethylene-sensitive flowers such as carnation and petunia.

Genes associated with, for example, membrane degradation, detoxification of lipid peroxides and ROS, protein degradation and nitrogen redistribution were identified during senescence of *Arabidopsis* and *Brassica* leaves and immature broccoli florets. This supports the view that petal and leaf senescence proceed through similar processes.

Methods to Delay Postharvest Flower Senescence

During commercial flower handling a main factor affecting flower life and quality is the temperature. Storage and transport at a low temperature severely slows down metabolic processes and greatly reduces ethylene production and sensitivity. Respiration rates of flowers increase exponentially with increasing temperature and, given the negative relation between respiration and vase life, it is important to keep the temperature as low as possible. Generally, in the temperature range between 0 and 10°C, Q_{10} values are between 3 and 5. However, in *Matthiola incana* and *Narcissus* a much higher Q_{10} value (around 7) was observed indicating that proper temperature control is crucial for quality maintenance in flower species. The effect of temperature on ethylene sensitivity is even more dramatic. Ethylene sensitivity of carnation flowers showed a Q_{10} value around 11 between 5 and 25°C. Consequently, most commercial flower species can be stored for periods up to 1 or 2 weeks at 1–4°C without negative side effects. Notable exceptions are orchid species and *Anthurium* varieties that suffer from such low temperatures and generally require a storage temperature of about 12°C.

In addition to temperature control, cut flowers may be treated after harvest with a diversity of chemical solutions. Apart from the inclusion of bactericides such as aluminum sulfate, chlorine, and quaternary ammonium compounds to inhibit bacterial growth, the vase water may be supplemented with chemicals to affect specific physiological processes. Ethylene-sensitive flower species are often treated with compounds that either block ethylene production (amino-oxyacetic acid, AOA) or ethylene perception (silver thiosulfate, STS). Such chemicals strongly reduce the effects of ethylene and may greatly prolong the vase life as both petal senescence and abscission are reduced. Of special interest is the recently introduced gaseous compound 1-methylcyclopropene (1-MCP). 1-MCP irreversibly binds to the ethylene receptor thereby preventing ethylene from turning the receptor off (ethylene receptor is a negative regulator). 1-MCP has recently been registered for use on cut flowers and potted plants in the United States and Canada. 1-MCP proved to be a very effective compound. Treatment of flowers with concentrations in the low $nl\,l^{-1}$ (parts per billion) range for only 4–6 h completely blocks the effect of $\mu l\,l^{-1}$ (parts per million) range ethylene concentrations. 1-MCP considerably prolongs flower life in a range of cut flower species (e.g., *Alstroemeria*, orchids, carnation, snapdragon (*Antirrhinum*), *Delphinium*, *Gypsophila*, *Lupinus*) and potted plants (e.g., *Begonia*, *Calceolaria*, *Campanula*, *Euphorbia pulcherrima*, geranium (*Pelargonium*), *Petunia*, *Rosa*), especially if ethylene is present in the atmosphere. It has repeatedly been observed that the effect of 1-MCP is less that that of an STS pretreatment. The reason for this is the possible breakdown of 1-MCP occupied receptors and production of new receptor protein. As STS stays present in the flower tissue it can occupy the newly formed receptors and offer prolonged protection against ethylene.

For ethylene-insensitive flower species, treatment solutions often contain specific combinations of some other plant growth regulators such as gibberellic acid and cytokinins. These plant hormones may prolong flower life, e.g., in tulip and iris, and may also be used to optimize flower opening, e.g., in rose and iris, and to counteract leaf yellowing, e.g., in *Alstroemeria*. To stimulate flower opening and proper development of flower buds, inclusion of sugar in the vase water, together with an effective bactericide, is recommended during the vase life.

Concluding Remarks

The overall picture of the sequence and relative importance of the different biochemical events during flower senescence is still lacking. In ethylene-sensitive flowers, flower life can be greatly extended by blocking production or perception of the senescence trigger, ethylene. This has led to development of very effective chemical treatments and molecular genetic strategies of commercial value. In flowers that exhibit an ethylene-insensitive petal senescence syndrome the basic senescence program may be similar to that in ethylene-induced senescence. However, the hormonal or developmental events that trigger senescence in ethylene-insensitive species have not been identified. Recent emphasis on basic cell death processes (PCD, apoptosis) and the use of gene arrays to study expression profiles of vast numbers of genes will surely lead to new concepts and treatments for prolonging the life of flowers.

List of Technical Nomenclature

1-methylcyclopropene	(1-MCP) An ethylene perception inhibitor; a gaseous hydrocarbon, thought to irreversibly block ethylene binding sites.
Aminooxyacetic acid	(AOA) An inhibitor of ACC synthase.
Q_{10}	Temperature coefficient: change in metabolic activity (fold) for each 10°C increase in temperature.
Silver thiosulfate	(STS) An ethylene perception inhibitor. The active ingredient is Ag^+, thought to irreversibly block ethylene binding site.
TUNEL	*In situ* terminal deoxynucleotidyl transferase-mediated dUTP nick end labeling of DNA 3′-OH groups. A way to show degraded DNA in intact nuclei.

See also: **Postharvest Physiology**: Genetic Engineering for Postharvest Quality; Postharvest Physiological Disorders of Fresh Crops. **Production Systems and Agronomy**: Commercial Flower Production Methodology. **Regulators of Growth**: Ethylene.

Further Reading

Borochov A and Woodson WR (1989) Physiology and biochemistry of petal senecence. *Horticultural Reviews* 11: 15–43.

Celikel FG and Reid MS (2002) Storage temperature affects the quality of cut flowers from the Asteraceae. *Horticultural Science* 37: 148–150.

Cornell University (2003) A summary of physiological processes or disorders in fruits, vegetables and ornamental products that are delayed or decreased, increased, or unaffected by application of 1-methylcyclopropene (1-MCP). www.hort.cornell.edu/department/faculty/watkins/ethylene/index.htm

Dangl JL, Dietrich RA, and Thomas H (2000) Senescence and programmed cell death. In: Buchanan BB, Gruissem W, and Jones RL (eds) *Biochemistry and Molecular Biology of Plants*, pp. 1044–1100. Rockville, MA: American Society of Plant Physiologists.

Have A ten and Woltering EJ (1997). Ethylene biosynthetic genes are differentially expressed during carnation (*Dianthus Caryophyllus* L.) flower senescence. *Plant Molecular Biology* 34, 89–97.

O'Neill SD (1997) Pollination regulation of flower development. *Annual Reviews of Plant Physiology and Plant Molecular Biology* 48: 547–574.

Reid MS and Wu M-J (1992) Ethylene and flower senescence. *Plant Growth Regulation* 11: 37–43.

Rubinstein B (2000) Regulation of cell death in flower petals. *Plant Molecular Biology* 44: 303–318.

Stead AD and van Doorn WG (1994) Strategies of flower senescence: a review. In: Scott RJ and Stead AD (eds) *Molecular and Cellular Aspects of Plant Reproduction*, pp. 215–238. Cambridge: Cambridge University Press.

Woltering EJ and de Vrije GJ (1995) Ethylene: a tiny molecule with great potential. *BioEssays* 17: 287–290.

Woltering EJ and van Doorn WG (1988) Role of ethylene in senescence of petals: morphological and taxonomical relationships. *Journal of Experimental Botany* 39: 1605–1616.

Woltering EJ, Have A ten, Larsen PB, and Woodson WR (1994) Ethylene biosynthetic genes and interorgan signalling during flower senescence. In: Scott RJ and Stead AD (eds) *Molecular and Cellular Aspects of Plant Reproduction*, pp. 285–307. Cambridge: Cambridge University Press.

Storage

J M DeLong and R K Prange, Agriculture and Agri-Food Canada, Kentville, Nova Scotia, Canada

A plan has been invented by which they (gourds) are preserved for food also... to last almost until the next crops are available. This method employs brine....

Pliny the Elder (77 AD) *Natural History*, vol. V, LXXIII

Pomegranates..., put them in a pot with no bottom, bury them in the ground,... so that no outside air can reach them; such fruits will be taken out not only sound, but even larger than they would ever be if they had hung on the tree.

Marcus Terrentius Varro (116–27 BC) *On Agriculture*, Book 1, LXII.

They that die by famine, die by inches.

Matthew Henry (1662–1714)

Pliny the Elder's and Varro's observations of first century agriculture point to an important fact in food production that has not changed since their day: unless a perishable commodity can be stored and maintained in a healthy state, its utility for human nutrition is sharply limited. The implications are obvious: as the art and science of storing foodstuffs is advanced, so also is the stability of the particular civilization. The histories of nations are often checkered with examples of how a periodic or long-term demise of a civilization is linked to outright famine or to the failure of storage practices to maintain the edibility of their main crops. Certainly, natural and political events are bound up in the decline of empires, but the loss of foodstuffs that would have otherwise sustained a population, heavily influence the demise of nations great and small.

Storing the Harvest – An Ancient Art

Without doubt, the ability to store harvested plant organs for extended periods of time has played a critical role in the development of agriculture (**Figures 1–3**). Simple baskets were used as early as 7000 BC to gather and store food until consumed.

Figure 1 Storing (A) and recording the storage (B) of raisons (*Vitis vinifera*) in Egypt about 2500 BC. Hieroglyphics from the tomb of Beni Hasan. Reproduced from Kays SJ (1997) *Postharvest Physiology of Perishable Plant Products.* Athens, GA: Exon Press: public domain.

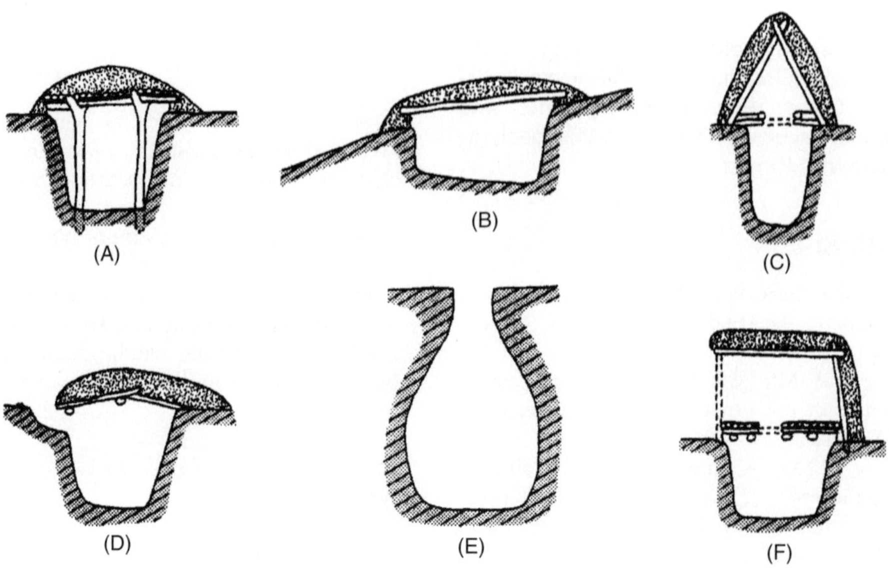

Figure 2 Examples of potato (A,B,D) and grain (C,E,F) storage pits dating from the Paleolithic period, 9000–7000 BC. Reproduced from Kays SJ (1997) *Postharvest Physiology of Perishable Plant Products.* Athens, GA: Exon Press: public domain.

Figure 3 A granary in Ostia, Italy around 600 AD. Courtesy of William P. Thayer.

Advances in technology such as fired clay pottery would enable primitive societies to store crops in concealed environments, creating a simple modified atmosphere. Middle Eastern artisans specialized in making pottery of numerous shapes and sizes for varied usage (ca. 4500 BC), while pre-Neolithic, Middle Eastern societies held grain in underground pits 9000–11000 years ago. Ancient Egyptians and Samarians are thought to have stored some of their crops in sealed limestone crypts to prolong storage life about 2500 BC. Pits were and are still used by primitive societies for storing various types of fruits and vegetables. Silos for long-term grain storage were used by the Romans and continued to be popular until well into the nineteenth century. Ancient (and modern) people often dried fruits such as apples (*Malus pumila*), apricots (*Prunus armeniaca*), figs (*Ficus* spp.), and grapes (*Vitis* spp.) to prolong the storage longevity of these perishable crops.

The Roman writer Varro (116–27 BC) described the construction of an underground pit for grain storage and how it was sealed following filling. It was recognized that entering a sealed storage area too quickly following opening could have disastrous consequences. Today, we know that respiration of stored grain decreases the oxygen (O_2) and elevates the carbon dioxide (CO_2) content of the storage room to levels that cannot support animal life. In essence, this ancient writer knew on a practical level certain fundamental principles of controlled or modified atmosphere technology.

Development of Storage Science

Some of the first experiments in which ripening of fruit was intentionally altered by changing the storage atmosphere were carried out in France by Jacques Berard around 1820. He demonstrated that fruit use O_2 and generate CO_2 while in storage, and if totally deprived of O_2, they did not ripen. Berard found that apples, apricots, peaches (*Prunus persica*), pears (*Pyrus communis*), and prunes (*Prunus domestica*) could be stored in an altered gas environment, and following their removal to ambient air and room temperature, were edible for extended periods.

Refrigeration has had a significant effect on the development of fruit and vegetable storage technology. The rationale for quickly lowering the temperature of harvested crops is essentially the same today as it was in antiquity: refrigeration greatly reduces food spoilage and waste. Although ice and snow have been used since the Roman era to preserve perishable foodstuffs, it was not until the 1800s that natural ice from the colder climes of the northern hemisphere was commonly used to refrigerate foods worldwide.

In the 1860s, Benjamin Nyce, an Ohio commercial storage operator, found that when he limited O_2 surrounding his fruit in an ice-refrigerated store, the produce was greatly improved. Nyce did not license the patent rights for his storage system despite strong commercial interest to implement the method, thus, it was not widely adopted. Ironically, Nyce's work resulted in a *de facto* prototype for modern controlled atmosphere (CA) storage, which, unfortunately, was not further developed for another 60 years.

In the early 1890s, the San Jose Fruit Packing Company experimented with augmented CO_2 levels as a preserver of fruit. In a railroad car experiment, a load of grapes, peaches, pears, persimmons (*Diospyros* spp.), and quinces (*Cydonia oblonga*) stored in elevated CO_2 and without refrigeration were in good condition following 11 days of travel, and held up well in the retail chain. In other early experiments with storage gas modification, elevated CO_2 was found to be helpful, but could not completely compensate for lack of refrigeration. Temperature control was/is fundamental for preservation of fruit and vegetable quality, with modification of the O_2 and CO_2 atmosphere being an important but secondary quality preservation method.

In the 1920s, Kidd and West, working at the Low Temperature Research Station in Cambridge, UK, conducted a number of experiments to determine the optimal atmospheric gas concentrations, i.e., lower O_2 and higher CO_2, for storing apples. It was known at that time that respiration fueled the generation of metabolic heat, which was deleterious to fruit quality in long-term storage. Therefore, one of the goals of

Table 1 Examples of climacteric and nonclimacteric fruits

Climacteric fruits	*Nonclimacteric fruits*
Apple (*Malus pumila*)	Blackberry (*Rubus* spp.)
Apricot (*Prunus armeniaca*)	Cherry (*Prunus avium*)
Avocado (*Persea americana*)	Cucumber (*Cucumis sativus*)
Banana (*Musa* spp.)	Date (*Phoenix dactylifera*)
Blueberry[a] (*Vaccinium* spp.)	Grape (*Vitis* spp.)
Fig (*Ficus* spp.)	Grapefruit (*Citrus paradisi*)
Kiwifruit (*Actinidia deliciosa*)	Lemon (*Citrus limon*)
Mango (*Mangifera indica*)	Olive (*Olea europaea*)
Muskmelon (*Cucumis melo*)	Orange (*Citrus sinensis*)
Nectarine (*Prunus persica*)	Pepper (*Capsicum* spp.)
Papaya (*Carica papaya*)	Pineapple (*Ananas comosus*)
Peach (*Prunus persica*)	Raspberry (*Rubus* spp.)
Pear (*Pyrus communis*)	Strawberry (*Fragaria* spp.)
Plum (*Prunus domestica*)	
Quince (*Cydonia oblonga*)	
Tomato (*Lycopersicon esculentum*)	

Adapted from Kader AA (2002) *Postharvest Technology of Horticultural Crops*, 3rd edn. Publication 3311. Oakland, California: University of California, Agriculture and Natural Resources and DeEll JR, Prange RK, and Peppelenbos HW (2003) Postharvest physiology of fresh fruits and vegetables. In: Chakraverty A, Mujumdar AS, Raghaven GSV and Ramaswamy HS (eds) *Handbook of Postharvest Technology. Cereals, Fruits, Vegetables, Tea, and Spices.* New York: Marcel Dekker Inc.
[a]Blueberry has been categorized as both a climacteric and nonclimacteric fruit.

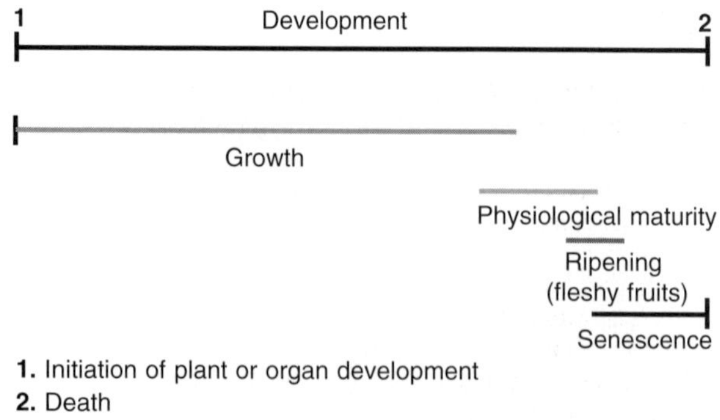

Figure 4 Stages of development for horticultural crops. Reproduced from Watada AE, Herner RC, Kader AA, Romani RJ, and Staby GL (1984) Terminology for the description of developmental stages of horticultural crops. *HortScience* 19: 20–21.

their research was to reduce the rate of storage respiration. Their studies led to assigning the term "climacteric" to indicate the burst of postharvest respiration in apples, which is now known to occur in many important fruit crops (**Table 1**). In subsequent years, Kidd and West also became increasingly aware of the role of ethylene in the climacteric rise in apples.

Following the seminal work of Kidd and West, CA storage became more common globally, although the UK was the first country to initially adopt commercial CA storage practices. This was soon followed by South Africa, the United States, Canada, Australia, New Zealand, Denmark, and the Netherlands. Today, CA storage is standard practice worldwide and is largely responsible for the high marketplace quality attained for many CA responsive fruits and vegetables. Interestingly, the challenge facing each major geographical region that adopted CA technology in the 1920s and 1930s is the same today: finding the optimal combination of temperature, O_2, and CO_2 for each species, cultivar, and even strain that leads to the highest quality retention for the longest duration. This goal is always tempered by the technology available and the economic viability of sustaining the storage environment for the desired length of time.

Influence of Preharvest Factors on Postharvest Quality

Preharvest factors greatly influence both the condition of the crop at harvest and the crop's storage and nutritive potential. The postharvest quality of fresh horticultural commodities markedly depends upon the quality attained at the time of harvest. Many preharvest factors are known to affect storage quality, including: genotype and cultivar selection; stage of maturity when harvested; climatic conditions such as temperature, light intensity, and rainfall amounts; soil texture and fertility; fertilizer type and application rates; disease and insect pressure; and growth regulator and pesticide application. Evaluating which preharvest conditions exert the most influence on postharvest fruit quality is difficult as they interact during the growing season and can change in degree of influence from year to year. Nonetheless, the goal in managing preharvest factors is ultimately to harvest the crop at the highest degree of quality and to sustain that quality throughout the storage period (*see* **Postharvest Physiology**: Postharvest Physiological Disorders of Fresh Crops).

Principles of Metabolism for Harvested Products

Senescence and Ripening

Physiological development of plant organs consists of separate, yet overlapping, stages including: growth, maturity, ripening (fleshy fruits), and senescence (**Figure 4**). From a postharvest perspective, ripening and senescence are particularly important as crops are harvested and stored within these phases of development.

Table 2 Changes that may occur during the ripening of fleshy fruits

Abscission (detachment of fruit from parent plant)
Carbohydrate conversion (starch hydrolyzed into sugars, e.g., fructose, glucose, sucrose)
Color changes (loss of chlorophyll, increase in anthocyanin, and carotenoid pigments)
Generation of flavor and aroma volatile compounds increase
Organic acid content decreases
Respiration rate and ethylene evolution increases
Seed maturation
Softening (changes in pectic substances and cell wall carbohydrate composition)
Wax deposition on skin

Adapted from Wills R, McGlasson B, Graham D, and Joyce D (1998) *Postharvest: An Introduction to the Physiology and Handling of Fruit, Vegetables and Ornamentals*, 4th edn. New York: CAB International.

Senescence is a highly ordered and complex sequence of physicochemical events in the developmental life cycle of a plant, whereby catabolism (degradation) exceeds anabolism (synthesis) leading ultimately to cell death; it occurs at the cellular, tissue, organ, and, eventually, organism level. During senescence, new metabolic pathways are activated or upregulated (e.g., starch degradation) and others are turned off or downregulated (e.g., chlorophyll synthesis). During advanced senescence, a progressive loss of regulatory control within the cell occurs due to the inability of the cell to maintain biochemical homeostasis, and involves events such as disruption of membrane integrity, declines in cytokinin concentration, and loss of anabolic enzyme and organelle function.

Many commodities are harvested during the process of ripening, which overlaps the onset of senescence or occurs as a stage within senescence (**Figure 4**). Ripening (usually restricted to fleshy fruits) is a developmental stage of coordinated, highly complex, and irreversible physicochemical events, which turn a physiologically mature, but inedible fruit into a valuable food commodity. Changes that can occur during ripening are highlighted in **Table 2**. Although "when to harvest" a particular crop is an arbitrary decision, the optimum stage of ripeness can be generally defined as that stage that promotes the highest economic quality for the longest period of time.

Respiration

In general, the science of postharvest physiology focuses on understanding and controlling respiratory metabolism and ethylene generation (**Figure 5**).

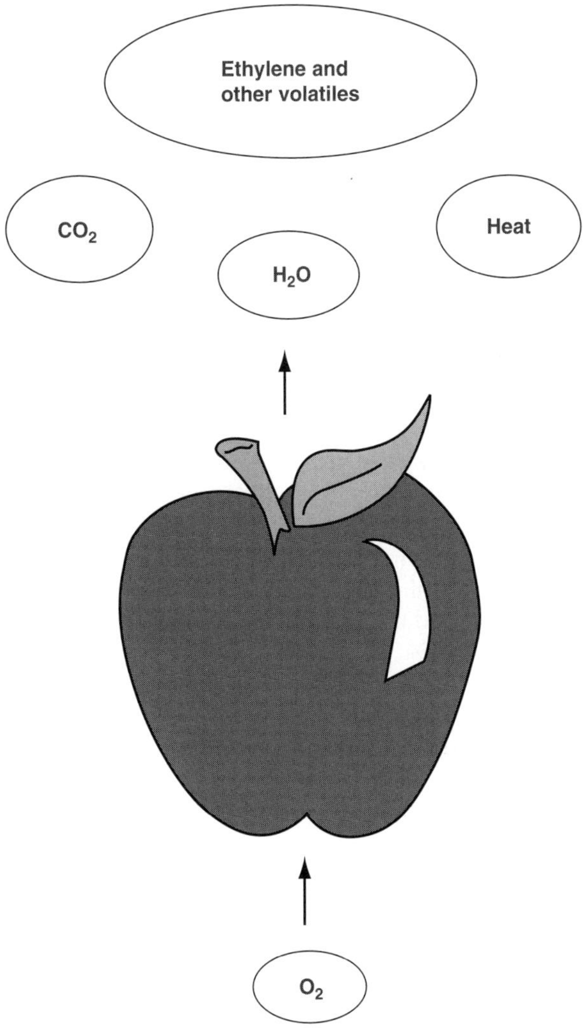

Figure 5 Simplified biochemistry of a stored apple. Following harvest, fruit and vegetables consume O_2 and generate CO_2, ethylene, organic volatile compounds, H_2O, and heat.

As plant organs are harvested, they no longer have access to water and nutrients previously provided by the root system, and are unable to synthesize carbohydrates due to leaf loss. Since the detached organ is living, substrates are necessary for metabolism, which are drawn from tissue stores of carbohydrates, lipids, and proteins contained within the organ at the time of harvest. These catabolic processes occurring within the cell are collectively termed respiration, although from a purely biochemical perspective, respiration is confined to the cytosol (glycolysis) and the mitochondria (tricarboxylic acid or citric acid cycle, and electron transport). Postharvest metabolism often leads to loss of the harvested organ's economic value due to the effects of respiration and senescence-related tissue degradation. In some cases (e.g., pear), a stimulation of respiration is required after harvest to achieve the degree of ripening, e.g., fruit softening, aroma, and flavor and texture development, associated with optimal eating quality. Nonetheless, slowing the rate of respiration generally prolongs the postharvest utilization and, hence, economic value of many horticultural commodities.

In modern crop storage facilities, respiration is controlled largely through the use of low temperatures, low O_2, and high CO_2 levels (**Table 3**). Low O_2 levels mediate a reduction in metabolic rate largely through the reduced activity of oxidase enzymes such as cytochrome oxidase, polyphenol oxidase, ascorbic acid oxidase, and glycolytic oxidase. High CO_2 levels change the activity of respiratory enzymes, can uncouple oxidative phosphorylation reactions, and inhibit ethylene biosynthesis and action. Besides altering the metabolism of the plant organ directly, low O_2 and high CO_2 atmospheres inhibit the growth of microorganisms, resulting in reduced storage rot and decay. In some cases, high CO_2 levels (e.g., 15 kPa or greater) can increase respiration by facilitating the formation of tricarboxylic acid cycle intermediates or damage tissue by increasing succinate levels.

Ethylene

Ethylene (C_2H_4) is a chemically simple but physiologically potent plant hormone that is produced upon wounding, detachment of plant organs during harvest, or exposure to various stresses during ripening and senescence. Research has shown that fruit ripening is either ethylene dependent or independent; hence, ethylene is a partial coordinator of ripening events, including induction of cell wall degrading enzymes, chlorophyll degradation, up-regulation of carotenoid biosynthesis, and respiratory metabolism. Although the biochemical pathway of ethylene synthesis is separate from that of respiration, its generation usually increases or decreases proportionately with the respiration rate. Since exposure to ethylene often increases the rate of respiration and senescence and thus the perishability of the crop, the suppression of *in vivo* ethylene generation is a major postharvest concern. Fortunately, the same storage measures that slow respiration, i.e., low temperatures, low O_2, and high CO_2, also suppress ethylene evolution.

The Climacteric

In the fruit of some plant species (**Table 1**), a concomitant burst of ethylene and respiration (usually measured as CO_2 evolution) occurs at or just prior to the onset of horticultural maturity or optimal ripeness (**Figure 6**). Termed the "respiratory climacteric" by Kidd and West in the 1920s, it can occur while the fruit is attached or detached from the parent plant, and provides a biochemical indicator of the stage of maturity. Concomitant with or immediately following the climacteric, an irreversible ethylene induced increase occurs in the expression of genes that control ripening related physiology, such as changes in fruit texture, color, and taste associated with optimal eating quality. As a result, climacteric fruits often progress through the various stages of ripening more rapidly than nonclimacteric fruit. Some nonclimacteric fruit (e.g., strawberry (*Fragaria* spp.), orange (*Citrus sinensis*)) do not advance further physiologically once the organ is detached from the parent plant. Therefore, knowledge of the optimum preharvest ripeness stage is critical for both climacteric and nonclimacteric species to ensure highest eating quality at harvest or after a storage period. For many climacteric fruits, harvesting for storage purposes (versus immediate consumption) just prior to the climacteric peak usually ensures an acceptable degree of ripeness as well as a high degree of keeping quality during the storage and poststorage periods. If harvest follows the climacteric, many fruits (even those held in ideal storage conditions) will not maintain optimal eating quality due to irreversible biochemical events that advance organ development into more advanced senescence. Although the arbitrary categorization of fruits into climacteric and nonclimacteric types has promoted research into the mechanism and meaning of the respiratory burst in climacteric fruits, the exact role of the climacteric is still unknown.

Modern Postharvest Storage

Storage

Goals Storing harvested crops is done for practical as well as economic reasons. For the former reason, storage allows access to the produce over a longer time period, which reduces waste and increases the crop's practical utility; for the latter reason, controlling the entry of produce into the marketplace helps to maximize economic returns. These goals are accomplished by: slowing the rate of the crop's metabolic activity; impeding the growth and proliferation of microorganisms that cause cellular damage and tissue degradation; and arresting moisture loss, which causes product shriveling and wilting. Adequate refrigeration, set at the lowest temperature that promotes the highest quality retention without freezing or inducing low-temperature damage, is critical to successful product storage (**Table 3**). The concomitant establishment of a CA

Table 3 Temperature, O_2, CO_2, and relative humidity ranges for maximizing storage quality and duration for various fruits and vegetables (A), and floral crops (B)

Plant	*Temperature range (°C)*	*O_2 (kPa)*	*CO_2 (kPa)*	*Relative humidity (%)*
(A)				
Fruit				
Apple	0–5	1–3	1–5	90–95
Apricot	0–5	2–3	2–3	90–95
Avocado	5–13	2–5	3–10	85–90
Banana	12–16	2–5	2–5	90–95
Cantaloupe (*Cucumis melo*)	3–7	3–5	10–15	90–95
Cherry	0–5	3–10	10–15	90–95
Grape	0–5	2–5	1–3	90–95
Grapefruit	10–15	3–10	5–10	85–90
Kiwifruit	0–5	1–2	3–5	90–95
Lemon	10–15	5–10	0–10	85–90
Mango	10–15	3–7	5–8	85–90
Olive	5–10	2–3	0–1	85–90
Orange	3–10	5–10	0–5	85–90
Peach (and nectarine)	0–5	1–2	3–5	90–95
Pear	0–5	1–3	0–3	90–95
Pineapple	8–13	2–5	5–10	85–90
Plum	0–5	1–2	0–5	90–95
Strawberry	0–5	5–10	15–20	90–95
Vegetable				
Artichokes (*Cynara scolymus*)	0	2–3	3–5	95–100
Asparagus (*Asparagus officinalis*)	0–5	–	5–12	95–100
Beans (green, snap, wax; *Phaseolus vulgaris*)	5–10	2–3	4–7	90–95
Broccoli (*Brassica oleracea italica*)	0	1–2	5–10	95–100
Brussel sprouts (*Brassica oleracea gemmifera*)	0	1–2	5–7	95–100
Cabbage (*Brassica oleracea capitata*)	0	2–5	3–7	95–100
Carrots[a] (*Daucus carota sativus*)	0	–	–	98–100
Cauliflower (*Brassica oleracea botrytis*)	0	2–5	2–5	95–98
Corn (sweet) (*Zea mays*; maize)	0	2–4	5–10	95–98
Cucumbers (*Cucumis sativus*)	4–12	3–5	0–5	85–100
Leeks (*Allium porrum*)	0	1–2	2–5	95–100
Lettuce (*Lactuca sativa*)	0	2–5	0	98–100
Onions (mature bulbs; *Allium cepa*)	0	1–3	5–10	65–70
Pepper (bell and chili)	5–12	2–5	2–5	95–98
Potatoes[a] (*Solanum tuberosum*)	4–12	–	–	90–98
Tomatoes (firm ripe)	8–12	3–5	3–5	90
(B)				
Floral crops				
Alstroemeria (*Alstroemeria* spp.)	0–4			
Anthurium (*Anthurium* spp.)	12.5–15.5			
Bird-of-Paradise (*Strelitzia reginae*)	7–10			
Carnation (*Dianthus caryophyllus*)	0–7			
Chrysanthemum (*Chrysanthemum* spp.)	−0.5–8			
Delphinium (*Delphinium* spp.)	0–4.5			
Freesia (*Freesia* spp.)	0–4			
Gypsophila (*Gypsophila paniculata*)	0–4.5			
Iris (*Iris* spp.)	−0.5–4			
Liatris (*Liatris spicata*)	0–5			
Lily (*Lilium* spp.)	0–4.5			
Lisianthus (*Eustoma grandiflora*)	1			
Narcissus (*Narcissus* spp.)	0–2			
Orchid (*Cattleya* spp.; *Phalaenopsis* spp.; *Paphiopedilum* spp.; *Phragmipedium* spp.)	0.5–15			
Rose (*Rosa* spp.)	0–4			
Snapdragon (*Antirrhinum majus*)	−1–5			
Statice (*Limonium* spp.)	1.5–4			
Tulip (*Tulipa* spp.)	−0.5–2			

In general, floral crops do not last well in CA conditions. With the exception of carnations, floral crops should not be stored for more than 2–3 weeks. The relative humidity recommendation for most cut flowers is 90–95%. Floral storage reference, adapted from Wills R, McGlasson B, Graham D, and Joyce D (1998) *Postharvest: An Introduction to the Physiology and Handling of Fruit, Vegetables and Ornamentals*, 4th edn. New York: CAB International. Fruit and vegetable crop storage recommendations adapted from Kader AA (2002) *Postharvest Technology of Horticultural Crops*, 3rd edn. Publication 3311. Oakland, California: University of California, Agriculture and Natural Resources.

[a] A reduction in O_2 and increase in CO_2 for storage is not recommended for these vegetables.

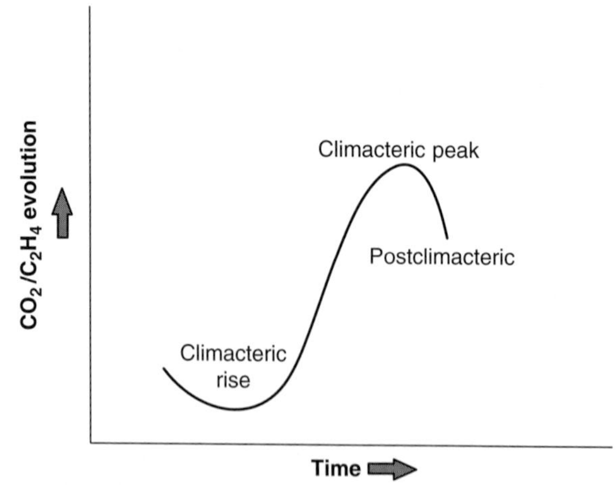

Figure 6 Respiration related CO_2 and C_2H_4 evolution over time in climacteric fruit. Reproduced from Watada AE, Herner RC, Kader AA, Romani RJ and Staby GL (1984) Terminology for the description of developmental stages of horticultural crops. *HortScience* 19: 20–21.

environment (for CA responsive crops) increases quality retention by further slowing respiration rates. Moisture loss is decreased by reducing the temperature difference between the produce and the air surrounding it, and by maintaining high humidity in the storage room.

Attaining these goals requires careful consideration of many factors when utilizing or constructing a building for the storage of harvested crops, including: the nature of plant product(s) to be stored; overall size based on anticipated stored tonnage; building materials; the number and dimensions of individual rooms; cooling room requirements; the type of refrigeration system with or without CA capacity; temperature and gas monitoring equipment; and grading and packing space, and related equipment requirements.

Storage rooms The size of a storage facility is based upon the maximum volume of produce to be stored at any one time. Building size is also influenced by the necessity to load the produce into its allotted space within 1 or 2 days of harvest, and the space needed to empty the room and grade and pack the product for market. While more costly, dividing a large store into several smaller rooms prevents breaking the storage atmosphere for those crops that are not immediately destined for the marketplace. Building size may also be influenced by cooling or precooling requirements before the crop is enclosed in a refrigerated or CA room. Precooling is necessary for removal of field heat at, and immediately following, harvest. Without cooling prior to storage, the demand on the refrigeration system greatly increases, and the field heat present is likely to cause some degree of crop quality loss before it is dissipated. Where available, a separate precooling room can greatly facilitate the removal of field heat and significantly lessen the refrigeration capacity needed to maintain low storage temperatures. Thus, the crop is cooled quickly to the lowest acceptable temperature resulting in a higher degree of product quality for the marketplace.

Controlled atmosphere environments A CA room will be required to establish O_2 and CO_2 levels that range from 1 to 5 kPa in most cases. Attention to structural details coupled with quality workmanship is therefore critical in achieving an air-tight room that can attain and maintain these O_2 and CO_2 concentrations. CA storage construction can be accomplished with many different materials: European CA rooms are mainly built with metal-faced insulating panels locked together with trademarked locking systems. Air-tightness is then maintained by taping and coating with plastic paints. In North America, CA buildings are often made with lumber and plywood or with concrete blocks or tilt-up concrete walls. The room is then coated with polyurethane foam insulation and covered with a fire retardant (**Figure** 7). The goal is structural stability coupled with air-tightness. A major point of leakage from CA rooms is the main entry door. Ideally, there should be only one door per room that is sealed with an airtight gasket system or with petroleum jelly. An access hatch or a double or triple-glazed removable window in the door will permit product viewing and serve as a point of entry for sample removal (**Figure** 7). Extreme caution must be used when entering a CA room as the low O_2 levels will not sustain human life. If entrance into the room is necessary, the use of a self-contained breathing apparatus is required, and the involvement of two qualified individuals is advisable as a safety measure.

Although there are several methods to reduce O_2 levels, the most widely employed is nitrogen gas flushing via a nitrogen generator, which displaces O_2 from the storage room thus reducing its concentration. The main advantages of nitrogen flushing are its nonexplosive chemical stability, and the process does not introduce ethylene or other gases into the storage atmosphere, which are potentially injurious to the crop. Other systems include: external burners that remove O_2 via combustion of natural gas or propane, molecular separators that operate on the principle of molecular segregation by filters or selective membranes, and ammonia cracking in which anhydrous ammonia is split at high temperatures into nitrogen and hydrogen gas.

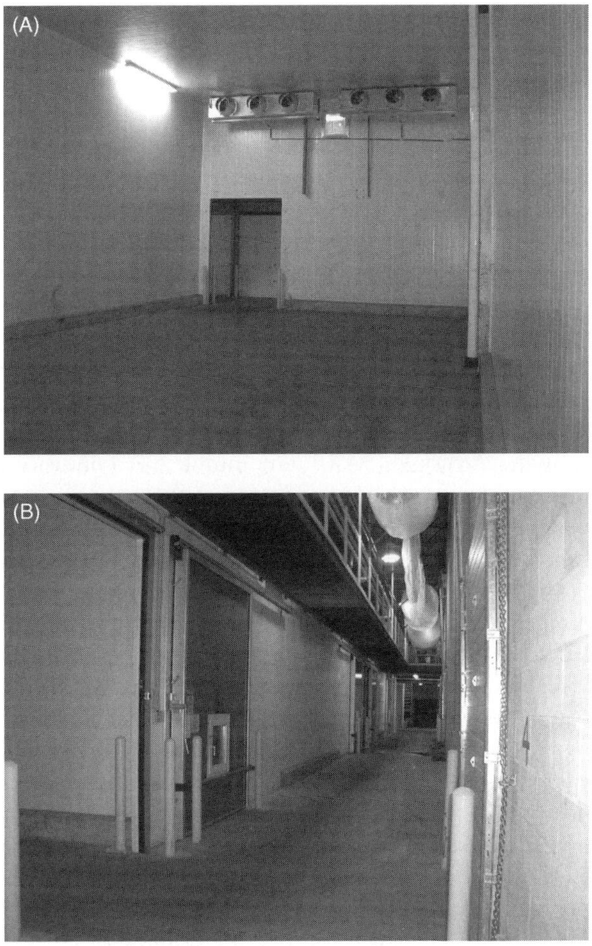

Figure 7 An empty controlled atmosphere storage room for apples with a 750-bin capacity and cooling fans placed on the door wall. The walls consist of metal sheeting on the front and back sides with polyurethane foam insulation in the middle. The open entry doorway measures 3.12×2.54 m. (A) A concourse area connecting several modern apple CA storage rooms with the viewing/sample removal window on the main doors. The large breather bags above the walkway help maintain constant pressure in the CA room during weather changes and prevent release of the CA gases to the outside air (B). Photographs were taken at Scotian Gold Cooperative Ltd., Kentville, Nova Scotia, Canada.

Carbon dioxide removal Although it may be beneficial to increase CO_2 levels in a CA environment, a CO_2 scrubbing system is often a requirement for many CA facilities as horticultural commodities vary markedly in CO_2 sensitivity (**Table 3**). A simple method for CO_2 removal is the use of hydrated lime in which the airflow from the CA room is directed over bags of lime that are externally connected to the store. The CO_2 in the airflow is absorbed by the lime and the purified air is returned to the CA room. The lime has to be changed periodically as it becomes saturated with CO_2 and, at that point, is unable to effectively absorb any additional CO_2 from the airstream. Bags of hydrated lime can also be placed in the storage room to absorb CO_2 directly from the atmosphere. The use of activated charcoal is another popular CO_2 scrubbing method, in which air from the CA store is passed over a bed of charcoal absorbent resulting in CO_2 removal. The CO_2 laden charcoal is then reactivated by passing outside air over the bed, which removes the CO_2 exhausting it to the outside air. This system has low operating costs and requires the absorbent to be changed about every 5 years.

Ethylene removal Although a CA environment with low O_2 and high CO_2 reduces both ethylene production by the stored crop and sensitivity to its effects, ethylene levels as low as 0.1 parts per million (ppm) can promote ripening and senescence of many crops. Removal of ethylene from storage rooms can be accomplished through an activated charcoal CO_2 scrubbing system. Nitrogen flushing or aeration of the store with outside air (when O_2 levels become too low) will also remove ethylene. When storage room air is circulated through aluminum silicate beads coated with potassium permanganate ($KMnO_4$), ethylene is effectively removed. $KMnO_4$, however, is a strong oxidizing agent and must be carefully handled to avoid injury. The beads must be replaced when they become saturated as indicated by a color change from purple to brown. Ethylene can also be reduced by catalytic oxidizers that operate on the principle of ethylene degradation via reaction with O_2 in the presence of a catalyst such as platinized asbestos. Exposure of the storage room air to ultraviolet radiation will remove ethylene as the airstream is passed by the lamps. However, the ozone or pre-ozone intermediates generated by ultraviolet exposure must then be removed by filtration of the air going back into the storage as they may be toxic to the stored produce.

In recent years, antiethylene compounds, e.g., ReTain® (aminoethoxyvinylglycine) and SmartFresh™ (methylcyclopropene), which are applied during the growing season or at the beginning of the storage period on some fruit crops, have been shown to markedly reduce the quantity of ethylene generated by the commodity over the holding period (particularly within a CA environment). Thus, the purging frequency of the storage room for high ethylene levels may be reduced for crops on which these products are registered.

Superatmospheric oxygen Oxygen levels above 21 kPa (ambient atmosphere) in the store have been shown to reduce respiration rates and ethylene evolution in some fruits and vegetables. The effectiveness of this treatment is influenced by the commodity, its maturity and stage of ripeness, the concentration of the O_2 regime imposed, the levels of CO_2 and ethylene in the storage room, and the temperature and duration of the storage period. For example, exposure of "Spartan" apples to 100 kPa O_2 (100%) at 1°C for 12 days results in reduced respiration, higher firmness, and less browning when sliced, compared with air stored fruit held at 1°C for 14 days.

Grapefruit (*Citrus paradisi*) respiration is stimulated by exposure to 80 kPa O_2 and 14°C, while apricots and cherries (*Prunus avium*) held under 30, 50, 75, and 100 kPa O_2 show no appreciable changes in respiration. It appears that longer exposure to elevated O_2 is generally injurious to fruits and vegetables, while short-term exposure to high O_2 followed by a return to a standard storage regime can lead to improved quality. Superatmospheric O_2 can also limit the growth of bacterial and fungal pathogens, particularly when combined with elevated CO_2 (15–20 kPa) levels.

As superatmospheric O_2 is highly combustible, utmost care must be taken when packaging or flushing with elevated O_2 systems. Research and practical experience with superatmospheric O_2 is only in its nascent stage. It is not known how elevated O_2 alters basic respiration and ethylene biochemistry in general, nor how it influences the storage quality of a broad spectrum of fruits and vegetables specifically. However, where the benefits of superatmospheric O_2 include improved storage quality and reduced pathogenic infection, its use on responsive crops will probably become more commonplace.

Monitoring of the storage atmosphere Without reliable and properly operating temperature, O_2, CO_2, and ethylene monitoring equipment, all of the effort and expense of establishing an ideal storage environment will fail in achieving the highest degree of crop quality over a long-term storage period. There are many equipment options for measuring atmospheric gases and temperature; the choice for each will be a compromise of cost per unit and the number of units required, precision of measurement demanded, long-term functional reliability, and availability and service options in a geographical area, to name a few. It is imperative that the storage facility be equipped with the necessary number and strategic placement of these measurement devices so that temperature and atmosphere conditions of the whole store can be consistently and reliably monitored over time.

Modified Atmosphere Packing

Modified atmosphere packaging (MAP) refers to the use of plastic or polymeric films that enclose fresh produce resulting in respiration and transpiration based changes in the atmosphere surrounding the commodity (**Figure 8**). The main goal of MAP is to create a gaseous equilibrium between the produce and the sealed atmosphere resulting in lower O_2 and higher CO_2 levels, which prolong fruit and vegetable quality. MAP is particularly useful for those commodities that have high respiration and ethylene generation rates, lose water rapidly, show browning on cut surfaces, or are susceptible to microbial growth. Although MAP can mimic CA conditions to some degree, the present state of the technology is unable to precisely control the O_2 and CO_2 levels of the sealed atmosphere over longer periods of time. As with CA, successful application of MAP must be integrated with adequate refrigeration in order to slow the rate of senescence.

Table 4 lists many of the films used in MAP technology. The permeability of the plastic film to gases and water vapor varies with the type and thickness of the film, but most types hold the produce in good condition for 2–4 weeks. In general, plastic film is four- to sixfold more permeable to CO_2 than to O_2. Recent developments in film manufacture permit a more precise matching of films with low, medium, or high respiration rate commodities as a means to establish more ideal modified atmospheres.

In addition to benefits similar to those attained in CA storage as O_2 and CO_2 levels are altered, MAP also prevents dehydration and wilting by arresting water loss from the atmosphere surrounding the crop. However, the increase in water vapor and respiration related heat generation may encourage proliferation of fungal and bacterial pathogens. Hence, packing films are often perforated with very small holes called microperforations to allow the outward movement of water vapor and heat, which reduces the incidence of rot. Perforated bags can alter the gaseous atmosphere of the pack and may prevent the establishment of ideal O_2 and CO_2 concentrations.

It may be desirable to change the atmosphere within the MAP environment more rapidly than can be accomplished by natural respiration alone. Flushing the pack with nitrogen or an appropriate combination of O_2, CO_2, and nitrogen can rapidly establish a more ideal storage atmosphere as well as remove any ethylene present. Removal of the pack air volume with a vacuum pump can also facilitate rapid establishment of CA-like conditions. MAP is also used to preserve prepared fruit and vegetables,

Figure 8 Minimally processed fruit (A) and prepared salad (B) sealed in plastic containers to create a modified atmosphere and kept on ice to reduce the rate of senescence.

which is collectively referred to as minimal processing (**Figure 8**).

An Eye to the Future

The commercial quality of any crop is determined by the interaction of pre- and postharvest factors encountered during the growing season or those imposed by the field or storage manager(s). A thorough understanding of the physiological and biochemical changes occurring in detached plant organs is critical for maintaining those organs in postharvest environments that arrest respiration and senescence related tissue degradation. Future

Table 4 Plastic films used for MAP of fruits and vegetables

Cellulose acetate
Ethylene vinyl acetate copolymers
Ethylene vinyl alcohol copolymers
Enhanced polyethylene resins
High-density polyethylene
Ionomer
Linear low density polyethylene
Low density polyethylene
Medium density polyethylene
Ultralow-density polyethylene
Polybutylene
Polyethylene terephthalate
Polyolefin
Polypropylene
Polystyrene
Polyvinyl butyral
Polyvinyl chloride
Polyvinylidene chloride

Adapted from Thompson AK (1998) *Controlled atmosphere storage of fruits and vegetables*. Chapter 6. *Modified Atmosphere Packaging*, pp. 95–116. New York: CAB International.

advances in perishable commodity storage will probably occur in such areas as: MAP technology; the use of consumer acceptable antitranspirant/protective coatings and antioxidant applications prior to storage; dynamic and nondestructive sensing systems to detect the condition of stored commodities (e.g., chlorophyll fluorescence, volatile emission monitoring, and digital imaging analysis) permitting alteration of the atmosphere for maximizing quality retention; and the use of novel gases such as nitric and nitrous oxide to arrest deterioration, to name a few. One of the most pressing research needs is to find alternatives for chemical treatments that historically have been successfully used to prevent oxidation, tuber and bulb sprouting, and pathogen related quality loss. The search for viable solutions to this need in particular will entail integration of three major areas: (1) genetic modification through traditional and molecular approaches; (2) manipulation of the environments in which the plant is grown and stored; and (3) application of treatments in the pre- and postharvest phases that maximize crop quality and nutritive value. This is a challenging task, which, in principle, has been with humanity from antiquity:

> The Italian seems to have had two things particularly in view in his farming: whether the land would yield a fair return for the investment in money and labor, and whether the situation was healthful or not.
>
> Marcus Terrentius Varro (116–27 BC) *On Agriculture*, Book I.

List of Technical Nomenclature

Climacteric The sudden rise in respiration in some fruit (e.g., apples) associated with optimal ripeness or maturity.

Controlled atmosphere Strict and continuous regulation of O_2 and CO_2 levels and temperature in the storage room.

Ethylene A natural gaseous hydrocarbon, produced by all plants, which can promote ripening and senescence.

Modified atmosphere Partial regulation of O_2 and CO_2 levels for prolonging the storage quality of fruits and vegetables.

Modified atmosphere packaging (MAP) Application of plastic films to fruits and vegetables to reduce O_2 and H_2O loss and increase CO_2 levels, resulting in prolonged storage life and improved product quality (MAP is often used with lower temperatures to further slow product deterioration).

Respiration Cellular biochemistry responsible for deterioration of harvested fruits and vegetables involving uptake of O_2 and generation of CO_2, organic volatile compounds, heat and H_2O.

Senescence The collective physiology and biochemistry responsible for the degradation and death of fruits and vegetables.

Superatmospheric oxygen A storage environment with O_2 levels >21 kPa.

See also: **Postharvest Physiology**: Genetic Engineering for Postharvest Quality; Postharvest Physiological Disorders of Fresh Crops; Ripening; Senescence, Flowers; Transport. **Regulators of Growth**: Ethylene.

Further Reading

Kader AA (2002) *Postharvest Technology of Horticultural Crops*, 3rd edn. Publication 3311. Oakland, California: University of California, Agriculture and Natural Resources.

Kader AA and Ben-Yehoshua S (2000) Effects of superatmospheric oxygen levels on postharvest physiology and quality of fresh fruits and vegetables. *Postharvest Biology and Technology* 20: 1–13.

Kays SJ (1997) *Postharvest Physiology of Perishable Plant Products*. Athens, GA: Exon Press.

Knee M (ed.) (2002) *Fruit Quality and its Biological Basis*. Boca Raton, FL: CRC Press.

Proceedings of the Colloquium (1997) Effect of pre-harvest factors on postharvest fruit quality. *HortScience* 32: 806–835.

Wills R, McGlasson B, Graham D, and Joyce D (1998) *Postharvest: An Introduction to the Physiology and Handling of Fruit, Vegetables and Ornamentals*, 4th edn. New York: CAB International.

Workshop Proceedings (2000) Modified atmosphere packaging 2000 – toward 2000 and beyond. *HortTechnology* 10: 481–510.

Transport

C B Watkins, Cornell University, Ithaca, NY, USA

Introduction

Few fresh crops are consumed at the point of production, and therefore they must be transported to markets for purchase by the consumer. The transport requirements can range from simple, involving short distances to a nearby local market, to sophisticated, involving long distances internationally by land, sea, or air. The modern marketplace, particularly in developed countries, involves the sourcing and therefore transport of fresh fruit, vegetable, and ornamental crops, including fresh cut products, from around the world. The mode of transport, the market demand, and therefore the price obtained by the grower/shipper for the crop, together with the requirements necessary to meet the level of quality to satisfy the consumer, greatly affects the economics of transport. The economics for any given crop in the marketplace can change quickly according to production location and transport costs. A market niche providing high financial returns that can justify the high costs of airfreight can be lost to development of production in an area closer to the market, thereby allowing use of cheaper transport methods.

Regardless of the production region, the same principles for maintaining fresh crop quality apply to transport of fresh crops as for on-site storage. After harvest, the objective of all fresh crop industries is to maintain quality by:

1. Reducing metabolic rates that result in undesirable changes in color, composition, texture, flavor and nutritional status, and growth (sprouting, rooting).
2. Reducing water loss that can result in loss of saleable weight, reduction in appearance factors, such as wilting and shriveling, and textural quality factors of softening, limpness, and loss of crispness.
3. Minimizing bruising, friction damage, and other mechanical injuries.
4. Reducing spoilage caused by pathogen infection, especially of damaged or wounded tissues.
5. Prevention of physiological disorders, especially freezing and chilling injuries.

These objectives are met by harvesting the crop at optimum maturity or quality, handling the crop carefully and rapidly to avoid mechanical injury and minimize deterioration, applying protective chemical preservatives (antioxidants, fungicides), heat, and modified atmosphere (MA) or controlled atmosphere (CA) storage treatments, providing protective containers and packaging, and ensuring appropriate relative humidity for the crop. Some crops have specialized requirements, e.g., pulsing of cut flowers with silver thiosulfate to inhibit ethylene action, carbohydrates necessary for opening, and biocides to inhibit bacterial growth. The fresh crop should be cooled quickly to remove field heat, and, if necessary, again after packaging. Proper refrigeration should be applied during storage, transport, and marketing. Good sanitation practices also must be maintained throughout these processes as well as during harvesting. Outbreaks of food-borne illness can have a devastating effect in the market; both for whole and fresh cut products.

While the factors involved in storage and transport are essentially the same, the relative importance of each will depend on the specific crop and production area. In some cases, the time in transit can represent a significant part of the storage life for a product, e.g., strawberries (*Fragaria x ananassa*), cherries (*Prunus avium*), and cut flowers, and therefore the transit time and the storage environment can greatly affect product marketability. In contrast, for products such as apple (*Malus pumila*) transported from the Southern Hemisphere to Europe or bulb crops from Europe to North America, transit times of even as much as 4 weeks still represent a relatively short time for a crop that can be stored for up to 12 months. Because of the large number of possible scenarios for managing a crop before and during transport, this article will present general themes of importance. For the purposes of this article, many of the examples of factors that affect physiology of horticultural products are for fruits and vegetables, but the principles apply to all harvested crops.

Pretransport Considerations

Packaging

Packages have four primary functions: containment, protection, convenience, and communication. The product may be graded for size and quality, and packaged manually or automatically depending on the crop and market requirements. Prepackage treatments can include application of waxes for appearance and/or reduction of water loss from the product.

1. Containment. Containment is a basic requirement for movement of a product from one point to another. The package used is a function of the growing region and market, and ranges from very simple open containers to highly designed packages involving technologies such as modified atmosphere (MA) generating films.
2. Protection. Packages provide protection for the product against environmental factors such as dust and water, as well as bruising, both impact and compression types, and friction injuries that can occur during handling and transport. Cartons must have stacking strength and durability to prevent collapse or crushing when they are on pallets, especially under high relative humidity conditions. Trays, cups, or other protective devices may be contained within the package to further prevent movement and/or contact between crop items. However, whatever the packaging used, it cannot protect the product against careless and rough handling, and maintaining product quality requires monitoring of all handling steps before, during, and after transport to prevent abuse.
3. Convenience. Crops are packaged in sizes that are convenient for the consumer, i.e., apportioned into appropriate sizes for market. The economies of scale required to transport crops over long distances may mean that larger containers are used. The product may be removed from the container and placed directly on display for purchase as single units, or repackaged in a distribution warehouse. Smaller packages (primary package) may also be contained in larger packages (secondary package), e.g., clamshell packages of strawberries, or cut produce held in MA packaging (MAP), within cartons. The secondary packages may be unitized into a tertiary package such as a plastic pallet shroud. Package size is affected by regulations designed to ensure that package weights are manageable by both sexes. There is also a trend toward use of smaller cartons for display purposes in retail markets, which reduce handling and therefore the potential for product damage. Apples exported from the USA to the UK, for example, are commonly packaged in two layer cartons, which are placed directly on display for the consumer, rather than in 20-kg cartons that require repackaging, or handling of the product on to shelf display.
4. Communication. Though of lesser importance to maintenance of crop quality, providing information about the product is an important function of packaging. In addition to the advertising factor of package marking, information is provided about gross and net package weight, unit size of the product, as well as declaration of the use of any postharvest treatments, e.g., antioxidant use or waxing of apples, which may be required by state or national regulations.

An important secondary function of packaging for fresh crops is to provide ventilation for rapid cooling of produce and/or insulation against temperature changes. The relevance of each depends on the crop, the cooling method utilized, and the subsequent requirements for temperature control.

Cooling Requirements

The physiology of the crop affects the cooling requirements and methods that can be used to cool the product. The general goal, however, is to reduce field heat rapidly, or to recool the product after it has been packaged, to minimize respiration rates (**Figure 1**) and associated heat production. Some crops with especially high respiration rates, such as asparagus (*Asparagus officinalis*), broccoli (*Brassica oleracea* var. *italica*), ripe tomatoes (*Lycopersicon esculentum*), leafy vegetables, avocados (*Persea americana*), mangoes (*Mangifera indica*), stonefruit, and strawberries, must be cooled rapidly to reduce their metabolic rates. Other products with lower respiration rates may have less critical requirements for rapid cooling, but in general should be cooled rapidly to ensure long-term storability. It has been shown for apple fruit, for example, that 1-day at 20°C can result in a loss of 1 week of storage life. Cooling also reduces ethylene production rates and sensitivity to its presence, decreases water loss, decay development, and the severity of crop responses to injury.

However, not all crops should be rapidly cooled, as some require curing at warmer temperatures. In onions, curing is used to dry the neck and outer scales. Potatoes are cured to develop wound periderms over damaged surfaces, and curing can reduce water loss and heal injuries. Curing is usually done in the field or in curing rooms, but may occur during transit, in the case of new crop potatoes. Distinct from curing, specific requirements may exist, such as for the banana (*Musa sapientum*), which requires special ripening treatments and therefore is not precooled. The final storage and/or transit temperature for a crop will be a function of its

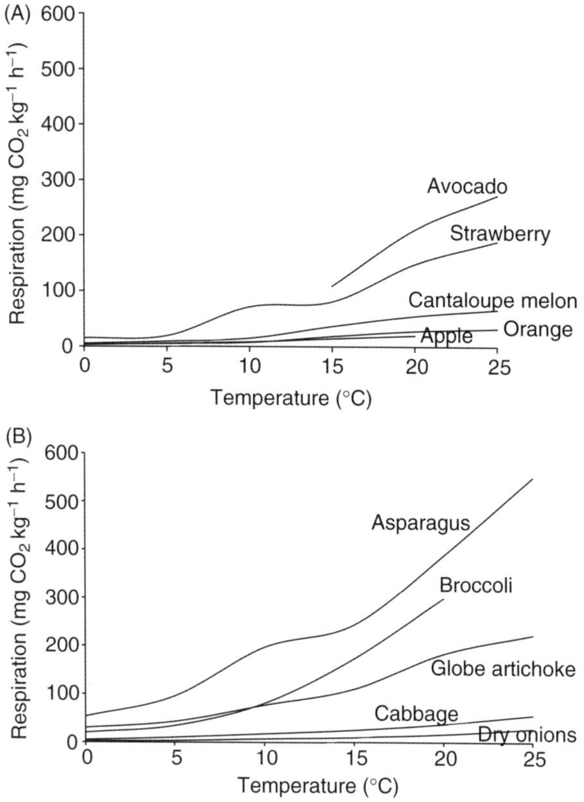

Figure 1 Respiration rates of selected fruit (A) and vegetables (B) at different storage temperatures.

susceptibility to freezing and chilling injury (see below).

Types of Cooling

The main methods used for cooling of fresh crops are room cooling, forced-air cooling, hydrocooling, forced-air evaporative cooling, package icing, and vacuum cooling. Detailed descriptions of the operation of each method are available from many sources. In general, the cooling method selected will be a function of economy, convenience, and personal preference, but the product type (**Table 1**) may greatly influence the type of cooling selected. Crops vary in their susceptibility to contact with water or ice. For example, strawberry fruit can be susceptible to decay if they become water saturated.

1. Room cooling (syn. passive cooling). Field or shipping containers are placed in a cold room, and cooling of the crop occurs passively by heat exchange between the product and the cold air. The rate of cooling is affected by the heat of respiration of the product, the refrigeration capacity, the final temperature desired, and the airflow around the product. Cooling rates can be maximized by ensuring that containers are stacked in patterns that maximize contact between the cold air and the product, by attention to stacking patterns in a room and ventilation of containers. Frequently, the cooling capacity of the refrigeration equipment is adequate to maintain cold temperatures of previously cooled crops, and not sufficient to ensure rapid cooling. For some crops, e.g., apples, bins of fruit harvested each day may be placed in several rooms overnight to increase cooling rates, and moved to the final storage room prior to receipt of the next day's harvest. Passive cooling is the most widely used method of cooling, but has the disadvantages of being slow and resulting in excessive water loss from the product. It is often assumed that crops such as the apple are less affected by slower room cooling. However, slow cooling rates of apple fruit kept in stacked cartons during passive cooling can result in greater softening during transit.
2. Forced-air cooling (syn. pressure cooling). Field or shipping containers are stacked in patterns so that cooling air is forced through, rather than around, the individual container. Operationally, the method requires a slight pressure gradient, usually developed with a fan, to force cold air through the containers. Vent holes should be placed in the direction that the air will move, and packing material that will interfere with free movement of air through the container minimized. Direct contact of cold air with the product results in much faster cooling, usually 0.1 to 0.25 the time that room cooling takes to cool a product. The rate of cooling is controlled by the volume of air passing over the product. The economics of cooling are affected by the fan speed and number of containers being cooled. Water loss from the product is less than that found for room cooling, but high relative humidity is required to minimize desiccation of strawberries, grape (*Vitis* spp.) stems, and other products susceptible to drying.
3. Hydrocooling. The temperature of the crop is reduced by cold water flowing over the product surface, either by flooding, spraying, or immersion of the product in an agitated bath of chilled water. The rate of internal cooling is related to the size and shape, and thermal conductivity of the product being cooled. The method is simple, economic and effective, avoids water loss, though it is generally suited to fruit and vegetables in bulk rather than packaged. Several vegetables are hydrocooled after packaging, however. The product, and its package if applicable, must be tolerant of wetting, and not damaged by the water flow or by the sanitizing agent, typically

Table 1 Cooling methods recommended for selected fresh crops

Crop	*Cooling method*					
	Room	*Forced-air*	*Hydro-cooling*	*Vacuum*	*Package icing*	*Forced-air evaporative cooling*
Fruit						
Apples (*Malus pumila*), pears (*Pyrus communis*)	x	x	x			
Lemons (*Citrus jambhiri*), Oranges (*Citrus sinensis*)	x					
Apricots (*Prunus armeniaca*), grapes (*Vitus vinifera*), strawberries (*Fragaria x ananassa*)	x	x				
Vegetables						
Cabbage (*Brassica oleracea capitata*), cauliflower (*Brassica oleracea botrytis*), beans (*Phaseolus vulgaris*), mushrooms (*Agaricus bisporus*)	x	x		x		
Dry onions (*Allium cepa*)	x	x				
Garlic (*Allium sativum*)	x					
Iceburg lettuce (*Lactuca sativa* var. *capitata*)	x			x		
Leaf lettuce (*Lactuca sativa*)	x	x	x	x		
Sweet corn (*Zea mays*)	x		x	x	x	
Cucumbers (*Cucumis sativus*), eggplant (*Solanum melongena*), tomatoes (*Lycopersicon esculentum*)	x	x				x
Artichokes (*Cynara scolumus*)	x		x		x	
Asparagus (*Asparagus officinalis*)	x		x			
Peas (*Pisum sativum*)	x	x		x	x	
Ornamentals						
Cut flowers	x	x				
Potted plants	x					

chlorine. The product must be transferred to a cold room quickly to avoid rewarming.

4. Forced-air evaporative cooling. The principle of operation is the same as forced-air cooling, but the air is cooled with an evaporative cooler rather than by mechanical refrigeration. It is possible to obtain air with temperatures a few degrees above the outside wet bulb temperature and at more than 90% relative humidity. Evaporative cooling is adequate for some chilling-sensitive products if shipped to local or regional markets.
5. Package icing. Products are cooled by filling packed containers with ice, either as finely crushed ice, flake ice, or a slurry of ice and water. The amount of ice required is a function of the product's weight. More expensive and water-tolerant packages are required, but the method has the advantage of maintaining high relative humidity around the product. However, the product must not be harmed by ice or by exposure to prolonged cold and wet conditions.
6. Vacuum cooling. The temperature of the crop is reduced by rapid evaporation of water, which is the primary refrigerant, from the product under low pressure. The pressure in the chamber is lowered to the saturation point corresponding to the lowest required temperature of the water. The method is best suited to vegetables with a high surface area to volume ratio, e.g., iceberg lettuce (*Lactuca sativa* var. *capitata*). Other leafy vegetables may be suitable, but because the weight loss associated with vacuum cooling is about 1%, many vegetables, and especially fruit, are less suited for cooling by this method.

1-Methylcyclopropene (1-MCP)

1-MCP is a new commercial compound that inhibits ethylene action and can maintain the quality of many crops. 1-MCP is commercially available as Ethylbloc^{TM} for use on ornamentals in North America, and as SmartFresh^{TM} for use on fruit, vegetables worldwide, and ornamentals outside of North America. 1-MCP has been registered in the USA and several other countries, and registration is pending in Europe and elsewhere. 1-MCP is already used extensively for maintaining quality of cut flowers during storage and transport. Its commercial use on other products has thus far been limited to apples but 1-MCP has the potential to markedly affect transport of horticultural crops around the world.

Loading Processes

Most transit vehicles have sufficient refrigeration capacity to maintain, rather than cool, fresh crops. To minimize warming of the crop and therefore avoid losing benefits of cooling it is necessary to ensure fast and efficient loading. The product should be cooled properly before loading into the container or other transit vehicle. It is desirable to refrigerate the loading dock and load assembly areas. The containers should be checked for operation before use. In addition to calibration of temperature controllers, refrigeration, and other mechanical requirements, they should be inspected for cleanliness, and should look and smell clean. Freedom from toxic materials that might be present from previous nonfood cargoes or from bacterial contamination should be ensured. If loads are certified as organic or as genetically modified, organism free additional steps may be required to prevent any contamination from previously transported products. The containers should be cooled to carrying temperature before loading, but the refrigeration units turned off before opening the doors for loading to prevent moisture condensing on the evaporator coil.

Ideally, a container is filled with uniform packaged product that has been unitized into larger units, and loaded into transit vehicles in such a way as to allow adequate airflow for temperature control. Calibrated temperature recording devices should be placed with each load.

In-Transit Preservation

During transport the primary goal of the shipper is to maintain the optimum temperature appropriate for the fresh crop. Relative humidity, although also important to product quality, is often a secondary consideration and is less often controlled. The use of MA and controlled atmosphere (CA) regimes during transit has increased, particularly during the last decade.

Temperature

The primary means of maintaining the quality of harvested fresh crops during transport is to reduce metabolic rates by use of low temperatures in the transit container or vehicle. The ideal temperatures for a range of fresh crops are presented in **Table 3** of **Postharvest Physiology**: Postharvest Physiological Disorders of Fresh Crops. As for on-site storage, the optimum temperature for transport can vary among fresh crops, especially susceptibility of many tropical and subtropical crops to chilling injury if stored between −1.5 and 15°C. Susceptibility of fresh crops to freezing injury must also be considered as the recommended storage temperatures of each crop is only 1–3°C above its freezing point.

Temperature control during transit can be by top-ice where crushed ice is placed on top of the load, or by ice contained within each package. These packages may be transported without further refrigeration. Control of temperature in an insulated vehicle with outside air also has been used, but is the least dependable. More typically, mechanical refrigeration is used for maintaining temperature during transport. Mechanical refrigeration not only cools the container or vehicle, but also provides heat to maintain appropriate temperatures during subfreezing transit conditions. Airfreight systems have little refrigeration capacity and are considered separately below.

The components of transport systems for good temperature control are:

1. An insulated body for the transit container or vehicle of sufficient strength to resist shock, vibration, flexing, and other movement, with qualities of low density, low thermal conductivity, and low moisture permeability and retention. The floor must be insulated, water tight, and be able to support forklift trucks.
2. Air circulation, to ensure adequate air movement through and around the loads. Air circulation is generally via overhead (or top air) delivery for over-the-road vehicles. In overhead delivery, conditioned air flows into the space between the cargo and vehicle ceiling. Bottom-air delivery systems are typically used in sea-going containers.
3. Temperature control based on use of automatically controlled thermostats to control the refrigeration system. Temperature of the air rather than the product is measured, and often in return-air units. In some newer units, the temperature of the supply stream in the discharge air duct is measured, reducing the risk of freezing of the fresh crops exposed directly to the airflow of conditioned air.

Relative Humidity

Relative humidity is a major factor in preventing moisture loss for some crops, and therefore maintaining their quality (**Table 2**). Humidity in both iced and mechanically refrigerated cars and trailers is usually high, but can be difficult to achieve during long transport periods. Large evaporator coils on the refrigeration unit will improve control of high relative humidity.

Air Exchange

An air exchange system is utilized to ventilate the load with outside air mostly in marine-based containers. Air exchange can prevent depletion of oxygen, and/or reduce undesirable concentrations of carbon dioxide, ethylene, or offensive odors. Hydrated lime and potassium permanganate also have been used to reduce carbon dioxide and ethylene concentrations, respectively.

Table 2 Optimum relative humidities for selected fruits and vegetables

Relative humidity (%)	*Fresh crop*
50–75	Pumpkins (*Cucurbita maxima*)
60–70	Chili peppers (*Capsicum annuum*)
65–70	Bulb onions, garlic, jicama (*Pachyrhizus erosus*)
70–75	Squash – hard (*Cucurbita maxima*)
75	Coconuts (*Cocos nucifera*)
85	Grapes
85–90	Carambola (*Averrhoa carambola*), celery (*Apium graveolens*), figs (*Ficus carica*), grapefruit (*Citrus paradisi*), lemons, mangoes (*Mangifera indica*), sweet corn, bell peppers (*Capsicum annuum*), sweet potatoes (*Ipomoea batatas*), tomatoes (*Lycopersicon esculentum*)
85–100	Watercress (*Lepidium sativum*)
90	Cantaloupe (*Cucumis melo*), feijoas (*Feijoa sellowiana*), loquats (*Eriobotrya japonica*), potatoes (*Solanum tuberosum*), watermelon (*Citrullus lanatus*)
90–95	Apples, apricots, bananas (*Musa* spp.), blueberries (*Vaccinium ashei*), cherries (*Prunus avium*), cranberries (*Vaccinium macrocarpon*), cucumbers, kiwifruit (*Actinidia deliciosa*), peaches (*Prunus persica*), pears, strawberries
90–98	Cauliflower
95	Beans, mushrooms, squash – soft (*Cucurbita pepo*)
95–100	Asparagus, broccoli (*Brassica oleracea*), Brussels sprouts (*Brassica oleracea*), leeks (*Allium ampeloprasum*), green onions (*Allium cepa x A. fistulosum*), spinach (*Spinacia oleracea*)
98–100	Beets (root) (*Beta vulgaris*), cabbage, carrots – topped (*Daucus carota*), lettuce, rutabagas (*Brassica napobrassica*)

Note that these relative humidities apply to each crop at its optimal storage temperature.

Modified and Controlled Atmospheres

MA and CA storage in which oxygen concentrations are lowered while those of carbon dioxide are increased can be used during transport to maintain crop quality. However, the potential for benefit and its application varies greatly by crop (**Table 3**), and economically, CA transport remains a niche market with relatively low volumes of high value commodities such as avocado, stone fruit, pears (*Pyrus communis*), mangoes (*M. indica*), and asparagus.

In general, the use of CA transport for many crops is uneconomical, e.g., in 1996, CA shipments represented only 2.5% of the total USA imports and exports volume combined. The economics of CA transport may become more favorable, however, in the future. Several types of CA containers are available, each with advantages and disadvantages. The largest supplier is TransFRESH, but others include Sabroe/Freshtainer, Carrier Transicold, Mitsubishi, Isolcell, and CONAIR-plus. TransFRESH uses a Tetrol CA system in which the desired atmosphere is injected into highly leak-proof containers, and the atmosphere maintained by introduction of fresh air and operation of a carbon dioxide scrubber. More sophisticated systems with greater atmosphere control employ a pressure swing absorber (PSA) or nitrogen membranes to generate nitrogen, along with carbon dioxide and ethylene scrubbers.

Control systems that generate MA/CA conditions in reefer ship holds also have been developed for apple and banana trades. CA storage has become accepted practice, and the general fresh produce carriers are able to offer CA capacity. However, ship-hold CA requires a single large shipment of product from one single destination to a single or perhaps two receiving ports, and container shipping is more convenient for smaller volumes of produce.

Use of MAP systems has been investigated for many fresh crops. MAs are usually obtained passively by product respiration within a film bag, the final atmosphere obtained being a function of film permeability to carbon dioxide and oxygen, film area, product weight and volume, and temperature. Successful development of a MA requires not only films that produce appropriate concentrations of oxygen and carbon dioxide to maintain quality, but avoidance of injurious gaseous concentrations when

Table 3 Recommended MA or CA conditions during transport of selected fruits and vegetables

Crop	*CA range*		*Potential for benefit*[a]	*Remarks*
	% O_2	% CO_2		
Fruit				
Apple	1–3	1–5	A	About 50% of production is stored under CA
Apricot	2–3	2–3	C	No commercial use
Cherry, sweet	3–10	10–15	B	Some commercial use
Fig	5–10	15–20	B	Limited commercial use
Grape	2–5	1–3	C	Incompatible with SO_2 fumigation
Kiwifruit	1–2	3–5	A	Some commercial use: C_2H_4 must be maintained below 20 ppb
Nectarine (*Prunus persica*)	1–2	3–5	B	Limited commercial use
Peach	1–2	3–5	B	Limited commercial use
Pear, Asian (*Pyrus bretschneideri*)	2–4	0–1	B	Limited commercial use
Pear, European	1–3	0–3	A	Some commercial use
Persimmon (*Diospyros kaki*)	3–5	5–8	B	Limited commercial use
Plum and Prune (*Prunus americana*)	1–2	0–5	B	Limited commercial use
Raspberry (*Rubus idaeus*) and other cane berries	5–10	15–20	A	Increasing use during transport
Strawberry	5–10	15–20	A	Increasing use during transport
Nuts and dried fruits	0–1	0–100	A	Effective insect control method
Avocado (*Persea americana*)	2–5	3–10	B	Limited commercial use
Banana	2–5	2–5	A	Some commercial use during transport
Grapefruit	3–10	3–10	C	No commercial use
Lemon	5–10	0–10	B	No commercial use
Lime (*Citrus limettioides*)	5–10	0–10	B	No commercial use
Olive (*Olea europaea*)	2–3	0–1	C	No commercial use
Orange	5–10	0–5	C	No commercial use
Mango	3–5	5–10	C	Limited commercial use
Papaya (*Carica papaya*)	3–5	5–10	C	No commercial use
Pineapple (*Ananas comosus*)	2–5	5–10	C	No commercial use
Vegetables				
Artichokes	2–3	2–3	B	No commercial use
Asparagus	Air	5–10	A	Limited commercial use
Beans, snap	2–3	4–7	C	Potential for use by processors
Beets	None		D	98–100% RH is best
Broccoli	1–2	5–10	A	Limited commercial use
Brussels sprouts	1–2	5–7	B	No commercial use
Cabbage	2–3	3–6	A	Some commercial use for long-term storage of certain cultivars
Cantaloupes	3–5	10–15	B	Limited commercial use
Carrots	None		D	98–100% RH is best
Cauliflower	2–3	2–5	C	No commercial use
Celery	1–4	0–5	B	Limited commercial use in mixed loads with lettuce
Corn, sweet	2–4	5–10	B	Limited commercial use
Cucumbers	3–5	0	C	No commercial use
Honeydews (*Cucumis melo*)	3–5	0	C	No commercial use
Leeks	1–2	3–5	B	No commercial use
Lettuce	1–3	0	B	Some commercial use with 2–3% CO_2 added
Mushrooms	Air	10–15	C	Limited commercial use
Okra (*Abelmoschus esculentus*)	3–5	0	C	No commercial use; 5–10% CO_2 is beneficial at 5–8°C
Onions, dry	1–2	0–5	B	No commercial use; 75% RH
Onions, green	1–2	10–20	C	Limited commercial use
Peppers, bell	3–5	0	C	Limited commercial use
Peppers, chili	3–5	0	C	No commercial use; 10–15% CO_2 is beneficial at 5–8°C
Potatoes	None		D	No commercial use
Radish (*Raphanus sativus*)	None		D	98–100% RH is best
Spinach	Air	10–20	B	No commercial use
Tomatoes				
mature-green	3–5	0–3	B	Limited commercial use
partially ripe	3–5	0–5	B	Limited commercial use

[a] A, excellent; B, good; C, fair; D, slight or none.

Adapted from Kader AA (ed.) (2002) Postharvest Technology of Horticultural Crops, 3rd edn, with permission.

the product is exposed to different temperatures throughout storage, transport, and marketing. However, fermentation resulting from temperature abuse can occur under commercial conditions, and MAP is therefore not commonly used for whole products. Exceptions exist, e.g., the strawberry, which tolerates and responds positively to elevated carbon dioxide injection into sealed plastic bags surrounding each pallet of fruit to a final concentration of approximately 15%.

MAP is used extensively to extend the storage life of fresh cut (minimally processed) products, and has become the mainstay of the industry. Fresh cut products, by definition, involve the disruption of plant tissues, and MAP serves to reduce water loss, cut surface browning, respiration rates, ethylene biosynthesis and action, and microbial growth. Desirable atmospheres may be different for fresh cut products than for whole products because deleterious responses, e.g., browning, may be prevented by atmospheres that would be unsuitable for the whole product. Cutting of products can reduce the boundaries to gas exchange between the product and its storage environment. Because of the generally shorter transport/storage periods involved, the fresh cut industry also can take advantage of the short-term tolerance of tissues to atmospheres that would cause damage over more extended time periods. Nevertheless, successful transport of fresh cut products requires close attention to temperature requirements, not only to prevent fermentation, but also to minimize the risk of microbial contamination.

MA conditions can develop inadvertently during transport of fresh crops, when air circulation is restricted. Development of carbon dioxide injury can cause losses to apple fruit in ship holds and in containers, and can be aggravated by use of cartons with limited ventilation. Blockages of air passages, such as floor drains, in tightly sealed containers or vehicles by debris or ice is common during winter shipments of broccoli, rapini (*Brassica campestris*), and Brussels sprouts (*Brassica oleracea* var. *gemmifera*) that are shipped with package ice and under top-ice.

Transportation Modes

Trucks, Trailers, and Containers

Both unrefrigerated and refrigerated trucks are used for short-haul wholesale delivery from distribution centers to supermarkets and other outlets. In some cases, open bed trucks with or without tarpaulins to protect the loads are used. Loss of temperature control at this point can result in loss of crop quality associated with moisture condensation and, subsequently, greater risk of decay and deterioration. Long-distance transportation requires refrigerated vehicles such as refrigerated semitrailers. The utilization of vehicles that are used for intermodal transport, i.e., the movement of containers via water, rail, and road, is common. Intermodel containers can provide door-to-door service with minimal handling of produce, uninterrupted temperature, and atmosphere control, except where quarantine inspections are required. These containers may have CA capacity (see above).

Sea Transport

Although containerization is an important component of the international sea shipment of fresh crops, about 50% of reefer cargo of perishable products is still transported by bulk and palletized. The transport period can be close to the storage life required for maximum product quality. Therefore, the same criteria important for land based storage apply to ship cargoes.

Air Transport

Air freight is the only means to transport highly perishable products from many parts of the world to distant markets, but of all transport strategies tends to be the most difficult to ensure a continuous cool chain. Ground handling can amount to more than 70% of the time taken to transport products from the shipper to the receiver. Many airports are not well equipped for maintaining cold temperatures. It is essential to ensure fast movement of product pickup and delivery. Load palletization and improved load handling systems can minimize delays. However, indirect flights involving changing planes and off-loading where priorities are given to passenger luggage can result in serious losses of temperature control.

A study of asparagus temperatures in transit from New Zealand to Japan illustrates how temperature abuses may occur during transit (**Figure 2**). At harvest, temperatures of the asparagus were lowered rapidly by hydrocooling and the spears then placed into cold storage. Product temperatures increased during trucking to the airport because of inadequate insulation, but were then lowered again by forced air-cooling at the freight forwarder. After several days, the asparagus was delivered to the airport where it was kept for 3–4 h outside before loading on the plane. In the first flight leg to Singapore, product temperatures increased, probably because of a higher hold temperature required for animals or other

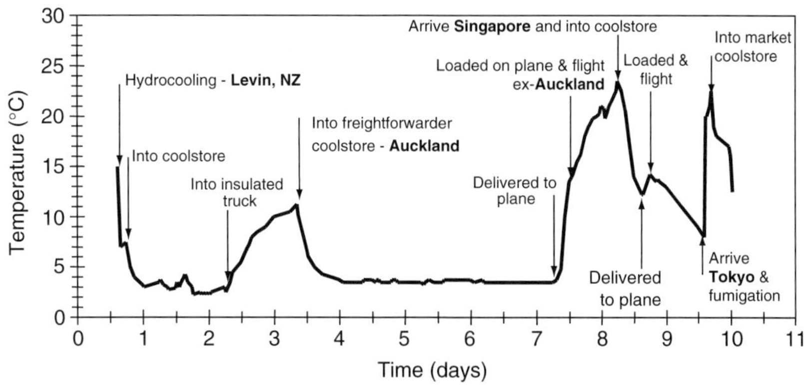

Figure 2 Temperature profile for asparagus harvested and packed in Levin, New Zealand and transported by road and air to Japan via Singapore. Reproduced with permission from Bollen AF, Brash DW, and Bycroft BL (1998) Air-freight coolchain improvements using insulation and supplemental cooling. *Applied Engineering in Agriculture* 14: 49–53.

freight. After unloading, product temperatures were again decreased by cold storage, increased slightly while loaded into the second plane, and cooled during flight. The asparagus were immediately fumigated upon arrival in Japan (although delays also can occur at this point), and then cooled in the market cold store. It has been demonstrated that heat unit accumulation is linearly correlated to asparagus shelf-life, and therefore a wide range of heat unit accumulation from 370 to 1160 °C-hours will result in considerable variation of shelf-life and quality.

Shipping packages used for air transport are usually the same as those used for land and sea transport. Containers built on pallets and shaped to make maximum use of the interior plane volume are in use. Most containers are unrefrigerated and provide minimal air circulation. Control of relative humidity is not available, and can be as low as 4% in jets at high altitude. Insulated airfreight containers for product shipment have not proven popular because of their bulk and higher delivery charges. Pallets of product, however, may be wrapped in insulating foil wrap, with or without supplemental cooling with ice or dry ice.

Transport of Mixed Commodities

Transport of commodities preferably involves single product types within a transit vehicle, but it is common for several products to be transported together, especially during shorter term transit situations. However, the transport of mixed loads results in conflicts between the optimum stacking of different package types as well as meeting the differing optimum storage regimes for each product. These factors need to be considered before such mixed products are transported as failure to recognize the significance of each may lead to rapid product deterioration.

1. Stacking of products within the transit vehicle. Mixed load transportation often involves use of different size and shape packages for each commodity. Variations in loading patterns can result in blockages in air circulation, and therefore less efficient temperature control. Special care is required to avoid such blockages, and use of pallets or racks to ensure adequate airflow along the transit vehicle floor.
2. Temperature. Crops with incompatible storage temperatures should not be stored together. However, because chilling injury is a function of the time that the product is kept at low temperature, chilling sensitive crops are sometimes transported for short distances at the suboptimum temperatures suitable for chilling resistant crops. The alternative of transporting chilling resistant crops at higher temperatures to avoid injury of chilling susceptible crops also can result in compromised quality because of more rapid deterioration of the chilling resistant crops.
3. Moisture compatibility. High relative humidity is recommended for most fresh crops to prevent loss of marketable weight and quality, but some crops such as garlic and dry onions require intermediate humidity levels (**Table 2**). Failure to maintain the relative humidity appropriate to the crops in the mixed load can result in unacceptable water loss from some products. In general, however, it is difficult to maintain high relative humidity during transport.
4. Ethylene production and sensitivity compatibility. Crops can vary greatly in their sensitivity to ethylene (**Table 4**). Typically, the fruit listed in **Table 4** have moderate to high ethylene production rates, and sensitivity to ethylene is expressed as accelerated ripening and/or decay. Most of the vegetables listed have low ethylene production, and symptoms of exposure to the gas include

Table 4 Examples of ethylene sensitive fruits, vegetables, and flowers

Fruits	*Vegetables*	*Flowers*
Apple	Beans	Anemone, windflower (*Anemone* spp.)
Apricot	Carrot	Bouvardia (*Bouvardia* spp.)
Guava (*Psidium guajava*)	Celery	Carnations (*Dianthus caryophyllus*)
Jackfruit (*Artocarpus heterophyllus*)	Collards (*Brassica oleracea*) and kale (*Brassica oleracea*)	Daffodils (*Narcissus* spp.)
Kiwifruit	Cucumber	Delphinium (*Delphinium, Consolida* spp.)
Lychee (*Litchi chinensis*)	Eggplant	Gypsophila (*Gypsophila* spp.)
Mango	Basil (*Ocimum basilicum*), chives (*Allium schoenoprasum*), dill (*Anethum graveolens*), parsley (*Petroselinum hortense*)	Holly, Ilex (*Ilex* spp.)
Mangosteen (*Garcinia mangostana*)	Lettuce	Lily (*Lilium* spp.)
Mamey sapote (*Pouteria sapota*)	Mushrooms	Rose (some cultivars) (*Rosa* spp.)
Squash	Green onion	Snapdragon (*Antirrhinum majus*)
Persimmon	Parsnips (*Pastinaca sativa*)	
Quince (*Cydonia oblonga*)	Pear	
Rambutan (*Nephelium lappaceum*)	Spinach	
Sweetsop (*Annona squamosa*)	Tomato (mature green)	
	Watermelon	

Table 5 Fresh crops that absorb odors from other crops and therefore should not be transported together

Odor producing crop	*Odor absorbing crop*
Apple, pear	Cabbage, carrots, celery, figs, onions, potatoes
Avocado	Pineapple
Onion, carrot	Celery
Ginger (*Zingiber officinale*)	Eggplant
Leek	Fig, grapes
Onion	Apple, celery, pears, citrus
Green onion	Fig, grapes, mushrooms, rhubarb (*Rheum rhaponticum*), corn
Pepper	Bean, pineapple, avocado
Carrot	Celery
Grapes, fumigated with sulfur dioxide	Other fruits and vegetables

toughening, skin yellowing, accelerated ripening and senescence, and bitterness. The incidence of the ethylene related disorder russet spotting in lettuce (*Lactuca sativa*) has been shown to be about three times greater in mixed loads than in straight loads during truck shipments. Care must be taken therefore to avoid mixed containers of crops where ethylene producers will cause deterioration of other crops. The use of 1-MCP may have a major impact on crop compatibilities.

5. Odor. Some crops absorb odors produced by other crops (**Table 5**), resulting in lower market appeal. In mixed loads, therefore, these crops must be kept separated from each other.

Summary

Successful transport of any fresh crop is primarily a function of the ability of the storage operator, shipper, and retail marketer to minimize the deterioration rate of a crop. Attention to the crop physiology and all aspects of the transport operation is required to produce high-quality crops in the marketplace. When all optimum conditions have been met, it is not uncommon for a crop being transported long distances to be of superior quality to that transported relatively short distances. Product losses during transport can be related to preharvest factors such as poor cultivar selection, adverse growing conditions, and pesticide or fertilizer damage, rough handling during harvesting and transport to the packinghouse; delays in cooling, packing equipment damage, or transport of crop after storage; and at the end of its storage potential, loading of warm produce, improper loading, excessive vibration or refrigeration equipment failure, or improper mixing of produce in the load. The importance of each factor will be a function of the quality of the product at the time of shipping, the storage potential of that product, and the transport time required.

List of Technical Nomenclature

Controlled atmosphere (CA) storage Storages in which atmospheres of oxygen, carbon dioxide, and nitrogen (and sometimes other gases) are controlled by external control systems.

Ethylene A plant growth regulator that controls the way that fruit grow and develop.

Fresh cut (syn. minimally processed) products Processed products that have been prepared without application of complete preservation procedures (heat sterilization, freezing, etc.), retaining their unprocessed, fresh-like properties and quality, and are used for convenient consumption.

Intermodal transport Transport that involves a combination of road, rail, and sea.

Modified atmosphere (MA) storage Similar to CA storage in that atmospheres of oxygen, carbon dioxide, and nitrogen are altered from normal atmospheric levels, but the atmosphere is not controlled. MAs are produced by plastic films or wax around products, the final atmosphere being a function of weight and volume of product, temperature, permeability characteristics of the film or wax, and film surface area.

See also: **Postharvest Physiology**: Postharvest Physiological Disorders of Fresh Crops; Ripening; Seed Storage; Storage.

Further Reading

Anon (1998) *American Society of Heating, Refrigerating and Air-Conditioning Engineers' Handbook*. Atlanta, Georgia: The American Society of Heating, Refrigerating and Air-Conditioning Engineers.

Bollen AF, Brash DW, and Bycroft BL (1998) Air-freight coolchain improvements using insulation and supplemental cooling. *Applied Engineering in Agriculture* 14: 49–53.

Gross KC, Wang CY, and Saltveit ME (2003) *The Commercial Storage of Fruits, Vegetables, and Florist and Nursery Stocks*. United States Department of Agriculture Handbook, No. 66. Washington DC. United States Department of Agriculture.

Heap R, Kierstan M, and Ford G (1998) *Food Transportation*. London: Blackie Academic and Professional.

Kader AA (ed.) (2002) *Postharvest Technology of Horticultural Crops*, 3rd edn. Publication 3311. Oakland, California: University of California, Division of Agriculture and Natural Resources.

Thompson JF, Brecht PE, Hinsch T, and Kader AA (2000) *Marine Container Transport of Chilled Perishable Produce*. Publication 21595. Oakland, California: University of California, Division of Agriculture and Natural Resources.

Watkins CB and Miller WB (2003) *A Summary of Physiological Processes or Disorders in Fruits, Vegetables and Ornamental Products that are Delayed or Decreased, Increased, or Unaffected by Application of 1-Methylcyclopropene (1-MCP)*. http://www.hort.cornell.edu/department/faculty/watkins/ethylene/index.htm

Postharvest Physiological Disorders of Fresh Crops

C B Watkins, Cornell University, Ithaca, NY, USA

Introduction

The harvest of any crop, whether fruit, vegetable, or flower, is associated with mechanical stresses, and because the plant part is separated from the parent, removal from a supply of water, nutrients, hormones, and energy. When harvested, the plant part has an altered ability to respond to stresses in the environment. Stress interrupts, restricts, or accelerates normal metabolic processes in an adverse or negative manner, and therefore is usually considered as potentially injurious to any plant system. However, consideration of postharvest systems is complicated as many storage regimens beneficially utilize stress conditions such as temperature and atmosphere modification to maximize storage potential of fresh crops. As highlighted by Kays, from a postharvest physiologist's position, stress is an external factor that will result in undesirable changes *only* if the plant or plant part is exposed to it for a sufficient duration or sufficient intensity. The postharvest period, in this context, can be seen as a time of stress management.

Postharvest physiological disorders are defined as those disorders that occur in fresh crops after harvest resulting from altered metabolism in response to imposition of stresses, and that are manifested as visible symptoms of cell death in the susceptible plant part. Physiological disorders are distinct from the many other undesirable postharvest changes in quality, such as water loss (wilting), loss of chlorophyll, softening, and other ripening related events associated with normal senescence, which affect fresh crop storage potential and thus marketability. The definition also excludes a number of direct postharvest injuries that can occur as a result of mechanical damage (e.g., bruising and other browning reactions associated with harvesting and handling), freezing damage, and exposure to gases or chemical solutions (e.g., ammonia leaks in cold storage, or skin damage from salts and antioxidants used in postharvest treatments). Pathological disorders are also distinct, but it is not uncommon for diseases to be associated with physiological disorders, especially as secondary infections.

Postharvest physiological disorders, by definition, develop after harvest. However, an overlap between

Table 1 Major physiological disorders of *Malus domestica* (apple) fruit

Postharvest factor	*Physiological disorder*
Nutrition	Bitter pit, senescent breakdown, watercore breakdown, Jonathan spot, lenticel blotch
Low temperature	Low temperature breakdown, superficial scald, soft scald, coreflush, or brown core
Atmosphere (lowered oxygen and/or elevated carbon dioxide)	Low oxygen injury, external carbon dioxide injury, internal carbon dioxide injury

pre- and postharvest expression can exist, and some disorders, especially those that are nutrition-related, may develop in fresh crops both on the parent plant and after harvest. In addition, it is important to note that preharvest factors, including climate, maturity at harvest, nutrition, and field management methods, can markedly affect susceptibility and tolerance of fresh crops to postharvest stresses. Therefore, considerable variation in the severity and timing of disorder expression can be observed even in the same crop, both within and among apparently similar lots of the same cultivar or strain.

A wide range of physiological disorders has been identified in harvested fresh crops. Those affecting the major commercial crops have been relatively well defined, although understanding of the biochemical basis for their development is often incomplete. The literature is most extensive for fresh crops that are stored for long periods, and of all crops, the apple (*Malus* × *domestica*) probably has the widest range of commercially significant physiological disorders (**Table 1**). The etiology of some disorders, particularly on minor crops, is often less complete. This article focuses on disorders associated with calcium deficiency, chilling injury, carbon dioxide and oxygen, and ethylene, and on hypotheses for disorder development.

Disorders Associated with Calcium Deficiency

Most information on nutrition and physiological disorders has centered on calcium. In addition to those described for apple (**Table 1**), a wide range of disorders associated with calcium deficiency are recognized (**Table 2**). Foliar symptoms of calcium deficiency are rarely observed, but disorders of fruits, storage roots, tubers, and compact leafy vegetables have been well described. These disorders can vary from localized symptoms in the case of bitter pit in apples or cavity spot of carrots (*Daucus carota*) to extensive cell death in disorders such as senescent breakdown of apples.

Table 2 Physiological disorders associated with calcium deficiency

Crop	*Physiological disorder*
Apium graveolus (celery)	Black heart
Brassica oleracea Group Gemmifera (Brussels sprouts)	Internal browning
Brassica oleracea Group Capitata (cabbage)	Internal tipburn
Capsicum annuum (pepper)	Blossom end rot
Daucus carota (carrot)	Cavity spot
Fragaria × *ananasa* (strawberry)	Tipburn
Lactuca sativa (lettuce)	Tipburn
Lycopersicon esculentum (tomato)	Black seed, blossom end rot
Manifera indica (mango)	Soft nose
Persea americana (avocado)	End spot
Phaseolus vulgaris (snap or green beans)	Hypocotyl necrosis
Pyrus communis (pear)	Cork spot, bitter pit
Solanum tuberosum (potato)	Spout failure, tipburn, internal rust spot
Chicory (*Cichorium intybus*)	Blackheart, tipburn
Parsnip (*Pastinaca sativa*)	Cavity spot

The soil solution usually provides an adequate supply of calcium to the plant, and therefore calcium deficiencies normally result from inefficient calcium distribution rather than limited calcium uptake. Calcium is moved to a greater extent in the xylem than in the phloem. A high growth rate of susceptible tissues can increase the occurrence of calcium deficient disorders, and calcium movement into fast-growing tissues, such as leaves, is much greater than into low-transpiring organs, such as fruits and storage organs. Water shortage or irregular water supply can also reduce calcium translocation into these organs. In contrast, magnesium and potassium translocation is less affected, resulting in a high magnesium plus potassium to calcium ratio in the tissue.

The role of calcium in plant tissues is assumed to be primarily in the maintenance of membrane structure and function, and of cell wall structure. Calcium stabilizes membranes by bridging phosphate and carboxylate groups of phospholipids and proteins at membrane surfaces. Calcium-deficiency disorders are probably related to impaired selective permeability of the membrane to many solutes, membrane disorganization, and loss of many membrane functions, and eventually loss of compartmentation in the cell. An involvement of calcium in cell walls also seems critical as the cell wall structure involves calcium cross-linkage, especially with pectin components. Tissues with low calcium may have weakened cell wall and middle lamella regions. Other possible biological functions of calcium include effects on enzymes and calcium–phytohormone interactions. However, very few enzymes are known to be activated by calcium, and *in vivo* free calcium in the cytoplasm and chloroplasts must be maintained to avoid enzyme inhibition, precipitation of phosphate, and competition with magnesium for binding sites.

Many factors that affect calcium distribution may also influence occurrence of calcium related disorders. These include humidity, root pressure, and phytohormone activity. Low calcium does not always result in development of physiological disorders. In apple, for example, bitter pit incidence during storage can be extremely high in early harvested fruit, but decline to a minimum as fruit reach physiological maturity, even though calcium concentrations in the tissue remain similar. As fruit mature further, the incidence of senescent breakdown during storage can increase. Therefore, the type and severity of the calcium related deficiency can vary greatly according to the maturity of the fruit at harvest, indicating the critical involvement of unknown factors in susceptibility of tissues to injury.

Other nutrients such as nitrogen can exacerbate calcium deficiency symptoms. The mechanism of nitrogen action may be preharvest by reducing calcium uptake, especially when the ammonium form of nitrogen is present, stimulating crop yields, organ size, and vegetative growth, or postharvest by increasing metabolic rates of the fresh crop. A high magnesium and/or potassium ratio to calcium may also result in greater susceptibility to calcium deficiency. Magnesium and potassium may replace calcium in the plasma membrane, but cannot replace its role in membrane stabilization, resulting in membrane dysfunction.

Commercial occurrence of calcium-deficient disorders appears related to increased yields achieved by modern agricultural processes. Many cultural practices for amelioration of these disorders have been investigated, but results have sometimes been inconclusive because of the wide range of interactions that can affect calcium uptake in the field. The most satisfactory methods for control of calcium-related physiological disorders involve direct application of calcium to the calcium deficient plant part. This has been achieved most successfully for the apple, for which many industries use routine calcium sprays and/or postharvest calcium drenches to reduce bitter pit and senescent breakdown of fruit during storage. Damage, particularly to skin tissues can occur however, if excess calcium concentrations are applied, and this technology is not suitable for many fresh crops.

It should be noted that the incidence of several physiological disorders, not necessarily classified at calcium deficiencies, are affected by calcium. In general, plant tissues with low calcium status are more susceptible to development of disorders. These include some of the chilling and atmosphere related disorders discussed below. Also, disorders, such as watercore development in apple, which result from dysfunctions in carbohydrate loading into overmature fruit, may be ameliorated by high calcium. However, they are not calcium deficient disorders *per se*.

Chilling Injury

Chilling injury can be defined as the visual manifestation of cellular dysfunction in crops exposed to chilling temperatures, usually in the range of -1.5–15°C. The lower temperature of this range for a given crop or plant part is a function of the freezing point. This injury is distinct from freezing injury that results from the direct cellular damage associated with ice formation. Because chilling injury occurs in crops growing in the field and is a major limitation to geographical distribution of many tropical and subtropical species, much literature exists on whole plant chilling injury. However, this article focuses on postharvest related chilling injury that occurs during storage, transport, and market distribution, and in home refrigerators.

Decreasing storage temperatures generally decrease metabolic rates of horticultural crops, and therefore temperature management is the most utilized technology for prolonging crop storage periods. However, many crops are susceptible to chilling injury, thus limiting application of this technology to maintain quality. Different patterns of response are observed when the maintenance of quality of chilling resistant and chilling sensitive crops over time at different temperatures is compared (**Figure 1**). The maximum storage period for a chilling resistant crop increases with lower temperature, being maximal at temperatures just above the freezing point. In contrast, the storage life of a chilling sensitive crop reaches a maximum at some higher temperature, depending on the commodity (see **Table 3** for examples), declining rapidly as temperatures are lowered. Lower storage temperatures may result in maintenance of desirable attributes such as firmness or greenness, but these positive attributes are outweighed by symptoms of chilling injury.

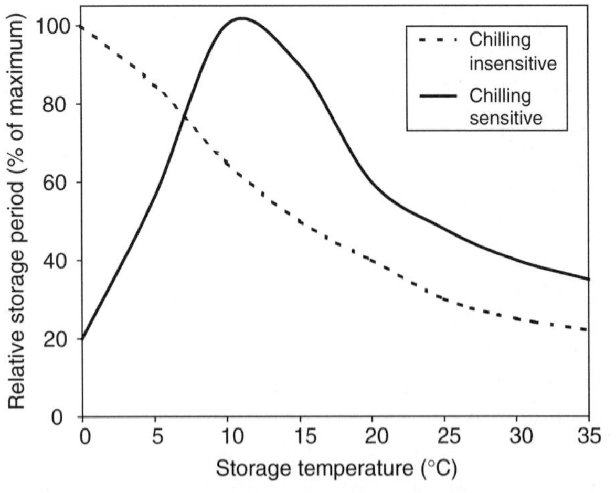

Figure 1 Stylized version of the relative changes in quality at different storage temperatures for chilling sensitive and chilling insensitive crops.

Chilling injury of harvested fresh crops is mostly observed in tropical and subtropical species that comprise the majority of recognized chilling sensitive crops (**Table 3**). Some definitions of chilling injury specifically incorporate the tropical and subtropical nature of the crop, and there is some controversy as to whether certain low temperature disorders occurring in temperate crops are manifestations of chilling injury. However, the physiological and biochemical mechanisms of injury probably are identical in all susceptible commodities, but the rate at which these changes occur differs dramatically. In tropical crops, for example, injury can occur within days of exposure to cold storage, whereas in a temperate crop such as the apple, chilling injuries occur only after several months. In addition, the mechanisms of chilling injury are likely to be similar in preharvest and postharvest situations, except

Table 3 Optimum storage temperatures for perishable products

Temperature	*Fruits*	*Vegetables*	*Flowers*
0°C or below (but above freezing point)	Apples (most cultivars), apricots, most berries (except cranberries), cherries, dates, figs, grapes, kiwifruit, loquats, nectarines, peaches, pears, persimmons, plums, prunes, quinces	Artichokes, asparagus, beans, beets, broccoli, Brussels sprouts, cabbage, carrots, cauliflower, celeriac, celery, chard, chicory, collards, corn, endive, escarole, garlic, leafy greens, horseradish, kale, kohlrabi, leeks, lettuce, mushrooms, onions, parsley, parsnips, green peas, radishes, rutabagas, salsify, spinach, turnips, watercress	Carnation, chrysanthemum, iris, lily-of-the-valley, dry rose, sweet pea, tulip
0–2°C	Apples (some cultivars), oranges	Asparagus, fullslip cantaloupe, water chestnuts	Allium, aster, bouvardia, crocus, freesia, gardenia, gerbera, hyacinth, narcissus, cymbidium orchid, ranunculus, rose in preservative
2–7°C	Apples (some cultivars), avocados (ripe), cranberries, guavas, oranges, pomegranates, tangerines, mandarins	Green beans, lima beans, cassava, 3/4 slip cantaloupe, southern peas, summer squash, tamarillos	Acacia, alstromeria, anemone, aster, bird-of-paradise, buddleia, calendula, calla, candytuft, columbine, cornflower, dahlia, daisy, delphinium, gerbera, gladiolus, gypsophilia, heather, lily, lupin, marigolds cymbidium orchid, poppy, phlox, primrose, protea, snapdragon, statice, stephanotis, stock, strawflower, sweet william, violet, zinnia, florists' greens
7–13°C	Avocados, carambolas, lemons, limes, papayas, passion fruit, pineapples	Cucumbers, eggplant (aubergine), muskmelons (casaba, crenshaw, honeydew), watermelon, okra, sweet peppers, pumpkins, winter squash, summer squash, taro, ripe tomatoes	Bird-of-paradise, heliconia, cattleya orchid, sweet william
13°C and above	Bananas, grapefruit, mango, plantain	Ginger, jicama, watermelon, sweet potatoes, green tomatoes	Anthurium, ginger, vanda orchid, poinsettia

Reproduced with permission from Kasmire RF and Thompson JF (1992) Selecting a cooling method. In: Kader AA (ed.) *Postharvest Technology of Horticultural Crops*, pp. 63–68. University of California Publication 3311. Oakland: University of California.

for the interaction with photosynthetic processes in the field.

The sensitivity of horticultural crops to chilling injury is affected by many factors including the species, cultivar, plant part, and morphology at the time of exposure to low temperatures. The physiological condition of the crop can also be important, especially the ripening status or relationship to the climacteric when exposure occurs. Some differences may be more apparent than real, e.g., ripe fruit may appear more resistant to chilling injury than unripe fruit, but only because alterations of already completed aspects of ripening can no longer occur. Preharvest factors, such as temperatures during the growing season and other field/orchard management techniques that influence the carbohydrate and mineral status of the crop, can also impact the crops susceptibility to chilling injury.

Chilling injury can be separated into a variety of visual symptoms, only the first four being physiological disorders *sensu stricto*:

1. surface lesions – pitting, large sunken areas, and discoloration;
2. water soaking of tissues;
3. internal discoloration (browning) of pulp, vascular strands, and seeds;
4. breakdown of tissues;
5. failure of fruits to ripen normally;
6. an accelerated rate of senescence;
7. increased susceptibility to decay;
8. shortened storage or shelf-life;
9. compositional changes; and
10. loss of growth (sprouting) capacit.

Table 4 illustrates variability of symptom development among different crops.

Table 4 Expressions of chilling injury in selected fresh crops

Crop	*Physiological disorder*
Aegle marmelos (bael)	Brown spots on skin
Actinidia deliciosa var. *deliciosa* (kiwifruit)	Low temperature breakdown
Ananas comosus (pineapple)	Internal browning (blackheart)
Annona spp. (custard apple, soursop)	Skin darkening, pulp discoloration
Asparagus officinalis (asparagus)	Discoloration and softening of stem tips
Capsicum annuum (pepper)	Sunken, water-soaked sheet pitting, darkening of seed
Carica papaya (papaya)	Skin scald, hard areas in pulp around the vascular bundles, water soaking of the tissue
Citrullus lanatus (watermelon)	Brown staining of the rind, rind pitting
Citrus sinensis (sweet orange); *C. limon* (lemon)	Rind pitting, browning of albedo or carpellary membranes between segments, scalding or staining
Cucumis melo (honeydew and cantaloupe melons)	Reddish-tan discoloration of surface and subsurface, sunken lesions, pitting
Cucumis sativus (cucumber)	Surface pitting (dark colored, watery blemishes), tissue collapse
Cyphomandra betacea (tree tomato)	Scald
Diospyros kaki (oriental persimmon)	Internal watery breakdown
Garcinia mangostana (mangosteen)	Hardening and browning of cortex
Gladiolus, chrysanthemum	Petal abscission ("shattering")
Hibiscus esculentus (okra)	Pitting, surface discoloration (blackening along the ridges), bronzing
Ipomoea batatas (sweet potato)	Brown to black internal discoloration, internal breakdown
Lycopersicon esculentum (tomato)	Pitting, blotchy color development, water soaking, and softening
Mangifera indica (mango)	Grayish scald-like discoloration of skin, skin pitting
Musa spp. (banana)	Dull skin color, skin blackening, hardening of central plancenta, subepidermal brown streaking, appearance of water soaked areas
Nephelium lappaceum (rambutan)	Darkening of the exocarp and spinterns
Olea europea (olive)	Tannish to brown discoloration of the flesh near or adjacent to the pit, which can progress to the skin
Persea americana (avocado)	Brown or gray flesh discoloration, scalding, skin pitting
Phaeseolus vulgaris (snap or green beans)	Pitting, rusty-brown streaks, spots
Prunus armeniaca (apricot)	Pitting, browning of peel
Prunus persica (peach)	Wooliness, flesh browning near the pit
Pyrus communis (pear)	Mealiness, superficial scald
Solanun melongena (eggplant)	Surface pitting, browning of seeds and vascular bundles
Solanum tuberosum (potato)	Mahogany browning (Chippewa and Sebago), sweetening
Vaccinium macrocarpon (cranberry)	Low temperature breakdown, rubbery texture, red flesh

Symptoms of chilling injury expressed as physiological disorders vary greatly according to plant type, and can be variable even within a plant type. Many of these symptoms, however, suggest that major changes take place at the membrane level, resulting in loss of membrane integrity, leakage of solutes, loss of cellular compartmentation. Occurrence of physiological disorders may be one of the later symptoms of chilling injury. Chilling injury symptoms may be visible at chilling temperatures, but usually occur after transfer of the crop to warmer temperatures. Differences in chilling sensitivity among varieties of the same species can differ greatly according to the storage temperature and the physiological dysfunction being evaluated.

Temperature/exposure time interactions exist in the development of chilling injury, and dysfunctions induced by chilling temperatures can be repaired upon transfer of the crop to nonchilling temperatures before permanent injury has occurred. This observation has led to inclusion of primary and secondary components in development of hypotheses of chilling injury development (**Figure 2**). The primary response to direct chilling stress is thought to be physical in nature, centered on the cell membranes. Criticisms of the membrane lipid hypothesis exist, and the possibilities of alternative transducers, such as cytosolic calcium, mechanisms mediated by cold labile protein, and structural changes in the cytoskeleton have been proposed. The secondary events include the multitude of metabolic processes that are adversely affected as a consequence of the primary event and lead to visible symptoms and cell death. The subdivision between primary and secondary events is not arbitrary as it is proposed that it allows the time dependent secondary events ("effects") to be conceptually separated from the more instantaneous primary event ("cause").

Chilling injury of crops can be alleviated by the following postharvest methods, although results can be variable between and within species:

1. temperature conditioning – pre-exposure of the product to temperatures slightly higher than chilling;
2. intermittent warming – interruption of cold storage with short periods of warm temperature;
3. chemical treatments – these include antioxidants, antitranspirants, fungicides, calcium, membrane stabilizers and glycerol homologs, and chemicals that modify fatty acid composition;
4. plant growth regulators – treatment with auxins, abscisic acid, and ethylene; and
5. modified atmosphere storage – increasing carbon dioxide and lowering oxygen partial pressures in the storage atmosphere either by controlled conditions or by use of film packaging, but results are greatly affected by timing and duration of the treatment as well as by commodity. In some cases, elevated carbon dioxide can interact negatively with chilling temperatures, resulting in increased injury to the crop.

Injuries Associated with Oxygen and Carbon Dioxide

The effects of controlled atmosphere storage and modified atmosphere packaging on the storage potential of fresh crops has been widely studied, and their postharvest tolerances to low oxygen and

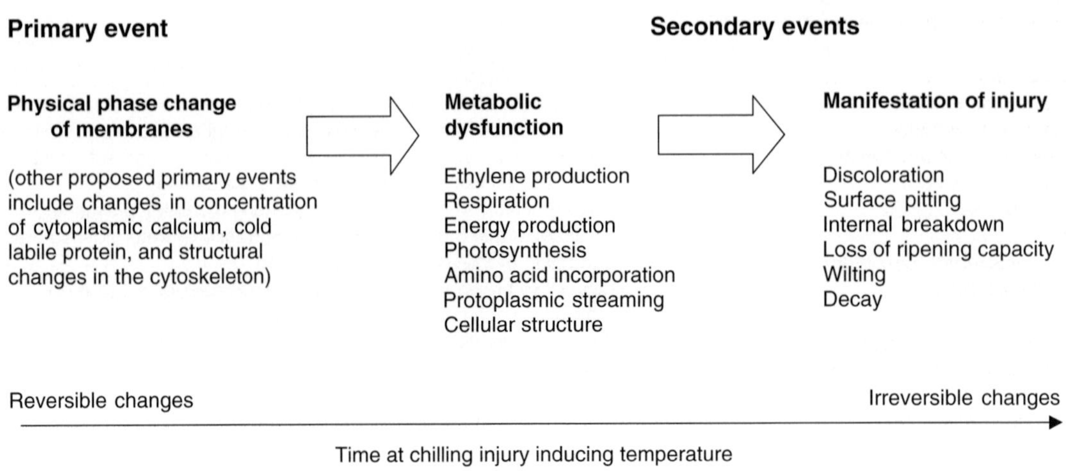

Figure 2 A schematic representation of responses of sensitive plants to chilling stress. Modified from Wang CY (1982) Physiological and biochemical responses of plants to chilling stress. *HortScience* 17: 173–186, and Raison JK and Orr GR (1990) Proposals for a better understanding of the molecular basis of chilling injury. In: Wang CY (ed.) *Chilling Injury of Horticultural Crops*. Boca Raton, Florida: CRC Press.

elevated carbon dioxide established. The goal of this research has been to find the atmosphere (and temperature) combinations that will result in maximum storage periods without injury. When a fresh crop is subjected to atmospheres outside safe limits at any temperature/time combination, damage may be manifested as irregular ripening, initiation and/or aggravation of certain physiological disorders, development of off-flavors, and increased susceptibility to decay. The tolerances of fresh crops to modified storage atmospheres are affected by metabolic and physical factors, and can vary greatly among species, cultivars and strains, organ types, and developmental stages. Time after harvest before exposure to elevated carbon dioxide, as well as treatment with antioxidants such as diphenylamine, can reduce susceptibility of fruit to carbon dioxide injury. Physical effects associated with skin thickness and permeance influence gas exchange between the outside and inside of the crop, and therefore their susceptibility to injury. These effects also impact tolerances of cut or minimally processed products to modified atmospheres compared with those of whole products.

Lower limits of oxygen range from $\leq 0.5\%$ (broccoli (*Brassica oleracea* var. *italica*), lettuce (*Lactuca sativa*), mushrooms (*Agaricus bisporus*)) to 14% (orange (*Citrus sinensis*) sections), while upper limits of carbon dioxide range from 1% (onion (*Allium cepa*)) to 30% (cherimoya (*Annona cherimola*)). Examples of physiological disorders are shown in **Table 5**, but in most crops, tolerance limits are indicated by development of off-flavors. Many other symptoms of negative responses to modified atmosphere are similar to those described for chilling injury, including softening, failure to ripen, or increased susceptibility to decay.

Understanding of the biochemical bases for development of these disorders is incomplete. Both lowered oxygen and elevated carbon appear to directly affect respiration and associated metabolic pathways of glycolysis, fermentation, the TCA cycle, and the mitochondrial respiratory chain, as well as pathways involved in secondary metabolism such as production of ethylene, pigments, phenolics, and volatiles. Increased carbon flux through the fermentation pathway by activation of pyruvate decarboxylase and alcohol dehydrogenase activities is a common feature of fresh crops exposed to anaerobic conditions, but direct evidence for injury by acetaldehyde and ethanol accumulations has not been demonstrated. Fresh crops exposed to high carbon dioxide, but usually not low oxygen, show high accumulations of succinate, which may be related to inhibition of succinate dehydrogenase by carbon dioxide. Succinate accumulation may be toxic to plant cells and it was proposed many years ago to be responsible for carbon dioxide injury of apple fruit. Recent evidence with tissues of different susceptibility to carbon dioxide has not supported this view. It is likely that effects of damaging levels of oxygen and carbon dioxide on fresh crops are due to progressive failure to maintain energy balance in the cell and metabolic cell function in response to accumulation of several toxic compounds.

Ethylene

Ethylene is a plant growth regulator that directly and indirectly affects plant metabolism. Many aspects of

Table 5 Expressions of physiological injury of selected fresh crops occurring in response to decreased oxygen and increased carbon dioxide concentrations in the storage atmosphere

Crop	*Decreased oxygen concentration*	*Increased carbon dioxide concentration*
Agaricus bisporus (mushroom)		Surface pitting, browning
Asparagus officinalis (asparagus)	Discoloration	Increased sensitivity to chilling injury, pitting
Brassica oleracea Group Gemmifera (Brussels sprouts)	Internal discoloration	
Capsicum annuum (pepper)	Breakdown	Calyx discoloration, internal browning
Carica papaya (papaya)		Aggravation of chilling injury
Cucumis sativus (cucumber)	Breakdown, increased chilling injury	Chilling injury, surface discoloration and pitting
Cynara scolymus (artichoke)	Blackening of inner bracts	Stimulated papus development and receptacle browning
Diospyros kaki (oriental persimmon)		Brown discoloration
Lactuca sativa (lettuce)	Breakdown at center	Brown stain
Lycopersicon esculentum (tomato)	Discoloration	Discoloration
Mangifera indica (mango)	Skin discoloration	Grayish flesh color
Olea europea (olive)		Aggravation of chilling injury
Persea americana (avocado)	Internal flesh browning	Skin browning
Prunus armeniaca (apricot)		Flesh browning
Prunus persica (peach)	Skin browning	Flesh browning
Pyrus communis (pear)	Internal flesh browning	Brownheart, coreflush, internal browning

the ripening process are initiated and coordinated by ethylene in climacteric crops and, commercially, temperature control and atmosphere modification are used to slow down its production and action. Nonclimacteric crops do not exhibit an increase in ethylene production during ripening and senescence, but ethylene can induce deleterious effects on chlorophyll degradation, abscission, texture, and flavor in these crops. Ethylene, often at low concentrations, has also been implicated in the occurrence of the following physiological disorders:

1. Russet spotting of lettuce. Well defined, localized spot-like lesions occur either in the epidermis or in the mesophyll, and, with increasing severity, vascular tissue discoloration and mesophyll cell collapse.
2. Inrolling of petals, fading, wilting, and abscission of many flowers, including closing of open flowers ("sleepiness").
3. Gummosis, bud necrosis, and flower blasting in flowering bulbs. Other effects can include inhibition of shoot and root elongation, and abscission of flower buds and leaves.
4. Epinastic responses (downward curvature of leaves) in ornamental plants.
5. Development of white inclusions in core kiwifruit (*Actinidia deliciosa* var. *deliciosa*) exposed to ethylene under CA conditions.
6. Aggravation of physiological disorders such as rusty brown discoloration and carbon dioxide induced brown stain in lettuce.
7. Development of bitterness in carrots, associated with isocoumarin accumulation.

Commercial Implications of Physiological Disorders

A combination of experimental data and empirical information derived from research and industry experience has identified the required preharvest nutrition status and the tolerance limits of temperature and atmosphere necessary to avoid development of physiological disorders in storage for most fresh crops. These tolerances together with changes of other quality factors prescribe the storage conditions and the maximum storage period available to each horticultural industry for marketing of most cultivars and strains. These limits, especially susceptibility of tropical and subtropical crops to chilling injury, impose major limitations on their availability in the market place. In addition, factors such as unusual climatic conditions, e.g., cooler than normal growing seasons, and changes in industry procedures can still result in unexpected losses of product value.

List of Technical Nomenclature

Chilling sensitive Used to describe plants or plant parts that suffer damage and ultimately die when exposed to chilling temperatures.

Climacteric A period in the development of certain fruit that exhibit a large increase in respiration accompanied by ethylene production during ripening.

Ethylene Plant growth regulator that controls the way that plants grow and develop; especially important in ripening of climacteric fruit.

Ripening The final stage of maturation during which the fruit develops optimum appearance and eating quality.

Senescence The process following maturation and ripening in which plant tissues grow old and die.

See also: **Postharvest Physiology**: Ripening; Storage; Transport. **Regulators of Growth**: Ethylene.

Further Reading

Bangerth F (1979) Calcium-related physiological disorders of plants. *Annual Reviews of Phytopathology* 17: 97–122.

Beaudry RM (2000) Responses of horticultural commodities to low oxygen: limits to the expanded use of modified atmosphere packaging. *HortTechnology* 10: 491–500.

Calderon M and Barkai-Golan R (eds) (1990) *Food Preservation by Modified Atmospheres.* Boca Raton, Florida: CRC Press.

Hakim A, Purvis AC, and Mullinex BG (1999) Differences in chilling sensitivity of cucumber varieties depends on storage temperature and the physiological dysfunction evaluated. *Postharvest Biology and Technology* 17: 97–104.

Kader AA (1985) Ethylene-induced senescence and physiological disorders in harvested horticultural crops. *HortScience* 20: 54–57.

Kasmire RF and Thompson JF (1992) Selecting a cooling method. In: Kader AA (ed.) *Postharvest Technology of Horticultural Crops*, pp. 63–68. University of California Publication 3311. Oakland: University of California.

Kays SJ (1997) *Postharvest Physiology of Perishable Plant Products.* Athens, Georgia: Exon Press.

Kirkby EA and Pilbeam DJ (1984) Calcium as a plant nutrient. *Plant, Cell and Environment* 7: 397–405.

Snowden AL (1990) *A Color Atlas of Post-Harvest Diseases and Disorders of Fruits and Vegetables Vol. 1: General Introduction and Fruits.* Boca Raton, Florida: CRC Press.

Wang CY (1982) Physiological and biochemical responses of plants to chilling stress. *HortScience* 17: 173–186.

Wang CY (1990) *Chilling Injury of Horticultural Crops.* Boca Raton, Florida: CRC Press.
Watkins CB (2000) Responses of horticultural commodities to high carbon dioxide as related to modified atmosphere packaging. *HortTechnology* 10: 501–506.

Genetic Engineering for Postharvest Quality

A Bernadac, A Latché, Z Li, M Bouzayen and J-C Pech, UMR 990 INRA-ENSAT, Toulouse, France

Introduction and Definitions

The appearance of fresh fruit and vegetables has long been the major criterion in making purchasing decisions by consumers, but there is nowadays growing emphasis on sensory properties such as texture, taste, and aroma and on nutritive value. From the quality standpoint, postharvest physiology studies are intended to elucidate the mechanisms governing the accumulation of metabolites responsible for quality attributes (e.g., aromas, vitamins, and antioxidant contents) as well as the mechanisms causing the deterioration of quality (overripening, softening, senescence, physiological disorders, or pathogen attacks). Postharvest technology is devoted to the development of methods that minimize deterioration of produce between harvest and consumption.

Historically, breeding has mainly been directed towards improving yield, uniformity, pest resistance, and storage qualities. In many cases these improvements have not been associated with better sensory quality. For instance, the increased storability and transportability of tomatoes (*Lycopersicon esculentum*) has been accompanied by a loss of flavor. Although quality traits exist in the germplasm of many fruit and vegetables, they have not been extensively exploited so far. It therefore becomes a challenge for breeders to introduce quality and nutritional traits without altering the agronomic properties of the crop. The route leading to the combination of acceptable agronomic traits with sensory and nutritional quality via traditional genetics is quite long. High flavor genotypes of tomato exist, but they are very small in size so that the introduction of this character into commercial varieties will require a long and complex breeding program. With recent genetic engineering developments, it has become possible to consider orienting approaches to restoring desirable consumer traits to cultivars with excellent agronomic properties.

The tremendous advances made in recent years in isolating and characterizing genes involved in either the overall regulation of the ripening–senescence process (biosynthesis and action of the plant hormone ethylene) or in individual pathways (cell wall degradation, chlorophyll breakdown, aroma volatiles production, synthesis of pigments, etc.) have opened up a wide range of opportunities for the control of quality traits through biotechnological methods (**Figure 1**).

Reducing Postharvest Deterioration and Increasing Shelf-Life

Control of Ethylene Biosynthesis and Action

The plant hormone ethylene regulates many aspects of plant development, including the senescence of plant organs and fruit ripening. Reducing ethylene biosynthesis and action represents a means to slow down the postharvest deterioration of horticultural products thus limiting postharvest losses and extending the shelf-life.

The ethylene biosynthetic pathway includes (1) the generation of *S*-adenosyl methionine (SAM) from the methionine cycle, (2) the conversion of SAM into 1-aminocyclopropane-1-carboxylic acid (ACC) and (3) the oxidation of ACC into ethylene. The two major genes involved in ethylene biosynthesis, *ACC synthase* and *ACC oxidase*, have been used for genetic manipulation, mainly in tomato. Sense or antisense suppression of ACC synthase of tomato has resulted in a very strong inhibition of ethylene production (around 99% inhibition at the peak of ethylene production). The development of red color was inhibited as well as softening and aroma production. The ripening-inhibited phenotype could be completely reversed by continuous treatment with ethylene for 6 days so as to become indistinguishable from naturally ripened fruit. Using antisense constructs of the other gene, *ACC oxidase* (*ACO*), strong reduction of ethylene production has also been achieved (97%). However the 3% residual ethylene was too high to inhibit the onset of color changes and softening of the flesh during ripening but it was low enough to slow down the rate of pigment accumulation and softening of the flesh at the overripening stage. In addition, the alterations occurring in the ripening phenotype were significant after detachment only. Cantaloupe melons (*Cucumis melo cantalupensis*) of the Charentais type is the first fruit where very high (>99.5%) inhibition of ethylene production has been achieved using antisense *ACO* gene. In these

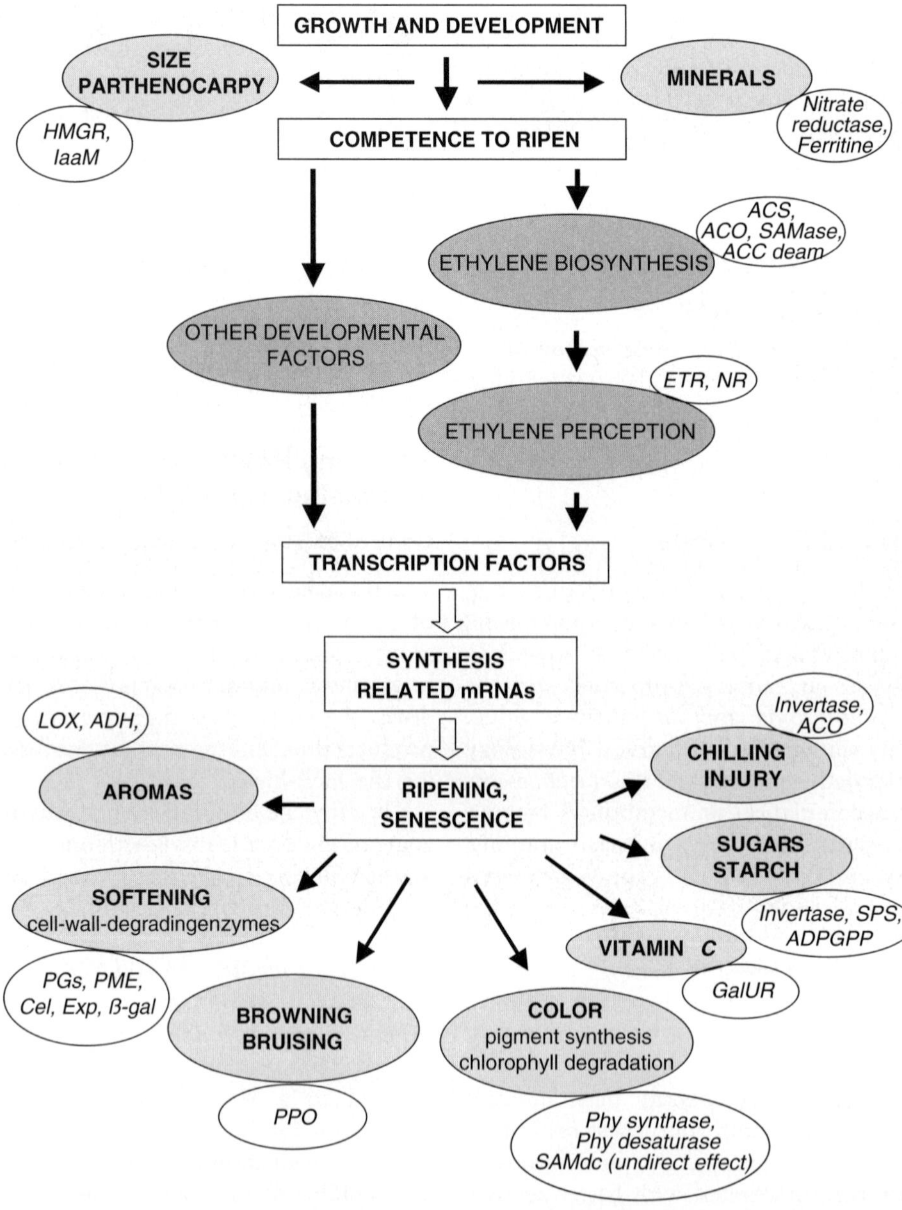

Figure 1 A scheme of the molecular mechanisms of fruit ripening and senescence showing the genes that have been targeted to the biotechnological improvement of postharvest quality. HMGR, hydroxymethylglutaryl coenzyme reductase; IaaM: indolacetamide monoxygenase; ACS, ACC synthase; ACO, ACC oxidase; SAMase, *S*-adenosyl methionine hydrolase; SAMdc, SAM decarboxylase; ACC deam, ACC deaminase; ETR and NR, "Ethylene Response" and "Never Ripe" ethylene receptors; SPS, sucrose phosphate synthase; ADPGPP, ADP glucose pyrophosphorylase; Phy, phytoene; PPO, polyphenol oxidase; PG, polygalacturonase; PME, pectine methyl esterase; Cel, cellulase; β-gal: β-galactosidase; Exp, expansin; LOX, lipoxygenase; ADH, alcohol dehydrogenase; GalUR, D-galacturonic acid reductase.

conditions, both attached and detached antisense fruit exhibited the same phenotype including inhibition of rind yellowing, peduncle detachment, part of flesh softening (**Figure 2**), aroma volatiles production, and the respiratory climacteric. However, some ethylene-independent pathways remained unaltered resulting in similar coloration of the flesh and accumulation of sugars and organic acids.

A number of transgenic lines of broccoli (*Brassica oleracea* var. *italica*) containing a tomato antisense *ACO* gene have been generated from hairy roots via *Agrobacterium rhizogenes*-mediated transformation. All of them showed reduced ethylene production in the florets 98 h after harvest. However about two-thirds of the lines exhibited hairy root-induced morphological changes as a result of *rol* loci expression. Among the lines with normal phenotype, two showed less postharvest yellowing. These data show that, similarly to ornamental flowers such as carnation (*Dianthus caryophyllus*), the senescence of

Figure 2 (see color plate 23) Some characteristics of ethylene-suppressed antisense ACC oxidase charentais cantaloupe melons (Antisense ACO, right) as compared to the nontransformed wild-type (WT, left). Note that ethylene-suppressed fruit do not develop the peduncular abscission zone (A); retain a green color of the rind on the vine (A) and after harvest (B); keep a firm flesh structure upon extended shelf storage (C), and do not develop chilling injury symptoms upon storage at low temperature (D). The color of the flesh is not affected by ethylene suppression (C).

broccoli florets, can be controlled by antisense *ACC oxidase* gene. Ethylene suppression through manipulation of ethylene biosynthesis genes has been achieved in other species. Sense or antisense ACC synthase and ACC oxidase apples (*Malus pumila*), as well as antisense ACC oxidase plums (*Prunus domestica*) are under field tests. Effects of ethylene suppression in apple are similar to those described in tomato and melon (inhibition of aromas production, of softening and of coloration).

Another approach to inhibit ethylene production consists in lowering the ethylene precursors by expressing heterologous genes. The T3 bacteriophage-derived *S*-adenosylmethionine hydrolase (SAMase) is capable of depleting SAM, a precursor of ethylene upstream in the biosynthetic pathway. The *SAMase* gene, driven by the E8 promoter restricting expression to the ripening fruit, has been expressed in tomatoes and melons. Ethylene synthesis was not reduced by more than 80% in either case, which was not enough to profoundly affect ripening. However in some lines of tomato, the period from pollination to ripening was reduced as well as the rate of ripening from the breaker stage to full red. In the melon, SAMase transgenic fruit exhibited more uniform ripening in the field and higher sugar content compared with control fruit. The higher sugar content of transgenic fruit was due to additional time on the vine associated with a delay in the formation of the slip. *ACC deaminase* is a bacterial gene encoding a protein capable of limiting the availability of ACC by transforming it into ammonia and α-ketobutyric acid. Tomato fruit expressing *ACC deaminase* and producing up to 85% less ethylene than control fruit ripen more slowly when removed from the vine early in ripening.

In contrast, fruit that remain attached to the plant ripen much more rapidly, exhibiting little delay relative to control fruit.

Taken together, these data show that the various components of the fruit ripening process have different sensitivities to ethylene and that some pathways are ethylene-independent and therefore not affected by ethylene suppression (**Table 1**). Significant alteration of the ripening process and shelf-life requires very strong inhibition of ethylene synthesis. When this can be achieved it becomes possible to set up a new postharvest handling procedure in which fruit can be harvested at full development with no risk of overripening, stored for the desired period of time, and allowed to ripen on command using exogenous ethylene. When residual ethylene remains above a threshold value, fruit can exhibit a slow-ripening character associated with a reduction of overripening. The slow ripening character however may appear, at least in the tomato, on detached fruit only. Fruit detachment induces a strong reduction of internal ethylene due to the diffusion of ethylene out of the stem scar. The postharvest strategy in this situation is to harvest fruit at the mature green or breaker stage and let the ripening process proceed at a slow rate.

Manipulation of ethylene synthesis can be achieved through genes exerting regulatory functions. One gene designated *E8*, related to ACC oxidase and a member of FeII dioxygenases, is supposed to be involved in the negative feedback regulation of ethylene biosynthesis. An antisense construct of this gene stimulates ethylene production during ripening of tomato fruit detached from the plant at, or well prior to, the onset of ripening, but the effects on ripening and quality have not been reported.

Ethylene acts through a receptor and a transduction pathway. Tomato contains a family of ethylene receptors named *LeETR1* to *5* and *NR*. A semidominant mutation of the *NR* gene is responsible for the inability to ripen of the *Never ripe* (*Nr*) natural mutant of tomato. Transgenic tomatoes expressing the *ETR1-1* mutant cDNA exhibited altered ethylene response, including ripening, indistinguishable from the *Nr* mutant. Slowing down the expression of *LeETR4*, a gene highly expressed in fruit but weakly in vegetative tissues, results in increased sensitivity to ethylene, with early ripening (up to 11 days earlier than wild-type) and more rapid development of color, indicating that *LeETR4* is a negative regulator of the ethylene transduction pathway. Antisense *NR*, on the contrary had a normal phenotype, but, in this case, reduction in *NR* expression resulted in a functional compensation consisting in an increased expression of *LeETR4*.

Biotechnological Control of Chilling Injury

Most fruit and vegetables of tropical and subtropical origin develop chilling injury at low nonfreezing

Table 1 Effects, on postharvest quality, of modifying the expression of genes from the ethylene biosynthesis pathway

Gene	*Type of modification*	*Plant species*	*Ripening processes*	*Changes*[a]
ACS	Antisense or cosuppression	Tomato	Ethylene production	↘ (99%)
			Pigment formation	↘
			Chlorophyll degradation	↘
			Firmness	↗
			Aroma production	↘
ACO	Antisense	Tomato	Ethylene production	↘ (97%)
			Acidity	↗
			Pigment formation	↘
			Cracking and damage	↘
	Antisense	Melon	Ethylene production	↘ (99.5%)
			Rind yellowing	↘
			Climacteric respiration	↘
			Peduncle abscission	↘
			Flesh softening	↘
			Aroma	↘
			Chilling injury	↘
	Antisense	Broccoli	Ethylene production	↘ (90%)
			Respiration	↘
SAMase	Overexpression	Tomato	Ethylene production	↘ (80%)
	Overexpression	Melon	Ripening duration	↗
			Sugar content	↗
			Peduncular abscission	↘
ACC deaminase	Overexpression	Tomato	Ethylene production	↘ (85%)

[a]Arrows indicate increase (↗), inhibition (↘), or no effect (→) on ripening or quality attributes.

temperatures which limits their storability and causes significant postharvest losses. Changes in membrane structure and composition as well as oxidative damage are considered as primary events of chilling injury. Transgenic tomato seedlings in which catalase has been inhibited by antisense RNA show higher susceptibility to chilling injury. It is not known whether antisense catalase fruit were more sensitive to chilling injury and whether overexpression of catalase would result in higher tolerance. Ethylene is considered as a mediator in the stress response. Other studies carried out with the melon indicated that ethylene-suppressed fruit containing an antisense *ACO* gene did not develop the characteristic pitting and browning of the rind associated with chilling injury when stored at low temperature (e.g., 2°C for 3 weeks) and upon rewarming to room temperature (**Figure 2**). Ethylene treatment of antisense ACO fruit restored the chilling sensitivity phenotype.

In potato (*Solanum tuberosum*), low temperatures result in sweetening due to the accumulation of hexoses. This causes sensory changes and browning reactions upon cooking and frying that are deleterious for the technological properties of potatoes. Potato plants overexpressing a putative vacuolar homolog of a tobacco (*Nicotiana tabacum*) cell wall invertase inhibitor under the control of the CaMV 35S promoter exhibited up to 75% reduction in cold-induced hexose accumulation. The processing quality of tubers was greatly improved without changing starch quality or quantity and tuber number and weight. Minimizing sugar accumulation at low temperature was also achieved by overexpressing an *Escherichia coli GlgC16* gene which encodes a starch-synthesizing enzyme, ADPglucose pyrophosphorylase, resulting in higher starch content and lower reducing sugars.

Reducing Softening and Cell Wall Degradation

Excessive softening is one of the major deteriorative process of fruit that limits transportation and shelf-life. Considerable effort has been devoted to inhibiting the expression of cell-wall-degrading genes in tomato with the aim of slowing down cell wall degradation and fruit softening. Endopolygalacturonase (*PG*) antisense or sense cosuppressed plants showed low levels of PG activity and reduced pectin degradation. However, no changes in firmness were detected, although fruit were less susceptible to cracking, splitting, mechanical damage, and postharvest fungal diseases. PG-suppressed tomatoes have been commercialized for a few years for the fresh market in the USA and for processed tomatoes in the USA and UK. The advantage for processed tomatoes is higher viscosity and the opportunity to process at lower temperature. However, for a number of commercial and marketing reasons, including GM-free "own brands" these processed tomatoes have been withdrawn from the market.

Reduction of pectin methyl esterase (PME) activity (<10% of control fruit) due to the expression of a *PME* antisense gene resulted not only in increased pectin methylesterification, but also in a higher molecular weight of pectins probably related to a lower capacity of polygalacturonase to hydrolyze the highly methylated pectins. Fruit quality was improved with a higher soluble solids content, but no other ripening processes were affected. Tomato juice prepared from low-PME tomatoes also exhibited improved qualities, including higher soluble solids, viscosity, and consistency. Combining antisense *PG* and *PME* has led to improved properties as compared with the antisense genes taken separately.

Tomato plants were transformed separately with antisense endo-1,4-β-glucanases (cellulase) *Cel 1* and *Cel 2* transgenes under the control of the 35S promoter. In both cases, although mRNA for each gene was reduced up to 95% in the ripe pericarp, fruit softening, measured using stress-relaxation analysis, was unaffected as compared with control fruit as was ethylene production. Suppression of both activities together may be necessary to observe any effect on softening or fruit quality characteristics (**Table 2**).

The enzyme β-galactosidase (β-gal) is capable of releasing galactosyl residues from the cell wall and is considered as participating to cell wall dissociation. Downregulation of β-gal activity is complicated by the fact that 6 β-gal genes (*TBG1* to *6*) are expressed during tomato fruit ripening. Fruit in which *TBG1* expression was reduced by 90% have been generated that underwent no changes in cell wall composition and texture. Downregulation of *TBG3* mRNA also resulted in no difference in firmness as compared with control fruit although there was less release of galactosyl residues during ripening and slower deterioration during long-term storage. The only successful decrease in tomato fruit softening (up to 40% less than controls) has so far been obtained by downregulating *TBG4* gene. *TBG3* gene expression was also reduced in antisense *TBG4* fruit, indicating a possible cooperation of the two genes, with, however, *TBG4* playing the major role in softening (**Table 2**).

Expansins are proteins that lack hydrolase or transglycosylase activity but cause cell wall loosening *in vitro*. Suppression to 3% of wild-type levels of the LeExp1 (*L. esculentum* expansin 1) protein,

Table 2 Effects of modifying the expression of genes participating in cell-wall degradation on tomato postharvest quality

Gene	*Type of modification*	*Ripening processes*	*Changes*[a]
PG	Antisense or cosuppression	Pectin degradation	↘
		Mechanical damage	↘
		Fungal diseases	↗
		Viscosity for processing	↗
PME	Antisense	Soluble solids content	↗
Cel (*Egase*)	Antisense	Ripening and firmness	→
Exp	Antisense	Firmness	↗
β-gal			
-*TBG1*	Antisense	Firmness	→
		Galactosyl residues	↘
-*TBG3*	Antisense	Firmness	→
		Galactosyl residues	↘
		Postharvest deterioration	↘
-*TBG4*	Antisense	Firmness	↗
		Galactosyl residues	↘

[a]Arrows indicate increase (↗), inhibition (↘), or no effect (→) on ripening or quality attributes.

which is highly and specifically expressed in ripening tomato fruit, resulted in significantly higher firmness throughout ripening. It also inhibited polyuronide depolymerization late in ripening but did not prevent the breakdown of structurally important hemicelluloses. In contrast, fruit overexpressing LeEXp1 protein were much softer than controls and exhibited precocious and extensive depolymerization of structural hemicelluloses but no alteration of polyuronide depolymerization. This indicates that LeExp1-mediated relaxation of the cell wall structure precedes and enhances polyuronide and hemicellulose depolymerization probably through a better access of the hydrolytic enzymes to their substrates. Alteration of expansin and Β-gal (*TBG4*) gene expression has so far been the only means to control fruit softening (**Table 2**).

Reduction of Browning and Bruising

Postharvest enzymatic browning is deleterious to whole and freshly cut fruit and vegetables. It corresponds to the oxidation of phenolic compounds following cell disruption by bruising, wounding, pathogen attacks, or physiological disorders. Antisense inhibition of polyphenol oxidase (PPO) using tuber-specific promoter abolished browning and strongly reduced bruising in potato sensitivity without any apparent side effects. Applications to other crops, such as pineapple (*Ananas comosus*), are awaited.

Biotechnological Control of Flavor

Increasing Sugar Content and Sweetness

Sugar content represents an important sensory parameter. Sucrose is the major photoassimilate transported from the leaves to the fruit and growing parts of the plant. Since most fruit accumulate glucose and fructose, sucrose cleavage takes place through the action of invertase (β-fructosidase) and sucrose synthase, which catalyzes the reversible hydrolysis of sucrose. Sucrose accumulation may favor high sugar levels because it contributes half the osmolarity of the equivalent in glucose and fructose and it is theoretically less accessible to respiratory loss than hexose sugars, which is beneficial to postharvest quality. For all these reasons, attempts have been made to increase the sucrose content of fruit through downregulating genes encoding enzymes capable of sucrose hydrolysis.

Transgenic tomatoes expressing a constitutive (35S promoter) antisense invertase gene exhibited normal growth, and fruit had increased sucrose and decreased hexose sugar concentrations. However, sucrose-accumulating fruit were around 30% smaller and had unexpected higher respiratory and ethylene production rates. Using the E8 fruit-specific promoter did not alter the levels of sucrose probably because sugars tend to accumulate prior to ripening when the E8 promoter is inactive. Using an antisense sucrose synthase cDNA under the control of the 2A11 fruit-specific promoter, sucrose synthase could be reduced by up to 99% but starch or sugar accumulation was not affected. The same result was found using the constitutive 35S promoter, although in that case there was a reduction in fruit set and fruit growth. Phenotypic alterations were also observed in carrots (*Daucus carota*) with silenced sucrose synthase.

Another strategy to increase the sucrose content in fruit or storage organs is to overexpress sucrose phosphate synthase (SPS), an enzyme participating in sucrose synthesis. Tomato plants expressing a corn

(*Zea mays*; maize) SPS showed a sucrose synthase activity increased by 27% in fruits. Seventy percent more sucrose was unloaded into 20-day-old fruit and the sucrose turnover was increased. However, despite the higher SPS activity, the total soluble sugar content remained unchanged. Due to interactions between the enzymes involved in the carbohydrate metabolism of the fruit, it seems difficult to obtain an improved sugar content by modifying a single gene. Therefore the biotechnological control of sugar content in fruit and vegetables is a goal yet to be reached.

Proteins such as monellin and thaumatin present in tropical fruit exhibit strong sweetening capacities. A gene encoding monellin has been introduced in tomatoes and lettuce (*Lactuca sativa*) using the 35S and E8 promoters, but the level of the protein was too low to detect changes in flavor. However, increased sweetness of fruit and vegetables using the monellin and thaumatin genes (either native or genetically modified) has been reported in several patents.

The starch content of fruit and vegetables could be included in the quality parameters since it modifies to the dry matter content and enhances processing quality. Expression of an ADPglucose pyrophosphorylase (ADPGPP) from *E. coli* (*GlgC16*) using tuber-specific promoters has resulted, in some experiments, in an increase of starch content (>20%) in potatoes, less accumulation of reducing sugars, and less browning due to the Maillard reaction during cooking. Other attempts to express the same gene in potato has led to an increase in starch turnover but not in starch content. In tomato, *GlgC16* expression under the control of a fruit-specific promoter led to an increase of the total solids content of 20–30% in fruit. Increasing the amylopectin/amylose ratio results in better technological starches in terms of gelatinization, retrogradation, and viscosity. This has been achieved in potatoes using a glycogen synthase from *E. coli*.

Aroma Volatiles

Few genes directly involved in aroma volatile production have been isolated in fruit and only three of them have been used for genetic manipulation.

Lipoxygenase (LOX) catalyzes the hydroperoxidation of polyunsaturated fatty acids which are precursors for the C6 aldehydes hexanal and hexenal. Reduction of LOX activity in tomato in a fruit-specific manner did not result in any significant changes in C6 aldehydes, indicating either that very low levels of LOX are sufficient for the generation of the aldehydes, or that a specific isoform of LOX not affected by the transformation is responsible for the production of these compounds. Alteration of the fatty acid composition of tomato fruit by overexpression of a yeast desaturase gene increased the concentrations of unsaturated and saturated fatty acids that are substrates for LOX activity. Transformed fruit had higher hexanal and hexanol production, but also higher levels of linoleic acid peroxidation products such as *cis*-3-hexenal and *trans*-2-hexenal and subsequently *cis*-3-hexenol via alcohol dehydrogenase. Alcohol dehydrogenase (ADH) reduces hexanal and *cis*-3-hexanal to hexanol and *cis*-2-hexenol, respectively. Tomato fruit in which *ADH* gene expression has been either downregulated or overexpressed exhibited corresponding lower and higher levels of hexanol and *cis*-3-hexenol. Fruit with increased levels of the alcohols were rated higher in ripe fruit flavor by a sensory panel. Genes involved in the production of esters (alcohol acyl transferase) have been cloned in strawberry (*Fragaria x ananassa*) and melon.

The production of aroma volatiles is influenced by the biotechnological control of other fruit attributes. Downregulated PG, PME, and PG+PME fruit occasionally exhibited a reduction in flavor volatiles production possibly related to a lower release and increased binding of volatiles to cell wall structures. Tomato fruit in which ACS was inhibited produced lesser amounts of aroma volatiles with the exception of some compounds. On the contrary aroma volatiles production was little affected in antisense *ACO* tomatoes except for a few compounds. These alterations are probably due to different levels of residual ethylene. In antisense *ACO* melons with very strong inhibition of ethylene production, aroma volatiles, primarily esters were almost completely inhibited in the homozygous line and to a lesser extent, but still strongly, in F_1 hybrids. Complete recovery was observed upon treating with exogenous ethylene. Tomatoes containing the antisense *E8* gene produced more ethylene and aroma volatiles. Downregulation of ethylene synthesis or a reduction of its action to extend shelf-life through biotechnology or breeding therefore results in lower production of aroma volatiles. Inhibition of the expression of a phytoene synthase gene in tomato not only resulted in the absence of red color, but also in lowered levels of all but some methanol and lipid-derived volatiles. In the tomato a number of aroma volatiles arise from the carotenoid pathway.

Control of Nutritional Compounds

Carotenoids

Carotenoids form one of the most important classes of plant pigments and play a crucial role in defining the quality parameters of fruits and vegetables in terms of color, antioxidant, and vitaminic activities for humans. A number of genes encoding enzymes of carotenoid biosynthesis have been cloned from bacteria, fungi, and plants, providing tools for genetic manipulation. Phytoene synthase (Psy), the first step in the carotenoid pathway, converting geranylgeranyl diphosphate (GGDP) into phytoene, has been silenced in tomato fruit, resulting in the inhibition of carotenoid biosynthesis, without apparent altered phenotypes of the plant except sensitivity to photobleaching. However, increasing the carotenoid content through manipulation of *Psy* was unsuccessful: constitutive expression of the cDNA in transgenic tomato plants led to dwarfism due to redirecting of GGDP from the gibberellin pathway into carotenoids, which reduced the levels of these hormones in the plants. Transferring a bacterial carotenoid gene (*crtI*) encoding the enzyme phytoene desaturase, which converts phytoene into lycopene to tomato plants, resulted in a twofold increase in provitamin A (β-carotene) content. However, the total carotenoid content was decreased due to the fact that the increase in β-carotene was at the expense of the lycopene.

Other genes such as phytoene-phytofluene desaturase and genes involved in the synthesis of xanthophylls have been cloned and represent potential targets for carotenoid biosynthesis modification in plants.

The gene responsible for the high-pigment mutant of tomato, *hp-2*, has been isolated by positional cloning. It encodes a nuclear protein of unknown function which is now available for enhancing pigment accumulation in tomatoes via biotechnological means.

The overexpression in tomato of the yeast *S*-adenosylmethionine decarboxylase gene (*ySAMdc*) fused with a ripening-inducible E8 promoter increased the level of the polyamines spermidine and spermine during ripening. These transgenic fruit presented a threefold increase in lycopene content.

Vitamin C

Fruit and vegetables represent the major sources of L-ascorbic acid (vitamin C) in human nutrition. The biosynthesis of vitamin C in plants includes two pathways leading to the immediate precursor L-galactono-1,4-lactone. One of these pathways proceeds via D-galacturonic acid. A gene encoding D-galacturonic acid reductase (Ga1UR) involved in the conversion of D-galacturonic acid into D-galactonic acid which is then concerted into L-galactono-1,4-lactone has been isolated from strawberry. Overexpression of this gene in *Arabidopsis* enhanced vitamin C content two- to threefold, thus opening the possibility to enhance vitamin C content of edible plants.

Control of Minerals (Nitrates and Iron)

A high nitrate content in leafy vegetables is an undesirable trait as nitrate is reduced to nitrite during digestion and can be converted into carcinogenic nitrosamines. The transfer of a nitrate reductase gene (*nia*) has been considered as an approach in reducing the nitrate content of vegetables. A chimeric nitrate reductase gene of tobacco has been stably expressed in transgenic lettuce. The level of nitrate was reduced but not sufficiently to reach very low levels, specially in older leaves. In addition, phenotypic alterations have been observed such as chlorosis, dwarfing, and early flowering. Further studies are needed to render this strategy fully applicable at the commercial level.

Increasing the iron content of vegetables can have health benefits. An increase in the iron content of lettuce of 1.2 to 1.7 times has been achieved by expressing a cDNA of soybean (*Glycine max*) ferritine in lettuce via *A. tumefaciens* transformation. In addition, transgenic lettuce had higher photosynthetic and growth rates, which represent interesting agronomic characters for commercial applications.

Biotechnological Control of Morphological Parameters

Fruit size represents an important commercial parameter. Overexpression of 3-hydroxy-3-methylglutaryl coenzyme reductase (HMGR), by enhancing the biosynthesis of mevalonate, is reported to increase cell division in fruit and therefore the final fruit size. Mevalonate is an essential compound in isoprenoid synthesis whose end products are important in stimulating cell division (cytokinins, abscissic acid, gibberellins and brassinosteroids).

The absence of seeds is a desirable trait in many fruit crops. However, the parthenocarpic trait is polygenic and often results in reduced fruit set and fruit size, so that it makes breeding programs cumbersome. Biotechnological methods have been considered as an alternative. Transgenic eggplants (*Solanum melongena*, aubergine) and tomatoes expressing the coding region of the *iaaM* gene from *Pseudomonas syringae* driven by an ovule-specific promoter show parthenocarpic development. The *iaaM* gene codes for an indolacetamide monoxygenase that converts tryptophan to indolacetamide, a precursor of the plant hormone auxin. Transgenic plants produced seedless fruit of marketable size with fruit sensory quality apparently unaffected as compared to control seeded fruit. In addition, transgenic plants are capable of setting fruit in adverse conditions.

Organ-Targeted Gene Expression

Expressing target genes in fruit only represents one of the major challenges for the development of the biotechnological control of fruit quality. Only few fruit-specific promoters are presently available, including the promoter of the *2A11* gene of unknown function which is expressed in young ovaries and fruit throughout development, and the promoter of the PG gene, which is expressed in a ripening-specific manner at rather late stages of ripening. The promoters of the *E8* and *E4* genes of unknown function have also been considered as ripening-specific. Expression of both genes is abundant in ripening fruit and is not detected in leaf, root, stem, or unripe fruit. They are both activated by ethylene. However, while ethylene stimulates *E8* expression only in fruit, it promotes *E4* expression in many plant organs. The promoters of two genes involved in carotenoid biosynthesis of pepper (*Capsicum annuum*) (*ccs* and *fib*) have been shown to direct gene expression in a fruit specific manner, not only in pepper, a nonclimacteric fruit, but also in tomato. The *fib* promoter has proven exceptionally strong in expressing GUS activity in ripening tomato fruit starting at the late immature-green stage. Other organ-specific promoters exist, such as tuber-specific promoters in the potato.

Conclusion

In conclusion, biotechnology-driven methods to slow down climacteric fruit ripening have been enabled by advances in our understanding of the molecular basis of these processes. The improvement of sensory quality has also become feasible through insights into the molecular basis of processes such as softening, aroma volatile production, and pigment accumulation. To date, these approaches have been restricted to climacteric fruit, such as tomato and to a lesser extent melon. However, the techniques have recently been validated in other climacteric fruit species such as apple, plum, and papaya (*Carica papaya*). Undoubtedly, the number of species engineered for postharvest quality will increase in the near future and will include nonclimacteric fruit, such as strawberries. There will also be a diversification of the number of target traits for improvement, including an increased focus on improving nutritional composition. These approaches are fully complementary to conventional breeding as desirable sensory and nutritional attributes can be rapidly transfered into conventionally bred elite lines. In the end, the generation of commercially successful fruit and vegetable cultivars with improved engineered postharvest traits will largely depend on the cooperation between biotechnologists, postharvest physiologists, and breeders.

List of Technical Nomenclature

β-Carotene	A compound responsible for the orange color of the fruit; corresponds to provitamin A.
Carotenoids	One of the classes of compounds responsible for the orange coloration of fruit.
Cell wall	External compartment of the cell responsible for the firmness of the tissues and corresponding essentially to a matrix of polysaccharides and proteins.
Cellulose	One of the groups of cell wall polysaccharides.
Ethylene	One of the five plant hormones controlling various aspects of plant development, particularly fruit ripening and senescence of plant organs.
Ethylene receptor	Protein capable of binding ethylene and corresponding to the first step of the signal transduction pathway leading to gene expression and physiological responses.
Hemicellulose	One of the groups of cell wall polysaccharides.
Inhibition of gene expression	Can be achieved by insertion of a cDNA corresponding to the target gene in sense (cosuppression) or antisense orientation.
Lycopene	A compound responsible for the red coloration of fruit.
Mature green, breaker stages	Stages of development at which fruit have gained competence to ripen and show the first change in color, respectively.
Polyuronides	One of the groups of cell wall polysaccharides.
Promoter	Nontranscribed DNA element capable of driving gene expression. The 35S promoter isolated from the cauliflower mosaic virus is the most widely used in biotechnology.

Transcription factors	Nuclear proteins that bind to the promoter for inducing (or repressing) gene expression.
Transgene	Additional gene (cDNA) originating from the same species or from other species (plants, animals, bacteria, viruses, etc.) generally transferred via *Agrobacterium* or biolistics to generate a transgenic plant.
Xanthophylls	A class of compounds responsible for yellow coloration in fruit.

See also: **Ethics and Biosafety**: Development and Commercialization of Genetically Modified Plants; Plant Genetic Engineering, Food Safety Issues. **Genetic Modification**: Transformation, General Principles. **Postharvest Physiology**: Postharvest Physiological Disorders of Fresh Crops; Ripening; Storage. **Regulators of Growth**: Ethylene. **Tissue Culture and Plant Breeding**: Regeneration of Fruit and Ornamental Trees via Cell and Tissue Culture.

Further Reading

Baldwin EA, Scott JW, Shewmaker CK, and Schuch W (2000) Flavor trivia and tomato aroma: biochemistry and possible mechanisms for control of important aroma components. *HortScience* 35: 1013–1021.

Cockshull KE, Gray D, Seymour GB, and Thomas B (1998) *Genetic and Environmental Manipulation of Horticultural Crops*. Wallingford: CAB International.

Gray JE, Picton S, Giovannoni JJ, and Grierson D (1994) The use of transgenic and naturally occurring mutants to understand and manipulate tomato ripening. *Plant, Cell and Environment* 17: 557–571.

Gutterson NC (1993) Molecular breeding for color, flavor and fragrance. *Scientia Horticulturae* 55: 141–160.

Kanellis AK, Chang C, Klee H, *et al.* (1999) *Biology and Biotechnology of the Plant Hormone Ethylene.*, Vol. 2. Dordrecht: Kluwer Academic.

Lelièvre JM, Latché A, Jones B, Bouzayen M, and Pech JC (1997) Ethylene and fruit ripening. *Physiologia Plantarum* 101: 727–739.

Pech JC, Bouzayen M, Latché A, *et al.* (2001) Physiological, biochemical and molecular aspects of ethylene biosynthesis and action. In: Bartz JA and Bretch JK (eds) *Postharvest Physiology and Pathology of Vegetables*, pp. 247–286. New York: Marcel Dekker.

Salamini F (2000) Biotechnology and molecular biology in practical horticulture. *Acta Horticulturae* 520: 17–42.

Valpuesta V (2002) *Fruit and Vegetable Biotechnology.* Cambridge: Woodhead Publishing.

Van den Berg H, Faulks R, Granado HF, *et al.* (2000) The potential for the improvement of carotenoid levels in foods and the likely systemic effects. *Journal of the Science of Food and Agriculture* 80: 880–912.

Seed Storage

J D Klein, ARO-Volcani Center, Bet Dagan, Israel

History and Purpose of Seed Storage

Seed storage has been an integral part of civilization since humans began to cultivate land in the Middle East 10 000 years ago. The archeological evidence includes stone lined pits and clay pots that contain the remains of stored seeds. One of the earliest written descriptions of seed storage is in the book of Genesis in the Bible, in which Joseph gathers grain from bountiful harvests for storage against a coming famine in Egypt. Although the Bible does not describe the storage facilities, the pyramids are constructed in a way that maintains low humidity and uniform temperature, two important factors in maintaining the viability and vigor of most stored seeds. Nonetheless, stories of viable 3000-year-old seeds being found in the pyramids must be regarded as apocryphal. The oldest viable seeds ever found were those of the sacred lotus (*Nelumbo nucifera*) found in a dried lake bed in Pulantien, China. Seedlings grew from these seeds, which were radiocarbon-dated to more than 1000 years old.

As is the case with the pyramids, the dried Chinese lake bottom provided uniform conditions of relatively low temperature and humidity. In addition, the hard pericarp of the lotus seed provides a barrier to water and gas transport between the outer environment and the seed. This virtually ideal combination of factors diminishes the possibility of membrane damage by hydration or oxidation, and accounts for the exceptional storage life of the sacred lotus seeds.

The major aims of seed storage are to conserve food value and viability. Seeds destined for food processing (milling, malting, oil extraction) are usually stored for no more than 2 years, while seeds stored for subsequent sowing are usually stored for no more than 3 or 4 years. Conservation of seeds for preservation of genetic diversity is carried out in seed banks, where facilities are designed to maintain seed viability for tens and even hundreds of years.

Factors Affecting Postharvest Quality of Seeds

Both biotic and abiotic factors can affect seed quality in storage. Infestations by insects, fungi, or bacteria can reduce germinability, as well as reduce the food value of the seeds by depleting their carbohydrate or protein reserves. In extreme cases, such infestations can result in the presence of toxic substances in the seeds; examples include ergot, caused by an alkaloid secreted by the fungus *Claviceps purpurea* in rye (*Secale cereale*) or wheat (*Triticum* spp.), and aflatoxin, caused by *Aspergillus flavus* or *A. parasiticus* in corn (*Zea mays*; maize) or peanut (*Arachis hypogaea*). Sometimes, seeds do not reach maturity because of climatic factors, use of inappropriate cultivars, or a combination of the two. Such seeds may lack oligosaccharides or other physiologically important compounds such as late embryogenesis abundant (LEA) proteins, which are considered important in maintaining membrane and organelle integrity as seeds dry and are subsequently stored. The storage quality of mature seeds can be reduced by excessive use of desiccating chemicals on the parent plant before harvest. Postharvest seed treatments can also affect seed storage quality, especially if seeds are stored for some time after treatment before sowing. Farmers in developing countries frequently treat seeds with vegetable oils to prevent insect attack in storage. However, these oils can be absorbed into seed tissues, where short chain (C9 and shorter) fatty acids in the oils can inhibit gibberellin induced amylolysis, and thus lead to a reduction in germination. Heat treatments, such as soaking cabbage (*Brassica oleracea* var. *capitata*) seeds at 50°C for 20 min to remove *Xanthomonas*, can also cause accelerated ageing of the seeds, resulting in a reduction in seed viability and seedling vigor.

Determining Seed Quality during Storage

Seed quality is usually examined during the course of storage, especially in seed banks where decisions must be taken as to when to renew the genetic material so that it may be conserved in optimum condition. The most accurate method of checking quality is to germinate a representative seed sample. In addition to being destructive, this method may be quite time-consuming, as germination of seeds can take anywhere from a few days to as much as several months, depending on seed type and degree of dormancy.

Many methods for analyzing seed quality during storage are based on seed physiology. These methods may be more rapid than a germination test, but they are just as destructive, and ultimately they are only correlative with germination results rather than directly diagnostic of seed quality. In the tetrazolium test, seeds or embryos are imbibed in a colorless tetrazolium chloride solution. If the seeds are at all viable, imbibition induces dehydrogenase activity, which releases hydrogen ions that reduce the tetrazolium to formazan, which stains living cells red. Although correct interpretation of the results can require extensive expertise, the results are available within 48 h. Flow cytometry can be used to determine the cell cycle status of embryo nuclei in dry seed. The G2/G1 ratio indicates the relative proportion of nuclei that have completed synthesis of new DNA and which are about to undergo mitosis again (G2) in comparison to those nuclei that are in the initial stage of cell growth (G1). Higher G2/G1 ratios are found in more vigorous seeds. The conductivity of seed leachate after 18–24 h of imbibition is considered to be proportional to the degree of cell membrane integrity, and is therefore inversely proportional to seed quality. Conductivity measurements are not always destructive, since, in principle, they can be carried out several times on the same seeds, which are redried after each test.

Among the nondestructive methods for analysis of seed quality are chlorophyll fluorescence, X-ray analysis of embryo quality, and measurement of volatiles emanating from the seeds. Chlorophyll fluorescence, especially in *Brassica* seeds, is inversely proportional to the degree of seed maturity, which in turn is proportional to seed germinability and potential longevity in storage. Measurements of chlorophyll fluorescence are very rapid (milliseconds/seed). Embryo size, which frequently correlates with resulting seedling quality, can be examined *in situ* by X-ray photography. The procedure can be automated to provide rapid results. Although metabolic activity in stored seeds is very slow, measuring the quantity and quality of volatiles produced by the seeds can give an indication of seed storage status. Ageing seeds frequently begin anaerobic respiration, and produce enhanced quantities of acetaldehyde and ethanol, which can be detected nondestructively. Lipid peroxidation of membranes in ageing seeds can be detected by increased amounts of volatiles, such as ethylene or methanol. Methanol can also be the result of pectin methyl esterase activity in seeds, which can be a sign of cell wall degradation associated with reduced seed quality.

Temperature and Moisture Effects

Whole Seed

Metabolic activity of all organisms slows with decreasing temperature, and resistance to low temperature injury is associated generally either with low water content or with high solute concentrations that prevent the freezing of free water. The vast majority (up to 80%) of all seeds are considered "orthodox." Such seeds are able to withstand drying to less than 10% moisture content at maturity, and, consequently, can withstand subfreezing temperatures. "Recalcitrant" seeds are those that tolerate only slight desiccation after harvest, if at all, and consequently are usually sensitive to chilling temperatures as well. There is a range of desiccation and temperature tolerances of recalcitrant seeds, such that some authorities have defined an "intermediate" seed storage characteristic, although it is probably better to consider desiccation tolerance as a continuum across all seed varieties. In general, recalcitrant seeds do not dry on the parent plant, tend to be large, and originate in aquatic regions, tropical rainforests, or climax forests in temperate zones. Recalcitrant seeds frequently can germinate immediately after harvest, since they do not seem to require the abscisic acid mediated signal that occurs during drying of orthodox seeds. For further information on acquisition of desiccation tolerance and initiation of germination *see* **Seed Development**: Onset of Desiccation Tolerance; Germination.

Most authorities consider moisture to be a more important factor than temperature in seed storage. Moisture can easily be quantified as a percentage of the total weight of the seed, but such measurements are destructive. The effect of moisture content on seed longevity in storage varies across species and even across cultivars. Water activity (A_w) expresses the energy status of the water, or the water potential of the seeds, and thus reflects the availability of water to participate in chemical and physical processes. It is measured (nondestructively) by determining the relative humidity of the air that is in moisture and temperature equilibrium with the seeds. The relationship between water activity and longevity is linear, and is similar from species to species. Water content at the same water activity can vary considerably across species. For example, an A_w of 0.3 equals a water content of 4.5% for oily rape seeds (*Brassica napus*) and 9.3% for nonoily wheat seeds. In oil seeds, minimal amounts of water are absorbed by the oil fraction, and the greater the oil content, the lower the total seed moisture level required to reach an A_w that can cause heating and loss of oil quality.

Cellular Level

There are both minimum and maximum levels of water in the seed that are critical for seed longevity. In most recalcitrant seeds, the mechanisms for protection of protein structure and membrane integrity in the event of water loss are largely absent, so that seeds cannot withstand drying without losing viability. As water loss increases, so too does the concentration of cytoplasmic components that can interact with protein surfaces. In orthodox seeds, these components can include proline or oligosaccharides such as raffinose or stachyose, as well as chaperonins such as LEA (dehydrin) or low molecular weight heat shock proteins that can form a water associated shell around the protein. As the seed dries further, sugar molecules can replace water at the protein surface, stabilizing the folded protein in a so-called glassy matrix against the time when rehydration occurs. These oligosaccharides are also hypothesized to protect drying cytoplasmic membranes from undergoing fusion, by replacing water molecules in the hydration layer of the membrane, keeping the phospholipids spaced properly in a state where they can be rehydrated without causing excessive membrane leakage that would otherwise result in cell death.

Paradoxically, bean breeders consciously try to produce germplasm with seeds that have minimal contents of raffinose-series oligosaccharides, since these sugars are implicated in producing flatulence (so-called "fartogenic compounds"). The resulting low-raffinose seeds do not stay viable as long in storage as do seeds with higher raffinose content.

Without proper spacing of the membrane components, ultimately proton gradients in the mitochondria can collapse, leading to production of free radicals. Antioxidant enzymes such as catalase, ascorbic acid peroxidase, superoxide dismutase, and glutathione reductase can protect cells in this case, as can molecular antioxidants such as glutathione, ascorbate, and certain phenols. At the lowest levels of hydration (corresponding to less than −150 MPa or 0.35 A_w), water is bound chiefly to ionic sites, and even orthodox seeds die, albeit slowly.

Control in Storage

Although it is clear that controlling the interaction of temperature and moisture is critical for successful seed storage, it is not certain that combining the lowest levels of each will result in synergistic enhancements in seed storage quality. In fact, the best storage moisture level is probably lower at high temperatures than at low temperatures. Determining this level is an inherently destructive procedure, which must be performed for each species when determining moisture and temperature conditions for storage. Factors such as oil content may lead to wide differences in calculated storage regimes. This has led to the adoption of drying orthodox seeds to equilibrium with a known relative humidity inside the drying container instead of measuring moisture content directly. Seeds can be stored at −18°C in standard freezers or even at −150 or −196°C in nitrogen vapor or in liquid nitrogen, respectively, without destructive moisture determinations (**Figure 1**). The latter temperatures can preserve seeds for tens and probably hundreds of years. In the absence of cooling facilities, seeds can be dried to very low moisture contents using vacuum or desiccants, sealed under vacuum and kept at ambient temperatures for

Figure 1 Seed storage in liquid nitrogen tanks. Photograph by Stephen Ausmus, USDA/Agricultural Research Service.

a few years, depending on variety. This method is not recommended for seed banks, however.

List of Technical Nomenclature

Orthodox seed	A seed that can usually be dried to very low moisture levels and can withstand freezing once desiccated.
Recalcitrant seed	A seed that dies if dried below a critical moisture level, and cannot tolerate low temperatures; usually short lived and difficult to store.
Water activity (A_w)	Expresses the energy status of the water, or the water potential of the seeds, and thus reflects the availability of water to participate in chemical and physical processes.

See also: **Biodiversity and Conservation**: Seed Banks. **Seed Development**: Germination. **Seed Dormancy**: Development of Dormancy.

Further Reading

Hong TD, Linnington S, and Ellis RH (1996) *Seed Storage Behaviour: A Compendium*. Handbooks for Genebanks No. 4. Rome: International Plant Genetic Resources Institute.

Vertucci CW and Farrant JM (1995) Acquisition and loss of desiccation tolerance. In: Kigel J and Galili G (eds) *Seed Development and Germination*, pp. 237–271. New York: Marcel Dekker.

PRIMARY PRODUCTS

Contents

Cellulose

M C Jarvis, Glasgow University, Glasgow, UK

Cellulose Structure

The primary structure of cellulose could hardly be simpler: it is a linear, β(1,4′)-linked polymer of glucose (**Figure 1**). In their native state, the polymer chains of cellulose are some 10^4 monomer units in length and are noncovalently bonded together into ordered strands called microfibrils. Microfibrils are the narrowest units of cellulose observed by microscopy.

Because cellulose is the most abundant of all biological substances and because its primary structure is so simple, it might be expected that we

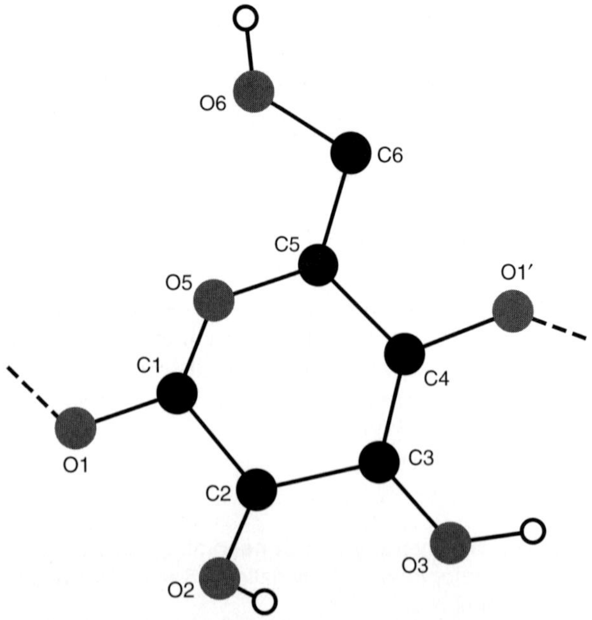

Figure 1 Numbering system used for carbon and oxygen atoms in a glucosyl unit of cellulose.

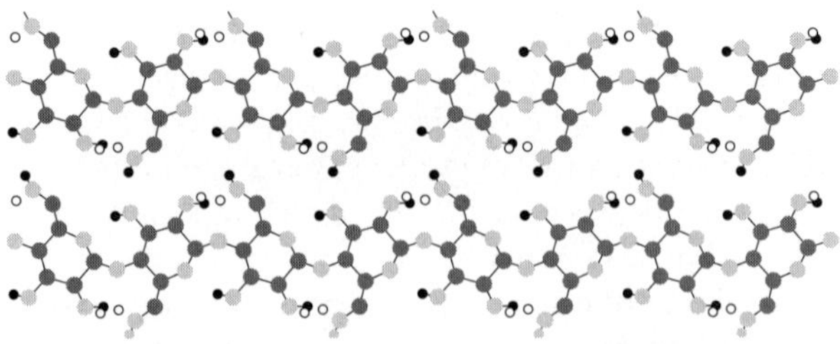

Figure 2 Two adjacent chains of cellulose 1β. The unit cell of the cellulose 1β lattice contains two chains, one at its origin and the other at its center. Those shown here are origin chains. Origin and center chains form alternating sheets within the lattice. The principal positions of the hydroxyl protons are black, but each proton can also occupy an alternative position (white circles).

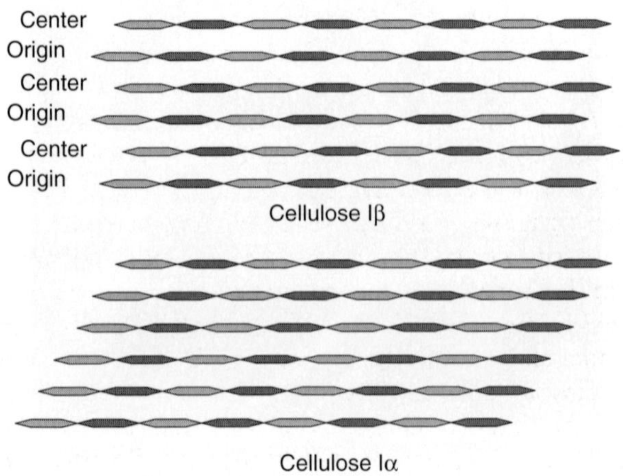

Figure 3 Comparison of the cellulose 1β and cellulose 1α structures, viewed side-on to the sheets of chains and with alternate glucosyl residues in each chain colored light and dark for clarity.

would understand how cellulose microfibrils form. In reality we are only now beginning to understand their supramolecular structure and to appreciate how it emerges from the mechanism of cellulose biosynthesis.

Native cellulose contains two crystalline forms, Iα and Iβ, in proportions that depend on the organism in which it is found. Cellulose Iα and Iβ were first distinguished by solid-state nuclear magnetic resonance (NMR). The main structural details of both crystal forms have been known for some time from modeling studies based on crystallographic analysis of the large, rectangular, highly crystalline microfibrils formed by certain algal and tunicate (animal) celluloses, which can contain more than 1000 chains.

The full structure of the monoclinic Iβ form has recently been published. It consists of parallel sheets of chains held together edge-to-edge by a complex system of multicentred hydrogen bonds (**Figure 2**). These involve O-2 and O-3 on one side of each glucosyl residue and O-6 on the other. The detailed hydrogen-bonding pattern of the triclinic Iα form is not yet published but it is known to differ from cellulose Iβ by longitudinal displacement of one sheet of chains relative to the next so that in principle the Iα–Iβ transition can be induced by bending. In celluloses from different algal sources the Iα and Iβ lattices have been shown by microdiffraction to occupy separate zones of a single microfibril, alternating either longitudinally or laterally (**Figure 3**).

Cellulose from higher plants is obviously much more important in practical terms but is much less understood, because although there is NMR evidence for the presence of the Iα and Iβ forms – or something very much like these – the crystalline units typically contain fewer than 20 chains and are too small for effective crystallography. Indeed they are so small that it is difficult to envisage how regular crystalline packing in any sense is possible, but the spectroscopic data indicate that the structures do have a considerable degree of order. Most current structural information has come from solid-state NMR experiments, which are highly effective for discerning the chain conformations in, for example, cellulose Iα and Iβ but less effective for determining how the chains are packed together.

In the primary cell walls of the soft, growing parts of plants the microfibrils visible by electron microscopy average 8–10 nm in diameter. However from spectroscopic and diffraction experiments the ultimate units from which primary-wall cellulose is constructed, best termed crystallites, are thinner still; some 2–3 nm in diameter, as estimated from the width of their diffraction peaks and the ratio ($\sim 2:3$)

of interior to surface chains measured by solid-state NMR. Within the larger structure of the cellulose microfibrils from the primary cell walls of onions (*Allium cepa*), a useful model system, each crystallite retains its identity: its surface chains remain identifiable by NMR. The crystalline units in wood cellulose are similar, although perhaps not identical, in size to those of primary cell walls. In contrast the textile cellulose from cotton (*Gossypium* spp.), flax (*Linum usitatissimum*), and ramie (*Boehmeria nivea*) have single-crystallite microfibrils with only about 30% of their chains at the surface, suggesting dimensions of 4–5 nm consistent with diffraction measurements.

Using a combination of solid-state NMR, Fourier transform infrared (FTIR) microscopy and molecular modeling, it has recently been shown that many of the crystal-surface chains of cellulose from higher plants are in a quite different conformation from any other form of cellulose that has been described so far. They lack the twin hydrogen bonds between successive glucose residues that are characteristic of cellulose Iα and Iβ, and have much more potential for hydrogen bonding to water, to hemicelluloses within the cell wall, and to other cellulosic fibers in paper and cotton. The hydroxyl groups of the surface chains are also more reactive chemically than those of the interior chains, which is relevant to the industrial preparation of substituted celluloses and in the dyeing of textiles.

Function in Plants

Cellulose is the principal contributor to the mechanical strength of living plants. Likewise, it is responsible for the mechanical properties of useful vegetable materials like wood, paper, and cotton. Cellulose typically comprises a quarter to one-half of the dry mass of plant cell walls, the remainder being made up of noncellulosic polysaccharides and, in wood, the phenolic polymer lignin.

Plant cell walls provide strength to living plants in two quite separate ways. In soft plant tissues like leaves and the stems of flowers, each cell is inflated by osmotic (turgor) pressure inside it so that the cell wall is stretched like the skin of a balloon. Plant tissues like these have only primary cell wall layers, which are permanently in tension and are quite flexible and elastic. Compressive stresses, like on the inside curve of a stem bending in the wind, are carried by turgor pressure. This is a very efficient way to build a plant, in the sense that it requires only a small investment of material in the polymers like cellulose that form the cell wall.

However, a plant relying on these principles cannot be more than 1–2 m tall, or the turgor pressure that is required becomes more than the cells can sustain. In trees, both tensile and compressive stresses are carried by wood. Wood must be stiff to withstand compression without buckling. Its stiffness is provided by thick, cellulose-rich secondary cell walls that, in most cases, are dead and have no turgor pressure within them. In this way a tree 100 m in height can withstand the stresses of wind and snow, but the investment of material in wood cell walls is prodigious and requires decades or centuries of photosynthesis by a huge canopy of foliage. That is why cellulose, largely in wood, is the most abundant material in the biosphere and why growing forests are the most important of the accessible sinks for atmospheric carbon dioxide.

Plant cell walls are often likened to manmade composite materials such as glassfiber. More accurately they should be described as nanocomposites, since the cellulose microfibrils that form the reinforcing phase are only nanometers in width rather than micrometers as in most synthetic composites. This difference in scale makes plant cell walls stronger than might be expected from a knowledge of the behavior of synthetic composites, for two reasons. First, the high surface area of the cellulose microfibrils means that fracture rarely involves failure of the adhesion between them and the matrix. Second, the microfibrils themselves are less likely to contain crystalline flaws that could initiate fracture.

However those aspects of theory that deal with the elastic properties of composite materials can be applied quite straightforwardly to plant cell walls. In particular, because cellulose is a very stiff material with an elastic modulus of about 170 GPa, the tensile properties of the plant cell wall depend strongly on the microfibril orientation within it. So the tensile modulus of wood, a crucial property in timber quality of the construction industry, is closely linked to the mean microfibril angle within the wood cell walls. Similarly the anisotropic growth of plant cells is achieved by the deposition of cellulose microfibrils in a predominant orientation normal to the direction of growth. This simple principle, linked intimately to the orientation of cell division planes, leads to all the diverse complexity of plant growth and form.

Biosynthesis

Synthesis of cellulose is the function of a complex of enzymes located at the surface of plant cells. Most of these enzymes span across the thickness of the plasma membrane. The raw material for cellulose synthesis is sucrose, from which uridine diphosphate glucose (UDP-glucose) is synthesized prior to the incorporation of the glucosyl residue into the growing cellulose chain. The key enzymes responsible for elongating cellulose chains are called cellulose synthases. There are approximately a dozen genes encoding cellulose synthases in the *Arabidopsis* genome. Some of these *Arabidopsis* cellulose synthases (CesAs) are responsible for cellulose synthesis in the primary walls of growing cells, while others are involved in secondary wall synthesis in woody tissues. It appears that at least two, probably three, distinct CesA proteins are needed to assemble a cellulose microfibril in any one type of cell wall, but how they cooperate is uncertain; for example, whether the two types of cellulose chain resembling the Iα and Iβ crystalline forms are the products of different CesA enzymes. Null mutants are known in each of the genes encoding CesA proteins of secondary wall cellulose synthesis: where these mutants have been characterized, cellulose is apparently not synthesized at all in the affected secondary wall layer.

At high magnification in transmission electron microscopy (TEM) of plant cell surfaces, the tracks of cellulose microfibrils may be seen emerging from protein bodies incorporated in the plasma membrane. These protein bodies are called "rosettes" because of their regular hexameric appearance on the inner face of the membrane, and they presumably represent the assemblage of proteins engaged in cellulose synthesis.

A number of enzymes other than CesAs are required for the synthesis of cellulose, and at least some of these are likely to form part of the rosettes. One of these enzymes is a membrane-bound cellulase encoded by the *KOR* gene in *Arabidopsis*. It is not clear why a cellulase should be needed for the synthesis of cellulose, but a detailed study of the effects of cellulose synthesis inhibitors in cotton has yielded evidence that sterol oligoglucosides play some role in the process, either as primers for chain initiation or as intermediates in chain elongation. Trimming off the sterol moiety would be one possible function for a cellulase.

The cotton model has also been used to demonstrate that enzymes involved in sucrose utilization are closely linked to the sites at the plasma membrane where cellulose is synthesized, and these may be associated with the rosette assemblages. For a number of other enzymes a more shadowy role in cellulose synthesis is suggested by mutant phenotypes that resemble those in which *CESA* genes are defective.

The manner in which cellulose microfibrils become correctly oriented around the cell is intimately tied to their biosynthesis, but it is not at all clear how the connection is achieved. It was once thought that cellulose microfibrils were laid down parallel to oriented arrays of microtubules in the underlying cytoplasm. While there are many occasions when microfibril and microtubule orientations are correlated, enough exceptions have been found to make it clear that microfibril orientation can be controlled without any involvement of a microtubule template. Genetic intervention to control microfibril orientation would be very useful for both timber and fiber crops, but the control mechanisms are not yet well enough understood to make this possible.

Cellulose Products: Paper

Paper and pulpwood together rank amongst the most important commodities in world trade, after oil and food. Paper is certainly not pure cellulose, but cellulose is its one indispensable component and the value of wood for papermaking depends on, amongst other things, its cellulose content.

Paper consists of fiber cells, generally from wood but sometimes from cereal straw or other nonwoody plant materials. The fiber cells are felted together and normally are held by noncovalent forces only, to make up the fabric of the paper. Coating and filling materials such as starch and kaolinite are normally added and, in the case of specialized papers, may constitute a large proportion of the total mass.

Coniferous timber (softwood) has longer fiber cells, typically 3–5 mm, than angiosperm timbers (hardwoods) or cereal straw. Softwood cells (tracheids) are also more uniform anatomically than the complex mixture of vessels, fibers (in the strict anatomical sense), and tracheids that make up hardwood tissue. It is sometimes said that softwoods are preferable for papermaking, but this is not strictly true. The long fibers of softwoods provide strength at the expense of smoothness, and most papers are made from a mixture of softwood and hardwood pulps in proportions that depend on the relative importance of strength and smoothness for the end use of the paper in question.

Making wood pulp means separating the wood cells. In the living tree the wood cells are held together by lignin, the main constituent of the middle lamella between one xylem cell wall and another. In mechanical pulping the wood chips are ground in such a way that the relatively brittle middle lamella lignin is shattered, while the tougher walls of the fiber cells remain intact. The lignin is not removed, only disintegrated, so papers made from mechanical pulp turn yellow with age as the lignin is photochemically oxidized.

There are various forms of chemical pulping that involve alkaline or acid hydrolysis of the lignin, with simultaneous reduction or substitution to prevent the hydrolyzed lignin fragments from repolymerizing.

Chemical pulping removes some but not all of the lignin, and the residual lignin after Kraft pulping is strongly colored so that Kraft papers are suitable only for packaging. To produce light-stable white paper it is necessary to use a pulping process like the acid sulfite process, which leaves the residual lignin less strongly colored, and to add a bleaching step. For environmental reasons, chlorine-based bleaching processes have been largely replaced by processes based on active oxygen.

Adhesion between fiber cells in paper depends on their surface properties. The cell surfaces are relatively hydrophobic and do not adhere well unless they are first damaged, by teasing out bundles of cellulose microfibrils so that these entangle and bind to one another. This process is called beating. It can entail some damage to cellulose within the cell walls also, with consequent loss of strength.

During acid-based pulping and oxidative bleaching, degradation of cellulose is possible and this may further reduce the strength of the resulting paper. If particularly durable, white paper is required, wood is not necessarily the best raw material. Nonwoody cells of dicotyledonous plants are in general attached to one another by pectin rather than lignin, and may be easier to separate. Flax, for example, is widely used in the specialized paper used to make banknotes. Flax has very long cellulosic fibers that can be separated without the harsh chemical treatments needed to remove lignin.

Despite predictions of the "paperless office" paper consumption is still rising, although recycling is beginning to offset the demand for pulpwood. Paper production consumes the equivalent of about 4 billion trees per year, and since paper is a short-lived product relative to furniture or constructional timber the net effect of rising paper consumption is to increase the flux of carbon into the atmosphere.

Textiles

Natural Cellulose

Despite the ubiquity of cellulose in plants, it is not normally found in a pure enough form, nor in long enough and fine enough cellular structures, to meet the requirements of a textile fiber. The cellulosic fibers of commerce are all isolated from plant parts that are in some way unusual.

Cotton is not only the most important but also the most extreme example, consisting of the grossly elongated seed hair cells of *Gossypium* spp. The hair cells have an outer waxy cuticle to retain water, a very thin primary wall, and a massively thickened secondary wall that consists essentially of cellulose, with only a small admixture of other polysaccharides like glucomannans. Most importantly, the seed hairs are single cells some centimeters long, so no cell separation is needed during the industrial isolation of cotton. Dewaxing, however, is required to remove waxy materials from the cuticle and provide absorbency. Compared with other natural cellulosic fibers, cotton has its microfibrils oriented in a less longitudinal direction and its tensile modulus is therefore somewhat lower.

Cotton is an intensively produced crop. More than half of the cotton acreage in the United States is now sown to genetically modified varieties. Intense competition on world markets keeps prices low enough to compete with synthetic fibers.

The majority of the other textile fibers are isolated from what are called bast tissues in the stems of plants. Anatomically these correspond to phloem fiber bundles, and they are located at the outer edge of the vascular ring or, where a true vascular cambium is present, immediately outside it. Thus lime (*Tilia* spp.) is called basswood in North America because when the bark is stripped off it peels along the line of the vascular cambium leaving a thin layer of bast fiber, phloem fiber cells, on its inner face. Lime trees have been exploited from prehistory to provide cellulose-rich textile fiber.

The best-known bast fiber is linen, which consists of the phloem fibers of the flax plant. Flax is the most ancient cultivated fiber, domesticated in the Near East almost as early as wheat (*Triticum* spp.) and barley (*Hordeum vulgare*). For most of the length of a flax stem, a layer of phloem fiber bundles lies just under the epidermis. The term "fiber" can be confusing when applied to bast fibers like flax. The fiber bundles can be seen with the naked eye, and consist of fiber cells some millimeters in length and about 50 μm in diameter, bonded to one another by the pectic middle lamella. The linen fibers of commerce are neither the phloem bundles nor the individual fiber cells. They consist of aggregates of cells intermediate in size between these extremes, and are separated out from the flax stem by a form of biotechnology that originated in the Neolithic.

Newly harvested flax stems are retted by microbial action, either in damp sheaves in the field or under anaerobic conditions in water. Microbial enzymes decompose the intercellular pectins faster than the largely cellulosic secondary cell walls of the fiber cells, leading to partial cell separation. A series of mechanical beating and combing operations then separates out the strong, pliable commercial fibers from one another and from the soft outer tissues and the woody, brittle underlying xylem. The yield is not high and the retting process requires high levels of experience, judgment, and luck with the weather, since the microbial ecology of the flax stems is not controlled except by its physical environment. These factors make linen an expensive fiber to produce, but its strength, smoothness, and luster make it competitive as a valued textile fiber for high-quality niche markets and the fashion industry.

Other bast fibers of some importance in international trade include jute (*Corchorus* spp.), hemp (*Cannabis sativa*), and ramie, which are isolated by processes similar to linen but vary in coarseness according to the diameter of the commercial fibers and the degree to which they are held together and stiffened by lignin: jute, for example, is quite highly lignified and this renders it coarser but more durable than other fibers. There are also a number of leaf fibers, associated with the parallel leaf veins of monocotyledonous plants. An example is sisal, a fiber once used for ropes, isolated from the leaves of *Agave sisalana*.

Other than cotton, the only common seed hair used as a commercial fiber is kapok, from the oriental tree *Ceiba pentandra*. Coir is one commercial fiber that is not strictly a seed hair, nor derived from either leaves or stems. Coir is a light, elastic, highly lignified fiber from coconut (*Cocos nucifera*) husks.

Across the world a wide range of plants are used locally as sources of cellulosic textile, paper, or rope fibers, without reaching international significance. Examples are pina, a beautifully fine, white textile fiber made from pineapple (*Ananas comosus*) leaves in the Phillipines; the bast fiber from nettles (*Urtica dioica*), which is quite like linen and substituted for it during both World Wars in Europe; and New Zealand flax (*Phormium tenax*), from a huge agave-like plant, which was of vital importance to the Maori economy before the arrival of Europeans.

Regenerated Cellulose

Rayon, developed in France at the end of the nineteenth century, was the first "synthetic" fiber. It is not strictly a synthetic material because the cellulose polymer remains intact throughout the processing of the raw material, high-purity wood pulp. However, the crystalline structure of the native cellulose is dismantled, allowing the cellulose to be dissolved and extruded into fine artificial fibers with a different chain packing arrangement, that of cellulose II (**Figure 4**).

To make rayon, wood pulp is processed to reduce the lignin and hemicellulose levels well below those used in papermaking. This dissolving pulp is steeped in concentrated sodium hydroxide, pressed, shredded and allowed to oxidize in contact with air. It is then treated with gaseous carbon disulfide to introduce xanthate (-OCS_2^-) substituents into the less crystalline parts of the cellulose microfibrils. The bulky xanthate groups disrupt the native crystal structure and allow the formation of a viscous colloidal solution called viscose.

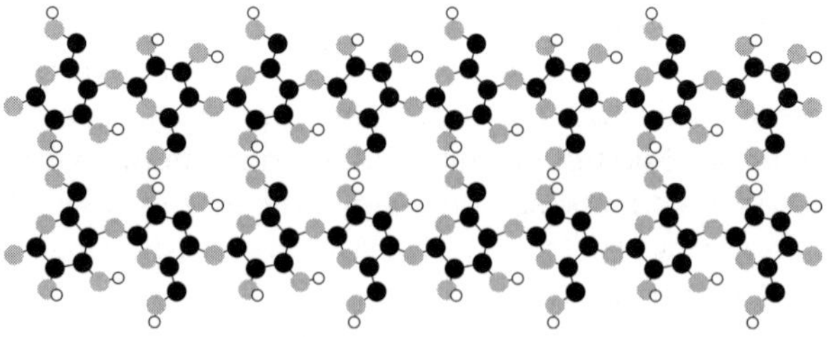

Figure 4 Two adjacent chains of cellulose II. As with cellulose 1β the unit cell of the cellulose II lattice contains two chains, one at the origin and the other (shown here) at the center. Origin and center chains form alternating sheets within the lattice as in cellulose 1β but in cellulose II the two sheets of chains run in opposite directions, i.e. the structure is antiparallel.

The viscose solution is then extruded through a spinneret into sulfuric acid, which decomposes the xanthate groups and allows the regenerated cellulose II to precipitate under tension as fine fibers. These are called "viscose rayon" from the nature of the dissolution processes. Cellulose II can also be made by dissolving wood pulp in cuprammonium solution and reprecipitating it as fibers; or by converting cellulose to cellulose acetate (see below), which dissolves in organic solvents, and then precipitating it by deacetylation. "Cuprammonium rayon" is produced in smaller quantities than viscose rayon and is used to make relatively fine fabrics.

The tensile strength of viscose rayon, in the dry condition, can be increased by crosslinking the cellulose xanthate chains with zinc salts during regeneration. This process leads to "high tenacity rayon," which is used in tires and belting. Rayon shrinks and loses much of its strength in water, so that rayon textiles must normally be dry-cleaned. This is a considerable disadvantage and surface treatment, to stabilize the fabric against moisture, can used to produce "high wet modulus rayon."

Rayon is a versatile fabric sharing many advantages with cotton. It is soft, comfortable, absorbent, and breathable, takes dyes easily and is derived from renewable raw materials. Wood is a cheaper raw material than any textile fiber but the cost of producing dissolving pulp and converting it into rayon leads to fiber costs that exceed that of cotton, so rayon is not a cheap alternative.

Modified Celluloses

The insolubility and mechanical strength of cellulose depend on the lateral association of the polymer chains by hydrogen bonding, particularly through O-2 and O-6 of the glucosyl monomer units.

Converting these hydroxyl groups to esters or ethers obstructs the formation of interchain hydrogen bonds and drastically changes the properties of cellulose. So, for example, acetylation or nitration can be used to make clear films for packaging, while cellulose substituted with carboxymethyl or hydroxyethyl groups gives viscous solutions in water.

The chemistry of cellulose modification is relatively simple but efficient. High-grade dissolving wood pulp similar to that used for rayon manufacture is swollen in sodium hydroxide solution and esterified with the appropriate acid chloride or anhydride. The degree and pattern of substitution strongly influence the physical properties of the product and are tightly controlled.

Modified celluloses, especially the water-soluble carboxymethyl and hydroxyalkyl forms, are used in an extraordinarily wide range of products – essentially, wherever the viscosity of an aqueous solution has to be modified. This includes foods, beverages, animal feeds, cosmetics, adhesives, paints and inks, paper, plasterboard, and oil drilling fluids.

Future Prospects

If paper is included, cellulose is the most important industrial feedstock after oil. Being a renewable resource its utilization seems likely to increase. Yet cellulose has a remarkably low public and scientific profile. This is probably because it is utilized mainly through industries with high output and relatively low added value, which do not support high levels of investment in science. Also the production of fiber crops and wood has attracted less research investment than agriculture, for a variety of reasons: most timber species can scarcely be called domesticated.

We can expect all this to change. Pressure to reduce dependence on oil-based raw materials is likely to focus public attention and research resources on cellulose production and utilization. With higher-value products like textiles and modified celluloses at the forefront, cellulose is likely to become increasingly prominent as a basic raw material.

See also: **Abiotic Stresses**: Mechanical Stress and Wind Damage. **Cell Walls and Fibers**: Cell Walls; Fiber Formation; Wood Growth and Development. **Ethics and Biosafety**: Development and Commercialization of Genetically Modified Plants. **Genetic Modification of Primary Metabolism**: Biopolymers; Wood Quality. **Growth and Development**: Cell Growth. **Production Systems and Agronomy**: Agricultural Crops. **Secondary Products**: Lignin. **Tissue Culture and Plant Breeding**: Clonal Forestry; Clonal Propagation, Forest Trees.

Further Reading

Anonymous (2002) *Walden's Paper Handbook*, 3rd edn. Ramsey: Walden-Mott Corporation.

Atalla RH and VanderHart DL (1999) The role of solid state C-13 NMR spectroscopy in studies of the nature of native celluloses. *Solid State Nuclear Magnetic Resonance* 15: 1–19.

Baskin TI (2001) On the alignment of cellulose microfibrils by cortical microtubules: a review and a model. *Protoplasma* 215: 150–171.

Debeaufort F, Quezada-Gallo JA, and Voilley A (1998) Edible films and coatings: tomorrow's packagings. *Critical Reviews in Food Science* 38: 299–313.

Delmer DP and Haigler CH (2002) The regulation of metabolic flux to cellulose, a major sink for carbon in plants. *Metabolic Engineering* 4: 22–28.

Edgar KJ, Buchanan CM, Debenham JS, *et al.* (2001) Advances in cellulose ester performance and application. *Progress in Polymer Science* 26: 1605–1688.

Eichhorn SJ, Baillie CA, Zafeiropoulos N, *et al.* (2001) Review: Current international research into cellulosic fibres and composites. *Journal of Materials Science* 36: 2107–2131.

Hirami M (1996) New developments in the rayon industry. *Journal of Macromolecular Science Pure and Applied* 33: 1825–1835.

O'Sullivan AC (1997) Cellulose: the structure slowly unravels. *Cellulose* 4: 173–207.

Wilkins TA, Rajasekaran K, and Anderson DM (2000) Cotton biotechnology. *Critical Reviews in Plant Science* 19: 511–567.

Williamson RE, Burn JE, and Hocart CH (2002) Towards the mechanism of cellulose synthesis. *Trends in Plant Science* 7: 461–467.

Amino Acids

P J Lea, Lancaster University, Lancaster, UK
R A Azevedo, University of São Paulo, Piracicaba, Brazil

Introduction

Amino acids are the major nitrogen containing compounds of plants and are the building-blocks of proteins. There are 20 different amino acids normally incorporated into proteins, although they may be subject to alteration afterwards, by enzyme reactions such as phosphorylation, methylation and acetylation. In addition to those in protein, over 300 additional amino acids have been isolated and characterized from plants. With the arrival of even more sensitive analytical techniques, this number is likely to increase. These amino acids may be present in low concentrations and play a vital part as an intermediate in a biosynthetic pathway, e.g., ornithine, homoserine, or cystathionine. In contrast they may act as a major storage form of nitrogen, e.g., canavanine in the seed of *Canavalia ensiformis*, or may be formed in high amounts in response to an external stress, e.g., γ-aminobutyrate. It is possible that some of these nonprotein amino acids may serve as insecticidal or fungicidal agents.

Humans and monogastric animals are not able to synthesize the following nine amino acids: lysine, threonine, methionine, phenylalanine, tryptophan, isoleucine, leucine, valine, and histidine. These are termed "essential" and must be supplied in the diet. They are however able to convert methionine to cysteine and phenylalanine to tyrosine, provided they are available. Plants are able to synthesize all 20 protein amino acids and these may be classified into "families" by the pathways that are involved. These pathways are frequently subject to very complex and tight regulation to prevent the wastage of energy and the key nutrients carbon, nitrogen and sulfur. It is interesting to note that the enzymes involved in the synthesis of the 'essential' amino acids are normally located within the chloroplasts of the leaves or the plastids of the nonphotosynthetic organs, such as the roots or seeds.

The Glutamate Family

The inorganic form of nitrogen ultimately available to plants for assimilation into amino acids is ammonia, which is normally present in equilibrium with the ammonium ion. Ammonia is formed through nitrate reduction in roots and shoots, through the fixation of atmospheric nitrogen by root or stem nodules, or may be taken up directly. Ammonia may also be generated internally by a large number of metabolic pathways such as photorespiration, phenylpropanoid synthesis, and the breakdown of nitrogen transport compounds.

Glutamine synthetase (GS) catalyzes the ATP-dependent conversion of glutamine into glutamate utilizing ammonia as a substrate (**Figure 1**). Two major isoforms of the GS enzyme exist: cytosolic GS (GS1), occurring in the cytoplasm of leaves and nonphotosynthetic organs, and chloroplastic GS (GS2), which is present in the chloroplasts of photosynthetic tissues and the plastids of roots or etiolated plants. In all plant species studied, chloroplastic GS2 is encoded by one nuclear gene per haploid genome. However, studies on a wide range of species have shown that cytoplasmic GS1 is encoded by a complex multigene family, which varies from three to six genes, which are expressed in different tissues and at different stages of development of the plant.

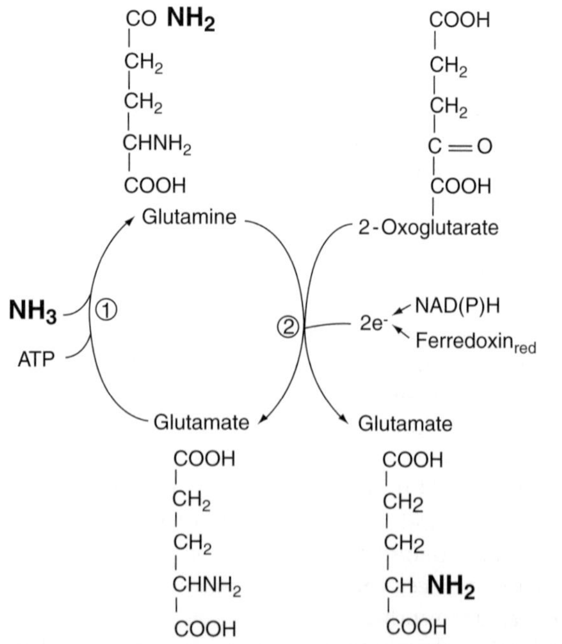

Figure 1 The glutamate synthase cycle. Enzymes: 1, glutamine synthetase; 2, glutamate synthase.

GS is inhibited by a range of compounds that include phosphinothricin, which is synthesized by a number of *Streptomyces* species. Phosphinothricin is a potent herbicide and is sold commercially as Glufosinate and Basta. The inhibition of GS causes a buildup of ammonia and a dramatic fall in the concentration of all amino acids, which perturbs a range of metabolic pathways including photorespiration. Some *Streptomyces* species also synthesize an enzyme phosphinothricin acetyl transferase, which detoxifies phosphinothricin by blocking the amino group and thus preventing it from binding to the active site. The phosphinothricin acetyl transferase gene is now used widely as a selective marker and also for the production of herbicide-resistant crops.

Following the incorporation of the ammonia into the amide position of glutamine, the nitrogen may be transferred to asparagine or used for the synthesis of the nucleotides required for DNA synthesis. However it is essential that the majority of the nitrogen is transferred to the α-amino group of glutamate. This reaction is carried out by glutamine 2-oxoglutarate amidotransferase (glutamate synthase, GOGAT), which catalyzes the reductant-dependent conversion of glutamine and 2-oxoglutarate to two molecules of glutamate (**Figure 1**). Two different forms of glutamate synthase are present in higher plants, one using ferredoxin as a source of reductant and the other NADH. Ferredoxin-glutamate synthase is predominant in the leaf, where it is located totally in the chloroplast. There is now evidence that there are two genes encoding the enzyme in higher plants and that a second form is present in the roots. In addition the NADH-glutamate synthase is present in high activity in nonphotosynthetic tissue, in particular in root nodules. It is still not clear whether there are one or two genes encoding NADH-glutamate synthase in plants.

Taken together, the operation of the two enzymes GS and glutamate synthase (termed the glutamate synthase cycle) allows the rapid and efficient assimilation of ammonia into the α-amino position. When the reaction takes place in the chloroplast, it is light driven and is a true photosynthetic reaction. The 2-oxoglutarate is made available by isocitrate dehydrogenase, and two distinct forms, using either NAD or NADP, have been shown to operate in both the mitochondria and cytoplasm. A further enzyme glutamate dehydrogenase has also been proposed to operate in the direction of ammonia assimilation and glutamate synthesis. The majority of evidence now indicates that glutamate dehydrogenase serves to generate 2-oxoglutarate, when there is a shortage of available carbon. It is possible that the balance between glutamine : glutamate or glutamine : 2-oxoglutarate may act as a signal to indicate the availability of nitrogen within the plant.

The nitrogen in the α-amino position of glutamate may now be redistributed to form other amino acids, prior to incorporation into protein (**Figure 2**). These reactions are carried out by aminotransferases (transaminases), the major forms of which synthesize alanine from pyruvate and aspartate from oxaloacetate. Provided the correct 2-oxo acid is available, all the remaining amino acids can then be formed. Distinct forms of these enzymes have been shown to be present in the chloroplast, mitochondria, peroxisomes, and cytoplasm. The reactions appear not to be regulated, but are dependent upon the availability of the substrates.

The synthesis of arginine represents a convenient mechanism of storing nitrogen, as each molecule contains four nitrogen atoms. Seeds frequently contain high concentrations in the protein and the soluble form is used for transport and storage, particularly in trees. Plants use the acetylated pathway for arginine synthesis, in which glutamate is first converted to ornithine, the presence of the *N*-acetyl group prevents the spontaneous cyclization of the semialdehyde (**Figure 3**). The first two enzymes, *N*-acetylglutamate synthase and *N*-acetylglutamate kinase, are both subject to feedback inhibition regulation by arginine. The conversion of ornithine to arginine follows the same pathway as found in mammals in the Krebs–Heinseleit urea cycle. Arginine and ornithine are the precursors of the synthesis of the polyamines, putrescine, spermidine, and spermine. These nitrogen-rich compounds accumulate in many tissues and play a role in plant differentiation, including flowering and root nodulation as well as stress resistance.

Proline, as well as being a major constituent of seed proteins, plays an important role as an

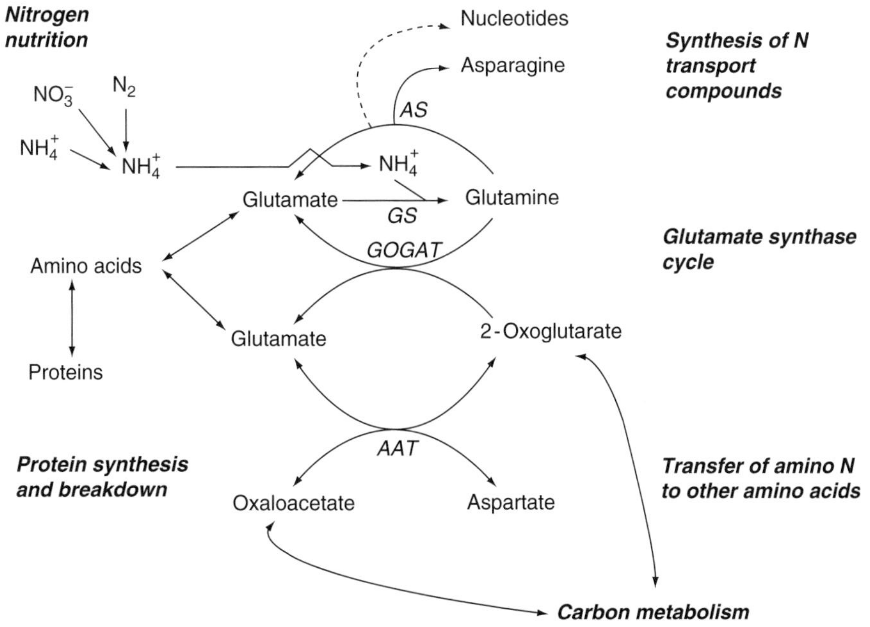

Figure 2 Enzyme reactions important in the assimilation of ammonia. GS, glutamine synthetase; GOGAT, glutamate synthase; AS, asparagine synthetase; AAT, aspartate aminotransferase. Adapted with permission from Lea PJ and Miflin BJ (2003) Glutamate synthase and the synthesis of glutamate in plants. *Plant Physiology and Biochemistry* 41 (in press).

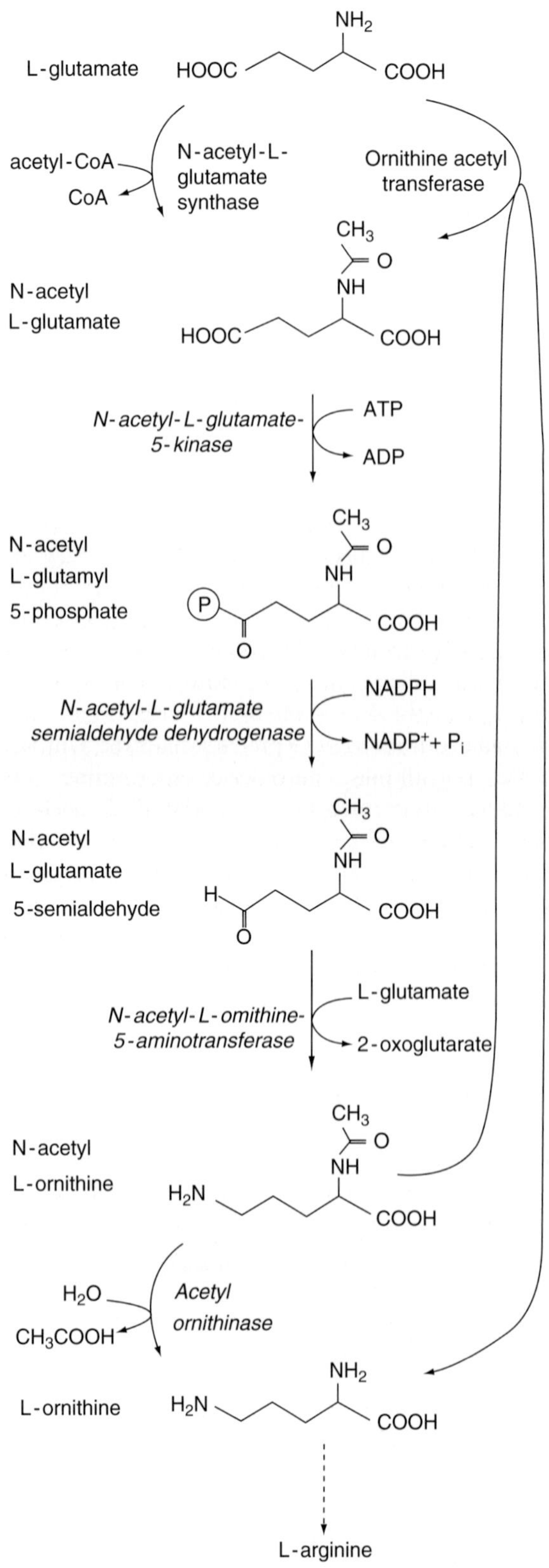

Figure 3 The acetylated pathway of ornithine and arginine synthesis. Adapted with permission from Ramón-Maiques S, Marina S, Gil-Ortiz F, Fita I, and Rubio V (2002) Structure of acetylglutamate kinase, a key enzyme for arginine biosynthesis and a prototype for the amino acid kinase enzyme family, during catalysis. *Structure* 10: 329–342.

osmoprotectant in resistance to drought stress and as an energy supply in pollen tube growth and nitrogen fixation in the legume root nodule. The first enzyme involved in the conversion of glutamate to proline is a fusion of γ-glutamyl kinase and glutamate semialdehyde dehydrogenase termed Δ^1-pyrroline-5-carboxylate synthase (P5CS). The cyclized pyrroline-5-carboxylate (P5C) is then reduced to form proline by P5C reductase (**Figure 4**). Under certain circumstances, ornithine may also act as a precursor of proline, via δ-ornithine aminotransferase. The kinase activity of P5CS is subject to feedback inhibition regulation by proline. During drought stress or high salt conditions the rate of proline synthesis is increased by a stimulation of the expression of the P5CS gene and also by a reduction in the expression of the gene encoding the enzyme proline dehydrogenase which catalyzes the first step of proline breakdown. A number of attempts have been made to obtain plants that are resistant to drought stress by manipulating their capacity to synthesize proline. Transgenic plants overexpressing normal and mutant P5CS and also δ-ornithine aminotransferase have been shown to accumulate proline and to be more resistant, although the differences have not proved to be large. In addition there is evidence that proline may have a detrimental effect on growth under nonstressed conditions.

The Aspartate Family

As indicated above, asparagine (**Figure 2**) is synthesized directly by the transfer of the amide nitrogen of glutamine to aspartate. Asparagine is a major transport compound for nitrogen and also a storage compound during times of nitrogen excess. The metabolism of asparagine by asparaginase forms aspartate and ammonia, which must be reassimilated through the glutamate synthase cycle.

Aspartate is the common precursor for the synthesis of the essential amino acids threonine, lysine, methionine, and isoleucine, which will be considered in a later section. Due to the low concentrations of lysine and threonine in cereals, research has focused on obtaining a thorough understanding of the genetic and biochemical control of each branch of the aspartate pathway (**Figure 5A, B**). Initial biochemical studies and more recent gene sequence evidence have established that all the steps apart from the final methylation of homocysteine take place in the chloroplast/plastid.

Three distinct isoenzymes of aspartate kinase (AK), which catalyzes the phosphorylation of aspartate, have been detected in plants, two lysine-sensitive and one which is threonine-sensitive. The

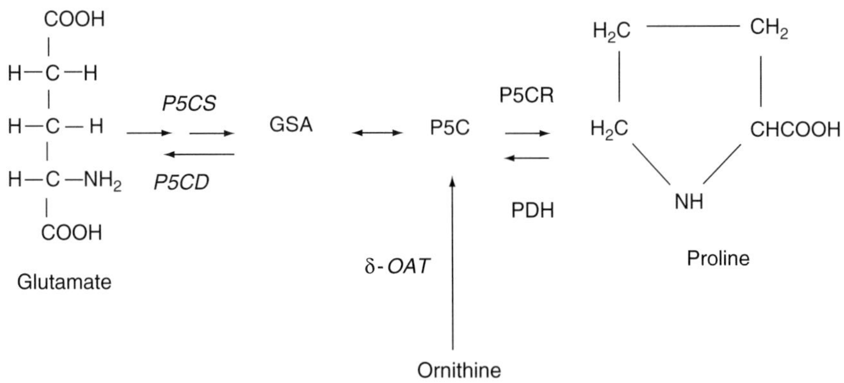

Figure 4 The pathways of proline synthesis in plants. Enzymes: P5CS, Δ^1-pyrroline-5-carboxylate synthetase; P5CR, Δ^1-pyrroline-5-carboxylate reductase; PDH, proline dehydrogenase; P5CD, Δ^1-pyrroline-5-carboxylate dehydrogenase; δ-OAT, δ-ornithine aminotransferase.

lysine-sensitive AK isoenzyme, which normally accounts for 60–80% of total AK activity in plants, is involved in the overall regulation of the pathway and can be synergistically inhibited by *S*-adenosylmethionine. The threonine-sensitive AK isoenzyme has been shown to be part of a bifunctional polypeptide also containing the threonine-sensitive homoserine dehydrogenase isoenzyme. The final enzyme required for the synthesis of threonine, threonine synthase (TS), is strongly activated by *S*-adenosylmethionine.

For the lysine branch of the pathway, the enzyme dihydrodipicolinate synthase (DHDPS) has been shown to be the key regulatory enzyme. Lysine is a potent inhibitor of DHDPS enzyme activity, and unlike AK, only one form of the enzyme has been observed in plants. It would be predicted that cystathionine-γ-synthase should be the key regulatory enzyme in the synthesis of methionine, as it competes with TS for the substrate O-phosphoserine; however, it is not subject to feedback inhibition. Recent information from transgenic plants containing elevated and decreased levels of both threonine synthase and cystathionine γ-synthase has indicated that there is a complex interaction between the two enzymes, involving both posttranscriptional and posttranslational controls. In this way the supply of carbon and sulfur is maintained to methionine, which is an essential precursor of ethylene and polyamines and is also an important donator of methyl groups.

There have been numerous attempts to produce plants containing elevated concentrations of threonine, lysine, and methionine, using both mutant selection and transgenic approaches. Plants containing an AK isoenzyme that is insensitive to lysine all accumulate soluble threonine, with little effect on the other amino acids. Plants that contain a DHDPS that is insensitive to lysine accumulate very high concentrations of lysine in the leaves, to a point that it is detrimental to growth. Expression of lysine-insensitive DHDPS in the seeds of canola (*Brassica napus*), leads to a greater than 100-fold increase in the concentration of soluble lysine, which caused the total lysine in the seed to be increased from 6 to 12%. In similar transgenic soybean (*Glycine max*), the total lysine concentration in the seed was increased over fivefold from 5.9 to 34%. These are clearly important results and it remains to be seen whether the high seed lysine lines will be able to produce adequate yields when grown under field conditions. The methionine concentration in plants has been increased by the overexpression of mutant forms of cystathionine γ-synthase; however, by far the greatest increases in methionine have been generated by reducing the activity of TS using an antisense gene. Unfortunately, in these plants there was clear evidence that the excess methionine was metabolized to ethylene, dimethylsulfide, and carbon disulfide.

Due to the requirement of obtaining cereal seeds with a high concentration of lysine, the pathway of lysine degradation has recently been examined in more detail (**Figure 6**). Initial studies of lysine catabolism in plants were carried out using ^{14}C-lysine in which the radioactivity was incorporated into glutamate and aminoadipate semialdehyde. This reaction is catalyzed by two enzymes, lysine α-ketoglutarate reductase (LKR), also known as lysine 2-oxoglutarate reductase (LOR), which catalyzes the formation of saccharopine, and saccharopine dehydrogenase (SDH) which hydrolyzes saccharopine into glutamate and aminoadipate semialdehyde. Most of the activity is present on a bifunctional polypeptide, although monofunctional forms of both enzymes have also been detected. It was first thought that the enzymes were restricted to the endosperm tissue of cereals, but more recent studies have indicated that they are expressed in other tissues,

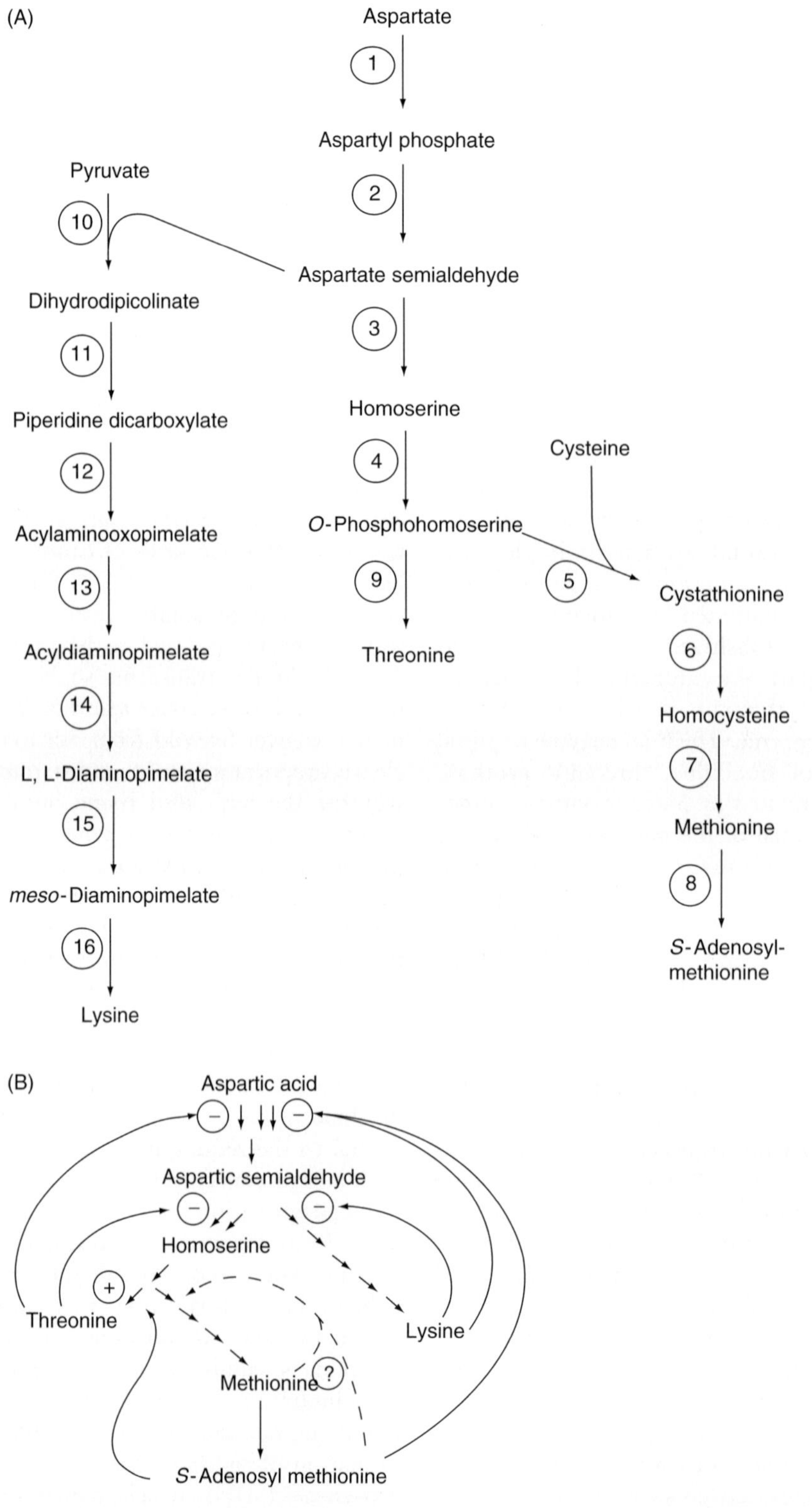

Figure 5 (A) Aspartate-derived biosynthesis pathways leading to the synthesis of methionine, threonine, and lysine. Enzymes: 1, aspartate kinase; 2, aspartate semialdehyde dehydrogenase; 3, homoserine dehydrogenase; 4, homoserine kinase; 5, cystathionine γ-synthase; 6, cystathione β-lyase; 7, methionine synthase; 8, *S*-adenosylmethionine synthetase; 9, threonine synthase; 10, dihydrodipicolinate synthase; 11, dihydrodipicolinate reductase; 12, piperidine dicarboxyate acylase; 13, acyldiaminopimelate aminotransferase; 14, acyldiaminopimelate deacylase; 15, diaminopimelate epimerase; 16, diaminopimelate decarboxylase. Adapted with permission from Azevedo RA, Arruda P, Turner WL, and Lea PJ (1997) The biosynthesis and metabolism of the aspartate derived amino acids in higher plants. *Phytochemistry* 46: 395–419. (B) The regulation of the aspartate pathway by end-product metabolites. $-$, enzyme inhibition; $+$, enzyme stimulation; ?, regulatory mechanism not clarified.

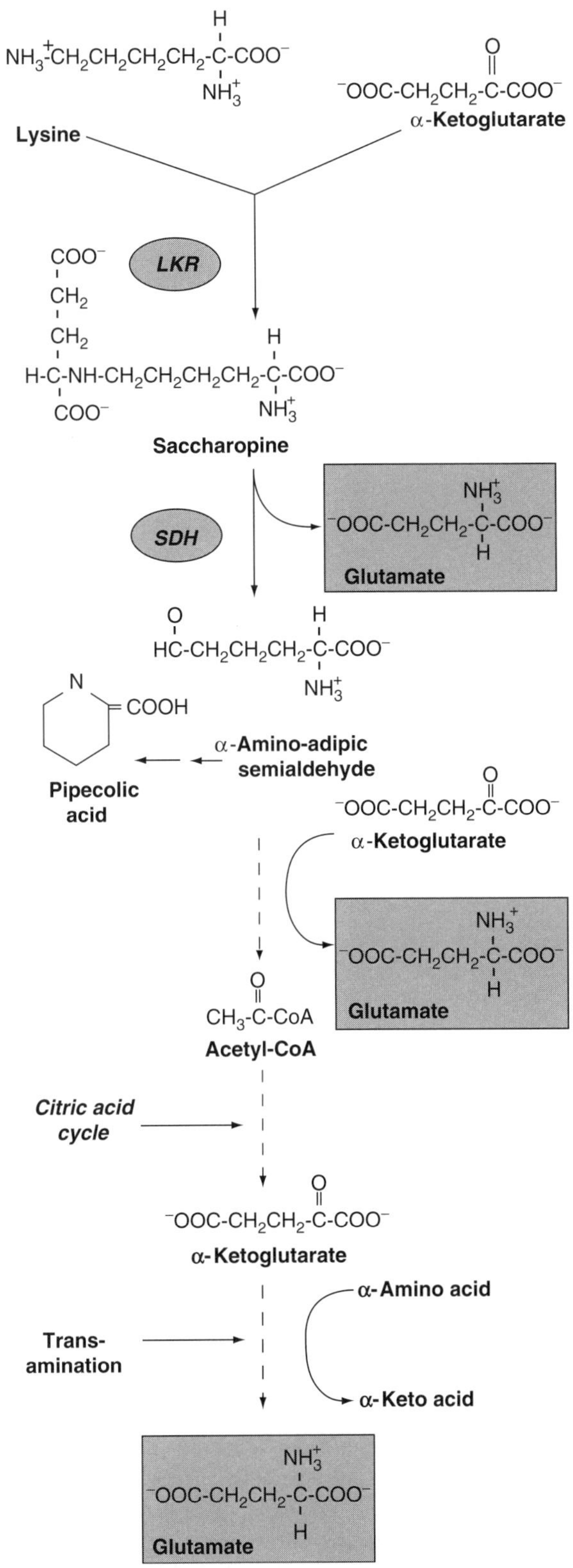

Figure 6 The lysine catabolism pathway of plants and metabolites derived from it. Broken arrows represent non-specified enzymatic reactions. Adapted with permission from Galili G, Tang G, Zhu X, and Gakiere B (2001) Lysine catabolism: a stress and development superregulated metabolic pathway. *Current Opinion in Plant Biology* 4: 261–266.

particularly during osmotic stress. Biochemical characterization of the LKR–SDH bifunctional enzyme has shown that the LKR, but not SDH, activity is stimulated by lysine, via a complex Ca^{2+}-dependent phosphorylation cascade. In the opaque-2 mutant of corn (*Zea mays*; maize), the seed has been shown to contain higher concentrations of lysine. Biochemical and molecular analyses have shown that the *opaque-2* gene, which encodes a transcriptional activation factor, regulates the activity of LKR in corn endosperms. In the opaque-2 mutant, LKR enzyme activity, protein, and mRNA have been shown to be reduced with the expression pattern markedly modified during grain development. Elevated concentrations of lysine have also been detected in the mature seeds of a knockout mutant of *Arabidopsis thaliana* which lacked the expression of the gene encoding the bifunctional LKR–SDH enzyme.

Glycine, Serine, and Cysteine

Glycine and serine, as well as being constituents of proteins, serve as precursors of phospholipids and purines and are the major source of one-carbon units in plant cells. However, in the leaves of temperate C_3 plants, the two amino acids play a key role in the process of photorespiration. Due to the oxygenase activity of RuBP carboxylase, the first enzyme involved in CO_2 assimilation, the two-carbon compound glyoxylate is transaminated by either glutamate or alanine to form glycine in the peroxisome. The glycine is transported into the mitochondria, where two molecules are converted into one molecule of serine, CO_2, and ammonia, with the concomitant production of NADH. The CO_2 released is lost to the atmosphere and has a severe effect on crop yield; the ammonia is reassimilated by the glutamate synthase cycle (**Figure 2**). Two enzymes are required for this reaction, glycine decarboxylase (GDC) and serine hydroxymethyltransferase (SHMT) (**Figure 7**). Such is the flux through the glycine to serine conversion (at least five times that of the standard tricarboxylic acid cycle), that the two enzymes account for more than 50% of the protein in the leaf mitochondria and are present at concentrations of 200 mg ml^{-1}.

The GDC enzyme is made up of four different polypeptides, P, H, T, and L. The first glycine molecule is decarboxylated by the P protein and the methylamine group is attached to a S atom on the flexible arm of the H protein. The T protein catalyzes the liberation of ammonia and the transfer of the methylene group to tetrahydrofolate, which is then available for SHMT. The L protein catalyzes the

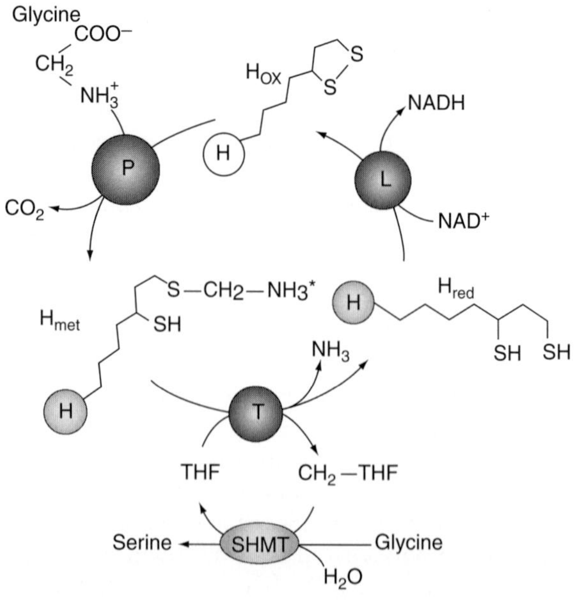

Figure 7 Outline of the reactions involved in oxidative decarboxylation and deamination of glycine in plant mitochondria. P, H, T, and L are the protein components of the glycine-decarboxylase multienzyme system. SHMT: serine hydroxymethyltransferase is involved in the conversion of CH_2-tetrahydrofolate to tetrahydrofolate at the expense of a second molecule of glycine. Adapted with permission from Douce R, Bourguignon J, Neuburger M, and Rébeillé F (2001) The glycine decarboxylase system: a fascinating complex. *Trends in Plant Science* 6: 167–176.

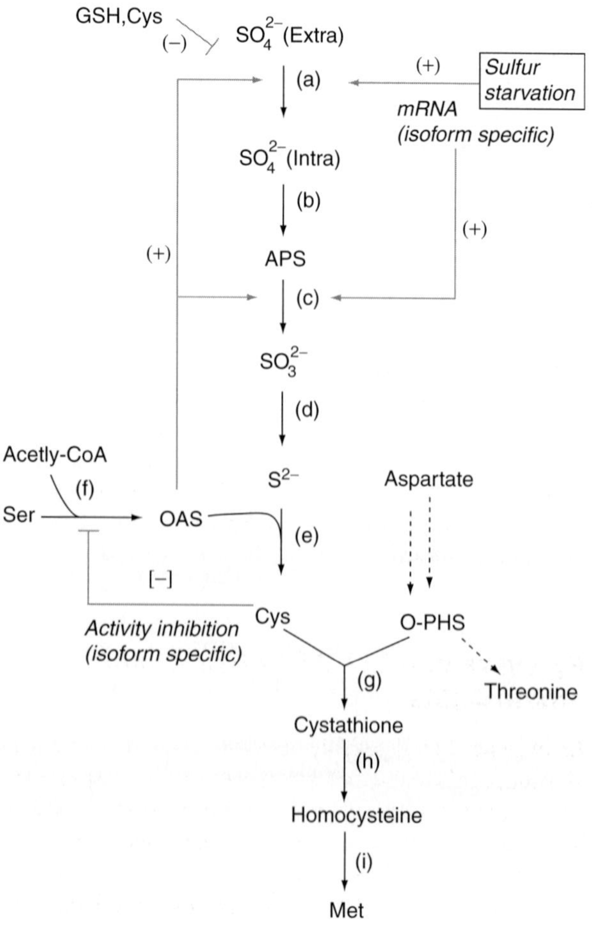

Figure 8 The pathway of sulfate reduction and conversion to cysteine and methionine. The enzymes involved in the pathway are: (a), sulfate transporter; (b), ATP sulfurylase; (c), APS reductase; (d), sulfite reductase; (e), *O*-acetylserine (thiol)-lyase; (f), serine acetyltransferase; (g), cystathione γ-synthase; (h), cystathione β-lyase; (i), methionine synthase. *O*-PHS, *O*-phosphohomoserine. Arrow and (+) indicates induction. Bar and (−) indicates repression; bar and [−] indicates inhibition. Adapted with permission from Saito K (2000) Regulation of sulfate transport and synthesis of sulfur-containing amino acids. *Current Opinion in Plant Biology* 3: 188–195.

synthesis of NADH using the reduced H protein as a source of reductant (**Figure** 7).

Serine and glycine can also be synthesized by a separate pathway, starting with phosphoglycerate and employing phosphorylated intermediates. It is assumed that this pathway operates in nonphotosynthetic tissues. However, there is strong evidence that although the photorespiration pathway in leaves is supposed to be a closed cycle, a considerable amount of glycine and serine can "escape." The two amino acids can be transported to other organs and used as precursors of cysteine and glutathione.

Cysteine, as well as being a key amino acid regulating the structure and function of proteins, is also the most important constituent of glutathione, a molecule which is essential for the prevention of damaging oxidation reactions within the cell. Glutathione is a tripeptide comprised of glutamate, cysteine, and glycine and is synthesized in the cytoplasm and chloroplast. The sulfate required for both the synthesis of cysteine and methionine (**Figure** 8) is first taken up by the roots and may be transported to the leaves, where it enters the chloroplast. The sulfate is activated by ATP sulfurylase by converting it to adenosine 5′-phosphosulfate (APS). The enzymatic step involved in the conversion of APS to sulfite has been the subject of debate for a long time. Recent data confirm that this reaction is catalyzed by a single enzyme called APS reductase. This enzyme is a GSH-dependent reductase, which used to be called APS sulfotransferase. The conversion of sulfite to sulfide requires a six-electron transfer catalyzed by ferredoxin-dependent sulfite reductase, in a similar manner to nitrite reductase. The carbon skeleton of cysteine is derived from serine via O-acetylserine by the action of serine acetyltransferase. This enzyme is located in the chloroplast, mitochondria, and cytoplasm, although only the cytoplasmic enzyme is feedback inhibited by μmol concentrations of cysteine. This inhibition of serine acetyltransferase can be reduced following calcium-dependent phosphorylation of the enzyme protein. The genes encoding sulfate uptake and reduction are induced during periods of sulfate starvation. O-acetylserine has been shown to be a positive regulator of this effect, whilst glutathione and cysteine cause repression.

The sulfide is finally inserted into cysteine by O-acetylserine (thiol)-lyase (also known as cysteine synthase), which is also located in the same three compartments as serine acetyltransferase. In the chloroplast, the two enzymes form a complex, although the O-acetylserine (thiol)-lyase is normally present in excess. Transgenic plants containing elevated activities of the sulfate-reducing enzymes and serine acetyltransferase have been shown to accumulate cysteine and glutathione. These plants have a greater capacity to resist oxidative stress induced by paraquat and hydrogen peroxide.

Pyruvate and the Branched Chain Amino Acids

As indicated in **Figure 2**, pyruvate is the precursor of alanine, following transamination by glutamate, catalyzed by alanine aminotransferase. Pyruvate is however also the starting-point for the synthesis of the branched chain amino acids, isoleucine, leucine, and valine (**Figure 9**), the reactions of which take place in the chloroplast/plastid. For leucine and valine, the carbon is derived solely from two molecules of pyruvate and the key regulatory enzyme is acetolactate synthase, also known as acetohydroxyacid synthase (see below). Leucine and valine act as feedback regulators, by binding synergistically on two separate domains of the enzyme protein, in order to inhibit activity. Leucine also inhibits the first enzyme unique to its own synthesis, isopropylmalate synthase. A branched-chain amino acid-aminotransferase catalyzes the final step in the pathway that produces leucine, valine, and isoleucine. In plants, the reactions are reversible from glutamate and are also probably involved in the degradation of the amino acids.

The carbon for isoleucine is derived from both threonine and pyruvate. Threonine, which is synthesized via the aspartate pathway shown above, is deaminated to form 2-ketobutyrate and ammonia by threonine deaminase (**Figure 9**). The biosynthetic isoenzyme is present in young actively growing tissue and is feedback inhibited by isoleucine. A degradative isoenzyme is present in old senescing tissue and is able to utilize serine and threonine as substrates but is not subject to feedback inhibition. One molecule of 2-ketobutyrate and one molecule of pyruvate are then able to act as a substrate for acetohydroxyacid synthase, which is the same enzyme as acetolactate synthase, used in the synthesis of leucine and valine

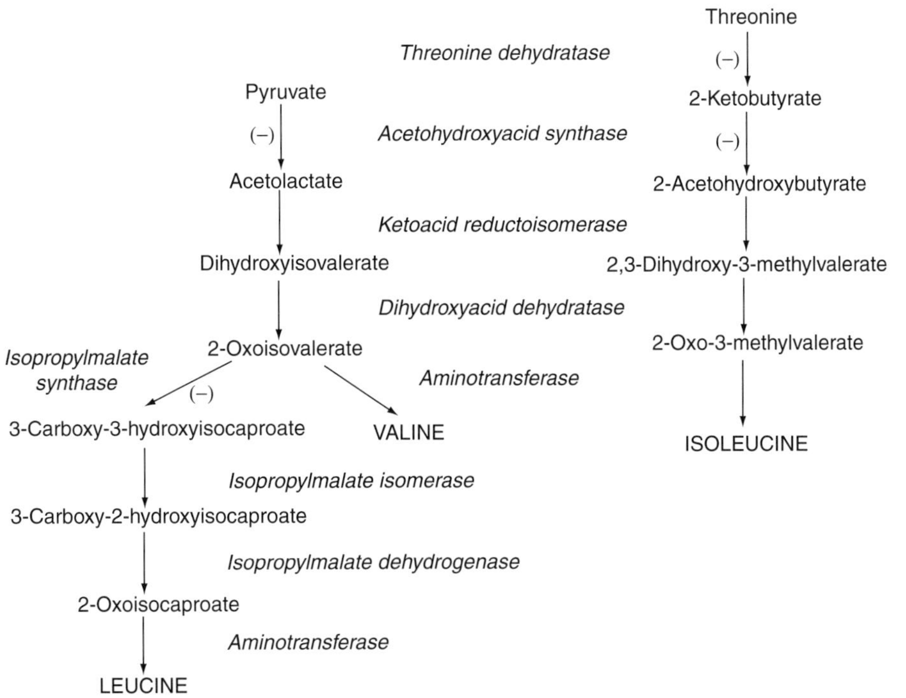

Figure 9 The pathway of the synthesis of isoleucine, leucine, and valine from pyruvate. (−) indicates the biosynthetic steps at which the pathway is regulated by feedback inhibition of the enzyme involved. Adapted with permission from Singh BK and Shaner DL (1995) Biosynthesis of the branch chain amino acids: from test tube to the field. *Plant Cell* 7: 935–944.

(see above). Isoleucine is able to act as a feedback inhibitor of the enzyme, although to a lesser extent than valine and leucine.

Several related classes of herbicides are known to inhibit acetolactate synthase/acetohydroxyacid synthase, such as the sulfonylureas, imidazolinones, triazolopyrimidine sulfonanilides, and pyrimidinyl oxybenzoates. These compounds are amongst the most advanced herbicides used in agriculture, as they exhibit extremely low toxicity to mammals and high efficacy, resulting in very low application rates. The herbicidal compounds are unusual in that they have no obvious chemical similarity to the substrates, cofactors, or allosteric inhibitors, but are able to bind to the evolutionary vestige of a quinone binding site. Herbicide-resistant weed species have appeared at many geographical locations and have been attributed to single point mutations, which can occur at multiple sites in the acetolactate synthase gene, resulting in a variable pattern of cross-resistance between the classes of inhibitors. Transgenic plants containing mutant inhibitor-insensitive forms of acetolactate synthase have been shown to be resistant to a range of the herbicides.

The Aromatic Amino Acids

The aromatic amino acids phenylalanine, tyrosine, and tryptophan are the precursors of a tremendous range of vitally important constituents of plants. These include, to mention just a few, amines, alkaloids, phenols, phenylpropanes, xanthones, stilbenes, flavonoids, melanins, and lignins. All of these compounds contain at least one aromatic ring and are ultimately derived from chorismate, which is synthesized via the shikimate pathway (**Figure 10**). The pathway involves seven steps commencing with phospho-*enol*-pyruvate and erythrose 4-phosphate. It is estimated that 20% of the fixed carbon in a plant passes down this pathway. In plants, all the steps are catalyzed by separate enzyme proteins, except the third and fourth reactions which are carried out by a bifunctional enzyme 3-dehydroquinate dehydratase-shikimate NADP oxidoreductase. All the enzymes have been shown to be located in the chloroplast/plastid, but there is also evidence of the presence of a second pathway in the cytoplasm, but this has not been confirmed. Somewhat surprisingly the first enzyme, 3-deoxy-D-*arabino*-heptulosonate 7-phosphate (DAHP) synthase is not subject to feedback inhibition by any of the end-product amino acids, although arogenate (a later metabolite) is a potent inhibitor.

The penultimate step in the shikimate pathway is catalyzed by 5-*enol*-pyruvylshikimate-3-phosphate synthase (EPSPS). This enzyme is the target for one of the most widely used herbicides, glyphosate, which is sold commercially as Roundup. The precise mechanism by which glyphosate inhibits EPSPS is still the subject of debate, but it would appear that it is able to bind to a complex of the enzyme and shikimate-3-phosphate, near to but not at the active site. Following application of the herbicide, the synthesis of chorismate and thus the three aromatic

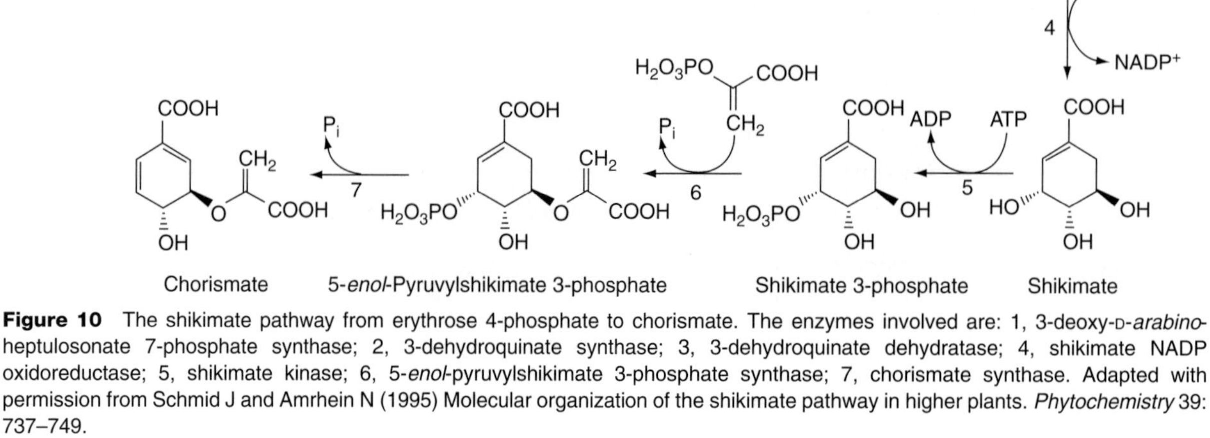

Figure 10 The shikimate pathway from erythrose 4-phosphate to chorismate. The enzymes involved are: 1, 3-deoxy-D-*arabino*-heptulosonate 7-phosphate synthase; 2, 3-dehydroquinate synthase; 3, 3-dehydroquinate dehydratase; 4, shikimate NADP oxidoreductase; 5, shikimate kinase; 6, 5-*enol*-pyruvylshikimate 3-phosphate synthase; 7, chorismate synthase. Adapted with permission from Schmid J and Amrhein N (1995) Molecular organization of the shikimate pathway in higher plants. *Phytochemistry* 39: 737–749.

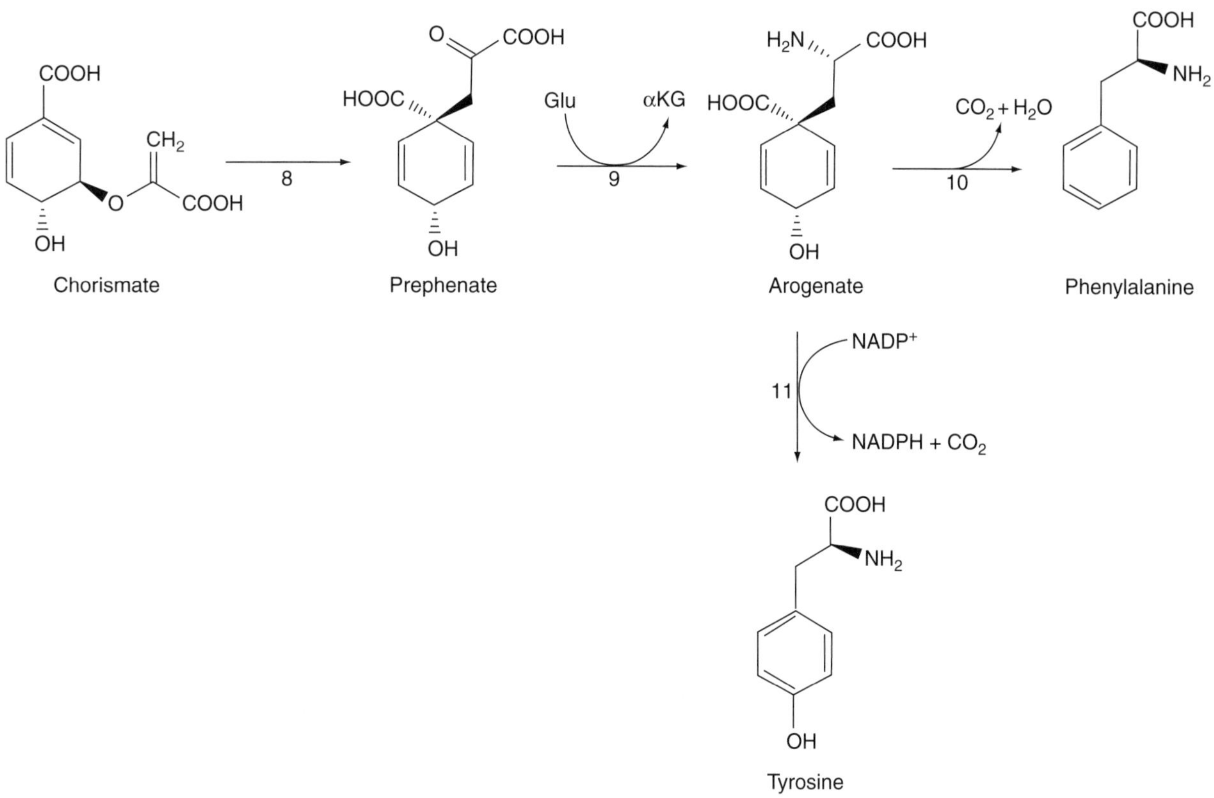

Figure 11 The pathway from chorismate to phenylalanine and tyrosine. The enzymes involved are: 8, chorismate mutase; 9, prephenate aminotransferase; 10, arogenate dehydratase; 11, arogenate dehydrogenase. Adapted with permission from Schmid J and Amrhein N (1995) Molecular organization of the shikimate pathway in higher plants. *Phytochemistry* 39: 737–749.

acids is prevented and there is a considerable buildup of shikimate and shikimate-3-phosphate. A range of mutant bacteria have been selected that contain a glyphosate-resistant form of EPSPS. The corresponding resistant EPSPS genes have been isolated and used to construct glyphosate-resistant crop plants that are now being grown commercially.

The pathway from chorismate to phenylalanine and tyrosine is less well characterized and does not appear to be the same as that found in microorganisms. It has now become clear that following the action of chorismate mutase, prephenate is first transaminated to form arogenate, which can then be converted to either tyrosine or phenylalanine (**Figure 11**). Two forms of chorismate mutase have been detected in higher plants, which exhibit different characteristics. The chloroplastic form is subject to inhibition by phenylanine and tyrosine and activation by tryptophan, whilst the cytoplasmic form is not regulated. In addition, arogenate dehydrogenase and arogenate dehydratase (which are both located solely in the chloroplast) are subject to strong inhibition by μmolar concentrations of tyrosine and phenylalanine, respectively.

Tryptophan is a precursor of a range of compounds that can act as growth regulators, chemical defense agents, and attractants for pollinators. In addition, the indole alkaloids vincristine and vinblastine have been shown to have considerable therapeutic activity in the treatment of cancer. The pathway from chorismate to tryptophan follows that previously demonstrated for microorganisms (**Figure 12**), and appears to be exclusively located in the chloroplast/plastid. A range of mutants of *A. thaliana* have been isolated by Last and colleagues, confirming the role of individual enzymes. The first enzyme in the pathway anthranilate synthase is exceedingly sensitive to feedback inhibition by tryptophan in the range of 1–5 μmol. Mutants containing an anthranilate synthase enzyme less sensitive to inhibition have been shown to accumulate tryptophan. An elevated flux through the pathway was stimulated following wounding or attack by *Pseudomonas syringae*, by increasing the expression of anthranilate synthase, which presumably allowed for an increased rate of synthesis of indole alkaloids as defense compounds.

Histidine

Histidine was described by Miflin in 1980 as the "Cinderella" of all of the amino acids and it has

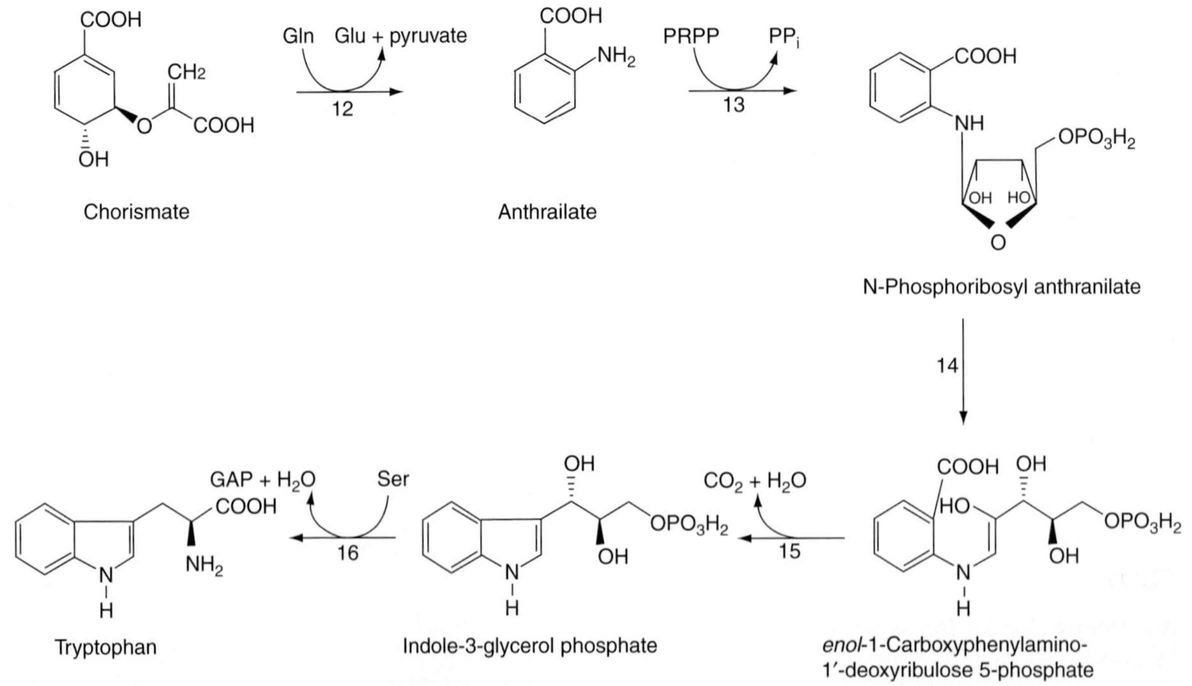

Figure 12 The pathway from chorismate to tryptophan. The enzymes involved are: 12, anthranilate synthase; 13, anthranilate phosphoribosyl transferase; 14, phosphoribosyl anthranilate isomerase; 15, indole-3-glycerol phosphate synthase; 16, tryptophan synthase. Adapted with permission from Schmid J and Amrhein N (1995) Molecular organization of the shikimate pathway in higher plants. *Phytochemistry* 39: 737–749.

changed little since. It is considered to be an ancient molecule as it is derived from an RNA metabolite and it regularly plays a key role in the active sites of enzymes. In plants, high concentrations of histidine are formed in species that hyperaccumulate metals. It is generally assumed that the pathway follows that known for some considerable time in bacteria and which starts with 5-ribosyl-α-1-pyrophosphate and is regulated by feedback inhibition of the first enzyme, ATP phosphoribosyltransferase. The pathway aroused considerable interest within agrochemical companies in the 1990s, when it became known that potent herbicidal compounds were found to inhibit ATP phosphoribosyltransferase, imidazole glycerol phosphate dehydratase, and histidinol dehydrogenase.

Histidinol dehydrogenase from cabbage (*Brassica oleracea* var. *capitata*) and imidazole glycerol phosphate dehydratase from wheat germ (*Triticum aestivum*) have been purified to apparent homogeneity and the corresponding complementary DNAs (cDNAs) cloned. In addition, more recently, cDNA clones have been obtained for ATP phosphoribosyltransferase, the bifunctional phosphoribosyl -ATP pyrophosphohydrolase/phosphoribosyl-AMP cyclohydrolase, *N*′-[(5′-phosphoribosyl)-formimino]-5-aminoimidazole-4-carboxamide ribonucleotide isomerase, and the bifunctional glutamine amidotransferase/cyclase and histidinolphosphate aminotransferase.

See also: **Genetic Modification, Applications**: Cell Factories. **Genetic Modification of Secondary Metabolism**: Alkaloids; Wood Quality. **Nutrition**: Mineral Uptake; Nitrogen Fixation. **Regulators of Growth**: Polyamines. **Secondary Products**: Lignin. **Water Relations of Plants**: Salt Stress. **Weeds**: Herbicide Resistance.

Further Reading

Azevedo RA, Arruda P, Turner WL, and Lea PJ (1997) The biosynthesis and metabolism of the aspartate derived amino acids in higher plants. *Phytochemistry* 46: 395–419.

Delauney AJ and Verma DPS (1993) Proline biosynthesis and osmoregulation in plants. *Plant Journal* 4: 215–223.

Douce R, Bourguignon J, Neuburger M, and Rébeillé F (2001) The glycine decarboxylase system: a fascinating complex. *Trends in Plant Science* 6: 167–176.

Falco SC, Guida T, Locke M, *et al.* (1995) Transgenic canola and soybean seeds with increased lysine. *Biotechnology* 13: 577–582.

Galili G, Tang G, Zhu X, and Gakiere B (2001) Lysine catabolism: a stress and development superregulated metabolic pathway. *Current Opinion in Plant Biology* 4: 261–266.

Lea PJ and Miflin BJ (2003) Glutamate synthase and the synthesis of glutamate in plants. *Plant Physiology and Biochemistry* 41 (in press).

Lea PJ and Morot-Gaudry J-F (2001) *Plant Nitrogen*. Berlin: Springer-Verlag.

Ramón-Maiques S, Marina S, Gil-Ortiz F, Fita I, and Rubio V (2002) Structure of acetylglutamate kinase, a key enzyme for arginine biosynthesis and a prototype for the amino acid kinase enzyme family, during catalysis. *Structure* 10: 329–342.
Saito K (2000) Regulation of sulfate transport and synthesis of sulfur-containing amino acids. *Current Opinion in Plant Biology* 3: 188–195.
Schmid J and Amrhein N (1995) Molecular organization of the shikimate pathway in higher plants. *Phytochemistry* 39: 737–749.
Singh BK (1999) *Plant Amino Acids: Biochemistry and Biotechnology.* New York: Marcel Dekker.
Singh BK and Shaner DL (1995) Biosynthesis of the branch chain amino acids: from test tube to the field. *Plant Cell* 7: 935–944.

Oils

A J Kinney, DuPont Experimental Station, Wilmington, DE, USA

Introduction

Oils are an important form of carbon storage in many plants and can contribute up to 80% of the total dry matter of a seed. Oils consist mainly of triacylglycerols (TAGs, **Figure 1**), which accumulate during the maturation phase of the seed in the embryo, the endosperm, or in both places. When the seed germinates the oil is used as an energy source.

The handful of common oilseed plants grown as crops, such as soybean (*Glycine max*), canola (oilseed rape; *Brassica napus*), cotton (*Gossypium* spp.), sunflower (*Helianthus* spp.), flax (*Linum usitatissimum*), and corn (*Zea mays*; maize) contain TAGs with the same acyl groups that are found in membrane lipids (**Figures 2** and **3**). These are mainly palmitate (16:0, that is, 16 carbons with no double bonds), stearate (18:0), oleate (18:1), linoleate (18:2), and linolenate (18:3). In contrast, many nondomesticated oilseeds contain substantial amounts of unusual acyl chains in their seed storage lipids (**Table 1**). More than 400 of these unusual fatty acids have been identified, including medium (C8–14) and very long (C20–24) chain saturated fatty acids, acyl chains containing hydoxylated, epoxidated, acetylenated and methylated functional groups, as well as fatty acids with conjugated double bonds (**Figure 4**).

TAGs are made from an esterification reaction, catalyzed by an enzyme called diacylglycerol acyltransferase (DGAT), of a third acyl group with a diacylglycerol (DAG). DAGs are also the substrate for membrane lipid synthesis, and thus all of the enzymatic reactions involved in oil biosynthesis, with the exception of DGAT, are analogous to those involved in membrane synthesis (*see* **Genetic Modification of Primary Metabolism**: Acyl Lipids). However, the individual enzymes catalyzing these reactions are often diverged, either in their substrate specificity or in their product, from the equivalent enzymes involved in membrane lipid biosynthesis.

Over 80 million metric tonnes of plant oils are produced annually for commercial purposes. The usefulness of a given plant oil depends upon its acyl composition. For example, the number and positions of double bonds determine melting range and heat stability of the oil, whereas a particular functional group might make a modified fatty acid suitable for industrial purposes.

In many cases, it has been possible to transfer the genes encoding diverged fatty acid biosynthetic enzymes from nondomesticated plants to commercial oilseed crops in order to produce industrially useful or nutritionally enhanced vegetable oils. It has also been possible to manipulate the endogenous lipid gene expression pattern in oilseed crops to change

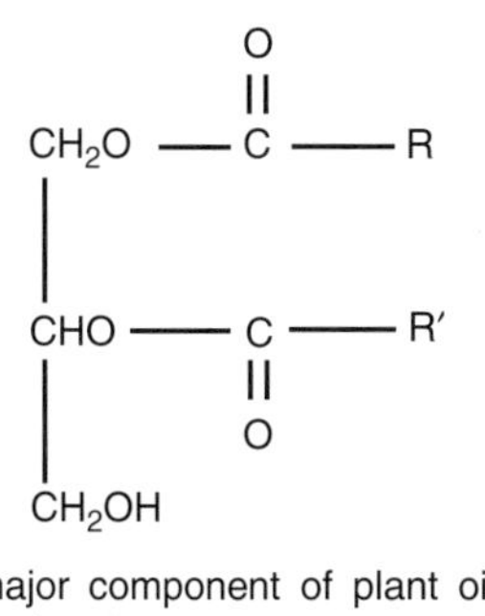

Figure 1 The major component of plant oil is triacylglycerol (TAG). The R and R′ residues are fatty acyl groups, examples of which are shown in **Figures 2–4**.

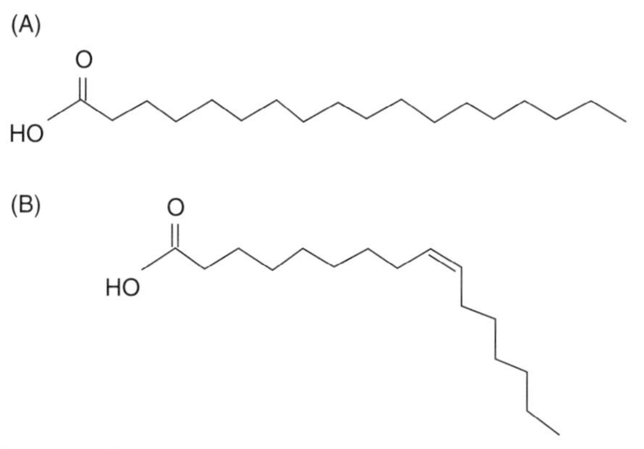

Figure 2 Structures of some of the acyl groups found in plant triacylglycerols. Examples of (A) a saturated fatty acid (stearic acid) and (B) a monounsaturated fatty acid (oleic acid).

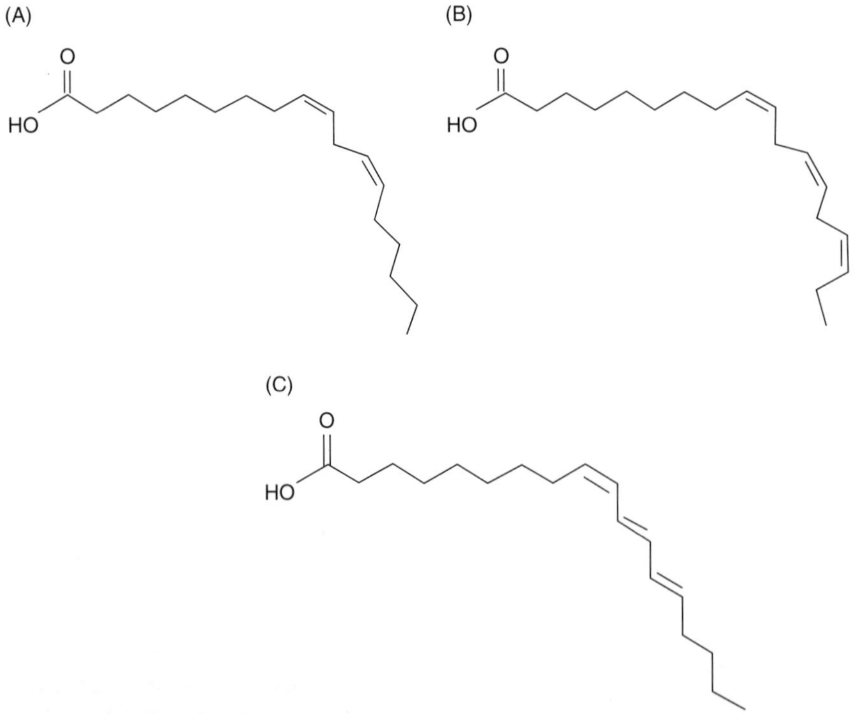

Figure 3 Structures of some of the acyl groups found in plant triacylglycerols. Examples of polyunsaturated fatty acids: (A) linoleic acid, (B) linolenic acid, and (C) eleostearic acid, a "conjugated" fatty acid.

Table 1 Some examples of unusual fatty acids found in the seed oils of various plant species (in most of these examples, the unusual fatty acid represents 50% or more of the total fatty acid content)

Plant species	*Unusual fatty acid in oil*	*Fatty acid type*
Cuphea hookeriana	Caprylic	Medium chain (C8)
Cuphea lanceolata	Capric	Medium chain (C10)
Ulmus americana	Capric	Medium chain (C10)
Cuphea wrightii	Lauric	Medium chain (C12)
Cuphea palustris	Myristic	Medium chain (C14)
Daucus carota	Petroselenic	Δ^6-Monounsaturated (C18)
Limnanthes alba	Eicosenoic	Δ^5-Monounsaturated (C20)
Oenothera biennis	γ-Linolenic	Δ^6-Polyunsaturated (C18)
Vernonia galemensis	Vernolic	12-Epoxy (C18)
Euphorbia lagascae	Vernolic	12-Epoxy (C18)
Ricinus communis	Ricinoleic	12-Hydroxy (C18)
Lesquerella fendleri	Lesquerolic	14-Hydroxy (C20)
Crepis alpina	Crepenynic	12-Acetylenic (C18)
Licania rigida	Licanic	4-Keto (C18)
Momordica charantia	Eleostearic	12-Conjugated (C18:3)
Impatiens balsamica	Parinaric	12-Conjugated (C18:4)
Calendula officianalis	Calendic	9-Conjugated (C18)
Sterculia foetida	Sterculic	9-Cyclopropene (C19)
Dichapetalum toxicarum	Fluorooleic	ω-Fluoro

the ratios of the existing fatty acids, and produce oils with enhanced functionality or nutritional value.

The Synthesis of Saturated and Monounsaturated Acyl Groups in the Plastid

Saturated acyl chains, up to C18, are synthesized in the plastids of oilseed cells by stepwise condensation of C2 units from malonyl-ACP to a growing acyl-ACP chain (**Figure** 5). Each elongation step is actually a cycle of four reactions, condensation followed by reduction, dehydration, and a second reduction. The rate of the elongation, and the substrate elongated, is controlled by the initial condensation reaction of the cycle. This condensation reaction is catalyzed by a β-ketoacyl synthase (KAS) enzyme. The first condensation of the growing chain, that of acetyl coenzyme A (acetyl CoA) with malonyl-ACP to form 3-ketobutyryl-ACP and CO_2, is catalyzed by an enzyme called KAS III. Subsequent condensations to C16 are catalyzed by a KAS I and the final C16 to C18 condensation by KAS II.

Plastidial KAS I and II enzymes are dimers of 46-kD and 50-kD polypeptides that are encoded by two closely related KAS A and KAS B genes. KAS A encodes a polypeptide with a relative mass of 46 kD, and a homodimer of this protein has KAS I activity.

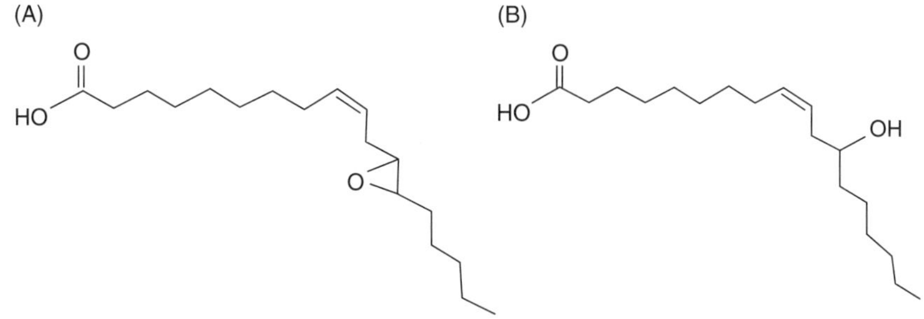

Figure 4 Structures of some of the acyl groups found in plant triacylglycerols. Examples of fatty acids with functional groups other than double bonds are (A) vernolic acid, an epoxy fatty acid, and (B) ricinoleic acid, a hydroxy fatty acid.

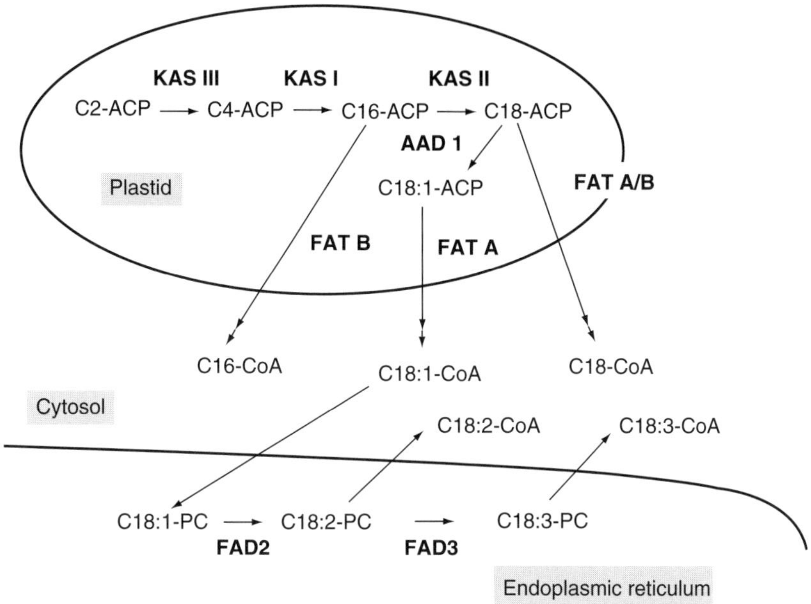

Figure 5 Key reactions generating the fatty acyl groups in common plant oils. Oil (triacylglycerol) is assembled from the acyl CoA molecules in the cytosol; thus, changing the acyl CoA pool will change the acyl composition of the oil. The key enzymes are β-ketoacylsynthases 1–3 (KAS 1–3); fatty acyl ACP thioesterases A & B (FAT A & B); acyl ACP desaturase 1 (AAD 1); fatty acid desaturases 2 & 3 (FAD2 & 3). Diverged forms of these key enzymes are responsible for the wide variation in the fatty acyl composition of less common plant oils. For example, a diverged form of FAT B leads to the accumulation of shorter chain acyl CoAs in the cytosol, whereas diverged forms of FAD2 introduce functional groups other than double bonds into the acyl chain, such as hydroxyl, epoxy, or acetylene groups. ACP, acyl carrier protein; CoA, coenzyme A; PC, phosphatidylcholine.

KAS B encodes a polypeptide with a relative mass of 50 kD, and a heterodimer of KAS A and KAS B gene products has KAS II activity.

Oils from a number of *Cuphea* species are rich in C8, C10, and C14 fatty acids. These species have a diverged form of seed KAS A, which results in a KAS I homodimer with increased specificity toward medium chain acyl-ACPs. This diverged KAS I has been referred to as KAS IV to distinguish it from the broader specificity C4–C14 KAS I.

In most plants, 18:0-ACP is desaturated in the plastid to δ-9 18:1-ACP, a reaction catalyzed by an acyl-ACP desaturase (the designation δ-9 means the double bond is nine carbons from the *delta* or carboxyl end of the fatty acid). This desaturase is encoded by a number of AAD genes (in soybean there are at least three AADs). Suppression of any one of these genes usually results in plant oils with an increased stearate content (20–30% of the total fatty acids). These modified oils may be useful as butter substitutes.

The family of soluble, plastidal, acyl-ACP desaturases (*AAD* gene family) includes a number of other members that vary from the standard δ-9 18:0-ACP desaturase in their substrate specificity, regio-specificity, or both. Examples include a δ-4 desaturase from carrot (*Daucus carota sativa*) and coriander (*Coriandrum sativum*), and a δ-6 16:0-ACP desaturase from *Thunbergia alata*, and a δ-9 16:0-ACP desaturase from *Doxantha* spp. and *Asclepias syriaca*. These species thus contain oils that have monounsaturated fatty acids with double bonds in positions other than the ninth carbon from the carboxyl group. Seeds of carrot and coriander, for example, contain oils rich in petroselenic acid (δ-6 18:1).

Expression of these novel AAD genes in other plants often results in the novel monounsaturated fatty acid showing up in the transgenic plant oil. The concentrations, however, are usually much less than those found in the species from which the gene was isolated. This is probably because additional diverged fatty acid enzymes are also needed, such as the thioesterase genes described below.

Not all unusual monounsaturated fatty acids are the result of variant AAD genes, however. For example, the oils of many *Limnanthes* species and a number of gymnosperm plants all contain 20-carbon monounsaturated fatty acids with a double bond at the δ-5 position, such as eicosa-5-enoic acid (δ-5 20:1). These fatty acids are the result of an extraplastidal, membrane-associated desaturase that inserts a double bond at the δ-5 position of 20:0-CoA. The gene encoding this enzyme is closely related to the genes encoding yeast, cyanobacterial, and mammalian δ-9 18:0-CoA desaturases.

Acyl chains are released from the plastid into the cytosol by the action of acyl-ACP thioesterases (**Figure 5**). There are two distinct gene families that encode these thioesterase enzymes: FAT A and FAT B. FAT A encoded enzymes hydrolyze predominantly 18:1-ACP with minor activities toward 18:0-ACP and 16:0-ACP. FAT B encoded enzymes usually hydrolyze saturated C14–C18 ACPs, preferentially 16:0-ACP, but they will also hydrolyze 18:1-ACP. Many plants that produce medium-chain saturated fatty acids have diverged forms of FAT B. The production of medium-chain (C8–C12) fatty acids in seeds such as *Cuphea*, described above, involves FAT B enzymes that have shorter chain length specificities than the long-chain FAT B enzymes found in all plants. Diverged FAT B genes have also been cloned from other medium-chain producing species, such as California bay laurel (*Umbellularia californica*), coconut (*Cocos nucifera*), and elm (*Ulmus* spp.). Expression of these FAT Bs in canola and other oilseeds results in substantial accumulation of medium-chain fatty acids in their oil. There is one reported example of a diverged FAT A that has high specificity toward 16:0-ACP rather than 18:1-ACP. This is found in mangosteen (*Garcinia mangostana*), which contains more than 50% stearate in its seed oil.

The Further Modification of Acyl Groups outside the Plastid

Once fatty acids leave the plastid they become esterified to coenzyme A (CoA); the acyl CoAs are the substrates for TAG biosynthesis (**Figure 5**). Before they are incorporated into oil, acyl groups may be further modified, either directly as acyl CoA or after transfer to another molecule.

Part of the 18:1-CoA pool is usually transferred to the membrane phospholipid, phosphatidylcholine (PC). While it is esterified to PC, 18:1 can be further desaturated to 18:2 and 18:3 by two specialized microsomal membrane associated desaturases, FAD2 (δ-12) and FAD3 (ω-6). The polyunsaturated acyl chains are then either re-esterified to CoA and become part of the cytosolic acyl CoA pool for TAG formation, or the PC itself is converted directly to DAG, which is then used to synthesize TAG. Both pathways to TAG appear to operate in all oilseeds. The result of either pathway is that TAG becomes enriched with 18:2 and 18:3, polyunsaturated fatty acids with two or three methylene interrupted double bonds (which are carbon–carbon double bonds separated in an acyl chain by a carbon atom containing two hydrogens). There are, however, a wide variety of oilseed plants that have TAG containing fatty acids with other types of functional groups or nonmethylene interrupted double bonds (two carbon–carbon double bonds separated in an acyl chain by a carbon–carbon single bond). Most of these unusual acyl groups are the result of the activity of diverged desaturases, members of the FAD2 gene family. These diverged desaturases include fatty acid 12-hydroxylases from *Ricinius communis* and *Lesquerella fendleri,* fatty acid 12-epoxygenases from *Vernonia galemensis* and *Crepis palaestina*, and a fatty acid 12-acetylenase from *Crepis alpina*.

The oils of these species are rich in the corresponding fatty acid product of these diverged desaturases. Some plant oils are rich in conjugated fatty acids, that is fatty acids with nonmethylene interrupted double bonds. For example, the oil from *Momordica charantia* is rich in α-eleostearic acid (18:3 $\delta^{9\text{-}cis,\ 11\text{-}trans,\ 13\text{-}trans}$), that from *Impatiens balsamica* in α-parinaric acid (18:4 $\delta^{9\text{-}cis,\ 11\text{-}trans,\ 13\text{-}trans,\ 15\text{-}cis}$), and that from *Calendula officinalis* in calendic acid (18:3 $\delta^{8\text{-}trans,\ 10\text{-}trans,\ 12\text{-}cis}$). FAD2 related conjugases, which synthesize conjugated double bonds, have been identified from all of these species. The *Calendula* conjugase is especially interesting since it represents the first example of divergence from a δ-12 regiospecificity for a FAD2-like gene product, since the $\delta^{8\text{-}trans/10\text{-}trans}$ conjugation is the result of action by the *Calendula* conjugase on the δ-9 double bond of linoleoyl-PC.

It has been suggested that small differences in active site geometry are responsible for the different functional outcomes of FAD2 desaturases and diverged FAD2 enzymes. For example, the *Lesquerella* hydroxylase enzyme is bifunctional, with some δ-12 desaturase activity in addition to hydroxylase activity, and it has been demonstrated in transgenic plants that only a few amino acids of a *FAD* encoded polypeptide control its ability to catalyze desaturation versus hydroxylation. Thus, small variations in the amino acid composition of these key enzymes have resulted in large variations in the fatty acid profile of the world's oilseeds.

Addition of functional groups are not the only possible modification of cytosolicic acyl CoAs. Many plant oils contain very long chain, unsaturated fatty acids (VLCFAs) of 20–26 carbons that are the result of the elongation of oleoyl CoA by a membrane-bound elongation complex. A long chain condensing enzyme (β-ketoacyl CoA synthase) has been purified from jojoba (*Simmondsia chinensis)* seeds, which have oil rich in C20–24 monounsaturated acyl chains, and the corresponding cDNA cloned. A related gene (*FAE*1) has been cloned from *Arabidopsis* by insertional mutagenesis.

It has become accepted, therefore, that a single condensing enzyme controls the elongation of 18-, 20-, and 22-carbon acyl CoAs in very long-chain FA producing seeds. In support of this is the observation that a single *FAE*1 homolog from *Limnanthes* catalyzes the formation of 20- and 22-carbon saturated fatty acids when expressed in soybean seeds.

There is one known angiosperm that accumulates liquid waxes instead of TAG in its seeds, the desert shrub jojoba. These liquid waxes comprise linear esters of monounsaturated C20, C22, and C24 fatty acids and fatty alcohols. Waxes are the direct result of cytosolicic 18:1-CoA elongation by a microsomal elongation complex analogous to that involved in VLCFA synthesis. The condensing enzyme involved is a member of the FAE/KCS gene family described above, and the fatty alcohol component of waxes is the result of a microsomal fatty acyl reductase.

Triacylglycerol Assembly and Oil Body Formation

Acyl CoAs are esterified to a glycerol backbone by membrane-bound acyltransferases. The initial reaction, transfer of the acyl chain from CoA to the *sn*-1 position of glycerol 3-phosphate to form lysophosphatidic acid, is catalyzed by glycerol 3-phosphate acyltransferase (GPAT). Very little is known about this enzyme. The second reaction, transfer of the acyl chain from the CoA ester to *sn*-2 to form phosphatidic acid, is catalyzed by lysophosphatidic acid acyltransferase (LPAT). Interestingly, there are two separate families of genes that encode plant LPATs, termed class A and class B. Both classes are expressed in the seeds of plants that accumulate unusual fatty acids, whereas common oil seeds, such as canola and corn, appear only to the express class A LPATs in their seeds. Thus, it is likely that class B LPATs are involved in the incorporation of unusual fatty acids into plant oils. The phosphate group of phosphatidic acid is then removed by phosphatidic acid phosphatase to form diacylglycerol (DAG). A final acyl group is then transferred to DAG in a reaction catalyzed by diacylglycerol acyltransferase (DGAT).

As for LPAT, there appear to be two separate gene families encoding DGAT enzymes. One type of DGAT is part of the larger acyl CoA:cholesterol acyltransferase gene family, but the other DGAT family is not related to any other known genes.

There is also an alternative pathway for TAG formation in plants, which involves the transfer of an acyl chain from a membrane phospholipid, phosphatidylcholine (PC), to DAG. This reaction is catalyzed by an enzyme called phospholipid:diacylglycerol acyltransferase, PDAT. The significance of this pathway in bulk oil biosynthesis is unclear. In oilseeds with exotic oils, this pathway may be involved in the incorporation of unusual fatty acids into TAG, since PDATs from species that produce unusual fatty acids have increased specificity toward the particular unusual fatty acid produced by that plant.

Oil is stored in the mature seed in the form of oil bodies. These are pools of TAG surrounded by a single monolayer membrane, produced from the budding of the outer ER membrane. The oil body membrane contains proteins known as oleosins that stabilize the oil body during desiccation of the seed.

The Genetic Modification of Plant Oils for Food and Industrial Uses

Plant breeders have performed germplasm selection and mutation breeding of oilseed crops for modified fatty acid composition for over 25 years. Mutants with variations in oleic, linoleic, and linolenic acids have been produced in all the major oilseed crops. In addition, mutants of soybean, canola, and sunflower have been produced with variations in palmitic and stearic acids. Indeed, mutation breeding has produced a number of commercially successful crops, such as low linolenic acid soybeans and high oleic acid sunflowers.

Knowledge of the pathways described above, and of the genes encoding the enzymes that catalyze them, has led to the advent of genetically modified (GM) oilseeds. By manipulating one or more of the above steps in developing seeds it has been possible to alter the relative abundance of the five major fatty acids at will. By introducing diverged versions of existing genes it has also been possible to produce oils containing unusual fatty acids in common oilseeds. Seed specific promoters are used to overexpress a gene or to downregulate an endogenous gene in a pathway, and this ensures that the growth, yield, and environmental sensitivity of the plant remain unaffected.

The first examples of these GM crops included high-laurate canola (50% C12 fatty acids) and high-oleic acid soybeans (85% oleic acid). Expressing a diverged FAT B complementary DNA (cDNA) from California bay laurel produced the high-laurate canola. The high-oleic acid soybeans were produced by suppressing expression of the endogenous soybean *FAD2-1* gene. The GM canola has found use, among other things, as a primary raw material for soap making, and the GM soybean as a highly oxidatively stable *trans* fatty acid free cooking oil.

Recently, quite exotic oils have been produced in common oilseed crops by expressing genes from nondomesticated oilseeds. For example, expression of the *Ricinius* and *Lesquerella* diverged *FAD2* cDNAs in canola resulted in oils containing 20% hydroxylated fatty acids, and expression of the *Vernonia* epoxygenase in soybean resulted in oils containing about 20% 12-epoxy fatty acids. In both these cases, there was also a concomitant increase in oleic acid, something that is often observed when diverged *FAD2*s are expressed in heterologous plants. Increases in oleic acid content, along with the appearance of conjugated fatty acids, were also observed when *FAD2* related conjugase cDNAs from *Mormordica charantia* and *I. balsamica* were expressed in soybean.

This increase in oleic acid is probably a sign that expression of a diverged enzyme in domestic crops causes some kind of imbalance in oil biosynthesis. Interestingly, increases in oil oleic acid were not observed when a *Calendula* conjugase cDNA was expressed in soybean, even though the soybeans were producing over 20% calendic acid. This is a sign that calendic acid is being metabolized normally by these soybeans. In addition, GM soybeans expressing the *Limnanthes* δ-5 desaturase and 16:0 condensing enzyme have also been produced, and these beans contained about 15% eicosenoic acid (δ-5 20:1) in their oil with no increased oleic acid.

The soybeans containing the *Calendula* and *Limnanthes* cDNAs promise to be the first of many new domestic oilseed crops producing exotic oils for a wide variety of nonfood uses.

List of Technical Nomenclature

AAD	Acyl ACP desaturase. EC 1.14.19.2 acyl-[acyl-carrier-protein] desaturase
ACP	Acyl carrier protein.
CoA	Coenzyme A.
DAG	Diacylglycerol.
DGAT	Diacylglycerol acyltransferase. EC 2.3.1.20 diacylglycerol O-acyltransferase
ER	Endoplasmic reticulum.
FAD	Fatty acid desaturase. EC 1.3.1.35 phosphatidylcholine desaturase
FAE	Fatty acid elongase (a misleading but standard name for a gene encoding a fatty acyl CoA condensing enzyme, KCS).
FAT	Fatty acid thioesterase. EC 3.1.2.21 dodecanoyl-[acyl-carrier protein] hydrolase; EC 3.1.2.22 palmitoyl-[acyl-carrier protein] hydrolase; EC 3.1.2.14 oleoyl-[acyl-carrier protein] hydrolase
GPAT	Glycerol 3-phosphate acyltransferase. EC 2.3.1.15 glycerol-3-phosphate O-acyltransferase
KAS	Ketoacyl ACP synthase. EC 2.3.1.41 3-oxoacyl-[acyl-carrier-protein] synthase
KCS	Ketoacyl CoA synthase. EC 2.3.1.16 acetyl-CoA C-acyltransferase
LPAT	Lysophosphatidic acid acyltransferase. EC 2.3.1.52 2-acylglycerol-3-phosphate O-acyltransferase
PC	Phosphatidylcholine.
PDAT	Phospholipid diacylglycerol acyltransferase. EC 2.3.1.158 phospholipid: diacylglycerol acyltransferase
sn-	Stereospecifically numbered. When a glycerol containing molecule, such as TAG, is drawn as a Fischer projection (with the secondary hydroxyl group to the left of the central carbon) the carbon atoms are designated *sn*-1, *sn*-2 and *sn*-3 counting from top to bottom.

TAG	Triacylglycerol.
VLCFA	Very long-chain fatty acid.

See also: **Genetic Modification of Primary Metabolism**: Acyl Lipids. **Secondary Products**: Epicuticular Waxes; Wax Pathways. **Seed Development**: Oil Deposition.

Further Reading

Badami RC and Patel KB (1981) Structure and occurrence of unusual fatty acids in minor seed oils. *Progress in Lipid Research* 19: 119–153.

Frentzen M (1998) Acyltransferases: from basic science to modified seed oils. *Fett/Lipid* 100: 161–166.

Gunstone FD, Harwood JL, and Padley FB (1994) *The Lipid Handbook*. London: Chapman & Hall.

Harwood JL (1996) Recent advances in the biosynthesis of plant fatty acids. *Biochimica et Biophysica Acta* 130: 7–56.

Huang A (1996) Oleosins and oil bodies in seeds and other organs. *Plant Physiology* 110: 1055–1061.

Kinney AJ (1997) Genetic engineering of oilseeds for desired traits. In: Setlow JK (ed.) *Genetic Engineering*., vol. 19, pp. 149–166. New York: Plenum Press.

Mekhedov S, Ilárduya OM, and Ohlrogge J (2000) Toward a functional catalog of the plant genome: a survey of genes for lipid biosynthesis. *Plant Physiology* 122: 389–401.

Moore TS Jr. (1993) *Lipid Metabolism in Plants*. Boca Raton: CRC Press.

Napier J, Sayanova O, Sperling P, and Heinz E (1999) A growing family of cytochrome b_5-domain fusion proteins. *Trends in Plant Science* 4: 2–4.

Shanklin J and Cahoon EB (1998) Desaturation and related modifications of fatty acids. *Annual Review of Plant Physiology and Plant Molecular Biology* 49: 611–641.

Stauffer CE (1996) *Fats and Oils*. Minnesota: Eagan Press.

Voelker TA (1996) Plant acyl-ACP thioesterases: chain-length determining enzymes in plant fatty acid biosynthesis. In: Setlow JK (ed.) *Genetic Engineering*, vol. 18, pp. 111–133. New York: Plenum Press.

Voelker TA and Kinney AJ (2001) Variations in the biosynthesis of seed storage lipids. *Annual Review of Plant Physiology and Plant Molecular Biology* 52: 335–361.

Proteins

P R Shewry, Rothamsted Research, Harpenden, UK

Introduction

Analysis of the recently completed genome sequences of rice (*Oryza sativa*) and *Arabidopsis* (a "model" plant species related to cultivated crucifers) shows the presence of about 25 500 and 50 000 genes, respectively, each encoding a protein. However, this number must be treated as a minimum, as both species were selected on the basis of their diploid chromosome constitution and small genome size, and genes encoding small proteins are not identified by the computer programs currently used to analyze the sequences. Many cultivated species have vastly bigger genomes, which may be associated with polyploidy. For example, the genome of hexaploid bread wheat (*Triticum aestivum*) comprises about 16×10^{12} base pairs of DNA compared with 430×10^6 base pairs for rice. Although much of the additional DNA in wheat is not transcribed (i.e., noncoding), the number of expressed genes, and hence proteins, is certainly greater than in rice.

Many of the proteins predicted from genome sequences have never been identified by direct biochemical studies and many may well remain undetected. This is because they are present in low amounts and/or transiently in specific organs or tissues. This is illustrated by studies of wheat grain development, in which about 4500–8000 expressed genes but less than 2000 individual proteins have been identified.

This article will not attempt to catalog the whole range of plant proteins but will focus on the quantitatively major types of proteins present in seeds and other harvested plant organs and their role in determining nutritional, quality, and processing properties. First, an historical background to plant protein analysis and classification is briefly provided.

Classification of Plant Proteins

Plant proteins were among the earliest proteins to be studied, with the gluten proteins of wheat being described as early as 1745 by Jacopo Beccari of the University of Bologna, Italy. Subsequent work led to the concept of classifying plant proteins on the basis of their solubility (or extractability from plant tissues) in a series of solvents, a concept formalized by Thomas Burr Osborne working at the Connecticut Agricultural Research Station (USA) between 1886 and 1928. Osborne studied the proteins of over 30 types of plant seed and divided them into four groups based on their solubility in water (albumins), dilute saline (globulins), alcohol/water mixtures (prolamins), and dilute alkali (glutelins). The terms albumin and globulin remain widely used today for proteins of plant and animal origin, while the term prolamin is applied only to a specialized type of protein in cereal seeds. However, glutelins are no

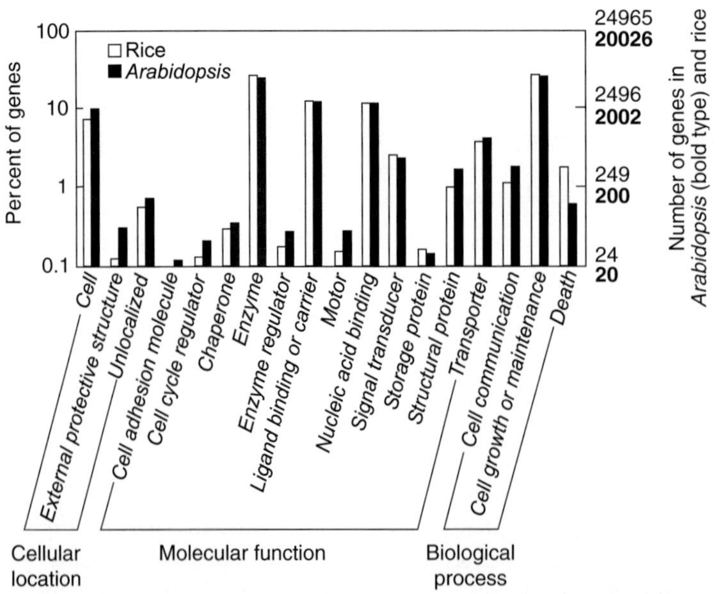

Figure 1 Functional classification of proteins encoded by genes predicted in rice and *Arabidopsis thaliana*. In this classification, "biological process," "cellular location," and "molecular function" are treated as independent attributes. Only 36.3% of the 25 426 predicted genes for *A. thaliana* are classified. For rice, only 20.4% of the 53 398 complete predictions, with both initial and terminal exons, could be classified. Reproduced with permission from Yu J, Hu SN, Wang J, *et al.* (2002) A draft sequence of the rice genome (*Oryza sativa* L. ssp *indica*) *Science* 296: 79–92.

longer regarded as a valid group and the term has rightly fallen into disuse.

The classification into "Osborne fractions" is still widely used by cereal chemists but most biochemists now prefer to classify proteins on other criteria, notably function and structural and evolutionary relationships. Thus, seed proteins are frequently divided into those with structural, metabolic, storage, and protective roles. Structural and metabolic proteins include enzymes and proteins present in cell structures (membranes, cell walls, etc.), while protective proteins include components with diverse properties (e.g., enzyme inhibition, chitin binding, surface activity), which together contribute to broad spectrum resistance to microbial pathogens (fungi, oomycetes, bacteria) and invertebrate pests. However, these groups of proteins have limited impact on the utilization of the seed when compared with the storage proteins, which account for half or more of the total seed proteins. The major, and often only, role of these proteins is to provide a store of nutrients for seed germination and seedling growth and their properties determine both the nutritional quality and processing properties.

A more detailed classification of the proteins encoded by the rice and *Arabidopsis* genomes has been attempted, by ascribing putative functions to gene products based on relationships to characterized proteins (mostly from other organisms). This allowed about 36% and 20% of the *Arabidopsis* and rice gene products, respectively, to be assigned to classes as shown in **Figure 1**. It is notable that several major classes comprise components that can be expected to be present in low amounts or transiently, for example, proteins involved in signal transduction and cell communication.

The classification of proteins into families has only been possible over the past two decades due to an explosion in the number of available sequences and, to a lesser extent, three-dimensional structures. This has allowed the construction of databases with free public access, such as SCOP for protein 3D structures and Pfam for plant protein families defined on amino acid sequence relationships. For example, Pfam currently divides some plant protein sequences into over 3000 families, which vary greatly in size.

Seed Storage Proteins

Seed storage proteins form the most widely studied plant proteins due to their abundance and impact on seed utilization. They show immense diversity in structure and properties among and within species. Nevertheless, the vast majority fall into only four groups defined initially on the basis of their solubility properties and sedimentation coefficients ($s_{20,w}$ values, a measurement of molecular mass determined by sedimentation ultracentrifugation). The most widely distributed seed proteins are globulins, which form the major storage protein fraction in dicotyledonous plants and also some cereals (oats, *Avena sativa*, and rice). Two types of storage globulin can be recognized, both of which were initially characterized in detail from legume seeds. These are the 11S "legumins" and 7S "vicilins." However, in reality, the sedimentation coefficients vary among species, from about 7–8S for vicilins and 11–13S for legumins. The differences in mass reflect the fact that 7S globulins are essentially trimeric proteins of molecular mass of about 150 000–190 000 and the 11S globulins hexamers of mass of about 300 000–450 000. The 7S globulin subunits typically have molecular masses of about 50 000 but posttranslational proteolysis and glycosylation can give rise to subunit chains with a wide range of masses. The 11S globulin subunits are always posttranslationally processed to give acidic (typical mass about 40 000) and basic (typical mass about 20 000) chains, which remain linked by a single disulfide bond. Glycosylation occurs only rarely.

Although the 7S and 11S globulin subunits have little or no similarity in their amino acid sequences, they do adopt almost identical three-dimensional structures and both are considered to belong to a larger superfamily of plant proteins called cupins (the latin name for a small cask, reflecting the fact that all members are based on a core β-barrell structure).

The similar structures of 7S and 11S globulins may facilitate their packing together in the same protein bodies, as both types of protein are stored in some species (see **Table 1**). However, in other species, either one or both types of globulin are stored with either prolamins (cereals) or 2S albumins.

Prolamins and 2S albumins also belong to a single superfamily of proteins, with most members containing a conserved motif of cysteine residues present in a tightly folded α-helical domain. However, the 2S albumins and prolamins have little other similarity. The 2S albumins occur widely in dicotyledonous seeds but have not so far been characterized from monocotyledonous plants, although their presence in fern spores demonstrates an origin predating that of monocotyledonous and dicotyledonous plants. They are typically synthesized as a preproproteins that is posttranslationally processed to give chains of about 8000 and 4000, which are linked by two disulfide bonds. In contrast, prolamins vary widely in mass (from about 10 000 to 100 000) and structure, with most containing repeated sequences (i.e., short sequences of amino acids that are repeated to form tandem or interspersed arrays). They may also form polymers with molecular masses exceeding 1 million. Both 2S albumin and prolamin fractions may contain forms that are rich in methionine, notably the 2S albumins of Brazil nut (*Bertholletia excelsa*) and

Table 1 Types of major storage proteins present in seed crops and their trivial names

Species	*Protein type*	*Trivial name*
Cereals		
Wheat	Prolamin	Gliadin/glutenin
Barley	Prolamin	Hordein
Rye	Prolamin	Secalin
Corn	Prolamin	Zein
Sorghum	Prolamin	Kafirin
Oats	Prolamin	Avenin
	11S globulin	–
Rice	Prolamin	–
	11S globulin	Glutelin
Legumes		
Soybean	7S globulin	β-Conglycinin
	11S globulin	Glycinin
Broad bean (*Vicia faba*)/pea (*Pisum sativum*)	7S globulin	Vicilin
	11S globulin	Legumin
French bean (*Phaseolus vulgaris*)	7S globulin	Phaseolin
Peanut	7S globulin	Conarachin
	11S globulin	Arachin
	2S albumin	–
Oilseeds		
Cottonseed	11S globulin	Gossypin
	7S globulin	Congossypin
	2S albumin	–
Sunflower	11S globulin	Helianthinin
	2S albumin	–
Oilseed rape (*Brassica napus*; canola)	11S globulin	Cruciferin
	2S albumin	Napin

sunflower (*Helianthus annuus*) and the β-zeins and δ-zeins of corn (*Zea mays*; maize).

The characteristics of albumin, globulin, and prolamin storage proteins are summarized in **Table 2**.

Tuber Proteins

Plant tubers are the second most important harvested organs after seeds, with five crops accounting for most of the world production. These are cassava (*Manihot esculenta*), potato (*Solanum tuberosum*), sweet potato (*Ipomoea batatas*), yams (*Discorea* spp.) and taro (see below), the latter two crops comprising several related species. These tubers all contain high amounts of starch with relatively low contents (varying from about 1–10% dry weight) of protein. Nevertheless, all except cassava contain specialized types of storage proteins. However, in contrast to seeds, these proteins do not fall into clearly defined families. The stem-derived tubers of potato contain glycoproteins of molecular mass of about 40 000, which account for about 40% of the total soluble protein. These proteins, called patatin,

Table 2 Summary of the characteristics of albumin, globulin and prolamin storage proteins

2S Albumins
Soluble in water
Molecular mass (M_r) typically ~ 10 000–15 000.
Processed posttranslationally to give large and small subunits
Two intrachain bonds within the large subunit and two interchain disulfide bonds
Not glycosylated
Specific components are rich in methionine
7S Globulins
Soluble in dilute salt solutions
Typically trimeric proteins of M_r 150 000–190 000.
Subunit M_r varies from ~ 40 000–80 000 but is typically ~ 50 000
Subunits may be proteolytically processed and glycosylated
Contain little or no cysteine and methionine
11S Globulins
Soluble in dilute salt solutions
Typically hexameric proteins of M_r 300 000–450 000. Subunits typically of M_r ~ 60 000 are posttranslationally processed to give M_r 40 000 (acidic) and M_r 20 000 (basic) chains associated by one interchain disulfide bond
Low in cysteine and methionine
Rarely glycosylated
Prolamins
Vary widely in structure, with subunit M_r ranging from ~ 10 000–100 000
Include monomeric forms and high M_r polymers stabilized by interchain disulfide bonds
Presence of repeated sequences and regions (domains) rich in specific amino acids results in unusual amino acid compositions.
Rich in proline and glutamine, poor in lysine and, in some cases, tryptophan, threonine, and methionine
No glycosylation or proteolytic processing
Soluble in alcohol/water mixtures when native and/or reduced

Reproduced with permission from Shewry PR (2000) Seed proteins. In: Black M and Bewley JD (eds) *Seed Technology and its Biological Basis*, pp. 42–84. Sheffield: Blackwell Science Ltd and Sheffield Academic Press.

exhibit hydrolase activity on a range of substrates, including acyl lipids (phospholipids, glycolipids, mono- and diacylglycerols) and β-1,3-glucans. The galactolipase activity may also play a role in inhibiting the growth of insect pests.

The root tubers of sweet potato also contain a single major storage protein fraction, sporamin, accounting for about 80% of the total protein. Sporamin comprises two major groups of proteins with similar masses (about 25 000) and over 80% sequence identity. They also show high sequence homology with members of the Kunitz family of proteinase inhibitors and inhibit trypsin *in vitro*, and have been reported to have antioxidant and radical scavenging activity.

Yams comprise several species in which the tuber appears to be derived from a swollen hypocotyl. They contain a storage protein fraction, termed dioscorin, which accounts for over 80% of the total tuber protein. Two closely related groups of proteins with molecular masses of about 28 000–29 000 are present. They are homologous to α-carbonic anhydrases from various species and also exhibit carbonic anhydrase activity. Again, they have been reported to inhibit trypsin, although only weakly, and to act as antioxidants and radical scavengers.

Taro comprises several related species of aroid, with the corm or underground stem being consumed. The most widely consumed species, *Colocasia esculenta*, produces edible corms that accumulate two types of storage protein. Trypsin inhibitors (molecular mass approx. 22 000 and 24 000) of the Kunitz family comprise about 40% of the total soluble proteins, with a further 40% being a second protein called tarin. Tarin again comprises several isoforms with molecular masses of about 12 500 which may form a dimer or tetramer and is a mannose-binding lectin.

It is clear, therefore, that tuber storage proteins differ in their structural and evolutionary relationships. Furthermore, they have biological activities that suggest they contribute to defense mechanisms to pests and pathogens or to resistance to abiotic stresses. Despite their low levels, they can make a significant contribution to diet in less developed countries.

Proteins and Seed Utilization

Nutritional Quality

Animals require a source of dietary protein to enable them to grow and replace proteins lost by breakdown. This is because they are only able to synthesize about half of the 20 amino acids commonly found in proteins; the remaining amino acids are required in the diet and thus termed "essential." If only one of these amino acids is limiting in the diet the other amino acids will be broken down, leading to poor feed conversion and nitrogen excretion. The only exception to this rule is ruminants, as the microflora present in the rumen can synthesize all 20 amino acids, providing the animal with a balanced diet. The amino acids essential for human nutrition are cysteine, methionine, tyrosine, phenylalanine, lysine, leucine, isoleucine, threonine, tryptophan, valine, and histidine (for children not adults). Of these, cysteine and tyrosine are not truly essential but can

Table 3 The essential amino acid compositions of selected cereals, legumes, and oilseeds in relation to the World Health Organization (WHO) Recommended Levels

	Legumes		*Cereals*		*Oilseeds*		*WHO*
	Broad bean	*Soybean*	*Wheat*	*Corn*	*Sunflower*	*Oilseed rape*	
Cysteine	0.8	1.3	2.6	3.1	1.5	2.9 }	3.5
Methionine	0.7	1.3	1.3	2.0	1.9	2.1 }	
Lysine	6.5	6.4	2.0	3.5	3.6	5.9	5.5
Isoleucine	4.0	4.5	3.6	3.6	4.3	4.7	4.0
Leucine	7.1	7.8	6.7	11.6	6.4	7.3	7.0
Phenylalanine	4.3	4.9	5.1	4.9	4.4	4.1 }	6.0
Tyrosine	3.2	3.1	2.6	2.3	1.9	3.0 }	
Threonine	3.4	3.9	2.7	3.9	3.7	5.0	4.0
Tryptophan	nd	1.3	1.1	0.9	1.4	1.2	1.0
Valine	4.4	4.8	3.7	4.9	5.1	6.0	5.0
Histidine	2.4	2.5	2.2	3.2	2.3	3.6	

Values are from a range of sources but all are expressed on similar wt% age basis.

be synthesized from methionine and phenylalanine, respectively. Hence, combined values for these pairs of amino acids are usually given.

Many seeds are deficient in one or more of the essential amino acids, which largely results from their low levels in the major groups of storage proteins that are present (**Table 3**). For example, the 7S and 11S globulins are low in sulfur-containing amino acids (cysteine and methionine), resulting in legume seeds being deficient in these amino acids. Similarly, the low levels of lysine and threonine in prolamins result in deficiency in cereal grains, with tryptophan also being deficient in corn. These deficiencies can be overcome by mixing seeds of different types (notably cereal and legume seeds) or by fortifying with amino acids produced by fermentation. An alternative approach is to improve the amino acid composition of the seed, either by exploiting natural variation, inducing further variation by mutagenesis, or genetic engineering.

The first two approaches led in the 1960s and 1970s to the identification of a number of spontaneous and induced high-lysine mutants of cereals, initially in corn and subsequently in barley (*Hordeum vulgare*) and sorghum (*Sorghum bicolor*). The best known of these are the *opaque-2* and *floury-2* mutants of corn, both of which were identified in the mid 1960s by visual examination of corn germplasm collections. Similarly, the barley mutant Hiproly, which was identified in germplasm from Ethiopia, has also been studied in some detail.

With the exception of Hiproly, all high-lysine mutants have decreased proportions of lysine-poor prolamins and compensatory increases in other, more lysine-rich, protein fractions. Hiproly is an exception in that the high-lysine phenotype results from specific increases in several lysine-rich proteins, most notably, two inhibitors of chymotrypsin called CI-1 (9.5 g % lysine) and CI-2 (11.5 g % lysine). Despite their ability to inhibit chymotrypsin and some other serine proteinases (notably subtilisin) these proteins do not appear to have antinutritional properties. Engineered forms of CI-2 with even higher levels of nutritionally essential amino acids are currently being used to improve the nutritional quality of transgenic cereals.

The low contents of sulfur-containing amino acids in legume seeds have led to attempts to improve the nutritional quality by expression of sulfur-rich proteins from other species. Attention has focused on methionine-rich 2S albumins from brazil nut and sunflower. It is also possible to increase the content of nutritionally essential amino acids by increasing the pool sizes of free amino acids, by uncoupling the feedback mechanisms that normally regulate their synthesis. In fact, it may be necessary to combine the expression of nutritionally enhanced storage proteins with modified amino acid biosynthesis in order to provide adequate levels of essential amino acids for protein synthesis.

Processing Quality

Seed proteins do not only contribute to satisfying the nutritional requirements of humankind, but they also confer functional properties that are crucial for food processing. Two crops are particularly important in this respect, wheat and soybean (*Glycine max*).

Wheat is one of the three major cereals that dominate world agricultural production, the other two being rice and corn. However, wheat is unique in that a high proportion of the total world production is consumed by humans in the form of processed foods, such as breads, pasta and noodles, cakes and biscuits, etc. The ability to make these products is determined by the prolamin storage proteins of the grain. When flour is mixed with water to form dough these

Table 4 Functional properties of soybean seed proteins of importance in food applications

General property	*Specific functional attribute*
Organoleptic	Color, flavor, odor
Kinesthetic	Texture, mouthfeel, smoothness, grittiness, turbidity, chewiness
Hydration	Wettability, water absorption, water-holding capacity, swelling, solubility, thickening, gelling, syneresis
Surface	Emulsification, foaming (aeration, whipping), protein-lipid film formation, lipid-binding, flavor-binding
Structural and rheological	Viscosity, elasticity, adhesiveness, cohesiveness, stickiness, dough formation, aggregation, gelation, network formation, extrudability, texturizability, fiber formation
Other	Compatibility with other food components, enzymatic activity, antioxidant properties

Reproduced with permission from Kinsella JE (1979) Functional properties of soy proteins. *Journal of the American Oil Chemists Society* 56: 242–258 © 1979 American Association for the Advancement of Science.

proteins come together to form a continuous viscoelastic network that entraps carbon dioxide released during fermentation to give leavened bread, and also confers cohesive properties to other food products. Highly elastic (strong) doughs are required to make bread and there is considerable interest in increasing the dough strength of higher yielding but low-quality wheats to allow their use for breadmaking.

The gluten proteins comprise a mixture of monomeric proteins (gliadins) and subunits that are assembled into high molecular mass (above 1 million) polymers stabilized by interchain disulfide bonds. Dough strength is determined by the amount, size, and properties of these polymers, which are in turn largely dependent on one group of glutenin proteins called the high molecular weight (HMW) subunits. Allelic variation in the composition of the HMW subunits is associated with differences in dough strength and, consequently, plant breeders routinely select for "quality associated" subunits. In addition, genetic engineering has been used to add additional genes for HMW subunits in order to increase their total amount and change their composition. This has resulted in increases in strength but has also led to the development of new types of wheat with "ultra-strong" properties.

Soybean proteins are highly pervasive in the food industry, being used to confer a range of properties including texture, water holding, emulsification, foaming, viscosity, and gelation (**Table 4**). As with wheat, these properties are determined by the storage proteins, the 11S glycinin and 7S β-conglycinin globulins. Individual 11S and 7S globulin subunits differ in their functional properties and it is possible to improve the gelation properties (to form harder gels) and emulsification activity by varying the proportions of different subunits. It is also possible to improve the properties of individual subunits by deleting short sequences from the N- and C-terminal domains or inserting several adjacent methionine residues into specific variable regions. The latter also improves the nutritional quality for monogastric animals.

Plant Food Proteins as Allergens

Allergenic reactions may occur in response to plant proteins present in pollen and foods, being characterized by the generation of IgE antibodies. Seven foods are responsible for the majority of food allergies, four of which derive from plants: cereals, peanut, soybean, and tree nuts. We do not currently understand why some proteins are allergenic while closely related proteins are not, but this topic is of considerable interest for two reasons. Firstly, there is a perception and also scientific evidence that the incidence of food allergy in industrialized countries is increasing. Secondly, there is concern that the introduction of novel foods, and particularly transgenic crops, into the diet will result in the inadvertent introduction of new allergens. Consequently, several decision trees for prediction of allergenicity have been developed, based on a combination of approaches (**Figure 2**).

The first approach is to determine whether the novel protein belongs to a family that contains other characterized allergens. This approach is limited by our current knowledge of protein allergens but it is already clear that the major ones fall into a relatively small number of protein families. The most important of these is the cereal prolamin superfamily, which contains several groups of allergenic proteins: nonspecific lipid transfer proteins (e.g., in peach (*Prunus persica*), apple (*Malus pumila*), corn, barley), 2S albumins (in oriental and yellow mustards, sesame (*Sesamum indicum*), Brazil nut, castor bean (*Ricinus communis*) cottonseed (*Gossypium hirsutum*), peanut (*Arachis hypogaea*)), inhibitors of α-amylase and trypsin (dietary allergens in wheat and rice, respiratory allergens in barley and wheat), and prolamins (in wheat). The second major family is the cupins, with 7S and 11S globulins being major allergens in soybean, peanut, and walnut (*Juglans regia*). Other plant food allergens are thiol proteinases (in kiwi fruit (*Actinidia chinensis*), fig (*Fiscus carica*), papaya (*Carica papaya*) soybean, pineapple (*Ananas comosus*))

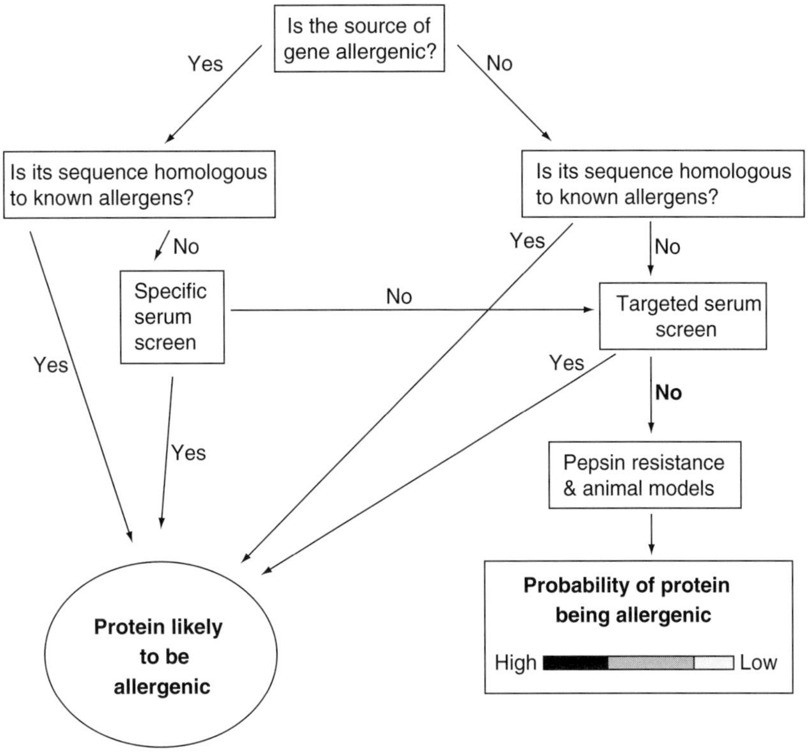

Figure 2 Modified decision tree for assessment of the allergenic potential of foods derived from biotechnology based on that developed by the FAO/WHO Expert Consultation. The "specific serum screen" uses sera known to be allergic to the source of the novel protein. The "targeted serum screen" is against sera from patients known to be allergenic to related plants. For example, a recombinant cereal protein should be tested against sera from patients allergic to grasses and other monocotyledonous species. Reproduced with permission from Mills ENC, Madsen C, Shewry PR, and Wichers HJ (2001) Plant food protein allergens – the role of structure and function in allergenic potential. *Food Allergy and Intolerance* 2: 194–209.

or are related to protective proteins (notably in rubber latex) or pollen proteins (profilins and Bet v 1 homologs).

In the second approach, the novel protein can be subjected to screens against sera from patients with well-characterized allergies, selected based on homology with characterized allergens and the species of origin.

Finally, it may be possible to predict allergenicity based on the characteristics of the protein itself. The most important common characteristic identified so far is high stability, particularly to thermal inactivation during food preparation and peptic digestion in the stomach. These predictive methods are not foolproof but should contribute to a precautionary approach to the introduction of novel foods and products.

The identification of allergic proteins and motifs (epitopes) may also facilitate the development of hypoallergenic foods, either by screening germplasm collections for natural variation in protein composition, mutagenesis, or genetic engineering. The latter could include downregulation of allergen gene expression or the expression of enzymes to modify allergen structure.

List of Technical Nomenclature

Albumins	Proteins that are soluble in water.
Essential amino acids	Amino acids that cannot be synthesized by animals and must be provided in the diet.
β-Conglycinin	7S globulin storage protein of soybean.
Cupins	protein superfamily characterized by a core β-barrel structure.
Dioscorin	storage protein of yam tubers.
Gliadins	Monomeric gluten proteins of wheat.
Globulins	Proteins that are soluble in dilute saline (0.5–1.0 mol l^{-1} NaCl).
Gluten	Wheat storage proteins (prolamins) that form a viscoelastic network in dough.
Glutenins	Polymeric gluten proteins of wheat.
Glycinin	An 11S globulin storage protein of soybean.
Glycosylation	The addition of sugar molecules to proteins, by attachment to serine, threonine, or asparagine residues.

Patatin	storage protein of potato tubers.
Prolamins	Proteins that are soluble in alcohol/water mixtures (e.g., 50% (v/v) propan-l-ol).
$s_{20,w}$	Sedimentation coefficient determined by sedimentation ultracentrifugation.
Sporamin	storage protein of sweet potato tubers.
Tarin	storage protein of taro tubers.
Zein	prolamin storage protein of corn.

See also: **Genetic Modification of Primary Metabolism**: Proteins; **Seed Development**: Storage Protein Deposition.

Further Reading

Bewley JD and Black M (eds) (1994) *Seeds, Physiology of Development and Germination*, 2nd edn. New York and London: Plenum Press.

Black M and Bewley JD (eds) (2000) *Seed Technology and its Biological Basis*. Sheffield: Blackwell Science Ltd and Sheffield Academic Press.

Mills ENC and Shewry PR (eds) (2003) *Plant Food Protein Allergens*. Oxford: Blackwell Scientific.

Owen MRL and Pen J (eds) (1996) *Transgenic Plants: A Production System for Industrial and Pharmaceutical Proteins*. Chichester: John Wiley and Sons.

Shewry PR (1995) Plant storage proteins. *Biological Reviews* 70: 375–426.

Shewry PR and Casey R (eds) (1999) *Seed Proteins*. Dordrecht: Kluwer.

Shewry PR, Halford NG, Tatham AS, *et al.* (2003) The high molecular weight subunits of wheat glutenin and their role in determining wheat processing properties. In: *Advances in Food and Nutrition Research*, vol. 45. Amsterdam: Elsevier Science.

Starch

S M Smith, Institute of Cell and Molecular Biology, University of Edinburgh, UK
K Denyer, John Innes Centre, Norwich, UK

Starch in Plants

Chemical Structure

Starch is comprised principally of two glucose polymers, amylose and amylopectin. Both polymers are made up of D-glucose units in cyclic pyranose form, linked through glycosidic bonds between carbons 1 and 4 of adjacent glucose monomers (**Figure 1**). The asymmetric carbon 1 is in α rather than β configuration. Such molecules are therefore referred to as α-1,4-D-glucans. α-Glycosidic linkages form relatively flexible, open-helical polymers, in contrast to β-glycosidic linkages which form much straighter molecules suited to structural fibrils such as cellulose. The glycosidic linkages protect the aldehyde groups of glucose molecules against oxidation, so that only the terminal glucose retaining the aldehyde group is readily oxidized. This is referred to as the reducing end of the glucan chain whereas the other end is nonreducing (**Figure 1**).

Amylose is principally a linear polymer with a degree of polymerization (DP) of between several hundred and a few thousand glucose residues, depending upon biological source (**Table 1**). Amylose may also contain infrequent α,1-6-linked branches. The branches also consist of α-1,4-D-glucan chains. Amylopectin is a much larger molecule and is highly branched (**Table 1**). Clusters of α-1,6-linked branches commonly occur at regular intervals of 25–30 residues, and the branches may themselves be branched. Branch chain lengths typically vary between 6 and 25 or more glucose residues. A proposed structure for amylopectin is shown in **Figure 2** (*see* **Seed Development**: Starch Synthesis).

Some starches contain small amounts of covalently-bound monoester phosphates, mostly at the C6 position of glucose residues, but with a significant amount at the C3 position. Most of the phosphorylation occurs on the amylopectin fraction rather than amylose. The extent of such phosphorylation is different in starches of different biological origin. Potato (*Solanum tuberosum*) tuber starch is one of the richest in phosphate containing up to 0.1% by dry weight, corresponding to phosphorylation of about 1 in 300 glucosyl residues (**Table 1**). Phosphorylation influences the physicochemical properties of the isolated starch which makes it better suited for some applications (see below).

Starch Synthesis and Organization into Granules

Starch is synthesized and accumulates in subcellular organelles of the plastid family: typically chloroplasts in photosynthetic tissues and amyloplasts in storage organs and other nonphotosynthetic cells. Key enzymes of starch synthesis are as follow:

1. ADPglucose pyrophosphorylase (AGPase; EC 2.7.7.27) which catalyzes the following reaction:

$$\text{glucose-1-phosphate} + \text{ATP} \leftrightarrow \text{ADPglucose} + \text{PP}_i$$

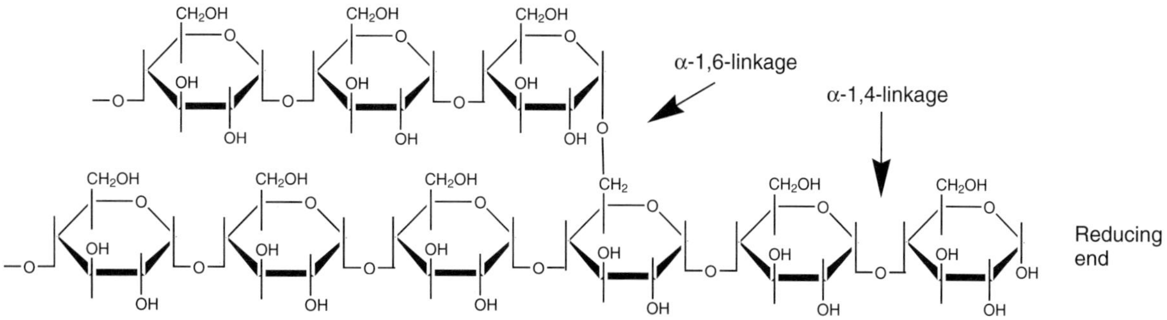

Figure 1 The glycosidic bonds present in amylose and amylopectin. Amylose contains very few α-1,6-links while amylopectin contains many.

Table 1 Comparison of some properties of amylose and amylopectin

	Amylose	*Amylopectin*
Proportion in starch	20–30%	70–80%
General structure	Essentially linear	Highly branched
Frequency of α-1,6 glucosidic bonds	0.1%	5%
Average chain length (glucosyl units)	10^2–10^3	20–30
Degree of polymerization	0.5–6×10^3	0.2–2×10^6
Mass	10^5–10^6	5×10^7–5×10^8
Physical character	Amorphous or helical	Semicrystalline
Phosphorylation	Trace	<0.3% of glucose in some plants
Solubility in water	Limited	Soluble
Stability in aqueous solution	Retrogrades	Relatively stable
Color with iodine	Dark blue	Red–brown
λ max of iodine complex	618 nm	550 nm
Inclusion molecules	Fatty acids, alcohols	None

2. Starch synthase (SS; EC 2.4.1.21) which transfers glucose from ADPglucose to the nonreducing end of an α-glucan chain:

$$(\alpha\text{-}1,4\text{-D-glucan})_n + \text{ADP-glucose} \rightarrow (\alpha\text{-}1,4\text{-D-glucan})_{n+1} + \text{ADP}$$

3. Starch-branching enzyme (SBE; EC 2.4.1.18) is involved in amylopectin synthesis. It cleaves an α-1,4-glucosidic bond and transfers the released glucan chain to an acceptor α-1,4-glucan by means of an α-1,6-glycosidic bond.

$$(\alpha\text{-}1,4\text{-D-glucan})_n \leftrightarrow (\alpha\text{-}1,4\text{-D-glucan})_{n-x}\text{-}1,6\text{-}\alpha\text{-}(\alpha\text{-}1,4\text{-D-glucan})_x$$

Other enzymes are involved in the creation of the final starch structure and in phosphorylation (*see* **Seed Development:** Starch Synthesis). The starch is organized into insoluble granules with distinctive morphology and sizes characteristic of the biological source. The granules typically vary in length or diameter from 1 μm to 100 μm. There may be one or multiple granules in each plastid. Starch commonly comprises about 75–80% amylopectin and 20–25% amylose, although leaf starches may have appreciably less amylose. The packing of amylose and amylopectin into the granule is still the subject of much research (*see* **Seed Development:** Starch Synthesis).

Occurrence and Function in Plants

Starch serves as a reserve for carbon and energy during plant growth and development. Its polymeric and insoluble nature enables large amounts of carbon to be stored without affecting the osmotic potential of the cell. During photosynthesis, fixed carbon can be exported from the chloroplast as triose phosphate, a 3-carbon sugar phosphate, for conversion into sucrose in the cytosol and subsequent export to sink tissues. As the requirements of such sink tissues become satisfied, progressively more fixed carbon is retained in the chloroplast and converted to starch, which serves as a temporary store. During the night, this starch can be broken down for export from the chloroplast, converted to sucrose, and exported to sink tissues. Leaf starch is thus transitory in nature.

Sink tissues utilize the imported sucrose firstly to satisfy their immediate energy and carbon requirements associated with growth and development. The additional sucrose is converted usually to hexose phosphates, imported into amyloplasts, and there converted to starch. The starch that accumulates in seeds, tubers, some roots, or the bark of trees is referred to as storage starch since it remains for relatively long periods of time. When the seed germinates or the tuber sprouts, the storage starch is broken down to provide the carbon and energy for new growth, until photosynthesis can be established. Thus, storage starch is analogous to the money we transfer to an investment account for use at an

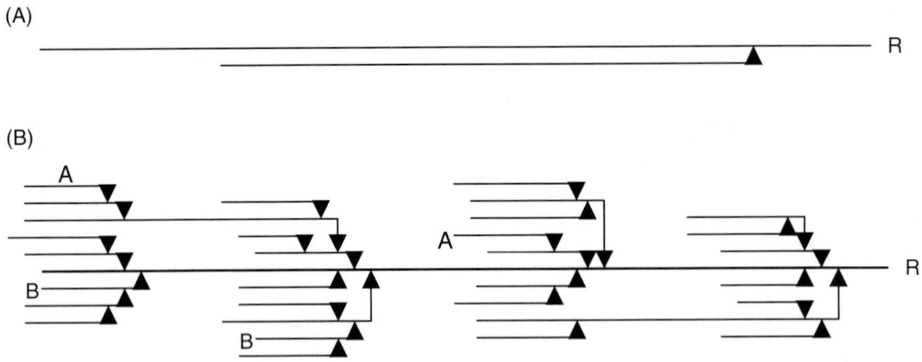

Figure 2 Representations of (A) amylose and (B) amylopectin structures. Lines are α-1,4-linked glucan chains. Arrowheads represent α-1,6-links. R, reducing end. A chains are those which are not branched, whereas B chains are themselves branched. The A and B chains shown are in the region of 12–15 glucose residues in length, and the distance between clusters of branch points is about 9 nm.

appropriate time in the future, while transitory starch is more analogous to the money that we keep in a current bank account or carry in our pockets, since it meets immediate requirements and fluctuates rapidly in amount according to circumstances.

Starch is also found in cells of the root cap, where it is thought to serve a role in gravity perception. Because of the high specific gravity of starch granules, the amyloplasts tend to sediment in the cell, displacing other organelles. It is hypothesized that this provides a mechanical stimulus which is then somehow transmitted to the cells of the root apical meristem, leading to a pattern of cell division and elongation which directs root growth towards the gravitational field. While there is strong evidence for this model, it is also found that starchless mutants are still able to partially perceive and respond to gravity.

The occurrence of starch in plant tissues is readily observed by iodine staining. Iodine interchelates into the amylose helix, resulting in intense dark blue staining. Amylopectin stains reddish-brown with iodine. Plant tissues may first be heated in ethanol to remove any chlorophyll, and are then typically treated with an acidic aqueous solution of iodine and potassium iodide, which immediately produces a blue–black coloration if starch is present.

Isolation and Properties of Starch from Plants

Isolation, Analysis, and Quantitation

Starch granules can readily be isolated due to their characteristic sizes and very high specific gravity (approximately 1.5 g cm^{-3}). Typically plant material is homogenized in aqueous buffer, the large cell and tissue debris removed by filtration through muslin or cheese cloth, and the filtrate further filtered through nylon mesh of defined pore size. The granules can be collected from the filtrate by centrifugation at low speed (e.g., 3000 *g* for 5 min), and washed in dilute detergent solution (e.g., 0.2% w/v sodium dodecyl sulfate) or treated with proteases to remove extraneous proteins, or in organic solvent (e.g., acetone) to remove lipids, and finally in water. The isolation of granules in this way provides material suitable for the analysis of starch composition and structure, including amylose and amylopectin content and chain-length distribution.

Starch granules can be solubilized by boiling in water or dimethylsulfoxide for 1 or more hours. The starch polymers can be precipitated from solution with ethanol (typically 80% v/v) which provides a further purification step. The proportions of amylose and amylopectin can be determined in several ways, e.g., by treatment with iodine solution, since the absorption spectra maxima for the iodine-amylose and iodine-amylopectin complexes are different (618 nm and 550 nm, respectively). Measurement of the absorbancies at these two wavelengths allows the proportions to be calculated. Alternatively, since amylose and amylopectin have very different masses, they can be separated by size-exclusion chromatography, using for example Sepharose™ eluted with NaOH solution (**Figure 3**). Other methods include selective precipitation of amylose with thymol or of amylopectin with the lectin concanavalin A. Branch chain-length determination first requires complete debranching of the amylopectin using an enzyme that cleaves α-1-6-glycosidic bonds (debranching enzyme, DBE). The products can then be separated by high-performance anion-exchange chromatography or reacted with charged fluorescent tags prior to separation on polyacrylamide gels, and the chain lengths and relative amounts can be quantified (**Figure 4**).

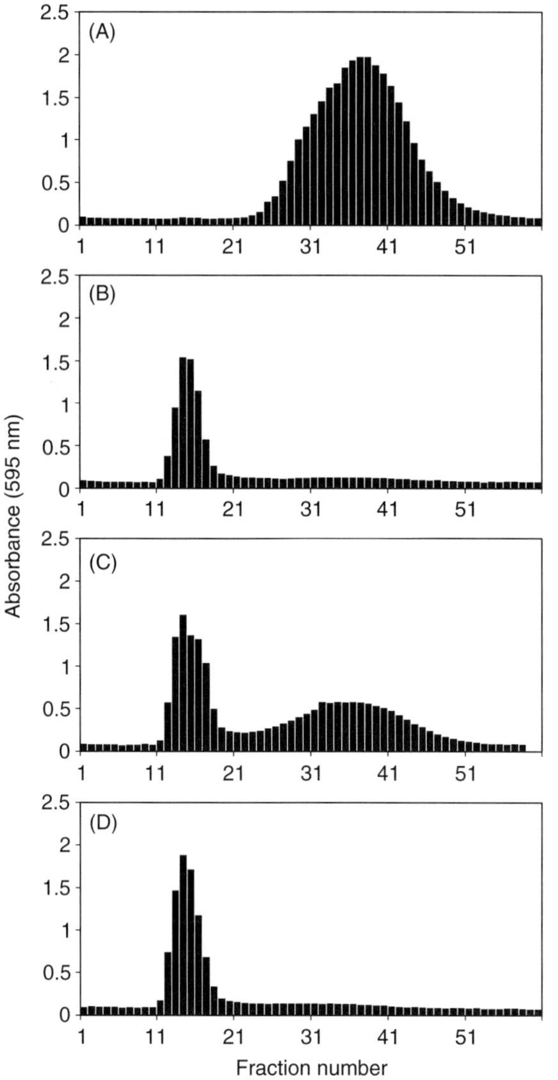

Figure 3 The elution profiles of starch from a column of Sepharose CL-2B. Samples of dissolved starch were fractionated according to molecular size by gel-permeation chromatography. Glucan was detected in the fractions by adding iodine solution and measuring the absorbance at 595 nm. Note that the relative peak areas do not reflect the relative amounts of amylose and amylopectin since these polymers have different iodine-binding capacities. (A) pea (*Pisum sativum*) amylose, (B) pea amylopectin, (C) wild-type pea starch, (D) amylose-free pea starch.

Quantitative isolation of starch from plant material may be achieved by first homogenizing and extracting the plant tissue several times in hot 80% (v/v) ethanol. This serves to remove all the soluble sugars and other polar solutes while leaving the starch in the insoluble fraction. The starch can then be dissolved by suspending the residue in water and boiling for several hours or autoclaving. The starch content of the solution is determined following complete hydrolysis to glucose by enzymes which are specific for α,1-4-glucans, thus avoiding any

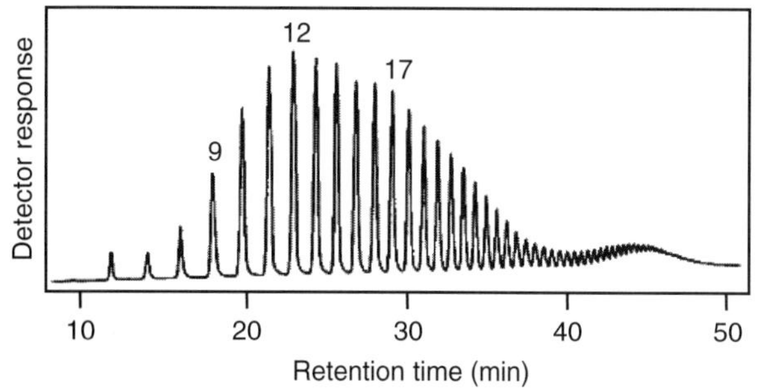

Figure 4 The chain-length profile of amylopectin from rice endosperm. Debranched amylopectin was analyzed using high-performance anion-exchange chromatography equipped with a pulsed amperometric detector. Data from Umemoto T, Yano M, Satoh H, Shomura A, and Nakamura Y (2002) Mapping of a gene responsible for the difference in amylopectic structure between japonica-type and indica-type rice varieties. *Theoretical and Applied Genetics* 104: 1–8.

contribution from cell wall material. This is typically achieved by treatment with a mixture of amyloglucosidase (which cleaves α-1,4 and α-1,6 glycosidic bonds) and α-amylase (an endoamylase that cleaves α-1,4 glycosidic bonds). The glucose content is then assayed enzymically, and compared to control samples which have not been treated with amylolytic enzymes.

Physicochemical Properties of Isolated Starch

Solubilization of isolated starch granules in water requires heat to bring about the absorption of water and swelling of the granules. During this process, crystallinity is lost from the granules, and soluble molecules are released into solution. The diffusion of starch molecules is particularly slow, so mechanical stirring is also required to achieve complete solubilization. For a typical wild-type starch the combined process of heating and stirring leads to an increase in the viscosity of the solution as hydration and swelling proceed, but then as complete dissolution is achieved, probably aided by fragmentation of large molecules, the viscosity of the solution falls again. The temperature at which swelling and dissolution occur, and the associated changes in viscosity, are characteristic for particular starches. In solution, the amylose and amylopectin molecules are hydrogen-bonded to water molecules rather than to themselves, and helices are disrupted. At this stage starch can be mixed with other materials, fractionated or chemically modified for various purposes (see below).

As a hot solution of starch is cooled, and if the concentration of starch is sufficiently high, gel formation occurs. In the gel, a complex network of hydrated molecules is held together by hydrogen bonding. During subsequent storage, particularly in the cold, the amylose and amylopectin molecules undergo slow reassociation, displacing some of the water molecules in a process called syneresis. This can be followed by recrystallization of the starch, a process called retrogradation. The result of these processes is precipitation, increased opacity, and stiffness of the gel. Reheating of the gel can reverse the syneresis and recrystallization processes to an extent. Retrogradation is clearly undesirable for many applications in food and industry. It is caused principally by the amylose fraction of starch, so amylose-free starch (see below) may be used in those cases where retrogradation is particularly undesirable. Alternatively, chemically modified starches are employed, such that the chemical groups added to the glucan polymers interfere with reassociation (see below).

Starch in the Diet

Starch in Food

The bulk of carbohydrate in the human diet takes the form of starch. Most of this starch is that occurring naturally in the plant products we eat. Cereal grains are the major source of starch in the human diet. It is estimated that cultivated cereal grain alone containing more than 10^9 tonnes of starch is produced each year worldwide. Corn (*Zea mays*; maize), rice (*Oryza sativa*), wheat (*Triticum aestivum*), barley (*Hordeum vulgare*), and sorghum (*Sorghum* spp.) are the main sources of this starch (**Table 2**). While much of the rice that is produced is consumed simply after boiling, other cereals are more often processed to produce, for example, bread, pasta, noodles, breakfast cereals, and other foods. The starch in cereal grains also fuels fermentations to produce beer and spirits. We also obtain a significant amount of starch from the consumption of pulses (peas and beans), tubers, and root vegetables and small amounts from other vegetables and fruit. Increasing amounts of starch in the human diet now come from processed starch added to convenience foods (see below).

The Behavior of Starch in Cooking

As explained above, when starch is heated in water it undergoes processes of hydration, swelling, and eventually solubilization. These processes occur during the cooking of starchy foods, but the extent of these changes depends upon the amount of water available and the cooking time. For example, pasta is cooked for a relatively short period of time and bread is made with a limited amount of water. Thus the starch is not fully hydrated or solubilized, and together with other ingredients that may affect the behavior of the starch, this imparts upon a particular food its distinctive texture and quality. In other foods, much more extensive swelling and solubilization of the starch content may be achieved, such as in custard, sauces, and puddings. Once such foods are cooled, they form starch gels. Retrogradation can be a problem for foods that have been cooked then chilled, even if starch solubilization is far from complete. Bread soon acquires a stale texture, brought about in part by retrogradation of the starch content. Brief reheating of such bread can reverse the retrogradation and "freshen" it. In chilled puddings, syneresis and retrogradation can lead to separation of gel and water, making an unappealing mess, but this can be reversed by brief reheating and mixing. The use of amylose-free or chemically modified starches in chilled food reduces the extent of retrogradation and provides a more stable product, while use of amylose-free starch in frozen foods reduces formation of slurries upon thawing.

The Nutritional and Health Value of Starch

The rate at which starch is digested in the gut to provide glucose for absorption into the circulatory system is an active area of research. It is important because elevated levels of blood glucose and insulin have been associated with diabetes, coronary heart disease, promotion of some cancers, and aging. There is a widely held belief that starch is digested

Table 2 Major starch crops

Common name	*Scientific name*	*Storage tissue*	*Major producers*
Maize or corn	*Zea mays*	Endosperm	North America, Europe, South America
Wheat	*Triticum aestivum*	Endosperm	North America, Europe
Sorghum, milo, or durra	*Sorghum bicolor*	Endosperm	Central America, Africa, Asia
Rice	*Oryza sativa*	Endosperm	Japan, Asia
Barley	*Hordeum vulgare*	Endosperm	Europe, North America
Oats	*Avena sativa*	Endosperm	Europe, North America
Potato	*Solanum tuberosum*	Tuber	Europe, North America
Manioc, tapioca, cassava	*Manihot esculenta*	Storage root	South America, Africa, Asia
Sago palm	*Metroxylon sagu*	Stem	South East Asia
Pea	*Pisum sativum*	Embryo	Europe, USA

relatively slowly in the gut resulting in a progressive absorption of glucose into the circulatory system. However this will depend on the form of starch being ingested. Increasingly the starch which we eat differs from that of the ancient human diet, since more processed foods are consumed.

Starch is digested primarily in the small intestine. However, some starch (estimated within the range 3–30%) may resist digestion in the small intestine and pass into the large intestine (colon). This is defined as resistant starch (RS) and is considered by some to have beneficial effects similar to some forms of dietary fiber (plant cell walls). RS may be further digested in the colon, and the resulting glucose fermented by microbes, with the production of acetic, lactic, butyric, and other short-chain fatty acids. These acids can be absorbed, transported in the blood system to the liver and peripheral tissues, and metabolized to provide energy to the host. It is estimated that in rats, short-chain fatty acids from starch fermentation can provide 5–30% of daily energy intake. Short-chain fatty acids have also been implicated as a protective factor against colon cancer (although other studies implicate RS in enhanced tumor formation in rodents). Furthermore, fermentation of RS in the colon is reported to reduce the amount of fecal ammonia, which is potentially beneficial to health.

RS can be classified into four types according to the underlying basis for its resistance:

- type 1, starch physically inaccessible, such as in whole cereal grains
- type 2, starch from plants producing granules that are naturally resistant to digestion
- type 3, retrograded and crystallized starch caused by processing or cooking
- type 4, chemically modified starches.

Unripe banana (*Musa sapientum*) is a source of much type 2 RS, but ripe bananas contain little starch. Raw potato contains a very high proportion of type 2 RS. Upon cooking the starch becomes readily digestible in the small intestine, but in cooked and chilled potato, retrogradation leads to formation of some type 3 RS. Type 3 RS is thought to be caused largely by retrogradation of the amylose content of the starch. Thus, high-amylose starches may be beneficial by producing RS, whereas processed amylose-free starches may produce little RS. A goal for food technologists is to acquire starches with desirable processing and storage characteristics, which are digested slowly in the small intestine, which may include some RS, and which are suitably palatable.

Applications for Extracted Starch

Overview

Starch is extracted for use in the food industry and in many others (e.g., manufacturing, chemical, engineering, environmental, and health care). The food sector alone uses about 70% of extracted starch. The wide range of applications for extracted starch is indicated in **Table 3**. Important features of starch which result in many of its applications are its ability to form gels, to thicken, and to bind. Ancient Egyptians (about 4000 BC) first noted the adhesive properties of starch and used it to bond papyrus strips together. The Chinese (AD 312) first exploited the water-repellent and strengthening properties of starch to produce resistance to ink in paper. Now starch is used extensively as an adhesive and coating material. It is also used as an important additive to a wide range of foods. Much is processed either chemically or physically to give polymers with a wide range of functions, but the majority is hydrolyzed to dextrins and glucose either for direct use in foods, or as a feedstock for many different fermentation processes. Thus, in addition to being relatively easy and inexpensive to produce, starch has great utility and versatility (**Figure 5**). The sources of starch and the way in which it is processed for different functions are described below.

Sources of Starch for Extraction

Cereals are the major source of starch for extraction and processing. It is estimated that at least 2×10^7 tonnes of cereal starch are extracted, which together with the starch extracted from other crops, gives an annual production of about 30 million tonnes. The United States, the European Union, and Japan are the largest producers. The greatest mass is produced from corn but other crops are major sources of starch for extraction (**Table 2**). Because different plants contain starches with different physical and chemical characteristics, including granule sizes, impurities (lipid and protein in

Table 3 Uses of extracted starch

Industry	*Uses*
Food	Sweeteners, sauces, soups, dressings, bakery, dairy, drinks, ice cream, lubricants
Materials	Paper, board, glues, plastics, rubber, concrete, biodegradable plastics
Chemicals	Cyclodextrins, feedstock for fermentations, alcohols, acids, fine chemicals
Environment	Sewage and water treatment
Health care	Pharmaceuticals, cosmetics, disposables

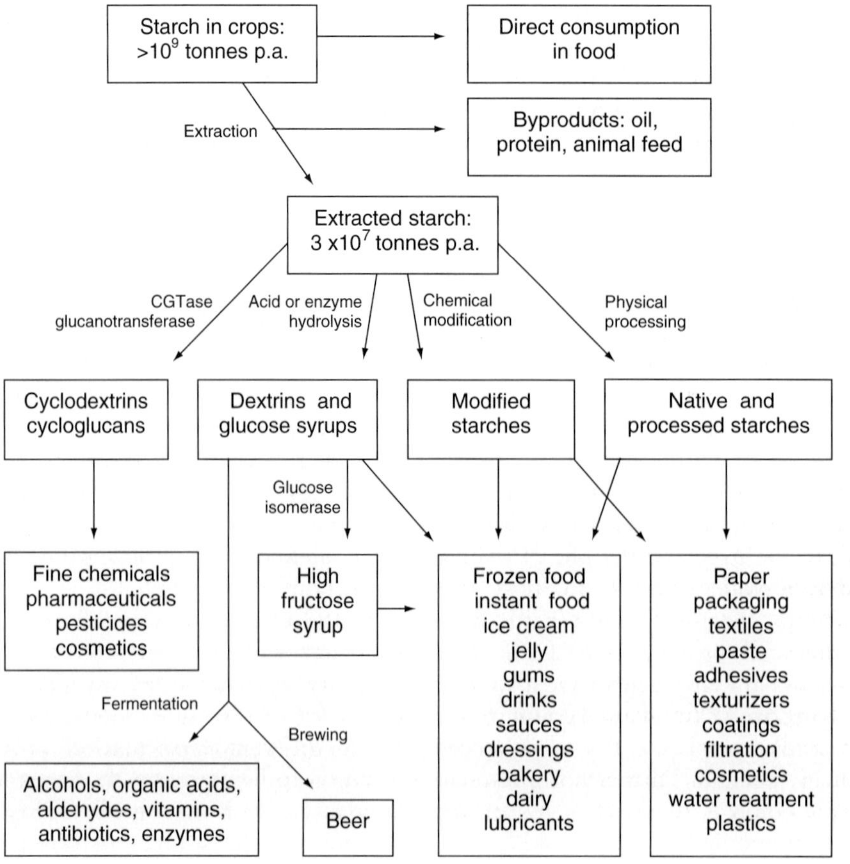

Figure 5 Summary of the processing and uses of starch.

particular), and level of phosphorylation, their properties are suited to different applications. Furthermore, mutant varieties producing modified starch have been bred either through selection of naturally arising variants or of variants arising from mutagenesis programs. The most important variant is "waxy" corn which derives its name from the waxy appearance of its seeds. The starch in this mutant lacks amylose due to the absence of granule-bound starch synthase I (GBSSI), the enzyme responsible for its synthesis. Subsequently mutants producing amylose-free starch have been developed for rice, barley, and sorghum. Transgenic potato plants lacking GBSSI have been produced using antisense gene technology to silence the GBSSI gene, and produce low-amylose starch. Another important mutant is the high-amylose mutant of corn ("amylomaize") which produces starch containing more than 60% amylose. It is mutated in a gene called amylose extender (*ae*) coding for a starch-branching enzyme.

Industrial Extraction of Starch

Starch is isolated on an industrial scale from seeds and tubers in the form of starch granules, since their physical properties (size and density in particular) allow them to be readily separated from other components. Cereal seeds comprise about 70% starch and only 15% water, so that extraction may begin with a dry milling process. Wheat is often dry-milled for extraction of starch, but corn is usually wet-milled. Potato tubers and root crops comprise almost 80% water and less than 20% starch, so that wet processing is required. In either case, the separation of starch granules from contaminating fiber (cell walls) and solutes involves a combination of sieving, filtration, washing, and centrifugation techniques. Ultimately the starch is dried, unless it is to be used directly for hydrolysis, chemical modification, or paper sizing. The byproducts from cereal processing include protein and oil, which have appreciable value. The solid wastes from cereal and potato processing can be used as animal feed and the liquid waste from potato processing can be used as a fertilizer.

Hydrolysis to Produce Maltodextrins and Glucose

Most of the extracted starch is hydrolyzed to products comprising varying amounts of malto-oligosaccharides (maltodextrins) and glucose. In some processes, hot acid (e.g., hydrochloric) is used first to bring about partial hydrolysis (achieving thinning or liquefaction), followed by neutralization and enzymic treatment to obtain products of the desired size. In other cases, enzyme hydrolysis alone is employed. The enzymes used are usually thermostable α-amylases (endoamylases) of bacterial or fungal origin, and fungal glucoamylases (exoamylases). Enzymes have the advantage over acid hydrolysis of not creating unwanted reaction products, of being more controllable, and using less energy. The maltodextrin and glucose syrups created in this way can be used in many further processes. Some are added directly to foods (confectionery, ice cream, jams, jellies, frozen desserts, cakes, chewing gum, canned fruit). They can increase sweetness, viscosity, moisture retention, color, and taste stability. Corn syrup is also added as "wort syrup" to some beer fermentations. Syrups are also used extensively as fermentation feedstocks for a wide range of chemical engineering processes. These include production of ethanol (for use as fuel or fuel supplement for cars in some parts of the world), other alcohols, aldehydes, organic acids, vitamins, antibiotics, enzymes, and single-cell protein. Finally, an increasing amount of glucose syrup is converted to fructose using the bacterial enzyme glucose isomerase. High-fructose corn syrup is added to increasing numbers of food products and soft drinks, providing sweetness with lower calorific value than sucrose or glucose. High-fructose corn syrup is both sweeter and less expensive that sucrose. In terms of global amounts used, amylases and glucose isomerase are two of the three most important enzymes in industry (protease being the third).

Chemical and Physical Modification of Starches

Starch can be modified in many different ways to make it suitable for an extremely wide range of industrial and food applications. Chemical modification often involves oxidation, esterification, or etherification. Each glucosyl residue has three hydroxyl groups available for substitution but usually no more than one is substituted and the majority of glucosyl residues remain unmodified. Chemical modification changes the gelatinization and cooking properties of starch, decreases retrogradation and increases water-holding capacity, and can change hydrophilicity or hydrophobicity. It is important in the manufacture of films and adhesives (**Table 4**). Intermolecular crosslinking of starch ("di-starch") is achieved by reaction with bifunctional reagents.

Table 4 Some modified starches, their properties, and uses

Modification	*E number*	*Treatment*	*Examples of industrial use*	*Examples of use in food*	*Benefits*
Pregelatinized	E1400	Moisture and heat	Oil drilling, mining	Coatings, pie fillings, instant products	Soluble in cold water
Hydrolyzed ("acid-thinned")	E1401	Acid	Textiles, laundry	Jellies, gums	Low hot paste viscosity, high gel viscosity
Oxidized	E1404	Hypochlorite	Textiles, paper, spray starch, adhesive	Thickener, gravy, sauces, jellies	Increased clarity, reduced retrogradation
Esterified	E1421	Acetic anhydride	Paper, textiles, film, packaging	Frozen foods, instant foods	Increased clarity, reduced retrogradation, forms films/fibers
Monophosphate	E1410	Phosphoric acid	Paper, textiles, metal refining	Frozen foods, baby foods	Stability to freeze–thaw cycles
Hydroxyalkyl ether	E1440	Propyleneoxide	Paper, textiles	Pie fillings, salad dressing	Increased clarity, increased stability
Di-starch phosphate (crosslinked)	E1412	Phosphorus oxychloride	Paper, metal sequestration	Many frozen and canned foods	Increased stability to freeze–thaw, pH, heat, and shear
Di-starch adipate (crosslinked)	E1422	Adipic anhydride	Paper	Many frozen and canned foods	Increased stability to freeze–thaw, pH, heat, and shear

Such starches undergo reduced swelling during gel formation and are more stable to physical forces (**Table 4**). In a further modification, the reactive groups of starch can be used to cross link it to other polymers such as cellulose, creating a covalently linked coating or copolymer. Physical modification of starch may involve treatments with heat, pressure, water, and/or mechanical disruption to create products with altered structure or molecular weight, for use in food and nonfood products.

Production of Cycloglucans

Cyclodextrins are cyclic glucans made up of 6, 7, or 8 glucose residues and are produced by the action of bacterial cyclodextrin glucanotransferase (CGTase) on starch. Such cyclodextrins have a hydrophobic interior and hydrophilic exterior. They have the ability to form inclusion complexes with a wide range of small organic molecules including ethylene, alcohols, organic acids, naphthols, steroids, and some vitamins. Such complexes provide the means to increase the solubility, stability, or reactivity of guest molecules, and to isolate one molecule from a mixture. They therefore have a wide range of uses in the formulation of medicines, pesticides, cosmetics, and foods, and are used in analytical chemistry. Several thousand tonnes are produced annually. More recently, related glucanotransferases have been shown to produce much larger cycloglucans (cycloamylose) from amylose, and relatively large complex cyclic products from amylopectin. The latter have desirable pasting properties and clarity which makes them useful for addition to confectionery and convenience foods. Other products include a starch derivative which can form a thermoreversible gel with the potential to replace gelatin.

Starch Plastics

The drive to replace plastics made from petroleum with products that are biodegradable and made from a renewable raw material has led to the development of plastics made from starch. Early products consisted of starch added to conventional plastics, but these were not fully biodegradable. Now plastics made solely from starch are produced and are set to increase in their range of applications and share of the market. It is estimated that in the European Union alone, the market for such plastics could be 10^6 tonnes per annum (comparing with 32×10^6 tonnes of petroleum-based plastics), with a value of 2×10^9 Euro, and providing 20 000 new jobs.

These new thermoplastic starches (TPS) are produced by mixing starch with plasticizer, and subjecting it to a combination of heat (120–220°C) and mechanical energy provided by high pressure. This melts the crystalline starch structure and creates a new structure in which the amylose and amylopectin molecules are held together through hydrogen bonding and van der Waals forces with the plasticizer. The plastic is formed by extrusion, compression or injection-molding to create products of the desired form. Starches of different origin can be used, including high-amylose and amylose-free starches, giving plastics with different properties. Similarly, different plasticizers can be used, including glycerol, glycol, sorbitol, polyethylene glycol, glucose, dextrins, and sucrose. Thus, a very wide range of plastics with different properties can be created, ranging from glassy to rubbery in nature. Products are available under trade names such as BIOPLAST and BIOFLEX. TPS is employed to make capsules for drug delivery, coatings, films, rubbish bags, flower pots, agricultural mulches, clay pigeons, nappy liners, disposable clothing, cups, cutlery, golf tees, packing materials, and more. Aging (changes in the structure leading ultimately to degradation) of the plastic begins as soon as it is made and continues over a period of months or years, depending upon the plastic and the environment.

The Future

The role of starch is changing from a cheap bulk ingredient to an increasingly important functional product with many applications. It is renewable, biodegradable, very versatile, beneficial to health and largely benign to the environment. The fundamental science of starch structure and synthesis is still an active area of investigation, and may lead to new ways to change its properties or increase the amount of starch made by plants. Biological ways to increase starch functionality could complement or bypass some currently used chemical and physical modification methods. Conventional mutagenesis and plant-breeding programs will probably be the most productive approach for the immediate future but genetic modification, possibly making use of enzymes that normally function in the metabolism of other glucans (e.g., glycogen), offers the potential to increase the range of starches made by plants in the future. Starch-based polymers may be increasingly developed to reduce dependence on petroleum-derived products or to create novel polymers.

List of Technical Nomenclature

Amylase An enzyme that hydrolyzes glycosidic bonds in starch.

Amylopectin	A polymer of α-1,4-linked glucose residues with many α-1,6-linked branches.
Amylose	A linear polymer of glucose residues linked through α-1,4-glycosidic bonds.
Cyclodextrin	A cyclic dextrin of 6-8 glucose residues.
Dextrin	A low molecular weight α-1,4-glucan.
Di-starch	Chemically modified starch containing intermolecular crosslinks.
Endoamylase	An amylase that hydrolyzes glycosidic bonds within α-1,4-glucans.
Exoamylase	An amylase hydrolyzing glycosidic bonds from one end of α-1,4-glucans.
Glucan	A polymer made up of glucose residues linked by glycosidic bonds.
Glycosidic bond	Ester bond between two sugars.
Maltodextrin	A low molecular weight α-1,4-glucan.
Plastid	A type of organelle of which chloroplasts and amyloplasts are examples.
Resistant starch	Starch which is not digested in the small intestine.
Retrogradation	Recrystallization of starch molecules upon storage of a starch gel.
Starch	Complex polymer comprising amylose and amylopectin.
Syneresis	Separation of water from starch gel during storage or freeze–thaw cycles.
Waxy starch	Starch lacking amylose.

See also: **Photosynthesis and Partitioning**: Primary Products of Photosynthesis, Sucrose and other Soluble Carbohydrates. **Seed Development**: Starch Synthesis.

Further Reading

Frazier PJ, Richmond P, and Donald AM (eds) (1997) *Starch Structure and Functionality*. Cambridge: Royal Society of Chemistry.

Galliard T (ed.) (1987) *Starch: Properties and Potential*. Chichester: John Wiley.

Lörcks J (1998) Properties and applications of compostable starch-based plastic material. *Polymer Degradation and Stability* 59: 245–249.

Meuser F, Manners DJ, and Seibel W (eds) (1995) *Progress in Plant Polymeric Carbohydrate Research*. Hamburg: Behr's Verlag.

Sivak MN and Preiss J (1998) *Starch: Basic Science to Biotechnology*. London: Academic Press.

Thompson DB (2000) Strategies for the manufacture of resistant starches. *Trends in Food Science and Technology* 11: 245–253.

Wang TL, Bogracheva TY, and Hedley CJ (1998) Starch: as simple as A, B, C? *Journal of Experimental Botany* 49: 481–502.

PRODUCTION SYSTEMS AND AGRONOMY

Contents

Agricultural Crops

G Russell, The University of Edinburgh, Edinburgh, UK

Introduction

According to the Merriam-Webster online dictionary, agronomy is a branch of agriculture dealing with field crop production and soil management. It is concerned with the production of food for humans and livestock, and of fiber and material for industrial processing. Agronomists are either people who provide advice to farmers or those who carry out the underpinning research, and the subject of agronomy has close links with the whole range of plant sciences including ecology. Many of the principles apply to grassland and to vegetable crops as much as to agricultural crops.

Cropping is as old as agriculture itself but the early part of its development was very much based on empiricism and on belief in divine intervention. However, knowledge was quickly gathered by experimentation and disseminated along the trade routes. By the twelfth century, there was an agronomic center of excellence at Seville in Spain where Ibn al Awam wrote the key medieval treatise on agricultural practices in a Mediterranean climate. His book described the cultivation of many species of crop and fruit tree as well as discussing suitability of soils, irrigation, manures, grafting, and crop diseases. In the nineteenth century, crop production was put on a firm scientific footing, particularly in terms of crop nutrient requirements. More recently, a greater understanding of the processes resulting in crop yield and the quality of the produce has built on discoveries in soil science, genetics, microbiology, crop physiology, and ecology.

The aim of farmers in cash-based production systems is generally to maximize the value of produce obtained per unit of input cost. The value of the produce depends both on the crop yield and on its quality.

Farming Systems

Although there are situations in which a single crop is grown year after year in the same field, it is normal for crops to be grown as part of a system where different crops follow each other in a rotational sequence that helps preserve fertility while controlling weeds, pests, and diseases. A species can be successfully cropped continuously where there is no significant risk of soilborne pathogens building up and there is an extended period of dry or cold weather between crops to limit the carry-over of foliar disease or weeds from one crop to the next. Issues that need to be considered include the interactions between the component crops, and the timing of harvest and planting. However, the particular rotation employed depends strongly on the climate and soil and on the prevailing economic conditions (**Figure 1**). Crops are only viable in a rotation when they can reach maturity within the growing season set by availability of water or adequate temperatures, and the need to harvest before the next crop. Rotations fell out of fashion

Figure 1 Crops of banana, wheat, and rice growing in Bangladesh. Photograph courtesy of Dr M R Turner.

in Western Europe when fertilizers and other agrochemicals were used to intensify farming but have returned to favor due to pressures of cost on the one hand and worries about pollution on the other. For the latter reason there has been an upsurge in interest in organic farming in many developed countries, although a good understanding of the underlying biological processes is needed for success. However, organic and high external input systems represent the extremes of a range of possibilities. Integrated farming systems have now been developed which rely heavily on rotations of crops as a means of taking advantage of biological means of enhancing soil fertility and controlling weeds, pests, and diseases while minimizing the inputs of fertilizer and pesticides without compromising profitability.

Yield Elaboration

Plants capture carbon dioxide from the atmosphere, and water and nitrogen from the soil, and use the energy in solar radiation to transform them into plant tissues with particular properties. The harvested yield can be considered as the product of aboveground biomass and a harvest index which can exceed 0.50 in cereals like wheat. The root weight is usually excluded in these calculations as it is generally a small proportion of the mature biomass. The physiology of these processes is now reasonably well understood and rapid progress is being made on understanding the genetic basis of it. However, there are still large gaps between being able to specify the genotype of an individual plant and being able to predict its performance in particular environments as part of a population of plants.

Resource Capture

Plants can be considered as a canopy of leaves, which intercept light and take up carbon dioxide, and a root system, which takes up water, nitrogen and other nutrients, connected by a transport system. However, it is often the storage organs that are harvested. These may be reproductive organs, e.g., in barley (*Hordeum vulgare*), stems, e.g., in sugar cane (*Saccharum officinarum*), or roots, e.g., sugar beet (*Beta vulgaris*). In most situations, the crop growth rate is proportional to the amount of solar radiation intercepted by the leaf canopy and thus total biomass depends on the integral of intercepted radiation over time. Thus the duration of the crop can be an important indicator of yield. Light absorption depends on the area of green leaf per unit area of land surface (leaf area index) and on the orientation of the leaves. The availability of water and nitrogen strongly influences the development and survival of the leaf canopy and thus influences the amount of solar radiation that is absorbed. In environments characterized by high irradiances for significant parts of the day, crop growth rate can be limited by the rate of uptake of carbon dioxide, particularly in C_3 species. Resource capture depends not only on the attributes of individual plants but also on their spatial distribution and population density. It also depends on the seasonal availability of the resources. Outside the tropics, there is a seasonal cycle of solar radiation receipts that is complicated by varying degrees of cloudiness. Water availability is related to rainfall unless irrigation is carried out or plant roots are able to access water from the water table. Nitrogen availability also varies seasonally as the rate of mineralization of soil organic matter depends on soil temperature (it is insignificant below about 7°C) and the presence of adequate but not an excessive amount of water. Most intensive cropping systems involve the application of nitrogen fertilizer in the form of ammonium nitrate or urea. A rapidly growing crop can withdraw most of the available water and nitrogen in the rooting zone. However, unlike cations such as calcium and to a lesser extent ammonium, nitrate is not held on the soil particles and so is easily leached whenever water flows through the soil profile and out of the rooting zone.

Resource Transformation

The effectiveness with which the resources are converted into biomass has a large impact on crop yield. Light use efficiency (LUE), defined as the ratio of mature biomass to cumulative intercepted solar radiation, was originally considered to be rather constant. However, more recent work has shown that there are differences between species. For example, C_4 plants such as corn (*Zea mays*) tend to show higher values than C_3 plants such as wheat (*Triticum aestivum*) that lose carbon dioxide through photorespiration while leguminous plants tend to have low values of LUE due to the energy costs of supporting the nitrogen-fixing symbionts in root nodules. LUE also tends to decline as crops approach maturity because of losses of biomass due to senescence or the action of phytophagous insects. Water is often the limiting factor for growth in dry regions and so the water efficiency (WUE) of crops is of considerable practical importance. It is defined as the mature biomass divided by the amount of water transpired by the crop or evaporated from the soil or leaves. The interpretation of WUE is complicated whenever significant amounts of water are evaporated from the soil surface and so transpiration efficiency is often a better measure of crop performance although more difficult to measure. There are

...nificant differences in transpiration efficiency between species and even between cultivars. Nitrogen use efficiency is often calculated as the ratio of mature biomass to the amount of nitrogen applied in fertilizer and the input of nitrogen is thus seriously underestimated in any system where soil-derived nitrogen makes up an important part of the total nitrogen. Nevertheless it can be used to compare cultivars and production systems, both of which have an effect on it. Note that although these efficiencies are in general use they are not true efficiencies as they are not dimensionless.

Increasingly, farmers have to produce crops that achieve particular quality standards. These include chemical composition, structural composition, and uniformity. Some quality aspects are under close genetic control, for example, protein composition in wheat, while others such as nitrogen concentration in barley grains depend on the environment and on the production practices employed by the farmer.

Management Interventions

Much knowledge about the best decisions to make in particular circumstances has arisen empirically. However, this knowledge has been backed up and systematized by extensive experimentation and field trials. In environments that are constraining but predictable, farmers have rather few decisions to make once the farming system has been chosen. In other systems, decisions have to be made during the entire growing cycle with interventions being made according to the state of the crop or the soil. The goal of agriculture is generally to maximize profitability by optimizing inputs while not exposing the farmer to excessive risk of failure and the skill of the farmer comes in taking general principles and modifying them to take account of the particular conditions in each field or even within each field.

Farming System

Choosing the farming system involves choosing the crops that are to be grown on the farm, the order in which they can be grown, and the level of inputs that will be applied. Many other decisions are made implicitly as necessary consequences of these decisions.

Genotype

Effective plant breeding has contributed immensely to increasing crop yields around the world. In many countries there are many cultivars for a farmer to choose from and carrying out comparative cultivar trials in a range of environments is an important agronomic activity (**Figure 2**). The choice of cultivar is often not simple, as compromises have to be made between yield, quality, and resistance to disease. Once the end use has been decided, however, the number of

Figure 2 (see color plate 24) Field trials of oilseed rape cultivars near Cambridge, UK. Photograph courtesy of Dr M R Turner.

possibilities is reduced although matching a cultivar to a particular environment is still complicated by the uncertainty of weather and outbreaks of pests and diseases. A recent development that may ultimately reduce this problem has been the use of mixtures of varieties in the same field. These are the modern equivalent of the landraces, i.e., mixtures of genotypes, employed in early agriculture and still used in some parts of the world today.

Establishment

Most agricultural crops are annuals and have to be established anew each year from seed or from vegetative organs such as potato tubers (*Solanum tuberosum*). The farmer has control over the type of material sown or planted, the date of planting, and the spacing of the individuals. These are key decisions as a well-established crop is often able to compete effectively with weeds and an appropriate population density is needed to ensure that the crop can reach maturity before running out of resources such as water. Not all plants survive to maturity and the farmer needs to estimate the losses so the seed rate can be increased to compensate.

Physical and Chemical Environment

In the beginnings of agriculture, the main interventions were designed to modify the physical environment. There are three main areas where the environment for crops can be modified. First, the nutritional status of the soil can be improved by adding lime to raise the pH or by adding fertilizer to bring the status of nitrogen, phosphorus, potassium, sulfur or the micronutrients closer to the optimum for the crops. Second, the water status of the soil can be altered. In many parts of the world, agriculture would be impossible without irrigation, which has become technically very sophisticated in areas of intensive agriculture. In other areas, the problem is an excess of water, which can be overcome by appropriate underdrainage. Irrigation is not the only solution to a shortage of water and dryland agriculture makes use of techniques such as short-duration crops and summer fallows to make the best use of a limited resource. It may also be possible to increase the rooting depth of crops or increase the water-holding capacity of soil by increasing the organic matter content. Finally, cultivation is used to improve the soil structure for plant growth by plowing and harrowing, either of the surface layers or at depth by subsoiling. Mechanical cultivations are energy intensive and much research has taken place into "no-till" and minimal cultivation methods.

Biological Environment

It is easy for someone familiar with modern farming systems to underestimate the potential effect of weeds, pests, and diseases on crop yield because these are all largely kept under control. However, in many parts of the world these factors may result in actual yields being only half of the attainable yield and major outbreaks can lead to a complete loss of crops. Initially, control was largely by rotation of crops or by mechanical means such as plowing. Pesticides based on sulfur or copper have been known for over a century but it is really only in the past 50 years that research into crop protection chemicals and breeding for resistance have had a large impact on agriculture. Now there is a greater emphasis on integrated pest management which uses ecological knowledge to reduce the amount of active ingredient required for control.

Sustainability

Crop yields around the world are still increasing as a result of the adoption of better cultivars and management together with the availability of appropriate inputs. However, questions are now being asked about the long-term sustainability of cropping. There are many definitions of sustainability but a key criterion must be the continuation of production without depleting the natural capital of the soil. This criterion can only be used if particular input levels are assumed. In fact, few agricultural systems would be sustainable if all nutrients, water, and energy had to be obtained from the farm itself or by recycling from the communities supplied from the farm. While it is possible to argue about what is sustainable, lack of sustainability is often much more obvious. Large-scale loss of soil by erosion or salinization, or lack of irrigation water due to overabstraction are clear signs of unsustainability. In the longer term, there is a concern about shortages of phosphorus as few cropping systems recycle phosphorus effectively. In market economies, the first sign of unsustainability may be loss of profitability due to either reduced yields or a need to increase inputs. In an increasingly crowded planet, there are also impacts of agriculture on other people and on wildlife, and systems may become unsustainable due to legislative or fiscal pressure designed to reduce pollution from agriculture or encourage biodiversity.

Research Areas

The study of molecular genetics holds out the promise of providing benefits to cropping. However, the gains may be less from the development of genetically modified (GM) crops than from a better understanding of the underlying biological processes so that genotypes can be matched more effectively to environment. At the other end of the spectrum, there has been considerable research into ecological aspects of cropping with the development of integrated and organic farming systems. In between comes the constant battle against weeds, pests, and diseases together with attempts to understand better the factors affecting uptake and use of water and nitrogen. Climate change is another major research topic. However, farmers have proved very adaptable in the past even on timescales similar to those suggested in the most recent climate change scenarios.

See also: **Biodiversity and Conservation**: Plant Diversity, Conservation and Use. **Crop Improvement**: Plant Breeding, Practice. **Genetic Modification**: Gene Cloning, General Principles. **Growth and Development**: Canopy Architecture; Field Crops. **Integrated Pest Management**: Principles. **Photosynthesis and Partitioning**: C_3 Plants; C_4 Plants. **Production Systems and Agronomy**: Grassland; Horticultural Crops, Protected; Multicropping. **Weeds**: Weed Technology and Control.

Further Reading

Azam-Ali SN and Squire GR (2001) *Principles of Tropical Agronomy*. Wallingford: CAB International.

Lægrid M, Bøckman OC, and Kaarstad O (1999) *Agriculture, Fertilizers and the Environment*. Wallingford: CAB International.

Lefroy EC and Hobbs RJ (2000) *Agriculture as a Mimic of Natural Ecosystems*. Current Plant Science and Biotechnology in Agriculture vol. 37. Dordrecht: Kluwer Academic.

Loomis RS and Connor DJ (1992) *Crop Ecology: Productivity and Management in Agricultural Systems*. Cambridge: Cambridge University Press.

Pretty J (2002) *Argi-Culture: Reconnecting People, Land and Nature*. London: Earthscan.

Grassland

M O Humphreys, IGER, Aberystwyth, UK

The Importance of Grassland

The world's grasslands whether unmanaged or managed and whether natural or sown are important in:

- providing a primary feed resource for livestock
- maintaining soil fertility
- protecting and conserving soil and water resources
- providing a habitat for wildlife (plants and animals)
- contributing to the landscape.

Table 1 Forage grasses of economic importance

Temperate – cool season	*Tropical – warm season*
Perennial ryegrass (*Lolium perenne*)	Bermudagrass (*Cynodon* spp.)[a]
Italian/annual ryegrass (*Lolium multiflorum*)	Zoysiagrass (*Zoysia* sp.)[a]
Tall fescue (*Festuca arundinacea*)	*Paspalum* spp.[a]
Meadow fescue (*Festuca pratensis*)	*Pennisetum* sp.
Orchardgrass/cocksfoot (*Dactylis glomerata*)	Sudangrass (*Sorghum bicolor*)
Timothy (*Phleum* sp.)	Buffelgrass (*Cenchrus ciliaris*)
Bromegrass/prairie grass (*Bromus* sp.)	*Brachiaria* spp.
Wheatgrass (*Agropyron* sp.)	*Heteropogon* sp.
Wild rye (*Elymus* sp.)	*Panicum* sp.
Meadow foxtail (*Alopecurus pratensis*)	Forage corn (*Zea mays*)
Reed canarygrass (*Phalaris arundinacea)*	Bamboos
Yorkshire fog (*Holcus lanatus*)	*Dichanthium* sp.
Sweet vernal grass (*Anthoxanthum odoratum*)	Rhodesgrass (*Chloris gayana*)
(*Deschampsia* sp.)[a]	*Eragrostis* sp.
Kentucky bluegrass/smooth-stalked meadowgrass (*Poa pratensis*)[a]	*Setaria* sp.
Bentgrasses (*Agrostis* spp.)[a]	*Spartina* sp.
Fine-leaved fescues (*Festuca rubra* sp.)[a]	*Miscanthus* sp.
Alkali grass (*Puccinellia* sp.)[a]	*Digitaria* sp.
	Andropogon gayanus

[a] Also has nonagricultural value.

Due to diminishing destruction of forests the global area of grassland has stabilized since the early 1990s at about 34 million km^2, which is about 26% of the world's total land area. Grasses usually predominate but herbaceous legumes, trees, and shrubs may also occur. Grasslands have long been associated with human agricultural activity and the main species of forage grasses and legumes having economic importance are shown in **Tables 1** and **2**. Cereals are also annual grasses but they are not considered here.

Evolution of Grassland

In Africa, grasslands have evolved with human intervention from about 2 million years ago as fire was used as a hunting tool and trees were destroyed. The close association between grazing animals and

Table 2 Forage legumes of economic importance in agriculture

Temperate – cool season	*Tropical – warm season*
White clover (*Trifolium repens*)[a]	Egyptian clover/berseem (*Trifolium alexandrium*)
Red clover (*Trifolium pratense*)	*Leucaena* spp.
Subterranean clover (*Trifolium subterraneum*)	*Sylosanthes* spp.
Alsike clover (*Trifolium hybridum*)	*Centrosema* spp.
Kura clover (*Trifolium ambiguum*)	*Macroptilium* spp.
Sweet clovers (*Melilotus* sp.)	*Desmodium* spp.
Strawberry clover (*Trifolium fragiferum*)	
Crimson clover (*Trifolium incarnatum*)	
Vetches (*Vicia* spp.)[a]	
Sulla (*Hedysarum coronarium*)	
Sanfoin (*Onobrychi viciifolia*)	
Galega (*Galega orientalis*)	
Alfalfa/lucerne (*Medicago sativa*)	
Bird's-foot trefoil (*Lotus corniculatus*)[a]	
Lupins (*Lupinus* sp.)[a]	
Soybean (*Glycine max*)	

[a] Also has nonagricultural value.

grassland controlled the evolution of the African savannas, which have produced many of the world's superior tropical grass species. The natural grasslands of Australia and Latin America evolved in the virtual absence of grazing animals until relatively recent times and the grasslands of India are the product of forest clearing only since 5000 BC. Therefore it is not surprising that Australia and Latin America are particularly sterile with regard to agriculturally useful grasses, although Latin America is a primary source of herbaceous legumes.

Few temperate grasslands exist today which are not managed either for domestic livestock production or wildlife. The agricultural grasslands of Europe are dominated by a few major groups of grass species such as the cocksfoots (orchardgrasses) and the ryegrass–fescues. Ancestors of these important species are still found in the forest zones of Southern Europe and North Africa. Following the retreat of the glaciers after the last ice age, these species evolved through intercrossing and natural selection and spread from the Mediterranean Basin into northern and western Europe as forests were cleared and grazing agriculture developed. In prehistoric Europe species of wild cattle were widely distributed in grazed open forests. By about 400 AD they were virtually exterminated through hunting and destruction of their natural habitat except for isolated areas particularly in eastern Europe. However, before the wild species disappeared, domesticated breeds had become an integral part of human activity. Fossils of domesticated cattle found in Macedonia (Greece) date from more than 9000 years BC. During Roman times animal husbandry became more important and spread through Europe as people migrated. In the Middle Ages cattle were mainly kept as working animals (to pull carts and plows) and for their meat, although milk production also developed in parts of northwestern Europe. Grazing with cattle, sheep, goats, and pigs was integrated with growing crops. Crop production helped to provide animal feed, particularly during the winter period, whilst the animals provided manure for the crops. Only in mountainous areas and cold climates, where cropping was not possible, did cattle, sheep, or reindeer remain the only products of extensive farming systems. Later in the twentieth century farms became more specialized in terms of animal and crop production and grazing-based farming (mainly cattle and sheep) tended to separate from landless farming (pigs and poultry). With increasing interest in organic production systems, mixed farming is now re-emerging in some parts of Europe.

Grassland and Livestock Production

The primary role of grassland is to provide forage for livestock. It is estimated that 24 million km^2 of tropical and temperate grasslands and savannas produce 19 billion t of green leaf dry matter per year. This contains nutrients and energy which herbivores must process to be of benefit to humankind. In western Europe about 17% (0.6 million km^2) of the total land area consists of permanent grassland compared to 26% for the world as a whole. In the United Kingdom grasslands, ranging from short-term pastures to permanent rough grazing, currently comprise about 64% of the agricultural land area and underpin nearly 40% of agricultural output.

Temperate grasslands support the major share of the world's milk and beef production, maintaining about 32% of the world's cattle which produce over 80% of the world's cow milk and over 70% of its beef and veal. In the United States about 61% of the nutrients consumed by dairy cattle is supplied by forages while beef cattle, sheep and goats, and horses obtain 83%, 91%, and 72% of nutrients from forages. In the United Kingdom an estimated 35% of the total cost of ruminant production is the cost of feed and 75% of the feed requirements are presently obtained from grass and forage, although this varies from around 60% of the feed for dairy cows up to 90% for sheep.

Pastoral and ranching livestock production systems depend on the availability of natural feed resources. The strategy for traditional pastoralists is to optimize livestock production on an area basis as investment in livestock numbers is important. Livestock and their products are virtually the only output and often climatic stresses (drought and/or cold) result in fragmented growing seasons, which lead to few options other than pasture production. The levels of available nutrients in most of this natural grassland is determined by natural soil fertility level. As well as these ecological constraints, there may be social constraints which define farming systems, such as conflicts between pastoral and crop production; restrictive systems of land tenure; subsistence versus commercial production; labor-intensive versus mechanized approaches; and the operation of market forces. Pastoral systems are often managed in traditional ways including nomadism and seasonal livestock movement. With improvements in agricultural practice (e.g., irrigation) traditional pastoralists may be replaced by farmers taking over traditional dry-season grazing lands for crop production. Crop residues may take priority over pasture for livestock feeding. In ranching systems grassland resources are utilized by pasture and herd management which may include the use of external supplements (e.g., urea, molasses, and mineral licks) and the introduction of oversown legumes which improve feed quality and soil fertility. There is little other control over land use. Availability of water for livestock is often a major constraint resulting in local land degradation around water sources. Grazing with a range of animal species may improve biological and economic efficiency due to differential grazing resulting in more efficient use of grazing resources and an increase in production efficiency.

More highly developed pasture farming intensifies grassland use with greater management of inputs, including fertilizers and improved forage species, and the use of supplementary feeding to alleviate seasonal constraints on production. In mixed farming systems crops are integrated with livestock and manure is used to maintain soil fertility together with crop rotations involving leguminous forages. Generally production levels increase with the development of more intensive management practices.

World milk and meat production is increasing steadily although total grassland area remains constant (**Figures 1** and **2**). There are very significant recent increases in production in some developing countries such as the increased output of milk in India (**Figure 3**). These data demonstrate an increasing world demand for animal feed. Some of this demand can be met from the use of cereal and pulse based concentrates but such inputs are often too costly for sustainable livestock systems. Crop residues may also be used but availability is often sporadic and generally nutritive value tends to be poor. However

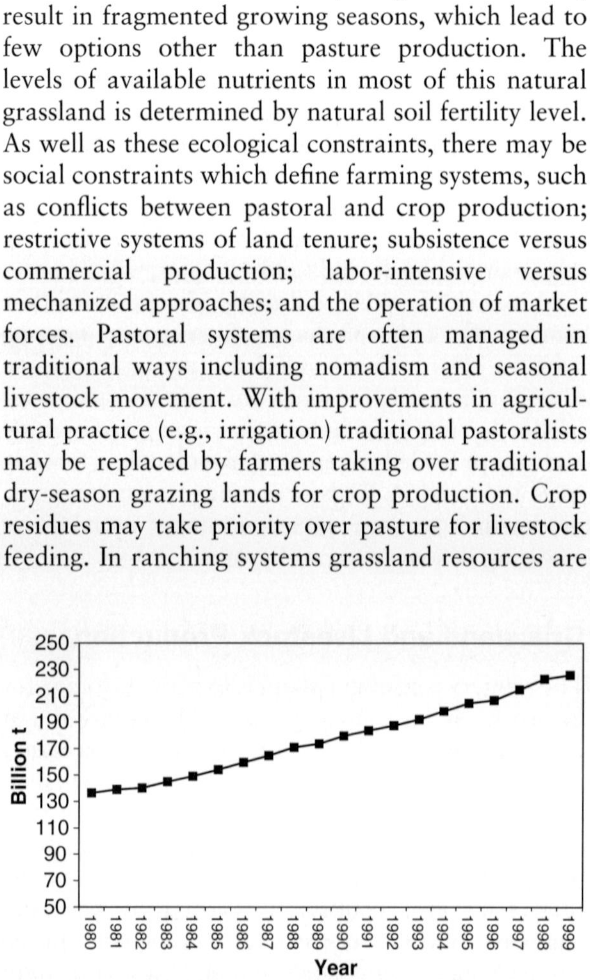

Figure 1 Meat, total world production. (Data from FAO Production Statistics.)

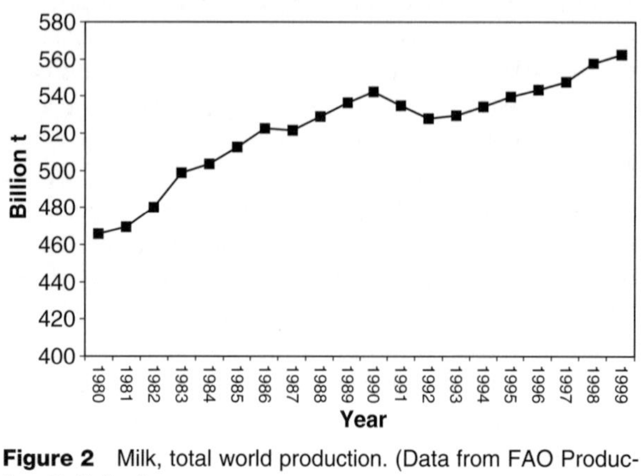

Figure 2 Milk, total world production. (Data from FAO Production Statistics.)

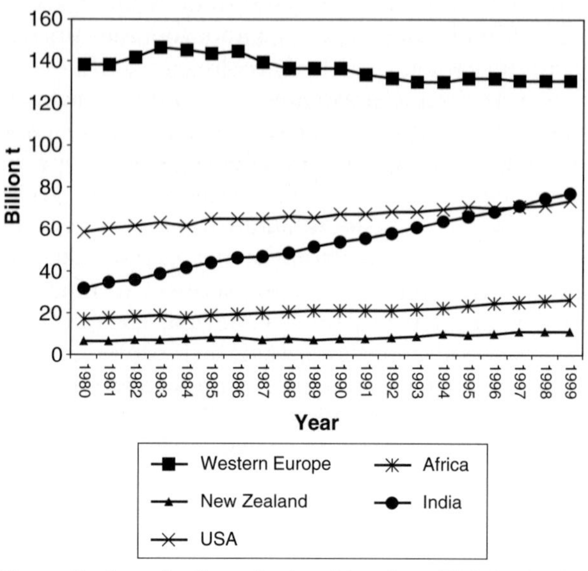

Figure 3 Annual milk production. (Data from FAO Production Statistics.)

with appropriate management there is considerable potential to make more and better use of grassland in a range of livestock production systems. Controlled use of improved grassland use is most well developed in temperate regions and this is reflected in what follows in this chapter. However tropical pastures have considerable potential and considerable progress is being made in their utilization.

Livestock and Pasture Management

There are two major components in the management of livestock production:

1. Herd management – which relates timing of livestock production and reproduction to the availability of feed supplies and market demands.
2. Sward management – which matches animal production with potential pasture production.

In Europe stocking practices have changed considerably during agricultural history. Initially farmers were less concerned with maximizing animal output as animals were kept mostly for subsistence and manure production. In modern times the livestock products of greatest value are milk, meat, and wool for national and international markets, and manure is often regarded as an undesirable byproduct. Stocking management has become a refined art, in which herbage accumulation and stocking density determine herbage allowance for livestock. This complex production system is manipulated by the farmer aiming to optimize animal production, whilst maintaining the quality and production potential of the sward. Farmers often rely on experience, but simple measurements of, for example, sward height, can be beneficial.

Livestock managers can choose between various combinations of stocking methods and intensities to manipulate sward and animal performance. Grazing can be regulated in relation to the time, method, and intensity of stocking in order to meet animal production targets and/or manipulate sward structure. Choice of variety and species composition is also important and forage grass and legume breeders have an important role in helping to ease management problems and increase the efficiency and sustainability of production. Forage availability and quality must be matched with animal requirements for energy and nutrients, both in the short term and over years. To solve problems of the seasonality of forage growth and herd requirements, the manager must use appropriate varieties and species. Forage grazed during the growing season must be rationed to allow conservation for feeding during periods of shortage. Stocking management also affects sward composition because defoliation, treading, and fouling at different intensities have differential effects on grassland species. Stocking management must be simple, flexible, and based on principles relating to herbage production, animal behavior, and market considerations. There are two main methods of stocking: continuous and rotational. Rotational grazing systems may vary considerably in relation to numbers of paddocks and the duration of grazing and resting periods and stocking rates can vary widely in both systems.

The Impact of Grazing

It is inherently cheaper to graze rather than to cut and feed forage. Grazed grass is about half the cost of grass silage and one-third the cost of concentrates in the United Kingdom. The advantages of stall-feeding are that the animals spend less energy on walking (20% of intake) and that nutrient losses can be reduced. Grazing animals remove forage selectively, both spatially and temporally, whereas cutting is instantaneous and removes all the available herbage above the cutting height. In heterogeneous swards selective intake during grazing can benefit animal performance. The ability to select is reduced when cut herbage is fed. Grazing animals return a large proportion of the ingested nutrients to the sward, whereas in a cutting system these are removed. However, animals return nutrients unevenly to the soil in patches or areas of high concentrations and they may also cause soil compaction. Larger-scale heterogeneity occurs when animals deposit excreta regularly in certain areas, e.g., near gates and watering points. For this reason a better distribution of recycled nutrients can be expected from rotational grazing in small paddocks than from continuous grazing in large paddocks. Thus harvesting method and grazing management have a large impact on grasslands in various ways.

Because cattle, sheep, goats, and horses are natural grazers, it follows that from an animal welfare point of view grazing is also preferred over housing. If dairy cows have a choice, they spend 80–99% of their lying-down time in the pasture rather than in a cubicle housing system. Housing animals is mostly a matter of convenience to feed and milk cows, particularly in winter. Grazing animals cannot always be fed according to their requirements for their production levels. Therefore, they are fed special forage, such as corn or silage, or concentrates to supplement grazed forage. Supplementary feeding is more conveniently carried out in feedlots or indoors than in the paddock. However, livestock grazing in pastures also plays an important role in maintaining biodiversity and enhancing landscape diversity.

Grass Preservation

Fodder conservation is necessary in most parts of Europe to maintain feed supply during winter. Over the past 50 years conservation has become more specialized and highly mechanized and there is a better understanding of the microbial and chemical processes involved. Traditionally haymaking was the only method, but this was gradually replaced by wet-silage making, which in turn has been replaced by wilted-silage making. This progression has reduced labor input and improved feeding value. Silage is produced by the controlled fermentation of sugars in plant material by lactic acid bacteria. Fructans is the main fermentable storage sugar in temperate grasses whereas legumes accumulate starch. Successful silage also depends on readily available sugars for good fermentation and a minimum of 3.7% sugar is required in fresh grass for good-quality silage without the use of additives. Legumes with their low water-soluble carbohydrate content and high buffering capacity are more difficult to ensile satisfactorily compared to grasses. Compared to temperate species, tropical grasses generally have a low available carbohydrate and protein content and poor digestibility. Although silage is not important in tropical areas considerable improvement in livestock production could be obtained by improving the nutritive value of sun-dried standing hay crops.

Interest in improving the quality of conserved grass in western Europe developed from the 1960s onwards as links between digestibility, maturity of herbage at cutting, and voluntary intake were clarified. Earlier cutting was adopted to optimize nutritive value rather than maximizing yield of dry matter, and this ethos eventually led to the dominance of silage over hay. Although grass is still the principal crop for silage, the ensiling of corn (*Zea mays*, maize) and whole-crop cereals has increased in recent years. The digestibility pattern of cereals is different from that of grasses since the declining quality of the leaf and stem is balanced to some extent by an increasing proportion of highly digestible grain. Corn ensiles well but corn silage is prone to aerobic deterioration, as is silage made from whole-crop cereals. Legumes, unless wilted or treated with an effective additive, tend to suffer extensive protein degradation during ensilage, although red clover appears to cause fewer problems in this respect.

The Value of Forage Legumes

Legumes are an essential part of sustainable farming systems because they supply their own nitrogen, enrich the N content of the soil, and have a high nutritive value. There are historical records of successful use of legumes in crop rotations dating from Roman times. Legumes can be grown as crops which are harvested and fed or in mixtures with grasses in grazed swards where they reduce the need for N fertilizer. The main forage legumes for temperate climates are species of *Trifolium* and *Medicago*. The United States grows about 13 million ha of alfalfa/lucerne (*M. sativa*), 5 million ha of white clover (*T. repens*), and 6 million ha of red clover (*T. pratense*). Other countries with large areas of forage legumes are Australia with 6 million ha of white clover and 17 million ha of subterranean clover and New Zealand with 9 million ha of white clover. New Zealand also exports about 4400 t of white clover seed each year. The main grassland and forage legumes in western Europe are white clover, red clover, and alfalfa. The productivity of grasslands containing legumes generally reflects the amount of N fixation, which has been estimated at between 65 and 280 kg ha^{-1} year^{-1} in western Europe. Animal production from white clover based pastures in Europe can be 80% and that from alfalfa and red clover can be 100% of that obtained from grass pastures fertilized with 400 kg N ha^{-1} year^{-1}.

White clover has superior nutritive value to ryegrass (*Lolium* spp.) in terms of protein content and digestibility which declines more slowly with increasing maturity. Animals fed pure clover or increasing proportions of clover in mixtures with grass produce more milk and liveweight gain than those fed pure perennial ryegrass. White clover has a place in permanent and ley pastures in farming systems which operate without or at low levels of N input. These exist in extensive farming systems for beef and sheep production in hill country, e.g., in central Europe and the United Kingdom and in so-called ecological or organic farming systems. Recent breeding of forage legumes has improved spring growth, pasture persistency, and resistance to pests and diseases. White clover is also useful in pastures in environmentally sensitive areas (e.g., for water harvesting) and areas designated for nature conservation in combination with grassland farming.

Red clover is better adapted to acid soils than alfalfa and is mainly used in Europe for silage and artificial drying. It has a high feeding value and sheep and cattle often prefer it to grass silage. Some alfalfa varieties are also well adapted to the climate of western Europe. Alfalfa has a lower cell wall content, and a higher cell soluble content (including crude protein, minerals) than ryegrass. Despite a lower organic matter digestibility compared to ryegrass, intake of alfalfa is 20–30% higher than ryegrass which has been attributed to a faster rate of digestion.

If legumes are to be used efficiently in sustainable grassland systems, their nitrogen-fixing potential must be optimized by using the most efficient combinations of plant and rhizobium genotypes which are effective in specific field situations. For most legumes in the developed world, selection of host genotypes has generally been on N-fertilized soils which tended to favor efficient use of soil N, rather than N fixation. Evidence suggests that all legumes are good scavengers of soil N and use this in preference to or in addition to fixed N. To optimize N fixation, selection should be carried out on low N soils. This approach has been adopted in the developing world with notable success, for example, in Brazil for soybean (*Glycine max*).

Nonagricultural Use of Grassland

Grassland can have recreational, functional, or landscape value apart from agricultural production. Intensively managed turf is widely used in many developed countries for sport and recreational activities. The turfgrass area in the United States is estimated to be at least 10 million ha and sales of turfgrass seed are second only to corn. In the United Kingdom amenity grassland is widely scattered over about 4% of the land surface. About half is seminatural grassland and the remainder is sown and maintained on a regular basis. It comprises areas differing widely in management and use, including nature reserves, parks, domestic lawns, road verges, airfields, dam-faces, reclaimed industrial wasteland and sports areas such as football pitches, golf courses, and school playing fields. On average mowing accounts for about 60% of the maintenance costs and fertilizer use for about 15%. The remainder is spent on the control of weeds, pests, and diseases and on renovation costs. Managers of amenity grassland must try to minimize input costs without risking sward quality, value, and longevity. Over the past 30 years considerable advances have been made in breeding slow-growing, hard-wearing grasses for intensively used areas and persistent fine-leaved grasses with good tolerance of pests and diseases; climatic and edaphic stresses; and attractive color and appearance throughout the year. This has helped to ease management problems and reduce costs. In more natural grassland greater understanding of ecological interactions is helping managers to maintain high levels of biodiversity while retaining important aspects of landscape appearance and protection.

Forage Grass and Legume Breeding

Domestication of forage crops is relatively recent with serious breeding to improve landraces and wild species ecotypes commencing early in the twentieth century. The collection and exploitation of natural variation from ecotypes and landraces has been important in improving temperate grasses and white clover for Europe; alfalfa for North America and the Mediterranean; timothy (*Phleum pratense*) for Scandinavia and many warm-season grasses for the United States. As breeders attempt to build a greater range of improvements into their cultivars, the possibilities of finding suitable combinations of traits within ecotypes become limited and new gene pools must be created by intercrossing diverse genetic resources. Wide crossing has been used very effectively in the improvement of ryegrasses for consistency of seasonal yield and nitrogen use efficiency; bermudagrass (*Cynodon dactylon*) for winterhardiness and digestibility; and *Agropyron* sp. and alfalfa/lucerne for environmental stress tolerance.

In western Europe reseeded swards containing improved modern ryegrass varieties can produce dry matter yields approaching 15 t ha^{-1} over a grazing season with an average grass production of 60 kg $ha^{-1}day^{-1}$. At a stocking rate of around two cows $ha^{-1}$and with good grazing management, daily intakes of 15 kg dry matter per cow are reasonably attained and should be sufficient for a daily milk yield of 18 liters. The maximum yield potential of ryegrass in temperate regions is estimated to be 29 t dry matter ha^{-1} for a grazing season – around 100 kg $ha^{-1}day^{-1}$. The best ryegrass varieties currently available in the United Kingdom can produce annual yields of 17 t of dry matter and there is clear potential to increase this by conventional breeding to 25 t. Forage crop breeders have also made significant advances in improving the energy value of grasses through increased digestibility and sugar content. Feeding high-sugar grasses, which accumulate 10–15% more sugars in leaves and stems, has improved milk production and milk protein content in dairy cows, liveweight gain in beef and sheep and reduced excreted N waste. Many tropical forage species are virtually undeveloped in terms of improvements in nutritive quality. They present breeders with considerable challenges to improve digestible energy and protein without reducing environmental stress tolerance and resistance to pests and diseases. Increases in cattle liveweight gains of between 35% and 60% have been achieved by improving digestibility in *C. dactylon* and *Setaria sphacelata*. Increased leafiness and a dwarfing gene in *Pennisetum americanum* increased cattle liveweight gain by 20%. Breeding programs have also been successful in increasing the mineral content of forage and in removing anti-quality factors such as alkaloids in reed canarygrass (*Phalaris arundinacea*). These objectives are important to improving animal health and welfare and hence productivity. For example, lambs grazing a low alkaloid, tryptamine-carboline-free reed canarygrass had up to 95% more liveweight gain per day compared to those grazing commercial cultivars.

Conclusions: The Value of Grassland

The agricultural benefits of grassland must be considered primarily in relation to how well livestock production is supported. The distribution of grassland yield within a year is more relevant in determining the number of animals that can be sustained by a grassland system (i.e., its carrying capacity) than the total annual yield. Equally important is the feeding value of forage which depends on protein/energy balance as well as overall digestibility. Increasing the content of soluble sugars in grasses improves digestibility, intake, efficiency of digestion, and N utilization by ruminants. Because many forage crops are perennial, they also have value in providing a stable ground cover which helps to prevent soil erosion and protects soil water against pollution (industrial and agricultural) and also excessive evaporation resulting in salinity problems. High yield may be of secondary importance in this context and may even be disadvantageous.

Grass is a complex crop and its agricultural value must be assessed in terms of the quantity and quality of livestock production (milk, meat, and wool) and also environmental and economic sustainability. In temperate grassland areas there is increasing emphasis on sustainability in farming systems rather than maximizing output at any cost. A key element in the efficiency of these systems is to optimize the protein/energy balance in ruminant feed and minimize input costs with on-farm produced feed assuming greater importance. The composition of forage coming from grasslands must be given the same attention as other livestock feeds. Grasses are rich in energy based on their carbohydrate content whilst legumes are very rich in protein. Grassland is perceived as the most natural way to feed animals and the survival of livestock producers in developed temperate regions will be linked to production from fewer, better-quality animals giving a better-quality product. In some tropical grasslands the utilization of herbage is poorly developed with little control of stocking rates leading to overgrazing and with little regard to grassland improvement. However in Australia and parts of Latin America, such as Brazil, considerable progress has been made in improving tropical pasture utilization. Considerable potential remains in other areas to improve synchrony between grassland growth and animal requirements in ways which sustain the basic resources of vegetation, soils, and animals.

List of Technical Nomenclature

Digestibility The fraction of green plant material which can be digested by ruminant animals.

Edaphic Relating to the physical, chemical, and biological characteristics of the soil.

Forage Grasses and herbaceous plants eaten by grazing animals.

Herbivore Any animal that feeds on plants.

Legume Nitrogen-fixing plants usually producing seed in pods and including clovers.

Rhizobia Nitrogen-fixing bacteria which coexist with leguminous plants in root nodules.

Rumen A complex stomach containing microorganisms which ferment plant material.

Ruminant Any animal, such as cattle and sheep, that has a complex stomach for the digestion of plants.

Silage Green plant material preserved by fermentation with lactic acid bacteria.

See also: **Crop Improvement**: Plant Breeding, Practice; Plant Breeding, Principles. **Nutrition**: Mineral Uptake; Nitrogen Fixation. **Photosynthesis and Partitioning**: Primary Products of Photosynthesis, Sucrose and other Soluble Carbohydrates. **Plants and the Environment**: Land Reclamation and Remediation, Principles and Practice. **Postharvest Physiology**: Senescence, Leaves. **Production Systems and Agronomy**: Agricultural Crops; Organic Farming. **Root Development**: The Rhizosphere and its Microorganisms. **Secondary Products**: Lignin. **Water Relations of Plants**: Basic Water Relations.

Further Reading

Bredenkamp GJ, Spada F, and Kazmierczak E (2002) On the origin of northern and southern hemisphere grasslands. *Plant Ecology* 163: 209–229.

Chapman GP (1996) *The Biology of Grasses*. Wallingford: CAB International.

Food and Agricultural Organization (2003) http://www.fao.org/ag/AGP/AGPC/doc/pasture/pasture.htm

Hodgson J and Illius AW (1996) *The Ecology and Management of Grazing System*. Wallingford: CAB International.

Humphreys LR (1991) *Tropical Pasture Utilization*. Cambridge: Cambridge University Press.
Loch DS and Ferguson JE (eds) (1999) *Forage Seed Production, vol. 2. Tropical and Subtropical Species*. Wallingford: CAB International.
Long SP, Jones MB, and Roberts MJ (eds) (1992) *Primary Productivity of Grass Ecosystems of the Tropics and Sub-Tropics*. London: Chapman & Hall.
Whitehead DC (1995) *Grassland Nitrogen*. Wallingford: CAB International.

Horticultural Crops, Protected

K E Cockshull, Horticulture Research International, Warwick, UK

Introduction

Modern systems of protected cropping offer some of the best examples of the ways in which plant science has been applied to the production of crops. In the most advanced of these systems, almost all aspects both of the aerial and the root environments around the crop are capable of being controlled and manipulated. With this measure of environmental control, it is feasible to maximize crop productivity throughout the cropping season and to force ornamental crops to produce flowers for specific occasions.

The major obstacle to protected cropping is that green plants are dependent upon light for their growth and development, and so the system must either allow solar radiation to reach the crop or it must provide radiant energy from an alternative source. Only radiant energy within a waveband spanning 400–700 nm is absorbed by chlorophyll and other pigments within the leaves, and utilized to create simple sugars from carbon dioxide (CO_2) from the atmosphere and hydrogen from water. This primary process is called photosynthesis, and the radiant energy that drives it is photosynthetically active radiation (PAR). The sugars are either converted into a storage product or are used almost immediately to provide the energy and the building blocks by which all other organic compounds required by the plants are synthesized. In theory, the cropping system could supply the required chemical energy directly, but this has not yet been achieved with green plants. By contrast, the mushroom does not require light for growth, and it has become a most important protected crop in many countries. However, its systems of production are so different from those of green plants that it will not be considered here.

The intention of this article is to show how knowledge of the plant and physical sciences has been applied to develop protected cropping systems, and to illustrate this by reference to the systems developed for two contrasting crops: salad tomatoes (*Lycopersicon esculentum*), and chrysanthemum (*Chrysanthemum morifolium*) and grown as a pot plant.

Basic Features

At the simplest level, protected cropping involves no more than the provision of protection from wind, rain, hail, and snow using a simple, transparent, or translucent canopy supported by props. Structures of this kind can lessen crop damage and thereby increase marketable yields. Indeed, they may be the only protection needed in regions where low ambient temperatures are unlikely to occur, but where they do occur, the merit of protection is enhanced if the system can raise and control the temperature of the air at a value above ambient. The usual source of energy for this purpose is a hydrocarbon fuel such as natural gas. Although the fuel can be used to heat air directly, a more uniform distribution of heat is achieved if the fuel is burnt in a boiler and then either hot water or steam is pumped around the protective structure. The advantage of raising the air temperature is that it regulates the rate at which the plant can synthesize new organic compounds from sugars and thus regulates the rate of formation of new cells. The latter process is especially important in relation to the initiation and expansion of new leaves because the productivity of many field crops is directly related to the quantity of solar radiation intercepted and absorbed by their leaves. Hence, productivity is usually proportional to the area of leaf surface per unit area of land, i.e., leaf area index (LAI). The same is true for protected crops and as the processes leading to the production of new leaf surface are enhanced in a heated environment, such crops have a higher LAI than crops planted at the same time, but in the field, and they grow faster.

The rates of most other plant processes are also accelerated by increased temperature, and crops held in warm, protected environments usually mature earlier than their contemporaries in the field. Protection is thus a means of accelerating growth, of advancing the natural cropping season, and of extending it when ambient temperatures begin to fall again in autumn. In northern regions, protection also provides a means of conserving those plants that would die if the temperature fell below 0°C.

Table 1 The transmission (%) of photosynthetically active radiation (400–700 nm) both direct (perpendicular to the material) and diffuse, and the transmission of infrared (IR) long-wave radiation (5000–14000 nm) by various glasshouse cladding materials

Material	*Thickness (mm)*	*Transmission (%)*		
		IR	*PAR*	
			Diffuse	*Direct*
Single glass	4	0	83	89
Double glass	2 × 3	0	72	82
Polyethylene (UV stabilized)	0.2	56	–	89–92
PMMA acrylic (double web)	16	–	75	84
Polycarbonate (double web)	10	–	65	74
Ethyl vinyl acetate	0.18	22	–	91
Teflon (fluorinated ethylene propylene)	0.05	57	–	96

Data from Waaijenberg D (1995) Greenhouse construction and equipment. In: Bakker JC, Bot GPA, Challa H, and Van de Braak NJ (eds) *Greenhouse Climate Control*, p. 170. Wageningen: Wageningen Pers.

Structures

The classic examples of frost sensitive plants are oranges (*Citrus sinensis*) and lemons (*Citrus limon*). Their fruits and fragrant flowers were once highly prized, and considerable money and ingenuity was invested to develop protected environments in which they could be safely conserved over winter in northern Europe. Although the earliest conservatories had just a few small panes of glass in their walls, later advances in materials and engineering made it feasible to construct buildings that could be clad almost entirely in glass. The advantages of glass as a cladding material are that it is rigid and is now available in thin sheets that are highly transparent to PAR (**Table 1**). Nevertheless, the structure required to support the glass cladding does obstruct light, and so the whole glasshouse may transmit less than 70% of the solar radiation falling upon it. The proportion transmitted will depend upon losses from reflection (related to the angle of incidence of the solar radiation), the thickness of the glass, its chemical composition, and the amount of dirt on its surfaces. The proportion that reaches the crop also depends upon the number and size of pipes and other obstructions installed between the crop and the glass (**Figure 1**).

Figure 1 (see color plate 25) Inside a modern "Venlo" glasshouse typical of those erected in northern Europe. Note the large glass panes, the light weight structure, and the ridge mounted ventilators. Photograph courtesy of Dalsem Horticultural Projects B.V.

The objects within the glasshouse, including the soil, absorb the transmitted radiant energy, are warmed by it, and then transfer energy to the surrounding air, but loss to the outside by conduction and convection is restricted by the closed nature of the protective structure. The warm objects also lose energy by radiation, but the especial merit of glass is that it absorbs long-wave radiation between 5 and 10 μm, i.e., the waveband emitted by warm objects. Consequently, a proportion of the radiated energy is returned from the cladding, either by conduction, convection, or re-radiation, and raises the temperature of the air within the glasshouse yet further.

This "greenhouse effect" is most valuable when ambient temperatures are low, but it is a hindrance when temperatures within the glasshouse become superoptimal for growth and development. In northern Europe, excessively hot air is exhausted from glasshouses by opening ventilators hinged from the roof ridge, while in North America, more active systems are frequently employed. One such system uses electric fans to extract hot air through one wall of the glasshouse while cooler air is drawn in through vents in the opposite, external wall. In the USA, especially, the incoming air may also be passed over a moist surface where it is further cooled by providing the latent heat required for evaporating the water. When the solar irradiance becomes too high, shading can be applied to the outside of the glasshouse to minimize the temperature gain. Systems of automated shading have been devised that draw shade materials over the glasshouse when either the internal air temperature or the external irradiance rises above some threshold level. An external system must be robust enough to withstand strong winds, and so it is more common to find shade screens installed within the glasshouse. In this position, they protect the crop from excessive radiation, but the internal air temperature will still rise because the radiant energy has already entered the glasshouse. A disadvantage of all shades is that they reduce the PAR transmitted to the crop and so reduce the potential for growth.

The thermal insulation properties of the relatively thin sheets of glass that are used to clad glasshouses are not high and growers seek improved insulation when energy costs rise. Improvements can be achieved by installing a second or even a third layer of cladding, and these additional layers can be of film plastic. However, there will inevitably be a further reduction in light transmission and the value of the consequent loss of yield may outweigh the value of the fuel saved (**Table 1**). In addition, as the enclosed air is no longer in contact with a cold external surface, its humidity will rise and may also lead to reduced yields. Often, the most cost-effective approach is to place an insulating cover over the crop at night and then to withdraw it and store it by day in such a way as to minimize any light loss. These covers are often referred to as thermal screens or thermal blankets.

Although glass has many advantages as a cladding material, fiberglass is more popular in some regions of the USA and, worldwide, film plastics are the most popular alternative (**Table 2**). Their principal advantage is that they are highly transparent to PAR (**Table 1**) and cost less than glass. They are also lighter in weight, require less support, and can simply be drawn over metal hoops to create walk-in tunnels. They can also be used to create "mulches" and even low tunnels to advance strawberry (*Fragaria* spp.) and vegetable crops in the field. Their main disadvantage is that they have a relatively short useful life and the costs of disposal can be high.

The first polyethylene films to be made did not absorb long-wave, infrared radiation, and so were not as thermally efficient as glass. However, additives are now routinely incorporated into the films to make them absorb in this waveband. Yet other additives are incorporated to discourage breakdown of the film under the influence of ultraviolet (UV) radiation and to discourage water droplets from forming and remaining on the inner surface of the film where they reduce light transmission. Many modern films are laminates and have the advantage that the additives can be incorporated into the layer where they will be most effective and long lasting. Indeed, the biggest advantage of plastic films may yet prove to be the ease with which their transmissivity in specific wavebands can be modified by incorporating appropriate compounds, as required. For example, removing the blue and UV wavebands that affect sporulation of saprophytic and parasitic fungi can reduce crop losses, while reducing transmission between 700 and 800 nm can modify stem extension. Finally, as infrared radiation in the 800–3000 nm waveband constitutes about one half of the energy of incident solar radiation, any additive that absorbs or reflects a high proportion of this waveband will greatly assist temperature control in summer.

Glasshouses with single spans greater than 12 m (40 feet) were once quite common in northern Europe, but new installations now consist of a

Table 2 Estimates of the area (ha) of greenhouses in different countries clad with either glass or plastic

Country	*Cladding*			*Total*
	Glass	*Plastic*	*Unspecified*	
Algeria			5000	5000
Belgium	542	2535		3077
China			600 000	600 000
Colombia			4500	4500
France	3500	4300		7800
Germany	3700	1100		4800
Greece	100	3470		3570
Israel			1400	1400
Italy	3750	25 000		28 750
Japan	2000	40 000		42 000
Morocco			1600	1600
Netherlands	10 000	100		10 100
Portugal	100	1400		1500
Romania	1450	4050		5500
South Korea		5107		5107
Spain	180	25 000		25 180
Turkey			9800	9800
UK	2400	380		2780
USA	4378	3182		7560

The cladding is described as "unspecified" where the original source of information did not differentiate between glass and plastics. Data from various years, mainly in the 1990s, taken from Enoch HZ and Enoch Y (1998) The history and geography of the greenhouse. In: Stanhill G and Enoch HZ (eds) *Greenhouse Ecosystems*, p. 10. Amsterdam: Elsevier.

Figure 2 A large glasshouse complex erected in The Netherlands by Dalsem Horticultural Projects B.V.

number of interconnected bays, each much less than 12 m wide. The standard bay width in Europe is 3.2 m (10 feet 6 in), although wider sizes are made, and each bay has a central ridge, giving a saw-tooth profile (**Figure 2**). In the USA and Canada, on the other hand, the roof is more likely to be covered with an inflated twin-skin film plastic, giving a rounded, barrel-shaped profile, and bay widths are less standardized. Gutter heights of at least 4.0 m (13 feet) are now standard on new glasshouses in most countries.

Control Systems

Modern control systems are most likely to be computer based and to have the ability to collate records of the external and internal environments, and to control all the systems available for modifying the environment. Such systems can produce both stable environments and dynamically changing ones, in which, for example, the change in temperature from day to night and vice versa can be progressive rather than abrupt. Many crop plants can integrate fluctuations in temperature over a range of at least 10°C and over a time span of at least 24 h, and dynamic control algorithms are now available that minimize energy consumption while sustaining a given average temperature. Software is also available that allows the setting for a controllable factor to be varied in relation to changes in an uncontrollable one, e.g., night temperature might be modulated according to the amount of solar radiation received in the day. Furthermore, one environmental factor might be altered dynamically so as to optimize the margin between costs and anticipated returns. In the "speaking plant" approach, environmental control is based directly upon the responses of representative plants or their parts, e.g., to maximize canopy photosynthesis or to regulate leaf temperature.

With the root environment, irrigation can be controlled by measuring evaporation or by estimating it from environmental records. In hydroponic systems, the pH and the electrical conductivity (EC) of the circulating solution can be continually monitored and controlled at values that have been established as optimal for the crop. The uptake of nutrients can be monitored and the addition of nutrients controlled, although, as yet, reliable electrodes are not available for all of the mineral elements in solution. Nutrient uptake is seriously hampered at low solution temperatures, and the temperature of any added irrigation water or nutrient solution should be kept above about 10°C. With cucumber (*Cucumis sativus*), it is desirable to control the root environment at an even higher temperature. The method of irrigation and its control depends upon whether the crop is grown in soil, soilless compost, water culture, or rockwool. For tomato, rockwool is supplied as slabs of a fibrous material wrapped in polyethylene. It has proved to be a most useful substrate as its physical properties are consistent, it is inert, sterile, and well aerated, and yet it also provides a reservoir of water and nutrients.

The Protected Environment

The installation of sophisticated computer control systems in greenhouses enables almost all aspects of a crop's aerial and root environments to be regulated, and provides a means by which crop productivity can be maximized. As plant growth is driven by photosynthesis, it is strongly influenced both by the carbon dioxide (CO_2) concentration of the surrounding air and by the amount of light intercepted by the plant's leaves. While it is not possible to modify the atmospheric concentration of CO_2 in the field, it is feasible to do so within the protected environment of a modern glasshouse where the leakage rate can be less than 0.5 air changes per hour when the ventilators are closed. Consequently, CO_2 from almost any convenient source can be added to raise the atmospheric concentration above the normal ambient one of about 350 vpm. In northern Europe, the exhaust gases from natural gas fired boilers or cogeneration engines have proved to be a cost-effective source of additional CO_2, even though the gases have first to be cooled and the concentrations of any pollutants reduced to safe levels. The cooled and scrubbed gases are usually introduced to the glasshouse at floor level so that the additional CO_2 can move upward through the leaf canopy by diffusion and convection. Simple gas analyzers are used to control the CO_2 concentration in the glasshouse at between 800 and 1200 vpm CO_2 because the net photosynthesis of most glasshouse crops approaches a maximal rate at concentrations in this range. Once ventilation begins, it is no longer either feasible or economic to maintain such high CO_2 concentrations. The problem then is that crop photosynthesis within the confines of a glasshouse can deplete the CO_2 concentration below even the normal ambient level and crop yield will suffer. For this reason, it can be economic to add CO_2 so as to maintain a concentration at or above 350 vpm, even with the ventilators open.

Yield of many glasshouse crops is also directly proportional to the amount of light intercepted by their leaves. Hence, to maximize this it is important to clean both surfaces of the glasshouse cladding material regularly and to manipulate crop density so that as much light as possible is absorbed by green leaves. Because growth and yield are so strongly correlated with the amount of light reaching a crop, it may be economically worthwhile to give additional light from electric lamps, especially if it is considered essential to sustain the required continuity of supply and product quality of the crop through the winter. The color of the light emitted by the lamps is not of great importance because most wavelengths of the visible spectrum can be used in photosynthesis. What is critical, however, is the efficiency with which a lamp converts electrical energy into PAR. The high-pressure sodium discharge lamp is particularly efficient in this regard, and so is widely used for supplementary lighting.

Atmospheric humidity is also an important factor in the protected environment because the spread of many diseases is influenced by humidity, usually by high humidity. High humidity also induces a number of physiological disorders caused by localized calcium deficiencies arising from a lack of transpiration and associated calcium transport. The glasshouse ventilators can be opened to allow the warm humid air to escape and be replaced by cooler air, but this utilizes more energy. Similarly, keeping heating pipes at a minimum temperature at all times will heat the glasshouse air and lower its relative humidity, but at a cost.

Water stress is a frequent cause of reduced crop yield in the field but in protected environments it can be avoided by growing plants in hydroponic systems, such as NFT (see earlier) or rockwool. The slabs of rockwool are normally wrapped in plastic so as to isolate them from the soil and to minimize water loss. The use of computer controlled irrigation systems allows water to be applied at the required volume and frequency to offset exactly the loss of water through evaporation. When combined with a nutrient fertilization system, stock solutions of nutrients can be added to the irrigation water at the required dosage in order to offset the uptake of specific ions by the crop. The nutrient solution is usually delivered to rockwool slabs through fine capillary tubes, and a similar system can be used to apply nutrient solutions to crops growing in other substrates. The use of hydroponic systems has almost eliminated the occurrence of severe water and nutrient stresses and the incidence of soilborne diseases in glasshouse crops, with the result that the productivity of these crops has dramatically increased.

Cultivation Systems for Protected Cropping

Salad Tomatoes

The special feature of the production of tomato (*L. esculentum*) under protection, particularly in northern Europe and Canada, is that each plant remains in continuous production for up to 8 months or more at a time. Also, as a new truss or cluster of flowers is initiated following the formation of every third leaf, a mature plant carries leaves, flowers, and fruits at all stages of their development. A wide range of cultivars producing fruit of various sizes, shapes, colors, and even biochemical composition is available, but this section will deal only with the protected production of salad tomatoes, i.e., round, red fruit intended for immediate consumption. Tomato plants cannot withstand frost and, as any temperature below 10°C adversely affects both pollen production and fruit set, fruiting plants can be produced only in heated greenhouses for much of the year in northern latitudes. In those regions, however, cropping cannot usually be profitably sustained throughout the winter because there is insufficient sunlight to produce enough fruit of adequate quality.

Normally, seed is sown toward the end of the year and the seedlings transplanted into small rockwool cubes and grown at close spacing in a propagation house. At this spacing, it is economically worthwhile to augment sunlight using high-pressure sodium lamps delivering about 25 μmol m^{-2} s^{-1} PAR for up to 18 h per day. When more widely spaced, however, as in the growing house, it is not usual to augment sunlight, even though fruit yield is directly proportional to the quantity of radiant energy incident upon the crop.

Once flower buds are visible, the young plants are transferred to the growing house, set out in pairs of rows separated by a path, and bumblebees are introduced to ensure pollination. Each plant is grown as a single-stem cordon, i.e., any shoots arising from the leaf axils are removed, and such a plant will grow more than 10 m (33 feet) in length and produce more than 30 fruit clusters (trusses) before cropping ceases in October. The stem and its fruit are supported on twine wound around a bobbin that is hooked onto a wire. The wire runs the length of each row and is attached to the glasshouse structure at gutter height. When the plant grows above the wire, twine is unwound from the bobbin and the bobbin is then moved along the wire. This causes the top of the plant to be lowered below the wire while the base of the stem trails along the ground. As the gutter heights of modern glasshouses are in excess of 3.5 m (11 feet 5 in), these operations have to be performed from a mobile, elevated platform.

The training system ensures that the young leaves with the highest photosynthetic potential are placed in the most favorable light environment near to the roof of the glasshouse. Yield and photosynthesis are closely correlated, and about 70% of all assimilates are partitioned to fruits in round tomatoes. Raising the atmospheric CO_2 concentration around the crop to increase net photosynthesis can increase yields, and the increase, which is approximately linear between 350 and 700 vpm CO_2, approaches an asymptote above 1000 vpm CO_2. The CO_2 can be obtained either from containers of the liquefied gas or by burning a hydrocarbon fuel such as natural gas. Even the exhaust gases from boilers or cogeneration units can be utilized, although they must first be cooled and have pollutants removed. Enrichment to 1000 vpm CO_2 is economically worthwhile when the ventilators are closed in winter, and when the ventilators are open in summer, photosynthesis can quickly deplete the atmospheric CO_2 concentration from 350 vpm to less than 300 vpm. At such times, it can be economical to add the gas to the glasshouse air to avoid depletion and, if the cost of the gas is low, it might be economical to enrich above ambient.

The crop is usually grown at an average 24-h air temperature of about 18.0°C, although the actual day and night temperature settings will be varied at different times of year. The atmospheric vapor pressure deficit (humidity) should be maintained at more than 0.3 kPa when the daytime irradiance is low, as this will sustain transpiration and so lessen the chance of calcium deficiency in the leaves. Cropping in northern Europe may continue until late October before the crop is removed and the house prepared for the next crop. Supply is usually switched to more southern latitudes during winter but, by late spring, it becomes impossible to maintain fruit quality in these southern regions as fruit set and uniformity of fruit ripening are both adversely affected at temperatures above 25°C.

Hydroponic cultivation systems enable the plants to be isolated from soilborne diseases, and the systems allow control over the supply of nutrients. One very successful hydroponic system is the "nutrient film technique" (NFT) in which plants are placed in shallow troughs that are sloped to create a 1:100 gradient along which a thin film of nutrient solution runs. The "spent" nutrient solution is collected in a central storage tank where its volume, pH, and electrical conductivity (EC) can be monitored and reset to desired values, before it is recirculated. In winter, it is considered desirable to limit cell enlargement, leaf size, and thus the vigor of vegetative growth by raising the EC of the nutrient using NaCl. Another very successful system is based on rockwool slabs. The young plants are placed over holes cut in the top of the polyethylene wrapper and their roots grow into the rockwool. An irrigation control computer supplies a standard volume of nutrient solution per plant every time a given quantity of solar radiation has been accumulated that day, and the solution is delivered to the plants via drippers. The polyethylene sleeve around the slab is slit along one side to allow any excess solution to run off, while a small reservoir of solution collects at the base of the slab. The run-off can be collected, but it should be sterilized before it is recirculated.

Markets frequently require fruit to be uniform in size, and to achieve this it is important that the supply of assimilate is matched by the presence of a suitable number of sinks for assimilates, i.e., developing fruit. The current preference in the UK is for fruit weighing 70 g (2.5 oz). This can be attained for much of the year if the crop is planted at a low density, e.g., 8000 plants acre^{-1}, and then, as the daily light integral increases, the shoot density is increased to as much as 16 000 shoots acre^{-1} by allowing plants to retain one side shoot.

Chrysanthemum as a Flowering Pot Plant

Unlike salad tomatoes, the marketable product is the entire plant together with the pot in which it is growing (**Figure** 3), and most growers are usually producing flowering pot plants in every week of the year. Furthermore, the production system is organized more like a factory production line with pots, compost, and young plants entering it, and flowering pot plants leaving it, having occupied the least possible space for the shortest possible time commensurate with producing a product of a specified quality. Other differences arise because many cultivars of chrysanthemum are chimeras, and so the plant is normally propagated vegetatively. Propagation is usually left to specialist companies who maintain healthy, vegetative stock plants and supply either rooted or unrooted stem cuttings, as required.

Another major difference from tomatoes is that flowering in chrysanthemum is controlled by daylength, and flower initiation and development generally occur most rapidly when the daylength is less than 12 h. The principal controlling factor is actually the duration of uninterrupted darkness, which has to exceed 12 h. Thus, to produce flowers in summer, blackout covers have to be drawn over the plants every afternoon so as to shorten the day and give 12 h of darkness; this mimicking of short-day conditions is required between the spring and autumn equinoxes. By contrast, to keep plants vegetative between the autumn and the spring equinoxes, it is necessary to mimic long-day conditions. This can be done by switching on domestic light bulbs for a relatively short time at night to ensure that the plants do not receive more than 10 h of continuous darkness. Today, the standard procedure is to employ a "night-break treatment," whereby the lamps might be turned on for up to 4 h once 8 h of darkness has elapsed. This treatment requires a PFD of only 2.5 μmol m^{-2} s^{-1} PAR at plant height, and the lamps do not have to be on continuously but can be cycled on and off during the "night-break" period.

Figure 3 The pot chrysanthemum cultivar 'Cedar Falls' bred by FGB.

At the start of the production line, the pots are filled with proprietary soilless compost to which the required minor mineral elements have been added. Then, depending upon the size of the pot and the size of finished plant that is required, up to five unrooted cuttings are inserted in the compost. The filled pots are placed on heated floors where the temperature can be controlled to give successful rooting and the cuttings are given long days, at least until rooting has occurred. Once rooted, the tip of each cutting is removed in order to remove apical dominance and encourage branching. When ready, the pots are respaced and grown either on concrete floors or on benches. Hydroponic irrigation systems are not easy to use when the whole plant and its pot comprise the marketable product; therefore, the pots are usually irrigated using either an ebb-and-flood sub-irrigation system or a piped distribution system with one or more drippers per pot.

All cultivars are classified in terms of the time taken to marketing from the onset of short days when grown under standard conditions. Thus, the date of marketing of those plants can be scheduled and short-day treatment can begin at the appropriate time. Flowering is more responsive to the average 24-h temperature than either the day or night temperatures alone, and the optimum for most cultivars lies between 18°C and 20°C. Flowering pot plants are produced year-round, and to produce plants of high quality in the poorer light conditions of winter, it is usual to boost photosynthesis by enriching the atmospheric CO_2 concentration to 1000 vpm, using any of the methods described for tomato above. However, the principal factor controlling winter quality is the daily solar radiation integral, and it is economically viable to use supplementary lighting at a PFD of about 75 $\mu mol\,m^{-2}\,s^{-1}$ PAR, from high-pressure sodium lamps to improve the quality of this crop. It is especially viable to use it during the first few weeks of short-day treatment as it assists uniform flower initiation. In general, the cultivars that are grown as flowering pot plants are naturally compact, however, synthetic plant growth regulators are used to give fine control over plant height. In addition, stem extension is strongly influenced by the actual temperature of the day and night periods, and chrysanthemum plants are more compact when the temperature is low at the beginning of the daylight period.

The flowering of poinsettia (*Euphorbia pulcherrima*) and kalanchoe (*Kalanchoe* spp.) is also controlled by daylength, and the systems for producing these species as flowering pot plants are very similar to those described above.

See also: **Growth and Development**: Leaf Development. **Production Systems and Agronomy**: Commercial Flower Production Methodology;

Further Reading

Atherton JG and Rudich J (eds) (1986) *The Tomato Crop*. London: Chapman & Hall.

Bakker JC, Bot GPA, Challa H, and Van de Braak NJ (eds) (1995) *Greenhouse Climate Control*. Wageningen: Wageningen Pers.

Cockshull KE (1985) *Chrysanthemum morifolium*. In: Halevy AH (ed.) *CRC Handbook of Flowering*, vol. 2, pp. 238–257. Boca Raton: CRC Press.

Cockshull KE (1992) Crop environments. *Acta Horticulturae* 312: 77–85.

Kamp PGH and Timmerman GJ (2002) *Computerised Environmental Control in Greenhouses*, 2nd edn. Ede: IPC Plant.Dier.

Larson RA (ed.) (1992) *Introduction to Floriculture*. San Diego: Academic Press.

Stanhill G and Enoch HZ (eds) (1998) *Greenhouse Ecosystems*. Amsterdam: Elsevier.

Orchard Crops

A D Webster, Maidstone, Kent, UK

Introduction

For the purposes of this article, orchard crops are taken to include those fruit species that are usually grown in areas experiencing temperate climatic conditions. However, it should be noted that these species are increasingly grown in areas with much hotter climatic conditions. In these areas, high altitude and/or specialized management treatments to aid dormancy breaking are employed. Traditionally, orchards consist of fruit trees of large stature (e.g., 10–15 m) planted in rows at wide spacings (e.g., up to 10 m apart). However, where methods of controlling the strong inherent vigor of the fruit species are available, trees are often grown at very close between-tree spacings. Nonetheless, for the purposes of this article, these are still considered to be orchards.

The most important fruit tree species and their origins are listed below, followed by a brief description of fruit tree propagation and some of the agronomic principles used in their culture.

Fruit Tree Species

All of the fruit tree species grown outside in the temperate climatic regions of the world are members of the family Rosaceae. These species and their origins are shown in **Table 1**. The most widely grown species are the apple and pear, which are known as pome fruits on account of their typical fruit structure, which is derived from several fused carpels surrounded by accessory floral tissue, the hypanthium. The hypanthium is the flesh of the fruit and this surrounds the inner pericarp, which is papery, contains several seeds and is often referred to as the core. Usually, the fruits have five fused carpels. Another less common pome (or pip) fruit is the quince, and other species occasionally grown for their fruits, such as *Amelanchier, Sorbus* and *Crataegus*, also belong within the pome fruits.

Stone fruits (peaches, nectarines, apricots, plums, and cherries) form a distinctive fruit type known as a drupe, which is fleshy and contains, at maturity, a single seed. The component fruit parts are the thin skin or outer wall (exocarp), the fleshy portion (mesocarp), and a woody stone (endocarp) that encloses and protects the seed. The fruits of the olive (*Olea europea*), a nonrosaceous species more adapted to Mediterranean climatic conditions, are also drupes, although they are not classified within the stone fruits. Other less commonly grown fruits, such as the mulberry (*Morus nigra*) are classed as multiple fruits in that they derive from the ovaries of several flowers, closely clustered within the inflorescence.

All of the aforementioned rosaceous species of pome and stone fruits are deciduous and require a dormant period each year, prior to which their

Table 1 Tree species of Rosaceae grown commercially for fruit production

Fruit type	*Botanic name*	*Origin and remarks*
Apple	*Malus pumila*	A natural hybrid of several *Malus* species originating from Kazakhstan and Russian central Asia
Apricot	*Prunus armeniaca*	Northern and northeastern China
Bullace, mirabelle and damson	*Prunus insititia*	Southern Europe and western Asia
Cherry (sweet)	*Prunus avium*	Eastern Europe and western Asia (Caspian and Black Sea regions)
Cherry (sour or tart)	*Prunus cerasus*	Eastern Europe and western Asia (Caspian and Black Sea regions)
Medlar	*Mespilus germanica*	Europe
Myrobalan	*Prunus cerasifera*	Southern Europe and western Asia
Peach (and nectarine)	*Prunus persica*	China
Pear (European or common)	*Pyrus communis*	Western and southwestern China
Pear (Asian or Chinese)	*Pyrus pyrifolia*	Western and southwestern China
Plum (European)	*Prunus domestica*	Probably a natural hybrid between the Myrobalan plum and the sloe (*P. spinosa*); southern Europe and western Asia
Plum (Japanese)	*Prunus salicina*	China
Quince	*Cydonia oblonga*	Iran

leaves are shed. In order for these deciduous fruit trees to recommence growth, they require a period of winter chilling, which may need to be supplemented or imposed artificially in areas where winters of sufficient severity are not experienced. To break dormancy, uniformly buds of fruit trees must be subjected to an accumulation of what are known as chilling units. There are various models used for calculating chilling unit requirements, but most are based on the accumulation of hours at temperatures above freezing but less than 12°C. Cultivars differ in their chilling requirements; some peaches suitable for subtropical climatic conditions need less than 200 h to break dormancy, whilst certain sweet cherry cultivars may need in excess of 1600 h.

Propagation of Fruit Trees

Pome and stone fruit tree species are almost all heterozygous and do not breed true-to-type when propagated from seed. With most fruit species, selected cultivars must be reproduced using asexual (vegetative) methods. Most fruit scion cultivars are very difficult to propagate from conventional cuttings or layers and the most reliable method of propagation is by budding or grafting a scion onto a genetically distinct rooted plant; this plant, which forms the roots and basal trunk of the tree, is called the rootstock.

Advances in propagation science, such as the development of improved macro- and micropropagation techniques, have, in recent years, made it much easier to propagate scion cultivars on their own roots, without the need for a rootstock. However, for most of the rosaceous fruit crops, rootstocks provide much more than a means of propagating the scion, conferring many favorable attributes to the composite stion (scion + rootstock) tree (see below). It is for this reason that rootstocks continue to be used and their popularity increases throughout the fruit production regions of the world.

Seed Propagation

Rootstocks are propagated either from seed or asexually using cutting or layering techniques. Successful propagation from seeds requires that the after-ripening and dormancy requirements of the seeds are satisfied prior to any attempt being made to germinate them. Seeds of all of the fruit-bearing species of Rosaceae show this requirement. The main requirement is for cold temperatures in order to satisfy the seed's chilling requirement and break dormancy. Dormancy breaking is usually achieved by placing the seeds in a moist medium (sand or vermiculaite) and subjecting them to temperatures just above freezing point for 6 to 12 weeks. This process is known as stratification. Part of this requirement for stratification may be achieved and the length of time needed for its completion reduced by soaking the seeds in solutions of gibberellic acid.

Once stratification is completed and the dormancy requirement satisfied, the rootstock seeds are planted in the nursery for germination. As with most other seeds, they require adequate water and oxygen, together with temperatures between 15°C and 25°C for optimum germination and growth.

Vegetative Propagation

Both rootstocks and scions are propagated using vegetative techniques. The selected rootstocks for pome and stone fruits, which affect the growth and survival of the scion cultivar in many beneficial respects (see below), are all propagated asexually (vegetatively). Propagation of these rootstocks is either by layering (stooling) or by cutting techniques. With layering, the propagule remains attached to the maternal stock plant until it is rooted; with cuttings, the propagule is severed from the stock plant prior to rooting. Success in layering and the closely related technique of stooling requires that the stock plants are severely pruned back to ground level each winter. As the new shoots begin to grow in the spring, soil or some other medium is heaped around the shoot bases but not so deep as to cover the shoot tip. This "earthing up" operation is repeated several times during the growing season and serves to partially exclude light and etiolate the shoot bases. Etiolated shoots produce roots much more freely than non-etiolated shoots. In the darkened conditions created, there is less production of secondary thickening tissues and, provided that water and oxygen are present in the surrounding medium, root initiation is encouraged. After leaf fall in late autumn, the earthing up medium is removed from the bases of the stock plants and the rooted shoots (often called "liners") are cut away and subsequently planted in a nursery prior to budding or grafting in the subsequent summer.

A few rootstock cultivars, especially those for stone fruits and quince rootstocks for pears, are propagated from cuttings. These are usually hardwood (winter cuttings) or softwood (summer) cuttings, although occasionally semihardwood or root cuttings may also be used. The techniques used are similar to those used for most other plant species. Cuttings are produced on stock plant hedges, which are severely pruned each winter. Nodal cuttings cut as near to their point of origin on the stock plant as possible and preferably derived from positions as close to the base of the stock as possible root best. Winter cuttings root better if taken in autumn or spring rather than in mid-winter. Summer cuttings require the creation of a favorable environment (mist or fogging techniques) to alleviate cutting desiccation during the process of root formation. For improved rooting, the bases of both summer and winter cuttings are treated with auxin (usually powders or liquid solutions of indolylbutyric acid) prior to insertion in the rooting medium. Stock plant etiolation or blanching (etiolation of just the shoot bases on the stock plant using black tape bands) can be used to improve the rooting of recalcitrant clones.

Micropropagation techniques have also been researched and developed for the propagation of many fruit tree rootstocks and scions. However, to date, these have not gained significant commercial acceptance. Problems of apparent "rejuvenation" of the micropropagated plants have been experienced, leading to delayed cropping on micropropagated scions and increased suckering on micropropagated rootstocks.

Budding and Grafting of Scions

The majority of fruit cultivars are propagated by budding or grafting the scion cultivar onto a rootstock. Budding is undertaken in summer, at a time when the buds have ripened sufficiently to survive their excision and transfer to the rootstock and when the sap flow within the rootstock is optimal for bud insertion. The buds used are lateral buds formed on shoots developed in the current season. These are nongrowing quiescent buds found in the leaf axils of the mid-portions of the extension shoots. They are cut individually from these shoots together with a small sliver of the surrounding stem tissues, including epidermis, phloem, cambium, and xylem tissues. A notch or T-shaped cut of similar dimension to the bud sliver is then cut out of or into the side of the rootstock stem at approximately 10–30 cm above soil level. The bud is then inserted into the cut on the rootstock and the two parts bound together with degradable tapes, to prevent desiccation and aid healing. Where necessary (if nondegradable), the bud ties are cut, once the union between the bud and the rootstock has healed. Efficient healing will only occur if the cambia of the two components (bud and rootstock) are brought into close contact with each other and if desiccation is prevented.

In the first summer and autumn, the bud remains quiescent and no outgrowth is made. In the subsequent winter, the rootstock is cut off just above the position of the bud and in the spring the bud of the scion begins its growth into a young tree. The trees may be lifted and sold following 1 or 2 years growth in the nursery.

The principles of grafting are similar to those of budding, but the size of the scion component is larger and the timing of the operation differs. Scion grafts comprise pieces of stem cut from 1-year-old shoots during the winter months; these usually contain two or three lateral buds. These are attached to the rootstocks using various types of grafts, the most common of which is called a whip and tongue graft. As with budding, the main objectives are to join the cambia of the two components and to prevent desiccation. Occasionally, a technique known as framework grafting is used where fruit growers wish to change an orchard from one scion cultivar to another, without the need to plant new young trees.

Choice of Rootstock

For most of the subtropical and tropical fruit species, where rootstocks are only used to provide a means of propagation, it is common to use seedling-raised rootstocks of the same species as the scion. Seedling-raised rootstocks are still utilized for propagation of some peaches and apricots, but growers are increasingly turning to clonal (vegetative) rootstocks for propagating most pome and stone fruits.

A critical consideration when using either seedling or clonally propagated rootstocks is their health status. It is particularly important that rootstocks and scions used in fruit tree propagation are guaranteed free of infection from damaging viral, phytoplasm, fungal, and bacterial diseases. Such infections influence graft compatibility and may reduce tree growth, cropping efficiency, and fruit quality.

The rootstocks used for propagating apples, pears, and all of the stone fruit species are many and varied. Whilst with apple the rootstock species is usually the same as that of the scion (*Malus pumila*), this is not the case with most of the other pome and stone fruit species. Rootstocks for pears are most frequently from another genus, the quince, whilst with stone fruits a wide range of *Prunus* species and their hybrids are used as rootstocks. With these crops, rootstock choice is influenced by climatic and edaphic conditions at the proposed orchard site and the tree management strategy planned by the grower. Some of the many characteristics provided by rootstocks are listed in **Table 2**.

Graft incompatibility is occasionally experienced between scions and rootstocks, leading to decline in tree growth and eventual tree death. This problem may be exacerbated by infection with plant viruses and phytoplasms. The beneficial attributes of some rootstocks, such as the quinces for pears, are so great that, where incompatibility occurs, graft-compatible interstems are inserted between the rootstock and scion to overcome this problem. Research is only just beginning to explain some of the reasons for variable graft compatibility between quinces and pear scion cultivars.

Table 2 Beneficial attributes conferred from selected rootstocks to scions budded/grafted onto them

Resistance to soil fungi
Resistance to soil pests
Resistance/tolerance to mice/voles
Resistance/tolerance to high soil pH
Resistance/tolerance to drought
Tolerance to wet soils and temporary anaerobis
Tolerance to severe winter cold
Control of tree vigor
Control of tree habit and branching
Induction of improved precocity and abundance of cropping
Induction of improved fruit size and quality

Control of Tree Shoot Growth

Shoot growth on many fruit trees is often excessively vigorous and erect, resulting in trees that are too high to facilitate easy management. Accurate targeting of the trees with agrochemical sprays is very difficult on very tall trees and this can result in unwanted spray drift of pesticides into the environment. The labor costs of pruning, fruit thinning, and picking are much higher for tall, vigorous trees than for smaller trees; the ideal tree is one that can be managed without the need for steps or ladders. Fruit production is often delayed on very vigorous trees and this reduced yield precocity results in poor returns on investment in the orchard. Finally, in countries where light levels are not intrinsically high, light levels in the basal and central parts of the branch canopy of very large trees are frequently too low, leading to reduced production of flowers and reduced fruit quality. Fruit growers use a combination of strategies to control the vigor and branching habit of their trees and also to manipulate the balance between shoot growth and fruiting.

Genetic Control of Tree Vigor

Scions with reduced vigor or compact habit The easiest and most natural (environmentally sensitive) method for controlling tree vigor and habit is via the scion cultivar itself. Several species, such as the peach and the morello (sour) cherry, are naturally small trees and can often be grown on quite invigorating rootstocks, yet still produce relatively small, manageable trees. In contrast, species such as the European (common) pears and sweet cherries are trees of the forest margin in the wild and naturally produce very large and unmanageable trees. Species such as the apple, plum, and apricot are intermediate in vigor but, are nonetheless, usually too vigorous when grown on their own roots or on invigorating rootstocks to be amenable to modern systems of production.

Many attempts have been made to either breed or select from natural populations scion cultivars with reduced vigor. "Sports" or mutants of the apple cultivars Red Delicious and MacIntosh, which exhibit reduced vigor and compact habit, have proved popular with growers. Most of these "sports" have been found as spontaneous mutants in orchards of the standard cultivars. Attempts have been made to simulate this process of natural mutation using techniques based on irradiation of scions or their treatment with chemical mutagens. Compact clones of the apple cultivar Bramley's Seedling have been produced and, more recently, compact clones of the pear cultivars Conference and Williams bon Chretien have been produced using irradiation breeding techniques. Unfortunately, many of the mutants produced, both those found growing naturally and those induced, are chimeras. As such they often prove unstable and may revert back to the vigorous parent at any time after planting in the orchard. Nevertheless, a few of the more stable clones, which are most probably periclinal chimeras, have proved of value and are grown in commerce.

In recent years, more information has become available on the genes that control shoot vigor in trees and scientists have begun to use genetic engineering techniques in attempts to control tree vigor. Most of this work has focused upon influencing the genes that control the production of gibberellins in the shoot internodes. Gibberellins are known to be responsible for internode extension and hence contribute greatly to the extension growth made each season by actively growing shoots. This strategy has some similarities with the use of many plant growth retarding chemical sprays, which also inhibit the production of active gibberellins in the tree (see below). Although several dwarfed transgenic mutants have been produced using this gene manipulation strategy, none have yet been fully tested in the orchard. Apart from the current concerns of consumers regarding genetically manipulated food plants, there are also some potential questions regarding the efficiency of the technique. Unless the reduction in active gibberellins is very tissue specific (i.e., only within the internodes), many other vital processes within the tree that rely on gibberellins for their success (e.g., fruit set) may also be affected. It will probably be essential to employ tissue-specific promoters with the gene manipulation technology if the technique is to succeed and produce a practical solution to vigor control.

Conventional breeding techniques (i.e., hybridization) have also been used occasionally to produce trees of reduced size and habit. Apple trees with a columnar habit have been produced by crosses made between popular cultivars and a mutant of the cultivar MacIntosh discovered as a single branch in a British Columbian orchard. Unusually, this mutation is controlled by a dominant gene and crosses have produced a wide range of apple cultivars with columnar habit and with fruits of differing sizes, color, and internal quality. The unusual habit of these trees, which produce almost no branches, is thought to be associated with changes in the metabolism of cytokinins within them. Peach trees of compact or semicolumnar habit have also been produced and their commercial potential is currently being evaluated.

Dwarfing rootstocks Much of the necessary control of tree vigor with apple, pear, sweet cherry, and plum trees is achieved using dwarfing rootstocks. The dwarfing rootstocks used are shown in **Table 3**. Vigor control has not proved to be of such high priority with peaches, nectarines, and apricots and few suitable dwarfing rootstocks have been developed. Rootstocks offering effective control of scion vigor are not available for subtropical and tropical fruit tree species, although with citrus crops, use of the closely related *Poncirus trifoliata* does result in some control of tree vigor. Many of the subtropical fruit species (e.g., the avocado, *Persea americana*) produce very large trees and growers would welcome the production of dwarfing rootstocks.

Research conducted intermittently over the last 60 years has focused on trying to understand how dwarfing rootstocks bring about their effect on scion vigor. The rootstock's roots and stem piece (shank), as well as the graft union between the scion and the rootstock, have all been implicated in this dwarfing effect. In all fruit species, the roots of the rootstock play a major role in bringing about any dwarfing influence upon the scion. However, with apples and pears (not stone fruits) the rootstock shank (the part of the tree trunk consisting of rootstock tissues) also causes part of the dwarfing. With dwarfing apple and quince rootstocks, increasing the amount of rootstock shank present on a tree (e.g., by budding higher above ground level) increases the dwarfing effect on the scion. Moreover, using a piece of dwarfing rootstock stem as an

Table 3 Types of deciduous fruits and the genera and species used as dwarfing rootstocks

Fruit species	*Genera and species used as rootstocks to control scion vigor*
Apple (*Malus pumila*)	*Malus pumila* clones Paradise (M.8) and Jaune de Metz (M.9) and their offspring
Pear (*Pyrus communis*)	Mostly clones of the quince (*Cydonia oblonga*). Also, a few clones of *Pyrus communis* (e.g., Pyrodwarf)
Sweet cherry (*Prunus avium*)	Selected clones of related *Prunus* species, such as *P. fruticosa, P. canescens, P. cerasus,* and *P. incisa*, as well as bispecific hybrids between these and other species
Plum (*Prunus domestica*)	Clones of *P. insititia* and complex hybrids between several related plum species
Apricot (*Prunus armeniaca*)	A few clones of *Prunus insititia* and several complex hybrids between other *Prunus* species

intermediary, grafted between an invigorating rootstock and a scion (i.e., as an interstock/interstem), also causes significant dwarfing of the scion. The graft union may also play a supplementary role in the dwarfing effect, as research has shown that it can significantly influence the transport (both acrotonic and basitonic) via either the xylem or phloem of water, minerals, photosynthates, and plant hormones.

The mechanisms involved in rootstock dwarfing of scions are poorly understood, and over the last 70 years theories based on disturbances to transport of water, mineral nutrients, and phenolics have all been proposed and then, following further research, subsequently rejected. More recent research in the 1990s has suggested that most dwarfing rootstocks significantly disturb the basipetal transport of auxins to the root system and the acropetal transport of cytokinins from the roots to the shoots. The balance between the concentrations of auxin and abscisic acid in the phloem of the tree also seem to be altered when a dwarfing rootstock is used. It is possible, therefore, that dwarfing rootstocks bring about their effects on the scion primarily by disturbing the subtle balances of hormones produced and distributed within the tree. Much more research will be needed before this enigma is resolved.

It is important to note, however, that dwarfing rootstocks dwarf scions in a rather different way to that achieved using either genetic manipulation or sprays with plant growth regulating chemicals, both of which influence gibberellin biosynthesis. With the latter techniques, all dwarfing is the result of shortened internodes. Although some dwarfing rootstocks also shorten scion internode length slightly, this is not the main method of dwarfing. Most dwarfing rootstocks also reduce scion vigor by inducing slower seasonal shoot elongation, and more rapid termination of shoot growth in late summer or early autumn. Many also induce branches to grow less vertically and more horizontally, which in itself results in slower extension growth (see below). Methods of dwarfing that reduce gibberellin-induced internode extension are considered to be less desirable than the dwarfing achieved by rootstocks. The mutual shading of leaves caused by internode shortening is thought to be detrimental to photosynthesis and to fruit coloring.

Vigor Control by Shoot Pruning and Training

The most traditional method of controlling fruit tree vigor is by pruning and/or training the growth of the shoots/branches. Unfortunately, pruning provides only a transient control of tree size. Severely pruned trees invariably respond by vigorous compensatory regrowth in the subsequent season and reduced production of fruits, due to removal of a high proportion of the floral buds and to increased competition between the strong new extension shoots and the young fruitlets for assimilates and other tree resources.

Mature fruit trees tend to maintain a relatively constant ratio between their shoot length and root length. Disturbing this ratio, for instance by shoot pruning, stimulates the tree to produce compensatory growth to restore the ratio. For this reason, severe pruning is not recommended as a method of controlling tree size unless accompanied by root pruning (see below).

The rate of extension shoot growth (branch growth) of most fruit tree species is greater on shoots with vertical orientation and reduced on shoots/branches with horizontal or downward pointing orientation. For this reason, training tree branches toward the horizontal is a much more effective method of controlling tree vigor than pruning. It is still not fully understood whether it is the branch orientation *per se* or the physical act of bending/twisting the branches to these orientations that causes the effects; it is probably a combination of both.

Occasionally, trunk or branch girdling has been used to help reduce tree vigor. This entails cutting a narrow groove almost all the way round the trunk in the spring. Phloem transport is temporarily inhibited by this treatment (which is also known as bark ringing) and growth is slightly reduced.

Vigor Control by Root Pruning and Root Restriction

Horticulturists have used root pruning as a means of fruit tree vigor control for many centuries. In the walled gardens of the Victorian era, trees such as the inherently vigorous sweet cherry were kept in check by pruning their roots on an annual or biennial basis. Occasionally, the whole tree would be lifted from the soil, its roots pruned, and then replanted into fresh, fertile soil. As labor became more expensive, the technique was abandoned and has only become of interest again as machines capable of pruning roots have been developed.

The success of the technique relies on the fact that the tree strives to maintain a fairly constant ratio between its root and shoot growth (see shoot growth above). When the roots of a tree are severely pruned it endeavors to restore this ratio by making more root growth but less shoot growth in the months following pruning. One problem with the technique is that it can weaken tree anchorage and lead to tree collapse in gales and it also has some deleterious effects on fruit size (see below).

Growing plants within containers restricts their root growth and hence also reduces shoot growth, and this, together with root pruning, is the basis for the production of ornamental bonsai trees. Growing fruit trees within conventional containers is rarely economical and fruit size and quality are often reduced. However, attempts have been made to plant trees within root restriction membranes buried in the orchard soil. These membranes, made of precision woven polymers, inhibit the growth of roots, but allow the passage of water and dissolved nutrients through the membrane pores from the surrounding soil. Trials with such root restriction membranes have demonstrated that the shoot growth of apple, pear, and sweet cherry trees can be significantly reduced after several years in the orchard. Research has shown that although some of the observed growth reduction is attributable to reduction in the supply of water and mineral nutrients, part of it is in some way directly associated with the restriction of root growth. It has been suggested that root restriction may cause a reduction in the root supply of cytokinins to the aerial parts of the tree, which, in turn, reduces shoot growth.

Vigor Control by Withholding Water Supply

In fruit growing regions where summer rainfall is minimal and soils have low water-holding capacity, shoot growth can be controlled by withholding supplies of irrigation water, either to the whole or only to part of the root system. The technique of "regulated deficit irrigation" has been used commercially in Australia and New Zealand for control of fruit tree vigor. Much of the normal irrigation water is withheld from the tree during the early spring, at the time when shoot growth is most vigorous. This seems to have no deleterious effects on fruit set. Once the fruits begin to swell irrigation is applied in abundance and the tree yields and fruit quality are not negatively influenced. The technique aids control of tree vigor, whilst making more efficient use of valuable water reserves. Recent research suggests that plant growth can be partially controlled by drying only a proportion of the root system.

Vigor Control Using Chemical Bioregulators

Since the 1960s, fruit growers have had access to various synthetic chemical plant bioregulators, which they have used for controlling the excessive vigor of trees. Almost all of these bring about their action by inhibiting the production of gibberellins by the trees. Several of these inhibitors, such as cyclocel and paclobutrazol, inhibit the gibberellin biosynthetic pathway at an early stage, prior to the production of any gibberellins. Others such as daminozide and prohexadione-Ca operate much later in the biosynthetic chain, preventing the production of active gibberellins from inactive ones. All of them cause shortening of the shoot internodes and hence reduction in tree vigor.

Chemical bioregulators differ greatly in their persistence both in the tree and in the surrounding soil. Cyclocel, which until recently was used extensively for growth control in pear trees grown in many parts of the world (not the UK), has long persistence in the tree. This results in undesirable residue levels in the fruits and the effect may persist for several years after completion of spraying. For this reason, cyclocel is no longer used for growth control of fruit trees in Europe. Paclobutrazol, which is sprayed onto the foliage of apples and pears, but more usually applied to the root systems of stone fruits, has long persistence in both the tree and soil. Fortunately, however, it is preferentially translocated between the shoots and roots and no detectable residues are found in harvested fruits. Daminozide is of less persistence than paclobutrazol and cyclocel but controversy concerning residues found in baby foods made from USA apples resulted in its withdrawal from use. The newest chemical to be released for growth control of fruit trees is prohexadione-CA. This has very short persistence in the tree and soil and several sprays per season may be needed to effect good growth control. It also has interesting side benefits in its action in reducing the incidence of the damaging bacterial pathogen fire blight (*Erwinia amylovera*).

Control of Tree Cropping

The objective of most fruit growers is to produce the highest yields possible of fruits that are of a size, color, and internal quality desired by the markets. Competition in the world fruit market is severe and growing trees that are aesthetically very pleasing, but which are relatively unproductive, is not a viable option economically. Therefore, fruit growers seek methods of improving their yields and the market acceptability of the fruits produced.

Most deciduous fruit trees are partially or fully self-sterile and require pollen from another compatible cultivar in order to set seed and produce fruits. This means that suitable sources of pollen must be planted within the orchard, at relatively close proximity to the trees of the commercial cultivar. Moreover, the pollen is moved from flower to flower by bees and other insects, and environmental conditions within the orchard must be favorable for this activity to be effective. This entails encouraging wild bees by sympathetic maintenance of headlands and banks around the orchard or the introduction of honey or mason bees to the orchard. For the bees to pollinate efficiently, the orchard must not be too cold and well sheltered from winds (see section on "Climate control" below).

Genetic Control of Cropping

The most obvious method of increasing tree yields is to choose a cultivar that yields abundantly. However, the major fruit markets are usually slow to adopt new cultivars and prefer to retain those popular with the consumer. Cultivars exhibiting high yields usually produce abundant flowers, and preferably this flowering should start only 2 or 3 years after planting the trees in the orchard. The flowers must be attractive to bees and other insects and have ovules that are viable and of reasonable longevity. This should ensure that fruits can be set, even if pollination occurs several days after flower opening (anthesis); the latter is known as having a long effective pollination period. Finally, shoot growth on the trees should not be so vigorous as to compete unfavorably with the fruitlets, so resulting in their abscission. Using apple as an example, the cultivars Braeburn, Golden Delicious, and Gala fit most of these criteria, whereas the traditional UK cultivar Cox's Orange Pippin (COP) does not. COP produces fewer flowers and the flowers are of poorer quality (ability to set) than most other popular commercial apple cultivars. Production of a clonal sport of COP, which is self-fertile, has helped increase yields in seasons unfavorable to bee activity and pollen transfer.

The appearance of fruit, particularly its size and color, has become particularly important to the multiple retailers and their consumers over the last 20 years. This has prompted researchers to seek clonal mutants of the major cultivars of apples and pears that have increased fruit size or improved skin color. The focus of this work has mainly been in apples and pears, where clones with increased red color (apple) or increased skin russeting (pear) have been selected.

Rootstocks, in particular dwarfing rootstocks, can have a very beneficial influence on fruit tree cropping. It would seem that by initiating the termination of shoot extension growth quite early in the summer, dwarfing rootstocks aid the partitioning of assimilates (photosynthates) to the sites of floral bud initiation and development. Flower buds on rosaceous fruit species are initiated and partially developed in the season prior to flowering and this earlier termination in the competing shoot growth leads to the production of more flower buds of improved setting ability. In the following spring, this can result in improved fruit set and yields. This hypothesis is somewhat simplistic, however, as there are rootstocks that dwarf sweet cherry scions, but which reduce rather than increase flower numbers and quality. Moreover, there are a few quite invigorating rootstocks for apple, which also induce the production of abundant flowering and fruit set.

Rootstocks may also influence fruit quality. Dwarfing rootstocks, which result in an open tree canopy with good exposure to light of all the fruits, invariably produce fruits of better color and sweetness. A few rootstocks, such as the apple rootstock M.9 and the plum rootstocks "Ferciana" and "Plumina" also seem to have a direct beneficial effect on fruit size, even after account has been taken of and corrections made for the crop yield per unit tree size (i.e., yield efficiency).

Climate Control and its Influence on Cropping

Fruit set and the associated yields are greatly influenced by climatic conditions. The most significant effect is at the time of flowering in the spring, when frosts can damage blossoms and cold windy weather can significantly inhibit the activity of the pollen vectors (bees and other insects) as well as reducing the germination of pollen on the stigma and its growth down the style into the ovary. Deciduous fruits mostly require temperatures of 15–25°C if pollination and flower fertilization is to be optimal. At temperatures lower than 10°C, there is minimal bee activity and poor pollination and flower fertilization. Wind speeds greater than $16\,\mathrm{km\,h^{-1}}$ also significantly inhibit the activity of honey bees, whilst rain during blossoming can result in the reduced viability or even death of the pollen.

Fruit growers endeavor to overcome these constraints on fruit set and production by providing shelter in the form of windbreaks within and around the orchard. It is important, however, to ensure that in orchards sited on sloping sites, the windbreak does not impede the natural flow of cold air down the slope, or this will create what is known as a frost pocket. During radiation frosts, long wave radiation is lost to the upper air and replaced by cold air at and above ground level; it is this cold air that causes the frost damage.

Where frost at blossom time is a regular occurrence, growers may install frost protection systems. The most popular of these are overhead irrigation systems that spray water onto the trees during the frost event. The heat released as the water freezes (the latent heat of fusion) protects the flowers from the worst frost damage. It is vital to ensure that the water supply continues throughout the frost period or damage may be increased rather than reduced. Other frost protection systems rely on large fans, which pull down the hot air escaping from the orchard surfaces. Others generate hot air, which is released around the trees.

Flower initiation and development on fruit trees is also influenced by climatic conditions. Adequate exposure to light is essential for flower initiation on branches and rootstock/pruning/training systems that facilitate good light exposure to the whole of the tree branch canopy are the ideal. Autumn weather conditions are also important and research has demonstrated that yields of pear trees can be markedly influenced by autumn temperatures, which in some way influence bud acclimation and quality. Warmer than average temperatures in early spring can negatively influence the development of flowers of apple cultivars such as Cox's Orange Pippin by, in some way, influencing ovule longevity at the time of flowering.

Tree Pruning and Training and its Effect on Tree Cropping

Severe pruning is deleterious to cropping for two reasons; firstly, it removes many of the floral buds; and, secondly, it stimulates strong shoot growth in the spring, which competes with the young fruitlets often triggering their abscission.

Training branches toward the horizontal and away from the vertical orientation increases the production of flower buds and these flowers usually have a greater potential to set fruit. Some of these benefits may be explained by the reduced competition from shoot growth on the branches, as horizontal training also reduces the vigor of shoot growth.

Chemical Bioregulators and Their Effects on Tree Cropping

Sprays of chemical bioregulators are used in fruit production for both increasing yields and increasing fruit quality. In seasons when frost damage is severe, sprays of gibberellic acid are used to induce the production and retention of parthenocarpic (seedless) fruits by the pear cultivar Conference. Sprays of quite low concentration (5–10 $mg\,l^{-1}$), shortly after the frost damage, can induce the set and retention of significant crops of seedless fruits, although fruit shape is often less pyriform and more oblong than normal. Although sprays of gibberellins (GA_{4+7} or GA_3) together with auxins (1-naphthalenacetic acid, NAA) have shown promise in stimulating the production of parthenocarpic fruits of other fruit species, such as plums, sweet cherries, and apples, these techniques have failed to gain commercial acceptance.

Several weeks after pollination and fruit set (fertilization) many fruitlets usually abscind from the trees. Unlike the first wave of abscission, which comprises unfertilized flowers, the fruitlets dropping at this stage are all fertilized. The causes for this natural fruit drop are not fully understood, but it is greatly enhanced when initial flower quality is poor and weather conditions following fruit set are unfavorable (cold and cloudy weather). Current research suggests that hormone-controlled assimilate supply to the fruitlets may be implicated in this wave of fruitlet abscission. Although sprays containing mixtures of gibberellins and auxins can reduce this fruit drop, they have proved too unreliable and expensive to be used widely in commerce.

With a few cultivars, a final phase of fruit abscission may occur just before harvest and this can prove most damaging in terms of loss of yield. Auxin sprays (e.g., NAA) are used in some countries to reduce the severity of this last period of fruit drop. It has been suggested that the gibberellins and/or auxins supplied in the sprays supplement or substitute for natural hormones produced by the fruits and so aid their ability to sustain growth and attract supplies of assimilates. It is also argued that they inhibit the formation of the abscission layer between the fruit stalk and its attachment to the tree at the fruiting spur.

Many fruit cultivars set too many fruits and if the natural mid-season fruit drop is insufficient too many of these are retained through until harvest. This results in fruits that are much too small for the commercial markets. Excessive fruit set and retention may also inhibit the production of floral buds for the following season. Floral buds begin to be initiated only 6–8 weeks following fruit set in the spring of the season prior to flowering. This inhibition of floral initiation is partly due to the seeds in the fruits and the gibberellins they produce; excessive supply of gibberellins, either naturally produced from fruits or sprayed onto the trees causes abortion of the young floral primordial and poor flower production in the subsequent spring. Where this problem goes unchecked it may lead to the development of a biennial sequence of cropping with the production of abundant small fruits in one year and almost no fruits in the subsequent year.

Growers require, therefore, moderate fruit set and retention on their trees, so as to improve final fruit size and quality and also avoid the establishment of a biennial sequence of cropping. One method of achieving this objective is to either reduce the flower numbers or prevent a proportion of them from setting. Inhibition of flower production using sprays of gibberellins applied in the late summer of the previous season has shown promise on peach cultivars, but is considered too risky and inconsistent for general use. Various chemicals, such as ammonium thiosulfate, endothallic acid, and pelargonic acid can be sprayed onto the flowers, causing desiccation of the vital organs (stigma, style, and pollen) and prevention of fruit set. This strategy is used with both apples and plums for flower thinning. Physical or mechanical removal of flowers has, to date, proved either prohibitively expensive or inefficient.

Most growers prefer to delay decisions on adjustment of crop abundance on the trees until the risk of frost has passed (i.e., several weeks after flowering). A more favored strategy is, therefore, fruitlet thinning, which is the stimulation of an increase in the natural mid-season fruitlet abscission. Auxins (e.g., NAA) and ethylene-releasing chemicals (e.g., ethephon) have been tested for this purpose and are approved for use in some countries. For many years, the carbamate insecticide carbaryl has also been used to thin the fruitlets of apples, although this use is now declining following the withdrawal of this product from use in Europe. More recently, sprays of benzyladenine have shown promise as fruitlet-thinners for use on apples and pears. The mode of action of these fruitlet thinning chemicals is still not fully understood. Current theories suggest that they induce transient stress in the trees, stimulate ethylene production close to the abscission layers, and/or inhibit the translocation of assimilates to the fruitlets.

Postharvest Storage of Fruits

Markets for tree fruits increasingly demand supplies throughout the year. Although production in the two hemispheres can go some way toward satisfying this demand, there is also a need to be able to store fruits for variable lengths of time so as to ensure consistent market supplies. Stone fruits are much more perishable than pome fruits and their storage is usually limited to a maximum of 5–6 weeks. Some cultivars of apples and pears can, however, be stored for 6 months or more.

The prime objective in fruit storage is to slow down the processes of ripening whilst keeping the fruits free of diseases and of good texture, juiciness, and flavor. Most fruit storage is carried out in cold rooms held at temperatures just above freezing. However, with some of the stone fruits slightly higher temperatures are needed if chilling injury is to be avoided. Cool temperatures alone are only suitable for relatively short-term storage of fruits. Where longer storage is required of apples and pears it is necessary to manipulate the store atmosphere in what is known as controlled atmosphere (CA) storage. This entails allowing the levels of carbon dioxide to increase, whilst lowering the levels of oxygen in the store. As apples, pears, and many other types of fruits ripen they produce ethylene in an autocatalytic process; if allowed to proceed unchecked ripening is very rapid. Attempts have been made to reduce this ethylene effect on fruit ripening by removing the gas from the cold stores or inhibiting its production in the fruits. Sprays of aminoethoxyvinylglycine (AVG), a chemical known to inhibit ethylene biosynthesis in the plant, have also been used on orchard apple trees to delay and reduce the rate of ethylene production by the fruits.

List of Technical Nomenclature

Acrotony (acrotonic)	Translocation of substances within the plant in an upward direction.
Anthesis	The stage of flowering when the blossom first opens.
Auxins	Natural and synthetic growth hormones found in or applied to fruit trees and other plants.
Basitony (basitonic)	Translocation of substances within the plant in a downward direction.
Benzyladenine (BA)	A synthetic chemical with cytokinin-like activity, which is used for aiding fruitlet abscission and branch induction on young trees.
Blanching	Excluding light from the basal portions of stems, so as to aid the induction of rooting.
Carbaryl	A carbamate insecticide that also has fruitlet thinning activity on apples.
Carpel	The part of the fruit that encloses the seed.
Chilling unit	An hour unit of temperature that when accumulated aids the breaking of winter dormancy.
Clone	A group of plants originating from the same individual by asexual methods of propagation.
Cyclocel	A synthetic chemical applied to plants to inhibit their growth.

Cytokinins Natural and synthetic growth hormones found in or applied to fruit trees and other plants.

Daminozide A synthetic chemical applied to plants to inhibit their growth.

Drupe A fruit derived entirely from the ovary, with a single seed, a stony endocarp, and a fleshy mesocarp.

Etiolation The growing of plants in semidarkness.

Endothallic acid A synthetic chemical used as a flower desiccant for crop reduction (thinning).

Hypanthium Accessory floral tissues including the receptacle from which the fleshy parts of pome fruits develop.

Internode The part of the young stem between leaf nodes.

Interstock (interstem) A species or cultivar, distinct from the rootstock and scion, which is used to form part of the trunk by grafting between the two aforementioned components.

Layering An asexual technique of vegetative propagation.

Micropropagation A technique of propagation carried out primarily *in vitro*, employing cultures of shoots that are encouraged to proliferate and/or root when growing on liquid or solid media under controlled environmental conditions.

Misting (fogging) Techniques used in the propagation of softwood (summer) cuttings, where fine droplets of water or fog are applied regularly to prevent desiccation and cool the cuttings.

Paclobutrazol A synthetic chemical applied to plants to inhibit their growth.

Parthenocarpy The production of fruits without seeds.

Pelargonic acid A synthetic chemical used as a flower desiccant for crop reduction (thinning).

Pome A fleshy type of fruit formed by the fusion of ovaries, calyx cup, and floral tube that has several carpels containing seeds.

Prohexadione-Ca A synthetic chemical applied to plants to inhibit their growth.

Rootstock The root and basal stem part of a fruit tree, which is a distinct genus, species, or cultivar from the scion, to which it is grafted.

Scion The fruiting part of a tree.

Stion The composite tree comprising rootstock (interstock) and scion.

Stratification A technique used to aid dormancy breaking of seeds of Rosaceae.

See also: **Postharvest Physiology**: Ripening; Storage.- **Tissue Culture and Plant Breeding**: Clonal Propagation, Forest Trees; Regeneration of Fruit and Ornamental Trees via Cell and Tissue Culture.

Further Reading

Jackson DI and Looney NE (1999) *Temperate and Subtropical Fruit Production*, 2nd edn. Wallingford: CAB International.

Wertheim SJ (1998) *Rootstock Guide: Apple, Pear, Cherry, European Plum*. Wilhelminadorp: Fruit Research Institute.

Westwood MN (1993) *Temperate Zone Pomology, Physiology and Culture*, 3rd edn. Oregon: Timber Press.

Organic Farming

A M Litterick and C A Watson, Crop and Soil Research Group, SAC, Aberdeen, UK

Introduction

Interest in organic farming is currently at an unparalleled level in the Western world. Farmers, growers, environmental campaigners, consumers, and policy-makers are increasingly concerned about the problems created by conventional, industrialized agriculture and are asking whether organic agriculture can provide opportunities to address some of these problems or improve aspects of our food and farming systems. A strong organic market has developed in many Western countries in response to consumer demand, and these markets continue to grow as the quality improves and the quantity of available organic produce increases.

Recent Growth of the Organic Food and Farming Industry

In 1985, only 0.1% of the total EU agricultural area was farmed organically. By 2001, that percentage had risen to over 2%. Extrapolating this rate of growth would lead to 30% of Western European agriculture being organic by 2010 (**Figure 1**). In the UK, the area of in-conversion and organic land

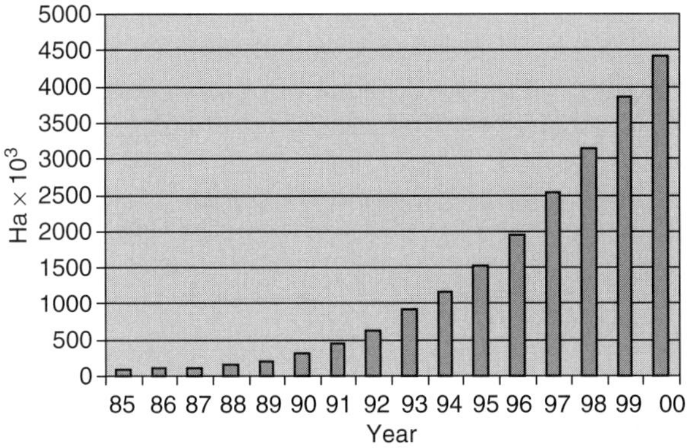

Figure 1 Expansion in the area of organic land in Europe (includes all 15 EU countries, all four European Free Trade Association countries, and all nine Accession countries). Data from 2000 are not yet complete. Figures based on data supplied by N Lampkin, Welsh Institute of Rural Studies, University of Wales, Aberystwyth.

totalled 4.3% in April 2002, with the highest number of producers located in Scotland, Wales, and the northwest and south of England.

Growth in interest and activity seems to be related to both the current financial pressures on agriculture and changing public perceptions of the agricultural industries. The difficult financial climate in agriculture is a significant driver for the increased interest in organic production, since current farm-gate prices for organic produce can be considerably higher than those for conventional produce. There is a strong perception amongst a significant proportion of the UK population that organic food is better, safer, or healthier and they buy organic food for these reasons. At present there is very little evidence to support such beliefs. There is, however, increasing evidence that some of the practices associated with conventional farming are creating environmental problems (mainly in terms of pollution and loss of biodiversity), and many people buy organic food because they feel that it is better for the environment. There is little doubt that the organic food and farming industry has moved beyond providing for a niche market: UK organic food sales for the year ending April 2002 were valued at £920 million. The European organic retail market has an estimated turnover of £6 billion and an increasing percentage (currently 3%) of the food sold in Europe is organic.

The Origins and Philosophy of Organic Farming

Organic agriculture is much more than the production of food without synthetic pesticides and fertilizers. Organic farming as practiced today has its origins in the work of a number of pioneers who were active in the early twentieth century. Key figures including Lady Eve Balfour, Mr Friend Sykes and Dr Albert Howard believed in the "rule of return," which states that "soil health and fertility must be maintained by encouraging the presence of humus." Mining the soil nutrients through intensive cropping followed by their replacement with artificial fertilizers was thought to lead to soil exhaustion and to crops vulnerable to pests and diseases. Such organic pioneers felt that soil required replenishment with organic matter, which they recognized as being a living substance, alive and teeming with organisms. These early organic voices had little evidence to prove that there was any harm in the use of synthetic fertilizers, and pesticides were little used at that time. The Soil Association was formed in 1946 by Eve Balfour and others to draw together evidence for the benefits of obeying the rule of return. It was not until the publication of Rachel Carson's book '*Silent Spring*' in 1962 that the potential dangers of pesticide use were clearly demonstrated. The organic movement grew slowly until the 1980s when awareness of the problems resulting from conventional agriculture began to increase. Rapid expansion of the industry began in the early to mid 1980s (depending on country) and continues today.

Organic agriculture is the only form of sustainable agriculture which is defined and regulated in law. The issue is confused somewhat because this form of agriculture is known under a number of different names, including biological, ecological, and biodynamic agriculture. The term "biological" is generally favored throughout mainland Europe, whereas Britain and the United States prefer "organic." Biodynamic farming is a specialized form of organic agriculture, which is part of a whole philosophy that encompasses education, art, nutrition, and religion as well as agriculture.

The principles and practices that lie behind these different names are largely similar. All forms of organic agriculture are based on the concept of the soil as a living system and all rely on the enhancement of natural cycles and the maintenance of links between soil, plant, animal, and humans. Successful organic agriculture depends on the fact that every farming practice affects many aspects of the farming system as a whole. This concept (**Figure 2**) is often described as "holistic." In other words, one component cannot be altered or removed from the system without positively or adversely affecting other components. In terms of management, organic systems require high inputs since detailed long-term planning is required to optimize the natural farm cycles.

The original definition of organic agriculture published by the United States Department of

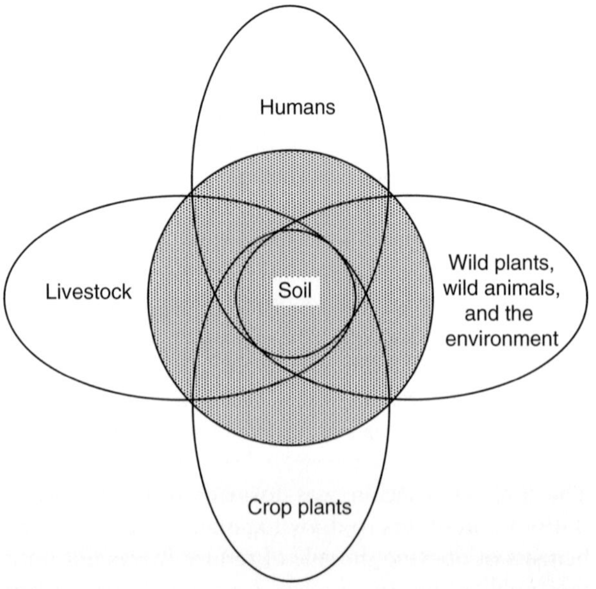

Figure 2 Links between the main components of organic farming systems.

Agriculture provides a useful summary of the most important practices:

> Organic farming is a production system which avoids or largely excludes the use of synthetically compounded fertilizers, pesticides, growth regulators and livestock feed additives. To the maximum extent feasible, organic farming systems rely on crop rotations, crop residues, animal manures, green manures, off-farm organic wastes and aspects of biological pest control to maintain soil productivity and tilth, to supply plant nutrients and to control insects, weeds and other pests.

The principles and practices which underpin organic agriculture have been defined in more detail by the International Federation of Organic Agriculture Movements (IFOAM) as:

- to produce food of high nutritional quality in sufficient quantity
- to interact in a constructive and life-enhancing way with natural systems and cycles
- to consider the wider ecological and social impact of the organic production and processing system
- to encourage and enhance biological cycles within the farming system, involving microorganisms, soil flora and fauna, plants, and animals
- to develop a valuable and sustainable aquatic ecosystem
- to maintain and increase the long-term fertility of soils
- to maintain the genetic diversity of the production system and its surroundings including the protection of plant and wildlife habitats
- to promote the healthy use and proper care of water, water resources, and all life therein
- to use as far as possible renewable resources in locally organized production systems
- to create a harmonious balance between crop production and animal husbandry
- to work as much as possible within a closed system with regard to organic matter and nutrient elements
- to give all livestock conditions of life with due considerations for the basic aspects of their innate behavior
- to minimize all forms of pollution
- to process organic products using renewable resources
- to produce fully biodegradable organic products
- to produce textiles which are long-lasting and of good quality
- to allow everyone involved in organic production and processing a quality of life which meets their basic needs and allows an adequate return and satisfaction from their work including a safe working environment
- to progress toward an entire production, processing, and distribution chain which is both socially just and ecologically responsible.

Regulation of Organic Food Production and Processing

In order to sell organic products, every farmer or processor must register with an approved certification body, comply rigorously with the standards of that body, and submit to regular inspections of his/her business premises and records. Land must undergo a minimum 2-year conversion period, during which the farmer complies fully with the standards of his/her chosen certification body, before it can be registered as organic. The organic standards of certification bodies in European Community countries differ from one another, but all must adhere to a set of basic minimum standards set out in Council Regulation (EEC) Nos. 2092/91 and 1804/99. The standards outline recommended practices in organic farming and food production and describe permitted, restricted, and prohibited inputs and practices for crop and livestock nutrition, animal welfare, pest, disease and weed control, and food processing. Each country has appointed an organization that audits and inspects certification bodies within that country to ensure that their inspection and certification procedures are adequate. In the UK, the United Kingdom Register of Organic Food Standards fulfils that function. Standards in countries outside the European Community must adhere to the minimum standards laid down by IFOAM.

There are currently 12 certification bodies in the UK, including the Soil Association, Scottish Organic Producers Association, Organic Farmers and Growers, and the Biodynamic Agricultural Association. Some of these certification bodies impose further, stricter regulations than the minimum standards set out in Regulations 2092/91 and 1804/99 on their producers.

Organic Crop Production

Crop Rotations

Since synthetic fertilizers and pesticides are prohibited in organic systems, modern organic crop production is radically different to its conventional counterpart. Balanced crop rotations are critical to the success of organic systems, but this is no longer the case in conventional agriculture, where chemical alternatives have allowed the intensification and specialization that have taken place over the last century. Here, economic pressures have led to the simplification of cropping systems and substitution of cultural methods of weed, pest, and disease control and fertility management with synthetic chemical alternatives. The reliance on crop rotations in conventional agriculture is now much reduced and continuous monocultures and short rotations without a fertility-building component are common in temperate parts of the developed world.

Crop rotations provide opportunities for organic farmers to prevent and control weeds, pests, and diseases, to provide nutrition for crops, and to minimize nutrient losses. Such opportunities exist because crops differ in their physiology, root architecture, nutrient demands, and susceptibility to pests, diseases, and weed pressure. In all organic rotations, there will be a fertility-building phase and a cash crop, or income-generating phase. In terms of time, the way in which these are partitioned will depend on climate, soil type, topography, available markets, and the presence/absence of livestock on the farm. The principles of rotation design are described in greater detail in **Table 1**.

The majority of organic farming enterprises in the UK and Europe are on traditional mixed (crop and animal) units which produce a range of cereals, fodder crops, and livestock. The success of these systems depends on ley–arable rotations, where nitrogen fixed by legumes (typically clover, *Trifolium* spp.) during the ley phase supports both animal production and a subsequent exploitative arable cropping phase. Management of crops and animals within the rotation aims to optimize the health and quality of the soil and the health and productivity of the crops and animals which depend on it. In addition to the traditional mixed farms, there are increasing numbers of specialized organic units throughout Europe, which are producing horticultural crops. Many of these farms are run without livestock and therefore operate "stockless" rotations. Such farmers may choose to operate stockless units because they have limited experience of working with livestock or because there is no history of livestock farming in their area and heavy investment in equipment, fencing, and infrastructure would be required to begin animal production.

Table 1 Key principles of crop rotation design

Principle	*Rationale*
Rotate deep- and shallow-rooting crops	Improve soil structure, aeration, water-holding capacity, and drainage
Alternate crops with large and small root biomass	High biomass crops increases the organic matter remaining in the soil for soil microbial and macrofaunal populations
Rotate N_2-fixing and N-demanding crops	Attempt to meet farm's N demands from within the system
Alternate weed-susceptible and weed-suppressing crops	Interrupt weed life cycle to reduce populations
Grow crops with different pest and disease susceptibilities	Break pest and disease life cycles, reduce host plant presence in rotation
Grow catch crops and green manures and undersow crops	Maintain soil cover to protect for erosion and leaching
Alternate autumn- and spring-sown crops	Combat weeds and distribute workload
Use appropriate crops, suited to climate and soils	Crops will suffer less stress and will therefore be less susceptible to pest and disease attack
Balance forage and cash crops	To make rotation economically as well as ecologically viable
Balance labor requirements and availability	Prevent over- and understaffing
Good timing of tillage operations	Balance positive results of tillage such as weed control with negative impacts such as disruption of macrofaunal activities and decreases in soil organic matter

Reporduced with permission from Robson MC, Watson CA, Fowler SM, *et al.* (2002) The agronomic and economic potential of break crops for ley/arable rotations in temperate organic agriculture. *Advances in Agronomy* 77: 369–427.

Table 2 Examples of different types of rotations

Typical ley/arable rotations[a]	
Example A	*Example B*
2–3-year short-term grazed ley (red clover or alfalfa on calcareous soils)	4–5-year grazed grass/clover ley
Wheat (or potatoes), then green manure[b]	Winter wheat
Potatoes/roots	Winter oats or barley
Wheat	2-year grazed red clover/Italian ryegrass ley
Rye or oats (undersown)	Winter or spring wheat
	Cereal/grain legume mix (or cereal undersown with ley)
Typical stockless arable rotations	
Example A	*Example B*
2-year red clover	1-year red clover cut and left as mulch
Wheat	Potatoes
Oats	Winter wheat
Beans	Spring beans
Wheat (composted manure may be applied)	Spring wheat (undersown) (composted manure may be applied)
Oats	Barley (undersown)
Typical stockless horticultural rotations	
Example A	*Example B*
2-year red clover	2-year red clover
Potatoes	Early potatoes with lettuce catch crop[c]
Carrots	Winter barley (composted manure may be applied)
Spring barley	Calabrese/cabbage
Winter wheat (undersown)	Spring oats (undersown)

[a] These examples are taken from Lampkin N (1990) *Organic farming.* Ipswich: Farming Press.
[b] Green manure: a crop such as grazing rye (*Secale cereale*) or vetch (*Vicia sativa*) which is grown for a limited period, then plowed in. Used to minimize nutrient losses and erosion from bare soil between main crops, to improve soil structure, or to help control pests or weeds.
[c] Catch crop: a crop which is grown following the main crop of the season. Main advantages include the minimization of nutrient losses and increased revenue.

Most organic farmers operate crop rotations throughout the whole farm, but some specialist growers of long-term crops (e.g., tree fruits and grapevines) have to import fertility from local sources and rely on wild plant understories and borders to improve plant (and thereby wild animal) biodiversity. Examples of stocked and stockless rotations are shown in **Table 2**.

Soil and Nutrient Management

It is recognized that plant and animal health and productivity in organic systems are heavily dependent on soil health and quality. Soil care and management is therefore central to the success of organic rotations. The farmer aims to choose a balance of crops for the rotation that will help to maintain and enhance soil nutrient content, soil nutrient cycling, soil structure, and soil organic matter content. Since synthetic fertilizers are prohibited in organic agriculture, the farmer often has to deal with a scarcity of readily available nutrients, so a primary aim of nutrient management in organic systems is to optimize the use of on-farm resources and minimize losses. The main nutrient inputs, outputs and flows on a mixed organic farm are illustrated in **Figure 3**.

The supply and management of N is more complex in organic than in conventional agriculture and organic farmers face a major challenge in that they must synchronize the availability of N mineralized from crop residues and animal manures with crop demand. Organic crops in traditional mixed farms rely mainly on symbiotically fixed N (from grass/clover leys and other legumes), and optimal use should therefore be made of crops that can help build soil fertility. Organic crops also obtain N from grazing animals (dung), imported composts and manures, and from atmospheric deposition.

Ensuring an adequate supply of P, K, and trace elements can also be difficult in organic systems (particularly on stockless farms) due to the restrictions on acceptable nutrient sources in the organic farming regulations. For example no genetically modified crops, crop residues, or animal products are permitted on organic farms. Organic materials such as farmyard manure and some types of compost are valuable sources of P and K. Various naturally occurring P, K, and trace element fertilizers including rock phosphate, basic slag, potassium sulfate, and seaweed extracts are approved for restricted use (i.e., they can be used when essential).

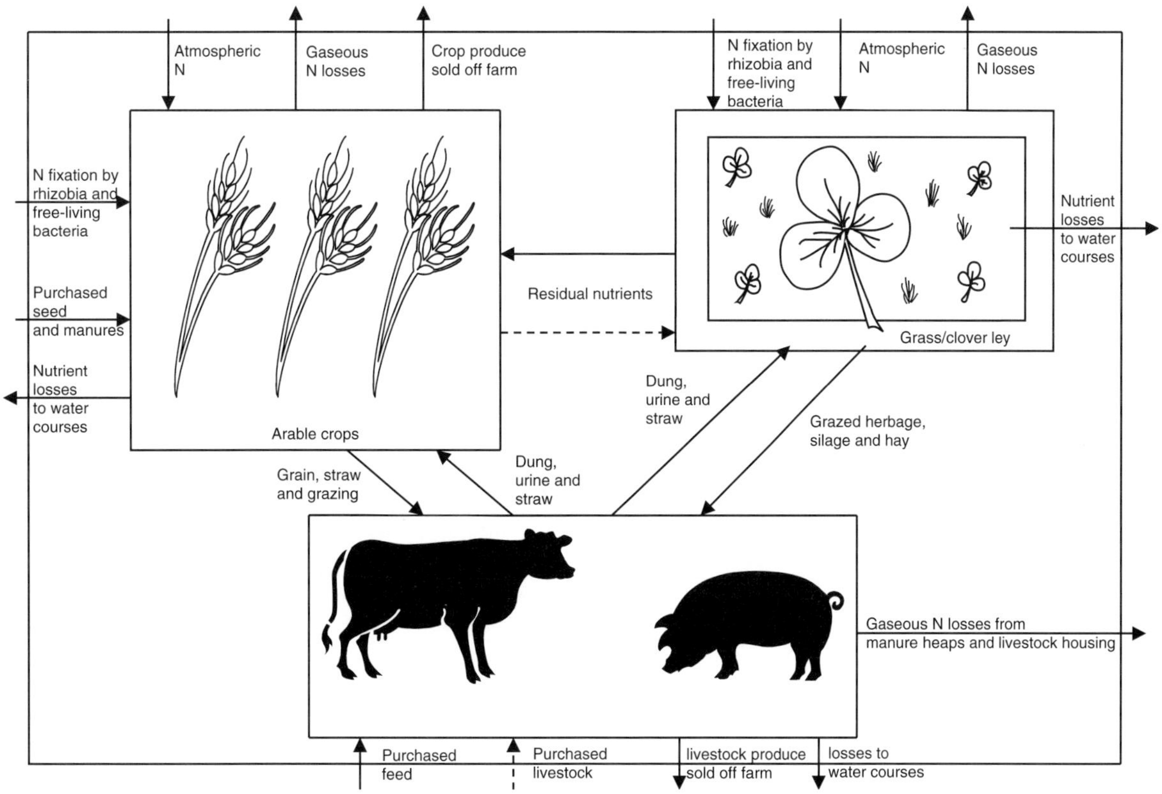

Figure 3 Nutrient inputs, outputs, and flows on a mixed organic farm.

An important feature of organic systems is that crop residues are returned to the soil (directly or indirectly) after harvest. There are three main advantages from this:

1. Nutrient losses from the system are minimized.
2. Deep-rooted crops such as alfalfa (*Medicago* spp. lucerne) and hemp (*Cannabis sativa*) can extract nutrients from depths inaccessible to other crops. Some of these nutrients eventually become available in the upper soil layers following microbial decomposition.
3. Soil organic matter (which is important due to its stabilizing effects on soil structure and because it acts as a reservoir for soil nutrients) levels are maintained or enhanced.

Nutrients can be lost from the system through leaching, soil erosion, volatilization, and denitrification. Management practices have a significant impact on nutrient losses, particularly N, from organic (and conventional) systems, and an understanding of the effects of such practices is therefore crucial if losses are to be minimized. Factors such as the choice of crops and cropping sequence, type of compost/manure/slurry, storage method, time of application (within the rotation and within the year), type and timing of cultivations and soil type are known to affect the nature and magnitude of nutrient losses from the farm. For example, nitrate leaching and losses of other nutrients including P and K are particularly severe when the soil is left bare for long periods. The practice prescribed by the organic regulations of balanced rotations with judicious use of catch crops and green manures provides an opportunity to significantly reduce losses of nutrients, in particular nitrate.

Control of Pests, Diseases, and Weeds

Since synthetic pesticides and some natural pesticides are prohibited in organic agriculture, farmers rely on a combination of cultural and management strategies to prevent and control pests, diseases, and weeds. Reactive treatments for pests and disease tend to be of little use in organic systems and are only allowed where their use can be shown to be essential to prevent crop loss.

Weeds are often regarded as the main problem for organic farmers, due to their competition with crop plants for light, water, and nutrients, and their potential to interefere with harvest machinery. However, they are also recognized as having positive roles in organic systems. The aim is therefore not to eliminate weeds altogether, but to maintain them at a manageable level to ensure that crop productivity or farm profitability is not compromised. Weeds can supply ground cover to an otherwise bare soil, reducing the risk of erosion, leaching, and soil crusting. The additional roots in the soil contribute to biological activity and soil structure. Weed species increase the plant diversity within the cropping system, and provide habitats for a wider range of insects and other invertebrates. Weeds can also act as a reservoir for beneficial mycorrhizal fungi, which naturally occur on most crop species. Plant roots infected with mycorrhizal fungi have been shown to take up more soil nutrients and have greater resistance to root pathogens. Weed control strategies can be successfully targeted towards key problem species, such as couchgrass (*Elymus repens*) or dock (*Rumex* spp.), and key growth stages in order to minimize the impact of weeds on crop yield and quality. The main weed control strategies in organic farming combine cultural techniques including the use of weed suppressing crops in the rotation with direct thermal or mechanical methods.

There is some scientific evidence to show that pests and diseases are generally not a significant problem in organic systems. It has been suggested that plants grown in fertile soils with high microbial activity and balanced nutrition are better able to resist pest and disease attack. Numbers of beneficial, parasitic, and predatory fungi and insects tend to be higher on organic farms due to positive environmental management and lack of pesticides. However major pest and disease damage is sometimes seen in organic crops, particularly in high-value vegetable and fruit crops, for example, root flies on brassicas and late blight (*Phytophthora infestans*) on potatoes (*Solanum tuberosum*). Cultural or husbandry techniques are of prime importance in the prevention and control of pests and diseases in organic systems. Examples of such techniques include: late sowing (to avoid the first generation of carrot fly, *Psila rosae*), use of resistant varieties, use of fleece covers to prevent insect pest attack on brassicas, and irrigation to minimize incidence of common scab (*Streptomyces scabies*) on potatoes. Pest and disease control in organic systems is achieved through a combination of such cultural and husbandry techniques with permitted treatments, including the use of steam, natural plant extracts, and biological control agents.

Organic Livestock Production

Livestock husbandry on organic farms depends on three major principles:

1. Animal production systems must conform to the highest welfare standards.
2. Livestock must be fed in a way suited to their physiology, using food produced either on-farm or sourced locally where possible.
3. Veterinary treatment must avoid routine prophylactic drug use. Animal health should be maintained and enhanced through good preventative husbandry and appropriate animal welfare, housing, and feeding systems.

The high welfare standards enforced under the organic regulations aim to ensure that livestock health and product quality (meat or livestock products such as milk and eggs) are optimized. Prolonged tethering, excessive confinement in cages or enclosed buildings, high stocking rates, and inappropriate bedding/flooring have all been shown to produce ill-health in animals and are therefore prohibited in organic systems. Organic livestock farming is a land-related activity. Livestock must therefore have access to a free-range area and the number of animals per unit of area must be limited to ensure integrated management of livestock and crop production on the farm, so minimizing any form of pollution, in particular of the soil and groundwater. The number of livestock must be closely related to the area available to avoid problems of overgrazing and erosion and to allow manure to be spread at acceptable rates to avoid environmental damage. The stocking rate must be adjusted to ensure that no more than 170 kg ha^{-1} N is applied per year.

Feeding regimes must be suited to the physiology of the animal. Excessive feeding of concentrates, which can cause animal health problems, is banned in organic farming. Instead, organic farmers aim to ensure that their livestock is fed in such a way that the potential and limitations of the animal's digestive system are realized. In order to ensure that the farm system as a whole is as balanced as possible, all forage required by ruminant livestock should be produced on-farm. Organic livestock should ideally be fed 100% organically grown feed of good quality. Up to 30% of the feed formula of rations on average may comprise feed from in-conversion farms, although when the in-conversion feeds come from part of the farm in question, this percentage can be increased to 60%. There is a current EU derogation (in place until 24 August 2005) which allows organic livestock to be fed a limited proportion of conventional feedstuffs when the farmer is unable to source sufficient organic feed. The maximum permitted percentage of conventional feed authorized per year is 10% in the case of herbivores and 20% for other species. These figures should be calculated annually as a percentage of the dry matter of feed from agricultural origin. The maximum percentage authorized of conventional feeds in the daily ration must be 25% calculated as a percentage of the dry matter.

In many modern conventional livestock systems, there is an almost total reliance on the routine use of veterinary drugs including antibiotics, anthelmintics, vaccines, growth promoting hormones, and trace elements. Organic systems seek to remove the need for such drugs by creating a system which optimizes animal health. Certain veterinary drugs are permitted in organic systems, but only when necessary. Withdrawal periods (between drug use and sale of livestock products or meat from treated animals) are generally longer than in conventional systems.

Organic Farming and the Environment

The widespread practice of intensive conventional agriculture has, in some cases, resulted in pollution of lakes, rivers, and seas, and occasionally soils. Soil structure and soil organic matter levels have declined in some poorly managed farms. Many intensive conventional farms have fewer trees, hedgerows, ponds, and bogs, and this loss of habitat has in turn resulted in a dramatic decline in the numbers of wild plant and animal species. Organic agriculture will provide one means to enable us to start to clean up and improve our environment.

Soil management is a high priority within organic systems. Good soil management practices and the resultant increase in organic matter levels will help to prevent soil erosion and drought sensitivity, which are becoming serious problems in the UK, particularly in the light sandy soils in the southeast of England. The regular addition of farmyard manure and composts to the soil is known to aid structure. This practice encourages earthworms and other soil fauna, which provide food for larger animals and also improve soil structure.

Contamination of soil and pollution of water from pesticides is almost completely absent in organic systems, since there are very few permitted pesticides and their use is severely restricted. The level of leaching and therefore water pollution from a farm will depend not only on the management system, but also on the crops grown, the topography, soil type(s), and the weather patterns. Several studies have shown that pollution of water by fertilizers, particularly nitrates and phosphates, is greatly reduced on land under organic management. The UK Environment Agency has stated that, to date, the low take-up of organic farming (in the UK) has meant that it has had a limited impact on water quality improvement strategies. However, it acknowledges that there are large potential benefits in pollution reduction to be gained if significant numbers of farmers convert to organic systems. It is worth noting that some recognized organic farming practices (e.g., plowing-in of grass–clover leys and storage of farmyard manure) can result in heavy losses of N (nitrate) and to a lesser extent K, particularly in wet weather. Organic agriculture does not provide perfect solutions to the potential problems of pollution and fertilizer loss. As with all farming systems, careful management is required to ensure that farming operations relating to manure preparation and use and crop nutrition are carried out with regard to weather, soil type, and the system as a whole.

A number of environmental organizations have published information on the decline in the numbers of wildlife species on intensively farmed land. English Nature, the government body responsible for the natural environment in England, reported that organic farming "provided a real opportunity for reversing some of the catastrophic declines in farmland biodiversity which have been seen in recent decades." There is a great deal of evidence that higher numbers and a greater range of species of birds, invertebrates, and wild plants occur on organic farms. Following research, the British Trust for Ornithology reported that "an expansion of organic farming would be a valuable component of any strategy seeking to enhance biodiversity on agricultural land."

The Countryside Commission of England and Wales considered that, in lowland areas, the management practices of mixed organic farms had noticeable positive effects on the aesthetic quality of the surrounding landscape. Traditional features including smaller fields, hedgerows, woodland, wetland, and ponds were being appropriately maintained or reintroduced. Organic farming provides the reasons to return these features to farmland, thereby making the countryside a more attractive place in which to live and work.

Organic farming uses more land per unit output than conventional farming. Some critics argue that conventional farming is therefore a better option in terms of the environment, because a greater proportion of land will be available for natural habitat creation off-farm.

Organic Food

The aim of organic farming is to produce nutritious food that is wholesome and fresh and has been subjected to minimal processing. A limited range of natural additives and processing aids (of plant and animal origin) are permitted in organic food, but they must only be used if it can be proved that each one is essential for the production of the food. Synthetic colorants, preservatives, and processing aids and those produced by genetic engineering are prohibited in organic food processing.

Consumers of organic food widely perceive it to be nutritionally better and tastier. In fact there has been little robust scientific work carried out to prove or disprove this. Differences have often been shown between organic and conventional foods, but there is little proof that these represent differences in food quality. Higher dry matter levels and lower pesticide levels tend to be found in organic fruit and vegetables, although conventionally produced cereals have been found to be better suited to modern baking requirements.

There is increasing concern over food safety following a number of European "food scares." The recent problem of bovine spongiform encephalopathy (BSE) in cattle (which would not have occurred if conventional agriculture had followed organic farming standards concerning the use of animal proteins in ruminant feeds) has raised further questions concerning the safety of conventionally produced foods. Because synthetic pesticides are not permitted, organically produced foods contain fewer (or no) residues of agrochemicals.

There are proven benefits from eating greater quantities of fresh fruit and vegetables, whether conventionally or organically produced. Claims have been made about the positive health benefits of organic vegetables which may be due to the presence of increased levels of secondary metabolites which can protect against certain cancers, but these are mainly unsubstantiated.

The limited research results available suggest that it is likely that there are health benefits from the consumption of organic food, but extensive research is needed to prove it.

Future of the Industry

The demand-led expansion of organic farming is continuing throughout much of Europe, although there is evidence that it is beginning to slow in some countries. The sector is constantly changing at present, with an increasing involvement by the multiple retailers and large farming companies and increased consumer interest in locally produced food and farmers' markets in countries such as the UK, Denmark, and Germany. European governments largely recognize the increasing importance of the organic farming sector and are increasingly investing in advice and support for organic farmers and in relevant research and development. Development is constrained in some countries by the lack of trained and/or experienced staff, but colleges and universities throughout Europe are now responding to the demand for training and education in organic farming through the provision of courses at all levels.

See also: **Diseases**: Plant Pathology, Principles. **Nutrition**: Nitrogen Fixation. **Production Systems and Agronomy**: Agricultural Crops; Horticultural Crops, Protected; Grassland. **Weeds**: Weed Biology; Weed Competition.

Further Reading

Conford P (2001) *The Origins of Organic Farming*. Edinburgh: Floris Books.

Foster C and Lampkin N (2000) *Organic and In-Conversion Land Area, Holdings, Livestock and Crop Production in Europe*. EU FAIR3-CT96-1794 Final Report. Aberystwyth: Welsh Institute of Rural Studies.

Hohenheim, University of (1999) *Organic Farming in Europe: Economics and Policy*, vols. 1-6. Hohenheim: University of Hohenheim.

International Federation of Organic Agriculture Movements (2003) http://www.ifoam.org

Lampkin N (1990) *Organic Farming*. Ipswich: Farming Press.

Lampkin N and Padel S (1994) *The Economics of Organic Farming*. Wallingford: CAB International.

Robson MC, Watson CA, Fowler SM, *et al.* (2002) The agronomic and economic potential of break crops for ley/arable rotations in temperate organic agriculture. *Advances in Agronomy* 77: 369–427.

Soil Association (2000) *The Organic Food and Farming Report*. Bristol: Soil Association.

Stockdale EA, Lampkin NH, Hovi M, *et al.* (2001) Agronomic and environmental implications of organic farming systems. *Advances in Agronomy* 70: 261–327.

UKROFS (2001) *United Kingdom Register of Organic Food Standards*. Reference OB4. London: United Kingdom Register of Organic Food Standards.

Woese K, Lange D, Boess C, and Bogl KW (1997) A comparison of organically and conventionally grown foods: results of a review of the relevant literature. *Journal of the Science of Food and Agriculture* 74: 281–293.

Commercial Flower Production Methodology

W B Miller, Cornell University, Ithaca, NY, USA

Introduction

Within agriculture, the production of greenhouse and flower ("floriculture") crops is one of the most intensive management of plants possible. Due to the use of a regulated environment (a greenhouse), nearly complete control of plant growth and development is possible. Thus, it becomes biologically possible to grow any plant at any time of the year anywhere in the world … even out of this world, as seen by NASA's interest in developing superintensive plant growth systems for long-term space flight. The key words in the above statement are "biologically possible," as the commercial flower industry must pay close attention to the cost and profitability of any procedures they use to grow their crops.

Objectives of Commercial Flower Production

The basic objective in commercial flower production is to produce high-quality products for consumption by the flower- and plant-loving public. This very often involves manipulation of floral induction, initiation and development, as in "out-of-season" production of flowers such as chrysanthemums (*Dendranthema × grandiflora*), poinsettia (*Euphorbia pulcherrima*), Easter lily (*Lilium longiflorum*), or other bulbous crops such as tulip or hyacinth. Central to this objective is distinguishing between plant growth (essentially dry weight gain) and plant development (processes leading to alterations in the vegetative body of the plant and manipulation of floral development). Very simplistically, the life of a commercial flower crop may be divided into four phases: (1) propagation, (2) vegetative development, leading to a plant with sufficient structure and stature to support floral development, (3) floral induction, initiation, and development to anthesis (or marketing), and (4) the postharvest phase that encompasses packaging, transportation, marketing, and use by the final consumer.

Products under consideration include cut flowers and foliage that are used in floral arrangements, flowering or nonflowering pot plants as used for indoor decoration, and "bedding plants" that are mainly destined for use in gardens and for outdoor decoration in the warmer months. A key concept in commercial flower production is that greenhouses are expensive to operate (e.g., winter heating); thus the area a crop occupies and time it takes to grow are both crucially important to the floriculturist. Thus, adoption of techniques based on plant physiology must be economically feasible, for example by:

- reducing time to grow the flower or plant to maturity
- reducing labor by improving uniformity, and allowing adoption of automation
- improving quality and thereby allowing a higher price (and hopefully profit) to be earned.

Propagation

The plants grown by commercial floriculturists may arise from seed or through vegetative propagation techniques (cuttings, division, layering, tissue culture, etc.). The ornamental plant breeding and seed production industries are highly developed, sophisticated, and worldwide in their scope. Breeding objectives vary enormously by crop, but in general characteristics such as flower color, plant form and size, productivity (in the case of perennial or semiperennial crops such as roses), earliness to bloom, fragrance, disease resistance, minimized plant-to-plant variation, seed production potential, enhanced postharvest performance (delayed leaf or flower senescence) are all crucial. In the case of vegetatively propagated crops, ease of propagation and perhaps amenability to *in vitro* culture are important.

Physiological techniques used by the seed industry are many. Technologies such as pregerminated and partially germinated seed offer the potential for more rapidly germinating and uniformly growing seedlings. The seedling industry has developed to the extent that seedlings are produced in small cells ("plugs") that may have soil volumes of less than $1\,cm^3$ per plant. Producers typically utilize a range of physiological and horticultural techniques to control seedling growth and development, such as growth retardants, nutrient stress, photoperiod manipulation, and temperature, all for the goal of producing the highest quality seeding for "growing-on." With such an intensity of production, highly sophisticated robots are now used for seed sowing, rougeing (removal of weak or defective seedlings and replacement with healthy ones), and transplanting into a final container for further growth. Efficient use of such machinery is predicated on a skillful blending of horticulture and plant physiology to obtain uniform plant growth.

Temperature, Light, Nutrition and Gases

As with all plants, the basic factors of temperature, light, water, nutrition, and gases are crucial for optimum growth (dry weight gain). Many commercial flower producers have the ability to increase dry weight gain by increasing photosynthesis through the use of assimilation lights (high pressure sodium, metal halide, etc.). Coupled to increases in dry weight, forcers are also interested in the coordinated morphological development of the crop, usually culminating in a flower. Thus, the environmental factors leading to floral induction, primordial initiation, and flower development are also important to understand, as well as the concept that each phase generally has its own optimum environmental parameters.

Temperature

As with any plant, temperature is a crucial environmental factor for floriculture production. Temperature plays a central role in determining crop growth rate (metabolism and respiration), plant size, response to other environmental cues (e.g., photoperiod), and, ultimately, the "quality" of the final plant or flower.

Horticultural considerations Professional growers are keenly aware of temperatures within their greenhouses. Temperature effects vary by crop. Some are "cool crops," that is, they grow well, and have best quality under "low" temperatures (e.g., 10–15°C night temperature). Examples are *Primula*, *Calceolaria*, *Cineraria*, *Viola*, and many spring bulb crops. Many crops fall into the "warm crop" category, and are best grown with night temperatures typically between 18 and 20°C and possibly above (poinsettia, many foliage plants). The majority are "moderate" temperature requiring, and grow well at temperatures between these extremes, often at 16–18°C night temperatures. Note that the absolute range of temperatures of these categories only spans about 10°C.

Furthermore, it is important to understand the temperature limits defining these categories are fluid, and crops in reality have rather complex temperature needs that vary throughout their life cycle. For example, crops are commonly "started warm" and "finished cool." For example, poinsettia is often grown in its early stages at temperatures of 23–25°C to promote rapid root growth and vegetative development prior to decapitation ("pinching") to promote lateral branching. At the end of the crop, temperatures may be reduced to 15–16°C to "harden" plants and intensify bract color prior to sale. Similar temperature manipulations exist for nearly all floriculture crops. These manipulations are taken to an extreme with crops such as Easter lily where plant development is nearly fully controlled by temperature, and the sales window is exclusively for the Easter holiday. In such cases, growers are constantly adjusting temperatures up or down by a few degrees, trying to "time" the crop so that it becomes saleable at the right moment.

In the 1930s and 1940s, massive investigations were carried out on effects of temperature on plant growth. This work led to the conclusion that most plants "grow better" (that is, taller, more robust) when night temperatures are somewhat cooler than day temperatures. As this is the normal situation in nature, this comes as no surprise. In a greenhouse, however, it is possible to reverse this situation and literally grow plants with night temperatures higher than the day temperature. Such a condition is known in the industry as "negative DIF," and has become important for those greenhouse growers interested in keeping plants shorter than normal, without the use of chemical growth regulators. Because warmer night temperatures increase respiration and needlessly waste carbohydrates. Other temperature "shock" techniques, such as a cold temperature pulse at sunrise, have been developed, and are as effective as negative DIF.

Dormancy and vernalization In many flower crops, dormancy development and remission are important. Controlled flowering of woody stems (lilac (*Syringa vulgaris*), *Buddleia*, *Hydrangea*) and forcing of bulbous crops (such as tulip (*Tulipa*), daffodil (*Narcissus*), hyacinth (*Hyacinthus*), and *Iris*) require detailed knowledge of dormancy development, and, more importantly, removal of dormancy at a specified point in time to allow forcing for the specified market. Generally, prolonged cold periods (especially for bulbs) are suitable for dormancy remission and vernalization in these crops. There are examples where long days can substitute for much of the cold required for vernalization, e.g., Easter lily. As an example of the difference in a biologically relevant dormancy-breaking treatment and one suitable horticulturally may be illustrated with the tulip. Many tulips flower adequately if given 10–12 weeks of cold. Horticulturally, however, many tulips grown as cut flowers need much more cold than this (16–18 weeks) to promote greater stem elongation leading to a better-quality flower.

Light

Quantity Above about 35°N latitude, short photoperiods, low solar angle, and cloudy weather combine to cause low light levels in the winter months. Since this period is important for flower production, many greenhouse producers use supplemental assimilation lighting to increase photosynthetic irradiance and dry weight gain. Research has indicated that in many crops, it is possible to deliver this supplemental lighting in a variety of ways (lower flux for a longer time vs. higher flux for a shorter time, etc.). General experience indicates large increases in both quantity and quality of production. The greenhouse lighting industry is sophisticated, and makes use of computer programs to optimize placement of lamps for best efficiency.

Duration (photoperiod) Light duration and control of the "photoperiod" is a crucial application of plant physiology in greenhouse crop production. Within about a decade of the original work on photoperiodism (i.e., early 1930s), the industry had adopted (with great assistance from university advisors and researchers) photoperiod control as a means of regulating the flowering of photoperiodic crops, especially of chrysanthemum (a short-day plant). In the industry, long nights (short days) are provided by manual or mechanical deployment of black cloth that excludes enough sunlight that plants perceive it as being "dark." Typically, growers deploy black cloth in the very late afternoon to minimize heat buildup under the cloth. Such heat can lead to "heat delay" whereby plants are much slower to flower under the supraoptimal temperature. Often, growers will remove the black cloth after natural nightfall, allow the heat to escape, then redeploy if necessary before sunrise. The exact duration of "night" needed depends on the critical photoperiod, and is crop specific. In practice, black-clothing to provide a dark period of about 14–16 hours is common.

Long days are provided by illuminating fluorescent or more commonly, incandescent lamps. The minimum irradiance used is typically 10 ft-candles (about 0.1–0.2 μmol m^{-2} s^{-1} photosynthetically active radiation) applied for 4 h from 10 p.m to 2 a.m. These times are very flexible, however, and in large installations crops may be lit for a few minutes each half-hour in a "rolling" fashion ("cyclic lighting") to allow a greater area of plants to be lit for the same size of electrical service.

Light quality and photomorphogenesis An emerging technology in floriculture is the development of photoselective films for use as greenhouse coverings. Experimental films have been developed that alter the ratio of transmitted red to far-red light. Specifically, films transmitting light enriched in red compared to far-red wavelengths cause shorter, more compact growth mimicking plants treated with growth retardants. While still in the research phase, photoselective films hold promise for growers of potted and bedding crops.

A related concept is to exclude the far-red light that naturally occurs at sunset each day. A good example is with Easter lily, where a significant degree of height reduction (desirable) can be obtained by covering the plants with black cloth about 30–45 min prior to sunset. Since this procedure eliminates the end-of-day far-red pulse, plants stay significantly shorter and compact as compared to plants that are exposed to the normal twilight.

Nutrition

Nutritional physiology and the roles of plant nutrients *per se* are not appreciably different for flower crops grown in greenhouses. What is unique is that many flower crops are grown in soilless substrates that are based on peat moss, coir (a coconut hull fiber), or composted and partially degraded tree bark combined with materials such as vermiculite (expanded mica), perlite (an expanded volcanic rock), sand, and others. When used in relatively shallow containers (pots, trays), these substrates provide the physical properties (large pores) leading to sufficient air-filled porosity upon draining to "container capacity" after watering. On the other hand, these materials suffer in their ability to retain nutrients, as they tend to have a fairly low cation exchange capacity (CEC) on a volume basis.

To compensate for substrates that provide adequate drainage and air-filled porosity but suboptimal chemical properties, flower crops are often grown under luxury fertility levels. Floriculture crops are usually grown with liquid fertilizer applied at each irrigation, often multiple times per day. Commonly, fertilizer in a ratio of 3:1:3, or 2:1:2 (N:P:K) is applied, with concentrations of 5–20 mmol l^{-1} N and K. Typically, a trace element preparation containing all the minor elements is included. In floriculture crop systems, most Ca is supplied by preplant liming, or perhaps during growth with $Ca(NO_3)_2$.

Implicit in the production of a plant in a container (for example pots or trays) is the ability to rapidly change the nutritional status of the root zone. Flower producers have the ability to monitor substrate and foliar nutrient levels, and if they fall above or below norms established for the crop, corrective steps may be taken. An estimate of the gross nutrient level in substrate may be obtained by determining the electrical conductivity (EC) of a specified mixture of the substrate and distilled water. If nutrients have accumulated, the EC increases, as the nutrients themselves are ions and therefore conduct electricity. Low substrate nutrient status is indicated by a low EC. Such tests are easy, fast, and economical to perform on-site for both flower growers and plant physiologists.

Gases

In greenhouse production, CO_2 and ethylene are the only two gases of major concern in so far as they are the only ones that can be economically altered to enhance or control plant growth and development.

Carbon dioxide In greenhouses, carbon dioxide concentrations are often increased to about 1000 ppm, usually in combination with supplemental assimilation lighting (see above) to increase photosynthesis, net dry weight gain, and total productivity. Depending on the crop, dramatic increases in productivity may be obtained with assimilation lighting and supplemental CO_2. The CO_2 is usually generated via specialized combustion units, or diluted from tanks of pure liquid CO_2. There are many practical considerations in the use of CO_2 including cost of the equipment, and the additional management concerns. For example, adding CO_2 to the greenhouse atmosphere is pointless during warm, high-light times of the year, when ventilation fans are running to keep the greenhouse from overheating. In such a case, any supplemental CO_2 is wasted. It is worth pointing out that the term "greenhouse gas" as applied to global warming is a misnomer, and that greenhouses *per se* have no effect whatsoever on global climate.

Ethylene Ethylene is almost always a problem in flower production (but see below, in "Plant Growth Substances"). Its presence in a commercial greenhouse is generally through one of the following sources:

- as a combustion contaminant (trucks, machinery, malfunctioning gas-fired greenhouse heaters)
- fungal activity on decaying plant material
- the presence of ripening fruits, vegetables, or senescing flowers emanating ethylene.

Of these, ethylene contamination from malfunctioning gas heaters is commonly seen. The only solution is to identify the problem and make corrective repairs. While there are no practical ways to scrub the greenhouse atmosphere of ethylene, it is easily removed by ventilation.

Detrimental ethylene effects on floriculture crops are many and include flower and leaf abscission, epinasty, flower and bud abortion, and senescence.

Plant Growth Substances

Basic investigations in plant physiology continue to have an enormous impact on the greenhouse industry, and the range of commercial plant growth substances available to the greenhouse industry is unique in horticulture. Within the Unites States, there are legally available and registered products in four major classes of growth substances: gibberellin, auxin, cytokinin, and ethylene. Another important group is the anti-gibberellins, products that are crucial in many crops grown in pots (containers). A listing of the main products available in North America is given in **Table 1**.

Gibberellin

Gibberellins find use in a wide range of crops, and differences exist between uses for gibberellic acid (GA_3) and other gibberellins such as GA_{4+7}. In

Table 1 The major growth substances used in commercial flower production

Chemical and form	*Commercial product*	*Main physiological response or commercial use*
Gibberellins		
Gibberellin A3	ProGibb	Dormancy remission, anti-leaf senescence
Gibberellins A4 + 7	Fascination (1:1 ratio of GA_{4+7} and benzyladenine)	Anti-leaf senescence, lateral branching
Anti-gibberellins		
Ancymidol	A-Rest	Height control (restriction of internode elongation)
Chlormequat chloride	Cycocel	Height control (restriction of internode elongation)
Daminozide	B-Nine	Height control (restriction of internode elongation)
Paclobutrazole	Bonzi	Height control (restriction of internode elongation)
Uniconazole	Sumagic	Height control (restriction of internode elongation)
Flurprimidol	Topflor	Height control (restriction of internode elongation)
Ethylene		
Ethephon	Florel, Ethrel	Flowering in bromeliads, lateral branching or planned flower abortion ("chemical pinching")
Anti-ethylene agents		
1-Methylcyclopropene	EthylBloc, SmartFresh	Anti-senescence, especially in ethylene-sensitive flowers in the presence of exogeneous ethylene
Auxins		
IBA, IAA, NAA	Various ("rooting powders")	Induction of adventitious roots in vegetative propagation
Cytokinins		
Benzyladenine	As a component of Fascination (see above)	Anti-leaf senescence, lateral branching

commercial flower production, gibberellic acid is used to substitute for cold in dormancy-breaking in plants such as *Azalea*, *Camellia*, and hydrangea, to improve flowering uniformity in *Cyclamen* (although much of this use has now been superseded by more uniformly flowering cultivars), for production of standards (tree-forms) of plants such as *Fuchsia*, to accelerate and increase flowering (*Spathiphyllum*, statice (*Limonium*)), altering inflorescence form (chrysanthemum), and reduction of chlorophyll breakdown (maintaining higher-quality foliage) in certain liliaceous crops such as *Alstroemeria*.

Other gibberellins, notably GA_{4+7}, have recently become registered for use in North America, with the major application being maintenance of chlorophyll and increased flower life in *Lilium*. Research has indicated that GA_{4+7} is much more effective than gibberellic acid in this regard, and since 2000, GA_{4+7} has become a major tool for greenhouse producers to produce higher-quality lily crops for the flower consuming public.

Anti-gibberellins (Growth Retardants)

Because of the inherently difficult task of growing plants in overall proportion to their pot, anti-GAs play an important role in flower crop production. The major growth retardants (the term "growth regulator" is also applied to these chemicals) are given in **Table 1**. The generalized mode of action of these materials is to block endogenous gibberellin synthesis, leading to reduced internode length with, typically, no effect on the number of nodes. Plants treated with growth retardants are often darker green and, less commonly, may show increased drought and cold resistance. The choice of which product to use is dictated by legal considerations, effectiveness on the crop of interest, the choice of application method (as a foliar spray, a direct substrate ("drench") application, or as a preplant soak or dip), and, of course, cost.

Since the growth retardants act by inhibiting gibberellin biosynthesis, the main effect is a reduction in cell length, but no major effect on cell division. Thus, it is crucial to understand that use of these products must precede the grand phase of growth, thus the decision to treat a plant with a growth retardant must be made before the plant is "too tall." A range of predictive tools, including plant growth models, have been developed to assist growers in making appropriate management decisions.

Growth retardants may have serious detrimental effects on plant growth including severely reducing bract size in poinsettia, severely delaying anthesis (in many crops), causing transient leaf chlorosis (e.g., chlormequat chloride on geraniums (*Pelargonium*)), and growth regulatory effects lasting much longer than needed or desired (e.g., failure of summer annuals to grow properly once planted in the garden). These problems are mainly horticultural in nature and are often related to improper use of the growth retardant.

Ethylene and Ethylene-Releasing Chemicals

Because of its close association with senescence, ethylene is generally considered detrimental in floriculture. There are many specific instances, however, where it is a useful tool. Ethephon (an ethylene-releasing chemical) is used to great effect to inhibit excessive stem elongation in bulbous crops such as daffodil and hyacinth. In the last 10 years, ethephon products have become increasingly popular and useful for increasing branching and inhibiting premature flowering on a wide range of floricultural crops.

Ethylene and ethephon are also useful for floral induction in bromeliads and to increase the flowering percentage and earliness of certain bulbous crops, notably the Dutch iris (*Iris* × *hollandica*).

Anti-ethylene Chemicals

In the early 1970s it was discovered that the silver ion (Ag^{+}) is a potent inhibitor of ethylene action. It is, however, more or less immobile in plants, and could not be easily applied to cut flowers, where ethylene-induced senescence is a central, but negative feature of flower development. Shortly thereafter, the concept of complexing the silver was introduced, and a complexed molecule, silver thiosulfate (STS), was introduced in the late 1970s. STS could be applied as a foliar spray to many ethylene-sensitive plants, but more importantly, was "pulsed" into the flower by placing cut stems into an STS solution and allowing the stem to draw in the STS. Since STS is an inhibitor of ethylene action it confers protection against exogenous, contaminating ethylene as might be found in and around flower handling, distribution, and marketing locations. Unfortunately, its toxicity (silver is a heavy metal) precluded its widespread use in the United States, although legal formulations do exist in Europe.

As of the late 1990s the ring hydrocarbon 1-methylcyclopropene (1-MCP) promises to profoundly affect floral postharvest handling. 1-MCP is a gaseous and essentially irreversible inhibitor of ethylene binding. Research in the last 10 years has indicated that 1-MCP often has no effects in ethylene-free air, but can remarkably reduce senescence as induced by subsequent ethylene exposure.

While 1-MCP is a gas, the commercial product is currently formulated as a white powder. When mixed in water, the 1-MCP gas is released typically into rooms or reasonably airtight chambers. At the risk of great simplification, exposure to $1\,\mu l\,l^{-1}$ 1-MCP for 8–10 h at ~20°C is an effective dose for most flowers. While 1-MCP binding appears to be irreversible, sensitivity to ethylene often slowly returns, presumably due to the *de novo* synthesis of new ethylene receptors in the flower.

Auxins

There are numerous auxin formulations available to the commercial flower producer, and their main use is to stimulate root formation in vegetative cuttings. Considering the hundreds of millions of vegetative cuttings of chrysanthemum and carnation (*Dianthus caryophyllus*) (two major cutting-propagated cut flowers), and the similarly massive use on a wide range of individually lesser pot plant crops, auxins are probably the most widely used growth substance in floriculture.

The specific use of auxin during propagation is a highly refined horticultural art. It may be applied as a powder or sprayed, dipped, or soaked as a liquid. Major propagation companies have an excellent proprietary understanding of the nuances of auxin application as affected by cultivar, time of year, and undoubtedly other variables.

Cytokinins

To the author's knowledge, the only registered cytokinin in North America is Fascination, in which benzyladenine is essentially an "inert ingredient" as far as its labeled use for *Lilium* postharvest quality is concerned. Experimentally, such a product can find use as a branching stimulant in a range of crops.

Abscisic Acid

To the author's knowledge, there are no commercial uses for abscisic acid (ABA) or ABA analogs in floriculture. Potential uses are easily imagined, however. One possibility would be to force stomatal closure to reduce water requirement in indoor landscape plantings, another might be to impose dormancy on a plant at any time during the year. Such a technique could greatly expand the locales where certain dormancy-requiring plants could be grown, as the environmental clues leading to dormancy (commonly short days and cooler temperatures) could be replaced by a hormone application.

Conclusions

Successful commercial flower production is a high synthesis of plant physiology, horticultural science, personnel and facilities management, marketing, and economics. The ability to regulate growth, development, and flowering are unparalleled in agriculture and allow sophisticated technologies to be adapted to flower production.

See also: **Flowering and Reproduction**: Flower Development. **Postharvest Physiology**: Senescence, Flowers; Senescence, Leaves. **Production Systems and Agronomy**: Horticultural Crops, Protected; Nursery Stock and Houseplant Production. **Regulators of Growth**: Cytokinins; Ethylene; Gibberellins; Juvenility; Photomorphogenesis. **Seed Development**: Germination. **Seed Dormancy**: Preharvest Sprouting.

Further Reading

Bunt AC (1988) *Media and Mixes for Container Grown Plants*. Dordrecht: Kluwer.

DeHertogh A and LeNard M (eds) (1993) *The Physiology of Flower Bulbs*. Amsterdam: Elsevier.

Dole JM and Wilkins HF (1999) *Floriculture: Principles and Species*. Englewood Cliffs: Prentice-Hall.

Hannan JJ (1998) *Greenhouses: Advanced Technology for Protected Horticulture*. Boca Raton: CRC Press.

Hartmann HT, Kester D, Davies F, and Geneve R (2002) *Hartman and Kester's Plant Propagation: Principles and Practices*. Englewood Cliffs: Prentice-Hall.

Larson RA (ed.) (1992) *Introduction to Floriculture*. New York: Academic Press.

Maki SL, Rajapakse S, Ballard RE, and Rajapakse NC (2002) Role of gibberellins in chrysanthemum growth under far red light-deficient greenhouse environments. *Journal of the American Society for Horticultural Science* 127: 639–643.

Mastalerz JW (1977) *The Greenhouse Environment*. New York: Wiley.

Nelson PV (1991) *Greenhouse Operation and Management*, 6th edn. Englewood Cliffs: Prentice-Hall.

Ranwala AP and Miller WB (2000) Preventive mechanisms of gibberellin4 + 7 and light on dark low temperature-induced leaf senescence in *Lilium* cv. Stargazer. *Postharvest Biology and Technology* 19: 85–92.

Rees AR (1992) *Ornamental Bulbs, Corms and Tubers*. Wallingford: CAB International.

Watkins CB and Miller WB (2003) A summary of physiological processes or disorders in fruits, vegetables and ornamental products that are delayed or decreased, increased or unaffected by application of 1-methylcyclopropene (1-MCP). http://www.hort.cornell.edu/department/faculty/watkins/ethylene/index.htm

Nursery Stock and Houseplant Production

R W F Cameron and M R Emmett, University of Reading, Reading, UK

Introduction

The production of ornamental plants for use in homes, gardens, and outdoor amenity areas is a commercial discipline that can be further divided into specialist sectors:

1. "Nursery stock" (or, in the US, "nursery") production is the propagation and cultivation of young plants for outdoor amenity use. This sector originates from nurserymen who specialized in the propagation and care of young fruit trees and bush fruit, prior to these being sold and planted as field crops. With the introduction of new plant species to western Europe in the nineteenth and early twentieth centuries, these nurserymen broadened their skills to produce a range of tree, shrub, and herbaceous plant species to meet the growing demand for ornamental garden plants. This led to the term "hardy ornamental nursery stock" being coined to describe those woody and herbaceous plants that could survive all year round in the temperate climates of Europe and North America. In order to meet customers' demands for quality and availability, nursery stock production is increasingly utilizing protected structures of one form or another (**Table 1**).
2. "House plant" production can be traced back to the fashion for conservatories and glasshouses attached to large country houses in Europe and North America in the eighteenth and nineteenth centuries. By the early twentieth century, the more popular varieties were being produced in protected structures by "market gardeners." As such, the traditional cultivation had much in common with that of cut flowers or even tender market garden crops such as tomato (*Lycopersicon esculentum*), which relied on glasshouses for their production in cooler climates.
3. "Bedding plants" are plants grown for the purpose of planting out in flower beds; in Europe, this term is usually only applied to annual plants or short-lived perennials. Their commercial production has similar origins to house plants, as they generally benefit from being grown under protected conditions prior to planting out.

By the end of the twentieth century, the concept of the "patio plant" became established, blurring the distinction between house and garden plants. Patio plants are often presented in containers, and placed on brick or stone patios adjacent to the house itself. In such circumstances, the hard landscape can absorb and reradiate thermal energy and enable the outdoor survival of species that are more vulnerable to cold and wet conditions. Growers traditionally associated with nursery stock, bedding and house plants may all supply plants to this growing market.

Development of the Global Market in Ornamental Plants

Details of the production and consumption of ornamental plants are collated annually by several national governments including the United Kingdom, United States, and Netherlands. The highly developed

Table 1 Percentage of UK hardy nursery stock plants grown under protection (glass or polythene structures)

Plant type	*Field grown area under protection (%)*	*Container grown plants under protection (%)*	*Plugs/cuttings under protection (%)*
Fruit trees, bushes, and plants	1.2	52.3	32.8
Trees intended for forestry use	0.4	0.1	0.0
Conifers	0.4	0.0	0.0
Other trees	3.7	8.5	0.4
Roses	1.7	8.7	0.0
Climbers	40.0	55.4	75.0
Rhododendrons	20.7	46.8	0.0
Shrubs	11.6	40.8	58.1
Herbaceous perennials	16.3	35.1	68.4
Heathers	14.5	22.6	80.0
Alpines	12.6	32.6	72.9
Herbs	9.9	58.8	69.0

Reproduced with permission from Anon (2003) *Hardy Nursery Stock Survey: Production during 2001.* York, UK: Department for Environment, Food and Rural Affairs – National Statistics (http://statistics.defra.gov.uk/esg/statnot/hns.pdf).

marketing and logistics of the Dutch auction markets has meant that the Netherlands has become a hub of the international trade in ornamental plants (particularly throughout Europe). In 2001, exports of garden and pot plants through the Dutch auctions alone totalled €1.3 billion. Data from the USA suggests that the wholesale value of the nursery and houseplant industry is US$6.9 billion per annum. Where once plants were produced on nurseries relatively close to the point of sale, today, certain species, even as quite large specimens, are transported across the globe to reach their market place.

Production Systems

Nursery production can be divided into a number of phases relating to the age or type of plant under cultivation. These phases are likely to comprise:

1. stock plant management or seed production;
2. propagation;
3. a weaning and growing-on stage often carried out under protection; and
4. a growing-on period after plants have either been potted-on into larger containers (container stage) or planted in a field to reach marketable size (field grown stage).

It should be noted that not all crops will necessarily follow this format; for example, some trees and rose (*Rosa* spp.) bushes are produced by budding or grafting techniques, which do not require protected propagation environments. The above phases may occur on different nurseries. Globally, the trend in recent years has been toward nurseries becoming specialist in either stock plant (phase 1), young plant (phases 2–3), or finished plant (phase 4) production.

Propagation

Propagation may be through sexual (i.e., seed) or asexual (i.e., vegetative propagation) means.

Propagation by Seed

Seed is the main method by which plants reproduce in nature, and it is widely used in the commercial propagation of many tree species, herbaceous perennials, and bedding plants. Sexual reproduction involves the development of a zygote with chromosomes from both the male and female parents, and hence the resultant seeds can be genetically variable. There is no guarantee, therefore, that a particular trait in a parent will necessarily be transferred on to the progeny. Many ornamental plants, however, have been selected because they possess an unusual or distinctive trait, and nurserymen are keen to retain this. For this reason, these plants tend to be reproduced asexually to ensure "trueness to type," i.e., that the progeny are genetically identical (clone) to the parent plant.

In annual flowering plants, however, where seed can be produced rapidly, genetic variation can be reduced over a number of generations by self-pollinating and careful selection of the progeny to retain the desired characteristics to produce homozygous or "inbred lines." Two different homozygous lines can then be crossed to produce plants with relatively stable characteristics. Such plants are generally more vigorous (hybrid vigor) than either of their parental homozygous lines and are known as F_1 hybrids.

Many woody plant species produce seeds that are in a physiologically dormant state and will not germinate without pretreatment. Such treatments may include the physical abrasion of the seed coat (scarification) to allow water and gases to permeate through to the embryo and/or stratification techniques to overcome dormancy. This may include temperature treatment (e.g., chilling at $\leqslant 5°C$), exposure to light, or growth regulating chemicals (e.g., gibberellic acid).

Vegetative Propagation

Cuttings Propagation of plants by cuttings is the most effective way of producing clonal varieties of ornamental trees and shrubs, and is the most common technique used in nursery production. The vast majority of cuttings are propagated through the use of leafy (softwood) cuttings, usually collected in summer, although some species can also be rooted as dormant leafless (hardwood) cuttings in winter. Successful rooting depends strongly on genotype, size and type of cutting selected, physiological aspects, and the propagation environment employed. Although somewhat of a generalization, rooting is easiest and most rapid in herbaceous plant species, intermediate in shrubs, and most challenging (or slow) in trees species. It is often considered that physiological condition can influence rooting and that tree species, particularly, need to be in a juvenile phase to optimize rooting. Rate or vigor of shoot growth may also be important, however, as studies with some shrub species show that optimum rooting coincides with the period of maximum shoot growth (**Figure 1**). This may relate to the production of endogenous substances, such as auxins (e.g., 3-indole-acetic acid, IAA), that promote adventitious rooting, but the precise hormonal and molecular mechanisms controlling root initiation remain undetermined. External application of auxin can

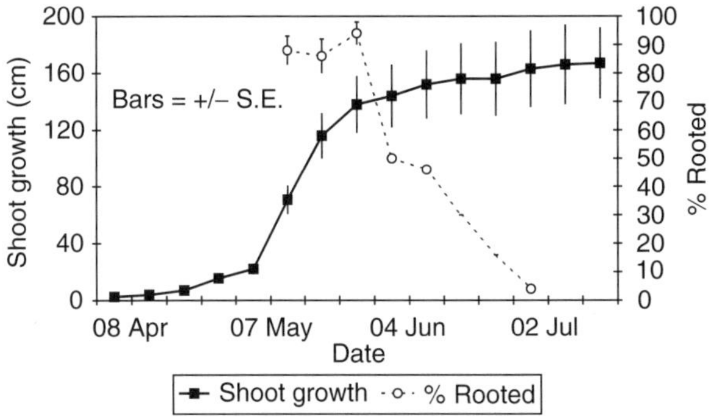

Figure 1 The relationship between shoot growth on the stock plant and percentage of cuttings that root in *Syringa vulgaris* cv. Madame Lemoine. Reproduced with permission from Cameron RWF, Ford Y-Y, Harrison-Murray R, Hand P, and Blake P (2000) *Elucidating and improving the role of auxin at all stages in the propagation of HNS cuttings. Final Report HH1209HN.* London: Department for Environment, Food and Rural Affairs (ex Ministry of Agriculture, Fisheries and Food).

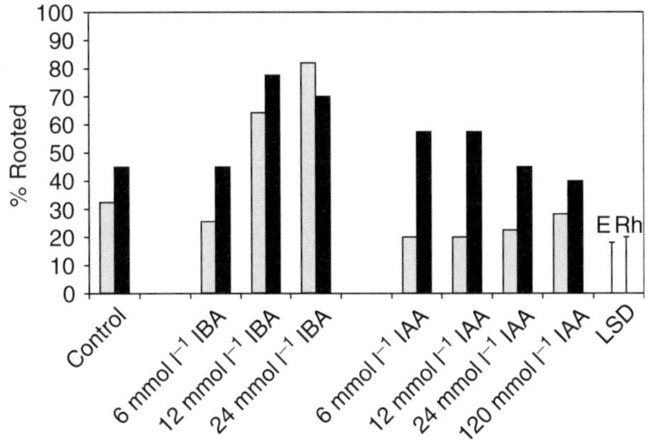

Figure 2 The effect of applied auxin type and concentration on rooting of *Elaeagnus pungens* cv. Maculata (shaded) and *Rhododendron* cv. Mrs G.W. Leak (solid) cuttings. (IBA = Indole-3-butyric acid, IAA = 3-indole acetic acid, LSD: least significant difference). Reproduced with permission from Cameron RWF, Ford Y-Y, Harrison-Murray R, Hand P, and Blake P (2000) *Elucidating and improving the role of auxin at all stages in the propagation of HNS cuttings. Final Report HH1209HN.* London: Department for Environment, Food and Rural Affairs (ex Ministry of Agriculture, Fisheries and Food).

improve rooting, and "rooting" dips and powders are commonly applied to the base of cuttings in commercial propagation. The form and concentration of auxin, however, can determine rooting percentage (**Figure 2**) and number of roots induced.

Water stress in the leafy cutting will reduce rooting potential; therefore, handling procedures and propagation environments are required to minimize water loss from the detached shoot. During collection from stock plants and preparation, cuttings should be kept wet and cool throughout. Similarly, propagation environments should be designed to maintain high relative humidity and avoid excessive temperatures to reduce transpiration, yet provide sufficient irradiance to allow photosynthesis to continue. The provision of polythene covers placed over the cuttings and appropriate levels of shading can allow many species to root, although more sophisticated mist and fog systems may be required to enable rooting of slower or more difficult species, or to allow larger cuttings to be used. When fog or mist is applied to cuttings then the medium that cuttings are inserted into needs to be free-draining to provide aeration and avoid waterlogging. Open structured media, such as 50% peat 50% perlite, are frequently used, but even these may become saturated with water unless the media is placed in contact with a substrate such as fine sand that helps draw water away through capillary action.

For most species, root initiation is optimized at a temperature of 18–25°C, and basal heat is often supplied in propagation environments through electrical cables or hot water pipes. Under ideal conditions, cuttings can form adventitious roots within 2–8 weeks, although some species can be slower. In some conifers, for example, which tend to be propagated in autumn or winter, root initiation may only occur after 12 weeks or more. Owing to propagation chambers being warm, humid environments, cuttings need to be reacclimatized (weaned) to natural atmospheric humidity and temperature before being grown-on. In species such as *Corylus maxima*, there is evidence to suggest that stomata may become dysfunctional after prolonged periods at high humidity, and cuttings fail to regulate their water status effectively after direct placement back to lower humidities. Therefore, to minimize excessive tissue desiccation, nurserymen will wean their cuttings by either progressively rolling up the sides of the polythene tents reducing the frequency of misting over a number of days.

Grafting Grafting is the physical union of two separate pieces of plant tissue. It involves the fusion of a stem section with one or more dormant buds present (scion) onto an existing plant with a rootsystem already present (rootstock). The graft is designed to match the vascular cambiums of the two pieces of tissue, encourage parenchyma cell production, and eventually form new xylem and phloem vessels between the two genotypes. Grafting is feasible between plant species, however, certain combinations never fuse effectively and these are termed "incompatible." To overcome such situations, a third genotype (which is compatible with both scion and rootstock) can be inserted between them to form an interstock. Although grafting is labor intensive, it can propagate plants that are difficult to propagate by other techniques, and can confer advantages associated with certain rootstocks, such as increased vigor or disease resistance. There are various forms of grafting based on how the scion and rootstock physically interlock. Bud grafting or budding is one such specialized technique where only a single bud of scion wood is used. Compared to other grafting techniques, it is relatively quick and hence frequently used in the large field-scale production of roses (T-budding) and ornamental trees (chip budding).

Although seed, cuttings, and grafting are the most commonly used techniques in nursery stock and houseplant production, a range of other techniques are practiced. **Table 2** summarizes the main techniques employed.

Management of Vegetative Stock Material

In some cases, cuttings are harvested from the current production crop. Although this is a potentially efficient process, there is the risk of transferring pests, diseases, and mutations into successive generations of cropping. Additionally, a growing schedule aimed at producing a high-quality finished crop does not necessarily yield good-quality cuttings. Consequently, the maintenance of stock plants to provide material for vegetative propagation is often undertaken by specialist nurseries.

There are notable concentrations of specialist stock plant nurseries in high altitude equatorial regions (such as Costa Rica and parts of Kenya) where the warm climate, high light levels, and relatively constant daylength enable suitable material to be produced all year round.

Table 2 The main propagation techniques used in the production of hardy nursery stock and house plants

Propagation technique	*Procedures*	*Plant types*
Seed	Germination rates can be affected by the level of development (maturity), storage conditions (viability), and seed dormancy	Trees, shrubs, herbaceous perennials, annuals. Rootstocks of roses and other species
Apomictic seed	Asexual seed – formation of a zygote without meiosis. Seeds have same genotype as mother plant	Not widely used; some members of the Rosaceae and *Citrus* spp.
Somatic embryogenesis	Embryos and seeds can be generated from cell suspension cultures	
Softwood stem cuttings	Various forms depending on the species, degree of lignification (soft, semi-ripe, etc.), and whether the stem is cut below a node (nodal) or between nodes (internodal)	Very wide range of tree, shrub, and herbaceous plant species
Hardwood stem cuttings	Leafless dormant stems are cut and propagated during autumn and winter	Bush fruits and some tree and shrub spp. (e.g., *Salix, Cornus*)
Root cuttings	In some species, adventitious shoots can generate from cut sections of root	*Papaver, Chaenomeles*
Leaf cuttings	Roots and new shoots can generate from the veins of cut leaf sections or from leaf bud cuttings, which are an entire leaf with a short piece of stem and axillary bud attached	*Saintpaulia, Begonia, Sedum, Lilium*
Micropropagation	The use of asceptic *in vitro* tissue culture techniques to produce large numbers of young plants. Comprises separate shoot multiplication, root induction, and weaning stages	Useful for introducing new cultivars onto the market
Layering	Bending attached stem over and covering with soil encourages root formation	*Rubus, Vitis,* climbing plants
Air layering	Stems are wounded and then covered with damp moss or peat to encourage root formation	*Ficus, Monstera*
Stooling or mound layering	New shoot growth is covered progressively with soil encouraging roots to form. Entire stem is removed to yield rooted cutting	Fruit rootstocks
Division	Physically splitting the crown, tuber, or rhizome to provide two or more progeny	Herbaceous perennials
Grafting	The fusion of the desired genotype scion onto a rootstock. Various forms include whip, side-tongue, side-veneer, cleft, wedge and bark grafting	Various tree and shrub spp., particularly slow-growing cultivars
Budding	A form of grafting, but using only a single bud of scion wood. Various methods include chip, T-, Inverted T-, patch and I- budding	Field production of roses, fruit, and ornamental trees

On a stock plant nursery, high standards of hygiene are essential. To maintain stock material in a virus free state, the stock plants may be derived from tissue culture and periodically tested for key diseases. Some countries operate certification schemes for stock material tested according to specific protocols.

Plug Cell Technology

The majority of plants propagated from seeds and cuttings are propagated in a tray comprised of modules commonly known as plug cells or plugs. Each plant is grown in a separate cell or plug, which allows easy removal from the tray at the time of potting. Various sizes of tray exist, but many production systems are standardized on a tray of dimensions 305 mm × 525 mm (length × width). Industry standardization on this (and similar) sizes has enabled the development of robotic handling systems capable of transferring individual plugs from propagation trays into larger containers and many other complex operations.

In the case of nursery stock, following propagation in plug trays, the crop is often transferred to an intermediate tray or container before it reaches its final pot. This intermediate module is often described as a "liner."

Field and Container Production of Nursery Stock

Traditionally, most nursery stock crops were planted in fields after the propagation stage. These were then grown to a marketable stage and lifted to be sold, either as a "bare-root" plant or potted into a container (containerized). This method of production is still widely used, especially for trees, roses, herbaceous perennials, and plants destined for civic landscaping projects. Since the 1970s, however, many growers have changed to a system where plants are grown entirely in containers until the point of sale (container grown). Container grown systems offer considerable advantages in product handling and presentation. Many aspects of container production are common to nursery stock, house, and bedding plants, and these are discussed below. Aspects relating to field production, such as soil preparation, will have much in common with other field systems (e.g., top fruit production).

Growing Media

For plants grown in containers, the growing medium needs to be free-draining, yet retain adequate supplies of moisture and nutrients. As such, *Sphagnum* peat provides an ideal substrate for many nursery stock species, due to its structure, with macropores providing aeration whilst the finer micropores adsorb and retain water effectively. It also encompasses a high cation exchange capacity, and thus is effective at absorbing excess nutrient ions and then releasing these slowly as the growing plant requires them. Even the disadvantages associated with peat, i.e., low pH and an inability to rewet after drying, can be readily overcome by adding $CaCO_3$ or $MgCO_3$, and by using wetting agents (surfactants) and/or frequent irrigation, respectively. In the UK and other countries, however, there is pressure to reduce the volume of peat extracted from peat bogs due to environmental issues, and alternative growing substrates are being sought. These may include chopped pine bark, composted garden and municipal waste (green waste), or by-products from other industries that utilize organic materials (e.g., sawdust, paper, or even coconut fibers (coir)). Providing consistent, high-quality alternative products in sufficient volume, however, remains problematic.

Crop Nutrition

Many nurserymen now rely on controlled release fertilizers incorporated into the growing medium to provide balanced nutrition to the developing plant. These are resin-coated granules that incorporate the primary plant nutrients. They release ions into the rhizosphere progressively over a number of months, and provide a continuous low concentration of nutrients to the plant roots. When a more specific growth response is required or a deficiency has developed in a crop, then nurserymen may rely on a liquid solution (liquid feed) of one or more elements to provide an instantaneous effect.

Irrigation

Consideration should be given to the quality of irrigation water used in nursery and houseplant production. Although the requirements for lower residue and microbial contamination may not be as stringent as for some edible crops, such as salads, it is desirable to use water of a reliable and definable quality. High salt concentrations or herbicide residues can be phytotoxic to sensitive crops, and even residues of calcium carbonate left on the foliage after irrigation can reduce the marketable value of an ornamental crop. Where water sources have high nitrate or phosphate levels, consideration should be given to reduce fertilizer rates to account for this.

Owing to container grown plants having only a limited volume of growing medium available there is a requirement for frequent irrigation. Depending on the weather conditions, container crops may require two or more irrigation periods per day. The majority of outdoor container plants are irrigated via overhead sprinklers, although other systems, such as drip irrigation or flood and drain, are utilized depending on the crop grown and the infrastructure of the nursery. Water use (evapotranspiration) in a crop is strongly determined by weather conditions, particularly temperature, relative humidity, and wind speed. As these may vary significantly from one day to the next, so does the volume of irrigation required to bring the growing medium back to a fully wetted condition (container capacity). Adjusting irrigation clocks manually on a daily basis is relatively inefficient, and the nursery industry is keen to adopt automated control systems that match irrigation supply to crop demand. This, combined with more effective application systems, is likely to reduce water consumption, reduce crop losses from over- or underwatering, and improve crop quality through growth regulation, enhanced uniformity, and reduced weed competition.

Nurseries are also attempting to improve their water use efficiency by the use of recycling or recirculation of water. Run-off from container beds and glasshouse roofs is collected in reservoirs, and after filtration and treatment can be reapplied to the crops. Recycled water requires treatment to avoid the spread of plant pathogens, and a variety of techniques can be employed, for example, exposure to chlorine, peroxyacetic acid, ozone, ultraviolet radiation, or allowing the water to perculate through a sand filter (slow sand filtration). This latter system allows the development of a microbial flora in the sand that is antagonistic to the main plant pathogen species.

Pruning, Flower Induction, and Scheduling

Private homes and gardens are a significant market for ornamental plants, the landscape industry being the other primary consumer. Retailers desire a uniform product that meets a number of specifications, such as size, number of flowers, container size, and freedom from disease, pests and weeds. They may also require that an individual crop is supplied at a particular period, or even in a number of consecutive batches in an attempt to extend the period in which it would normally be marketed (scheduled production). This provides a number of challenges for the nursery industry. Young plants may be pruned regularly to enhance the number of branches present and improve form. To meet specific sales dates the crops may need to be held back through the use of cold stores or a delayed potting date, or advanced (forced) by using higher temperatures to attain the final height or flowering requirement at an earlier date.

Growers of protected crops may manipulate the photoperiod in order to schedule flowering. In the case of potted chrysanthemums (*Chrysanthemum morifolium*), a short-day plant, all year round production is achieved through extending the photoperiod with artificial lighting during the early vegetative stage of the crop, and then creating short photoperiods using blackouts to stimulate floral development in the later stages. Poinsettia (*Euphorbia pulcherima*), a Christmas house plant, is treated in a similar way. The majority of nursery stock species are long-day plants.

Growth Regulation

Many ornamentals are grown under protected conditions to assist in maintaining schedules, crop protection programs, and physical cleanliness. The disadvantage of protected environments is that the relatively higher temperatures and lower light conditions can contribute to a "stretched" habit due to longer internodes. This problem is compounded by the fact that, in general, customers prefer a more compact habit.

Growers often refer to treatments that help maintain a compact habit as "growth regulation," but it should be noted that such treatments often only affect plant height and have limited impact on dry matter accumulation. A range of techniques are employed including:

1. the use of restricted irrigation regimes;
2. the use of chemical growth regulators (e.g., paclobutrazol and daminozide);
3. managing the temperature regime so that there is a minimum differential between day and night temperatures (a treatment known as "DIF"); and
4. pinching and trimming of shoot tips.

Changing the spectral composition of light by altering the ratio of red/far red irradiance can also reduce internode extension in some crops. Special polythene films (spectral filters), which alter the light spectrum in which the crop is grown and hence regulate its growth, are likely to be used more widely in future.

Table 3 The influence of rooting and overwintering environments on the survival of cuttings from "difficult to produce" shrub species *Magnolia x soulangeana, Cotinus coggygria* cv. Royal Purple, and *Rhododendron* cv. Coccineum Speciosum

Rooting environment	*Winter environment*	*Survival (%)*		
		Magnolia	*Cotinus*	*Rhododendron*
Fog	Well-ventilated tunnel	95	72	79
	Enclosed tunnel	95	75	62
	Cold store	95	71	95
Mist	Well-ventilated tunnel	67	55	20
	Enclosed tunnel	55	43	8
	Cold store	69	71	46
Least significant difference (LSD) ($P=0.05$)		9	14	12

Cuttings were placed in either "wet fog" or intermittent mist environments to root, then finally placed in either a side-ventilated tunnel, an enclosed tunnel with poor ventilation, or a cold store at 2°C for the winter period following propagation.
Reproduced with permission from Cameron RWF and Judd HL (2000) *Development of effective overwintering strategies for rooted cuttings and young liners. Final Report HDC Project HNS 84*, pp. 44. East Malling: Horticultural Development Council.

Biotic and Abiotic Stress

Nursery production comprises a wide range of plant species, but individual numbers of any one crop on an individual holding may be relatively small. Therefore, the normal problems associated with monocropping are infrequent in this sector. Nevertheless, pest species such as slugs, vine weevil (*Otiorhynchus* spp.), thrips, red spider mite (e.g. *Tetranychus* spp.), and various aphid species can be problematic across a range of species, especially where plants are grouped closely under protection. Where feasible and economically viable, nurseries have implemented biological control techniques (introduction of natural predators) to control pest populations. With the exception of very specific crop routines, the application of pesticides in heavy doses is rarely required. Likewise, the main effects of fungal disease, such as with *Botrytis cinerea* and powdery mildew (e.g., *Sphaerotheca* spp.), can be ameliorated through good management practices and environmental control.

Abiotic factors can impinge on the production of some ornamental species. Many broadleaf evergreen plant species, e.g. *Rhododendron* sp., and golden leaf forms of many ornamentals are susceptible to injury from high irradiance levels and high temperatures. Injury can manifest itself as lesions on the leaf margins or chlorosis in the leaf center (sun-scalding), and such plants are normally cultivated under shading to minimize damage.

Younger, less well-developed plants are particularly prone to the extremes of temperature. A young rooted cutting may be susceptible to cold stress, whereas an older specimen may be tolerant of the same sub-zero temperature. Rooted cuttings are often overwintered in plug trays before being potted-on. Care is required to ensure that environmental conditions are appropriate to maximize survival rates. **Table 3** demonstrates how different factors during the propagation and overwintering stages may interact to determine plant survival.

Container plants also are more susceptible to temperature extremes due to the limited thermal buffering offered by the small volume of growing medium. For this reason, nursery stock growers in the colder regions of North America and Europe, may stack or bury their container specimens every winter and insulate them with snow or other materials to minimize the risk of frost injury to the roots.

List of Technical Nomenclature

Abiotic	Relating to environmental influence.
Adventitious roots	Roots not derived from the first true root from a seedling (radicle), i.e., can be initiated from stem tissues such as vascular cambium.
Amenity	Relating to ornamental or recreation purposes.
Auxin	A plant growth regulator involved in cell development.
Biotic	A biological influence.
Clone	A plant genetically identical to the parent.
Gibberellin	A plant growth regulator involved in cell development.
Habit	The size and shape of a plant.
Herbaceous	Nonwoody perennial plants.

Liner	A young plant that requires further development prior to sale.
Perlite	A light-weight, chemically inert growing medium.
Progeny	Offspring.
Run-off	Excess water that is not retained in the container.
Transpiration	Movement of water through the plant / cutting from the roots or stem base to the atmosphere via the leaves.
Weaning	Acclimatization required before a plant is moved from one environment to another.

See also: **Growth and Development**: Leaf Development. **Production Systems and Agronomy**: Horticultural Crops, Protected. **Root Development**: Root Growth and Development. **Seed Dormancy**: Development of Dormancy; Seed Production; Seed Quality; Seed Treatments. **Tissue Culture**: Clonal Propagation, *In Vitro*. **Tissue Culture and Plant Breeding**: Clonal Propagation, Forest Trees; Regeneration of Fruit and Ornamental Trees via Cell and Tissue Culture; Somatic Hybridization.

Further Reading

Armitage AM (1994) *Ornamental Bedding Plants.* Wallingford: CAB International.

Cameron RWF, Harrison-Murray RS, Ford YY, and Judd H (2001) Ornamental shrubs: Effects of stock plant management on the rooting and establishment of cuttings. *Journal of Horticultural Science and Biotechnology* 76: 489–496.

Cameron RWF, Harrison-Murray RS, Fordham M, *et al.* (2003) Rooting cuttings of *Syringa vulgaris* cv. Charles Joly and *Corylus avellana* cv. Aurea: The influence of stock plant pruning and shoot growth. *Tree Structure and Function* (in press).

Davis TD, Haissig BE, and Sankhla N (eds) (1988) *Adventitious Root Formation in Cuttings.* Portland, Oregon: Dioscorides Press.

Hartman HT, Kester D, Davies F, and Geneve R (2001) *Plant Propagation, Principles and Practice*, 7th edn. New York: Prentice Hall.

Lamb K, Kelly J, and Bowbrick P (1995) *Nursery Stock Manual: Grower Manual 1.* London: Grower Books.

Larson RA (ed.) (1992) *Introduction to Floriculture*, 2nd edn. London: Academic Press.

Macdonald B (1989) *Practical Woody Plant Propagation*, 2nd edn. Portland, Oregon: Timber Press.

Netherlands Census Data on Crop Areas and Values. www.minlnv.nl/international

United Kingdom Census Data on Crop Areas and Values. www.defra.gov.uk

United States Census Data on Crop Areas and Values. www.nass.usda.gov

Plantation Crops and Plantations

K V Peter and A Kurian, Kerala Agricultural University, Vellanikkara, India
V L Chopra, National Research Centre on Plant Biotechnology, New Delhi, India

Introduction

Plantation crops are high-value commercial crops which play a vital role in the agricultural economy and export trade of many developing and developed countries (**Table 1**). Tea (*Thea sinensis*), coffee (*Coffea* spp.), cocoa (*Theobroma cacao*), rubber (*Hevea brasiliensis*), coconut (*Cocos nucifera*), arecanut (*Areca catechu*, betel nut), oil palm (*Elaeis guineensis*), cashew (*Anacardium occidentale*), and cardamom (*Elettaria cardamomum*) are the major crops grown on a plantation scale. These crops are perennial in nature and the cultivation is labor-intensive, generating considerable employment opportunity during on-farm operations and off-farm processing activities. The most obvious characteristic is that plantations are monocultures, the same species occupying large contiguous areas usually in even-aged stands. Exceptions are coconut, arecanut, cocoa, and coffee, which are grown either as a mono crop or as a multiple crop. Cultivation of these crops is mostly confined to the tropical region between 20° N and 20° S of the equator. Tea, cardamom, and coffee prefer subtropical conditions (**Table 2**).

Plantation crops serve a variety of human needs not only as sources of food, oil, and industrial raw material but also flavoring materials. Coconut is the most versatile among 2700 species of the palms, rightly called "the tree of heaven," it provides edible and industrial oils, protein-rich milk, invigorating water, timber, fiber, roofing and matting material, activated carbon and shell powder (**Figure 1**). Oil palm, recognized as the highest and cheapest edible oil-yielding crop, is unique in producing two types of oil, palm oil and palm kernel oil, used for culinary and industrial purposes. Being rich in β-carotene, palm oil contributes substantially to nutritional and energy requirement in many places. Arecanut is chewed as a stimulant and forms an essential requisite for several religious and social ceremonies. Cardamom, known as "the queen of spices," is a major oriental spice. Cashew yields highly delicious and nutritive nut, often referred as "wonder nut." Cocoa yields cocoa beans which are the source of

Table 1 Plantation crops and the countries where they are grown

Crop	*Center of origin*	*Countries where grown*
Coconut (*Cocos nucifera*)	South East Asia or Pacific Islands	India, Indonesia, Philippines, Sri Lanka, Malaysia, Thailand, Papua New Guinea, Fiji, Samoa, Zanzibar, Solomon Islands
Arecanut (*Areca catechu*)	Cochin China, Malay Peninsula, East Indies	India, East Africa, Madagascar, Zanzibar, Sri Lanka, Pakistan, Malaysia, Bangladesh, Indonesia, China, Philippines, Fiji
Oil palm (*Elaeis guineensis*)	Guinea coast of West Africa	Malaysia, Indonesia, Nigeria, Thailand, Ivory Coast, Colombia, Papua New Guinea
Rubber (*Hevea brasiliensis*)	Amazon River basin of South America	Indonesia, Thailand, Malaysia, China, India
Coffee (*Coffea* spp.)	Africa	Brazil, Vietnam, Costa Rica, Indonesia, Philippines, India
Tea (*Thea sinensis*)	Irrawaddy River basin of Southeast Asia	India, China, Sri Lanka, Kenya, Russia, Indonesia
Cocoa (*Theobroma cacao*)	Upper Amazon basin in South America	Côte d' Ivoire, Ghana, Indonesia, Brazil, Nigeria, Cameroon, Malaysia, Ecuador, India
Cashew (*Anacardium occidentale*)	Brazil	India, Brazil, Vietnam, Tanzania, Mozambique, Indonesia
Cardamom (*Elettaria cardamomum*)	Western Ghats of India	India, Guatemala, Tanzania, Sri Lanka, El Salvador, Vietnam, Laos, Cambodia, Papua New Guinea

Table 2 Soil and climatic requirements of plantation crops

Crop	*Altitude (m)*	*Temperature (°C)*	*Rainfall (cm year^{-1})*	*Soil type*	*Soil pH*
Coconut	Up to 1000	27±5–7	100–300	Laterite, coastal sandy, alluvial	5.0–8.0
Arecanut	Up to 1000	10–40	150–500	Laterite, red loam, alluvial	Acidic to neutral
Oil palm	Up to 300	15–32	200	Loamy soils	4.0–6.0
Rubber	Up to 450	21–35	200–300	Variety of soils	4.0–6.5
Tea	700–2500	20–27	150	Sandy loam to silty loam	3.2–6.2
Coffee	1000–1500	15–30	100–250	Red laterites	4.5–6.5
Cocoa	Up to 900	10–38	100–150	Well-drained rich soils	Acidic to neutral
Cashew	Up to 1000	15–35	100–200	Red sandy loams or coastal sand	5.6–7.0
Cardamom	600–1200	10–35	150–400	Clay loam	Acidic

Figure 1 A coconut tree (*Cocos nucifera*) in fruit.

cocoa butter and cocoa mass used for making chocolate and confectionery. Tea is the most popular and the cheapest beverage in the world, consumed by two-thirds of the world's population. Coffee is the second important beverage, whose dried beans roasted, ground, and brewed make a stimulating and refreshing drink.

Plant Growth and Development

Coconut, arecanut, and oil palm, members of the palm family Arecaceae, grow with a single stem, a crown of paripinnate leaves, and an adventitious root system. The palms are monoecious in nature and the inflorescence is a spadix. Cashew, coffee, tea, cocoa, and rubber are trees with a well-developed tap root and spreading lateral root system. Except rubber, all other species bear simple leaves and are evergreen. In rubber, the trifoliate leaves are arranged in groups or storeys, a cluster of spirally arranged leaves emerging from each storey. The leaflets bear extra floral nectaries. A rubber tree of more than 3 or 4 years shows wintering behavior, in which the mature leaves are shed rendering the tree wholly or partly leafless for a short period. The bark of the rubber tree contains laticifers occurring in concentric cylinders which are interconnected. The laticifers

Table 3 Growth characteristics of plantation crops

Crop	*Family*	*Height (m)*	*Type of inflorescence/flower*	*Pollination*	*Economic produce*
Coconut	Arecaceae	20–30	Spadix/monoecious	Cross	Fruit (nut)
Arecanut	Arecaceae	30	Spadix/monoecious	Cross	Fruit (nut)
Oil palm	Arecaceae	20–30	Spadix/monoecious	Cross	Fruit (nut)
Rubber	Euphorbiaceae	30	Panicle/monoecious	Cross	Latex
Tea	Camelliaceae	9 Pruned to low bush	Single or clusters/ hermaphrodite	Cross	Leaf
Coffee	Rubiaceae	5	Condensed cyme/ hermaphrodite	Self in *Coffea arabica* Cross in *Coffea canephora*	Seeds (beans)
Cocoa	Sterculiaceae	6–14	Dichasial cyme cauliflorus/ hermaphrodite	Cross	Seeds (beans)
Cashew	Anacardiaceae	12	Panicle/andromonoecious male and hermaphrodite	Cross	Fruit (nut)
Cardamom	Zingiberaceae	2–5	Panicle/hermaphrodite	Cross	Fruit (capsule)

contain latex which is extracted through controlled excision of bark called tapping. Coffee and cocoa show dimorphic branching with orthotrophic and plagiotrophic shoots. The point at which plagiotrophic shoots are produced in cocoa is called the jorquette. Tea develops two types of shoots, aperiodic and periodic. Aperiodic shoots arise from pruned bush frame and growth is continuous. Periodic shoots develop from axillary buds of aperiodic shoots showing rhythmic growth passing through alternate phases of growth and dormancy. Cardamom is a perennial with subterranean rhizomes. The palms and rubber plants are monoecious, producing separate male and female flowers, while in oil palm male and female inflorescences are separate. The cashew bears male and hermaphrodite flowers in the same inflorescence, whereas tea, coffee, cocoa, and cardamom have hermaphrodite flowers with male and female sex organs in a single flower. All the species except *Coffea arabica* are cross-pollinated by insects or wind and the seedling progenies are highly heterozygous. In the majority of plantation crops, fruits are the main economic produce except tea and rubber where leaf and latex respectively form the economic produce (**Table 3**).

Figure 2 A plantation of tea (*Thea sinensis*), showing terrace cultivation.

Production Systems and Agronomy

Plantation crops in general require high management input in terms of tillage, fertilizers, irrigation, and pesticides. Because of their perennial nature and large size, plantation crops are planted in large pits providing wider spacing. Crops such as tea and rubber grown in undulating and sloping terrain are planted in contour terraces to prevent soil erosion (**Figure 2**). Coconut and arecanut offer opportunities for intercropping other economic species whereas cocoa and coffee form compatible intercrops. Plantation crops are heavy feeders of nutrient elements. Heavy manuring of organic and inorganic fertilizers in split doses is used to maintain the vegetative reproductive balance in species valued for their fruits, to keep them in vegetative phase as in tea, and to enhance the yield of latex in rubber. Micronutrient application is used in cases where deficiencies are found. Planting of shade trees and shade regulation are practiced in plantations of tea, coffee, cardamom, and cocoa which are shade-loving species. Leguminous cover crops are established in the interspaces of rubber and oil palm plantations (**Figure 3**) to conserve soil and to improve soil structure and fertility (**Table 4**).

Figure 3 Rubber (*Hevea brasiliensis*) is planted in widely spaced rows, allowing space for a leguminous crop in between.

Pruning and training are regular operations for tea, coffee, and cocoa. In tea, young plants are trained to form a low spreading bush through a series of operations including centering, tipping, bending, and pegging. Pruning helps to maintain the plants in permanent vegetative phase and to stimulate growth of young shoots. In coffee, pruning helps to build strong woody framework, promotes production of berry-bearing wood, and maintains production efficiency through crop regulation. Pruning helps to restrict the height of the jorquette to a convenient height in cocoa trees.

Propagation

Because plantation species are perennial crops with a long juvenile phase and heterozygous nature, much care must be given to the production and selection of planting material. The different stages in the production of quality planting materials include selection of mother plants and seeds, and the adoption of proper nursery techniques. Lack of clonal propagation techniques in palm species makes the selection procedures very stringent. Selection of seed nuts is of utmost importance in palm species as the performance of the new progeny can be evaluated only several years after planting. In coconut and arecanut a series of selections is made at different stages – seed garden, mother palms, seed nuts, and seedlings – following selection criteria and eliminating poor-quality offspring. In tea, cocoa, and rubber, polyclonal seeds are used in cases where seed

Table 4 Production systems of plantation crops

Crop	*Spacing (m)*	*Pit size*	*Manures and fertilizers (year^{-1})*
Coconut	7.5 × 7.5	0.75–1.2 m^3	15–25 kg FYM palm^{-1}, NPK 0.34 to 1.0:0.17 to 0.50:0.68 to 2.0 kg palm^{-1}, dolomite 1 kg palm^{-1}, $MgSO_4$ 0.5 kg palm^{-1}
Arecanut	2.7 × 2.7	80 cm^3	12 kg FYM palm^{-1}, NPK 100:40:140 g palm^{-1}
Oil palm	9.0 × 9.0	60 cm^3	50 kg FYM palm^{-1}, NPK 1.2:0.6:1.2 kg palm^{-1}, $MgSO_4$ 175 g palm^{-1}
Rubber	420–445 plants ha^{-1}	75–90 cm^3	10:10:10 NPK mixture 300 kg ha^{-1}, 50 kg ha^{-1} $MgSO_4$ if yellowing occurs
Tea	0.75–1.35	45–60 cm^3	NPK 90–180:20–50:50–180 kg ha^{-1}, $ZnSO_4$ 2–20 kg ha^{-1}, $MnSO_4$, borax each at 4–6 kg ha^{-1}
Coffee	1.2–5.0	45 cm^3	NPK 140:90:120 kg ha^{-1}
Cocoa	2.7–3.0	50 cm^3	NPK 100:40:140 g plant^{-1}, dolomite 100 g plant^{-1}
Cashew	7.5–10.0	90 cm^3	NPK 725:320:750 g plant^{-1}
Cardamom	1.2–3.0	90 × 90 × 45 cm	NPK 75:75:150 kg ha^{-1} for rainfed crop, 125:125:250 kg ha^{-1} for irrigated crop

FYM, farmyard manure; NPK, nitrogen:phosphorus:potassium.

propagation is adopted. In cardamom, seed propagation has the advantage of not spreading katte, a viral disease. The seeds are subjected to scarification with 20% nitric acid for 10 min before sowing to enhance germination. The selected seeds pass through a primary nursery for a shorter period and a fairly lengthy secondary nursery. Seedlings 12 to 18 months old are preferred for planting in the main field except in cashew and cocoa where 3–6-month-old seedlings are used.

In view of the high variability exhibited by seedling progenies, vegetative propagation is preferred for large-scale planting of plantation crops wherever clonal propagation is feasible. It has the added advantage of precocious bearing. In coffee, a single-noded green wood cutting with leaves clipped and dipped in 5000 ppm indole butyric acid (IBA) strikes roots in 3 to 4 months. In tea, single-node semihard wood cuttings with a leaf dipped in 200 ppm naphthalene acidic acid (NAA) can be rooted easily. Patch budding is a successful method of clonal propagation of rubber and cocoa whereas in cashew it is softwood grafting. In cardamom rhizomes with two or three shoots ensure early bearing.

See also: **Primary Products**: Oils. **Production Systems and Agronomy**: Oil Palm and Coconut; Rubber. **Secondary Products**: Rubber Production.

Further Reading

Banerjee B (1993) *Tea Production and Processing*. New Delhi: Oxford University Press and IBH Publishing Co.

Bavappa KVA, Nair MK, and Kumar TP (1982) *The Arecanut Palm*. Kasaragod: Central Plantation Crops Research Institute.

Chadha KL and Rethinam P (1994) *Advances in Horticulture*, vols. 9 and 10. New Delhi: Malhotra Publishing House.

Clifford MN and Wilson KC (1985) *Coffee*. Westport: AVI Publishing Co.

Hartley CWS (1977) *The Oil Palm*. Harlow: Longman.

Mandal RC (1992) *Cashew Production and Processing Technology*. Bikanir: Agrobotanical Publishers.

Peter KV (2002) *Plantation Crops*. New Delhi: National Book Trust of India.

Purseglove JW, Brown EG, Green CL, and Robbins SRJ (1981) *Spices*, vol. 2. New York: Longman Inc.

Sethuraj MN and Mathew NM (1992) *Natural Rubber: Biology, Cultivation and Technology*. Amsterdam: Elsevier.

Thampan PK (1993) *Handbook on Coconut-Palm*. New Delhi: Oxford University Press and IBH Publishing Co.

Wood GAR and Lass RA (1985) *Cocoa*, 4th edn. New York: Longman Inc.

Oil Palm and Coconut

B S Jalani, A Kushairi and S C Cheah, Malaysian Palm Oil Board, Kuala Lumpur, Malaysia

Introduction

Perennial crops such as oil palm (*Elaeis guineensis*) and coconut (*Cocos nucifera*) have played and will continue to play important roles in the agriculture of the humid and semihumid tropics where they are cultivated on a large scale in both plantations and smallholdings for food and nonfood industrial purposes.

The seed of the modern oil palm industry was sown during the previous century when four seedlings from West Africa, the center of origin and diversity, were planted in Bogor, Indonesia in 1848. Oil palm was initially planted as ornamental and the first commercial plantings in Southeast Asia started in 1911. The crop diversification policy in Malaysia in the late 1950s, which was later followed by Indonesia, stimulated the rapid expansion and modernization of the industry.

On the other hand, coconut is endemic to Malesia, and later spread to many coastal tropical and semitropical areas. Coconut is the "tree of life" which supplies the basic necessities of food, fuel, and shelter for rural people in many of these areas.

This article deals with the achievements and limitations of improvement and biotechnology of these two important food and industrial crops in bridging the yield gap from the wild state to the theoretical limit.

Oil Palm

Origin and Distribution

The oil palm genus *Elaeis* (Arecaceae) has two species, *E. guineensis* ($2n = 32$) and *E. oleifera* ($2n = 32$). The oil palm of commerce (sometimes called the African oil palm), *E. guineensis*, is indigenous to the wild and semiwild groves of tropical and subtropical West Africa distributed from Senegal south to Angola and Zaire.

There is a secondary distribution of oil palm growing in semiwild groves on the Atlantic coast of Brazil, predominantly in Bahia, but it is believed that they were brought to Brazil from Africa by slave traders in the seventeenth century.

Elaeis oleifera is native to tropical areas of Central and South America, growing wild with very little

Figure 1 Young oil palm with fruit bunches. Upper insert: Mature bunch; lower insert: cross-section of mature fruit showing separate sources of palm oil from the mesocarp and palm kernel oil from the endosperm or kernel.

large-scale planting compared to *E. guineensis*. This is partly due to the very low yield despite the high quality of the oil. However, *E. oleifera* may play a very important role in the future with respect to interspecific hybridization and biotechnology.

Biology

Oil palm is an unbranched monoecious plant, with male and female flowers in separate inflorescences (**Figure 1**). It grows to a height of 30–40 m and can live to more than 150 years. It is a feather palm with its stem topped by 35 to 60 pinnate fronds borne on a columnar stem. It has an adventitious root system.

Oil palms produce about two to three fronds per month. Separate male and female inflorescences are borne on the same palm in alternate cycles and oil palm is essentially a cross-pollinated (allogamous) plant. The duration of each floral cycle is dependent on both genetic and environmental factors. The flowers are bisexual in origin, but in "males," the stigmas are suppressed while in "females," the stamens are underdeveloped. Each male inflorescence produces a large quantity of pollen, averaging 30 g.

Each female inflorescence may have from 700 to 1500 flowers, which are receptive for 36 to 48 h after anthesis. The flowers within an inflorescence open sporadically, taking up to 4 weeks between the initial and the last flushes. Normally, most flowers are receptive (up to 82%) on the second day after anthesis, which would be the most suitable time for pollination.

Fruits

The fruit consists of epicarp, mesocarp, endocarp, embryo, and endosperm. The epicarp consists of a thin layer of parenchymatous cells. The mesocarp, the primary oil storage of the fruit, consists of parenchymatous cells containing oil droplets. The endocarp or shell consists of stone cells. The testa cells have brown contents. The endosperm cells contain oil droplets while the aleurone grains have crystalloids. The kernel (endosperm), being the secondary oil storage of the fruit, is oval in shape and about 3 cm in length.

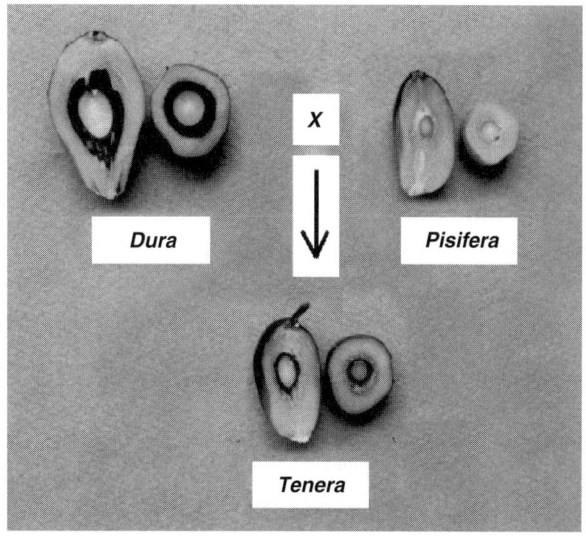

Figure 2 Different types of oil palm fruits: *dura* × *pisifera* = *tenera* as commercial D × P planting materials.

Fruits of *E. guineensis* are classified into three distinct forms based on shell thickness (**Figure 2**): (1) *dura*, homozygous dominant ($Sh^+ Sh^+$), which has a thick shell (2–8 mm); (2) *pisifera*, homozygous recessive ($Sh^- Sh^-$), which is shell-less; and (3) *tenera*, heterozygous hybrid ($Sh^+ Sh^-$) resulting from a *dura* × *pisifera* (D × P) cross, which has a thin shell (0.5–4 mm). However, there still exists, within each fruit form, considerable variation in shell thickness under apparent polygenic control.

The ratio of mesocarp to bunch of the *dura* is 25–65%, *tenera* 75–85%, and *pisifera* 95%. The *pisifera* has lower kernel content, while the *dura* and *tenera* are similar (5–10%). The *pisifera* is generally female sterile. Occasionally, there may be fertile *pisiferas* but the fruit set is generally low and oil to bunch ratio seldom exceeds that of *tenera*. Thus, *pisifera* is not exploited as planting material. It is however, used as the male parent in *dura* × *pisifera* (D × P) crosses to produce *tenera* hybrids.

Fruit color (nigrescens, virescens, and albescens) can also be used as a marker to distinguish between types of *E. guineensis* but it makes no significant contribution to oil yield.

Development of *tenera*

One of the major factors in yield improvement has come from the exploitation of genes affecting shell thickness. This discovery was the cornerstone which led to the commercial utilization of *dura* × *pisifera* (D × P) or *tenera* planting materials. There is no appreciable differences in fresh fruit bunch (FFB) yield between *tenera* and *dura* but the oil to bunch ratio (O/B) of the *tenera* (24%) is higher than that of the *dura* (18%), giving an approximately 30% higher oil yield (**Table 1**).

The hybrid nature of *tenera* requires the separate selection of *dura* and *pisifera* populations. Reciprocal recurrent selection and modified recurrent selection methods are usually used in improvement programs. Extensive progeny testing is essential in the selection of *pisiferas* for good general combining ability and/or specific combining ability with *duras*. The selected *pisiferas* are used as male parents to combine with selected *dura* female parents for the production of commercial quantities of D × P (*tenera*) seeds.

Breeding Populations

The primary objectives in oil palm breeding are:

- increased oil yield per hectare
- improvement of oil quality
- reduced height increment rate
- pest and disease tolerance
- improvement of physiological traits (bunch index, total dry matter, and bunch dry matter)
- exploitation of genotype × environment interaction.

Table 1 Average bunch composition (%) of oil palm *dura* and *tenera*

Components	*Dura* ($sh^+ sh^+$)	*Tenera* ($sh^+ sh^-$)
Fruit/bunch	60	60
Mesocarp/fruit	60	80
Shell/fruit	30	10
Kernel/fruit	10	10
Oil/wet mesocarp	50	50
Oil/bunch	18	24

Other traits of interest include high early yield, kernel yield, yield profile, carotene level, vitamin E content, and other minor components.

In oil palm, breeding and selection is synonymous with the developments of four Deli *dura* seedlings introduced to Bogor in 1848. Currently 90% of global D × P seed production is based on the Deli *dura* as the female parent. This is not surprising as the Deli *duras* are early bearing and superior in bunch and fruit characteristics.

It is interesting to note that, although the palms have very narrow genetic base and were initially grown as ornamentals, by chance the Deli *dura* had a valuable genotype and is still considered the best *dura* for breeding and seed production. The *pisiferas* which are used as male parents also originated from a limited number of palms from Africa.

dura and *tenera*/*pisifera* Breeding Populations

Modern *dura* breeding populations are almost exclusively derived from the Deli *dura* population. Some other notable *duras* are Serdang, Elmina, Gunung Melayu, Dabou, Lofindi, Dami, Coto, and NIFOR. There have been introgressions amongst these *dura* and other populations that has resulted in some overlap between different breeding programs.

Early research programs in Africa concentrated on breeding and improvement of *tenera* populations and the improvement of *tenera*/*pisifera* in Southeast Asia was largely based on the African elite *pisiferas* such as AVROS, Yangambi, La Me', and Ulu Remis.

The introgression of different breeding populations in seed production, especially in most common sources of introgressions such as Deli *dura* with Yangambi *pisifera* and others, has led to the "derived *pisifera*." The most notable derived *pisiferas* are Dumpy-AVROS, URT-AVROS, and URT-Cameroon. The development of these derived *pisiferas* sometimes reflect differences in emphasis of the breeding programs. For examples, in Ulu Remis *pisifera* the emphasis is on improvements in yield, while Dumpy-AVROS *pisifera* emphasizes both yield and short palm height.

Selection of Parental Palms

Reciprocal recurrent selection or modified recurrent selection are used to maintain and select independently separate lines of *duras* and *teneras*. The procedure is also able to avoid inbreeding effects and to allow some characteristics to be selected in *dura* and *tenera* lines to overcome problems of negatively correlated traits. For example, bunch number and average bunch weight are negatively correlated and it is not possible to maximize both in the same palm except through high selection pressure over a long period. Usually, average bunch weight, fruit to bunch, kernel to bunch, and shell to fruit are emphasized in the selection of *dura* parents, while bunch number, mean fruit weight, and mesocarp to fruit are stressed in *tenera* lines. Selection for oil is emphasized in both parental lines.

Different descendent lines of Deli *dura* have been developed through selfing and sib-mating. Some of these superior Deli *dura* lines were introgressed with other *duras*. Family selection followed by individual palm selection is commonly practiced to overcome low heritability estimates of some characters. For example, heritability of bunch yield and its components is low, while bunch quality traits such as mesocarp and oil contents of the fruits have higher estimates.

Selection of *pisifera* is indirect and troublesome compared to *dura* selection because their fruit production is so poor. The strategy is to evaluate or progeny test the *pisiferas* on the basis of actual testcrosses with selected *duras* as well as using *tenera* with *dura* crosses. Thus, the evaluation of *pisiferas* in progeny test of D × P, D × T, or T × D is an integral part of breeding scheme such as the reciprocal recurrent selection and modified recurrent selection. Breeding designs and specific mating systems such as North Carolina Model I, diallel or simply in biparental combinations, are used to generate these progenies. In-depth scrutiny of genetic variance components, heritability estimates, and combining abilities follow these mating designs. In the North Carolina Model I mating design, one *pisifera* is crossed with several, usually three to five *duras*, and the mean performance over all crosses is taken as a measure of the breeding value of the *pisiferas*, which allows the exploitation of general combining ability for the production of commercial seeds.

Due to the large demand for seeds coupled with long breeding cycle and other testing requirements (e.g., multilocation tests in differing areas), it is not possible to confine seed production only to those crosses between the *duras* and *pisiferas* that have undergone progeny testing. The normal procedure in commercial production is to select the *dura* and *pisifera* independently and only selected *dura* would be combined with any selected *pisifera*.

Important Traits

There are several components of oil yield but the most important are fresh fruit bunch and fruit quality characters. Fresh fruit bunch yield is highly correlated to the weight per bunch and number of bunches per palm. Bunch number and average bunch weight and fruit to bunch ratio and fruit weight have negative correlations. Positive relationships are shown between fresh fruit bunch and its components, mean fruit weight and mesocarp to fruit, bunch number, and oil yield.

Individual palms are selected from superior families on the basis of the mesocarp to fruit, shell to fruit, oil to dry mesocarp, and oil to bunch. Other characters such as fruit to bunch and kernel contents are given special attention to ensure that the traits are at reasonable levels.

Performance of Oil Palm Worldwide

Oil palm is unique in that it produces two types of oil, palm oil from the mesocarp and palm kernel oil from the endosperm (kernel). This can cause confusion as the general public thinks that these two oils are the same. The fatty acid composition of palm oil is significantly different from that of palm kernel oil and coconut oil (**Table 2**). Some of the similarities between palm kernel oil and coconut oil are due to their derivation from the endosperm.

The theoretical oil yield from oil palm is estimated at 18.6 tonnes ha^{-1} $year^{-1}$ (**Table 3**), thus making it the highest oil-bearing crop of any cultivated species.

There have been tremendous improvements in oil yield and its components in the last 100 years. The average oil yield from the wild and semiwild groves is about 0.2 tonnes ha^{-1} $year^{-1}$ (**Table 3**).

In properly managed growing areas, the yield is much higher, but is dependent on inputs (e.g.,

Table 2 Fatty acid composition (%) and iodine value (%) of palm oil, palm kernel oil, and coconut oil

Composition	*Palm oil*	*Palm kernel oil*	*Coconut oil*
$C_{6:0}$	–	0.3	0.5
$C_{8:0}$	–	4.2	8.0
$C_{10:0}$	–	3.7	6.4
$C_{12:0}$	0.2	48.7	47.3
$C_{14:0}$	1.1	15.6	17.6
$C_{16:0}$	44.0	7.5	8.4
$C_{16:1}$	0.1	–	–
$C_{18:0}$	4.5	1.8	2.8
$C_{18:1}$	39.2	14.8	5.8
$C_{18:2}$	10.1	2.6	1.8
$C_{18:3}$	0.4	–	–
$C_{20:0}$	0.4	–	–
Others	–	0.8	1.4
Iodine value	53.3	17.8	8.0

planting materials, agronomic practices) and the environment (e.g., soils, rainfall distribution). In the four largest palm oil-producing countries (**Table 3**), the average oil yield ranges from 3.9 to 2.1 tonnes ha^{-1} $year^{-1}$. A good commercial plantation in Malaysia is capable of producing 5.5 tonnes ha^{-1} $year^{-1}$ of oil yield. The yield is, however, dependent on the ownership of the crop which reflects the levels of inputs. Independent smallholdings produce an average oil yield of 3.3 tonnes ha^{-1} $year^{-1}$, organized smallholdings average 3.8 tonnes ha^{-1} $year^{-1}$, and private plantations yield 5.5 tonnes ha^{-1} $year^{-1}$. The average oil yields of the best plots, the best progenies, and the best individuals are 8.6, 11.5, and 14.9 tonnes ha^{-1} $year^{-1}$, respectively (**Table 3**) and the best individual yield is almost 80% of the theoretical yield.

Table 3 Oil yield of oil palm on the basis of total dry matter production

Yield class	*Total dry matter production tonnes ha^{-1} $year^{-1}$*	*Oil yield tonnes ha^{-1} $year^{-1}$*
Maximum theoretical yield	44	18.6
Best experimental yield:		
Best individual palms		14.9
Best selected progenies		11.5
Best plots	39	8.6
Good commercial yield (Malaysia)	32	5.5
Average national yields:		
Colombia		3.9
Malaysia	20	3.7
Indonesia	19	3.4
Nigeria	16	2.1
Wild and semiwild groves (Africa)		0.2

Improvements in Oil Yield

The performance of the unselected original progenies of the Deli *dura* was quite spectacular, with oil yield of 2.8 tonnes ha^{-1} $year^{-1}$ (**Table 4**). The oil yields from the unselected, first, and fourth cycle selection of Deli *dura* were 2.7, 3.4, and 4.5 tonnes ha^{-1} $year^{-1}$, an average improvement of 0.43 tonnes per selection cycle. These *duras* were used as commercial planting materials from 1910 until early 1960s.

In the early 1960s in Malaysia, *tenera* ($D \times P$) replaced the *dura* as commercial planting material and oil yield increased to 4.9 tonnes ha^{-1} $year^{-1}$. Subsequent plantings with improved selections has resulted in further improvements of oil yield (**Table 4**).

Improvement through Agronomic Practices

The substantial increase in yield has been obtained through a combination of selective breeding and improved agronomic practices and factory efficiency. There was a gross improvement of oil yield from 1.3 tonnes ha^{-1} $year^{-1}$ in 1951 to 5.34 tonnes ha^{-1} $year^{-1}$ in 1991. Changes in fertilizer application, improved nursery practices and soil/water management, and improved factory practices have all contributed to this improvement.

The reduction in the period to first fruiting from 36 to 30 months after field planting together with other traits, such as tolerance to diseases (*Ganoderma* and *Fusarium*) and drought should help to further increase productivity.

The Role of Introduced Pollinating Insects

Prior to 1981, it was normal practice for oil palm plantations in Malaysia to conduct assisted pollination to palms up to 10 years old to improve fruit set. The results were costly and not satisfactory. The

Table 4 Yield performance of oil palm planting materials

Materials	*Year of planting*	*Fresh fruit bunch tonnes ha^{-1} $year^{-1}$*	*Oil to bunch (%)*	*Projected oil yield tonnes ha^{-1} $year^{-1}$*
Dura				
F_1 Bogor (unselected)	1878	16.5	17.6	2.8
Tg Morawa (unselected)	1885	16.0	17.4	2.7
Elmina (1st selection cycle)	1933	20.1	17.0	3.4
OPRS (4th selection cycle)	1969	24.8	18.3	4.5
Tenera				
DD × CI/UA/SP	1962	22.6	21.9	4.9
DD × AVROS	1964	31.0	23.5	7.3
DD × AVROS	1968	31.1	22.1	6.9
DD × AVROS	1970	31.6	24.2	7.6
DD × Dumpy-AVROS	1979	33.3	25.8	8.9
DD × Yangambi	1988	35.1	26.0	9.5
DD × Yangambi/Nigeria	1992	35.5	31.5	11.2

introduction of the pollinating weevil *Elaedobius kamerunicus* from Africa was made in 1982 with great success and the weevil has spread throughout the country and to Indonesia, the Philippines, and Thailand. Effective pollination has resulted in higher fruit set, bigger kernels, and bigger bunches, and the weevil has not had any detrimental effects on the local insects or environment. The introduction of the weevil has increased bunch weight by 28%, the flower to fruit ratio by 7%, and the kernel to fruit ratio by 17%. The oil yield increase due to pollination by the weevil amounted to 1.2%.

Development of Dwarf Palms

There is great interest in breeding dwarf palms as these are easier to harvest. Several dwarfing genotypes are available; a notable one is Dumpy E206. This has been introgressed into *pisifera* and yields are comparable to *pisifera* but with 30% shorter plants. The Malaysian Palm Oil Board germplasm collection from Africa contains other important sources of dwarf palms for breeding and there are a number of ongoing introgression programs using these materials. Early results of progeny testing showed that the height increment of the progenies averaged 45 cm $year^{-1}$ which is 40% less than the current commercial D × P materials.

Breeding for High Unsaturation Oil (High Iodine Value)

Palm oil has an iodine value of 53 and is semisolid at room temperature. On fractionation, it can be split into 55% liquid olein and 45% solid stearin. Olein fetches a higher market price and one objective in oil palm breeding is to further increase the proportion of olein to stearin. There are two approaches to achieve this objective: conventional breeding and biotechnology (see below). The screening of germplasm has identified plants with iodine values that exceed 60 which have been used to produce high iodine value selections for commercial planting.

The iodine value of *E. oleifera* is very high, up to 80. The interspecific hybrids between *E. oleifera* and *E. guineensis* have an iodine value of 65 but the hybrids have oil yield too low for commercial exploitation.

Breeding for High Lauric Oil

Palm kernel oil is an important source of lauric oil which is in great demand for lauric acid with the expansion of oleochemical industry worldwide. The supply of lauric oil from coconut is dwindling, partly due to its low yield and declining cultivation and palm kernel oil is one way to increase supply.

Table 5 Composition (%) of carotenoids of palm oil derived from *E. guineensis*, *E. oleifera*, and their hybrids

Type	E. guineensis	E. oleifera	E. oleifera × E. guineensis *hybrid*
Phytoene	1.27	1.12	1.83
cis β-carotene	0.68	0.48	0.38
Phytofluene	0.06	Trace	Trace
β-Carotene	56.02	54.08	60.53
α-Carotene	35.06	40.38	32.78
cis α-Carotene	2.49	2.30	1.37
ζ-Carotene	0.69	0.36	1.13
γ-Carotene	0.33	0.08	0.23
δ-Carotene	0.83	0.09	0.24
Neurosporene	0.29	0.04	0.23
β-Zeacarotene	0.74	0.57	1.03
α-Zeacarotene	0.23	0.43	0.35
Lycopene	1.30	0.07	0.05
Total ($mg\,l^{-1}$)	500–700	4300–4600	1250–1800

Screening of germplasm has revealed a considerable level of variation for kernel to bunch which could be the basis for exploitation of elite materials with high kernel content. High-yielding elite *dura* palms with high kernel to bunch contents have been introgressed to produce new planting materials with high kernel to bunch (over 10%) as compared to the current commercial planting materials with 6% kernel to bunch. Planting materials with high kernel contents will soon be released for commercial planting.

High Carotenoid and Vitamin E Content

Palm oil contains very high level of carotenes and vitamin E which are very useful for pharmaceutical and nutraceutical purposes. There is considerable variation within and between species of oil palm and their hybrids (**Table 5**). *E. guineensis* has a lower carotene content (500–700 mg l^{-1}) than *E. oleifera* (4300–4600 mg l^{-1}) while their hybrids are intermediate (1250–1800 mg l^{-1}) between the two parents. The vitamin E content also follows a similar pattern (**Table 6**).

Breeding for Resistance to Pests and Diseases

The oil palm industry in Southeast Asia is relatively free of major pests and diseases. Basal stem rot caused by *Ganoderma boninense* is probably the only major disease. Only recently, a technique has been developed enabling the screening of germplasm and existing materials to select resistant palms for breeding purposes.

In Africa and some parts of Central and South America, vascular wilt caused by *Fusarium oxysporium* is quite prevalent, causing death to both young and old palms. Tolerant genetic materials together with effective nursery screening technique for *F. oxysporium* should be able to produce planting materials resistant to this disease.

The expansion of oil palm cultivation in Central and South America has been hampered by serious diseases such as "sudden wilt" or *Marchitez*, recently identified basal stem rot and bud rot. According to one opinion, bud rot ("pudricion de cogollo" or "amarelecimento fatal") is caused by biotic factors which are yet to be identified. As a result interspecific hybrids of *E. oleifera* and *E. guineensis*, which show some form of tolerance to bud rot, have been planted in some areas in Central and South America. Although showing some form of tolerance, the interspecific hybrids are slowly being affected by bud rot. Another opinion, especially gained through experience of a similar problem in Malaysia and Indonesia, holds that bud rot is caused by abiotic factors, namely poor soil/water management, and inadequate and unbalanced major and minor nutrients.

Breeding for Physiological Traits and Environmental Adaptation

To maximize yield it is desirable to have site-specific planting materials based on the knowledge that genotypes differ in their response to environmental factors such as soil, rainfall, temperature, humidity, and agronomic practices.

There is a need therefore, to breed for more efficient palms with high harvest index (HI) for site-specific environment. There is sufficient genetic variability from the available germplasm to breed for short efficient palms for increased planting density and high harvest or bunch index (BI).

Table 6 Composition (%) of tocopherol and tocotrienols of palm oil derived from *E. guineensis*, *E. oleifera*, and their hybrids

Type	E. guineensis	E. oleifera	E. oleifera × E. guineensis *hybrid*
α-Tocopherol	21	15	19
α-Tocotrienol	23	27	28
γ-Tocotrienol	45	54	42
δ-Tocotrienol	11	4	11
Total (mg l^{-1})	600–1000	700–1500	600–1600

Role of Germplasm

Almost 90% of the world production of oil palm seed comes from descendants of Deli *dura* which can be traced to the four Bogor palms. Realizing the importance of broadening the oil palm genetic base, many expeditions have been mounted by different groups to collect materials from the wild and semiwild groves in West Africa.

Biotechnology

Capitalizing on the palm's high productivity, the techniques of biotechnology can be employed to divert the palm's biological resources towards the production of novel, high-value products. Biotechnology is being applied to complement and enhance improvement of the palm through conventional breeding.

Genetic engineering Opportunities abound in the area of genetic engineering for oil palm by diverting the intermediates of oil synthesis to the production of industrial chemicals (bioplastics, polymers), nutraceuticals, or even pharmaceuticals. With the knowledge of the pathways involved in fatty acid biosynthesis, manipulating oil composition through genetic engineering has been an attractive goal. In oil palm, attempts are being made to produce a more liquid oil by increasing the oleic acid content (iodine value 70). The pathway that needs to be manipulated to achieve this objective is illustrated in **Figure 3**.

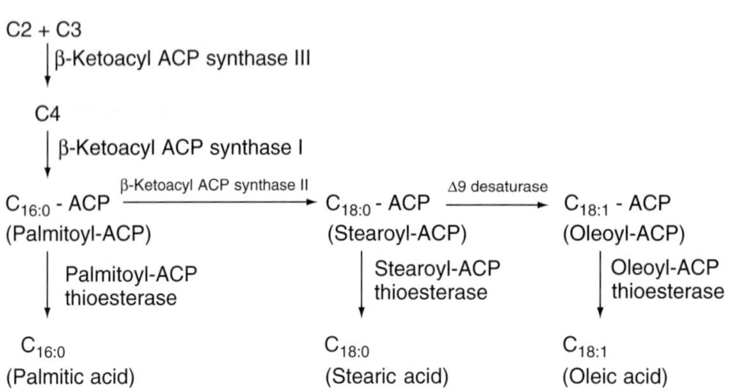

Figure 3 Pathway to achieving high oleic oil content in palm oil.

Genomics The economic value of genomic studies in a perennial crop such as oil palm lies in the ability to identify genetic markers for early selection. The use of DNA markers could greatly reduce the number of breeding cycles. Marker-assisted selection requires the availability of genetic linkage maps and genetic markers associated with the trait of interest. Genetic linkage maps for the location of genes which are potentially useful as markers in oil palm have been constructed. Quantitative trait loci (QTL) affecting carotene and oleic acid content in palm oil have been mapped. In addition, other genomic approaches are being applied to oil palm, including the generation of expressed sequence tags (ESTs) from various tissues and the use of the DNA microarray for profiling gene expression.

Tissue culture There is a need to augment the seed production in oil palm with clonal material produced via tissue culture. *In vitro* micropropagation is an attractive route to the production of oil palm planting materials. Recent field trials have indicated that increase in yield is achievable through this process. The efficiency of production through culture can be further enhanced with the use of liquid suspension cultures.

Coconut

Origin and Distribution

Coconut (*Cocos nucifera*; Arecaceae) is widely distributed in tropical regions of the world, ranging between 20° N and 20° S latitude. The fruit can float and endure long journeys in seawater. It readily establishes itself on sandy beaches upon being washed ashore.

Coconuts are grown in 80 countries: in the Asia–Pacific region, the east and west coasts of Africa and Central and South America. Coconut is the "tree of life" as it plays a vital role, in addition to giving food, drink and shelter, in the economies of many small island countries in the Caribbean, Indian, and Pacific Oceans. Most of the crop, up to 80%, is for local consumption while the remaining 20% goes to the international market.

Biology

Coconut is a perennial monocotyledon ($2n = 32$). It produces adventitious roots at the base of the stem with a prominent bole as it grows older (**Figure 4**). The palm has a single terminal meristem producing 30–45 fronds on the top of its stem. Generally there are three types of coconut, namely tall, dwarf, and their hybrids.

Figure 4 Young coconut palm showing high yield, precocity and early bearing.

Flowering occurs 3 to 10 years after planting. The inflorescence is a spath and spadix type, with both male and female flowers. Coconut being allogamous with very fertile pollen, cross-pollination is very common with hybrids showing hybrid vigor.

The coconut fruit is a fibrous drupe with very thin epicarp, very thick, fibrous mesocarp surrounding a seed consisting of endocarp (shell) which has three "eyes" containing the embryo, fleshy endosperm, and water. The fruit normally matures 12 months after fertilization.

Coconut has a long life span, over 80 years, and grows to a height of 30 meters depending on the environment. In plantations, palms are usually replaced when they reach 60 years.

Uses of Coconut Palm

Coconuts have a variety of uses. The leaves provide fodder, shelter, and materials for making baskets and other traditional products. The husks are used as fuel, planting substrate, ropes, and mats. The endocarp (shell) is used as fuel and for making handicraft curios. The trunk is utilized as building materials, shingles, flooring, furniture, curios, and fuel. Both young and mature coconuts provide a refreshing, healthy drink.

The endosperm, kernel, or meat is the main commercial commodity of the coconut industry. While in a fresh condition, the meat is processed to produce coconut skim milk and desiccated shredded coconut. After drying, the meat is called copra (5% moisture content) which upon further processing yields coconut oil and palm kernel cake. Copra, coconut oil, and palm kernel cake are traded in the international market.

Breeding and Yield Performance

Breeding objectives Copra production is dependent on the number of nuts per palm and the quality and quantity of the meat in the nut. The average oil yield of a coconut grove is about 0.5 tonnes ha^{-1} $year^{-1}$, which is very low compared to palm oil and palm kernel oil. However, coconut oil has its own niche and specialty functions. In order to be competitive and sustainable, the primary objective in coconut breeding is therefore to increase yield. Other important attributes are early bearing, resistance to pests and diseases, and tolerance to abiotic stress such as drought, salinity, wind, and poor management.

Table 7 Heritability estimates of important traits in coconut

Trait	*Dwarf*	*Tall I*	*Tall II*	*Hybrid*
Number of bunches	0.08	0.65	0.11	0.13
Number of nuts per palm	0.49	0.37	0.14	0.30
Copra per nut	0.70	0.78	0.40	0.57
Copra per palm	0.09	0.17	0.35	0.27

Hybridization followed by progeny testing and selection is the most common procedure. However, the emphasis is now on hybrids of dwarf × tall and its reciprocal to exploit the advantages of dwarf coconut (precocity, dwarfness, high nut number) and that of tall coconut (high copra content). The best hybrids are further improved by reciprocal recurrent selection to exploit the traits with good combining abilities and high heritability. Copra to nut has very high heritability, while number of nuts and copra production have fairly high heritability estimates (**Table 7**), thus improving yield in coconut should be reasonably easy.

Germplasm Palm height and fruit color are used to classify the different types of coconut: tall and dwarf for height and red, yellow, and green for the color of the immature fruits. The varieties are named after their origin, color and/or height, such as Malayan Yellow Dwarf, Brazilian Green Dwarf, Fijian Tall, Cameroon Red Dwarf, Tall, West African Tall, Tagnanan Tall, and so on. The Coconut Genetic Resources Database contains over 1300 coconut accessions in various countries in South Asia, Southeast Asia, South Pacific, and the Latin America–Caribbean region.

Yield improvement There has been impressive yield improvement in coconut (**Table 8**). For examples, there were differences in copra yield between Malayan Yellow Dwarf and Malayan Tall of 1.16 and 1.54 tonnes ha^{-1} $year^{-1}$ as "old varieties" and 2.24 and 2.30 tonnes ha^{-1} $year^{-1}$ as "modern varieties" respectively (**Table 8**).

Table 8 Copra yield from different types of coconut

Types	*Number of nuts ha^{-1} $year^{-1}$*	*Copra tonnes ha^{-1} $year^{-1}$*	*Oil tonnes ha^{-1} $year^{-1}$*
Old varieties planted in the 1940s and early 1950s			
Malayan Yellow Dwarf	–	1.16	0.75
Malayan Tall	–	1.54	1.00
Hybrid	–	3.28	2.13
Average high yielding varieties	–	2.25	1.46
High yielding blocks	–	2.29	1.49
Modern varieties			
Malayan Green Dwarf	–	2.24	1.45
Malayan Red Dwarf	–	2.15	1.40
Malayan Yellow Dwarf	–	2.45	1.60
Malayan Tall	8855	2.30	1.56
Hybrid			
MAWA	28819	6.06	4.09
MATAG	28945	7.24	4.78

MAWA, hybrid of Malayan Yellow Dwarf × West African Tall; MATAG, hybrid of Malayan Yellow or Red Dwarf × Tagnanan Tall from the Philippines.

Table 9 Precocity and early bearing of new coconut varieties

Years after planting	Number of nuts ha^{-1}		
	MAWA	MATAG	Malayan Tall
3	–	900	–
4	16 200	21 292	–
5	14 300	26 293	1 150
6	20 500	27 880	3 850

However, upon crossing with elite germplasm materials, the hybrids have very high yield. The MAWA (hybrid of Malayan Yellow Dwarf × West African Tall) and MATAG (hybrid of Malayan Yellow or Red Dwarf × Tagnanan Tall from the Philippines) are yielding 6.06 and 7.24 tonnes ha^{-1} $year^{-1}$ of copra or 3.94 and 4.73 tonnes ha^{-1} $year^{-1}$ of coconut oil respectively (**Table 8**). These hybrids, especially MATAG, are also early-bearing (**Table 9**).

Breeding for Local Adaptation and Resistance to Pests and Diseases

In addition to low yield and high cost of production, the expansion of coconut cultivation in many countries are hampered by several pests and diseases. The lethal yellowing-like diseases (LYD) have devastated huge areas in many countries in the Latin America–Caribbean, Africa, and Pacific regions. The disease is thought to be caused by phytoplasmas and is probably the most important threat to global coconut cultivation.

Almost all of the imported germplasm and hybrids succumb to these diseases but a number of local varieties and a few imported selections show some degree of tolerance. Progenies of these types have been selected for further screening to lower disease incidence and slow rates of spread.

Other diseases include stem bleeding caused by *Ceratocystis paradoxa*, bud rot caused by *Phytopthora palmivora*, basal stem rot caused by *G. bonineense* and *G. lucidum,* and red ring disease caused by the red ring nematode *Rhadinaphelenchus cocophilus*. The notable pests are black headed caterpillar (*Opisina arenosella*), red palm weevil (*Rhynchophorous ferrugineus*), rhinoceros beetle (*Oryctes rhinoceros*), termites (*Odontotermes obesus*), and whitefly (*Aleurodicus pulvinatus*).

Biotechnology

The application of biotechnology to coconut is to facilitate the breeding programs in yield improvement and crop protection, especially lethal yellowing-like diseases. Various DNA marker technologies are used for the evaluation of biodiversity in coconut germplasm, genomic analysis, mapping population, linkage groups, and indexing of pathogens of importance to coconut. It is now possible, using molecular markers, to distinguish coconut populations from different origins. This will help in the management and utilization of germplasm. In addition the indexing of coconut pathogens by molecular markers would help to facilitate the safe movement of germplasm.

In vitro multiplication of high-yielding hybrids and individuals with desirable attributes such as resistance to lethal yellowing-like diseases and tolerance to abiotic stress is a long-term hope for the production of homogeneous planting materials. Attempts in somatic embryogenesis show that coconut is a highly recalcitrant species in *in vitro* culture.

Certain figures and tables are reproduced with kind permission of the Malaysian Palm Oil Board (MPOB).

List of Technical Nomenclature

Latex	A viscous mixture produced in the bark of some tropical trees from which rubber is derived.

See also: **Diseases**: Breeding for Disease Resistance. **Flowering and Reproduction**: Pollination. **Primary Products**: Oils. **Production Systems and Agronomy**: Plantation Crops and Plantations. **Secondary Products**: Rubber Production.

Further Reading

Basiron Y, Jalani BS, and Chan KW (eds) (2000) *Advances in Oil Palm Research*, vol. 1. Bangi, Selangor, Malaysia: Malaysian Palm Oil Board.

Corley RHV and Tinker PBH (2003) *Oil Palm*, 4th edn. Ames: Iowa State University Press.

Davidson L (1991) Management for efficient cost-effective and productive oil palm plantations. *Proceedings of PORIM International Palm Oil Conference* 9–14 September 1991, Kuala Lumpur, Malaysia, pp. 153–167.

Lee CH and Toh PY (1991) Yield performance of Golden Hope D × P planting materials. *Planter* 47: 317–324.

Mandal RC (1998) *Coconut Production and Protection Technology*. New Delhi: Vadams eBooks Ltd. 168pp.

Ohler JG (1984) *Coconut, Tree of Life*. Rome: Food and Agriculture Organization.

Ohler JG (1999) *Modern Coconut Management*. Rome: Food and Agriculture Organization.

Sharma M, Singh G, Padman CVR, and Toh, TS (2000) MAWA and MATAG: two high yielding hybrids for revitalization of the Malaysian coconut industry. *Proceedings of the International Planters Conference*, 17–20 May 2000, Malaysia, pp. 605–624.

Woodroof JG (1979) *Coconuts: Production, Processing and Products*, 2nd edn. Westport: AVI Publishing Co.

Rubber

B S Jalani, Malaysian Palm Oil Board, Kuala Lumpur, Malaysia
O Ramli, Malaysian Rubber Board, Selangor, Malaysia

Introduction

There are many latex-bearing dicotyledonous species from the humid and wet tropical and subtropical regions of the world, but none is as productive as the Para rubber tree, *Hevea brasiliensis,* which is indigenous to the Amazon basin of South America. The native populations of this area exploited the rubber trees, including trading the latex to enhance the basic necessities of their subsistence nomadic and seminomadic life.

Serious domestication and modernization of the rubber industry started less than 130 years ago. A Briton, Sir Henry Wickham, collected rubber seeds from the Amazon basin in 1876 and some of these seeds were dispatched to Southeast Asia in 1877. The surviving seeds became the nucleus of the modern rubber industry. Rubber is now grown in many parts of South and Southeast Asia, Africa, and South America. Rubber is therefore, perhaps the youngest of the major domesticated crops in the world.

The remarkable success of rubber development was due to the systematic exploitation of agricultural research. The most important single component was breeding and selection, which was started in Indonesia in 1910–20, and a few years later in Malaysia. This led to the rubber "boom" in Southeast Asia which has continued until the present day. Currently, Malaysia, Indonesia, Thailand, and Vietnam account to almost 80% of the world's natural rubber production.

Origin and Distribution

Hevea (Euphorbiaceae) has two subgenera with 10 species. Subgenus *Hevea* has nine species, including *H. brasiliensis*, and subgenus *Microphyllae* has only one species. *H. brasiliensis* is the only species of commercial importance. Other species may play an important role in crop improvement and biotechnology in the future with expanded objectives from solely on latex to exploitation of timber woods and other products such as nutraceuticals and pharmaceuticals.

All the species occur in the wild of the whole humid tropic Amazon basin and parts of the uplands, covering Brazil, Bolivia, Colombia, Ecuador, French Guiana, Guyana, Peru, Suriname, and Venezuela. The Rio Negro region of the Amazon basin in Brazil is considered as the center of genetic diversity of the genus *Hevea.*

Biology

All *Hevea* species are diploids with $2n=36$. Interspecific hybridization occurs in nature. All species contain latex in all parts of the plant. There are other features common to all the species (**Figure 1**). The leaves are trifoliate with long petioles giving rise to pinnately veined leaflets. Separate male and female flowers are borne on the same inflorescences with females at the ends of the main branches of the panicles. Each flower has a five-lobed calyx borne on the basal disc of five free or united glands, without petals. The male flower has between five and ten stamens with their filaments united into a column with sessile anthers. The ovary is three-celled with one ovule per cell. Pollination is mainly by insects, almost entirely by midges and to a lesser extent by thrips. Fertilization occurs within 24 h after pollination. The fruit has a trilobular capsule, usually containing three seeds, which usually dehisces explosively upon ripening.

The seeds are large, ovoid in shape with a slightly flattened ventral surface and the testa is hard and shiny, brown or gray–brown with numerous darker mottles or streaks on the dorsal surface. An integument lines the inner surface of the testa and encloses the embryo, cotyledon, and endosperm. The seeds are recalcitrant and lose viability and deteriorate rapidly on storage unless special precautions are taken. Germination is hypogeal and mostly occurs within 3–25 days after sowing.

Growth and Wintering

The stem growth or elongation is discontinuous and is characterized by tiers or whorls of leaves. After germination, the first tier of leaves appears and is quickly followed by rapid elongation of the stem, then a rest period in which scale leaves develop around a terminal bud. The terminal bud then opens a new whorl of leaves, followed by rapid elongation of the stem and another rest period. By repetition of this sequence, leaves are produced in whorls separated by lengths of bare stem. Although elongation of the stem is intermittent, the girth increases continuously.

From 3 or 4 years onwards, rubber trees are subject to wintering, i.e., annual shedding of senescent leaves which renders the trees leafless for a short period. Within 2 weeks, this is followed by the

Figure 1 Rubber tree, *Hevea brasiliensis* (A) Growth of latex–timber clone (left) with bigger bole and latex clone. (B) Trifoliate leaves. (C) Inflorescence. (D) Young fruit. (E) Mature fruit after dehiscence. (F) Tapping rubber tree for latex production.

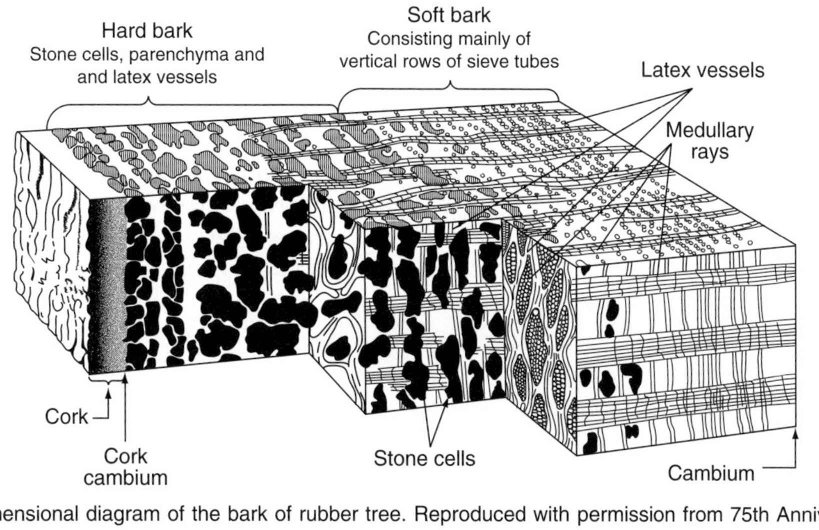

Figure 2 Three-dimensional diagram of the bark of rubber tree. Reproduced with permission from 75th Anniversary of the rubber research institute of Malaysia (1925–2000): Milestones in rubber research. The Malaysian Rubber Board.

burst of terminal buds and expansion of new leaves a week later. Latex production falls slightly during the leaf fall and more markedly during refoliation.

Bark and Latex Vessel System

Bark

The trunk of a mature rubber tree has a central wood cylinder surrounded by bark. The two tissues are separated by a thin layer of vascular cambium which produces xylem (wood) on the inside and phloem on the outside.

The virgin bark has three concentric layers (**Figure 2**). The outermost layer, the periderm, consists of phellum (cork cells) and an inner phellogen (cork cambium). The middle layer, the hard bark, contains mostly disorganized and nonfunctional sieve tubes and laticifers or latex vessels. Soft bark, the innermost layer, is mainly composed of concentric cylinders of sieve tubes and phloem parenchyma alternating with thinner cylinders of laticifers.

Latex Vessel System

The rubber-containing latex is formed and stored in the laticifers which are differentiated by the cambium at fairly regular intervals, normally at the rate of 1.5–2.5 rings per annum. They are fairly close to each other and are interconnected, thus forming a cylindrical meshwork of tubes interweaving between the medullary rays. The number of latex vessel rings increases with the age of the tree.

The latex vessels are aligned at an angle to the vertical axis of the trunk and generally incline from lower left to upper right as one faces the tree. This is the reason for the tapping cuts to be made to slope from high left to low right at an angle of about 30°, as this cuts more vessels and gives a higher yield than if the cut is made in the opposite direction.

The amount of rubber-containing tissue per unit volume of bark is clearly an important determinant of yield. This depends on the number of latex vessels rings, the number of vessels per millimeter of ring, and the mean diameter of the vessels. The number of latex vessel rings is positively and significantly correlated with yield.

Bark Renewal

The paring away of the bark during tapping stimulates the tree to form wound tissue. New cork cells on the outside and cork parenchyma on the inside are formed by differentiation of newly formed phellogen. The vascular cambium is activated to produce new phloem, including rings of latex vessels, at an accelerated rate. As a result, during the first 6 months after tapping, bark regeneration is particularly fast. After about 8 years, the regeneration has progressed sufficiently for the renewed bark to be tapped. Usually the renewed bark has a greater number of latex vessel rings than that found in the virgin bark.

Tapped Latex

The latex that flows out upon tapping the bark consists of the contents of latex vessel rings at different stages of development, and small amounts of materials from other types of cell in the phloem. The most common materials are the rubber particles with smaller amounts of lutoids and Frey–Wyssling particles. The rubber hydrocarbon molecule consists of long chain monomer isoprene C_5H_8. The contents of tapped latex consist of 36% rubber, 60% water, and 4% other components such as acetone extract (lipids, resins, and waxes), carbohydrates, and ash (**Table 1**).

Table 1 Composition of tapped latex

Component	*Content (%)*
Rubber	35.62
Acetone extract (lipids, resins, waxes)	1.65
Proteins	2.03
Carbohydrates	0.34
Ash	0.70
Water	59.62

Reproduced with permission from Morris JE (1989) Processing and Marketing. In: Webster CC and Baulkwill W (eds) (1989) Rubber. Oxford: Blackwell Scientific Publishing.

Narrow Genetic Base

The rubber breeding material used in Asia has a very narrow genetic base, having originated from seeds collected by Sir Henry Wickham in a very small area of Brazil. The initial breeding base came from about 2000 seedlings from the Wickham collection imported into Asia in 1877. Most of these seedlings went to Sri Lanka, with much smaller numbers going to Malaya and Indonesia. There were some other introductions to Southeast Asia in later years but due to their poor yield they have not been used in subsequent breeding programs. Hence, rubber breeding in Asia (and in Africa) is based on limited germplasm.

The intensive use of directional selection for high yield, phenotypic assortative mating, and the extensive use of clonal propagation in the early breeding programs have aggravated the problem of the narrow genetic base. As a result, some undesirable features have become apparent which can be illustrated by the following:

1. Most of the clones bred to date in Malaysia (and other Southeast Asian countries) can be traced to seven "primary clones." Some inbreeding has therefore featured in the early breeding programs. Some of these primary clones are themselves related, so inbreeding may be even greater.
2. There is a diminishing return obtained from breeding effort. There was a substantial yield increase in the early phases of breeding, but this has diminished in later phases (see below).
3. Genetic erosion seems to have occurred as materials selected in one region may be found to be less adaptable in another. For example, Southeast Asian selections are susceptible to South American leaf blight (SALB) which could be due to lack of resistance in the original Wickham collection or subsequent loss through genetic erosion following selection in the absence of the disease.

Breeding Objectives

The objectives of breeding programs need to be carefully planned to take into account technological advance, socioeconomic factors, and the demands of the industry; this planning is all the more difficult because *Hevea* has a long generation time (about 30 years). In general, there are two primary objectives:

1. To breed clones with a genetically high yield potential with dry latex as the main product.
2. To achieve high vigor in order to reduce the unproductive juvenile period.

After opening, there are two competing sinks for photosynthate, namely latex offtake and tree growth. Growth rate tends to decline under tapping and only vigorously growing trees are able to sustain maximum latex yield for many years.

Secondary objectives include:

1. Good girthing on tapping to sustain yield and reduce wind damage losses through trunk snap.
2. Thick virgin bark to minimize wounding incidence known to affect yield productivity on later panels.
3. Good bark renewal to enhance economic tapping life.
4. General resistance to major diseases to ensure better growth and yield and also to minimize the risk of losses as in severe cases of SALB attack.
5. Tolerance to wind to minimize losses (ensuring good tapping stand throughout economic life); but this may be associated with branching habit, tree height, long latex flow or low plugging index leading to adverse partition and hence imbalance between crown and trunk components, wood properties, etc.
6. Tolerance to drought to ensure higher yield.

Tolerance to wind and major diseases are considered most important amongst these secondary characters.

More recently several novel secondary characters have been added to the breeding objectives to meet new demand of the industry and to take advantage of the rapid developments in biotechnology. These are:

1. Good response to simulation coupled to low intensity tapping.
2. Good growth form for high-quality wood production.
3. Production of proteins, nutraceuticals, and pharmaceuticals.

Past *Hevea* Breeding

In Southeast Asia, *Hevea* breeding started during the early part of the last century. Mass selection, or what is better known as "mother tree selection" or "ortet selection," was used and early breeders screened hundreds of thousands of seedling trees planted on rubber estates. This was then followed by selection and multiplication of the selected trees. After evaluation, the best-performing trees were cloned for further evaluation in field trials. Rapid progress was achieved, especially in Indonesia, between 1919 to 1926 and a number of popular clones such as Tjir, PR, GT, BD, AVROS, and LCB series were selected. In Malaysia, selection of outstanding mother trees started in 1928 resulting in clones such as PB, Pil, GI, Ch, and Lun N series. Some of these clones are still being planted today in some rubber-growing countries.

Mother tree selection was followed by empirical crosses with the best-performing ortet clones in various combination to produce another generation of progenies for selection. This empirical approach has produced many successful clonal progenies such as RRIM 600, RRIM 623, PB 5/51, and PB 28/59 from Malaysia and PR 255 and PR 261 from Indonesia.

The conventional breeding system of cyclical 'generation-wise assortative mating' (GAM) ensued empirical crosses a few decades later. Under this system, the best genotypes in one generation were used as parents for the next cycle of breeding. One disadvantage of this method is that these new clones can be traced back to only a few parents, resulting in slow genetic advance and some inbreeding depression. New clonal material from GAM has superior performance compared to older clones probably because of the heterogeneity in the initial breeding stock. The various phases of *Hevea* breeding using Malaysia as an example are shown in **Table 2**.

Improvement of Latex Yield

The primary economic product of *Hevea* is latex measured as dry weight. **Table 3** shows the increased yield over time in Malaysia. The initial introductions from the wild had a very low yield, averaging $250\,kg\,ha^{-1}\,year^{-1}$ and in early or pilot planting

Table 2 Phases of *Hevea* breeding in Malaysia

Phase	*Year*	*Clones produced*
I	1928–1931	RRIM 500 series
II	1937–1941	RRIM 600 series
III	1947–1958	RRIM 700 series
IV	1959–1965	RRIM 800 series
V	1966–1973	RRIM 900 series
VI	1974–still continuing	RRIM 2000 series

Table 3 Improvement of dry latex yield in *H. brasiliensis* in Malaysia

Period	*Phase of selection*	*Materials*	*Girth at 5th year (cm)*	*Dry latex yield ($kg\,ha^{-1}\,year^{-1}$)*
1876	Germplasm	Wild	—	250
1900s	Pilot	Unselected	—	550
1920s	Early	Pil B 84	—	1175
1928–31	Phase I	RRIM 501	—	1264
		RRIM 513	—	1433
1937–41	Phase II	RRIM 600	41.3	1573
1947–58	Phase III	RRIM 712	41.2	1658
1959–65	Phase IV	RRIM 803	46.7	2000
1966–73	Phase V	RRIM 936	45.7	2146
1974–present	Phase VI	RRIM 2027	61.0	3036
Theoretical yield				9000

from unselected materials this was increased to 550 $kg\,ha^{-1}\,year^{-1}$. The initial selections in the 1920s yielded 1175 $kg\,ha^{-1}\,year^{-1}$ of dry latex.

In Malaysia, rubber research and development started in 1925 and selection from the first phase (1928–31) was able to produce clonal materials that were released in the 1950s with the dry latex yield of 1264 $kg\,ha^{-1}\,year^{-1}$. This was later followed by the same series of clonal materials with average dry latex yield of 1433 $kg\,ha^{-1}\,year^{-1}$. The second phase of breeding, from 1937 to 1941, produced new clones with the dry latex yield averaging 1573 $kg\,ha^{-1}\,year^{-1}$. This material was released in the 1960s. There was a further yield improvement from the third phase of breeding cycle (1947–58) with yields of 1658 $kg\,ha^{-1}\,year^{-1}$ from clonal materials released in the 1970s.

There was a further progress in the 1980s resulting from the 1959–65 breeding cycle (fourth phase) with clones producing 2000 $kg\,ha^{-1}\,year^{-1}$ of dry latex, while the fifth phase of breeding cycle (1966–73) attained 2146 $kg\,ha^{-1}\,year^{-1}$. The modern clones of the sixth breeding phase (1974 – present) can yield over 3000 $kg\,ha^{-1}\,year^{-1}$. In addition, these modern clones also have good timber characteristics (see below).

Breeding for Latex–Timber Clones

The commercialization of rubber wood (heveawood) in the 1970s has changed the *Hevea* breeding strategy. Heveawood has been accepted as an alternative timber to the natural forest species for manufacturing wood-based products and there has been a remarkable increase in demand for heveawood. The supply comes from the felling of 25–30-year-old trees which are replaced by replanting. Heveawood production is now a very important aspect of the rubber growing industry.

This has led to a shift in the breeding paradigm to concentrate on latex yield as well as high timber volume. The new types of planting material are known as 'latex–timber' clones. These show, in addition to high latex yield, good growth form, rapid growth, a long straight bole and good canopy wood (**Figure 1**). The total wood volume consists of the bole (clear trunk) volume and the canopy wood volume. The bole wood is the premium wood for sawn timber and furniture while the canopy wood is used for chips and medium-density fiberboard. The performance of these clones is shown in **Table 4**. It is possible that in the future there will be a further shift from latex–timber to timber–latex clones, with the implication that timber will be the primary product with latex playing a secondary role.

Breeding for Disease Resistance

Rubber trees are susceptible to diseases caused by numerous pathogens, for example, *Oidium*, *Colletotrichum*, *Fusicoccum*, and *Corynespora*. SALB has been responsible for the failure of rubber plantations in South America and none of the Asian and African clones is tolerant to this disease.

The emphasis in *Hevea* breeding should be directed at getting horizontal or field resistance as opposed to immune or vertical resistance. Valuable sources of genes conferring resistance to different diseases are available from the germplasm recently collected from the wild. These genes from wild sources need to be identified and their disease resistance genes introgressed into the modern high-yielding cultivars.

Breeding for Local Adaptation

Due to the increased demand for rubber and the extension of cultivation of other more profitable

Table 4 Comparative performance of latex and latex–timber clones

Planting materials	*Mean girth at 5th year (cm)*	*Dry latex yield* ($kg\,ha^{-1}\,year^{-1}$)	*Wood volume*			*Total wood volume* (m^3)
			Age (years)	*Clear bole* (m^3)	*Canopy* (m^3)	
Latex clone						
RRIM 600	41.3	1388	15	0.16	0.21	0.37
Latex–timber clones						
RRIM 911	49.5	1632	22	0.46	0.69	1.15
RRIM 921	50.3	1488	22	0.63	0.68	1.21
RRIM 2001	53.4	2850	17	0.41	0.82	1.23
RRIM 2008	60.0	2686	14	0.33	0.99	1.32
RRIM 2014	55.9	2007	14	0.53	0.80	1.33
RRIM 2025	47.4	2700	14	0.63	1.20	1.87
RRIM 2027	61.0	3036	16	0.60	0.70	1.30

crops in traditional rubber-growing areas, the cultivation of rubber has moved to new areas where conditions are often marginal. Locally adapted material is therefore required, and in well-organized growing countries, planting is carried out using "environmax" recommendation, i.e. clones are planted in the most suitable sites to enable them to perform to their full potential. In the longer term specific breeding is required for local adaptation to specific conditions. Breeding elsewhere may actually result in dysgenic change in the opposite direction to that intended. This could be one reason for the non-performance of many clones in exchange programs.

Role of Germplasm

The Needs for a New Genetic Base

Despite the narrow genetic base of the initial *Hevea* population, spectacular yield improvement was achieved in the earlier phases of breeding programs. The genetic gain, however, has not been sustained in the later phases of breeding, probably due to inbreeding depression and genetic erosion following directional selection for high yield, phenotypic selection for high yield, phenotypic assortative mating, and the extensive use of clonal vegetative propagation leading to further erosion of the genetic variability.

In addition to adverse effects on long-term yield improvement, the narrow genetic base could also increase vulnerability to various diseases. SALB is the greatest single threat to the rubber cultivation, not only in Asia and Africa but also in South America. Screening tests for resistance to SALB have shown that none of the oriental clones is resistant to the disease. Fortunately, the absence of this disease has enabled rubber cultivation to flourish elsewhere, especially in the East.

Germplasm Collections

In view of the importance of broadening the genetic base, SALB-resistant clones of *H. brasiliensis* and other wild species (*H. benthamiana* and *H. spruceana*) were introduced to Malaysia in 1951–52. A total of 513 clones or "genotypes" were screened for SALB resistance and other traits. Only one clone, IAN 873, was found to be high yielding, vigorous, and reasonably tolerant to SALB and *Colletotrichum*. This clone was subsequently recommended for planting, in 1975.

In 1966, there was another introduction of 1010 seeds from six species and one subspecies but only 76 survived due to poor germination. These included *H. benthamiana*, *H. guianensis*, *H. pauciflora*, *H. rigidifolia*, *H. spruceana*, and *H. pauciflora* var. *coriacea*.

In 1981, a multinational expedition by the International Rubber Research and Development Board succeeded in collecting a total of 64 736 seeds and 1533 meters of budwood from 194 presumably high-yielding ortet trees of mainly *H. brasiliensis* and a small number of *H. benthamiana* and their interspecific hybrids. In 1995, Malaysia conducted another expedition to collect seeds of various species of *Hevea* from the Amazon basin. A total of 4.7 tonnes of seeds were collected, out of which 3.5 tonnes were retained in Brazil and 1.2 tonnes were sent to Malaysia. The collection comprised seeds of *H. brasiliensis*, *H. guianensis*, *H. spruceana*, *H. rigidifolia*, *H. pauciflora*, *H. benthamiana*, and their interspecific hybrids.

There were also multilateral exchanges of clones between rubber-growing countries in Asia in 1954 and 1974. These clones, however, are mainly derived from Wickham materials and hence cannot be considered as new germplasm for widening the *Hevea* genetic base.

Impact of New Genetic Resources

These new genetic resources have widened the *Hevea* genetic base, including the availability of SALB-resistance genes and other genetics variations to create a greater impact on future *Hevea* improvement. Some of the newly recommended clones, RRIM 900 and RRIM 2000 series (**Tables 3** and **4**) are derived from crosses between oriental and occidental clones.

These introductions have also been important for heveawood breeding. Fast-growing genotypes with long boles from these introductions have been introgressed to latex–timber clones, namely RRIM 900 and RRIM 2000 (**Tables 3** and **4**). A number of genotypes of *H. brasiliensis* and other *Hevea* species such as *H. pauciflora*, *H. guianensis*, and *H. nitida*, though low latex yielders, are potential materials for the establishment of rubber forest projects for timber production.

Biotechnology

The aim of biotechnology techniques in *Hevea* is to insert into its genome genes that will (1) enhance the rubber biosynthesis and increase latex vessel density thus improving the production of rubber; (2) increase wood volume to augment timber output; and (3) encode proteins of commercial value.

Increased Rubber Content in Latex

The rubber content in latex varies from 40% in trees that are left untapped to 28% in trees that are tapped intensively or that have been subjected to sustained yield stimulation. Trees that undergo ethylene-based yield stimulation resulting in long flow duration usually have a low rubber content in the latex. Therefore, there is a mounting need to increase the rate of rubber biosynthesis (**Figure 3**) with the aim of raising the rubber content in latex harvested from ethylene-stimulated trees.

The latex C-serum contains a proteinaceous stimulator of rubber biosynthesis, while several sequenced *Hevea* genes show homology to the eukaryotic initiation factor 5A (eIF5A) of *Medicago sativa* which encodes closely related soluble proteins that stimulate the incorporation of isopentenyl diphosphate into rubber. A total of 17 isolated complementary DNAs (cDNAs) showed 90% identity between them, and they could also be grouped into seven isoforms of eIF5A proteins. The native protein, which is slightly smaller due to posttranslational modifications, is called the rubber biosynthesis stimulator (RBS). *In vitro* studies show that both native protein (RBS) and its recombinant counterparts stimulate rubber biosynthesis.

Hevein: Latex-Specific Gene Promoter

Although the CaMV 35S constitutive promoter facilitates strong GUS (gene encoding β-glucuronidase) expression in latex, its recombinant proteins are unselectively expressed in all tissues making it less useful in *Hevea* transgenic work as the proteins could be toxic or deleterious to the plant. It is therefore, pertinent to isolate promoters that facilitate latex vessel-specific expression. An upstream sequence has been identified as the hevein promoter which directs the expression of hevein to the organelles called lutoids in the latex. The 5′-flanking region of hevein (1.3 kb) has been isolated by genome walking and transcriptional site determined by primer extension analysis. The sequences of the gene promoters revealed the consensus eukaryotic TATA and CAAT box sequences.

Modification of Anatomy to Improve Latex–Timber Production

In *Hevea*, the secondary phloem region of the bark is interspersed by latex vessel system. Hence phloem differentiation may have implications on the latex vessel density which has strong bearing on rubber productivity. It is envisaged that homeotic and structural genes influence wood and phloem/laticifer development. This led to the construction of cDNA libraries from tissue representing the cambium and early differentiating cambial tissue both adjacent to xylem and phloem.

Expressed Sequence Tags

A catalog of genes expressed in latex or expressed sequence tags (ESTs) by sequencing random cDNA clones derived from latex would provide useful tools for profiling latex gene expression, identifying new genes, and generating a pool of genes for future studies. About 80% of plasmid preparations of 2370 clones from the latex cDNA library contain cDNA inserts of >0.5 kb. Using basic local alignment search tool (BLAST) analysis, about 78% of the ESTs matched with known sequences while the remaining 22% were classified as genes of no known function. A substantial proportion (8%) of the ESTs of known function consists of two biosynthesis-related proteins, rubber elongation factor (REF) and a small rubber particle protein.

Shortening the Breeding Cycle

Rubber as a perennial crop species requires a long breeding and selection cycle. This has prompted

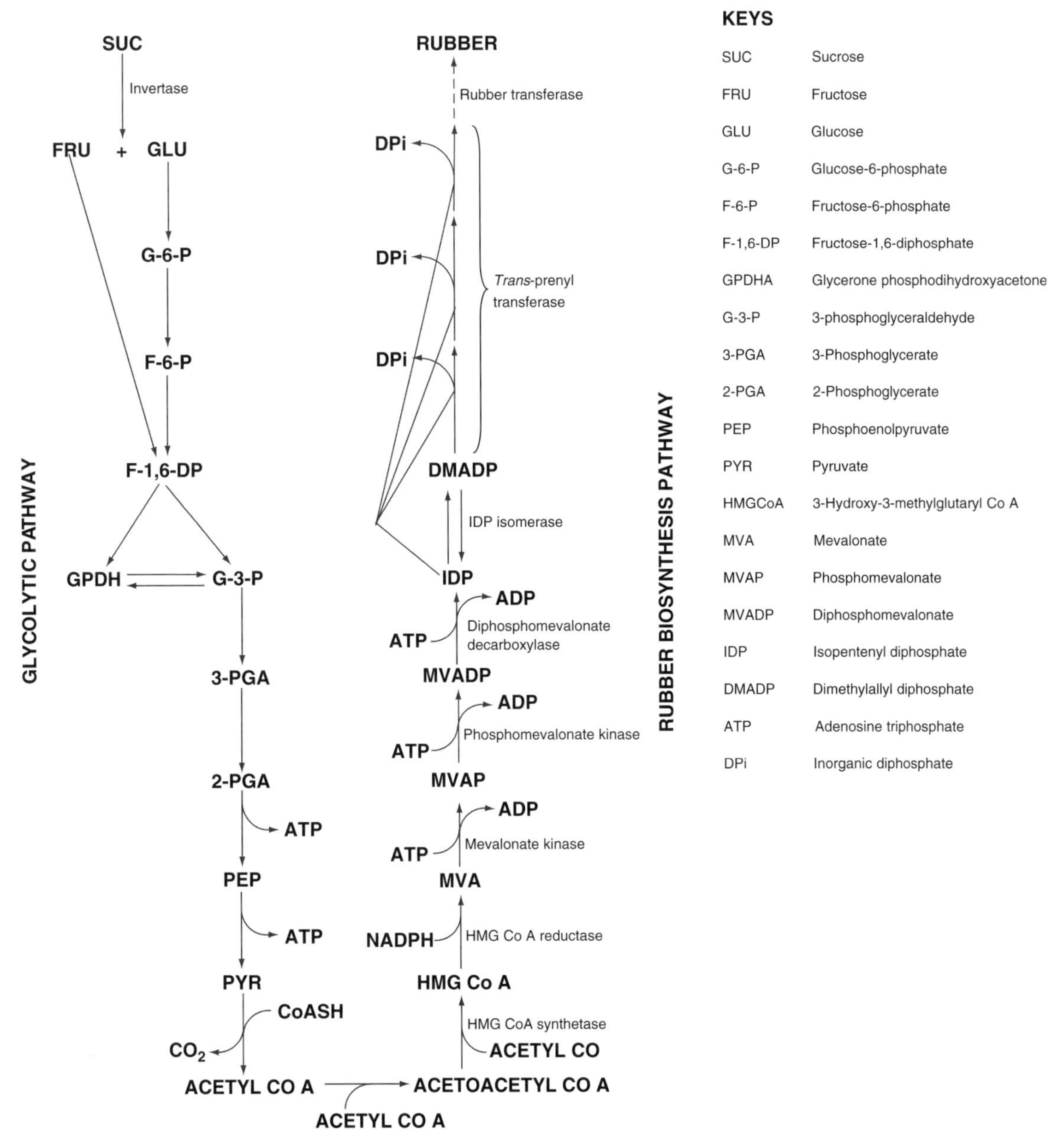

Figure 3 Steps in the biosynthesis of rubber. Reproduced with permission from Yusof F and Chow KS (2003) The Biosynthesis of Rubber in *Hevea brasiliensis*. Monograph No. 4. The Malaysian Rubber Board.

investigations into the possibility of shortening the cycle, including development of early selection techniques and shortening the testing cycle. The early selection techniques use parameters in the seedlings or immature rubber trees that are strongly associated with yield of mature trees. Direct methods of predicting the yield of mature trees include tapping young plants for a certain period of time and obtaining the amount of latex from the leaflets or petiollules.

Indirect methods to predict the potential yield of the mature trees include latex vessel system in the bark plugging index and photosynthetic rate. Most of these methods have limitations and are not consistently related with yield or mature trees due to influence of environments. The other alternative method for early selection and shortening the breeding cycle is by using the molecular techniques which have been successfully applied to many important economic crops. These techniques have much to offer as new tools to assist in the rubber breeding and selection. They include the selection of parents for crossing, culling of progenies, validation of genetic origins, identification of parentage, protection of new elite breeding materials, confirmation of relatedness, phylogenetic and conservation studies, as well as the construction of a genetic map for linkage to quantitative trait loci for marker assisted selection.

List of Technical Nomenclature

Copra	Dried flesh of coconut fruits consisting of endosperm and embryo tissues.
Palm kernel oil	Oil derived from the kernel (seed) of palm fruits that enriched in lauric acid.
Palm oil	Oil derived from the mesocarp of palm fruits that contains mostly stearic and oleic acids.

See also: **Biodiversity and Conservation**: Germplasm Conservation. **Diseases**: Breeding for Disease Resistance. **Secondary Products**: Rubber Production.

Further Reading

Faridah Y and Chow KS (2003) *The Biosynthesis of Rubber in* Hevea brasiliensis. Monograph No. 4: 27 pp. Kuala Lumpur: Malaysian Rubber Board.

Masahuling B and Ramli O (2000) Breeding superior rubber clones for latex and timber. *Bulletin of the Genetics Society of Malaysia* 6: 6–7.

Masahuling B, Othman H, and Ramli O (1996) Evaluation, utilization and conservation of Hevea germplasm. In: Quah SC, Kiew R, Bujang I, *et al.* (eds) *Underutilized Tropical Plant Genetic Resources: Conservation and Utilization*, pp. 285–298. Selangor: Universiti Pertanian Malaysia.

Ong EL, Chan H, Yeang HY, Lim CL, Fauziah AR, and Subki I (2000) *Milestones in Rubber Research.* 75th Anniversary of the Rubber Research Institute of Malaysia (1925–2000). Kuala Lumpur: Malaysian Rubber Board.

Ramli O and Masahuling B (1996) *Hevea* genetic resources in RRIM. *Bulletin of the Genetics Society of Malaysia* 2: 18–20.

Ramli O, Masahuling B, Md Zain AA, *et al.* (1999) Newer rubber planting materials for wood production. *Malaysia Forester* 62: 130–144.

Simmonds NW (1989) Rubber breeding. In: Webster CC and Baulkwill WJ (eds) *Rubber*, pp. 85–124. Singapore: Longman Scientific and Technical.

Tan H (1987) Strategies in rubber tree breeding. In: Abbott AJ and Atkin RK (eds) *Improving Vegetatively Propagated Crops*, pp. 27–62. London: Academic Press.

Tan H (1995) Genotype × environmental interaction studies in rubber (*Hevea*) clones. *Journal of Natural Rubber Research* 10: 63–76.

Tan H and Tan AM (1996) Genetic studies of leaf disease resistance in *Hevea*. *Journal of Natural Rubber Research* 11: 108–114.

Webster CC and Paardekooper EC (1989) The botany of the rubber tree. In: Webster CC and Baulkwill WJ (eds) *Rubber*, pp. 57–84. Singapore: Longman Scientific and Technical.

Yeang HY (2002) *Advances and Challenges in* Hevea Biotechnology. International Rubber Research Development Board Workshop on Breeding, Agronomy and Socio-Economics, 28 August–7 September 2000. 5 pp. Malaysia Indonesia.

Zain M, Ramli O, Masahuling B, and Ong SH (1997) *RRIM 2000 series clones – Characteristics and Descriptions*. Kuala Lumpur: Malaysian Rubber Board.

Multicropping

S N Azam-Ali, University of Nottingham, Nottingham, UK

Nomenclature

Multicropping is probably the oldest form of agriculture and is still widespread, particularly in tropical and low-input production systems. In principle, sequential cropping, where a second crop is grown after the harvest of a first crop, is a simple form of multicropping. However, most multicrops involve the cultivation of two or more species on the same piece of land where the growth cycles of different species overlap for at least part of their duration. Such multicrops may take the form of intercrops, where growth and yield are confined to a single growing season, or agroforestry, where a perennial component, e.g., trees or shrubs, occupies some of the land over a series of growing seasons. **Table 1** describes various types of cropping system ranging from the simplicity and uniformity of monocrops, where similar plants grow alongside each other simultaneously and sequentially from one season to the next, through to the multiplicity and heterogeneity of agroforestry systems where diverse plant types coexist within and between seasons. These complex systems may include dual canopy crops where high-value plantation crops such as oil palm (*Elaeis guineensis*), coconut (*Cocos nucifera*), cocoa (*Theobroma cacoa*), or rubber (*Hevea brasiliensis*) are interplanted with annual species. Here, a lower, shorter-duration crop complements the upper canopy of the plantation species. Alternatively, they may be multipurpose home gardens containing a woody component alongside vegetables, staples, and fruits.

The choice of species for multicropping varies with geographical location, level of input, and local preference. The actual combinations of particular crops and their planting sequences and arrangements

Table 1 Different types of cropping system

System	*Definition*	*Remarks*
Monocropping	The continuous growing of the same species on a piece of land over a sequence of growing seasons	Risk of residual transfer of pests, diseases, and weeds from one season to the next
Ratooning	The residual stumps of a crop are allowed to regrow to produce a second crop	Cheap, low-input second crop with shorter growing period than sequential cropping. Yields of second crop are often poor and unreliable, often because of pest and disease transfer. Restricted to crops such as sorghum, millet, rice, and pigeonpea
Sole crop	Crop composed of individual plants of the same variety of one species	Interactions occur between plants that are all at the same stage of growth. Management of inputs, control measures, and irrigation can be synchronized to meet the temporal and spatial demands of the crop. Mechanization is possible
Sequential cropping	Crops are grown one after the other with no overlapping phase	Usually involves an additional crop after the harvest of the main single rainy season crop. Potential residual effects in the reservoir of nutrients, weeds, or pests and diseases with implications for the succeeding crop or crops. Requires a relatively long potential growing period (over 180–200 days). The growth of the second crop is often risky because of the limited amount of residual moisture. Typically includes legumes or oilseeds after paddy rice or rainfed cereals
Multicropping	Two or more species growing on the same piece of land, where at least part of the growth cycles of different species overlap	Plants of different species are close enough to allow interactions between them, usually at different stages of their growth
Intercropping	A multicrop composed of two or more annual crops grown simultaneously on a piece of land within a single growing season	Intercrops may retain the same population as the sole crops that they are composed of in which each plant of one species is replaced by a single plant of one or more other species ("replacement intercrop") or additional plants of one or more species may be fitted into an existing population of sole crops ("additive intercrop")
Mixed intercrop	Multicrops of randomly arranged plants of different species	Such crops may be planted as replacement or additive intercrops where the grower requires an additional output from a second species without affecting the yield of the main crop
Row intercrop	Intercrops grown in structured arrangements of different species arranged in alternating blocks of narrow rows	Interspecies interactions occur between rows and intrarow competition occurs within rows. Individual species may be sown, fertilized, and harvested separately
Strip intercrop	Intercrops where blocked rows of each species are grown separately	Strips are wide enough for some plants within each block to behave as if in a sole crop
Relay cropping	The interplanting of one species with a second species before the first crop reaches maturity	The very short overlap means competition between species is minimal. The second crop may have better soil moisture conditions because of earlier sowing and protection from the first crop. Sowing of second crop before harvest of first crop reduces labor peaks at harvest. There may be damage to seedlings of second crop at harvest of first. Examples include paddy rice and legumes, cereals and legumes, cereals and other cereals
Agroforestry	Land use systems in which woody perennials (trees, shrubs) are grown in association with herbaceous plants (annual or perennial crops, or pastures) and/or livestock	Capture and use of resources is distributed in space and time. Plants within the system make demands on resources at different times ("temporal complementarity") or use resources more efficiently at any one time ("spatial complementarity"). In agroforestry systems the woody component/s may provide more than one product (e.g., timber, fuelwood, fodder, fruit) and/or "services" (shelter, shade, soil or water conservation)
Alley cropping	A specific form of agroforestry in which trees and shrubs are established in hedgerows within arable cropped land	The woody component is always planted in rows but the spacing between rows depends on the species mix and topography. Sometimes known as "hedgerow intercropping"

are determined primarily by the total length of the growing season, the architectural dimensions and duration of each species, and the suitability of different crops to particular environments. In areas of low rainfall, early maturing and drought-tolerant species such as millet (*Pennisetum glaucum*) and sorghum (*Sorghum* spp.) are usually evident. In areas of higher rainfall, various combinations of cereals and legumes with different maturities are used. For example, combinations of rice (*Oryza sativa*) and other cereals or legumes may be found in higher-rainfall areas with a single extended rainy season such as southeast Asia. Typical species combinations in regions of limited rainfall include maize (*Zea mays*), sorghum, or millet mixed with beans, soybean (*Glycine max*), groundnut (*Arachis hypogaea*) or pulses such as cowpea (*Vigna unguiculata*), pigeonpea (*Cajanus cajan*), or chickpea (*Cicer arietinum*). Economic or social value, ease of cultivation, nutrient requirements, planting density and spacing, variety and type of plant, and other cultural factors influence the choice and arrangement of different species.

The Ecology of Multicropping

In ecological terms, multicropping is often more stable than sole or monocropping, particularly where inputs such as fertilizers, pest and disease control, and irrigation are limited or unavailable. In many temperate, intensive systems, sole crops are grown so that external inputs can be timed to meet the needs of a particular species. In this way, maximum productivity can be achieved within the limits set by the cost-effectiveness of inputs, the genetic potential of the crop, and the physical potential of its environment. In such circumstances, multicrops are rare. In less intensive systems, there is more likelihood of encountering mixed cropping. Here, the lack of inputs, such as nutrients and water, means that resources must be sequestered from within the system itself and/or the spatial and temporal demand for the maximum yield of a particular species must be balanced against the optimal performance of the whole system. The prevalence of multicrops becomes greater as we move from relatively benign to more hostile environments and in many dry, tropical regions multicropping is often the dominant system of cultivation over large tracts of land.

As well as the greater potential stability of multicropping systems, it is also possible that certain combinations of multicrops may produce a greater yield than that achieved by their constituent sole crops grown under similar conditions. Reported yield advantages vary but typically range between 25% and 40%, particularly for cereal/legume intercrops where an improvement in the yield fraction of the cereal component may more than compensate for the yield loss incurred by the legume component. However, farmers' preference for multicrops are often unrelated to the possibility that they may provide greater combined yields *per se* than their constituent sole crops. Often, multicropping is chosen to reduce the risk of complete crop failure or to ensure a more stable range of yields in variable environments. The unreliability of annual crop yields increases with the inherent variability of the environment. Much of this annual variability is reduced when crops are grown together because environmental, pest, or disease pressures do not equally affect different species. Pest and disease effects on one species may be reduced because specialized pathogens may settle on nonhost component species of the multicrop. This "fly-paper effect" encourages overall stability by the additional compensatory growth of the resistant species in the space left by diseased individuals. For example, virus diseases may spread more rapidly through adjacent plants of the same species than when these plants are separated by dissimilar, nonsusceptible, plants of another species. Insect vectors are also discouraged from attacking plants that are intermingled with other species than when a single species is grown in a block.

Multicrops containing species with different times to full canopy resist the establishment of weeds and may protect the soil against erosion and wind damage. Similarly the late planting of a second crop into one that is reaching maturity (relay cropping) is common in climates with a distinct wet and dry season where the rainy season cannot support the growth of two full-season crops. In addition to usefully using soil moisture that would otherwise be unavailable the following year, relay cropping can stabilize soil aggregates and discourage weeds at the end of the first growing period. By using crops of varying durations and by adjusting the sowing date of relay crops into established crops, both the demand for natural resources, e.g., nutrients and water, and human resources for sowing and harvesting are spread across the season.

Multicropping is usually suitable where human labor is available and land is not. To be economically viable, the demands of the multicrop should be less or the yields more than when the species are grown as sole crops on the same land area. Because multicrops are composed of species of different maturities with different peak requirements for specific resources the maximum demand for each resource is less than when the whole land area is occupied with identical plants of the same species. **Table 2** summarizes the major advantages of multicropping systems and **Table 3** lists their major disadvantages.

Resource Capture by Multicrops

Plants grow in two media (soil and air) and two dimensions (space and time). This means that roots must compete for nutrients and water and leaves must simultaneously intercept light and absorb CO_2

Table 2 The major advantages of multicropping

Insurance against complete crop failure
Increased protection against soil erosion
Distributes demands for labor and inputs during the growing season
Distributes storage requirements over the year
Provides options to grow a number of plants of different species within the main crop
Possibility of using resources more efficiently
Possibility of nitrogen transfer from legumes to associated nonlegumes
Discourages the spread of diseases and pests between susceptible individuals

Table 3 The major disadvantages of multicropping

Allelopathic effects mean that the combination of some species may be harmful
Difficult to mechanize planting and harvesting
Difficult to apply fertilizers and control measures to specific component species
Risk of damage to plants during removal or introduction of new species into the multicrop
Land may be occupied by the duration of the longest-growing species

for plants to achieve optimal growth and yield. The ability of individual plants to occupy a given space has implications for their use of resources to grow (i.e., changes in the area, weight, or size of different organs) at different developmental stages in the life of the crop. Most crops are grown primarily for some part of their total productivity, usually the reproductive component. The relationship between reproductive yield and overall growth is not always simple and may vary with environment and crop duration. For the optimal growth and development of a multicrop, individual plants of different species must simultaneously "capture" and "convert" resources such that the combined performance of the whole system is greater than the maximal performance of an individual species. A complicating feature of multicrops is that for much of their lives plants must compete for resources with neighbors of both the same and at least one different species. This competition occurs in both space and time and has different characteristics above and below ground. In intercrops, intra-(i.e., within) species competition usually precedes inter-(i.e., between) species competition for light. The early life of a plant within an intercrop is therefore similar to that of a comparable sole crop. However, differences in size and growth rate of adjacent plants, e.g., a short legume growing adjacent to a tall cereal, mean that competition between plants may occur earlier than in the comparable sole crop. In a typical cereal/legume intercrop, individual legume plants become more crowded and cereal plant less crowded earlier in the season than their sole counterparts.

As well as differences in air and soil temperatures, humidity, wind speed, and CO_2, competition within multicrops occurs for the relative capture and use of light, water, and nutrients by each component species of a multicrop compared with their sole counterparts.

Capture and Use of Solar Energy

In a multicrop, lower-lying neighbors of a different species may usefully employ light transmitted through a dominant upper canopy species. The idealized pattern of light distribution is often achieved by using species with vertically orientated leaves in the uppermost layers and species with more horizontal leaves at the bottom of the canopy. This arrangement is best demonstrated by intercrops composed of a tall C_4 species such as millet, sorghum, and maize, overlying a C_3 legume such as cowpea, common bean, or groundnut. In such cases, the high intensities of radiation often found in tropical environments can be usefully intercepted by the cereal component allowing the penetration of lower intensities of radiation that can now be used because they are below the saturating irradiance for C_3 species.

Most multicrops include species with differing durations. Here, there is a temporal element of reduced competition for light. In cereal/legume intercrops, the cereal is often harvested before the legume. This means that the legume plants in the intercrop continue to grow without competition from the taller cereal. The light intercepted by the legume plants remaining in the intercrop is different to that in a comparable sole crop because positions previously occupied by cereal plants are now missing and adjacent legumes can occupy some or all of this space.

To calculate the efficiency with which crop systems convert solar energy into dry matter, we must divide the final dry weight of the crop in $g\,m^{-2}$ (including grain, leaves, stems, roots, etc.) by the amount of radiation intercepted during the season ($MJ\ m^{-2}$). Not all plant organs have the same energy content per unit of dry weight so we need to account for the energy value ($MJ\ g^{-1}$) of the various lipid, protein, and carbohydrate fractions. By adjusting for the energy values of the different plant compounds we can calculate the biological efficiency in terms of MJ of productivity per MJ of radiation intercepted.

Capture and Use of Water

In a multicrop, any interactions between the roots of adjacent plants of different species depend on their lateral and horizontal distribution and planting arrangement. The root systems of many species can be modified by the distribution of resources in the soil. The abilities of roots of different species to extend into variable soils have implications for the amount of water extracted by each species within a multicrop relative to that extracted by a sole crop.

In principle, multicropping allows a more complementary rate of extraction of water throughout the soil in space and time than in sole crops where each plant must compete in a similar location as its neighbors. The greatest complementarity should therefore occur in multicrops composed of species that differ substantially in root distributions and durations, e.g., agroforestry.

Capture and Use of Nutrients

Soil and plant type, fertilizers, previous crops, and arrangement of species and plants all influence how species compete for nutrients within a multicrop. Because nutrients are taken up from the soil solution, the soil water balance of a multicrop relative to its sole counterparts influences the capture of specific nutrients. Uptake rates vary with plant age and maximum nutrient demand for one species within a multicrop may not coincide with that of another. The balance between the amount of nutrients captured by each species within mixed crops also depends on the overall uptake of nutrients by the whole system. Where the rooting depth of each species within a multicrop is similar, total nutrient availability may also be similar. In climates where rainfall is less than soil moisture storage during the season, supply precedes demand for nutrients and timing of demand may not be critical to availability. However, in the humid tropics, where rainfall exceeds evaporation, earlier mineralization means that nutrients may be washed into deeper soil layers beyond crop roots. In these circumstances, deep-rooted components, e.g., perennial woody species, can intercept nutrients as they move beyond the root front of crops. In agroforestry systems, deep-rooted perennials can act as "nutrient pumps" that supply nutrients to the soil surface, particularly through the decomposition of leaves.

There is the potential to interplant nonlegumes with nitrogen-fixing legumes to make fixed N_2 available to adjacent nonlegumes via root/nodule exudation and/or vesicular fungi or through better uptake of N released by the legume. However, these processes appear to contribute little to the nitrogen economy of the nonlegume component during the same growing season.

Calculating the Performance of Multicrops

Methods to compare the performance of multicrops and their constituent sole crops include the Land Equivalent Ratio (LER), the Area–Time Equivalency Ratio (ATER), and the Crop Performance Ratio (CPR).

Land Equivalent Ratio (LER)

The Land Equivalent Ratio (LER) calculates the relative land area required from sole crops to provide the same yield from a multicrop composed of the same species. For a multicrop composed of two species, *i* and *j*:

$$\mathrm{LER} = (Y_{ij}/Y_{ii}) + (Y_{ji}/Y_{jj}) \quad [1]$$

where Y is the yield per unit area, Y_{ii} and Y_{jj} are sole crop yields of the component crops i and j, and Y_{ij} and Y_{ji} are the respective multicrop yields. The partial LER values, L_i and L_j, express the ratios of the yields of crops i and j when grown as multicrops, relative to their equivalent sole crops. Thus,

$$L_i = (Y_{ij}/Y_{ii}) \quad [2]$$

$$L_j = (Y_{ji}/Y_{jj}) \quad [3]$$

The total LER of a multicrop is the sum of its two partial land equivalent ratios (i.e., $\mathrm{LER} = L_i + L_j$). When $\mathrm{LER} = 1$ or less there is no net advantage to multicropping. An $\mathrm{LER} > 1$ means that more land is required to produce the same yield from sole crops of each component than with a comparable multicrop. The LER has become the most widely used index for assessing the relative performance of various forms of multicrop, in particular intercrops. However, the use of the LER concept takes no account of the relative durations of each species within the multicrop or sole crop systems. This means that published estimates of LER may exaggerate the advantages of multicropping because any land left unused after the harvest of the shorter duration sole crop is not included in the calculations.

Area–Time Equivalency Ratio (ATER)

The Area–Time Equivalency Ratio (ATER) calculates yield as a function of both land area and time by including the duration of land occupied by a multicrop and each component sole crop. This is done by redefining Y as the quantity q per unit of time, t, so that $Y = q/at$. An area time equivalency ratio can then be defined as

$$\mathrm{ATER} = \sum_{c=1}^{n} Y_c^i / Y_c^m \quad [4]$$

So, productivity by species c in the intercrop can be expressed in terms of both area and time (where $Y_c^i = q_c^i/at^i$, where t^i is the intercrop duration). Similarly, the productivity of species c as a sole crop can also be expressed in terms of area and time ($Y_c^m = q_c^m/at^{mi}$, where t^{mc} is the sole crop duration). Equation [4] can then be rearranged to give:

$$\text{ATER} = \sum_{c=1}^{n}(q_c^i/at^i)/(q_c^m/at_c^m) \qquad [5]$$

and

$$\text{ATER} = \sum_{c=1}^{n}\frac{(at_c^m)(q_c^i/q_c^m)}{at^i} \qquad [6]$$

For each crop c, $(at_c^m)(q_c^i/q_c^m)$ is the area × time, i.e., hectare.days (ha.d) required by sole crops to produce the quantity of c achieved in the multicrop. When all species in the multicrop are included in the calculation, ATER gives the ratio of area.time required in sole cropping to area.time used by the multicrop to produce the same quantities of all the component crops, i.e.,

$$\text{ATER} = \sum_{c=1}^{n} at_c^m/at^i = at^m/at^i \qquad [7]$$

ATER can now be presented in a form similar to LER, i.e.,

$$\text{ATER} = [(y_1^i/y_1^m)(t_1^m) + (y_2^i/y_2^m)(t_2^m) + \ldots(y_n^i/y_n^m)(t_n^m)]/t^i \qquad [8]$$

When time is included, apparent advantages calculated on the basis of LER often disappear and the use of the ATER suggests that multicrops and sole crops of the same species often use land with similar efficiency.

Crop Performance Ratio

Neither LER nor ATER identifies the resource-capturing principles responsible for any measured differences between multicrops and sole crops nor do they measure the biological efficiencies in terms of biomass or yield produced relative to energy captured during the season. A knowledge of the energy equivalents of different plant compounds is important in evaluating multicrops which are often composed of plant products with substantially different composition. By referring all calculations to the energy equivalent of biomass (MJ) we can compare biological outputs of different species composed of organs with different energy values.

The concept of a Crop Performance Ratio (CPR) can be used to:

1. Calculate multicrop productivity (adjusted for energy) per unit of land relative to that required from the same area by sole crops sown in the same proportions as the multicrop.
2. Account for the effect of differences in crop duration for each species on the performance of the multicrop and sole crops.
3. Compare the capture or conversion of resources such as solar radiation, water, or nutrients by intercrops and sole crops on the same basis as that for dry matter or yield.

For an intercrop composed of two species (a and b), the Crop Performance Ratio CPR_a for species a is given by:

$$\text{CPR}_a = \frac{Q_{ia}}{P_{ia}Q_{sa}} \qquad [9]$$

where Q_{ia} and Q_{sa} are its productivity per unit area (MJ m^{-2}) in the multicrop and sole crop respectively and P_{ia} is the proportional sown area of species a in the multicrop. A similar calculation can be used for species b so that for both species together:

$$\text{CPR}_{ab} = \frac{Q_{ia} + Q_{ib}}{P_{ia}Q_{sa} + P_{ib}Q_{sb}} \qquad [10]$$

Because their sown proportions multiply the sole crop values in the multicrop, this calculates their "expected" productivity if unit land area had been sown with sole crops in the same proportions as in the multicrop. Again a CPR > 1 indicates a multicrop advantage and a CPR < 1 a multicrop disadvantage. The CPR concept can be used to compare the capture or use of any resource, e.g., radiation, transpiration, or nutrients by a multicrop with its constituent species.

For multicrops composed of species with different durations we need to account for the durations, t_a and t_b, for species a and b respectively. For the sole crops the rate R (MJ d^{-1}) of productivity is given by:

$$R_{sa} = \frac{Q_{sa}}{t_a} \text{ and } R_{sb} = \frac{Q_{sb}}{t_b} \qquad [11]$$

For the same species in the multicrop, the rates of productivity are given by:

$$R_{ia} = \frac{Q_{ia}}{t_a} \text{ and } R_{ib} = \frac{Q_{ib}}{t_a} \qquad [12]$$

Now, the time-adjusted Crop Performance Ratio (CPRT) for species a is given by:

$$\text{CPRT}_a = \frac{R_{ia}}{P_{ia}R_{sa}} = \text{CPR}_a \qquad [13]$$

and for species b is:

$$\text{CPRT}_b = \frac{R_{ib}}{P_{ib}R_{sb}} = \text{CPR}_b\left(\frac{t_b}{t_a}\right) \quad [14]$$

For the multicrop, the overall Crop Performance Ratio (CPRT_{ab}) is given by:

$$\text{CPRT}_{ab} = \frac{(Q_{ia} + Q_{ib})t_b}{P_{ia}Q_{sa}t_b + P_{ib}Q_{sb}t_a} \quad [15]$$

Conclusions

Despite their history and importance to the sustenance, livelihoods, and welfare of much of the world, it remains unclear whether, and under which circumstances, various forms of multicropping may be more biologically efficient than their constituent sole crops. Expressed on a daily average over a whole growing cycle, multicropping and sole cropping probably have similar efficiencies in their capture and conversion of resources. This does not mean that multicropping is not worthwhile – the occupancy of land is usually related to the number of harvests, i.e., crops grown per year rather productivity per day and on this basis multicrops consistently outperform sole crops. Similarly, reproductive or economic yields are rarely the only criteria on which to advocate any particular cropping system, particularly in the seasonally arid tropics. A major consideration is how much risk is involved in cultivating a particular cropping system. Multicrops, in which species are grown for varying durations and different purposes, are inherently less risky than sole crops and provide some degree of insurance against adverse biological and physical conditions through their complementary sharing of resources in space and time.

See also: **Growth and Development**: Canopy Architecture; Field Crops. **Photosynthesis and Partitioning**: C_3 Plants; C_4 Plants. **Production Systems and Agronomy**: Agricultural Crops.

Further Reading

Azam-Ali SN and Squire GR (2002) *Principles of Tropical Agronomy.* Wallingford: CAB International.

Francis CA (ed.) (1986) *Multiple Cropping Systems.* New York: Macmillan.

Hiebsch CK and McCollum RE (1987) Area-×-Time Equivalency Ratio: a method for evaluating the productivity of intercrops. *Agronomy Journal* 79: 15–22.

Ong CK and Huxley P (eds) (1996) *Tree–Crop Interactions: A Physiological Approach.* Wallingford: CAB International.

Sinoquet H and Cruz P (eds) (1995) *The Ecophysiology of Tropical Intercropping*, Proceedings of the INRA conference, Guadeloupe, French Antilles, December 1993. Paris: INRA.

Vandermeer J (1989) *The Ecology of Intercropping.* Cambridge: Cambridge University Press.

Willey RW (1985) Evaluation and presentation of intercropping advantages. *Experimental Agriculture* 21: 119–133.

R

REGULATORS OF GROWTH

Contents

Auxins

R M Napier, Horticulture Research International, Wellesbourne, UK

Introduction

Auxins are natural plant hormones, mobile biochemicals which are transported from cell to cell and organ to organ to signal and coordinate growth and development. The experiments of Charles Darwin on phototropic bending by grass coleoptiles in the 1880s led him to conclude that plants contained a mobile "influence" mediating growth. In the 1920s Fritz Went collected diffusates from coleoptiles and showed that the "influence" induced bending when applied to unstimulated coleoptiles. Shortly after, in 1934, Kogl identified indole-3-acetic acid (IAA) as the active compound.

The main natural auxin in all plants is IAA and it is synthesized primarily in dividing meristematic cells. Auxin is an absolute requirement for plant cell division, but through directed and nondirected transport it also mediates control over many other key events during development. These include asymmetric growth during phototropism and gravitropism, apical dominance, fruit set and development, abscission, both lateral and adventitious root initiation, phyllotaxy, and vascular patterning. It is not surprising, therefore, that auxin concentrations in the plant are carefully controlled, that there is a sophisticated machinery for polar auxin transport, and that some microorganisms have capitalized on their own ability to synthesize auxins in order to perturb a host plant's development and exploit the disturbance for either pathogenesis or symbiosis. It is also clear that addition of auxin to plants will affect growth, and synthetic auxins are used widely in a range of agricultural and horticulture practices. The largest single market for auxins is as selective herbicides, but auxins are also applied as agrochemicals on fruit crops and for root induction during clonal propagation of nursery stocks.

Auxins

In addition to IAA a small number of naturally occurring, related compounds have auxin activity: 4-chloroindole-3-acetic acid, phenylacetic acid and indole-3-butyric acid (IBA) being the best characterized. Indole-3-butyric acid is probably converted to IAA by β-oxidation in plant cell peroxisomes, making the active form of this compound IAA.

A much more heterologous group of synthetic compounds are found to have auxin activity. Some of the most widely used are listed in **Table 1**.

IAA Synthesis and Breakdown

Measurement

Detailed measurements of auxin synthesis and of free auxin concentrations are becoming routine. Currently the technique of choice is gas chromatography coupled to mass spectrometry which is highly sensitive. Unfortunately, these exacting analytical

Table 1 Auxins and auxin transport inhibitors

	Formula	*Abbreviation*	*Uses*
Auxin			
Indole-3-acetic acid	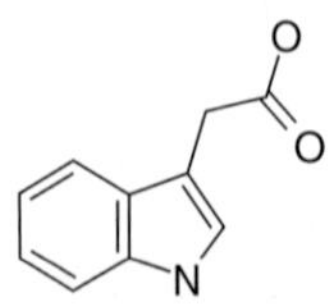	IAA	Natural auxin
Indole-3-butyric acid		IBA	Natural auxin. Rooting compound
2,4-dichlorophenoxyacetic acid	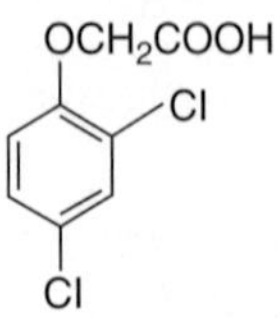	2,4-D	Selective herbicide
2,4,5-trichlorophenoxyacetic acid	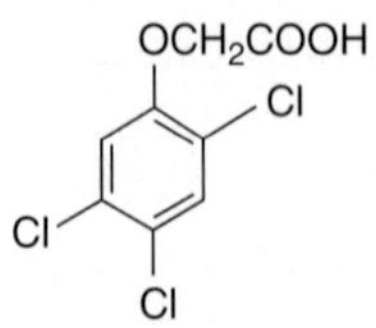	2,4,5-T	Selective herbicide. Brushwood killer
Chloramben	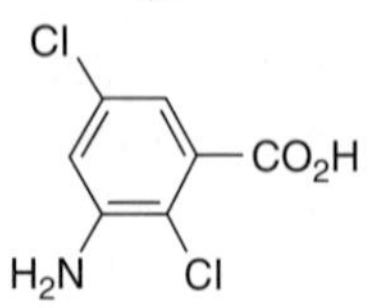	Amiben	Selective herbicide, preemergence
Picloram	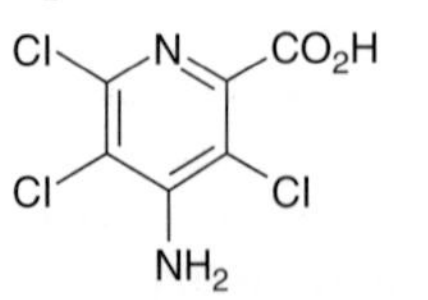	Tordon, Grazon	Selective herbicide
Triclopyr	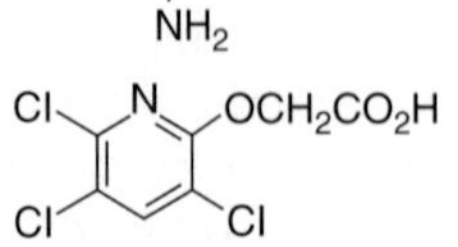	Garlon	Selective herbicide, Orchard fruit anti-abscission agent
Quinclorac	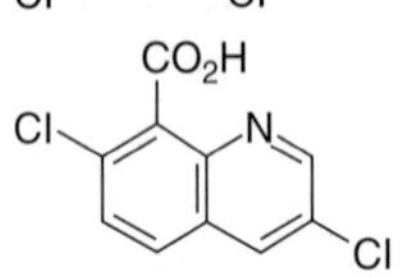		Herbicide
Auxin transport inhibitors			
1-naphthylphthalamic acid	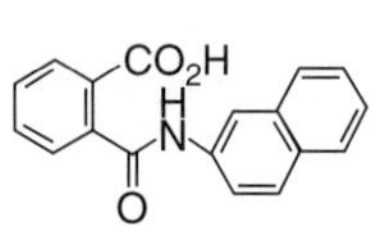	NPA	Auxin efflux inhibitor
2,3,5-triiodobenzoic acid	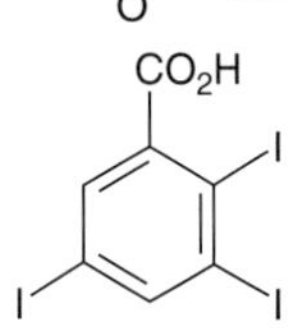	TIBA	Efflux inhibitor and weak auxin

techniques are available in very few laboratories. Protocols are also developing for parallel analyses of a number of plant hormones to give data about antagonistic and complementary signals from the same small samples. Only auxin is dealt with here.

Auxin Conjugates and Storage

For many years it was thought that IAA was synthesized in apical meristems. However, recent evidence suggests that auxin is, or can be, produced throughout the plant, including the roots. Furthermore, not all IAA production is from *de novo* synthesis. In germinating seeds IAA is produced from breakdown of stored forms of the hormone, conjugates of amino acids, proteins, and sugars. Broadly, monocotyledonous plants (cereals and grasses) accumulate sugar conjugates, dicotyledonous plants accumulate amino acid conjugates. Hydrolysis of these conjugates during germination precedes or coincides with the start of root extension. Large fluxes of IAA released from conjugates stored in the kernal mediate extension in cereal coleoptiles, but this auxin moves in the vascular strands to the coleoptile tip and is then distributed back down to the elongating cells by the polar auxin transport system (see below).

De Novo Synthesis

Stored IAA reserves are finite and soon after seedling emergence *de novo* IAA synthesis commences. The amino acid tryptophan is the precursor for a set of parallel IAA biosynthetic pathways, none of which is fully mapped or understood at the genetic or protein levels. A few days later a further biosynthetic pathway is activated, a tryptophan-independent pathway. The intermediates on this pathway have not been identified, but tryptophan is not one of them. It seems likely that this tryptophan-independent pathway remains the constitutive biosynthetic source of IAA throughout the rest of a plant's life. However, when a plant is stressed, such as by wounding, or during major development events like germination, then IAA synthesis is stepped up by switching on the tryptophan-dependent pathways.

Very young leaves, less than 0.5 mm long in *Arabidopsis*, produce large amounts of IAA. These early high concentrations and high synthetic capacity help drive leaf expansion, but as the leaf expands, synthesis and concentrations fall. Concentrations of free IAA were found to be 100-fold lower in the expanded leaf lamina. In fully-grown *Arabidopsis* plants the highest concentrations of IAA are found in expanding fruiting bodies, the siliques (**Figure 1**). It is likely that much of this is synthesized by seeds during embryo development (see below). The inflorescence stalk contains more IAA than most other parts of the plant, but it is not clear if this is synthesized *in situ* or if it is in transit.

Deactivation

Control of auxin action can be mediated by removing IAA as well as by synthesis. Auxins are removed by oxidation and by conjugation into inactive storage compounds. Some enzymes involved in these processes are known, but their regulation – genetic or biochemical – is understood poorly.

Homeostasis

The balance of synthesis, breakdown, conjugation, and transport is regulated rigorously to give auxin homeostasis. Some of the fine measurements of IAA concentration have illustrated that leaf expansion, for example, is reduced by both increases and decreases of auxin. Clearly, changes in auxin concentration are important as plants respond to stimuli, but homeostasis is critical for both optimal development and to keep the system primed for stimulatory responses. At present, the feedback mechanisms for homeostasis are not well characterized.

Nonplant Auxin Synthesis

The auxin IAA is made by certain microorganisms as well as by plants, but it appears that only plants can control auxin concentrations to give auxin homeostasis. Additionally, it appears that only plants have auxin receptors and respond to auxin as a hormone, although there is some evidence that plant nematodes can sense auxin and use this as a signal to move towards a new host. Bacteria such as the pathogens *Pseudomonas syringiae* and *Agrobacterium tumefaciens* synthesize copious amounts of auxin from trytophan, making use of a trytophan-dependent pathway not exploited by plants. Tryptophan monooxygenase converts tryptophan to indoleacetamide, which is converted by a hydrolase to IAA. Such auxin contributes to the pathogenic responses induced in plant hosts, such as in the crown gall tumors caused by agrobacterial infections. Likewise there is evidence to suggest that symbiotic mycorrhizal fungi produce auxin and that this auxin then contributes to the changing morphology of the root system in these associations.

Auxin Transport

All plant hormones are moved around the plant in the vasculature where, once loaded, they move passively as passenger molecules. Auxin is special in that plants also have a polar transport system to

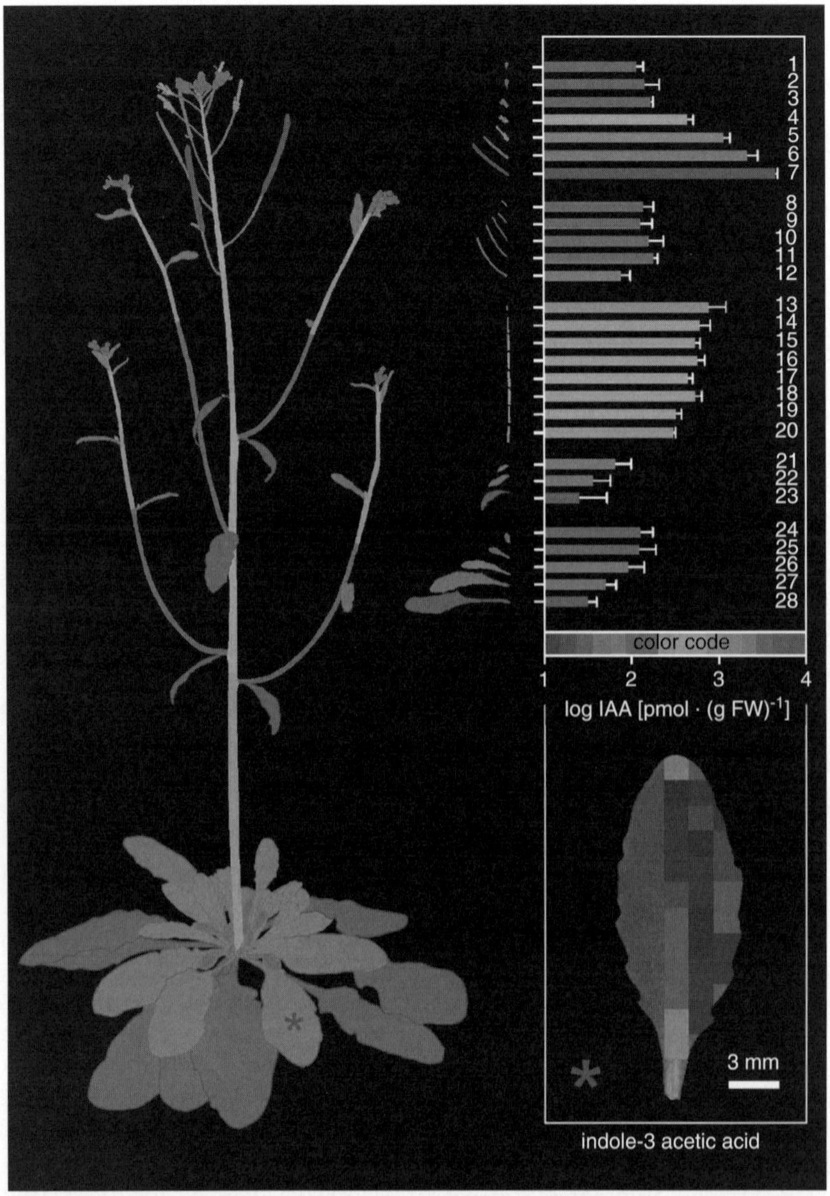

Figure 1 (see color plate 26) Auxin distribution in a mature *Arabidopsis* plant. The distribution of free IAA was measured using gas chromatography/mass spectrometry. Concentrations are illustrated by color-coding the organs, the scale is shown to the right. Reproduced with permission from Müller A, Düchting P, and Weiler EW (2002) A multiplex GC-MS/MS technique for the sensitive and quantitive single-run analysis of acidic phytohormones and related compounds, and its application to *Arabidopsis thaliana*. *Planta* 216: 44–56.

move IAA vectorally. Polar auxin transport requires energy and contributes to many of the important auxin-dependent responses described below, such as apical dominance and gravitropism.

The consequence of polar auxin transport is that certain cells (or tissues) receive extra auxin, an auxin stimulus. The donor cells might or might not become depleted of auxin, depending on whether or not they are replenished by new synthesis or by influx from other cells. Polar transport functions at the cellular level and is now believed to be effected by the combined activities of both auxin influx proteins and auxin efflux proteins. Candidate proteins and the genes encoding them have been described, both having been identified from screening mutant populations and the application of molecular genetics. The influx carrier gene, now known as *AUX1*, was isolated from seedlings insensitive to the auxin 2,4-D. The AUX1 protein is a member of the amino acid transporter protein superfamily. It is an integral membrane protein spanning the lipid bilayer with 10 or 11 helices joined by short loops of more hydrophilic amino acids.

The first efflux carrier gene was identified from plants with an extreme flower morphology. Pin-like organs formed with no development of flower parts.

This morphology gave rise to the name *PIN* for the gene family and there are more than ten *PIN* genes in *Arabidopsis*. Individual members of the *PIN* gene family are expressed in a defined set of cells or tissues, but family members appear to be expressed in most parts of the plant. The efflux proteins act to pump auxin out of cells to effect auxin gradients and confer directional information. The proteins themselves sit in the plasma membrane, span the membrane 10 times, but are not related to *AUX1*. They are localized asymmetrically around the cell such that efflux activity induces polar auxin pumping (**Figure 2**).

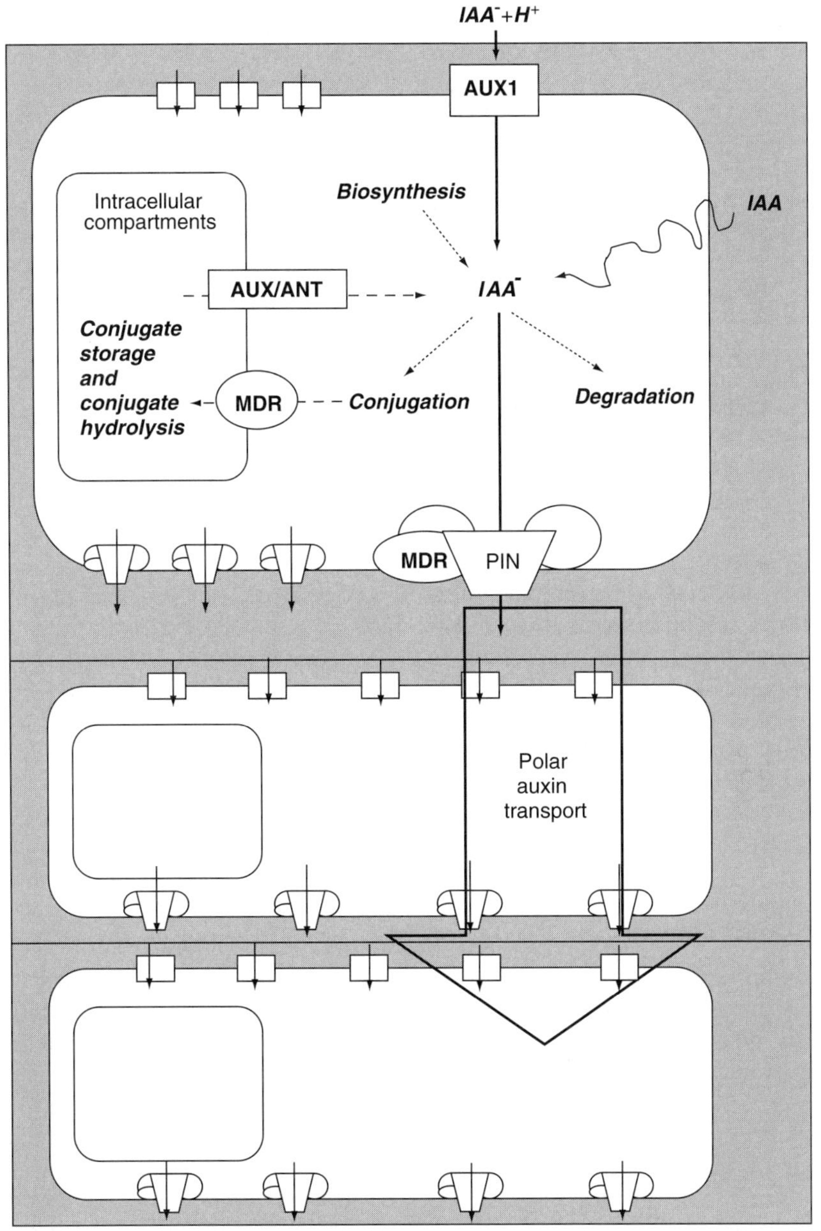

Figure 2 Auxin in the plant cell and polar transport. Auxin enters plant cells by both diffusion as uncharged IAA (top right) and by carrier-mediated uptake. The uptake carrier AUX1 is indicated as a box. The acidic cell wall space provides protons for IAA^- cotransport down the proton gradient across the plasma membrane into the cytoplasm. The auxin pool in the cytoplasm will be the result of biosynthesis, degradation, conjugation, and transport. Both free IAA and IAA conjugates are likely to be partitioned within cells, although the identity of compartments used for storage remains unclear. Polar auxin transport is driven by efflux of IAA through a complex of proteins that includes PIN. The activity of the efflux complex is regulated by continual secretion of components of the complex. The PIN complex and AUX1 protein are targeted to specific (and opposite) faces of the cell to give direction to auxin flow. The targeting mechanism is not understood. The net effect of efflux at the base of each cell and accumulation by the adjacent top face of the next cell in each file is polar auxin transport. Additional proteins implicated in auxin transport and compartmentation are named ANT (neutral amino acid transporter) and MDR (a member of the multidrug resistance family of transporters).

The speed of polar auxin transport is generally measured at about 10–20 mm h^{-1}. Under circumstances in which auxin might be moved through few cells to effect a gradient for differential growth (such as for gravitropism in roots or coleoptiles) and for small plants, such a transport rate seems sustainable. However, the passage of an auxin stimulus from the apex of a tree to its roots presents a challenge of a different magnitude. Observations have been made of a wave of auxin release travelling in advance of, and more rapidly than, net polar auxin transport. The mechanisms for such a wave have still to be identified, but merit further research.

It is not correct to surmise that all auxin movement is through the polar transport system. Considerable quantities of free IAA and of auxin conjugates are carried in the vasculature, particularly the phloem. Indeed, the relatively high pH in phloem acts to concentrate free IAA and unloading from the phloem will require the action of carrier proteins like PIN and AUX1, although the mechanisms regulating unloading sites remain to be identified.

Many of the auxin-driven responses considered below fail if auxin is absent or in excess, or if polar auxin transport is defective. It is, perhaps, not surprising that hormone delivery is as important as the hormone itself. As a result, auxin physiology has benefited from the use of drugs that act specifically to inhibit polar auxin transport. Of these, naphthylphthalamic acid (NPA) (**Table 1**) is most specific and has been used widely. Triiodobenzoic acid (TIBA) has also been used, but has additional activities. Recently, 1-naphthoxyacetic acid has been identified as a useful tool to inhibit uptake.

Vascular Patterning

Auxin transport inhibitors have dramatic effects on the development of the vasculature in developing tissues. Auxin both determines sites of vascular differentiation and promotes vascular connectivity. Most work has been done in leaves and using *Arabidopsis*, in which auxin transport inhibitors lead to confinement of the vasculature to the leaf margins. The zone of cell division in leaves lies at the margins and auxin synthesis in very young leaves is extremely active, as noted above. It seems likely that auxin from the margins induces vascular development and the auxin gradient from edge to center of the leaf lays down the direction for early patterning. As the vasculature forms it also generates an auxin sink. The new phloem exports IAA, reinforcing local gradients, but before and during vascular development polar transport provides directionality.

The pattern of primary and secondary vascular strands is set very early in leaf development and once set is not changed by auxin stimuli. In addition to synthesis at the leaf margins it has been suggested that specialized leaf cells in the lamina, such as glandular hydathodes, provide auxin and help determine tertiary patterning. It can be seen that combining these various axes and gradients of auxin, along with other stimuli, gives a complex set of developmental cues which result in diverse, yet reproducible, venation patterns.

Embryonic Patterning

Just as polar flow of auxin is essential for cell axis formation in the vasculature, it plays the same, key role in establishment of the plant body axis in young embryos. In the same way as auxin transport inhibitors induce errant patterns of development in leaves, they prevent the establishment of bilateral symmetry in embryos. Auxin responses during embryogenesis have also been well studied using morphogenic *Arabidopsis* mutants. One known as *MONOPTEROS* shows the same embryonic failure to establish a polar body axis as seen after NPA treatment and the *MONOPTEROS* gene turns out to code for an auxin-regulated transcription factor, a protein controlling gene expression. Mechanisms of auxin action and the control of gene expression are considered below.

Organ Patterning

Evidence is accumulating to show that auxin and its polar transport determine the apical–basal pattern of development in some floral tissues such as the gynoecium (the role of auxin in fruit development is discussed below). Perhaps the most widely observed outcome of auxin-induced patterning is in phyllotaxis, the regular arrangement of leaves on a stem. Once again, polar auxin transport inhibitors and *Arabidopsis* mutants defective in polar transport like *pin1-1* have helped illustrate that accumulation of IAA in cells at the side of the shoot apical meristem initiates organogenesis and determines the position of the next leaf primordium. Some delicate and beautiful experiments have been described by the group of Kulemeier in Switzerland. Microapplications of transport inhibitors or IAA to tomato (*Lycopersicon esculentum*) shoot apices have shown that, for leaves and flowers, auxin application determines primordium development in the radial dimension. Primordium development always arises at a fixed distance from the summit of the apical meristem in the apical–basal dimension (**Figure 3**).

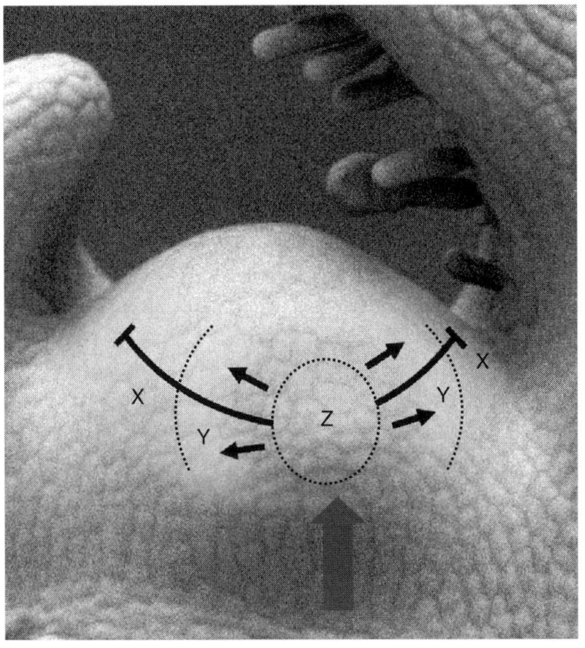

Figure 3 Auxin determines phyllotaxy. A lateral view of a tomato shoot apical meristem showing two leaf primordia, P3 to the right at the rear, P2 to the left at the rear, and P1, the area at the face of the apical dome (z). Auxin delivery that is dependent on polar auxin transport is indicated by the gray arrow. This auxin causes local expression of organ identity and outgrowth genes. Lateral cell–cell signaling leads to boundaries (y) and inhibition of organogenesis (x). Reproduced with permission of Elsevier Ltd. from Kuhlemeier C and Reinhardt D (2001) Auxin and Phyllotaxis. *Trends in Plant Science* 6: 187–189.

Apical Dominance and Branching

In addition to controlling the initiation of leaf primordia at shoot apices, it has been recognized for many years that auxin transported back from the shoot apex controls the outgrowth of side branches from axillary buds. Gardeners and horticulturalists make use of this apical dominance to control plant structure by pruning. By removal of the stem apex the source of auxin is removed and its inhibitory action on axillary buds is released, leading to side branch outgrowth and bushier plants. Consistent with these utilitarian observations, classical experiments by Thimann and Skoog showed that if auxin is applied to the pruned apical stump, bud outgrowth is inhibited. More recently, experimental plants generated with low auxin levels are bushier. However, apically derived auxin does not enter the axillary bud and it is unclear how the signal is transferred. Root-derived cytokinins are also involved in controlling bud break and other root–shoot signals are likely to be added to the list of controls before long.

Apical dominance is active in roots as well as stems, but for roots the contribution of additional specific signals is less clear. Root tip-derived auxin certainly inhibits lateral root initiation close to the apex; however, lateral root development is controlled by auxin at several stages. Further up the root behind the elongation zone lateral primordia form, but remain dormant. These primordia will not grow out unless the tip is removed (apical dominance) or until a second, shoot-derived, polar flux of auxin arrives. Not long after emergence a lateral root becomes auxin-autonomous.

Tropisms and Epinasty

Plants respond to the vectoral signals light and gravity by differential growth. Auxin gradients, generated by lateral (polar) auxin transport, have been implicated and the experiments of Cholodny and Went are common schoolroom lessons. Blocks of agar were used to collect endogenous signals from two sides of gravistimulated seedling stems. The blocks were then applied to unstimulated seedlings and bending was observed. Auxin applied in similar blocks was shown to give the same differential growth.

In stems, greater auxin concentrations enhance elongation growth; in roots, auxin inhibits growth. More recently, auxin measurements have shown that the side of stems expanding more rapidly contain a little more IAA, although the gradient across the stem is generally only of the order of two-to threefold. Auxin reporter genes have shown expression patterns corresponding to small auxin gradients (**Figure 4**), but even so it is argued that the magnitude of the gradient is insufficient to support the considerable differences in elongation rate across the same tissues. The Cholodny–Went hypothesis remains controversial. Recent data on the stimulus-induced redistribution of specific members of the PIN family in roots (e.g., PIN3), data from auxin measurements and observations from auxin response mutants all support the hypothesis. Other data, such as kinetic measurements of gravistimulated roots and experiments in which seedling shoots are bathed in excess IAA yet are still found able to respond by differential growth, argue against such a straightforward hypothesis. Either way, it is clear that auxin is a requirement for both differential growth and tropic growth. It might not be the only signal involved.

A similar phenomenon to tropic growth is epinasty, petiole bending induced by a range of nondirectional stimuli. The contribution auxin plays in epinasty is seen most clearly in plants treated with auxinic herbicides (see below). The top sides of the petioles extend excessively leading to curvature turning the face of the leaf away from light. Other hormones, in particular ethylene, are also involved in epinasty.

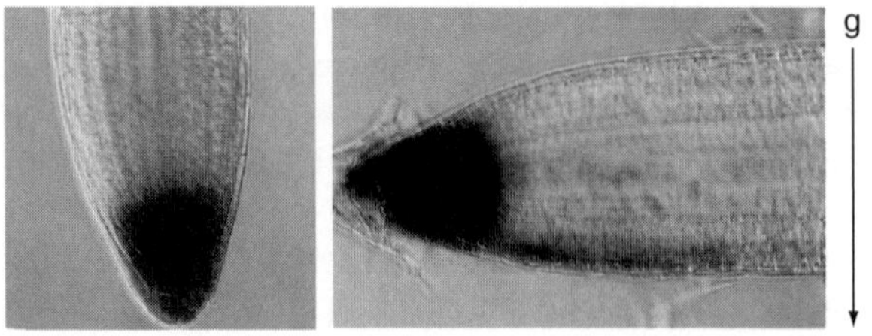

Figure 4 (see color plate 27) Auxin distribution after gravistimulation in roots. Expression of an auxin-responsive reporter gene (*DR5-GUS*) is observed by the accumulation of blueness. The vertical root tip is rich in auxin, but the cells behind the tip show little or no expression, indicating low auxin concentrations. However, if the root tip is turned through 90 degrees to lie horizontal, auxin is seen to accumulate in cells along the lower side in the root elongation zone. In roots, auxin inhibits cell elongation and so the cells on the upper side elongate to push the root tip back towards the vertical. Reformatted from the original; reproduced with permission from Rashotte AM, DeLong A, and Muday GK (2001) Genetic and chemical reductions in protein phosphatase activity alter auxin transport, gravity response, and lateral root growth. *Plant Cell* 13: 1683–1697. Copyrighted by the American Society of Plant Biologists.

Cell Division and Tissue Culture

So far, only the influence of auxin on tissue development has been considered. However, one of the most profound actions of auxin on plants is over the control of cell division. In the absence of auxin or at low concentrations plant cells will not divide. The cell cycle arrests at G_1. Cell expansion can proceed at low auxin concentrations, but only at elevated concentrations will the cell cycle advance past G_1 to G_2/M and division. During division cell expansion is minimal.

Amongst the most striking examples of auxin control over division is in cell and tissue culture for which auxin must be included in almost all media. Furthermore, auxin (and cytokinin) concentrations are manipulated to promote proliferative callus growth, regeneration, and root induction. For root induction auxin concentrations are raised; for regeneration auxin concentrations are reduced, cytokinins raised.

Secondary meristems are important for the generation of many plant tissues. Amongst the most prolific is the cambium which gives rise to secondary xylem and phloem and the tissues of the phellum (bark). Auxin concentrations have been measured in thin, sequential sections across the cambial region of pine trees, from xylem (wood) to bark. The highest concentration corresponded with cambial cells and it has been suggested that this zone of high auxin conveys positional information. In this way auxin is acting as a morphogen and the steepness of the declining gradient of auxin across cells adjacent to the cambium determines the radial width of differentiating zones of division and expansion. As for cell division (high) and cell expansion (low), there are concentration thresholds which combine with other positional stimuli to mediate the length of time cambial derivatives spend dividing, differentiating or laying down secondary walls. Consequently, much secondary growth is governed by auxin concentrations in, and delivery from, the cambium.

It has been reported that all cambial auxin is delivered by polar auxin transport, mostly from the expanding stem apices. Polar transport inhibitors deplete almost all the auxin from the cambial region. However, considering the speed of transport (above) it is recognized that there are additional sources and the dividing cambium might itself contribute to the pool.

Adventitious Rooting and Wound Responses

Adventitious roots form from stem tissues, generally as a result of damage or removal of the primary root system. Many plants do produce some adventitious roots naturally and they add valuable lateral support to tall stems in cereals, for example. However, addition of auxin to cut or damaged stems often induces a strong adventitious rooting response and the horticultural industries relying on clonal propagation make good use of this response, for ornamentals, trees, flowers, and general garden plants. Synthetic auxins IBA and 1-naphthylacetic acid (1-NAA) are both sold widely in hormone rooting powders and dips.

For lateral root primordia to form, cells of the root xylem pericycle divide and differentiate. For adventitious rooting a similar series of events is likely to take place in equivalent cells in the stem vasculature but the exact cells giving rise to the first division are less certain. New root primordia form at the side of the procambium, or within callus tissue that forms at the cut or wounded surface. The initiation of root primordia in stem tissues requires a redifferentiation response and, although auxin promotes this, the molecular or genetic mechanisms are unclear.

There are mutants of *Arabidopsis* that show precocious rooting along the seedling hypocotyl, *superroot 1* and *2* (*sur1* and *sur2*) for example. In both, auxin levels are elevated due to distinct effects on auxin biosynthetic pathways.

Wounding plant tissues gives rise to rapid cell division and callus formation. Callus is undifferentiated and apolar in its growth, but not only can new, polar organs arise from it (like adventitious roots), but adjacent cells can redifferentiate to reform vascular connectivity. The process is slow, commencing many days after damage and callus growth, but xylem elements start to form with a polar orientation and line up to re-establish the vasculature as fully formed, thickened xylem elements die. Auxin and polar auxin transport are essential for this response and it is likely that this polar redifferentiation occurs along a reforming apical–basal auxin gradient. This is not a general event, this redifferentiation occurs only in a narrow band of cells that starts and finishes at damaged vascular strands. A similar process is likely to take place during graft development so that vascular continuity is regained.

Fruit Growth

Auxin plays a vital role in all stages of reproductive growth. The role of auxin in gynoecium development and embryo polarity have been recorded above and some of the highest auxin concentrations have been found to be in developing fruit. In the early 1950s Nitsch showed that achene (seed) removal from strawberry (*Fragaria x ananassa*) receptacles inhibited receptacle enlargement. Replacing the achene with a supply of auxin maintained fruit growth. Cessation of auxin supply also led to ripening in this nonclimacteric fruit.

In climacteric fruit, ripening is mediated by ethylene not auxin. Nevertheless, in a wide range of fruit as well as in cereal seed-heads it is clear that the developing embryo and seed are rich auxin sources and that this auxin mediates many developmental responses in adjacent tissues. Indeed, parthenocarpy can be induced in many species by auxin application and this has been used to generate seed-free fruit by both topical application and, recently, by genetic engineering. The bacterial IAA biosynthetic gene *iaaM* coding for tryptophan monooxygenase was transformed into tomato, eggplant (*Solanum melongena*, aubergine), and other fruit under the control of a carpel-specific promoter. This gave rise to localized overproduction of IAA and induced seed-free fruit of marketable quality with the added advantage that production was no longer seasonal. The need for bees to pollinate in the glasshouse was removed, a problem during winter months. Some apple (*Malus pumila*) and cherry (*Prunus* spp.) crops are sprayed with auxin at flowering to induce fruit set, although there is generally a mix of hormones used.

Fruit crops, particularly *Citrus* spp., are sprayed with auxins to improve fruit size by promoting "June drop" but, more widely, orchards are sprayed after the drop to increase fruit size and hexose content and to delay fruit abscission. Addition of auxin delays the development of the abscission zone. Both exocarp and seeds are rich in IAA, but, as in strawberries, it is likely that as the testa matures around the seed it forms an impervious barrier, blocking auxin release, in the absence of auxin abscission proceeds. Leaf abscission is also initiated as auxin transport from the leaf declines.

Treatment of orchards to delay fruit abscission is becoming increasingly common to facilitate automated harvesting, maximizing the yield of ripe fruit collected in the minimum number of passes by harvesters. Lychee (*Litchi chinensis*) and nut crops are also treated with auxins to promote fruit size and stall abscission.

Herbicides

The largest commercial exploitation of any plant hormone has been in the use of synthetic auxins as selective herbicides. Around the world a number of populations of auxinic herbicide-tolerant weeds have arisen, an indication of the long time over which applications have been made. The first phenoxyacetic acid auxins, such as 2,4-D (**Table 1**), were developed in the 1940s and use has been extensive ever since. Farmers have exploited their selectivity for they have a much higher activity against most dicotyledonous plants than against cereals and this had led to widespread use on cereal crops, sugarcane and turf to control broadleaved weeds. Auxins were the active defoliants in "Agent Orange" and phenoxy auxins are still used to control brushwood.

The mechanisms of herbicidal action and the basis of selectivity between monocotyledons and dicotyledons is unclear. Induction of a massive synthesis of ethylene has been suggested to be the mechanism of activity, at least in part because a byproduct of ethylene overproduction is cyanide. However, recent work shows that inhibition of either ethylene synthesis or ethylene receptors does not prevent herbicidal activity. Therefore, although induction of the ethylene-synthesizing enzymes is a recognized effect of auxin application, ethylene remains a symptom of herbicidal auxin and is not the sole mediator.

Auxinic herbicides also induce abscisic acid (ABA) synthesis and this might, or might not, be a consequence of elevated ethylene synthesis. Ethylene and ABA act antagonistically in growth control. Elevated ABA will shut stomata and the consequent block on photosynthesis might be the source of excessive hydrogen peroxide production, as recorded in auxinic herbicide-treated plants. In addition to the damage done by peroxide itself, such reactive oxygen species will induce defense and senescence responses, giving rise to necrosis and death.

Saturating auxin clearly gives rise to a number of damaging symptoms, but genetic data from resistant biotypes, as well as the observations of monocotyledon/dicotyledon selectivity suggest that there is a single target site for these herbicides and this seems likely to be the auxin receptor.

Auxin Perception, Receptors, and Signaling

Perception and Receptors

Many techniques have been used in experiments to identify auxin receptors and many candidates have been examined in detail. Photoactive and radiolabeled auxins have been shown to bind to a diverse set of proteins, affinity purification has identified others, and genetic screens have yielded a vast amount of information on auxin transport proteins, signaling intermediates and on the transcriptional regulation of auxin-induced genes. Yet there is still scope for the discovery of novel receptors.

The first candidate auxin receptor was identified in the early 1970s in membrane fractions of *Zea* coleoptiles. It was characterized as a protein which bound active auxins and is referred to as auxin-binding protein 1 (ABP1). Purification and sequencing of ABP1 followed in the late 1980s. It is a soluble protein targeted to the endoplasmic reticulum. Experiments to test its function used microelectrodes to record ion fluxes across the plasma membrane and these showed that ABP1 was necessary for auxin to stimulate proton efflux and membrane hyperpolarization. However, it was also necessary for ABP1 to be present on the outer face of the plasma membrane, not just in the endoplasmic reticulum. Later, immunological and biochemical techniques confirmed a small population of ABP1 on the cell surface. Transformation to give plants overexpressing ABP1 confered a phenotype in which auxin-induced cell expansion is dependent on ABP1 and a homozygous null mutant (ABP1 knockout) was found to be embryo lethal. The three-dimensional crystal structure of ABP1 with and without auxin bound to it have been determined at high resolution and give the first images of a plant hormone recognition site. All these observations are consistent with ABP1 being an auxin receptor, but many auxin-regulated events have remained unlinked to ABP1 activity, most notably auxin-regulated cell division and auxin-regulated gene activation. Consequently, ABP1 is unlikely to be the only auxin receptor, but at present no other strong candidates are known.

Signaling

There is also a gap in our knowledge about signaling intermediates involved in the expected biochemical cascade from receptor activation to effectors. However, a great deal is known about some of the effectors, the complexes of genes and proteins which mediate auxin-regulated transcriptional activity. The first auxin-insensitive mutant for which the defective gene was cloned and sequenced was a mutant known as *axr1*. The sequence data suggested the gene product coded for an ortholog of a ubiquitin-activating enzyme and, therefore, the AXR1 protein was likely to be involved in protein degradation. More recently, many other components of the proteasome, an enzyme complex involved with protein degradation and turnover, have been shown to be involved in auxin-mediated transcriptional control.

The targets of proteasomal degradation appear to be auxin-regulated transcription factors. Biochemical studies discovered several families of protein strongly upregulated by auxin application to plant tissues. Once cloned and sequenced it became clear that some of these were transcription factors and as these were examined further, similar genes were identified giving rise to the *Aux/IAA* gene family. In addition, a large family of proteins was identified that bound to auxin response elements, DNA sequences located in the promoter regions of auxin-regulated genes. These are the auxin response factor (ARF) proteins. Both ARF and Aux/IAA proteins have DNA-binding domains and oligomerization domains and they form both homodimers and mixed dimers (**Figure 5**).

The current model for auxin-regulated gene expression is that at low auxin concentrations, mixed ARF-Aux/IAA dimers bind tightly to DNA repressing transcription. A rise in auxin concentration leads to dissociation of the dimers releasing the Aux/IAA protein, allowing transcription to proceed. The ARF remains bound to the promoter DNA and may associate with another ARF forming a homodimer. The dissociated Aux/IAA proteins become phosphorylated and substrates for rapid degradation by the ubiquitin–proteosome pathway. A fall in auxin concentration slows protolysis and newly transcribed Aux/IAAs reassemble with the ARFs to repress transcription again.

The defective gene in the embryo morphology mutant *MONOPTEROS*, mentioned above, is a member of the *Aux/IAA* gene family. Given the number of ARFs and Aux/IAAs and the large number of different, mixed dimers that can be made, it can be seen that this system can account for the great variety of different responses controlled by auxin. However, linking each response to a defined pair of transcriptional regulators will take considerable effort.

Summary

There are still gaps in our knowledge of auxin synthesis, signaling, and action, but a great deal has been learnt. There are many agriculturally useful

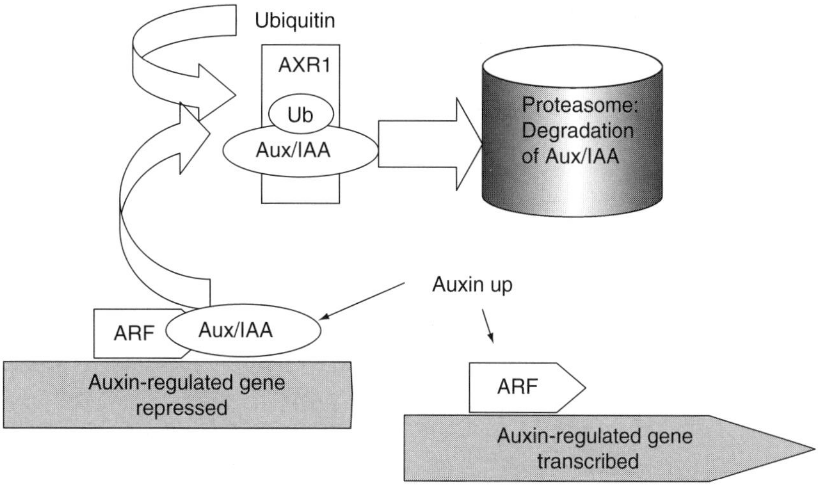

Figure 5 Schematic illustration of auxin-mediated gene activation. In low auxin concentrations ARF and Aux/IAA proteins are bound as a dimer to promoter regions of auxin-responsive genes. On addition of auxin, the dimers dissociate, leaving auxin response factor (ARF) bound and allowing transcription to proceed. The displaced Aux/IAA protein is ubiquitinated via AXR1 and then degraded in the 26S proteosome. Amongst the first genes to get transcribed after auxin addition are the *Aux/IAAs* and so if high auxin concentrations do not persist, the genes are re-repressed.

applications of both auxins and of the knowledge of auxin physiology and, as we learn more about the genetics and biochemistry of auxin, a still wider set of applications can be anticipated.

Acknowledgements

Funding has been provided by BBSRC and EC contract IC15CT980118.

See also: **Cell Walls and Fibers**: Wood Growth and Development. **Flowering and Reproduction**: Fertilization. **Genetic Modification**: Transformation, General Principles. **Growth and Development**: Cell Division and Differentiation; Cell Growth; Field Crops; Leaf Development; Molecular Biology of Development. **Postharvest Physiology**: Genetic Engineering for Postharvest Quality; Ripening; Senescence, Flowers; Senescence, Leaves. **Production Systems and Agronomy**: Orchard Crops. **Regulators of Growth**: Abscisic Acid; Cytokinins; Gibberellins. **Root Development**: Genetics of Primary Root Development; Lateral Root Initiation; Root Growth and Development. **Seed Development**: Embryogenesis. **Tissue Culture**: General Principles; Organogenesis; Somatic Embryogenesis. **Weeds**: Herbicide Resistance.

Further Reading

Berleth T, Mattsson J, and Hardtke CS (2000) Vascular continuity and auxin signals. *Trends in Plant Science* 5: 387–393.

Davies PJ (1995) *Plant Hormones: Physiology, Biochemistry and Molecular Biology*. New York: Kluwer Academic.

Friml J and Palme K (2002) Polar auxin transport: old questions and new concepts?. *Plant Molecular Biology* 49: 273–284.

Kuhlemeier C and Reinhardt D (2001) Auxin and phyllotaxis. *Trends in Plant Science* 6: 187–189.

Leyser O (2002) Molecular genetics of auxin signaling. *Annual Review of Plant Biology* 53: 377–398.

Ljung K, Hull AK, Kowalczyk M, *et al.* (2002) Biosynthesis, conjugation, catabolism and homeostasis of indole-3-acetic acid in *Arabidopsis thaliana*. *Plant Molecular Biology* 49: 249–272.

Zazimalova E and Napier RM (2003) Points of regulation for auxin action. *Plant Cell Reproduction* 21: 625–634.

Abscisic Acid

J A D Zeevaart, Michigan State University, East Lansing, MI, USA

Introduction

Abscisic acid (ABA) is a plant hormone that has multiple roles in plant growth and development. Although the name implies involvement in abscission, the role of ABA in this process is still not well established. However, it is now evident that ABA plays an important role in mitigating various stress responses, so that it has been called a stress hormone. Responses to ABA can be very rapid as well as long term. Although no receptors for ABA have been isolated, it appears that there is more than one ABA receptor. Evidence indicates that for some responses ABA is perceived on the external surface of the plasma membrane, whereas for other responses perception of ABA appears to be inside the cell.

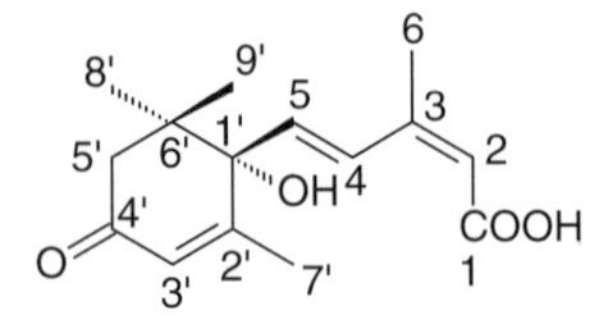

Figure 1 Structure of *S*-(+)-abscisic acid with numbering system.

History

Abscisic acid was first described as abscisin II in 1965 by Fred Addicott and associates at the University of California, Davis, USA, as an abscission-accelerating compound that had been isolated from young cotton fruits. At the same time, Philip Wareing and co-workers at the University of Wales, Aberystwyth, UK, isolated a substance called dormin, which was supposed to induce dormancy in woody species. After it was established that abscisin II and dormin are chemically identical, the name abscisic acid (ABA) was adopted.

Occurrence

ABA is ubiquitous in plants and has also been found in several fungi and in some cyanobacteria. The amounts of ABA in plants vary considerably. Leaves contain more ABA than roots. Most developing seeds and fruits are rich sources of ABA. The water status of plants is the main condition that regulates the ABA content of plants. Dehydration can increase the ABA level as much as 50- to 100-fold in a few hours. Conversely, rehydration of desiccated plants results in rapid degradation of this excess ABA.

Chemistry, Biosynthesis, and Degradation

ABA is a C_{15} organic acid with one asymmetric carbon atom at the C-1′ position (**Figure 1**). The naturally occurring form is *S*-(+)-ABA and the side chain has the 2-*cis*,4-*trans* configuration. Photo-isomerization to the 2-*trans*,4-*trans* isomer renders the compound biologically inactive. ABA is a sesquiterpenoid, which means that the molecule is built up of three isoprenoid units. Consequently, it was originally proposed that the biosynthesis of ABA involves the condensation of three molecules of isopentenyl diphosphate, the universal building-block of terpenoids. This pathway of ABA biosynthesis has indeed been found to operate in fungi. By contrast, in green plants ABA is a breakdown product of carotenoids, specifically of violaxanthin and neoxanthin. A pivotal step in the biosynthesis of ABA is the cleavage of a carotenoid molecule (C_{40}) into a C_{25} fragment and a C_{15} product, xanthoxin, which is further oxidized in two steps to ABA (**Figure 2**). The cleavage step takes place in chloroplasts, the organelles where carotenoids are localized. The rapid increase in ABA during dehydration is mostly due to increased expression of the enzyme 9-*cis*-epoxycarotenoid dioxygenase (NCED), catalyzing the cleavage reaction. Mutants of *Arabidopsis* deficient in ABA have been instrumental in elucidating the ABA biosynthetic pathway (**Figure 2**). The *ABA1* gene encodes the enzyme zeaxanthin epoxidase (ZEP) that converts zeaxanthin to violaxanthin. The next steps, isomerization of all-*trans*-violaxanthin to 9-*cis*-violaxantin and 9′-*cis*-neoxantin, are still poorly understood. The cleavage of these latter carotenoids is the first committed step in ABA biosynthesis. *Vp14* of maize (*Zea mays*) was the first gene to be isolated that encodes an NCED. The substrates for NCEDs must be in the 9-*cis* configuration in order to give the 2-*cis*,4-*trans* configuration of xanthoxin and ABA. (The 2,3 double bonds of xanthoxin and ABA correspond to the 9,10 double bonds of carotenoids; **Figure 2**). The last two conversions from xanthoxin to ABA are localized in the cytoplasm. *ABA2* encodes a unique short-chain alcohol dehydrogenase that converts xanthoxin to abscisic aldehyde. There are two mutants in *Arabidopsis* for the final step of ABA biosynthesis catalyzed by abscisic aldehyde oxidase (AAO3). This enzyme requires a molybdenum cofactor, which is lacking in the *aba3* mutant.

There are two different mechanisms by which plants degrade ABA. First, ABA can be oxidized at the 8′-position to give 8′-hydroxy-ABA. This compound is unstable and rearranges to phaseic acid (PA), which in turn is converted to dihydrophaseic acid (DPA). This compound accumulates to very high levels in seeds of some plants, such as bean (*Phaseolus vulgaris*). Second, ABA can be conjugated to glucose to give the glucose ester of ABA (ABA-GE) (**Figure 3**). In most plants, the oxidative pathway is the predominant one. The catabolites can also be conjugated to glucose: PA to PA-GE, whereas DPA is mainly conjugated at the 4′ position to the DPA-glucoside. Other minor catabolites of ABA that have been identified in some plants are the 1′,4′-*cis*- and *trans*-diols, 7′-hydroxy-ABA, and ABA-glucoside (**Figure 3**).

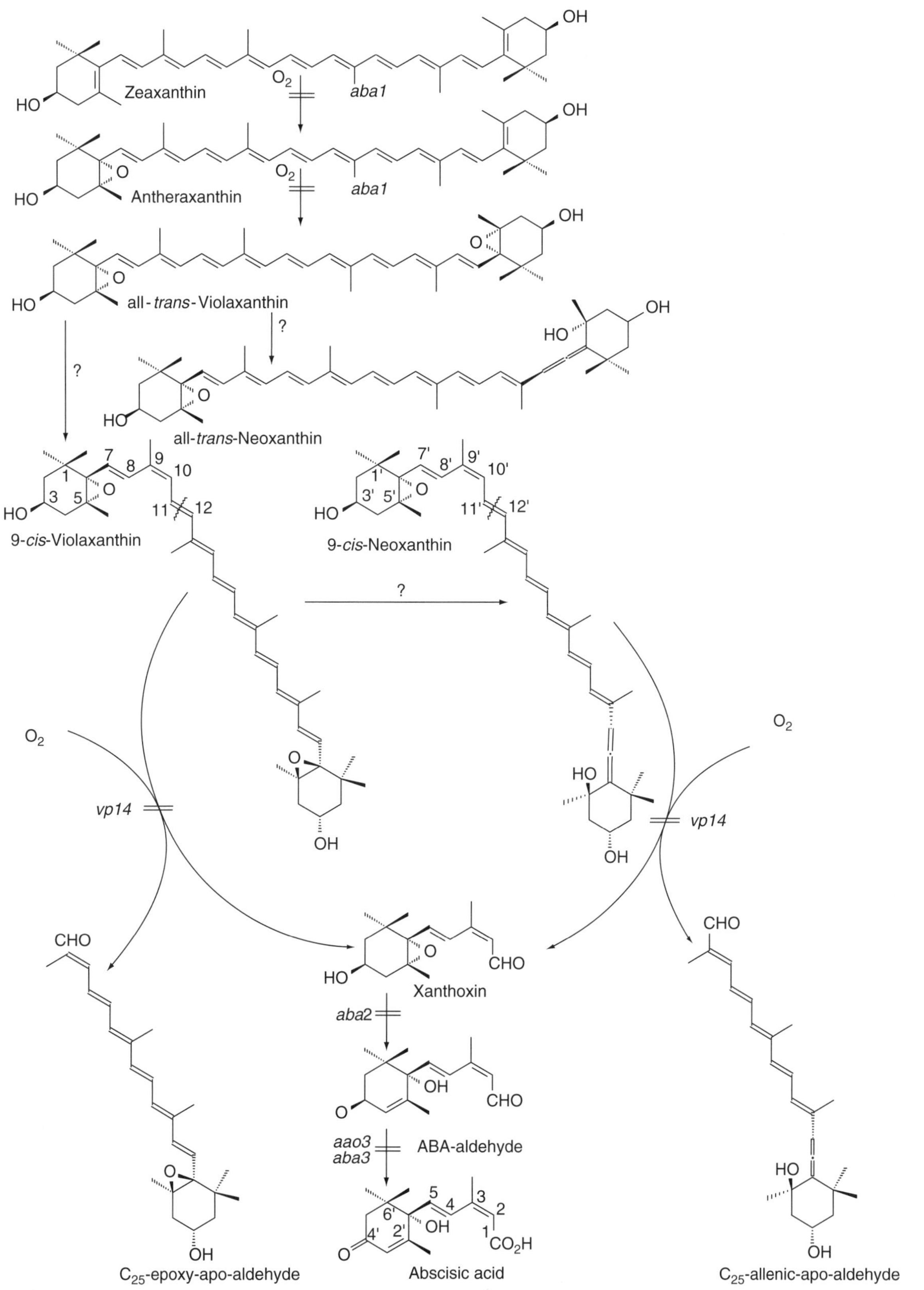

Figure 2 Biosynthetic pathway of ABA in higher plants. The metabolic blocks in the *aba1*, *aba2*, *aba3*, and *aao3* mutants of *Arabidopsis* and of *vp14* in maize (*Zea mays*) are indicated. Reproduced from Zeevart JAD (1999) Abscisic acid metabolism and its regulation. In: Hooykaas PJJ, Hall MA, and Libbenga KR (eds) *Biochemistry and Molecular Biology of Plant Hormones*, with permission of Elsevier.

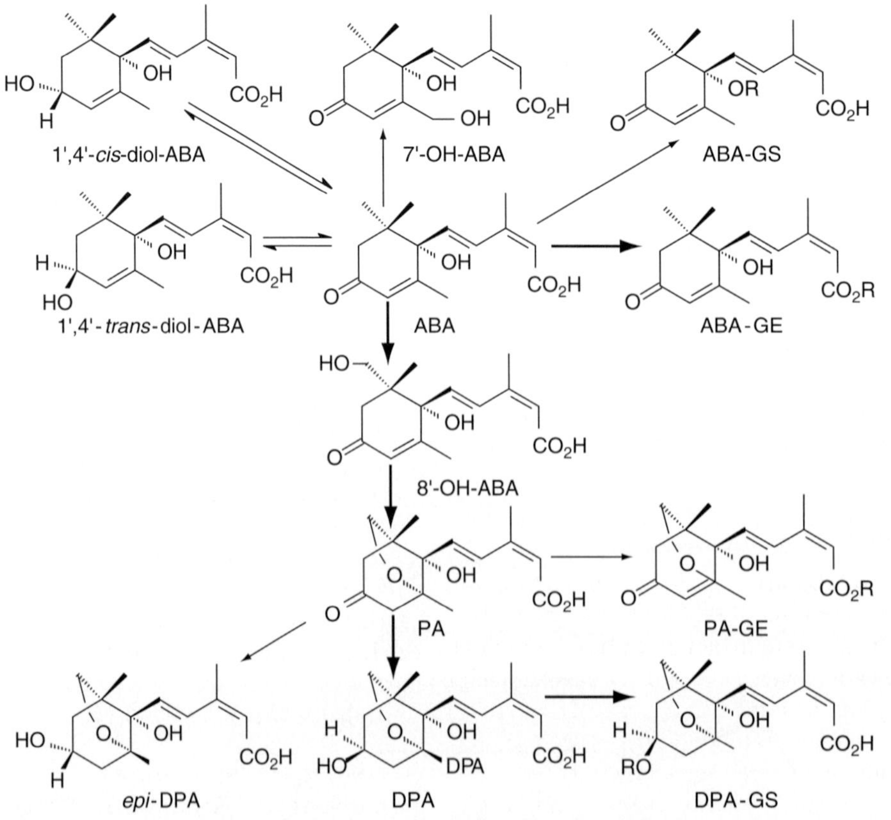

Figure 3 Catabolism of ABA in plants. The major pathways of oxidation and conjugation are indicated in bold arrows. R, glucose. Reproduced from Zeevart JAD (1999) Abscisic acid metabolism and its regulation. In: Hooykaas PJJ, Hall MA, and Libbenga KR (eds) *Biochemistry and Molecular Biology of Plant Hormones*, with permission of Elsevier.

Physiological Roles of ABA

Mutants

The roles of ABA in plant growth and development have been elucidated by various methods. Initially, ABA was applied to plants or plant parts to study its effect. In most instances, the effects were inhibitory, counteracting the effects of auxin or gibberellin. Much more informative has been the work with ABA-deficient or ABA-insensitive mutants. Mutants that are impaired in ABA biosynthesis are known in tomato (*Lycopersicon esculentum*) (*flacca*, *sitiens*, and *notabilis*) and in *Arabidopsis* (*aba1*, *aba2*, *aba3*, and *aao3*; see **Figure 2**). These mutants accumulate only a fraction of the ABA present in the corresponding wild-type plants. All ABA-deficient mutants have in common that their seeds are nondormant, the plants wilt readily, and developing seeds sprout without becoming dormant. All these expressions of ABA deficiency can be reversed by treatment with ABA. ABA-deficient mutants also show reduced growth. This has been ascribed to lack of turgor because of excessive transpiration. However, more recent work has demonstrated that this growth inhibition is at least in part due to increased ethylene production, which is suppressed by ABA in wild-type plants.

Water Relations

ABA plays a major role in ameliorating stress responses. When plants are water-stressed, ABA levels increase, which leads to rapid closure of stomata, the tiny pores in leaves through which gas exchange takes place (carbon dioxide in, water vapor out). Whereas growth of shoots ceases quickly during water stress, root growth continues during mild stress. Studies with mutants indicate that ABA is required for sustained root growth during water stress. Thus, acquisition of water is enhanced by sustained root growth, and water is conserved by closure of stomata and cessation of shoot growth. ABA also induces expression of genes encoding proteins (dehydrins) that protect macromolecules, membranes, and organelles from the harmful effects of dehydration. These proteins cross-protect against a number of stresses. For example, proteins induced by water stress will also protect against cold stress.

Embryo and Seed Development

Following fertilization, the embryo develops into a seed. About halfway through embryo development, the ABA content reaches its maximum and then gradually declines. At the same time, reserves accumulate either in the endosperm or in the cotyledons of the developing seed. ABA induces the synthesis of so-called late embryogenesis abundance (LEA) proteins as the seed becomes gradually desiccated. These proteins protect cellular structures and enzymes as the seed becomes desiccated and attains a dormant state. In mutants deficient in ABA, the embryo continues to develop into a seedling without entering the dormant state, a phenomenon known as vivipary.

Seed Germination

Seeds of many species are initially dormant and will not germinate when imbibed in water. ABA and gibberellin (GA) play important roles in maintaining dormancy or promoting germination, respectively. Seed dormancy has been studied extensively in *Arabidopsis* owing to the availability of hormone mutants. GA-deficient mutants do not germinate unless GA is added to the medium. On the other hand, ABA-deficient mutants germinate without an initial period of dormancy. The double mutant, deficient in both ABA and GA, germinates without addition of GA. This indicates that the main role of GA is to counteract the effects of ABA and that GA is not required *per se* for seed germination in *Arabidopsis*.

Antagonistic effects between GA and ABA also play a role in the germination of cereal grains. These seeds consist of an embryo and starchy endosperm surrounded by three or four layers of aleurone cells. Upon hydration, the embryo starts to produce GA, which induces the synthesis and/or secretion of digestive enzymes, such as α-amylase and protease, in the aleurone cells. This effect of GA can be counteracted by ABA. ABA also induces its own set of hydrophilic proteins, which protect enzymes and cellular structures. Thus, in case of water stress, ABA not only causes cessation of mobilization of reserves used for initial seedling growth, but it also produces proteins that protect the young seedling from dehydration. Perception of GA and ABA for induction and inhibition of α-amylase, respectively, is at the plasma membrane, whereas the receptor for the ABA stress response appears to be localized intracellularly.

Signaling

Although no specific ABA receptors have been identified, much has been learned about the intermediate steps between ABA perception and physiological responses. Among the rapid responses, guard cell responses are the best characterized. ABA first causes an increase in cytosolic Ca^{2+}, which leads to a depolarization of the plasmalemma and efflux of K^+ and anions, mostly malate. This causes water to move out of the guard cells, resulting in closure of the stomatal pore. The stomatal response can only be induced by *S*-(+)-ABA. Other responses are slower and can be induced by both (+)- and (−)-ABA. Many of these responses involve ABA-induced gene expression during stress or seed development. *Cis* elements have been identified for many ABA-induced genes. Some of the genes encode transcription factors that have their highest expression during the later stage of seed development. In two ABA-insensitive mutants, *abi1* and *abi2*, both seed development and various aspects of vegetative growth are affected. The wild-type genes of these mutants encode phosphatases, indicating that phosphorylation and dephosphorylation represent important steps in ABA signaling. However, the proteins targeted by these phosphatases are not yet known.

See also: **Seed Dormancy**: Development of Dormancy; Embryogenesis; Germination. **Water Relations of Plants**: Drought Stress; Stomata. **Regulators of Growth**: Ethylene; Gibberellins.

Further Reading

Addicott FT (ed.) (1983) *Abscisic Acid*. New York: Praeger.

Cutler AJ and Krochko JE (1999) Formation and breakdown of ABA. *Trends in Plant Science* 12: 472–478.

Davies WJ and Jones HG (eds) (1991) *Abscisic Acid: Physiology and Biochemistry*. Oxford: BIOS Scientific.

Finkelstein RR (2002) Abscisic acid. In: Taiz L and Zeiger E (eds) *Plant Physiology*, 3rd edn, pp. 539–558. Sunderland: Sinauer Associates.

Finkelstein RR, Gampala SSL, and Rock CD (2002) Abscisic acid signaling in seeds and seedlings. *Plant Cell* 14: S15–S45.

Rock CD (2000) Pathways to abscisic-regulated gene expression. *New Phytologist* 148: 357–396.

Schwartz SH, Qin X, and Zeevaart JAD (2003) Elucidation of the indirect pathway of abscisic acid biosynthesis by mutants, genes, and enzymes. *Plant Physiology* 131: 1591–1610.

Zeevaart JAD (1999) Abscisic acid metabolism and its regulation. In: Hooykaas PJJ, Hall MA, and Libbenga KR (eds) *Biochemistry and Molecular Biology of Plant Hormones*, pp. 189–207. Amsterdam: Elsevier.

Zeevaart JAD and Creelman RA (1988) Metabolism and physiology of abscisic acid. *Annual Review of Plant Physiology and Plant Molecular Biology* 39: 439–473.

Cytokinins

P E Jameson, Massey University, Palmerston North, New Zealand

Introduction

Cytokinins were originally discovered following the search for a compound that could promote cell division. It had been recognized in the 1930s that while addition of auxin to growth media enabled prolonged culture of plant tissue, cell division did not occur. Coconut (*Cocos nucifera*) milk provided a cell division-inducing factor as did autoclaved herring sperm DNA. Kinetin was isolated from the autoclaved herring sperm DNA and identified in 1955. Subsequently, kinetin was shown to be an artifact of the autoclaving process. However, when Skoog and Miller carried out tissue culture experiments on fully defined media containing kinetin and indoleacetic acid (IAA), they found that undifferentiated tobacco (*Nicotiana tabacum*) callus tissues were stimulated to form shoots or roots, depending on the relative levels of cytokinin and auxin. A high relative ratio of cytokinin:auxin led to shoot formation, whereas the opposite led to root formation. A more intermediate ratio led to undifferentiated growth (**Figure 1**A). It was suggested that the cytokinins promoted shoot organogenesis, inhibited root organogenesis, and promoted cell division.

In 1963, Stuart Letham extracted and identified the compound known as zeatin from corn (*Zea mays*; maize) kernels. This was the first naturally occurring cytokinin to be identified, and it has taken almost 40 years since then to obtain confirmation that cytokinins are indeed synthesized by the plant itself. Since Letham's discovery, the origin of cytokinins in plants has variously been suggested to arise from the breakdown of tRNA (which has cytokinin moieties attached to it), from surface dwelling microbes (many of which can secrete cytokinins into culture), or *in planta* from precursors via a cytokinin synthase enzyme (the activity of which could be detected in plant extracts, although the enzyme itself remained elusive).

In the intervening years, a wealth of information has been gained on cytokinin metabolites and metabolism, but little at the genome level. However, in the last few years, several highly significant breakthroughs have occurred in cytokinin research. In 1999, a number of genes coding for metabolic enzymes were isolated, and in 2001, not only was a family of genes coding for the key cytokinin biosynthetic enzyme isolated, but also the gene coding for a cytokinin receptor was identified.

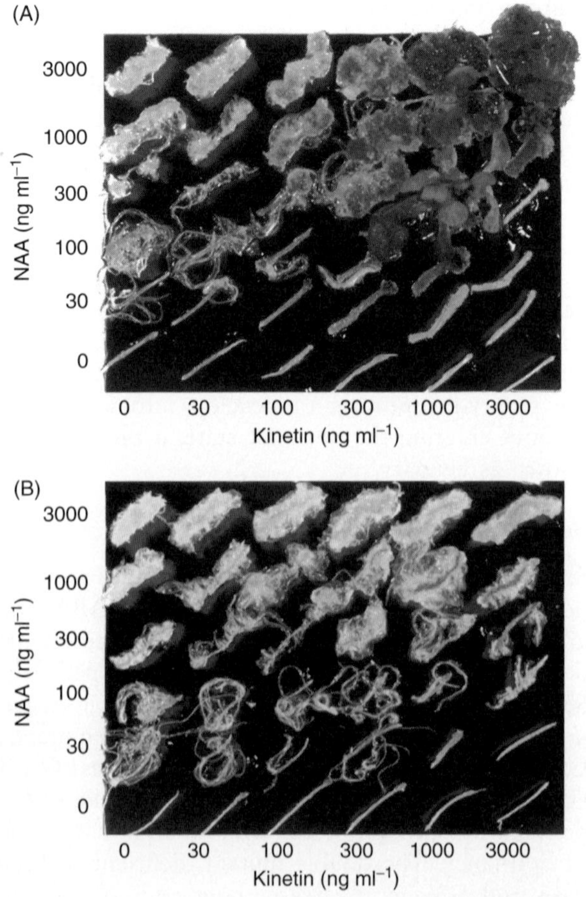

Figure 1 (see color plate 28) Callus growth, and shoot and root organogenesis of wild-type *Arabidopsis* and the cytokinin-resistant mutant *cre1-1* in media containing different cytokinin and auxin concentrations. Hypocotyl segments were excised and cultured on media containing different levels of kinetin and NAA. Wild-type explants (A) proliferated rapidly, turned green and produced shoots in the presence of high concentrations of cytokinins; the *cre1-1* explants (B) did not. Reproduced with permission from *Nature* from Inoue T, Higuchi M, Hashimoto Y, *et al.* (2001) Identification of *CRE1* as a cytokinin receptor from *Arabidopsis*. *Nature* 409: 1060–1063: © 2001 Macmillan Publishers Ltd.

It is now apparent that plants have the ability to biosynthesize their own cytokinins via an isopentenyl transferase (IPT) – known commonly as cytokinin synthase – and that cytokinin molecules are detected by a receptor of similar type to the ethylene receptor.

Cytokinin Structure

A cytokinin was originally defined as a compound that, in the presence of auxin, induced cell division in a suitable assay material grown on a defined medium. There now appear to be three categories of cytokinins: the isoprenoid type, exemplified by zeatin, which is the most abundant (**Figure 2**); the

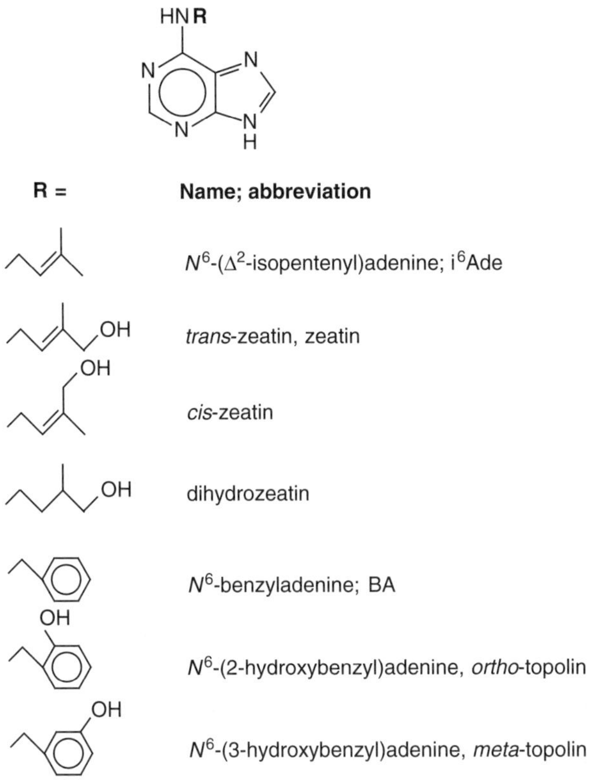

Figure 2 Structures of natural adenine and synthetic phenylurea cytokinins. Reproduced with permission from Mok DWS and Mok MC (2001) Cytokinin metabolism and action. *Annual Review of Plant Physiology and Plant Molecular Biology* 52: 89–118: © 2001 by Annual Reviews.

naturally occurring adenine derivatives with aromatic substituents; and the synthetic diphenylurea type (**Figure 2**).

Almost all of the cytokinins identified so far exist as the free base (e.g., zeatin), as the riboside (e.g., zeatin riboside), and as the nucleotide (zeatin nucleotide). In plant tissues, these forms appear to be readily interconvertible, so it has been difficult to determine which forms are biologically active.

Biosynthesis

Following the extraction of the so-called "free" cytokinins by methanol or ethanol solutions, alkaline hydrolysis of the plant tissue releases cytokinin molecules attached to particular tRNAs. There has been some debate over several decades as to whether the "free" cytokinins originate by *de novo* synthesis or are released following the breakdown of tRNA.

tRNA as a Source of Endogenous Cytokinins

The cytokinins released from tRNA were often found to be structurally different from those found as "free" forms, and, significantly, only the *cis*-isomer of zeatin, and not the *trans*-isomer, was released from tRNA. The *cis*-isomer possesses significantly lower activity in bioassays, but can be present in plants at relatively high levels. A *cis-trans* isomerase also exists, so *cis*-zeatin could be converted to the active *trans*-zeatin. Recently, a novel gene was cloned, the enzyme of which was shown to specifically glucosylate *cis*-zeatin, indicating that there are possibly *trans*- and *cis*-specific genes involved in cytokinin homeostasis. However, the slow turnover rate of tRNA is still considered to be insufficient to account for the amount of cytokinin present in plants, and the breakdown of tRNA is neither regulated nor sufficiently localized to provide the spatial and temporal control required of cytokinin biosynthesis.

De novo Biosynthesis

The most significant recent advance in cytokinin research has been the confirmation of *de novo* biosynthesis of cytokinins by plants. For many years, it was believed that the key step in cytokinin biosynthesis in plants would be the same as that in bacteria, in which the 5-carbon side chain (the isopentenyl group) is attached to the N^6 of adenosine monophosphate (AMP) by a DMAPP (dimethylallypyrophosphate):AMP isopentenyl transferase (IPT) (**Figure 3**). While enzyme activity capable of carrying out this reaction was first detected in slime mold, and subsequently in *Agrobacterium tumefaciens*, it was never purified to homogeneity from plant extracts, nor were the plant genes cloned. However, several bacterial *ipt* genes were cloned during the late 1980s/early 1990s.

Utilizing the newly completed *Arabidopsis* genome and the sequences of *ipt* genes from several bacteria, two Japanese groups independently identified a total of nine potential isopentenyl transferase sequences, seven of which (AtIPT1 and AtIPT3–8) were subsequently shown to code for DMAPP:AMP isopentenyl transferase activity in transformed *Escherichia coli*. Overexpression of *AtIPT4* in *Arabidopsis* induced cytokinin responses in callus culture in the absence of added cytokinins. Furthermore, it was shown that the preferred substrates for purified AtIPT4 were ATP and ADP, and not AMP (the preferred substrate for the *A. tumefaciens* IPT enzyme) (**Figure 3**). The "hypothetical side chain" shown in **Figure 3** could arise if a hydroxylated derivative of DMAPP provided the side chain.

Sites of Biosynthesis

While cytokinins have been detected in all parts of the plants analyzed, traditionally the main sites of biosynthesis were considered to be in the roots.

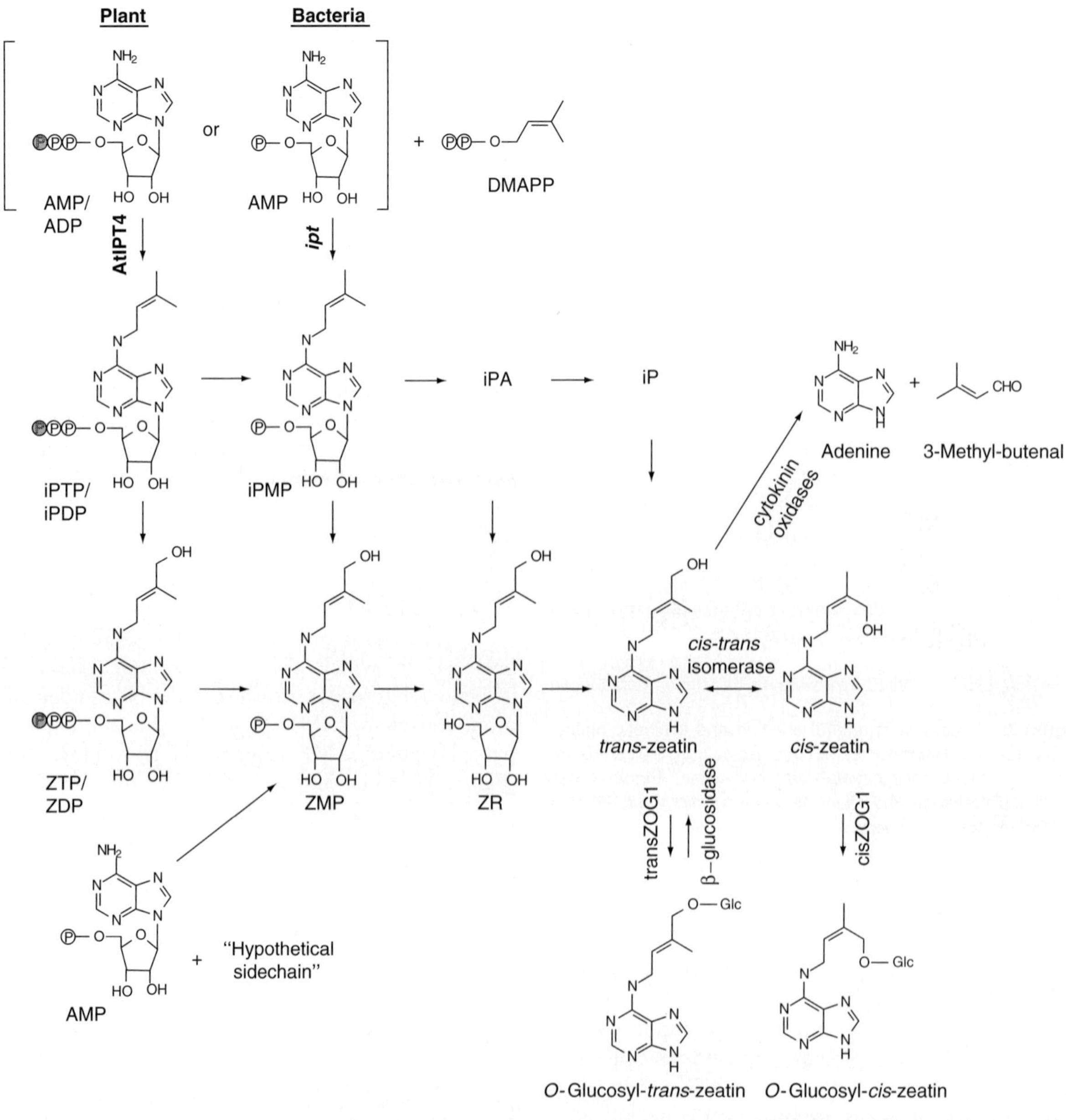

Figure 3 Proposed biosynthetic and metabolic pathway for cytokinins. Left, the proposed biosynthesis of zeatin tri-/diphosphate in *Arabidopsis*. Both ADP and ATP are likely substrates for the plant IPT enzyme, and these and their di- and triphosphate derivatives are indicated together (e.g., ATP/ADP). The biosynthesis of cytokinins in bacteria (e.g., *A. tumefaciens*) is compared next to it. Right, several possible modifications and the degradation of zeatin. Reproduced with permission from Haberer G and Kieber J (2002) Cytokinins: new insights into a classic phytohormone. *Plant Physiology* 128: 354–362: © 2002 The American Society of Plant Biologists.

Developing seeds, cambial tissue, and the shoot apex were also thought to be sites of synthesis, and possibly the leaves. Now, with a family of *ipt* genes available, detailed expression studies with reporter genes will be able to determine precisely the sites of synthesis. Preliminary reports indicate that *ipt promoter-GUS* constructs express in root tips, developing seeds, and at the base of leaves, lateral buds, and siliques of *Arabidopsis*, with some members of the gene family showing expression only in seeds. Thus, it would appear that the results from classical biochemical and physiological studies are supported by those at the level of gene expression.

Perception and Signal Transduction

In order to cause a biological response, the active form of the cytokinin molecule must be detected by a receptor protein. Following such detection (referred to as perception), a signal transduction pathway is activated, and target genes will be up- or down-regulated downstream from this. **Figure 4** depicts a model of the proposed cytokinin signal transduction pathway in *Arabidopsis*.

The identification of one of the cytokinin receptors, shown to be encoded by the *Arabidopsis CRE1* (cytokinin response 1) gene, relied on finding mutants that did not exhibit classic responses to cytokinins. Hypocotyl segments were excised and cultured on media containing different levels of kinetin and NAA (naphthalene acetic acid). Wild-type explants proliferated rapidly, turning green and producing shoots in response to high concentrations of cytokinin (and auxin) (**Figure 1**A), while explants from the *cre1-1* mutant did not (**Figure 1**B). Root growth of the *cre1-1* mutants was also resistant to cytokinin, and roots continued to elongate at concentrations of cytokinin that inhibited root growth of wild-type *Arabidopsis*. The *CRE1* gene was shown to encode a histidine kinase.

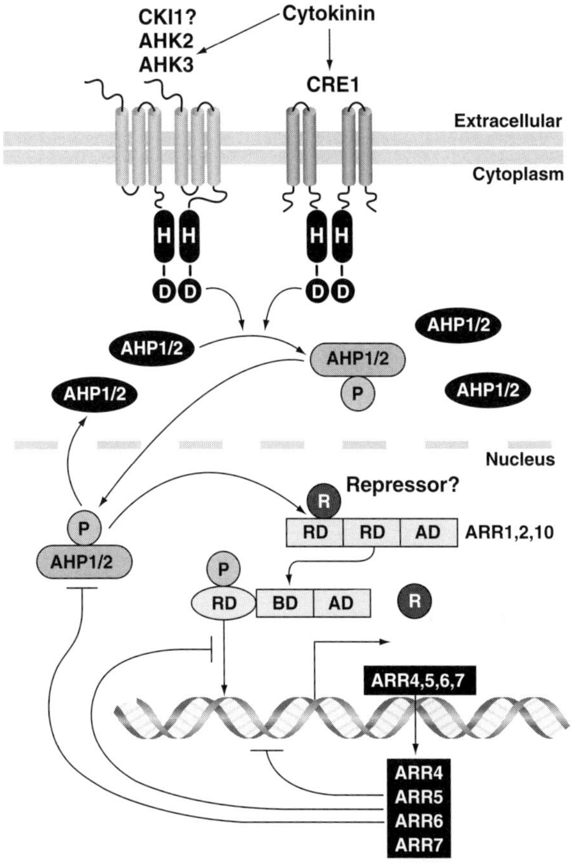

Figure 4 Model of cytokinin perception and signal transduction pathway in *Arabidopsis*. Cytokinin signal is perceived by multiple histidine protein kinases at the plasma membrane. After perception of the cytokinin signal, these histidine protein kinases initiate a signaling cascade through the phosphorelay that results in the nuclear translocation of AHPs from the cytosol. Activated AHPs interact with sequestered ARRs or ARR complexes in the nucleus, transfer the phosphate to the receiver domain of their cognate B-type ARRs, and, in turn, release the transcription activator ARRs from putative repressors in the nucleus. The dephosphorylated AHP shuttles back to the cytosol, where it can be rephosphorylated. The liberated ARRs bind to multiple *cis* elements in the promoters of target genes. The activation of the transcription repressor *ARR*s as cytokinin primary response genes provides a negative feedback mechanism. RD, response domain; BD, DNA binding domain; AD, transcription activation domain; R, putative repressor. Reproduced with permission from Sheen J (2002) Phosphorelay and transcription control in cytokinin signal transduction. *Science* 296: 1650–1651: © 2002 by American Association for the Advancement of Science.

When the *CRE1* gene was expressed in a yeast strain deficient in the *SLN1* gene (which encodes an osmosensing histidine kinase), the signal transduction pathway was activated by *trans*-zeatin, isopentenyladenine (iP, i^6Ade), benzyladenine (BA), and thidiazuron (TDZ), all of which are cytokinins that have been shown to be physiologically active in various bioassays. Significantly, the *CRE1* receptor did not respond to *cis*-zeatin, or to IAA, abscisic acid, or gibberellic acid, confirming its action as a specific cytokinin receptor molecule, and also confirming that the *cis*-zeatin isomer is unlikely to be biologically active *per se*.

Perception of the cytokinin molecule is only the first step in the signal transduction pathway, which culminates in gene activation. In the current model, histidine phosphotransfer proteins (AHPs) transmit the signal from the histidine protein kinase receptors to nuclear response regulators (ARRs), which can activate or repress transcription (**Figure 4**). In *cre1-1* mutants, two particular ARRs were downregulated, suggesting that they are specific components of the cytokinin signal transduction pathway. Indeed, when ARR2 was overexpressed in *Arabidopsis*, this was in itself sufficient to promote shoot formation *in vitro* in the absence of cytokinin and to delay leaf senescence, thus confirming the role of nuclear response regulators in the signal transduction pathway.

Metabolism

When applied to plant tissues, cytokinins are rapidly metabolized. Different plants, and indeed even different tissues within organs of a plant, have been shown to follow particular metabolic patterns. However, ultimately the cytokinins are either broken down by side chain cleavage, or are conjugated into storage or inactive forms (**Figure 5**).

Side Chain Cleavage and Cytokinin Oxidase

Side chain cleavage results in irreversible loss of activity, and is the consequence of the activity of

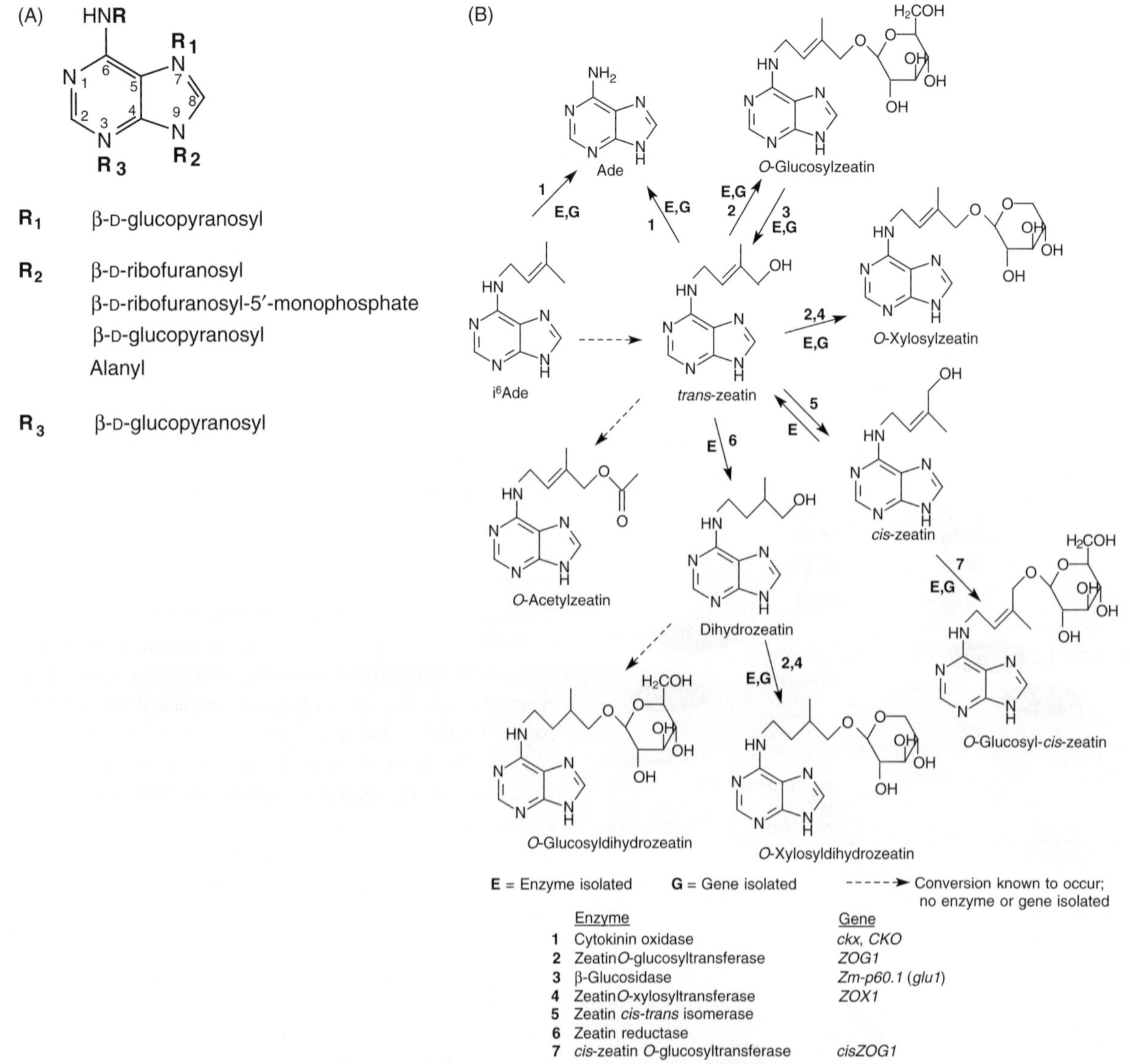

Figure 5 (A) Modifications of the adenine ring. (B) Enzymes (E) and genes (G) involved in the modification of the N^6-isoprenoid side chain. Reproduced with permission from Mok DWS and Mok MC (2001) Cytokinin metabolism and action. *Annual Review of Plant Physiology and Plant Molecular Biology* 52: 89–118: © 2001 by Annual Reviews.

cytokinin oxidase. Cytokinin oxidase is particularly active against cytokinin molecules such as zeatin and isopentenyladenine and their ribosides (molecules with a double bond in their side chain), but not against the dihydroderivatives (no such double bond), or cytokinins with an O-glucosyl moiety on the side chain. A particular gene family appears responsible for cytokinin oxidase production, and it is becoming apparent that cytokinin oxidase expression and activity is high in specific regions where endogenous cytokinin content is also elevated, and/or cytokinin synthase is expressed. Transgenic plants with elevated cytokinin content also exhibit an increase in cytokinin oxidase activity.

So far, there appear to be at least three genes coding for functional cytokinin oxidase enzymes in *Arabidopsis*. As more experiments are carried out using the promoters from these particular cytokinin oxidase genes linked to reporter genes such as GUS, we should begin to see the subtle effects exerted by cytokinin oxidase in regulating cytokinin levels. Such regulation is likely to be found in specific sites of the plant, such as the site of root initiation, where the presence of too much cytokinin might inhibit growth. The consequences of overexpression of a cytokinin oxidase gene on growth and development are significant and are described below.

Conjugation

Conjugation of the side chain and of the purine molecule occurs, leading to deactivation or inactivation of the cytokinin (**Figure 5**). Conjugation of the side chain is predominantly through the addition of a glycosyl grouping (glucose or xylose). Genes responsible for addition of *O*-glycosyl moieties to the side chain have been isolated. Such conjugation tends to deactivate the molecule, but metabolic studies indicate that an *O*-glucosyl moiety can be removed by β-glucosidase activity, thus releasing active cytokinin. Glucosylation of cytokinin has been shown to occur at positions 3, 7, and 9 of the purine moiety, and alanine has also been shown at position 9 in lupins (*Lupinus* sp.) and other members of the Fabaceae. Such conjugates are not considered biologically active *per se*.

Glycosylation can also occur on the ribose moiety. Indeed, a family of novel cytokinin metabolites was recently identified in *Pinus radiata*, in which the ribose of ZR, DZR, and iPA was not only glycosylated, but phosphorylated forms of these conjugates also exist. It will be interesting to determine whether such novel cytokinins are perceived by cytokinin receptor molecules.

Analysis of Cytokinins

Careful characterization of endogenous cytokinins has always been problematic because of the low levels and similar chemical structures of many of the different forms. Traditionally, cytokinins were characterized by their ability to cause cell division in a callus bioassay in the presence of auxin. Several different bioassays were used to determine cytokinin levels, usually based on the promotion of callus growth, the ability of cytokinins to delay senescence in excised leaf sections, or the induction of biosynthesis of betacyanin in *Amaranthus* leaf disks in the dark. Today, sophisticated mass spectrometry methods are employed to identify and quantify cytokinin levels, but ultimately the only way of assessing whether a newly identified cytokinin is biologically active is through bioassay.

While mass spectrometric methods provide identification, and quantification if stable isotopes are used as internal standards, many laboratories employ high performance liquid chromatography (HPLC) procedures to separate individual cytokinins followed by immunoassay (either radioimmunoassay (RIA) or enzyme linked immunosorbent assay (ELISA)), especially when assaying multiple samples. Antibodies are available that cross-react with cytokinins with a hydroxylated side chain (Z-type) or with a non-hydroxylated side chain (iP-type). *O*-Glucosides must be treated with β-glucosidase before detection is possible, while it is common to treat the nucleotides with alkaline phosphatase, followed by HPLC prior to immunoassay. Cytokinins with aromatic side chains will not be detected with the antibodies used to detect Z-type and iP-type cytokinins, and novel antibodies have been raised for detection of these compounds.

However, analyses of cytokinin are often poorly performed with the various forms not separated and the immunoassays frequently misused. In contradiction to original claims, immunoassays are not suitable for assay of crude samples, because of nonspecific cross-reactivity with the antibodies. Even now they are used to detect pooled mixtures of zeatin-like (i.e., Z, ZR, DZ, DZR) and isopentenyl adenine-like compounds (iP, iPA), with the researchers omitting to take into account differential cross-reactivity. Frequently, researchers have omitted to preserve the nucleotides during extraction (breakdown of which will add to the pool of ribosides), or to isolate the glucosides (*O*-glucosides will not be detected by the common antibodies used, which cross-react to the side chain). Arguably, the most significant misinterpretations of data appear in the seaweed extract industry, where gross overestimates of cytokinin content have been obtained from misinterpreted bioassay data, and from gas chromatographic data where other compounds co-chromatographing with cytokinins enhanced the peak area.

Cytokinins and Plant Development

Experiments based on exogenously applied cytokinin and/or correlation of endogenous levels with a developmental event have led to the cytokinins being implicated in many aspects of plant growth and development. This review will focus on the promotion of cell division, shoot organogenesis, the release of apical dominance, the inhibition of root growth, and the delay of senescence, which are features of particular horticultural/agronomic significance.

In the early 1960s, callus bioassays were used to show that cell division-inducing compounds were elevated in developing fruit and seeds. Subsequently, tight positive correlations between cytokinins, such as zeatin, and cell division have been shown in cereals (particularly wheat (*Triticum* spp.), corn (*Zea mays*; maize) and rice (*Oryza sativa*)) and fruit (such as plum (*Prunus domestica*), apple (*Malus pumila*), and kiwifruit (*Actinidia deliciosa*)), again leading to the suggestion of a key role for cytokinins in cell division. Recently, application of cytokinin has been shown to upregulate key components of the cell cycle, thus implicating a direct role for cytokinin in cell division.

Application of cytokinins to axillary buds of pea (*Pisum sativum*) has been known for many years to release the inhibited buds from apical dominance (imposed either by the apical bud itself or by application of IAA to a decapitated shoot). Correlations between cytokinin levels and outgrowth of axillary buds have been shown for various branching and nonbranching mutants, with the nonbranching mutants usually showing lower levels of cytokinins. However, work with the *ramosus* mutant of pea would suggest that root-exported cytokinin is not the only signal regulating release of axillary branches in intact pea.

Numerous experiments have implicated cytokinins in the delay of senescence, for example, cytokinins applied to senescing leaves led to a regreening in some instances, the immersion of petioles of cut leaves in solutions of kinetin delayed senescence, and subjecting plants to various forms of stress that lead to leaf senescence decreased the cytokinin levels in xylem sap (see below). In addition, the cytokinin biosynthetic capacity of tobacco leaves was found to be reduced at the onset of senescence, and endogenous cytokinins showed a decline in active forms at the onset and during senescence, although some senescing leaves have also been found to contain unusually high levels of cytokinin O-glucosides.

Many of the classical studies described above have been supported by observations of transgenic plants overexpressing a bacterial cytokinin synthase (*ipt*) gene. Such plants frequently showed the release of axillary buds from apical dominance and delayed senescence of leaves. However, high endogenous cytokinin content also led to inhibition of root growth (**Figure 6**), and so rooted transgenic plants could not be obtained. The initial experiments with transgenic plants used constitutive promoters so that cytokinin production was not controlled in either a temporal- or spatial (i.e., organ or tissue)-specific manner. Heat shock and copper-inducible promoters provided some temporal control (and again confirmed that plants with elevated cytokinin levels exhibit reduced apical dominance and delayed senescence), but more valuable developmental information is being obtained from controlled overexpression of the *ipt* gene from the use of fruit- and seed-specific promoters, and from senescence-inducible promoters. For example, a chimeric *vicilin-ipt* gene was introduced into tobacco, and fertile plants were regenerated. The vicilin promoter specifies embryo-specific expression in tobacco. Transgenic plants were morphologically normal and cytokinin levels were specifically enhanced only in developing seeds. Here, cell division was enhanced and an increase in the dry weight of the *vicilin-ipt* seeds was recorded. Such data lend support to the early correlative studies linking cytokinins with cell division in developing seeds and fruit.

Figure 6 (see color plate 29) Tobacco transformed with the bacterial *ipt* gene. Note release of axillary buds and inhibition of root growth in lines overexpressing the *ipt* gene. Reproduced with permission from McKenzie MJ, Mett V, Reynolds PHS, and Jameson PE (1998) Controlled cytokinin production in transgenic tobacco using a copper-inducible promoter. *Plant Physiology* 116: 969–977: © The American Society of Plant Biologists.

Tight temporal control of senescence was obtained when the promoter of a senescence associated gene (*SAG12*) was linked to the *ipt* gene. Expression of SAG genes increases at the onset of senescence. When the chimeric *SAG12-ipt* gene was introduced into tobacco, as the leaves began to senesce the SAG12 promoter was induced, activating the *ipt* gene and leading to the delay of senescence.

The transgenic experiments outlined above all utilize the enhanced production of cytokinin. To confirm the role of a plant hormone in a certain developmental response, experiments with plants with decreased levels of a particular hormone provide valuable contrasting information to that obtained with plants overexpressing a particular hormone. Unfortunately, unlike gibberellin and abscisic acid biosynthetic mutants, true cytokinin deficient mutants have not been found, probably because they would be lethal. However, with the recent isolation of genes coding for cytokinin oxidase, the enzyme responsible for side chain removal and deactivation of cytokinins, plants with reduced endogenous cytokinin content have been produced. The results confirmed that the cytokinins have key functions in meristem activity, with opposing roles in shoots and roots. The transgenic plants overexpressing cytokinin oxidase showed retarded shoot development, but enhanced growth of the root system, as a consequence of alterations in cell proliferation in the apical meristem.

Particularly useful hormone mutants are those defective in their perception of a hormone. Such mutants have been especially useful in ethylene research, and have very recently been utilized in cytokinin research. The *cre1* mutants are relatively insensitive to exogenous cytokinins, so that more cytokinin is required to promote callus proliferation, shoot formation, and to inhibit root growth than is required by the wild-type, which is further support for a key role for cytokinin in these processes.

Cytokinins as Root-to-Shoot Signaling Molecules

Classical definitions of plant hormones required proof of translocation from a site of synthesis to a site of action. While such a requirement is no longer normally included in the definition, there has been some debate as to whether cytokinins do function as long-distance signals from root to shoot, or are paracrine signals, acting at the site where they are produced.

A number of early correlative studies showed that various forms of mineral or water stress led to senescence and also a decreased level of cytokinin in the xylem sap. Such data were interpreted as showing a causal link between root produced cytokinins and the onset of senescence in the leaves. A substantial amount of work has been carried out using radiolabeled cytokinins applied to derooted plants or supplied via a wick into the xylem. The studies showed that cytokinins do move rapidly in the xylem and spread to most parts of the plant, with the majority accumulating in the expanding leaves at the end of the transpiration stream. In addition, lateral movement from xylem to phloem has been shown in lupin. However, conflicting results have been reported in transgenic plants. In one instance of controlled overexpression of cytokinin in the roots (by tetracycline derepression of the *ipt* gene), additional cytokinin was not translocated to the shoots (no release of apically inhibited axillary buds or delay of leaf senescence). In another instance of controlled overexpression of cytokinin in the roots (*ipt* gene controlled by a copper-inducible promoter), both release of apical dominance and delay of leaf senescence were recorded. Moreover, recent work with decapitated chick peas supports the idea of cytokinins as root-to-shoot signaling molecules by showing that xylem sap cytokinins represented one source of active cytokinins appearing in lateral buds after shoot decapitation.

Cytokinins as a Signal of Mineral Nutrient Deficiency

Recent work supports a role for cytokinin as a root-to-shoot signal communicating nitrogen availability. Marked changes in the cytokinin profile of roots occurred during the recovery of corn roots from nitrogen starvation, followed by an increase in the concentration of cytokinin in the xylem sap and in the leaves (**Figure 7**), supporting the suggestion that the cytokinins act as root-to-shoot signaling molecules.

Rootstock Effects May be Mediated by Cytokinin

Rootstocks are widely used to improve yield in many tree crops. The mechanism(s) by which this occurs has variously been suggested to involve nutrients, water movement, the graft union itself, and changes in hormone concentration. While all the classical plant hormones have been implicated, attention continues to be focused on the cytokinins. Recent work with apple rootstocks supported earlier work with sweet cherry (*Prunus avium*).

The cytokinin content of the apple shoot sap differed with the type of rootstock used, with higher shoot sap cytokinin levels associated with the more invigorating rootstocks, and lower levels associated

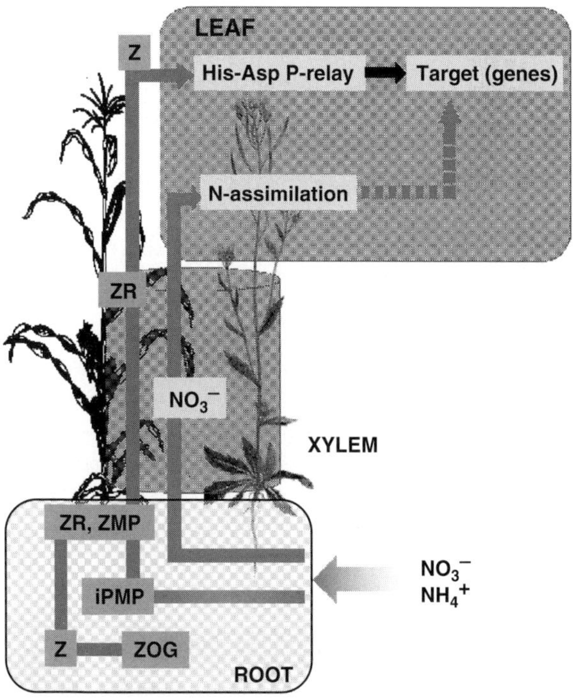

Figure 7 Current scheme of multiple routes communicating nitrogen availability of root to the shoots. Reproduced with permission from Takei K, Takahashi T, Sugiyama T, Yamaya T and Sakakibara H (2002) Multiple routes communicating nitrogen availability from roots to shoots: a signal transduction pathway mediated by cytokinin. *Journal of Experimental Botany* 53: 971–977.

with the more dwarfing rootstocks. Consequently, cytokinins may, at least in part, be a contributing factor in the response of shoots to certain rootstocks.

Cytokinin Production by Microbes

Because so many of the plant associated microbes produce cytokinin molecules in culture, and because (until 2001) cytokinin metabolic genes had been identified but not biosynthetic ones, there were suggestions that the "free" cytokinins found in plants were produced exclusively by microbial symbionts of plants. However, with the identification of the cytokinin synthase gene family in *Arabidopsis* and cytokinin synthases in *Petunia* and broccoli (*Brassica oleracea* var. *italica*), such sources of cytokinin no longer need to be invoked. However, many of the disease symptoms caused by phytopathogenic microbes appear to reflect the action of plant hormones, such as the "green islands" associated with some bacterial, fungal, and viral infections, as well as the tumors or gall formations, and the fasciation diseases, associated with bacterial infection.

Gall Forming Bacteria

The best characterized cytokinin producing plant pathogen is *Agrobacterium tumefaciens*. Crown galls formed by *A. tumefaciens* contain high levels of both cytokinins and auxins. In the late 1950s, it was shown that gall tissue freed of the bacterium could be maintained *in vitro* on hormone-free media. This observation led eventually to the profound discovery that a portion of bacterial DNA (the T-DNA portion of the T_i plasmid) is transferred and actually incorporated into the plant genome, whereupon the transferred auxin genes and the *ipt* gene are activated. If the auxin genes in the T-DNA are inactivated, a "shooty gall" results, whereas if the cytokinin biosynthesis gene is inactivated "rooty galls" result (**Figure 8**). Such observations confirmed the original suggestion by Skoog's laboratory that the relative levels of cytokinin and auxin control organogenesis. Certain *A. tumefaciens* strains have an additional cytokinin synthase gene (*tzs*, also located on the T_i plasmid), and it is this gene, or the *ipt* gene, that has been used in transformation experiments to overexpress cytokinin synthase activity *in planta*.

The strategies used by other gall forming bacteria differ from *A. tumefaciens*. Both *Pseudomonas syringae* pv *savastanoi* (which causes olive and oleander knot) and *Erwinia herbicola* pv *gypsophilae* (which causes galls on gypsophila) release copious quantities of cytokinin into the culture medium. Both have plasmids carrying cytokinin synthase genes (and auxin biosynthetic genes, but not necessarily on the same plasmid). However, the presence of the bacterium is required if the gall is to be maintained on hormone-free medium, and no gene transfer is invoked.

The strategy used by *Rhodococcus fascians* to induce shooty gall formation and fasciations remains unsolved. A cytokinin synthase gene has been detected and the plasmid on which it is located is necessary for pathogenicity. In contrast to the other gall-forming microbes, infection by *R. fascians* does not lead to elevated levels of cytokinins in diseased tissue, and, indeed, may lead to decreased levels. However, elevated auxin levels have been reported. Thus, we have shooty galls with the opposite ratio of auxin to cytokinin than might be expected from other studies.

Phytopathogenic Fungi

Considering the vast range of phytopathogenic fungi, relatively little work has been conducted on these

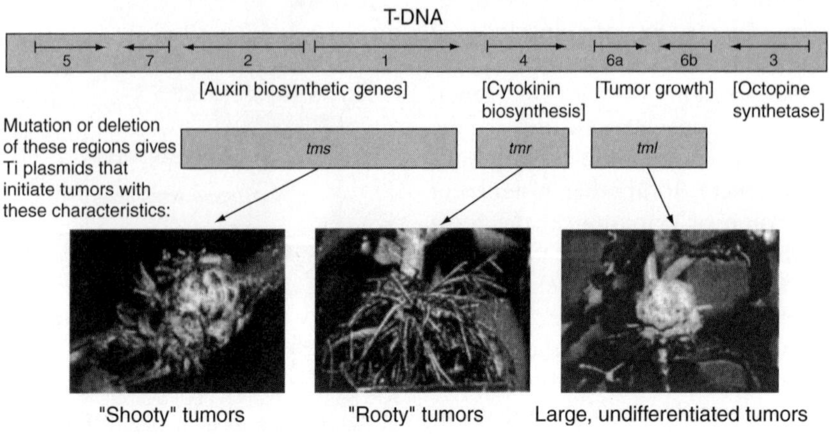

Figure 8 Gall formation induced by *A. tumefaciens*. Note the classic gall formation is modified by rooty galls when the *ipt* gene is deleted or to shooty galls when the auxin biosynthetic genes are deleted. Figure courtesy of Professor Roy Morris.

pathogens. Elevated levels of *trans*-zeatin may be involved in clubroot formation by *Plasmodiophora brassicae*, and in green island formation by biotrophic and hemibiotrophic fungi.

Phytopathogenic Viruses

Three independent research groups have shown that the metabolism of cytokinins is significantly altered in three different host–virus combinations. Infection led to a decrease of active forms and an increase in conjugated forms of cytokinin. Cytokinin application inhibited virus replication, and prevented the virus-induced decrease in free radical scavenging enzymes. These data have led to the suggestion that the decrease in cytokinin is necessary to provide an environment of free radicals for viral replication.

Horticultural and Agronomic Uses of Cytokinins

Tissue Culture

Since organogenesis was shown to be controlled in defined medium by manipulating the relative levels of cytokinin and auxin, the predominant commercial use of cytokinins has been in tissue culture, where a relatively higher ratio of cytokinin to auxin is used to promote the multiplication of shoots. However, transfer of shoots to cytokinin-free medium is usually required at the stage of root initiation, to avoid the inhibitory effect of cytokinin on root initiation and elongation.

Kinetin and 6-BAP are the most frequently used cytokinins in commercial tissue culture laboratories. While kinetin has never been shown to occur naturally, 6-BAP is a naturally occurring cytokinin belonging to the group of cytokinins with an aromatic side chain – the group that also includes *ortho*- and *meta*-topolin (**Figure 1**). Thidiazuron, one of the synthetic diphenylurea cytokinins (**Figure 1**), has also been used, often for the more recalcitrant woody species.

Promotion of Fruit Growth by Diphenylurea-Type Cytokinins

Application of diphenylurea (DPU) compounds such as CPPU (N-phenyl-N′-(2-chloro-4-pyridyl)urea) causes quite extraordinary increases in fruit size of kiwifruit, apples, grapes (*Vitis* spp.), etc. To obtain size increases without shape distortion, the compounds must be applied at a quite specific time in fruit development. The size increase may be due to enhanced cell division and/or cell expansion.

There is a debate concerning the mode of action of the DPU-type compounds, and whether they can be classed as cytokinins. It has been shown that DPU-type cytokinins strongly inhibit cytokinin oxidase *in vitro*, and it was postulated that their activity may be due to a cytokinin preserving effect. However, kiwifruit treated with CPPU showed decreased levels of cytokinin, particularly zeatin, suggesting a feedback inhibition on biosynthesis in the presence of CPPU. As mentioned above, it was recently shown that thidiazuron binds to the cytokinin receptor, CRE1, and can trigger the downstream signal cascade, providing evidence that the diphenylurea compounds function as active cytokinin molecules.

Seaweeds and Seaweed Extracts

Throughout the world, extracts containing seaweeds, and frequently other components, are claimed to have growth enhancing effects, with the cytokinin (often along with other ingredients) claimed as the growth promoting compound. While proof of efficacy in the field is frequently lacking, there is no doubt that seaweeds (especially the rapidly growing brown kelps) contain a full complement of cytokinins, as do some of the extracts. However, the methodology used to assess cytokinin-like activity in seaweed extracts has frequently led to overestimates of the true cytokinin content.

Transgenic Plants/Genetic Engineering

Attempts are being made to harness the potentially useful effects of cytokinin in promoting cell division, in shoot organogenesis, and in delaying senescence by genetically engineering crop plants with the cytokinin synthase gene. Up until now such work has used the *ipt* genes from *Agrobacterium*. Targeting the gene is essential if the detrimental side-effects of overexpression are to be avoided. Seed-specific promoters linked to the *ipt* gene are being used to promote cell division in the seed (i.e., to increase sink capacity), while several groups are focusing on the SAG12 promoter to increase cytokinin levels, and to delay senescence in leaves (i.e., to increase source longevity), or in harvested tissues such as broccoli heads to increase postharvest shelf-life. Overexpression of cytokinin is also being investigated as a means to promote shoot formation in recalcitrant species in tissue culture. With the isolation of plant *IPT* genes, the promoters of these genes may enable highly refined targeting of particular plant *IPT* genes.

Concluding Statement

In an article published in the 1980s, the cytokinin was described as "the plant hormone without a role." This statement can comfortably be laid to rest as genomics based research refines, but overtly supports, the roles that were suggested for the cytokinins several decades ago.

List of Technical Nomenclature

Abscisic acid	A plant hormone that controls the way plants grow and develop.
Apical dominance	The influence exerted by a terminal bud in suppressing the growth of lateral (auxillary) buds.
Auxin	A plant hormone that controls the way plants grow and develop.
Axillary buds/ lateral buds	Buds that occur in the axils of leaves.
Betacyanin	A plant pigment.
Bioassay	A "biological assay": utilizes the response of a plant tissue or organ to detect a biologically active molecule such as a plant hormone.
Callus	Undifferentiated cells that form on cut surfaces of dicotyledonous plants.
Cambial tissue	A layer of actively dividing cells.
Cell cycle	The different stages a cell passes through from one division to the next.
Chimeric gene	A gene constructed of several components derived from different genes (e.g., promoter from one gene and a coding region from another).
***cis*-isomer**	On the side chain of the cytokinin molecule, the groups either side of the double bond lie on the same side.
Conjugation	In biochemistry this refers to the addition of a molecule (e.g., a sugar) to another molecule (e.g., a cytokinin) which may then change its properties.
Constitutive promoter	A promoter that causes a gene to be expressed continuously.
Cytokinin	A plant hormone that controls the way plants grow and develop.
Embryo	The young plant in a seed before seed germination.
Endogenous	Occurring within.
Ethylene	A plant hormone that controls the way plants grow and develop.
Exogenous	From outside.
Explant	Often refers to a small section of plant placed in tissue culture.
Expression studies	Such studies monitor the activity (expression) of genes by measuring mRNA levels.
Gas chromatography	A method of separating components in a mixture, but may not have the power to separate all components.
Gene	A sequence of DNA that codes for (usually) a protein.
Genetic engineering	The manipulation of genetic material for practical purposes; involves the splicing of new genetic material into a genome.
Gibberellin	A plant hormone that controls the way plants grow and develop.
GUS	A reporter gene; if the gene is expressed and the tissue is supplied with the appropriate substrate, the tissue will turn blue.
Histidine kinase	A receptor molecule.
HPLC	A method of separating components in a mixture, but may not have the power to separate all components.
Hypocotyl	Part of seedling stem below the cotyledons.
IAA	A naturally occurring auxin.
Immunoassay	An assay based on antibody–antigen recognition; in the case of cytokinins, antibodies that recognize certain cytokinin molecules (the antigen).
Inducible promoter	A promoter that is turned on by some external cue such as light or copper (e.g., a copper-inducible promoter).
Kinetin	A synthetic cytokinin.
Mass spectrometry	A technique that enables a molecule to be identified.
Meristem	Growing point at tip of shoot (shoot apex) or root (root apex).
Organogenesis	The formation of organs (e.g., shoots or roots) from undifferentiated cells.
Petiole	Leaf stalk.
Phloem	The conducting tissue involved in conducting sugars and other compounds.
Phytopathogenic	A pathogen (virus, bacterium, or fungus) that infects plants.
Promoter	The portion of a gene that determines whether the gene is expressed or not.
Receptor	A molecule whose function is to detect and promulgate a signal.
Reporter genes	Such genes produce a product that can be visualized (e.g., a blue color or fluorescence); they are used to observe where and when the promoter of a particular gene is active.

SAG gene	A senescence associated gene whose expression either increases or decreases at the onset of senescence.
Senescence	The controlled process that leads to the deterioration and eventual death of a tissue, organ, or whole plant.
Shoot or root apex	Growing point at tip of shoot (shoot apex) or root (root apex).
Signal transduction	The process by which a cell converts a signal into a response. Signal transduction starts with the receptor responding to a specific signal (e.g., cytokinin), followed by pathway transduction of that signal along a chain of molecules and ending with (in the case of cytokinins) a change in gene expression.
Siliques	Seed pods such as those found on *Arabidopsis*.
Transgenic plant	A plant whose genome contains new DNA from a plant or another organism.
trans-isomer	On the side chain of the cytokinin molecule, the groups either side of the double bond lie on opposite sides.
Transpiration stream	The flow of water through the xylem from the roots to the leaves.
tRNA	Transfer RNA molecules recognize amino acids and guide them to the ribosome for protein synthesis.
Xylem	The water and mineral conducting tissue.
Zeatin	A naturally occurring cytokinin.

See also: **Genetic Modification**: Transformation, General Principles. **Growth and Development**: Cell Division and Differentiation. **Postharvest Physiology**: Senescence, Flowers; Senescence, Leaves. **Production Systems and Agronomy**: Horticultural Crops, Protected. **Regulators of Growth**: Auxins. **Tissue Culture**: Organogenesis.

Further Reading

Haberer G and Kieber JJ (2002) Cytokinins. New insights into a classic phytohormone. *Plant Physiology* 128: 354–362.

Jameson PE (2000) Cytokinins and auxins in plant–pathogen interactions – An overview. *Plant Growth Regulation* 32: 369–380.

Jameson PE (1993) Plant hormones in the algae. In: Round FE and Chapman DJ (eds) *Progress in Phycological Research,* vol. 9. pp. 239–279. Bristol: Biopress Ltd.

Jameson PE, Zhang H, and Lewis DH (2000) Cytokinins: extraction, separation and analysis. *Methods in Molecular Biology* 141: 101–121.

Lohrmann J and Harter K (2002) Plant two-component signaling systems and the role of response regulators. *Plant Physiology* 128: 363–369.

Mok DWS and Mok MC (eds) (1994) *Cytokinins: Chemistry, Activity, and Function.* Boca Raton: CRC Press. New data has overtaken some of the information in this book, but for details of classical studies, particularly focusing on function, the reader is directed to Chapter 5: Cytokinins as phytohormones – sites of biosynthesis, translocation, and function of translocated cytokinin (DS Letham); Chapter 9: Cytokinin metabolism and compartmentation (PE Jameson); and Chapter 12: Cytokinins and plant development (DWS Mok and RC Martin).

Mok DWS and Mok MC (2001) Cytokinin metabolism and action. *Annual Review of Plant Physiology and Plant Molecular Biology* 52: 89–118.

Sakakibara H and Takei K (2002) Identification of cytokinin biosynthesis genes in *Arabidopsis*: a breakthrough for understanding the metabolic pathway and the regulation in higher plants. *Journal of Plant Growth Regulation* 21: 17–23.

Sheen J (2002) Phosphorelay and transcription control in cytokinin signal transduction. *Science* 296: 1650–1652.

Takei K, Takahashi T, Sugiyama T, Yamaya T, and Sakakibara H (2002) Multiple routes communicating nitrogen availability from roots to shoots: a signal transduction pathway mediated by cytokinin. *Journal of Experimental Botany* 53: 971–977.

Gibberellins

P Hedden, Rothamsted Research, Harpenden, UK

Definition and Discovery

Gibberellin, commonly abbreviated to GA, is a member of a group of naturally occurring tetracyclic diterpenoid carboxylic acids, most of which possess the *ent*-gibberellane (C_{20}) or *ent*-20-norgibberellane (C_{19}) carbon skeletons. Currently 126 different GAs have been identified from higher plants, fungi, or bacteria. Some GAs possess biological activity in higher plants and a limited number of these act as endogenous regulators of plant growth and development, mediating developmental and environmental signals. GAs with this phytohormonal activity possess a C_{19} skeleton, with a C19–C10 lactone, a 7-carboxylic acid group and, for stimulation of vegetative growth, a 3β-hydroxy function, as exemplified by two of the most common endogenous GA growth regulators, GA_1 and GA_4 (**Figure 1**). Many GAs possess biological activity when applied to plants due to metabolism to the active forms by the plant. This has been shown clearly using dwarf mutants in which different steps in the biosynthetic pathway to GAs are blocked. Only biosynthetic intermediates beyond the block, which therefore can be converted to the active form, show activity when applied to such mutants.

Gibberellins were discovered in Japan early in the twentieth century as the causative agent for the overgrowth symptoms in rice plants suffering from the "bakanae" (silly seedling) disease. The disease was shown to be due to infection with the phytopathogenic fungus *Gibberella fujikuroi*, from which an active principle was extracted that reproduced the bakanae symptoms when applied to rice plants. The term "gibberellin" was first used in 1935 by Teijiro Yabuta for this active substance, from which two crystalline solids, named gibberellin A and gibberellin B, were eventually obtained. Both gibberellin A and B were subsequently shown to be impure and three compounds, named gibberellin A_1, A_2, and A_3, were separated from gibberellin A. The composition of gibberellin B is unclear. After World War II the results of the Japanese research on gibberellins were available outside Japan and stimulated work on these compounds, particularly in the USA and UK. Groups at the US Department of Agriculture laboratories in Peoria, Illinois, and at the ICI Akers Laboratory in the UK independently isolated a pure, highly biologically active compound from cultures of *G. fujikuroi* and agreed on the name gibberellic acid. The structure of gibberellic acid, which proved to be identical with gibberellin A_3, was proposed by the British group in 1956 (**Figure 1**).

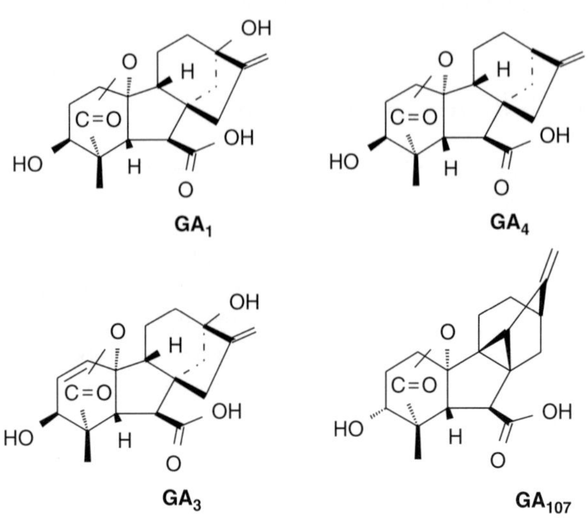

Figure 1 Examples of C_{19}–GA structures.

The ability of gibberellic acid to restore growth to dwarf mutants of pea (*Pisum sativum*) and maize (*Zea mays*), and to induce bolting in rosette plants, prompted the suggestion that GAs might be endogenous growth regulators in higher plants. This proposal was strengthened by the extraction from higher plants of substances with GA-like activity and confirmed in 1958 by the chemical identification of gibberellin A_1 from immature seeds of runner bean (*Phaseolus coccineus*). Subsequently, many new gibberellins were identified in higher plants as well as in *G. fujikuroi*, and the practice of naming them according to the original Japanese system of gibberellin A_{1-n} in chronological order of discovery was used in several laboratories. The use of the abbreviation GA_n for gibberellin A_n was introduced some time later and is now common practice.

Occurrence and Distribution

Gibberellins are thought to be ubiquitous in higher plants, but their occurrence in lower plants is uncertain and has not been systematically investigated. In shoots of higher plants, GAs are usually most abundant in growing tissues, where they are present at concentrations in the range 10^{-11}–10^{-8} g g^{-1} fresh weight, depending on the tissue, species, and nature of the GA. The most widespread active GA is GA_1, although its 13-deoxy analog, GA_4 is usually also present, but at lower concentrations. In some species, such as *Arabidopsis thaliana* and members of the Cucurbitaceae, GA_4 is the predominant form and indeed has more growth-stimulating activity than GA_1 in these species. Although at one time thought to be only a fungal product, GA_3 is now known to be produced by some plant species, albeit usually as a relatively minor component. The bioactive GAs are always accompanied by their biosynthetic precursors and catabolites (see below), which may be present at much higher concentrations than the active forms. Roots also produce GAs, which are normally present at lower concentrations than in shoots. Reproductive tissues generally have a higher GA content than vegetative organs, with particularly large amounts in anthers and pollen, and in developing seeds. Concentrations greater than 10^{-5} g g^{-1} fresh weight are found in the endosperm and/or immature cotyledons of some species. Developing seeds often contain more structurally diverse GAs than are found in other tissues. For example, apple (*Malus pumila*) seeds contain more than 20 different GAs, some of which are highly functionalized on rings C and D, including those containing a cyclopropane ring due to the formation of a C9–C15 bond, as in GA_{107} (**Figure 1**).

Gibberellins are produced by a number of fungal species, including *G. fujikuroi*, *Sphaceloma manihoticola*, *Phaeospheria* sp., and *Neurospora crassa*. The three first species excrete large amounts ($mg\,l^{-1}$) of GAs, the major GA produced in each case being GA_3, GA_4, and GA_1, respectively. Cultures of *G. fujikuroi* are used for the commercial production of GA_3 and also of mixtures of GA_4 and GA_7, which accumulate in certain strains. *Neurospora crassa* produces low ($10^{-9}\,g\,g^{-1}$ fresh mycelium) amounts of GA_3, which is not excreted. Gibberellins are not known to have a physiological function in fungi, but effects on the host may facilitate infection and improve the nutrient supply. Certain endophytic bacteria, principally *Rhizobium phaseoli*, *Azospirillum lipoferum*, and *A. brasilense*, also produce GAs, which may modify the growth of the host plant. Gibberellin-related compounds are synthesized in the gametophytes of certain fern species and induce the formation of antheridia (sperm-forming organs). These compounds, known as antheridiogens, are C_{19}-GAs modified by methylation of the carboxyl group and/or rearrangement of the C/D ring structure, and include 9,15-cycloGAs, such as those found in apple seeds.

Biosynthesis

Gibberellins are formed from *trans*-geranylgeranyl diphosphate (GGPP) via the hydrocarbon *ent*-kaurene, which is converted to GAs by a series of oxidative reactions. The pathway in vegetative tissues of higher plants is shown in **Figure 2**. Conversion of GGPP to *ent*-kaurene occurs in plastids and proceeds in two steps via *ent*-copalyldiphosphate, requiring the action of two enzymes, *ent*-copalyldiphosphate synthase (CPS) and *ent*-kaurene synthase (KS). Formation of the bioactive GAs from *ent*-kaurene occurs outside the plastid, the initial reactions being catalyzed by cytochrome P450-dependent monooxygenases on membranes, while the later steps require the activity of soluble 2-oxoglutarate-dependent dioxygenases. The cytochrome P450-mediated reactions include the conversion of *ent*-kaurene to *ent*-kaurenoic acid by *ent*-kaurene oxidase (KO), and the formation of GA_{12} in three steps from *ent*-kaurenoic acid, catalyzed by *ent*-kaurenoic acid oxidase (KAO). A third monooxygenase, GA 13-hydroxylase (GA13ox), converts GA_{12} to GA_{53}; these GAs are metabolized to GA_9 and GA_{20}, respectively, by the dioxygenase, GA 20-oxidase (GA20ox). The 20-oxidase reactions comprise at least three steps, involving conversion of C20 from a methyl group via the alcohol to an aldehyde, from which it is lost in the formation of the lactone function in the C_{19}–GA skeleton. The biological active products, GA_4 and GA_1, are formed from GA_9 and GA_{20}, respectively, by 3β-hydroxylation, which is catalyzed by another dioxygenase, GA 3-oxidase (GA3ox). Some GA 3-oxidases also produce GA_3 or its 13-deoxy analog GA_7, as byproducts in two-step reactions via 2,3-didehydro intermediates. Another group of dioxygenases, GA 2-oxidases (GA2ox), introduce a 2β-hydroxy group, resulting in the formation of inactive products. Thus, they catalyze a catabolic reaction, which is important in regulating the concentration of the active GA species. Gibberellins commonly form conjugates, particularly with glucose, and these may take the form of glucosyl ethers, in which glucose is coupled to a hydroxyl group, or glucosyl esters, in which the linkage is through the 7-carboxylic acid group. Glucosyl ether conjugates of 2β-hydroxy GAs, in which the glucose is coupled through the 2β-hydroxyl group, are particularly common and may serve to sequester these GA catabolites in the vacuole. Glucosyl ester formation is reversible and in some cases may function to produce a reserve from which GAs can be generated rapidly, for example, in seed germination.

Gibberellin biosynthesis in *G. fujikuroi* differs from that in higher plants in several respects: a single enzyme (CPS/KS) catalyzes both steps in the formation of *ent*-kaurene from GGPP, 3β-hydroxylation occurs early in the pathway as one of the reactions catalyzed by GA_{14} synthase, a highly multifunctional cytochrome P450 monoxygenase that converts *ent*-kaurenoic acid to GA_{14} (3β-hydroxy GA_{12}), and 13-hydroxylation is the final step in the pathway to GA_3. Furthermore, removal of C20 is accomplished by a cytochrome P450 rather than a dioxygenase. The genes for the GA-biosynthetic enzymes are clustered in *G. fujikuroi*, whereas they are dispersed throughout the genome of the higher plant *A. thaliana*.

Distribution and Regulation of GA Biosynthesis

Sites of GA biosynthesis in plants have been investigated by examining the tissue distribution of enzyme activity for specific steps, either in homogenates or intact tissues, and, more recently, by determining the pattern of expression of the genes that encode the biosynthetic enzymes. Highest levels of GA biosynthesis occur in growing tissues. Development control is particularly important for the first committed step of the pathway, catalyzed by CPS, which, in *A. thaliana*, is expressed in growing

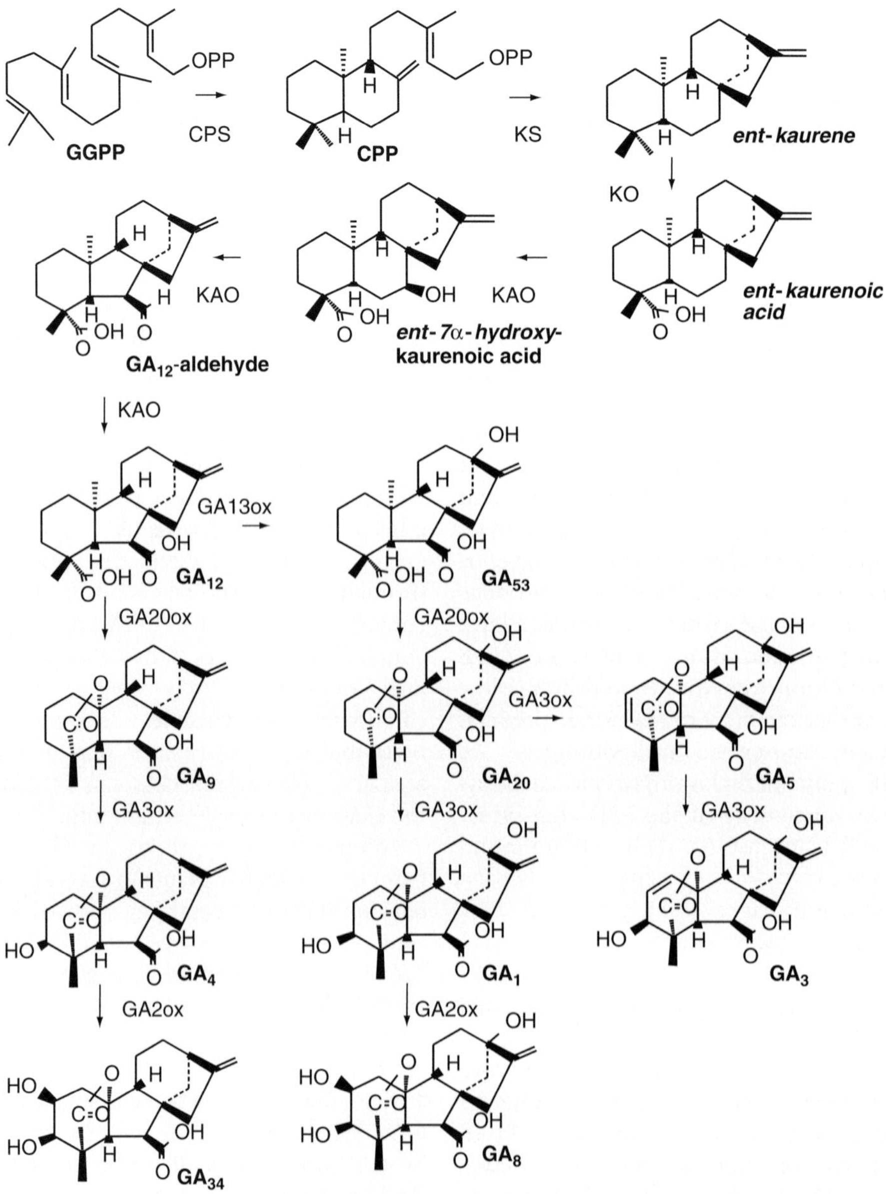

Figure 2 Generalized GA-biosynthetic pathway for vegetative tissues in higher plants, showing intermediates and enzymes.

tissues and vascular elements, consistent with *ent*-kaurene synthesis occurring in proplastids, rather than in mature chloroplasts. Enzymes responsible for early steps of the pathway are encoded by one or a small number of genes, whereas later enzymes, particularly the dioxygenases, are encoded by gene families. The GA-biosynthetic genes in *A. thaliana* are listed in **Table 1**. *GA20ox* and *GA3ox* gene family members show differential tissue-specific patterns of expression and may thus be involved in the regulation of different developmental processes, such as seed germination, leaf and stem expansion, pollen development, and fruit growth.

The dioxygenase genes are important sites of regulation by endogenous and environmental signals. In a homeostatic process, GAs regulate their own synthesis by suppressing the expression of *GA20ox* and *GA3ox* genes and enhancing *GA2ox* expression. In contrast, auxin stimulates GA biosynthesis in at least some circumstances by upregulating expression of *GA20ox* and/or *GA3ox* genes and suppressing *GA2ox* expression. In both these cases, the principal site of regulation, *GA20ox* or *GA3ox*, may differ depending on the species. Many responses of plants to changes in light quantity, quality, or duration (photoperiod) are mediated by GAs. Such changes may modify the concentration of the hormone through altered biosynthesis, or may change the amplitude of the response to GA. In long-day rosette plants, induction of stem elongation after exposure

Table 1 *Arabidopsis* genes involved in GA biosynthesis and mutants

Gene	*Number of genes in family*	*Mutant*
CPS	1	ga1
KS	1	ga2
KO	1	ga3
KAO	2	
GA20ox	5	ga5 (*GA20ox1*)
GA3ox	4	ga4 (*GA3ox1*)
GA2ox	$6^a + 2^b$	

[a] One gene contains a large insertion and is apparently not expressed, i.e., is a pseudogene.
[b] Two *AtGA2ox* genes are distantly related to the others and are specific for C_{20}-GAs.

to long days is due to both enhanced GA production, as a result of increased expression of *GA20ox*, and higher sensitivity to GA. Conversely, tuber formation in potato (*Solanum tuberosum*) in short days is the result of reduced biosynthesis of GAs, which inhibit tuberization. There is evidence for a diurnal fluctuation in *GA20ox* gene expression, increasing during the light period and decreasing during the dark, that would allow the GA content to reach higher levels in long days. The photoreceptor for regulation of GA biosynthesis by photoperiod is probably phytochrome, which is also involved in the induction of seed germination by red light. In this case, the light stimulates expression of *GA3*ox genes and thereby enables the production of the active GAs from their non-3β-hydroxylated precursors.

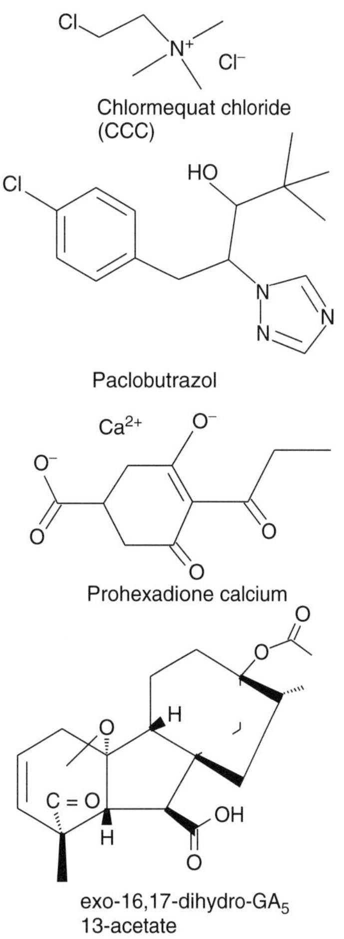

Figure 3 Structures of inhibitors of GA biosynthesis that act as growth retardants.

Chemical Regulation of GA Biosynthesis

Chemical inhibitors of GA biosynthesis are used extensively in agriculture and horticulture as growth retardants, primarily to control plant stature (see below). Three classes of inhibitors have been developed, differing in their site of action in the GA-biosynthetic pathway. Examples of each type are shown in **Figure 3**. The "onium" inhibitors, exemplified by chlormequat chloride and AMO-1618, contain a permanent positive charge and inhibit the conversion of GGPP to CPP by CPS. They are thought to act by mimicking a charged transition state in the cyclization. Their high water solubility allows for ease of application and efficient transport within the plants. However, they are relatively inefficient inhibitors, particularly chlormequat chloride, which is the most widely used retardant, mainly on cereals. Potential side effects of these compounds are inhibition of epoxy squalene cyclization and transmethylation reactions in steroid biosynthesis.

A large group of compounds with a nitrogen-containing heterocyclic ring target *ent*-kaurene oxidase. Prominent members of this group are the triazoles, paclobutrazol, and uniconazole. The heterocyclic compounds inhibit cytochrome P450 monooxygenases, with highest affinity for methyl hydroxylases. Thus, they are active only against *ent*-kaurene oxidase in the GA-biosynthetic pathway, but some of these inhibitors have activity also against sterol 14-demethylase and abscisic acid 8′-hydroxylase. They are highly efficient inhibitors that function at low concentrations, but have low water solubility and are transported poorly in the plant, primarily in the xylem.

The third group of retardants are the acylcyclohexanediones, such as prohexadione, which act on the dioxygenases of GA-biosynthesis, with highest activity against GA 3β-hydroxylases. They are also effective against 2-oxidases, but have relatively little activity against GA 20-oxidases. They act as mimics of 2-oxoglutarate and block binding of this cosubstrate to the enzyme active site. Although less efficient as inhibitors, they have higher water solubility than the nitrogen-containing heterocyclic compounds and are transported more readily throughout the plant. As side effects, they may inhibit other 2-oxoglutarate-dependent dioxygenases. For example, inhibition of flavanone 3β-hydroxylase can result in reduced anthocyanin production and therefore loss of color in flowers and fruit. A new group of retardants that is still being developed for commercial application comprises 16,17-dihydroGAs, such as exo-16,17-dihydroGA$_5$ 13-acetate. These compounds, which are thought to inhibit GA 3β-hydroxylation, are extremely potent growth retardants on members of the Poaceae, but are much less effective on dicotyledonous plants.

Mode of Action

Physiological Function

Knowledge of the physiological roles of GA in plants has come from observations of GA-deficient mutants and the effects of applying GAs to such mutants and to normal plants. It is clear from such experiments that GAs are important regulators of growth throughout the life cycle of the plant (**Figure 4**). Gibberellins are necessary for germination of seeds, in which they stimulate the growth of the embryo. Additionally, in cereal seeds GAs produced in the embryo and scutellum induce the synthesis of hydrolases (amylases, proteases, and nucleases) in the aleurone cell layers and their secretion into the endosperm, where they break down macromolecular reserves for utilization by the growing embryo. They also induce the release of minerals from the aleurone. In tomato (*Lycopersicon esculentum*) seeds, and possibly in seeds of other dicotyledonous plants, GA-induced hydrolases break down the endosperm, which would otherwise form a physical barrier to radicle emergence. Elongation growth in seedling tissues (hypocotyl, epicotyl, coleoptile, and mesocotyl), stem internodes, and roots is dependent on GA. Stimulation of internode growth by GA can be the result of both increased rates of cell division and enhanced longitudinal cell growth, although, in some species, such as rice (*Oryza sativa*), it is due only to increased cell expansion. It has been shown in several species that GA-induced cell elongation is partly the result of a change in cell growth polarity, from a random to longitudinal orientation. Thus, GAs induce transverse alignment of cortical microtubules, which mediate the transverse deposition of cell-wall microfibrils. In addition, GAs promote cell enlargement, probably by loosening the cell wall and changing the turgor pressure within the cell. Leaf expansion is dependent on GA, as is leaf shape. Thus, in some species, most notably tomato, treatment with GA causes a reduction in the degree of leaf dissection. Furthermore, the transition from juvenile to adult leaf shape is mediated by GA in a number of species. The formation and degree of branching of trichomes on the surface of leaves is also mediated by GAs.

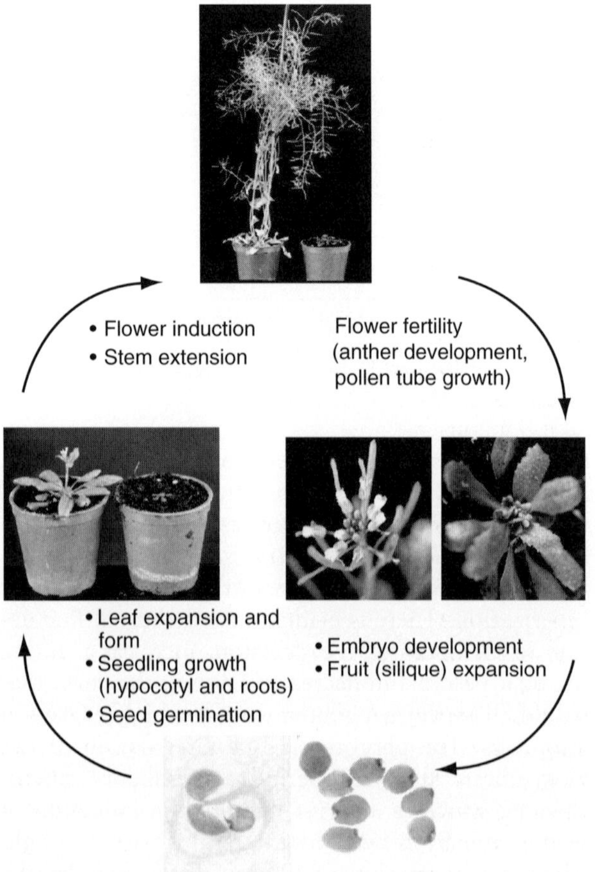

Figure 4 Developmental processes in which GA participates during the life cycle of *Arabidopsis*, illustrated by a comparison of Landsberg *erecta* (left) with the GA-deficient *ga3-1* mutant at different developmental stages. Photographs courtesy of Dr A. Phillips.

Application of GAs induces bolting in rosette plants, even in the absence of inductive conditions (long days and/or exposure to low temperature). In *Arabidopsis*, which is a facultative long-day plant, the stem extension associated with bolting has an absolute requirement for GA, whereas flower induction occurs in the absence of GAs in long days (although it is delayed), but is absolutely dependent on GAs in short days. There are thus at least two signaling pathways, one GA-dependent and the other GA-independent, to flower induction in *Arabidopsis*. The GA-independent pathway operates only in long days. In addition, exposure to long days enhances GA biosynthesis and signaling and, therefore, stimulates the GA-dependent pathway. An involvement for GAs in the induction of flowering by exposure to low temperatures (vernalization) has not been clearly established, although in at least one case, that of *Thlaspi arvense*, vernalization has been shown to enhance GA biosynthesis. Application of GAs induces cone formation in juvenile conifers that are not normally competent to do so. GA_3 is effective with members of the Cupressaceae and Taxodiaceae, but not with the Pinaceae, which require 13-deoxy GAs, such as GA_4.

Gibberellins are required for the normal development of flowers and, in particular, are necessary for anther development and pollination in many species. Thus, GA-deficient mutants are often male sterile. Such mutants may also have abnormal development of and lack of pigmentation in petals, the normal source of GAs for which are the anthers. Gibberellins affect sex expression in monoecious and dioecious species, promoting the formation of male flowers in *Cannabis sativa* and members of the Cucurbitaceae. However, GA-deficiency in maize often results in the formation of anthers in the ears. Gibberellins are necessary for normal fruit set and growth, and for seed development. In some species, such as pea, application of GAs to unpollinated ovaries or to seedless fruit will promote normal, but parthenocarpic fruit growth, while in other species auxins and cyokinins may also be required. Although seeds are often very rich in GAs, there is little evidence to suggest that seed-derived GAs stimulate fruit growth directly. Rather, seeds may provide other signals, such as auxin, that induce GA production within the fruit.

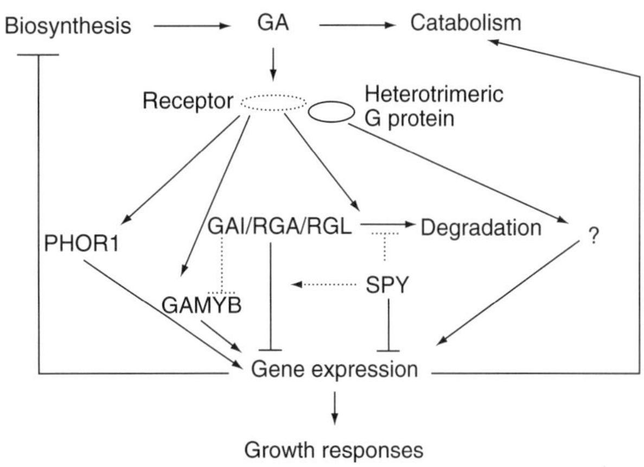

Figure 5 Gibberellin signaling pathways showing the identified components and their possible relationship. GA receptors, which are thought to be present on the external face of the plasma membrane, have not been identified.

Signal Transduction Pathways

A number of components of GA signal transduction pathways have been identified as a result of analyzing mutants with altered responsiveness to GAs, but, in most cases, it is still not clear how these components interact. A summary of the current understanding of GA signaling is presented in **Figure 5**. Although no receptors for GAs have been identified, there is evidence from work with cereal aleurones that GAs are perceived on the outer face of the cell membrane in this tissue. Further work with cereals indicates that GAs signal via heterotrimeric G proteins, both in the aleurone and vegetative shoot, but this is not the only GA signaling pathway and it may have minor or no significance in dicotyledonous plants. Several components of GA signaling are negative regulators that suppress GA responses in the absence of GAs, which function by inhibiting the action of these suppressors. *Arabidopsis* contains a family of five proteins with overlapping functions, comprising GAI, RGA, and three RGA-like (RGL) proteins, that belong to the GRAS protein superfamily. Cereals appear to have only a single example of these proteins: RHT in wheat (*Triticum aestivum*), DWARF8 in maize, SLN1 in barley (*Hordeum vulgare*), and SLR1 in rice. Overexpression of these proteins, even in an unrelated species, causes semidwarfism and, in rice, has been shown to inhibit α-amylase production. These proteins contain a highly conserved sequence near the N-terminus, known as the DELLA domain, which is absent from other GRAS proteins; deletion of this domain causes the protein to be no longer subject to inactivation by GA, producing a dwarf phenotype that cannot be normalized by GA application. Such gain-of-function deletion mutations have been found in several species, including wheat, in which *Rht* alleles were used in the production of high-yielding, lodging resistant, semi-dwarf varieties that enabled the "green revolution." In common with other GRAS proteins, GAI/RGA etc are localized in the nucleus, where they may function as dimers in regulating transcription. GA is thought to act by inducing their degradation by specific proteases in an ubiquitin-mediated process. The DELLA domain is the recognition sequence that is required for GA-induced degradation.

Another negative regulator of GA signaling is SPINDLY (SPY), of which there are two homologs in *Arabidopsis*. Loss-of-function mutations in SPY result in an overgrowth phenotype. The N-terminal part of SPY contains 10 tetratricopeptide repeats, which are involved in protein–protein interactions, while the C-terminus functions as an *N*-acetylglucosamine transferase. Transfer of *N*-acetylglucosamine (GlcNAc) onto serine or threonine residues by SPY is thought to compete with phosphorylation and thereby regulate the activity of the protein substrate. The substrates for SPY have not been identified, but may include proteins involved in GA signaling. However, the function of SPY is unlikely to be specific for this process. Further negative (SHI) and positive (PHOR1, SLY, and PKL) regulators of GA responses have been described, but their relationship with other components is not completely clear. PHOR1, which was isolated from potato, is localized in the cytosol in the absence of GA, which induces its transport into the nucleus, where it functions as a positive effector of GA responses.

The endpoint of GA signaling is a change in the level of expression of specific genes, the products of which bring about the developmental changes that are characteristic of GA responses. For example, induction of α-amylase gene expression in cereal aleurones results in the breakdown of starch in the endosperm, and induction of *LEAFY* (*LFY*) expression in the *Arabidopsis* shoot apex causes the transition from a vegetative to a floral apical meristem. Elements within the promoters of these genes confer GA responsiveness and have been shown to bind to a class of transcription factors, known as GAMYB. It has been demonstrated in barley aleurones that GA signaling enhances GAMYB expression and it may also bring about posttranslational activation of the protein. Details of how GAMYB expression is regulated by GA signaling are not yet known, but correlative evidence suggests the involvement of cGMP in this process and the GRAS protein SLN1 (in barley) is thought to act upstream of GAMYB by repressing its transcription and/or activation.

Another consequence of GA action in the cereal aleurone is an increase in cytosolic Ca^{2+} concentration as well as enhanced production of the Ca^{2+}-binding protein, calmodulin, and of an endoplasmic reticulum-localized Ca^{2+}-ATPase. A general role for Ca^{2+}/calmodulin in GA signaling has not been defined, but in the aleurone Ca^{2+} is required for the synthesis and secretion of α-amylase, which is a metalloenzyme.

Applications for Gibberellins and Inhibitors of Gibberellin Biosynthesis

Many of the physiological processes in which GAs participate, such as seed germination, internode elongation, flower development, and fruit set and expansion, are important in agriculture and horticulture. Consequently, controlling these processes by manipulating GA content has long been part of crop husbandry. Depending on the application, GA content may be increased by applying GAs, or decreased by application of chemical inhibitors of GA biosynthesis. In some cases, natural semidwarf varieties that have mutations affecting GA biosynthesis, such as *le* pea, which has a mutation in a GA 3β-hydroxylase gene, or GA signal transduction, such as *Rht* wheat, are in common use. Genetic modification of GA biosynthesis and signal transduction can also be used to introduce beneficial traits into crop species and could provide an alternative to chemical treatments.

Gibberellins in Commercial Use

Gibberellins are produced commercially by fermentation of *G. fujikuroi*, the most important and widely used product being GA_3 (gibberellic acid), although GA_4, usually in combination with GA_7, is produced for particular applications from some strains of the fungus. The biggest use of GA_3 worldwide is on seedless grape (*Vitis vinifera*) varieties in order to increase berry size and modify their shape. Applications at full bloom cause the berries to elongate, while an additional prebloom application of GA_3 increases the length of the rachis, producing a looser cluster that is less susceptible to fungal infection, and decreases pollen viability and ovule fertility, so reducing fruit numbers and increasing the weight of remaining fruit. Gibberellins are used on a number of top fruit species to improve fruit set, size, shape, and skin quality. Fruit set in citrus, particularly in mandarin oranges, which set few seeds, can be improved by application of GA_3. Later application of GA_3 to citrus fruit is used to delay ripening and to improve the appearance of the rind. Poor pollination in apple and pear (*Pyrus communis*) due to adverse weather conditions at flowering time can result in low yields. In these circumstances fruit can be induced to grow parthenocarpically by application of GA_3; in the case of apples, the GA needs to be used in combination with an auxin and cytokinin to obtain optimal fruit growth. Mixtures of GA_4/GA_7 and benzyladenine, a cytokinin, are applied routinely to apple trees just after flowering to modify fruit shape, by increasing fruit length relative to the diameter. These treatments also reduce the incidence of russeting, a browning of the skin that decreases the value of the fruit. GA_4 is the more active GA component and preparations containing GA_4 virtually free of GA_7 are commercially available.

A major use of GA_3 is in the production of malt for brewing and other purposes. Malt is obtained by the hydrolysis of polysaccharides and proteins in the barley endosperm during germination of the grain in the steeping process. Addition of GA to the steep to induce the production and secretion of α-amylase and other hydrolytic enzymes from the aleurone shortens the malting time by 2–3 days, increases yield, and produces a more uniform quality. Numerous minor uses for GAs include sugar production from sugar cane (*Saccharum officinarum*), seed production in rosette species, releasing dormancy in potato tubers and cone production in conifers. In this last application breeding programs can be shortened dramatically by inducing cone formation in juvenile trees.

Inhibitors of Gibberellin Biosynthesis

The main agronomic function of GA-biosynthesis inhibitors is to control stem growth. In terms of value, growth retardants are the most important group of plant growth regulators, with the use of chlormequat chloride (CCC) to control lodging in wheat being by far the largest application for such compounds. Although almost all commercial wheat varieties are semidwarf due to the presence of *Rht* genes, depending on soil and climatic conditions, it is often necessary to control stem growth further with CCC. As well as providing improved stem stability, the dwarf habit results in higher grain yields due to a greater proportion of assimilate being directed to the grain. Other cereals respond little to CCC so that combinations of CCC, or the related compound mepiquat chloride, with ethephon (which releases ethylene) are used to control height. Other types of inhibitors, such as prohexadione and the 16,17-dihydroGAs, are being developed for use on cereals and may become important in the future. Inabenfide, a pyridine-type *ent*-kaurene oxidase inhibitor, is used commercially to prevent lodging in rice. After CCC, mepiquat chloride is the most important retardant, with its major application on cotton to control excessive vegetative growth.

Orchard and amenity trees are commonly treated with growth retardants to restrict shoot growth. The modern high-density fruit orchards necessitate rigorous control of growth, while pruning is highly labor-intensive and, therefore, expensive. Reduced shoot growth also provides higher fruit yields due to the redirection of assimilates. A number of growth retardants have been used on trees. Daminozide, an inhibitor of GA 3β-hydroxylase, was commonly used on apple trees, but has been discontinued due to fears about its safety. Triazoles, such as paclobutrazol, are used on fruit trees in some countries, but the persistency of some of these compounds has restricted their use. Other targets for growth retardants are turf grasses and ornamental pot plants, on which a broad range of different inhibitors are used.

See also: **Growth and Development**: Cell Division and Differentiation; Cell Growth; Field Crops; Leaf Development; Molecular Biology of Development. **Flowering and Reproduction**: Flower Development. **Regulators of Growth**: Juvenility; Photomorphogenesis; Phytochromes and other Photoreceptors; Vernalization. **Root Development**: Root Growth and Development.

Further Reading

Hedden P and Kamiya Y (1997) Gibberellin biosynthesis: enzymes, genes and their regulation. *Annual Review of Plant Physiology and Plant Molecular Biology* 48: 431–460.

Hedden P and Phillips AL (2000) Gibberellin metabolism: new insights revealed by the genes. *Trends in Plant Science* 5: 523–530.

Hooley R (1994) Gibberellins: preception, transduction and responses. *Plant Molecular Biology* 26: 1529–1555.

Olszewski N, Sun T-P, and Gubler F (2002) Gibberellin signaling: biosynthesis, catabolism and response pathways. *The Plant Cell* 14: S61–80.

Phinney BO (1983) The history of gibberellins. In: Crozier A (ed.) *The Biochemistry and Physiology of Gibberellins*, pp. 19–52. New York: Praeger.

Rademacher W (2000) Growth retardants: effects on gibberellin biosynthesis and other metabolic pathways. *Annual Review of Plant Physiology and Plant Molecular Biology* 51: 501–531.

Richards DE, King KE, Ait-Ali T, and Harberd NP (2001) How gibberellin regulates plant growth and development: a molecular genetic analysis of gibberellin signalling. *Annual Review of Plant Physiology and Plant Molecular Biology* 52: 67–88.

Yamaguchi S and Kamiya Y (2000) Gibberellin biosynthesis: its regulation by endogenous and environmental signals. *Plant and Cell Physiology* 41: 251–257.

Ethylene

G E Schaller, University of New Hampshire, Durham, NH, USA

Introduction and History

Ethylene (C_2H_4) is a simple gaseous hydrocarbon that has profound effects upon plant growth and development. The phrase "One bad apple spoils the whole bunch" encapsulates a great deal of information about ethylene, the hidden protagonist in the events described: (1) that ethylene is biosynthesized by the apple; (2) that its biosynthesis is regulated; (3) that it can modify plant development; and (4) that its effect has economic repercussions. Although most popularly associated with ripening, ethylene plays a role throughout the entire life of the plant. Ethylene is a regulator of seed germination, seedling growth, leaf and petal abscission, organ senescence, stress responses, and pathogen responses. The complete biosynthetic pathway for ethylene in plants has been deduced and an understanding is starting to emerge as to mechanisms by which plants regulate ethylene production. Through mutational studies using the model plant *Arabidopsis*, components in the signal transduction pathway have also been identified, including ethylene receptors and transcription factors. The mechanisms for ethylene biosynthesis, perception, and signal transduction are conserved in plants. As a result, methods developed to regulate ethylene responses can be applied to many agriculturally important plant species.

Ethylene was one of the first plant hormones to be discovered, its discovery hastened by a form of air pollution peculiar to the nineteenth and early twentieth centuries. At that time, illuminating gas was used for lighting and gas lines ran throughout towns and cities. Leaks from the pipelines resulted in premature senescence and abscission in nearby vegetation, sometimes seriously damaging trees and greenhouse plants. Dimitry Neljubov identified ethylene as the "active" component in illuminating gas and published his results in 1901. In the following years, a variety of physiological effects of ethylene were cataloged and early on some of these were realized to be of commercial importance. For example, a patent was taken out on the use of ethylene to color citrus fruit in 1923. Although the utility of ethylene was recognized, it was not until the 1930s that plants were demonstrated to produce ethylene themselves, thereby establishing ethylene as an endogenous regulator of plant growth and development.

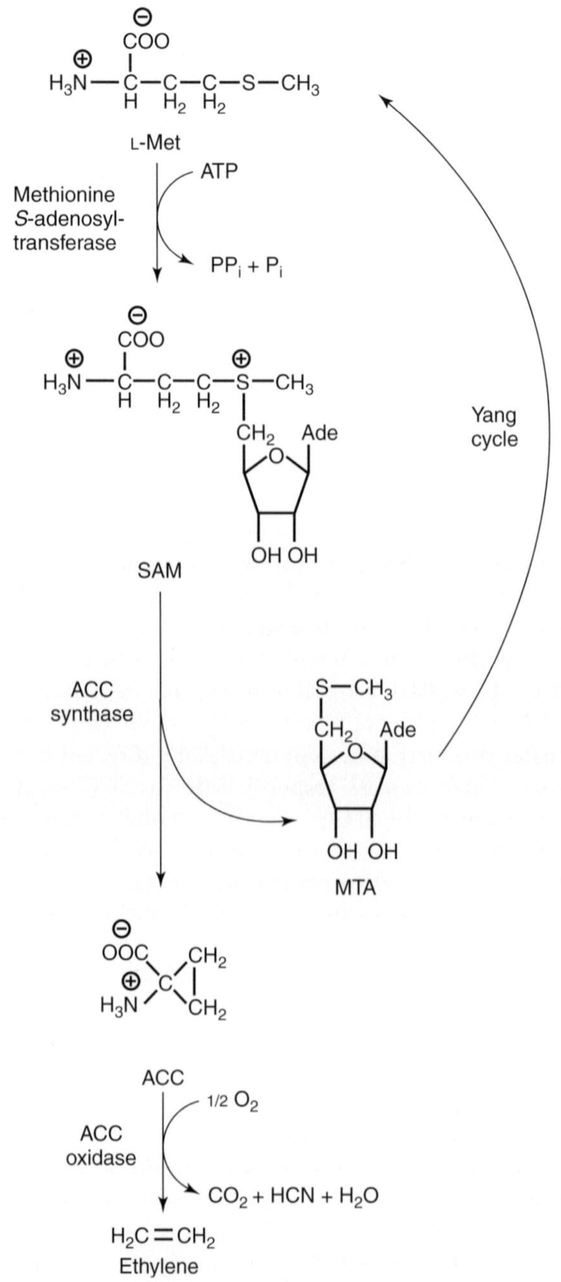

Figure 1 Pathway for ethylene biosynthesis. Chemical structures and the enzymes catalyzing each step are shown. Abbreviations used are L-Met (L-Methionine), SAM (*S*-adenosylmethionine), ACC (1-aminocyclopropane-1-carboxylic acid), and MTA (5′-methylthioadenosine).

Biosynthesis of Ethylene

The major steps in the biosynthetic pathway for ethylene in plants were deduced during the 1960s and 1970s. Ethylene is synthesized in three steps from the cellular pools of L-methionine (**Figure 1**). In the first step, methionine is converted into *S*-adenosylmethionine (SAM) by the enzyme methionine adenosyltransferase. This step is not specific to the pathway for ethylene biosynthesis as SAM is also used in methylation reactions and polyamine biosynthesis. In the second step, SAM is converted to 1-aminocyclopropane-1-carboxylic acid (ACC) by the enzyme ACC synthase. The other product of this reaction, 5′-methylthioadenosine, is recycled back to methionine in a sequence of steps commonly referred to as the Yang cycle after its discoverer. The Yang cycle allows for formation of ethylene even when methionine levels are low. In the third step, ACC is converted to ethylene by action of the enzyme ACC oxidase, which is also called ethylene-forming enzyme.

The production of ethylene is carefully regulated and varies in response to internal and external factors. Among the internal factors that can regulate ethylene production is genetic background. Crop cultivars may emit different amounts of ethylene and this often correlates with differences in shelf-life. Rates of ethylene production also vary with development, high rates being frequently associated with cell division, fruit ripening, and organ senescence. Other plant hormones, in particular auxin, have also been demonstrated to induce ethylene biosynthesis. External factors such as light, gravity, nutritional factors, physical contacts, and the gases oxygen and carbon dioxide have all been shown to affect ethylene production. Of particular interest has been the finding that ethylene biosynthesis increases in injured cells. This "stress ethylene" can be induced by abiotic stresses such as cold, high temperature, flooding, radiation, bending, and wounding. Stress ethylene can also be induced by biotic stresses associated with infections and insect damage.

The biosynthesis of ACC from SAM is the primary control point in ethylene biosynthesis, and the activity of ACC synthase is thus the principal regulator of ethylene levels in the plant. In all plants examined, ACC synthase is encoded by a gene family, which raises the possibility that different isoforms of the enzyme can be regulated in independent fashion. Regulation of ACC synthase has been demonstrated at the transcriptional level, with expression of the genes found to vary temporally, spatially, and in response to environmental stimuli such as wounding, anaerobiosis, and ozone. Hormones also regulate expression, one example being the *Arabidopsis* ACS4 isoform whose expression is stimulated by auxin. Regulation also occurs at the posttranscriptional level. Stability of a tomato (*Lycopersicon esculentum*) ACC synthase is regulated by phosphorylation. In addition, genetic and biochemical data indicate that the C-terminus of ACC synthase is an inhibitory domain, and it has been proposed that modification of this domain could regulate activity of the enzyme.

Like ACC synthase, ACC oxidase is encoded by a gene family. In general, the level of ACC oxidase activity is not considered limiting to the formation of ethylene. However, there is evidence that expression levels of ACC-oxidase isoforms are differentially regulated. Increasing the level of ACC-oxidase activity may be particularly important under conditions of high ethylene production. For instance, the gene encoding ACC oxidase was originally identified as a result of its increased expression during tomato ripening. In addition, expression of the gene encoding one *Arabidopsis* isoform is induced by ethylene, indicating that the plant responds to ethylene by increasing its capacity to synthesize the hormone.

Signaling by Ethylene

Only in the past decade have significant inroads been made into elucidating the manner by which plants perceive and transduce the ethylene signal. An important contribution to our understanding of ethylene signal transduction has come from the isolation of mutants that affect ethylene responses in *Arabidopsis*. These mutations fall into two main classes: (1) mutations that render a plant insensitive to ethylene; and (2) those that result in a constitutive ethylene response. From this work, a genetically defined pathway for signal transduction has been established (**Figure 2**). Even though there are many ethylene responses in the plant, all those examined make use of the same initial signal transduction pathway. Identification of the gene products associated with the mutants has also been edifying. The ethylene signal transduction pathway combines components of disparate evolutionary origin: some of the protein motifs are found in bacteria, others in eukaryotes only, and others are specific to plants. By making use of the information gained from *Arabidopsis*, similar proteins have been identified in other plant species indicating that the mechanism of ethylene signaling is conserved.

Perception of ethylene is mediated by a family of membrane-associated receptors. The ethylene-binding site of the receptors is located within the N-terminal transmembrane domains that are also responsible for membrane association. The binding site contains a copper cofactor that is required for the high-affinity ethylene binding that receptors display. Dominant ethylene-insensitive mutations arise from changes in amino acids within the region of the receptor involved in ethylene binding. For example, the *etr1-1* mutation arises from a change in one of the amino acids thought to be required for coordinating the copper cofactor, thereby preventing the binding of ethylene to the receptor. Ethylene receptors are evolutionarily related to bacterial sensor proteins. Like the bacterial sensors, the ethylene receptors form dimers and contain histidine kinase domains. The role of histidine kinase activity in signaling by the receptors is unknown. Histidine kinase activity has been demonstrated for the ethylene receptor ETR1, but some ethylene receptors contain diverged histidine kinase domains that are predicted to lack enzymatic activity. Genes encoding ethylene receptors have now been identified in crops such as tomato, musk melon (*Cucumis melo* (reticulatus group)), cucumber (*Cucumis sativus*), mango (*Mangifera indica*), peach (*Prunus persica*), broccoli (*Brassica oleracea* var *italica*), grapes (*Vitis* spp.), rice (*Oryza sativa*), tobacco (*Nicotiana*

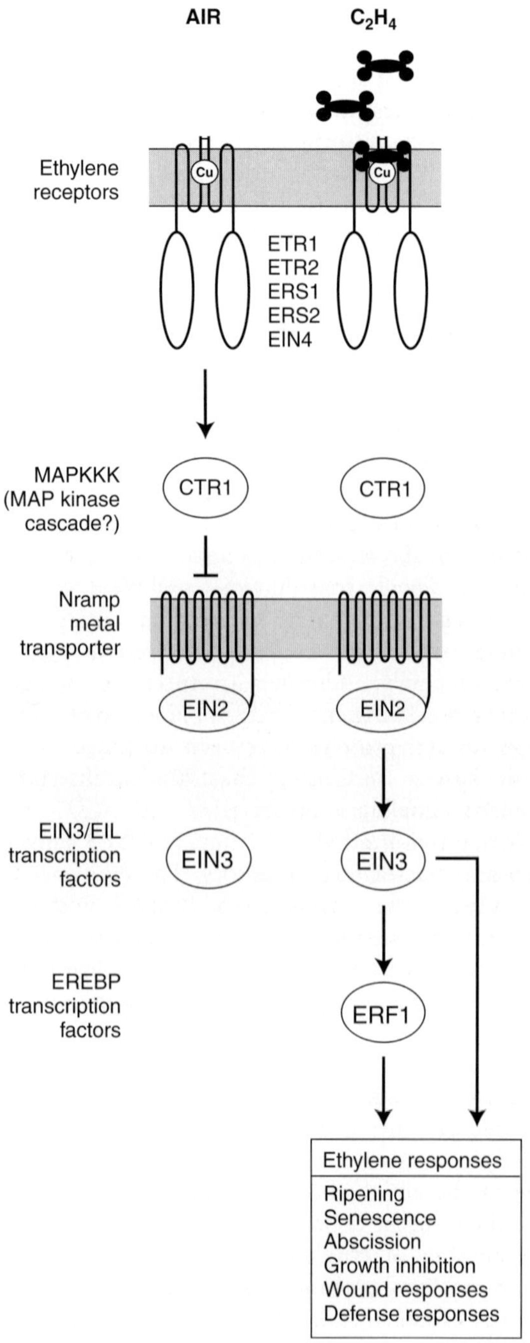

Figure 2 Pathway for ethylene signal transduction. Proteins from *Arabidopsis* are shown along with their predicted order of action based on genetic analysis. Not included in this pathway are genes defined by mutations but whose gene products have not been determined. The predicted transmembrane structure is shown for the ethylene receptors and EIN2. Soluble protein domains are indicated by circles. In air, ethylene receptors activate CTR1, which then suppresses ethylene responses. In ethylene, the receptors are inactivated, leading to inactivation of CTR1. As a result, EIN2 is activated and stimulates the EIN3/EIL family of transcription factors. Among the genes transcriptionally induced by EIN3 is *ERF1*, which encodes a member of the EREBP family of transcription factors. Both families of transcription factors are involved in regulating ethylene responses.

tabacum), geranium (*Pelargonium* spp.) and rose (*Rosa* spp.).

The CTR1 and EIN2 proteins of *Arabidopsis* have been shown to act in the same primary response pathway as the ethylene receptors. CTR1 belongs to the Raf family of protein serine/threonine kinases that initiate MAP kinase cascades in eukaryotes. Thus, the involvement of CTR1 in ethylene signal transduction indicates that other members of the MAP kinase cascade may also participate in signaling. CTR1 acts downstream of the ethylene receptors and has been shown to physically interact with the receptors. Loss-of-function mutations in CTR1 result in constitutive ethylene responses. EIN2 acts downstream of CTR1 and is an integral membrane protein with similarity to the Nramp family of metal ion transporters. Loss-of-function mutations in EIN2 result in ethylene insensitivity. Overexpression of the cytosolic C-terminal domain of EIN2 induces some ethylene responses.

Two distinct families of transcription factors have been implicated in mediating changes in gene expression in response to ethylene. Members of the EIN3/EIL family of transcription factors are directly activated by the ethylene signal transduction system, and are required for ethylene-dependent gene induction. Loss-of-function mutations in *EIN3* render a plant ethylene-insensitive. Among the genes regulated by the EIN3/EIL family are those for a second set of transcription factors: the ethylene response element binding proteins (EREBPs). The EREBPs are a large family of related transcription factors that are unique to plants, some of which mediate the effects of ethylene by binding to a promoter element known as the GCC box.

An important aspect of ethylene signal transduction is that the pathway is negatively regulated. In air, the receptors are apparently in an "on" state and suppress ethylene responses. Binding of ethylene to the receptors turns them "off," relieving the suppression on downstream components in the pathway and resulting in the induction of ethylene responses. This model for negative regulation results in the counter-intuitive situation in which a decrease in the number of receptors is predicted to induce ethylene responses, while an increase in the number of receptors is predicted to suppress ethylene responses. Consistent with this model, loss-of-function mutations in the ethylene receptors result in activation of ethylene responses in the plant. Interestingly, expression of some members of the ethylene receptor family is induced by ethylene, indicating that dynamic changes in ethylene receptor levels may play a role in regulating ethylene signal transduction.

Ethylene Responses in Plants

Ethylene binds with high affinity to its receptors and the signal is transduced through a series of conserved proteins to regulate downstream responses. However, even though the same initial signal transduction pathway is used, the ultimate responses to ethylene are quite variable (**Figure 2**). Ethylene responses are dependent upon developmental, environmental and tissue-specific factors, and vary considerably in their kinetics. One of the most sensitive responses in plants is the inhibition of seedling and root growth, which can occur within 15 min of ethylene application and is responsive to concentrations as low as $0.1\,\mu l\,l^{-1}$ (ppm) ethylene, saturation of the response occurring at $10\,\mu l\,l^{-1}$. Most ethylene effects are reversible and, upon removal of ethylene, seedling growth returns to its original rate within 15 min to 1 h. On the other hand, some ethylene responses may take days to become apparent, such as the stimulatory effect of ethylene on senescence or the initiation of adventitious roots. In order to have some specificity in the type of response initiated, the ethylene signal transduction pathway must interact with and be regulated by other pathways. Transcription mediated by the large EREBP family of transcription factors represents a potential target for such regulation.

Ethylene is commonly considered an "aging" hormone due to its stimulatory effects on ripening, senescence and abscission. Fruit is termed climacteric if it ripens in response to ethylene (e.g., apple (*Malus pumila*), banana (*Musa sapientum*), cantaloupe melon (*C. melo cantalupensis*), tomato) and non-climacteric if it fails to respond to ethylene (e.g., bell pepper (*Capsicum frutescens*), cherry (*Prunus avium*), grape, strawberry (*Fragaria ananassa*)) when harvested at commercial maturity. In general, climacteric fruit must reach a certain stage of maturity before becoming sensitive to ethylene. Climacteric fruit increase their ethylene production and/or sensitivity during the ripening process. The ability of ethylene to accelerate the senescence of leaves and flowers is also dependent upon the type of plant and the maturity of the tissue. However, in all cases, leaves and flowers will eventually senesce even in the absence of ethylene. For example, the leaves of ethylene-insensitive mutants of *Arabidopsis* still senesce but do so more slowly than leaves of wild-type plants. Abscission is restricted to zones of competent cells found at the base of leaves, floral organs, and fruits. As with senescence, ethylene may accelerate the process of abscission, but abscission will occur in the absence of ethylene.

Ethylene regulates many aspects of plant growth. It promotes seed germination, thereby countering the effects of abscisic acid, and stimulates hook formation in dark-grown seedlings, which serves to protect the delicate apical meristem as the germinating seedling penetrates the soil. Ethylene stimulates the production of adventitious roots and root hairs and promotes flowering in some plant species, notably bromeliads such as pineapple. Furthermore, ethylene is capable of either inhibiting or promoting cell division and elongation, with promotion occurring primarily in aquatic plants such as rice. The changes in cell growth are facilitated by changes in cell wall biosynthesis, as reflected by changes in the levels of cell wall hydroxyproline (a component of hydroxyproline-rich glycoproteins) and the direction that cellulose microfibrils are deposited.

Ethylene plays general and specific roles in plant stress responses. It is produced in response to just about any abiotic stress, whether the stress arises from a change in temperatute or physical wounding of the plant tissue. In response to flooding, ethylene mediates the formation of aerenchyma, gas-filled channels that are thought to increase the availability of oxygen to submerged tissues. The role of ethylene in mediating the plant responses to biotic stresses, such as those that arise from pathogen infection, is complex. Ethylene may be produced as a result of lesions formed during infection. However, there are also pathogen-specific effects, with the response of the plant to the pathogen modulated by a complex interplay between the ethylene, jasmonic acid, and salicylic acid signaling pathways. Ethylene-insensitive tomato plants showed greater resistance to some pathogens. However, an ethylene-insensitive tobacco plant showed increased susceptibility to normally nonpathogenic fungi of the soil. The biotic stress-induced ethylene can serve a variety of purposes. It can lead to senescence and abscission of damaged organs, increased lignification, and the production of enzymes such as glucanases and chitinases that degrade fungal cell walls.

Several examples of how other plant hormones may modulate ethylene responses have already been given, but these represent the tip of the iceberg. Many plant hormones stimulate ethylene production, such that ethylene can be considered a second messenger for the primary hormone. This has been reported for auxin, ABA, cytokinins, gibberellins, and brassinosteroids. Historically, the role of ethylene as a second messenger led to some confusion in defining what the primary responses were for the plant hormones. This was particularly true in the case of auxin, many of whose effects are mediated by ethylene, a notable exception being the auxin-induced cell elongation. Cytokinin is often thought of as a plant hormone that counteracts the effects of ethylene, owing to its delaying effect on senescence; but in dark-grown seedlings, cytokinin treatment induces ethylene production leading to changes in cell growth. The stimulation of ethylene production by other plant hormones can be synergistic, indicating that different mechanisms may be employed by these hormones to regulate ethylene production.

Manipulation of Ethylene Responses

There is a long history of ethylene responses being manipulated for agricultural purposes. Egyptian bas-reliefs from two millennia ago depict the practice of fig gashing. Wounding of the figs results in the release of stress ethylene, which accelerates the ripening process. Wood smoke has been used in many cultures to stimulate flowering or fruit ripening, ethylene being the active component in the smoke. However, it was not until the twentieth century that the physiological basis for these effects was uncovered. Our increased understanding of the role of ethylene in plants, along with mechanisms involved in its production and signal transduction, allow for targeted manipulation of ethylene responses (**Figure 3**).

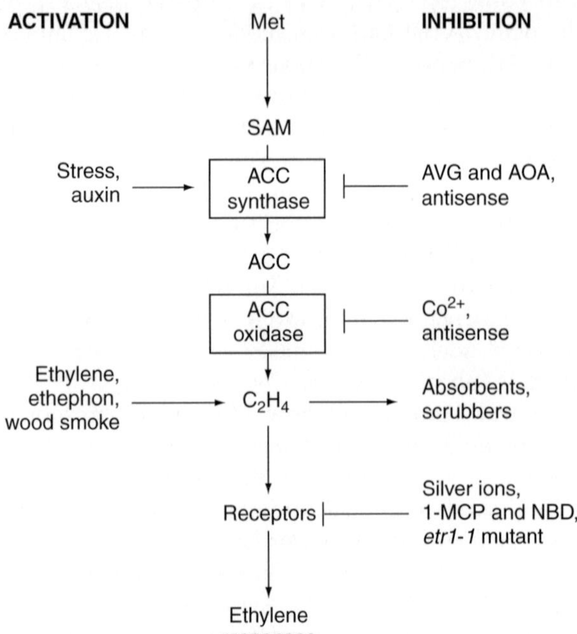

Figure 3 Manipulation of ethylene responses. An abbreviated version of the pathway for ethylene biosynthesis and perception is shown. Key biosynthetic enzymes are boxed. Ethylene responses can be activated by inducing the plant to produce ethylene or by adding sources of ethylene. Ethylene responses can be inhibited by removing ethylene, by inhibiting key proteins with chemical antagonists or by modifying proteins through transgenic methods. Met, L-methionine.

Chemical Strategies to Induce Ethylene Responses

A number of chemical compounds have been identified that release ethylene, and the ease with which these nongaseous compounds can be manipulated makes them the most convenient ethylene source for agricultural purposes. Chief among these compounds is ethephon, the abbreviated name of 2-chloroethanephosphonic acid. Ethephon is a liquid that is stable below pH 3.5 but which releases ethylene when the pH is raised. Ethephon was developed for use as a plant growth regulator in the late 1960s and is sold under the brand names Cerone, Ethrel, Florel, and Prep. Ethephon may be the most widely used plant growth regulator in agriculture. The ethylene precursor ACC can be readily converted to ethylene by plants, but its use as an ethylene source is primarily confined to research.

The ability of ethylene to stimulate plant maturation processes such as ripening and senescence has been of particular use in agriculture. Application can either be preharvest or postharvest. Historically, one of the primary uses of ethylene has been to enhance the ripening of fruit such as tomatoes, bananas and honeydew melons (*C. melo* (inodorous group)). Currently, of greatest economic importance is its use in the processed tomato market. Ethylene is used to enhance the ripening of tomatoes so that a higher percentage of the crop is ready for mechanical harvest. It is also used in the treatment of citrus fruits to eliminate green patches on the skin, although the rest of the fruit does not show changes in ripening in response to ethylene. Ethylene's ability to stimulate the maturation of tobacco leaves makes it possible to reduce the number of harvestings from seven to three.

The ability of ethylene to act as a growth retardant has resulted in widespread use on cereal crops such as barley (*Hordeum vulgare*), wheat (*Triticum aestivum*), oats (*Avena sativa*), and rye (*Secale cereale*) in the United States and Europe. Treatment results in shorter plants with stiffer stems and thus reduces lodging, the permanent bending of the plant to the ground such as can happen in heavy rainfall. Lodging can prevent proper development of the cereal and makes harvesting the grain difficult.

Almost every action discovered for ethylene has found some agricultural use. The ability of ethylene to facilitate abscission has been exploited with fruit trees such as apple, cherry and plum (*Prunus domestica*), where ethylene can be used prior to mechanical harvesting. The ability of ethylene to induce flowering has proved particularly useful with Bromeliads. Field-grown pineapples (*Ananas comosus*) are treated with ethylene to stimulate uniform flowering in the crop and thus uniformly timed fruit production. Ethylene is also used with ornamental Bromeliads to stimulate flowering at any time of the year. The ability of ethylene to promote dehiscence has been applied to the harvesting of cotton. Ethylene treatment enhances opening of the cotton bolls and also improves the response of the cotton plants to the defoliants that are applied prior to harvest. Ethylene has been found to stimulate latex flow in rubber trees. Its application increases the duration of the flow and allows for shorter cuts to be made into the bark of the tree.

Chemical Strategies to Inhibit Ethylene Responses

Inhibition of ethylene responses also plays an important role in agriculture. This can be accomplished by using compounds that slow down plant metabolism, thus delaying the onset of ethylene-mediated processes. In controlled atmosphere fruit storage, ripening is delayed by manipulating levels of oxygen and carbon dioxide. Carbon dioxide appears to act as a mild toxicant by reducing cellular metabolism.

Reducing the levels of ethylene present in the plant can also prevent activation of ethylene responses. At the simplest level, the plant-produced ethylene is removed, a practice that is of particular importance during storage of horticultural products. For this purpose various absorbents, e.g., brominated charcoal and potassium permanganate, and scrubbers are used. The air in closed storage rooms can be circulated through oxidizing scrubbers that contain metal catalysts. Ethylene levels can also be reduced by inhibiting the enzymes required for ethylene biosynthesis. Activity of the biosynthetic enzyme ACC synthase is inhibited by AVG (aminoethylvinylglycine) and AOA (aminooxyacetic acid). Neither AVG nor AOA is registered for commercial use owing to concerns over potential toxicity, but these compounds are used in research. Activity of the enzyme ACC oxidase is inhibited by cobalt ions (Co^{2+}).

Compounds have been discovered that are capable of inhibiting the action of ethylene within the plant. These compounds target the plant ethylene receptors rather than downstream components in the signal transduction pathway. Silver was found to be an inhibitor of ethylene action in 1976 and has since been commercialized in the form of silver thiosulfate (STS). The primary use of silver has been in the floral industry to inhibit prebloom abscission and to extend the vase life of cut flowers such as carnations (*Dianthus caryophyllus*). Silver ions apparently function by replacing the copper required for ethylene binding by the receptors. Antagonists of ethylene binding have also been discovered. In particular, strained cycloalkenes such as 2,5-norbornadiene and *trans*-cyclooctene are competitors for the same binding site as ethylene, but are unable to induce ethylene responses. These compounds have not been commercialized owing to their unpleasant odor and possible carcinogenicity. However, the cyclo-olefine 1-methylcyclopropene (1-MCP) has recently been approved for commercial use (licensed by Rohm and Haas Co. and by BioTechnologies for Horticulture, Inc.). 1-MCP is a potent competitive antagonist for ethylene binding with greater affinity for the binding site than ethylene itself. It has no detectable odor and no known toxicities. 1-MCP is effective at $nl\ l^{-1}$ concentrations. A single treatment can double the vase life of cut carnations and retard ethylene responses in bananas and green tomatoes for more than a week.

Transgenic Modification of Ethylene Responses

The identification of genes encoding proteins required for ethylene biosynthesis and signal transduction provides a nonchemical alternative to the modification of ethylene responses in the plant. Signaling through the ethylene pathway can be altered by suppressing, overexpressing or using mutant versions of these proteins. In all cases, due to the pleiotropic effects of ethylene on plant growth and development, it will be necessary to evaluate how modifying the desired trait affects the overall viability of the plant. The use of tissue-specific or inducible promoters should help minimize undesirable side-effects.

The first demonstrations of the transgenic approach came after the cloning of the ethylene biosynthetic genes encoding ACC synthase and ACC oxidase from tomato plants. Antisense technology was used to reduce expression of these genes in tomato. In both cases, it was possible to reduce ethylene production by more than 95% in the fruit. As a consequence, ripening of the transgenic fruits was suppressed. An alternative approach to reduce ethylene biosynthesis has also been demonstrated. Rather than inhibit the native biosynthetic genes by antisense technology, the gene encoding a bacterial ACC deaminase was introduced into tomato, its expression serving to deplete the endogenous ACC pool. One potential advantage of reducing biosynthesis, rather than perception, of ethylene is that the process can be reversed by subsequent treatment with ethylene.

The identification of genes involved in ethylene perception provides a new set of targets for transgenic modification. The ethylene receptor mutant *etr1-1* should be of particular use owing to the dominant nature of this mutation. It is capable of rendering a plant ethylene insensitive in a background containing wild-type copies of itself as well as wild-type copies of other ethylene receptors. The *etr1-1* mutation from *Arabidopsis* can also be used to generate ethylene insensitivity in other plant species. When transgenically expressed in tomato, the *etr1-1* mutation delayed ripening and senescence of the fruit. When expressed in petunia (*Petunia* spp.), the *etr1-1* mutation delayed senescence and abscission of flowers following pollination, more than doubling the flower-life. Different mutant versions of the receptor show varying degrees of ethylene insensitivity, thus refined modulation of ethylene responses should be possible with transgenic plants.

List of Technical Nomenclature

1-MCP	1-Methylcyclopropene.
ACC	1-Aminocyclopropane-1-carboxylic acid.
AOA	Aminooxyacetic acid.
AVG	Aminoethylvinylgylcine.
EREBP	Ethylene response element binding protein.
L-Met	L-Methionine.
MAP kinase	Mitogen-activated protein kinase.
MTA	5′-Methylthioadenosine.
NBD	2,5-Norbornadiene.
ppm	Parts per million.
SAM	*S*-Adenosylmethionine.
STS	Silver thiosulfate.

See also: **Abiotic Stresses**: Mechanical Stress and Wind Damage. **Diseases**: Bacterial Diseases; Fungal Diseases. **Growth and Development**: Cell Growth. **Postharvest Physiology**: Ripening; Senescence, Flowers; Senescence, Leaves. **Regulators of Growth**: Cytokinins; Gibberellins. **Root Development**: Genetics of Primary Root Development.

Further Reading

Abeles FB, Morgan PW, and Saltveit ME Jr. (1992) *Ethylene in Plant Biology*, 2nd edn. San Diego: Academic Press.

Bleecker AB (1999) Ethylene perception and signaling: an evolutionary perspective. *Trends in Plant Science* 4: 269–274.

Bleeker AB and Kende H (2000) Ethylene: a gaseous signal molecule in plants. *Annual Review of Cell and Developmental Biology* 16: 1–18.

Johnson PR and Ecker JR (1998) The ethylene gas signal transduction pathway: a molecular perspective. *Annual Review of Genetics* 32: 227–254.

Kende H (1993) Ethylene biosynthesis. *Annual Review of Plant Physiology and Plant Molecular Biology* 44: 283–307.

Mattoo AK and Suttle JC (1991) *The Plant Hormone Ethylene*. Boca Raton: CRC Press.

Schaller GE (2000) Histidine kinases and the role of two-component systems in plants. *Advances in Botanical Research* 32: 109–148.

Sisler EC and Serek M (1999) Compounds controlling the ethylene receptor. *Botanical Bulletin of Academia Sinica* 40: 1–7.

Brassinosteroids

S D Clouse, North Carolina State University, Raleigh, NC, USA

Overview

Plant hormones are essential regulators of plant development beginning with seed germination and culminating in whole-plant senescence. Until recently it was generally believed that the five classes of compounds comprising abscisic acid, auxin, cytokinins, ethylene, and gibberellins could account for most or all of the growth regulatory effects of plant hormones. During the past 5 years, however, application of molecular genetics and biochemical analysis to a number of dwarf mutants has revealed that another group of compounds, termed brassinosteroids (BRs), are as critical in the normal development of a plant as are the classical plant hormones. This article provides a brief history of BR research, summarizes the structure, natural occurrence, and biosynthesis of BRs, examines physiological responses to BRs along with practical agricultural applications, and surveys molecular approaches to understanding the mode of action of BRs in promoting elongation, division, and differentiation of cells in multiple developmental programs.

History of Research in Brassinosteroids

Discovery at the US Department of Agriculture

BR research developed from work beginning in the 1940s at the US Department of Agriculture (USDA) by J. W. Mitchell and colleagues in which they used organic extracts of pollen from over 60 species to search for new compounds with plant growth-promoting properties. An extract from *Brassica napus* pollen proved to have marked effects on cell elongation and division in bioassays using bean second internodes and promoted overall growth of radishes, leafy vegetables, and potatoes when young seedlings were sprayed in greenhouse experiments. Based on these results, USDA researchers mounted a major effort to identify the active component of the *B. napus* extract (which had been termed "brassins"), and from 227 kg of *Brassica* pollen milligram quantities of a novel plant steroid were obtained. This compound, which was named "brassinolide," accounted for all of the brassin activity observed in a wide variety of bioassays.

Brassinolide (BL) structure was determined by single crystal X-ray analysis and was found to be a polyhydroxylated derivative of 5α-cholestane (**Figure 1**). Thus, plants were shown to produce a growth-promoting steroid with structural similarity to cholesterol-derived animal steroid hormones such as androgens, estrogens, and corticosteroids from vertebrates, and ecdysteroids from insects and crustacea. Shortly after the publication of the BL structure in 1979, several synthetic routes to BL and its analogs were designed resulting in a more ready supply of BRs for bioassays and field experiments to define the physiological roles of BRs and their potential application in increasing crop yields. Many of the early studies focused on exogenous application of BRs to various species of intact plants or isolated plant organs and observation of the growth response. These laboratory studies consistently showed that BRs strongly promoted cell elongation in a wide array of plants, often mimicking the effects of auxin, but in some cases showing unique properties. Field studies included both horticultural and agronomic crops and often involved spraying with 24-epibrassinolide, a stereoisomer of BL that was more efficiently synthesized than BL itself. Results from field experiments were more variable than bioassays, sometimes showing profound increases in yields after spraying with BR, while other experiments showed no effect of BR treatment.

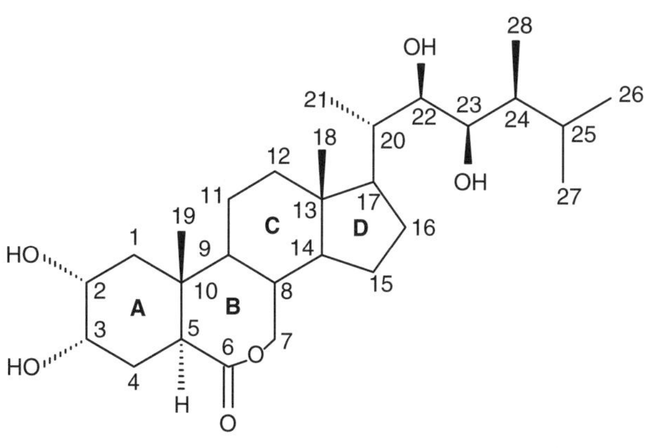

Figure 1 The structure of brassinolide, a commonly occurring brassinosteroids with high biological activity.

Early Work on Brassinosteroids in Japan

Besides the USDA, Japanese scientists were also heavily involved in early BR research. Work in the 1960s at Nagoya University showed that a methanol extract of insect galls of *Distylium racemosum* had extremely high activity in the rice lamina inclination assay, which had been established originally as an auxin bioassay. After publication of the BL structure, it was found that pure BL had the same activity in the rice lamina assay and castasterone (identical to BL except for a ketone vs. lactone at C6) was identified as one of the active components of the *D. racemosum* extract. The rice lamina inclination assay became established as one of the most sensitive and specific bioassays for BR activity and continues to be used widely. During the 1980s several Japanese groups applied biochemical and biophysical instrumentation to measuring endogenous BR levels in a wide spectrum of plants and unraveling the biosynthetic pathway to BL, primarily in cell suspension cultures of *Catharanthus roseus*. This work showed that BRs are not confined to *Brassica* species but appear to be ubiquitous in the plant kingdom, and that the membrane sterol campesterol is the biosynthetic progenitor of BL.

Molecular Genetics Applied to Brassinosteroid Research

Molecular analysis of BR action began in the early 1990s with the cloning of the first BR-regulated genes and the identification of a BR-insensitive mutant in *Arabidopsis thaliana*. The year 1996 was pivotal in BR research, when several reports of BR biosynthetic mutants appeared, including the critical observation that the extreme dwarf phenotype exhibited by these mutants could be rescued only by BR, and not by any other growth-promoting compound. Thus, the prediction by many of the early BR researchers that these compounds were a new class of plant hormones was confirmed and BRs were widely accepted by plant scientists as essential regulators of normal plant growth and development. The last 5 years have seen an increasing number of laboratories become involved in BR research and significant advances have been made in understanding BR signal transduction and the complexities of BR biosynthesis.

Structure and Natural Occurrence of Brassinosteroids

Figure 1 shows the chemical structure of BL, the first and most active BR characterized. Using nuclear magnetic resonance, mass spectrometry, and X-ray crystallography, BL was established to be (22*R*,23*R*,-24*S*)-2α,3α,22,23-tetrahydroxy-24-methyl-B-homo-7-oxa-5α-cholestan-6-one. To date, 40 free BRs and four conjugates have been fully characterized from plants, which differ from BL by variations in the substitutions at C2 and C3 in the A ring, the presence of a lactone, ketone, or deoxo function at C6 in the B ring, the stereochemistry of the hydroxyl groups in the side chain, and the presence or absence of a methyl (methylene) or ethyl (ethylene) group at C24. The conjugates consist of glycosylated, meristylated, and laurylated derivatives of the hydroxyls in ring A or in the side chain. Many of the known BRs have proven to be biosynthetic precursors or metabolic products of BL, although castasterone, the immediate precursor of BL, has been shown in some plants to have biological activity independent of its conversion to BL. Numerous structure–function assays have been undertaken to ascertain the biological activity, and the likely affinity for the BR receptor, of the various BRs. Optimal activity results from the configuration found in BL, namely a lactone function at C6/C7, *cis*-vicinal hydroxyls at C2 and C3, *R* configuration of the hydroxyls at C22 and C23, and a methyl substitution at C24.

BRs occur endogenously at quite low levels, that vary among plant tissues. Pollen, the original source of BL, and immature seeds show the highest levels (1–100 ng g^{-1} fw), while shoots and leaves usually have lower amounts (0.01–0.1 ng g^{-1} fw). Bioassay results suggest that root tissue contains BRs but they have yet to be identified and quantified rigorously from this tissue. Cultured crown gall cells of *C. roseus* have levels of BL and castasterone (30 ng g^{-1} fw) equivalent to that of pollen. In general, young growing tissues contain higher levels of BRs than mature tissues which is consistent with the observation that BRs show greater physiological effect on immature than older tissue. BRs have been found in all plants examined to date and characterized in tissues of at least 44 plant species, including 37 angiosperms (28 dicots and nine monocots), five gymnosperms, one alga, and one pteridophyte.

Brassinosteroid Biosynthesis and BR-Deficient Mutants

BRs are ultimately derived from mevalonate via the isoprenoid pathway, which also produces gibberellins and abscisic acid. However BRs are unique among plant hormones in being sterol derivatives. The BL biosynthetic pathway can be divided into the cycloartenol to campesterol segment which produces the bulk membrane sterols campesterol, sitosterol, and stigmasterol (**Figure 2**), and the campesterol to BL segment which is unique to BR biosynthesis (**Figure 3**). Details of the pathway have been provided by feeding of labeled intermediates and monitoring their fate in cell suspension cultures and whole plants, and by measuring endogenous sterol and BR levels by mass spectrometry in *Arabidopsis*, pea (*Pisum sativum*), and tomato (*Lycopersicon esculentum*) mutants blocked in various steps of the respective pathways.

Cycloartenol to Campesterol

Mutants with blocks very early in the sterol biosynthetic pathway such as *smt*1 and *fackel* have some characteristics of BR-deficient mutants in adult plants, but show unique defects in embryogenesis not found in sterol or BR biosynthetic mutants later in the pathway. Moreover, these mutants, while retaining some sensitivity to BR, are not rescued to wild-type by BR treatment, as are the downstream mutants. Mutants later in the sterol pathway have also been identified. The *Arabidopsis* mutant, *ste*1 accumulates Δ^7 sterols with a decrease in the corresponding Δ^5 sterols, and a null allele of *ste*1, *dwarf7* (*dwf7*) has many of the phenotypic features of BR-deficient mutants (although not as severe), which are rescued by BR treatment. The *DWF7* gene showed significant sequence conservation with yeast C5 desaturases, and feeding experiments with labeled intermediates revealed that *dwf7* accumulated episterol while 5-dehydroepisterol was not detected. Thus, *dwf7*/*ste*1 are lesions in the gene encoding the Δ^7-C5-desaturase that converts episterol to 5-dehydroepisterol. Similar types of analyses showed the *Arabidopsis dwf5* mutant is a lesion in the gene encoding the Δ^7 sterol reductase required for the conversion of 5-dehydroepisterol to 24-methylenecholesterol and that the allelic mutants *dwarf1*/*dimunito*/*cabbage*1 (*dwf*1/*dim*/*cbb*1) of *Arabidopsis* are defective in the conversion of 24-methylenecholesterol to campesterol. These mutants share many of the phenotypic characteristics of *dwf7*, including intermediate dwarfism and rescue to wild-type by BR treatment. The *lkb* mutant of pea is also blocked in the C24 reduction of 24-methylenecholesterol to campesterol.

Campesterol to Brassinolide

The terminal pathway from campesterol to BL proceeds via a series of side chain hydroxylations

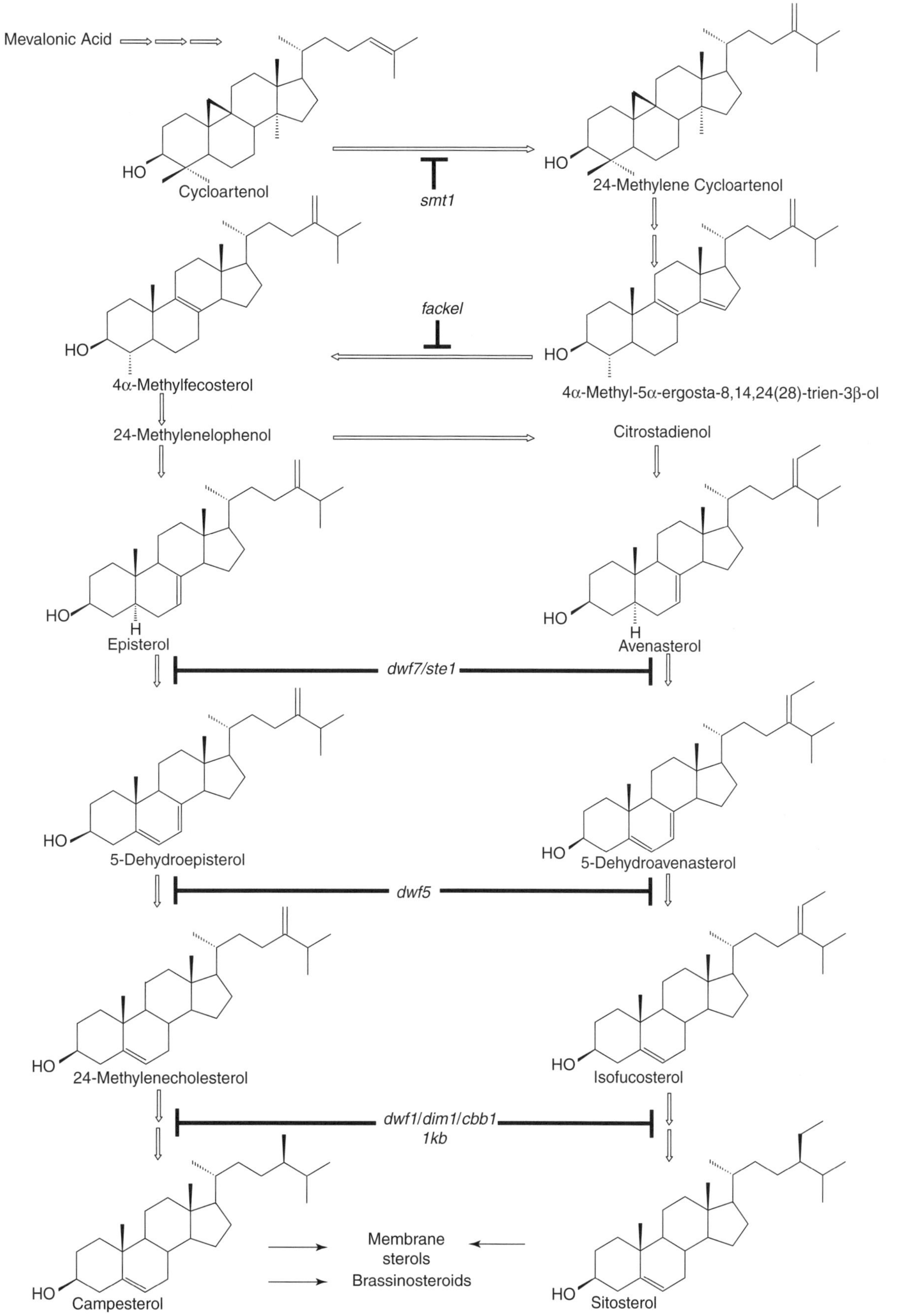

Figure 2 The biosynthetic pathway to campesterol, the sterol precursor of brassinolide, showing known mutants in *Arabidopsis* and pea that block conversion of intermediates at the steps indicated.

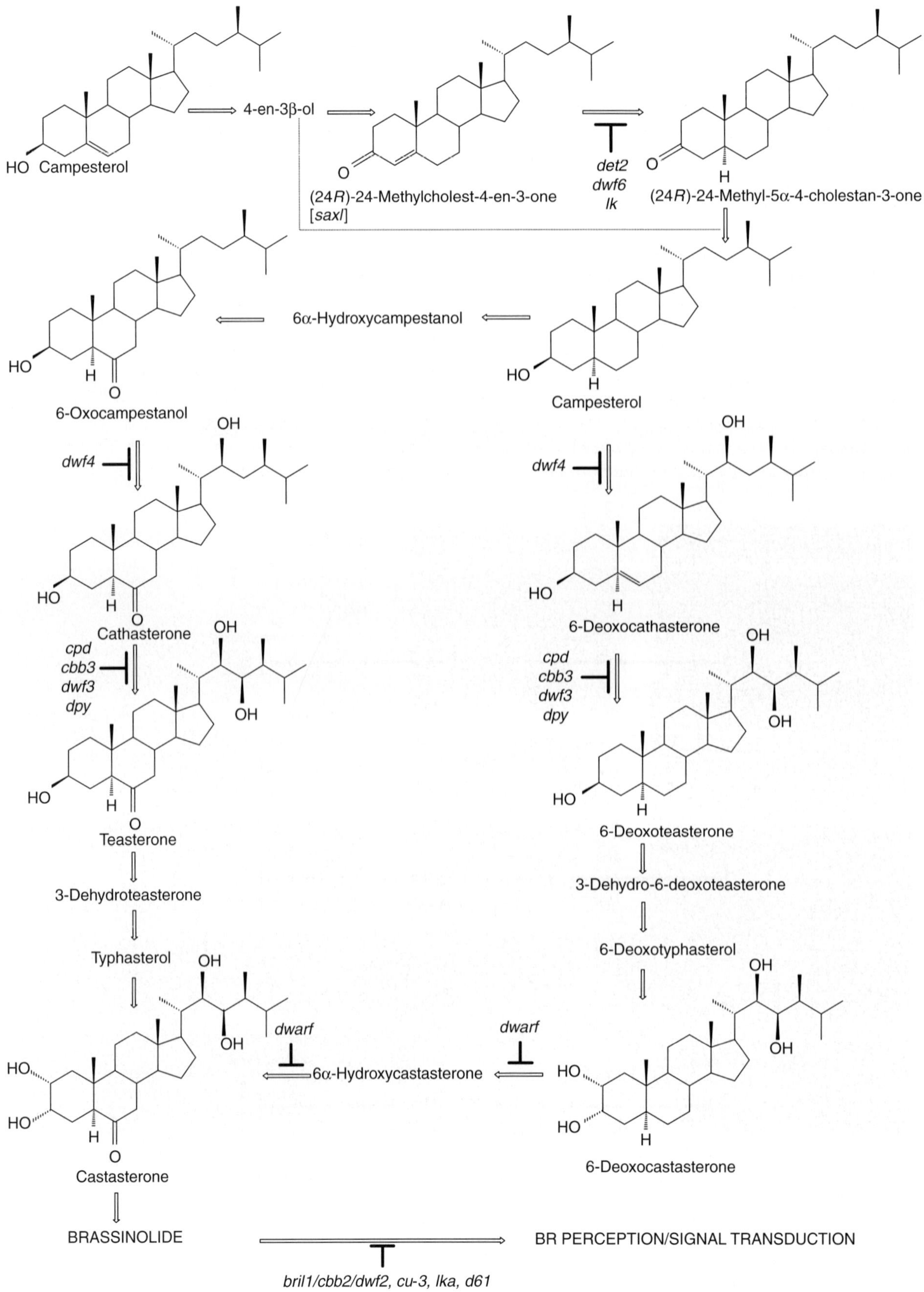

Figure 3 The biosynthetic pathway from campesterol to brassinolide showing known mutants in *Arabidopsis*, pea, tomato, and rice that block the conversion of intermediates at the steps indicated, or prevent BR perception and/or signal transduction.

combined with ring reductions, oxidations and epimerizations (**Figure** 3). Campesterol was predicted to be the plant sterol progenitor of brassinolide based on side chain structure; and the relative biological activities, co-occurrence, and molecular structure of teasterone, typhasterol, and castasterone suggested that BL was synthesized from campesterol by way of these intermediates. Analysis by mass spectrometry demonstrated that 6-deoxo forms of teasterone, typhasterol, and castasterone were found to co-occur with 6-oxo forms in many plant species, including *Arabidopsis*, and parallel pathways to BL termed early and late C6 oxidation were demonstrated in several species. Recent results have shown that conversion of 6-deoxo to oxo forms can occur at several steps in the pathway, and thus BL synthesis proceeds more via a metabolic grid rather than two parallel pathways that converge only at castasterone. Interestingly, in tomato castasterone biosynthesis appears to proceed only via the 6-deoxo pathway.

Mutants in several steps of this segment of the pathway have been identified, including *constitutive photomorphogenesis and dwarfism* (*cpd*), *dwarf4* (*dwf4*), *sax1*, and *de-etiolated2* (*det2*) of *Arabidopsis*, *lk* of pea, and the *dumpy* (*dpy*) and *dwarf* mutants of tomato. These mutants show a characteristic phenotype in the light, including dwarf stature, dark green, rounded and downward-curling leaves, prolonged lifespan, reduced fertility and apical dominance, and altered vascular development (**Figure 4**). In the dark, *Arabidopsis* mutants exhibit some of the features of light-grown plants including shortened hypocotyls and open cotyledons. Tomato and pea BR mutant do not show this de-etiolated effect to the extent seen in the *Arabidopsis* mutants. All of the phenotypic alterations exhibited by BR-deficient mutants are rescued to wild-type by exogenous application of BR.

The *DET2* gene encodes a sterol 5α-reductase required for the reduction of (24*R*)-24-methylchol-

Figure 4 (see color plate 30) Phenotype of BR dwarf mutants. (A) Comparison of several 5-week-old BR-deficient or BR-insensitive (*bri1*) mutants of *Arabidopsis* with a wild-type *Arabidopsis* plant of the same age. See the text for a description of each mutant. (B) Close-up of an 8-week-old *bri1* mutant of *Arabidopsis* showing dark green, curled leaves and cabbage-like phenotype. (C) The *dpy* mutant of tomato (left) shows a dwarf phenotype with curled leaves and suppression of axillary shoots. When treated with 0.1 μM brassinolide twice weekly, the mutant is rescued to wild-type phenotype (right). (D) Phenotype of the *curl-3* tomato BR-insensitive mutant. Note the similarity to *bri1* (B). Adapted with permission from Sakurai A, Yokota T, and Clouse SD (1999) *Brassinosteroids: Steroidal Plant Hormones*. Tokyo: Springer-Verlag and Koka CV, Cerny RE, Gardner RG, *et al.* (2000) A putative role for the tomato genes dumpy and curl-3 in brassinosteroid biosynthesis and response. *Plant Physiology* 122: 8–98.

est-4-en-3-one to (24*R*)-24-methyl-5α-cholestan-3-one during the four-step conversion of campesterol to campestanol. The functional equivalence of plant and mammalian 5α-reductases was demonstrated by the fact that transformation of the mutant with human 5α-reductases (driven by the CaMV 35S promoter) rescued the Det2 phenotype to wild-type without BR application, and was prevented by specific inhibitors of the mammalian 5α-reductases. Moreover, recombinant DET2 protein expressed in human embryonic kidney cells was able to reduce several 3-oxo,$\Delta^{4,5}$ mammalian steroids, including testosterone and progesterone. The *lk* mutant of pea acts at the same step in BL biosynthesis as *det2*. The *Arabidopsis sax1* mutant blocks BL synthesis in a step preceding *det2*, and feeding experiments suggest that SAX1 acts in a putative side pathway involving conversion of campersterol to 6-deoxocathasterone via 22-hydroxylated intermediates.

DWF4 is a cytochrome P-450 steroid hydroxylase that is responsible for C22 hydroxylation of the side chain of campestanol and 6-oxocampestanol to form 6-deoxocathasterone and cathasterone, respectively. The dramatically reduced endogenous levels of cathasterone vs. 6-oxocampestanol coupled with the much higher activity of cathasterone in bioassays suggest that the DWF4 reaction may be a rate-limiting step in BL biosynthesis. CPD is also a cytochrome P-450 hydroxylase that converts cathasterone and 6-deoxocathasterone to teasterone and 6-deoxoteasterone by C23 hydroxylation. The naturally occurring *dpy* mutant of tomato is an intermediate dwarf with severely altered leaf morphology, including the downward-curling and dark green color typical of the *Arabidopsis* BR mutants. Spraying with BL rescues the mutant phenotype to wild-type (**Figure 4**) and feeding studies coupled with measurement of endogenous BR levels show that *dpy* is blocked at the same step in BL biosynthesis as *cpd*. Another cytochrome P-450, the DWARF protein of tomato, has been shown by feeding experiments in mutant plants and yeast to catalyze the conversion of 6-deoxocastasterone to castasterone via the intermediate 6α-hydroxycastasterone.

Regulation of Brassinosteroid Levels in Plants

Regulation of the activity of BR biosynthetic enzymes and the expression of the corresponding genes may be a mechanism by which plants control the levels of BR in various tissues, with a resultant modulation of growth. For example, expression of the *CPD* gene is confined to cotyledons and leaf primordia in etiolated seedlings and is detectable in the adaxial parenchyma of expanding leaves in light-grown plants. Expression of *CPD* is not affected by ethylene, gibberellin, cytokinin, jasmonic acid, or salicylic acid, but is specifically downregulated by BRs in both dark and light, suggesting a feedback regulatory loop. Differential regulation of the early and late C6 oxidation pathway has also been suggested. Based on feeding experiments in *Arabidopsis*, it has been proposed that 6-deoxo intermediates of the late C6 oxidation pathway are more active in the light, while 6-oxo intermediates of the early C6 oxidation pathway are more active in the dark. In contrast to *Arabidopsis*, work in tomato showed that late C6 oxidation intermediates were present, but early C6 oxidation intermediates were not detected. Moreover, 6-deoxo intermediates were more effective in rescuing the *dpy* mutant than 6-oxo intermediates in both the light and the dark.

Metabolism of BL may be another method by which active BR levels are controlled in plants. The use of labeled BRs and cell suspension cultures has demonstrated glucosyl, sulfate, and acyl conjugates of BRs as well as hydroxylation at C25 and C26 of the side chain. Overexpression of the *Arabidopsis BAS1* gene (which encodes a cytochrome P-450 hydroxylase) results in inactivation of BL by C26 hydroxylation and creates a phenotype similar to BR-deficient mutants. Furthermore, a steroid sulfotransferase from *B. napus* inactives BRs by *O*-sulfonation of the C22 hydroxyl.

Physiological Responses and Molecular Mechanisms

Cell Elongation

Application of exogenous BR at nanomolar to micromolar levels causes pronounced elongation of hypocotyls, epicotyls, and peduncles of dicots, as well as coleoptiles and mesocotyls of monocots. The extreme dwarfism of BR-deficient and BR-insensitive mutants confirms the promotive role of BRs in stem elongation in intact plants, and microscopic examination of cell files in various organs shows that mutant cells are shorter than the corresponding wild-type cells. Young vegetative tissue is particularly responsive to BRs, and while both BR and auxin promote elongation, their kinetics are quite different. Auxin generally shows a very short lag time of 10–15 min between application and the onset of elongation, with maximum rates of elongation reached within 30–45 min while BR has lag times of at least 45 min with elongation rates continuing to increase for several hours. The ability of BR to promote elongation in auxin-insensitive mutants and in the presence of inhibitors of auxin action suggests an independence of BR and auxin action, although synergisms between BR and auxins do occur in many systems. Additive effects on elongation are often seen with gibberellins, while inhibitory effects of cytokinins, abscisic acid, and ethylene on BR-induced elongation have been described. It is intriguing why plants have three different signals – BR, auxin, and gibberellins – controlling the same physiological process of elongation. Differences in the temporal and spatial expression of genes in response to these three hormones may be of critical importance.

How do BRs promote cell expansion? One plausible mechanism is that BRs increase the level of activity of wall-modifying enzymes. The plant cell wall forms a highly crosslinked, rigid matrix that opposes cell expansion and differentiation. In order for elongation and other morphogenetic processes to occur, the cell wall must be modified, i.e., by wall relaxation or loosening and by incorporation of new polymers into the extending wall to maintain wall integrity. Several proteins with possible roles in cell wall modification processes have been identified, including glucanases, xyloglucan endotransglycosylases (XETs), and expansins. One scenario of wall extension proposes that expansins are primarily responsible for wall relaxation, but glucanases and XETs affect the extent of expansin activity by altering the viscosity of the hemicellulose matrix. Furthermore, XETs may function to incorporate new xyloglucan into the growing wall. It has been clearly shown that BRs alter the biophysical properties of plant cell walls and also increase the abundance of mRNA transcripts encoding wall-modifying proteins, particularly XETs and expansins. BR-regulated XET genes have been cloned in soybean (*Glycine max*), tomato, rice (*Oryza sativa*), and *Arabidopsis*. BR-induced expansion is accompanied by proton extrusion and hyperpolarization of cell membranes and there is some evidence that modulation of ATPase activity by BR may be involved. BRs have recently been shown to regulate the flow of water into cells via aquaporins and have previously been found to alter the cortical microtubule orientation from longitudinal to transverse, both of which effects may be involved in cell expansion.

Cell Division and Differentiation

BRs promote cell division in cultured parenchyma cells of *Helianthus tuberosus* and in Chinese cabbage (*Brassica pekinensis*) protoplasts. BL also accelerates the rate of cell division in petunia protoplasts in the presence of auxin and cytokinin but can not take the place of either hormone. In soybean cell suspension cultures, BL treatment hastened the onset of mitosis and increased transcript levels for genes encoding histone H_2B and high mobility group 1 proteins, both of which are associated with DNA replication during cell division. BRs also increased expression of the cyclin D3 in *Arabidopsis* cell suspension cultures and could substitute for cytokinin in the maintenance of these cultures.

A good deal of evidence supports the role of BRs in promotion of vascular differentiation. Nanomolar levels of BL stimulate tracheid formation in *H. tuberosus* explants and isolated mesophyll cells of *Zinnia elegans*. In the zinnia system, BRs also regulate the expression of several genes associated with xylem formation. The intense expression of a BR-regulated XET in paratracheary parenchyma cells surrounding vessel elements in soybean epicotyls also suggests a role for BRs, and XETs, in xylem formation. The modification of cambial division seen in BR-deficient mutants also implicates endogenous BRs in xylem differentiation *in vivo*, and it is relevant that BRs have been identified in cambial scrapings of *Pinus silvestris*.

Other Effects

Other physiological effects attributed to BRs include promotion of gravitropism in roots; induction of extracellular invertase associated with assimilate partitioning; modulation of stress responses including a protective effect against plant injury from chilling, drought and salt stress; promotion of pollen tube elongation; acceleration of senescence and a variety of effects on reproductive biology.

BR-Insensitive Mutants and Signal Transduction

To identify components of the BR signal transduction pathway, a screen for BR insensitivity based on the ability of mutant *Arabidopsis* seedlings to elongate roots in the presence of BR concentrations inhibitory to wild-type root elongation was undertaken. This analysis uncovered the mutant *brassinosteroid insensitive* 1 (*bri*1), that exhibits a pleiotropic phenotypic including severely dwarfed stature, reduced apical dominance, delayed flowering and senescence, male sterility, and nearly complete insensitivity to BRs in a variety of assays. Cloning and sequencing of *BRI*1 revealed that this gene encoded a leucine-rich repeat receptor kinase. BRI1 exhibits homology with plant and animal receptor kinases and leucine-rich repeat proteins in three major domains including the putative extracellular ligand-binding domain, the single membrane-spanning domain, and the cytoplasmic kinase domain. Several alleles of *BRI*1 with identical phenotype have been isolated in independent screens and sequence analysis of several mutant alleles confirmed that the putative ligand-binding domain and kinase domain are essential for *in vivo* function. BR-insensitive dwarf mutants have also been identified in pea, tomato, and rice.

Plant receptor-like kinases function in many diverse physiological processes including growth and development, embryogenesis, fertilization, abscission, disease resistance, and response to light. Animal receptor kinases dimerize upon ligand binding and autophosphorylate intracellular residues. This activates the kinase domain which phosphorylates intracellular targets such as G proteins, phospholipase C, and transcription factors, thus amplifying and proliferating the initial signal. Ligand-dependent dimerization followed by autophosphorylation is believed to be a conserved mechanism in plants, although no complete analysis of any plant receptor kinase has yet been accomplished. Preliminary biochemical analysis of the BRI1 kinase domain *in vitro* has shown that it is an active kinase that autophosphorylates on multiple Ser and Thr residues.

Recently it has been shown that BRI1 is localized in the plasma membrane and is ubiquitously expressed in all organs of young, growing *Arabidopsis* plants, and that the extracellular domain is capable of perceiving BRs. A BRI1/green fluorescent protein shows increased binding of tritiated BR in the plasma membranes compared to wild-type plants, and the BR-binding activity can be precipitated by antibodies to green fluorescent protein. These results show that BRI1 is either the BR receptor or part of a BR receptor complex including other steroid-binding proteins. This is quite significant, given that very few ligands for plant receptor kinases have been identified. Moreover, it illustrates a clear difference between plant and animal steroid signal transduction. In the latter, intracellular steroid receptors that migrate to the nucleus upon ligand binding predominate.

Several critical questions need to be addressed to further understand how BRI1 functions in BR signaling pathways.

1. Does BRI1 bind BR directly or indirectly in the presence of a sterol binding peptide?
2. Does BRI1 undergo homo- or heterodimerization?
3. What are the specific amino acids of the kinase domain that undergo autophosphorylation *in vivo*?
4. How does this autophosphorylation affect kinase activity and binding of intracellular substrates for the BRI1-KD?

Several research groups are currently addressing these problems which will help bridge the gap between BR perception at the cell surface and the growing number of BR-regulated genes that have been identified.

Practical Applications to Agriculture

Spraying of BRs on fields has resulted in variable results but promising increases in yields of wheat, rice, watermelon (*Citrullus vulgaris*), and potatoes (*Solanum tuberosum*) have been reported. In China, 28-homoBL has been registered as a plant growth regulator to increase yields in tobacco (*Nicotiana tabacum*), sugar cane (*Saccharum officinarum*), rape seeds (*Brassica napus*), tea (*Camellia sinensis*), and certain fruits, while in Russia and Belarus, 24-epiBL has been registered on potato, tomato, cucumber (*Cucumis sativus*), pepper (*Capsicum* spp.), and barley. Inconsistent results have been attributed to the physiological status of the crop (with stressed crops showing greater response to BR than those grown under optimal conditions) and the instability of BRs under field conditions. To address the latter, Japanese researchers have developed a BR analog called TS303 which is stable in the field and has shown promising results in increasing rice yields, enhancing fruit set in tomatoes and grapes (*Vitis vinifera*), and preventing cold damage in pears (*Pyrus communis*).

The dramatic phenotype of BR-deficient and insensitive mutants is clear evidence of the critical importance of BR activity in the growth and development of plants and it is likely that greater understanding of the molecular mechanisms of BR action could have practical impact on generating transgenic crop plants of many species with altered growth properties. In fact, preliminary results indicate that overexpression of BRI1 or DET2 can lead to larger plants than the wild-type control. Conversely, antisense inhibition of these genes leads to dwarfism, which may be desirable in many agriculture contexts, for example the ornamental nursery industry. Alternatively, the recent discovery of the BR biosynthesis inhibitor, brassinazole (a compound analogous in function to the gibberellic acid inhibitor paclobutrazol), allows a chemical approach to growth control that does not require generation of transgenic plants.

See also: **Cell Walls and Fibers**: Cell Walls. **Growth and Development**: Cell Division and Differentiation; Cell Growth; Field Crops; Molecular Biology of Development. **Regulators of Growth**: Abscisic Acid; Cytokinins; Ethylene; Gibberellins; Jasmonates; Photomorphogenesis; Phytochromes and other Photoreceptors.

Further Reading

Altmann T (1999) Molecular physiology of brassinosteroids revealed by the analysis of mutants. *Planta* 208: 1–11.

Asami T and Yoshida S (1999) Brassinosteroid biosynthesis inhibitors. *Trends in Plant Science* 4: 348–353.

Clouse SD (2000) Plant development: a role for sterols in embryogenesis. *Current Biology* 10: R601–604.

Clouse SD and Sasse JM (1998) Brassinosteroids: essential regulators of plant growth and development. *Annual Reviews of Plant Physiology and Plant Molecular Biology* 49: 427–451.

Cutler HG, Yokota T, and Adam G (1991) *Brassinosteroids: Chemistry, Bioactivity, and Applications.* Washington, DC: American Chemical Society.

Fujioka S and Sakurai A (1997) Biosynthesis and metabolism of brassinosteroids. *Physiologia Plantarum* 100: 710–715.

Khripach V, Zhabinskii V, and De Groot A (2000) Twenty years of brassinosteroids: steroidal plant hormones warrant better crops for the twenty-first century. *Annals of Botany* 86: 441–447.

Mandava NB (1988) Plant growth-promoting brassinosteroids. *Annual Reviews of Plant Physiology and Plant Molecular Biology* 39: 23–52.

Sakurai A, Yokota T, and Clouse SD (1999) *Brassinosteroids: Steroidal Plant Hormones.* Tokyo: Springer-Verlag.

Sasse JM (1997) Recent progress in brassinosteroid research. *Physiologia Plantarum* 100: 696–701.

Schumacher K and Chory J (2000) Brassinosteroid signal transduction: still casting the actors. *Current Opinion in Plant Biology* 3: 79–84.

Yokota T (1997) The structure, biosynthesis and function of brassinosteroids. *Trends in Plant Science* 2: 137–143.

Jasmonates

G J Loake and P L Nurmberg, University of Edinburgh, Edinburgh, UK

Introduction

Jasmonates are naturally occurring fatty acids that have been identified in a wide variety of plant species (**Figure 1**). Interestingly, their structure is similar to that of animal eicosinoids, lipid based signals produced during tissue inflammation. Jasmonic acid (JA) was first identified from cultures of the fungus *Lasiodiplodia theobromae*. Interest from the perfume industry drove the early studies on methyl-jasmonate (Me-JA), due to its volatile nature and presence in the essential oils of jasmine (*Jasminum grandiforum*) and other plant species. Plant physiologists began to study jasmonates when it was discovered that JA and Me-JA in *Artemisia* leaves and stems functioned as a senescence promoting substance. Furthermore, these compounds were found to inhibit growth in several plant species. Intensive studies over the past 20 years have provided significant insights into the physiological role of jasmonates and their biosynthesis and metabolism. Jasmonates are now widely regarded as endogenous plant growth substances, which play key roles in both plant development and responses to abiotic and biotic stimuli. The concentration of JA is dependent on tissue type, development stage, and environment cues. The highest concentrations are found in flowers and reproductive tissues, and the lowest levels are present in roots and mature leaves.

Jasmonate Biosynthesis

Jasmonates are synthesized through the octadecanoid pathway (**Figure 2**). The precursor for the biosynthesis of jasmonates is linolenic acid. Lipid analysis of plant cell membranes has revealed this fatty acid to be present in large quantities as a component of glycerides. In contrast, negligible amounts of free fatty acids are found within the cytoplasm. Therefore, the first step in JA biosynthesis is the release of the unsaturated fatty acid precursor, linolenic acid, from a cellular membrane. The identity of this membrane, however, is presently unclear. The concentration of JA rapidly increases in response to environmental stimuli. Hence, the mechanism controlling the release of linolenic acid constitutes a pivotal regulatory control point. Linolenic acid is characteristically found attached at the sn-2 position of plant membrane phospholipids. Recently, a gene, designated *DAD1*, has been identified that encodes the phospholipase A_1 activity responsible for the release of linolenic acid from cell membranes.

The released linolenic acid is oxidized by the action of lipoxygenase (LOX). The expression of genes encoding LOX isoforms are activated by a number of environmental stimuli, preceding increases in enzyme activity. LOX catalyzes the incorporation of molecular oxygen into polyunsaturated fatty acids. In the case of JA biosynthesis, a specific 13-LOX incorporates oxygen into the fatty acid at the n-13 position of linolenic acid. The product of the reaction is 13-hydroperoxylinolenic acid, a conjugated hydroperoxydiene. Lipoxygenases are widely distributed among plant organs. Seeds are a particularly rich source of LOX activity, which presumably reflects the huge mobilization of lipid reserves following germination. Vegetative lipoxygenases are found in the cotyledons of germinating seedlings and in young leaves and stems. The

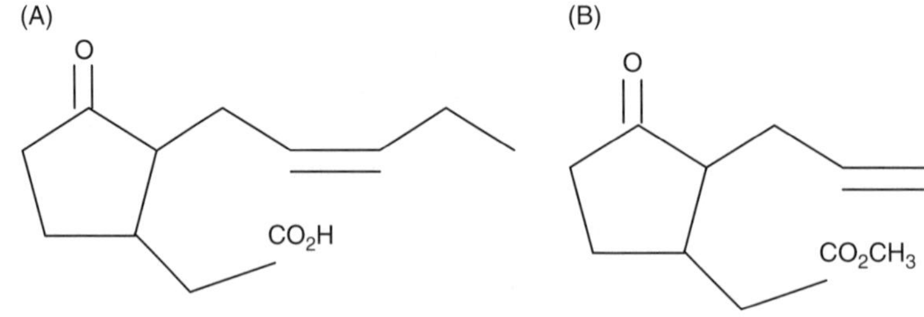

Figure 1 Structure of jasmonic acid (A) and methyl jasmonate (B).

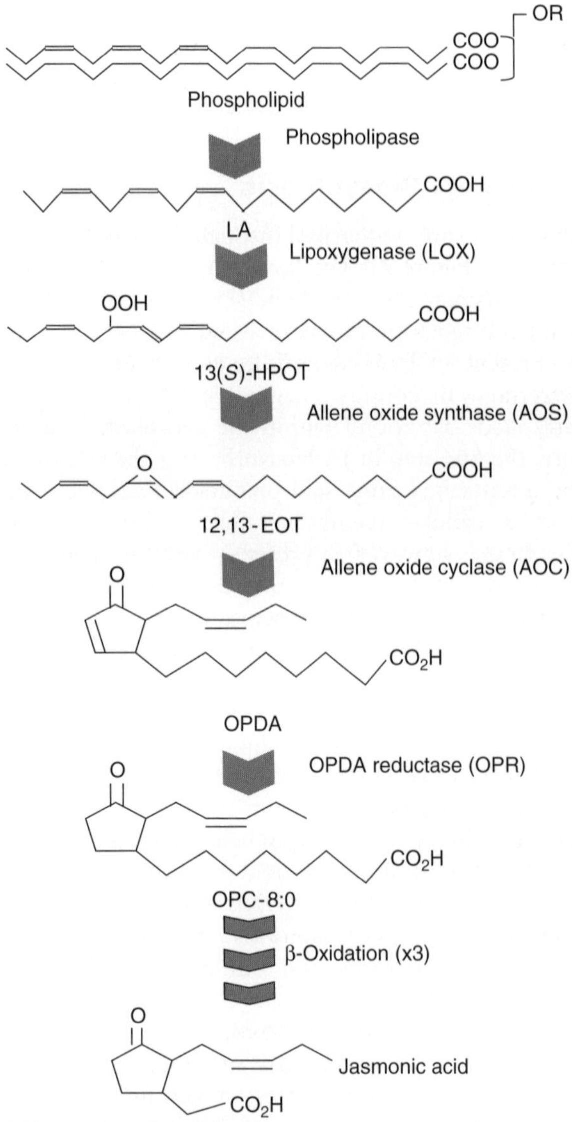

Figure 2 The biosynthesis of jasmonic acid and related jasmonates. See text for details.

location of LOX activity in a variety of plant organ and cell types suggests a high division of labor among plant lipoxygenases.

The generated 13-hydroperoxylinolenic acid is then dehydrated by the action of allene oxide synthase (AOS). The gene encoding this activity is rapidly induced in response to JA, suggesting jasmonate biosynthesis is under the control of a positive regulatory loop. This intermediate (12,13-epoxy linolenic acid) is rapidly cyclized to the cyclopentenone 12-oxo-phytodienoic acid (OPDA) by the action of allene oxide cyclase (AOC). This activity was first detected in flax (*Linum usitatissimum*) seed by Zimmerman and Feng in 1978. These investigators reported the formation of a cyclic fatty acid from the reaction of linolenic acid with flax seed extract. This enzyme (allene oxide synthase, AOS) is found throughout the plant kingdom. AOC activity is present in a variety of tissues including the leaves of developing seedlings and ripening fruits.

The enzyme OPDA reductase catalyzes the reduction of the double bond in the cyclopentenone ring of OPDA to form the cyclopentanone 3-oxo-2-(2′-pentenyl)-cyclopentane-1-octanoic acid. It is thought NADPH provides the necessary reductant. The identification of intermediates in the pathway having hexanoic and butanoic acid side chains led to the proposal that chain shortening to produce the 12-carbon product JA occurs by three cycles of β-oxidation. The final product, (1R,2S)-3-oxo-2-(2′-pentenyl)cyclopentane acetic acid (jasmonic acid) retains the same *cis* stereo configuration of the side chains as the parent 12-oxo-PDA. The *cis* form of JA is easily isomerized to the *trans* form; indeed, originally, JA was the term given to denote the *trans* form of the compound, which had unknowingly isomerized from the *cis* form during its isolation. Recently, the gene encoding the methyl-transferase activity responsible for the production of Me-JA from JA has been identified. This gene, designated *JMT*, encodes an *S*-adenosyl-methionine:jasmonic acid carboxyl methyltransferase. The expression of the JMT gene, like other genes encoding enzymes of jasmonate biosynthesis, is rapidly induced by Me-JA. Although most attention has been focused on the role of JA and Me-JA as signals in plant development and responses to the environment, other intermediates in the octadecanoid pathway may also have biological activity.

Jasmonates and Plant Development

JA and Me-JA are powerful inhibitors of germination and plant growth. The application of Me-JA to *Avena* and *Lactuca* seeds strongly inhibits germination. Moreover, exogenous application of relatively high concentrations of JA (10^{-3} mol l^{-1}), to rice (*Oryza sativa*), lettuce (*Lactuca sativa*), and wheat (*Triticum* spp.) seedling induces a growth inhibition of over 90%. In this context, the physiological events induced by JA resemble those of the plant hormone abscisic acid (ABA), although ABA is biologically active at significantly lower concentrations. Me-JA at 10^{-6} mol l^{-1} is also a powerful inhibitor of primary root growth in the model plant *Arabidopsis thaliana*. This phenomenon facilitated the development of a rapid screen to identify Me-JA resistant mutants in this species. The characterization of these mutants has provided significant insights into the functions of jasmonates in plant growth and development. Pollen germination is also influenced by JA. In *Impatiens*, 10^{-5} mol l^{-1} JA actively inhibited pollen tube growth. In contrast, however, *Lilium* pollen tube growth was strongly stimulated by 10^{-7} mol l^{-1} JA.

JA has also been shown to cue the initiation of leaf senescence, which results in a rapid degradation of chlorophyll and ribulose 1,5-bisphosphate carboxylase (RuBPCase) and a concomitant decrease in photosynthetic activity. These events follow an increase in protease and peroxidase activities. The expression of genes encoding proteins in photosynthetic carbon assimilation are suppressed by Me-JA. In both the cytoplasmic and plastidic compartments of barley leaf cells, Me-JA affects various steps in the formation of chloroplast proteins, such as the small and large subunits of ribulose-1-5-bisphosphate carboxylase (rubisco), the light harvesting chlorophyll a/b binding proteins, and the 65- and 68-kD proteins of photosystem II. Interestingly, the induction of leaf senescence by JA can be counteracted by the application of cytokinins, which are known to delay the onset of senescence in leaf tissue.

JA has also been shown to promote the biosynthesis of ethylene during fruit ripening in tomato (*Lycopersicon esculentum*). In this context, it has been demonstrated to increase the activity of 1-aminocyclopropane-1-carboxylic acid (ACC) synthase and ACC oxidase, key enzymes in the biosynthesis of ethylene. Whether jasmonates play a more general role in fruit ripening remains to be determined. There is also some evidence that JA and Me-JA are involved in responses to mechanical stimuli, known as thigmotropism. *Bryonia dioica* produce tendrils that coil around objects, thereby providing support for subsequent plant growth. A touch stimulus applied to the ventral side of the tendril promotes coiling. In contrast, a stimulus to the dorsal side inhibits this response. The application of JA and octadecanoid pathway intermediates induces the coiling response. Lignification of a zone of cells termed the Bianconi plate, in addition to proliferation of a layer of collenchyma cells around this plate, is promoted during the coiling response. These developmental changes are also activated by JA.

A physiological role for JA in tuber development has also been proposed. JA may elaborate tuber development by disrupting intracellular microtubules, and facilitating radial cell expansion, which precedes the development of the tuber. Moreover, in potato (*Solanum tuberosum*), the tuber inducing stimulus is thought to be a jasmonate family member. Even low levels of tuberonic acid-β-glucoside provides a powerful signal for tuber formation. This compound is synthesized in the leaves and transported downwards to the stolons. It is interesting that the β-glucoside conjugate rather than the free acid is the most active compound. Conjugated fatty acids may therefore also be important signaling compounds in plant cells.

Role of Jasmonates in Plant Reproduction

The requirement for jasmonates in anther dehiscence was uncovered via the analysis of *A. thaliana* mutants. A mutant that was found to be insensitive to growth inhibition by Me-JA, designated *coi1*, exhibited a male sterile phenotype. Further analysis revealed that this was due to a failure in anther dehiscence, i.e., the release of developed pollen from the anthers, even after the flowers had opened. This discovery suggested that Me-JA played an essential role in anther dehiscence. Further support for this came from the analysis of a triple mutant defective in the activity of three fatty acid desaturases, which are required for the production of the jasmonate precursor, linolenic acid. A massive reduction in linolenic acid content in this mutant resulted in reduced jasmonate biosynthesis. Like the *coi1* mutant, this triple mutant was also male-sterile, due to a defect in anther dehiscence. Furthermore, a delayed dehiscence phenotype was also found in a mutant, designated *opr3*, in which the activity of OPDA reductase, a key enzyme in jasmonate biosynthesis, was abolished in flower organs. Therefore, the characterization of male-sterile *A. thaliana* mutants uncovered a key role for jasmonates in anther dehiscence. The function of jasmonates in pollen development, however, is likely to be complex.

Careful analysis of these JA deficient mutants has shown that they also exhibit a reduction in pollen viability. Moreover, the length of the anther filament in some of these mutants is also reduced. Therefore, jasmonates may be required for multiple steps during pollen development rather than just anther dehiscence. In this context, perhaps the most important mutant to be uncovered in this series is *dad1*, which has defects in anther dehiscence, pollen maturation, and flower bud development. The *DAD1* gene was found to encode a phospholipase A1 (PLA1) lipolytic enzyme. Addition of linolenic acid to this mutant restored the wild-type phenotype. Thus, *DAD1* was found to encode the lipolytic activity responsible for the release of linolenic acid, the precursor required for jasmonate biosynthesis, from plant cellular membranes. This was an important discovery. Interestingly, *DAD1* gene expression was restricted to the filaments and was engaged immediately preceding flower opening. Furthermore, the DAD1 protein was found to be targeted to the chloroplasts. This is in agreement with some of the other enzymes integral to JA biosynthesis, which have also been shown to accumulate in the chloroplasts. The overexpression of this gene in transgenic *Arabidopsis* plants resulted in some plants being devoid of green pigmentation, and which died before the production of true leaves. Hence, the overexpression of DAD1 presumably results in the extensive hydrolysis of membrane phospholipids, resulting in the complete destruction of chloroplasts.

In angiosperms, force is required to break the stomium, a collection of specialized epidermal cells that join anther cell walls. The targeted destruction of the stomium resulted in anthers that failed to dehisce; thus, dehiscence requires functional stomium cells. The force to break the stomium is supplied by dehydration and shrinkage. Concomitantly, the liquid surrounding the pollen grains is absorbed resulting in their desiccation. Careful observation of the *dad1* mutant revealed that the dessication step was blocked, resulting from a blockage in water transport to the vascular tissue (**Figure 3**). A similar phenomenon was found for the *opr3* and *coi1* mutants; thus, water transport in anthers is regulated by JA. Filament expansion is largely driven by water transported from the dehydrating anthers. Moreover, petal cells are also dependent on water uptake for expansion; this increase in size also drives flower opening. These processes could also be controlled by water transport regulated by localized JA accumulation. Thus, JA could integrate pollen maturation, anther dehiscence, and flower opening, resulting in effective self-pollination of *Arabidopsis* flowers.

Jasmonates and Insect Defense

Insect herbivores grazing on plant leaves cause mechanical injury to the leaves, which cues the activation of a plethora of inducible defense responses. One such response is the transcriptional activation of insecticidal genes, for example, those encoding serine proteinase inhibitors (*PIN*), which perturb the digestive system of feeding insects. This reprograming of gene expression occurs not only in leaves under direct attack, but also in challenged systemic leaves. Hence, both local and long range systemic signals are generated, which engage defense gene expression throughout the plant. Furthermore, intra- and interspecific airborne signals may also be released that activate anti-insect defenses in neighboring plants (**Figure 4**). The exogenous application

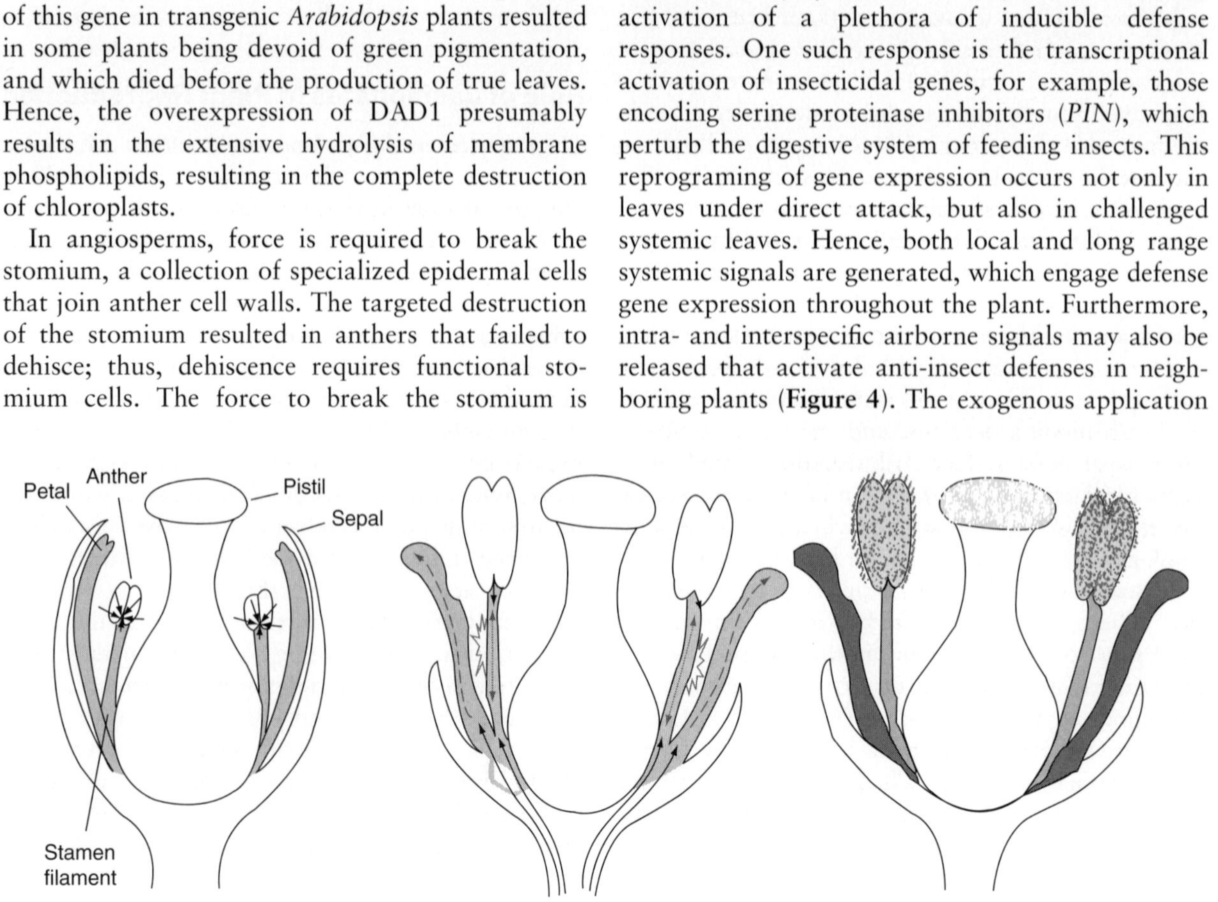

Figure 3 The role of JA in plant reproduction. The shaded areas represent organs that actively take up water and elongate in response to JA. Elongation of organs represented by arrows. Shading represents organs that actively take up water, resulting in their elongation in response to JA.

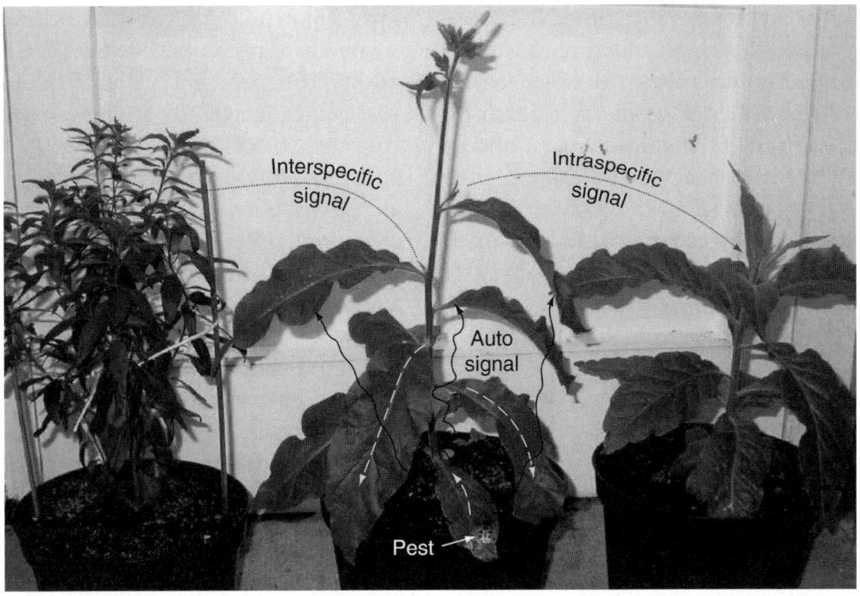

Figure 4 Volatile Me-JA functions as an airborne signal to cue the deployment of defense responses against insect herbivores. The dashed arrows represent endogenous signaling, solid arrows represent liberation of volatile jasmonate, and dotted arrows represent intra- and interspecific defense signaling cued by JA accumulation in response to herbivorous insect attack.

of JA onto the leaf surface was found to activate the expression of *PIN* and other defense related genes. This suggested that JA may be a key signaling molecule in the activation of defense related genes in response to insect attack. In this context, various intermediates of JA biosynthesis, such as linolenic acid and OPDA, were also found to activate defense related gene expression. The specificity of this response was confirmed by the failure of stearic and arachidonic acids to elaborate this response, demonstrating that it was not a general effect of fatty acids *per se*. Furthermore, a structural specificity of hydroperoxides and linolenic acid was shown. Only 13(*S*)-hydroperoxylinolenic acid, the intermediate in JA biosynthesis, and not 13(*S*)-hydroperoxylinoleic acid, a structurally similar fatty acid, could engage defense related gene expression. In a similar manner, linolenic acid was shown to induce *pin* gene expression, but the structural analog γ-linolenic did not. These results suggested that the applied JA biosynthetic intermediates were being converted directly to JA within the unwounded leaf. Hence, the enzymes responsible for JA biosynthesis may therefore be continually present, at least at a basal level, within leaf tissue. Thus, the rate determining step in jasmonate biosynthesis is probably the release of the fatty acid precursor linolenic acid, following the activation of *DAD1* encoded phospholipase A1 activity in response to insect feeding.

Confirmation of a function for jasmonates in insect defense was provided by the observation that mechanical injury resulted in a massive and transient accumulation of JA and Me-JA in leaf tissue, which preceded the transcriptional activation of defense related genes. Further evidence that JA was an essential cue in the establishment of insect resistance was provided by an elegant series of experiments employing mammalian anti-inflammatory drugs. The cyclooxygenase inhibitors aspirin and propyl gallate and the lipoxygenase inhibitors SHAM and ZK139817 can effectively block the accumulation of JA following mechanical wounding. Importantly, when JA biosynthesis is blocked by application of these enzyme inhibitors and the plant is subsequently injured, no defense related gene expression occurred in the leaves. Furthermore, the exogenous application of JA complemented the impact of these agents on JA biosynthesis. These experiments provided compelling evidence that JA is an essential component of the insect resistance signaling pathway. Recent evidence has suggested that JA signaling involves an E3 ubiquitin ligase mediated pathway, reviewed by Turner and coworkers in 2002.

Genetic studies have further underscored a role for jasmonates in insect defense. The *def1* mutant of tomato fails to engage defense related gene expression in response to exogenous JA application. Importantly, this mutant exhibited increased susceptibility to tobacco hornworm (*Manduca sexta*) larvae. Furthermore, the *Arabidopsis* triple mutant (*fad3 fad7 fad8*), which is compromised in the production of linolenic acid, the precursor for jasmonate, was found to show increased susceptibility to *Bradysia impatiens*, the common fungal gnat, compared to wild-type plants. Thus, either insensitivity toward JA or an inability to synthesize this molecule resulted in increased susceptibility to insect attack. JA is not, however, the only octadecanoid derived signal involved in insect resistance. Somewhat surprisingly, the *Arabidopsis opr3* mutant, defective in OPDA reductase activity and hence JA production, was found to be no more susceptible to fungal gnat larvae than wild-type plants. Thus, an octadecanoid intermediate that accumulates upstream of OPDA reductase activity may also cue defense responses directed against insect feeding. In this context, the exogenous application of OPDA has been shown to engage the expression of a plethora of defense related genes. Thus, OPDA, JA, and possibly other as yet unidentified octadeconoid derived signals may function in combination to convey effective insect resistance.

Function of Jasmonates in Disease Resistance

The first indication that jasmonates might be involved in the establishment of disease resistance against microbial pathogens came from studies employing plant cell cultures. These experiments demonstrated that the production of antimicrobial phytoalexins by these cell cultures was cued by the addition of Me-JA, suggesting that Me-JA may be an important signal in the deployment of this key plant defense mechanism. Furthermore, JA was shown to induce the expression of genes encoding basic pathogenesis related proteins, which are known to possess antimicrobial activity. Similarly, exogenous JA application also leads to the accumulation of small antifungal peptides known as defensins. Therefore, exogenously applied jasmonates can orchestrate the deployment of a number of pivotal plant defense responses. Importantly, JA accumulates in both local and systemic tissues following pathogen infection, suggesting that JA functions as an endogenous signal during the establishment of disease resistance. The exogenous application of Me-JA to *Arabidopsis* plants induced resistance against the necrotrophic fungal pathogens *Alternaria brassicicola* and *Botrytis cinerea*. These pathogens require dead plant cells to complete their life cycle. Hence, jasmonate signaling alone is sufficient to establish disease resistance, at least against some classes of pathogens.

Arabidopsis mutants exhibiting insensitivity to JA and mutants compromised in SA signaling have provided additional evidence that jasmonate signaling is important for resistance against necrotrophic pathogens. The JA insensitive *coi1* mutant did not express the defensin gene, *PDF1.2*, following *A. brassicicola* infection, and exhibited enhanced susceptibility to both *B. cinerea and A. brassicicola*. Me-JA dependent expression of the antifungal protein Thi2.1 was also not visualized in the *coi1* mutant. Additionally, another JA insensitive mutant, designated *jar1*, was extremely susceptible to root rot caused by a number of *Pythium* species, compared to wild-type plants. In contrast, none of these mutants were more susceptible to biotrophic pathogens, such as *Peronospora parasitica*. Conversely, *Arabidopsis* mutants compromised in salicylic acid (SA) dependent signaling were more susceptible to biotrophic, but not necrotrophic, pathogens. The SA based pathway is often required for local resistance against biotrophic pathogens, which feed on living plant cells. Moreover, SA is essential for the development of systemic acquired resistance (SAR).

Recently it has emerged that plants possess the capacity to prioritize their defense responses. In this context, evidence for cross-talk between the jasmonate and salicylic acid (SA) defense signaling pathways has been uncovered. Transgenic plants silenced for expression of phenylalanine ammonia lyase (PAL), a key enzyme in phenylpropanoid metabolism that may be required for SA biosynthesis, showed reduced disease resistance to tobacco mosaic virus (TMV), but enhanced grazing induced resistance to larvae of the insect herbivore *Heliothis virescens*. In contrast, transgenic plants overexpressing PAL exhibited enhanced disease resistance to TMV, but decreased resistance to *H. virescens*. These observations suggested that while plants can deploy either effective jasmonate or SA based defense responses independently, they cannot be deployed in combination. Thus, plants prioritize the activation of given sets of defense responses that are targeted against specific classes of pathogen.

Jasmonates also play an important role in the establishment of induced systemic resistance (ISR). In *Arabidopsis* this defense mechanism is cued by the growth of *Pseudomonas fluorescens* in the rhizosphere, the environment surrounding the roots. ISR conveys resistance against a broad spectrum of microbial pathogens. This resistance mechanism occurs independently of SA, because transgenic *Arabidopsis* plants depleted in SA, while compromised in SAR, still establish ISR. However, *Arabidopsis* mutants that are insensitive to JA fail to develop ISR. Thus, jasmonates may also play an important function in this defense mechanism. In addition, *Arabidopsis* mutants insensitive to the gaseous plant hormone ethylene (ET) also fail to develop ISR. Hence, JA and ET function in combination to establish ISR. Interestingly, to date, the development of ISR has not been shown to correlate with the expression of defense qresponse genes. Thus, the molecular basis of this defense mechanism remains undetermined. Rather than leading to constitutive defense gene expression, as observed in SAR, perhaps root colonization by *P. fluorescens* leads to a super-induction of defense gene expression in response to virulent pathogens, which is sufficient to convey effective resistance.

Jasmonate Conjugation

Specific conjugates have been detected with glucose introduced at carbons 11 or 12, suggesting the presence of a JA selective glucosyltransferase. Moreover, conjugates of dihydro-JA with a number of other molecules, including the hydrophobic amino acids leucine, isoleucine, and valine, have also been detected. Conjugation is a characteristic of many signal molecules including the plant hormones auxin, cytokinin, gibberellin, and SA. The rapid metabolism of plant signal molecules is obviously crucial to maintain their transient nature. Rapid release of free JA from glucose conjugated forms, by the action of a JA selective glucosidase, may play a key functional role in the augmentation of *de novo* JA biosynthesis in response to environmental cues. Therefore, specific glucosyltransferase and glucosidase enzymes may play central roles in the modulation of jasmonate signaling.

Surface-to-Air Signaling

The plant kingdom is rich in natural chemicals, many of which are volatile. Plants have been shown to release a bouquet of volatiles during attack from insect herbivores. These chemicals are thought to elicit multiple responses including: deterring insect feeding, plant–plant communication, and the recruitment of the herbivorous predators. One of the best studied volatiles is Me-JA, which may play an important role in plant–plant signaling. The levels of Me-JA can be increased *in planta* by the over expression of the *JMT* gene, which encodes a jasmonic acid carboxyl methyl transferase. Interplant airborne signaling by Me-JA has been well documented. The release of volatile Me-JA from plants under herbivore attack may cue the deployment of defense responses in distant plants of the same species to prepare them for an impending insect challenge. Many and perhaps most plants, however, may not produce sufficient amounts of Me-JA for this task. But some flowers, such as jasmine, in addition to the leaves of several *Artemisia* species, for example, sagebrush, have been shown to undertake airborne interplant signaling. Placing tomato and sagebrush plants together in a sealed container results in the accumulation of insecticidal proteins. This response was found to be dependent on the release of Me-JA. Furthermore, field experiments have demonstrated that damaged sagebrush plants release significant amounts of Me-JA, which can be detected by gas traps at a significant distance from the damaged plants. Importantly, tomato plants growing adjacent to damaged sagebrush exhibited a striking reduction in damage resulting from insect herbivores. Blocking air contact between sagebrush and the adjacent tobacco plants abolished this reduction. Thus, airborne Me-JA may cue the deployment of insect defense responses in tomato by interspecies communication.

Future Applications

It is now apparent that jasmonates undertake multiple functions in the growth and development of plants and their responses to environmental stimuli. Insights into the molecular basis of jasmonate function is providing opportunities for the rational design of advantageous agricultural traits. For example, JA has been shown to play a central role in pollen development. In some contexts, the generation of male-sterile crops is an important agricultural target. This could be of significant benefit in some crop species where self-fertilization would ideally be abolished, facilitating the production of F_1 hybrids. Either the inhibition of jasmonate synthesis or the development of jasmonate insensitivity in the male reproductive organs, during flower development, would lead to male-sterility, thereby preventing self-fertilization. Thus, the manipulation of jasmonate action during flower development may represent a valid target to facilitate the production of hybrid crops. In addition to their roles in plant development, jasmonates are also important in plant disease resistance. The overexpression of the *JMT* gene, as described above, in *Arabidopsis* leads to the enhanced accumulation of Me-JA; thus, JMT activity may represent a key rate determining step in Me-JA biosynthesis. These transgenic plants exhibited striking resistance against the necrotrophic pathogen *B. cineria*. Hence, manipulating endogenous Me-JA levels may provide an important mechanism to generate necrotroph resistant crops. Research into the molecular basis of jasmonate action is therefore not only providing novel insights into plant growth and development, but is also revealing potential strategies for the design of crop plants.

List of Technical Nomenclature

Anther	The structure containing pollen.
Biotroph	A pathogen requiring living plant cells.
Cytokinin	A plant hormone controlling development.
Dehiscence	A mechanism of pollen release from the anther.
Filament	The shaft supporting the anther.
Methyl-transferase	An enzyme that transfers a methyl group.
Necrotroph	A pathogen that feeds on dead plant cells.
Octadecanoid pathway	A biochemical pathway that produces jasmonates.
Plastids	Organelles; for example, the chloroplast.
Proteinase inhibitors	Substances that inhibit the activity of enzymes that digest protein.
Senescence	Develomentally programed death.
Thigmotropism	Responses to mechanical stimuli.
Turgor	Osmotic pressure.

See also: **Flowering and Reproduction**: Reproductive Strategies. **Regulators of Growth**: Abscisic Acid; Ethylene.

Further Reading

Creelman RA and Mullet JE (1997) Biosynthesis and action of jasmonates in plants. *Annual Review Plant of Physiology and Plant Molecular Biology* 48: 355–387.

Farmer EE (2001) Surface-to-air signals. *Nature* 411: 854–856.

Reymond P and Farmer EE (1998) Jasmonate and salicylate as global signals for defense gene expression. *Current Opinion in Plant Biology* 1: 404–411.

Thaler JS (1999) Jasmonate-inducible plant defences cause increased parasitism of herbivores. *Nature* 399: 686–688.

Turner JG, Ellis C, and Devoto A (2002) The jasmonate signal pathway. *The Plant Cell* 14 (Supplement): S153–S164.

Vick BA and Zimmerman DC (1980) Oxidation systems for modification of fatty acids: the lipoxygenase pathway. In: Stumpf PK and Conn EE (eds) *The Biochemistry of Plants: a Comprehensive Treatise*. London: Academic Press.

Polyamines

A F Tiburcio, T Altabella and A Cordeiro, Universitat de Barcelona, Barcelona, Spain

Introduction

Polyamines (PAs) are aliphatic nitrogen-containing compounds of low molecular weight and polycationic nature. PA levels in plants depend on their biosynthesis, conjugation, degradation, transport, and conversion to other metabolites, such as alkaloids. The genes encoding most of the enzymes involved in PA metabolism have been cloned recently. This has enabled studies on enzyme regulation and the manipulation of PA metabolism through the generation of transgenic plants. Many lines of research, such as treatments with exogenous PAs, the quantification of the endogenous PA levels and the correspondent enzyme activities, together with the application of specific inhibitors and studies with mutant and transgenic plants, show that PAs are involved in the regulation of many aspects of plant development. These include cell division, morphogenesis, senescence, stress responses, and secondary metabolite production. In most of these processes, PAs act in concert with other hormonal and environmental signals. The mechanism(s) of action of PAs is (are) still largely unknown. Some of these actions might be related to their polycationic nature, enabling interactions with polyanionic macromolecules and thus modulating their biological functions. The interaction of PAs with most plant hormones suggests that the former act as plant growth regulators. It is assumed that the union to a specific receptor (not yet characterized) would initiate the chain of events leading to the differential expression of specific genes. This hypothesis is strengthened by the effects of PAs on the expression of several genes, as well as by their interaction with different signaling pathways. The study of PA response defective mutants should enable us to clarify the mechanisms of perception and signal transduction of these plant growth regulators in the near future.

Polyamines Are Aliphatic Polycations with Structural and Growth Regulator Functions

Polyamines are ubiquitously found in living organisms. The major forms are the diamine putrescine (Put), the triamine spermidine (Spd), and the tetraamine spermine (Spm) (**Figure 1**). Due to the

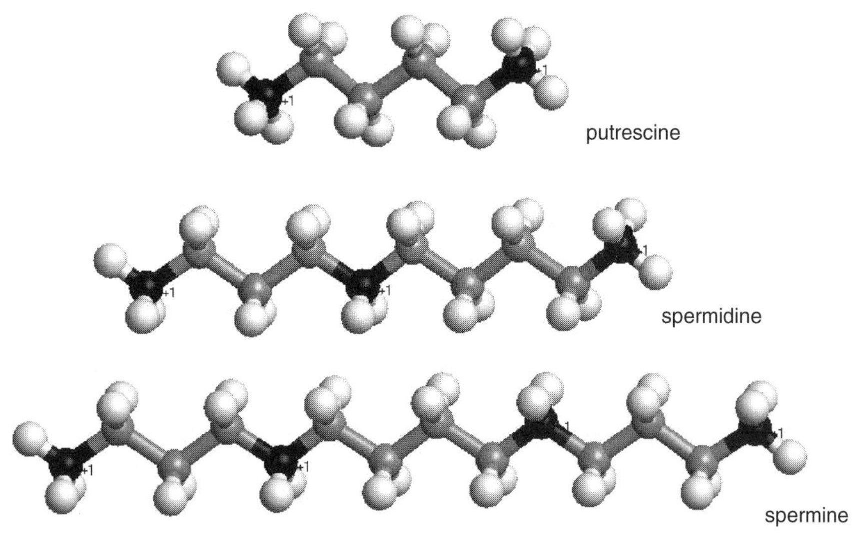

Figure 1 Molecular structure of the common polyamines. Nitrogen atoms are represented in black, carbon in gray and hydrogen in white.

presence of amino groups, these molecules are positively charged at physiological pH values, acting as polycations, with charges distributed at fixed lengths along a flexible carbon chain.

PAs play a structural role attributed to the electrostatic interactions between their positively charged amino groups and a variety of negatively charged molecules. In this way, PAs interact with nucleic acids, phospholipids, and proteins, affecting their structure and some of their biological properties. Recently, it has been shown that PAs modulate protein synthesis by affecting the secondary structure of messenger RNA (mRNA), transfer RNA (tRNA), and ribosomal RNA (rRNA).

PAs are essential for growth and development of several organisms. In mammals, the gene for ornithine decarboxylase (ODC, EC 4.1.1.17) (**Figure 2**), the main enzyme responsible for Put biosynthesis in animals, is a cellular protooncogene. It has been demonstrated that malignant transformation induced by several carcinogens or oncogenes is associated with the constitutive activation of the ODC gene.

For years there has been a debate on the regulatory function of PAs in plants, since they have to be present at much higher concentrations than phytohormones to affect a given physiological process. This is probably due to the dual function (structural and regulatory) of PAs in contrast to the classical hormones, which only play a regulatory role. Polyamines are now recognized as a new group of plant growth regulators. In contrast to animal hormones, which exert very specific physiological effects, plant hormone functions overlap broadly. Throughout this article, we will see how PAs affect most of the growth and development phases in

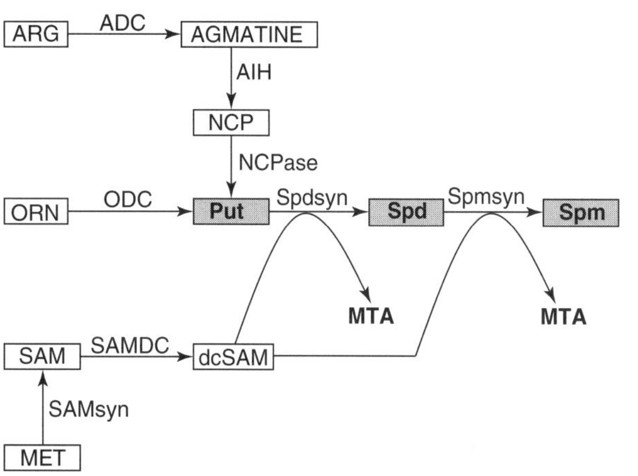

Figure 2 The polyamine biosynthetic pathway in plants. ARG, arginine; ADC, arginine decarboxylase; AIH, agmatine iminohydrolase; dcSAM, decarboxylated SAM; MET, methionine; MTA, methylthioadenosine; NCP, *N*-carbamoylputrescine; NCPase, NCP amidohydrolase; ORN, ornithine; ODC, ornithine decarboxylase; Put, putrescine; SAM, *S*-adenosylmethionine; SAMDC, SAM decarboxylase; SAMsyn, SAM synthase; Spd, spermidine; Spdsyn, spermidine synthase; Spm, spermine; Spmsyn, spermine synthase.

plants, interacting both positively and negatively with other groups of hormones.

Polyamine Biosynthesis

In higher plants, Put is formed through two biosynthetic routes. One of them includes the reaction catalyzed by ODC (**Figure 2**), whereas the other starts with the decarboxylation of arginine by arginine decarboxylase (ADC, EC 4.1.1.19), to form agmatine, which in turn is converted to Put through a series of intermediates (**Figure 2**). The genes for both

ODC and ADC have been cloned from a variety of plants. The expression patterns of the corresponding mRNAs indicate that ODC is expressed mainly in dividing tissues, whereas ADC is expressed during cell elongation and stress situations.

The conversion of Put to Spd occurs through the addition of an aminopropyl group. This group derives from methionine, which is initially converted to *S*-adenosylmethionine (SAM), and, subsequently, into decarboxylated SAM, in reactions catalyzed by SAM synthase (EC 2.5.1.6) and SAM decarboxylase (SAMDC, EC 4.1.1.50), respectively (**Figure 2**). Spd synthase (EC 2.5.1.16) catalyzes the transfer of the aminopropyl group from decarboxylated SAM to Put to form Spd.

Another aminopropyl group, also derived from decarboxylated SAM, is necessary for the conversion of Spd into Spm, in a reaction catalyzed by Spm synthase (EC 2.5.1.22). The other product of the aminopropyl transfer reactions, 5′-methylthioadenosine (**Figure 2**), is recycled back into SAM. Genes for SAMDC, Spd, and Spm synthases have also been cloned from a number of plants.

A particular characteristic of plants is that SAM, aside from participating in numerous transmethylation reactions, as it does in other organisms, is also the precursor for the biosynthesis of ethylene (**Figure 3**).

Inhibitors of PA biosynthetic enzymes have been characterized and have proven very useful in the study of this metabolism (**Table 1**).

Polyamine Biosynthetic Enzymes Are Regulated at the Transcriptional, Translational, and Posttranslational levels

ADC is subject to transcriptional, translational, and posttranslational regulation in various plant species. Often, a lack of correlation exists between transcript levels and ADC enzyme activity. In some plants, it was shown that ADC is synthesized as a precursor peptide that undergoes posttranslational modifications to generate the active protein. For example, oat (*Avena sativa*) ADC is synthesized as a precursor that is specifically processed into two peptides. In *Arabidopsis thaliana* under osmotic stress, one of the two existing forms of ADC is activated, suggesting that the same species may contain isoenzymes with differing functions. Recently, the regulation of ADC at the translational level by a short open reading frame in its 5′ untranslated region has been described in clover (*Trifolium* spp.).

Unlike ADCs, plant ODCs are inhibited by Put, Spd, and Spm. In mammals, there is a protein named antizyme, which interacts with the N-terminal region of ODC, causing a conformational change of the enzyme and its rapid degradation. An ODC inhibitor of proteinaceous nature has also been identified in barley (*Hordeum vulgare*). In this case, it was suggested that rather than initiating a degradative process, this protein may be involved in the regulation of ODC activity.

Plant SAMDC synthesis seems to be inhibited by Spd at the transcriptional level. Furthermore, SAMDC is reported to be synthesized as a preprotein, which is autoprocessed to two subunits that constitute the active form of the enzyme.

Polyamine Biosynthetic Enzymes Are Compartmentalized

Using cell fractionation experiments and enzyme activity measurements or autoradiographic methods with labeled specific inhibitors, it has been possible to localize the activities of PA biosynthetic enzymes in some cellular compartments (**Table 2**). However, due to the limitations of these techniques, the data obtained should be considered with some caution. Clearer and more reliable information may be obtained from immunological studies.

Recently, antibodies against plant ADCs became available. Polyclonal antibodies against plant ADCs

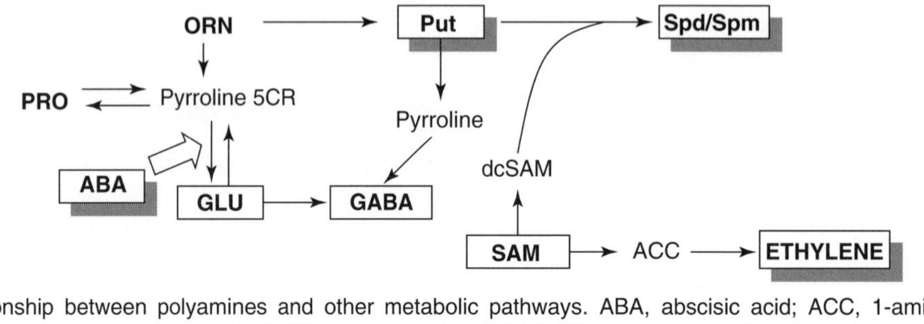

Figure 3 Relationship between polyamines and other metabolic pathways. ABA, abscisic acid; ACC, 1-aminocyclopropane-1-carboxylic acid; GABA, γ-aminobutyric acid; dcSAM, decarboxylated SAM; GLU, glutamic acid; ORN, ornithine; PRO, proline; Put, putrescine; Pyrroline 5CR, pyrroline-5-carboxylate; SAM, *S*-adenosylmethionine; Spd, spermidine; Spm, spermine.

Table 1 Inhibitors of polyamine biosynthetic enzymes

Enzyme	*Inhibitor*	*Structure*	*Mechanism*
Arginine decarboxylase (ADC)	L-canavanine	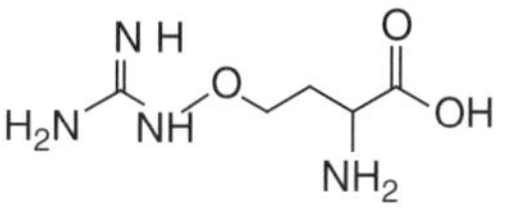	Reversible and competitive
	DL-α-methylarginine	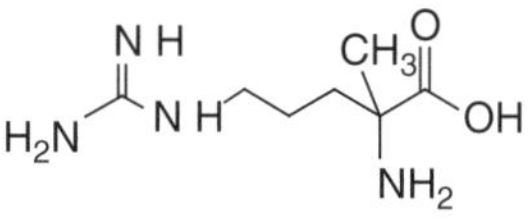	Reversible and competitive
	DL-α-difluoromethylarginine (DFMA)	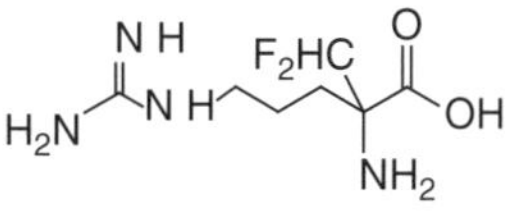	Irreversible
Ornithine decarboxylase (ODC)	L-canaline	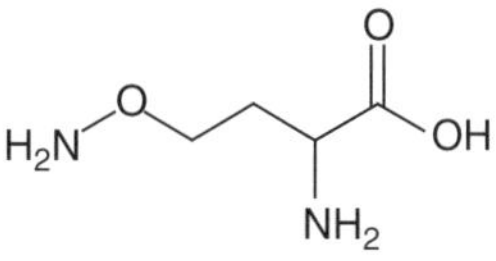	Reversible and competitive
	DL-α-methylornithine	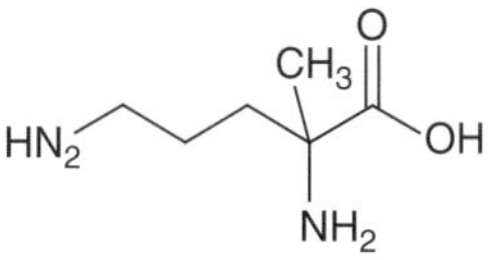	Reversible and competitive
	DL-α-difluoromethylornithine (DFMO)	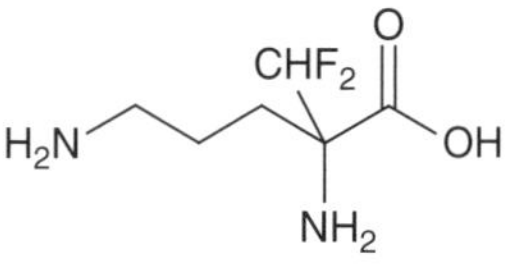	Irreversible
S-Adenosylmethionine decarboxylase (SAMDC)	Methylglyoxal bis(guanylhydrazone) (MGBG)	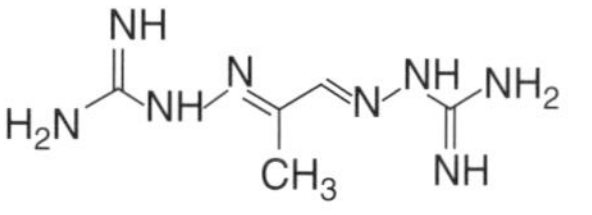	Reversible and competitive
Spermidine synthase	Cyclohexylamine (CHA)	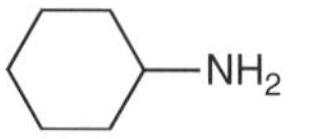	Reversible and competitive

Table 2 Subcellular localization of polyamine biosynthetic enzymes by biochemical and immunological methods

Enzyme	*Localization*	*Methods*
Arginine decarboxylase (ADC)	Chloroplasts, mitochondria, cytoplasm, nucleus	Cellular fractionation–enzyme activity/ immunological analysis Labeled inhibitors (^{14}C or ^{3}H)
Ornithine decarboxylase (ODC)	Vacuole, nucleus, cytoplasm	Cellular fractionation – enzyme activity Labeled inhibitors (^{14}C or ^{3}H)
S-Adenosylmethionine decarboxylase (SAMDC)	Cytoplasm	Cellular fractionation – enzyme activity
Spermidine synthase		

have been used to demonstrate that oat ADC is localized in chloroplasts, associated with thylakoid membranes, whereas in tobacco (*Nicotiana tabacum*) ADC is present in chloroplasts and nucleus.

Degradation (Catabolism) of Polyamines

Amino oxidases are a class of copper-dependent enzymes that include diamino oxidases (DAO, EC 1.4.3.6), highly specific for diamines, as well as PA oxidases (PAO, EC 1.5.3.3), which oxidize Spd and Spm on their secondary amino groups. The reaction products of DAO action on Put are pyrroline, H_2O_2, and ammonium, whereas PAO produces pyrroline and 1,5-diazabicyclononane from Spd and Spm, respectively, together with diaminopropane (Dap) and H_2O_2 (**Figure 4**). Dap may be converted to β-alanine, while pyrroline may be catabolized to γ-aminobutyric acid (GABA). GABA is then transaminated and oxidized to succinic acid, which is

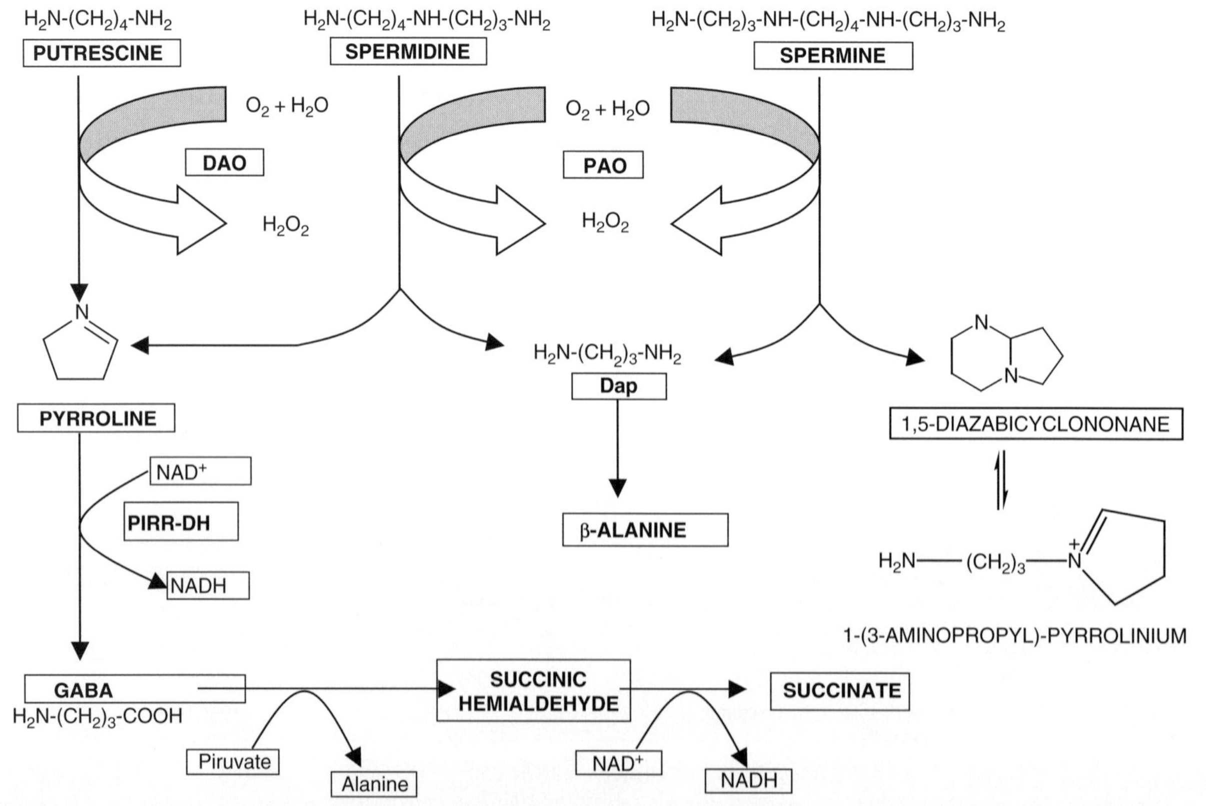

Figure 4 Polyamine degradation by amine oxidases. DAO, diamine oxidase; Dap, diaminopropane; GABA, γ-aminobutyric acid; PAO, polyamine oxidase; PYRR-DH, pyrroline dehydrogenase.

incorporated into the Krebs cycle, thus recycling the carbon and nitrogen coming from Put and Spd (**Figure 4**). Recently, both DAO and PAO have been cloned from a number of plant species.

Polyamines Form Conjugates with Differing Molecules

In nature, PAs occur not only as free bases but also as conjugates, forming part of peptides (e.g., glutathionylspermidine), amino acids (e.g., putreamine), or antibiotics (e.g., bleomycin). PAs can conjugate with acetic acid, forming acetyl derivatives, which are widespread in the animal kingdom. In plants, however, the hydroxycinnamic acid amides are the most common PA conjugates. Thus, in some plants, Put, agmatine, Spd, and Spm form amide bonds with one or two molecules of coumaric, caffeic, and/or ferulic acids, giving rise to hydroxycinnamoyl acid amides (**Figure 5**).

Posttranslational covalent linkage of PAs to proteins is catalyzed by a class of enzymes known as transglutaminases (TGases, EC 2.3.2.13). Similar to TGases of animal origin, most plant TGases play structural functions. Nevertheless, a new possible function in the photosynthetic process has also been proposed, since TGases catalyze the union of PAs to chloroplast proteins.

Polyamines Are Precursors for Secondary Metabolites

Put is the precursor not only of Spd and Spm, but also of pyrrolidine alkaloids. These include tobacco alkaloids such as nicotine and nornicotine. Furthermore, Spd and Spm can form conjugates with fatty acids or cynnamic acids, giving rise to macrocyclic alkaloids, as is the case of the Spd-derived *Cannabis* alkaloids.

For many years, it was assumed that Put used in the biosynthesis of nicotine and other pyrrolidine alkaloids came from the decarboxylation of ornithine via ODC. However, studies on tobacco callus have demonstrated that ADC, and not ODC, is the most important source of Put for alkaloid biosynthesis.

The flux of the diamine pools can be driven to "primary" or "secondary" pathways by means of genetic manipulation. In the case of Put, this is of significance because this diamine is a precursor not only of nicotine, but also of other medicinally important compounds such as, for example, tropane alkaloids.

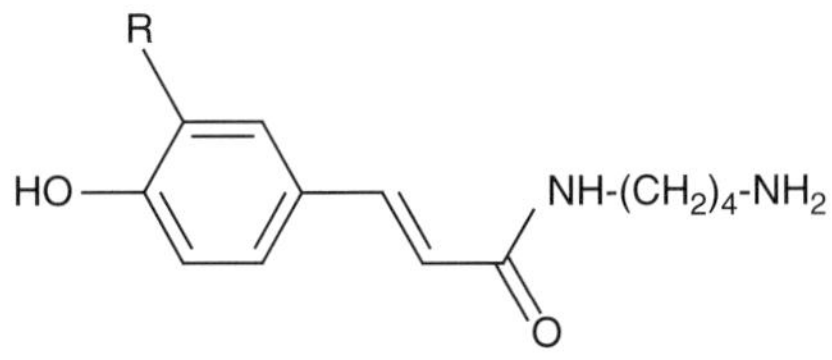

R=H Coumaroylputrescine
R=OH Caffeoylputrescine
R=OCH_3 Feruloylputrescine

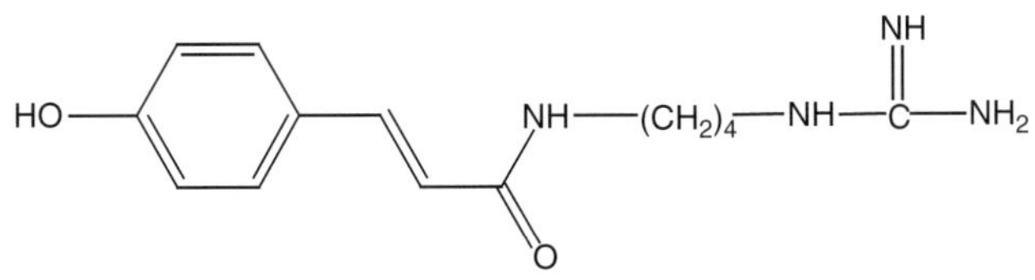

Coumaroylagmatine

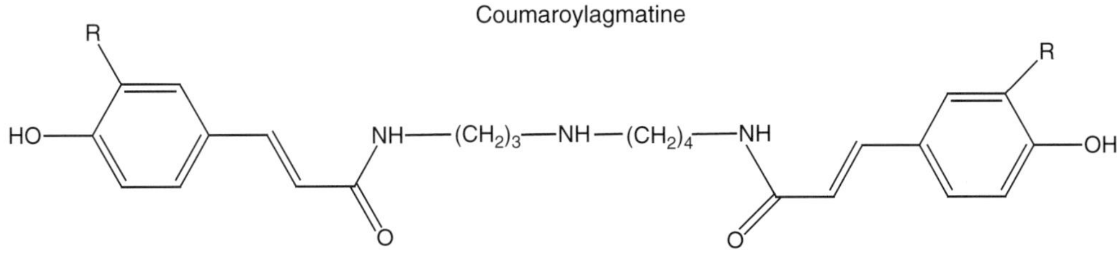

R=H Dicoumaroylspermidine
R=OCH_3 Diferuloylspermidine

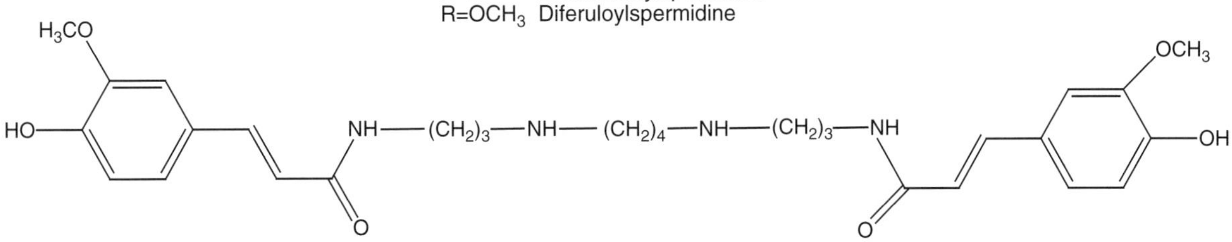

Diferuloylspermine

Figure 5 Chemical structure of some polyamine–cinnamic acid conjugates.

Polyamines Are Transported throughout the Plant

In general, cells are equipped with efficient transport systems for the uptake of external PAs. It is not known whether there are individual transporters for the various PAs or only a single one, capable of transporting all the PAs. In some systems there seem to be several PA transporters, whereas in others only a single transporter appears to exist. In plants, the studies on PA transport are more difficult than in animals due to cell walls, which can sequester exogenously supplied PAs. However, there is evidence to suggest that PA transport occurs as an active bidirectional process. PA import depends on ATP and on the Ca^{2+} concentration, being stimulated by auxins and cytokinins.

Long-distance transport is a common feature of hormone action. It is therefore important to determine how PAs are translocated inside the plant. Studies on long-distance transport show the existence of a nonpolar PA translocation system within plants. These compounds may be absorbed by the root system or synthesized in different parts of the plant and transported, via xylem and/or phloem, to other parts of the plant. Put, for example, is rapidly absorbed via roots against a concentration gradient in tomato (*Lycopersicon esculentum*) and corn (*Zea mays*; maize) seedlings and translocated, as such or as its metabolites, to the upper parts of the plant. The translocation, which is dependent on both temperature and relative humidity, seems to be related to plant transpiration.

Receptors and Signaling Pathways

Specific receptors for Put and Spd have been identified in the plasma membrane of *Escherichia coli*. In plants, several Spd-binding proteins solubilized from zucchini (*Cucurbita pepo*) plasma membranes have been characterized, but the question about the existence of PA receptors is still controversial. In animal cells, it has been proposed that PAs control the basal level of second messengers and modulate extracellular signals transduced through G protein-coupled receptors. In plants, the information on the role of PAs in signal transduction is scanty. It is known that PAs influence the activity of specific protein kinases. PAs may also play a role in plant gene regulation. Furthermore, it has been suggested that PAs, like auxins and cytokinins, may use the phosphatidylinositol pathway for the transduction of signals between plant cells.

Polyamines Regulate Cell Division

When slices of the parenchyma tissue of *Helianthus tuberosum* were excised from a dormant tuber and grown without exogenous hormones the cells did not divide, while the presence of PAs (Spm, Spd, Put, or cadaverine), supplied to the medium, actively stimulated cell division. This growth stimulation was similar to that observed by treatment with auxins. Further work has shown that a correlation exists between PA levels and cell division.

Sugar beet (*Beta vulgaris*) cells grown in suspension culture were used to investigate the regulation of PA levels during the transition from a quiescent to a proliferating state. During this transition there was an increase in DNA synthesis which was accompanied by a dramatic increase in cellular PA content. The increase in PAs was preceded by a rapid but transient increase in ODC and SAMDC activities. Inhibition of SAMDC activity by methylglyoxal bis(guanylhydrazone) (MGBG) (**Table 1**) reduced cell division. Inhibitors of ODC and ADC activities individually had little effect on cell division, but in combination led to a reduction in cell division. The transition of cells from G2 to mitosis is known to depend on the activity of a multiprotein complex, which includes cyclins. The expression of a mitotic cyclin in quiescent sugar beet cells in suspension culture was induced both by PAs (especially Put) and by their precursors arginine and ornithine. This indicates that PAs have a role in controlling gene expression, and, more specifically, in controlling cell division. Moreover, the observed correlation between peaks of casein kinase II activity and peaks of PA levels in synchronizable tobacco cells suggests that interactions with PAs might be important regulators of that kinase activity during the cell cycle.

Polyamines Are Involved in Plant Morphogenesis

Embryogenesis

The first morphogenetic effect of PAs noted in higher plants was on embryoid formation in cultures of carrot (*Daucus carota*). In the presence of high levels of auxin, these cultures grow as calli, forming no embryoids. When auxin is removed, ADC activity and Spd content rise, and embryogenesis occurs. If the increases in ADC activity and Spd titer are prevented by the inhibitor dl-α-difluoromethylarginine (DFMA) (**Table 1**), embryogenesis only occurs when Spd is added. It is likely that this effect is truly on morphogenesis and not on general growth processes, since nonembryogenic lines of carrot fail to show the rise in ADC activity or any dependence on Spd. Carrot cells grown in the presence of auxin are not embryogenic, and produce more ethylene and less PAs, presumably by switching the fate of SAM, which is a common precursor of both substances (**Figure 3**). The evidence suggests that the morphogenetic potential to produce embryoids is elicited by PAs and inhibited by ethylene. This hypothesis is strengthened by the observation that several chemical inhibitors of PA production reduce embryoid formation, and especially by the increased somatic embryogenesis observed in a transgenic carrot cell culture with higher Put content due to the overexpression of mouse ODC. It remains to be determined if the interaction between ethylene and PAs occurs at the level of their biosynthesis or of signaling pathways (**Table 3**).

Flower Induction

Thin cell layer (TCL) explants of flowering tobacco plants cultured *in vitro* can be made to differentiate vegetative buds, floral buds, or roots depending on the conditions of the culture. During vegetative and floral bud differentiation, there is an increase in ADC activity and a decrease of ODC activity. Application of DFMA inhibits bud initiation, while the inhibitor of ODC dl-α-difluoromethylornithine (DFMO) (**Table 1**) does not. Endogenous PAs increase during bud development, and Spd levels in floral buds are much higher than in vegetative buds. When Spd formation is inhibited by cyclohexylamine (CHA, an inhibitor of Spd synthase activity) (**Table 1**), flowering decreases in parallel with the decline of Spd levels, and this effect is reversible by applied Spd.

A similar effect of Spd on the flowering of *A. thaliana* has also been demonstrated. Spd levels in *Arabidopsis* flowers are higher than in vegetative tissues. CHA inhibits flowering, while Spd

Table 3 Processes of plant growth and development in which polyamines are involved

Physiological process	*Increases in*	*Key enzyme(s)*[a]	*Inhibition by*[b]	*Interactions with phytohormones*	*Possible mechanisms of action*
Cell division	Polyamines	ODC SAMDC	DFMO MGBG	Auxins	Induction of mitotic cyclins
Somatic embryogenesis	Spermidine	ADC	DFMA	Auxins Ethylene	Interaction with ethylene biosynthesis and/or signaling pathway
Floral induction	Spermidine	ADC	DFMA	Auxins Cytokinins	Interaction with molecules of the "floral stimulus"
Floral development	Polyamines	ODC	DFMO	Unknown	Interaction with floral identity genes?
Root formation	Putrescine	ADC	DFMA	Auxins Cytokinins	Interaction with auxin signaling?
Tuber formation	Polyamines	SAMDC	MGBG	Ethylene	Interaction with ethylene biosynthesis and/or signaling pathway
Senescence and fruit ripening	Spermidine, spermine	ADC	DFMA	Ethylene	Inhibition of lipid peroxidation Interaction with ethylene biosynthesis and/or signaling pathway
Stress	Polyamines	ADC	DFMA	Ethylene Abscisic acid	Free radical scavenging Interaction with abscisic acid signaling pathway Production of phytoalexins, induction of pathogenesis-related proteins
Secondary metabolism	Putrescine	ADC	DFMA	Auxins	Control of putrescine methyltransferase?

[a] ADC, arginine decarboxylase; ODC, ornithine decarboxylase; SAMDC, *S*-adenosylmethionine decarboxylase.
[b] DFMA, DL-α-difluoromethylarginine; DFMO, DL-α-difluoromethylornithine; MGBG, methylglyoxal bis(guanylhydrazone).

application promotes flowering under noninductive conditions (short day). Furthermore, in a late-flowering *Arabidopsis* mutant, the addition of Spd significantly promotes flowering. These results show that Spd plays a role in floral induction (**Table 3**).

Flower Development

While PAs formed via ADC are involved in floral initiation, there is also evidence that PAs derived from ODC are required for flower development (**Table 3**). Thus, transgenic tobacco plants overexpressing mouse ODC had abnormal flower morphology. These results are in accordance with the observations in a male sterile stamenless tomato mutant line, which showed abnormally high levels of PAs in all floral organs. Exogenously applied PAs often caused abnormal floral development, and conversely, some inhibitors of PA biosynthesis caused normal pollen development in the mutant. The abnormally high PA levels may contribute to the pattern of abnormal stamen development.

Rooting

It is well established that root differentiation in the whole plant and stem segments cultured *in vitro* proceeds in general from perivascular cells. In contrast, in TCL explants, composed only of epidermal, subepidermal, and cortical cells, roots can differentiate from these superficial cells when the explants are cultured *in vitro*. Therefore, this system is useful for the study of root initiation, since it indeed reflects organ initiation in the strict sense. When the inhibitors DFMA and DFMO were used, root formation was inhibited in both cases, but PA levels were only decreased by DFMA. When explants exposed to DFMA were transferred to a medium lacking the inhibitor but containing Put, they were able to differentiate roots, whereas DFMO-mediated root inhibition could not be reversed by exogenous Put (**Table 3**).

Tuber Formation

The role of polyamines in tuber formation has been studied by transgenic approaches. In antisense SAMDC potato plants (*Solanum tuberosum*), a reduction in the SAMDC transcript level was accompanied by a range of stunted phenotypes with highly branched stems, short internodes, and small and chlorotic leaves, as well as inhibited root growth. Additionally, these antisense potato plants did not flower, and produced very small and elongated tubers. The characteristics of these antisense plants correlated with reduced levels of SAMDC transcript and enzyme activity, PA content, and increased rate of ethylene evolution. Although these results indicate that PAs play a role in tuber formation, further work is required in order to understand fully the relationship between PAs and ethylene in this process (**Table 3**).

Polyamines Act as Antisenescence Agents

Treatment with Spd or Spm prevents the loss of chlorophyll, stabilizes the molecular composition of the thylakoid membranes, and delays senescence in dark-incubated oat leaves. The inhibition of lipid peroxidation may be one of the mechanisms responsible for the antisenescence effects of PAs. This inhibition of lipid peroxidation by Spm was suggested to be due to direct binding of Spm to microsomal phospholipids. Another possible molecular mechanism by which Spm may inhibit lipid peroxidation could be related to the ability of this PA to form a ternary complex with Fe^{2+}, changing its susceptibility to autoxidation and thus its ability to generate free oxygen radicals.

Transgenic tomato plants overexpressing yeast SAMDC under the control of an ethylene- and ripening-inducible promoter showed delayed leaf senescence, as well as delayed fruit ripening. The transgenic tomato lines are being used to study the interactive roles of PAs and ethylene in the regulation of plant senescence (**Table 3**).

Polyamines and Stress

The first observation that Put levels increase in plants subjected to potassium deficiency dates back to the 1950s. Since then, many studies have shown that other kinds of abiotic stress, such as osmotic stress, acid stress, high NH_4^+ concentrations, and atmospheric contaminants (e.g., SO_2 or Cd^{2+}), as well as low temperatures also lead to Put increases. This response is mainly mediated by the ADC pathway. However, stress induced Put accumulation is not common in all plants. For instance in tobacco leaves under osmotic stress, Put levels and ADC activity decrease, whereas Spd and Spm levels increase. This behavior has been observed in other plant systems, suggesting that Spd and Spm may also contribute to stress resistance. It is noteworthy that excessive Put accumulations may have deleterious effects on plants. For instance, transgenic tobacco plants overexpressing oat ADC generated in our laboratory displayed chlorotic and necrotic lesions that correlated with the levels of Put increase. This has also been observed more recently in *Arabidopsis*. A more direct involvement of PAs in stress responses has been observed in carrot cell cultures overexpressing mouse ODC.

Under osmotic stress, there is a correlation between the increases in PA levels and the accumulation of proline. In *A. thaliana*, the expression of pyrroline-5-carboxylate synthase (one of the enzymes responsible for the proline increase) is induced by osmotic stress. The expression of pyrroline-5-carboxylate synthase is regulated at the transcriptional level by an abscisic acid dependent pathway. Since PA metabolism is connected to proline synthesis (**Figure 3**), it is possible that the interaction between that amino acid and PAs occurs through the abscisic acid signaling pathway (**Table 3**).

High ozone levels induce a number of biochemical alterations in plants, some of which are defense mechanisms against oxidative stress. Such mechanisms include increases in the levels of ascorbic acid, peroxidases, phenolic compounds, ethylene, and PAs. When tobacco and tomato plants are fed with PAs, there is a decrease in leaf damage caused by ozone. Furthermore, ADC activity and Spd levels increase in ozone-treated oat leaves before the first lesions become visible. DFMA treatment does not allow the increase in ADC activity, leading to more severe lesions. These results suggest that PAs have a protective role against ozone. The formation of superoxide radicals is inhibited *in vitro* by PA application. It is thus possible that PAs act as free radical scavengers.

PAs have also been implicated in biotic stress responses, with roles in the resistance against viral, bacterial, or fungal attacks in plants. A number of mechanisms for PA actions in plant–pathogen interaction have been proposed. For instance, in cereals, fungal and bacterial pathogens elicit the production of PA conjugates, which may function as phytoalexins, whereas in tobacco Spm induces the expression of pathogenesis-related proteins and the resistance against tobacco mosaic virus infection.

The involvement of PAs in the responses to such a broad variety of stress processes may be the starting-point for many interesting agronomic applications, especially by manipulating PA metabolism with transgenic plants.

Outlook

Although the PA biosynthetic pathways in plants are now well established, further efforts are needed to clarify other aspects of PA metabolism, including subcellular localization and organization of PA metabolic enzymes, PA degradation, conjugation, and transport. Most of the genes encoding PA biosynthetic enzymes in plants have recently been isolated and antibodies against some of the enzymes are available. These tools have been used to initiate the study of the regulatory mechanisms controlling cellular PA levels. However, future work is needed to analyze the regulation of PA gene expression. With the availability of the cloned genes, the effects of the manipulation of PA levels on cell division, morphogenesis, senescence, stress, and secondary metabolism have been studied using transgenic approaches. This has increased our understanding of how PAs may act, providing clues for advantageous biotechnological applications. Furthermore, studies with transgenic plants are being complemented with genetic approaches. Future experiments should be designed to identify the components of the PA signaling pathway(s) in plants. In addition, the possible crosstalk with differing signaling pathways should also be investigated in order to unravel PA interactions with other plant hormones and environmental stimuli. We expect that a combination of molecular and genetic approaches, together with biochemical and cell biology studies will provide important insights into the mechanisms of PA action in plants.

List of Technical Nomenclature

ADC	Arginine decarboxylase.
CHA	Cyclohexylamine.
DAO	Diamine oxidase.
Dap	Diaminopropane.
DFMA	dl-α-Difluoromethylarginine.
DFMO	dl-α-Difluoromethylornithine.
MGBG	Methylglyoxal bis(guanylhydrazone).
ODC	Ornithine decarboxylase.
PA(s)	Polyamine(s).
PAO	Polyamine oxidase.
Put	Putrescine.
SAM	*S*-adenosylmethionine.
SAMDC	*S*-adenosylmethionine decarboxylase.
Spd	Spermidine.
Spm	Spermine.
TCL	Thin cell layer.
TGase(s)	Transglutaminase(s).
GABA	γ-Aminobutyric acid.

See also: **Abiotic Stresses**: Cold Stress; Free Radicals, Oxidative Stress and Antioxidants; Mechanical Stress and Wind Damage. **Flowering and Reproduction**: Flower Development. **Growth and Development**: Leaf Development; Molecular Biology of Development. **Postharvest Physiology**: Ripening; Senescence, Flowers; Senescence, Leaves. **Primary Products**: Amino Acids. **Root Development**: Root Growth and Development. **Seed Development**: Embryogenesis; Germination. **Tissue Culture**: Organogenesis; Somatic Embryogenesis. **Water Relations of Plants**: Drought Stress.

Acknowledgments

This work was supported by CICYT-BIO99-453. AC is supported by a FCT, Portugal, studentship (BPD/3579/2000).

Further Reading

Bouchereau A, Aziz A, Larher F, and Martin-Tanguy J (1999) Polyamines and environmental challenges: recent development. *Plant Science* 140: 103–125.

Cohen SS (1998) *A Guide to the Polyamines*. New York: Oxford University Press.

Evans PT and Malmberg RL (1989) Do polyamines have roles in plant development? *Annual Review of Plant Physiology and Plant Molecular Biology* 40: 235–269.

Galston AW and Kaur-Sawhney R (1995) Polyamines as endogenous growth regulators. In: Davies PJ (ed.) *Plant Hormones: Physiology, Biochemistry and Molecular Biology*, 2nd edn, pp. 158–178. Dordrecht: Kluwer Academic.

Galston AW, Kaur-Sawhney R, Altabella T, and Tiburcio AF (1997) Plant polyamines in reproductive activity and response to abiotic stress. *Botanica Acta* 110: 197–207.

Kumar A, Altabella T, Taylor MA, and Tiburcio AF (1997) Recent advances in polyamine research. *Trends in Plant Science* 2: 124–130.

Kumar A and Minocha SC (1998) Transgenic manipulation of polyamine metabolism. In: Lindsey K (ed.) *Transgenic Plant Research*, pp. 187–199. Harwood: Academic Publishers.

Malmberg RL, Watson MB, Galloway GL, and Yu W (1998) Molecular genetic analyses of plant polyamines. *Critical Reviews of Plant Science* 17: 199–224.

Tiburcio AF, Altabella T, Borrell A, and Masgrau C (1997) Polyamine metabolism and its regulation. *Physiologia Plantarum* 100: 664–674.

Walden R, Cordeiro A, and Tiburcio AF (1997) Polyamines: small molecules triggering pathways in plant growth and development. *Plant Physiology* 113: 1009–1013.

Juvenility

S R Adams, Horticulture Research International, Wellesbourne, UK

Introduction

Juvenility is the early phase of plant growth, from seed germination, during which flowering cannot be induced. It is important as it affects when a plant can become reproductive. The length of the juvenile phase can vary greatly from a few days in some herbaceous species to 30–40 years in beech (*Fagus sylvatica*) trees (**Table 1**). In a few species, there is no

Table 1 The approximate length of the juvenile phase for a range of plant species

Plant	*Common name*	*Approximate duration of juvenile phase*
Herbaceous		
Arabidopsis thaliana	Thale cress	4 days
Glycine max	Soyabean	11–32 days
Petunia x *hybrida*	Petunia	12–36 days
Oryza sativa	Rice	14–45 days
Brassica oleracea var. *botrytis*	Cauliflower	2–8 weeks
Brassica oleracea var. *italica*	Broccoli	5 weeks
Brassica oleracea var. *gemmifera*	Brussels sprout	7–11 weeks
Woody		
Vitis spp.	N/A	1 year
Prunus spp.	N/A	2–8 years
Malus spp.	N/A	4–8 years
Citrus spp.	N/A	5–8 years
Hedera helix	English ivy	5–10 years
Fraxinus excelsior	Ash	15–20 years
Acer pseudoplantanus	Sycamore maple	15–20 years
Quercus robur	English oak	25–30 years
Fagus sylvatica	European beech	30–40 years

juvenile phase; flower primordia can form within the seed. Having completed the juvenile phase, plants may be described as "ripe to flower." Once mature, the adult condition is highly stable, and reversion to the juvenile state is rare. The transition from juvenile to adult may also be associated with changes in leaf shape and thickness, pigmentation, thorniness, and rooting ability. However, these changes are often less distinct in herbaceous plants than in woody species.

Importance of Juvenility

By having a juvenile phase, plants avoid the low seed yields that would occur if they were to flower precociously when still small and with limited photosynthetic capacity. Delaying flowering and fruiting also has the advantage that more resources can be partitioned for vegetative growth, enabling young plants to compete for light. The length of the juvenile phase is very important for scheduling crop production, and in determining the time needed for breeding programs. The problem is worse when breeding woody species, which tend to have a longer juvenile phase, and because of this, more research has been carried out on woody than on herbaceous species. It may be commercially desirable to breed some species of plants with a short juvenile phase, although for other species that suffer from premature floral initiation, selection for a longer juvenile phase may prove beneficial.

Characteristics of Juvenile Plants

The potential or ability to flower is the only consistent criterion that can be used to distinguish between plants that are juvenile and those that are adult. Nonflowering plants are not necessarily juvenile; they might be capable of flowering but have not been exposed to inductive conditions. In some species (particularly woody plants), the phase change from juvenile to adult may be associated with other characteristics: leaves may change in shape, thickness and cuticular characteristics, longevity in autumn, and phyllotaxy. Other characteristics associated with the stage of development may relate to pigmentation, thorniness, branching, bark characteristics, rooting ability, disease resistance, tracheid width and length, cold resistance, and the way in which photosynthates are partitioned to the main stem or branches. For example, English ivy (*Hedera helix*) is well known for its different juvenile and adult forms (**Table 2**).

Different parts of a plant can exhibit different maturity characteristics, particularly in woody species. The lower and inner parts of plants may retain juvenile characteristics while the upper and peripheral parts are adult. This can be seen by assessing where flowering occurs, as well as by looking at the characteristics of cuttings. For example, in ivy, cuttings taken from the base of plants tend to exhibit juvenile characteristics while those from the top of the same plant may show adult characteristics. Similarly, when grafting birch, scions from the top of adult plants flower much sooner than those taken from the base. Other characteristics affected by juvenility, such as rooting, thorniness, anthocyanin formation, and leaf retention, have been seen to differ between the upper and lower parts of a plant.

Quantifying the Length of the Juvenile Phase

The juvenile phase ends when plants attain the ability to flower. However, first flowering is often used to indicate the time when plants become adult. Plants are often incorrectly assumed to be juvenile during the intervening period when plants are waiting for inductive conditions, or when flower induction or evocation is occurring. One way of quantifying when plants actually attain the ability to flower is to transfer batches at regular intervals from a noninductive to an inductive environment for flowering. Plants transferred from the noninductive environment before the end of the juvenile phase will flower at the same time (and for terminal flowering species, have the same leaf number) as those grown

Table 2 Characteristics of juvenile and adult English ivy (*Hedera helix*)

Character	*Juvenile*	*Adult*
Flowers	Absent	Present
Leaves	3- or 5-lobed palmate	Entire, ovulate
Phyllotaxis	Alternate	Sprial
Growth habit	Climbs or ground cover	Erect, shrub form
Aerial roots	Present	Absent
Pigmentation	Anthocyanin present	Anthocyanin absent

Adapted from Thomas B and Vince-Prue D (1984) Juvenility, photoperiodism and vernalization. In: Wilkins MB (ed.) *Advanced Plant Physiology*, pp. 408–439. London: Pitman and Hackett WP, Cordero RE, and Srinivasan C (1987) Apical meristem characteristics and activity in relation to juvenility in *Hedera*. In: Atherton JG (ed.) *Manipulation of Flowering*, pp. 93–99. London: Butterworths.

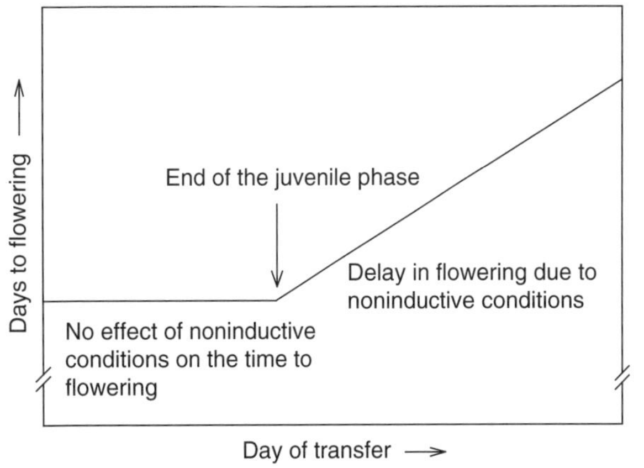

Figure 1 Assessing the length of the juvenile phase by transferring plants at regular intervals from noninductive to inductive conditions.

continually in the inductive environment (**Figure 1**). Flowering will be delayed in plants that remain under noninductive conditions after the juvenile phase has ended. The duration of the juvenile phase can be estimated by examining the data visually, using regression analysis, or through fitting more complex flowering models. Experiments of this type work well, although care is needed when interpreting the results if the different environments are likely to have a direct effect on the length of the juvenile phase. Other workers have assessed the effect of exposing plants to a short period of highly inductive conditions at differing times.

While morphological changes, such as leaf shape, may be associated with the change from juvenile to adult, most of these characteristics change gradually. There is often no distinct morphological change associated with the transition from juvenile to adult making it difficult, particularly in herbaceous plants, to use this as an indicator of juvenility. However, for a few species, a specific character such as thorniness, anthocyanin formation, phyllotaxy, or leaf shape may provide a useful indication. In woody species, plant vigor may also change with plant age (termed "ageing"), but this should be considered as independent from the change from juvenile to adult. Although not juvenile, young vegetatively propagated trees undergo changes associated with "ageing", exhibiting a loss of vigor as they develop.

Factors Affecting the Length of the Juvenile Phase

The length of the juvenile phase is genetically controlled; however, as with most genetic characters, there are interactions with the environment. Various workers have shortened the length of the juvenile phase through breeding. For example, by crossing precocious birch (*Betula* spp.) parents, seedlings were obtained that flowered in 2 years compared with the normal 5–6 years. However, the length of the juvenile phase is not consistent for a given genotype. Instead, it is affected by various environmental factors, and so chronological time (or the number of dormancy cycles) does not necessarily provide a good indication of when juvenility will end.

In general, conditions that promote vegetative growth have been found to hasten the end of the juvenile phase. Camellia (*Camellia japonica*) seedlings flowered within 2 years compared with the normal 4–5 years when grown under continuous light, at a minimum temperature of 18°C, and with regular feeding. A similar response has been observed in birch. Light integral and temperature have also been found to affect the length of the juvenile phase in a number of herbaceous species. In *Petunia* spp., shading plants resulted in the length of the juvenile phase being prolonged from 23 to 36 days. In another experiment, the length of the juvenile phase in *Petunia* was shortest at 22°C (12 days); higher or lower temperatures delayed the end of the juvenile phase (18 days at 29°C and 21 days at 14°C). In apple (*Malus* spp.) and pear (*Pyrus communis*) seedlings, inverse correlations were found between stem diameter and the length of the juvenile phase.

In some woody plants, girdling or ringing the stem by removing bark, scoring, or strangulation with wire has been used to hasten flowering. However, it is not clear whether this really is due to an effect on juvenility, as these treatments tend to be more effective on slightly older plants. The response is thought to be due to the shoot retaining a higher proportion of the assimilates produced. Conversely, removing plant material through pruning tends to delay flowering and may also prolong juvenility. In apples, at least, there is evidence to indicate that flowering can be hastened by 1–3 years by grafting onto dwarfing rootstocks (e.g., M8 or M9), although this advantage diminishes under optimal conditions for growth.

Physiological Basis for Juvenility

There is a very poor understanding of the mechanisms involved in regulating phase change. The results of many experiments have been inconclusive or even appear to contradict each other. It could well be that different species are juvenile for different physiological reasons. It has been suggested that plants may need to attain either a given size, assimilate availability, critical node number, or critical distance between the meristem and roots for juvenility to end. However, the response appears to be more complex. In blackcurrent (*Ribes nigrum*), a close correlation was found between plant size and the ability to induce flowering. However, when cuttings were repeatedly taken before plants had reached a critical size, plants of the third cutting cycle flowered. It was suggested that a minimum number of cell divisions may need to occur in the shoot apex to end juvenility. However, some plants can be maintained in a juvenile state by continually taking cuttings.

In the case of photoperiodic plants, response to an inductive floral stimulus requires that the leaves are competent to respond to photoperiod and produce the floral stimulus, and that the meristems have the capability to respond to such a stimulus. Juvenility could, therefore, be a property of either the plant or of the meristem. When apices from juvenile herbaceous plants are grafted onto adult plants, they sometimes flower sooner than they would otherwise, indicating that the meristems are competent to respond. Accordingly, it has been suggested that it is more likely that the leaves of juvenile plants are incapable of producing the floral stimulus, or that a signal from the roots maintains the plants in a vegetative state. A different response has been obtained for some woody species, although again the results have often been conflicting. Robinson and Wareing showed that there was little response when seedling scions were grafted onto adult trees; the phase change of the scion material (from juvenile to adult) was not accelerated. However, young mango (*Mangifera indica*) seedlings that were approach-grafted to adult shoots flowered early, providing that the adult shoots were girdled and the leaves were removed from the seedlings. It could be that these treatments increased the translocation of assimilates from the adult shoot to the seedling, and hence enabled transmission of the floral stimulus. Work conducted on ivy has shown that while grafting an adult scion onto the juvenile rootstock can cause the adult partner to revert to a juvenile form, it has not been possible to induce flowering of juvenile scions by grafting them onto adult plants. Therefore, in some plants, both apex and plant characteristics may be important in determining when juvenility ends.

Once the juvenile phase has ended, the adult phase is relatively stable. Asexual propagation of adult plant material does not normally cause reversion to the juvenile phase. In contrast, however, plants naturally revert to a juvenile state following sexual (and apomictic) reproduction. *In vitro* manipulation can also be used to prevent adult plants from giving juvenile offspring. Embryoids from root callus of adult ginseng (*Parax ginseng*) formed flowers when subcultured for 1 month on a medium containing GA_4; plants are normally juvenile for 3 years when raised from seed. While rare, reversion from the adult to juvenile phase has been reported in some woody plants following low nitrogen, low light, and high temperature treatments.

Effects of Growth Regulators

It has been suggested that growth regulators may control maturation, and that ratios rather than absolute concentrations may be critical. While various growth regulators have been shown to affect juvenility, it is unclear whether changes in their concentrations directly control maturation. Their action could be indirect, for example, by affecting partitioning or mobilization of assimilates.

Gibberellins

It has been suggested that gibberellins are important in maturation. When applied exogenously, gibberellins have caused adult ivy, blackwood (*Acacia melanoxylon*), citrus, pear, plum (*Prunus domestica*), almond (*Prunus amygdalus*), and apricot (*Prunus armeniaca*) to revert to a juvenile form. Also ancymidol, an inhibitor of gibberellin synthesis, has been shown to prevent ivy from reverting to a juvenile state when exposed to low light. Similarly, the growth regulator paclobutrazol, which acts by inhibiting GA biosynthesis, has been observed to promote flowering. However, the growth regulator chlormequat when used in conjunction with stem girdling has also been shown to promote flowering, despite an increase in GA-like substances. A different response appears to occur with conifers. In some Cupressaceae and Taxodiaceae species,

GA_1, GA_3, GA_4, GA_7, and GA_9 were shown to be very effective in promoting early flowering. GA_{13} and GA_{24} were less effective, and GA_5 was almost ineffective, due to poor leaf penetration. While GA_{4+7} promotes flowering in most Pinaceae, it is questionable whether it affects maturation. The ratio of male to female strobili increases with age, whereas GA_{4+7} is usually more effective at promoting female strobili. Furthermore, while gibberellins can promote flowering, their application does not appear to cause a true phase change, as young plants tend to revert to the nonflowering state when treatments cease.

Auxins

Work on Brussels sprouts (*Brassica oleracea* var. *gemmifera*) suggests that auxins can promote flowering and shorten the length of the juvenile phase. Auxins may act by stimulating differentiation of the vascular system, thus increasing the supply of nutrients and hormones to developing organs. However, the effect of applying auxins is not always consistent; for example, applying auxins to blackcurrant inhibited flowering.

Abscisic Acid

It has been reported that juvenile ivy leaves contain five times more abscisic acid (ABA) than adult leaves. However, it has been suggested that the GA/ABA ratio may be more important than the absolute concentration, and that ABA may act by regulating GA levels. The addition of ABA can prevent the rejuvenation of ivy induced by GA_3. ABA has been reported to encourage flowering in blackcurrant; however, it is doubtful that it induced phase change, as some of the control plants also flowered.

Cytokinins

The potential role of cytokinins is unclear. Cytokinins can induce apex rejuvenation of adult Douglas fir (*Pseudotsuga menziesii*) and Norway spruce (*Picea abies*), and zeatin (a natural cytokinin) has been reported to promote flowering in apple. Similarly, applying PBA (a synthetic cytokinin) has been shown to induce flowering in 4-week-old grape (*Vitis* spp.) seedlings, which would not normally be expected to flower for 3–5 years.

Ethylene

Ethephon, an ethylene releasing agent used to induce fruit drop (as a control of fruit size), tends to increase flower bud initiation in the following season. Ethephon has also been shown to induce mango seedlings to flower 3 years earlier than normal. Furthermore, when day lily (*Hemerocallis* spp.) plantlets were cultured, an increased ethylene level was recorded around the time juvenility was thought to have ended.

Genetic Basis for Juvenility

While it is clear that the length of the juvenile phase is genetically controlled, the precise genes involved and the way in which they interact is not yet understood. Work on apples indicates that the inheritance of juvenile characteristics is additive, indicating multigenic factors governing this phase. A number of genes involved in the flowering process have been identified in pea (*Pisum sativum*). The gene *Sn* is thought to control the production of an inhibitor in leaves that decreases with age, while the gene *Lf* increases the number of inductive cycles needed for induction in young plants and affects the minimum leaf number. The *Hr* gene also delays flowering by increasing the number of inductive cycles required, although it has no effect in the first 3 weeks. The *Veg* gene, when homozygous recessive, prevents flowering under favorable conditions. It has been suggested that some of these genes may be involved in determining the length of the juvenile phase.

Recent work, primarily on *Arabidopsis* and *Antirrhinum*, has revealed key genes involved in the flowering process. Mutation in the *CENTRORADIALIS* (*CEN*) gene in *Antirrhinum* or its homolog *TERMINAL FLOWER1* (*TFL1*) in *Arabidopsis* results in terminal flowering of the inflorescence. However, as well as having a terminal flower, the recessive mutant *tfl1* in *Arabidopsis* shows earlier flower commitment. Furthermore, when *CEN* is overexpressed in tobacco (*Nicotiana tabacum*) the plants flower later. It is possible, therefore, that this gene affects the length of the juvenile phase. The gene *HASTY* may also have a role in controlling the length of the juvenile phase in *Arabidopsis*, as it appears to reduce the competence of shoots to respond to *LEAFY* (*LFY*) and *APETALA1* (*AP1*), both of which are involved in floral meristem identity. When juvenile *Citrus* seedlings were transformed to constitutively express *LFY* and *AP1*, the length of the juvenile phase was shortened. Plants flowered early and had fewer juvenile traits; these characteristics were maintained in the offspring of the transgenic plants. However, similar responses have not been observed in *Arabidopsis*. It is possible that in *Citrus*, *AP1* interacts with other MADS box proteins involved in the regulation of phase change. *LFY* may also regulate the expression of these proteins.

While numerous genes affecting flowering time have been identified, it is often unclear which of these do so by affecting the length of the juvenile phase.

List of Technical Nomenclature

Adult	The phase in which plants have the ability to flower.
Apomictic	Reproduction involving specialized generative tissues, but not dependent on fertilization.
Embryoids	A mass of plant tissue that resembles an embryo.
Evocation	Changes at the apex that lead to flowering.
Induction	Changes in leaves that trigger flowering.
Juvenile	The phase of development when flowering cannot be induced.
Photoperiod	Duration of the light period in a day.
Photosynthates	Carbohydrates formed as a result of photosynthesis.
Phyllotaxy	The arrangement of leaves on a stem.
Strobilus	The cone of a gymnosperm.

See also: **Flowering and Reproduction**: Flower Development; Reproductive Strategies. **Regulators of Growth**: Vernalization. **Seed Development**: Physiology of Maturation.

Further Reading

Adams SR, Pearson S, and Hadley P (2001) Improving quantitative flowering models through a better understanding of the phases of photoperiod sensitivity. *Journal of Experimental Botany* 52: 655–662.

Bernier G, Kinet JM, and Sachs RM (1981) *The Physiology of Flowering*, Vol. 1: *The Initiation of Flowers*. Boca Raton: CRC Press.

Greenwood MS (1995) Juvenility and maturation in conifers: current concepts. *Tree Physiology* 15: 433–438.

Hackett WP (1985) Juvenility, maturation, and rejuvenation in woody plants. *Horticultural Reviews* 7: 109–155.

Hackett WP, Cordero RE, and Srinivasan C (1987) Apical meristem characteristics and activity in relation to juvenility in *Hedera*. In: Atherton JG (ed.) *Manipulation of Flowering*, pp. 93–99. London: Butterworths.

Haffner V, Enjalric F, Lardet L, and Carron MP (1991) Maturation of woody plants: a review of metabolic and genomic aspects. *Annales des Sciences Forestiares* 48: 615–630.

Meilan R (1997) Floral induction in woody angiosperms. *New Forests* 14: 179–202.

Siraj Ali YS, Tayama HK, Prince TL, and Carver SA (1990) Identification of the developmental phases in poinsettia. *Journal of the American Society for Horticultural Science* 115: 728–731.

Thomas B and Vince-Prue D (1984) Juvenility, photoperiodism and vernalization. In: Wilkins MB (ed.) *Advanced Plant Physiology*, pp. 408–439. London: Pitman.

Waring PF (1987) Juvenility and cell determination. In: Atherton JG (ed.) *Manipulation of Flowering*, pp. 83–92. London: Butterworths.

Zimmerman RH (1972) Juvenility and flowering in woody plants: A review. *HortScience* 75: 447–455.

Zimmerman RH (1976) Symposium on juvenility in woody perennials. *Acta Horticulturae* 56: 1–317.

Vernalization

M Piñeiro and J M Martínez-Zapater, Centro Nacional de Biotechnología, CSIC, Madrid, Spain

Definition and Historic Background

Vernalization (a term derived from the Latin word "vernalis," meaning "related to" or "occurring in spring") is defined as the acquisition or acceleration of the ability to flower by a prolonged chilling treatment. Once a plant has become vernalized, this state is stable throughout subsequent mitotic divisions for long periods of time. However, this vernalized state is not inherited by the progeny of a vernalized plant, implying that the vernalized state is reset and must be acquired *de novo* each generation. The promotion of flowering by an extended period of low temperature is a powerful adaptation for spring-flowering plants in temperate climates. Together with photoperiod (day length), temperature is one of the environmental factors with a predictable pattern of seasonal variation, and both factors are used by plant species to fine tune their flowering time and ensure reproductive success. **Figure 1** shows the effect of vernalization on the flowering time of *Arabidopsis thaliana*.

The phenomenon of floral induction in response to low temperature has been known for two centuries as a result of early agricultural experience, and appears in reports before the 1830s. During the second half of the nineteenth century and the early years of the twentieth century, the phenomenon was studied systematically and several reports described how a prolonged exposure to cold (winter) promoted flowering in various rosette-forming biennials and cereals. At the beginning of the twentieth century,

Figure 1 Effect of vernalization on the flowering time of *A. thaliana*. Flowering is accelerated in vernalized plants (+V) compared with nonvernalized plants (−V).

Klebs observed that sugar beet and other biennials remained vegetative for years unless they were exposed to near-freezing temperatures. However, he discussed his results on promotion of flowering by low temperatures and photoperiod responses in terms of photoassimilate availability. It was Gassner who, between 1910 and 1918, studied the floral response to low temperatures, and introduced the concept of the requirement of a cold treatment for flowering. The term vernalization was formulated for the first time by Lysenko in Russia (1929). He observed that the seeds of winter cereals could be desiccated after exposure to low temperature and then stored in the vernalized condition. This condition was maintained during storage, and sowing the seeds resulted in plants that flowered as spring varieties. He coined the term "jarovization" (spring cereals in Russian are called jarovoe, derived from the root word "jar," god of spring), and translated it into English as vernalization. The term vernalization implies that winter varieties of plants are made to behave as spring or summer varieties by exposure to cold.

Types of Vernalization Response

The ability of plants to perceive and respond to low temperatures provides them with a means of adapting their development to the environment, optimizing their chances of self-perpetuation. In this way, different species display different vernalization requirements for flowering and commonly genetic variation for the vernalization requirement exists within species. Some plant species show an absolute (qualitative) requirement for vernalization; flowering will only occur after a prolonged exposure to cold. Other species display a facultative (quantitative) vernalization response; these plants will flower earlier when they have been exposed to low temperatures than in the absence of a cold treatment, but the chilling treatment is not essential for flowering. Sugar beet (*Beta vulgaris*) and biennial varieties of henbane (*Hyoscyamus niger*) are examples of the first type of vernalization response (obligate requirement). Winter varieties of cereal species such as wheat (*Triticum* spp.), rye (*Secale cereale*), and barley (*Hordeum vulgare*) provide examples of the facultative requirement for vernalization. Furthermore, different vernalization requirements are combined with specific photoperiodic requirements in each species (see **Table 1**).

Factors Influencing the Vernalization Response

The exposure of plants to low temperatures accelerates flowering in vernalization-requiring species. However, the magnitude of the response to the chilling treatment is highly influenced by several factors, such as the temperature and duration of the period of cold exposure, the developmental stage of the plant, and the photoperiodic conditions during and after the exposure to low temperature.

Temperature and duration of the cold treatment There is a wide range of optimum temperatures for vernalization. In *Secale cereale* cv Petkus, temperatures between 1 and 7 °C are effective, although temperatures higher or lower than these can still have some effect in accelerating the time of flowering (**Figure 2**). Within this range, every species has a characteristic vernalization temperature.

The optimum duration of cold exposure required for maximal acceleration of flowering is also species-specific. Below the optimum, longer cold treatments will lead to larger reductions in flowering time, while above the optimum, longer exposures do not have further effects. In some species, very short treatments can have a substantial effect on the time of flowering. In *S. cereale* cv Petkus a 4-day treatment is enough to cause a detectable change in flowering time; in celery, 8 days of cold produces an effect, although the maximum effect can only be obtained after more than 1 month. However, in species such as henbane, more than 3 months of cold treatment are required for maximal promotion of flowering, with shorter periods of cold resulting in only partial acceleration of flowering. The effect of the duration of the chilling treatment on the flowering time of *A. thaliana* is shown in **Figure 3**.

Table 1 Examples of combinations of vernalization and photoperiod requirements for flowering in different species

Photoperiodic requirement	*Vernalization requirement*		
	Absolute (qualitative)	*Facultative (quantitative)*	*None*
Absolute LD (qualitative)	*Hyoscyamus niger* (biennial), *Campanula medium*, *Oenothera biennis*, *Beta vulgaris*	*Dianthus arenarius*, *Dianthus gallicus*, *Oenothera stricta*	*Hyoscyamus niger* (annual), *Dianthus superbus*, *Oenothera acaulis*, *Scabiosa ukranica*
Facultative LD (quantitative)	*Dianthus barbatus*, *Scabiosa succisa*, *Leucanthemum cobennense*	*Dianthus caryophyllus*, *Scabiosa conescens*, *Sinapis alba*, *Arabidopsis thaliana*, Winter cereals	*Dianthus prolifer*, *Nigella damascena*, *Scrophularia arguta*, Spring cereals
Day neutral	*Scrophularia vernalis*, *Euphorbia lathyris*, *Saxifraga rotundifolia*, *Lunaria biennis*	*Geum intermedium*, *Pisum sativum* (some varieties), *Daucus carota*	*Scrophularia peregrina*, *Euphorbia peplus*, *Senecio vulgaris*, *Lunaria annus*
SD	*Chrysanthemum morifolium* (depending on different varieties)		

LD, long days; SD, short days.

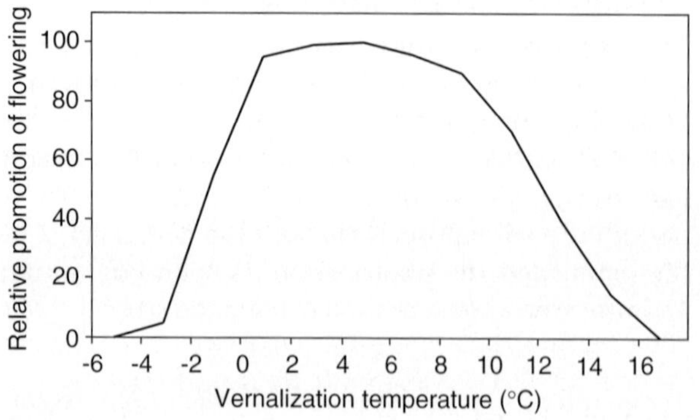

Figure 2 Effect of different vernalization temperatures on the promotion of flowering in *S. cereale* cv Petkus. Adapted with permission of Springer-Verlag from Purvis ON (1961) The physiological analysis of vernalization. In: Ruhland W (ed.) *Encyclopedia of Plant Physiology*, pp. 76–122. Berlin: Springer-Verlag.

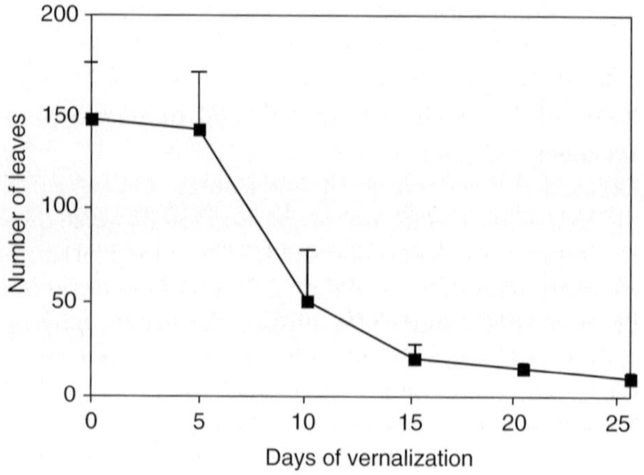

Figure 3 Effect of the duration of the vernalization treatment on the flowering time, expressed as number of leaves produced before flowering, of *A. thaliana* (ecotype *Sf-2*). Adapted with permission of Springer-Verlag from Lee I, Bleecker A, and Amasino R (1993) Analysis of naturally occurring late flowering in *Arabidopsis thaliana*. *Molecular and General Genetics* 237: 171–176.

Vernalization can be reversed in some species, such as rye and henbane, by a short exposure to high temperatures (usually between 30 and 40 °C), provided it takes place within a few days of the end of the chilling treatment. This response is known as devernalization. Normally, a period of 3–5 days in normal growth conditions is enough to stabilize the vernalized state of the plant and make the devernalization treatment completely ineffective.

Developmental stage of the plant Another factor that can influence the effectiveness of the vernalization response is the developmental stage of the plant. Commonly, herbaceous plants can acquire the vernalized state as imbibed seeds or seedlings; however, many biennial species require the development of rosette leaves before they are able to respond to vernalization. In plants that respond to chilling treatment as moistened seeds or during the seedling stage, the sensitivity to vernalization usually decreases as the plant grows older, whereas in biennial species this sensitivity increases with age. Metabolic activity seems to be a prerequisite for effective vernalization, since dry seeds are not susceptible to cold treatment. In cereals, imbibed seeds are responsive to vernalization treatment, even when hydrated to levels that are not sufficient for germination. Developmental requirements are very different among species. In cereals, seedlings are as equally responsive to treatment as mature plants. However, biennials, such as *H. niger*, will not respond to a chilling treatment before 10 days of growth, while sugar beet requires the formation of rosette leaves and responds to cold only when $2\frac{1}{2}$–3 months old. Finally, in cases such as *Lunaria biennis* or *A. thaliana*, the requirement for vernalization decreases with increasing age of the plant.

Photoperiod The vernalization response is influenced by photoperiod in two different ways. First, photoperiodic conditions before or during the exposure to low temperatures can affect the acquisition of the vernalized state. Second, the magnitude of the response to the cold treatment can be affected by photoperiod after the vernalization requirement has been satisfied. In the first case, day length can substantially modify the vernalization requirements for many plant species. The duration of the period of exposure to low temperatures required to promote flowering in these species can be decreased or even eliminated by growth under certain photoperiodic conditions. This is the case for winter varieties of crop cereals: these varieties display a quantitative vernalization requirement, but exposure to short days can reduce or even completely replace their vernalization requirement. In a similar way, the vernalization of *Campanula medium*, a biennial with an absolute requirement, is completely replaceable by a period of growth under short days. Other examples of replacement of vernalization by short days are species such as *Succisa pratensis*, *Salvia pratensis*, and *Coreopsis grandiflora*. Growth under long-day conditions can also substitute for vernalization in certain species such as *Silene armeria* or members of the legume family.

The second effect is related to the existence in many plant species of specific photoperiodic requirements after exposure to low temperature. Very frequently, species susceptible to vernalization show a subsequent requirement for long days. This situation mimics seasonal changes in temperate climates where the end of the period of low temperatures (winter) is associated with an increase in day length (spring). However, this is not a universal rule and some plant species that require vernalization are completely day-neutral or even require a period of growth in short days to initiate flowering (see **Table 1**). Biennial varieties of henbane (*H. niger*) and sugar beet display an absolute requirement for long days after vernalization. Winter cereals behave as quantitative long-day plants after their vernalization requirement has been satisfied. However, species such as *Scrophularia vernalis* or *Cardamine pratensis* will flower irrespective of day length once they have been vernalized. Finally, some varieties of chrysanthemum require a period of growth under short days after the chilling treatment to induce flowering.

The Physiology of Vernalization

Site of Cold Perception

Since plants can respond to a cold treatment, there must be specific organ(s) capable of perceiving low temperatures. Which organs are involved in the perception of cold has long remained an open question. Classically, the problem was addressed by cooling only certain parts of the plant and keeping the rest at normal growth temperatures, and subsequently monitoring which of the cooling treatments were effective in promoting flowering. Experiments in which only the shoot apex, leaves, or roots were exposed to cold indicated that the shoot apical meristem was responsible for the perception of cold. In plants like henbane, beet (*Beta* spp.) or chrysanthemum, the vernalization response is observed only when the shoot apical meristem is cooled, but not when roots or leaves are exposed to low temperatures. In cereal seeds, the embryos or even those portions of them containing dividing cells, but not the endosperm, can become vernalized providing further support to the notion that meristems can perceive low temperatures. However, experiments carried out with *Lunaria biennis* and *Thlaspi arvense* revealed that this hypothesis was too simple. In these experiments, plants were regenerated from leaves of vernalized or nonvernalized plants. Shoots regenerated from nonvernalized leaves remained vegetative, while plants regenerated from leaves that had been exposed to a low temperature treatment behaved as vernalized plants, producing flowering shoots. These results suggested that the shoot apical meristem is not the only site for cold perception, and organs other than the apex can also perceive the cold.

The Role of Hormones in the Vernalization Response

Initial grafting experiments suggested the existence of a graft-transmissible molecule, named vernalin, which was produced upon vernalization. This molecule was thought to be responsible for the observed induction of flowering in a nonvernalized biennial henbane plant after grafting with a vernalized henbane, or even a cold-treated biennial of a different species. However, these initial results on the graft-transmission of the vernalized state could not be reproduced in many other species, making it impossible to generalize about the existence of such a molecule. In henbane and many other biennial plants, the vernalization requirement for flowering can be substituted by treatment of the plants with the hormone, gibberellic acid (GA). The fact that nonvernalized biennial plants, such as henbane, can be induced to flower by treatment with gibberellin led to the proposal that vernalin could be a form of gibberellin. Furthermore, vernalization of plants such as *Raphanus sativus* (radish) and *T. arvense* leads to changes in GA metabolism in the apex of cold-treated plants. This suggests that changes in GA biosynthesis could mediate the promotion of flowering associated with vernalization. However, the application of GA cannot replace the vernalization requirement of wheat and other winter cereals or vernalization-requiring varieties of *Pisum*. In *A. thaliana*, GA treatments decrease the flowering time of both wild-type ecotypes and late-flowering mutants. However, the acceleration of flowering observed in these late-flowering mutants after the application of GA is the same in vernalization-sensitive and vernalization-insensitive late-flowering mutants. Finally, *Arabidopsis* mutants deficient in GA biosynthesis still display a regular vernalization response, suggesting that GA deficiency does not compromise the ability of *Arabidopsis* to perceive and respond to vernalization. Therefore, although GA can substitute for the vernalization requirements of many biennial plants, the role of gibberellins in the vernalization response needs to be clarified.

Genetic Basis for the Vernalization Response

Crosses between annual and biennial or spring and winter varieties of the same species have allowed the identification of loci responsible for the vernalization requirement of these varieties. Typically, a small number of loci responsible for a dominant or recessive behavior have been identified in different species. As shown in **Table 2**, the difference in vernalization requirement for flowering of annual and biennial varieties of *H. niger* is due to a single locus that responsible for the dominant biennial flowering habit. In other cases, such as sugar beet, the biennial growth habit is recessive and is also determined by a single locus. Genetic analyses performed in diploid wheat (*Triticum monococcum*) and barley (*H. vulgare*) have shown that in both species two loci act antagonistically to determine the vernalization requirement. Dominant alleles of *Vrn-Am1* in wheat and *Vrn-H1* in barley promote spring growth habit (effectively reducing the vernalization requirement), whereas dominant alleles of *Vrn-Am2* (wheat) and *Vrn-H2* (barley) increase the vernalization requirement, thus promoting the winter growth habit.

The vernalization requirement for flowering in *A. thaliana*, like that of wheat and barley, can be either recessive or dominant. Mutations affecting loci in the so-called autonomous floral promotion pathway (*fca*, *fpa*, *fve*, *fy*, *ld*) confer a recessive vernalization-responsive late-flowering phenotype. Furthermore, although *Arabidopsis* lines normally used in the laboratory are early flowering and behave as spring varieties, many naturally occurring populations display a dominant requirement for vernalization and behave as winter annuals. Crosses between early-flowering and late-flowering vernalization-responsive ecotypes of *Arabidopsis* have demonstrated that the synergistic action of two loci is responsible for the dominant winter growth habit of those ecotypes that behave as winter annuals. These two loci are called *FLOWERING LOCUS C* (*FLC*) and *FRIGIDA* (*FRI*), and dominant alleles of *FRI* and *FLC* interact to delay flowering in the absence of chilling treatment. Loss-of-function mutations in any of these genes eliminate the late-flowering phenotype conferred by these dominant alleles. In conclusion, genetic analyses performed to date in several species seem to indicate that a reduced number of loci confer the vernalization requirement in a dominant or recessive fashion.

Molecular Genetics of Vernalization

Vernalization remains one of the most poorly understood plant developmental processes at the molecular

Table 2 Genes conferring vernalization requirement in different species

Species	*Number of loci*	*Type of requirement*
Hyoscyamus niger (henbane)	1	Dominant
Beta saccharifera (sugar beet)	1	Recessive
Triticum monococcum (wheat)	2	1 dominant (*Vrn-Am2*) and 1 recessive (*Vrn-Am1*)
Hordeum vulgare (barley)	2	1 dominant (*Vrn-H2*) and 1 recessive (*Vrn-H1*)
	2	Dominant[a] (*FRI* and *FLC*)
Arabidopsis thaliana	5	Recessive[b] (*fca, fpa, fve, fy, ld*)

[a] Loci responsible for vernalization requirements in natural populations.
[b] Loci identified as recessive mutations uncovering a vernalization requirement.

level. All the information available has been obtained using *A. thaliana* as a model species. Three features can be distinguished in the molecular mechanism of the vernalization process: (1) the factors responsible for generating the vernalization requirement; (2) the vernalization pathway, or how the cold stimulus is perceived and a signal that accelerates flowering is generated; and (3) the molecular mechanism allowing the maintenance of the vernalized state through cell division.

Molecular Determinants of the Vernalization Requirement

As mentioned above, the vernalization requirement of *Arabidopsis* results from the interaction between the *FRI* and *FLC* loci. The *FLC* gene encodes a MADS-box-containing transcription factor (a class of plant transcriptional regulators involved in the control of different aspects of development). Dominant alleles of *FLC* encode a protein that seems to play a role in repressing flowering. The flowering time in *Arabidopsis* appears to be inversely related to the expression level of *FLC*. The *FRI* gene encodes a protein without significant similarity to other proteins of known function, and that seems to be involved in the positive regulation of *FLC* expression. Active alleles of *FRI* act to cause an increase in the level of the FLC product, which is accompanied by a delay in flowering. Vernalization has the opposite effect, and prolonged exposure to cold results in a decrease in *FLC* expression and an acceleration of flowering. Spring ecotypes of *Arabidopsis* carry inactive alleles of *FRI*. In these ecotypes, mutations affecting loci in the autonomous floral promotion pathway (*fca*, *fpa*, *fve*, *fy*, *ld*) also result in an increase in *FLC* transcript level. This can explain the recessive requirement for vernalization and the late-flowering phenotype caused by these mutations. Therefore, the level of FLC product is a key component in generating the vernalization requirement in *Arabidopsis* (**Figure 4**). In fact, by increasing the number of genomic copies of the *FLC* gene in winter annual (thus containing *FRI*) ecotypes of *Arabidopsis*, plants can be generated that display an absolute requirement for vernalization and behave as biennial varieties. In *Brassica* species that display a quantitative response to vernalization, the two main quantitative trait loci (QTL) responsible for this response are collinear with the regions of the *Arabidopsis* genome where *FLC* and *FRI* reside. In addition, the transcript levels of the putative ortholog of *FLC* in biennial brassicas are also reduced by vernalization. It remains an open question as to whether these conclusions can be applied to other species.

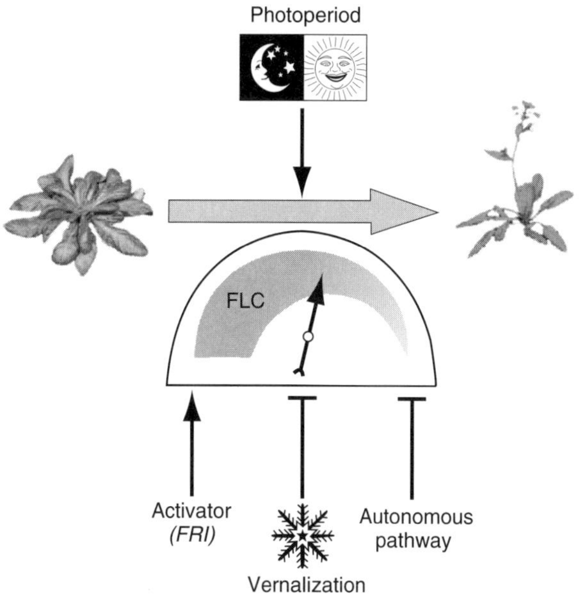

Figure 4 Schematic model showing the role of *FLC* in the control of flowering time in *A. thaliana*. *FRI*, vernalization, and the activity of the autonomous floral promotion pathway modulate the levels of the FLC product, which acts to repress flowering.

Transduction of the Low Temperature Signal

The elements required for the perception of the cold stimulus and the components of the signal transduction pathway responsible for the vernalization response remain obscure, even in the model species *Arabidopsis*. The selection of *Arabidopsis* mutations that reduce or eliminate the vernalization response of late-flowering mutants has allowed the identification of at least three loci, *VRN1–3*, which could participate in the vernalization pathway. Among these mutations, *vrn2* prevents the suppression of *FLC* expression by cold treatment. The *VRN2* gene encodes a Polycomb group protein; these types of factors have been associated with the maintenance of chromatin in a transcriptionally inactive configuration through cell division.

Molecular Basis for the Mitotic Stability of the Vernalized State

Once the vernalized state has been achieved, it is stable for the rest of the life cycle of the plant. Consistent with its role in establishing the vernalization requirement of *Arabidopsis*, the "off" state of the *FLC* gene is also stable throughout subsequent mitotic divisions following the cold treatment. This stability of the vernalized state throughout mitotic divisions and the resetting of this state in each generation is reminiscent of phenomena related to epigenetic switching of gene expression involving chromatin remodeling. DNA methylation is indeed one of the mechanisms that can contribute to the epigenetic control of transcription, and a possible role for DNA methylation in the establishment of the vernalized state has been suggested. Vernalization treatments seem to reduce the level of DNA methylation in *Nicotiana tabaccum* and *A. thaliana*. Furthermore, the level of DNA methylation can be reduced using demethylating agents or, in the case of *Arabidopsis*, using genetically modified lines with reduced expression levels of DNA methyltransferases. These reductions in DNA methylation can have an effect on flowering that is analogous to vernalization in *Thlaspi arvense*, *Arabidopsis*, and winter wheat varieties. However, in *Arabidopsis* DNA, demethylation and vernalization do not substitute for each other but have an additive effect on the promotion of flowering, suggesting that they have independent effects on flowering time. Although it is tempting to think that DNA methylation might be at least one of the components involved in promoting the epigenetic changes associated with the acquisition of the vernalized state, changes in the methylation state of a particular gene following vernalization have not been described so far.

Other Cold-Responsive Processes

Besides flowering, other plant developmental processes display a response to cold analogous to the vernalization response. Cold is required in many cases to break a specific resting period, a state in which the growth and metabolism of organs and primordia are suspended. Examples of these developmental processes are germination, break of bud dormancy in deciduous species, or break of growth arrest in bulbous, tuberous and other species.

Seed germination in many plant species depends on a prolonged period of exposure to cold. The seeds of species such as *Acer saccharum*, *Cormus florida*, *Malus* spp., and *Rosa* spp., among others, undergo a period of dormancy after desiccation, and germination occurs only after the seeds have been exposed to a period of low temperatures. This chilling treatment is known as stratification and facilitates the germination of the seeds under favorable environmental conditions.

Plants such as deciduous species are seasonally exposed to extreme environmental conditions such as chilling or freezing that can damage or even kill the organism. These species have evolved to undergo periods of dormancy that prevent the damage by exhibiting dormant buds, which are extremely resistant. These buds are initiated during the growing season, and contain the meristems or primordia that will give rise to vegetative or floral organs during the next growing season. Once inclement weather conditions have passed, bud dormancy is broken and growth resumed. In many cases, restoration of metabolic activity and cell division is dependent on a period of cold exposure. Bulbous species such as tulip (*Tulipa* spp.), hyacinth (*Hyacinthus* spp.), and crocus (*Crocus* spp.) also show low temperature requirements that are often related to the breaking of a resting period and induction of stem elongation.

In contrast to vernalization, in which the effect of cold-promotion of flowering is delayed in time, in these other cold-responsive processes, low temperatures have a direct effect on pre-existing primordia. Therefore, cold-mediated release of growth arrest does not seem to require the establishment of a mitotically stable memory of cold exposure, such as that described for vernalization. However, it is conceivable that the mechanism of cold perception and the signaling cascade(s) triggered by the exposure to low temperature could be similar in all the above cold-responsive processes. For instance, GA application can significantly reduce the duration of the cold treatment required for germination in many species, and can also accelerate the break of bud dormancy in some tree species. Furthermore, as described for vernalization, the interaction with photoperiod can on some occasions modulate the time of bud dormancy breaking: the chilling treatment is sometimes necessary but not sufficient to restore the growth of overwintering buds, and long days following the exposure to cold are required.

Relevance of Vernalization for Agriculture

Vernalization has been interpreted as an adaptation of plants from temperate climates to seasonal changes. Knowledge of the vernalization requirements for different crop varieties is crucial to maximize their yield. In fact, farmers intuitively had the notion of vernalization before it was formulated or studied systematically. For example, it was known that winter varieties of cereals had to be planted before the end of the winter if they were to set fruit before the next cold season; however, spring cereals flowered early after sowing in spring. The importance of the management of cold treatment is illustrated by the findings of Lysenko on the vernalization of imbibed cereal seeds (see above). These findings were crucial to prevent frost damage of winter varieties of cereals in extremely cold regions of Russia, enabling vernalized seeds of these winter varieties to be mechanically sown early in spring and behave as spring varieties. Although the vernalization requirements of many crop species have been selected against many horticultural and winter-flowering ornamental crops, such as strawberries (*Fragaria* spp.) or primroses (*Primula* spp.), still display strong vernalization requirements that have to be considered when growing these crops. For example, the propagation of strawberry plants in climates with low temperatures during autumn and subsequent planting in warm regions with temperate winters is a common horticultural practice to obtain an early strawberry crop with substantially increased market value.

List of Technical Nomenclature

Annual plant	A species or variety with a life cycle that is completed within 1 year.
Apex	The tip of the shoot, containing the meristem and young organ primordia.
Autonomous pathway	In *Arabidopsis*, a genetic pathway promoting flowering independently of photoperiod and vernalization.
Biennial plant	A species or variety with a life cycle spanning 2 years, with flowering taking place during the second year.
DNA methylation	Addition of methyl groups to specific sites of the DNA molecule; associated with gene inactivation.
Dormancy	A period of quiescence, inactive state.
Epigenetic	When applied to transcriptional regulation, stable alteration of gene function not due to changes in the DNA sequence of the gene.
Gibberellin	A plant growth molecule that controls the development of plants.
MADS-box proteins	Transcriptional regulators involved in the control of different plant developmental processes and containing conserved DNA-binding domains.
Orthologous	Refers to genes with the same ancestral origin found in different species.
Photoperiod	Duration of the daily light period.
Vernalin	A hypothetical growth molecule that promotes flowering upon vernalization.

See also: **Flowering and Reproduction**: Flower Development; Reproductive Strategies. **Regulators of Growth**: Gibberellins

Further Reading

Bernier G, Kinet JM, and Sachs RM (1981) *The physiology of flowering, Vol. 2*. Boca Raton: CRC Press.

Chouard P (1960) Vernalisation and its relations to dormancy. *Annual Review of Plant Physiology* 11: 191–238.

Lang A (1965) Physiology of flower initiation. In: Ruhland W (ed.) *Encyclopedia of Plant Physiology*, pp. 1380–1536. Berlin: Springer-Verlag.

Michaels SD and Amasino RM (2000) Memories of winter: vernalization and the competence to flower. *Plant Cell and Environment* 23: 1145–1153.

Napp-Zinn K (1984) Light and vernalisation. In: Vince-Prue D, Thomas B, and Cockshull KE (eds) *Light and the Flowering Process*, pp. 75–88. London: Academic Press.

Napp-Zinn K (1987) Vernalization – environmental and genetic regulation. In: Atherton JG (ed.) *Manipulation of Flowering*, pp. 123–132. London: Butterworth.

Purvis ON (1961) The physiological analysis of vernalization. In: Ruhland W (ed.) *Encyclopedia of Plant Physiology*, pp. 76–122. Berlin: Springer-Verlag.

Phytochromes and Other Photoreceptors

R E Kendrick, Wageningen University, Wageningen, The Netherlands
J L Weller, University of Tasmania, Hobart, Australia

Introduction

The light absorbed by plants spans a broad spectral range. This covers the entire spectrum visible to the human eye and extends beyond it to include both shorter (UV) and longer (far-red) wavelengths. Plants contain many compounds that are capable of absorbing light. Some of these, such as phenylpropanoids and flavonoids, which are products of secondary metabolism, are used for UV-screening. Others are photoreceptor proteins, which have acquired light-absorbing properties by incorporation of a nonprotein group termed a chromophore. Although the protein part of the molecule (the apoprotein) does not absorb light in the visible region of the spectrum, the protein can have a strong influence on the absorption range of the chromophore.

For some chromoproteins, light absorption is not their primary function, but is merely an indirect consequence of their structure (e.g., compounds involved in electron transport processes such as cytochromes and flavins). In other cases, the function of the protein is to trap light and convert its energy to a form useful for the plant through photosynthesis. These are often called light-harvesting proteins. A third class of chromoproteins regulates plant development in response to light. These are called sensory photoreceptors and essentially provide information to the plant about its light environment, giving it the capacity to respond to any changes. It is these photoreceptors that are the topic of this article (**Table 1**).

Phytochrome

Phytochrome (from Greek root words meaning "plant" and "color") is the photoreceptor involved in regulating many photomorphogenic processes in plants. It is one of a subset of photoreceptor proteins that are photochromic. Photochromic photoreceptors exist in two different forms, with different light-absorbing properties, and can be reversibly interconverted between these forms by consecutive absorption of light in the appropriate wavebands. In the case of phytochrome, the two forms have maximal absorption in the red and far-red region of the spectrum, and are referred to as Pr and Pfr, respectively. The basis for this transformation is a *cis-trans* isomerization within the chromophore.

Discovery of Phytochrome

The existence of phytochrome was first proposed by Harry A. Borthwick, Stirling B. Hendricks and coworkers at the US Department of Agriculture in the early 1950s. Their work showed that many plant responses were most effectively induced by red light and that this induction could be cancelled by immediate exposure to far-red light. These observations led to the suggestion of a pigment with two interconvertible forms – a red light-absorbing form and a far-red light-absorbing form. A pigment exhibiting this photochromic property was first detected in 1959 using sensitive dual-wavelength spectrophotometry. Phytochrome was subsequently purified using this spectrophotometric assay. Since phytochrome is present at a low concentration (10^{-6} mol l^{-1}) and chlorophyll also absorbs strongly in the red region of the spectrum, detection of phytochrome by *in vivo* spectrophotometry is restricted to tissues low in chlorophyll and usually to tissues grown in complete darkness, which contain no chlorophyll at all.

Table 1 Summary of the main plant photoreceptors

Type	*Main absorption*	*Responses*
UV-B photoreceptors	UV-B	Flavonoid biosynthesis
Cryptochromes	UV-A, blue and green	Photomorphogenesis
Phototropin	UV-A and blue	Phototropism
Zeaxanthin	Blue	Stomatal movement
Phytochromes	Red and far-red	Photomorphogenesis

The Phytochrome Chromophore

In remarkable agreement with predictions based on action spectra, the phytochrome chromophore phytochromobilin is a member of a group of compounds called tetrapyrroles, which include the chlorophylls, the heme- in hemoglobin and the bile pigments found in the gall bladder of mammals. However, unlike chlorophylls, which are cyclic, phytochromobilin is a linear molecule more closely related to the chromophores of light-harvesting proteins in cyanobacteria and red algae. It is attached via pyrrole ring 1 to a cysteine residue in the amino-half of the phytochrome apoprotein. This process is referred to as photoreceptor assembly and takes place spontaneously when both components are present. Light absorbed by the chromophore results in its isomerization. This, in turn, results in a conformational change of the protein that initiates the signal transduction chain(s) that leads to the terminal responses.

The Phytochrome Family

There are now known to be several different forms of the phytochrome apoprotein, all of around 120–130 kDa, which are encoded by a small gene family. In *Arabidopsis*, the best-studied model plant, these different forms of the apoprotein have been designated as PHYA, PHYB, PHYC, PHYD, and PHYE, and have counterparts in other species. An evolutionary tree constructed for phytochrome sequences from several plant species suggests that they fall into four subfamilies: phyA, phyB (including *Arabidopsis* phyD), phyE and phyC (**Figure 1**). It is likely that the common angiosperm ancestor contained representatives of these four phytochromes, with both duplication and/or loss of members taking place during evolution. For example, monocotyledons so far examined appear to have fewer members as a result of less diversity in the phyB subfamily. In other taxa, duplications have accumulated. In some cases, recent duplications have occurred independently in different taxa, as in the case of the phyB subfamily in *Arabidopsis* and tomato (*Lycopersicon esculentum*). This has created some confusion for phytochrome nomenclature, since the system was originally based on the five phytochromes present in *Arabidopsis*, and individual members of the phyB subfamily in other species (e.g., tomato phyB1 and phyB2) are technically not homologous with either of the two closely related phyB-type phytochromes in *Arabidopsis* (phyB and phyD).

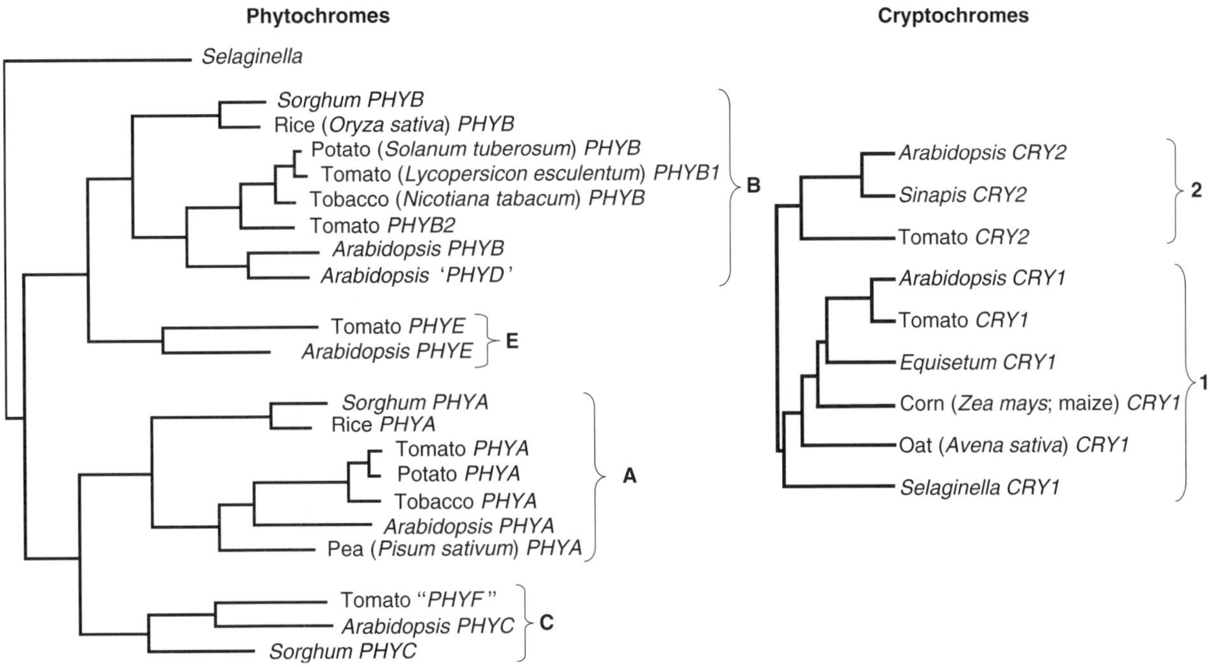

Figure 1 Family trees for phytochromes and cryptochromes from a range of plant species based on the sequences of their genes.

Among the members of the phytochrome family, phyA is unique in a number of ways. Firstly, it is present at a much higher level in dark-grown plants, in the order of 50 times more abundant than the next most abundant phytochrome. Secondly, its level is strongly reduced by light, owing to strong down-regulation of *PHYA* gene expression, and to rapid breakdown of the phyA protein after conversion to its Pfr form. These distinct properties of phyA are reflected in its unique physiological roles, which are described in more detail below.

Localization and Primary Action

In general, the physicochemical properties of phytochrome have been extensively studied for phyA and phyB only, both of which form dimers in solution, through affinity of binding sites in the C-terminal parts of their molecules (**Figure 2**). Early studies suggested that phytochrome was a cytoplasmic protein. In addition, the observation of a number of rapid phytochrome responses influencing parameters related to membranes, and the observation of regular arrangements of phytochrome in certain lower plants, both suggested that it might show some affinity for membranes. However, phytochrome was also clearly shown to regulate expression of nuclear genes, necessitating a relay of information to the nucleus. Recent observations have shown that both phyA and phyB are cytosolic when synthesized in darkness, but migrate to the nucleus after activation by light.

Although the nature of the primary action of phytochrome has been debated for many years, it is now generally accepted that phytochrome has serine/threonine protein kinase activity, which is modified by conformational changes of the protein upon photoconversion of Pr to Pfr. However, it is still not clear whether this is the only primary action of phytochrome, or whether all phytochromes have this property. One cytoplasmic target of phytochrome kinase activity, called phytochrome kinase substrate 1 (PKS1), has been conclusively identified. Phytochrome has also been shown to interact with proteins in the nucleus and to control the activity of transcription factors, which bind to light-responsive elements within the promoters of certain light-regulated genes. One protein, called phytochrome-interacting factor 3 (PIF3) associates with both phytochrome and known light-responsive elements in the promoters of target genes suggesting that there may be a rather close coupling between light absorption and the switching on and off of gene expression.

Blue Light-Absorbing Photoreceptors

Several distinct classes of photoreceptors are now known to mediate developmental responses to blue light. These include the *cryptochrome* and

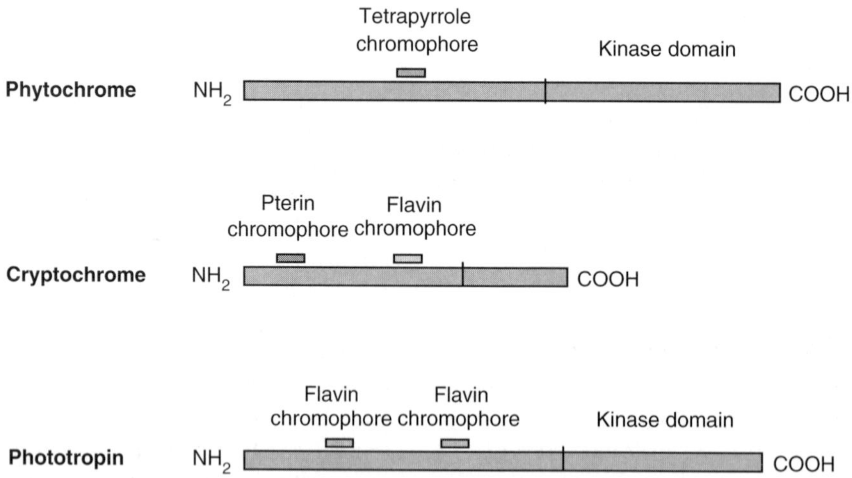

Figure 2 Generic structure of the phytochromes, cryptochromes, and phototropin photoreceptors.

phototropin families and a third, as yet uncharacterized, photoreceptor system.

The Cryptochromes

The contribution of a blue light-specific photoreceptor system to de-etiolation was first inferred from observations that blue light could promote de-etiolation, even in plants grown under continuous red light (i.e., under conditions that are saturating for phytochrome activity). In addition, mutants strongly deficient in phytochrome chromophore synthesis (and hence in the activities of all phytochromes) retain substantial responsiveness to blue light. The primary blue light photoreceptor in de-etiolation, cryptochrome 1 (cry1), was originally identified by Margaret Ahmad and Anthony R. Cashmore at the University of Pennsylvania, after cloning of the defective gene in an *Arabidopsis* mutant showing impaired de-etiolation responses to blue light. A second member of the cryptochrome family, cry2, was identified by its homology to cry1.

The cryptochrome apoproteins are around 75 kDa in molecular mass and have two distinct parts (**Figure 2**). The N-terminal half shows similarity to enzymes called photolyases, which are activated by light to repair certain kinds of damage to DNA. This part of the molecule binds two chromophores – a flavin (flavin adenine dinucleotide or FAD) and a pterin. The C-terminal halves of cry1 and cry2 show only a relatively low degree of similarity to each other and to other known proteins, but have been demonstrated to induce de-etiolation independently of light exposure. The mechanism of action of cry1 and cry2 appears to involve light-mediated redox reactions and communication between the two halves of the molecule. Both cry1 and cry2 have been shown to be localized in the nucleus where they can interact with phytochrome and serve as a substrate for its kinase activity. However, action spectra for a number of blue light-induced phenomena involving changes in ion fluxes across cell membranes suggest that cryptochromes could also be involved in light-driven redox reactions outside the nucleus.

Phototropin

Phototropin is a photoreceptor that is essential for the phototropic bending of seedling parts (e.g., hypocotyls and coleoptiles) in response to irradiation with unilateral blue light. It was first identified by Winslow Briggs and colleagues at the Carnegie Institute of Washington as a 120-kDa membrane protein, which was rapidly phosphorylated in response to blue light. The phototropin apoprotein is encoded by the *nonphototropic hypocotyl 1* (*NPH1*) gene and binds two flavin mononucleotide (FMN) chromophores. It contains conserved regions also found in flavin-binding proteins important for sensing light, oxygen, or voltage in many organisms from bacteria to mammals. It appears that as a result of light absorption the photoreceptor phosphorylates itself, and this leads to a change in its conformation. However, it is not yet understood how this could lead to the change in the distribution of growth regulators, which is also known to play a role in tropic bending.

A Third Blue Light Photoreceptor System

Xanthophylls are yellow-brown pigments that absorb blue light. One in particular, zeaxanthin, has long been considered as a potential candidate for the chromophore of an additional blue light photoreceptor. Mutants blocking zeaxanthin biosynthesis have been identified and have been shown to lack normal blue light-induced stomatal opening, although other typical cryptochrome- and phototropin-regulated responses remain normal. There is still debate about the existence of a zeaxanthin-based photoreceptor, since zeaxanthin is also an intermediate in the biosynthesis of abscisic acid, a key hormone involved in the stomatal closure mechanism.

UV-B-Absorbing Photoreceptors

Short-wavelength UV light has many damaging effects on biological systems and it is thought that plants have the capacity to prevent damage of exposure to UV-B by production of screening pigments or compounds such as flavonoids, which can quench the damaging radicals produced as a result of UV-B exposure. Diverse physiological experiments have also predicted the existence of a number of UV-B photoreceptors in plants, although none have yet been characterized. It may be that there is no single photoreceptor for UV-B and the responses to UV-B are generic in nature, induced by the presence of the damaging radicals, irrespective of their origin. The physiological importance of UV-B damage has been of particular interest in recent times, in view of the higher levels of UV-B reaching the surface of the earth as a result of the depletion of the ozone layer.

Genetic Investigations of Photoreceptor Function

One effective way to study the function of a photoreceptor has been to identify plants that lack that specific photoreceptor as a result of mutation in the gene encoding its apoprotein. Screening under monochromatic light for defective photoresponses has generally identified these mutants. In the case of cryptochrome and phototropin, isolation of mutants was essential for conclusive identification of the photoreceptor. Mutants for all but one (phyC) of the eight known photomorphogenic photoreceptors in *Arabidopsis* have now been identified. These types of screens have also identified mutants that are defective in phytochrome chromophore synthesis. Where no mutation is available in a known gene, antisense techniques have been applied in an attempt to reveal a photoreceptor function. Overexpression studies have also been conducted. Mutants with defective light responses, but normal photoreceptor content, are presumed to have defects in cellular components that are necessary for normal photoreceptor function (signal transduction chain mutants).

Photoreceptor Mutants

Studies with type-specific photoreceptor mutants have shown that both phyA and phyB play a role in the induction of germination. During de-etiolation, nearly all photoreceptors have been shown to contribute to some extent, i.e., a high degree of genetic redundancy exists, reflecting the high selection pressure for photon detection. The B-type phytochrome family members play a dominant role in shade avoidance and near-neighbor detection. In the case of flower induction, it appears that Pfr in the phyB pool has an inhibitory effect. In contrast, phyA and cryptochromes have a promotive effect on flowering. This clearly demonstrates a role for phyA in light-grown plants, despite it being light labile. Studies of type-specific phytochrome and cryptochrome mutants have revealed that members of both families can provide input signals to the biological clock.

Chromophore Mutants

Some of the first photomorphogenic mutants, which were selected for their partial inability to de-etiolate in white light, were subsequently shown to have defects in genes that encode enzymes involved in phytochrome chromophore biosynthesis. As expected on the basis of the studies with type-specific phytochrome mutants, chromophore mutants exhibit loss of responsiveness both to far-red (owing to a reduction in phyA) and red light (owing to a reduction in phyB and other phytochromes). However, in later stages of development, the chromophore-deficient mutants in many cases recover a degree of photomorphogenic responsiveness. This was originally interpreted as leakiness of the mutations, but may also be explained by more recent observations that small gene families encode some of the enzymes necessary for chromophore biosynthesis. While such mutants are not suitable for addressing the roles of individual phytochromes, they have been used as phytochrome-deficient systems for micro-injection studies with pharmacological agents in an attempt to understand phytochrome signal transduction. In a transgenic approach, such mutants have been phenocopied by overexpressing a mammalian gene in plants, which reduces a precursor of the phytochrome chromophore to an inactive product.

Transgenic Approaches

In heterologous systems, such as the overexpression of phyA from a monocotyledon (e.g., oat, *Avena sativa*) in a dicotyledonous species, the overexpressed gene can behave in a different way to the endogenous phyA gene. The oat phyA produced is much more stable than the endogenous phyA in its Pfr form, having important implications for the predicted wavelength responsiveness in the far-red, which is partially determined by its kinetics of destruction. Overexpression can even result in interference with the action of an endogenous phytochrome gene leading to a negative-dominant phenotype.

Light Signal-Transduction Mutants

Genetic analysis has revealed one class of recessive mutants that constitutively exhibits aspects of photomorphogenesis in complete darkness. Since these are recessive mutants, which implies they are loss-of-function mutants, it has led to the hypothesis that the default development of a plant is to develop as a light-grown plant; therefore, it follows that the growth strategy of etiolation must result from repression of this default pattern. Analysis of these mutants has revealed that many proteins are involved that are nuclear localized in the dark. Excitation of phytochrome appears to result in suppression of these repressive components. These events are components of the signal transduction chain closely associated with the regulation of gene expression. The high degree of redundancy between photoreceptors during de-etiolation almost certainly points to a common mode of signal transduction. However, where there is specificity, either temporal or spatial, the development of unique pathways must be anticipated. This aside, common elements of signal transduction, such as changes in calcium metabolism and guanine triphosphate (GTP)-protein activation may still be involved in many responses.

Photoreceptor Coaction and Interaction

It appears that many processes are regulated by multiple photoreceptors acting in concert. There are examples of both synergistic and antagonistic interactions. A good example of interaction is in the case of phototropism. Although blue light is essential for the growth curvature, pre-irradiation of a seedling with red light (acting via phytochrome) modifies the sensitivity of the seedling to the subsequent blue light. Within an individual gene family we refer to both overlapping and discrete functions of individual family members (**Table 2**). At the stage of seedling establishment, detection of any photons irrespective of their wavelength conveys advantage. This means that all phytochromes and cryptochromes appear to work in concert to achieve de-etiolation. However, closer studies have revealed some complex interactions. For example, phyA and phyB stimulate de-etiolation in the red and far-red region of the spectrum, respectively, but the presence of phyA reduces the effectiveness of phyB in red light. In addition, there is a complex developmental pattern of responsiveness. This could reflect the temporal and spatial expression of the photoreceptor genes or the state of the capacity of the system to respond (competence) to input signals following activation of a photoreceptor.

Modes of Photoreceptor Action

Physiological studies have identified different types of responses to light. A pulse of light can induce some responses, and these are often termed inductive responses. Other responses require longer periods of continuous irradiation and often show an increased response with increase in irradiance. Thus, it appears that photoreceptors can in essence function either as a switch or a photon counter. It is difficult to generalize about photoreceptor action, since their individual mechanisms of action may be quite diverse. However, analysis of phytochrome responses has revealed several modes of action.

The Low Fluence Response (LFR)

This is the classical phytochrome response, which is induced by a pulse of red light and nullified by a subsequent pulse of far-red light. The precise response to Pfr can either be a graded response or a response at a sharp threshold, but in both cases the

Table 2 The overlapping and discrete roles played by members of photoreceptor families in *Arabidopsis*

Response	*Phytochromes*					*Cryptochromes*		*Phototropin*
	phyA	*phyB*	*phyC*	*phyD*	*phyE*	*cry1*	*cry2*	*nph1*
Seed germination	+	+				+?	+?	
De-etiolation	+	+	+?	+	+	+	+	
Shade avoidance	(+)	+		+	+			
Flowering	+	+				+	+	
Photoperiodism	+	+						
Circadian timing	+	+				+	+	
Phototropism	(+)	(+)				(+)	(+)	+

+, active; (+), modifier; +?, probable; blank, unknown.

level of Pfr present after far-red light is insufficient to lead to a response. If an increasing period of darkness is introduced between a red pulse and a far-red pulse there is an escape from reversibility. Such experiments provide information about the kinetics of phytochrome action. Analysis of mutants reveals that B-type phytochromes play an important role in this response mode of phytochrome action.

Very Low Fluence Response (VLFR)

Action spectra for these responses coincide with the absorption spectrum of the Pr form of phytochrome. They are incredibly sensitive and can be saturated by very low fluences of red light. It has been calculated that only one in a million molecules of Pr have to be excited. This means that these responses can be elicited by the green light often used as safelight for photomorphogenic studies and even by the low level of conversion of Pr to Pfr occurring in response to far-red light. Therefore, these responses can only be studied by growing plants in absolute darkness and they are obviously not reversible by far-red light. Examples of processes showing this kind of response are the stimulation of seed germination and regulation of gene expression. Analysis of mutants has revealed that phyA is active in this mode of phytochrome action.

High Irradiance Responses (HIR)

In contrast, other responses require continuous irradiation for maximal effect and show an increase in response to irradiance. Such responses cannot be explained on the basis of the absence or presence of Pfr, but reflect the cycling between the two forms. This implies that phytochrome can act as a photon counter. Both phytochromes and cryptochromes can act as irradiance sensors. In the case of de-etiolation, far-red light has been shown to be very effective in de-etiolation and is consequently called the far-red-HIR. Physiological analysis predicted that this was a labile phytochrome and type-specific mutants showed that it is the phyA pool that is involved in this response. Red-HIR responses have also been demonstrated. Under continuous irradiation with red light, the phyA pool is rapidly depleted and the phyB-type phytochrome family members act as irradiance sensors. By definition, where the cryptochromes are involved in photon counting it can be described as a blue-HIR response.

List of Technical Nomenclature

Apoprotein	The protein part of a chromoprotein.
Chromophore	The part of a photoreceptor that gives it the capacity to absorb light.
Cryptochrome	A blue/UV-A-absorbing photomorphogenic photoreceptor.
Etiolation	The strategy of dark growth.
Photochromic	A compound that exists in two forms that are interconvertible by light.
Photomorphogenesis	The responses of plants to the light environment about them.
Photoreceptor	A protein with an attached chromophore that absorbs light.
Phototropin	The blue/UV-A-absorbing photoreceptor involved in phototropism.
Phytochrome	A red/far-red-absorbing photomorphogenic photoreceptor.
Skotomorphogenesis	The strategy of dark growth or etiolation.

See also: **Regulators of Growth**: Circadian Rhythms; Photomorphogenesis.

Further Reading

Kendrick RE and Kronenberg GHM (eds) (1994) *Photomorphogenesis in Plants*, 2nd edn, p. 828. Dordrecht: Kluwer Academic Publishers.

Lin C (2000) Photoreceptors and regulation of flowering time. *Plant Physiology* 123: 39–50.

Neff MM, Fankhauser C, and Chory J (2000) Light: an indicator of time and place. *Genes and Development* 14: 257–271.

Sage LC (1992) *Pigment of the Imagination. A History of Phytochrome Research*. San Diego: Academic Press.

Smith H (2000) Phytochromes and light signal perception by plants – an emerging synthesis. *Nature* 407: 585–591.

Photomorphogenesis

R E Kendrick, Wageningen University, Wageningen, The Netherlands
J L Weller, University of Tasmania, Hobart, Australia

Introduction

Life on earth depends on the energy emitted as a result of nuclear fusion in the sun. This solar energy, which takes the form of electromagnetic radiation (**Table 1**), is initially fixed in the biosphere by the process of photosynthesis in plants and microorganisms. Photosynthesis is the main source of energy

Table 1 The visible spectrum and its flanking regions

Region	*Wavelength (nm)*
UV	
UV-C	<280
UV-B	280–320
UV-A	320–400
Visible	
Blue	400–480
Green	480–550
Yellow	550–590
Orange	590–650
Red	650–700
Near infra-red	
Far-red	700–800

Table 2 List of responses throughout the life cycle of a plant, which are collectively grouped under the title plant photomorphogenesis

Regulation of seed germination
Seedling establishment (de-etiolation)
Responses to UV-B
Phototropism
Sun-tracking of flowers and leaves
Shade avoidance and near neighbor detection
Entrainment of the circadian clock
Photoperiodic processes such as induction of flowering and dormancy mechanisms

for the green plant, and the efficient use of the available photosynthetic radiation is the key strategy for plant survival. It is therefore not surprising that plants have evolved photosynthetic pigments that together can harvest light over the broad band of available radiation. The main photosynthetic pigments, the chlorophylls, are present at high concentration and absorb across a range of wavelengths from 400 to 700 nm. Within this range, their main absorption peaks are in the blue (BL) and red (R) regions of the spectrum. The less effective absorption by chlorophylls in the green region of the spectrum results in a preferential reflection and transmission of green light, which gives leaves their typical green color. Since light is the energy source for plants, they essentially compete with each other for access to it.

Most green plants are sedentary and, unlike most animals, are unable to move in search of a more energy rich environment. Instead, plants can modify the way they grow, so that they can use the available energy in a more efficient way. Any modification of development that increases the plant's ability to compete for light will provide a clear advantage, which can be acted upon by natural selection. A number of different developmental strategies have evolved, and any individual plant species may exhibit different strategies in different situations. For example, the plant may increase the efficiency of its light capture by modifying its photosynthetic structures. This may include modification of the chloroplasts, and of the shape, size, and thickness of leaves (i.e., they show shade tolerance). Other species compete directly for access to light by outgrowing their neighbors (i.e., they show shade avoidance). Arrangement of leaves may also be modified to minimize self-shading. Yet other species avoid competition by shifting their period of vegetative growth to a time of year when fewer other species are present. In addition, plants often exhibit a growth bending toward the light (phototropism). These growth changes in response to light, which provide a clear selective advantage, can be clearly demonstrated by time-lapse photography.

Another important developmental response to light is seen predominantly in seed plants. With the development of the seed habit it became necessary for seeds to germinate in a moist environment, often below the soil surface. This has resulted in the development of a seedling growth strategy in darkness or under very low light levels in which normal development of leaves is repressed and stem elongation is stimulated. Upon emergence into light, elongation dramatically slows down, and leaves begin to develop before the limited food reserves in the seed are exhausted.

Collectively, modifications of plant development in response to light such as these are known as "photomorphogenesis" (from the Greek words "photo" meaning light and "morphogenesis" meaning development of form). **Table 2** lists a number of processes that are encompassed by this term. In the related phenomenon of "photoperiodism," processes such as flowering, bud outgrowth, and storage organ formation are regulated by daylength, rather than by just the quantity and quality of light. Finally, in addition to these developmental responses to light, some plants also show daily movements in response to light. Examples are the phenomena of sun tracking and sleep movements of leaves and flowers.

Photoreceptors in Photomorphogenesis

Developmental responses to light do not depend on the highly abundant light harvesting pigments of photosynthesis, but on distinct photoreceptors, which provide information to the plant about the light conditions under which it is growing. However, given the importance of photosynthesis for plant growth, we could expect that wavelengths of light most appropriate to serve as developmental signals would also be those that are best for photosynthesis.

It is now known that there are several different families of photomorphogenic photoreceptors, which do predominantly absorb light in the R and BL regions of the spectrum. These include the phytochromes, which absorb principally in the R and far-red (FR) light regions, and several other photoreceptors that absorb in the BL/UV region, including the cryptochromes (*see* **Regulators of Growth:** Phytochromes and other Photoreceptors). These photoreceptors also vary in their sensitivity to other variables in the light environment, such as irradiance and duration.

Thus, it seems that selective pressures have favored the development of a relatively sophisticated light sensing system, which allows the plant to respond in different ways to different competitive situations throughout its life cycle.

Action Spectra

Unlike in animals, plant photoreceptors are not organized into specialized structures, but are distributed throughout the plant. Also, unlike the light harvesting pigments, photomorphogenic photoreceptors are generally present at very low levels, which means that they are not obvious to the naked eye. Their existence was first inferred from physiological experiments that determined the relative effectiveness of different wavelengths of light in the induction of a given photomorphogenic response. This relationship is known as an "action spectrum."

The light absorbing properties of many biological pigments derive from the interaction between the protein and an attached chemical group called a chromophore. Different chromophore groups have characteristic light absorbing properties, which determine their colors. Comparison of the action spectrum for a photomorphogenic response with absorption spectra of known pigments can therefore provide useful information about the general nature of the photoreceptor or photoreceptors regulating the response in question. In some cases, predictions made on the basis of action spectra have been remarkably accurate, as in the case of the discovery of phytochrome. However, the level of resolution in the measurement of action spectra is usually fairly low, relative to that of absorption spectroscopy. Action spectra are therefore of limited value in distinguishing the action of photoreceptors, which have only minor differences in their absorption spectra.

Germination

A dormant seed is essentially in a state of suspended animation and must make the critical (all or none) decision of whether or not to germinate. This is particularly critical for small seeds with limited food reserves. It is clear that for small seeds buried in soil, it is a selective advantage to remain dormant until brought close to the soil surface (**Figure 1**). Otherwise, the germinating seedling risks expending all the energy reserve of the seed before reaching the full light conditions under which photosynthesis can begin.

Light is known to regulate seed germination in several different ways, depending on the species. In some species, dormant seeds can be induced to germinate by incredibly small amounts of light. In these cases, the response does not depend on the wavelength of the light. It has been shown that very sensitive responses such as this are regulated by phytochrome A (phyA), a member of the phytochrome photoreceptor family (*see* **Regulators of Growth:** Phytochromes and other Photoreceptors). Such extreme sensitivity to light may represent a very early warning to buried seeds that they are near the soil surface.

In other cases, the induction of germination requires higher amounts of light, and shows wavelength dependence more characteristic of phytochrome action. A typical example is provided by dormant lettuce (*Lactuca sativa*) seeds, which germinate if exposed to a few minutes of R after imbibition, but do not germinate if maintained in darkness. An irradiation with FR given immediately after the R negates the effect of the R. This R/FR reversibility of germination can be repeated many times, and the germination response is only dependent on the final irradiation. Observations such as this have led to the idea that phytochrome operates as a molecular switch, and is activated when it is converted from the R-absorbing form (Pr) to the FR-absorbing form (Pfr).

In this response, as in others discussed below, phytochrome clearly detects the presence or absence of R. The significance of being able to monitor R is obvious when it is considered that photosynthetic pigments have an absorption maximum in the R. High levels of R are therefore a good indicator of an environment productive for photosynthesis, and linking germination to R is therefore a sensible strategy.

But what is the significance of FR to the plant? The answer to this question appears to lie in the changing spectral quality of light in the natural environment. Under a dense vegetation canopy, the light is depleted in the R region of the spectrum, but is still relatively rich in longer FR wavelengths. A low R : FR photon ratio is thus a specific indicator of vegetational shade. Phytochrome, specifically phytochrome B, gives a

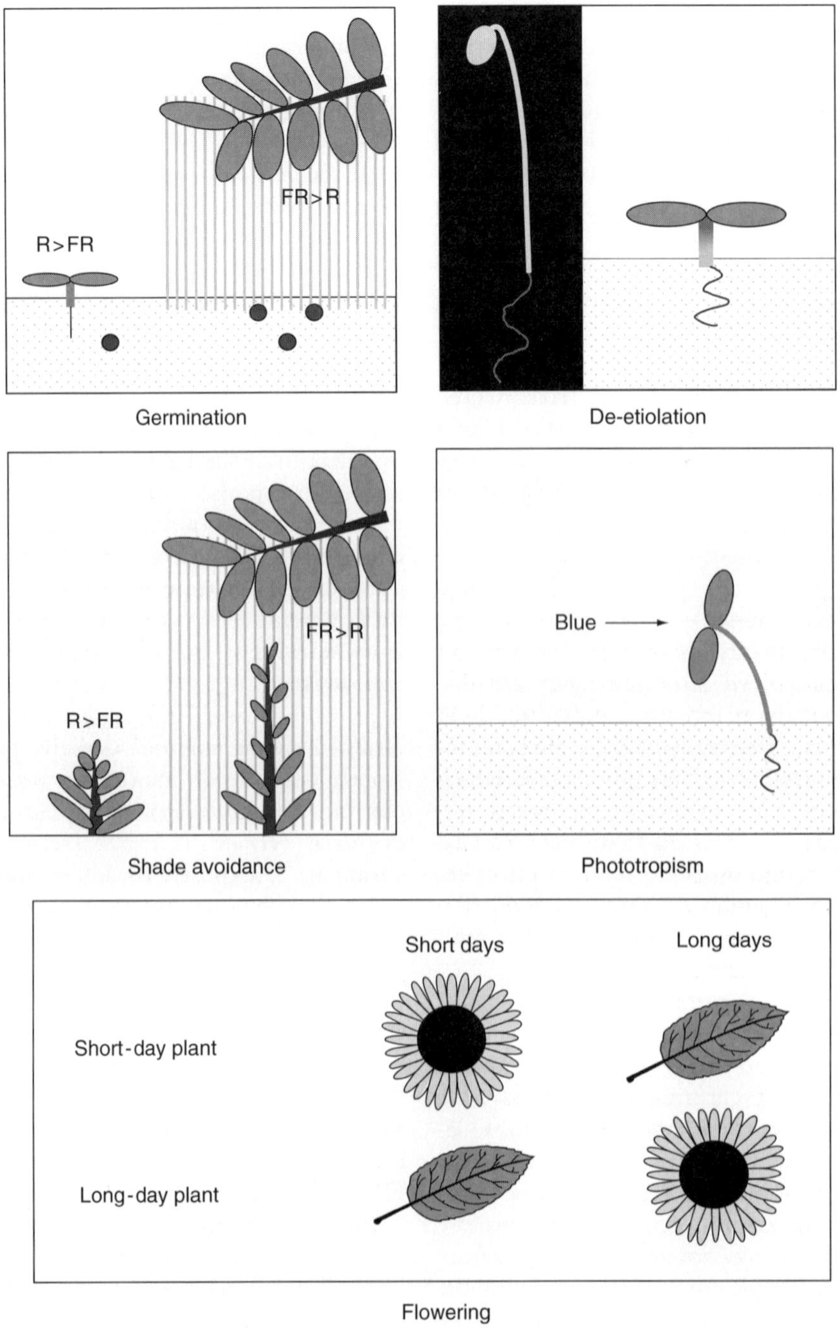

Figure 1 Examples of photomorphogenic processes exhibited by plants throughout their life cycle.

seed the ability to monitor this ratio, and allows it to make an "informed decision" about whether or not to germinate (**Figure 1**).

Phytochrome has also been implicated in other responses in which seed germination is actively inhibited by light. It is possible that these different modes of light responsiveness for germination may coexist in many species. Their relative prominence may also explain differences in the specific light requirements for germination among species.

Skotomorphogenesis (Etiolation) and Seedling Establishment (De-etiolation)

Because the germinating seed has only a limited supply of energy, it is important that the emerging seedling reaches the light and becomes a self-sufficient photosynthetic green plant as quickly as possible. Seedlings germinating under the soil initially direct their energy into rapid, negatively geotropic elongation growth, and do not waste energy in producing leaves. This strategy of growth has been called skotomorphogenesis or etiolation, and is generally seen wherever seedlings are germinated in darkness (**Figure 1**). The bean sprouts (*Phaseolus aureus*) eaten in Chinese restaurants provide a typical example of an etiolated seedling. These seedlings have an elongated hypocotyl and poorly developed primary leaves, which are devoid of chlorophyll. In addition, the shoot apex is protected by the formation of a characteristic hook structure.

Once exposed to light, the whole developmental strategy of the plant rapidly changes. Stem elongation ceases, the apical hook opens, and the primary leaves expand to become photosynthetic. Protective pigments such as anthocyanin may also be produced. This process of seedling establishment in the light environment is called de-etiolation. Underlying the visible changes in the seedling are profound changes in the expression of many genes, e.g., the formation of the photosynthetically competent plastids. The genes that encode ribulose bisphosphate carboxylase/oxygenase, the key enzyme involved in the fixation of carbon dioxide in photosynthesis and the most abundant protein on earth, are inactive in etiolated seedlings, but rapidly become active after exposure to light. This is a tightly coordinated process involving the integrated action of both plastid and nuclear genomes. Inhibition of elongation growth results from a modification of cell wall properties, such as deposition of the cellulose microfibrils in the secondary cell wall. These processes could be mediated by changes in plant hormone metabolism and action. De-etiolation of the aerial parts of the plant is accompanied by enhanced growth and branching of roots below the soil surface.

Both inhibition of elongation growth and expansion of leaves can result from photoconversion of Pr to Pfr by R, in the classical low fluence response mode of phytochrome action. However, other wavelengths are also clearly effective, particularly in the BL and FR regions of the spectrum. Unlike the response to R, which can be induced by a single pulse of light, responses to BL and FR generally require continuous exposure at high irradiance, and have therefore been termed high irradiance responses. Mutants deficient in specific phytochromes have shown that the FR high irradiance response is regulated by phyA, whereas R/FR-reversible responses are predominantly regulated by phyB. The basis for the differences in wavelength and irradiance requirements of these two phytochromes has yet to be satisfactorily explained.

Phytochrome also absorbs BL and is also partially responsible for de-etiolation responses to this waveband. However, the existence of other, specific BL-absorbing photoreceptors for de-etiolation was postulated for a long time. Progress in identifying photoreceptors responsible for BL-mediated phenomena has been hampered by the fact that action spectra for BL-induced responses have several peaks of action in the BL/UV-A region of the spectrum. Flavins and carotenoids were long considered as two candidate pigments, but neither shows the full absorption profile predicted by action spectra. It is now known that the primary photoreceptors for BL-mediated photomorphogenic responses, the cryptochromes, each carry two chromophores: a pterin and a flavin.

The stage of seedling establishment is a critical step in the life cycle of every plant, hence the very strong selection pressure for the use of multiple photoreceptors and cross-talk between them to signal this transition. It is obvious that at low light levels below the soil surface no individual photoreceptor will be saturated, and having a higher photoreceptor concentration would confer selective advantage. This would clearly favor the duplication of the phytochrome gene family in seed plants. This gives the seedling an early warning of proximity to the soil surface. Analysis of mutants for most photoreceptors involved in de-etiolation has shown that heterozygotes exhibit a partial loss of response particularly under low irradiances. This is termed haplo-insufficiency, and implies that light responses are limited by the photoreceptor abundance. During the evolution of seed plants on earth, colonization of the land brought with it unique problems. Photosynthetic organisms by design are adapted for optimal performance in the light environment. Studies of recessive mutants that de-etiolate in darkness demonstrated that the de-etiolation arises as a result of suppression of a series of repressors. Once activated, the photoreceptors relieve this suppression resulting in the plant adopting its default strategy. While mutants deficient in individual photoreceptors survive in the laboratory, due to genetic redundancy between them, it is predicted that they would be severely disadvantaged in the natural environment. Such a seedling would have to expend more energy before perceiving the light environment, decreasing survival prospects. It is ironic that the R/FR reversible phyA, the most abundant phytochrome in a dark-grown seedling and the first detected spectrophotometrically, probably plays little role in detection of the R:FR photon ratio. It can be considered as an antenna phytochrome, hence, the relatively high concentration present.

Shade Avoidance and Near Neighbor Detection

Even after it is established in the light and becomes fully photosynthetic, a seedling remains in competition for the available photosynthetically active radiation. Under highly competitive conditions where the degree of shading by other larger plants is high, seedlings are able to modulate their growth to resume a light seeking strategy (**Figure 1**). This is referred to as the shade avoidance syndrome, and is in effect a partial re-etiolation (**Table 3**). As described above in "Germination," the fact that FR is transmitted through leaves, whereas R is absorbed, makes the R : FR photon ratio a good indicator of the degree of shading. Under laboratory conditions, the addition of FR to the background white light results in dramatic effects on growth and development, despite the fact that it cannot be used for photosynthesis. Harry Smith and his colleagues at Leicester University, UK have shown that in many species the growth rate is related to the R : FR ratio in a negative way; in other words, the lower the ratio the greater the growth rate. Changes in the R : FR ratio have been shown to modulate growth rate with a lag of only 15 min, indicating that the plant continuously monitors its surroundings. Sensitivity to R : FR is sufficiently high that, in some cases, plants can even sense the quality of reflected light, and thus can detect the presence of adjacent plants even before direct shading occurs. This may explain the observation that photoreceptors are present in the surface layers of young growing tissues, since in the interior of green stems, the R : FR photon ratio decreases as a result of absorption of R by chlorophyll. The B-type phytochrome family members play a dominant role in the way plants respond to changes in R : FR photon ratio as a result of direct shade of other plants or reflectance from neighbors. Plants continue to use phytochrome as a detector of vegetational shade throughout their period of vegetative growth. Part of the shade avoidance syndrome is an increased apical dominance and acceleration of flowering. This acceleration of reproduction and seed production is a clear survival strategy for a plant that finds itself in deep shade.

Table 3 The shade avoidance syndrome

Parameter	*Response*
Extension growth	Accelerated rapid internode extension with reduced specific stem weight
Leaf development	Retarded, resulting in thinner leaves with a smaller leaf area that exhibit early senescence; plastid development and chlorophyll retarded; reduction of anthocyanin levels in immature leaves
Apical dominance	Strengthened, branching inhibited
Flowering	Accelerated
Assimilate distribution	Markedly changed storage deposition; leaf and root sinks reduced at the expense of stems

Photoperiodism

Another parameter of the light environment that changes is the daily duration of light and darkness, dependent on the time of year. This can be used as a reliable signal or indicator of seasonal change, such as the onset of winter. This phenomenon of photoperiodism was first characterized in detail in plants by Whiteman W Garner and Harry Allard, but later shown in insects, birds, and mammals. Both plants and animals have evolved pigment systems that can provide information about the duration of light and darkness, enabling them to anticipate the seasons. In animals, examples of photoperiodism are the onset of bird migration and initiation of sexual behavior. In plants, the most widely studied photoperiodic response is associated with reproduction: flowering and the development of dormancy mechanisms, such as the production of winter buds in deciduous trees and the formation of bulbs and corms. Daylength is the most constant parameter of the environment that can predict the time of year. Temperature can show considerable fluctuations, and therefore it is perhaps not surprising that living organisms have locked on to the parameter of photoperiod and are not fooled by conditions such as a few warm days in early spring. An example of man's interference with photoperiodism in plants can be seen in autumn in situations where deciduous trees grow close to street lights. On branches adjacent to the light, the leaves remain green and leaf fall is delayed as a result of the artificially long days typical of summer.

Although an oversimplification, the flowering responses of plants can for convenience be split into three groups: short-day plants (e.g., *Chrysanthemum* ssp., Japanese morning glory (*Pharbitis nil*)) that flower as days grow shorter in autumn (**Figure 1**); long-day plants (e.g., many grasses and cereals) that flower as days grow longer in late spring and summer; and day-length-indifferent plants (e.g., roses (*Rosa* spp.) and tomato (*Lycopersicon esculentum*)) that given a suitable temperature will flower at any time of the year. Such photoperiodic responses demonstrate that plants have the capacity for time measurement.

Research has shown that the plants effectively measure the length of darkness in each daily sequence. In Japanese morning glory, a short-day plant, flowering is readily induced under a daily cycle of 10 h light and 14 h darkness, whereas plants remain entirely vegetative when the daily light period is extended to 16 h. In fact, it is not necessary for this extension to be made with continuous light, and a short light break of a few minutes given at approximately the middle of the 14-h night period is equally effective in preventing flowering. In many long-day plants, a light break in the night under short days has the same effect as long days. If the effectiveness of different wavelengths of light is investigated it is found that R is most effective and that FR can negate its effect. Experiments have shown that phytochromes present in the leaves perceive the light signal, and this results in the release of a chemical signal that is transported to the apical meristem of the plants where it changes the fate of the cells produced.

Photoperiodism has been used to human advantage. Since the daylength is detected by phytochrome, only low levels of energy are required to artificially extend the day. Light breaks are also effective in switching development at the expense of little energy. Daylength can be shortened by simply covering plants with a black plastic sheet to shorten naturally long days. Using these simple procedures it is possible for horticulturists to manipulate the growth of plants such as Chrysanthemum, a flower traditionally typical of autumn, to flower at any predetermined date throughout the year.

Phototropism

The phenomenon of phototropism, or bending of plants toward the light, was first described a long time before the discovery of phytochrome. Charles Darwin wrote a very clear account of it in his book "The Power of Plant Movement," published in 1880. Action spectra subsequently showed that the photoreceptor absorbs in the BL and UV-A region of the spectrum, but the favored candidates for the photoreceptors, flavins and carotenoids, both lacked some of the predicted absorption bands in the action spectrum, which meant that the photoreceptor remained elusive for many years. Eventually, a genetic approach enabled the photoreceptor to be identified. The isolation of aphototropic mutants (i.e., mutants lacking the capacity to exhibit phototropism) in *Arabidopsis* showed that phototropism can be genetically separated from phytochrome and cryptochrome action. However, excitation of both cryptochrome and phytochrome can modify the phototropic response to BL. Recently, Winslow R. Briggs and colleagues at the Carnegie Institute of Washington at Stanford, USA cloned the gene impaired in one group of aphototropic mutants, and showed that the gene product, which they named phototropin, had clear characteristics of a BL photoreceptor. The phototropin protein is related to sensory proteins in other organisms from bacteria to mammals, which respond to light, oxygen, or voltage. Phototropin has two flavin chromophores.

Light Signal Transduction

Light responses are very diverse, and occur over a wide range of time scales. Any explanation of light signaling must take into account this diversity. Considering the rapidity of some phytochrome responses, Stirling B. Hendricks and Harry A. Borthwick suggested in 1967 that the primary mechanism of phytochrome action might involve changes in membrane permeability. However, as phytochrome shows little affinity for membranes, it is not clear how it can influence membrane related processes. In addition, phytochrome present in the cytosol in the dark clearly influences gene transcription, which occurs in the nucleus upon irradiation. Both phytochromes and cryptochromes are now known to move to the nucleus upon exposure to activating light, suggesting the possibility of a relatively direct link between photoreceptors and the modification of gene expression.

The phytochrome molecule itself has kinase activity, which is different in its two forms. Researchers are currently looking for molecules that may act as molecular partners or substrates for phytochrome. Both nuclear and cytosolic phytochrome interacting proteins have now been identified, suggesting multiple sites of phytochrome action. Transcription factors, as well as other key proteins are involved in phytochrome signal transduction. Other studies have implicated signaling pathways similar to those observed in animals and microorganisms. Activation of a guanine triphosphate (GTP) binding protein, modification of phosphatidylinositol metabolism, involvement of cyclic guanine monophosphate (GMP), changes in cytoplasmic calcium concentration, activation of calmodulin, and induction of kinase activity have all been implicated in phytochrome signal transduction. There is even evidence for different transduction chains being induced by phytochrome within a single cell.

Studies of signal transduction in cryptochrome responses have implicated the following elements:

GTP binding proteins, membrane depolarizations, redox chains, and electron transport, as well as calcium/calmodulin cascades. Ultimately, these processes could mediate changes in plant hormone metabolism (by modification of synthesis, degradation, their effectiveness, or the availability of their receptors), resulting in changes in growth and development.

In the case of phototropin, the photoreceptor is a serine/threonine kinase, which upon excitation with BL autophosphorylates itself. How the modified photoreceptor sets in motion the chain of events leading to differential growth is still not understood, although calcium metabolism again appears to play a role. A great deal of research over the years has implicated a central role for the plant hormone auxin in growth curvatures of plant organs. In the case of the oat (*Avena sativa*) coleoptile, excitation of the photoreceptor in the tip, close to the point of auxin synthesis, results in asymmetric distribution of auxin as it moves downward to the elongating zone. This results in cell wall plasticity being maintained for a longer period on the shaded side resulting in differential growth and curvature toward the light.

Biotechnological Modification of Photomorphogenesis

Knowledge of photomorphogenic processes, such as the photoperiodic control of flowering, has already led to efficient crop production strategies in photoperiodic species. A continued increase in fundamental knowledge of photomorphogenesis should enable us to modify the architecture and response of plants to our advantage. Most crops are grown in dense stands, and this leads to shade avoidance responses at the expense of crop yield. At face value, the suppression of shade avoidance offers a good prospect to increase yield in monocultures or increase the diversity of crops that can be grown in the shade of other plants. In contrast, an elongated, shade avoiding morphology is actually desirable in some fiber crops, and is usually achieved by high-density sowing and a relatively inefficient use of seed. Constitutively shade avoiding plants could then be grown at lower densities.

List of Technical Nomenclature

Absorption spectrum	A graph indicating the relative absorption of different wavelengths of light by a substance.
Action spectrum	A graph indicating the relative responsiveness of different wavelengths of light in eliciting a standard response.
Chromophore	The constituent of a proteinaceous photoreceptor that gives it the capacity to absorb visible light.
Cryptochromes	A family of blue/UV-A photoreceptors.
Far-red light	Light in the waveband 700–800 nm.
Photomorphogenesis	Responses of plants that occur as a result of processing information about their light environment.
Photoperiod	The length of the daily light period, which varies with the time of year.
Photoreceptor	A chromoprotein that absorbs light and is used in energy transfer or information processing.
Phototropin	The blue/UV-A photoreceptor involved in phototropism.
Phytochromes	A family of photochromic photoreceptors that absorb red and far-red light.
Seedling establishment (de-etiolation)	The transition of an etiolated dark-grown seedling into a green autotrophic plant.
Skotomorphogenesis (etiolation)	The strategy of dark growth that leads to etiolation.

See also: **Regulators of Growth**: Auxins; Circadian Rhythms; Photoperiodism; Phytochromes and other Photoreceptors.

Further Reading

Balleré CL (1999) Keeping up with the neighbours: phytochrome sensing and other signalling mechanisms. *Trends in Plant Science* 4: 97–102.

Kendrick RE and Kronenberg GHM (eds) (1994) *Photomorphogenesis in Plants*, 2nd edn. Dordrecht: Kluwer Academic Publishers.

Neff MM, Fankhauser C, and Chory J (2000) Light: an indicator of time and place. *Genes & Development* 14: 257–271.

Sage LC (1992) *Pigment of the Imagination. A History of Phytochrome Research*. San Diego: Academic Press.

Smith H (1998) Engineering a phytochrome gene. In: Cockshull KE, Gray D, Seymore GB, and Thomas B (eds) *Genetic and Environmental Control of Horticultural Crops*, pp. 175–190. Oxford: CABI Publishing.

Smith H (2000) Phytochromes and light signal perception by plants – an emerging synthesis. *Nature* 407: 585–591.

Thomas B and Vince-Prue D (1997) *Photoperiodism in Plants*. London: Academic Press.

ISBN 0-12-227050-9

ENCYCLOPEDIA
OF
APPLIED PLANT SCIENCES

ENCYCLOPEDIA OF APPLIED PLANT SCIENCES

Editor-in-Chief

BRIAN THOMAS

Editors

DENIS J MURPHY

BRIAN G MURRAY

ELSEVIER
ACADEMIC
PRESS

Amsterdam Boston Heidelberg London New York Oxford
Paris San Diego San Francisco Singapore Sydney Tokyo

Elsevier Ltd., The Boulevard, Langford Lane, Kidlington, Oxford, OX5 1GB, UK

Elsevier Inc., 525 B Street, Suite 1900, San Diego, CA 92101-4495, USA

First edition 2003

Library of Congress Control Number: 2003109025

A catalogue record for this book is available from the British Library

ISBN: 0-12-227050-9 (set)

This book is printed on acid-free paper

Printed in the United Kingdom

EDITORIAL ADVISORY BOARD

FOREWORD

I often tell visitors and policy makers who want to learn about current research in the plant sciences and agriculture that 'This is not your father's plant science'. As amply evidenced in these three volumes of the Encyclopedia of Applied Sciences, plant biology is undergoing a significant change in emphasis. The change engages far greater numbers of plant scientists in genetics, molecular biology, and biotechnology than ever before. This is not surprising given the growing amount of information on genome structure and functional genomics and the tools of genetic transformation of plants. These changes makes it possible for laboratory scientists to have a far greater impact on agriculture production and products than in any other period of time in the long history of applied plant sciences. The future promises to bring even more changes that bring sustainability to production agriculture and greater variability to its products.

These volumes contain chapters from world authorities in a range of topics and carry information that will inform the advanced reader, as well as the newcomer to the field, about the latest advances in the expanding fields of plant science. The reader will learn about the use of transgenic plants with altered metabolism that enhances accumulation of proteins and carbohydrates, alkaloids and terpenoids, and other components. Other chapters report the state of the science and recent breakthroughs in genetics and biochemistry of photosynthesis, plant growth and development, and plant nutrition. There is also a much needed update on the many experimental approaches to develop transgenic plants with novel characteristics, including enhanced resistance to pests and pathogens; biofortification and enhanced food safety; altering seed biology; reducing ripening times and post-harvest losses; bioremediation, and so on. The advances that have been made in the field are truly remarkable and are a testament to the many opportunities that modern genetics and biotechnology have brought to plant sciences.

However, while the advances have been great the informed reader will realize that very few of the discoveries reported in these chapters have yet found their way into commercialization. Furthermore, there has been a disturbing trend in recent years to slow the delivery of new biotechnology products to the marketplace. Governmental regulatory agencies in the US and the EU, as well as countries around the world, have shown reluctance to grant approval of new products, or to approve for sale products that have gained regulatory approval in another country.

At this juncture the products that make it to the marketplace are those for which significant profit can be accurately predicted: to date, the direct beneficiaries of the few approved products are the farmers. It remains to be seen whether subsequent products will be limited to plants whose traits benefit the farmer/producer, or whether products that have more obvious 'consumer benefit' will be approved for release. And, given the troubling trend of governments to expand regulatory oversight, and therefore cost of bringing a product to market, it also remains to be seen whether public sector institutions will or will not be a major player in developing genetically enhanced crops, or whether this opportunity will be left to major private sector companies.

Nevertheless, the quality of plant science research continues to improve, enhancing our knowledge of plant biology, and creating more and more opportunities to make contributions to sustainably produce sufficient amounts of nutritious and safe foods to feed an ever-growing population. And, to bring plant made

pharmaceuticals and other specialty products to market. The concerns aside, I remain optimistic and hope that that you share this optimism after reading these volumes.

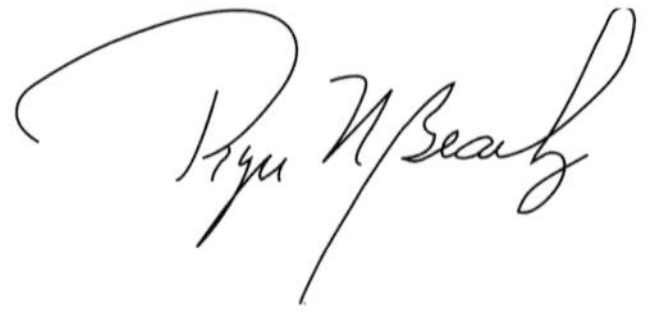

Roger N. Beachy, Ph.D.
President
Donald Danforth Plant Science Center
St. Louis, Missouri, USA

INTRODUCTION

Plants have a huge impact upon human existence. All of our food and much of our other energy needs are ultimately derived from the products of plant photosynthesis. Plants also have their impact through the social and economic dimensions of human existence. All agriculture, which provides us with our food, is ultimately based on plant production, be it directly through arable crops such as cereals, horticultural crops including fruits and vegetables or indirectly through pasture, forage and plant derived animal feedstuffs. Industrial crops such as fiber crops or woodlands and forests are a vital source of renewable materials for clothing, building and recreation, that we take for granted in everyday life. Flowers and ornamental plants delight us with their beauty, fragrance and their seemingly infinite variety.

Societies and economies all over the world have been able to gain increasing benefits from plants as our scientific understanding of their properties and potential has increased rapidly over recent decades. The range and complexity of this new information about plants is now potentially bewildering. A real challenge is how can such accumulated knowledge be made available to those taking an interest in the area for the first time. Equally how can a specialist in one aspect of plant applications gain access to information from wider or complementary areas of the subject? This is where we hope that this work will be of the most value.

The Encyclopedia of Applied Plant Sciences provides a friendly and comprehensive portal to the huge amount of knowledge that is available on plants and how we make use of them. The Editors, assisted by a distinguished Editorial Board have commissioned articles from an unparalleled group of experts across the plant sciences to provide introductory overviews and summaries of the state-of-the-art for their areas. Each article is sufficient for the reader to gain an understanding of the essentials of the topic covered and also provides an entry to the substantial amount of literature that lies beneath the articles. Rigorous editing and comprehensive indexing by the publishers ensures that the articles are understandable and that information can be located as needed. Articles covering the breadth of plant science are grouped under clearly described entries that will help the reader in finding not only the article they require but also articles on closely related topics.

Wherever possible authors have tried to show how the underlying science and information links through to applications in practical situations. Plant science undoubtedly has an exciting future and in this Encyclopedia we signpost this through a range of entries on the potential of the new genetic technologies, both in addressing traditional challenges and in making novel contributions to agriculture and other applications of plant science.

Putting the encyclopedia together has been a long and involved project. We are grateful for the help and support from the staff at Elsevier, the expertise and time given by the editorial board and the huge amount of work by the authors of the articles. We believe that the end result is a unique assemblage of the key information in this important field and are privileged to have been involved in the project. The value in the work is down to the quality of the contributors. Omissions or lack of balance are entirely the responsibility of the Editors. We hope that the work proves to be not only of practical value but can act as a catalyst for readers to gain entry to the exciting world of plant science.

Brian Thomas
Denis Murphy
Brian Murray

GUIDE TO USE OF THE ENCYCLOPEDIA

Structure of the Encyclopedia

The material in the Encyclopedia is arranged as a series of entries in alphabetical order. Some entries comprise a single article, whilst the majority of entries are on more diverse subjects and consist of several articles that deal with various aspects of the topic. In the latter case the articles are arranged in a logical sequence within an entry.

To help you realize the full potential of the material in the Encyclopedia we have provided three features to help you find the topic of your choice.

1. Contents List

Your first point of reference will probably be the contents list. The complete contents list appearing in each volume will provide you with both the volume number and the page number of the entry. On the opening page of an entry a contents list is provided so that the full details of the articles within the entry are immediately available.

Alternatively you may choose to browse through a volume using the alphabetical order of the entries as your guide. To assist you in identifying your location within the Encyclopedia a running headline indicates the current entry and the current article within that entry.

You will find "dummy entries" where obvious synonyms exist for entries or where we have grouped together related topics. Dummy entries appear in both the contents list and the body of the text. For example, a dummy entry appears for Arable Crops which directs you to Production Systems and Agronomy: Agricultural Crops where the material is located.

Example

If you were attempting to locate material on Biotechnology via the contents list.

BIOTECHNOLOGY *see* DISEASES: Genetic Modification of Disease Resistance, Fungal Pathogens; Genetic Modification of Disease Resistance, Viral Pathogens. GENETIC MODIFICATION: Gene Cloning, General Principles; Insertional and Transposon Mutagenesis; Transformation, General Principles; Transformation in Dicotyledons; Transformation in Monocotyledons; Transformation in Plastids; Transgene Stability and Inheritance. GENETIC MODIFICATION, APPLICATIONS: Flower Color; Nitrogen Fixation; Oxidative Stress; Cell Factories; Molecular Farming; Plantibodies. GROWTH AND DEVELOPMENT: Control of Gene Expression, Posttranscriptional Regulation.

At the appropriate location in the contents list, the page numbers for these articles are given.

If you were trying to locate the material by browsing through the text and you looked up Biotechnology then the following information would be provided.

Biotechnology *see* **Diseases**: Genetic Modification of Disease Resistance, Fungal Pathogens; Genetic Modification of Disease Resistance, Viral Pathogens. **Genetic Modification**: Gene Cloning, General Principles; Insertional and Transposon Mutagenesis; Transformation, General Principles; Transformation in Dicotyledons; Transformation in Monocotyledons; Transformation in Plastids; Transgene Stability and Inheritance. **Genetic Modification, Applications**: Flower Color; Nitrogen Fixation; Oxidative Stress; Cell Factories; Molecular Farming; Plantibodies. **Growth and Development**: Control of Gene Expression, Posttranscriptional Regulation.

Alternatively if you were looking up Genetic Modification the following information would be provided.

GENETIC MODIFICATION

Contents

2. Cross References

All of the articles in the Encyclopedia have been extensively cross referenced. The cross references, which appear at the end of an article, have been provided at three levels:

i. To indicate if a topic is discussed in greater detail elsewhere

ROOT DEVELOPMENT/Root Growth and Development.
See also: **Growth and Development**: Cell Division and Differentiation Cell Growth. **Nutrition**: Ion Transport; Mineral Uptake; Nitrogen Fixation. **Root Development**: Genetics of Primary Root Development; Lateral Root Initiation; The Rhizosphere and its Microorganisms. **Water Relations of Plants**: Phloem; Xylem.

ii. To draw the reader's attention to parallel discussions in other articles

ROOT DEVELOPMENT/Root Growth and Development.
See also: **Growth and Development**: Cell Division and Differentiation; Cell Growth. **Nutrition**: Ion Transport; Mineral Uptake; Nitrogen Fixation. **Root Development**: Genetics of Primary Root Development; Lateral Root Initiation; The Rhizosphere and its Microorganisms. **Water Relations of Plants**: Phloem; Xylem.

iii. To indicate material that broadens the discussion

ROOT DEVELOPMENT/Root Growth and Development.
See also: **Growth and Development**: Cell Division and Differentiation; Cell Growth. **Nutrition**: Ion Transport; Mineral Uptake; Nitrogen Fixation. **Root Development**: Genetics of Primary Root Development; Lateral Root Initiation; The Rhizosphere and its Microorganisms. **Water Relations of Plants**: Phloem; Xylem.

3. Index

The index will provide you with the volume number and page number of where the material is to be located, and the index entries differentiate between material that is a whole article, is part of an article or is data presented in a table. Detailed notes are provided on the opening page of the index.

4. Color Plates

The color figures for each volume have been grouped together in a plate section. The location of this section is cited in the contents list. Color versions of black and white figures are cited in figure captions within individual articles.

5. Contributors

A full list of contributors appears at the beginning of each volume.

CONTRIBUTORS

W W Adams III
University of Colorado
Boulder, CO, USA

S R Adams
Horticulture Research International
Wellesbourne
UK

G N Agrios
University of Florida
Gainesville
FL, USA

B S Ahloowalia
International Atomic Energy Agency
Vienna, Austria

G Alibert
INP-ENSAT
Auzeville, France

T Altabella
Universitat de Barcelona
Barcelona, Spain

A Altman
The Hebrew University of Jerusalem
Rhovot, Israel

J L Araus
University of Barcelona
Barcelona, Spain

S Arcioni
Institute of Plant Genetics
Perugia
Italy

A E Arthur
John Innes Centre
Norwich, UK

S N Azam-Ali
University of Nottingham
Nottingham, UK

R A Azevedo
University of São Paulo
Piracicaba, Brazil

M Bacon
University of Lancaster
Lancaster, UK

M Båga
University of Saskatchewan
Saskatoon, SK, Canada

P Barcelo
DuPont Wheat Transformation Laboratory
Barcelona, Spain

H S Bariana
University of Sydney
Camden, NSW, Australia

A R Barr
University of Adelaide
Glen Osmond, SA, Australia

L Bastiaans
Wageningen University
Wageningen, The Netherlands

M H Behboudian
Massey University
Palmerston North, New Zealand

J N B Bell
Imperial College London
Ascot, UK

G D Bending
Horticulture Research International
Wellesbourne, UK

R L Benech-Arnold
IFEVA
Buenos Aires, Argentina

L R Benjamin
Rothamsted Research
Harpenden, UK

P Berjak
University of Natal
Durban, South Africa

A Bernadac
UMR 990 INRA-ENSAT
Toulouse, France

I J Bingham
Scottish Agricultural College
Aberdeen, UK

D Blakesley
Horticulture Research International
West Malling, UK

J M Bonga
Canadian Forest Service – Atlantic
Forestry Center
Fredericton, NB, Canada

A-M Boudet
Université Paul Sabatier
Toulouse, France

M Bouzayen
UMR 990 INRA-ENSAT
Toulouse, France

J Brown
University of Idaho
Moscow, ID, USA

V Buchanan-Wollaston
Horticulture Research International
Wellesbourne, UK

M A S Burton
University of Hertfordshire
Hatfield, UK

P D S Caligari
Universidad de Talca
Talca, Chile

R W F Cameron
University of Reading
Reading, UK

S Cameron
Canadian Forestry Service
Fredericton, NB, Canada

M Cannell
Syngenta, Jealotts Hill International Research Centre
Bracknell, UK

M Cappadocia
Université de Montréal
Montréal, Canada

J S Carmichael
University of North Dakota
Grand Forks, ND, USA

A C Cassells
National University of Ireland
Cork, Ireland

K J Chalmers
University of Adelaide
Glen Osmond, SA, Australia

B Chanson
Université Montpellier II
Montpellier, France

D Chávez
University of Illinois
Chicago, IL, USA

S C Cheah
Malaysian Palm Oil Board
Kuala Lumpur, Malaysia

R N Chibbar
National Research Council Canada
Saskatoon, SK, Canada

V L Chopra
National Research Centre on
Plant Biotechnology
New Delhi, India

L P Christensen
Danish Institute of Agricultural Sciences
Aarslev, Denmark

P Christou
Fraunhofer Institute of Molecular Biology and
Applied Ecology
Schmallenberg, Germany

S D Clouse
North Carolina State University
Raleigh, NC, USA

K E Cockshull
Horticulture Research International
Wellesbourne, UK

J I Cooper
NERC Centre for Ecology and Hydrology
Oxford, UK

A Cordeiro
Universitat de Barcelona
Barcelona, Spain

C Darley
University of York
York, UK

E Dayan
Agricultural Research Organization
Hanegev, Israel

J Degenhardt
Max Planck Institute for Chemical Ecology
Jena, Germany

G-J de Klerk
Center for Plant Tissue Culture Research
Lisse, The Netherlands

J M DeLong
Agriculture and Agri-Food Canada
Kentville, Nova Scotia, Canada

B Demmig-Adams
University of Colorado
Boulder, CO, USA

N Dengler
University of Toronto
Toronto, ON, Canada

K Denyer
John Innes Centre
Norwich, UK

M Dow
National University of Ireland
Cork, Ireland

J G Dubrovsky
Universidad Nacional Autónoma de Mexico
Cuernavaca Morelos, Mexico

E M Duffy
National University of Ireland
Cork, Ireland

J M Dunwell
University of Reading
Reading, UK

G Dutfield
Queen Mary College
London, UK

P Elomaa
University of Helsinki
Helsinki, Finland

M R Emmett
University of Reading
Reading, UK

D E Evans
Oxford Brookes University
Oxford, UK

J D Everard
DuPont Experimental Station
Wilmington, DE, USA

W E Finch-Savage
Horticulture Research International
Wellesbourne, UK

K Fisher
Virginia Polytechnic Institute and State University
Blacksburg, VA, USA

M E Foley
USDA
Agricultural Research Service
Fargo, ND, USA

B V Ford-Lloyd
University of Birmingham
Birmingham, UK

C H Foyer
Rothamsted Research
Harpenden, UK

J B Free
University of Wales
Cardiff, UK

C Freeman
University of Wales
Bangor, UK

S C Fry
The University of Edinburgh
Edinburgh, UK

M D Gale
John Innes Centre
Norwich, UK

G Galiba
Agricultural Research Institute of the Hungarian Academy of Sciences
Martonvásár, Hungary

J Gershenzon
Max Planck Institute for Chemical Ecology
Jena, Germany

K G Gilbert
University of Bristol
Bristol, UK

T A Gorshkova
KIBB, RAS
Kazan, Russia

K S Gould
University of Auckland
Auckland
New Zealand

J Grace
University of Edinburgh
Edinburgh, UK

D J Gray
University of Florida
Apopka, FL, USA

J Gril
Université Montpellier II
Montpellier, France

K Groten
Rothamsted Research
Harpenden, UK

X Gu
North Dakota State University
Fargo, ND, USA

N D Hagan
CSIRO Plant Industry
Canberra, Australia

D L Hallahan
Central Research and Development
E I DuPont de Nemours & Company
Wilmington, DE, USA

J B Harborne[†]
University of Reading
Reading, UK

F Hay
Royal Botanic Gardens Kew
Ardingly, UK

P Hedden
Rothamsted Research
Harpenden, UK

J S Heslop-Harrison
University of Leicester
Leicester, UK

T J V Higgins
CSIRO Plant Industry
Canberra, Australia

D F Hildebrand
University of Kentucky
Lexington, KY, USA

H W M Hilhorst
Wageningen University
Wageningen
The Netherlands

S J Hiscock
University of Bristol
Bristol, UK

P J Holloway
The Long Ashton Research Station
Bristol, UK

R Hull
John Innes Centre
Norwich, UK

M O Humphreys
IGER
Aberystwyth, UK

R Hunt
University of Sheffield
Sheffield, UK

[†]Deceased

B S Jalani
Malaysian Palm Oil Board
Kuala Lumpur, Malaysia

P E Jameson
Massey University
Palmerston North, New Zealand

M F Jardinaud
Biotechnology and Plant Breeding, ENSAT
Auzeville, France

M C Jarvis
Glasgow University
Glasgow, UK

P P Jauhar
United States Department of Agriculture – Agricultural Research Service
Fargo, ND, USA

C E Jeffree
University of Edinburgh
Edinburgh, UK

R Johnson
Colby College
Waterville, ME, USA

D L Jones
University of Wales
Bangor, UK

H D Jones
Rothamsted Research
Harpenden, UK

L Jones
University of York
York, UK

V G Kakani
Mississippi State University
MS, USA

J Kallerhoff
INP-ENSAT
Auzeville, France

J Kang
University of Toronto
Toronto, ON, Canada

K J Kasha
University of Guelph
Guelph, ON, Canada

S P Kell
University of Birmingham
Birmingham, UK

R E Kendrick
Wageningen University
Wageningen, The Netherlands

A R Kermode
Simon Fraser University
Burnaby, British Columbia, Canada

A D Kinghorn
University of Illinois
Chicago, IL, USA

A J Kinney
DuPont Experimental Station
Wilmington, DE, USA

J D Klein
ARO-Volcani Center
Bet Dagan, Israel

R M D Koebner
John Innes Centre
Norwich, UK

A Kohli
John Innes Centre
Norwich, UK

M J Kropff
Wageningen University
Wageningen, The Netherlands

K Kunert
University of Pretoria
Pretoria, South Africa

A Kurian
Kerala Agricultural University
Vellanikkara, India

A Kushairi
Malaysian Palm Oil Board
Kuala Lumpur, Malaysia

M Labra
University of Milan
Milan, Italy

J Langridge[†]
Adelaide University
Glen Osmond, SA, Australia

P Langridge
Adelaide University
Glen Osmond, SA, Australia

E Larsen
Danish Institute of Agricultural Sciences
Aarslev, Denmark

A Latché
UMR 990 INRA-ENSAT
Toulouse, France

D W Lawlor
Rothamsted Research
Harpenden, UK

P A Lazzeri
DuPont Wheat Transformation Laboratory
Barcelona, Spain

P J Lea
Lancaster University
Lancaster, UK

R C Leegood
University of Sheffield
Sheffield, UK

R A Leigh
University of Cambridge
Cambridge, UK

Z Li
UMR 990 INRA-ENSAT
Toulouse, France

S E Lid
Agricultural University of Norway
Ås, Norway

A M Litterick
Crop and Soil Research Group, SAC
Aberdeen, UK

R E Litz
University of Florida
Homestead, FL, USA

[†]Deceased

G J Loake
University of Edinburgh
Edinburgh, UK

U Lüttge
Darmstadt University of Technology
Darmstadt, Germany

P Maliga
Rutgers University
Piscataway, NJ, USA

J Maluszynska
University of Silesia
Katowice, Poland

M Maluszynski
IAEA
Vienna, Austria

T Marks
Horticulture Research International
West Malling, UK

J M Martínez-Zapater
Centro Nacional de Biotechnología, CSIC
Madrid, Spain

G Matthews
Imperial College of Science, Technology and Medicine
Ascot, UK

N Maxted
University of Birmingham
Birmingham, UK

M R McAinsh
Lancaster University
Lancaster, UK

S McQueen-Mason
University of York
York, UK

J Memelink
Leiden University
Leiden, The Netherlands

M Mencuccini
University of Edinburgh
Edinburgh, UK

K Meyer
DuPont Central Research and Development
Wilmington, DE, USA

W B Miller
Cornell University
Ithaca, NY, USA

D J Miralles
University of Buenos Aires
Buenos Aires, Argentina

D J Murphy
University of Glamorgan
Cardiff, UK

B G Murray
University of Auckland
Auckland, New Zealand

D Nadella
North Dakota State University
Fargo, ND, USA

R M Napier
Horticulture Research International
Wellesbourne, UK

R E L Naylor
Trelareg Consultants
Banchory, Scotland, UK

P Newell
Formerly of The University of
the South Pacific
Suva, Fiji Islands

W E Newton
Virginia Polytechnic Institute
and State University
Blacksburg, VA, USA

P S Nobel
University of California
Los Angeles, CA, USA

P L Nurmberg
University of Edinburgh
Edinburgh, UK

A M Nuutila
VTT Biotechnology
Espoo, Finland

E O' Connor
Bio-Renewables Ltd
Ely, UK

C E Offler
University of Newcastle
New South Wales, Australia

K-M Oksman-Caldentey
VTT Biotechnology
Espoo, Finland

O-A Olsen
Pioneer Hi-Bred International
Johnston, IA, USA

J L Osborne
Rothamsted Research
Harpenden, UK

N W Pammenter
University of Natal
Durban, South Africa

Y S Park
Canadian Forest Service – Atlantic
Forestry Center
Fredericton, NB, Canada

J W Patrick
The University of Newcastle
New South Wales, Australia

M J Paul
Rothamsted Research
Harpenden, UK

J-C Pech
UMR 990 INRA-ENSAT
Toulouse, France

K V Peter
Kerala Agricultural University
Vellanikkara, India

M Petitprez
Biotechnology and Plant Breeding, ENSAT
Auzeville, France

P W B Phillips
University of Saskatchewan
Saskatoon, SK, Canada

M Piñeiro
Centro Nacional de Biotechnología, CSIC
Madrid, Spain

A H Pickering
Massey University
Palmerston North, New Zealand

M Pletsch
Universidade Federal de Alagoas
Maceió, Brazil

K M Plummer
University of Auckland
Auckland, New Zealand

Y Poirier
University of Lausanne
Lausanne, Switzerland

S Pond
Canadian Forestry Service
Fredericton, NB, Canada

S A Power
Imperial College London
Ascot, UK

R K Prange
Agriculture and Agri-Food Canada
Kentville, Nova Scotia, Canada

P V V Prasad
University of Florida
Gainesville, FL, USA

F Pupilli
Institute of Plant Genetics
Perugia, Italy

C B Purrington
Swarthmore College
PA, USA

A S Raghavendra
University of Hyderabad
Hyderabad, India

O Ramli
Malaysian Rubber Board
Selangor, Malaysia

K R Reddy
Mississippi State University
MS, USA

M J Reiss
Institute of Education
University of London, UK

P Roberts
University of York
York, UK

D Robinson
University of Aberdeen
Aberdeen, UK

T L Rost
University of California
Davis, CA, USA

E C Rowe
Wageningen University
Wageningen, The Netherlands

G Russell
The University of Edinburgh
Edinburgh, UK

P R Ryan
CSIRO Plant Industry
Canberra, Australia

F Sala
University of Milan
Milan, Italy

R A Sánchez
IFEVA
Buenos Aires, Argentina

S Sanogo
New Mexico State University
Las Cruces, NM, USA

G E Schaller
University of New Hampshire
Durham, NH, USA

P J Sharp
Plant Breeding Institute
University of Sydney, Australia

T Shepherd
Scottish Crop Research Institute
Dundee, UK

P R Shewry
Rothamsted Research
Harpenden, UK

K Shinozaki
RIKEN Tsukuba Institute
Tsukuba, Japan

N W Simmonds†
University of Edinburgh
Edinburgh, UK

L Skøt
Institute of Grassland and Environmental Research
Aberystwyth, UK

G A Slafer
University of Buenos Aires, Buenos Aires
Argentina

H Slater
Lancaster University
Lancaster, UK

S M Smith
Institute of Cell and Molecular Biology
University of Edinburgh, UK

D Sparkes
University of Nottingham
Loughborough, UK

E Stoger
RWTH Aachen
Aachen, Germany

R Straughan
University of Reading
Reading, UK

J Sutka
Agricultural Research Institute of the Hungarian Academy of Sciences
Martonvásár, Hungary

I Szarejko
University of Silesia
Katowice, Poland

D A Tabah
University of Bristol
Bristol, UK

A G Taylor
Cornell University
Geneva, NY, USA

†Deceased

T H Teeri
University of Helsinki
Helsinki, Finland

B Thibaut
Université Montpellier II
Montpellier, France

B Thomas
Horticulture Research International
Wellesbourne, UK

J M G Thomas
University of Florida
Gainesville, FL, USA

G A Thompson
University of Arkansas
Little Rock, AR, USA

A F Tiburcio
Universitat de Barcelona
Barcelona, Spain

K Tsunewaki
Fukui Prefectural University
Fukui, Japan

R M Twyman
University of York
York, UK

A J E van Bel
Justus-Liebig-University
Giessen, Germany

J E G van Dam
WUR, Wageningen
The Netherlands

B Veit
AgResearch
Palmerston North, New Zealand

R Verpoorte
Leiden University
Leiden, The Netherlands

J A Walsh
Horticulture Research International
Wellesbourne, UK

H L Wang
University of Arkansas
Little Rock, AR, USA

C B Watkins
Cornell University
Ithaca, NY, USA

C A Watson
Crop and Soil Research Group, SAC
Aberdeen, UK

A D Webster
Maidstone, Kent, UK

J L Weller
Univeristy of Tasmania
Hobart, Australia

P J White
Horticulture Research International
Wellesbourne, UK

E J Woltering
Wageningen University
Wageningen, The Netherlands

X B Yang
Iowa State University
Ames, IA, USA

K Yu
University of Kentucky
Lexington, KY, USA

J A D Zeevaart
Michigan State University
East Lansing, MI
USA

R L Zimdahl
Colorado State University
Fort Collins, CO
USA

M Ziv
The Hebrew University of Jerusalem
Rohovot
Israel

CONTENTS

VOLUME 1

C

D

E

F

G

Volume 2

REGULATORS OF GROWTH

Photoperiodism

B Thomas, Horticulture Research International Wellesbourne, Warwick, UK

Definitions and Discovery

"Photoperiodism" is defined as "The response to changes in daylength that enables plants (or any other living organisms) to adapt to seasonal changes in their environment." During the spring and autumn periods, daylength alone is an ambiguous indicator of the time of year but the progressive changes in daylength provide a certain environmental signal for the passage of the seasons. In order to locate the time of year accurately, a timekeeping mechanism in photoperiodism operates with precision and must be insensitive to unpredictable variations in the environment such as temperature. The precision of daylength sensing must also accommodate the fact that the rate of change of daylength varies during the year, being fastest in spring and autumn. The seasonal range and rate of change of daylength is lower in the tropics than at higher latitudes. In lower latitudes, timekeeping needs to be more precise than at higher latitudes in order to locate a seasonal event with the same degree of accuracy.

That plants responded to the daily duration of light was proposed independently by Julien Tournois and Hans Klebs at the beginning of the twentieth century. It was the American physiologists Garner and Allard, however, who first saw clearly that flowering and many other responses in plants could be accelerated either by long days (LD) or short days (SD), depending on the plant. They were led to their discoveries by studies on soybean (*Glycine max*), in which flowering tended to occur at the same time in the field, irrespective of planting date and Maryland Mammoth cultivar of tobacco (*Nicotiana tabacum*), that grew large, while failing to flower in summer, while plants growing in pots under glass flowered while still quite small in winter and early spring. These results suggested a seasonal response, and after eliminating temperature and light intensity as causal factors Garner and Allard concluded that the tobacco and soya plants would only flower if the duration of the daily light period was sufficiently short.

Garner and Allard introduced the terms photoperiod and photoperiodism and classified plants into the photoperiodic groups we use today. Short-day plants (SDP) are those that flower or in which flowering is accelerated by days which are shorter than a "critical" daylength. Long-day plants (LDP) are plants that flower or in which flowering is accelerated when the daily light period exceeds a critical daylength. Plants that flower at the same time irrespective of the photoperiodic conditions are called day-neutral plants (DNP). Plants that respond to daylength can be further subdivided into obligate (or qualitative) types, where a particular daylength is essential for flowering, or facultative (or quantitative) types, where a particular daylength accelerates but is not essential for flowering. Many important crop species are potentially photoperiodic; for example, many cereals such as wheat (*Triticum aestivum*) and barley (*Hordeum vulgare*) are LDP, while SDP include rice (*Oryza sativa*) and soybean (see **Table 1**).

The length of the day and the night are mutually linked within the 24-h daily cycle. Photoperiodic responses could therefore be theoretically determined by either the length of the day or the length of the night. Classic experiments with SDP *Xanthium* revealed that flowering only occurred if the night length was greater than 8.5 h, irrespective of the relative durations of light and darkness in the experimental cycle. When the night was sufficiently long, flowering occurred even when the accompanying light periods were long; in contrast, short light periods did not cause flowering if they were coupled with short nights. The conclusion was that, for SDP, the nightlength is clearly the decisive factor. If a long night is interrupted by a short period of light (e.g., 30 min) near the middle, SDP respond as if they had been exposed to a long day. For LDP, such night breaks are only effective if given in combination with daylengths that are just longer than those needed to permit flowering. Also, in LDP, unlike in SDP, the amount and spectral composition of the light given during the day period, especially in the latter part, has a large effect on flowering. Where the response to daylength depends primarily on the length of the dark period, the plants are called dark-dominant, and conversely, if the light period is the main influence, they are called light-dominant. In general, most SDP are dark-dominant and most LDP are light-dominant. The responses of dark-dominant SDP and light-dominant LDP to combinations of light and darkness are shown in **Figure 1**.

Table 1 Examples of long-day plants (LDP) and short-day plants (SDP), including some important crop species

SDP		*LDP*	
Qualitative or absolute	*Quantitative*	*Qualitative or absolute*	*Quantitatitve*
Coffea arabica	*Cannabis sativa*	*Arabidopsis thaliana*	*Allium ampeloprasum*
Fragaria × *ananassa*	*Cucumis sativus* (some cvs)	*Avena sativa* (spring strains)	*Antirrhinum majus*
Glycine max cv. Biloxi	*Glycine max*	*Brassica juncea, B. carinata, B. pekinensis*	*Arabidopsis thaliana*
Hibiscus cannabinus, H. esculentus	*Gossypium* spp.	*Delphinium elatum* (garden hybrids)	*Dianthus carthusianorum* cv Napoleon III, *D. caryophyllus* (glasshouse carnation)
Humulus japonicus, H. lupulus	*Helianthus annuus, H. tuberosus*	*Fuchsia hybrida*	*Eustoma grandiflorum*
Kalanchoë blossfeldiana	*Nicotiana tabacum*	*Hyoscyamus niger* (annual)	*Hemerocallis fulva*
Nicotiana tabacum cv. Maryland Mammoth	*Oryza sativa*	*Jasminium grandiflorum*	*Linum usitatissimum*
Oryza sativa	*Rhododendron spp. (florist's azalea)*	*Lemna gibba, L. minor*	*Hordeum vulgare* (spring cvs)
Perilla (red)	*Ricinus communis*	*Lolium temulentum* cv. Ceres	*Lolium temulentum* Ba 3081
Pharbitis nil	*Rosa gallica, R. rugosa*	*Nicotiana sylvestris*	*Medicago sativa*
Xanthium strumarium	*Sesamum indicum*	*Papaver somniferum*	*Petunia hybrida*
	Sorghum bicolor, S. halepense	*Silene armeria, S. coeli-rosa*	*Pisum sativum*
	Zea mays	*Trifolium pratense* (English Montmorency)	*Secale cereale* (spring cvs)
			Solanum tuberosum
			Triticum aestivum (spring cvs)
			Vicia faba

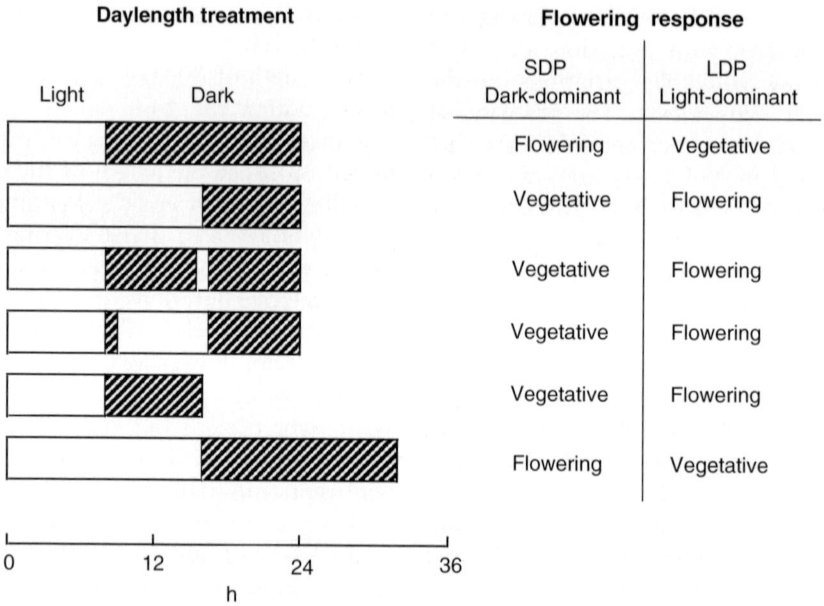

Figure 1 The typical flowering responses of long-day plants (LDP) and short-day plants (SDP) to combinations of light and dark periods of different lengths.

Occurrence

The ability to detect seasonal change and respond to it confers a selective advantage on plants because it provides a means of anticipating, and consequently preventing, the adverse effects of a particular seasonal environment. For example, at high latitudes short days in the autumn precede the low temperatures of winter. The shortening autumn days act as a signal to induce bud dormancy and cold hardiness, enabling perennial plants to survive the unfavorable winter environment. Summer conditions can also be unfavorable and bud dormancy is induced by the long-day conditions in some desert species of plant. Photoperiodic plants are common, even in tropical latitudes where the seasonal daylength changes are small and daylength is used to synchronize reproductive or other activities with seasonal events such as dry or rainy periods. Coincident flowering in members of a population increases the chances of outbreeding and hence genetic recombination. For this reason, synchronization of floral initiation through photoperiodic sensitivity can confer advantages independently of whether reproduction is matched with a particular favorable environment. A further potential benefit of photoperiodic responses is that they can enable organisms to occupy an ecological niche in space and time. For example, a response to short days can enable a woodland plant to flower and seed before the dense leaf canopy is formed in the spring.

Processes Controlled by Photoperiodism

Flowering is the best-known and most intensively studied example of photoperiodism in plants, but it is by no means the only aspect that is regulated by the length of the day. A second type of response, which is widespread, is that which enables organisms to avoid unfavorable conditions at particular times of the year through the establishment of a quiescent or dormant phase. For plants, the onset of bud dormancy and frost tolerance is frequently established in response to the shortening days of the autumn. Tuberization, which is also associated with dormancy, is usually a response to short days, but an important exception to this pattern is bulbing in onions, which is driven by long days. Although other aspects of plant development, including growth, seed germination, and pigmentation have frequently been shown to be modified by daylength in experimental situations, the evidence that this is a response to the duration of light and darkness, rather than the amount of light, is in most cases inconclusive.

Daylength Detection and Measurement

In order to respond to photoperiod, plants must be able to "measure" the length of the day. This capability must include at least two components. Firstly, there must be a means of light detection in order to distinguish between day and night. Secondly, there must be an internal timekeeper, which is relatively insensitive to environmental factors such as temperature. A common feature of photoperiodism appears to be that daylength perception is a separate process from the response to photoperiod. When either the leaves or shoot tips of photoperiodically sensitive plants are exposed to different daylengths, flowering depends on the daylength given to the leaves and not to the apex. Equivalent results are found for vegetative responses such as tuberization and bulbing. In several instances, leaves from plants that have been given a daylength treatment which initiates flowering have been grafted to plants that have not been exposed to permissive daylengths, with the result being flowering in the receptor plants. Daylength therefore is perceived in leaves and results in a localized change in the properties of that leaf. Flowering then occurs as a result of a signal transmitted from the leaves to the apex. The change in the leaf is termed "induction," while the response at the apex leading to the initiation of flowering is sometimes called "evocation."

Photoreceptors

In plants, light is detected by photoreceptor molecules that mediate the wide range of developmental responses shown to light quality and quantity (*see* **Regulators of Growth**: Phytochromes and other Photoreceptors). Photoreceptors that have been shown to be involved in photoperiodism fall into two groups, the phytochromes and the cryptochromes. Phytochromes are a family of chromoproteins characterized by sensitivity to the red and far-red parts of the spectrum. Up to five different phytochromes, PHYA to PHYE, can be distinguished in any one plant species, based on the DNA sequence coding for the protein part of the molecule. Studies have shown that different phytochromes and cryptochromes have different properties and have different roles in the photoperiodic process (**Table 2**).

Phytochrome A appears to be required for the detection of long days in LDP. Phytochrome B, on the other hand, appears to have a general role in establishing the sensitivity of both LDP and SDP to daylength, but it is not yet known whether it is a part of the daylength-measurement system. The cryptochromes are flavoproteins that mediate plant responses to blue light. Two members of the family (CRY1 and CRY2) have been found and these are active in promoting flowering by light in LDP members of the Brassicaceae (**Table 2**). In all other families tested, promotion of flowering in LDP and SDP is mediated only by phytochromes.

Time Measurement

The biological clock needed for photoperiodic time measurement, in most cases seems to be based on endogenous circadian rhythms in light sensitivity.

Table 2 Roles of different photoreceptors in flowering in long-day plants (LDP) and short-day plants (SDP)

	LDP	*SDP*
Phytochrome A	Promotes	?
Phytochrome B	Inhibits	Inhibits
Phytochrome C	Neutral	?
Phytochrome D	Inhibits	?
Phytochrome E	Inhibits	?
Cryptochrome 1	Promotes (Crucifers)	None
Cryptochrome 2	Promotes (Crucifers)	None

None, No SDP species selectively sensitive to blue light has been found; ?, unknown.

Evidence for the involvement of circadian rhythms in both SDP and LDP has been obtained using two classic experimental strategies. One is the "Bunsow protocol" in which a fixed light period is followed by a long night, which is systematically interrupted with a light pulse at different times. A series of peaks and troughs in the flowering response to these treatments, with a period of approximately 24 h, indicates the operation of circadian rhythm in light sensitivity (**Figure 2**). The second is the "Nanda–Hamner protocol" in which the light period is held constant and the total length of the light–dark cycle is systematically varied by extending the dark period. In these "resonance" experiments photoperiodic responses are shown to some combinations of light and darkness and not to others, and permissive cycle lengths are separated by about 24 h.

External Coincidence

The potential role of circadian rhythms in light sensitivity in daylength measurement was first postulated by Bunning. He suggested that the circadian clock consisted of two half cycles, one "photophil" and the other "scotophil." When light was received in the scotophilic phase, the daily cycle was perceived as a long day, but the absence of light during the scotophilic phase produced a short-day response. This idea was subsequently developed to form the "external coincidence" model. In this model, the scotophil phase is restricted to a narrower time band, called the photoinducible phase, where light acts to affect photoperiodic induction. In SDP the circadian rhythm is entrained by the dusk signal, so that the photoinducible phase always occurs at about the same time in darkness (**Figure 2**). The progress of the rhythm in darkness can be followed by light interruptions, which show a rhythm with a period of approximately 24 h. Light received during the photoinducible phase (dawn or a night-break) prevents the short-day response. Thus it is easy to think that the circadian rhythm in light sensitivity provides the timing base that allows the length of the dark period to be distinguished by the plant.

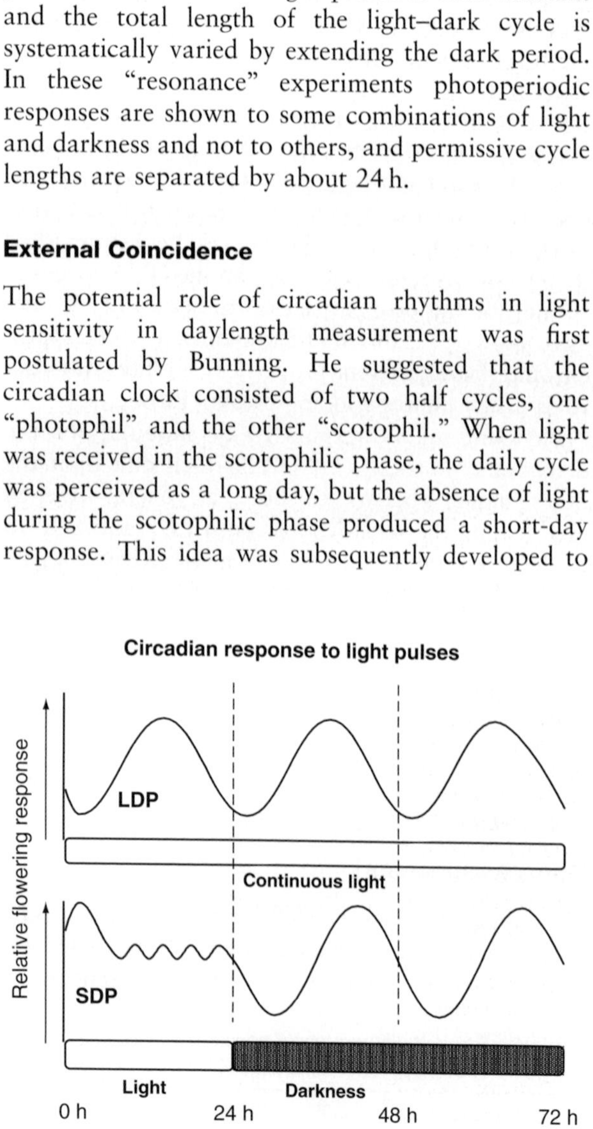

Figure 2 Responses of LDP and SDP to light pulses. For LDP, plants respond to far-red light added to a background of low levels of white or red light. The phase of the rhythm of the response is set at the beginning of the light period. For SDP, there is a rhythm of response to red light pulses in darkness. The phase of the rhythm is set at the beginning of the dark period by apparently suspending a rhythm initiated at the beginning of the previous light period. In this way, the first trough is always the same time from the beginning of the dark period, enabling the length of the dark period to be measured.

For LDP or light-dominant plants the situation is not quite as straightforward. Because sensitivity to night-breaks is much reduced, it is very difficult to do night-break experiments. Rhythms in response to a night-break have been described in the LDP but the pattern of response varies with the duration of the experimental dark period. In this, LDP differ from SDP where the night-break rhythm is independent of the length of the dark period, the phase normally being set at the end of the preceding photoperiod. In contrast, LDP show a circadian rhythm in response to adding far-red light to a background of white fluorescent or red light. This rhythm runs during the light period and the phase is set by the dark-to-light transition (or dawn).

Photoperiodic sensitivity in plants appears to be consistent with external coincidence models where light and darkness set the phase of a rhythm in light sensitivity and whether a particular light/dark treatment is seen as a long day or short day depends on where light falls during the progress of this phased rhythm. The phase of the rhythm is established in SDP (dark-dominant plants) at the light-to-dark transition (dusk) and the light will prevent the short-day response if it falls during the photoinducible phase of the dark period. If light is not received during this phase, the cycle is seen as a short day. In LDP (light-dominant plants) the rhythm is established at the beginning of the light period (dawn) and continues to run in continuous light. For a long-day response, light has to fall on the rhythm during the photoinducible phase, which occurs towards the end of the long day. It is worth noting that not all researchers are agreed that external coincidence models and circadian rhythms are the means by which plants detect daylength. An argument has been made for semidian (12 h) rather than circadian (24 h) rhythms being the time base for photoperiodism, although evidence is restricted to only one species and there is disagreement on the interpretation of the data.

Internal Coincidence

An alternative to external coincidence for photoperiodic mechanisms in plants is the internal coincidence model. In this case, there are two rhythms. One is initiated at the beginning of the light period, the other being initiated at the end of the light period or beginning of the dark period, and the phase relationship of these rhythms determines whether induction takes place or not. This model is extremely difficult to test and its complexity is generally thought to be unnecessary because in the simplest model systems such as *Pharbitis*, it is possible to explain the observations on the basis of a single rhythm and an external coincidence model. However, it is certainly worth noting that the dark-dominant and light-dominant types of response require the two types of rhythms (i.e., dusk-phased and dawn-phased, respectively) which have been claimed to be required for internal coincidence models. There is good evidence that both types of mechanism can exist and control the same process in different cultivars of the same species. It is possible to envisage situations where both mechanisms may operate together, either as a fail-safe, should one mechanism be disabled, or to establish a dual-key system, where both need to be satisfied to obtain a response.

Other Photoperiodic Responses

Dormancy

A period of dormancy during which growth temporarily ceases is a feature of the annual growth cycle of many plants. Such growth inactivity can be imposed directly by the environment (by low winter temperatures, for example) or, for the lateral buds, may be a consequence of apical dominance by the terminal bud. However, many woody plants also show a period of true dormancy when growth is not resumed under favorable environmental conditions, nor when the influence of other buds is removed. In some plants, dormancy involves only a suspension of growth without the development of any specialized structures. In other cases specialized resting organs are formed. The resting buds of woody plants develop when the cessation of normal internode extension leads to a telescoping of the shoot. In some species, the lack of elongation alone leads directly to the formation of a resting bud because the overlapping stipules of the unexpanded leaf primordia form the protective bud scales (e.g., in *Betula*). In other species, the protective outer leaf scales are modified leaves. In general, the onset of dormancy in woody plants is promoted by short days and inhibited by long days. Examples of woody plants in which dormancy is affected by photoperiod are shown in **Table 3**.

Bulbing and Tuberization

Many higher plants are capable of vegetative (i.e., nonsexual) means for propagating or perpetuating themselves through the formation of resting structures that have both a storage and reproductive function. Control of the formation of storage organs by photoperiod may be either quantitative or qualitative. Vegetative storage organs arise by lateral swellings of a number of different tissues including stems (tubers, corms), roots (tuberous roots or root-tubers), and leaves (bulbs). Once formed, frequently in response to daylength, storage organs enter into a state of dormancy during which they are more resistant to unfavorable environments such as water stress and high or low temperatures. In some cases plants propagate vegetatively without passing through a dormant storage phase, such as in strawberry (*Fragaria x ananassa*), where daughter plants are produced on runners, which are formed by extreme elongation of internodes in what is otherwise a rosette plant. Most photoperiodically

Table 3 Effects of daylength on the induction of dormancy in some common trees and shrubs

1. Dormancy induced by short days, prevented by long days
Acer palmatum, A. rubrum
Albizia julibrissin
Betula pubescens
Catalpa bignonioides
Cercidiphyllum japonicum
Cercis canadensis
Cornus alba, C. florida, C. nuttallii
Fagus sylvatica
Larix decidua
Liriodendron tulipifera
Platanus occidentalis
Populus balsamifera (= tacamahaca), P. robusta
Rhus typhina
Robinia pseudacacia
Tsuga canadensis
Ulmus americana
Weigela florida
2. Dormancy delayed but not prevented by long days
Acer pseudoplatanus
Aesculus hippocastanum
Liquidambar styraciflua
Paulownia tomentosa
3. Induction of dormancy not influenced by daylength
Fraxinus excelsior
Fuchsia boliviana
Malus (apple)
Pyracantha coccinea
Sorbus aucuparia

induced storage organs are favored by exposure to SD (**Table 4**). However, the formation of bulbs in the genus *Allium* is favored by LD, as is runner formation in strawberries.

Where tuberization, almost always a short-day response, has been studied, the reaction to light and darkness, sensitivity to night breaks, and spectral sensitivity are very similar to those for flowering in SDP. In the most direct test for photoperiodic regulation a brief exposure to red light prevents the inductive effect of a long night on tuberization in several species. Perception of daylength is accomplished by young leaves that are most sensitive when just fully expanded. Similarly, if we take bulbing in onions (*Allium cepa*), which is a long-day response, we find that the physiology of the response to light and daylength is very similar to flowering in LDP. Thus the stimulation of bulbing is promoted by wavelengths in the far-red part of the spectrum when given in the latter part of a long day, and this is mediated by phytochrome A. With both tuberization and bulbing, daylength perception takes place in the leaves and the response occurs elsewhere in the organism, but the effects of light and the chronobiology of the response is almost identical to the analogous daylength promotion of flowering.

Table 4 Influence of photoperiod on the formation of storage organs

Species	*Storage organ*
Favored by short days	
Apios tuberosa	Root tubers
Begonia evansiana	Aerial stem tubers
Begonia socotrana	Aerial stem tubers
Begonia tuberhybrida	
cv. Camelliaflora	Underground stem tubers
cv. Multiflora	Underground and aerial stem tubers
Dahlia hybrida	Root tubers
Dioscorea alata	Root tubers
Dioscorea divaricata	Aerial axillary stem tubers
Helianthus tuberosus	Underground stem tubers
Oxalis sp.	Underground stem tubers
Phaseolus multiflorus	Root tubers
Solanum andigena	Underground stem tubers
Solanum demissum	Underground stem tubers
Solanum tuberosum	Underground stem tubers
Trientalis borealis	Underground stem tubers
Ullucus tuberosus	Underground stem tubers
Favored by long days	
Allium ascolonium	Bulbs
Allium cepa	Bulbs
Allium proliferum, A. sativum	Underground and aerial bulbs
Brodiaea laxa	Corms

Florigens and Other Signals

The perception of the daylength signal occurs primarily in the leaves and, from grafting experiments, it is known that induced leaves are independently capable of generating over many days or even weeks. At the other extreme, favorable cycles must be continued until the apex has become recognizably floral, indicating the need for a continual supply of a floral stimulus (presumably from the leaf) if flower development is to be sustained. An intact and functioning nutrient transport system is a requirement for communication between the sites of perception in the leaves and the sites of response, suggesting that the signals are chemical, rather than biophysical in nature. Daylength response is lost when the pathway is disrupted by removing the source leaf, inhibiting transport by localized heat or cold treatments applied at intermediate points in the transport path or stem girdling. Also, when leaves taken from an induced plant are grafted to a receptor plant, promotion of flowering or other responses only occur when a graft union has developed.

The identity of the flowering signals is one of the major unanswered questions of classical plant physiology. A number of ideas abound, and they fall into three main groups.

Florigens and Antiflorigens

It was proposed by Chailakhyan more than half a century ago that the signal passing between leaves and response sites is a specific flowering hormone, which he proposed to call "florigen." This idea was based on a series of experiments showing that grafting of leaves from one donor species to a separate receptor species could cause flowering (**Table 5**).

This strongly suggested the participation of common signals in different species. Other grafting experiments suggested that other substances inhibiting flowering may be involved; the appropriate daylength would then lead to removal of an "antiflorigen" rather than (or in addition to) synthesis of a floral hormone. After 60 years of experimental effort, no convincing candidates for florigen or antiflorigen have been isolated.

Plant Growth Substances

The second possibility is that the flowering signals are combinations of the known growth substances, such as auxins, gibberellins, and cytokinins (*see* **Regulators of Growth**: Gibberellins). Daylength has been found to cause changes in the content of endogenous growth substances in a wide range of photoperiodic species. However, for most of the

Table 5 Examples of successful transfer of flowering stimulus between species or genera following grafting

Donor	*Response type*	*Receptor*	*Response type*
Intraspecific grafts			
Glycine max cv Agate	DNP	*G. max* cv Biloxi	SDP
Chenopodium rubrum 60° 47′ N	SDP[a]	*C. rubrum* 34° 90′ N	SDP
Pisum sativum various genetic lines	DNP or LDP[a]	*P. sativum* line G	LDP
Interspecific grafts			
Gossypium hirsutum	DNP	*G. davidsonii*	SDP
Nicotiana tabacum cv Delcrest	DNP	*N. sylvestris*	LDP
Intergeneric grafts			
Blitum virgatum	LDP	*Chenopodium rubrum*	SDP
Chenopodium polyspermum	SDP	*Blitum capitatum*	LDP
Cucumis sativus	DNP	*Sicyos angulatus*	SDP
Centauria cyanus	LDP	*Xanthium strumarium*	SDP

DNP, day-neutral plants; LDP, long-day plants; SDP, short-day plants.
[a] Facultative photoperiodic response.

known growth substances, changes in endogenous levels or a response to exogenously applied growth substances is limited to isolated individuals, with no evidence for a more general involvement. The exception to this is for gibberellins. It has been known for a long time that applied gibberellin (GA) can substitute for photoperiodic induction in many plants. In particular, GAs cause flowering in many LDP which grow as rosettes in short daylength; in these plants, the flowering response (either to GA or long daylength) is accompanied by elongation of the flowering stem. There are also a few qualitative SDP where GA application can evoke flowering in noninductive daylengths. Unfortunately, any simple interpretation of the action of GA in the control of flowering is ruled out because exceptions, or opposite responses, can be listed for all of the generalizations made above.

Multifactorial Signals

The third idea is that there are no specific flower-promoting substances, but that the signal is multi-component, including carbohydrates and growth hormones, possibly acting in sequence. The main evidence for this comes from the range of environmental conditions that affect flowering and the detection of changes in the levels of several metabolic components being transported from leaves into the apex at a very early stage after induction.

However, the existence of a leaf-generated signal in plants where induction is achieved through exposure to the appropriate photoperiod can hardly be questioned and the identity of this signal is presently unknown. Moreover, it is evident from grafting experiments that, at least in some cases, the resultant transmissible product from leaves is the same stimulus (or is functionally equivalent), despite the fact that induction may be achieved via different means.

Photoperiod Genes

Sensitivity to daylength frequently has an important impact on flowering time in the field. Flowering time is one of the most important traits in plant breeding, and plant breeders have identified a number of daylength response genes. Examples of these are the *Ppd* genes of wheat, the *Sn*, *Dne*, and *PpD* genes in pea (*Pisum sativum*), the *ea* genes in barley, and the *E* genes in soybean. In the face of the complex and sometimes contradictory physiology of photoperiodism, the application of physiological and molecular genetics to the photoperiodic control of flowering is the most likely route to understanding the process in the future. The large number of genetic mutants in *Arabidopsis*, a qualitative LDP, indicate that there are almost a hundred genes that can contribute to flowering time, although most of these are probably not strictly daylength-response genes. Two important genes that have been isolated from mutants impaired in daylength response are (1) *CO* (*constans*), a gene that is expressed in a circadian manner and required for floral initiation in long days, and (2) *LHY*, a gene that is required for the response to daylength and other aspects of circadian rhythmicity. These genes are almost certainly part of the induction and daylength measurement systems respectively. Mutants impaired in photoreceptor content or sensitivity also affect daylength-dependent flowering. Isolation of further genes from models such as *Arabidopsis* and their homologs in crop species will open up new opportunities for modifying the daylength responses of plants by targeted breeding and genetic manipulation.

Applications of Photoperiodism

Although it is difficult to quantify the benefits of photoperiodism to farmers, horticulturists, and plant breeders, they have been estimated to amount to billions of dollars. Testing plants for their photoperiodic requirements is standard practice in breeding programs, thus avoiding a major reason for crop failures. Plant breeders can manipulate daylength to obtain multiple generations of plants per year. They can obtain seed from plants that fail to reproduce out of doors because the days are too long, by growing them in shortened days in the greenhouse. Also, new crosses between varieties are possible by using daylength to synchronize flowering in plants that normally flower at different times.

Photoperiodism has had a particular impact on horticulture. Daylength manipulation by the use of blackouts or supplementary lighting to promote or enhance flowering has been used for a wide range of ornamental species. In the case of the major horticultural crop *Chrysanthemum*, daylength-sensitive cultivars of this SDP have been deliberately selected. This allows plants to be maintained in a vegetative state or brought to flowering as required by the grower through varying daylength. Production of poinsettia (*Euphorbia pulcherrima*), the most valuable ornamental crop in the United States, is also dependent on daylength management. Timing of production for a specific market, e.g., Christmas, Easter, or Mother's Day, is very important in floriculture as there is a price premium at these times. Daylength regulation makes this possible for many crops, including begonias, Christmas cactus, chrysanthemums, and poinsettias.

See also: **Flowering and Reproduction**: Reproductive Strategies. **Production Systems and Agronomy**: Commercial Flower Production Methodology; **Regulators of Growth**: Circadian Rhythms; Gibberellins; Phytochromes and other Photoreceptors; Vernalization.

Further Reading

Jackson SD and Thomas B (1997) Photoreceptors and signals in the photoperiodic control of development. *Plant Cell and Environment* 20: 790–795.

Lumsden PJ (1991) Circadian rhythms and phytochrome. *Annual Review of Plant Physiology and Plant Molecular Biology* 42: 351–371.

Lumsden PJ and Millar AJ (1998) *Biological Rhythms and Photoperiodism in Plants*. Oxford: BIOS Scientific.

Sage LC (1992) *Pigment of the Imagination: A History of Phytochrome Research*. San Diego: Academic Press.

Thomas B and Vince-Prue D (1997) *Photoperiodism in Plants*. London: Academic Press.

Circadian Rhythms

U Lüttge, Darmstadt University of Technology, Darmstadt, Germany

Overview

The awareness of rhythmicities determining human life and the life around us dates back to ancient cultures, in Egypt and in Europe (stone henges), and in the classical Greek and Latin civilizations. Rhythms that are independent of the natural light/dark (LD, or also DL) cycle, functioning as an external driver, and that run freely under constant environmental conditions, including constant light (LL) or constant darkness (DD), are called free running or endogenous rhythms, i.e., the driver lies inside the organism.

The discovery of endogenous movements in plants is attributed to de Mairan (1729) who showed that day/night movements of leaves of *Mimosa pudica* continued in DD. An elegant example of this phenomenon is the flower clock designed by Linné in 1745, where 12 groups of plants that open and close their flowers at different times of the day were planted in a circle, and used to tell the time.

Wilhelm Pfeffer (1845–1920), one of the founders of modern plant physiology, began to study leaf movements systematically. He showed that energy is required for the motor driving the leaf movements. The rhythm is strongly dampened in DD, it dissipates as respiration consumes energy reserves, and is sustained in LL where photosynthesis continuously provides energy.

Bioluminescence of the unicellular marine dinoflagellate *Gonyaulax polyedra* also provides an excellent illustration of endogenous rhythmicity. Large populations of these microalgae begin to produce bioluminescence when they are stimulated by a boat passing over the sea. This response to mechanical stimulation can be simulated in the laboratory. It is rhythmic in LD, i.e., diurnally, and continues endogenously in both DD and LL. The amplitude is given by the activity of the underlying metabolic reactions, as it is shown to be dependent on temperature (**Figure 1**). In *G. polyedra*, as it is a marine organism, the amplitude is highest at around 20°C, and rapidly decreases at temperatures below 16°C and above 24°C. Like the leaf movements, stimulated bioluminescence of *G. polyedra* is also rapidly dampened in DD, due to the lack of supply of photosynthetic substrates and energy for the biochemical reaction of bioluminescence, i.e., the driver is exhausted in DD and the system is dissipative. In LL, long-lasting, undampened oscillations are recorded.

Phenomenologically there is a vast variety of rhythms in organisms, which can have any period length. Man is familiar with his own rhythms ranging from periods of ca. 10^{-3} s of nervous actions, 10^0 s of heartbeat, 10^0–10^1 s of respiration, ca. min of circulation and smooth muscle activities, day of various diurnal activities, month of menstrual cycle to years. Similarly, in plants there are electrical rhythmicities (seconds to minutes), and many other rhythms shorter than diurnal rhythms (ultradian), e.g., cytoplasmic streaming, some leaf movements, tendril movements, stomatal guard cell movements, or longer than diurnal rhythms (infradian), e.g., tidal and (semi-)lunar rhythms (egg production of the brown alga *Dictyota dichotoma*), annual activities of photosynthesis, growth and development, and the 30–40-year rhythms of bamboo flowering. A basic distinction can be made between period lengths related to natural environmental rhythms (diurnal, tidal, lunar, annual) and period lengths unrelated to any regular environmental events.

Among the former, diurnal rhythms are the most prominent phenomena. In plants and other organisms, the external driver is very often the light period/dark period rhythm of our environment, and in this case we refer to the rhythms as diurnal. (Note the ambiguity using the term "day," which may refer to the entire 24-h period of the environmental rhythm or to the light period only as a contrast to "night.") Under continuous external conditions, the period length of the free-running rhythm is never exactly the 24 h of the solar day, but is always somewhat shorter or longer, i.e., approximately ("circa") a day ("diem"), or a "circadian" rhythm. Diurnal rhythms are exactly 24 h because they are carried along, or entrained, by the natural day/night rhythm of 24 h. When they are released from this external driver, i.e. from entrainment, the internal driver comes through, with the circadian endogenous period. Thus, the fact that a rhythm is circadian, is strong evidence for its endogenous nature, where no hidden external driver is involved. The quest for the underlying "biological clock," has become a rapidly expanding field of scientific inquiry termed chronobiology.

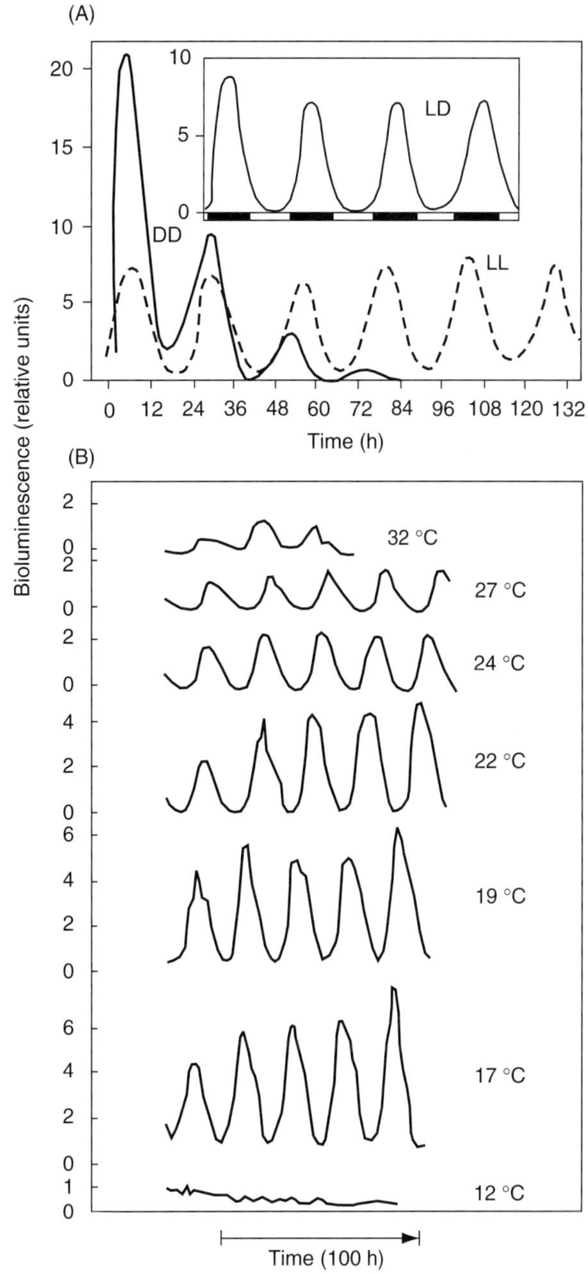

Figure 1 Rhythms of mechanically stimulated bioluminescence of the unicellular dinoflagellate *Gonyaulax polyedra*, as studied by B. Sweeney and her research team. In LD there is continuous diurnal rhythmicity and it is also free-running in both DD and LL, where there is strong dampening in DD after 3–4 periods and continuous oscillation with constant amplitude in LL for many periods (A). The amplitude in LL is strongly dependent on temperature (B). (Part (A) reproduced with permission from Lüttge U, Kluge M, and Bauer G (2002) *Botanik*, 4th edn. Weinheim: Wiley-VCH.)

Definitions and Terminology

In the preceding section, technical terms, such as amplitude and period of rhythms have already been used, but these need to be defined precisely.

The output of a large clock on the wall is the rhythmic movement of its hands. Its periodic driver is its pendulum. The simple geometry behind the swinging of a pendulum (**Figure 2A**) allows us to deduce and define the terminology needed to discuss rhythms.

An undampened pendulum is continuously and rhythmically swinging back and forth between points A and C and with an angle φ, via points B and D, as represented in **Figure 2A**. These points, and any other point the pendulum can take, are phases of the swinging movement. Thus, we can draw a phase-space

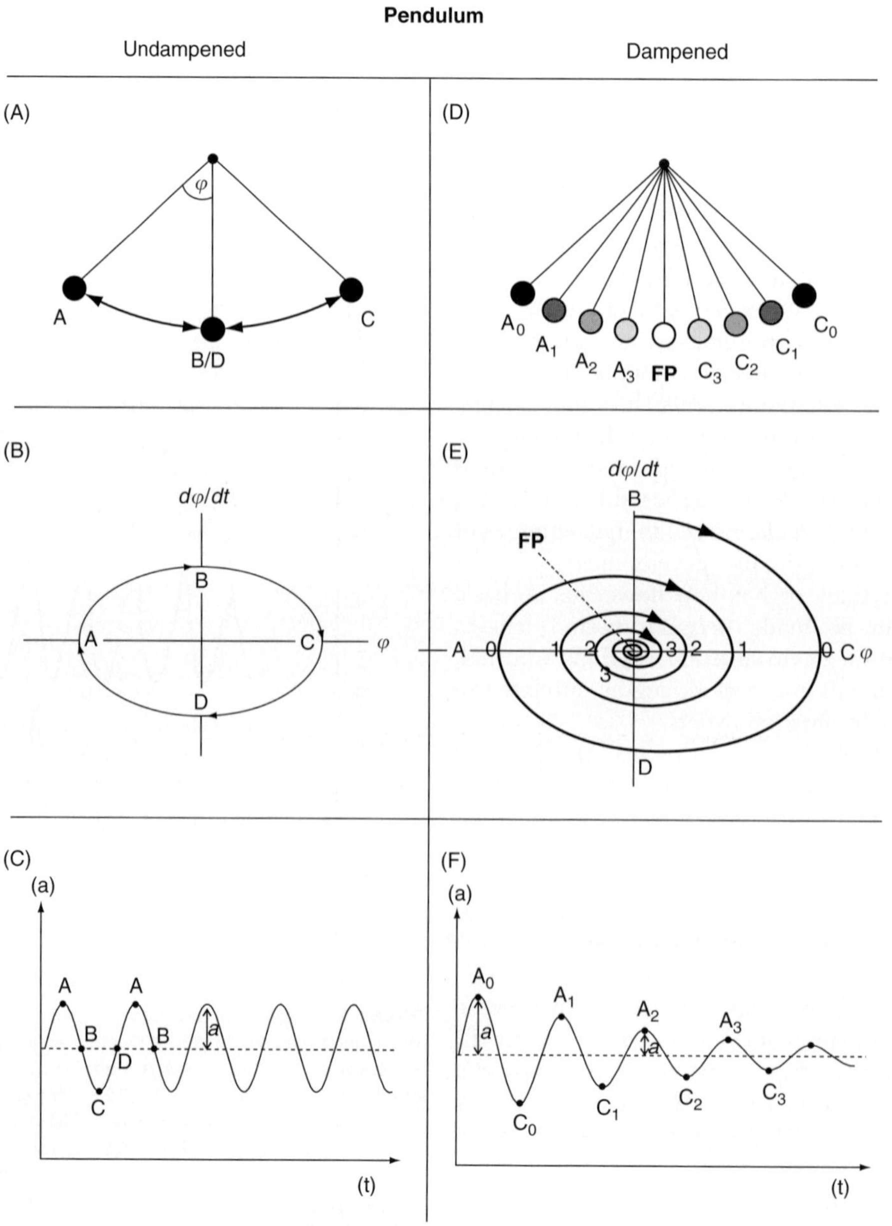

Figure 2 Derivation of basic features of rhythmic oscillations from consideration of an undampened (A–C) and a dampened (D–F) pendulum. (A) Periodically driven pendulum (e.g., by its weights in a large wall clock) swinging without dampening between points A and C through points B and D with a maximum angle of φ. (B) Phase-space diagram, with the angle φ largest at points A/C and the change of φ with time, t, ($d\varphi/dt$) largest at points B/D, showing an elliptic closed cycle, a stable trajectory, or path of the pendulum. (C) Progression in time (time-series), a regular sine-wave in this case, where points A-B-C-D are phases, and the amplitude a corresponding to the angle φ in (A) is largest at points A and C (with different directions of a) and zero at points B and D. (D) Dampened pendulum, where the angle φ gets increasingly smaller from points A_0/C_0 to A_3/C_3 and the pendulum eventually moves into a stable fixed point, FP. (E) Trajectory of the dampened pendulum in phase-space of φ and $d\varphi/dt$ spiraling into the fixed point. (F) Time series of the dampened pendulum with decreasing amplitude a.

diagram, where φ is the angle of the pendulum and $d\varphi/dt$ is its time derivative, i.e., the angular velocity (**Figure 2B**). $d\varphi/dt$ is highest at points B/D and zero at A and C. This phase-space diagram depicts the path or trajectory of the tip of the pendulum. For the undampened pendulum it is a closed cycle. The progression in time would be depicted by a regular sine-wave with a constant amplitude (a) (**Figure 2C**; compare with LL in **Figure 1**). If we had measured data points representing the line of the sine-wave or trajectory, we would call this a time-series measurement. For a dampened oscillation (**Figure 2F**; compare with DD in **Figure 1**), the angle φ would get increasingly smaller, and points A and C would get increasingly closer to each other (A_o to A_3, C_o to C_3) as time progresses (**Figure 2D**). In phase-space, we do not see a cycle but a spiraling into an equilibrium, which in the language of dynamical systems is often called a fixed point (FP), where the pendulum then stays (**Figure 2E**), and in the time-series we would see strong dampening indicated by a progressively decreasing amplitude (a) (**Figure 2F**).

Figure 3A reiterates the definition of the term phase. The time required to return from a given phase to the same phase again (i.e., from A to A, from B to B, etc.) is the period length (p) and $1/p$ is the frequency. The phase can be shifted by external signals or control parameters, called *zeitgeber*, without changing period length or amplitude (**Figure 4**).

Basic Operational Elements in Biological Oscillations

Any system, even if composed only of a very small number of elements, can move on a limit cycle, which means it oscillates regularly as long as it has (1) feedback loop(s) and/or (2) a beat oscillator or hysteresis switch. What we see in circadian rhythmicity are overt rhythms, which are somehow controlled by external control parameters, which we can

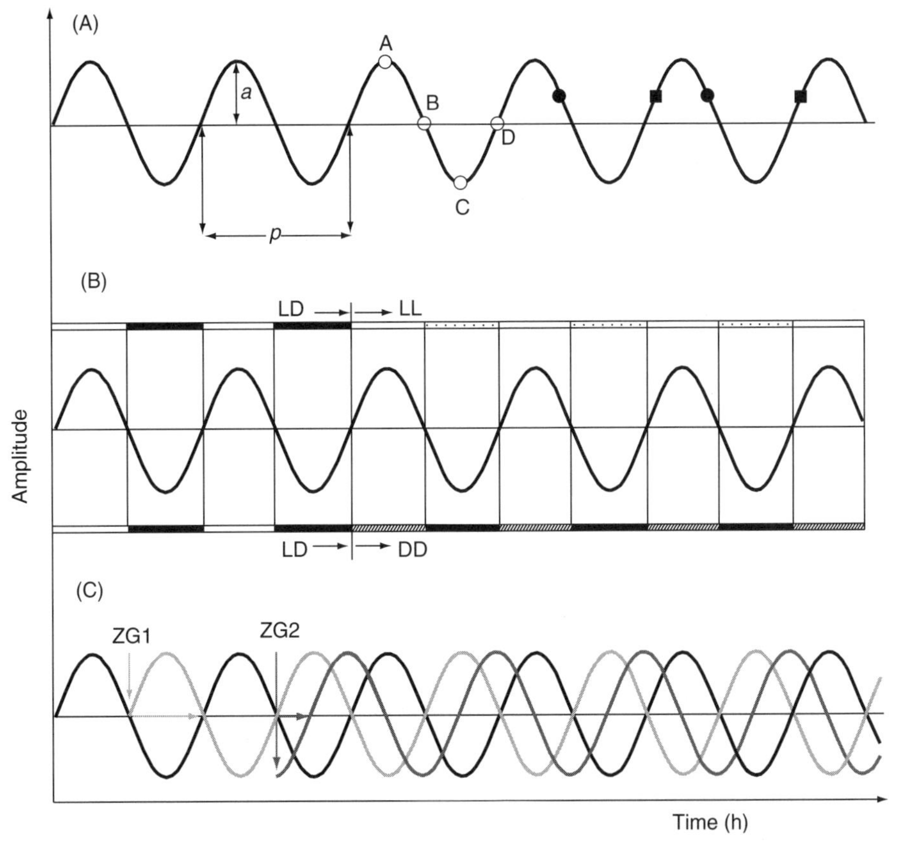

Figure 3 More terminology. (A) Phases, e.g., A, B, C, D (as shown in **Figure 2**) or solid circles and squares; amplitude a; period p, i.e., the time needed to return from a given phase to the same phase again. $1/p$ is the frequency. (B) Conventional denotation of diurnal rhythms (LD or DL) in external changes of light (L) and darkness (D) and endogenous rhythms in continuous darkness (DD) or light (LL). Dark and light bars indicate actual darkness and illumination, respectively. Dotted bars in LL indicate the time of the dark period experienced by the system before application of LL; the so-called subjective dark period. Hatched bars in DD indicate the time of the light period experienced before application of DD; the so-called subjective light period. (C) Phase-shifting. An external control parameter, e.g., light, temperature, etc., may function as a *zeitgeber* (ZG) determining the phase of the rhythm (setting the watch). In the examples shown ZG1 shifted the phase by 180° and ZG2 by 90°, i.e., 1/2 and 1/4 period, respectively.

also see or measure and manipulate. They are connected via essentially three black boxes: an input pathway, a beat oscillator or hysteresis switch, and an output pathway (**Figure 4**). All the basic elements, control parameters, overt rhythms, input and output pathways, and the oscillator are further connected by (positive and/or negative) feedback loops.

The input and output pathways are not necessarily linear, but may constitute branching and anastomosing networks. One given oscillator may require the action of different control parameters at the same time and thus different input pathways, and one given oscillator may also determine different overt output rhythms operating different output pathways. Oscillators themselves may also be complex network structures. The whole system then is a meta-network of coupled individual networks.

The control parameters may feed into the input networks via opening and closing of a gate. This gate itself may oscillate and may be under circadian control. This can be studied by phase shifting (**Figure 3C**; see also **Figure 5**). An external control parameter may set the time, i.e., it may function as a *zeitgeber*; however, its action may not be the same at all circadian times. Such gating is revealed by studying phase-response curves (**Figure 5**). The control parameter – in nature, but mainly also in experiments – itself may be rhythmic. This allows further investigation of entrainment. If periods of input oscillations are not too far away from the endogenous circadian period, the overt rhythm follows the external rhythms and the system is entrained (**Figure 6B**). If the external period is too short or too long the system escapes entrainment and the overt time series may either show the circadian period (**Figure 6A**) or become arrhythmic (**Figure 6C**).

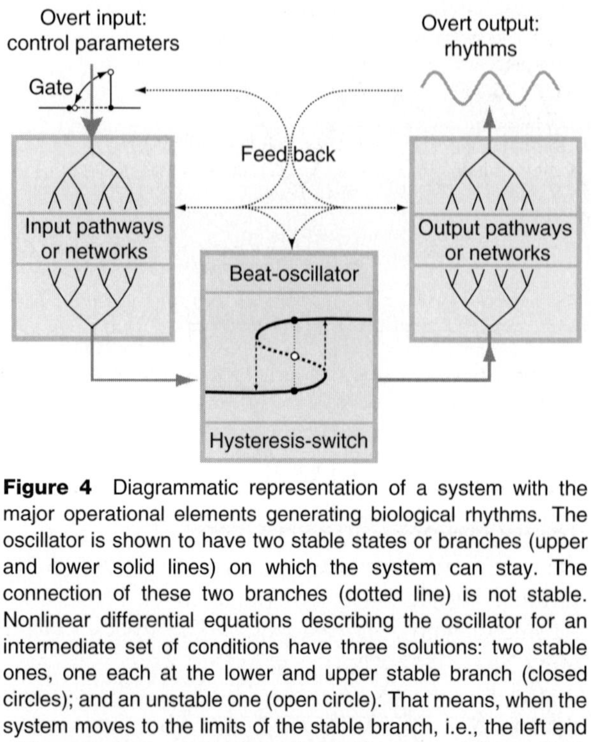

Figure 4 Diagrammatic representation of a system with the major operational elements generating biological rhythms. The oscillator is shown to have two stable states or branches (upper and lower solid lines) on which the system can stay. The connection of these two branches (dotted line) is not stable. Nonlinear differential equations describing the oscillator for an intermediate set of conditions have three solutions: two stable ones, one each at the lower and upper stable branch (closed circles); and an unstable one (open circle). That means, when the system moves to the limits of the stable branch, i.e., the left end of the upper one or the right end of the lower one, it jumps down to the lower one or up to the upper one (arrows): It operates as a hysteresis switch. This model of an oscillator is generic, i.e., applicable to very different systems, for example, technical systems, enzymes, biomembranes, etc. The oscillator receives information of control parameters via input pathways or networks. Entry of control parameter information may be gated. The oscillator operates output pathways or networks, which release overt output rhythms. All elements, gates, input and output networks, overt output rhythms, and the oscillator, are connected via positive and negative feedback loops. The system is a coupled meta-network.

Overt Rhythms

The output rhythms are the time series we observe (**Table 1**). We can see circadian rhythms at all levels of scaling from organs, to cells and membranes, and transcript levels (mRNA) and enzymes. Major rhythms of organs studied are the leaf movements mediated by the motor organs of the pulvini and induction of flowering. Flower induction is light regulated via a circadian oscillator, which has a light sensitive phase. In whole cells movements, the cell

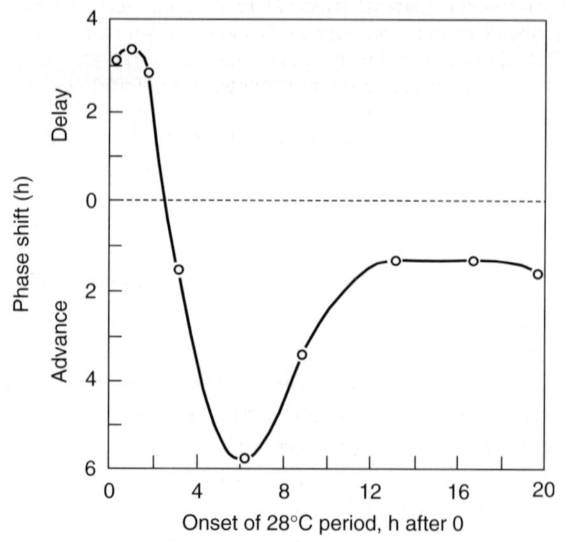

Figure 5 A classical experiment by I. Moser. Phase-response curve of circadian leaf movements of *Phaseolus coccineus* when the temperature was raised from 20°C to 28°C for 4 h at time 0 (abscissa), which refers to the maximum night position of the leaves. The final phase shift may be a delay of more than 3 h or an advance of almost 6 h, and any length of time in between depending on the time (abscissa) of the circadian rhythm when the temperature change to 28°C was applied. Reproduced with permission from Bünning E (1973) *The Physiological Clock*. New York, Heidelberg, Berlin: Springer-Verlag.

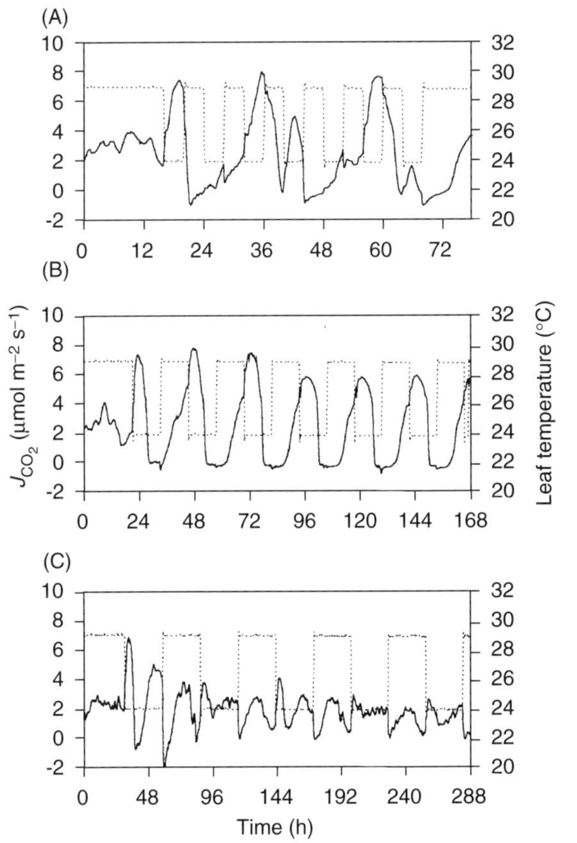

Figure 6 Response of the CO_2 exchange rhythm (solid lines) in the Crassulacean acid metabolism (CAM) plant *Kalanchoë daigremontiana* in LL to externally imposed temperature rhythms (dotted lines). External temperature rhythms were rectangular symmetric changes between 24°C and 29°C. With a short period of 8 h of the imposed rhythm, the circadian period came through (A) with a 24-h period of the imposed rhythm there was exact entrainment (B), and with a long period of 56 h of the imposed rhythm the time series became arrhythmic (C). Reproduced with permission from Lüttge U, Grams TEE, Hechler B, Blasius B, and Beck F (1996) Frequency resonances of the circadian rhythm of CAM under external temperature rhythms of varied period lengths in continuous light. *Botanica Acta* 109: 422–426.

cycle and overt oscillations of metabolic networks receive most attention. While membranes may provide a structural basis of the oscillators themselves, various membrane activities are also seen as circadian output phenomena. A vast number of enzymes show circadian oscillations in their transcript (mRNA) and protein levels and activities; a selection of the most interesting and well studied ones is listed in **Table 1**. Since light is the foremost external control parameter of circadian rhythmicity in plants, it may not be surprising that a very large number of circadially oscillating enzymes are directly or indirectly (photorespiration, crassulacean acid metabolism) related to phytosynthesis. In conclusion, the phenomenology of overt circadially rhythmic output is so vast and immensely diverse that it appears that almost any aspect of plant life may be under circadian control.

Table 1 Examples of overt circadian rhythmic output phenomena at different scaling levels of organization

Scaling level	*Circadian output*
Organs	
Roots	Transport
Leaves (pulvini)	Movements
Shoots	Growth, induction of flowering
Cells	Stomatal guard cell movements; taxis, i.e., movements of unicellular organisms; cell cycle and cell division; metabolic networks, e.g., C_3 and C_4 photosynthesis, crassulacean acid metabolism (CAM), respiration, energy metabolism
Membranes	Electrical potential; H^+-transporting ATPase activity; K^+-transport; membrane components (lipids)
Enzymes	
Photosynthesis	Ribulose-bis-phosphate carboxylase/oxygenase small and large subunit; chlorophyll a/b-binding; ATP-synthase; chlorophyll synthesis enzymes
Photorespiration	Several enzymes
Crassulacean acid metabolism	Several enzymes
Carbohydrate metabolism	Several enzymes
Bioluminescence	Luciferase binding protein
N-metabolism	Nitrate reductase; nitrate transporter; nitrogenase
Ethylene synthesis	1-Aminocyclopropane-1-carboxylic acid oxidase

External Control Parameters

The external control parameters reveal themselves in their action in entrainment and as *zeitgebers* in phase-shifting. While some rhythms in animals, e.g., locomotor activity in the fruitfly *Drosophila*, and in fungi, e.g., conidiation, only occur in darkness, circadian rhythms in plants are usually strongly dampened in DD and need light. Light does act as a control parameter via photosynthesis, where external CO_2 concentration may be another control parameter. Signaling functions are exerted by red and blue light as major control parameters in plants. Another very important external control parameter is temperature. In contrast to the amplitude, which may strongly depend on temperature (**Figure 1**), circadian rhythms are always temperature compensated in that the circadian period is temperature independent over a large range of temperatures. Sometimes there is a slight overcompensation where the rhythm slows down a little (slightly increased period length) rather than accelerating at increased temperature. The control parameters of light intensity and temperature may also determine thresholds delineating windows or ranges of parameter intensities within which a system is rhythmic and beyond which it is arrhythmic. This was observed for light intensity and temperature in overt rhythms as different as bioluminescence of *Gonyanlax polyedra* and the CAM cycle of *Kalanchoë daigremontiana* (*see* **Photosynthesis**: CAM Plants).

Output and Input Networks

Naturally the three "black boxes" pose larger problems of unraveling. Output networks in fact largely remain black boxes. We know much more about possible input networks because these are largely identical to the signaling chains and networks we have in the regulation of physiology, growth, and development, at least with respect to the various receptors and messengers involved (**Table 2**). While receptors (in chronobiology also called *zeitnehmer*) for the control parameters red light and blue light, i.e., phytochromes and cryptochromes, are known and studied and characterized at the molecular level, no temperature receptor has been detected. Both light and temperature signals then feed into signal transduction networks with the well-known primary and secondary messengers, e.g., all the known groups of phytohormones, calcium ions, phytoinositides, and also cyclic AMP.

Oscillators

In mammals, including man, the suprachiasmatic nucleus of the hypothalamus in the brain has a central role in rhythmicity. In *Drosophila*, the *PER* ("period") and *TIM* ("timeless") genes and in the fungus *Neurospora*, the *FRQ* ("frequency") gene appear to have "master gene" functions in a hierarchy of clock-controlled genes (CCGs), just as maternal master genes head a hierarchical sequence of gap genes, pair rule genes, homeotic genes, and segment polarity genes in the fruitfly's development. From these observations, the concept arose that there should be one central oscillator in all organisms.

Table 2 Receptors and messengers involved in input networks of circadian rhythms in plants

Receptors
Light
phytochromes (red light)
cryptochromes (blue light)
Temperature
no receptor known
Messengers
Phytohormones:
auxin(s)
gibberellic acid
abscisic acid
kinetin(s)
ethylene
Calcium ions
Phytoinositides (inositol phosphates of plants)
Cyclic-AMP

However, there are many examples where different overt rhythms occur side by side and apparently are independent of each other. A list of examples for plants is given in **Table 3**, which shows that different types of overt output rhythms may occur in the same organ, tissue, or even cell. These may be independent of each other or get coupled or synchronized. Of course, this leaves the question open whether these are activities of one oscillator operating different output networks or are due to different oscillators. The different oscillations of two catalase genes in the same cells of *Arabidopsis* and the outputs of class 1 and class 2 genes in the cells of *Synechococcus* (**Figure 7**) appear to make the ruling of a single oscillator rather unlikely.

Another possibility is that in a plant organ each cell individually contains a copy of the same oscillator governing the same output rhythm. These autonomous oscillators can desynchronize and synchronize, respectively, at times. This was shown for adjacent stomata in leaves and for CAM in small patches of cells in a given leaf of *Kalanchoë* (*see* **Photosynthesis**: CAM Plants).

Thus, the one central oscillator appears to be just a figment of our imagination, an idea rather than a solid fact. **Figure 4** shows that even a given system for one given overt output already is a super network of coupled networks, and network integration is moved to the next level of higher complexity if we consider coupling and synchronization of different oscillators. This implies that another idea, which must be abandoned in most cases, is that of linear chains of deterministic unidimensional cause–effect linkages of signaling and regulation systems. We are mainly dealing with network-like systems. This greater complexity has the advantage of conferring robustness, adaptive plasticity, and flexibility to the respective systems. If networks are open there may still be some kind of hierarchy. If networks are closed in themselves, hierarchy is lost, although knots may still be differentiated in being more central or more peripheral in network architecture.

Network theory may help us to stay on top of the bewildering phenomenology of signaling and regulation elements. However, we can and must ask what is

Table 3 Different overt rhythms in one cell, tissue, or organ

Organisms	*Organ, tissue, cell*	*Overt circadian rhythms*
Pharbitis nil, Angiosperm	Shoot system	Leaf movements, photoperiodic flowering
Arabidopsis thaliana, Angiosperm	Shoot system	Leaf movements, hypocotyl elongation
N. tabacum, Angiosperm	Different tissues	Cytoplasmic calcium levels
Phaseolus, Angiosperm	Leaves	Opening and closing movements of stomata, pulvinar movements
Angiosperm	Leaves	Opening and closing movements of stomata, photosynthesis
Angiosperm	Leaves	Movements of adjacent individual stomata
Kalanchoë, Angiosperm	Leaves	Crassulacean acid metabolism (CAM), nitrate reductase
Kalanchoë, Angiosperm	Leaves	Crassulacean acid metabolism (CAM) in adjacent patches of cells
Arabidopsis thaliana, Angiosperm	Same cell	Two catalase genes, light harvesting complex (*LHCb*) promoter gene
Chlamydomonas reinhardtii, Unicellular Chlorophyceae	Same cell	Phototaxis, chemotaxis, stickiness to glass, cell division, ammonia uptake
Euglena, Unicellular Euglenophyceae	Same cell	Photosynthesis, gravitaxis
Gonyaulax polyedra, Unicellular Dinophyceae	Same cell	Photosynthesis, bioluminescence, mitosis
Synechococcus elongatus, Unicellular prokaryotic Cyanophyta	Same cell	Class 1 and class 2 genes

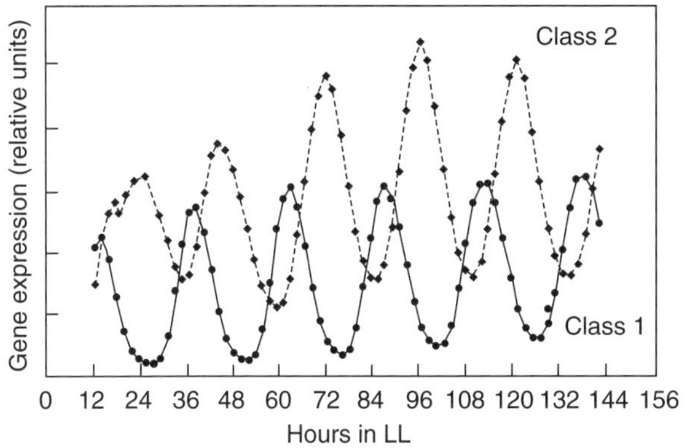

Figure 7 Different phase shifted oscillations of expression of a class 1 (•) and a class 2 (♦) gene in *Synechococcus elongatus* as recorded by the group of Susan S. Golden measuring bioluminescence of luciferase reporter strains, a new method allowing continuous registration of time series of gene activities. Reproduced with permission from Min H and Golden S (2000) A new circadian class 2 gene, *opcA*, whose product is important for reductant production at night in *Synechococcus elongatus* PCC 7942. *Journal of Bacteriology* 182: 6214–6222.

the possible nature of individual oscillators. As already noted above, any system with nonlinear dynamics can potentially become an oscillator. The beat oscillator depicted in **Figure 4** can be an inorganic system of physics, civil engineering, or inorganic chemistry. In organisms, it can be a gene system, but also a posttranslational system independent of the actual operation of genes. An example of the latter is the tonoplast membrane hysteresis switch operating in circadian rhythmicity of the CAM cycle (*see* **Photosynthesis**: CAM Plants). At the molecular level circadian oscillations occur at all points on the

Table 4 Most widely used organisms in chronobiology – so-called model organisms

Organism	*Organization*	*Taxonomy*
Drosophila melanogaster	Eukaryotic, multicellular	Insecta
Neurospora crassa	Eukaryotic, mycelial	Ascomycota
Synechococcus elongatus PCC 7942	Prokaryotic, unicellular	Cyanobacteria
Gonyaulax polyedra	Eukaryotic, unicellular	Dinophyceae
Chlamydomonas rheinhardtii	Eukaryotic, unicelullar	Chlorophyta
Arabidopsis thaliana	Eukaryotic, multicellular	Angiosperms
Kalanchoë fedtschenkoi *Kalanchoë daigremontiana*	Eukaryotic, multicellular	

way from the gene to the protein, i.e., in transcription, transcript abundance, translation, and posttranslational processing.

Approaches of study used are: (1) genetics, with mutagenesis and molecular biology for the molecular oscillators; (2) biochemistry, biophysics, and physiology for the posttranslational oscillators; and (3) theoretical physics of nonlinear dynamics, including modeling and computer simulations. The most intensely studied model systems in the kingdoms of animals, fungi, and plants are listed in **Table 4**.

Perhaps the closest advance in reconstruction of a molecular oscillator is achieved by the group of Susan S. Golden for the cyanobacterium *Synechococcus elongatus*. The core is a three-gene-cluster

KAI A, *B* and *C* ("kai" is Japanese for cycle), where *KAIC* encodes an ATP binding peptide. A *CIKA* gene encodes a circadian input kinase regulating a bilin-type of phytochrome-related photoreceptor. A group of factors at the output side was also identified. Evidently, this oscillator is not the result of a single gene activity, but a regulation network.

For higher plants, the greatest molecular advance of looking into the black boxes is made for blue light controlled rhythmicity in *Arabidopsis thaliana*. The entire genome of *A. thaliana* is now sequenced. A large number of genes have been identified that participate in rhythmicity. Some of them can be assigned to the black boxes of **Figure 4** and the feedback pathways, but many questions remain unanswered (**Figure 8**). This highlights the problem of how difficult it is to work out interconnections within and between networks, and to determine sequences of events. While we have been able to glance into the black boxes, much remains to be done regarding deciphering the networks. For the aspect of integration of networks, it is noteworthy that some outputs of the *Arabidopsis* circadian system depicted in **Figure 8** again are gene oscillations. Outputs are also growth and morphology, as may be concluded from the names of some of the genes involved, e.g., light mediated development gene (*DET1*), constitutive photomorphogenetic gene (*COP1*), early flowering gene (*ELF3*), and late elongated hypocotyl gene (*LHY*).

There are cases where oscillator and output appear to be uncoupled, e.g., in DD. In plants, continuous circadian rhythmicity in DD is rare, as most circadian rhythms run in LL. As mentioned above, this is due to energy dissipation and a lack of photosynthetic energy supply in DD. While the overt output rhythms cease oscillating, the oscillator itself may continue. A special *ZGT* gene (Chinese zhong guang tiaokong meaning clock and light controlled), which couples an oscillator with outputs, was discovered in tobacco (*Nicotiana tabacum*).

Consequences of Short-Term Fluctuations

Besides molecular coupling in the output pathway, e.g., via the *ZGT* gene, there are also completely different and purely physical interactions determining the output. Let us consider a system consisting of the essential elements of circadian oscillations depicted in **Figure 4**. If we add short term fluctuations, e.g., stochastic noise, to its dynamics, this may have different consequences. Such dynamics always exist in biological systems. Then, first the system must be robust under the influence of the fluctuations. Robustness must be given by the organization of the system, and it is in fact inherent in the network structure, as pointed out above. Second, with meta networks consisting of many identical subsystems, the subnetworks will differ in fluctuations. Under certain conditions, this may lead to a loss of synchronization between these elements, as shown experimentally, for instance, in CAM leaves (*see*

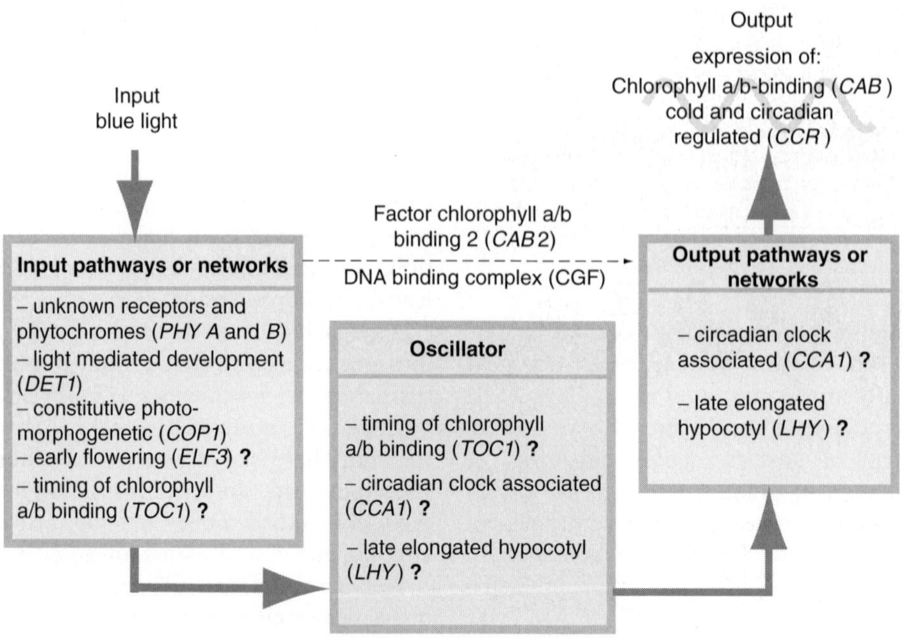

Figure 8 A glance into the black boxes of **Figure 4**: evidence for genes involved in circadian rhythmicity in *Arabidopsis thaliana* related to the scheme of **Figure 4**.

Photosynthesis: CAM Plants). Third, fluctuations may cause stochastic resonance. If we consider an oscillation where the peaks never reach a threshold level above which overt rhythmicity would be seen (**Figure 9A**), stochastic noise superimposed to the oscillation may lead to a rhythmic output if noise has the appropriate intensity. Weak noise has no effect (**Figure 9B**), and strong noise completely overrides the oscillator (**Figure 9C**). With medium noise, the peaks of the oscillator rhythmically pass the threshold level and a rhythmic output is seen (**Figure 9D**). This phenomenon is called stochastic resonance.

Stochastic resonance has been observed in biological systems, and this includes applications in medicine. An experimental proof for stochastic resonance in the circadian rhythmicity of plants is lacking to date. However, simulations with a model that excellently reproduces many other facets and

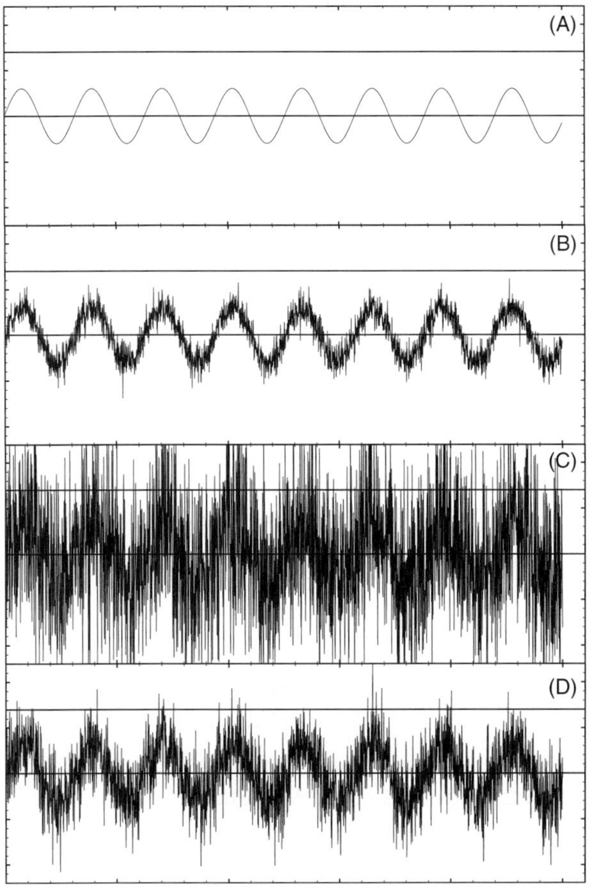

Figure 9 Schematic simulation of stochastic resonance. An oscillation (sine wave) only shows overt output when it passes an upper ceiling indicated by the upper horizontal line. (A) The sine wave without added noise does not pass the ceiling and there is no rhythmic output. (B) With low noise intensity added there is no output just like in (A). (C) With strong noise intensity added the sine wave is totally overridden and no rhythm is observed. (D) With appropriate intermediate noise intensity the sine wave rhythmically passes the ceiling and an overt rhythm is obtained.

properties of the endogenous CAM cycle are highly suggestive. Using the CAM model with the tonoplast membrane beat oscillator (*see* **Photosynthesis**: CAM Plants) Freder Beck and colleagues have simulated stochastic resonance. At too high a temperature, the system moves directly into a fixed point with low vacuolar malate content. At too low a temperature, another fixed point is reached with high vacuolar malate content. At intermediate temperatures, the system may oscillate, i.e., display a stable limit cycle (**Figure 10A**). This behavior is also observed in intact leaves *in planta*. At a temperature close to the upper ceiling, the oscillation is strongly dampened and it spirals into a fixed point (**Figure 10A**). If a noise term is added to the nonlinear differential equations running the simulated oscillations, the trajectories in the first three cases (temperature too high, too low, or intermediate) become less smooth, but nothing changes in principle. In the case of the temperature close to the critical upper ceiling, noise prevents the system from remaining in the fixed point to which the dampened oscillation spiraled in. It is thrown by noise-kicks stochastically out of the fixed point into nearby limit cycles, from where it returns to the fixed point after a full cycle, so that a quasiperiodic output with almost regular oscillations is obtained at this temperature (**Figure 10B**). The time series simulation of **Figure 11** gives a different graphical representation of this phenomenon, and also demonstrates the strong dependence of the signal/noise relation on noise intensity alluded to above (**Figure 9**).

Biologically Adaptive Value of Endogenous Clocks

Overt circadian rhythmicity notably is an experimental artifact. It cannot occur under natural conditions, where constant environmental conditions with constant light or darkness never occur. Nevertheless, a biologically adaptive value of endogenous clocks is often discussed. In the natural day/night rhythm, it is thought to allow internal timing of physiological processes, and to provide readiness, alertness, and preparedness to the changing conditions of day and night. However, there is little solid evidence that this indeed is of adaptive value.

So far, there appears to be only one conclusive experiment by the group of Golden with the prokaryote cyanobacterium *Synechococcus elongatus* demonstrating a competitive advantage of circadian rhythmicity. They obtained mutants and strains with a short circadian period of 23 h, a medium period of 25 h (wild-type), and a long period of 30 h. When grown separately in pure culture, all strains

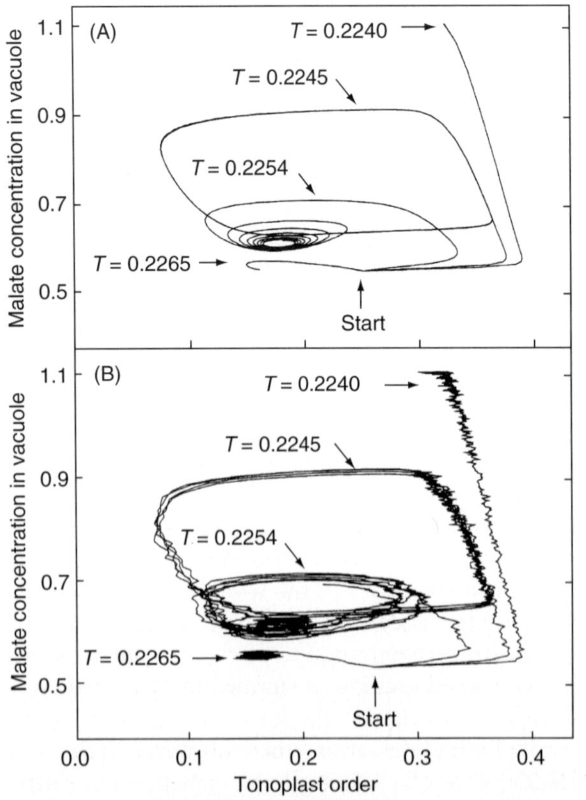

Figure 10 Simulation of trajectories in the phase plane of vacuolar malate levels (ordinate) and the state of order of the tonoplast membrane (abscissa) in CAM. All trajectories start at the same point ("Start"). All numbers given are relative numbers of the mathematical simulation model. (A) Trajectories without noise. For the scaled low relative temperature (0.2240), the trajectory runs into the upper fixed point with high vacuolar malate content. For the higher temperatures, the trajectories reach the lower fixed point with low vacuolar malate content on a direct path (0.2265) or on a spiral path (0.2254). At intermediate temperatures (e.g., 0.2245), limit cycle oscillations occur. Temperature regulates the distance of fixed points to limit cycles. (B) Trajectories as in (A) with a noise component added. Trajectories at the temperatures 0.2240, 0.2245, and 0.2265 are not affected in principle. However, the trajectory, which in (A) without noise spiraled into the fixed point is now thrown by noise-kicks stochastically into a nearby limit cycle from where it returns to the fixed point after a full cycle, so that now almost regular oscillations are obtained at this temperature. Reproduced with permission of Springer-Verlag from Lüttge U (2000) The tonoplast functioning as the master switch for circadian regulation of crassulacean acid metabolism. *Planta* 211: 761–769.

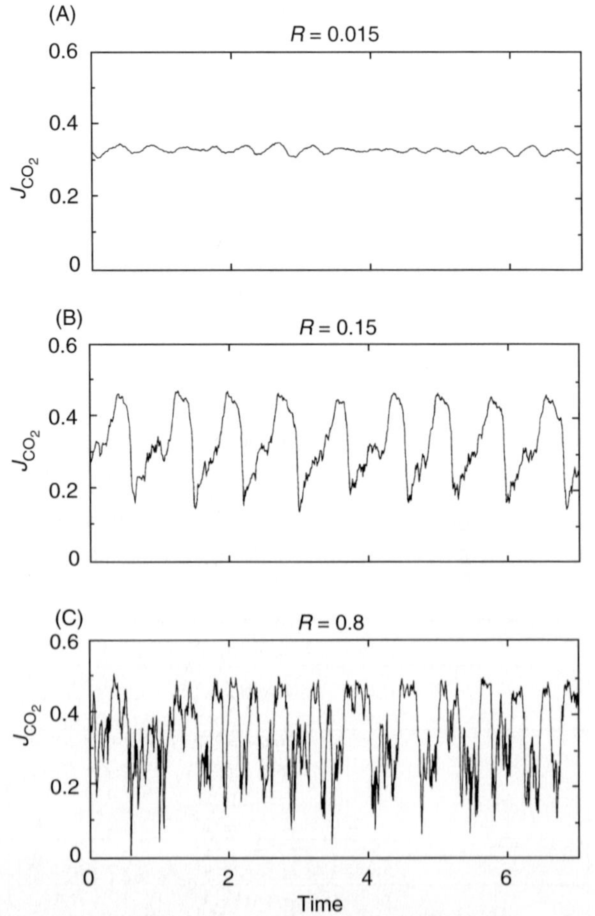

Figure 11 Time series of net CO_2 exchange in CAM simulated at various levels of noise (*R*) increasing from top to bottom (A to C). All numbers given are relative numbers of mathematical simulation. These simulations with the CAM model (see also **Figure 10**) illustrate the signal/noise relationships explained by the schematic simulations in **Figure 9** by using an operational model of a real biological system, i.e., the CAM cycle. Reproduced with permission of Springer-Verlag from Lüttge U (2000) The tonoplast functioning as the master switch for circadian regulation of crassulacean acid metabolism. *Planta* 211: 761–769.

performed similarly well under normal DL or in LL. If they were mixed, however, the strain having an endogenous period closest to an applied external period rapidly outcompeted the other two strains, i.e., the short-period strain (23 h) won at L:D of 11:11 h, the wild-type strain (25 h) at L:D of 12:12 h, and the long-period strain (30 h) at L:D of 15:15 h.

Conversely, simulations of photosynthesis of higher plants by Williams and Gorton comparing a linear and a rhythmic model suggested that rhythmicity does not have advantages for CO_2 acquisition in the field. Thus, photoperiodic timing may be a side-product of evolution of life under the continuous entrainment by the rhythm of days and nights, which may have forced the time constants of the turnover of elements in signaling and regulation networks to evolve in a way that such networks display circadian oscillations. It is intriguing to ask whether the message of the experiment with *Synechococcus elongatus* can be applied to other more highly organized organisms and higher plants assuming a general advantage of biological clocks for fitness.

List of Technical Nomenclature

Amplitude	Maximum extent of oscillation.
Clock controlled gene (CCG)	A gene showing circadian rhythmicity of expression.
Control parameter	An external or environmental factor interacting with rhythmicity.
Crassulacean acid metabolism (CAM)	Ecophysiological modification of photosynthesis with nocturnal CO_2 fixation.
Dampening	Decrease of amplitude in time series of rhythms.
Day	(1) Solar day of 24 h, or (2) light period of the solar day.
Desynchronization	Individual elements or oscillators operating independently and out of phase with each other.
Dissipative	System loses ("dissipates") energy due to contact with environment.
Diurnal	Day/night cycle of 24 h.
Entrainment	System carried along by an external rhythm, following the imposed external rhythm.
Feedback	Positive or negative, where downstream elements or events either stimulate or inhibit upstream elements or events.
Frequency	Number of oscillations per unit of time.
Hysteresis switch	Change between different loops of pathways connecting two states of a system.
Limit cycle	Stable trajectory of an oscillating system, regular stable oscillation.
Network	A complex arrangement of interconnected elements.
Nonlinear dynamics	Discipline in theoretical physics dealing with nonlinear cause–effect relations in time and space.
Oscillator	Putative basic structure generating rhythmicity.
Period	Time required by an oscillation to return from a given phase to the same phase again.
Phase	Any point in the cycle of an oscillation; time during oscillation measured in units of the period.
Phase response	Response to phase shifting parameters at different phases.
Phase space	Diagram consisting of the dynamical variables of the system (such as phase and phase velocity).
Phase shifting	Setting the clock by an external control parameter.
Rhythm	
Circadian	*Circa diem*, i.e., approximately 24 h.
Endogenous	In the absence of any oscillation of external parameters or conditions.
Free running	As endogenous.
Overt	Observable output of endogenous oscillator.
Simulation	Computer aided output of the mathematical model of a system.
Stochastic resonance	Rhythmic output of a system under the ordering influence of noise.
Synchronization	Individual elements or oscillators coupled, producing coordinated oscillations.
Time series	Measurements of an overt output over time.
Trajectory	Path of a system in phase space.
Zeitgeber	An external control parameter.
Zeitnehmer	A receptor for an external control parameter.

See also: **Flowering and Reproduction**: Flower Development. **Growth and Development**: Canopy Architecture; Cell Division and Differentiation; Cell Growth; Control of Gene Expression, Post Transcriptional Regulation; Control of Gene Expression, Regulation of Transcription; Molecular Biology of Development. **Photosynthesis and Partitioning**: C_3 Plants; C_4 Plants; CAM Plants. **Regulators of Growth**: Abscisic Acid; Auxins; Cytokinins; Ethylene; Gibberellins; Photomorphogenesis; Photoperiodism; Phytochromes and other Photoreceptors. **Seed Development**: Embryogenesis.

Further Reading

Barak S, Tobin EM, Andronis C, Sugano S, and Green RM (2000) All in good time: the *Arabidopsis* circadian clock. *Trends in Plant Science* 5: 517–522.

Bünning E (1973) *The Physiological Clock: Circadian Rhythms and Biological Chronometry*, 3rd edn. Berlin, Heidelberg: Springer-Verlag.

Edmunds LN and Tamponnet C (1990) Oscillator control of cell divison cycles in *Euglena:* role of calcium in circadian timekeeping. In O'Day DH (ed.) *Calcium as an Intracellular Messenger in Eucaryotic Microbes*, pp. 97–123. Washington DC: American Society for Microbiology.

Lüttge U (2000) The tonoplast functioning as the master switch for circadian regulation of crassulacean acid metabolism. *Planta* 211: 761–769.

Lüttge U (2002) Circadian rhythmicity. *Progress in Botany* 64: 277–319.
McClung CR (2000) Circadian rhythms in plants: a millenial view. *Physiologia Plantarum* 109: 359–371.
Millar AJ (1999) Biological clocks in *Arabidopsis thaliana*. *New Phytologist* 141: 175–197.
Somers DE (1999) The physiology and molecular bases of the plant circadian clock. *Plant Physiology* 121: 9–19.
Sweeney BM (1987) *Rhythmic Phenomena in Plants*. San Diego: Academic Press.
Watts DJ (1999) *Small Worlds: The Dynamics of Networks between Order and Randomness*. Princeton: Princeton University Press.
Winfree AT (1990) *The Geometry of Biological Time*. Berlin, Heidelberg, New York: Springer-Verlag.

ROOT DEVELOPMENT

Contents

Genetics of Primary Root Development

R M Twyman, University of York, York, UK

Introduction

The roots of a mature plant provide support and facilitate the uptake of nutrients and water. Although different plants show tremendous variation in root morphology, the basic anatomical elements of the root are the same, consisting of dermal tissue (epidermis), ground tissue (cortex, endodermis, pericycle), and vascular tissue (phloem, xylem). These concentric cell layers are derived from the embryonic epidermis (L1), ground (L2), and vascular (L3) tissues. When the embryo germinates, the root that emerges is called the radicle (primary root). In many plants, this diversifies to form an extensive, branching network of lateral roots in the soil. In gymnosperms, however, the primary root remains dominant and is known as the taproot. The root system originates from a small population of cells, the root apical meristem (RAM), which forms at the basal end of the plant embryo.

Roots are tip-growing structures. The root grows in length due to the proliferation of self-renewing stem cells in the RAM, which is found just behind the root cap. Cells leaving the RAM are deposited in files (representing the radial cell layers) and go through successive stages of elongation and differentiation. The young root can thus be divided into four longitudinal zones: the cap, the division zone, the elongation zone, and the differentiation zone (**Figure 1**). The RAM does not produce lateral organs in the same way as the shoot meristem. Instead, lateral roots may arise from the differentiated part of the main root. However, the RAM does continuously supply cells to the root cap, which is worn away by abrasion as the root pushes through the soil.

The genetic basis of root development has been extensively studied in *Arabidopsis* because it provides several very tractable examples of pattern formation and morphogenesis. In particular, because the cell divisions in the RAM are stereotypical, the *Arabidopsis* root provides a valuable experimental model for the study of cell lineage and cell–cell

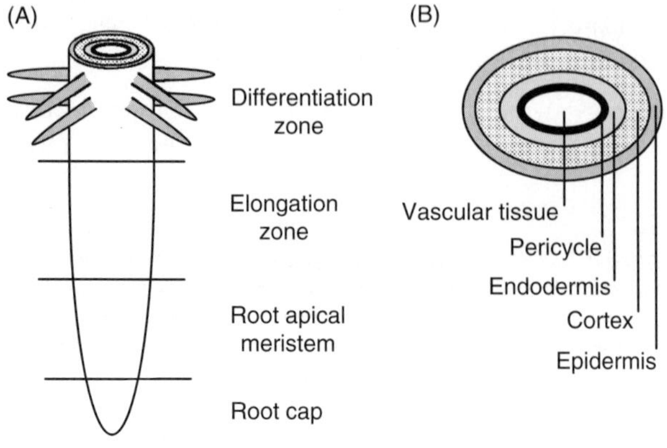

Figure 1 General structure of the *Arabidopsis* root. (a) Longitudinal organization. (b) Radial organization.

interactions in plant development. This article discusses the specification of the RAM in the *Arabidopsis* embryo, the role of the RAM in root patterning, and the specification of cell fates during root differentiation.

Specification of the Root Apical Meristem

The origins of the RAM can be traced back to the very first cell division in the plant embryo, when the elongated zygote divides asymmetrically to produce a small apical cell (which gives rise to most of the embryo) and a larger basal cell (which forms the basal part of the embryo and the suspensor). The apical cell undergoes a series of stereotyped divisions to produce the embryo, while the basal cell divides several times in the same plane to produce the suspensor. The embryonic root is derived from the lower tier of embryonic cells in the globular stage embryo and the upper cell of the suspensor, which is called the hypophysis (**Figure 2**).

Axial patterning depends on the expression of a series of genes in zones along the apical–basal axis. Mutations in these genes generate patterning defects that abolish parts of the axis and their derived structures. In the case of root development, the most important genes are those that specify basal parts of the axis. For example, the *MONOPTEROUS* (*MP*) gene is expressed in the bottom tier of cells of the proembryo, and the *mp* mutant has no hypocotyl, radicle, or RAM. Similarly, the *HOBBIT* (*HBT*) gene is required for the formation of the hypophysis, and *hbt* mutants do not correctly form a RAM. The product of the *MP* gene is a transcription factor that binds to auxin-responsive elements. The role of auxin in root development has been revealed by the analysis of genes that act upstream of *MP*. For example, the *GNOM* (*GN*) gene, which is expressed just after fertilization, encodes a protein required for polarized auxin transport. In normal embryos, auxin accumulates at the future basal pole because auxin efflux pumps are localized specifically at the basal ends of each cell. In *gn* mutants, there is no precise localization of these efflux pumps, auxin is evenly distributed, the zygote divides symmetrically rather than forming apical and basal cells of different sizes, and the basal pole of the embryo is not specified.

Once the RAM is specified, root growth is dependent on its continuous proliferation. This does not occur in mutants lacking functional *ROOT MERISTEMLESS1* (*RTM1* or *RTM2*) genes, while cell proliferation in the shoot meristem is unaffected.

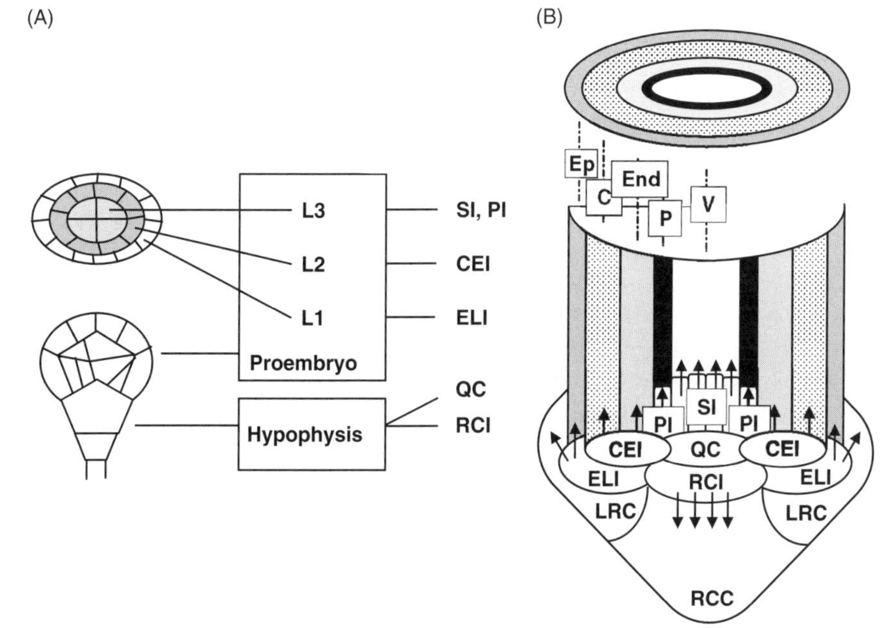

Figure 2 (A) Origins of the cell types in the root. Schematic diagrams on the left show the globular proembryo in transverse (upper) and longitudinal (lower) section. The origins of the three radial cell layers L1, L2, and L3 are shown in the transverse section. The longitudinal section is labeled proembryo and hypophysis. The letters on the right show the cells in the promeristem that are derived from these embryonic structures. SI, stele initials; PI, pericycle initials; CEI, cortical/endodermal initials; ELI, epidermal/lateral root cap initials; QC, quiescent center; RCI, root cap columella initials. (B) Organization and activity of the root apical meristem. The different cell types in the file meristem and mature root are shown in different shades and are labeled as follows: Ep, epidermis; C, cortex; End, endodermis; P, pericycle; V, vasculature. The cells of the root cap are labeled as follows: LRC, lateral root cap; RCC, root cap columella. They are derived from the initial cells of the promeristem, which are identified using the same abbreviations listed for (A) above.

Conversely, overproliferation of cells is the defect in *fass* mutants, which are characterized by fat, barrel-like roots with extra cell layers.

Establishing and Maintaining Radial Organization

Origins of the Radial Cell Layers in the Root

Radial patterning is established early in development so that by the early heart stage the plant embryo clearly comprises three layers, known as L1, L2, and L3. Mutants such as *knolle* and *keul*, which perturb embryonic radial pattering, disrupt the organization of cells in the root and shoot while the axial patterning discussed above is preserved. Roots contain four distinct concentric layers of cells surrounding the vasculature, so root development must involve a series of periclinal divisions that diversify the basic three layers of the embryo. In normal embryos, the L1 (epidermis) layer of the embryo gives rise to the epidermal and lateral root cap initials, which generate the lateral root cap and the outer epidermal cell layer. The L2 (ground) layer gives rise to the endodermal/cortical initials, which generate the endodermis and cortex layers. Finally, the L3 (stele) layer gives rise to the stele initials, which generate the vasculature and pith. The hypophysis (terminal suspensor cell) gives rise to the stem cells of the promeristem (the quiescent center) and the root cap columella initials (**Figure 2**).

Maintaining the Organization of Radial Cell Layers

Once the initial cells of the RAM have been established, continual proliferation and differentiation of these cell types allows the root to grow and lay down files of cells organized in the correct concentric pattern. The root of *Arabidopsis* is particularly interesting because the cell divisions are stereotyped and largely invariant. For example, each L2-derived cortical/endodermal initial cell (CE initial) divides transversely to produce a replacement initial cell and one daughter cell. The daughter cell then goes on to divide periclinally to produce one cortical cell and one endodermal cell, which enter the file meristem. The process is reiterated indefinitely.

Two types of experiment have been used to investigate the molecular basis of these stereotyped divisions: cell ablation and genetic analysis. Ablation experiments have shown that, while the cell divisions follow a defined pattern, they are regulated largely by cell–cell contacts rather than depending on lineage. For example, the ablation of one CE initial cell is followed by the unexpected division of an adjacent pericycle initial cell, so that the CE initial is replaced

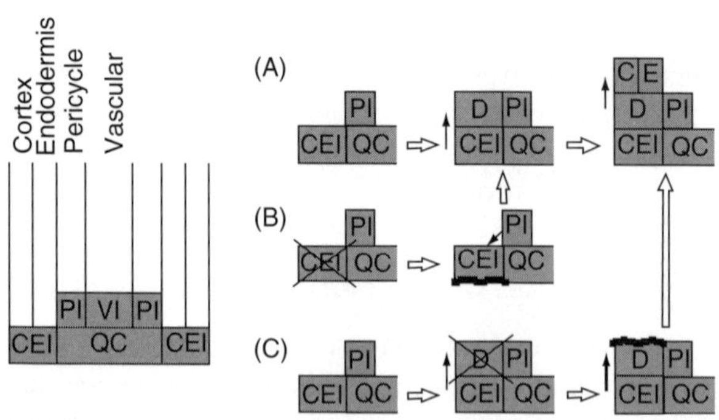

Figure 3 Laser ablation studies in root development. Left panel shows schematic representation of the promeristem (CEI, cortical/endodermal initials; QC, quiescent center; PI, pericycle initial; SI, stele initials). Corresponding cell layers in the file meristem are shown above. (A) In normal development, each CEI gives rise to a daughter cell (D), which divides to form cortical and endodermis cells (C, E) adding to existing files. (B) If the CEI cell is ablated, an unexpected division of the PI cell replaces it. (C) If the daughter cell is ablated, the CEI undergoes an extra division to replace it. Wavy lines indicate squashed remains of ablated cells. Reproduced with permission from Twyman RM (2000) *Instant Notes in Developmental Biology*. BIOS Scientific Publishers, Oxford.

by a pericycle daughter cell (**Figure 3**). The replacement cell then behaves in the same manner as a CE initial. If root development were lineage dependent, the pericycle initial would define a separate lineage (i.e., it would only be capable of producing pericycle cells) and such a replacement would not occur. If the daughter cell of a CE initial is ablated, the underlying CE initial cell divides again to replace it. However, if three or more daughter cells lying above the CE initial are ablated, the CE initial becomes unable to divide asymmetrically (**Figure 3**). This shows that cell behavior is determined by signals from adjacent cells in the same file. Recently, it has been shown that these signals may be constrained within layers of cells by channeling through plasmodesmata.

Several mutants have been identified that disrupt the radial organization of the *Arabidopsis* root. *SHORT ROOT* (*SHR*), *SCARECROW* (*SCR*), and *PINOCCHIO* (*PIC*) are required for the periclinal division of CE daughter cells and the formation of separate cortical and endodermal layers. In *shr* mutants, the undivided daughter cells differentiate as cortical cells, so the root lacks an endodermal layer. The opposite effect is seen in *pic* mutants. In *scr* mutants, the daughter cells express both cortical and endodermal markers and appear to have an intermediate character. Double mutants of *shr* and *scr* show the *shr* phenotype, suggesting that *SHR* acts upstream of *SCR* in the pathway that controls periclinal cell division. Both genes encode transcription factors so SHR may directly or indirectly activate

SCR. Recently, the SHR protein has been shown to move between cells in order to regulate the *SCR* gene.

Tropism in Root Growth

Root growth shows characteristic tropisms including positive geotropism (growth in the direction of gravity, also described as positive gravitropism) and avoidance of obstacles (sometimes described as negative thigmotropism). However, in a medium that is free of obstructions, the roots of most plants do not grow straight downwards. Most roots demonstrate a phenomenon called circumnutation, which is characterized by a circular or elliptical movement around the growth direction, so the root effectively grows downwards in waves and coils rather than in a linear fashion. In many plants, these waves and coils progressively straighten out so very little evidence of circumnutation remains. In *Arabidopsis*, however, they are preserved in the mature root system and this allows the molecular processes involved in circumnutation to be investigated by the analysis of mutants. Interestingly, *Arabidopsis* roots growing in space still undergo circumnutation even in the absence of gravity. Mutations that show abnormal circumnutation have been identified in *Arabidopsis*, e.g., *spr1*, *spr2*, *spr3*, *spir1*, *spir2*, *tor1*, *tor2*, and *tor3*. The functions of many of these genes remain to be elucidated, although it is suggested they may help to organize microtubules in the cell and therefore define the cell's symmetry.

Gravitropism is based on the redistribution of auxin in the elongation zone of the developing root. If the root is not growing vertically downwards, then auxin accumulates in the lower parts of the root, inhibiting cell elongation and causing the root to bend in the direction of gravity. Since only the root tip is sensitive to tropisms, this response requires a mechanism to sense gravity at the root tip and a signaling pathway to connect the root tip and elongation zone. Sensing may be mediated by starch-rich statoliths in the root cap columella cells, which respond to gravity by falling to the lowest point in the cell. Alternatively, or in addition, the columella cells may sense gravity by the displacement of the protoplast within the cell wall. Gravitropic sensing is relayed to the responding cells in the elongation zone of the root by signaling pathways that involve calcium and pH changes. Mutations that disrupt any of these processes can generate agravitropic plants.

Mutations that disrupt or abolish the gravitropic response have been identified in several plants. In *Arabidopsis*, mutants that do not accumulate starch in the root cap and therefore lack statoliths can still sense gravity, although the gravitropic response is weakened. Other mutants that show weak or abnormal gravitropic responses represent genes that encode components of the gravitropism signal transduction pathway (e.g., *RHA1*) or the auxin transport pathway (e.g., *AUX1*, *AGR1*).

Root Differentiation

Root hairs are formed only in the zone of differentiation. The root epidermis in *Arabidopsis* comprises 16–22 longitudinal files of cells. Each epidermal cell can adopt one of two alternative fates, becoming either a trichoblast (hair cell) or an atrichoblast (nonhair cell). However, all the cells in a single file adopt the same fate, and these fates appear to be specified during the later stages of embryogenesis. Generally, files of trichoblasts are separated by one or two files of atrichoblasts and ethylene signaling may play an important role in determining cell fates. Mutants have been identified that lack root hairs or have too many root hairs. These define genes involved in the specification of epidermal fates and the differentiation process itself.

WEREWOLF (*WER*) and *GLABRA2* (*GL2*) are positive regulators of atrichoblast differentiation and loss-of-function mutants are characterized by extra root hairs. WER and GL2 are both transcription factors, and WER is thought to act upstream of GL2 because *GL2* expression is abolished in *wer* mutants. *GL2* is expressed preferentially in atrichoblasts and acts relatively late in the differentiation process to suppress the formation of root hairs. In *gl2* mutants, it is still possible to identify atrichoblast cells even though they have hairs, because they are similar to nonhair cells in most other respects. Loss of function *caprice* mutants have fewer root hairs than normal, indicating that *CAPRICE* (*CPC*) is a positive regulator of trichoblast differentiation. Interestingly, both WER and CPC have MYB-like domains which suggest they interact. The MYB domain is a DNA binding domain found in several eukaryotic transcription factors. Whereas WER is a functional transcription factor, CPC lacks a transactivation domain. Therefore, CPC may repress atrichoblast differentiation by sequestering WER into inactive complexes. The differentiation of epidermal cells could depend on the balance of *WER* and *CPC* expression, with high ratios of *WER* favoring atrichoblast differentiation and high ratios of *CPC* favoring trichoblast differentiation. In agreement with this, *cpc wer* double mutants show intermediate numbers of root hairs. Another gene that promotes atrichoblast differentiation is *TRANSPARENT TESTA GLABRA* (*TTG*) and this appears to act

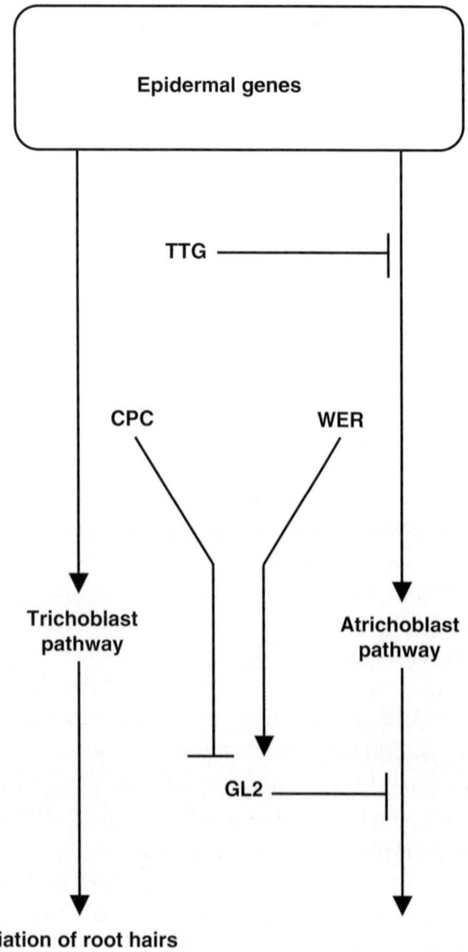

Figure 4 Summary of genes involved in the differentiation of root hair and nonhair cells. *TTG* operates early in the pathway to promote trichoblast differentiation. The expression of *GL2*, which inhibits the formation of root hair cells in the atrichoblast lineage, is controlled by the competitive influences of *WER* (positive) and *CPC* (negative).

independently of WER and CPC. In *ttg* mutants nearly all epidermal cells develop as trichoblasts and the effect is much stronger than that seen in *gl2* mutants. The complete conversion of atrichoblasts to trichoblasts in *ttg* mutants suggests that *TTG* acts at an earlier stage that *GL2* and is involved in the determination of cell fates rather than the mechanics of differentiation (**Figure 4**).

Applications of Root Development

Most of the known genes affecting root development have been identified in *Arabidopsis*, but genes and mutant phenotypes characterized in other species suggest that similar developmental mechanisms are involved in many plants. For example, the tobacco (*Nicotiana tabacum*) *hyp2* and tomato (*Lycopersicon esculentum*) *gib1* mutants each have an additional layer of cortical cells in the root, and a component of the auxin efflux carrier required for patterning the embryo and specifying the embryonic root has been isolated from *Brassica juncea*. In other species, novel genes with roles in root development have been cloned, e.g., the rice (*Oryza sativa*) *OSTF1* gene, encoding a root-specific transcription factor, and the rice *Roc-1* gene, encoding a transcription factor that may have an important role in the specification of the protoderm layer. Root development has an important impact on many aspects of plant physiology and performance, including nutrient uptake and utilization, interactions with pests and pathogens, and tolerance of extreme weather and other abiotic stresses. An understanding of the genetic basis of root development will provide the basis for rational strategies for plant improvement, based on the introduction or modification of genes controlling specific aspects of root growth and morphogenesis.

List of Technical Nomenclature

Ablation	Targeted destruction of a particular cell, e.g., with a laser beam.
Atrichoblast	Epidermal nonhair cell in the root.
Auxin	A plant hormone that controls growth and development.
Axial pattern	The organization of cells along the shoot–root axis of the embryo or seedling.
Circumnutation	A circular or elliptical movement around the growth direction.
Differentiation	Cell specialization during development.
File meristem	The part of the root apical meristem containing organized cells whose fate has been specified, and which are waiting to enter the elongation zone.
Geotropism	Growth in response to gravity. Positive geotropism is downward growth.
Gravitropism	Growth in response to gravity. Positive gravitropism is downward growth.
Hypophysis	The terminal cell of the suspensor, which gives rise to part of the root.
Initial cells	Stem cells that give rise to particular differentiated cell types, e.g., CE initials give rise to cortex and endodermal cells.
Lateral organs	Organs periodically derived from the shoot meristem, such as leaves.
Lineage	A cell's line of descent.

Morphogenesis	The formation of structures and shapes, e.g., cylindrical roots vs. flat leaves.
Pattern formation	The organization of cells in development, e.g., the formation of radial cell layers.
Periclinal division	A cell division in which one of the daughter cells enters a different cell layer.
Promeristem	The part of the root apical meristem containing the quiescent center and the initial cells (stem cells).
Radial pattern	The organization of cells in concentric layers in the root or shoot.
Root apical meristem	The population of proliferating cells that gives rise to the root, comprising the promeristem and the file meristem.
Statolith	A particle of inorganic matter whose movement in response to gravity can be detected.
Stem cells	Self-renewing cells. Cells that divide to produce a replacement stem cell and a daughter cell that goes on to differentiate.
Suspensor	The structure connecting the plant embryo to the ovule.
Thigmotropism	Growth in response to contact with an object. Negative thigmotropism is the avoidance of an obstruction.
Trichoblast	Epidermal hair cell in the root.

See also: **Growth and Development**: Molecular Biology of Development. **Root Development**: Lateral Root Initiation; Root Growth and Development; The Rhizosphere and its Microorganisms.

Further Reading

Costa S and Dolan L (2000) Development of the root pole and cell patterning in *Arabidopsis* roots. *Current Opinion in Genetics and Development* 10: 405–409.

Dolan L, Janmaat K, Willemsen V, *et al.* (1993) Cellular organization of the *Arabidopsis* root. *Development* 119: 71–84.

Howell SA (ed.) (1998) Root development. In: *Molecular Genetics of Plant Development*, pp. 289–311. Cambridge: Cambridge University Press.

Migliaccio F and Piconese S (2001) Spiralization and tropisms in *Arabidopsis* roots. *Trends in Plant Science* 6: 561–565.

Paquette AJ and Benfey PN (2001) Axis formation and polarity in plants. *Current Opinion in Genetics and Development* 11: 405–409.

Scheres B and Berleth T (1998) Root development: new meanings for root canals. *Current Opinion in Plant Biology* 1: 32–36.

Lateral Root Initiation

J G Dubrovsky, Universidad Nacional Autónoma de Mexico, Cuernavaca Morelos, Mexico
T L Rost, University of California, Davis, CA, USA

Introduction

During embryogenesis, the first root in the embryo axis is called the radicle. When it emerges during germination it becomes the primary root. Lateral roots (sometimes called secondary roots) will then form from the primary root. Roots that develop on other parts of a plant, such as the stem, or leaves, are called adventitious roots. Roots are heterotrophic organs, meaning that their nutrition depends significantly on photosynthates produced by leaves. Photosynthesis in turn depends significantly on the water and mineral compounds taken up by the roots. Because plants do not move, root growth and new root formation are the only way for a plant to explore the underground area around the plant. In this process, plants constantly produce new roots to fulfill the important functions of water and mineral uptake and transport. Many lateral roots do not have a long lifespan so that root turnover is a common phenomenon.

Plants can produce a prodigious number of lateral roots. For example, a 4-month wheat plant (*Triticum aestivum*) can form 14 million live root tips. The shape of a root system and its success in absorption and uptake of water and minerals is dependent on an efficient and effective mechanism to make new lateral roots. Thus, root initiation is a most essential component of root system development.

This article outlines the most important aspects of lateral root primordium (LRP) initiation, namely when and where it starts, what tissues participate, and what role cell proliferation plays during LRP initiation and early development.

Ontogenetic, Structural, and Temporal Aspects

When Does Initiation Start?

LRP initiation usually starts in the primary root after seed germination. However, in some plants LRP initiation actually starts during the development of the radicle in the embryo axis; cucumber *(Cucumis sativum)* is an example. In the radicle of mature cucumber embryos, LRPs are well developed and each consists of a few hundred cells. These primordia are initiated during embryo development opposite prospective protoxylem strands (**Figure 1**). In these species, lateral root emergence occurs within two to three days of germination.

In peas (*Pisum sativum*), preformed lateral roots also exist, but they cannot be observed unless the seeds are treated with a cell division inhibitor such as colchicine. Once treated, the emerging radicle can be seen to have up to 10 preformed lateral root primordia.

Where Does Initiation Start?

In most plants the first visible signs of lateral root initiation occur in the differentiation zone of the primary root; this is basal to the zone of cell elongation. This usually takes place within the first few centimeters from the root tip, but varies between species or cultivars. In corn (*Zea mays*, maize) roots, in some cultivars, the first initiation events can take place within the first centimeter proximal to the zone of cell elongation, whereas in other cultivars, it can be 6–8 cm away from the elongation zone. The location along the root also depends on the rate of root growth; initiation sites are closer to the root tip in slower growing roots. Generally, under the same conditions in the same species and cultivar, the distance from the root tip to the location of LRP initiation is relatively stable. In plants of some groups, the first initiation events actually happen in the root apical meristem where most cells are cycling. The most striking examples are ferns, e.g., *Marsilea* and *Ceratopteris*, where LRP initiation takes place within the first 100 μm from the apical cell. In roots of angiosperm plants, LRP initiation within the meristem takes place in some water plants, such as water hyacinth (*Eichhornia crassipes*), water lettuce (*Pistia stratiotes*), or in some Cucurbitaceae (pumpkin, *Cucurbita pepo*, and giant pumpkin, *Cucurbita maxima*), and banana (*Musa acuminata*). Some examples of early LRP initiation along the root are shown in **Table 1**.

Circumferentially, the lateral root (LR) initiation site is related to the structure of the vascular cylinder. Commonly, in ferns (Pterophyta) and dicotyledonous plants it takes place opposite the protoxylem poles. In monocotyledonous plants, characterized with polyarch roots, LRP initiation usually takes place opposite the protophloem. Generally LRPs are initiated in an acropetal direction from the more mature part of the root toward the root apex; thus the youngest primordium is formed closest to the tip.

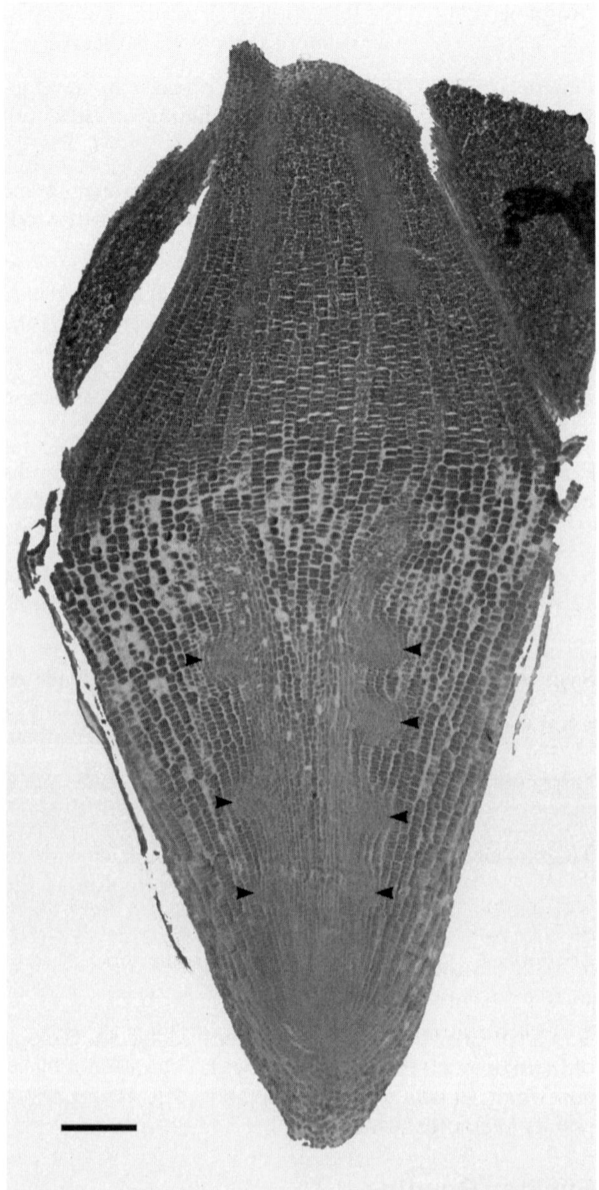

Figure 1 (see color plate 31) Mature embryo of cucumber plant (*Cucumus sativum* cv Muromsky) 12 h after the start of seed imbibition. At this stage no cell division occurred. The 7-μm thick section was made in the plane of the cotyledons. Arrowheads indicate embryonic LRPs in the radicle that are located along the prospective protoxylem strands. In the upper part of the embryo, shoot apical meristem, leaf primordium and parts of cotyledons (to the right and to the left from the embryo) can be seen. Scale = 200 μm. Photograph courtesy of Dr J.G. Dubrovsky.

What Tissues Initiate Lateral Root Primordia?

The first land plants had dichotomous branching. In this type of branching, all tissues of the parent axis divided at the tip to form two equal branches. In living primitive plants, this type of branching can be found in Lycophyta and some other lower vascular plants. Monopodial root branching of endogenous origin developed in more advanced plants. Some transitional stages from dichotomous to monopodial branching can be found in such contemporary plants as *Lycopodium clavatum* and *Selaginella lyallii*. In monopodial branching, the parent root is the main organ axis from which lateral organs emerge. During lateral root development the new primordia are

Table 1 Some characteristics of LR initiation and the spatial distribution of LRP on the parent root in five species

Species		*Relation of the initiation site to the proximal meristem border*	*Spatial arrangements of lateral roots*[a]
Ceratopteris thalictroides	70	*Acropetally*[b]	*Regular*[d]
Eichornia crassipes	300	*Acropetally*	*Tendency to be regular*
Victoria trickeri	680	*Slightly basipetally*[c]	*Tendency to be regular*
Cucurbita maxima	1000	*Acropetally*	*Regular and irregular*
Arachis hypogaea	1050	*Slightly basipetally*	*Tendency to be regular*

Based on data from Mallory TE, Chiang S-H, Cutter EG, and Gifford EM (1970) Sequence and pattern of root formation in five selected species. *American Journal of Botany* 57: 800–809.

[a]The relative position of the proximal meristem border level was considered as that at which the protophloem tissue has differentiated or where the cells of the parent root tissues are still dividing.

[b]Acropetally to the proximal meristem border means LRP initiation within the root apical meristem.

[c]Basipetally to the proximal meristem border means LRP initiation outside the root apical meristem.

[d]Regular refers to a relatively repeated pattern of LRP formation.

located basal to the root tip, so that the growth of the main root axis and formation of an LR are developmentally separate events.

In fern roots LRP are initiated in the endodermis. This tissue represents the internal layer of the cortical tissue complex. In the division Sphenophyta (genus *Equisetum*), in gymnosperms and angiosperms, roots are initiated in the pericycle, the most external tissue of the vascular cylinder or stele. The pericycle alone is responsible for LRP initiation, but tissues like the endodermis can participate in primordium development. The pericycle is commonly a uniseriate tissue, but in roots of plants in some families (Fabaceae and Cucurbitaceae) it can be biseriate or multiseriate throughout the perimeter of vascular tissues, or can be bi- or multiseriate only opposite the protoxylem and uniseriate in between (**Figure 2**). During LRP initiation, in many species, the endodermis and even the cortex sometimes contribute to forming a temporary root cap-like structure that has a protective function for developing LRP. In some plants, the tissues of the vascular cylinder in the vicinity of the initiated primordium are also activated and participate in the formation of vascular connections between the parent and lateral root.

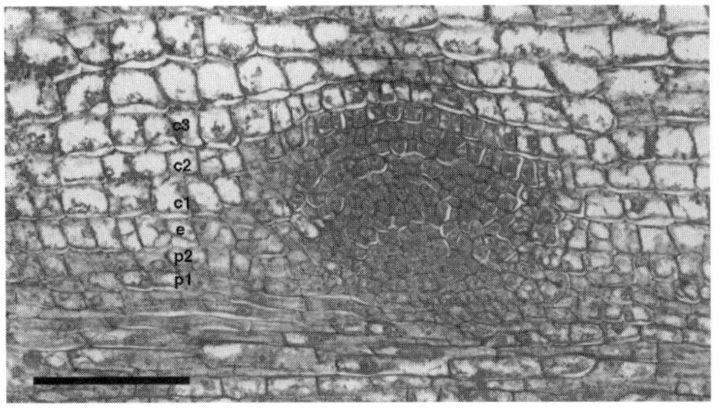

Figure 2 (see color plate 32) The most basal (see **Figure 1**) embryonic LRP in cucumber plants (*Cucumus sativum* cv Muromsky) 18 h after the start of seed imbibition. The 7 μm section was made in the same plane as in **Figure 1**. Note that two layers of pericycle (p1 and p2), endodermis (e), and three layers of the cortex (c1 to c3) participate in LRP formation. Procambium tissues, located below the pericycle layers, participate in the formation of future vascular connections between the parent root and the lateral root. Scale = 100 μm. From Dubrovsky JG (1986) Origin of tissues of embryonic lateral root in the cucumber, tissue interactions, and positional control in development. *Ontogenez* 17: 176–189. (English translation from Russian appeared in *Soviet Journal of Developmental Biology*, Consultant Bureau, New York 17: 119–128); reproduced with the permission of *Ontogenez, Russian Journal of Developmental Biology*.

An example of participation of the endodermis and the cortex in LRP formation is the cucumber (**Figure 2**). Up to three or four cortical layers located near the developing primordium participate in its formation, whereas cells of the same cell files located outside the primordium elongate and do not contribute to lateral root formation.

Cell Proliferation during Lateral Root Initiation and Development

Pericycle cells in the protoxylem pole express genes needed for continued cell cycling, whereas those outside the phloem do not. For example, a gene for cyclin-dependent kinase, cdc2, a key molecule in cell cycle regulation, is expressed in protoxylem associated pericycle cells within the differentiation zone, where cell proliferation generally does not occur. This demonstrates that these pericycle cells are competent for cell division. Cyclin dependent kinases are activated by cyclins D (during G1 to S transition) and B (during G2 to M transition) and their expression was found in protoxylem-associated single pericycle cells. Indeed, in *Arabidopsis thaliana* many mitotic figures can be found in pericycle cells located opposite the protoxylem poles and only some divisions result in LRP initiation. Cells that initiate primordia are sometimes called founder cells. The number of founder cells varies in different taxa. In ferns, one endodermal founder cell is needed for each LRP. In

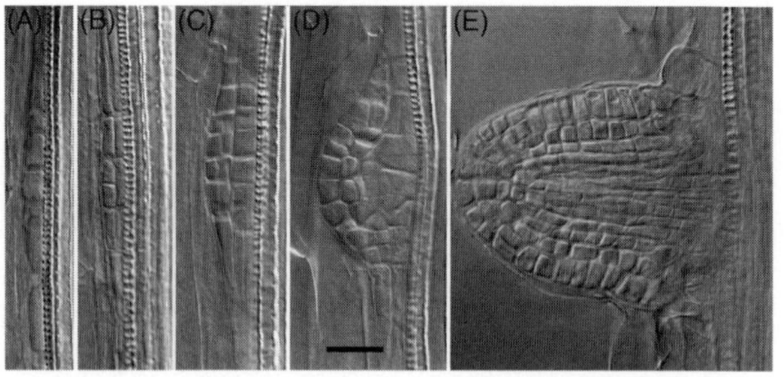

Figure 3 Initiation of LRP and stages of its development in *Arabidopsis thaliana.* (A) First anticlinal divisions in pericycle founder cells, the youngest stage of LRP initiation. (B) The first set of periclinal divisions in the LRP. (C) The second set of periclinal divisions in the LRP. (D) Youngest LRP stage that can be developed autonomously, with features of advanced histogenesis. (E) Recently emerged lateral root. Scale = 20 μm. Photograph courtesy of Dr J.G. Dubrovsky and Dr T.L. Rost.

angiosperms, the minimum number of founder cells giving rise to a primordium has not been extensively studied. It has been estimated to be between three and 162 in different species. Founder cells and their descendants commonly have a relatively short cell-division cycle resulting in rapid formation of LRPs. The duration of the cell cycle in different species is between 3 and 8 h in the first stages of LRP development. When a primordium has formed, cells divide more slowly finally attaining cell cycle times similar to cells in the primary root apical meristem.

The first divisions of founder cells in many species are anticlinal divisions (new cell walls are perpendicular to the parent root surface) (**Figure 3A**). Then cells enlarge radially and start periclinal divisions (new cell walls are parallel to the parent root surface) (**Figure 3B, C**). As a result, a lateral root primordium of a few hundred cells is formed (**Figure 3D**) that soon emerges on the surface of the parent root and becomes a lateral root (**Figure 3E**). Generally, primordium development consists of two phases, the early phase of primordium initiation and a later phase of lateral root meristem formation. The first phase is dependent on the parent root. The second phase is autonomous, and if the primordium is experimentally excised from the parent root, it can still form a lateral root *in vitro*. The processes of cell proliferation and histogenesis of the LRP are tightly coordinated. This has been proven with the analysis of differential gene expression during primordium formation. A collection of transgenic *Arabidopsis* lines has been produced which carried tissue-specific promoters fused with the β-glucoronidase (GUS) reporter gene (a gene encoding the enzyme whose activity can be detected histochemically); from the first moments of initiation, a highly ordered series of cell divisions generated tissues of developing LRs. Tissue identities acquired during development can be followed from the first periclinal divisions in the developing primordium.

Regulation and Control of Lateral Root Initiation

Hormonal Control of LRP Initiation

Thousands of studies have been conducted to determine the relative roles of plant growth regulators (hormones) on lateral root initiation and development. There is general agreement that there are at least three specific steps in lateral root development: initiation, organization of the LRP meristem, and LR emergence. Each of these steps is most likely regulated independently. Many of the published studies use seedling roots and often the experiments involve excising the root tip and placing it in different culture environments. This is done so that specific hormones and other treatments can be applied under controlled conditions.

It is generally agreed that the two most important hormones involved in lateral root development are auxin and cytokinin. Basically, auxin is provided by the cotyledons in the seedling and later by the shoot. Auxin travels acropetally (toward the root tip) and then back, basipetally (toward the base of the root). Cytokinin travels basipetally. Hypothetically, at some level from the root tip, cytokinin and auxin establish a balance that becomes optimal for LRP initiation.

The support for this generalization is based on experiments with seedlings of several different plant species; pea, lettuce (*Lactuca sativa*), radish (*Raphanus sativus*), and *Arabidopsis* are examples. Lateral roots are initiated in an acropetal series, meaning that they are initiated first near the base of the seedling root and that new laterals are induced in series toward the tip of the growing root. This process is regulated by auxin from the shoot and cotyledons. If the cotyledons are removed, new lateral roots will not be initiated; this can be reversed by adding auxin to replace the cotyledons. The only way to change the acropetal initiation series is to excise the root and treat it with exogenous auxin. In that kind of experiment, it has been seen that new lateral roots are initiated between existing laterals; this rarely happens during normal growth.

Lateral root initiation is a developmentally timed event. At some point, along the axis of the root, the pericycle cells become competent to divide and initiate an LRP. This requires that the proper hormone composition and environmental cues are exactly correct. If for some reason one of these things is not correct, initiation will not occur. If, for example, pea roots are grown at temperatures above 25°C, the root will continue to slowly elongate, but lateral roots will not be initiated. If roots are transferred to an ideal temperature lateral roots will be initiated, but only in the new growth. This suggests that lateral root initiation in peas is temperature sensitive, and that in intact roots if the LRP is not initiated at a specific developmental time, then it may not be initiated.

Another interesting observation is that in pea seedlings grown in the dark only about 10 lateral roots will form. These laterals are developed from the primordia initiated in the radicle during embryogenesis. If the seedlings are placed in the light new laterals will form as the root system develops, suggesting that the development of the shoot system is necessary for the development of new LR.

Cytokinin is known to control cell division, and it is necessary for LR initiation. If the root tip of a seedling primary root is removed, more LRs will be initiated than in control roots with the tip intact. If cytokinin is added in place of the root tip the number of LRs is decreased. This suggests that cytokinin acts to moderate the number of initiation sites. Cytokinin comes from the root tip, it moves basally and acts along with auxin to initiate only a certain number of lateral roots.

If the tip of the root is removed and placed in culture, in the presence of auxin at the base of the root in an agar block, the number of LR initiation sites is multiplied. In fact in some cases a fasciation occurs where the laterals are so close to each other that a continuous sheet of primordia forms. This suggests that without cytokinin no limits on initiation events occur and without auxin movement much less sites are stimulated.

Genes Involved in LRP Initiation

Currently, it is not known how hormones activate certain pericycle cells to induce LRP formation. As was mentioned above, auxin is a major hormonal factor affecting LRP initiation. If genes responsible for auxin signaling are mutated, the resulting plants will be deficient in LR initiation. For example, in *Arabidopsis thaliana* auxin-resistant mutants *axr1* and *aux1* have reduced numbers of LRs. Respective proteins are involved in protein degradation and polar auxin transport, respectively. Conversely, when mutants overproduce auxin, as in *superroot* (*sur*) or *aberrant lateral formation* (*alf1-1*), increased LRP initiation occurs. The respective genes (*AXR1*, *AUX1*, *ALF1-1*) thus affect LR initiation. The most rapid response to auxin at the gene level is the expression of genes from the *Aux/IAA* family. Some of these genes are involved in regulation of LRP development. *SHORT HYPOCOTYL 2/IAA3*, *IAA 28*, *SOLITARY ROOT/IAA 14* are required for LRP development. Respective mutants have fewer lateral roots, or none at all.

Independently of auxin-related genes affecting LRP initiation, a few other genes were discovered that are auxin-autonomous and are important for the process of LRP initiation itself. The first such gene characterized was *ALF4-1* (*aberrant lateral root formation 4-1*). Plants deficient in this gene (*alf4-1* mutants) are abnormal in LRP initiation and cannot form lateral roots at all. When these plants are treated with auxin, the LRP initiation does not occur except for a few primordia forming near the root tip, but roots remain sensitive to the hormone since primary root growth is inhibited. *ALF4-1* has been cloned and encodes a protein with unknown function. A gene called *LRP1* is expressed from early stages of LRP development and involved in primordium initiation. *LATERAL ROOT INITIATION 1* (*LIN1*) gene is involved in coordination of LRP initiation with external nutrients or internal metabolic status. *lin1* mutant is able to initiate lateral roots under conditions of high sucrose to nitrogen ratio in the growth medium whereas in wild-type plants under these conditions LRP initiation is completely inhibited.

Genes responsible not only specifically for lateral root development, but for normal root development are also needed for LRP initiation. For example, in *rml (rootmeristemless)* mutants of *Arabidopsis thaliana*, a normal radicle was formed during embryogenesis, but postembryonic development, including cell proliferation, was impaired. LRP initiation occurs, but the primordial meristem is arrested. RML genes were shown to be important for activation of cell proliferation in the root meristem. This gene encodes γ-glutamylcysteine synthetase, an enzyme involved in the biosynthesis of glutathione (γ-glutamylcysteinyl glycine, an antioxidant and reducing compound important for redox homeostasis in cells). *HOBBIT* is another gene that affects general root development, including LRP initiation. The primary root cannot grow normally because founder cells are not properly specified. Few other genes affecting pattern formation and thus root growth are known to be required for LRP initiation, but the overall process of genetic control and gene interaction is yet to be understood.

The described genes responsible for lateral root initiation were all discovered in a model plant, *Arabidopsis thaliana*. Only a few genes with similar functions are known in other plants. Among them are *DIAGEOTROPICA* (*DGT*) in tomato (*Lycopersicon esculentum*), *LATERAL ROOTLESS 1 (LRT1)* in corn, and *HYPERNODULATION ABERRANT ROOT FORMATION 1 (HAR1)* in *Lotus japonicus*.

Problems of Spatial Regulation of Root Initiation

Assuming all initiated primordia develop into roots, the spatial distribution of lateral roots on a parent root reflects the pattern of LRP initiation. In many plants the pattern is difficult to follow. However, in other plants some tendency to regularity has been described. Frequently laterals are initiated opposite the protoxylem poles. When LRPs are initiated nonrandomly, some spatial pattern can be observed as in the examples in **Table 1**. Regular spacing or the tendency to such spacing indicates that there is some coordination between LRP initiation along individual protoxylem strands and between strands within a parent root. Generally, new primordia are initiated in the acropetal pattern. Hypothetically, the spatial coordination of initiation could be regulated by signals from an initiated primordium to the site of future LRP initiation. How this signaling mechanism works is yet to be discovered. In some polyarch roots, as in banana plants, each successive LRP is initiated circumferentially at an angle 30° or greater. This indicates the existence of some kind of negative control (inhibition of initiation in close proximity) of LRP initiation.

Effects of Chemicals on Lateral Root Initiation

After seed germination, LR initiation starts relatively rapidly. It has been found that LR initiation and subsequent formation is very sensitive to stress factors and is an excellent bioassay. More than 1000 different compounds have been tested on root growth of corn and cucumber plants in a search for active cytostatics (compounds that inhibit cell division). Inhibitors of protein and nucleic acid synthesis, nonspecific nonelectrolytes (alcohols, ethers, ketones, and phenolics), heavy metal salts, platinum and other noble metal complexes, some herbicides, and other compounds were tested. A simple rule emerged: if the compound is not a cytostatic, LRP initiation takes place and if a compound stops cell division (cytostatic activity), LRP initiation is inhibited. Even such poisonous compounds as thallium and mercury salts did not completely inhibit LRP initiation. These experiments showed that lateral root development is resistant to many factors and implies root plasticity and adaptability.

Alternate Strategies for LRP Initiation

Proteoid and Proteoid-like Roots

Proteoid roots, found in members of the Proteaceae, have peculiar root development. LRPs are initiated very close one to another and produce clusters of laterals that are determinate and relatively short lived. In some species, 280 to 1000 LRPs per centimeter of parent root can be initiated. A few other plant families also form proteoid-like roots. All species with proteoid and proteoid-like roots can grow in poor soils. They are usually not mycorrhizal (fungal symbiosis), but are able to mobilize soil phosphorus, by exudation of nutrient-solubilizing compounds and by increased root surface area. The initiation of LRPs in clusters can be induced by absence of phosphorus, and in some species, of iron, in nutrient media or in soil. The most important applied aspect of proteoid roots is in members of the Fabaceae, for example in *Lupinus albus*. Citrate, malate, and succinate are major organic anions exuded by proteoid-like roots that ameliorate soils. Auxin is considered as the most important signal inducing clustered primordia, but more work is needed to understand how LRP initiation is controlled in proteoid and proteoid-like roots.

Hairy Roots

A soil bacterium *Agrobacterium rhizogenes* causes "hairy root" disease in some dicotyledonous plants. A bacterial Ri (root inducing) plasmid integrates its transfer-DNA (T-DNA) into the plant genome (DNA). As a result, rapidly growing and intensely branched (hairy) adventitious roots are developed. When roots induced by the bacteria in aseptic conditions are maintained *in vitro* on a nutrient medium, the transformed root tissue can grow for years. Such cultures have been established for several hundred species. In many species, especially from Solanaceae, Asteraceae, and other families, roots synthesize secondary metabolites important for pharmacology (for example, alkaloids, and others). Intensive growth and rapid LR initiation make hairy root culture a major potential commercial source of important rootborne compounds. Genes of the Ri plasmid inserted in the root-cell DNA caused increased sensitivity to auxin and subsequently greater proliferation of lateral roots. Paradoxically, root initiation itself in hairy roots is understudied and detailed analysis of this important process is yet to be done.

Lateral Root Initiation in Desert Plants

To cope with drastic environments, desert plants have elaborated many adaptive features, some of which are expressed in root development. In some Cactaceae from the Sonoran Desert, primary roots grow only for 2 or 3 days after germination (a determinate root growth pattern). Apparently, when cells of the primary root meristem stop proliferating, a signal is sent to start LRP initiation. This mechanism represents a case of physiological root tip decapitation. Many lateral roots in turn have determinate growth and induce new LRPs. As a result, a compact root system is formed that is sufficient for water and mineral uptake. Another interesting feature is the rapid formation of LRs after a period of drought. In adult plants of *Opuntia ficus-indica* and many other species, when water becomes available after a drought, LRs are formed rapidly and can grow several millimeters in less than 24 h (these usually short-lived roots are called "rain roots"). Rain roots are an important strategy to rapidly increase the root surface area, and water uptake efficiency.

Future Perspectives

Many problems related to our understanding of LR development are still open questions. We are still far from designing crop plant root systems that can grow faster or function more efficiently. To be able to do so, we have to better understand the control points for LR initiation and development. Gene hunting and uncovering the functions of already known genes in combination with structural–physiological analysis is the best approach to analyze how LR initiation is controlled and regulated in plants.

Acknowledgments

JGD was supported by CONACyT, Mexico (31832-N) and PAPIIT UNAM (IN210202-3), Mexico.

See also: **Regulators of Growth**: Cytokinins. **Root Development**: Genetics of Primary Root Development; Root Growth and Development.

Further Reading

Altman A and Waisel Y (1997) *Biology of Root Formation and Development*. New York: Plenum Press.

Casimiro I, Beeckman T, Graham N, *et al.* (2003) Dissecting Arabidopsis lateral root development. *Trends in Plants Science* 8: 165–171.

Clowes FAL (1961) *Apical Meristems*. Philadelphia: FA Davis.

Dubrovsky JG (1986) Origin of tissues of embryonic lateral root in the cucumber, tissue interactions, and positional control in development. *Ontogenez* 17: 176–189. (English translation from Russian appeared in *Soviet Journal of Developmental Biology*, Consultant Bureau, New York 17: 119–128).

Dubrovsky JG, Rost TL, Colón-Carmona A, and Doerner PW (2001) Early primordium morphogenesis during lateral root initiation in *Arabidopsis thaliana*. *Planta* 214: 30–36.

Flores HE, Lynch JP, and Eissenstat D (eds) (1997) *Radical Biology: Advance and Perspectives on the Function of Plant Roots*. Rockville: American Society of Plant Physiologists.

Fukaki H, Tameda S, Masuda H, and Tasaka M (2002) Lateral root formation is blocked by a gain-of-function mutation in the *SOLITARY-ROOT/IAA14* gene of *Arabidopsis*. *Plant Journal* 29: 153–168.

Guttenberg H von (1968) *Der Primäre Bau der Angiospermenwurzel*. Berlin: Gebruder Borntraeger.

Ivanov VB (1994) Root growth responses to chemicals. *Soviet Science Review, series D, Physicochemical Biology* 13: 1–70.

Jackson MB (ed.) (1986) *New Root Formation in Plants and Cuttings*. Dordrecht: Martinus Nijhoff.

Malamy JE and Benfey PN (1997) Down and out in *Arabidopsis*: the formation of lateral roots. *Trends in Plant Science* 2: 390–396.

Malamy JE and Ryan KS (2001) Environmental regulation of lateral root initiation in *Arabidopsis*. *Plant Physiology* 127: 899–909.

Mallory TE, Chiang S-H, Cutter EG, and Gifford EM (1970) Sequence and pattern of root formation in five selected species. *American Journal of Botany* 57: 800–809.

McCully ME (1975) The development of lateral roots. In: Torrey JG and Clarkson DT (eds) *The Development and Function of Roots*, pp. 105–124. New York: Academic Press.

Rogg LE, Lasswell J, and Bartel B (2001) A gain-of-function mutation in *IAA28* suppresses lateral root development. *Plant Cell* 13: 465–480.

Sussex IM, Godoy JA, Kerk N, *et al.* (1995) Cellular and molecular events in a newly organizing lateral root meristem. *Philosophical Transactions of the Royal Society of London B* 350: 39–43.

Tieghem P van and Douliot H (1886) Origine des radicelles chez les légumineuses et les cucurbitacées. *Bulletin de la Société Botanique de France* 8: 494–501.

Van den Berg C, Weisbeek P, and Scheres B (1998) Cell fate and cell differentiation status in the *Arabidopsis* root. *Planta* 205: 483–491.

Watt M and Evans JR (1999) Proteoid roots: physiology and development. *Plant Physiology* 121: 317–323.

Mycorrhizae

E M Duffy and A C Cassells, National University of Ireland, Cork, Ireland

Mycorrhizae and Symbiosis

The plant root surface and the area influenced by root exudates (the rhizosphere) support a rich and diverse microflora and microfauna. In return for nutrients from the plant, approximately 40% of the dry matter produced by a plant may be released into the rhizosphere, where free-living organisms maintain soil fertility and, in natural ecosystems, provide a biological buffer against plant pathogens and soilborne pests. These organisms include plant growth promoting rhizosphere bacteria (PGPRs) that produce an array of antimicrobial compounds in the soil, and may also promote plant growth and stimulate plant defense mechanisms. There also exist organisms that feed within the tissue of the root: these include microorganisms that are pathogenic, that is, damage the plant, or are symbiotic, where mutual benefits are conferred. The former includes a wide range of fungal, bacterial, and nematode plant pathogens; the latter includes nitrogen fixing bacteria and mycorrhizal fungi. The term "mycorrhizae" is translated as "root fungus." The fungal partner in a mycorrhiza (plural mycorrhizae or mycorrhizas) generally derives its energy from the plant and in return increases the capability of the root system to collect minerals and water from within and outside the rhizosphere as the fungal mycelium can extend into the surrounding soil. Mycorrhizal fungi especially aid the plant by harvesting phosphate and other immobile minerals in deficient soils. Mycorrhizae may beneficially influence the rhizophere microflora. The fungus also protects the plant against biotic and abiotic stresses, possibly by stimulating the plant's stress responses. Under these circumstances, the relationship between the mycorrhizal fungus and its host plant is symbiotic (mutually beneficial). In mycorrhizae, there is a fine line between mutualism and parasitism by the fungal partner, as is seen in attempts to develop commercial mycorrhizal inoculants. The responses range from beneficial effects, to no effect, to pathogenicity. This poses the question of what is the distinction between mycorrhizal fungi and biotrophic fungi that live on the plant without causing disease symptoms, indeed the latter may benefit the plant by "immunizing" or inducing resistance mechanisms that act against pathogens and pests.

Origin and Evolution of Mycorrhizae

Plant fossil records indicate the presence of endosymbiotic fungi associated with roots going back to the early Devonian period, and it has been hypothesized that mycorrhization was involved in the colonization of the land by plants. This is supported by fossil evidence of the relationship since Paleozoic times. Zygomycete fungi form mycorrhizal associations with the roots/rhizoids of approximately 80% of land plants including Bryophytes (mosses) and Pteridophytes (ferns). In contrast to the relatively small number of endomycorrhizal (intracellular) fungal species associated with a very large and diverse number of plant species, is the large number of ectomycorrhizal (intercellular) fungal species (estimated at 5000) associated with relatively few plant species (**Table 1**). This has led to the hypothesis that ectomycorrhizae are a more recent development in plant evolution. Some plant families, e.g., the Brassicaceae, which produce toxic glucosinolides and isothiocyanates, and the Caryophyllaceae (e.g., carnations (*Dianthus caryophyllus*)), which have a modified root cap, do not form mycorrhizal associations, and this supports the hypothesis that these families may have evolved mycorrhizal independence, indicating a general trend toward reduced mycorrhizal dependence.

Types of Mycorrhizae

Seven types of mycorrhizae are distinguished based on the host plant, the fungus, whether fungal hyphae penetrate the host cells, the structure of the fungus in the host cells, and the form of mycelium associated with the root surface (**Table 1** and **Figures 1** and **2**). Ectomycorrhizae are common on approximately 2000 species of trees and shrubs of the Pinaceae (pines), Fagaceae (beech), Betulaceae (birch), and Myrtaceae (myrtles), which grow in subartic and temperate regions. The ectendomycorrhizal fungi display characteristics of both ectomycorrhizal and arbuscular mycorrhizal fungi, and are found on pines and larches. The arbuscular mycorrhizae are the most abundant endomycorrhizal fungi, occurring on approximately 25000 plant species, including most temperate and tropical crop species and tropical trees. The arbutoid types are found on woody plants of genera in the Arbutoideae and Pyrolaceae. The monotropoid mycorrhizae are found on parasitic plants in the Monotropaceae where they connect the non-photosynthetic host to neighboring trees, which acts as a source of carbohydrates. Three types of mycorrhizae are associated with the Ericales (heathers, rhododendrons, and vaccinia) and with genera in the Epacridaceae and Empatreaceae, which grow on nutrient poor, acid soils. Finally, there are the orchidaceous mycorrhizae, which are universal and obligate on orchids.

The Fungi in Mycorrhizae

The fungi that form ectomycorrhizal associations with plants are members of the Eumycotina ("true fungi"); most are basidiomycetes and include members of 25 families, e.g., Agaricales and Russulales. Some are members of the ascomycetes, e.g., of the Tuberales. The exception to this is the two species of *Endogone* of the Zygomycota, which are normally

Table 1 The main characteristics of mycorrhizae

	Ectomycorrhizae	*Endomycorrhizae*					
		Ectendomycorrhizae	*Arbuscular mycorrhizae*	*Arbutoid mycorrhizae*	*Monotropoid mycorrhizae*	*Ericoid mycorrhizae*	*Orchid mycorrhizae*
Fungal sheath present	+	+ or −	−	+	+	−	−
Hyphae enter cells	−	+	+	+	+	+	+
Hartig net formed	+	+	−	+	+	−	−
Hyphal coils in cells	−	+	+	+	−	+	+
Vesicles in tissues	−	−	+ or −	−	−	−	−
Achlorophyllous host plant	−	−	− or +	− or +	+	−	+
Fungal taxon	Basidiomycetes Ascomycetes Zygomycetes	Basidiomycetes Ascomycetes	Zygomycetes	Basidiomycetes	Basidiomycetes	Ascomycetes Basidiomycetes	Basidiomycetes
Host plant taxon	Gymnosperms Angiosperms	Gymnosperms Angiosperms	Bryophytes Pteridophytes Gymnosperms Angiosperms	Ericales	Monotropaceae	Ericales	Orchidaceae

The two most important economic types, the ectomycorrhizae and the arbuscular mycorrhizae are highlighted for contrast. The endomycorrhizae consist of six subgroups and are distinguished by the occurrence of penetration of cells of the host plant. Note that arbuscular mycorrhizae are formed on the roots of all plant taxons; the rare exception to mycorrhizal competence is shown in some higher plant families, which are considered to have become mycorrhizal independent. Mycorrhizal associations with non-Zygomycete fungi may have evolved later than with Zygomycetes.

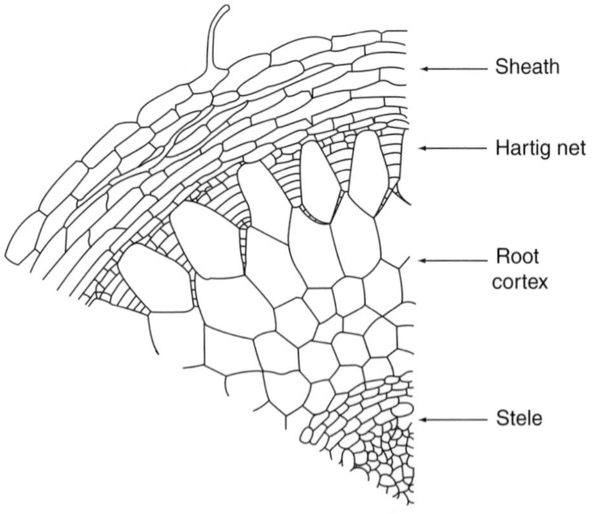

Figure 1 Illustration of a typical ectomycorrhizae. The sheath is composed of a dense mat of fungal mycelium of one or more isolates. The mycelium may expand out into the surrounding soil depending on the soil fertility, soil moisture content, and other factors. In the Hartig net, the epidermal cells of the root are physically separated by fungal mycelium. The mycelium grows intercellularly in the cortical cells, but does not penetrate the stele of the root (based on Isaac S (1992) *Fungal–Plant Interactions.* London: Chapman & Hall).

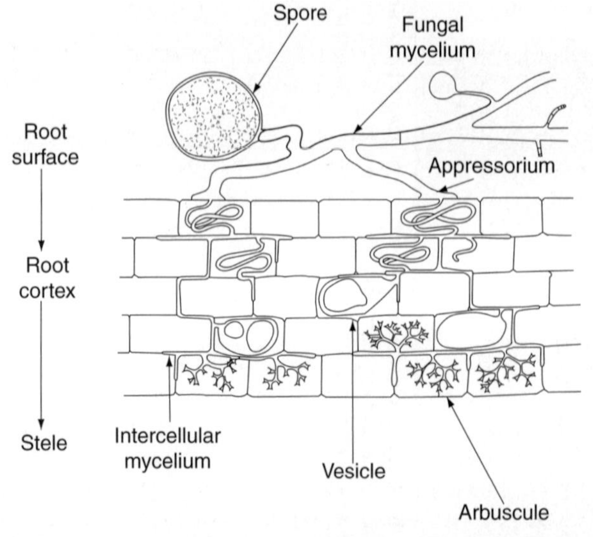

Figure 2 Illustration of a root infection by an (vesicular) arbuscular mycorrhizal fungus. The fungus penetrates the cell wall using appressoria, the mycelium grows intercellularly and forms vesicles and arbuscules by penetrating the cell walls. The penetrated cells are not killed on penetration; the plasmalemma invaginates around the fungal structure, and mineral, metabolite, and water exchange occurs across the intact membranes (based on Isaac S (1992) *Fungal–Plant Interactions.* London: Chapman & Hall).

associated with arbuscular mycorrhizae, but can also form ectomycorrhizal associations. Many of the ectomycorrhizal fungi can be cultivated on artificial media, but in some cases with difficulty.

The fungi that form arbuscular mycorrhizae have been grouped in the Endogonaceae (*Endogone* spp.) of the Endogonales and in three families of the Glomales, of the Zygomycota. The families (and genera) are Glomaceae (*Glomus* spp.), Acaulosporaceae (*Acaulospora* and *Entrophospora* spp.), and Gigasporaceae (*Gigaspora* and *Sclerocystis* spp.). Recent results from genomic analysis are likely to result in a revision of these taxonomic groupings. Arbuscular mycorrhizae are noncultivable on artificial media (see "Production of Mycorrhizal Inoculum" below).

The Ericaceous mycorrhizae are formed with ascomycete and basidiomycete fungi while the orchidaceous mycorrhizae are basidiomycetes, commonly from the genus *Rhizoctonia*, species of which are also pathogenic on plant roots.

Mycorrhizal Symbiosis

Studies using radioactive carbon dioxide have shown that ^{14}C-labeled sugars synthesized in the leaves of the host plant pass into the roots and accumulate in the fungal mycelium, where they are converted to the fungal storage carbohydrate trehalose. It is likely that this carbohydrate flow is unidirectional as most higher plants cannot metabolize trehalose. The exception is in monotropoid and orchidaceous mycorrhizae where the partner plants can utilize trehalose. Similar studies using ^{15}N have shown that mycorrhizal fungi can pass nitrogen to the host plant. This is important where the turnover of organic matter is low, as in the forests at high altitude or in peat. The higher fungi associated with ectomycorrhizae, e.g., *Rhizoctonia*, are capable of breaking down lignified plant tissues and proteins and utilizing amino acids and other organic nitrogen sources. Ectomycorrhizae are also known to produce plant growth regulators, which may explain their effects on root morphology and longevity.

It is often considered that the greatest benefit to plants from mycorrhization is the increase in phosphate supply. The availability of phosphorus is low in many soils and may be present in insoluble forms. Mycorrhizae, by extending the mineral acquisition range, play a critical role in solubilizing and acquiring phosphorus, which is transferred to the plant in the form of phosphates. Protection against drought stress is another important benefit of mycorrhization, given the increasing desertification in many undeveloped areas of the world and the predicted consequences of global warming for desertification in important crop growing regions of Europe. The mycelial network established in mycorrhizae greatly extends the soil penetration away from the immediate rhizosphere, and, together with the improved mineral nutrition of mycorrhized plants, is considered to be responsible for the improved drought resistance.

Mycorrhizae in Ecosystems

While colonizing relatively few plant taxa (see **Table 1**), the importance of the ectomycorrhizae and the ectendomycorrhizae lies in that their host plants are the dominant species in some of the world's most important ecosystems, including boreal, temperate, and subtropical forests. They appear to be restricted to dicotyledonous species. The fungal species involved are very diverse, but contain many of the advanced basidiomycete fungi that are capable of lignin decomposition, and, in many cases, the mycorrhizae contain more than one fungal species. A characteristic of ectomycorrhizae and ectendomycorrhizae is the ability of the fungal partners to permeate the F and H horizons of the forest floor, and to contribute to rapid breakdown of organic matter, including lignified material, and to nutrient cycling. It has been estimated that the mycorrhizal fungi account for 15% of the net primary production in forest ecosystems. Their presence is reflected in the mushroom flora associated with specific forest ecosystems. In the tropics, ectomycorrhizae are characteristic of plants growing at high altitude on poor soils.

(Vesicular) arbuscular fungi (VAM, syn. AMF) are the largest group of the endomycorrhizae, that is, of those mycorrhizae that enter the cells of the host plant. They are very widespread in natural vegetation and in areas of traditional farming where they increase the immobile mineral acquisition potential and water absorbing surface for the host plant. They may also contribute to "biological buffering" associated with good soil fertility, which is important in preventing the build up of pathogen inoculum. The most marked effects of AM fungal presence are generally seen in stressed environments, for example, the highest percentage of mycorrhizal root colonization occurs in older, fixed sand dunes. Mycorrhizal plants occur at all successional stages of dune development, although nonmycorrhizal plants have been observed in fertile patches at the high tide line where nutrient accumulation occurs. It has been hypothesized that mycorrhizae are important in this type of ecosystem due to low phosphorus availability, hence the importance of the presence of mycorrhizae for plant establishment and growth. In addition to this, mycorrhizal hyphae help to bind sand grains and improve ecosystem stability. Studies utilizing ^{32}P have confirmed this showing an increase in the concentration of phosphate in plant tissues relating to increased biomass production. The general view regarding phosphate acquisition is that AMF infected roots have advantageous geometry and better distribution than nonmycorrhizal roots to acquire phosphorus from transitory localized and diluted sources of the nutrient.

Similarly, desert and arid ecosystems are very fragile and subject to constant disturbance, and in this situation, mycorrhizae play an essential role in sustaining the vegetation cover. AMF can also be used in the reclamation of damaged soils (e.g., heavy metal contamination), and are known to increase tolerance to other extreme conditions, such as chilling and drought. Positive tolerance can be attributed to changes in the plant's rate of photosynthesis, or levels of carbohydrates and proteins. It has been argued that the improved tolerance to biotic and abiotic stresses is a result of the more balanced nutritional status, as induced by mycorrhizae, but recently studies are beginning to focus more on mycorrhizal induced physiological changes, and how these contribute to improved tolerance/resistance under different stress conditions.

Mycorrhizal Fungal Colonization and Biotic Stress Resistance

In contrast to the "gene for gene" specificity shown in many plant–pathogen interactions, AMF symbiosis is essentially nonspecific, a single species of fungus being capable of colonizing many plant species, which is in contrast to the narrow host ranges of biotrophic fungal pathogens. This type of association has evolved as an example of extreme compatibility where mechanisms have evolved to coordinate cellular development of both organisms to achieve a functional mutualistic state.

The first step in mycorrhizal colonization is probably chemical recognition by the mycorrhizal fungus of the presence of a potential host plant. Whether there is any specificity in the chemical recognition is arguable, at least for AMF, as most have a wide host range. The most striking modifications induced during resistance to pathogen attack are the formation by the host cell of wall appositions and papillae at the site of penetration attempts. This is generally characterized by primary and secondary plant wall components, such as callose, phenolics, proteins, or silicon, which all contribute to reinforcement of the cell wall and thus resistance. In contrast, in the case of the AMF colonization, roots of most host plants show very little reaction at the cytological level to the development of appressoria or infection hyphae, and no significant modifications occur in the walls of epidermal cells in contact with these primary infection structures. The lack of, or weak, host response in outer root cortex cells to AMF contrasts with the extensive modifications that are induced by arbuscule development in the parenchyma cells of the inner cortex. In order for the fungus to become integrated into the host's root system, it is now accepted that the symbiosis must be the result of molecular communication between the plant and fungus through the exchange of recognition and response signals. Mycorrhizal fungi, particularly the Zygomycetes, have a long association with plant roots, and may have developed unique mechanisms of interaction with the plant defense systems, including strategies to attenuate plant defenses.

The development of antibodies, molecular probes, and stains for various cellular components has assisted in dissecting the molecular interface of the two interacting organisms. Xyloglucans, β-(1,4)-glucans, hydroxyproline rich glycoproteins, and nonesterified polygalacturonans have all been found to be localized at the interface. The presence of a mixture of primary cell wall components at this interface indicates that the plant cells that contain arbuscules maintain their ability to synthesize and secrete cell wall material, and the failure of these components to assemble further has been suggested to be the result of weak lytic activity from the fungus. The molecular changes that occur in root cells to enable formation and functioning of the mycorrhizae are currently being researched, and have focused on the host defense mechanism. Most literature on this topic concurs that defense genes show minor transient increases following initial colonization, but these are insignificant in comparison with the upregulation following an attack by a pathogenic fungus. Plants possess a broad spectrum of basic defense mechanisms, pre-established or induced, which render them resistant to most potential colonizers (nonhost resistance). Inducible defense involves a complex series of plant processes that are triggered as a general response to attack by a pathogen; namely, specific or nonspecific pathogen recognition, signal transduction, and the activation of transcription factors leading to expression of defense proteins and antimicrobial compounds. Major inducible plant defense responses include cell wall modifications, such as those described above, enhancement of secondary metabolism, and the accumulation of antimicrobial proteins (including "pathogenesis related proteins"). Their rapid activation during incompatible interactions leads to cytological and molecular modifications that are associated with resistance. Of particular interest is the hypothesis that defense mechanisms are sensitized in mycorrhizal plants and, as a consequence, defenses are rapidly upregulated on pathogen attack. This may explain the resistance shown by artificially mycorrhized plants to pathogens. Both the speed and scale of the defense response, in the appropriate tissue location, are believed to be critical factors in overcoming pathogen attack.

Determination of Mycorrhizal Dependence

Mycorrhizal dependence is a measure of the benefits of mycorrhization. It is important to note that the benefits of mycorrhization are strongly influenced by the fertility of the soil in which the mycorrhizae are grown. The first step in this analysis is confirmation that the mycorrhizal inoculation has been successful. This is based on decolorizing the roots and examining them for the presence of the AMF diagnostic structures, namely, vesicles, arbuscules, and spores, as appropriate. It is also essential to confirm the persistence of the mycorrhizal fungus by repeating this screening at the end of the trial. DNA probes are now available for some AMF and, increasingly, these will facilitate confirmation of mycorrhization colonization and its persistence. Confirming the presence of mycorrhizal fungi in herbaceous crops is relatively straightforward; however, doing this for woody species with deep roots has posed sampling problems. Tests that confirm whether the control plants have been mycorrhized by resident mycorrhizal fungi in the field must also be carried out. In commercial inoculant trials, determination of the beneficial effects is usually based on measuring the increase in dry matter of mycorrhizal plants versus non-mycorrhizal or naturally mycorrhized plants. This is expressed quantitatively, as relative field mycorrhizal dependence (RFMD) in the formula:

RFMD (%) = dry weight of mycorrhizal plant − dry weight of nonmycorrhizal plant ÷ dry weight of the mycorrhizal plant × 100

In the case of crop plants, it is more meaningful to measure the effects of crop yield, and where the crop is graded, e.g., strawberry (*Fragaria* spp.), fruit size distribution and fruit grades, as appropriate, are taken into consideration.

Production of Mycorrhizal Inoculum

The production of ectomycorrhizal inoculum, in general, follows the following stages: (1) isolation of the microorganism in pure culture; (2) scaling up of the cultures in fermentation vessels; (3) a further scaling up by growth on sterilized seedlings or cooked cereal grains; and (4) formulation of the product. In addition to ensuring that no contamination occurs during the multiplication, it is important that the product have an adequate shelf-life to facilitate marketing and distribution. *Pisolithus tinctorius* inoculum is produced on a commercial scale by culture in 1.5 l jars containing sterilized vermiculite and peat imbibed with liquid culture medium. The substrate is dried to 10–20% moisture. This inoculum is added to sterilized nursery soil for inoculation of seedlings.

The production of inoculum of arbuscular mycorrhizal fungi is complicated as the latter cannot yet be cultured and so production traditionally has been on the roots of a propagation host plant in sterilized soil or peat mixes; alternatively production has been in soilless (hydroponic) systems, in sand or expanded clay, or in nutrient films. The general problem with these systems is that they are not aseptic, and there is the risk that the inoculum may contain inoculum of pathogens. To avoid this it is important to use a propagation host that is not susceptible to the major diseases of the target crop. The process begins with the selection of an appropriate isolate(s), ideally one(s) adapted to the soil in which the crop will be grown (see above). This inoculum is used to inoculate surface-sterilized seed of the propagation host, which is planted in sterile substrate and maintained in isolated conditions, with routine examination for pathogens. Pesticides and fungicides that are compatible with the AMF isolate can be used. As with ectomycorrhizal inoculum, formulation of the inoculum is important to optimize shelf-life. In the case of *Glomus* spp., which produce large numbers of vesicles in the root tissue, infected roots can be chopped up and dried to reduce water content, and used as inoculum. A method that is reported to be successful is the use of soilless culture in expanded clay, where the AMF may penetrate the clay particles, which are then air-dried and can be ground up to give particles of the desired size range. The particles can be introduced into the field beside the seedlings with standard fertilizer application equipment. AMF spores can be surface sterilized and used as inoculum for aseptic plant tissue cultures. Productivity of spore production is reported to be high in these systems, and advances in plant tissue bioreactors may allow economic scaling up in the future. For legal reasons, associated with product claims and labeling, commercial mycorrhizal inoculants are commonly referred to as "biofertilizers."

Ectomycorrhizae

The surface of roots infected with ectomycorrhizal fungi is sheathed or mantled with mycelium, which penetrates the root, sometimes forming a "Hartig net" amongst the enlarged epidermal cells (**Table 1** and **Figure 1**). The mycelium penetrates the root tissue, but is excluded from the stele and does not penetrate the root cells. The main root continues to grow normally, however, the lateral roots thicken, may be pigmented (yellow), and have an altered branching pattern, which may be characteristic of the specific mycorrhizal partner fungus. The fungus may constitute of the order of 25% of the dry weight of the root. There is limited hyphal outgrowth into the surrounding soil.

Ectomycorrhizae in Forestry

The forestry industry plants billions of seedlings every year. Traditionally, they were planted in soil naturally infected with mycorrhizal spores using a ratio of forest soil to substrate of 1:10. An alternative method is to inoculate seedlings and to plant these into the nursery beds with noninoculated plants. A problem with these methods is that the inoculum may also carry pathogens and pests, and the mycorrhizal mycelium may not make contact with surrounding noninoculated seedlings. In the 1980s and 1990s, in order to overcome these criticisms, attempts have been made to produce inoculum from pure cultures of ectomycorrhizal fungal isolates (see "Production of Mycorrhizal Inoculum"). While various isolates have been screened, those using *P. tinctorius* (Pt) and *Thelephora terrestris* have attracted most commercial attention. Pt inoculation has been shown to improve the performance of a wide range of ectomycorrhizal forestry species on several marginal sites affected by extreme acidity and heavy metal pollution.

Endomycorrhizae

A characteristic of endomycorrhizae is penetration of the root cell by the mycorrhizal fungus. Ectendo-, arbutoid, and monotropoid mycorrhizae are similar to ectomycorrhizae in that they form a sheath and Hartig net, but there is also penetration of the root cells. Ericoid and orchid mycorrhizae do not form a sheath or Hartig net. The arbuscular mycorrhizal fungi (AMF) lack a fungal sheath and Hartig net, but produce characteristic arbuscules and also may produce characteristic vesicles (**Table 1** and **Figure 2**). AMF can produce a vast network of fine mycelium extending into the soil beyond the rhizosphere.

AMF in Agriculture and Horticulture

Mycorrhizae may be absent in high-input agricultural systems or the strains present may have been weakened by selection for ability to grow in fertilized soils. Specifically, they can be suppressed by applications of phosphate fertilizer, and while they can survive as introduced inoculants on the planting material, they may not sporulate in chemically fertilized soils and may not survive intercropping periods in temperate zones, especially in rotations involving nonmycorrhizal plants, such as brassicas and sugar beet. The cost of reintroduction under these circumstances may be uneconomic. Nevertheless, AMF have been the subject of intensive research in recent years in the search for more sustainable crop production strategies, and beneficial results have been reported for inoculation of tropical crops, such as banana (*Musa sapientum*) and pineapple (*Ananas comosus*), and for temperate crops, including strawberry, corn (*Zea mays*; maize), and potato (*Solanum tuberosum*). For the many reports claiming beneficial effects, there are as many reporting no effects or even adverse effects. This may be due to difficulty in developing suitable inoculants due to the strong interaction between the crop genotype, the AMF strain, and the environment. Commercial AMF inoculants are either single isolates, often aggressive colonizing strains of *Glomus* spp., or mixed isolates of *Glomus* spp., or native field isolates from the soil in which the mycorrhized crop is to be grown. A number of commercial companies are now selling AMF inoculants, but the cost per plant is high, at approximately 10 US cents. Considering that the cost of a transplanted seedling is of the order of 5 US cents, it is difficult to convince growers to pay so much for the "promise" of improved plant performance.

AMF in Micropropagation

Microplants are, like transplanted seedlings, amenable to mycorrhizal inoculation during the establishment stage. The aim in "biotizing" plants (inoculating with beneficial microorganisms) is to develop a beneficial rhizosphere prior to plant exposure to any type of stress. If mycorrhizal fungi are to be added to a production system, the strategy should be that they have the greatest chance possible of becoming the main colonizers of the plants; hence, they are generally added to sterilized soils where there is no competition from indigenous organisms. Inoculation into unsterilized soils can also be particularly beneficial if highly efficient strains with rapid infection rates are used.

The transfer of *in vitro* plantlets to *extra vitrum* conditions, that is, at the acclimatization stage, is one of the most critical stages in the micropropagation process, and can result in high plant mortality rates, a consequence of the poor physiological quality of microplants. They can suffer from desiccation, both due to thin cuticles and malfunctional stomata. Furthermore, microplants are grown in an aseptic environment, and thus colonization of their roots can improve establishment and subsequent growth, by preparing them better to cope with the *extra vitrum* environment when normally saprophytic organisms can become facultative pathogens. Inoculation of microplants before transplanting (at acclimatization) has the possibility of increasing crop uniformity, reducing transplant/microplant mortality, and improving growth. From an economic point of view, the amount of initial inoculum needed for preinoculation is very low compared with field inoculation, e.g., 31 kg hectare^{-1} compared to 250–2000 kg hectare^{-1}.

Studies have focused on mycorrhizal introduction during acclimatization of tissue cultured plants. To extrapolate from this, it has been suggested that growth and survival could be further improved if AM fungi could be introduced at an even earlier stage, e.g., *in vitro*, and further develops the prospect of *in vitro* weaning. Examples of crops raised from microplants that have been inoculated with AMF include: sour orange (*Citrus aurantium*; Seville orange) with *Glomus fasiculatus*, apple (*Malus pumila*) microplants with *Glomus intraradices*, and grapevine (*Vitis* spp.) microplants with many *Glomus* spp. Lists also include strawberry, potato, ginger (*Zingiber officinale*), pineapple, and coffee (*Coffea* spp.).

AMF in Revegetation

As discussed above, AMF inoculum is expensive and so its exploitation has been restricted to niche applications, usually publicly funded projects. These include revegetation of amenity areas, landscaping of roads and motorways, and in ecological projects aimed at establishing vegetation on polluted sites and stabilizing soil and sand dunes in drought prone or coastal areas.

List of Technical Nomenclature

Arbuscule — Branched haustorial structure of certain endomycorrhizal fungi that forms within living cells of the root.

Biotroph — An organism that can only live and reproduce on another living organism.

F horizon	Fragmented leaf litter and fine organic matter less than 2 mm.
H horizon	Organic matter less than 2 mm interwoven with fine roots and associated with some leaf residues.
Inoculum	Microorganism or its parts, capable of colonizing the host plant.
Rhizosphere	The area of the soil immediately surrounding the plant root.
Vesicle	Swollen unbranched haustorial structure formed by some endomycorrhizal fungi in the root cells.

See also: **Diseases**: Bacterial Diseases; Fungal Diseases. **Nutrition**: Aluminum Toxicity; Deficiency Diseases, Principles; Mineral Uptake. **Plants and the Environment**: Amenity Horticulture; Global Warming Effects; Land Reclamation and Remediation, Principles and Practice. **Production Systems and Agronomy**: Agricultural Crops; Grassland; Horticultural Crops, Protected; Nursery Stock and Houseplant Production; Orchard Crops; Organic Farming. **Regulators of Growth**: Jasmonates. **Root Development**: Root Growth and Development. **Tissue Culture**: Clonal Propagation, *In Vitro*; General Principles. **Tissue Culture and Plant Breeding**: Clonal Propagation, Forest Trees. **Water Relations of Plants**: Drought Stress; Salt Stress.

Further Reading

Cassells AC, Doyle BM, and Curry RF (eds) (2000) Methods and markers for quality assurance in micropropagation. *Acta Horticulturae* 530.

Douds DD and Millner DD (1999) Biodiversity of arbuscular mycorrhizal fungi in agroecosystems. *Agriculture Ecosystems & Environment* 74: 77–93.

Douds DD and Kapulnik Y (eds) (2000) *Arbuscular Mycorrhizas: Physiology and Function*. Dordrecht: Kluwer Academic Publishers.

Gianinazzi S, Schuepp H, Haselwandter K, *et al.* (eds) (2002) *Mycorrhizal Technology in Agriculture*. Basel: Birkhauser Verlag.

Isaac S (1992) *Fungal–Plant Interactions*. London: Chapman & Hall.

Khan AG, Kuek C, Chaudhry TM, *et al.* (2000) Role of plants, mycorrhizae and phytochelators in heavy metal contaminated land remediation. *Chemosphere* 41: 197–207.

Mukerji KG, Chamola PBP, and Singh J (eds) (2000) *Mycorrhizal Biology*. New York: Kluwer Academic/ Plenum Publishers.

Varma A and Hock B (eds) (1999) *Mycorrhiza: Structure, Function, Molecular Biology and Biotechnology*. Berlin: Springer-Verlag.

Wilkinson DM (2001) Mycorrhizal evolution. *Trends in Ecology and Evolution* 16: 64–65.

Root Growth and Development

I J Bingham, Scottish Agricultural College, Aberdeen, UK
D Robinson, University of Aberdeen, Aberdeen, UK

Primary Structure and Function

The principal functions of the roots of terrestrial plants are the acquisition of water and nutrients and anchorage of the plant. Whilst the morphology of a root system in the field can vary widely with genotype and soil conditions, the basic structure of individual root axes is remarkably consistent. The demands of life in the soil have constrained the design of a root during evolution.

To access water and nutrients, a root must extend through the soil by deforming or displacing soil particles ahead of the root tip. Growth of cells near the tip generates the axial force required to overcome the mechanical resistance of the soil. A conical root tip, radial symmetry in cross-section, and a small diameter facilitate penetration of soil pores. Efficient uptake of water and nutrients relies on close contact being maintained between the root surface and water-holding pores and ion-exchange surfaces. Absorbed material must then be delivered to the shoot, sometimes over considerable distances (e.g., several meters), at a rate commensurate with shoot demand. At the same time the root must be supplied with assimilates from the shoot to fuel the growth and activity of its tissues. These long-distance transport requirements can be met only by the differentiation of specialized cells with low axial resistance.

This article discusses how roots are constructed to meet these various demands. The discussion is confined to angiosperms, as they form the largest and most economically important group of plants.

The primary structure of a root is shown in **Figure 1**. It consists of an apical meristem, protected by the root cap, followed by a zone of rapid cell expansion and a zone of maturation. Growth results from both the division and enlargement of cells. Cell division provides a continuous supply of cells for expansion, but does not itself contribute directly to root extension. It is by the irreversible elongation of cells that the root extends. Cell elongation occurs as part of the overall cell division cycle within the meristem, and also after a cell leaves the meristem and enters the zone of rapid cell expansion. Thus, it is the absence of cell division in the latter that distinguishes these two zones. Together the apical meristem and

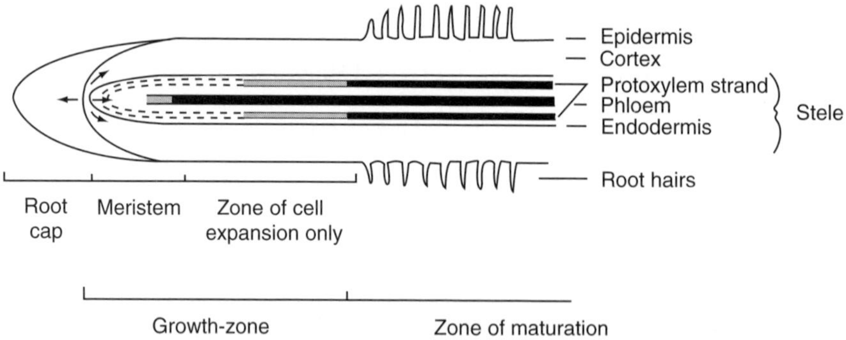

Figure 1 Diagram of a longitudinal section of the apex of a typical root (not to scale) showing the primary root structure and main regions of development. Arrows indicate the direction of cell division and displacement in the meristem. The relative positions of maturation of some vascular cells are shown. Bold shading indicates mature cells; light shading and broken lines represent immature cells becoming discernible morphologically. Cells mature closer to the apex in slow-growing roots. The size of the meristem and zone of rapid cell expansion vary with species and growth conditions. In corn (maize) grown under good conditions, the meristem may extend for 1–1.5 mm behind the quiescent center, and the entire growth-zone for 10–12 mm. In *Arabidopsis* typical values are 0.7 and 2.5 mm respectively (depending on the age of the root). Redrawn from a diagram originally published in: Esau K (1941) Phloem anatomy of tobacco affected with curly top and mosaic. *Hilgardia* 13: 437–490, with permission from University of California.

zone of rapid cell expansion make up the root growth-zone. Although it is cell expansion that directly determines root growth, changes in the rate of cell production can result in changes in growth rate by altering the number of cells expanding at a given time.

When viewed in cross-section, a mature root exhibiting primary growth consists of several discrete tissue layers (**Figure 2**). The outer layer of cells is the epidermis. Root hairs form as outgrowths from epidermal cells. The cortex is composed of one (e.g., in *Arabidopsis thaliana*) or more layers of parenchyma cells. Typically these are highly vacuolated with large intercellular spaces. In many species, the outer layers of the cortex develop into a hypodermis. Here the cells have thickened walls and few intercellular spaces. If the hypodermis develops a hydrophobic Casparian band it becomes known as the exodermis. The endodermis is a single cell layer that separates the cortex from the stele, which contains the vascular tissue. As with the exodermis, the endodermis possesses a Casparian band when mature.

The stele consists of the pericycle, xylem, and phloem. The pericycle is usually a single layer of parenchyma cells lying just inside the endodermis, although in many monocotyledons several cell layers may be present. The xylem is arranged in poles, alternating with strands of phloem. The number and arrangement of xylem poles differ between species. In the center of the root, xylem may differentiate to form a solid core (e.g., *Ranunculus* sp.), or parenchymatous pith may develop instead (e.g., *Zea mays*). Dicotyledonous roots may be distinguished from those of monocotyledons by the presence of a cambium located between the phloem and xylem. Divisions in these cells give rise to the secondary growth found in most dicotyledon roots, but absent from most monocots.

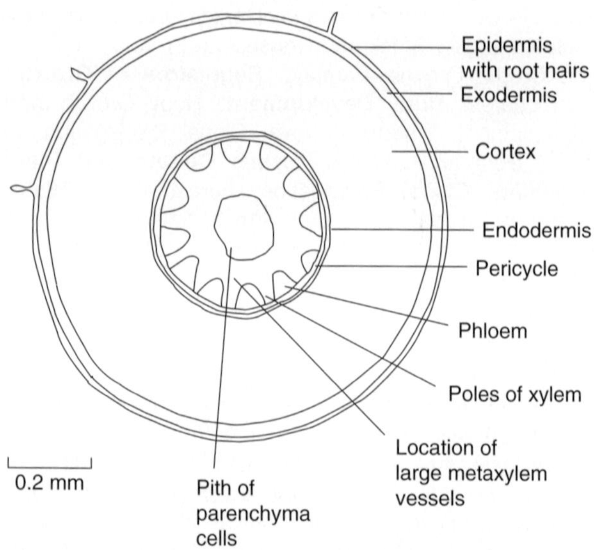

Figure 2 Diagram of a transverse section through the mature region of a *Zea mays* primary root showing the location of tissues and the arrangement of vascular strands. Adapted from: Peterson CA (1988) Endodermal Casparian bands: their significance for ion uptake by roots. *Physiologia Plantarum* 72: 204–208, with permission from Munksgaard International Publishers Ltd.

Apical Meristem

The meristem consists of groups of cells known as initials, and their derivatives. The initials remain close to the boundary between the root apex and root cap, in the promeristem, and divide continuously. The daughter cells (those arising from division of the initial) may themselves undergo several cycles of division. With each new division, and associated expansion, the earliest-formed daughter cells are displaced further away from the apex and ultimately out of the meristem. The plane of division relative to the root axis is critical to establishing the three-dimensional structure of the root. Most divisions are transverse to the long axis of the root and produce extensive files of cells. Some divisions occur parallel to the long axis (radial and tangential) and increase the number of cell files in a tissue.

Tracing cell files back towards the apex in histological preparations allows a tentative identification of their initials to be made. Apical organization in a root can be described as "closed" or "open." In species exhibiting closed organization (e.g., *Z. mays, A. thaliana*), distinct tiers of initials can be identified. Obvious tiers are absent in species whose root apices have open organization (e.g., *Trifolium repens*) and the initials for the cortex and root cap are not easily distinguished. In roots with closed and open organization, the initials are arranged around a group of relatively inactive cells known as the quiescent center.

Classical anatomy and radioisotope labeling of DNA synthesis have provided a wealth of information regarding the cellular organization and the rates and location of cell production in the roots of diverse species. But it was the adoption of *A. thaliana* as the model species for investigations of root development in the late 1980s that significantly advanced our understanding of the control of these processes. *Arabidopsis* has many advantages over other species for the investigation of root development. Its roots are narrow (120 μm diameter) and almost transparent. The cortex comprises only one cell layer. Numerous *Arabidopsis* mutants defective in some aspect of root growth or development have been identified which, when allied with the genomic information now available for this species, allows root development to be studied at the molecular genetic level.

Arabidopsis has three tiers of initials giving rise to (1) the stele, (2) the endodermis and cortex, and (3) the columella of the root cap and the peripheral cells of the root cap and epidermis (**Figure 3**). Molecular marking techniques have confirmed these cell lineages, originally proposed from histological examination. For example, reporter gene constructs, which are specific for particular types of differentiated cells (e.g., vascular and root cap cells), have been found expressed in their meristematic initials. In roots older than a week, some reorganization of the apex occurs. A periclinal division in the cortical initials creates separate initials for the cortex and endodermis. The cellular organization of the root is set during the development of the embryo. Mutations in several genes expressed during embryogenesis disrupt the eventual organization of the meristem. Examples include hobbit (*hbt*) and shortroot (*shr*). Hobbit mutants fail to develop a functional meristem, while shortroot mutants lack an endodermis. Molecular and biochemical analysis suggests that SHR encodes a putative transcription factor, whilst HOBBIT codes for a protein with homology to a subunit of the anaphase-promoting complex of the cell cycle.

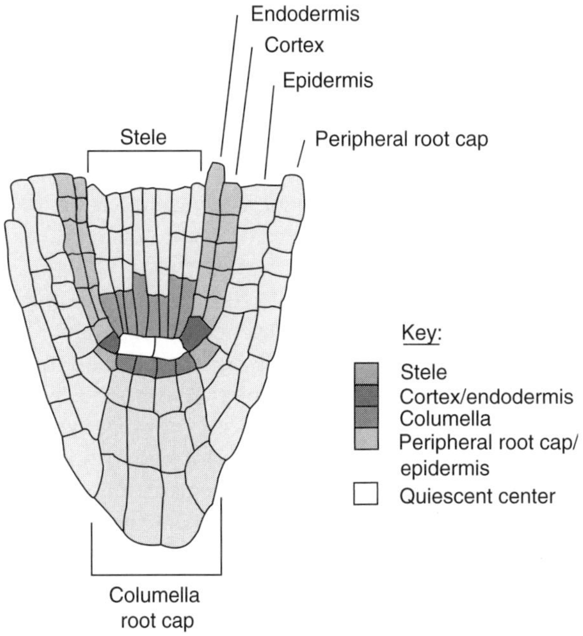

Figure 3 (see color plate 33) Organization of the root meristem of *Arabidopsis* showing four groups of initials arranged in three tiers surrounding the quiescent center. For the same shade, initials are shown by bold shading, and their daughter cells by light shading. Redrawn from Figure 2c, Van den Berg C, Weisbeek P, Scheres B (1998) Cell fate and cell differentiation status in the *Arabidopsis* root. *Planta* 205, 483–491, Copyright Springer-Verlag with permission.

Once the organization of the meristem has been established, the cellular pattern is sustained during subsequent growth. The consistent cell lineage patterns that are observed suggest that the fate of a cell is determined by its initial, and is inflexible. However, the developmental fate of a cell depends on its position within the meristem, and its fate can be altered after its production by the initial. In experiments where single cortical initials were destroyed by laser ablation, pericycle initial cells divided to fill the cavity and subsequently developed into cortical and endodermal cells rather than pericycle cells. Also when a cortical cell was isolated from its derivatives, its differentiation was disrupted. These and other experiments indicate that the older daughter cells within a cell file pass information to the more apical cells influencing their differentiation. Thus, events specifying cell fate during embryogenesis must be followed by position-dependent signals to ensure that the correct pattern of development occurs during root growth.

For the first few days after germination, the extension rate and length of apical meristem of a primary root may increase. Eventually, if the growth conditions are uniform, the meristem operates in a steady state. Its size and cell number remain relatively constant and the rate of cell production equals the rate of cell displacement out of the meristem. DNA labeling has shown that it takes about 12–25 h for a cell to complete one cycle of division. Mitotic cells of the stele have longer cycle times than those of the cortex and so tend to be larger than neighboring cortical cells because of the greater period for expansion between divisions. They also undergo fewer cycles before passing out of the meristem. In *Z. mays*, the last divisions in the xylem vessel files occur much closer (e.g., 0.24 mm) to the root cap boundary, compared with cells of the cortex (e.g., 1.22 mm). Even within a single file, individual cells can have different cycle times. Consequently, cells within the meristem do not divide synchronously.

The quiescent center is a population of cells with very low rates of mitotic activity. A small fraction of the cells may divide, but with long cycle times. The extent of cell production depends on the type of meristem. In open meristems the rates tend to be greater than those in closed meristems. In each case the dividing cells are located towards the periphery of the quiescent center. For a long time the function of the quiescent center remained obscure. It is thought that it may serve as a population of stem cells to replace initials at the end of their proliferative lifespan, or in the recovery of the meristem after injury. Laser ablation experiments have shown that the quiescent center of *Arabidopsis* inhibits the differentiation of surrounding cells. This is essential for maintaining the cell production capability of the meristem. When the quiescent center cell contacting a columella initial was ablated, the initial displayed a more advanced state of differentiation normally associated only with its daughter cells. Similar effects were observed in cortical initials when their adjacent quiescent center cell was ablated, suggesting that in *Arabidopsis* the quiescent center may control the balance between cell proliferation and cell differentiation.

Cell Expansion

The biophysics of cell expansion has been studied extensively. The generally accepted model is one in which load-bearing bonds within the cell wall are broken causing the walls to relax. This lowers the turgor pressure and establishes a water potential gradient across the cell membrane. As water enters the cell, the turgor pressure increases causing the cell walls to extend irreversibly. The rate at which a cell expands is a function of its turgor pressure and cell wall properties. To sustain a given rate of expansion, the solute and water fluxes into the cell must be sufficient to maintain its turgor pressure and keep pace with wall loosening.

A root is made up of a number of cell files. It is the sum of the extension of the individual cells in each file that constitutes root extension. As cells in adjacent files cannot move across each other, all files must extend at the same rate. Cell expansion occurs predominantly in the longitudinal direction, although some radial expansion may take place close to the apex. This increases the root's diameter.

The relative rate of root extension is not uniform along the length of the growth-zone. For example in *Z. mays*, it increases with distance from the root apex, reaching a maximum at about 4 mm, before decreasing to zero at the basal end of the growth-zone (**Figure 4**). Under conditions conducive to fast growth rates, water and solute fluxes into expanding cells are nonlimiting. Furthermore, measurements of turgor pressure made using a single-cell pressure probe have shown that turgor is relatively uniform

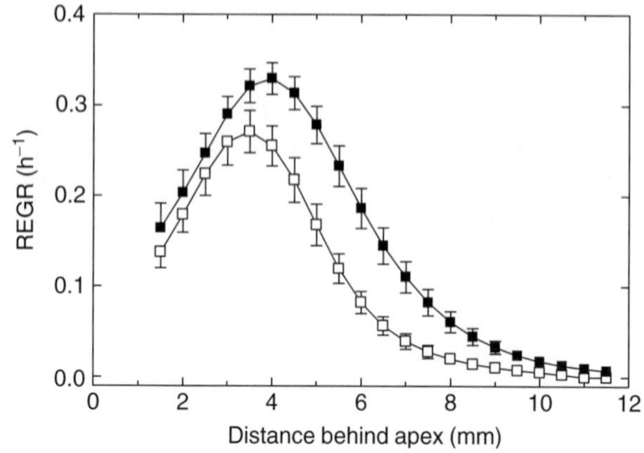

Figure 4 Relative elemental growth rate (REGR; sometimes referred to as local growth rate or strain) along the growth-zone of a primary root of young maize seedlings. Closed symbols represent control roots. Open symbols are for roots from seeds artificially aged prior to germination and shows a lower maximum REGR and shorter growth-zone after aging. After Bingham IJ and Merritt G (1999) *Seed Science and Technology* 27, 151–162, reproduced with permission from International Seed Testing Association.

along the growth-zone. This implies that changes in relative extension rate along the growth-zone are controlled by changes in cell wall properties. In the region of increasing extension rate there is a progressive loosening of cell walls, while in the region of decreasing growth rate there is a gradual tightening, until the cell ceases elongating altogether. This probably involves changes in cell wall extensibility, in the minimum turgor pressure required for growth (sometimes called the yield threshold), or both.

The growth rate of roots is reduced under adverse soil conditions such as low water potential or excess salinity. This response is often associated with a shortening of the growth-zone and a reduction in the maximum relative rate of extension. Extension rate in the apical region of the growth-zone is usually less affected. For example, *Z. mays* roots can continue elongating at low soil water potentials, albeit at a slower rate, because enhanced wall loosening in the apical 2–3 mm enables cells in this region to extend despite a reduction in turgor pressure. Abscisic acid (a plant growth regulator) has been implicated in signaling this increased wall loosening. The biochemical basis of wall loosening and tightening in extending roots is not fully resolved. The orientation of cellulose microfibrils in the wall is believed to control the anisotropy of cell growth, but there is little evidence to suggest that it regulates extension rate along the growth-zone.

Cell wall loosening and tightening probably result from, respectively, cleavage and formation of cross-links between cell wall polymers. The activities of several groups of enzymes have been measured in the growth-zone of roots showing different extension rates. These groups include the expansins, trans-glycosylases (e.g., xyloglucan endotransglycosylase, XET), and peroxidases. The first two groups have been linked with wall loosening and the last with wall tightening. Changes in activity correlate with some, but not all aspects of wall loosening and tightening. The current view is that several processes are likely to be involved in regulating cell wall properties along the growth-zone.

Cells in the growth-zone accumulate inorganic and organic solutes to maintain their turgor pressure during elongation. Organic solutes are translocated into roots from photosynthetic tissues by the phloem whose mature sieve tubes extend well into the growth-zone. Several lines of evidence indicate that unloading of solutes from the phloem and their radial transport into the root cortex occurs mostly via the symplast within the elongation zone. The most compelling evidence comes from studies using membrane-impermeant fluorescent dyes (**Figure 5**).

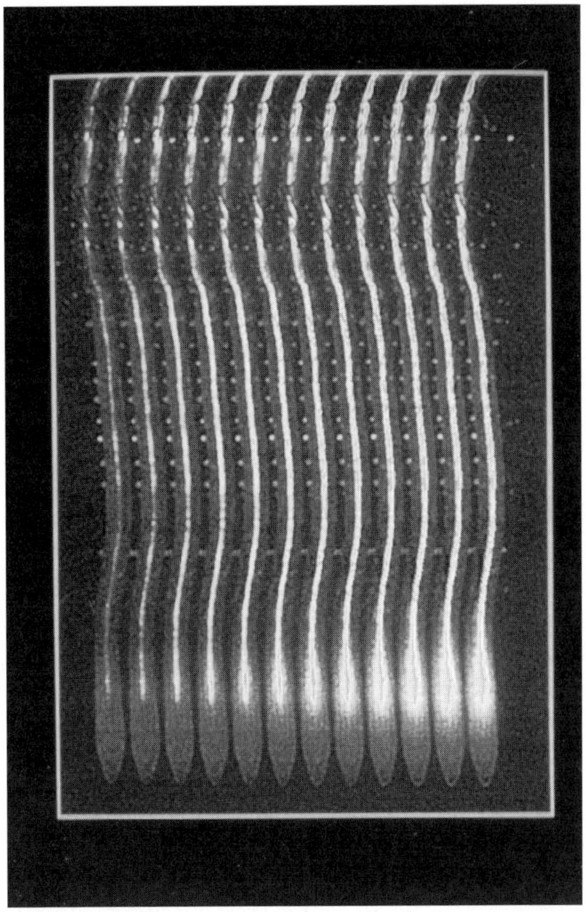

Figure 5 (see color plate 34) False-color confocal laser scanning microscope images showing the progression of phloem unloading at the tip of a primary root of *Arabidopsis thaliana*. A membrane impermeant fluorescent dye (5(6) carboxyfluorescein) introduced into the phloem of the leaf was translocated in the protophloem to the growing root tip, from where it moved radially towards the epidermis and acropetally towards the root apex. Most intense unloading was in the zone of cell expansion. Images were taken at 3 min intervals; blue, low concentration; yellow, high concentration. Images first appeared in Oparka KJ, Prior DAM, and Wright KM (1995) Symplastic communication between primary and developing lateral roots of *Arabidopsis thaliana*. *Journal of Experimental Botany* 46: 187–197, reproduced with permission of Oxford University Press and K Oparka.

Such evidence also indicates that intense phloem unloading precedes the elongation and emergence of lateral roots from the pericycle of the parent root, and that there is limited symplastic communication between the root apex and root cap. The pathway for inorganic solutes is less certain. They may be taken up from the soil directly into the growth-zone, or perhaps more likely, they may be absorbed by the mature regions of the root and delivered to the growth-zone via the stele.

In a growing root, cell division and expansion must be coordinated. However, cell division can be inhibited (e.g., by using mitotic inhibitors such as colchicine) without initially affecting expansion, demonstrating that these processes can be partly uncoupled. Understanding the molecular co-regulation of cell division and expansion in roots promises to be one of the key areas of future research.

Cell Differentiation and Maturation

Using molecular markers, it has been shown that cell initials in the promeristem exhibit gene expression specific to cell types, suggesting that differentiation has already started in these cells. Differentiation then progresses as the daughter cells are displaced through the meristem.

Root Cap

Most roots are terminated at their apical end by a conical collection of living parenchyma cells, the root cap. The root cap is derived from cell divisions in the apical meristem that displace daughter cells forward, in the direction of root growth. In species that have closed apical meristems, the root cap originates from its own set of initials distinct from those of the cortex. In a few species (e.g., *A. thaliana*) initials that generate the peripheral layers of the root cap also generate the epidermis (**Figure 2**). In most, however, the epidermis is derived from the cortical initials. The histological origin of root cap cells is less obvious in species that have open apical meristems (e.g., *Allium cepa*).

The root cap's main function is to protect the underlying cells of the apical meristem from damage as the growing root is forced between soil particles. The root cap secretes a carbohydrate-rich mucilage, "mucigel," that acts as a lubricant, reducing friction between root and soil. Mucilage gives young roots their typical slimy appearance and texture. Soil particles can become embedded in the mucigel, forming a water-retentive sheath around the sub-apical part of the root. This helps to maintain the root's hydraulic connection to the bulk soil and, therefore, to its main external reservoir of water and nutrients.

As the root penetrates the soil, cells of the root cap progressively slough off and are replaced from certain cells in the apical meristem. The loss of root cap cells is part of the process of "rhizodeposition" whereby materials derived from living plants enter the soil's organic matter. The loss of cells from the root cap was once considered to involve only dead cells. We now know, however, that many of the cells lost from the root cap are alive and continue to function for a short time when detached from the root. These "border cells" ensheath the root tip, trapped by the root's coating of mucigel. Their rates of production vary between species and in response to various internal and external stimuli. Intriguingly, the metabolic activities of border cells can differ significantly from those of attached root cap cells. This has led to the suggestion that border cells may be sources of chemical signals that specifically attract or deter soilborne microbes towards or away from the root. Border cells are also believed to lubricate the root to reduce friction as it penetrates through the soil.

Some root cap cells (columella cells; **Figure 3**) contain conspicuous starch grains known as statoliths. Statoliths have long been implicated in the gravitropic response of roots. The idea is that gravity pulls the dense statoliths to the bottom of the root cap cells in which they are contained, allowing the root to detect its orientation with respect to the gravitational field and to initiate appropriate growth responses to it. Evidence obtained for *Arabidopsis* using a combination of hypergravity (growing plants in a high-speed centrifuge), microgravity (growing plants in Earth orbit), starch-deficient mutants, and laser and genetic ablation of statolith-containing cells has shown that the amount of statolith sedimentation is correlated with the strength of the gravitational field and of the gravitropic response. Since the gravitropic response of the root involves differential expansion of the upper- and lower-most files of cells in the growth zone, some form of signal must be transmitted from the root cap to the elongation zone. The nature of the signal(s) is unknown but is the subject of current research and debate.

Vascular Differentiation

In general, the first vascular cells (i.e., closest to the apex) that can be distinguished morphologically are the immature metaxylem elements. These appear larger and more vacuolated than the neighboring cells. The metaxylem elements are followed by the protoxylem elements and the sieve tubes of the protophloem. In contrast, the first cells to mature are the phloem sieve tubes. Typically, this occurs close to the proximal margin of the apical meristem and near the start of the zone of rapid cell expansion (**Figure 1**).

Recent studies of field-grown *Z. mays* and *Glycine max* roots have cast doubt on the classical interpretation that the large metaxylem vessels mature (i.e., the cytoplasm and end walls disappear and the vessel opens for conduction) within a few centimeters of the root tip. Cryomicroanalysis of inorganic solute accumulation and examination of hand sections, which is less disruptive than traditional techniques involving embedding, has shown that the large vessels are not open for conduction until at least 15 cm or more behind the root tip. This has raised important questions about the role of young tissues of primary roots in the uptake and long-distance transport of water. It is difficult to imagine that they contribute to water capture by mature plants when their large vessels are closed. Although the narrow (early) metaxylem and protoxylem vessels open closer to the root tip and so can transport water, they have a much greater axial resistance to water flow (since, from Poiseuille's law, resistance is inversely proportional to the fourth power of vessel radius).

Endodermis and Exodermis

The roots of all plants possess an endodermis. The Casparian band is formed from the deposition of suberin and lignin in the radial and transverse cell walls. This occurs at or near the end of the growth-zone where the protoxylem vessels mature and become conductive. These are known as state I cells. Further from the root tip there may be an additional layer of suberin deposited over the inner surface of all endodermal cell walls (state II). This is found in many, but not all, species. In monocotyledons a third stage of development occurs (state III) in which an extra cellulose wall is laid down under the suberin lamellae and is typically thickest on the inner tangential wall. Endodermal cells are connected to the pericycle and cortex by plasmodesmata, providing a symplastic pathway across the root. Development of the state II and state III endodermis is asynchronous and occurs first in cells aligned with the phloem. Those that remain in state I are known as passage cells. In some species they persist, but in others (e.g., *Triticum aestivum, Hordeum vulgare*) they eventually mature into state II and state III cells.

Of the angiosperm species studied to date, 91% have been found to possess an exodermis. About 3% possess a hypodermis with no Casparian band; this group includes legumes and some grass species (e.g., *T. aestivum, H. vulgare, Avena sativa*). The remainder lack a hypodermis altogether. The Casparian band in the exodermis tends to develop further from the root tip than that of the endodermis, but is comparable in structure. As with the endodermis, maturation from state I to states II and III can occur. In some species (e.g., *Z. mays, Helianthus annuus*) the exodermis is uniform, where all cells are in state II. In others (e.g., *A. cepa)* the exodermis is dimorphic, with long cells in state II and short cells where maturation to state II is delayed. The development of the exodermis is influenced by environmental conditions including water availability.

It is widely accepted that the lipophilic nature of the Casparian bands and suberin lamellae in the endodermis and exodermis restrict the flow of water and solutes through the apoplast. This enables the root to exercise some metabolic control over the transport of nutrients to the xylem. If the apoplastic pathway is completely restricted in either the exodermis or endodermis, then solutes must cross a membrane *en route* to the xylem. At the same time the apoplastic restriction helps prevent loss of water, nutrients and other solutes to the soil. However, the extent of this apoplastic restriction has been questioned. Evidence now suggests that the endodermis and exodermis represent a barrier of variable resistance and some apoplastic flow of water and solutes through them is possible. The extent of suberization and lignification is influenced by growth conditions, as is the chemical composition of the suberin: aliphatic suberin is less hydrophobic than aromatic.

Root Hairs

Conspicuous on subapical root zones of most herbaceous plants, root hairs are tubular protrusions of epidermal cells into the surrounding soil. Each root hair arises from a single cell and grows from the tip. In some species (e.g., *A. thaliana*) only certain cells of the root epidermis (trichoblasts) can form root hairs. In other species, all epidermal cells have the potential to form hairs. Root hairs are typically between 5 and 20 μm in diameter, 80 to 1500 μm long, and can attain densities in excess of $1000\,mm^{-2}$.

For an apparently simple process, root hair development involves a plethora of genetic controls. These have been investigated most extensively in *Arabidopsis*, which has become a model species for the study of root hair growth; indeed, root hairs are themselves used to investigate the genetic, molecular, and positional regulation of cell fate in plants. Four groups of root hair morphogenesis genes have been identified, those involved in hair initiation, the outward swelling of the trichoblast, the transition from swelling to tip growth, and tip growth itself. Between two and eleven genes are associated with each group, many of which have multiple functions. Their regulation is complex, as revealed by the analysis of the many *Arabidopsis* mutants that express abnormal root hair development. Environmental factors also influence root hair development. Shortages of nitrogen and phosphorus often increase root hair length, density or both.

The root hairs of legumes are the main sites of entry of N_2-fixing bacteria such as *Rhizobium* sp. The bacteria are encouraged to accumulate at the surface of the root hair by chemical attractants secreted by the plant. The bacteria secrete NOD factors (lipochito-oligosaccharides) that stimulate the plant to form more root hairs (and bacterial entry points). The root hairs are induced to curl and to bind with the bacteria via lectins. The bacteria enter the root hair by means of an infection thread that penetrates into the cortex where the root nodule is eventually formed.

The precise functions of root hairs remain enigmatic. It is often claimed that root hairs, because of their enormous surface area, must be the main sites of water and nutrient absorption, and are crucial in anchorage. Yet experiments often reveal root hairs to have limited effects in these roles. For example, the forces needed to uproot individual *Arabidopsis* plants are almost identical whether those plants have root hairs or not; the strength and number of lateral roots contribute far more to resisting uprooting. Phosphorus uptake per plant is usually not in proportion to the surface area provided by root hairs, especially if the plant is heavily colonized by arbuscular mycorrhizal fungi. There is no doubt that root hairs are physiologically active, but there is little experimental evidence that they are essential for uptake or anchorage. Root hairs are short-lived, with a lifespan in the order of days. Consequently, mature parts of roots are usually hairless.

It is possible (although again the evidence is scanty) that root hairs help to maintain hydraulic continuity between the root and the soil's water-filled pores. Root–soil contact is vital for water and nutrient uptake. Root hairs help trap soil particles and mucilage close to the root's surface and minimize the risk that the root loses contact with the bulk soil, as can occur in strongly structured soils if roots are confined mainly to wormholes, cracks, and macropores. They may also be important for locally anchoring the root behind the growth-zone, so that during extension the root can overcome the mechanical resistance of the soil without lifting the plant upwards.

Secondary Growth

Species differ widely in the amount and characteristics of secondary growth in their roots. Secondary growth begins with cell divisions in the vascular cambium. Divisions also occur in the pericycle. The vascular cambium and the dividing pericycle form a sleeve of cells that continue to divide to generate secondary phloem and xylem tissue. As the vascular cylinder expands radially, the cortex and endodermis may be disrupted and eventually sloughed. The outer surface of the root is then made up of cork layers produced by a cork cambium derived ultimately from the pericycle.

Because of these structural changes, roots with secondary growth are unlikely to function in water and nutrient uptake. They will, however, be important for the long-distance transport of resources to and from fine roots, and for anchorage. Species (e.g., *Convolvulus arvensis, Ranunculus* sp.) that have little secondary growth may retain a functional root cortex.

Tissue Degeneration

Part of normal root development in most species is the gradual degeneration of specific tissue types. This can occur even in quite young tissue. In the roots of some grass species, including *T. aestivum, H. vulgare*, and *Lolium perenne*, this involves the progressive loss of most of the cortex. The outer surface of such roots may then be formed by the innermost layer of cortical cells and the endodermis. Whether this loss of cortex impairs water and nutrient uptake significantly will depend on the physiological integrity of the remaining cells and on the extent to which root–soil contact is affected. This degeneration may be less evident in other species, such as *Z. mays*, in which it is usual for only the epidermis to be lost.

A particularly important type of tissue degeneration is the formation of aerenchyma; large, continuous gas-filled spaces within the cortex. Aerenchyma can arise from the collapse or separation of adjoining files of cortical cells. The resulting spaces act as low-resistance, long-distance pathways by which oxygen can be transported to the root apex. Aerenchyma is common in roots of wetland species (e.g., *Oryza sativa*), but it can also develop in many others in response to temporary waterlogging.

See also: **Growth and Development**: Cell Division and Differentiation; Cell Growth. **Nutrition**: Ion Transport; Mineral Uptake; Nitrogen Fixation. **Root Development**: Genetics of Primary Root Development; Lateral Root Initiation; The Rhizosphere and its Microorganisms. **Water Relations of Plants**: Phloem; Xylem.

Further Reading

Clarkson DT (1996) Root structure and ion uptake. In: Waisel Y, Eshel A, and Kafkafi U (eds) *Plant Roots: The Hidden Half*, pp. 483–510. New York: Marcel Dekker.

Esau K (1977) *Anatomy of Seed Plants*, 2nd edn. Chichester: John Wiley & Sons.

Foreman J and Dolan L (2001) Root hairs as a model system for studying plant cell growth. *Annals of Botany* 88: 1–7.
Hose E, Clarkson DT, Steudle E, Schreiber L, and Hartung W (2001) The exodermis: a variable apoplastic barrier. *Journal of Experimental Botany* 52: 2245–2264.
McCully ME (1999) Roots in soil: unearthing the complexities of roots and their rhizospheres. *Annual Review of Plant Physiology and Molecular Biology* 50: 695–718.
Oparka KJ, Prior DAM, and Wright KM (1995) Symplastic communication between primary and developing lateral roots of *Arabidopsis thaliana*. *Journal of Experimental Botany* 46: 187–197.
Peterson CA and Enstone DE (1996) Functions of passage cells in the endodermis and exodermis of roots. *Physiologia Plantarum* 97: 592–598.
Pritchard J (1994) The control of cell expansion in roots. *New Phytologist* 127: 3–26.
Van den Berg C, Weisbeek P, and Scheres B (1998) Cell fate and cell differentiation status in the *Arabidopsis* root. *Planta* 205: 483–491.
Webster PL and McLeod RD (1996) The root apical meristem and its margins. In: Waisel Y, Eshel A, and Kafkafi U (eds) *Plant Roots: The Hidden Half*, pp. 51–76. New York: Marcel Dekker.
White PJ (2001) The pathways of calcium movement to the xylem. *Journal of Experimental Botany* 52: 891–899.
Wu Y and Cosgrove DJ (2000) Adaptation of roots to low water potentials by changes in cell wall extensibility and cell wall proteins. *Journal of Experimental Botany* 51: 1543–1553.

The Rhizosphere and its Microorganisms

G D Bending, Horticulture Research International, Warwick, UK

Introduction

The term rhizosphere was first used in 1904 by Hiltner to describe the zone of intense bacterial activity in the soil surrounding legume roots (from Greek; "rhiza" meaning root and "sphere" meaning field of influence). In more recent years, the meaning of the term has broadened so that it now represents the generic volume of soil influenced by the root. Significant quantities of plant metabolites are lost from the root into the surrounding soil. Carbon loss from the root has a great impact on microbial communities, the population sizes and activities of which are elevated in the rhizosphere relative to the surrounding soil. These organisms interact with the plant and each other in a variety of positive, negative, and neutral ways. Plant–microbe rhizosphere interactions can influence plant growth, nutrition and development, susceptibility to disease, and the dynamics of nutrients, heavy metals and xenobiotics. As such, they have considerable potential for biotechnological exploitation.

Definitions

A number of distinct regions can be recognized within the rhizosphere (**Figure 1**). These include the area within the tissues of the plant root (the endorhizosphere), the root surface (the rhizoplane), and the soil adjacent to the root (the ectorhizosphere). In plants colonized by mycorrhizal fungi, the ectorhizosphere has an additional component, the mycorrhizosphere, which can extend for a considerable distance from the root.

Rhizosphere Carbon Flows

Amounts and Composition of Rhizodeposition

The driving force for the development and maintenance of the rhizosphere is the deposition of carbon by the plant root, with four categories of such "rhizodeposits" being recognized:

- "Diffusates," including sugars, amino acids, and hormones. These compounds are released

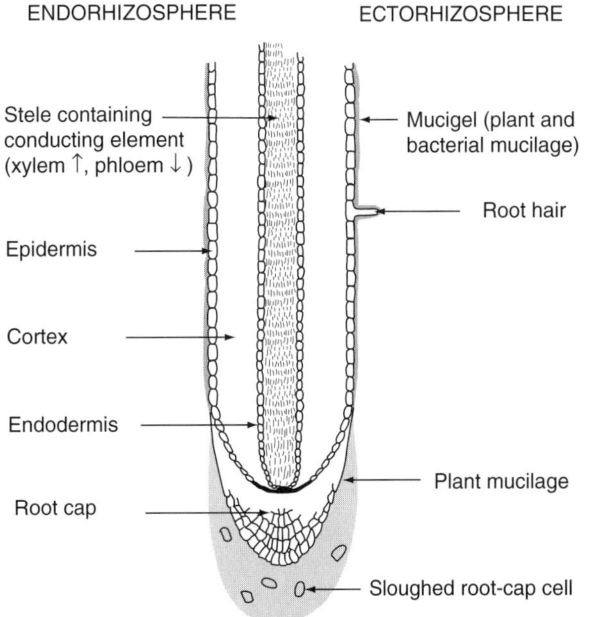

Figure 1 Structure of the rhizosphere. Reproduced from Lynch JM (1983) The spermosphere and rhizosphere. In: *Soil Biotechnology – Microbiological Factors in Crop Productivity*. Oxford: Blackwell Publishing Ltd.

passively from the root down the steep electrochemical gradient that exists for them between the cytoplasm of root cells and the soil solution.

- Secretions such as polysaccharides, enzymes, siderophores, organic acids, and allelochemicals. These compounds are actively released from roots. In the case of organic acids, anion channels are important in the secretion process, while for polysaccharides and phenolics, secretion may involve vesicles.
- Root debris sloughed off during growth through soil, including root caps, epidermal cells, and whole roots.
- Excretions of waste products including alcohols and CO_2.

The quantity and composition of rhizodeposits varies enormously and is controlled by a number of interacting factors. These include plant genotype, age and health, environmental variables such as light intensity, temperature, soil nutrient status and pH, and the presence of rhizosphere microbial populations. Most exudation occurs from the zone just behind the tip of primary and lateral roots, with further quantities lost from the zone of elongation and the junction of primary and lateral roots in some species.

Root carbon (C) budgets have been determined for a number of plant species in laboratory investigations using ^{14}C tracer techniques. While these methods are effective at quantifying fluxes, they are not able to distinguish root from microbial respiration, and therefore to quantify absolute amounts of C lost as rhizodeposits. Using tracer techniques, the percent net fixed photosynthate translocated to the roots of wheat (*Triticum* spp.), corn (*Zea mays*; maize), tomato (*Lycopersicon esculentum*), and pea (*Pisum sativum*) has been estimated to be between 30% and 60%. When respiration (arising from both plant roots and rhizosphere organisms) is measured, the amount of C lost from the roots is between 40% and 80% of photosynthate. Generally, between 10% and 30% of the C lost from the roots is recovered as rhizodeposits.

Much less is known about rhizodeposition from mature plants. In long-term experiments with wheat and white mustard (*Sinapis alba*), the amounts of C recovered in root biomass at crop harvest were 20% and 20–35%, respectively, of that which had been translocated to the root. With root respiration estimated at 30% of the C translocated to the root, the amount of material lost via rhizodeposition was estimated to be 0.5 and 2.8 $t\,C\,ha^{-1}\,year^{-1}$ for wheat and mustard, respectively. In field experiments with wheat, the proportion of fixed C transferred to the root declined from 50% after 42 days to 2% at 154 days. Total amounts of C lost via rhizodeposition, including soil respiration and root residues, were estimated at between 1.2 and 1.9 $t\,C\,ha^{-1}\,year^{-1}$, representing 15% of C fixed by the plant. In the case of forest trees, 40–73% of net fixed C can be transferred to the roots, and losses of C, including that mediated through mycorrhizal fungi, has been calculated at 5.8–7.5 $t\,C\,ha^{-1}\,year^{-1}$. When bark was removed to stop the flow of assimilate to the roots, it was found that for forest trees, current assimilate provides energy for over 50% of soil respiration.

Advances in reporter gene technology could help to improve characterization of rhizosphere C fluxes. Bacteria harboring a marker gene reporting on single C sources have the potential to provide quantitative information on the spatial and temporal dynamics of rhizosphere C flow. However, application of the technology has been limited to date.

Significant quantities of nitrogen (N) are also lost from the plant as rhizodeposits. In barley (*Hordeum vulgare*) and pea, rhizodeposition in 7–14-week-old plants has been estimated to account for between 32%–71% and 15%–48%, respectively, of the below-ground N budget, amounting to 20% and 7%, respectively, of total plant N at maturity.

Direct Effects of Root Exudates

A direct function has been unequivocally demonstrated for only a small number of rhizodeposit components. Most of these functions relate to nutrient mobilization.

Nutrient Mobilization

Plants secrete a variety of extracellular phosphorus (P)-mobilizing enzymes, the induction of which is triggered by P deficiency. Most roots can produce acid phosphatase, which hydrolyzes inorganic P from organic phosphomonoesters, which form 30–80% of the soil P reserve in agricultural soils. In lupins (*Lupinus* spp.), acid phosphatase production can be increased by up to 20 times under P stress.

Similarly, many plant species exude organic anions such as citric and malic acids in response to deficiency of several nutrients, including P, potassium (K), iron (Fe), and manganese (Mn) (**Table 1**). These compounds lower rhizosphere pH, increasing availability of $H_2PO_4^-$ and a number of micronutrients, including Mn^{2+}, $Fe^{2+/3+}$, and Zn^{2+}. Inorganic anions, including HCO_3^-, and H^+ are also exuded from plant roots. Flow of these ions is largely controlled by cation/anion balance, which is affected

Table 1 Effect of nutrient status on organic acid production by selected plant species

Plant species	*Nutrient*	*Root organic acid content (nmol g⁻¹ fw root)*	
		Malate	*Citrate*
Brassica napus[a]	+P	3350	1100
	−P	21700	8100
Hordeum vulgare[b]	−Al	2686	1888
	+Al	8358	1666
Zea mays[a]	−K	23000	12000
	+K	33000	19700
Phaseolus vulgaris[a]	+Fe	18000	800
	−Fe	68000	5100

Adapted from Jones DL (1998) Organic acids in the rihzosphere – a critical review. *Plant and Soil* 205: 25–44. Reproduced with kind permission of Kluwer Academic Publishers.
fw, fresh weight.
[a] + Represents nutrient-sufficient media; − represents nutrient-deficient media.
[b] + Represents a toxic concentration of Al; − represents no Al.

by mineral N source, with NH_4^+ uptake inducing H^+ extrusion. However, many plants enhance H^+ excretion under Fe or P deficiency. H^+ release can result in rhizosphere acidification, enhancing the availability of $H_2PO_4^-$, Mn^{2+}, $Fe^{2+/3+}$, and other cations in the rhizosphere.

Many plant species produce compounds to aid solubilization of $Fe^{2+/3+}$ so that availability for plant uptake is increased. In dicotyledons, many of these compounds are phenolic in nature. For example, in tomato, Fe deficiency induces secretion of caffeic acid, which increases solubility of Fe minerals. In grasses and cereals, Fe deficiency promotes the secretion of siderophores, which are hydroxy- and amino-substituted iminocarboxylic acids with highly effective capacities to solubilize Fe minerals. Exudation of siderophores can increase by up to 20 times on the imposition of Fe deficiency in cereals.

Detoxification

Organic acids play a role in mediating detoxification of metals by plants. Aluminum (Al) toxicity can limit crop production in many areas. In several crop plants, including wheat and corn, Al^{3+} tolerance is the result of increased exudation of malate or citrate in response to the metal (**Table 1**). The organic acids complex Al^{3+} rendering it nontoxic. The exudation of organic acids allows the plant to grow at higher Al^{3+} concentrations.

Water Acquisition

Mucilage, consisting of polysaccharides containing hexose and pentose sugars and uronic acids, is secreted by root cells as the root grows through the soil. Additional mucilage is secreted by rhizosphere microbes. On contact with the soil, mucilage forms a gel, which has a number of beneficial properties. The gel binds soil particles and microbes together with the root to form a "rhizosheath." As the moisture content of soil falls following plant uptake, the mucilage dries. Since it binds the soil and root together, the gel ensures that gaps do not form as the soil shrinks, so that hydraulic conductivity is maintained. Other possible benefits of mucilages include aiding lubrication as roots move through the soil and the absorption of ions, including $Fe^{2+/3+}$, Ca^{2+}, and $H_2PO_4^-$. Further, the binding of soil particles together by root-derived mucilage promotes soil stability, increasing resistance to erosion. The rhizosheath may also prevent the incursion of pathogens and herbivores.

Signal Molecules

A variety of specific secondary chemicals secreted into the rhizosphere by the plant can have an impact on the growth of other plants, and beneficial and pathogenic microbes. Some compounds exuded by plant roots may suppress the growth of other plant species, an interaction described as "allelopathy." Many of these interactions have only been demonstrated in laboratory studies and a conclusive demonstration that the compound is able to exert effects at concentrations existing in the field is lacking. The best-known allelopathic effect is that of the black walnut tree (*Juglans nigra*), which has been reported to kill other plants at a distance of 25 m from the tree, via production of the highly toxic quinone juglone (5-hydroxy-1,4-napthoquinone). A number of allelopathic interactions have the potential for exploitation in agricultural systems. For example, in Africa, intercropping grain crops with the legume *Desmodium uncinatum* is highly effective at suppressing infection of the crops by the root parasite *Striga hermonthica*.

Plants produce a variety of signals that influence the activities of specific rhizosphere microbial communities. In the case of root nodule bacteria and mycorrhizal fungi, plants produce flavonoids, which can induce root infection and establishment of the symbiosis (see below). Plant roots can also secrete molecules that interfere with or mimic bacterial N-acylhomoserine lactones, which are signal molecules used by Gram-negative bacteria to regulate population size. These compounds could play an important role in regulating the dynamics of rhizosphere bacteria, including plant-growth promoting, biocontrol, and pathogenic strains.

Microbial Interactions in the Rhizosphere

Carbon in the rhizosphere has a stimulatory effect on the soil microbiota, with greatly enhanced populations in the rhizosphere relative to the bulk soil. Using plate-counting methods, the rhizosphere/bulk soil (r/s) ratio of bacteria, actinomycetes, and fungi is usually in the ranges 2–20, 5–10, and 10–20, respectively. In young roots, rhizosphere bacterial communities are dominated by species with fast growth rates and capacities to utilize simple substrates (r strategists). As roots mature, there is a shift in dominance to bacterial communities with relatively slow growth rates and the capacity to degrade more complex substrates (K strategists). Molecular analysis of bacterial community structure has indicated that rhizosphere soil may be phylogenetically less diverse and dominated by a smaller number of strains, relative to bulk soil. Faunal communities are also more abundant in the rhizosphere relative to bulk soil, including groups feeding directly on plant roots and microbial communities. The stimulation of microbial communities in the rhizosphere provides a variety of benefits to the host plant, including enhanced nutrient solubilization, growth stimulation, protection from infection by root pathogens, and the development of mutualistic symbioses.

Plant Growth-Promoting Bacteria and Fungi

A number of free-living rhizosphere bacteria and fungi enhance plant growth directly. Such plant growth-promoting microbes can act by a variety of mechanisms, including nitrogen fixation, the solubilization of nutrients, including $H_2PO_4^-$ and $Fe^{2+/3+}$, the production of phytohormones, and reduction of metal toxicity. In some circumstances, nitrogen fixation by free-living rhizosphere bacteria can make significant contributions to plant N, which can amount to 1.5–20% in some rice (*Oryza sativa*) genotypes. Plant growth-promoting rhizobacteria are typically effective colonizers of plant roots, with chemotaxis to root exudates and motility playing a key role.

Diseases and Pathogen Suppression

Rhizodeposition also stimulates the germination of pathogen propagules and directed growth toward the root, which can lead to disease. Soil-borne pathogens fall into two broad groupings: nectrotrophic and biotrophic pathogens. Nectrotrophic pathogens, including *Fusarium*, *Verticillium*, and *Pythium* rapidly kill all or part of the host following their entry through plant roots. For nectrotrophic fungal pathogens with broad host range, such as *Pythium* and *Fusarium*, plant exudate components, including sugars and amino acids, stimulate propagule germination and growth toward the root. Subsequent infection occurs through wounds or breaks in the root surface. For those pathogens with limited host ranges, propagule germination stimulants can be compounds specific to the host family, such as organic sulfur (S) compounds in the case of the interaction of *Sclerotium cepivorum* with *Allium* spp. Nectrotrophic pathogens characteristically have wide host ranges and attack young, debilitated, or senescing tissues. Host cells are rapidly killed by cytolytic enzymes or toxins.

Biotrophic pathogens, such as *Plasmodiophora brassicae*, have a narrow host range and infect plants directly or through natural openings. Such pathogens initially grow within host tissues without cell death, although they can cause changes to root structure and physiology. Specialized parasitic structures called haustoria are formed within host cells, through which nutrients are drawn from the host.

The rhizosphere also supports populations of bacteria that may have negative effects on plant growth and development without infection of root tissues. The mechanisms by which such deleterious rhizosphere bacteria operate include the production of phytotoxins and phytohormones, competition for nutrients, and the inhibition of mycorrhizal fungi.

The rhizosphere microbial community has some capacity to protect the host plant against pathogens. Although many agrochemicals are available to control root disease, concerns over the environmental and human-health implications of pesticide use, and the desire for sustainable agricultural systems, have been a driving force behind great interest in developing biological disease control based on rhizosphere antagonists of pathogens. A number of biocontrol agents are available as commercial formulations for use in agriculture and horticulture.

There are many examples of bacteria that can suppress growth of pathogenic fungi in the rhizosphere. Effective colonization of the root is a key factor determining the ability of these bacteria to exert biocontrol. A number of these bacteria produce antifungal metabolites, including antibiotics (**Table 2**), extracellular enzymes, and hydrogen cyanide (HCN). Competition between rhizosphere bacteria and fungal pathogens for nutrients has also been identified as a biocontrol mechanism. For example, the sequestration of Fe^{3+} by bacterial siderophores and chelators can limit availability of the nutrient to pathogens, restricting their growth through the rhizosphere. Exposure of roots to nonpathogenic rhizosphere bacteria, including strains of *Bacillus* spp. and *Pseudomonas* spp., can induce resistance of host plants to some pathogenic fungi. Several mechanisms have been implicated in induced resistance, including

Table 2 Selected rhizosphere disease biocontrol agents and the antibiotic mechanisms involved

Biocontrol agent	*Antibiotic compound*	*Pathogen/disease*	*Host*
Agrobacterium radiobacter	Agrocin 84	*Agrobacterium tumefaciens* (crown gall)	Stone fruit (drapes), roses (*Rosa* spp.)
Pseudomonas fluorescens 2-79	Phenazine-1-carboxylate	*Gaeumannomyces graminis* var *tritici* (take-all)	Wheat
Pseudomonas fluorescens F113	2,4-Diacetylphloroglucinol	*Pythium ultimum, Fusarium oxysporum, Phoma beta, Rhizopus stolonifera* (damping-off)	Sugar beet (*Beta vulgaris*)
Pseudomonas fluorescens CHAO	HCN, 2,4-diacetylphloroglucinol, pyoluteorin	*Thielviopsis basicola* (black root rot)	Tobacco (*Nicotiana tabacum*)
Pseudomonas fluorescens Pf-5	Pyrrolnitrin, pyoluteorin	*Pythium ultimum, Rhizoctonia solani* (damping-off)	Cotton (*Gossypium* spp.)

Adapted from Brimecombe MJ, de Leij FA, and Lynch JM (2001) The effect of root exudates on rhizosphere microbial populations. In: Pinton R, Varanini Z, and Nannipieri P (eds) *The Rhizosphere. Biochemistry and Organic Substances at the Soil–Plant Interface*, pp. 95–140. New York: Marcel Dekker.

enhanced production of phytoalexins, stress-related proteins and degradative enzymes, and the strengthening of epidermal cells.

Many rhizosphere fungi, including mycorrhizal fungi (see below) are able to suppress soil-borne plant pathogens. Fungi have the advantage over bacterial biocontrol agents in that they are generally more effective at spreading through the soil and rhizosphere. A variety of mechanisms are involved in control of fungal pathogens by rhizosphere fungi, including competition for nutrients, antibiotic production, and induced resistance. In addition, many fungi are able to parasitize spores, sclerotia, or hyphae of other fungi, resulting in biocontrol. Mycoparasitism is initiated by host sensing, which is generally followed by directed growth toward it, recognition, penetration, and degradation. Production of a number of degradative enzymes, including chitinases, proteases, and glucanases is involved in the biocontrol process.

Mutualistic Interactions

A significant proportion of plant assimilate is directed to specific symbiotic microbes that inhabit the plant root. These mycorrhizal and nitrogen-fixing mutualistic symbiotic associations between plant roots and soil microbes have profound influences on plant growth and development. In both symbioses, the plant provides the microbe with carbon, in return for which the microbial partner provides the host with nutrients.

Mycorrhizae Mycorrhizae (from German; "myco" meaning fungus, "rhiz" meaning root) are mutualistic symbioses between plant roots and soil fungi that involve bidirectional flow of nutrients; carbon from the plant host to the fungus and nutrients from the fungus to the plant. Over 80% of land plants are able to form mycorrhizae, and for these plants, mycorrhizae, not roots, represent the primary nutrient-absorbing organs. A number of plant families are predominantly nonmycorrhizal, including the Brassicaceae, Chenopodiaceae, Caryophyllaceae, Juncaceae, and Cyperaceae. Nonmycorrhizal plants are more abundant in harsh habitats, such as saline and arid soils.

Between 4% and 20% of net photosynthate can be transferred from the plant to its fungus partner. Fungal hyphae radiate into the soil from mycorrhizal roots and improve the effectiveness of nutrient uptake relative to nonmycorrhizal plants. This can arise from a number of mechanisms:

- Fungal hyphae extend beyond the area of nutrient depletion surrounding the root.
- Fungal hyphae greatly increase the surface area for absorption of nutrients relative to nonmycorrhizal roots.
- Hyphae are able to extend into soil pores that are too small for roots to enter.
- Some mycorrhizal fungi can access forms of N and P that are unavailable to nonmycorrhizal plants, particularly organic forms of these nutrients.

Additionally, mycorrhizal fungi are able to provide protection to the host plant against root and shoot pathogens, with a number of mechanisms proposed. These include antibiotic production, induced resistance, competition for root infection sites, and providing a physical barrier to infection.

A single plant root may be colonized by many different mycorrhizal fungi; mycorrhizal fungi generally have low specificity and are able to colonize a variety of different plant species. There is clearly niche differentiation among mycorrhizal fungi and great variation between fungal species in terms of the benefits they are able to provide their host. The mycelial network in the soil can link plants of the same and different species and there is some evidence that carbon can be transferred between individual plants via this network. The mycorrhizosphere can have a distinct bacterial community structure relative to the rhizosphere and bulk soil. Mycorrhizosphere inhabitants include intrahyphal bacteria in ectomycorrhizal fungi and intraspore bacteria in some arbuscular fungi. Some mycorrhizosphere bacteria can promote mycorrhiza formation, although the mechanisms involved are unclear. There are several categories of mycorrhiza, which are defined on the basis of structure and morphology.

Arbuscular mycorrhizae are the commonest group, and are found on angiosperms, gymnosperms, pteridophytes, and bryophytes. The fungus partners are members of the Zygomycetes in the order Glomales. Fungal hyphae undertake intracellular penetration of root cortical cells to form arbuscules, through which exchange of nutrients and carbon occurs. The mycelium of arbuscular fungi spreads into the soil where it differentiates to form hyphae with roles in colonization of new roots and nutrient uptake. The extramatrical mycelium of arbuscular mycorrhizal fungi (AMF) is particularly noted for uptake and translocation to the host of ions with low diffusion coefficients that are relatively immobile in the soil solution, such as P, Zn^{2+}, and Cu^{2+}. However, AMF may also have a role in providing mineral forms of N, K^+, and other nutrients to the host. There is evidence that AMF could play a role in the tolerance of some plants to heavy metal contamination, with the development of metal tolerance by the fungi and binding of metals to polyphosphate within fungal hyphae implicated. The diversity of AMF communities is critical to the maintenance of plant diversity, productivity, and nutrient uptake in natural ecosystems. Many farming practices, including fertilizer application, cultivation, and fumigation can have deleterious impacts on AMF communities, which are known to be less diverse and abundant in conventional agricultural systems relative to organically managed and seminatural areas. Commercial inocula of arbuscular mycorrhizal fungi are available for use in degraded habitats and agricultural systems, although application of these products has been relatively limited to date.

Ectomycorrhizae are found almost entirely on woody perennials, including members of the Pinaceae, Betulaceae, Fagaceae, and Dipterocarpaceae. Over 2000 fungus species are known to be capable of forming ectomycorrhizae, with most being basidiomycetes or ascomycetes. Ectomycorrhizal fungi and their hosts are mostly generalists, capable of forming symbioses with a range of partners. The fungus forms a mantle of hyphae on the outside of the plant root, which extends from the root into the surrounding soil. Hyphae penetrate into the cortex where they ramify between cells to form a structure called the Hartig net, through which materials are exchanged between the symbionts. The primary function of the fungal mycelium is absorption of nutrients from the soil and the translocation of these materials to the host. Ectomycorrhizal fungi have been shown to translocate mineral forms of N, P, and micronutrients from the soil to the host. Some ectomycorrhizal fungi typical of temperate and boreal forests are able to produce a suite of extracellular enzymes that mobilize organic forms of N and P, which are otherwise unavailable to the host plant. In some countries, an inoculum of selected ectomycorrhizal fungi is used in nurseries to enhance the growth and survival of seedlings.

Nitrogen fixation There are many different symbiotic associations between nitrogen fixing bacteria and plant roots. The most significant of these for agriculture is the Fabaceae–*Rhizobium* spp./*Bradyrhizobium* sp. root nodule symbioses. The bacteria persist in a dormant or saprophytic state in the soil before infecting a suitable root via the root hair. Legume roots exude various flavonoid and isoflavonoid molecules that induce expression of *nod* (nodulation) genes by such rhizobial bacteria. This results in the formation by the bacterium of lipo-oligosaccharide Nod factors, the precise structure of which determines the host range and specificity of *Rhizobium* spp.

Once inside the plant, an infection thread, consisting of an infolded plasma membrane, forms around the bacterium. Bacteria multiply inside the infection thread, which grows into the inner cortical cells, where bacteria are released into the cytoplasm. Here, root cells are induced to divide, resulting in formation of a root nodule containing enlarged nonmotile bacterial cells. The host plant provides the bacteria with carbohydrates. In return, rhizobial bacteria fix nitrogen from the atmosphere into NH_4^+, via nitrogenase. The NH_4^+ is converted into amides or ureides, which are passed to the plant xylem. Simultaneous infection by arbuscular mycorrhizal fungi stimulates the Fabaceae–*Rhizobium* symbiosis, enhancing the number and size of nodules, thereby increasing nitrogen fixation. In Fabaceae field crops, the amount of nitrogen fixed in nodules can be as much as $120\,kg\,N\,ha^{-1}$ in a growing season. Seed and soil inoculants of rhizobia are used for many crops, including soybean (*Glycine max*), bean

(*Phaseolus vulgaris*), peanut (*Arachis hypogea*), and clover (*Trifolium* spp.).

Recent work on root nodule bacteria has demonstrated that rhizobia are not restricted to *Rhizobium*/*Bradyrhizobium*, with nitrogen-fixing strains of *Ralstonia*, *Burkholderia*, and *Methylobacterium* isolated from nodules of some tropical Fabaceae. The common characteristic among rhizobia appears to be the possession of *nod* genes. A variety of bacterial genes involved in symbiotic nitrogen fixation have been characterized. In many rhizobia, these genes are located on plasmids, and in several instances *nif* (nitrogen fixing), *nod*, and other genes involved in nodule formation and functioning are clustered together. The genomes of a number of rhizobial species have been sequenced and genomic sequencing of the model host *Lotus japonicus* is in progress. These developments should allow progress in understanding the molecular mechanisms underlying the functioning of the root nodule symbiosis.

Phytoremediation

It is well established that biodegradation of numerous xenobiotics, including pesticides, polyaromatic hydrocarbons, and oils, is enhanced in the rhizosphere relative to nonrhizosphere soil. Most of the enhanced biodegradation can be attributed directly to enhanced cometabolic degradation resulting from the elevated microbial community surrounding roots. However, the production of surfactants and other metabolites by plant roots and their associated microbial communities could also play a role in enhancing the solubilization of xenobiotics, increasing their availability for biodegradation.

In the case of sites contaminated with heavy metals and radionuclides, phytoextraction has been proposed as a potential remediation mechanism. Solubilization of the contaminants by root and microbe-derived organic acids and siderophores results in the uptake of the metal into the shoot biomass of hyperaccumulator species, which can then be removed from the site.

List of Technical Nomenclature

Allelopathy	The deleterious impact of one plant on another.
Ectorhizosphere	The region of the rhizosphere external to the root.
Endorhizosphere	The region of the rhizosphere within the root.
Mycorrhiza	The mutualistic symbiosis between specialized soil fungi and plant roots.
Mycorrhizosphere	The rhizosphere region within and around structures of mycorrhizal fungi.
Rhizodeposition	Exudation of carbon by the plant root.

See also: **Diseases**: Fungal Diseases. **Nutrition**: Mineral Uptake; Nitrogen Fixation. **Photosynthesis and Partitioning**: Sources and Sinks. **Plants and the Environment**: Phytoremediation. **Root Development**: Mycorrhizae; Root Growth and Development.

Further Reading

Grayston SJ, Vaughan D, and Jones D (1996) Rhizosphere carbon flow in trees, in comparison with annual plants: the importance of root exudation and its impact on microbial activity and nutrient availability. *Applied Soil Ecology* 5: 29–56.

Jones DL (1998) Organic acids in the rhizosphere – a critical review. *Plant and Soil* 205: 25–44.

Lugtenberg BJJ, Dekkers L, and Bloemberg GV (2001) Molecular determinants of rhizosphere colonization by *Pseudomonas*. *Annual Review of Phytopathology* 39: 461–490.

Lynch JM (ed.) (1987) *The Rhizosphere*. Chichester: Wiley Interscience.

McGrath SP, Zhao FJ, and Lombi E (2002) Phytoremediation of metals, metalloids and radionuclides. *Advances in Agronomy* 75: 1–56.

Nehl DB, Allen SJ, and Brown JF (1996) Deleterious rhizosphere bacteria: an integrating perspective. *Applied Soil Ecology* 5: 1–20.

Pinton R, Varanini Z, and Nannipieri P (eds) (2001) *The Rhizosphere: Biochemistry and Organic Substances at the Soil–Plant Interface*. New York: Marcel Dekker.

Powlson D, Bateman G, Davies K, Gaunt J, Hirsch P, and Barlow P (eds) (2001) Proceedings of the millenium conference on rhizosphere interactions. *Plant and Soil* 232, 1–233.

Rengel Z (2002) Breeding for better symbiosis. *Plant and Soil* 245: 147–162.

Smith SE and Read DJ (1997) *Mycorrhizal Symbiosis*, 2nd edn. London: Academic Press.

Whipps JM (2001) Microbial interactions and biocontrol in the rhizosphere. *Journal of Experimental Botany* 52: 487–511.

S

SECONDARY PRODUCTS

Contents

Lignin

K Meyer, DuPont Central Research and Development, Wilmington, DE, USA

Introduction

Lignin is an important constituent of the plant secondary cell wall. As such, it is an essential contributor to the rigidity and hydrophobic nature of specialized plant cells that provide mechanical support and carry out water transport. The presence of lignin has important implications for human utilization of plant-derived feedstocks. Environmental impact and energy requirements of processes for isolation of cellulose from plant material are strongly influenced by both quantity and quality of the lignin constituent. Moreover, lignin has an impact on the suitability of plant material as feedstock for microbial fermentation and animal feed. Modifications of lignin deposition in genetically modified plants have been the focus of efforts bridging the fields of plant physiology, biochemistry, and biotechnology. This article will outline physiological and cell biological aspects of lignification, followed by a brief overview of the chemical properties of the lignin polymer. Finally, a more detailed description of the enzymatic machinery that channels carbon to the lignin monomers will be provided.

Properties of Lignin

Lignin is a hydrophobic, three-dimensional plant polymer of very high, hitherto unknown, molecular weight, comprised of aromatic molecules with propene side chains, the monolignols. The three commonly found monolignols that build the bulk of the polymer are *p*-coumaryl alcohol, coniferyl alcohol, and sinapyl alcohol (**Figure 1A**). The ability to synthesize monolignols and polymerize them oxidatively into lignin is thought to have evolved with the colonization of land by plants. Lignins are found in pteridophytes (ferns) and all higher plants (gymnosperms and dicotyledonous as well as monocotyledonous angiosperms). **Figure 2** illustrates the distribution of the lignin polymer in the stem of the herbaceous angiosperm *Arabidopsis thaliana* as revealed by histochemical staining. Lignin is found in the extracellular matrix of the xylem cells that function in water transport. Lignification confers compressive strength to the walls that allow the water-conducting vessel elements to withstand the negative pressure associated with water transport facilitated by transpiration. In addition to this, lignin has a central role in waterproofing the walls of these elements. Lignin is also found in the sclerified parenchyma or fiber tissue. This tissue has a central role in providing the stem with the mechanical strength required for an upright growth habit. In woody plants, where the vascular tissue is formed by a ring of meristematic cells (the cambium), lignin is

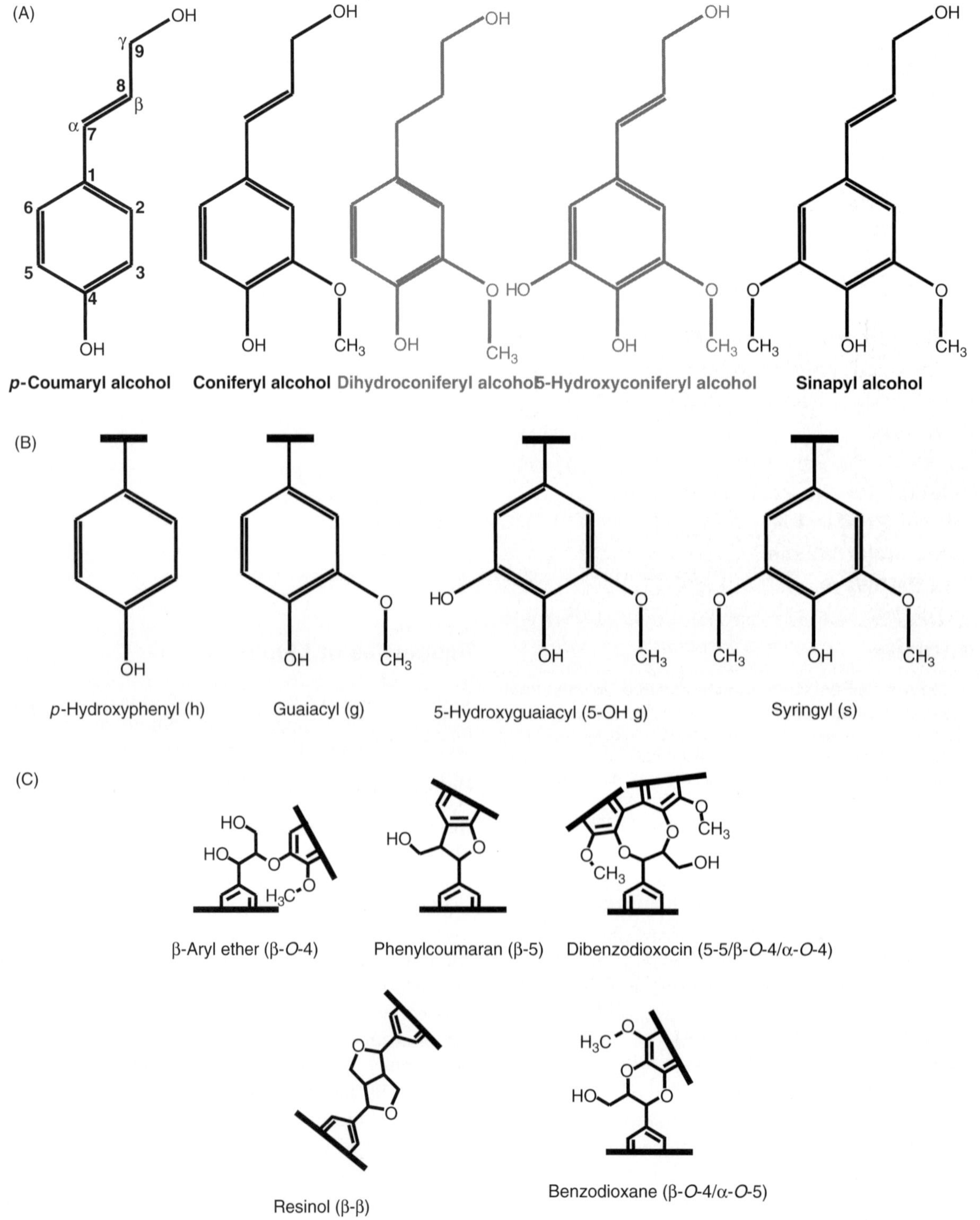

Figure 1 (A) Monolignols. The numbering of the carbon atoms of the monolignol molecule and specific labeling of the carbon atoms of the propene side chain are indicated. Monolignols highlighted in gray are minor constituents of the monolignol pool in wild-type plants but become a significant constituent of the monolignol pool in transgenic plants with altered monolignol biosynthesis. (B) Lignin monomers. (C) Interunit linkages in the lignin polymer.

abundant in the walls of the vessel elements and the surrounding fiber cells of the angiosperms and the homologous tracheids and fiber cells of the gymnosperms.

The extracellular assembly of the lignin polymer succeeds the deposition of the primary cell wall, and in the case of vessel elements and tracheids, is ultimately followed by cell autolysis. There is a

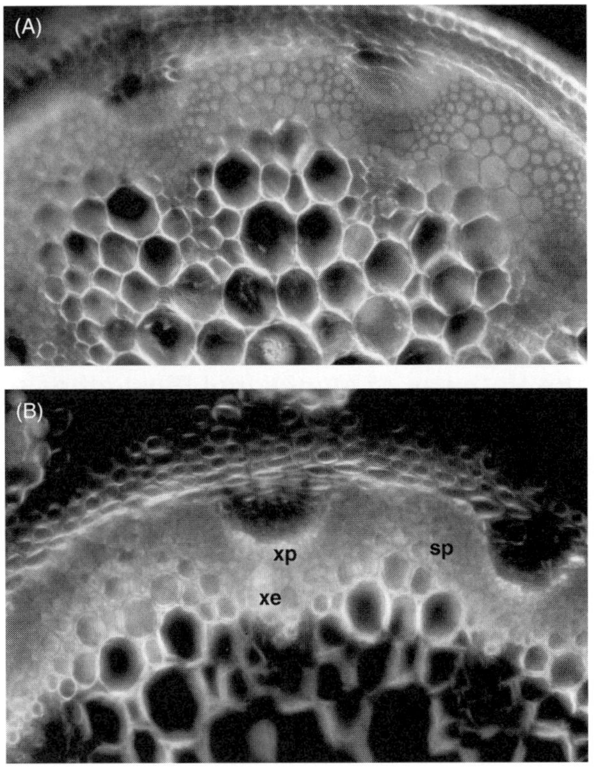

Figure 2 (see color plate 35) Distribution and monomeric composition of lignin in a herbaceous plant as revealed by histochemical staining. (A) Cross-section of an *Arabidopsis* stem stained with phloroglucinol. Presence of aldeyhyde end groups in lignin produce a characteristic pink color. (B) Cross-section of an *Arabidopsis* stem stained with the mäule reagent. This reagent gives a yellow color with guaiacyl lignins (g lignins) and a red color in the presence of syringyl residues (g/s lignins). xe, xylem elements; xp, xylem parenchyma; sp, sclerified parenchyma.

general pattern of spaciotemporal lignin deposition, which is conserved in most lignifying cells of angiosperm and gymnosperms. Lignification starts at the middle lamella in the S1 region of the primary cell wall at postulated lignin initiation sites after completion of the S2 region of the secondary cell wall. The lignin polymer permeates the matrix polysaccharides of primary and cellulose fibrils and xylans of the secondary cell wall. Finally, lignification proceeds through the S2 and S3 regions of the secondary cell wall.

Methods of Lignin Analysis

The complexity of the lignin polymer and its natural resistance to chemical degradation dictates that there is no single analytical method that can provide a complete picture of monomer composition, as well as the nature and relative abundance of particular monomer linkages. The current picture of the lignin polymer and its variability among different plants, tissues, cells, and even on the subcellular level within a lignified cell wall has been attained through various methods. These include histochemical staining (**Figure 2B**), oxidative degradation, microautoradiography using radiolabeled lignin precursors, and, more recently, immunohistochemistry performed using antibodies raised against lignin epitopes created *in vitro*. Moreover, recent advances in the determination of the relative abundance of interunit linkages of isolated lignin polymers stems from the application of NMR techniques.

Definition of Lignin

A clear definition of what constitutes lignin is still hotly debated at this point in time. Recent controversy has been triggered by the unusual monomeric composition of lignins in plants with perturbations of the monolignol pathway (discussed in more detail below) and the ensuing question of whether these unusual monomers are truly part of the lignin polymer. The following operational definition of lignin was recently proposed by the research groups of John Ralph and Ron Sederoff: "If a phenolic molecule derived from the monolignol-specific pathway couples via dehydrogenative polymerization with a growing aromatic polymer or oligomer of phenylpropanoids and is further integrated into the polymer via endwise coupling with new monolignols or monomers, the resulting polymer constitutes lignin."

Heterogeneity of Lignin Monomer Composition

There are significant differences in the lignin monomer composition of gymnosperms (i.e., softwoods) and angiosperms (i.e., hardwoods and lignins of herbaceous plants). With very few exceptions, gymnosperm lignins are comprised only of monomers derived from *p*-coumaryl alcohol and coniferyl alcohol (referred to as guaiacyl (g)- type lignins), whereas angiosperm lignins are assembled from all three monolignols (referred to as guaiacyl/syringyl (g/s)- type lignins) (**Figure 1B**). Lignin monomer composition also varies dramatically between cells of different tissue types (**Figure 2B**) and even within the wall of a given cell. For example, during the development of a tracheid cell, a lignin rich in *p*-hydroxyphenyl residues is first deposited in the cell corners at the middle lamella during the early stages of lignification, whereas at later stages of lignification in the S2 and S3 layers constituting the secondary cell wall, a lignin assembled from coniferyl alcohol monomers is deposited. A predominant tissue-specific pattern of lignin deposition in the softwood of angiosperm trees consists of guaiacyl-type lignins in the cell walls of vessel elements and guaiacyl/syringyl-type lignins in the cell walls of fiber and xylem parenchyma cells. Thus, in certain hardwood species, g/s ratios can vary more than fifty-fold, ranging from 88:12 in cell walls of vessel elements to 12:88 in fiber cell walls.

Lignin Polymerization and Interunit Linkages

Apoplastic polymerization of monolignols is considered to be a spontaneous process that succeeds the one electron oxidation of the monomers catalyzed by peroxidases and/or oxidases (discussed in more detail below). The reactivity of the monolignol and its ability to form various linkages in the polymerization process is determined by the substitution pattern of the aromatic ring. **Figure 1C** shows common interunit linkages found in lignin. In the native lignin polymer, the monomer residues are linked to one another in as many as 10 different ways through C–O bonds, comprised of alkyl aryl and diaryl ether structures and C–C bonds including alkyl and/or aryl moieties. Surprisingly, the β-aryl ether linkage (β-O-4) makes up more than 50% of the interunit linkages found in guaiacyl- as well as guaiacyl/syringyl-type lignins. Otherwise, phenylcoumaran (β-5), dibenzodioxocin (5-5/β-O-4,α-O-4), and 1,2-diarylpropane (β-1) linkages together constitute the second most prominent types of interunit linkages. Phenylcoumaran and dibenzodioxocin linkages clearly require the presence of coniferyl alcohol-derived monolignol residues. Hence, these interunit linkages are more abundant in gymnosperm lignins and angiosperm lignins that lack syringyl monomers.

Lignins can be polymerized *in vitro* using the monolignols and a suitable oxidizing enzyme (see below). There is a significant difference in the relative abundance of the various interunit linkages between lignins synthesized *in vitro* and lignin isolated from plant material. As mentioned above, the latter shows a strong predominance of certain β-aryl ether (β-O-4) linkages. A template polymerization mechanism has been proposed to account for this, in which the lignin polymer grows in a linear, albeit slightly branched, fashion from postulated lignification initiation sites in the primary cell wall. According to this model, the predominant reaction of early lignification *in vivo* is the addition of monolignols to a growing polymer chain. The lignification initiation sites have been proposed to consist of certain polysaccharide epitopes, i.e., hydroxycinnamic acid esterified to the cell wall, arrays of "dirigent" cell wall glycoproteins, and calcium enriched sites, either alone or in combination. The direction of lignification in the secondary cell wall has been shown to be influenced by the direction of cellulose microfibrils and the hemicellulose network of the cell wall. The hemicellulose fraction forms the interface between cellulose and lignin by embedding the cellulose microfibrils and forming covalent linkages to lignin. Another important contributor to the difference in interunit linkages between lignins generated *in vitro* and those isolated from plant sources can be attributed to the exquisite control that the lignifying cell exerts over monolignol supply. Thus, at a given point in the lignification process *in vivo*, the carefully controlled balance of certain monolignols provided by the cell may promote the formation of only certain interunit linkages.

The Pathway of Monolignol Biosynthesis

Monolignols are synthesized through the phenylpropanoid pathway from phenylalanine (and tyrosine in monocotyledonous angiosperms, i.e., grasses). RNA blot experiments, expression of reporter gene constructs, and immunolocalization studies of enzymes of the phenylpropanoid pathway suggest that the bulk of monolignols is produced in the cells that undergo lignification. Hence there is only a limited transfer of monolignols from neighboring xylem or ray parenchyma cells to tracheids or vessel elements, presumably at later stages of cell differentiation to sustain lignification after the water-conducting cells have undergone autolysis. Moreover, there is a close correlation between the expression pattern of enzymes of the phenylpropanoid pathway and those of the shikimate pathway; this suggests that monolignols are formed in nonphotosynthetic lignifying cells from erythrose 4-phosphate (an intermediate of the pentose phosphate pathway) and phosphoenolpyruvate (an intermediate of glycolysis), thereby suggesting a close correlation between lignin deposition and carbohydrate metabolism. Thus, the carbon currency that feeds lignin biosynthesis seems to be sucrose, which is translocated from the source leaves via the phloem. Despite the importance of lignin as a major sink for photosynthetically fixed carbon (about a third of the woody biomass of the world is comprised of lignin), we are at a very early stage in understanding the principles and mechanisms regulating the flow of carbon from sucrose to monolignols. A first step toward elucidation of these mechanisms has been made recently by demonstrating that the activity of both a plastid-localized transketolase and a phospenolpyruvate-phosphate translocator have a strong influence on the accumulation of phenylpropanoids, including lignin.

The following will summarize the sequence of interconversions that link primary metabolism (i.e., biosynthesis of aromatic amino acids) to lignin deposition. The focus of the summary will be on the description of the properties of the enzymatic machinery. This involves a description of subcellular localization, substrate specificity, and kinetic properties of the enzymes and tissue specificity of their expression. Moreover, the effects of altered expression of the respective proteins on lignin properties will be summarized. Prior knowledge of the primary features of the lignin polymer allows one to infer some of the properties of the biosynthetic machinery providing lignin monomers. For instance, it must be a high-flux pathway, as is reflected by the high catalytic center activity of the respective enzymes. Furthermore, the dynamics of the lignification process outlined above suggest that the biosynthetic machinery must have built-in mechanisms that facilitate precise control of the hydroxylation and methylation pattern of the aromatic monomers produced at a given stage of differentiation.

The current picture of the monolignol pathway depicted in **Figure 3** was formed after several recent and very significant revisions, including the cloning of the gene encoding the 3-hydroxylase enzyme that generates the 3,4-hydroxylated caffeoyl intermediate. Characterization of the kinetic properties and substrate specificity of this enzyme revealed that shikimate and quinate esters of the 4-hydroxylated coumaryl moiety constitute the preferred substrate of the 3-hydroxylase enzyme. Thus, 3-hydroxylation of the aromatic ring requires two enzymatic steps: the transfer of the acyl moiety from coenzyme A (CoA) to quinic or shikimic acid and the hydroxylation of the aromatic ring. From these results, the monolignol pathway may be dissected into segments based on the state of oxidation of the involved intermediates. Early interconversions (i.e., introduction of the 4-hydroxyl group) occur at the level of the free acid (cinnamic acid). Subsequent introduction of the 3-hydroxyl group and its methylation takes place at the level of the hydroxycinnamic acid esters (hydroxycinnamoyl CoA thioesters and shikimate and quinate esters). The final introduction of the 5-hydroxyl group and its methylation takes place after the hydroxycinnamoyl esters have been reduced to hydroxycinnamyl aldehydes and alcohols.

Enzymes of Monolignol Biosynthesis

Phenylalanine/tyrosine ammonia lyases (PAL, TAL) (E.C.4.3.1.5) The PAL enzymes catalyze the one-step deamination of phenylalanine to cinnamic acid at the entry point into the pathway. The active PAL

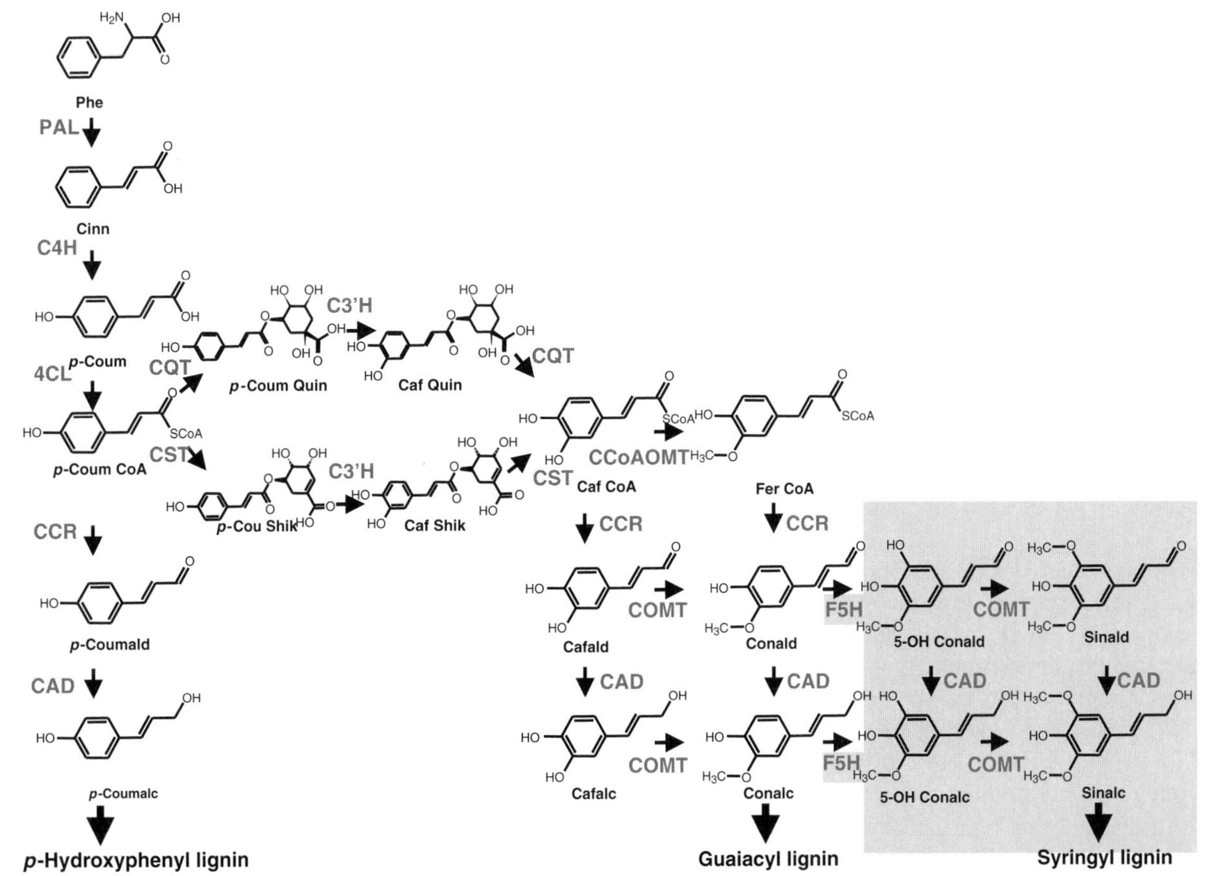

Figure 3 The biosynthetic pathway to monolignols. Interconversions (indicated by shaded background) are absent in most gymnosperms.

enzymes are operationally soluble cytoplasmic homotetramers of subunits that range in size from 60 to 80 kDa. Recombinantly produced PAL enzymes from plants show Km values for phenylalanine in the low $\mu mol\,l^{-1}$ range (10–30 $\mu mol\,l^{-1}$) and catalytic center activities of 10–30 s^{-1}. The PAL enzyme(s) of grasses also have significant activity toward tyrosine. The extent to which tyrosine ammonia lyase (TAL) activity contributes to monolignol biosynthesis in monocotyledonous plants is unknown. PAL enzymes are encoded by a small family of closely related genes, implying redundancy at this step. The spatiotemporal expression pattern of PAL genes is closely correlated with monolignol biosynthesis. Significant reductions in PAL activity have been generated in transgenic plants through cosuppression or antisense. These reductions are accompanied by a significant decrease in lignin content.

Cinnamate 4-hydroxylase (C4H) (E.C.1.14.13.11) C4H catalyzes the 4-hydroxylation of the aromatic ring of cinnamic acid. C4H is a cytochrome P450-dependent monoxygenase encoded by a single gene in most plants. It is a membrane-bound 58-kDa heme-containing protein that interacts closely with a cytochrome P450 reductase that is essential for catalysis. There is a surprisingly high level of sequence conservation of primary amino acid sequences (greater than 80%) of C4H proteins from divergent plant species. C4H proteins, heterologously produced in yeast microsomes, have Km values ranging between 2 and 10 $\mu mol\,l^{-1}$ for cinnamic acid and turnover numbers ranging from 5 to 8 s^{-1}. The central role of 4-hydroxylation in monolignol synthesis and the lack of molecular and biochemical redundancy at the level of C4H catalysis is indicated by the strong impact that is exerted by disruptions in C4H expression and/or function on monolignol production and lignin deposition. Like PAL, the spatiotemporal pattern of C4H gene expression is found to be closely correlated with monolignol biosynthesis, i.e., high levels of C4H expression are observed in lignifying tissues.

Hydroxycinnamyl CoA ligases (4CL) (E.C.6.2.1.12) 4CL enzymes are operationally soluble, monomeric enzymes of 60-kDa molecular weight belonging to the class of adenylate-forming CoA ligases. There is clearly redundancy at the level of 4CL enzyme activity, both in gymnosperms and angiosperms. 4CL proteins of the angiosperms belong to two groups of evolutionary divergent sequences. For example, in *Arabidopsis* there are three distinct 4CL proteins that share only 60% sequence identity. Km values ranging from 10 to 25 $\mu mol\,l^{-1}$ and turnover numbers of 5–15 s^{-1} have been determined for hydroxycinnamic acid substrates and recombinantly produced 4CL proteins. A close correlation exists between monolignol synthesis and the expression of 4CL enzymes. Transgene-mediated down-regulation of 4CL activity is accompanied by moderate reductions in lignin quantity, presumably as a result of functional redundancy.

***p*-Hydroxycinnamoyl CoA:shikimate/quinate *p*-hydroxycinnamoyl transferases (CST, CQT)** Although these enzyme activities have been demonstrated in a wide array of plant species, including those that do not accumulate soluble shikimate or quinate esters of hydroxycinnamic acids as soluble secondary products, the central role of these enzymes in monolignol formation went unrecognized until the recent discovery of the substrate specificity of the enzyme introducing the 3-hydroxyl group of the aromatic ring (see below). The soluble enzymes have been partially purified from various plant sources; however, no report could be found at this point for the cloning of the corresponding genes. CST and CQT enzymes catalyze both forward and reverse reactions. They form shikimate or quinate esters from the hydroxycinnamoyl CoA thioesters and free shikimate and quinate as part of the forward reaction, and hydroxycinnamoyl CoA thioesters from free CoA and hydroxycinnamoyl shikimate and quinate esters as part of the reverse reaction. Km values of partially purified CST enzymes have been reported to be in the low $mmol\,l^{-1}$ range for the hydroxycinnamoyl CoA thioesters, the low $\mu mol\,l^{-1}$ range for the acyl acceptors in the forward reaction, and in the low $\mu mol\,l^{-1}$ range for both substrates in the reverse reaction. Clearly one of the next milestones in the study of plant monolignol biosynthesis will be the isolation of the CST and CQT genes, the determination of their tissue-specific expression and analysis of substrate specificity, and kinetic properties of the recombinantly produced enzymes.

Hydroxycinnamoyl shikimate/quinate 3-hydroxylase (C3′H) C3′H was only very recently cloned from *Arabidopsis*. Its expression pattern in this plant is virtually identical to that of C4H, the other "early" hydroxylase in the monolignol pathway. Like C4H, the enzyme is an ER-localized cytochrome P-450-dependent monoxygenase of 58 kDa. The enzyme was expressed in yeast microsomes to determine its kinetic properties and substrate specificity. The heterologously produced enzyme shows Km values and turnover numbers of 7 $\mu mol\,l^{-1}$ and 10 s^{-1} and 18 $\mu mol\,l^{-1}$ and 6 s^{-1} for coumaryl shikimate and coumaryl quinate, respectively, indicating that the shikimate ester is the preferred substrate. There is no molecular or biochemical redundancy at this biosynthetic step in *Arabidopsis*, currently being the only plant in which this enzyme has been described at the biochemical and molecular level. Disruption of the C3′H gene in an *Arabidopsis* mutant results in complete blockage of monolignol production after introduction of the 4-hydroxyl group. Plants that lack C3′H activity produce lignin that is entirely comprised of *p*-hydroxyphenyl residues.

Caffeoyl CoA *O*-methyltransferase (CCoAOMT) (E.C.2.1.1.104) Once produced by the reverse reaction of CST or CQT, caffeoyl CoA can have one of two fates (see **Figure 3**). Caffeoyl CoA can be reduced to caffeyl aldehyde (discussed below) or it can be methylated at the 3-hydroxy group to give feruloyl CoA. The latter reaction is catalyzed by *S*-adenosylmethionine-dependent *O*-methyltransferase activities that exhibit a high level of specificity for hydroxycinnamoyl CoA thioesters. Enzymes of this class are encoded by small gene families in plants. They are operationally soluble homodimers comprised of 27–30 kDa subunits. Km values of caffeoyl CoA O-methyltransferase (CCoAOMT) for caffeoyl CoA of $5\,\mu mol\,l^{-1}$ and catalytic center activities ranging from $2\,min^{-1}$ to $2\,s^{-1}$ have been reported. Sequences encoding the catalytic subunits of CCoAOMTs share remarkable levels of sequence conservation (greater than 80% sequence identity) in diverse plant species. Disruption of this enzymatic step leads to reduced lignin deposition, which has a somewhat more pronounced effect on guaiacyl than syringyl lignin deposition. This finding has led some researchers to suggest that methylation of caffeoyl CoA seems to be part of a coniferylalcohol-specific "metabolic channel" of the monolignol pathway.

Hydroxycinnamoyl CoA reductases (CCR) (E.C.1.2.1.44) These enzymes reduce hydroxycinnamoyl CoA thioesters to hydroxycinnamylaldehydes. They are monomeric, operationally soluble 38-kDa enzymes that belong to the class of nicotinamide adenine dinucleotide phosphate (NADP)-dependent oxidoreductases. Two CCR genes that share around 80% sequence identity have been isolated from most plant species studied so far. Complete disruption of one CCR gene (*IRX4*) in *Arabidopsis* by mutation leads to a reduction in lignin deposition of more than 50%, indicating that there is some, albeit limited, redundancy at the CCR step. Km values and turnover numbers for the four hydroxycinnamoyl CoA substrates of the enzymes ranging from 28 to $45\,\mu mol\,l^{-1}$ and 3 to $12\,min^{-1}$ have been reported.

Hydoxycinnamyl alcohol dehydrogenases (CAD) (E.C.1.1.1.195) CAD activities in plants reduce hydroxylated or methoxylated cinnamaldehydes to the corresponding alcohols. These enzymes belong to the large family of plant alcohol dehydrogenases. The first class of plant CAD enzymes that was described at the molecular level is referred to as the CAD2 family of hydoxycinnamyl alcohol dehydrogenases. Proteins of this class are operationally soluble dimeric enzymes with molecular weights of 38 kDa per monomer that exert activity toward a wide range of cinnamaldehydes with the lowest Km (4 μmol) for coniferylaldehyde. Members of this protein family in angiosperms and gymnosperms share close to 60% sequence identity. The expression pattern of the CAD2 protein agrees well with the important role of this enzyme in monolignol biosynthesis.

The other distinct class of CAD enzymes that may be involved in lignin biosynthesis is the CAD1 class of plant alcohol dehydrogenases. These are soluble, monomeric 35-kDa proteins that exert activity toward coniferylaldehyde and 4-coumaryl aldehyde.

Disruption of CAD2 activity in angiosperms does not affect lignin content. Lignin quality, however, is affected through the incorporation of monomers derived from coniferyl and sinapyl aldehyde into the lignin polymer. The most dramatic reduction in CAD activity has been achieved in a mutant gymnosperm tree. Less than 1% of wild-type CAD activity in this plant is accompanied by the incorporation of dihydroconiferyl alcohol monolignols (**Figure 1A**) into the lignin polymer.

Hydroxycinnamyl alcohol/aldehyde *O*-methyltransferases (COMT) (E.C.2.1.1.68) These enzymes catalyze the methylation of both the 3- and 5-hydroxyl groups of monolignol intermediates (i.e., hydroxycinnamic acids, alcohols, and aldehydes) *in vitro*. They are operationally soluble homodimers comprised of 40 kDa subunits that are encoded by small gene families in angiosperms. Like CCoAOMT, these enzymes show signature sequences of *S*-adenosylmethionine-dependent O-methyl transferases. However, sequence similarity of these proteins to CCoAOMT does not extend significantly beyond these regions. Km values and catalytic center activities of these enzymes for caffeoyl and 5-hydroxyconiferyl aldehyde and alcohol ranging between 2 and $12\,\mu mol\,l^{-1}$ and $0.2–0.8\,s^{-1}$ have been reported. The kinetic parameters for aldehyde and alcohol substrates are significantly better than those found for the corresponding free acids. Activity of COMT enzymes toward caffeylaldehyde and caffeyl alcohol may suggest that another metabolic route to coniferyl alcohol may be operating in parallel to that acting at the level of CoA thioesters (discussed above and see **Figure 3**). This route would be initiated by the CCR- (and CAD)- catalyzed reduction of caffeoyl CoA to caffeylaldehyde (and caffeyl alcohol), followed by the COMT-mediated methylation yielding coniferylaldehyde (and coniferyl alcohol). Transgenic downregulation experiments of COMT activity suggest that the more important role of COMT activity is the methylation of 5-hydroxyconiferylaldehyde and alcohol, which is part of sinapyl alcohol synthesis. Disruption of COMT activity has little influence on lignin quantity. However, in angiosperms it is accompanied by complete disruption of sinapyl alcohol production and syringyl lignin deposition. This observation is in complete agreement with the substrate preference profile of the hydroxylase introducing the 5-hydroxy group (F5H), which is part of sinapyl alcohol production (discussed below). The resulting lignin also contains 5-hydroxyguaiacyl monomeric units (**Figure 1B**) derived from the incorporation of 5-hydroxyconiferyl alcohol by lignin. The polymerization of lignins with 5-hydroxyconiferyl alcohol monomers (**Figure 1A**) is accompanied by the occurrence of unusually high levels of benzodioxane linkages, which are the product of radical coupling of 5-hydroxyguaiacyl residues (**Figure 1C**).

Ferulate 5-hydroxylase (F5H) The F5H hydroxylase is part of the monolignol pathway branch operating at the level of coniferylaldehyde and alcohol that leads to sinapyl alcohol production. Like other hydroxylases of the monolignol pathway, the enzyme is present as a single gene in the angiosperms that have been studied so far. F5H genes from angiosperms share more than 60% sequence identity. The F5H protein shows the typical signature sequences of cytochrome P-450-dependent monoxygenases with an N-terminal membrane anchor. Although the enzyme accepts ferulate as a substrate ($Km = 1\,mmol\,l^{-1}$), its preference for coniferylaldehyde and alcohol is pronounced with Km values of 1 and $3\,\mu mol\,l^{-1}$, respectively. At $3–4\,min^{-1}$, the turnover number of this enzyme is significantly lower than that of C4H and C3′H. Studies in transgenic plants demonstrate that tissue specificity of syringyl lignin deposition in angiosperms (see above) is controlled at the level of ferulate 5-hydroxylase expression. Expression of reporter genes under the control of the ferulate 5-hydroxylase promoter demonstrates that F5H expression coincides with deposition of syringyl-rich lignins (Clint Chapple, Max O Ruegger, unpublished observations). Using an F5H overexpression strategy, a lignin has been generated in some transgenic lines that contains less than 3% guaiacyl residues and about 10% 5-hydroxyguaiacyl residues, with the remainder entirely consisting of syringyl residues. This lignin is significantly less condensed than the typical angiosperm g/s-type lignin, i.e., devoid of carbon–carbon linkages that involve the C5 carbon of the aromatic ring. More specifically, the lignin lacks the phenylcoumaran (β-5) and dibenzodioxocin linkages (5-5/β-O-4,α-O-4) (see **Figure 1C**) of g-type and g/s-type lignins. A more linear, less branched lignin, highly enriched in β-aryl ether linkages is envisaged to exist and seems to function normally in these transgenic plants.

In summary, high levels of redundancy at most steps of the monolignol pathway are obvious. This redundancy, however, clearly does not extend to the three hydroxylases of the pathway. Thus, it is tempting to speculate that the single gene nature, especially of C3′H and F5H, was preserved during evolution because it is part of an intricate mechanism in which a central determinant of the future properties of the lignin polymer, its monomeric composition, is controlled at the level of transcription of a single gene.

Monolignol Conjugation

Hydroxycinnamoylalcohol glucosyltransferases (GT) (E.C.2.4.1.111) Coniferin and syringin, the 4-O-glucosides of coniferyl and sinapyl alcohol are considered the "transport" forms of monolignols. These glucosides are rarely detected in lignifying tissue, perhaps indicating that they are rapidly turned over or it may question their involvement in the sequence of reactions that lead to lignin deposition. The occurrence of these conjugates in some lignifying tissues may merely indicate their storage or detoxification at certain stages of cell or tissue differentiation when lignin deposition cannot keep pace with monolignol production. Very recently, two glucosyltransferase proteins have been identified from *Arabidopsis* that exert complementary substrate specificity profiles for coniferyl and sinapyl alcohol. These are soluble, monomeric 53-kDa proteins that are members of the large family of plant glucosyltransferases. The enzymes share 85% sequence identity, exhibit Km values for the two substrates below $300\,\mu mol\,l^{-1}$ and have turnover numbers of $6–8\,s^{-1}$.

Intracellular Transport of Monolignols to Sites of Polymerization

It is reasonable to assume that the monolignol synthesis machinery described above is cytosolic and somehow organized around the three cytochrome P-450-dependent monoxygenases of the pathway that are inserted into the cytosolic face of the endoplasmic reticulum (ER). Lignin polymerization takes place in the apoplast. Thus, a mechanism must exist that facilitates the cytoplasmic transport and secretion of monolignols into the apoplastic space. To date, the biochemical or molecular properties of components of this transport and secretion machinery have not been described. Some early studies, conducted using radiolabeled monolignol precursors and microautoradiography of monolignol-synthesizing tissues, describe the existence of ER- and Golgi-derived cytoplasmic, monolignol-containing vesicles.

Extracellular Polymerization

No single enzyme has been identified at the molecular level that has been shown to be both necessary and sufficient for lignin polymerization *in vivo*. Both laccases and peroxidases can be used to polymerize monolignols *in vitro*. Up to now, genetic approaches, as well as transgenic downregulation approaches using suitable peroxidase or laccase candidate genes, have failed to produce plants with disrupted lignin deposition. The difficulty in elucidating the role of individual laccases and peroxidases in lignin polymerization may lie in the high levels of redundancy at this step. Both laccases and peroxidases belong to large divergent gene families in plants.

Laccases (E.C.1.10.3.2) Laccases are members of the family of *p*-diphenol:O_2 oxidoreductases. These are copper-containing enzymes that reduce both atoms of molecular oxygen to H_2O (i.e., use molecular oxygen as the electron acceptor). Laccase proteins that are strong candidates for oxidizing enzymes in lignin polymerization have recently been described at the molecular and biochemical level. The genes are predominantly expressed in lignifying tissues. A targeting signal in the primary amino acid sequence of these proteins may suggest their secretion to the apoplastic space. Activity staining using coniferyl aldehyde (in the presence of catalase to eliminate peroxidase activity) provided further evidence for the involvement of laccase in lignin formation. Moreover, kinetic properties and substrate specificity of a 90-kDa laccase purified from developing xylem further substantiated evidence for the involvement of this enzyme in lignin deposition. Km values ranging from 2 to 25 $\mu mol\,l^{-1}$ and turnover numbers of 8–10 s^{-1} were reported for monolignol substrates.

Peroxidases (E.C.1.11.1.7) Peroxidases (hydrogen peroxide oxidoreductases) are heme-containing metalloproteins that use hydrogen peroxide (H_2O_2) as the electron-accepting cosubstrate. The evidence for their involvement in apoplastic oxidation of monolignols is as strong as that available for the involvement of laccases. Anionic peroxidase enzymes that are candidates for lignin oxidation show Km values for monolignols of 10–20 $\mu mol\,l^{-1}$. A strong correlation between H_2O_2 generation and lignin deposition has been observed in several plants. Like laccases, peroxidases that have an implied role in monolignol oxidation are glycoproteins with a putative signal peptide for apoplastic targeting.

Outlook

Future challenges in lignin research can be grouped into four main areas. First, the enzymes catalyzing the acyl transfer reactions in the synthesis of coumaryl shikimate and quinate intermediates of monolignol synthesis need to be identified at the molecular level and their biochemical properties need to be determined. Second, components of the monolignol transport and excretion machinery need to be identified. Third, the nature of the oxidizing catalyst(s) in lignin polymerization needs to be clarified. Finally, the components that integrate vascular development in plants with biosynthetic activity of the monolignol pathway need to be determined. Significant progress has recently been made in this area with the functional dissection of promoters of monolignol-specific genes and the description of the corresponding transcription factors. Moreover, additional proteinaceous and hormonal components controlling vascular development in plants are being discovered. The description of this field of research, however, was beyond the scope of this article.

It is evident that the field of lignin biochemistry has entered a phase that allows for implementation of several metabolic engineering strategies leading to reduced lignin quantity in genetically modified plants. The impact of this reduction on agricultural performance has yet to be determined. An alternative, safer strategy entails the alteration of lignin quality, by reducing the presence of interunit linkages, which are difficult to cleave. It has been demonstrated that this can be accomplished by shifting the monolignol supply to highly methoxylated monomers, which have fewer opportunities for intramolecular crosslinks.

List of Technical Nomenclature

Catalytic center activity The unit for catalytic power of multimeric enzymatic catalyst expressed as μmol of product formed per second per μmol of active enzyme complex, i.e., per active site.

Guaiacyl residue A monomeric lignin component derived from coniferyl alcohol.

Km	The unit for affinity of an enzyme for a particular substrate, expressed as substrate concentration at which 50% of maximum velocity is achieved.
Syringyl residue	A monomeric lignin component derived from sinapyl alcohol.
Tracheid	A specialized water-conducting cell of gymnosperm wood (softwood).
Turnover number	The unit for catalytic power of monomeric enzymatic catalyst expressed as μmol of product formed per second per μmol of enzyme.
Vessel element	A specialized water-conducting cell of angiosperm wood (hardwood).
Xylem	Water-conducting plant tissue.

See also: **Cell Walls and Fibers**: Cell Walls; Fiber Formation. **Growth and Development**: Cells. **Primary Products**: Cellulose.

Further Reading

Davin LB and Lewis NG (2000) Dirigent proteins and dirigent sites explain the mystery of radical precursor coupling in lignan and lignin biosynthesis. *Plant Physiology* 123: 453–461.

Dixon RA, Chen F, Guo D, and Parvathi K (2001) The biosynthesis of monolignols: a metabolic grid, or independent pathways to guaiacyl and syringyl units? *Phytochemistry* 57: 1069–1084.

Donaldson LA (2001) Lignification and lignin topochemistry – an ultrastructural view. *Phytochemistry* 57: 859–873.

Herrmann KM (1995) The shikimate pathway as entry to aromatic secondary metabolism. *Plant Physiology* 107: 7–12.

Lewis NG (1999) A 20th century roller coaster ride: a short account of lignification. *Current Opinion in Plant Biology* 2: 153–162.

Lewis NG and Yamamoto E (1990) Lignin: occurrence, biogenesis and biodegradation. *Annual Review of Plant Physiology and Molecular Biology* 41: 455–496.

Lewis NG, Davin LB, and Sarkanen S (1999) The nature and function of lignins. In: Barton DHR, Nakanishi K, and Meth-Cohn O (eds) *Comprehensive Natural Products Chemistry*, pp. 618–745. London: Elsevier Science.

Mellerowicz EJ, Baucher M, Sundberg B, *et al.* (2001) Unraveling cell wall formation in the woody dicot stem. *Plant Molecular Biology* 47: 239–274.

Plomion C, Leprovost G, and Stokes A (2001) Wood formation in trees. *Plant Physiology* 127: 1513–1523.

Ralph J, Lapierre C, Mariata JM, *et al.* (2001) Elucidation of new structures in lignins of CAD- and COMT-deficient plants by NMR. *Phytochemistry* 57: 993–1003.

Roberts K and McCann MC (2000) Xylogenesis: the birth of a corpse. *Current Opinion of Plant Biology* 3: 517–522.

Schmid J and Amrhein N (1995) Molecular organization of the shikimate pathway in higher plants. *Phytochemistry* 4: 737–749.

Schoch G, Goepfert S, and Morant M (2001) CYP98A3 from *Arabidopsis thaliana* is a 3′-hydroxylase of phenolic esters, a missing link in the phenylpropanoid pathway. *Journal of Biological Chemistry* 276(39): 36566–36574.

Sederoff RR, Mackay JJ, and Ralph J (1999) Unexpected variation in lignin. *Current Opinion in Plant Biology* 2: 45–152.

Antifeedant Substances in Plants

C B Purrington, Swarthmore College, PA, USA

Introduction

Although the literature on antifeedancy is dispersed rather thinly under a variety of cognates and related terms (e.g., antifeeding compound, feeding deterrent, feeding rejectant, feeding suppressant, feeding inhibitor, gustatory repellent, phagodepressant), it is unified by the hope that antifeedant sprays might someday become an important pest management tool. In particular, it is hoped that antifeedants might lack the dangers associated with the use of natural or engineered pesticides, both of which frequently have detectable if not substantial toxicity toward other animals, such as ourselves. Unfortunately, most of the literature on this topic has suggested that antifeedants are simply not effective, or not reliably so, against many of the important pests of common crops. Furthermore, insects are capable of habituating to the presence of antifeedants after a given amount of time. For this and several other reasons, the marketing of specific antifeedants has had little success.

Although it is easy to be pessimistic about the commercial viability of antifeedants, research on these compounds is producing valuable insight into the ecology and evolution of plant resistance strategies, as well as the corresponding suite of adaptations in herbivores. Knowledge of the presence and diversity of antifeedant compounds in plants is especially important for theories on herbivore host breadth. Early theories on host plant selection highlighted the importance of secondary compounds as feeding cues (i.e., feeding stimulants), but it is now generally believed that antifeedants (and the related repellents, which act on olfactory cells) are even more important in determining host breadth of herbivorous pests. Oddly, however, the word antifeedant has not caught on outside of the field of crop protection and insect neurophysiology. In most recent books and review articles on the ecology and evolution of plant resistance, the word "antifeedant" is never mentioned, and is seemingly displaced by an overemphasis on plant toxins, or rather on the toxic characteristics of compounds that may have other effects such as deterrency. Therefore, a secondary goal of reviewing antifeedants here is to argue for the importance of the term in studies of plant/herbivore ecology and evolution.

Although in this article the emphasis is on antifeedant compounds, it should be noted that many morphological structures (e.g., wooliness, spinescence, coloration, etc.) have similar effects on herbivore feeding, and these features were undoubtedly viewed by breeders as antifeedants long before antifeedant chemicals were identified in plant tissues.

Definitions of Antifeedants

Most researchers define antifeedants as those chemicals that have antifeedant properties at low concentration, and that act on very specific sensory cells (antifeedant receptors) in the pest. The neurons associated with these antifeedant receptors either prevent insect feeding (feeding deterrent effect) or cause cessation or slowing of further feeding (feeding suppressant effect). Another mode of action of some antifeedants is through an apparent ability to block the function of a herbivore's feeding-stimulant receptors, or an ability to bind directly to its normal feeding cues, such as sugars and amino acids. An example of this mechanism is the action of DEET (*N*, *N*-diethyl-*m*-toluamide) in repellent lotions, which deter blood-sucking arthropods by blocking their ability to perceive feeding stimulants. The very potent antifeedant azadirachtin acts in a similar way by reducing the sensitivity of sugar-sensing cells in herbivorous insects and thus causing the insects to incorrectly assess nutritional adequacy of treated host leaves. Only careful electrophysiological studies on the two categories of receptors (feeding deterrent versus feeding stimulant) can determine whether the putative antifeedant has a direct effect on antifeedant receptors, a distinction that is important when asking certain research questions. Another type of "false" antifeedant category is those compounds that act nonspecifically on all gustatory sensilla (e.g., through immediate and general cell toxicity) are also not considered true antifeedants. The finer points of categorizing antifeedants, however, are not usually an issue because data are rarely available on the precise electrophysiological mode of action of antifeedant compounds or extracts.

Although antifeedants can belong to a variety of chemical classes, the majority are alkaloid, flavonoid, and terpene secondary compounds. Some of the more commonly known antifeedants are presented in Table 1. In general, antifeedants are often bitter (when tasted by humans), although this is certainly not diagnostic of an antifeedant, and they are usually not involved in primary metabolism (there are exceptions, such as some sugars).

Identifying Antifeedants through Electrophysiology

When studying a particular antifeedant, it is important to determine early on whether the compound's

Table 1 Examples of naturally occurring antifeedants and the herbivores on which they are active

Antifeedant	*Chemical class*	*Source plant (family)*	*Herbivore*
Azadirachtin	Limonoid	*Azadiractha indica* (Meliaceae)	*Schistocerca gregaria* (Orthoptera)
Caffeine	Alkaloid	*Coffea arabica* (Rubiaceae)	*Danaus plexippus* (Leipidoptera)
Capsaicin	Alkaloid	*Solanum capsicum* (Solanaceae)	*Leptinotarsa decemlineata* (Coleoptera)
Cocaine	Alkaloid	*Erythroxylum coca* (Erythroxylaceae)	*Leptinotarsa decemlineata* (Coleoptera)
Gymnemic acid	Saponin	*Gymnema sylvestre* (Asclepiadaceae)	*Homo sapiens*
Nicotine	Alkaloid	*Nicotiana tobacum* (Solanaceae)	*Schistocerca gregaria* (Orthoptera)
Phlorizin	Phenolic	*Malus domestica* (Pomoideae)	*Myzus persicae* (Homoptera)
Strychnine	Alkaloid	*Strychnos nuxvomica* (Loganiaceae)	*Pieris brassicae* (Lepidoptera)
Tannic acid	Tannin	*Medicago sativa* (Fabaceae)	*Hypera postica* (Coleoptera)
Warburganal	Sesquiterpenoid	*Warburgia salutaris* (Canellaceae)	*Spodoptera exempta* (Lepidoptera)

effect on feeding is simply due to its repellent qualities and not to the activation of receptors involved in taste (gustatory receptors). A simple way of eliminating the olfactory component of a feeding response is to remove or disable the olfactory organs and then compare the responses of disabled and control subjects to a putative compound. When the locations of olfactory and gustatory sensilla are known, however, it is possible to obtain very useful electrophysiological data. These data can establish both that olfactory neurons are not being activated (ruling out a repellency activity) when exposed to the putative antifeedant and that gustatory receptors are transmitting signals for cessation of feeding. In some cases, it is even possible to use electrophysiological data to identify individual gustatory sensilla that are involved with a specific antifeedant or class of antifeedant. With insects, these cells can be located on one or more of a variety of structures including buccal surfaces, mouthparts, tarsi, and antennae. For caterpillars, sensilla located on the maxilla are especially important. Unfortunately, because electrophysiological measurements are extremely hard to obtain, even when receptor locations are known (knowledge that applies to only a few of the important insect crop pest species), they are infrequently gathered. More importantly, the nature of collecting such data makes electrophysiology an inefficient tool for mass screening of putative antifeedants, which is a major emphasis in antifeedant research.

Another technique for establishing antifeedant action is through the use of *in vitro* binding assays, in which the strength of antifeedant/receptor interactions are used as indicators of the antifeedant potency. These experiments necessarily require isolation of receptor molecules from the antifeedant-sensing cells, and thus only well-studied organisms, usually insects, are involved in this methodology.

Identifying Antifeedants with Feeding Arenas

A far simpler and thus much more common technique for testing putative chemicals and plant extracts for antifeedant properties is the choice arena (**Figure 1A**). This technique uses dishes or cages in which a single herbivore (insect, rabbit, etc.) is presented with unadulterated host plant tissue, as well as host tissue that has been sprayed with the test compound. The antifeedant index is calculated by measuring for each arena the relative amount of each type of disk consumed. In general, an antifeedant is considered effective if, on average, 95% of the treated leaf is left uneaten. As mentioned above, the calculated effect might include strict antifeedant

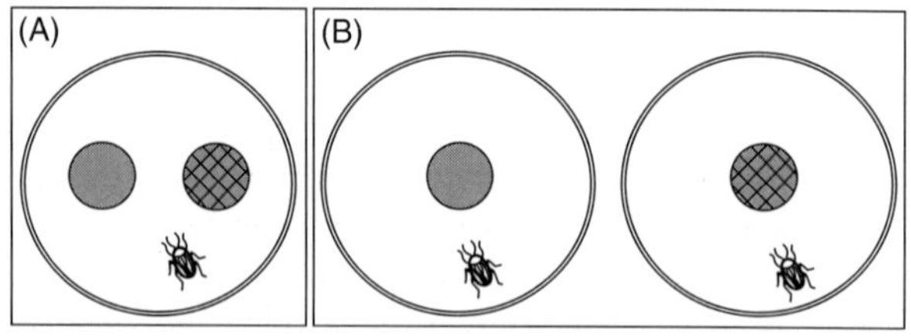

Figure 1 Choice (A) and nonchoice (B) feeding arenas can be used to estimate the antifeedant effect of test compounds (hatched circles) relative to untreated host material (shaded circles).

effects (those due to the activation of antifeedant receptors), as well as reduced feeding due to the repellent qualities of the tested compound or extract. If the experimenter wishes to separate these effects, the herbivore's olfactory function must be impaired in some way so that only gustatory contributions to the experimental outcome are measured.

A popular variant of this design is to replace the host plant tissue with disks fashioned out of stem pith, artificial diet, filter paper, agar, or glass fiber mat that have been soaked in a feeding stimulant, although this methodology is considerably less meaningful. In particular, using artificial disks probably overvalues the deterrent effect of tested compound because the strength of the applied feeding stimulant can rarely be as strong as that provided by unadulterated, preferred host tissue. Despite these drawbacks associated with artificial disks, use of living material in bioassays has a very important and often overlooked handicap of its own. When living host material is used in bioassays, the phytotoxic effects of putative antifeedants must also be considered. Because many antifeedants are toxins, it is not surprising that some of them are toxic to plants, and thus treated plants may develop an antifeedant quality that is only indirectly caused by the applied compound. For example, plants often produce gamma-aminobutyrate in response to various types of stress, and this compound has antifeedant properties against a variety of tested insects. An additional indirect mechanism of action is via a simple reduction in nutritional content, such as that resulting from a cessation of photosynthesis caused by a test compound that is lethal to plant cells, and which could therefore easily change repellent, antifeedant, and stimulant qualities of plant tissues. For the reasons just described, many of the more careful studies on antifeedants also include control tissue that has been treated with the solvent used to extract or resuspend the antifeedant compound, because many common solvents can stress plant tissue and so decrease their acceptability to insects.

Another common experimental methodology is to provide the herbivore with the control and test options singly, thus creating a nonchoice arena (**Figure 1B**). Nonchoice assays are particularly useful in assessing whether an antifeedant is an absolute deterrent, evident when an herbivore's feeding is completely and permanently inhibited when presented with antifeedant-treated host tissue or artificial disks. Although the nonchoice scenario is often accused of being biologically irrelevant, it is very appropriate when testing the efficacy of antifeedant compounds for use in monocultures, where the pests might not have alternative hosts for miles. Similarly, if the pest of interest does not display significant mobility and host selection in nature, it is experimentally inappropriate to facilitate such movement within a test arena with choices available. In this scenario, data from choice arenas would overestimate the true effectiveness of the putative antifeedant substance.

As might be expected, collection of data from both choice and nonchoice arenas yields a much more satisfying conclusion about the effectiveness of an antifeedant. Specifically, if the antifeedant effect tends to diminish when assessed with a nonchoice design, the compound might be viewed as a rather poor candidate for use in agricultural settings. The more careful studies will also typically include arenas with disks that have different concentrations of the "stimulant" and the "antifeedant," because manipulation of these amounts can have very important effects on the antifeedant index.

An important complication in interpreting data from both choice and nonchoice arenas is that toxicity effects (antibiosis) can inflate the antifeedant index, which should strictly be a measure of the chemical's effects on the feeding behavior of the insect. For example, if ingestion of the test disk causes rapid and irreversible muscle paralysis, failure to continue feeding might appear, incorrectly, to be due to antifeedant effects. A useful methodology for estimating the antibiosis component of antifeedant action is to deliver the compound directly to the herbivore's gut, and then estimate growth rate or survivorship. Of course, many compounds (most notably azadirachtin, which is discussed below) have both antifeedant and toxic activities, so researchers should, ideally, report the LD_{50} (dosage causing 50% mortality) for compounds together with an index of antifeedancy.

Another important consideration in antifeedant research is whether herbivores respond differently to antifeedant compounds over time. One such effect, called habituation or desensitization, occurs when the deterrent effect of a compound on a test insect diminishes over time. Therefore, choice and nonchoice trials of short duration might dramatically overestimate the true antifeedant properties of a compound. For herbivores that are polyphagous (i.e., not specializing on a given plant taxa), habituation to antifeedants is particularly rapid. Because much antifeedant research is directed at developing crop protection chemicals, the prevalence of habituation in such trials has diminished the early hopes that antifeedants would become popular commercial products.

Antifeedants Can Be Feedants to Other Pests

For those wishing to make antifeedants commercially viable, a constant source of disappointment is the fact that some antifeedants, despite being very active against many herbivore species, are potent feeding stimulants to a few economically important pests. For example, gossypol, a dimeric sesquiterpene found in cotton (*Gossypium* spp.), has antifeedant action toward generalist insect pests (e.g., *Heliothis* spp. and *Epicauta* spp.) but acts as a feeding stimulant for the boll weevil (*Anthonomous grandis*), a specialist on this species. In the end, it might be necessary to treat crops with both "specialist-targeted" and "generalist-targeted" antifeedants to achieve full protection of a given crop species. Because the vast majority of research on antifeedants is performed with generalist Lepidopteran larvae (*Spodoptera* spp., *Heliothis* spp., and *Pieris* spp.) and Orthopterans (*Locusta migratoria* and *Schistocera gregaria*), our knowledge of antifeedants affecting specialists is limited.

Evolution of Antifeedants

Because most identified antifeedants are toxic, it is not particularly surprising that herbivores have evolved gustatory senses that can perceive such toxins. The response of many insect species to specific antifeedants, therefore, may be explained as an adaptation to avoid further toxicity. In other words, genetic variants in a herbivore population that cannot perceive an existing or new plant toxin tend to have low fitness, and will not contribute many offspring to the next generation. However, there are numerous examples showing that chemicals (e.g., strychnine) can have strong antifeedant effects even though the insect species has probably had no prior encounter with the chemical throughout evolutionary time. In these instances, it is necessary to invoke the evolution of a generalized antifeedant receptor. Most hypotheses on the evolution of antifeedant sensitivity suggest that deterrent sensilla were originally sensitive to a wide array of compounds. Indeed, at first they were not even involved in deterrent activity, but rather in general gustatory identification of food. It is assumed that over evolutionary time these cells gradually lost their sensitivity to certain compounds (e.g., water, sugars, amino acids, etc.) but retained their sensitivity to other classes of compounds. Coupled with this restriction of sensitivity is the development of an antifeeding neural activity for these receptors. For some herbivores, these antifeedant receptors still identify large numbers of compounds as antifeedants.

Why do some plants apparently lack strong antifeedants? Even for antifeedants that are active at very low concentrations, it is likely that there is a metabolic cost involved in their production and storage. Therefore, protection comes at some fitness cost, and plant species that lack such antifeedants might enjoy high levels of growth and reproduction, at least in areas where herbivore pressures are absent or minimal. Like other types of chemicals involved in plant resistance, antifeedant content (e.g., phenolics, hydrogen cyanide, cucurbitacins) of some plants can be increased by herbivory. Induction thereby increases resistance when insects are present, but minimizes the fitness costs of producing a resistance phenotype when pests are absent or infrequent.

When considering the adaptive benefit of toxic secondary compounds, selection might reasonably favor those compounds that are easily identified by existing deterrent cells in herbivorous species. Plants possessing compounds that are not easily perceived by pests would therefore be subject to the occasional massive loss of tissue that occurs before the herbivore ceases feeding. Therefore, one could make the prediction that natural selection imposed by herbivore sampling (ingestion of small amounts of tissue) could add to the selective advantage of those toxic compounds that can be perceived by common pests.

Commercialization of Plant-Derived Antifeedants

The ultimate goal of antifeedant research is to isolate or synthesize a compound that functions as an absolute antifeedant, i.e., one that completely inhibits herbivores from ever attempting to feed again, but is at the same time nontoxic to even the targeted pest species (and therefore unlikely to be toxic to other animals such as ourselves). But if that antifeedant is not acutely toxic, evolution of countermeasures in the pest populations are likely to rapidly occur, and fix, as soon as mutations appear that diminish the ability of sensory cells from perceiving the antifeedant.

To date, only azadirachtin (**Figure 2**) derived from the neem tree (*Azadirachta indica*) has received any commercial success; this is marketed under a variety of trade names, e.g., Margosan-O, Azatin, Bioneem, Neemesis, Neemgard. Azadirachtin probably owes much of its success to its multiple modes of action. In addition to its much-touted antifeedant qualities, it is repellent and also toxic to many insects, in part due to its deleterious effects on insect growth regulation and maintenance of circadian rhythm. Azadirachtin is also an "indirect" antifeedant due to its toxic effects on treated plants, which thus become less advantageous hosts for herbivores. Resistance to azadirachtin, therefore, is regarded as being unlikely to develop rapidly in natural populations because it acts at many different points in the life cycle of affected insects. But despite statements that insects will never be able to mount countermeasures, laboratory selection experiments (with aphids) have shown that the antifeedant and toxic effects on insects can diminish after 40 generations. Interestingly, aphids exposed to a blend of neem compounds, rather than just to azadirachtin, did not develop resistance, indicating that synergistic effects of antifeedants are important in plant resistance.

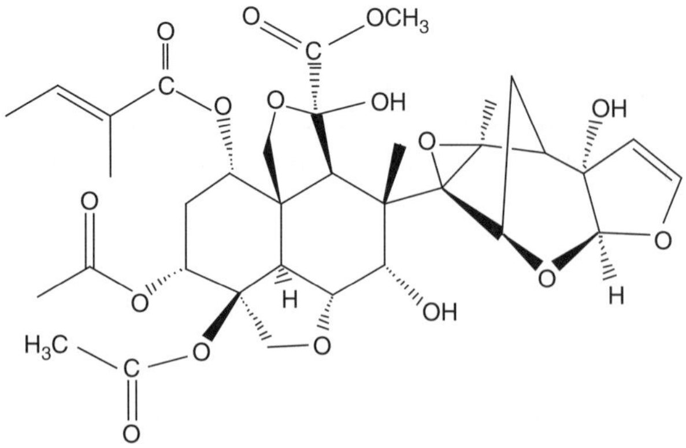

Figure 2 Chemical structure of azadirachtin, a potent antifeedant isolated from the neem tree.

In general, there are several reasons why antifeedant research has not generated successful products. First, as mentioned above, potent antifeedants that naturally occur are likely to be toxins also, and thus are unlikely to be viewed as "safe" in the eyes of regulatory agencies or the consuming public. Azadirachtin, for example, has toxic effects on certain mollusks, nematodes, fungi, protozoa, and even viruses. Although many of these groups can be pestiferous, the wide range of activity also warrants concerns about additional, untested toxicity against neutral or beneficial organisms in sprayed fields. Second, because toxic secondary compounds often exist in plant vacuoles and other cell macrostructures, spraying extracted antifeedant chemicals on plant leaf surfaces (even species that normally produce the antifeedant) may cause plant stress or mortality. Third, if an antifeedant is to be successfully commercialized, it is important to use choice trials of long duration to rule out the possibility that insects will habituate to its presence. Most trials do not last for the entire period in which feeding takes place, and thus promising antifeedants are often advanced to field tests only to reveal that herbivores habituate rather quickly to their presence. Fourth, attempts at cloning antifeedant genes and incorporating them into transgenic plants is unlikely to be easy, because many of the secondary products involved in antifeedancy are synthesized by the sequential action of multiple genes; to date, there have been few successes of reliably transferring whole biosynthetic pathways into plants that may lack the necessary precursors. Finally, as stated earlier, antifeedants identified to date do not have entirely predictable effects on all herbivores; in contrast, many insect toxins are reliably toxic against the majority of insect targets.

See also: **Pests**: Genetic Modification of Pest Resistance, Insect Pests. **Secondary Products**: Phytoalexins, Natural Plant Protection.

Further Reading

Chapman RF (1974) The chemical inhibition of feeding by phytophagous insects: a review. *Bulletin of Entomological Research* 64: 339–363.

Chyb S, Eichenseer H, Hollister B, Mullin CA, and Frazier JL (1995) Identification of sensilla involved in taste mediation in adult western corn rootworm (*Diabrotica virgifera virgifera* LeConte). *Journal of Chemical Ecology* 21: 313–329.

Dethier VG (1947) *Chemical Insect Attractants and Repellents*. Philadelphia: The Blakiston Company.

Mordue (Luntz) AJ and Blackwell A (1993) Azadirachtin: an update. *Journal of Insect Physiology* 39: 903–924.

Norris DM (1986) Anti-feeding compounds. In: Kato T, Krämer DW, Kuck K-H, and Morris DM (eds) *Chemistry of Plant Protection: Sterol Biosynthesis, Inhibitors and Anti-Feeding Compounds*, pp. 97–146. New York: Springer-Verlag.

Schoonhoven LM (1982) Biological aspects of antifeedants. *Entomologia Experimentia et Applica* 31: 57–69.

Pharmaceuticals, Plant Drugs

A D Kinghorn and D Chávez, University of Illinois, Chicago, IL, USA

Historical Perspective and Background

For millennia, humankind has used whole, powdered, and extracted plants for the treatment or alleviation of many types of diseases. Gradually, from about 200 years ago, it became possible to purify the pharmacologically active organic compounds produced by plants as new drugs, to the extent that drugs from the so-called higher plants (gymnosperms and angiosperms) became the most widely used medicines available. This trend continued until 50 or 60 years ago, when antibiotics from microbial sources and new synthetic drugs began to be introduced into drug therapy. In the nineteenth century, medicinally important compounds such as atropine, caffeine, cocaine, morphine, and quinine were the first plant drugs to be obtained in pure form. Additional drugs from plant sources such as digitoxin, digoxin, paclitaxel, reserpine, tubocurarine, vinblastine, and vincristine continued to be introduced in the twentieth century. It has been estimated that, worldwide, about 120 drugs of plant origin are on the market. Most of these drugs were discovered in the laboratories of major pharmaceutical companies, although academic, governmental, and private research institutes have made major contributions.

The majority of our plant drugs were discovered as a result of following up on folkloric observations of medicinal uses of plants in temperate or tropical countries. For example, digitoxin was isolated from the purple foxglove (*Digitalis purpurea*). An English physician in the late eighteenth century noted that some of his patients were using this plant to treat dropsy (edema). Later on, digitoxin was purified and structurally determined, and is still in use as a reliable therapy to treat congestive heart failure (a cause of edema). An excellent and rather remarkable example of a drug obtained from the tropical rain forest is the above-mentioned tubocurarine, which was isolated from *Chondrodendrum tomentosum* (Menispermaceae) used in poisonous blowgun darts in South America to paralyze small animals, and then later introduced into Western medicine as a skeletal muscle relaxant for use in surgery. The use of tubocurarine has now been largely superseded by synthetic drugs of related structure. **Figure 1** shows the structures of digitoxin (a steroidal glycoside) and tubocurarine (a complex bisbenzylisoquinoline alkaloid; shown as its chloride salt).

Even today, about three-quarters of the world's population rely on traditional medicine for their primary medical care, mainly in developing countries. Consequently, a major part of their drug therapy is derived from locally available medicinal plants. In some countries, the documentation and laboratory and clinical investigation of the medicinal flora are particularly well developed, most particularly in the People's Republic of China, where the

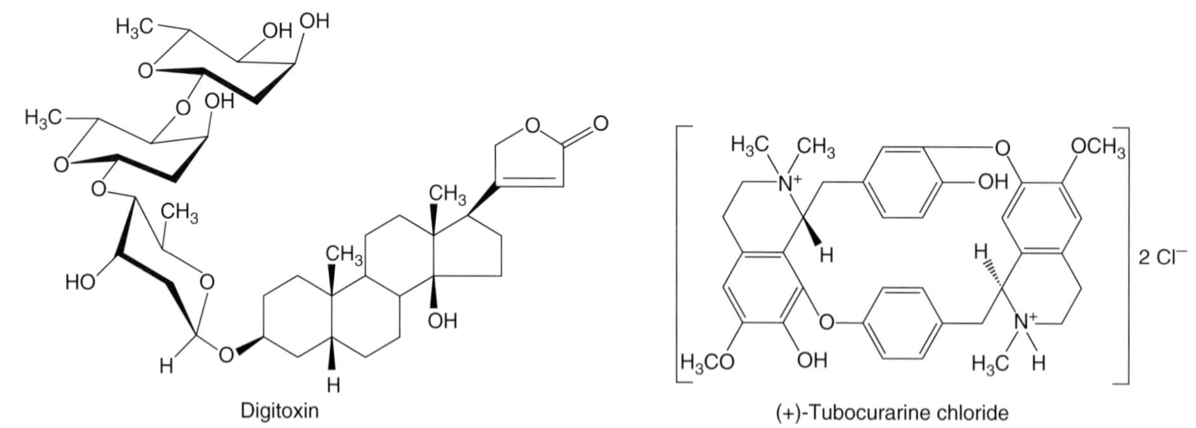

Figure 1 Structures of digitoxin and (+)-tubocurarine chloride.

detailed study of several thousand herbal products used in traditional Chinese medicine has long been ongoing. About 500 medicinal plants are in common usage in the People's Republic of China, and research on some of these has led to promising new drugs. Another particularly well-developed system of traditional medicine in which treatment with herbal products is featured prominently is the Ayurvedic system of the Indian subcontinent.

There are an estimated 250 000 species of gymnosperms and angiosperms, but these occur in a disproportionate manner in the world's tropical rainforests, which occupy only about 7–8% of the earth's land surface and contain half to two-thirds of the known higher plants. The types of plant constituents used as drugs are classified as "secondary metabolites," which are found in only specific organisms or related groups of organisms, often at high concentration levels. Although their functions in the plant are not completely understood, many of these substances have been found to serve a beneficial role, such as acting to deter predators or to stimulate pollination. It is thought that tropical rainforest plants (as opposed to temperate species) are particularly prone to produce specialized secondary metabolites, due to factors such as high humidity, along with generally elevated temperatures, a continuous growing season, and the considerable species density encountered where they grow. It has been estimated that nearly 50 drugs have been developed from tropical plants, including quinine, reserpine, scopolamine, tubocurarine, vinblastine, and vincristine. Since only a relatively small percentage of these plants has ever been subjected to detailed biological or chemical investigation in the laboratory, there are perhaps hundreds more drugs that await discovery from the plants of the tropical rainforests. However, many authors have written to express their concern at the increasing destruction of the tropical rainforests, leading to an alarming loss of biodiversity, which will greatly reduce opportunities for future drug discovery in this new century and beyond.

Pharmacology is defined as the "scientific study of medicinal drugs and their effects on the body." There are many plant secondary metabolites from plants which have proven to act at specific pharmacological targets, and so have provided a more complete idea of exactly how drugs act on animals and humans in the laboratory or clinic, and more recently at a cellular level. For example, morphine is a major alkaloid (nitrogenous base) found in the opium poppy (*Papaver somniferum*), and in fact was the first compound of this plant to be purified, in 1805. Opium has a history of use dating back thousands of years, and the Greeks used this as a cathartic, hypnotic, and narcotic. Morphine is a potent analgesic, and has been used as a standard pain-killing substance in experiments with laboratory animals, having subsequently been found to bind selectively to the mu (μ) receptor, one of three opioid receptor subtypes (with the other two being kappa (κ) and delta (δ)). A second example of a useful "pharmacological tool" has been obtained from croton oil, the seed oil of *Croton tiglium* (Euphorbiaceae), a plant formerly used as a severe purgative. Later on, it was found that when croton oil was applied to the backs of mice repeatedly following a single subthreshold dose of a chemical carcinogen, papillomas resulted. Laborious phytochemical work eventually led to the isolation and structural characterization of the novel compound 12-O-tetradecanoylphorbol 13-acetate (TPA) as the most abundant croton oil skin irritant and "tumor-promoting" constituent. While TPA has no human or veterinary drug use, it has been found to interact specifically with protein kinase C, and can be used in biochemical experiments to specifically probe this important phosphorylating enzyme. **Figure 2** shows

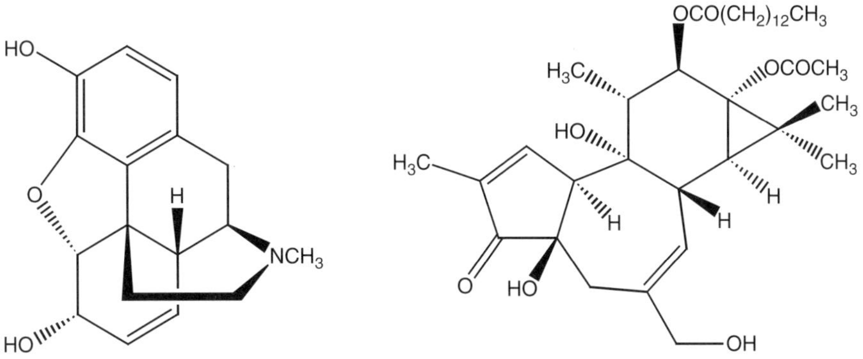

Morphine   12-*O*-Tetradecanoylphorbol 13-acetate

Figure 2 Structures of morphine and 12-*O*-tetradecanoylphorbol 13-acetate.

the structures of morphine (a morphinan isoquinoline alkaloid) and 12-*O*-tetradecanoylphorbol 13-acetate (a phorbol ester).

Unfortunately, it is often found that many promising drug candidates from plants and other natural sources often have certain undesirable characteristics, for example, in being relatively insoluble in water, or low heat stability. Alternatively, it may be possible to reduce the inherent toxicity of a natural compound. It is often possible to use such substances as "lead compounds," so that they can be optimized by chemical synthetic methods, to afford commercially viable drugs. The drug morphine (**Figure 2**) has probably served as a "chemical template" to a greater extent than any other plant product, in an effort to maintain or improve on its potent analgesic activity, but also to curtail its side effects. There are many clinically useful analgesics modeled on the molecule of morphine, and retaining the particular structural features of morphine required to exhibit potent painkilling activity. One example is the structurally simpler molecule levorphanol, which is a more potent painkiller than morphine, and also causes less emesis and constipation. However, the addiction liability of levorphanol is as great as that of morphine. The structure of levorphanol is shown in **Figure 3** (this synthetic drug is used as its tartrate salt).

Examples of a few of the plant drugs and their therapeutic uses are shown in **Table 1**. It should be noted that the majority of the compounds listed are alkaloids. Compounds in this class form water-soluble salts when treated with acids, and as such are in a convenient form for laboratory evaluation and clinical testing. Glycosides such as digitoxin and digoxin are also to some extent water soluble. The lack of solubility in water of other classes of plant secondary metabolites such as the terpenoids and phenols has restricted their development as clinically useful drugs.

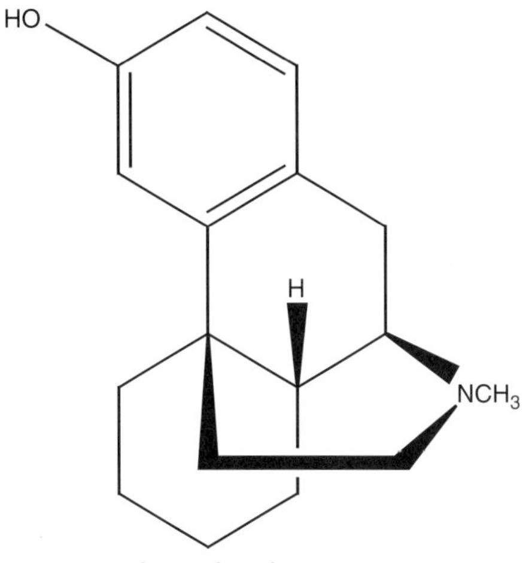

Levorphanol

Figure 3 Structure of the analgesic synthetic morphine derivative levorphanol.

Examples of Recently Introduced Plant-Derived Drugs

Anticancer Agents

While there are many types of anticancer agents on the market, it is not always realized how many of these are derived from higher plants. For example, two of the newer types of anticancer agent are the semisynthetic derivatives etoposide and teniposide, which are based on a plant compound of the lignan class, podophyllotoxin. The latter compound was first discovered as a major constituent of the poisonous plant American mandrake (*Podophyllum peltatum*), which grows in the eastern half of the United States. When podophyllotoxin was tested for antineoplastic activity in tumored mice, this compound proved to be too toxic for further development. Etoposide and teniposide both have a polar modified sugar unit introduced into the molecule when compared with their parent, podophyllotoxin.

Table 1 Examples of therapeutic agents isolated from plants

Drug name	*Plant of origin (family)*	*Therapeutic activity*
Alkaloids		
Atropine	*Atropa belladonna* (Solanaceae)	Antisialogogue
Cocaine	*Erythroxylum coca* (Erythroxylaceae)	Local anesthetic
Codeine	*Papaver somniferum* (Papaveraceae)	Analgesic; antitussive
Morphine	*Papaver somniferum* (Papaveraceae)	Analgesic
Physostigmine	*Physostigma venenosum* (Fabaceae)	Cholinergic used in glaucoma
Pilocarpine	*Pilocarpus jaborandi* (Rutaceae)	Cholinergic used in glaucoma
Quinine	*Cinchona officinalis* (Rubiaceae)	Antimalarial
Quinidine	*Cinchona officinalis* (Rubiaceae)	Antiarrhythmic
Scopolamine	*Datura metel* (Solanaceae)	Prevention of motion sickness
Reserpine	*Rauvolfia serpentina* (Apocynaceae)	Antihypertensive; psychotropic
Tubocurarine	*Chondrodendron tomentosum* (Menispermaceae)	Skeletal muscle relaxant, used as an adjuvant in surgery
Vinblastine	*Catharanthus roseus* (Apocynaceae)	Anticancer agent
Vincristine	*Catharanthus roseus* (Apocynaceae)	Anticancer agent
Steroidal glycosides		
Ditoxin	*Digitalis purpurea* (Scrophulariaceae)	Cardiotonic
Ditoxin	*Digitalis lanata* (Scrophulariaceae)	Cardiotonic

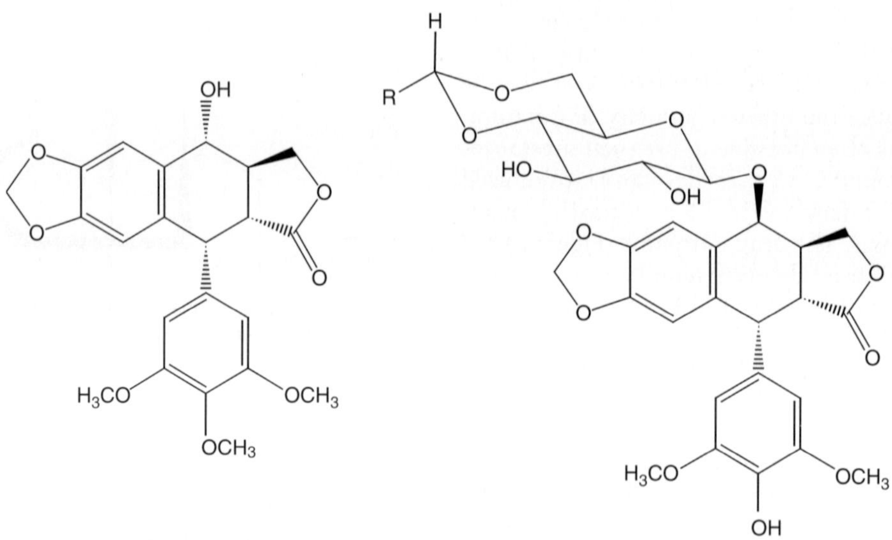

Podophyllotoxin

		R
	Etoposide	CH_3
	Teniposide	thienyl

Figure 4 Structures of the lignan, podophyllotoxin, and its semisynthetic derivatives etoposide and teniposide.

It is of interest that the resultant minor structural differences between these semisynthetic compounds led to different clinical uses. Etoposide is used for the treatment of small-cell lung cancer and testicular cancers, while teniposide is employed in combination chemotherapy in refractory childhood acute lymphoblastic leukemia. The structures of podophyllotoxin and its derivatives etoposide and teniposide are shown in **Figure 4**.

A second example of a plant-derived anticancer agent is paclitaxel (originally known in the scientific literature as taxol), a nitrogen-containing diterpenoid which was first obtained as a constituent of the bark of the Pacific yew (*Taxus brevifolia*). Since this is a small, slow-growing tree, great concern was expressed initially, while paclitaxel was being developed, as to how it would be possible to obtain sufficient of the drug to meet the clinical demand without wiping out the populations of *T. brevifolia* in the Pacific Northwest region of the United States and Canada. Several alternative approaches towards rectifying this problem were proposed, and finally

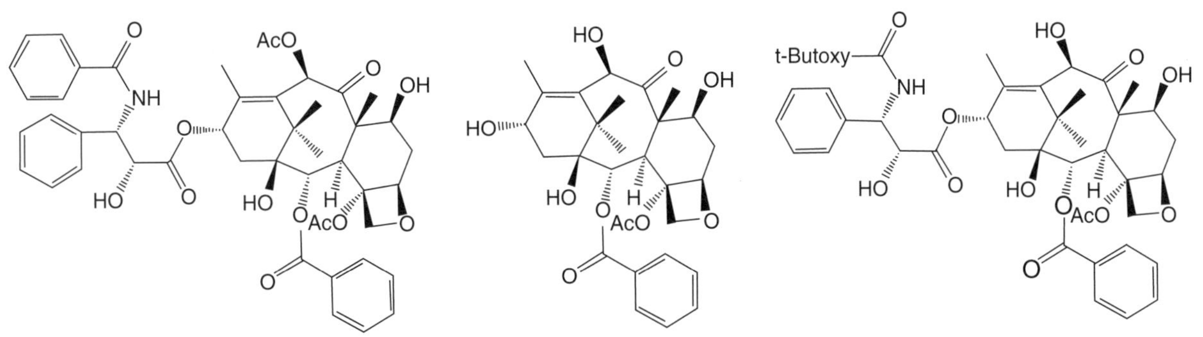

Paclitaxel 10-Deacetylbaccatin III Docetaxel

Figure 5 Structures of paclitaxel, from *Taxus brevifolia*, and its analogs 10-deacetylbaccatin and docetaxel.

the situation was resolved by generating the diterpenoid portion (10-deacetylbaccatin III) of the paclitaxel molecule by growing widely available ornamental *Taxus* species (such as the European yew, *T. baccata*) in greenhouses. The nitrogen-containing side chain is then incorporated onto 10-deacetylbaccatin III semisynthetically to afford the intact molecule of paclitaxel. Initially, paclitaxel was used clinically to treat ovarian cancer resistant to other forms of chemotherapy. However, the spectrum of use of this anticancer drug has gradually increased, to include the treatment of breast cancer and AIDS-related Kaposi's sarcoma. A second taxane anticancer drug structurally related to paclitaxel is the semisynthetic compound docetaxel, which is indicated for advanced or metastatic breast cancer after the failure of prior chemotherapy. The structures of paclitaxel, 10-deacetylbaccatin III, and docetaxel are shown in **Figure 5**.

Two members of a third group of newly introduced anticancer agents of plant origin are based structurally on the quinoline alkaloid camptothecin. This parent compound was isolated initially from the Chinese plant *Camptotheca acuminata* (family Nyssaceae), and exhibited poor solubility in its initial clinical trials. The semisynthetic derivatives topotecan and irinotecan are water soluble, and indicated for patients with metastatic ovarian cancer and metastatic carcinoma of the colon or rectum, respectively. These compounds are anticancer agents of the topoisomerase I inhibitor class. The structures of camptothecin, topotecan, and irinotecan are shown in **Figure 6**.

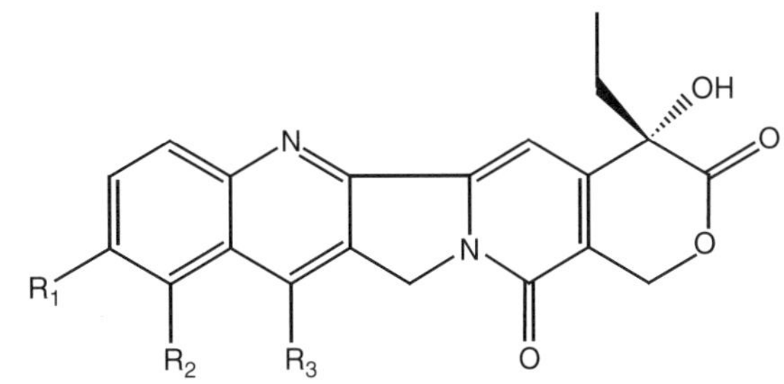

	R_1	R_2	R_3
Camptothecin	H	H	H
Topotecan	OH	$CH_2N(CH_3)_2$	H
Irinotecan	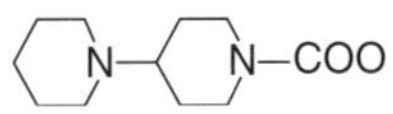	H	CH_2CH_3

Figure 6 Structures of the quinoline alkaloid camptothecin and its analogs topotecan and irinotecan.

Antimalarial Agents

Despite many attempts towards its eradication, infection of malaria by the blood parasite *Plasmodium falciparum* remains a formidable problem in tropical areas of the world. The sesquiterpene lactone artemisinin (also known as "qinghaosu"), isolated from the Chinese medicinal plant *Artemisia annua* (Asteraceae), has become the prototype member of an important new class of antimalarial agents. This compound, a sesquiterpene lactone with an unusual endoperoxide functional group in the molecule, is only sparingly water soluble. However, when the carbonyl group of artemisinin is reduced to a secondary alcohol and a hemisuccinate salt is formed, the resultant compound is sodium artesunate, which is used clinically in the People's Republic of China and certain countries in Southeast Asia to treat falciparum malaria. An alternative strategy to improve on the parent compound is to produce the ether, artemether, which has been developed recently by a major pharmaceutical company to treat malaria in combination chemotherapy. The structures of artemisinin, sodium artesunate, and artemether are shown in **Figure 7**.

Topical Pain-Reliever

The phenolic amide capsaicin has recently been introduced as a topical pain reliever in the United States, and is used in creams particularly for the

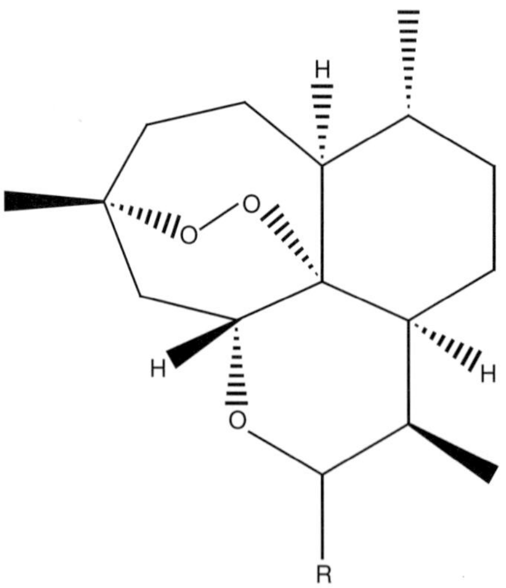

	R
Artemisinin	=O
Sodium artesunate	β-$OCOCH_2CH_2CO_2Na$
Artemether	β-OCH_3

Figure 7 Structures of the sesquiterpene lactone artemisinin and its analogs sodium artesunate and artemether.

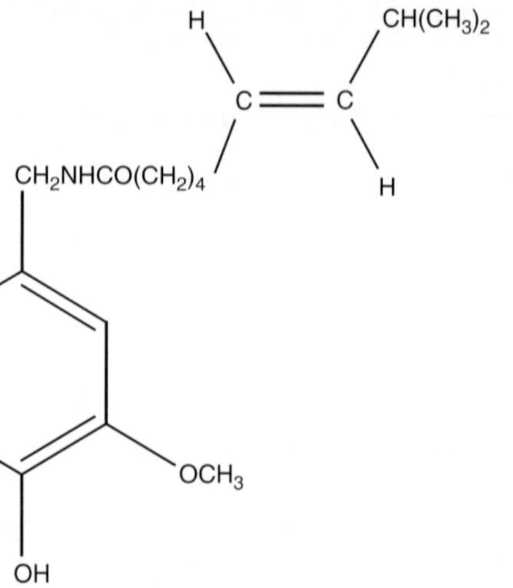

Capsaicin

Figure 8 Structure of capsaicin.

alleviation of pain from rheumatoid arthritis, osteoarthritis, and neuralgia. The compound is extracted from the cayenne pepper (*Capsicum frutescens*) and other similar commercial varieties. Capsaicin, the vanillyl amide of isodecenoic acid, is an extremely pungent compound, and is used in low concentrations (0.025–0.075%) in over-the-counter formulations. Capsaicin is known to act by causing a depletion of substance P, a neuropeptide, from local sensory C-type nerve fibers. When substance P is depleted, this affects the transmission of pain impulses from the peripheral nerves to the spinal cord. The structure of capsaicin is shown in **Figure 8**.

Use of Herbal Remedies

In the course of the last decade, there has been a major increase in interest in the use of phytomedicines ("herbal remedies") in Western countries, among both the general public and health professionals. These materials are purchased from pharmacies or health food stores in the crude form, either as powders, tablets, capsules, or as various types of extract. Many have become extremely popular, and for several years *Echinacea* spp., garlic (*Allium sativum*), *Ginkgo bilboa*, ginseng (*Panax* spp.), and St. John's wort (*Hypericum perforatum*), have been among the major phytomedicines available. In the United States such products are marketed as dietary supplements, and so are not regarded as drugs and do not require a prescription for their sale. To present just two examples of "herbal remedies" here, information is given on the constituents of two of these new phytomedicines, ginkgo and feverfew.

Ginkgo

Commercial ginkgo products are produced from the dried leaves of *Ginkgo biloba*, a tree which is native to East Asia and has been described as a "living fossil," because of its ancient taxonomic lineage. In Germany and other countries in Europe where ginkgo products are regulated as drugs (and not dietary supplements), uses include the treatment of headache, hearing loss, mood disturbance, short-term memory loss, and peripheral vascular disease. Several commercial ginkgo products are produced in the form of a chemically standardized extract prepared containing 6% terpenes and 24% flavone glycosides. While flavone glycosides are found in many plants, the ginkgo terpenes are most unusual, and constitute a mixture of the diterpene lactone derivatives ginkgolides A, B, C, J and M, and the related sesquiterpene lactone bilobalide. The structures of the ginkgolides and bilobalide are shown in **Figure 9**. It appears that the therapeutic effects of ginkgo are due to a mixture of constituents, with there being definite biological activities associated with both the ginkgolides and the flavonoids. A number of clinical trials have been performed already while others are ongoing, and, although some of these are controversial, the efficacy of ginkgo has generally been supported in such investigations.

Feverfew

The herbal remedy feverfew consists of the leaves of *Tanacetum parthenium*, and has attracted much

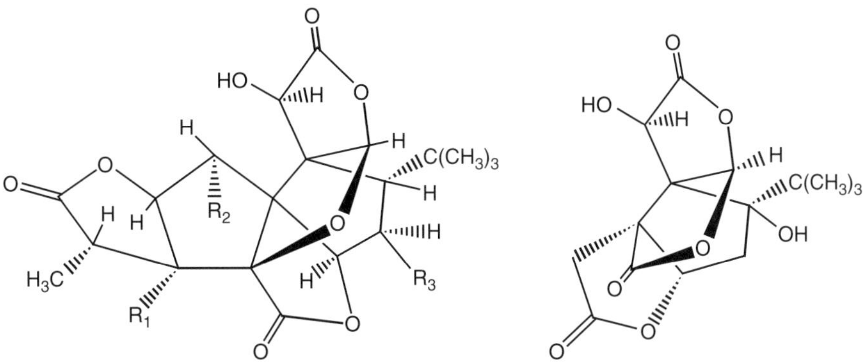

Bilobalide

Ginkgolide	R_1	R_2	R_3
A	OH	H	H
B	OH	OH	H
C	OH	OH	OH
J	OH	H	OH
M	H	OH	OH

Figure 9 Structures of the diterpene lactones ginkgolides A, B, C, J, and M and the sesquiterpene lactone bilobalide.

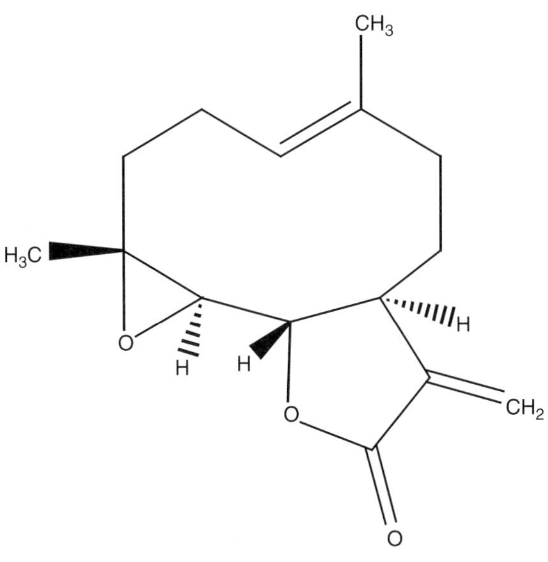

Parthenolide

Figure 10 Structure of the sesquiterpene lactone parthenolide.

interest because of its effects in reducing the severity and frequency of migraine headaches. The plant is commonly chewed fresh or in a freeze-dried form, and the perceived activity has been associated with the presence of the sesquiterpene lactone, parthenolide. The structure of this compound is shown in **Figure 10**. Possibly parthenolide acts together with other constituents of feverfew leaves to elicit the antimigraine activity. The efficacy of feverfew has been substantiated by certain clinical trials.

Future Perspective

Plant-derived drugs have been a mainstay of therapy in Western medicine for about 200 years and are extremely important in systems of traditional medicine in developing countries. The increasing interest in herbal remedies in Western Europe and North America by both lay persons and health care professionals further attests to the importance of plant medicinal products. Since plant extracts are frequently included in drug discovery screening regimens in the pharmaceutical industry, it may be anticipated that additional plant-derived small-molecular-weight organic molecules will be introduced as single-entity drugs in the future, although probably more of these will be semisynthetic substances rather than their naturally occurring form. Plant natural products can be manipulated chemically in the laboratory to have improved efficacy, less toxicity, and/or enhanced water solubility and stability.

List of Technical Nomenclature

Alkaloid A nitrogenous base capable of forming salts with acids.

Analgesic A drug that reduces pain without producing anesthesia or loss of consciousness.

Antiarrhythmic A drug that counters the loss of rhythm of the heart muscle.

Anticancer	A drug effective in treating tumors in clinical trials.
Antihypertensive	A drug that reduces the blood pressure of persons suffering from high blood pressure.
Antineoplastic	A compound that can reduce the size of tumors in experimental animals.
Antisialogogue	A drug used for the control of secretions (e.g., bronchial, salivary).
Cardiotonic	A drug exerting a favorable effect on the action of the heart.
Cathartic	A drug promoting evacuation of bowels; purgative.
Cholinergic	Relating to nerve cells or fibers that employ acetylcholine as their neurotransmitter.
Diterpene	A compound composed of four isoprene (C_5H_8) units.
Edema	An abnormal accumulation of fluid in cells, tissues, or cavities of the body, resulting in swelling.
Emesis	Vomiting.
Flavonoid	A compound consisting of two C_6 groups (substituted benzene rings) connected by a three-carbon aliphatic chain.
Glaucoma	Disease of the eye involving increased ocular pressure leading to defects in the field of vision.
Glycosides	Compounds which yield one or more sugars among the products of hydrolysis.
Hypnotic	A drug causing sleep.
Papilloma	A benign tumor of the skin or mucous membrane.
Psychotropic	A drug affecting the mind (used in mental illnesses).
Sesquiterpene	A compound based on three isoprene (C_5H_8) units.
Steroid	A compound containing a tetracyclic steroidal nucleus.
Terpenoid	A compound based on isoprene (C_5H_8) building-blocks (normally two, three, four, or six, as in mono-, sesqui-, di-, and triterpenoids).

Further Reading

Cox PA (2000) Will tribal knowledge survive the millennium? *Science* 287: 44–45.

Dewick PM (2002) *Medicinal Natural Products: A Biosynthetic Approach*, 2nd edn. Chichester: John Wiley.

Evans WC (2002) *Trease and Evans' Pharmacognosy*, 15th edn. London: Ballière Tindall.

Huang KC (1993) *The Pharmacology of Chinese Herbs*. Boca Raton: CRC Press.

Kinghorn AD and Balandrin MF (eds) (1993) *Human Medicinal Agents from Plants*. Symposium Series No. 534. Washington, DC: American Chemical Society.

Newman DJ, Cragg GM, and Snader KM (2000) The influence of natural products upon drug discovery. *Natural Products Reports* 17: 215–234.

Robbers JE and Tyler VE (1999) *Tyler's Herbs of Choice*. Binghampton, NY: Haworth Herbal Press.

Robbers JE, Speedie MR, and Tyler VE (1996) *Pharmacognosy and Pharmacobiotechnology*. Baltimore: Williams & Wilkins.

Soejarto DD and Farnsworth NR (1989) Tropical rainforests: potential source of new drugs? *Perspectives in Biological Medicine* 32: 244–256.

Phytoalexins, Natural Plant Protection

J B Harborne,[†] University of Reading, Reading, UK

The Phytoalexin Concept

Ideas of a chemical basis to disease resistance in plants date back many years. Thus, Marshall Ward in 1905 developed a toxin theory with the view that there were compounds in plants capable of inhibiting the growth of pathogenic fungi. These early ideas about disease resistance were developed throughout the 1920s and 1930s with the proposals that plant polyphenols of various types were responsible for combating infection. Polyphenols were recognized to be antimicrobial in activity and were known to be widely present in plant tissues. They became the most likely agents for plant protection. However, experimental evidence in support of these ideas was generally inconclusive.

The major breakthrough in the study of chemically based disease resistance mechanisms came with the enunciation of the phytoalexin hypothesis by Muller and Börger in 1941. The key feature of this theory was the novel idea that chemical substances capable of protecting plants from disease are not present in the healthy plant but are formed *de novo* upon infection (**Table 1**). Such compounds capable of

[†] Deceased.

Table 1 The major tenets of the phytoalexin concept

1	A phytoalexin is a compound which inhibits the development of the fungus in hypersensitive tissues and is formed or activated only when the host plants come in contact with the parasite
2	The defense reaction occurs only in living cells
3	The inhibitory agent is a discrete chemical substance, a product of the host cell
4	The phytoalexin is nonspecific in its toxicity towards fungi; however, fungal species may be differentially sensitive to it
5	The basic response in both resistant and susceptible cells is the same, the basis of differentiation between resistant and susceptible hosts being the speed of formation of the phytoalexin
6	The defense reaction is confined to the tissue colonized by the fungus and its immediate neighborhood
7	The resistant state is not inherited; it is developed after the fungus has attempted infection. The sensitivity of the host cell which determines the speed of the host reaction is specific and genotypically determined

Table 2 Phytoalexins of some common crop plants

Plant species	*Phytoalexins*	
	Chemical type(s)	*Example*
Apple (*Malus pumila*)	Biphenyl	Aucuparin
Broad bean (*Vicia faba*)	Polyacetylene	Wyerone
Cabbage (*Brassica campestris*)	Indole	Cyclobrassinin
French bean (*Phaseolus vulgaris*)	Pterocarpan	Phaseollin
Grape vine (*Vitis vinifera*)	Stilbene	Viniferin A
Parsley (*Petroselinum crispum*)	Furanocoumarin	Bergapten
Pea (*Pisum sativum*)	Pterocarpan	Pisatin
Peanut (*Arachis hypogaea*)	Stilbene	Resveratrol
Pepper (*Capsicum annuum*)	Sesquiterpene	Capsidiol
Potato (*Solanum tuberosum*)	Sesquiterpene	Rishitin
Rice (*Oryza sativa*)	Diterpenoid	Momilactone A
	Flavanone	Sakuranetin
Sorghum (*Sorghum vulgare*)	3-Deoxyanthocyanin	Luteolinidin

warding off (Greek, "alexos") disease organisms from the plant (Greek, "phytos") were termed phytoalexins. This novel theory developed from experiments carried out by these two workers on resistant factors in the potato (*Solanum tuberosum*) to the blight disease *Phytophthora infestans*. But they were unsuccessful at that time in isolating a potato substance with the properties of a phytoalexin.

It remained for Cruickshank and Perrin in 1960 to be the first to characterize a phytoalexin, namely pisatin, a pterocarpan from the pea plant (*Pisum sativum*). It could be isolated in quantity from empty pea pods inoculated with the conidia of the brown rot fungus (*Monilinia fructicola*), and it was undetectable in control material inoculated with water. Furthermore, it proved to be significantly fungitoxic against *Monilinia fructicola* and against a range of 10 other fungi tested. In fact, it fitted the phytoalexin hypothesis perfectly and obeyed all the tenets originally proposed by Muller and Börger (**Table 1**).

Following this first success, investigations for phytoalexins were carried out on most of the major crop plants, including potato, tobacco (*Nicotiana tabacum*), broad bean (*Vicia faba*), pepper (*Capsicum annuum*), and peanut (*Arachis hypogaea*) with similar results. New phytoalexin structures, characteristic for that plant species or the plant family to which it belongs, were uncovered (**Table 2**). Surveys for phytoalexin induction were extended to wild plant species. In one extensive survey, over 500 species of the family Fabaceae were found to respond positively in this way to fungal inoculation. At the latest count, some 300 chemical structures have been characterized as phytoalexins, following induction experiments with about 900 plant species belonging variously to 40 plant families.

The phytoalexin response is now recognized as a disease resistance mechanism widespread in plants. In fact, it is the rule rather than the exception, that phytoalexins will be induced. However, a major question that still needs answering is how effective are phytoalexins in plant protection in the field. For example, continuing experiments with phytoalexin production indicate that phytoalexins may be deactivated following further metabolism by the enzymes present in a particular pathogenic fungus. In other cases, such as that of potato blight, it is apparent that the production of phytoalexins is not directly related to the well-known *R* gene resistance of that crop plant. Hence, attention in recent years has concentrated on the signaling systems through which phytoalexins are elicited, and on the relative contribution of phytoalexins to the many newly recognized resistance mechanisms in plants. In the present account of phytoalexins, it is intended to discuss the chemistry of phytoalexins, their distribution and further metabolism, their antimicrobial nature and their significant contribution to disease resistance in most crop plants.

Chemical Variation in Phytoalexin Production

A considerable array of different secondary metabolites have been characterized as phytoalexins in plants following elicitation. Almost every class and subclass of natural product is represented – phenolics, alkaloids, terpenoids, polyacetylenes, and so on. Some examples are shown in **Table 2**. Some typical structures are illustrated in **Figure 1**.

In general, the type of phytoalexin produced is typical of the family to which that plant belongs. Thus, in the Solanaceae, the phytoalexins are universally sesquiterpenoids. Typical are rishitin, the potato phytoalexin and capsidiol, a phytoalexin of the pepper plant (**Figure 1**). Similar sesquiterpenoids have been encountered in other Solanaceae studied, including tomato (*Lycopersicon esculentum*) and tobacco. In the Brassicaceae, indole-based sulfur compounds are formed, such as camalexin, the phytoalexin of *Arabidopsis thaliana*. Again, in the Rosaceae, the phytoalexins are dibenzofurans or have closely related biphenyl structures (e.g., aucuparin).

In the Fabaceae, a large family which has been investigated intensively, the phytoalexins are typically isoflavonoids and can be subdivided into isoflavones, isoflavanones, pterocarpans, or isoflavans. Representative structures are the pterocarpans pisatin from pea and phaseollin from the French bean (*Phaseolus vulgaris*). Similar isoflavonoids occur widely in the family, but occasionally unrelated structures occur. For example, the stilbene resveratrol is a phytoalexin produced by the peanut (or groundnut).

In some plants, only one or two main phytoalexins are synthesized. This is true of *Arabidopsis thaliana*, where only camalexin (**Figure 1**) is produced and of the pea where pisatin and maackiain are the major products induced. In other species, however, the

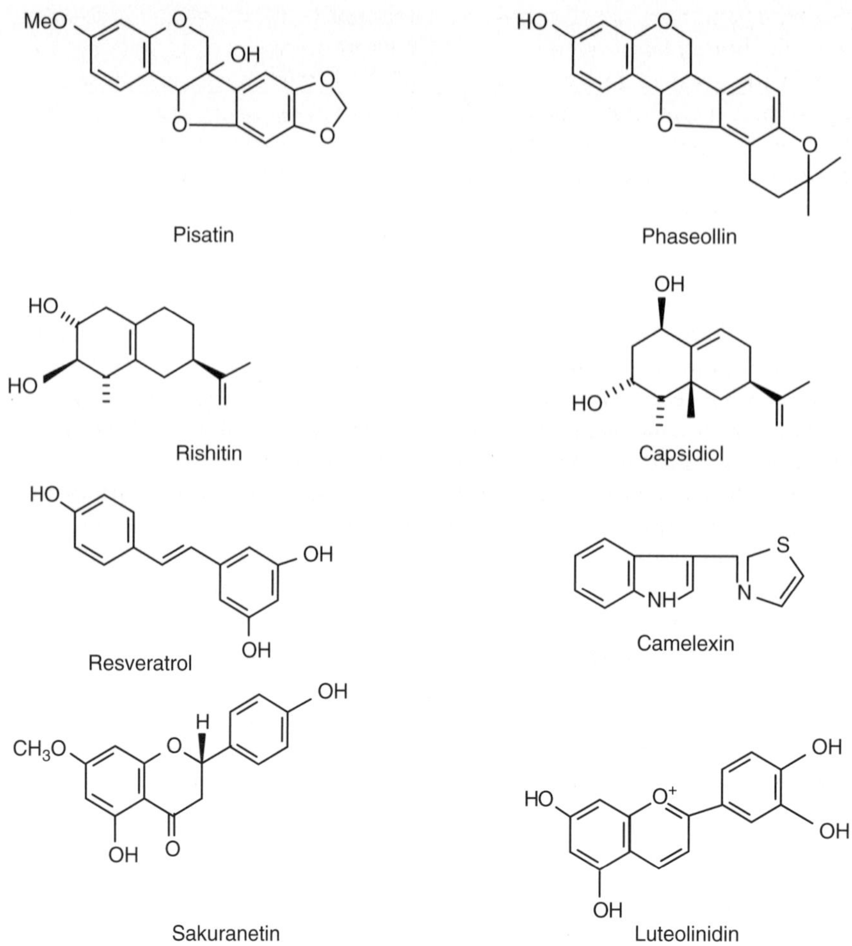

Figure 1 Structures of some representative phytoalexins.

phytoalexin response can be complex. In the cabbage (*Brassica oleracea*), nine phytoalexins with indole-sulfur structures are produced, while in rice, no fewer than 15 phytoalexins may be formed: 14 diterpenoids and one flavanone. Again, in the potato, between 15 and 20 sesquiterpenoids may be induced. Finally, in the carnation (*Dianthus caryophyllus*), about 30 anthranilic acid-derived phytoalexins are formed.

This wide range of phenolic, terpenoid, and nitrogen-based molecules are all capable of killing fungi or at least severely inhibiting fungal growth. This fungitoxicity of phytoalexins is effective against a range of plant fungi, although some are more sensitive than others to a particular phytoalexin structure. In terms of antimicrobial activity, there does not appear to be any obvious structure–activity relationships. Nevertheless, there is one chemical feature that phytoalexins have in common. There is a significant degree of lipophilicity present. It is notable that phytoalexins generally lack glycosidic attachment in spite of the fact that most constitutive secondary metabolites in plants are glycosides. Furthermore, in most aromatic structures, free hydroxyls are masked by O-methylation (see for example pisatin, **Figure 1**) and prenyl groups may be present to increase the lipophilic properties. This lipophilicity present in the great majority of phytoalexins may be important to permit movement of the toxic agent to the site of infection, so that it can interact with the fungal spore germination in the waxy environment of the leaf surface.

The Experimental Approach to Phytoalexin Synthesis

For a chemical substance to be regarded as a phytoalexin, its formation has to be induced *de novo* in healthy plant tissue following inoculation or infection with microorganisms. One simple procedure for this is the drop-diffusate technique (**Figure 2**). Leaves or leaflets are floated on water in the light and droplets of a spore suspension of a nonpathogenic fungus is placed on the upper leaf surface. A nonpathogenic fungus is used, since if an organism known to be pathogenic on that species were to be used, it might destroy the phytoalexin as rapidly as it is formed. A second batch of leaves is set up as a water control. After 48 h, the droplets are collected and examined. In a positive experiment, these droplets will contain high levels of phytoalexins, in relatively pure solution free from other plant constituents. The absence of any chemical in the control droplets confirms the *de novo* synthesis in the experimental droplets. Indeed, if antifungal material appears in the control droplets, this is good evidence that constitutive antifungal agents are present and are being leached out of the leaf into the droplets.

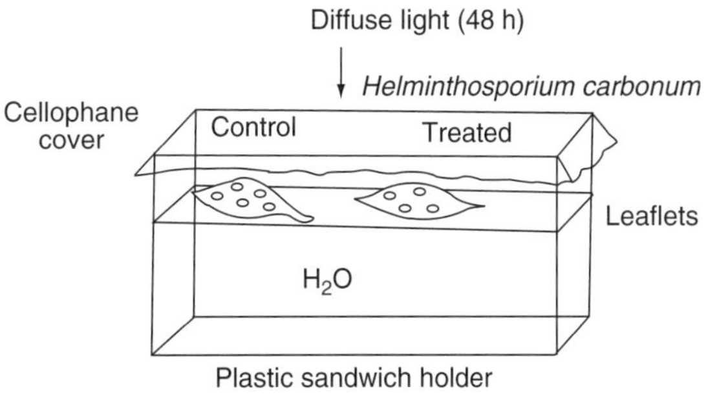

Figure 2 Drop diffusate technique for phytoalexin induction. Reproduced with permission from Harborne JB (1993) *Introduction to Ecological Biochemistry*, 4th edn., p. 278. London: Academic Press, with permission from Elsevier Ltd.

In this leaf droplet system, fungal spores start germinating within the droplets in 1–2 h and the resulting germ tubes then penetrate the host cells. As a consequence of this "trigger," the plant responds rapidly by the massive synthesis of phytoalexin at the point of infection. These phytoalexins synthesized within the leaf are "pulled out" onto the leaf surface where the fungal invasion is occurring and hence there is the accumulation of material in the droplets. It is possible that this drop-diffusate technique may not provide a complete picture of phytoalexin synthesis, since certain substances may not diffuse from the leaf into the overlying droplets. However, tests have shown that for most species the same compounds are present both in the leaf and in the droplets. Furthermore, the phytoalexins are usually phytotoxic as well as fungitoxic and hence there is a greater likelihood that the phytoalexins are exported from the leaf. A sign of a positive phytoalexin response in many plants is the subsequent necrosis of the plant tissue where the droplets have been sitting. While the drop-diffusate is usually applied to leaf tissue, it may be applied with suitable modification to the induction of phytoalexins in stem, hypocotyl, or seed.

A necessary sequel to the drop-diffusate experiment is a test on the droplets for fungitoxicity. This is carried out simply by thin layer chromatography (TLC) bioassay in which the two extracts (phytoalexin extract and water control) are developed on a TLC plate in a solvent system such as ether–hexane (3:1). After drying, the plate is sprayed with a fungal spore suspension of *Cladosporium herbarum* or *Botrytis cinerea* and incubated for up to 5 days at 25–30°C. During this time, the fungus grows over the whole plate except where the fungitoxic zones are present; these areas appear as white spots on a grey background. The final stage in the investigation is the chemical identification of the compounds that are present.

Phytoalexin Synthesis and Disease Resistance

The phytoalexin response is not confined to warding off fungal pathogens. While fungal inoculation usually gives the most reproducible results, infections with bacteria, e.g., *Pseudomonas syringae*, and with viruses, e.g., tobacco mosaic virus, have also resulted in *de novo* synthesis of phytoalexins. Stressing the plant may also induce such synthesis to some extent and a range of abiotic agents (e.g., dilute cupric chloride, the enzyme pectinase) have been employed to mimic the natural process. Nevertheless, in spite of these various conditions which can set off the phytoalexin response, it is still regarded as an important mechanism for resisting a variety of fungal invasions in the natural ecosystem.

When fungal attack occurs, it is now recognized that an elicitation takes place at a very early stage in the infection. Several kinds of elicitors have been encountered: either oligosaccharide fragments broken away from the plant or fungal cell walls; or else proteinaceous materials. Elicitation involves the *activation* of DNA regions in the nuclear genome of the plant cell, the synthesis of messenger RNA, and the subsequent production of the enzymes required for phytoalexin synthesis. In addition to macromolecular elicitors, it is apparent that certain messenger molecules can act as signals to alert this defense response. One such signal is salicylic acid, which can function in a volatile form as methyl salicylate. Another effective signal molecule is jasmonic acid, which can also be present in volatile form as methyl jasmonate. Application of either of these signals to healthy plant tissue can initiate the production of phytoalexins. It has even been proposed that the methyl jasmonate released from one infected plant can enable a neighboring plant to activate its disease resistance mechanisms.

The continued synthesis of phytoalexin during the first few days following infection is essential for maintaining disease resistance. For example, in the pepper plant (*Capsicum annuum*), the sesquiterpenoid phytoalexin capsidiol is produced continuously from 3 h after infection until at least 6 days later. The amount of capsidiol synthesis is critical and in a resistant cultivar, $1204\,\mu g\,ml^{-1}$ of capsidiol is required to maintain resistance. Capsidiol is in fact only fungistatic at low concentrations (e.g., $3.75\,mmol\,l^{-1}$) and only becomes fungitoxic at levels about $5\,mmol\,l^{-1}$. Susceptible cultivars are unable to produce sufficient capsidiol to ward off disease and thus succumb to infection.

The difference between resistance and susceptibility towards fungal invasion may be determined not only by the ability of the plant to produce sufficient phytoalexin but may also involve a subtle switch in the pathway of phytoalexin synthesis. Such a switch occurs in sorghum (*Sorghum* spp.) plants infected by anthracnose, *Colletotrichum sublineatum*. In a compatible susceptible sorghum cultivar, two 3-deoxyanthocyanins are present: apigeninidin and apigeninidin 5-(caffeoylarabinoside). However, in an incompatible resistant cultivar, two further pigments are produced: luteolinidin and 5-methylluteolinidin. These new pigments are several-fold more fungitoxic than the original two phytoalexins. Additionally, successful defence depends on increased synthesis of phytoalexin and the resistant cultivar, at 96 h after infection, has twice the amount of material than the susceptible form.

One of the advantages of following the synthesis of phytoalexins that are intensely colored, as in sorghum, is that it is then possible to see where the phytoalexins are localized. Indeed, it has been found that synthesis in sorghum is taking place in subcellular inclusions within the host epidermal cells and it is restricted to the first cells that come under fungal attack.

The Further Metabolism of Phytoalexins

In some plants such as the pepper (see above), phytoalexin synthesis in quantity may be sufficient to protect it against a particular pathogen. However, as a result of the coevolution between plants and their microbial parasites, some fungi have developed the ability to detoxify phytoalexins as rapidly as they are formed. Thus, they become pathogenic on that particular host plant. Detoxification involves the further metabolism of the phytoalexin to produce inactive products. Such further metabolism has been investigated in the pea plant, from which the first fully identified phytoalexin – pisatin – was obtained.

One of the important pathogens of pea is the organism *Nectria haematococca*, which exists in nature as a range of different strains, varying in their ability to infect pea plants. Pathogenic strains have the ability to detoxify pisatin. This is achieved by the presence of an enzyme, pisatin demethylase, which removes the O-methyl group from pisatin, converting it to 6a-hydroxymaackiain (**Figure 3**), which is no longer fungitoxic. Other strains of *N. haematococca* either have low levels of this enzyme or lack the enzyme altogether and hence are nonpathogenic. Thus, the ability of *N. haematococca* to successfully infect the pea plant depends on the ability to produce sufficient pisatin demethylase to inactivate the pisatin as soon as it is formed.

What is true for *N. haematococca* seems to operate for other pea pathogens and their ability generally to

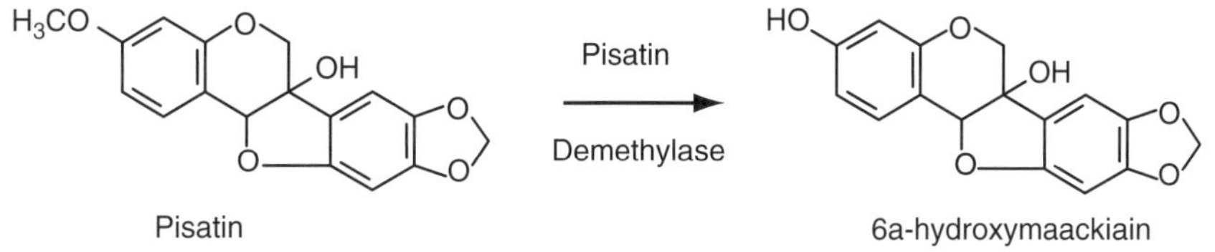

Figure 3 Further metabolism of pisatin by pathogenic strains of *Nectria haematococca*.

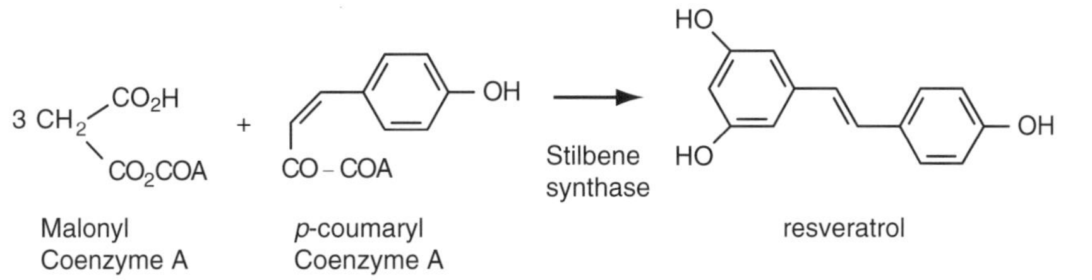

Figure 4 One-step biosynthesis of the phytoalexin resveratrol.

infect peas depends on their capacity to produce this key enzyme, pisatin demethylase. It is even possible to convert a nonpathogen of pea such as *Cochliobolus heterostrophus* by genetic engineering into a pathogen. This is done by transferring the gene for pisatin demethylase synthesis from *Nectria* to *Cochliobolus*. The transformed organism expresses the demethylase enzyme and is then capable of producing infected lesions on the pea plant.

What is true for pisatin also extends to the phytoalexins produced by other plants. Many phytoalexins are capable of undergoing single-step enzymic metabolism to yield inactive products. Thus, the potato phytoalexin lubimin is oxidized to the inactive isolubimin by *Gibberella pulicaris*. And the parsnip (*Pastinaca sativa*) phytoalexin xanthotoxin is detoxified by ring opening to the corresponding benzofuran carboxylic acid by the same microorganism.

Genetic Transfer of Phytoalexin Synthesis

The best evidence to date that phytoalexins are effective in providing plants with disease resistance has been obtained by foreign phytoalexin expression in a novel plant. Stilbene phytoalexins such as resveratrol require only a single condensing enzyme, stilbene synthase, to link the precursors malonyl-CoA and *p*-coumaryl-CoA for their synthesis (**Figure 4**). Pathogen resistance has thus been engineered into tobacco plants by the transfer of stilbene synthase genes from grapevine (*Vitis vinifera*), a resveratrol-producer. The regenerated tobacco plant proved to have improved resistance to infection by *Botrytis cinerea*. This could be traced immediately to the rapid synthesis of up to 40 μg resveratrol g^{-1} fresh wt. A similar experiment carried out with stilbene synthase transfer to the rice plant was also successful. A third experiment with tomato was slightly more variable. Here, resveratrol was again produced following elicitation of the transformed plant, but it only provided useful resistance to infection by potato blight. Other organisms such as *Botrytis cinerea* elicited resveratrol synthesis, but not in sufficient amount to prevent infection. Overall, the results obtained by this technique are distinctly promising and this approach may have important practical benefits when applied to a range of crop plants.

Phytoalexin Toxicity

Although phytoalexins are most frequently thought of as antifungal agents, it is true that they can be toxic to other classes of living organisms. Thus, viruses, bacteria, higher plants, and animals are all susceptible to the action of individual phytoalexins. Cell death as well as nonlethal inhibitions occur, indicating the biocidal as well as biostatic capacities of these agents. Fears have been expressed, for example, about the potential toxicity of xanthotoxin, a furanocoumarin phytoalexin which can accumulate in diseased celery tissue and which could adversely affect fieldworkers harvesting the celery (*Apium graveolens*) crop.

Fungitoxicity has been recorded for most phytoalexins when tested against a range of microorganisms and there is a considerable range of sensitivities to a given substance. Fungitoxicity is expressed as an ED_{50} value, the median effective dose required for the inhibition of mycelial growth of a named fungus. Some examples of the variations recorded for two Rosaceae phytoalexins are given in **Table 3**.

Table 3 Inhibition of spore germination by dibenzofuran phytoalexins

Phytoalexin	ED_{50}($\mu g\,ml^{-1}$) against		
	Alternaria alternata	*Botrytis cinerea*	*Fusarium culmorum*
α-Cotonofuran	27	35	14
β-Pyrurfuran	96	58	20

Table 4 ED_{50} values of phytoalexins against *Pyricularia oryzae*

Phytoalexin	*Versus spore germination*	*Versus hyphal growth*
Momilactone A	31.4–314[a]	—
Oryzalexin D	30	—
Oryzalexin E	62.5	—
Oryzalexin S	20	—
Sakuranetin	15	35
Naringenin		360

[a] Values obtained for inhibition of germ tube growth

While no overall structure–activity relationships can be established for all phytoalexins, there is no doubt that within certain subclasses, it is possible to discern structural features which enhance activity. For example, the flavanone naringenin is formed during phytoalexin induction in the rice plant, but it has only one tenth of the activity (with an ED_{50} of 360 ppm) compared to the main phytoalexin sakuranetin (its 7-methyl ether), which shows an ED_{50} of 35 ppm, against *Pyricularia oryzae* (**Table 4**). These data for the rice phytoalexins (**Table 4**) also indicate that when more than one class of phytoalexin is produced, there can be significant variations in their relative toxicities.

Phytoalexins are generally regarded as multisite toxicants and capable of affecting a variety of physiological and biochemical pathways. Phytoalexins with isoflavonoid structures such as pisatin are capable, for example, of causing membrane dysfunction and it is possible that the site of action of certain phytoalexins may be at the cell membrane. Alternatively, it has been proposed that isoflavonoid derivatives function as uncouplers of oxidative phosphorylation in the fungal cell. If this is so, then the relative acidity and number of free phenolic groups could determine the antifungal potency.

The Practical Application of Phytoalexin Research

The practical value of knowledge about phytoalexins in terms of protecting agricultural or horticultural crops from disease has still to be demonstrated. There is, however, no reason why eventually some practical advantage may be gained by enhancing the phytoalexin response of these plants. In developing practical applications, the following four points should be borne in mind.

1. Phytoalexins are only one barrier in a multilayered defense system within higher plants. Constitutive chemicals, particularly those at the leaf surface, may provide an alternative barrier. There are also many macromolecular barriers which may be constitutive (e.g., lignification of cells) or induced (e.g., pathogenesis related (PR) proteins). It is not always possible to determine during infection which are the most effective defense systems.
2. Phytoalexins are generally less toxic than synthetic fungicides, such as benomyl. This may not matter too much *in vivo*, since there is a considerable concentration of toxin at the point of attack. On the other hand, use of phytoalexins as a spray to protect a crop is unlikely to succeed for the same reason and also because of the relative lability of phytoalexins under field conditions.
3. Research on genetic engineering of crop plants to introduce new genes for phytoalexin synthesis has been successful. However, this approach requires significant refining before it becomes a practical procedure.
4. A large variety of stimulants to phytoalexin synthesis have been identified. Methyl jasmonate could be used as a vapor on glasshouse crops to alert plants to defend themselves. However, it is going to be prohibitively expensive to try out the same procedure in the field. Applied research has concentrated recently on finding a suitable synthetic chemical capable of continually inducing phytoalexin synthesis in plants. In the short term, such an approach is particularly promising and may be successful before very long.

List of Technical Nomenclature

Biphenyl	Two phenyl groups, linked by a C–C bond.
3-Deoxyanthocyanin	A rare class of anthocyanin pigment, lacking a 3-hydroxyl substituent.
Dibenzofuran	An aromatic phenolic structure.
Diterpenoid	Formed from four isoprene units.
Flavanone	Formed in the flavonoid pathway by isomerization of a chalcone.
Furanocoumarin	A coumarin with a fused furan ring attached.

Indole	An aromatic alkaloid with a nitrogen atom in a five-membered ring.
Polyacetylene	An aliphatic hydrocarbon with one or more acetylenic bonds.
Pterocarpan	A subclass of isoflavonoid.
Sesquiterpenoid	Formed from three isoprene units.
Stilbene	A phenolic with a C_6–C_2–C_6 skeleton.

See also: **Diseases**: Breeding for Disease Resistance; Fungal Diseases; Genetic Modification of Disease Resistance, Fungal Pathogens. **Regulators of Growth**: Jasmonates. **Secondary Products**: Allergens.

Further Reading

Bailey JA and Mansfield JW (1982) *Phytoalexins*. Glasgow: Blackie.

Daniel M and Purkayasthe RP (1994) *Handbook of Phytoalexin Metabolism and Action*. New York: Marcel Dekker.

Hammerschmidt R (1999) Phytoalexins: what have we learned after 60 years? *Annual Review of Plant Pathology* 37: 285–306.

Harborne JB (1999) The comparative biochemistry of phytoalexin induction in plants. *Biochemical Systematics and Ecology* 27: 335–367.

Kuc J (1995) Phytoalexins, stress metabolism and disease resistance in plants. *Annual Review of Plant Pathology* 33: 275–297.

Mansfield JW (2000) Antimicrobial compounds and resistance. In: Sluserenko A, Fraser RSS, and van Loon L-C (eds) *Mechanisms of Resistance to Plant Diseases*, pp. 325–370. Dordrecht: Kluwer Academic.

Allergens

L P Christensen and E Larsen, Danish Institute of Agricultural Sciences, Aarslev, Denmark

Introduction

Plants have evolved sophisticated biochemical processes in which unique, diverse, and biologically active secondary plant products are produced. Many of these biologically active substances often have beneficial properties for the plants, such as protection against attack from fungi, bacteria, and herbivores. In some cases, these substances are beneficial to humans as well and may be used medicinally; in other cases, they are toxic to humans or can be responsible for various types of dermatitis. Dermatitis caused by plants can result from contact with living, damaged, or processed plant materials, and may manifest itself in a person immediately, a few hours after contact, or may not occur until the second or subsequent exposure.

Chemical, pharmacological, and dermatological studies of dermatitis induced by natural products have enabled the various plants and plant products to be classified into four major groups: the contact allergens, the phototoxic compounds, the chemical irritants, and the mechanical irritants. However, the most frequent type of plant-induced dermatitis is caused by the contact allergens and, consequently, this article will focus on these compounds.

Contact Allergens

Allergic Contact Dermatitis (ACD)

The most frequent type of skin inflammation condition induced by plant products is allergic contact dermatitis (ACD), which shows the typical features of eczema. In ACD, skin contact with the sensitizing substance (allergen) is required for the eczema to develop and progress. Contact allergens from plants are typically secondary metabolites with molecular weights of below 700. Allergenic secondary metabolites are often referred to as haptens, although this term is also used for allergenic synthetic compounds. Once an individual has become well sensitized, often only minute amounts of the allergen may be needed to elicit an allergic reaction. The skin sensitization reaction to a contact allergen is a multistep process with two principal stages, (1) sensitization and (2) elicitation:

(1) Sensitization. A state of sensitization to the contact allergen is induced. This may happen on the first exposure to the contact allergen or after several exposures.
(2) Elicitation. A sensitization response is elicited. This may happen when a subject who is sensitive to a given contact allergen is exposed to, or challenged with, the same contact allergen.

To cause skin sensitization, the hapten must penetrate into the epidermis beneath the stratum corneum, where it binds covalently to soluble or cell-bound carrier proteins to generate the antigen (hapten–protein complex) (**Figure 1**). A specific clone of lymphocytes in the immune system then recognizes the antigen. The lymphocytes migrate from the skin via lymphatics to regional lymph nodes where they proliferate, differentiate, and disseminate via blood vessels throughout the body (propagation phase). The individual is now sensitized. When the sensitized individual re-encounters the allergen, an eczematous response develops at the site of contact (elicitation phase).

ACD induced by plants can take many forms ranging from a dry scaly erythema to a severe papular or vesicular inflammation with edema. The molecular mechanisms that decide whether an individual will become sensitized or not are still largely unknown. However, susceptibility to sensitization by contact allergens is, in most cases, genetically determined and is related to race, sex, and possibly age. Sensitization is also dependent on the sensitizing potential of the allergen, the quantity, and method of exposure. Consequently, not all individuals are affected similarly.

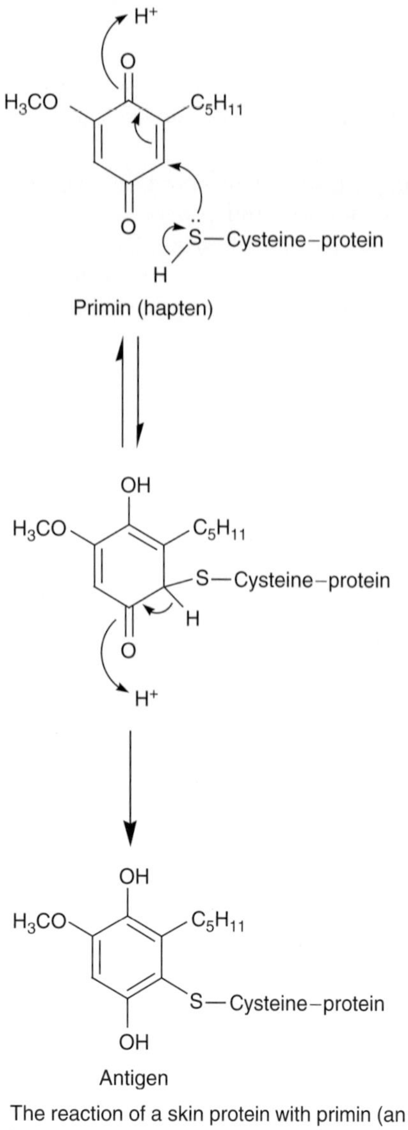

Figure 1 The reaction of a skin protein with primin (an example of a hapten) to form a protein-hapten complex (antigen).

Some of the most potent plant contact allergens are the "urushiols" found in poison ivy (*Toxicodendron radicans*) and poison oak (*Toxicodendron diversilobum*). Very strong contact allergens like the urushiols may induce sensitization immediately after the first exposure. After sensitization has occurred, renewed contact with the allergen(s) will elicit an immune response that normally peaks after a delay of 24–72 h, depending on the allergen(s) and the degree of exposure. As the immune response in ACD is delayed, it is often referred to as a delayed-type hypersensitivity or type-IV hypersensitivity in contrast to the immediate (type I) hypersensitivity reactions that are usually caused by macromolecules such as proteins. For weak contact allergens, sensitization first occurs after a prolonged and repeated exposure to them. However, once sensitization has occurred, subsequent exposure to a small amount of allergen may induce ACD. The course of the allergic reaction is then similar to that produced by more potent sensitizers. Many cases of occupational dermatitis are actually due to prolonged exposure to weak contact allergens.

An allergen may also elicit ACD in an individual already sensitized to a different but structurally related compound. In such a case, the pair of compounds are said to be cross-reactive. Cross-reactivity is very relevant to human sensitization. Individuals who have been sensitized by poison ivy or poison oak also react strongly to, for example, Japanese lacquer tree (*Toxicodendron vernicifluum*) and mango (*Mangifera indica*), which contain structurally related urushiols and resorcinol derivatives, respectively. Furthermore, compounds that are structurally related to an allergen may not themselves be sensitizers, and yet they may be capable of producing dermatitis due to cross-reactivity. Such secondary allergens are known to be elicitors. The naphtoquinone lapachol is an example of a secondary allergen in Teak (*Tectona grandis*) where the primary sensitizer is the closely related deoxylapachol (**Figure 2**).

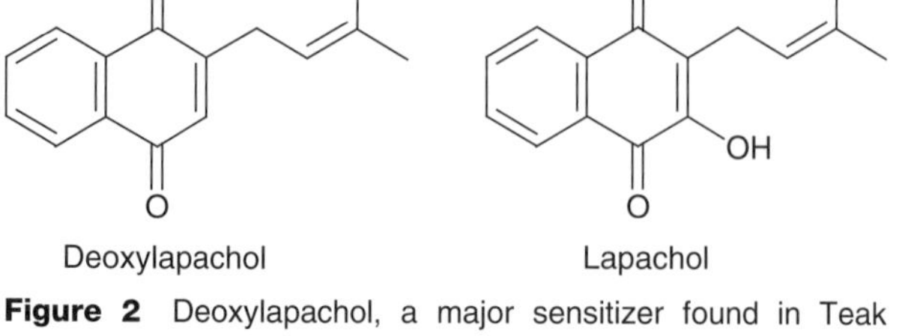

Figure 2 Deoxylapachol, a major sensitizer found in Teak (*Tectona grandis*, Verbenaceae). Lapachol is only a secondary allergen (see text).

Isolation and Identification of Plant Allergens

Plant allergens can often be identified from an offending plant by means of patch testing on a sensitized individual if the allergen(s) in the offending plant is known. However, if the allergen(s) is not known, various operations are required to isolate and identify the allergen(s). This is a multistep, and often long, process that consists of three main stages: (1) preparation of the plant extract; (2) isolation and purification of the allergen(s) from the plant extract; and (3) identification of the chemical structure of the allergen(s). Following each stage of purification, which is done by various chromatographic techniques, the different fractions must be tested on the sensitized individual in order to follow the biological activity through the purification process. Finally, when the pure allergen(s) is obtained, it is identified by sophisticated spectroscopic methods.

Types of Plant Allergens

Sesquiterpene lactones (SQLs) Allergic contact dermatitis to plants of the Asteraceae (syn. Compositae) family, which constitute one of the largest flowering plant families in the world, is one of the major causes of dermatoses in Western Europe. For example, the genera *Dendranthema* and *Tanacetum*, with their numerous species and cultivars, are one of the most common causes of ACD among florists and horticulturists. Also, medicinal plants such as *Saussurea lappa* (costus root) and *Arnica* species, as well as their preparations (e.g., tincture of arnica and costus oil), are responsible for many cases of ACD. More than 200 species of the approximately 25 000 different Asteraceae species have caused ACD. The main allergens in this plant family are sesquiterpene lactones (SQLs), which are cyclic C_{15} terpenoids containing a lactone ring (**Figure 3**). SQLs are widely distributed in Asteraceae and so far over 1800 different SQLs have been isolated from this family. In addition to Asteraceae, SQLs have also been found in members of several plant families including Acanthaceae, Amaranthaceae, Apiaceae (syn. Umbelliferae), Aristolochiaceae, Lamiaceae, Lauraceae, Magnoliaceae, and in a genus of liverworts of the family Jubulaceae where they also are the main cause of ACD (**Table 1**). SQLs are usually concentrated in fragile glandular trichomes and also on pollen grains, and are therefore easily spread with the wind. Consequently, airborne dust particles containing

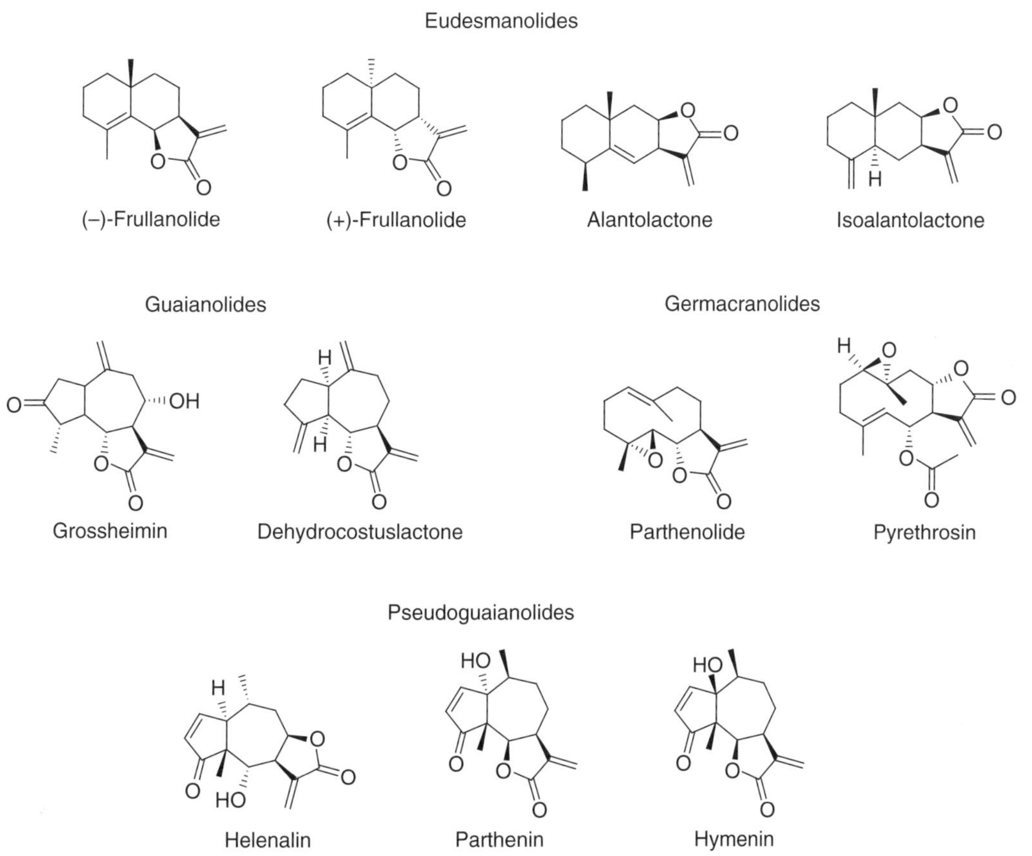

Figure 3 Examples of allergenic sesquiterpene lactones classified according to their carbocyclic skeleton.

Table 1 Examples of common plant source(s) of allergenic sesquiterpene lactones (SQLs)

Family/genus/species	*Common name(s) of allergenic species*[a]	*Examples of allergenic SQLs*
Asteraceae		
Achillea spp.	Yarrow, milfoil (= *A. millefolium*)	α-Peroxyachifolide
Ambrosia spp.	Western ragweed (= *A. psilostachya*)	Ambrosin, coronopilin, damsin, parthenin, parthenolide, psilostachyin A, B and C
Anthemis cotula	Dog fennel, stinking may-weed	Anthecotulide
Arnica spp.	Arnica (= e.g., *A. montana*)	Carabron, helenalin, xanthalongin
Artemisia spp.	Mugwort (= *A. vulgaris*)	Arteglasin A, cumambrin A and B, 8-deoxycumambrin, ludovicin A, B and C
Chamaemelum nobile	Sweet (Roman) chamomile	3-Dehydronobilin[b], 1,10-epoxy-nobilin[b], isohydroxynobilin[b], nobilin[b]
Chamomilla recutita	Wild chamomile, German chamomile	Anthecotulide
Cichorium intybus	Chicory, escarole, witloof	Lactucin, lactupicrin
Cnicus benedictus	Blessed thistle	Cnicin
Cynara scolymus	Artichoke	Cynaropicrin, grossheimin
Dendranthema spp.	Chrysanthemum (= e.g., *D. indicum* and its cultivars)	Arteglasin A
Dittrichia viscosa	Sticky elecampane, stinkwort	Deacetoxyxanthinin, inuviscolide
Eupatorium cannabinum	Hemp agrimony	Eupatoriopicrin
Gaillardia spp.	Blanket flower (= *G. pulchella*)	Helenalin, spathulin
Helenium spp.	Sneezeweed, false sunflower (= *H. autumnale*)	Alternilin, autumnolide, florilenalin, helenalin, isotenulin, tenulin
Helianthus annuus	Sunflower	4,5-Didehydroniveusin A, niveusin B and C
Inula spp.	Elecampane (= *I. helenium*)	Alantolactone, isoalantolactone
Iva spp.	Marshelder (= *I. xanthifolia*)	Coronopilin, ivalin, ivasperin, parthenin
Lactuca sativa	Lettuce	Lactucin, lactupicrin
Parthenium spp.	Congress grass, carrot weed, Indian ragwort (India) (= *P. hysterophorus*)	Ambrosin, coronopilin, guayulin A and B, hymenin, parthenin, tetraneurin E
Saussurea lappa	Costus root	Costunolide, dehydrocostuslactone
Tanacetum spp.	Feverfew (= *T. parthenium*)	Parthenolide, pyrethrosin, reynosin
Taraxacum officinale	Dandelion	Taraxinic acid-1′-*O*-β-glucopyranoside
Telekia speciosa	Large yellow ox-eye	Alantolactone
Jubulaceae (liverworts)		
Frullania spp.	Frullania (= e.g., *F. dilatata*)	Costunolide, (−)- and (+)-frullanolide
Lauraceae		
Laurus nobilis	Sweet bay, bay leaf, noble laurel	Costunolide, dehydrocostuslactone, laurenobiolide
Magnoliaceae		
Magnolia spp.	Magnolia (= *M. grandiflora*)	Costunolide, parthenolide

Adapted from Evans FJ and Schmidt RJ (1980) Plants and plant products that induce contact dermatitis. *Planta Medica* 38, 289–316. Hausen BM (1988) *Allergiepflanzen – Pflanzenallergene: Handbuch und Atlas der Allergie-Induzierenden Wild- und Kulturpflanzen. Kontaktallergene.* Landsberg: Ecomed Verlagsgesellschaft mbH. Lovell CR (1993) *Plants and the Skin.* Oxford: Blackwell Scientific Publications.

[a] If more than one plant species from a genus is known to be allergenic and contains sensitizing SQLs, the common name(s) of the selected plant species is followed by its botanical name in parentheses.

[b] Potential allergen but not biologically tested.

SQLs have been suspected as a cause of airborne contact dermatitis. ACD to plants containing SQLs accounts for close to 50% of all reported cases of plant dermatitis. In nearly all cases where SQLs are skin sensitizers, they contain the α-methylene-γ-butyrolactone entity, a powerful electrophile, as part of their structure. This moiety is known to react readily with amino groups of lysine residues in skin proteins, forming antigens. This indicates that the α-methylene-γ-butyrolactone moiety is fundamental for the allergenic activity of SQLs (**Figure 4**).

SQLs can be classified according to their carbocyclic skeleton. Allergenic SQLs have primarily been found within the four major classes or structural groups of SQLs: germacranolides, guaianolides, eudesmanolides, and pseudoguaianolides (**Figure 3**). Cross-reactions are frequent between SQLs inside a structural group, such as between the eudesmanolides alantolactone and isoalantolactone, whereas cross-reactions seem to be rare between SQLs belonging to different groups. SQLs of the same structural group occur in unrelated families and even the same sensitizing SQLs are often found in more than one plant species (**Table 1**). As a result, individuals who become sensitized to one plant species may also be allergic to other species. For example, the allergenic SQLs of *Frullania* species (liverworts), primarily made up of the eudesmanolides (+)-frullanolide and (−)-frullanolide (**Figure 3**), cross-react with related SQLs of higher plants. A striking case of noncross-reactivity is provided by the Asteraceae species *Parthenium hysterophorus*. A population of this species grows as a weed (Indian ragwort) in India where it is a notorious cause of ACD. The active component is the SQL parthenin. However, in another population of the same species that grows in South America, parthenin is replaced by its diastereomer hymenin (**Figure 3**). Individuals sensitized to the Indian strain of *P. hysterophorus* show sensitization responses when patch tested with parthenin, but not with hymenin. Lack of cross-sensitivity between these epimers has also been demonstrated in studies on guinea pigs.

More than 3000 SQLs have been isolated from plants, of which approximately 100 have been able to cause ACD. At least a further 100 SQLs are potential contact allergens by virtue of their structural features but have not yet been biologically tested. Examples of plant sources for allergenic SQLs are given in **Table 1**.

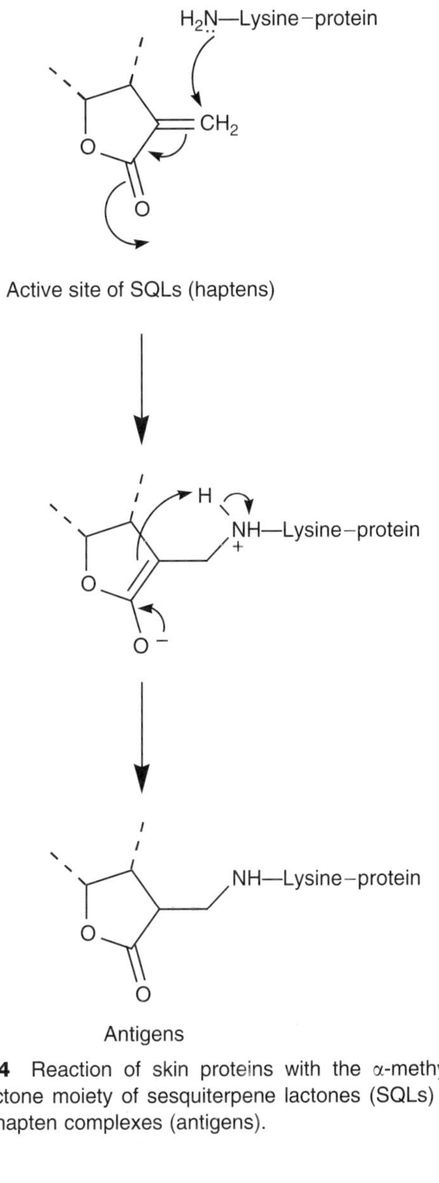

Figure 4 Reaction of skin proteins with the α-methylene-γ-butyrolactone moiety of sesquiterpene lactones (SQLs) to form protein–hapten complexes (antigens).

Tulipalin A and its precursors Contact dermatitis from *Alstroemeria* species (Alstroemeriaceae) and tulips (Liliaceae) is well known in the horticultural industry. The condition evoked is known as "tulip fingers" and gives rise to eczematous dermatitis of the hands, which may spread to the forearms, face, and neck. Patients with "tulip fingers" typically present fissuring, erythema, scaling, and tenderness of the fingers and fingertips. The causative agents have been identified as tuliposides and free α-methylene-γ-butyrolactone (tulipalin A), which have also been found in other genera of Alstroemeriaceae and Liliaceae (see **Figure 5** and **Table 2**). Despite the structural relationship between tulipalin A and SQLs, there is no evidence of any cross-reactivity between them. Likewise, people sensitive to tulips and *Alstroemeria* are not normally sensitive to species of the Asteraceae family, and vice versa. Tulipalin A is formed from tuliposides by enzymatic hydrolysis (**Figure 5**). It has antibiotic properties and is released in large amounts upon fungal attack or after excision of plant parts, but is also present *in situ*, although in relatively small amounts compared to tuliposides.

Quinones and their precursors A very common cause of ACD in central and northern America is

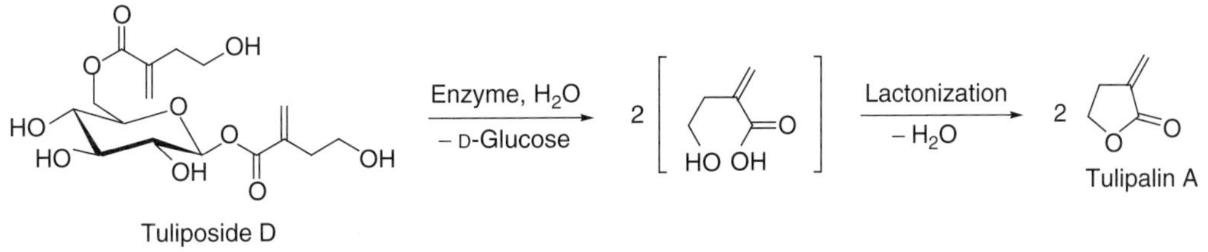

Figure 5 The formation of tulipalin A (α-methylene-γ-butyrolactone) from tuliposides (e.g., tuliposide D).

Table 2 Distribution of tuliposides and tulipalin A in the plant families Alstroemeriaceae and Liliaceae

Family/genus	*Tuliposides and tulipalin A*
Alstroemeriaceae	
Alstroemeria spp. (Peruvian lily)[a]	Detected in all species investigated (>30) including several cultivars
Bomarea spp.	Detected in all species investigated (6)
Liliaceae	
Erythronium spp.	Detected in all species investigated (4) including a few cultivars
Gagea spp.	Detected in nearly all species investigated (5)
Tulipa spp. (tulip)[a]	Detected in all species investigated (>50) including many cultivars

Adapted from Christensen LP, Kristiansen K, and Ørgaard M (2000) Alstroemeriaceae. In: Avalos J and Maibach HI (eds) *Dermatologic Botany. Dermatology: Clinical and Basic Science Series*, pp. 273–310. Boca Raton: CRC Press.
[a]Common name in parentheses.

exposure to *Toxicodendron* species, especially poison ivy and poison oak. The allergenic compounds of poison ivy and poison oak and related species of the Anacardiaceae family are the urushiols, a mixture of 3-alkyl and 3-alkenyl catechols with side chain lengths of usually 15 or 17 carbon atoms. The number of double bonds in the 3-alkenyl catechols vary between one and three, and are normally situated at positions 8, 11, and/or 14 of the alkenyl side chain. Furthermore, the urushiols from different Anacardiaceae species differ in the relative proportions of the individual components. In poison ivy, the alkyl and alkenyl groups are, for example, predominantly of C_{15} chain length, whereas in poison oak the chain length is predominantly C_{17}. Allergens that are very closely related to the urushiols have been isolated from the fruit pulp of *Ginkgo biloba* (Ginkgoaceae) and from species of the Proteaceae family. Some of these allergens occur also in members of the Anacardiaceae family (see **Figure 6** and **Table 3**).

The 3-alkyl and 3-alkenyl catechols are readily oxidized to the corresponding orthoquinones, which are highly reactive toward nucleophilic sites in proteins, and there is a substantial body of evidence suggesting that these quinones are the actual haptens reacting with skin proteins to form antigens (**Figure 7**). The urushiols are in this context regarded as prohaptens as they need to be oxidized to their more reactive analog.

The "phacelioids" are a group of quinones and phenols that are structurally related to the urushiols and some of which have highly sensitizing properties (**Figure 8**). The phacelioids have been found in *Phacelia* and related species of the Hydrophyllaceae family and they are responsible for the allergenic properties of these plants (**Table 3**). Geranylbenzoquinone, from *Phacelia* species, is, like many other quinones, a very potent allergen. It is about 30 times stronger than the corresponding geranylhydroquinone that is also a powerful allergen. These findings are consistent with the theory that phenolic contact

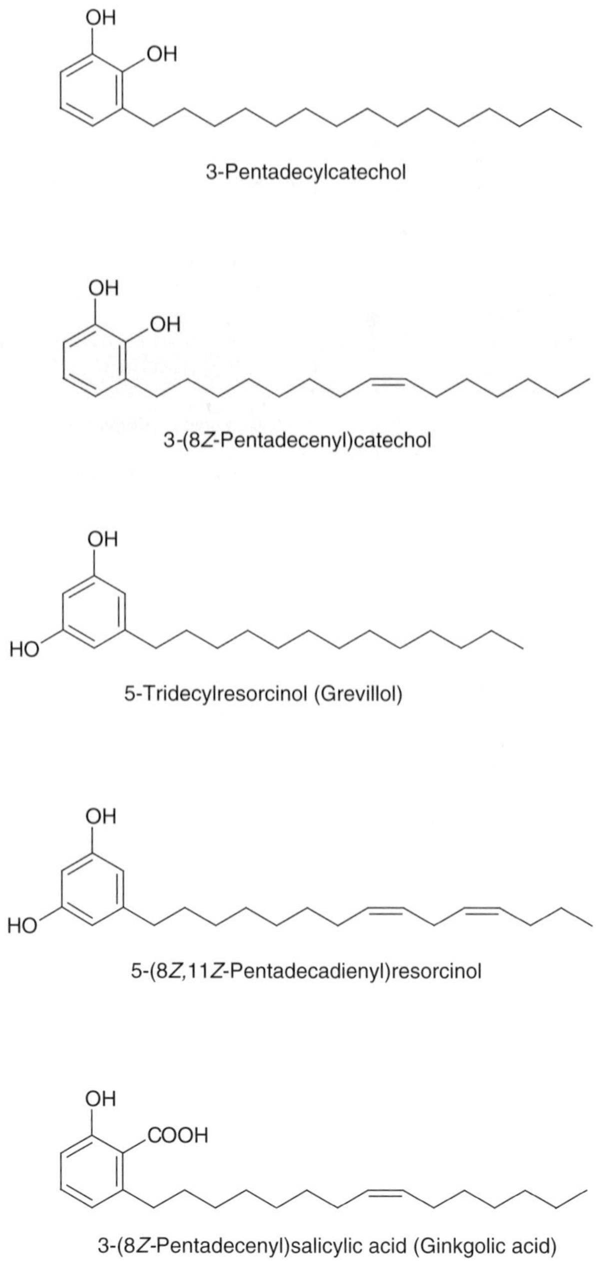

Figure 6 Examples of urushiols (3-pentadecylcatechol and 3-(8*Z*-pentadecenyl)catechol) and structurally related allergens.

Table 3 Examples of common plant sources of allergenic quinones and their precursors

Family/genus/species	*Common name(s) of selected species*	*Quinones and their precursors*
Anacardiaceae		
Anacardium occidentale	Cashew nut	C_{15}-Resorcinols (e.g., 5-(8*Z*,11*Z*-pentadecadienyl)resorcinol), anacardic acid, ginkgolic acid
Mangifera spp.	Mango (= *M. indica*)	C_{17}-Resorcinols (e.g., 5-(2*Z*-heptadecenyl)-resorcinol)
Metopium spp.	Chechém negro (= *M. brownei*)	C_{15}- and C_{17}-Urushiols (e.g., 3-pentadecyl-catechol)
Semecarpus spp.	Indian marking-nut tree, Dhobi's nut (= *S. anacardium*)	C_{15}- and C_{17}-Urushiols (e.g., 3-(8*Z*-heptadecenyl)-catechol)
Smodingium argutum	African poison ivy	C_{15}- and C_{17}-Urushiols (e.g., 3-penta- and 3-hepta-decylcatechol)
Toxicodendron spp.	Poison ivy (= *T. radicans*), Poison oak (= *T. diversilobum*)	C_{15}- and C_{17}-Urushiols (e.g., 3-pentadecylcatechol, 3-(8*Z*-pentadenyl)catechol, 3-heptadecylcatechol)
Araceae		
Philodendron scandens subsp. *oxycardium*	Sweetheart vine	5-(8*Z*,11*Z*,14*Z*-heptadecatrienyl)resorcinol
Bignoniaceae		
Paratecoma peroba	Peroba do Campos, Peroba amarella	Lapachol and related quinones
Tabebuia spp.	Lapacho (used for several species)	Deoxylapachol, lapachol and related quinones
Cupressaceae		
Calocedrus spp.	Incense cedar (= *C. decurrens*)	Thymoquinone
Thuja spp.	Western red cedar (= *T. plicata*)	Thymoquinone
Ebenaceae		
Diospyros spp.	African ebony (= *D. crassifolia*), Macassar (= *D. celebica*)	Macassar quinone and related quinones
Fabaceae		
Acacia spp.	Australian blackwood (= *A. melanoxylon*)	Acamelin, 2,6-dimethoxy-1,4-benzoquinone, melacacidin
Bowdichia spp.	Sucupira (= *B. nitida*)	Bowdichione, 2,6-dimethoxy-1,4-benzoquinone
Dalbergia spp.	Cocobolo (= *D. retusa*), Brazilian rosewood (= *D. nigra*)	Dalbergiones (e.g., (*R*)- and (*S*)-4-methoxydalbergione, (*S*)-4,4′-dimethoxydalbergione, (*S*)-4′-hydroxy-4-methoxydalbergione)
Machaerium scleroxylon	Pao ferro, Caviuna vermelha, "Santos rosewood"	(*R*)-3,4-Dimethoxydalbergione
Gesneriaceae		
Streptocarpus spp.	Cape primrose (= *Streptocarpus* hybrids and species)	2,6-Dimethoxy-1,4-benzoquinone
Ginkgoaceae		
Ginkgo biloba	Ginkgo, Maidenhair tree	Ginkgolic acid and related compounds
Hydrophyllaceae		
Phacelia spp.	Desert heliotrope, scorpion weed (= *P. crenulata*)	Phacelioids (e.g., geranylbenzoquinone, geranyl-hydroquinone, 1-oxofarnesylhydroquinone, geranylgeranylhydroquinone)
Turricula parryi	No common name	Phacelioids (e.g., turriculoic acid-B, turriculol-E, turriculol-F)
Wigandia spp.	No common names	Phacelioids (e.g., 2,3-dimethoxygeranyl-benzoquinone, farnesylhydroquinone, wigandol)
Orchidaceae		
Cymbidium spp.	No common name	2,6-Dimethoxy-1,4-benzoquinone
Cypripedium calceolus	Lady's slipper	Cypripedin
Primulaceae		
Primula spp.	Poison primrose (= *P. obconica*)	Primin (found in >8 species) and primetin (found in >14 species)
Proteaceae		
Grevillea spp.	Silk oak (= *G. robusta*)	5-Tridecylresorcinol (grevillol) and related resorcinols

Continued

Table 3 Continued

Family/genus/species	*Common name(s) of selected species*	*Quinones and their precursors*
Sterculiaceae		
Mansonia altissima	African black walnut, Mansonia, Bété	Mansonone A
Verbenaceae		
Tectonia grandis	Teak	Deoxylapachol, lapachol

Adapted from Aregullin M and Rodriguez E (2000) Hydrophyllaceae. In: Avalos J and Maibach HI (eds) *Dermatologic Botany. Dermatology: Clinical and Basic Science Series*, pp. 187–199. Boca Raton: CRC Press, Christensen LP (2000) Primulaceae. In: Avalos J and Maibach HI (eds) *Dermatologic Botany. Dermatology: Clinical and Basic Science Series*, pp. 201–235. Boca Raton: CRC Press, Evans FJ and Schmidt RJ (1980) Plants and plant products that induce contact dermatitis. *Planta Medica* 38, 289–316, Guin JD, Beaman JH, and Baer H (2000) Toxic Anacardiaceae. In: Avalos J and Maibach HI (eds) *Dermatologic Botany. Dermatology: Clinical and Basic Science Series*, pp. 85–142. Boca Raton: CRC Press, Hausen BM (1988) *Allergiepflanzen – Pflanzenallergene: Handbuch und Atlas der Allergie-Induzierenden Wild- und Kulturpflanzen. Kontaktallergene*. Landsberg: Ecomed Verlagsgesellschaft mbH, and Lovell CR (1993) *Plants and the Skin*. Oxford: Blackwell Scientific Publications.

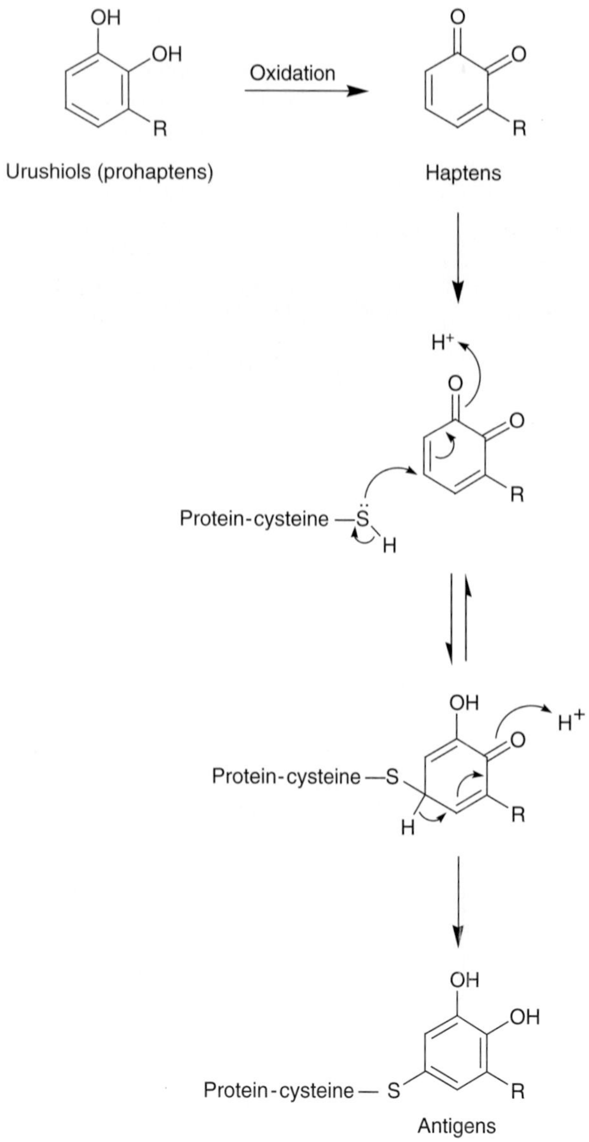

Figure 7 Possible mechanism for the reaction of skin proteins with urushiols (prohaptens) to form protein–hapten complexes (antigens). R = 3-akyl or 3-alkenyl side chain.

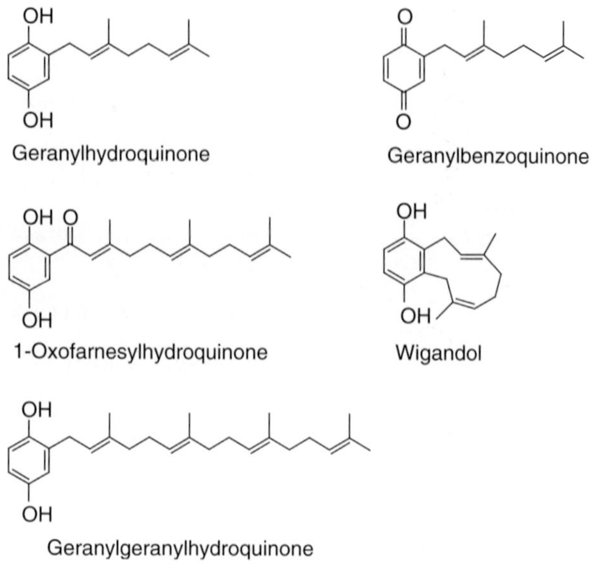

Figure 8 Chemical structures of some allergenic hydroquinones and quinones (phacelioids) from the Hydrophyllaceae family.

allergens require oxidation to a quinone in order to gain sensitizing activity, as in the case of the allergenic urushiols.

One of the common causes of natural product-induced dermatitis in Europe has been the quinone primin (**Figure 9**) from *Primula obconica* (poison primrose) of the Primulaceae family. *Primula obconica* is widely used as an indoor ornamental plant, especially in northern Europe. Primin is one of the most sensitizing substances isolated from plants so far and has also been found in a few other *Primula* species and in the unrelated *Miconia* species of the Melastomataceae family (**Table 3**). In *P. obconica*, primin is situated in trichomes on the surface of the plant, and it has been shown that this strong contact allergen can be released directly into the air; consequently, this allergen may cause airborne

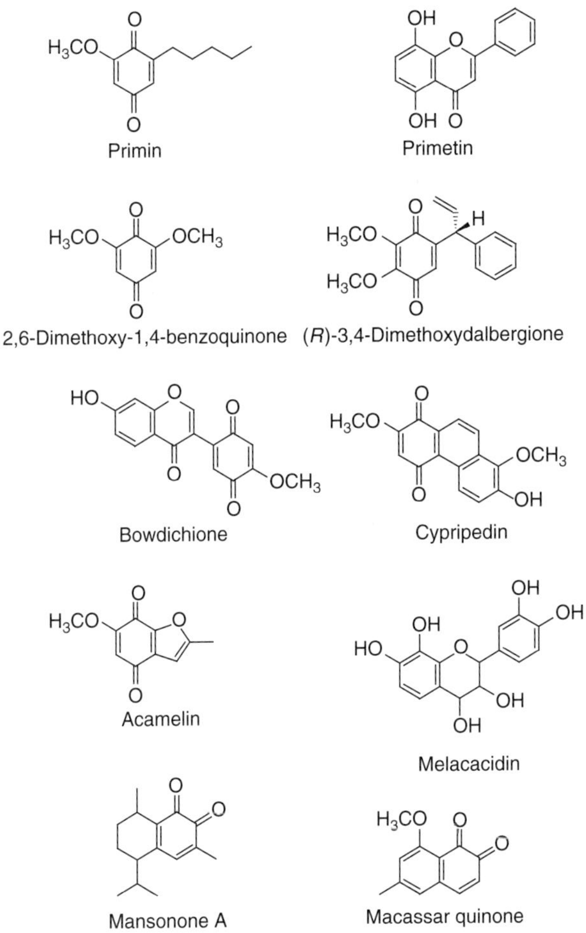

Figure 9 Examples of allergenic quinones and related phenols found in various sensitizing plant species (see **Table 4**).

contact dermatitis in highly sensitized individuals. However, primin is not the only allergen present in members of the genus *Primula*. For example, the main sensitizer in *Primula mistassinica* has been identified as the dihydroxyflavone primetin, which has also been found in several other *Primula* species (**Figure 9** and **Table 3**). The sensitizing properties of primetin are probably due to its uncommon arrangement of hydroxy groups that easily can be oxidized on the skin to the corresponding sensitizing primetinquinone.

Brazilian rosewood (*Dalbergia nigra*) and the heartwoods of other *Dalbergia* species, as well as related species such as *Machaerium scleroxylon* (Pao ferro) of the Fabaceae family, are used to make high-class furniture and cabinetwork, knife handles, wooden jewelry, and musical instruments, and can cause occupational contact dermatitis (especially musical instrument manufacture). The responsible allergens are quinones that belong to the class of neoflavonoids, named dalbergiones, of which the strongest sensitizer is (*R*)-3,4-dimethoxydalbergione (see **Table 3** and **Figure 9**). The sensitizers found in Australian blackwood (*Acasia melanoxylon*), also from the Fabaceae family, have been identified as the quinones acamelin and 2,6-dimethoxy-1,4-benzoquinone, and the hexahydroxyflavan melacacidin. 2,6-Dimethoxy-1,4-benzoquinone has also been found to be the causative agent in the heartwoods of the genus *Bowdichia* (Fabaceae), together with the isoflavone quinone bowdichione (**Figure 9**). Besides hardwoods, 2,6-dimethoxy-1,4-benzoquinone has been found to be the main allergen in *Streptocarpus* hybrids and species (Gesneriaceae), and is one of the few quinones that appears to be relatively widespread in nature (see **Table 3**). Other examples of quinone sensitizers from hardwoods are the orthoquinones: macassar quinone and related quinones from *Diospyros* species (Ebenaceae), commonly known as Ebony, and mansonone A from *Mansonia altissima* (Sterculiaceae), also known as Mansonia or Bété (see **Table 3** and **Figure 9**). Teak from the Verbenaceae family is one of the most valuable timbers in the world, but is also a common cause of ACD. As mentioned above, the allergens involved have been identified as the naphtoquinones deoxylapachol and lapachol (**Figure 2**), which are also common in many species of the Bignoniaceae family (see **Table 3**).

In the Orchid family (Orchidaceae), one of the largest families in the plant kingdom, species from several genera, including *Cypripedium*, *Paphiopedilum*, and *Cymbidium*, are known to cause ACD. The allergens responsible are quinones and the main sensitizers have so far only been identified in *Cypripedium calceolus* and in *Cymbidium* species as cypripedin and 2,6-dimethoxy-1,4-benzoquinone, respectively (**Figure 9**).

Polyacetylenes and other allergens The polyacetylene falcarinol is, like many other highly biologically active compounds, very reactive. For example, it reacts easily with amino acids in skin proteins forming antigens. Falcarinol is widespread in the Apiaceae and Araliaceae families and has been identified as the major allergen in common plant species such as English ivy (*Hedera helix* subsp. *helix*, Araliaceae), carrot (*Daucus carota*, Apiaceae), and parsley (*Petroselinum crispum*, Apiaceae) (see **Table 4** and **Figure 10**). The high concentrations of falcarinol in some plant species such as English ivy sometimes result in irritant reactions. The concentration of falcarinol is increased in, for example, carrot roots when they are infected with the fungus *Fusarium*, suggesting that falcarinol may have antifungal action. Another example of a strong sensitizing polyacetylene is (*R*)-octadeca-9*Z*,17-diene-12,14-diyne-1,16-diol (**Figure 10**). This

Table 4 Examples of some nonquinoid contact allergens and their botanical sources (see **Figures 10** and **12**)

Plant allergens	*Plant source(s)*	*Common name(s)*
Allicin, diallyl disulfide	*Allium sativum* (Alliaceae)	Garlic
Allyl isothiocyanate	*Brassica nigra* (Brassicaceae) and may also be present in other Brassicaceae species	Black mustard
Chlorophorin	*Chlorophora excelsa* (Moraceae)	African teak, iroko, kambala
Coleon O	*Coleus* spp. (Lamiaceae)	Coleus
Cryptopleurine	*Cryptocarya pleurosperma* (Lauraceae)	Poison walnut
Falcarinol	Widely distributed in the Apiaceae and Araliaceae families	Carrot, parsley (Apiaceae); English ivy, umbrella tree, ginseng (Araliaceae)[a]
Hydrangenol	*Hydrangea* spp. (Hydrangeaceae)	Hydrangea
Masonin	*Narcissus* spp. and hybrids (Amaryllidaceae)	Daffodil, narcissus
Oxyayanin A	*Distemonanthus benthamianus* (Fabaceae)	Ayan, Nigerian satinwood
Sesamin	*Sesamum indicum* (Pedaliaceae)	Sesame
(+)-Usnic acid	Widely distributed in lichens	Lichens

Adapted from Evans FJ and Schmidt RJ (1980) Plants and plant products that induce contact dermatitis. *Planta Medica* 38, 289–316, Hausen BM (1988) *Allergiepflanzen – Pflanzenallergene: Handbuch und Atlas der Allergie-Induzierenden Wild- und Kulturpflanzen. Kontaktallergene.* Landsberg: Ecomed Verlagsgesellschaft mbH, and Lovell CR (1993) *Plants and the Skin.* Oxford: Blackwell Scientific Publications.

[a]Examples of common species from the Apiaceae and Araliaceae containing falcarinol.

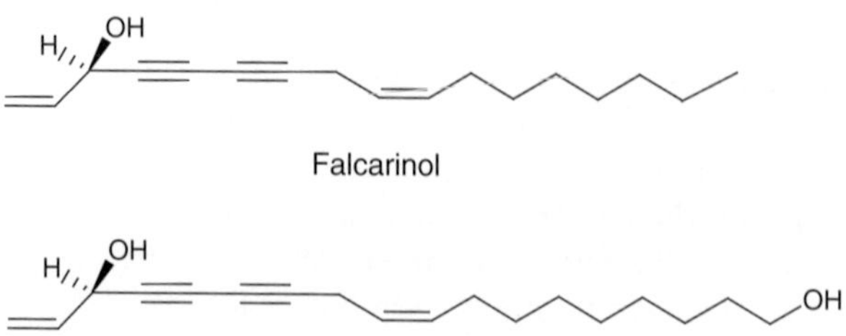

Figure 10 Chemical structures of falcarinol and (*R*)-octadeca-9*Z*,17-diene-12,14-diyne-1,16-diol, examples of highly sensitizing polyacetylenes in the Apiaceae and Araliaceae plant families.

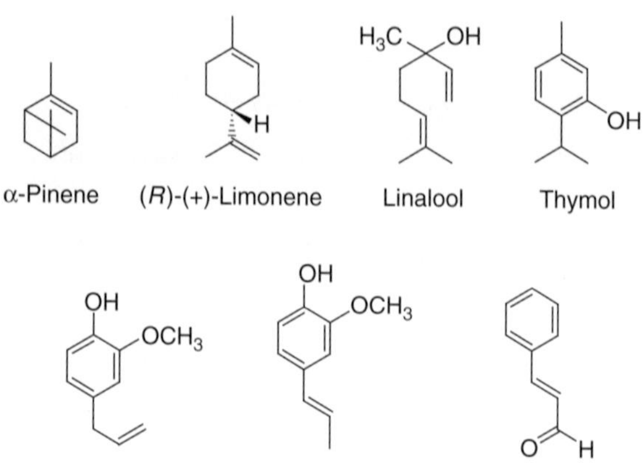

Figure 11 Examples of some allergenic volatile compounds common in various essential oil preparations.

polyacetylene, which show cross-reactivity to falcarinol, is the major allergen in the Japanese species *Dendropanax trifidus* (Japanese name kakuremino) of the Araliaceae family.

The essential oils are complex mixtures of substances containing terpenoids and phenylpropanoids with low boiling points, i.e., phenols, esters, alcohols, and ethers. In general, volatile oils have low allergenic potential and may, in some cases, be weak irritants as well. Examples of some common allergenic constituents of essential oils are given in **Figure 11**.

Several other secondary plant metabolites are known to give rise to allergic reactions. These substances belong to various chemical classes, including alkaloids, flavonoids, stilbenes, lignans, diterpenes, isocoumarins, and the lichen compounds (**Figure 12**). Examples of some miscellaneous allergens are listed in **Table 4**, together with their botanical sources.

A special group of allergenic substances are those produced by fungal endophytes. Fungal endophytes infect many grass species and often confer benefits to the hosts such as reduced herbivory by insects and animals owing to their production of toxic and allergenic substances. In some parts of the world, e.g., Australia and New Zealand, facial eczema of sheep caused by compounds produced by fungal endophytes is a very big problem. The causative agents have been shown to be composed of toxic and allergenic alkaloids, such as the sporidesmins.

Phototoxic Compounds

The term phytophotodermatitis describes the skin eruption induced by contact with plants followed by exposure to sun. The condition is caused by plants belonging to a small number of families, notably Apiaceae, Rutaceae, and Moraceae. Children at play and gardeners are particularly at risk. Phototoxic

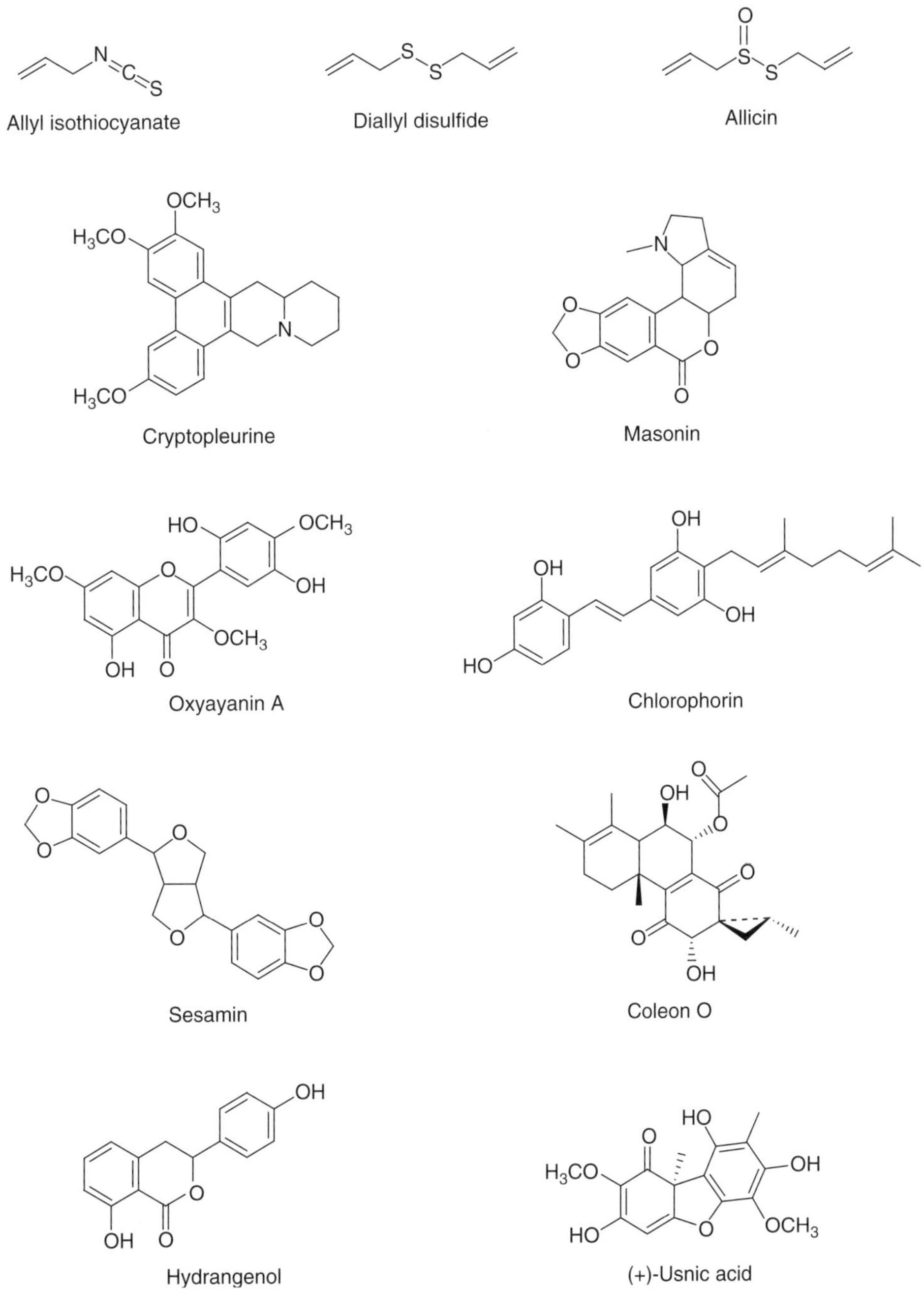

Figure 12 Chemical structures of some nonquinoid allergens from various plant sources (see **Table 4**).

reactions are often distinctive, although they are sometimes misdiagnosed. Typically a streaky, often bullous, erythematous eruption occurs on light-exposed areas, which may sometimes resemble that of sunburn. The lesions are painful rather than itchy and leave a characteristic gray-brown area of hyperpigmentation, which may persist for several months. This type of reaction is due to phototoxic compounds that act as primary irritants. Since no allergic mechanism is involved, phototoxic reactions can occur in any individual given sufficient exposure to the phototoxins and long-wave ultraviolet radiation (UV-A). Psoralens (furocoumarins) appear to be responsible for the majority of phototoxic reactions in humans caused by plants. Common phototoxic psoralens are shown in **Figure 13**. Although other phototoxic substances (e.g., α-terthienyl and phenylheptatriyne) (**Figure 14**) have been isolated from members of the Asteraceae family, they are probably not of clinical relevance. The same is true for the phototoxic compound hypericin (**Figure 14**) from St. John's wort (*Hypericum perforatum*,

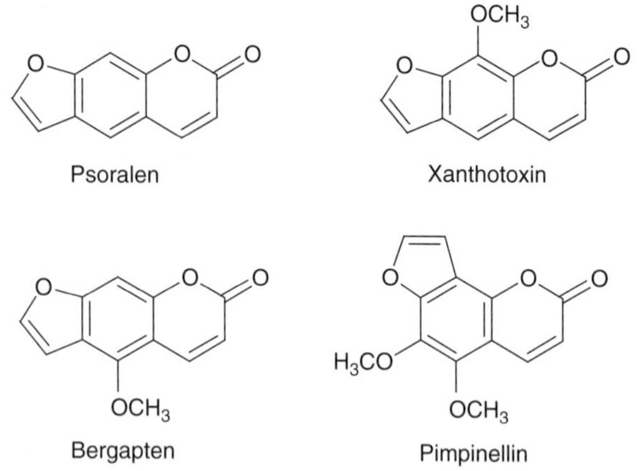

Figure 13 Chemical structures of some common photosensitizing psoralens.

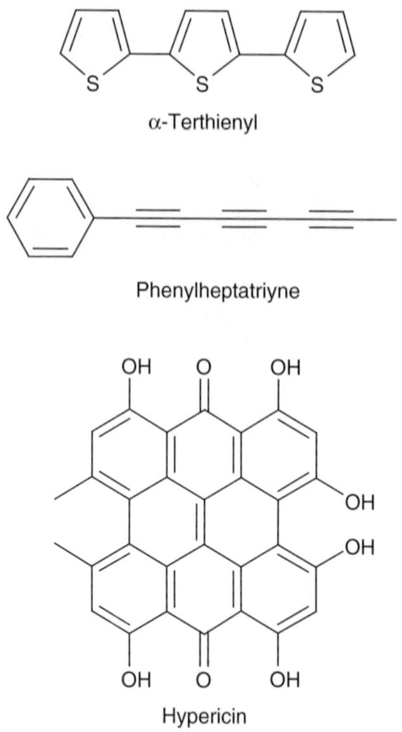

Figure 14 Phototoxic compounds from the Asteraceae (α-terthienyl and phenylheptatriyne) and from *Hypericum* species (hypericin) of the Hypericaceae family, which may not be of clinical relevance.

Hypericaceae) and other *Hypericum* species. However, hypericin can cause severe phototoxicity in grazing animals. When hypericin is absorbed by the animal and enters the peripheral circulation, subsequent exposure to sunlight will cause severe sunburn and other damage such as serious necrosis of the skin followed by infection and starvation.

The phototoxic action of the psoralens is probably due to their ability to react with nucleobases in DNA under the influence of UV-A radiation. On further exposure to UV-A radiation, these adducts react with neighboring nucleobases giving rise to crosslinks in the DNA (**Figure 15**). The resulting interstrand crosslinks in DNA lead to the characteristic acute bullous lesions. The mechanism for the hyperpigmentation is less clear and may occur independently of the acute inflammatory reaction. Although no allergic mechanism is involved in phototoxic reactions, psoralens are occasional sensitizers. Photoallergic reactions to plants occur rarely and have only been reported a few times for psoralens. The major allergenic psoralens appear to be the relatively poorly phototoxic compounds sphondin and isobergapten (**Figure 16**).

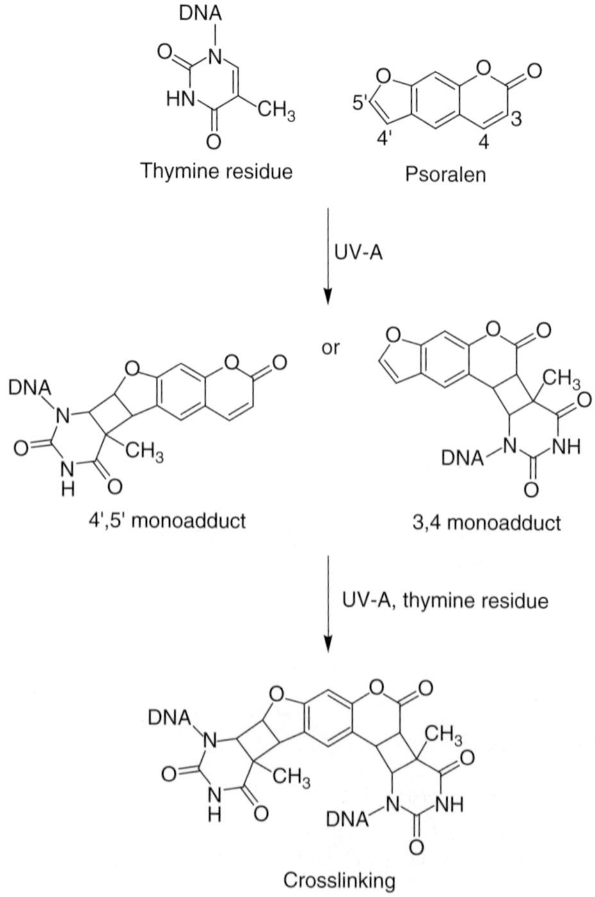

Figure 15 Possible mechanism for the reaction of psoralens with nucleobases in DNA (e.g., thymine residues) under the influence of UV-A radiation resulting in the formation of interstrand crosslinks in DNA, and hence leading to photodermatitis.

The psoralens have only limited distribution in the plant kingdom and are found primarily within the families of Apiaceae, Moraceae, Rosaceae, Rutaceae, and Fabaceae. The distribution of psoralens varies widely between plant species and may even vary within a species according to geographical location

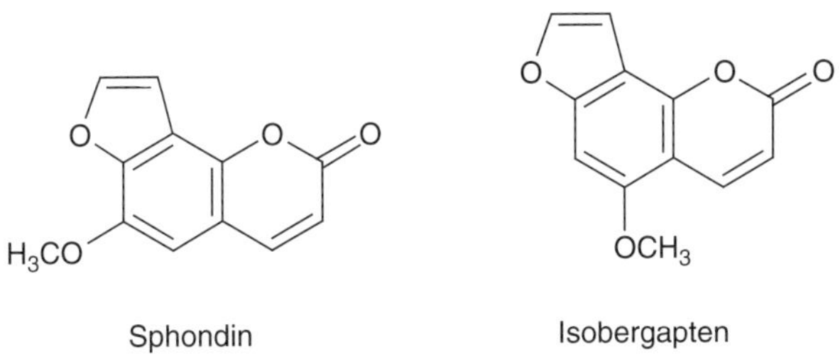

Figure 16 Sphondin and isobergapten are examples of allergenic psoralens. These compounds are poorly phototoxic.

and climate. The content of psoralens with antifungal properties might increase upon fungal attack. It has been shown that the development of disease-resistant cultivars of vegetables such as celery (*Apium graveolens*) and parsnip (*Pastinaca sativa*) of the Apiaceae family have resulted in increased concentrations of psoralens in these plants and, hence, a greater risk for photodermatitis. Some of the most common psoralen-containing plant species found to elicit photodermatitis are given in **Table 5**.

Table 5 Plant species containing psoralens (furocoumarins) and reported to cause photodermatitis

Family/genus/species	*Common name(s)*
Apiaceae	
Ammi majus	Bishop's weed, Queen Anne's lace
Anethum graveolens	Dill
Angelica archangelica	Angelica
A. sylvestris	Wild angelica
Anthriscus cereifolium	Garden chervil
A. sylvestris	Cow parsley, wild chervil
Apium graveolens	Celery
Daucus carota	Carrot
Foeniculum vulgare	Fennel
Heracleum laciniatum[a]	Tromsø palm
H. mantegazzianum	Giant Russian hogweed, wild rhubarb
H. sphondylium	European cow parsnip, hogweed
Levisticum officinale	Lovage
Pastinaca sativa[a]	Parsnip
Petroselinum crispum	Parsley
Peucedanum ostruthium[a]	Masterwort
Fabaceae	
Psoralea corylifolia	Bavchi, scurf pea
Moraceae	
Ficus carica	Fig
Rutaceae	
Citrus aurantifolia[a]	Lime
C. bergamia	Bergamot
C. limetta	Sweet lemon
C. limon	Lemon
C. paradisi	Grapefruit
C. sinensis	Sweet orange
Dictamnus albus	Burning bush, gas plant
Pelea anisata	Mokihana
Ruta chalepensis[a]	Frinked rue
R. graveolens	Rue

Adapted from Hausen BM (1988) *Allergiepflanzen – Pflanzenallergene: Handbuch und Atlas der Allergie-Induzierenden Wild- und Kulturpflanzen. Kontaktallergene.* Landsberg: Ecomed Verlagsgesellschaft mbH and Lovell CR (1993) *Plants and the Skin.* Oxford: Blackwell Scientific Publications.
[a]Other species from this genus than those shown in this table contain phototoxic psoralens and have been reported to cause photodermatitis.

Irritants

Plants can irritate the skin by mechanical and/or chemical means and the term irritant contact dermatitis is used to describe a cutaneous inflammatory response to a chemical and/or physical injury. In irritant contact dermatitis, no immunological response is mediated, so potentially anyone can be affected and the severity of the response depends on the barrier function of the individual's skin as well as on the potency and duration of the irritant stimulus.

Chemical Irritants

The skin's reaction to chemical irritants occurs shortly after contact. Unlike allergenic substances, chemical irritants will affect all subjects on the first exposure if the concentration is sufficient. However, strong contact allergens, such as primin, falcarinol, and the urushiols (see **Figures 6, 9** and **10**), can, in some cases, also give rise to irritant contact dermatitis in high concentrations. The mechanism by which chemical irritants exert their effect is not completely understood. In general, it is believed that inflammation of the skin arises in response to tissue damage.

Plants with stinging hairs are very common causes of irritant contact dermatitis. The plants are characterized by the presence of hypodermic syringe-like emergences that are capable of penetrating the skin and injecting small amounts of irritating chemicals which produce a relatively mild skin irritation (itching and urticaria). In the case of the common stinging nettle (*Urtica dioica*) and other Urticaceae species, the irritants consist of several pro-inflammatory mediators, including histamine, acetylcholine, and serotonin (**Figure 17**). Plants with stinging hairs have also been recorded in the families of Euphorbiaceae, Loasaceae, and Hydrophyllaceae.

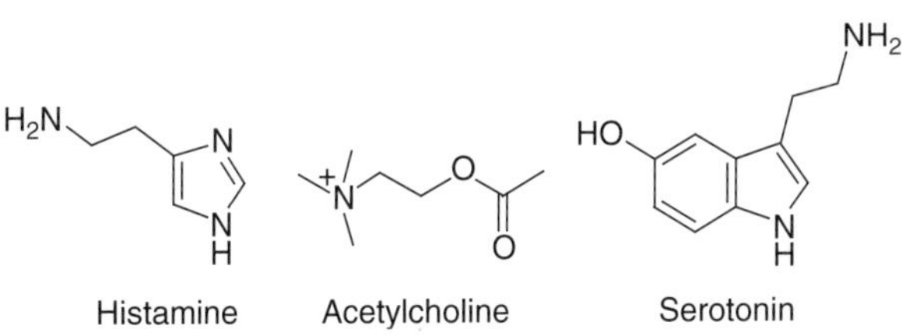

Figure 17 Chemical irritants from the common stinging nettle (*Urtica dioica*) and other species of the Urticaceae family.

Several members of the Ranunculaceae family cause dermatitis when crushed on the skin. The irritant properties of the plants have been ascribed to protoanemonin, which has been isolated from several *Ranunculus* and *Pulsatilla* species. Protoanemonin is produced in damaged plant tissues by enzymatic cleavage of the glucoside ranunculin (**Figure 18**).

Capsaicin and related capsaicinoids are responsible for the pungent taste of, for example, chilli peppers (*Capsicum annuum*, Solanaceae) and are powerful skin irritants. Capsaicin induces erythema but no blistering in mammalian skin. The gingerols (e.g., 6-gingerol) and shogaols (e.g., 6-shogaol) of ginger (*Zingiber officinale*, Zingiberaceae) are capsaicin-like compounds and are found in the rhizome of this rubefacient species (**Figure 19**). The spicy taste of many Brassicaceae (syn. Cruciferae) species, including mustard (*Brassica nigra*) and horseradish (*Armoracia lapathifolia*), is due to irritant isothiocyanates, such as allyl isothiocyanate (**Figure 12**) and benzyl isothiocyanate (**Figure 19**), which are produced by decomposition of glucosinolate glycosides when the plants are damaged.

Many species of the Euphorbiaceae and Thymelaceae families contain a highly irritant sap or latex. The irritant principles of the Euphorbiaceae are tigliane polyl esters (tetracyclic diterpenoids). Some genera, including *Hippomane*, produce daphnane (tricyclic diterpenoid) and ingenane orthoesters (**Figure 20**), also found in members of the daphne family (Thymelaceae).

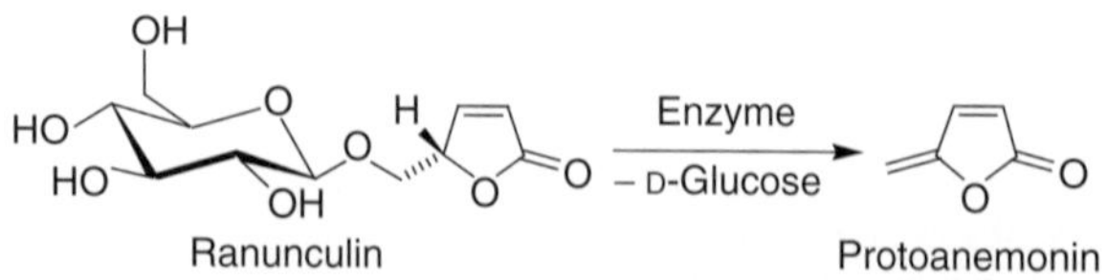

Figure 18 The formation of the unsaturated lactone protoanemonin in members of the Ranunculaceae family by means of simple enzyme-induced cleavage of the glucoside ranunculin.

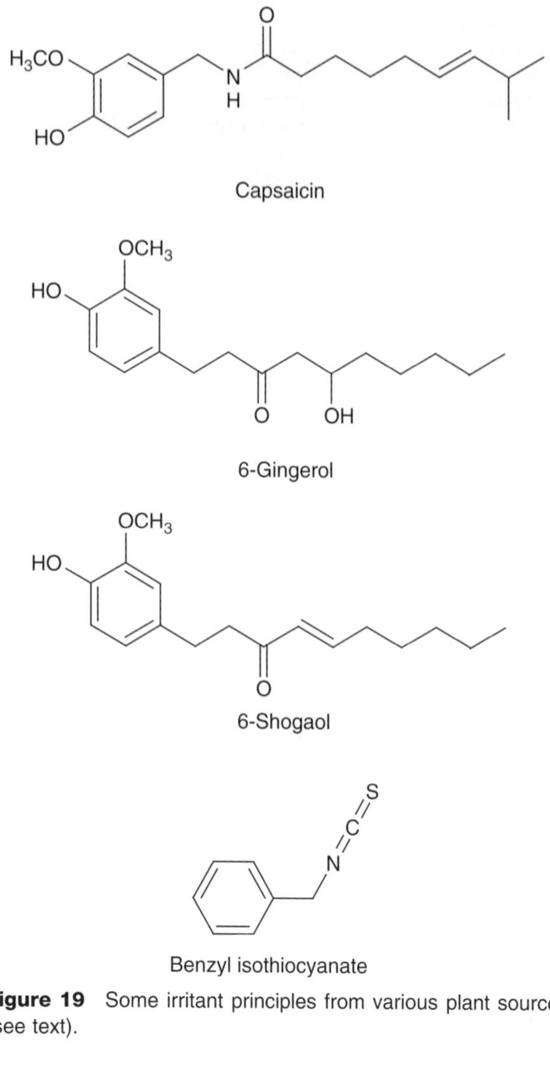

Figure 19 Some irritant principles from various plant sources (see text).

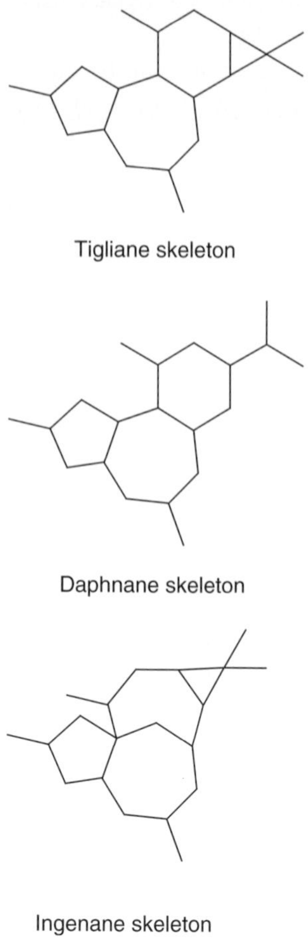

Figure 20 The irritant principles of the Euphorbiaceae and Thymelaceae are esters of closely related diterpene polyols based on the tigliane, daphnane, and ingenane hydrocarbon skeletons.

Mechanical Irritants

Many plant species are able to cause an inflammatory response by mechanical damage of the skin. The mechanical damage can be caused by thorns (roses, Rosaceae), spines and glochids (cacti, especially *Opuntia* species, Cactaceae), stiff hairs (several members of the borage family, Boraginaceae), T-shaped hairs (American dogwoods, Cornaceae), irritant fibers (rose hips, Rosaceae; tulip bulbs, Liliaceae), and by means of minute needle-like calcium oxalate crystals that are present in the tissues of many plants. These crystals, thought to be waste products of cellular metabolism, may themselves give rise to irritant contact dermatitis or may act, in part, by enhancing the penetration of toxins through the stratum corneum. Calcium oxalate crystals, which are present in the dry outer scales of daffodil (*Narcissus* species) and hyacinth (*Hyacinthus* species) bulbs are, for example, partly responsible for the condition known as lily rash and hyacinth itch. An allergic sensitivity may co-occur, especially in *Narcissus* species where the main allergen seems in be masonin (see Figure 12).

List of Technical Nomenclature

Allergic contact dermatitis (ACD)	A specific immunological response to an external substance such as secondary plant products.
Antigen	A substance that stimulates the production of antibodies.
Contact allergens	Compounds that are able to penetrate into the epidermis beneath the stratum corneum by contact and to induce an immunological response.
Edema	Effusion of serous fluid into body cavities or into the interstices of cells in tissue spaces.
Emergences	Outgrowths, as a prickle, on the surface of plants.
Epimers	Diastereomers that have a different configuration at only one chiral center.
Erythema	Superficial dermatitis characterized by redness in patches.
Glandular trichomes	Small hairs placed on the surface of plants containing or bearing glands.
Glochids	Tufts of shorter hooked hairs.
Hapten	Nonprotein chemical compounds with allergenic properties.
Hypersensitivity	An allergic response to a substance.
Irritant contact dermatitis	A cutaneous inflammatory response to a physical and/or chemical injury.
Necrosis	The death of a piece of animal tissue.
Nucleobases	Nitrogenous bases that are building blocks in, for example, DNA.
Occupational dermatitis	Dermatitis due wholly or partly to plant contact in the work place.
Papular inflammation	Inflamed swelling of the skin, characterized by a small, solid, rounded, raised area on the skin.
Phototoxic compound	A compound that becomes toxic under UV-A light.
Phytophotodermatitis	Cutaneous reaction on the skin induced by contact with plants followed by exposure to sun.
Prohapten	A substance that after metabolism becomes a hapten.
Vesicular inflammation	Inflamed area of the skin covered with small vesicles containing fluid.

See also: **Secondary Products**: Antifeedant Substances in Plants; Epicuticular Waxes; Pharmaceuticals, Plant Drugs; Phytoalexins, Natural Plant Protection.

Further Reading

Aregullin M and Rodriguez E (2000) Hydrophyllaceae. In: Avalos J and Maibach HI (eds) *Dermatologic Botany. Dermatology: Clinical and Basic Science Series*, pp. 187–199. Boca Raton: CRC Press.

Christensen LP (2000) Primulaceae. In: Avalos J and Maibach HI (eds) *Dermatologic Botany. Dermatology: Clinical and Basic Science Series*, pp. 201–235. Boca Raton: CRC Press.

Christensen LP, Kristiansen K, and Ørgaard M (2000) Alstroemeriaceae. In: Avalos J and Maibach HI (eds) *Dermatologic Botany. Dermatology: Clinical and Basic Science Series*, pp. 273–310. Boca Raton: CRC Press.

Christensen LP and Larsen E (2000) Direct emission of the allergen primin from intact *Primula obconica* plants. *Contact Dermatitis* 42: 149–153.

Evans FJ and Schmidt RJ (1980) Plants and plant products that induce contact dermatitis. *Planta Medica* 38: 289–316.

Gäfvert E and Karlberg A-T (1998) Isolation and identification of contact allergens. In: Lepoittevin J-P, Basketter DA, Goossens A, and Karlberg A-T (eds) *Allergic Contact Dermatitis. The Molecular Basis*, pp. 43–67. Berlin: Springer-Verlag.

Guin JD, Beaman JH, and Baer H (2000) Toxic Anacardiaceae. In: Avalos J and Maibach HI (eds) *Dermatologic Botany. Dermatology: Clinical and Basic Science Series*, pp. 85–142. Boca Raton: CRC Press.

Hausen BM (1988) *Allergiepflanzen – Pflanzenallergene: Handbuch und Atlas der Allergie-Induzierenden Wild- und Kulturpflanzen. Kontaktallergene.* Landsberg: Ecomed Verlagsgesellschaft mbH.
Lovell CR (1993) *Plants and the Skin.* Oxford: Blackwell Scientific Publications.
Lovell CR (1997) Phytodermatitis. *Clinics in Dermatology* 15: 607–613.
Lovell CR (2000) Phytophotodermatoses. In: Avalos J and Maibach HI (eds) *Dermatologic Botany. Dermatology: Clinical and Basic Science Series*, pp. 51–65. Boca Raton: CRC Press.
Mitchell J and Rook A (1979) *Botanical Dermatology. Plants and Plant Products Injurious to the Skin.* Vancouver: Greengrass.
Paulsen E (1992) Compositae dermatitis: a survey. *Contact Dermatitis* 26: 76–86.
Roberts DW and Lepoittevin J-P (1998) Hapten-protein interactions. In: Lepoittevin J-P, Basketter DA, Goossens A, and Karlberg A-T (eds) *Allergic Contact Dermatitis. The Molecular Basis*, pp. 81–111. Berlin: Springer-Verlag.

Dyes

K G Gilbert, University of Bristol, Bristol, UK

Introduction

Today, most dyes used in the commercial dyeing industry are synthesized from the byproducts of fossil fuels, e.g., aniline and other aromatic derivatives. However, until the end of the nineteenth century, all colors used for dyeing came from natural products, as this was the only available source. These natural dyes were widely used and traded. Of the sources available for the production of natural dyes, many were plants and these were widely grown as crops in traditional rotation systems. Unfortunately, since the synthetic chemical revolution, most of the plants used for the production of dyes have ceased to be cultivated and scant information is available about the use of them for the production of color. Although natural dyes were, at one time, relied upon for all colored textiles, they are rarely used today and a misconception that natural dyes produce pale washed out pastels and shades of brown and beige has crept in. This is not the case, and, in reality, many strong and vibrant shades can be produced, which are largely comparable with, and often surpass, their synthetic counterparts.

The commercial dyeing industry's main arguments for not using natural dyes are that they require a two-step dyeing process (pretreatment with a mordant before dyeing) and that the colors are not useful for dyeing modern synthetic fibers. In the light of modern chemistry, both of these problems could be solved by chemically altering the dye molecules to adapt them for direct dyeing and fixing to synthetic fibers. At present, this is not economically viable because synthetic dyes are readily available and much cheaper than natural products. Synthetic dye production is heavily reliant on the byproducts of the oil industry and, therefore, can only be produced while the oil resources last. This is a contentious point, since nobody really knows how long the oil reserves will last, with forecasts citing anything from 30–1000 years. However, it would be useful to have alternative sources in place in case the more pessimistic predictions are correct.

Plants Used for Dye Production

Many different plants have been used for the production of colors through the ages and in different geographical regions. Several texts have been devoted to listing such plants. However, of the many plants recorded, there are only a few that were commonly used, and these are best listed for their production of specific-colored compounds.

The most commonly used dyestuff is indigo (dark blue) and while it is now produced largely by synthetic means, it has never been replaced by a satisfactory synthetic alternative. The synthetic product is exactly the same, chemically, as the plant-derived product and, as such, the synthetic and natural product can be readily interchanged in the dyeing process. Natural indigo can be derived from a wide variety of plant species, genera, and families, representing many different geographical regions, ranging from temperate (e.g., *Isatis* spp.) to subtropical and tropical (e.g., *Indigofera* spp.). Other species have also been used for large-scale production of indigo, e.g., *Polygonum tinctorium* in Japan, China, and Russia. Yellow dyestuff is mainly found in the form of flavonoids, which can be found in many common species, but the richest sources include weld (*Reseda luteola)* and golden rod (*Solidago* spp.), although other species, such as quercitin and persian berries, have also been used commercially. Red dyestuffs are mainly derived from anthraquinones, the richest plant source being madder (*Rubia tinctorum*). Natural dyes can also be found in the form of anthocyanins and carotenoids, but these tend to be unreliable in terms of fastness to light and washing. Therefore, their usage is generally confined to use as food dyes and colorings, where fastness is less important. From the three primary colors given above, many different colors can be obtained by mixing the dyes, e.g., green was produced by overdyeing weld-dyed cloths with indigo (probably the basis for Lincoln green).

Indigo

Known as the "King of Colors," indigo is the deep blue dye used to dye denim jeans. Its importance in world terms is immense and has been for hundreds of years. Even today, indigo contributes around one-tenth of the total international dye market.

The cultivation of indigo plants and the extraction of indigo from them was a worldwide trade, which supported many communities. Although indigo could be made from plants in large quantities, the decline of the natural indigo trade was largely because the process was not well understood. Quality and purity could neither be controlled nor guaranteed. In this context, it was perhaps unfortunate that the tropical indigo producers in Bengal did not start their research into the commercial production of natural indigo until the synthetic chemists were well on the way to producing synthetic indigo. It has been suggested that, had this been the case, the production of synthetic indigo may not have been so important and natural indigo may have survived and perhaps still have been used in large quantities today. Quality and supply were also a problem with most other natural dyes and, consequently, in the 1830s, chemists began to develop methods for producing synthetic coloring materials. The first coloring matter to be made in this way was mauveine, by William Perkin in 1856, and in 1880 a German chemist, Adolf von Baeyer, patented the production of synthetic indigo. It was several years before indigo could be produced on a commercial scale, but within 10 years of the first commercial synthetic indigo production, the natural indigo trade was, in commercial terms, dead. Little research has been done on natural indigo and its sources since that time, even though it is a totally renewable resource. Small amounts of naturally derived indigo are still produced, mainly in the Asian sub-continent, from tropical *Indigofera* species. This source provides enough indigo to satisfy the demands of small-scale craft dyers, supplying exclusive markets. However, there are still no standards in place to guarantee quality.

The Chemistry of Indigo

Indigo is an aromatic compound derived from two molecules of indoxyl joined at position 7 on each of the indoxyl rings (**Figure 1D**). It is practically insoluble in water and many other common solvents, but there are some solvents that will dissolve indigo to some extent (**Table 1**). However, indigo can be

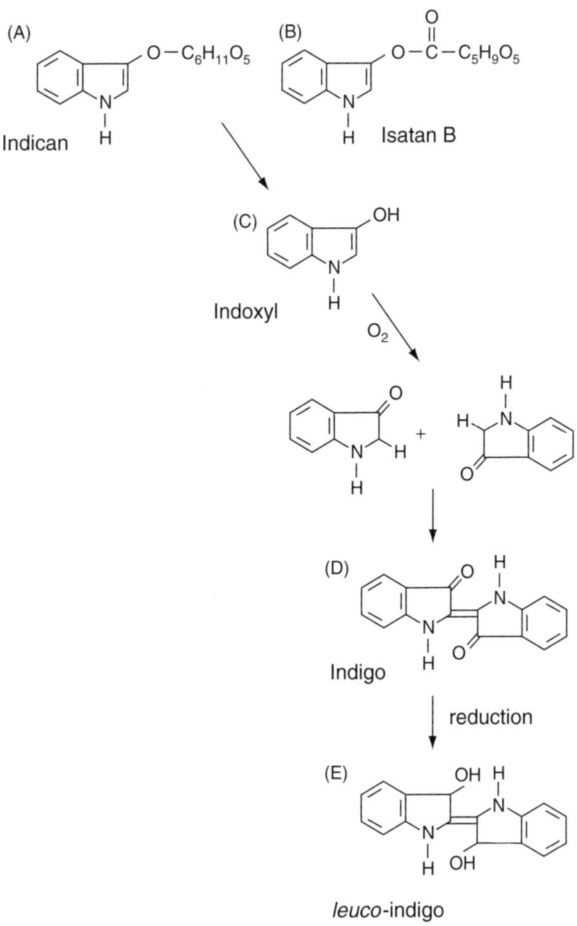

Figure 1 Indigo production from the precursors indican (found in all indigo producers) and isatan B (found only in *Isatis* spp.) and reduction of indigo to its colorless (soluble) form in the dye vat.

reduced to a "leuco-" (colorless) form, which is soluble in alkaline water (**Figure 1E**). This forms the basis for dyeing with indigo. In the insoluble form, indigo is of little use as a dye, as it cannot be fixed to cloth; however, in the reduced form, it is absorbed by cloth. On reoxidation in the air, the insoluble indigo particles are retained in the fibers. This is why the textile does not immediately appear blue when the cloth is removed from an indigo vat, but rather a yellowish green, which deepens in color on contact with the air, finally assuming a deep blue color after several minutes. Once formed, indigo is a very stable molecule and remains so for many centuries. Lightfastness is also very good; in fabrics found in stately homes dating back to medieval times, indigo-dyed yarns remain blue, while most other colors fade. In fact, some of the blue coloring seen on such textiles was originally green, but the yellow component, which is highly susceptible to light degradation, has faded leaving only the blue dye behind.

The major problem in indigo-dyed fabrics is poor rub-fastness, which can be seen clearly in denim dyed

Table 1 Solubility of indigo in various solvents

Solvent	*Solubility*	*Initial color*	*Color after 24 h*
Xylene	i	–	–
Toluene	δ–i	Pale blue	Pale blue
Butan-1-ol	δ	Pale blue	Pale blue
HCl (dilute)	i	–	–
Ethanol	δ	Pale brown	Pale brown
Methanol	δ	Brown "haze"	Brown
Water	δ–i	Pale blue	Pale blue
Acetone	δ	Purple	Purple
Chloroform	s	Blue	Blue
1,1,1-Trichloroethane	s	Blue	Blue
1,1,2,2-Tetrachloroethane	s	Blue	Colorless
Ethyl acetate	s	Blue	Purple
Tetrahydrofuran	s	Purple	Red
Pyridine	s	Blue	Red
Dimethyl sulfoxide	s	Blue	Red/brown
N-Methyl,2-pyrrolidone	s	Blue	Red
Ammonia solution	δ	Brown	Green

s, soluble; i, insoluble; δ, slightly soluble.

with indigo and it was probably this property that saved indigo as a contemporary dyestuff. Jeans became popular mainly because of the uneven fading, which meant that no two pairs were ever the same. This gave the clothing a unique quality, which was adopted as the dress-code of the nonconformist youth of the late 1960s. The origin of jeans came about during the height of the synthetic dye revolution in the mid-nineteenth century when Levi Strauss, a Bavarian tailor, marketed the indigo-dyed, heavy-duty, cotton drill trousers, which we know today as denim jeans, for wear by the gold miners of California. The fabric was produced in Nîmes in France (*bleu de Nîmes* — blue denim) and was used to produce hard-wearing work wear in Genoa, where the sailors used the heavy serge cloth to make their clothes withstand the rigors of life at sea (*bleu de Gênes* — blue jeans).

Isatis tinctoria (Woad)

The earliest known source of indigo is woad (*Isatis tinctoria*), a biennial species, previously a native of southeast Russia, that has become indigenous to northern Europe. It is thought that the Ancient Britons used woad to dye their bodies to make them look more fearsome in battle, as reported by Pliny in AD 44. In northern Europe, woad was the traditional source of indigo. The wealth created by trading this commodity was immense and some of the infrastructure that arose from it is still apparent today in the old trading centers around Europe, especially Toulouse and Erfurt. Even though the dye derived from woad was said to yield a lower quality product than most of the tropical and subtropical species, the woad trade dominated the European indigo market up until the seventeenth century. The economic dependence on indigo trading was so great that dyers and traders in Europe were bound by law to use woad-derived indigo and were not allowed to deal in the tropical product. The penalties for breaking this law were severe and, in some places, punishable by death. In the seventeenth century, woad ceased to be grown in Western Europe as a source of indigo and it was replaced by the tropical indigo trade from Dutch and Portuguese operatives in India and Spanish, French, and English colonies in the Americas. From 1897 onwards, the natural indigo trade worldwide was reduced considerably by the production of synthetic indigo in Germany. However, cultivation of woad continued in England until the beginning of the nineteenth century and it was used for the dyeing of uniforms for the police and armed forces.

The genus *Isatis* contains around 30 species, most of which produce indigo. However, only *Isatis tinctoria* was grown commercially in Europe for indigo production. Woad belongs to the same family as the brassicas (oilseed rape, *Brassica napus*, *Arabidopsis*, cabbages, etc.) and has many similarities with them. In its normal growing cycle, seed is set in May and is ripe by late July or early August. The seeds drop and germinate immediately, while the ground is still warm. After germination, the young plantlet appears to stop growing for a short period, when it produces a taproot up to 1 m long. Once established, it forms a small rosette of leaves, which protects the meristematic tissue during the winter. In the following spring, the plant produces a long stem, bearing yellow flowers on short racemes. The seeds form in a purple/brown siliqua or pod, each containing one or two seeds. When the seeds ripen in late summer, the cycle starts again.

To make a dye from woad, it is necessary to use fresh leaf material. In its natural life cycle, the leaf-producing period is very short and occurs during the winter when light quality is poor. Therefore, woad is treated as a biennial, sown in early spring and harvested throughout the summer during its vegetative phase.

Woad produces two indoxyl-forming substances in its leaves, indican and isatan B, which, when exposed to the air, form the blue compound indigo (**Figure 1A** and **B**). Although the chemistry involved was not well understood, it was known that the plants could be used to make a blue dye. Consequently, a complex series of procedures was devised to produce the grey-blue powder known as "woad" or "couched woad," which contained the mysterious blue coloring matter. This involved harvesting the leaves from the field, macerating them to a pulp, and then making woad balls from the sticky leafmass. The balls (about 12.5 cm diameter) were allowed to dry in the sun, where they reduced in size to about 7 cm and went hard, before being broken into small pieces. The pieces were heaped into piles, watered liberally, and allowed to compost in the air for several weeks (couching). During this time, the piles were continuously turned to maintain aerobic conditions. Once the material had converted to a dark clay-like substance, it was thoroughly dried, collected, and packed tightly into barrels. The couched woad was thought to improve in "dyeing power" if stored, roughly doubling its efficacy as a dye over a period of 4 years. There is much confusion between the names woad and indigo, woad is the name of the plant, indigo is the dye that it contains. Nowadays, indigo precursors are extracted by steeping fresh leaves in water. The liquor obtained is made alkaline and then aerated to produce indigo as a pure compound, free of the plant material found in couched woad. This method of steeping in water is similar to that used for production of tropical indigo, but because of the different climate in Europe, this method was not obvious until the chemistry was better understood.

As a consequence of the decline in its use, there has been no attempt to breed woad to improve indigo precursor content or crop architecture. However, current research has shown that modern techniques of extraction can produce yields of indigo comparable to those obtained from more exotic species. Consequently, it is possible that natural indigo derived from woad may become economically important to European agriculture again. Recent research has shown that indigo precursor production in woad is susceptible to environmental conditions, particularly light quality and quantity, and that some plants are better suited to particular climates. Changes in environmental conditions (especially light) can account for a ten-fold difference in overall indigo yield. This is evident both among plants derived from different areas of Europe and within specific plant populations. In fact, when the genetic make-up of different plants was studied using modern DNA fingerprinting techniques, they revealed a high degree of genetic diversity among plants selected from different regions. This probably came about because different communities selected plants they considered most suitable for their region and growing regime. This diversity provides an excellent gene pool, which is being studied to understand how the genetic make-up of a particular plant is related to specific characteristics. From this, markers can be identified for particular traits of interest, which will be of use for selection of plants for use in a program of classical breeding to develop plant varieties best suited to modern husbandry techniques, commercial indigo production, and climate in a particular region. It is hoped that this will make natural indigo production as a commercial venture an even more attractive proposition.

Yellow Plant Dyes

Yellow compounds can be obtained from most plants, including naturally occurring species, and cultivated plants. Therefore, in many cases the source of yellow dyes was dictated by what was available locally but, in commercial terms, the quality and quantity of the dyes produced was not sufficient to justify growing plants solely for dyestuffs. However, there is a small group of plants that produce sufficient quantities of yellow dyes, in the form of flavonoids, to be considered of economic importance: *Reseda luteola* (weld or dyer's rocket), *Reseda lutea* (dyer's greenweed), *Genista tinctoria* (dyer's broom), and *Solidago* spp. (golden rod). Several other natural yellow dyes are produced from plants, including quercitin, which is the most commonly found yellow dyestuff. However, the main source of yellow dye is quercitron, isolated from the inner bark of *Quercus tinctoria*. Other sources include old fustic from dyer's mulberry (*Chlorophora tinctoria*), young fustic from Venetian or dyer's sumach (*Cottinus coggygria*), and Persian berries (*Rhamnus frangula*), but these were not generally grown as crops. Saffron, the very expensive dye derived from the stigma of *Crocus sativus*, is well known but little used in textile dyeing.

The flavonoids form a large group of aromatic chemicals, although not all of them are useful as dyes. Flavonoids are mordant dyes, which means that fabrics require pretreatment with a metal salt (e.g., Alum), making the dyeing process more complicated than that employed for modern synthetic dyes. The principle flavonoid coloring matter is luteolin (**Figure 2A**), which produces the most vibrant and light-fast dye. However, several other flavonoids from different plant sources were in common use, e.g., apigenin, quercitin, berberine, emodin, and rhamnetin (**Figure 2**).

Reseda luteola—Weld and Other Sources of Flavonoids

The most commonly used plant for the production of the flavonoid yellow dye luteolin was weld. It was also used with indigo to produce fast green dyes, which are uncommon in plants. Weld is said to be the

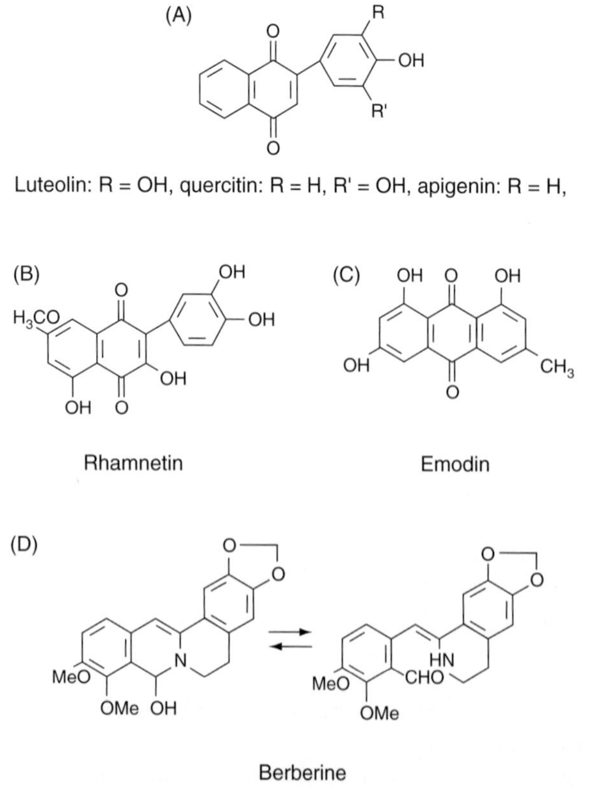

Figure 2 Flavonoids found commonly in plants.

oldest European dyestuff known. It was used by the Gauls and other nations North of the Alps and in North Africa in the time of Julius Caesar, particularly in conjunction with indigo for green dyes.

Weld is a biennial or polycarpic perennial, which attains a height of up to 1.5 m. The coloring matter is concentrated in the flowers and seeds. Weld is perhaps the easiest dyeplant to grow as a crop because it requires little or no fertilizer. In fact, the quality of the dyestuff obtained from weld is improved if plants are grown on impoverished soils. It produces a small rosette of leaves in the first year and, after winter vernalization, flowers in the spring of the following year. In Europe, an autumn sown crop was favored, but in the UK, plants were usually sown in spring. Plants were harvested in the same way as a hay crop, gathered up into sheaves and sold directly to the end-user. This method is still favored today, although an annual variety of weld, which is more suited to modern farming techniques, is usually used. Once dried, the dye is extracted from the plant material for use.

Weld contains yellow coloring matter mainly in the form of luteolin and apigenin. Luteolin is present in the form of the free aglycone and as luteolin glucosides. Several methods of extraction have been tried including using hot water and alcohol. However, of the methods tried, none are exhaustive and substantial amounts of the dye are left in the plant material. This is also the case when plant material is used directly in the dye vat and it may be that the useful components are exhausted and the remaining compounds are merely waste products.

Red Plant Dyes

Prior to the age of indigo dyeing, red dyes were considered to be the most valuable and dyers producing red cloth were held in high esteem. It was not until the quality of blue dyes improved during the middle ages that the taste for red diminished and indigo dyeing took its place. It was at this time that the color of the clothes of the Virgin Mary began to be depicted as blue and dignitaries started to wear blue rather than red robes. Indeed, in the thirteenth century churches of Thuringia, the devils on the stained glass windows are blue rather than red or black, as had been previously used.

Red textile dyes derived from plants are usually in the form of anthraquinones. As with the yellow flavonoids, they require the cloth to be mordanted prior to dyeing. Anthraquinones occur in many different plants, but possibly the most important red plant dyestuff was obtained from the root of the madder plant (*Rubia tinctorum*).

Rubia tinctorum — Dyer's Madder

Rubia tinctorum is a perennial plant related to lady's bedstraw and goosegrass or cleavers, which are common wild flowers in Europe. It produces an array of fast colors from its roots, ranging from pale pinks and violets, through deep reds to oranges and browns. This is uncommon, as red dyes are scarce both from natural and synthetic sources, and as such madder is considered a valuable commodity. The most significant of the madder dyes is alizarin (**Figure 3A**). A less important dye, pseudopurpurin, is often found in madder-dyed textiles and appears to be coextracted with alizarin (**Figure 3B**).

Madder is thought to be a native species of Persia and the eastern Mediterranean, although it has now become naturalized in Europe. It was used by all the ancient civilizations including the Egyptians, Persians, and Indians and latterly the ancient Greeks and Romans. There is also archaeological evidence that it was used by the Vikings. During the Crusades, madder was introduced into Italy and France and was later cultivated by the Moors in Spain. From here it was exported north and was grown widely throughout Mediterranean Europe. By the mid-seventeenth century it had become an important

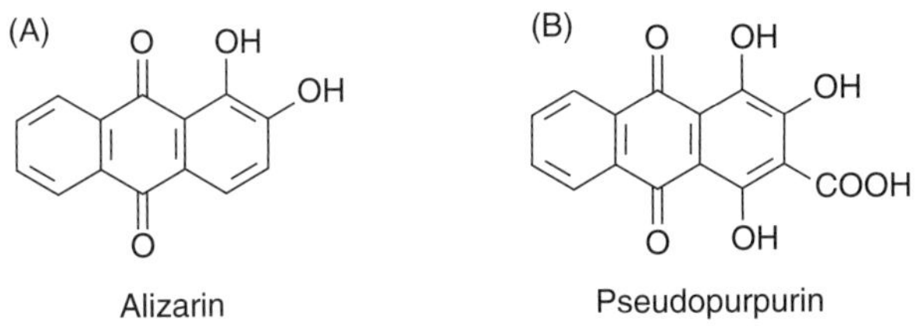

Figure 3 Anthraquinones alizarin and pseudopurpurin. The most useful constituents for dyeing found in madder (*Rubia tinctorium*).

dye crop. Its cultivation declined during the French Revolution and the Napoleonic wars, but its importance regained impetus during the mid-nineteenth century, before it was largely replaced by synthetic alizarin in the early twentieth century.

The importance of madder should not be overlooked, in the middle ages it was used to produce the bright red clothes for common people. This red color, produced using the more expensive insect and mollusk dyes, was previously the preserve of the upper classes. As such, it was used only for the dyeing of clothes for special occasions, particularly weddings. Madder was also used in conjunction with indigo and woad for the dyeing of military uniforms, generally for peace-time festive parades, e.g., the British Redcoats. However, until 1915, the French Military wore madder-dyed breeches as part of their combat uniform, as it was good camouflage for bloodstains. This practice continued until the First World War when they changed the color to blue, as it made them less easy targets as accuracy in warfare improved.

Traditionally, madder is propagated from root cuttings, taken manually from established plants in late April/early May. This is probably a consequence of poor seed set and germination in temperate climates. However, if the plants grown for seed are maintained with a good nutritional status, flowering and seedset are good and the seeds can be drilled directly into the ground. Once propagated the plants are left, without further cultivation, for 2–3 years for the roots to grow, then they are harvested, washed, dried, powdered, and stored. As with couched woad it is thought that dye quality improves with age. However, if they are not stored properly and kept dry, spoilage occurs.

Other Plant Sources of Red Dyes

Although madder was considered the most important source of red dye, several other species have been used. Brazilwood was used in Europe to dye leather for bookbinding in the thirteenth century. The major dye constituent, brasilein, is derived from the auto-oxidation of its precursor brasilin, which is found in the rasped wood of the soluble redwoods of *Caesalpinia* spp. Logwood, a close relative of Brazilwood, comes from the tree *Haematoxylon campechianum* and contains haematin as its major coloring matter.

Indian madder, Chay root or Chay-aver (*Oldenlandia umbellata*), a small shrub or herb of 30–50 cm in height, which is a native of southern India, was used in India, Sri Lanka, and northern Burma, although it produces less than a quarter of the dye obtained from madder. Other plants related to madder, e.g., *Rubia cordifolia* and *Rubia munjista*, were used in some parts of India; these were known colloquially as "Munjeet" or "Manjeet."

Galium tinctorium, *Galium mullugo* (great ladies bedstraw or wild madder), *Galium verum* (yellow ladies bedstraw), and *Galium aperine* (goosegrass or cleavers) also produce anthraquinones, but they are considered inferior to alizarin from madder. Several other species that produce red dyes are known, but generally, they are not considered to be viable contributors to the natural dye market.

See also: **Genetic Modification, Applications**: Flower Color.

Further Reading

Balfour-Paul J (1997) *Indigo in the Arab World*, 1st edn. London: Curzon Press.

Balfour-Paul J (1998) *Indigo*: Richmond: British Museum Press.

Chenciner R (2000) *Madder Red, A History of Luxury and Trade*. Richmond: Curzon Press.

Dean J (1999) *Wild Colour*. London: Mitchell Beazley, Octopus Publishing.

Delamare F and Guineau B (1999) Colour: *Making and Using Dyes and Pigments*. London: Thames and Hudson.

Dunn ST (1905) *Alien Flora of Britain*, 1st edn. London: West, Newman & Co.

Goodwin J (1982) *A Dyer's Manual*. London: Pelham Books.

Hurry JB (1930) *The Woad Plant and its Dye*. Reprinted in 1973. Clifton, NJ: Augustus M. Kelley.

Mills JS and White R (1994) *The Organic Chemistry of Museum Objects*, 2nd edn. Oxford: Butterworth Heinemann.

Perkin AG and Everest AE (1918) *The Natural Organic Chemistry of Colouring Matters*, 1st edn. London: Longmans Green and Co.

Sandberg G (1989) *Indigo Textiles, Technique and History*. London: A & C Black.

Seefelder M (1994) *Indigo in Culture, Science and Technology*. Landsberg, Germany: Ecomed.

Stoker KG (1997) *The Cultivation of Woad (Isatis tinctoria) for Production of Natural Indigo: Agronomy, Extraction and Biochemical Aspects*. PhD Thesis, University of Bristol.

Volatiles for the Perfume Industry

D J Murphy, University of Glamorgan, Cardiff, UK

Introduction

The various industries that produce perfumes and related odorants rely to a great extent on plant volatiles for many of their most important products. The global perfumes industry is immensely lucrative – estimates of the annual trade in perfumes in France alone are well in excess of US$25 billion, much of which is plant-derived. In addition to traditional scents and perfumes, which continue to grow in use for both men and women, there are many additional markets that utilize plant volatiles as a key selling point. These include the essential oils and new higher-value scented cleansing products, such as soaps and creams. Together, these trends ensure a buoyant market for plant volatiles.

Diversity of Volatiles

Plants accumulate a bewildering diversity of volatile compounds that serve many important physiological roles. Some volatiles act as deterrents to potential pests while others may attract predators of pest species. For example, as caterpillars graze on certain plants, the damaged plant tissues release specific volatiles, which attract parasitic wasps that then paralyze and lay their eggs in the caterpillars. Plant volatiles may also attract pollinators or other beneficial animals. In some cases, plants even attract insects by mimicking their sex pheromones in what has been termed a "sexual swindle." In the case of *Ophyris* orchids, the cuticle of the flower contains a standard mixture of straight chain saturated and unsaturated hydrocarbons of 21–29 carbons chain length. However, the relative proportions of the hydrocarbons closely mimic that found in sex pheromones of pollinators like male bees. The male bees repeatedly attempt to copulate with the orchid flowers, hence ensuring an efficient transfer of pollen.

Aroma Compounds

Typical aroma compounds produced by plants include the following:

- the largest group is undoubtedly the ubiquitous lower isoprenoids (mono- and sesquiterpenoids)
- aliphatic straight-chained hydrocarbons and aldehydes that are found in many fruits and flowers
- highly odoriferous derivatives of benzoic acid and phenylpropane that are especially common in the Apiaceae
- a wide range of sulfur- and nitrogen-containing volatiles.

Many plant volatiles are (coincidentally) attractive to humans, ranging from the relatively simple hexenals and hexenyl acetates, etc., that are the main signature from mown grass, to the extremely complex bouquets produced by flowers like jasmine (*Jasminum* spp.) and honeysuckle (*Lonicera* spp). Some of the most important and widely used plant volatiles used in the fragrance and perfume industry are the products of so-called "essential oils." The term essential oils is not truly accurate as many of the volatiles are alcohol-soluble and are not true oils at all, but this traditional term remains in widespread use. Essential oils are very volatile fragrances that easily evaporate and are very different in character from true oils such as olive oil or safflower oil. Essential oils are complex mixtures of plant produced chemicals, for example essential lavender oil contains more than 50 different compounds.

The active ingredients of most essential oils occur in minute quantities in plants – rose (*Rosa* spp.) petals yield just 0.2% w/w essential oil. This means that the extracts must be highly concentrated in order to be marketed for human use. One feature shared by all volatile fragrances is that they all readily bind to receptors in olfactory neurons. Another important character is that they evaporate at room temperature. Indeed, the fragrances used in perfumes are classified on a scale from 1 to 100, according to how readily they dissipate. Pure essential oils are expensive and must be obtained from reputable sources. Adulteration of essential oils is very common. For example, the finest natural French lavender (*Lavandula* spp.) oils harvested in the Haute Provence contains up to 70% linalyl acetate as its most potent fragrance. However, while many other lavender oils from France may have even higher levels of linalyl acetate, some of these oils are fortified with synthetic products and may have no traces of natural lavender. Sandalwood (*Santalum album*) oil can be adulterated with diverse oils such as castor, palm, and linseed.

List of Technical Nomenclature

Pheromone	A volatile hormone secreted by animals (especially insects), that influences the behavior of other animals of the same species.

See also: **Genetic Modification of Secondary Metabolism**: Terpenoids. **Photosynthesis and Partitioning**: Primary Products of Photosynthesis, Sucrose and other Soluble Carbohydrates. **Primary Products**: Oils.

Further Reading

Hay RKM and Waterman PG (eds) (1993) *Volatile Oil Crops: their biology, biochemistry and production.* New York: John Wiley.

Lewinsohn E, Ziv-Raz II, Dudai N, *et al.* (2000) Biosynthesis of estragole and methyl-eugenol in sweet basil (*Ocimum basilicum* L): developmental and chemotypic association of allylphenol O-methyltransferase activities. *Plant Science* 160: 27–35.

Pichersky E and Gershenzon J (2000) Plant odyssey: volatile substances. *Biofutur* 224: 22–26.

Pichersky E and Gershenzon J (2002) The formation and function of plant volatiles: perfumes for pollinator attraction and defense. *Current Opinion in Plant Biology* 5: 237–243.

Schiestl FP, Ayasse M, Paulus HF, *et al.* (1999) Orchid pollination by sexual swindle. *Nature* 399: 421–422.

Secondini O (1991) *Handbook of Perfumes and Flavors.* New York: Chemical Publishing Co.

Teranishi R, Buttery RG, and Sugisawa H (eds) (1993) *Bioactive Volatile Compounds from Plants*, ACS Symposium vol. 525. Washington: American Chemical Society.

Tronson D (2001) The odour, the animal and the plant. *Molecules* 6: 104–116.

Rubber Production

D L Hallahan, Central Research & Development, EI DuPont de Nemours & Company, Wilmington, DE, USA

Definitions

The term 'rubber' encompasses any material that is highly elastic, i.e. a material that can be stretched without breaking and that returns to its original length on removal of the stretching force. Natural rubbers are produced by many plant species, often (though not always) as a constituent of latex. A plant latex is defined simply as a milky fluid present in laticifers, or latex ducts, which seeps out on wounding (tapping). Natural rubbers belong to the class of natural products termed polyisoprenoids (polymers of isoprene).

History

The elastic properties of the rubber that could be extracted from certain plants were known and appreciated by the ancient civilizations of Central and South America long before the arrival of Europeans. Aztec murals reveal that rubber was collected as a tribute from subject peoples and used in religious ceremonies. After its "discovery" by Europeans as a novelty material for making balls, it began to be used in the manufacture of waterproof clothing and footwear for the European market. Since no method of preserving latex was known at the time, proofing of clothing and leather was carried out locally and the product exported.

The first challenge to be met to enable the large-scale use of rubber was to identify an efficient solvent for the material. In 1818, James Syme discovered that coal tar naphtha could be used to this end, and this discovery was exploited by Charles Mackintosh for the fabrication of a rubber waterproof layer sandwiched between two layers of fabric. However, the unstabilized natural material remained generally unsatisfactory because changes in temperature resulted in changes in its texture and elasticity. It was soft and tacky when hot, and hard and brittle when cold. Interest in natural rubber as a commodity was waning when, in 1839 (and after some 5 years of effort during which he and his family were largely destitute), Charles Goodyear discovered that a combination of heat and sulfur somehow stabilized rubber whilst retaining its desirable properties. Later termed "vulcanization", the perfected process opened the door for exploitation of natural rubber on a commercial scale.

After the discovery of vulcanization, rubber would be inextricably linked to modern industrial development, and particularly to transportation. North American imports of natural rubber more than doubled every decade of the nineteenth century after 1840, to over 15000 tons (representing around half the total world production) by 1890. A sad and sobering consequence of this expansion in demand is the fact that, in the latter half of the nineteenth and early part of the twentieth centuries, it was satisfied largely by slave labor. Millions of people living in "colonial" territories of Amazonia in South America and The Congo in Africa were exploited unmercifully in the collection of "wild rubber."

Large-scale rubber collection in Amazonia began near the Brazilian Atlantic port city of Pará, and expanded westwards into the amazon basin. A number of rubber-producing trees were exploited initially, including various *Sapium* species (yielding a latex called *caucho blanco*), *Castilloa* (*Castilla*) species (yielding *caucho negro*), and *Hevea* species. *Hevea brasiliensis*, yielding the high-quality rubber termed *jebe fino* or *siringa fina*, was to be found only in a specific region near the Putumayo river, a tributary of the Amazon. The related species, *Hevea guianensis*, yielded a lower-quality rubber termed *jebe débil*. *Hevea* trees could only be tapped for about 6 months a year in Amazonia, because of their preponderance in low-lying, flooded areas. Although, initially, rubber could be obtained year-round from *Castilloa* trees (*C. elastica*, *C. ulei*) on dryer land, rubber extraction from these trees involved their harvesting and destruction, and the supply was quickly exhausted. As early as 1857, Thomas Hancock (the British inventor of a process for mastication of rubber and eventual holder of a British patent for vulcanization based largely on observations of Goodyear's vulcanized material) speculated on the desirability of developing plantations of rubber-producing trees. During the 1870s, Kew Gardens in London received three collections of seeds of *Hevea brasiliensis* from South America. Only one of these appears to have supplied the hevea germplasm subsequently exploited in Asian plantations. This was supplied by Henry Wickham, then living in the Rio Tapajos region of Upper Amazonia, in 1876. Of the 70 000 seeds he brought to Kew, 2 700 germinated, and a large proportion of the resulting seedlings were then dispatched to Sri Lanka, with a few sent to Indonesia and Singapore. It was largely from the collection at the Heneratgoda Botanic Gardens in Colombo, Sri Lanka, that the hevea rubber industry in Asia is believed to have developed.

Sources of Natural Rubber

Many plant species (estimates put the number known at around 2000) produce rubber, either within latex or other tissue. As indicated above, the latex tapped from the tree *H. brasiliensis* is today the only significant commercial source of natural rubber. However, rubbers from other species have been used commercially in the past and continue to be studied in order to better understand the physiology and biochemistry of natural rubber, as well as its physical constitution and properties. Indian laurel (*Ficus elastica*) is a well-known household plant that produces a rubber-containing latex. Gutta percha latex is obtained from trees of the *Sapotaceae* family (*Palaquim gutta* and *P. oblongifolia*) in the Malaysian peninsula. The rubber of this species is present in a viscous, grayish latex, which exudes slowly from cuts in the bark and rapidly turns brown after exposure to the air. Balata latex is obtained from the tropical American tree *Mimusops balata*, as white or reddish exudates. Chicle is obtained from the saprodilla tree *Archras zapote*, again a tropical American species. The central American tree *Castilla elastica* yields *caucho negro* rubber as indicated above, and another Brazilian species, *Manihot glazovii*, yields a material known as ceara rubber. The dandelion species kok-saghyz (*Taraxacum kok-saghyz*), from Kazakhstan, and krim-saghyz (*T. megalorhizon*), found in the Crimea and throughout the Mediterranean region, also yield a high-quality rubber in their roots. *Taraxacum kok-saghyz*, as well as the nonlatex-producing American desert shrub guayule (*Parthenium argentatum*) were investigated as potential alternative sources of natural rubber during World War II, and interest in guayule remains strong to this day. In this species, rubber is produced seasonally within parenchymatous cells of the stem and root, and its isolation requires harvesting of the plant and maceration of the tissue.

Commercial Production of Natural Rubber

Natural rubber, despite the development of many synthetic polymer alternatives, remains a high-volume commodity material, along with other plant products, such as sugar, starch, vegetable oil, wood pulp, and cotton. Currently, some 6 810 000 tonnes of natural rubber are produced annually, with 90% of this production from Asian countries (data from International Rubber Study Group website). Today, all of this material comes from the tapping of hevea trees, which remains entirely a manual process. The website of the International Rubber Research and Development Board provides a good current overview of rubber production (http://www.irrdb.com/irrdb/NaturalRubber/).

Various tapping systems have been employed over time, but all current systems involve the continual reopening of a single cut in the bark of the trunk (**Figure 1**). As the bark is continually shaved away from the lower edge of the cut, a characteristic tapping panel develops on the trunk, with the bark regenerating from the top of the panel downwards.

Figure 1 (see color plate 36) A rubber tree, *H. brasiliensis*, being tapped for latex. Courtesy of Linatex Inc., Yately, Hampshire, UK.

Latex is directed from the cut into a collection vessel (which may be earthenware, aluminum, glass or even a coconut shell; **Figure 1**) by a small spout worked into the bark. Using such methodology, a single person can tap around 500–600 trees in 3–4 h. The yield of latex is dependent on a variety of factors, including the number of times the cut has been opened, rate of bark consumption, angle and direction of the cut, and the time at which the cut is made. Above all, latex yield (and quality) is dependent on the clone of hevea employed. Early yields of rubber were of the order of 250 kg $ha^{-1}year^{-1}$, but selective breeding has increased yields to the point that now over 2000 kg $ha^{-1}year^{-1}$ may be obtained. Stimulants such as copper sulfate or phenoxyacetic acid derivatives may also be employed to increase yield, by prolonging latex flow.

A key aspect of rubber production, as alluded to earlier, is the preservation of the latex. The latex collected after tapping, if left untreated, will undergo spontaneous coagulation and eventually putrefaction. Shortterm preservation may be achieved by the addition of anticoagulants such as ammonia, which despite having drawbacks (which have led to considerable research into alternative preservation systems) remains the predominant technique in use today. Once stabilized in this manner, the natural rubber component of the latex may be concentrated by evaporation, creaming, or centrifugation. This is essential as the dry rubber content of latex varies between 30 and 40% by weight, making it uneconomical to transport this material. Increasing the dry rubber content to 60% not only addresses this economic issue but also (except in the case of evaporation) serves to increase the purity of the product. The longest established method of processing this material further involves its conversion into ribbed smoked sheets (RSS). This is achieved by first diluting the latex in water with an antioxidant (usually sodium bisulfite) and an antimicrobial agent (usually paranitrophenol), and allowing it to coagulate in pans, troughs or tanks after addition of acetic or formic acid. The coagulum is then sheeted, by passing several times through rollers, the final rolling imparting the ribs that aid in drying. Drying may then be carried out in a smoke house. If a hot air chamber is used then the sheets are referred to as air-dried sheets (ADS). Over time, many variations of this basic process have been introduced, leading to a variety of grades of the final product. A number of technically specified rubbers (TSR) are described further in the "Further Reading" section below. A variety of crepe rubber grades are also produced, either from latex or field-coagulated rubber. Pale latex crepe and sole crepe are high-purity, light-colored rubbers prepared under controlled conditions, which include a bleaching step.

Most rubbers are compounded (formulated appropriate to their end use) by maceration, softening, addition of additives (antioxidants, fillers, etc.), and vulcanization. Natural rubber is a viscoelastic material, i.e., one exhibiting liquid (viscous) and solid (elastic) properties. Vulcanization, by introducing crosslinks, reduces the flow properties and makes the material more elastic. The vulcanization process involves the crosslinking of the polymer chains of the rubber; generally, sulfur is used, although other substances may be employed. Sulfur vulcanization employs heat, sulfur, activators (such as zinc oxide and stearic acid), and accelerators (such as sulfenamides and thiazoles) to form an active sulfurating agent that abstracts hydrogen atoms from the rubber molecules to form polysulfidic crosslinks.

The Nature of Natural Rubber and Latex

In 1826, Faraday elucidated the basic chemical composition of natural rubber and, in 1879, Bouchardat identified the simple 5-carbon molecule isoprene ($CH_2=C(CH_3)-CH=CH_2$) as its basic constituent. Many subsequent studies have shown that the rubber hydrocarbon of *H. brasiliensis* latex is predominantly a polymer of *cis*-1,4-polyisoprene. The presence of a small proportion of *trans*-1,4 bonds has recently been proven by NMR analysis, based on assignments derived from studying the rubber of *Lactarius volemus*. This arises because their synthesis requires a short-chain allylic diphosphate initiator (see below). This is the situation for the majority of the rubbers mentioned above. The rubbers gutta percha, balata and chicle, however, contain primarily *trans*-1,4 bonds.

One of the most remarkable features of natural rubbers is their high molecular mass, which makes these molecules the largest of all known isoprenoid compounds. The molecular weight distributions of rubbers isolated from *H. brasiliensis*, *P. argentatum*, and *T. kok-saghyz* are presented in **Figure 2**. This analysis shows the rubbers of hevea and guayule having a weight average molecular weight (Mw) of ca. 1.4×10^6, whilst the rubber extracted from *T. kok-saghyz* is even larger, with Mw of 2.18×10^6. The latter figure implies that a polyisoprenoid chains of *T. kok-saghyz* with molecular weight of 2.18×10^6 would contain some 32 000 isoprene monomers, although this may be an overestimate as the rubber used in this analysis was not treated to remove bound proteins or lipids. A rubber may be operationally defined as a polyisoprenoid containing

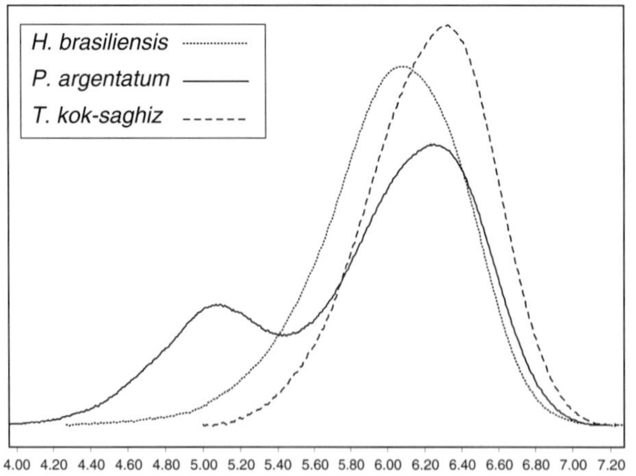

Figure 2 Molecular weight distributions of rubbers isolated from *H. brasiliensis, P. argentatum,* and *T. kok-saghiz.* Analysis of rubbers by gel permeation chromatography shows that Hevea rubber has a weight average molecular weight (Mw) of 1.44×10^6 and polydispersity (Mw/Mn) of 5.71. The corresponding values for *P. argentatum* are 1.47×10^6 and 2.3, and for *T. kok-saghiz* are 2.18×10^6 and 1.8. Analyses performed by the Corporate Center for Analytical Science, The DuPont Company.

more than about 500 isoprene units (or molecular weight >30 000), although clearly this is a lower limit on molecules that may contain tens of thousands of isoprene units.

The physical properties of natural rubbers derive from their stereochemistry, degree of crosslinking and the proportion and composition of the non-rubber ingredients coextracted from the latex. One of the key properties of natural rubber as a material is its crystallization behavior. Stretching of natural rubber, like other polymers with a highly stereo-regular structure, results in rapid crystallization as the polymer molecules order themselves. This phenomenon is the basis of the high tensile strength and crack growth resistance of natural rubber. Crystallization causes an increase in the stress rubber can withstand on straining, and the stress at break (known as the "green strength") is an important measure of rubber quality. That rapid crystallization cannot be due to the stereochemistry and/or length of the polyisoprenoid chain alone is clear, since synthetic *cis*-1,4-polyisoprene exhibits a significantly lower rate of crystallization than the natural material. Recent analyses have shown that the fatty acid content of the rubber is a key factor in determining crystallinity under strain. Unvulcanized rubber that had both proteins and free fatty acids removed exhibited reduced green strength. Transesterification (to remove fatty acids linked to the terminal –OH groups of the polyisoprene chain; see below) reduced this to a level equivalent to synthetic *cis*-1,4-polyisoprene.

Most natural rubbers are isolated from latex, a milky fluid that flows from plants on wounding. Latex is not a secretion but consists of the cytoplasm of a network of specialized cells termed laticifers. Two types of laticifers have been described. In almost all the Euphorbiaceae, for example, a simple non-articulated laticiferous system is formed by the apical intrusive growth of individual cells. A compound, articulated laticiferous system is formed in *H. brasiliensis* and *T. kok-saghyz*, where a series of simple or branched cells with perforated end walls form a reticulated network. The laticiferous system of *H. brasiliensis* has been the most thoroughly studied, and readers are referred to the volume edited by Sethuraj and Mathew for further reading.

In latex-producing and nonlatex-producing plants, rubber is compartmentalized in subcellular particles present in the cytosol of the rubber-producing cells. The average diameter of these particles varies considerably among species, with a range of between 0.01 μm and 15.0 μm. Almost nothing is known about the ontogeny of these particles, although it is believed they derive from the rough endoplasmic reticulum. The particles are formed of a homogenous rubber core surrounded by a monolayer proteolipid membrane, analogous to the oil storage bodies of oilseed plants. The protein and lipid composition of the outer membrane varies considerably between species and, in some, can be very complex. Isolated rubber particles from *H. brasiliensis*, for example, contain a large number of diverse proteins, but by far the most abundant proteins are low molecular weight molecules termed "small rubber particle protein" (SRPP) and "rubber elongation factor" (REF). In guayule rubber particles, by contrast, the predominant protein is a cytochrome P450 enzyme, allene oxide synthase, involved in lipid metabolism rather than rubber biosynthesis.

Related Polyisoprenoids

Whilst very many plant species produce rubber, all organisms contain polyisoprenoids of lower molecular weight. Low molecular weight rubbers are present in many species, and still smaller polyisoprenoids (polyprenols) have been identified in at least some tissues of many different plant species. Dolichols, or α-dihydropolyprenols, are ubiquitous as they play a key role in transglycosylation of proteins, acting as carriers of sugars. As is the case for rubber, all of these molecules are synthesized from isopentenyl diphosphate (IPP) and contain either two or three *trans* units at the ω-end of the chain (**Figure 3**).

Short-chain polyprenols have been identified in very many plant species. In all cases, a mixture of

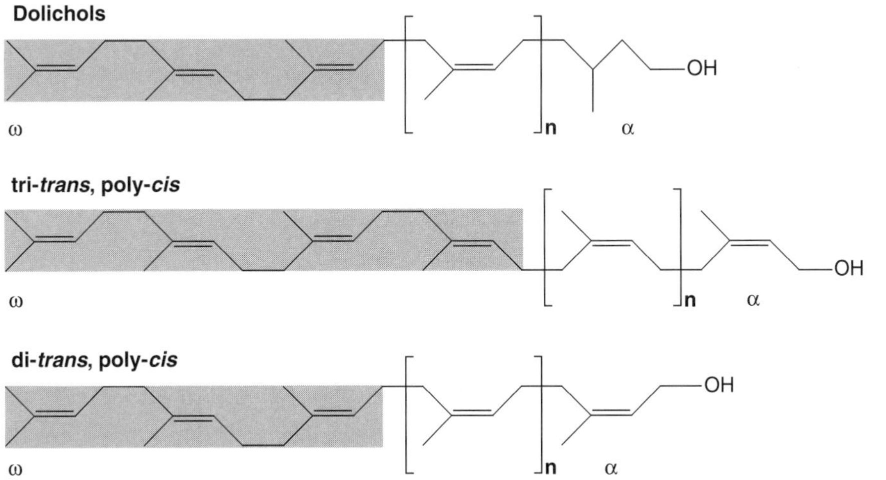

Figure 3 Structures of plant polyprenols. The structures of dolichols ($n = 12$–18), tri-*trans*, poly-*cis* polyprenols (e.g., ficaprenols, $n = 10$–12) and di-*trans*, poly-*cis* polyprenols (e.g., betulaprenols, $n = 6$–9) are shown. The part of the molecules formed from the allylic terpenoid diphosphate initiators are shown against a shaded background whilst the remainder of the molecules are formed by the activity of *cis*-prenyltransferases catalyzing sequential additions of IPP. The ω- and α-ends of the molecules are indicated.

chain length distribution is found when these compounds are isolated from plants. Plant polyprenols are typified by the betulaprenols and ficaprenols isolated from *Betula verrucosa* (silver birch) and *F. elastica* (Indian laurel), respectively. Betulaprenol-type polyprenols have two *trans* units at the ω-end, and are isolated as a mixture of molecules of varying chain length. This natural mixture contains prenols of 6, 7, 8, and 9 isoprene units (C30–C45), and these polyprenols are found only in the wood of this species, not in leaves. Ficaprenol-type polyprenols have three *trans* units at the ω-end and again are isolated as a mixture (10–12 isoprene units; C50–C60), in this case from leaves. Dolichols are larger molecules (16–22 isoprene units; C80–C110), and even longer polyprenols have been reported from, for example, *Potentilla* and *Rosa* spp. These species usually contain two groups of polyprenols: one where the dominant chain length is C95 or C100 and the second with C120 or C125 as the dominant forms. Some *Potentilla* spp. contain a third grouping of 38–40 isoprene units (C190–200).

Low molecular weight rubbers have been characterized from many plants (e.g. *Pycnanthemum incanum*, *Solidago altissima*, and various *Helianthus* spp.), as well as the fungus *Lactarius volemus*. These molecules have been useful as "models" for the larger rubbers, as it has been possible to carry out detailed NMR studies of these materials to determine, for example, the nature of stereochemistry at the ω-end. The fungal rubber showed a unimodal molecular weight distribution (centered around 2.4×10^4) and was shown to possess two *trans* units at the ω-end. Interestingly, the lowest molecular weight fraction of this rubber contained two dominant forms of polyprenols: C105 and C160.

In the context of polyisoprenoids as a whole, natural rubber may be viewed as one extreme of a continuum of molecules of increasing molecular weight. The fundamental common feature of *cis*-polyisoprenoids, including natural rubber, appears to be their possession of a short stretch of *trans* units at the ω-end. Whilst the rubbers of *P. argentatum* and *T. kok-saghiz* have a unimodal molecular weight distribution, that of *H. brasiliensis* has a bimodal distribution analogous to that of the polyprenols of *Potentilla* and *Rosa* spp. (**Figure 2**). Differences in the pattern of molecular weight distribution of *H. brasiliensis* rubber have been observed in different clones of this species; thus, this characteristic appears to be under genetic control. One might reasonably speculate as to whether distinct polymerase (prenyltransferase) enzymes are responsible for synthesizing different size classes of polyisoprenoids in those plants producing polyisoprenoids with multimodal molecular weight distributions.

The Biosynthesis of Isopentenyl Diphosphate

As indicated above, all polyprenol synthesis is based on the simple five-carbon molecule isopentenyl diphosphate (IPP). Two pathways for the synthesis of this molecule are known to occur in plants and bacteria. In bacteria and plant plastids, a pathway involving glyceraldehyde 3-phosphate and pyruvate as precursors occurs, whilst in the cytosol of plant cells acetate and mevalonic acid are intermediates.

The latter pathway, the acetate/mevalonate pathway, was discovered before the former, and all subsequent investigations have indicated that it is this pathway that is responsible for the generation of IPP in hevea latex.

In the acetate/mevalonate pathway, acetate is converted to IPP by the sequential action of six enzymes (**Figure 4**). Studies of the physiology of hevea latex have shown that acetate is an effective precursor for rubber synthesis, and demonstrated the presence of all of the enzymatic activities necessary for its conversion to IPP. Recent molecular biological approaches, in particular expressed sequence tag (EST) analyses, have yielded clones for many genes expressed in hevea latex. Of the minimum of six genes encoding enzymes of the core acetate/mevalonate pathway to IPP in hevea, only HMG-CoA reductase and HMG-CoA synthase had been cloned prior to the analysis of latex EST libraries. Two cDNAs, encoding enzymes termed HMGR1 and HMGR2, were isolated using a heterologous hybridization probe, and genomic southern blotting confirmed the presence of at least two genes for HMG-CoA reductase in hevea. Subsequent investigations revealed a third gene, termed HMGR3. An EST homologous with HMGR1 was also identified in a hevea latex library. Analysis of the expression of HMGR1 and HMGR3 indicated that whilst HMGR3 performed a "housekeeping" role, HMGR1 was most probably the isoform present in laticifers and involved in rubber biosynthesis. Of the other four genes of the acetate/mevalonate pathway, hevea homologs have been identified only as ESTs, and no expression data is available at this point. Now that all the genes for IPP biosynthesis in hevea are available, however, detailed analysis of their expression is likely to yield insight into the regulation of this pathway in latex.

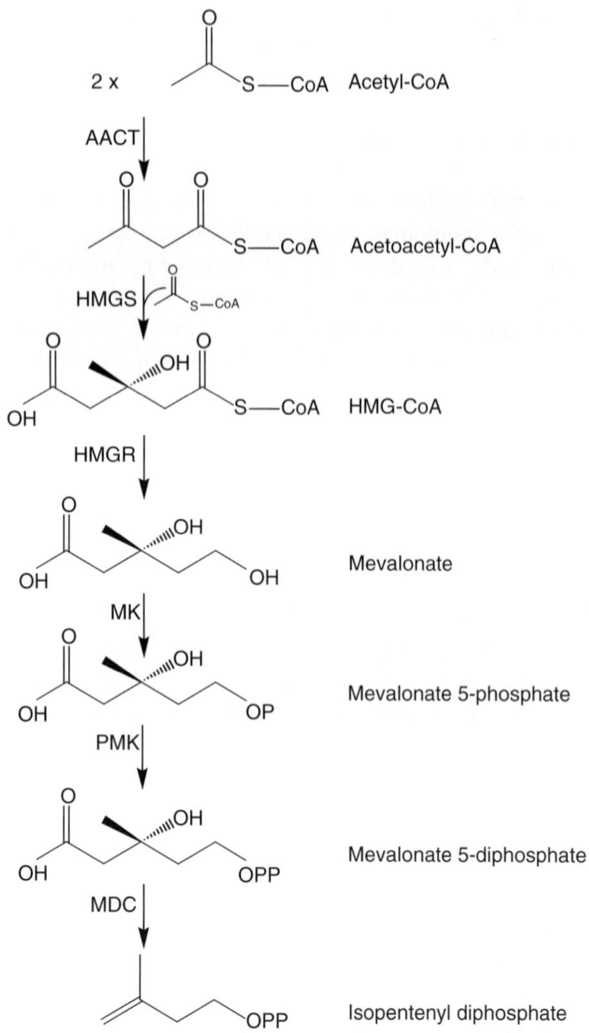

Figure 4 The acetate/mevalonate pathway of isopentenyl diphosphate synthesis. The enzymes involved are acetyl-CoA acetyltransferase (AACT), HMG-CoA synthase (HMGS), HMG-CoA reductase (HMGR), mevalonate kinase (MK), phosphomevalonate kinase (PMK), and mevalonate-5-phosphate decarboxylase (MDC). The letter "P" denotes a phosphate moiety.

The Biosynthesis of Rubber

Practically nothing is known about the biosynthesis of *trans*-1,4-rubbers such as chicle, balata, and gutta percha, since attention has focused on understanding the biosynthesis of the economically more valuable *cis*-1,4 rubber of hevea. Polyisoprenoids are synthesized by enzymes termed prenyltransferases, using isopentenyl diphosphate (IPP) as the "monomer" (**Figure 5**). This polymerization, at least in *cis*-1,4-polyisoprenoids, requires a short allylic diphosphate initiator molecule. The enzyme isopentenyl diphosphate isomerase (IPPI) catalyzes the interconversion of IPP to the allylic diphosphate isomer, dimethylallyl diphosphate (DMAPP). DMAPP can serve to initiate the synthesis of new rubber *in vitro* in isolated rubber particles supplied with IPP, and specific inhibition of this enzyme also inhibits rubber synthesis *in vitro*. Two distinct clones of this IPPI have recently been isolated from hevea latex and shown to be catalytically active following expression in *E. coli*. When *in vitro* rubber biosynthesis assays were carried out in the absence of initiating allylic diphosphates, addition of active IPPI was found to stimulate rubber synthesis. Condensation of IPP and DMAPP, catalyzed by *trans*-prenyltransferases, yields the additional allylic diphosphates geranyl diphosphate (GPP), farnesyl diphosphate (FPP), and geranylgeranyl diphosphate (GGPP) which serve a similar role as initiators in rubber biosynthesis (**Figure 5**). Although many prenyltransferases are known that synthesize these short

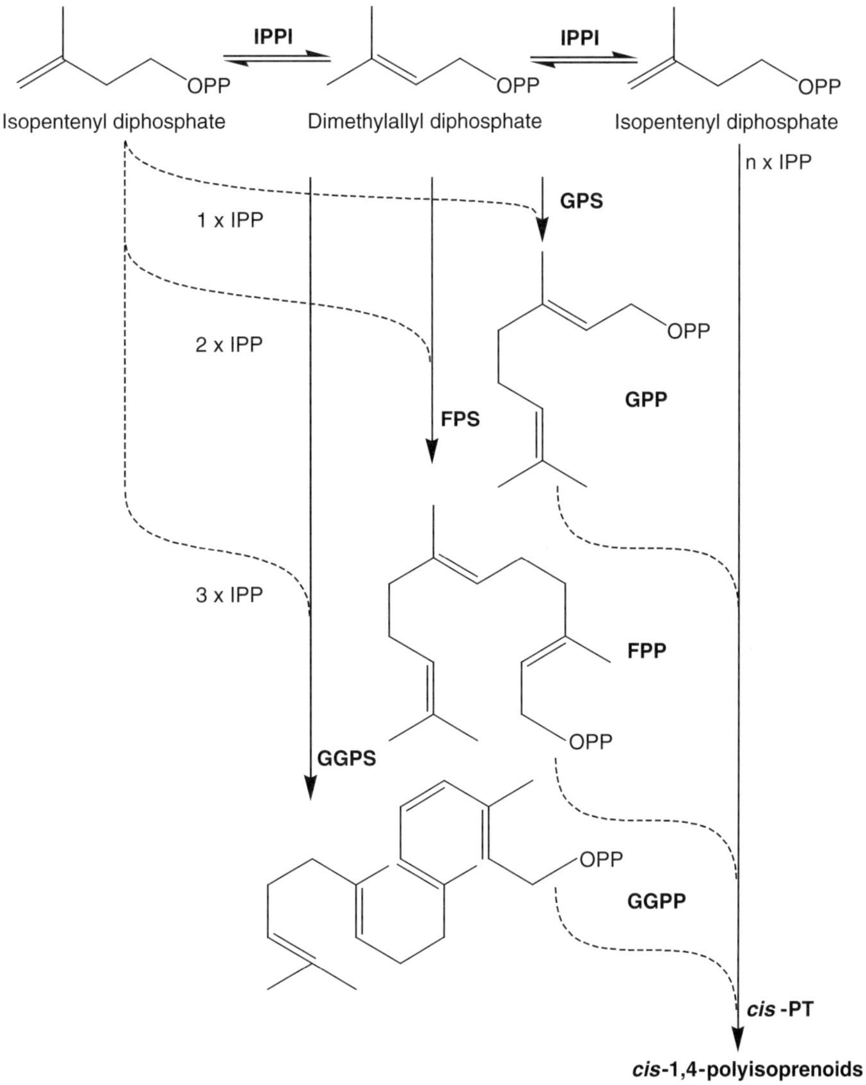

Figure 5 Isomerization of IPP and subsequent synthesis of allylic diphosphates and *cis*-polyisoprenoids. Isopentenyl diphosphate (IPP) and dimethylallyl diphosphate (DMAPP) are interconverted by the action of the enzyme IPP isomerase (IPPI). Condensation of DMAPP and IPP yields the allylic terpenoid diphosphates geranyl diphosphate (GPP), farnesyl diphosphate (FPP), and geranylgeranyl diphosphate (GGPP). These condensations are catalyzed by the *trans*-prenyltransferases geranyl diphosphate synthase (GPS), farnesyl diphosphate synthase (FPS), and geranylgeranyl diphosphate synthase (GGPS), respectively. All of these allylic diphosphates (including DMAPP) may serve as initiators of *cis*-polyisoprenoid biosynthesis *in vitro*. However, in *H. brasiliensis*, FPP is believed to be the principal initiating species for rubber biosynthesis, which is catalyzed by a *cis*-prenyltransferase enzyme, the rubber synthase or rubber transferase.

trans-polyprenyl diphosphates, no example of a *trans*-prenyltransferase capable of synthesizing longer chain molecules has been identified.

The known *trans*-prenyltransferase enzymes require magnesium ions and have been shown to generate a carbocation radical by dephosphorylation of bound allylic diphosphates, which reacts with an incoming molecule of IPP resulting in chain extension. Comparison of the primary amino acid sequences of *trans*-prenyltransferases indicates that they share a common ancestor, and this gene family is closely related in primary structure to the terpenoid cyclases. Both the *trans*-prenyltransferases and terpenoid cyclases share a common catalytic mechanism, although the cyclases act on the products of *trans*-prenyltransferase activity to generate a variety of terpenoid structures. Since the *trans*-prenyltransferases all share a common catalytic mechanism, it would appear that chain length is determined by the structure of individual enzymes. This conclusion has been supported by mutagenesis studies, where the primary structure of these enzymes is selectively altered. Modification of several *trans*-prenyltransferases in this manner has yielded forms that are capable of synthesizing longer polyprenol diphosphates than the original enzyme. It is believed that this "deregulation" results from elimination of an amino acid residue at the base of a "cavity," which in the wild-type enzyme blocks further extension of the chain. This interpretation is supported by structural analysis of crystallized *trans*-prenyltransferases and cyclases, which shows in all cases a cavity formed by 10–12 mostly antiparallel α-helices termed the "terpenoid synthase fold." Even the mutated enzymes, however, cannot synthesize polyprenol diphosphates longer than 30–50 carbons.

The key enzyme in rubber biosynthesis is, of course, that responsible for the synthesis of the rubber chain. This enzyme, termed rubber transferase or rubber synthase, can be analyzed biochemically using intact rubber particles, on or in which the enzyme is intimately bound. However, all attempts to isolate this activity from these particles have (to date) proved unsuccessful. Thus, until recently, this enzyme, arguably the most significant in rubber synthesis, has proved the most elusive.

As mentioned above, it has long been known that *trans*-prenyltransferases were also required for *cis*-1,4-polyisoprene synthesis, since studies *in vitro* had shown that such rubbers required a short *trans*-polyprenyl diphosphate initiator (DMAPP, GPP, FPP, or GGPP) before further chain elongation could occur. It was assumed that the enzyme(s) catalyzing chain elongation using these initiators would be similar in nature to the *trans*-prenyltransferases, because the reactions catalyzed by these enzymes differed only in stereochemistry (**Figure 6**). However, the characterization of a bacterial *cis*-prenyltransferase, synthesizing undecaprenyl diphosphate, revealed a *cis*-prenyltransferase with a unique primary and secondary structure, which is very distinct from the *trans*-prenyltransferase family of enzymes. Subsequently, many bacterial and plant *cis*-prenyltransferases have been identified, which are all closely related in primary structure. All of these enzymes require magnesium ions for activity, and although the details of their catalytic mechanism have not been elucidated, it is likely that it proceeds via carbocation radical generation as in the *trans*-prenyltransferases, despite their different structures.

The three-dimensional structures of two bacterial *cis*-prenyltransferase enzymes, undecaprenyl diphosphate synthases from *Micrococcus luteus* and *E. coli*, have now been solved. In addition to differing dramatically in primary structure from the *trans*-prenyltransferases, the three-dimensional structure of these *cis*-prenyltransferases is classified into a completely different structural family. Although in neither case was a substrate or analog present in the crystallized proteins, the resulting structures have allowed speculation as to the nature of catalysis and the active sites of these enzymes. Mutagenesis studies, such as those conducted for the *trans*-prenyltransferases, have also been reported for these *cis*-prenyltransferases. It would appear from these studies that an elongated hydrophobic tunnel, lined at the bottom by large amino acid residues, is most likely the site of elongation of the *cis*-polyisoprenoid product. Chain termination may occur in a manner analogous to that of the *trans*-prenyltransferases, in that a bulky hydrophobic amino acid at the bottom of the tunnel limits elongation.

Analyses of prenyltransferases catalyzing the synthesis of short polyisoprenoids (*trans*- or *cis*-) are very useful in developing our understanding of the fundamental mechanisms involved in polyisoprenoid

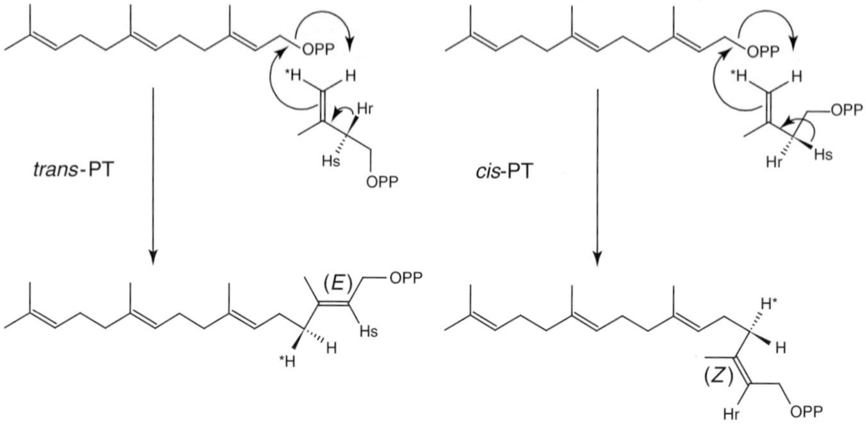

Figure 6 Synthesis of polyisoprenoids by *cis*- and *trans*-prenyltransferases. Ionization of an allylic diphosphate initiator (in this case FPP) in the active site of the prenyltransferase results in the formation of a carbocation, which attacks the double bond of incoming IPP from the *Si* face. In the case of *trans*-prenyltransferases, the *proR*-H at the C-2 position of IPP is then removed to form a new allylic double bond; in the case of *cis*-prenyltransferases, the *proS*-H is removed.

biosynthesis. However, it is likely to be difficult to extrapolate from these enzymes to those responsible for the synthesis of natural rubbers, i.e. molecules thousands of isoprenoid units long. For example, it is inherent in the models proposed for the action of bacterial undecaprenyl diphosphate synthases that product is released back out of the top of a hydrophobic pocket once chain extension has been completed. It is difficult to imagine such a mechanism operating with the enzymes responsible for natural rubber biosynthesis. Despite such caveats, knowledge of the primary sequences of the bacterial enzymes has allowed identification of related enzymes in plants, and several distinct isoforms of *cis*-prenyltransferase have been cloned from *H. brasiliensis* latex. No direct evidence for an involvement of these gene products in natural rubber biosynthesis is available. However, the fact that these genes were isolated as cDNAs prepared from latex RNA may be of some significance in this regard. It would seem unlikely that these hevea gene products are responsible for the synthesis of short-chain polyisoprenoids, since it is known that such molecules are absent from *H. brasiliensis* latex. It will be interesting to see if one or other of these isoforms might be involved in natural rubber biosynthesis as studies of these genes continue, or indeed whether a completely distinct prenyltransferase remains to be identified as the elusive rubber synthase.

List of Technical Nomenclature

Carbocation A cation containing an even number of electrons, with a significant proportion of the excess positive charge located on one or more carbon atoms.

Cyclase An enzyme that catalyzes the formation of a cyclic molecule from one or more acyclic precursors.

Laticifer Secretory cells of plants that produce a milky white watery latex.

Monomer A molecule that can undergo polymerization thereby contributing constitutional units to the essential structure of a macromolecule.

Number average molecular weight (Mn) The arithmetic mean of the distribution of number of molecular weight species in a distribution curve showing relative amount by number (mole fraction).

Polydispersity A measure of the breadth of the distribution of polymer chain sizes, generally taken to be the ratio of Mw/Mn.

Polymer A molecule of high relative molecular mass, the structure of which essentially comprises the multiple repetition of units derived from molecules of low relative molecular mass.

Polymerase An enzyme that catalyzes the formation of macromolecules or polymers.

Prenol An alcohol possessing the general formula $H–[CH_2–C(CH_3)=CH–CH_2]_n–OH$ in which the carbon skeleton is composed of one or more isoprene units.

Polyprenol A subgroup of prenols in which n in the general formula $H\text{-}[CH_2–C(CH_3)=CH–CH_2]_n\text{-}OH$ is >4.

Vulcanization A process for crosslinking rubber in order to improve its material properties (e.g., decreased plastic flow, diminished surface tackiness, increased tensile strength, increased solvent resistance).

Weight average molecular weight (Mw) The arithmetic mean of the distribution of molecular weight species in a distribution curve showing relative amount by weight (weight fraction).

See also: **Production Systems and Agronomy**: Oil Palm and Coconut.

Further Reading

Blackley DC (1997) *Polymer Latices: Science and Technology; Types of Latices, 2nd edn., Vol. 2.* London: Chapman & Hall.

D'Auzac J, Jacob J-L, and Chrestin H (eds) (1988) *Physiology of Rubber Tree Latex.* Boca Raton, FL: CRC Press.

Han K-H, Shin DH, Yang G, *et al.* (2000) Genes expressed in the latex of *Hevea brasiliensis. Tree Physiology* 20: 503–510.

International Rubber Research and Development Board Website; http://www.rudb.com/irrdb/NaturalRubber/ (2003).

International Rubber Study Group Website; http://www.rubberstudy.com (2003).

Morawetz H (2000) History of rubber research. *Rubber Chemistry and Technology* 73: 405–426.

Morel ED (1906) *Red Rubber.* New York: The Nassau Print.

Sethuraj MR and Mathew NM (1992) *Natural Rubber: Biology, Cultivation and Technology. Developments in Crop Science* 23. Amsterdam: Elsevier.

Stanfield ME (1998) *Red Rubber, Bleeding Trees.* New Mexico: University of New Mexico Press.

Wang K and Ohnuma S-I (1999) Chain-length determination mechanism of isoprenyl diphosphate synthases and implications for molecular evolution. *Trends in Biochemical Science* 24: 445–451.

Epicuticular Waxes

P J Holloway, The Long Ashton Research Station, Bristol, UK
C E Jeffree, University of Edinburgh, Edinburgh, UK

Definitions

In the generic sense, the terms wax or waxy are used to describe the physical characteristics of certain classes or mixtures of organic substances. Such materials are inert solids or semisolids with melting points in the range 35–100°C. They are insoluble in water but soluble in apolar solvents, such as chloroform, benzene, and diethyl ether. In addition, most waxes will form hard, glossy films on polishing; these coatings have much lower free surface energies than glass, for example, and are not easily wetted by water.

A wide variety of natural products fitting this general description can be obtained from plant, animal, and mineral sources, although nowadays many commercial waxes are manufactured synthetically. However, waxes cannot be defined unequivocally in simple chemical terms because they often differ markedly from one another in composition. Nevertheless, many natural waxes consist of complex mixtures of long-chain nonpolar compounds, with the molecular weights of individual constituents commonly ranging from about 200 to 600, but occasionally exceeding 1000; these values correspond with chain lengths of between about 20 and 60 carbon atoms.

Occurrence

Waxes are omnipresent in extant higher plant species, as well as mosses and ferns, implying that these compounds appeared early in the evolution of land plants, despite the fact that they are poorly preserved in the macrofossil record. Most plant waxes are found in intimate association with the cuticle, the thin continuous nonliving "skin" that covers the aerial surfaces of all plants. These cuticular waxes are synthesized exclusively by the underlying epidermal cells (*see* **Secondary Products:** Wax Pathways), as are the unique series of C_{16} and C_{18} hydroxy and epoxy alkanoic acids, which interesterify to form the characteristic polyester cutin matrices of cuticular membranes themselves. Wax constituents are subsequently secreted not only onto the surface of the cuticle but also become embedded within it. The former are usually referred to as "epicuticular" or "superficial", but may sometimes be called "surface lipids." "Intracuticular" is used to denote waxes present within the cuticle, but these will not be dealt with in this article. The primary physiological functions of the cuticle complex are to waterproof (liquid water) the plant surface and to limit water loss (water vapor) from internal tissues.

Isolation and Analysis

The extraction of epicuticular waxes from plant material is straightforward and usually involves brief exposure of the surfaces of detached intact organs, such as leaves and fruits, to an appropriate organic solvent; chloroform is often the solvent of choice. According to the level of refinement required for the investigation, this may be achieved by:

- total immersion of individual plant parts for about 30 s in a large volume of solvent;
- running a fine stream of solvent over the selected surfaces and collecting the effluent;
- wiping selected regions of the surface with extractive-free cotton wool swabs soaked in solvent, followed by further extraction of the swabs in the same solvent

In each case, solvent washings are combined, then evaporated, and the solid residue obtained dried to constant weight. Overexposure to solvents can cause leaching of intracuticular waxes and, if severe, may result in removal of internal lipids and pigments from the specimens.

Such extraction procedures are often combined with gravimetric or planimetric measurements of the plant samples in order to obtain a quantitative assessment of the mean epicuticular wax content per unit weight or area. The amount of wax on leaf surfaces may be as little as 1 μg cm^{-2} on some temperate species, but can reach as much as several mg cm^{-2} on a number of tropical palm species; in the latter case, these wax layers are so thick that they can be removed by mechanical methods. The wax deposits on the fruits of a given species are invariably heavier than those on the corresponding leaves, which in turn may carry different amounts of wax on their adaxial and abaxial surfaces. The effects of solvent washing on a plant surface can also be monitored by scanning electron microscopy (SEM) in order to confirm the efficiency of removal of epicuticular deposits.

A summary of typical analytical procedures used in the determination of the composition of epicuticular waxes is given in **Figure 1**. Thin-layer chromatography (TLC) on silica gel provides useful qualitative information about the various classes of constituents present by comparisons of R_F values observed with those of appropriate reference compounds, together with spray reagents for specific functional groups in some cases. TLC is also valuable as a preparative method for isolating fractions corresponding with individual wax classes; these can then be examined in more detail by other chromatographic or spectroscopic techniques.

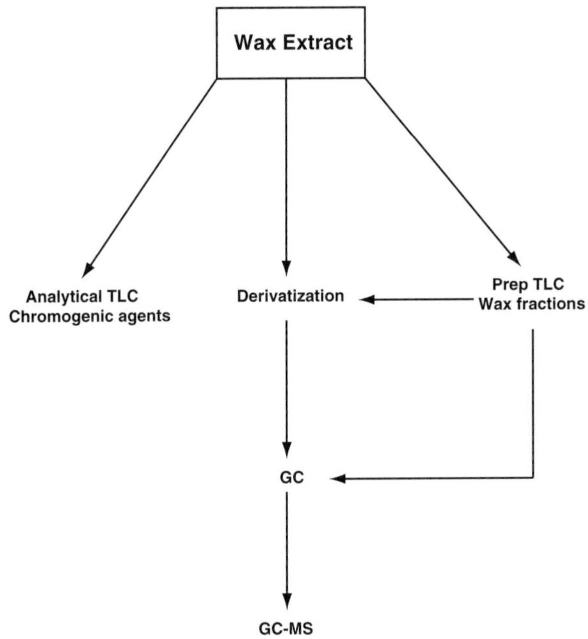

Figure 1 Summary of analytical procedures most commonly used to determine the qualitative and quantitative composition of plant epicuticular waxes. Adsorption thin-layer chromatography (TLC) fractionates a wax into its constitute classes according to their polarity (**Tables 1** and **2**). Individual compounds are separated from one another by gas–liquid chromatography (GC) and identified by mass spectrometry (MS).

Gas–liquid chromatography (GC) with flame ionization detection (FID) is used to identify and quantify each constituent in a wax, although preliminary derivatization of any free hydroxyl or carboxylic acid groups present is normally carried out; the former are usually analyzed as the corresponding trimethylsilyl ethers, the latter as methyl esters. Stationary phases capable of operating at temperatures well above 300°C are necessary in order to achieve complete separations; short-length, bonded, fused silica capillary columns in conjunction with both temperature and gas flow programing are used routinely nowadays. Because such columns also offer high resolution, they can be used as a one-step method of analysis for a total wax extract. Compounds are often characterized first from their retention indices following coinjection with a series of known *n*-alkanes, whereas quantitation is achieved by subsequent addition of an appropriate internal standard and calculation of response factors with reference to model compounds for each class of constituent. Final identification of compounds is made by mass spectrometry (MS) from their molecular weights and diagnostic fragmentation patterns; some examples are shown in **Figures 2–5**. The most powerful analytical tool available to the wax chemist today is still GC coupled to MS (GC-MS). However, if single constituents can be isolated from a wax, confirmatory structural information may be obtained

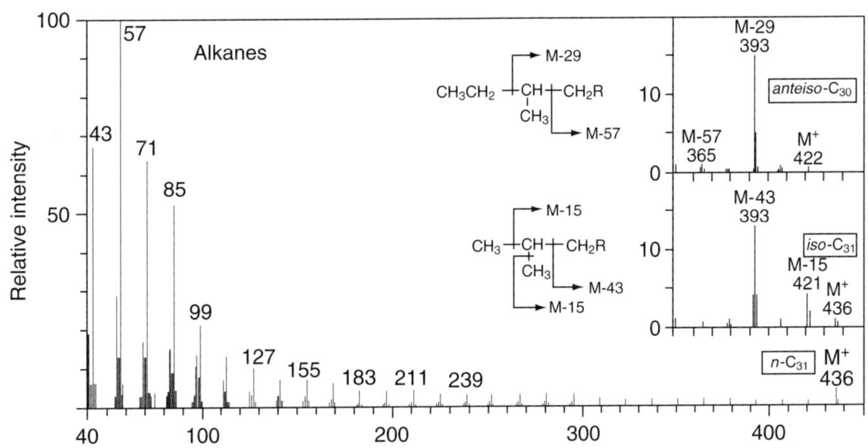

Figure 2 Mass spectra of *n*-hentriacontane (*n*-C_{31}), 2-methyltriacontane (*iso*-C_{31}) and 3-methylnonacosane (*anteiso*-C_{30}) obtained from a GC-MS analysis of the alkane fraction isolated from the epicuticular wax of tomato fruits. Fragmentation schemes showing the diagnostic ions for the two types of branched (*br*) homologs are also included; only the high ends of the spectra are shown, the lower ends being similar to that of the *n*-homolog.

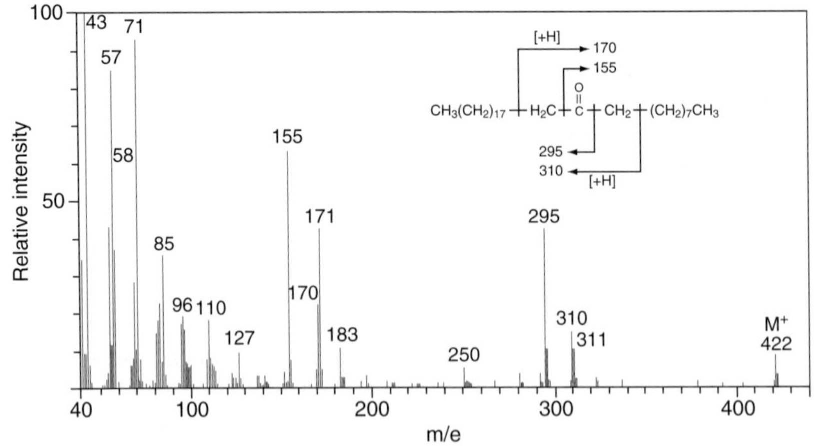

Figure 3 GC-MS of the C_{29} ketone fraction isolated from the epicuticular wax of plum (*Prunus domestica*) fruits. The fragmentation scheme for the major component, *n*-10-nonacosane, is shown.

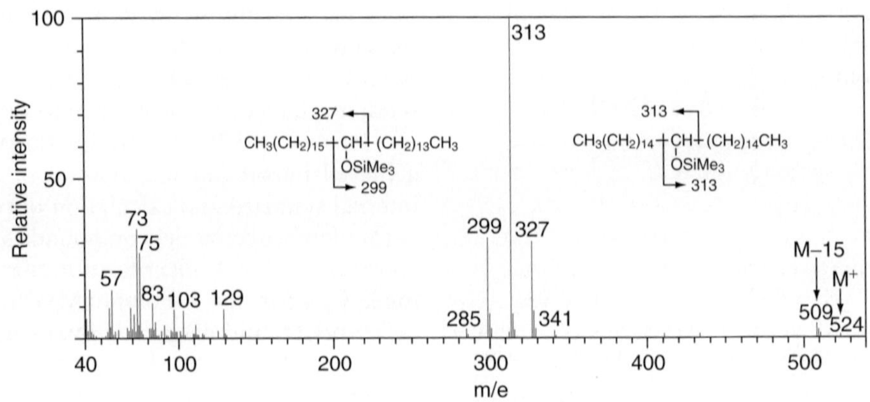

Figure 4 GC-MS of the C_{31} secondary alcohol fraction isolated from the epicuticular wax of pea leaves, after conversion to the corresponding trimethylsilyl (–$OSiMe_3$) derivatives. The fragmentation schemes for the major components, *n*-16-(trimethylsiloxy)hentriacontane and *n*-15-(trimethylsiloxy)hentriacontane, are shown.

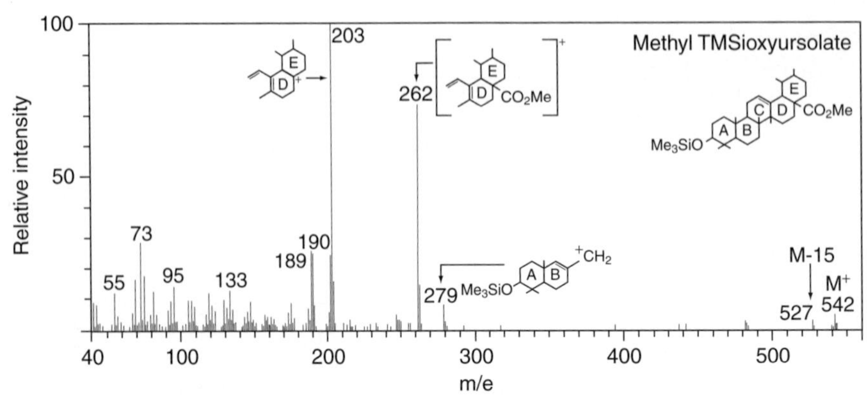

Figure 5 GC-MS of ursolic (3β-hydroxyurs-12-ene-28-oic) acid isolated from the epicuticular wax of apple leaves, after conversion to the corresponding trimethylsilyl (–$OSiMe_3$) ether methyl ester (–CO_2Me) derivative. The origin of the diagnostic retro-Diels-Alder rearrangement ion (*m/z* 262) is shown, together with those of other important fragmentation ions.

by nuclear magnetic resonance (NMR). High-performance liquid chromatography (HPLC) has found only limited use for wax analysis, mainly due to the difficulty in detecting eluted compounds.

Composition

The chemistry of the wax secreted onto the surfaces of plants shows a great diversity in composition and complexity, with major differences not only between species, but also sometimes even between different regions on the same species. The adaxial and abaxial surfaces of pea (*Pisum sativum*) leaves, and the sheaths and blades of the flag leaves of barley (*Hordeum vulgare*), are striking examples of the latter. On the other hand, the composition of waxes from disparate species can sometimes be remarkably similar, as in the case of the leaves of cabbage (*Brassica oleracea*) and *Clarkia elegans* or of the leaves of tulip (*Tulipa gesneriana*) and the needles of Sitka spruce (*Picea sitchensis*). Nevertheless, wax biosynthesis and its ultimate deposition on the cuticle surface are under strict genetic regulation. This factor is highlighted by the existence of several spontaneous or induced subglaucous or "waxless" mutants of a number of crop species (peas, brassicas, corn (*Zea mays*; maize), sorghum (*Sorghum bicolor*), and barley) and recently in *Arabidopsis*; these plants show drastic alterations in the composition, amount, or microstructure of their foliar waxes, compared with the normal form.

It is difficult to classify plant waxes chemically, but they can be divided roughly into those that contain predominantly long-chain aliphatic constituents and others that possess large amounts of cyclic constituents, especially triterpenoids; inevitably, intermediate types with both categories of compound are also known. In exceptional cases, as on some ferns, the "wax" or "farina" found on the fronds is composed almost entirely of crystalline lipophilic flavonoids.

Many aliphatic compounds have been identified in epicuticular waxes and these belong to various classes, which, in turn, invariably show either a predominantly odd or even carbon number; some representative examples are given in **Tables 1** and **2**, respectively. Plant species differ from one another in the numbers of these classes present and their relative amounts, as well as in the homolog content within each class. Homologs differ from one another by one $-CH_2-$ group. Positional isomerism is commonly encountered in classes that possess one or more functional group, such as hydroxy and oxo, in mid-chain positions. Methyl-branching at terminal positions may sometimes be detected in certain classes, especially alkanes and esterified primary alcohols and alkanoic acids.

A comprehensive survey of wax chemistry is outside the scope of this article, but it is worth recording here, for reference, the main aliphatic constituents present in the leaf waxes of some familiar plant species. High proportions of alkanes are found on the leaves of peas and brassicas, aldehydes on sugar cane (*Saccharum officinarum*) leaves and lemon (*Citrus limon*) fruits, ketones on leek (*Allium porrum*) leaves, primary alcohols on the leaves of many grasses and cereals, symmetric secondary alcohols on the leaves of brassicas and *C. elegans*, asymmetric secondary alcohols on some mosses, the needles of many gymnosperms and the leaves of carnations (*Dianthus* spp.), and β-diketones on the leaves of many eucalypts, box (*Buxus sempervirens*), and barley. Although there are some notable exceptions, such as esters, the most common homologs in an odd-numbered class are generally C_{29}, C_{31}, and C_{33}, compared with C_{26}, C_{28}, and C_{30} for an even-numbered class. However, attempts to utilize wax chemistry as a chemotaxonomic marker have only met with limited success.

Plant waxes that are rich in cyclic constituents are less common than their predominantly aliphatic counterparts. They are usually found coating the surfaces of fruits or isolated from leaves that exhibit a glossy finish, such as *Rhododendron* species and cherry laurel (*Prunus laurocerasus*). Pentacyclic triterpenoids and sterols (tetracyclic triterpenoids) are the two main classes of cyclic constituents that

Table 1 Principal classes of aliphatic constituents with a predominantly odd carbon number found in plant epicuticular waxes

Class[a]	*Common examples*
Alkanes	*n*-Nonacosane *n*-Hentriacontane
Ketones	*n*-15-Nonacosanone *n*-16-Hentriacontanone
Secondary alcohols	*n*-10-Nonacosanol *n*-15-Nonacosanol
β-Diketones	*n*-14, 16-Hentriacontandione *n*-16, 18-Tritriacontandione

[a] Arranged in increasing order of polarity.

Table 2 Principal classes of aliphatic constituents with a predominantly even carbon number found in plant epicuticular waxes

Class[a]	*Common examples*
Esters	*n*-Octadecyl hexacosanoate *n*-Hexadecyl octacosanoate
Aldehydes	*n*-Hexacosanal *n*-Octacosanal
Primary alcohols	*n*-Hexacosanol *n*-Octacosanol
Acids	*n*-Hexacosanoic *n*-Octacosanoic

[a] Arranged in increasing order of polarity.

occur most frequently, although the latter are more commonly detected as minor components. Pentacyclics with ursane and oleanane skeletons are the most abundant classes. Ursolic acid and its analogs are the major constituents of apple (*Malus pumila*) fruit waxes, whereas oleanolic acid and its derivatives dominate in those from olives (*Olea uropaea*) and grapes (*Vitis vinifera*); α- and β-amyrins, also of the oleanane series, can be detected in many leaf waxes. The ubiquitous plant sterols, β-sitosterol, campesterol, and stigmasterol, are often secreted onto the surface of foliage.

It should be noted that esterification is a widespread mechanism for the formation of wax components. This may occur with any aliphatic or cyclic constituent containing hydroxyl or carboxylic acid groups. Some examples are esters of sterols with alkanoic acids $>C_{16}$ and triterpenols with acetic acid. Very long-chain esters, known as estolides, are found in many conifer waxes. These compounds are formed by the linear interesterification of various ω-hydroxyalkanoic acids (e.g., *n*-16-hydroxyhexadecanoic), α,ω-alkanediols (e.g., *n*-hexadecane-1,16-diol), alkanoic acids and, sometimes, primary alcohols; the corresponding molecular weights range from about 800 up to 1500. Such compounds might provide the "missing" link between the cuticular lipids, wax, and cutin.

Physical Form

The fine structure of epicuticular wax layers *in situ* has been studied extensively over the last 35 years or so with the aid of the scanning electron microscope (SEM), using both coated and uncoated, fresh or dried specimens. Indeed, Barthlott and his group have cataloged over 10 000 species and proposed a classification scheme for use in the systematics of angiosperm families. Epicuticular waxes show a great diversity in ultrastructure, which can range from an extremely thin amorphous film to complex layers of microcrystals with different shapes and sizes, (0.1 μm–20 μm). Common crystal types observed are hollow tubes, simple or decorated platelets, rodlets, filaments, and ribbons, which are generally flattened and exhibit different degrees of adornment. Wax granules having an indeterminate shape may also be present; these may represent truncated tubes, plates, or cuboid crystals. Although these different crystal types often appear to be distributed randomly over the cuticle surface, they can sometimes be found in regular arrangements or definite patterns. For example, on *Convallaria majalis* leaves, wax plates standing on their edges form parallel arrays over large areas and also appear to radiate from the stomatal apertures. On the other hand, platelets arranged in rosettes appear to be a feature of the epicuticular wax layers in the order Fabales. In the Strelitziacea, patterning extends to the formation of remarkable wax chimneys surrounding the stomatal complexes. The walls of these tubes are constructed from fused co-aligned wax crystals that appear to arise from a well-defined ring of cuticle at the base. Such patterning effects indicate that there is precise control over the development of wax morphology in some species, but how this is achieved is not known.

If many wax crystallites occur on the cuticle surface, they scatter short wavelength light and, collectively, are responsible for the bluish matt "bloom" visible on many leaves, stems, and fruits. Cogent examples are the fruits of plums and grapes, the leaves of brassicas, cereals, Eucalypts, carnations, and the bloodroot (*Sanguisorba canadensis*), the latter celebrated by M.C. Esher in his lithograph, "Dewdrop." Smooth glossy leaves, like *Rhododendron* spp. and cherry laurel, are usually characterized by the presence of a dense amorphous covering of epicuticular wax. Sometimes, the epicuticular wax layer may feel greasy to the touch, as can be detected on some cultivars of apple fruit.

Relationships between Composition and Ultrastructure

Tube-like crystals of epicuticular wax occur widely in the plant kingdom and chemical analyses have revealed the presence of two main types of constituent in the corresponding isolated waxes. The first group contains substantial amounts of asymmetrical secondary alcohols, in which *n*-10-nonacosanol and closely related positional isomers predominate; β-diketones are the main components of the second type. Familiar examples of species containing these compounds have already been cited above. The chemical distinctions between the two types of tube are also supported by some morphological differences. Thus, those of the secondary alcohol class are usually shorter and wider (0.3–1.1 μm long by 0.1–0.2 μm wide) than those of the β-diketone class (2.0–3.0 μm long by 0.2–0.3 μm wide).

Tubular crystals with morphologies different from those described above are also known. For example, on leaves of brassicas, *C. elegans*, and *Centranthus ruber* there are large wax tubes (up to 8.0 μm long by 1.0 μm wide) that are transversely ridged and taper gradually from base to apex. These waxes all contain mixtures of alkanes and symmetrical secondary alcohols and ketones. On the other hand, the tubules that are found on the surface of *Cerinthe minor* leaves are associated with the occurrence of δ-lactones (1,5-alkanolides) in the wax.

Other correlations between chemistry and ultrastructure have also been made. Platelets are often formed on plant surfaces by epicuticular waxes that contain high proportions of primary alcohols (see above for examples). Similar associations between aldehydes and the formation of wax filaments, and between symmetrical ketones and ridged rodlets, have been observed on disparate groups of plant species. It should also be noted that wax fine structure can differ markedly according to the position on the same plant. There are β-diketone-rich tubes on the abaxial surfaces of the flag leaves of wheat (*Triticum* spp.) in contrast to primary alcohol-rich plates on the corresponding adaxial surfaces, whereas, in barley, the former predominate on the leaf sheaths and the latter on both surfaces of the leaves.

Recrystallization experiments *in vitro* using volatile organic solvents can be used to confirm possible crystal structure–chemistry relationships for epicuticular waxes. Thus, total waxes, individual class fractions isolated from them and, in some cases, pure synthetic compounds can be compared in a model system and monitored by SEM. However, the results that have been obtained from such investigations fall into three broad categories:

- Good agreement between the crystalline form of the epicuticular wax *in situ*, that of the total wax after recrystallization *in vitro*, and that of one of the components of the wax after purification and recrystallization *in vitro*.
- Good agreement between the crystalline form of the epicuticular wax *in situ* and that of the total wax after recrystallization *in vitro*, but not that of any of the components of the wax after purification and recrystallization *in vitro*.
- No agreement between the crystalline form of the epicuticular wax *in situ* and *in vitro* after recrystallization.

Examples in the first category include *n*-10-nonacosanol in many monocotyledons, dicotyledons, and gymnosperms, β-diketones and hydroxy-β-diketones in cereals and Eucalypts, and *n*-hexacosanol and *n*-octacosanol in grasses and legumes. Micrographs illustrating the morphology of epicuticular wax layers and crystal forms *in situ* and *in vivo* are shown in **Figures 6–13**.

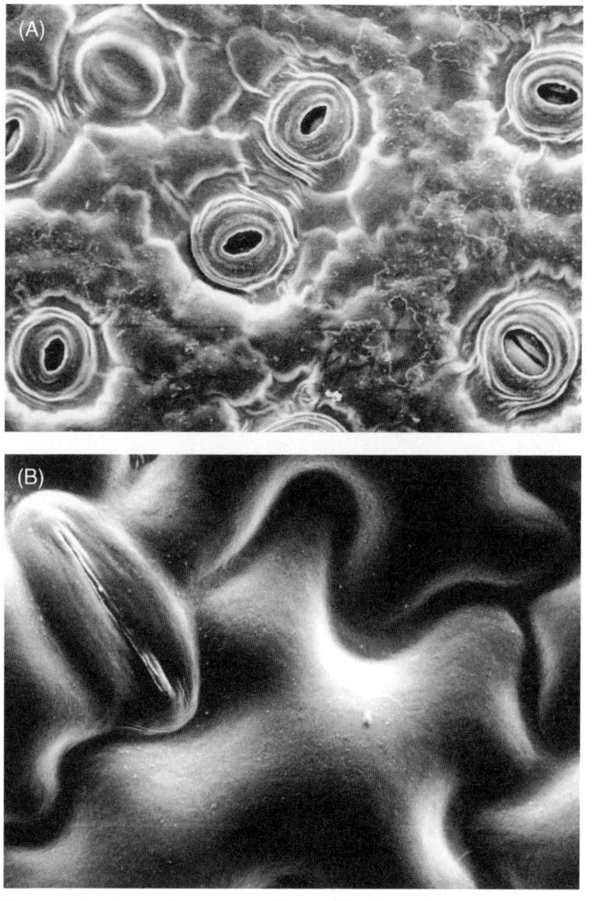

Figure 6 Amorphous wax films. (A) The abaxial surface of *Rhododendron ponticum* leaves (x550) is covered with a dense crust of triterpenoid-rich epicuticular wax, which thins out in the vicinity of the numerous stomata present. (B) The adaxial surface of field bean (*Vicia faba*) leaves (x880) is finely corrugated, replicating the convex shape of the underlying epidermal cells. The surface of the cuticle appears smooth and has an extremely thin coating (ca. 3 μg cm^{-2}) of wax.

Possible Mechanisms for Wax Secretion *In Planta*

The mechanism for transport of epicuticular wax constituents from the epidermis, through the cuticle and onto the plant surface has been the subject of much debate and experimentation since the nineteenth century, but remains an enigma to this day. For a long time, these materials were thought to be extruded in a liquid or semisolid form via pores; this hypothesis was based mainly on the visual interpretation of the large cylindrical wax structures observed with the light microscope on leaves such as those of *Strelitzia*, banana (*Musa* spp.), bamboo (*Arundinaria* spp.), and the nodes of sugar cane stems (*Saccharum officinarum*). However, successive detailed investigations by both light and electron microscopy have consistently failed to provide unequivocal evidence for the existence of either pores or any other form of discrete channel through which waxes might be able to migrate. The perceived problem for transport then is the fact that wax molecules are large, involatile, and most will be hard solids at physiological temperatures. Consequently, it is difficult to envisage

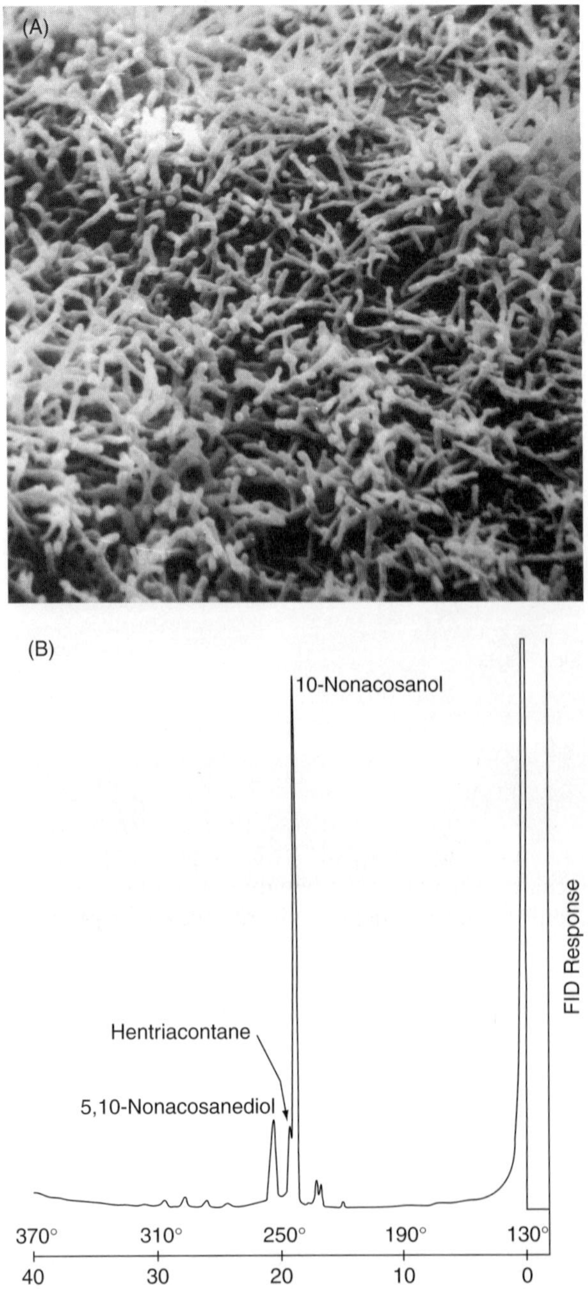

Figure 7 (A) The small tubular microcrystals of epicuticular wax found on the adaxial surface of *Aquilegia vulgaris* leaves (x5200) contain large amounts of *n*-10-nonacosanol. (B) GC analysis of total wax after silylation; Dexsil 300 column with flame ionization detection (FID).

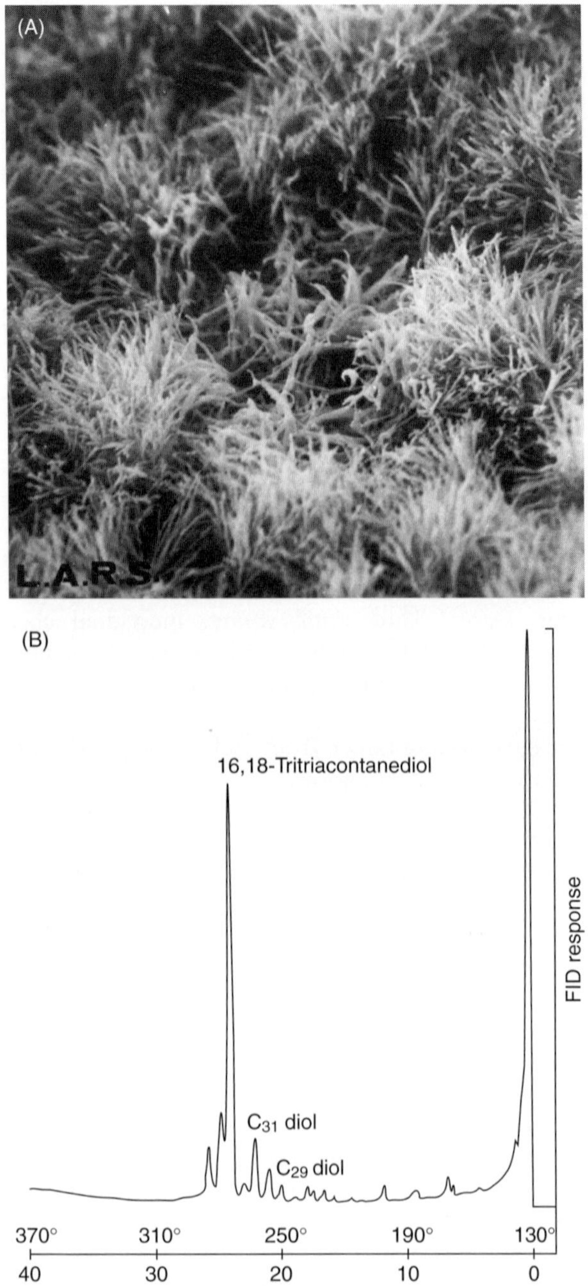

Figure 8 (A) The aggregations of long tubular microcrystals of epicuticular wax that occur on the adaxial surface of *Eucalyptus globulus* leaves (x2200); they are characterized by the presence of *n*-16,18-tritriacontanedione as the major component. (B) GC analysis of total wax after reduction with sodium borohyride to the corresponding diol and silylation; Dexsil column as **Figure 7**.

how they would be able to move in their native state directly through the complex architecture of either the cell wall or the cuticle. The *in vitro* experiments carried out with isolated waxes and their individual constituents suggest that a solvent or some other form of solubilizing agent might be instrumental in the transport process. This could facilitate the permeation of wax in a liquid form through the cuticle. Once on the surface and exposed to the atmosphere, the transport medium would then evaporate or dissociate, leaving the wax to form a crystalline or semicrystalline layer by self-association. To date, headspace analysis has not been able to detect the presence of volatile compounds specifically emitted from the surfaces of very glaucous plants and

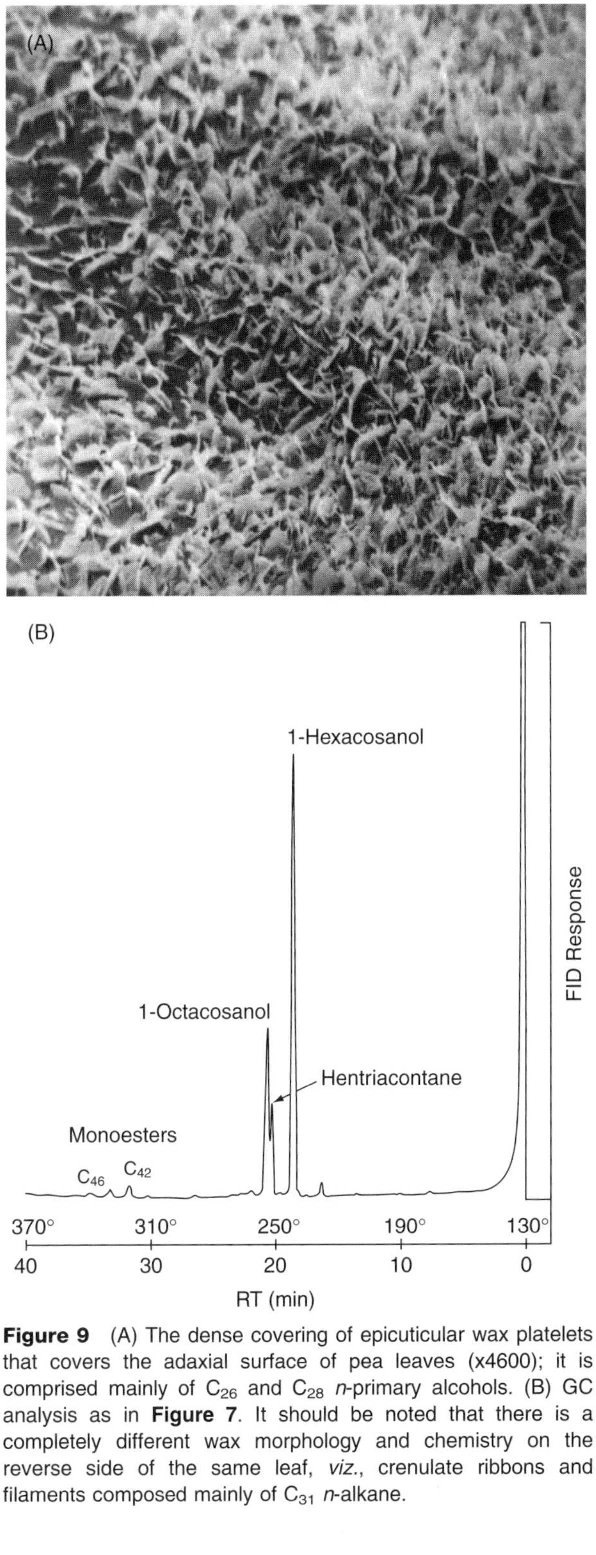

Figure 9 (A) The dense covering of epicuticular wax platelets that covers the adaxial surface of pea leaves (x4600); it is comprised mainly of C_{26} and C_{28} *n*-primary alcohols. (B) GC analysis as in **Figure 7**. It should be noted that there is a completely different wax morphology and chemistry on the reverse side of the same leaf, *viz.*, crenulate ribbons and filaments composed mainly of C_{31} *n*-alkane.

therefore potential candidates as wax solvents. However, acyl carrier proteins have been identified recently in the epicuticular waxes of broccoli (*B. oleracea* var. *italica*); such proteins might be able to provide the missing piece in the jigsaw puzzle.

Chemical Changes during Development

The amounts and composition of the epicuticular waxes on leaves and fruits may change substantially and continuously during the course of a plant's growth. Wax production is most rapid in the earlier stages of development, but it gradually declines as the organ reaches maturity. If damaged, wax layers can be regenerated but this ability is usually confined to the youngest tissues during their normal developmental phase of wax production. Detailed studies of the ontogeny of waxes on the surfaces of several species have been published and include the leaves of wheat, corn, peach (*Prunus persica*), beech (*Fagus sylvatica*), and lime (*Tilia tomentosa*), as well as tomato (*Lycopersicon esculentum*) and *Citrus* fruits. In most cases, major changes were detected in the relative proportions of the various wax classes identified, some of which were absent in the early stages, but that later became major constituents; the reverse also occurred in some species. In addition, the ratio between cyclic and aliphatic constituents, if both classes were present, fluctuated markedly. There was also a marked tendency for homolog chain-lengths of aliphatic classes to increase with the maturity of the wax layer. For example, in tomato, *n*- and *br*-alkanes were the main components of immature fruits, but these were replaced by pentacyclic triterpenols and flavonoids in mature ones; *br*-alkanes were not detected in the wax of ripe fruits and the chain-length distribution of the *n*-alkanes shifted more toward C_{31}. In beech, primary alcohols are dominant on juvenile leaves, alkanes and phenolic esters on intermediate stages, and aldehydes on mature leaves. The switch in the production of primary alcohols to alkyl esters during the development of corn leaves is also accompanied by a dramatic change in wax microstructure. It should be noted that some of the ontogenetic changes in wax composition that have been observed within one species are equivalent to the gross differences that are reported between different species.

Effects of Environment and other External Factors

Various abiotic factors have been shown to have a profound effect on the production, composition, and microstructure of the epicuticular waxes of a number of plant species. Thus, wax production is often stimulated by exposure to high levels of solar radiation, elevated temperatures, and low humidities; concurrent changes in wax class compositions and homolog distributions within individual classes may also be detected, reflecting, in turn, effects on

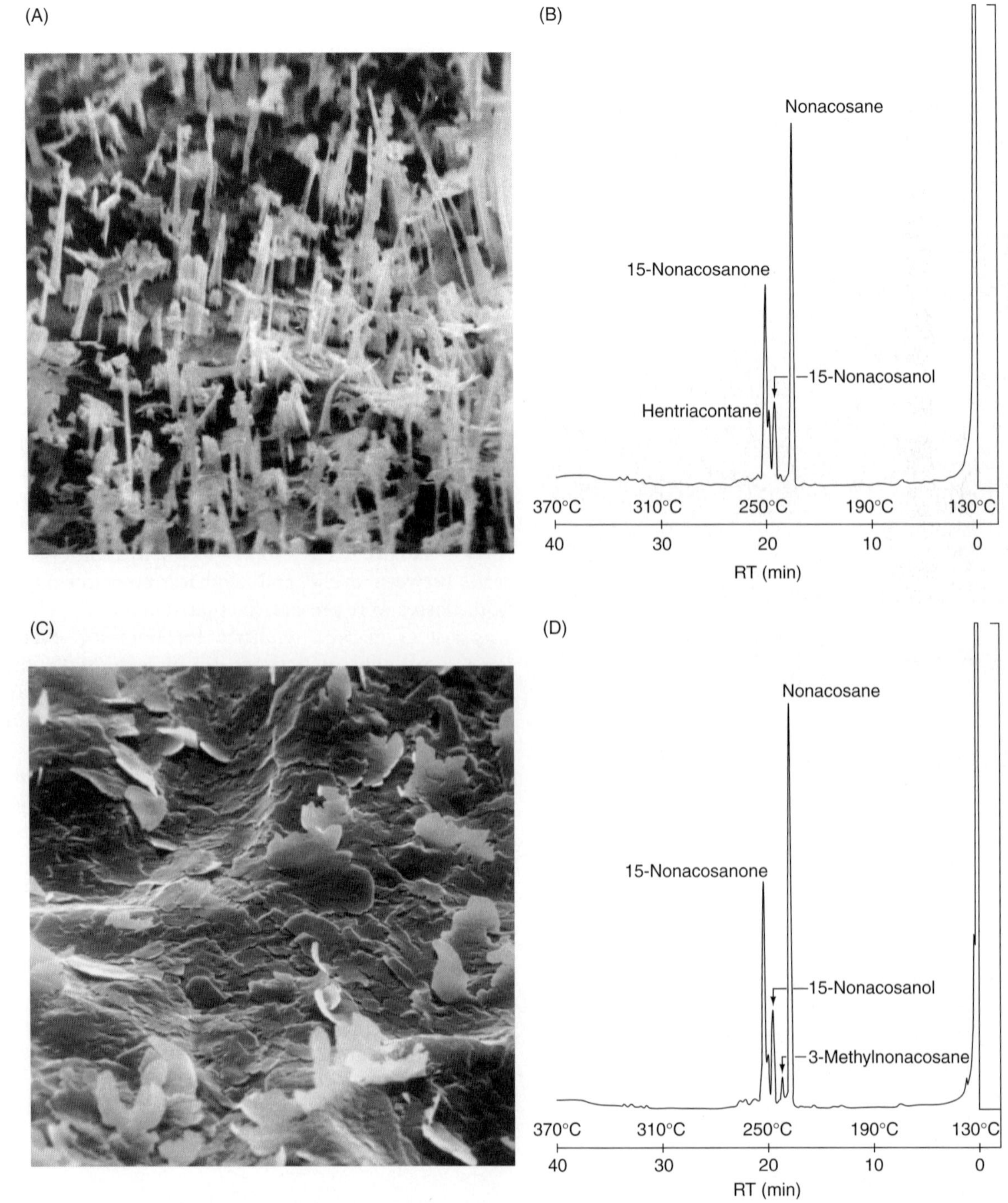

Figure 10 Effects of two mutations on the ultrastructure and composition of the epicuticular waxes on the abaxial leaf surfaces of oilseed rape (*Brassica napus*). The surface of Nilla normal plants shows the morphology (A, x4700) and chemistry (B) characteristic of most species in the cabbage family. Wax ultrastructure on the Nilla mutant (C, x5400) is drastically altered in comparison with the normal type, showing extensive plate-like formations covering and extending upwards from the cuticle surface. However, the only change that can be detected in wax composition (D) is the production of a small amount of branched (*br*) alkane. In the Rigo mutant (E), there are extensive modifications to both the crystal structure as well as the chemistry (F) of the epicuticular wax. Wax production in this mutant is partially inhibited and switches to the formation of *n*-aldehydes; this is manifested by the appearance of sparsely distributed short tubular microcrystals. RT, retention time. GC details for all as described in **Figure 7**.

fundamental biosynthetic processes. However, these chemical changes may not always be mirrored by modifications to the crystalline form of the resultant wax deposits. Environmental effects on wax ultrastructure are well documented for leaves on normal and mutant forms of *B. oleracea*.

(E)

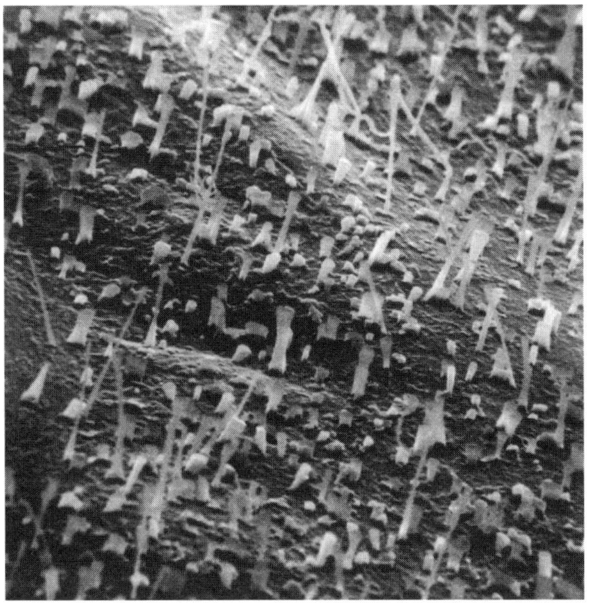

(F)

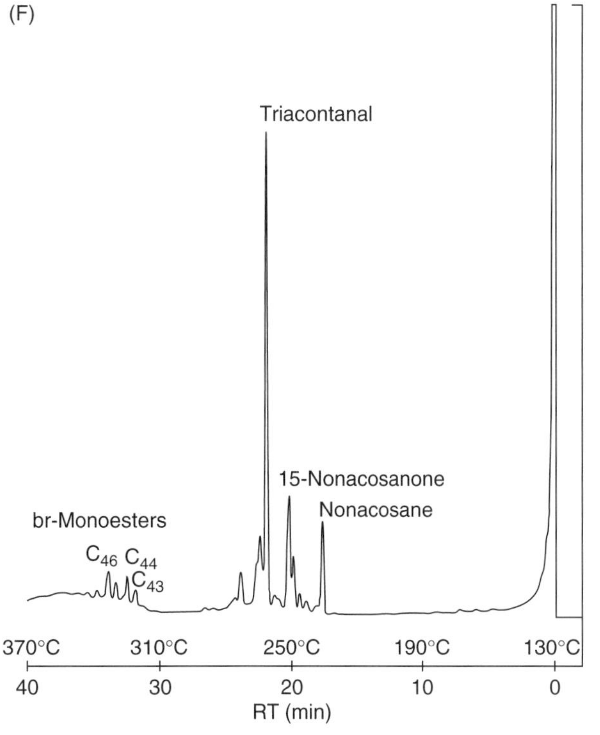

Figure 10 Continued.

Most wax constituents are resistant to chemical modification under normal environmental conditions. However, aldehydes, which are readily oxidized to the corresponding alkanoic acids, are exceptions; this transformation can be detected during the course of a growing season on the leaves of beech for example. The deterioration of waxes often observed on conifer needles is a symptom that is usually attributed to air pollution. Although the secondary alcohol tube waxes present can be oxidized by nitrogen dioxide (NO_2) *in vitro* to yield oxidized analogs, such as *n*-10-nonacosanone and alkanoic acids, that have planar rather than tubular crystal morphologies, the NO_2 concentrations necessary are well above those that occur even in a heavily polluted atmosphere. On the other hand, long-term exposure to atmospheric pollutants, such as acid rain, oxides of nitrogen (Nox), sulfur dioxide (SO_2), and ozone (O_3), probably has a direct effect on wax biosynthesis, especially that in the needles of conifer species, resulting in major changes in wax morphology and coverage. A number of herbicides, including trichloroacetic acid (TCA), ethyl *N*,*N*-dipropylthiocarbamate (EPTC), diallate, and ethofumesate, are specific inhibitors of cuticular lipid biosynthesis and are active in the vapor phase. The visible appearance of plants, such as peas and brassicas, treated with sublethal doses of these compounds gradually changes from glaucous to glossy, as wax production and crystallite formation are suppressed.

Epicuticular waxes, especially those of a finely crystalline nature, may be damaged or even removed by weathering from strong winds or heavy rains. However, in the majority of plant species, repairs to any wax layers that have been damaged can only take place during the period of organ expansion. Once expansion is completed, synthesis of wax ceases and the layers are then subjected to a variety of environmental hazards for the rest of the life of the organ. Thus, the epicuticular wax layers may accumulate detritus, become habitats for microorganisms, or suffer further abrasion and smoothing by mutual contact with adjacent organs on the same plant or its close neighbors. Ballistic impacts of precipitation and wind-blown particles are also known to disrupt epicuticular ultrastructure. All of these processes can alter the intrinsic nature of the wax layer, increasing its wettability and permeability, as well as reducing its reflectivity and anti-adhesive properties. Such effects are particularly important in evergreen species with long-lived leaves. In conifers, for example, needle surfaces may become so densely colonized by epiphytic fungi and algae within 2 or 3 years that the original wax layer is replaced almost entirely by a new surrogate surface with entirely different physicochemical properties.

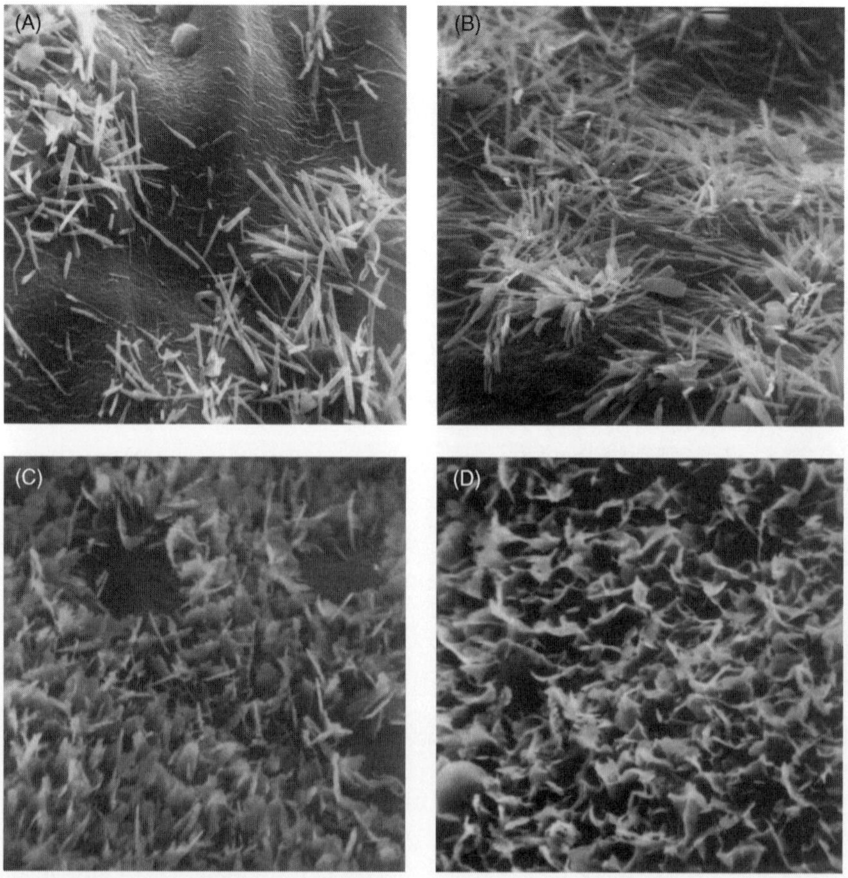

Figure 11 Epicuticular waxes that recrystallize in the same physical form as observed *in planta*. (A) Wax on the flag leaf sheath of barley *in situ* (x4500) and (B) the isolated wax *in vivo* (x4300). The tubular wax crystals are composed mainly of β-diketones and hydroxy-β-diketones. (C) Wax on the adaxial surface of a vegetative leaf of barley *in situ* (x9000) and (D) the isolated wax *in vivo* (x10500). *n*-Hexacosanol is the principal constituent of the wax platelets.

Role in Water Repellency

Epicuticular wax layers have a considerable influence on the wettability of plant surfaces in contrast to the corresponding intracuticular waxes, which control their permeability to water. The degree of surface wetness is functionally important to plants in several ways. For example, highly hydrophobic surfaces discourage the adhesion of dirt and other particulate contaminants falling in rainwater. Because most of these extraneous materials are more hydrophilic than waxes, they adhere preferentially to water droplets, which can then roll off the plant surface. On the other hand, the presence of water films on leaves impedes the transport of carbon dioxide and, ultimately, inhibits photosynthesis. In addition, excess water on the surface may infiltrate through open stomatal pores and damage the underlying mesophyll cells, providing, in turn, a medium for the growth of invading bacteria and fungi. Plant surfaces are also the targets for applications of agrochemicals and hydrophobicity plays an important part in the efficiency of deposition, retention, and spreading of impacting spray droplets. Most pesticide formulations contain added surfactants (surface-active agents) in order to reduce the surface tension of the spray liquid and thereby improve their performance on difficult-to-wet crops and weeds. Water has an equilibrium surface tension of $72\,\mathrm{mN\,m^{-1}}$, but addition of an appropriate organosilicone surfactant can decrease this to about $20\,\mathrm{mN\,m^{-1}}$, thus, enabling complete wetting of all plant target surfaces.

It is well known that plant waxes are hydrophobic substances; this property is attributable to the orientation of their molecules in the solid state. X-ray and electron diffraction studies of aliphatic constituents reveal that their long chains are arranged perpendicularly or inclined to certain planes, in individual monomolecular layers. Long-chain alkanoic acids and primary alcohols are dimeric in the solid state, their adjacent polar end groups being sandwiched between the hydrocarbon chains. The net result of these molecular orientations is that low free surface energy methyl groups are exposed at an air–solid

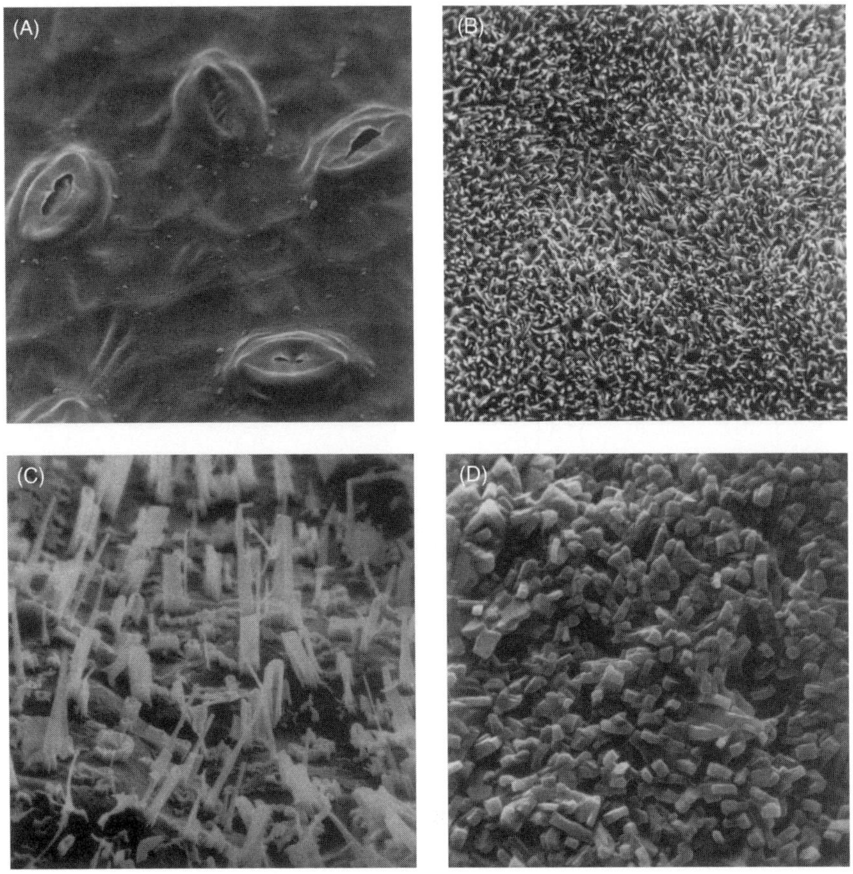

Figure 12 Epicuticular waxes that do not recrystallize in the same form as observed *in planta*. (A) On the abaxial surface of peach leaf, the wax layer is essentially amorphous (x1350); (B) after extraction and recrystallization it forms dense masses of platelets (x2900). The main constituents of this wax are triterpenoid acids and *n*-alkanes. (C) Characteristic tubular forms of epicuticular wax occur on the adaxial surface of Brussels sprout (*Brassica oleracea* var. *gemmifera*) leaves (x5600), in marked contrast to the short stubby rodlets that appear following recrystallization of the isolated wax (D, x6600). The wax from this species is predominantly a mixture of *n*-nonacosane, *n*-15-nonacosanone, and *n*-15-nonacosanol.

interface. Maximum hydrophobicity occurs with *n*-alkanes, where the terminal methyl groups are arranged in the tightest possible packing. The situation is less clear with the molecular composition of the surfaces of solid cyclic wax components but they tend to be more wettable than their aliphatic counterparts.

The relative wettabilities of isolated epicuticular waxes, as well as plant surfaces, can be quantified by the contact angles formed by water, *viz.*, the angle between the solid surface and a tangent drawn to the liquid surface where it meets the solid (**Figure 14**). Such measurements provide an inverse measure of the adhesion between a solid surface and a liquid, with complete wetting = 0° and repellency = 180°. When isolated and reconstituted as a smooth layer, most waxes and single wax components show contact angles only between about 95° and 110°, the latter being recorded for *n*-alkanes. However, the surfaces of plants exhibit a much wider range of wettabilities with contact angles being as low as 45° up to 150° and above; the important modifying factor present on plants is microroughness. If a surface has a contact angle less than 90°, this suggests that it has an incomplete covering of wax or possesses other surface features such as corrugations or trichomes that enhance wetting. It is well known that the hydrophobicity of a waxy surface is increased by roughening it, and this provides an explanation for plant surfaces with contact angles greater than 110°. The most water-repellent plant surfaces known are those that have well-defined microcrystalline epicuticular wax deposits. The effective surface areas of these layers are several orders of magnitude greater than that of a corresponding smooth layer and, in turn, cause the entrapment of air films between water droplets and the cuticle surface; the contact angle between air and water is 180°. The contribution of a wax layer to water repellency can be readily demonstrated by the observation of a substantial decrease in contact angle after its removal from the plant surface by gentle

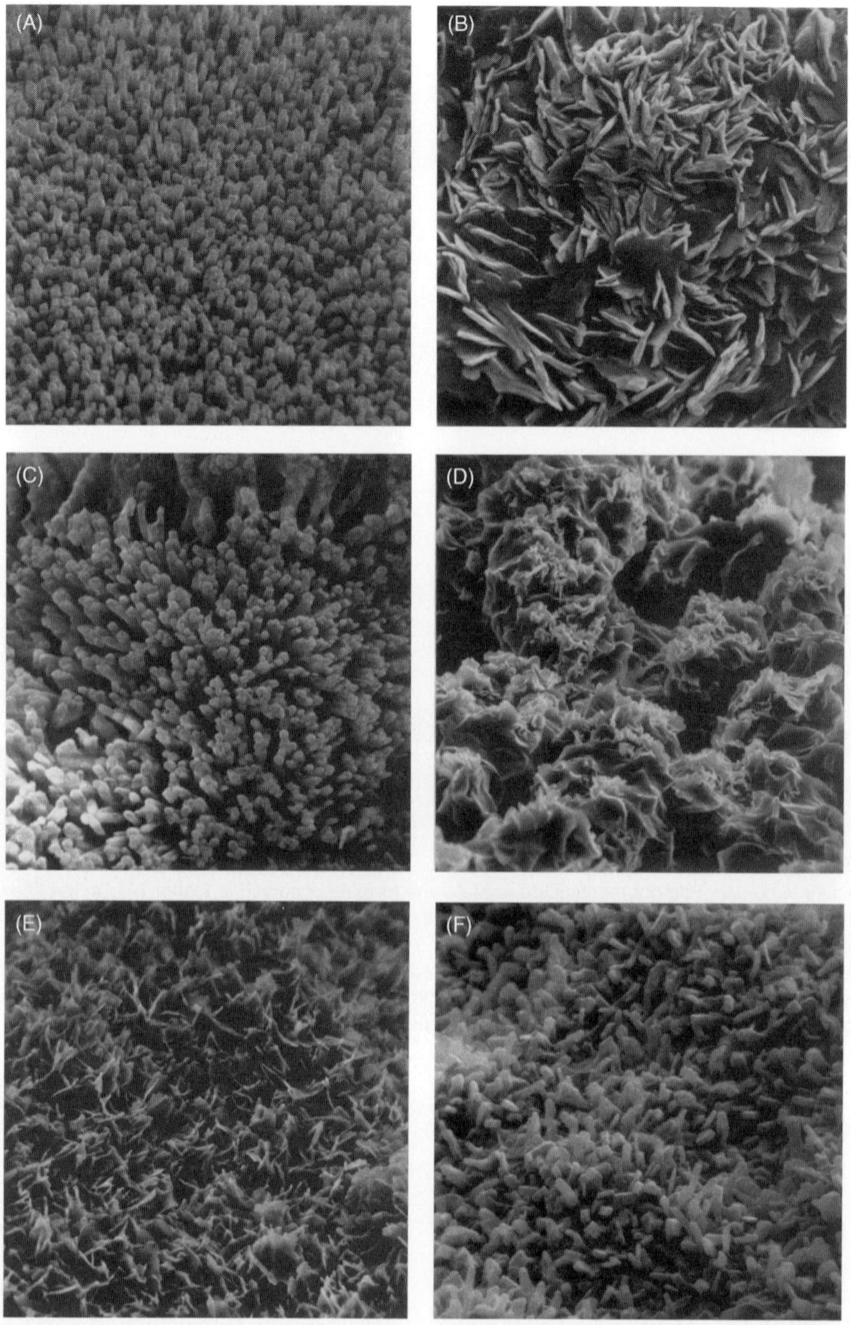

Figure 13 Isolated wax fractions and pure compounds after recrystallization. (A) Purified *n*-10-nonacosanol (x5000) and (B) synthetic *n*-15-nonacosanol (x5500); the asymmetrical secondary alcohol (A) produces tubes, as opposed to plates for the corresponding symmetrical compound (B). (C) The *n*-secondary alcohol fraction isolated from the epicuticular wax of greater celandine (*Chelodonium majus*) leaves recrystallizes in the same tubular form (x6500) as the wax *in situ* and *in vivo*. However, without the secondary alcohol component, the isolated wax recrystallizes as large aggregates of crenulate plates (D, x1800). Recrystallized *n*-primary alcohols (E, x6200) and *n*-aldehydes (F, x6100) isolated from the leaf wax of pea, showing platelet and rodlet types of crystal morphology, respectively.

abrasion or solvent washing. In addition, it is possible to calculate the fraction of air–liquid interface for a difficult-to-wet surface from the contact angle recorded on the surface and that obtained from a smooth layer of the wax isolated from the same surface.

It should be noted, in passing, that some of the most water-repellent leaves known in nature, such as those of the floating ferns *Salvinia* and *Azolla*, and the lotus *Nelumbo*, are extremely rough. At the 100-μm scale, there are trichomes and papillate epidermal cells, with ridged reticulate patterning on

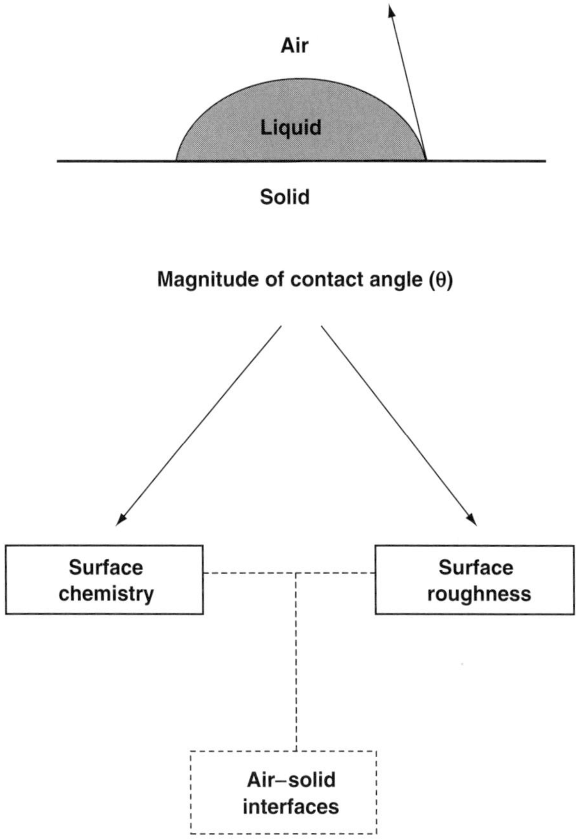

Figure 14 Summary of the main physicochemical factors governing the wettability of plant surfaces. The liquid droplet shown exhibits a contact angle of about 80° on the surface of the solid.

the cell surface discernible at the 10-μm level, together with wax crystallites 1 μm in size. Many insect wings and bird feathers also display an array of chemical and macroscopic/microscopic features that contribute to their highly water-repellent nature – the proverbial water off a duck's back.

Possible Functions in Pest and Disease Resistance

It is generally agreed that the intrinsic chemical and morphological features of plant surfaces are important signals influencing the behavior of phytophagous and entomophagous insects, as well as the infectivity of fungal pathogens. However, attempts to establish correlations between susceptibility or resistance of plant species to damage by these biological agents, with either the composition or fine structure of the corresponding epicuticular waxes, have met with limited success. Most of the work has been done using cultivars or mutant forms of important crop species, particularly *B. oleracea* and *Z. mays*; nevertheless, much of the published information is either circumstantial or anecdotal. Part of the problem is thought to arise from complications caused by the presence of other stimulatory or inhibitory organic compounds, e.g., sugar esters, amino acids, and volatile diterpenoids, also secreted onto the cuticle surface or even dissolving in the layer of epicuticular wax. This is especially important for leaves that possess glandular trichomes, such as tobacco (*Nicotiana tabacum*) and potato (*Solanum tuberosum*), or vesicles, as in the case of lupin (*Lupinus albus*).

In some plant species, epicuticular waxes may be instrumental in providing some of the visual, tactile, or chemical signals necessary for the initial attachment and subsequent movement, feeding, and oviposition of herbivorous insects, as well as other insects that prey on them. For example, microcrystalline waxes, particularly if they occur in a flat plate-like form, may be exfoliated by insects walking over the surface and thus could deter their attachment due to the lack of adhesion. Another possible physical mechanism for defense or attraction involves the degree of scattering of incident light by the waxy blooms of glaucous species; this effectively alters the visible appearance of the surface to prospective insect pests and may influence host selection. Some insects, such as locusts, are known to palpate the foliage of plants prior to feeding, and bait tests have confirmed that most of the commonly found aliphatic epicuticular wax constituents stimulate acceptance behavior, whereas others, such as the pentacyclic triterpenoids, α- and β-amyrin, act as deterrents. The effects on growth and reproduction of ingesting the epicuticular wax from *Z. mays* leaves and classes of constituents isolated from it have been examined in detail for the fall armyworm (*Spodoptera frugiperda*) using artificial diets. In most cases, a significant increase in the growth rates of larvae was observed.

There have been few direct attempts to examine the influence of epicuticular waxes on the developmental sequences leading from spore adhesion to germination and infection by plant pathogenic fungi. It is known that surface hydrophobicity is an important factor for some pathogens, but not for others, although many spores will germinate quite readily on artificial surfaces, such as glass and cellophane. However, the success or failure to develop viable infection structures subsequently may be dependent on the recognition of specific chemical or morphological clues. A cogent example of this phenomenon is *Erysiphe graminis* on ryegrass (*Lolium* spp.) leaves, which develops normally on the adaxial surface but not on the corresponding abaxial surface. The wax on the adaxial surface is rich in primary alcohols and plate-like in form, whereas that on the abaxial surface occurs as thin amorphous sheets composed mainly of aldehydes. This could provide a possible explanation for the resistance of the latter surface to infection by the powdery mildew, especially since germlings of this fungus developed normally when the wax layer was removed.

Economic Importance

Waxes obtained from a number of vegetable sources have had a long history in commerce worldwide and the reader is referred to the comprehensive treatise by Warth. They have found uses in the food, cosmetic, candle, rubber, adhesives, paper, printing, and electrical industries, but are most highly valued as essential ingredients in the manufacture of high-quality polishes for a wide variety of materials. Although many of the traditional vegetable waxes are no longer produced commercially, having been replaced by synthetic or semisynthetic substitutes, there are three that are still of economic importance. The most valuable is carnauba isolated from the leaves of a Brazilian palm (*Copernicia cerifera*); it is one of the hardest waxes known with a melting point of 85°C and is composed mainly of very long-chain alkyl esters. Another Brazilian palm wax, ouricuri, is used as a substitute for carnauba and is obtained from the leaves of *Syagrus coronata*, *Cocos coronata*, or *Attalea excelsa*. Finally, there is candelilla from Mexico, the wax isolated from the shrubs *Euphorbia antisiphilitica* and *E. cerifera*. It differs in composition from the two types of palm waxes in having a high content of *n*-alkanes, with C_{31} as the dominant homolog. Japan wax extracted from the berries of *Toxicodendron* spp. is also still on the market, but it is not a true wax because it is a glycerol-based vegetable tallow.

List of Technical Nomenclature

Functional groups	Unbranched = *n*-(normal); branched = *br* (—CH(CH_3)—), e.g., *iso*-(2-methyl), *anteiso*-(3-methyl), etc.; alkane (—CH_3) = -ane; unsaturation (—HC=CH—) = -ene; alcohol (-OH) = -ol or hydroxy-; ketone (—CO—) = -one or oxo-; aldehyde (—CHO) = -al; acid (—COOH) = -oic acid; ester (—CO–O—) = -yl for alcohol moiety then -oate for acid moiety.
Primary carbon number and systematic stem name	10, decan-; 20, icosan-; 30, triacontan-; 40, tetracontan-.
Prefixes for intermediate homologs	+1 = hen-, +2 = do-, +3 = tri-, +4 = tetra-, +5 = penta-, +6 = hexa-, +7 = hepta-, +8 = octa-, and +9 = nona-.

See also: **Diseases**: Bacterial Diseases; Fungal Diseases. **Growth and Development**: Leaf Development. **Secondary Products**: Wax Pathways. **Water Relations of Plants**: Basic Water Relations.

Further Reading

Barthlott W, Neinhuis C, Cutler D, *et al.* (1998) Classification and terminology of plant epicuticular waxes. *Botanical Journal of Linnean Society* 126: 237–260.

Cutler DF, Alvin KL, and Price CE (eds) (1982) *The Plant Cuticle*. London: Academic Press.

Gould RF (ed.) (1964) *Contact Angle, Wettability and Adhesion*. Washington DC: American Chemical Society.

Hamilton RJ (ed.) (1995) *Waxes: Chemistry, Molecular Biology and Functions*. Dundee: Oily Press.

Harwood JL and Boyer J (eds) (1990) *Lipids, Membranes and Aspects of Photobiology*. London: Academic Press.

Holloway PJ (1970) Surface factors affecting the wetting of leaves. *Pesticide Science* 1: 156–163.

Holloway PJ, Rees RT, and Stock D (eds) (1994) *Interactions between Adjuvants, Agrochemicals and Target Organisms*. Berlin: Springer-Verlag.

Juniper BE and Jeffree CE (1983) *Plant Surfaces*. London: Edward Arnold.

Juniper BE and Southwood R (1986) *Insects and Plant Surfaces*. London: Edward Arnold.

Kerstiens G (ed.) (1996) *Plant Cuticles: an Integrated Functional Approach*. Oxford: BIOS Scientific Publishers.

Kolattukudy PE (ed.) (1976) *Chemistry and Biochemistry of Natural Waxes*. Amsterdam: Elsevier.

Mangold HK (ed.) (1984) *CRC Handbook of Chromatography, Lipids.*, vol. 1. Boca Raton: CRC Press.

Martin JT and Juniper BE (1970) *The Cuticles of Plants*. London: Edward Arnold.

Percy KE, Cape JN, Jagels R, and Simpson CJ (eds) (1994) *Air Pollutants and the Leaf Cuticle*. Berlin: Springer-Verlag.

Warth AH (1956) *The Chemistry and Technology of Waxes*. New York: Reinhold Publishing.

Wax Pathways

T Shepherd, Scottish Crop Research Institute, Dundee, UK

Introduction

The major components in plant waxes are formed by a condensation–elongation process in which long acyl chains are synthesized by sequential elongation of a C_2 primer derived from acetyl-CoA with C_2 units derived from malonyl-CoA. Chains of up to C_{16} and

C_{18} are formed in the synthesis *de novo* and are then extended to C_{30} or higher by a second elongation system. Modification of acyl chains give rise to the wax constituents which include fatty acids, aldehydes, alkanes, primary alcohols, alkyl esters, secondary alcohols, ketones, and various polyoxygenated compounds. Numerous techniques are used to work out the pathways of wax synthesis. Some of the most useful approaches involve blockage of the biosynthetic sequence at one or more points by use of inhibitors (chemical and photoperiod) and by mutation, resulting in accumulation of measurable levels of intermediates. Tracer studies can demonstrate incorporation of isotopically labeled precursors into reaction products and purified enzymes may be used with labeled substrates *in vitro* to validate specific steps of a biosynthetic pathway. More recently, molecular biology techniques have been used to identify and clone cDNA encoding specific enzymes and proteins. The cDNA can be transferred to model organisms such as *Arabidopsis*, *Escherichia coli*, and yeast for further investigation and characterization of the cDNA gene products. Wax mutants are usually characterized by changes in the coverage, morphology and composition of the wax, which often present visually as a loss of the normal glaucous waxy bloom, giving a glossy appearance. Such mutants have been termed *eceriferum* (*cer*) in *Arabidopsis* and barley (*Hordeum vulgare*) and *glossy* (*gl*) in corn (*Zea mays*; maize), and various *Brassica* species. Some mutants lacking the truly glossy phenotype are termed subglaucous.

Formation of Malonyl-CoA

Formation of malonyl-CoA from acetyl-CoA is the first step in wax synthesis. The reaction is catalyzed by acetyl-CoA carboxylase (ACCase) which contains three functional activities, biotin carboxylate carrier protein (BCCP), biotin carboxylase (BC), and carboxyltransferase (CT). The BCCP protein contains a covalently bound molecule of biotin to which carbon dioxide derived from bicarbonate is added to form N-1′ carboxybiotin-BCCP. Subsequently the CO_2 is transferred to acetyl-CoA to form malonyl-CoA. These reactions are catalyzed by BC and CT respectively (**Figure 1**). Biotin is attached to BCCP by a flexible arm 1.6 nm long which may aid movement of biotin between BC and CT.

Two types of ACCase occur in plants, one in plastids, the other in the cytosol. In many species, including dicotyledons but excluding the Poaceae, plastidial ACCase occurs in a heteromeric form as dissociable multisubunits of BCCP, BC and nonidentical α- and β- subunits of CT. In the Poaceae, plastidial ACCase occurs in a multifunctional nondissociable homomeric form and in all plant species studied to date, cytosolic ACCase is also homomeric. Plastidial ACCase supplies malonyl-CoA for the synthesis *de novo* whereas cytosolic ACCase supplies malonyl-CoA for further elongation of acyl chains

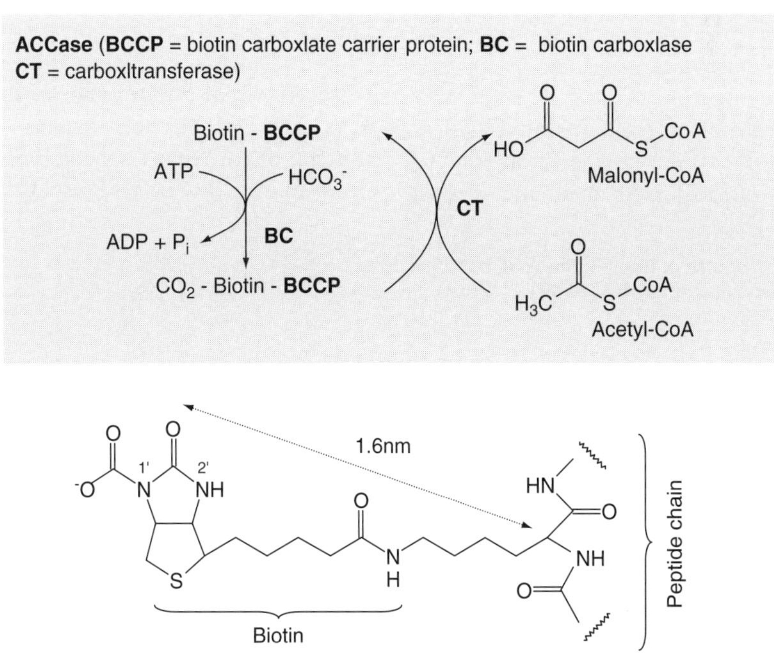

Figure 1 Synthesis of malonyl CoA by the action of acetyl-CoA carboxlase (ACCase); structure of the carboxybiotin prosthetic group of carboxybiotin-BCCP.

and for synthesis of polketides and flavonoids. ACCase therefore has a key role in regulation of acyl chain production. Activation of ACCase involves light-stimulated reduction of the enzyme by thioredoxin, and in pea (*Pisum sativum*), the CT β-subunit is phosphorylated, and its activity may be regulated by a process of phosphorylation and dephosphorylation.

Genes that encode some or all of the ACCase subunits from several species have been identified, cloned, and their gene products characterized. For example, in *Arabidopsis*, the *CAC1*, *CAC2*, *CAC3*, and plastidic *accD* genes encode the BCCP, BC, α-CT, and β-CT subunits of heteromeric ACCase respectively, whereas the *acc1* gene encodes the cytosolic homomeric ACCase.

Acyl Chain Formation: Synthesis *De Novo*

The synthesis *de novo* of long acyl chains occurs in plastids, and is catalyzed by an enzyme complex known collectively as fatty acid synthase (FAS). The complex includes acyl carrier protein (ACP), which forms a thioester link to acyl groups via a pantetheine 4′-phosphate prosthetic group (**Figure 2**). This flexible side chain of ACP is thought to act as a lever arm, which aids movement of the extending acyl chain between the other enzyme components of FAS. Multiple tissue-specific isoforms of ACP exist, and in *Arabidopsis* expression of the leaf-specific isoform of ACP is stimulated by light.

Formation of acyl chains is initiated by transfer of an acetyl group from acetyl-CoA to a cysteine thiol of the condensing enzyme 3-ketoacyl-ACP synthase (KAS), another key component of FAS, to serve as the C_2 primer for elongation (**Figure 3**, step 1). A malonyl group is transferred from malonyl-CoA to the pantetheine thiol of ACP to serve as the C_2 elongating substrate (step 2). KAS then catalyzes the decarboxylation of malonyl-ACP to form a carbanion, which undergoes a Claisen condensation reaction with the acyl carbonyl group of the KAS-bound acetyl primer (step 3). The net effect is transfer of the acyl (acetyl) group to C-2 of malonyl-ACP to produce acetoacetyl-ACP and CO_2, and regeneration of the KAS cysteine thiol. The CO_2 released contains the carbon atom added to acetyl-CoA during the formation of malonyl-CoA. There then follows a three reaction sequence of reduction, dehydration, and reduction. Stereospecific reduction of 3-ketoacyl-ACP to D-3(*R*)-hydroxyacyl-ACP by NADPH is catalyzed by 3-ketoacyl-ACP reductase (step 4). Dehydration of D-3(*R*)-hydroxyacyl-ACP to the corresponding Δ2*trans* (*E*)-unsaturated acyl-ACP is catalyzed by 3-hydroxyacyl-ACP dehydrase (step 5). Reduction of the Δ2*trans* (*E*)-unsaturated acyl-ACP to the corresponding C_4 acyl-ACP by NADPH is catalyzed by (*E*)-2, 3 enoyl-ACP reductase (step 6) which completes the first cycle of elongation. To start the next cycle, the C_4 acyl group is transferred from ACP to the KAS cysteine thiol, freeing ACP to accept another malonyl group from malonyl-CoA (step 7). The cycle repeats a further six times to produce palmitoyl-ACP (16:0-ACP), then one further time to produce stearoyl-ACP (18:0-ACP). Introduction of a double bond between C-9 and C-10 may occur when the acyl chain is 16 or 18 carbons long by the action of acyl-ACP desaturase.

Formation of the carbon–carbon bond during the Claisen condensation (**Figure 3**, step 3) is a key event during chain elongation. Consequently, this reaction, catalyzed by the condensing enzyme KAS, has been subject to much detailed investigation. The effect of inhibitors such as cerulenin and arsenite on fatty acid synthesis reveal the presence of multiple isoforms of KAS. Three elongation systems operate in plastids, and the chain length specificities of the condensing enzymes are shown in **Figure 4**. Production of short

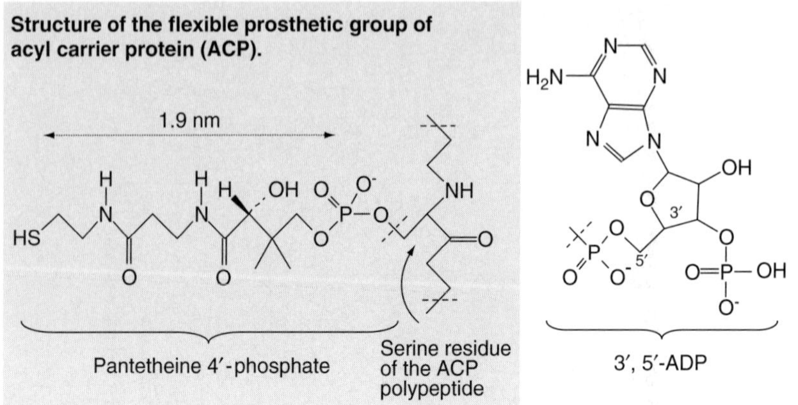

Figure 2 Structures of acyl carrier protein and coenzyme A. In coenzyme A, pantetheine 4′-phosphate is linked to 3′, 5′ADP rather than the serine residue of the ACP polypeptide.

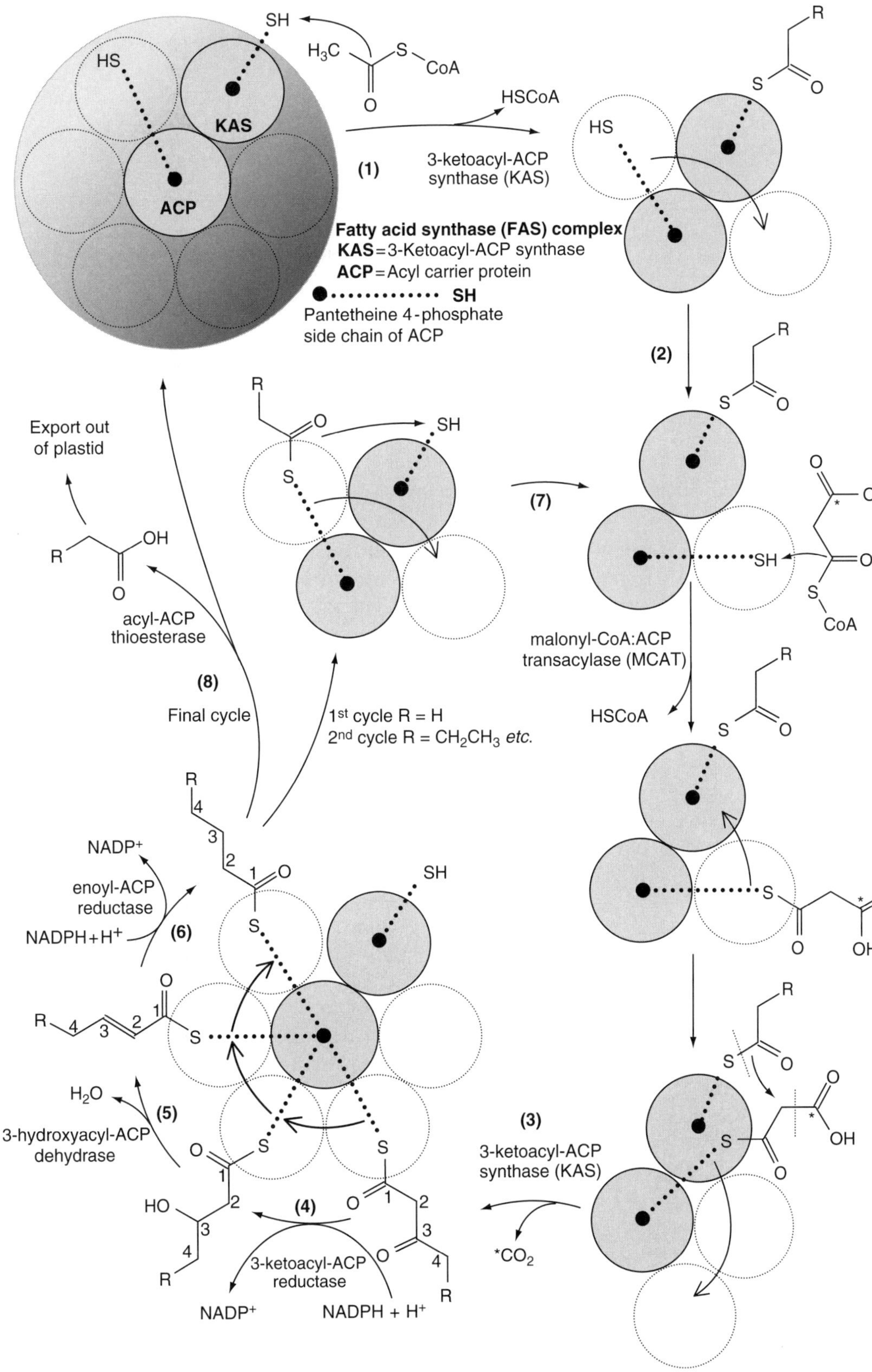

Figure 3 Synthesis *de novo* of acyl chains by fatty acid synthase (FAS) in plastids. The spatial arrangement shown for the different enzyme components of FAS is representational and does not indicate a definitive structure.

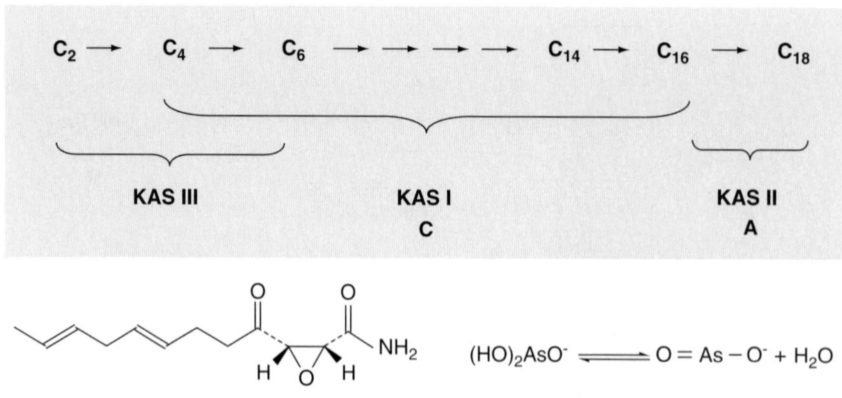

Cerulenin (C) Arsenite (A)

Figure 4 Specificities of 3-ketoacyl-ACP synthase (KAS) isoforms involved with acyl chain elongation in fatty acid synthesis. Cerulenin (C) and arsenite (A) are chemicals which inhibit elongation at the steps shown. Cerulenin binds to the active site cysteine residue.

acyl chains (C_4 and perhaps C_6) involves KAS III, further elongation to C_{14} and C_{16} involves KAS I, and the final elongation to C_{18} is catalyzed by KAS II. Genes encoding all three KAS isoforms have been cloned and expressed sequence tag (EST) clones for each KAS have been identified from a variety of plant species. There is extensive sequence homology between KAS I and II, whereas KAS III appears to be more closely related to a different condensing enzyme, polyketide chalcone synthase (CHS).

Models for catalysis by KAS I, II, and III have been proposed, based on crystallographic studies of enzyme–substrate interactions in *Escherichia coli* (**Figure** 5). Each KAS isoform has a reactive site cysteine (Cys) residue to which acyl groups are transferred from ACP or CoA (steps 1, 2). Hydrogen bonds formed between the C-1 acyl oxygen of the incoming thioester and the backbone NH groups of the Cys residue and an adjacent amino acid (glycine, Gly, in KAS III; phenylalanine, Phe, in KAS I) are thought to facilitate the reaction. Hydrogen bonding increases the electrophilicity of the acyl group to nucleophilic attack by the Cys thiol (steps 1, 2). Carbanion formation by decarboxylation of malonyl-ACP is catalyzed by two basic residues close to the active site Cys (steps 3, 4). In KAS I and II these are both histidines (His), whereas in KAS III there is a His and an asparagine (Asn). Proton abstraction by His initiates the decarboxylation and the resulting carbanion is stabilized in an enolate conformation by the second His (KAS I, II) or Asn (KAS III). The condensation reaction involves attack by the carbanion on the C-1 acyl of the Cys thioester (steps 5, 6), a process which is also assisted by hydrogen bonding.

Free ACP formed during acyl transfer from acyl-ACP to KAS inhibits the acyl transfer process and the inhibition is reversed by acylation of ACP, whereas ACP has little inhibitory effect on either KAS II or KAS III (steps 1, 2). However, C_{16}-ACP acts as a feedback inhibitor of KAS III, therefore overproduction of acyl chains by KAS inhibits initiation of new acyl chains by KAS III.

In most plants the major products of FAS are even carbon-number saturated straight (*n*-)acyl chains. The enzyme complex can, however, utilize alternative primers, which result in formation of structurally modified acyl chains. Use of propionyl-CoA (C_3) results in formation of odd-carbon *n*-acyl chains as minor components. In some plants such as the Brassicaceae, branched primers give rise to branched acyl chains. Even-carbon *iso*-branched (*i*-) chains come from use of isobutyryl-CoA (C_4), derived from the amino acid valine. Similarly, odd-carbon *iso*- (*i*-) and *anteiso*- (*a*-) chains arise from the use of C_5 primers most probably derived from leucine and isoleucine respectively (**Figure** 6). Internally branched acyl chains are not common in plant waxes, but use of methylmalonyl-CoA as the elongation unit in place of malonyl-CoA results in introduction of methyl branches within the chain, generally at odd-carbon positions (**Figure** 6).

Termination of acyl-ACP chain extension in the plastid can occur via two mechanisms, which determine the subcellular distribution of acyl chains. Some are incorporated into plastidial lipids, entering the requisite pathways by the action of plastidial glycerol-3-phospate acyltransferase (G3PAT). The remainder, including those destined for incorporation into wax components, are hydrolyzed by thioesterases and the fatty acids produced are transported across the plastidial membrane (**Figure** 3, step 8). Free acids may then be converted to acyl-CoA derivatives by an acyl-CoA synthase at the outer plastidial membrane. However, recent evidence

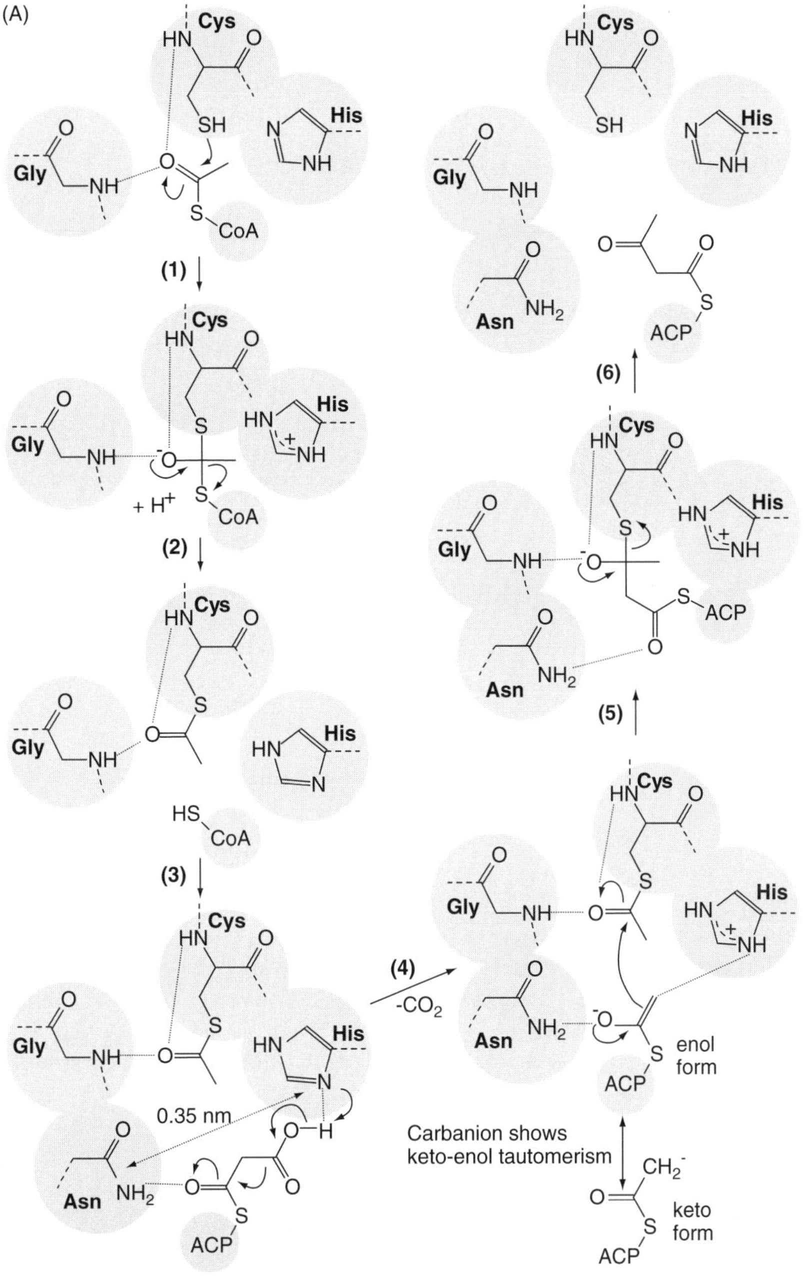

Figure 5 Mode of action of (A) 3-ketoacyl-ACP synthase (KAS) isoform III and (B) 3-ketoacyl-ACP synthase (KAS) isoforms I and II. Amino acid residues shown include parts of the active site cysteine (Cys), catalytic histidine (His) and asparagine (Asn) residues, and adjacent phenylalanine (Phe) and glycine (Gly). Hydrogen bonds are indicated by: · · · · · .

suggests that acyl groups incorporated into phosphatidyl choline (PC) or another glycerolipid are more likely to provide primers for subsequent elongation. In most plants at least two groups of acyl-CoA thioesterases are active, which differ in their specificity for acyl chains. One group (FAT A) has a high specificity for unsaturated substrates, particularly oleoyl-ACP (Δ9 18:1-ACP), whereas the other (FAT B) has a higher relative affinity for saturated acyl-ACP. Studies with leek (*Allium porrum*) have shown that the activity of FAT B, particularly to stearoyl-ACP (18:0-ACP), is expressed predominantly in the epidermis. Acyl chains transported out of the plastid may then be substrates for synthesis of cellular complex lipids, cutin, suberin and waxes.

Acyl Chain Synthesis: Elongation

Acyl chains destined for incorporation into plant waxes are lengthened further by a series of

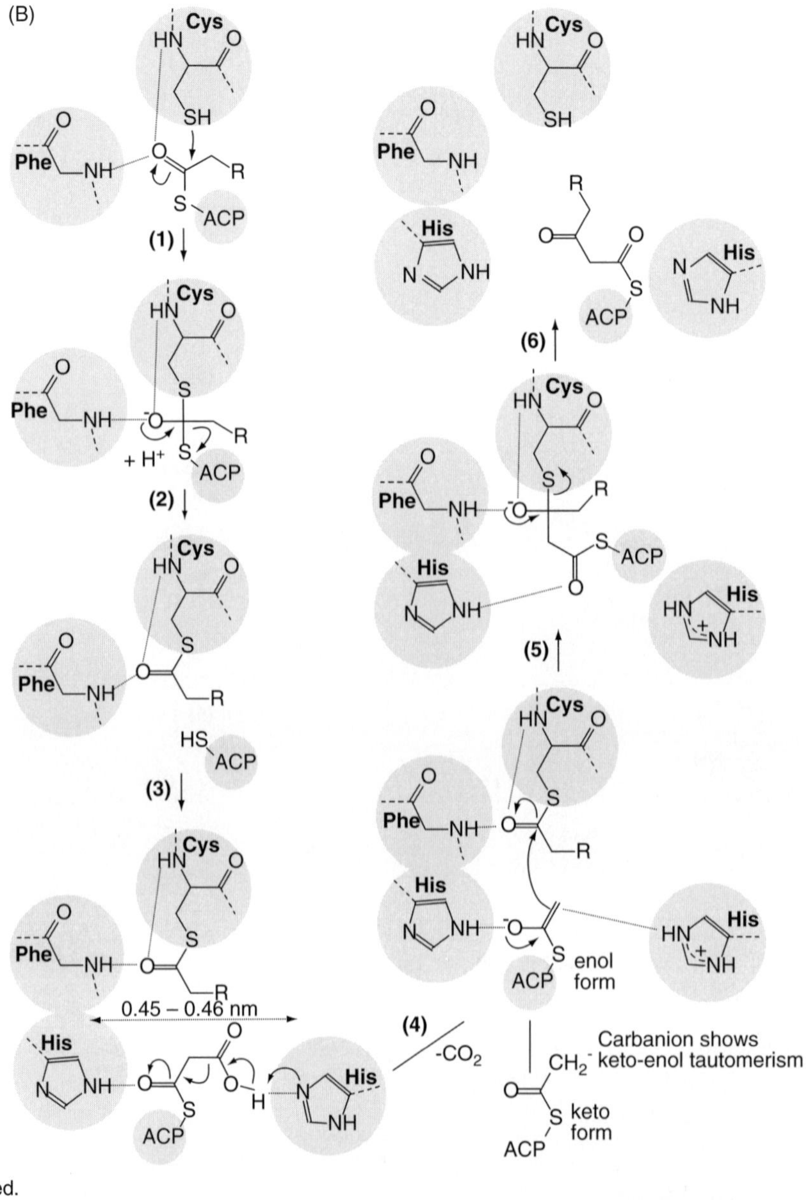

Figure 5 Continued.

extraplastidial ATP-dependent elongation systems associated with the endoplasmic reticulum, Golgi apparatus, and plasma membrane of epidermal cells. Chains are extended by C_2 units derived from malonyl-CoA in sequences of condensation, reduction, dehydration, and reduction, similar to those catalyzed by plastidial FAS. Collectively referred to as fatty acyl elongases (FAE), the enzyme systems consist of 3-ketoacyl-CoA synthase (KCS), 3-ketoacyl-CoA reductase, 3(*R*)-hydroxyacyl-CoA dehydrase, and (*E*)-2,3 enoyl-CoA reductase. The major difference between FAS and FAE is that in the latter the extending acyl chain is carried by CoA rather than ACP and malonyl-CoA is the C_2 donor, rather than malonyl-ACP.

Acyl Chain Modification

The products of elongation are transformed into the major wax components by enzyme complexes associated with the elongation systems. There are two distinct sequences called the decarbonylative (**Figure** 7) and the reductive (**Figure** 8) pathways respectively, to highlight the key reactions in each.

Decarbonylation Pathways

The initial step in this pathway involves reduction of acyl-CoAs to intermediate aldehydes by acyl-CoA reductase using NADH as the hydrogen source (**Figure** 7). Alkanes are formed by decarbonylation of the aldehydes and the one carbon reduction in

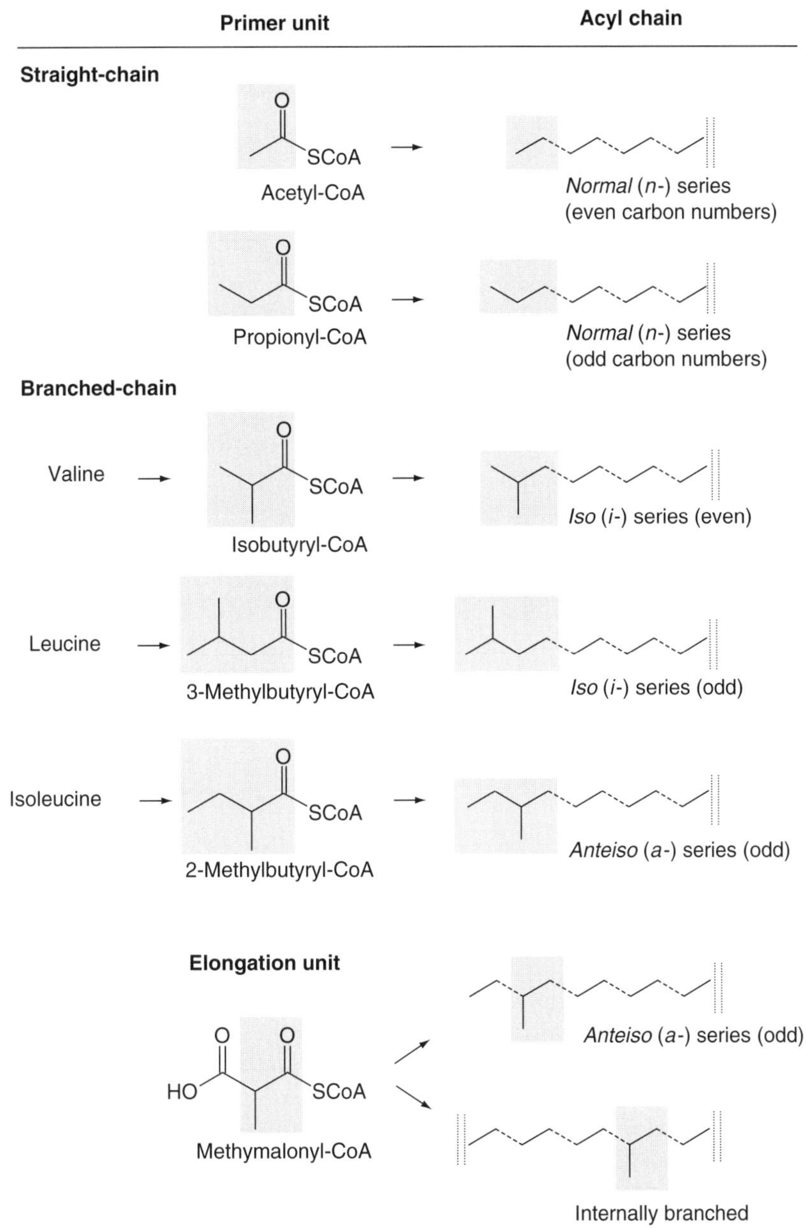

Figure 6 Alternative primer molecules and branched elongation units utilized by fatty acid synthase (FAS) for synthesis of acyl chains.

chain length gives the characteristic odd-carbon *n*-alkanes associated with plant waxes. Formation of carbon monoxide during decarbonylation has been verified *in vitro* using preparations obtained from pea microsomes by sucrose density gradient centrifugation. Radiolabeled [1-^{3}H, 1-^{14}C]-C_{18} aldehyde and [2-^{3}H, 2-^{14}C]-C_{18} fatty acid were converted into the [1-^{3}H]-C_{17} alkane, with concomitant production of an equivalent amount of ^{14}CO. The carbon monoxide was trapped as a carbonyl adduct with the rhodium complex $RhCl(PPh_3)_3$ (PPh_3 = tri-triphenylphosphine). An alternative mechanism involving decarboxylation would have produced $^{14}CO_2$; however, this byproduct could not be detected. The fate of the carbon monoxide released *in vivo* remains unknown.

Secondary alcohols are formed via the action of a hydroxylase, believed to be a cytochrome P-450 dependent mixed function oxidase (or monoxygenase) using NADPH as a coenzyme. Hydroxylation is stereospecific, a single enantiomer is formed when an oxygen atom is inserted between the carbon and one of the hydrogens of an acyl methylene (CH_2) group without reversal of configuration. Oxidation of the alcohol by NAD^+ gives the corresponding ketone. Secondary alcohols, ketones, and alkanes have

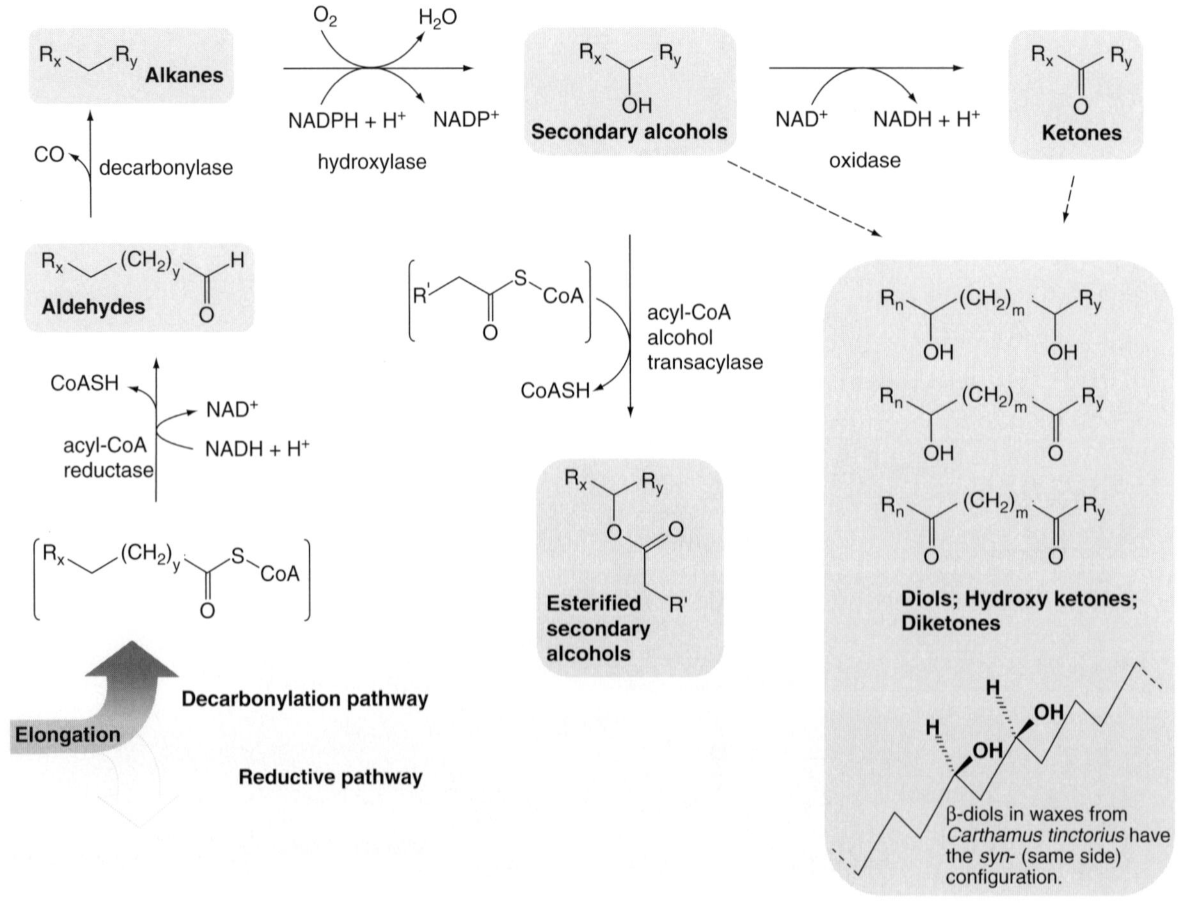

Figure 7 Reactions and products of the decarbonylation pathway.

similar distributions of homologs when all three classes are found together. The site of hydroxylation varies with species and plant age from mid-chain to locations closer to an alkyl terminus, although the 10-hydroxy isomer is often predominant (**Figure 9**). Formation of several positional isomers with the hydroxyl at one or more carbons to either side of the predominant isomer indicates that hydroxylation is not always under tight control. In contrast, ketone formation appears to be more specific, with a single positional isomer predominant in many species, although in some such as brassicas and *Rubus idaeus*, several positional isomers are formed. Secondary alcohols usually occur in the free form, but they may also be esterified to fatty acids. Additional hydroxylation and oxidation can yield diols, hydroxy ketones, and diketones (**Figure 10**), and there is great variation in the site of insertion of the second oxygen. In conifers this occurs from three to six carbons to either side of C-10, whereas in brassicas and *Carmathus tinctorius* substitution occurs at one or two carbons from the initial midpoint insertion giving a series of dioxygenated compounds with α- and β-substitution patterns (**Figure 10**). The formation of *syn* β-diols in *C. tinctorius* (**Figure 7**) suggests that a single regiospecific hydroxylase is responsible for introduction of both hydroxyl functions.

A variation on the decarbonylation pathway is thought to be the source of β-diketones and related compounds and also alkan-2-ol derivatives found in waxes from barley and other cereals and grasses. In this sequence, 3-keto acyl-CoA intermediates derived from the action of 3-ketoacyl-CoA elongase feed into two subsequent pathways. Some are transformed into methylketones by the action of a decarboxylase and then into alkan-2-ols by methylketone reductase (**Figure 11**, path 1). Esterification can then yield alkan-2-ol esters. Further reaction may involve insertion of an oxo group into the alkan-2-ol moiety.

Alternatively, the 3-ketoacyl intermediates undergo two further condensation steps to produce a polyketide (**Figure 11**, path 2). After a sequence of reduction, dehydration, and reduction and then more full cycles of elongation, a β-diketone is produced following decarbonylation or decarboxylation. Additional reactions can lead to formation of hydroxy-β-diketones and oxo-β-diketones. In barley three domains within a multifunctional polypeptide

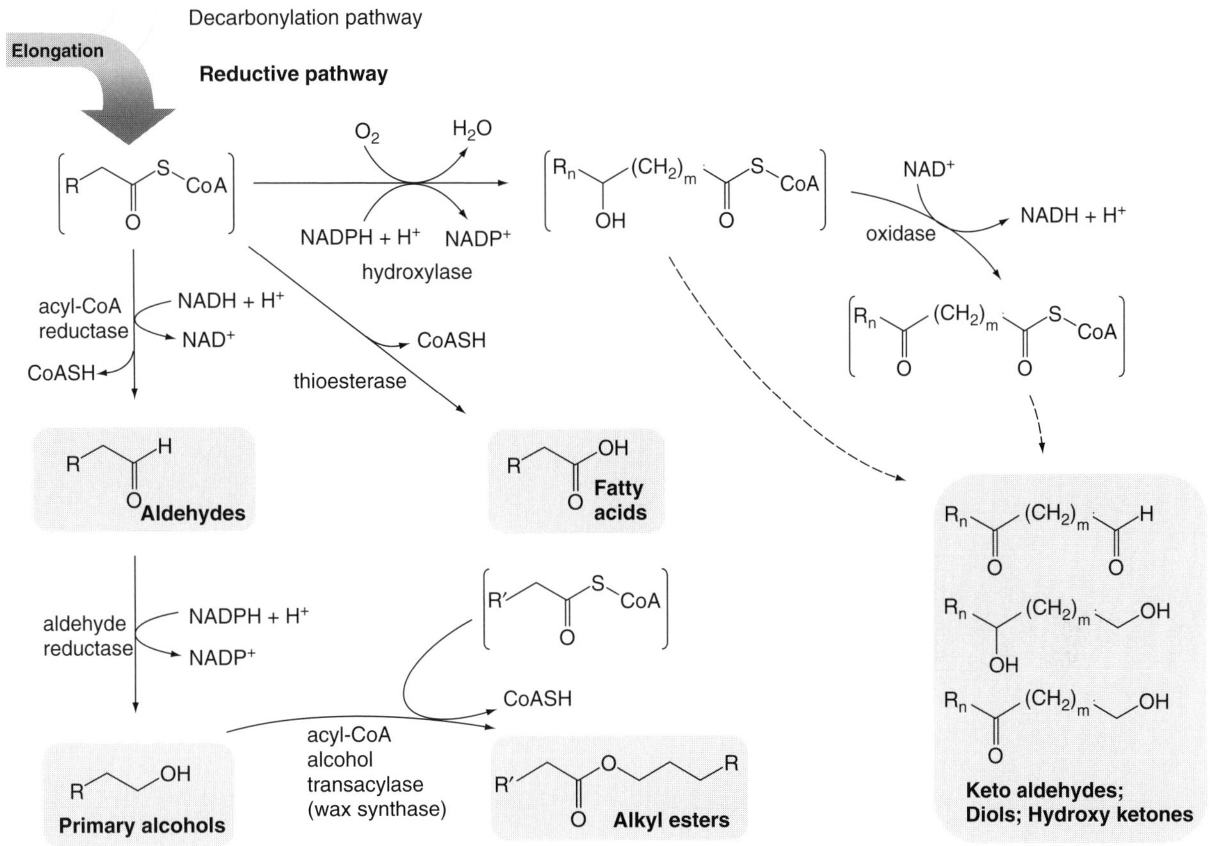

Figure 8 Reactions and products of the reductive pathway.

known as *cer-cqu*, the product of the *cer-cqu* gene, are involved at different points in the pathway. The condensation (KCS) activity of 3-ketoacyl elongase is carried out by the *q* domain, the *u* domain is associated with hydroxylase activity, and the *c* domain is involved with elongation in an as-yet undetermined way. The barley β-keto elongase uses C_{14} and C_{16} 3-oxoacyl-CoA for synthesis of esterified alkan-2-ols, whereas C_{18} 3-oxoacyl-CoA is used for synthesis of β-diketones. Labeled C_{12}, C_{14}, and C_{16} fatty acids are incorporated into diketones, but the C_{18} acid is not incorporated, which shows that the oxo groups are formed during elongation and not subsequently. The linkage between both branches of the pathway is demonstrated by the effect of chemical inhibitors and mutation. For example a mutation in *cer-c* blocks formation of β-diketones, but increases formation of esterified alkan-2-ols. Both branches of the pathway are not always active simultaneously and it is more common to find β-diketones as the lone product, although in *Sorghum bicolor* only esterified alkan-2-ols are formed. It is of note that alkan-2-ols are also found as components of the polymers cutin and suberin.

The position of the retained β-diketo group in β-diketones (**Figure 12**) and hydroxy β-diketones (**Figure 13**) varies with species, although the 14, 16 diones are most common. Stereospecific hydroxylation can occur to either side of the β-diketone functional group, possibly due to the β-diketone substrate assuming different orientations. The formation of equivalent hydroxyl substituents with both *R* and *S* stereochemistries indicates different stereospecific hydroxylase enzymes can be active (**Figure 14**).

The Reductive Pathways

This sequence also involves formation of an intermediate aldehyde from acyl-CoA via the action of acyl-CoA reductase, followed by reduction of the aldehyde to a primary alcohol by aldehyde reductase, using NADPH (**Figure 8**). In some plants (e.g., pea, jojoba (*Simmondsia californica*)), a single enzyme carries out both reductions, and the aldehyde intermediate remains bound. Acyl-CoA can also furnish free fatty acid following hydrolysis. Several mechanisms for synthesis of long chain esters are possible, although it is unclear in which form the acid moiety is

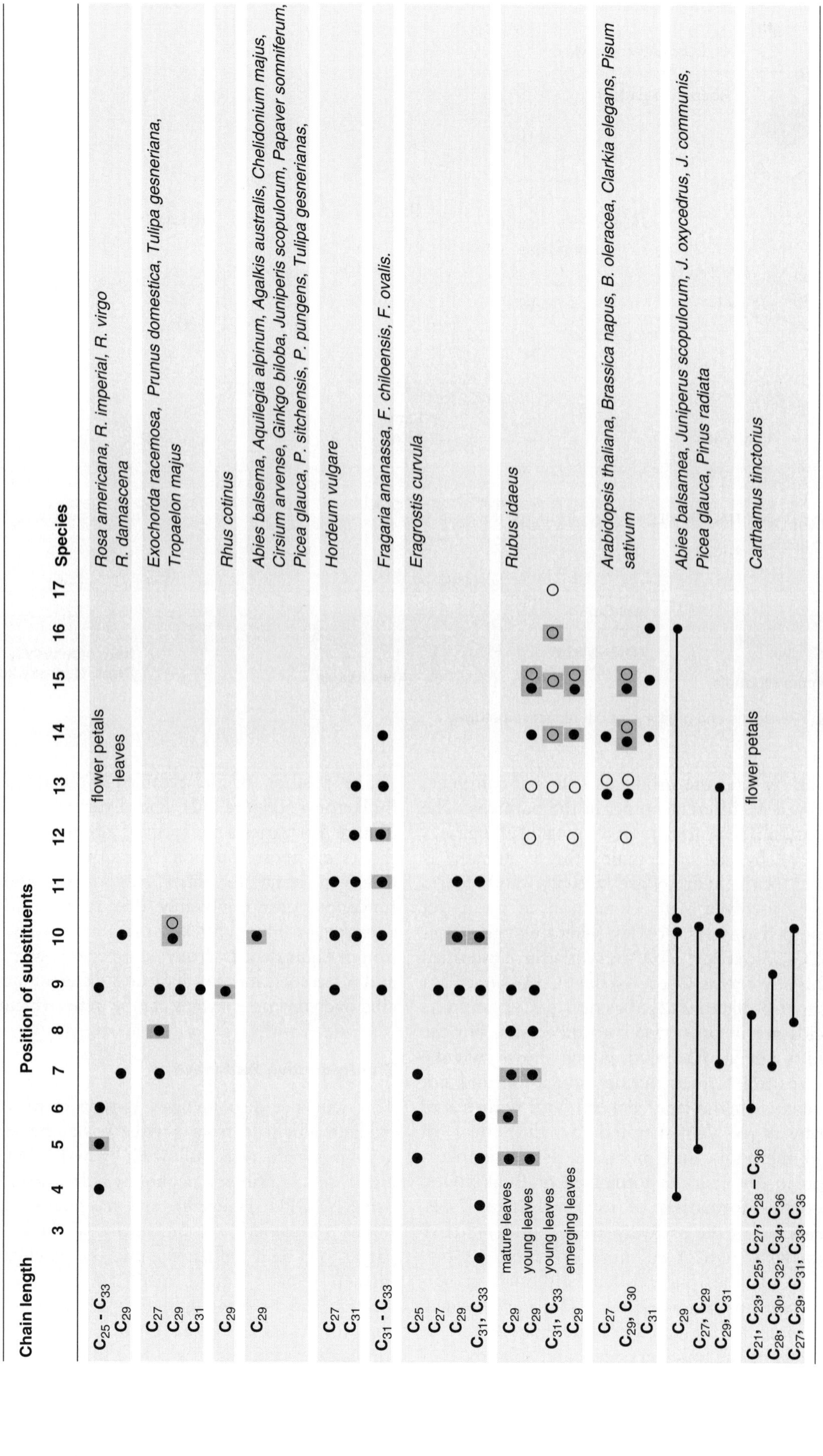

Figure 9 Position of hydroxyl and oxo substituents in: • Secondary alcohols, ○ ketones, and •—• diols; ▒ indicates predominant isomers.

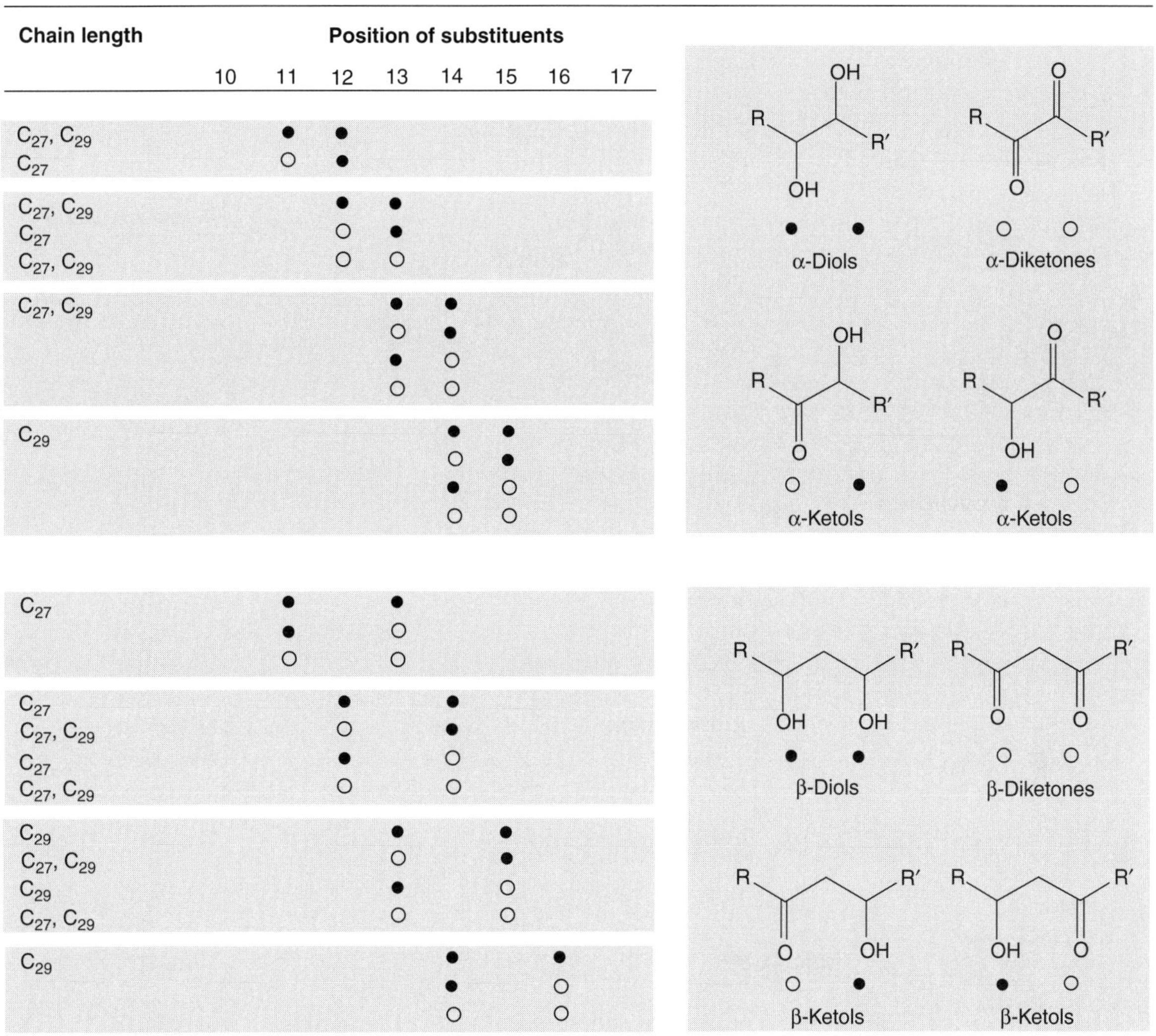

Figure 10 Position of hydroxyl and oxo substituents in α-diols, α-ketols, α-diketones, β-diols, β-ketols, and β-diketones in *Brassica* cuticular waxes.

introduced. These include direct esterification of acids with alcohols and transfer of acyl groups from phospholipids or acyl-CoA to alcohols. Acyl-CoA alcohol transacylase (wax synthase) has been identified from several plants, and purified from jojoba. In some species the ester composition is consistent with a random combination of acid with alcohol. In many others, however, esterification is nonrandom and often shows a preference for short acid–long alcohol combinations although combinations with mid chain lengths may be most abundant. Oxygenated substituents can be introduced by hydroxylation and oxidation of the acyl chain, probably at the acyl-CoA stage, to produce ketoaldehydes, keto alcohols, diols, and hydroxy fatty acids. The latter two groups may include ω-hydroxy compounds in which hydroxlation has occurred at the terminal methyl group (ω-position). Some of these compounds occur in the free form and as esters, others are only found as components of simple polymeric esters known as estolides found in wax from conifers (**Figure 15**) or as components of cutin and suberin.

Products of the reductive pathways consist predominantly of even-carbon homologs when derived from *n*-acyl chains, although odd-carbon compounds may also be present. Branched-chain primary alcohols, acids, and esters are found in the wax from some plant species such as the Brassicaceae. These are formed by analogous pathways using branched acyl-CoA chains as substrates.

A Multiplicity of Elongation Systems

A schematic representation of chain elongation and the channeling of acyl intermediates into the main products of the reductive and decarbonylation

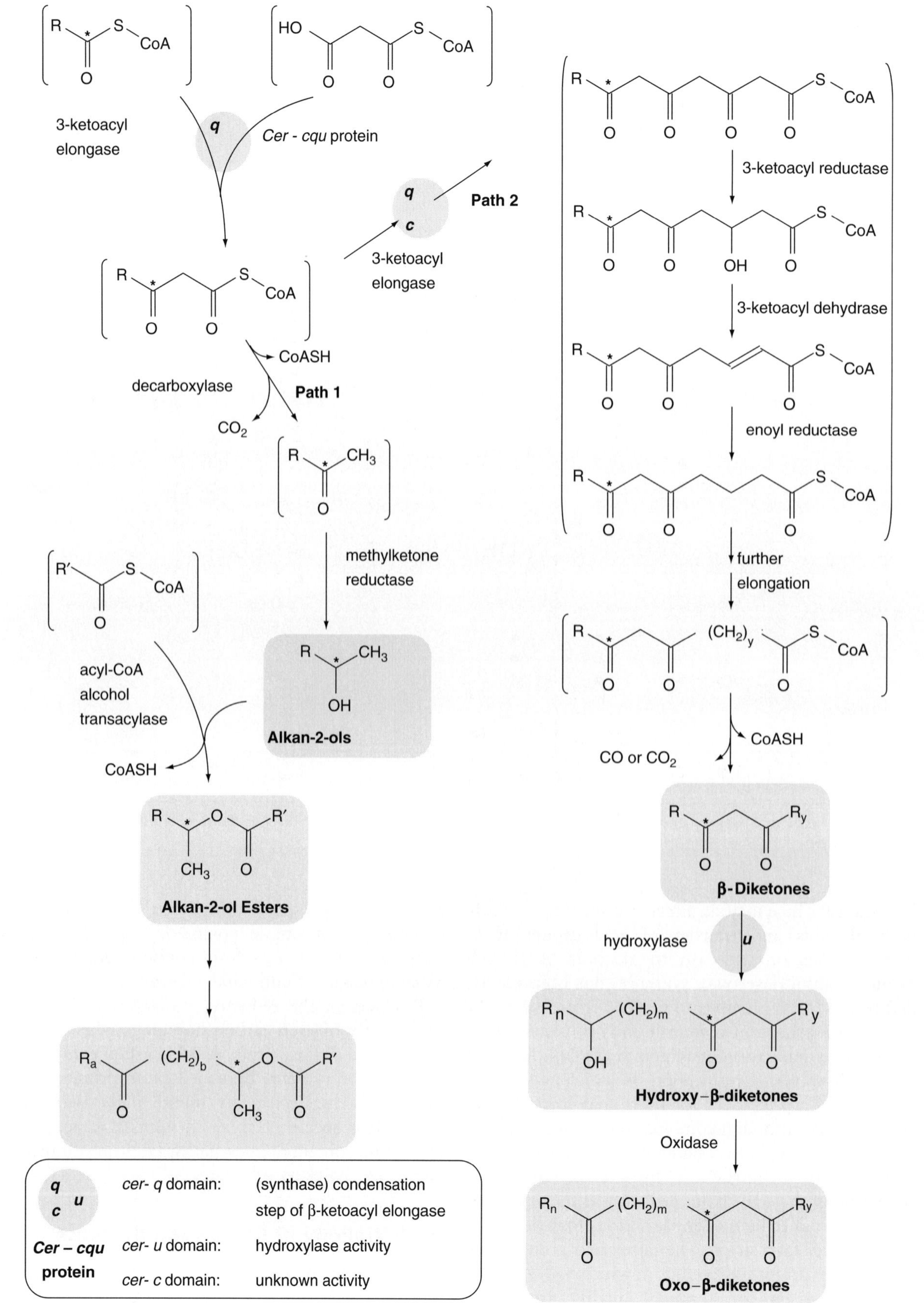

Figure 11 Formation of β-diketones and alkan-2-ols.

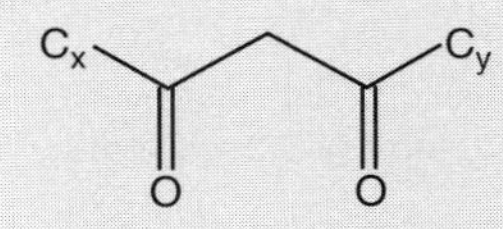

β-diketones

$C_x = CH_3(CH_2)_{x-1}$

$C_y = (CH_2)_{y-1}CH_3$

dione	6, 8 C_{29}	8, 10 C_{31}	10, 12 C_{33}	8, 10 C_{29}	10, 12 C_{31}	12, 14 C_{33}	10, 12 C_{29}	12, 14 C_{31}	14, 16 C_{33}	12, 14 C_{29}	14, 16 C_{31}	16, 18 C_{33}	14, 16 C_{29}	
y		21			19			17			15		13	
x	5	7	9	7	9	11	9	11	13	11	13	15	13	**Species**
	●	●	●											*Buxus sempervirens*
				●	●						●			*Rhododendron baileyi, R. racemosum, R. hemitrichotium*
					●	●		●						*Andropogon halli, A. scoparius; Eragrostis curvula; Festuca glauca, F. ovina; Panicum virgatum*
						●	●	●	●	●				*Dianthus caryophyllus*
									●	●	●	●	●	*Hordeum vulgare*
											●			*Avena sativa; Agropyron elongatum, A. intermedium, A. smithii; Elymus cinerus; Secale cereale; Triticum aestivum; Triticale hexaploide*
										●	●			*Eucalyptus risdoni*
												●		*Acacia* spp.; *Eucalyptus* spp.

Figure 12 Position of diketo (oxo) substituents in β-diketones.

Hydroxy (oxo) dione	5-OH 14, 16 C_{31}	5-oxo 14, 16 C_{31}	8-OH 14, 16 C_{31}	9-OH 14, 16 C_{31}	5-OH 12, 14 C_{31}	5(R)-OH 12, 14 C_{33}	5(S)-OH 12, 14 C_{33}	5-oxo 12, 14 C_{33}	5-OH 14, 16 C_{33}	25(S)-OH 14, 16 C_{31}	25-oxo 14, 16 C_{31}	26(R)-OH 14, 16 C_{31}
	(1)	(3)	(1)	(1)	(1)	(1)	(1)	(3)	(1)	(2)	(4)	(2)
n	4		7	8	4	4			4	13		13
m	8		5	4	6	6			8	8		9
y	15		15	15	17	19			17	6		5
Species												
Poa ampla	●	○			●				●			
Triticum aestivum			●	●								
Andropogon halli, A. scoparius					●	●			●			
Festuca glauca, F. ovina; Panicum virgatum						●						
Eragrostis curvula						●		○				
Agropyron smithii; Avena sativa; Elymus cinerus; Hordeum vulgare; Secale cereale; Triticale hexaploide; Triticum durum										●		
Agropyron intermedium										●	○	
Agropyron elongatum										●	○	●

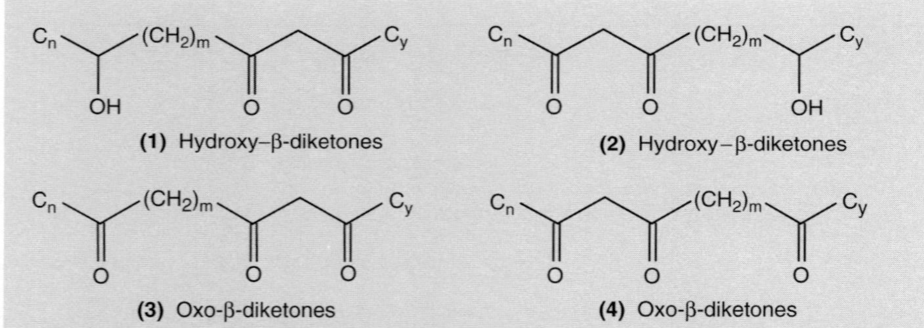

$C_n = CH_3(CH_2)_{n-1}$ $C_y = (CH_2)_{y-1}CH_3$

Figure 13 Position of hydroxyl and oxo substituents in hydroxy- and oxo-β-diketones.

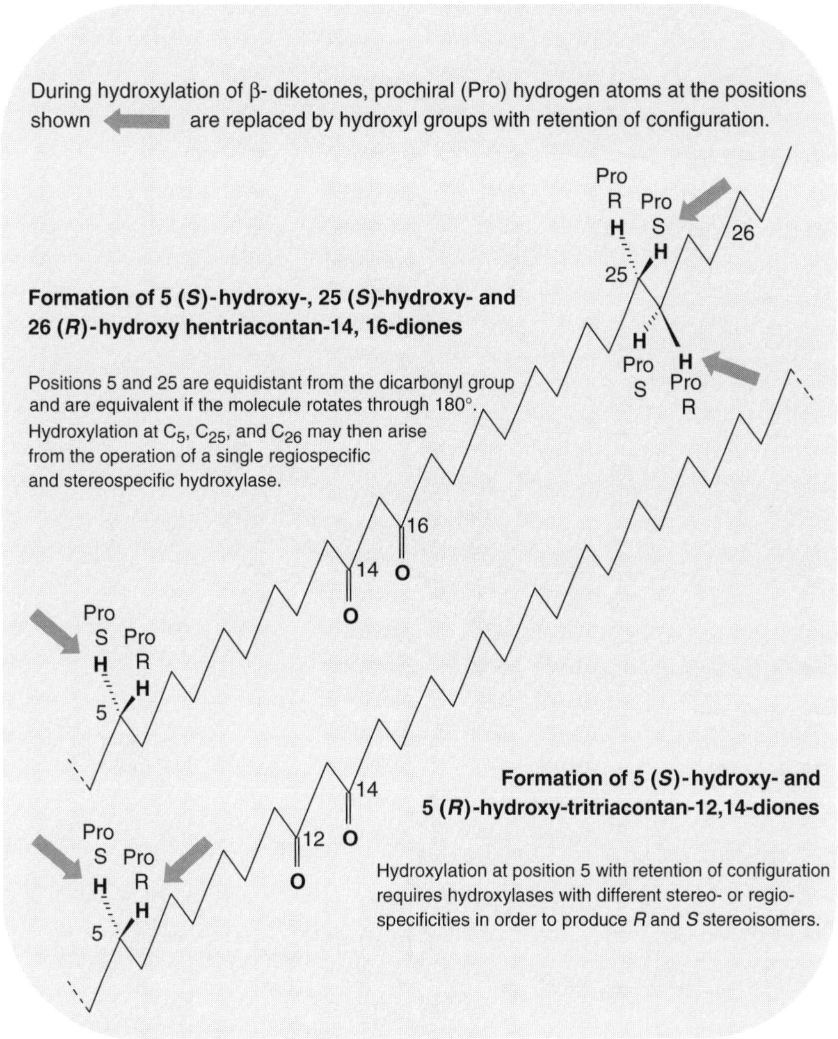

Figure 14 Regio- and stereospecificity of hydroxylation of β-diketones.

pathways is shown in **Figure 16**. The chain lengths shown for each class of compound are those found most frequently in plant waxes. Distributions for acids and alcohols include those found in wax esters. Many of the wax *cer, gl*, and other mutants used to investigate wax biosynthesis are defective at specific points in the elongation sequence, resulting in accumulation of shorter products than normally found for wild-type plants. Similar effects can be induced by chemical inhibitors or by variation in photoperiod. Blockages in elongation occur between C_{16} and C_{18}, C_{20} and C_{22}, and at each step between C_{24} and C_{32} (**Figure 16**), depending on plant species. Such findings support the existence of sequential elongation systems with multiple isoforms of elongase, or of elongase subunits, with specificity for a particular step or sequence of steps in elongation. Some mutants also display defects in the associated pathways, and the resultant changes in the distribution of wax components, particularly for oxygenated compounds such as secondary alcohols, ketones, and β-diketones, are often responsible for the phenotypic classification of the mutation.

Numerous genes involved in the elongation and associated pathways have been cloned, and their gene products studied. However, few have been assigned definite identities or functions. The *Arabidopsis cer1* gene encodes a fatty acid decarbonylase, maize *gl8* encodes a 3-ketoacyl reductase, and similar sequences have been found in barley, leek, and *Arabidopsis*. An alcohol-producing fatty acyl-CoA reductase has been identified as the product of the *TAA1a* gene from bread wheat. This enzyme is responsible for formation of the alcohol moieties of wax esters present in the tapetum tissues of pollen. Genes encoding 3-ketoacyl-CoA synthase (KCS) have been identified from seeds of *Arabidopsis*, *Brassica*, and jojoba. The *Arabidopsis kcs-1*, *CUT1*, *fdh* (fiddlehead), and *hic* (high carbon dioxide) genes are also thought to encode KCS in leaves and stems.

Substitution	1, 15 C_{30}	1, 11 C_{28}	1, 13 C_{30}	1, 15 C_{32}	1, 17 C_{34}	1, 19 C_{36}	1, 9 C_{28}	1, 11 C_{30}	1, 13 C_{32}	1, 15 C_{34}	1, 17 C_{36}	1, 9 C_{30}	1, 11 C_{32}	1, 13 C_{34}	1, 15 C_{36}	1, 7 C_{30}	1, 9 C_{32}	1, 11 C_{34}	1, 16 C_{26}	
n	15	17	17	17	17	17	19	19	19	19	19	21	21	21	21	23	23	23	10	
m	13	9	11	13	15	17	7	9	11	13	15	7	9	11	13	5	7	9	14	
(1)	●	●	●	●	●	●	●	●	●	●	●	●	●	●	●	●	●	●		*Osmundia regalis*
(2) a	○	○	○	○	○	○	○	○	○	○	○	○	○	○	○	○	○	○		*O. regalis*
b		16					18	16	18				16	18				18		
(3)		●						●					●							*O. regalis*
																			●	*Avena sativa*
(4)		■	■	■	■	■	■	■	■	■	■			In estolides						*Juniperus scopulorum*

[a] as free alcohols [b] as esters with palmitic (16) and oleic (18) acids

		C_9	C_{10}	C_{11}	C_{12}	C_{13}	C_{14}	C_{15}	C_{16}	C_{24}	
m α, ω – diols		9	10	11	12		14		16		
m ω – fatty acids					11	12	13	14	15	23	
(5)	As diesters	●	●	●	●						*Triticum aestivum*
	In estolides						●		●		*Juniperus scopulorum*
(6)	In estolides				■						*J. sabina*
									■		*J. communis*
										■	*J. rigida*
					■	■	■	■	■		*J. scopulorum*

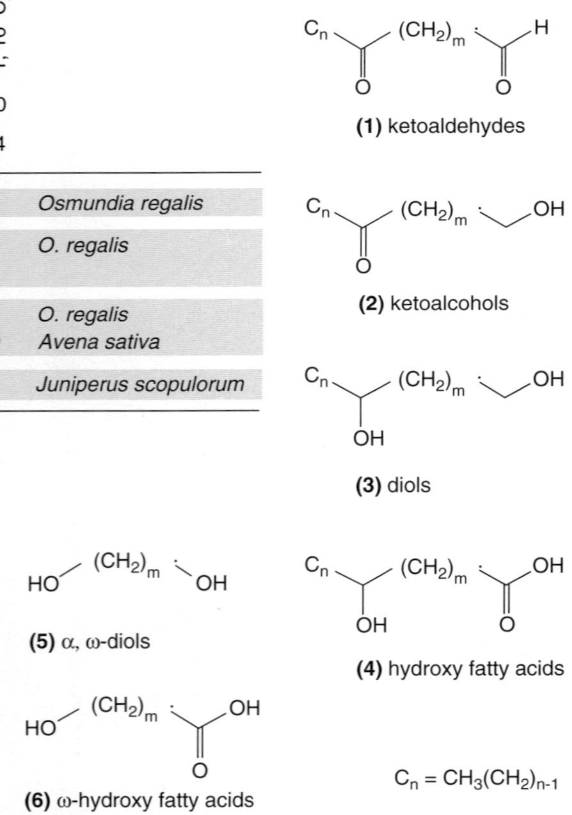

Figure 15 Position of hydroxyl and oxo substituents in ketoaldehydes, ketoalcohols, diols, and hydroxy fatty acids.

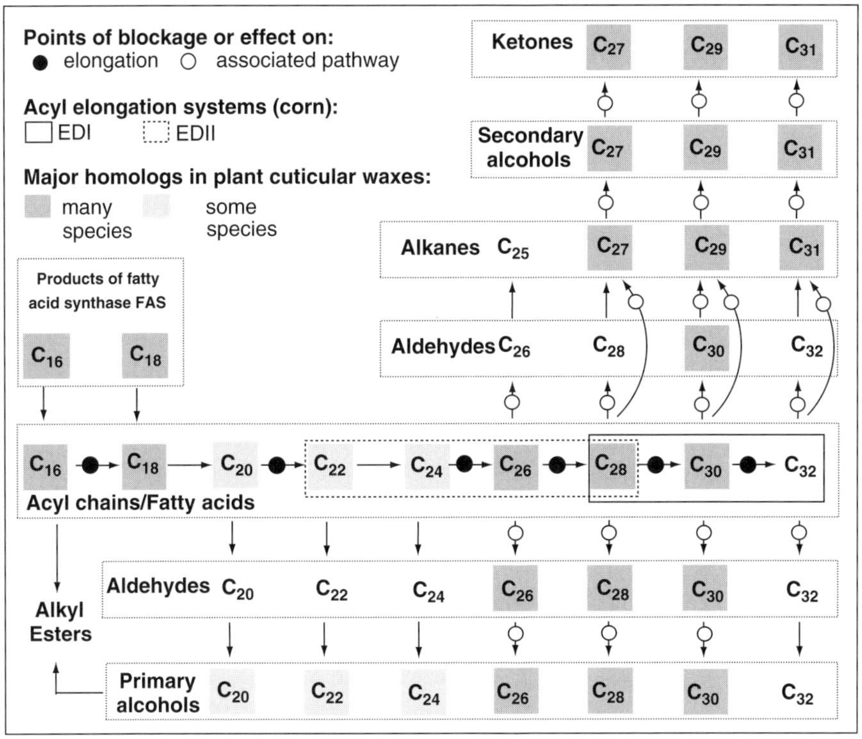

Step	Inhibitors/ Mutation	Plant
Elongation		
C_{16} → C_{18}	Cerulenin	
C_{20} → C_{22}	Arsenite	Barley
	2-ME	Barley
	Cerulenin	Leek
C_{24} → C_{26}	*kcs-1*[c]	*Arabidopsis thaliana*
C_{26} → C_{28}	Cyanide	Barley
Photoperiod		Barley
	gl3	Narrow stem kale
	cer6, kcs-1[c]	*A. thaliana*
C_{28} → C_{30}	TCA	*Brassica oleracea*, barley
	gl1, gl2, gl3	Brussels sprout
	gl5	Cauliflower
	gl3[c]	Corn
	cer2[c], *cer6*[c], *kcs-1*[c]	*A. thaliana*
C_{30} → C_{32}	TCA	Barley
	gl2[c], *gl4*[c], *gl16*	Corn
	wa	Pea

Step	Inhibitors/ Mutation	Plant
Associated Pathways		
Acyl → Aldehyde		
C_{26}, C_{28}, C_{30}, C_{32}	*gl5*[c], *gl11*[c]	Corn
	cer8	*A. thaliana*
Aldehyde → Alcohol		
C_{26}, C_{28}, C_{30}	*cer4*[c]	*A. thaliana*
	gl2	Brussels sprout
Aldehyde → Alkane		
C_{30}	*cer1*[c]	*A. thaliana*
	gl5[c]	Corn
Acyl → Alkane		
C_{28}, C_{30}, C_{32}	*gl1, gl2, gl3*	Brussels sprout
	cer2[c]	*A. thaliana*
Alkane → Secondary Alcohol		
C_{27}, C_{29}, C_{31}	*gl1, gl2, gl3*	Brussels sprout
Secondary Alcohol → Ketone		
C_{27}, C_{29}, C_{31}	*gl1, gl2, gl3*	Brussels sprout
Acyl → Fatty acid		
	cer3[c]	*A. thaliana*

2-ME: 2-Mercaptoethanol

TCA: Trichloroacetic acid

[c]genes have been cloned

Figure 16 Schematic representation of acyl chain elongation and the channelling of acyl intermediates into the associated decarbonylative and reductive pathways. Numbers indicated by C_n refer to the length of the acyl chain. The effects of chemical inhibitors and genetic mutation on elongation and the associated pathways are indicated for a range of plant species. Elongation and chemical transformations are defective at the points shown.

The *kcs*-1 and *CUT1* mutation effects elongation at each step between C_{24} and C_{30}, with reduced levels of the corresponding C_{26}–C_{30} aldehydes and primary alcohols (*kcs-1*, *CUT1*) and C_{29} alkanes, secondary alcohols, and ketones (*CUT1*). However, for *kcs-1* there is no complete blockage in chain length extension of any component, suggesting that there is considerable redundancy in elongase activity and chain length specificity, whereas the main effect of *CUT-1* occurs at the C_{24}–C_{26} step. Several gene products (*Arabidopsis cer2* and *cer3*, corn *gl2* and *gl7*) are nuclear associated and appear to have regulatory functions while others are probably involved with supply of precursors or substrate to various steps in the reaction sequences (*Arabidopsis cer8*, *cer9*, and *cer19*; corn *gl1* and *gl8*).

There is abundant evidence for separate but parallel elongation systems linked with some of the different associated pathways. The products of the reductive and decarbonylative pathways often differ greatly in length although the individual reactions of the associated pathways do not appear to be sensitive to substrate chain length. For example, the major primary alcohols are often much shorter than the corresponding alkanes. However, if both pathways share a pool of precursor aldehydes from a common elongation system, then the distributions of the alkanes and alcohols should be similar. These observations point to a degree of separation in the elongation systems that feed into the reductive and decarbonylation pathways respectively. In barley the differential effects of the *cer-cqu* mutation, treatment with inhibitors or stimulants and incorporation of C_{18} fatty acid on the production of alkanes, alcohols, and β-diketones indicate that at least two parallel elongation systems operate. One extends acyl chains for production of alkanes and primary alcohols via the associated pathways, the other elongates 3-ketoacyl chains for production of β-diketones (**Figure 17**). Formation of very long chain alcohols, aldehydes, acids, and alkanes in corn and barley is attributed mainly to an elongation system, ED-I, active at the seedling stage. A second system, ED-II, active over the whole of the plant's life, is primarily responsible for production of esters, although some ED-I products may also be synthesized. These systems also differ in the extent of elongation (**Figure 16**). In wheat, a system similar to corn ED-I functions at the seedling stage, whereas in mature plants the 3-ketoacyl elongation system becomes active, producing β-dicarbonyl compounds. Branched *iso-* and *anteiso-* constituents are particularly abundant in free and esterified fatty acids and alcohols from *Brassica* waxes (**Figure 18**). In contrast, the other products of the associated pathways, including longer chain acids, aldehydes, alkanes, secondary alcohols, and ketones, do not have branched components under normal conditions. This specificity for *n-* compounds is defective in some *Brassica* mutants (gl_4, gl_6, and gl_{Nilla}), resulting in the formation of branched alkanes and secondary alcohols. In wax esters from kale (*Brassica oleracea acephala*) and swede (*B. napobrassica*) there is an apparent preference for doubly branched and unbranched combinations of acid with alcohol over singly branched combinations. Taken together these observations suggest the operation of

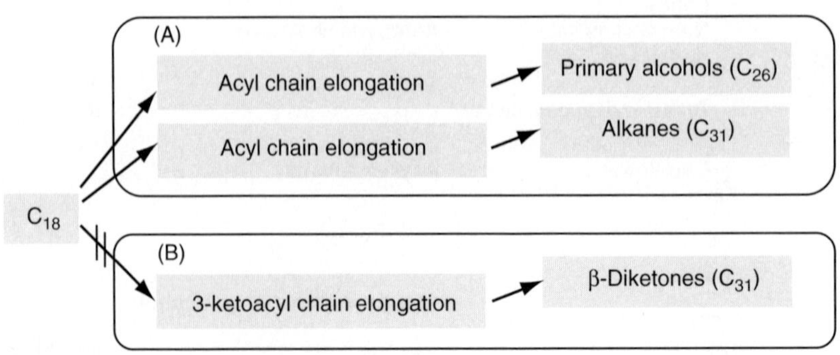

Elongation Products	*cer-cqu* mutation	Cyanide	2-Mercaptoethanol	C_{18} Fatty acid
Primary alcohols	No effect			Incorporated
Alkanes	No effect	Stimulated	Inhibited	Incorporated
β-Diketones	Impaired synthesis	Inhibited	No effect	Not incorporated

Figure 17 Effects of mutation, chemical treatment, and incorporation of C_{18} precursors on the distribution of constituents in barley cuticular wax. Two or more parallel elongation systems function producing (A) primary alcohols and alkanes and (B) β-diketones.

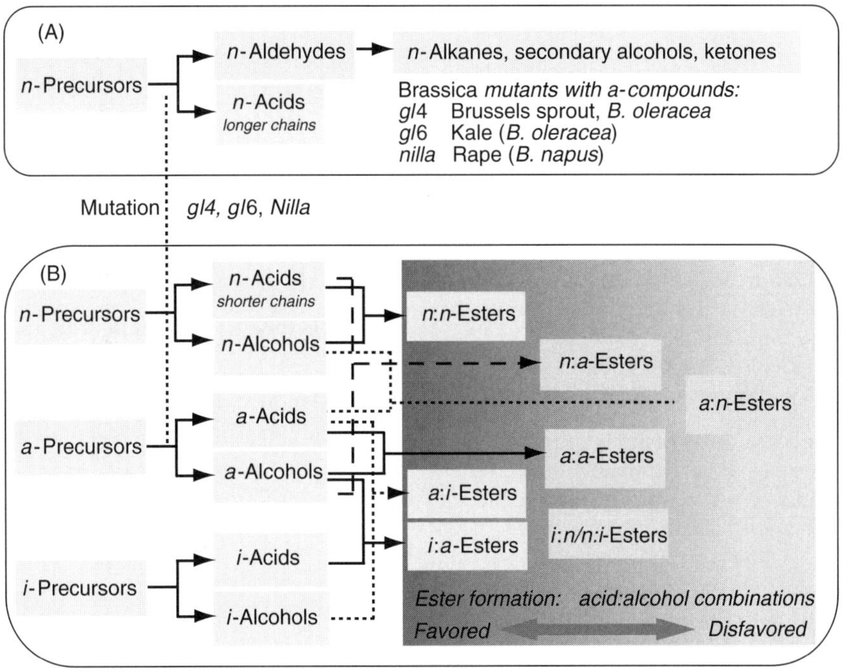

Figure 18 Distribution of straight-chain and branched-chain constituents in brassica cuticular waxes. At least two parallel elongation systems are believed to function. One system (A) produces straight-chain components, which are mainly derived from the associated decarbonylation pathways. The other system (B) produces straight-chain and branched-chain components which are derived from the associated reductive pathways. *n*-, straight chain; *a*-, *anteiso*-branched chain; *i*-, iso-branched chain.

two or more parallel elongation systems and associated pathways with specificities for normal straight-chain (*n*-) and branched substrates and also different esterification systems with primary specificities for branched and unbranched substrates, but which display a degree of cross over.

Deposition of Wax

One of the least understood aspects of wax biosynthesis is how the wax is secreted. Individual components must move from the sites of synthesis (endoplasmic reticulum, Golgi apparatus, and plasma membrane), through the cell wall, pectin layer, and cuticle and onto the cuticle surface. Transportation to the plasma membrane may occur within intracellular vesicles, perhaps derived from the endoplasmic reticulum, Golgi apparatus, and plasma membrane. In sorghum the local density of vesicles increases adjacent to the site of wax secretion when wax production is induced by light. There is little evidence for macroscopic pits, pores, or secretory structures within the cell wall and cuticle, therefore a form of diffusion through microscopic spaces and channels is the most likely mechanism for deposition. Diffusion may be facilitated by the presence of small lipid transport proteins (LTPs) associated with the cell wall. Due to their hydrophobic properties, LTPs probably have the ability to move from the outer layer of the plasma membrane, through the cell wall and on to the cuticle surface.

Future Developments

A major objective for the future is to identify and clone the genes for all the enzymes involved in wax synthesis, and progress will undoubtedly accelerate with further application of state-of-the art molecular genetics techniques. Recently, complementary DNAs (cDNA) encoding acyl-CoA reductase and wax synthase from jojoba and KCS from *Lunaria annua* have been expressed in *Arabidopsis*. This resulted in the switching of lipid biosynthesis from production of triglycerides to formation of wax esters. The modification of plant cuticular layers by alteration of the chemical composition of the wax is of considerable interest and potential benefit. The physiochemical characteristics of the wax play a key role in determining how plants interact with their environment, for example in relation to problems such as disease resistance, drought tolerance and the uptake of foliarly applied chemicals.

Acknowledgments

The author would like to acknowledge the support of the Scottish Executive Environment and Rural Affairs Department.

List of Technical Nomenclature

Alkenol Compound containing a double bond and a hydroxyl substituent.

Biotin A bicyclic molecule and one of the B vitamins, which can act as a carrier of the carboxyl group (carbon dioxide) in carboxylation reactions.

Chiral Spatial arrangement of atoms within a molecule or structure that cannot be superimposed on its mirror image, commonly a carbon atom to which four different atoms or groups are attached.

Claisen condensation Condensation reaction in which removal of an α-hydrogen from an ester produces a mesomerically stabilized carbanion, which in turn reacts with the carbonyl group of another ester or ketone. The products of the reaction are β-ketoesters or β-diketones.

Coenzyme A Complex molecule which forms a thioester linkage with acids to form acyl-CoA derivatives.

Complementary (or copy) DNA (cDNA) A length of single-stranded DNA consisting of a complementary copy of a segment of messenger RNA (mRNA) extracted from cells. The cDNA is made *in vitro* on the RNA template, using reverse transcriptase.

Condensation Chemical reaction joining two or more molecules in which bond formation (e.g., carbon–carbon, carbon–sulfur) is accompanied by the elimination of a molecule of water or an other simple molecule.

Cytosol Liquid portion of cytoplasm.

Decarbonylative pathway Sequence of reactions in which alkanes are formed from aldehydes by loss of carbon monoxide (decarbonylation).

Diketone Compound with two keto (–CO–) functional groups.

Diol Compound with two hydroxyl (–OH) functional groups.

Electrophilicity Reactivity of an electrophile. An electrophile is an electron-deficient reactive center which participates in bond forming reactions by receiving electrons, e.g., from a nucleophile.

Enantiomer One of a pair of chemical structures in which the spatial arrangements of atoms are mirror images of each other.

Expressed sequence tag (EST) Small segment of DNA used for sequence matching with other (known) DNA sequences.

Heteromeric Consisting of separate sub units with specific function.

Homolog Member of a series of compounds of the same chemical type which differ only in the lengths of the carbon chains (C_n). The difference corresponds to different numbers of methylene (CH_2) groups within the molecules.

Homomeric Consisting of a multifunction unit.

Hydroxyketone (ketol) Compound with the keto (–CO–) and hydroxyl (–OH) functional groups.

Hydroxylation Reaction where a hydroxyl group (OH) is introduced into an alkyl chain.

Isomers Compounds of similar type with the same chemical formulae (same numbers of different types of atoms) but which differ in the spatial arrangement of atoms. Examples include stereoisomers (enantiomers) and positional isomers.

Ketoaldehyde Compound with the keto (–CO–) and aldehyde (–CHO) functional groups.

Nucleophile An electron-rich reactive center which participates in bond-forming reactions by donating electrons, e.g., to an electrophile.

Plastid Membrane-surrounded organelle within a plant cell, such as a chloroplast, which is distinct from mitochondria. Plastids are the site of formation of acyl chains by the synthesis *de novo*.

Polyketide A chain with methylene (CH_2) interrupted keto (oxo) functional groups (—$CH_2COCH_2COCH_2CO$—).

Positional isomers Isomers which differ only in the location of specific atoms of functional groups within the molecule.

Prochiral Atom or grouping within a nonchiral molecule which if replaced by another atom or group will result in formation of a chiral molecule.

Reductive pathway Sequence of reactions in which aldehydes and primary alcohols are formed from acyl-CoA by reduction.

Stereospecific Property or process relating to or involving a particular spatial arrangement of atoms in a molecule.

Thioester Sulfur-containing ester (RCOSR′) formed by a condensation reaction between an acid RCO_2H and a thiol R′SH.

Thioredoxin Protein containing a dithiol functional group which can carry out reduction or oxidation reactions.

See also: **Diseases**: Bacterial Diseases; Breeding for Disease Resistance; Fungal Diseases; Viral Diseases. **Genetic Modification**: Gene Cloning, General Principles; Insertional and Transposon Mutagenesis; Transformation, General Principles. **Genetic Modification of Primary Metabolism**: Acyl Lipids. **Growth and Development**: Leaf Development. **Primary Products**: Oils. **Secondary Products**: Epicuticular Waxes.

Further Reading

Bianchi G (1995) Plant waxes. In: Hamilton RJ (ed.) *Waxes: Chemistry, Molecular Biology and Functions*, pp. 175–222. Dundee: The Oily Press.

Gülz PG (1994) Epicuticular leaf waxes in the evolution of the plant kingdom. *Journal of Plant Physiology* 143: 453–464.

Gunstone FD and Herslöf BG (2000) *Lipid Glossary 2*. Bridgewater: The Oily Press.

Holroyd GH, Hetherington AM, and Gray JE (2002) A role for the cuticular waxes in the environmental control of stomatal development. *New Phytologist* 153: 433–439.

Harwood JL (1994) Lipid metabolism. In: Gunstone FD, Harwood JL, and Padley FB (eds) *The Lipid Handbook*, pp. 605–664. London: Chapman & Hall.

Kolattukudy PE (1996) Biosynthetic pathways of cutin and waxes, and their sensitivity to environmental stress. In: Kersteins G (ed.) *Plant Cuticles: An Integrated Functional Approach*, pp. 83–108. Oxford: BIOS Scientific.

Kolattukudy PE, Croteau R, and Buckner JS (1976) Biochemistry of plant waxes. In: Kolattukudy PE (ed.) *Chemistry and Biochemistry of Natural Waxes*, pp. 289–347. Amsterdam: Elsevier.

Kunst L and Samuels AL (2003) Biosynthesis and secretion of plant cuticular wax. *Progress in Lipid Research* 42: 51–80.

Lemieux B (1996) Molecular genetics of epicuticular wax biosythesis. *Trends in Plant Science* 1: 312–318.

Post-Beittenmiller D (1996) Biochemistry and molecular biology of wax production in plants. *Annual Review of Plant Physiology and Plant Molecular Biology* 47: 405–430.

Schultz DJ and Ohlrogge JB (2002) Metabolic engineering of fatty acid biosynthesis. In: Min Kuo T and Gardner HW (eds) *Lipid Biotechnology*, pp. 1–25. New York: Marcel Dekker.

Tulloch AP (1976) Chemistry of waxes of higher plants. In: Kolattukudy PE (ed.) *Chemistry and Biochemistry of Natural Waxes*, pp. 235–287. Amsterdam: Elsevier.

Von Wettstein-Knowles P (1993) Waxes, cutin and suberin. In: Moore TS Jr. (ed.) *Lipid Metabolism in Plants*, pp. 127–166. Boca Raton: CRC Press.

Von Wettstein-Knowles P (1995) Biosynthesis and genetics of waxes. In: Hamilton RJ (ed.) *Waxes: Chemistry, Molecular Biology and Functions*, pp. 91–129. Dundee: The Oily Press.

SEED DEVELOPMENT

Contents

Embryogenesis

M F Jardinaud and M Petitprez, Biotechnology and Plant Breeding, ENSAT, Auzeville, France

Introduction

The flowering plant life cycle is divided into haploid and diploid generations. The gametophytic (haploid) phase begins after meiosis with two kinds of differentiated organs: the pollen grain (male) and the embryo sac (female). Fusion of products of these two haploid structures determines zygote formation, the starting point of the sporophytic (diploid) phase.

In flowering plants (angiosperms), two fertilizations occur. One of the two sperm cells of the pollen grain fertilizes the egg cell to produce the zygote, the other sperm cell merges with the central cell of embryo sac leading to endosperm differentiation. This is the first step in seed development. The endosperm will provide nutrients for embryo development and in some cases support the later seedling.

Angiosperms are divided into two subdivisions, dicotyledons and monocotyledons, representing two types of body organization with either one or two cotyledons in the seed. In dicotyledons, the seedling, in which two cotyledons are symmetrically positioned across the axis, is composed of a shoot apex, a hypocotyl, an embryonic root, and in some cases an epicotyl (**Figure 1**). The embryos of monocotyledons differ from the former in three aspects: (1) they bear only one cotyledon (the scutellum in cereals), (2) in addition to embryonic root meristem, roots can be derived from the lower part of epicotyl, (3) the shoot apex is well developed in the mature seed. Nevertheless, embryogenesis in monocotyledons and dicotyledons is similar up to the 8-celled stage. Typical development of a dicotyledonous embryo (**Figure 2**) goes through four successive stages: globular, heart-shaped, torpedo-shaped, and cotyledonary stages, followed by periods of storage accumulation, seed desiccation, and quiescence.

The mature embryo is organized along a longitudinal axis with cell proliferation occurring mostly at the two ends of this apicobasal axis from two populations of stem cells: the shoot and the root meristems. From the zygote to the seedling, embryogenesis serves to build the apicobasal pattern, to differentiate the primary tissue types, and in many cases to develop storage specialized tissue essential for germination and seedling development.

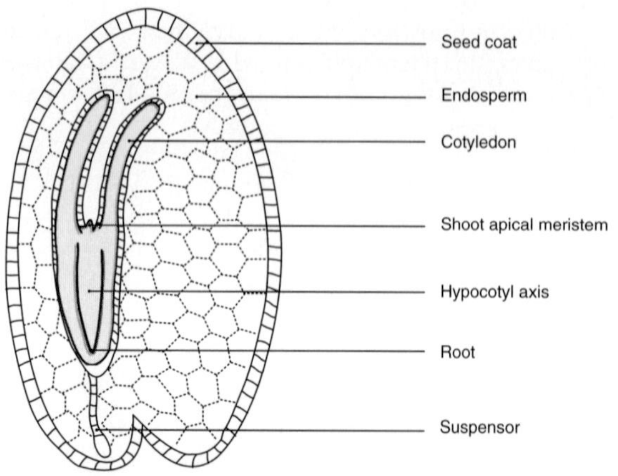

Figure 1 Schematic representation of a maturing dicotyledonous seed embryo (e.g., *Arabidopsis*). The torpedo stage embryo is surrounded by endosperm.

The Shoot–Root Axis in Early Embryogenesis

Although there is considerable variation in how embryos in different genera form, the overall trends in early zygotic divisions and developmental patterns are remarkably similar. Here we describe as an example the *Arabidopsis* pattern of embryo development, one of the most well studied, to illustrate dicotyledonous embryogenesis.

The First Division: Asymmetric Cleavage of the Zygote

During the life cycle, a single cell, the zygote, produces a multicellular organism with many different specialized cell types. To generate such different cell types requires cell divisions in which daughter cells have

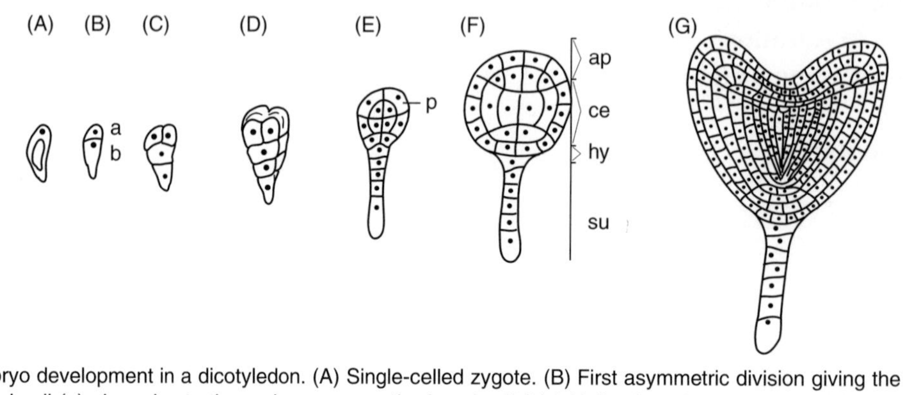

Figure 2 Embryo development in a dicotyledon. (A) Single-celled zygote. (B) First asymmetric division giving the 2-celled embryo. (C, D) The apical cell (a) gives rise to the embryo proper, the basal cell (b) initially gives the suspensor. (E) Octant stage embryo: periclinal divisions give rise to the protoderm (p). (F) The globular embryo comprises the apical part (ap), the central part (ce) deriving from the apical first cell, and the hypophysis (hy) and suspensor (su) deriving from the first basal cell. (G) Heart-shaped stage embryo.

different fates. Such divisions, called asymmetric cell divisions, cleaving daughter cells containing different cellular components are the result of an early polarization of the zygote. In flowering plants, the embryo sac and the egg cell already display polarity along the micropylar–chalazal axis. The embryonic polarity is invariant in all angiosperms: the apical tip towards the chalaza and the basal tip towards the micropyle. Identification of the determinants of this early polarization is still open. First of all, the 1-celled zygote (**Figure 3**) elongates along the apical–basal axis, and randomly arranged cortical microtubules are reorganized in a parallel array perpendicular to this axis. Next, these microtubules form a ring around the preprophasic cell, the location of which points out the future division plane. The nucleus migrates towards the upper part of the cell while the vacuole mainly fills the lower part. Such movements of organelles could be related to dynamics of microtubules and actin microfilaments as indicated by zygotic polarization in the brown alga *Fucus*. In this organism, the zygote's axis is determined by localized secretion of wall components able to bind to transmembrane proteins anchoring cytoskeletal components like actin and microtubules. In angiosperms the zygote divides asymmetrically into a small apical cell and a large vacuolated basal cell. The apical daughter cell will give rise to the major part of the later embryo whereas the basal daughter cell will form mainly the suspensor but also contribute to the embryo. Many genetic studies involving *Arabidopsis* mutants have tried to answer the question: Which factors determine different cell fate of the two daughter cells? Size difference between the two daughter cells cannot explain the difference in fate because in a significant number of flowering plants, this size difference is absent or opposite to that observed in *Arabidopsis*. As mentioned above, the apical–basal axis of the embryo is

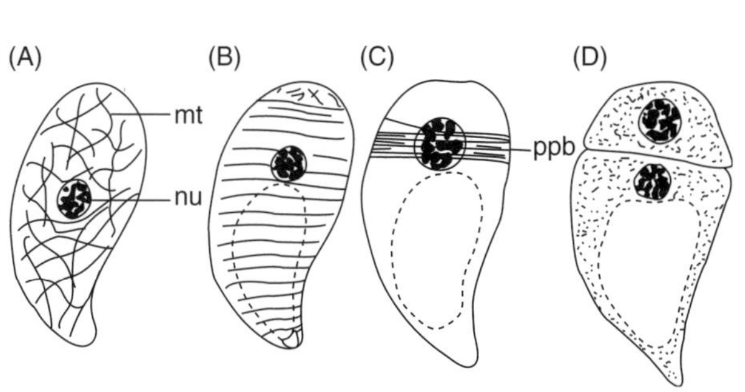

Figure 3 First asymmetric division. (A) In the zygote the cortical microtubules (mt) are randomly arranged and the nucleus (nu) is in the central part. (B) The vacuole develops in the basal part, microtubules arrange in a parallel orientation, and the nucleus migrates towards the upper part. (C) Preprophasic stage: the cortical microtubules form a preprophasic band (ppb) as a ring around the nucleus, indicating the future division site (D).

aligned with the micropylar–chalazal axis of the embryo sac, suggesting that positional cues in the maternal tissues could determine the embryo polarity axis. Nevertheless, cell culture studies reveal that plant embryos can also develop outside the maternal tissue (in *in vitro* culture), and to date, very little genetic evidence argues in favor of a maternal control of zygotic polarity. Mutational analyses have been performed to isolate *Arabidopsis* mutants with an altered embryonic pattern in order to identify genes that control early embryogenesis. Among them, the *GNOM* gene is a good candidate for contribution in axis fixation. It encodes a protein homologous to a guanosine diphosphate–guanosine triphosphate (GDP–GTP) exchange factor that acts on a G-type protein involved in vesicle transport to the Golgi network. Such a vesicle transport mediated by the actin cytoskeleton could serve to fix the axis through secreted cell wall components, possibly arabinogalactane proteins (AGPs).

Another interesting mutant is the *twin* mutant deleted in the *TWN* gene. The apical cell aborts and the embryo develops from the basal cell. This suggests that the two daughter cells could express the capacity to form an embryo but, in normal conditions, the apical cell suppresses the potential of the basal cell.

However, other plants display much more variation in the first division and nevertheless develop normal zygotic embryo. For instance, in rice, the zygote elongates first after fertilization and then divides unequally and transversally. The first division of the smaller terminal cell is often oblique, but may also be transverse or longitudinal. It can be concluded that a correct asymmetrical first division is not an essential requirement for zygotic embryogenesis. More important seems to be an asymmetrical distribution of cytoplasmic determinants such as messenger RNA (mRNA) in determining the embryo polarity.

From Two Cells to Globular Embryo

Cell lineages derived from the apical cell will progressively differentiate to form cotyledons, shoot meristem, hypocotyl axis, and embryonic root (**Figure 4**). By contrast, the basal cell will divide four times according to the same orientation as the first division plane of the zygote and thus specify two kinds of organs: (1) the hypophysis which will give rise to the quiescent center of the root meristem and the columella root cap, (2) the suspensor, a 7–10-cell highly differentiated organ, anchoring the embryo to the embryo sac which possibly functions as an "umbilical cord" conveying nutrients from the maternal tissues to the developing embryo. The suspensor senesces after the heart-shaped stage and disappears in the mature seed.

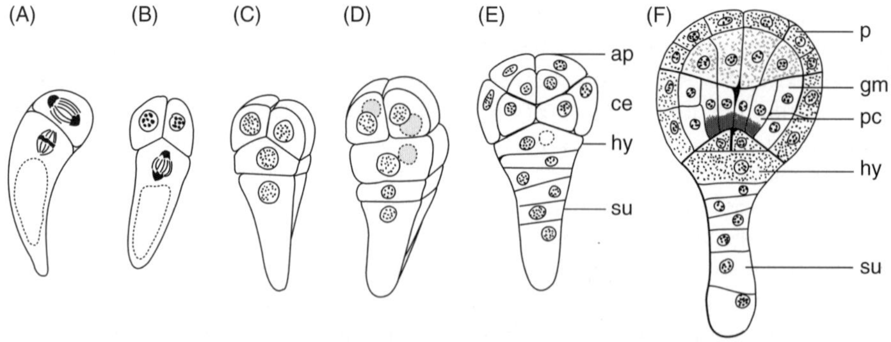

Figure 4 (see color plate 37) From the first division to the globular embryo. (A, B, C, D) First three waves of division; (E) octant stage; (F) the globular-stage embryo is limited by protoderm (p), the central part (ce) is differentiated in an external ground meristem (gm) and an inner procambium (pc). At this stage gene expression of *ML1* (blue), *WUS* (yellow), *PIN1* (red) and *LEC1* (green) are restricted to cell specific zones. Hy, hypophysis; su, suspensor.

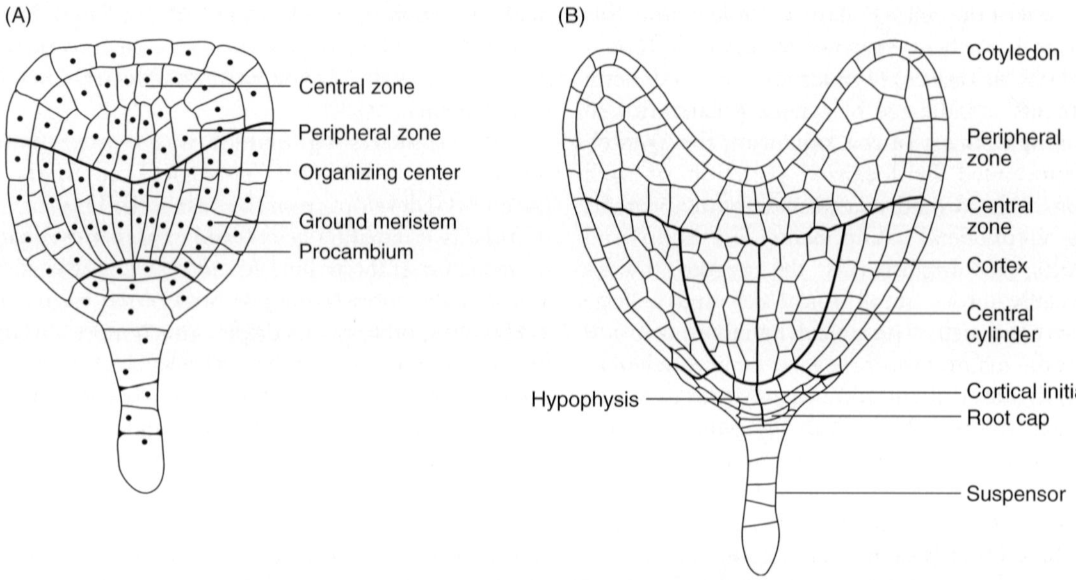

Figure 5 (A) Late globular transition and (B) heart-shaped embryos. In the late globular embryo (A) the apical tier contains an organizing center summated by a central zone and surrounded by the peripheral region. The central zone of the proper embryo is formed by the ground meristem and the inner procambium. In the heart-shaped stage embryo (B) the basic body organization is in place: cotyledons develop on both sides of shoot apical meristem and the root is derived from central part and development of the hypophysis.

At the same time, the apical cell also divides four times, but with each division plane being in an orthogonal direction in regard to the previous one. The apical cell adopts the classical mechanism of spindle orientation rotating from one division to the next. The third division wave in the apical cell lineage is periclinal (**Figure 4E**) and gives rise to the protoderm, an outer cell layer of the proembryo. At the next stage, the globular embryo, the three primordial tissue layers are differentiated: protoderm, ground meristem, and procambium. The spatial organization of these three tissues consists of three concentric layers leading to a new radial axis of symmetry of the globular embryo. This axial symmetry is reinforced after the globular stage, with the differentiation of outer protoderm in epidermis and the lateral root cap, and the middle layer of ground meristem giving rise to cortex and endoderm, whereas the inner procambium forms vascular primordium and pericycle (**Figure 4F**). The root meristem forms the basal end.

Transition from Globular to Heart-Shaped: Appearance of the Final Body Plan

The root meristem comes from two distinct cell types: the upper tier of stem cells comes from the basal cells of the proembryo, while the lower tier derives from the hypophysis whose development is induced by the proembryo (**Figure 5A**). The root meristem is a layered organized stem cell system. Its core, the quiescent center, consists of four nondividing cells between two files of stem cells. Daughter cells generated by the upper tier extend the cell files of the root. The role of the quiescent center is to maintain, by local interactions, the undifferentiated status of the adjacent stem cells.

Later on, opposite to the root meristem, a cell mass differentiates giving the shoot apical meristem. The primary shoot meristem is a part of a complex organization of the shoot apex arranged in three zones (**Figure 5B**): (1) the central zone harbors slowly dividing stem cells at the summit and an organizing center beneath maintaining the stem cells, (2) all around is a ring-shaped peripheral region, where the daughter stem cells descended from the organizing center divide faster and primordia are initiated, (3) the rib zone consists of fast-dividing cells that contribute to the inner tissue of the stem. The cells in the peripheral region are competent to give rise to primordia, but cotyledons or leaf primordia are only initiated at specific sites according to the phyllotactic pattern. Biochemical and genetic evidence indicates that local auxin accumulation contributes to the initiation of cotyledon primordia.

In the transition from globular stage to heart-shaped stage the embryo exchanges the radial symmetry for a bilateral organization when the cotyledons are first formed from two lateral domains at the apical end. At this stage, the body plan and the main tissue layers of the mature embryo have been determined. The cotyledons increase in size by cell division and cell elongation events. Cortex and endoderm cells differentiate and accumulate storage proteins and lipids that will later be used to support seedling growth.

Following this description of the sequence of cellular and tissular events leading to the development of the embryo, a question arises: What are the determinants and the genes involved in signaling and cell fate determination?

Genes in Embryo Development

Early Embryogenesis

From the first division of the zygote, the two daughter cells exhibit differential gene expression. Only the apical cell expresses *ML1* gene, a homeobox gene whose expression will first concern all of the embryo proper (at the 2- to 8-cell stage) and later will be restricted to protoderm of the central and apical part of the globular embryo (**Figure 4F**). Another homeobox gene, *WUS*, is expressed early in the inner four apical cells of the 16-cell embryo. Later on, its expression is restricted to cells of the organizing center, which will give rise to the shoot meristem, and close to the vascular primordium. *WUS* could maintain the pluripotent capacity of the shoot meristem precursor cells. At the transition from 16- to 32-cell stage, inner cells of the central part of the embryo express the *PIN1* gene and accumulate PIN1 protein, an auxin carrier, in their basal plasma membrane, thus inducing an apical–basal polarity.

Both the central and basal domains of the early embryo contribute to root apical meristem formation, and signaling between the two regions may be critical. The *HBT* gene is required for correct hypophysis cell formation; it could either specify the basal region, or control the division program through which the hypophysis produces root meristem. Hypophysis development also depends on the *BDL* gene implicated in auxin-mediated apical–basal patterning. An auxin-dependent process appears to mediate the cell fate of the hypophysis, although it is not known whether auxin itself is the signal. Another gene, *LEC1*, is required for the maintenance of suspensor cell fate. *LEC1* encodes a subunit of the CCAAT box-binding transcription factor, and is expressed exclusively during embryogenesis and plays a central role in controlling both early and late embryogenesis.

SERK1 is a gene that encodes a leucine-rich repeat transmembrane receptor kinase. It is expressed in ovules prior to fertilization and transiently during early embryo development. Its expression persists in all the cells of the embryo proper until the heart-shaped stage and disappears at later stages. *SERK1* could be a component of an embryogenesis signaling pathway, able to recognize an extracellular ligand to switch on the embryogenesis program.

During mid-embryogenesis, the *SCR* gene is transcribed in the ground tissue of the heart-shaped embryo, and after ground cell division, in the endodermis and quiescent center only. The *SCR* gene encodes a putative transcription factor whose mechanism of action is not yet unraveled, but plays an important function in controlling the radial pattern of cell types in embryo development.

From Radial to Bilateral Symmetry

The transition from radial to bilateral organization of the apical embryo is related to differential expression of genes involved in the initiation of primordia. The *STM* homeobox gene is expressed at the late globular stage in the central region of the embryo apex by only a limited set of cells within the apex, precluding cells with a shoot meristem fate from participating in organ formation. *STM* encodes a transcription factor that could maintain the undifferentiated state of meristematic cells by repressing primordia-promoting genes like *AS1*.

In the central part of the apical meristem, the stem cells of the organizing center express the *CLV3* gene encoding a small protein ligand that activates a membrane-bound receptor kinase CLV1, repressing the *WUS* gene and allowing the cells to differentiate and enter the flanking peripheral zone. The *ANT* gene, encoding a transcription factor, is expressed in the peripheral zone of the apical meristem, and maintains the proliferative cell state during organ development. **Figure 6** depicts the pattern of expression of these genes according to the development of the apical region of the embryo.

Two other genes have been shown to be expressed in the vascular precursor cells: *ZLL* is expressed just below the meristem primordia and is required to maintain meristem cell identity within the apex. The *MP* gene is required for correct cell axialization and development of vascular files. *MP* codes for a protein similar to an auxin response factor (ARF1), a transcription factor bearing nuclear localization sequences, and a DNA-binding domain.

Three kinds of genes have to date been identified as acting in the developmental process of embryogenesis: "general" genes mostly involved in cell division or cell elongation control, genes involved in auxin transport or auxin response, and transcription factor coding genes. It is not surprising that these latter are the most represented, because transcriptional control of gene expression is a key event in a morphogenetic process. The longitudinal axis of the embryo contains three nonoverlapping transcriptional territories: apical, central, and basal. Each tissue layer has a distinct transcriptional program established in a territory-specific manner. This suggests that a prepattern of different transcriptional regulatory domains has been established in the globular embryo before the morphogenetic events leading to differentiation of cotyledons at the heart-shaped stage. Such a prepattern may be related to cell–cell signaling involving not only diffusible molecules but also positional cues via cell wall components.

In rice, different homeobox genes are expressed during early embryogenesis. *KN2* and *KN3* are expressed in the region where the shoot apical meristem is organizing. KN3 expression marks the boundaries of embryonic organs following formation of the shoot apical meristem suggesting its involvement in patterning organ positions. The *TF1* gene, encoding a GL2 homeodomain protein, is strongly expressed in the rice embryo at the globular stage and preferentially localized to the protoderm at 3 to 6 days after fertilization; such an expression pattern is closely related to *ML1* gene expression in the *Arabidopsis* embryo.

Hormones in Control of Embryo Development

Auxin could play a major role in providing positional information for coordination of embryo patterning from the globular stage onwards. Auxin is a highly

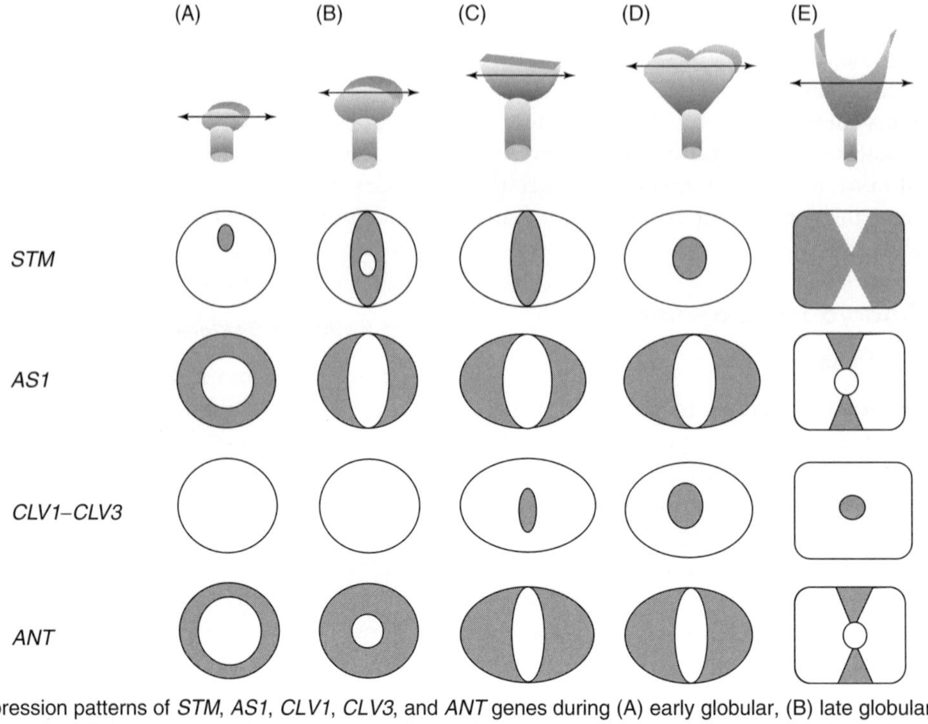

Figure 6 Expression patterns of *STM, AS1, CLV1, CLV3*, and *ANT* genes during (A) early globular, (B) late globular, (C) transition, (D) heart-shaped, and (E) torpedo-shaped stages of embryogenesis. On the schematic representation of apical sections, the gray zone indicates the expression area of the corresponding gene.

diffusible molecule occurring in both active and conjugated (inactive) forms in tissues. Shoot meristems and leaf primordia are considered as the main sites of synthesis of auxins. Influx and efflux carrier proteins take part in a polarized basipetal transport of auxin along the embryonic axis. Preventing auxin transport results in high auxin concentrations near the source and relatively lower concentrations elsewhere. High concentrations of auxin promote tissue outgrowth resulting in cotyledon expansion, whereas low levels of auxin or auxin-induced signaling between the cotyledons is related to lack of tissue outgrowth. Auxin is involved in promoting the development of lateral primordia and determining their radial position. Asymmetric auxin distribution determines polarity and pattern in the embryo. In wheat, the distribution of auxin changes during embryo development. The shift from radial to bilateral symmetry is correlated with a redistribution of auxin in the embryo, associated with the onset of an active directed auxin transport from the basal pole towards the shoot apical meristem.

The *PIN1* gene encodes a putative transport protein of the efflux carrier complex. *PIN1* expression becomes polarized at the mid-globular stage (100 cells), before the two cotyledons have started to develop. Beside *PIN1*, the *MONOPTEROS* (*MP*) gene encodes an auxin response factor, which is a transcription factor likely to bind to promoter elements of auxin inducible genes. The *MP* gene has an early function in the establishment of body and vascular strands in the embryo. *PINOID*, which encodes a serine-threonine kinase, is first expressed at the globular stage and is proposed to negatively regulate auxin signaling. Both gene functions seem to be required for proper transport and coordinated localization of auxin. Recent work based on the precise localization of auxin correlated the auxin position with cell differentiation events in the root meristem. An auxin peak was observed in the root meristem and modifications of the position of this peak were associated with corresponding changes in the distal cell fates. This reveals a dual function of auxin as a positional signal within individual meristem besides its role as a long-range signal promoting tissue outgrowth. Auxin has long been known to regulate a wide variety of plant responses, but recent molecular genetic findings directly implicate this hormone in the organization of cell patterns and morphogenesis. It is plausible that auxins act as intercellular messengers in patterning processes, and that auxin-mediated long-distance signaling could simultaneously integrate morphogenesis.

Sterols play essential roles in the development of eukaryotic organisms. In plants, bulk sterols like sitosterol and campesterol are integral membrane components, but also serve as biosynthetic precursors of brassinosteroids. It is not surprising that mutations in genes encoding sterol biosynthetic enzymes have a dramatic effect on plant development. In fact, all the mutants having defects in the sterol biosynthetic pathway revealed alterations in embryo development. Among the cloned genes, *FACKEL*, encoding a sterol C14 reductase, is expressed early in the globular embryo and controls the orientation of cell division plane. The *SMT1* gene encodes an enzyme responsible for C24 alkylation of the sterol side chain. In the *smt1* or *fackel* mutants, basic cell division is not affected but in the embryo, cell expansion defects are observed until the globular stage. Compounds of the sterol biosynthetic pathway may play a hormonal-like role in controlling embryogenesis and position-dependent cell fate.

Seed Development

Seed development is a pivotal process in the life cycle of angiosperms. It is initiated by the double fertilization which leads to the development of the embryo and the endosperm. Among angiosperms, monocotyledonous and dicotyledonous seed development have important similarities and differences. In dicotyledons, the embryo is normally the major part of the mature seed, and the endosperm tissues are often ephemeral. In contrast, in monocotyledons, endosperm tissues in many cases constitute the bulk of the seed. Embryogenesis provides a bridge between the gametophytic phase and the differentiation of the sporophytic plant through shoot and root meristems. As such, the plant embryo must establish the polarized plant body plan and enable the young plant to survive the adverse conditions of vegetative rest in a conserved state for many years. Although genetic studies in *Arabidopsis* have begun to reveal a few genes involved in pattern formation, cell differentiation, and organ development, the problem remains to decipher the signaling networks that take place very early in embryogenic regions and their spatiotemporal regulation mechanisms.

List of Technical Nomenclature

Actin microfilaments	Filamentous strands which are a component of the cell cytoskeleton.
Auxin	Plant growth regulator that controls plant development by promoting cell division and elongation.
Brassinosteroid	Plant growth regulator that controls plant development by stimulating stem elongation, inhibiting root growth and development.

Chalazal pole The site of an ovule opposite the micropyle, and adjacent to the stalk region.

Columella The central part of a root cap in which the parenchyma cells are arranged in a series of columns.

Cytoskeleton A network of fibers in the matrix of cells that provides a supporting framework for organelles, anchors the cell membrane, and facilitates cellular traffic. The major components of cytoskeleton are the microfilaments (of actin) and microtubules (of tubulin).

Embryo sac One cell in each ovule undergoes meiosis resulting in four haploid megaspores. One of these survives and its nucleus divides three times, producing a single large cell which contains eight haploid nuclei and is called the embryo sac.

Endosperm Typically a $3n$ tissue in the seeds of angiosperms formed by the fusion of a sperm nucleus with the two polar nuclei in an embryo sac. The endosperm serves as a food substance for the early growing embryo.

Gametophyte That plant generation which gives rise to the gametes by means of mitosis. Typically haploid.

Homeobox The homeobox sequence codes for a protein domain that is involved in binding to DNA. Homeoboxes appear to determine when particular groups of genes are expressed during development.

Hypophysis The top cell of a suspensor which gives rise to the development of the root in the embryo of angiosperms.

Micropyle The opening in the integument of an ovule through which the pollen tube may pass and enter the embryo sac.

Microtubules Components of the cell cytoskeleton mainly constituted of tubulin. These structures include a cortical array in quiescent cell and the mitotic spindle in dividing cells.

Periclinal division A plane of division or cell wall establishment which is parallel with the surface of the organ.

Phyllotactic pattern The arrangement of leaves on a plant stem.

Pollen grain A mature microspore in a seed plant with a distinctive cell wall and containing a sperm cell.

Primordium A group of undifferentiated cells that represents the initial stages in development of a plant organ.

Quiescence State of a cell that is not at that time undergoing cell division cycles.

Spindle An aggregation of microtubules that aid in the movement of chromosomes during mitosis or meiosis.

Sporophyte The diploid phase of the life cycle of plants that gives rise to the production of spores by means of meiosis. In higher plants it is the dominant phase of the life cycle.

Stem cell Cell that gives rise to a lineage of cells by producing, upon division, two dissimilar daughters, one replacing the original stem cell, the other differentiating further.

Sterol Any 17-carbon tetracyclic structure, with additional alcohol groups and side chains. The best-known examples are cholesterol (in animal) and sitosterol (in plant).

Zygote The diploid cell produced after an egg cell is fertilized. Beginning of the new sporophyte plant.

See also: **Flowering and Reproduction**: Fertilization. **Growth and Development**: Cell Division and Differentiation. **Regulators of Growth**: Brassinosteroids. **Root Development**: Root Growth and Development. **Seed Development**: Endosperm Differentiation; Nutrient Loading of Seeds; Physiology of Maturation; Seed Production; Seed Quality. **Tissue Culture**: Artificial Seeds; Somatic Embryogenesis.

Further Reading

Chaudhury AM, Koltunow A, Payne T, *et al.* (2001) Control of early seed development. *Annual Review of Cell Developmental Biology* 17: 677–699.

Goldberg RB, de Paive G, and Yadegari R (1994) Plant embryogenesis: zygote to seed. *Science* 266: 605–614.

Harada JJ (1999) Signaling in plant embryogenesis. *Current Opinion in Plant Biology* 2: 23–27.

Jürgens G (2001) Apical–basal pattern formation in *Arabidopsis* embryogenesis. *European Molecular Biology Organization Journal* 20: 3609–3616.

Paquette AJ and Benfey PN (2001) Axis formation and polarity in plants. *Current Opinion in Genetics and Development* 11: 405–409.

Souter M and Lindsey K (2000) Polarity and signalling in plant embryogenesis. *Journal of Experimental Botany* 51: 971–983.

Endosperm Differentiation

S E Lid, Agricultural University of Norway, Ås, Norway
O-A Olsen, Pioneer Hi-Bred International, Johnston, IA, USA

Introduction

The cereal grain was an object of interest to the early plant anatomists, and outlines of the main developmental stages of the endosperm date back to the last part of the nineteenth century. In spite of this fact, insight into the processes underlying endosperm development is fairly new, and has progressed significantly only within the last ten years or so. Factors that contributed to this slow progress include the fragility and inaccessibility of the endosperm coenocyte, as well as the simple composition of the endosperm, which compared to embryos, might have appeared less challenging to plant anatomists and embryologists.

A problem that still remains unsolved today is the evolutionary origin of the angiosperm endosperm, but the possible explanations that are being discussed include the endosperm as an evolved altruistic twin embryo or a continued development of the megagametophyte triggered by fertilization. A combination of novel insight into the evolution of angiosperms and improved techniques for characterizing developing endosperm is needed to solve this problem. This article focuses on the developmental biology of nuclear endosperm, which is the most common type of angiosperm endosperm, including in the cereal species.

The Megagametophyte and the Central Cell

The megagametophyte, or embryo sac, is situated within the body of nucellus parenchyma cells, which are covered by the integuments (**Figure 1A**). The embryo sac, representing the gametophytic generation of the plant, is the result of a process which starts with the differentiation of a nucellus parenchyma cell into a megaspore mother cell. The megaspore mother cell in turn subsequently undergoes meiosis to give rise to four megaspores. Of the four haploid cells resulting from this female meiosis in corn (*Zea mays*; maize), one cell persists to give rise to the first cell of the embryo sac. During development of the corn embryo sac, which displays the so-called polygonum type of development, three mitotic divisions give rise to two groups of four nuclei, one at the end closest to the embryo (the

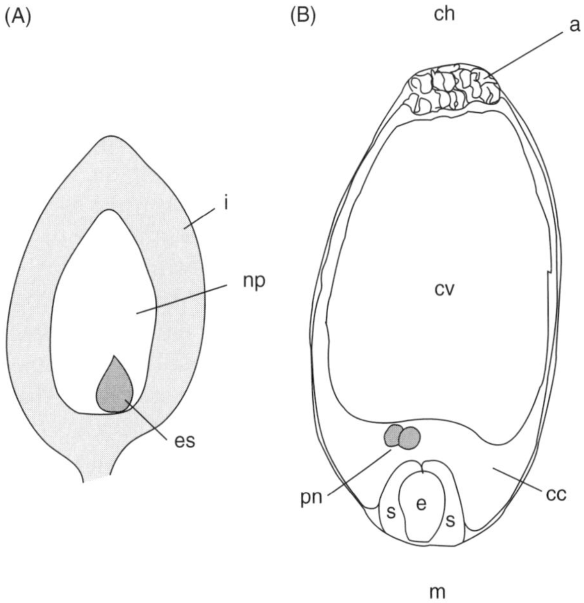

Figure 1 The embryo sac. This and the following figures are based on corn (maize). The embryo sac is situated within the body of nucellus parenchyma which are covered by integuments (A). The mature embryo sac consists of four cell types; the egg cell and the two synergids at the micropylar end, the antipodals at the chalazal end and the central cell containing the two polar nuclei and a large central vacuole (B). es, embryo sac; np, nucellus parenchyma; i, integuments; e, egg cell; s, synergids; a, antipodals; cc, central cell; pn, polar nuclei; cv, central vacuole; ch, chalazal end; m, micropylar end.

micropylar end) and one in the opposite (chalazal) end of the embryo sac. One nucleus from each side then moves to the center of the embryo sac, where they remain in contact. Fusion of these two polar nuclei does not take place until fertilization. The two nuclei plus the cytoplasm of the mid portion of the embryo sac later form the central cell (**Figure 1B**). The remaining nuclei have different fates; one of the micropylar nuclei becomes the egg cell, the two others the synergids, whereas the three chalazal nuclei undergo repeated mitosis forming about 20 to 40 antipodal cells. The cellularization process of the embryo sac is mediated by the formation of radial microtubular systems formed on the nuclear envelopes. Little is known about the factors involved in embryo sac differentiation, but genetic evidence clearly demonstrate the involvement of genes expressed both in the embryo sac itself as well as in the surrounding maternal tissues.

The endosperm is formed from the central cell, which after fertilization becomes triploid ($3n$). Features of the central cell that are important for endosperm development include the large central vacuole and the location of the bulk of the central cell cytoplasm containing the polar nuclei in the distal part close to the embryo (**Figure 1B**).

The Endosperm Coenocyte

During the double fertilization process in angiosperms, one of the two male gametes carried by the pollen tube fuses with the egg cell nucleus to form the diploid zygote, and the other with the diploid central cell nucleus to form the endosperm. In some species, like corn, the male gamete first fuses with one of the two polar nuclei, the product subsequently fusing with the second polar nucleus.

As stated above, the events in early endosperm formation differ dramatically from the common mode of development in somatic tissues. Rather than being unique, however, recent evidence suggests that this mode of development is shared by endosperm, megagametophytes, and pollen grains, thus representing an integral part of the life cycle of plant cells. As outlined below, the events in endosperm development are likely to be the result of a few key regulatory steps in the control of the cell cycle and the microtubular (MT) cytoskeletal arrays compared to somatic cells, rather than involving novel processes not active in somatic tissues.

After fertilization, mitotic division of the primary endosperm nucleus ensues quickly, in corn within 3 to 5 h. Already during the first mitotic divisions, the distinguishing feature of nuclear endosperm becomes apparent, namely the occurrence of mitosis without the default cytokinesis that accompanies mitoses in somatic cells. As a result, the central cell becomes multinucleate, and is referred to as the endosperm coenocyte. The mechanism suppressing cytokinesis is unknown, but observations both in barley (*Hordeum vulgare*) and wheat (*Triticum* spp.) show that a phragmoplast is partly formed; in wheat endosperm this phargmoplast is even partially functional. In barley, the remains of the nonfunctional phragmoplast are seen as a cluster of tubulin on the nuclear envelope of the sister nuclei after completion of each division.

Recent data from cell cycle research show, not surprisingly, a tight linkage between the molecules that regulate progression through the cell cycle and the MT cytoskeletal arrays. One example of this is that a key player in cell cycle progression, mitogen activated protein (MAP) kinases, are present in phragmoplasts. This or similar molecules are therefore candidates for regulators of the phragmoplast function in endosperm.

Two other important phenomena occur during the syncytial stage of endosperm development, i.e., determination of the plane of mitosis and migration of endosperm nuclei in a predictable pattern within the former central cell. The experimental evidence for both of these events comes from corn research carried out over the last 50 years, and involve studies of corn grain color sectors after somatic mutations of genes encoding enzymes involved in anthocyanin biosynthesis at different times after the double fertilization. As confirmed by direct microscopy observations, the corn primary endosperm nucleus divides in a plane perpendicular to the longitudinal axis of the embryo sac, such that each daughter nucleus is positioned on each side of the embryo as viewed from the abgerminal side of the grain (**Figure 2A–D**). The next two divisions are at perpendicular planes relative to each other, the eight nuclei resulting from the three first divisions positioning themselves in a circle in the

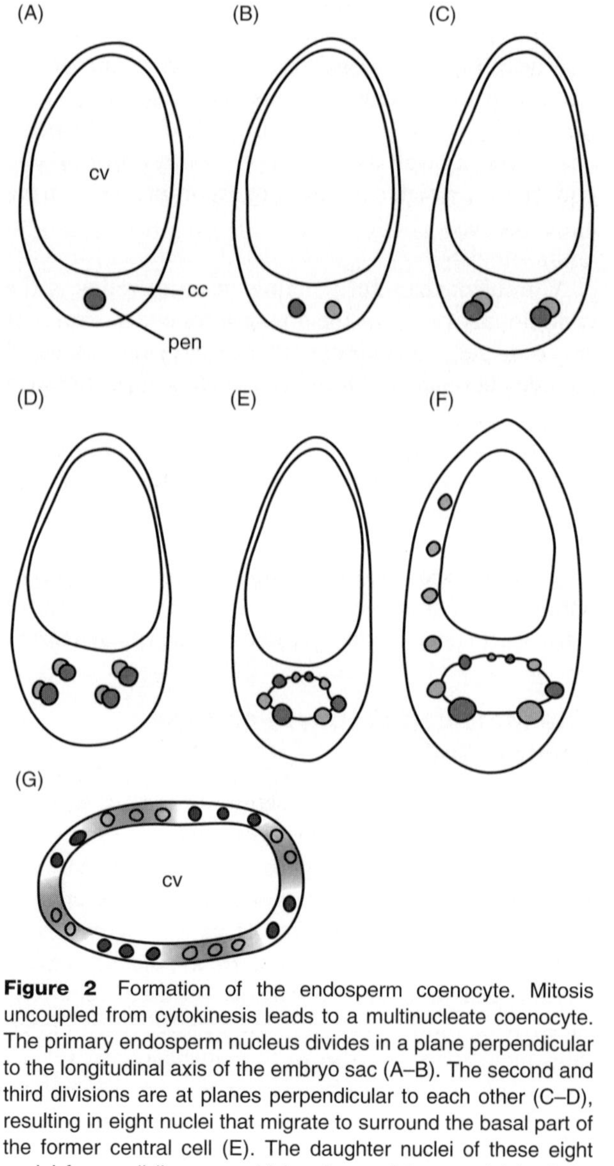

Figure 2 Formation of the endosperm coenocyte. Mitosis uncoupled from cytokinesis leads to a multinucleate coenocyte. The primary endosperm nucleus divides in a plane perpendicular to the longitudinal axis of the embryo sac (A–B). The second and third divisions are at planes perpendicular to each other (C–D), resulting in eight nuclei that migrate to surround the basal part of the former central cell (E). The daughter nuclei of these eight nuclei form cell lineages which sub populate one-eighth of the cytoplasm of the central cell in a proximal–distal pattern. (F) Only one of the eight cell files is drawn. Subsequent divisions lead to a multinucleate coenocyte, shown in a transverse view (G). pen, primary endosperm nucleus; cc, central cell; cv, central vacuole.

proximal cytoplasm of the former central cell. During the ensuing stage, the pattern of migration for each nucleus becomes restricted, such that the daughter cells of the eight nuclei form cell lineages each of which populates one-eighth of the cytoplasm of the central cell in a proximal–distal pattern (**Figure 2E–F**). After the completion of endosperm cellularization, these subpopulations of cells each gives rise to a part of the endosperm with a shape corresponding to a sector of an orange (**Figure 2G**). The function of this orderly pattern of development is unknown, but the predictable planes and early migration pattern may be significant for the early events in endosperm differentiation (see below).

The number of mitotic divisions during the coenocytic stage appears to be under genetic control, and is terminated by a mitotic arrest, at least in barley. In corn, three or four rounds of mitosis occur before this arrest, giving rise to 256–512 endosperm nuclei. Whether or not a similar arrest occurs in other cereal species has been difficult to establish by direct observations due to the fact that the duration of the coenocytic stage in rice (*Oryza sativa*), wheat, and corn is only 4 days, compared to 6 days in barley. However, based on the highly conserved nature of the events leading to cell wall formation in the endosperm coenocyte, it seems reasonable to assume that a similar mitotic arrest is of general occurrence. The mechanism involved in mitotic arrest is unknown, as is also the stage of the cell cycle in which the arrest occurs. In plants, gene products involved in cell cycle arrest both in G1 and in G2 have been described, representing possible mechanisms for the control of cell cycle arrest in endosperm.

Recently, studies in *Arabidopsis*, which also have a polygonum type of embryo sac and a nuclear mode of endosperm development, have shed light on the regulation of the initial steps in endosperm formation. In this system, mutations in the *medea*, *fis2*, and *fie* genes lead to endosperm development in the absence of fertilization, suggesting that their role is to delay endosperm development until after fertilization. The sequence of the *medea* and the *fie* genes shares similarity with a class of genes known to regulate chromatin structure in *Drosophila*, the so called *polycomb* genes. The implications of this finding, which is assumed to be valid also for the grass species, are twofold. First, it shows that the male nucleus is not a prerequisite for endosperm development. Second, it suggests that the chromatin of endosperm nuclei is configured in an endosperm specific way, facilitating the expression of the genes that drive endosperm development. A complete understanding of the processes underlying endosperm development must eventually also include an understanding of these processes, which set the stage for the post-fertilization events leading to endosperm formation.

Endosperm Cellularization

During the mitotic hiatus in barley, and in the nuclear endosperm of all other species investigated so far, the endosperm coenocyte is subdivided into cellular compartments via a highly conserved pathway involving radial microtubular systems (RMS). During this process, which starts immediately after the onset of the mitotic hiatus, microtubuli assemble on the envelope of all endosperm nuclei (**Figure 3A**). At first, microtubuli from each nucleus overlap, but soon clear interzones form between adjacent RMS, forming what have been termed cytoplasmic phragmoplasts in barley (**Figure 3B**). Although the anticlinal walls that subdivide the endosperm coenocyte into cellular compartments are formed along the whole length of the cytoplasmic phragmoplast, the part of the wall closest to the former central wall first becomes visible in the light microscope. For this reason, before it was shown that cytoplasmic phragmoplasts were involved in anticlinal cell wall formation, the first anticlinal walls were referred to as "free growing" by some authors. The result of the activity of the cytoplasmic phragmoplast is that each

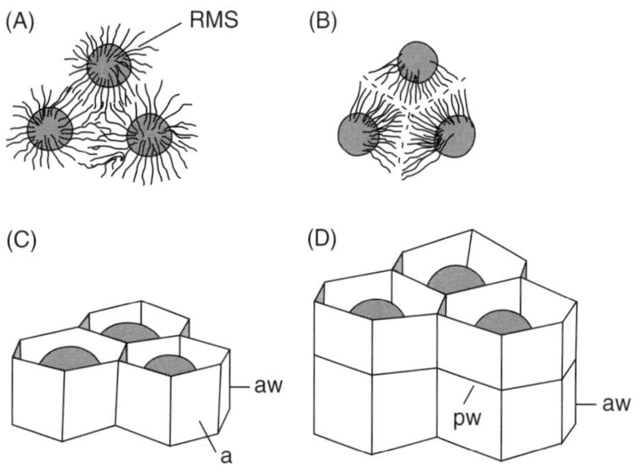

Figure 3 Endosperm cellularization. After the mitotic hiatus, the endosperm coenocyte is subdivided into cellular compartments by formation of radial microtubular systems (RMS) protruding from each nucleus (for simplicity, only three nuclei are shown). Initially, microtubuli between neighboring nuclei in the syncytium overlap (A), before distinct interzones between adjacent RMS are formed (B). The interzones between RMS, referred to as cytoplasmic phragmoplasts, facilitate anticlinal cell wall deposition, which leads to formation of alveoli, cellular compartments left open towards the interior central vacuole (C). The first periclinal mitosis is linked to the formation of a functional interzonal phragmoplast depositing a periclinal cell wall which cuts off an outer layer of cells and an inner cell layer of open-ended wall tubes similar to the first layer of endosperm alveoli (D). a, alveoli; aw, anticlinal cell wall; pw, periclinal cell wall.

nucleus and the surrounding cytoplasm of the endosperm coenocyte is encased in a cylinder of cell walls, or an alveolus, with its opening towards the central vacuole (**Figure 3C**).

The radial microtubuli and cytoplasmic phragmoplasts appearing in early endosperm development differ from MT cytoskeletal arrays forming during the somatic cell cycle. The possibility is, however, that they do not represent entirely new processes, but rather variations over the same theme displayed in somatic cells. Currently, one model that explains how polymerization of MTs occurs in plants is that initiation and extension of microtubuli occur at nuclear envelopes, and that MTs are translocated from this site to other cellular locations where they function. If this holds true, radial microtubular arrays may be the result of a block of MT transport away from nuclei.

The cytoplasmic phragmoplasts that form the anticinal cell walls bear strong resemblance to interzonal phragmoplast, both representing arrays of MT with opposing (plus) polarity. Recent research in several species, including *Arabidopsis*, have identified several similarities between the two types of phragmoplasts, including the Knolle syntaxin, supporting the view that cytoplasmic phragmoplasts contain most if not all of the basic elements present in interzonal phragmoplasts. Further research is needed to identify additional components in both interzonal and cytoplasmic phragmoplasts, hopefully leading to a better understanding of the mechanism involved in the initiation of anticlinal cell wall deposition in endosperm.

The next phase in endosperm cellularization occurs when the cell cycle arrest is lifted, which occurs synchronously throughout the entire coenocyte. In barley, as well as in all other species with nuclear endosperm examined so far, the plane of this mitosis is periclinal, i.e., parallel to the former central cell wall. Importantly, this mitosis is linked to the formation of functional interzonal phragmoplasts depositing periclinal cell walls as in meristematic cells. This cuts off an outer layer of cells and an inner layer of cell wall tubes similar to the first layer of endosperm alveoli (**Figure 3D**). Completion of endosperm cellularization involves repeated rounds of the same process, periclinal divisions of nuclei in the innermost alveoli, and centripetal extension of the anticlinal walls facilitated by nuclear MT arrays. The result of this process is that anticlinal walls from all sides of the endosperm completely invade the former central vacuole, completing the process of endosperm cellularization. This stage is reached at 6 DAP (days after pollination) in barley, and around 4 DAP in wheat, corn, and rice.

Endosperm Cell Fate Specification and Differentiation

The cereal endosperm consists of four major cell types, the starchy endosperm, the aleurone layer, the transfer cells, and the cells of the embryo-surrounding region. Similar to many other plant systems, the specification of endosperm cell fate seems to be determined by positional cues. These cues, yet to be identified, are assumed to be perceived by cellular receptors similar to those known from animal systems, activating the relevant genes specifying the different endosperm cell types. In the early endosperm, four developmental domains, corresponding to the four cell types, are believed to be defined by positional signaling. Of these, two cover the perimeter of the endosperm; the first is over the ventral vasculature corresponding to the pedicel in corn and to the nucellar projection in barley and wheat, the second is the remainder of the perimeter of the endosperm. The third zone is represented by the lining of the embryo cavity of the endosperm, in which the embryo develops. The fourth, and largest, developmental zone is the entire interior of the endosperm.

The Aleurone Layer

Of the four endosperm cell types, the underlying mechanism for cell fate specification is best understood for the corn aleurone layer. Two corn mutants have been instrumental in elucidating these mechanisms. The first of these mutants is *crinkly4*, which in homozygous condition leads to patches of endosperm lacking aleurone cells and instead carries starchy endosperm cells in peripheral positions. The second mutant is *dek1*, which in homozygous condition leads to a total absence of aleurone cells. *Crinkly4* encodes a receptor-like protein kinase with similarity to the tumor necrosis factor receptor, a family of receptors that is central in inflammatory responses in mammals. The cloning of *Crinkly4* represented a major advance in endosperm research, strongly suggesting the possibility that aleurone cells were specified by a localized, but yet unidentified ligand. Further support for the involvement of positional specification of aleurone cell fate has recently come from studies of unstable alleles of *dek1*. In one part of these studies, *dek1/dek1/dek1::Mu* endosperms, lacking aleurone cells, were studied. On rare occasions, the Mu transposon excised, creating a *dek1/dek1/Dek1* cell. Remarkably, even late in grain development, this somatic reversion led to a redifferentiation from a starchy endosperm cell identity to an aleurone cell identity in the periphery of the endosperm. In a second set of experiments, using *Ds* (*Dissociation*) induced chromosome breakage, endosperms in which somatic mutations led to a loss of *Dek1* function were created, resulting in the respecification from aleurone to starchy endosperm cell fate in the peripheral position of the endosperm. These studies are in agreement with studies in wheat, where a redifferentiation of inner daughter cells from periclinal divisions of aleurone cells to starchy endosperm cell fate was reported several decades ago. The *Dek1* gene was recently cloned and shown to encode a predicted membrane anchored calpain-like cystein proteinase, however, the function of DEK1 in aleurone signaling is yet to be defined. A general model for endosperm cell fate specification is presented in **Figure 4**. Interestingly, barley has three layers of aleurone cells, probably reflecting either a difference in the distribution of the proposed ligand, or a different sensitivity of the signal transduction system that transmits the signal to the nucleus. Interestingly, corn lines with multilayered aleurone have also been reported, resembling the three-layered aleurone of barley.

Although the exact time point for the specification of aleurone cell fate is currently unknown, it has been proposed that aleurone cell fate is specified already when the first periclinal division takes place in the endosperm (**Figure 4A–B**). Support for a different fate in the two daughter cells of this division comes from barley, where the two daughter cells display dramatically differently MT cytoskeletal arrays. In the aleurone initials (blue nuclei, **Figure 4B**), the full complement of MT arrays of somatic meristems is present, including hooplike cortical arrays, preprophase bands (PPBs), and interzonal phragmoplasts. In contrast, in the starchy endosperm initials (red nuclei, **Figure 4B**), hooplike cortical arrays, and PPBs are missing. Most likely, this difference is reflected in the pattern of cell divisions in the two tissues; aleurone cell divisions are restricted to either the periclinal or anticlinal plane, whereas in the starchy endosperm, cell divisions occur at random planes. In somatic cells, PPBs predict the plane of division in the ensuing mitosis. Introduction of the PPB in aleurone cells may therefore be a mechanism to match the expansion of the aleurone layer with the growth of the inner mass of starchy endosperm cells that the aleurone layer encloses. In the aleurone layer, periclinal divisions contribute to the mass of starchy endosperm cells, as cells removed from the peripheral position redifferentiate to become starchy endosperm cells (see above). In corn, divisions in the aleurone layer are predominantly anticlinal from 20 DAP and onwards. Possibly, internal pressure from the expanding starchy endosperm cell mass regulates the rate of cell divisions in the aleurone layer via the PPB, initially periclinally and at later stages anticlinally.

(A) (B) (C) (D) (E)

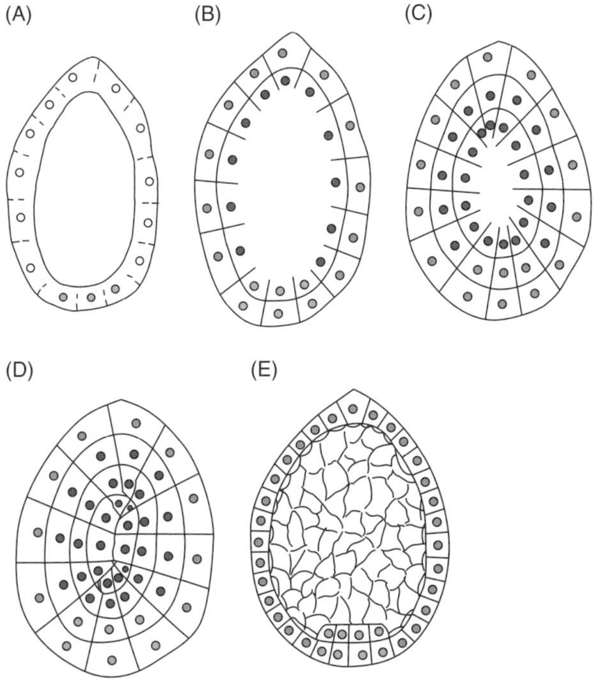

Figure 4 A model for endosperm cell fate specification. Transfer cell fate specification is believed to occur by positional signaling already at the coenocytic stage in the basal domain (green nuclei, A). The remainder of the nuclei have the potential to assume the aleurone or the starchy endosperm cell fate. After the first periclinal division, the peripheral cell layer acquires aleurone identity (blue) by positional signaling. The inner cells and their descendants assume starchy endosperm identity (shown as red, B–C). After completion of the cellularization process (D), mitotic divisions of starchy endosperm cells are at random orientation, leading to a loss of the cell file pattern established during the cellularization process (E).

Transfer Cells

For the second peripheral developmental domain of the endosperm, the zone over the ventral vascular bundle of the grain, mutant data are lacking. However, based on *in situ* hybridization experiments, the *END1* (*Endosperm1*) transcript from barley is detected in the cytoplasm of the endosperm coenocyte over the vascular bundle, thus indicating that this developmental domain is established already at the coenocytic stage (**Figure 4A**). Interestingly, the restricted distribution of the *END1* transcript in the coenocyte also suggests a mechanism for gene expression involving positional signaling.

One possibility, yet to be explored, is that the mechanisms involved in the specification of aleurone and transfer cell fates are very similar, and that the two pathways share one or more components.

The hallmark of the endosperm transfer cells is their very thick and elaborate cell walls, facilitating the uptake of sucrose and other nutrients from the maternal plant tissues. Marker transcripts for corn transfer cells include transcripts such as *Betl1-4*, which may have a role in plant defense. The soluble acid invertase *Incw2* from corn also represents a molecular marker for the endosperm transfer cell layer in corn. Two mutants perturbed in transfer cell function are *miniature-1* and *reduced grain filling1*. Recent studies, based on the promoter upstream regions of the *BETL* genes in barley using electrophoretic mobility shift assays, aim at identifying transcription factors activating the *BETL* genes.

The Embryo-Surrounding Region

The third developmental domain of the endosperm consists of the dense cells that line the cavity in the proximal part of the endosperm, commonly referred to as the embryo-surrounding region (ESR). The exact role of the ESR is unknown, but it possibly functions in nutrient uptake or protection of the embryo from invading pathogens. Little information is available on the mechanism underlying ESR cell fate specification. Marker transcripts for this region are *Esr1–4*. The function of these transcripts is unknown. In most species examined, endosperm cellularization is initiated in the vicinity of the embryo, suggesting that this developmental domain is also specified at an early developmental stage. The observation that corn endosperms that spontaneously fail to form embryos also appear to lack *Esr* transcripts suggests that ESR expression depends on communication between the endosperm and the embryo. On the other hand, embryoless corn mutants form a normal embryo cavity, indicating that the formation of an embryo cavity is an intrinsic part of the endosperm developmental program. Studies using *Esr* promoters to isolate transcription factors active early in ERS differentiation should yield valuable information on the mechanisms underlying ESR cell fate specification and differentiation.

The Starchy Endosperm

The internal part of the endosperm consists of a single developmental domain in which the cells assume starchy endosperm cell fate (**Figure 4A–E**). In spite of the fact that starchy endosperm cells are one of the most intensively studied cell types, very little is known about the molecular basis for how the fate of these cells is specified. Based on the *crinkly4* and *dek1* phenotypes, in which the missing aleurone cells are replaced by starchy endosperm cells, endosperm cells appear to take on the starchy endosperm fate unless modified by signals specifying alternative cell fates. After completion of the cellularization phase (**Figure 4D**), mitotic divisions of starchy endosperm cells are at random orientation, leading to a loss of the cell file pattern established during the cellularization process (**Figure 4E**). Therefore, defining the lack of PPBs as a feature of the starchy endosperm developmental program, at least in barley, the developmental domain resulting in the differentiation of starchy endosperm cells is established at the time of completion of the cellularization process. Currently, no clues exist about the molecular basis for starchy endosperm cell fate specification. Shortly after the completion of the celullarization process, starch grains accumulate, and 5–8 days later, prolamin storage proteins appear.

The basis for the high rate of starch accumulation in the starchy endosperm relative to other tissues is not known in detail, but high influx of sucrose may be part of the explanation. Other contributing factors may be the presence of isoforms of ADPglucose pyrophosphorylase as seen in barley, which are highly active in the absence of 3-PGA, an allosteric activator of the enzyme in vegetative tissues. Recently, this isoform has been shown to be localized to the cytosol, rather than in the plastids containing the enzyme in photosynthetically active tissues.

The prolamin storage proteins, e.g., zeins in corn and hordeins in barley, are differentially produced only in starchy endosperm cells, representing the most commonly used developmental markers for this tissue. Intensive research into the molecular biology of storage protein genes has revealed that transcription is regulated by the binding of transcription factors to two common promoter upstream motifs found in several genes. The first of these is the so-called prolamin-box (P-box), for which protein factors have been identified. The second is a binding site 20 bp downstream of the P-box in the 22-kDa zein gene promoter which binds the basic leucine zipper protein Opaque2, an endosperm specific transcriptional activator of zein gene expression. The activity of these transcriptional activators leads to prolamin gene expression around 10 DAP, which in corn is approximately 6 days after completion of the cellularization phase. No information has yet been published on the mechanisms involved in the transcriptional activation of *Opaque-2*. Mutants lacking starchy endosperm cells but containing aleurone cells should yield valuable information about the initial events required for starchy endosperm cell fate specification.

List of Technical Nomenclature

Alveolus	An elongated open-ended (towards the interior) chamber formed around each nucleus during endosperm cellularization.
Angiosperm	Group of plants whose seeds are born within an ovary (flowering plants).

Anticlinal Alignment of the cell division plane at right angles to the outer surface.

Antipodal cells Three or more cells located in the upper part of the mature embryo sac.

Central cell The larger cell of the embryo sac, containing the polar nuclei and a large central vacuole.

Chalaza The basal region of ovule where the stalk unites with the integuments and the nucellus.

Coenocyte The multinucleated mass of protoplasm resulting from mitotic activity of the primary endosperm nucleus; equivalent of *syncytium*.

Cytokinesis The division of a cell's cytoplasm as opposed to its nucleus.

Cytoskeleton The intercellular network of microtubules, actin microfilaments, and intermediate filaments.

Endosperm The nutritive tissue surrounding and nourishing the embryo in seed plants.

Integument Cell layer or layers covering the nucellus and ovule, ultimately forming the seed coat.

Megagametophyte The female gametophyte, or embryo sac, located within the ovule of seed plants.

Micropyle A minute canal in the seed coat formed by extension of the integuments, through which the pollen tube enters upon fertilization.

Microtubuli (MT) One of the types of protein filaments of the cytoskeleton.

Mitogen activated protein kinase (MAP kinases) A protein involved in mediating phosphorylation signals in MAP kinase cell signaling pathways.

Nucellar projection The equivalent of pedicel in barley and wheat.

Nucellus The tissue surrounding the ovule in a seed of flowering plants, itself surrounded by one or two integuments.

Pedicel The stalk of individual flower of an inflorescence containing the ventral vasculature of the seed.

Periclinal Alignment of the cell division plane parallel to the outer surface.

Phragmoplast A spindle-shaped system of microtubules arising between daughter nuclei at telophase, within which the cell plate is formed in plants.

Polar nuclei The two nuclei that migrate to the center of the embryo sac prior to the double fertilization in flowering plants.

Preprophase band (PPB) Cytoskeletal structure formed around the cell inside the plasma membrane predicting the plane of cell plate formation during cytokinesis.

Primary endosperm nucleus The second product of double fertilization in addition to the zygote.

Radial microtubular systems (RMS) The microtubuli protruding radially from individual nuclei during endosperm cellularization.

Syncytium A multinucleated mass of protoplasm.

Synergids Two short-lived cells located in the basal part close to the egg cell in mature embryo sacs of flowering plants.

Tubulin Protein building block of microtubules.

See also: **Seed Development**: Embryogenesis; Nutrient Loading of Seeds; Oil Deposition; Physiology of Maturation; Seed Quality; Starch Synthesis; Storage Protein Deposition.

Further Reading

Becraft PW, Brown RC, Lemmon BE, Olsen O-A, and Opsahl-Ferstad H-G (2000) Developmental biology of endosperm development. In: Bhojwani SS and Soh SS (eds) *Current Trends in the Embryology of Angiosperms*, pp. 353–374. Dordrecht: Kluwer Academic.

Berger F (1999) Endosperm development. *Current Opinion in Plant Biology* 2: 28–32.

Brown R, Lemmon BE, Nguyen H, and Olsen O-A (1999) Development of endosperm in *Arabidopsis thaliana*. *Sexual Plant Reproduction* 12: 32–42.

Brown RC, Lemmon BE, and Olsen O-A (1994) Endosperm development in barley: microtubule involvement in the morphogenetic pathway. *Plant Cell* 6: 1241–1252.

DeManson DA (1997) Endosperm structure and development. In: Larkins BA and Vasil IK (eds) *Cellular and Molecular Biology of Plant Seed Development*, pp. 73–115. Dordrecht: Kluwer Academic.

Lid SE, Gruis D, Jong R, *et al.* (2002) The *defective Kernell* (*dek1*) gene required for aleurone cell development in the endosperm of maize grains encodes a membrane protein of the calpain gene superfamily. *Proceedings of the National Academy of Sciences, USA* 99: 5460–5465.

Lopes MA and Larkins BA (1993) Endosperm origin, development, and function. *Plant Cell* 5: 1383–1399.

Olsen O-A (2001) Endosperm development: Cellularization and cell fate specification. *Annual Review of Plant Physiology and Plant Molecular Biology* 52: 233–267.

Olsen O-A, Brown R, and Lemmon BE (1995) Pattern and process of wall formation in developing endosperm. *BioEssays* 17: 803–812.

Olsen OA, Linnestad C, and Nichols SE (1999) Developmental biology of the cereal endosperm. *Trends in Plant Science* 4: 253–257.

Otegui M and Staehelin A (2000) Syncytial-type cell plates: a novel kind of cell plate involved in endosperm cellularization in *Arabidopsis*. *Plant Cell* 12: 933–947.

Nutrient Loading of Seeds

J W Patrick and C E Offler, University of Newcastle, New South Wales, Australia

A J E van Bel, Justus-Liebig-University, Giessen, Germany

Introduction

Seeds are the reproductive propagules of gymnosperms and angiosperms. They are formed by maturation of the ovule following fertilization and are comprised of an embryo and nutritive tissue (filial generation comprised of endosperm or cotyledons) enclosed by a seed coat (maternal generation). This structure affords the new generation the adaptive advantages of protection from desiccation and an initial source of nutrients on germination regulated by developmental controls responding to favorable environmental cues. For angiosperms, nutrient storage occurs either in the endosperm (endospermic seeds such as cereals – see **Figure 1**) or in two large cotyledons (seed leaves attached to the embryonic axis) following consumption of the endosperm by the developing embryo (nonendospermic seeds such as grain legumes – see **Figure 1**). Seeds of today's grain crops, cereals, and grain legumes, have been bred to optimize nutrient storage to maximize yield.

During seed development, the seed coat is the delivery route for nutrients to support growth of, and storage in, the embryo and endosperm (**Figure 1**). The filial tissues, which are symplasmically isolated from the maternal tissues, collectively function as a compartment committed to storage of nutrients released from the seed coat (**Figure 1**). However, this is not the only role of nutrient loading in seed development. It is now apparent that imported nutrients also act as signals regulating developmental progression of filial cells from division to expansion and storage.

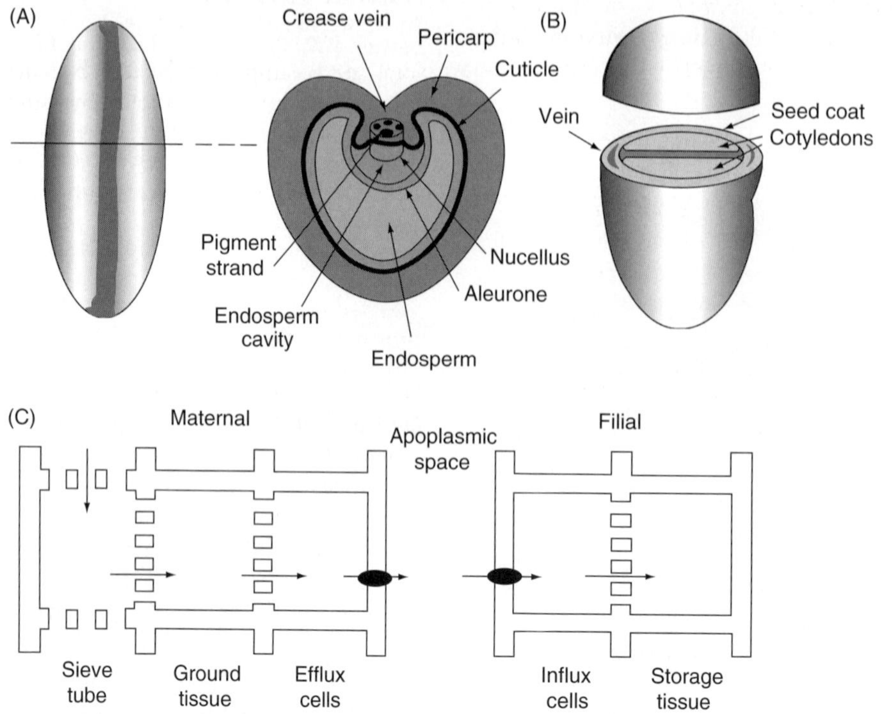

Figure 1 The anatomy of (A) wheat (endospermic) and (B) bean (nonendospermic) seeds together with (C) a diagrammatic representation of a common cellular pathway of nutrient loading in seeds. Nutrients are imported from the parent into maternal seed tissues through the phloem embedded in ground tissues that form a symplasmic compartment. The imported nutrients are effluxed from specialized maternal seed coat cells into an apoplasmic space. Specialized filial cells, located at the maternal/filial interface, support influx of nutrients from the apoplasmic space. Thereafter, the retrieved nutrients are moved symplasmically to cellular sites of growth and storage. Adapted with permission of CSIRO Publishing, from Patrick JW and Offler CE (1995) Post-sieve element transport in developing seeds. *Australian Journal of Plant Physiology* 22: 681–702.

In most sink organs (net importers of nutrients), experimental access to transport processes moving nutrients from the importing vascular cells to storage cells (postsieve element transport and see **Figure 1**) is thwarted by the intimate anatomical connections between these cells. The morphology of developing seeds, and especially those of cereals and grain legumes (see **Figure 1**), obviates this experimental impasse. In particular, the absence of symplasmic linkage between maternal and filial generations permits independent investigation of nutrient transport and metabolism in these two organs. As a consequence, considerable progress is being made in understanding nutrient loading in developing seeds.

Seed Growth and Storage

After setting, seeds enter a prestorage phase of development dominated by cell division prior to a storage phase when most of their dry matter is accumulated (**Figure 2**). Initially, cell division, cell expansion, and storage overlap sequentially. During this time, absolute growth rates of developing seeds increase to a plateau reached coincidentally with attainment of final seed volume (**Figure 2**). Thereafter, absolute growth rate remains constant for the remainder of seed fill (storage), before declining precipitously as seed maturity is approached (**Figure 2**).

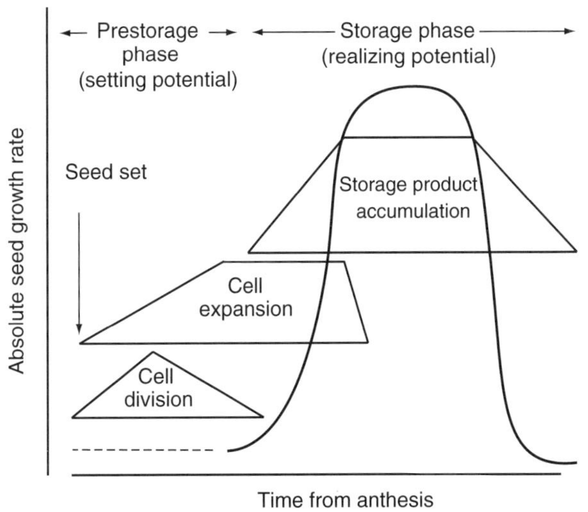

Figure 2 Diagrammatic representation of the key events in seed development in relation to absolute growth rate (dry weight gain per unit time, solid line) as a measure of seed loading of nutrients from seed set to cessation of nutrient import. Key developmental events of cell division, expansion, and storage product accumulation within the filial tissues are depicted as rates. Divergent lines denote periods of increasing rate and convergent lines denote decreasing rates.

Seeds are strong nutrient sinks. A measure of the sink strength of seeds is seen in retrospective analyses of crop yield. Increases in seed yield have resulted from a greater diversion of plant biomass into seeds rather than through an increase in overall crop biomass. This conclusion is based on strong positive correlations between seed yields and ratios of seed to shoot biomass (harvest index) across cultivars of endospermic and nonendospermic seed species. Thus, yield increase has come from greater seed biomass gain at the expense of other vegetative organs that compete for the finite nutrient pool.

Final seed biomass is a function of both filling rate and duration. These two determinants of yield are related inversely. In most circumstances, fill rates are the most important regulator of final seed biomass. Fill rate (sink strength) is a function of seed cell number (sink size) and activities of seed physiological processes (sink activity). The potential size that a seed may achieve at maturity is positively related to cell number and often inversely related to the number of seeds developing on the plant. Both seed and cell numbers are set during the prestorage phase of seed development.

Nutrients for Seed Loading Come from a Number of Sources through the Phloem

Seed set is supported largely by current photoassimilates (carbon) produced by subtending leaves (**Figure 3**) with secondary contributions from vegetative parts of inflorescences (cereals) and fleshy pod walls (grain legumes). Nitrogen for seed loading is derived largely from senescing source leaves resulting in a negative relationship between supplies of nitrogen and photoassimilates. As a consequence, carbon supply for continued seed loading in monocarpic species relies increasingly on nonstructural carbohydrates remobilized from stems and inflorescences (**Figure 3**). For example, as the photosynthetic rate of the flag leaf in wheat (*Triticum* spp.) drops, grain loading relies on remobilization of stem reserves. Under environmental stress conditions that lower leaf photosynthesis, dependence on remobilized carbon reserves increases and may contribute upward of 60% of final seed biomass.

Most nutrients are imported into seeds exclusively through the phloem (**Figures 1** and **3**) as shown by comparing solute composition of xylem and phloem saps with seed nutrient budgets. Calcium is an exception and its import pathway is uncertain. Entry through the xylem is excluded during the storage phase of seed development as excess water, imported through the phloem, is recycled via the xylem back to vegetative organs (**Figure 3**). Analyses of phloem saps

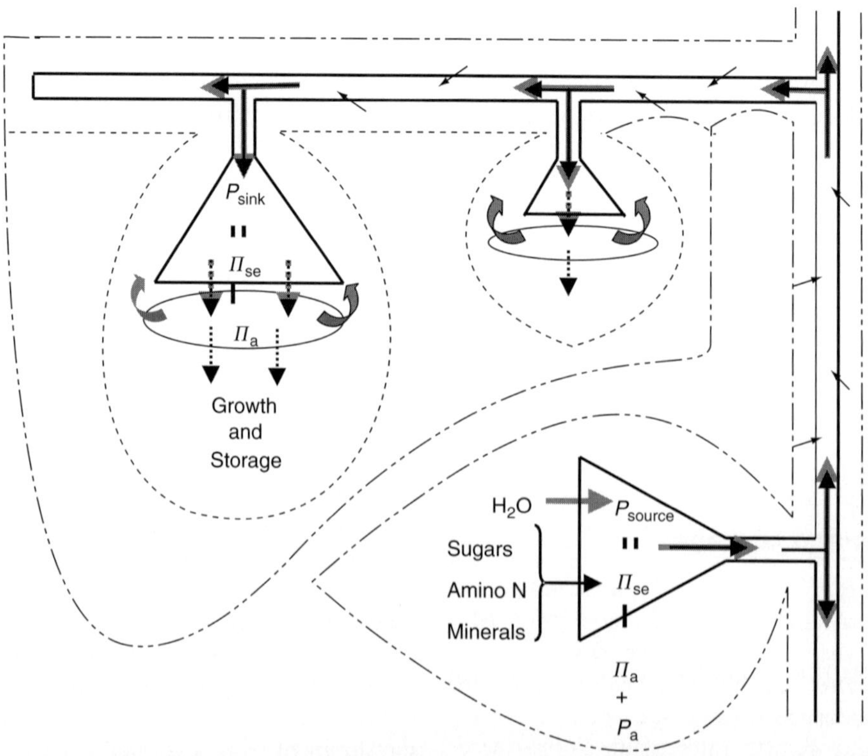

Figure 3 Pressure flow hypothesis of phloem transport highlights the key transport events contributing to seed loading of nutrients. Sugars, amino nitrogen compounds, and minerals are loaded into leaf minor veins (⟶) and result in high osmotic pressures (Π_{se}). The large difference in osmotic pressures between minor veins (Π_{se}) and surrounding leaf apoplasm (Π_a) drives water uptake (⟶). This water flow generates a high hydrostatic pressure (P_{source}) in the veins that is unaffected by variations in tension (P_a) caused by leaf transpiration. Nutrients, remobilized from stores located along the phloem path, provide an additional source of nutrients loaded into the phloem (⟶). On reaching the seed coat, nutrients are unloaded from the phloem into the seed apoplasmic space (····▸) from whence they are taken up into the filial tissues (····▸) to support growth and storage processes. The nutrient efflux lowers osmotic pressures in the phloem (Π_{se}). This, in combination with high osmotic pressures in the apoplasmic space (P_a), lowers phloem hydrostatic pressure (P_{sink}) to values below those in leaf minor veins (P_{source}). The resulting difference in hydrostatic pressures drives a mass flow of solution containing the loaded nutrients from leaf to seed (⟶). Phloem water, in excess of requirements for volume growth by seeds, is returned to the parent plant through the xylem (↻).

demonstrate that sucrose is the principal imported sugar. Nitrogen is imported in organic form and, in grain legumes, is present in phloem saps at equimolar concentrations to sucrose. Phloem sap contains a range of mineral elements with potassium predominating.

Seed Loading – Source or Sink Limited?

Seed loading may be limited by nutrient supply (source limited) or by the capacity of the developing seed to utilize available nutrients (sink limited). Manipulating production of, and intersink competition for, photoassimilates has been used to assess source- or sink-limitation (see **Table 1**). These studies demonstrate that the answer depends upon the phase of seed development. Seed set and cell division (prestorage) phases are acutely sensitive to alterations in supplies of photoassimilates and amino nitrogen compounds (**Table 1**). Sensitivity of seed set and size to photoassimilate supply is elegantly demonstrated by introgression of *Rht* dwarfing genes into genomes of modern wheat cultivars. Expression of these genes reduces stem height and thus competition for photoassimilates during the critical period of seed set and prestorage development when seed number (per inflorescence) and cell number (per seed), respectively, are determined.

In contrast, seed filling is less dependent upon current photoassimilate supplies (**Table 1**). Any shortfall in photoassimilate supplies is buffered by remobilization of stored reserves. The premise that growth of filling seeds is sink-limited is supported by other independent observations. For instance, *in vivo* absolute growth rates of cotyledons are not exceeded when they are supplied with sucrose *in vitro* at concentrations that are saturating for their growth rate. In contrast, supply of amino nitrogen compounds limits rates of seed storage protein accumulation as indicated by supplementary nitrogen doses

Table 1 Summary of experimental observations used to deduce whether seed growth is source- or sink-limited for photoassimilates and amino nitrogen compounds at key stages of development

Seed development phase	*Limitation*	*Experimental evidence*
Seed set/cell division	Photoassimilate and amino nitrogen supply	Negative: shading, defoliation; positive: CO_2 fertilization, sink removal, *Rht* dwarfing genes
Seed filling	Sink-limited for photoassimilates	Photoassimilate supply independent. *In vivo* and *in vitro* growth rates similar
	Source-limited for amino nitrogen	Increased *in vivo* or *in vitro* supplies enhance seed storage protein levels

enhancing storage protein levels. However, these dose–response relationships are not proportionate, indicating colimitation by sink processes. Cessation of seed growth appears to result from anatomical or metabolic changes in seeds rather than from a lack of nutrient supply.

Control Points of Seed Loading

Seeds Influence their Nutrient Supplies

Photosynthetic rates of source leaves often closely match changes in biomass gains by developing seeds they supply and particularly so under conditions of sink-limited growth. This relationship results from reversible feedback of sink demand on leaf photosynthesis. It is attributed to sucrose-regulated expression of genes encoding proteins engaged in photosynthesis and phloem loading. In addition, plant hormones may also play a role in this phenomenon. In contrast, demand for seed nitrogen has no influence on remobilization of reserves.

Münch Pressure Flow Identifies Control Points in the Phloem

The pressure flow hypothesis of phloem transport provides a powerful tool to identify key control points of seed loading (see **Figure 3**). The hypothesis describes import rate (R_f) of a nutrient as:

$$R_f = L_p(P_{source} - P_{sink})AC \quad [1]$$

Differences in hydrostatic pressure (P) between source (leaf minor veins) and sink ends of the phloem path, modulated by its hydraulic conductivity (L_p) and cross-sectional area (A), determines volume flow of water carrying nutrients at certain concentrations (C). At quasi water equilibrium, water potential (Ψ) of sieve element sap (se) approximates that of the surrounding apoplasm (a) such that:

$$P_{se} - \Pi_{se} = P_a - \Pi_a \quad [2]$$

and, as a consequence:

$$P_{se} = (\Pi_{se} - \Pi_a) + P_a \quad [3]$$

where Π is sap osmotic pressure and P hydrostatic pressure.

Hydrostatic pressures in sieve elements of source leaves (P_{source}) are largely set by Π_{se} resulting from phloem loading of major osmotic species into leaf minor veins (**Figure 3**). Relative contributions of solute species to phloem sap osmolality, in descending order of significance, are sugars (ca. 50% and mostly sucrose) and potassium (interchangeable with sucrose), amino compounds (ca. 15% and rise during seed filling), keto acids and inorganic compounds, and mineral elements. Variations in evaporative loss altering leaf P_a (**Figure 3**) does not appear to impact on P_{source} as phloem loading is turgor regulated.

Phloem sap concentration (C) of each transported nutrient is determined by phloem loading of nutrients in source leaves and along the axial path following remobilization of reserves (**Figure 3**). Indeed, import rates into developing seeds of minor osmotic species, such as mineral elements and individual amino nitrogen compounds, rely solely on sieve element loading. Source-limitation of amino nitrogen supply points to their phloem sap concentrations and hence their phloem loading being decisive in determining amino nitrogen compound levels imported into developing seeds, a conclusion confirmed by comparative studies with genotypic variation in seed storage protein levels.

Hydraulic conductances ($L_p \times A$) of differentiating axial phloem paths may impose limitations on nutrient supply during seed set and cell division phases of seed development. Here, importing seed tissues are composed primarily of dividing cells to which nutrients are delivered symplasmically in procambial vascular strands over distances of up to 2 mm. Hydraulic conductivities and cross-sectional areas of these strands are considerably less than those of the interconnecting sieve elements and, as such, determine path conductance (eqn [1]). A hint that path conductance could be a key factor in regulating nutrient supply to young developing seeds is the observed increase in seed set and size following exposure to atmospheres enriched in carbon dioxide (**Table 1**). Source limitation is likely to result from impedance of nutrient delivery to the developing seed sinks rather than from a lowered production of photoassimilates. The latter is far in excess of the small sink demand at seed set. Procambial strand conductances are likely to be determined by transport properties of their interconnecting plasmodesmata. Size exclusion limits of plasmodesmata are dynamic properties that are highly regulated in ways that are just beginning to be understood.

In contrast, sink-limitation for photoassimilates during seed fill (**Table 1**) implies that the conductive capacity of the vascular system (eqn [1]) does not impose any limitation on photoassimilate supply. A conclusion verified by the discovery that exudation rates from excised peduncles exceed seed growth rates by two-fold and that grain growth rates are unaffected by 50% removal of stem cross-sections supporting nutrient flows to developing seeds. Thus, seed control of import rates must be exercised through regulating hydrostatic pressures in sieve elements unloading nutrients into developing seeds (i.e., P_{sink} – eqn [1]; **Figure 3**). In this context, seed water potentials and turgor pressures are independent of the water relations of vegetative organs, and are controlled by seeds themselves. This has the consequence of biasing partitioning of phloem nutrients to developing seeds under conditions of water stress. Combined with remobilization of stem and inflorescence reserves, this outcome can result in maintenance of seed yields under drought conditions encountered during the latter part of seed fill. P_a has not been measured in developing seeds but is likely to exert minimal influence on P_{se} relative to Π_a (**Figure 3**). This has been shown by reproducing seed import rates into attached seed coat halves with solution osmolalities (Π_a) that match those of seed apoplasmic saps.

Assimilate Loading in Prestorage Phase Seeds

The expression of enzymes (invertases) that hydrolyze sucrose to glucose and fructose is a feature of seeds during their prestorage phase that may or may not be sustained throughout the remainder of seed development. In tropical grasses, an extracellular invertase is expressed in juxtaposing maternal and filial cells throughout seed development. The resulting hexoses function as developmental signals, as graphically illustrated by the compromised structural development of seed mutants of corn (*Zea mays*, maize) with perturbed invertase function. In addition, hydrolysis to hexoses in the seed apoplasm generates a steep transmembrane sucrose gradient that favors its release by facilitated diffusion from maternal tissue to the seed apoplasm (**Figures 3** and **4**).

Prestorage phase seeds of grain legumes and temperate cereals express genes encoding soluble and extracellular invertases in their maternal tissues. An extracellular invertase localized to seed coats of grain legumes may enhance unloading by facilitated diffusion. Expression of this extracellular invertase co-occurs with that of a hexose transporter expressed exclusively in abaxial epidermal cells of

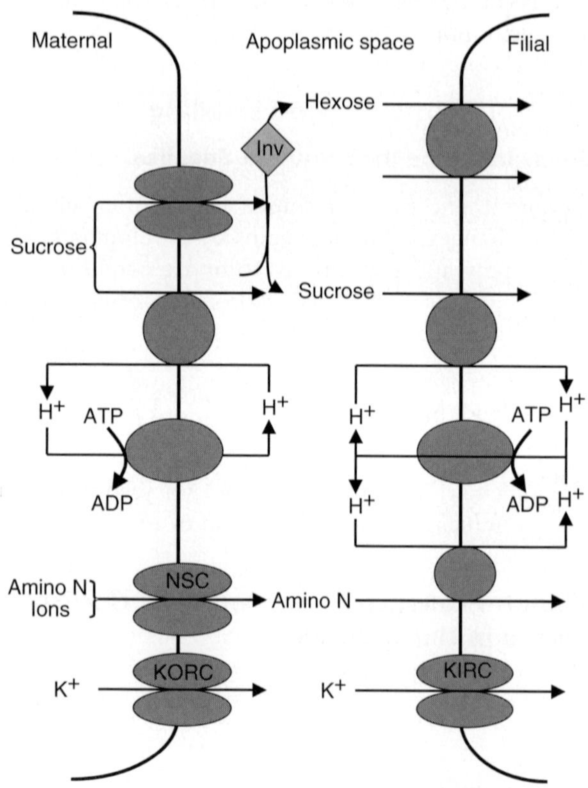

Figure 4 Membrane transport of nutrients between maternal and filial seed tissues occurs through channels (channel symbol) or carriers (carrier symbol). Efflux from maternal cells through channels has been deduced for sucrose, amino nitrogen compounds, and potassium. Amino nitrogen compounds, potassium, and other ions may be effluxed through nonselective channels (NSC) or, for potassium, through potassium outwardly rectifying channels (KORC). In some species, a component of sucrose efflux could be in antiport with protons returning down their electrochemical gradients generated by proton-pumping H^+/ATPases. Influx of sucrose, hexose, and amino nitrogen compounds into filial cells is carrier-mediated in symport with protons vectorally displaced to the apoplasmic space by proton-pumping H^+/ATPases. Potassium influx may occur through potassium inward rectifying channels (KIRC).

the developing cotyledons. Physiological studies of prestorage phase *Vicia* cotyledons have revealed the presence of two putative hexose transporters. One of these transporters exhibits kinetic properties that account for biomass gain of prestorage phase cotyledons at *in vivo* apoplasmic concentrations of hexose (100 mmol l^{-1}). In addition, hexoses have a signaling role as suggested by concurrence of high hexose concentrations with mitotically active regions of developing cotyledons and hexose induction of cell division in cultured cotyledons. Apoplasmic levels of hexoses decline when expanding cotyledons crush the innermost layer of seed coat cells expressing extracellular invertases.

For young *Vicia* cotyledons and barley (*Hordeum vulgare*) endosperm, sucrose symporter activity increases in concert with cotyledon biomass gains and is matched by a commensurate increase in *SUT* transcript levels. The latter event in *Vicia* cotyledons results from a progressive recruitment of abaxial cotyledon epidermal cells to express SUTs in a basipetal wave initiated at their pole most distal from the funiculus. Expression occurs coincidentally with the cotyledon epidermal cells contacting seed coat tissues as cotyledons expand outward. The switch to sucrose symporter activity coincides with loss of extracellular invertase activity from seed coats (see above) coupled with sucrose replacing hexoses as the predominant sugar in the seed apoplasm. The signal(s) initiating SUT expression are yet to be determined, but indirect evidence points to apoplasmic sugars. In this context, substrate derepression appears to be responsible for activation of amino acid transporters following depletion of endosperm reserves.

Sucrose symporter expression is coupled with transfer cell development in both broad bean (*Vicia faba*) and barley. Differentiation of these cells at the maternal/filial interface enhances membrane transport of nutrients, which is a common feature of many seeds, and occurs in one or both of maternal and filial tissues. The specialized transfer cells are characterized by an elaborately invaginated secondary wall that is often polarized to the direction of solute flow and a dense organelle-rich cytoplasm. The contours of the invaginated wall ingrowths are closely followed by the plasma membrane resulting in amplifications of membrane surface area of up to 20-fold. Sucrose and amino acid transporters are colocalized with H^+/ATPases to wall ingrowth plasma membranes. As for SUT expression, signals initiating the development of seed transfer cells are unknown, but hexoses are strongly implicated as a component of the signaling pathway.

Assimilate Loading during the Storage Phase of Seed Development

Symplasmic Transport in Maternal Tissues

Nutrients are imported into maternal tissues by a simple vascular system, usually comprised of a single vein that extends partially or fully along the seed axis. Maternal vascular systems in some species are elaborated by repeated branching. These modifications have been found not to influence rates of nutrient import. The ready movement of membrane-impermeable fluorescent molecules from the phloem into ground tissues demonstrates that sieve elements of maternal vascular systems interconnect symplasmically with the surrounding cells (**Figure 1**). Pore sizes of plasmodesmata interconnecting sieve elements and adjacent parenchyma are large, allowing for rapid passage of nutrients. Current evidence suggests that nutrients are moved from importing sieve elements to the surrounding vascular parenchyma cells by bulk flow of phloem sap. Subsequent transport is by diffusion through plasmodesmata interconnecting the cells of the ground tissues to those involved in membrane transport of nutrients to the seed apoplasm (**Figure 1**).

Symplasmic transport is probably rate limited by conductivities of interconnecting plasmodesmata. In this context, plasmodesmata located at the sieve element/vascular parenchyma interfaces are considered to be key sites of control of seed loading as suggested by their relatively low frequency and large drops in sucrose concentrations (200–400 mmol l^{-1}) across the cell interfaces they interconnect. In addition to delivering nutrients, bulk flow of sap from the phloem provides a mechanism to meet water requirements for volume growth in expanding seeds.

Transport through a symplasmic compartment allows for imported nutrients to accumulate in short-term storage pools and to undergo metabolic interconversion. The former functions to buffer variations in rates of nutrient import and the latter to modify imported nutrients to meet the nutritional requirements of filial tissues. For instance, the extensive postsieve element pathways in grain legume seeds, comprised of highly vacuolate cells, are estimated to meet embryo demand for 4–12 h. Starch turnover also buffers sucrose levels. Bulk flow through the phloem means that developing seeds exert little control over rates at which minor osmotic species are imported. Thus, short-term storage of these nutrients, coupled with metabolic interconversions, is critical to meet nutritional requirements of the filial tissues. In this context, imported amides undergo considerable metabolism prior to being released to the seed apoplasm. For other nutrients, such as micronutrients, only limited remobilization in maternal tissues may take place thus placing strong control over levels reaching filial seed tissues.

Membrane Transport to the Seed Apoplasm

Nutrient concentration differences at sites of symplasmic/apoplasmic exchange in maternal seed tissues provide driving forces for efflux by facilitated diffusion through carriers or channels (**Figure 4**). However, in some grain legume species, an energy-dependent sucrose efflux appears to be coupled in antiport with protons (**Figure 4**). Sucrose transport may need to be energized to generate the high osmotic concentrations in the seed apoplasm and to rapidly regulate adjustments in apoplasmic sap osmotic pressures. Photoassimilate release appears to occur from specialized cells located at or near the maternal/filial interface (**Figure 1**).

Less is known about release of amino nitrogen compounds, but for pea (*Pisum sativum*) seed coats their release may occur through nonselective pores (**Figure 4**). A nonselective ion channel has been detected in release cells of French bean (*Phaseolus vulgaris*) seed coats that is permeable to a wide range of electrolytes including large organic ions such as glutamate. In addition, potassium channels have been detected in protoplasts derived from release cells of bean seed coats (**Figure 4**).

Transporter and channel activity in maternal seed tissues result in characteristically high levels of nutrients being deposited in seed apoplasms, and combined these reach osmotic concentrations of 300–400 mOsmol. Major osmolytes in seed apoplasms are sugars (sucrose or hexose depending upon whether an extracellular invertase is present), amino nitrogen compounds, and potassium (**Figure 4**). The spectrum of amino nitrogen compounds changes during seed development. Following a decrease in osmotic concentrations from prestorage phase seed, apoplasmic nutrient levels remain steady across seed fill and during the diurnal cycle. This suggests a tight regulation of apoplasmic nutrient levels through integration of nutrient transport to and from the seed apoplasm (see below).

Membrane Transport from the Seed Apoplasm

During storage product accumulation, sucrose is taken up from seed apoplasms of grain legumes and temperate cereals. In tropical cereals, it undergoes extracellular hydrolysis and is accumulated as free hexoses (**Figure 4**). Sucrose influx has biochemical and biophysical characteristics consistent with sucrose/H^+ symport in seeds of grain legumes and temperate cereals (**Figure 4**). Sucrose/H^+ symporter (SUT) presence in seeds has been confirmed by expression of SUT genes in broad bean (*VfSUT1*), pea (*PsSUT1*), rice (*Oryza sativa*; *OsSUT1*), barley (*HvSUT1*), and wheat (*TaSUT1*). The mechanism of sugar uptake by endosperm of tropical cereals is less certain, but may involve hexose/H^+ symport (**Figure 4**).

High resistances are encountered for nutrient entry into the apoplasm of filial tissues. Thus, it is not surprising to find that sucrose symporters are colocalized with H^+-ATPases to the plasma membranes of cells bounding this interface (**Figure 1**). These transport cells are characterized by a dense cytoplasm, lack of storage product accumulation and, in some species, have differentiated into transfer cells with wall ingrowths polarized toward the maternal tissues. Accumulated photoassimilates are moved symplasmically through high frequencies of plasmodesmata for storage in endosperm cells or cotyledon storage parenchyma cells (**Figure 1**).

A saturable low-affinity and high-capacity amino acid uptake system becomes apparent as storage product accumulation progresses (**Figure 4**). In this context, a number of low-affinity and broad-specificity amino acid transporters (AAPs) have been cloned from seed cDNA libraries of *Arabidopsis*, pea, and broad bean. The AAPs are well suited for transporting the wide range of amino acids and amides released from seed coats. Their coordinate expression with storage protein accumulation is indicative of a primary role in supplying substrate for storage protein biosynthesis. Potassium channels have been detected in French bean (*P. vulgaris*) cotyledons (**Figure 4**).

Integration of Phloem Import, Seed Loading, and Storage Functions

Integration between these processes could be achieved by feed-forward and feedback signals operating along the seed-loading transport pathway to regulate plasmodesmal and membrane transport with metabolic accumulation in the filial tissues. Activity of metabolic machinery responsible for nutrient accumulation appears to be regulated through feed-forward control by nutrients (sugar; nitrogen) acting as signals to regulate gene expression. Quantitative estimates of plasmodesmal numbers and cross-sectional areas suggest that symplasmic transport is unlikely to rate limit nutrient transport by diffusion in filial tissues. However, activities of amino acid and sucrose transporters appear to be regulated by substrate feedback derepression providing a positive link with metabolic sequestration (**Figure 5**). A key role for sucrose/H^+

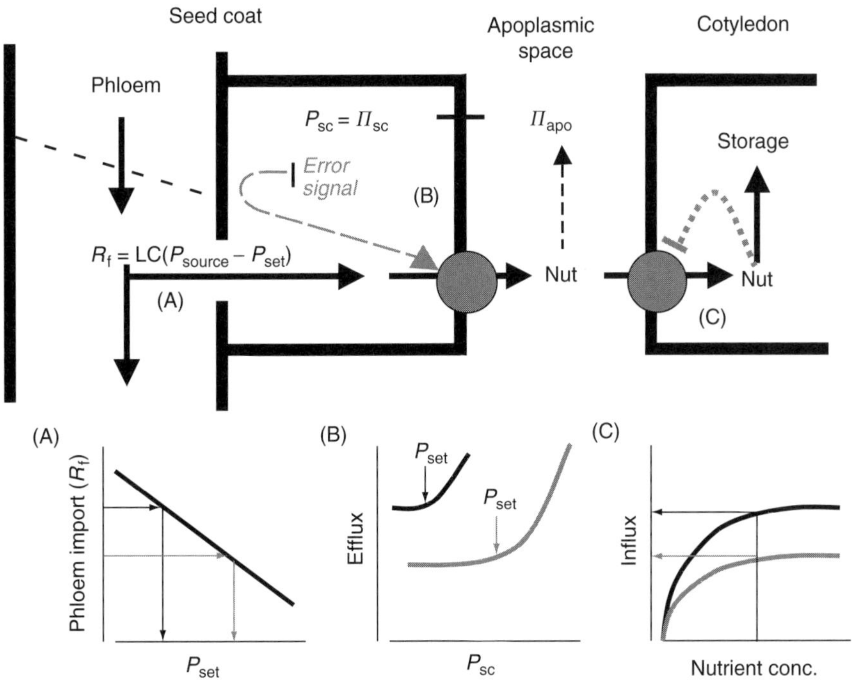

Figure 5 A mechanistic model describing the integration of nutrient transport to and within developing legume seeds. The import of nutrients through the phloem (A) and efflux from the seed coat to the seed apoplasm (B) is mediated by a turgor-dependent efflux mechanism (regulated by a turgor-homeostat) and uptake of nutrients by the cotyledons (C). Metabolically induced changes in nutrient pool sizes in the cotyledon symplasm could regulate the activities of transporters located in the plasma membranes of the cotyledon dermal cell complex ((C) noting increase from light to dark curve) by derepression of transporter gene expression (broken line). Nutrients account for the osmotic content of the apoplasm solution and their pool size is relatively small (turnover time <1 h). Thus, an increased rate of nutrient withdrawal from the seed apoplasm by the cotyledons will result in an immediate decrease in the apoplasm osmotic concentration. Since the osmotic difference between the seed apoplasm (Π_{apo}) and coat (Π_{sc}) is relatively small (0.1–0.2 MPa), an incremental decrease in the osmolality of the apoplasm solution will elicit an amplified increase in the turgor pressure of the seed coat (P_{sc}, (B)). Such displacement of coat turgor above the turgor set point (P_{set}) results in an error signal ((B) and see broken line), the magnitude of which causes an immediate compensatory increase in assimilate efflux (B). The altered nutrient efflux acts to maintain the apoplasm osmolality independent of the enhanced flux through it (C). Consequently, the increased potential for nutrient uptake by the cotyledons can be fully realized ((C) and note dark curve). In the short-term (minutes), the other consequence of the turgor homeostat is that seed coat turgor and hence phloem import by bulk flow driven by the turgor difference between the sieve tubes and unloading cells ($P_{st} - P_{sc}$, (A)) is maintained. Under conditions where cotyledon demand is sustained, the turgor set point of the seed coat adjusts downward (B). The turgor adjustment results from decreases in the osmotic concentration of the coat symplasm, whilst that of the seed apoplasm is homeostatically maintained (C). The decrease in the turgor set point of the efflux cells serves to enhance the pressure difference between these cells and the importing sieve elements. As a result, the rate of phloem import into the seed coat is increased to a level commensurate with the new rates of efflux from the seed coat and uptake by the cotyledons ((A) and note change from light to dark arrows). L, hydraulic conductance; C, nutrient concentration. Adapted with permission of CSIRO Publishing, from Patrick JW and Offler CE (1995) Post-sieve element transport in developing seeds. *Australian Journal of Plant Physiology* 22: 681–702.

symporters in storage product accumulation has been confirmed by overexpression of StSUT1 in pea cotyledons enhancing their growth rates.

Symplasmic and membrane transport are arranged in series in maternal seed tissues (**Figure 1**). Plasmodesmal conductances, particularly for those at sieve element/vascular parenchyma interfaces could play significant roles in regulating sieve element unloading in wheat grains and indeed nutrient supply to the filial tissues. For developing seeds of grain legumes, transport control appears to be shared between symplasmic and membrane transport with regulation of plasmodesmal conductances not understood. However, there is some evidence to suggest that phloem import into, and nutrient release from, maternal tissues are regulated by feedback control driven by filial nutrient demand signaled through shifts in pool sizes of apoplasmic nutrients (**Figure 5**). Alterations in rates of nutrient withdrawal are rapidly detected by changes in their apoplasmic pool size as these are small with high turnover rates. Small differences in nutrient concentrations between maternal symplasm and apoplasm ensure that small changes in apoplasmic nutrient levels cause large changes in their transmembrane concentration (and osmotic) differences.

Release by facilitated diffusion from maternal seed tissues provides a mechanism specific for each nutrient to respond to filial demand. However, for major osmolytes, a decline in their apoplasmic pool sizes translates into significant shifts in osmotic concentration that would result in elevated turgor pressures of maternal cells thus slowing phloem import (**Figure** 5). Grains of temperate cereals avert this outcome by the presence of apoplasmic barriers separating their endosperm cavities from maternal phloem. Under these conditions, turgor alterations in the nucellar projection transfer cells will not influence pressures in the crease phloem. This is consistent with phloem import being unaffected when endosperm cavities of attached wheat grains are perfused with solutions of differing osmolalities. An alternative strategy is based around turgor-dependent release of nutrients from maternal tissues of tropical cereals and grain legumes ensuring turgor maintenance and hence phloem import rates (**Figure** 5).

Recirculation of Phloem Water and Not Loaded Nutrients

Sites at which phloem-imported water is recycled and the export routes followed have not been resolved in developing seeds. However, it is clear that loaded nutrients are not returned to the parent plant in recycled water or by any other means. This outcome is considered to be achieved by strategically positioned semipermeable apoplasmic barriers in maternal seed tissues and/or efficient retrieval mechanisms located along xylem export routes. Examples of the latter include diminished xylem paths (funicles of grain legumes) or complete discontinuities (pedicels of wheat grains), with closely juxtaposed xylem and phloem elements with or without intervening transfer cells. These structural modifications probably contribute to the independence of seed water relations from the vegetative plant body; a feature central to regulating phloem import (see above).

Conclusions

Substantial progress is being made in understanding membrane and, to a lesser extent, plasmodesmal transport processes contributing to seed loading. For membrane transport, the exciting frontiers are to isolate and functionally characterize membrane proteins responsible for nutrient release to the seed apoplasm. Of equal importance is to understand regulation of plasmodesmal transport of nutrients in the maternal seed tissues. Knowledge about recycling of phloem water is fragmentary and would repay further study, as it is feasible that this process could rate limit nutrient import through the phloem.

As the information base on seed loading grows, opportunities to modify the process will become technically feasible. For instance, genetic manipulation of plasmodesmal function would offer opportunities to enhance symplasmic conductance to amplify photoassimilate supplies at seed set to reduce rates of flower, fruit, and seed abortion. This provides considerable potential for improvement of yields of grain crops. The opportunity currently exists for overexpression of transporter proteins to increase rates of seed fill and possibly final seed size.

List of Technical Nomenclature

Apoplasm	The extracellular spaces accessible to solutes, largely comprising the fluid-filled spaces in cell walls and surface films of intercellular spaces.
Embryo	A young plant developed in the seed before germination.
Endosperm	Triploid nutritive tissue in seeds resulting from fusion of a sperm nucleus with the two polar nuclei of the ovule.
Filial	The next generation resulting from fertilization of male and female gametes.
Hydraulic conductivity	Capacity of a structure to support mass flow of water per unit cross-sectional area for a given hydrostatic pressure difference between the ends of the transport pathway.
Osmotic pressure	Potential pressure that can be developed as a result of separating pure water from a solution containing dissolved solutes; at low solute concentrations, osmotic pressure is a direct function of the combined solute concentration.
Ovule	The female reproductive structure that on fertilization forms the seed.
Plasmodesmata	The plasma membrane-lined micropores linking adjacent cells.
Sink	An organ/tissue that is a net importer of a specified nutrient.
Source	An organ/tissue that is a net exporter of a specified nutrient.
Symplasm	The cytoplasmic continuity of adjoining cells interconnected by plasmodesmata.
Turgor pressure	The positive hydrostatic pressure in plant cells resulting from the cell walls resisting volume change brought about by water uptake.

Water potential Free energy state of water in a system relative to pure water at standard temperature and pressure measured in pressure units derived from the chemical potential of water divided by the molal volume of water.

See also: **Flowering and Reproduction**: Flower Development. **Photosynthesis and Partitioning**: Sources and Sinks. **Seed Development**: Artificial Seeds; Embryogenesis; Endosperm Differentiation; Seed Production; Starch Synthesis; Storage Protein Deposition. **Water Relations of Plants**: Basic Water Relations; Phloem.

Further Reading

Patrick JW and Offler CE (1995) Post-sieve element transport in developing seeds. *Australian Journal of Plant Physiology* 22: 681–702.

Patrick JW and Offler CE (2001) Compartmentation of transport and transfer events in developing seeds. *Journal of Experimental Botany* 52: 551–564.

Weber H, Heim U, Golombek S, Borisjuk L, and Wobus U (1998) Assimilate uptake and the regulation of seed development. *Seed Science Research* 8: 331–345.

Wobus U and Weber H (1999) Sugars as signal molecules in plant seed development. *Biological Chemistry* 380: 937–944.

Starch Synthesis

K Denyer, John Innes Centre, Norwich, UK
S M Smith, University of Edinburgh, Edinburgh, UK

The Physiological Importance of Starch in Seeds

Starch is accumulated in the seed as a store of carbon to be used to fuel the postgerminative growth and development of the young plant until it becomes fully autotrophic. Seeds also store carbon in other forms, most commonly as oil and protein. The seeds of different species store different proportions of these various storage compounds. Some species, such as the grasses (e.g., corn, wheat, rice) and grain legumes (e.g., peas and some beans), have seeds with a very high proportion of starch. For example, starch accounts for approximately 70% of the dry weight of a mature wheat (*Triticum aestivum*) seed and 50% of that of a pea (*Pisum sativum*) seed. However, there is evidence to show that the amount of starch accumulated by such starch-storing seeds may be greater than is strictly required for postgerminative growth. First, many seeds will germinate and grow successfully even if a large part of the storage tissue is damaged or removed. Second, the *rug3* mutant of pea accumulates almost no starch in its seeds but these seeds still germinate and grow successfully and produce normal plants. Third, many cultivated varieties of cereals and grain legumes have seeds that are larger than those of their wild relatives. This is due to plant breeding where, as part of the effort to increase yield, plants with larger seeds have been selected. In other words, much of the starch in these seeds is there for our benefit.

Species that have seeds with high proportions of oil at maturity may accumulate starch transiently during seed development. For example, *Brassica napus* and *Arabidopsis thaliana* accumulate starch during the early phase of seed development when oil accumulation is beginning. Later in development, this starch is broken down so that there is essentially no starch in the mature seeds. It has been postulated that the temporary reserve of starch in these seeds provides a store of carbon that can be converted into oil later in development.

Different species store starch in different parts of the seed. In a cereal grain, which is a single-seeded fruit, starch accumulates in the cells of the developing endosperm. In other seeds, such as pea seeds, the main storage tissues are the cotyledons of the embryo. The endosperm in pea seeds exists, but only transiently, very early in seed development and it does not accumulate starch. Starch may also accumulate in the maternal tissues closely associated with the seed. For example, cells in the pericarp of a cereal grain (one of the maternal tissues surrounding the embryo and endosperm) accumulate starch but only early in seed development. The pericarp of a mature cereal grain contains essentially no starch. Similarly, starch accumulates transiently in the seed coats of developing pea seeds.

The Structure of Starch Granules

Starch accumulates in the form of granules, which vary enormously in shape and size between species. Some species even have different sorts of starch granule within the same cell. For example, in the cells of the endosperm of wheat and related species, there are both large, disk-shape granules and smaller, spherical granules. The number of starch granules in a plastid (the organelle in which starch is synthesized; see below) also varies from one (e.g., pea embryos) to several hundreds (e.g., grains of Inca wheat, *Chenopodium quinoa*). There is a wide range of variation in the number of granules per plastid even between related species. For example, within the grasses, oat (*Avena sativa*) and rice (*Oryza sativa*) endosperms have compound starch granules made up of several separately initiated granules whereas corn (*Zea mays*; maize) endosperm typically has single starch granules per plastid.

Despite large variations in granule morphology, the internal organization of starch granules is less variable (**Figure 1**). Rings of hard (semicrystalline, not readily digested by amylase) and soft (relatively disorganized, easily digested by amylase) material can be seen in starch granules when these are observed under the light microscope or when fractured granules are etched by treatment with α-amylase and viewed by scanning electron microscopy (SEM). What we see as rings are actually concentric, three-dimensional shells. These shells, often referred to as growth rings, are 100 nm or more apart. It has been suggested that one growth ring forms per day although there is no direct evidence for this.

High-magnification microscopy of various kinds (SEM, transmission electron microscopy (TEM) and atomic force microscopy (AFM)) has shown that growth rings appear to be composed of elongated structures known as blocklets that are stacked on top of one another. These occur in both the hard and soft layers of the growth ring, although the blocklets in the hard, crystalline layers are generally larger than those in the soft layers. The diameter of the blocklets varies between 20 and 500 nm depending on the source of the starch. Studies of partially degraded starch granules suggest that molecules of the major glucose polymer in starch (amylopectin; see below) are organized into "superclusters" and these may relate to the blocklets observed by microscopy.

Repeating layers separated by a distance of approximately 9 nm are seen within blocklets in high-resolution TEM and repeats of this length are also seen in small angle X-ray scattering profiles. This repeat distance is remarkably constant regardless of the species from which the starch was extracted and despite variations in the composition of starch from different mutants. This structural periodicity is thought to be due to the molecular structure of the major starch polymer, amylopectin.

Amylopectin is a large, multiply branched glucose polymer. Details of its chemical structure are given elsewhere in this volume (*see* **Primary Products: Starch**). Most of the branches of amylopectin consist of short chains of glucose (containing approximately 6–20 glucose units). These short chains are arranged in clusters within the polymer with longer chains of glucose extending between and joining adjacent clusters (**Figure 1**). Inside a hydrated starch granule, amylopectin exists in a semicrystalline state. Within the blocklets or superclusters of amylopectin, adjacent amylopectin molecules are thought to lie parallel to one another with their clusters of short chains in register. The chains within the clusters lie approximately along the radius of the starch granule with their nonreducing ends towards the granule surface.

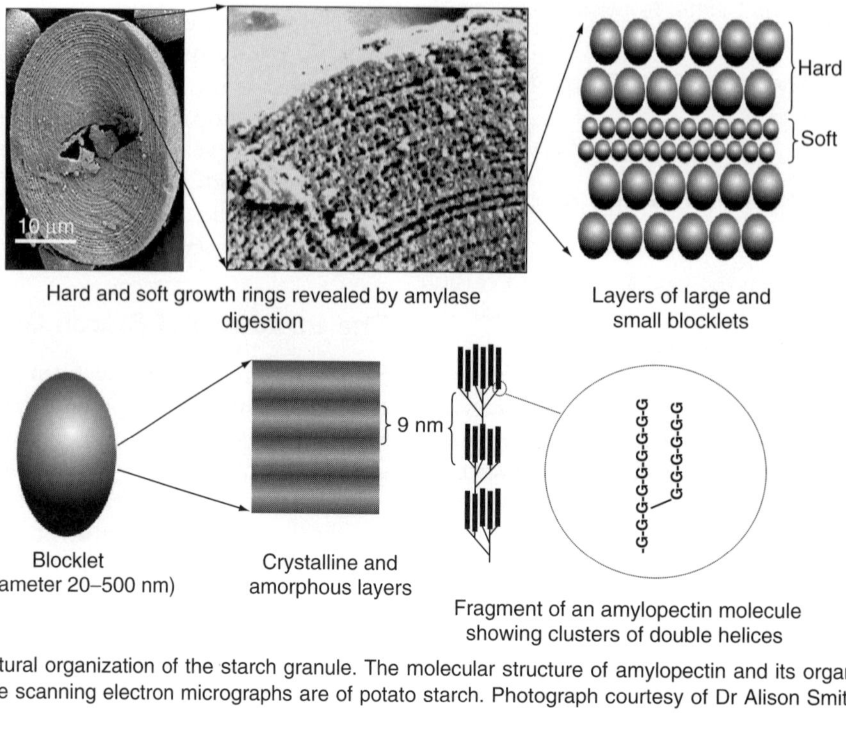

Figure 1 The structural organization of the starch granule. The molecular structure of amylopectin and its organization within the granule is shown. The scanning electron micrographs are of potato starch. Photograph courtesy of Dr Alison Smith.

Pairs of short chains form double helices which pack together to form crystalline arrays. Between the crystalline layers are noncrystalline (amorphous) regions containing the branch points that join together the chains within the clusters. The 9 nm repeats are explained by the alternating crystalline and amorphous layers in the amylopectin molecule.

The physical properties of the semicrystalline amylopectin molecule resemble those of a synthetic side-chain liquid-crystal polymer. Models using this concept have helped to explain the physical properties of starch and the way in which starch granules gelatinize when heated in different solvents. The relevance of this to the cooking properties of starch is discussed elsewhere (*see* **Primary Products**: Starch).

Two types of amylopectin crystals called A-type and B-type have been found in plant starches (**Table 1**). The chains within the amylopectin molecule in A-type crystals are more densely packed than in the B-types. Granules contain only A-type crystals (A-type starch, found in, e.g., most normal cereal seeds), only B-type (B-type starch, found in, e.g., seeds of high-amylose corn mutants) or both types of crystals (C-type starch, found in, e.g., pea seeds). The type of crystal depends upon the arrangement of the branch points within the amylopectin molecules and this is reflected in the chain-length profile of the amylopectin. Starches with amylopectin with relatively short average chain lengths have A-type crystals. In these crystals, the branch points are found in both the crystalline and noncrystalline regions of the amylopectin clusters. B-type starches have amylopectin with longer average chain lengths and most of the branch points are within the noncrystalline region of the clusters. The differences between the amylopectin molecules that form A-type and B-type crystals are very subtle. For example, the differences between them in the average chain lengths can be as little as one glucose unit. However, the nature of the crystals has a profound effect on properties of starch such as its susceptibility to hydrolysis by α-amylase.

Starch granules also contain a second sort of glucose polymer called amylose. Amylose is a relatively infrequently branched molecule compared to amylopectin and is smaller in mass (*see* **Primary Products**: Starch). Amylose molecules are found throughout the starch granule. Those molecules in the center of the starch granule are larger and more concentrated than the amylose molecules at the periphery of the granule. Amylose is not crystalline in the granule even though purified amylose readily crystallizes from solution. However, it is not clear whether amylose exists only in the amorphous regions between blocklets or whether it is also dispersed amongst the amylopectin molecules within the semicrystalline blocklets.

In the seeds of all wild plants, the ratio of amylose to amylopectin is fixed within fairly narrow limits, between approximately 75% amylopectin:25% amylose and 66% amylopectin:33% amylose. However, mutants with very different ratios of amylose to amylopectin exist. For some of these mutants, the distinction between amylose and amylopectin becomes blurred. For example, abnormal amylopectin molecules with fewer than normal branches take on some of the physical characteristics of amylose molecules. Such starches are sometimes referred to, perhaps misleadingly, as "high-amylose" starches.

There are also mutants that have no amylose in their starch granules. They are sometimes referred to as "waxy" mutants since the original corn mutant lacking amylose has seeds with a waxy appearance. These zero-amylose starches exist in a number of different species and in all cases, the mutants have a near-normal amount of starch with apparently normal internal and external granule structure. This shows that amylose is not required for starch granule synthesis or structural integrity. Its role is therefore unclear. However, the fact that nature has not exploited more fully the possible variation in the ratio of amylose to amylopectin suggests that there is some as yet unknown advantage to having both types of polymer in the normally observed ratio in a wild-type plant.

How Starch is Synthesized

The Precursors and their Utilization

Starch in seeds is synthesized from carbohydrate (usually sucrose) imported from the photosynthetically active parts of the plant. The sucrose is transferred from the cells of the maternal tissues to the cells of the offspring via the apoplast (extracellular space). After import into the cells of the

Table 1 Types of amylopectin crystals

Starch type	*Crystal type*	*Amylopectin chains*	*Example*
A	A	Short	Cereal seed
B	B	Long	Potato, high-amylose corn
C	A + B	Both	Pea seed

developing seed, some of the imported sucrose may be temporarily stored but most is broken down and converted into hexose phosphates. This pool of metabolites in the cytosol provides carbon for many biochemical pathways in the developing seed. In the cells of the storage tissues, it provides the carbon building-blocks for the synthesis of starch and other storage products, for the synthesis of structural materials in the seed such as cell walls and membranes, and for respiration which provides the energy and reducing power for these synthetic processes.

Starch is synthesized inside organelles called plastids. Those occurring in storage tissues are usually referred to as amyloplasts while those in green tissues are chloroplasts. Green seeds like those of oilseed rape (*B. napus*) or pea contain small amounts of chlorophyll in the plastids of their storage tissues. However, these plastids do not receive enough light through the walls of the seed and the fruit to support photosynthetic carbon fixation to any significant degree. Thus plastids of seed storage tissues rely on a supply of carbohydrates from the cytosol from which to synthesize starch.

The metabolites imported into plastids to support starch synthesis vary from species to species. In most species, it is a hexose phosphate, either glucose 1-phosphate or glucose 6-phosphate. Inside the plastids, glucose 1-phosphate is converted into a high-energy form of glucose called adenosine 5′-diphosphoglucose (ADPG) by the enzyme ADPG pyrophosphorylase. In the direction of ADPG synthesis, this reaction also consumes ATP and generates pyrophosphate. It is the glucose moiety of the ADPG molecule that is used to build the chains of glucose of which the starch polymers are composed. Starch synthase (SS) catalyzes the transfer of the glucosyl units to the growing polymer chains.

In developing pea seeds, plastids specifically import glucose 6-phosphate and not glucose 1-phosphate to support starch synthesis (**Figure 2**). We know this because there is a mutant of pea that lacks plastidial phosphoglucomutase (PGM), the enzyme that converts glucose 6-phosphate to glucose 1-phosphate. The mutant is unable to synthesize starch despite the fact that glucose 1-phosphate is present outside the plastids. In the endosperm of all grasses (including our commonly cultivated cereals), some hexose phosphate is imported into plastids to support starch synthesis but the major imported metabolite is ADPG that is made in the cytosol. Evidence for this comes from mutants that lack the cytosolic enzyme responsible for the synthesis of ADPG in the cytosol or lack the transporter protein capable of importing ADPG across the plastid envelope. As far as we know, only grasses can make ADPG both inside and outside the plastids and they do so only in the cells of the endosperm. The reasons for this difference between species in the source of ADPG for starch synthesis are not presently understood.

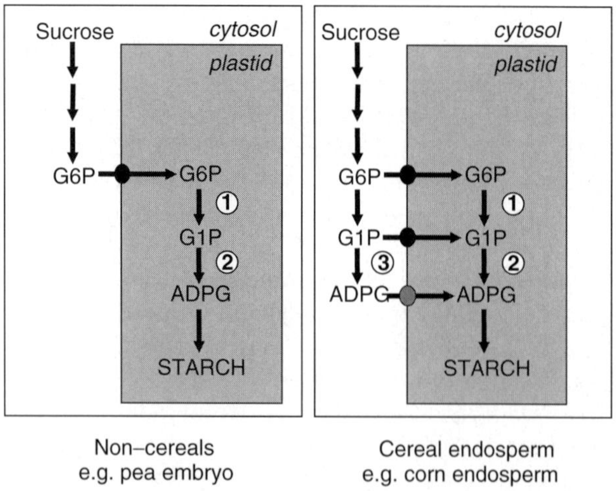

Figure 2 The pathway of conversion of sucrose to starch in typical cells from cereal and noncereal seeds. G6P, glucose 6-phosphate; G1P, glucose 1-phosphate; ADPG, ADPglucose. The following enzymes are shown: 1, plastidial phosphoglucomutase; 2, plastidial ADPG pyrophosphorylase; 3, cytosolic ADPG pyrophosphorylase. The black ovals represent transporters in the inner plastidial envelope.

Growth of the Starch Granule

Starch granules are synthesized within the stroma of the plastid. Although a rudimentary thylakoid membrane system has been observed in some seed plastids, it is not known to have a role in starch synthesis. Most of the enzymes involved in starch synthesis are soluble, stromal enzymes although a subset of these enzymes is also found inside the starch granules themselves. Early experiments using radioactively labeled precursors showed that synthesis occurred by deposition of material on the surface of the granules. However, exactly how starch granules are initiated and grow is still not known. Starch polymers could be synthesized entirely in a soluble form and then deposited on the granule surface where they could partly crystallize. Alternatively, starch polymers protruding from the granule surface and already partly crystalline could be further elongated. Both of these processes may occur to some extent. As well as growth at the surface, it has been shown recently that granules are permeable to small metabolites and some synthesis and/or further elaboration of starch polymers occurs inside the granules by the granule-bound enzymes (see “The Synthesis of Amylose” below).

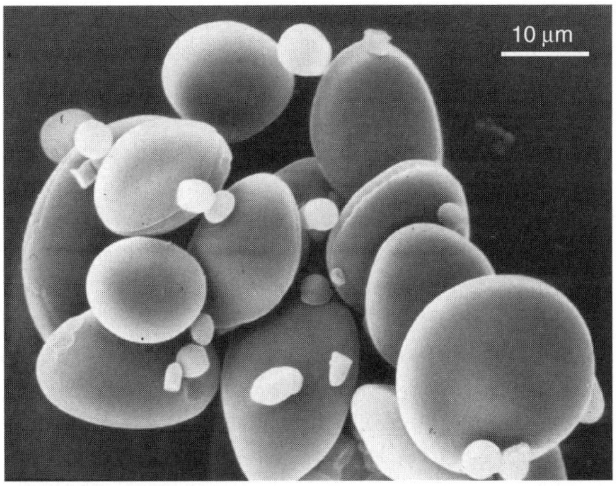

Figure 3 Starch granules from wheat seeds observed using a scanning electron microscope. Note the two types of granules: large flattened disk-shaped A-granules, each with an equatorial groove, and smaller, near-spherical B-granules.

The deposition of material on the granule surface in a uniform manner would result in a spherical starch granule. Such granules occur but the granules of many species are far from spherical in shape which suggests that the site of granule growth is regulated in some way. The growth of the A-type starch granules in the seeds of wheat and related species provides an example of directed deposition of material on the granule surface (**Figure 3**). These granules grow predominantly from two ridges around the equator, resulting in a flattened, disk-shaped granule with an equatorial groove. How the location of granule growth is determined is not understood. Within the stroma of the plastid there are internal membranes, tubules, and cytoskeletal structures. Some or all of these may be involved in the spatial coordination of polymer synthesis and/or its deposition.

The Synthesis of Amylopectin

A minimum of two types of enzyme are required to build a branched glucose polymer such as amylopectin. First, a SS that can elongate the polymer by the sequential addition of single glucose moieties to its nonreducing ends is needed. Second, a starch branching-enzyme (SBE) is needed that can cleave the α-1,4 linkage between glucose units in a chain and then add the severed chain via an α-1,6 linkage to another part of the molecule (or to a different molecule). Both of these types of enzymes exist in plants in multiple isoforms. There are five isoforms of SS encoded in the *Arabidopsis thaliana* genome that represent five distinct classes of isoform (**Figure 4A**). These are granule-bound starch synthase (GBSS-I), SS-I, SS-II, SS-III, and a fifth isoform that is

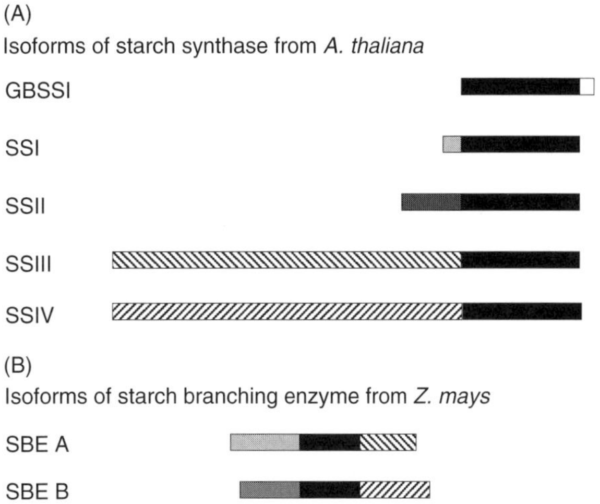

Figure 4 The primary structures of starch synthase and branching enzymes. The length of the bars indicates the relative number of amino acids in the mature proteins. For each enzyme, the major structural domains are indicated by areas of different shading. The most highly conserved domains in both enzymes are shown in black although these differ between SS and SBE. Domains that do not share appreciable sequence similarity are shaded differently. The conserved (black) domains in the starch synthases are responsible for catalytic activity. In the starch branching enzymes, the domains shown in black contain four highly conserved regions that are present in all members of the α-amylase superfamily of enzymes.

sometimes referred to as SS-V but which is referred to here as SS-IV. Evidence, based principally on comparisons of amino acid sequences, suggests that all five classes of SS isoform are probably present in all plants. Similarly, there are two classes of starch-branching enzyme (SBE A and SBE B) that probably also exist in all plants (**Figure 4B**).

The conservation between species of these classes of SS and SBE suggests that each has a unique role in starch synthesis. Evidence is accumulating from studies of mutant and transgenic plants and from studies of the kinetic properties of the isolated enzymes that supports this idea. For example, the role of GBSS-I in amylose synthesis (see below) is clearly unique amongst the isoforms of SS and is conserved between species. Comparison of amylopectin chain-length profiles of transgenic potatoes (*Solanum tuberosum*) with reductions in the amount of either SS-II or SS-III also supports the idea that these isoforms make distinct contributions to the synthesis of starch. Biochemical studies of isolated isoforms of SBE show that these differ significantly in their substrate specificities, with those of the SBE B class preferring long-chain amylose as a substrate. Thus, the A and B forms of SBE probably also have different roles.

In addition to the multiple classes of isoform of SS and SBE, some species have more than one gene encoding isoforms of a particular class. In corn there are two forms of SS-II. SS-IIa is expressed most strongly in the endosperm and SS-IIb in the leaves. There are also multiple forms of SBE-II (**Table 2**) and GBSS-I in different species that are expressed in different tissues. SBE-IIb in barley (*Hordeum vulgare*) is endosperm specific but SBE-IIa is also expressed in other parts of the plant. GBSS-Ia is expressed strongly in pea embryos but very little in pea pods whereas GBSS-Ib is expressed in pods but not very much in embryos. In wheat, GBSS-I is expressed in the endosperm and in pollen grains and a second isoform of GBSS-I, encoded by another gene, is expressed in the pericarp but not in the endosperm or pollen. It is expected that these different, tissue-specific isoforms have broadly similar roles in starch synthesis but subtle differences in their activities cannot be excluded.

Different isoforms of SS and SBE are expressed at different times during seed development. For example, the two isoforms of SBE in developing pea, barley, and wheat seeds are differentially expressed. In these species, the SBE B isoform (also known as SBE-II in pea, SBE-I in barley and as WBE-I in wheat; see **Table 2**) is preferentially expressed late in development whereas the SBE A isoform accounts for most of the SBE expression in young seeds. In developing pea seeds, GBSS-I is expressed later in development than SS-II. These different patterns of expression probably account for differences in the physical and chemical properties of starches from immature and mature seeds. For example, the amylose content of starch (as a percentage of the mass of starch) increases gradually throughout seed development in several species. In pea seeds (and probably in other seeds too), the initiation of starch synthesis occurs in different cells at different times reflecting the different developmental ages of cells in the embryo. This indicates that there are spatial as well as temporal differences in expression of starch biosynthetic enzymes in developing seeds.

Even though there are multiple isoforms of SS and SBE with different roles, it is thought likely that other enzymes are required to construct a molecule as complex as amylopectin. There are a number of what are traditionally regarded as starch "degrading" enzymes present in plastids during starch synthesis. Some of these could modify the starch polymers during their synthesis and may actually be required to synthesize a crystallization-competent amylopectin molecule. For example, studies of mutants of several species that lack isoamylase, a starch-debranching enzyme that cleaves the α-1,6 bonds in amylopectin, have shown that this enzyme is required for normal starch synthesis (**Table 3**). In the absence of isoamylase, reduced amounts of starch often in the form of more numerous, smaller granules together with a soluble form of starch called phytoglycogen are commonly observed. Various theories based on the phenotypes of such mutants have been proposed to explain the role of isoamylase in starch synthesis. It could take part in the determination of the structure of amylopectin by trimming away some of the branches made by SBE. It could help to destroy soluble glucans that would otherwise accumulate at the expense of starch granules. It could be involved in the initiation of starch granules and glucan particles and influence the number of these that form per plastid. These ideas are not mutually exclusive and further investigation may reveal other potential roles for isoamylase and the other starch "degrading" enzymes.

Despite the progress being made in understanding the roles of the different isoforms of SS and SBE, and

Table 2 Isoforms of starch branching enzyme (SBE)

Species	*SBE A*	*SBE B*
Pea	SBE-I	SBE-II
Barley	SBE-IIa and b	SBE-I
Wheat	WBE-II	WBE-I
Corn	BEII	BEI
Potato	SBE A	SBE B
Rice	RBE-IIa and b (≡RBE4 and 3)	RBE-I (≡RBE1)

Table 3 Some commercially important mutants

Species	*Mutation*	*Enzyme affected*	*Phenotype*
Corn	*amylose extender (ae)*	SBE	High amylose
	shrunken 2 (sh2)	AGPase	Reduced starch, sweet
	waxy (wx)	GBSS-I	Amylose-free
	sugary (su1)	Isoamylase	Reduced starch, sweet
Pea	*rugosus (r)*	SBE	Reduced starch, sweet
	rugosus (rb)	AGPase	Reduced starch, sweet
Rice	*alk, gel(t), acl(t)*	SS-IIa	Low gelling temperature
	waxy (wx)	GBSS-I	Amylose-free
Wheat	*waxy*	GBSS-I	Reduced amylose
Barley	*waxy (wx)*	GBSS-I	Reduced amylose

of other enzymes involved in starch synthesis, we still do not understand how the amylopectin molecule is constructed. One of the biggest outstanding mysteries is the nonuniform branching of the amylopectin molecule. The branch points are periodically clustered such that regions of high branch-point density are separated by regions with relatively few branch points. This gives rise to the clusters of chains within the amylopectin molecule that, in turn, are responsible for many of the physical properties of starch and distinguish starch from other, similar glucans such as the glycogens made by animals and bacteria. Whether the properties of the branching enzymes or the starch synthases, or other, as yet unknown proteins are responsible for the formation of clusters remains to be discovered.

The Synthesis of Amylose

Amylose is synthesized by GBSS-I. This enzyme operates inside the starch granule rather than in the stroma like the amylopectin-synthesizing enzymes. It uses small malto-oligosaccharides as the initial substrate for amylose synthesis and elongates these using ADPG as the glucose donor. Both small malto-oligosaccharides and ADPG diffuse into the granule to the site of action of GBSS-I. Evidence from transgenic plants with reduced activities of GBSS-I shows that GBSS-I can synthesize amylose deep within the granule. However, in wild-type plants, amylose synthesis probably occurs mostly just underneath the granule surface within a framework of the most recently deposited amylopectin.

Some of the unique properties of GBSS-I which allow this isoform of SS, but not others, to synthesize amylose are known. The enzyme is able to add several glucose moieties to the growing amylose molecule before releasing the chain and rebinding to another. This processive elongation, as it is called, does not occur with other isoforms of SS that have been investigated. SS-II from pea seeds for example, is able to add a single glucose moiety only at each encounter with the glucan substrate. Almost all of the GBSS-I protein in a wild-type seed is found tightly associated with starch granules. This contrasts with other isoforms of SS that exist either partly or completely in the soluble phase of the plastid. The parts of the GBSS-I protein that are responsible for unique properties such as the processive mode of action and tight binding to starch granules have not yet been identified.

Amylose is synthesized at the expense of amylopectin. Mutants that lack GBSS-I, and therefore have no amylose, have instead more than the normal amount of amylopectin per gram dry weight. This suggests that there is competition between the amylose- and amylopectin-synthesizing pathways. This could operate at the level of competition for the common substrate, ADPG, or it could be via some other mechanism such as competition for space within the granule. Currently there is no clear evidence for these or other possibilities. However, there are circumstances when the availability of ADPG is known to influence the relative amounts of amylose and amylopectin that are synthesized. In mutants that have a reduced capacity to synthesize ADPG, the relative amylose content is lower than in wild-type plants. This is thought to be due to the poor affinity of GBSS-I for ADPG compared to the affinity of the other starch synthase isoforms. Thus, GBSS-I is relatively more affected by a decline in ADPG concentration than are the amylopectin-synthesizing enzymes.

With respect to the competition for space, GBSS-I accounts for most of the granule-bound starch synthase activity and may normally outcompete the more minor, granule-bound amylopectin-synthesizing enzymes, rapidly filling the available space with amylose. However, when GBSS-I is absent, these minor enzymes may be able to fill the space with amylopectin instead. This competition for limited space may also account for the upper limit (approximately 33%) on the amount of amylose synthesized in wild-type starch granules.

The Economic Importance of Starch and of Genetic Variation

The starchy seeds of cereals and grain legumes are of great economic importance. They form a major part of the human diet and are also fed in bulk to our domesticated animals. As well as whole seeds being used as food for humans (for example boiled rice or flour for bread), starch extracted from cereal grains, particularly corn, is used as a food ingredient – usually after post-extraction modification of some kind. Extracted starch is also used by many non-food industries (e.g., in the manufacture of paper and corrugated cardboard) (*see* **Primary Products**: Starch).

The high value of starch-containing seeds has fueled much research on seed biology and has prompted the selection and study of an array of mutants with altered seed development and/or composition. Many of these mutants are scientific curiosities but some have become cultivated varieties (**Table 3**). Sweetcorn varieties of corn, for example, are mutants that lack isoamylase (see "Synthesis of Amylopectin" above). The seeds accumulate less than the normal amount of starch and larger than normal amounts of sucrose. The latter makes the seeds more palatable to sweet-toothed humans.

Super-sweetcorn varieties are double mutants that lack the enzyme that makes ADPG (ADPG pyrophosphorylase) as well as lacking isoamylase (a starch debranching enzyme). They have even more sucrose and less starch than sweetcorn. Other sweet seeds, such as garden peas, are also starch mutants. The peas most commonly grown for the frozen food market are wrinkled-seeded mutants that lack one isoform of SBE and have a lower than normal starch content. Other sweet, low-starch varieties of peas that are grown commercially lack ADPG pyrophosphorylase.

Mutants that lack or have reduced amounts of amylose are also cultivated. For example, waxy maize varieties lack GBSS-I and have starch that is 100% amylopectin. The starch extracted from these varieties is valued because it has superior pasting properties. Similarly, partial waxy mutants of wheat, with lower than normal GBSS-I activity, are grown very largely because flour from these seeds makes the best quality Japanese udon noodles. High-amylose corn mutants are also cultivated, although at the moment on a relatively small scale. The starch from these varieties is prized for its suitability to make for example, films, adhesives, and biodegradable plastics.

Often, starch mutants have a low starch content and this yield penalty may limit the commercial success of a mutant variety even though the starch extracted from it has very valuable properties. However, steady increases in the yields of mutant varieties through breeding and selection have been achieved. Thus it may be possible to commercially cultivate other seeds with altered and interesting starch compositions in the future.

List of Technical Nomenclature

Amylase	An enzyme which hydrolyzes α-1,4-glycosidic bonds in starch.
Amylopectin	Polymer of α-1,4-linked glucose residues with many α-1,6-linked branches.
Amylose	Linear polymer of glucose residues linked through α-1,4-glycosidic bonds.
Cereal	A crop-plant member of the grass family (e.g., maize, wheat, and rice).
Cotyledon	A seed leaf specialized for food storage.
Endosperm	Triploid storage tissue of developing seeds, persistent in cereal seeds.
Isoform	One of a group of enzymes that all catalyze the same reaction.
Legume	A member of the Fabaceae, e.g., peas and beans.
Plastid	An organelle, some forms of which synthesize starch (e.g., chloroplast).
Seed	Embryo-containing structure formed from the ovule following fertilization.
Starch	Complex glucose polymer comprising amylose and amylopectin.
Waxy starch	Starch lacking amylose.

See also: **Primary Products**: Starch.

Further Reading

Denyer K, Johnson P, Zeeman S, and Smith AM (2001) The control of amylose synthesis. *Journal of Plant Physiology* 158: 479–487.

Donald AM (2001) Plasticization and self-assembly in the starch granule. *Cereal Chemistry* 78: 307–314.

Emes MJ, Tetlow IJ, and Bowsher CG (2001) Transport of metabolites into amyloplasts during starch synthesis. In: Barsby TL, Donald AM, and Frazier PJ (eds) *Starch: Advances in Structure and Function*, pp. 138–143. Cambridge: Royal Society of Chemistry.

Gallant DJ, Bouchet B, and Baldwin EJF (1997) Microscopy of starch: evidence of a new level of granule organisation. *Carbohydrate Polymers* 32: 177–191.

Kossmann J and Lloyd J (2000) Understanding and influencing starch biochemistry. *Critical Reviews in Plant Science* 19: 171–225.

Myers AM, Morell MK, James MG, and Ball SG (2000) Recent progress towards understanding the biosynthesis of the amylopectin crystal. *Plant Physiology* 122: 989–997.

Preiss J and Sivak MN (1998) Biochemistry, molecular biology and regulation of starch synthesis. *Genetic Engineering* 20: 177–223.

Slattery CJ, Kavakli IH, and Okita TW (2000) Engineering starch for increased quantity and quality. *Trends in Plant Science* 5: 291–298.

Smith AM (1999) Making starch. *Current Opinions in Plant Biology* 2: 223–229.

Thompson DB (2000) On the non-random nature of amylopectin branching. *Carbohydrate Polymers* 43: 223–239.

Oil Deposition

D J Murphy, University of Glamorgan, Cardiff, UK

Introduction

Seeds are the most common and widespread sites for the accumulation of storage oils in plants. These oils are deposited during seed maturation in the form of cytosolic oil bodies. It is likely that all seeds accumulate at least some cytosolic oil bodies and even non-oilseed species, such as peas (*Pisum sativum*), contain about 2% of their seed weight as lipid. True oilseeds accumulate oil bodies as one of their major storage reserves in amounts ranging from about 20% seed weight in soybean (*Glycine max*) to 42% in oilseed rape (*Brassica napus*), and as high as 76% in some of the larger seeded nuts. Oil bodies are found in several different types of seed tissue. In albuminous dicotyledons, such as castor bean (*Ricinus communis*) and coriander (*Coriandrum sativum*), oil bodies accumulate mainly in the endosperm. In exalbuminous dicotyledons, such as oilseed rape and sunflower (*Helianthus annuus*), oil bodies are found mainly in the embryo cotyledons and axis. Finally, in monocotyledons, such as cereals, oil bodies may be located in the embryo scutellum and/or in the aleurone layer.

Seed oil bodies are organelles for the long-term storage of neutral lipids that are eventually mobilized after germination. The lipid storing cells of seeds tend to contain numerous small oil bodies. In seeds that undergo desiccation (so-called orthodox seeds), a single cell may contain from dozens to hundreds of oil bodies of only about 0.5–1.0 μm diameter. In contrast, many nondesiccating (i.e., recalcitrant) seeds contain numerous larger oil bodies of 2–10 μm diameter. The vast majority of seeds store lipid in the form of triacylglycerols (TAGs) of various acyl chain lengths and functionality. The functionality of a fatty acid means the extent to which it is modified from the standard saturated acyl chain structure. Typical fatty acid modifications include acyl chain desaturation, hydroxylation, epoxidation, or acetylenation. To date, there is only one known plant species, the desert shrub *Jojoba chinensis*, that does not store its oil as TAG; it stores a long-chain liquid wax instead.

Seed oils are of tremendous economic importance with an annual global traded market of some 100 million tonnes that is worth more than US$60 billion. There is great interest from many companies in the biotechnological manipulation of oilseeds to produce novel fatty acids either for food use or as industrial feedstocks. Indeed, one of the first transgenic crops with a modified seed composition to be approved for commercial cultivation was a lauric-oil oilseed rape first cultivated on a large scale in 1995.

Oil Accumulation

Seed oil bodies have been recognized as distinctive lipid storing organelles since the early 1960s, but the mechanism of their biogenesis has always been controversial, and even today there remain large gaps in our understanding of this process. Most investigators would now support an endoplasmic reticulum origin of seed oil bodies but, as in animal and fungal systems of storage lipid accumulation, there are many unclear and puzzling aspects of this mechanism. For example, although there are a few reasonable images of direct physical connections between oil bodies and endoplasmic reticulum, it is surprisingly difficult to observe this phenomenon under the microscope, even during the stage of oilseed development when there is maximum oil body biogenesis and oil bodies make up as much as half of total cell volume. There have been many reports that the most active fraction in TAG biosynthesis is the microsomal membranes, which are normally equated with the endoplasmic reticulum. However, the microsomal fraction is made up of a mixed population of membrane vesicles, many of which are derived from the endoplasmic reticulum, but with some originating from the plasmalemma, Golgi, or from fragments of other lysed membranes, such as the plastid envelope. Even the endoplasmic reticulum membrane itself cannot be considered as a single entity; rather it is a complex three-dimensional network of tubules and sheets that is in constant flux. Staehelin has described no fewer than 16 different types of morphologically distinct endoplasmic reticulum domains in plants, each of which is more or less specialized for a particular function, ranging from vacuole formation to oil body biogenesis. The existence of specific endoplasmic reticulum domains for oil body formation can explain much of the apparently conflicting biochemical data relating to TAG biosynthesis as discussed below.

The activity of TAG biosynthetic enzymes, such as acyltransferases, has been reported in high-speed membrane pellets, soluble fractions, and even oil bodies. Other enzymes contributing to TAG formation, including fatty acid elongases, hydroxylases, and acyltransferases, were reported by Whitfield and coworkers and Richards and coworkers to be present in two discrete fractions from sucrose density gradients. The denser fraction contained most endoplasmic reticulum marker enzymes including those of phospholipid formation, while the lighter fraction contained high specific activities of the enzymes of TAG biosynthesis from glycerol 3-phosphate. The latter fraction, termed the low-density membrane (LDM) fraction, has a density of only $1.05\,g\,ml^{-1}$ and a TAG/phospholipid ratio of 2.93, whereas the values for the bulk endoplasmic reticulum fraction were 1.10 and $0.18\,g\,ml^{-1}$, respectively. It has been proposed that the low-density membrane fraction is derived from a TAG synthesizing domain of the endoplasmic reticulum. This domain would be enriched in TAG biosynthetic enzymes that could function in a concerted manner to channel acyl groups toward TAG formation rather than to membrane lipid production. This is particularly important when the acyl groups destined for TAG formation are incompatible with membrane bilayer organization. For example, the short to medium chain (C8–C14) acyl groups that are found in many types of agriculturally important seeds, including oil palm (*Elaeis guineensis*) and coconut (*Cocos nucifera*), can act as membrane destabilizing detergents. Indeed, dodecanoic acid (C12 lauric acid), which can constitute over 90% of some seed oils, is the major component of most domestic detergents, including liquid soaps, washing-up liquids, and shampoos. One of the sodium salts of lauric acid is sodium dodecoylsulfate (SDS), which is a key chemical used in many research laboratories, e.g., for protein solubilization and separation. If such short or medium chain fatty acids are allowed to accumulate in a normal bilayer membrane, they will first destabilize it and finally will completely solubilize and dissolve the membrane. Therefore, species that accumulate high levels of short to medium chain storage lipids must have rigorous mechanisms to ensure their channeling in the endoplasmic reticulum membrane toward the formation of TAGs and away from membrane lipids.

The fact that low-density membrane domains are highly enriched in TAG suggests that, following its formation, TAG tends to accumulate at its site of synthesis, rather than equilibrating with the bulk pool of membrane lipids. Phospholipid membranes can only accommodate up to 3%mol/mol TAG without disrupting the bilayer organization. Higher levels of TAG cause a phase separation, whereby the TAG forms a separate globular domain between the bilayer leaflets. Direct evidence for such globular TAG domains was recently reported by Lacey and coworkers from cell-free extracts of sunflower seeds, where 1H-NMR spectroscopy revealed that TAG molecules synthesized in microsomal membranes were tumbling isotropically in a similar manner to TAG in oil bodies, rather than having the more motion-restricted spectra expected if they were intercalated into the lipid bilayer. Together, these data imply that TAGs are formed in specialized domains of the endoplasmic reticulum. Apparently, the newly formed TAGs are not free to disperse laterally throughout the endoplasmic reticulum membrane system, and their localized accumulation soon results in a phase separation to form microdomains of TAG molecules sandwiched between the bilayer leaflets. Mountford and Wright have proposed the existence of analogous structures in the plasma membranes of some animal cells. It has been estimated that the diameters of such structures are only about 25 nm, which may explain why they have yet to be observed directly in either animals or plants. Since they all copurify in the low-density membrane fraction, these globular TAG microdomains in the endoplasmic reticulum apparently remain adjacent to the rest of the TAG biosynthetic apparatus, which implies that they are constrained or tethered within this region of endoplasmic reticulum.

Clearly, when such a complex membrane system as the endoplasmic reticulum, with all of its many subdomains, is disrupted by cell fragmentation, a highly heterogeneous population of vesicles is created. Some of these vesicles will contain varying amounts of TAG, plus its associated biosynthetic enzymes, and will therefore have different buoyant densities, possibly even colocalizing with the soluble fraction, e.g., following nonequilibrium discontinuous density gradient centrifugation. This may explain the various reports of TAG formation in these different subcellular fractions from seeds. In addition to being found in membrane and soluble fractions, there have been persistent reports of the presence of TAG biosynthetic enzymes on oil bodies in developing seeds.

There are at least four possible explanations for the presence of TAG biosynthetic enzymes on seed oil bodies. Firstly, the enzymes may be on membranous appendages of the endoplasmic reticulum that remain attached to the oil bodies following their budding off from the endoplasmic reticulum proper. Such appendages may facilitate refusion of oil bodies with the endoplasmic reticulum for the further metabolism of oil body TAG, e.g., by desaturases or transacylases during sunflower seed development, as has been reported by Mancha and Stymne in 1997. Secondly, the enzymes may be localized on the oil body itself. This would require them to insert into a phospholipid monolayer rather than a bilayer while retaining their biological activity, which seems less likely. Thirdly, the TAG enzymes could be on small fragments of membrane that originally formed continuities between endoplasmic reticulum and oil bodies of the kind that has occasionally been observed in ultrastructural studies. Shearing of these connecting regions of membrane during cell disruption could leave some of the membrane fragments, together with their complement of enzymes, attached to the oil bodies. The fourth possibility is that there is never any physiological connection of TAG enzymes with oil bodies, and that all such associations are artefactual, i.e., the proteins may adhere to the oil bodies following homogenization of the tissue and the resultant disruption of the endomembrane system. This would imply that there are no stable continuities between the endoplasmic reticulum and oil bodies, and that the latter are released from the membrane free of any residual endoplasmic reticulum proteins, with the exception (in desiccation tolerant seeds only) of their unique structural proteins – the oleosins and caleosins. At present, it is not possible to distinguish between these different possibilities and models of oil body biogenesis therefore remain speculative. Several models for the formation of oil bodies from the endoplasmic reticulum membrane in various organisms are discussed in recent reviews by Murphy and Vance in 1999, Zweytick and coworkers in 2000, and by Murphy in 2001.

Most plants in temperate and subtropical climates, including nearly all of the major annual crops, produce seeds that have undergone desiccation as a normal part of their development. Such seeds often contain as little as 5–10% moisture and are able to survive over the winter, or even for many years, in this dry and dormant state. Desiccation tolerant, or "orthodox," seeds tend to contain many small storage oil bodies with diameters in the range of 0.5–2.0 μm. Such small oil bodies differ from the larger oil bodies in many oily fruits or recalcitrant seeds in that they are surrounded by an outer coat that is primarily made up of a novel class of proteins called "oleosins," plus a monolayer of phospholipid.

Oleosins

Oleosins are a class of relatively low molecular weight alkaline proteins of about 15–24 kDa that accumulate on the surfaces of oil bodies. Oleosins are found especially (and perhaps exclusively) in desiccation tolerant seeds. The structure, biosynthesis, and function of oleosins have been the subject of several reviews, including those of Moloney in 1999, Tai and coworkers in 2002, Frandsen and coworkers in 2001, and Murphy in 2001. In addition to their biological interest, oleosins have recently been used in a biotechnological context as vehicles for the overexpression in transgenic plants of a variety of fusion proteins including β-glucuronidase (GUS), xylanase, and hirudin. The presence of proteins in isolated and washed oil bodies from peanut (*Arachis hypogaea*) seeds was recognized by Jacks and colleagues as long ago as 1967. It is now believed that oleosin genes are present in all higher plant genomes, as well as in several lower plants such as gymnosperms, but that oleosin proteins are not necessarily accumulated in all oil storing tissues of plants. Oleosins are relatively abundant proteins in many seeds that accumulate large quantities of storage oil bodies. For example, oilseeds such as oilseed rape and sunflower (*Helianthus annuus*) may contain 40–45% w/w storage oil, which itself contains 1–5% w/w oleosin. In such seeds, oleosins can make up as much as 10% total cell protein.

Well over a decade after the first oleosin gene was sequenced, the function of these unique and often very abundant proteins remains to be fully explained. In the past, three possible roles have commonly been ascribed to oleosins: firstly, they are involved in the biogenesis of oil bodies in seeds; secondly, they maintain the small size and exceptional stability of many seed oil bodies; and thirdly, they are involved in oil body mobilization, and possibly act as lipase binding sites. It is unlikely that oleosins are essential for either oil body biogenesis or mobilization, for the following reasons. Both nondesiccating oilseeds and oil bearing fruits are fully capable of accumulating large quantities of cytosolic oil bodies in the absence of oleosins. There have been relatively few studies of oil body biogenesis in such tissues, but the available data suggest that TAG biosynthesis on the endoplasmic reticulum is followed by the formation of cytosolic oil bodies of 1–5 μm diameter in the same way as occurs in oleosin containing tissues. This would appear to rule out any requirement for oleosins in oil body biogenesis *per se*. By the same token, oleosin deficient seeds, including cocoa (*Theobroma cacao*) and red oak (*Quercus rubra*), are fully able to break down their storage TAG despite the absence of oleosins, which suggests that oleosins are also not required for any of the processes associated with oil body mobilization in seeds.

The only difference between oil bodies in oleosin expressing tissues and those in nonexpressing tissues is that the former tend to be smaller and are more stable to dehydration/rehydration. This may provide a clue to the true function of oleosins in seeds. As described in more detail below, Leprince and coworkers have observed that the oleosin deficient oil bodies of desiccation sensitive seeds are unstable to drying and rehydration *in vivo*. It is important to point out that the oil bodies retained their integrity during the drying phase of the cycle and only coalesced following rehydration of the seed tissues. This rehydration induced oil body coalescence led to a catastrophic and fatal phase inversion in the cells of such tissues. Severe disruption of oil body integrity has also been reported in lettuce (*Lactuca sativa*) seeds subjected to cycles of freezing and thawing, which cause a combination of hydration and mechanical stresses. So a possible function of oleosins is to enable seeds to undergo extensive dehydration (often followed by a prolonged period of dormancy), and then allow them to suffer the rigors of rehydration while still being able to store over 50% of their weight as oil bodies. This is by no means a trivial adaptation as it would allow oleosin-containing seeds to colonize new habitats away from the moist tropical and subtropical regions. The importance of oil body integrity in contributing to seed viability is demonstrated by observations that one of the earliest markers for deterioration in stored seeds is the coalescence of their oil bodies. This coalescence is presumably caused by damage to the oleosin layer that normally stabilizes such oil bodies, leading to the irreversible loss of seed viability. Whether oleosin proteins accumulate in their seeds or not, oleosin genes appear to be present in all plant species that have been analyzed to date. It is therefore possible that oleosins have other, as yet unknown, functions, e.g., in tropical plants that do not accumulate seed oleosins.

In summary, there is no evidence that oleosins are required for either the biogenesis or mobilization of oil bodies in plants. In those seeds where they are expressed, oleosins are synthesized in conjunction with oil bodies, although even here their presence is not necessarily a prerequisite for oil body accumulation. The relative rates of oleosin and (TAG+ phospholipid) accumulation probably determine the final size of the small oil bodies found in oleosin expressing seeds (but even in the absence of oleosins, oil bodies of 1–5 μm can still accumulate). The one function that does seem to be definitely associated with oleosins is to provide a stabilizing outer coating to storage oil bodies in those seeds that normally undergo desiccation and rehydration prior to development of a new seedling. Such oilseeds invest huge metabolic resources, often amounting to over 10% total cell protein, in the accumulation of oleosins. This considerable investment demonstrates the importance of oleosins for seeds. It is suggested that oleosins may be required for lipid storing seeds to withstand dehydration/rehydration and therefore be capable of delaying their germination until conditions are more appropriate. Oleosins may have additional and as yet unknown roles, but the available evidence to date is consistent with a single role that is related to the acquisition of desiccation tolerance by oilseeds.

List of Technical Nomenclature

Albuminous seed A seed containing a persistent endosperm in which much of its storage products may be located.

Aleurone layer A layer of endosperm enclosing the embryo that is found in some plant seeds; It is often rich in storage protein bodies and oil bodies.

Endoplasmic reticulum A complex network of interconnected tubules, vesicles, and sacs that are found in all eukaryotic cells; it can serve many specialized functions in the cell, including protein synthesis, sequestration of calcium, production of neutral lipids, storage and production of glycogen, and insertion of membrane proteins.

Endosperm Nutritive tissue found in all developing seeds but absent in mature exalbuminous seeds.

Exalbuminous A seed lacking any appreciable endosperm that accumulates most of its storage products in cotyledons.

Organelles Specialized intracellular structures found in eukaryotes. Examples found in plants include the nucleus, endoplasmic reticulum, mitochondria, and plastids.

Scutellum (literally "shield") The single cotyledon in cereals, which forms a digestive organ for the endosperm. Sugars are transferred from the endosperm to the embryo through the scutellum.

Triacylglycerol (TAG) A common form of storage lipid found in most organisms. A TAG molecule is made up of a single glycerol backbone esterified to three fatty acyl residues.

See also: **Genetic Modification of Primary Metabolism**: Acyl Lipids. **Primary Products**: Oils. **Seed Development**: Germination; Onset of Desiccation Tolerance; Physiology of Maturation.

Further Reading

Frandsen GI, Mundy J, and Tzen JT (2001) Oil bodies and their associated proteins, oleosin and caleosin. *Physiologia Plantarum* 112: 301–307.

Jacks TJ, Yatsu LY, and Altschul AM (1967) Spheresome membranes from peanut, *Arachis hypogaea*. *Plant Physiology* 42: 585–597.

Lacey DJ, Beaudoin F, Dempsey CE, Shewry PR, and Napier JA (1999) The accumulation of triacylglycerols within the endoplasmic reticulum of developing seeds of *Helianthus annuus*. *Plant Journal* 17: 397–405.

Mancha M and Stymne S (1997) Remodelling of triacylglycerols in microsomal preparations from developing castor bean (*Ricinus communis* L.) endosperm. *Planta* 203: 51–57.

Moloney MM (1999) Seed oleosins. In: Shewry PR and Casey R (eds) *Seed Proteins*, pp. 781–806. Dordrecht: Kluwer.
Mountford CE and Wright LC (1988) Organization of lipids in the plasma membranes of malignant and stimulated cells: a new model. *Trends in Biochemical Sciences* 13: 172–177.
Murphy DJ (2001) The biogenesis and functions of lipid bodies in animals, plants, and microorganisms. *Progress in Lipid Research* 40: 325–438.
Murphy DJ and Vance J (1999) Mechanisms of lipid-body formation. *Trends in Biochemical Sciences* 24: 109–115.
Richards D, Taylor R, and Murphy DJ (1993) Localization and possible substrate specificity of the oleate hydroxylase of developing castor bean seeds. *Plant Physiology and Biochemistry* 31: 89–94.
Staehelin LA (1997) The plant ER: a dynamic organelle composed of a large number of discrete functional domains. *Plant Journal* 11: 1151–1165.
Tai SS, Chen MC, Peng CC, and Tzen JT (2002) Gene family of oleosin isoforms and their structural stabilization in sesame seed oil bodies. *Bioscience, Biotechnology and Biochemistry* 66: 2146–2153.
Whitfield HV, Murphy DJ, and Hills MJ (1993) Subcellular localisation of fatty acid elongase in developing seeds of *Lunaria annua* and *Brassica napus*. *Phytochemistry* 32: 255–258.
Zweytick D, Athenstaedt K, and Daum G (2000) Intracellular lipid particles of eukaryotic cells. *Biochimica et Biophysica Acta* 1498: 1–20.

Storage Protein Deposition

D J Murphy, University of Glamorgan, Cardiff, UK

Seed proteins provide more than half of the global intake of dietary protein by human populations, and are also major nutrients for many domesticated animals. All seeds contain some protein as a nitrogen reserve for the emerging seedling after germination, and some seeds contain as much as 40% of their dry weight as protein. In addition to their important roles as macronutrients, seed proteins also contribute to the processing properties of food products derived from seeds. One of the best examples of this is the role of the cereal proteins in the production of flours and their conversion to different types of breads.

In the future, there are plans to engineer seed proteins in crops such as wheat in order to improve their breadmaking performance (*see* **Primary Products**: Amino Acids). Another important target for seed protein modification is to enhance their nutritional quality by increasing the levels of essential amino acids, especially methionine, lysine, and tryptophan, which are relatively deficient in most seed proteins (*see* **Primary Products**: Proteins; Amino Acids).

Finally, some seed proteins can be troublesome allergens, e.g., in peanuts, or can contribute to conditions like coeliac disease, which is caused by an intolerance to the prolamin family of cereal proteins. Efforts to remove or otherwise manipulate undesirable proteins, either by conventional breeding or by GM techniques, are now underway.

See also: **Primary Products**: Amino Acids; Proteins. **Seed Development**: Physiology of Maturation; Starch Synthesis.

Further Reading

Bewley JD and Black M (1995) *Seeds: Physiology of Development and Germination*, 2nd edn. New York: Plenum Press.
Habben JE and Larkins B (1995) Genetic modification of seed proteins. *Current Opinion in Biotechnology* 6: 171–174.
Larkins BA and Vasil I (eds) (1997) *Cellular and Molecular Biology of Plant Seed Development*. Dordrecht: Kluwer.
Matsuoka K and Bednarek SY (1998) Protein transport within the plant cell endomembrane system: an update. *Current Opinion in Plant Biology* 1: 463–469.
Muntz K (1998) Deposition of storage proteins. *Plant Molecular Biology* 38: 77–99.
Shewry PR and Casey R (eds) (1999) *Seed Proteins*. Dordrecht: Kluwer.
Shewry PR and Gutteridge S (eds) (1992) *Plant Protein Engineering*. Cambridge: Cambridge University Press.

Physiology of Maturation

A R Kermode, Simon Fraser University, Burnaby, British Columbia, Canada

Introduction

Seeds control survival and reproductive capacity and therefore occupy a critical position in the life history of higher plants. The successful establishment of the new plant, both temporally and spatially, as well as the vigor of the young seedling, is largely determined by the physiological and biochemical features of the seed. Consider some of the key processes that occur during seed maturation. These include: (1) the accumulation of food reserves that will ultimately allow the seedling to survive before it commences photosynthesis and autotrophic growth; (2) the synthesis of protective compounds that allow the seed to withstand some degree of water loss and to survive for long periods under adverse environmental conditions; (3) hardening of the protective layers of the seed (the seed coat), and intracellular accumulation of antimicrobial compounds, permitting survival of seeds under adverse environmental conditions and conferring some resistance to pathogens, insects, and herbivores; and (4) development of dormancy in some species, which effectively distributes seed germination in time and space. The following article will outline some of the key processes that occur during seed maturation and the control of some of these processes by plant hormones and other factors. The reliance of humans on seeds depends on germination (for plant cultivation); moreover, seeds such as those of cereals and legumes are themselves major food sources or sources of industrial products whose importance lies in the macromolecules (proteins, starch, and oils) accumulated during development and maturation. Some of the modern approaches that promise to elucidate the genes and encoded proteins that underlie the biochemical and physiological basis of seed maturation are discussed below.

An Overview of Embryo and Seed Development

The mature seed is comprised of three parts: an embryonic root (radicle) and shoot, food reserve tissues or organs (cotyledons, endosperm, or megagametophyte), and a protective structure, the seed coat (testa). In cereals, most of the food reserves of the seed are accumulated in the endosperm (**Figure 1A**), a tissue that is not part of the embryo. In legumes, such as the common bean (*Phaseolus vulgaris*) or soybean (*Glycine max*), the endosperm is resorbed during development and it is the cotyledons that store the bulk of the seed's reserves (**Figure 1B**). In seeds of conifer species, there is no endosperm; instead, the embryo develops within the megagametophyte, a haploid maternal tissue that becomes the major storage organ of the seed (**Figure 1C**). The seed tissues surrounding the embryo play a major role in the ultimate survival of a seedling. They provide a unique physical and chemical environment that is critical for promoting maturation events and for preventing embryos from prematurely switching to a germination program.

The development of most seeds can be divided conveniently into three confluent stages (**Figure 2**). During histodifferentiation, the single-celled zygote undergoes extensive mitotic division, and the resultant cells differentiate to form the basic body plan of the embryo (the axis and cotyledons); concurrently, there is the formation of the triploid endosperm or haploid megagametophyte. Thereafter, cell division ceases during the seed expansion (maturation) stage and there is cell expansion. As shown in **Figure 2**, as well as the induction of expression of genes for cell expansion, many genes are expressed that encode enzymes and other proteins required for the synthesis of reserves. Thus, a key process during seed maturation is the deposition of reserves (normally proteins along with lipids or carbohydrates), primarily in the storage tissues (i.e., cotyledons, endosperm, or megagametophyte). Finally, the development of most seeds is terminated by some degree of drying (maturation drying), which results in a gradual reduction in metabolism as water is lost from seed tissues and the embryo passes into a metabolically inactive, or quiescent state. In preparation for these final events of maturation, the embryo and other seed tissues express genes at earlier stages that specify products important for survival in the dry state.

In many wild plant species, seed maturation is accompanied by the imposition of dormancy, a condition that has a genetic basis and is also influenced by the environment during seed development (e.g., the degree of dormancy). The term dormancy refers to the ability of seeds to delay germination until the environmental conditions are appropriate for survival of the seedling. Dormancy is beneficial for wild species by timing seedling emergence to occur in the proper season, distributing seed germination over many years (in some instances), and reducing competition among offspring. In contrast to quiescent seeds, seeds that are dormant at maturity must receive signals or conditions that allow them to break or terminate dormancy in order to commence germination. Seeds of many conifer species (e.g., yellow cedar (*Chamaecyparis nootkatensis*) and western white pine (*Pinus monticola*)) exhibit profound dormancy at the time of seed shedding or dispersal, and require several months in the moist state (generally at low or fluctuating temperatures) before they are able to germinate. Most domesticated crops show very little dormancy.

Reserve Deposition during Seed Maturation

After germination, the reserves that have been accumulated in the storage organs of the seed are broken down and the resulting energy, carbon, and nitrogen will support seedling establishment until the plant becomes self-sufficient by commencing photosynthesis. As shown in **Table 1**, the seeds of different species accumulate different types and amounts of reserves in their storage organs (endosperm,

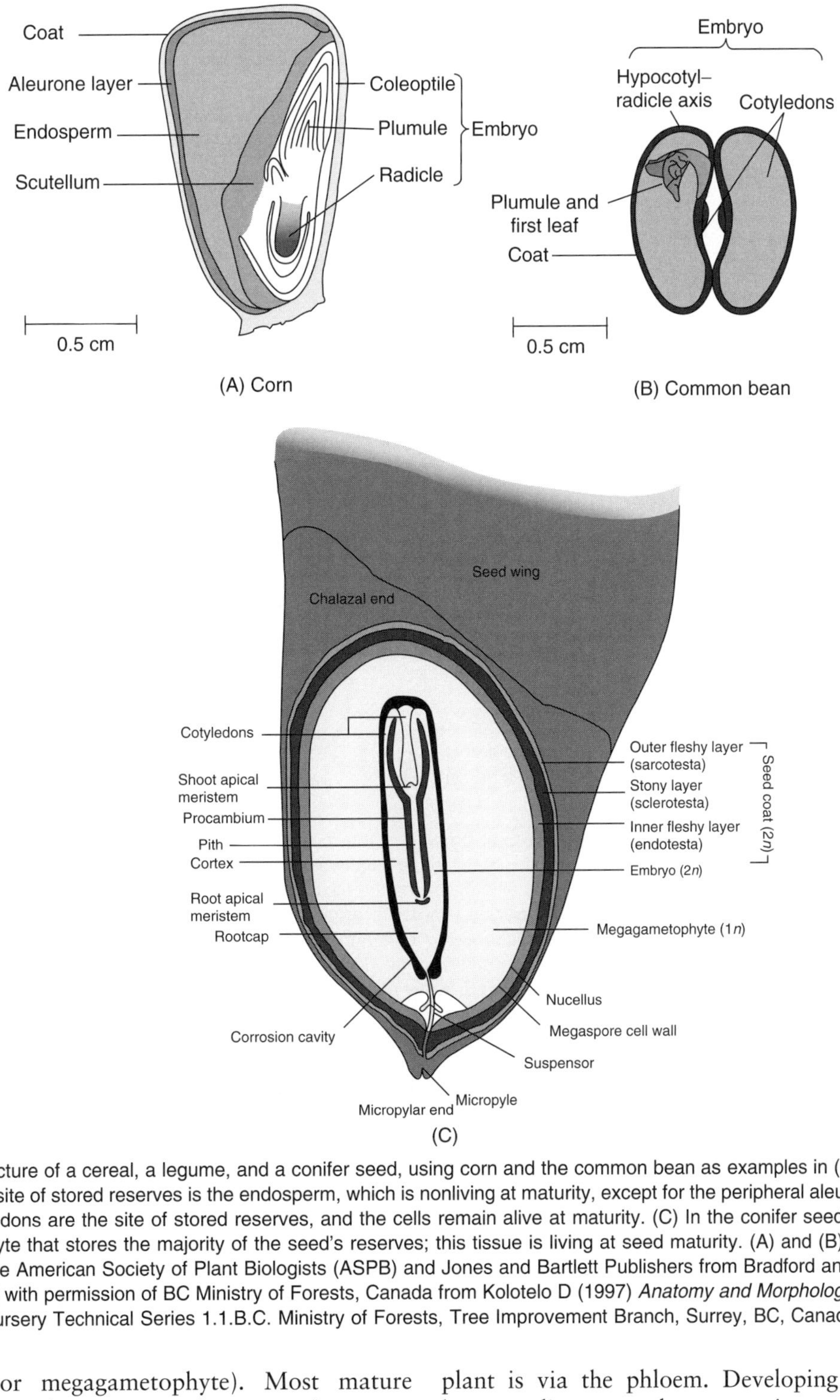

Figure 1 Structure of a cereal, a legume, and a conifer seed, using corn and the common bean as examples in (A) and (B). (A) In corn, the major site of stored reserves is the endosperm, which is nonliving at maturity, except for the peripheral aleurone layer. (B) In bean, the cotyledons are the site of stored reserves, and the cells remain alive at maturity. (C) In the conifer seed, it is the haploid megagametophyte that stores the majority of the seed's reserves; this tissue is living at seed maturity. (A) and (B) reproduced with permission of the American Society of Plant Biologists (ASPB) and Jones and Bartlett Publishers from Bradford and Bewley (2003). (C) Reproduced with permission of BC Ministry of Forests, Canada from Kolotelo D (1997) *Anatomy and Morphology of Conifer Tree Seed.* Forest Nursery Technical Series 1.1.B.C. Ministry of Forests, Tree Improvement Branch, Surrey, BC, Canada.

cotyledons, or megagametophyte). Most mature seeds contain at least two or three stored reserves in appreciable quantities.

The food reserves are synthesized *in situ* from photoassimilate translocated to the developing seed from the parent plant. Sucrose is the form of sugar translocated to the developing seed; the major pathway for its long-distance transport from the sites of photosynthesis within vegetative parts of the plant is via the phloem. Developing cereal grains have no direct vascular connections with the parent plant, and a short-distance transport mechanism operates to move assimilates from the vascular tissues to the region of reserve accumulation, the endosperm. In temperate cereals, such as wheat (*Triticum* spp.) and barley (*Hordeum vulgare*), assimilates are supplied via the vascular tissue running in the furrow (crease), and must first pass

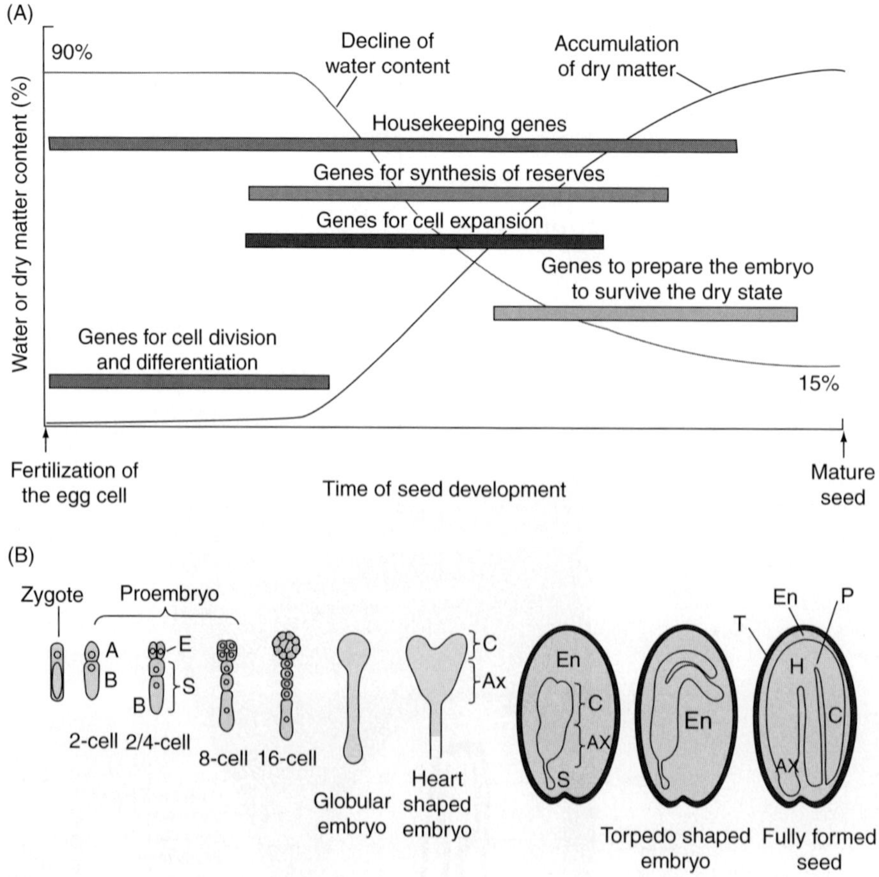

Figure 2 The stages of seed development. (A) General scheme of the major events occurring during seed development. The different sets of genes that are expressed at the various times throughout development are indicated by the horizontal bars. After the histodifferentiation stage (the stage of cell division and differentiation), the next stage (expansion) is characterized by an increase in dry weight as storage reserves (protein, starch, and oil) are laid down, displacing water in the storage tissues. The final stage is maturation drying; in many seeds there is a final decline in water content to 12–15%. (B) Embryogenesis, using oilseed rape as an example. After fertilization to form the zygote, mitosis occurs to produce an apical (A) and a basal (B) cell. The latter forms the suspensor (S), which facilitates development of the embryo (E) by serving as a conduit for nutrients. The suspensor later degenerates during the final stages of seed maturation. The apical cell undergoes extensive cell division and tissue formation until it forms a torpedo shaped embryo enclosed in a nutritive endosperm. Reserve deposition then occurs, with the endosperm and parent plant supplying nutrients for the formation of proteins and oils in the cotyledons. The supply of nutrients from the parent plant for reserve synthesis arrives via the phloem to the seed coat, from which the nutrients diffuse inward to the developing embryo. The endosperm is completely resorbed by the time the embryo has completed reserve accumulation. A, apical cell; Ax, axis; B, basal cell; C, cotyledons; E, embryo; En, endosperm; H, hypocotyl; P, plumule (shoot apex); S, suspensor; T, testa (seed coat). (Reproduced with permission of the American Society of Plant Biologists (ASPB) and Jones and Bartlett Publishers from Bradford and Bewley, 2003.)

through the funiculus-chalazal region, then through the nucellar projection, and finally through the aleurone layer before entering the starchy endosperm (**Figure 3**). In some legumes (e.g., certain cultivars of soybean and field and garden pea (*Pisum arvens* and *Pisum sativum* respectively)), the translocated sucrose produced by photosynthesis in the leaves and pods may be stored temporarily as starch in the pod prior to remobilization and transfer to the developing seed.

Storage Protein Synthesis and Deposition

The storage proteins of seeds are generally classed on the basis of their solubility characteristics and their sedimentation coefficient (**Table 2**). In most seeds, the syntheses of the various storage proteins are initiated at similar, but not identical, times during development.

Storage protein synthesis and deposition in dicotyledonous seeds The process of storage protein synthesis and deposition in seeds of dicotyledonous species can be illustrated using the cotyledons of the pea seed as an example (**Figure 4**). During early developmental stages (e.g., ~5 days after flowering), the storage parenchyma cells of the cotelydons contain a single large vacuole that occupies the majority of the cells' volume. At this stage, the cells appear similar to

Table 1 Types and amounts of storage reserves accumulated by seeds of different species[a]

	Average composition (%)			
	Protein	*Oil*	*Carbohydrate*[b]	*Major storage organ*
Cereals				
Barley	12	3[c]	76	Endosperm
Dent corn (maize)	10	5	80	Endosperm
Oats	13	8	66	Endosperm
Rye (*Secale cereale*)	12	2	76	Endosperm
Wheat	12	2	75	Endosperm
Legumes				
Broad bean (*Vicia faba*)	23	1	56	Cotyledons
Garden Pea	25	6	52	Cotyledons
Peanut (*Arachis hypogaea*)	31	48	12	Cotyledons
Soybean	37	17	26	Cotyledons
Other				
Castor bean (*Ricinus communis*)	18	64	Negligible	Endosperm
Oil palm (*Elaeis guineensis*)	9	49	28	Endosperm
Oilseed rape (canola; *Brassica napus*)	21	48	19	Cotyledons
Pine (*Pinus* spp.)	35	48	6	Megagametophyte

[a] Reproduced with permission of Kluwer from Bewley JD and Black M (1994) *Seeds. Physiology of Development and Germination*, 2nd edn. New York: Plenum Press.
[b] Mainly starch.
[c] In cereals, oils are stored within the scutellum, an embryonic tissue.

those of vegetative tissues. However, by later stages (e.g., 12–14 days after flowering), the vacuoles begin to subdivide forming smaller vacuoles and protein deposition within the vacuoles has commenced at this time, with accumulation occurring first at the periphery of the vacuoles (**Figure 4**). By day 20, there are numerous small vacuoles filled with storage protein forming distinct and discrete bodies. At seed maturity, these protein storing organelles of the cotyledonary storage parenchyma cells are referred to as protein bodies or protein storage vacuoles. The latter term is used in part to differentiate between these organelles of dicotyledonous seeds (that originate from vacuoles) versus some of the protein storing organelles of cereals grains that originate from the ER (see subsequent section).

Where are storage proteins synthesized within storage parenchyma cells, and how are they transported intracellularly to reach the protein body/vacuole? Synthesis generally occurs on the rough ER, and there is vectorial transport to the protein storing organelles. An example can be illustrated for the globulin storage protein, glycinin, which is a 12S legumin-type storage protein. The glycinin holoprotein is a protein complex, a hexamer comprised of six subunits. Each individual subunit in turn is comprised of an acidic and basic polypeptide, which are synthesized sequentially on the same mRNA template (**Figure 5**). The initial translation product is a preproprotein and contains the signal peptide at the N-terminus, the acidic and basic polypeptides joined by a short linker, and a short (penta-) peptide at the C-terminus (**Figure 5**). Following translocation of the precursor glycinin from the cytosol, across the ER membrane into the lumen of the ER (a process mediated by the N-terminal signal peptide), the protein undergoes a number of posttranslational processes. These occur after cleavage of the signal peptide (a cotranslational event, which generates the proprotein) and, collectively, they are important for the stability of the storage protein and for its competence to be transported out of the ER. The posttranslational modifications include:

1. cleavage of the C-terminal pentapeptide;
2. folding and assembly of the proprotein into the correct (three-dimensional) tertiary and quaternary structure. For glycinin, this involves the formation of trimers in the ER; after arrival at the protein storage vacuole, hexamers are formed; and
3. formation of disulfide bridges, in which the ER-localized enzyme protein disulfide isomerase uses the –SH groups of the cysteine residues of the acidic and basic subunits to create intermolecular S-S bonds (**Figure 5**).

Provided these processes occur in a correct manner, the proprotein is ultimately transported to protein storage vacuoles. This occurs via vesicles that bud from the transitional elements of the ER, delivering the protein to the Golgi complex. Following transit through the various compartments of the Golgi

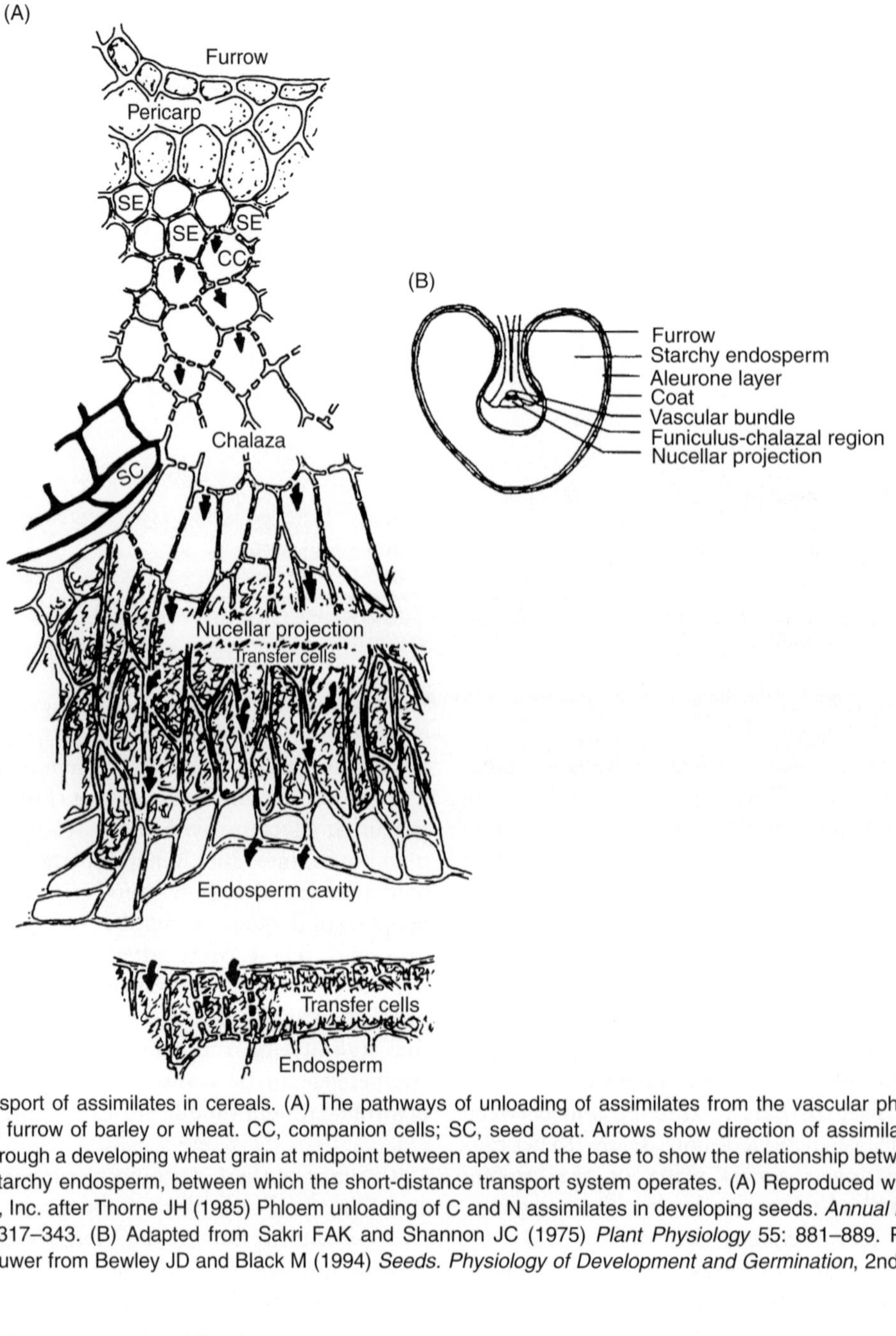

Figure 3 Transport of assimilates in cereals. (A) The pathways of unloading of assimilates from the vascular phloem (SE, sieve elements) in the furrow of barley or wheat. CC, companion cells; SC, seed coat. Arrows show direction of assimilate transport. (B) Cross-section through a developing wheat grain at midpoint between apex and the base to show the relationship between the vascular tissue and the starchy endosperm, between which the short-distance transport system operates. (A) Reproduced with permission of Annual Reviews, Inc. after Thorne JH (1985) Phloem unloading of C and N assimilates in developing seeds. *Annual Reviews of Plant Physiology* 36: 317–343. (B) Adapted from Sakri FAK and Shannon JC (1975) *Plant Physiology* 55: 881–889. Reproduced with permission of Kluwer from Bewley JD and Black M (1994) *Seeds. Physiology of Development and Germination*, 2nd edn. New York: Plenum Press.

complex, vesicles bud from the trans-Golgi network to deliver proglycinin to the protein storage vacuole. Upon arriving at its final destination, the protein undergoes the final steps in its maturation, including removal of the linker sequence (a process effected by a vacuolar cysteine endoprotease). As a result, the acidic and basic polypeptides are joined only by disulfide bridges, and assembly of the trimers into a hexamer yields the mature holoprotein, which is stabilized by surface charge interactions (**Figure 5**).

Although other globulin-type storage proteins do not have the same features as preproglycinin (for example, in other legumes, the linker and the C-terminal pentapeptide may be absent from legumin-type storage proteins), the above example demonstrates some of the principles of storage protein synthesis. For example, generally the processes of signal peptide cleavage, protein folding and assembly, creation of disulfide bridges, and glycosylation (if present; see below) must occur in the correct manner before storage proteins are competent for transport out of the ER. Many of these processes are facilitated by proteins that are permanent ER residents (molecular chaperones), such as

Table 2 Classification of seed storage proteins[a]

Protein type	*Solubility characteristics*
Albumins	Water or dilute buffers at neutral pH
Globulins (7S vicilins and 11S legumins)	Salt solutions
Prolamins	Aqueous alcohols (70–90%)
Glutelins	Dilute acid or alkaline solutions

[a] After Osborne TB (1924) The vegetable proteins. *Monographs in Biochemistry*. London: Longmans, Green & Co.

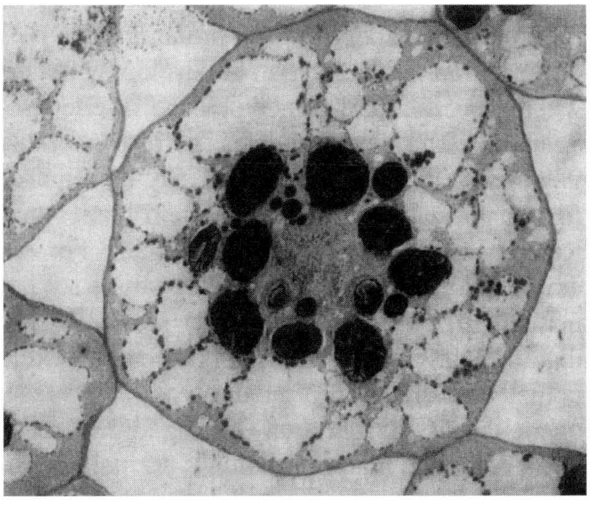

Figure 4 Storage protein deposition in storage parenchyma cells of pea seed. Electron micrograph of a cell from a developing pea cotyledon. Numerous protein storage vacuoles are visible. Storage protein is being accumulated as electron dense deposits at the periphery of the protein storage vacuoles. At the end of seed development, the protein storage vacuoles will be completely filled with protein. The nucleus is at the center, surrounded by large, electron dense starch grains. Reproduced with permission of John Wiley and Sons, Inc. from Vitale A and Chrispeels MJ (1992) Sorting of proteins to the vacuoles of plant cells. *Bioessays* 14: 151–160.

binding protein (BiP), which promotes correct folding and assembly of the storage protein polypeptides and protein disulfide isomerase, which forges disulfide bonds (when present). If these processes do not occur in the correct manner, molecular chaperones (especially BiP) are important for retaining misfolded and misassembled proteins, thus preventing their export from the ER. Restrictions in the transport of proteins out of the ER can be viewed as a late stage in a series of mechanisms utilized by the cell to ensure faithful expression of genetic information at the level of functional and correctly localized proteins. In plants, the ER may be a major site of degradation of some of the proteins that fail to assume their functional conformation (whether or not they are associated with BiP or a BiP-like protein); other sites of degradation may be the vacuole and the cytosol.

Glycosylation is a modification that many seed storage proteins undergo *en route* to their target organelle. Many undergo *N*-linked glycosylation, in which the glycan is linked to the amide nitrogen of an asparagine residue. This modification occurs cotranslationally, as a result of the nascent storage protein containing a tripeptide consensus signal (i.e., asparagine-X-serine/threonine, in which X is any amino acid except proline). *N*-linked glycosylation occurs in the ER lumen and involves a dolichol lipid intermediate and ER glycosyltransferase. The latter enzyme transfers sugars *en bloc* from the dolichol intermediate to asparagine residues at sites on the polypeptide where the consensus signal is present. Generally, while still in the ER, the glycans may be trimmed, a process that is now recognized to be critical for normal seed development. In their mature state, storage proteins that are modified in this manner often have both high mannose and complex *N*-linked glycans (**Figure 6**), the latter arising as a result of "maturation" of the original *N*-linked glycan when the storage protein transits through the Golgi complex *en route* to the protein storage vacuole. *N*-linked glycans do not play a role in targeting of storage proteins to the protein storage vacuole; however, they appear to play a fundamental role in promoting correct folding and, as a consequence, enhance protein stability (e.g., via protection against proteolysis).

As intimated above, a late posttranslational modification of storage proteins (often completed within the protein storage vacuole) is proteolytic processing. In some cases, the polypeptide is simply cleaved into two or three smaller polypeptides, all of which remain together in the oligomer; in others, a domain is lost (e.g., from the N-terminus, C-terminus, or from within the polypeptide). The degree of processing is also variable; it may be complete or only partial, with some polypeptides remaining unprocessed. Interestingly, the assembly of glycinin trimers into hexamers (**Figure 5**) *in vitro* requires the proteolytic cleavage of proglycinin into its basic and acidic subunits. Thus proteolytic cleavage may be a potential regulatory step in the pathway leading to glycinin deposition. **Figure 7** shows some of the modifications that take place during the synthesis, transport, and deposition of other types of storage proteins found within seeds of dicotyledonous species (e.g., the 7S vicilins of the globulin family of storage proteins and the 2S albumins; **Table 2**).

An important event in the deposition of seed storage proteins is their specific targeting from the trans-Golgi network to the protein storage vacuole. This process can be summarized as follows:

1. The types of targeting signals on seed storage proteins are diverse. Some are discrete signals

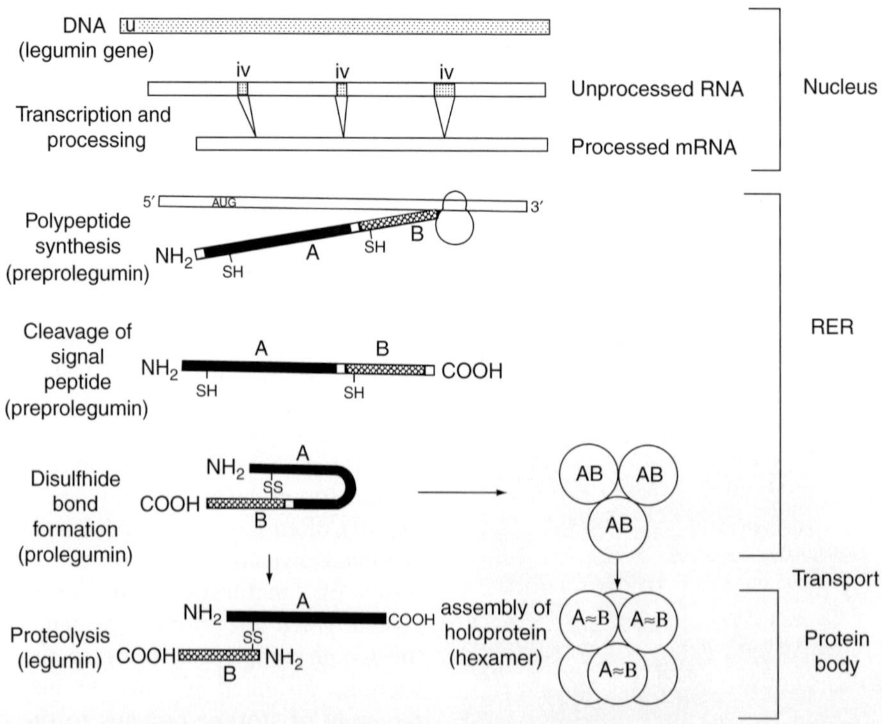

Figure 5 Synthesis and processing of soybean glycinin, a 12S legumin type storage protein. Following transcription, the mRNA encoding glycinin is transported from the nucleus to the rough endoplasmic reticulum (RER) where it is translated. The mRNAs for the acidic (A) and basic (B) subunits are joined by sequences encoding a 4-amino-acid linker on the primary translation product (preprolegumin). The mRNA also encodes the signal peptide at the N-terminal end and a pentapeptide at the C-terminal end. The primary product (prolegumin) is processed to yield the A and B subunits joined by the linker sequence and disulfide bonds (S-S); the signal peptide and the pentapeptide are cleaved off. In the final step of processing, the mature protein is formed as the linker sequence is removed, and the acidic and basic subunits are joined only by the disulfide bonding. The mature acidic subunit contains 278 amino acids (Mr ~ 40 000) and the basic subunit 180 amino acids (Mr ~ 20 000). Assembly of the subunits within the protein bodies yields the mature hexameric legumin holoprotein. Reproduced with permission of Kluwer from Bewley JD and Black M (1994) *Seeds. Physiology of Development and Germination*, 2nd edn. New York: Plenum Press.

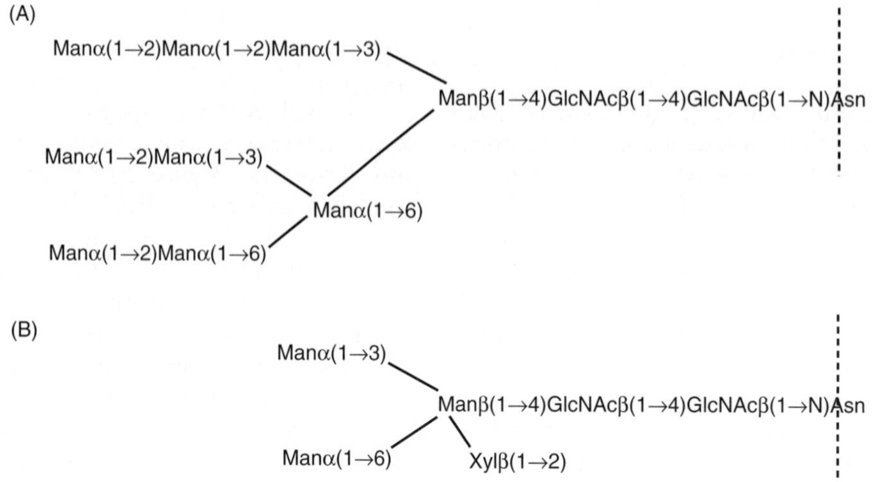

Figure 6 The *N*-linked glycans of seed proteins. (A) A typical high mannose glycan (soybean agglutinin). (B) Probable structure of one version of a modified or complex glycan of phaseolin. Reproduced with permission of Kluwer from Bewley JD and Black M (1994) *Seeds. Physiology of Development and Germination*, 2nd edn. New York: Plenum Press.

(contiguous stretches of amino acids at either the C-terminus, N-terminus, or occur internally within the polypeptide) (**Figure 8**). Other sorting signals may be three-dimensional "signal patches;" these are conformation-dependent signals and involve noncontiguous stretches of amino

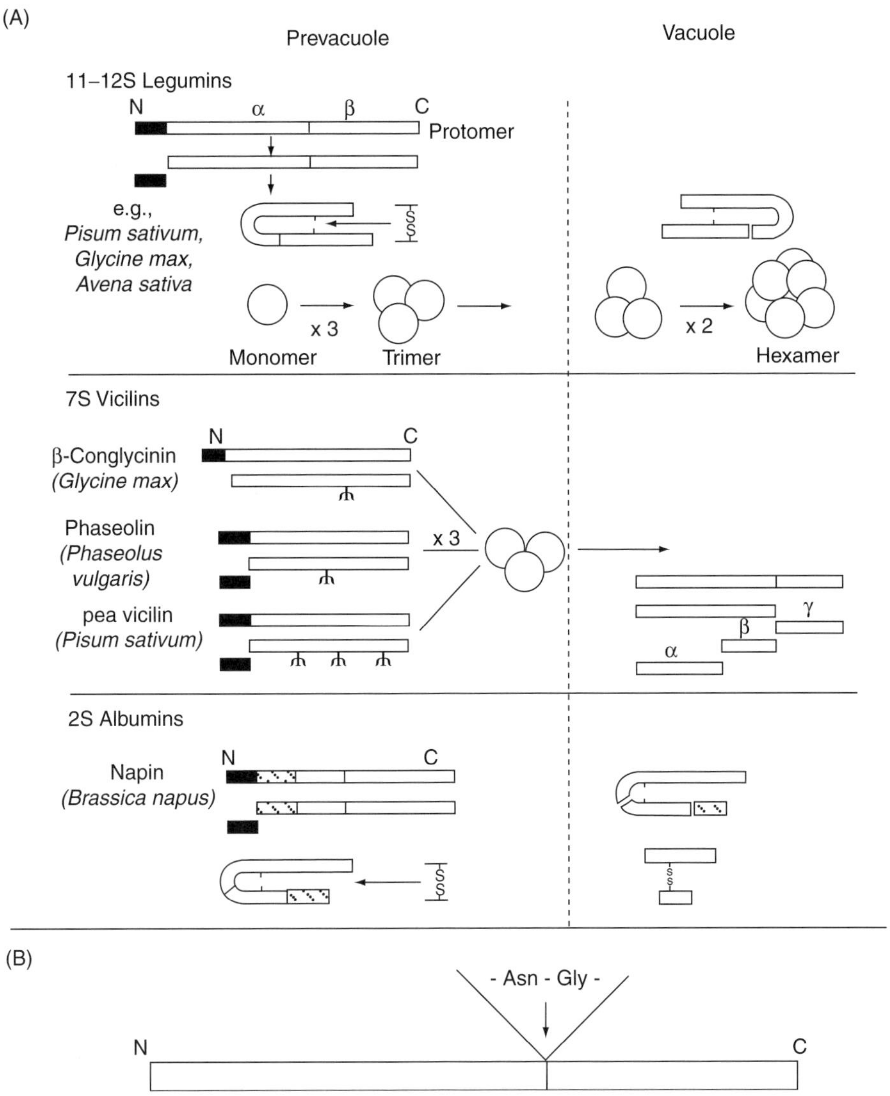

Figure 7 (A) Synthesis and processing of storage proteins of dicotyledonous seeds (2S albumins and 7S vicilins). Signal peptides are shown in solid black; propeptides are shown by hatched boxes and glycosylation sites are shown by inverted symbols. Processing events that take place *en route* to the vacuole are shown on the left; those occurring within the vacuole are shown on the right. (B) The proteolytic cleavage site used by vacuolar endoproteases, in which there is cleavage of the peptide bond on the C-terminal side of an asparagine (Asn) residue of the protein. The other amino acid is often, but not always, glycine. Reproduced with permission of Academic Press (Elsevier) from Srivastava LM (2001) *Plant Growth and Development*. San Diego: Academic Press.

acid sequences that are brought together by the folding of the protein to form a specific three-dimensional domain on the surface of the protein.

2. Cells of developing seed storage tissues may contain two vacuole types (protein storage vacuoles and lytic vacuoles) that differ in their luminal and membrane composition; sorting of proteins for targeting to these two types of vacuoles is proposed to occur at the trans-Golgi network, but involves different mechanisms and different vesicle types.
3. The targeting mechanism may involve a specific vacuolar sorting receptor that segregates proteins into clathrin coated vesicles at the trans-Golgi network. Another mechanism involves physical aggregation of storage polypeptides (involving protein–protein interactions); both the acidification of transport vesicles as well as the generally acidic pI of storage proteins may aid in the aggregation process.

Storage protein synthesis and deposition in seeds of cereals In seeds of some cereals (e.g. corn (*Zea mays*; maize), rice (*Oryza sativa*), and *Sorghum* spp.), there is direct formation of protein bodies from the rough ER. The prolamins (alcohol-soluble storage proteins abundant within the endosperm of

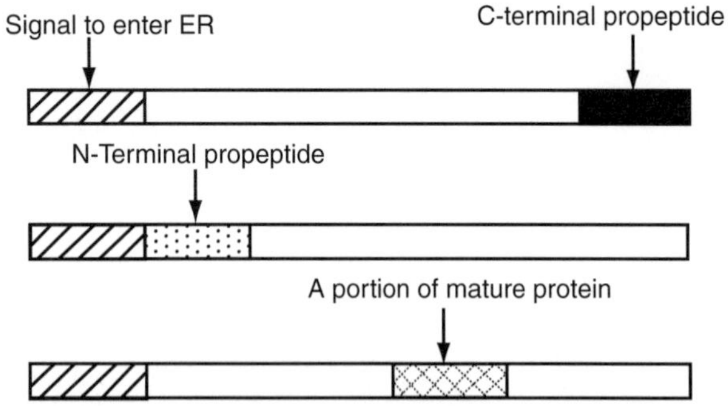

Figure 8 Types of vacuolar targeting signals on seed storage proteins. Shows three different regions in which the vacuolar sorting and targeting information occur in plant proteins: in the cleaved C-terminal or N-terminal propeptide, or internally within the mature protein. The signal peptide for insertion in the ER is shown for each protein. Reproduced with permission of Elsevier from Gal S and Raikhel NV (1993) Protein sorting in the endomembrane system of plant cells. *Current Opinions in Cell Biology* 5: 636–640.

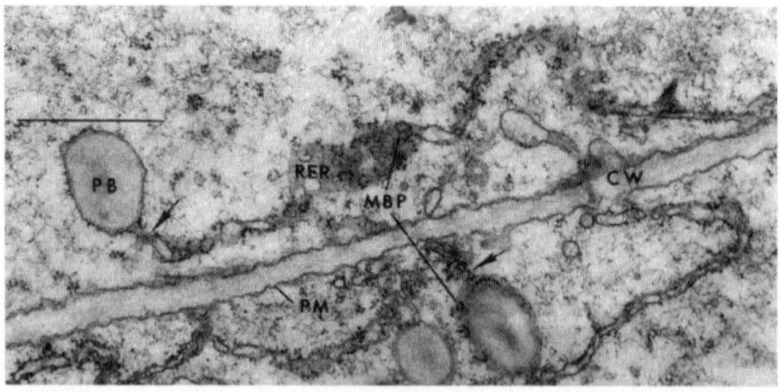

Figure 9 Direct protein body formation from the rough endoplasmic reticulum (RER). Electron micrograph of a developing cell of the starchy endosperm of corn to illustrate the formation of protein bodies (PB) forming from RER and the proximity of membrane bound polyribosomes (MBP). Note the continuity between the protein body membranes and extended RER cisternae (arrow). CW, cell wall. Reproduced with permission of the American Society of Plant Physiologists from Larkins BA and Hurkman WJ (1978) Synthesis and deposition of zein in protein bodies of maize endosperm. *Plant Physiology* 62: 256–263.

cereal grains; **Table 2**) are retained within the lumen of the rough ER to form protein deposits that remain surrounded by ER membrane (**Figure 9**). In contrast, prolamins of other cereals, such as wheat, barley and oats (*Avena sativa*) accumulate (along with globulins) in vacuole derived protein bodies, and in most cases transit through the Golgi complex to arrive there. Rice seeds accumulate two types of storage proteins: the globulin-like glutelins and the prolamins. The former are transported to vacuoles via the Golgi complex, while the latter aggregate within the ER lumen. These different pathways exhibited by the two storage protein types appear to be facilitated by a differential targeting of their respective mRNAs onto distinct rough ER membranes (cisternal ER and protein body ER).

Lipid Synthesis and Deposition

Seed storage oils (lipids) are generally triacylglycerols, i.e., esters of glycerol and fatty acids. The esterification of glycerol with fatty acids occurs in the cytoplasm, and newly synthesized triacylglycerols accumulate within specialized organelles referred to as oil bodies. These organelles are unique in that they are delimited by only a half-unit membrane; the size of oil bodies is species dependent and ranges between 0.6 and 2.0 μm in diameter. Generally, these organelles contain an integral membrane protein, oleosin, which may serve to maintain oil body integrity during the desiccation stage of seed development and/or serve as a docking site for lipases during postgerminative reserve mobilization. Lipid synthesis occurs on specific domains of smooth ER where the enzymes for triacylglycerol synthesis are located. The newly synthesized triacylglycerols accumulate between the two leaflets of the bilayer that comprise the ER membrane; once the ER membrane reaches a certain size, the lipid droplet buds off carrying the single leaflet with it (**Figure 10A**). Oleosins are synthesized concomitantly on the rough ER; they are targeted to the monolayer as a result of a highly

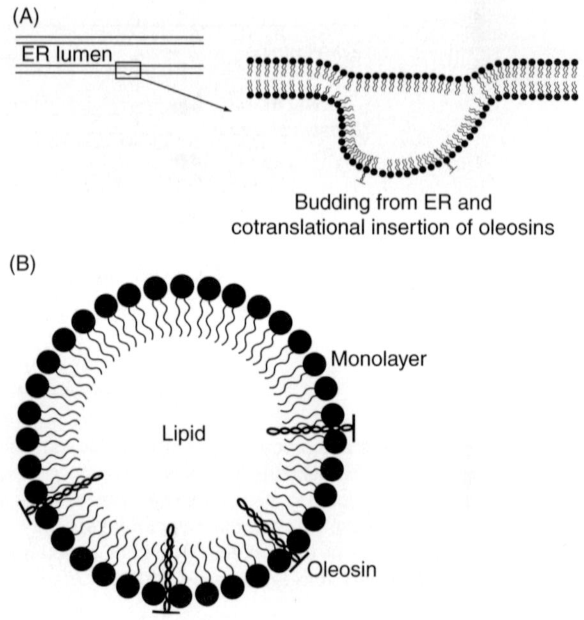

Figure 10 Model showing the formation of oil (lipid) bodies in seeds. (A) Synthesis and deposition of triacylglycerols and their accumulation between the two leaflets of the bilayer comprising the ER membrane. (B) Targeting of oleosin to the oil body. Reproduced with permission of Academic Press (Elsevier) from Srivastava LM (2001) *Plant Growth and Development.* San Diego: Academic Press.

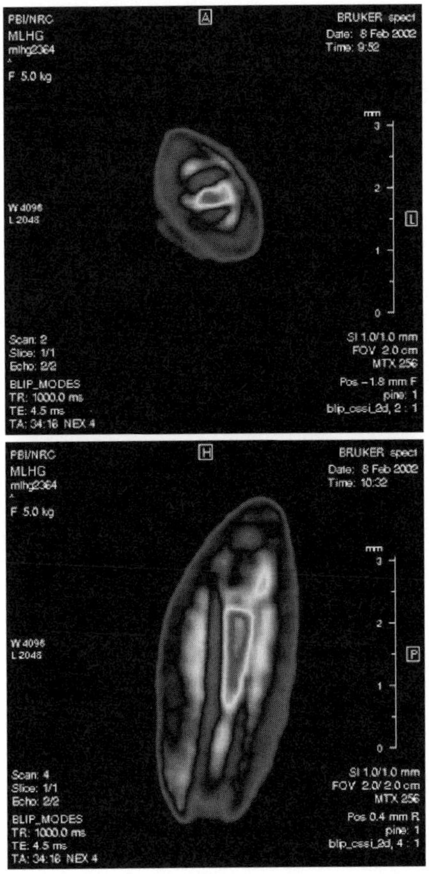

Figure 11 (see color plate 38) NMR showing distribution of lipid in a western white pine (*Pinus morticola*) seed shortly after imbibition. Photograph supplied courtesy of A. Feurtado and M. Grawel, Simon Fraser University and Plant Biotechnology Institute, Canada.

conserved internal hydrophobic domain (**Figure 10B**; note that **Figure 10B** does not accurately represent the orientation of oleosin, in which its central hydrophobic domain is embedded within the interior of the oil body, and the N-terminal and C-terminal regions of the protein extend beyond the surface of the oil body).

NMR is useful for examining the overall distribution of oils (lipids) in seeds and for examining gross changes in its accumulation during seed development and decline following postgerminative reserve mobilization (**Figure 11**).

Starch Synthesis and Deposition

Starch, a mixture of two structurally different glucose polymers (amylose and amylopectin), is synthesized and stored as starch granules contained within amyloplasts, special starch-storing plastids. In cereals, starch is accumulated mainly in the cells of the starchy endosperm, which are dead at seed maturity, and the starch grains are tightly appressed against protein bodies. In seeds of dicotyledonous species (e.g., pea and the common bean), starch is synthesized in the storage parenchyma cells of the embryonic cotyledons (and in the root-shoot axis) – cells that are living at maturity. Three major enzymes participate in starch synthesis: (1) ADP glucose pyrophosphorylase (which catalyzes the synthesis of ADP-glucose from glucose-1-phosphate and ATP); (2) starch synthase (which transfers the glucosyl unit from ADP-glucose to the nonreducing end of a pre-existing α-1,4-glucan primer); and (3) the starch branching enzyme (which hydrolyzes an α-1,4 bond and attaches the released α-1,4 glucan segment to the same or another glucose chain by an α-1,6-linkage, thereby producing branch points in the amylopectin polymer). **Figure 12** shows the biosynthetic steps in starch biosynthesis and a schematic representation of the levels of organization within the starch granule.

In certain mutants, there are defects in the biosynthetic pathway for starch (**Table 3**). In certain mutants of corn, this leads to the accumulation of sucrose and accounts for the sweetness of their seeds (referred to as sweetcorn).

Other Seed Reserves: Cell Wall Hemicelluloses and Phytin

Hemicelluloses, such as mannans, galactomannans, and arabinoxylans, are often deposited as food reserves in the cell walls of storage tissues, such as the endosperm or perisperm. In some seeds, such as those of fenugreek (*Trigonella foenum-graecum*), the walls become so thickened that they occlude the lumen of cells; as a result these cells are dead at maturity. In other seeds, such as those of date palm (*Phoenix dactylifera*), the cells that accumulate these reserves remain living at maturity. Seeds also accumulate mineral stores, often in the form of phytin, which is a mixed salt of phytic acid (myo-inositol hexaphosphoric acid). Phytin is generally deposited inside protein bodies, sometimes in inclusions known as globoids. Some aspects of the synthesis of phytin are not well understood; fusion of phytin containing vesicles (possibly ER derived) with the membrane of the protein storage vacuole allows phytin to be sequestered within this organelle.

Development of the Seed Coat

A key process that can have a major impact on seed viability and vigor is the development and hardening of the seed coat (testa), which involves many structural and biochemical changes. These include lignification of the outer cell walls of the seed coat

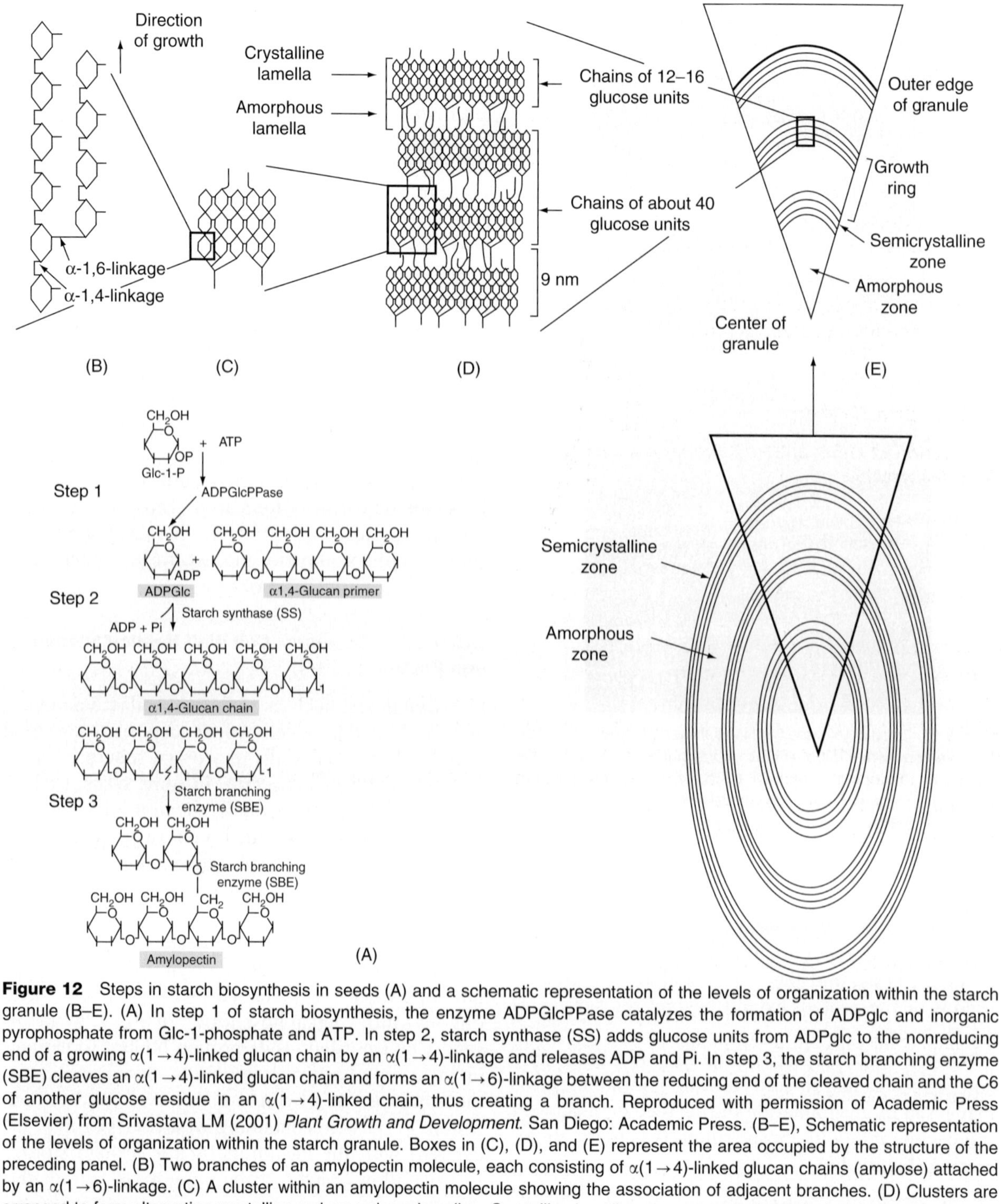

Figure 12 Steps in starch biosynthesis in seeds (A) and a schematic representation of the levels of organization within the starch granule (B–E). (A) In step 1 of starch biosynthesis, the enzyme ADPGlcPPase catalyzes the formation of ADPglc and inorganic pyrophosphate from Glc-1-phosphate and ATP. In step 2, starch synthase (SS) adds glucose units from ADPglc to the nonreducing end of a growing α(1→4)-linked glucan chain by an α(1→4)-linkage and releases ADP and Pi. In step 3, the starch branching enzyme (SBE) cleaves an α(1→4)-linked glucan chain and forms an α(1→6)-linkage between the reducing end of the cleaved chain and the C6 of another glucose residue in an α(1→4)-linked chain, thus creating a branch. Reproduced with permission of Academic Press (Elsevier) from Srivastava LM (2001) *Plant Growth and Development.* San Diego: Academic Press. (B–E), Schematic representation of the levels of organization within the starch granule. Boxes in (C), (D), and (E) represent the area occupied by the structure of the preceding panel. (B) Two branches of an amylopectin molecule, each consisting of α(1→4)-linked glucan chains (amylose) attached by an α(1→6)-linkage. (C) A cluster within an amylopectin molecule showing the association of adjacent branches. (D) Clusters are arranged to form alternating crystalline and amorphous lamellae. Crystalline lamellae are more closely packed and represent a more ordered arrangement of amylose and amylopectin chains than amorphous lamellae; the two together form the semicrystalline zone. (E) Slice through a starch granule showing alternating zones of semicrystalline and amorphous zones. B–E reproduced with permission from Elsevier Press from Smith AM (1999) Making starch. *Current Opinions in Plant Biology* 2: 223–229.

and the deposition of cuticle and waxy substances. Seed coat thickness is a factor that has recently been implicated in the control of seed dormancy, and regulation of testa thickness by hormones such as abscisic acid is being investigated.

Accumulation of Compounds with a Defense Role

Seeds generally accumulate a number of compounds that have a defense role. Some of these are deposited

Table 3 Seed mutants with defects in the starch biosynthesis pathway

Enzyme affected	*Mutant*	*Component affected*	*Plant*
ADPGlcPPase	*Shrunken2*	Large subunit	Corn
	brittle2	Small subunit	Corn
	Rugosub	Large subunit	Pea
SS	*Waxy*	One of GBSS	Corn, rice, potato (*Solanum tuberosum*)
SBE	*sbe1 and sbe2*	Two SBEs (in family A)	Corn
	Rugosus	SBE (family A)	Pea

SS, starch synthase; SBE, starch branching enzyme; GBSS, granule-bound starch synthase. Reproduced with permission of Academic Press (Elsevier) from Srivastava LM (2001) *Plant Growth and Development*. San Diego: Academic Press.

within protein storage vacuoles along with storage proteins and include lectins, protease inhibitors, and α-amylase inhibitors.

Maturation Drying

Acquisition of Desiccation Tolerance during Seed Maturation

The maturation drying period (desiccation) in many seeds is initiated by the severing of the vascular supply from the parent plant to the seed (funiculus detachment) and senescence of the pod or capsule. Similarly, pectic substances in the lumina of xylem elements of the rachis of wheat and barley (laid down during the final stages of grain maturation) may lead to the progressive dehydration of the ear by cutting off its water supply. Whatever the route of water loss by seeds, the expression of genes that specify products important for the acquisition of desiccation tolerance occurs earlier, during the maturation (expansion) phase of seed development (**Figure 2**). During very early development, seeds are generally intolerant of drying, but they later undergo a transition to a desiccation tolerant state at a particular time. Generally, this occurs around mid-development for the so-called orthodox seeds.

The majority of seeds are referred to as "orthodox," in which desiccation occurs as a preprogramed and final stage in their development (**Figure 2**). Seeds of the orthodox type and other desiccation tolerant structures such as spores and pollen are unique in the degree of water loss tolerated; as much as 90–95% of the original water is removed during their development. In this dehydrated state, the seed can survive the vagaries of the environment, and, unless dormant, will resume full metabolic activity, growth, and development when conditions conducive to germination are provided. Some of the mechanisms underlying desiccation tolerance of seeds, including some of the putative protective molecules and cellular and subcellular changes, are noted in **Table 4**.

Orthodox seeds can be stored for long periods under conventional conditions, i.e., in the dry state and at low temperature. Recalcitrant seeds, on the other hand, do not undergo maturation drying, nor are they capable of withstanding water loss of the magnitude of that experienced by orthodox seeds. The seeds are shed at relatively high moisture contents and are highly susceptible to desiccation injury; in order to remain viable, they must not undergo any substantial change in moisture. They are not storable under conditions suitable for orthodox seeds, and even when stored under moist conditions, their viability is frequently brief and only rarely exceeds a few months. Thus, the terms orthodox and recalcitrant have been used to describe the storage behavior of seeds. A category intermediate between orthodox and recalcitrant is now recognized (e.g., coffee, *Coffea* spp.), in which seeds survive desiccation but become damaged during dry storage at low temperatures (0°C and −20°C). It is important to note, however, that the situation is in fact more complex, and there is a gradual continuum of desiccation tolerance across orthodox and recalcitrant species.

Desiccation Acts as a "Switch" to Terminate Development and Promote the Transition to a Germination and Growth Program

The mature, quiescent seed is able to commence germination immediately when hydrated under the appropriate conditions. There is substantial evidence to suggest that it is the final stage of seed development – maturation drying – that enables seeds to "switch off" or downregulate the transcription of genes associated with seed development (e.g., storage protein genes). Moreover, desiccation promotes a normal transition to a germination and postgerminative program, and its molecular role in this transition has been elucidated (see "Further Reading"). An interesting aspect of control by desiccation is its ability to overcome some of the constraints of the seed environment that maintain embryos in a developmental mode (see below).

Control of Seed Maturation

Elucidation of some of the important events that control seed maturation is undoubtedly one of the

Table 4 Possible components of desiccation tolerance in seeds and their protective action

Site or process affected	*Protective component*	*Protective action*	*Possible mode of protection*
Membranes	Carbohydrates: sucrose plus raffinose and/or stachyose	Prevent changes in selective permeability due to lateral phase separation of phospholipids in the bilayer and the phase transition from liquid crystalline to gel	Hydroxyl groups of sucrose replace water on hydrophilic (polar) end groups of membrane phospholipids; oligosaccharide (raffinose and/or stachyose) inhibits sucrose crystallization during drying, preventing loss of its protective potential
	Lipid-soluble antioxidants (e.g., tocopherols)	As above; prevent de-esterification of membrane phospholipid and free fatty acid accumulation	Scavenging activity increases resistance to free radical mediated desiccation injury
Structure/metabolism	Reserves: carbohydrates, lipids, proteins	Prevent "whole scale" mechanical disruption of cellular components	Critical level of reserves in vacuoles/storage bodies confers mechanical strength to whole cell
		Prevent loss of tightly bound ("vital") water necessary for structural and functional integrity of biomolecules	Water binding capacity of cells enhanced with increased number of sorption sites
	Hydrophilic, denaturation resistant proteins (e.g., LEAs, other desiccation inducible polypeptides)	As above	Water binding capacity of cells enhanced with increased number and strength of sorption sites; native conformation of protective molecules maintained throughout drying; bind ions and thereby counteract damaging effects of increasing ionic strengths of cytosol during drying
	"Repair" proteins, proteases, ubiquitin and extension protein, HSP/molecular chaperones, some LEAs	Rapid re-establishment of structural and metabolic integrity following imbibition	Efficient repair of membranes and other cellular components restores normal functioning; aid proteins in recovering their native conformation; degradation of damaged or denatured proteins

Reproduced with permission of CRC Press from Kermode AR (1990) Regulatory mechanisms involved in the transition from seed development to germination. *Critical Reviews in Plant Sciences* 9: 155–195.

most exciting areas of seed biology. Key interactions between different hormone signaling pathways and other regulatory cues provided by the seed environment are involved. Collectively, these are important for controlling most of the processes that ultimately determine seedling vigor – protein and oil reserve accumulation, regulation of precocious germination, induction of dormancy, and development of stress tolerance.

Embryos excised from developing seeds at certain stages often germinate when placed on water or on a minimal culture medium. Thus, even though they have not completed their development, embryos germinate precociously, but often fail to establish viable seedlings. Therefore, an important role of the seed environment is to maintain embryos in a developmental mode until they are fully formed and have accumulated sufficient reserves to permit successful germination and subsequent seedling establishment. Studies to elucidate the mechanisms whereby the seed environment retains the embryo in a developmental mode and suppresses germination have implicated several regulatory factors, including abscisic acid (ABA) and restricted water uptake. One component of the chemical environment surrounding the developing embryo is the osmotic environment. This specialized environment maintains a highly negative osmotic potential during development, which in turn is thought to suppress precocious germination of embryos by restricting water uptake.

As noted above, a large component of growth or maturation of the developing embryo and seed storage tissues is the laying down of reserve materials. Vigor is a measure of a seedling's ability to emerge and become established over a range of environmental conditions. This in turn is largely dependent upon earlier events during seed development – in particular, the stable accumulation of

Figure 13 (see color plate 39) Vivipary in mangroves (A) and in mutant (*vp-1*) corn seeds (B). Red mangrove (*R. mangle*) plants grow in intertidal waters. The seeds germinate and grow to lengths of 20–25 cm while still on the parent plant. Note a cluster of seedlings with long sword-like roots. Eventually the seedlings fall and become planted in the mud below. (A) Reproduced with permission of Academic Press (Elsevier) from Srivastava LM (2001) *Plant Growth and Development*. San Diego: Academic Press. (B) Reproduced with permission of Cell Press from McCarty DR, Hattori T, Carson CB, Vasil V, Lazar M, and Vasil IK (1991) The *Viviparous-1* developmental gene of maize encodes a novel transcriptional activator. *Cell* 66: 895–905.

reserves (protein and lipid or starch) within the storage parenchyma cells of the embryo and seed storage organs. A role for ABA has been established in reserve deposition, particularly in promoting the synthesis of storage proteins and lipids.

Continued axial growth of the embryo (germination) very rarely takes place within the (maturing) seed on the plant. Two exceptions are (1) vivipary and (2) preharvest sprouting of some cultivars of wheat and barley (and several other species). Suggestive evidence for a role of ABA in the control of germination of the developing seed is provided by cases of vivipary. This condition is characterized by germination of the embryo within the fruit on the parent plant. There is an uninterrupted progression from embryogenesis to germination with little or no cessation of growth (quiescence) and, in most cases, little or no dehydration. Relative insensitivity to ABA appears to be the case in mangroves (*Rhizophora mangle*; **Figure 13A**), where vivipary is a normal occurrence.

Preharvest sprouting (unlike vivipary) occurs in cereal grains at physiological maturity when there is sufficient rain before harvest (**Figure 14**). Serious losses to grain quality occur, particularly in wheat; damage occurs as the hydrolytic enzymes associated with postgerminative reserve mobilization degrade the starch and protein in the grain. Again, a relative insensitivity to ABA appears to underlie the sprouting susceptibility that characterizes some of the wheat, barley, and sorghum cultivars that are particularly prone to this phenomenon.

Figure 14 Preharvest sprouting in wheat. This phenomenon refers to germination of grains on the parent plant at physiological maturity, and is prevalent in climates characterized by cool, damp autumns or heavy rainfall around the time of harvest. The three ears on the left have sprouted; they are from plants of cultivar Boxer that were grown under cool, moist conditions during the later stages of grain maturation. The three ears on the right are from plants of the same variety grown under identical conditions, but show no visible signs of sprouting, indicating intrapopulation variability. Reproduced with permission of Academic Press (Elsevier) from Srivastava LM (2001) *Plant Growth and Development*. San Diego: Academic Press.

Mutants of corn (**Figure 13B**) and *Arabidopsis* that are either deficient in ABA or exhibit a relative insensitivity to ABA (so-called "response mutants") have been invaluable for elucidating the role of ABA and for defining the components of the ABA signaling pathways that control maturation events and interact with other signaling pathways. Screens to identify suppressors of ABA signaling mutants have also uncovered interesting evidence for cross-talk between hormone signaling pathways. These and other approaches have led to the following conclusions regarding the control of maturation:

1. ABA, as well as promoting storage reserve synthesis and deposition, plays a role in the acquisition of desiccation tolerance during seed development. For example, ABA has a promotive role in the accumulation of desiccation protectants (including soluble sugars and LEAs – Late Embryogenesis Abundant proteins). This is shown in **Figure 15**, where a severe mutant of *Arabidopsis* (exhibiting relative ABA insensitivity; *abi3-6*) produces seeds that are intolerant of desiccation. Seed viability is not altered in ABA-deficient (*aba*) and the less severe ABA-insensitive (*abi-3)* mutants of *Arabidopsis*, yet seeds of double mutants exhibiting these two traits do not undergo desiccation on the parent plant, are intolerant of artificial desiccation, and fail to produce some of the late abundant proteins (LEAs). These double-mutant seeds accumulate only low amounts of the major storage proteins, and are deficient in several low molecular weight polypeptides, both soluble and bound, some of which are heat stable. During development (14–20 days after pollination), the low amounts of various maturation specific proteins are degraded.
2. ABA plays a key role in dormancy inception during seed development. Mutants with defects in ABA signaling pathways have reduced dormancy, which is generally accompanied by disruption of seed maturation and precocious expression of germinative/postgerminative genes.
3. Characterization of the ABA insensitive mutants of corn and *Arabidopsis* led to the cloning of novel transcription factors. Lesions in the genes encoding these factors underlie some of the phenotypic characteristics of these mutants that are disrupted in seed maturation. For example, the *VP1* gene of cereals and the *ABI3* gene of *Arabidopsis* encode novel transcription activators, which play a role in the expression of ABA responsive genes during seed development (e.g., storage protein and *LEA* genes). The promoter elements of target genes under control of these transcription factors (those that mediate regulation by ABI-3 and VP-1) have been identified. These include TACGTGTC (an ABA responsive element, ABRE), a small sequence

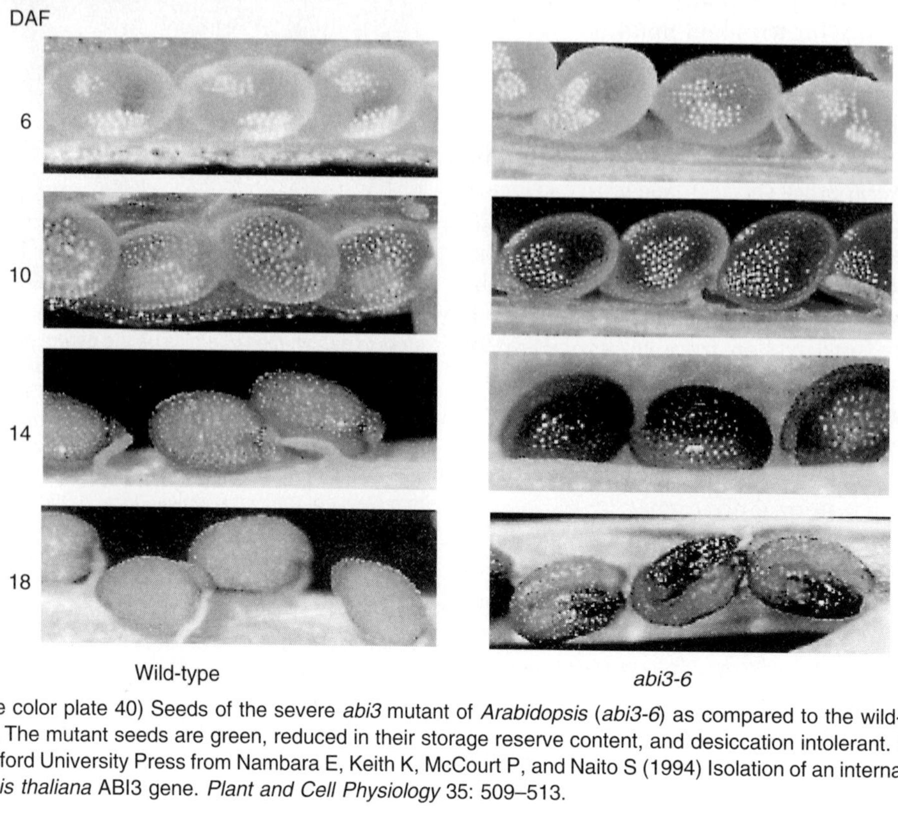

Figure 15 (see color plate 40) Seeds of the severe *abi3* mutant of *Arabidopsis* (*abi3-6*) as compared to the wild-type. DAF, days after fertilization. The mutant seeds are green, reduced in their storage reserve content, and desiccation intolerant. Reproduced with permission of Oxford University Press from Nambara E, Keith K, McCourt P, and Naito S (1994) Isolation of an internal deletion mutant of the *Arabidopsis thaliana* ABI3 gene. *Plant and Cell Physiology* 35: 509–513.

located just downstream of the ABRE, and a quantitative element (the sph box/RY repeat), conserved in many seed specific gene promoters.

4. Redundancy in developmental and physiological response pathways is the rule rather than the exception in seeds and is exemplified by the emerging and mounting evidence for cross-talk between developmental and hormonal response pathways. These studies have determined that there are key interactions between the ABA, ethylene, and gibberellin (GA) signaling pathways in controlling seed maturation and the inception of dormancy. For example, the locus *ERA3* of *Arabidopsis* (in which mutations lead to phenotypes characterized by enhanced responsiveness to ABA and prolonged seed dormancy) is allelic to the *ethylene insensitive2* (*EIN2*) locus. The *EIN2* gene encodes a novel integral membrane protein with unknown molecular function. Mutations in this pathway lead to an overaccumulation of ABA suggesting that *ERA3* (*EIN2*) is a negative regulator of ABA synthesis. In addition, mutations that decrease ethylene sensitivity (such as *ein2*) increase the sensitivity of seeds to ABA. Conversely, seeds of mutants with a constitutive ethylene response (*ctr1* mutants) show reduced responsiveness to ABA, similar to the ABA response of wild-type seed in the presence of the ethylene precursor ACC. Like ethylene, GA appears to act as an antagonist to ABA function during seed development and germination. Interestingly, when an ABA deficient mutant of corn (*vp-5*) is manipulated either genetically or via biosynthesis inhibitors to induce GA deficiency during early seed development, vivipary is suppressed in developing kernels and the seeds acquire desiccation tolerance and storage longevity. Major accumulation of GA_1 and GA_3 occurs in wild-type corn kernels, just prior to a peak in ABA content during development. It is speculated that these GAs induce a developmental program that leads to vivipary in the absence of normal amounts of ABA, and that a reduction of GAs re-establishes an ABA/GA ratio appropriate for suppression of germination and induction of maturation. Induction of GA deficiency does not suppress vivipary in *vp-1* mutant kernels, suggesting that VP1 (see above) acts downstream of both GA and ABA in programing seed development.
5. Recessive mutants of *Arabidopis* with lesions at the *FUSCA3* (*FUS3*) and *Leafy Cotyledon* (*LEC1*) gene loci lead to various abnormalities during mid-embryogenesis and late embryogenesis, including loss of dormancy and failure to acquire desiccation tolerance. FUS3 and LEC1 modulate the abundance of ABI3 protein in seeds, and synergistic interactions between the three proteins (ABI3, FUS3, and LEC1) are thought to control various key events, including accumulation of chorophyll and anthocyanins, sensitivity to ABA, and expression of individual members of the 12S storage protein gene family. Interestingly, part of FUS3 (a continuous stretch of 100 amino acids) shows significant similarity to the B3 domain of ABI3 and VP-1 proteins, a domain that interacts with the RY *cis* promoter motif of several seed proteins. Thus, both FUS3 and ABI3 may be essential components of a regulatory network acting in concert through the RY-promoter element to control gene expression during late embryogenesis and seed development.
6. ABI5, a member of the family of basic leucine zipper transcription factors, regulates a subset of *LEA* genes during seed development and in vegetative tissues in the presence of ABA. It has been implicated in ABA regulated gene transcription mediated by ABI3/VP1 proteins, and direct physical interactions between ABI3 and ABI5 (mediated by the B1 domain of ABI3) have recently been demonstrated. ABI5 itself may play a key role during postgerminative stages of seedling development; more specifically it reinduces seedling "quiescence" under adverse environmental conditions. There are also interesting interactions between hormone signaling pathways and sugar sensing. During germination and early postgerminative seedling development/growth, sugars can repress nutrient mobilization, hypocotyl elongation, cotyledon greening and expansion, and shoot development. High sugar accumulation during early seedling development may elicit a protective mechanism (i.e., developmental arrest), as it probably reflects poor growth conditions at a stage when the young seedling is highly vulnerable.

Figure 16 shows some of the seed maturation processes that are modulated by ABA and other hormones, and **Figure 17** illustrates some of the factors that interact with ABI3 to control maturation events and prevent a premature switching to a germination and growth program.

Novel Approaches to Elucidate the Biochemical and Physiological Basis of Seed Maturation

It is presently an exciting time to undertake the challenges of understanding the biochemical, physiological, and genetic components that control seed

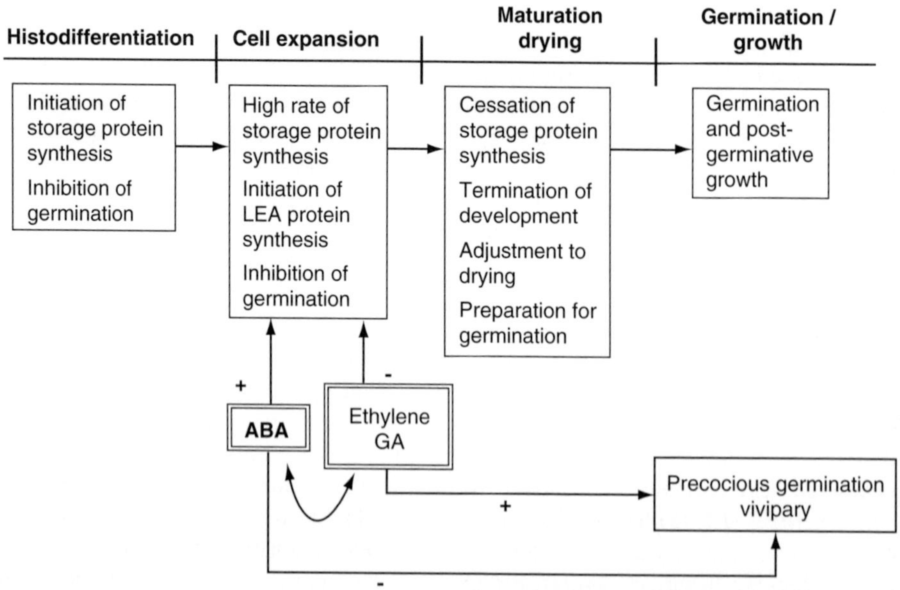

Figure 16 Events during the development and germination/growth of seeds that are affected by ABA and other hormones. Desiccation tolerance is generally acquired by seeds around mid-maturation, when LEA proteins and other protective substances are synthesized. Induction of a subset of LEA proteins also occurs following the transition to germination and growth, i.e., in seedlings and plant vegetative tissues when they are subjected to water deficit related stresses. Reproduced with permission from Marcel Dekker, based on Kermode AR (1995) Regulatory mechanisms in the transition from seed development to germination: interactions between the embryo and the seed environment. In: Galili G and Kigel J (eds) *Seed Development and Germination*, pp. 273–332. New York: Marcel Dekker.

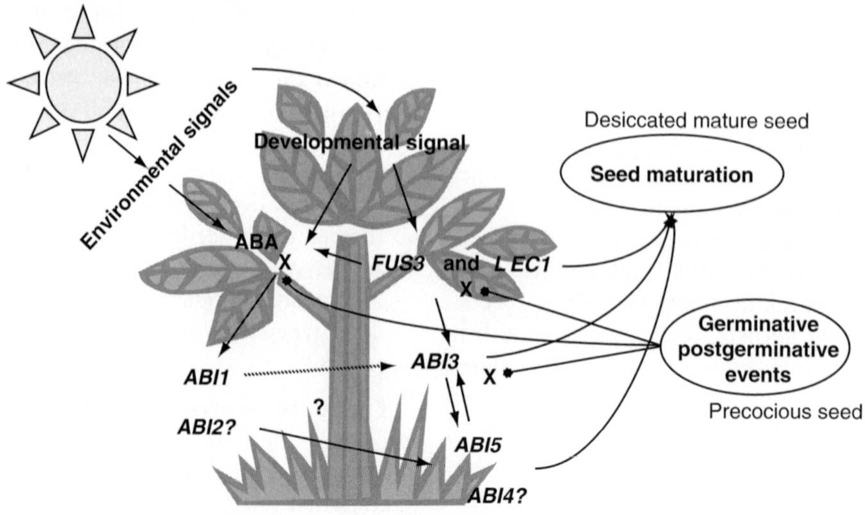

Figure 17 Factors that interact with ABI3 to control maturation events and prevent a premature switching to a germination and growth program. The actions of FUS3 and LEC1 are shown upstream of ABI3; this is based on the findings that FUS3 and LEC1 can regulate the amount of ABI3 protein and, generally, defects within these genes cause more widespread phenotypic consequences on developmental processes than those caused by *ABI3* gene mutations. Adapted with permission of Elsevier from Bonetta D and McCourt P (1998) Genetic analysis of signal transduction. *Trends in Plant Science* 6: 231–235.

maturation. Several novel approaches are now available including the yeast one- and two-hybrid systems for examining protein–protein interactions, and the use of differential display reverse transcription-PCR, gene and enhancer trap tagging, and microarrays to identify maturation specific genes. Proteomics approaches could yield invaluable information concerning posttranslational controls involved in hormone signaling pathways that control maturation. These and similar research avenues may yield more decisive results than the traditional approaches.

Finally, it is noteworthy that the control of seed maturation is a complex and multifaceted property. There are clearly a multitude of genes the expression of which mediate events beyond histodifferentiation to control reserve synthesis and deposition, inception of dormancy, acquisition of stress tolerance, and other processes that ultimately control seed viability and seedling vigor.

See also: **Seed Development**: Endosperm Differentiation; Seed Production; Seed Quality; Starch Synthesis. **Seed Dormancy**: Development of Dormancy; Genetics of Dormancy; Preharvest Sprouting.

Further Reading

Bewley JD and Black M (1994) *Seeds. Physiology of Development and Germination*, 2nd edn. New York: Plenum Press.

Bradford K and Bewley JD (2003) Seeds. Biology, technology, and role in agriculture. In: Chrispeels MJ and Sadava DE (eds) *Plants, Genes, and Crop Biotechnology*, 2nd edn, pp. 212–239. American Society of Plant Biologists (ASPB) and ASPB Education Foundation. Sudbury: Jones and Bartlett Publishers.

Finkelstein RR, Gampala SSL, and Rock CD (2002) Abscisic acid signaling in seeds and seedlings. *The Plant Cell (Supplement)*: S15–S15.

Gazzarina S and McCourt P (2001) Genetic interactions between ABA, ethylene and sugar signaling pathways. *Current Opinion in Plant Biology* 4: 387–391.

Kermode AR (1990) Regulatory mechanisms involved in the transition from seed development to germination. *Critical Reviews in Plant Sciences* 9: 155–195.

Kermode AR (1995) Regulatory mechanisms in the transition from seed development to germination: interactions between the embryo and the seed environment. In: Galili G and Kigel J (eds) *Seed Development and Germination*, pp. 273–332. New York: Marcel Dekker.

Kermode AR and Bewley JD (1999) Synthesis, processing and deposition of seed proteins: The pathway of protein synthesis and deposition in the cell. In: Shewry PR and Casey R (eds) *Seed Proteins*, pp. 807–841. Dordrecht: Kluwer Academic Publishers.

Kermode AR and Finch-Savage W (2002) Desiccation sensitivity in orthodox and recalcitrant seeds in relation to development. In: Black M and Pritchard H (eds) *Desiccation and Plant Survival*, pp. 149–184. Wallingford: CAB International.

Srivastava LM (2001) *Plant Growth and Development*. San Diego: Academic Press.

Vertucci CW and Farrant JM (1995) Acquisition and loss of desiccation tolerance. In: Galili G and Kigel J (eds) *Seed Development and Germination*, pp. 701–746. New York: Marcel Dekker.

Onset of Desiccation Tolerance

W E Finch-Savage, Horticulture Research International, Wellesbourne, UK

Introduction

During development, the seeds of many species progressively gain the ability to withstand significant moisture loss. At maturity, they can remain viable after as much as 90–95% of their water has been removed. In this dehydrated state, the seed can survive for long periods and, unless dormant, will resume full metabolic activity, growth, and development when conditions are suitable for germination. There is a clear quantitative relationship between seed moisture content and longevity in storage, which has great importance for mankind in the supply and distribution of seeds for agriculture and, more recently, in the conservation of plant genetic resources. This article summarizes the changes that occur as seeds acquire desiccation tolerance during their development, and outlines current understanding of the underlying mechanisms involved.

Desiccation Tolerance Differs between Species

The development of most angiosperm seeds is often considered to be divided into three stages (**Figure 1**). During the initial histodifferentiation stage, the single-celled zygote undergoes extensive mitotic division, and the resultant cells differentiate to form the embryo. At the same time the triploid endosperm or haploid megagametophyte is formed. In the second stage, cell division ceases during seed expansion and deposition of reserves. In most cases, seeds progressively gain desiccation tolerance during this stage, although, at the end of development, seeds from different species can differ in their capacity to withstand water loss. In the majority of species, the development of seeds is terminated in the third stage by some degree of drying (maturation drying), which results in a gradual reduction in metabolism as water is lost from seed tissues and the embryo passes into a metabolically inactive, or quiescent, state. In these so-called "orthodox" seeds, desiccation occurs as a preprogramed and final stage in their development. A second type of seeds (recalcitrant), on the other hand, does not undergo maturation drying, nor are they capable of withstanding the extent of water loss tolerated by orthodox seeds.

Recalcitrant seeds are shed at relatively high moisture contents and can be highly susceptible to

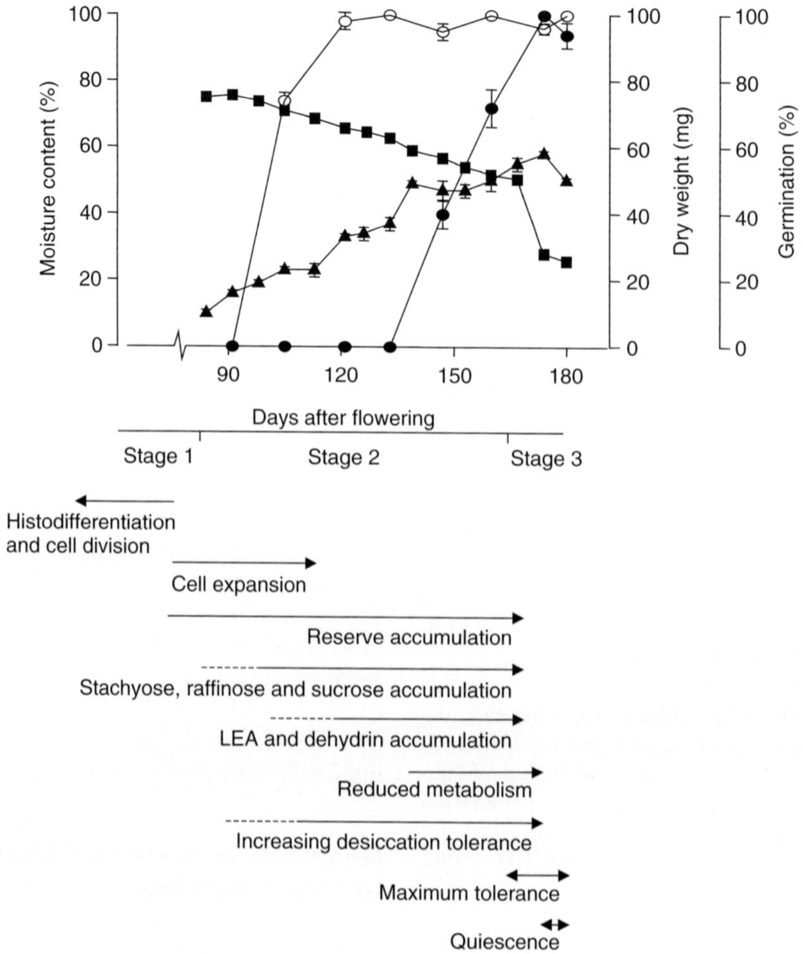

Figure 1 An example of orthodox seed development in *Acer platanoides* (Norway maple). Changes in dry weight (▲), percentage moisture content (■) and germination (●, ○), before (○) and after (●) drying to 10% moisture content. Germination after drying to this moisture content indicates the acquisition of full desiccation tolerance. Tolerance to this level of desiccation develops gradually during reserve accumulation. Other changes that are thought to be desiccation-tolerance-related (see text) are also shown. Seeds of this species are dormant throughout development and require low temperature treatment to induce germination. Data are redrawn from Greggains V, Finch-Savage WE, Quick WP, and Atherton NM (2000) Putative desiccation tolerance mechanisms in orthodox and recalcitrant seeds of the genus *Acer*. *Seed Science Research* 10: 317–327.

desiccation injury. In contrast, orthodox seeds can be dried and stored for long periods in the dry state and at low temperature. Recalcitrant seeds do not survive these conditions and even when stored under moist conditions, their viability is frequently brief and only rarely exceeds a few months. The terms orthodox and recalcitrant refer to this difference in storage potential rather than their ability to withstand desiccation. The association of these terms with desiccation tolerance has been misleading, since there is no clear division into two categories on the basis of tolerance and a gradual continuum of desiccation tolerance exists across orthodox and recalcitrant species.

The moisture contents to which seeds of different species can survive dehydration at the end of development, ranges from just less than that of vegetative tissues to almost complete tolerance. A recent study has shown a continuous scale of desiccation tolerance across 64 orthodox and recalcitrant species, with critical water contents for survival ranging from 0.1 to $>1.2\,\mathrm{g\,g^{-1}}$ dry weight. Even in genetically related species there can be wide variation in desiccation tolerance at shedding, for example, within the genera *Acer*, *Citrus*, *Coffea*, *Dipterocarpus*, *Hopea*, and *Shorea*.

Development of Orthodox Seeds

The three phases of seed development described above are marked by distinctive changes in fresh weight, dry weight, and water content (**Figure 1**). During histodifferentiation and early cell expansion, there is a rapid increase in whole seed fresh weight and water content. Generally, the amount of water in the seed remains relatively stable as dry weight rapidly increases during the later part of the seed expansion phase of development. There is an associated decline in percentage water content as reserves are deposited. Then, as the seed undergoes maturation drying and approaches quiescence, there is a period of water loss, resulting in a rapid decline in percentage seed water content.

Little is known about the mechanism and route of water loss from seeds. Some studies suggest that water is lost primarily by evaporation from the surface of surrounding seed structures. Another suggestion is that water moves from the seed to the parent plant by a metabolically active process. However, in soybean (*Glycine max*) and castor bean (*Ricinus communis*), the desiccation period is thought to be initiated by the severing of the vascular supply to the seed, which suggests that water loss from the seed to the parent plant is an unlikely cause. Similarly, the accumulation of pectic substances in the lumina of xylem elements of the rachis of wheat (*Triticum* spp.) and barley (*Hordeum vulgare*) may reduce water supply to the ear causing dehydration.

During very early development, seeds are generally intolerant of drying and then later become desiccation tolerant (**Figure 1**). In many cases, drying of seeds at a desiccation tolerant stage of their development promotes germination upon rehydration. For example, seeds of *Phaseolus vulgaris* (french bean) become desiccation-tolerant about 26 DAP (days after pollination), approximately half-way through development. Between 26 and 32 DAP an increasing proportion of seeds can be induced to germinate, whereas those dried at 22 DAP fail to germinate when rehydrated owing to a lack of cellular and metabolic integrity. After 32 DAP, seeds are both tolerant of desiccation and germinability can be induced by drying. The transition to a desiccation-tolerant state approximately midway through development is also characteristic of seeds from other commonly studied species such as soybean, corn (*Zea mays*; maize) and barley. Therefore, desiccation tolerance can be gained well before the completion of other major developmental events, such as reserve deposition and the start of maturation drying. In other species, such as *Acer platanoides*, illustrated in **Figure 1**, desiccation tolerance is initiated later and full tolerance of all seeds is not achieved until maturation drying. Once dried, the seeds remain desiccation-tolerant during subsequent imbibition, but progressively lose tolerance of re-imposed desiccation at some stage during germination. At any one time during development, the susceptibility to desiccation injury can differ between seed tissues and this difference may remain at maturity, although this is more commonplace in recalcitrant seeds.

Seed Development Differs in Recalcitrant Species

With the exception of highly desiccation-sensitive species, the physiology of seed development, following initial histodifferentiation, has strong similarities across the recalcitrant species studied so far and is similar to that of orthodox seeds before they reach maximum dry weight (mass maturity) and maturation drying. During development, tolerance of desiccation increases throughout reserve accumulation as percentage moisture content decreases in most recalcitrant seeds as it does in orthodox species. These changes are associated with a reduction in vacuolar volume and the appearance of a semi-viscous state. However, in general, recalcitrant seeds at shedding have had almost no net loss of water; they can still be accumulating dry weight, they retain active metabolism, remain desiccation-sensitive, and have no requirement for desiccation to stimulate subsequent germination. There appears to be no clear end point to recalcitrant seed development and seeds are shed at different moisture contents in different years on the same plant. Tolerance tends to increase to a maximum near the time of shedding as moisture content declines and so those shed with the lowest moisture content are most tolerant to desiccation.

Estimates of Desiccation Tolerance Can Vary

The impact of drying rate is species- and developmental stage-specific; it is, therefore, necessary to define the rate of water loss when reporting the extent of desiccation tolerance of seeds during development. For example, seeds of several legumes are unable to withstand rapid drying when harvested at early stages of development, but survive slow drying. In general, tolerance of rapid drying in orthodox seeds occurs only at or near the completion of reserve deposition just after the onset of natural drying. Gradual water loss may allow protective changes to occur and hence increase the seed's resistance to desiccation injury, whereas, rapid drying may not allow time for these protective changes to take place so that considerable disruption to cellular membranes and internal structures could occur. Repair cannot take place during drying itself when the seed reaches a critical dry (and quiescent) state before these processes can be initiated. In addition, repair during subsequent imbibition may be limited because the influx of water can be too rapid for the weakened cell to withstand.

In contrast to these general findings with orthodox seeds, recalcitrant seeds can appear more desiccation tolerant when rapidly dried because the extent of damage is time dependent, and thus rapid drying minimizes damage during drying to any given moisture content. Temperature is also a factor in the survival of recalcitrant seeds during drying because low temperatures are not tolerated in many tropical species. A further complicating factor in some recalcitrant seeds is that the onset of metabolism linked to germination has been observed close to the time of harvest, or natural shedding, and this will progressively increase sensitivity to drying.

The Acquisition of Desiccation Tolerance

The increase in desiccation tolerance as water content declines, with the addition of seed reserves and the apparent continuous range of critical moisture contents observed across species, suggests that desiccation tolerance is a quantitative feature. However, it is more accurate to express the degree of desiccation tolerance in terms of water potential rather than moisture content, as this reflects the amount of water available to the cytoplasm. When data are presented in this way, there is convincing evidence that tolerance appears to be acquired in discrete water potential steps during development. Each step is thought to have a different water activity with associated metabolic processes, and may be related to specific desiccation stresses and discrete patterns of gene expression. Thus, the attainment of different levels of desiccation tolerance during development may be determined by progressive acquisition of a number of tolerance mechanisms or by sufficient accumulation of desiccation protectants. For example, it appears that desiccation sensitivity in recalcitrant seeds can occur at a minimum of two levels. In one case, the removal of freezable water (water that is not bound or structure associated) is tolerated and minimum survivable moisture content coincides with the quantity of nonfreezable (matrix-bound) water in the tissue. This suggests that recalcitrant seeds, unlike orthodox seeds, require this bound water for the maintenance of membrane integrity. In the second case, associated with more sensitive recalcitrant species, viability is lost as freezable water is removed. Seeds of these species may lack mechanisms that stabilize membranes as water is removed and may experience mechanical damage resulting from the reduction in cell volume. Other mechanisms are also likely to be limiting, such as the provision of adequate protection against free radical attack resulting from metabolic disruption. In all cases, solute leakage precedes viability loss during desiccation, suggesting that significant membrane damage has occurred.

Mechanisms Underlying the Acquisition of Desiccation Tolerance

Studies on the effects of desiccation on sensitive tissues do not distinguish between the causes of desiccation intolerance and changes during the death of cells caused by desiccation. Nevertheless, it is clear that the integrity of membranes in seeds is of crucial importance to the maintenance of viability during drying and subsequent imbibition. While drastic changes to membranes have been observed upon their drying *in vitro*, the evidence suggests that membranes in desiccation-tolerant seeds appear to be protected. It is likely that protection involves a number of metabolic and/or structural adjustments. In addition, cells must also be capable of repairing damage that is likely to be caused by desiccation.

Although in some species, desiccation tolerance can be acquired well before the completion of reserve deposition, it is possible that the accumulation of a critical amount of reserves is required before the seed can withstand desiccation. Highly vacuolated cells, which contain little reserve material, may undergo severe mechanical disruption during water loss, but this damage may be limited by the presence of sufficient reserves. In addition, the accumulation of certain proteins and sugars are thought to play a more direct role in the acquisition of desiccation tolerance. However, it is likely that the underlying basis of desiccation tolerance is diverse and is not simply restricted to the synthesis of these specific proteins and sugars.

Proteins and Desiccation Tolerance

An abundant group of hydrophilic proteins that are temporarily regulated during seed development has been associated with desiccation tolerance. The genes encoding these late-embryogenesis-abundant (LEA) proteins, have highly coordinated regulation so that their corresponding mRNAs peak just prior to, or during, desiccation. Dehydrins are a subset of the LEA proteins that exhibit some common structural features, suggesting a protective function during desiccation. Dehydrin genes are responsive to both developmental and environmental cues and are transcribed in response to desiccation. Dehydrin proteins also accumulate during seed development and, in response to seed drying, in a number of desiccation-sensitive recalcitrant species. However, dehydrin proteins are not detected in mature undried axes of a range of recalcitrant tropical wetland species. These species would not normally be exposed to significant drying and are considered to be particularly sensitive to desiccation. Desiccation sensitivity may therefore be due in part to an inability to accumulate a sufficient quantity of dehydrins or other LEA proteins.

Generally, the mRNAs encoding LEA proteins are detected in embryos from around mid-development. The high levels of abscisic acid (ABA) observed at this stage of development are thought to induce the accumulation of these polypeptides and hence prepare the embryo for desiccation or possible cellular disruption upon subsequent rehydration. Further evidence for a regulatory role for ABA has been provided by comparisons of wild-type and ABA mutants. However, the role of ABA in this and other aspects of the acquisition of desiccation tolerance remains unclear at present.

Specific small heatshock proteins (HSPs) of the cytosolic classes (I and II) accumulate in seeds of several plant species including recalcitrant species. In *Arabidopsis* and other species, class I small HSPs are first detected during mid-maturation and become most abundant in dry seeds, which is consistent with a role in desiccation tolerance. In some seeds (e.g., *Arabidopsis*), the proteins decline rapidly during germination, while in others (e.g., sunflower, *Helianthus* spp.), they persist.

The presence of dehydrins and small HSPs in recalcitrant seeds supports the view that their presence alone is not sufficient to confer desiccation tolerance. However, the converse might be true, i.e., their absence may imply an inability to tolerate desiccation and the absence of specific protective proteins cannot be ruled out as the cause of desiccation sensitivity. For example, the membranes surrounding oil bodies of seeds contain integral proteins, called oleosins, which may maintain the integrity of these organelles during desiccation and subsequent imbibition. Oleosins are present in the membranes of oil bodies in desiccation-tolerant seeds, but are absent, or their amount is diminished, in desiccation-sensitive seeds.

Sugars and Desiccation Tolerance

Drying and rehydration of model membrane systems usually results in the fusion of vesicles and loss of the ability to transport calcium. However, when disaccharides, such as trehalose, are present in concentrations equivalent to those in desiccation-tolerant organisms, functional vesicles are preserved. Membrane fusion during desiccation may therefore be prevented as a result of the sugars' hydroxyl groups interacting (i.e., forming hydrogen bonds) with the polar head groups of phospholipids and functional groups of proteins. Thus, the sugars are thought to alter physical properties of dry membranes so that they resemble fully hydrated ones. In seeds, sucrose and raffinose are much more common than trehalose and they have been suggested to fulfill this role.

Orthodox seeds accumulate considerable amounts of sugars throughout maturation, which may be important in the acquisition of a desiccation-tolerant state, but the evidence can be contradictory. For example, stachyose accumulates in immature soybean seeds subjected to slow drying, but does not increase significantly when seeds are maintained at high humidity. However, an increase in the amount of raffinose is not correlated with the acquisition of desiccation tolerance of wheat embryos. Studies with recalcitrant seeds show that there is no clear link between the presence of sugars and the level of desiccation tolerance in seeds. However, the lack of a clear relationship with recalcitrant seeds is perhaps not surprising since the proposed mechanisms for the involvement of sugars in desiccation tolerance operate at moisture contents below those at which they can survive. For example, sugars may protect the cell during severe desiccation by glass formation; in the presence of sugars a supersaturated liquid is produced with the mechanical properties of a solid. Only sugar mixtures equivalent in concentration and composition to those of desiccation-tolerant embryos are able to form glass at ambient temperature. It is suggested that glass formation prevents cellular collapse during desiccation and promotes a state of metabolic quiescence by restricting diffusion of substrates and products within cells. There is also some evidence that the protective effect of sugars extends to preventing irreversible changes to proteins.

Metabolic Imbalance and Desiccation Sensitivity

Recalcitrant seeds, perhaps because they remain moist, show the absence of substantial developmental arrest and maintain active metabolism throughout development to the time of shedding. In contrast, the metabolic activity of orthodox seeds is thought to decline in a programed way before or during the early stages of maturation drying, so that seeds are shed in a quiescent state. This organized decline in metabolic activity, which presumably has a role in protection against desiccation damage, does not occur in species with the most sensitive seeds and is not completed in those species with more tolerant recalcitrant seeds. These seeds, like those of orthodox species in the early stages of development, lose viability at high water potentials (ca. -3–-11 MPa) when they are metabolically active, respiration is measurable, and presumably membranes are still hydrated. However, at this level of hydration, it is thought that metabolism becomes "unregulated" and repair processes become inoperative. It is possible that this is responsible for the lipid peroxidation and free radical activity that has been associated with seed viability loss in several recalcitrant species during desiccation. However, to date, it has been difficult to tell whether this reported accumulation of free radicals is a cause or a consequence of viability loss. Nevertheless, it is reasonable to speculate that membrane damage and viability loss during drying of these seeds may result from unregulated metabolism enhanced by inadequate protection by free radical scavengers.

In orthodox seeds, the transition to desiccation sensitivity following germination of pea (*Pisum sativum*) and cucumber (*Cucumis sativus*) seeds is accompanied by a desiccation-induced imbalance of metabolism (i.e., increased emission of CO_2 and fermentation products) in the radicle that precedes loss of membrane integrity. Fermentation products are also released during desiccation of desiccation-sensitive recalcitrant seeds during development. Products resulting from this imbalanced metabolism (e.g., acetaldehyde) disturb the phase behavior of phospholipid vesicles and may therefore enhance the membrane damage induced by dehydration.

Summary

Desiccation tolerance is established through a programed sequence of events during the later stages of seed development and is then lost progressively during subsequent germination. Tolerance is clearly a complex property involving mechanisms for cellular protection, and also for cellular repair during rehydration. Desiccation tolerance appears to be acquired in discrete water potential steps during development that have different water activities and associated metabolic processes, which may be related to specific desiccation stresses and discrete patterns of gene expression. Thus, the attainment of different levels of desiccation tolerance may be determined by progressive acquisition of specific tolerance mechanisms or by sufficient accumulation of desiccation protectants. Proteins such as the dehydrins appear to have a protective role and, at very low moisture contents, sugars may play a key protective role by functionally replacing water and thus stabilizing membranes and other sensitive systems.

See also: **Biodiversity and Conservation**: Recalcitrance. **Postharvest Physiology**: Seed Storage. **Regulators of Growth**: Abscisic Acid. **Seed Development**: Embryogenesis; Physiology of Maturation.

Further Reading

Bewley JD and Black M (1994) *Seeds. Physiology of Development and Germination*, 2nd edn. New York: Plenum Press.

Black M and Pritchard HW (2002) *Desiccation and Survival in Plants: Drying without Dying*. Wallingford: CAB International.

Hoekstra FA, Golovina EA, and Buitink J (2001) Mechanisms of plant desiccation tolerance. *Trends in Plant Science* 6: 431–438.

Kermode AR (1995) Regulatory mechanisms in the transition from seed development to germination: interactions between the embryo and the seed environment. In: Galili G and Kigel J (eds) *Seed Development and Germination*, pp. 273–332. New York: Marcel Dekker.

Kermode AR (1997) Approaches to elucidate the basis of desiccation-tolerance in seeds. *Seed Science Research* 7: 75–95.

Kermode AR and Finch-Savage WE (2002) Desiccation sensitivity in orthodox and recalcitrant seeds in relation to development. In: Black M and Pritchard HW (eds) *Desiccation and Survival in Plants: Drying without Dying*, pp. 149–184. Wallingford: CAB International.

Pammenter NW and Berjak P (1999) A review of recalcitrant seed physiology in relation to desiccation-tolerance mechanisms. *Seed Science Research* 9: 13–37.

Vertucci CW and Farrant JM (1995) Acquisition and loss of desiccation tolerance. In: Galili G and Kigel J (eds) *Seed Development and Germination*, pp. 701–746. New York: Marcel Dekker.

Seed Quality

A G Taylor, Cornell University, Geneva, NY, USA

Introduction

Seed quality is an important factor in the agronomic life cycle of plants and must be given careful attention to ensure successful continuation of the genome from generation to generation. High-quality seeds provide farmers the potential for an optimal plant stand maximizing crop yield. Seed quality is determined during development on the mother plant and must be maintained to the time of sowing. This discussion will focus primarily on how seed quality relates to postharvest physiology and seed aging.

Several conditions or factors can lower the quality of seeds. The complete loss of quality refers to the inability of seeds to resume active growth after hydration. Events that lead to the loss of seed quality may be assessed on two levels: by whole-plant (seedling) response and by specific physiological (both biochemical and biophysical) changes. This article will address selected parameters of both groups and describe methodologies and their efficacy in evaluating seed quality. This discussion is limited to the evaluation of seeds that are desiccation tolerant (*see* **Seed Development:** Onset of Desiccation Tolerance), and would germinate if provided a favorable environment of water, temperature, and oxygen (*see* **Seed Dormancy:** Development of Dormancy).

Three Categories of Seed Quality

Seed quality may be grouped into three categories, each having several components. These three categories are relatively independent, but each contributes to the overall concept of seed quality. The importance of their respective components will depend on the species, application and the severity of a problem. The categories are: (1) description, (2) hygiene, and (3) performance potential. The descriptive category concerns the purity of a seed lot. The presence of other crop seeds, weeds, inert matter, and other cultivars that would lower quality are components of this category. Detection of other cultivars of the same species requires either growing plants to observe off-types or DNA fingerprinting technology. Other contaminants can be detected by visual inspection. Seed weight and seed lot uniformity are additional components of the descriptive category.

The hygienic category pertains to biological contaminants that may injure the seed or organisms that adversely affect seed health, such as fungi, bacteria or viruses, insects, mites, and noxious weed seeds. Specific pathological testing is required to quantify seed health problems, while visual inspection can detect the presence of insects, mites, and weed seeds. Methods to test each of these components is beyond the scope of this article, but the reader is referred to books and monographs in the Further Reading section. The remainder of this article will address the last category, potential performance, and will focus on the germination and vigor of a seed lot.

Overview and Relative Sensitivity of Selected Parameters

The loss of seed quality is attributed to aging or deterioration during seed storage. Factors that directly affect the rate of seed aging are temperature, relative humidity (which determines seed moisture content), and time (*see* **Postharvest Physiology:** Seed Storage). The objective of this article is to present techniques to measure or assess seed quality.

A number of events occur with the loss of seed quality before the loss of viability. In addition, aging can continue past the loss of viability (*postmortem* events). A number of events occurring during aging are expressed at the whole-plant and/or biochemical/biophysical levels. Selected parameters that are associated with the loss of seed quality are shown in **Figure 1**. Those parameters at the top of the figure

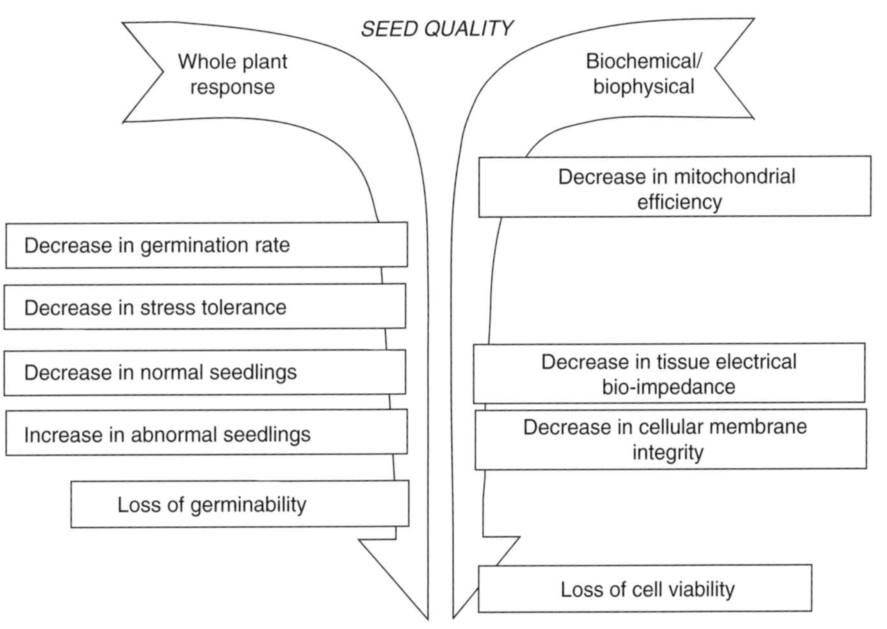

Figure 1 The relative sensitivity of selected whole-plant and physiological events to aging during storage. Those parameters in the upper portion of the figure are most sensitive or are early indicators of aging, while seeds become nongerminable and eventually all cells die at the bottom of the figure. (Illustration by D H Paine).

are most sensitive and can be considered to be early indicators of seed aging. The decline in seed quality is accompanied by changes in both whole-plant and physiological responses until seeds do not germinate, and ultimately, all cells die. This illustration provides a framework to describe methods used to assess seed quality. In addition, applications and limitations of each method are presented.

Several parameters are described in this article to illustrate whole-plant and physiological responses. Nondestructive tests in which seeds are termed "germinable" or "nongerminable" use methods that allow subsequent germination on the same seed. Seeds aged under specific conditions to obtain a range of seed qualities are used for destructive testing methods. Since the time period for seeds to age is dependent on the temperature and seed moisture content, the terms "time-of-aging" or "aging," expressed as a percentage, refer to a percent of the total time it takes for all seeds in the sample to become nonviable. This calculation is used to compare the relative sensitivity of a parameter to aging and to make comparisons between parameters.

The loss of seed quality is a challenge to quantify because germination does not decline in a linear manner with aging (**Figure 2**). In this example the total percentage of germinated seedlings of nonaged seeds was close to 100%, and showed only an initial slight decline until a critical time (about 40%), then decreased sharply. This illustrates the concept that the relationship between viability and time is an important characteristic of seed quality. High-quality seeds remain germinable longer than low-quality seeds independent of their environment. In general, stored seeds do not die immediately but maintain quality without an appreciable loss of germination for a period of time.

Little difference can be measured in the viability of samples between 0% and 25% aging making it difficult to determine the degree of aging within this range (**Figure 2**). Because the standard germination test lacks sensitivity to these early aging events, vigor tests have been developed to better forecast field sowing potential. Vigor as defined by the Association of Official Seed Analysts is "those seed properties, which determine the potential for rapid, uniform emergence and development of normal seedlings under a wide range of field conditions." This operational definition not only encompasses the concept of field seedling performance but also includes the robustness of germination. Many vigor tests have been developed to provide this supplemental information on a seed lot. Germination rate and stress tests are examples of whole-plant responses, while mitochondrial efficiency reflects vigor at the physiological level and can be used to detect early stages of aging.

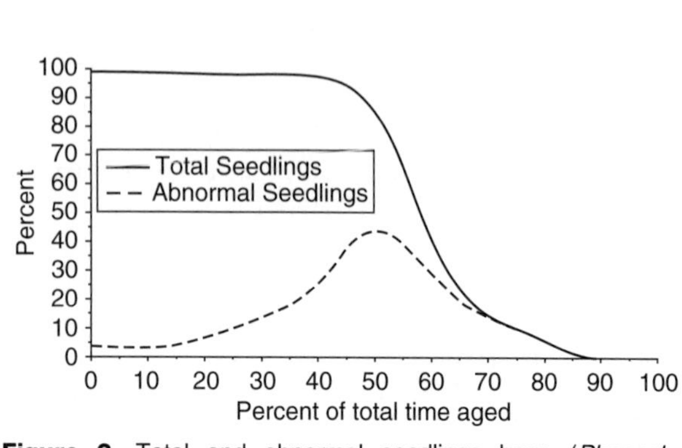

Figure 2 Total and abnormal seedlings bean (*Phaseolus vulgaris*) illustrating seed aging in storage. All seeds were nongerminable at 100%.

Whole-Plant (Seedling) Responses

The most widely used, practical methods to assess seed quality are based on whole-plant responses. The whole-plant response measures germination, which is of paramount interest to farmers or consumers of seeds. These germination tests do not require sophisticated instrumentation or apparatus, are generally inexpensive to perform, and many samples can be tested concurrently. A sample of seeds is tested in a destructive manner, and the results are considered representative of that particular seed lot.

Germination Test

The standard germination test is the only test accepted in commerce to assess the germination of a seed lot. Germination test conditions and procedures have been published by two recognized organizations: International Seed Testing Association (ISTA) and in the United States and Canada, the Association of Official Seed Analysts (AOSA). The testing rules are grouped by seed types: agricultural and vegetable, tree and shrub, and flower, spice, herb and medicinal species. (References are provided in the Further Reading section for those interested in germination test procedures for a particular species.)

Criteria used to evaluate seedlings are important to this discussion. Visible germination (radicle emergence) is one criterion used by seed physiologists to denote that seed has successfully germinated. The seed analyst, following the ISTA or AOSA rules, evaluates seedlings and classifies them as normal or abnormal. Normal seedlings are those that have all constituent parts including roots, shoots, and leaves. Abnormal seedlings are those seeds that germinate but lack the necessary seedling structures or development of these structures to be classified as normal seedlings. In commerce, only normal seedlings are used to record germination percentage on the label.

Total germinated seedlings and abnormal seedlings are illustrated for snap beans (*Phaseolus vulgaris*) with aging (**Figure 2**). Normal seedlings can be obtained by subtraction (total seedlings – abnormal seedlings). During seed aging, an increase in abnormal seedlings occurs prior to the loss of viability (**Figure 1**). In other words, the normal seedling class passes through an abnormal growth phase prior to the loss of viability.

Germination Rate

A decline in germination rate is the most sensitive indicator of aging at the whole-plant level (**Figure 1**). Lettuce (*Lactuca sativa*) seeds exposed to mild aging conditions revealed a reduction in both the speed and uniformity of germination (**Figure 3**). A 14-h lag in the T50 (time for 50% of the seeds to germinate) was measured for nonaged and aged samples (0% vs. 25% of total aging). However, there were no differences in the percentage total germinated seedlings with any of the aging treatments. Thus, germination rate provides an index of the early stages of aging, prior to a measurable decrease in the total germination percentage. Early seedling growth is another sensitive indicator of aging and may be more applicable than germination rate in some species.

In practice, germination rate can be recorded in a time course manner or at a particular reference time (e.g., 1, 2 or 3 day). Data acquisition and processing can be automated with computerized imaging systems. However, a nonaged reference sample is needed to compare with the test sample as germination rate is influenced by environmental conditions of the test and cultivar. In addition, the same test conditions can be used that are used for a standard germination test. In this manner, two pieces of information can be obtained on the seed lot over the course of one test: standard germination (percentage of normal seedlings) and germination rate.

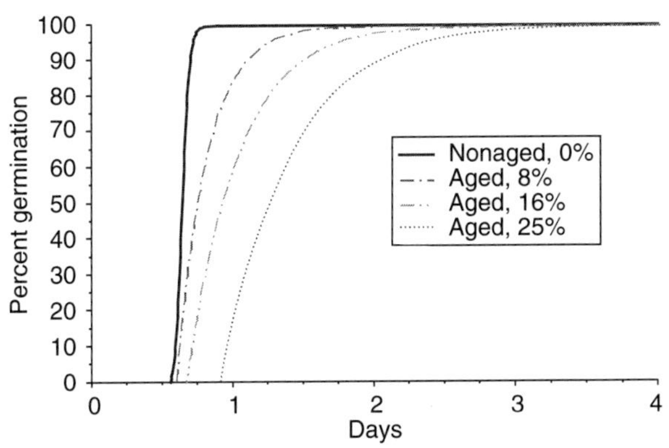

Figure 3 Germination rate of lettuce (*Lactuca sativa*) seeds affected by mild aging conditions, and prior to a reduction in final germination. Original data used with permission from Hacisalihoglu G, Taylor AG, Paine DH, Hilderbrand MB and Kahn AA (1999) Embryo elongation and germination rates as sensitive indicators of lettuce seed quality: Priming and aging studies. *HortScience* 34: 1240–1243.

Stress Tolerance

A decrease in stress tolerance is also a more sensitive indicator of seed quality than the standard germination test (**Figure 1**). Stress tests are the most common group of vigor tests that are routinely conducted by the seed industry and in other seed-testing laboratories. Though a wide range of test conditions could be employed, the two most common are: (1) rapid aging followed by a standard germination test, and (2) testing the seed in a cold, wet condition to mimic an early planting environment.

The premise for rapid aging tests is to determine the germination of a lot before and after aging. In the following example, two seed lots are contrasted using this test method: lot A with an initial 5% aging and the lot B at 20% aging (**Figure 2**). A germination test would show little differences between the two lots. If both lots were rapidly aged to add 30 percentage points, the total accumulated aging time would then be 35% and 55%, respectively. The rapid aging would result in profound differences in the viability of the two lots, thus indicating differences in seed quality.

To age seeds rapidly in preparation for the test above, seed moisture content must be increased either by vapor hydration or by the controlled addition of liquid water. Vapor-phase hydration may be accomplished over water in a near 100% relative humidity environment (accelerated aging) or may be accomplished by using a saturated salt to provide a given relative humidity (saturated salt aging). The addition of a known amount of water to a given sample of seeds (controlled deterioration) with known moisture content can be used to increase the seed moisture content to a desired level (*see* **Seed Development**: Seed Treatments). These techniques to increase seed moisture content are accompanied by exposure to warm temperatures (generally 35–45°C) for a specified period of time (usually several days).

Stress tolerance may also be tested directly by applying stress to the seeds, removing the stress(es), and then later measuring germination. The two environmental stresses most commonly used are suboptimal temperatures and wet conditions. Both of these stresses are employed during the "cold test": seeds are exposed to a cold, wet condition for a period of time, transferred to a temperature optimal for germination, and then the seedlings evaluated. There are several variables in a cold test including the temperature, the medium water conditions, and duration of test (both pre- and poststress conditions). Cold tests may use a field soil, a soilless medium, or other medium. Cold tests are widely used for corn (*Zea mays*, maize) and other crop seeds, and techniques have been established for routine testing.

Physiological Responses

A myriad of biochemical and biophysical events occur at the cell and subcellular levels during aging in storage, and physiological changes of interest are those that are correlated with some aspect of seed quality. Physiological changes are rarely observed in low moisture seeds in storage, so seeds must be hydrated to express the event. This may require only a short-term soak in water or may require controlled hydration, but seeds are analyzed prior to the completion of germination (*see* **Seed Development:** Seed Treatments). Physiological responses tend to be consequences or symptoms of the aging process rather than the cause of aging.

Methods used to detect physiological changes associated with seed quality should have the following characteristics: (1) they provide objective data, (2) their results are not confounded by seed morphology or composition, (3) they may be performed on many samples, and (4) they can be conducted in a short period of time. No one test has all of the desired characteristics; however, each test requires less time to perform than the standard germination test. Collectively, physiological tests are diagnostic tools that can be used to supplement information from the standard germination test and can provide quick results on a seed lot. The applications and limitations of selected tests follow.

Cell Viability

Seeds must possess viable embryo cells to germinate (**Figure 1**). Many methods are available to assess cell viability including the use of vital stains. The most widely used vital stain in seed testing is tetrazolium chloride, originally developed in the 1940s. Dehydrogenase activity within cells is closely correlated with viability, and its measurement is the basis for this test. An aqueous solution of tetrazolium chloride is colorless but when reduced by cellular dehydrogenase activity forms a water-insoluble, red compound, formazan. Tetrazolium is a topical stain and can be used to visualize tissue viability. Seeds must first be imbibed in water to become metabolically active. However, tetrazolium solution does not readily enter most seeds, and seeds must be pierced or cut to allow access of the solution to the embryo. Data on staining is subjective, and considerable experience is needed to interpret the staining patterns accurately. Despite these limitations, tetrazolium testing has been developed for many species and is especially useful for seeds that are viable, but dormant.

Cell Membrane Integrity

Cell membranes maintain solutes such as sugars, amino acids, and electrolytes compartmentalized within the cells. The integrity of cell membranes is lost with aging, resulting in leakage of cellular constituents into the intracellular spaces and then into the germination environment. Seeds must be imbibed for cells to become turgid, and subsequent leakage occurs by passive diffusion.

A number of compounds leak from seeds and may be used as an indirect measure of membrane integrity. Leakage of electrolytes, amino acids, and sinapine from cabbage (*Brassica oleracea*) seeds is illustrated (**Figure 4**). Leakage kinetics differ by compound and by seed quality (germinable vs. nongerminable). Amino acid leakage is more accurate in differentiating germinable from nongerminable seeds than the electrolyte method. This can be attributed to the location of amino acids and electrolytes. Amino acids and electrolytes are intracellular, while electrolytes also present in the seed covering tissue. Therefore, a large background level of electrolytes is detected from high-quality seeds, which is not related to membrane integrity. Sinapine, an alkaloid compound found exclusively in seeds of Brassicaceae, exhibits a little leakage from all seeds during the first few hours of hydration after which only the nongerminable seeds will continue to leak.

Leakage appears to have tremendous potential as a rapid seed-quality test because the data is objective, and many samples can be tested either in bulk or on a single-seed basis. Unfortunately, there are major limitations for wide-scale use of leakage tests for most plant species. The inner layer of the seed coat contains a semipermeable layer that restricts or prevents the diffusion of many solutes. This semipermeable layer allows water and gas diffusion as the seeds readily germinate, and is composed of cutin or suberin. Thus, the semipermeable layer is a morphological barrier that confounds the interpretation of leakage tests as solutes cannot diffuse from the embryo to the environment. In practice, the use of the electrolyte leakage test is restricted to large-seeded legumes, *Brassica*, and a few other species

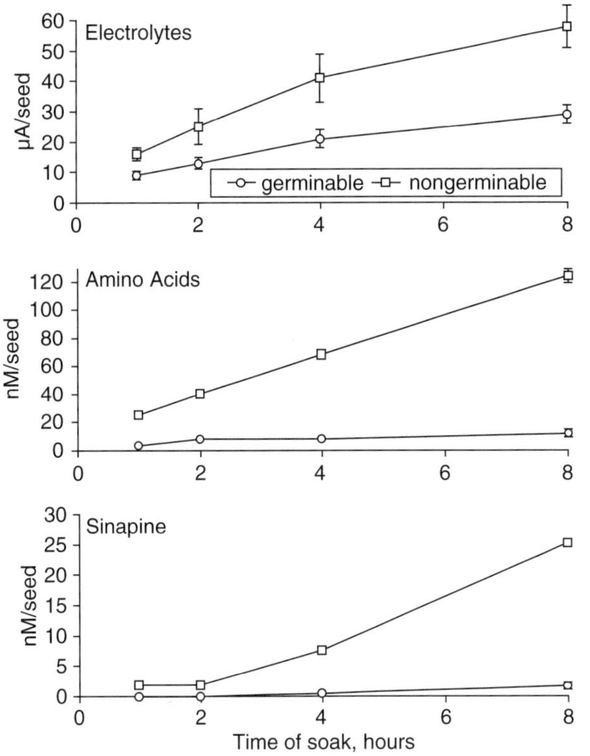

Figure 4 Leakage is illustrated from germinable and nongerminable seeds of cabbage (*Brassica oleracea* cv. Capitata). Original data used with permission from Hill HJ, Taylor AG and Xuang XL (1988) Seed viability determinations in cabbage utilizing sinapine leakage and electrical conductivity measurements. *Journal of Experimental Botany* 39:1439–1447; and Taylor AG, Lee SS, Beresniewicz MB and Paine DH (1995) Amino acid leakage from aged vegetable seeds. *Seed Science and Technology* 23: 113–122.

that lack a semipermeable layer. Leakage, however, is not a sensitive test and should not be considered as a vigor test (**Figure 1**).

Electrical Bio-Impedance

Electrical impedance spectroscopy or bio-impedance spectroscopy has been developed for medical diagnostics, and has application in seed-quality testing. Electrical impedance spectroscopy is a biophysical test in which electrical properties of the seed that change with aging are detected. Briefly, seeds are exposed to alternating current and the capacitive and resistive components of the seed tissue are measured.

Sophisticated instrumentation is needed to analyze the electrical properties of seeds. An impedance/gain-phase analyzer interfaced with computer was used to generate impedance data from germinable and nongerminable *Phaseolus vulgaris* seeds (**Figure 5**). This test provides a complete spectrum of data obtained over a wide range of frequencies and can be modeled by several mathematical parameters to further assess seed quality. These models may also provide biophysical insights to further explain the mechanisms behind seed aging.

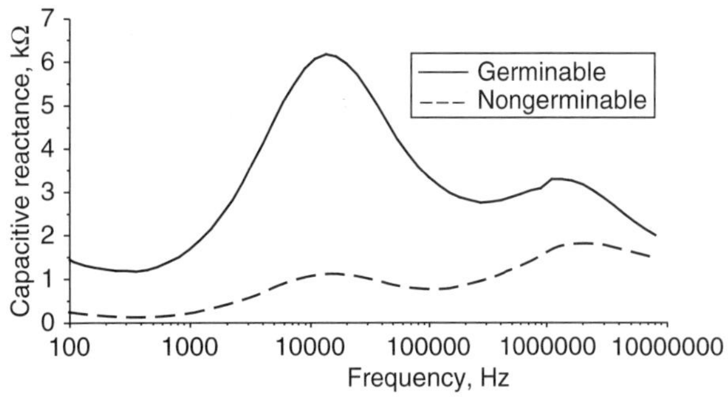

Figure 5 Electrical impedance spectra from germinable and nongerminable seeds of beans (*Phaseolus vulgaris*) reveal differences between seed qualities.

Seed moisture content is critical to differentiate seed quality and to obtain reproducible data. Seeds must be partially imbibed to a moisture content insufficient for the completion of germination. The corresponding water potential of the partially hydrated seed is about −5 MPa (*see* **Seed Development:** Seed Treatments). The method is nondestructive and the seed can be dried and stored or used for another purpose. Unlike leakage tests, the seed coat does not alter the impedance spectra. Electrical impedance spectroscopy research is relatively recent and further work is needed to adapt and expand this technology for seed-quality assessment.

Mitochondrial Efficiency

Resumption of respiratory activity and the functionality of the mitochondria are essential to provide energy to the germinating seed. A convenient method to measure respiration rate is to measure a product or byproduct of respiration. A decrease in mitochondrial efficiency is related to ethanol production (the byproduct of anaerobic fermentation), and is an early indicator of seed aging (**Figure 1**). A sensitive test can be enhanced by a test that employs the ratio of two measurements in which one response increases with aging, and the other decreases.

During hydration under aerobic conditions ethanol production increases as seeds age, whereas under anaerobic conditions (hypoxic), ethanol production decreases as seeds age. The ratio of ethanol production of the anaerobic to aerobic ratio (ANA) results in an extremely early indicator of seed aging. For example, *Brassica* seeds aged under mild conditions exhibited only a slight decline in a germination test between 0% and 38% time-of-aging (**Figure 6**). In contrast, the ANA exhibited a substantial drop between 0% and 25% time-of-aging. Thus the ANA ethanol test is extremely sensitive to the effects of aging on respiration and mitochondrial efficiency.

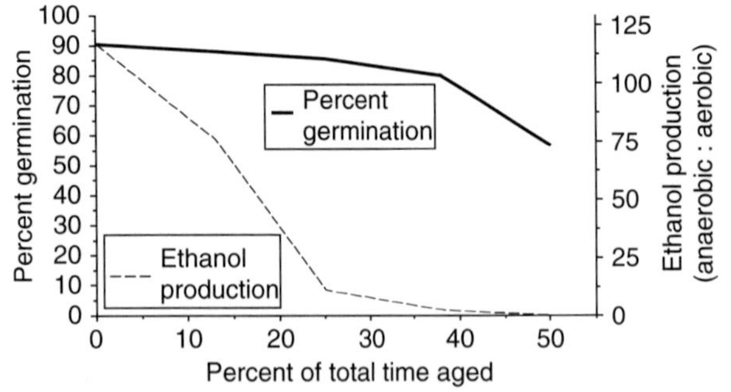

Figure 6 The effects of aging of cabbage (*Brassica oleracea* cv Capitata) seeds on percentage germination and ethanol production ratio (anaerobic : aerobic). Original data used with permission from Taylor AG, Johnson CF, Kataki PK, and Obendorf RL (1999) Ethanol production by hydrated seeds: a high resolution index of seed quality. *Acta Horticulturae* 504: 153–160.

Further research and development is needed to expand the application of this technology.

Conclusions

The performance potential is an important characteristic of seed quality, and many techniques can be used to estimate it. The standard germination test results provides a basis for the measurement of seed quality and is recognized internationally. Other whole-plant tests, especially stress tests, are routinely used in seed testing laboratories to provide supplemental information (vigor assessment) of crop species. Selected physiological methods that analyze biochemical (tetrazolium vital staining and ethanol production related to mitochondrial efficiency) and biophysical (electrolyte leakage and electrical impedance spectroscopy) events can be associated with the degree of aging. In summary, a number of tests can be used in standard practice or as specialized tests for research.

The relative sensitivity of seeds to aging events provides a framework to relate the whole-plant responses to physiological deterioration (**Figure 1**). Other or newly developed physiological methods can be added to this diagram as markers or indexes of seed quality. The value of new physiological tests should depend upon their ability to provide objective data, provide results that are not confounded by seed morphology or composition, and provide a way to conduct tests on many samples in a short time period. A test that can be conducted on a single-seed basis and is nondestructive would also be of great benefit.

List of Technical Nomenclature

Abnormal seedlings Seedlings that lack the necessary structures or development of these structures to be classified as normal.

Cell membrane integrity Essential to compartmentalize cellular solutes and may be measured by leakage tests.

Cell viability A prerequisite for germination that may be measured with vital stains.

Electrical bio-impedance A method in which seeds are exposed to alternating current over a range of frequencies. The capacitive reactive and resistive components of the seed tissue are measured and related to seed quality. Also termed electrical impedance spectroscopy.

Germination rate Germination speed and uniformity of a seed lot. A decrease in germination rate is the most sensitive indicator of aging at the whole-plant level.

Germination test A standardized procedure for testing a particular species and classifying the seedlings as normal or abnormal.

Mitochondrial efficiency The ability of the mitochondria to respire substrates to form energy needed for germination.

Normal seedlings Seedlings that have all constituent parts including roots, shoots, and leaves. The percentage of seedlings with this classification is used for labeling the commercial seed lots.

Seed aging Deterioration of seeds in storage; the aging rate for a particular seed lot is determined by the temperature and relative humidity, which determines moisture content of the seed and time it will take to lose germinability.

Seed quality A broad term that includes three categories of a seed lot: description, hygiene, and potential performance.

Stress test Tests to measure tolerance to various stresses are the most common group of vigor tests. Examples include rapid-aging test and the cold test.

Vigor Germination robustness. An important concept to describe the early stages of aging prior to the loss of germination.

See also: **Postharvest Physiology**: Seed Storage. **Seed Development**: Seed Treatments; Onset of Desiccation Tolerance. **Seed Dormancy**: Development of Dormancy.

Further Reading

Association of Official Seed Analysts (1983) *Seed Vigor Testing Handbook*. Las Cruces, NM: Association of Official Seed Analysts.

Association of Official Seed Analysts (2000) *Rules for Testing Seeds*. Proceedings of the Association of Official Seed Analysts.
Basra AS (ed.) (1995) *Seed Quality*. New York: Food Products Press.
Copeland LO and McDonald MB (2001) *Seed Science and Technology*, 4th edn. Boston: Kluwer Academic.
Hampton JG (2002) What is seed quality? *Seed Science and Technology* 30: 1–10.
International Seed Testing Association (1999) *International Rules for Seed Testing*. *Seed Science and Technology* 27 (Supplement).
Repo T, Paine DH, and Taylor AG (2002) Electrical impedance spectroscopy in relation to seed viability and moisture content. *Seed Science Research* 12: 17–29.
Taylor AG (1997) Seed storage, germination and quality. In: Wien HC (ed.) *The Physiology of Vegetable Crops*, pp. 1–36. Wallingford: CAB International.
Taylor AG, Johnson CF, Kataki PK, and Obendorf RL (1999) Ethanol production by hydrated seeds: a high resolution index of seed quality. *Acta Horticulturae* 504: 153–160.

Seed Treatments

A G Taylor, Cornell University, Geneva, NY, USA

Introduction

Seeds are the cornerstone of agriculture as they are the propagules that insure plant genetic survival from one generation to the next. Seeds of most species are safely dried after harvest from the mother plant and can be stored until the time of sowing. Desiccation tolerance gives seeds their unique ability to be stored as a low moisture content dispersal unit with relatively long shelf-life. This allows access for the manipulation of the seed's physical, chemical, or biological environment.

Seed treatments is a broad term and may not only be indicative of application of certain compounds or processes, but also the subjection of seeds to various energy forms such as heat, radiation, or magnetism. The original meaning of seed treatments is the application of a pesticide (referred throughout this article as plant protectants) to control pests. The term seed enhancements is used by both academia and industry, and better defines the scope of this article. Seed enhancements are postharvest methods that improve germination or seedling growth or facilitate the planting of seeds and the delivery of other materials required at time of sowing. Seed enhancements include two major categories: (1) the manipulation of the physiological status of seeds by hydration and (2) the application of other agents or materials, hereafter referred to as coating methods. This article will address both categories (in reverse order): coating methods and hydration methods. Technologies will be presented discussing the methodology, applications, and commercial utilization, including the role and application of water for each topic. Selected methods within each category are presented in order from no to increasing levels of water application.

This article will focus on seeds that are the sexual propagules from annual, biennial, or perennial plants. However, asexual propagules are also used in agriculture and found in nature. The most common asexual "seed" is the tuber from potatoes (*Solanum tuberosum*). Seed potatoes may also be treated with chemical or biological agents. The seed treatment technology is comparable to planter box or spraying of liquid treatments as described under "Coating Methods." Seed potatoes are not amenable to many treatments described in this chapter as they have high water content.

Water Status of Seeds

The application of seed treatments may alter the seed moisture content or may be used to hydrate the seed to a desired level. Therefore, an overview of water in seeds is needed before discussing particular seed treatments. Seeds' water status may be expressed in terms describing its concentration or in terms of water relations. Moisture content, based on fresh weight (fw), is commonly used to express water concentration. Moisture content, MC, is calculated as:

$$MC = W \cdot (S + W)^{-1}$$

where W is the weight of the water in the seed and S is the weight of the dry seed. The MC of seeds ranges from low in storage to high in germination (for example: from 6 to 55 g $H_2O \cdot g^{-1}$ fw, respectively).

Water movement is passive in seeds, moving from high to low concentrations. Consequently, dry seeds can take water vapor from humid air. Seed MC comes into equilibrium with the relative humidity (RH) of the air and is a relationship described by a moisture isotherm (**Figure 1**). The type of water binding varies with the RH, and three regions are shown for snap bean (*Phaseolus vulgaris*) in this isotherm. The three regions of water binding are observed: region I, <20% RH, region II, 25–65% RH, and region III, 70–97% RH. Seed moisture content increases dramatically as the RH increases from 70% to 97%. To provide perspective, most

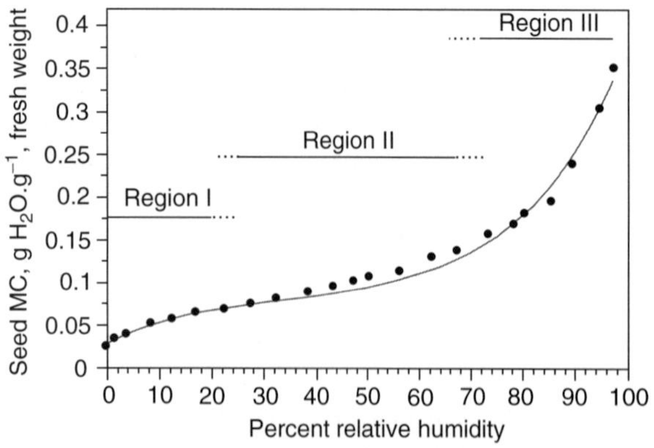

Figure 1 Moisture isotherm of bean (*Phaseolus vulgaris*) seeds revealing three regions of water binding. Original data used with permission from Taylor AG (1997) Seed storage, germination and quality. In: Wien HC (ed.) *The Physiology of Vegetable Crops*, pp. 1–36. Wallingford: CAB International.

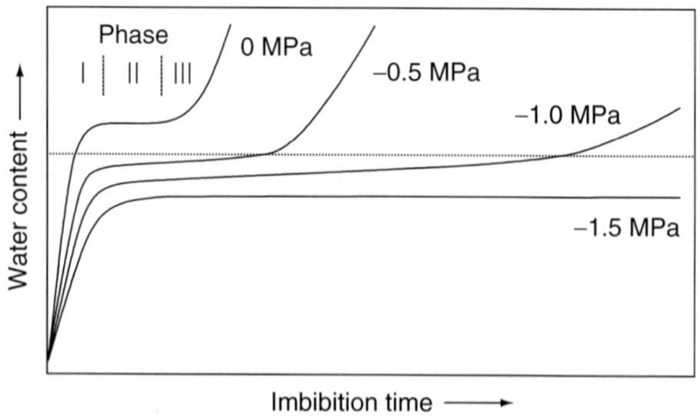

Figure 2 Time courses of water content at water potentials 0, −0.5, −1.0 and −1.5 MPa. Three phases, imbibition (I), activation (II) and growth (III) are shown for 0 MPa curve. Reprinted with permission from Bradford KJ (1995) Water relations in seed germination. In: Kiegel J and Galili G (eds) *Seed Development and Germination*, pp. 351–396. New York: Marcel Dekker.

seeds in agriculture are stored in region II, and commercial seed lots are commonly maintained in equilibrium at 40–50% RH. Seed aging or deterioration is accelerated in region III.

Water uptake that precedes germination can also be described in terms of water relations. Water potential (Ψ) is used to quantify the energy status of water, which is measured in MPa. The Ψ of pure water is 0 MPa, while a decrease in Ψ is shown by more negative values. The components of the total water potential are the algebraic sums of the matric, osmotic, and pressure potentials. Seeds in storage at 50% RH have a Ψ of about −100 MPa. This very negative Ψ is due to the large matric component. Water uptake curves are shown for seeds placed in environments of 0, −0.5, −1.0 or −1.5 MPa (**Figure 2**). Seeds imbibe water rapidly in phase I germination due to the large matric potential. Imbibition time depends on water availability, contact surface area to the water supply, seed coat permeability, and species. An increase in seed volume (swelling) is concurrent with an increase in moisture content. Phase II occurs after seeds are imbibed, and is accompanied by metabolic activity. Phase III commences at the onset of visible germination (growth of radicle emergence through the seed covering). Both osmotic and pressure potentials are responsible for the total water relations in phases II and III. Visible germination (phase III) can first be achieved at 0 MPa, while a longer period of time is required for lower water potentials (−0.5 and −1.0 MPa) (**Figure 2**). There is insufficient water potential for seeds to complete germination at −1.5 MPa so seeds remain in phase II. Thus, seed moisture content can be regulated depending on the water potential of the imbibing environment.

Coating Methods

Coating methods includes techniques for the application of compounds to the surface or external portions of seeds with minimal influence on the physiology. These techniques do not significantly alter or increase the seed moisture content; however, during pelleting the moisture status may briefly increase. All coating methods described below are used commercially, and techniques employed depend upon the crop value and the particular application needs.

Planter Box

Mixing seeds with a dry powder containing a particular active ingredient is the simplest coating method used. In agriculture, this on-farm operation is known as a planter or hopper box application, and is mostly used for application of plant protectants. Water is not required for this coating method. The major disadvantage of this method is that dry powders do not adhere well onto the seed surface, causing lack of uniform dosage. In addition, worker safety is a concern since the operator can be exposed to the powder (dust) by inhalation and skin contact. The planter box method is more appropriate for formulations that have a short shelf-life or that may be incompatible with other compounds already applied to the seed. For example, biological seed treatments containing bacteria or fungi as the active ingredient may have poor survival rates if applied with water or with chemical treatments. The control of a sporadic pest during the planting season is another application. A seed lot may be nontreated or treated only with a fungicide, but an insect pest that attacks seeds and seedlings may also be present in the field. A planter box application of a labeled insecticide may be included to control this transient insect. The trend is for less use of this method due to the disadvantages described.

Conventional Liquid Treatments

The objective of liquid application is to apply a small quantity of material uniformly over seeds. The major principles include the exact metering of liquid formulation with known seed weight, uniform liquid distribution (usually by atomization) onto the seed mass, and finally, mixing. A wide range of materials can be applied with this method, and chemical plant protectants have the greatest commercial use. Chemical plant protectants are commonly formulated to be soluble, dispersed, or suspended in water, in which water acts as a carrier or diluent for the active ingredient. In general, the total liquid applied with active ingredient is less than 1% of the seed weight. Therefore, the change in seed moisture content is negligible after treating. If additional liquid is used, a drying powder may be added during treatment. Conventional liquid treatments procedure can not only be conducted on small samples, but also has been scaled-up for large commercial applications capable of treating tonnes per hour; thus this technique can be performed either on a batch or a continuous basis. For chemical plant protectants, a dye or colorant is added along with an adhesive or binder. This method is widely used for most crop species due to application uniformity, short application time, low material cost, and the fact that no additional energy is needed for drying.

Film Coating

Film coating is a technique originally developed for the pharmaceutical and confectionery industries. A liquid formulation that includes a film-forming polymer is sprayed at a highly controlled rate onto a tumbling mass of seeds over time to achieve a uniform deposition of materials. Formulations may also contain plasticizers, colorants, and other ingredients and are commercially available as ready-to-use (RTU) formulations or as dry powders that can be dispersed in water. Generally, the increase in seed weight from film coating is small (0.5% to 5.0%), so that the original size and shape of seed is not obscured (**Figure 3**). The aqueous film-forming formulation must be dried during spraying to avoid agglomeration. For example, if a film coating formulation has 25% solids, and a weight increase of 5% is desired, then 200 g total of film-forming suspension is needed to treat 1 kg of seed. In this scenario, 150 g of water must be removed by drying during the coating operation to result in a deposition of 50 g of coating. The rate of film application is limited by the effectiveness of the drying air being forced through the seed mass and must be closely controlled. To facilitate drying, perforated pans have been used for film coating (**Figure 4**). Fluidized bed technology has been employed, but is limited to small-scale applications. The benefits of film coating include uniform placement of materials in the correct dosage, increased worker safety (an essentially dust-free environment), enhanced appearance due to the addition of pigments, and an excellent delivery for a number of plant protectants at rates higher than those allowed using slurry techniques. Film coating is a value-added treatment, and is used primarily for high-value seeds such as vegetable and flower seeds.

Pelleting

Seed pelleting, along with coating and encrusting, consists of the application of solid particle fillers with a binder or adhesive to form a more or less

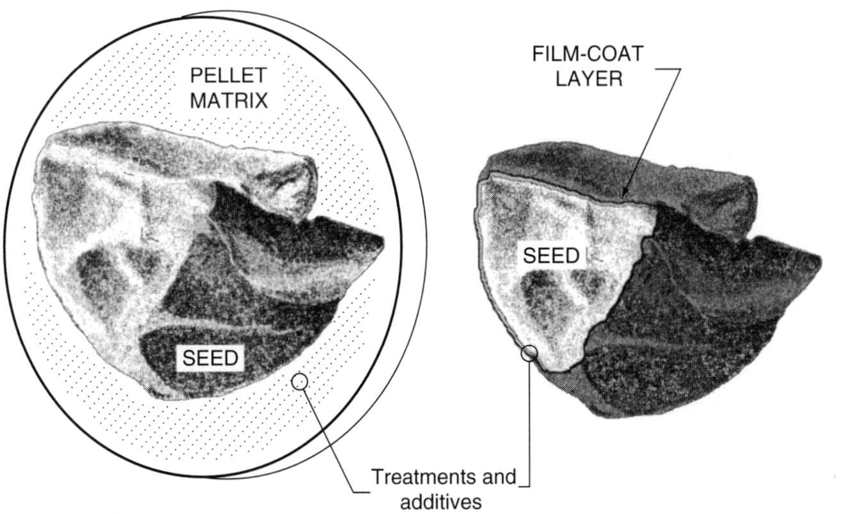

Figure 3 Illustration of a pelleted (left) and film coated (right) onion (*Allium cepa*) seed. Both the pellet matrix and film coating layer can serve as a delivery system for plant protectants. Illustration courtesy of D.H. Paine.

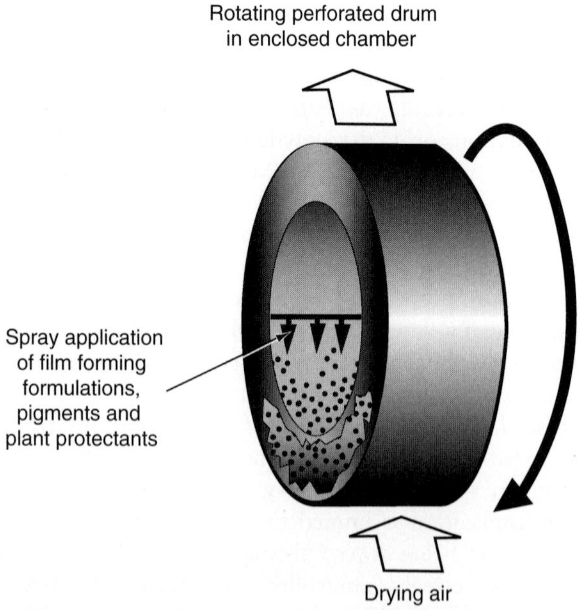

Figure 4 Film coating of seeds in a side-vented drum system. Polymeric film coating formulation is sprayed onto seeds over time with concurrent drying. Adapted with permission from Halmer P (1999) Commercial seed treatment technology. In: Black M and Bewley JD (eds) *Seed Technology and its Biological Basis*, pp. 257–286. Sheffield: Sheffield Academic Press.

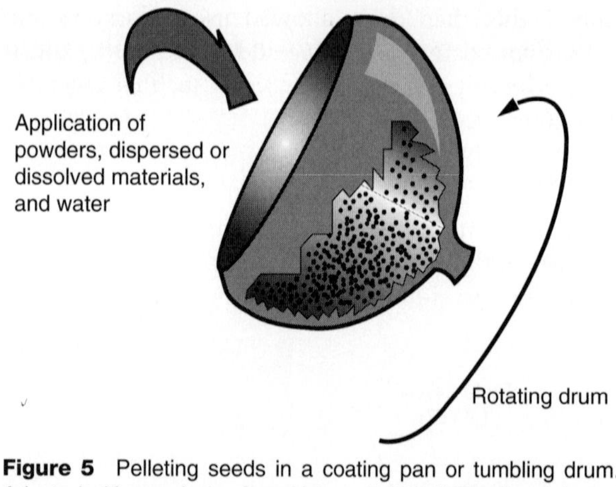

Figure 5 Pelleting seeds in a coating pan or tumbling drum. Adapted with permission from Halmer P (1999) Commercial seed treatment technology. In: Black M and Bewley JD (eds) *Seed Technology and its Biological Basis*, pp. 257–286. Sheffield: Sheffield Academic Press.

spherically shaped dispersal unit. Consequently, the shape of the seed is obscured (**Figure** 3). The weight increase varies by species and final desired product, but is usually 2- to 50- or more fold (200% to ⩾5000% increase) for pelleting, while coating and encrusting are "mini" pellets ranging from 0.2- to 2-fold (20% to 200% increase) the original seed weight. In this process, seeds are generally pelleted on a batch basis in a coating pan or tumbling drum (**Figure** 5). In addition, rotary pan technology has been developed that can rapidly pellet seeds compared to conventional coating pans. Pellets are sized during and at the end of the process to insure uniformity of final pellet size. Most materials used are proprietary; however, a dry formulation or blend consisting of diatomaceous earth with finely ground polyvinyl alcohol are examples of a filler and binder, respectively. Seeds are tumbled with repeated applications of the coating formulation followed by intermittent spraying of seeds with water to activate the binder and cause the formation of the pellet around each seed. Unlike film coating, pelleting is a wet process, and the pellets are not dried until the completion of coating. Plant protectants can be layered in a zone during the formation of the pellet; this procedure is known as pellet loading. In this manner, the active ingredients are physically separated from contact with the seed surface, but available during germination and early seedling growth. Film coating polymers can also be added after pelleting to provide a dust-free and cosmetic finish. The pellets improve the plantability of the seeds, and the pelleting process serves as a delivery system for plant protectants needed to control pests. Pelleting allows for precision placement of small and irregularly shaped seeds during mechanized planting operations. In addition, the ballistics of a seed is improved by the pelleting, which facilitates aerial sowing of selected species. Pelleting is routinely performed by the seed industry on high-value, small-seeded horticultural and flower seeds, and selected species for aerial sowing.

Hydration Methods

The objective of most hydration methods is to enhance physiological activity of seeds. Methods are presented in which the seed moisture content becomes the major variable that is regulated. Several of these methods are used commercially for particular species or sowing conditions, while others have applications in research or specialized uses.

Solvent Infusion

Many compounds including plant growth regulators (PGR) and hormones are not soluble in water. Methods are needed to apply these materials and to permit them to diffuse into seeds. Solvent infusion is a method in which seeds are treated by soaking in an organic solvent with an active ingredient in solution. Acetone and dichloromethane (methylene chloride) are used for this application after which seeds are ventilated to remove all residual solvents. Water is not used, and mixing water with organic solvents may be lethal to seeds. Thus, the initial seed moisture content should be low to prevent injury. Solvent infusion is a soak method without water added, as the physiological enhancement of the PGR occurs during germination. Solvent infusion has limited commercial application largely due to the expense and logistics of handling organic solvents. Moreover, proper disposal of the used solvents must be considered for large-scale use.

Humidification

The initial seed moisture content can have an influence on the susceptibility to mechanical injury and hydration damage, especially in large-seeded legumes. Seeds are subjected to mechanical stresses during harvesting, handling, and sowing. Rapid water uptake also causes stress during imbibition, particularly in cold, wet soil conditions. Both mechanical and hydration injury result in a decreased germination. Seeds are most susceptible with low moisture content (region I) and the lower portions of region II (**Figure 1**). Increasing the seed moisture content to the upper portion of region II or region III decreases the susceptibility to these stresses. The seed moisture content can be increased using humidified air circulation. Though seed moisture content can be increased to the upper regions of region III by humidification, water vapor diffusion is much slower than liquid water imbibition. Care is needed to avoid condensation, and prolonged exposure can result in microbial growth on the seed surface. Humidification and other methods are used to achieve desired seed moisture content for the storage of large-seeded legumes.

Priming

Seed priming refers to several different techniques used to advance germination by controlled hydration, but to a level insufficient for the completion of germination. During priming, seeds are able to imbibe or partially imbibe water and achieve elevated seed moisture content in phase II (**Figure 2**). Time of treatment ranges from less than 1 day to several days, and cool temperatures are commonly used (15–20°C). Since seeds have not completed germination, they remain desiccation-tolerant and can be dried again for long-term storage. All priming techniques rely on the controlled uptake of water to achieve a critical moisture content that will activate metabolic activity in a controlled environment. Three priming techniques are employed: liquid systems, solid particulate systems, and controlled hydration. Liquid systems regulate the water potential of the imbibing solution with solutions of large molecular weight polyethylene glycol (PEG 8000) or inorganic salts. In solid particulate systems, seeds are mixed with a solid particulate material such as a fine-grade vermiculite and water. Seeds, due to their negative matric potentials, are able to imbibe water from the solid material and achieve a desired water potential or seed moisture content. The proportion of seed, solid particulate, and water is critical since an excess amount of water will permit the seeds to complete germination. Controlled hydration is based on the addition of a known amount of water to a seed sample of known weight and initial seed moisture content. The weight of water needed to increase a given seed weight to a desired moisture content can be calculated:

$$W_a = S_i \cdot (MC_f - MC_i) \cdot (1 - MC_f)^{-1}$$

where W_a is the weight of water to add to the initial weight of seed S_i, MC_i is the initial seed sample moisture content, and MC_f is the final seed moisture. Seed MC is expressed as g $H_2O \cdot g^{-1}$ fw. Water can be added over time to allow seeds to slowly achieve the desired seed MC.

The benefits of priming include increased germination rate and percentage seedling emergence (especially under suboptimal temperatures for germination) and decreased photosensitivity. Seeds primed by the seed industry are dried to their original moisture for storage. Priming reduces shelf-life, so primed seeds should be stored under ideal conditions and sown soon after treatment. Priming has been used most extensively on high-value vegetable and flower seeds.

Soaking and Pregermination

Noncontrolled water uptake includes those methods in which water is freely available to seeds ($\Psi = 0$ MPa). Soaking seeds in water, also known as steeping, allows imbibition and seeds to enter phase II; however, an oxygen supply is needed for most species to complete germination. Hence, water is aerated to increase the dissolved oxygen in solution. Since water is not limited, seeds will germinate in time, or the process must be arrested at a specific time to prevent the onset of phase III. It may be desirable for seeds to complete germination (phase III), and specialized methods can then be used to deliver the seeds or carefully handle the germinated seeds.

Pregerminated seeds may be sown in a viscous gel that suspends the seeds and prevents damage to the germinated seeds. This technique is termed gel seeding or fluid drilling. For other applications, individual seeds may be encapsulated in a gel coating that prevents dehydration. Hydroxyethyl cellulose polymers have been used to produce a thixotropic gel for fluid drilling, while gel capsules are made of alginate. These gels may also serve as a carrier for materials required at time of sowing including plant protectants, nutrients, plant growth regulators, and activated carbon. Finally, methods to carefully dry germinated seeds so that they may be stored and sown with conventional planters have been achieved from recent advances on desiccation tolerance mechanisms. The use of pregerminated seeds allows for a more rapid and synchronous seedling establishment and the ability to remove nongerminated seeds prior to desiccation. Soaking and hydration have found specialized applications for flower seeds and other high-value crops in greenhouse production.

Applications

Chemical plant protectants are the largest group of compounds used commercially as seed treatments. Older materials are generally broad spectrum, while new compounds have specific modes of action for controlling pests. Also, new chemistry materials are being developed that are effective at very low dosage rates. To that end, precision is needed in metering these materials uniformly over the seed mass to obtain a target dosage on a per-seed basis. Plant protectants applied as seed treatments are also challenging traditional in-furrow applications of liquids or granules. It is common that 1 kg of active material is applied per hectare as an in-furrow treatment. In comparison, the same material can be applied effectively as a seed treatment with a reduction of over 75% in pesticide usage. Generally, high loading rates are needed for these plant protectants (up to 10% of the seed weight), which require either pelleting or film coating techniques to apply.

Biologicals, fertilizers, and nutrients are other examples of seed treatments. The major biological used commercially is *Rhizobium* inoculation of legume seeds. Coating techniques are used to apply and to preserve the viability of the bacteria for efficient nitrogen fixation. The addition of lime and molybdenum may be used for regulating pH and providing micronutrients to foster the symbiotic relation. Other biologicals used to control soilborne diseases and to replace chemical seed treatments have been tested. In reality, biological and chemical treatments are generally applied together for a growing season of protection. Chemical treatments provide reliable short-term control of pests at time of sowing and are effective under a wide range of soil and environmental conditions. Biologicals, particularly those organisms that can colonize the growing root system, provide long-term protection, as the biological is able to proliferate.

Nutrients or fertilizers can be applied by pelleting technologies, and other additives may be included to alter coating pH to favor uptake of a particular macro- or micronutrient. High concentrations of fertilizers in close proximity to the seeds may decrease osmotic potential (salt effect), and may retard or prevent germination. Application of fertilizers as a seed treatment is restricted to marginal lands that have poor soil fertility or if a particular nutrient is deficient early in the growing season as generally large quantities of nutrients are required for sustained plant growth.

Film coating and, in particular, pelleting may result in a physical barrier around the seed. It is important that water uptake and gas exchange occur in these coated seeds to insure optimal germination. A restriction of oxygen uptake from pelleted seeds has resulted in the use of oxygen-liberating compounds to be incorporated into the pellet. Calcium peroxide may be applied as a dry powder during pelleting, which forms hydrogen peroxide after sowing to release oxygen to the germinating seeds. This technique is important for pellets sown in irrigated fields or in crops, for example, rice (*Oryza sativa*), that are sown into wet soil conditions.

Finally, coatings or additives to the coatings may be used to regulate water availability at time of sowing. Hydrophilic compounds have been studied to hold water around the germinating seed until it becomes established. In contrast, hydrophobic or other polymers have been used to retard or delay water movement into the seeds. Large-seeded legumes are susceptible to hydration damage when sown in a cold, wet soil (see "Humidification" section). Coatings that retard water uptake can improve germination and survival under these stress conditions. Polymer coatings have also been developed that delay germination by several days. In this manner, seedling emergence times can be adjusted for different species or varieties sown at the same time. These coatings have applications in multicropping systems or in hybrid seed production in which the parental lines may have different maturity, for example, corn (*Zea mays*, maize). Success of the hydrophobic coatings is dependent on complete and uniform application of the polymers, and film-coating technologies are well suited to achieve this goal.

Conclusions

This article focused on treatments that are initially applied to dry seeds in storage in which the benefit or enhancement is realized during germination or at time of sowing. Another approach is for the seed treatments to be effective in storage. Plant protectants may be applied after harvest to control pests active during storage. Application of antioxidants and other compounds has been explored to improve longevity by retarding physiological changes that result with seed aging. Physical treatments such as heat, radiation, or magnetism were also another area not addressed. Though research has been performed with these treatments, results are inconsistent or inconclusive. Additional research is needed to better understand the mechanism of these physical treatments on seeds before commercial utilization.

In summary, seeds act as a delivery system for precise and uniform application of compounds. Plant protectants applied onto seeds provide a system to control pests in the target zone with minimal environmental contamination. Film coating reduces worker exposure to harsh compounds by producing a dust-free product. Physiological methods enhance performance under stressful conditions, thus providing greater insurance for crop establishment. Several methods can be integrated into a seed treatment system; for example, a current commercial practice is to first prime and then pellet seeds. The techniques outlined in the article provide a basis to develop future treatments and strategies. The importance of water and water content always needs to be considered as these strategies are envisioned. In conclusion, seed treatments have tremendous potential to enhance seed and seedling performance.

List of Technical Nomenclature

Conventional liquid treatment	Application of small amounts of materials in water resulting in uniform distribution with minimal change in seed moisture content.
Film coating	Application of a more or less continuous deposition of materials without obscuring the seed's size or shape, and resulting in minimal product dust-off.
Humidification	Application of humidified air circulation that results in increased seed moisture content.
Pelleting	Application of solid particles as a filler with binder or adhesive resulting in a more or less spherical shaped dispersal unit.
Planter box	Application of dry formulations to seeds; generally an on-farm operation conducted prior to sowing.
Priming	Controlled application of water or water potential to allow seeds to imbibe, and advance physiologically, but prevent the completion of germination.
Seed enhancements	Postharvest methods that improve germination, seedling growth, or facilitate the planting of seeds and the delivery of other materials required at time of sowing.
Seed treatments	Any technique in which seeds are subjected to a compound, process or to various energy forms.
Soaking/pregermination	Application of water to advance germination; soaking is generally of short-term duration to partially or fully imbibe seeds, while pregermination requires an aerated soak to result in complete germination (sprouting).
Solvent infusion	Application of materials dissolved or dispersed in an organic solvent, followed by removal (evaporation) of the solvent.

See also: **Postharvest Physiology**: Seed Storage. **Seed Development**: Artificial Seeds; Germination; Seed Production. **Tissue Culture**: Artificial Seeds.

Further Reading

Bradford KJ (1995) Water relations in seed germination. In: Kiegel J and Galili G (eds) *Seed Development and Germination*, pp. 351–396. New York: Marcel Dekker.

Clayton PB (1993) Seed treatment. In: Matthews GA and Hislop EC (eds) *Application Technology for Crop Production*, pp. 329–349. Wallingford: CAB International.

Halmer P (1999) Commercial seed treatment technology. In: Black M and Bewley JD (eds) *Seed Technology and its Biological Basis*, pp. 257–286. Sheffield: Sheffield Academic Press.

Hill HJ (1999) Recent developments in seed technology. *Journal of New Seeds* 1: 105–112.

Jeffs KA (ed.) (1986) *Seed Treatment*. London: British Crop Protection Council.

Khan AA (1992) Preplant physiological seed conditioning. In: Janick J (ed.) *Horticultural Reviews*, pp. 131–181. New York: John Wiley.

Martin TJ (ed.) (1988) *Applications to Seeds and Soil*. London: British Crop Protection Council.

Martin TJ (ed.) (1994) *Seed Treatment: Progress and Prospects*. London: British Crop Protection Council.

Pill WG (1995) Low water potential and presowing germination treatments to improve seed quality. In: Basra AS (ed.) *Seed Quality: Basic Mechanisms and Agricultural Implications*, pp. 319–360. New York: Haworth Press.

Scott JM (1989) Seed coatings and treatments and their effects on plant establishment. *Advances in Agronomy* 42: 43–83.

Taylor AG (1997) Seed storage, germination and quality. In: Wien HC (ed.) *The Physiology of Vegetable Crops*, pp. 1–36. Wallingford: CAB International.
Taylor AG and Harman GE (1990) Concepts and technologies of selected seed treatments. *Annual Review of Phytopathology* 28: 321–339.
Taylor AG, Allen PS, Bennett MA, *et al.* (1998) Seed enhancements. *Seed Science Research* 8: 245–256.
Taylor AG, Eckenrode CJ, and Straub RW (2001) Seed coating technologies and treatments for onions: challenges and progress. *HortScience* 36: 199–205.

Germination

R Johnson, Colby College, Waterville, ME, USA

Introduction

Germination is the process that changes a quiescent seed into a growing seedling. This conversion of a dry dispersal unit into a new plant is one of the most important and intriguing events in plant development. As defined by seed scientists, germination begins with imbibition (uptake of water) by a seed and ends when the embryonic axis begins to elongate. In most species, the end of germination is marked by the protrusion of the embryonic root, called the radicle, from the seed (**Figure 1**). In cereal grains, the coleorhiza also elongates and initially covers the radicle as it protrudes through the seed coat. Soon after germination, the radicle breaks through the coleorhiza as well. The process of germination, while conceptually simple, involves many complex developmental and physiological events. Imbibition of the seed results in the rehydration of tissues that had undergone severe dehydration during the later stages of seed development. Metabolic processes such as respiration and protein synthesis must be restarted. Repairs need to be made to membranes damaged by the dehydration and subsequent rehydration. Hydraulic pressure must develop in cells of the radicle tip resulting in cell expansion which will allow this structure to push against the tissues that surround and contain it. At the same time, loosening of the radicle cell walls and weakening of the surrounding structures is important in allowing the radicle to protrude from the seed. All of these processes must be properly controlled and coordinated so that germination will occur at the appropriate time, ensuring the survival of the embryo as a new plant.

Requirements for Germination

For germination to occur, all seeds require water, oxygen, and an appropriate temperature. Light is also an important factor that affects germination in many species. It is important to note, however, that many plants produce seeds that do not actually germinate even when they are placed into completely appropriate conditions. These viable seeds that do not germinate under normally suitable conditions are classified as dormant. A dormant seed will thus not germinate even when provided with a sufficient water supply, an atmosphere containing sufficient oxygen, and a temperature that is appropriate for the species. Only after dormancy is removed, or broken, may the seed germinate. Seed dormancy may be imposed by a variety of factors, either physical or physiological. Dormancy imposed by a seed coat that is impermeable to water may be broken by physical damage to the seed coat or by enzymatic digestion occurring in the gut of an animal. Physiologically imposed dormancy is a more complex phenomenon, and can be broken, depending on the species, by exposure to warm dry conditions (after-ripening), cool wet conditions (stratification), light, heat, chemicals in smoke, or other factors. Dormancy, and its eventual release after exposure to appropriate conditions, allows seeds to postpone germination until a later time when environmental conditions may be more suitable for survival of the seedling.

Most crop plants produce seeds that have little or no dormancy, and seeds that are dormant when produced can lose their dormancy as described above. Nondormant seeds that are capable of germinating if exposed to the appropriate environmental conditions but which are dry and thus metabolically inactive are referred to as quiescent seeds. The specific conditions that are required for germination of a quiescent seed can vary significantly between species.

Water

As germination begins with imbibition, water is clearly required. A seed must be exposed to moist soil so that it can take up sufficient water into its tissues to allow germination. The availability of water to a seed is generally viewed in terms of the water potential (Ψ, measured in MPa) of the soil or other surrounding medium. Experimental evidence indicating that seeds require exposure to water potentials above a critical threshold value has led to the development of the hydrotime model of germination. According to this model, the critical water potential (Ψ_b) is the highest water potential to which a seed can be exposed and still not germinate. Upon

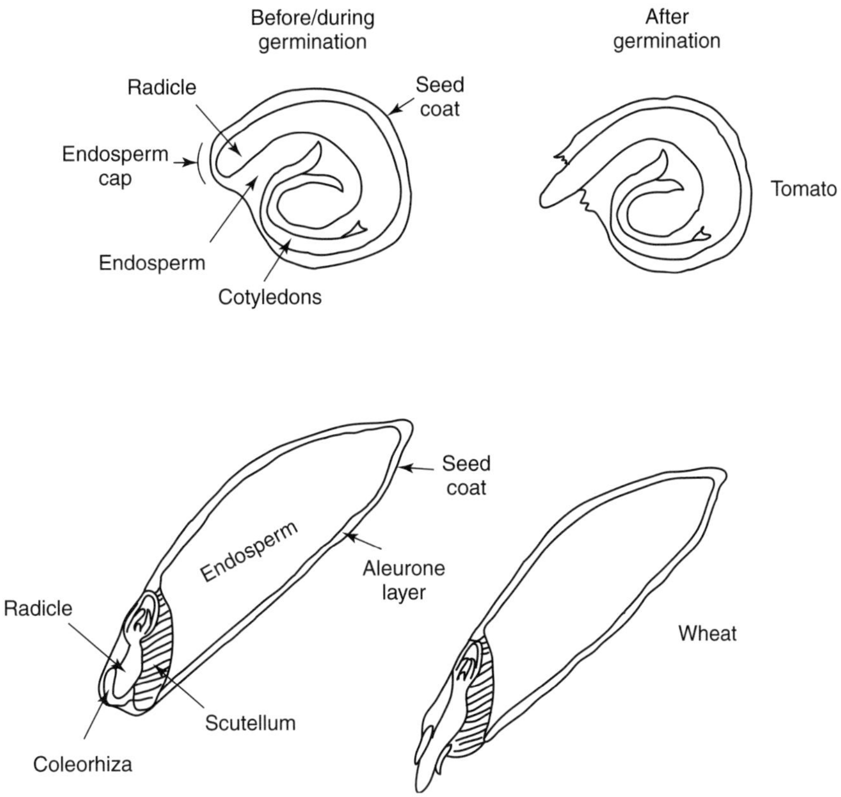

Figure 1 Typical features of seeds during and after germination. Tomato (*Lycopersicon esculentum*) is a representative dicot and wheat (*Triticum aestivum*) is representative of cereal grains.

exposure to water potentials above the threshold, seeds will complete germination after exposure to the required amount of accumulated hydrotime (θ_H, measured in MPa · h). Within a population of seeds of a particular species, the amount of hydrotime required to complete germination tends to be relatively constant, while considerable variability in the Ψ_b may exist among individual seeds. This variability in the threshold water potential can result in significant variability within a population of seeds in the amount of actual time required for germination.

Temperature

Seeds will germinate only within an appropriate temperature range. Corn (*Zea mays*; maize) kernels, for example, will germinate within a range of approximately 10–42 °C, and germinate optimally at 32–35 °C. Grains of wheat (*Triticum aestivum*), barley (*Hordeum vulgare*), and oats (*Avena sativa*) are able to germinate at considerably lower temperatures than corn. For these species germination can occur above approximately 3 °C and is maximal at 20–25 °C. Below their optimal temperature, seeds will germinate more slowly than they would at the optimal temperature. The amount of time required for a seed to germinate at suboptimal temperatures can be modeled in a similar manner to that described above for the hydrotime model. A seed will have a minimum temperature (T_b) below which it will not germinate. The seed will complete germination after exposure to a required amount of thermal time (θ_T, in degrees · h) above the minimum temperature. Experiments testing this model have indicated that among seeds within a population the T_b was relatively constant, while the amount of thermal time required for germination (θ_T) varied among seeds, resulting in different rates of germination for individual seeds even at the same temperature.

Oxygen and Light

As aerobic respiration is generally required for germination, seeds must be in the presence of oxygen. Seeds that are near the soil surface are normally exposed to a sufficient oxygen supply such that oxygen is not a limiting factor for germination, but in flooded soils seeds can indeed be well hydrated at the proper temperature and yet be unable to germinate due to a lack of oxygen. While such anaerobic conditions are inhibitory to the germination of most seeds, there are exceptions, such as rice (*Oryza sativa*) grains that are indeed able to germinate under anaerobic conditions in flooded fields.

The effect of light on germination (and dormancy) is complex. While seeds of most crop plants will germinate equally well in light or darkness, profound effects of light have been documented for many other species. Seeds of some plants require an exposure to light for dormancy breakage, while in other cases, exposure to light during imbibition inhibits germination. Many of these seed responses to light are mediated by phytochrome. In the case of Grand Rapids lettuce (*Lactuca sativa*), for example, low fluences of red light break the dormancy of hydrated seeds and this effect is photoreversible by far-red light. The inhibitory effect of light is a more direct effect on germination itself (rather than dormancy) as it requires high irradiance light during the final stages of germination.

Measurement of Germination

It is often valuable to measure the germination characteristics of a seed or a population of seeds. Two measurements that are commonly found in the literature are the germination capacity and the germination rate. The germination capacity is defined as the fraction of seeds in a given population or seed lot that is able to germinate within a specified period of time. For seeds that are to be used for agricultural sowing, at least 80% germination is usually desirable. The germination rate is the reciprocal of the amount of time required for germination to be completed, and is an indication of how fast the germination process occurs rather than of how many seeds in the population are able to germinate. The germination rate can be measured for an individual seed, and the rate for an entire population can be obtained by calculating the mean of the rates of the individual seeds.

Events during Germination

A quiescent (but nondormant) seed will begin to germinate upon exposure to the appropriate environmental conditions. Germination begins when the seed starts to take up water from the environment and is completed when the embryonic axis begins to elongate. Events occurring after radicle protrusion are not considered to be a part of germination, but are rather a part of early seedling growth. Germination itself includes a combination of biophysical, biochemical, and physiological processes. Physical factors involved in radicle protrusion include water uptake by radicle cells resulting in the buildup of turgor, cell wall loosening in the radicle cells that facilitates cell expansion under these conditions, and a weakening of the tissues adjacent to the radicle tip that allows radicle extension to occur.

Water Uptake

Uptake of water from the soil occurs down a water potential (Ψ) gradient. As a dry seed has an extremely low Ψ (-20 to -100 MPa), water uptake can begin to occur very rapidly from moist soil, which has a Ψ above -1 MPa, often approaching 0 MPa (the Ψ of pure water). The uptake of water by a seed occurs in a triphasic pattern (**Figure 2**). Initially, in phase I, water is taken up rapidly in response to the steep gradient of Ψ between the soil and the seed. In many species this water uptake takes place through the seed coat, but in species with impermeable seed coats uptake must occur through specific regions or structures designed for water uptake. Phase I results in the uptake of water equivalent to approximately one to three times the dry weight of the seed tissues, and raises the Ψ of the seed to around -1.5 to -1.0 MPa.

Phase II can be described as a lag phase in the water uptake process. At this point the seed is essentially fully hydrated and the cells of the embryo are under turgor. Further water uptake is limited by the fact that these turgid cells are prevented from expansion by the rigidity of their cell walls or by surrounding structures. The amount of time required for this phase to be completed will depend on a variety of factors, including the temperature and water potential of the seed, as discussed in the previous section. Phase III begins immediately upon protrusion of the embryo and the resulting release of constraints on cell expansion. Continued growth of the seedling and uptake of water into the growing

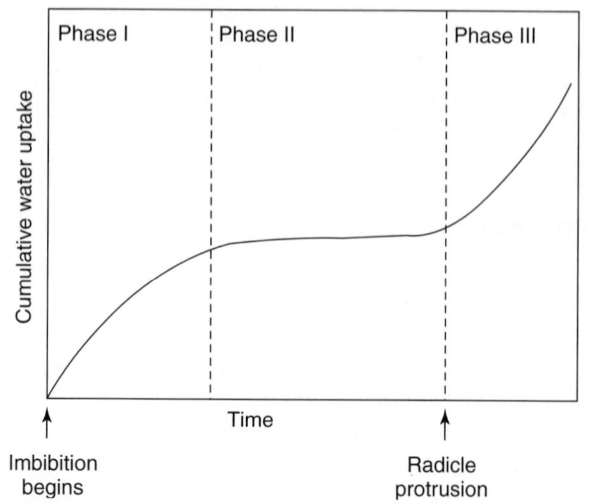

Figure 2 Uptake of water by seeds during and after germination. In phase I, beginning immediately upon imbibition, water uptake is rapid. In phase II, seed water content remains constant and no further fresh weight increase is observed. In phase III, which begins with radicle protrusion, elongation of the embryo results in further uptake of water.

tissues result in rapid and sustained water uptake after the completion of germination.

Metabolism

By the later part of phase I of water uptake, cells are sufficiently hydrated for metabolic reactions to proceed. Respiration generally begins within a few hours of the start of imbibition, including the reactions of glycolysis, the Krebs cycle and oxidative phosphorylation, and the pentose phosphate pathway. As mobilization of storage reserves (such as starch) is primarily a postgerminative event, the energy source for glycolysis during germination comes from sucrose and small oligosaccharides that are already present in the seed. Pyruvate produced from glycolysis is then transported to the mitochondria and converted to acetyl CoA. Krebs cycle reactions and oxidative phosphorylation in the mitochondria then use this acetyl CoA to produce ATP. In addition to glycolysis an alternative route for hexose breakdown, the pentose phosphate pathway, is also used in germinating seeds. This pathway produces NADPH and erythrose-4-phosphate, both of which are needed for biosynthetic reactions.

Another important early metabolic event during germination is protein synthesis. Polysomes form very rapidly upon seed hydration, and protein synthesis begins, utilizing stored mRNAs already present in the seed. These are mRNAs that were produced late during seed maturation and are thus available for immediate translation upon hydration and reconstitution of the protein synthesis machinery. Transcription of new mRNAs, however, also occurs very rapidly upon imbibition and translation from these new mRNAs can begin as early as 1 h after the start of imbibition. This rapid activation of respiration and protein synthesis during imbibition provides the energy and the newly produced proteins that are required for radicle protrusion.

Generation of Expansive Forces

Protrusion of the radicle through the constraining tissues that surround it requires the generation of expansive forces. Cells of the radicle will be rapidly hydrated during phase I of seed water uptake resulting in the formation of significant turgor. The fairly low solute potential (Ψ_s) within seed embryos (-1 to -2 MPa) is sufficient to result in very significant positive hydraulic pressure within the cells. If the cell walls of the radicle cells are sufficiently relaxed so that this pressure is not contained within the individual cells, then the radicle as a whole will exert a strong hydraulic pressure against the tissues that surround and contain it. As long as these surrounding structures are strong enough to resist the pressure, however, elongation of the radicle cannot occur and germination will not be completed.

Constraints on Radicle Elongation

For seeds of some species, such as rapeseed (*Brassica napus*) and lettuce, the endosperm and seed coat surrounding the radicle do not provide a sufficiently strong barrier to prevent its elongation. In this case, the primary restraint on elongation of the hydrated radicle comes from the walls of the radicle cells themselves. Once these cell walls are loosened, no further barrier exists, and radicle protrusion occurs. In many other species, structures such as the endosperm or seed coat provide a physical constraint that prevents radicle elongation.

In tomato (*Lycopersicon esculentum*) seeds the radicle tip is covered by an endosperm cap (**Figure 1**). The cell walls of the cap are sufficiently strong to prevent radicle elongation, and must be weakened before germination can be completed. A number of cell wall degrading enzymes, including β-mannanase, β-1,3-glucanase, and polygalacturonase increase during imbibition prior to radicle emergence. In addition to these hydrolytic enzymes, expansins, which act as wall-loosening factors by disrupting noncovalent linkages between polymers, also increase during imbibition. Apparently, the concerted effort of several different wall-loosening proteins is required to weaken the endosperm cap, allowing radicle emergence.

Hormonal Control of Germination

Germination (or lack thereof) is often the result of a balance between competing factors. In the case of tomato, a balance exists between the turgor-driven growth potential of the radicle and the physical resistance of the surrounding structures. This balance between growth and restraint must be properly regulated so that germination occurs only under the correct conditions and that events take place in a suitably controlled manner. It has been known for many years that the phytohormones gibberellin (GA) and abscisic acid (ABA) play significant roles in the regulation of seed germination. In a general sense, GA stimulates germination while ABA tends to inhibit germination, although the mechanisms by which these hormones interact with environmental factors and the ways that hormone levels are transduced into germination responses are still not entirely clear. Recent evidence also suggests that brassinosteroids play an important role in stimulating germination of *Arabidopsis* seeds.

Gibberellin

A large body of evidence exists demonstrating the importance of GA in seed germination. Much of this information comes from the study of mutants with altered germination characteristics (**Table 1**). For example, *ga1* mutants of *Arabidopsis* lack the ability to synthesize GA and these plants produce seeds that are unable to germinate without exogenous application of GA. Such results indicate that *Arabidopsis* seeds require endogenously produced GA for normal germination.

Dormant seeds by definition are unable to germinate when allowed to imbibe under normal conditions. Dormant seeds of many species, however, can be forced to germinate by exogenous application of GA. Whether exogenous GA stimulates germination through the same pathways that are relevant for stimulation of nondormant seeds by endogenous GA is not entirely clear, as there are some significant differences in both gene expression and growth patterns between normally germinating oat grains and dormant grains forced to germinate by addition of exogenous GA.

One very important mechanism by which GA exerts an effect on seed germination is through the stimulation of gene expression (**Table 2**). The classical effect of GA in stimulating α-amylase gene expression in cereal aleurone layers is critically important for the mobilization of starch reserves to provide energy for early seeding growth. This mobilization of storage reserves, however, is primarily a postgerminative event and is thus not likely to be important in the regulation of germination itself. In addition to α-amylase genes, GA has also been shown to stimulate expression of genes encoding proteins such as mannanase and expansins, which appear to be involved in cell wall weakening of radicle covering structures. In this way, GA stimulated gene expression may be much more directly related to the regulation of germination.

Abscisic Acid

During seed development, ABA plays a very important role in preventing precocious germination of the developing seeds and in the imposition of dormancy during embryogenesis. Corn mutants, such as *vp2*, that are deficient in ABA biosynthesis produce viviparous seeds which sprout on the maternal plant during seed development. Other corn mutants, such as *vp1*, with defects in ABA perception, show this phenotype as well (**Table 1**). The normal *VP2* gene encodes a protein required for one of the steps in ABA biosynthesis, while the normal *VP1* gene encodes a transcription factor that is required for the activation of a number of ABA-upregulated genes. Without the production of the VP1 transcription factor, expression of these genes in response to ABA is prevented and germination of the developing seeds fails to be suppressed.

The role of ABA in mature imbibing seeds is more complex. Although exogenous application of ABA usually inhibits germination, there is not a clear relationship between endogenous ABA and the ability of mature seeds to germinate. For example, ABA concentrations are similar in dormant and nondormant wheat embryos. The ability of an imbibing seed to inactivate ABA is much better correlated with nondormancy than is the amount of ABA in the seed before imbibition.

Table 1 Some examples of mutations affecting seed germination

Mutation	*Phenotype*[a]	*Gene product affected*
Arabidopsis thaliana		
abi1	Insensitivity to ABA	Protein phosphatase 2C
abi3	Insensitivity to ABA	VP1-like transcription factor
abi5	Insensitivity to ABA	bZIP transcription factor
era1	Hypersensitivity to ABA	Farnesyl transferase
ga1	Inability to synthesize GA	Copalyl diphosphate synthase
ga2	Inability to synthesize GA	*ent*-kaurene synthase
lec1	"Leafy" cotyledons, vivipary	Transcription factor
tt4	Pale weak seed coat, reduced GA requirement	
Lycopersicon esculentum **(tomato)**		
gib1	Inability to synthesize GA	
Zea mays **(corn)**		
vp1	Insensitivity to ABA, vivipary	Transcription factor
vp2	Inability to synthesize ABA, vivipary	Phytoene desaturase
vp7	Inability to synthesize ABA, vivipary	Lycopene synthase

[a] ABA, abscisic acid; GA, gibberellin.

Table 2 Some examples of genes expressed during seed germination

Name	*Gene product*	*Possible role*	*Effect of gibberellin*[a]	*Effect of abscisic acid*[a]
Arabidopsis thaliana				
AtEPR1	Extensin	Cell wall modification?	+	None
***Avena fatua* (wild oat)**				
AFN1	Glucosyl transferase	Signaling?	+	
AFN3	Glutathione peroxidase	Stress responses?	None	
***Hordeum vulgare* (barley)**				
HvPTR1	Peptide transporter	Peptide transport	None	None
***Lycopersicon esculentum* (tomato)**				
	Endo-β-mannanase	Cell wall weakening	+	None?
CEL55	Cellulase	Cell wall weakening	+	None
LeEXP4	Expansin	Cell wall weakening	+	None
GLUB	β-1,3-Glucanase	Cell wall weakening	+	–
LeXPG1	Polygalacturonase	Cell wall weakening		None
***Nicotiana tabacum* (tobacco)**				
βGLU1	β-1,3-Glucanase	Cell wall weakening	+	–
***Triticum aestivum* (wheat)**				
GF2.8	Germin	Pathogen defense		

[a] + indicates that the hormone stimulates gene expression, – indicates that the hormone inhibits expression of the gene, "none" indicates that the hormone has been shown to have no significant effect on expression of the gene.

A number of genes whose expression is stimulated by GA are suppressed by ABA (**Table 2**). In addition to the well-established role of ABA in suppressing GA-stimulated expression of α-amylase genes, ABA also has effects on genes likely to be more directly involved in the regulation of germination. For example, GA-stimulated expression of β-1,3-glucanase is suppressed by ABA in both tomato and tobacco (*Nicotiana tabacum*) seeds. In the case of rapeseed germination, where the barrier to radicle expansion comes from the radicle cell walls themselves, exogenous application of ABA prevents radicle emergence by inhibiting wall loosening within the radicle.

Molecular Biology of Germination

For any seed, the factors limiting embryo expansion must be regulated in an appropriate manner. Clearly the hormones GA and ABA play an important role in this regulation. As described above, a number of genes that may be important in germination are stimulated by GA or inhibited by ABA. In order to increase our understanding of the molecular biology of germination, efforts have been made to isolate these and other genes (**Table 2**). One strategy for identifying germination-related genes has been to search for genes that are expressed specifically during germination.

Some of the earliest germination-related genes to be identified were the appropriately named germin genes of cereal grains, which are expressed during germination and early seedling growth. The germin protein is an oxalate oxidase, which catalyzes the conversion of oxalate to two molecules of CO_2 and hydrogen peroxide. While associated with germination, germin probably does not play any role in controlling the germination process. Rather, it appears to protect the seed and seedling from pathogen attack. More recently, a number of genes have been identified that are specifically expressed in the embryo or the endosperm cap of germinating tomato seeds. Mannanase, glucanase, polygalacturonase, and expansin genes are all expressed in the endosperm cap and may have important roles in weakening this structure during germination.

A related approach used to identify germination-related genes has been to compare gene expression patterns during imbibition between genetically identical dormant and germinating seeds. In germinating wild oat (*Avena fatua*) grains, only a very small fraction of mRNAs increase in abundance specifically in the germinating grains, indicating that upregulation of only a few genes may be sufficient for germination. Genes specifically upregulated during germination (before radicle emergence) include those encoding glutathione peroxidase and a UDP-glucose glucosyltransferase. It is possible that glutathione peroxidase could play a role in stress responses and that the UDP-glucose glucosyltransferase might play a role in signaling that is important for radicle emergence.

Although considerable work remains to be done before we reach a clear understanding of the molecular biology and biochemistry of seed germination, much progress has already been made and this continues to be a very active area of research. The ability to study mutants lacking the function of specific genes and to follow expression patterns of genes during germination promises to yield exciting new information about the regulation of germination in the years to come.

Applied Aspects of Seed Germination

As the process that results in the conversion of a quiescent seed into a growing seedling, seed germination is clearly of utmost importance for agriculture. For sowing of crops it is desirable to have vigorous seeds in which germination is rapid and synchronous within the seed lot. For this reason seed lots to be sold for agricultural planting are carefully tested for their germination characteristics. On the other hand, germination at undesired times can result in significant agricultural losses. One such example is that of preharvest sprouting of cereal grains. Mature grains of wheat, barley, rice, and corn may germinate within the ear of the maternal plant if exposed to cool wet conditions before harvest. This results in the initiation of storage reserve breakdown within the grains, reducing the quality of the grains and the price that can be obtained for them.

The process of malting, important for the production of beer and a variety of food products (e.g., malt syrup, malt flour), is a case of carefully controlled germination under industrial conditions. Large batches of (typically barley) grains are allowed to germinate and to undergo a brief period of postgerminative seedling growth. Within the grains, α-amylase and other hydrolytic enzymes are produced, but the process is halted before significant breakdown of storage reserves begins. The grains are then dried and heated to produce a malt that can later be used for brewing or other purposes.

As production of practically everything we eat or drink relies upon seed germination in one form or another, it is difficult to overestimate the importance of this process. Further advancements in our understanding of germination should provide significant benefits in terms of increased agricultural productivity and reduced losses.

List of Technical Nomenclature

Abscisic acid	A plant hormone.
Dormancy	A state characterized by the inability to germinate under normally suitable conditions.
Endosperm	A storage tissue in seeds.
Germination capacity	The fraction (or percentage) of seeds in a population that are able to germinate.
Germination rate	The reciprocal of the amount of time required for germination to be completed.
Gibberellin	A plant hormone.
Imbibition	The uptake of water by seeds.
Phytochrome	A photoreceptor sensitive to red and far-red light.
Radicle	The embryonic root.
Vivipary	Premature germination of seeds on the maternal plant.
Water potential	A measure of the amount of free energy associated with water.

See also: **Cell Walls and Fibers**: Cell Walls. **Regulators of Growth**: Abscisic Acid; Gibberellins; Phytochromes and other Photoreceptors. **Seed Dormancy**: Development of Dormancy; Genetics of Dormancy. **Water Relations of Plants**: Basic Water Relations.

Further Reading

Bewley JD (1997) Seed germination and dormancy. *Plant Cell* 9: 1055–1066.

Bewley JD and Black M (1994) *Seeds: Physiology of Development and Germination*, 2nd edn. New York: Plenum Press.

Bradford KJ, Chen F, Cooley MB, *et al.* (2000) Gene expression prior to radicle emergence in imbibed tomato seeds. In: Black M, Bradford KJ, and Vásquez-Ramos J (eds) *Seed Biology: Advances and Applications*, pp. 231–251. Wallingford: CAB International.

Holdsworth M, Kurup S, and McKibben R (1999) Molecular and genetic mechanisms regulating the transition from embryo development to germination. *Trends in Plant Science* 4: 275–280.

Johnson R (2000) Characterization of germination-related genes in *Avena fatua* seeds. In: Black M, Bradford KJ, and Vásquez-Ramos J (eds) *Seed Biology: Advances and Applications*, pp. 253–259. Wallingford: CAB International.

Mayer AM and Poljakoff-Mayber A (1989) *The Germination of Seeds*, 4th edn. Oxford: Pergamon Press.

Welbaum GE, Bradford KJ, Yim K-O, Booth DT, and Oluoch MO (1998) Biophysical, physiological and biochemical processes regulating seed germination. *Seed Science Research* 8: 351–371.

Artificial Seeds

D J Gray, University of Florida, Apopka, FL, USA

Background

Conventional Seed

Production of most major agronomic crops utilizes seeds for stand establishment. This is due to their overall low cost and efficiency when compared to other methods of planting crops, such as the use of cuttings. With few exceptions, a seed is composed of a zygotic embryo, which is encased in nutritive tissues and several layers of protective coats. Due to the common use of seeds, we tend to take their special properties for granted. Seeds are extremely durable due to the protective coats and are desiccation tolerant, so that they can be dried and become quiescent. These qualities allow seeds to be stored *en masse* at room temperature and sown with relatively simple planting equipment. Due to their ability to become quiescent, seeds also are used for germplasm preservation in seed repositories. The primary disadvantage to the use of seed rests with the origin of the zygotic embryo, which is the result of sexual reproduction; the progeny of two parents. This has led to the development of often complex breeding programs from which inbred parental lines are developed. Such inbred lines are used to produce uniform hybrid progeny when crossed. For many crops, such as fruits, nuts, and certain ornamentals, it is not possible to produce a true-breeding seed from two parents due to genetic barriers to selfing. For other crops, such as forest trees, the generation time is too long to achieve rationally an inbred breeding program. Therefore, for such crops, propagation is accomplished either vegetatively by cuttings or the use of relatively low-quality open pollinated seed is tolerated.

Artificial Seed

For a number of applications, it would be advantageous to combine the efficiency aspects of seed with clonal plant production. The discovery of somatic embryogenesis in the 1950s provided a possible alternative to conventional seed. Somatic embryos are structurally equivalent to zygotic embryos, but are true clones, since they arise from the somatic cells of a single "parent." Somatic embryos differ from zygotic embryos in that they are produced via *in vitro* culture, develop without nutritive and protective seed coatings, and do not typically become quiescent. The field that seeks to use somatic embryos as functional seed is termed "artificial or synthetic seed technology." Thus, artificial seeds are defined from a practical standpoint as somatic embryos engineered to be of use in commercial plant production and germplasm preservation. The actual structural complexity of artificial seed depends on requirements of the specific crop application. Therefore, a functional artificial seed may or may not require a synthetic seed coat, be hydrated or dehydrated, quiescent or nonquiescent, depending on its intended usage.

Structure and Development of Conventional versus Artificial Seeds

Zygotic and Somatic Embryos

Zygotic and somatic embryos share many structural characteristics, with both typically passing through globular, torpedo, and cotyledonary stages for dicots or globular, scutellar, and coleoptilar stages for monocots. They also share a key useful developmental feature in that both are able to form complete plants without separate rooting and shoot development phases. Since somatic embryos develop from somatic cells, instead of zygotes, they differ in that the former can be used to produce duplicates of a single genotype.

Desiccation tolerance and quiescence Desiccation tolerance and onset of quiescence, which are key developmental mechanisms leading to the usefulness of zygotic embryos in conventional seed, are either missing or have been overlooked in somatic embryos. Zygotic embryos in conventional seed begin to lose water content and enter a resting period during maturation which is the major factor allowing them to be conveniently stored. Basic developmental mechanisms that induce or allow the resting phase differ among species and have been variously categorized. Generally, two broad categories of arrested growth, "quiescence" and "dormancy," are distinguished. Quiescence is a resting phase that can be reversed solely by the addition of water. Dormancy is a form of quiescence that requires factors in addition to water, such as cold or heat treatments, for resumption of growth to occur. Typically, quiescent seed can be kept for prolonged time periods without a noticeable loss of viability. Seed germplasm repositories utilize specific temperature/relative humidity regimes to optimize long-term viability.

In contrast to resting zygotic embryos in seeds, somatic embryos typically do not become quiescent; rather, they tend to continue to grow and typically either germinate, become disorganized into embryogenic tissue, or die. Lack of a resting phase in somatic embryos is a major drawback to their use as artificial seed. Somatic embryos with reversible, arrested growth (i.e., quiescence) will be needed in order to mimic seed storage and handling characteristics.

Dormancy, wherein embryo proliferation occurs but not germination and plant development, has been documented in somatic embryos of several species and may be unrecognized in those of others. Certain types of dormancy, such as those regulated by nonembryonic seed tissues, clearly cannot be a factor in existing embryogenic culture systems. However, mechanisms of quiescence and dormancy that are controlled by water availability, temperature, light, inhibitors, promoters, etc., may be functional in somatic embryos of species with corresponding zygotic embryo dormancy. Quiescence has been induced experimentally in somatic embryos of several species, usually by controlled dehydration, often combined with modification of culture nutrients (such as increase of carbohydrate sources) and plant growth regulators (such as addition of abscisic acid (ABA)).

To experimentally induce quiescence in (normally) rapidly growing somatic embryos, lowering of water content by dehydration is a logical approach since dehydration and rehydration cause arrest and resumption of growth in orthodox seeds. The first example of induced quiescence in somatic embryos was documented in orchardgrass (*Dactylis glomerata*). Somatic embryos were dehydrated in a 70% RH environment and survived up to 21 days at 23°C. The embryos became discolored, decreased in size, and their outer cell walls collapsed. Embryo water content was reduced from 83% to 13% within 24 h. This water content is similar to that of seeds maintained at 70% RH and is adequate for maintenance of viability during prolonged seed storage. During rehydration, the embryos regained their normal appearance and germinated. However, only mature embryos which contained starch and lipid storage compounds survived. Of mature dehydrated somatic embryos stored for 21 days, 4% produced plants after imbibition. This study convincingly demonstrated that quiescence could be induced to occur in somatic embryos.

Grape (*Vitis vinifera*) somatic embryos were used to first demonstrate quiescence in a dicotyledonous species. They were treated similarly to orchardgrass and quiescence was induced. In grape, genotypic differences in response were noted; those varieties that produced relatively well-developed somatic embryos in culture were most responsive. After 21 days, 34% of dehydrated somatic embryos from one genotype produced plants following imbibition. More recently, somatic embryos of grape, dehydrated to 25% of original moisture content, exhibited 90% viability after 42 months of storage at 6°C. Such somatic embryos were capable of germination and plant development. This study demonstrated that high germination percentages and long storage periods are possible.

Quiescence has also been observed in somatic embryos of corn (*Zea mays*, maize) and soybean (*Glycine max*). Approximately 90–100% selected, high-quality somatic embryos of alfalfa (*Medicago sativa*, lucerne), produced plants when treated with ABA before drying. After 1 year of dried storage without humidity control, 60% of the embryos germinated into seedlings. The ability of alfalfa embryos to survive dehydration was thought to be due to addition of ABA; this appeared to induce desiccation tolerance. Thus, ABA-treated somatic embryos of alfalfa mimicked typical quiescent seed. Further refinements to the system showed that $10\,\mu mol\,l^{-1}$ ABA applied to 14-day-old embryos was most effective. Slow drying was more effective than rapid drying.

In somatic embryos, timed application of ABA and controlled drying appear to coincide with the developing seed environment. Induction of desiccation tolerance in somatic embryos shows that environmental conditions regulate whether or not they assume the complex developmental pathways normally associated with those of zygotic embryos in seed.

Nutritive and Seed Coat Tissues

Tissues surrounding zygotic embryos in seed serve to provide nutrition and/or protection. A major distinction between zygotic and somatic embryos is that the latter develop without these tissues. In addition, they create barriers to regulate gas exchange. For example, protective tissues limit oxygen availability, which influences respiration and embryo development and prevents precocious germination. For somatic embryos, the culture vessel in which they are grown determines the available gas regime. Availability of oxygen and other gases depends on subculture interval, whether or not the vessel is tightly sealed, as well as gas usage and evolution by other associated tissues, such as subtending callus. This is a stark contrast to the highly regulated and developmentally controlled seed environment. In fact, immature zygotic embryos often do not develop normally if removed prematurely from the seed and placed onto culture medium; this reinforces the role of seed tissues in providing the proper environment for development.

Embryo nutrition: zygotic embryos Within seeds, the tissue that performs a nutritive function is the endosperm or female gametophyte in angiosperms and gymnosperms, respectively. Nucellar tissue also performs a nutritive function in many seeds. All substances required by the embryo must pass either through the suspensor or directly from the encasing nutritive tissues. Relative dependence on suspensor versus endosperm for nutrition varies between species. Depending on species, endosperm may or may not persist in the mature seed to provide nutrition during germination. The composition of storage substances also differs among species but consists predominantly of carbohydrates, lipids, and proteins. Endosperm may also function to regulate embryo water balance. Embryos initially develop with a suspensor connection; however, those with endospermic seeds, such as tobacco (*Nicotiana tabacum*), shift from suspensorial to endosperm nutrition at some point in their development. Such embryos are able to utilize nutrients from endosperm at an early stage. In nonendospermic seeds, such as cucurbits and legumes, endosperm either does not form in appreciable amounts or is exhausted during development so that the mature embryo must contain adequate storage reserves for germination.

Germination requires energy reserves stored either in endosperm (endospermic seeds) or the embryo itself (nonendospermic seeds). Storage reserves are mobilized during germination. The complex developmental mechanisms that lead to the accumulation then usage of storage compounds are partially responsible for the long-term viability and rapid germination qualities of seed.

Embryo nutrition: somatic embryos The ability of somatic embryos to develop in a manner that parallels that of zygotic embryos is surprising considering the significant differences in the environments in which they develop. Zygotic embryos of a given species are adapted to grow in very species-specific and complex nutrient environments. Whereas, with somatic embryos, there is no endosperm tissue to feed the developing embryo and modulate nutrient flow. Somatic embryos develop in a highly unorganized manner on or within callus cultures or immersed in liquid medium. In fact, only a few commonly available culture media have been used to produce somatic embryos from hundreds of different species. In retrospect, this is rather unexpected, considering the large range of nutrient regimes among various seed types that these reports parallel. Often, a suspensor is the only link between the embryo and growth medium. Interestingly, somatic embryos commonly develop perched above subtending callus or explant tissue on narrow suspensors without any opportunity for nutritive factors to be absorbed by the embryo body. This demonstrates that the suspensor can serve as the pathway to provide all needed nutrition and that endosperm is not necessary for embryo development. However, it is likely that these significant differences in the environments between zygotic and somatic embryos result in the irregularities noted with the latter.

Protective coatings: zygotic embryos Seed coats are composed of structurally complex lignified tissues that originated from the floral ovule and, for seed pods, the ovary. Seed coats variously fulfill a number of functions. Often, there are structured openings through the coats to permit the entrance of water to facilitate germination. Some seed coats possess accessory tissues that constitute "wings" or other structures, such as hinges that spring open with drying to facilitate dispersion. They provide protection from disease-causing organisms and mechanical stress. They also function to preserve dormancy in some species by a variety of mechanisms, including exclusion of water. The seed coat provides a limited amount of space in which the embryo and accessory tissue must develop; this often causes the embryo to become compressed and/or flattened into a shape and size distinct for a given species or variety. In comparison, corresponding somatic embryos tend to be larger and not flattened, with wider hypocotyls and fleshier cotyledons. The force exerted by the developing embryo within the seed coat also determines pressure potential, thus affecting osmotic potential and water content.

Protective coatings: somatic embryos Artificial seed coats have been demonstrated. However, mimicking certain of the complex functions of seed is not possible and probably not necessary. Specialized dispersal structures are not needed, since artificial seed will be handled with mechanical planting devices. Seed coat-induced pressure to alter morphology and water content of somatic embryos, while possibly preferable in producing high-quality analogs to conventional seed, may be difficult to achieve.

Artificial seed coats consisting of plastic resins, organic ingredients, or other materials have been demonstrated. However, calcium alginate, which forms a gel, has been used in most examples. Additives, including starches, sugars and antimicrobial substances like fungicides, have been incorporated into the coating. Functionally, such coatings then constitute an artificial endosperm and seed coat in one structure when used to encapsulate an embryo. Both hydrated and dehydrated artificial seed coats have been tested.

To date, most examples of artificial seed coats have not shown adequate utility. In nearly all instances, somatic embryos encapsulated in such coatings lose viability faster than embryos that are not encapsulated. Nevertheless, as described below, protective and/or nutritive coatings will be required for certain practical implementations of artificial seed technology; hence, research into their development is urgently needed.

The foregoing discussion of nutritive and seed coat tissues has underscored significant differences in physical environment that occur between somatic embryos and zygotic embryos. Both somatic and zygotic (in seed) embryos placed on an appropriate germination medium are connected to the medium by a liquid interface. However, only the somatic embryo is directly exposed to medium. The zygotic embryo is separated from medium by seed coatings and endosperm, although a direct channel to the outside may occur in some types of seed. Thus, somatic embryos develop and germinate in an environment unlike that of zygotic embryos. Certain aspects of the complex seed environment, such as presence of an array of polysaccharides, minerals, proper pH, etc., are obviously necessary for development of normally germinating somatic embryos and are provided in commonly used culture media. However, since zygotic embryos excised from seed and placed on nutrient medium often do not develop properly and tend to exhibit abnormalities common to somatic embryos, it is probable that additional modifications will be necessary. Some of the nutrient factors in seeds such as specific polysaccharides and amino acids can be readily incorporated into culture medium and have been used in a number of systems, including alfalfa and Norway spruce (*Picea abies*). However, simulating complex nutrient pulses and the precise physical environment created in seed will be more difficult.

Types of Artificial Seed

Depending on the specific application and production needs of a given crop, functional artificial seed can assume several different forms. The somatic embryo may be quiescent or nonquiescent and a protective coating may or may not be present. For example, uncoated nonquiescent somatic embryos could be used to produce those crops that are now laboriously micropropagated by tissue culture. Manpower reduction and, hence, cost reduction achieved by producing plants by somatic embryogenesis would be significant compared to that of micropropagation. Uncoated, quiescent (dried) somatic embryos would be useful for germplasm storage since they can be hand-stored in existing seed storage repositories. Cost of manipulating somatic embryos for such storage would be similar to that of seed but would constitute a clonal source of germplasm.

Nonquiescent somatic embryos placed in a hydrated encapsulation constitute a type of artificial seed that may be cost effective for certain field crops that pass through a greenhouse transplant stage such as carrot (*Daucus carota*), celery (*Apium graveolens*), seedless watermelon (*Citrullus vulgaris*), and other vegetables. However, if long-term storage is not necessary, artificial seed production and planting would have to be closely coordinated.

Artificial seeds consisting of quiescent encapsulated somatic embryos could be more useful than the nonquiescent form, since the artificial seed production phase could be separated from the plant production phase.

Dehydrated, quiescent somatic embryos encapsulated in artificial coatings are the form of artificial seed that most resembles conventional seed in storage and handling qualities. These consist of somatic embryos encased in artificial seed coat material, which then is dehydrated. Under these conditions, the somatic embryos become quiescent and the coating hardens. Theoretically, such artificial seeds are durable under common seed storage and handling conditions. Upon rehydration, the seed coat softens, allowing the somatic embryo to resume growth, enlarging and emerging from the encapsulation. This type of synthetic seed would be necessary for field planting using conventional technology. Unfortunately, there has not been sufficient development of the dehydrated, encapsulated form of artificial seed to allow commercialization. Since dehydrated uncoated somatic embryos can survive significantly longer storage periods, it appears that the materials currently used as artificial seed coats are detrimental to survival. Additional research is required to develop more acceptable coatings.

In addition to application of an artificial seed coat to facilitate planting, other methods of delivering somatic embryos to soil, such as fluid drilling, have been proposed.

Applications for Artificial Seeds

The particular usage of artificial seed for a given crop will depend on several factors, including specific production needs that are not met with existing systems. For artificial seed to be of commercial use, crop production must ultimately be more profitable than with existing systems. Various crops are discussed below to illustrate possible applications of different forms of artificial seed.

Ornamental Crops

Many ornamental crops already are propagated via tissue culture where per-plant production costs are relatively high due to manpower requirements. Micropropagation of shoot cultures is routinely used; this process requires multiple laborious handling steps to ultimately produce rooted plants. Use of artificial seed, in this form consisting of naked hydrated embryos, would alleviate much of the manpower requirements. Somatic embryos mass-produced in culture could be selected and placed directly into planting flats, resulting in rooted plants and thus eliminating several labor-intensive steps.

Fruit and Nut Crops

Most fruit and nut crops currently are economically propagated clonally from cuttings due to genetic self-incompatibility and long breeding cycles. While artificial seed could possibly be used to facilitate production, existing methods are cost effective and development costs of the technology may not be justified. Efficiency aspects of artificial seed may be further reduced if the resulting clones require grafting, as do many existing fruit and nut varieties. However, use of dehydrated, quiescent, uncoated forms of artificial seed could be of significant use in germplasm conservation, since they could be stored in conventional seed repositories. This would alleviate the expensive reliance on field gene banks, in which plants are kept in field repositories where they are subject to loss from environmental disasters.

Vegetable Crops

Artificial seed could provide a cost advantage to those vegetable crops that currently have high seed costs and high per-plant value. For example, many hybrid varieties are very expensive to produce. Some are produced as transplants in greenhouses before transfer to the field. For seedless watermelon, the combination of high seed cost, due to barriers to seed production, and low germination rates can result in transplants costing US$1.00 each. For seedless watermelon, artificial seed could reduce per-plant costs by circumventing barriers to seed production. Artificial seed, in this case naked or encapsulated hydrated somatic embryos, could be treated in the same manner as conventional seed in transplant production.

Hybrid Cereals

Another application of artificial seed would be to remove the reliance on inbred and male-sterile parental lines in hybrid seed production. For example, the hybrid corn industry relies on inbred parentals to produce uniform hybrid seed. Mass seed production is made possible by use of male-sterile lines as females. Resulting hybrid plants exhibit increases in yield and quality; however, development of true-breeding parental lines is extremely time-consuming and costly. As an alternative, artificial seed could be used to clone and mass-produce one outstanding hybrid, eliminating the most demanding parts of a hybrid program. For this usage, artificial seed would have to function as true analogs of conventional seed in planting and production so that existing equipment could be used.

Conifers in Forestry

Due to relatively low per-plant value (when value is amortized over production time) and high volume of plants needed, seed must be used to establish stands of conifers. Conifers are relatively unimproved genetically due to barriers associated with the long reproductive life cycle and extended time needed to evaluate progeny. Therefore improvement via conventional breeding is extremely slow. Ability to select existing outstanding individuals from the forest and clone them economically as artificial seed would allow rapid progress in tree quality to be achieved. The form of artificial seed most amenable to forest tree production would be dehydrated, quiescent somatic embryos, encapsulated in a protective coating. This would allow artificial seed to be utilized in conventional production systems.

Conclusion

To date, the greatest progress in implementation of artificial seed technology has been with conifers. Production systems in place today allow the efficient establishment in soil of thousands of trees at a time. However, development of similar systems for other crops has been slow. This may be due to the lack of a driving need for improvement, since existing systems tend to be profitable in light of the relatively large amount of investment capital needed to produce true artificial seed (i.e., commercially useful somatic embryos, as defined here). Advances in the state-of-the-art of embryogenic cell culture are continuing and it is likely that certain improvements eventually will actualize the use of artificial seed for additional crop applications.

List of Technical Nomenclature

Abscisic acid (ABA)	A plant growth regulator associated with embryo maturation and quiescence.
Artificial seed	A somatic embryo used in commercial plant production or germplasm preservation.

Artificial seed coat	Protective man-made material used to coat, protect and/or provide nutrition to somatic embryos.
Desiccation tolerance	Ability of embryos to survive dehydration.
Dormancy	Resting phase in embryos in which reversal requires factors in addition to water.
Endosperm	Nutritive tissue of zygotic origin, which encases and provides nutrition to zygotic embryos.
Quiescence	Resting phase of embryos usually reversed by addition of water.
Seed coat	Protective tissue of maternal origin surrounding zygotic embryos for protection.
Somatic embryo	Clonal embryo resulting from *in vitro* cell culture; the duplicate of one "parent."
Suspensor	Organ connecting developing embryo to subtending tissue.
Zygotic embryo	Embryo resulting from sexual reproduction; the progeny of two parents.

See also: **Crop Improvement**: Chromosome Engineering. **Postharvest Physiology**: Seed Storage. **Seed Development**: Embryogenesis; Germination; Seed Production. **Tissue Culture**: Somatic Embryogenesis.

Acknowledgments

This research was supported by the Florida Agricultural Experiment Station and a grant from The Florida Department of Agriculture and Consumer Services Viticulture Trust Fund and approved for publication as Journal Series No. R-09249.

Further Reading

Bewley JD and Black M (1985) *Seeds: Physiology of Development and Germination.* New York: Plenum Press.

Gray DJ (1987) Introduction to the symposium. In: Proceedings of the Symposium on synthetic seed technology for the mass cloning of crop plants: problems and perspectives. *HortScience* 22: 796.

Gray DJ and Purohit A (1991) Somatic embryogenesis and development of synthetic seed technology. *Critical Reviews in Plant Sciences* 10: 33–61.

Ragahavan V (1986) *Embryogenesis in Angiosperms.* New York: Cambridge University Press.

Trigiano RN and Gray DJ (2000) *Plant Tissue Culture Concepts and Laboratory Exercises*, 2nd edn. Boca Raton: CRC Press.

Seed Production

R E L Naylor, Trelareg Consultants, Banchory, Scotland, UK

Introduction

Seeds are a basic resource in agriculture. They represent the foundation and continuation of all crops. The grower requires that the seeds sown fulfill the basic requirement of being capable of producing a good plant stand. The seeds thus have to be alive and capable of germination without the delay imposed by prolonged dormancy. The producer of these seeds therefore has to pay particular attention to fostering all the biological processes involved in seed production and to minimize any detrimental influences on subsequent germination.

In some cases, seeds are still collected from the wild, albeit from specific locations. An example is forest trees in many of which geographic provenances or specific seed trees may be identified with the aim of restricting the genetic variability in the established plants. In field crops, the seeds may themselves be the agricultural product (e.g., cereals, mustard, coconut) in which case there may be additional quality criteria determined by the end use of what is then an agricultural product. Crops in which the seed is not the main product, e.g., crops produced for their leaves (lettuce, grass) or their stems (flax, sugar cane) or their roots (sugar beet, carrots) are likely to require a different agronomy if seed production is the aim of growing the crop. Even in clonally produced crops (e.g., potatoes) sexual reproduction and seed production are essential for the generation of new cultivars. The major focus of this article is on producing seed for sowing the next crop.

Seed production is generally preceded by pollination and starts with the fertilization of the ovule. It can be considered as ending when the seed is shed from the parent plant or is harvested. After-ripening, seed storage conditions, and handling all play a role in determining seed quality at planting. This article will summarize the biological processes involved in seed production and then use this to develop the principles of seed production agronomy. It will examine the impacts of various sorts of "stress" on seed production. The focus will be on conventional seeds; species with recalcitrant seeds will not be considered. Seed storage after harvest will not be considered.

The Biological Background

After fertilization of the ovule, seed development up to harvest involves many changes. The embryo grows and develops in complexity. Increase in the amount of storage material in or adjacent to the embryo leads to changes in seed weight. The water content of the seed decreases and its physiological state changes. The seed and fruit coats provide increasing protection for the seed.

Changes during Embryo Development

Between fertilization and harvest the fertilized ovule undergoes growth (increase in size, weight, and numbers of cells) and development (increase in complexity and organization). The first phase of seed development is one of rapid cell division. Increases in the size and number of cells requires water and assimilates. There are different patterns of embryo and seed development. The embryo of cereal seeds (strictly a fruit) has many cells and is highly developed. It bears a number of leaf primordia and a set of root initials and a rudimentary vascular strand. The embryo typically occupies about 25% of the volume of the cereal seed, the rest of the volume is the starchy endosperm. After reimbibition and at the onset of germination, the embryo grows rapidly by the extension of already existing cells. The embryo of legumes consists of two large cotyledons containing the storage reserves and an embryo axis. The cotyledons may compose 90% of the seed volume. The embryo of many species is relatively undeveloped. Seeds of members of the Apiaceae (carrots, parsnips) take a long time to germinate because, after imbibition, the embryo has to grow inside the seed before the radicle emerges from the seed coat.

Changes in Seed Mass

The second stage of seed development is the accumulation of reserves. After fertilization, the developing seed is a strong sink for assimilates. These may be the products of current photosynthesis, or have been produced earlier and stored and are now being retranslocated to the seed. In many crop species it is the products of current photosynthesis that provide the majority of seed mass. Because seeds do not generally have a vascular connection with the phloem, substances which accumulate in the seed must move apoplastically after unloading from the phloem. Import of stored assimilates may be important if weather conditions prevent high assimilation during grain filling.

The major storage carbohydrate in many seeds is starch. This is insoluble and has to be synthesized from simple soluble sugars (mostly sucrose) transported to the seed. Any impairment of the transpiration stream can therefore influence the rate of biomass accumulation in the seed. Thus water shortage or breakage of stems (lodging) can lead to some young seeds aborting or seeds being less well filled or maturing later. Starch is not the only storage carbohydrate – in peas some 40% of the storage carbohydrate may be hemicelluloses. In coffee and date seeds the hemicellulose is stored in the cell walls. Sucrose entering the seeds may also be converted to oils and fats. Cereal seeds may contain up to about 12% protein. The protein content of cereal seeds is achieved largely through redistribution of nitrogen accumulated before anthesis. The protein content of legume seeds may be about 20–30% (peas and beans). Oil is a main storage compound in species such as linseed, sunflower, and oilseed rape (canola).

Changes in Hydration

Immediately after fertilization, the water content of many seeds is often about 80%. As the seed increases in weight, the water content declines and the aim in most crops is to harvest under about 15% water content. Harvesting at higher moisture contents risks damaging those seeds which have not yet fully developed, so seed weight and quality may be low. For most species a water content of 10% or less is required for safe seed storage and so seeds are often dried if harvested at water content much above this.

Changes in the Seed Coat

Few crops have naked embryos. The major exceptions are the gymnosperms used as forest trees. Other types of plant have a seed coat (testa) and often a fruit coat (pericarp) with a range of complexity across the range of crop species. The seed and fruit coats are important because they provide physical protection for the seed. They restrict water loss and slow down the drying of the seed during maturation. Changes in the seed coat are both genetically and environmentally controlled and can influence how the seed behaves when planted. For example, wheat varieties with a white seed coat are generally nondormant and when seeds mature in moist conditions they are at risk of premature grain sprouting (the seeds germinate while still on the parent plant). Red wheat varieties have a short period of primary dormancy which generally prevents this. In legumes, a colored seed coat usually slows down water ingress during the early reimbibition of seeds planted to germinate and this is important in preventing imbibition damage restricting stand establishment.

The influence of weather during maturation is critical and interacts strongly with genotype. Irrespective of the color of the seed coat, wheat varieties are more likely to sprout when maturation conditions are cool and moist. Thus we can define the relative susceptibility to sprouting of a set of varieties but not predict from just a knowledge of the genotype, the actual extent of sprouting in any season or at any site.

Changes in Physiological State

During this time there are changes in the content of plant hormones which regulate embryo development, the accumulation of reserves, and seed coat development and these may be stored for later regulation of seed germination and early seedling growth. The "normal" development of seeds culminates in the production of a dry seed. Stored dry, this can retain viability for many years. Seeds harvested very early, before completing embryo development and accumulation of reserves, may be capable of germinating, but may not survive drying. Thus the progressive dehydration on the parent plant is important in determining seed quality.

Properties of the embryo or of the seed coat may determine whether the seed is able to germinate immediately or whether it is dormant. Hormonal control of the embryo is one mechanism. In legumes, "hard-seededness" may prevent germination because the maturation conditions lead to a very impermeable and rigid seed coat. Substances in seed coats may prevent oxygen diffusion to the embryo.

Principles of Agronomy for Seed Production

Agronomy is the set of practices describing "how" to grow crops as distinct from the underpinning crop science which provides the "why." However, the ways in which we grow crops have evolved pragmatically as growers have sought to improve food supply and the supply of viable healthy seeds for the next crop. The aim of agronomy is to provide good growing conditions (in as much as these are controllable by the crop manager) and to manipulate these so that the crop grows in such a way as to optimize the yield of the plant part required. For crops such as cereals, the seed is the product, so there may not need to be much adjustment of the agronomy to produce high-quality seed crops. In contrast, for many vegetable, fiber, and forage crops the agronomy for producing seeds may be quite different from that required for growing the main crop of leaves, stems, or roots. It may need to take account that selection of cultivars for enhanced leaf, stem or root production may have led to types which do not reproduce easily or synchronously.

The main features of any agronomy system for production of seeds are:

- choosing a good site
- choice of an appropriate cultivar
- selection of high-quality seeds
- creating a good seedbed to enhance seedling emergence and growth
- sowing an appropriate density of seeds
- adequate water
- sound crop nutrition
- sound crop protection
- good harvesting conditions.

Weather is something that the grower usually has to accept, except for protected crops where management of water availability, temperature, and sometimes daylength, is possible.

Clearly the amount of green leaf area and the capture of radiation input play a large part in determining the accumulation of carbohydrate. The management of canopy area is an important feature of modern agronomy systems and is usually achieved by judicious use of nitrogen. Plant growth regulators can be used to modify plant architecture to improve canopy structure, to prevent lodging in order to ensure the crop can be harvested, and to alter the reproductive structure. Levels of phosphorus and potassium are important in seed crop production systems. Phosphorus generally increases seed yield. Potash is important for flowering.

Seeds are bought in the expectation that they are of high quality. This means they have high purity (no contamination with weed seeds), high health (no contamination with seedborne pests or diseases), and high planting value (all seeds germinate), and are true to type (the bag contains what is stated on the label). To achieve the latter, seed crops should be planted with a degree of isolation from other plantings of the same species to prevent contamination by pollen. The recommended isolation distance depends on whether the species is predominantly cross-pollinated or self-pollinated. In wheat, rice, chick pea, lentil, beans, and tomato, which are self-pollinated, the recommendation is 5 m. For sorghum, cowpea, and eggplant which are semicrossed (by wind or insects) the recommendation is 100 m. In cross-pollinated crops such as sunflower, cabbage, carrot, cucumber, and onion the recommended isolation distance is 600 m.

Crop rotation is extremely important for commercial seed production for a number of reasons and long gaps between seed crops and the previous growing of the same crop are recommended. This helps to reduce the buildup of pests and diseases. In addition, the likelihood of contaminants of different varieties of the same species is reduced.

Any agrochemicals used for crop protection must not have any deleterious effect on seed production and quality. This may restrict the timing of application of particular products.

Impact of Stress

The main stresses on plants that are relevant to seed production are those which lead to a reduction of seed quantity or of seed quality. A restriction of photosynthesis and restriction of movement of assimilates from leaves to seeds will lead to a reduction in quantity. Environmental conditions during the phase of seed maturation influence seed quality at harvest. Conditions at harvest and the harvesting process can lead to damage of seeds and reduced quality.

Low photosynthesis may have a number of causes. Low light intensity sets an upper limit to the amount of energy which can be chemically bound in photosynthesis. Even when light intensities are high, photosynthetic rate may be restricted by low or high temperatures. A lack of water may lead to closure of stomata thus preventing gas exchange into the leaf and restricting photosynthesis. In addition to an effect on assimilation, restricted water supply limits the translocation of assimilates and nutrients to the developing seeds. A major effect of lodging is to injure the vascular system of the plant thus preventing any further seed filling.

Weeds influence production mainly by capturing light energy which we would prefer to go to the crop plants. International regulations do not permit trade in seed samples contaminated by weed seeds. Diseases of crop plants may reduce the number of seeds produced or their size. These effects can be achieved through reducing the green leaf area and thus restricting the amount of photosynthesis, by reducing root size and thereby restricting the uptake of water and nutrients, or by interfering with the translocation stream of the plant. Invertebrate pests achieve their impacts in similar manners.

Mitigation

In protected cropping systems, temperature and water availability are relatively simple to control. In field crops, irrigation may be an option for crops with high-value seeds but temperature control is not easy. Careful site selection can contribute to minimizing the likely impact of low or high temperatures.

Protection of crops against weeds, diseases, and pests is a well-developed technology with a range of options in many crops but few in others. The range of products available for application to crops is reducing as older compounds are not being reregistered and so their use will not be permitted. Nevertheless, high standards of crop protection are required for seed crops to achieve satisfactory levels of purity and seed health.

Harvesting

For most field crops there are two main alternatives for harvesting seeds. The plants may be cut and left to dry in the field in windrows, followed by picking up with a harvester. Alternatively the plants may be sprayed with a desiccant and when this has achieved its effect the crop harvested directly. Cutting and windrowing can lead to large numbers of seeds falling on the ground and not being recovered. This can also lead to problems of volunteer plants in the following crop. Desiccation can lead to higher seed recovery but must not be delayed otherwise the seeds may be shed. In some cases the seed needs to be extracted from the fruit. After the seeds are obtained they may require to be dried in order to store them safely. Much information is available on environmental conditions which provide safe storage environments for seeds.

Some Case Studies

In this section the seed production procedures for six crops are described. The crops have been selected to illustrate a variety of crop and seed types (**Table 1**) and the range of procedures used to generate high-quality seeds for planting.

Wheat

Wheat (*Triticum aestivum*) is a major food source for much of the world population and supplies large proportions of carbohydrate in the diet. The seed is the plant part which is consumed. Most varieties are self-fertilizing so seed for sowing can be obtained by saving some of the previous crop or by identifying crops with particularly good seed qualities of viability, purity, and health. Thus no special techniques are required for seed crops.

Wheat is grown in diverse environments around the world with many thousands of cultivars providing improved genotypes adapted to particular locations. Seed yields are highest where water and nutrients do not limit production. Seed quality is highest when seeds mature in dry conditions.

Table 1 Crops with different reproductive features

Crop	*Use*	*Plant part used*	*Family*	*Life cycle*	*Crop planting material*	*Pollination*
Wheat	Cereal	Seed	Poaceae	Annual	Seed	Self
Alfalfa	Forage legume	Leaves and stems	Fabaceae	Perennial	Seed	Cross
Brussels sprouts	Vegetable	Leaves	Brassicaceae	Annual	Seedlings	F_1 hybrid
Carrot	Vegetable	Root	Apiaceae	Biennial	Seed	F_1 hybrid
Potato	Vegetable	Tuber	Solanaceae	Perennial	Seed tubers	Self/cross
Rose	Ornamental	Flowers	Rosaceae	Woody perennial	Grafted plants	Cross

Maturation in damp conditions can lead to sprouting of the grain on the parent. Seed growth and maturation in warm conditions result in high yields whereas in hot dry conditions, seed weight may be low and seed numbers may be low because of abortion after fertilization or shedding during development.

The targets for wheat cropping systems are to provide the site and resources to permit establishment of a regular and adequate stand, development of a good population of fertile tillers, production of ears with a large number of grains, and adequate seed filling. Moisture stress can reduce seed yield in a number of ways depending on when it occurs. Tillering may be reduced thus limiting the population of fertile tillers. Later, moisture shortage during reproductive development can restrict the number of seeds initiated per ear. Although wheat is largely self-pollinated, there can be restriction of fertilization of ovules when a water shortage during flowering causes pollen sterility. Shortage of water during seed development leads to imperfect grain filling and low seed weight. Crops that lodge suffer a severe restriction of supply of assimilates to the developing grains. The agronomy of wheat involves decisions on sowing density, fertilizer amount and timing of applications, good crop protection, possibly use of plant growth regulators to manipulate plant form to increase fertile tillers and to protect against lodging. All these decisions will impact on seed quantity. The decisions on when to harvest can affect seed quality.

The options for crop protection in wheat are many, but they may not be accessible to all growers worldwide. Herbicides, fungicides, and insecticides, as well as plant growth regulators are available to protect the crop at all stages. Because the grains may be used for human consumption, there is considerable effort to produce grain samples that are free from contamination by invertebrate pests, plant diseases, and residues of agrochemicals.

The seeds ripen synchronously and are harvested in a single pass. Seed harvested with a combine should be between 20% and 8% moisture content because outside this range seeds may be damaged during the threshing stage. Swathing or cutting the crop can be done when the seeds are at 30% moisture. Seeds are usually safe for storage below 14% moisture content but may need to be carefully dried to get down to this moisture content.

The seed quality of wheat is defined in different ways for specific end uses. Thus for home feeding to animals, seeds high in protein are required. For baking, seed quality incorporates measures of the content of gluten and of the enzyme α-amylase. One common quality criterion is the absence of damaged grains so timing of harvesting and the correct setting of machinery are important.

Alfalfa

Alfalfa (*Medicago sativa*, lucerne) is a herbaceous perennial legume used primarily as a hay crop and has a high feeding value. In many regions it is a key component of cropping systems valued for its deep roots and its contribution to soil fertility because of nitrogen fixation, and it is established from seeds. Seeds shed from seed crops can survive in the soil for several years and so a gap of 4 years free of alfalfa is recommended for seed crops. Alfalfa is mainly cross-pollinated.

Environments that favour vegetative growth of alfalfa (for forage) are generally not those that favor reproductive growth. Thus, seed production of alfalfa is often located in areas away from the main areas of use and conducted by specialist seed growers. There is a strong environmental influence on seed yield with locations and seasons differing widely. Commercial seed yields are well below the theoretical potential. The main yield components influenced by the environment are the number of inflorescences per square meter and the proportion of flowers that set pods. The small seeds (weighing about 0.002 g) are borne within spiral pods.

Production of seed crops of alfalfa demands different management from pasture use. In pasture, animals may eat the young seed heads, and defoliation may reduce the green area available for photosynthesis. Plant spacing and sowing rate determine the onset and extent of interplant competition and influence the seed yield, although the results may not be consistent from one year to the next, nor between regions. Generally for seed crops, low sowing rates are advised (1 kg ha^{-1}) and close row spacings (15 cm) to be avoided. This permits much branching and maximum production of seed heads borne at a harvestable height and produces heavier seeds. In addition this crop structure permits good crop ventilation which minimizes disease development. Seed yield per plant can vary from 10 to 60 g.

Alfalfa is cross-pollinated, but in adverse conditions may be able to set seed after self-pollination. Nevertheless, cross-pollination leads to a greater quantity and higher quality of seeds. A key feature of alfalfa seed production is the provision of pollinators. Alfalfa flowers produce pollen and nectar which attracts many insects, particularly bees. Honeybees learn to extract nectar from the flowers without tripping them; thus ground bees and bumble bees are often more important pollinators. In districts where bees are absent, alfalfa seed crops often fail. In contrast to open field plots where the proportion of flowers producing seed was about 35%, this increased to 55% where bees were introduced. Seed yields in similar experiments were about 675 kg ha^{-1} in open field plots and about 1100 kg ha^{-1} in cages with bees. Obviously, it is not feasible to cage large areas for commercial seed production but this can be done on small plots for multiplication of small quantities of breeders' seed. Clearly also the use of insecticides in alfalfa seed crops may be disadvantageous. In most areas native bees are effective pollinators and may trip from 30% to over 90% of the flowers. In the northwest United States and Canada, the cornerstone of alfalfa seed production has been to introduce the alfalfa leaf-cutting bee. This species is a much more efficient pollinator than the honeybee. The technology of managing and rearing these leaf-cutter bees is well developed in western Canada.

In common with many legumes, alfalfa may produce seeds with a hard seed coat. This is a survival mechanism which can enable natural populations to survive stress, but is undesirable in cultivated types. Hard seeds may comprise 10–60% of a seed lot of alfalfa and therefore limit the stand establishment. The proportion of hard seeds reflects the environmental conditions during seed maturation. Samples with a high proportion of hard seeds may be scarified to encourage germination.

The flowers open over a long period and thus the seeds do not ripen synchronously. Harvest has to be timed to obtain the maximum amount of good-quality seed but inevitably some seeds will ripen and shed before harvest while others will be immature. Cutting and maturing the crop in windrows can allow additional seed maturation. Desiccants are sometimes used to allow maturation on the intact plant but harvest must then take place within a few days.

Seeds are dried after harvest to below 12% moisture. Specialized seed-cleaning equipment is used to remove weed seeds. Before sowing seeds are usually coated with appropriate strains of the nitrogen-fixing bacterium *Rhizobium*.

Brussels Sprouts

Brussels sprouts are a form of cultivated *Brassica oleracea* in which the lateral buds develop as small round vegetative buds. They are important seasonal vegetables in temperate regions. The inflorescence bears a large number of flowers which ripen over a long period of time. The varieties of Brussels sprouts readily cross-fertilize with each other and with other types within the species (cabbage, cauliflower, calabrese, kale, and fodder kale) and wild weedy brassicas. Thus, wide isolation distances are recommended (sometimes over 1 km) and good weed management. Modern varieties are mostly F_1 hybrids and so the seed is not true to type. Seed production exploits the strong self-sterility in the species. Commercial crops are mostly established by transplanting seedlings. This is usually done mechanically so high-quality seed is required to produce trays of modular transplants in which there are no blanks (i.e., 100% germination) and all seedlings are the same size.

For seed crops, seeds of the two parent varietal lines are sown in separate nursery beds in spring. Suitable parent plants are selected and transplanted to polytunnels in autumn. In spring the plants flower and blowflies are introduced to ensure pollination. The strong self-incompatibility and wide dispersal of pollen by the flies ensures the seeds are the products of cross-fertilization between the two varieties. This can be tested by growing a sample of seedlings and assessing their uniformity. This method will not work for the parent inbred lines which have to be maintained by hand-pollination at the flower bud stage which overcomes the self-incompatibility.

The seed pods are brittle and may shatter and shed seed before harvest. The decision on timing of harvest is therefore important in determining the quantity and also the quality of seeds. The crop is usually cut and windrowed for up to 2 weeks. The windrows are not turned in order to avoid pod shattering and seed shedding. The seed is then usually dried to reduce seed moisture content to below 10% (if stored in sacks) or 8% (for bulk storage).

Carrot

Carrots (*Daucus carota*) are grown worldwide as a vegetable. Carrots are biennial plants, the swollen roots are produced in the year of sowing and are harvested, stored, and consumed. To produce seeds, the roots must be left and after vernalizing, the plants then flower in the following year. The plant produces a tall stem with many branches bearing seeds. Not all the branches behave similarly. The primary umbel is the topmost umbel on the main stem and secondary and tertiary umbels are also formed. The primary umbel is the first to flower and produces heavier seeds with a higher germination percentage and faster germination rate. The embryo is very small relative to the size of the seed which weighs about 0.8 mg.

Many modern varieties are F_1 hybrids and display high yields and great uniformity of roots, but cannot be used themselves to produce true-to-type seed. The F_1 hybrids are produced through exploiting cytoplasmic male sterility. In order to produce new carrot seed the plant breeder has to maintain three inbred lines – the donor, the male sterile, and the restorer line which are planted in separate rows and may need to be planted at different times. Seed crops may be sown in late summer and left in the ground or may be sown earlier, the roots lifted, stored, and replanted in spring. Thus carrot seed crop management is complicated. It is a specialized activity which may involve sowing at higher densities than appropriate for root production in order to maximize the number and proportion of primary umbels. Seed growers may also harvest the different classes of umbels separately to grade the seed by size.

At harvest the rows of male plants are removed first. Hand-harvesting of separate umbel classes as they mature is one option. On a larger scale, mechanical harvesting is used and takes place when the primary umbels are mature. Usually the crop is cut, wind-rowed, and picked up and threshed after a few days. The seeds are cleaned and the spines removed. Seeds are usually stored at 7–10% moisture content.

Potato

The potato (*Solanum tuberosum*) is a major provider of starch in the diet of many people around the world. The tubers are eaten and it is generally tubers that are planted in fields. These "seed tubers" are grown by specialists, often in cooler areas where there is less risk of insects spreading disease. Cultivars are propagated clonally but this process cannot lead to new genetic combinations of characters to produce new varieties. For "true" potato seed to be produced, the plant must flower. Some economically important potato diseases are not seed-transmitted. There is some interest in developing the use of true potato seeds for growing crops so that the tubers can be retained for eating. This section will consider only the production of true potato seeds.

Cultivars of potato differ in the extent of flowering. Cultivars are usually highly heterozygous and so seeds from selfed berries on commercial crops usually show great segregation of characters for plant form and tuber characters. Specialist plant breeders aim to manage the combination of characters of the offspring and usually hand-pollinate selected parent lines.

The fertilized ovules develop inside a small fleshy fruit (with the same structure as a tomato) which is harvested, usually by hand, about 40 days after pollination. After a period of storage at ambient temperature, the fruits become soft and the seeds can be extracted by hand. Alternatively, with larger quantities, the fruits may be soaked and macerated and the debris floated off. The seeds are then dried in a thin layer before storage. Potato seeds are usually dormant for 6–12 months.

Roses

Ornamental roses (*Rosa* spp.) are an important horticultural product. The most common method of commercial propagation is to take a bud from a desired cultivar and graft it onto a species rose rootstock selected for its ability to produce a vigorous root system. An important rootstock species is *Rosa laxa*. Specialist rose breeders and growers either produce their own rootstocks or buy them in. The production of bulk supplies of *Rosa laxa* for rootstocks is an important market and so seed production is a specialist activity.

Rosa shrubs produce flowers and seeds at a young age. The seeds (achenes) are borne within a fleshy, berrylike hip that remains on the plant after ripening. The hips are usually hand-picked in autumn after they have turned red. The seeds are recovered by macerating the hips and recovering the seeds by flotation. The process can be aided by soaking the hips until the pulp ferments and then pushing the mass against a sieve. After extraction the seeds are spread in a thin layer and dried prior to storage.

Seeds collected shortly after ripening and not allowed to dry are less dormant than fully dried seeds. After drying the seeds are classed as deeply dormant and may require 3–9 months exposure to cold temperatures (stratification) before they will germinate.

Conclusions

The wide variety of seed structure among cultivated plants requires a range of methods for seed production. The aim is constant: to produce healthy seeds, free of contaminants, true to type, and giving high and rapid germination. The techniques to achieve this vary between annual, biennial, and perennial species and between crops grown for their seeds, leaves, stems, or roots.

See also: **Crop Improvement**: Chromosome Engineering; Plant Breeding, Practice. **Energy Crops**: Biomass Production. **Flowering and Reproduction**: Fertilization; Reproductive Strategies. **Growth and Development**: Canopy Architecture. **Postharvest Physiology**: Ripening; Seed Storage. **Seed Development**: Seed Quality.

Further Reading

Basra AS (ed.) (1995) *Seed Quality: Basic Mechanisms and Agricultural Implications*. New York: Food Products Press.

Bean EW (1977) *Principles of Herbage Seed Production*, 2nd edn. Aberystwyth: Welsh Plant Breeding Station.

Cromarty A (1982) Techniques for drying seeds. In: Dickie JB, Linington S, and Williams JT (eds) *Seed Management Techniques for Genebanks*, pp. 88–130. Kew: Royal Botanic Gardens.

Harrington JF (1960) Thumb rules of drying seeds. *Crops and Soils* 1: 16–17.

Hebblethwaite PD (ed.) (1980) *Seed Production*. London: Butterworths.

International Seed Testing Association (1999) International rules for seed testing. *Seed Science and Technology* 27 (Suppl.).

Kelly AF and George RAT (1998) *Encyclopaedia of Seed Production of World Crops*. Chichester: John Wiley & Sons.

McDonald MB and Copeland LO (1997) *Seed Production Principles and Practices*. New York: Chapman & Hall.

Nellist ME (1981) Predicting the viability of seeds heated with dry air. *Seed Science and Technology* 9: 439–455.

Soffe RJ (ed.) (1995) *The Agricultural Notebook*, 19th edn. Oxford: Blackwell Science.

SEED DORMANCY

Contents

Development of Dormancy

H W M Hilhorst, Wageningen University, Wageningen, The Netherlands

Definition and Ecological Relevance

Dormancy is a phenomenon that can be found among most known forms of life, including spores, larvae, cysts, seeds, bulbs, tubers, and even a number of higher animals. It is generally characterized by the virtual absence of metabolic activity and/or a lack of further development and growth. Dormancy is a trait that has been acquired during the evolution in a selection for the ability to survive adverse environmental conditions, such as cold and drought. It is not clear yet when dormancy became apparent during the time since seed-bearing plants first appeared. However, there is some consensus that the occurrence of dormancy is related to climatic changes during the Earth's history. This is corroborated by the observation that the number of plant species with dormancy generally increases with the geographical distance from the equator and, hence, with the amount of variation in precipitation and temperature. There is little doubt that the phenomenon of dormancy has significantly contributed to the development of new species and the successful dispersion of those already present.

Seed dormancy has been under study since the days of the Greek philosopher Theophrastus (372–287 BC), the "father of seed physiology." Very often dormancy is simply regarded as the absence of germination. However, absence of germination can have several causes. First, the seed can be dead. Second, the environment may be limiting to germination, and third, the cause may be in the seed or dispersal unit itself. With this in mind, a satisfactory definition can be formulated: "Dormancy is the absence of germination of a viable seed under otherwise favorable conditions." The ecological relevance of seed dormancy can be defined as an effective delay of germination to avoid germination and subsequent growth under unfavorable climatic conditions.

Categories of Dormancy

The numerous studies of seed dormancy have yielded a vast number of classifications and definitions. Of the many classifications of seed dormancy the one that was published by the Russian seed biologist Nikolaeva in the 1970s has been widely adopted. An additional classification of dormancy is based on the timing of its occurrence: primary dormancy indicates the type of dormancy that occurs prior to dispersal as part of the seed's developmental program, whereas secondary dormancy denotes the acquisition of dormancy in a mature hydrated seed as a result of the lack of proper conditions for germination. The essence of this latter classification is a distinction between absence of germination and dormancy. Within this context "absence of germination" is identical to the term "quiescence" which is commonly used to denote the absence of germination because of a lack of water. This is an important aspect in relation to the persistence of seed banks in the soil. When discussing the development of dormancy as part of the seed's developmental program it is important to distinguish between the different types (**Table 1**). A complicating factor is that in many species the observed dormancy is composed of different types. For example, many seeds possess a seed coat that poses a mechanical restraint to embryonic growth but that may also contain chemical inhibitors, such as phenolic compounds (mechanical and chemical dormancy). In addition, in the same seed, endosperm tissue may restrict embryo growth until it is degraded by hydrolyzing enzymes induced by embryonic factors (physiological/mechanical dormancy). There does not seem to be a preference among higher plant families or genera for a certain category or type of dormancy.

In order to explain the acquisition of dormancy during seed development it is thus mandatory to study all the possible aspects that may contribute to the overall effect of dormancy on seed behavior. Regrettably, most developmental studies have neglected this and have predominantly focused on physiological dormancy.

Dormancy is Part of the Developmental Program of the Seed

During the first half of seed development most seeds become germinable, marking the completion of histodifferentiation and morphogenesis of the embryo. Some species produce ripe seeds with underdeveloped or underdifferentiated embryos, such as *Heracleum* and celery (*Apium graveolens*). In these seeds, embryo development may be resumed upon rehydration of the seed after abscission and dispersal. Other species display so-called vivipary in which

Table 1 A classification of dormancy

Categories	*Groups*	*Caused by*	*Mechanism*
Primary dormancy (predispersal)	Exogenous (outside embryo)	Maternal tissues or endosperm	Inhibition of water uptake (physical dormancy) Mechanical restraint to embryo expansion and radicle protrusion (mechanical dormancy) Modification of gas exchange Prevention of leaching of inhibitors from embryo Supplying inhibitors to the endosperm (chemical dormancy)
	Endogenous (inside embryo)	1. Underdeveloped embryo (morphological dormancy) 2. Metabolic blocks (physiological dormancy) 3. Morphophysiological dormancy (1 and 2 combined)	Embryo in mature seed has to complete development prior to germination Physiological mechanisms largely unknown
	Combinational	Combination of exogenous and endogenous dormancy	
Secondary dormancy (postdispersal)		Metabolic blocks, induced in nondormant seeds when the germination environment is unfavorable	Physiological mechanisms largely unknown

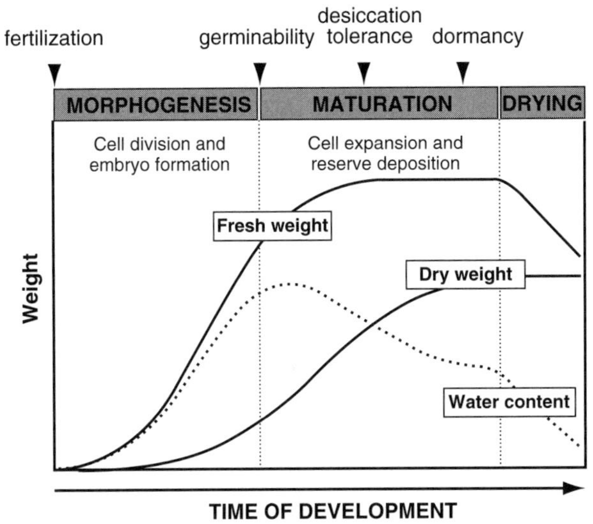

Figure 1 Schematic view of fresh and dry weight accumulation during seed development. The three major developmental phases and some important events are indicated above.

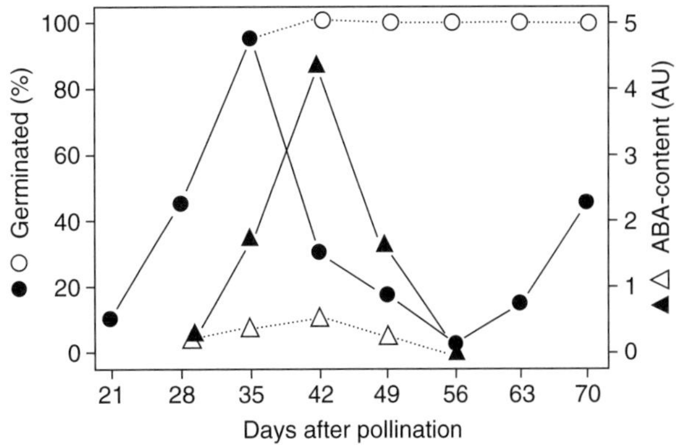

Figure 2 Accumulation of abscisic acid (ABA) and acquisition of germinability in developing wild-type (—) and ABA-deficient mutant (- - -) tomato seeds. (HWM Hilhorst, original data.)

seeds germinate in the fruit while still on the parent plant. An example is mangrove (*Rhizophora mangale*). However, the majority of higher plant species produce seed with fully developed embryos that do not germinate during the entire period on the parent plant. Seeds of many species also do not germinate after dispersal. Thus, germination is suppressed during development and a persistent block to germination may be present in the mature seed (primary dormancy).

Seed development can be divided into a number of distinct phases that are marked by morphological and physiological parameters (**Figure 1**). One is a period of morphogenesis during which the embryo's body plan is established through intensive cell divisions and the embryonic organs and tissues are formed. A following phase is a period of seed maturation. It includes the arrest of tissue and organ formation, the accumulation of nutrient reserves, and changes in embryo size and in fresh and dry weights. Finally, seeds undergo further drying to low moisture content, unless they develop in a wet environment, such as a fleshy fruit.

The Occurrence of Abscisic Acid during Seed Development

Developmental arrest and primary dormancy have long been associated with the presence of the plant hormone abscisic acid (ABA) during seed development. Characteristically, ABA levels rise during the first half of development and decline during late maturation, when seed water content decreases (**Figure 2**). Sometimes two ABA peaks occur, e.g., in bean (*Phaseolus vulgaris*) and pea (*Pisum sativum*). ABA has been detected in embryo, endosperm, and testa and in fruit tissues such as pericarp and locular tissue. ABA action has been related to a number of other developmental processes including storage protein synthesis, suppression of precocious germination, induction of late embryogenesis abundant proteins, and induction of desiccation tolerance. The use of ABA-deficient and ABA-responsiveness mutants has made a landmark contribution to the current hypotheses on the role of ABA in developmental processes, including the induction of primary dormancy.

ABA in the Induction of Primary Dormancy

ABA has long been associated with dormancy, mainly because the hormone could be detected in both developing and mature seeds while it was known that it was inhibitory to germination when applied exogenously. The availability of ABA-deficient and -responsiveness mutants, especially of *Arabidopsis thaliana*, tomato (*Lycopersicon esculentum*), and corn (*Zea mays*, maize) has clearly demonstrated that the absence of or insensitivity to ABA during seed development results in the formation of nondormant seeds. **Figure 2** shows the patterns of ABA content and germinability of wild-type and ABA-deficient (*sitiens* mutant) tomato (cv Moneymaker) seeds during development. ABA levels are up to 10 times higher in the wild-type seeds than in the mutant seeds. Wild-type tomato seeds may become completely nongerminable (dormant) during the maturation phase whereas ABA-deficient seeds retain full germinability. Thus, ABA deficiency during development is associated with the absence of primary dormancy in the mature seed. It was shown that this association only occurred if the embryo contained the dominant *Aba* allele. Crosses between wild-type and mutant plants indicated that maternal ABA (i.e., located in testa and fruit tissues) had no influence on dormancy. Apparently, the rise in ABA content during seed development is required to induce dormancy. Interestingly, seeds in overripe fruits of the *sitiens* mutant may germinate within the fruit (viviparous germination). This does not occur in the wild-type fruits, whereas in both genotypes ABA contents at the end of maturation are comparably low. Thus, ABA does more than directly inhibit germination. Manipulation of seed ABA content by genetic modification of tobacco (*Nicotiana tabacum*) has shown that overexpression of zeaxanthin oxidase, one of the enzymes of the ABA-synthetic pathway, results in more dormant phenotypes, whereas knocking-out of the gene encoding for this enzyme yields less dormant phenotypes.

Apart from ABA content, sensitivity to ABA may play a role in the expression of dormancy or inhibition of germination. In wheat (*Triticum aestivum*) and corn, several cultivars exhibit vivipary, also known as preharvest sprouting, under warm and humid conditions. These sprouting-susceptible cultivars have a reduced sensitivity to ABA. Analysis of mutants in *Arabidopsis* and corn displaying vivipary has led to the identification of ABA-responsive genes that are responsible for these phenotypic characteristics: *ABI3* and *VP1*, respectively. These homologous genes encode for transcription factors that play a role in the regulation of gene expression. These genes have been shown to be pivotal in seed development. They are involved in maintaining the developmental state in seeds and avoiding transition to the vegetative or growth stage. In other words, seeds of mutants lacking this gene will show characteristics of growth.

Dormancy is a Developmental Event

Mutant analysis has revealed a number of genotypes with altered dormancy characteristics, but which possess normal ABA contents throughout their development. Examples are the *abi3*, *leafy cotyledon* (*lec1*, *lec2*), and *fusca* (*fus3*) mutants in *Arabidopsis*. These mutants all possess phenotypes that are characteristic of the vegetative state, e.g., reduced tolerance to desiccation, active meristems, expression of germination related genes, and absence of dormancy. *LEC1* and *FUS3* loci probably regulate developmental arrest, as mutations in these genes cause a continuation of growth in immature embryos. ABA-controlled dormancy (via *ABA* and *ABI*) might represent a different mechanism preventing germination, which occurs later and is additive to the developmental arrest controlled by *LEC1* and *FUS3*. In *Arabidopsis* it was shown that *ABI3* is also active during vegetative quiescence processes in other parts of the plant. In conclusion, the induction of dormancy during development is only partially a response to ABA and should rather be considered a developmental event.

More mutants are known with altered dormancy characteristics but with normal ABA content and ABA sensitivity. Examples are the *aberrant testa shape* (*ats*) and *transparent testa* (*tt*) mutants of *A. thaliana*. The *ats* seed shape mutant has a reduced testa thickness and germinates faster and to higher levels than wild-type seeds. This mutant produces ovules in which the integuments do not develop properly. The *tt* mutants produce seeds with defects in the pigmentation of the testa. This testa modification may enhance the uptake of water and oxygen and the leaching of inhibitory substances from the seed. These examples are clear cases of altered physical and chemical dormancy in which modification of seed coat properties reduces the restraint to germination.

Dormancy and the Environment

Dormancy Depends on the Genome, the Environment, and Interactions between Genome and Environment

Dormancy has a hereditary component. This has been demonstrated for a large number of species, including many crops, such as rice (*Oryza sativa*), soybean (*Glycine max*), and barley (*Hordeum vulgare*) and weeds such as wild oats (*Avena fatua*). Classical genetic analysis has demonstrated that between one and 10 often recessive genes are related to dormancy, depending on the species. However, mutational and molecular analysis now indicates that the number of genes directly or indirectly involved in the regulation of dormancy is much higher. Also the degree of dormancy is hereditary. For example, in 10 genetically pure lines of *Avena fatua* seed dormancy ranges from 0% to nearly 100%. In many species the occurrence of dormancy is coupled to morphological or physiological characteristics. Examples are seed color and seed coat thickness but also herbicide resistance. An example of the latter is the lower degree of dormancy in atrazin-resistant biotypes of *Amaranthus* spp. as compared with the susceptible biotypes. Hybridization studies with dormant and nondormant pure lines of a species have revealed that seed dormancy may be controlled by the genotype of both the maternal and paternal parent as well as the embryo or a combination of the three, depending on the species.

There is a very strong influence of the environment on the acquisition of dormancy in developing seeds.

Environmental factors known to affect dormancy include: day length, light quality, temperature, soil moisture, and mineral nutrition. However, effects of these factors on the level of dormancy are ambiguous and depend on the species. Although exceptions are known, exposure to elevated temperatures during the second half of seed development (i.e., maturation phase) generally results in seeds with less dormancy. Similarly, water stress during seed development results in less dormancy in the majority of species studied. An explanation for the dormancy relieving effect of temperature and water stress is that these conditions may speed up after-ripening of the seeds while still on the mother plant. After-ripening of dry seeds effectively removes dormancy, especially at higher temperatures (see below).

Interactions between the environment and the genome appear to be very complex. It has been shown that environmental factors influence such seed characteristics as size, weight, and coat thickness. It is evident that this may have an impact on seed dormancy. To add to the complexity, position of seeds within a fruit and of fruits on the mother plant may have a profound effect on dormancy.

Dormancy and Stress Responses Overlap

The content of ABA responds to abiotic stresses, such as heat, cold, and drought. Exposure to stress will increase ABA contents, not only in seeds but also in other plant organs. For example, drought will increase ABA contents in leaves in which it induces closure of the stomata to prevent further water loss. In many studies that aimed at identifying dormancy-related genes, genes were found that responded to ABA but were also known to come to expression as a response to stress. Thus, we may address the question: what is the relationship between dormancy and stress? Obviously, if we consider the ecological relevance of dormancy, unfavorable periods for growth are generally stressful: too cold, too warm, or too dry. It is during such periods that seeds strive to maintain their dormancy, but also have to be equipped to tolerate the stressful conditions.

In *Avena fatua* pure dormancy lines are available that have been used for molecular genetic characterization of dormancy. Differential screening for gene expression of dormant and nondormant lines resulted in the isolation of several genes that were expressed in the dormant seeds, but not in the nondormant. However, if the nondormant seeds, and even seedlings, were stressed by exposure to elevated temperatures the "dormancy genes" came back to expression. From this we may conclude that dormancy and stress responses overlap and that control by ABA is the common denominator.

The Breaking of Dormancy

The breaking of primary (physiological) dormancy is usually achieved through prolonged periods of dry storage, often called after-ripening. This process may commence when seeds are still on the plant and are low in moisture, typically between 5% and 15% on a fresh-weight basis. There is a clear relationship between the period of after-ripening required to break dormancy and the ambient temperature. Higher temperatures will shorten the after-ripening period. At present, not much is known about the processes that occur in dry seeds. The previously mentioned dormancy genes of *A. fatua* are reduced in their expression upon after-ripening. Apparently, RNA-degrading enzymes can be active in a dry environment. However, water distribution in seeds may be uneven. This leaves the possibility of the existence of seed parts with higher amounts of water in which chemical reactions are more likely to occur. Generally, the sensitivity to gibberellins (GAs) increases during after-ripening. The gibberellins are a group of hormones of which some are absolutely required for germinative growth and are, as such, antagonists of ABA.

Another condition that may be very effective in the breaking of dormancy is cold. Exposure of fully hydrated seeds to cold conditions, generally below 15°C, will gradually relieve dormancy. During this cold imbibition, or prechilling, inhibitory compounds, including ABA, may leak out and hence enable germination. As in after-ripening the cold exposure increases the sensitivity to GAs.

The conditions that are conducive to the breaking of dormancy can be related to the natural environment of the seed. In temperate regions the summer is the growth season and seeds will germinate in spring, following the cold winter months during which dormancy is broken. Alternatively, in hot and dry climates, the cooler wet season is most favorable for growth and thus, the hot and dry summer months will provide the conditions to break dormancy.

It needs to be stressed that the aforementioned conditions are generally effective in breaking physiological dormancy. Other types of dormancy may require other conditions. For example, seeds with thick and hard seed coats or fruit tissues may require extended periods in the soil to enable the degradation of the tissues that surround the embryo. It is not known whether microorganisms play a role in this process. Also passage through the digestive tract of animals is known to degrade these tissues. A high

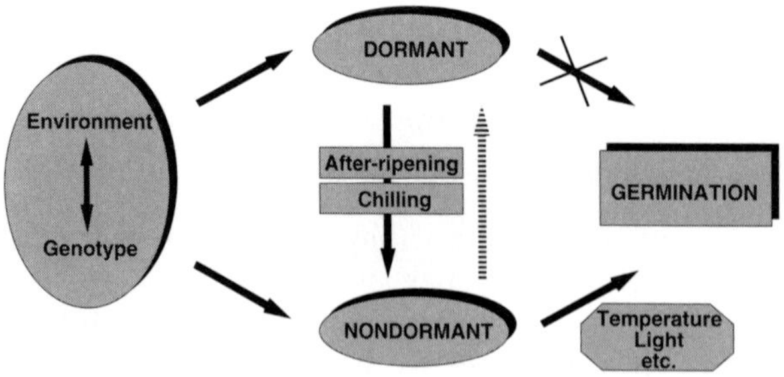

Figure 3 Schematic overview of transitions between dormant and germinative states and environmental factors involved. The cross indicates a nonpermissive transition. The dashed arrow indicates a transition as a result of the absence of factors that are required for germination.

level of adaptation to specific environments can be found in species that respond to fire. Many of them contain specific structures in their water-impermeable seed coats that open up in response to extremely high temperatures and thus allow uptake of water. In other cases the smoke may relieve dormancy by as yet unknown chemical reactions with seed coat components.

Dormancy induction and relief are quantitatively related. A deep dormancy will require longer periods of higher/lower temperatures to break dormancy by after-ripening and cold, respectively. In addition, the environment of the developing seed will largely determine the depth of dormancy. These relationships are summarized in **Figure 3**.

The Breaking of Primary Dormancy is a Prerequisite for Germination

The breaking of dormancy is accompanied by an increase in sensitivity to light and other germination stimulants. Furthermore, the range of temperatures over which germination can occur widens (**Figure 4**). In other words, seeds become sensitive to their environment during the process of dormancy relief. If the required set of conditions for germination is not (fully) available, germination will not take place and seeds may become dormant again (**Figure 3**). Thus, the transition from the dormant to nondormant state is reversible. Application of exogenous GAs may short-circuit this transition and induce germination in dormant seeds. This shows that the formation of GAs is a crucial step in the induction of germination. However, seedlings resulting from this treatment often display a stunted growth. Apparently, dormancy breakage also releases other factors that are required for normal growth. There is a vast amount of information on a wide array of chemicals that break dormancy or stimulate germination, including alcohols, organic acids, nitrate and re-

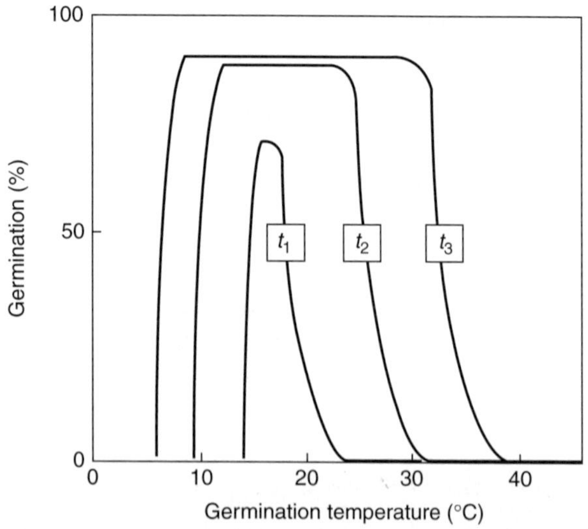

Figure 4 Effect of different times ($t_3 > t_2 > t_1$) of after-ripening on the germination temperature window. Prolonged after-ripening opens the window, indicating relief of dormancy.

spiratory inhibitors. As yet, the action mechanisms of these componds are unknown.

Dormancy is a Component of Seed Quality

In many modern crops dormancy has disappeared during domestication of their wild progenitors. Also breeding programs selecting for improved germination and emergence traits may have contributed to this. However, in many crops dormancy may reappear when germination and seedling growth occur under stress conditions, for example in the field. There is good evidence that the depth of dormancy may be expressed in terms of stress tolerance. A good example of this is the occurrence of so-called thermodormancy in lettuce (*Lactuca sativa*). Lettuce seed germination is often inhibited at elevated temperatures. A dormancy-breaking treatment may increase the upper temperature limit considerably (**Figure 4**). This is a common practice in enabling growth of lettuce in hot conditions. In other words, such a treatment makes lettuce seeds more tolerant to higher temperatures. It is also known that seeds from the essentially nondormant ABA-deficient *sitiens* mutant of tomato have a better ability to overcome osmotic stress during germination than the wild-type seeds. Yet, germination of both genotypes on water and under optimal temperature is similarly high. Thus, the assessment of the depth of dormancy can be improved and made more discriminating by execution of the test under suboptimal conditions. Alternatively, testing for germination speed rather than total germination is usually a more sensitive means to reveal differences in dormancy. Seed batches of the same species may germinate at 100% but may show differences in speed of germination which is indicative of (small) differences in depth of dormancy.

Dormancy in commercial seed is particularly common in species that are still relatively close to their wild progenitors and because of this successful production is often severely hampered. These include seeds from many tree and flower species but also true potato (*Solanum tuberosum*) seed is notorious for its dormancy. Among these species different types of dormancy may occur that each require a specific treatment for relief.

Prospects of Seed Dormancy Research

Genetic analysis of dormancy mutants has firmly established the crucial role of ABA in the acquisition of seed dormancy during development. Advanced genetic and mutant analysis are identifying additional genes. It is not known yet whether these genes with unknown functions are downstream targets of ABA or whether they have an impact on seed dormancy in an independent way. The molecular identification of these genes will be of pivotal importance, as well as the identification of more target genes. Transcriptomics and proteomics approaches will be most efficient tools in accomplishing this.

List of Technical Nomenclature

Abiotic	Not related to living organisms.
Abscisic acid	(ABA) Plant growth substance that controls growth and development.
Abscission	Becoming detached, e.g., leaf, seed.
Dormancy	Quiescence, inactivity, arrest of growth.
Genotype	Genetic constitution of an individual or group.
Gibberellins	(GA) Plant growth substance that controls growth and development.
Maturation	Ripening of a fully developed speed.
Morphological	Relating to form.
Mutant	Genetically changed, usually in one gene.
Phenotype	Set of observable characteristics of an individual or group.
Physiological	Relating to function.
Proteomics	Technique to study expression of many proteins at the same time.
Testa	Seed coat.
Transcription factor	Protein that controls transcription of DNA into mRNA.
Transcriptomics	Technique to study expression of many genes at the same time.
Vegetative	Nonsexual.

See also: **Crop Improvement**: Mutation Techniques. **Flowering and Reproduction**: Reproductive Strategies. **Growth and Development**: Molecular Biology of Development. **Postharvest Physiology**: Seed Storage. **Regulators of Growth**: Abscisic Acid; Gibberellins. **Seed Development**: Embryogenesis; Germination; Seed Production; Seed Quality. **Seed Dormancy**: Genetics of Dormancy; Preharvest Sprouting.

Further Reading

Baskin CC and Baskin JM (1998) *Seeds: Ecology, Biogeography, and Evolution of Dormancy and Germination.* San Diego: Academic Press.

Bewley JD and Black M (1994) *Seeds: Physiology of Development and Germination*, 2nd edn. New York: Plenum Press.

Hilhorst HWM (1995) A critical update on seed dormancy. 1. Primary dormancy. *Seed Science Research* 5: 61–73.

Hilhorst HWM and Toorop PE (1997) A review on dormancy, germinability and germination in crop and weed seeds. *Advances in Agronomy* 61: 111–166.

Koornneef M, Bentsink L, and Hilhorst HWM (2002) Seed dormancy and germination. *Current Opinion in Plant Biology* 5: 33–36.

Viémont J-D and Crabbé J (2000) *Dormancy in Plants: From Whole Plant Behavior to Cellular Control.* Wallingford: CAB International.

Yadegari R and Goldberg R (1997) Embryogenesis in dicotyledonous plants. In: Larkins BA and Vasil IK (eds) *Cellular and Molecular Biology of Plant Seed Development*, pp. 3–52. Dordrecht: Kluwer Academic.

Genetics of Dormancy

D Nadella and X Gu, North Dakota State University, Fargo, ND, USA
M E Foley, USDA, Agricultural Research Service, Fargo, ND, USA

Introduction

Seed dormancy is the temporary suspension of growth during the developmental phase of a plant resulting in the failure of germination. Dormancy is an essential adaptive trait that promotes the survival of wild species by distributing their germination over time and space. In cultivated cereal grains, dormancy confers tolerance to preharvest sprouting of the mature grains on the mother panicle under wet or humid conditions. The dormant seeds transform into a germinable state by a process of afterripening. Dormancy is enforced through the seed covering structures and/or the embryo. In general, dormancy is a quantitative trait with large influences of environment and of genotype × environment interaction on its expression. Dormancy is mainly governed by nuclear factors and can be the dominant or the recessive form of the trait. Depending on the species or accession, and the amount of afterripening undergone, dormancy may be under the control of major genes or polygenes. Molecular marker techniques have been helpful in locating the quantitative trait loci (QTLs) governing dormancy in many grasses. The genes and QTLs governing dormancy are conserved in different grass species through evolution. This article summarizes the genetic basis of dormancy in important cereal grains including the effects of environmental factors and the genotype × environment interaction. The concept of colinearity of grass genomes in locating the putative orthologous genes and QTLs underlying dormancy is described. The role of some mutants in understanding the mechanism of dormancy is discussed in cereals and in the dicotyledonous model species *Arabidopsis thaliana*. Applications of the knowledge on genetic basis of dormancy for improved crop production and efficient weed management are outlined.

Definitions

Seed dormancy is the temporary failure of a viable seed to germinate after a specific length of time under a particular set of environmental conditions that normally allow germination after the dormant state has been terminated naturally or artificially. Afterripening denotes the transition of dormant seeds to a less dormant or completely nondormant state, and it occurs over some period of time through exposure of mature seeds to certain environmental conditions. Afterripening does not cause an abrupt change from a dormant to a fully germinable state. Rather, seeds in a population become more responsive to a range of conditions that promote and less responsive to conditions that restrict germination resulting in a gradual increase in the proportion of seeds that germinate until the entire population becomes nondormant. The mechanism by which afterripening removes dormancy is unknown, but the environmental conditions that bring about afterripening vary with species. For example, many grasses such as rice (*Oryza sativa*), wild oat (*Avena fatua*), and barley (*Hordeum vulgare*) normally require afterripening under warm, dry conditions, whereas species like *Arabidopsis* (*Arabidopsis thaliana*) require cool, moist conditions referred to as stratification or chilling. Dormancy can be reinduced in nondormant seeds under unfavorable conditions (**Figure 1**). Thus, primary dormancy occurs in seeds dispersed directly from the mother plant and secondary dormancy is induced in partially or fully afterripened seeds.

There is no direct way to measure the degree of dormancy or amount of afterripening a seed has incurred. Thus, dormancy in a population of seeds is measured in terms of the delayed onset and reduced rate of germination with respect to nondormant or fully afterripened seeds from the same population under the same germination conditions. Physiologically, dormancy has often been understood as a balance between growth-inhibiting and growth-promoting substances or the result of the suppression of metabolic processes. Some studies suggest individual enzymatic or stocastic nonenzymatic reactions are the basis for afterripening. Dormancy and afterripening are interrelated processes. In the absence of a clear understanding of the underlying mechanisms for these processes, the term "germinability" or the germination capacity is used. Germinability denotes the tendency of seeds in a population for immediate, intermediate, or much delayed germination due to

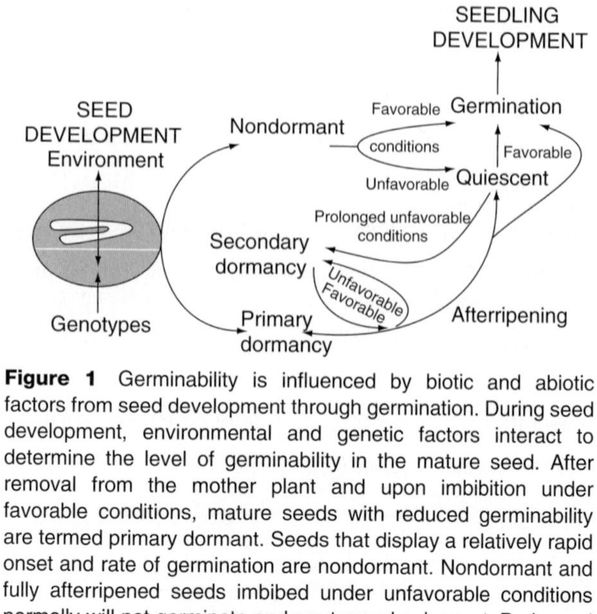

Figure 1 Germinability is influenced by biotic and abiotic factors from seed development through germination. During seed development, environmental and genetic factors interact to determine the level of germinability in the mature seed. After removal from the mother plant and upon imbibition under favorable conditions, mature seeds with reduced germinability are termed primary dormant. Seeds that display a relatively rapid onset and rate of germination are nondormant. Nondormant and fully afterripened seeds imbibed under unfavorable conditions normally will not germinate and are termed quiescent. Prolonged periods under unfavorable conditions for germination can result in the induction of secondary dormancy. Seeds with primary and secondary dormancy require afterripening under certain environmental conditions to achieve a state of nondormancy. Annual dormancy-nondormancy cycles account for some of the periodicity of germination in the soil seedbank.

internal factors, when the prevailing environmental conditions are favorable for germination of the species. Germinability takes into account the relative nature of dormancy and afterripening as it relates to germination.

Seed dormancy is not a universal phenomenon. Lack of dormancy in rice, wheat (*Triticum aestivum*) and barley seeds results in preharvest sprouting (PHS) when conditions after dry-down but before harvest are wet or humid and seeds germinate in the head. Preharvest sprouting reduces seed yield and quality. Seed dormancy and resistance to preharvest sprouting are directly correlated. Some mutations affecting seed development such as viviparous 1 (*vp1*) of corn (*Zea mays;* maize) cause immature embryos to germinate precociously, without completing their development program. This is termed precocious germination or vivipary.

Structural Components of a Seed and their Genetic Constitution

The major morphological structures of a mature seed are the hull, pericarp, testa, endosperm, and embryo. Technically, a grass seed is a fruit called a caryopsis or grain. The pericarp and testa are often tightly fused and are referred to as the seed coat or covering structures. Dormancy can be 'embryo-imposed', where control of dormancy resides within the embryo, "coat-imposed", which is enforced by seed covering structures such as the endosperm, pericarp/testa and, hull, or dormancy can be imposed by both embryo and seed coat. Seed covering structures may impose dormancy by being impermeable to water, gas exchange and leaching of germination inhibitors from the seed, or by providing mechanical resistance against protrusion of the radicle. Many possible physiological mechanisms for embryo-imposed dormancy have been hypothesized but direct proof is lacking. Seed components are of either maternal or zygotic origin and vary in their genetic constitution and ploidy levels. The hull is derived from lemma and palea, and the pericarp is derived from the ovary wall of the floret from the maternal parent. The testa is derived from the integuments of the female gametophyte. The aforementioned structures are of maternal genetic constitution and are diploid. The embryo results from fertilization between the male and female gamete; it has the genetic constitution of a hybrid and is diploid. The triploid endosperm is the result of fusion of two polar nuclei of the female gametophyte with one male gamete. The seed or caryopsis is of the same generation for the embryo and endosperm, but of the previous generation for the hull and pericarp/testa. In general, the generation of a seed is designated with the generation of the embryo.

Seed components may act both independently and in conjunction/interaction with other components in the germination process and thus in dormancy. The components involved in dormancy vary with the period of afterripening of seeds. The effect of individual seed components on dormancy can be determined at different stages of afterripening of seed, either by germination of each component separately, or by germination of an intact caryopsis and apportioning the effects genetically to different components. The role of individual components on germinability of seeds can be determined by germination studies using various treatments: (1) intact seed (hull, pericarp/testa, endosperm, and embryo); (2) dehulled seed (pericarp/testa, endosperm, and embryo); (3) dehulled seed with pericarp/testa removed (embryo and endosperm); and (4) excised embryos (embryo). Using this approach, hull- and pericarp/testa-imposed dormancy were detected in the rice cultivar N22 at 10 days after harvest (**Table 1**). By 40 days after harvest following exposure to warm, dry conditions, only the hull affected germinability. However, one disadvantage with the individual components approach is the effect of a component on dormancy in isolation may not be

Table 1 Effect of hull and pericarp at different days after harvest in some rice cultivars

Cultivar	*Treatment*	*Germination (%)*					
		Day 10	*Day 20*	*Day 30*	*Day 40*	*Day 50*	*Day 60*
N22	Hull intact	0	10	20	46	78	98
N22	Dehulled	0	15	76	98		
N22	Pericarp scraped	96	100				
N22	Pericarp scraped and enclosed in hull	0	21	48	66	86	98
IR13429–287–3	Hull intact	0	16	24	74	100	96
IR13429–287–3	Dehulled	38	82	98			
Mahsuri	Hull intact	81	100				

Adapted from Seshu DV and Sorrells ME (1986) Genetic studies on seed dormancy in rice. In: *Rice Genetics: Proceedings of the International Rice Genetics Symposium*, pp. 369–382. Los Banos, Philippines: International Rice Research Institute.

the same as in an intact caryopsis. That is, while an excised embryo may not show dormancy, it may express dormancy through an interaction with other components in an intact caryopsis. Genetic effects of coat-imposed factors can be detected as differences in germinability between reciprocal F_1s after ruling out cytoplasmic effects, which are also maternally inherited. The reciprocal differences include the effect of endosperm (based on dosage differences between the female and male gametes). To differentiate between cytoplasmic effects and coat-imposed factors, segregation for dormancy needs to be evaluated in the F_3 generation (of the embryo). While segregation in the F_3 generation confirms the genetic effect of seed-covering structures, the absence of segregation indicates that observed reciprocal differences in the F_1 generation are due to cytoplasmic factors. The effect of the embryo on dormancy in a genotype can be detected from the germination behavior of the F_1 hybrids compared to that of the recessive female parent. For example, a role of the embryo on dormancy is inferred in rice cultivars T1926 and T2105 by the delayed germination of F_1 grains from crosses involving these parents compared to the nondormant female parent Ptb10. In wheat cultivar RL 4137, the embryo effect is inferred by the intermediate dormancy of F_1 seeds from crosses involving this cultivar with each of five nondormant female parents. However, determination of individual seed component effects through genetic means should also be attempted cautiously as complex inheritance patterns involving the interaction of both coat and embryo factors may affect the expectations. For example, reciprocal F_1 differences of germinability are not observed when there are duplicate recessive factors for dormancy, one of which is from the seed coat. However, the involvement of coat-factors is indicated by delayed segregation in the F_3 generation.

The major seed component contributing to germinability varies in different species and even in individual genotypes within a species. Coat-imposed dormancy is generally thought to account for low germinability in dormant cereal grains because there is little evidence of dormancy due to the excised embryos in most cases. However, embryo dormancy is well documented in at least one genotype of wild oat, namely M73, and in a nondomesticated species of rice (*O. sativa f. spontanea*). Genetic studies also indicate embryo dormancy in many cultivars of cereals. Dormancy in Asian cultivated rice (*O. sativa*) is generally imposed by the hull, pericarp/testa or both with a stronger and more prolonged influence by the hull. Red color of the pericarp is known to confer dormancy in some wild strains of rice. In cultivated hexaploid wheat, resistance to PHS has long been associated with the red color of the pericarp. Coat factors other than the red pericarp color also influence dormancy as in some white-grained wheats. The genetic effect of both embryo and coat-imposed factors are reported for some genotypes of red- and white-grained wheats. In wild oat, the pericarp/testa is the predominant factor contributing to the long-term dormancy in most of the genotypes. Both pericarp and embryo dormancy exist in the strongly dormant line M73.

Inheritance of Dormancy

The inheritance of seed dormancy has been investigated in terms of delayed onset and reduced rate of germination in weed species like wild oat, whereas it has been studied in terms of both delayed germination and resistance to preharvest sprouting in domesticated field crops such as barley, rice, and wheat. The variation for dormancy is under genetic control with strong influences of the environment and the interaction of genotype and environment (G $\times$ E).

Environmental Influences and Heritability

The onset, maintenance and removal of seed dormancy is influenced by environmental factors reducing the genetic component of the variation, i.e., heritability. Heritability is the proportion of the total phenotypic variation of a population that is attributable to the effects of genes. Heritability is expressed in two ways; broad sense heritability (h^2b) is the ratio of total genetic variation including the additive, dominance and interaction effects to the total phenotypic variation; narrow sense heritability (h^2n) measures the proportion of the variation that is due to "additive effects" of genes in a population. Estimates of heritability for dormancy vary in different species and in different populations (**Table 2**). In wild oat, heritability for dormancy is about 50% under natural field conditions, but a higher estimate is reported in another population (M73/SH430) grown under controlled environmental conditions. The heritability estimates for preharvest sprouting under field conditions range from low to high in wheat and high in sorghum and barley.

The environmental conditions under which the plants are grown, such as, temperature, soil moisture, photoperiod, etc., profoundly influence the degree and duration of dormancy. Temperature is the most important of all these factors. Temperature influences dormancy: (1) during seed maturation on the maternal plant; (2) at the time of imbibition; and (3) the rate of afterripening. Dormancy levels

Table 2 Estimates of heritability for dormancy in different species of grasses

Name of species and cross/population	*Heritability (h^2)*	SE/CI [a]	*Reference*
Wild oat *(Avena fatua)*			
AN, ST, SF[b]	0.51–0.56	0.03–0.08	Jana and Naylor (1980)[g]
M73/SH430	0.75	(0.64–0.83)	Fennimore *et al.* (1998)[c,h]
Wheat *(Triticum aestivum)*			
KS75216/Clark's Cream	0.44	0.16–0.63	Upadhyay and Paulsen (1988)[d,i]
KS82W404/Clark's Cream	0.47	0.2–0.6	
KS82W439/Clark's Cream	0.42	0.12–0.62	
White Chief/Clark's Cream	0.56	0.33–0.71	
Houser/Clark's Cream	0.10[e]		Paterson and Sorrells (1990)[d,j]
NY18/Clark's Cream	0.84	(0.78–0.88)	Anderson *et al.* (1993)[c,k]
NY18/NY10	0.56	(0.46–0.64)	
Barley (*Hordeum vulgare*)			
Bode/Agneta	0.69–0.80		Buraas and Skinnes (1984)[c,l]
Sorghum (*Sorghum bicolor*)			
IS 9530/Redland B2	0.63		Lijavetzky *et al.* (2000)[f,m]

[a] Standard errors or confidence intervals;
[b] site names of natural populations;
[c] heritability in narrow-sense;
[d] heritability in broad-sense;
[e] standard-unit heritability;
[f] heritability calculated based on F_3/F_2 parent-offspring regression. F_3, families from 10 extreme resistant and susceptible F_2 individuals were used.
[g] Jana S and Naylor JM (1980) Dormancy studies in seed of *Avena fatua*. II. Heritability for seed dormancy. *Canadian Journal of Botany* 58: 9–93.
[h] Fennimore SA, Nyquist WE, Shaner GE, Myers SP, and Foley ME (1998) Temperature response in wild oat (*Avena fatua* L.) generations segregating for seed dormancy. *Heredity* 81: 674–682.
[i] Upadhyay MP and Paulsen GM (1988) Heritabilities and genetic variation for preharvest sprouting in progenies of Clark's Cream white winter wheat. *Euphytica* 38: 93–100.
[j] Paterson AH and Sorrells ME (1990) Inheritance of grain dormancy in white-kernelled wheat. *Crop Science* 30: 25–30.
[k] Anderson JA, Sorrells ME, and Tanksley SD (1993) RFLP analysis of genomic regions associated with resistance to preharvest sprouting in wheat. *Crop Science* 33: 453–459.
[l] Buraas T and Skinnes H (1984) Genetic investigations on seed dormancy in barley. *Hereditas* 101: 235–244.
[m] Lijavetzky D, Martinez MC, Carrari F, and Hopp HE (2000) QTL analysis and mapping of pre-harvest sprouting resistance in sorghum. *Euphytica* 112: 125–135.

decrease with increasing maternal plant growth temperature in most of the grass species like wild oat, wheat, and rice. The time required to reduce dormancy by afterripening under warm, dry conditions decreases with increasing temperature in rice and wild oats. The level of dormancy increases with increasing germination temperature over a range of 4–24 °C (**Figure 2**) in dormant wild oat lines in contrast to the nondormant lines. A lower level of dormancy at low germination temperatures was also reported in wheat. A maturation temperature by germination temperature interaction showed a significant influence on the level of dormancy in both wild oat and wheat. Other environmental factors such as high soil moisture and short day-length conditions during the maternal plant growth also increase dormancy in wild oat. Because of the great influence of environmental conditions on dormancy, extreme care should be taken in maintaining the constant conditions during germination experiments. Apart from the influence of the environment, the $G \times E$ interactions have also been important components of variation for dormancy in many grasses like wild oat, rice, barley, and wheat.

Genetic Component

Genetic variation for dormancy exists both within and across species. For example, the duration of dormancy varies widely in natural populations of wild oat grown under similar conditions. Some genotypes of wild oat germinate immediately after seed dehiscence, while others take weeks to many months of afterripening to achieve their full germination potential. In general, nondomesticated species possess higher levels of dormancy compared to cultivated species. Some nondomesticated weedy strains of rice have dormancy levels about five times

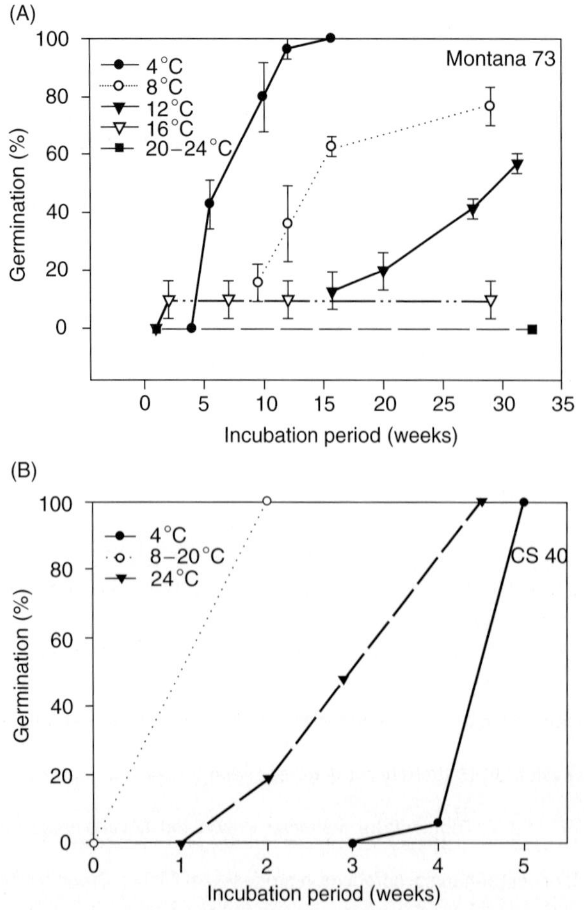

Figure 2 Influence of temperature on seed germination of two pure lines of *Avena fatua*: A Montana 73 and B CS40. Germination is suppressed with increasing temperatures in the dormant line Montana 73 in contrast to the nondormant line CS40. Adapted with permission from Naylor JM and Fedec P (1978) Dormancy studies in seed of *Avena fatua*. 8. Genetic diversity affecting response to temperature. *Canadian Journal of Botany* 56: 2224–2229.

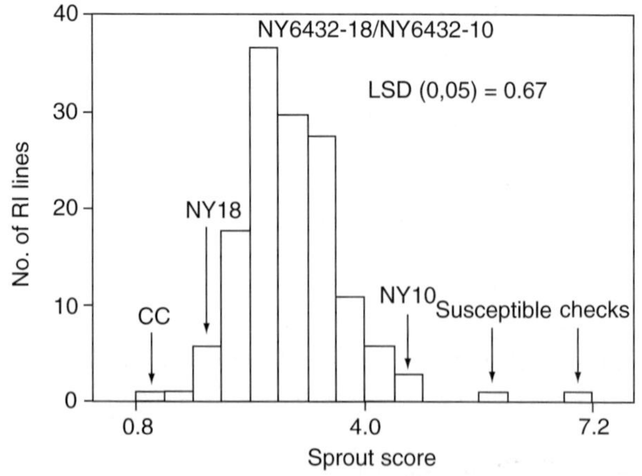

Figure 3 Phenotypic distribution for preharvest sprout scores of the F_5-derived recombinant inbred (RI) lines from the cross of two white-grained wheats (NY18/NY10). NY18 is moderately resistant and NY10 has low levels of resistance to PHS. The PHS scores are averages over seven environments. The RI lines show continuous variation ranging in their scores from 0.8 (resistant to PHS) to 7.2 (susceptible to PHS). Clark's Cream is a hard white winter wheat that shows a high level of dormancy and the Geneva and Houser lines are susceptible checks. Adapted from Anderson JA, Sorrells ME, and Tanksley SD (1993) RFLP analysis of genomic regions associated with resistance to preharvest sprouting in wheat. *Crop Science* 33: 453–459.

higher than cultivated species. Dormancy is stronger in African cultivated rice (*O. glaberrima*) than in Asian cultivated rice. Seeds of the japonica subspecies of Asian cultivated rice generally have a shorter duration of dormancy than the tropical indica subspecies, and those of javanica have a moderate degree of dormancy. In most modern cultivars of bread wheat, seed dormancy lasts for only a few days, although there are varieties in which dormancy can last for several months.

The mode of inheritance of dormancy varies within and among different species. Dormancy is controlled by recessive factors in some grasses like wild oat and barley, whereas dormancy is dominant or partially dominant in rice. Resistance to preharvest sprouting in sorghum is partially dominant. There are red-grained and white-grained dormant wheat varieties whose mode of inheritance differs. Even within the white-grained wheats, the mode of inheritance differs from genotype to genotype. Dormancy is dominant in crosses involving the dormant white wheat Clark's Cream, whereas dormancy is governed by recessive factors in some other white wheats such as Kenya 321 and AUS 1408. Dormancy in red-grained wheat is associated with the red color of the pericarp and is dominant. In some crosses between red- and white-colored wheat, both dominance for dormancy associated with the red color of the pericarp and partial dominance for embryo dormancy are reported. In crosses involving dormant and nondormant red-grained parents (red × red), dormancy is partially dominant.

The inheritance of dormancy can be under the control of one or a few major genes, or under the control of polygenes. However, the phenotypic distribution in populations of seeds segregating for dormancy may be uni-, bi- or tri-modal in many cases (**Figure 3**). That is, the distribution is not exactly discrete even when governed by major genes because of the large influence of the environment and/or the role of modifier genes. Because of the complexity, the genetic models for dormancy proposed in the F_2 generation of a cross need to be confirmed using backcrosses to both of the parents and by progeny testing in advanced generations (e.g., F_3). Dormancy is essentially a quantitative trait displaying continuous phenotypic variation and thus locating individual loci governing the trait has been difficult. Hence, quantitative genetic methods are employed to calculate the magnitudes of the additive, dominance, and interaction effects of the trait, to test whether simple additive dominance models are adequate to explain the variation in a cross or complex epistatic interactions are also operating, and to estimate an approximate number of loci segregating for the trait in that cross.

The genetic control of dormancy (major genes or polygenes) in a population depends on the parental combination, the seed component under investigation and the duration of afterripening. The inheritance of dormancy is also influenced by the way in which the phenotypic data are represented and environmental conditions such as temperature under which germination takes place are evaluated. The individuals have been grouped into dormant and nondormant classes based on a certain percentage of germination in some studies whereas in others, the rate of germination over a period of time is measured. In the F_2 generation of a wild oat population (M73/CS40), germination data fit to a monogenic ratio with dominance for dormancy when the seed is classified into those that germinated and those that do not by 2 weeks. However, if the rate of germination over a period of 15 weeks is the measure, then nondormancy is partially dominant being controlled by at least two genes. When seeds of the cross M73/SH430 were germinated at 20 °C, nondormancy was partially dominant controlled by two genes; in another study, where seeds were germinated at 15 °C nondormancy showed complete dominance under the control of three genes.

Dormancy is mainly influenced by nuclear genes in many grasses such as rice, wheat, barley, and wild oat. Dormancy measured at different periods of afterripening displayed continuous variation in segregating populations from some crosses in rice and was inferred as governed by multiple genes. A monogenic (3:1) or a digenic (13:3) segregation was inferred for dormancy in intact seeds from some other crosses when the F_2 population was classified into groups of dormant and nondormant individuals based on their germination per cent. Hull-imposed dormancy in rice displayed a monogenic ratio of 3:1 in crosses involving the dormant parent N22 and a digenic complementary ratio of 9:7 in those involving the cultivar IR13429-287-3, after the pericarp/testa-imposed dormancy was removed by afterripening. The role of a few modifiers was suggested in both of these crosses. Also, in wheat, dormancy exhibits both major gene and polygenic inheritance. In some white-grained wheats, dormancy seems to be governed by duplicate recessive factors displaying a 15:1 segregation ratio of nondormant:dormant individuals (plants with less than 50% dormancy classified as nondormant and others as dormant) in the F_3 generation, whereas dormancy is governed by a single recessive gene in a cross between dormant white *T. sphaerococcum* and a nondormant cultivar of *T. aestivum*. A single gene *Phs* controlled dormancy in a cross between two red-grained cultivars, Boxer and Soleil, with varying levels of dormancy. Dormancy in red-grained wheats is inherited as a pleiotropic effect of the dominant *R* loci for the red pericarp color. There are three homoeologous R loci located on group 3 chromosomes of hexaploid wheat. Complex polygenic inheritance has also been reported for preharvest sprouting in wheat for some white × white crosses. Simple additive-dominance models were adequate to explain the inheritance of germination in three sets of these crosses, while one cross, Houser/Clarks Cream, also showed dominant × dominant epistasis. In barley, a continuous variation under the control of several recessive genes was reported for preharvest sprouting resistance in F_8 lines of the cross Agneta/Bode. In sorghum, a simpler genetic control for preharvest sprouting resistance was suggested.

Three nuclear factors of more or less equal strength segregate in populations derived from crossing a nondormant cultivated oat (*A. sativa*) with a dormant wild oat. At least three factors with epistatic interactions seem to be involved in the inheritance of dormancy in wild oat. In an earlier study using F_1, F_2 and backcross generations derived from different dormant and nondormant parents, dormancy was determined to be a quantitative trait controlled by at least two genes. The locus *EE* from the nondormant parents SH430 and CS40 governs early germination and the loci *L1* and *L2* from the dormant parents AN127 and M73 respectively, govern dormancy or late germination. The locus *L1* exhibited epistatic interaction to the locus *E*. In a more recent study, a 3 locus model was proposed by classifying lines from different generations of the cross M73/SH430 into nondormant and dormant caryopses based on days to germination. The loci *G1* and *G2* from SH430 promote early germination, and the locus *D* from the dormant parent M73 promotes late germination. The expression of locus *D* for late germination is dependent upon the alleles present at loci *G1* and *G2*. A caryopsis will be nondormant if it has at least one copy of either *G1* or *G2*, or at least one copy of both G1 and G2 alleles regardless of the genotype at *D* locus. All the individuals with genotype dd will be nondormant irrespective of alleles at *G1* or *G2*. The F_7 RI lines showed a segregation of 7:1 nondormant to dormant and the only dormant types in the F_7 generation are the dormant parental type *g1g1g2g2DD*. A simple additive dominance model seems to be satisfactory at a germination temperature of 15 °C, whereas epistatic gene action is detected at 20 °C by quantitative genetic analysis of the same population. About six factors were estimated to be segregating between the dormant and nondormant parents.

Quantitative Trait Loci

Locating the individual genes that govern quantitative traits like dormancy is difficult owing to the absence of discrete phenotypic segregation. Rather, the inheritance of such a trait is traced to the chromosomal segments governing the trait's expression. Regions of the genome associated with expression of a quantitative trait are known as quantitative trait loci (QTLs). Molecular markers abundantly distributed throughout the genome have been useful to mark the chromosomal location of QTLs governing a quantitative trait. Molecular markers are pieces of DNA that are tightly linked to the trait of interest. There are different molecular marker techniques, such as restriction fragment length polymorphisms (RFLPs), randomly amplified polymorphic DNAs (RAPDs), simple sequence repeats (SSRs) and amplified fragment length polymorphisms (AFLPs). Morphological markers, biochemical markers, e.g., isozymes, and DNA markers have been used to mark individual loci for dormancy in rice, wheat, barley, and wild oats. Studies in wheat, barley, and sorghum indicate epistatic interactions among dormancy QTLs; many QTLs that are specific to individual environments, while a few QTLs of large effects that are expressed over a range of environments. Hence, it is important to examine dormancy QTLs in different genetic backgrounds and across a range of environments to understand their contribution to the trait in a particular species.

QTLs for dormancy in rice were identified in the backcross inbred lines (BILs) derived from the cross Nipponbare/Kasalath and a set of recombinant inbred lines (RILs) from the cross Peikuh/W1944. The parents of both the crosses are of cultivated rice, except W1944, which belongs to a wild species (*O. rufipogan*). In the BIL population, five putative dormancy QTLs linked to RFLP markers together explained 48.2% of the phenotypic variation for germination of freshly harvested seeds. A Kasalath-derived locus marked by the RFLP marker C1488 on chromosome 3 accounted for the largest phenotypic effect (22.5%). In the RIL population, a total of 17 QTLs on seven chromosomes with phenotypic effects ranging from 7.3% to 21.8% were detected in different germination conditions. Of the 17 QTLs, 12, 2, and 3 QTLs were identified intact, dehulled, and both intact and dehulled seeds, respectively. In wheat, at least six different chromosomal regions are associated with resistance to preharvest sprouting in the white-kernelled × white-kernelled cross NY6432-18/Clarks Cream. Of these, three unlinked RFLP markers, BCD1434 (1AS), CDO431 (1BS) and CDO64 (2S), together explained 32% of the phenotypic variation. The three loci showed a highly significant three-way interaction. In another population derived from the cross NY18/NY10, four QTLs accounted for 51% of the phenotypic variation.

Tolerance to preharvest sprouting in the cross of a red-kernelled × white-kernelled wheat was mapped to chromosomes 6B and 7D. In wild oat, two RAPD markers are linked to germinability. The loci marked by OPX-06 and OPT-04 explained 12.6% and 6.8% of the phenotypic variation, respectively, in the F_2 population of the cross M73/SH430. In barley, the major QTL *SD1* (of the four QTLs SD1–SD4) is located on the long arm of chromosome 7 marked by the RFLP marker PSR128 in a cross between dormant Steptoe and nondormant Morex. The locus explains approximately 50% of the phenotypic variation for dormancy and is epistatic to loci *SD2* and *SD4*. In sorghum, two unlinked QTLs on chromosomes *E* and *F* contribute 83% of the total phenotypic variation for resistance to preharvest sprouting across two environments. There is an epistatic interaction between these loci.

Genome Colinearity and Orthologous Dormancy Loci

Comparative mapping for grass genomes has revealed conservation of gene order or 'colinearity' within a chromosomal segment between species. Colinearity, which is often referred to as synteny, has been reported for *R* loci that impart red pericarp color and the *Vp1* locus, which prevents the precocious germination of immature embryos. Putative orthologous loci affecting seed germinability or dormancy have been identified for corn, barley and wheat indicating that the location and function of some loci have been conserved through evolution. The relative positions of orthologs of the corn *Vp1* gene are conserved in rice and wheat on chromosome 1 and chromosome 3, respectively. Furthermore, both wheat and rice have been reported to carry *Vp1* orthologs in loose genetic linkage with *R* loci and the *Rd* locus respectively, controlling seed coat pigments. The QTL for resistance to preharvest sprouting in wheat located on the short arm of chromosome 1 (marked by BCD 1434) corresponds to a QTL on the homoeologous chromosome 5S of barley. The major QTL *SD1* on barley chromosome 7 marked by PSR128 may be orthologous to a locus for resistance to preharvest sprouting in wheat on chromosome 5DL. This locus, which is marked by BCD1874 in wheat, is conserved in the tribe *Triticeae,* in oat, and is closer to the predicted location of the viviparous locus *Vp9* on corn chromosome 7.

Mutants Affecting Dormancy and Germination

Mutants affecting dormancy are rare in nature and may contribute little to the natural variation for the dormancy trait. However, some natural and induced mutants that have altered seed morphology, or that have altered functions with regard to germinability compared to the wild-type seed, provide insight into pathways, signals, and mechanisms underlying dormancy and germination. Mutants deficient in or insensitive to hormones, such as gibberellins (GA), and abscisic acid (ABA), seed coat mutants of *Arabidopsis* and viviparous mutants in cereals have been especially well characterized.

The plant hormones ABA and GA are thought to be involved in the inhibition and promotion of germination, respectively. The *aba* mutants of *Arabidopsis* that are inhibited in the biosynthesis of ABA have reduced or no seed dormancy and a reduced requirement for light during germination. In contrast, the *Arabidopsis abi* mutants produce normal levels of ABA but are insensitive to ABA. Research into the germinability of *aba*, *abi*, and *aba/abi* double mutants and wild-type seeds suggests that the level of embryonic ABA during seed development and maturation plays a major role in the induction, and perhaps the maintenance, of dormancy. The *ga1* mutants of *Arabidopsis* are defective in gibberellin biosynthesis and show reduced germination. Research conducted on *Arabidopsis ga1* mutants, mutants insensitive to gibberellins (*gai*), *ga1/gai* double mutants, and wild-type seeds strongly indicates that gibberellins play a major role in the induction of germination. Embryos of the *ga1* seeds readily germinate when the testa layer is removed indicating that the GA requirement for germination may be imposed by the testa. Mature *Arabidopsis* seeds are characterized by a brown testa owing to the presence of condensed tannins. Seed coat mutants showing altered flavonoid pigmentation represented by *transparent testa* (*tt*) and mutants with altered testa structure represented by *aberrant testa shape* (*ats*) have been useful in understanding the role of the testa in seed dormancy and germination. A reduction or absence of phenolic compounds in the testa of pigmentation mutants appears to improve the seed coat permeability, resulting in a faster onset and higher rate of germination compared to wild-type seeds.

Precocious germination of the corn *vp1* mutant is brought about by a reduced sensitivity to the hormone ABA. The mutation has various pleiotropic effects on seed development including a defect in anthocyanin biosynthesis in the aleurone layer of the endosperm. Molecular analysis of the corn *Vp1* gene and protein have provided few clues as to its function; however, it seems to be a transcription factor required for seed maturation-related gene expression. The *ABI3* gene in *Arabidopsis* functions in seed maturation and dormancy and is orthologous to *Vp1*. *Vp1* orthologs have also been identified in rice, wheat and wild oat. The correlation between *afVp1* expression and embryo dormancy in imbibed seeds suggests that the gene could play an important role in seed dormancy in wild oat. The existence of other viviparous mutants in rice such as *riv1*, *riv2* and *enl1* suggests that there are different genes regulating precocious germination. These rice mutants are ABA-insensitive and affect precocious germination without obvious pleiotropic effects on seed development. The rice *enl1* mutant lacks an endosperm and may be useful for investigating the role of the endosperm, and embryo and endosperm interactions on seed development, dormancy, and germination.

Implications for Dormancy in Agriculture

Breeding for Resistance to Preharvest Sprouting

Dormancy plays an important role in the survival of many plant species by distributing germination over time and space. The process of domestication with selection for rapid and uniform germination has led to non- and less-dormant genotypes. Nondormancy is not always a desirable trait because an insufficient level of dormancy in some crops leads to preharvest sprouting. On the other hand, while a certain level of dormancy is desirable to impart resistance to preharvest sprouting in rice, barley, and wheat, a higher level of dormancy may lead to poor stand establishment owing to slow and inconsistent germination, may adversely affect the malting of barley, and may lead to weed problems in the subsequent crop. Thus, breeding for predictable and low levels of dormancy is an important objective for some crop improvement programs.

Weed Management

Variability in germination as a result of seed dormancy is the primary reason for the persistence of many weeds. Genotypes with varying levels of dormancy and seeds in various stages of afterripening in natural populations ensure that a proportion of the population will survive to reproduce. Environmental influences and G × E interactions add to the variability in weed seed germination. In addition to conferring the phenotypic plasticity in primary dormant seeds, environmental factors such as temperature and moisture play a role in the induction of secondary dormancy when conditions are unfavorable for germination (**Figure 1**). Primary and secondary dormancy are important factors influencing the population dynamics of many weeds. Weed ecologists modeling the population dynamics of weeds rely on environmental parameters to predict weed seedling emergence. Since it is clear from estimates of heritability that dormancy is influenced by both environmental and genetic factors, adding a genetic and G × E component to these models should improve their accuracy and increase the efficiency of current weed management strategies. A detailed knowledge of the genetic and environmental factors underlying dormancy and identification and cloning of dormancy alleles from various plants is critical for the management of crops and weeds in the twenty-first century.

List of Technical Nomenclature

Additive effects	Difference between the homozygotes (*AA*, *aa*) of a locus
Diploid	An organism with two sets of genetic material, consisting of paired chromosomes, one from each parent.
Dominance	The masking effect of an allele of a gene over its alternative allele in the heterozygous condition.
Epistasis	Dominance of a gene over a nonallelic gene.
F_1	The first generation of a cross.
F_2 and F_3	The second and third generations, respectively, of a cross obtained by self-fertilization of the F_1 and F_2 individuals, respectively.
Genotype	The genetic constitution of an organism.
Homoeologous loci	Loci in different genomes in polyploid species that have arisen from a common ancestor.
Incomplete dominance	Two alleles of a gene resulting in contrasting phenotypes, but their heterozygote is intermediate between the two phenotypes.
Mutant	A variant individual arising owing to a sudden heritable variation in a gene or in chromosome structure.
Orthologous loci	Genes that arose from a common ancestor and are conserved in different species.
Phenotype	Appearance of an individual as contrasted with its genetic make-up or genotype.
Recessive	The member of an allelic pair whose expression is masked by the presence of the other member of a homologous chromosome.
Reciprocal crosses	Crosses in which the sources of male and female gametes are reversed.
Triploid	An organism with three basic sets of chromosomes.

See also: **Seed Development**: Germination.

Further Reading

Anderson JA, Sorrells ME, and Tanksley SD (1993) RFLP analysis of genomic regions associated with resistance to preharvest sprouting in wheat. *Crop Science* 33: 453–459.

Baskin CC and Baskin JM (1998) *Seeds-Ecology, Biogeography, and Evolution of Dormancy and Germination.* San Diego, CA: Academic Press.

Bench-Arnold RL, Sanchez RA, Forcella F, Kruk BC, and Ghersa CM (2000) Environmental control of dormancy in weed seed banks in soil. *Field Crops Research* 67: 105–122.

Bewley JD (1997) Seed germination and dormancy. *The Plant Cell* 9: 1055–1066.

Debeaujon I, Leon-Kloosterziel KM, and Koornneef M (2000) Influence of the testa on seed dormancy, germination, and longevity in Arabidopsis. *Plant Physiology* 122: 403–413.

Fennimore SA, Nyquist WE, Shaner GE, Doerge RW, and Foley ME (1999) A genetic model and molecular markers for wild oat (*Avena fatua* L.) seed dormancy. *Theoretical and Applied Genetics* 99: 711–718.

Foley ME (2001) Seed dormancy: an update on terminology, physiological genetics, and quantitative trait loci regulating germinability. *Weed Science* 49: 305–317.

Foley ME and Fennimore SA (1998) Genetic basis for seed dormancy. *Seed Science Research* 8: 173–182.

Gu X, Chen Z, and Foley ME (2003) Inheritance of seed dormancy in weedy rice. *Crop Science* 43: 835–843.

Han F, Ullrich SE, Clancy JA, and Romagosa I (1999) Inheritance and fine mapping of a major barley seed dormancy QTL. *Plant Science* 143: 113–118.

Koornneef M and Karssen CM (1994) Seed dormancy and germination. In: Koornneef M and Karssen CM (eds)

Arabidopsis, pp. 313–334. Cold Springs: Cold Springs Harbor Laboratory Press.
Lin SY, Sasaki T, and Yano M (1998) Mapping quantitative trait loci controlling seed dormancy and heading date in rice, *Oryza sativa* L., using backcross inbred lines. *Theoretical and Applied Genetics* 96: 997–1003.
Simpson GM (1990) *Seed Dormancy in Grasses.* New York: Cambridge University Press.

Preharvest Sprouting

R L Benech-Arnold and R A Sánchez, IFEVA, Buenos Aires, Argentina

Definitions, Causes and Predisposing Factors

Dormancy can be defined as the failure to germinate because of some internal block of the seed that prevents completion of the germination process. While the adaptive significance of dormancy is quite evident for plants living in the "wild," it has always been seen as a complication in seeds from plants that are grown as crops. Indeed, no farmer would tolerate uncertainty on the emergence of a crop because of the possible existence of remnant dormancy in the seed lot that is used to generate a new crop. For that reason, crops that originally must have had dormancy have been selected heavily against this trait throughout their domestication process. In some cases, this selection against dormancy has gone too far and the seeds are germinable even prior to crop harvest. This situation, combined with rainy or damp conditions prevailing during the last stages of maturation, may lead to germination in the mother plant, a phenomenon that is better known as preharvest sprouting. Though its occurrence has been reported for several cultivated plants, preharvest sprouting is predominantly a feature of cereal crops, i.e., wheat (*Triticum aestivum*), sorghum, barley (*Hordeum vulgore*), oats (*Avena sativa*), rye (*Secale cereale*), rice (*Oryza sativa*), and even millets (*Setaria italica*). Therefore, the cereal grain is used as the "model system" for discussing the different aspects of this phenomenon dealt with in this article.

Inception of dormancy occurs very early in cereals. Embryos are usually fully germinable from the early stages of development, i.e., 15–20 days after pollination (DAP), if isolated from the entire grain and incubated in water. The entire grain, however, reaches an ability to germinate well after it has been acquired by the embryo. This seed coat (endosperm plus pericarp)-imposed dormancy is the barrier preventing untimely germination and its duration depends on the genotype and on the environment experienced during maturation and beyond. Therefore, though cases of embryo dormancy have been reported for barley and other cereal crops, preharvest sprouting-susceptible cultivars are usually those whose coat-imposed dormancy is terminated well before harvest maturity. For this reason, breeding programs that select for genotypes that lose their dormancy a few weeks after crop harvest should be an obvious step toward avoiding preharvest sprouting; however, due to the paucity of knowledge on the genetic, physiological and environmental control of dormancy, it is very difficult to adjust the timing of dormancy loss to a precise and narrow "time window." In malting barley, for example, breeders are compelled to work on a "knife's edge." The malting process itself requires grain germination; hence, a low dormancy level at harvest is a desirable characteristic because the grain can be malted immediately after crop harvest, thus avoiding the costs and deterioration resulting from grain storage until dormancy is terminated.

In addition to the combination of low dormancy and damp weather prior to crop harvest, the confluence of other environmental conditions might be required for preharvest sprouting to occur. Seeds that present full dormancy do not germinate at any temperatures. This is also the case for dormant cereal grains: in summer cereals, such as sorghum, dormancy is not expressed at high temperatures (i.e., 30 °C), while in winter cereals, such as wheat and barley, dormancy is not expressed at low temperatures (i.e., 10 °C or lower). It should be emphasized that the depressed germination that occurs as temperatures exceed or fall below certain values (depending on whether it is a winter or summer cereal) is truly an expression of dormancy and not an inevitable effect of temperature on germination, for it does not take place in isolated embryos or in grains that have after-ripened. The higher the dormancy level, the narrower the thermal range within which dormancy is not expressed. This lack of expression of dormancy at, for example, low temperatures in grains from winter cereals implies that in years when damp conditions are combined with low air temperatures around harvest time, both resistant (high dormancy) and susceptible (low dormancy) cultivars might be expected to sprout.

The environment experienced by the mother plant during seed development may sometimes modulate the speed with which seeds are released from dormancy and thus the sprouting behavior of a crop. Among the different factors acting on the mother plant, temperature appears to be responsible for year-to-year variation in grain dormancy within the same genotype. A "time window" was identified within the grain-filling period of barley, with sensitivity to temperature for the determination of the speed of dormancy release after physiological maturity. This "time window" was found to range from 300 °C days to 350 °C days after heading (accumulated over a base temperature of 5.5 °C). A positive linear relationship was established between the average temperature perceived by the crop during this "time window" and the germination index of the grains 12 days after physiological maturity (PM), which is approximately half-way between physiological and harvest maturity (HM). The established relationship shows that the higher the temperature experienced during the sensitivity "time window," the faster the rate with which grains will be released from dormancy after physiological maturity and, consequently, the lower the dormancy level prior to crop harvest.

In summary, an early termination of dormancy within seed development is instrumental in preharvest sprouting susceptibility. Some conditions predispose the crop to suffer from this adversity. For example, in barley, high temperatures during seed development followed by low temperatures (and obviously damp conditions) prior to crop harvest are excellent conditions for preharvest sprouting. A scheme summarizing the causes of and predisposing factors for sprouting is presented in **Figure 1**.

Consequences of Preharvest Sprouting

Preharvest sprouting has a wide range of consequences, all of which are adverse. In terms of seed viability these consequences range from the immediate loss of seed viability upon subsequent desiccation when the embryo has grown beyond the point at which it loses its desiccation tolerance, to a dramatic reduction in seed longevity when embryo growth has not gone that far. Initiation of the germination process triggers the synthesis of enzymes that promote reserve mobilization. During germination, the growing embryo synthesizes and secretes gibberellins into the starchy endosperm. Gibberellins diffuse into the aleurone layer and induce the synthesis of α-amylase. This endoamylase starts the degradation of the starchy endosperm and is the only enzyme that can hydrolyze raw starch. α-Amylase in cereals is commonly divided into two types: an endogenous late maturity, green, or low pI (isoelectric point) group; and a germination or high pI group that is associated with sprouting. In addition to α-amylase, other enzymes participate in the

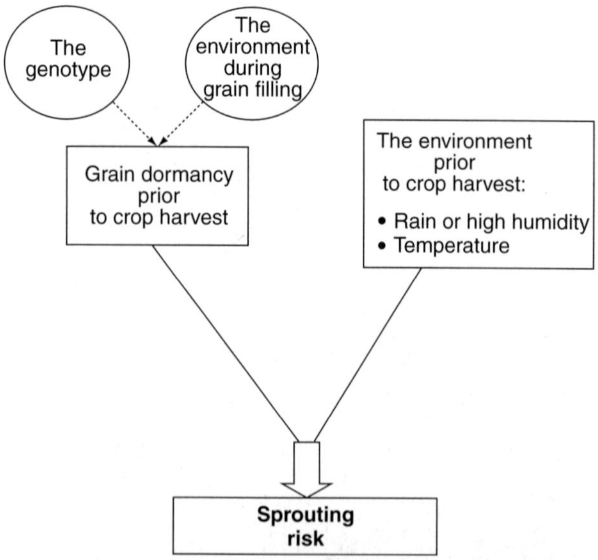

Figure 1 Summary of the factors that determine the risk level for a crop to suffer from preharvest sprouting. Reproduced with permission of CABI Publishing from Benech-Arnold RL, Giallorenzi MC, Frank J and Rodriguez V (1999) Termination of hull-imposed dormancy in barley is correlated with changes in embryonic ABA content and sensitivity. *Seed Science Research* 9: 39–47.

hydrolysis of starch in the endosperm to simple sugars for use by the embryo, namely, debranching enzymes (R-enzyme, pullulanase), isoamylase, exoamylase (β-amylase), and α-glucosidase (maltase). Proteolytic enzymes function during sprouting to mobilize nitrogen for the embryo and seedling, and to release bound or inactive enzymes. β-Amylase, debranching enzymes, and probably other enzymes are complexed with proteins during maturation and then freed by proteolysis during germination. Many types of proteolytic enzymes – endopeptidases, carboxypeptidases, aminopeptidases, etc. – are associated with sprouting.

The consequences of this hydrolysis of reserves depend on the types of products for which the cereal is intended and on the processing methods used. Breads baked from hard wheats are affected more than most products by preharvest sprouting of the grain. Production of the bread is complicated by extreme stickiness of the dough. Sprouting weakens the dough strength, decreases the amylograph peak viscosities, and results in poor handling and machining properties. Stickiness of dough is usually attributed to extensive enzymic hydrolysis of damaged starch and altered rheological properties to proteolytic enzymes. Preharvest sprouting also affects both the processing and quality of the many kinds of noodles that are made from wheat. In dry noodles, high α-amylase weakens the dough so that the noodles cannot support their own weight and break

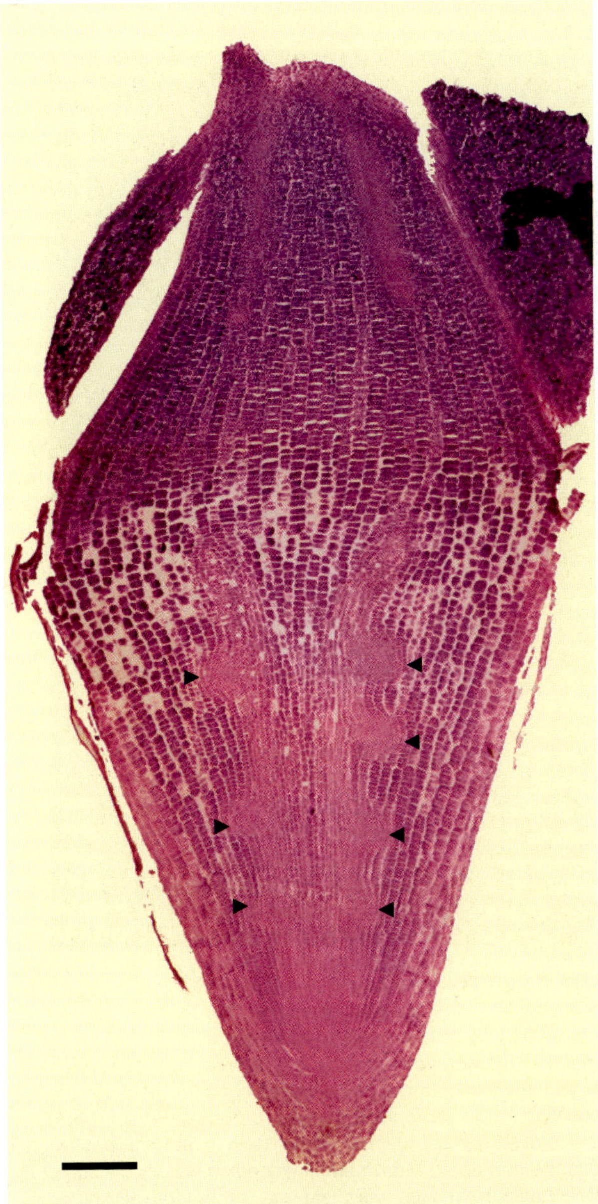

Plate 31 Root Development Mature embryo of cucumber plant (*Cucumus sativum* cv Muromsky) 12 h after the start of seed imbibition. At this stage no cell division occurred. The 7-μm thick section was made in the plane of the cotyledons. Arrowheads indicate embryonic LRPs in the radicle that are located along the prospective protoxylem strands. In the upper part of the embryo, shoot apical meristem, leaf primordium and parts of cotyledons (to the right and to the left from the embryo) can be seen. Scale = 200 μm. Photograph courtesy of Dr J.G. Dubrovsky.

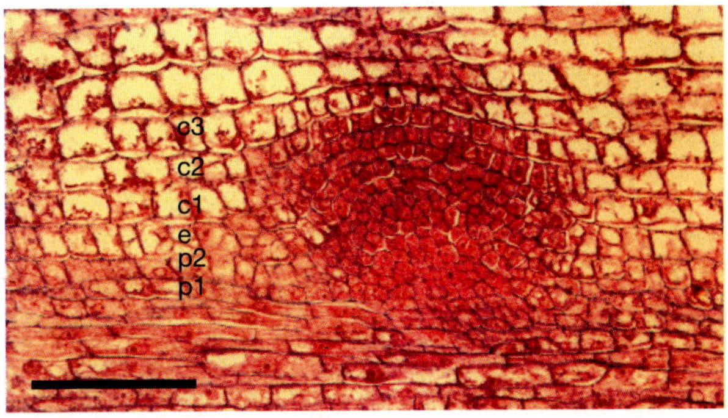

Plate 32 Root Development The most basal (see **Figure 1**) embryonic LRP in cucumber plants (*Cucumus sativum* cv Muromsky) 18 h after the start of seed imbibition. The 7 μm section was made in the same plane as in **Figure 1**. Note that two layers of pericycle (p1 and p2), endodermis (e), and three layers of the cortex (c1 to c3) participate in LRP formation. Procambium tissues, located below the pericycle layers, participate in the formation of future vascular connections between the parent root and the lateral root. Scale = 100 μm. From Dubrovsky JG (1986) Origin of tissues of embryonic lateral root in the cucumber, tissues interactions, and positional control in development. *Ontogenez* 17: 176-189. (English translation from Russian appeared in *Soviet Journal of Development Biology*, Consultant Bureau, New York 17: 119-128); reproduced with the permission of *Ontogenez, Russian Journal of Development Biology.*

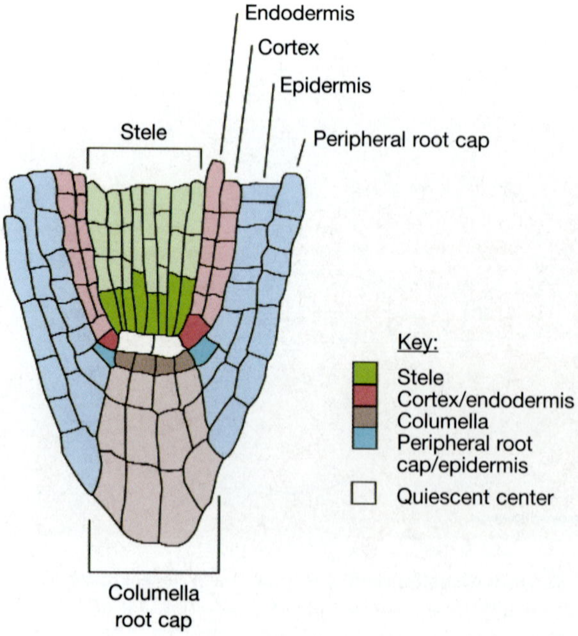

Plate 33 Root Development Organization of the root meristem of *Arabidopsis* showing four groups of initials arranged in three tiers surrounding the quiescent center. For the same shade, initials are shown by bold shading, and their daughter cells by light shading. Redrawn from Figure 2c, Van den Berg C, Weisbeek P, Scheres B (1998) Cell fate and cell differentiation status in the *Arabidopsis* root. *Planta* 205, 483-491, Copyright Springer-Verlag with permission.

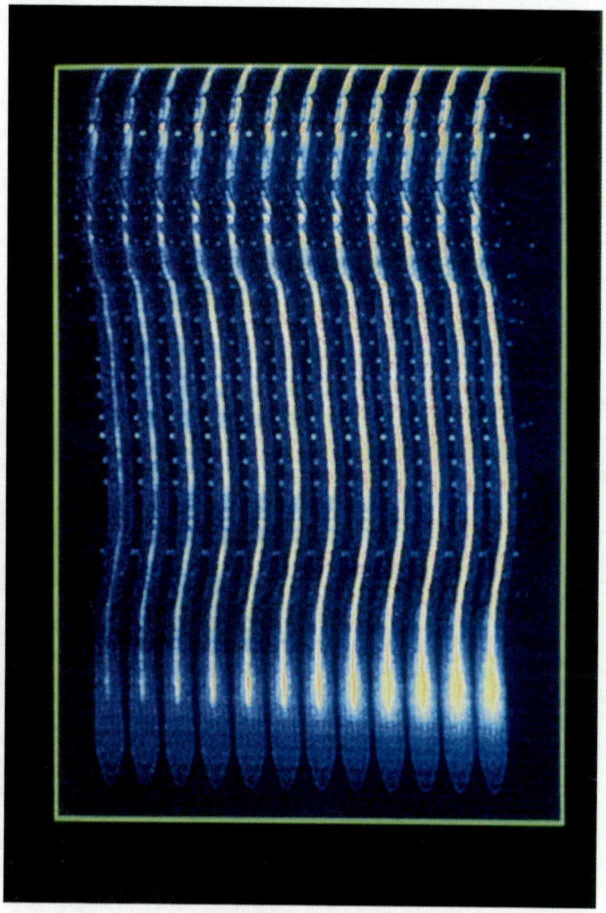

Plate 34 Root Development False-color confocal laser scanning microscope images showing the progression of phloem unloading at the tip of a primary root of *Arabidopsis thaliana*. A membrane impermeant fluorescent dye (5(6) carboxyfluorescein) introduced into the phloem of the leaf was translocated in the protophloem to the growing root tip, from where it moved radially towards the epidermis and acropetally towards the root apex. Most intense unloading was in the zone of cell expansion. Images were taken at 3 min intervals; blue, low concentration; yellow, high concentration. Images first appeared in Oparka KJ, Prior DAM, and Wright KM (1995) Symplastic communication between primary and developing lateral roots of *Arabidopsis thaliana*. *Journal of Experimental Botany* 46: 187-197, reproduced with permission of Oxford University Press and K Oparka.

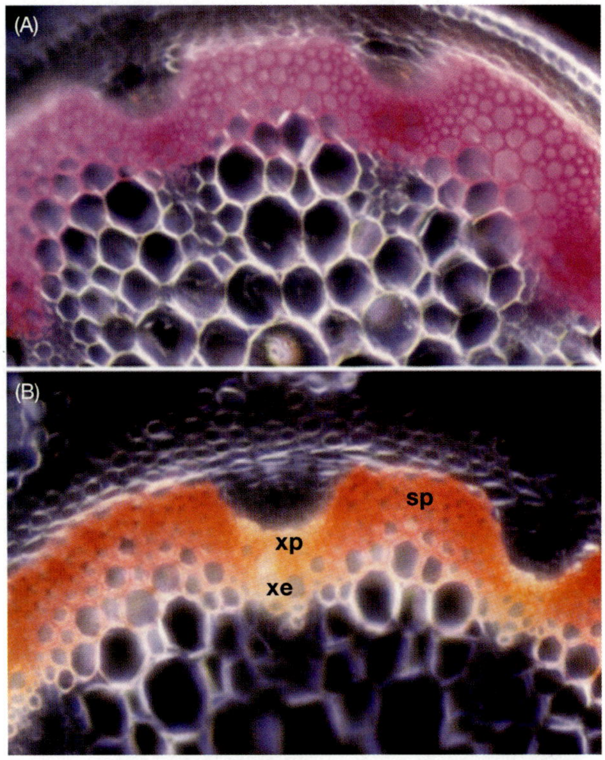

Plate 35 Secondary Products Distribution and monomeric composition of lignin in a herbaceous plant as revealed by histochemical staining. (A) Cross-section of an *Arabidopsis* stem stained with phloroglucinol. Presence of aldeyhyde end groups in lignin produce a characteristic pink color. (B) Cross-section of an *Arabidopsis* stem stained with the mäule reagent. This reagent gives a yellow color with guaiacyl lignins (g lignins) and a red color in the presence of syringyl residues (g/s lignins). Xe, xylem elements; xp, xylem parenchyma; sp, sclerified parenchyma.

Plate 36 Secondary Products A rubber tree, *H. brasiliensis*, being tapped for latex. Courtesy of Linatex Inc., Yately, Hampshire, UK.

Plate 37 Seed Development From the first division to the globular embryo. (A, B, C, D) First three waves of division; (E) octant stage; (F) the globular stage embryo is limited by protoderm (p), the central part (ce) is differentiated in an external ground meristem (gm) and an inner procambium (pc). At this stage gene expression of *ML1* (blue), *WUS* (yellow), *PIN1* (red) and *LEC1* (green) are restricted to cell specific zones. Hy, hypophysis; su, suspensor.

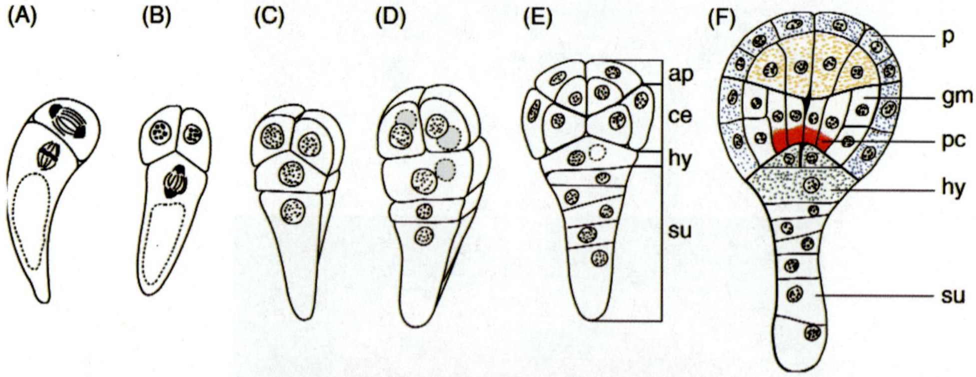

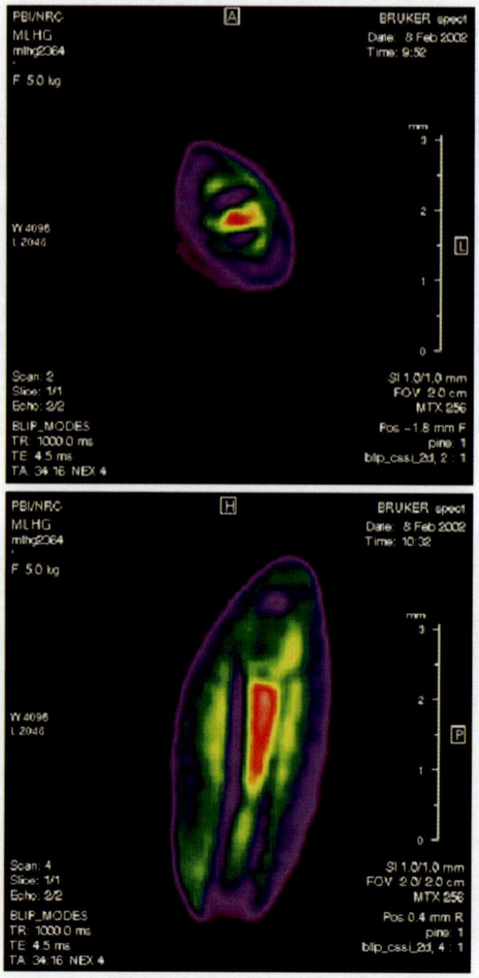

Plate 38 Seed Development NMR showing distribution of lipid in a western white pine (*Pinus morticola*) seed shortly after imbibition. Photograph supplied courtesy of A. Feurtado and M. Grawel, Simon Fraser University and Plant Biotechnology Institute, Canada.

Plate 39 Seed Development Vivipary in mangroves (A) and in mutant (*vp-1*) corn seeds (B). Red mangrove (*R. mangle*) plants grow in intertidal waters. The seeds germinate and grow to lengths of 20-25cm while still on the parent plant. Note a cluster of seedlings with long sword-like roots. Eventually the seedlings fall and become planted in the mud below. (A) Reproduced with permission of Academic Press (Elsevier) from Srivastava LM (2001) *Plant Growth and Development*. San Diego: Academic Press. (B) Reproduced with permission of Cell Press from McCarty DR, Hattori T, Carson CB, Vasil V, Lazar M, and Vasil IK (1991) The *Viviparous-1* developmental gene of maize encodes a novel transcriptional activator. *Cell* 66: 895-905.

Plate 40 Seed Development Seeds of the severe *abi3* mutant of *Arabidopsis* (*abi3-6*) as compared to the wild-type. DAF, days after fertilization. The mutant seeds are green, reduced in their storage reserve content, and desiccation intolerant. Reproduced with permission of Oxford University Press from Nambara E, Keith K, McCourt P, and Naito S (1994) Isolation of an internal deletion mutant of the *Arabidopsis thaliana* ABI3 gene. *Plant and Cell Physiology* 35: 509-513.

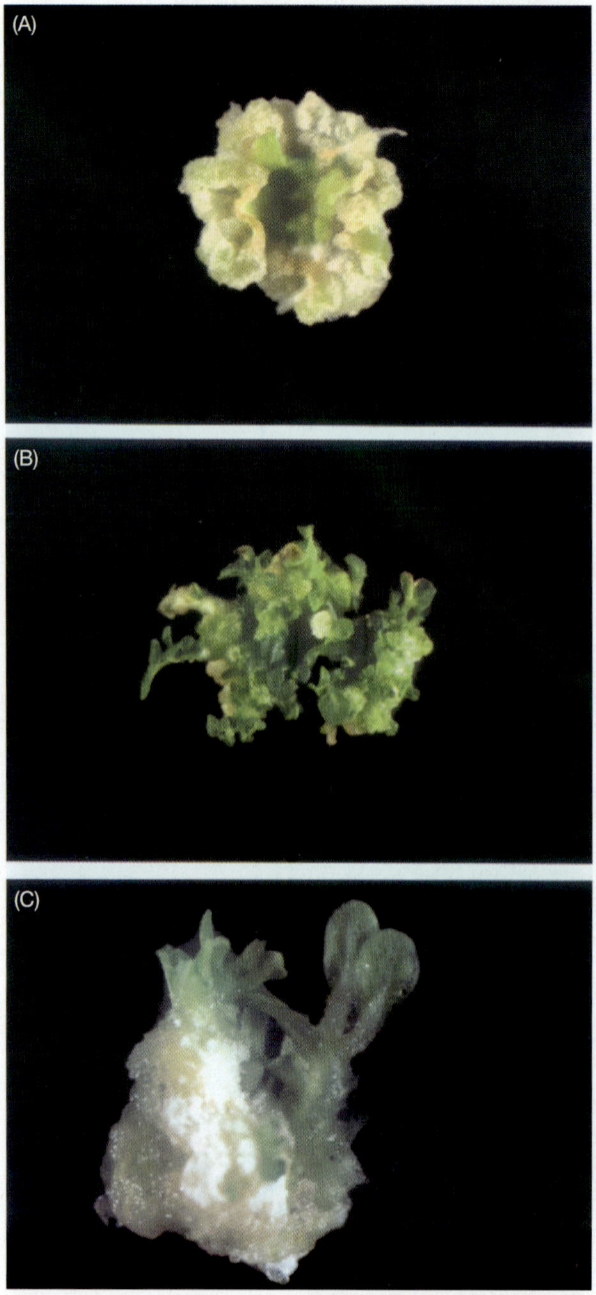

Plate 41 Tissue Culture (A) Callus formation on a tobacco leaf explant in the presence of 1 ppm of each 6-benzylaminopurine (BA) and 1-naphthaleneacetic acid (NAA), after 21 days in culture. (B) Direct bud formation on a tobacco leaf explant in the presence of 2 ppm BA and 0.5 ppm NAA after 24 days in culture. (C) Indirect bud formation from callus on a tobacco leaf explant in the presence of 2 ppm BA and 1 ppm NAA after 30 days in culture.

Plate 42 Tissue Culture Alginate-encapsulated embryos of Norway spruce. Reprinted with permission from Gupta PK, Timmis R, Pullman G, *et al*. (1991) Development of an embryogenic system for automated propagation of forest trees. In: Vasil IK (ed.) *Scale-Up and Automation in Plant Propagation,* pp. 75-93. New York: Academic Press.

Plate 43 Tissue Culture and Plant Breeding Tomato plants differentiating from callus cells grown from cotyledons after 6 weeks of culture in a synthetic medium. At this stage, plantlets can be excised, rooted, transferred to soil in the greenhouse, and grown to the flowering stage.

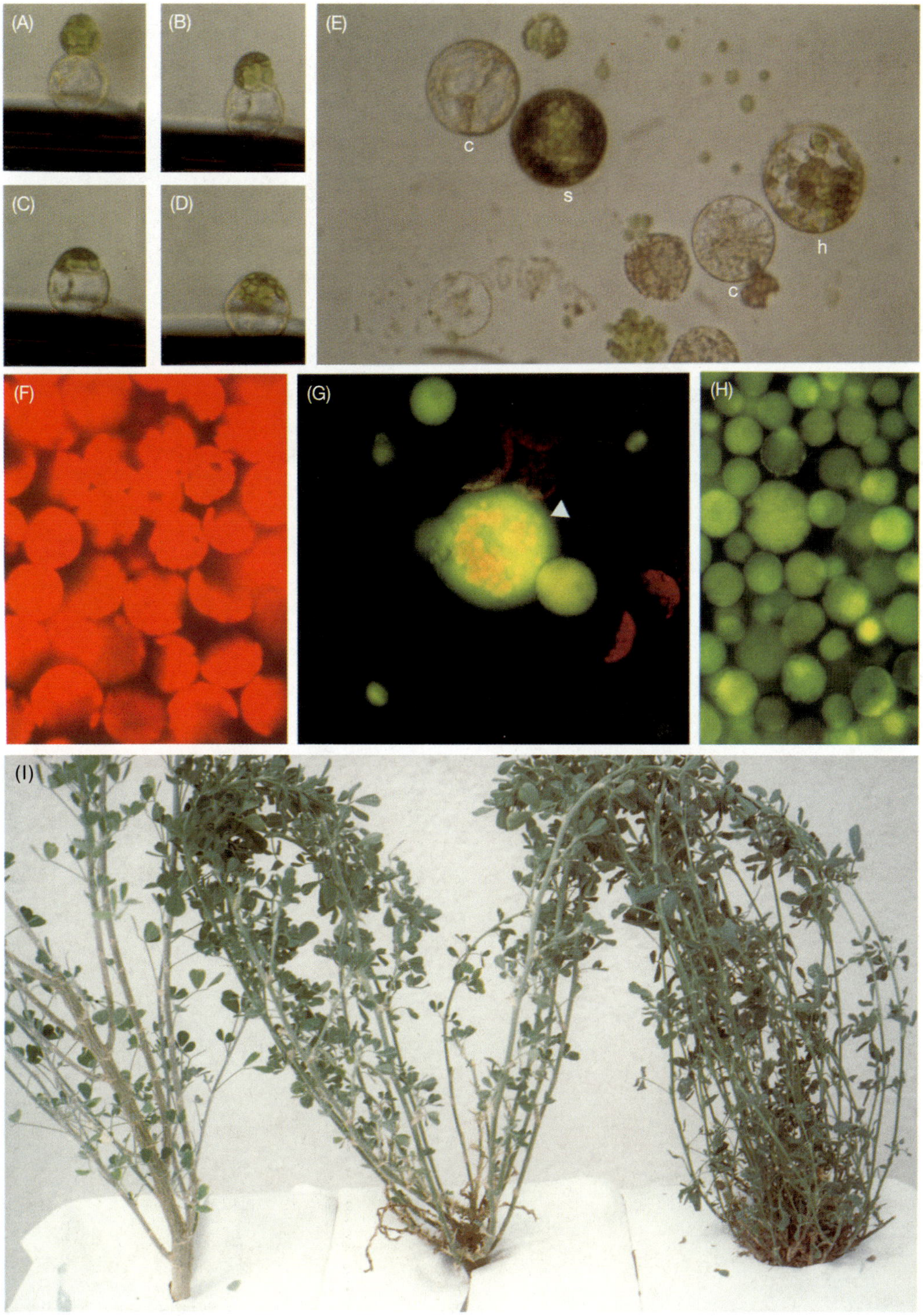

Plate 44 Tissue Culture and Plant Breeding (A-D) The different phases of electrofusion of two protoplasts. (A) The protoplasts are aligned by the application of the AC field. (B and C) The DC pulses induce the breakdown of membranes and (D) The heterokaryon is formed, but the cytoplasms are not mixed yet. (E) Electrofusion products: h – heterokaryon from mesophyll and callus protoplasts; c and s – homokaryons of callus and mesophyll protoplasts, respectively. (F) Mesophyll protoplasts observed under UV light. (G) The arrowhead indicates a heterokaryon derived form mesophyll and callus protoplasts, observed under UV light. (H) Callus protoplasts treated with FITC and observed under UV light. (I) Somatic hybrid plants (middle) between *M. arborea* (left) and *M. sativa* (right) Arcioni S, Damiani F, Mariani A, and Pupilli F (1997). Somatic hybridization and embryo rescue for the introduction of wild germplasm. In: Mckersie BD and Brown DC (eds) *Biotechnology and the Improvement of Forage Legumes.* pp. 61-89. Oxford, UK, New York: CAB International.

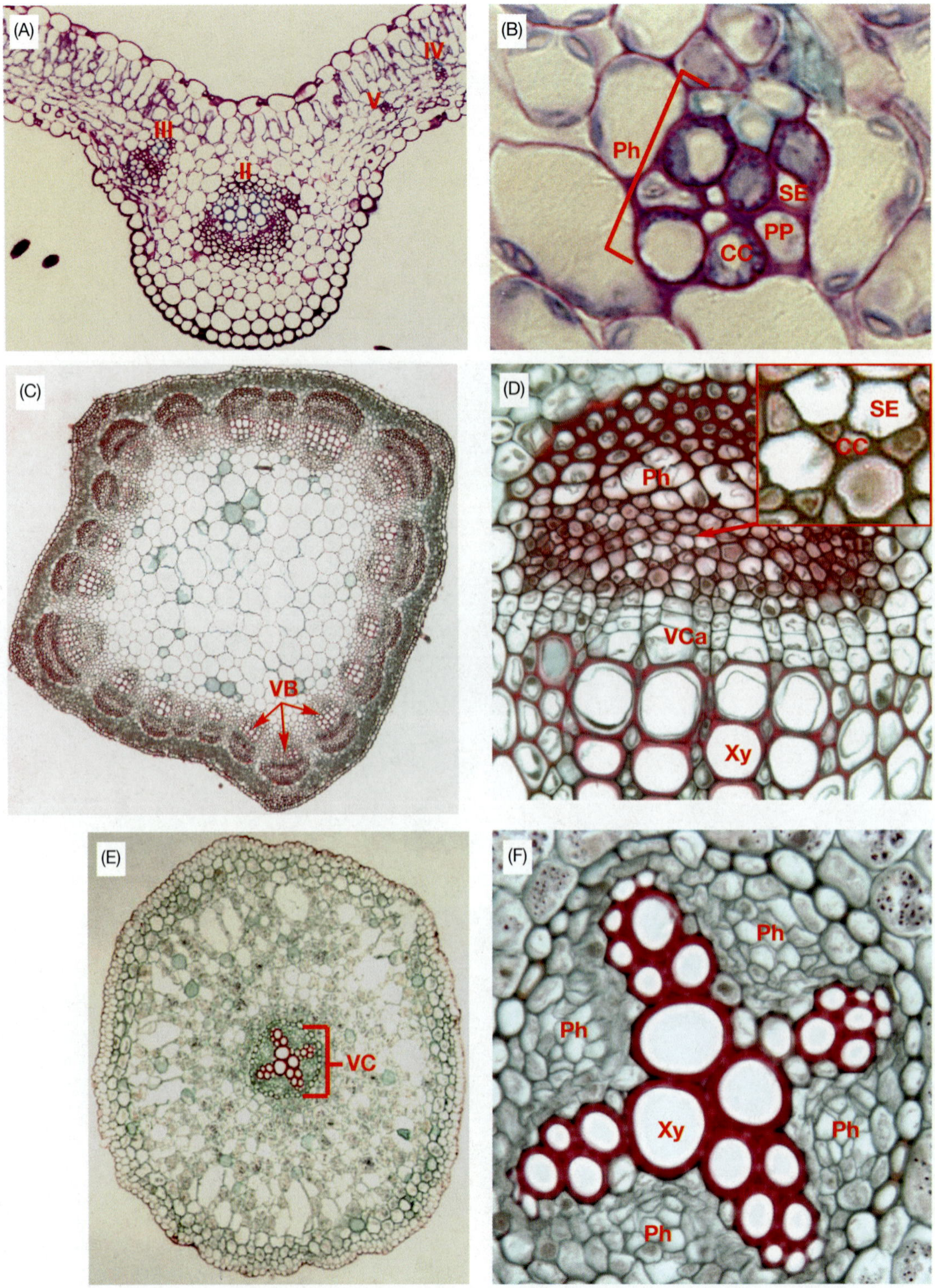

Plate 45 Water Relations of Plants Light micrographs showing the generalized vascular structure in leaves, stems, and roots of many dicotyledonous plants. (A) Cross-section of a leaf showing II, III, IV, and V order veins (x40). (B) Collecting phloem in the IV order minor vein containing the transfer cell type of companion cells: note the cell wall ingrowths (x630). (C) Cross-section of a stem shows the vascular bundles and secondary development of phloem and xylem (x40). (D) Transport phloem in the vascular bundle of the stem contains large sieve elements and small companion cells (x400). (E) Cross-section of a root shows the vascular cylinder located in the center (x40). (F) Phloem and xylem alternate in the vascular cylinder (x400). CC, companion cell; Ph, phloem; PP, phloem parenchyma; SE, sieve element; VCa, vascular cambium; VB, vascular bundle; VC, vascular cylinder; Xy, xylem.

during the dehydration process. Malt from barley is generally used for controlled, uniform amylosis in most fermentation processes. Preharvest sprouting of the barley usually lowers conversion of the malt and extractability of fermentable material and increases the growth of molds.

In addition to the complications outlined above, preharvest sprouting reduces grain yield as a result of carbohydrate respiration, which, at the same time, generates a favorable environment for saprophytic attack.

Physiological Control of Sprouting

As mentioned above, the primary cause for sprouting susceptibility is a low dormancy prior to crop harvest. Consequently, much effort has been channeled into trying to understand the mechanisms through which the duration of dormancy is differentially controlled in genotypes with contrasting behavior to sprouting. The role of plant hormones such as abscisic acid or gibberellins in the control of seed dormancy and germination has long been recognized. Therefore, these plant growth regulators have been suspected from the beginning to be behind the sprouting behavior of cereals (see below). However, other substances have been implicated also in the physiological control of sprouting. The glumes (bracts) of wheat, for instance, contain an unknown inhibitor that delays sprouting and is simply inherited. Similarly, the well-known resistance of red wheats to preharvest sprouting relative to white wheats has been attributed to precursors of the pigment phlobaphene in the testa layer of the former. These compounds (catechin- and tannin-like materials) are known to inhibit germination and occur in lower amounts in white wheats than in red wheats, in which they decline during afterripening to permit germination. Tannin-containing sorghums, though less palatable than white sorghums, are very resistant to sprouting. Pigments in the seed coat may be part of a two-factor system that inhibits germination directly or by interfering with gaseous exchange.

Another aspect that should be considered is the rate of absorption of moisture by kernels. Indeed, water for germination is usually available for limited periods during most sprouting episodes; thus, a high rate of water absorption or the possibility of remaining moistened for a longer time after damp conditions have ended, might contribute to sprouting susceptibility. Absorption of moisture by kernels is influenced by morphology of the inflorescence and characteristics of the seed coat. Imbibition is increased by features associated with awns in wheat and is affected by waxiness, pubescence, and angle of the inflorescence in barley. Also implicated in the control of the rate of water absorption are grain hardness, color, restriction by the seed coat, thickness of the testa and other layers, size, and surface-to-volume ratio.

Hormonal Regulation of Dormancy in the Developing Cereal Grain

The Role of Abscisic Acid (ABA)

Research into the mechanisms of dormancy in the developing seeds of many species suggests a strong involvement of the phytohormone and germination inhibitor abscisic acid (ABA). Indeed, ABA-deficient or ABA-insensitive mutants of *Arabidopsis* and corn (*Zea mays*; maize) precociously germinate and application of the ABA-synthesis inhibitor fluridone has been shown to reduce dormancy in developing seeds of some species. In cereals, the imposition of dormancy on the embryo by the structures that surround it has been suggested to be mediated by the high levels of endogenous ABA existing in the embryos during grain development. ABA content in embryos is usually low until 15 DAP (days after pollination). From then onwards, ABA content increases coinciding with the acquisition of the capacity of the embryo to germinate if isolated from the rest of the grain; hence, one possibility is that precocious germination would be prevented by the surrounding structures from impeding ABA leaching outside the embryo. ABA content has been reported to peak at around physiological maturity and to decline thereafter. However, in contrast to what might have been expected, no correlations have been found between ABA levels in the embryo during seed development and susceptibility or resistance to sprouting. Although inhibition of ABA synthesis (either genetically or using chemicals) has been shown to accelerate the termination of dormancy, susceptible genotypes do not have lower ABA levels during grain development than resistant ones. However, barley appears to be an exception to the general rule and therefore deserves special attention.

In barley, glumellae (the hull) adhering to the caryopsis represents a further constraint for embryo germination in addition to that already imposed by endosperm plus pericarp. This differs from wheat and sorghum where embryo germination is restricted only by endosperm and pericarp. Following the dynamics of the release from dormancy imposed by the different structures surrounding the embryo in grains from a sprouting-susceptible barley cultivar (B1215) and a moderately sprouting-resistant one (Quilmes Palomar), it can be seen that, as expected, embryos from both varieties are able to germinate precociously from early stages of development if excised from the entire grain (**Figure 2A**). In both cultivars, dormancy imposed by endosperm plus testa/pericarp is steadily overcome at a similar rate throughout development (**Figure 2B**). However, although caryopses presented low dormancy from well before physiological maturity (PM, defined as the moment when the grain has attained maximum dry weight), the presence of the hull prevents grain germination prior to that stage. Hull-imposed dormancy started to be removed from PM onwards, with a rate that is different depending on the cultivar: in B1215 grains this restriction is removed abruptly, while in Quilmes Palomar grains, the removal occurs at a lower rate (**Figure 2C**).

Dormancy imposed by the hull is a common feature in grasses and it has been suggested that a high polyphenol oxidase activity existing in the barley hull results in oxygen deprivation for the embryo. The way in which oxygen influences germination of dormant seeds is largely unknown but it has been hypothesized that oxygen concentration might determine the rate at which germination inhibitors are catabolized. This proposition is strongly supported by the fact that the dormancy breaking effect of a strong oxidant like hydrogen peroxide is through a reduction in the endogenous level of the germination inhibitor abscisic acid (ABA). How then can this mechanism operate differentially throughout development and between genotypes presenting contrasting sprouting behavior? Both in the sprouting-susceptible and sprouting-resistant barley cultivars, ABA content in the embryo is usually similar until PM, and maximum ABA content occurs prior to PM (**Figure 3**). However, immediately after PM, a dramatic reduction in embryonic ABA content takes place in embryos from the sprouting-susceptible B1215, coinciding with the abrupt termination of hull-imposed dormancy that takes place in these grains after PM (**Figure 2C**); in contrast, in Quilmes Palomar embryos, ABA content is kept at high levels for longer (i.e., until 43 DAP).

In the light of these results and within the frame of the proposition that the hull impedes embryo germination because it interferes with ABA oxidation (or metabolism) through oxygen deprivation, it could be argued that release from hull-imposed dormancy occurs because a high concentration of oxygen is unnecessary when germination inhibitors (i.e., ABA) are no longer present. However, different sprouting behavior among genotypes is better explained by differences in embryo responsiveness to ABA. The sensitivity of an embryo to ABA is measured as the embryo's capacity to overcome the inhibitory action of a certain concentration of the hormone. In other words, higher concentrations of ABA are required to

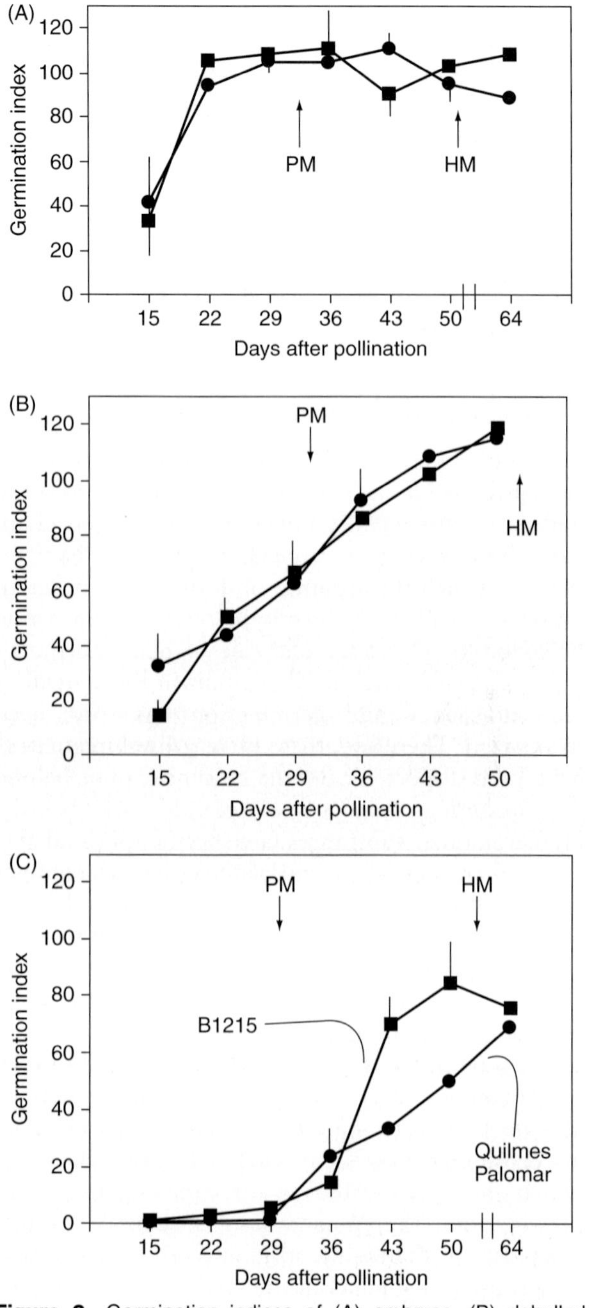

Figure 2 Germination indices of (A) embryos, (B) dehulled caryopses and (C) grains from a sprouting-susceptible (B1215, ■) and a moderately sprouting-resistant (Quilmes Palomar, ●) barley cultivar, harvested on different days after pollination and incubated at 20°C for 12 days. PM and HM are the points at which the crops had reached physiological and harvest maturity, respectively. Reproduced with permission of CABI Publishing from Benech-Arnold RL, Giallorenzi MC, Frank J and Rodriguez V (1999) Termination of hull-imposed dormancy in barley is correlated with changes in embryonic ABA content and sensitivity. *Seed Science Research* 9: 39–47.

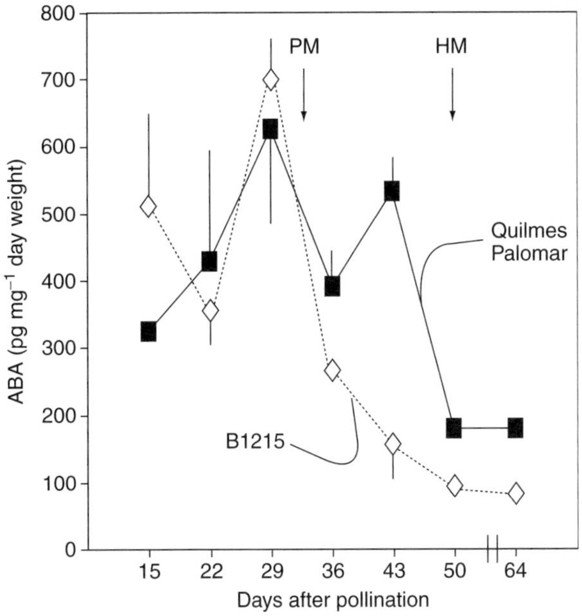

Figure 3 Abscisic acid (ABA) content (expressed as picograms of ABA per milligram of dry weight) in embryos from a sprouting-susceptible (B1215, ◇) and sprouting-resistant (Quilmes Palomar, ■) barley cultivar, harvested on different days after pollination. PM and HM are the moments the crops had reached physiological and harvest maturity, respectively. (Reproduced with permission of CABI Publishing from Benech-Arnold RL, Giallorenzi MC, Frank J and Rodriguez V (1999) Termination of hull-imposed dormancy in barley is correlated with changes in embryonic ABA content and sensitivity. *Seed Science Research* 9: 39–47.)

block germination of isolated embryos from sprouting-susceptible cultivars than those required to inhibit embryo germination of resistant ones. In most cases, differences in the sensitivity of embryos to ABA between sprouting-resistant and sprouting-susceptible cultivars are maintained throughout the whole developmental period. The nature of the low sensitivity to ABA observed in embryos from sprouting-susceptible genotypes remains unclear but it might be due either to a high rate of degradation of the hormone in the outside walls of the embryo or to alterations in the ABA transduction pathway.

The Role of Gibberellins (GA)

The central role of gibberellins (GAs) in promoting seed germination was suggested decades ago. Since then, the identification of GA-deficient mutants of *Arabidopsis* and tomato (*Lycopersicon esculentum*) seeds that will not germinate unless exogenously supplied with GAs has confirmed this hypothesis. It has been proposed that endogenous GAs control germination through two processes: a decrease in the mechanical resistance of the tissues surrounding the embryo and promotion of the growth potential of the embryo (*see* **Seed Development**: Germination), thus antagonizing the effect of ABA. But in addition to its role as germination promoter, it has been demonstrated that the pattern of exit from dormancy in developing cereal grains can be altered by inhibiting GA synthesis, suggesting that this pattern (and consequently the sprouting behavior of the crop) depends on the extent to which ABA action, as a dormancy imposer, is counterbalanced by the effect of GAs. Indeed, application of paclobutrazol, an inhibitor of GA synthesis, almost immediately after anthesis of a sprouting-susceptible sorghum variety, results in a pattern of exit from dormancy that resembles the characteristic pattern of a sprouting-resistant variety, despite the fact that sprouting-resistant genotypes do not have lower GA levels during development (see also for ABA, above). Nevertheless, it could be predicted from this information that lowering GA content through genetic means should result in genotypes with an extended dormancy and, consequently, greater resistance to sprouting.

Breeding for Resistance to Sprouting: Genetic and Molecular Aspects of Sprouting Resistance

Though resistance to sprouting is strongly associated with red grain color in wheat or with the presence of tannins in the testa in sorghum, much of the effort to improve the trait has focused on white wheat or tanninless sorghums. This is understandable since using both red wheat and tannin-containing sorghum would result in sprouting control at the expense of grain quality. Most studies find that resistance of white wheat to preharvest sprouting is a quantitative trait, though other studies indicate that dormancy is controlled by one or two recessive genes. Wheat was found to have three QTLs (quantitative trait loci) that explained more than 80% of the total phenotypic variance in seed dormancy. A major QTL was located on the long arm of chromosome 4A and two minor QTL were on chromosomes 4B and 4D. In sorghum, two unlinked QTLs, *phsE* and *phsF*, were found to influence preharvest sprouting in an F_2 population generated by crossing a sprouting-susceptible variety with a sprouting-resistant one, together accounting for 53% of the phenotypic variance in preharvest sprouting.

Early genetic investigations revealed that seed dormancy in Scandinavian barleys was governed by several recessive, nucleoplasmic loci with high heritability. Genetic control of barley seed dormancy has also been studied by means of QTL mapping. A saturated molecular marker linkage map based on the six-row Steptoe/Morex (S/M) mapping population has been developed and extensively used for QTL analysis by the North American Barley Genome Mapping Project. Steptoe is a six-row feed barley with high levels of dormancy and Morex is a six-row malting type that does not express dormancy. Four regions of the barley genome on chromosomes 1 (7 H), 4 (4 H), and 7 (5 H) were associated with most of the differential genotypic expression for dormancy in the S/M cross. They were designated SD1 to SD4 (SD for seed dormancy) and accounted for approximately 50, 15, 5, and 5% of the phenotypic differences, respectively, in germination following several postharvest periods. Interestingly, comparative maps suggested a homologus relationship between the major QTL of wheat and barley gene SD4.

Breeding for resistance to sprouting is obviously the most effective and economic means of dealing with this problem. As mentioned above, breeding practices often work against dormancy of cereals; however, it should not be very difficult to obtain genotypes with an extended duration of dormancy and, consequently, with resistance to sprouting. Either traditional or marker-assisted breeding could lead to resistant genotypes. However, the possibility of moving release from dormancy to a precise and narrow "time window," as in the case of barley, might require a more complex approach. Studies linking molecular biology with physiology (i.e., identification of candidate genes), appear to be a promising means of achieving such an objective. For example, if genes controlling sensitivity to hormones (either ABA or GAs) are finally identified and their participation in the control of dormancy is eventually known, then efforts should be directed to understanding the regulation of those genes.

The gene *Vp1*, for example, encodes a transcription factor whose involvement in the control of embryo sensitivity to ABA has been shown since the isolation of corn *vp1* mutants that are insensitive to ABA and present viviparity. Preharvest sprouting in cereals is very similar phenotypically to the *vp1* mutation in corn, raising the interesting possibility that preharvest sprouting in cereals is caused, in part, by the physiological disruption of the *Vp1* function. Genes homologous to *Vp1* from barley and other gramineae, such as rice, sorghum and *Avena fatua*, have been cloned and sequenced and, in some cases, close correlations between expression of the gene and dormancy were found; however, in other cases, such an association was not detected. Manipulation of *Vp1* function in those cases where a relationship with sprouting behavior has been found could represent an opportunity for adjusting timing of dormancy release.

Protein kinases often act in the transduction of external signals and could play a role in the effects of environmental conditions on expression of dormancy. For that reason, a protein kinase mRNA (PKABA1) that accumulates in mature wheat seed embryos and that is responsive to applied ABA was cloned and its expression was analyzed during imbibition of dormant and nondormant wheat seeds. When dormant seeds are imbibed, embryonic PKABA1 mRNA levels remain high for as long as the seeds are dormant, while they decline and disappear in embryos of germinating seeds. The role of this kinase in dormant seeds is currently under investigation, but a potential role of phosphorylation-dependent responses in maintenance of seed dormancy is also supported by recent characterization of the *abi1* mutant of *Arabidopsis* (an ABA-insensitive with no dormancy). G-protein coupled receptors can participate in hormone-dependent signaling cascades affecting germination-related genes. Recently, it has been shown that overexpression of GCR1, a G-protein-coupled receptor gene, decreases seed dormancy in *Arabidopsis*. Whether expression of this type of gene is related to dormancy depth in cereal seeds is not yet known, although it is an interesting possibility.

In summary, both genetic and molecular biology studies could be of use in the search for cultivars that are resistant to preharvest sprouting without having a long-lasting dormancy. However, complementarity with physiological studies is important if such a goal is to be achieved. The most profitable genetic investigations would be those that, for example, through QTLs analysis, highlight the participation of genes with known physiological function. If the resulting phenotype happens to correlate well with some characteristic of that gene (i.e., differences between phenotypes in terms of gene expression timing, sequence, regulation, etc.) then the possibilities for manipulating the system are great.

List of Technical Nomenclature

α-Amlyase An enzyme that starts the degradation of the starchy endosperm in the germinating grain.

Abscisic acid (ABA) A plant growth substance, equivalent to a hormone, that controls several growth and developmental processes and inhibits germination.

°C days Units for measuring thermal time, a variable that combines temperature and time effects on plant development.

Dormancy Failure to germinate because of some internal block of the seed that prevents the completion of the germination process.

Gibberellins (GA)	A plant growth substance, equivalent to a hormone, that controls several growth and developmental processes and promotes germination.
Harvest maturity (HM)	When the developing seed has undergone desiccation and has reached between 10% and 12% moisture content.
Physiological maturity (PM)	When the developing seed has acquired its maximum dry weight.
Quantitative trait	A trait with continuous phenotypic variation.
Quantitative trait loci (QTL)	Set of genes or alleles controlling a quantitative trait.
***Vp1* (*Viviparous-1*)**	A gene that, among other processes, controls embryo sensitivity to ABA.

See also: **Regulators of Growth**: Abscisic Acid; Gibberellins. **Seed Dormancy**: Development of Dormancy; Genetics of Dormancy; **Seed Development**: Germination; Physiology of Maturation.

Further Reading

Benech-Arnold RL (2002) Bases of pre-harvest sprouting resistance in barley: physiology, molecular biology and environmental control of dormancy in the barley grain. In: Slafer GA, Molina-Cano JL, Araus JL, Savin R, and Romagosa I (eds) *Barley Science. Recent Advances From Molecular Biology to Agronomy of Yield and Quality*, pp. 481–502. New York: Food Product Press.

Black M (1991) Involvement of ABA in the physiology QJ;of developing and mature seeds. In: Davies WJ (ed.) *Abscisic Acid Physiology and Biochemistry*, pp. 99–124. Oxford: Bios Scientific Publishers Ltd.

Jacobsen JV, Gubler F, and Chandler PM (1995) Gibberellin action in germinated cereal grains. In: Davies P (ed.) *Plan Hormones. Physiology, Biochemistry and Molecular Biology*, pp. 246–271. Dordrecht: Kluwer Academic Publishers.

McCarty DR, Hattori T, Carson CB, *et al.* (1991) The *viviparous-1* developmental gene of maize encodes a novel transcriptional activator. *Cell* 66: 895–905.

Paulsen GM and Auld AS (2002) Preharvest sprouting of cereals. In: Benech-Arnold RL and Sánchez RA (eds) *Seed Physiology: Applications to Agriculture*. New York: Food Product Press.

Steinbach HS, Benech-Arnold RL, and Sánchez RA (1997) Hormonal regulation of dormancy in developing Sorghum seeds. *Plant Physiology* 113: 149–154.

T

TISSUE CULTURE

Contents

General Principles

M Ziv and A Altman, The Hebrew University of Jerusalem, Rehovot, Israel

Introduction

Plant cell and tissue culture techniques involve the culture of cells, protoplasts, tissues, and organs isolated aseptically, on a defined sterile medium under controlled environmental conditions.

Plant *in vitro* technologies are used mainly for the regeneration of organs or somatic embryos for propagation, virus elimination, the transformation and generation of transgenic plants for improvement of plant traits, and for germplasm storage. Additionally, tissue cultures and cell suspension cultures under controlled conditions – especially defined medium composition – are valuable as model systems for basic studies of cell physiology and metabolism. The ability to control metabolic pathways of *in vitro* cultured cells and tissues resulted also in a major practical application, i.e., the production of secondary metabolites – phytochemicals for the food and pharmaceutical industries. Propagation in tissue culture (micropropagation) is used to develop high-quality pathogen-free plants, selected genotypes, or transformed cloned plants. Propagation in tissue culture depends on cell competence and totipotency, that is, the ability of plant cells to re-express their genetic potential, to undergo dedifferentiation and redifferentiation, and to regenerate new plants. The pathway of regeneration may include several steps, expressed either in organogenesis – unipolar organ formation, or somatic embryogenesis – bipolar embryonic axis formation. Micropropagation has an advantage over traditional plant propagation, due to its potential rapid, large-scale multiplication of new genotypes and the generation of pathogen-free plants. Plants for micropropagation are selected for their unique agricultural traits such as disease and insect resistance, product quality, and their ability to tolerate environmental stress conditions.

Plant transformation and somatic cell genetics, making use of tissue culture methods for the generation of new crop plants, have contributed greatly to plant breeding and crop improvement. The outstanding progress and remarkable achievements in plant tissue culture techniques and molecular biology, including embryo rescue, haploid production, somatic cells hybridization, and the introduction of selected genetic traits to cells, are the major *in vitro* systems that enabled production of novel plant crops. Transformation – the introduction of DNA to plant cells – was achieved by using bacterial or viral transformation vectors and mediated gene transfer through electroporation and the biolistic method (i.e., bombardment of cells with microprojectiles coated with the desired DNA). Regeneration of haploid cells and plants from isolated microspores is another important contribution to the production of homozygous offspring for further nonconventional breeding. The isolation of protoplasts provided a system for somatic hybridization through protoplast fusion, which overcomes interspecific crossing barriers. The culture of very young embryos allowed embryo rescue techniques used to overcome embryo abortion in incompatible crossings of several species. Progress in transgenic plant production has been reported for vegetable, flower, and fruit crops as well as for forest trees.

Success in tissue culture depends greatly on the nutrient media, the culture condition, and the plant material used as a source for explants. The nutrient medium consists of minerals, vitamins, carbohydrates, and growth regulators, mainly auxins and cytokinins, at a specific level and ratio. In most cases the isolated plant explants are cultured on a semisolid medium containing a solidifying gelling agent, usually agar. However, liquid cultures, either a floating raft support or in flasks on a rotary shaker, have also been found suitable for growth and a morphogenic response of several plant species. Liquid cultures are used for cell suspension cultures and for propagation in large-scale bioreactor cultures.

Temperature and light quality and duration have either an enhancing or an inhibiting effect on the response of plant tissue *in vitro* and need to be adapted to each plant or tissue requirements. In general, cultures are kept in the dark or under low light intensities.

The physiological status of the tissue has a paramount role in the response of isolated cultured explants. Usually juvenile and young actively growing tissues respond better to hormonal signals in the culture medium than older tissues. Various types of plant cultures are used *in vitro* – shoot tip meristems, buds, roots, callus, cell suspensions, protoplasts, pollen, ovule and embryo cultures – and each type is cultured on a specific nutrient medium and under specific culture conditions.

This article provides a general review of the principles of plant cell and tissue culture and the readers are referred to the appropriate articles in the sections dealing with various growth and developmental aspects.

Laboratory Facilities, Media, and Plant Material

Tissue Culture Facilities and Equipment

The facilities of tissue culture laboratories are organized into three major areas. One area is used for medium preparation and dispensing of nonaseptic media to containers before autoclaving, and for initial plant preparation. The area should contain an autoclave, either a horizontal or an upright model, an analytical balance, pH meter, magnetic stirrers, a hot plate or microwave oven to dissolve the agar, and various membranes for filter sterilization. The second area is used for explant isolation, sterilization, and planting, and for continuous transfer of new or established cultures. The most essential item in this area is a laminar flow hood with a horizontal or vertical air flow, used for the preparation of sterile media components and to dispense the autoclaved or filter sterilized medium to sterile containers as well as for plant explant isolation and subculture. The third area is used for the maintenance of the cultures in growth chambers under controlled temperature, and light quality and duration as well as humidity as specified for each plant culture.

Medium Preparation

Selection of the nutritional components and the level of plant hormones determine the success of growth and development of the cultured cells, tissues, or organs *in vitro*. There is a large variety of formulations and information published in manuals and laboratory protocol books on various media suitable for cells, protoplasts, tissues, and organs.

In general, most nutrient media consist of macro- and microelements, a carbon source, vitamins, amino acids, and growth regulators. Other organic supplements such as plant extracts are added to specific species and for specific purposes. The inorganic salts contain nitrogen as ammonium and nitrate, phosphorus, potassium, magnesium, calcium, and sulfur. The essential microelements include manganese, boron, zinc, copper, molybdenum, and sometimes cobalt. The iron is added in chelated form as ferrous-EDTA to prevent precipitation.

Concentrations of the various components of the most popular media formulated for tobacco callus and for other formulated media appear in **Table 1**. The use of stock solution can greatly simplify the procedure of media preparation. The chemicals are prepared in stock solutions 10 or 100 times the required final concentration, dissolved in distilled water, and used for preparation of 1 or 2 liters by adding the required aliquot of a specific component in the final concentration requirement.

The carbohydrates used are usually sucrose or glucose, whereas fructose is used only for very specific species requirement. Most media contain the sugar alcohol myo-inositol, which was found to be beneficial during cell division and cell wall synthesis.

The growth regulators used in tissue culture belong to four classes of plant hormones – auxins, cytokinins, gibberellins, and abscisic acid as well as growth retardants – and are also prepared as stock

Table 1 Various media used for tissue culture regeneration and proliferation (in mg l^{-1})

Media components	*Murashige T and Skoog F (1962)*	*Gamborg OL et al. (1968)*	*McCowan B and Lloyd GB (1982)*	*White RR (1954)*	*Schenk RV and Hildebrandt AC (1972)*
$(NH_4)NO_3^6$	1650		400		
$(NH_4)_2SO_4$		134			
$(NH_4)H_2PO_4$					300
KNO_3	1900	2500		80	2500
$Ca(NO_3)_2 \cdot 4H_2O$			556	208.4	
$CaCl_2 \cdot 2H_2O$	440	150	96		200
Na_2SO_4	370	250	370	360	400
KH_2PO_4				200	
$NaH_2PO_4 \cdot H_2O$	170		170		
KCl		130.5		16.5	
K_2SO_4			990	65	
$FeSO_4 \cdot 7H_2O$	27.8	27.8	27.8		15
Na_2EDTA	37.26	37.3	33.6	2.5	20
$MnSO_4 \cdot 4H_2O$	22.3				
$MnSO_4 \cdot H_2O$		10	22.3	3.8	10
$ZnSO_4 \cdot 7H_2O$	8.6	2.0	8.6	3.0	1.0
H_3BO_3	6.2	3.0	6.2		5.0
KI	0.83	0.75		0.75	1.0
$NaMoO_3$				0.001	
$Na_2MoO_4 \cdot 2H_2O$	0.25	0.25	0.25		0.1
$CuSO_4 \cdot 5H_2O$	0.025	0.025	0.25	0.001	0.2
$CoCl_2 \cdot 6H_2O$	0.025	0.025			0.1
Myo-inositol	100	100	100		1000
Nicotinic acid	0.5	1.0	0.5	0.5	5.0
Pyridoxine · HCl	0.5	1.0	0.5	0.1	0.5
Thiamine · HCl	0.1	10.0	1.0	0.1	5.0
Glycine	2.0		2.0	3.0	
Sucrose	30000	20000	20000	20000	30000

solutions (**Table 2**). However, inclusion of gibberellins and abscisic acid is relatively limited, and they are only added to cultures when enhanced elongation is required or for the precocious germination and desiccation of encapsulated somatic embryos, respectively. They are dissolved in ethanol, potassium hydroxide, or dimethyl sulfoxide and stored in a refrigerator. They can be divided into small vials for longer storage periods in $-18°C$. Measured aliquots are added to attain a final concentration in 1 or 2 liters of prepared medium. The growth hormones are used at a concentration of 10^{-5} to 10^{-9} mol l^{-1}, depending on the plant tissue and culture purpose. Various tissues may require only auxin, or cytokinin, or both, or neither.

The organic supplements, vitamins such as thiamine, pyridoxine, nicotinic acid, and amino acids such as glycine, are prepared as stock solutions and used according to the required final concentration of the prepared medium. The pH is usually adjusted to 5.6–5.8 with 0.5 mol hydrogen chloride or sodium hydroxide. Agar in the range of 0.6–1% or Gelrite 0.1–0.3% is added to the medium to obtain a semisolid medium after pH adjustment. Prior to autoclaving the medium is heated gently to below boiling point to dissolve the agar and subsequently dispensed to the culture containers.

The media are steam sterilized by autoclaving at 120°C for 15–20 min, cooled to about 40°C, and dispensed to test tubes, Petri dishes, jars, culture boxes, or any other disposable containers. Heat-labile substances are sterilized, by using membrane filters of 0.45 or 0.2 μm pore size, and added to the medium after autoclaving and cooling and before dispensing to the culture containers.

Other organic supplements such as casein hydrolyzate, fruit, and seed extracts are sometimes added to the medium according to a specific formulation requirement. If heat labile, they can be filter sterilized through membrane filters and added aseptically after autoclaving.

Types of Cells, Tissues, and Organs for Explant Isolation

The type of cells, tissues, or organs used in culture is determined by the genotype of the plant and its availability, its physiological and developmental state, and the purpose of the established culture. The source can include shoots, meristem tips, leaves,

Table 2 Growth regulators and the concentration range used in plant tissue culture

Growth regulators	*Abbreviation*	*Concentration* (μ mol l^{-1})
Auxins		
Indole-3-acetic acid	IAA	5.0–15.0
1-Naphthaleneacetic acid	NAA	0.5–50.0
2,4-Dichlorophenoxyacetic acid	2,4-D	0.05–25.0
Indole-3-butyric acid	IBA	0.5–50.0
p-Chlorophenoxyacetic acid	4-CPA; CPA	5.0–50.0
Indole-3-acetyl-L-aspartic acid	IAA-L-aspartic acid	0.05–25.0
Cytokinins		
Zeatin (6-[4-hyroxy-3-methyl-but-2-enylamino] purine *trans* mixed isomers)	Z	0.05–25.0
t-Zeatin riboside	ZR	0.05–25.0
6-benzylamino-9-[2-tetrahydropyranyl]9H purine	BPA	0.05–25.0
Kinetine (6 furfurylaminopurine)	KN	0.5–25.0
6-Benzylaminopurine (N^6-benzyladenine)	BAP or BA	0.5–25.0
N-phenyl-N′ 1,2,3-trihydroxybenzene	TDZ	0.1–15.0
Other growth regulators		
Gibberellic acid	GA_3	0.05–25.0
Abscisic acid	ABA	0.5–50.0
Jasmonic acid (3-oxo-2-[2′-*pn*-tenyl]-cyclopentaneacetic acid	JA	0.05–15.0
Paclobutrazol	PAC	1.0–10.0
Ancymidol	ANC	1.0–10.0

stems, roots, flower organs, seeds, seedlings, and embryos.

Shoot apical meristems of 0.2–0.5 mm are used for virus-free clonal propagation, for direct genetic transformation, and for long-term germplasm storage (cryopreservation). The meristems, isolated from the terminal buds aseptically, are cultured on agar-solidified media with low levels of growth hormones to eliminate callus formation.

Potentially, terminal and axillary buds are the most suitable explant to initiate plant cultures, especially when isolated from young actively growing shoots. The composition of the medium and the level of growth hormones may influence the successful survival and growth of explants after isolation. Changes in the growth hormones balance are required for further proliferation of isolated plant shoots.

Leaves are used as an explant source for callus production, for regeneration of buds directly or indirectly through callus (**Figure 1**). They were the first organs used to isolate plant protoplasts from the mesophyll tissue. Young actively growing leaves can regenerate buds directly at the cut surface of the explant in media with a high cytokinin to auxin ratio (**Figure 2**).

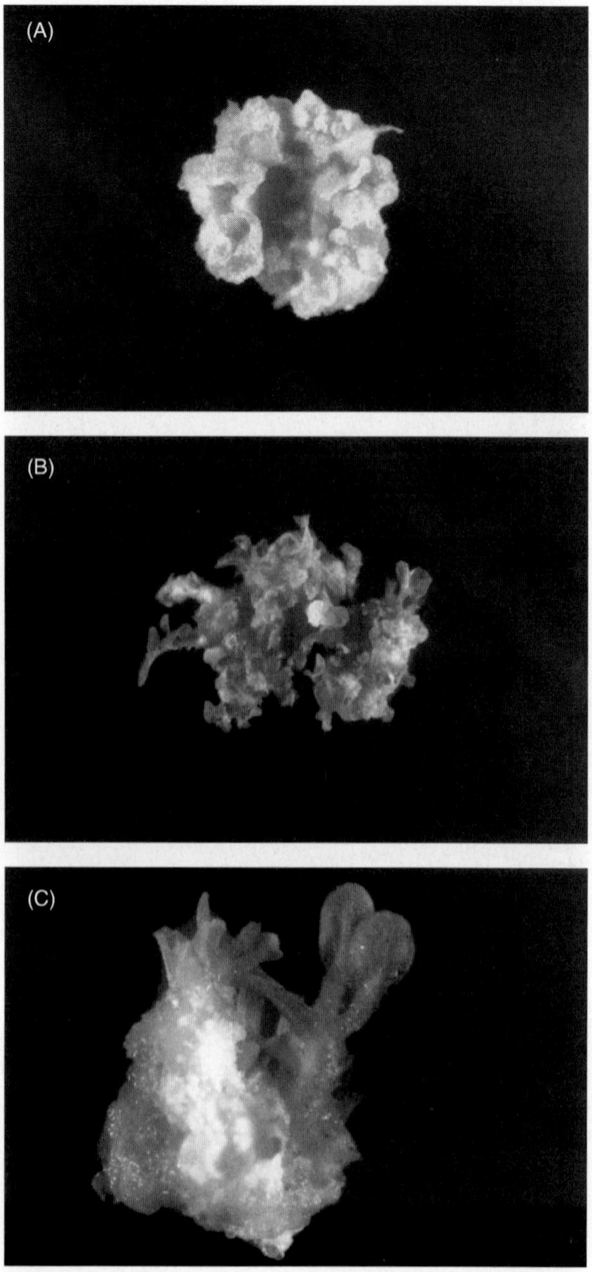

Figure 1 (see color plate 41) (A) Callus formation on a tobacco leaf explant in the presence of 1 ppm of each 6-benzylaminopurine (BA) and 1-naphthaleneacetic acid (NAA), after 21 days in culture. (B) Direct bud formation on a tobacco leaf explant in the presence of 2 ppm BA and 0.5 ppm NAA after 24 days in culture. (C) Indirect bud formation from callus on a tobacco leaf explant in the presence of 2 ppm BA and 1 ppm NAA after 30 days in culture.

Stems are also a major explant source used for callus production from the pith parenchyma cells, or for bud regeneration from the vascular tissue.

Figure 2 Bud regeneration from the basal cut surface of *Ornithogalum dubium* leaf explant in the presence of 1 ppm kinetine (KN) and 0.1 ppm NAA.

Figure 3 Adventituous bud formation from a poplar root explant cultured in liquid medium on a shaker in the presence of BA and NAA after 28 days.

Tobacco (*Nicotiana tabacum*) stems were one of the first organs used for callus production. Other tissues from stems include epidermis and subepidermis tissues, the cambium, and cortex tissues, and their use depend on the specific plant species involved.

Roots isolated from seedlings or from aseptic plants are very easy to culture, with the root tip being the most responsive section. In most plants, however, roots are not responsive to culture manipulation for the induction of organs or somatic embryos, but there are some exceptions, as was shown in poplar (*Populus* spp.) (**Figure 3**). Once transformed by *Agrobacterium rhizogenes*, roots of several species form a root biomass and are used mainly for phytochemical production in liquid cultures in bioreactors.

Flowers are used mainly as a source for the isolation of anthers, pollen grain, ovaries, ovules, and immature embryos. Anthers and ovules are commonly used for haploid plant production and for *in vitro* fertilization. Petals and sepals from young flowers were also used as an explant source.

In plants producing storage organs such as bulbs, tubers, or corms, the tissue of the storage organ, the terminal or the axillary buds are isolated as explants for clonal propagation in plants that are soil contaminated or virus infested through continuous vegetative propagation. Scales in many bulbous plants are easily induced to regenerate buds as was first reported in *Lilium*, and the pith parenchyma cells from tubers are used for the initiation of callus cultures.

Seeds and seedling are also used as an explant source or to obtain aseptic embryonal axis tissue for the isolation of hypocotyls, cotyledons, and epicotyls.

Plant Material and Explant Establishment

The physiological state of the plants from which the fragments are isolated, the organ from which the explant is excised, the type of explant, and the pretreatment of the plant organ or the tissue determine the success of the explant response in culture. Actively growing young tissues usually respond best in culture.

The organ or tissue is removed from the seedling or from the plant stem, root or leaf in a laminar hood, under aseptic conditions. It is washed in running water, surface sterilized by immersion in 70% ethanol for 30–60 s, followed by immersion in a disinfectant containing 1.5–3% active sodium hypochlorite, or with a milder solution of 9–10% calcium hypochlorite. The duration of sterilization varies between 10–30 min depending on the plant material and the type and concentration of the disinfectant. Often a wetting agent such as Tween 20 is added to the disinfectant to a final concentration of 0.01% (1–2 drops l^{-1}). After surface sterilization the plant material is rinsed in sterile distilled water three or four times in the laminar hood.

The size of explants varies from 0.1–0.2 cm for apical meristems, up to 0.5–1.0 cm for leaf sections or disks from stem, tuber, or root tissue. Usually, explants are cultured with their normal polar direction, but are sometimes inverted for better contact of the tissue with the medium. In explants isolated from leaves, the lower epidermis is placed in contact with the medium since its wax content is lower and it has a higher number of stomata that are potentially more competent than normal epidermal cells. After excision, the initial phase of plant culture is carried out in small containers with single explants on agar-solidified medium, to control and prevent the spread of contamination.

After the establishment phase in culture, viable responding tissues that show initiation of cell division and growth are subcultured to larger containers or to liquid shake cultures for further growth and proliferation.

Isolated explants and plant cultures at various phases and stages of development are kept in growth chambers under controlled temperature (21–25°C), either in the dark or in a 16/8 light–dark photoperiod with a photon flux density of 60–80 micromoles.

The physiological state of the plant material, which varies according to the age of the plant and the timing of isolation, is a major determinant in the success of the explant response in culture. The physiological state depends on whether the tissue is juvenile or mature, active or dormant, and at the vegetative or the reproductive stage. These factors determine the internal hormonal balance of the tissue and affect the plant material response with cell division initiation on the cut surface of the explant and at advanced stages of dedifferentiation.

The internal hormonal balance determines the level of added external hormones and the types of regenerating organs – roots, leaves, meristematic centers, flower organs, etc. The physiological state of plant material can be manipulated prior to explant isolation by growing or storing plants under specific controlled environmental conditions, mainly temperature and photoperiod. In recalcitrant plants, pretreatment of the source plant material can enhance their response to hormonal signals in the medium.

Callus and Cell Suspension Cultures

Induction of growth in an isolated explant takes place at the peripheral layers of the wounded cut surface and involves activation of the cells in preparation for cell division. The inner cell layers, underneath the damaged cells, dedifferentiate and return to the meristematic state. Removal of an explant and wounding the tissue was shown to induce an initial response in the cells associated with ethylene evolution and with an enhanced metabolic activity and cell division. Increased metabolic and mitotic activity result in the formation of callus tissue made of parenchymatic cells. Cell division was shown to be closely related to auxin addition, with 2,4-dichlorophenoxy acetic acid (2,4-D), one of the most effective auxins for callus induction in many plant species. The cells continue to divide, forming several layers of new cells that surround the inner older cells nearest to the original explant tissue as can be seen in a carrot (*Daucus carota*) root explant (**Figure 4**).

Callus tissue from the original explant can be continuously subcultured to fresh agar or liquid medium. The callus can be maintained on the same medium or transferred to a new medium with a different ratio of auxins to cytokinins to promote redifferentiation, formation of meristematic centers and somatic embryos, or organ development under appropriate manipulation of the medium.

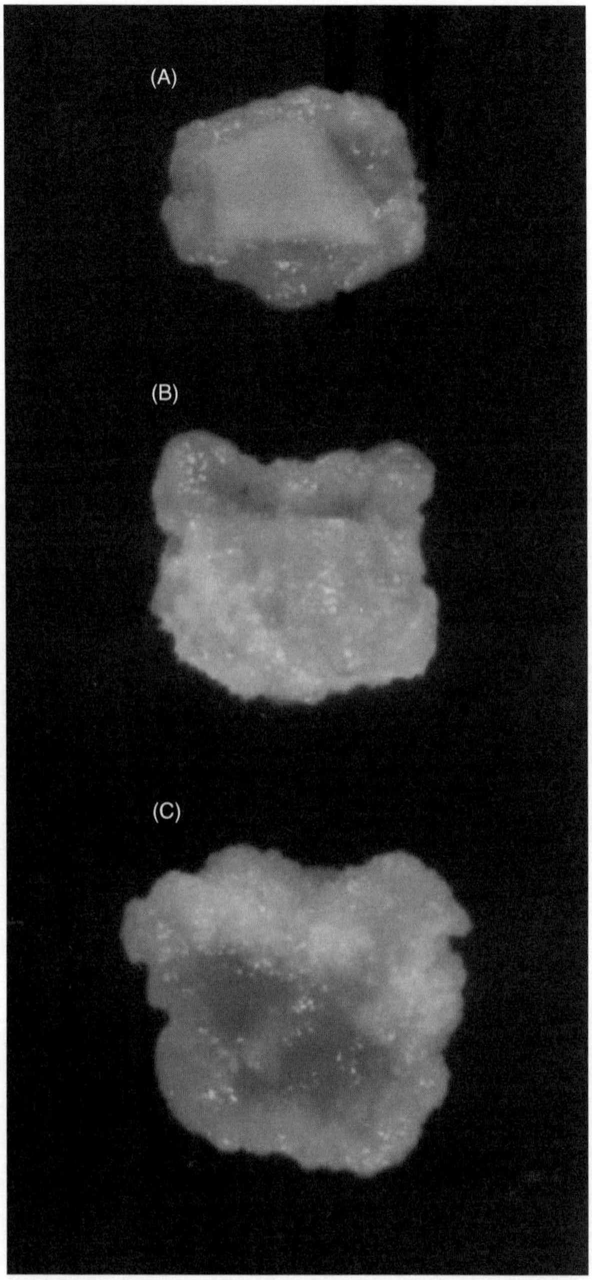

Figure 4 Stages in callus formation in a carrot root explant in the presence of 2 ppm 2,4-D and 0.5 ppm BA after (A) 10, (B) 18, and (C) 26 days in culture.

Callus fragments can be transferred to liquid media in flasks on rotary shakers. Cell suspension cultures are used for production of secondary metabolites and the induction of somatic embryogenesis or organogenesis from meristematic callus clusters.

The continuous agitation promotes the breakdown of the larger clumps of callus into smaller ones, forming a suspension of small aggregates and singulated cells. Serial filtering and sizing of the cell suspension provides a more homogeneous starting culture and can be a better source for selection of cell lines having a more rapid growth rate, efficient metabolite production, or specific selected traits.

Cells in suspension undergoing redifferentiation to somatic embryos are first induced to form pro-embryogenic clusters by auxin, often without or with very low levels of cytokinin. Upon removal or lowering of the auxin level the proembryogenic clusters develop somatic embryos, as was first shown in carrot cell suspensions and later in many other plant species. Callus tissue kept in the dark remains white or yellow, but in the light it becomes green and develops chloroplasts. Often, in long-term cultures, callus tissue may lose the ability to develop chloroplasts and chlorophyll and remain etiolated.

Plant Protoplasts and Somatic Cell Hybridization

Protoplasts are isolated cells produced by removing the surrounding cell wall either by mechanical means or by the use of cell wall degrading enzymes. Removal of the cell wall leaves the protoplast surrounded by the plasmalemma membrane. Protoplasts are isolated mainly from mesophyll cells of leaves, from callus or suspension cell cultures and from pollen grains. Isolation and maintenance of the plasma integrity is mandatory for protoplast viability and isolation efficiency, and is carried out in a specific isolation medium. The initial stage of isolation requires plasmolysis of the cells by a suitable osmoticum, and in enzymatic isolation, a mixture of cell wall-degrading enzymes. The solution contains mannitol as an osmoticum and a mixture of cellulases, hemicellulases, and pectinases.

Following isolation the protoplasts are washed with a decreasing osmoticum solution, centrifuged, and resuspended in a cell wall-regenerating medium. Isolated intact protoplasts undergo cell wall regeneration followed by rapid cell division and formation of small cell colonies. The cell colonies are subcultured to a new agar medium to form a callus, or to liquid medium for the establishment of cell suspension cultures. These can be further manipulated by growth hormones to regenerate organs or somatic embryos that are originated from single cells in the proembryonic clusters.

Although no longer a major field in tissue culture, protoplasts research served for over 30 years to study cell wall development, and was used mainly for somatic cell hybridization. Somatic cell hybridization aimed at overcoming crossing barriers and incompatibilities in breeding programs for the improvement of plant crops.

Fusion of protoplasts requires an optimal medium and culture condition to overcome the negatively charged membrane's repulsion. It is carried out in a positively charged medium or by electrofusion in alternating electrical fields. The media for fusion of protoplasts includes polyethylene glycol, a high level of Ca^{2+}, and a pH ranging from 5.6 to 6.6. Following fusion the protoplasts are washed, resuspended in a medium with a lowered osmoticum, and induced to undergo cell wall regeneration and cell division.

Protoplast fusion can be used for cytoplasmic hybrids (cybrids) formation, when two protoplasts fuse one with an enucleated cytoplasm and the other contains the nuclear genetic material from one parent only.

Selection of somatic hybrids depends on the types of fused protoplasts, the use of various markers, or on fluorescence labeling. Somatic hybrid cell colonies can be further induced to form organs or somatic embryos by subculture to regeneration media.

Anther and Pollen Culture

Tissue culture methods provide a suitable system for the production of haploid plants, using microspores, uninucleated pollen cells, or young anthers for culture. The pollen gametophyte is a totipotent cell and can be induced to divide and form haploid somatic cells *in vitro* inside the intact anther or after isolation. The cell colonies can be used for the production of doubled haploid homozygous lines after the doubling of the chromosomes and are used for breeding programs. *In vitro* regenerated doubled haploid plants are used for production of isogenic lines in breeding programs of several crop plants.

The use of isolated microspores proved to be a better system for haploid production since anther tissue cells often begin to divide and cannot be separated from the microspore-derived haploid cells. Haploid plants have been produced from several plant genera, from microspores isolated at the optimal developmental stage, which were induced to divide, form callus, and undergo regeneration through organogenesis or somatic embryogenesis. The culture medium used is similar to the basic medium used in tissue culture, with auxins, cytokinins, and organic additives. Often amino acids and activated charcoal (0.25–0.5%) are added according to the specific requirements of the plant species to enhance cell division.

The physiological state of the donor plant, as determined by environmental and intrinsic factors, affects the microspores' response to hormonal signals in culture and can determine their morphogenetic pathway. Cold pretreatment of flowers or the stamens in the Solanaceae and Graminaceae was found to improve the response of the isolated microspores to culture conditions and to increase the efficiency of callus formation and regeneration.

Morphogenesis and Organization Aspects in Culture

The potential of almost all cells to express totipotency and regenerate an entire plant is one of the fundamental assets of plant tissue culture. Isolated cells and tissues undergo dedifferentiation and redifferentiation through manipulation of various factors in the culture medium. Cells undergoing dedifferentiation return to a ground stage and can be induced to redifferentiate to new cells, tissues, organs, and a whole plant.

Plant regeneration can follow one of two pathways, organogenesis and somatic embryogenesis. In many plant species both pathways can be induced by the manipulation of the media components. In some species only organ formation was observed without somatic embryo induction. The development of a unipolar structure after dedifferentiation culminates in the formation of an organ, a shoot, root, leaves, or flowers. Organ formation directly on the explant is often associated with vascular and connective tissue that exists in the original plant fragment.

In a medium with equal amounts of auxin and cytokinin, callus formation is observed. Lowering the cytokinin in relation to auxin level induces root formation while increasing cytokinin to auxin ratio usually leads to the initiation of shoot primordia. Somatic embryogenesis occurs when a bipolar meristem develops into a proembryo and a somatic embryo with an embryonal axis, consisting of root and shoot meristems. The continued development of proembryos through the various embryonal stages is similar to the development of zygotic embryos. Somatic embryos were shown to originate from a single but not an isolated cell. These cells are often located in a small cluster of cells connected to neighboring cells by plasmodesmata.

Formation of both roots and shoots on the same explant is often observed in long-term cultures with the changes in hormonal ratio can be seen in *Narcissus* (**Figure 5**).

The morphogenetic pathway, either organogenesis or somatic embryogenesis, is determined by the presence of competent cells capable of expressing totipotency. Interaction between intrinsic inductive signals, present in the isolated tissue, and extrinsic signals in the medium and the microenvironment, determine the organization and morphogenesis of the induced cells. Removal of existing gradients within the tissue after isolation and the formation of new gradients are part of the induction and determination phase and provide the signals for the specific morphogenetic pattern. The two pathways that occur, directly on the isolated tissue or indirectly from the callus tissue, are presently used commercially for micropropagation of herbaceous and woody plant species.

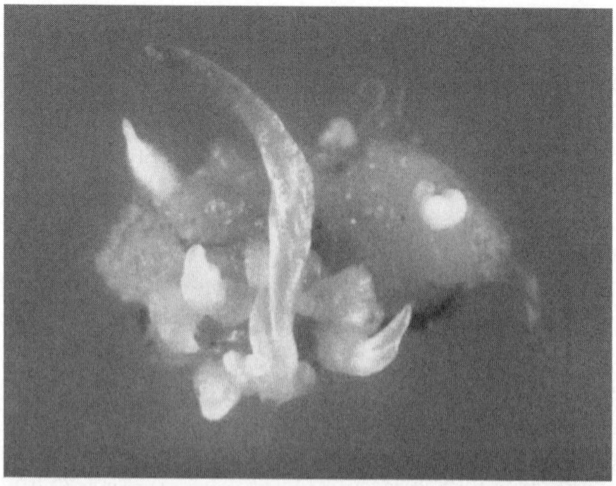

Figure 5 Development of buds and a root from a *Narcissus* bulb scale explant in response to changes in hormonal ratio in the medium after 35 days in culture.

Applications of Plant Tissue Culture

Micropropagation

Micropropagation is the first and major commercial application of tissue culture techniques. It is currently used for a large variety of herbaceous and woody plant species including forest trees, through enhanced axillary bud formation, organogenesis, and/or somatic embryogenesis.

In vitro propagation of pathogen-free, elite, selected, or recalcitrant genotypes is carried out in four distinct stages: (1) explant establishment stage, (2) regeneration and proliferation stage, (3) acclimation and rooting stage, and (4) the final transplanting *ex vitro*.

The initial stage involves the introduction of aseptic plant material to prevent contamination. The plant material is surface sterilized and cultured singly in small culture containers, usually test tubes containing the selected medium with a low level of growth regulators. Virus elimination, carried out at this stage, involves virus indexing, thermotherapy, and chemotherapy using the meristematic dome tissue. Shoot tips and axillary buds are the most common explants used at the initiation stage of micropropagation since they contain existing buds. In other types of explants – leaves, roots, or stem tissue – the regeneration involves adventitious bud formation through dedifferentiation and redifferentiation and requires careful examination of the genetic stability of the cultures to prevent somaclonal variation. Direct adventitious bud formation from meristematic clusters, seen in fern leaves (**Figure 6**), is a more stable pathway and the buds can be subcultured for four or five generations with limited variation.

At the second stage the established explants are subcultured to a medium with high cytokinin and a lower auxin level to enhance bud production and shoot proliferation. This stage is repeated in several cycles to increase the shoot biomass, in either agar or liquid cultures. Large-scale liquid cultures have been used for the proliferation of several ornamental, vegetable and forest plants to overcome manual handling and introduce automation technologies. Liquid medium in 2–5-liter bioreactors enhanced proliferation in several plant species, resulting in a plant biomass 5–10-fold of the initial inoculum after 30 days in culture (**Figure 7**). The problem of hyperhydricity (abnormal shoot and leaf structure) is encountered mainly in liquid cultures, but also in agar media. This is manifested in glossy vitreous leaves (**Figure 8**) lacking normal mesophyll

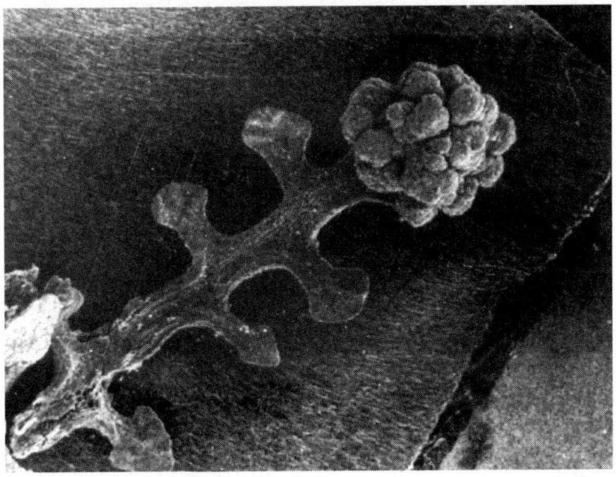

Figure 6 Meristematic bud clusters on a fern leaf cultured in liquid medium in the presence of 2 ppm KN after 24 days.

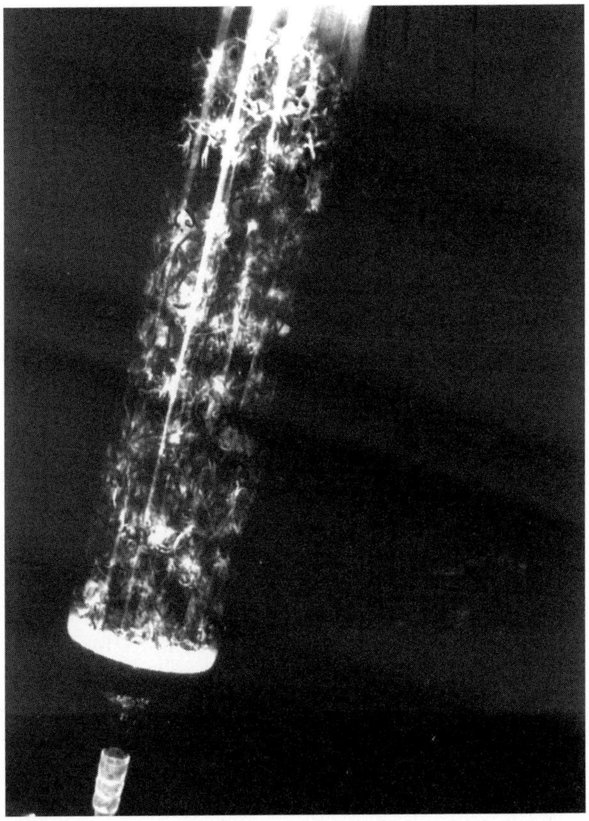

Figure 7 Bioreactor culture of *Ornithogalum dubium* after 24 days in culture.

Figure 8 Carnation (*Dianthus caryophyllus*) shoots with (a) hyperhydric and (b) normal leaves. Reproduced with permission from Ziu M (1986) *In vitro* hardening and acclimation of tissue cultured plants. In: Witheres LA and Alderson PG (eds) *Plant Tissue Culture and its Agricultural Application*, pp. 187–196. London: Butterworths.

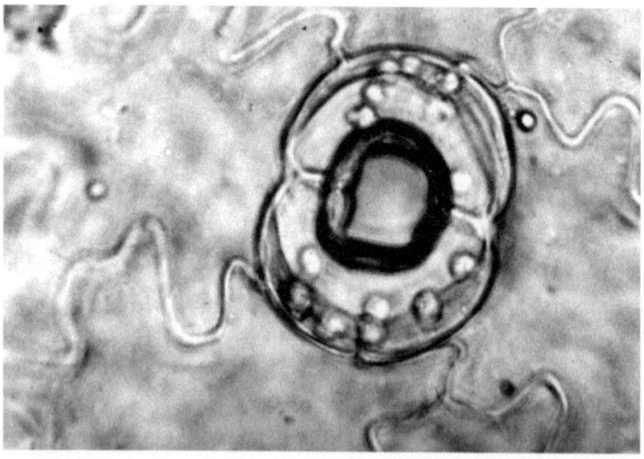

Figure 9 An abnormal stoma with a very large pore from a hyperhydric carnation leaf. Reproduced with permission from Ziv M (1991) Quality of micropropagated plants – vitrification. *In vitro Cell Biology* 27: 64–69.

development, with distorted cuticlar and stomatal structure (**Figure 9**).

At the third stage the shoots are acclimated on a hardening medium, usually without or with a low level of cytokinin, and with lower sucrose and higher auxin levels for root induction. The regenerated plants are hardened for transplanting *ex vitro* (**Figure 10**).

At the fourth stage the plants are removed from the medium, washed, transplanted to an aseptic soil mixture, and grown under mist in a controlled temperature and humidity environment, to prevent leaf desiccation. After establishment the plants are exposed to an environment with higher light intensity and lower humidity for normal growth.

Figure 10 Carnation plant with roots on a hardening–rooting medium in the presence of 1ppm indole-3-acetyl-L-aspartic acid (IBA) after 21 days in culture. Reproduced with permission from Ziu M (1986) *In vitro* hardening and acclimation of tissue cultured plants. In: Witheres LA and Alderson PG (eds) *Plant Tissue Culture and its Agricultural Application*, pp. 187–196. London: Butterworths.

Plant Transformation and Genetic Manipulation *In Vitro*

Transgenic plant production, using tissue culture methods to introduce new genes into plants, was first established in tobacco protoplasts and several other Solanaceae species that served as model plants. Transformation, using various organs, has since been extended to almost all plant genera both dicotyledons and monocotyledons. Genetic modification, using vector-dependent or vector-independent methods, has been achieved in vegetable, ornamental and fruit tree crops, and in forest plant species.

The two major technologies, *Agrobacterium*-mediated DNA transfer and microprojectile accelerated particle bombardment, are based on the use of selectable markers, reporter genes, and various molecular biology technologies to identify the transformed plants.

The *Agrobacterium tumefaciens* tumor-inducing (Ti) plasmid is the most common and widely used method for plant transformation. The plasmid is first disarmed of the tumor-inducing T-DNA and then cloned with the selected foreign genes and one or several markers. Various plant explants from leaves, stems, roots, embryos, apical meristems, and seed hypocotyls are co-cultivated with the bacteria and incubated for 24–72 h. The explants are then transferred to a selection and regeneration medium for 3–4 weeks. The regenerated buds or shoots are transferred to a selection medium and the surviving transformants are tested for the presence of the insertion, an antibiotic or a herbicide marker, or tested for a reporter gene. In addition, the transformation event is verified by Southern blot analysis or polymerase chain reaction (PCR), used to ensure the presence of the introduced genes. Transformed plants originating from buds are then subcultured to a rooting medium before they are transferred to the greenhouse. Transformed somatic embryos are hardened prior to transplanting.

The second, more efficient, and advantageous method for transformation makes use of particle-mediated bombardment with a microprojectile gun.

The particle gun has successfully transformed both dicotyledons and monocotyledons. The apparatus accelerates metal particles coated with DNA into plant cells by using meristematic tissue or embryogenic callus. The bombarded tissue is transferred to a regeneration medium with a selectable marker to test the targeted plants for the insertion of the genes by the above-described methods.

Tissue culture techniques have contributed vastly to transformation and genetic improvement of plants by expanding the potential of nonconventional breeding for the introduction of new traits into crop plants.

Genetic selection and improvement of plants *in vitro* has also benefited from somaclonal variation, either spontaneous due to genetic instability in cultured cells, or induced by selective pressure. Somaclones isolated to date include stress and disease resistant mutants, and early and higher yielding crops and grains with higher protein content. These traits were found to be heritable, but were often either stable or unstable after one or more generation. Understanding of somaclonal variation will contribute to potential use of *in vitro* selection for improved agronomic traits. Somaclonal variation can sometimes interfere in clonal propogation, due to off types regeneration.

Secondary Metabolite Production

Plants are a source of innumerable phytochemicals used in the pharmaceutical, food, and cosmetic industries. The potential of plant cells in culture for the production of secondary metabolites for medicinal and food purposes was first established in cell suspension and callus cell aggregates in shake cultures or in bioreactors. However, the level of most of the secondary metabolites produced *in vitro* is low compared to their level *in vivo*. Improving metabolite production from cell suspension cultures by selection of high-yielding cell lines has been achieved in several species. The use of organs, hairy roots, or bud clusters, being organized and differentiated structures as in organs producing metabolites in intact plants, was established recently as an alternative more efficient culture system. The cells or organs are cultivated in liquid medium agitated in shake cultures or by aeration mixing in impeller or airlift bioreactors.

The production of secondary metabolites in cell cultures, as in intact plants, is usually compartmented in the vacuole or in cell organs, requires the destruction of the tissue for the extraction of the product, and makes the system costly, with the eventual loss of the biomass. Metabolites diffusing into the medium, or enabled excretion of the products, provide a more economical *in vitro* system. Following entrapment and immobilization of cells in a resin, foam, hollow fiber gels or a polymer, and increasing cell membrane permeation, the metabolite is induced to diffuse into the medium. The extraction is conducted either when the medium is changed or under continuous flow conditions.

Cell cultures are also applied for biotransformation using a specific precursor and enzymes associated with a specific phase of a metabolic cycle for the production of a target molecule. This is usually applied to metabolic reaction that is biologically but not chemically feasible.

Germplasm Storage

Tissue culture methods are utilized for genetic conservation of rare, unique, endangered, and selected plant material and for controlled genetic stability. These cultures are preserved as cells, calluses, embryos, buds, shoot tips, ovules, and anthers, stored, and regrown again. *In vitro* conservation can be either under growth-limiting conditions for short-term storage or cryopreservation for long-term storage.

The maintenance and continuous transfer of cultures for an extended period requires the conservation of stock cultures. Short-term conservation can be maintained under limiting growth conditions, including low temperatures and growth inhibiting conditions. Cultures kept at 5–8°C and sealed to prevent dehydration of the tissue have a low metabolic activity, reduced cell division, and slow growth, and can be kept for about 8 to 12 months.

Growth-limiting conditions in a medium with a high osmoticum, usually 3–5% mannitol or growth inhibitors including abscisic acid, cycocel, paclobutrazol, or other growth retardants, slow proliferation and growth of the tissue. Limiting growth conditions can be imposed by a combination of both reduced temperature and inhibiting growth regulators.

Storage of germplasm in liquid nitrogen at ultralow temperatures (i.e., cryopreservation) has been achieved successfully for the long-term conservation of meristems and shoot tips, calluses, cell suspensions, and embryos.

Cryopreservation involves cooling of the plant material inside a small tube of 2 ml capacity with a cryoprotectant solution of 0.5–1 mol dimethyl sulfoxide, 1–1.5 mol ethylene glycol or propylene glycol, and 1–2 mol sorbitol or sucrose, depending on the plant tissue. The cooling, in a stepwise freezing procedure of 0.5–1°C min^{-1}, should reach a subfreezing temperature of -40°C, prior to freezing in liquid nitrogen at $-196^\circ C$. The frozen cultures are stored at -80 to $-100^\circ C$ or in liquid nitrogen to prevent ice formation and damage to the tissue.

The cryopreserved tissue is thawed rapidly to prevent ice formation, by first dipping the plant tissue in a $40^\circ C$ sterile water bath. After thawing the tissue is washed gently from the cryoprotectant by incubation for a short period in a suitable medium at low temperature, or after several subcultures in a semisolid medium. It is then transferred to a semisolid agar medium for regrowth and further regeneration and proliferation.

Conclusion

The integration of plant cell culture and molecular biology provides remarkable advancements to the understanding of plant growth, development, and regeneration and to the production of transgenic plants with improved agronomic traits.

Tissue culture biotechnology has become an economically viable enterprise for plant micropropagation through improved culture media, advanced culture methods, and protocols for automated scaled-up propagation. This is complemented by the production of transformed new plants resistant to biotic and abiotic stress and with better yields, and by the use of cell or organ cultures for phytochemical production, all contributing to the development of an advanced plant biotechnology.

List of Technical Nomenclature

Acclimatization	The hardening of a micropropagated plantlet, a shoot, or a seedling during the transition from tissue culture to *ex vitro* condition.
Adventitious	The development of new organs, in shoots, roots, leaves, or callus, from nonexisting meristems.
Callus	A disorganized tissue arising from a wounded explant upon isolation or from a group of proliferating cells, usually made up of parenchyma cells.
Competence	The ability of a cell or a group of cells to be induced to redifferentiate and dedifferentiate.
Cybrid	A cytoplasmic hybrid, originating from the fusion of a whole protoplast with an enucleated or irradiated cytoplasm.
Dedifferentiation	Changes in a cell or a group of cells from a differentiated state to a nondifferentiated phase.
Explant	An isolated plant organ, a tissue, or cells used to initiate an *in vitro* culture.
Hyperhydricity	Hydrated tissue and abnormal morphogenesis in tissue cultured plants.
Meristemoid	A group of meristematic cells arising in callus, or any other cultured tissue capable of developing into somatic embryos, roots, or shoots.
Organogenesis	The formation of organs in tissue culture.
Redifferentiation	Changes in dedifferentiated cells in response to a morphogenetic signal.
Somatic embryogenesis	Development of embryoids arising from somatic cells either directly on an explant or from callus tissue.
Subculture	The transfer of a plant, organ, tissue, or cells to a new culture medium.
Totipotency	The potential of fully differentiated competent cells to re-express their genetic information, to divide, differentiate, and develop into a whole new organism.

See also: **Genetic Modification**: Gene Cloning, General Principles; Transformation in Monocotyledons; Transformation in Plastids; Transformation, General Principles; Transgene Stability and Inheritance. **Production Systems and Agronomy**: Plantation Crops and Plantations. **Tissue Culture**: Artificial Seeds; Clonal Propagation, *In Vitro*; Micropropagation; Organogenesis; Secondary Metabolism in Plant Cell Cultures; Somatic Embryogenesis. **Tissue Culture and Plant Breeding**: Clonal Propagation, Forest Trees; Regeneration of Fruit and Ornamental Trees via Cell and Tissue Culture; Somaclonal Variation.

Further Reading

Aitken-Christie J, Kozai T, and Smith MAL (1995) *Automation and Environmental Control in Plant Tissue Culture*. Dordrecht: Kluwer Academic.

Bhojwani SS and Razdan MK (1996) *Plant Tissue Culture: Theory and Practice*. Amsterdam: Elsevier.

Dixon RA (1985) *Plant Cell Culture: A Practical Approach*. Oxford: IRL Press.

Gamborg OL and Phillips GC (1995) *Plant Cell, Tissue and Organ Culture Fundamental Methods*. New York: Springer-Verlag.

George EF (1993) *Plant Propagation by Tissue Culture*, 2nd edn. Andover: Exegetics.

Lindsey K (1991) *Plant Tissue Culture Manual, Fundamentals and Applications*. Dordrecht: Kluwer Academic.

Lumsden PJ, Nicholas JR, and Davies WJ (1994) *Physiology, Growth and Development of Plants in Culture.* Dordrecht: Kluwer Academic.

Soh WY and Bhojwani SS (1999) *Morphogenesis in Plant Tissue Cultures.* Dordrecht: Kluwer Academic.

Street HE (1977) *Plant Tissue and Cell Culture*, 2nd edn. Oxford: Blackwell Scientific Publications.

Vasil IK and Thorpe T (1994) *Plant Cell and Tissue Culture.* Dordrecht: Kluwer Academic.

Micropropagation

A C Cassells, National University of Ireland, Cork, Ireland

Introduction

The term "micropropagation" refers to the large-scale multiplication of plants by bud proliferation or bud induction on plant culture media *in vitro* (**Figure 1**). It is implicit that when micropropagation is used commercially as an alternative to vegetative propagation, the plants produced should be true-to-type, free of pathogens, and free of microbial contaminants. The latter are environmental organisms that enter the cultures on or in the explant at establishment of the tissue culture, or enter during the serial subculture steps in multiplication. Vegetative multiplication of plants in the field is a slow process that is dependent on plant growth cycles, requires labor for pesticide treatments, irrigation, etc., and is at risk from crop infection or loss due to abiotic and biotic stresses, e.g., drought and disease, respectively. While growth can be accelerated and biotic and abiotic stress better managed by cultivation in glasshouses, the process is still stock-dependent and relatively expensive, as reflected by selling prices of 5 US cents for unrooted chrysanthemum cuttings and up to 50 US cents or more for cuttings of woody plants. Commercial micropropagation is a substitution industry in the production of vegetative planting material (propagules). It has the advantage over *in vivo* multiplication that it is independent of seasonality. It can be used to multiply disease-free plants with little risk of pathogen re-infection. However, it currently has high labor costs, with a propagule selling price for bare-rooted

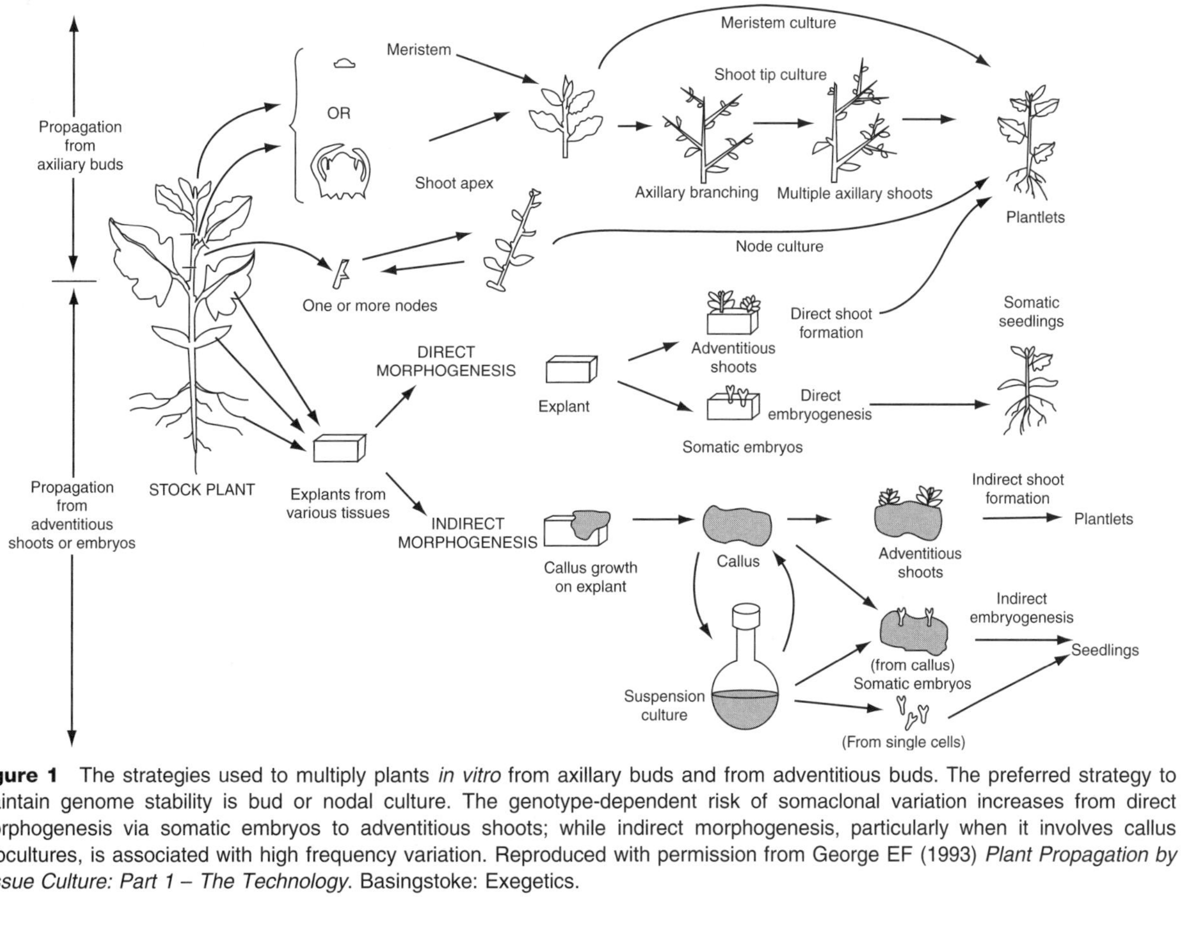

Figure 1 The strategies used to multiply plants *in vitro* from axillary buds and from adventitious buds. The preferred strategy to maintain genome stability is bud or nodal culture. The genotype-dependent risk of somaclonal variation increases from direct morphogenesis via somatic embryos to adventitious shoots; while indirect morphogenesis, particularly when it involves callus subcultures, is associated with high frequency variation. Reproduced with permission from George EF (1993) *Plant Propagation by Tissue Culture: Part 1 – The Technology.* Basingstoke: Exegetics.

microplants of ca. 15 US cents. The key feature of micropropagation is the ability to multiply rapidly from single elite plants, e.g., disease-free individuals and new breeders' lines, compared with conventional methods while maintaining high health status. Automation is currently being investigated in an attempt to reduce the labor component of micropropagation, which amounts to approximately 60% of the total costs.

The Micropropagation Laboratory

In addition to administrative offices, staff facilities, and storage areas, the micropropagation laboratory consists of five functional areas: (1) the culture vessel washing and media preparation area; (2) the explant preparation area; (3) the subculture area; (4) the growth rooms; and (5) the shipment area. Many micropropagators also have glasshouse facilities for the establishment and growing on of microplants. In area (1) are located the equipment for washing of vessels for recycling, for the preparation and dispensing of media, and for autoclaving the media and contaminated cultures from production before disposal. This area contains, in addition to autoclaves, the equipment for water purification (commonly stills), agar dispensers, pH meters, and balances for weighing the media components. Preparations of basal media in powder form can be purchased, but many micropropagation companies prepare media in-house to reduce costs. Area (2) is used to prepare explants for introduction into culture. A dissecting microscope is required to isolate the smaller apical meristem tips used for pathogen elimination. This is the area where plant tissue and explants are surface sterilized prior to introduction to the culture medium under aseptic conditions. Usually this involves immersing the explant in 80% aqueous ethanol for 30 s, followed by immersion in a 20% aqueous dilution of household bleach for 20 min, followed by several rinses in distilled water. The exposed ends of the explant are trimmed. Area (3) contains the laminar flow cabinets where the surface-sterilized explants are placed in the prepared media-containing vessels under aseptic conditions. It is also where aseptic cultures are subdivided and cultured in the mass propagation stages. Cultures are maintained in temperature-controlled growth rooms in the range 20–28°C, with a set photoperiod (usually a 16-h day length), under lights of defined photosynthetic flux (PFF) in the range 30–100 $\mu mol\, m^{-2} s^{-1}$. Subculture cycles are commonly carried out at 4–6-week intervals. Cultures are inspected regularly for signs of contamination and may be tested for subliminal (latent) contamination by placing pieces of the tissues on appropriate microbial media. When the desired numbers of plants have been obtained, the culture vessels are transferred to the shipment area where they may be despatched to the nursery, or the microshoots or microplants are removed from the vessels, washed free of medium and substrate, e.g., agar, and packed as microshoots or bare-rooted microplants for establishment or shipment.

Plant Tissue Culture Media

The development of plant tissue culture media has been based on mineral analyses of plant tissues and, particularly, of the liquid endosperm of coconut, *Cocos nucifera* (coconut water, coconut milk). The most universal media formulation is that of Murashige and Skoog published in 1960, which consists of salts of minerals and trace elements: myo-inositol and glycine and the vitamins thiamine, nicotinic acid, and pyridoxine. The latter may not be synthesized in explants until organogenesis occurs. To this basal medium at pH 5.8 are added sucrose as a growth substrate, agar to solidify the medium, and plant growth regulators. In 1958, Skoog and Miller showed that the responses of tobacco (*Nicotiana tabacum*) pith explants in culture were dependent on the concentrations and ratios of the plant growth regulators auxin and cytokinin. In most protocol studies, the basal medium of Murashige and Skoog is used and the hormonal balance is determined following their principle based on factorial combinations of auxins and cytokinins. There are two general exceptions to this approach; some species, e.g., *Rhododendron* spp. or woody plants, have their own specialized basal media; and, in some cases, hormones have to be employed sequentially, firstly, to induce competence in the explant, and then for expression of competence.

The main auxins used in tissue culture (in order of increasing potency) are: 3-indoleacetic acid (IAA), 3-indolebutyric acid (IBA), 1-naphthalene acetic acid (NAA), and 2,4-dichlorophenoxyacetic acid (2,4-D). The principal cytokinins used (in order of increasing potency) are: 6-furfuryl aminopurine (kinetin), 6-benzylaminopurine (BA), N6-(2-isopentyl)adenine (2-iP), and 4-hydroxy-3-methyl-*trans*-2-butenyl-aminopurine (Zeatin). Several anti-auxins are used in plant tissue culture, principally to reduce endogenous explant auxin, e.g., 2,3,5-tri-iodobenzoic acid. Some specific uses of plant growth regulators include that of jasmonic acid to induce tubers in potato (*Solanum tuberosum*) *in vitro* and of abscisic acid to induce dormancy in somatic embryos. Gibberellins are used to increase stem elongation prior to *in vitro* rooting. A more recently introduced compound with cytokinin-like properties, thidiazuron (TDZ), is very potent in inducing somatic embryos.

The Culture Environment

Two aspects of the physical environment are important in micropropagation, namely, the gaseous exchange properties of the culture vessel, and the light spectrum and energy levels. It has been demonstrated that for stomatal function (and avoidance of hyperhydricity), it is important that the microshoots develop a transpiration stream *in vitro*. This results in calcium uptake necessary for stomatal function and as a counter ion to auxin in basipetal auxin movement. A transpiration stream can be achieved by bottom cooling of the cultures (1–2°C lower than the growth room temperature) by running water through the shelves or by using porous lids. Gaseous exchange is also important to vent ethylene to the atmosphere (ethylene is associated with *in vitro* growth abnormalities such as hyperhydricity) and to facilitate photosynthesis (see below).

The spectral qualities of the growth room lights (red and blue light components) and the light energy available for photosynthesis (the photosynthetic photon flux, PFF) are also important. Low PFF ca. $30\,\mu mol\,m^{-2}\,s^{-1}$ may be enough in the heterotrophic phase of culture, but to increase photosynthetic competence or for photoautotrophic culture, high PFF is required. Photoautotrophic culture is carried out with elevated growth room CO_2 at 1–2000 ppm and a PFF of $100\,\mu mol\,m^{-2}\,s^{-1}$ or higher, in vessels that allow gaseous exchange with the ambient atmosphere.

Production of Pathogen-Free Plants

Where available, stock plants for micropropagation should be obtained from certified disease-free sources, or should be sent to the appropriate agencies for pathogen testing and elimination. This is not always possible for micropropagators who work with a wide range of plant species, the diseases of which may be poorly documented. Under these circumstances, plants should be observed for any symptoms of disease, and reference made to the literature on diseases of the crop or of related crops (see American Phytopathological Society in the "Further Reading"). The stock plants may be tested for known diseases of the crop or of related crops following European and Mediterranean Plant Protection Organization (EPPO) guidelines (see "Further Reading"), with the caveat that most serological diagnostics are strain-specific. Broad spectrum (non-strain-specific) DNA diagnostics are being developed, but are not yet in commercial production. Inoculation of indicator plants is a valuable backup to validate serological tests and to detect uncommon pathogen strains. If disease-free stock plants are not available, heat therapy may be attempted to eliminate pathogens, and, less commonly, antibiotics are used to reduce bacterial contamination before meristem excision. In general, meristem excision is a valuable tool in eliminating both pathogens and endophytic (internal) microorganisms in plants.

Antibiotics and the antiviral chemical ribavirin have been used successfully *in vitro* to eliminate bacteria and viruses, respectively. Such antimicrobial compounds can be used more effectively *in vitro* than *in vivo*, as it is possible to maintain inhibitory concentrations in the former where the chemicals are not "diluted" by plant growth. It should be borne in mind that most antimicrobial compounds used in tissue culture are biostatic rather than biocidal, and that microbial elimination is based on the excision of new growth from the explant; carryover of infected tissues must be avoided. This can be achieved by serial subculture of new shoots on the antimicrobial compound. However, the *in vitro* culture should be screened for contaminants and plants grown from the culture tested in the approved manner (see EPPO guidelines) before mass clonal propagation, especially of important crops, is carried out. The danger is that undetected microbial contaminants will be mass propagated *in vitro* and may either overrun the cultures (quite often this happens when cultures are transferred to lower strength medium for rooting or when the hormones in the medium are changed) to become "vitropaths." Such contaminants may subliminally reduce plant performance *in vitro* and following establishment, or they may become pathogenic to the crop in the field, or be a source of inoculum for disease in other crops.

Micropropagation Strategies

There are two basic strategies used to multiply plants *in vitro*: (1) proliferation of preformed buds; or (2) bud induction termed adventitious regeneration (**Figure 1**). Plant cell totipotency, namely, the ability of the plant cell to regenerate into a plant, is the basis of adventitious regeneration. A problem associated with the latter is variability in the progeny plants from some genotypes, which is a possible reflection of the fluidity of the plant genome. Cells of tissues in plants may be genetically variable due to endopolyploidization and amitosis in the plant tissues, or in

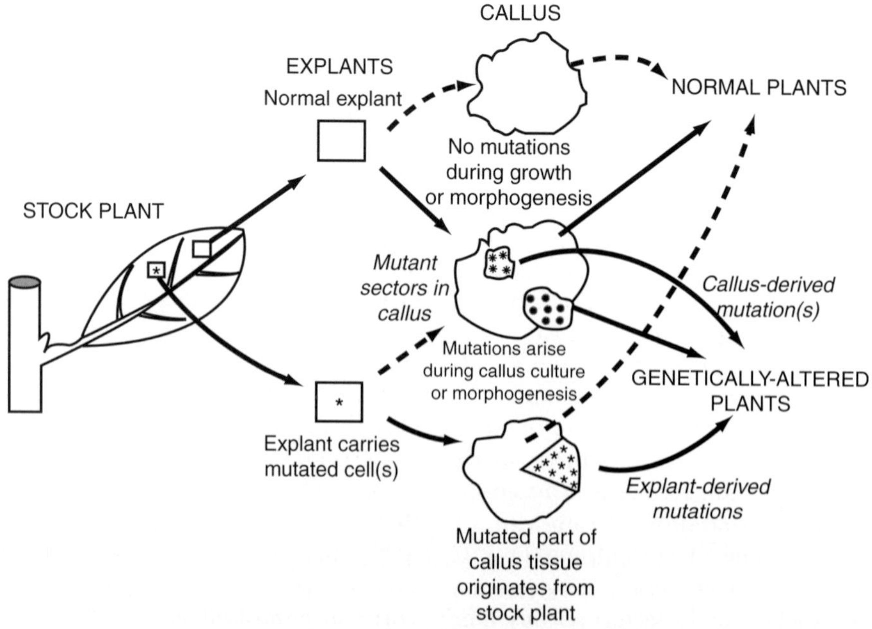

Figure 2 The origins of somaclonal variation. Variation may exist in the cells of the explant or arise *de novo* during callus formation. Adventitious shoots may be chimeral. Reproduced with permission from George EF (1993) *Plant Propagation by Tissue Culture: Part 1 – The Technology.* Basingstoke: Exegetics.

callus formation on the explant (**Figure 2**); this variability is not expressed in the germ cells but can be expressed where the mutant cells are totipotent in culture. These strategies, including direct and indirect organogenesis and somatic embryogenesis, are shown in **Figure 1**. Genotype and callus proliferation before organogenesis are related to variability in unstable genotypes. The potential of somatic embryogenesis as a basis for artificial seed production and cost reduction in micropropagation is discussed briefly below.

The Stages in Micropropagation

The micropropagation process is conveniently divided into stages (based on Murashige, and Debergh and Maene; see **Figure 3**) (**Figure 4**).

Stage 0

The importance of plant selection cannot be overemphasized in micropropagation. Two elements of stock plant selection are important, namely, the genetic quality and freedom from pathogens. Selection for genetic quality is based on the choice of, where possible, at least five uniform individual plants whose performances have been determined. Micropropagators should also be familiar with the legal position regarding plant breeders' rights and plant patents. Selection for freedom from disease is more problematic. The best advice is to obtain stock from a competent certification authority. If disease indexing is to be carried out by the micropropagator, it should be based on approved diagnostic strategies as described by the EPPO or another official certification body.

When preparing for micropropagation, it is important to ensure that stock plants do not become infected in the glasshouse. This was the original reason for Debergh and Maene designating this preparatory stage 0. Plant-holding glasshouses should be insect-proofed and plant handling should be minimized. Bottom watering should be used and fertilized applications minimized to discourage rapid plant growth.

Stage 1

In this stage, the aim is to establish an aseptic culture of the plant. The preferred approach is to establish the smallest apical tip explant that will survive. Usually, the apical dome and first pair of leaf primordia are excised with the aid of a dissecting microscope and established on the plant tissue culture medium in a laminar flow cabinet. The design of the medium used for this step follows the principles described above (see: "Further Reading" for extensive tables of media). It can be useful to include casein hydrolysate to encourage the expression of bacterial contaminants. Antioxidants

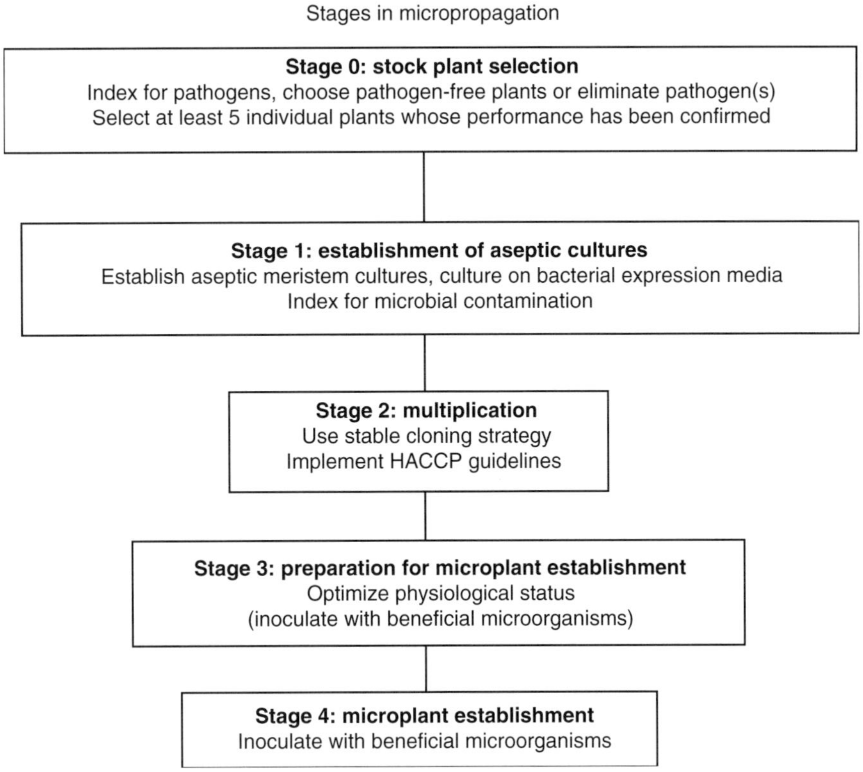

Figure 3 The stages in micropropagation and the associated genetic and health quality assurance actions.

Figure 4 The influence of the culture vessel atmosphere on microplant morphology. Left: control potato microplant culture with unrestricted gaseous exchange with the atmosphere. Center: growth in a vessel with a lid that is semipermeable to ethylene; note microplant stem elongation and suppression of leaf expansion. Right: growth in an airtight vessel; note the formation of microtubers from the nodal buds.

and phenol-binding agents, e.g., polyvinylpyrrolidone, can be incorporated also. In stage 1, the aim is to establish a single plant from each meristem explant, avoiding any adventitious regenerants that may be genetically unstable. These individual plants can then be indexed for bacterial contamination by taking the basal tissue at subculture and placing it on bacteriological media. When the culture test for bacteria is negative, the process can advance to stage 2.

Stage 2

Stage 2 is the mass multiplication stage. Here, a stable cloning strategy should be followed avoiding callusing and adventitious regeneration where the genotype is unstable *in vitro*. For most species, bud proliferation is used with the exception of crops such as *Begonia* and *Saintpaulia*, which are vegetatively propagated *in vivo* from leaf cuttings and are stably regenerated *in vitro* from leaf explants. It is also essential that laboratory and growth room contamination with microorganisms and microarthropods is avoided. To achieve this, hazard analysis and critical point control (HACCP) should be employed. The cultures should be routinely visually inspected for contamination, and contaminated cultures should be autoclaved before disposal. Cultures should also be routinely checked for latent microorganisms by culture indexing for nonexpressed contamination. Various species have been proposed as markers for process failure, e.g., heat-resistant *Bacillus* species indicate failure of the media and instrument sterilization, *Penicillium* indicates problems at media pouring. *Pseudomonas* indicates failure to decontaminate the explant, *Rhodotorula* indicates contaminated air, and *Staphylococcus* indicates poor aseptic technique.

Stage 3

This stage is concerned with the preparation of plants for return to the environment. Multiplication by bud proliferation or adventitious regeneration involves microshoot production; in many cases, the microshoots may self-root, while in other cases, it may be necessary to induce rooting by reducing or eliminating cytokinin and manipulating the auxin concentration in the medium. Sometimes, microshoots are rooted *in vivo* in the nursery. Another consideration at this stage is the physiological quality of the microshoots/microplants. In the most extreme cases, the shoots may show signs of the "hyperhydricity syndrome"; this is expressed as hyperhydrated tissues, shoot tip necrosis and loss of apical dominance, and other physiological defects. Aberrant *in vitro* physiology may result in heavy losses or serious growth checks at microplant establishment *in vivo*, where the *in vitro*-formed leaves may not be photosynthetically competent and the stomata nonfunctional. Similarly, the *in vitro* roots may have impaired function. Several strategies have been used to improve the weaning of microplants, including transferring the microplants in stage 3 to a high sucrose medium, reducing the sucrose content of the medium to encourage photosynthetic competence, or transferring to a photoautotrophic sugar-free medium. The high-sucrose strategy treats the microplants like seedlings and builds up their food reserves to enable redevelopment of functional organs at establishment. The low-sucrose strategy is aimed at reducing sucrose inhibition of photosynthetic competence, so the plants become mixotrophic *in vitro* (that is, photosynthesis partially contributes to their energy requirement for growth). Photo-autotrophic culture is a form of 'micro' hydroponics.

Stage 4

In this stage, the plants are established in the glasshouse. The main problems are the impairment of physiological functions discussed above and lack of disease resistance. The former can be addressed by establishing microplants in high humidity and low light to reduce desiccation and photo-bleaching of poorly developed photosynthetic apparatus, respectively. Damping-off due to pathogen attack is more problematic, as microplants may be sensitive to the chemicals used to control such diseases in seedlings. Resistance to pathogen attack and improved drought stress have been achieved by inoculation *in vitro* and *in vivo* with arbuscular mycorrhizal fungi. Beneficial effects have also been reported following inoculation with growth-promoting biocontrol bacteria *in vitro* and *in vivo*.

Automation of Micropropagation

Micropropagation is very labor intensive. When carried out in high labor-cost regions, such as the USA and European Union countries, micropropagules (microplants, microtubers, etc.) compete only with propagules whose selling price is greater than ca. 15 US cents. One strategy to expand the market for microplants is to produce them in lower labor-cost economies, but this has resulted in problems such as lack of production infrastructure. An alternative has been to automate production, but in heterotrophic culture it has been difficult to control contamination problems associated with scaling-up of culture vessels, e.g., microshoot or bud cluster production in bioreactors. As an alternative to automating bud and shoot cultures, considerable effort has focused on the development of artificial seed based on encapsulation of somatic embryos. Somatic embryogenesis can be achieved in many crops and, in principle, it has been shown that somatic embryos can be encapsulated in artificial endosperm and given a protective seed coat. The artificial endosperm is essentially the plant tissue culture medium, and the seed coat may be made of carbohydrates or gelatin, materials that dissolve in the presence of water. The main difficulty with artificial seed has been providing shelf-life, and achieving synchronous germination.

Quality Control in Micropropagation

The elements of quality control that are important in micropropagation are genetic and epigenetic quality, physiological quality, and health quality. Variability in adventitiously regenerated plants, called somaclonal variation, includes polyploidy, aneuploidy, chromosome breakage and rearrangement, point mutations, and changes in DNA methylation. Given that a stable cloning strategy has been employed, then variability in the progeny should not be a problem. However, even with cloning based on precocious axillary bud proliferation, genetic instability may arise when the bud clusters give rise to adventitious shots through callusing of the explant. Using nodal culture, epigenetic changes may arise resulting, for example, in a delay in flowing, which have serious economic consequences. More immediately obvious are the consequences of poor physiological quality (see above). Both epigenetic changes and poor physiological quality are consequences of adverse media, culture vessel, and genotype interactions.

Trends in Micropropagation

To remain competitive, micropropagation must continue to be technologically innovative, to reduce production costs and maintain and improve quality. While there is the prospect that machine vision systems for automated cutting of nodes in subculture may reduce costs, precision cutting of bud clusters for a uniform product is a more remote prospect due to the irregular shapes involved. There is also the labor requirement to sort, pack, and establish the microplants, which, unless also automated, has high costs. A better prospect for automation of microplants appears to be the production of artificial seed, where all stages are more amenable to automation, and the product (the "synseed") has the potential for handling by existing seed planters. However, as mentioned above, good shelf-life and synchronous development are the exception rather than the norm for somatic embryos of most crops.

The potential for quality improvement of microplants to produce a value-added propagule seems more attainable than price reduction. *In vitro* weaning and biotization *in vitro* and *in vivo* offer considerable potential to improve microplant performance, both at establishment and postestablishment in the field. There is also great potential in the development of DNA diagnostics, which offer both strain-specific and broad spectrum screening for multiple pathogens in a single test. Again, subject to cost, these would have potential in ensuring the health status of microplants. Micropropagation is established as the preferred method for the mass multiplication of elite individual plants.

List of Technical Nomenclature

Autotrophic culture Syn. photoautotrophic culture; the energy for growth is provided by the ambient lighting. In plant tissue culture, this is commonly fluorescent lighting.

Epigenetic Variation that is related to differences in the regulation of gene expression as opposed to mutation.

Explant The part of the plant that is cultured *in vitro*.

Heterotrophic culture The energy for growth is provided by carbon compounds in the medium. The main carbon source is usually sucrose.

Hyperhydricity Syn. vitrification: refers to a syndrome of aberrations in the morphology and physiology of plants *in vitro* and is initially characterized by the hyperhydration of affected tissues.

Somaclonal variation Phenotypic variation, either genetic or epigenetic, displayed in plants adventitiously regenerated *in vitro*.

See also: **Biodiversity and Conservation**: Germplasm Conservation. **Diseases**: Bacterial Diseases; Fungal Diseases; Viral Diseases. **Regulators of Growth**: Abscisic Acid; Auxins; Brassinosteroids; Cytokinins; Ethylene; Gibberellins; Jasmonates; Juvenility; Photomorphogenesis. **Seed Development**: Artificial Seeds. **Tissue Culture and Plant Breeding**: Clonal Propagation, Forest Trees; Somaclonal Variation. **Tissue Culture**: Artificial Seeds; Clonal Propagation, *In Vitro*; General Principles; Organogenesis; Somatic Embryogenesis. **Water Relations of Plants**: Drought Stress; Uptake, Loss and Control.

Further Reading

American Phytopathological Society http://www.apsnet.org.

Cassells AC, Doyle BM, and Curry RF (eds) (2000) *Methods and Markers for Quality Assurance in Micropropagation*: *Acta Horticulturae*, vol. 530. Leuven: International Society for Horticultural Science.

Debergh PC and Zimmerman RH (eds) (1991) *Micropropagation: Technology and Application*. Dordrecht: Kluwer.

European and Mediterranean Plant Protection Organization (EPPO) *Guidelines*. http://www.eppo.org.
George EF (1993) *Plant Propagation by Tissue Culture: Part 1 – The Technology*. Basingstoke: Exegetics.
George EF (1996) *Plant Propagation by Tissue Culture: Part 2 – In Practice*. Basingstoke: Exegetics.
Trigiano RN and Gray DJ (eds) (2000) *Plant Tissue Culture Concepts and Laboratory Exercises*. 2nd edn. Roca Baton: CRC Press.

Clonal Propagation, *In Vitro*

B S Ahloowalia, International Atomic Energy Agency, Vienna, Austria

Introduction

Clonal propagation of plants refers to the production of genetically identical plants through nonsexual methods. Many plants are propagated from vegetative parts such as bud grafts, slips, stem cuttings, bulbs, tubers, and runners. For example, date palm (*Phoenix dactylifera*) trees are propagated from offshoots and rooting them in soil. Yet in other plants, leaves develop adventitious buds on the margins and leaf veins, which on sprouting develop into new plants. The plants produced in this manner are exact copies or clones of the mother plant, and genetically identical unless a mutation occurs in the cells and tissues of the original plant.

Vegetative propagation is used for the multiplication of many important crops, shrubs, and trees. These include crops such as cassava (*Manihot esculenta*), potato (*Solanum tuberosum*), sweet potato (*Ipomoea batatas*), ginger (*Zingibar officinale*), sugar cane (*Saccharum* spp.), taro yam (*Colocasia esculenta),* turmeric (*Curcuma domestica*), and yam (*Dioscorea* spp.). Several ornamentals, e.g., azalea (*Rhododendron* spp.), carnation (*Dianthus* spp.), chrysanthemum (*Chrysanthemum* spp.), jasmine (*Jasminum officinale*), roses (*Rosa hybrida* and *chinensis*), and tulips (*Tulipa* spp.), and many fruits such as banana and plantain (*Musa* spp.), pineapple (*Ananas comosus),* passion fruit (*Passiflora edulis*), and kiwi fruit (*Actinidia chinensis*), are also propagated from vegetative parts. In many other fruits, such as apple (*Malus domestica*), citrus (*Citrus sinensis*, *C. paradisi*, and several other *Citrus* species), lychee (*Litchi chinensis*), mango (*Mangifera indica*), and pear (*Pyrus communis*), shoots and buds are grafted to produce clones of the original mother tree. Bud sports or spontaneous mutations are also propagated in this manner to obtain new varieties.

In Vitro Propagation

In vitro propagation is very much like multiplication of plants from vegetative parts except that plants are produced by culturing tiny pieces on a synthetic medium in a suitable container instead of soil in a pot. The *in vitro* propagation of plants is initiated by culturing pieces of stems, leaves, apical and axillary buds, immature seeds, embryos, and cotyledons. The basis of *in vitro* propagation is rapid multiplication of plants by culturing tiny pieces on media under controlled environment and aseptic (germ-free) conditions. Since strict aseptic conditions have to be maintained in the production of *in vitro* cultured plants, the culture and transfer of cuttings and plants is carried out under germ-free conditions in laminar-flow cabinets, and the containers in which plants are cultured are kept tightly sealed. The laminar-flow cabinets are designed to produce sterile air. The instruments such as scalpels, scissors, and forceps used for cutting and transfer of plant pieces are sterilized by alcohol dip and flaming or dry heat. Likewise, the media are also sterilized.

Most culture media contain basic major and minor salts, vitamins, sugar, plant growth regulators, and agar with pH adjusted between 5.6 and 5.8. The major salts provide nutrients such as nitrogen, phosphorus, potassium, calcium, and magnesium, and minor salts elements such as copper, manganese, molybdenum, sulfur, and zinc, which along with vitamins are essential for the growth and differentiation of plant cells and tissues. The sugar in the culture medium provides carbon for photosynthesis. However, the composition of media (particularly the type and amount of sugar and growth regulators) is changed according to different plant species. Several commercial companies sell ready-made basal media in powdered form. Time saving, accuracy, and standardization of composition are the main advantages for using such media. The ready-made basal media are without growth regulators and sugars.

The amounts of various ingredients, particularly those of plant growth regulators, are changed to suit the proliferation and growth of different types of plant and sometimes even of different varieties. The media and glass containers are sterilized, usually with steam in a pressure cooker or an autoclave. Many laboratories use glass jars such as jam jars, which can be used repeatedly by washing and autoclaving or disposable plastic containers such as Watson Module^{TM} cups which come presterilized. Tiny pieces of plants or tissues, which have been surface sterilized, are then transferred to the cooled, solidified medium, and kept under controlled conditions of light, day length, humidity, and temperature to provide summer-like conditions for shoot proliferation, root formation, and plant growth. The controlled environment conditions for culture and growth allow plant propagation on a year round basis, in a small space and production of high-quality disease- and insect-free plants in large numbers. This process is called micropropagation. This technique is particularly suitable for large-scale multiplication of plants propagated from vegetative parts, especially ornamentals, such as African violet (*Saintpaulia ionantha*), chrysanthemums, roses, *Streptocarpus*, carnation, and *Gerbera*. Many crops such as banana and plantain, cassava, garlic (*Allium sativum*), ginger, pineapple, potato, sweet potato, and yams (*Colocasia esculenta* and *Dioscorea* spp.) can also be micropropagated. Conventionally these plants are propagated from cuttings, corms, bulbs, and tubers, and repeated multiplication in soil leads to infection with viruses, fungi, bacteria, and nematodes. In contrast, the micropropagated plants are free from pathogens and of high quality.

Process of *In Vitro* Cloning

The production of plants through micropropagation is a four-step process: (1) explant culture, (2) shoot proliferation, (3) rooting, and (4) plant hardening and transfer to soil.

Explant Initiation

Tiny pieces (explants) between 2 and 10 mm long, taken usually from various plant parts (organs) such as shoot tips, stem nodes, and leaves, are the starting material to initiate plant micropropagation. Initiation of explants is the very first step in micropropagation. A good clean explant, once established in an aseptic condition, can be multiplied several times; hence, explant initiation in an aseptic condition should be regarded as a critical step in micropropagation. Very often, explants fail to establish and grow, not due to the lack of a suitable medium but because of contamination from bacteria and fungi.

The explants can also be obtained as small pieces from stems, apical leaves and axillary buds, immature whole seeds, or their parts such as embryos and cotyledons and hypocotyledons, and in some cases even roots, inner core and scales from the bulbs, floral buds, and floral parts. The tiny pieces are surface sterilized to get rid of the fungi, bacteria, and insects. To achieve surface sterilization, the explants are first washed in sterile water, rinsed in ethanol, and then immersed in solutions of chemicals with chlorine base. Calcium or sodium hypochlorite-based solutions, 1–3% (w/v) are usually used for soft herbaceous materials, such as leaves and shoots. A 5–7% solution of Domestos^{TM} disinfectant (which contains 10.5% w/v sodium hypochlorite, 0.3% sodium carbonate, 10.0% sodium chloride, and 0.5% w/v sodium hydroxide, and a patented thickener) is also effective for surface sterilization of many explants. Other surface sterilants used include mercuric chloride (though its use should be avoided as far as possible, since it is highly toxic) and potassium permanganate. The explants are rewashed three to four times with sterile distilled water after sterilization with chemicals.

The explants produce shoots or roots and even somatic embryos, depending upon the chemicals in the medium called plant growth regulators or phytohormones. There are two sets of growth regulators that are most critical in the process of micropropagation: the auxins such as indole-3-acetic acid (IAA), indole-3-butyric acid (IBA), naphthalene acetic acid (NAA), and cytokinins such as kinetin (6-(furfurylamino) purine), 6-benzyl aminopurine (BA or BAP), zeatin ((E)-2-methyl-4-(1*H*-purin-6-ylamino)-2-buten-1-ol), and 2iP (6-(rr-dimethylallylamino) purine). The relative ratio of cytokinin to that of auxin in the culture medium determines the pathway the explants take. By using relatively high concentrations of auxins (particularly the synthetic auxin-like compound 2,4-D and (2,4-dichlorophenoxyacetic acid), an explant can be turned into a lump or a mass of cells – a callus. The callus is then proliferated by culturing on a series of media containing different types and varying amounts of growth regulators that make the cells differentiate into adventitious buds, shoot primordia, and even somatic embryos that resemble sexual embryos. This process of obtaining plants from cells through callus cultures is called organogenesis and that of embryos as somatic embryogenesis. The pathway of cloning plants through callus culture often leads to the production of plants, a proportion of which do not resemble the parent, and are often genetically different from the parent. Many such types of plants originate from somaclonal variation that involves changes in the structure or function of DNA.

In practice, a majority of plants are micropropagated from explants taken from shoot tips and nodal cuttings with apical or axillary buds. Pieces taken from these parts are used for establishing mother cultures that are used as a source of further cuttings to multiply plants. Located in the shoot tip are a group of cells called the shoot meristem. The shoot meristem cells are capable of division and making more meristems or adventitious buds that give rise to multiple shoots.

Shoot Proliferation

During this stage there is a rapid numerical increase and growth of structures destined to form shoots. Once explants are established, they are cultured on media that promote multiple shoot formation by forming either adventitious buds or shoot primordia (also known as stage II). To achieve high rate of shoot meristem proliferation, explants are cultured on media containing relatively high cytokinin and low auxin. So good is the shoot proliferation capacity in a small space that as many as 150–200 orchid buds can be produced on 20 ml medium in one 7-cm tall plastic cup container with 5 cm diameter base. Efficient multiplication invariably requires growth regulators in the medium.

Rooting

The shoots produced from stage II are rooted by further culture on rooting medium. In most plants, root formation is promoted by adding auxin to the subsequent medium. Often, shoots produce roots on media without growth regulators or with extremely low (0.01–1 mg l^{-1}) auxins such as IBA or IAA. In case of cloning through somatic embryos, shoot and root formation occur simultaneously. Such embryos germinate on media without any growth regulators, although some require culture on media with gibberellins and/or desiccation for maturity.

Plant Hardening and Transfer to Soil

The tiny *in vitro* produced plants (microplants) have to be acclimatized to the natural conditions of air (carbon dioxide), humidity, and temperature, a process called weaning or hardening. Usually the microplants are cultured under high humidity and their stomata function in a manner not suitable for plant survival under low humidity and variable temperature and high light intensity. A gradual adjustment to natural conditions is achieved by keeping microplants under shade, and infrequent watering during the first 1 to 3 weeks. In practice this is achieved by keeping such plants under plastic covers that allow sufficient air circulation, provide diffused light, and maintain relatively high humidity. In some cases, it helps to grow the plants in vermiculite and irrigate them with a low-concentration nutrient solution. The hardened plants are then grown in compost or soil for a few weeks and then transferred to large pots or field.

It is better to transfer the microplants as bunches rather than as separated single plants in the initial stages of establishment in soil. Bunch planting allows better survival and root growth. The soil should be low in nutrient level and well aerated at this stage. Other growth conditions such as temperature, light, and moisture need to be adequate but not excessive. The micropropagated plants should not be exposed to direct sunlight for the first 5 to 6 days. Frequent and overwatering should be avoided during the first 14 days of soil transfer. Each plant has its own specific requirement of humidity and temperature after transfer to soil. For example, strawberry (*Fragaria x ananassa*) microplants prefer low temperature (12–16°C) and low relative humidity, whereas those of orchids (Orchidaceae) prefer high temperature (20–22°C) and high humidity. Microplants usually are slow to grow in the beginning; however, once established in soil, *in vitro* propagated plants grow like normal plants, and usually outperform the conventionally propagated plants in vegetative growth.

New Industry

Micropropagation allows propagation of disease-free, high-quality plants, tubers and bulbs. During the past 30 years, a whole new industry based on *in vitro* plant propagation has developed. Worldwide millions of plants of fruits such as banana and plantain, citrus, pineapple, strawberry, and crops such as sugar cane, potato, sweet potato, and cassava are multiplied through micropropagation. Among ornamentals, species of *Anthurium*, *Gerbera*, orchids (*Cymbidium*, *Dendrobium*, *Oncidium*), roses, and many more are produced through micropropagation. Plants micropropagated in Asia where labor is relatively cheap are shipped to Europe and the United States as microplants (orchids), microtubers (e.g., potato), and microbulbs (e.g., daffodils (*Narcissus pseudonarcissus*) and tulips). The sophistication in micropropagation ranges from low-tech in used liquor bottles to high-tech bioreactors. In large-scale plant micropropagation laboratories, it is common to make several liters of medium in bulk and dispense exact amount in thousands of containers using mechanical devices such as a peristaltic pump.

A wide variety of containers from reusable milk bottles, used liquor bottles, jam jars, glass test-tubes, plastic Petri dishes, and Magenta™ vessels to disposable plastic cups such as Watson Modules™ (**Figure 1**) are used for commercial-scale culture of plants. Each type of vessel has its advantages and limitations. The reusable glass vessels are cheap to buy but require high labor and energy costs for bulk cleaning and sterilization. Disposable plastic materials such as Watson Modules come presterilized, save labor costs, allow stacking of culture thus freeing

Figure 1 Potato microplants cultured on medium in a Watson Module™ container. Reproduced by permission of Teagasc-Agriculture and Food Development Authority from Ahloowalia BS (1994) Mini-tubers for seed potato production. *Farm and Food* 4: 4–6.

Figure 2 Micropropagated plants of potato after 10 (left) and 21 days (right) of transfer to soil in Watson Module™ containers. Reproduced by permission of Teagasc-Agriculture and Food Development Authority from Ahloowalia BS (1994) Mini-tubers for seed potato production. *Farm and Food* 4: 4–6.

space in culture rooms, and can be turned into pots for soil growing of microplants (**Figure 2**); however, they cannot be reused for plant culture.

Future Developments

Many innovations are taking place in the technology of cloning plants through *in vitro* culture. These include the automated containers, "bioreactors," in which cells rather than plants are cultured as suspension in the liquid medium. With the help of computers, the medium is maintained and adjusted at the desired temperature, pH, oxygen, and CO_2 content, and replenished automatically. Plant cells maintained in this manner produce somatic embryos that can be matured on solid medium. Such systems are being developed for the multiplication of trees. However, many cells undergo changes in their genetic makeup after a short duration of culture that leads to the production of plants that are completely different from the parent. The procedures for plant production from bioreactor-cultured cells are as yet limited to a few experimental plants only. Many problems of forming plant shoots and embryos and their maturation and germination into microplants, and transfer to soil, still remain to be solved.

List of Technical Nomenclature

Callus A cluster of undifferentiated plant cells that have the capacity to regenerate a whole plant in some species.

Cell Cells are the building blocks of all living organisms. A cell is the smallest structural unit of a living organism that is able to grow and reproduce independently.

Cell culture A technique of growing cells in laboratory conditions.

Clonal propagation Asexual multiplication of plants that are genetically alike and originate from a single individual or an explant.

Clone A genetic replica of an organism obtained through non-sexual (no fertilization) reproduction process.

Explant A tiny piece of a plant or group of cells used for culture on artificial medium for growth or maintenance.

In vitro In a test-tube or other laboratory apparatus. (Opposite of *in vivo* – in the living organisms.)

***In vitro* culture** Storage and growing of living material in tissue culture.

Microplants Plants produced through *in vitro* culture or micropropagation.

Micropropagation Multiplication of plants from tiny pieces on synthetic medium under controlled environment and aseptic conditions.

Organogenesis Production of organs, such as adventitious buds, primordial shoots, and roots from the cells and callus cultures.

Plant propagation Production of plants from vegetative parts such as cuttings and slips without sexual generation.

Shoot meristem A group of special cells in the apical region of the shoots, which are capable of dividing and give rise to axillary buds, lateral shoots, and stem.

Somaclonal variation Variation that originates in tissue cultures. It may be physiological, genetic, or epigenetic.

Somatic embryo Embryos of nonsexual origin in plants obtained from cultures of cells and calluses.

Tissue culture A technique in which portions of a plant (or animal) are grown on a synthetic culture medium.

Vegetative propagation Reproduction of plants using a nonsexual process involving the culture of plant parts such as stem and leaf cuttings.

See also: **Production Systems and Agronomy**: Nursery Stock and Houseplant Production. **Root Development**: Genetics of Primary Root Development; Root Growth and Development. **Tissue Culture and Plant Breeding**: Regeneration of Fruit and Ornamental Trees via Cell and Tissue Culture.

Further Reading

Ahloowalia BS (1995) Watson Module: A new tool for plant micropropagation. *BioLink* 2: 17.

Ahloowalia BS (2000) High-tech propagation of horticultural plants. In: Chadha KL *et al.* (eds) *Biotechnology in Horticulture and Plantation Crops*, pp. 26–37. New Delhi: Malhotra Publishing House.

Debergh PC and Zimmerman RH (1991) *Micropropagation: Technology and Application*. Dordrecht: Kluwer Academic Publishers.

Dixon RA (1985) *Plant Cell Culture: A Practical Approach*. Oxford: IRL Press.

Evans DA, Sharp WE, Ammirato PV, and Yamada Y (1983) *Handbook of Plant Cell Culture, vol. 1, Techniques for Propagation and Breeding*. New York: MacMillan.

George EF, Puttock DJM, and George HJ (1987) *Plant Culture Media, vol. 1, Formulations and Uses*. Edington: Exgenetics Ltd.

Kyte L (1987) *Plants from Test Tubes: An Introduction to Micropropagation*. Portland: Timber Press.

Murashige T (1974) Plant propagation through tissue cultures. *Annual Review of Plant Physiology* 255: 135–166.

Thorpe TA and Harry IS (1997) Application of tissue culture to horticulture. In: Altman A and Ziv M (eds) *Horticultural Biotechnology: In Vitro Culture and Breeding*, pp. 39–49. Leuven: Acta Horticulturae.

Torres K (1988) *Tissue Culture Techniques for Horticultural Crops*. New York: Van Nostrand Reinhold.

Wetter LR and Constabel F (1982) *Plant Tissue Culture Methods*. Saskatoon: National Research Council of Canada.

Organogenesis

G-J de Klerk, Center for Plant Tissue Culture Research, Lisse, The Netherlands

Introduction

In plants, differentiated somatic cells may reinitiate the ontogenetic program: when given the proper stimuli, they develop into adventitious meristems. These new meristems generate adventitious roots, shoots, or embryos. The formation of adventitious roots and shoots is referred to as (adventitious) "organogenesis," whereas "somatic embryogenesis" denotes the formation of adventitious embryos. *In vitro*, adventitious organogenesis may occur at very high frequencies, and is one of the corner stones of biotechnological breeding and propagation methods. Biotechnological breeding techniques such as genetic engineering and haploid production involve adventitious regeneration of a complete plant from somatic cells. Many micropropagation protocols involve the formation of adventitious shoots from, for example, leaf or scale fragments. Before planting *ex vitro*, microcuttings are treated with auxin to obtain adventitious roots. Regeneration of somatic embryos from cell suspensions will – if broadly applicable – revolutionize plant propagation. This article deals with adventitious organogenesis *in vitro* from the scientific and practical points of view.

Occurrence of Organogenesis

Formation of Plant Meristems

An organism begins its existence as a fertilized egg cell, the zygote. From this single, morphologically simple cell, the embryo proper and the suspensor are formed following a series of cell divisions. In the embryo proper, the shoot and root apical meristem arise in a polar manner. In postembryonic development, the shoot apical meristem produces the stem, leaves, axillary meristems, and flowers, while the root apical meristem generates the primary root, but no lateral organs. During the plant's life, differentiated cells may produce adventitious meristems; a large part of the plant's body, in particular much of the root system, is derived from adventitious meristems. New root meristems that produce lateral roots are formed from certain cells in the root pericycle. Adventitious roots may regenerate from stems, usually from cells in between the vascular tissues, and this happens frequently in many species during natural development. In monocotyledons, the initial root derived from the embryo radicle usually aborts early on, and new roots arise adventitiously from the stem. In some species, leaves, shoots, embryos, or inflorescences can occur on leaves that have originated from adventitious meristems, a phenomenon known as epiphylly. Well-documented examples that display this phenomenon are *Bryophyllum* and *Kalanchoë*.

Certain environmental conditions may also induce the formation of adventitious meristems. Flooding, for example, results in the formation of adventitious roots from stems of *Rumex palustris*. When plants have been damaged, they replace or repair the missing parts. Excised branches regenerate adventitious roots relatively easily, and this property has been exploited since ancient times for the vegetative propagation of elite plants. Loss of the shoot apical meristem occasionally results in the formation of an adventitious meristem; however, it is usually an axillary meristem that resumes growth. After removal of the root tip, a lateral root is formed. A root from which the shoot has been removed may produce a new, adventitious shoot, usually from pericycle cells. In this case, the inducing signal is likely to be cytokinin, which is synthesized in the root tips. Leaves can generate complete plants; Hagemann reported in 1931 that out of 1196 species, 308 formed buds and roots from excised leaves.

Organogenesis in Tissue Culture

In tissue culture, adventitious shoot and root formation can occur very abundantly. In 1957, Skoog and Miller reported their classical experiments on the regeneration of roots and shoots from tobacco (*Nicotiana tabacum*) pith cultured *in vitro*. In their experiments, plant hormones of the auxin and cytokinin type played a pivotal role. Roots were formed at a high, shoots at a low, and callus at an intermediate auxin:cytokinin ratio. This concept has been repeatedly verified in many species, and recent confirmation has come from transgenic plants with increased cytokinin or auxin synthesis. In tissue culture, virtually all plant species may regenerate adventitious roots and shoots if suitable plant tissue is used and the appropriate conditions are established. *In vitro*, adventitious organogenesis occurs at a high rate because excised plant tissues can survive *in vitro* for a protracted period (provided there are sufficient nutrients, wilting does not occur, and microorganisms are absent), and because the hormones auxin and cytokinin can be easily applied.

Organogenesis is a Developmental Process

Concepts

Cell–cell interactions have been shown to play a major role in the embryonic development of animals. In experiments, in which excised embryonic tissues were transplanted from one individual to another, it was found that groups of cells influenced the differentiation of other groups of cells, a phenomenon known as induction. In the exciting study published in 1924 by Mangold and Spemann, a piece of blastopore lip from a newt early gastrula was transferred to another embryo (also an early gastrula). The graft, which was placed at a different site in the host to that in the donor, induced the formation of an almost complete secondary embryo. Thus, the grafted piece of lip regulated the direction of development in the adjacent tissues. It was also found that only cells in a particular state were able to respond to the signal; this particular state of reactivity is referred to as competence. The signal caused cells to become determined with respect to their future development. This is a gradual process, occurring over various cell generations. When the fate of cells has been determined, the signal is no longer required, and the cells will follow their fate even when grafted to another region. The formation of specialized cells in an organ from determined cells is referred to as differentiation.

In higher animals, morphogenetic processes that have been terminated can be reawakened to replace lost organs, and is known as regeneration. The renewal of a limb in salamanders is a well-known example. In the case of vertebrate limb regeneration, it is believed that the first step consists of cell divisions beneath the wound surface. A blastema is formed consisting of cells that supposedly have dedifferentiated to some extent.

The term organogenesis refers to the formation of organs, so in the plant's postembryonic development, e.g., to the formation of leaves and lateral roots. However, in plant tissue culture, the term is now being used in a narrow sense for the formation of roots and shoots from tissues without pre-existing meristems, so it is equal to adventitious organogenesis.

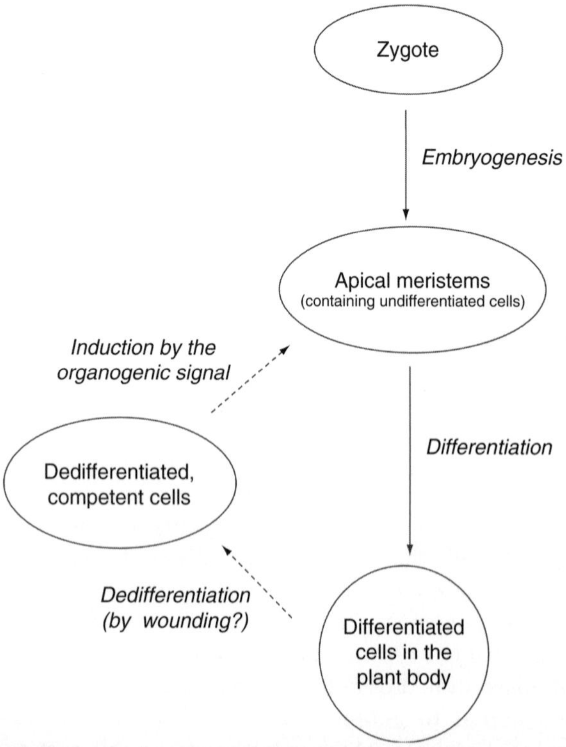

Figure 1 Breakdown of developmental processes into successive steps. Adventitious organogenesis involves dedifferentiation, followed by induction and differentiation.

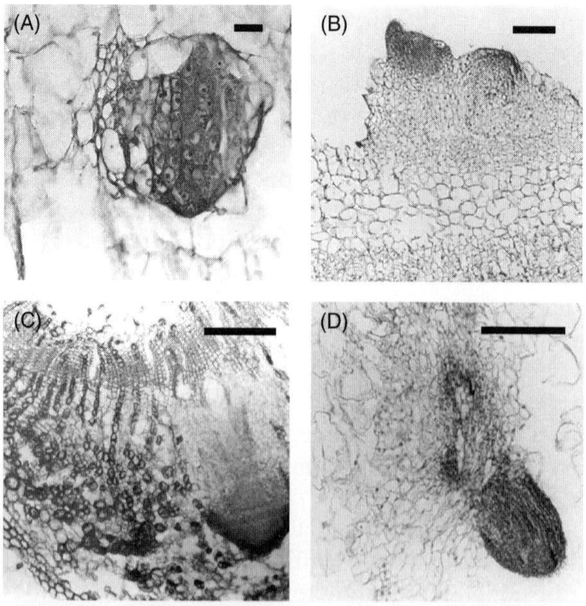

Figure 2 Formation of adventitious meristems either directly (A, C) or indirectly with an intermediate callus phase (B, D). (A) Adventitious shoot meristem in tomato cotyledons; (B) shoot primordia regenerated from thin cell layers of tobacco; (C) root primordium in a *Prunus avium* microcutting; (D) root primordium regenerated from tobacco pith explants. Scale bar = 200 μm in (B, C and D) and 20 μm in (A). The photographs were kindly provided by M. M. Altamura, University La Sapienza, Rome.

Timing of the Successive Steps in Organogenesis in Plants

Initially, researchers used histological methods to identify the successive phases in organogenesis. They distinguished a phase in which there are no visible changes ("induction period") followed by a period in which cell divisions are manifest ("initiation period"). This is then followed by the "expression period," during which a root or shoot meristem grows out. The major limitation of this approach is that histological observations provide little information about the physiological and developmental state of cells.

Christianson and colleagues adopted an alternative approach in which the response to the organogenic signal is the essential feature. By transferring tissues between media with different hormonal compositions, they recognized an initial phase in which the tissue develops the competence to respond to the organogenic signal; this period was labeled the "dedifferentiation phase." Then, in the "induction phase," cytokinin and auxin exert their shoot- and root-inducing action, respectively. After that, in the "differentiation phase," the signals are no longer required; the fate of the cells has been determined and shoots or roots are formed. A summary of the developmental process is shown in **Figure 1**.

This approach has proved fruitful in studies on adventitious shoot (*Convolvulus*, *Nicotiana*) and root (*Malus*) formation. A typical example is the two-step procedure in *Arabidopsis*, in which first callus is induced and from this callus shoots. The occurrence of a dedifferentiation phase is also obvious in indirect root formation. First, many cell divisions occur leading to the formation of a callus, from which root primordia are formed. When the newly formed meristems seem to be formed directly from cells in the initial tissue, transfer studies indicate that the sensitivity to the inducing stimulus is small during the initial period. **Figure 2** shows examples of direct and indirect root and shoot formation.

Control of Organogenesis

Some Characteristics of Plant Hormones

After uptake, applied hormones are rapidly deactivated; only a minor fraction (<5%) is present in the active, nonmetabolized form. A remarkable exception to this is ethylene. The endogenous level of this gaseous hormone usually remains low because it easily diffuses out of tissues into the atmosphere. However, ethylene may accumulate in submerged tissues because of the low solubility of gases in water.

The uptake of various hormones from the medium and their rate of deactivation may be very different, for example, auxin is taken up much faster than cytokinin. It should be noted that applied hormones may alter the metabolism of endogenous hormones, for example, the application of auxin enhances ethylene synthesis in plants. **Table 1** presents an overview of the three hormones that have major roles in organogenesis.

Dedifferentiation Phase

Table 2 reviews the three main phases of adventitious organogenesis. Perhaps the most crucial facet of organogenesis is that a particular cell or group of cells deviates completely from the normal developmental pathway. In about 1960, Steward suggested that one of the requirements for adventitious organogenesis is the physical isolation of cells from the maternal tissues. This supposedly removes the restrictions necessary for proper functioning of cells in the whole plant. Skoog stressed the involvement of gradients of physiologically active substances. It has also been proposed that adventitious organogenesis depends on wounding. When a plant tissue is wounded, cells are triggered to initiate a healing process, and this involves a process of dedifferentiation and redifferentiation during which cells may become temporarily competent to respond to the organogenic signal. An argument in favor of this

Table 1 Characteristics of the plant growth regulators cytokinin, auxin, and ethylene, which have been shown to play an essential role during adventitious organ formation

Compounds	*Inhibition*	*Promotion*
Auxin • Indole-3-acetic acid (IAA) • Indole-3-butyric acid (IBA) • 1-Naphthaleneacetic acid (NAA) • 2,4-Dichlorophenoxyacetic acid (2,4-D) • Dicamba	• 2,3,4-triiodobenzoic acid (TIBA) and 1-*N*-naphthylphthalamic acid (NPA) inhibit polar auxin transport • *p*-Chlorophenoxyisobutyric acid (PCIB) is a genuine anti-auxin competing with auxin for the auxin receptor	• Antioxidants such as phenolic compounds inhibit oxidative destruction of some auxins (IAA, IBA)
Cytokinin • Zeatin (Z) • Zeatinriboside (ZR) • Isopentenyladenine (iP) • Isopentenyladenosine (iPa) • 6-Benzylaminopurine (BAP) • Thidiazuron (TDZ)	• Lovastatin and simvastatin Inhibit cytokinin synthesis, although very unspecifically	
Ethylene • Ethylene	• Aminoethoxyvinylglycine (AVG) inhibits ethylene synthesis • Silver ions inhibit ethylene action; is preferably added as silverthiosulfate (STS)	• 1-aminocyclopropane-1-carboxylic acid (ACC), a direct precursor of ethylene, is rapidly converted to ethylene within the plant

Table 2 Characteristics of the successive steps in adventitious organ formation

Phase	*Characteristics*	*Promotion*	*Inhibition*
Dedifferentiation phase	Cells develop competence to respond to the organogenic signal	Wounding related compounds (ethylene, methyljasmonate)	Oxidative stress related to wounding
Induction phase	Cells respond to the organogenic signal and become determined to either root or shoot formation	Organogenic signals are promotive (auxin for root induction, cytokinin for shoot induction)	Auxin inhibits shoot induction; cytokinin inhibits root induction; ethylene is inhibitory (by disturbing the establishment of polarity?)
Differentiation phase	Cells do not require the organogenic signal anymore; the organ is being formed		Concentrations of the organogenic stimulus that are promotive in the induction phase are now inhibitory

hypothesis is that compounds known to be involved in the wounding reaction favor adventitious organogenesis, e.g., ethylene (**Figure 3**) and jasmonate. Furthermore, adventitious organogenesis usually occurs only after wounding. Interestingly, nonwounded *R. palustris* shoots that form adventitious roots after flooding first show an essential accumulation of ethylene. Root meristems may also be formed without previous wounding in very young, only slightly differentiated stem tissue.

In addition to the wounding reaction per se, wounding is also essential for other reasons. First, a transverse cut interrupts the active downward (from the apex to the base) transport of auxin. As a result, auxin accumulates at the basal wounding surface, and therefore organogenesis often occurs at this site. Second, the epidermis is quite impermeable to inorganic and organic nutrients, and plant hormones. When, for example, excised apple (*Malus pumila*) microcuttings are treated with auxin, uptake via the epidermis is negligible and almost all uptake is via the cut surface. So as a result of wounding, the uptake of medium components is enormously enhanced. Third, wounding involves stress from which the tissue suffers. Compounds that protect the tissue against the stress, e.g., antioxidants such as ascorbic acid or certain phenolic compounds, may thereby enhance adventitious organogenesis.

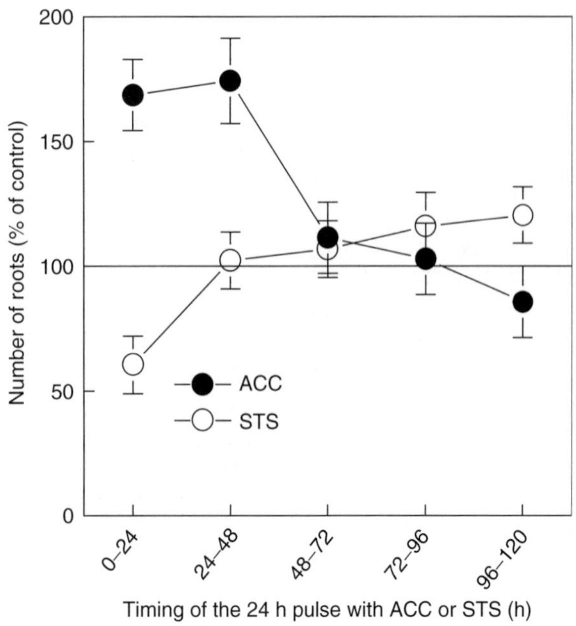

Figure 3 Effect of 24 h pulses with STS (an inhibitor of ethylene action) and ACC (a direct precursor of ethylene) during the rooting treatment of apple. Note that during the first 2 days, ACC promotes and STS inhibits rooting. This is during the dedifferentiation phase.

The differential requirements for plant hormones during adventitious organogenesis have hardly been explored. It has been reported that an otherwise inactive auxin, such as phenylacetic acid, and the anti-auxin *p*-chlorophenoxyisobutyric acid are promotive during the initial dedifferentiation phase, having no or even an inhibitory effect after that. It is possible that this promotion is caused by auxin-induced ethylene synthesis. During the dedifferentiation phase, sufficiently high levels of cytokinin are also required; lovastatin, a compound that blocks cytokinin synthesis, specifically inhibits adventitious root formation in apple microcuttings when applied during the first phase. This is not surprising because of the involvement of cytokinin in cell division.

Induction and Differentiation Phases

In the induction phase, auxin and cytokinin should be supplemented at the appropriate concentrations. Christianson's transfer experiments demonstrated that during induction, low levels of auxin and high levels of cytokinin are required for adventitious shoot formation. For root regeneration, high auxin and low cytokinin should be supplied during the induction phase. Ethylene inhibits adventitious root formation during the induction phase, probably because it interferes with the establishment of polarity. In the third phase, during which the meristem generates the organ, the hormonal composition of the medium is less critical; however, the high concentrations of exogenous hormones required for induction are usually inhibitory.

Endogenous Factors

Adventitious shoots and roots can be induced from virtually all types of tissues, but to varying degrees according to the rules of thumb shown in **Table 3**. A major effect is exerted by the genotype, although the ability to regenerate is probably determined by only a few genes. The genetic component is unlikely to involve sensitivity to hormones, but it is more likely to involve the maintenance of morphogenetic competence.

Table 3 Rules of thumb for selecting tissues for adventitious organ formation

The capability to regenerate decreases with:

- the extent of differentiation (meristematic cells regenerate more easily than fully differentiated cells)
- the ontogenetic age (juvenile tissues are more capable of adventitious organogenesis than adult tissues)
- the physiological age (recently formed tissue has the highest regeneration performance)
- the dormancy state (dormant tissues are less able to regenerate than nondormant ones)

When adventitious shoots and roots develop from the same organ, they usually originate from different tissues. In cultured carrot (*Daucus carota*) petioles, for example, application of indoleacetic acid (IAA) results in the formation of adventitious shoots, roots, and embryos. Roots originate from cells near the vascular bundles, subepidermal cells give rise to somatic embryos, and shoots are formed from large parenchyma cells. This suggests that cells may be capable of only one of the regenerative pathways. It is not known whether this is because cells are genuinely incapable of following certain regenerative pathways (the cells are not totipotent), or whether the position of a cell within the explant results in differences in the microenvironment that are favorable for one of the pathways (e.g., because of a hormonal gradient within the explant). If, as the data suggest, the former explanation is true, then to achieve adventitious organogenesis in recalcitrant species, it is more important to select explants with the appropriate types of cells than to develop highly sophisticated media.

Nonhormonal Environmental Factors

For mineral nutrition, most researchers use Murashige and Skoog's formulation (MS) or derivatives of it. MS was initially developed for tobacco callus growth, but it is satisfactory for shoot and root growth in many species. The mineral composition of MS is similar to that of well-growing, healthy plants. It might be the case that meristematic tissues have a specific mineral composition, and thus require another mineral component in the medium. A reduction in the strength of the medium (e.g., half the concentration of all nutrients) is believed to be promotive in many species, and, as a rule, the MS concentration is halved for adventitious rooting *in vitro*. Furthermore, the ratio of NH_4:NO_3 and the amount of nitrogen may play a role. Casein hydrolysate is sometimes added as organic nitrogen source, and may enhance regeneration. Sucrose is usually added at a concentration of 2–4%. Occasionally, the type of carbohydrate added is critical for adventitious organogenesis. Explants are usually cultured on medium solidified with agar, agarose, or gelrite. The solidifying agent may have a huge effect on adventitious organogenesis.

The pH of the media influences adventitious organogenesis, but this is unlikely to be a direct effect. The pH in the medium determines the uptake of many compounds, the solubility, the activity of enzymes secreted into the medium, the activity of antibiotics, etc. A large change in the pH within the tissues has a devastating effect. The buffering capacity of tissue culture media is low, and it has been observed that the pH of media, originally set at pH 5.5, usually changes considerably during culture. A buffer such as Tris (tris-(hydroxymethyl)aminomethane) is very toxic. The only buffer that is well suitable is MES (2-(*N*-morpholino)ethanesulfonic acid), although at 10 mmol l^{-1}, the commonly used concentration, MES is also inhibitory.

Various other factors have been shown to influence adventitious organogenesis, namely, light, temperature, electrical currents, and the gaseous environment. The reason may be trivial, such as photo-oxidation of some auxins (in particular IAA), or genuine.

Molecular Biology and Biochemistry

Determinations in Explants

Many researchers have carried out biochemical and molecular determinations in explants from which roots or shoots are regenerating. It is very doubtful, however, whether these studies tell us much about adventitious organogenesis because of the very small proportion of cells directly involved in regeneration. For example, in a model system to study root regeneration using 1-mm stem slices cut from apple microcuttings, at 72 h small meristems had been formed that made up far less than 1% of the volume of a slice. Obviously, the biochemical and molecular characteristics of the meristems or their founder cells are swamped by those of the other cells. It should also be noted that within a tissue cells may be very different from one another. A notable example concerns the accumulation of cytokinin in certain cells. A second interesting example is the occurrence of short-lived starch grains which are formed during the very early stages of regeneration of shoots, roots and embryos in the founder cells from which regeneration occurs. The latest very sensitive molecular techniques such as microarrays may be successful in picking up changes in gene expression early in the regeneration process in extracts of explants. Such studies could be accompanied by microscopical studies (*in situ* hybridization), to establish whether the response occurs in the founder cells only.

An involvement of DNA methylation in developmental processes in plants can be deduced from the finding that plants with a reduced level of methylation display many phenotypic abnormalities. Methylation also plays a major role in the development of mammals; for example, early in mammalian embryogenesis global demethylation occurs and after that remethylation. In *Arabidopsis* and tomato (*Lycopersicon esculentum*), methylation levels in young seedlings are approximately 20% lower than in mature leaves, whereas the greatest methylation occurs in seeds. It is possible that hypermethylation during embryogenesis is followed by demethylation during germination and subsequent slow remethylation. Methylation also depends on the type of tissue involved; cells of different tissues of the carrot plant, for example, are characterized by specific methylation patterns. In adventitious organ formation, azacytidin, a drug that causes hypomethylation leads to reduced shoot and root regeneration.

Mutant Studies

In plant developmental biology, the study of mutants is becoming increasingly important; only a few of these studies are directly relevant to adventitious organogenesis. In tobacco, an auxin insensitive mutant, *rac*, has been isolated, which forms callus tissue in response to auxin application, instead of adventitious roots. This suggests that *rac* mutants are unable to respond to the rhizogenic stimulus exerted by auxin during the induction phase. In *Arabidopsis*, adventitious organogenesis is achieved in a two-step method. First, roots are cultured with a 2,4-dichlorophenoxyacetic acid (2,4-D) containing medium, which apparently results in sufficient dedifferentiation. Then, a shoot-inducing medium containing cytokinin is applied. Three *Arabidopsis* mutants, *srd1*, *srd2*, and *srd3*, originally isolated as temperature sensitive mutants with defects in shoot formation, have been examined with respect to their time of action. The mutant genes *srd2* and *srd3* were active during the first phase in which callus formation is induced. These mutants did not form adventitious shoots, but adventitious root formation was not impaired. The third mutant gene *srd1* acted later during the phase in which cytokinin was added and inhibited shoot but not root formation. Such temperature sensitive mutants are very elegant tools for dissecting the process of organogenesis. The *hoc* mutant of *Arabidopsis* regenerates shoots from root explants on hormone free medium.

Recently, much research has been initiated on genes that have regulatory roles in the formation and maintenance of the apical root and shoot meristems. These studies may have a large impact on the understanding of adventitious organogenesis. The *SHOOTMERISTEMLESS* (*STM*) gene and its ortholog *Knotted1* (*Kn1*) function to maintain undifferentiated cells in the shoot apical meristem. In wild-type embryos, the earliest *STM* transcript is found in one or two cells of the late globular stage embryo. Seeds that are homozygous recessive (*stm/stm*) form only a root, hypocotyl, and cotyledons after germination and the apical meristem is absent. Calluses from these incomplete seedlings are unable to regenerate shoots, showing that the *STM* gene is required in adventitious organogenesis. Transgenic plants with overexpression of *Kn1* spontaneously form adventitious meristems. The overall appearance of plants overexpressing the *STM* gene resembles that of plants transformed with cytokinin synthesizing genes, and, indeed, *STM* overexpressing plants have increased levels of cytokinin. In tomato, overexpression of the homeobox gene *LeT6* results in spontaneous production of adventitious meristems.

Many other interesting genes have been reported. *ZWILLE* (*ZLL*), for example, is required to maintain the meristem cell identity within the apex. The *CLAVATA* (*CLV*) gene products promote differentiation of cells and/or repress cell division. Thus, cells mutated in one of the *CLV* loci accumulate undifferentiated cells in the meristems. The *WUSCHEL* (*WUS*) gene occurs very early in the apical region, and whereas *CLV* and *STM* are expressed from the heart and globular stage onward, *WUS* is expressed as early as the 16-cell stage. It is possible that *WUS* maintains the pluripotent capacity of the shoot meristem precursor cells.

In *Arabidopsis*, mutants occur that are incapable of forming lateral roots. One of the genes involved, the *LRP1*, is also expressed early in adventitious root formation. In *Arabidopsis*, the *ROOT MERISTEMLESS1* (*RML1*) gene has been identified. *RML1* mutants have an embryonic root, but fail to initiate cell division upon germination. Lateral roots and adventitious roots initiated from callus show similar defects. The *RML1* gene encodes the first enzyme of glutathione (GSH) synthesis, indicating that this enzyme is essential for the initiation and maintenance of cell division in roots.

Concluding Remarks

From the applied point of view, there are two major problems relating to organogenesis: the recalcitrance to regenerate and the increased incidence of mutations in plants originating from adventitious organogenesis. The latter phenomenon is referred to as somaclonal variation and is dealt with elsewhere (*see* **Tissue Culture and Plant Breeding:** Somaclonal Variation). With respect to the recalcitrance to organogenesis, recent progress has been made by the identification of the successive phases of adventitious organogenesis (that display large differences in hormonal requirements), and by the beneficial effects of hormone-like compounds other than auxin and cytokinin, in particular, wound-related compounds. Knowledge regarding the molecular regulation in shoot and root apical meristems and in lateral root meristems is rapidly increasing. Genes such as those mentioned above are essential for the functioning of the apical meristems, and are probably active from very early on in the induction phase. Furthermore, as these genes control the transition to a more differentiated state, they might determine the regenerative capacity of somatic cells. When these findings on molecular regulation are applied to studies on adventitious organogenesis, the understanding of this process will be improved and new tools for its control may be developed.

List of Technical Nomenclature

Adventitious meristem	Meristem that is not directly derived from an apical meristem, but originates from somatic cells.
Apical meristem	The meristem at the tips of stems and roots.
Auxin	A type of plant growth regulator that induces adventitious root formation; some auxins promote the formation of somatic embryos; auxins inhibit the outgrowth of axillary buds.
Axillary meristem	Meristem in the axil of a leaf; the outgrowth of most axillary meristems is inhibited by the apical meristem.
Competence	Ability of cells or tissues to respond to an inducing signal.
Cytokinin	A type of plant growth regulator that induces adventitious shoot formation; cytokinins promote the outgrowth of axillary buds.
Dedifferentiation	Return of a specialized cell to a more undifferentiated state.
Demethylation	The process by which methyl groups are removed from genomic DNA.
Determination	The commitment of cells to a particular developmental pathway even though there may be no morphological features that reveal this determination.
Differentiation	The process by which cells become specialized, e.g., become vascular cells.
Explant	Excised small organs or pieces of tissue that are transferred to an *in vitro* nutrient medium.
Induction	The process by which a signal exerts its effect.
Meristem	A group of undifferentiated cells that divide and replenish themselves, while at the same time producing cells that undergo differentiation.
Methylation	The process by which methyl groups are added to certain nucleotides in genomic DNA; methylated DNA is not easily transcribed.
Organogenesis	The process of formation of an organ, e.g., a leaf; in tissue culture, "organogenesis" is used in a narrow sense and refers only to adventitious root or shoot formation.
Parenchyma	A thin-walled, relatively little specialized cell type; may carry out various functions.
Regeneration	Repair or replacement of a body part following loss or injury.
Somatic cells	Any cell other than the cells in the meristems.
Somatic embryo	An embryo originating from a somatic cell formed in a process of dedifferentiation and redifferentiation.
Totipotency	The capacity of a single cell to form a complete organism.

See also: **Growth and Development**: Cell Division and Differentiation; Molecular Biology of Development. **Regulators of Growth**: Auxins; Brassinosteroids; Cytokinins; Ethylene. **Root Development**: Root Growth and Development. **Tissue Culture**: Clonal Propagation, *In Vitro*; Micropropagation. **Tissue Culture and Plant Breeding**: Somaclonal Variation.

Further Reading

Christianson ML (1987) Causal events in morphogenesis. In: Green CE, Somers DA, Hackett WP, and Biesboer DO (eds) *Plant Tissue and Cell Culture*, pp. 45–55. New York: Liss.

De Klerk GJ (1990) How to measure somaclonal variation: a review. *Acta Botanica Neerlandica* 39: 129–144.

De Klerk GJ, Van Der Krieken W, and De Jong J (1999) The formation of adventitious roots: new concepts, new possibilities. *In Vitro Cellular and Developmental Biology* 35: 189–199.

Joy IVRW and Thorpe TA (1999) Shoot morphogenesis: structure, physiology, biochemistry and molecular biology. In: Soh W-H and Bhojwani SS (eds) *Morphogenesis in Plant Tissue Cultures*, pp. 171–214. Dordrecht: Kluwer Academic Publishers.

Souter M and Lindsey K (2000) Polarity and signalling in plant embryogenesis. *Journal of Experimental Botany* 51: 971–983.

Sugiyama M (1999) Organogenesis in vitro. *Current Opinion in Plant Biology* 2: 61–64.

Wolpert L (1998) *Principles of Development*. Oxford: Oxford University Press.

Somatic Embryogenesis

J Kallerhoff and G Alibert, INP-ENSAT, Auzeville, France

Definition

Sexual fertilization involves the fusion of a pollen grain and an ovule, to form a zygotic embryo which matures into a seed bearing randomly the genetic characters of both parents. Using *in vitro* culture techniques, an embryo can also be obtained from female (gynogenesis) or male (androgenesis) gametes, with the regenerated plants containing then only one set of parental chromosomes. In contrast to zygotic embryos formed exclusively from gametal cells, somatic embryos derive from nonreproductive (somatic) cells. When these somatic tissues are placed in liquid or solid culture media, following specific conditions of subculture, somatic embryos are induced and can be cultured over long periods of time, provided they are subcultured on a regular basis. Somatic embryogenesis exploits totipotency, which is the characteristic capacity of a plant cell, within a given tissue performing a defined function, to alter its genetic program towards a dedifferentiation process. This involves the loss of its primary function, followed by the acquisition of a new mode of division, which leads to the formation of a new plant.

History

The first credible mention of the production of somatic embryos *in vitro* is attributed to H. Waris in the late 1950s who showed that when *in vitro* cultured seedlings of carrot (*Daucus carota*) or *Oenanthe* were supplemented with high concentrations of some amino acids such as glycine, new plants could be produced from root tips after undergoing morphological steps similar to those of zygotic embryo development. Waris called this phenomenon "neomorphs." It is however now admitted that F.C. Steward and J. Reinert, independently in 1958, were the first to demonstrate the ability of cell suspension cultures of carrot to form somatic embryos. Several authors have since demonstrated that this property was neither specific to carrot nor to the cell suspensions, but that it could be extended to most of the higher plants examined. Indeed, depending on the plant and on the environmental culture conditions, different organs can be used as the starting material for somatic embryo formation. Experimental models with carrot, alfalfa (*Medicago sativa*, lucerne), *Arabidopsis*, or corn (*Zea mays*; maize), have elucidated the main steps involved in the basic mechanisms underlying somatic embryogenesis. This article will focus mainly on somatic embryogenesis in angiosperms, whilst pointing out whenever necessary differences and essential characteristic features of the less-studied case of gymnosperms.

Somatic Embryogenesis versus Zygotic Embryogenesis

Zygotic Embryogenesis

The understanding of zygotic embryo development in angiosperms has to a major extent been unraveled by G. Jurgens on the model plant *Arabidopsis thaliana*. Plant zygotes are produced from the double fertilization event occurring inside the embryo sac: one of the sperm cells of the pollen grain fuses with the oosphere (the female gamete) (**Figure 1A**) to give the diploid zygote while the other sperm cell fuses with the two nuclei of the central cell to give the triploid endosperm. Subsequently, the zygote undergoes an asymmetrical division leading to two cells that have different fates. The smaller cell gives rise to the embryo while the larger one forms the suspensor, a transitory structure that allows exchange of nutrients and molecular signals between the maternal tissues and the embryo. The asymmetrical division of the zygote determines the apicobasal pattern of the embryo development; the root meristem will be formed from the hypophysal cells located in the upper part of the suspensor, while the shoot apical meristem will appear on the opposite side. The development of the embryo proceeds through well-defined stages, the most prominent ones being the globular, the heart-shaped, and the cotyledonary stages (**Figure 1A**). The globular phase is characterized by the formation of the protoderm, which is the precursor of the epidermis of the new plant, whilst the heart-shaped and cotyledonary stages lead respectively to the emergence of the root meristem just above the suspensor, and to the shoot meristems between the cotyledons. Thereafter, the embryo enters the maturation phase during which reserves accumulate in the cotyledons or in the endosperm. Grain filling is followed by the set-up of the two teguments and seed formation is completed with the drying phase and dormancy.

Somatic Embryogenesis

Somatic embryo development follows the same pattern as that of zygotic embryo with some

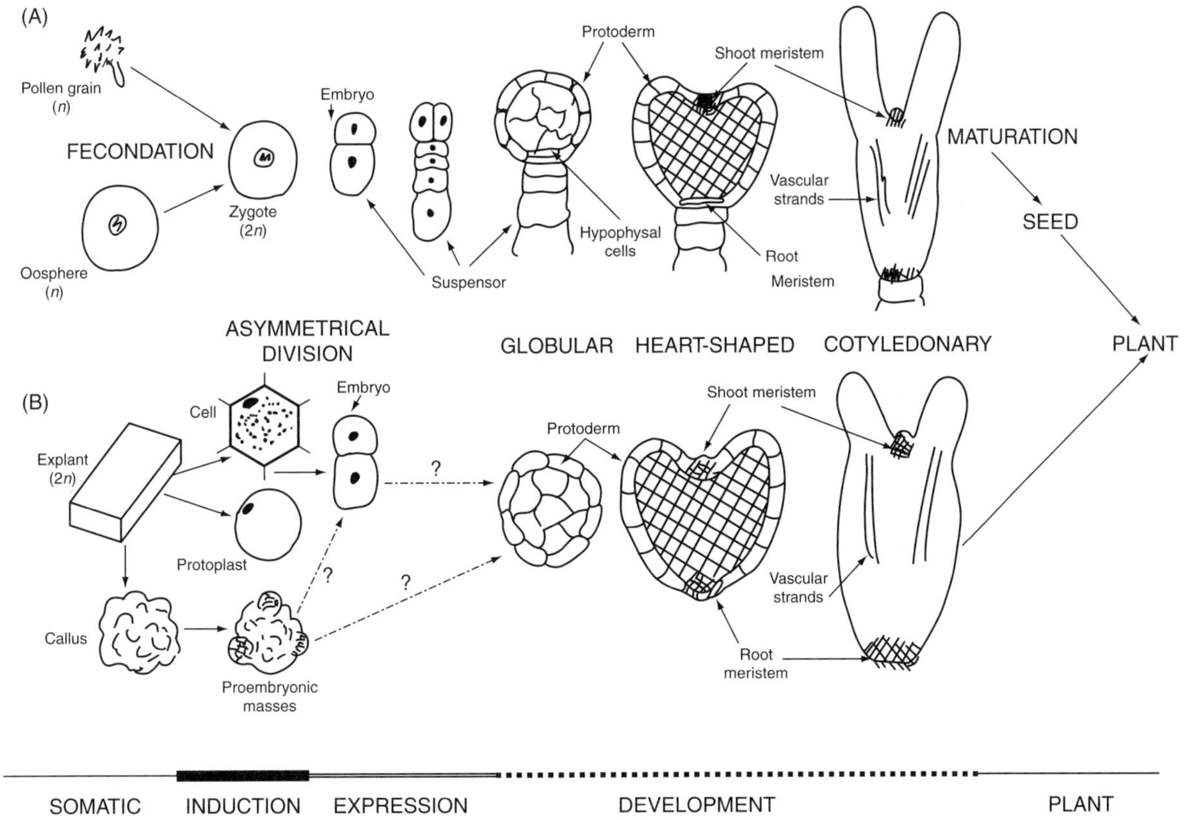

Figure 1 Zygotic embryogenesis (A) versus somatic embryogenesis (B). The drawings are not drawn to scale.

particularities (**Figure 1B**). However, several differences between the two are observed. One major difference is the presence of an induction phase which constitutes the prerequisite step for the reorientation of the somatic program, with a shift to the dedifferentiation one. During this phase, specific treatments given to the plant material lead to the acquisition of an "embryogenic competence" in some cells, hence the general term "proembryogenic cells."

Depending on the plant species, a stimulus applied directly to selected organs leads to some cells acquiring "embryogenic competence." Leaves of *Cichorium* hybrid 474 treated by glycerol to induce an osmotic shock, combined with an incubation at 35°C, give rise to the seclusion of newly formed cells, with a high nucleus : cytoplasmic ratio, and an increased deposit of callosic material in the cell wall. This type of somatic embryogenesis is called "direct" because it develops without any intervening callus phase. Direct somatic embryogenesis can also be induced on protoplasts, as is the case in alfalfa or sunflower (*Helianthus annuus*). The stress resulting from the wall degradation during protoplast preparation is certainly, together with growth regulator treatments, a major factor in the acquisition of the embryogenic competence. In carrot, the most commonly studied model, disorganized callus first produced from somatic tissues can be propagated over long periods before somatic embryogenesis is induced. Treatment of these unorganized callus cultures with high auxin concentrations gives rise to small densely aggregated cells called proembryogenic masses (PEMs), which are the first morphological markers of the embryogenic expression program. Development of embryos from proembryogenic material is more generally achieved by transferring the explants or the PEMs alone, to a medium devoid of auxins. Somatic embryos then follow the same successive morphological developmental steps as described for zygotic embryogenesis: globular, heart-shaped, and cotyledonary (**Figure 2**).

Another general difference between zygotic and somatic embryos in angiosperms is the absence of the suspensor in somatic embryos. In gymnosperms, however, somatic embryos do carry a suspensor and in vigorously shaken embryogenic suspension cultures, they result from polyembryogeny of the primary somatic embryos. Moreover, in angiosperms, somatic embryos generally lack a maturation phase and are able to germinate directly to form a whole plant. In conifers, high osmotic pressure or media containing abscisic acid (ABA) are often used to trigger maturation of somatic embryos. In some

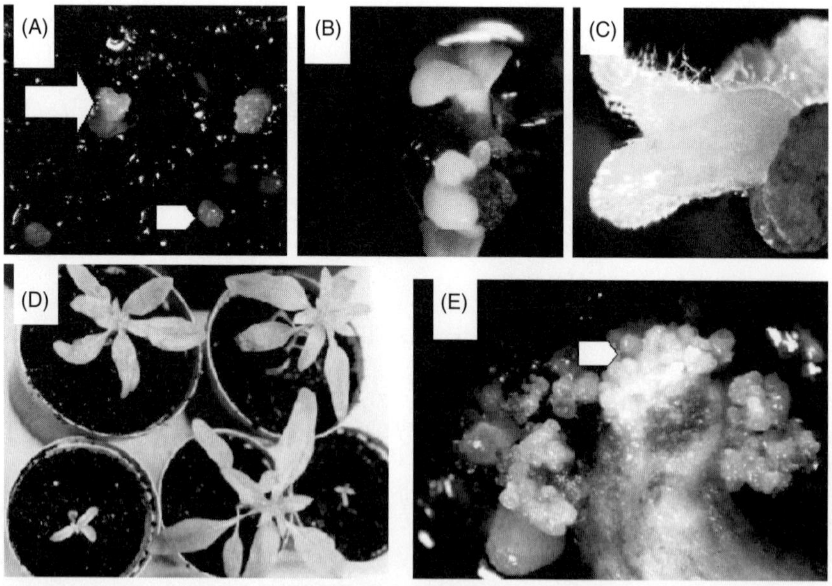

Figure 2 Somatic embryogenesis in sunflower (*Helianthus annuus*). A, globular stage (small arrow) and heart-shaped stage (large arrow) somatic embryos emerging on 8-day-old seedling hypocotyl epidermal layer; B, cotyledonary stage; C, germinating somatic embryo; D, plant regenerated from somatic embryos; E, cycling somatic embryogenesis. Secondary embryos at the globular stage (arrow) are formed on the hypocotyl of a primary somatic embryo.

cases, a desiccation phase is required before the embryo is able to germinate to form the whole plant.

Developmental Events in Somatic Embryogenesis

The Initial Material

Several starting materials such as chopped leaves, hypocotyls, cotyledons, or embryogenic axes of immature or mature embryos are used for the production of somatic embryos. However, zygotic immature embryos seem to be the best, and are specially suited to conifers. In the case of carrot, alfalfa, or *Arabidopsis*, previous establishment of well growing cell suspension cultures is the method of choice for all-year availability of source material required for embryo production.

As presented above, somatic embryo pattern can be divided into three steps: induction, expression, and maturation. We will describe each of these steps in the light of results recently obtained on much-studied models, such as carrot, alfalfa, *Arabidopsis*, and sunflower.

The Induction Phase

The aim of the induction phase is to allow cells of the studied material to complete a transition phase from a somatic state to a proembryogenic one. Competent cells are those which give rise to embryos after transfer to conditions in which the exogenous applied stimuli are removed. The competence is in most cases induced by a hormonal treatment, generally auxin (mainly 2,4-D) in the presence or not of a cytokinin, following a stress which could be wounding, osmotic shock, or high-temperature treatment.

Several indirect evidences have involved chitin-like molecules (similar to the *Nod* factors that act in the differentiation of nodules in leguminous plants) in somatic embryogenic competence acquisition in carrot cells. It has indeed been shown that endochitinase enzymes extracellular protein 3 (EP3) are secreted into the medium of embryogenic culture cell suspensions. The molecules formed through endochitinase action would possess hormone-like properties, as somatic embryogenic capacity is reinitiated in temperature-permissive carrot mutants *ts11* lacking this property. Despite many efforts, no substrate molecule for these endochitinases has to date been identified in plants. It has nevertheless been reported that EP3 genes coding these enzymes are not expressed in the embryogenic cells, but rather in other cells of the suspension. These results point to the existence of cell-to-cell communication between the different types of cells within the suspensions, which could reflect the degree of responsiveness to initiation of somatic embryogenesis. More recently, a class of compounds called arabino galactan (AGPs) proteins was isolated from a suspension culture medium of embryogenic cells. These AGPs induced embryogenic competence when added to non-embryogenic cultures of carrot cell lines. Moreover, the existence of a particular class of soluble AGPs containing *N*-acetylglucosamine and having the ability to be cleaved by endochitinases has led to the proposition that AGPs could be candidate substrate molecules of the EP3 endochitinases. Thus, in addition to hormones and stresses, signal molecules could assume the exchange of messages between cells within a cell population or a tissue, and play a central role in the acquisition of embryogenic competence of the material.

Among a tissue or a cell suspension culture, a low proportion of cells (only 1–2% in a carrot cell suspension culture) becomes competent. According to several authors, these cells can be considered as somatic embryo arrested in their development at a very early stage. Indeed the messenger RNA (mRNA) content of these cells shows striking similarities with those of well-developed embryos. Moreover embryogenic clusters in carrot show polarity distribution of DNA synthesis capacity and activated calmodulin.

Identification of biochemical markers associated with the acquisition of the embryogenic competence would be very useful for the identification of the responding material. It has been shown, by cell tracking, using promotor driven luciferase gene that expression of the *somatic embryo receptor-like kinase* (*SERK*) gene coincides with expression of competent cells. Moreover it has also been shown that the endogenous level of ABA was several times higher in competent cell clusters than in nonembryogenic cells. These two characters could therefore be retained as markers of the embryogenic state of competence. Thus, with the exception of the role of stresses and the involvement of some signal molecules, the mechanism of somatic embryogenic competence acquisition remains as yet largely unknown.

The Expression Phase

Following induction, some cells enter the embryo developmental program after the stimuli have been removed. The pattern of the expression phase is characterized by two successive processes: the acquisition of first apicobasal polarity and then radial and bilateral polarity (**Figure 1B**).

Apicobasal polarity fixation Apicobasal polarity acquisition of the zygote seems a prerequisite for embryo development in animals and plants. Morphologically, apicobasal polarity is clearly visible in zygotic embryogenesis (**Figure 1A**), where the zygote divides generally asymmetrically into two cells bearing different sizes, contents, and fates. In somatic embryogenesis, polarity acquisition is not as clear as in its zygotic counterpart, either because the embryogenic cells are embedded among nonembryogenic ones, and thus difficult to observe, or the polarity process occurs through tiny modifications within the embryogenic cells, and is therefore hard to identify. In alfalfa protoplasts, transfer from a medium containing a high auxin level to a medium without auxin leads to asymmetrical divisions at the origin of somatic embryos. In sunflower, embedding protoplasts into agarose matrix and culturing them in the presence of 2,4-D leads to asymmetrical division and to the formation of embryoid structures (embryo-like structures). In carrot, transfer of isolated cells pretreated with auxin to a medium lacking the hormone results in asymmetrical divisions, giving rise to embryos. The origin of the embryos from single cells was confirmed in carrot by video cell tracking, but the heterogeneity of the carrot cell suspension cultures or of the starting explant material makes it hard to study the initial events occurring prior to the formation of the somatic embryos. However, owing to the precision of the zygotic embryo pattern formation program, these first stages could be very similar in both cases.

In zygotic embryogenesis, the polar axis is determined by the first division and fixes the apicobasal formation of the embryo. In somatic embryos, it is not clear whether this axis is really initiated at such an early step or if it is acquired later, at the globular stage following unknown mechanisms. Experiments involving protoplasts in alfalfa and in sunflower point towards a mechanism similar to those described in zygotic embryo formation.

Radial and bilateral polarity development The appearance of the globular embryo is the most constant morphological event detected at the onset of the somatic embryogenic program. The definition of the apicobasal axis is difficult to detect during early stages of globular embryo formation, as the suspensor is often absent. The formation of the protoderm (a mono cell layer around the embryo formed by anticlinal divisions) is absolutely essential for completion of this determinate stage. It will later form the epidermis of the entire embryo. In many cases, the protoderm is only partially formed or even absent, leading to aborted embryos. It is generally assumed that the radial pattern proceeds through positional information mediated by cell-to-cell interactions.

Following the globular stage, the embryo enlarges rapidly and follows the same morphological pattern as that described for zygotic embryo: heart-shaped and cotyledonary stages. As in zygotic embryos, polar auxin transport is assumed to be involved in the bilateralization of the globular embryo to heart-shaped and cotyledonary stages. Indeed, the addition of inhibitors of the polar auxin transport such as 2,3,5-triiodobenzoic acid (TIBA) to culture medium leads to the formation of abnormal structures with fused cotyledons while the formation of the shoot and the root meristems are not altered. Thus, mechanisms other than polar auxin transport seem to be involved in the process. The pertaining question yet to be answered is the stage at which the bilateral polarity is acquired: is it before the globular stage or at a late globular stage? One important feature of the heart-shaped and cotyledonary stages is the initiation of the shoot and the root meristems and the emergence of the cotyledons. In somatic embryogenesis, the lack of germinating capacity of many embryos can be explained by defects occurring at this stage.

Formation of abnormal embryos is a general feature observed in somatic embryogenesis. Many of the morphological forms obtained are very similar to that obtained in zygotic development mutants produced in *Arabidopsis*. For example somatic globular embryos lacking a normal protoderm formation are very similar to *knolle* mutant embryos. In some cases, the apical somatic embryo meristem can be only partially organized or even lacking, as observed for the *shoot-meristemless* mutants. The cotyledons can be either lacking or on the contrary be multiple and fused as presented in **Figure 3** for soybean (*Glycine max*), a pattern encountered in the *gurke* or *gnom* mutant. Lastly, root formation is often difficult in somatic embryos, a characteristic undoubtedly linked to the absence of a suspensor and to some alterations in the hypophysal cell group, as in the mutant *hobbit*. As induction of somatic embryogenesis requires in most cases high concentrations of hormones (mainly auxins), such abnormalities could be related to a detrimental effect of the hormonal balance inside the cells.

The Maturation Phase

The induction and expression phases represent the embryogenesis process, *sensu stricto*, during which all the organs of the embryo are established. In zygotic embryos, the maturation phase concerns events linked with synthesis and accumulation of reserves, desiccation, formation of teguments, and dormancy. Generally in somatic embryos from angiosperms, these two later steps are not required and after a rapid growing step, the embryos can germinate directly to form plantlets. Storage proteins of the same type as that of their zygotic counterparts are found in somatic embryos but they accumulate to a lesser extent. Moreover, it seems that the synthesis of these compounds starts at an early stage of the embryo formation, while in zygotic embryos they appear mainly after the completion of the structural organ formation. Similarly, late embryogenic abundant (LEA) proteins which are involved in desiccation resistance of the seeds are not synthesized as abundantly in somatic embryos as in zygotic embryos, and seem to be present at all stages of the embryo development.

In conifers, the somatic embryogenic pattern is very similar to that of the zygotic one. For maturation, immature somatic embryo suspension cultures must be transferred to a medium containing ABA and a high osmolarity. Following this transfer, storage proteins and lipids accumulate in the embryos with a composition similar to the zygotic embryo content. Finally, drying the embryos greatly improves their germination potential. LEA proteins accumulate during this period. Following imbibition somatic embryos can develop to seedlings.

Cyclic Embryogenesis

Cyclic embryogenesis is defined as the capacity of somatic embryos to develop into primary embryos, induced on explants without an intervening callus phase. Furthermore tertiary embryos can be produced on the secondary embryos and so on, depending on the culture medium composition and the extent of subcultures. Cyclic somatic embryogenesis is well documented in alfalfa. In this case, somatic embryogenic callus can be induced on leaf or petioles of selected genotypes by high concentrations

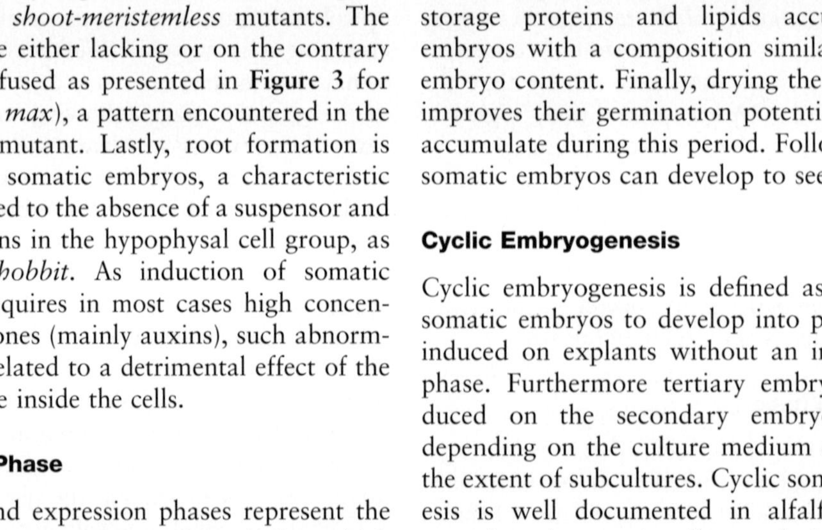

Figure 3 Somatic embryogenesis in soybean (*Glycine max*). A, globular stage (small arrow) and heart-shaped stage (large arrow) somatic embryos formed on cotyledons of immature zygotic embryo; B, somatic embryo at the cotyledonary stage, detached from the explant (somatic embryogenesis was induced by α-naphthalene acetic acid (NAA)); C, abnormal somatic embryo (fingerlike cotyledons) appearing after induction with 2,4-D.

of auxin and cytokinin. Upon transfer of the embryos to a medium free of growth regulators, new somatic embryos are repeatedly formed on the older ones in a cycle of about 30 days. Somatic embryos can then either be set for plant production or for further somatic embryo production.

Similar patterns with some variations in the technical protocol have been also described for plants as different as carrot, sunflower, Japanese maple (*Acer*), etc. (**Figure 3**). This system allows mass production of embryos through exponential propagation.

Somatic Embryogenesis as a Model System for Understanding Zygotic Embryogenesis

Somatic embryogenesis is a powerful model system to be used when studying mechanisms involved in zygotic embryo formation. The young zygotic embryo is minute and its development occurs deeply embedded within maternal tissues. Moreover, dissection of ovaries to achieve removal of a large number of young embryos requires much effort. Most of the available results on zygotic embryo development concern morphological descriptions recorded through microscopy observations. Molecular and biochemical analyses are largely hampered by physical inaccessibility and by the technical difficulty of extracting solely embryos without tissues issued from maternal origin. In contrast, somatic embryos can be obtained on a large scale, from very early stages of development and sometimes with a more or less synchronous pattern. This characteristic makes somatic embryos useful for molecular studies. The ability to identify mutants in carrot, *Arabidopsis,* and corn has led to the characterization of genes specifically involved in developmental regulation. Developmental stage-specific molecular markers involved in determining events during the differentiation of cells have also been identified. The use of *Arabidopsis* leafy cotyledon mutants (*LEC1* and *LEC2*) has allowed the identification of control sequences required for the establishment of the state of embryonic competence. Furthermore, the isolation of the *BABY BOOM* (*BBM*) gene from immature pollen grains of oilseed rape (*Brassica napus*) suggests a role for this gene in promoting cell proliferation and morphogenesis during embryogenesis.

Somatic Embryogenesis and Plant Breeding

Propagation of Cloned Lines

Based on the early work of Steward on carrot, plants can now routinely be produced through somatic embryogenesis. **Table 1** indicates a list of species where this technique has been successfully applied. Woody as well as nonwoody plants can be propagated from somatic material giving rise to clones of selected genotypes of high economical value. The superiority of this *in vitro* propagation technique over others such as microcuttings, bud culture, or organogenesis is its efficiency, associated with the high conformity of the produced clones respective to the donor parent. This conformity is due to the unicellular origin of the somatic embryo in contrast to the pluricellular origin of organogenic buds, which arise from the structural reorganization of dedifferentiated somatic cells into meristems, which thereafter develop into shoots.

In gymnosperms, efficient somatic embryogenesis can only be induced in immature zygotic embryos. Thus because each zygotic embryo carries a specific genome, plants obtained from several immature embryos are genetically different from each other and different from the donor parent. So the plants issued from the somatic embryos may not carry in their genome all the selected agronomically important traits of the parent. A solution to this problem has been found by cryopreservation in liquid nitrogen, in the presence or absence of a cryoprotectant such as glycerol, allowing the conservation of immature embryos during months and sometimes over several years without loss of plant regeneration potential. Long-term conservation can

Table 1 A nonexhaustive list of species successfully regenerated into entire plants by somatic embryogenesis

Woody plants	*Nonwoody plants*	
	Monocotyledons	*Dicotyledons*
Abies balsamea	*Asparagus*	*Antirrhinum majus*
Albizia lebbeck	*Dactylis glomerata*	*Arabidopsis thaliana*
Citrus spp.	*Iris germanica*	*Arachis hypogea*
Cocos nucifera	*Musa* spp.	*Begonia*
Coffea canefora	*Oryza sativa*	*Cicer arietinum*
Cycas	*Panicum maximum*	*Cichorium intybus*
Eucalyptus sutens	*Saccharum officinarum*	*Cucumis sativus*
Eonymus europaeus	*Sorghum bicolor*	*Daucus carota*
Feijoa sellowima	*Triticum aestivum*	*Gossypium*
Picea glauca	*Zea mays*	*Helianthus annuus*
Pseudotsuga menzesii		*Lycopersicon esculentum*
Quercus robur		*Medicago sativa*
Sequoia sempervirens		*Pelargonium*
Sorbus domestica		*Pisum sativum*
Populus		*Ranunculus asiaticus*
Vitis vinifera		*Rosa hybrida*
		Solanum sarrachoides
		Vigna unguiculata

be also used in order to preserve genotypes which are being field-trialed. Following agronomic evaluation of the clones, elite genotypes can then be removed from cryogenic storage and used for mass propagation. This technique offers several other advantages:

1. Multiple somatic embryos coming from one immature embryo can be produced from subsequent production runs, then pooled together and stored until sufficient numbers have been produced. Culture batches can be set continuously throughout the year.
2. The cryopreserved embryos can be recovered when necessary and can be made to germinate synchronously to provide plants of uniform age and size during the suitable growing season.

Several species such as pine (*Pinus* spp.), Norway spruce (*Picea abies*), white spruce (*Picea glauca*), black spruce (*Picea mariana*), and Caribbean pine (*Pinus caribaea*) have been successfully produced by this method.

Artificial Seed Production

Mass propagation of somatic embryos in the field is not practical as they are devoid of teguments and are very fragile to handle. The embryos are therefore coated with gels enriched with nutrients, growth hormones, or pesticides to provide them with physical protection, and to improve conservation and germination capacities. Encapsulation into beads of sodium alginate (2–3%) seems to give the best results. This method has been applied to Norway spruce and loblolly pine (*Pinus taeda*) with good success. Refined methods involve the use of polyethylene oxide plastic resin or a mixture of wheat (*Triticum aestivum*) starch and kappa-carrageenan, which favor gas exchanges and/or nutrients with the embryos. Several plants as diverse as carrot, mango (*Mangifera indica*), papaya (*Carica papaya*), or horseradish (*Armoracia lapathifolia*) have been treated with this technology with experimental success. However, the economical viability of synthetic seed technology still remains to be proven.

Genetic Engineering of Plants

One of the prerequisites of most gene transfer technologies is responsiveness *in vitro* of the starting explants. The principle consists in transferring and integrating foreign gene(s), generally into the cell nuclear genome. If this cell can be oriented to adopt dedifferentiation followed by an embryogenetic program, an entire plant harboring and expressing the transgene in all its cells can then be produced. As somatic embryos arise in most cases from single cells to give rise to plants with high genetic conformity relative to the donor parent, somatic embryogenesis represents the technique of choice to produce transgenic plants.

Gene transfer can be mediated by agrobacteria or by direct gene transfer using electroporation or particle bombardment. Graminaceous plants are less sensitive to *A. tumefaciens* than dicotyledonous plants and, as such, most of transgenic wheat, corn, *Sorghum*, barley (*Hordeum vulgare*), banana (*Musa sapientum*), sugar cane (*Saccharum officinarum*), etc. have been obtained following particle bombardment of embryogenic cell suspension cultures. Thus somatic embryogenesis has gained in importance in recent years and is now integrated as being part of the most important revolution in the history of plant breeding.

List of Technical Nomenclature

Callus	A mass of nonorganized cells obtained from tissues or organs cultured *in vitro* generally in presence of auxin.
Calmodulin	A protein that binds calcium. The calcium–calmodulin complex activates other enzymatic proteins in a cascade manner.
Cell suspension culture	Callus under vigorous agitation in a culture medium dissociate into individual cells (or cell clumps). The cells can divide in presence of the required hormones.
Cryopreservation	A technique to decrease the metabolic activity of an organism by a step-by-step cooling from the ambient to the liquid nitrogen temperature.
Dormancy	A period during which a seed reduces its metabolic activity after having lost part of its water content.
Electroporation	A technique to introduce molecules (mainly genes) inside cells by the mean of electric currents which induce "pores" on the plasma membrane.
Explant	A piece of organ or tissue used for *in vitro* culture.
Hormones	Plant growth factors: organic molecules that control the plant growth and development.
Imbibition	A period during which a dormant seed absorbs water in order to reinitiate its metabolism.

Meristem	A mass of undifferentiated cells present in the shoot buds and the apical root at the origin of the formation and growth of the shoots and the roots.
Mutant	An organism in which one or more character(s) has been changed at the gene level by physical or chemical treatment.
Particle bombardment	A technique to introduce genes into cells by gun bombardment using gold or tungsten microparticles, coated with the required DNA.
Polarity	In a cell, the establishment of two poles characterized by the segregation at each pole of different components of the cell.
Promoter	In chromosomes, a noncoding sequence of nucleotides which controls the expression of a gene.
Somatic	The somatic cells constitute the body of the plants or animals. They generally contains two sets of chromosomes each coming from one of the two parents.

See also: **Production Systems and Agronomy**: Nursery Stock and Houseplant Production; Plantation Crops and Plantations. **Seed Development**: Artificial Seeds; Embryogenesis. **Tissue Culture and Plant Breeding**: Clonal Propagation, Forest Trees; Regeneration of Fruit and Ornamental Trees via Cell and Tissue Culture. **Tissue Culture**: Artificial Seeds; Clonal Propagation, *In Vitro*; General Principles; Micropropagation; Organogenesis.

Further Reading

Dodeman VL, Ducreux G, and Kreis M (1997) Zygotic embryogenesis versus somatic embryogenesis. *Journal of Experimental Botany* 48: 1493–1509.

Dudits D, Gyorgyey J, Bogre L, and Bako L (1995) Molecular biology of somatic embryogenesis. In: Thorpe TA (ed.) In vitro *Embryogenesis in Plants*, pp. 267–308. Dordrecht: Kluwer Academic.

Mordhorst A, Toonen MAJ, and de Vries S (1997) Plant embryogenesis. *Critical Reviews in Plant Sciences* 16: 535–576.

Schmidt EdDL, de Jong AJ, and de Vries S (1994) Signal molecules involved in plant embryogenesis. *Plant Molecular Biology* 26: 1305–1313.

Sorvari S, Toldi O, Ahanen K, *et al.* (1997) Using polysaccharides and galactomannans as gelling agents in capsule formation of artificial seeds. *Journal of the American Society for Horticultural Science* 122: 878–883.

Zimmerman JL (1993) Somatic embryogenesis: a model for early development in higher plants. *Plant Cell* 5: 1411–1423.

Artificial Seeds

S Pond and S Cameron, Canadian Forestry Service, Fredericton, NB, Canada

Definition

Strictly speaking, the definition of an artificial seed (also called a synthetic seed or synseed, seed analog, or manufactured seed) is a somatic embryo inside a coating; it may be sown in the same manner as a conventional seed. Somatic (nonzygotic) embryos are bipolar (having both root and shoot meristems), arise from somatic (nonsexual) cells, and are genetically identical. Zygotic embryos are a result of sexual reproduction (fusion of gametes) and are not genetically identical making them less useful for the production of artificial seeds. The somatic embryo may be hydrated or desiccated; the coating may be water impermeable or water soluble and may also enclose nutrients (artificial endosperm) and other additives deemed necessary (e.g., mycorrhizal fungi, fungicides, and/or bacteriocides). The term "artificial seed" may also refer to unencapsulated (naked) somatic embryos (either hydrated or desiccated). Due to the low success rate and/or high cost of somatic embryo production in some species, buds, shoots, bulbs, or other meristematic tissues that can produce a whole plant may also be encapsulated. These are also considered to be artificial seeds. The type of artificial seed produced and its economic feasibility vary greatly among species, depending on the ease and cost of natural reproduction (i.e., by seed or by vegetative propagation) and the state of current technology in somatic embryogenesis for that species. The decision to use artificial seed technology depends on the plant species, the need for improvements in production efficiency, or other justification for the development and production costs associated with artificial seed.

Role of Natural Seed and Artificial Seed in Plant Propagation

The seed is the delivery system in sexually reproducing plants that passes on genetic material from one generation to the next. A natural seed develops from a fertilized ovule and consists of a zygotic embryo with a nutritive supply surrounded by a protective coat. The seed structures surrounding the embryo protect it during storage and sowing, and nourish it during germination. They also regulate gas exchange (which influences embryo respiration), and embryo development, and prevent precocious germination. The majority of seeds from temperate species tolerate desiccation – i.e., they can survive without water for long periods of time – and in this anhydrous state can survive extremely adverse conditions for many years. They can also be kept in frozen storage or in liquid nitrogen for long-term germplasm storage. These dried orthodox seeds can be mechanically planted directly in the field in a cost-efficient manner. Automated machinery is readily available to handle conventional seed. These attributes make seeds the most feasible and cost-efficient means of storing and propagating plants.

However, there are drawbacks to the use of natural seeds as the sole units of propagation. Many plants are poor seed producers and require supplemental methods of propagation to decrease production costs and supply sufficient quantities of plant material to meet demands. In other cases, the plants do not breed true. Sexual reproduction recombines specific genes, making it impossible to preserve unique gene combinations. Inbred strains also often result in decreased vigor and lower seed productivity. In order to produce an exact copy or clone of the original plant, these species or varieties must be vegetatively propagated. This has been done very successfully in many plant species. Many tropical species have recalcitrant seeds (i.e., they are sensitive to chilling or desiccation and cannot be stored) and thus require alternative methods to store and propagate the germplasm. Diseases are intrinsic in seeds of some species, so other means of propagation are needed to produce disease-free plants. In some species with long reproductive cycles (gymnosperms being a prime example) where the time from germination to first flowering is measured in years, production of natural seed from a desired cross is a very slow process.

Vegetative reproduction through tissue culture can circumvent the above problems. Large numbers of identical clones can be produced in a short period of time and a cost-effective manner; sterile or unstable genotypes can be multiplied efficiently; and tissue can be cryopreserved to permit long-term storage of germplasm. Tissue culture is used to produce disease-free plants, and genetically engineered plants are initially multiplied using *in vitro* methods.

There are a number of ways to vegetatively propagate plants through tissue culture, such as organogenesis and meristem culture. However, the most promising method for the development of encapsulable units suitable for artificial seed is somatic embryogenesis (SE). Somatic embryogenesis mimics the zygotic embryogenesis that occurs in a natural seed. Superficially similar stages are produced in most species in somatic and zygotic embryogenesis, each resulting in an embryo (either somatic or zygotic, respectively) with shoot and root apices ready to germinate. Somatic embryogenesis has the potential to economically produce unlimited numbers of identical somatic embryos that can be germinated into crops of identical plants. However, these embryos are naked. The tissue culture nutrient medium and not the mother plant provides nutrition. Availability of oxygen and other gases is regulated, often poorly, by the culture environment and not by the protective layers of the seed coat. As a result, developmental problems may occur during maturation of the somatic embryos because of nonoptimal culture methods. These problems must be addressed before embryo production and artificial seed production can be automated.

Naked somatic embryos can be used as artificial seeds, but they are not quiescent and they lack the nutritive and protective layers present in a natural seed. The embryo epidermis is unlikely to have protective waxy layers. Therefore, embryos are germinated under laboratory conditions (*in vitro*) until they become photoautotrophic, after which they are gradually hardened off. They are then transplanted to a greenhouse where they are further acclimatized, at which point they can be treated as normal seedlings. This process is labor intensive and costly. If these naked embryos were encapsulated with nutritive material and a protective coating, they would hypothetically be protected during handling and germination, and would receive nutrients and/or other factors to enhance germination. Thus, a coating is seen as a way to germinate artificial seed in less demanding conditions, and reduce handling costs. Desiccation of the embryos prior to encapsulation can induce quiescence, allowing them to be stored without loss of vigor. The ultimate goal in the production of artificial seeds is to produce a coated propagule as cost efficiently as natural seed. Ideally, the propagule could be stored for long periods of time without germinating or losing vigor. It would germinate and produce a whole viable plant when transferred to conditions conducive to germination; also, it would be strong enough to withstand the stress of mechanical seeding equipment. However, many technical and biological barriers still prevent attainment of these goals.

History

In 1902, J. Gottleib Haberlandt predicted that artificial embryos would be successfully cultivated from vegetative cells. In 1958, Stewart and Reinhart were the first researchers generally acknowledged to have proved this prediction by independently discovering SE in carrot (*Daucus carota*) plants. Earlier reports by Levine in 1947 and Curtis and Nichol in 1948 describe structures, in carrot and orchid (Orchidaceae) respectively, that may have been somatic embryos.

At a conference in 1977, Toshio Murashige first publicly addressed the concept of artificial seeds and, since then, different approaches have been actively pursued to make it a viable technology, with varying levels of success. In the mid-1970s, Keith Walker's group at Monsanto worked on alfalfa (*Medicago sativa*, lucerne), soybean (*Glycine max*), and other vegetable crops while Robert Lawrence's group at Union Carbide began working on forest trees, celery (*Apium graveolens*), and lettuce (*Lactuca sativa*) using fluid drilling and polyoxyethylene for seed tapes or sheets. In the 1980s, Redenbaugh and coworkers developed sodium alginate hydrogels for single embryo artificial seeds and were able to produce hydrated artificial seeds of alfalfa and celery, although they had very low conversion rates (7–10%, respectively). Work on encapsulation of hydrated embryos continues today. Research is ongoing at Weyerhauser Ltd as part of their integrated Miniplug® system on a sophisticated botanic seed analog for conifer somatic embryos which incorporates a perfluorocarbon-oxygenated nutrient medium separated from the embryo by a porous membrane, and covered with a rigid waterproof seed coat with a weak spot for root emergence.

In the 1980s, work also started on desiccation of embryos for use as artificial seed. In 1985, Gray and Conger reported plant recovery (4%) from desiccated quiescent orchard grass (*Dactylis glomerata*) somatic embryos. This was repeated with grape (*Vitis vinifera*) somatic embryos in 1987 (equilibrated to 13% water content, stored for 21 days with 34% germinating and converting after imbibition). McKersie's group in 1989 kept alfalfa somatic embryos desiccated for 1 year with a 60% survival rate after imbibition. Compton and coworkers were able to germinate plants from dried corn and soybean embryos in 1992 at about a 60% success rate. Attree and colleagues in 1991 and 1992 recovered viable plants from 81% of interior spruce embryos after dry storage for 14 days (using embryos matured in a medium with polyethylene glycol as the osmoticant) and, in 1999, patented his desiccation method and an encapsulation method which uses a melted nonhydrated water-soluble compound (polyethylene glycol) to coat the desiccated embryos and yield a hardened capsule when solidified. When planted, the capsule gradually dissolves after exposure to water.

The journal and patent literature is currently filled with accounts of attempts to commercialize artificial seed technology for fruits and plantation crops, vegetable crops, ornamental crops, spice, medicinal/aromatic plants, and trees. There have been many modest successes published in the development of desiccated artificial seed but, in most cases, too many biological and/or technological problems remain to use the technology on a commercial basis.

Feasibility of Producing Artificial Seeds

As discussed previously, there are a number of possible artificial seed systems. The type of artificial seed produced, based on its economic feasibility, will vary greatly among species, depending on the ease and cost of natural reproduction (i.e., by seed or by vegetative propagation), the state of current SE technology for that species, the need for improvements in production efficiency, or other justification for the development and production costs associated with artificial seed and the end use of the artificial seed (e.g., storage, greenhouse germination, field germination). As stated by Murashige and Street in the 1970s, and as has been proved consistently in the literature, the major stumbling block in establishing artificial seed production as a viable technology is a lack of understanding of the SE process and an inability to consistently produce high-quality propagules that can germinate in a soil environment with an acceptably high success rate.

The development of the ideal viable, quiescent, low-cost artificial seed was described as a 10-step process by Redenbaugh in 1987. Some steps generally apply to more than one species whereas other steps may be species-specific. The latter are inevitably the most demanding in terms of development, and are noted as such.

1. Selection of the crop based on technological and/or commercial potential (species-specific).
2. Establishment of an SE system (species-specific).
3. Optimization of the clonal production system (optimizing protocols to synchronize and maximize the development of normal mature embryos capable of conversion to normal plants) (species-specific).
4. Automation of embryo production.
5. Posttreatment of mature embryos to induce quiescence (which may involve a desiccation step).
6. Development of an encapsulation and coating system.
7. Optimization and automation of the encapsulation system.

8. Optimization of conversion requirements for greenhouse and field growth (watering, fertilizer, transplantation, etc.).
9. Identification and control of any pest and disease problems that may be unique to artificial seeds.
10. Determination of the economic feasibility of using the artificial seed delivery system for a specific crop compared with other propagation methods (cost–benefit analysis of encapsulation versus other options).

Methods of Producing Artificial Seeds

Hydrated Somatic Embryos

Naked embryos Naked, hydrated, artificial seeds have the lowest development cost of all the artificial seed systems using somatic embryos, but have the highest handling and germination costs (**Figure 1**). They require a well-established, cost-efficient SE system capable of routinely producing large numbers of high-quality somatic embryos showing good conversion percentages. Hydrated embryos are nonquiescent and must be germinated immediately after production or stored under moist conditions for only very short periods of time. They are fragile and, in most cases, require careful, individual handling that increases the cost per seed. They are germinated *in vitro* under well-controlled conditions and transferred manually through various hardening-off procedures before transplantation to the greenhouse.

This method may be applied to high-value ornamental crops where the high end-value of the crop justifies the high cost per plant. An example of a crop that uses this method of propagation is Colorado blue spruce (*Picea pungens*) in which only a small percentage of trees have good color. They are selected from seed-grown nursery stock at 3–7 years of age for color and form. The parent trees are insufficiently juvenile by then to be propagated by any means other than by grafting which results in a cost to the consumer of US$60.00 per tree. Propagating blue spruce trees using SE allows a portion of the SE tissue from each clone to be frozen while the remainder is used to produce somatic embryos. The embryos are germinated and planted, after which the seedlings are observed for several years for development of the desired blue color. The tissue from the clones that produce the best color can then be thawed and used for large-scale production of naked, hydrated somatic embryos at a lower cost than grafted trees.

Although costly, mass propagation using naked embryos has the benefit of circumventing the biological and technological limitations involved in coating and/or drying somatic embryos. Machinery has been developed to automate planting of *in vitro* germinated somatic embryos into sterile plugs for greenhouse rearing (Fast Plant® developed by Silvagen, Halifax, Nova Scotia and currently used in interior spruce) that decreases handling costs.

Encapsulated embryos Hydrated mature somatic embryos can be encapsulated in a hydrogel (**Table 1**). The majority of the published reports discuss this type of artificial seed (**Figure 2**). The coating should be nontoxic and offer protection from mechanical damage while providing water and nutrients to the embryo during storage and sowing. However, there are problems associated with the use of hydrogel capsules. In some species, embryo vitrification due to aberrant water relations is a problem. Gas exchange through the gel may inhibit growth, and nutrients can leach from the beads. Excessive gel firmness may inhibit germination. The gel provides an excellent nutrient medium for microorganism growth and,

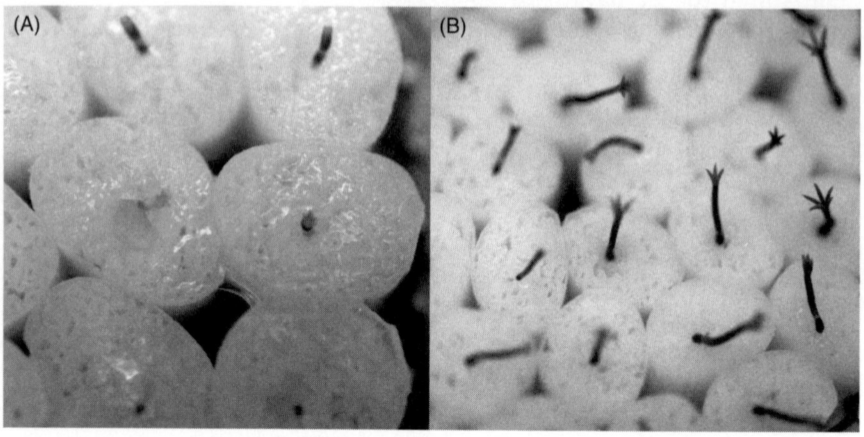

Figure 1 Germination of hydrated, naked white spruce embryos on sorbarods: (A) at 3 days; (B) at 14 days. Reproduced with permission from Maruyama E, Kinoshita I, Ishii K, *et al* (1997) Alginate-encapsulated technology for the propogation of tropical trees: *Cedrela odorata* L., *Guazuma crinita* Mart., and *Jacaranda mimosaefolia* D. Don. *Silvae Genetica* 46: 17–23.

Table 1 Coating substances and additives for somatic embryos

Hydrogels[a]	*Additives*	*Dry coatings (or coatings applied hydrated and then dried)*
Sodium alginate	Fertilizer	Polyethylene glycol 1000–4000
Potassium alginate	Nutrient salts (N,P,K)	Polyethylene oxide homopolymers
Guar gum	Cytokinins	Acrylic copolymer containing carboxyl group
Agar	Humic acids	Potassium propenoate-propenamide copolymers
Agarose	α-Keto acids	Carboxyl methylized cold soluble swelling starch
Amylase	Gibberellic $acid_3$	Synthetic trioctachedral smectite
Pectin	Gibberellic $acid_{4/7}$	Synthetic sodium-magnesium lithium silicate
Dextran	Pesticides (ethazol, thiophantenethyl, fenaminosulf, ethazol, benomyl, chloroneb, captan, metalaxyl)	Starch + synthetic polymer of acrylamide
Gelatin	Biocontrolagents (Trichoderma harsianum, T. kininoii, Rhizobium)	Sodium acrylate
Starch	Sucrose	
Amylopectin	Activated charcoal	
Mannan		
Modified cellulose (e.g., carboxymethyl cellulose, polyacrylamide)		
Tragacanth gum		
Carrageenan with locust bean gum		
Polysaccharides and galactomannans + starch		
Sodium pectate		

[a] Singly or in combination

Figure 2 (see color plate 42) Alginate-encapsulated embryos of Norway spruce. Reprinted with permission from Gupta PK, Timmis R, Pullman G, *et al.* (1991) Development of an embryogenic system for automated propagation of forest trees. In: Vasil IK (ed.) *Scale-Up and Automation in Plant Propagation*, pp. 75–93. New York: Academic Press.

therefore, requires sterilization (which affects gel characteristics) and planting into a sterile environment or the addition of additives to control microbial growth. The gel capsules are also susceptible to dehydration and require specialized care during handling, sowing, and germination (e.g., at the pre-greenhouse transplant stage). Hydrated capsules can survive only short-term storage under cold, moist conditions ($\sim$1 month). The capsules are fragile and incompatible with most conventional seeding machines, requiring development of machinery engineered specifically for gel-encapsulated embryos. Singulation may be a handling problem because uncoated capsules tend to be sticky.

Attempts are being made to improve gel bead characteristics to address some of these problems. The hardness of the gel can be lowered using special alginate formulations, with or without the addition of substances that interfere with the hardening reaction. Adding inactive substances (inert granules, powders, air bubbles) may prevent the formation of a continuous gel matrix. Sakamoto and coworkers developed self-breaking beads that swell gradually after being sown in humid and low electrolytic conditions until they split and break open. Inert oxygen-carrying compounds such as perfluorocarbons and silicone have been added to the gel matrix to address the problem of oxygen availability. The gel matrix (or artificial endosperm) may contain additives (**Table 1**). Mamiya's group used sodium alginate with slow-release sucrose capsules to encapsulate asparagus (*Asparagus officinalis*) somatic embryos. The slow-release sucrose worked well, but as has been seen in other species, total covering of the embryo by the gel matrix interfered with normal germination of the embryo.

To protect the hydrated beads from desiccation, various coating strategies have been proposed. Some examples are: membrane-like coatings with wax and resin; polyorganosiloxanes; a thin membrane of alginate with a liquid center; a self-breaking, liquid-center capsule; and a dried water-soluble polymer coating (Elvax 4260). Tackiness can be minimized by rolling the capsules in a suitable powder (Tullanox).

A hydrated gel with or without additives can be used as an artificial endosperm and be surrounded by a rigid manufactured seed coat. The seed coat can be made of cellulosic material, glass, plastic, a cured polymer resin, paraffin, wax, varnish, etc. usually with an orifice covered with a thin film layer to allow protrusion of the radicle during germination (as described in several published patents). These hydrated encapsulated artificial seeds are suitable for plants that have a high unit value (such as vegetable transplants), as they are relatively expensive to develop and have a high unit cost per seed.

Germinating embryos also may be planted using fluid drilling. Fluid drilling was originally developed as a crop establishment technique using a thickened gel matrix (in contrast to the solidified matrix used in encapsulated embryos). Germinated seeds are incorporated into a protective gel mixture that is then continuously extruded into the seedbed. The use of the gel with the addition of sucrose, fertilizers, plant growth hormones, pesticides, microorganisms, etc. results in greater and more uniform seedling emergence with earlier and greater yields, and allows bulk handling of many small plantlets. Gels for fluid drilling include: Laponite (magnesium silicate clay), Liqua-gel (potassium starch acrylamide), Terrasorb (starch or synthetic copolymer), Hydrozorb 30 (potassium acrylate), N-gel (various cellulose based materials – Natrosol). Natrosol has the highest oxygen diffusion rate.

Strict control of seedbed conditions is critical to allow seedlings to establish. This method has been applied to germinating carrot somatic embryos. However, very low conversion percentages were reported. There are still biological and technical problems to be overcome regarding normal development of embryos, interaction of embryos with the gel mixture (e.g., oxygen diffusion rate and level of nitrogen source), and definition of seedbed conditions for optimum germination. The method remains quite labor intensive as it requires controlled germination of the plants before they are fluid drilled.

Desiccated Somatic Embryos

Naked embryos Desiccation induces quiescence in somatic embryos, providing more handling flexibility in large-scale production systems. Dehydrated, naked, quiescent somatic embryos may be frozen for indeterminate germplasm storage or stored at room temperature for shorter periods of time. Desiccation allows somatic embryos of seasonal crops (e.g., vegetables for spring planting, trees for reforestation) to be produced year-round, stored, and distributed for germination as required. Desiccation has the added advantage of improving germination. Mild desiccation has been shown to cause a developmental switch from storage reserve synthesis to storage reserve catabolism, yielding more vigorous seedlings that exhibit more rapid and uniform development.

Unencapsulated desiccated embryos have the disadvantage of being brittle (especially the cotyledons if they are reflexed, or spread open), requiring extra care in their handling to prevent mechanical damage. Gentle handling and germination under controlled conditions in the laboratory or greenhouse are required, thus increasing the cost per seed. Overall, development costs are lower because no encapsulation is required. Not only must SE technology be capable of producing low-cost, high-quality somatic embryos, but the embryos must also be capable of withstanding desiccation and rehydration without suffering damage. Therefore, both optimized maturation and postmaturation protocols must be defined that increase storage product reserves in the embryos and increase desiccation tolerance. Currently this is the major stumbling block in the development of this technology in many species. Desiccation tolerance is the result of several mechanisms working together to limit water loss, protect cell

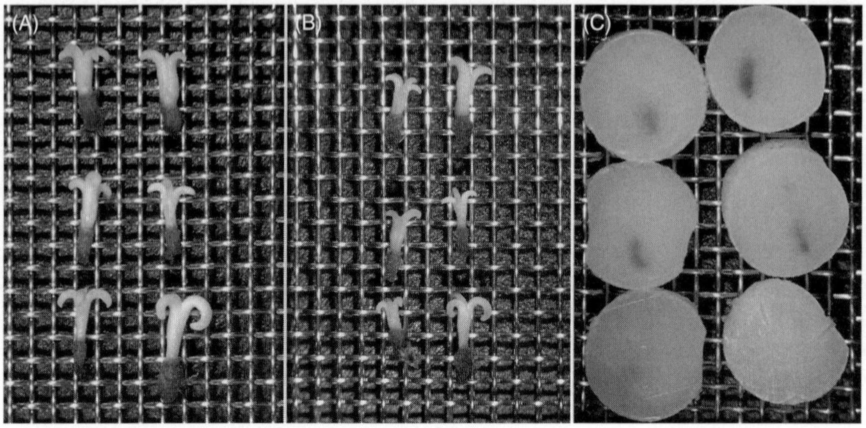

Figure 3 Nonaqueous encapsulation of white spruce. (A) Desiccated, cold-stored embryos; (B) embryos after rapid dehydration to $<10\%$ moisture content; (C) embryos in discs of polyethylene glycol 1000.

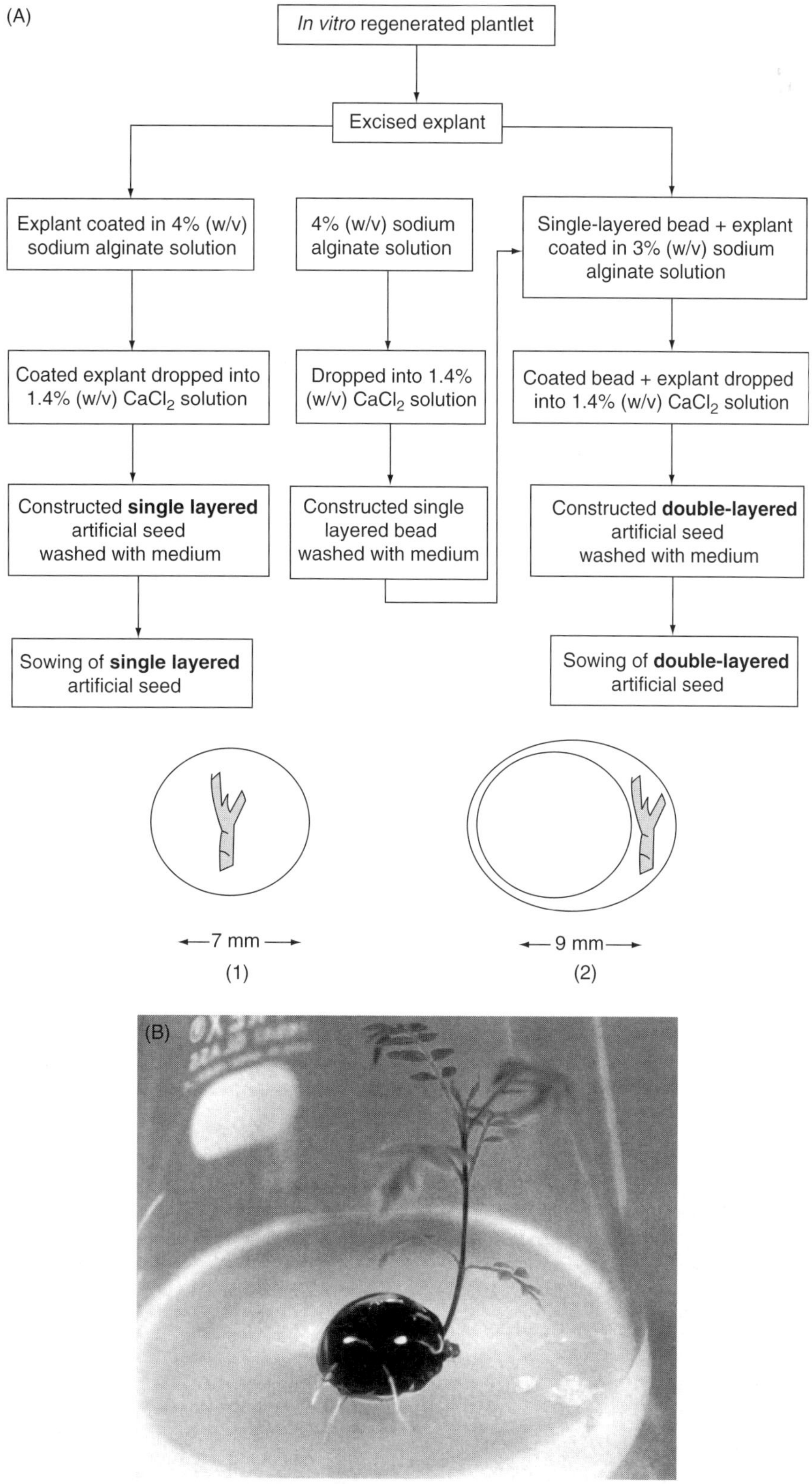

Figure 4 Double-layer encapsulation of meristems in alginate. (A) Schematic of the process; (B) germination of a *Jacaranda mimosaefolia* meristem from a double-layered bead containing charcoal. Reprinted with permission from Maruyama E, Kinoshita I, Ishii K, *et al.* (1997) Alginate-encapsulated technology for the propagation of tropical forest trees: *Cedrela odorata* L., *Guazuma crinita* Mart., and *Jacaranda mimosaefolia* D. Don. *Silvae Genetica* 46: 17–23.

membranes from damage during desiccation, and repair mechanisms that correct minor damage that has occurred during rehydration. Methods that increase the amount of time the embryos spend in the maturation process provide embryos with larger storage reserves and increased desiccation tolerance. Exposure to exogenous abscisic acid (ABA) and sublethal stress is important in promoting desiccation tolerance. Maturation media with high osmotic potential induced by either increased levels of permeating osmoticants (e.g., sucrose, mannitol), nonpermeating osmoticants (e.g., polyethylene glycol 4000) or higher gel strength media (to limit water availability), induce water stress, delay germination, increase storage reserve production, and increase desiccation tolerance. Other sublethal stresses (e.g., low temperature, nutrient deprivation) can have similar effects on desiccation tolerance. Properly pretreated embryos remain viable when they are rapidly desiccated to less than 15% moisture content.

Encapsulated embryos Coated desiccated embryos represent the ideal form of artificial seed. All the advantages of desiccated embryos are present, with the added advantage of encapsulation for handling. Encapsulated embryos can be tailored to the required size and shape necessary for use with conventional seed planting equipment. Desiccated somatic embryos destined for storage and/or field sowing must have abundant storage reserves and high conversion percentages (i.e., the percentage of embryos capable of converting into normal actively growing plants). Like their naked counterparts, they require as starting material quiescent somatic embryos capable of withstanding desiccation and rehydration and thus capable of surviving periods of cold, frozen, or room-temperature storage. The desiccated embryos are coated with a protective layer to protect them from mechanical damage during seed handling and may incorporate a nutritive layer to nourish and protect them from desiccation during the early stages of germination (**Figure 3**). The coating(s) must be nontoxic, nonaqueous (so they can be applied without rehydrating the embryos), and either meltable at a relatively low temperature (so the embryo does not suffer thermal damage during coating) or dry and attached to the embryos with an adhesive. The coating needs to dissolve readily in water after sowing to allow unimpeded germination of the embryo.

As with other methods, large numbers of normal mature somatic embryos must be generated in a cost-efficient manner to make nonhydrated artificial seeds competitive in price with natural seeds. As well, there should be some economic rationale for using artificial seed, such as the production of improved stock in conifers or in other species where the reproduction rate is slow, sporadic, or not true-to-type. Development costs are high in this type of artificial seed because rigorous protocols are needed to produce embryos able to withstand desiccation, as well as nonaqueous encapsulation media plus automated coating technology. Operational costs of actual seed production should be low once the development phase is past.

Table 2 Examples of artificial seeds in horticultural and plantation crops

Type of artificial seed	*Coating*
Hydrated naked somatic embryos	
Over 300 species form somatic embryos many of which could be used for artificial seed e.g., Colorado blue spruce	—
Hydrated encapsulated somatic embryos	
Alfalfa, asparagus, bamboo, *Brassica*, *Camellia japonica*, caraway, carrot, celery, cinnamon, coffee, *Coptis chinensis*, eggplant, geranium, ginger, horseradish, lettuce, lily, papaya, rice, sandalwood, tangerine, spruce (Norway, white, black)	Alginate
Dried, naked somatic embryos	
Alfalfa, carrot, grape, orchardgrass, spruce	—
Dried, encapsulated somatic embryos	
Brassica, carrot, papaya, interior spruce	Polyox, calcium alginate, sodium alginate with cellulose derivative A, Polyethylene glycol 1000
Other meristematic tissue, encapsulated	Alginate
1. Axillary buds	
Mulberry, kiwifruit, blackberry, raspberry, swamp cabbage	
2. Adventitious buds	
Vanilla, olive	
3. Shoot tips	
Banana, cardamon, apple, *Solanum* sp., oak, birch, tropical tree species	
4. Protocorms	
Orchid, *Cymbiidium giganteum, Dendrobium wardianum*	
5. Nodal segments	
Dianthus, Solanum sp.	

Other Meristematic Tissue

Axillary buds, shoot tips, and other meristematic tissue In cases where an SE system has not yet been developed due to the difficulty of producing embryogenic tissue, or problems with somatic embryo maturation, organogenically derived tissue can be used as the propagule. Propagules such as tissue culture-propagated axillary buds, adventitious shoots, shoot tips, protocorms, or hairy roots with shoots, are encapsulated in a manner similar to the hydrated somatic embryos, usually in a sodium alginate solution with nutrients, antibiotics, and often with activated charcoal (**Figure 4**). The propagules can then be planted on soil or other substrates such as cotton, filter paper, or soilrite. This method has been used successfully on a small scale in many horticultural and forestry crops (**Table 2**).

Conclusion

After 25 years of investigation, artificial seed technology remains a work in progress. Successful commercialization of any of the naked or coated embryo systems still requires substantially more knowledge of the basic mechanisms behind SE. Some problems to be overcome are: loss of embryogenic potential with age of the SE culture; asynchronous development and lack of mature embryo uniformity; precocious germination; structural anomalies; lack of desiccation tolerance; need for a sugar source for germination (heterotrophism); and low conversion percentages. In most species, these problems are caused by inappropriate cultural conditions and not factors intrinsic to SE. In the few species where substantial SE cultural advances have been achieved, researchers are well on their way to developing artificial seed.

Bulk production methods for economically producing tens or hundreds of thousands of propagules are rapidly evolving and being improved. The goal is a reliable, cost-effective SE system to synchronize the production of large numbers of high-quality somatic embryos showing high conversion percentages without encapsulation, and capable of surviving the stresses associated with encapsulation and sowing. The capability to enter a quiescent state upon desiccation and be stored in this dried state would be advantageous, as would the ability to be germinated in nonsterile conditions in the laboratory or greenhouse/field. Desiccation is expected to reduce handling costs after encapsulation, because the embryos can be handled like natural seed.

Aside from the costs to develop the encapsulable units for artificial seeds, additional investments are necessary to develop encapsulation methods and machinery for seed handling both during production and sowing. In the absence of front-end development costs, embryos will be fragile, incurring high handling expenses and the use of specialized greenhouses or laboratory space for germination and conditioning, thereby restricting use of the technology to only the highest-value crops due to the high unit cost. Nonetheless, for some crops (i.e., where SE systems are difficult to develop, or where synchronization, maturation, or germination problems persist), sowing naked propagules or encapsulation of hydrated tissue (e.g., shoots, buds, bulbs) may be the only alternative. After the initial expenditure, it is anticipated that artificial seeds can be produced quite cheaply by automating both embryo production and encapsulation.

Only after reliable, species-specific methods of bulk somatic embryo production tailored for artificial seed use have been devised can universal encapsulation and automation systems be developed. Although some progress has been made in demonstrating the feasibility of artificial seed and implementation has been successfully accomplished on a small scale, commercial use of artificial seed is still more a concept than a reality.

List of Technical Nomenclature

Conversion percentage The percentage of somatic embryos that "convert" to growing plantlets with normal elongating hypocotyls and roots.

Encapsulable unit An embryo, somatic embryo, bud, adventitious shoot, etc. that can be encapsulated, but which must have a high conversion percentage in its unencapsulated form and show a tolerance to the stresses associated with the encapsulation and sowing.

Heterotrophic The type of plant that requires a source of carbohydrates as its carbon source.

In vitro From the Latin meaning "in glass," referring to plant material that is cultured in aseptic conditions on artificial nutrient medium.

Photoautotrophic The type of plant that uses CO_2 as its carbon source (by using photosynthesis) and does not require a source of carbohydrates (e.g., sugars) in the growing medium as a carbon source.

Propagule A plant unit or structure (e.g., somatic embryo, protocorm, shoot tip) that will give rise to new plants.

Somatic embryo An embryo produced from somatic cells or tissues and not as a result of fertilization of an egg cell.

Somatic embryogenesis The process whereby somatic cells give rise to bipolar structures called somatic embryos.

Zygotic embryo A bipolar structure produced as a result of fertilization of an egg cell.

See also: **Biodiversity and Conservation**: Seed Banks. **Postharvest Physiology**: Seed Storage. **Seed Development**: Seed Production. **Tissue Culture and Plant Breeding**: Clonal Propagation, Forest Trees; Regeneration of Fruit and Ornamental Trees via Cell and Tissue Culture. **Tissue Culture**: Somatic Embryogenesis.

Further Reading

Attree SM and Fowke LC (1999) *Maturation, Desiccation and Encapsulation of Gymnosperm Somatic Embryos.* United States Patent No. 5,985,667.

Cervelli R and Senaratna T (1995) Economic aspects of somatic embryogenesis. In: Aitken-Christie J, Kozai T, and Smith MAL (eds) *Automation and Environmental Control in Plant Tissue Culture*, pp. 29–64. Boston: Kluwer Academic Publishers.

Gray DJ, Compton ME, Harrell RC, and Cantliffe DJ (1995) Somatic embryogenesis and the technology of synthetic seed. In: Bajaj YPS (ed.) *Biotechnology in Agriculture and Forestry*, Vol. 30, *Somatic Embryogenesis and Synthetic Seed I*, pp. 126–151. Berlin: Springer-Verlag.

Gupta PK, Timmis R, Pullman G, *et al.* (1991) Development of an embryogenic system for automated propagation of forest trees. In: Vasil IK (ed.) *Scale-Up and Automation in Plant Propagation*, pp. 75–93. New York: Academic Press.

Hay IE and Charest PJ (1999) Somatic embryo germination and desiccation tolerance in conifers. In: Jain SM, Gupta PK, and Newton RJ (eds) *Somatic Embryogenesis in Woody Plants*, Vol. 4, pp. 61–96. Boston: Kluwer Academic Publishers.

Kitto SL, Pill WG, and Molloy DM (1995) Fluid drilling as a delivery system for somatic embryo-derived plantlets. In: Bajaj YPS (ed.) *Biotechnology in Agriculture and Forestry*, Vol. 30, *Somatic Embryogenesis and Synthetic Seed I*, pp. 181–192. Berlin: Springer-Verlag.

Maruyama E, Kinoshita I, Ishii K, *et al.* (1997) Alginate-encapsulated technology for the propagation of tropical forest trees: Cedrela odorata L., *Guazuma crinita* Mart., and *Jacaranda mimosaefolia* D. Don. *Silvae Genetica* 46: 17–23.

McKersie BD, van Acker S, and Lai F-M (1995) Role of maturation and desiccation of somatic embryos in the production of dry artificial seed. In: Bajaj YPS (ed.) *Biotechnology in Agriculture and Forestry*, Vol. 30, *Somatic Embryogenesis and Synthetic Seed I*, pp. 152–169. Berlin: Springer-Verlag.

Redenbaugh K (ed.) (1993) *Synseeds: Applications of Synthetic Seeds to Crop Improvement.* Boca Raton: CRC Press.

Redenbaugh K, Viss P, Slade D, and Fujii J-A (1987) Scale-up: artificial seeds. In: Green CE, Sommers DA, Hackett WP, and Biesboer DD (eds) *Plant Tissue and Cell Culture*, pp. 473–493. New York: Alan R. Liss.

Redenbaugh K, Fujii J-A, and Slade D (1991) Synthetic seed technology. In: Vasil IK (ed.) *Scale-Up and Automation in Plant Propagation*, pp. 35–74. New York: Academic Press.

Roberts DR, Webster FB, Cyr DR, *et al.* (1995) A delivery system for naked somatic embryos of interior spruce. In: Aitken-Christie J, Kozai T, and Smith MAL (eds) *Automation and Environmental Control in Plant Tissue Culture*, pp. 245–256. Boston: Kluwer Academic Publishers.

Sakamoto Y, Onishi N, and Hirosawa T (1995) Delivery systems for tissue culture by encapsulation. In: Aitken-Christie J, Kozai T, and Smith MAL (eds) *Automation and Environmental Control in Plant Tissue Culture*, pp. 215–244. Boston: Kluwer Academic Publishers.

Timmis R (1998) Bioprocessing for tree production in the forest industry: conifer somatic embryogenesis. *Biotechnology Progress* 14: 156–166.

Uozumi N and Kobayashi T (1995) Artificial seed production through encapsulation of hairy root and shoot tips. In: Bajaj YPS (ed.) *Biotechnology in Agriculture and Forestry*, Vol. 30, *Somatic Embryogenesis and Synthetic Seed I*, pp. 170–180. Berlin: Springer-Verlag.

Secondary Metabolism in Plant Cell Cultures

A M Nuutila and K-M Oksman-Caldentey, VTT Biotechnology, Espoo, Finland

Introduction

Plant cell cultures have been investigated for more than 30 years as a potential source of secondary metabolites to rival extraction processes based on whole plant material. However, success so far has been limited. Many questions still remain unanswered, with a particular need for enhanced productivities in cell cultures and simplified and improved process technology. Metabolite engineering might offer new possibilities to enhance the production in cultured cells. The connection between the expression of secondary metabolism and the

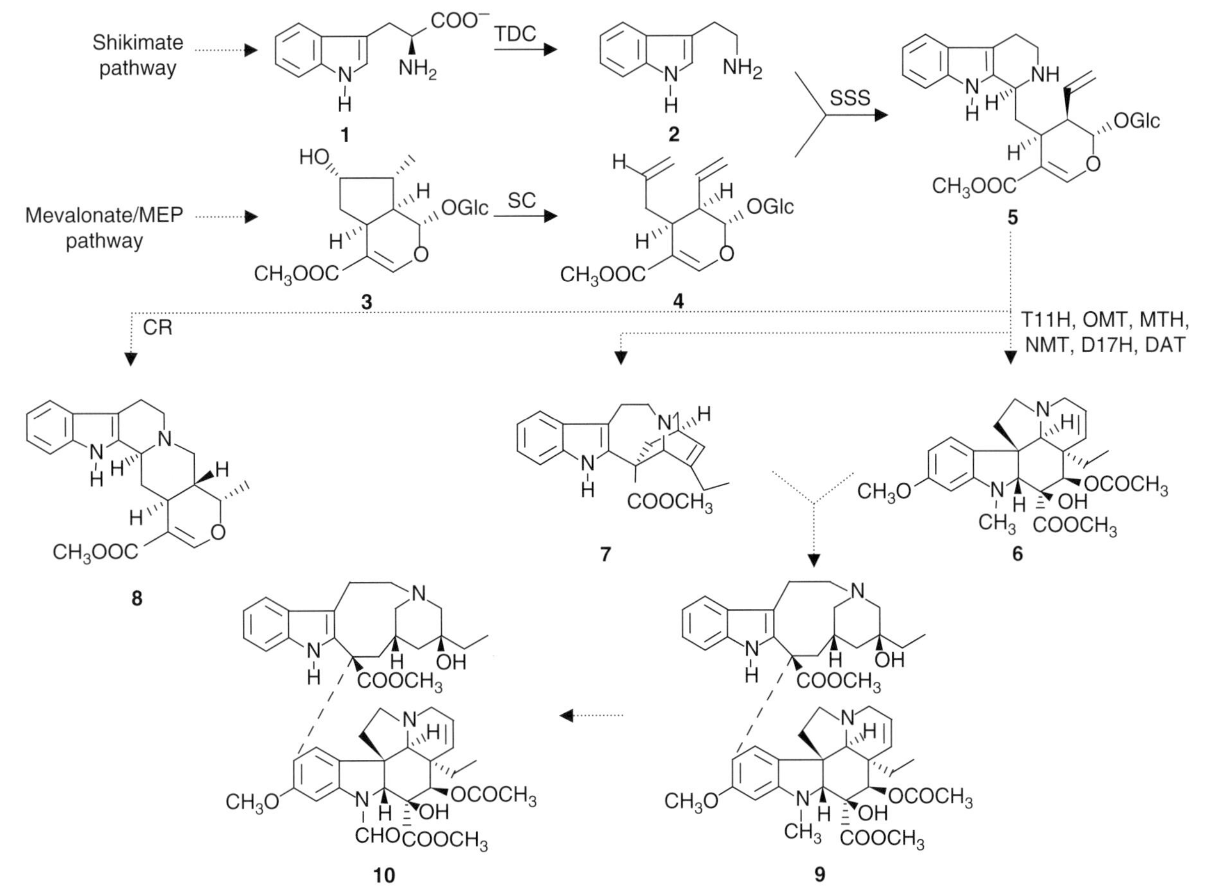

Figure 1 The biosynthesis route of terpenoid indole alkaloids in *C. roseus*. The broken line indicates that more than one enzymatic step is involved. **1**, Tryptophan; **2**, tryptamine; **3**, loganin; **4**, secologanin; **5**, strictosidine; **6**, vindoline; **7**, catharanthine; **8**, ajmalicine; **9**, vinblastine; **10**, vincristine.

differentiation program has been shown to exist, and the occurrence of certain pathways of secondary metabolism can depend on the general development or primary differentiation of cells or organs. Many plant secondary metabolites are commercially valuable and are used as pharmaceuticals in medicine. However, for commercial utilization, the production of secondary metabolites by plant cell cultures must be economically viable and easy to scale up.

Plant Secondary Metabolites

Plant secondary metabolites can be defined as compounds that have no recognized role in the maintenance of fundamental life processes in the plants that synthesize them. These compounds may serve to deter potential herbivores or pathogens, to discourage competing plant species, to attract pollinators or symbionts, or to further the interests of the plant in other ways. They are produced in specific parts of plants at defined developmental stages. The amounts are often low (less than 1% of dry weight) and highly variable. Furthermore, many plants containing interesting compounds, for example, those used as pharmaceuticals, are difficult to grow or are becoming endangered as a result of overharvesting. Since chemical synthesis is often not economically viable owing to their complex structures, the production of plant compounds in cell cultures remains an attractive alternative.

Secondary metabolites comprise large and diverse groups of chemicals that are specific for certain plant species; these are generally stored in cell vacuoles. Based on their biosynthetic origins, plant metabolites can be divided into three major groups: the terpenoids, the alkaloids, and the phenylpropanoids (including the allied phenolic compounds). The terpenoids are derived from the five-carbon precursor, isopentenyl diphosphate (IPP), the alkaloids are biosynthesized from amino acids, and the phenylpropanoids are biosynthesized by either the shikimic acid pathway or the malonate/acetate pathway. The biosynthetic pathways are long, complicated, and still largely unknown. One of the best-characterized pathways is that of terpenoid indole alkaloids (**Figure 1**).

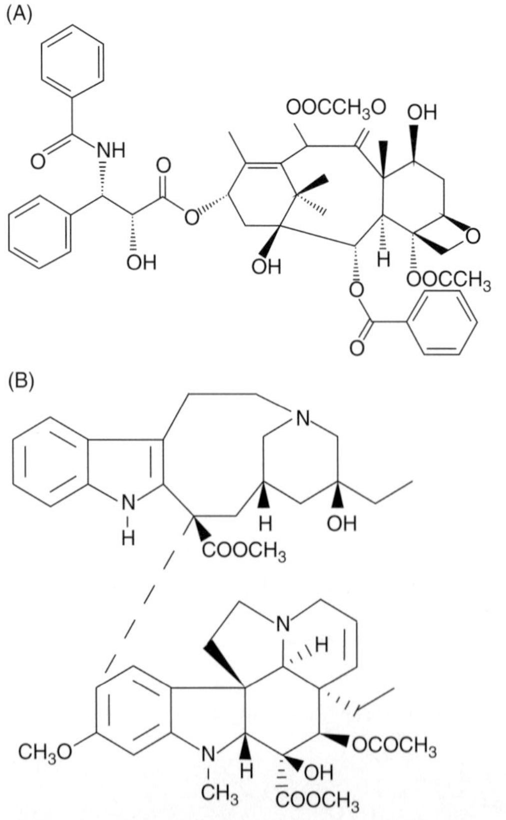

Figure 2 Chemical structures of (A) paclitaxel and (B) vinblastine found in Pacific yew trees (*T. brevifolia*) in Madagascar periwinkle (*C. roseus*), respectively.

In addition to their intrinsic value to the plant, many secondary metabolites (also called phytochemicals) are commercially valuable and are used as pharmaceuticals in medicine. Today, about 25% of all prescription medicines are derived from plants and, in the case of anticancer drug discovery and design, the percentage is even higher. Such compounds include vinblastine (**Figure 2**), vincristine, paclitaxel (Taxol^{TM}) (**Figure 2**), camphothecin and the semisynthetic drugs etoposide, etopophos, and teniposide, which are chemical derivatives of the plant lignan podophyllotoxin. A number of phytochemicals are also used in foods as colors or flavors and as fragrances in perfumes.

Secondary Metabolism in Plant Cell Cultures

Despite the fact that secondary metabolism in plant cell cultures has been studied for more than 30 years, progress has been somewhat disappointing, with only a very limited number of successes in the market place. These successes include shikonin production by cell suspension cultures of *Lithospermum erythrorhizon* and berberine production by cell cultures of *Coptis japonica*. Rosmarinic acid production by cell cultures of *Coleus blumeii* has been carried out on a large scale, and sanguinarine, which has market potential in oral hygiene products, has been produced by cell cultures of *Papaver somniferum*. An example of an expensive new drug, the production of which is now partially based on plant cell cultures, is paclitaxel, an antineoplastic drug, which was originally extracted from the bark of 50–60-year-old Pacific yew trees (*Taxus brevifolia*).

The connection between the expression of secondary metabolism and the differentiation program has been pointed out repeatedly. In higher plants, the occurrence of certain pathways of secondary metabolism can depend on the general development of the organism and/or on the development of single organs, tissues, and particular specialized cells. Synthesis and accumulation can be endogenously controlled by development-dependent differentiation processes, and/or can be regulated by exogenous factors such as light, temperature, nutrition, and wounding. The ability of tissues or cells to synthesize a distinct secondary compound is a manifestation of primary differentiation. This primary differentiation determines the specificity of the cellular response to exogenous factors.

Selection of High-Producing Cell Lines

The genetic variability in tissue culture-derived plants or cell culture clones has been documented in several studies, and has been named "somaclonal variation." It can arise either from genetic or epigenetic changes. Cryptic chromosomal rearrangements, such as chromosome breakage, translocation, deletion, and inversion, are to a large extent responsible for the genetic variation in plant cell cultures. Similar changes can occur at the gene level. Somaclonal variation has been widely used as a breeding tool in the search for agriculturally interesting traits. Cultured plant cells are heterogeneous and it is therefore possible to select cells with respect to a particular desired property. The selection of cell clones with high yields of secondary metabolites has been used in order to obtain high-producing plant cell lines. A good example is the selection of periwinkle (*Catharanthus roseus*) with high yields of vindoline, ajmalicine, and serpentine. Also, a number of other studies has demonstrated the increased steroidal and tropane alkaloid levels based on selection. However, continuous selection is often necessary to maintain the high-producing cell lines, otherwise the production level will decline over time. All the achievements so far have been based on the selection criteria mentioned above and on the optimization of the culture conditions, indicating a very empirical approach to achieving high levels of production.

Effect of Culture Medium

The production of secondary metabolites in cultured cells can be affected by culture conditions (i.e., exogenous factors): medium composition, gases, pH, temperature, light and inoculum size, as well as the culture system employed. The presence of sucrose in the medium is known to enhance secondary metabolite production in many cases, but other carbohydrates have also been used. Basal salts also affect secondary metabolism, with phosphate and nitrogen being the key factors. High levels of phosphate have been found to enhance cell growth, but they have a negative effect on secondary metabolite formation. It has been observed that phosphate decreases the level of alkaloids in cell cultures of tobacco (*Nicotiana tabacum*) and periwinkle by reducing the activities of enzymes that channel primary metabolites into secondary biosynthetic pathways. Additionally, it has been reported that phosphatase activities in *Hollarrhena dysenteria* cell cultures were reduced by the presence of phosphate in the medium. This resulted in severe inhibition of alkaloid synthesis. In the case of nitrogen, the presence of nitrate and ammonium ions in the medium results in a rapid increase of biomass, but secondary product formation is inhibited. Product accumulation can be strongly enhanced by reducing or deleting mineral nitrogen content. Moreover, the balance between nitrate and ammonium ions in the medium can drastically affect the release of secondary metabolites into the medium. Since secondary products are usually stored in the vacuoles of the cells, their release into the medium would facilitate the harvesting process.

Increased productivity can also be achieved by changing the hormones or growth regulators and their concentrations in the medium. Since a certain level of differentiation is a prerequisite for secondary product formation, the addition of auxin or cytokinin has an effect on secondary metabolite production. In order to achieve good production, the type and concentration of auxin and cytokinin used should support cell division and differentiation. In many plant cell cultures, the synthetic auxin 2,4-dichlorophenoxyacetic acid (2,4-D) has been found to inhibit secondary metabolite formation; therefore, the natural auxins naphthaleneacetic acid (NAA) and indoleacetic acid (IAA) are mainly used in combination with different cytokinins.

Since the optimal conditions for product synthesis are not often favorable to growth, maximum production has been achieved, in some cases, using a two-stage system. In the first stage, large amounts of biomass are produced as quickly as possible. In the second stage, the conditions favor product formation, not growth (**Figure 3**). An example of an operational two-stage process is the production of shikonin by cell cultures of *L. erythrorhizon* used by the Mitsui Company (Mobara, Japan).

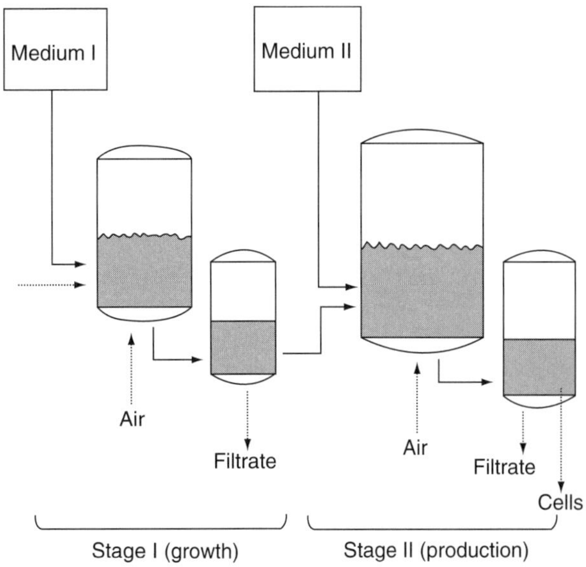

Figure 3 A two-stage production system. In the first stage, biomass is produced in a growth medium; in the second stage, the product formation is induced using a production medium.

Elicitation

Elicitors are stress agents that induce the accumulation of antibiotically active secondary metabolites (phytoalexins) in plants, as well as in cultivated plant cells. The accumulation of secondary products is regarded as a part of the defense system of the intact plant. Elicitors have also been shown to induce a range of other plant secondary metabolites. The production of these compounds is a dynamic defense response exhibited by plant cells when challenged by an elicitor.

The most commonly used biotic elicitors are fungal homogenates from families such as *Phytophora*, *Aspergillus*, and *Alternaria*, but abiotic elicitors, e.g., inorganic salts of cadmium, copper, and vanadyl, have also been used. The possibility of using chitosan as an elicitor has been intensively studied. Chitosan is widely distributed in nature and can be extracted, for example, from crab shells. The activity of chitosan depends strongly on the degree of polymerization and its N-acetylation. Chitosan has also been shown to have an effect on differentiated cells such as hairy roots, which are usually rather resistant to the commonly used elicitors. Although the use of elicitors does not directly increase secondary metabolite production, they may increase cell permeability. The economic viability of the production process depends in part on the capacity of the cells to secrete the desired metabolite into the surrounding medium (see above). However, increased permeability of the plant membranes has often been connected with the loss of viability of the plant cells.

Another interesting group of elicitors is the jasmonates. Jasmonates are plant hormones that have a dual effect on plant growth and development. They are strongly involved in the signal transduction cascades that are activated by stress situations, such as wounding and pathogen attack, and can thus initiate the plant defense system in order to increase the accumulation of secondary metabolites. Jasmonic acid and methyl jasmonate are both widely used to stimulate secondary metabolite production, e.g., nicotine alkaloids or several phenylpropanoids, in plant cells.

Gaseous Regime

The gaseous regime is critical for both growth and production in cell cultures. The beneficial effect of carbon dioxide on the production of secondary metabolites has been well established. However, only a few data are available on the effect of dissolved oxygen. There have been reports on reduced alkaloid production, e.g., in *Berberis wilsonae* cell cultures, the alkaloid production was reduced when the dissolved oxygen concentration was below 50% of air saturation. Ethylene has also been shown to have an effect on cell culture growth and secondary metabolite production, but the data are not consistent: in some cell cultures (e.g., *Thalictrum rugosum* and *Coffea arabica*), ethylene has stimulated secondary metabolite production, but in others (e.g., *C. roseus*), production has been impeded.

Secondary Metabolism in Hairy Root Cultures

High levels of secondary metabolite production, dependent on primary differentiation of cells, is often lost in undifferentiated cell suspension cultures during long periods of subculturing. This is mainly due to the instability of the cell lines. Therefore, production of secondary metabolites in differentiated tissue cultures, i.e., hairy root cultures, has gained much interest in recent years. The soil-borne pathogen *Agrobacterium rhizogenes* is able to transfer part of its DNA (T-DNA), carried on a large plasmid, to the genome of the host plant. Integration and expression of T-DNA genes in the host plant cell lead to the development of transformed roots, which can be excised and grown *in vitro* as hairy root cultures. The hairy roots are fast growing, laterally highly branched, and are able to grow in hormone-free medium. The advantages are that they are genetically and biochemically stable and produce high levels of secondary metabolites, characteristic to the host plant. The secondary metabolite production is comparable to that of the intact plant, and has been reported to be stable for years, in contrast to other types of plant cell culture. The productivity is highly linked to cell differentiation: production is clearly decreasing when the roots start to form callus and reappearing when the roots are allowed to redifferentiate. Compared to conventional root cultures, hairy roots possess higher growth rates, which are comparable to those of suspension cell cultures.

Hundreds of plant species, including many medicinal plants, have been successfully transformed via hairy roots. Most of the studies have been made with alkaloid-producing hairy roots. The highest levels so far have been obtained in hairy roots of tropane alkaloid-producing *Duboisia* plants, exceeding $30\,mg\,g^{-1}$. Several physical and chemical factors have been found to influence the growth and productivity of hairy root cultures. However, they are not as easily modified by changing the culture conditions as in the case of undifferentiated plant cells.

Several patent applications have been filed with the specific aim of using root cultures, transformed with *A. rhizogenes*, for the production of phytochemicals. This is an indication of the commercial potential of processes based on the use of differentiated roots grown *in vitro* for production of important secondary metabolites.

Metabolic Engineering

Metabolic engineering has been successfully employed in microorganisms for the increased production of pharmaceuticals and for the production of new compounds, e.g., antibiotics. Although efficient methods for gene cloning, transformation, and regeneration of transgenic plants are now available, the progress in improving medicinal plants has so far been relatively slow. This is due to the fact that surprisingly little is known about how secondary metabolites are formed in plants, how the biosynthetic pathways are regulated, and which genes are directing the formation of the various metabolites. The majority of plant genes involved in primary metabolism have now been identified, as a result of large-scale DNA sequencing projects, but only very limited numbers of genes involved in secondary metabolism have been identified.

Berberine was the first alkaloid for which each biosynthetic enzyme was identified and the respective genes characterized from the primary metabolite precursors to the end product, and this was achieved using *B. wilsoniae* cell suspension cultures. Good progress has also been made toward understanding the biosynthesis of morphine in opium poppy (*Papaver somniferum*) and tropane alkaloids (e.g., scopolamine) in *Hyoscyamus niger*. In addition, the elucidation of enzymatic synthesis of some other alkaloids is nearly complete: ajmaline, vindoline, corydaline, macarpine, and berbamunine. Even though considerable progress has been made in unraveling certain alkaloid biosynthetic pathways, the pathways leading to lignans, terpenoids, (iso)flavonoids, and coumarins are still for the most part unknown. However, the rapid development of genomic, proteomic, and metabolomic tools will create great opportunities to engineer the often complex pathways of plant secondary metabolites, and thus increase the productivity of these important compounds.

Different approaches can be applied in order to obtain high levels of desired secondary metabolites in transgenic plant cells. In the case of rate-limiting enzymes, the flux can be increased by introduction of the encoding gene regulated by a strong promoter or by introduction of a heterologous gene coding for an enzyme with a similar function. It is also possible to reduce the flux toward competitive pathways and/or catabolism, and increase the number of productive cells. In addition to metabolic engineering of existing biosynthetic routes, novel products could be obtained from plants by genetic engineering, for example, secondary metabolites of great value and possible new lead compounds for the pharmaceutical industry. If this is the case, secondary metabolite production in plant cell or tissue culture may become a viable industrial approach.

Production of Plant Secondary Metabolites on a Large Scale

For production of secondary metabolites on an industrial scale, shake flask cultures of plant cells are not adequate. Bioreactors can be used on a large scale and they facilitate the control of culture conditions to a greater extent than shake flasks. Plant cells differ from microbial cells in terms of their large size, rigid cell wall, extensive vacuoles, and slow metabolism. The morphology of the various plant cell culture types, i.e., suspension cultures, hairy root cultures, organogenic cultures, or embryogenic cultures, is also more varied than that of microbial cultures. The choice of bioreactor is determined by many factors, including the scale, the mixing mechanism, the method of aeration, the morphology of the cell culture, and the resistance of the cells to the shear stress generated in the bioreactor. In many cases, a successful bioreactor design involves a balance of potentially conflicting factors. Although a sufficient supply of oxygen is necessary, care must be taken to ensure that the aeration does not strip off carbon dioxide or other volatiles and thus reduce growth or production.

Most of the bioreactors used to grow plant cells are derived from reactors used for microbial cultivation, e.g., the classical stirred tank and air-lift reactors, but several novel types of bioreactor have also been developed (**Figure 4**, **Table 1**), including the rotating drum bioreactor, which has been used to cultivate *L. erythrorhizon* up to volumes of 1000 l, and, more recently, the wave bioreactor. However, the largest reactors (up to 75 000 l) used are the conventional stirred tank reactors.

Table 1 Bioreactor types used to cultivate plant cell, tissue, and organ cultures

Culture type	*Reactor type*	*Size range (l)*
Cell suspension culture	Stirred tank	1–75 000
	Air-lift or bubble column	1–1500
	Rotating drum	2–1000
Hairy root culture	Stirred tank	1–20
	Modified stirred tank	1–14
	Air-lift or bubble column	1–13
	Droplet reactor	2–500
	Mist reactor	1–35
	Wave reactor	1
Embryogenic and organogenic cultures	Stirred tank	1–10
	Air-lift or bubble column	0.6–5

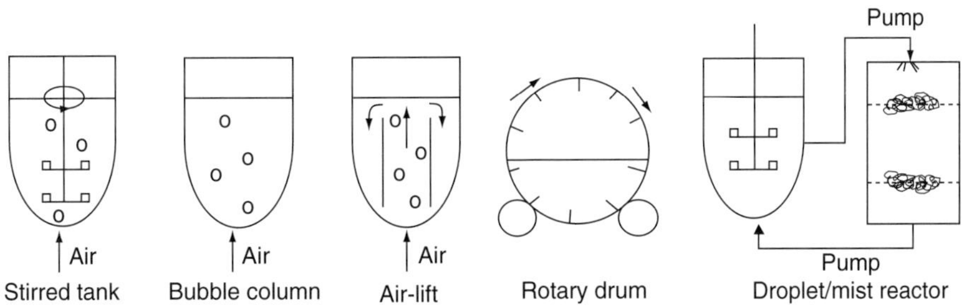

Figure 4 Bioreactors used for various plant cell cultures.

The morphology of hairy root cultures demands that special measures must be taken to ensure a good level of growth of these cultures in bioreactors. Hairy roots can be very susceptible to mechanical stress, and root disruption and loss of specific morphology have been reported. However, hairy roots of some species have been successfully grown in conventional stirred tank reactors. Improved growth characteristics have been achieved with a novel configuration: an impeller-mixed vessel with a mesh to separate the roots from the stirrer. To avoid shear stress, air-lift vessels ranging from 1 to 13 l have been successfully used with hairy root cultures. Trickling-film, droplet, and nutrient mist reactors have also been developed for use with hairy root cultures. In these reactors, the roots are in contact with air most of the time and the medium is sprayed over the roots. The results so far have been rather promising, as growth of cultures in these systems is good. The new wave bioreactor has also been used for hairy root cultures, currently with a working volume of 100 l.

The large-scale cultivation of plant cells and hairy roots in bioreactors is now possible for many cell lines. However, no single bioreactor design or process format is optimal for all culture lines; the design must be chosen with regard to the characteristics of the particular cell line used.

Commercial Utilization of Secondary Metabolites from Plant Cells

In addition to production technology that is suitable on a large scale, a highly productive cell line is also necessary for commercial success. Some of the production processes that have reached pilot scale are listed in **Table 2**. Several researchers have attempted to define the cost of producing secondary metabolites using plant cell biotechnology. The final cost is not easy to calculate and it depends on many factors: the time needed for the production of biomass and product, the volumetric proportion of biomass in vessels, cell morphology, specific productivity, and market size. Using the current technology status, the levels of production achieved would result in a selling price of approximately US$5000 per kg, which is not economically feasible. However, with further advances in the production technology, i.e., more highly productive cells, product secretion, cell reuse through perfusion strategy, lower cost nutrients, lower capital costs, and improved efficiencies in manufacturing and other operations, the product selling price could be reduced significantly (approximately US$2000 per kg). At these reduced price levels, commercial production of pharmaceuticals by cell cultures would be economically feasible. It is predicted that such technological advances could be achieved in the next 10 years.

Paclitaxel from Plant Cell Cultures

The antineoplastic drug paclitaxel (Taxol™) is an example of an expensive new drug, the production of which is already partially based on plant cell cultures. Paclitaxel was originally extracted from the bark of 50–60-year-old Pacific yew trees (*Taxus brevifolia*). Today, callus and cell suspension cultures of *Taxus* species are used as a potential alternative source of paclitaxel and related taxanes. Early attempts to induce callus from *T. brevifolia* were not successful, but, in recent years, cell cultures of many *Taxus* species have been induced. For example, *T. cuspidata* cultures produce paclitaxel in amounts equivalent to those found in the bark of the intact plant (0.01–0.02% w/w). Callus cultures from *T. baccata*, *T. brevifolia*, *T. cuspidata*, and *T.* x *media* have also been established. The amount of paclitaxel produced from these cultures ranges from 0.0001% to 0.013% (w/w). *T. brevifolia* and *T. baccata* can also be transformed by *Agrobacterium tumefaciens* and the transgenic cell lines maintained their ability to produce paclitaxel. *T. brevifolia* and *T.* x *media*

Table 2 Plant cell-based production processes that have reached pilot scale

Product	*Species*	*End-use*	*Producer*
Arbutin	*Datura inoxia*	Cosmetic	Mitsui Chemicals (Japan)
Berberin	*Coptis japonica* *Thalictrum minus*	Pharmaceutical	Mitsui Chemicals (Japan)
Catharantine	*Catharanthus roseus*	Pharmaceutical	Mitsui Chemicals (Japan)
Echinaceae polysaccharides	*E. purpurea* *E. augustifolia*	Pharmaceutical	Diversa (Germany)
Geraniol	*Geramineae* spp.	Pharmaceutical	Mitsui Chemicals (Japan)
Ginseng biomass	*Panax ginseng*	Pharmaceutical	Nitto Denko (Japan)
Rosmarinic acid	*Coleus blumei*	Pharmaceutical	Nattermann (Germany)
Scopolamine	*Duboisia* spp.	Pharmaceutical	Mitsui Chemicals (Japan)
Shikonin	*Lithospermum erythrorhizon*	Pharmaceutical, cosmetic	Mitsui Chemicals (Japan)
Taxol	*Taxus brevifolia* *T. chinensis*	Pharmaceutical	Mitsui Chemicals (Japan) Phyton Catalytic (USA/Germany)

embryos and immature embryos have also been cultured *in vitro* and the embryo-derived seedlings contain paclitaxel, 10-deacetylbaccatin III, and other taxanes. The process of producing paclitaxel from cell cultures has been patented, and patents have been applied for by several companies as well as research groups.

List of Technical Nomenclature

2,4-D	2,4-Dichlorophenoxyacetic acid.
IAA	Indoleacetic acid.
NAA	Naphthaleneacetic acid.

See also: **Secondary Products**: Pharmaceuticals, Plant Drugs. **Tissue Culture**: General Principles.

Further Reading

Croteau R, Kutchan TM, and Lewis NG (2000) Natural products (secondary metabolites). In: Buchanan B, Gruissem W, and Jones R (eds) *Biochemistry and Molecular Biology of Plants*, pp. 1250–1318. Rockville, MD: ASPP.

Eibl R and Eibl D (2002) Bioreactors for plant cell and tissue cultures. In: Oksman-Caldentey K-M and Barz WH (eds) *Plant Biotechnology and Transgenic Plants*, pp. 163–199. New York: Marcel Dekker.

Goldstein WE (1999) Economic considerations for food ingredients produced by plant cell and tissue culture. In: Fu T-J, Singh G, and Curtis WR (eds) *Plant Cell and Tissue Culture for the Production of Food Ingredients*, pp. 195–213. New York: Kluwer/Plenum.

Oksman-Caldentey K-M and Hiltunen R (1996) Transgenic crops for improved pharmaceutical products. *Field Crops Research* 45: 57–69.

Saito K and Mizukami H (2002) Plant cell cultures as producers of secondary compounds. In: Oksman-Caldentey K-M and Barz WH (eds) *Plant Biotechnology and Transgenic Plants*, pp. 77–109. New York: Marcel Dekker.

Sevón N and Oksman-Caldentey K-M (2002) *Agrobacterium rhizogenes*-mediated transformation: root cultures as a source of alkaloids. *Planta Medica* 68: 859–868.

Verpoorte R and Alfermann AW (eds) (2000) *Metabolic Engineering of Secondary Metabolism.* Dordrecht: Kluwer.

Verpoorte R, van der Heijden R, and Schripsema J (1993) Plant cell biotechnology for the production of alkaloids: present status and prospects. *Journal of Natural Products* 56: 186–207.

Verpoorte R, van der Heijden R, ten Hoopen HJG, and Memelink J (1999) Metabolic engineering of plant secondary metabolite pathways for the production of fine chemicals. *Biotechnology Letters* 21: 467–479.

TISSUE CULTURE AND PLANT BREEDING

Contents

Clonal Propagation, Forest Trees

J M Bonga and Y S Park, Canadian Forest Service – Atlantic Forestry Center, Fredericton, NB, Canada

Introduction

Currently, most commercial forestry is practised in natural forests in the Northern Hemisphere and in high rainfall tropical areas. However, it is projected that production from these forests is insufficient to meet the future demand for wood. Furthermore, it is expected that progressively more of the natural forests will be designated as protected ecological reserves and as recreational forests. Therefore, production on the remaining commercial sites will have to be increased greatly if the production demands are to be met. To improve forest productivity, one may expect that many of the commercial woodlands will increasingly be managed by planting of the best genetic stock currently available. In addition, future crops will have to be adapted to projected environmental changes, such as global warming and increased drought and soil salinity. Therefore, efficient techniques are required that will ensure rapid genetic improvement of planting stock for deployment in plantation forests.

Most commercial forest tree species have not been domesticated to the same degree as agricultural crops. With many annual agricultural crops, production has been increased greatly over that of their wild ancestors as a result of breeding, often over centuries. Similarly, breeding has greatly improved the adaptation of the crops to environmental and pathological stresses. Breeding of forest trees on a similarly intensive scale has never been an option owing to the long breeding cycle of trees. Therefore, most forest tree planting stock currently in use is still close to its unimproved wild ancestors. Clearly, if domestication as intensive as that practised with agricultural crops could be achieved with trees, huge gains in productivity and in adaptation to changing environmental conditions would be possible.

Cloning of commercially important trees is an integral part of producing and deploying genetically improved trees. Cloning, in contrast to breeding, does not alter the genetic composition of its offspring but produces genetically identical copies of improved genetic material. Various forms of cloning have been used in forestry, the traditional methods being grafting and rooting of cuttings. In addition to these, advanced methods for *in vitro* cloning of trees have been developed over the last few decades, offering the potential for efficient clonal propagation and greatly increased forest productivity.

Traditional Cloning

Grafting

In grafting, a shoot cut from a donor tree (called a scion) is attached to the stem of another tree (called the rootstock) in such a way that the cambial layers of both are placed in contact with each other so that they can fuse and thus form a union between the scion and rootstock. This is accomplished by making matching slanting cuts on both the scion and rootstock and wrapping the cambial layers together with a thin rubber band until a graft union is formed. Once the graft union is healed, there is no exchange of genetic material between the scion and rootstock, each maintaining its genotype; however they will influence each other owing to differences in growth rate or redistribution of nutrients to different parts of the plant.

Grafting is impractical for producing large numbers of plants for reforestation, but is widely used in fruit and ornamental tree propagation. For most commercially important forest tree species, grafting, although not used for mass-propagation of planting stock, is widely employed in establishing seed orchards, which are used for producing genetically improved seeds. Such clonal seed orchards are established by grafting shoot cuttings from superior genotypes (plus trees), which were selected either from wild populations or from a breeding program, onto juvenile rootstocks. Thus, several grafted individuals are produced for each of the selected superior trees. In these clonal orchards, wind pollination occurs among the superior genotypes, thus resulting in the production of genetically improved seeds that are used for reforestation. These grafted individuals are also used to facilitate breeding and to preserve breeding material in clonal archives.

Rooting of Cuttings

The traditional method for mass production of clonal planting stock is by rooting of cuttings. It is accomplished by inserting severed shoots into a rooting medium and permitting them to develop adventitious roots. For easy-to-root species, such as poplars and willows, the rooted cuttings can be produced in nursery beds, but the most common rooting environment is the greenhouse. As the cuttings have no effective means to replace moisture loss through evapotranspiration, maintenance of a high relative humidity, using a misting and fog irrigation system, is generally required.

Rooting of cuttings works well when the donor plants are juvenile seedlings. It is effective with mature trees only when these produce juvenile sprouts, such as stump and root sprouts, for use as cuttings. For a few forest tree species, notably within the genera *Populus*, *Cryptomeria*, and *Eucalyptus*, rooting of cuttings obtained from superior mature trees is a practical propagation method for producing genetically improved planting stock.

Unfortunately, for many tree species, in particular the conifers, rooting of cuttings from mature trees is not possible. Therefore, for these species, cloning is limited to young plants. Although these plants are too young to have demonstrated their full genetic potential, cloning by rooting of cuttings of seedlings raised from controlled crossings of superior trees in breeding orchards can result in substantial genetic gain. In some conifer breeding programs, in particular for *Picea*, *Larix*, and *Chamaecyparis*, hedging has been used to delay maturation of the stock (donor) plants. For example, a yellow-cedar hedge can produce cuttings capable of rooting for about 13 years, whereas unhedged seedlings will do so for only 5 years.

There are additional advantages in rooting of cuttings, particularly, when dealing with species that are poor seed producers. In this case, rooting of cuttings of young seedlings can produce a large number of propagules from a small quantity of seed. The quantity of genetically improved seed produced by breeding is often too small to be used directly for reforestation, which makes mass-clonal propagation of seedlings derived from these seeds an attractive proposition.

In vitro Cloning

In vitro cloning is a technology that has developed rapidly in the last few decades. In conjunction with traditional cloning methods, it is increasingly finding important functions in genetic improvement programs. In *in vitro* culture, a small piece of tissue (called an explant) is removed aseptically from a donor plant and placed on a sterile nutrient medium that supplies all the nutrients, including minerals, carbohydrates, vitamins, and microelements, required for growth of the tissue. Because both the explant and nutrient medium are free of microorganisms, the tissue can grow without interference from them. This is an advantage not enjoyed by traditional rooting of cuttings, where application of organic compounds can stimulate growth of unwanted microbes. By manipulating the levels of plant growth regulators and other growth-regulating chemicals *in vitro*, growth of the tissue can be directed into unorganized callus formation or organized growth such as the development of shoots, roots, or embryos.

Through *in vitro* clonal propagation, plants can be regenerated from normal diploid tissues, or from genetically unique cells such as haploid or triploid cells, or from genetically modified cells, thus creating a wide variety of new genotypes.

Organogenesis

In organogenesis, new plants arise from axillary or adventitious buds *in vitro*, which elongate into shoots and, subsequently, form adventitious roots. Organogenesis can often be achieved with tissues from seedlings and sometimes from mature trees. Organogenesis has been effective in propagating some species, for example, several members of the *Eucalyptus* and *Populus* genera, and one conifer, *Pinus radiata*. However, propagation by organogenesis has not been effective for most other tree species. Either the propagation rates are too low to be commercially effective or the propagules are not true-to-type, i.e., their growth habits differ from that of their donors. A problem that often arises with plants multiplied through organogenesis is precocious flowering. This and other genetic instability problems are not necessarily intrinsically associated with organogenesis; often they are the result of the technology not yet having been properly optimized for the species.

Somatic Embryogenesis

Initiation of somatic embryogenesis (SE) *in vitro* is a recent development and is now available for some conifers and hardwoods. The technology has been greatly improved over the past two decades and, for a few conifers, has reached a level that is sufficiently advanced for implementation in industrial reforestation programs.

In conifer SE, zygotic embryos, or megagametophytes that contain zygotic embryos, are removed from immature seeds and placed on a tissue culture medium that, for most species, contains 2,4-dichlorophenoxyacetic acid (2,4-D). This potent growth regulator causes the zygotic embryo to produce a rapidly growing, translucent tissue mass composed of many small, immature somatic embryos with suspensors. These somatic embryos are clones of the original zygotic embryo. Once enough of this embryo-suspensor mass has been produced, the tissue is transferred to a medium free of 2,4-D but containing instead the growth-inhibiting hormone abscisic acid (ABA) and a solute that increases the osmotic potential of the medium. This allows the many immature embryos in the embryo-suspensor mass to develop into mature embryos that will germinate and develop into young plants ready for planting (**Figure 1**).

SE has many advantages over organogenesis. In organogenesis, shoots have to be rooted before a plant is obtained. As somatic embryos have both a shoot and a root meristem, a separate rooting step is not required. Furthermore, SE often forms propagules faster and in larger numbers per explant than organogenesis. Another advantage of SE is that somatic embryos can be encapsulated, thus forming artificial seed for large-scale planting; thus the propagation process can be automated. However, the most important advantage of SE is that the embryogenic tissue can be maintained in liquid nitrogen (cryopreservation) at a temperature ranging from -120 to $-196\,^{\circ}\mathrm{C}$ without killing it. In general, axillary or adventitious buds formed by organogenesis cannot be stored in this manner.

The ability to store embryogenic tissue without change in genetic makeup or loss of regenerative capacity offers a significant opportunity to develop genetically improved clone lines. In the past, with the exception of a few hardwood species, the development of forest tree clone lines equivalent to crop "varieties" was not possible because genetically identical genotypes could not be produced over time. With

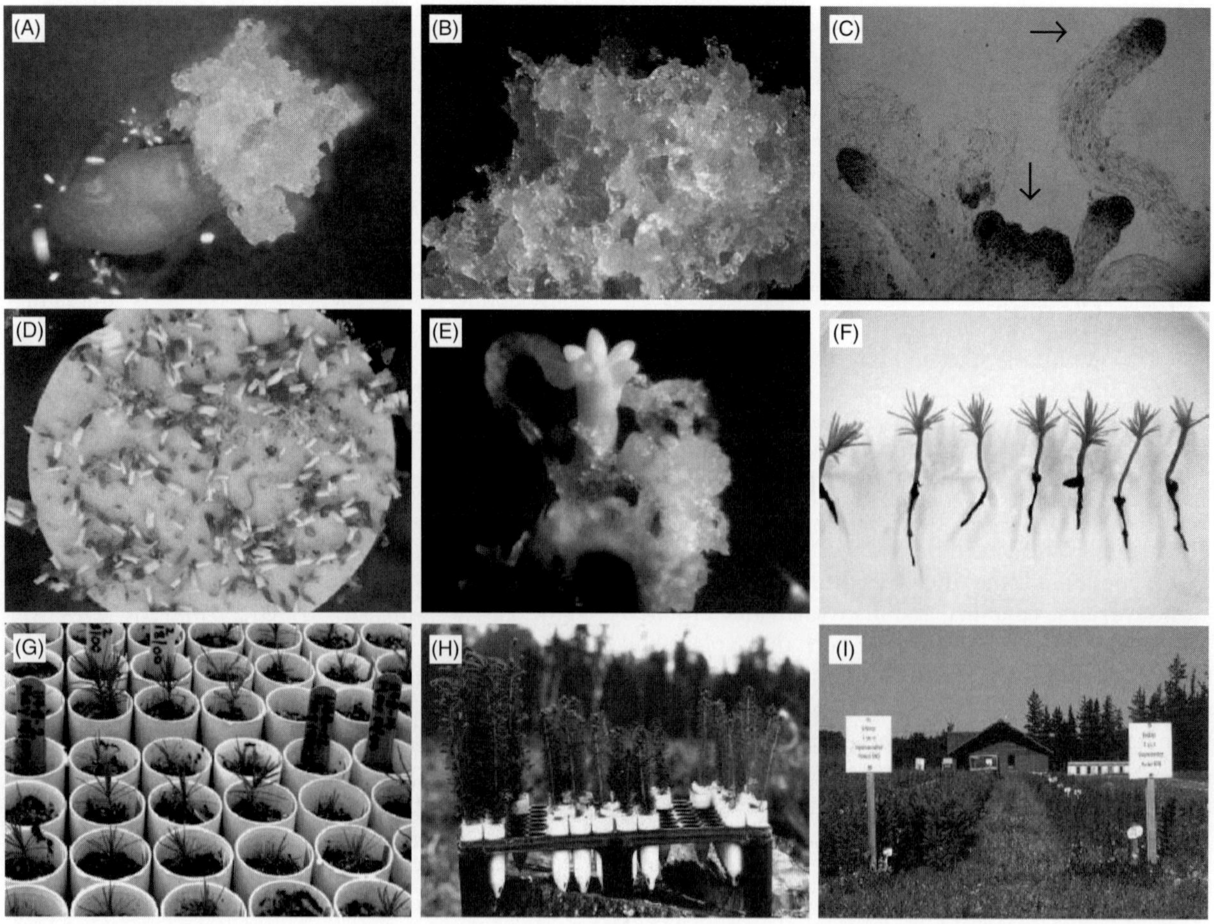

Figure 1 Clonal propagation by somatic embryogenesis: (A) extrusion of embryogenic tissue from a megagametophyte explant of eastern white pine (*Pinus strobus*); (B) proliferation of embryogenic tissue in white spruce (*Picea glauca*); (C) a microscopic view of embryogenic tissue showing somatic embryo heads, each with a suspensor (→), and cleaving somatic embryos (↓), thus forming more embryos; (D) mature somatic embryos of eastern white pine; (E) a mature somatic embryo of jack pine (*Pinus banksiana*) showing well-formed cotyledons; (F) germinated somatic embryos of eastern white pine at the time of transplanting in a greenhouse; (G) greenhouse culture of somatic embryo-derived eastern white pine; (H) four clones of somatic embryo-derived plants of white spruce ready for field testing showing within-clone uniformity and among-clone variability; and (I) clonal nursery testing of white spruce.

SE and cryopreservation, it is now possible to select the best clone lines (clonal varieties) and to produce a large number of genetically identical trees from these clone lines consistently over time. This is accomplished by dividing each embryo-suspensor tissue mass into two portions, one of which is cryopreserved while the other is propagated to produce clonal plants for testing in the field. Once field testing has shown which are the best-performing clones, the corresponding embryogenic tissue can be retrieved from liquid nitrogen storage, thawed and used for propagating plants for commercial plantations.

The use of tested clonal varieties in commercial forestry (i.e., clonal forestry) will result in much more genetic gain than can be obtained by conventional tree breeding. The actual development of clonal varieties may involve controlled crossbreeding of proven superior trees and selection of SE clones developed from seed obtained by this crossbreeding. Multigeneration tree breeding, in conjunction with production of SE clones and the periodic genetic evaluation of these, plays an integral part in developing clonal varieties. The clonal varieties thus developed will use indigenous genes in the most favorable combinations without introducing foreign genes.

Encapsulation of somatic embryos into artificial seed is an area of current research. Even though commercially viable systems for artificial seed production or an automated embryo planting system are still lacking, SE can be effectively used in the development of superior clonal varieties. Although production of planting stock entirely by the SE process is possible for several commercially important conifers, albeit at a high cost, only a few plants of these clonal varieties need be propagated by SE.

These few can easily be multiplied by serial rooting of cuttings at a cost of about 1.5–2 times the cost of producing seed-propagated planting stock.

Advanced Cloning Techniques

In vitro cloning techniques offer the opportunity to clone plants from selected cells that have a genetic makeup that is different from that of the normal diploid cell in the plant body. Other cells that can be used for cloning plants with an unusual genetic makeup are those contained in immature embryos that will abort if left to mature in the seed. Plants can also be regenerated from cells modified by fusion of protoplasts or genetic engineering and then be multiplied by traditional or *in vitro* cloning. This technology is still largely experimental and not yet ready for commercialization.

Regeneration of Trees from Cells with Differing Ploidies

The plant body is mostly composed of diploid cells. In addition, it contains haploid and, in the case of angiosperms, triploid cells. The haploid cells are in the pollen of all plants and, in the case of gymnosperms, in the nutritive tissue (megagametophyte) of the seed. In angiosperms, the endosperm is triploid. Haploid or doubled haploid and triploid plants can be produced through *in vitro* cloning, using the available haploid and triploid cells as starting material.

The aim of haploid cell culture is to obtain homozygous diploid plants for controlled hybridization to capture hybrid vigor in a reproducible fashion. In agricultural crops, homozygous plants are obtained through controlled inbreeding over several generations. The controlled crossing of two homozygous lines will produce a single hybrid genotype, a method that is used for producing a hybrid variety. This process is not practical for trees because the breeding cycle is too long. However, homozygous lines can be obtained quickly if haploid cells are turned into diploid ones *in vitro* and if plants are regenerated from these. This technique has been successful with some cereal crops and with a number of hardwood tree species but not yet with conifers. During the culture of haploid cells *in vitro*, the chromosome number of the cells is doubled by colchicine treatment, producing a "doubled haploid". Plants cloned from the doubled haploid cells will be homozygous and, if fertile, can be used for controlled breeding between doubled haploid plants of different genotypes. For a few hardwood tree species, triploid trees have been cloned from triploid endosperm cells *in vitro*.

Embryo Rescue

Sometimes, immature embryos with an unusual genetic makeup are formed, which unfortunately, will abort if left to mature in the seed. Such abortion occurs when there is a genetic incompatibility between the developing embryo and the maternal tissue surrounding it. Even though not viable within the seed, these embryos can develop normally and germinate properly if they are removed from the seed and grown to maturity *in vitro*. This process is called embryo rescue. Embryo rescue has the capacity to create genotypes that are not obtainable by breeding. By cloning *in vitro* or by traditional rooting of cuttings, the new rescued genotype can then be mass-produced. Embryo rescue has been achieved for a number of tree species.

Protoplast Hybridization

Recombination of genes by sexual reproduction is limited by sexual compatibility of two partners. The sexual incompatibilities within and between species are not always due to genes or chromosomes being incompatible but to such factors as male sterility, seedless plants, inhibition of pollen germination or tube growth, or endosperm failure. Such limitations can sometimes be overcome by fusion of protoplasts. Protoplasts are formed by enzymatic removal of cell walls from cells. These cell-wall-free cells fuse easily when brought into contact with each other. After fusion, a new cell wall is formed and plants are regenerated from the new cell. When two haploid protoplasts are fused, one derived from one parent and one from the other, the fusion product is diploid and the plants cloned from these are diploid. This process is very similar to sexual breeding except that often parents can be hybridized that are sexually incompatible. Fusion of diploid protoplasts creates tetraploid plants, and fusion of a haploid and a diploid protoplast produces triploid plants. Protoplast fusion between sexually incompatible partners and subsequent multiplication of plants by *in vitro* cloning has been achieved for only a few tree species, primarily those of the genus *Citrus*.

Genetic Transformation

Most genetic transformation of plants involves *in vitro* technology, in particular, the use of embryogenic cultures. Cells that regenerate somatic embryos are genetically modified either by infection with *Agrobacterium* or bombardment with tungsten or gold microprojectiles, both being used to carry the foreign genes into the embryogenic cells. The advantage of genetic engineering over conventional breeding is that genes that are completely foreign to the breeding population can be introduced, thus, desired genes can be added to already improved genotypes. For example, genes for insect or disease resistance or genes providing tolerance to salt, heavy metals, or drought can be added to embryogenic cultures that are being cloned *in vitro*. However, not all cells are genetically transformed during the transformation process; therefore, transformed cells must be separated from nontransformed cells. This is achieved by inclusion of a marker gene into the gene construct that carries the desired foreign gene. For example, a gene making cells resistant to the antibiotic kanamycin is often used for this purpose. By exposing the genetically transformed tissue to high levels of kanamycin, the nontransformed cells are killed while the transformed cells will divide and form somatic embryos.

Clonal Forestry

In a broad sense, the term "clonal forestry" refers to any use of clonal propagules in forest management. However, in a more restrictive sense, "clonal forestry" refers to the use of tested clones in plantations.

Clonal forestry has been practised mainly with angiosperm trees; recently, however, clonal forestry with conifer species has generated keen interest. Primarily, this is owing to advancements in traditional as well as in *in vitro* vegetative propagation techniques. There are many advantages to clonal forestry, particularly for the deployment of tested clones: (1) the amount of genetic improvement is much greater than is possible through conventional tree breeding; (2) it offers the flexibility to introduce rapidly suitable clones to meet changing product goals, site conditions, and other variables; and (3) it offers the ability to manage diversity in clonal plantations.

In order to practise clonal forestry, an effective mass cloning method must be available. As mentioned previously, the main traditional cloning technique is rooting of cuttings. This method has been particularly effective and cost efficient for hardwood species such as poplar. In conifer species, however, mass propagation of rooted cuttings is generally possible only with seedlings up to about 5 years of age. This is a serious limitation because, by the time the genetic superiority of a clone line has been determined through lengthy genetic tests, the donor plant has become too old for further mass-cloning by rooting of cuttings.

The recent advancements in SE technology have made it possible to circumvent this problem, at least for some conifer species. For most spruce and a few larch and pine species, up to 60% of the seeds will form embryogenic cultures and, of these, about 80% will form clonal plants that are plantable in the field. These rates are high enough for industrial application, particularly for developing high-value clonal varieties when used in combination with cryopreservation as outlined earlier.

Despite the advantages of clonal forestry, there are public concerns. The main concern is that a narrow genetic base may make clonal plantations vulnerable to diseases and insects, leading to plantation failure. However, in contrast to highly domesticated agricultural crop plants, most forest tree populations have a large range of genetic variability for pest resistance characteristics. Therefore, individuals that show a high level of pest resistance can be selected for cloning and deployment in clonal forestry.

There is, however, a risk that susceptible genotypes will unknowingly be deployed in the clonal plantation. In general, it is assumed that the more genotypes that are deployed in a clonal plantation, the lower the risk. However, increasing the number of clones in a plantation will result in a reduction in the maximum amount of wood produced. Thus, a balancing act is required and hence the question arises "What is a safe number of clones in a clonal plantation?" This is a difficult question, because the pest–host system is complex, and building a model is difficult when dealing with totally unknown diseases and insects. However, using various approaches to quantify this question, scientists generally agree that planting 15–25 clones in a plantation should be sufficient for protection, yet still confer the benefits of clonal forestry. Some general considerations in determining the number of clones that should be used in clonal plantations are as follows: (1) if the species is short-lived or short-rotation, a lower number of clones may be used because the exposure period to potential risk is reduced; (2) a lower number may be acceptable if forest management systems are intensive and include pest control measures; and (3) the more well-known a clone, the more acceptable is its extensive use.

Once the appropriate number of clones has been decided, a deployment strategy must take into consideration the configuration of deployed clones. Such a configuration can consist of clones in a random mixture or in a mosaic structure. Alternatively, mixed planting of a clonal mixture and sexually produced seedlings obtained from low-cost seed orchard seeds can be considered. For example, in a plantation, 40% of the plants can be a mixture of the best clones identified from genetic tests, and the remaining 60% of plants can be propagated from seed orchard seed. Apart from reducing the cost of planting stock, as the clonal stock is more expensive, it will increase the initial diversity of the plantation.

Typically, in forests in temperate climates, where rotations are 40–60 years, about 40% of the plantation basal area is commercially thinned at about one-half rotation age, leaving superior quality trees for final harvesting. It is likely that the remaining crop trees will be tested clones as well as the best trees propagated from seed orchard seeds. This will maintain acceptable diversity and produce high-value plantations.

As implied, clonal forestry must be based on tree breeding and genetic testing. Through tree breeding, progressively improved trees are produced in each generation. Genetic testing of clones obtained at each progressive breeding cycle will produce further improved clones. Therefore, the composition of clonal mixtures in subsequent clonal plantations will change over time. Furthermore, evaluation of genetic tests at regular intervals will lead to continually revised clonal compositions that are available for plantations. Thus, the diversity of clonal plantations can also be managed through time.

The cost of producing clonal trees is higher than that of seed-propagated trees. Therefore, clonal forestry is likely to be practised only on the most productive commercial sites to take advantage of specific characteristics of the tested clones. Thus, clonal forests will represent only a portion of plantation forests. On most reforestation sites, natural regeneration and other forest management methods that foster sustainable forestry can be expected to remain the main course of reforestation.

List of Technical Nomenclature

Adventitious An organ (a root or a shoot) or an embryo arising *in vitro* from tissues from which they do not normally arise on the intact plant.

Axillary A shoot arising in the leaf–stem axil.

Cambial layer A thin layer of tissue between the bark and the wood of a tree. It forms new bark and wood tissue during the growing season.

Clone Plants derived asexually from a single parent plant. Clones are all genetically identical to the parent plant.

Diploid The cell contains two sets of chromosomes, one haploid set of chromosomes from one parent and another haploid set from the other parent.

Endosperm The triploid nutritive tissue surrounding an angiosperm embryo.

Explant A small piece of tissue or organ aseptically excised from a donor plant for *in vitro* culture.

Haploid The cell contains only half the diploid number of chromosomes. This is found mainly in gamete cells such as pollen, sperm, and unfertilized eggs.

Homozygous The two haploid sets of chromosomes in the cell are identical. Plants with such cells form pure breeding lines.

In vitro Culture of tissues or cells aseptically on a nutrient medium inside plastic or glass containers.

Megagametophyte The haploid nutritive tissue surrounding a conifer embryo.

Meristem An area of active cell division.

Organogenesis The formation of adventitious shoots and roots *in vitro*.

Protoplast A cell minus its cell wall.

SE See somatic embryogenesis.

Somatic embryogenesis Adventitious embryos arising from cells or tissues *in vitro*. Whereas, in organogenesis, clonal plants are formed by first producing adventitious shoots that are subsequently rooted, somatic embryos contain both a shoot and a root meristem and a separate rooting step is not needed.

Suspensor An organ attached to an embryo. In the maturing seed, it pushes the embryo into a nutrient-filled cavity in the surrounding tissue that feeds the embryo.

Triploid The cell contains three haploid chromosome sets.

See also: **Crop Improvement**: Double Haploid Production; Hybridization and Plant Breeding; Plant Breeding, Principles. **Seed Development**: Artificial Seeds. **Tissue Culture and Plant Breeding**: Somatic Hybridization.

Further Reading

Ahuja MR and Libby WJ (eds) (1993) *Clonal Forestry I: Genetics and Biotechnology*. Berlin: Springer-Verlag.

Ahuja MR and Libby WJ (eds) (1993) *Clonal Forestry II: Conservation and Application*. Berlin: Springer-Verlag.

Bonga JM and von Aderkas P (1992) In Vitro *Culture of Trees*. Dordrecht: Kluwer Academic Publishers.

Park YS, Barrett JD, and Bonga JM (1998) Application of somatic embryogenesis in high-value clonal forestry: Deployment, genetic control, and stability of

cryopreserved clones. *In Vitro Cellular and Developmental Biology – Plant* 34: 231–239.
Park YS, Bonga JM, and Mullin TJ (1998) Clonal forestry. In: Mandal AK and Gibson GL (eds) *Forest Genetics and Tree Breeding*, pp. 143–167. New Delhi: CBS Publishers.
Tzfira T, Zuker A, and Altman A (1998) Forest-tree biotechnology: genetic transformation and its application to future forests. *Trends in Biotechnology* 16: 439–446.

Clonal Forestry

D Blakesley and T Marks, Horticulture Research International, West Malling, UK

Introduction

The management and exploitation of forest resources is a major global issue and one where current advances in plant biotechnology can be used to effect in crops that hitherto have received limited directed genetic improvement. Most tree regeneration is by seed from natural stands, where no selection, breeding, or management has been practised. Consequently, yields are highly variable and afforestation is on the basis of adaptive survival and not crop requirements. Increasingly, better genotypes comprising traits for improved productivity and disease resistance are being identified, and procedures have been developed to facilitate planting of clonal forests to exploit the high genetic potential in plantation trees. Biotechnological methods used in selective breeding and genetic modification will facilitate rapid utilization of these new genotypes in forestry, minimizing pressure on existing valuable resources in natural forests.

Global Forestry and Wood Production

It is estimated that in the year 2000, forests covered a total area of 3869 M hectares (ha), amounting to some 30% of the earth's land mass (Food and Agriculture Organization of the United Nations, FAO). From these forests, 14.6 Mha are removed each year, primarily in the tropics (approx. 11 Mha), for firewood (biomass), fodder, construction, paper, and processing (e.g., fiberboard). A total of 5.2 Mha of both natural and plantation forests are replanted each year and there is, additionally, conversion of 1.5 Mha of natural forest to plantation status. The total area grown as managed plantations amounted to 187 Mha, of which 62% was planted in Asia and much of the rest in South America; 4.5 Mha of these plantations are replanted annually, although only 3 Mha are deemed successful.

The volume of wood harvested, however, is perhaps the more relevant statistic. In the year 2000, this was estimated to be 3.4 billion m^3, or nearly 1 m^3 ha^{-1} of total forest, although as these data are difficult to ascertain, it could be much higher (FAO). Considering that much of the global forest is relatively unproductive, with extraction rates for natural forest being as low as 2 m^3 ha^{-1} $year^{-1}$, there is a great need to increase the production rates of plantation forests to alleviate current rates of deforestation. The need to conserve natural forests has been well argued, for both biodiversity and maintenance of the geophysical features of the environments they cover. Fenning and Gershenzen argue that to serve many of the diverse needs of tree extraction, production rates in natural forests are too slow, and plantations of biotechnologically improved trees provide the most viable option to produce a sustainable future resource.

Growth is faster in the tropics, where plantations of short rotation species and hybrids of *Acacia* and *Eucalyptus* have been used with great success (e.g., in Indonesia and Brazil, respectively). In the temperate and boreal zones, conifers with longer rotation times have been planted, although *Populus* and *Salix*, with relatively short rotation times, have also been productive, particularly in the USA, where 20 000–30 000 ha are planted annually. In stark contrast to the major agricultural and horticultural crops, forest trees grown in plantations have only been "improved" for a few generations. They are genetically and phenotypically similar to the "wild" species and the majority are still grown from seed. Breeding has often focused upon improving yield and form, using production methods based upon seed orchards comprising vegetatively propagated trees with proven field performance. This has produced seed of a much higher performance than that occurring in natural stands (e.g., *Picea sitchensis* (Sitka spruce), UK). However, as orchards are prone to invasion by pollen from other stands, genetic gain achieved in progeny seed is not as high as the potential in parental material. To exploit the potential of better-performing genotypes and avoid dilution from inferior pollen, superior quality trees can be vegetatively propagated using two approaches. Progeny from seed orchards can produce relatively small numbers of cloned individuals ("bulking") and clones have been produced from proven families. This technique generally uses cuttings taken from hedges grown from seedlings and is most frequently used with difficult-to-propagate conifers. This is the approach adopted for Sitka spruce in the UK. Alternatively, genetically superior individuals are identified and large numbers of clones are produced from these, either through a similar hedging process, or through tissue culture systems such as axillary shoot culture or somatic embryogenesis. Cloning individuals of proven genotypic and phenotypic quality can have a dramatic effect upon the genetic gain within a plantation (**Figure 1**). More recently, the process of selection has benefited from the use of marker-assisted breeding techniques, based on the use of molecular markers such as restriction fragment length polymorphisms (RFLPs), randomly amplified polymorphic DNAs (RAPDs), amplified fragment length polymorphisms (AFLPs), and microsatellites. These have been applied to a range of species (e.g., *Populus*) to identify and target particular traits.

Clonal Forests and the Role of Biotechnology

The application of clonal forestry within plantations is likely to open up new opportunities for improving tree productivity, although this is not a new concept and dates back hundreds of years in Asia. Clones of *Cryptomeria japonica* have been planted as small monoclonal blocks adapted to defined locations in Japan. The selection of appropriate *Cunninghamia lanceolata* (Chinese fir) clones in China has raised mean production levels up to six times that of unselected seed-derived trees (in one clone to $1170\,m^3\,ha^{-1}$ at 39 years). Such gains in productivity, set against limited problems of pest or disease across stands of restricted genetic diversity, argue well for the use of clonal trees.

One of the major problems in establishing a clonal production system for a particular genotype, however, is maturation. Rooting ability, and hence the propagation capacity, is often lost as a tree matures. Juvenile material is normally required to initiate cutting or tissue culture production. Furthermore, the field performance of many genotypes will not have been fully assessed, so information on this is required before large-scale production is viable. This may take up to 10 years, during which time access to juvenile material must be retained. How this is addressed depends upon the species and starting material selected as suitable for cloning. When controlled crosses are used in species such as *Pinus radiata*, immature embryos can be removed and used to initiate somatic embryogenesis. When embryogenic callus has been produced this can be used either to induce embryos for field trialling, or can be cryopreserved in a juvenile responsive state in liquid nitrogen to await trial results. The frozen material is later thawed and used in the production and germination of somatic embryos. Alternatively, inducing axillary shoot cultures of mature genotypes has been achieved in several species, including *Prunus avium* and *Fraxinus excelsior*, where a change in their physiology maintains an easy-to-root vigorous state. Other species do not suffer from this problem and *Populus*, *Salix*, and many genotypes of *Eucalyptus* can be readily propagated by cuttings or

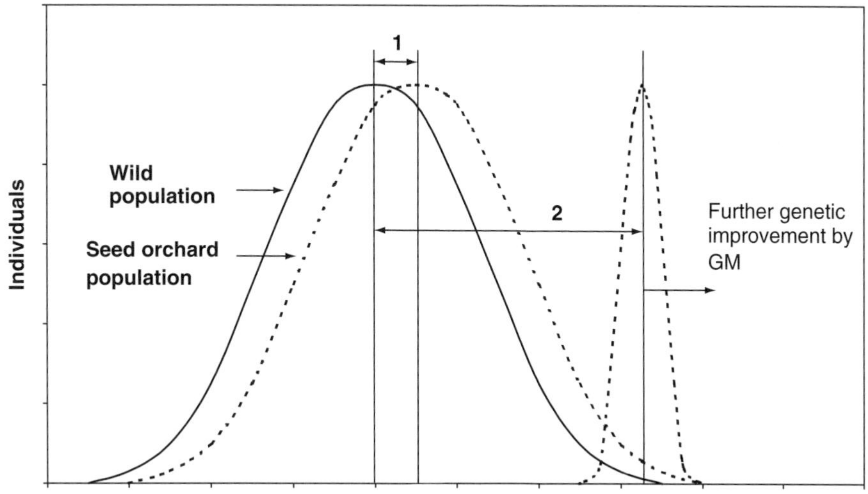

Figure 1 The genetic gain achieved by conventional breeding of wild populations to obtain improved genotypes is either limited (1), or requires repeated breeding cycles to provide a significant gain in the quality of selected traits. Cloning of trees of proven value from either the wild or conventionally improved populations offers a much larger genetic gain (2). The introduction of specific traits using GM technology could increase this yet further.

tissue culture to produce propagules for field planting. Poplar (*Populus*), especially, is amenable to gene transformation and has become a model for both the development of transgenic techniques and for the introduction of genes of agronomic importance.

Assuming that genetic modification (GM) becomes socially and politically acceptable, it could make an enormous contribution to tree improvement programs, by allowing breeders to overcome some of the critical barriers that have hindered improvement in the past. These include very long generation times, self-incompatibility, and the necessity for repeatedly backcrossing to isolate rare alleles. Many of the traits that are of interest to tree breeders, such as delay or inhibition of sexual reproduction and tolerance of domestication, are deleterious to plants in the natural environment. Examples of agronomic traits, such as herbicide resistance and insect resistance, and production efficiency traits, such as wood properties, which are of interest to tree breeders, are considered below.

Application of Genetic Modification

Herbicide Resistance

Herbicide resistance in poplars was the first example of GM in forest trees, being achieved in the late 1980s. It now represents the most developed GM system in forest trees. Several types of herbicide tolerance have been introduced into poplars, including Roundup®, which has received considerable attention from Strauss and his coworkers in the USA. Herbicide resistance is highly effective in poplars, giving superior weed control as young trees compete for water, light, and nutrients during their establishment phase. This ultimately provides the forester with higher survival rates and increased productivity. Positive effects on the environment may result from reduced water consumption and reduced soil erosion owing to herbicide substitution for tillage. However, the effects of increased herbicide use on biodiversity in the environment is a more complex issue, which is controversial and unproven. Biodiversity levels within plantations may suffer directly from the elimination of weeds and their associated insect fauna, while increased productivity may reduce tree extraction pressure on the wider landscape, leading to more positive effects on biodiversity.

Insect Resistance

Chemical toxins are usually employed to control insect pests on forest trees, but they are expensive and can have a significant environmental impact. In contrast to herbicide application, most of the applied pesticide does not make contact with the target insects. Consequently, there are both strong economic and environmental benefits to be gained from transgenic trees with enhanced pest resistance, particularly if this obviates the necessity for pesticide application. Endotoxins from the bacterium *Bacillus thuringiensis* (Bt) are particularly potent in trees, and have been widely employed in GM studies. Low avian and mammalian toxicity is an important attribute of these toxins. However, the problem of insect resistance to Bt will require careful analysis and research. Control of Bt expression in time and space, through tissue or wound-specific promoters, or plastid transformation, may be necessary. In addition, management practices within GM plantations may help to extend the effectiveness of a particular toxin. The strategic planting of nontransgenic trees, for example, may delay the acquisition of resistance by providing refuges or "refugia" for the maintenance of a population of nonresistant insect pests.

Lignin and Wood Properties

A very large number of genes are known to be involved in wood production. Consequently, the application of conventional breeding techniques is impractical, whereas the potential for genetically modifying wood properties is considerable. Strategies for downregulating lignin are of considerable interest. It is now possible to reduce lignin content or alter its structure thus making it easier to extract, hence, facilitating pulping. For example, suppression of cinnamyl alcohol dehydrogenase (CAD) in poplar has been shown to cause both a modest reduction in lignin content and a change in lignin crosslinking, which eases pulping. An antisense gene for the xylem-specific 4-coumarate:coenzyme A ligase has also been reported to repress lignin content by up to 45% in transgenic poplar. Processing of trees with reduced lignin content would require fewer chemicals, fibers would suffer less damage during lignin extraction, and damage to the environment would be reduced. Conversely, elevated lignin levels might enhance trees grown for firewood or bioenergy production.

Transgene Containment

Stable transgenic trees are important, not just to maintain the genetic characteristics of the original genotype, but also to contain the transgene(s). A major concern expressed about GM technology is the potential escape of genetically modified trees into the wider environment, where they might affect biodiversity and compete with wild, natural populations of the same, or related, species. Parallels have been drawn with the use of potentially invasive exotic plantation species. For example, fast-growing exotic *Acacia* species are being planted outside their native ranges for pulp and biomass production. Invasion of native South African woodlands by introduced *Acacia mearnsii* is becoming a serious problem. It is worth noting, however, that in the case of exotics, thousands of genes are being introduced into the environment, in comparison to one or two transgenes that encode highly specific traits. Nevertheless, a method of restricting gene flow through engineered sterility mechanisms to contain transgenes is essential, particularly for traits such as herbicide tolerance and insect resistance, which have the potential to increase the fitness of a species and hence its natural competitiveness.

Many options are currently being explored for inducing floral sterility in forest trees. One of the most promising is the introduction through nuclear transformation of floral meristem-specific cell-lysing toxins, which cause cell death. An alternative approach is to target the plastid genome (plastome) for foreign DNA integration rather than the more conventionally targeted nuclear genome. The benefits of this technology are the production of transgenic plants that have a specific region of plasmid DNA incorporated at a precise, prechosen site within the plastid genome, and in which high levels of transgene expression can be achieved. Additionally, a major attraction for plastid transformation technology is that it offers an opportunity to restrict transgene flow through pollen, since the plastome is maternally inherited in most plant species. Thus, transgenic pollen is not produced and released into the environment, so avoiding potential introgression of transgenes into neighboring plants. Whatever the sterility technology employed, one important issue facing tree breeders today is confidence in its robustness.

Although there have been significant advances in nuclear transformation of poplar in the USA, there have yet to be sufficiently long-term field trials in a range of different environments that would be likely to satisfy regulatory authorities in many parts of the world. A second concern is the impact of transgene products on the wider ecosystem. Horizontal gene flow could confer a selective evolutionary advantage to existing weed species. Consumption of transgenic toxins by aphid pests, for example, could accumulate higher in the food chain, initially in beneficial predator species. More research will need to be carried out into the persistence of more stable toxins, particularly if they enter the food chain or accumulate in the soil. A third concern is resistant biotypes, for example, the evolution of pests harboring resistance to toxins expressed in transgenic trees. It is almost inevitable that this will occur, but the timescale is poorly understood. There are contingencies for this, involving both pest management and careful use of the available toxin gene pool. A good example of this is the Bt family of toxins, the various members of which have an extensive toxicological diversity including different target insect specificities and different rates of environmental persistence.

Genetic modification will almost certainly increase in complexity, with the incorporation of suites of genes that produce more complex molecules, or affect more than one trait. If the efficacy of this technology is proven, and the ecological impact of transgenic trees is confined within plantations, then they should not present major environmental concerns.

Clone Numbers and Managing Risks

Clonal forestry is still in its infancy in comparison to other plantation crops such as fruits or oil palms. One of the most important potential threats to the environment posed by clonal forestry is the use of single or small numbers of clones, planted to increase uniformity, yields, and production efficiency. Restricted diversity is often viewed as an issue of clone deployment within the plantation, rather than a threat to the wider environment *per se*. Various schemes have been devised to reduce risk to the plantation from pests, disease, or climate change, incorporating the use of different numbers of clones or mosaics of small blocks of different clones. Mixed planting with seedlings has also been suggested, to reduce initial planting costs and to provide an early return upon capital investment when the stand is thinned. Managing this in practice will be confounded by the long rotation time of some hardwood species, which will have been selected on the basis of their performance over only a relatively short period of time.

Surprisingly, even the deployment of a small number of clones can represent a broader genetic base within a plantation than one comprised entirely of seedlings. Seeds may have been collected from few parent trees within a natural population or an existing plantation. They may represent just a fraction of the genetic diversity contained within those populations. In contrast, clonal forests may comprise a broad range of heterozygous clones of known genetic make-up. Even if plantation managers are confident that their clonal deployment strategy insures against environmental risks, there is another issue which is of concern, both to tree breeders and conservationists; the legacy of clonal stands, which are left to reproduce "naturally", in the absence of healthy, diverse populations of "wild" trees.

Conservation of Forest Tree Genetic Resources

The primary task of gene conservation is the maintenance of the evolutionary adaptive potential of a species, community, and ecosystem. This applies equally to natural forests in protected areas and forests viewed as repositories of species and genotypes with economic potential. Efforts to conserve and enhance forest genetic resources for use in clonal forestry must include the management of protected forest areas, productive forests, and breeding populations. Biotechnology will play an increasingly important role in programs that involve tissue culture of elite genotypes. The production of transgenic tree species, as well as efforts toward molecular breeding, have further emphasized the need for safe and reliable protocols. Long-term storage of somatic embryos and transgenic lines is necessary, for example, during field evaluation and selection of different genotypes. The duration of storage depends on the species and the trait being assessed, but may last for many years. There may also be a role for biotechnology through *ex situ* conservation of rare or endangered genetic resources.

Ex Situ Conservation Methods

Ex situ conservation methods for vegetatively propagated woody species traditionally relied on field or glasshouse collections. Such collections are expensive to maintain, are susceptible to environmental changes, and do not present a safe long-term option. Consequently, some 25 years ago, alternative *in vitro* approaches were considered. Shoot tip culture methods were modified to slow down growth rates and hence reduce maintenance costs. This was achieved most conveniently by employing reduced temperatures and was carried out initially with herbaceous crops. One major advantage of restricted growth techniques is that they utilize the same basic facilities as micropropagation. Cultures can be readily switched to rapid multiplication regimes when required. Restricted growth techniques have been used for a number of temperate woody species, including apple (*Malus pumila*) and pear (*Pyrus communis*). However, they do not alleviate some of the main problems associated with the culture of plant tissue *in vitro*, such as high costs of labor and space, and risks of somaclonal variation.

Cryopreservation

Cryopreservation techniques, which were developed for use with animal semen and blood cells, offered an alternative low cost and safe method of storage, which was thought to reduce the risk of somaclonal variation. In theory, it should be possible to store cells and tissues indefinitely at the temperature of liquid nitrogen (−196°C). The main technical problems with cryopreservation of plant tissues derive from damage caused by ice crystals formed in vacuolated cells. Within certain limits, these problems can be minimized by the use of cryoprotectant compounds in combination with controlled cooling conditions and/or procedures that remove freezable water from cells. The first successful cryopreservation protocols for meristems were achieved with rapid freezing procedures, but these were improved by the development of more reliable two-step freezing protocols. In the latter, plant tissue is dehydrated by slow cooling, typically to −40°C, before it is plunged into liquid nitrogen.

More recent strategies avoid the problem of ice crystal formation by inducing the process of vitrification, where ice nucleation is inhibited and water undergoes a change of state from a liquid to a noncrystalline "glass". Under cellular conditions, it is difficult to achieve this transition because it almost certainly occurs at temperatures in the region of −100°C and ice crystals are usually formed before that temperature is reached during the cooling process. The glass transition temperature can be raised by introducing high concentrations of solutes into the cells so that a point is reached at which the transition occurs before ice nucleation occurs. The so-called "vitrification" technique employs high concentrations of nontoxic compounds with suitable physical properties, for example, cocktails of cryoprotectant compounds such as DMSO (dimethyl sulphoxide), glycerol, and polyethylene glycol. Dereuddre and his coworkers also pioneered a vitrification protocol that does not require freeze-induced cell dehydration, the so-called encapsulation-dehydration technique. Meristems or embryos are protected by encapsulation in alginate beads, precultured on a medium with high levels of sucrose, and dehydrated. Dehydration to moisture contents in the range 15–25% ensures that a glass transition occurs in both the alginate matrix and the plant tissue upon freezing in liquid nitrogen. Building on early successes with herbaceous plants, there have been significant advances in cryopreservation of forest tree species using vitrification protocols, including shoot tips of eucalyptus (*Eucalyptus* spp.), willow (*Salix* spp.), pear, apple, and mulberry (*Morus* spp.), and somatic embryos of walnut (*Juglans* spp.). Significant progress has also been made with conifers, and some tree breeders routinely cryopreserve embryogenic callus (*Pinas radiata*) and somatic embryos (*Picea abies*) to conserve multiplication and regenerative capacities of embryogenic lines, transgenic or not. However, the genetic fidelity of embryogenic tissue lines stored in this way should be routinely monitored and the cryopreserved material evaluated in the field.

The Future

Making significant changes to the productivity of plantation forests by conventional means will always be a long-term process because of the generally long rotation times prevalent in forest trees, whereas global demands upon this resource necessitate the use of processes that will achieve production targets in the short term. To make any serious inroad into the annual reduction in the global forest, serious consideration needs to be given to maximizing the planting of high-yielding short rotation clones. Although some of these may be identified from conventionally produced trees, it is likely that many will need to be produced by biotechnological means, either because of the time taken to introduce new genes, or the impracticality of achieving the desired function (e.g., altered lignin biosynthesis). Also, practical mechanisms, whereby field assessment of clones can be optimized and the desired characteristics of the clones proved to be robust in the field, will need to be more widely utilized. The advancement of technological developments is perhaps not the issue driving the deployment of clonal forests, but rather the acceptance of the concept of clonal forests and the underpinning biotechnology that supports it. Societal acceptance, as well as that of policy makers, is crucial in fulfilling the potential that new biotechnologically improved clonal plantations can offer.

List of Technical Nomenclature

Axillary shoot culture	Induced branching of shoots by culture *in vitro*.
Biodiversity	Variety of species in a natural community or habitat.
Cryopreservation	Long-term storage (preservation) of tissue by freezing in liquid nitrogen.
Cryoprotectant	A chemical that helps to protect tissues from freezing during cryopreservation.
Cuttings	Shoots used to vegetatively propagate individuals.
Downregulation	Reduced expression of a gene.
Genetic modification (GM)	Modification of the genome of an organism by manipulation of its DNA.
Genotype	The genetic constitution of an organism.
Lignin	A complex noncarbohydrate polymer that is deposited in the cellulose of plant cell walls during secondary thickening, providing strength.
Marker-assisted breeding	Procedure to link genetic markers to phenotype characteristics.
Phenotype	The observed characteristics of an individual.
Plantation	Trees grown under cultivation.
Plastid genome	A genome belonging to a membrane-bound cellular organelle.
Rotation time	The time taken for tree crop to reach maturity.
Somaclonal variation	Expression of genetic variation in vegetative tissues.
Somatic embryogenesis	Induction of somatic embryos by culture *in vitro*.
Transgenic	Organism whose genome incorporates genes from another species.
Vegetative propagation	Reproduction of an individual by nonsexual means.

See also: **Biodiversity and Conservation**: Germplasm Conservation. **Crop Improvement**: Molecular Markers. **Ethics and Biosafety**: Plant Genetic Engineering, Ecological Issues. **Genetic Modification**: Transformation, General Principles; Transformation in Dicotyledons; Transformation in Monocotyledons. **Genetic Modification of Secondary Metabolism**: Wood Quality. **Pests**: Genetic Modification of Pest Resistance, Insect Pests. **Secondary Products**: Lignin. **Tissue Culture**: Clonal Propagation, *In Vitro*; Micropropagation; Somatic Embryogenesis. **Tissue Culture and Plant Breeding**: Clonal Propagation, Forest Trees. **Weeds**: Herbicide Resistance.

Further Reading

Fenning TM and Gershenzon J (2002) Where will the wood come from? Plantation forests and the role of biotechnology. *Trends in Biotechnology* 20: 291–296.
Global Forest Resources Assessment 2000 (2002) *Main Report, FAO Forestry Paper 140*. Rome: FAO, p. 479.
Mathews JH and Campbell MM (2000) The advantages and disadvantages of the application of genetic

engineering to forest trees: a discussion. *Forestry* 73: 371–380.
Merkle SA and Dean JFD (2000) Forest tree biotechnology. *Current Opinion in Biotechnology* 11: 298–302.
Strauss SH, DiFazio SP, and Meilan R (2001) Genetically modified poplars in context. *The Forestry Chronicle* 77: 271–279.
Wu L, Hu XS, and Han YF (2000) Molecular genetics and developmental physiology: implications for designing better forest crops. *Critical Reviews in Plant Sciences* 19: 377–393.

Regeneration of Fruit and Ornamental Trees via Cell and Tissue Culture

R E Litz, University of Florida, Homestead, FL, USA

Introduction

Trees, irrespective of whether they are angiosperm or gymnosperm species, are very heterogeneous plants. Accordingly, trees that have been derived from seedling populations of a known maternal parent can be extremely variable for most horticultural traits. Trees and woody shrubs that have considerable horticultural value are therefore generally propagated vegetatively in order to preserve their unique phenotype. For example, the prized mangoes (*Mangifera indica*) of India are clones that were selected from outstanding seedling trees of the Lakh Bagh, a large orchard of seedling mango trees that was established near Lucknow at the time of the Mogul emperor Akbar the Great in the sixteenth century. The introduction of vegetative propagation methods, e.g., grafting, into India by the Portuguese a few years earlier made it possible for the superior mango tree selections to be spread throughout India.

For many tree species, their long juvenile period has precluded extensive breeding programs that involve controlled pollinations and backcrosses. This is due to the commitments of land, labor, and time that are necessary to evaluate heterogeneous segregating seedling populations. As a result, most tree selections have resulted from random pollinations, and the outstanding trees have then been propagated vegetatively. The difficulties implicit in breeding of tree species have resulted in the relative absence of genetic information about these plants.

Production of many fruit and ornamental trees is restricted to relatively few horticultural selections. Some extreme examples include the commercial production of kiwifruit (*Actinidia deliciosa*), which is restricted to the single variety "Hayward," and mangosteen (*Garcinia mangostana*), which is known as a female tree that reproduces asexually from polyembryonic seeds. There has consequently been some erosion of genetic diversity within some species. This is particularly apparent among vegetatively propagated plantation crops that are grown under monoculture conditions in the tropics and subtropics, e.g., papaya. Clonally propagated plants are under intense disease and insect pressure under these conditions, and production can fail in the absence of sufficient genetic diversity.

In Vitro Regeneration of Fruit and Ornamental Trees

The recovery of plants by stimulating the development of existing meristems or through the induction of new meristems or somatic embryos under *in vitro* conditions is the basis for the micropropagation industry and the production of genetically modified plants using transformation or mutation induction procedures, respectively. Enhancing the efficiency of vegetative propagation using *in vitro* procedures began to have an impact on the foliage and cut flower industry several decades ago. These micropropagation procedures enabled the production of elite selections of certain types of herbaceous ornamental species at low cost and with great speed and efficiency. The commercial application of the same cloning procedures to fruit and ornamental trees has not achieved the same level of commercial success. There are a number of reasons for this slower pace of innovation.

Most ornamental species are herbaceous, and changes that are associated with phase change, i.e., the transition from the juvenile to the mature stage, tend to be rather insignificant. On the other hand, woody plants undergo quite profound changes in morphology during phase transition. Changes in morphology are associated with distinct alterations in gene expression that affect cell and tissue determination, e.g., alterations in secondary metabolism in the mature phase that include lignification and changes that affect organization and development, such as the floral organs, etc. Since elite trees can only be identified when they are in their mature phase, it is often very difficult to stimulate the growth of their meristematic tissues at this late stage of their development and to reprogram somatic cells, because cells and organs within mature phase plants are so highly determined.

The Plant Growth Medium and Growing Conditions

Stock plants of elite angiosperm trees are either field-grown specimens or small vegetatively propagated plants that are hard pruned and maintained in a controlled environment such as a greenhouse or a growth chamber. The maintenance of stock plants under controlled conditions is preferred since it is easier to optimize growth of the plant and to control systemic fungal and bacterial infections. Tissue dissected from an elite specimen tree in order to establish *in vitro* cultures (known as the "explant") first must be surface-disinfected in order to cleanse the tissue of superficial bacterial and fungal contaminants. This is generally accomplished using a dilute sodium hypochlorite or mercuric chloride solution, followed by thorough rinsing with sterile water before its introduction onto a semi-solid sterile plant growth medium. The optimum plant growth medium formulation must be determined empirically for each plant species; however, increasingly, formulations consisting of relatively low major salts concentrations are considered to be the most appropriate for use with most woody plants. Some examples of low salt formulations include Woody Plant Medium, Andersons Rhododendron Medium, Murashige & Skoog Medium that has been modified to contain half or quarter strength major salts, B5 medium, etc. The basal medium is supplemented with vitamins, organic acids, sucrose, and a gelling agent – agar, gellan gum, or an agar : gellan gum mixture. The pH of the plant growth medium is fairly well buffered, although the pH is adjusted to 5.7 prior to sterilization at 1.1 kg cm^{-2} for 15–20 min. Growth conditions are usually 25 °C and, if light is required, it is usually 60–80 μmol s^{-1} m^{-2} provided by gro lux or cool white fluorescent lights with a 16-h photoperiod.

In Vitro Regeneration from Existing Meristems (Shoot Tip/Nodal Culture)

The standard micropropagation practice for regenerating plants *in vitro* (i.e., shoot tip or nodal culture) involves the initiation of developmental growth from apical buds (containing the apical meristem) and nodal tissues (containing axillary meristems) that have been removed from a stock plant, surface sterilized, and cultured on sterile plant growth medium. The rationale for this approach is that the integrity of the meristem can be retained in this manner. A prerequisite for *in vitro* cloning of trees by shoot tip and nodal culture is that their shoot apices must be wholly or at least partially rejuvenated during the establishment phase. Rejuvenation requires the expression of embryonic or juvenile phase developmental genes.

Shoot tip/nodal culture has four recognized developmental stages: establishment (of the explants), proliferation (of axillary buds), shoot formation (by re-establishing apical dominance), and rooting of the shoots. Many protocols have indicated a particular growth medium that has been formulated for establishment of the excised apical and lateral buds of a particular woody species. The establishment medium optimizes recovery and initial growth of the explant, and is supplemented with an auxin, e.g., indole-3-acetic acid (IAA), indole-3-butyric acid (IBA), α-naphthaleneacetic acid (NAA), and either a cytokinin, e.g., 6-benzylaminopurine (BAP), kinetin, zeatin, 6-(γ, γ-dimethylallylamino)-purine (2-iP), or a plant growth regulator such as thidiazuron (TDZ), whose action is similar to that of cytokinins. Explants of woody plants can be especially difficult to establish *in vitro* owing to the accumulation of oxidative products in the plant growth medium, which are toxic to the explant. Death of explanted tissue can be avoided by frequent subculture during the first few days and/or by incorporating an antioxidant in the plant growth medium. Following an empirically determined period on establishment medium, the tissue is subcultured onto another medium formulation that stimulates proliferation of the axillary or nodal meristems. This is achieved by suppressing apical dominance with a strong cytokinin. The proliferation of lateral meristems is normally allowed to continue for a fixed number of subcultures; at the end of each cycle the individual lateral buds are dissected from the proliferating axillary buds and recultured onto proliferation medium. Prolonged maintenance on proliferation medium can cause physiological disorders, such as hyperhydricity, which is caused by excessive water accumulation in the tissue, and habituation, which is a result of acquired insensitivity to exogenously applied plant growth regulators.

Apical dominance can be restored to the meristems by subculturing lateral buds onto medium without cytokinin. In order to stimulate rooting of the shoots that have developed *in vitro*, exposure to a potent rooting auxin, e.g., IBA or NAA, is usually necessary. Different rooting approaches have been adopted with varying degrees of success, depending on the species: (1) induction of *in vitro* rooting and development on medium with low auxin concentrations; (2) induction of rooting *in vitro* with a brief exposure to a high auxin concentration followed by root development *ex vitro*; and (3) root induction by dipping in high auxin concentration and development *ex vitro*. In general, options (2) and (3) have been the most effective for many woody plants.

Table 1 Regeneration of elite fruit and ornamental trees by shoot tip and nodal culture

Plant family	*Species*	*Common name*
Annonaceae	*Annona muricata*	Soursop
Ericaceae	*Vaccinium* spp.	Blueberry
	Rhododendron spp.	Rhododendron
Moraceae	*Artocarpus* spp.	Breadfruit, jakfruit
	Ficus carica	Fig
	Ficus spp.	Ornamental Ficus
	Morus spp.	Mulberry
Myrtaceae	*Psidium guajaba*	Guava
Rosaceae	*Crataegus pinnatifida*	Hawthorne
	Malus x domesticum	Apple
	Prunus spp.	Almond, cherry, peach, nectarine, plum
	Rosa spp.	Rose
	Rubus spp.	Cane fruit

Application of Shoot Tip and Nodal Culture to Fruit and Ornamental Trees Shoot tip and nodal culture is the standard commercial micropropagation procedure, and has had its greatest application as an alternative commercial method for propagating plant species. Its efficacy is dependent upon its cost-effectiveness in comparison with traditional cloning procedures. Although relatively few tree species can be propagated by this procedure, shoot tip or nodal culture has great potential utility for rapidly propagating certain fruit and ornamental tree species, including scion and rootstock selections. Among fruit and ornamental species, particular success involving shoot tip and nodal culture has been achieved among angiosperm tree species within the Ericaceae, Moraceae, and Rosaceae families (**Table 1**). There have been no successful reports of shoot tip and nodal culture involving any elite gymnosperm species selection.

In Vitro* Regeneration of Fruit and Ornamental Trees *De Novo Plants can be regenerated from cell and tissue cultures in which there are no pre-existing meristems. These regeneration pathways are less likely to be used for propagating plants, because somatic mutations occur more commonly during rapid cell cycling, thereby altering the genetic integrity of elite selections. However, there are a few notable exceptions; date palm (*Phoenix dactylifera*) and oil palm (*Elaeis guineensis*) have both been propagated by somatic embryogenesis. On the other hand, the application of genetic engineering strategies to horticulturally important woody plants is dependent on the ability to regenerate an elite selection from single cells.

Most field-grown horticultural and agronomic crops are seed propagated. In order to define a cell-to-whole plant regeneration protocol for such crops, it is customary to utilize immature zygotic embryos or the hypocotyl and cotyledons of newly germinated seedlings as explants. Horticulturally important trees, however, can be vegetatively propagated selections that might be several hundred years old (see above). *De novo* regeneration strategies involving the appropriate explant and plant growth medium must consequently be established empirically.

Most cells within a plant are highly determined or specialized following the embryonic or immediate postembryonic (seedling) stage. Therefore, regeneration of plants from cells of elite (mature phase) trees generally involves redetermination of highly differentiated cell types. Induction refers to a change in the developmental potential or competence of a cell, i.e., a change from one highly differentiated state to another highly differentiated state. Alternatively, in certain angiosperm tree species, it is rarely possible to locate cells within the organism that already have the competence to differentiate as whole plants. Cells that have morphogenic competence possess certain general morphological characteristics; they are small, isodiametric in shape, have a prominent nucleus, and are richly cytoplasmic.

Two regeneration pathways have been defined that involve induction from differentiated cells: organogenesis and somatic embryogenesis. Organogenesis refers to the induction of *de novo* meristematic regions, either the apical meristem or the root meristem, from somatic tissues. Somatic embryogenesis refers to the induction of *de novo* embryos from somatic tissues. Somatic embryogenesis has been clearly demonstrated to have a single-cell origin; however, organogenesis apparently can have either a single-cell or multiple-cell origin, although the former is more characteristic. It is almost certain that organogenesis and somatic embryogenesis represent the same regeneration pattern, but under different epigenetic control. The embryogenic pattern of development occurs when competent cells have been completely rejuvenated, so that embryonic genes are expressed. Organogenesis occurs from competent cells in which many embryo-specific genes are not expressed, e.g., development of the cotyledons. Whether an induced cell follows an organogenic or embryogenic pathway is determined by several genetic and epigenetic factors, including taxon, genotype, source of the explant (organ or tissue), stage of development of the plant and of the tissue at the time of excision of the explant, plant growth medium composition, etc. Two patterns of induction are recognized in organogenesis and somatic embryogenesis: permissive induction and directive induction.

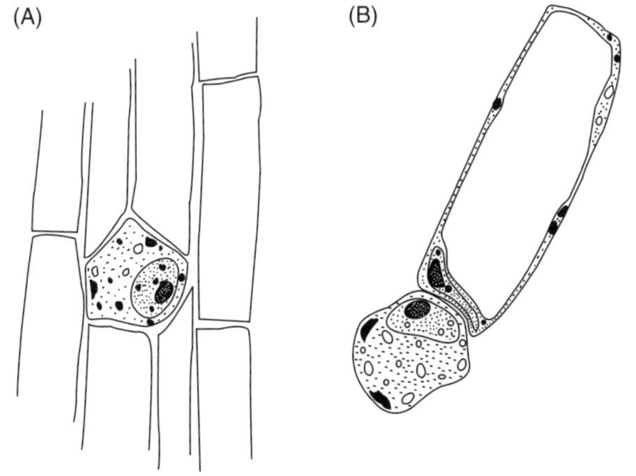

Figure 1 Diagrammatic representation of origin of morphogenically competent cells. (A) Competent cells within the explanted tissue and (B) following an unequal cell division.

Permissive induction The morphogenic potential of competent cells already present within a tissue before their excision and explanting can be released as a result of their removal from surrounding plant tissue. It is implicit with permissive induction that morphogenically competent cells are present within certain plant tissues even before excision (**Figure 1**). Prior to their excision and explanting, the morphogenic potential of these cells would most probably have been suppressed by the tissue environment. For example, many tropical and subtropical angiosperm tree species have polyembryonic seeds, in which adventitious embryos develop from morphogenically competent cells within the nucellus or the inner integument, e.g., mango, citrus, *Syzygium* spp., mangosteen (*Garcinia mangotana*), etc. Within a few angiosperm tree species, different ecogeographical races may have either polyembryonic or monoembryonic seeds, i.e., containing only a zygotic embryo. Removal of the nucellus from either type of seed and explanting it to an *in vitro* environment can stimulate the development of embryos from morphogenically competent cells within the nucellus.

Directive induction More generally, it is necessary to stimulate limited division of cells within excised tissue before an induction event can occur. Directive induction involves a change in competence of cells within an explant, unlike permissive induction. Limited cell proliferation can result in the formation of a callus mass, consisting of dedifferentiated cells (i.e., of altered competence) or it is possible that only a single cell division may be involved. Acquisition of morphogenic competence is accompanied by altered cell morphology, i.e., an asymmetric cell division can occur spontaneously, resulting in the formation of two unequally sized daughter cells (**Figure 1**). One (morphogenically competent) daughter cell is small with an isodiametric shape, a prominent nucleus, and is richly cytoplasmic. The other, highly differentiated, daughter cell is large with a prominent vacuole, but would have no morphogenic competence. The acquisition of morphogenic competence associated with an asymmetric cell division is considered to be homologous with an asymmetric division of the zygote to form two daughter cells that become the embryo and the suspensor. The most critical factors for *in vitro* directive induction include genotype, source, and stage of development of the explant, plant growth medium formulation, and plant growth regulator composition of the medium. Genotype can strongly affect induction, and some of the most important woody fruit and ornamental cultivars are not highly responsive *in vitro*.

Table 2 Regeneration of elite fruit and ornamental trees by organogenesis

Plant family	*Species*	*Common name*
Anacardiaceae	*Mangifera indica*	Mango
Moraceae	*Artocarpus heterophyllus*	Jakfruit
	Morus spp.	Mulberry
Oleaceae	*Olea sativa*	Olive
Oxalidaceae	*Averrhoa carambola*	Carambola
Rosaceae	*Pyrus communis*	Pear
Rutaceae	*Citrus* spp.	Various citrus

Organogenesis and Somatic Embryogenesis of Fruit and Ornamental Trees

Development from a competent, i.e., induced, cell occurs by permissive induction in the absence of the stimulus for directive induction. The expression of development can be as adventitious root or shoot meristems (organogenesis) or as adventitious embryos (somatic embryogenesis) (see **Tables 2** and **3**, respectively). Plant growth medium formulations that are effective for shoot tip and nodal culture of woody plants are also appropriate for induction of organogenic and embryogenic cultures. However, the growth medium supplements and growing conditions are significantly different for each *de novo* pathway.

Organogenesis Organogenic cultures of woody plants are induced by a combination of an auxin (usually NAA) and a cytokinin (e.g., kinetin, BAP, 2iP). The presence of the auxin is necessary to stimulate rapid cell division within the explant; the cytokinin also stimulates cell division, but is essential for organization of the apical meristem. (The auxin : cytokinin ratio can determine the organization of shoot (high cytokinin to auxin ratio) and root

Table 3 Regeneration of elite fruit and ornamental trees by somatic embryogenesis

Plant family	*Species*	*Common name*
Angiosperms		
Anacardiaceae	*Mangifera indica*	Mango
	Anacardium occidentale	Cashew
Hippocastanaceae	*Aesculus hippocastnum*	Horse chestnut
Lauraceae	*Persea americana*	Avocado
Myrtaceae	*Myrciaria cauliflora*	Jaboticaba
	Syzygium spp.	Rose apple, Malay apple
Oleacea	*Olea sativa*	Olive
Rosaceae	*Eriobotrya japonica*	Loquat
	Malus x domesticum	Apple
	Rosa spp.	Rose
Rubiaceae	*Coffea* spp.	Coffee
Rutaceae	*Citrus* spp.	Various citrus
Sapindaceae	*Dimocarpus longan*	Longan
Sterculiaceae	*Theobroma cacao*	Cacao
Gymnosperms		
Zamiaceae	*Ceratozamia euryphyllidia*	NA
	Ceratozamia hildae	NA

Figure 2 Shoot organogenesis from cultured leaves of the Andean fruit naranjilla or lulo (*Solanum quitoense*), a perennial shrub.

(high auxin to cytokinin ratio) meristems.) With many woody angiosperms, a necessary prerequisite for organogenesis has been the use of leaves from *in vitro* shoots that develop from shoot tip and nodal cultures. Incubation in a 16 h light photoperiod is important, and (shoot) organogenic callus cultures are generally green. Once adventitious apical buds have appeared on the callus (**Figure 2**), they can be removed and transferred onto proliferation medium in order to stimulate lateral bud proliferation (see shoot tip and nodal culture) or they can be transferred onto a growth medium without plant growth regulators to stimulate apical dominance (see shoot tip and nodal culture). Shoots that have been derived from organogenic cultures can be rooted in the same manner as shoots derived from those tip and nodal cultures.

Somatic embryogenesis There are four distinct stages in somatic embryogenesis that are easily recognized in systems that have been optimized: (1) induction; (2) maintenance; (3) development/maturation; and (4) germination/conversion. Permissive induction of embryogenic cultures from competent cells within excised tissues of woody plants normally requires explanting onto a rather simple plant growth basal medium, as is the case with nucellar cultures of citrus. Directive induction of embryogenic cultures, on the other hand, requires the presence in the plant growth medium of a strong auxin, e.g., 2,4-dichlorophenoxyacetic acid (2,4-D), picloram, dicamba, etc., to stimulate rapid cell division. In addition, medium supplements for embryogenic cultures of woody plant species usually include relatively high sucrose concentrations ($30–90\,g\,l^{-1}$) and a source of reduced organic nitrogen. Once embryogenic cultures have been induced, they are technically no longer callus cultures, although they continue to be referred to as such in much of the literature. Embryogenic cultures are characteristically friable and watery in appearance, and can be white, light brown, or honey-colored, depending on the species. A source of light is usually not required for induction and maintenance of embryogenic cultures.

Very shortly after cultures are recognized as being embryogenic, they undergo a rapid change in appearance while still on induction medium. The cultures, still relatively friable, consist of smooth globular structures of various diameters, cells, and small cell aggregates (**Figure 3**). The embryogenic cells in cultures of angiosperm trees organize as "proembryonal masses" (PEMs) of various stages of development and disorder. PEMs are proembryos that have lost their ability to organize and develop as single embryos in the presence of the primary induction agent (strong auxin), but enlarge and produce embryogenic cells and secondary globular embryos and PEMs from the protoderm. Embryogenic cultures of gymnosperms consist of proembryonic cells and proembryos that develop to the precotyledonary stage. Secondary somatic embryos develop from the meristematic region of precotyledonary somatic embryos. Proliferation of embryogenic cultures on induction medium is normally referred to as the maintenance stage.

Embryogenic cultures of angiosperm trees in maintenance medium can have two genotype-specific

Figure 3 Diagrammatic representation of a typical embryogenic culture. Different sizes of globular proembryonal masses (PEMS) develop from a transient callus in the presence of the induction agent (e.g., 2,4-D). In the continuous presence of 2,4-D, secondary PEMs and embryos develop from the periphery of large PEMs. In the absence of the inducing agent, small globular embryos can develop individually as single embryos.

Figure 4 Germinating somatic embryos derived from leaves of an elite selection of the longan (*Dimocarpus longan*), a subtropical fruit tree of Southeast Asia.

responses: (1) proliferation of PEMs in induction medium and (2) proliferation and development of cotyledonary embryos (from the hypocotyl) in induction medium. The latter type of maintenance culture is less sensitive to auxin. Embryogenic cultures cannot be maintained under inductive conditions indefinitely and gradually lose competence to regenerate. Proembryonal masses of many plant genotypes can proliferate on semisolid induction medium; however, a few genotypes can proliferate much more efficiently as suspension cultures in liquid plant growth medium consisting of induction formulation. Growth of embryogenic cultures in suspension is highly genotype-specific. Embryogenic suspension cultures consist of an initial inoculum of an embryogenic culture in liquid induction medium, which are maintained on a rotary shaker in darkness or subdued light. The rapid proliferation of embryogenic cultures in suspension is the basis for automation of somatic embryogenesis in bioreactors. Maintenance of embryogenic cultures is perhaps the most critical stage of somatic embryogenesis, because this is the stage at which proliferation of the culture is optimized, and this is the stage that is utilized for genetic transformation, for *in vitro*-induced mutation and selection, and for cryopreservation. Therefore, it is essential to optimize the maintenance stage.

Auxin is not only the primary induction agent for embryogenic cultures, but also inhibits the subsequent development of somatic embryos of many genotypes. Transfer of embryogenic cultures from the maintenance medium containing strong auxin stimulates somatic embryo development, and generally inhibits the continuing proliferation of the cultures. Development of somatic embryos to maturity is a distinct stage of somatic embryogenesis that is mediated by different growth parameters (**Figure 4**).

Development of angiosperm tree somatic embryos from PEMs can be affected by cytokinins, which stimulate organization of the apical meristem, and the gradual lowering of the osmolarity of the plant growth medium. The recovery of good quality somatic embryos is related to the basic seed physiology of the species, whether the seeds/embryos are recalcitrant (no dormancy and no desiccation tolerance) or orthodox (dormancy and desiccation tolerance) in their storage behavior. Under *in vitro* conditions, all somatic embryos develop as recalcitrant embryos since dormancy does not normally occur under these conditions. Quiescence of orthodox angiosperm tree somatic embryos must be imposed by the inclusion of agents that inhibit embryo growth and development in the plant growth medium; such agents include abscisic acid (ABA) and increased osmolarity of the growth medium. Unlike angiosperm tree somatic embryos, somatic embryos of many gymnosperm species are highly responsive to ABA and high osmolarity (provided by polyethylene glycol) for normal maturation (**Figure 5**). Somatic embryos can develop either in absolute darkness or under a light photoperiod; however, light is probably not essential for development. Mature somatic embryos "germinate" according to the sequence observed with zygotic embryos: elongation of the hypocotyl and growth of the radicle, followed by elongation of the epicotyl. Germination is light-sensitive. The recovery of viable plants from somatic embryos is referred to as conversion.

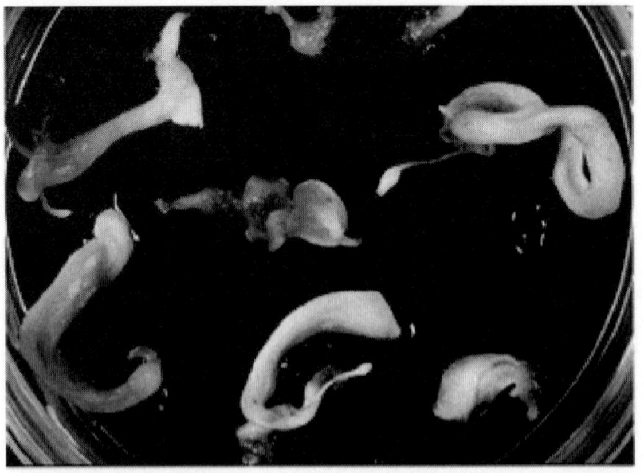

Figure 5 Somatic embryos derived from leaves of an elite gymnosperm, the cycad *Ceratozamia hildae*.

Somatic Cell Genetics of Fruit and Ornamental Trees

It is extremely tedious to consolidate interesting genetic traits in perennial plants using controlled backcrossing in conventional breeding, and the strategy must involve extensive field plantings. Because of the difficulties that are associated with the breeding of perennial plants, *in vitro* strategies that involve the use of shoot tip/nodal culture, organogenesis, and somatic embryogenesis have been adopted that would permit the genetic manipulation of existing cultivars.

Somaclonal Variation

Regeneration of plants from somatic tissues can result in the phenotypic expression of somatic mutations that have accumulated in these tissues or that have resulted from the *in vitro* cell cycle. (With some species, the *in vitro* cycle itself is apparently mutagenic.) Tissue culture-induced off-types with interesting horticultural traits have been recovered for a few fruit and ornamental trees, e.g., thornless cane fruit (*Rubus* spp.). Although somaclonal variation represents a potential source of genetic variability in clonally propagated crop plants, it is undirected and random. Therefore, most of the off-types have little or no horticultural interest. Commercial propagation of oil palm by somatic embryogenesis resulted in the wide-scale production of trees with distorted, i.e., mantled, fruit, and the affected trees were of no value. Although banana (*Musa* gp. AAA) is not a woody species, the production of off-types from shoot tip and nodal culture can exceed 5% even under optimal *in vitro* growth conditions that have been designed to minimize variation. Potentially useful banana somaclonal variants have been identified under field conditions, including selections having increased resistance to Panama disease caused by the fungal pathogen *Fusarium oxysporum*. Somaclonal variation within existing clones has considerable application for dessert and cooking banana improvement since most of these plants are sterile triploids, in which the existing genetic variation is due to somatic mutation.

Induced Mutations and Selection

The most problematic issues involved with identification of useful somaclonal variants are as follows: (1) the process is random and undirected; (2) a large number of uninteresting off-types are produced; and (3) plants must be observed and off-types identified under field conditions, which is expensive and laborious for tree species. Certain genotypes are amenable to *in vitro* selection at the single cell level; however, this is only effective if the action of the selective agent is the same at the single cell and whole plant levels and if the plant can be regenerated from single cells. Moreover, since many horticultural traits are conferred by more than a single gene, it is usually very difficult to select for such traits *in vitro*.

The number of mutations can be greatly increased by exposure of cell and tissue cultures to physical and chemical ionizing agents. This has been utilized quite effectively in conjunction with shoot tip and nodal cultures and increasingly with *de novo* regeneration systems. Chimeral meristematic zones result from exposure of plants to ionizing irradiation or chemical mutagens. Homohistonts can be isolated from chimeric tissue by shoot tip and nodal culture after approximately four subculture cycles. Regenerated plants can be screened under field conditions for horticultural utility. Embryogenic cultures can also be exposed to gamma-irradiation or chemical mutagens followed by screening with an effective selection agent. For example, certain plant pathogens produce one or more phytotoxic compounds during symptom development and, either alone or as a crude culture filtrate, can stimulate the host resistance response. Selection of *in vitro* cultures for the constitutive expression of the resistance response has been used to identify peach (*Prunus persica*) embryogenic cells with resistance to *Xanthomonas campestris*, mango cultures for resistance to *Colletotrichum gloeosporiodes* (the cause of anthracnose), grape (*Vitis* spp.) cultures for resistance to *Elsinoe ampelina* (the cause of anthracnose), date palm (*Phoenix dactylifera*) cultures for resistance to *F. oxysporum* (the cause of Bayoud disease), and citrus for resistance to *Phoma tracheiphila* (the cause of "mal secco" disease). Although *in vitro* mutagenesis and selection can be a powerful tool to address specific breeding objectives, there are relatively few chemical or physical

Table 4 Genetic transformation of woody fruit and ornamental plants

Family	*Species*	*Horticultural trait*	*Gene(s)*
Anacardiaceae	*Mangifera indica*	Fruit ripening	ACC oxidase in antisense
			ACC synthase in antisense
			Alternative oxidase in antisense
Lauraceae	*Persea americana*	Resistance to Phytophthora root rot	Glucanase
Rosaceae	*Malus x domesticum*	Fire blight resistance	HrpN
			AttacinE
	Prunus spp.	Plumpox virus resistance	PPV coat protein
Rutaceae	*Citrus aurantifolia*	Citrus tristeza virus resistance	CTV coat protein
Vitaceae	*Vitis vinifera*	Grapevine fanleaf virus resistance	GFLV coat protein
		Arabis mosaic virus resistance	ArMV coat protein
		Grapevine virus A and B resistance	GVA and GVB coat proteins
		Foliage and fruit disease	Chitinase

agents that can be utilized for *in vitro* selection, and this has limited its application. The use of induced mutation strategies is also limited by the large number and randomness of gene mutations that occur.

Genetic Transformation

As an alternative to induced mutation and selection for plant improvement, the targeting of specific horticultural traits using genetic transformation has considerable potential for vegetatively propagated trees, because the genetic integrity of the clone can be retained with the exception of the altered trait(s). Several fruit species have been transformed either with marker genes or with marker genes together with a gene for a useful horticultural trait (see **Table 4**); however, relatively few woody ornamental species have been transformed. Breeding objectives that have been addressed include the control of fruit ripening, and resistance to insect pests, viruses, and pathogenic fungi and bacteria. The implications of the use of single genes in perennial species for controlling continuously evolving insect pests and pathogens has not been assessed. For example, it is not known how targeted and nontargeted insect populations will be affected by long-term exposure to transgenic plants with insecticidal proteins. Therefore, determination of the effectiveness of this approach must await field observations.

Manipulating the Ploidy of Fruit and Ornamental Trees

Many woody species are naturally occurring polyploids, which can contribute to their high level of heterozygosity. It has been possible to manipulate the ploidy of a few angiosperm and gymnosperm woody species using *in vitro* procedures.

Table 5 Regeneration of haploid woody fruit and ornamentals

Family	*Species*	*Common name*
Angiosperms		
Annonaceae	*Annona squamosa*	Sugar apple
Caricaceae	*Carica papaya*	Papaya
Rosaceae	*Malus x domesticum*	Apple
Rutaceae	*Citrus* spp.	Various citrus
Sapindaceae	*Litchi chinensis*	Lychee
	Dimocarpus longan	Longan
Vitaceae	*Vitis vinifera*	Grapevine
Gymnosperms		
Zamiaceae	*Zamia* spp.	NA
	Ceratozamia spp.	NA

Haploidy

Recovery of haploid plants of a few woody species has been reported, although the regenerants have apparently not been utilized in a breeding program. One of the advantages of producing haploids of woody plants would be the recovery of homozygous diploids that could be used as male parents in controlled pollinations in any conventional breeding program. In this manner, it would be possible to limit the variation from one parent, and move more rapidly to cultivar release. Haploids of angiosperm trees have been produced *in vitro* in two ways: either by anther and microspore culture or from the ovary following pollination with irradiated pollen (see **Table 5**). Regeneration of haploids of angiosperm trees is normally via the embryogenic pathway. Haploid gymnosperms have been recovered *in vitro* by culturing the megagametophyte, the haploid storage organ of gymnosperm seeds, and regeneration can be via either the organogenic or the embryogenic pathway.

Triploidy

From a plant breeding perspective, the recovery of seedless selections of certain fruit crops can be very

Table 6 Regeneration of triploid fruit and ornamental trees

Family	*Species*	*Common name*
Annonaceae	*Annona squamosa*	Sugar apple
Passifloraceae	*Passiflora foetida*	NA
Rosaceae	*Eriobotrya japonica*	Loquat
	Pyrus communis	Pear
Rutaceae	*Citrus grandis*	Pummelo

important. Dessert bananas (*Musa* gp. AAA) and cooking bananas (*Musa* gps. AAB, ABB) are naturally occurring triploids, and almost all of the important selections of these crops are entirely seedless. The endosperm is a triploid nutritive tissue that occurs within the seed of angiosperms, and has been demonstrated to have morphogenic potential in a number of woody species (see **Table 6**). It is not clear whether selections that have been made from triploid regenerants could have horticultural interest; however, for fruits either with many seeds (guava; *Ps idium* spp.) or with large and hard seeds (*Annona* spp.), this approach has interesting possibilities. Regeneration has been via the organogenic pathway.

Tetraploidy

Production of tetraploid plants has been a standard breeding procedure for many years, and can be accomplished *ex vitro* by applying colchicine to the apical meristem. Genetic variation can also be enhanced by somatic hybridization, which involves the production of interspecific and intergeneric hybrids by forced fusion of enzymatically isolated protoplasts. This strategy has been utilized as a breeding tool for improving certain perennial fruit crops, and most of the somatic hybrids have involved the following forced hybridizations: (1) citrus interspecific crosses; (2) intergeneric crosses between citrus and related species in the genera *Atalantia*, *Clausena*, *Fortunella*, *Poncirus*, and *Severina*; and (3) interspecific crosses between passionfruit (*Passiflora edulis* f. *flavicarpa*) and other *Passiflora* spp. This appears to be a potentially useful strategy for developing new rootstock selections (e.g., citrus), although it has limited value for addressing scion breeding objectives owing to the great heterogeneity and polyploidy of recovered plants.

Conclusions

Efficient regeneration pathways for elite selections of horticulturally important woody plants from cell and tissue cultures provide a unique opportunity for improving many of these crops. Advanced micropropagation systems could have an impact on production of many tree crops, including species that are difficult to propagate vegetatively by traditional means. The application of traditional breeding methods to many of these species has been difficult owing to their long juvenile periods and the high cost of maintaining field plantings of segregating populations of seedling trees for many years. Until now, it has been impossible to respond to crop-threatening disease and pest situations of woody plants because of the time and costs involved. Consequently, impacted traditional cultivars are often replaced by less acceptable selections. Somatic cell genetics applied to threatened traditional selections is certain to have a major impact on improving perennial woody plants, not least because it will be possible to change an existing clonal selection by altering one or more genes that control a single horticultural trait. Thus, it will be possible to add significant value to an already highly prized selection.

List of Technical Nomenclature

Adventitious Development of organs or embryos from somatic tissues or callus.

Auxin A natural or synthetic plant growth substance that induces cell division and elongation. Also utilized to induce adventitious rooting.

Competence Developmental potential of a cell.

Cytokinin The natural or synthetic plant growth substances that induce cell division and adventitious shoot meristems. Suppress apical dominance.

De novo For the first time.

Dedifferentiation Change in the competence of cells, i.e., from differentiated to nondifferentiated states.

Differentiation Induction of cell and tissue types with a specific function.

Directive induction Change in the competence of a cell.

Explant Tissue excised from a plant in order to initiate a culture.

Ex vitro Nonsterile conditions.

Genetic transformation Introduction of foreign genes into cells and their expression.

Haploid A plant with half the normal chromosome number.

Induction Change in the competence of a cell or organ.

In vitro	Tissue growth in a container under sterile conditions.
Juvenility	The stage(s) of plant development before flowering.
Meristem	A group of actively dividing cells at the tip of roots and shoots and in the axillary region of stems, leaves and flowers.
Micropropagation	*In vitro* vegetative propagation of plants.
Morphogenesis	Differentiation of form.
Organogenesis	Induction and development of adventitious shoot or root meristems.
Permissive induction	Expression of a predetermined developmental potential.
Polyploid	Plants with multiples ($>2x$) of its base chromosome number.
Protoplast	A plant cell lacking its cell wall.
Somaclonal variation	Tissue culture-related variation.
Somatic embryogenesis	Induction and development of adventitious embryos.
Suspension culture	Growth of tissue and cell cultures in a liquid medium.

See also: **Crop Improvement**: Mutation Techniques; Plant Breeding, Practice; Plant Breeding, Principles. **Tissue Culture and Plant Breeding**: Somaclonal Variation; Somatic Hybridization.

Further Reading

Ammirato PV, Evans DA, Sharp WR, and Bajaj YPS (eds) (1990) *Handbook of Plant Cell Culture: 5 Ornamental Species.* New York: McGraw-Hill.

Debergh PC and Zimmerman RH (eds) (1991) *Micropropagation Technology and Application.* Dordrecht: Kluwer Academic Publishers.

Geneve RL, Preece JT, and Merkle SA (eds) (1997) *Biotechnology of Ornamental Plants.* Wallingford: CAB International.

Hammerschlag FA and Litz RE (eds) (1992) *Biotechnology of Perennial Fruit Crops.* Wallingford: CAB International.

Litz RE (1995) Fruit crop biotechnology. In: Parker SP (ed.) *Encyclopedia of Science and Technology*, pp. 62–64. New York: McGraw-Hill.

Litz RE (ed.) (in press) *Biotechnology of Fruit and Nut Crops.* Wallingford: CAB International.

Mohan Jain S, Brar DS, and Ahloowalia BS (eds) (1998) *Somaclonal Variation and Induced Mutations in Crop Improvement.* Dordrecht: Kluwer Academic Publishers.

Mohan Jain S, Sopory SK, and Veilleux RE (eds) (1996) In Vitro *Haploid Production in Higher Plants: Important Selected Plants.* Dordrecht: Kluwer Academic Publishers.

Pierek RLM (1997) In Vitro *Culture of Plants.* Dordrecht: Kluwer Academic Publishers.

Somaclonal Variation

F Sala and M Labra, University of Milan, Milan, Italy

The Discovery of Somaclonal Variation

Complete, or almost complete, genetic uniformity is a desirable trait, and frequently a must, in crop production. In agricultural practice, genotype stability maintenance is not considered a problem, once a favorable genetic constitution has been achieved by breeders; homozygous genomes are reproduced by inbreeding (e.g., in cereals), while heterozygous genomes are reproduced by cuttings (e.g., in fruit trees and in poplar, *Populus* spp.). This certainty was undermined 30 years ago when Alan Durrant, from the University College of Wales, subjected plants of a highly inbred flax (*Linum usitatissimum*, cv. Stormont Cirrus) cultivar to a number of different elevated concentrations of the nutrients potassium, nitrogen, and phosphorus. Although individual plants of an inbred cultivar are expected to be genetically homogeneous and morphologically identical, the resulting plants appeared morphologically different according to the growing conditions; their development was clearly influenced by nutrient concentration. But what was most surprising was that when seeds produced by these plants were germinated and grown under normal nutrient conditions, the progeny from any one nutrient treatment, although similar to each other, differed both from the original common parent cultivar and also from the progeny of plants exposed to different nutrient conditions. One group was large and highly branched, for example, whereas another was small and spindly. Apparently, these traits were inherited by their progeny! This behavior was reminiscent of the Lamarckian concept of evolution, i.e., changes acquired in response to an altered environment become hereditary. At first, Durrant's results did not meet with much acceptance, but, subsequently, Christopher Cullis at the John Innes Institute in Norwich, UK demonstrated that flax plants grown under the conditions previously used by Alan Durrant were characterized by changes in the analyzed DNA fraction, the repetitive DNA: a few of the sequences increased in number, while most showed a decrease. Overall, there was a 15% difference in total DNA content between the two groups with the most divergent DNA content. Cullis also verified that the repetitive DNA alterations arose during mitosis when the original cultivar was grown under nutrient stress: the DNA in the lower part of the flax stem had the original composition, while the altered pattern appeared in the younger upper region and in the flowers derived from it.

Further experimental observations showed that this type of DNA change is not limited to flax grown under nutrient stress. In fact, a number of investigators have noticed that when different plant species are exposed to different stresses, DNA variability can be induced. In general, the accumulated evidence has made it clear that the plant genome can change in a single generation and that the change is inherited in the sexual or clonal progeny.

The development of *in vitro* techniques for tissue, callus, or protoplast culture for biotechnological applications has shown that genomic abnormalities and phenotypic variation are frequent in regenerated plants. In 1981, Larkin and Scowcroft, CSIRO, Canberra, Australia, proposed the terms "somaclonal variation" and "somaclonal variant" to identify variant plants differentiated from cells and tissue grown in culture. Since then, somaclonal variation has been essentially associated with *in vitro* cell culture, although the earlier observations in flax, and more recent observations on heritable genomic changes in plants under different environmental abiotic or biotic stresses, now point to a more general concept. This concept proposes that plants not only tolerate genetic changes induced by abiotic and biotic stresses, but may have "generators of diversity" as well that act to create new genomic combinations within the plant or plant populations. Virginia Walbot and Christopher Cullis, Stanford University, USA, proposed that such "generators of diversity" serve to diversify the genetic constitution of the plant as part of its strategy to adapt to changing environments. Thus, regeneration from dedifferentiated plant material is only one of the stresses that cause somaclonal variation, other biotic and abiotic stresses may also promote genetic changes. The message that geneticists and molecular biologists are increasingly passing on to the plant breeders is that the plant genome is remarkably plastic.

How Frequent is Somaclonal Variation?

The above reported observations propose the existence of a direct correlation between stress and the extent of genomic changes. Although this has not yet been demonstrated conclusively, it is common practice to avoid conditions of stress if genomic uniformity is required in plant reproduction, or to employ stresses if induction and selection of new genotypes is desired. For example, the protocols employed in clonal propagation, where the aim is to produce millions of genetically uniform crop plants for agricultural use, are designed in order to avoid tissue culture in the dedifferentiated state. On the other hand, the protocols used for crop improvement through tissue culture largely depend on cell and tissue dedifferentiation in the presence of excess hormone concentrations.

Assessment of the mutation frequency in a plant population regenerated from tissue culture or subjected to different stresses is usually based on the observation of morphological and physiological changes, and more rarely on the analysis of the plant genome. Observation of morphological and physiological changes is sometimes possible, and sometimes not, while the detection of genomic changes, which are of diverse nature and distributed in the genome, depends on the tool used for DNA analysis. Thus, the way in which somaclonal variation is assessed makes a considerable difference to the evaluation of its frequency in a population; it depends on the traits being considered and does not include cryptic changes that are not easily detectable. As a consequence, it is not possible to compare directly the results for somaclonal variation in stressed plant populations obtained by different research teams. The more recent available data suggest that somaclonal variation may affect all plants of a stressed population, and that claims of absent or rare somaclonal variation in the published literature may have been affected by the investigation procedure.

What Triggers Somaclonal Variation?

Causative agents of somaclonal variation include abiotic stresses (plant cell dedifferentiation and differentiation, or plant growth under extreme salt, cold, and drought conditions) and biotic stresses (infection by parasites). It is useful if the inducing agent itself is included in the definition; thus, it is better to use the terms "tissue culture-induced variation" or "cold (salt, drought, etc.)-stress induced variation." However, the term "somaclonal variation" is now also commonly used to define plant variants induced by stresses other than those obtained *in vitro*.

The source of genetic variability in somaclones has been the subject of debate. The relative importance of (1) the genomic variability pre-existing in adult plant cells, and (2) the genomic changes subsequently induced by the stress are still under discussion. In fact, while it is well established that cells in adult plant tissues may not be genetically homogeneous, a large number of observations have shown that a large fraction of variability is directly induced by stress.

What are the Changes in Somaclonal Variants?

Somaclonal variation is characterized by genetic and epigenetic factors. Although a clear-cut boundary between these two factors cannot be drawn, we may define genetic factors as those that are related to changes in DNA. These include: (1) single gene mutations, which may be dominant, codominant, or recessive; (2) deletion/insertion; (3) changes in the number of certain genome fractions (e.g., microsatellite DNA regions); (4) transposon or retrotransposon integration or relocation; and (5) repeated duplication of the chromosome within the intact nuclear membrane (endoreduplication) without spindle formation, or other processes at the nuclear DNA level that might lead to polyploidy or aneuploidy.

Epigenetic factors are defined as those involving alteration of gene expression and are not based on either DNA sequence modification or alteration in the quantity of DNA. The main epigenetic change is a change in the methylation of different DNA sequences, including transposons or retrotransposons. Some epigenetic changes are stable for many sexual generations, while others are unstable or reversible, either somatically or through meiosis. Epigenetic aspects of somaclonal variation essentially include gene silencing or activation.

Homeotic mutations, which affect the activity of several genes, have also been described in somaclonal variants. These may be of the genetic type, if base changes are involved, or of the epigenetic type, if due to methylation. Changes in organellar DNA sequences have also been detected in somaclonal variants, including deletion or amplification of specific sequences.

The importance and role of the above factors in determining somaclonal variation in plants remains largely to be determined.

Molecular Tools for Detecting Genomic Changes in Somaclonal Populations

A number of different techniques are presently available for identifying genetic diversity among individual plants. Analysis of DNA sequences is usually used for classification of plants based on known evolutionary history, while techniques based on the analysis of random genomic sequences are preferred for identifying genomic changes in somaclonal variants. Since the development of polymerase chain reaction (PCR)-based methodologies, the preferred tools have been those based on the analysis of microsatellite regions with the random amplified microsatellite polymorphism (RAMP) approach, the random amplified DNA sequences with random amplified polymorphic DNA (RAPD), or the amplified fragment length polymorphism (AFLP) approach. Random changes in methylation may also be routinely analyzed by an AFLP-derived methodology, which makes use of restriction enzymes that are sensitive to methylation.

All of the above analytical procedures are used to assess the rate of DNA variation in somaclones. Usually, a direct correlation is found between the rate of DNA polymorphism in a somaclonal population and the rate of morphological changes. However, it should be noticed that, due to the random nature of DNA exploration, no direct association of molecular and morphological traits is usually found. A further time-consuming task is that of identifying the precise type and site of DNA variation. This has been attempted only in a few cases of particular interest.

As further research produces new insight into the organization and function of the plant genome, a better understanding of the molecular basis of somaclonal variants could be achieved in the near future.

Transmission to the Sexual Progeny

Somaclonal variation is most frequently inherited by the sexual progeny through meiosis and segregates following Mendelian rules. In a few cases, the variation is transmitted only to asexually propagated plants and is lost through meiosis, the sexual progeny then returning to the original phenotype.

Genetic Improvement of Crops through Somaclonal Variation

Traditionally, the main sources of variability for the breeder have been mutation and crosses with related or wild cultivars. Occasionally, wider crosses with sexually incompatible species have been successfully achieved. The genetic variability induced by somaclonal variation is undesirable in traditional breeding programs, but, if properly set up and controlled, it offers an additional and powerful tool to plant breeders.

Plant tissue culture is the most widely used technique for creating genetic variability through somaclonal variation. Tissue culture protocols are available for most crop species, although improvements are still required for many crops, especially cereals, woody plants, and tropical plants. *In vitro* techniques for the culture of protoplasts, leaves, stems, anthers, microspores, ovules, and embryos have been used, or are attempted with different species, to create new genetic variation from which to select useful traits.

Many factors have been found to play a role in influencing somaclonal variation by tissue culture, including: the genotype of the donor plant, the tissue source, the culture duration, the degree of departure from organized cell growth, and the use of growth regulators in the culture medium. However, despite the increase in our understanding of how these factors work, it is still not possible to control their effect completely and predict the outcome of a somaclonal breeding program. An important consideration is that, usually, in programs of plant genetic improvement through somaclonal variation, the problem is not the induction of variability through a round of dedifferentiation and differentiation in the *in vitro* culture, but the selection of the somaclones of interest. Dwarfism, sterility, and albinism in regenerated plants are clear indicators of the occurrence and frequency of variation, but the selection, for instance, of salt-resistant crop plants will require approaches for the early selection of promising genotypes from among the many regenerated plants. Furthermore, in some cases, somaclonal variation is manifested only in successive generations of regenerated plants as the characters concerned are recessive and are not observed in the primary regenerates. Thus, the analysis of F_1 and subsequent generations of seed families is frequently necessary.

Many new successful crop varieties have been produced by somaclonal variation. These include plants with desirable morphological traits, such as disease resistance, insect resistance, drought, cold, salt and metal tolerance, improvement of nutritional value, and many more. Some *in vitro*-induced mutants were similar to known spontaneous or induced mutations, while others were not. In some cases, these mutants carried desirable traits that could not be obtained with traditional breeding and mutagenesis. In other cases, attempts to exploit somaclonal variation have been unsuccessful because useful changes were not detected or were not novel, or were associated with other deleterious changes. Furthermore, in some cases, the changes were not stable after selfing or crossing. These are not common and the general lesson learnt from the analysis of somaclonal variation across crop species is that the selected mutation is most frequently stable and inherited through sexual generations.

Figure 1 (see color plate 43) Tomato plants differentiating from callus cells grown from cotyledons after 6 weeks of culture in a synthetic medium. At this stage, plantlets can be excised, rooted, transferred to soil in the greenhouse, and grown to the flowering stage.

Additional advantages of somaclonal variation for crop improvement are that it is cheaper than other methods of genetic manipulation, it is universally applicable, and it does not require "containment" procedures, as is the case for genetically modified plants.

In conclusion, crop improvement through somaclonal variation is a powerful tool but is not a panacea (**Figures 1** and **2**).

Somaclonal Variation in Transgenic Plants

Transgenic plants are expected to integrate and express one (or more) foreign gene(s) in an otherwise unmodified genomic environment. Indeed, while extensive molecular analysis has established that the foreign gene becomes permanently integrated into the chromosomal DNA of transgenic plants, much less attention has been paid to its expression. Nevertheless, evidence for the occurrence of morphological and agronomic changes in transgenic plants has been consistent. In fewer cases, there is evidence for molecular changes at the DNA level; within a transgenic population, most or all plants may bear detectable or cryptic somaclonal variation with a sound genetic basis. This is not surprising, since the presently used transformation procedures see cell culture as an obligatory step toward the differentiation of transgenic plants. This is the case for *Agrobacterium tumefaciens* infection, particle bombardment, and direct protoplast and cell

Figure 2 Normal (left) and dwarf (right) tomato plants differentiated from the *in vitro* culture of plants illustrated in **Figure 1**. The production of dwarf plants is a strong indication of the occurrence and extent of somaclonal variation.

transformation. Apparently, the only exception is represented by the approach presently used for the transformation of the model plant *Arabidopsis thaliana*, which is based on flower infection with *A. tumefaciens* followed by development of transgenic seeds. However, since even this treatment may represent a stress, there are sufficient reasons to expect that transgenic plants produced with this approach may show DNA changes due to somaclonal variation.

In our experience with transgenic rice (*Oryza sativa*), poplar, and sugarcane (*Saccharum officinarum*) populations transformed with different transformation protocols, DNA changes were always recorded when investigated with the appropriate molecular tools (**Figure 3**). A quantitative correlation between DNA changes and morphological and agronomic changes may be hypothesized, but cannot be easily demonstrated.

From an applied point of view, an important question is: is somaclonal variation in transgenic

Figure 3 Changes in leaf morphology observed in insect-resistant poplar (*Populus nigra*) plants produced by infecting leaves with *Agrobacterium tumefaciens* carrying a *Bacillus thuringiensis* toxin gene.

plants triggered exclusively by the stressful transformation treatment or is it influenced also by the integration of the foreign gene? Control experiments have been performed to try to verify this. Transformation in the presence or absence of the foreign gene has shown that the stressful transformation conditions play the major role in inducing DNA changes. Disruption of other genes by integration of the foreign gene cannot be ruled out, but should be considered a rare event.

How to Exploit Transgenic Plants Bearing Somaclonal Variation

Somaclonal variation can disturb both basic and applied studies on transgenic plants. The use of transgenic plants in order to assign a role to cloned genes of unknown function may be impaired by concomitant variant traits due to somaclonal variation. For example, in the agronomic and industrial exploitation of transgenic plants, transgenic elite cultivars might show undesirable changes compared with the original plant. For instance, when insect-resistant, transgenic sugarcane plants were produced, they were found to be morphologically identical, but the agronomic analysis of selected plants showed changes in some agronomic traits. Rare DNA changes were also observed by AFLP analysis.

Nevertheless, a large array of commercially exploitable transgenic plants has been produced. In fact, several different approaches can be applied to overcome the problem of somaclonal variation in transgenic plants. One may be that of using recurrent backcrossings to restore the original genotype while retaining the foreign gene. A second approach may be to utilize the best-performing transgenic plants, regardless of genomic changes. The former approach is particularly suitable for annual crops, such as rice, while the latter is suitable for plants that are commercially reproduced by cuttings, such as sugarcane or poplar. But these approaches are not appropriate for all cases. For instance, in the hypothetical case of transgenic olive trees, it would not be practical to wait years before verifying that somaclonal variation had not affected flavor, yield, ripening, or other industrial traits in selected trees.

Concluding Remarks

The Mendelian principle that genes have a fixed, linear order on the chromosome and that genetic loci are inherited unchanged from parent to offspring have been the cornerstones of genetics. However, it is now clear that this basic principle is frequently violated, and that stressful environmental conditions may give rise to rapid and heritable changes in the amount and organization of the genome.

Plants clearly have a higher degree of genetic instability and variability as compared to animals and this may be acceptable in plants because their development is less tightly regulated than animal development. This instability possibly reflects the need of plants to adapt to changing environmental conditions: higher animals have greater control over their environment, while plants are at the mercy of climate, weather, and sometimes of the breeder, who is responsible for exposing plant cells to *in vitro* culture or to other stresses. As a result of this, somaclonal variation has now been recognized as a powerful tool to reveal pre-existing variation and/or induce new variation in plants. Compared to induced mutagenesis, tissue culture may result in a higher frequency of mutations, and can produce mutants that otherwise could not be obtained with traditional chemical or physical mutagenesis.

The potential for crop improvement through somaclonal variation is presently underrated. One reason for this is the great enthusiasm that gene transfer into plants has created in the same period of time in which somaclonal variation was described. Indeed, compared to gene transfer, genomic changes in somaclonal variants are more unpredictable and less easily controlled. However, given the present atmosphere of uncertainty in the public's acceptance of transgenic plants, especially in the European community, and the long list of control measures that are exclusive to transgenic plants, it would be a good idea for somaclonal variation to receive more attention both from plant breeders and from the molecular biologist.

Further study of the molecular events underlining somaclonal variation is also desirable, since it will shed light on the mechanism by which plants react to stress in the natural environment. Understanding this phenomenon could be important in defining the cellular mechanisms acting in the process of evolution.

See also: **Crop Improvement**: Molecular Markers; Mutation Techniques. **Genetic Modification**: Transformation, General Principles. **Tissue Culture**: Micropropagation; Organogenesis; Somatic Embryogenesis.

Further Reading

Arencibia A, Carmona ER, Cornide MT, *et al.* (1999) Somaclonal variation in insect-resistant transgenic sugarcane (*Saccharum* hybrid) plants produced by cell electroporation. *Transgenic Research* 6: 1–12.

Cullis CA (1983) Environmentally induced changes in plants. *CRC Critical Reviews in Plant Sciences* 1: 117–131.

D'Amato F (1985) Cytogenetics of plant cell and tissue culture and their regenerates. *CRC Critical Reviews in Plant Sciences* 3: 73–112.

Durrant A (1962) The environmental induction of heritable changes in *Linum*. *Heredity* 17: 27–61.

Durrant A (1971) Induction and growth of flax genotrophs. *Heredity* 27: 277–298.

Kaeppler SM, Kaeppler HF, and Rhee Y (2000) Epigenetic aspects of somaclonal variation in plants. *Plant Molecular Biology* 43: 179–188.

Larkin PJ and Scowcroft WR (1981) Somaclonal variation: a novel source of variability from cell cultures for plant improvement. *Theoretical and Applied Genetics* 60: 197–214.

McClintock B (1984) The significance of responses of the genome to challenge. *Science* 226: 792–801.

Sala F, Arencibia A, Castiglione S, *et al.* (2000) Somaclonal variation in transgenic plants. *Acta Horticulturae* 530: 411–419.

Walbot V and Cullis CA (1983) The plasticity of the plant genome – is it a requirement for success? *Plant Molecular Biology Reports* 1: 3–11.

Wang G, Castiglione S, Chen Y, *et al.* (1996) Poplar (*Populus nigra* L.) plants transformed with a *Bacillus thuringiensis* toxin gene: Insecticidal activity and genome analysis. *Transgenic Research* 5: 1–13.

Wessler SR, Bureau TE, and White SE (1995) LTR-retrotransposons and MITEs: important players in the evolution of plant genomes. *Current Opinions in Genetics and Development* 5: 814–821.

Somatic Hybridization

S Arcioni and F Pupilli, Institute of Plant Genetics, Perugia, Italy

Introduction

In the plant kingdom, the genetic information present in two different genomes can be combined using sexual or asexual processes. For a long time, plant breeders have been tempted to cross very different species to create new forms. To overcome sexual barriers, *in vitro* fertilization and embryo or ovary culture are employed; however, these technical refinements have only been used on a relatively small scale because each combination between two species brings specific problems (*see* **Crop Improvement**: Hybridization and Plant Breeding). Somatic cell fusion, leading to the formation of viable cell hybrids, has been a tremendously powerful technique in human genetics and immunology, but the fusion of plant cells was not possible because of their rigid cellulose cell wall until the development of enzymatic techniques for the digestion of the cell wall, leading to the large-scale isolation of protoplasts by Cocking. Protoplast fusion provides an opportunity to synthesize hybrids between taxonomically distant parents beyond the limits of sexual compatibility, but it also creates cells with new genetic, nuclear as well as cytoplasmic, constitutions, which otherwise could not be obtained. Because angiosperms usually have a cytoplasmic genome that is maternally inherited, prior to the development of somatic hybridization, new cytoplasmic–nuclear combinations could only be produced sexually by the slow process of backcrossing.

Another advantage of protoplast fusion is that a set of genes responsible for specific useful traits can be transferred without knowing the molecular details of the genes that confer the traits. A classical example is the successful transfer of disease resistance genes in potato (*Solanum tuberosum*). However, beyond the almost unlimited possibilities to create new hybrids between sexually incompatible species, the question yet to be answered is whether protoplast fusion can really circumvent the genetic and biological barriers of sexual incompatibility. No fertile intergeneric somatic hybrids have been reported so far. Asymmetric fusion, i.e., the limited transfer of chromosomes or chromosome segments from one species to another, may be the solution. The ability to regenerate plants from intraspecific, interspecific, and even intergeneric hybrid cells also provides the opportunity to study the physiological interactions between organelles, and, in this respect, these hybrids can be considered as experimental material that simulates what has happened in nature over thousands of years of plant evolution. Cell hybridization might be exploited further for the mapping of genes and linkage analysis in somatic plant cells; this approach has been used with great success in human–rodent cell hybrids. The whole process of somatic hybridization involves several stresses for plant cells that may result in a variety of rearrangements of a genetic or epigenetic nature. Such rearrangements are the basis of new morphological variability, which, in some cases, is utilized for agricultural benefit. The aim of this article is to overview the somatic hybridization process in plants, including the methodological aspects of: (1) screening of suitable sources of protoplasts amenable to fusion; (2) setting up the experimental conditions for protoplast fusion, culture and selection of heterokaryons, and plant regeneration from selected calli; and (3) characterization of somatic hybrid plants. The production and characterization of somatic hybrid plants in alfalfa (*Medicago sativa*, lucerne) is used as a case study.

Need for Somatic Hybridization

In general, when somatic hybridization is to be used the primary objective is the transfer of useful agronomic traits from a species (donor) that is sexually incompatible with the target crop (recipient). For example, in alfalfa there are three main objectives: (1) to introduce traits for resistance to biotic and abiotic stresses; (2) to improve the feeding value; and (3) to increase the green biomass production. Few (or none) of these objectives can be achieved by sexual hybridization as alfalfa is cross-fertile with only a few species that are of little agronomic value. Cross incompatibility in alfalfa is mainly due to postfertilization barriers in that endosperm formation or function in the hybrid embryo is strongly repressed. Examples of wild *Medicago* species that could be the source of useful agronomic traits are listed in **Table 1**. As well as these traits, somatic hybridization can be used to introduce into alfalfa the genes for tannin synthesis that could render alfalfa forage amenable for grazing by ruminants and with higher feeding value. Finally, somatic hybridization could be used to improve the vigor and, thus, forage production through polyploidization by fusing the cells of superior genotypes selected at the diploid level. This would maximize the level of heterozygosity that in

Table 1 Sources of traits for stress resistances for alfalfa

Species	*Resistance to*	*Pathogen or insect attacker*	
		Common name	*Scientific name*
M. rigidula	Fungal disease	Anthracnose	*Colletotrichum trifolii*
M. suffruticosa	Fungal disease	Spring black stem	*Phoma medicaginis*
M. murex	Fungal disease	Common leaf spot	*Pseudopeziza medicaginis*
M. scutellata	Fungal disease	Sclerotinia crown	*Sclerotinia trifoliorum*
M. arborea	Bacterial disease	Bacterial wilt	*Corynebacterium insidiosum*
M. truncatula	Insect attack	Spotted alfalfa aphid	*Therioapis maculata*
M. blancheana	Insect attack	Alfalfa weevil	*Hypera postica*
M. noeana	Insect attack	Potato leafhopper	*Empoasca fabae*
M. disciformis	Insect attack	Alfalfa seed calchid	*Brucophagus roddi*
M. littoralis	Salinity		
M. polymorpha	Drought		

autotetraploid plants, such as alfalfa, is positively correlated with vigor.

To exploit the potential of somatic hybridization as a tool for gene transfer in alfalfa, the genome of this species has been combined with those of two closely related diploids, namely *M. coerulea* and *M. falcata*, and the distantly related tetraploid shrub *M. arborea*. These three parent combinations were chosen to analyze separately the genetic consequences of genetic hybridization between parents that differ in ploidy level and genetic background. Moreover, hexaploid somatic hybrids between alfalfa and both *M. coerulea* and *M. falcata* could be an interesting base for the constitution of highly heterozygous varieties that are able to reduce the inbreeding depression encountered by alfalfa when cultivated in evironments with poor pollinating conditions. Finally, *M. arborea* could be the source of genes for stress adaptation and winter forage production.

Screening of Suitable Sources of Protoplasts Amenable to Fusion

A variety of tissues, including roots of seedlings, hypocotyls, leaves, and cultured cell lines, have been used for protoplast isolation. Protoplasts suitable for fusion should be viable, perfectly spherical, able to undergo mitotic divisions after fusion, and regenerable. Another important feature is pigmentation, because the presence of chlorophyll in protoplasts of only one fusion partner can be utilized as a selection criterion to identify heterokaryons. As far as outcrossing species like alfalfa are concerned, the genotypes used as protoplast source should be previously selected for their capacity to regenerate plants from undifferentiated cell lines, and then be propagated vegetatively. This is because the regeneration capability is recognized to be under the control of a few rare alleles occurring within the heterogeneous populations of outcrossing species. The selected plants should be maintained in a controlled environment (12 h photoperiod, 216 $\mu mol^{-2} s^{-1}$, $20 \pm 1°C$, 80% relative humidity).

From the experimental point of view, there are three critical points for successful protoplast isolation: (1) protoplast source; (2) enzymatic mixtures; and (3) time of enzymatic digestion. **Tables 2 and 3.** list the tissues, the enzymatic mixtures, and experimental conditions for protoplast isolation in alfalfa and other *Medicago* spp.

Protoplast Source

To get viable protoplasts that are able to develop further after fusion, it is necessary to use explants or cell cultures that are actively growing. Protoplast sources include young leaves, cotyledons, and roots or cell suspensions at the log growth phase.

Enzymatic mixtures In many enzymatic mixtures, some or all of the three classes of enzymes commonly used for digesting the cell wall, namely cellulase, hemicellulase and pectinase, are present. The enzymatic activity and specificity of the different enzymes may vary according to either the species they are isolated from or the commercial supplier. Therefore, for each protoplast source, the concentration and composition of the proper enzymatic mixture should be adjusted empirically.

Time of enzymatic digestion In general, digestion for too long may damage or release protoplasts that are too fragile, so it is recommended to make the digestion time as short as possible. The use of desalted enzymes reduces the incubation period and enhances protoplast viability. In general, rootlets and peeled leaves require a shorter (4–6 h) incubation

Table 2 Experimental conditions for protoplast isolation from *Medicago* spp. involved in somatic hybridization

Species	*Source*	*Enzyme mixture*[a]	*Incubation*		
			Time (h)	*Temperature (°C)*	*Agitation (r.p.m.)*
M. sativa	Cotyledons	A	16	25	10–20
	Leaves	B	4–5	28	Occasional
	Cell lines	E	16–18	23	40
	Roots	A	16	25	60
M. coerulea	Leaves	F	4	28	–
	Cell lines	E	16	25	40
M. falcata	Leaves	b	16	26	–
	Cell lines	g	16–18	25	40
M. arborea	Leaves	c	6	28	10–20
	Cell lines	g	16	25	40
M. glutinosa	Cotyledons	d	16	25	30
M. varia	Cotyledons	d	16	25	30
M. polymorpha	Cotyledons	e	16	28	30
M. scutellata	Leaves	c	16–18	27	10–20
M. disciformis	Leaves	c	16–18	27	10–20
M. littoralis	Cell lines	g	16–18	25	40

[a]For composition of enzyme mixtures a–g see Table 3.

Table 3 Enzymatic solutions for protoplast isolation in *Medicago* spp.

Enzymes or chemicals	*Enzyme mixtures*[a]						
	A	*b*	*c*	*d*	*e*	*f*	*g*
Cellulase							
Onozuka R-10	1	0.5	1			5	
Driselase					1		2
Rhozyme HP-150	1	0.5	1	2	1		
Macerozyme R-10	0.5		0.5	0.03		1.25	
Pectolyase Y23	0.25						0.3
Cellulase Y-C					0.1		0.5
Pectinase		0.25					1.2
Meicelase				2			
Mannitol	0.15 mol l^{-1}	0.15 mol l^{-1}	0.15 mol l^{-1}	0.7 mol l^{-1}	0.55 mol l^{-1}	0.6 mol l^{-1}	0.55 mol l^{-1}
Sorbitol	0.15 mol l^{-1}	0.15 mol l^{-1}	0.15 mol l^{-1}				
MES[b]	1.5 mmol l^{-1}	1.5 mmol l^{-1}	1.5 mmol l^{-1}				
NaH_2PO_4	0.35 mmol l^{-1}	0.35 mmol l^{-1}	0.35 mmol l^{-1}				
PH	5.7	5.5	5.7	5.7	5.8	5.6	5.8
CPW[c] salt solution	–	–	–	+	+	+	+

[a]Unless otherwise specified, the concentrations are expressed as % (w/v).
[b]2-(*N*-morpholino)ethane sulfonic acid.
[c]Plasmolyzing solution of Frearson EM, Power JB, and Cocking EC (1973) Isolation, culture and regeneration of *Petunia* leaf protoplasts. *Developmental Biology* 33: 130–137.

time than cotyledons and cell suspensions (8–18 h). The enzyme treatments are usually carried out in the dark at temperatures ranging from 23°C to 28°C with gentle agitation.

Mannitol alone or in combination with sorbitol is added to all protoplast isolation solutions to avoid osmosis through the protoplast membrane. Following enzymatic digestion, the protoplast suspension should be purified from debris and undigested or partially digested cells, which could interfere with the fusion process. The first purification step is usually a filtration through a mesh sieve with a pore size of around 60 μm, which retains cell clumps and undigested tissues. Mesophyll and cotyledon protoplasts are centrifuged at low speed for a short time, and the pelletted protoplasts are clean enough for fusion. Root and cell suspension protoplasts need an additional purification step consisting of centrifugation in a sucrose solution (15–20%), after which the protoplasts float to the top of the tube because of their low density. Floated protoplasts are in general very clean and homogeneous in size, features that render them amenable to fusion.

Early experiments of protoplast fusions involving *M. sativa* cell suspension protoplasts and mesophyll protoplasts of other species failed due to the inability of hybrid calli to regenerate plants. It was concluded that cell suspension protoplasts have a key role in the development of hybrid calli, but have a negative influence on regeneration since these protoplasts do not usually regenerate plants. As a matter of fact, when highly regenerable mesophyll protoplasts of *M. sativa* were fused with callus protoplasts of *M. coerulea*, some hybrid calli regenerated fertile plants. This strategy was also effective for the other two parent combinations involving as fusion parents *M. sativa* mesophyll and *M. falcata* and *M. arborea* callus protoplasts.

Setting up the Experimental Conditions for Protoplast Fusion, Culture and Selection of Heterokaryons and Plant Regeneration from Selected Calli

Protoplast fusion can be induced by a number of treatments: (1) divalent cations (Ca^{2+}) in alkaline solutions; (2) water-soluble polymers; and (3) electric field pulses. Plant protoplasts are negatively charged and therefore spontaneous fusion does not take place. The neutralization of this negative charge by divalent cations such as Ca^{2+} in a solution of high pH can promote protoplast aggregation. In the areas where the protoplasts are in close contact, the membranes of the two cells start to coalesce, inducing cytoplasmic continuity, which broadens to the whole surface of membrane adhesion. The same effect of protoplast agglutination and membrane coalescence is obtained with polyethylene glycol (PEG), dextran, and polyvinyl alcohol (PVA). Many laboratories use PEG in combination with high pH and high Ca^{2+} concentration in the fusion buffer. However, PEG preparations are often toxic for plant cells. This problem can be overcome by deionizing PEG solutions or lowering their concentration. In the electric field mediated method, protoplasts are first brought in close contact by the application of a nonuniform alternating electric field (AC), followed by the application of continuous electric field pulses (DC) of high intensity for a short time, which leads to the breakdown of membranes at the points of adhesion, and, later, to cell fusion. **Figure 1A–D** shows the different phases of electrofusion of two protoplasts. This method has proved to be very efficient in terms of hybrid yield because it avoids the nonphysiological conditions imposed on protoplasts by PEG and/or high pH. Furthermore, by proper adjustment of the fusion parameters it is possible to maximize the number of binucleate fusion products.

In alfalfa, the optimal results in terms of highest percentage of binucleate interspecific fusion products (heterokaryons) and minimal cell damage were obtained by applying an AC field of 150 V cm^{-1} and 1–5 pulses of a DC field of 1500 V cm^{-1} of a duration of 35 μs each. The sensitivity of alfalfa protoplasts to the electric field depended on the protoplast source and their size; in particular, larger protoplasts required a lower voltage electric field than smaller ones to fuse, and rootlet protoplasts were less sensitive to the AC field than cell suspension protoplasts. Another important factor affecting fusion efficiency was the uniformity of the size of isolated protoplasts, and in homogeneous populations, a restricted range of optimal AC fields can induce almost all the protoplasts to undergo the fusion process.

Fusion experiments with protoplasts isolated from plants of two different species produce a suspension of homofusion products, heterokaryons, and unfused cells (**Figure 1E**). To avoid tedious and time consuming procedures for hybrid selection at the plant level, the development of methods to select heterokaryons in the fusion mixture is essential. One of the most commonly used strategies is to set up culture conditions that preferentially enhance the growth of hybrid cells compared to parental cells. An example of these strategies is based on the genetic complementation of auxotrophic mutants (i.e., genotypes that lack specific features that are necessary for growth), such as chlorophyll or nitrate reductase deficient mutants. Additionally, a combination of nuclear destruction by heavy irradiation and cytoplasmic metabolic inhibitors can provide a means for obtaining new nucleus–cytoplasm combinations. The combination in the same genotype of a dominantly expressed trait, such as a drug resistance, and a recessive trait, such as auxotrophy, gives rise to a double mutant called a universal hybridizer. When such a mutant is fused with any wild-type cell and cultured on a minimal medium supplemented with the drug to which the mutant is resistant, then only hybrids are able to grow, because the universal hybridizer is auxotrophic and the wild-type is susceptible to the drug. Similarly, resistance to two different drugs can be introduced in the parental lines through transformation, and then the hybrids can be selected by adding both drugs to the culture medium. When spontaneous or induced mutants are not available, and/or genetic transformation cannot be utilized, heterokaryon selection can be performed manually or automatically with the aid of specific devices. This method can also rescue new types of hybrids with various arrays of nuclear and cytoplasmic parental contribution that differ from those obtained in the presence of selective pressure or with inactivation treatments.

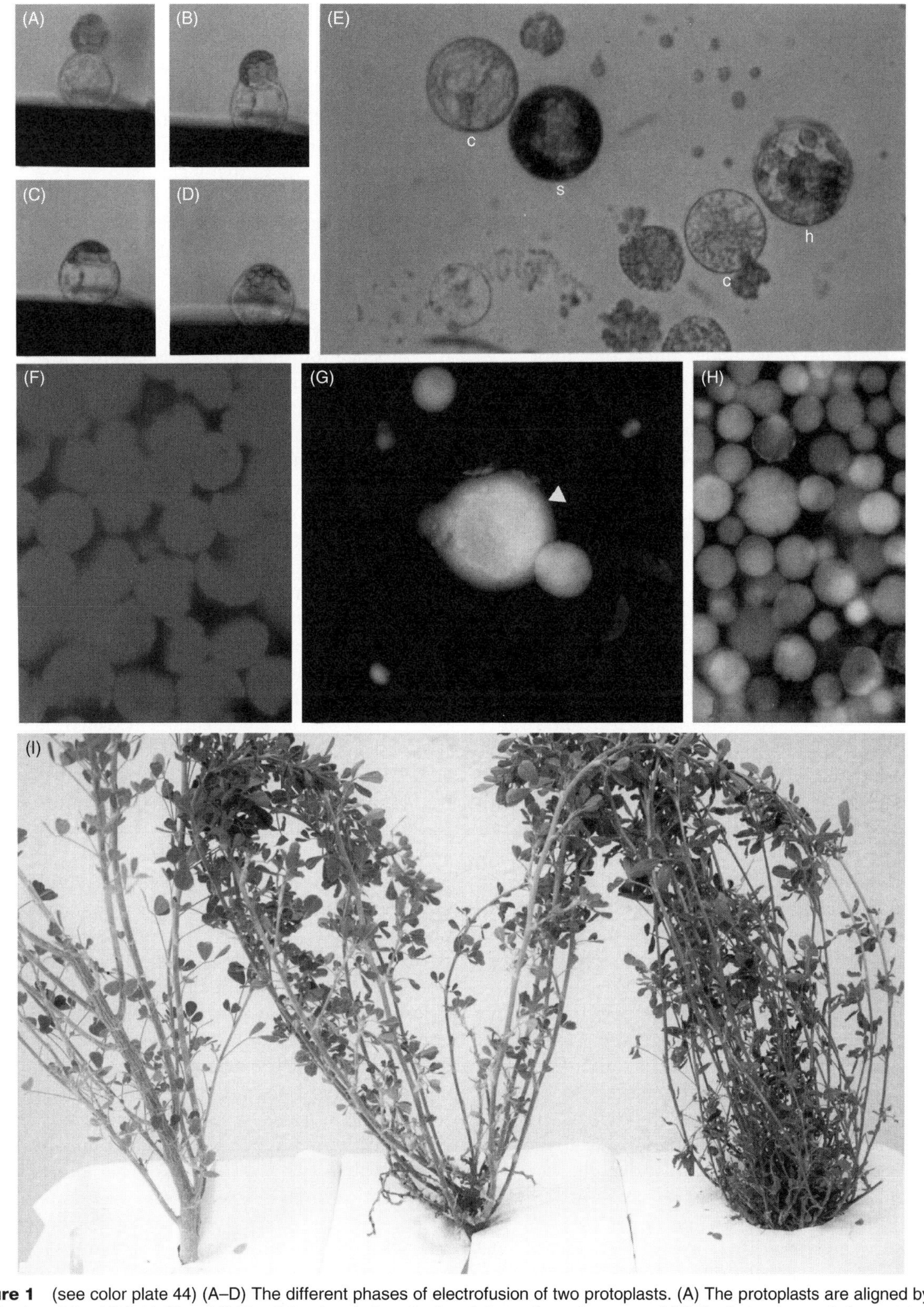

Figure 1 (see color plate 44) (A–D) The different phases of electrofusion of two protoplasts. (A) The protoplasts are aligned by the application of the AC field. (B and C) The DC pulses induce the breakdown of membranes and (D) The heterokaryon is formed, but the cytoplasms are not mixed yet. (E) Electrofusion products: h – heterokaryon from mesophyll and callus protoplasts; c and s – homokaryons of callus and mesophyll protoplasts, respectively. (F) Mesophyll protoplasts observed under UV light. (G) The arrowhead indicates a heterokaryon derived form mesophyll and callus protoplasts, observed under UV light. (H) Callus protoplasts treated with FITC and observed under UV light. (I) Somatic hybrid plants (middle) between *M. arborea* (left) and *M. sativa* (right) Arcioni S, Damiani F, Mariani A, and Pupilli F (1997). Somatic hybridization and embryo rescue for the introduction of wild germplasm. In: Mckersie BD and Brown DC (eds) *Biotechnology and the Improvement of Forage Legumes.* pp. 61–89. Oxford, UK, New York: CAB International.

Morphological differences such as the presence or absence of chloroplasts in mesophyll and cell line protoplasts, respectively, can be used to identify fusion products under the microscope. The physical isolation of heterokaryons with a micropipette is feasible, but tedious and labor consuming. Moreover, the hybrids must be recovered within a few hours of fusion and before the mixing of parental cytoplasms, otherwise the hybrid cells become undistinguishable from the parental cells. The addition of fluorescent dyes such as fluorescein isothiocyanate (FITC) or fluorescein diacetate (FDA) to the colorless protoplasts makes them fluoresce green–yellow under ultraviolet light (UV) (**Figure 1H**), while mesophyll protoplasts appeared red (**Figure 1F**); the hybrid cells, therefore, appear with red spots (**Figure 1G**). This treatment extends the time limit for hybrid identification by another 3–4 days; it reduces the risks of misinterpretations and allows the use of equipment such as a fluorescent activated cell sorter (FACS) for the automatic screening and isolation of differentially stained protoplasts.

Since neither antibiotic resistant, nor auxotrophic mutants, were available in *Medicago*, hybrid cells were selected on the basis of their color using the FITC staining procedure. Fusion products were embedded in semisolid medium 2–3 days after fusion, and the heterokaryon development was followed visually until the stage of 50–80-cell clusters, which could be picked up manually with the needle of a syringe. Prolonged culture of heterokaryons in close proximity to rapidly dividing parental protoplasts can be critical as it results in a higher percentage of dividing heterokaryons. Similarly, the surrounding of newly isolated hybrid calli with actively dividing nurse calli greatly enhanced their growth.

The culture of hybrid cells does not differ substantially from that of normal plant cells, and a culture medium with a high ratio of auxin/cytokinin growth regulators is initially required to enhance mitotic division of hybrid cells. When a callus is formed as a result of many divisions, the regeneration process is triggered by lowering the auxin to cytokinin ratio. In general, the regeneration of hybrid plants is more difficult than that of the corresponding parental genotypes. This is probably due to the number of stresses to which the hybrid cells are subjected (enzyme treatment for protoplast isolation, exposure to fusogenic agents, selective pressures with drugs of various nature). In addition, either polyploidization or the genetic distance of the parents contribute to prolong the regeneration phase of hybrid calli.

Even when the sources of parental protoplasts were properly selected, regeneration from hybrid calli in the genus *Medicago* was not easy. In fact, although *M. sativa* + *M. coerulea* somatic hybrids (S + C) followed the regeneration pathway of the *M. sativa* parental line, the other two somatic hybrid plants, *M. sativa* + *M. arborea* (S + A) and *M. sativa* + *M. falcata* (S + F) regenerated with extreme difficulty after more than 1 year of culture.

Even when the hybrid cells were subjected to selective pressure or physically isolated, the hybrid nature of the regenerated plants should be confirmed because escapes from the selective agent and misclassification of hybrid cells cannot be excluded *a priori*. Isozymes that are different forms of the same enzymatic activity are often polymorphic between parental lines of somatic hybrids, and are visualized as bands of different molecular weights in an electrophoretic pattern. Each band corresponds to a gene that can be taken as a marker of the presence of a determinate genome or a fraction of it in the hybrid genome. The general rule is that polymorphic bands must be simultaneously present in the hybrids. However, isozyme analysis provides only circumstantial evidence of hybridity; chimerism, i.e., the simultaneous presence of both parental cells in the hybrid plants that have not originated from a real fusion event, cannot be ruled out. If the parental lines have different isozymes and they are simultaneously present in the hybrids, the analysis of the number, size, and shape of the chromosomes of the hybrid cells can confirm hybridity and exclude chimerism.

With the advent of the DNA based technologies, a new generation of genetic (molecular) markers is now available allowing the contribution of each parental genome in the hybrid to be quantified at both cytoplasmic and nuclear levels. This is particularly useful for those hybrids in which one parent contributes poorly to the constitution of the hybrid genome (asymmetric somatic hybrids). The most commonly used markers are restriction fragment length polymorphisms (RFLPs), random amplified polymorphic DNA (RAPD), and amplified fragment length polymorphisms (AFLPs). They are all based on variation in DNA sequences, and therefore are able to reveal a great number of parent specific alleles that allow the nuclear contribution of each parent to the hybrid genome to be quantified. Similarly, the DNA of chloroplasts and mitochondria can be analyzed with these markers to investigate the cytoplasmic genetic constitutions of somatic hybrids.

Esterase isozymes and chromosome analyses were sufficient to confirm the hybridity of S + C and S + A because parental genomes were present symmetrically in the hybrids, and therefore it was relatively easy to find polymorphic bands between the parents that were present in the hybrids. Conversely, because the *M. sativa* parent contributed almost entirely to the S + F genome, a great number of RAPD markers was screened to find the few *M. falcata* specific alleles retained in the somatic hybrids.

Characterization of Somatic Hybrid Plants

Once the hybrid nature of the regenerated plants has been ascertained, a detailed study of their genetic make-up and field performance is required if these plants are to be included in breeding programs. Such analyses are carried out at molecular, cytogenetic, and morphological levels. The results of the characterization of the three *Medicago* hybrids S + C, S + A, and S + F are reported in this section.

Cytological

In general, polyploidization and hybridization result in a decrease of fertility in somatic hybrids compared to the parental lines. Such low fertility is often due to irregularities of chromosome pairing and segregation, resulting in gamete lethality. Chromosome counts showed that all parental chromosome sets are present in S + C, this hybrid having $2n = 48$ $(32 + 16)$ chromosomes, while in S + F $(2n = 33)$ and S + A $(2n = 56)$ 15 and 8 parental chromosomes were missing, respectively. Because chromosomes of the parental species are morphologically indistinguishable, it was not possible to determine the relative contribution of the two parents to the S + A and S + F chromosome sets. Although S + C and S + F were fertile, S + A was completely sterile, and no flowers were observed after 3 years of cultivation under field conditions. The analysis of microsporogenesis of S + C showed that the number and kind of meiotic abnormalities were not dramatically different from those observed in cultivated tetraploid alfalfa. In all metaphases analyzed of S + F, two chromosomes of unexpected length were observed, which could be the result of a translocation.

Molecular

The ability to monitor the fraction of a parent genome present in the hybrid during the recurrent cycles of backcrossing may speed up the elimination of the unnecessary genetic material conferred by the donor parent. A molecular characterization of the nucleus of the three somatic hybrids S + C, S + A, and S + F was carried out with different kinds of molecular markers. RFLP analysis of S + C showed that 30% of the bands specific to the *M. coerulea* parent were not present in the somatic hybrid genome. Since cytological observations excluded any loss of chromosomes from the hybrids, the lack of specific bands was probably the result of the small-scale rearrangements of DNA that occurred in *M. coerulea* during protoplast isolation and fusion. In the S + A RFLP pattern, the absence of approximately one-third of the *M. sativa* specific bands and half those of *M. arborea* indicated large-scale genome rearrangements, which in this case were due to the loss of parental chromosomes. In S + F, the analysis of random RAPD loci was carried out mainly to determine if *M. falcata* contributed to the hybrid genome, and it was shown that only a small number of *M. falcata* specific bands was present in the hybrid pattern, thus confirming the highly unbalanced parental contribution to the hybrid. In conclusion, chromosome counts and molecular analyses of random loci showed that both S + A and S + F underwent large-scale rearrangements, while small-scale rearrangements occurred in S + C.

The genes controlling the synthesis of ribosomal RNA (rRNA) are organized in repeated units formed by conserved coding regions interspersed with variable noncoding regions (IGS, inter genic spacers). Almost all the rDNA genes are located in the nucleolar organizing region (NOR) placed in specific chromosomal segments. As the structure of rDNA is fast evolving for both the number of repetitions of the basic units and length of the noncoding regions, the evolution of these genes is frequently studied in somatic hybrid plants in comparison with their parents. The analyses of the RFLP variations of IGS and of the number and organization of NORs through fluorescent *in situ* hybridization (FISH) of *Medicago* somatic hybrids showed that, whereas S + C retained all six parental nucleolar organizing regions (NORs) and all parental RFLPs, S + A retained five of six parental NORs while losing half of the *M. arborea* specific RFLPs, indicating that simple chromosome loss of one *M. arborea* NOR accounted for the RFLP losses. S + F underwent dramatic alterations as five of six parental NORs were retained and new rDNA RFLPs were created and amplified differentially among somaclonal variant plants.

Morphological

None of the analyses listed above, however, provide information on to what extent the somatic hybridization process acts on the gene expression of somatic hybrids.

To investigate how the genome rearrangements detected at cytological and molecular levels could have affected the field performance of *Medicago* somatic hybrids, a comparative morphological analysis of traits related to forage production was carried out. In general S+C and S+F showed a morphological resemblance to *M. sativa*, while S+A appeared intermediate between the parents. *Medicago arborea* differs from *M. sativa* for many traits, the most striking of them being *M. arborea*'s shrub habit with a lignified single stem, while *M. sativa* has a mostly herbaceous prostrated habit with many (up to 30) stems originating from the crown (**Figure 1I**). S+A showed 6–15 stems per plant, originating from a crown-like structure, and showed active growth in winter, like *M. arborea*. However, the most striking feature of S+A was the inability of these plants to flower.

In conclusion, the various extents of rearrangements detected in S+F and S+C did not affect greatly the morphophysiological features of the plants, whereas the chromosome elimination of both parents, or the incompatibility of the two distantly related parental genomes in S+A, determined macroscopic disturbances such as sterility.

Conclusions

Somatic hybridization was introduced more than 20 years ago as a technological aid to classical breeding to overcome the failure of wide hybridization to enrich the cultivated germplasm with useful traits. Most of the promises and challenges have not as yet been realized, and only a few examples of somatic hybrid plants in potato and *Brassica* spp. have been utilized as commercial varieties. The reasons for this are similar to those encountered with sexual crosses: the more genetically distant are the parents to be crossed the more chromosomally unstable and sterile are their progeny resulting, in many cases, in unsuitable material for breeding purposes. On the other hand, somatic hybrids expressing cytoplasmic male sterility are utilized in plant breeding, even though their practical use has not been fully exploited as yet. So has some 20 years of research on protoplast fusion in many leading laboratories of plant biotechnology worldwide been a total waste of time? Of course not. Although somatic hybrids do not have a direct breeding value, they constitute a unique biological resource to study how two distant genomes interact in the same individual. The study of interaction mechanisms of two different genomes in a new individual obtained in the absence of selection may provide new insights to clarify how modern plants have evolved over time, and perhaps to predict how they could evolve in the future. Moreover, on the basis of the experience of somatic hybridization, plant biotechnologists have abandoned the idea that a direct transfer of genes to a new genetic environment is sufficient for obtaining the expression of the desired trait. Early experiments on plant transformation, which were often unsuccessful, suggested that this might be the case. The continuous evolution of biomolecular techniques has clarified the molecular basis of traits of agronomic interest, and the advances in scientific knowledge have increased the possibilities of successful gene transfer. These new insights make the potential of genetic transformation for plant improvement infinitely higher than that showed by somatic hybridization. Thus, protoplast fusion can no longer be considered a suitable tool for transferring single or a few genes, but could be more efficiently utilized for producing new genetic variability derived from new random combinations of genes of the parental genomes forced to remain in the same nucleus.

Acknowledgments

Contribution n. 17 from the Institute of Plant Genetics Research Division of Perugia.

List of Technical Nomenclature

Autotetraploidy A tetraploid individual formed by four similar genomes.

Backcross A cross between an offspring and one of its parents.

Callus Undifferentiated plant cell line retaining the potentiality to regenerate a whole plant.

Epigenetics Relationship between genes and their phenotypic effect.

Inbreeding depression Decrease of vigour as a consequence of self-fertilization.

Isozymes Multiple forms of the same enzyme.

Outcrossing Tendency of an individual to cross-hybridize.

Somaclonal variation The appearence of new traits in plants regenerated from cell lines.

See also: **Crop Improvement**: Alien Cytoplasms; Hybridization and Plant Breeding; Plant Breeding, Principles. **Genetic Modification**: Transformation, General Principles. **Tissue Culture**: General Principles. **Tissue Culture and Plant Breeding**: Somaclonal Variation. **Tissue Culture**: Somatic Embryogenesis.

Further Reading

Alexander RG, Cocking EC, Jackson PJ, and Jett JH (1985) The characterization and isolation of plant heterokaryons by flow cytometry. *Protoplasma* 128: 52–58.

Bajaj YPS (1994) Somatic hybridization – a rich source of genetic variability. In: Bajaj YPS (ed.) *Biotechnology in Agriculture and Forestry vol. 27. Somatic Hybridization in Crop Improvement I*, pp. 3–32. Berlin, Heidelberg, New York: Springer Verlag.

Belliard G, Vedel F, and Pelletier G (1979) Mitochondrial recombination in cytoplasmic hybrids of *Nicotiana tabacum* by protoplast fusion. *Nature* 281: 401–403.

Carlson PS, Smith HH, and Dearing RD (1972) Parasexual hybridization. *Proceedings of the National Academy of Science USA* 69: 2292–2294.

Cocking EC (1978) Selection and somatic hybridization. In: Thorpe TA (ed.) *Frontiers of Plant Tissue Culture*, pp. 151–158. Calgary: University of Calgary Press.

Constabel F (1984) Fusion of protoplasts by polyethylene glycol (PEG). In: Vasil IK (ed.) *Cell Culture and Somatic Cell Genetics of Plants. Vol 1: Laboratory Procedures and Their Applications*, pp. 414–422. London: Academic Press.

Constabel F and Cutler AJ (1985) Protoplast fusion. In: Fowke LC and Constabel F (eds) *Plant Protoplasts*, pp. 54–65. Boca Raton: CRC Press.

Galbraith DW (1984) Selection of somatic hybrid cells by fluorescence-activated cell sorting. In: Vasil IK (ed.) *Cell Culture and Somatic Cell Genetics of Plants. Vol 1: Laboratory Procedures and Their Applications*, pp. 433–447. London: Academic Press.

Gaynor JJ (1986) Electrofusion of plant protoplasts. In: Evans DA, Sharp WR, and Ammirato PV (eds) *Handbook of Plant Cell Cultures. Techniques and Applications*, pp. 149–171. New York: Macmillian Publishing Company.

Melchers G, Sacristan MD, and Holder AA (1978) Somatic hybrid plants of potato and tomato regenerated from fused protoplasts. *Carlsberg Research Communications* 43: 203–218.

Pelletier G (1993) Somatic hybridization. In: Hayward MD, Bosemark NO, and Romagosa I (eds) *Plant Breeding, Principles and Prospects*, pp. 93–106. London: Chapman & Hall.

Power JB, Cummis SE, and Cocking EC (1970) Fusion of isolated plant protoplasts. *Nature* 225: 1016–1018.

U

Ultraviolet-B Responses *see* **Plants and the Environment**: Ozone Depletion.

W

WATER RELATIONS OF PLANTS

Contents

Basic Water Relations

P S Nobel, University of California, Los Angeles, CA, USA

Introduction

Water is crucial to plant growth and development. It is involved in essentially all plant metabolic processes, including acting as a reactant in photosynthesis, being essential in the movement of carbohydrates and nutrients throughout plants, and serving as the main ingredient of the nectar of flowers. Most of the water that evaporates on land is transpired by plants. This water has typically been taken up from the soil by plants – water enters their roots, moves along their stems in a special conducting tissue (the xylem), and is vaporized in their leaves before diffusing out to the surrounding air, where the water vapor concentration is generally much lower than in the leaves. The productivity and sustainability of many ecosystems, including agroecosystems, is controlled by water. For instance, in the 38% of the land area of the Earth that is arid ($\leqslant$250 mm annual rainfall) or semiarid ($\sim$250 to 500 or 600 mm annual rainfall, depending on seasonal rainfall distribution), the biomass of native plants is limited by their water acquisition. Furthermore, agriculture in such regions can depend on irrigation; e.g., water supplied by wells or canals in the Central Valley and the Imperial Valley of California is crucial for the intensive and highly productive cultivation of plants there. In semiarid parts of Africa, Australia, India, and other regions, rainfed agriculture depends on growing drought-tolerant crops.

Water Energetics

Liquid Water

As does any substance, water moves spontaneously toward regions of lower energy. For plant water relations, such a thermodynamic constraint is conveniently described by the water potential, Ψ, which represents the energy of water per unit volume (e.g., joules m^{-3}) and has the same units as pressure (newtons m^{-2} or pascals). Indeed, Ψ is usually expressed in megapascals (MPa) for plant water relations.

The tendency of water to move spontaneously toward lower energies means that it will tend to move toward regions with lower hydrostatic pressure, P. This principle underlies the flow of blood out of the left ventricle of the heart as its muscular walls contract and the flow of water out of a faucet into a kitchen sink. Water in plants also tends to move spontaneously toward regions of lower P, which means along the xylem from the root up the stem to the uppermost leaves and branches.

Water also tends to move toward regions of higher osmotic pressure, Π. For instance, water tends to flow into algal and other cells placed in distilled water, a process termed "osmosis." Based on the Van't Hoff relation for dilute ideal solutions, Π is directly and linearly dependent on the concentration of solutes, namely, $\Pi = RT\Sigma c_j$, where R is the international gas constant (8.31 J mol^{-1} K^{-1}), T is the absolute temperature in Kelvin units, and c_j is the concentration of solute species j (mol m^{-3}). Under certain circumstances, such as when the activity of water (a_w) is affected by the surfaces of colloids or other interfaces, as is the case in most plant cells, Π can be equated to $-RT/\bar{V}_w \ln a_w$, where $\bar{V}_w$ is the partial molal volume of water (about 18×10^{-6} m^3 mol^{-1}).

Everyday observations indicate that water tends to flow downhill, e.g., falling raindrops and the inevitable flow of rivers from highlands to the ocean. This gravitational component of Ψ is energetically expressed as $\rho_w gh$, where ρ_w is the density of water (about 1000 kg m^{-3}), g is the gravitational acceleration (about 9.8 m s^{-2}), and h (m) is the height above some reference level in the gravitational field. Using relationships among physical quantities, $\rho_w g$ can be shown to equal about 0.010 MPa m^{-1}. Consequently, substantial gravitational contributions to the water potential occur only if water moves over large distances – not a few micrometers across cells or a few millimeters across tissues, but rather a few meters up a plant.

Now that the three energetic components of the water potential have been described, Ψ can be presented as follows:

$$\Psi = P - \Pi + \rho_w gh \qquad [1]$$

Sometimes a term for matric pressure (or matric potential) is added on the right-hand side of eqn [1], but this is not necessary as interactions at interfaces can be accounted for by their effects on the local hydrostatic pressure or on the activity coefficient of water (which affects Π). Equation [1] is the basic energetic equation governing the water relations of plants and indicates that water spontaneously tends to flow toward lower P, higher Π, and lower h. Specifically, a difference in water potential between two locations leads to water flow, and $\Delta\Psi$ equals 0 at equilibrium. As indicated, the gravitational term is important only for describing water energy and movement over considerable distances (>1 m). For instance, as one moves 10 m vertically downward in a column of water at equilibrium ($\Delta\Psi = 0$), such as in a lake or ocean, P increases by 0.1 MPa (about 1 atmosphere) as the gravitational term similarly decreases ($\rho_w g$ in eqn. [1] is 0.01 MPa m^{-1}). In contrast, the focus at a cellular and tissue level is on the interaction between the osmotic and the hydrostatic components of Ψ.

Water Vapor

For air, the water potential of water vapor, Ψ_{wv} (MPa), depends on the relative humidity:

$$\Psi_{wv} = \frac{RT}{\bar{V}_w} \ln \frac{\text{relative humidity in \%}}{100} \qquad [2]$$

At 20°C, $RT/\bar{V}_w$ equals 135 MPa, so Ψ_{wv} can have large negative values. For instance, at 50% relative humidity (air that is half saturated with water vapor), Ψ_{wv} is (135 MPa) (ln 50/100) or (135 MPa) (−0.693) by eqn [2], which is −94 MPa, an extremely negative water potential. Because Ψ for plants is generally −0.5 to −5 MPa, water tends to move energetically downhill from the plants to the much drier atmosphere. Even in a rainstorm, the relative humidity of the air rarely exceeds 95%, in which case Ψ_{wv} is 135 ln 0.95 or −6.9 MPa. On the other hand, at night leaf temperatures can decrease substantially, leading to dew formation as the air adjacent to leaves or other surfaces becomes saturated with water vapor, which then condenses out (the ability of air to hold water vapor decreases rapidly with temperature, as air with 50% relative humidity at 20 °C exceeds water vapor saturation when cooled to 8 °C, leading to condensation).

Capillaries: Surface Tension

Water has a relatively high surface tension. This is exhibited at air–water interfaces, such as those that occur in soil or in the cell walls of plant cells. Surface tension is related to the hydrogen bonding based on electrostatic attraction between the positively charged atom of hydrogen on one water molecule and the negatively charged atom of oxygen on an adjacent water molecule. In particular, surface tension equals the force per unit length existing at the water surface, which is dimensionally equivalent to the energy per unit area (force/distance = force × distance/distance^2 = energy/area). Indeed, surface tension represents the energy needed to expand the surface area by breaking hydrogen bonds, which results when water molecules are brought from the interior to the surface.

For a cylindrical pore of radius r (m), the surface tension σ (N m^{-1} or Pa m) acts over a perimeter of $2\pi r$, leading to a force of $2\pi r\sigma \cos\alpha$, where α is the contact angle between the water surface and the solid at the air–water interface (**Figure 1**). This pulling or stretching of the water surface leads to a tension in

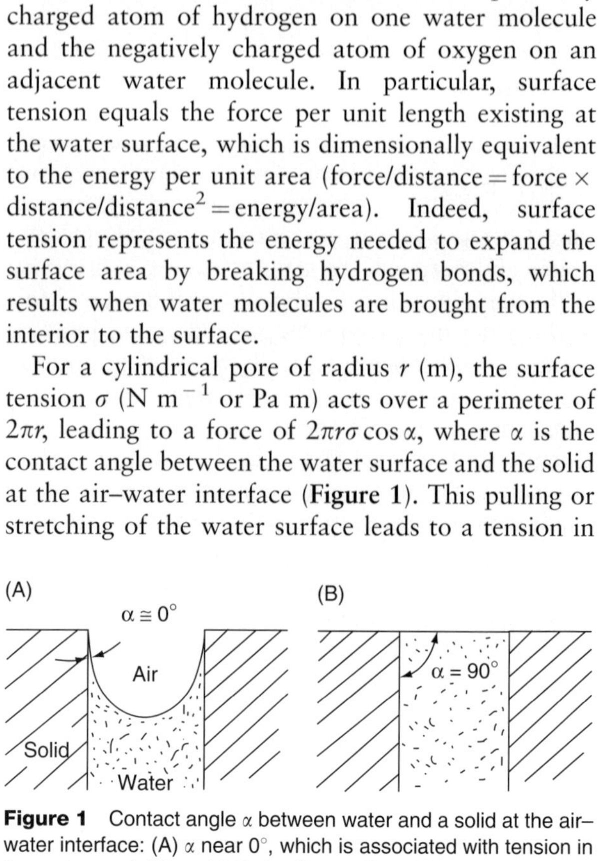

Figure 1 Contact angle α between water and a solid at the air–water interface: (A) α near 0°, which is associated with tension in the water; and (B) α of 90°, leading to $P = 0$ MPa.

the water, $-P$ (P is the pressure), which acts over an area πr^2. Thus, $2\pi r\sigma \cos\alpha$ equals $-P\pi r^2$, so

$$P = \frac{-2\sigma \cos\alpha}{r} \quad [3]$$

which is known as the Young and Laplace equation. (This equation can also be expressed as $P = -\sigma(1/r_1 + 1/r_2)$, where r_1 and r_2 are the radii of curvature in mutually perpendicular directions.)

In the soil, α can be near 0°(cos 0 = 1; **Figure 1A**) and the radius of curvature at the concave air–water interface between adjacent relatively small soil particles can be small, e.g., 1 μm. By eqn [3], P then equals $-(2)(7.28 \times 10^{-8}$ MPa m$)(1)/(1 \times 10^{-6}$ m) or -0.15 MPa. When soil water is available, it flows into the pores, leading to an increase in α and a raising of P toward 0 (eqn [3]). As the soil dries, its water content decreases and water is lost from the larger pores first; α is then near 0° only for smaller and smaller pores, leading to a decreasing P (eqn [3]) and hence decreasing the soil water potential Ψ (eqn [1]).

The Young and Laplace equation can also be used to estimate P in the cell wall pores. These pores are very fine – often only 5 to 10 nm in diameter – but water is generally available and flows in, leading to an α approaching 90° (at 90°, the water level is flush with the cell wall surface and $P = 0$ MPa) (**Figure 1B**). For a cell wall pore radius of 5 nm and α of 87° at the air–water interface, $P = -(2)(7.28 \times 10^{-8}$ MPa m$)(1)(\cos 87°)/(5 \times 10^{-9}$ m) or -1.5 MPa, a typical water potential for a leaf.

Water Flow

The flow or more properly the flux density (the latter term is recommended by the Système Internationale d'Unités) of water is proportional to the force causing the movement (the term flow often is used to describe the amount of substance moving per unit time and the flux density to describe the amount time^{-1} area^{-1}). The force can be represented by differences in Ψ or P with position. The volume flux density J_V, which is relatively easy to measure and appropriate if water is the main substance moving, can be expressed in units of m^3 m^{-2} s^{-1} or m s^{-1}. Indeed, J_V equals the average velocity of the fluid movement, which is often exploited in experimental studies. Different formulations are used to describe J_V in various parts of the soil–plant–atmosphere continuum.

In the soil, Darcy's law – formulated in 1856 – indicates that:

$$J_V = -L^{\text{soil}} \frac{\partial \Psi^{\text{soil}}}{\partial x} \cong -L^{\text{soil}} \frac{\partial P}{\partial x} \cong L^{\text{soil}} \frac{\Delta P^{\text{soil}}}{\Delta x^{\text{soil}}} \quad [4]$$

where L^{soil} (m^2 s^{-1} MPa^{-1}) is the soil hydraulic conductivity coefficient. Thus, water in the soil flows toward lower soil water potentials (ψ^{soil}), which in turn means toward lower soil hydrostatic pressures (P^{soil}; due to the absence of membranes in the soil, osmotic pressure differences do not directly influence the flow there). L^{soil} is sensitive to the soil water content, becoming extremely low in dry soil for which the water phases are small and widely separated from each other. This underlies the action of the soil crust, which dries out and hence leads to little water movement out of the soil during drought but which can have a high L^{soil} when wetted by rainfall, facilitating water entry into the soil.

At the root surface, the volume flux density can be represented by:

$$J_V = L_P(\Delta P - \sigma\Delta\Pi) \quad [5]$$

where L_P is the root hydraulic conductivity coefficient (m s^{-1} MPa^{-1}) and the reflection coefficient σ is a dimensionless parameter introduced by irreversible thermodynamics that quantifies the influence of solutes. In particular, σ varies from 1 for impenetrable solutes to 0 for freely penetrating solutes, so it represents the relative contribution of a particular solute to osmotic water flow. L_P can be broken down into a radial component across root tissues (L_R, m s^{-1} MPa^{-1}) and an axial hydraulic conductance per unit length along the root xylem (K_h, m^4 s^{-1} MPa^{-1}).

The flux density expressed per unit area along tubes, such as those in the xylem composed of dead, conducting cells with thick cell walls, was described by Hagen in 1839 and Poiseuille in 1940, leading to Poiseuille's law (also known as the Hagen–Poiseuille law):

$$J_V = \frac{-r^2}{8\eta} \frac{\partial P}{\partial x} \quad [6]$$

where η is the viscosity of the fluid (η for water at 20 °C is 1.00×10^{-3} Pa s) and r is the tube radius (when expressed per cylindrical tube, the volumetric flow rate is $-\pi r^4/8\eta$). Flow along the xylem generally does not limit L_P as much as L_R does, unless the xylem becomes blocked by air bubbles in the individual tubes, leading to cavitation.

Water vapor diffusion out of a leaf can be described by Fick's first law, which was enunciated in 1855:

$$J_{\text{wv}} = -D_{\text{wv}} \frac{\partial c_{\text{wv}}}{\partial x} = g_{\text{wv}}\Delta c_{\text{wv}} = g'_{\text{wv}}\Delta N_{\text{wv}} \quad [7]$$

where J_{wv} (mol m^{-2} s^{-1}) is the molar flux density of water vapor, D_{wv} is the diffusion coefficient for water vapor in air (2.4×10^{-5} m^2 s^{-1} at 20 °C), c_{wv} is the water vapor concentration in air (mol m^{-3}), g_{wv} and g'_{wv} are water vapor conductances (m s^{-1} and mol m^{-2} s^{-1}, respectively), and N_{wv} is the mole fraction of water vapor in air (0.023 for air saturated with water vapor at 20 °C). To convert J_{wv} to a volume flux density, the molar flux density needs to be multiplied by the partial molal volume of water: $J_{wv} \times \bar{V}_w = J_V$.

Parts of the Water Movement Pathway

Soil and Root–Soil Air Gap

Water readily moves through wet soil, so the drop in water potential over this part of the soil–plant–atmosphere continuum is relatively low under wet conditions. As the soil dries, L^{soil} in Darcy's law (eqn [4]) decreases, so a larger $\Delta\Psi^{soil}$ is then needed to maintain substantial water uptake by a root. For instance, L^{soil} for sandy loam can decrease by a factor of 10^6 as Ψ^{soil} goes from −0.01 MPa (very wet soil) to −10 MPa (very dry soil). As the soil dries, roots can become dehydrated and therefore shrink radially, especially young fleshy roots that have not developed extensively suberized or lignified tissues. Consequently, an air gap can develop between a root and the surrounding soil across which water must move as a vapor (cf. Fick's first law, eqn [7]). For instance, young roots of various desert succulents and other plants can shrink 20% in the radial direction as Ψ^{soil} decreases from −0.01 to −10 MPa. The developing air gap serves as an increasing resistance for water movement, which can help limit water loss from a plant to the drying soil, especially in the initial stages of drought when the root water potential is somewhat higher than Ψ^{soil}.

Root

Various cell types are involved for water moving from the root surface to the root xylem (**Figure 2**). Water entering young roots must first cross a layer of epidermal cells. Root hairs projecting from such cells increase the surface area for water and nutrient uptake, as do the hyphae of mycorrhizal fungi. Water then moves across the root cortex (**Figure 2**) either by moving in the cell walls (apoplastic pathway) or through the cells (symplastic pathway). The cells in the symplastic pathway are interconnected by plasmodesmata (pores through the plasma membrane that link the cytoplasm of one cell to that of an adjacent cell) and water can also move from cell to cell across other parts of the plasma membrane. Most cells of the endodermis have their radial walls impregnated and hence blocked with suberin (termed a Casparian strip), forcing water and solutes to move into the cytoplasm of endodermal cells. After crossing the endodermis, water can then enter the conducting cells of the xylem. For older roots of dicotyledons, layers of periderm whose cell walls often contain suberin and lignin form outside the vascular tissue, and the other outer cell layers can become disrupted. Primarily because of the suberin, radial movement of water across such older roots is reduced, especially under dry conditions.

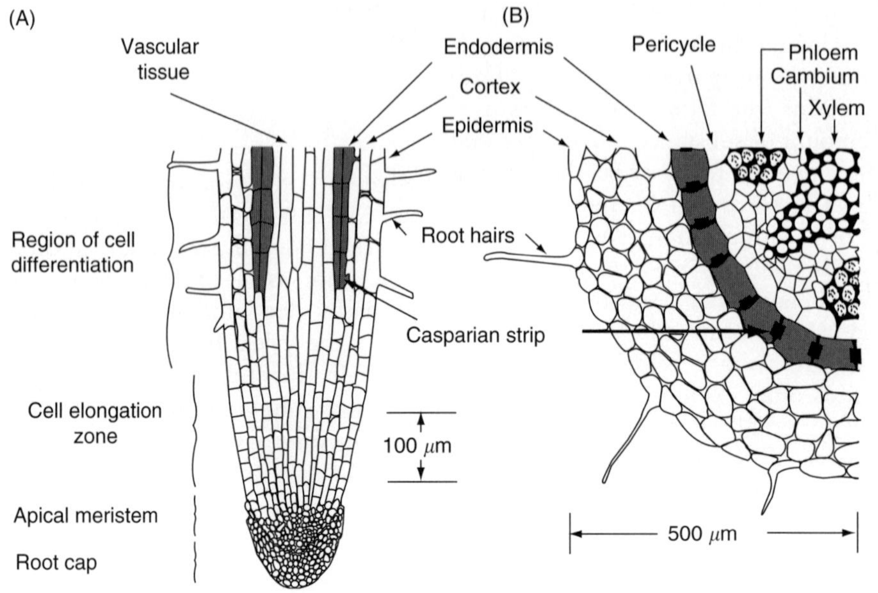

Figure 2 Longitudinal section (A) near the tip of a young root and (B) cross-section about 10 mm back from the tip. Reproduced by permission from Nobel PS (1999) *Physiochemical and Environmental Plant Physiology*, 2nd edn. San Diego: Academic Press.

Xylem

The xylem forms the plumbing of a plant, allowing water to travel along the roots, up the stem, and across the leaves. The conducting cells of the xylem – tracheids and vessel elements (also called vessel members) – have the remarkable feature that they perform their main physiological role when they are dead. The narrow tracheids that often taper at their ends and the generally shorter and wider, more highly derived, vessel elements consist of tubes of cell walls without any plasma membranes or cytoplasm. For tracheids, the cell wall is interrupted by pits in regions with only thin primary cell wall material and where sequential tracheids abut, facilitating water flow. Vessel elements line up, forming vessels, and most of the end cell walls between sequential elements can disappear during development, which greatly facilitates water flow. In particular, pits can provide about half of the resistance to water movement along the xylem, the rest being the frictional resistance involved in Poiseuille flow (eqn [6]), where the conductivity of an individual lumen is proportional to its radius raised to the fourth power. Thus wide, unblocked vessels account for a disproportionately large amount of the water flow along the xylem. During drought, as well as during alternating freeze–thaw cycles, air bubbles (embolisms) can block individual xylem vessels, thereby decreasing K_h.

Leaves

Water moves across leaves in the xylem, which for dicotyledons involves increasingly finer branches of veins (for monocotyledons, leaf veins tend to be parallel with less pronounced tapering). Water from along the xylem and from its terminal cells moves both symplastically and apoplastically across a leaf and evaporates from the cell walls. Therefore, the interior of a leaf becomes nearly saturated with water vapor, which then diffuses out of a leaf through the stomata toward air with a lower water vapor content (lower c_{wv}) in a process called transpiration. When the stomata are closed, a limited amount of water still escapes from the leaf through the waxy cuticle on the leaf surface; this water loss is known as cuticular transpiration.

Phloem

Although most water movement in a plant occurs in the xylem, water flow also occurs in the phloem, a process that requires an input of metabolic energy to increase the solute content and hence the osmotic pressure locally, leading to water entry by osmosis. Such water entry raises the local hydrostatic pressure, leading to pressure-driven flow along the phloem. The main function of the phloem is to take photosynthetic products from leaves or other sources and deliver them to regions of growth or storage (the organic compounds can later be moved out of the latter region in the phloem as well). However, certain young fruits have a higher (less negative) water potential than the stems on which they occur, so water cannot enter such fruits by flowing energetically downhill in the xylem. Indeed, the phloem can be a major pathway for water entry into young fruits of apple (*Malus pumila*), grape (*Vitis vinifera*), tomato (*Lycopersicon esculentum*), and various species of cacti (Cactaceae).

Summary: Cohesion–Tension Theory

The cohesion–tension theory for water flow in plants postulates that evaporation and loss of water by leaves creates a tension ($P<0$) in the xylem sap. Because water molecules are hydrogen-bonded together, the tension is sustained along the length of the xylem, causing water to be pulled up the plant. Putting water under tension due to leaf transpiration causes the conducting cells of the xylem and the other cells in the stem to contract slightly, leading to measurable decreases in stem diameter (described as a water capacitance effect that depends mainly on the changing water content of parenchyma cells). Such tension is experimentally difficult to measure – evidence from centrifuging plant parts and thereby increasing the forces indicates that large tensions can develop, whereas direct measurement with pressure probes inserted into specific cells does not reveal the large negative pressures expected, mainly for technical reasons. In any case, when the stomata close at night and transpiration therefore nearly ceases, the tension abates and the cells reswell, leading to a measurable increase in the diameter of stems of trees and other plants. When the stomates reopen the next day, water is initially lost from the leaves, causing them to become thinner (often by 5–10%) and a tension to develop in the leaf xylem. Water subsequently moves from living cells in the stem to the conducting xylem cells. Indeed, such capacitance effects cause water movement in the roots to lag behind transpiration from the leaves by a few minutes for small herbs to hours for large trees. Such reversible daily changes become progressively smaller during drought, when less water flows through the soil–plant–atmosphere continuum.

Once the soil attains a lower water potential than the plant, water tends to move out of plant roots to the soil, although such loss is reduced when certain cell layers in the roots become extensively suberized and lignified. Indeed, the inexorable laws of thermodynamics and the properties of plant tissues even describes uptake of water into a plant in regions where the soil is wet, movement along the root xylem, and then a loss of plant water back to the soil from roots in dry regions (often called "hydraulic lift" if the dry regions are nearer to the soil surface than are the wet regions). Water plays many roles in plants and in turn in all ecosystems, and its movement from region to region along a plant has long been described in terms of physical constraints and applicable equations. Yet only recently has the actual pathway been clarified and the role of water channels across membranes (aquaporins) been described, suggesting that many new discoveries are imminent in the crucial and experimentally active field of plant water relations.

List of Technical Nomenclature

a_w	Activity of water	
c_j	Concentration of solute j	mol m^{-3}
c_{wv}	Concentration of water vapor	mol m^3
g	Gravitational acceleration	~ 9.8 m s^{-2}
g_{wv}	Water vapor conductance	m s^{-1}
g'_{wv}	Water vapor conductance	mol m s^{-1}
h	Height above reference level	m
r	Radius	m
x	Distance	m
D_{wv}	Diffusion coefficient of water vapor	2.4×10^{-5} m^2 s^{-1} at 20°C.
J_V	Volume flux density	m^3 m^{-2} s^{-1} = m s^{-1}
J_{wv}	Molar flux density of water vapor	mol m^{-2} s^{-1}
K_h	Axial hydraulic conductance per unit length	m^4 s^{-1} MPa^{-1}
L_P	Root hydraulic conductivity coefficient	m s^{-1} MPa^{-1}
L_R	Radial hydraulic conductivity coefficient	m s^{-1} MPa^{-1}
L^{soil}	Soil hydraulic conductivity coefficient	m^2 s^{-1} MPa^{-1}
N_{wv}	Mole fraction of water vapor	
P	Hydrostatic pressure	MPa
R	International gas constant	8.31 J mol^{-1} K^{-1}
T	Temperature (absolute)	Kelvin units
$\bar{V}_w$	Partial molal volume of water	$\sim 18 \times 10^{-6}$ m^3 mol^{-1}
α	Contact angle	°
η	Viscosity	1.00×10^{-3} Pa s for water at 20°C
Π	Osmotic pressure	MPa
ρ_w	Density of water	~ 1000 kg m^{-3}
σ	Surface tension of water	7.28×10^{-8} MPa m at 20°C
σ	Reflection coefficient	
Ψ	Water potential	MPa
Ψ_{wv}	Water potential of water vapor in air	MPa

See also: **Nutrition**: Deficiency Diseases, Principles. **Water Relations of Plants**: Stomata; Uptake, Loss and Control; Xylem.

Further Reading

Boyer JS (1995) *Measuring the Water Status of Plants and Soils*. San Diego: Academic Press.
Kramer PJ and Boyer JS (1995) *Water Relations of Plants and Soils*. San Diego: Academic Press.
Nobel PS (1999) *Physiochemical and Environmental Plant Physiology*, 2nd edn. San Diego: Academic Press.
Nobel PS and Cui M (1992) Hydraulic conductances of the soil, the root–soil air gap, and the root: changes for desert succulents in drying soil. *Journal of Experimental Botany* 43: 319–326.
Nobel PS and De la Barrera E (2000) Carbon and water balances for young fruits of platyopuntias. *Physiologia Plantarum* 109: 160–166.
Salisbury FB and Ross CW (1992) *Plant Physiology*, 4th edn. Belmont: Wadsworth.
Sperry JS, Adler FR, Campbell GS, and Comstock JP (1998) Limitation of plant and water use by rhizosphere and xylem conductance: results from a model. *Plant, Cell and Environment* 21: 347–359.
Steudle E (2000) Water uptake by plant roots: an integration of views. *Plant and Soil* 226: 45–56.
Taiz L and Zeiger E (1998) *Plant Physiology*, 2nd edn. Sunderland: Sinauer Associates.
Tyree MT (1997) The Cohesion–tension theory of sap ascent: current controversies. *Journal of Experimental Botany* 48: 1753–1765.
Zimmerman MH (1983) *Xylem Structure and the Ascent of Sap*. Berlin: Springer-Verlag.

Xylem

M Mencuccini, University of Edinburgh, Edinburgh, UK

Introduction

This article will cover the most important elements of water flow in the xylem of plants. First, the fundamentals of xylem anatomy with regard to water transport are described. Second, water flow is described at two scales: the individual conduit and the entire plant. Third, the causes of xylem cavitation and embolism, its ecological consequences, and the strategies for recovery from embolism are discussed.

Definitions and Functional Xylem Anatomy

Xylem (from the Greek *xylon*, for wood) is a complex tissue present in all vascular plants (also called tracheophytes). The term wood is frequently used as a synonym for xylem in trees and shrubs. Xylem is made up of different cell types, which can be grouped under the two fundamental types of parenchyma and sclerenchyma cells. Parenchyma is made up of living cells with a variety of functions, some of which are still not well understood. The main ones are storage, vertical and radial transport of nutrients and carbohydrates, production of protective resins and other defensive compounds against external wounds, production of tyloses to block dysfunctional conduits, and possibly repository of dangerous catabolic waste (such as oils, tannins, gums, aromatic compounds, and resins) in the central part of the stem. Parenchyma cells are normally arranged in rays, to facilitate transport. Interestingly, their location is such that all xylem conduits are always in direct contact with at least one parenchyma cell.

The sclerenchyma is made up of dead cells with the fundamental functions of transport (of water, hormones, and nutrients), storage (of water), and mechanical support. The fundamental (and most ancient) sclerenchyma cell type is the tracheid, a thin elongated hollow cell with a lignified cell wall and a central lumen, through which water flows. The conducting lumen originates from the programmed self-destruction of the original cytoplasm and nucleus of the immature living cell. The cell wall is composed of several layers with different chemical and physical properties, which are classed as middle lamella (adjoining adjacent conduits), primary (essentially made of hydrophilic cellulose and hemicellulose polymers and pectin), and secondary wall (a lignified, hydrophobic layer directly lining the lumen).

Tracheids, which carry out both functions of transport and mechanical support, are present in all vascular plants but especially in conifers, bryophytes, and ferns, which are evolutionarily primitive. The more recent flowering plants (known as angiosperms) display instead distinct cell types for transport (vessels) and mechanical support (fibers), as well as tracheids. Vessels normally have much larger lumina than tracheids, and are made up of series of several hundreds or thousands of barrel-like segments (called vessel elements) stacked on top of one another. Fibers are shorter than vessels or tracheids, have small lumina and comparatively very thick walls. Lateral walls of both tracheids and vessel elements are heavily lignified by the secondary wall, but the extent of this thickening varies from species to species in both extent and shape (e.g., annular, spiraliform, scalariform, reticulated, or pitted) and provides fundamental taxonomic diagnostics. While many intermediate cell types exist in nature that are not easily classed into one type or the other, the fundamental anatomical distinction between tracheids and vessels is the lack of perforation in the secondary thickening of the end walls of the latter. However, between successive vessel elements of the same vessel, the perforation of the end wall is complete, and water can flow seamlessly from one vessel element to the one above.

The overall dimensions of tracheids, vessels, and fibers can be obtained by macerating wood in concentrated acidic solutions and using a light microscope, or by using more modern noninvasive techniques based on X-ray scanning. Tracheids are normally 1–5 mm long, although exceptional lengths (particularly in extinct species) of up to 1–3 cm have been reported. Their diameter (or width of their major axis, where the section is not circular) can vary between 10 and 80 μm. In the case of vessels, lengths can range from a few centimeters to several meters, whereas diameters can range from 20 μm up to hundreds of micrometers in ring porous species (see below), to half a millimeter in lianas.

Water flows from one tracheid (or vessel) to the adjacent one by means of little apertures in the secondary walls called pit pores. While pore structures can differ widely among flowering plants, in most modern conifers the bordered pit represents the predominant form of pitting. In a bordered pit, the secondary walls on both sides of the pit curve inside the lumina to form two arched structures around the pit, leaving only a central circular hole for water flow. Inside this almost spherical space, the two primary walls form the pit membrane, a mesh made up of miniscule pores, at whose center a bulging structure, called the torus, is located. The torus, whose diameter is larger than the size of the two central openings in the borders, has the function of closing the aperture in case of cavitation (see below) of one of the two adjoining conduits. The pores of the pit membrane are of the order of tens to hundreds of nanometers and, because of their very small size, represent a major resistance to water flow. Because the mesh network is partly made of pectins, it behaves much like a hydrogel, by shrinking and swelling in response to ion concentrations in the flowing solution. For instance, if a plant stem is sequentially perfused with dissociating solutions of KCl, NaCl, CaCl, or KNO_3, of greater and greater ionic concentration, systematic increases in flow rates can be demonstrated at constant hydrostatic pressures. Although the actual diameter reduction of the mesh network is minimal (as it is mostly composed of cellulose, which does not respond to changes in ion concentration), changes in hydraulic resistance of the pore channel can be noticeable, since hydraulic resistance is inversely related to the fourth power of the radius of a capillary (see below). This provides the plant with a means to partially regulate the hydraulic resistance of its network by controlling the ionic composition of the xylem sap.

Water Flow in the Xylem and the Ascent of Sap

The Plant Water Potential

In any part of the soil–plant–atmosphere continuum, the thermodynamic status of water can be described using the concept of water potential, symbolized by the Greek letter Ψ. Water potential is defined as the chemical energy of one mole of water (joules per mole) minus the chemical energy of the same mole of water at a reference state (pure free water at the same temperature and pressure and at a reference elevation) and divided by the partial molal volume of water. As such, it has units of MPa (megaPascals), i.e., pressure units. Water moves spontaneously only from a region of higher chemical potential to one of a lower chemical potential, and in so doing it releases free energy. Given this definition, Ψ assumes a value of zero when water is freely available, becomes negative when water is "scarce," and positive when water is under pressure. However, it must be remembered that this definition discounts the presence of atmospheric pressure: the absolute pressure of water is always $+0.1$ MPa higher than the value of its Ψ: for instance, when $\Psi = 0$, absolute pressure will be $+0.1$ MPa.

Because plants are continuously losing water from their leaves, most of the time plant water potentials are negative (i.e., the columns of water in the xylem are under "tension," much like a stretched rubber band).

Sap Flow Inside Individual Conduits

Water flow in the xylem can occur either in response to a pressure or a tension gradient. Inside individual conduits, the velocity profile of water flow approximates a parabolic distribution, according to Poiseuille's law, with no or little flow along the walls and maximum flow velocities at the center, where the effect of friction against the cell wall is minimal. The significance of Poiseuille's law for water movement in individual conduits can be demonstrated using techniques such as MRI (magnetic resonance imaging), which allows the flow of water inside individual conduits to be visualized. Because of this parabolic dependency of flow velocity and because of their much larger areas, large vessels have mass flow rates far in excess of those recorded in small vessels. As mentioned above, one expects the mass flow to increase with the fourth power of the conduit radius. For instance, maximum flow velocities can range from 1 to 3 $m\,h^{-1}$ in conifers, from 0.7 to 6 $m\,h^{-1}$ in diffuse-porous hardwoods, from 4.0 to 45 $m\,h^{-1}$ in ring-porous hardwoods, and up to 100 $m\,h^{-1}$ in herbaceous species. This correlates well with the differences in average conduit diameter across these broad taxonomic groups.

Sap Flow in Intact Plants

Sap flow rates in intact plants can be described using the Ohm's law, in analogy to electrical theory, as follows:

$$E = \frac{\Delta\Psi}{R} \quad [1]$$

where E, $\Delta\Psi$, and R indicate the water flow rate (in units of mass of water per second), the water potential difference driving the flow between two points conveniently chosen (for instance the inside of the root stele and the leaf xylem), and the xylem hydraulic resistance, respectively (see also **Figure 1**). The xylem hydraulic resistance represents the inevitable energy losses of the unit mass of flowing water from roots to leaves per unit time, owing to the cohesion (i.e., friction) between adjacent water molecules and the adhesion between water molecules and cell walls in these tiny capillaries. The above equation implies that a plant can maximize the flow of sap through its xylem by two means: by maximizing the water potential difference between roots and leaves; and by minimizing the hydraulic resistance between these two points (i.e., a more efficient xylem). This fundamental distinction gives rise to a number of very important hypotheses about plant behavior in relation to xylem physiology and ecology, which will be developed below. However, we first need to understand the concept of xylem vulnerability.

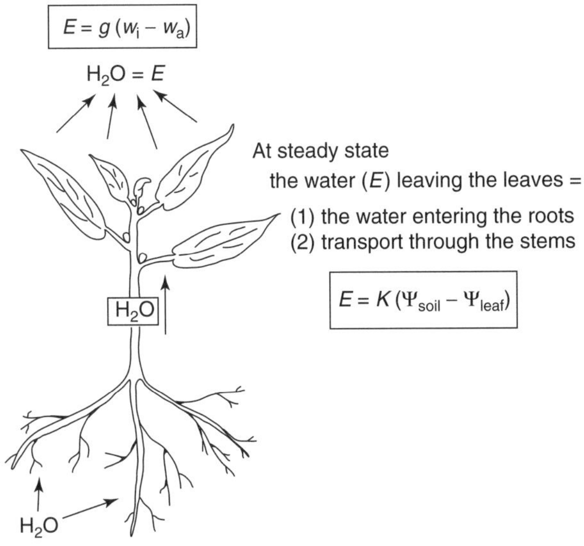

Figure 1 A diagrammatic representation of water flow in plants. Sap movement is driven by the losses occurring inside the leaves by evaporation (E). This decreases plant leaf potential (Ψ) relative to stem, root, and soil water potentials, causing sap to flow upward. The losses by evaporation are controlled by the degree of aperture of the stomata (symbolized here by g, stomatal conductance) and by the difference in water vapor pressure between air (W_a) and intercellular spaces (W_i) inside the leaf. Similarly, liquid water flow is driven by plant hydraulic conductance (K) and the difference in water potentials between the leaves and the soil at the root surface. At steady state, these two fluxes must be identical.

Vulnerability of Xylem Water Transport

Drought

Water transport occurs under negative water potentials, sometimes at extreme levels (down to about −15 MPa, i.e., about −150 atmospheres). This thermodynamic status of water, referred to as "metastable," is the reason why water flow in the xylem is vulnerable to formations of emboli, similar to that which occurs in human blood vessels during scuba diving. The tensions developing in the lignified xylem can be so intense that one can observe plant stems shrinking and swelling as a consequence of its fluctuations. The shrinkage is minimal (a few tens of micrometers); however, precision displacement transducers exist that can resolve these tiny changes.

It is likely that the evolution of xylem features when plants first colonized the land 400 million years ago also included the capacity to resist this enormous tension without imploding. Lignin serves beautifully as a very stiff and strong construction material, despite its relatively low density compared to materials such as steel. For instance, extant species that normally grow in drought-prone environments and have routinely high levels of tension in their xylem also tend to have high values of wood density. This correlation is likely to have resulted from the evolutionary need described above.

Pure water has a tensile strength well in excess of −15 MPa (sometimes, figures down to −100 MPa are reported). Therefore, under ideal conditions, it can resist tensions in excess of what plants normally experience. However, the stability of the water column is drastically reduced if an air–water interface develops locally. The interface can be created by the action of external agents (wind, insects, browsers, etc.), which can break the bark and the xylem and create an artificial air entry point, or by the presence of other dysfunctional conduits. The entry of air in the first conduit open to the outside does not necessarily cause widespread embolism: water is normally sucked out of the broken conduit into the adjacent functional ones, while the open conduit fills with air at atmospheric pressure. The retreat of the air–water meniscus stops at the pit, as it does not have enough energy to move through the very small pit pores. Energy to cause the meniscus to "air seed" the adjacent functional conduit is only created if the difference in pressure between the air phase and the water phase reaches a critical point (the "entry point pressure"). In nature, this is normally obtained if water tension reaches critical levels (either because the soil dries out or because the loss of water from the leaves is too fast, i.e., atmospheric drought). However, one can also experimentally manipulate the levels of air pressure on the other side of the meniscus and prove that this same phenomenon occurs under identical conditions. Therefore, the stability of xylem water under tension is completely dependent on the segregation between the water and air phases in different compartments.

Once "air seeding" has occurred, a small gas bubble penetrates inside the air seeded conduit. Upon retreat of the water from the air-seeded conduit, the gas bubble expands occupying the entire conduit, albeit at a pressure much lower than atmospheric, and probably close to full vacuum. This event is referred to as cavitation. Over time, gases diffuse from the liquid to the empty cavity and bring air pressure back up to atmospheric. The conduit is now embolized (**Figures** 2 and 3).

For anatomical reasons, each individual conduit has its own characteristic "entry-point pressure," i.e., the critical tension upon which it is air seeded. However, each woody stem is made of hundreds of thousands or even millions of different conduits in the same cross-section. Therefore, at the level of the entire plant, there is no single "entry-point pressure," and the phenomenon can be described only statistically for the population of conduits. Such synthetic descriptions of the susceptibility of xylem

Figure 2 A scanning electron micrograph of a cross-section of a stem of Asiatic dayflower (*Commelina communis*), a common herbaceous plant of the spiderwort family. The plant was subjected to a drought cycle and then quickly frozen in liquid nitrogen before being cut and mounted in the scanning electron microscope (SEM). The image clearly shows several cavitated vessels, indicated by the large elliptical black holes. Also visible are some intercellular spaces that emptied of water as a result of drought. They can be recognized because of their much smaller size, their irregular shape, and the tendency towards rhomboidal form.

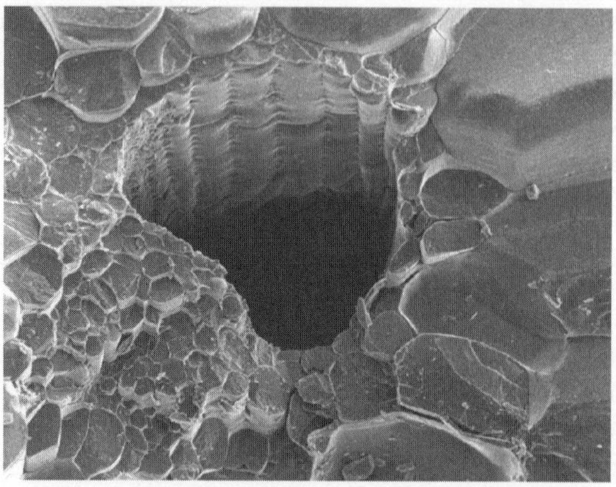

Figure 3 An SEM image at a greater magnification to that in Figure 2 of a single cavitated vessel of a *Commelina communis* stem. The internal architecture of the vessel is clearly visible. The rugosity of the secondary wall is mostly due to the shape of the adjacent cells. Some annular thickenings of the wall and a large number of crevices at the margins between adjacent parenchyma cells are evident. The irregularity of the vessel cross-section is an artifact of the technique employed to mount the specimen on the SEM.

to cavitation owing to drought are called vulnerability curves (**Figure 4**). Vulnerability curves are not static and may change over time. For instance, during plant growth, new conduits are produced with different anatomical features. This can impact on

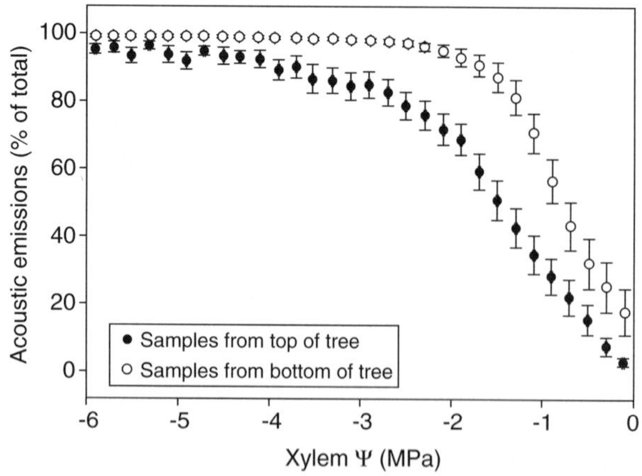

Figure 4 A vulnerability curve to cavitation for the xylem of a mature Scots pine tree (*Pinus sylvestris*). The curves were obtained in the laboratory from woody segments sampled from the tree trunk at different heights, corresponding to different nodes from the top. Xylem water potential was measured using *in situ* stem psychrometers, whereas the extent of cavitation was estimated using a technique based on counting the ultrasonic acoustic emissions (UAE) resulting from the sudden vibration of the tracheid walls following cavitation. A complete dehydration cycle would normally last 48 h. The curves start at the lower right-hand corner with fully saturated stems and no UAE. When the first tracheids begin to cavitate, a few UAE are recorded. The curves progressively accelerate as they move toward the upper left-hand corner, in correspondence with the most frequent tension when cavitation occurred, followed by a clear plateau. Xylem samples from the bottom of the tree were much more vulnerable to cavitation than those from the top, as more UAE could be recorded at less negative water potentials.

the xylem vulnerability. Also, in the short term, if several intense drought cycles occur sequentially, the xylem can progressively lose its resistance to cavitation. This xylem "fatigue" likely results from conditions of intense water tension, when the air–water menisci in the pit pores exert a very strong pulling force on the entire pit mesh, causing it to stretch irreversibly and break. Upon a subsequent cycle, the air-entry point pressure is sensibly lowered. It is unclear whether repair mechanisms exist for xylem fatigue or whether it leads to irreversible conduit dysfunction.

Freeze–thaw

There is a second process by which air emboli can develop in the xylem related to cycles of frost and thaw during autumn, winter, and early spring. This process is referred to as freeze–thaw cycle embolism. Contrary to situations of drought, there is no transient phase of development of a cavitated conduit before the embolus forms, and the mechanism does not involve air seeding from an adjacent dysfunctional conduit.

When the xylem sap freezes (normally at temperatures between -1.0 and $-5.0°C$; supercooling does not normally occur in the xylem sap as its solute concentration is low), gas is forced out of solution, as ice has much lower gas solubility than liquid water. Hence, on freezing, gas bubbles form in the xylem. The size of these bubbles varies depending on the speed of freezing. If freezing proceeds fast, many small bubbles form, scattered in the sap. If freezing proceeds slowly, bubbles coalesce, creating more critical conditions. Also, if conduits are large, there is a greater tendency for bubbles to coalesce, again creating critical conditions.

On thawing, all gases must redissolve if the conduit is to maintain its function. This can occur because the surface tension at the air–water interface tends to make the bubble compress and undergo pressure, facilitating redissolution into the liquid. However, this must occur very rapidly, as any development of water tension owing to evaporating conditions in the atmosphere will reverse this process and make the bubble expand, instead of helping its compression.

For many species, particularly the ring-porous angiosperms (so called because their very large vessels are not scattered randomly in cross-section, but form distinct annual rings produced during the early part of the growing season), xylem can be extremely sensitive to freeze–thaw cycles. For example, in oak (*Quercus* spp.) and ash (*Fraxinus* spp.), one freeze–thaw cycle in the autumn can be sufficient to embolize all the large vessels produced during the previous spring. However, in diffuse-porous angiosperms (so called because they lack the characteristic ring of very large vessels of the ring-porous species, e.g., maple (*Acer* spp.), birch (*Betula* spp.), etc.), several freeze–thaw cycles are required to completely embolize a woody stem. This number varies depending on factors such as the characteristic vessel diameters and the specific weather conditions during thawing. It is normally in the order of several tens of cycles. In tracheid-bearing plants, winter embolism is more limited (because of the small size of their conduits), with many species being almost completely immune from it.

One final point relates to the relationship between cell frost hardiness and resistance to xylem winter embolism. These are two quite different processes. Frost hardiness is an acclimation mechanism of the living cells to freezing temperatures. It develops in response to temperature reductions in the autumn to protect living cells from intracellular freezing. It is normally lost in the spring when temperatures start to increase. Winter embolism is largely a physical process, which is dependent on the xylem anatomical characteristics and weather conditions. Damage to living cells occurs in response to out-of-season cycles of thaw–freeze (in this order, i.e. damage occurs upon freezing). Winter embolism instead occurs in response to cycles of freeze–thaw (inverse order, i.e. damage occurs upon thawing). Winter embolism has been studied less intensively than cavitation by drought. Some of these conclusions may be modified when more information is available. For instance, some recent reports highlight a potential role for parenchyma cells in protecting xylem conduits from freeze–thaw cycle embolization in some species. In these species, winter embolism would only occur at temperatures below the threshold for frost hardiness of the parenchyma cells.

Ecological Consequences of Xylem Embolization

Xylem physiology plays a central part in influencing plant behavior. The first vascular plants blessed with a slender vascular strand of tracheids are reported for the Silurian period (about 415 million years ago), about the same time rudimentary stomata appeared (400–410 million years ago). Therefore, there is a strong link between the evolution of the water transport system and the control of water loss by stomata on the leaves. There are good ecological reasons for the existence of this coordination. For a plant to function, any loss of water by the leaves must readily be replaced by water coming from the soil through the stem. When a conduit embolizes, it is filled with air at atmospheric pressure. Under these conditions, unless the conduit is "repaired," it can no longer function to transport water to the leaves. Therefore, over time, embolization would deprive leaves of large numbers of functional conduits and would increase the likelihood of leaf water stress. We can express this concept using the notation of eqn (1) by saying that embolization increases the value of xylem hydraulic resistance (as the same amount of water flows through a reduced number of conduits, its viscous energy losses per unit time are greater). Alternatively, we can say that hydraulic resistance is itself a function of sap tension.

As mentioned earlier, plants have two basic strategies to minimize the limitations to water flow through their xylem, i.e., increase the water potential difference between roots and leaves, or reduce the hydraulic resistance between them. (*En passant*, it is worth noting that the degree of "adaptation" of a plant to its environment is not increased by maximizing xylem water flow *per se*. However, maximum xylem water flow is well correlated with maximum stomatal opening and, therefore, to minimal diffusional limitations to photosynthesis.)

Decreasing the hydraulic resistance is costly. Precious carbohydrates are employed to build hydraulic support structures instead of productive leaves. However, increasing the water potential difference can be a dangerous strategy, as already noted. Large levels of tension may lead to xylem embolism and increases in hydraulic resistance. In both cases, behavioral tradeoffs are apparent. In the first case, the alternative is between having a redundant hydraulic structure with limited diffusional limitation to photosynthesis (at the price of reduced leaf area construction, but with maximum performance of individual leaves), or accepting a certain degree of diffusional limitation to photosynthesis because of an underdesigned hydraulic system (with reduced photosynthetic performance of individual leaves but with the benefit of an increased leaf area). In the second case, the choice is between avoiding xylem embolism at the cost of a reduced water potential difference, or accepting that a certain fraction of the xylem system is always embolized, with the advantage of being able to maximize the water potential difference (and hence flow rate).

In each of these two cases, theoretically optimal solutions can be found, and significant differences emerge among different plant types. For instance, grass species have an inherent low cost of xylem construction (owing to their very small stature). In this case, ample hydraulic supply to the leaves may be an optimal strategy, and the extra allocation of carbohydrates to the xylem results in only minor reductions in leaf area growth. On the contrary, large trees have very large xylem construction costs (because of their massive height). While leaf area is roughly a two-dimensional object, xylem is a three-dimensional object. Hence, a unit reduction in leaf area growth has minor effects in increasing the xylem support structure. The optimal solution may lie in accepting an under-designed xylem system while leaf area growth is sustained.

How Frequently Does Xylem Embolism Occur?

Of the two tradeoffs mentioned in the previous section (i.e. relative redundancy of the hydraulic structure, and relative avoidance of xylem embolism), the first one is the direct result of the constraint of having to operate within a fixed carbon budget. The second one is subtler. It could be asked why plants cannot afford to have a hydraulic system that is immune from cavitation and embolism. After all, plant species differ widely in their vulnerability to cavitation, and some species have a very resistant xylem. For instance, while poplars and willows can lose 100% of their conductive capacity by a water potential of -1 to -1.5 MPa, some desert shrubs can endure water potentials of -5 to -10 MPa before even showing initial small conductance losses.

If some species can build a xylem resistant to embolism down to -5 or -10 MPa, why is this strategy not adopted by all species? The response to this question probably resides in the unavoidable compromise between efficiency in water transport and safety from embolism. Safety from embolism calls for building conduits with very small pit pores: the smaller the pit pores, the more negative the entry-point pressures for air seeding. On the other hand, very small pit pores are a hindrance to water transport: the smaller the pit pores, the greater the hydraulic resistance for water movement in the xylem. Across several tens of species, there is a significantly negative relationship between xylem conductivity and resistance to embolism. Low resistance to cavitation characterizes species whose xylem is very efficient in conducting water, and vice versa. Other tradeoffs may be present and more research is required on this important topic.

Of the multitude of publications on this topic, there has never been a case reported of a plant species showing no evidence of cavitation throughout a yearly cycle. Even continuously watered plants growing under favorable conditions in a greenhouse or a growth room do show some limited losses (frequently around 15–20%).

Coping with Embolism and Strategies for Refilling

Use-Once-and-Throw-Away Strategies

Plants have evolved a number of strategies to recover from cavitation and embolism. Without these strategies, embolism would be an entirely cumulative process, with no way back. This situation does occur in some species. For instance, we have already mentioned the case of ring-porous trees in relation to freeze–thaw cycle embolism. After the first frost, the large vessels of the previous year are effectively lost forever, as they are physically too large to be refilled during the following spring. It is apparently more effective for these species (e.g., oak, ash, chestnut (*Castanea* spp.), etc.) to build new vessels every year and accept the inevitable damage occurring late in the season. At the beginning of the growing season there are no large functional vessels present, so there must be a very tight chronological coupling between new xylem growth and new leaf growth, otherwise the new sprouts will soon die of dehydration. In fact, early flushing species statistically tend to have inherently low levels of winter embolization; in late-flushing species freeze–thaw cycle embolism is typically a significantly limiting factor and new xylem is required before leaf growth resumes.

Active Refilling Strategies

Active strategies for xylem refilling represent a more conservative use of the existing xylem, as each individual conduit can undergo several distinct drought cycles and still recover its function. The grapevine (*Vitis* spp.) symbolizes one strategy of "active" embolism repair. In winter, the xylem of grapevines is entirely cavitated (this is easily seen by their very low wood water content). At the time of bud flushing, the root system increases ion pumping in anticipation of the leaf requirements for nutrients and solutes. This results in the formation of a significant osmotic pressure in the root stele, as water follows the ions from the soil to the stele through a semipermeable membrane. The generated pressure can amount to 0.1 or even 0.2 MPa (i.e., 1 to 2 atmospheres), and results in the gradual rehydration of the entire xylem. Gas bubbles are literally expelled upward through the pit pores to the atmosphere.

Root pressure is developed not only by grapevines, but also by many other species. Birches and maples are the most notable examples, and this feature is exploited by man in the spring (exudation of maple and birch syrup). Many herbaceous species also develop root pressure on a daily basis, thereby providing a year-round effective strategy for xylem refilling. In these cases, bubbles are not physically expelled through the pit pores as in the grapevine, but are dissolved in the slowly flowing sap. As mentioned above, if the sap falls under even limited levels of pressure, the surface tension at the air–water interface tends to compress the bubbles and increase the gas pressure. This facilitates dissolution (**Figures 5** and **6**).

A form of localized stem pressure (in contrast to the root pressure mechanism just discussed) represents a second repair strategy. Currently, evidence for the formation of localized stem pressure is very limited and considerable disagreement exists as to its extent and even existence. At least for some species, there is evidence that refilling can occur even when the xylem sap is under high tension. Traditional physical theory predicts that, under these circumstances, refilling is impossible, as the sap will tend to be sucked away from the gas bubble, facilitating its expansion (instead of its compression). However, some authors have recently proposed that formation of localized pressure in cavitated conduits is physically possible even if the rest of the functional xylem is under tension. The proposed mechanism involves an active role of the adjacent parenchyma cells, which, on detection of an embolized conduit in their

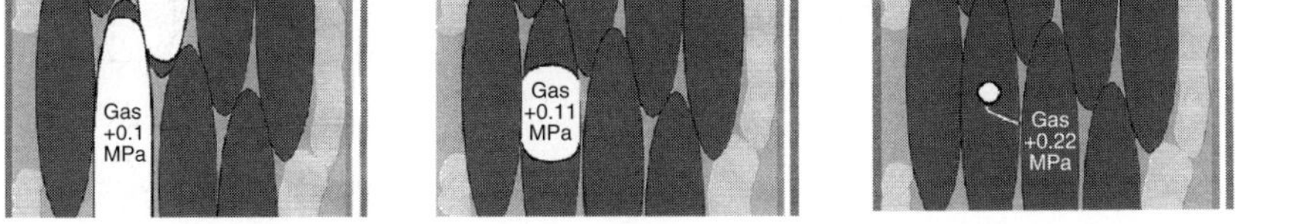

Figure 5 A diagrammatic representation of the refilling process for the common bean (*Phaseolus vulgaris*), which shows regular daily cycles of root pressure. (A) The typical situation during the day, while the plant transpires from its leaves. There are two embolized (white color) vessels at the center of the diagram, inside which the air pressure is assumed to be atmospheric (i.e., +0.1 MPa). They contain only a small amount of water in their terminal tapered ends. The rest of the vessels (dark color) are assumed to be functional and operating at a working tension of −1.0 MPa. After sunset, two conditions may occur. (B) The condition without root pressure. Assuming transpiration stops completely after dusk and the soil is entirely saturated, xylem water potential is in equilibrium with atmospheric pressure at a positive +0.1 MPa. The gas bubbles are now slightly compressed as a consequence of the surface tension of water. Hence, they are at a slightly higher pressure than water, which facilitates their dissolution in the static sap. (C) The condition of a xylem under hydrostatic pressure by the roots, amounting to an extra +0.1 MPa (i.e., an absolute value of xylem water potential of +0.2 MPa). Now the bubbles are compressed to a far greater degree and, are therefore under a much greater pressure. Their dissolution is much faster than in the previous case.

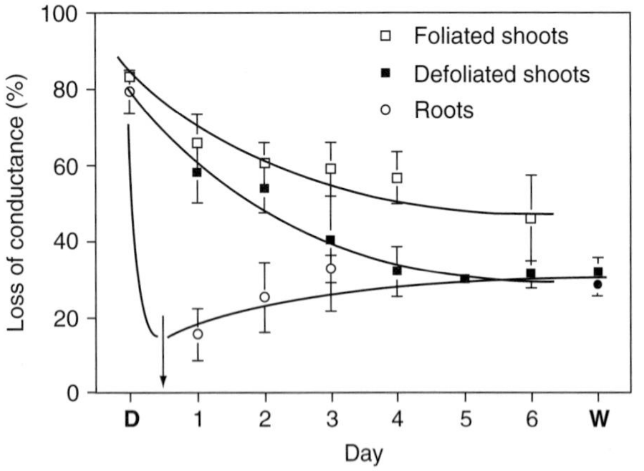

Figure 6 An experiment illustrating the importance of refilling for pinto bean (*Phaseolus vulgaris*) plants. The X-axis of the graph plots a drought sequence. The letters D and W indicate values for plants at the peak of the drought and for well-watered controls, respectively. The numbers 1–6 indicate the number of days since cessation of the drought cycle and irrigation was started again (indicated by the black arrow). The Y-axis plots the per cent loss of conductance due to embolism for each category. The roots of the plants refilled their embolized xylem overnight after irrigation (○), whereas the recovery of the foliated shoots from the same plants (□) was still incomplete after an entire week (compare the values of 6 with W). The recovery of the shoots occurred significantly faster if the stems were defoliated during the refilling process (■), and was completed after 4 days. Roots probably refill easily because, upon irrigation, they are surrounded by water-filled pores and absorb it from every side. Stems take longer to refill probably because it proceeds gradually upward from the base of the stem to the tips of the petioles. Defoliating the stems probably helps because it eliminates water tension in the xylem during the day, augmenting the effects of root pressure. (M Mencuccini and JP Comstock, unpublished data.)

vicinity, start pumping ions, or water directly, into the cavitated/embolized conduit. Anatomical features, such the hydrophobic nature of the secondary wall, the hydrophilic nature of the primary wall inside the pits, and the presence of specific water channels in the parenchyma membranes, would help the refilling of the conduit. This hypothesis is appealing, but conclusive proof is still lacking. Among other issues, the biochemical signal for the detection of a cavitated conduit adjacent to a parenchyma cell is not known.

Future Research in This Area

Most of the evidence in support of the theories discussed above comes from invasive measurements taken on plant segments in a laboratory environment. The emphasis of research is likely to gradually move away from destructive sampling toward non-invasive *in situ* techniques, such as NMR (nuclear magnetic resonance). Furthermore, there is a need for a better understanding of the physiological behavior of large trees, a system that has been largely ignored in the past in favor of more manageable grass or shrub species. The study of large trees creates massive logistical difficulties, which are currently approached by using construction cranes *in situ* or by building towers and walking platforms above the canopy. Also, sampling difficulties will always be present in this kind of research, unless new technologies are developed, such as portable CT or NMR scanners.

The physical mechanisms of xylem refilling are still poorly understood. Several aspects remain unexplored and research on this topic is urgently required. Significant limitations to this research are represented by the difficulty of isolating and visualizing *in vivo* the main areas where the core of the refilling process takes place, i.e., the inside of the pits.

Finally, we still have an incomplete picture of the variability in xylem hydraulic conductance and vulnerability to cavitation across the entire plant kingdom. Broad syntheses of the existing studies on the taxonomic and evolutionary aspects are likely to be very fruitful if focused on the relationships between xylem physiology and leaf physiology, particularly the carbon balance of leaves and phloem. For example, we understand the physiological implications of the differences in xylem anatomy between the major plant groups (e.g., bryophytes, conifers, angiosperms); however, our knowledge of the anatomical characteristics of stomata for these same groupings is extremely limited. Because of the tight ecological links between xylem water transport and leaf water losses, research in this area is likely to yield some very interesting insights into plant behavior and evolution.

Acknowledgment

The Natural Environment Research Council is acknowledged for its financial support of the research relevant to this article.

List of Technical Nomenclature

Air seeding	The major mechanism by which cavitation by drought occurs in the xylem of plants.
Entry-point pressure	The least negative tension required to force air through the pit membrane of a conduit.
Fiber	The unit with functions of mechanical support of the stem.

Hydraulic conductance	The inverse of the resistance.
Hydraulic resistance	The drop in sap water potential required to force a unit amount of water through a plant.
Hydrogel	A polymer with strong water-absorbing and swelling capacities.
Parenchyma cell	Living cell of the xylem.
Plant water potential	The relative chemical energy of the water.
Pit	The hole on the secondary wall of a conduit through which water flows to an adjacent conduit.
Pit pore	The pore in the pit membrane made up of the cell primary wall of a conduit.
Surface tension	The tension of a liquid when in contact with an air interface.
Torus	The central bulging inside the pits of conifers.
Tracheid	The water-conducting unit of the xylem of bryophytes, ferns, and conifers.
Vessel	The water-conducting unit of the xylem of angiosperms.
Vessel element	The basic unit of a vessel.
Water channel	A family of proteins of cell membranes specific for water transport in and out of the cell.
Xylem refilling	The process of recovery from cavitation and embolization.

See also: **Cell Walls and Fibers**: Fiber Formation; Wood Growth and Development. **Nutrition**: Ion Transport. **Primary Products**: Cellulose. **Secondary Products**: Lignin. **Water Relations of Plants**: Basic Water Relations; Drought Stress; Uptake, Loss and Control.

Further Reading

Cannell MGR and Dewar RC (1994) Carbon allocation in trees: a review of concepts for modelling. *Advances in Ecological Research* 25: 50–104.

Grace, J. (1993) Consequences of xylem cavitation for plant water deficits. In: J.A.C. Smith, H. Griffith (eds.). *Water Deficits: Plant Responses from Cell to Community*, pp. 109–128. Oxford: Bios Scientific Publishers.

Mencuccini M (2003) The ecological significance of long distance water transport: short-term regulation, long-term acclimation and the hydraulic costs of status across plant life forms. *Plant, Cell and Environment* 26: 163–182.

Raven JA (1993) The evolution of vascular plants in relation to quantitative functioning of dead water-conducting cells and stomata. *Biological Review* 68: 337–363.

Sperry JS, Hacke UG, Oren R, and Comstock JP (2002) Water deficits and hydraulic limits to leaf water supply. *Plant, Cell and Environment* 25: 251–263.

Tyree MT and Sperry JS (1989) Vulnerability of xylem to cavitation and embolism. *Annual Review of Plant Physiology and Plant Molecular Biology* 40: 19–38.

Zimmermann MH (1983) *Xylem Structure and the Ascent of Sap*: Berlin: Springer-Verlag.

Phloem

G A Thompson and H L Wang, University of Arkansas, Little Rock, AR, USA

Introduction

Plants emerging from an aquatic environment to live on land evolved several morphological adaptations that required the ability to efficiently utilize water. These adaptations included the evolution of roots to absorb water and nutrients from the soil, leaves that require water for structural integrity and photosynthesis, and shoots to serve as functional links between roots and aerial organs. Although the size of vascular plants varies greatly from the tiny millimeter long duckweeds (Lemnaceae) to the giant 100-m-tall redwood trees (*Sequoia sempervirens*), functionally similar vascular systems evolved to interconnect the different plant organs. Such vascular systems allow rapid and controlled long-distance translocation of water in which photoassimilates, nutrients, and macromolecules are transported throughout the plant. This transport system consists of two types of tissues, the xylem and phloem (**Figure 1**). The xylem transports water and minerals from roots to the aerial organs, while the phloem translocates water and direct or indirect products of photosynthesis from source tissues to the sink region where they are utilized (root and shoot apices and developing tissues) and/or stored (seeds, storage roots, or fruits). The internal structure of the conducting cells within both xylem and phloem is optimized for low-resistance rapid transport. Low resistance is achieved in vessel elements of the xylem by cellular apoptosis resulting in total autophagy of the cellular contents. In contrast, selective autophagy of certain organelles occurs in the living sieve elements of the phloem. In addition to assimilate transport, the importance of the phloem tissue in redistributing water and mineral nutrients, as well as the long-distance translocation

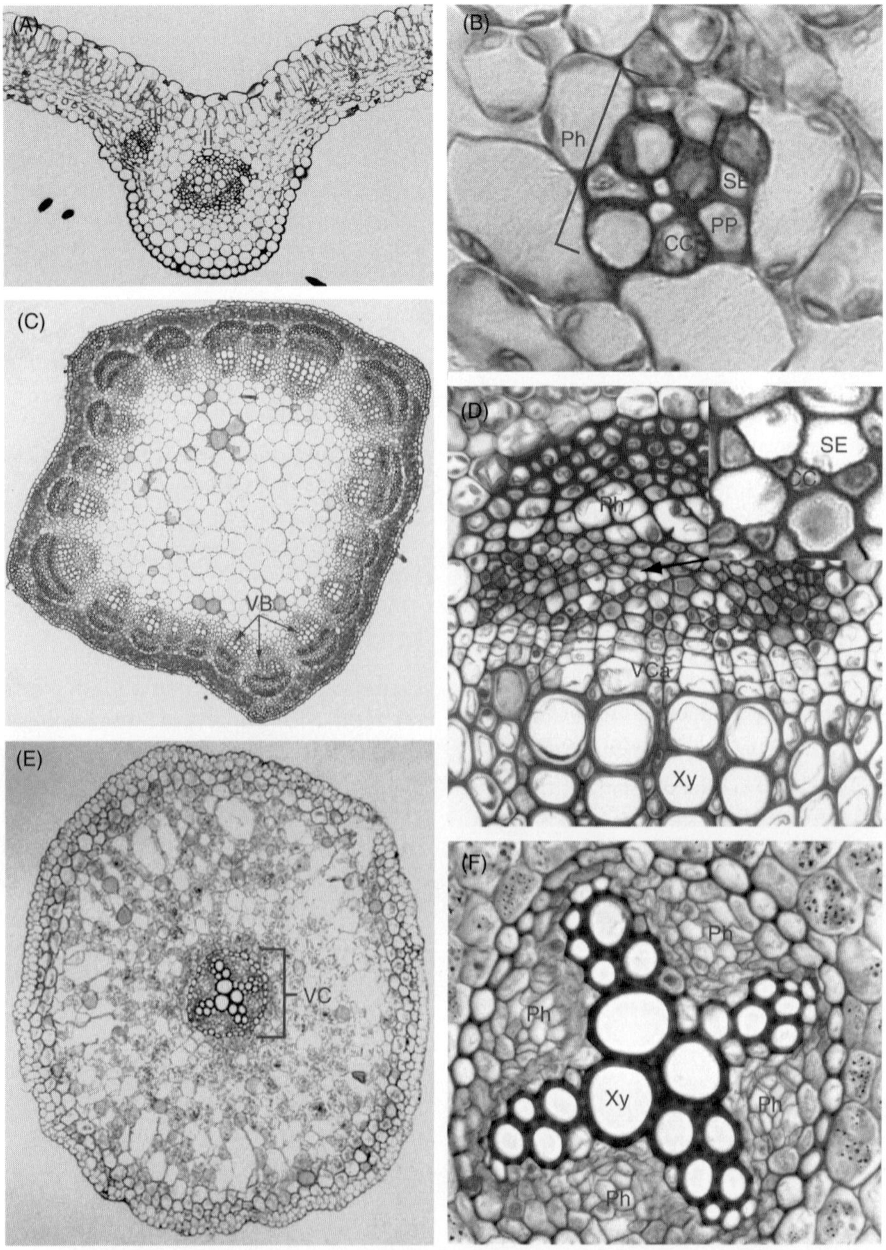

Figure 1 (see color plate 45) Light micrographs showing the generalized vascular structure in leaves, stems, and roots of many dicotyledonous plants. (A) Cross-section of a leaf showing II, III, IV, and V order veins (x40). (B) Collecting phloem in the IV order minor vein containing the transfer cell type of companion cells: note the cell wall ingrowths (x630). (C) Cross-section of a stem shows the vascular bundles and secondary development of phloem and xylem (x40). (D) Transport phloem in the vascular bundle of the stem contains large sieve elements and small companion cells (x400). (E) Cross-section of a root shows the vascular cylinder located in the center (x40). (F) Phloem and xylem alternate in the vascular cylinder (x400). CC, companion cell; Ph, phloem; PP, phloem parenchyma; SE, sieve element; VCa, vascular cambium; VB, vascular bundle; VC, vascular cylinder; Xy, xylem.

of physiological signals and macromolecules throughout the plant, is being recognized.

Structural Specialization for a Unique Transport Function

Phloem is a complex tissue composed of several cell types and is developmentally classified as either a primary or secondary tissue (**Figure 1**). Primary phloem differentiates from the procambial region of meristems and is functional in expanding organs. Secondary phloem originates from the vascular cambium, which occupies a lateral position in stems and roots, and can extend in the form of strips into leaves with secondary growth. Primary and secondary phloem contains similar cell types. Protophloem sieve elements are the first cells of vascular bundles to differentiate and become functional in young organs. Arranged in succession, these phloem elements differentiate at a distance of less than 1 mm from the apical meristem to provide the nutrition for continuing cell division and growth. From the beginning of seedling growth, phloem elongation keeps pace with the growth of the stems and organs and is guided by the orientation of procambial strands so that phloem extends throughout the plant body to its farthest ends. During secondary growth, the vascular cambia continuously adds phloem radially to the outside and xylem to the inside of the plant axis.

Sieve elements are the most highly specialized cells in the phloem due to their modified and metabolically restricted protoplasm and their connections with contiguous sieve elements through modified cell wall areas (sieve areas and sieve plates). Sieve plates, penetrated by large open sieve pores lined with plasma membrane, join individual sieve elements to form the sieve tube. To establish a low resistance pathway, terminally differentiated sieve elements undergo exceptional cytoplasmic reorganization to become a highly specialized living conducting system for the long-distance translocation of assimilates. Mature sieve elements are characteristically devoid of nuclei and ribosomes and are physiologically associated with companion cells for protein and metabolite exchange. Reorganization of the sieve element endomembrane system results from degeneration of the tonoplast (surrounding the vacuole) and dictyosomes (Golgi bodies), and structural changes in the endoplasmic reticulum forming the sieve element reticulum. The sieve element reticulum and the sieve element plasma membrane form a tightly coupled membrane complex held together by clamp-like structures that also anchor plastids and mitochondria. This complex delineates a thin layer of parietal cytoplasm that is separated from the transport stream located in the lumen of the sieve element.

An individual sieve element is physiologically associated with one or more companion cells (**Figure 1B, 1D, 1F**). Molecular trafficking occurs between the cells through specialized, branched plasmodesmata, thus, the two cell types function as a unit referred to as the sieve element–companion cell complex. Companion cells have large nuclei and nucleoli, numerous large mitochondria, abundant ribosomes, and a dense background cytoplasm, all of which are characteristic of metabolically active cells. Comprehensive physiological and molecular studies indicate at least three important roles for companion cells. They supply energy as ATP to the sieve elements, synthesize macromolecules that are required by the mature enucleate sieve elements, and are involved in assimilate loading in the source tissues and assimilate unloading in the sink tissues.

Phloem parenchyma cells adjacent to sieve elements or companion cells can also be involved in loading and unloading of solutes into and from the conducting elements. In some plant species, parenchyma cells further differentiate into transfer cells that are characterized by extensively invaginated plasma membrane and abundant mitochondria for intensive membrane transport of solutes. Phloem parenchyma cells also function as storage cells. Moreover, in both the primary and secondary phloem, it is common to find separated or clustered sclerenchyma cells (phloem fibers or sclereids), which serve to support and protect the phloem cells or exist as storage cells.

Phloem Translocation

In 1928, Mason and Maskell discovered that girdling the phloem of a tree blocked sugar transport through the sieve elements. This finding stimulated a period of extensive research to understand how sugars are translocated through the phloem. Beginning in the 1940s, a comprehensive view of phloem transport emerged, aided by the availability of radioactive isotopes to label sugars *in vivo*. The noninvasive technique of isotopically labeling sugars by supplying $^{14}CO_2$ to a source leaf proved invaluable to the initial understanding of phloem loading, long distance translocation, and phloem unloading. The use of radioactive tracers also allowed the rates of assimilate translocation to be accurately measured. The mass transfer rate, which is the quantity of material passing through sieve elements in a given cross-sectional area per unit time, ranges from 0.6 to $14.5\,g\,h^{-1}\,cm^{-2}$, whereas the velocity, which is the linear distance traveled per unit time, ranges from 0.3 to $1.5\,m\,h^{-1}$, a rate that is significantly greater than movement over long distances by simple diffusion.

Current theories explaining the mechanisms of assimilate translocation are based on the pressure-flow hypothesis first proposed by Ernst Münch in 1930. Münch designed a very simple model for the "mass" or "bulk" flow of assimilates based on the differential concentration of osmotically active solutes at separate locations in a continuous transport system. Water crossing a semipermeable membrane into a region of high solute concentration and out of a connected region of low solute concentration creates a turgor gradient in the pathway that passively transports solutes until equilibrium is achieved.

In plants, the pressure flow model of translocation links the xylem and phloem as an integrated system to maintain water movement through the phloem and prevent the system from coming to equilibrium (**Figure 2**). The generalized scheme for establishing and maintaining the turgor gradient in the phloem begins with active phloem loading of osmotically active solutes (primarily sugars) into the sieve elements against a concentration gradient. Plants use different strategies to load sugars against the concentration gradient and, in some woody plant species, phloem loading is an entirely passive process that does not involve high sugar concentrations in source tissues to generate the turgor gradient for mass flow. In most plants, the high concentration of sugars in the sieve element lowers the solute potential in the sieve elements causing a steep gradient of water potential across the plasma membrane. This stimulates the influx of water from the xylem vessel elements and high turgor develops in the sieve elements of source tissues. Unloading sugars from the sieve elements in sink tissue lowers the solute concentration in the sieve tube. The increase in solute potential results in water moving out of the sieve element and back into the xylem vessel elements, decreasing turgor in the sieve element. As a consequence, a high turgor gradient is established that serves to drive the mass flow of water containing assimilates and other solutes from source tissues to the sink tissues.

The rate of mass flow (Rm) can be expressed as:

$$\mathrm{Rm} = (AC)Lp(\psi_{\mathrm{p\ se\text{-}source}} - \psi_{\mathrm{p\ se\text{-}sink}}) \qquad [1]$$

where A is the cross-sectional area of the sieve elements, C is the assimilate concentration in the phloem, Lp is the hydraulic conductivity of the path, and ψ_{p} is the turgor potential of the sieve elements in the phloem of the source tissues ($\psi_{\mathrm{p\ se\text{-}source}}$) and sink tissues ($\psi_{\mathrm{p\ se\text{-}sink}}$). The turgor in the sieve elements at either location can be directly measured or calculated from the water potential (ψ_{w}), solute potential (ψ_{π}), and gravity potential (ψ_{g}):

$$\psi_{\mathrm{w}} = \psi_{\mathrm{p}} + \psi_{\pi} + \psi_{\mathrm{g}} \qquad [2]$$

The effects of gravity (ψ_{g}) on water movement at the cellular level is minimal and generally ignored. If the water potential of the sieve element sap ($\psi_{\mathrm{w\ se}}$) is in equilibrium with the water potential of the phloem apoplast ($\psi_{\mathrm{w\ apo}}$), and the apoplastic turgor potential ($\psi_{\mathrm{p\ apo}}$) is considered to be minimal, the turgor potential in the sieve elements will be:

$$\psi_{\mathrm{p\ se}} = \psi_{\pi\ \mathrm{apo}} - \psi_{\pi\ \mathrm{se}} \qquad [3]$$

At the source region in most plants, active accumulation of assimilates occurs across the cell walls surrounding a sieve element–companion cell complex. Accumulation is preceded by the prephloem transport of assimilates, starting in the photosynthetically active

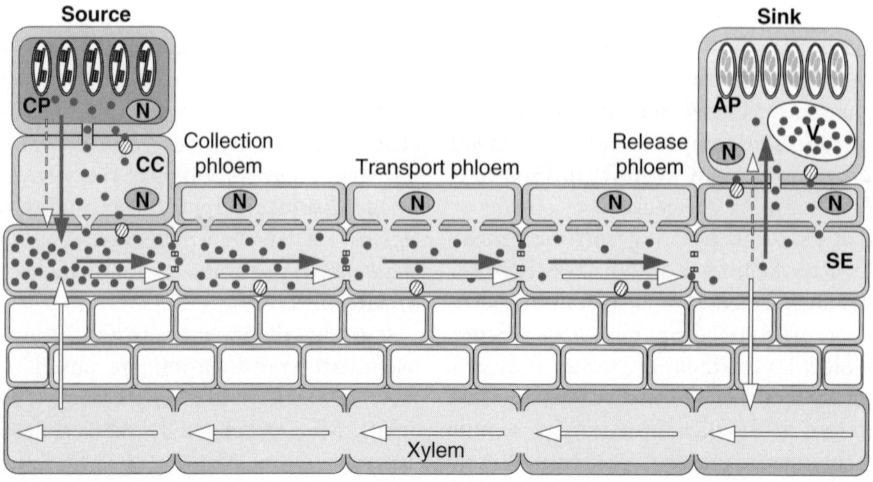

Figure 2 Sugars (small shaded dots) synthesized in photosynthetic cells of source tissues are loaded into the sieve elements (SE) of the collection phloem by either apoplastic or symplastic mechanisms (see **Figure 3A–C** for details). Sugar accumulation in the sieve element lowers the water potential causing water to enter from the xylem and surrounding cells, resulting in high turgor pressure and mass flow through the transport phloem. In the release phloem, sugars are unloaded from the sieve elements by either apoplastic or symplastic mechanisms (see **Figure 3D–F** for details). Sugar removal increases the water potential in the sieve element causing water to move into the xylem or to be utilized for growth. Large, hatched circles represent the locations of sugar transporters involved in apoplastic phloem loading in the collection phloem, sugar retrieval in the transport phloem, and apoplastic phloem unloading in the release phloem. CC, companion cell; N, nucleus; CP, chloroplasts; AP; amyloplasts; V, vacuole. Shaded arrows indicate the direction of assimilate flow and unshaded arrows indicate the direction of water flow.

mesophyll cells. Prephloem transport of assimilates continues through connecting mesophyll cells and eventually into the bundlesheath and phloem parenchyma cells that contact a sieve element–companion cell complex. Assimilates move into the sieve element–companion cell complex symplastically or are released into the apoplast to be actively loaded into the conducting elements. Accordingly, the concentration of solutes is increasing in the sieve elements while being reduced in the apoplast surrounding the conducting elements, generating a low solute potential in the sieve element ($\psi_{\pi\ se}$) and high solute potential in the apoplast ($\psi_{\pi\ apo}$). This creates a steep gradient of water potential across the plasma membrane driving the influx of water from the xylem and surrounding cells into the sieve element (see eqn [2]) and generates increased turgor potential ($\psi_{p\ se\text{-}source}$) in the sieve elements of source tissues (see eqn [3]).

In the sink region, removal of sugars from the sieve elements for utilization in growth or to be accumulated in storage compartments as osmotic solutes or polymers reduces the solute concentration and increases the water potential in the sieve elements. The increase in water potential causes water to move out of the sieve elements, reducing turgor in the sieve element ($\psi_{p\ se\text{-}sink}$). As a consequence, a high turgor gradient is established and drives the mass flow from the source tissues to the sink tissues. Phloem unloading occurs along the entire length of the axial phloem pathway as different sink tissues compete for assimilates. Rapid removal of assimilates into an individual sink decreases the turgor in the connecting sieve elements creating a stronger turgor gradient and, thus, increasing the mass flow of assimilates into this sink (see eqn [1]). In order to obtain sufficient photoassimilate, nutrients, and/or water, various cellular pathways and mechanisms evolved for efficient phloem unloading in different types of sinks, or at the different development stages in the same sink tissues. Transpiration rates are relatively low in most heterotrophic growth sinks and storage organs; thus, most of the water and nutrient demands for these tissues are met through the phloem pathway. However, the final destiny of water imported into sink tissues via the phloem varies. For example, phloem-imported water is used for cell growth and volume expansion in the elongation zone of plant roots, while in some storage sinks, where sugar is either utilized or stored, the water is recirculated to the xylem.

Strategies for Efficient Phloem Loading and Unloading

Within the plant, phloem is generally divided into three functional parts: the collection phloem (loading), the transport phloem (long distance movement), and the release phloem (unloading) (**Figure 2**). In general, sieve elements are smaller than the companion cells in the collection phloem (**Figure 1B**), whereas in the transport and the release phloem, the sieve elements are usually larger than the companion cells (**Figure 1D**), reflecting their respective roles. Other structural features in the phloem also reflect alternative mechanisms utilized in both the entry and exit of assimilates into and from the conducting cells.

Phloem Loading

The vascular system of a photosynthesizing leaf is composed of decreasing orders of vein sizes. The smallest or minor veins containing the assimilate-exporting sieve elements of the collection phloem have a close spatial relation to the photosynthetically active mesophyll cells (**Figure 1A, 1B**). Prephloem transport begins with the symplastic movement of assimilates by directed diffusion from the mesophyll cells to the bundle sheath and phloem parenchyma cells before entering the sieve element–companion cell complex. Companion cells of the minor veins are specialized for assimilate loading into the sieve elements for long-distance transport. Companion cell structure and extent of their symplastic connections with the surrounding parenchyma cells, as well as the configuration of minor veins, varies greatly among different species suggesting alternative strategies of phloem loading.

Apoplastic loading of sucrose In some plant families, the conducting elements are symplastically isolated due to the lack of functional plasmodesmata connecting the sieve element–companion cell complex to the surrounding cells (**Figure 3A**). Sucrose is released into the apoplast at the end of the prephloem pathway, requiring an energized mechanism to "actively transport" sugars across the plasma membrane barrier into the sieve element–companion cell complex to concentrate the sugars. These species often have transfer cell-type of companion cells adjacent to the bundle sheath and phloem parenchyma cells (**Figure 1B**). Transfer cell-type of companion cells with their finger-like cell wall projections enlarge the surface area of the plasma membrane, concomitantly increasing the number of membrane transporter proteins that are required for apoplastic loading. There is considerable genetic and molecular information elucidating the mechanism of the sucrose/proton cotransporter in phloem loading. Specific transporters have also been identified for active loading of amino acids and ions into phloem sieve elements. Moving sucrose from the apoplast

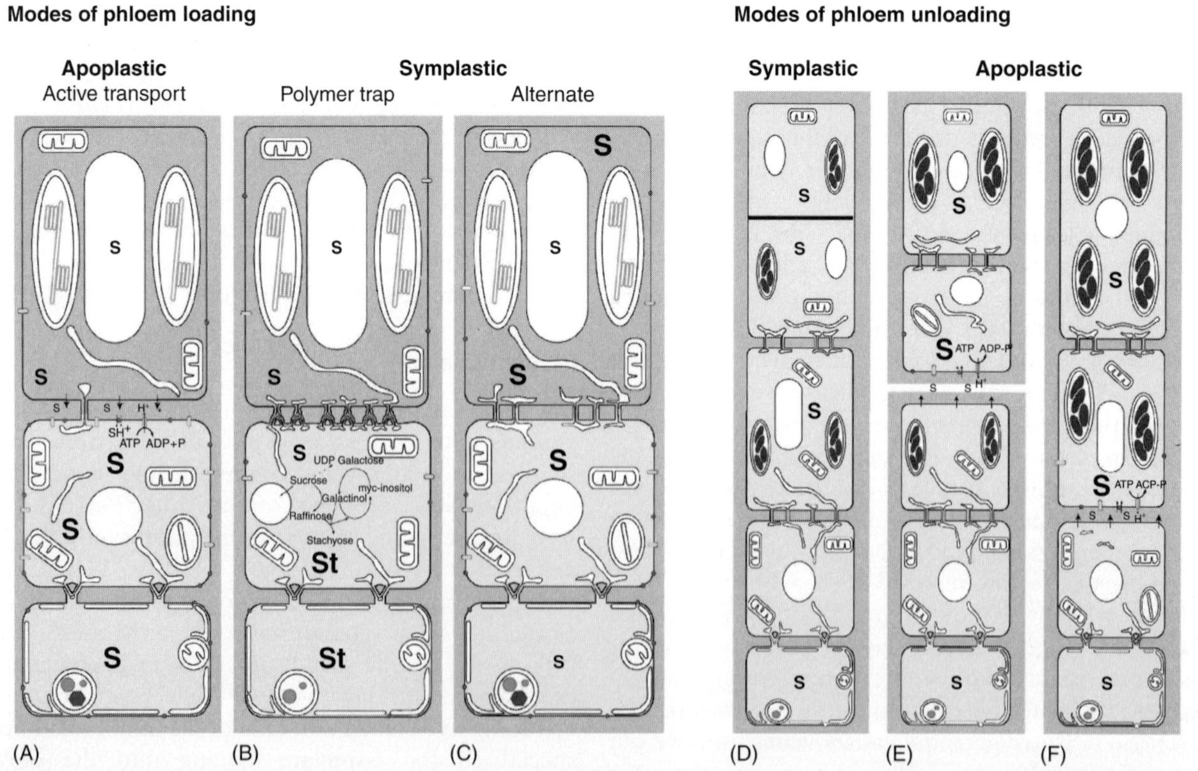

Figure 3 (A) Apoplastic loading by active transport occurs in plants where the conducting tissue is partially isolated from the surrounding cells. Sucrose (S) produced by photosynthesis in the mesophyll cells moves from cell to cell symplastically through the bundle sheath and phloem parenchyma, where it is released into the apoplast. Sucrose/proton cotransporters in the plasma membrane of the companion cells drive apoplastic loading. (B) Symplastic loading by "polymer-trapping" is suggested for plants translocating the sucrose-galactosyl polymers raffinose, stachyose, or verbascose. These larger oligosaccharides are synthesized from sucrose in intermediary-type companion cells and, due to their increased size, are trapped in the sieve element–companion cell complex. (C) Symplastic loading by alternate mechanisms involves sucrose moving down a concentration gradient from the mesophyll into the sieve element–companion cell complex of the phloem. This alternate mechanism of phloem loading appears to lack a sugar-concentrating step in the source tissue. (D) In most cases, sucrose unloading occurs down the concentration gradient and is symplastic. (E) Apoplastic unloading typically occurs across the physically separated maternal/filial interface in seeds, where assimilates leave the maternal symplast and are actively taken up in the filial tissue. (F) Unloading also involves an apoplastic step and active re-uptake (retrieval) of either sucrose or hexoses where sieve element–companion cell complexes are symplastically isolated. Reproduced with permission from Schulz A and Thompson GA (2002) Phloem structure and function. *Encyclopedia of Life Sciences* 14: 190–199. London: Nature Publishing Group.

into the companion cell is a thermodynamically unfavorable process that requires energy. The energy for apoplastic loading of sucrose is provided by plasma membrane-bound ATPases that hydrolyze ATP to pump protons from the cytoplasm into the apoplast. Apoplastic sucrose loading is inhibited by *p*-chloromercuribenzenesulfonic acid (PCMBS), a sulfhydryl group modifier that binds to the plasma membrane-located sucrose transporters. Sucrose accumulation in the apoplast following PCMBS treatment is diagnostic of plant species that utilize apoplastic phloem loading.

Symplastic loading of sugars Many plant species are insensitive to the inhibitory effects of PCMBS on phloem loading. Plants translocating raffinose or higher polymers of sucrose-galactosyl sugars (stachyose, verbascose) load assimilates symplastically into the collection phloem of mature leaves (**Figure 3B**). These plants also have structurally distinctive intermediary-type companion cells with numerous branched plasmodesmata linking them to bundle sheath cells, maintaining symplastic continuity with the prephloem pathway. The "polymer-trap hypothesis" provides an explanation of how sugars can accumulate to higher concentrations within the sieve elements than the surrounding cells of the prephloem pathway for osmotically driven translocation. Sucrose diffuses symplastically through the prephloem pathway into the intermediary cells. Within the intermediary cell, sucrose-galactosyl polymers (raffinose and stachyose) are synthesized from sucrose and galactose, resulting in sugar molecules that are too large to diffuse back through the narrow orifices of the plasmodesmata in the intermediary cells and, consequently, back into the cells of the prephloem pathway.

Some woody plant families use an "alternative" pathway to symplastically load sucrose without further modification through the prephloem pathway into the sieve element–companion cell complex (**Figure 3C**). In these taxa, the highest sucrose concentration is in the mesophyll cytoplasm rather than the sieve elements. Photoassimilates diffuse from mesophyll into the sieve elements along the concentration gradient, rather than against the concentration gradient observed in other species. Since the gradient extends from the assimilate-producing mesophyll cells to the sink tissues where sugar unloading has established the gradient, the term "phloem loading" does not apply.

Phloem Unloading

Phloem unloading also consists of two interconnected steps: assimilate unloading from sieve elements and the postsieve element movement of assimilates into the sink cells. Solutes move from sieve elements into sink tissues by symplastic, apoplastic, or combined cellular pathways, depending upon the tissue, its developmental stage, and the ultimate fate of the assimilates. In general, sugars are removed from the vicinity of the sieve elements to maintain the turgor gradient and the motive force for continued translocation. In growth sinks, such as root tips where assimilates are rapidly utilized, plasmodesmata interconnect sieve elements and the developing meristematic cells, allowing symplastic movement of solutes through layers of cells into the differentiating and growing root cells (**Figure 3D**).

In developing seed sinks, assimilates are utilized for cell division and growth in the early stages and accumulate in storage cells at the later stages (**Figure 3E**). In wheat (*Triticum* spp.) grains, the pathway and mechanism of phloem unloading changes and impacts the control of photoassimilate partitioning. The development of wheat grains can be divided into two phases, cell division and grain filling. The cell division phase is from the time of fertilization to the cessation of cell division within the endosperm when the cell numbers per grain are determined. Numerous plasmodesmata interconnect the sieve element–companion cell complexes to the chalaza cells and the chalaza cells to the nucellar projection cells (terminal cells of maternal tissue), suggesting a symplastic pathway. However, histochemical studies combined with fluorescent tracer experiments indicate that phloem unloading at this developmental stage is a combination of symplastic and apoplastic pathways. Assimilate supply from the photosynthesizing source leaf determines the extent of sink development (cell number and size in a grain). During the grain-filling phase, photoassimilates are transported and stored in the endosperm cells (**Figure 4**). An apoplastic barrier, composed of lignin, phenolics, and suberin, forms on the cell walls of chalazal cells (pigment strand cells) at the grain-filling stage. This apoplastic barrier forces the transported photosynthate through a symplastic pathway and blocks the leakage of assimilates released from the nucellar projection. Meanwhile, the nucellar projection cells progressively develop wall ingrowths, becoming transfer cells in which the increased plasma membrane surface area promotes apoplastic sugar unloading from the symplastic domain. Thus, phloem unloading into the developing grain is greatly promoted and serves as a strategy to realize the full sink potential under different environmental conditions. If the availability of assimilates derived directly from photosynthesis is limited, assimilates stored in the stem or old leaves remobilize to the grain.

In developing fruit sinks, such as tomato (*Lycopersicon esculentum*), shifts in phloem unloading from a symplastic to an apoplastic pathway is coupled with increasing activity of extracellular invertase (**Figure 3F**). During the early stages of tomato fruit development, dye tracer experiments showed that a membrane-impermeable dye moved symplastically from the sieve elements through the phloem parenchyma cells into the storage parenchyma cells. However, when the fruit reached the stage of cell expansion, symplastic movement only extended from sieve elements to the phloem parenchyma cells. In other experiments, ^{14}C-labeled sucrose imported from sieve elements was released from the phloem parenchyma cells into the apoplast where it was hydrolyzed by extracellular invertase into glucose and fructose. Hexose transporters located in the plasma membrane of storage parenchyma cells remove glucose and fructose from the apoplast and promote sucrose release from the phloem parenchyma cells. This is another strategy to lower the turgor pressure in the sieve element of the sink, promoting efficient phloem translocation.

Disrupting Vascular Continuity

Physical wounding that severs a portion or the entire vascular bundle disrupts vascular continuity and can severely alter water relations in the plant. Fortunately, damaged vascular bundles can be regenerated through the formation of wound-phloem and wound-xylem elements to rejoin the disconnected bundles. Within individual sieve elements, disruption

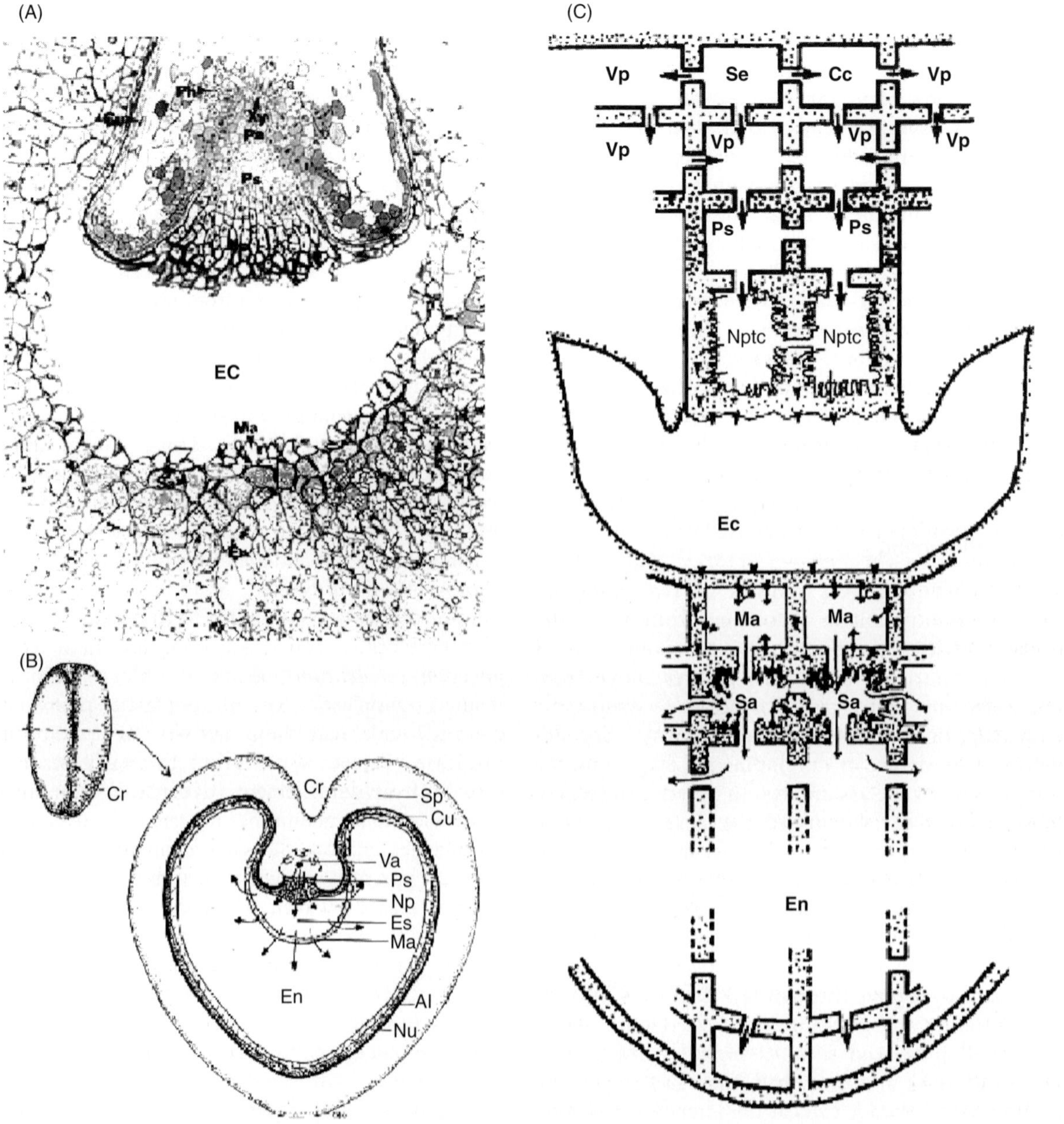

Figure 4 Cellular pathway of photosynthate transfer during the grain filling stage of wheat. (A) Light micrograph of a transverse section through the middle of the developing wheat grain. (B) Schematic representation of generalized assimilate movement and arrangement of tissues in the developing wheat grain. (C) Schematic of the detailed cellular pathway followed by assimilate transport in the developing grain during the grain-filling stage. Al, aleurone; Cr, crease; Cu, cuticular layer; Cc, companion cell; Ec, endosperm cavity; En, endosperm; Ma, modified aleurone layer; Np, nucellar projection; Nu, nucleus; Nptc, nucellar projection transfer cell; Pa, parenchyma cells; Ph, phloem; Ps, pigment strand; Sa, sub-aleurone cell; Se, sieve element; Sm, small endosperm; Sp, seed pericarp; Va, vascular bundle; Vp, vascular parenchyma; Xy, xylem; arrows show the direction of assimilate flow. Reproduced with permission from Wang HL, Patrick JW, and Offler CE (1993) A novel experimental system for studies of photosynthate transfer in the developing wheat grain. *Journal of Experimental Botany* 44: 1177–1184, Oxford University Press and Wang HL, Patrick JW, and Offler CE (1995) Assimilate delivery pathways in the developing wheat grain and their implications for control of carbon transport. Carbon partitioning and source–sink interactions in plants. *Current Topics in Plant Physiology* 13: 156–169, American Society of Plant Biologists.

of the phloem results in a sudden release of turgor, creating a hydraulic surge that initiates a series of reversible events to regain turgor and to reduce the loss of assimilates from disrupted sieve elements. The maintenance of turgor prevents the collapse of sieve elements and subsequent loss of that portion of the transport system until continuity is restored by the formation of wound phloem. In an intact vascular system, the structure of individual sieve elements is optimized to serve as a low-resistance pathway with sieve plates, containing unobstructed sieve pores of ~200–400 nm in diameter, joining individual elements in the sieve tube. Turgor can be rapidly re-established in damaged sieve tubes by the immediate accumulation of clotting proteins at sieve plates followed by the slower callose-mediated constriction of the sieve pores.

Depending upon the species, there appear to be alternative mechanisms that involve mostly physical interactions among pre-existing clotting proteins in the sieve elements. Classically, the phloem of most angiosperms contains proteinaceous structures, collectively called P-protein (phloem protein), that accumulate in differentiating and persist in translocating sieve elements. P-protein is initially deposited into ultrastructurally distinct polymorphous or crystalline bodies during sieve element differentiation. P-protein bodies either persist or, more often, disperse forming a network of filaments that appear to be immobilized through interactions with the appressed endomembrane system, allowing the assimilate stream to remain undisturbed during normal transport functions. Upon wounding, the filaments are released from their parietal position, accumulating at the downstream sieve plate and occluding the sieve pores. The polymerized aggregates and filaments in sieve elements of members of the Cucurbitaceae are composed of polymers of a 96-kDa phloem filament protein that appears to be covalently cross-linked with a 48-kDa dimeric phloem lectin. Grafting studies demonstrated that the subunits of both P-proteins are also freely mobile in the assimilate stream but the interrelationship between mobile P protein subunits and polymerized structures remains unresolved. A dynamic equilibrium between the polymerized and unpolymerized P-proteins could be essential for clotting the unusually large sieve pores (up to 1 μm in diameter) of cucurbits. Changes in the redox state of the phloem sap could be a mechanism to control gel-sol transitions between the two structural states and regulate the clotting function of these proteins.

Recent investigations using confocal laser scanning microscopy as a noninvasive technique to observe functioning sieve elements in broad bean (*Vicia faba:* Fabaceae) provides convincing evidence for the hypothesis that preformed structural proteins serve as a turgor-responsive clotting mechanism in sieve elements. Interestingly, members of the Fabaceae have multiple mechanisms, which might act in concert to control the flow of the assimilate stream in response to wounding. Like most angiosperms, the sieve elements in the phloem of broad bean contain filamentous P-protein and P plastids (plastids containing proteinaceous inclusions). An unusual feature of this taxon is the presence of elongated crystalloid P-protein bodies (typically one 10–30-μm crystal per sieve element), which are located within the sieve element adjacent to the sieve plates. Changes in turgor caused by the insertion of a microcapillary into a sieve element resulted in the formation of detached strands of filamentous P protein and bursting of the P plastids, releasing the protein contents into the assimilate stream where a large clot formed at the sieve plate. In addition, the crystalloid P-protein bodies undergo an extremely rapid and reversible transformation from the crystalline to an amorphous state, plugging the sieve plate in response to wound-induced changes in turgor. This structural change appears to require specific divalent cations (Ca^{+2}, Ba^{+2}, Sr^{+2} but not Mg^{+2}), which are either absent or highly regulated in the sieve element. Because the process is reversible, the crystalloid P-protein bodies appear to serve as an effective biological valve to regulate the assimilate flow in response to injury.

Sieve pore occlusion is completed by the deposition of wound callose, a β-1,3-glucan polysaccharide, between the plasma membrane and cell wall, which physically closes the protein-plugged sieve pores. In contrast to the immediate formation of protein clots, microscopically visible callose deposition occurs within minutes after wounding. Callose is synthesized by plasma membrane-localized callose synthase, which is activated by calcium fluxes to utilize substrate from the cytoplasm and deposit callose on the surface of the cell wall of the sieve pore, effectively constricting the plasma membrane of the pores. Thus, wound callose in sieve pores efficiently seals the severed sieve elements and is depolymerized following the formation of wound phloem to re-establish assimilate transport.

Other Substances that Flow Through the Phloem

The composition of phloem sap (assimilate stream) could provide additional details to understand the function and mechanism of phloem translocation and carbon–nitrogen partitioning throughout the plant. For most plant species, phloem sap contains ~15–25% dry matter as solutes with sucrose accounting for up to 90% of that amount. Other components of the phloem sap include free amino acids (6–25%), organic acids (such as citrate and malate), proteins, RNAs, hormones, and mineral nutrients (potassium, phosphorus, chloride, sulfur, magnesium, calcium, sodium, iron, zinc, copper, and boron).

Phloem plays a significant role in transporting mineral nutrients and nitrogenous compounds to the developing and storage sinks and in nutrient recycling. Mineral nutrients are generally acquired through roots and transported along the water flow in the xylem to the shoot, primarily to the source leaves. These xylem-imported nutrients then move from the xylem into the phloem and are subsequently translocated into sink organs. For example, nitrogen, in the form of asparagine, moves from the xylem to the phloem in minor veins of lupin (*Lupinus* spp.) leaves and is redistributed into other parts of the plant. In wheat, up to 20% of leaf iron and 70% of leaf zinc was remobilized to the developing grain. The transition mineral nutrients such as iron, zinc, copper, and manganese are probably chelated when transported in the phloem pathway. Four groups of molecules, including sugars, organic acids, small peptides, and proteins, are apparently involved in the translocation of metal nutrients in the phloem of different plant species. However, little is known about chelating compounds and the mechanisms involved in the phloem translocation of metal nutrients.

In addition to sugars and nutrients, a wide variety of molecules enter sieve elements and move with the photoassimilates in the osmotically driven translocation stream from source to sink tissues. Low molecular weight signaling molecules, such as hormones, are translocated in the phloem initiating physiologically amplified responses at a distance from their sites of synthesis. Peptide hormones, such as systemin, an 18 amino acid polypeptide inducer of proteinase inhibitor synthesis, appear to be transported throughout the plant in the phloem. Proteins and other macromolecules play significant roles in phloem function and long-distance signaling throughout the plant. The approximately 200 soluble proteins that are translocated in sieve elements are quickly being identified and will shortly provide a unified picture of their physiological roles in the phloem. Protein activation or modification systems and complete redox control systems appear to operate within the sieve element–companion cell complex in response to environmental cues. In addition to proteins, the long-distance transport of low levels of endogenous mRNA through the phloem appears to be a mechanism for gene-specific systemic acquired silencing.

The Continuing Journey to Understand Phloem Function

It is of fundamental importance to our overall comprehension of plant growth and productivity to understand the basic biochemical, physiological, and morphological events that occur in the phloem. The phloem has long been recognized as playing a central role in unifying the organs and tissues of vascular plants to distribute sugars, hormones, amino acids, and other organic compounds from the sites of synthesis or remobilization to the sites of use or storage. Modern agricultural practices have shown that the improvement of crop yield has been associated, not with an increase to total biomass production, but with a greater partitioning of available assimilates to the harvested organs. Comprehensive models are needed to understand the dynamic control of carbon–nitrogen partitioning and the ultimate effects on yield. An improved understanding of the translocation mechanisms for assimilates and nutrients in the phloem will also have an impact on the nutritional quality of food plants for humans and animals.

The living conducting elements of the phloem are well suited for a pivotal role in establishing and propagating long-distance signals to regulate plant metabolism and adaptation to environmental stresses. Although an increasing number of phloem-specific peptides, proteins, and RNAs are being identified, the physiological significance of these phloem-mobile macromolecules requires further investigation. Moreover, many plant viruses and viroids appear to have adopted the plant's endogenous machinery for long-distance transport to facilitate systemic infections. Restricting the capacity of viruses to enter and move in the phloem is a powerful strategy to minimize systemic infection and could provide a basis for developing virus-resistant plants. Another strategy to minimize viral infections is to control phloem-feeding insects that transmit the viruses from plant to plant. Phloem-feeding insects, such as aphids and whiteflies, have evolved a highly specialized method of feeding from sieve elements and are among the most destructive insects to worldwide agriculture. To reduce our reliance on chemical pesticides, control measures will increasingly depend on the exploitation of phloem-based genetic host resistance and an understanding of how features of the phloem can deter aphid infestation. Thus, understanding the mechanisms involved in phloem development, function, and regulation will provide novel and useful insights for crop improvement and optimizing agricultural management.

List of Technical Nomenclature

Apoplast The entity of extracellular space in the plant body, including the cell wall and

	the lumen of dead cells but excluding the air-filled intercellular spaces.
Companion cells	Specialized parenchyma cells in the phloem tissue of angiosperms that derive from an unequal division of the sieve element–companion cell mother cell and are functionally associated with sieve elements.
Phloem loading	The process where (photo)assimilates enter the sieve element–companion cell complex.
Phloem unloading	The process where (photo)assimilates exit from the sieve element–companion cell complex
Plasmodesmata (plural)	Cylindrical cell-to-cell connections, delimited by the plasma membrane, with a cytoplasmic and an endoplasmic reticulum pathway.
Sieve elements	Conducting cells of the phloem tissue that translocate photoassimilates from source to sink tissues.
Sieve plate	A part of the sieve element wall containing one or more clusters (= sieve areas) of sieve pores through which the protoplasts of adjacent sieve elements are interconnected.
Sieve pores	Highly specialized, dilated plasmodesmata between sieve elements.
Sink	Location in the plant where photoassimilates are utilized for growth or are stored.
Source	Location in the plant where photoassimilates are synthesized or are remobilized from storage organs.
Symplast	The entity of living protoplast in the plant body linked by plasmodesmata or sieve pores spanning the cell walls.
Transporter	A specific membrane protein responsible for transferring solutes across membranes.
Turgor	The pressure within the cell resulting from the movement of water into the cell.
Water potential	The algebraic sum of the solute potential and the pressure potential, or wall pressure; the potential energy of water.

See also: **Diseases**: Viral Diseases. **Nutrition**: Mineral Uptake. **Photosynthesis and Partitioning**: Sources and Sinks. **Postharvest Physiology**: Transport. **Seed Development**: Nutrient Loading of Seeds. **Water Relations of Plants**: Xylem.

Further Reading

Behnke H-D and Sjolund RD (eds) (1990) *Sieve Elements. Comparative Structure, Induction and Development.* Berlin: Springer-Verlag.

Daudet FA, Lacointe A, Gaudillere JP, and Cruiziat P (2002) Generalized munch coupling between sugar and water fluxes for modelling carbon allocation as affected by water status. *Journal of Theoretic Biology* 214: 481–498.

Knoblauch M, Peters WS, Ehlers K, and van Bel AJ (2001) Reversible calcium-regulated stopcocks in legume sieve tubes. *Plant Cell* 13: 1221–1230.

Martinelli LW and McCutchan JS (eds) (2000) Assimilate transport and partitioning: integration of structure, physiology and molecular Biology. *Australian Journal of Plant Physiology* 27.

Patrick JW (1997) Phloem unloading: sieve element unloading and post-sieve element transport. *Annual Review of Plant Physiology and Plant Molecular Biology* 48: 191–222.

Schulz A (1998) Phloem. Structure related to function. *Progress in Botany* 59: 429–475.

Thompson GA and Schulz A (1999) Macromolecular Trafficking in the Phloem. *Trends in Plant Science* 4: 354–360.

van Bel AJE and van Kesteren WJP (eds) (1999) *Plasmodesmata.* Heidelberg: Springer-Verlag.

van Bel AJ, Ehlers K, and Knoblauch M (2002) Sieve elements caught in the act. *Trends in Plant Sciences* 7: 126–132.

Wang HL, Patrick JW, and Offler CE (1995) Assimilate delivery pathways in the developing wheat grain and their implications for control of carbon transport. Carbon partitioning and source–sink interactions in plants. *Current Topics in Plant Physiology* 13: 156–169.

Stomata

M R McAinsh, Lancaster University, Lancaster, UK

Introduction

Stomata (singular stoma, from the Greek for "mouth") are pores on the surface of the leaves and other aerial parts of most higher plants that allow uptake of CO_2 for photosynthesis and the loss of water vapor from the transpiration stream. This gaseous exchange between the plant and the environment is regulated by changes in the size of the stomatal pore. These changes are brought about by alterations in the turgor of a pair of specialized cells, called guard cells, which surround the pore. The term stoma refers to the pore and the guard cell pair. The major function of stomata is to allow sufficient CO_2 to enter the leaf in order to maintain optimal photosynthesis under the prevailing environmental conditions whilst conserving as much water as possible. This requires guard cells to integrate information from a range of environmental signals (such as light intensity, temperature, CO_2 concentration, water availability, and humidity) when formulating the final size of the stomatal pore. Consequently, guard cells have been extensively studied as a model cell type in which to examine stimulus–response coupling in plants providing a detailed understanding of the mechanisms underlying the control of guard cell turgor and hence the size of the stomatal pore.

The Structure of Stomata and Mechanics of Stomatal Movements

The occurrence of stomata in plants can be traced back though the fossil records providing a comprehensive history of the stoma. For example, fossils of one of the earliest recorded vascular plants, *Cooksonia pertoni*, clearly show the presence of stomata. In extant plants, stomata can be found on the leaves of all vascular plants. They are also present in more primitive plants such as the mosses, liverworts, and hornworts.

In most herbaceous plants stomata are found on both the upper (adaxial) and lower (abaxial) surfaces of the leaves although there are usually fewer stomata on the upper leaf surface. These plants are referred to as amphistomatous. In other species, including most tree species, stomata are only present on the lower surface. Such plants are referred to as hypostomatous. In contrast, aquatic plants with floating leaves, such as water lilies, have stomata on the upper surface only and are referred to as epistomatous or hyperstomatous. In addition, some aquatic plants that grow in water and on land, such as the water fern *Marsilea*, have stomata on the upper surface of floating leaves but stomata on both surfaces of aerial leaves.

The number of stomata present on a leaf expressed on a unit area basis is referred to as the stomatal frequency or density. This can vary markedly within a single leaf, between plants and individuals of a species, and within a plant community. Stomatal frequency is also affected by environmental factors (including water availability, light intensity, temperature, and CO_2 concentration), leaf morphology, and the genetic background of the plant. Furthermore, stomatal frequency can be influenced by cell size; smaller guard cells are usually associated with higher stomatal frequencies. Therefore, stomatal numbers are often expressed on the basis of the total number of cells present in the leaf epidermis per unit area. This is called the stomatal index. Despite the problems associated with stomatal frequency measurements the total area of the leaf surface occupied by stomata (open to an aperture of 6 μm) has been estimated to range from 0.3% to 1.20% of the leaf surface area depending on the species.

There is considerable variation in the anatomy of stomata. It is possible, however, to distinguish two main types. One type has dumbbell-shaped guard cells with bulbous ends; the pore is a long slit between the two "handles" of the dumbbells (**Figure 1A**). In this type the guard cells are flanked by a pair of differentiated epidermal cells called subsidiary cells. The guard cells, subsidiary cells, and the pore are referred to as a stomatal complex. Stomata of this type are typical of grasses but are also found in a few other monocots. The other type has kidney bean-shaped guard cells that result in an elliptical stomatal pore (**Figure 1B**). These are found in dicot plants and nongrass monocots. Subsidiary cells are often absent in species with kidney bean-shaped guard cells, in which case the guard cells are surrounded by

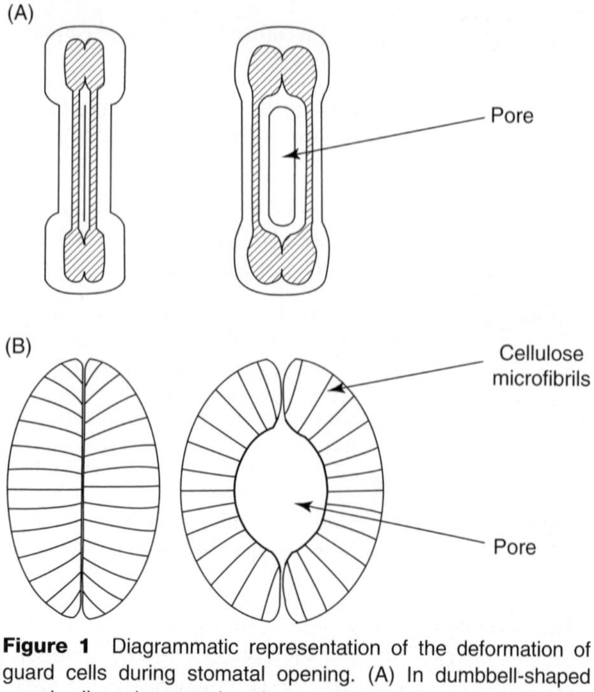

Figure 1 Diagrammatic representation of the deformation of guard cells during stomatal opening. (A) In dumbbell-shaped guard cells an increase in volume causes the thin-walled bulbous ends of the cells to swell pushing the middle portions of the cells apart and hence widening the slitlike pore. (B) In kidney bean-shaped guard cells the radial arrangement of cellulose microfibrils fanning out from the pore maintains the cell cross-sectional area constant whilst allowing some expansion of the cell walls in a lengthwise direction. Therefore, an increase in volume causes the cells to curve outward increasing the pore size.

ordinary epidermal cells, although the presence of subsidiary cells is not uncommon (**Figure 2**). Dumbbell-shaped and kidney bean-shaped guard cells are both relatively small cells that vary in size considerably between species. Guard cells vary from <10 to 80 μm in length and from a few micrometers to 50 μm in width although the latter is dictated largely by the size of the stomatal pore (see below).

Changes in the size of the stomatal pore result from alterations in guard cell turgor. The water potential of a guard cell (Ψ) is essentially equal to the turgor pressure (P) minus the osmotic pressure (Π):

$$\Psi = P - \Pi \ (\mathrm{Pa})$$

During stomatal opening the accumulation of solutes inside guard cells (see below) leads to an increase in osmotic pressure, which lowers the water potential causing the uptake of water and an increase in turgor pressure. Conversely, during stomatal closure the loss of solutes from guard cells leads to a decrease in the osmotic pressure, which raises the water potential causing the loss of water and a decrease in turgor pressure. This results in a reversible increase in guard cell volume of 40–100% depending on the species. The deformation of the guard cell walls caused by these volume changes is an essential feature of guard cell mechanics and stomatal movements.

The walls of guard cells are differentially thickened; there are regions that are up to 5 μm across compared to the 1–2 μm typical of epidermal cells. Kidney bean-shaped guard cells have very thick inner and outer (lateral) walls, a thin dorsal wall (the wall in contact with the subsidiary or epidermal cells), and a thickened ventral (pore) wall. They also exhibit an unusual pattern of the cellulose microfibrils that reinforce the cell wall. In ordinary cells with a cylindrical shape, the cellulose microfibrils are oriented transverse to the long axis of the cell. As a result, an increase in volume results in cell expansion in the direction of the long axis, since the cellulose reinforcement offers the least resistance at right angles to its orientation. In contrast, in kidney bean-shaped guard cells the cellulose microfibrils fan out radially from the pore (**Figure 1B**). This maintains the cross-sectional area of the guard cells constant whilst allowing some expansion of the guard cell walls in a lengthwise direction. Consequently, since the length of a stoma remains virtually constant during opening and closing an increase in guard cell volume causes cells to curve outward into the surrounding subsidiary and epidermal cells increasing the size of the stomatal pore. The pattern of wall thickening in dumbbell-shaped guard cells is very different to that of kidney bean-shaped cells. The walls of the bulbous ends are thin compared to the heavily thickened and relatively inflexible walls of the handles of the dumbbells in the middle portion of the guard cells (**Figure 1A**). An increase in cell volume causes the thin-walled bulbous ends of the guard cells to swell pushing the middle portions of the cells apart and hence widening the slitlike stomatal pore.

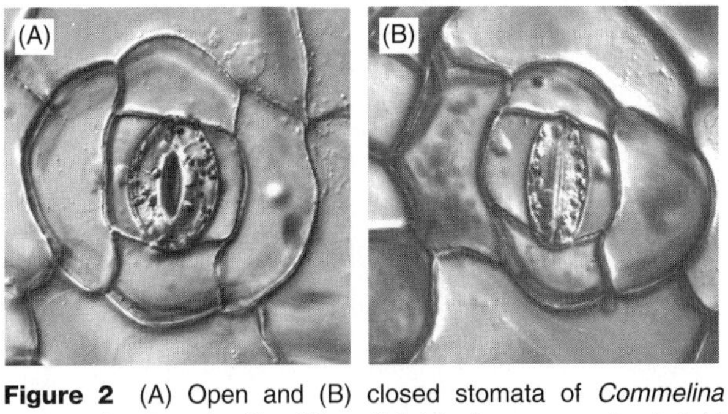

Figure 2 (A) Open and (B) closed stomata of *Commelina communis* viewed with differential interference contrast light microscopy. Each pore consists of two guard cells surrounding the pore and two flanking subsidiary cells. These are collectively referred to as a stomatal complex. Magnification × 400.

Ionic Relations of Stomatal Movements

The changes in guard cell osmotic pressure that underlie stomatal movements result from fluxes of anions (negative ions) and cations (positive ions) into or out of guard cells across the plasma membrane together with the synthesis or breakdown of osmotically active molecules within the cell. To avoid the potentially deleterious effects on cellular metabolism of high concentrations of salts in the cytoplasm ions move across the vacuolar membrane and are accumulated in the guard cell vacuole. This necessitates the generation of a compatible solute such as sucrose in the cytoplasm in order to increase the osmotic pressure of the cytoplasm. Potassium ions constitute the major cationic species involved in the control of stomatal movements and were first localized to guard cells by Macallum in 1905 using a histochemical test. However, it was not until almost half a century later that Japanese scientists reported a relationship between K^+ levels and stomatal aperture. The fluxes of K^+ during stomatal movements are accompanied by changes in the levels of anionic species present. This is essential in order to maintain the charge balance of the guard cell. The major anionic species involved are Cl^- and the inorganic ion malate^{2-}. During stomatal opening K^+, and to a lesser extent Cl^-, are taken up by guard cells and malate^{2-} is synthesized from starch. Conversely, during stomatal closure K^+ and Cl^- are lost from guard cells and malate^{2-} levels decrease.

Malate^{2-} may be released from guard cells, metabolized in the guard cell mitochondria, or channeled back into starch. In addition, although a significant proportion of the changes in osmotic pressure that occur during stomatal movements can be attributed to potassium salts, other osmotica may also play a role under certain conditions particularly when stomatal apertures are small or in the early stages of opening. The most likely candidate for this is sucrose.

The diffusion of ions across biological membranes occurs very slowly and could not account for the large and rapid fluxes of ions that take place during stomatal movements. Therefore, specific transport systems exist in both the plasma and vacuolar membranes of guard cells that provide a route for the movement of K^+ and its accompanying anions coupled to a favorable thermodynamic gradient. Importantly, the influx of ions into and the efflux of ions out of guard cells during stomatal movements occur through different sets of transport proteins.

Stomatal Opening

The generation of a favorable thermodynamic gradient for the uptake of ions during stomatal opening is an active process requiring the input of energy. Proton-pumping ATPases (H^+-ATPase) energize the guard cell plasma and vacuolar membranes establishing a proton motive force. This consists of a proton concentration (pH) gradient and an electrical gradient. Ions can either move passively through ion channels in response to the electrical component of the proton motive force or uptake can be driven by the pH gradient via a proton-coupled cotransport system.

K^+ uptake at the plasma membrane occurs passively through ion channels. The major plasma membrane K^+ uptake channel in guard cells is called the inward K^+ rectifier (K^+_{in}) due to the direction of the K^+ flux through the channel under physiological conditions (**Table 1**). This channel is activated by a shift in the plasma membrane potential (the electrical gradient across the membrane) to values more negative than -100 mV (hyperpolarization) when the net driving force for K^+ would typically be into the cell. Importantly, K^+_{in} is also inhibited by potential signaling intermediates (see below) including physiological increases in cytosolic free Ca^{2+} ($[Ca^{2+}]_{cyt}$), ranging from 100 nmol l^{-1} to 1.5 μmol l^{-1}, and cytoplasmic pH (pH_{cyt}).

The movement of Cl^- into guard cells of closed stomata is opposed by both the Cl^- concentration gradient (33 mmol l^{-1} in the apoplast, 50–120 mmol l^{-1} in the cytoplasm) and the plasma membrane potential. Therefore, it is likely that the accumulation of Cl^- by guard cells occurs via a cotransport system. Since stomatal opening is enhanced when the external pH is decreased this may take the form of a nH^+/Cl^- symporter although this remains to be confirmed.

Accumulation of K^+ by the guard cell vacuole cannot occur passively against an estimated vacuolar membrane potential with the cytoplasm -20 to -40 mV more negative than the vacuole. Therefore, K^+ uptake may be mediated by either a K^+/nH^+ antiport system or by a vacuolar membrane pyrophosphatase that is thought to pump H^+ and K^+ simultaneously into the vacuole. However, to date there is no evidence for a vacuolar membrane cation exchanger and only limited evidence for pyrophosphatase-dependent K^+ translocation at the guard cell vacuolar membrane. Similarly, although the predicted vacuolar membrane potential would be sufficient to drive a limited accumulation of anions passively via channels little is known about the anion channels that mediate the transport of Cl^- and malate^{2-} into the guard cell vacuole.

Table 1 Summary of plasma membrane ion channels involved in guard cell turgor regulation

Channel	*Permeability*	*Activation kinetics*	*Voltage dependency*	*Notes*
K^+_{in}	Highly selective for K^+	Rapid activation	Activated by hyperpolarization	Inhibited by increased $[Ca^{2+}]_{cyt}$ and pH_{cyt} Sensitive to $[K^+]_{ext}$ and $[K^+]_{cyt}$
K^+_{out}	Highly selective for K^+	Rapid activation	Activated by depolarization	Sensitive to pH_{cyt} Sensitive to $[K^+]_{ext}$
Stretch-activated channels	Anion channel: Cl^- Cation channel: K^+	Not determined	Weak or no voltage dependency	Activated by stretch
R-type anion channels	$NO_3^- > I^- > Br^- > Cl^- >$ malate^{2-}	Rapid activation	Strongly activated by depolarization	Activated by $[Ca^{2+}]_{cyt}$
S-type anion channels	$NO_3^- > I^- > Br^- > Cl^- >$ malate^{2-}	Slow activation	Weakly activated by depolarization	Activated by $[Ca^{2+}]_{cyt}$ Regulated by abscisic acid

Stomatal Closure

Three ion channels capable of supporting K^+ efflux from the vacuole to the cytoplasm during stomatal closure have been identified in the vacuolar membrane of guard cells. These are the fast-activating vacuolar (FV) channel, the vacuolar K^+ (VK) channel, and the slow vacuolar (SV) channel. The permeability to K^+, activation kinetics, and voltage dependency of these channels varies. They also exhibit differential sensitivity to $[Ca^{2+}]_{cyt}$ and pH_{cyt} (**Table 2**). It has been proposed that the VK and SV channels act in concert to mediate the prolonged efflux of K^+ from the guard cell vacuole during stomatal closure. Activation of the VK channel in response to stimulus-induced increases $[Ca^{2+}]_{cyt}$ (see below) would allow K^+ efflux into the cytoplasm shifting the vacuolar membrane potential to less negative potentials (depolarization) on the cytosolic side and hence activating the SV channel. The resultant efflux of Ca^{2+} (and K^+) from the vacuole through the SV channel would cause further depolarization of the vacuolar membrane potential and maintain the Ca^{2+} activation of both the VK and SV channels. This model requires a mechanism to allow the repolarization of the vacuolar membrane, possibly through continued activity of vacuolar H^+ pumping or via parallel anion efflux from the vacuole, in order to maintain the driving force for continued K^+ efflux. Although the efflux of anions from the guard cell vacuole may occur through specific ion channels these remain to be identified.

K^+ efflux from the cytoplasm requires depolarization of the plasma membrane potential. A shift in the membrane potential to values more positive than -100 mV results in inactivation of K^+_{in}. In addition, a second plasma membrane K^+ channel called the outward K^+ rectifier (K^+_{out}) due to the direction of the K^+ flux through the channel (**Table 1**) opens at membrane potentials more positive than the K^+ equilibrium potential allowing K^+ efflux from the cell. The potential at which K^+_{out} activates becomes more positive with increasing concentrations of external K^+. This allows the channel to open even in the face of the increases in apoplastic K^+ that occur during closure. Interestingly, K^+_{out} is insensitive to $[Ca^{2+}]_{cyt}$ but activated by both cytoplasmic and apoplastic alkalization. Stretch-activated K^+-permeable channels have also been identified in the plasma membrane of guard cells together with an ensemble of K^+ conductances collectively called leak conductances although the function of the latter remains to be determined.

Two main anion efflux channels have been identified in the guard cell plasma membrane. These are the rapid activating (R-type) and slow activating (S-type) anion channels (**Table 1**). R-type anion channels activate rapidly, spontaneously inactivate after only a few seconds, and are strongly activated by depolarization of the plasma membrane to between -10 and -80 mV. In contrast, S-type anion channels exhibit slow activation kinetics, show little time-dependent inactivation over a period of tens of seconds, and are only weakly activated by depolarization of the plasma membrane and remain open at the normal range of the resting membrane potentials in guard cells. In common with many ion channels identified in guard cells R- and S-type anion channels are both activated by increases in $[Ca^{2+}]_{cyt}$. The R-type anion channel is also pH sensitive.

Table 2 Summary of vacuolar membrane ion channels involved in guard cell turgor regulation

Channel	*Permeability*	*Activation kinetics*	*Voltage dependency*	*Notes*
Fast-activating vacuolar (FV)	P_K:P_{Cl} of 6:1 (P_{Cl} still significant)	Very fast activation	Large at positive potentials	Activated by $[Ca^{2+}]_{cyt}$ $<100\,nmol\,l^{-1}$ Activated by increasing pH_{cyt} Sensitive to vacuolar–cytosolic K^+ gradient
Vacuolar K^+(VK)	$K^+ > Rb^+ > NH_4^+ \gg Cs^+ \sim Na^+ \sim Li^+$	Instantaneously and time-dependent activation	Voltage-dependent	Activated by $[Ca^{2+}]_{cyt}$ (100–600 nmol l^{-1}) Inhibited by $[Ca^{2+}]_{cyt} > 600\,nmol\,l^{-1}$ Inhibited by increasing pH_{cyt}
Slow vacuolar (SV)	P_{Ca}:P_K of 3:1	Time-dependent activation	Open at cytosolic positive potentials	Activated by $[Ca^{2+}]_{cyt} > 600\,nmol\,l^{-1}$ Activated by increases in pH (in the range pH 6.0–8.0)

Stomatal Responses and Guard Cell Signal Transduction

Stomata are affected by a wide range of environmental variables including light (both quality and quantity), CO_2 concentration, temperature, water availability and humidity, and pollutants. In addition, factors such as the availability of mineral nutrients, fungal pathogens, and the biological clock regulating stomatal circadian rhythms can also exert a marked influence on stomatal behavior. Therefore guard cells are required to integrate signaling information from all of these sources when formulating the final size of the stomatal pore.

Ca^{2+} as an Intracellular Second Messenger

Ca^{2+} is a ubiquitous second messenger in both animals and plants. Increases in $[Ca^{2+}]_{cyt}$ (above a resting level of $100\,nmol\,l^{-1}$) play an important role in linking the perception of many extracellular stimuli to cellular responses. Stimulus-induced increases in $[Ca^{2+}]_{cyt}$ were first reported in higher plants in the early 1990s in guard cells in response to the plant hormone abscisic acid (ABA). The central role of Ca^{2+} in guard cell signal transduction is now firmly established. Increases in guard cell $[Ca^{2+}]_{cyt}$ have been observed in response to stimuli that cause stomatal opening and closure (**Table 3**) and many components of Ca^{2+}-based signal transduction pathways present in animal cells have also been identified in guard cells. In addition, a number of Ca^{2+}-permeable channels have been identified in guard cells which allow influx of Ca^{2+} across the plasma membrane or release of Ca^{2+} from intracellular stores (**Figure 3**). These include: in the plasma membrane, stretch-activated Ca^{2+}-selective channels, K^+_{in} (which exhibits a limited permeability to Ca^{2+}), nonselective voltage-gated Ca^{2+}-permeable channels, and H_2O_2- and hyperpolarization-activated Ca^{2+} channels; in the vacuolar membrane, SV channels, cyclic ADP-ribose (cADPR)-activated Ca^{2+}-permeable channels, and voltage-dependent Ca^{2+} channels (27 ps, 14 ps conductance). There is also good evidence for phospholipase C-inositol 1,4,5-trisphosphate (IP_3)-mediated Ca^{2+} mobilization. A novel lipid signaling molecule, sphingosine-1-phosphate, has also been implicated in the generation of increases in guard cell $[Ca^{2+}]_{cyt}$.

Table 3 Stimulus-induced increases in guard cell $[Ca^{2+}]_{cyt}$

Closuring stimuli	*Opening stimuli*
Abscisic acid (ABA)	Auxin
Elevated CO_2	Fusicoccin
Oxidative stress	
External Ca^{2+}	
Low external K^+	

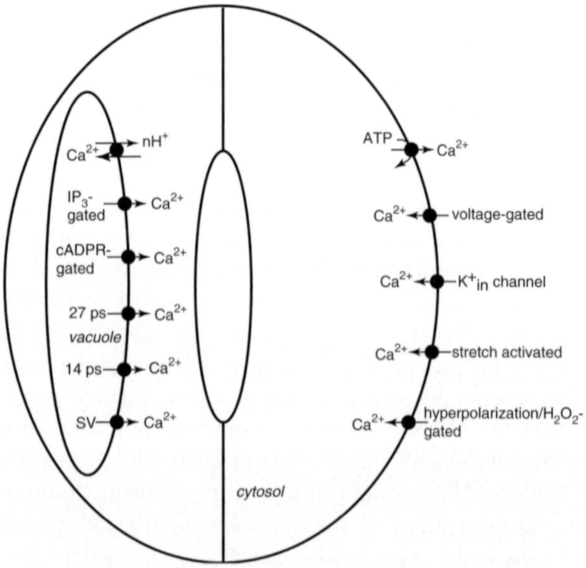

Figure 3 Ca^{2+}-permeable channels in the guard cell plasma and vacuolar membranes. Modified with permission from McAinsh MR, Brownlee C, and Hetherington AM (1997) Calcium ions as second messengers in guard cell signal transduction. *Physiologia Plantarum* 100: 16–29.

ABA Signal Transduction

Stomatal closure in response to the plant hormone ABA has been widely used to study the signaling pathways involved in the control of guard cell turgor, and hence stomatal aperture, at the cellular and molecular level. This has allowed models integrating the activities of ion channels in both the guard cell plasma and vacuolar membranes to be developed. Since the activity of many of the ion channels involved in guard cell turgor regulation are either directly or indirectly (through Ca^{2+}-dependent depolarization of the plasma membrane) regulated by $[Ca^{2+}]_{cyt}$ (see above) most models focus on $[Ca^{2+}]_{cyt}$ as a key signaling intermediate (**Figure 4**). However, this does not exclude the possibility of Ca^{2+}-independent signaling events. For example, growth temperature is known to affect profoundly the ability of ABA to induce a change in $[Ca^{2+}]_{cyt}$ without affecting ABA-induced turgor loss. In addition, pH_{cyt}, phospholipase D and phosphatidic acid, an ABA-activated and Ca^{2+}-independent serine-threonine protein kinase, and a farnesyltransferase have all been implicated in ABA signaling pathways. Taken together, this suggests that the activities of ion channels, which represent the final effectors in guard cell signal transduction, can be regulated by both Ca^{2+}-dependent and Ca^{2+}-independent signaling pathways.

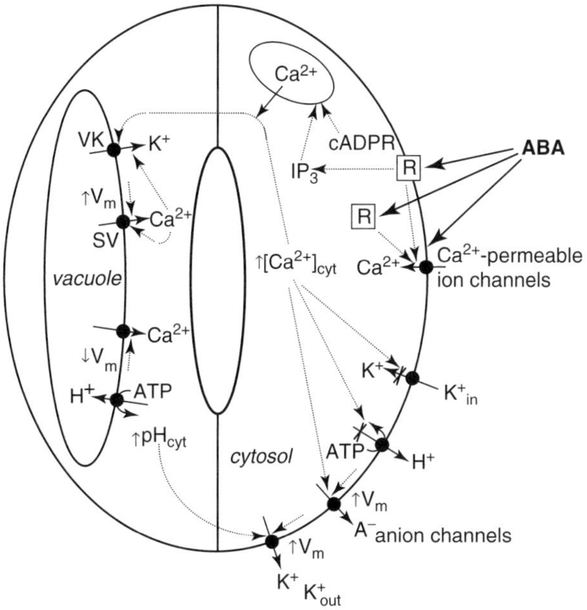

Figure 4 A simplified model of Ca^{2+}-dependent ABA-turgor signaling in guard cells. Perception of ABA by an extracellular or an intracellular receptor (R) results in an increase in $[Ca^{2+}]_{cyt}$ through Ca^{2+} influx and/or Ca^{2+} release from intracellular stores. This may occur via plasma membrane Ca^{2+}-permeable channels, cADPR- or IP_3-mediated Ca^{2+} release, or Ca^{2+}-induced Ca^{2+} release through the activation of VK channels causing depolarization of the vacuolar membrane potential and the activation of SV channels. In addition, activation of the vacuolar H^+-ATPase may increase pH_{cyt}, hyperpolarizing the vacuolar membrane potential, and activating vacuolar Ca^{2+} release channels. Abscisic acid (ABA)-induced increases in $[Ca^{2+}]_{cyt}$ inhibit both K^+_{in} channels and the plasma membrane H^+-ATPase, and promote the opening of plasma membrane anion channels. The resultant Ca^{2+}-dependent depolarization of the plasma membrane activates K^+_{out} channel. Efflux of cations and anions from the guard cell results in a reduction in guard cell turgor and hence closure of the stomatal pore. Modified with permission from Ward JM, Pei Z-M, and Schroeder JI (1995) Roles of ion channels in initiation of signal transduction in higher plants. *Plant Cell* 7: 833–844. Copyrighted by the American Society of Plant Biologists.

List of Technical Nomenclature

Activation	The conversion of a channel into a functional state.
Anion	A negatively charged ion.
Cation	A positively charged ion.
Depolarization	Change in membrane potential to a potential less negative than the resting potential; the opposite of hyperpolarization.
Hyperpolarization	Change in membrane potential to a potential more negative than the resting potential; the opposite of depolarization.
Membrane potential	The potential difference or voltage difference across a membrane.
Permeability	A characteristic of a channel pore that describes which ions are able to travel through the channel.
Second messenger	Intracellular metabolite or ion that couples an extracellular stimulus to a physiological response.
Selectivity	The affinity of a channel pore for different ions.
Signal transduction	The means whereby cells respond to a signal.
Stomatal frequency	Number of stomata present on a leaf expressed per unit area.
Stomatal index	Number of stomata present on a leaf expressed on the basis of the total number of cells present in the leaf epidermis per unit area.
Voltage dependence	Any channel characteristic that changes with membrane potential.

See also: **Photosynthesis and Partitioning**: C_4 Plants; CAM Plants; C_3 Plants. **Regulators of Growth**: Abscisic Acid; Circadian Rhythms. **Water Relations of Plants**: Basic Water Relations; Uptake, Loss and Control; Drought Stress.

Further Reading

Allen GJ and Sanders D (1997) Vacuolar ion channels of higher plants. *Advances in Botanical Research* 25: 217–252.

Assmann SM and Wang XQ (2001) From milliseconds to millions of years: guard cells and environmental responses. *Current Opinion in Plant Biology* 2: 421–428.

Blatt MR (2000) Cellular signalling and volume control in stomatal movements in plants. *Annual Review of Cell and Developmental Biology* 16: 221–241.

Hetherington AM (2001) Guard cell signalling. *Cell* 107: 711–714.

MacRobbie EAC (1998) Signal transduction and ion channels in guard cells. *Philosophical Transactions of the Royal Society of London B* 353: 1475–1488.

Mansfield TA, Hetherington AM, and Atkinson CJ (1990) Some current aspects of stomatal physiology. *Annual Review of Plant Physiology and Plant Molecular Biology* 41: 55–75.

McAinsh MR, Brownlee C, and Hetherington AM (1997) Calcium ions as second messengers in guard cell signal transduction. *Physiologia Plantarum* 100: 16–29.

Schroeder JI, Allen GJ, Hugouvieux V, Kwak JM, and Waner D (2001) Guard cell signal transduction. *Annual Review of Plant Physiology and Plant Molecular Biology* 52: 627–658.

Ward JM, Pei Z-M, and Schroeder JI (1995) Roles of ion channels in initiation of signal transduction in higher plants. *Plant Cell* 7: 833–844.
Willmer C and Fricker M (1996) *Stomata*. London: Chapman & Hall.

Uptake, Loss, and Control

M Bacon, University of Lancaster, Lancaster, UK

Water Uptake by Roots

The favorable water balance of a plant depends on the capacity of its roots to take up sufficient water. If water loss from the leaves is greater than uptake by the roots a water deficit may develop. Mild water deficits within a plant will have significant effects on growth and physiology, whereas severe deficit will lead to cell and, in the most severe of cases, plant death. Given this clear demonstration of how important water uptake is in sustaining plant life, it may seem surprising that relatively little is known about the processes that regulate root water uptake. The current understanding of water uptake into roots is presented and the highly sophisticated physiology involved is demonstrated below.

The Driving Forces for Uptake

The driving force for water uptake during the day in the vast majority of conditions is set up by the loss of water, via the stomata on the leaf surface. This loss of water generates a negative water pressure (often termed a "tension" or "potential") within the plant, emanating from the site of water loss. The cohesive and tensile forces of water mean that this negative pressure is propagated throughout the plant, generating a whole plant water potential gradient from the outermost cells of the root to the mesophyll cells of the substomatal cavity, underneath stomata. Some controversy does, however, surround the issue of the ability of a plant to generate such massive negative pressures (up to 10 MPa).

During the night, when transpiration is restricted by a reduction in uptake by stomatal closure, water uptake may still occur, at a reduced rate, via osmosis. The active uptake of nutrient ions into the cells and xylem vessels of the plant will generate an osmotic flow of water into the plant. This flow of water will then displace water within the xylem and generate a positive root pressure driven flow, in contrast to the negative pressure flow during the day. Positive root pressure will help a plant to recharge with water overnight. This is most visible in certain angiosperm species, which exhibit guttation, the phenomenon in which xylem sap is extruded onto the leaf surface in response to significant root pressures.

Pathways of Water Movement

On passing from soil to root, water will first enter the cell walls of the outermost root cells. The network of cell walls, termed the apoplast, offers one of three potential pathways of water movement from soil, through the root cortex and into the xylem vessel of the root, for mass transfer to the rest of the plant. The potential pathways of water movement through the root are summarized in **Figure 1**. In the apoplastic pathway, water, which is driven by the

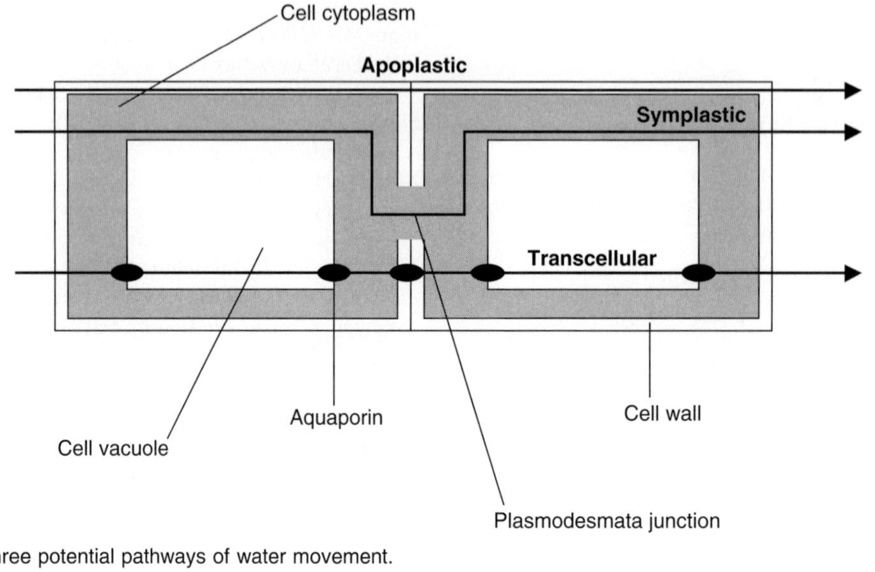

Figure 1 The three potential pathways of water movement.

negative xylem pressure, flows between cells in the cell wall matrix. The second pathway, the symplastic pathway, is a pathway in which water is absorbed by a cell and transported through cellular connections known as plasmodesmata, remaining in the cellular cytoplasm. The final possible pathway is the transcellular pathway, in which water moves across membranes (primarily the cell and vacuolar membranes) from cell to cell, due to the high permeability of the membranes. Unlike the other pathways, this pathway will not permit the passive movement of any significant quantities of nutrient ion.

Resistances to Water Transport

An endodermal layer (see **Figure 1**) acts as the primary barrier against direct apoplastic flow into the root xylem. The endodermis forms close to the root apex in a single layer of cells immediately adjacent to the stele, which lay down insoluble suberin and lignin in the walls to form a water impermeable apoplastic barrier, often visualized as Casparian bands. The functional significance of this structural feature is to force all water and dissolved nutrient ions and other solutes to enter cells before entering the xylem. In this way, the plant can exert active control over the ions and other solutes dissolved within the water entering the root.

A secondary barrier in mature regions of the roots of some species, growing under certain conditions, is the exodermis. Acting in a similar way to the Casparian band, but located in cortical cells close to the root epidermis, this layer acts as a barrier against water flow into the root. More importantly, however, its function is to prevent water flowing out of the root from these mature conducting regions of the root.

Which Pathway?

The long-held view in the literature is that the apoplastic pathway is the primary pathway of water movement under normal conditions. Water then moves symplastically on arriving at the Casparian band, before entering the xylem. The significance of the other pathways has long been a source of controversy. The discovery of water channels (aquaporins) to facilitate water transport across membranes has, however, refueled the discussion (see later).

The predominance of a particular pathway of water movement may change under different environmental conditions. Detailed measurements of water flow (hydraulic conductance) through roots have suggested that pathways other than the apoplastic pathway may have physiological significance under certain conditions. A "composite transport model" developed by Ernst Steudle over the last 10–15 years takes into account root structure and the potential pathways of water movement to explain why the hydraulic conductance of plant roots is variable and apparently under metabolic control. The model assumes that both the apoplastic and symplastic pathways have a functional role and interact with one another. The observed changes in root hydraulic conductance (or resistance to flow) is explained in terms of the driving forces behind water flow. During transpiring conditions, the water potential gradient between the water in the soil and the xylem vessel drives an apoplastic flow of water. Consequently, flow is determined by demand. However, under nontranspiring conditions, due to stomatal closure, the driving force for water uptake will switch from being driven by tension gradient to an osmotic gradient, generated by dissolved solutes as discussed earlier. As a result of this change in driving force, the pathway of water transport will move from predominantly apoplastic to predominantly symplastic flow.

The Apoplastic Bypass

The Casparian bands within the endodermis act to force water to move symplastically, and provide a single point control for access to the xylem stream. This control mechanism is well established in the literature, but has recently been challenged. The discovery that certain solutes appear to enter the xylem unchecked, and their quantities within the xylem stream are directly proportional to transpiration rate, suggests that a purely apoplastic pathway facilitated by some form of physiological or structural "bypass" of the symplastic pathway may exist. Indeed, the Casparian band may not be as effective as first thought in forcing water movement into a symplastic pathway. Various environmental conditions may alter the chemical composition and amounts of suberin and lignin within the developing Casparian bands, and compromise the impermeability of the layer to water.

Aquaporins

It is now evident that water flow across membranes is not passive, but is facilitated by specific water channels or "aquaporins" under metabolic control. Aquaporins are a highly conserved set of membrane integral proteins of around 30 kDa, containing six membrane domains that form the hub of the pore. The expression and activity of aquaporins is under metabolic and environmental control. Is regulation of aquaporin activity (i.e., membrane permeability) of physiological significance? Clearly, when water flow is predominantly symplastic, the answer is yes. As already discussed, symplastic flow will predominate under nontranspiring conditions and in older sections of roots that possess well suberized apoplasts (impermeable to water). The full significance of aquaporin expression and activity is still to be realized. Ongoing work investigating the role of plant growth regulators, environmental factors, and nutrients in controlling root hydraulic conductivity, all suggest a pivotal role for a metabolic regulation of water flow.

Control – The Sensing and Signaling of Soil Water Status

The control of water uptake is governed by sophisticated physiology, and can regulate uptake in response to demand. The long-held understanding that many such plant responses are purely biophysical is now called into question; a network of active control systems are in place to regulate water loss and govern uptake. An assessment of the current status of understanding of the regulation of water loss is appropriate in order to build a whole plant view of water uptake, loss, and control.

Water Loss

Although water loss occurs primarily through the leaf surface, it is important to remember that water can exit roots just as easily as it can enter, if the driving forces for water movement switch, as occurs under drought, when the soil water potential falls below that of the root. The "waterproofing" of roots as a result of increased suberin and lignin deposition has already been discussed as a means of preventing reverse flow of water and dissolved solutes via this pathway. Water will leave the leaf surface via the stomatal pores or the cuticle. While cuticular loss is measurable, it is generally low and relatively constant when compared to controlled water loss via the stomatal pores.

Stomatal water loss The majority of water loss from the leaf occurs via the stomata (pores on one or both sides of the leaf surface), the apertures of which are tightly regulated. When the stomatal pores are open, water evaporates from the cell walls of mesophyll into the intercellular spaces below the stomatal complex. This space is directly connected to the air outside the leaf. The driving force for evaporation is the water potential gradient between the apoplast of the mesophyll cell wall and the air. The level of connection between this air space and the atmosphere is regulated by the size of the stomatal aperture. Stomatal aperture size is regulated by changes in the turgor (water pressure) of the pair of guard cells that form the aperture (*see* **Water Relations of Plants**: Stomata). The rate of water loss realized is therefore highly dependent on the control of stomatal aperture. This variable resistance is the key determinant of water loss from the leaf, water uptake, and whole plant water status, as it determines the magnitude of the water potential gradients throughout the plant (see above). Stomatal and cuticular resistances to water loss are in series with a boundary layer resistance to form the three barriers to water loss at the leaf surface. Boundary layer resistance is a function of an unperturbed, water saturated layer of air close to the leaf surface. Such a resistance is highly dependent on the environment, particularly wind speed.

Cuticular water loss As already discussed, under most conditions cuticular water loss is minimal and is a function of the production and secretion of various waterproof waxes into and onto the leaf surface.

The Control of Water Loss

The control of water loss from plants is governed by a plant's ability to modulate stomatal aperture in response to the changing environment (*see* **Water Relations of Plants**: Stomata). Research over the past 20 years has established the existence of various root-to-shoot signaling mechanisms that allow the perceived water status of the soil to be communicated to the shoots.

Hydraulic Signaling

Until recently, the view that held sway in the literature was that changes in the water status of the soil would propagate a change in the water potential gradient in the entire plant from root cell to stomatal guard cell. With a change in water potential, guard cells would lose the turgor necessary to remain open, the aperture would close, and water loss would be minimized. While this may indeed still hold true, experiments in which the water status of a plant is maintained even though availability of soil water changes (by allowing the roots to dry the soil) do not support this original supposition. By splitting the root system of a plant, it is possible to supply a plant with sufficient water via one half of the root system. By withholding water from the other half of the root system, it is possible to create an experimental system in which roots are in contact with drying soil in the absence of any hydraulic changes. The original view in the literature suggests that the plant would not respond to this treatment. Surprisingly, even though the water status of the plant can be maintained, stomata close and the rate at which evaporative area is produced (leaf area) is inhibited. Removal of the roots in contact with drying soil removes this inhibition of stomatal opening and growth. This and other experiments have led plant physiologists to propose that the plant can perceive the water status of the soil and signal such information chemically in the water carrying xylem vessels to the shoot. Root-to-shoot signaling of soil water status using chemical signaling molecules would explain the frequently observed phenomenon of reduced rates of stomatal water loss in the absence of any change in plant water status. Of course, in nature, chemical and hydraulic regulation of growth occur together, with the relative importance of the two mechanisms varying between genotypes and with different soil conditions, and with the extent of the drought conditions. For many years now, the literature has concentrated on hydraulic limitations on plant functioning in drying soil, and so it is important to put the case for chemical regulation. It should not be forgotten, however, that the two systems depend upon each other for effective functioning, and that often distinctions between them are arbitrary and unrealistic.

But how can roots perceive the drying soil? This question remains unanswered. The root cells may perceive a change in the water content of the soil, the increased mechanical impedance of the soil, a change in the electrochemical properties of the soil, or a combination of these stimuli.

Chemical Signaling

Many chemical signals that move from roots to shoots in the transpiration stream may provide shoots with a measure of the access that roots have to soil water. The rate of stomatal water loss may then be modified as a function of the intensity of this signal. What are the candidate signals for stomatal pore and growth regulation?

Abscisic acid There is strong evidence from a number of different studies that the plant growth regulator abscisic acid (ABA) plays an important role in the regulation of stomatal behavior in plants. This hormone may also have a role in the root to shoot signaling process. However, it is not clear whether there is always enough ABA in the xylem stream to explain the stomatal and growth responses that are generated by changes in soil water availability. Recent experiments where xylem pH has been manipulated, and experiments with transgenic plants and mutants seem to support the case for a central role of ABA as a root-to-shoot signal that regulates stomatal aperture (*see* **Water Relations of Plants:** Stomata).

pH – an ABA dependent signal A well-watered plant contains a significant concentration of ABA in the transpiration stream, and delivery of this hormone into the leaf over the course of the day can be very substantial. If all of this ABA reached the site of action on the guard cells then the stomata would nearly always be fully closed. Clearly this is not the case, and leads us to suggest that most ABA delivered to the leaf in the xylem is intercepted in some way before it can reach the stomata. ABA is a weak acid and will thus distribute between the leaf compartments according to their pH. Relatively alkaline compartments will accumulate ABA, which then becomes trapped there. Under most circumstances, the symplast of the leaf is relatively alkaline when compared to the apoplast. In a well-watered plant, most xylem delivered ABA enters the symplast, and has no access to binding sites on the stomata. Small reductions in the water potentials of leaves can alkalinize the apoplast of the leaf, and the ABA trapped in the symplast may be released to gain access to sites of action around the plant.

A drought induced alkalinization of the xylem stream allows a greater proportion of a root sourced ABA signal to penetrate to sites of action within the leaf. Mild soil drying can increase xylem sap pH, and this change can be enough to limit stomatal conductance, even without the production of extra ABA. This is an important observation because it suggests that there is always enough ABA in the plant to close stomata, even when the plant is well supplied with water. The involvement of ABA in the pH responses of stomata using *flacca*, an ABA deficient mutant, confirms the central role of pH as an ABA dependent drought signal. This tomato (*Lycopersicon esculentum*) mutant possesses an inability to synthesize ABA in any great quantity. In control plants, an increase in pH within the xylem, and ultimately the apoplast surrounding the stomatal guard cells, causes stomata to close. Against a low ABA background, this pH response disappears, and is only restored when low concentrations of ABA are added to the xylem stream. These low concentrations of hormone are found in well-watered plants and do not, themselves, limit stomatal opening under well-watered conditions when the pH of the xylem is lower.

Very similar effects of ABA can be demonstrated in terms of limiting the rate of leaf production, and hence minimizing the rate of evaporative area production. ABA can limit leaf expansion at physiologically relevant concentrations in a pH dependent manner, identical to that seen with stomatal conductance.

Ethylene Recently, there has been considerable interest in the possibility that ethylene can regulate the growth of plants in response to changes in soil water availability. Evidence in the literature suggests that ethylene has little or no effect on stomatal behavior, but can act as an inhibitor of growth to limit evaporative area production, reducing water loss. While ethylene would act as a local signal to regulate growth, its precursor, 1-aminocyclopropane-1-carboxylic acid (ACC), may act as a root sourced signal transported in the xylem.

Interaction of signals Most work on root-to-shoot signaling has concentrated on the effects of single hormones. This has been necessary to generate enough precision to convince sceptics that hormonal signaling is indeed a reality. As more genetic tools and new technologies become available, the interactions between hormones that are undoubtedly important in this complex field can be investigated.

Application of Our Understanding

One of the most notable applications of our understanding of the control of water loss has emerged from the original experiments using plants with split root systems to generate chemical signals, while maintaining plant water status. The partial root zone drying technique (PRD), pioneered by Brian Loveys and colleagues in the Southern Australian vineyards, uses a variation of this experimental technique to reduce excessive transpirational water loss and vine vigor, while not compromising the plant's water status. By alternating the side of the root system that is allowed to dry at regular intervals, the viability of the whole of the root system is maintained (**Figure 2**). The chemical signals produced by the roots in contact with drying soil reduce transpiration and the rate of side shoot vigor, both common problems in wine grape production. Reduced vigor opens the canopy up to the sun and allows light to penetrate deep into the canopy and to the developing grapes. This, coupled with an apparent redirection of photosynthate away from excessive vegetative vigor toward reproductive growth, causes significant increases in grape quality with no yield penalty. Using the PRD technique dramatically enhances the subsequent quality of the wine produced. All of these benefits are joined by a significant reduction in the amount of water used to irrigate the crop, with consequent environmental benefits. Development of the technique is continuing in Australia and within Europe on a variety of fruiting crops, including tomatoes, olives (*Olea europaena*), corn (*Zea mays*; maize), cotton (*Gossypium* spp.), and *Citrus* spp.

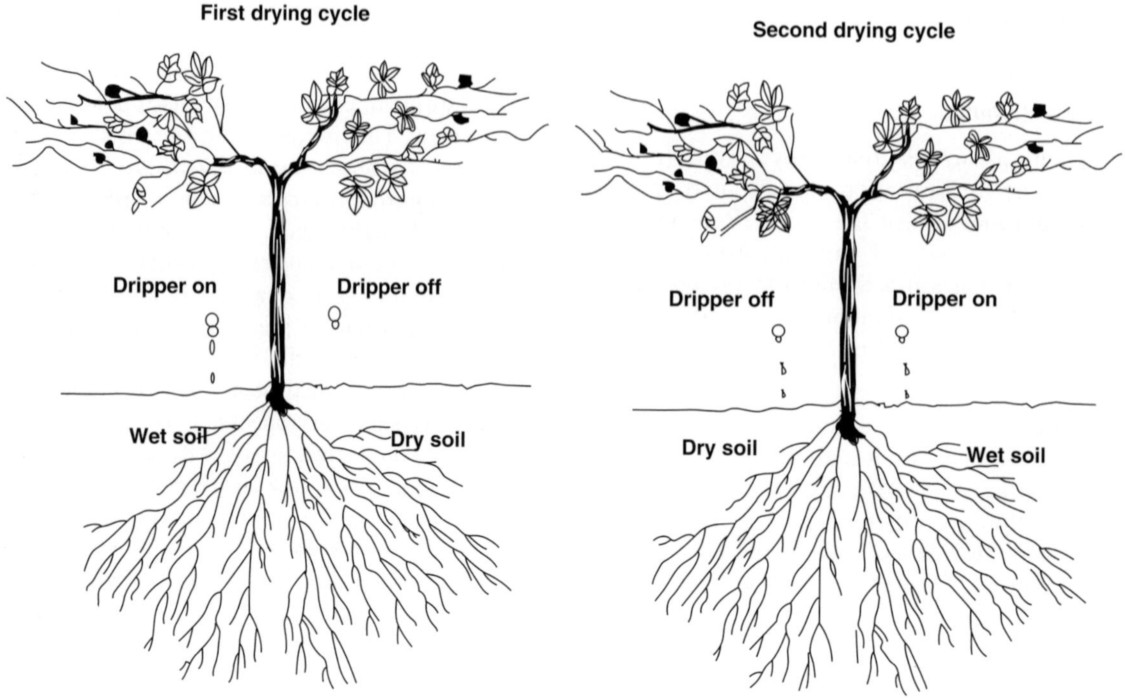

Figure 2 Schematic of the partial root zone (PRD) technique in which alternate halves of a root system are irrigated to sustain root drying without compromising water status.

List of Technical Nomenclature

ABA	Abscisic acid, one of the plant growth regulators.
Aquaporins	Proteins that mediate the transport of water across membranes.
Casparian strip (band)	The waterproofed band of apoplast close to the stele.
Cortex	The outer part of the root.
Epidermis	The outermost tissue layer.
Ethylene	A gaseous plant growth regulator.
Exodermis	Specialized tissue layer between root cortex and epidermis.
PRD	Partial root zone drying.
Stele	The inner part of the root.
Vacuolar	Pertaining to the cell vacuole.
Xylem	The water conducting vessels for long distance transport.

See also: **Regulators of Growth**: Abscisic Acid; Ethylene. **Water Relations of Plants**: Basic Water Relations; Drought Stress; Phloem; Xylem.

Further Reading

Davies WJ and Zhang J (1991) Root signals and the regulation of growth and development of plants in drying soil. *Annual Review of Plant Physiology and Plant Molecular Biology* 42: 55–76.

Davies WJ, Tardieu F, and Trejo C (1994) How do chemical signals work in plants that grow in drying soil? *Plant Physiology* 104: 309–314.

Davies WJ, Bacon MA, Thompson DS, Sobieh W, and Rodriguez L (2000) Regulation of leaf and fruit growth in plants growing in drying soil: exploitation of the plants' chemical signalling system and hydraulic architecture to increase the efficiency of water use in agriculture. *Journal of Experimental Botany* 51: 1617–1626.

Jackson MB (1993) Are plant hormones involved in root-to-shoot communication? *Advances in Botanical Research* 19: 103–187.

Steudle E (2000) Water uptake by plant roots: an integration of views. *Plant and Soil* 226: 45–56.

Steudle E (2000) Water uptake by roots: effects of water deficit. *Journal of Experimental Botany* 51: 1531–1542.

Stoll M, Loveys B, and Dry P (2000) Hormonal changes induced by partial rootzone drying of irrigated grapevine. *Journal of Experimental Botany* 51: 1627–1634.

Thompson DS, Wilkinson S, Bacon MA, and Davies WJ (1997) Multiple signals and mechanisms that regulate leaf growth and stomata behaviour during water deficit. *Physiologia Plantarum* 100: 303–313.

Tyerman SD, Bohnert HJ, Maurel C, Steudle E, and Smith JAC (1999) Plant aquaporins: their molecular biology, biophysics and significance for plant water relations. *Journal of Experimental Botany* 50: 1055–1071.

Wei C, Steudle E, and Tyree MT (1999) Water ascent in plants: do ongoing controversies have a sound basis? *Trends in Plant Science* 4: 372–375.

Wilkinson S (1999) pH as a stress signal. *Plant Growth Regulation* 29: 87–99.

Drought Stress

K Shinozaki, RIKEN Tsukuba Institute, Tsukuba, Japan

Physiological and Molecular Responses to Drought Stress

Water deficit has an adverse effect on plant growth. Therefore, drought stress is the most severe environmental stress for plant growth and crop production. Plants respond and adapt to drought stress in order to survive under water deficit conditions. Drought stress induces various biochemical and physiological responses in plants. Among them, stomata closure in response to water deficit is the best-known physiological plant response. Stomata closure decreases water loss from leaves. Accumulation of various small molecules, such as various sugars, sugar alcohols, and proline, is observed in various plants. These small molecules are thought to function in osmotic adjustment. A number of genes have been described that respond to drought stress at transcriptional level. Their gene products are thought to function in stress tolerance and response (**Figure 1**). Stress-inducible genes have been used to improve stress tolerance of plants by gene transfer. It is important to analyze functions of stress-inducible genes not only for further understanding of molecular mechanisms of stress tolerance and response of higher plants but also for improvement of stress tolerance of crops by gene manipulation.

The plant hormone abscisic acid (ABA) is produced under water-deficit conditions and plays important roles in response to and tolerance against dehydration. ABA functions in stomata closure in leaves. Most of the genes that have been studied to date are also induced by ABA. It appears that dehydration triggers the production of ABA, which, in turn, induces various genes. The existence has been described of ABA-independent as well as

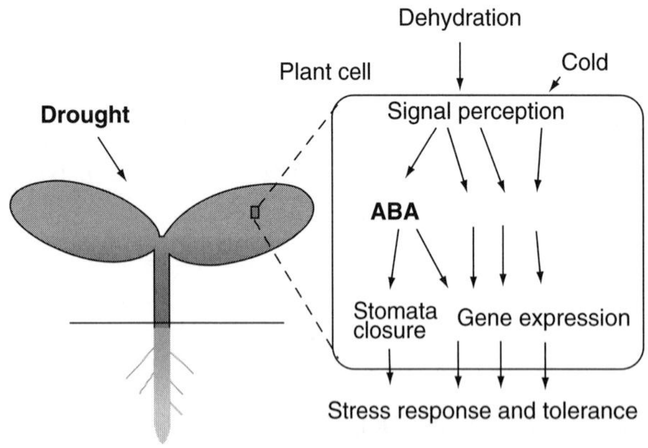

Figure 1 Physiological and molecular responses to drought stress in plant cells. Molecular and cellular responses to drought stress include perception of dehydration signal, signal transduction to cytoplasm and nucleus, gene expression, and responses and tolerance to drought stress. ABA, abscisic acid.

ABA-dependent signal transduction cascades between the initial signal of drought stress and the expression of specific genes. *Cis-* and *trans*-acting elements that function in ABA-independent and ABA-responsive gene expression by drought stress have been precisely analyzed. A variety of transcription factors have been shown to function in stress-responsive gene expression, which suggests complex regulatory mechanisms in the response to drought stress.

Complex mechanisms seem to be involved in gene expression and signal transduction in the response to drought stress. Details of molecular mechanisms of regulating plant genes to drought stress still remain to be solved concerning signal transduction cascades. Genetic and molecular approaches have been used to solve complex abiotic stress signal transduction pathways. In this article, molecular processes in drought stress response and tolerance are described. Functions of drought-inducible genes, regulation of their gene expression, and signal transduction pathways in drought stress response and tolerance are also discussed. The approaches to improve drought stress tolerance by gene transfer of stress-inducible genes are described, which is important for molecular breeding of drought stress tolerance.

Isolation of Drought-Inducible Genes

Many plant genes are induced by environmental stresses, such as drought, low temperature, and high salinity (**Figure 1**). These stress-inducible genes have been cloned using various methods, such as differential cloning, differential display, and so on. Using the polymerase chain reaction (PCR), many drought-inducible genes have been identified to encode proteins involved in transcription, signal transduction, and other regulatory processes. Using yeast transformation, plant genes that are involved in osmotic or high salinity stress tolerance have been cloned.

Recently, many stress-inducible genes have been identified using microarray technology. This technology has been developed extensively and now has become a powerful tool for global analysis of expression profiles of genes in response to various abiotic and biotic stresses and various treatments with hormones and chemicals. There are two major systems in microarray technology: (1) DNA chip-based technology arrays complementary DNA (cDNA) or oligonucleotides on glass slides at a density of up to 1000 genes cm^{-2}; (2) Affimetrix GenChip arrays gene-specific synthesized oligonucleotides. The arrayed sequences are hybridized simultaneously to a two-color fluorescently labeled cDNA probe pair from mRNA samples of plants that are stress-treated or untreated. This microarray technology allows large-scale comparative analysis of gene expression under various stress conditions.

Recently, expression profiles of genes under drought, cold and high-salinity stress conditions have been analyzed using microarrays and GenChip. Various kinds of transcripts have been reported to be increased after drought, cold, and high-salinity treatments. Many drought-inducible genes are also induced by salt stress and low temperature, which suggests the existence of similar mechanisms of stress responses. A strong relationship in the expression of these stress-responsive genes is observed. However, more significant cross-talk is observed between drought and high salinity responses than that between drought and cold responses. Several gene groups showing different expression profiles are identified by microarray analysis during stress treatment.

Function of Drought-Inducible Genes

Various genes respond to drought stress in various species, and functions of their gene products have been predicted from sequence homology with known proteins (**Figure 2**). Genes induced during drought stress conditions are thought to function not only in protecting cells from water deficit by the production of important metabolic proteins but also in the regulation of genes for signal transduction in the drought stress response. Thus, these gene products are classified into two groups. The first group includes proteins that probably function in stress tolerance; such as chaperones, late embryogenesis abundant (LEA) proteins, osmotin, antifreeze proteins, mRNA binding proteins, key enzymes for osmolytes biosynthesis, water channel proteins,

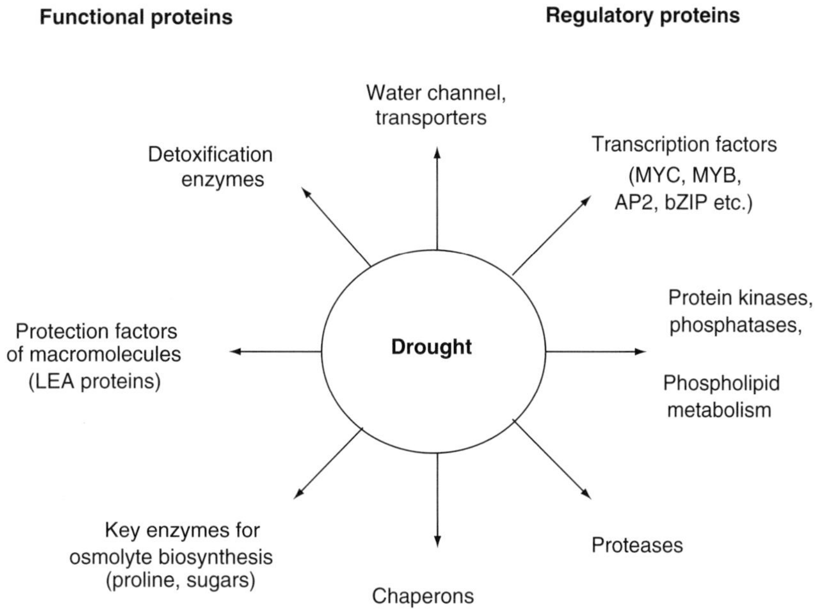

Figure 2 Functions of drought stress inducible genes in stress tolerance and response. Gene products are classified into two groups. The first group includes proteins that probably function in stress tolerance (functional proteins), and the second group contains protein factors involved in further regulation of signal transduction and gene expression that probably function in stress response (regulatory proteins).

sugar and proline transporters, detoxification enzymes, and various proteases. LEA proteins, chaperones, and mRNA-binding proteins have been analyzed biochemically and shown to be involved in protecting macromolecules like enzymes, lipids, and mRNAs from dehydration. Proline, glycine betaine, and sugars function as osmolytes and protect cells from dehydration. Key enzymes of several osmolytes have been cloned and analyzed biochemically. Water channel proteins, sugar transporters, and proline transporters are thought to function in transport of water, sugars, and proline, respectively, through plasma membranes and tonoplast to adjust osmotic pressure under stress conditions. Detoxification enzymes are involved in protection of cells from active oxygens. Proteases and nucleases are thought to be required for protein turnover and recycling of amino acids and nucleotides.

The second group contains protein factors involved in further regulation of signal transduction and gene expression that probably function in stress response, such as protein kinases, transcription factors, and enzymes in phospholipids metabolism (**Figure 2**). Genes for a variety of transcription factors that contain typical DNA-binding motifs have been demonstrated to be stress-inducible. These transcription factors function in further regulation of various functional genes under stress conditions. Various protein kinase genes are induced or upregulated by dehydration. Stress-inducible genes for protein phosphatases are also reported. These protein kinases and phosphatases may be involved in modification of functional proteins and regulatory proteins involved in stress signal transduction pathways. Phospholipids, such as inositol 1,4,5-triphosphate, diacylglycerol, and phosphatidic acid, are believed to be involved in the stress signaling process in plants. Enzymes involved in phospholipid metabolism of which genes are stress-inducible may play important roles in stress signaling as well. Existence of variety of drought-inducible genes suggests complex responses of plants to drought stress. Their gene products are involved in drought stress tolerance and stress responses.

Gene Expression in Response to Drought Stress

Dehydration triggers the production of ABA, which, in turn, not only causes stomata closure but also induces various genes. Many drought-inducible genes are also induced by exogenous ABA treatment, which indicates important roles of ABA in drought-responsive gene expression. However, many drought-inducible genes are induced in ABA-deficient (*aba*) and ABA-insensitive (*abi*) mutants, which indicates the existence of ABA-independent processes of gene expression. Northern blot analysis and promoter analysis have suggested the existence of ABA-independent as well as ABA-dependent transcriptional regulatory systems between the perception of drought-stress signal and the expression of specific genes. As shown in **Figure 3**, it is now hypothesized that at least four independent signal pathways function in the activation of stress-inducible genes under dehydration conditions: two are ABA-dependent (pathways I and II) and two are ABA-independent (pathways III and IV). There are at least two ABA-independent signaling pathways in response to cold stress (pathways IV and V). One of the ABA-dependent pathways requires protein biosynthesis (pathway I).

Most drought-inducible genes are upregulated by exogenous ABA treatment. The levels of endogenous ABA increase in many plants under drought and high salinity conditions. In one of the ABA-dependent pathways (pathway II), drought-inducible genes do not require protein biosynthesis for their expression. These dehydration-inducible genes contain potential ABA-responsive elements (ABREs; PyACGTGGC) in their promoter regions. ABRE functions as a *cis*-acting DNA element involved in ABA-regulated gene expression. ABRE was first identified in wheat (*Triticum aestivum*) *Em* and rice (*Oryza sativa*) *RAB* genes, and its DNA-binding protein EmBP1 was shown to encode a bZIP protein. Two cDNAs encoding DNA-binding proteins that specifically bind to the ABRE have been cloned by yeast one-hybrid screening and shown to contain the bZIP structure (AREB/ABF). These bZIP transcription factors regulate ABA-responsive gene expression downstream of *abi1/2* and *era1* mutations (pathway II).

Biosynthesis of novel protein factors is necessary for the expression of ABA-inducible genes in one of the two ABA-dependent pathways (pathway I). Promoters of drought-inducible genes contain conserved DNA-binding motifs, MYB and MYC. MYC and MYB transcription factors are thought to function in the regulation of ABA-inducible genes which respond to drought stress rather slowly after the production of ABA-inducible transcription factors (pathway I). Many stress- and ABA-inducible genes encoding various transcription factors have now been reported. These contain conserved DNA-binding motifs, such as zinc finger, homeobox, and WRKY motifs. These transcription factors are thought to function in the regulation of ABA inducible genes, which respond to drought stress rather slowly after the production of ABA-inducible transcription factors (pathway I).

ABA-Independent Gene Expression in Expression Response to Drought Stress

There are at least two ABA-independent regulatory systems in drought inducible gene expression. One of the ABA-independent pathways of drought stress response overlaps with that of cold stress response (pathway IV). A *cis*-acting element including

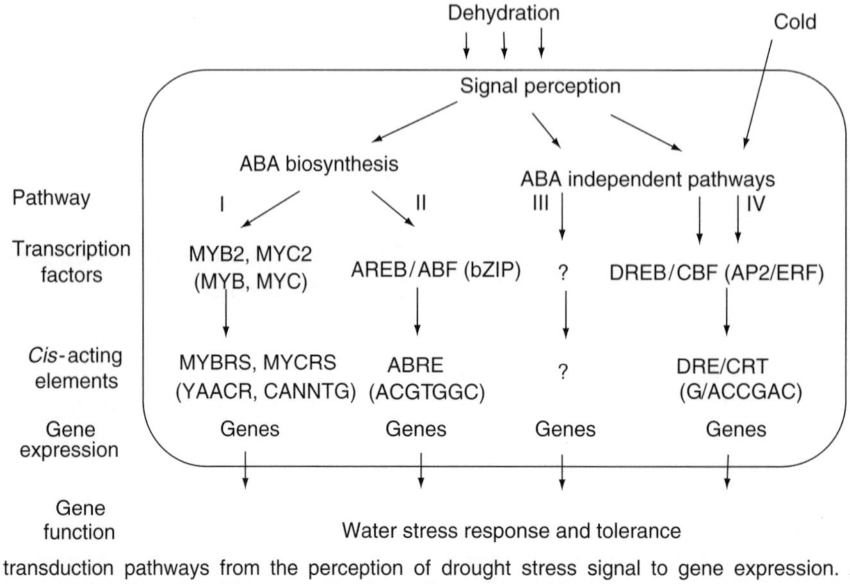

Figure 3 Signal transduction pathways from the perception of drought stress signal to gene expression. At least four signal transduction pathways exist (I–IV): two are ABA-dependent (I and II) and two are ABA-independent (III and IV). Protein biosynthesis is required in one of the ABA-dependent pathways (I). In another ABA-dependent pathway, ABRE functions as an ABA-responsive element and does not require protein biosynthesis (II). In one of the ABA-independent pathways, DRE/CRT is involved in the regulation of genes not only by drought and salt but also by cold stress (IV). Another ABA-independent pathway is controlled by drought and salt, but not by cold (III). ABF, ABRE binding factor; ABRE, ABA-responsive elements; AREB, ABRE binding protein; CRT; C-repeat; DRE, dehydration responsive element; DREB, DRE-binding protein; MYB, MYB transcription factor; MYBRS, MYB recognition site; MYC, MYC transcription factor; MYCRS, MYC recognition site.

A/GCCGAC, named the dehydration responsive element (DRE) and C-repeat (CRT), is essential for the regulation of the induction of many stress-inducible genes under drought, low-temperature, and high-salt stress conditions in an ABA-independent pathway. All the DRE/CRT binding proteins (DREBs and CBFs) contain a conserved DNA binding motif (AP2/ERF motif). Two independent families of DREB proteins, DREB1/CBF and DREB2, mainly function as *trans*-acting factors in two separate signal transduction pathways under low-temperature and dehydration conditions, respectively. Overproduction of the DREB1A and CBF1/DREB1B cDNAs driven by the 35S CaMV promoter in transgenic plants significantly improved stress tolerance to drought and freezing. Microarray analysis identified many genes with various functions as DREB/CBF target genes, which suggests over-expression of many genes function in drought stress tolerance.

In the ABA-independent pathways, there are several drought-inducible genes that do not respond to either cold or ABA treatment, suggesting that there is a specific pathway that functions in one of dehydration stress response (pathway III). These genes include *RD19* and *RD21* which encode different cysteine proteases, and *ERD1* which encodes a Clp protease regulatory subunit.

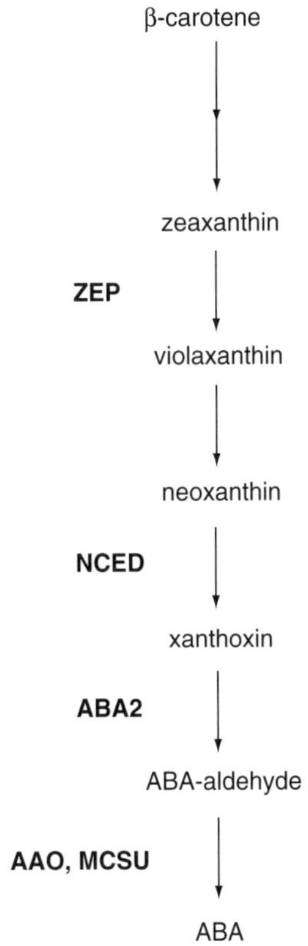

Figure 4 ABA biosynthesis pathway. Intermediates and enzymes involved in ABA biosynthesis are shown. ABA is synthesized from β-carotene via oxidative cleavage of neoxanthin. AAO, ABA aldehyde oxidase; ABA2, ABA deficient 2 (short chain alcohol dehydrogenase); MCSU, molybdenum cofactor sulfurase; NCED, 9-*cis*-epoxycarotenoid dioxygenese; ZEP, zeaxanthin epoxidase.

Drought Stress Response of ABA Biosynthesis

ABA is produced under drought-stress conditions *de novo*, which requires protein biosynthesis. As mentioned above, this process is important for drought-inducible gene expression. ABA is synthesized from β-carotene via zeaxanthin, neoxanthin, xanthoxin, and ABA-aldehyde (C40 pathway), and the step of conversion of neoxanthin to xanthoin is rate-limiting in the ABA biosynthesis under drought stress (**Figure 4**). Genetic evidence supports the C40 pathway, and biochemical studies suggest that the cleavage of neoxanthin is a key regulatory step in ABA biosynthesis.

Many ABA-deficient mutants that do not produce ABA have been isolated in various plants. Based on genetic analysis of ABA-deficient mutants, most of the enzymes involved in ABA biosynthesis have been identified, such as zeaxanthin epoxidase (ZEP; ABA1/LOS6 in *Arabidopsis*, ABA2 in tobacco (*Nicotiana tabacum*)), 9-*cis*-epoxycarotenoid dioxygenase (NCED; VP14 in corn (*Zea mays*; maize)), ABA aldehyde oxidase (AAO in *Arabidopsis*), molybdenum cofactor sulfurase (MCSU; ABA3/LOS5 in *Arabidopsis*). In various plants, NCED genes are significantly induced by drought stress but not by cold stress, which suggests that ABA mainly functions in osmotic stress responses. Transgenic plants that overexpress NCEDs accumulate ABA and are tolerant to drought stress. This also indicated a key role of NCED in ABA biosynthesis.

Genetic Analysis of Signal Transduction in Response to Dehydration

Genetic analysis of *Arabidopsis* mutants with the *RD29A* promoter::luciferase transgene suggests complex signaling pathways in drought, salt, and cold stress responses. There are various mutants that greatly expressed *RD29A* or repressed it in response to dehydration, high salinity, cold and ABA. It has been proposed that stress-signaling pathways, including ABA-independent and ABA-dependent pathways, are not completely independent. Compared with wild-type *RD29A-LUC* plants, mutants exhibited either a constitutive (*cos*), high (*hos*), or low (*los*) level of *RD29A-LUC* expression in response to various stress or ABA treatments. The characterization and cloning of some of the mutations have provided new insight into the mechanisms of stress and ABA signal transduction. The loci of the mutations, such as *fiery1* (*fry1*), *hos1*, *los1*, *los2*, *los5/aba3*, *los6/aba1*, and *sad1*, have been identified and their genes have been cloned by Zhu's group.

Signal Perception and Signal Transduction in Drought Stress Response

Signal transduction pathways from the sensing of dehydration signal or osmotic change to the expression of various genes, and the signaling molecules that function in stress signaling, have not been extensively studied in plants. Signal transduction pathways in drought stress response have been studied based on the knowledge in yeast and animal systems as shown in **Figure 5**. Two component systems function in sensing osmotic stress in bacteria and yeast. Plants as well as cyanobacteria contain many genes encoding sensor histidine kinases and response regulator homologs, which suggests the involvement of similar osmosensing mechanisms in higher plants. One of the histidine kinases, ATHK1, is shown to function as osmosensor in *Arabidopsis*. Other sensing mechanisms may function during drought stress responses, such as mechanical sensors of cytoskeletons and sensors of superoxides produced by stress. Stomata closure is well characterized as a model system in the responses of plant cells to dehydration stress and ABA treatment. During stomata closure, the level of cytoplasmic calcium is increased, which suggests that calcium functions as a second messenger in the osmotic stress response. Phospholipids play important roles in calcium signaling. However, the roles of calcium and phospholipids in osmotic stress signaling are still unelucidated.

ABA plays important roles in drought stress responses. ABA is involved in not only stomata closure but also induction of many genes. Several mutants in ABA signaling have been identified and their genes encode protein phosphatases (ABI1/2), farnesyl transferase (ERA1), protein kinases (OST1/SRK2E), and several proteins involved in mRNA processing (ABH1, SAD1). These suggest that protein modification and mRNA processing are involved in ABA signaling. However, various signaling molecules seem to be involved in ABA signaling, such as phospholipids and cyclic ADP ribose. MAP kinase cascades and calcium dependent protein kinase are suggested to be involved in drought stress response and ABA signaling. Complex signaling cascades are thought to function in molecular responses to drought stress. Molecular analysis of

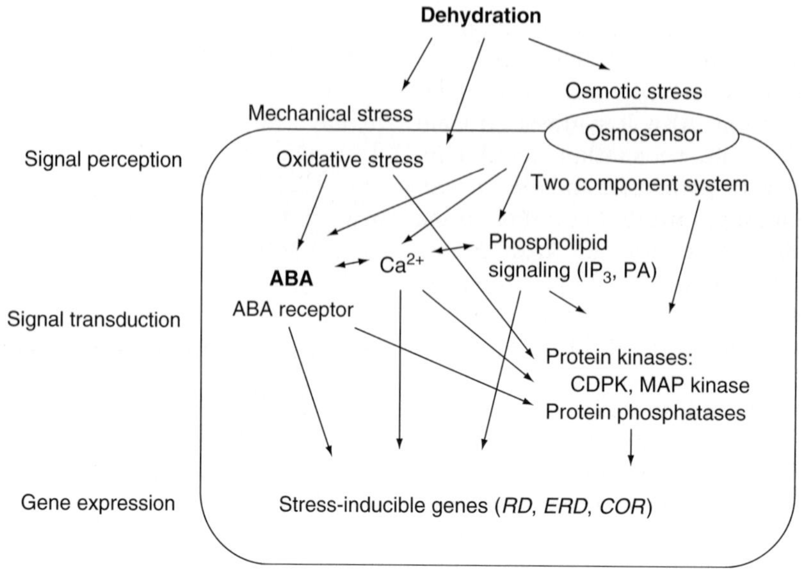

Figure 5 Second messengers and protein factors involved in the signal perception and the signal transduction in drought stress response. Two-component histidine kinase is thought to function as an osmosensor in plants. Calcium and phospholipids are most probable cellular second messengers of drought stress signal. Phosphorylation process functions in water stress and ABA signal transduction pathways. ABA plays important roles in the regulation of gene expression as well as physiological responses during water stress. CDPK, calcium dependent protein kinase; MAP, mitogen activated protein kinase.

the signaling process is mainly in progress based on genetics and gene cloning.

Improvement of Stress Tolerance Using Gene Transfer

Recently, several different approaches were attempted to improve stress tolerance of plants by gene transfer of stress-inducible genes. Genes for key enzymes involved in osmolyte biosynthesis and/or detoxification have been used for the improvement of stress tolerance in transgenic plants. Used genes for transformation were *Escherichia coli* mannitol 1-phosphate dehydrogenase for mannitol, Δ1-pyrroline-5-carboxylate synthetase and proline dehydrogenase for proline, *Arthrobacter globiformis* choline dehydrogenase for glycine betaine, and so on.

LEA proteins and detoxification enzymes were overexpressed in transgenic plants to produce stress tolerant phenotype of the plants, which indicates that their gene products really function in stress tolerance. In all these experiments, a single gene for a protective protein or an enzyme was overexpressed under the control of the CaMV 35S constitutive promoter in transgenic plants although a number of genes have been shown to function in environmental stress tolerance and response. In the next step, it is important to manipulate several genes to achieve strong stress tolerance for the application of this technology in the development of stress tolerant transgenic crops.

Overproduction of genes for stress-induced transcription factors in transgenic plants activated the expression of many target genes involved in stress tolerance under unstressed normal conditions and significantly improved stress tolerance to drought and freezing. Overproduction of the DREB1A and CBF1/DREB1B cDNAs driven by the 35S CaMV promoter in transgenic plants significantly improved stress tolerance to drought and freezing. However, the DREB1A transgenic plants revealed severe growth retardation under normal growth conditions. The DREB1A cDNA driven by the stress-inducible *RD29A* promoter was expressed at low level under unstressed control condition and strongly induced by dehydration, salt, and cold stresses. The *RD29A* promoter minimized negative effects on growth of plants, whereas the 35S CaMV promoter caused severe growth retardation under normal growth conditions. Moreover, this stress-inducible promoter enhanced tolerance to drought, salt, and freezing at higher levels than that of the 35S CaMV promoter. These results suggest that regulatory genes involved in stress response are important for the improvement of stress tolerance by gene transfer.

Perspectives

The total nucleotide sequences of the *Arabidopsis* and rice genomes were determined in 2000 and 2002, respectively. All the stress-inducible genes have been identified by systematic analysis of gene expression using microarray. However, signal transduction pathways have not been fully understood yet. In the next decade, it will be important to develop novel methods to analyze complex networks of stress responses of higher plants. Reverse genetic approaches as well as forward genetics become more important for understanding complex signaling processes in environmental stress responses. Efficient gene disruption method as well as transgenic approaches using RNAi and overexpressors will contribute for the precise understanding of molecular networks in drought stress response and tolerance. Global analysis of gene expression and protein–protein interactions during signal transduction is necessary for this purpose.

See also: **Plants and the Environment**: Global Warming Effects. **Water Relations of Plants**: Basic Water Relations.

Further Reading

Bray EA, Bailey-Serres J, and Weretilnyk E (2000) Response to abiotic stresses. In: Buchanan B, Gruissem W, and Jones R (eds) *Biochemistry and Molecular Biology of Plants*, pp. 1158–1203. Rockville, MD: American Society of Plant Biologists.

Ingram J and Bartels D (1996) Molecular basis of dehydration tolerance in plants. *Annual Review of Plant Physiology and Plant Molecular Biology* 147: 377–403.

Ishitani M, Xiong L, Stevenson B, and Zhu JK (1997) Genetic analysis of osmotic and cold stress signal transduction in *Arabidopsis*: interactions and convergence of abscisic acid-dependent and abscisic acid-independent pathways. *Plant Cell* 9: 1–16.

Kasuga M, Liu Q, Miura S, Yamaguchi-Shinozaki K, and Shinozaki K (1999) Improving plant drought, salt, and freezing tolerance by gene transfer of a single stress-inducible transcription factor. *Nature Biotechnology* 17: 287–291.

Seki M, Narusaka M, Ishida J, *et al.* (2002) Monitoring the expression profiles of 7000 *Arabidopsis* genes under drought, cold and high-salinity stresses using a full-length cDNA microarray. *Plant Journal* 31: 279–292.

Shinozaki K and Yamaguchi-Shinozaki K (2000) Molecular responses to dehydration and low temperature: difference and cross-talk between two stress signaling pathways. *Current Opinion in Plant Biology* 3: 217–223.

Xinog L, Schumaker KS, and Zhu JK (2000) Cell signaling during cold, drought, and salt stress. *Plant Cell* (Supplement): S165–183.

Salt Stress

D J Murphy, University of Glamorgan, Cardiff, UK

Introduction

Salts, and especially sodium chloride, are required for normal growth and development of all plants. However, most plants, and especially the major crops, will only tolerate a relatively narrow range of ambient salt concentrations before their performance is adversely affected. Too much salt in the growth medium, or substrate (e.g., soil, sand, or water) of a plant will result in symptoms of stress that may culminate in death. Salt stress is an increasing problem in many agricultural areas where overirrigation can lead to salt accumulation in the soil. Salt stress is also found in areas where rainfall is decreasing due to climate change. Salt stress in agricultural systems is only one of several interrelated land management problems faced by farmers. Others include soil erosion, compaction, acidity, and waterlogging. Many of these stresses are interrelated, and there is no straightforward panacea for any of them. There are two manifestations of excess salt content in soils, namely salinity and sodicity. Of these, salinity is the more prevalent and serious form of salt stress in most affected regions, although sodicity can be a particularly acute problem, for example in some parts of Australia. In saline soils, sodium is associated with chlorine, with which it forms a salt. The presence of salt in the soil reduces the availability of water to plants and at high enough concentrations can kill them. In contrast, in sodic soils, much of the chlorine has been washed away, leaving behind sodium ions attached to tiny clay particles in the soil. As a result, these clay particles lose their tendency to stick together when wet. This can result in unstable soils, which may erode or become impermeable to both water and roots.

Soil salinity and sodicity problems are common in arid and semiarid regions, where rainfall is insufficient to leach salts and excess sodium ions out of the rhizosphere. Some 6–10% of the total global land surface is covered with different types of salt-affected soils and as much as one-third of all agricultural land may be affected by some form of salt stress. At present, there are nearly 954 million ha of saline soils on the earth's surface. Such salt-affected soils are found throughout the world. More than 80 million-ha of such soils are in Africa, 50 million ha in Europe, 357 million ha in Australasia, nearly 147 million ha in the Americas, and about 320 million ha in South and Southeast Asia. The extent of areas affected by salt stress is increasing annually due to a combination of factors, such as poor irrigation, more use of marginal land for agriculture (as better land is swallowed up by urbanization), climate change, and greater pressure on scarce water resources for industry or for burgeoning human populations, especially in developing countries.

Clearly, salt stress is a widespread and increasingly serious challenge for global agriculture. Out of about 5000 crops that are cultivated throughout the world, only a few can survive with water that contains more than 1% salt. Falling crop yields due to salt stress often occur in regions with particularly high population growth such as South and Southeast Asia. Although global population is predicted to stabilize by the second half of the twenty-first century, it will still be necessary to feed an additional 3–4 billion people over the next 50 years. Therefore an important challenge for scientists will be to attempt to mitigate the adverse impact of salt stress on crop yields. In this brief review, the effects of salt stress on crops will be examined together with some of the mechanisms for withstanding salt stresses that have been developed by plants that are already adapted to saline habitats. Finally, the development of new varieties of salt-tolerant crops by both traditional breeding and transgenic approaches will be described.

Units

The salinity of soils is expressed quantitatively in various ways but the most common units are the amount (w/w) of sodium ions, measured in ppm, or the electrical conductivity, measured as Siemens cm^{-1} or $S m cm^{-1}$. A concentration of 1 ppm sodium ions is approximately $0.7\,\mu S m cm^{-1}$. The older unit of electrical conductivity that is still often found in the literature is $Mho\,cm^{-1}$ and $1\,S m cm^{-1}$ equals $1\,Mho\,cm^{-1}$ (Mho is a reciprocal ohm). Water pressure is expressed in pascals, where $1\,Pa = 1\,N\,m^{-2}$. One atmosphere equals approximately 1 bar or 0.1 MPa.

Crops and Salt Stress

Almost all of the major cultivated crops are susceptible to even fairly modest levels of salt stress. However, there are degrees of relative tolerance, with crops like sugar beet (*Beta vulgaris*), cotton (*Gossypium* spp.), and barley (*Hordeum vulgare*) showing the best performance in slightly salty soils, while grapes (*Vitis vinifera*), wheat (*Triticum aestivum*), and alfalfa (*Medicago sativa*, lucerne) have a lower tolerance and red clover (*Trifolium pratense*), beans

(*Phaseolus* spp.) and *Citrus* crops have little or no tolerance to salt. A more comprehensive list of crops that can tolerate some degree of salt stress is shown in **Table 1**.

A good example of the types of salt stress that affect crops can be seen in the case of Pakistan, parts of which are severely affected by salt accumulation in the soil. The hardest-hit region is the Indus plain, which covers almost 20 million ha and is serviced by some 60 000 km of irrigation canals. Almost one-third of this region is affected by the various types of salt stresses that are often present as a combination of salinity and sodicity. In total, 1.9 million ha are simply saline, while 1.9 million ha are permeable saline–sodic, 1.0 million ha are impermeable saline–sodic, and 0.03 million ha are sodic. Many of these problems of excess salt are caused by the intensive and continuous use of irrigation, which has raised the water table in the irrigated areas. The higher water table causes salts to be brought nearer the soil surface and there is insufficient rain to wash them away from the areas in which they accumulate. The aridity of the Indus region also leads to a high evapotranspiration rate and the consequent accumulation of salt on the soil surface. The area of salt-affected land in Pakistan is estimated to be increasing annually by some 40 000 h. Similar problems are found in other salt-affected countries around the world.

Saline soils can sometimes be improved by leaching the salts from the root zone of crops using controlled irrigation. Leaching is the process by which extra water is added to a field and allowed to soak through the soil and drain away underground. A common method of leaching is to pond the water in basins over the entire field. Sometimes the excess water is removed by pumping from wells. Permissible depths for groundwater tables vary according to the type of soil being irrigated. The amount of leaching water that enters the soil by surface flooding determines how much salt is removed from the soil. When water is leached through the soil, a surface depth of 15 cm of water for every 30 cm of plant root will leach out 50% of the salt while 60 cm of water per 30 cm of root zone leaches out 90% of the salt. However, leaching of salts by this sort of irrigation must be very carefully managed in order to avoid further additions to an already high water table. Productivity of saline soils can also be ameliorated by chemical treatments, such as the application of gypsum, sulfur, or sulfuric acid. More drastic and expensive treatments include scraping off the salt layer, levelling the land, deep plowing, subsoiling, and sanding (i.e., adding sand).

The major drawback of all these control and ameliorative methods is their expense. They are also often difficult to implement in a particular region as the exact nature of the optimal treatment can vary considerably from one area to another – sometimes even from one field to another. The use of an inappropriate treatment may exacerbate the problem and a "one size fits all" approach is rarely successful. Another factor is that salt stress in many regions has

Table 1 Crops and related species that can withstand moderate and high salinity. Note that the agronomic performance of these crops will still be adversely affected by salt stress and none of them can be classified as halophytes

Moderate salinity	*High salinity*
Some crops can grow in soils with electrical conductivity up to 15 000 μSm cm^{-1} (9600 ppm), providing the soil is fairly loamy and the underground water table is reasonably deep (2–3 m)	At high salinity (15 000–20 000 μSm cm^{-1}, 9600–12 800 ppm), with coarse-textured soil, where good-quality irrigation water is not available, the growth of most plants will be restricted and only a few salt tolerant plants like these can grow
Cereals: rice (*Oryza sativa*), sugar cane (*Saccharum* spp.), oat (*Avena sativa*), wheat (*Triticum aestivum* and *T. durum*), *Aegilops*, *Triticale*, sorghum (*Sorghum bicolor*), barley (*Hordeum vulgare*), corn (*Zea mays*, maize), pearl millet (*Penasitum thypdodeum*), rye (*Secale cereale*), *Triticale* (hybrid of rye × wheat cross)	Fruits: date palm (*Phoenix dactylifera*), coconut (*Cocos nucifera*)
Brassicas: mustard (*Sinapis alba*), rapeseed (*Brassica napus*)	Grasses: kallar grass (*Leptochloa fusca*), orchardgrass (*Dactylis glomerata*), bermudagrass (*Cynodon dactylon*), Rhodes grass (*Chloris gayana*)
Vegetables: spinach (*Spinacea oleracea*), sugar beet (*Beta vulgaris*), tomato (*Lycopersicon esculentum*), and carrot (*Daucus carota*)	Woody species: jojoba (*Simmondsia chinensis*), guava (*Psidium guajava*), jujube (*Zizyphus mauritiana*), mesquite (*Prosopis* spp.), mangroves (*Rhizophora* spp.), acacias (*Nilotica, Ampliceps, Stenophylla, Machonochieana*), *Atriplex* spp.
Fodder and forage crops: dhancha (*Sesbania sesban*), berseem (*Trifolium alexandrinum*), alfalfa (*Medicago sativa*, lucerne), *Melilotus* spp., e.g., sweet clover, honey clover, Indian clover, and white clover	
Fiber crops: cotton (*Gossypium hirsutum*), sunhemp (*Crotalaria juncea*), and kenaf (*Hibiscus cannabinus*)	
Fruits: fig (*Ficus carica*), grape (*Vitis vinifera*), pomegranate (*Punica granatum*), zizyphus (*Zizyphus jujuba*)	

multiple causes, often including climate-related factors. Therefore, various different ameliorative treatments may have to be applied to different extents over several successive seasons. This may raise technical and management challenges that make it difficult to implement such strategies at local levels. For these reasons, much attention has been focused on biological solutions to the problems raised by salt stress in agriculture. In particular, a great deal of useful information on salt tolerance can be obtained by studying plants that grow naturally in salty habitats, i.e., halophytes.

Halophytes

High levels of salt in the substrate of growing plants occur naturally in coastal estuaries and saltmarshes, and are also common in arid or desert environments, which often have significant salt accumulations in the soil. Halophytes are plants that are adapted to such habitats. Halophytes have acquired mechanisms to enable them to withstand the presence of levels of salt that would be lethal to nonhalophytic plants. However, the additional energy expenditure required to cope with salt stress often means that the growth rates of halophytes are lower than plants that do not grow in such salty environments.

Plants that are able to tolerate moderate to high concentrations of salt often have highly concentrated leaf sap. In such plants, the water potential of their leaf sap may reach -17 MPa, compared with a range of -1 to -3 MPa in normal plants. This enables these halophytes to absorb water osmotically from highly saline soils where the water potential may be as low as -15 MPa. Other halophytes are able to absorb and store water in swollen leaves, e.g., in succulents like *Mesembryanthemum crystallinum*. Some halophytes excrete excess salt either onto the leaf surface (e.g., the tamarisk or salt cedar, *Tamarix chinensis*) or in special salt glands (e.g., the desert saltbush, *Atriplex polycarpa*) as the water is selectively reabsorbed after deposition of the salt. In some cases, this may cause the surfaces of the plants to have a white sheen of salt crystals. The nature and mechanism of salt exclusion from halophytes is now being studied at the physiological and molecular levels.

Mangroves are important halophytes in tropical estuarine habitats. They exhibit several different types of mechanisms for coping with highly saline conditions. The black mangrove (*Avicennia germinans*) and the white mangrove (*Laguncularia racemosa)* are able to take up seawater through their roots, but then excrete excess salt through pores, or salt glands, located on the surface of leaves. A different strategy is employed by the red mangrove (*Rhizophora mangle*), which excludes the salt in seawater at the root–substratum interface. Mangroves are noted for their pneumatophores, erect roots that are an extension of the underground root system. Because these roots are exposed at least part of the day and not submerged under water, the root system can obtain oxygen in an otherwise anaerobic substrate.

As mentioned above, plants that tolerate salt and grow in saline environments often have high intracellular salt levels. A major component of the osmotic adjustment in these cells is accomplished by ion uptake. The utilization of inorganic ions for osmotic adjustment would suggest that salt-tolerant plants must be able to tolerate high levels of salts within their cells. However, enzymes extracted from these plants show the same high sensitivity to salt as is found in nonhalophytes. This suggests that, although halophytes can accumulate intracellular sodium ions, they are able to exclude these ions from the cytosol itself, i.e., the ions must be stored in an organelle like the vacuole. Plants can use three strategies for the maintenance of a low cytosolic sodium concentration, namely exclusion, compartmentation, or secretion. The mechanism of most interest to researchers trying to transfer salt tolerance from halophytes to crops is compartmentation. We now know that the compartmentation of sodium occurs due to the action of ATP-dependent vacuolar transporter proteins that move potentially harmful ions from the cytosol into the large central vacuole. These ions, in turn, act as an osmoticum within the vacuole to maintain water flow into the cell. The manipulation of these mechanisms in order to improve the salt tolerance of crop plants is a major goal of breeders and molecular biologists.

Breeding and Engineering for Improved Salt Tolerance

Historically, the breeding of salt-tolerant varieties of most of the major crop plants has not been very successful. To a great extent this is due to the fact that crop plants tend to be rather poorly tolerant of salt stress and therefore there is little or the natural variation that is the prerequisite for any breeding program. However, there are some crops that have a degree of tolerance for growth in soils and further breeding work may well yield results, particularly if modern biotechnological tools like marker-assisted selection and genomic profiling are deployed to assist the breeding program.

Although there is only limited variation in salt tolerance in breeding populations of most conventional crops, in some cases the existing variation may make it worthwhile to select germplasm with moderately enhanced performance on slightly saline soils. For example, the wild perennial *Glycine* species have been suggested as a potential source of germplasm to improve soybean for agronomic traits including salt tolerance. The annual wild soybean (*Glycine soja*) is a close relative of cultivated soybean (*G. max*) and has been effectively used in soybean breeding. Genetic variation in salt tolerance in natural populations of *G. soja* has been reported. Other studies indicated that although there is genetic variation in salt tolerance among individuals in *G. soja* populations, the measured salt tolerance can depend on the experimental conditions used. The results indicated that salinity and other environmental effects interact in several ways that may obscure inheritance studies of salt tolerance. It is therefore important that researchers have a basic understanding of salinity–environment interactions so that accurate assessments of salt tolerance can be made and reported. This is especially important where relatively small tolerance effects are being compared, as in these kinds of intraspecific crosses.

Apart from the slow progress in breeding salt-tolerant crop cultivars using intraspecific or near-relative crosses, there are three more recent approaches that may result in more promising results in the future. The first is to use wide crosses in order to introgress salt tolerance characters into a crop from another nonsexually compatible species, often with the assistance of tissue culture techniques and chemical treatments. Secondly, individual genes (rather than entire chromosomes or large segments thereof) can be copied from a salt-tolerant organism (not necessarily a plant) to the crop of interest to produce a transgenic variety. Thirdly, and most radically, is the possibility of domesticating true halophytes as mainstream crops. There are advantages and drawbacks for all three strategies as outlined below.

Wide Crosses

An example of a wide cross is the production of a hybrid between wheat and the sand couch (*Thinopyrum bessarabicum*), a wild grass found on the edges of the Black Sea. The sand couch can tolerate salt levels that are up to 70% of those found in seawater and is sufficiently closely related to wheat to form hybrids, although these are normally sterile. Hybrid plants were produced by hand-pollinating the stigmas of emasculated wheat plants with pollen from the couch grass. Since the newly formed hybrids contain two sets of nonpairing chromosomes they would normally be sterile. The primary hybrids were therefore treated with colchicine in order to effect chromosome doubling, hence restoring their fertility. The new hybrid species, containing both wheat and couch grass genomes, was named "Tritipyrum." While wheat is killed at less than half-strength seawater, Tritipyrum was able to survive and produce seed under such conditions. Tritipyrum may have another advantage over its parents in certain situations. In countries such as Iran, cattle are grazed on fields of wheat in winter. Because this new variety is also semiperennial, it continues to produce new growth for several months, so Tritipyrum could be grown on both good and salt-affected soils, serving a dual purpose as both a forage and a grain crop. Despite the promising preliminary results, the new crop variety is still a long way from the market. This illustrates one of the drawbacks of the hybrid approach – namely that it can take well over a decade to progress from the development of the initial salt-tolerant research material to a final commercial cultivar on the farm. It is possible that this timescale may be reduced in future as more precise methods of molecular marker-assisted selection, based on known traits, become cheaper and easier to deploy even in less well-equipped centers. Nevertheless, this approach remains daunting and unattractive to most commercial organizations and is therefore mostly undertaken in the public domain.

Genetic Engineering

The genetic engineering approach to improving salt tolerance in crops is based on observations that many halophytes accumulate sodium ions in the vacuoles of their leaf cells via a particularly efficient ATP-dependent transporter protein. Although there are many other physiological mechanisms that regulate salt tolerance in halophytes, the activity of this single enzyme appears to be one of the most important determinants of salt tolerance. This was shown most clearly when the *AtNHX* gene, encoding a vacuolar transporter from *Arabidopsis thaliana*, was introduced into the genome of rapeseed (*Brassica napus*) under the control of a highly active hybrid bacterial gene promoter. This resulted in the overexpression of the vacuolar transporter protein in the transgenic rapeseed plants. Growth of the transgenic plants in glasshouses under normal low-salt conditions was unaffected. However, while growth of wild-type plants was severely affected by the presence of 200 mmol NaCl (or 40% of the salt concentration of seawater) in the growth solution, the transgenic plants grew, flowered, and produced seeds with a normal composition. Note that in this case the introduced gene did not come from a halophyte but from a closely related species that was not salt tolerant. This indicates that salt tolerance may be achieved simply by overexpressing (using a viral promoter) an existing, or endogenous, gene(s), such as one encoding a vacuolar transporter, in a plant of interest, as recently demonstrated with *A. thaliana*.

It is possible that better levels of salt tolerance could be achieved by transferring a copy of a vacuolar transporter gene from a true halophyte to a major crop. This has now been done by producing transgenic rice (*Oryza sativa*) plants that express a $Na(+)/H(+)$ antiporter gene (*AgNHX1*) from *Atriplex gmelini*. The activity of the vacuolar-type $Na(+)/H(+)$ antiporter in the transgenic rice plants was eightfold higher than that in wild-type rice plants. Salt tolerance assays followed by nonstress treatments showed that the transgenic plants overexpressing *AgNHX1* could survive under conditions of 300 mmol NaCl for 3 days while the wild-type rice plants could not. These preliminary data are interesting in showing that some degree of salt tolerance is regulated by a single gene, but so far none of the transgenic plants has been tested under field conditions where there may be additional variables that affect other aspects of their physiology. For example, the metabolic cost of the increased enzyme activity may offset any yield gains from improved performance in saline conditions. Nevertheless, the results are promising enough to be followed up since, in some salt-prone regions, even a small increase in salt tolerance may be a very useful agronomic trait.

Domesticating Halophytes

A third approach that has been mooted as a way of producing salt-tolerant crops is the more radical suggestion of domesticating true halophytes, rather than trying to "force" conventional nonhalophyte crops into salt tolerance, whether by breeding or genetic engineering. The attractions of this strategy are obvious in that the plant is already completely adapted to salt stress. The drawbacks should be equally apparent, however. Although there are dozens of existing species of halophyte, few of them are particularly attractive as potential crops. Among some examples of possible halophyte crops are pickleweed (*Salicornia bigelovii*) as an oilseed or the four-wing saltbush (*Atriplex canescens*) as a forage crop. Halophytes such as these may have superior salt-tolerance traits but they are unlikely to possess many of the other key agronomic traits that relate to good field performance. Therefore, extensive and possibly lengthy selection programs will be required to optimize characters like disease resistance, product yield, nitrogen efficiency, etc. Although this may be a daunting prospect, the deployment of modern advanced breeding tools may considerably shorten such a domestication program. For example, in just a decade, huge progress has been made in domesticating the wild oilseed plant, meadowfoam (*Limnanthes alba*), into a potential commercial crop. Although meadowfoam is not a halophyte, this progress does demonstrate the feasibility of crop domestication strategies.

List of Technical Nomenclature

Halophyte A plant that is tolerant of moderate to high salt concentrations in its growth substrate.

Salinity Excessive concentration of salts in the growth medium, especially soils. The most common salts are sodium and chloride ions.

Sodicity A sodic soil has an excessive concentration of sodium (but not chloride) ions associated with the negatively charged clay particles.

Transgenic Describes an organism that contains gene(s) that have been inserted using recombinant DNA technology.

Vacuole Membrane-bound compartment(s) that can occupy as much as 80–90% of the volume of a plant cell.

See also: **Water Relations of Plants**: Basic Water Relations; Drought Stress.

Further Reading

Alam SM, Ansari R, and Athar Khan M (2000) Saline agriculture and Pakistan. *Pakistan Economist* 19: May 08–21, available on the web at: http://www.pakistaneconomist.com/issue2000/issue19&20/i&e3.htm

Apse MP and Blumwald E (2002) Engineering salt tolerance in plants. *Current Opinion in Biotechnology* 13: 146–150.

Flowers TJ and Yeo AR (1995) Breeding for salinity resistance in crop plants: Where next? *Australian Journal of Plant Physiology* 22: 875–884.

Foy CD (1984) Adaptation of plants to mineral stress in problem soils. *Ciba Foundation Symposium* 102: 20–39.

Kijne JW, Prathapar SA, Wopereis MCS, and Sahrawat KL (1998) *How to Manage Salinity in Irrigated Lands: A Selective Review with Particular Reference to Irrigation in Developing Countries*. Colombo: International Irrigation Management Institute.

Knapp SJ and Crane JM (1999) The development of meadowfoam as an industrial oilseed: breeding

challenges and germplasm resources. In *Proceedings of the Fourth National Symposium on New Crops*, pp. 225–233. New York: John Wiley.
Mass EV and Nieman RH (1978) Physiology of plant tolerance to salinity. In: Jung GA (ed.) *Crop Tolerance to Suboptimal Land Conditions*, pp. 277–299. Madison: ASA Special Publication 32.
Murphy DJ (2003) Biotechnology, its impact and future prospects. In: Barber J and Archer M (eds) *From Molecular to Global Photosynthesis*. London: Imperial College Press, in press.
Nieman RH and Shannon MC (1976) Screening plants for salinity tolerance. In: Wright MJ (ed.) *Plant Adaptation to Mineral Stress in Problem Soil*, pp. 359–367. Ithaca: Cornell University Press.
O'Leary JW and Glenn EP (1994) Global distribution and potential for halophytes. In: Squires VR and Ayoub A (eds) *Halophytes as a Resource for Livestock and Rehabilitation of Degraded Lands*, pp. 7–17. Amsterdam: Kluwer.
Shannon MC (1984) Breeding, selection and the genetics of salt tolerance. In: Staples R and Toenniessen GH (eds) *Salinity Tolerance in Plant Strategies for Crop Improvement*, pp. 231–254. New York: John Wiley.
Xiong L and Zhu JK (2002) Molecular and genetic aspects of plant responses to osmotic stress. *Plant Cell Environment* 25: 131–139.
Zhu JK (2001) Plant salt tolerance. *Trends in Plant Science* 6: 66–71.

Plant Responses to Waterlogging

D E Evans, Oxford Brookes University, Oxford, UK

Introduction

The chief consequence of waterlogging for higher plants is the development of oxygen deficiency, particularly in the root environment. This deficiency results from the fact that oxygen diffuses slowly in water compared to air. In freely drained soils, roots develop into gas filled spaces between soil particles. These pores in the soil structure connect to the surface, and therefore the atmosphere, where the oxygen concentration is 20.8 kPa. Oxygen concentrations similar to atmospheric can occur to a depth of more than 1 meter in the soil. Waterlogging results in the filling of these interstitial spaces within the soil and prevents oxygen diffusion resulting in hypoxia (partial oxygen deficiency) or anoxia (the almost total absence of oxygen). In addition to lack of availability of oxygen, tissue may be damaged by the accumulation of gaseous products that would otherwise freely diffuse from the root. In agriculture, waterlogging may result in reduced yield or the death of the crop; this is particularly noticeable early in the growing season where flooded areas of a field may result in failure of seeds to germinate or death of newly germinated seedlings.

Root growth is affected by even small decreases in available oxygen. The critical oxygen pressure (COP – the oxygen concentration at which growth or metabolism are inhibited) for corn (*Zea mays*; maize) at 25°C is 20 kPa – only just below the oxygen concentration of air. As aerobic metabolism generates a much higher ATP yield than anaerobic, tissue relying on anaerobic metabolism rapidly becomes depleted of ATP and of energy storage reserves. When oxygen concentrations decrease further, tissue damage will occur. Most plant tissues will die rapidly in anoxic conditions, with seedlings being particularly vulnerable. However, plant adaptations exist that permit some plants to survive hypoxia and in some instances anoxia in the soil environment.

Low oxygen concentrations detected by the tissue cause a rapid change in the free calcium concentration of the cytoplasm. This is central to the initiation of responses to low oxygen (see below) and almost certainly results from calcium release from mitochondria. Production of the gaseous hormone ethylene (ethene) and its precursor aminocyclopropane-1-carboxylic acid (ACC) is also of key importance. Ethylene induces changes in root structure and activity and ACC transport to the shoot induces further damage, like defoliation, often observed in waterlogging.

Two major classes of adaptations to hypoxia and anoxia exist: anatomical and biochemical. In the former case, the structure of the tissue is modified, to create spaces for gas diffusion and to minimize oxygen demand. Biochemical adaptations include the induction of pathways of anaerobic metabolism and enzymes for the removal of toxic waste.

Anatomical Adaptations to Waterlogging

The chief anatomical adaptation to waterlogging shown by plants is the formation of aerenchyma – tissue containing gas spaces. It is formed in the roots of wetland species like rice (*Oryza sativa*), and in some dryland species in adverse conditions. In some species, it is formed constitutively (i.e., is always present) whilst in others it is a result of abiotic stress, commonly hypoxia resulting from waterlogging. There are two types of aerenchyma: lysigenous and schizogenous. Lysigenous aerenchyma is formed when previously formed cells die within a tissue (e.g., the root cortex) to create a gas space. Lysigenous aerenchyma is found in rice, wheat (*Triticum aestivum*), barley (*Hordeum vulgare*), and corn. Schizogenous aerenchyma is formed when intercellular gas spaces form within a tissue as it develops and without cell death taking place. Spaces are formed by the differential growth of adjacent cells with cells separating from one another. Wetland species like *Rumex* show characteristic schizogenous aerenchyma, while some species (like *Saggitaria lancifolia)* show both schizogenous and lysigenous aerenchyma.

Lysigenous aerenchyma formation is initiated by the gaseous plant hormone ethylene (ethene) formed in hypoxic conditions. Lysigeny is the result of the activation of a cell death pathway. The first point of aerenchyma formation in corn and rice is the death of cells in the mid cortex of the root, a short distance behind the growing tip. Cell death in corn then progresses into surrounding cells. Studies of cell death in corn suggest that the pathway initiated resembles programmed cell death (apoptosis) in animal cells in a number of ways. However, one major difference is the need for the cell walls of the dying cells to be removed and this is achieved by the induction and release of cell wall degrading enzymes. At the end of the process, gas spaces are created behind the root tip that convey oxygen to the growing tissues. Removal of cells also reduces the demand for oxygen in that zone.

Biochemical Adaptations to Waterlogging

Anaerobic conditions created by waterlogging cause rapid changes in gene expression. Initially, normal protein synthesis stops and genes encoding polypeptides known as "transition polypeptides" are induced. Subsequently, a range of new proteins (anaerobic proteins, ANPs) are transcribed. In corn, for instance, this numbers about 20 proteins. Most of these are metabolic enzymes involved in establishing anaerobic metabolism as, in the absence of oxygen, the citric acid cycle and oxidative phosphorylation cannot function. Three pathways of anaerobic metabolism (fermentation) have been described in plants. Induction of pyruvate decarboxylase and alcohol dehydrogenase results in alcoholic fermentation, with the production of ethanol and carbon dioxide. Induction of lactate dehydrogenase results in lactic acid fermentation, with the production of lactate. The third pathway involves the production of alanine from glutamate and pyruvate. In corn, waterlogged roots first produce lactic acid and subsequently carry out alcoholic fermentation. The transition from lactate to alcoholic fermentation occurs as the cytosolic pH of the tissue decreases (cytoplasmic acidosis – one of the consequences of metabolism in low oxygen), favoring the activity of the latter pathway. Alanine fermentation is common in some (e.g., barley) but not all (e.g., corn) roots in waterlogging and alanine aminotransferase is induced by hypoxia in these species.

Of the 20 genes observed to be induced by anaerobosis in corn, many are directly involved in carbohydrate metabolism (alcohol dehydrogenase, aldolase, glyceraldehyde-3-phosphate dehydrogenase, phospho-hexose isomerase, pyruvate decarboxylase, and sucrose synthase). The importance of the induction of these enzymes in waterlogging is illustrated by the fact that corn mutants lacking alcohol dehydrogenase can only survive a few hours of submergence at 27°C when most corn genotypes will survive for 3 days. Other enzymes also induced by hypoxia are likely to be involved in the anatomical changes described above. These include enzymes of cell wall loosening and degradation like xyloglucan endotransglycosylase.

While the initiation of anaerobic fermentation results in survival until waterlogging ends and normal oxygen pressures are restored, anatomical changes provide a mechanism for long-term supply of oxygen to submerged tissues. In the extreme case of deep-water rice, which is exposed to flooding for more than a month during the growing season, a further characteristic permits survival. In this species, flooding induces shoot elongation at the rate of up to 250 mm per day, so that it is never submerged. Here, the combination of anatomical changes, modified metabolism and modified growth permit survival in extreme conditions.

Crop losses due to waterlogging world-wide are large; for instance, in Western Australia 2 million ha of land used for wheat is susceptible to waterlogging, with yield losses ranging from 15 to 50%. Estimated losses of wheat alone in Australia were Aus$300 million per annum in 2000. Alteration in rainfall, with an increased flooding as a result of global environment change are likely to significantly increase these losses. Increasing waterlogging tolerance using a combination of molecular genetic and plant breeding approaches to enhancing crop tolerance is currently an important goal.

List of Technical Nomenclature

Anoxia Absence of oxygen.

Hypoxia Deficiency of oxygen.

Aerenchyma Tissue containing gas spaces formed or present in plants often as a result of oxygen deficiency. Aerenchyma may be lysigenous – where spaces are formed by cell death – or schizogenous – where spaces are formed by the separation of cell walls.

See also: **Root Development**: Root Growth and Development. **Water Relations of Plants**: Basic Water Relations; Uptake, Loss and Control.

Further Reading

Dennis ES, Dolferus R, Ellis M, *et al.* (2000) Molecular strategies for improving waterlogging tolerance in plants. *Journal of Experimental Botany* 51: 89–97.

Drew MC (1997) Oxygen deficiency and root metabolism: injury and acclimation under hypoxia and anoxia. *Annual Review of Plant Physiology and Plant Molecular Biology* 48: 223–250.

Gunawardena A, Pearce DM, Jackson MB, Hawes CR, and Evans DE (2001) Characterization of programmed cell death during aerenchyma formation induced by ethylene or hypoxia in roots of maize (*Zea mays* L.). *Planta* 212: 205–214.

Jackson M and Armstrong W (1999) Formation of aerenchyma and the process of plant ventilation in relation to soil flooding and submergence. *Plant Biology* 1: 274–287.

Sachs MM, Subbaiah CC, and Saab IN (1996) Anaerobic gene expression and flooding tolerance in maize. *Journal of Experimental Botany* 47: 1–16.

Subbaiah CC and Sachs MM (2003) Molecular and cellular adaptations of maize to flooding stress. *Annals of Botany* 90: 119–127.

Vriezen WH, Zhou Z, and Van Der Straeten D (2003) Regulation of submergence induced shoot elongation in *Oryza sativa* L. *Annals of Botany* 91: 263–270.

WEEDS

Contents

Weed Biology

R E L Naylor, Trelareg Consultants, Banchory, Scotland, UK

Introduction

In this article, weeds are defined and the common features of their biology described. How weeds cause losses in crop yields and how those losses are achieved are considered and the role of weeds in arable field ecology is assessed. A brief outline of the options for managing weeds is given.

What is a Weed?

Weeds are plants we do not want. Many of us believe we can look at a piece of ground and identify which plants are the weeds. A simple definition of a weed is a plant growing where humans do not want it to grow. The grass growing on the lawn is desirable, but when it grows a few centimeters away in the flowerbed it is a weed. Thus, any plant species and any type of plant can be a weed. The moss in the lawn, bracken (*Pteridium aquilinum*) in grassland, wild oats (*Avena fatua*, an annual grass), or docks (*Rumex* spp., perennial dicotyledonous plants) in an arable field, and sycamore (*Acer pseudoplantus*) trees in a coniferous forest are all examples of weeds. Even a crop species can be a weed, for example, when small potatoes, *Solanum tuberosum*, are left behind in the soil after harvest they can sprout and grow in the following crop (usually termed volunteer weeds). The European Weed Science Society has adopted the following definition: "a weed is any plant or vegetation, excluding fungi, interfering with the objectives or requirements of people." A simpler definition from The Weed Science Society of America is "a plant growing where it is not desired."

Clearly then, to describe the biology of weeds might require an account of the biology of all types of plants. It is more useful to try to define what it is about particular species that makes them weedy and to try to identify general features that they have in common. Even this is a difficult task because there are many reasons why we do not want a plant to grow in a particular location. Why a plant is classed as a weed and what features of its biology make it a weed are considered in this article. The major commercial impact of weeds is as negative factors in cropping systems, so how weed biology influences crop losses and the extent of those losses are outlined briefly below. The beneficial effects of weeds are also discussed and a brief outline of the techniques available for managing weeds is given.

Why is a Plant a Weed?

The major reason for calling a plant a weed is because it leads to reduced crop yield, which is usually the result of the weed interfering with the crop's capture of light, water and nutrients. Some of the more problematic weeds are those that lead to large reductions in yield and have so far proved difficult to manage, such as wild oats (*A. fatua*) or cleavers (*Galium aparine*) in cereals. The management of weeds leads to increased costs of production. In addition to reduced crop quantity, weeds frequently reduce crop quality. This may be because the presence of contaminating seeds reduces the price that can be commanded for the crop seeds. The occurrence of unwanted species is not appreciated by the purchaser of vegetables. The occurrence of the poisonous berries of *Solanum nigrum* in peas (*Pisum sativum*) is particularly undesirable.

A mat of vegetation (weeds) at the base of a crop stand may increase the humidity by conserving moisture, resulting in delayed ripening and/or an increased level of moisture in the crop at harvest necessitating the cost of crop drying. The higher humidity can also lead to a higher incidence of disease, e.g. the presence of weeds in oilseed rape can increase levels of *Botrytis* in the crop. Those weed species that climb up crop plants interfere with harvesting by making operation of the harvester (man or machine) more difficult. Species that grow vigorously late in the season can also interfere with the harvesting of potatoes and sugar beet, *Beta vulgaris*.

A number of species may not lead directly to reductions in yield but may achieve this indirectly through their effects on crop health. Species that are related to crops and grow in fields may act as alternate or alternative hosts for crop pests and diseases. The disease club root (caused by *Plasmodiophora brassicae*) occurs in many members of the brassica family, including crops such as oilseed rape (*Brassica napus*, also known as canola), cabbages (*Brassica oleracea* var. *capitata*), and turnips (*Brassica rapa*). Weeds in the same plant family such as charlock (*Sinapis arvensis*) and shepherd's purse (*Capsella bursa-pastoris*) are important vectors of this disease. Red clover, *Trifolium pratense*, can provide an alternative source of the pea mosaic virus, which is then transmitted to crop plants by aphids. The common chickweed (*Stellaria media*) is the source of mosaic virus for a number of crop species. Many wild grasses harbor the disease ergot of cereals (caused by the fungus *Claviceps purpurea*) in which a cereal grain may be replaced by a hard sclerotium called an ergot. The sclerotia are poisonous and if they are harvested with the grain, the resulting flour, when consumed, can lead to the potentially fatal condition known as St Anthony's Fire. Some of the earliest legislation for quality control of flour relates to millers checking their grain for ergots. Domestic stock may be poisoned if they consume grasses with ergots. The weed fat hen (*Chenopodium album*) is a host for the black bean aphid, which damages broad and field beans. Many weed species harbor nematodes, which may infect crop species and cause significant yield loss.

Seeds or propagules (e.g., potato tubers) may be left in the soil from a previous crop, as a result of precocious seed shedding prior to harvest, a late harvest or inefficient harvesting. Many small seeds of cereals pass through the combine and are returned to the soil with the straw. Rarely are all potato tubers extracted from the soil. Such survivors, or volunteer crops, are weeds in subsequent crops and, in addition to interfering with crop growth, may act as a "disease bridge" from one season to another, which is implicated in the initiation of many rust (e.g., *Puccinia striiformis*) outbreaks in cereals and potato blight (*Phytophthora infestans*).

Relatively few plants are parasitic on crop plants but their occurrence can reduce yield greatly. In temperate climates, dodder (*Cuscuta* spp.) occasionally occurs. In tropical areas, the parasitic plant witchweed (species of *Striga*, mainly *S. hermonthica*) can infect the staple crops corn (*Zea mays*; maize) and sorghum (*Sorghum bicolor*) and drastically reduce yields to 10% of a healthy crop.

There are many species that have an impact on animal production and animal health. Thistles and similar plants with spines or thorns inhibit animal foraging and protect adjacent plants from grazing, thus effectively reducing the area available for grazing. A number of plants are poisonous to domestic stock (and sometimes humans) and stock need to be kept away from them or the plants need to be removed from areas where the animals graze. The yellow Compositae *Senecio jacobaea* is not only poisonous when fresh but remains so after drying and so hay made from infested pastures may be poisonous. The foxglove (*Digitalis purpurea*), the garden laburnum tree (*Laburnum anagyroides* and other species), and the garden shrub rhododendron (*Rhododendron ponticum* and other species) are all poisonous to humans as well as farm stock. Some plants contain aromatic compounds that can taint animal products, for example, wild garlic or ramsons (*Allium ursinum*) can impart an undesirable flavor to cows' milk. Ingestion of some plant species e.g., the giant hogweed (*Heracleum mantegazzianum*), may lead to photosensitization of humans or animals. Bracken (*P. aquilinum*) is thought to contain teratogens. Many plant species can act, often incidentally, as intermediate hosts or a vehicle for ingestion of animal pests and parasites.

In some situations, plants can be regarded as a safety hazard. For example, tall vegetation on the roadside can reduce vision on corners and is frequently mown to maintain visibility. Furthermore, scrub plants beneath power cables can be a potential fire hazard.

The presence of weed species can also have an effect on the wool industry, where the value of wool fleeces is based, in part, on the level of plant fragments in the wool. Plants that have seeds with hooks or barbs, e.g., the burr-marigolds (*Bidens* spp.), which can become trapped in the wool, can reduce wool quality.

In many parts of the world, aquatic weeds may prevent water flow. The plant mass blocks ditches and irrigation channels and transpires water that could have been delivered to crop plants. The water hyacinth (*Eichhornia crassipes*) is an important tropical aquatic weed. In continental Europe, Canadian pondweed (*Elodea canadensis*) was originally introduced as a garden pond plant, but it has colonized areas where it is not wanted and become a serious problem. The growth of some aquatic weeds may reduce the oxygen content of the water and lead to the death of fish.

Individual growers do not want neighbors to see that they have a weedy field, so pride is a factor in weed management decisions. There is some evidence that perhaps 30% of herbicide applications may be cosmetic in the sense that they could not be justified economically, i.e., the costs of the chemical and its application are not recouped in the value of the extra yield gained. However, such analyses are rather crude and ignore the benefit of dealing with weeds while the population is still small, in order to save losses or costs in subsequent crops. Thus, the costs and benefits should probably be assessed on a complete crop rotation and not on a single season basis.

What Features Make a Plant a Weed?

It is convenient to focus on the case of agricultural weeds and then to extend the concepts to other situations. Trying to describe the common biological features of all weeds is difficult because we could be describing the biological characteristics of all plants. An easier task is to consider the features of competitive weeds in arable crops and rotations. An interesting approach to this was made by H. G. Baker in 1965 who tried to define what would be the "design features" of an "ideal weed." He considered that the ideal weed would have no special requirements for germination so it would respond quickly to environmental germination cues. It would have discontinuous, self-controlled germination of seeds, which would provide some insurance for survival of the species at a site even if a whole cohort of seedlings was eliminated (e.g., by weed management). Great seed longevity would enable a soil seed bank to persist over time and rapid seedling growth would allow the young weed plant to grow and compete with the crop for the resources of light, water and nutrients. Annual weeds rely on the replacement of populations from seed each year and so an early transition to reproductive growth combined with the capacity to continue producing seed over an extended period would maximize the number of seeds produced. Many weeds are what have been termed "pioneer" species and, in the initial phases of a population invasion, the individuals may be far apart. In such circumstances, reproduction would be more likely to be successful if the plant was self-compatible, so that pollen from the same plant could fertilize the ovules. In established populations, the possibility of crossing exists and the ideal weed would have an easy method of cross-pollination, such as by wind rather than by, for example, a specialized insect visitor. The production of copious numbers of seed in favorable circumstances would ensure the ideal (annual) weed would more than replenish the soil seed banks and provide an opportunity for dispersal to new habitats. However, in agricultural systems, crop managers take measures to manage weeds and so the environment may not be conducive to the production of large numbers of seeds. In such adverse conditions, the ideal weed would maintain the production of at least some seeds. Because the environment (at the scale of size of the seed) is likely to differ from year to year as a result of crop rotation, the ideal weed would need to disperse well over long and short distances and not be dependent on a specialized means of dispersal.

H.G. Baker suggested that the ideal weed would have special means of competition. By this he meant the weed's ability to climb up crop plants and use them for support, the possession of spines that protect the weed from grazing and possibly allelopathy. Baker also considered the additional features that might be possessed by the ideal perennial weed. Obviously it would have vigorous vegetative reproduction, enabling weed growth to keep up with crop growth and providing consolidation of the species at that site. Vegetative reproduction and spread would be aided if the weed was able to split relatively easily, for example, by having brittle lower nodes or rhizomes. Soil cultivation would thus serve to spread fragments of the plant to new locations. This ability to fragment would be combined with the ability of the fragments to regenerate.

The putative features of an "ideal weed" given above can be used to judge why any particular species is a successful weed in some circumstances. Fortunately, few weed species possess all these features to a high degree, but many have at least one of these properties. Consideration of weed biology can, and indeed should, form the foundation of rational and integrated methods of weed management.

How Large are the Crop Losses?

Various authorities have estimated a global yield loss due to weeds to be about 10%. Greater losses may occur in low input tropical systems. In one survey, organic farmers in the USA identified weeds as more difficult to manage than pests and diseases. The precise impact of a weed infestation depends mainly on the crop species (and variety) being grown, which weed species are present, how many weeds there are, the relative time of emergence of the weeds and the crop plants, and the general crop vigor and health.

Not all weeds have identical impacts on yield. An informal survey of weed specialists in Europe in December 2001 asked for identification of the worst weeds (defined as those most widespread and/or most difficult to manage) in Britain, Denmark, Finland, France, Germany, Hungary, Italy, The Netherlands, Spain, and Switzerland. Those who replied identified ten annual grasses (*Alopecurus myosuroides*, *Apera spica-venti*, *Avena fatua*, *A. sterilis*, *Cynodon dactylon*, *Cyperus* spp., *Echinochloa crus-galli*, *Lolium rigidum*, *L. multiflorum* and *Sorghum halepense*), nine annual dicots (*Chenopodium album*, *Conyza canadensis*, *Galium aparine*, *Polygonum aviculare*, *P. persicaria*, *Bilderdykia convolvulus*, *Solanum nigrum*, *Stellaria media* and *Xanthium strumarium*), and four perennial species (*Convolvulus arvensis*, *Elytrigia repens*, *Rumex obtusifolius*, *Sonchus*/*Cirsium* spp. (perennial thistle species)). These species constitute the most important weed species in the majority of field, vegetable and plantation crops and are those that remain the most difficult to manage, in spite of advances in crop technology in the latter half of the twentieth century. Other weed species may occur locally or may be limited to specific crops in which they create a problem, but are otherwise not too difficult to manage, particularly if herbicides are incorporated into the weed management plan.

There is great variation in yield loss according to the type of crop, the season and the timing of the infestation. Early weed infestations can smother those crops that are slow to establish and reduce yield by as much as 90% (**Table 1**). In contrast, weed establishment late in the season in a fast-growing aggressive crop may have little impact on yield. This emphasizes the value of detailed observations of young crops together with early intervention.

The effect of a weed species on a crop obviously depends on the population density of the weed. The most useful relationship describes the effect of weed density on the percentage yield loss of the crop (**Figure 1**). This relationship is a rectangular hyperbola (an equation of the general form $y = 1/x$) and demonstrates an increasing impact on yield of increasing weed populations, but the proportional loss decreases at larger weed densities and reaches an asymptote. At this level the weeds are probably interfering with each others growth as much as with crop growth. The curve relates propotional yield loss (YL) to weed density (n) and is described by three parameters: the asymptotic proportion of crop yield lost at high weed densities (A); the proportion of crop yield lost per individual weed at low densities (I); and the threshold weed density below which no yield loss can be detected (T).

$$\mathrm{YL}(\%) = ((I - T) \times n)/(1 + ((I - T) \times n)/A)$$

The values of I, A and T can be relatively easily determined and then used in decision support systems for weed management. This approach is proving useful for many crops in which yield is measured per unit area. However, where the yield of an individual plant part is important (e.g., in size-graded vegetables such as carrots, *Daucus carota sativa*, or parsnips, *Pastinaca sativa*) then an alternative equation may be more useful. In this equation (**Figure 2**), the reciprocal of individual plant or plant part weight (w) is linearly related to weed density (n) in an equation of the form:

$$1/w = 1/(W + (c \times n))$$

where W is the maximum individual crop weight in the absence of competition and c is the slope of the

Table 1 Examples from a range of countries of comparisons of crop yield on weedy and weed-free plots

Crop	*Weed species present*	*Weed-free yield*	*Weedy yield*	*Yield loss due to weeds (%)*	*Response to weed management (%)*
Bread wheat (*Triticum aestivum*)	*Alopecurus myosuroides*	5.54	3.96	28.5	39.9
Bread wheat	*Galium aparine*	11.8	4.05	65.7	191.4
Durum wheat (*Triticum durum*)	*Phalaris minor*	6.1	5.6	8.2	8.9
Barley (*Hordeum vulgare*)	BLW	6.29	5.98	4.9	5.2
Rice (*Oryza sativa*)	BLW + G	3.82	3.03	20.7	26.1
Rice (*Oryza sativa*)	BLW + G	3.95	1.26	68.1	213.5
Corn (*Zea mays*)	BLW + G	1.87	1.596	14.7	17.2
Corn (*Zea mays*)	BLW + G	10.5	4.5	57.1	133.3
Sugarcane (*Saccharum officinarum*)	BLW + G	132.5	108.7	18.0	21.9
Cotton (*Gossypium hirsutum*)	BLW + G	865	715	17.3	21.0
Potato (*Solanum tuberosum*)	BLW	51.8	43.2	16.6	19.9
Sugarbeet (*Beta vulgaris*)	BLW	85.5	16.9	80.2	405.9
Broccoli (*Brassica oleracea italica*)	BLW + G	11.4	5.8	49.1	96.6
Kale (*Brassica oleracea acephala*)	BLW + G	25.4	15	40.9	69.3

BLG, a mixture of broad-leaved weeds; G, a mixture of grass weeds. Yield loss due to weeds (%) is the difference between weedy and weed-free plots expressed as a per cent of weed-free plots. Yield response due to weed management (%) is the difference between weedy and weed-free plots expressed as a per cent of weedy plots.
Based on data published in various volumes of Proceedings BCPC Weed Control Conference, Brighton (1989).

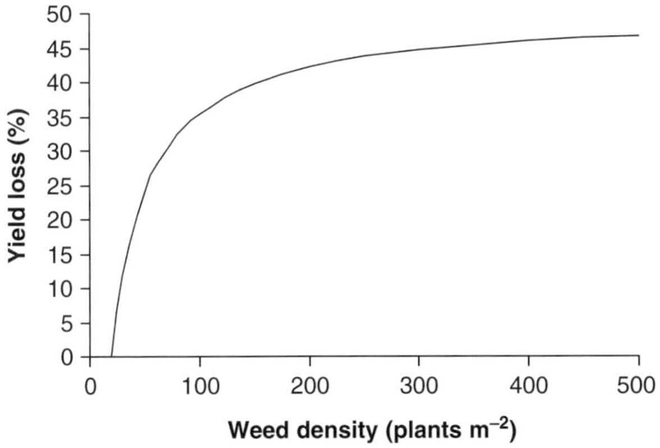

Figure 1 An asymptotic response of crop yield to weed density. The curve has parameters $A = 50\%$, $T = 20$ weeds m^{-2} and $I = 1.5\%$ weed^{-1}. See text for details.

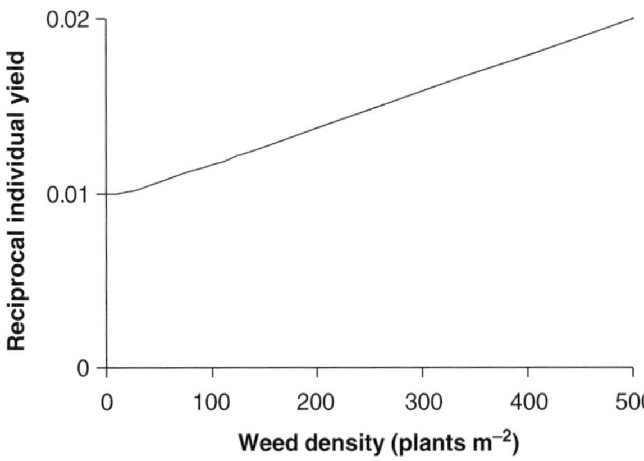

Figure 2 The response of individual plant yield to weed density. The line has parameters $W = 100\%$ and $T = 20$ weeds m^{-2}. See text for details.

relationship. In this case, there is not a constant impact per individual weed.

How are Crop Losses Achieved?

This topic is covered fully in the entry on **Weeds:** Weed Competition, but a few points deserve mention here. Although there have been many experiments to show the effects on crops of competition with weeds for light, water and nutrients, this is a simplistic approach to the study of interference of weeds with crop growth. Weeds and crops occupy the same space and exploit the light, water, and nutrients within that space, thus interfering with each others growth in an interactive and dynamic way. Different amounts of light capture lead to different assimilate reserves, which leads to differing investment in new leaf canopy and root systems, resulting in further resource "capture." As each individual grows, it modifies its own local environment and the environment of neighboring plants.

In temperate climates, weeds reduce yield mainly by shading crop plants and thereby reducing the amount of photosynthesis that the crop can undertake. The

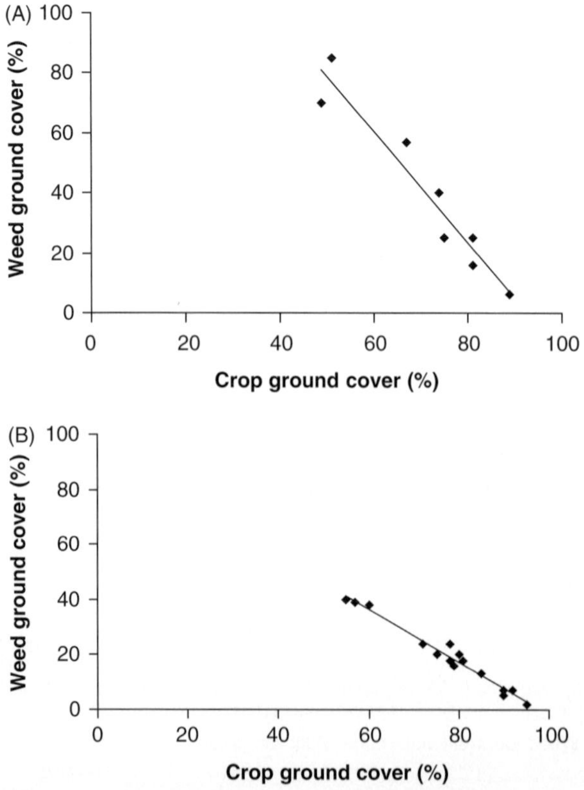

Figure 3 The relationship between crop leafiness (ground cover %) and weed cover (%) in a set of winter wheat varieties (A) and a set of spring varieties of wheat, barley, and oats (B). Based on data of Richards M (1989) Proceedings BCPC Weed Control Conference, Brighton.

shading effect is greater in fertile conditions when there is adequate moisture. The identification of so-called "competitive varieties" is important for organic systems in which herbicides are not used and for low input systems where herbicides may not be affordable (**Figure** 3). In semi-arid climates, the main effect of weeds is usually achieved through their capture of the scarce resource, water, less being available for the crop plant. Some weeds are very efficient at taking up water and can accentuate a water shortage in drought-susceptible crop species. The roots of many weeds are concentrated in the surface layers of the soil, in contrast to the more deeply rooted crop species, giving the weeds an advantage when scavenging for surface-applied water and soil nutrients. Because all nutrients are taken up in solution, water shortage limits nutrient uptake.

Weed Population Dynamics

The biology of annual weeds emphasizes reproduction. Seed production, dispersal, longevity, dormancy, and germination are important features of weed population dynamics (*see* **Weeds**: Weed Seed Biology). The weeds growing in a crop can be considered as merely the vehicle by which seeds in the soil produce further seeds for incorporation into the soil seed bank. The production of copious seeds provides a means of consolidation at an already invaded site and a means for invading new sites. Reduction of seed production in annual weeds is a priority for management. If reproduction is not prevented, the new seeds can germinate and invade subsequent crops. Similarly, for perennial weeds the emphasis of management needs to be on reducing spread through preventing sexual reproduction as well as limiting vegetative growth. An understanding of the population dynamics of weed species should enable management decisions to be taken on a rational basis rather than in the expectation, or hope, that the measures will be successful. Many of the models of weed populations are based on information on reproduction of a limited number of weed species and are used to compare the efficacy of particular management options over a long term. The models may be particularly useful in appraising the likely effectiveness of combinations of management decisions (concerning tillage, sowing time, crop type, stubble treatment, etc.) in alternative scenarios but inevitably can only be as good as the data on which the models are based. Models can be used to assess the efficacy of intervention at different stages in the life cycle of weeds. They may assist in decision-making on whether to use a cheaper herbicide that is less effective rather than a more effective but more expensive one. Models can help in deciding whether or not to treat the weed, i.e., to incorporate a weed "threshold" into management in order to prevent unnecessary or uneconomic (i.e., cosmetic) treatments. More sophisticated models could incorporate a consideration of the beneficial effects of weeds.

Many weed species in intensive agricultural systems show a scattered distribution, both within and between fields. Weed management by growers may lead to local extinction of a species. However, reinvasion can occur along with crop seeds, on farm machinery or by natural seed dispersal. The resulting small, perhaps transient, population in a landscape is termed a metapopulation. The dynamics of the metapopulation depends partly on the size of the local subpopulations and their extinction rate. The connectivity between patches is important. Both the dispersal rate and the distance between patches determine the reinvasion rate. These factors can lead to large-scale temporal and spatial variations on overall and local population sizes.

There is a difference in the population ecology of annual and perennial weeds. The logistic equation of population growth has two parameters: the intrinsic rate of increase, r, and the ceiling population or carrying-capacity, K. Annual weeds are generally "pioneer" species, colonizing open habitats and achieving their spread through seed dispersal. Individual plants may be short-lived and a large proportion of biomass is allocated to reproduction. They are typified as "r-selected species:" population size is often well below the carrying-capacity and is regulated by the physical effects of the environment on r. In contrast, K-selected species tend to occupy the closed vegetation of later stages of vegetation succession (or grassland), be long-lived, with a prolonged vegetative stage, and allocate a small proportion of biomass to reproduction. Population size is often close to the carrying-capacity and is determined by the influence of biotic factors on K. These two "strategies" represent the extremes and the behavior of an individual species is usually intermediate. However, the most important annual and perennial weeds seem to lie nearer the extremes than the center of the range.

Seed Dispersal

Seed dispersal by plants is a passive process in the sense that the plants are unable to choose the destination of their offspring. However, various features of the seed and plant can aid dispersal. Taller plants may disperse seeds over greater distances. Lighter seeds may be dispersed further by wind, but may have a lower chance of producing a new seedling because of their limited energy reserves. Heavier seeds fall nearer the parent plant, where there may be a greater chance of landing in a suitable microsite for germination. Even where plants have seeds with structures that enhance dispersal by wind, the majority of seeds fall near the parent plant. Adaptations for animal dispersal (externally or internally) may achieve greater distances of dispersal and achieve a distinctly more or distinctly less random pattern than wind dispersal. For a weed, the advantage of dispersal is the chance of migration to a new site. A disadvantage may be the absence of another plant with which to achieve cross-pollination.

(For features of "seed rain," the seed bank in the soil and the topics of seed dormancy and germination *see* **Weeds**: Weed Seed Biology.)

Weed Introduction

Dispersal of species over very great distances to new habitats is achieved by introduction. There are numerous cases of plants originally introduced as decorative or useful species becoming weeds. This has usually occurred because a species has been introduced to an environment where its natural diseases, pests, and predators are absent. Canadian Pondweed (*Elodea canadensis*) was probably introduced to the UK attached to timber imported for building railways; it quickly colonized the waterways where it hindered the passage of boats. The species also spread to other countries in Europe and by 1870 had reached France, Belgium, Germany, Russia, Hungary, Denmark, and Sweden. It appears not to have advanced beyond a latitude of about 66° N, probably because of the low temperatures. *Elodea canadensis* is still common in all of the areas that it colonized, but after about 5 years its growth slows down, and it stops being so troublesome. The reason is not fully understood.

The flat-stemmed cacti belonging to the genus *Opuntia*, commonly known as prickly pears, are used for human consumption and animal feed (the stems). They are widely cultivated in Latin America, the Mediterranean area, and the Middle East. In the early nineteenth century, a single plant of *Opuntia stricta* was introduced to Australia as an ornamental plant. The species soon became naturalized and any attempts to remove the cacti by plowing merely stimulated vegetative propagation. A new strain of spineless *Opuntia* was shipped to Australia in 1914 for use as forage; however, the plants produced viable seed and the offspring contained both the spiny and the spineless plants. Cattle and sheep do not eat the spiny plants. By 1925, prickly pear cacti were a serious invader of rangeland in eastern Australia rendering large areas unusable for grazing. The Australian government initiated a search for suitable biological control agents. The larvae of the moth *Cactoblastis cactorum* was introduced as the major control agent, and by 1933 almost 90% of the prickly pear plants had been eradicated.

The spiny European shrub gorse (*Ulex europaeus*) has been introduced to other countries where it competes very successfully with the native vegetation. It was introduced into New Zealand and sold by seed merchants and nurseries for private cultivation until the 1890s. It now covers more than 3% of the total land area of New Zealand, including significant areas of agricultural use and forest plantations. It is regarded as New Zealand's commonest and most costly weed. In North America, gorse was first introduced in Oregon, probably in the 1960s, and has now spread as far south as San Diego and as far north as coastal British Columbia. In addition to its role in the loss of productive agricultural land, it is undesirable because gorse thickets are potential sites for wildfires. Once established, gorse produces large amounts of litter, which accumulates, acidifies the soil and excludes plants that do not tolerate acid conditions. Thus, the species provides a threat to the natural biodiversity of an area.

It is rarely economical to control gorse on grazing land by using conventional means and much marginal pasture land is converting to gorse cover. In New Zealand, the number of viable seeds of *U. europaeus* in the top 5 cm of the soil appears to decline by 10% in about 10 years. The emphasizes the potential for regeneration of gorse populations from the seed bank.

Control of gorse in New Zealand using biological control agents began in 1989. Gorse spider mites (*Tetranychus lintearius*) have proved moderately successful, especially strains imported from Spain and Portugal. The use of biological control is unlikely to provide the degree of control achieved with herbicides. The aim instead is to provide a cost-effective and continuing reduction in the incidence of gorse. Spider mites can apparently reduce the density of gorse bushes by nearly 20%.

St John's Wort (*Hypericum perforatum*) was introduced to the USA in 1696 for its medicinal and ornamental properties. It has been a popular folk remedy for treating minor cuts, burns, and inflammation. The small black glandular dots on the foliage and petals contain a photodynamic yellow pigment called hypericin, which is said to have an anti-depressant effect in humans but can also photosensitize light-sensitive skin if consumed in high dosage. Animals that graze on St. John's Wort are also affected by the chemical constituents. Depression of the central nervous system and increased sensitivity to temperature change may make animals difficult to handle. Livestock usually avoid it unless food is scarce. St. John's Wort is a perennial plant with a horizontal spreading root system that reproduces by both seeds and runners. A single plant may produce 15 000–30 000 seeds per year, which can remain viable in the soil for up to 10 years. The plant is now found in most states of the USA and has become an important weed of rangelands. Its habitat in the USA is very different from that in Europe where it is mainly a hedgerow plant surviving in the shade. Chemical control of St John's Wort can be difficult due to its extensive root system and uptake of herbicides through the small waxy leaves is restricted. Several insects have been introduced to limit St John's Wort, including a moth that eats foliage and flowers (*Aplocera plagiata*), a root-boring beetle (*Agrilus hyperici*), a midge that forms galls in the leaf buds (*Zeuxidiplosis giardi*), and two foliage-eating beetles (*Chrysolina hyperici and C. quadrigemina*). The latter beetle, in particular, has proved very effective, but has cycles of good and less good control.

Beneficial Effects of Weeds

It may seem perverse to consider the beneficial effects of weeds. Nevertheless, weeds are a component of the arable field ecosystem and so contribute to their overall ecology. Agricultural areas generally have a reduced biodiversity; indeed, the extent of biodiversity has been used as an index of the intensity of management. Current European Union (EU) policy is to encourage farmland biodiversity through less intensive farming either by reducing the intensity of management or reducing the farmed area through set-aside. We can consider weeds as valuable indicators of biodiversity because of their role in providing food or shelter for animal species. Birds are particularly prominent animals and much of the decline in farmland birds has been associated with the reduction of crop weediness. The species composition of sown rotational grassland contains few species of grasses, mainly *Lolium* (ryegrass) species and *Festuca* (fescue) species. The seed mixtures sown may also contain red or white clover. This is very different from the much more complex mixtures sown in the first half of the twentieth century. A typical mixture then might have contained additional species of grasses (e.g., *Phleum pratense, Dactylis glomerata, Poa trivialis, Cynosurus cristatus*), legumes (e.g., *Trifolium incarnatum, Onobrychis viciifolia, Anthyllis vulneraria*) and many other species, which were included to improve the mineral nutrition of the sward (e.g., *Achillea millefolium, Plantago lanceolata, Potentilla anserina, Chicorium intybus, Symphytum officinale*). This more complex mixture was seen as a more efficient exploitation of the available niches in the habitat resulting in a more mineral-rich herbage. It would also provide a degree of insurance against one species failing.

Weeds can provide extra protection from erosion by water and wind. Bare soil is very prone to erosion by water and wind. The presence of a plant cover helps to reduce the momentum of impacting raindrops. It also reduces the windspeed at the surface of the soil thus providing protection from wind erosion. Plant roots help to bind the soil to reduce its ability to move.

There is considerable interest in designing weed management systems that allow populations of beneficial organisms to develop. Insects that pollinate crops (e.g., bumblebees) need a source of nectar and pollen when the crop is not in flower and weeds can provide this. Ground beetles, spiders, and hoverfly larvae are natural predators of aphids and weeds may provide a food source for them. This would reduce the reliance on aphicides. Furthermore, the weeds themselves may be more attractive to crop pests; it has been reported that aphid infestations on sugar beet are lower in weedy fields than in weed-free fields.

Weeds may also provide a valuable food source for other organisms that are valued by society, or the land manager. The acceptance of a certain level of weeds in a crop may allow rarer farmland birds to survive. Postponing weed control until later in crop growth may permit birds to rear chicks before the food supply is removed. Weeds in stubbles can provide winter feed for a range of organisms. Some weeds may actually be valuable plants despite the fact that they grow where we do not want them to. Many wild plants continue to be used as herbs for flavoring and others have medicinal properties. "Herbal" remedies are perceived by many as safer than synthesized pharmaceutical drugs, despite the latter having undergone a rigorous approval process. Many drugs have their origins in plant secondary metabolites and indeed some are still derived by extraction from deliberate plantings of such medicinal plants. *Digitalis purpurea* (foxglove) is a poisonous plant and a source of the glycosides digitalin and digitoxin, which are used as cardiac stimulants and vasoconstrictors.

There is increasing concern about the impact of farming on rural diversity. Although farmers may wish to minimize the negative effect of reduced crop yields, wider society may judge the beneficial effects of weed populations as food or refuge for other organisms are more important. Farmers may be asked to manage fields for conservation and only eliminate the particularly noxious weeds. Thus, weed management in temperate cereal fields could be targeted at *Alopecurus myosuroides* and *Galium aparine*, while leaving other weed species such as *Stellaria media* or members of the Polygonaceae to provide food for beneficial invertebrates and vertebrates. This selective management will provide a severe challenge for farmers and growers in the future. It results from changed perceptions of the reasons for managing weeds, but does not alter the reasonably well understood impact of weeds on crop yields. The management emphasis will need to change from regarding boundary areas as merely the strips of land in between the productive fields, to considering fields as areas in which we have modified the native vegetation.

Managing Weeds

Weed management is covered in more detail in the entry **Weeds**: Weed Technology and Control. A successful weed management strategy exploits our knowledge of weed biology to achieve crops with minimal weed populations, which suffer little interference from weed growth, at an economic cost.

Weeds, like crops, are living organisms that respond to their environment. We manipulate the environment to provide conditions conducive to the growth of crop plants and it is not surprising that other plants can exploit these conditions too. Growers must "fine-tune" their manipulation of plant environments to give the advantage to the crops over the weeds. Weed management is just one of the considerations in a multi-factorial decision-making process for the selection and definition of a crop management system. Many of the decisions taken by a grower influence the species composition and the density of weed infestations. The time of the previous harvest, management of the residues of the previous crop, type and timing of soil tillage, fertility management, adjustment of soil pH, varietal choice, use of clean seed, date of planting, sowing rate, row spacing, and crop nutrition are all decisions made about the crop that influence the development of weed populations.

An essential part of the decision-making process for weed management is an appraisal of whether management is necessary. Reduction of crop yield is the major justification for the effort and expense of weed management in arable crops, while in horticultural crops effects on crop quality are also important. In most agricultural systems there has been a reduction in manual weed removal. The mechanization of weed management requires a source of power and the ability to sow crops in rows so that the weeds in between the rows are easily identified and removed. In the second half of the twentieth century, the development of synthetic herbicides was a major advance leading to great improvements in food supply and food security around the world. The weed management systems being developed now exploit our knowledge of weed biology and integrate many aspects of crop management with any beneficial aspects of weeds, especially their contribution to biodiversity.

Assessing weed populations in the field can be problematic. The correct identification of weed species is important and has to be done at the seedling stage if intervention is to be effective. The distribution of weeds also tends to be patchy, so assessing weed density is not straightforward. The whole field needs to be inspected, not just the area near the gate, in order to determine the modal density. The implication of weed patchiness is that weed management measures should be localized rather than imposing an average overall treatment, which might be ineffective in high density patches and costly in low density areas.

Although it is feasible to develop a weed management strategy for a farm or individual field, the implementation of the strategy is more difficult. This is mainly because the weather plays an important part in determining the ability to work on the land and to perform particular activities. Thus, the strategy will always need to be flexible and incorporate a number of approaches in order to be effective. Weed managers need to appraise the range of techniques they use to manage weeds and to determine the benefits in extra yield to be attained for the cost of the weed management

Weeds in the Future

Changes in crop management and in climate are likely to alter the occurrence of weed species in fields. Climate change may result in the spread of "new" weeds from other areas either as a direct response or as a response to changes in cropping systems. Invasions of new species into areas will create new challenges for growers. The desire to reduce costs of production is leading to increased interest in early drilling and minimum tillage systems, but these are known to alter the weed flora that predominates. Reliance on the repeated use of single herbicides or single herbicide groups to control particular weeds has led to the evolution of herbicide-resistant weed biotypes, which have proved difficult to manage.

See also: **Growth and Development**: Field Crops. **Production Systems and Agronomy**: Agricultural Crops. **Seed Development**: Germination. **Weeds**: Weed Competition; Weed Seed Biology; Weed Technology and Control.

Further Reading

Brain P and Cousens R (1990) The effect of weed distribution on predictions of yield loss. *Journal of Applied Ecology* 27: 735–742.

Clarke JH (ed.) (1992) *Set-aside*, p. 283. Farnham: BCPC.

Clarke JH, Davies DHK, Dampney PMR, *et al.* (1996) Rotations and cropping systems. *Aspects of Applied Biology* 47: 482.

Cousens R and Mortimer AM (1995) *Dynamics of Weed Populations*. Cambridge: Cambridge University Press.

Firbank LG, Carter N, Darbyshire JF, and Potts GR (eds) (1991) *The Ecology of Temperate Cereal Fields*, p. 469. Oxford: Blackwell Scientific Publications.

Naylor REL (1993) Weed seed ecology: current difficulties and future directions. In: *British Crop Protection Council Conference* – Weeds, pp. 483–484. Farnham: BCPC.

Naylor REL (ed.) (2002) *Weed Management Handbook*, 9th edn. Oxford: Blackwell Science.

Radosevich S, Holt J, and Ghersa C (1997) *Weed Ecology: Implications for Management*, 2nd edn. New York: John Wiley & Sons.

Rao VS (2000) *Principles of Weed Science*, 2nd edn. Enfield, NH: Science Publishers Inc.

Zimdahl R (1999) *Fundamentals of Weed Science*, 2nd edn. San Diego: Academic Press.

Weed Competition

L Bastiaans and M J Kropff, Wageningen University, Wageningen, The Netherlands

Introduction

Competition

Competition in its broadest sense can be defined as the mutually adverse effect of organisms or species on one another. Interference competition refers to a direct interaction between two organisms in which one physically excludes the other from a portion of habitat, and hence from the resources that could be exploited there. Exploitation competition represents the indirect form of competition, in which the interaction between organisms or species is mediated through one or more shared resources. The specific situation of crop–weed competition is a typical example of exploitation competition, in which the focus is on the effect of resource capture by weeds on growth and production of the crop. Those resources of which supply cannot meet demand are of major interest, as they determine the attainable yield of the crop. If weeds capture such resources, crop growth will be reduced, resulting in yield loss.

Darwin identified the central role of competition in selection processes of organisms in general. Since then, competition has been regarded as one of the major forces behind the appearance and life history of plants, and the structure and dynamics of plant communities. Because of the important role of competition between plants in a wide range of ecosystems, competition between plants has been studied from different perspectives. In ecology, scientists have studied competition between plants to understand succession patterns of vegetation, to comprehend the diversity and stability of plant communities, and to help define management strategies for seminatural ecosystems. In forestry, the main thrust has been the optimization of tree stand densities and thinning practices to maximize wood production. In agricultural sciences, competition studies have focused on optimization of sowing densities and on the evaluation of intercropping systems, where the potential for yield advantage is confined in the ratio between intraspecific and interspecific competition. In crop–weed competition research, most emphasis has been on minimization of the adverse effect of weed species on crop production and on the development of predictive tools for yield loss assessment to rationalize the use of herbicides.

The Negative Impact of Weeds

Worldwide a 10% loss of agricultural production can be attributed to the competitive effect of weeds, in spite of intensive control of weeds in most agricultural systems. Without weed control, crops might be completely lost, and for this reason weed management is one of the key elements of most agricultural systems. In addition to the reduced productivity of a crop following the capture of limiting resources by weeds, the quality of the produce might be negatively affected, particularly if individual plant size is one of the quality characteristics. As weeds often appear in patches, they might also affect the uniformity of the produce. In some other instances, damage is not just related to yield reduction. The simple presence of weed seeds above legal purity standards, like for instance the presence of poisonous seed of black nightshade (*Solanum nigrum*) in peas to be processed for canning, dramatically reduces the relative value of the crop. Other weeds, such as *Galium aparine* (cleavers) in cereals, interfere with mechanical harvesting operations, resulting in a lower acreage being harvested per unit of time. Weeds might also affect the efficiency of the land use system causing opportunity costs as financial resources are allocated to weed control. In extreme situations, the labor requirement for weeding might restrict or determine the area growing under production. This is typically the case in situations where weed control is to a large extent based on hand weeding, like in many smallholder farms in the tropics and on organic farms. Furthermore, some arable weed species act as alternate or alternative host for pests and pathogens.

Other Mechanisms of Yield Reduction

Yield reduction does not necessarily have to result from competition only, as other mechanisms through which weeds can have a negative effect on crop production have been identified. Allelopathy is one such mechanism, in which plant interference is mediated by the addition of plant produced phytotoxins to the environment. Chemicals with allelopathic potential are present in virtually all plants and in most tissues, including leaves, stems, flowers, roots, seeds, and buds. Under appropriate conditions, these chemicals may be released into the environment, generally the rhizosphere, in sufficient quantities to affect neighboring plants. The negative effect of weeds on crop production through the release of allelochemicals is just one example of the role of allelopathy in agriculture. Replanting problems, autotoxicity, and problems with crop rotations have also been attributed to allelochemicals. Currently, a lot of attention is focused on exploring the possibility of utilizing allelopathic crops to inhibit weed growth.

Parasitism is yet another mechanism of plant interference, in which the weed plant derives resources directly from a host plant. *Orobanche* spp. (broomrapes) and *Striga* spp. (witchweeds) belong to the most troublesome parasitic weeds in agricultural production systems. Both weed species connect to the root system of their host and withdraw nutrients as well as carbohydrates. *Orobanche* is a root parasite of, for example, legumes, sunflower (*Helianthus* spp.), tobacco (*Nicotiana tabacum*), and tomato (*Lycopersicon esculentum*). It lacks chlorophyll and depends completely on its host throughout its entire life span. *Striga* is a parasite of major cereal food crops, such as corn (*Zea mays*; maize), *Sorghum* spp., and millet (*Setaria italica*), and threatens agriculture throughout large parts of Africa. It is a so-called hemiparasite as, after emergence, the shoot starts producing chlorophyll. Apart from withdrawal of nutrients and carbohydrates the parasite has a pathogenic effect on the host plant, responsible for up to 80% of its yield reducing effect.

Resource Competition

Light

In potential production situations (i.e., water and nutrients are not limiting) crop growth rate and productivity are mainly determined by light, temperature, and crop characteristics. The availability of light varies greatly during the day. Its intensity is relatively predictable following diurnal and seasonal schemes, although changes occur very rapidly due to clouds. Light interception by a crop canopy is determined by the leaf area index of the species, and the light absorption characteristics of the leaves. These last characteristics are related to leaf thickness and leaf angle distribution. The leaf angle distribution determines the amount of radiation absorbed per unit of leaf area. Planophile leaves (horizontally oriented) capture light with a higher efficiency than erectophile leaves (vertically oriented).

In mixed canopies, apart from leaf area index and light absorption characteristics, plant height is an important factor in the distribution of light over the competing species. If a leaf is positioned above another leaf it will absorb a larger amount of radiation. A strong correlation between plant height and competitive ability has been demonstrated for many crop species. However, there are often tradeoffs. In rice (*Oryza sativa*), for example, it has been shown that taller varieties are more competitive, but these taller varieties have a lower yield potential as a result of lodging and/or a reduced harvest index.

Light, as such, cannot be stored in the system. If the canopy does not intercept it, it will reach the soil surface and be lost for crop growth. Competition for light is therefore an instantaneous process of resource capture. Competition for light is also considered size-asymmetric, meaning that relative to their size, larger plants receive a disproportionate share of the light-resource. Plants that grow fast in early growth stages thus have a strong advantage and can build up a larger share in the canopy. It is for this reason that competitiveness is not only related to light absorption characteristics, but also to light use efficiency. Some weed species demonstrate a strong phenotypic plasticity with respect to height development. *Chenopodium album*, for example, can overtop a sugar beet (*Beta vulgaris*) crop even when the weeds emerge under a closing canopy. It has been demonstrated that the decrease in the red:far-red ratio as a result of preferential absorption of red light by plants stimulates stem extension.

Water

All plants require water for transpiration, and their requirements are determined by the amount of absorbed radiation, temperature, vapor pressure deficit, and species characteristics. If there is a shortage in water supply, transpiration requirements cannot be met by water uptake by the roots, causing closure of the stomata. Because CO_2 is also transported through the stomata, photosynthesis is then reduced as well. If the shortage of water persists, this might also have direct effects on morphological features. Leaf extension rate, for instance, tends to reduce as a result of water shortage. Among the factors that determine crop growth in rainfed environments are the seasonal water supply, soil water holding capacity, root development, water use efficiency, and the ability of the crop to endure or tolerate water stress.

Plants growing in a mixture in a water limited production situation compete for water and generally also for light. The importance of water competition is determined by the length, severeness, and timing of the drought period(s). Competition mechanisms for water differ principally from competition for light. Competition for light is a process of direct competition for resource capture, with an instantaneous nature: if the resource is not captured, it is lost because the resource is not stored in the system. In contrast, water can be stored in relevant quantities in the soil compartment of the system. In periods where soil moisture content is high enough for potential growth, all species in a mixture can meet their water requirements for transpiration during that period. This is often the case during the beginning of the growing season. However, the amount of available soil moisture will be reduced during the growing season, when rainfall and other processes that increase soil moisture content cannot meet the water losses by evaporation and drainage. Therefore, plant transpiration in a period when water is not limiting growth affects water availability later in the season, reflecting the indirect nature of competition for water. Especially when the life cycles of species in a mixture differ, it may happen that an early maturing species does not suffer from water stress itself, but increases the water stress effects for the later-maturing species by enhancing total water loss earlier in the season.

Species differ in their water use efficiency with respect to dry matter production, which affects competitive relationships. It is well known that C_4 species require less water per unit dry matter produced when compared to C_3 species. A more efficient species may have an advantage in a drought situation, but, following the indirect nature of competition for water, it is often the timing of drought periods that determine if that holds. It should be realized that the ability of a species to outcompete neighbors or control resource availability and the ability to survive stress are not necessarily the same phenomenon.

Nutrients

In many production situations, the supply of macronutrients (N,P,K) limits the growth of the crop or vegetation for at least part of the growing season. Various processes influence the availability of nutrients in the soil. Apart from inputs (e.g., fertilization) and losses (e.g., leaching, uptake by plants) to the relevant soil compartment, availability is sometimes determined by conversions within the system. Nitrogen, for instance, can be present in a mineral (nitrate, ammonium) form that can be taken up by plants, but also in an organic form that prevents uptake by plants. Mineralization and immobilization are the processes governing the transition from one state to the other.

Competition for nutrients is to a large extent comparable to competition for water, though there are also differences. Similar to the resource water, nutrients can be stored in the soil system. Uptake of nutrients early on in the season will therefore effect nutrient availability later in the season. In contrast to the resource water, nutrients are also stored in relevant quantities in the plant tissue of the crop or the vegetation. This gives an additional importance to the uptake of nutrients in the period when supply is still adequate, as the early uptake of nutrients will also influence the performance of the crops later on. This demonstrates the complexity and the dynamic character of competition for nutrients.

Quantification of Crop Yield Loss

Density Based Models

Weed density is often used as the explanatory variable in descriptive models that try to relate the level of weed infestation to the loss in crop yield resulting from competition:

$$YL = \frac{aN_w}{1 + aN_w} \quad [1]$$

where YL is relative yield loss, N_w is weed density or the number of weed plants per unit area, and a is the fractional yield loss per unit weed density as $N_w \rightarrow 0$, or the yield loss caused by the first weed added to a weed free crop. This equation is generally known as the hyperbolic yield loss-weed density equation (**Figure 1**, upper curve). The asymptotic shape of the curve illustrates that the yield loss caused by the first weed added to the crop has the strongest effect. For each additional weed, the effect on crop yield diminishes, owing to an increased intraspecific competition of the weeds.

In this one-parameter regression model, maximum fractional yield loss is 1 (or 100%) at high weed densities; meaning a complete loss of the crop. However, high weed density does not result in a complete crop failure in all situations. An extended version of the yield loss–weed density equation permits a maximum fractional yield loss of less than 1:

$$YL = \frac{aN_w}{1 + \frac{a}{m}N_w} \quad [2]$$

In this two-parameter model, m is the maximum relative yield loss at high weed densities. A statistical comparison of various equations showed that at the field level, this equation gave the best description of crop yield loss due to weeds for a range of crop–weed combinations. Parameters a and m are, however, not constant for a specific crop–weed combination, but vary strongly from site to site and from year to year. This indicates that, although the above equation has large ability for describing and summarizing the consequences of weed infestations in retrospect, its predictive ability is rather poor. Consequently, the potential for using early-season weed density counts as a basis for rationalization of herbicide use in decision-support systems is rather limited. The instability in parameters a and m and the concomitant variation in the yield loss-weed density function reflect that factors other than just weed density have an important effect on the size of the yield reduction resulting from competition by weeds. Environmental factors, such as weather conditions and resource availability, and crop density are among these factors. Moreover, the period between crop and weed emergence has been identified as a principal factor in this respect, and for that reason additional variables representing the difference in the period between crop and weed emergence have sometimes been added to the standard version of the yield loss-weed density equation.

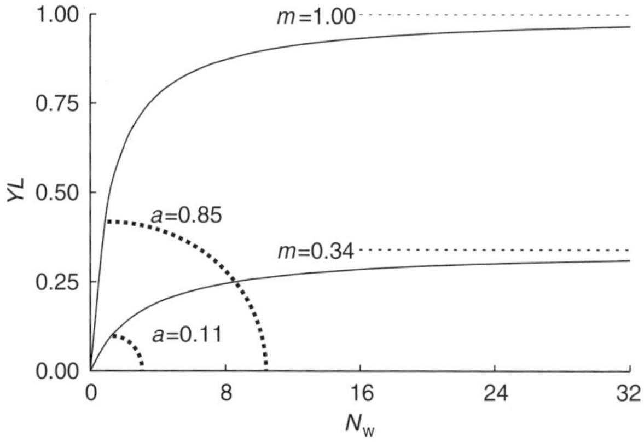

Figure 1 The hyperbolic yield loss-weed density curve (eqn [2]) illustrating the meaning of regression coefficients a and m. Curves represent the yield loss (YL) of tomatoes as a function of nightshade density (N_w) when field-seeded ($YL = 0.85N_w/(1+0.85N_w)$) or transplanted ($YL = 0.11N_w/(1+0.11\ N_w/0.34)$).

Relative Leaf Area Approach

The relative leaf area approach is an alternative descriptive regression model for early prediction of crop losses by weed competition. This approach stems from the observation that in simulation studies where weed density and the period between crop and weed emergence were varied, a close relationship existed between relative leaf area of the weeds shortly after crop emergence and final yield. This observation suggests that, rather than using weed density as such, it is better to weigh the density with the leaf area of the plants at the moment of observation. In this way the stronger competitive ability of early emerging and thus taller weed plants will be accounted for. This alternative model therefore relates yield loss to relative weed leaf area (L_w expressed as the share of the weed species in the total leaf area) shortly after crop emergence, using the relative damage coefficient, q, and the maximum fractional yield loss, m, as model parameters:

$$YL = \frac{qL_w}{1 + \left(\frac{q}{m} - 1\right)L_w} \qquad [3]$$

The relative damage coefficient, q, reflects the relative competitive ability between weed and crop after correction for their plant size. When crop and weed have identical physiological and morphological characteristics, q, approaches unity and a linear relation between relative leaf area of the weeds and yield loss is obtained (**Figure 2**). When the weed is a stronger competitor than the crop, the relative damage coefficient, q, will be larger than 1 and a convex curve is found above the diagonal line. When the crop is the strongest competitor, q will be smaller than one and a concave curve is found under the diagonal line. Because leaf area accounts for density and age of the weeds, this regression model accounts for the effect of weed density and the effect of the relative time of weed emergence.

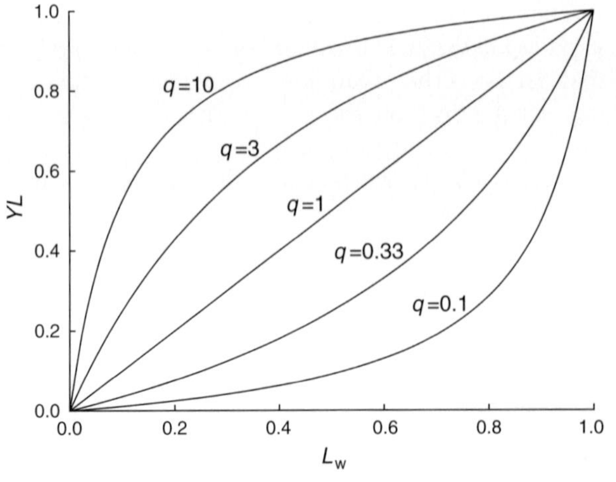

Figure 2 The relation between yield loss (YL) and relative leaf area of the weeds (L_w) according to eqn [3] at different values of parameter q.

Relative weed leaf area appeared to be superior over weed density as an explanatory variable in descriptive yield-loss models, especially if results from more than one site and year were simultaneously examined. Analysis of five sugar beet – *C. album* experiments, conducted in 3 years at different weed densities and with relative weed emergence ranging from 0 to 30 days after crop emergence, demonstrated that a general hyperbolic relationship between yield loss and weed density did not exist (**Figure 3**). However, the data of the competition experiments were accurately described by the yield loss-relative leaf area model. The value of q was found to be larger than 1, indicating that *C. album* was a stronger competitor than sugar beet.

Determination of the relative leaf area of the weeds is not always easy, limiting practical implementation of this model. For this reason, relative leaf area has sometimes been successfully replaced by relative weight of the weeds, leaving the general principle of the model unchanged. At the same time, with the rapid progress in the area of technology development for precision agriculture, it is likely that in the near future the leaf area assessment problem will be overcome.

Mechanistic Models

The two approaches discussed above are both descriptive: the models give a quantitative description of the consequences of the presence of weeds for the productivity of the crop, without considering the processes that determine competition. For a better understanding of competition, quantitative insight into the dynamics and mechanisms of competition is needed. Competition involves a complex of processes

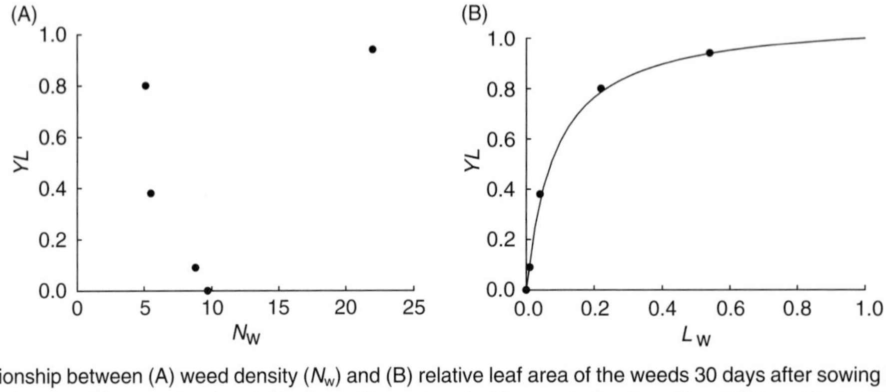

Figure 3 Relationship between (A) weed density (N_w) and (B) relative leaf area of the weeds 30 days after sowing (L_w) and yield loss (YL) for five experiments with sugar beet and *C. album*. Reproduced with permission from Kropff MJ and Lotz LAP (1992) Systems approaches to quantify crop–weed interactions and their application in weed management. *Agricultural Systems* 40: 265–282.

that together govern the capture, utilization, and distribution of resources over competing species. Ecophysiological competition models are developed to dynamically simulate the growth of all competing species, based on how morphological, physiological, and phenological processes are affected by environmental factors, such as resource availability, and vice versa. The models can be used to quantitatively analyze the crop–weed ecosystem and to identify the crucial processes, relevant environmental conditions, and important plant characteristics that together determine the outcome of competition in a specific environment. Examples of these are the use of models for determination of the critical period for weed control, or the identification of genetic traits that determine crop competitive ability.

In a crop–weed competition model, each species is represented by its own growth model, and, additionally, these growth models are linked by submodels that account for the distribution of resources over the competing species (**Figure 4**). Exactly how this distribution of resources is simulated is strongly related to the nature of the resource. Under favorable conditions, light is the main factor determining the growth rate of the crop and its associated weeds. Given the instantaneous nature of radiation, the distribution of light is mainly determined by the ability of both species to capture this resource. Incoming radiation is partly reflected by the canopy and the remaining fraction is potentially available for absorption by the mixed canopy. Radiation attenuates exponentially within a canopy as a function of cumulative leaf area index counted from the top downwards and light extinction coefficients. In most competition models, the leaf canopy is separated into a large number of leaf layers, and light extinction is then calculated for each layer, starting from the top. Within each layer, the leaf area of each species is weighted by its extinction coefficient, and the distribution of absorbed radiation over the competing species is then proportional to the species share in this effective leaf area. Water and nutrients are slightly different resources, as both can be stored in the soil. This requires that in the model, the actual supply has to be kept track of, like, for instance, through the introduction of a soil moisture balance, if it concerns water. Additionally, the demand of the species for these resources has to be determined. Transpiration is, for example, driven by the absorbed amount of radiation and the vapor pressure deficit within the canopy. The more light a species absorbs, the more water is required for transpiration.

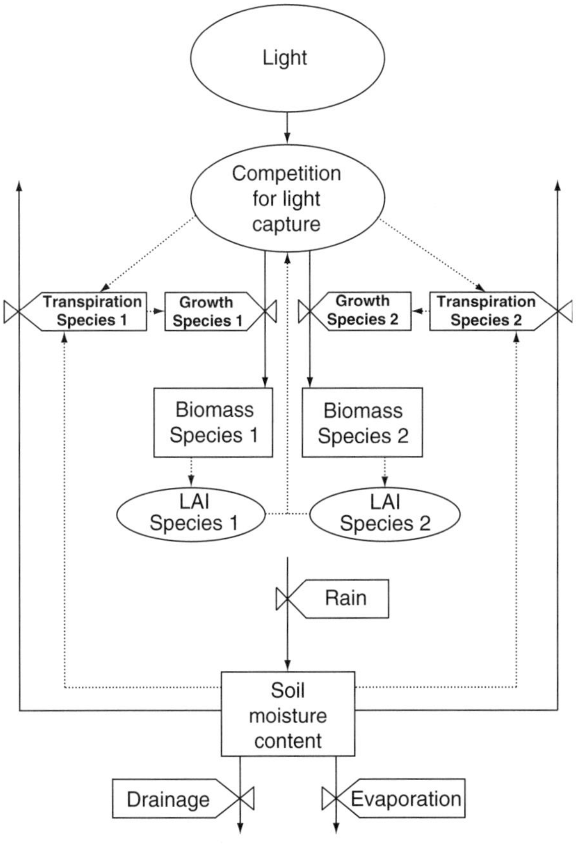

Figure 4 Simplified representation of the general structure of a crop–weed competition model accounting for competition for light and water. LAI, leaf area index. Reproduced with permission from Kropff MJ and van Laar HH (1993) *Modelling Crop–Weed Interactions.* Wallingford: CAB International.

In contrast to the descriptive models, the processes underlying competition (e.g., capture, utilization, and distribution of resources) are explicitly simulated in mechanistic crop–weed competition models. It is, however, a misconception to believe that comprehensive, mechanistic crop–weed competition models have better predictive ability than simple descriptive regression models. An increased complexity will be accompanied by an increased parameter and function requirement. Apart from the practical implications of this, each parameter estimate will have its own uncertainty, and these uncertainties and "errors" will accumulate in the final result, undermining the precision of the outcome of the simulation. For this reason, the models are not well suited for operational decision making. Their strength is their ability to assess the relative importance of various processes or traits in the competition process. One example of this is the use of crop–weed competition models in the development of more competitive crop cultivars. By conducting simulations and changing individual traits one at a time, the sensitivity of the model outcome for the specific trait can be determined, and this is a good indicator for the relative importance of that trait in the competition process.

Simulations under weed free conditions can be used to reveal the consequences of the various traits for yield potential, and as a result the model can be used for exploring the trade-off between competitive and yielding ability. Adaptations in general crop management might also favor the crop in its competition with weeds, and also in these situations the model can be used to explore the consequences of such management changes.

For the development of integrated weed management systems, particularly those that aim for a reduced reliance on herbicides, insight into crop–weed competition is essential, both for minimizing the detrimental effect of weeds in current crops and for the long-term management of weed populations. In the last decades, insight into crop–weed interactions, and particularly competition, has largely increased. The development of descriptive yield loss models resulted in a much better understanding of the harmful effect of weeds on crop productivity. From then on, the attention shifted toward insight into the processes that govern competition, and a much better appreciation of the complex and dynamic character of competition was obtained. Mechanistic models were developed to synthesize the existing knowledge, to determine the relative importance of various processes that are relevant in competition, and to identify gaps in current knowledge. The better understanding of crop–weed competition has great potential for the improvement of alternative strategic weed management systems.

List of Technical Nomenclature

Allelopathy	Interaction between plants in which interference is mediated through the release of phytotoxins to the environment.
Competition	The mutually adverse effect of organisms or species on one another.
Descriptive model	A regression model that relates a response variable to one or more independently determined explanatory variables.
Exploitation competition	An indirect form of competition, in which the interaction between organisms or species is mediated through one or more shared resources.
Interference competition	A direct interaction between two organisms in which one physically excludes the other from a portion of habitat.
Mechanistic model	A model that simulates the behavior of a system based on underlying processes.
Model	A simplified representation of a system.
Parasitism	Interaction in which one organism derives resources directly from a host.
System	A limited part of reality that contains interrelated elements.

See also: **Weeds**: Weed Biology; Weed Seed Biology; Weed Technology and Control.

Further Reading

Bastiaans L, Kropff MJ, Goudriaan J, and van Laar HH (2000) Design of weed management systems with a reduced reliance on herbicides poses new challenges and prerequisites for modelling crop–weed interactions. *Field Crops Research* 67: 161–179.

Cousens R (1985) An empirical model relating crop yield to weed and crop density and a statistical comparison with other models. *Journal of Agricultural Science* 105: 513–521.

Cousens R and Mortimer M (1995) *Dynamics of Weed Populations*. Cambridge: Cambridge University Press.

Kropff MJ and van Laar HH (1993) *Modelling Crop–Weed Interactions*. Wallingford: CAB International.

Kropff MJ and Spitters CJT (1991) A simple model of crop loss by weed competition from early observations on relative leaf area of the weeds. *Weed Research* 31: 97–105.

Liebman M, Mohler CL, and Staver CP (2001) *Ecological Management of Agricultural Weeds*. Cambridge: Cambridge University Press.

Weed Seed Biology

R E L Naylor, Trelareg Consultants, Banchory, Scotland, UK

Introduction

Although seed production and dispersal are important parts of the life cycle of any plant species, these aspects are particularly crucial for the survival of populations of annual weed species (**Figure 1**). Weed seed production is influenced by the pollination and compatibility systems of the species and by the opportunity for, and success of, seed development on the parent plant. Growing weed populations and the seed dispersal mechanisms of individual species influence the seed rain at a location. The balance between new seeds arriving in the seed rain, seed

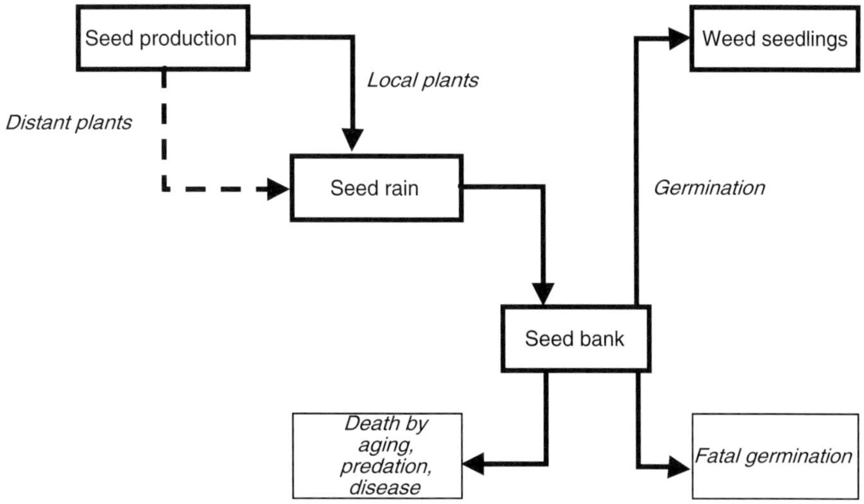

Figure 1 Weed seed dynamics.

death, and seed germination determines the dynamics of the weed seed bank in the soil. Genetic and environmental conditions each have a role to play in the development of seed dormancy and may influence seed size, shape, and physiological polymorphisms. Weed seed germination is a response to environmental cues and produces a new cohort of weed seedlings. These topics will receive attention in the following sections. The term seed will be used hereafter to indicate a complete propagule irrespective of whether the morphological structure is a seed or fruit.

Weed biology has implications for the ways in which weeds are managed in crops. Many features of crop management provide responses to the challenge presented by weeds and in particular by weed seed biology. Crop managers aim to minimize the seed rain and incorporation of seeds into a weed seed bank in the soil, because these seed banks are of great significance for future weed infestations. Crop managers also attempt to restrict the germination of weed seeds and the establishment of new weed populations in a growing crop.

Seed Production

Seed production by annual weeds is crucially important to their survival. It is no less important in perennial species where seed production as a consequence of sexual reproduction is important for the reassortment of genes and creation of new biological variability. Thus, although in annual species seed production takes place just once (but may be over an extended period), in perennial species seed production can take place on more than one occasion over an extended period of years. The seed is an important biological feature for dispersal of a species. In perennial species, vegetative reproduction may be important for consolidation and growth of a population at a single location. For perennials also, the production of seeds provides for new genetic individuals and the possibility of long-distance dispersal.

The production of flowers, fruits, and seeds can vary widely between related species, between populations of the same species at different sites, and between years at the same site. The number of seeds produced is a phenotypic character reflecting the interaction between genetic seed production potential and the environment experienced by the individual plant during its life. Local weather conditions and the crop management imposed by growers influence the number of plants achieving reproduction and the likelihood of pollination. Environmental conditions during seed development can influence the proportion of fertilized ovules which are retained, the amount of reserves they accumulate, and the degree of dormancy. Nevertheless, the most successful annual weeds generally have a large seed output which is maintained even in relatively adverse conditions.

For annual weeds, which possess many of the features of pioneer plants, the individual plants may be some considerable distance apart, more than the area over which pollen can easily be spread by the wind. Self-compatibility ensures that pollen from the same plant can fertilize ovules, thus ensuring some successful seed production at a site even when neighboring plants are scarce. In contrast, in perennial species, in which more energy is devoted to vegetative means of propagation, there are frequently mechanisms to ensure outcrossing by restricting the compatibility of pollen to fertilize ovules on the same plant.

One measure of the reproductive success of a weed is the number of seeds produced (**Table 1**). Different species have different capacities for seed production. In addition there is great phenotypic variation within species. Within one population, individual plants of blackgrass *(Alopecurus myosuroides)* produced from 4 to 54 inflorescences and bore a total of from 14 to 7614 seeds. This suggests that the potential for population growth is great.

The number of seeds produced by an individual plant or by a weed population is not the only measure of reproductive success. It may be better to take into account the success of the seeds in producing new seedlings at a later date, or the success of an individual plant in producing further plants. Sometimes the growing weed has been regarded merely as the mechanism whereby one seed produces further seeds.

Seed Size

Seed size is an important aspect of seed biology and represents the "investment" by the parent plant towards the success of individual seeds in producing a seedling. Generally large seeds are more successful in producing seedlings. They may also germinate earlier than smaller seeds, permitting more successful capture of environmental resources and therefore may grow bigger and faster than seedlings from later germinating small seeds. There is an evolutionary tradeoff between producing many small seeds, each with a small chance of survival, and fewer large seeds, most of which produce seedlings. Larger seeds, having greater storage reserves, are usually better able to emerge from depth than are small seeds of the same species. Of particular importance to weed species, larger seeds may be more successful at producing seedlings in hazardous environments, e.g., in arable fields where the grower takes measures to reduce weed infestations. A further consequence of seed size or mass is the potential for dispersal, sometimes aided by the development by maternal tissues on the outside of the seed of structures to aid dispersal (e.g., by wind or by animals).

Table 1 Examples of seed production in some weed species. The values (taken from a wide variety of sources) represent some of the highest values recorded for each species

Species	*Common name*	*Seeds* m^{-2}
Alopecurus myosuroides	Blackgrass	6500
Avena fatua	Wild oats	250
Bellis perennis	Daisy	1300
Bromus tectorum	Brome	14850
Capsella bursa-pastoris	Shepherd's purse	40000
Cardamine hirsuta	Hairy bitter-cress	640
Chamaenerion angustifolium	Rosebay willowherb	76000
Chenopodium album	Fat hen	72000
Cirsium arvense	Creeping thistle	5200
Convolvulus arvensis	Bindweed	2000
Elytrigia repens	Couch	1000
Galium aparine	Cleavers	3500
Matricaria discoidea	Pineapple weed	7000
Papaver rhoeas	Corn poppy	17000
Ranunculus repens	Creeping buttercup	100
Raphanus raphanistrum	Wild radish	2250
Rumex crispus	Curled dock	60000
Senecio jacobea	Ragwort	150000
Senecio vulgaris	Groundsel	1200
Spergula arvensis	Spurrey	10000
Stellaria media	Chickweed	15000
Veronica hederifolia	Ivy-leaved speedwell	45

It is assumed that all seeds which are produced by a plant are viable. This may not be the case. Samples of seed of blackgrass collected from different locations in UK had germinations from 8% to 42%. The proportion of seeds with a ripe caryopsis was 14–55%. In some cases the seed may be used as a vehicle for dispersal even if sexual reproduction has not taken place (apomixis). The seeds are thus clones of the parent and such seeds are termed apomictic. This occurs in, e.g., dandelions (*Taraxacum* spp). Apomixis may replace sexual reproduction in adverse climatic conditions or at the end of a growing season.

In many plant species, including weeds, there may be more than one type of seed on a parent plant. Such polymorphisms may be of size, color, shape, or of physiological state. Polymorphisms can be interpreted as adaptations by the species to ensure survival in an unpredictable environment. Polymorphic differences in seed size can be interpreted as meaning that large seeds, not being dispersed far, ensure survival at the locality (persistence) while small seeds have a greater chance of being dispersed further away, but have a lower chance of landing at a safe site. In some, but not all, species, larger seeds are more dormant. Populations or plants with apparently higher mean seed weight usually have a higher proportion of large seeds. Further implications of polymorphisms are considered below.

An important issue for crop managers is how to manage crops so that future weed populations are minimized. Clearly, it is important to manage existing weed populations so that no new seed is produced and shed to the soil. Weed control in crops not only serves to increase yield through reduced competition from weeds, but also contributes to reducing weed populations in future crops.

The Seed Rain

The seeds shed from weeds onto the surface of the soil are known as the seed rain. This comprises seeds of a mixture of species and is distributed unevenly over time and space. The species composition of the seed rain is rarely in the same proportions as the parent mixed weed population because different species have different seed outputs (**Table 1**). The composition of the seed rain may also not correlate with the composition of the seed bank. This may be because the existing seed bank is large compared to the seed rain or because of selective predation of seeds on the soil surface before they are incorporated into the soil. There is likely to be a greater correlation of the composition of the seed rain with assessments of the seed bank taken soon after seed shedding than with samples taken prior to the next seed rain season.

There are reports of propagules with structures favoring wind dispersal moving several kilometers. However, most seeds are shed around the parent plant with rapidly decreasing numbers of seeds being dispersed to great distances. This pattern tends to consolidate the occurrence of a species at a site where it has been able to grow and reproduce successfully, but also to provide the opportunity to spread to new, more distant sites. Even in species which appear to possess characters which enhance wind dispersal, most seeds may fall near the parent plant. Ragwort *(Senecio jacobaea)* seeds bear a pappus but fewer than 0.5% of seeds land more than 1 m away. In general, taller species have a greater distance of dispersal. In many agricultural systems, new weed infestations comprise single plants. Over time, these develop into patches of small populations.

In many arable situations the seed rain may be about 10 000 seeds m^{-2} shed to the surface of the soil. Incorporation into the soil is important because most seeds germinate better when covered by soil than on the soil surface. Seeds remaining on the soil surface are liable to be eaten by predators. Indeed the reduction of weed seeds at the soil surface consequent on the development of efficient weed management has been linked to the decline in farmland birds. Seeds on the surface may fall down cracks in the soil and this may be assisted by the bending or twisting of elongated structure (awns) on seeds of some species. Some animals may bury seeds in the soil (earthworms are particularly important for this) but a major mechanism for seed burial is soil management prior to sowing crops. The techniques of plowing and of surface cultivation are designed to bury crop residues and to create a new seed bed, but at the same time serve to bury weed seeds. Plowing to 20 cm tends to bury 90% of seeds deeply (to greater than 10 cm) where they may survive, dormant for a considerable time. Replowing, of course, tends to bring them to the surface again. Surface cultivation mixes seeds into the surface layers of the soil (typically 60% in the top 5 cm) where they may be more likely to experience the environmental cues for germination. Thus, choice of implement can have a great effect on the subsequent weed seedling population establishing and its composition as well as weed seed longevity.

Farm implements may move seeds horizontally on or in the soil and this may contribute to the spread of infestations throughout a field. Typical sequences of involving five operations of farm machinery in a field have been shown to move seeds horizontally up to 15 m. An individual operation moves different size and shapes of seeds by different amounts. Farm machinery can be an important vector for the dispersal of seeds because of seeds contained in soil attached to tyres or other parts. Farm machinery hygiene is important to prevent the longer-distance spread of weed species particularly when machinery is shared between a number of users.

Many seeds have structures which are interpreted as attracting animals which then become important dispersal vectors. Seeds with hooks or barbs can be dispersed on the surface of animals attached to their feathers or fur. Domestic stock or wild animals can disperse seeds.

The seed rain is important because it adds new fresh seeds to the seed bank. It is these seeds which are likely to have the greatest vigor and longevity. This is reflected in the old gardener's saying: "One year's seeding, seven years' weeding." Failure to prevent seed rain can lead to rapid and large increases in the weed seed bank. A field with a seed bank of 5500 seeds m^{-2}, after 2 years of set-aside with no weed control had 198 500 seeds m^{-2}.

Clearly, an imperative for weed management is to prevent the return of fresh weed seeds to the soil. This is achieved by attempting to kill or otherwise remove the weeds from the growing crop. Often when crops are harvested mechanically, weeds seeds are detached from the parent plant and deposited on the soil surface with the crop residues.

Mechanical systems within harvesting machinery which screen out the weed seeds and enable them to be removed elsewhere and destroyed can play an important part in the long-term reduction of weed populations. Burning crop residues after harvest has provided some destruction of weed seeds and can make an important contribution to the restriction of weed populations, e.g., of wild oat *(Avena fatua)*. However, burning of crop residues is not always permissible now.

Seeds as Food

Many animals eat seeds (a process termed "spermophagy") either while they are still on the parent plant or after they have been shed. Most seed heads of the thistle *Cirsium canescens* contained endophagous insects that damaged about 25–50% of seeds. Grazing of the inflorescences of butterbur (*Petasites hybridus*) by cattle and sheep has sometimes prevented any seed production. After seed shedding, seeds on or near the soil surface can provide important food for birds, small mammals, and invertebrates. The cereal grains are an important food for wild farmland birds, seeds of important grass weeds such as wild oat, blackgrass, and the brome-grasses (*Bromus* spp.) do not seem to be important food species for them. Farmland birds do, however, consume the seeds of some important dicotyledonous weed species, e.g., species of fat hen and relatives (*Chenopodium* spp.), knotgrasses and relatives (*Polygonum* spp.), chickweeds (*Stellaria* spp.) and others. Unfortunately, the seeds of cleavers (*Galium aparine*) do not seem to be an important food for birds.

Not all species are equally predated. Many spermophagous animals exhibit food preference. Frequently species with large seeds are preferred, possibly because less energy is used in foraging if larger items can be located and consumed. The distribution of seeds in the soil can itself influence the level of predation. Local patches of initially high seed density may attract high populations of predators whose activity can result in patches of low density of weed seeds.

The Seed Bank

The seed bank in the soil is of considerable significance to crop managers. Weed populations arise from the seed bank in the soil which thus represents a potential weed infestation. In his book, On The Origin of Species, Charles Darwin reported that from about 200 g (dry weight) of soil he counted 537 seedlings emerging over the next 6 months. This both emphasizes the size of the weed seed bank in the soil and indicates that not all seeds germinate together. Because weed seeds may have considerable longevity, so the shedding of seed in any one year may provide weed populations in crops for many years to follow: eliminating a weed population in any one year is unlikely to lead to continued absence of weeds thereafter. Because the seed bank in the soil has been derived from seed populations over many years, the species composition and spatial distribution are quite variable. The composition of the weed seed bank reflects previous cropping and weed population history and has been termed an "ecological memory."

Frequently the number of species in the weed seed bank is less than the number of species found in adjacent natural or seminatural habitats. This reflects the reduced biodiversity of agricultural land. Often the weed seed bank contains a higher proportion of annual species. However, although poor in species numbers, the seed density of weed seed banks is frequently greater than seed banks in adjacent natural and seminatural habitats, sometimes by several orders of magnitude. Again (not surprisingly) the majority of seeds are of annual species.

The size of the seed bank is affected by the history of the weed population and of the weed management in the particular field (**Table 2**). The total number of viable weed seeds in the soil is typically 30 000–80 000 seeds m^{-2} but may be double this. For comparison, a typical sowing rate for cereals might be about 350–500 seeds m^{-2}. Arable fields with a long history of effective weed management may contain "only" 3000 seeds m^{-2}. Individual weeds species often exceed this, e.g. blackgrass may have 5500 seeds m^{-2}. Organic systems, in which use of herbicides is not permitted, frequently have large weed seed banks in the soil and weeds are recognized as a major constraint in such systems. Nevertheless, large weed seed banks can develop even when herbicides are used.

The distribution of seeds in the seed bank is not even. There are more seeds near the soil surface because that is where the seed rain lands. Mixing takes place by soil management with farm machinery or by animals burying seeds. There is great horizontal spatial variability in both seed density and in species composition of the seed bank. The borders (headlands) of fields usually contain more weed seeds and grow more weeds than do the central portions of fields. Because most seeds are not dispersed far from the parent plant, the distribution is patchy, with a different typical patch size for different species. The general farming system may be considered as a

Table 2 Some estimates of the size of the weed seed bank under a range of crops grown either with or without herbicides. Values are the high recorded values, not typical values, for each crop/management category

Crop	*System*	*Seeds* m^{-2}
Cereals	Herbicide	17 000
Maize	Herbicide	130 000
Potatoes	Herbicide	40 000
Sugar beet	Herbicides	138 000
Cereals	Organic	74 000
Vegetables	Organic	48 800
Sown rotational grassland	Organic	20 000

Based on a wide variety of sources, including Radosevich S, Holt J, and Ghersa C (1997) *Weed Ecology: Implications for Management*, 2nd edn. New York: John Wiley.

response to soil physical conditions (soil type, wetness) and seems to influence the species composition of the seed bank. The management of the immediately preceding crop has a major influence on seed density. Soil chemical features seem to have little impact on seed banks. Both the vertical distribution and the horizontal patchiness have implications for weed managers.

As a broad generalization, a plant species may produce either long-lived seeds or seeds that germinate relatively soon after shedding. Clearly, the former behavior leads to the development of a persistent seed bank, with many of the seeds greater than 1 year old. Species with seeds that germinate soon after shedding have transient seed banks in which the seeds survive usually for less than 1 year and need constant replenishment if the population is to survive at that location. Although these represent extremes of species behavior, many of the most important arable weeds produce large persistent seed banks. For species of northwestern Europe, perennials generally have less persistent seeds than related annuals and biennials. Increased habitat disturbance (e.g., soil cultivation) tends to select for species with more persistent seed banks. Greater persistence of seeds is not consistently related to smaller seed size. Generally, species with long or flat seeds have less persistent seed banks.

The size of the seed bank is constantly changing. Fresh seeds are entered into the seed bank via the seed rain generally in late summer and autumn. Some weed species can produce seeds for long periods of the year, e.g., *Stellaria media* and *Poa annua*. Seeds of some species may germinate in autumn, others in spring and this germination reduces the size of the seed bank. In addition, predation and death of individual seeds contribute to a decline in the seed bank. In general, the weed seed bank is dominated by seeds less than 1 year old. Although older seeds buried deep in the soil are less likely to germinate, they nevertheless are important for the long-term survival of weed species at a particular site. Different species have different characteristic rates of decline in viability and hence in their longevity.

There are many causes of seed death (as opposed to their consumption as food). The oxygen supply to the seed is important to permit respiration. The restricted oxygen supply in cold wet soils may contribute to seed death, particularly of older, deteriorated seeds. Fungi may kill seeds indirectly through depriving them of oxygen.

In the absence of a fresh seed rain, the death or predation of seeds leads to a decline in the seed bank. Often a constant proportion of seeds survives each year so that a plot of the logarithm of seed numbers against time shows a negative linear slope. In the absence of soil disturbance, this may be a slow process, i.e., have a shallow slope. Crop managers are interested in accelerating this natural decline. Systems of soil cultivation, such as rotovation, which bring seeds to the surface, expose them to light and to fluctuating temperatures can promote removal of seeds from the seed bank by germination. So long as these seedlings are killed and not allowed to grow, reproduce, and provide a fresh seed rain, there will be a net decline in the size of the seed bank. In an experiment where seed rain was prevented, the seed bank in a field declined over 5 years from 20 000 to 12 000 m^{-2} when undisturbed, to 10 000 seeds m^{-2} when cultivated twice yearly, and to 7000 seeds m^{-2} when cultivated four times a year. The more occasions such cultivation is practiced, the greater the decline. The stale seed bed is a common and useful technique of weed management in organic cropping systems. This involves preparing a seed bed for the crop well in advance of sowing so as to encourage weed seed germination, followed by killing of the weeds before preparing the seed bed again and sowing the crop. An alternative is never to cultivate the soil so as not to bring buried seeds to the surface. Eventually, the deeply buried seed will die and the surface layers will become depleted of weed seeds as they germinate. Crucial to the success of this minimum tillage approach is the prevention of any fresh seed rain.

Germination, Dormancy, and Longevity

Seed germination is the first stage of the transition from a dormant (but live) seed to an autotrophic independent growing plant. Precise germination requirements have been evaluated and quantified for some crop species but less attention has been paid to weed species. There have been numerous laboratory studies which are of limited relevance to understanding what influences weed seed germination in the field because they have investigated only single or constant factors.

When weed seeds are shed many do not germinate immediately: they are dormant. There are many different mechanisms of dormancy, all of which serve to limit the time of germination of weed seeds. Freshly shed seeds may not be able to germinate (**Figure 2**). This is termed primary dormancy and may be due to one or more dormancy mechanisms. In many members of the Ranunculaceae (including the buttercups) and the Apiaceae (including cow parsley (*Anthriscus sylvestris*) the dormancy may be morphological because the embryo is not fully developed. In these species, germination is delayed because the

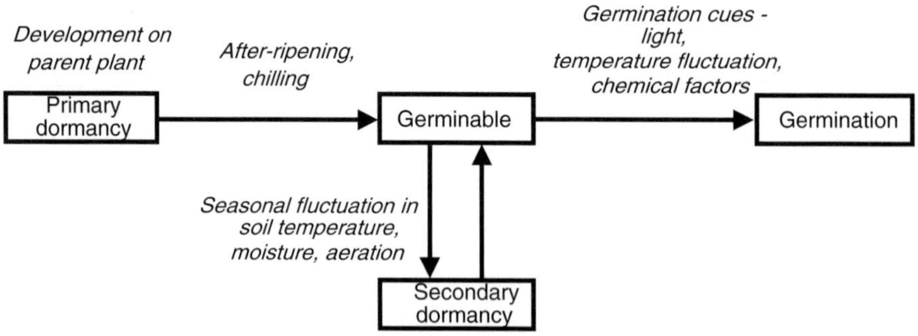

Figure 2 Scheme of dormancy changes in seeds.

embryo has to grow and differentiate before germination can occur. This is also why umbelliferous crops such as carrots and parsnips are termed "slow germinators." Dormancy may also be mechanical (sometimes termed the straitjacket effect) and caused by a very rigid hard seed coat. In commercial seeds, scarification or cracking the seed coat can help germination. In weed seeds, this might be achieved by weathering of the seed coat in the soil or passage through the gut of an animal. Seed coats may also impose a physical dormancy by restricting the access of water or oxygen into the seed. In the Fabaceae (e.g., smooth tare (*Vicia tetrasperma*)) this is often linked to seed coat color: darker seeds have more anthocyanin in the seed coat and this restricts water access. Finally, dormancy may be physiological, linked to endogenous hormone levels within the seed.

In addition to different species having different dormancy mechanisms and different lengths of dormant period, there are also differences within species, particularly between populations. For example, when plants from 50 populations of *Poa annua* in the north of Scotland were grown in the glasshouse, the seeds had different germinabilities (7–100%) and different mean germination times (4–28 days). Germinability and mean germination time were not associated. Although the maturation environment can influence seed dormancy, this was not the case here where all the seeds matured in the same (glasshouse) conditions. This suggests that the populations had different responses to the maturation environment which resulted in different dormancy depths. In microspecies of *Taraxacum* (dandelions) there are distinct differences in the response of fresh seeds to temperature and light and in survival in the soil.

The position of the seed on the parent plant often influences dormancy. This is notable in umbellifers where seeds on the primary umbel are larger, heavier, less dormant, and less viable than seeds on secondary or tertiary umbels. In the Asteraceae genus *Xanthium* (cockleburs) the seeds are borne and dispersed in pairs: the upper seed is more dormant and germinates about 1 year later than the other. Similarly in wild oats (*Avena fatua* and *A. ludoviciana*), the seeds borne at different positions within the spikelets have different germination requirements.

Seed polymorphism also influences germination. In the weed fat hen (*Chenopodium album*) as in many related species, the seeds are polymorphic: they may be brown or black and have smooth or reticulate seed coats. In UK most fat hen seeds are black and have thick seed coats. These seeds need to experience a cold period or encounter a nitrate-rich soil solution before they will germinate. Up to 5% of seeds may be large and brown, have thin coats, and germinate earlier than the black seeds. These seeds are the first to ripen on the parent plant. In spurrey (*Spergula arvensis*) seeds with a papillate seed coat germinate better than nonpapillate forms at low temperatures; the reverse occurs at high temperatures. These seed coat morphs also differ in their response to water availability.

The occurrence of seed size and dormancy polymorphisms in a species make it difficult to ascribe a particular set of conditions as germination promoting. A treatment which apparently promotes germination of say 55% of seeds is not necessarily characteristic of the species: it may actually promote germination in 100% of one morph and inhibit it in other morphs. Because the maturation environment can change the proportion of different morphs and alter the depth of dormancy, the conditions that promote germination may not always be constant from year to year. This makes it difficult to predict the germination of individual weed species. Seed dormancy is ecologically important and of agronomic significance because it provides "dispersal over time."

Because each species has its own characteristic germination requirements, different weed species emerge at different times. Soil management, which is designed to create a good seed bed for crops seeds also does this for weed seeds in the soil. The timing of cultivation and sowing can therefore have a profound effect on which species are killed by cultivation and which emerge with the crop. Clearly the species composition of the established weed population can depend on the soil and crop management decisions taken and may differ from year to year.

Naturally, weed seedlings which emerge before or as the crop seedlings emerge have the potential not only to have a great competitive effect on the crop but also to produce large numbers of seeds. Weed seedlings that emerge late, into an already established crop, suffer from competition from the crop and their seed production is restricted. This emphasizes the importance of controlling weeds in a young crop for long-term weed management.

Primary dormancy is gradually lost, and the seed is then capable of germinating if the correct environmental cues are perceived by the seed. If not, the seed remains ungerminated and eventually may re-enter a period of secondary dormancy. Seeds that are buried deeply in the soil may progress through repeated cycles of secondary dormancy and release. The main cues for germination, which are common to all plant species, include fluctuating temperature, light, and soil aeration. These conditions occur near to the soil surface and so seeds that are buried deep in the soil tend not to germinate. It is generally the seeds in the 2–5 cm nearest the surface that produce the seedlings which establish the next weed population. Seeds on the surface can be damaged by exposure to high temperatures.

Over time, the germinability and vigor of seeds declines and ultimately the seed dies. Low seed vigor may lead to the process of germination taking longer. Different species have different characteristic seed longevities. In some species the seeds survive for only about 1 year. In other species longevity of seeds can be considerable with the records of survival of seeds in the soil of over 100 years. In addition, seeds may be eaten by soil organisms or attacked by fungi or other soil microorganisms which lead to rotting. Two classical seed burial experiments in the United States showed that at least one seed germinated from seeds of 3 out of 20 species placed in the soil 100 years previously (Beal's experiment in Michigan) or from 36 of 107 species after 39 years (Duval's experiment in Virginia). These figures represent the extremes of seed survival; the majority of seeds last less long. Nevertheless, following deep plowing after a prolonged period of grass, there are frequent observations of flushes of weeds comprising species not seen in fields since the previous arable crop.

Many investigations of soil seed banks show declines in the seed bank in excess of the number of seeds germinating. This may be due to aging followed by death, by disease, or by predation. Some seeds, particularly those deep in the soil, may germinate but not emerge and produce established seedlings. This may be the fate of numerous small seeds which do not have large reserves to fuel germination and growth. Such "fatal germination" is useful in depleting the weed seed bank and soil.

Modeling and Predicting Germination

There are numerous observations of the responses of weed species to changes in crop management, some of which are instituted to restrict the growth and spread of weeds. Although these observations are useful for the crop manager, they do not provide a sound basis on which to anticipate the likely effects of changes in crop management on weed populations. Central to modeling the dynamics of weed populations is a detailed and quantitative consideration of the fluxes between various categories (seeds, seedlings plants), various states (live, dormant, dead), and various locations (near the soil surface, deeply buried). This core section needs to be preceded by a section containing quantitative descriptions of the influence of competition from the crop species and variety and of weed management measures on seed production and the seed rain. It must then be followed by information on the removal of individuals from the seed bank through death or as germinating seedlings and the survival of seedlings to reproduction. Application of such a model in a recursive way can provide useful "scenario testing" for assessing the implications of modified crop production systems on weed populations in the medium to long term. Examples of the use of this approach have suggested that crop rotation can greatly reduce the populations of weed species with relatively short-lived seeds (and therefore with transient seed banks) but the effect on species with longer-lived seeds (and hence with more persistent seed banks) was less. Seed survival in the soil over the winter seems to be an important determinant of the overall population dynamics with seed production and plant survival of less importance. Nevertheless, populations of weeds within crops can be efficiently and effectively reduced (almost eliminated) using chemical herbicides with the effect of drastically reducing the seed rain.

The incorporation of weed population models into decision support systems could provide a useful tool for crop managers. The combination of "expert" scientific knowledge, with "best practice" guidelines and encyclopedic information on specific products and procedures has the potential to facilitate technology transfer into farm practice. The ideal system would be able to incorporate local weather forecasts to predict the efficacy of the procedure selected.

Conclusions

Weed populations in crops are dependent on seed invasion for their initial generation and often on seed production and survival for their continued existence at a site. Weed seed populations comprise a mixture of species and seed ages. The species composition and density of the weed seed bank bear a loose relationship to the weed vegetation in previous and in subsequent years. The dynamics of weed seed populations is complex and influenced by abiotic factors (weather), biotic factors (seed predators) and by crop management procedures. Weed seed populations are important because they impose a continued need for weed management over a series of years.

See also: **Production Systems and Agronomy**: Agricultural Crops; Organic Farming. **Seed Development**: Germination; Seed Production; Seed Quality. **Seed Dormancy**: Development of Dormancy. **Weeds**: Herbicide Resistance; Weed Biology; Weed Competition; Weed Technology and Control.

Further Reading

Bunce R and Howard D (eds) (1990) *Species Dispersal in Agricultural Habitats*: London: Pinter.

Cousens R and Mortimer AM (1995) *Dynamics of Weed Populations*. Cambridge: Cambridge University Press.

Fenner M (ed.) (2000) *Seeds: The Ecology of Regeneration in Plant Communities*, 2nd edn. Wallingford: CAB International.

Harper JL (1977) *Population Biology of Plants*. London: Academic Press.

Leck M, Parker V, and Simpson R (eds) (1989) *Ecology of Soil Seed Banks*. San Diego: Academic Press.

Naylor REL (ed.) (2002) *Weed Management Handbook*, 9th edn. Oxford: Blackwell Science.

Radosevich S, Holt J, and Ghersa C (1997) *Weed Ecology: Implications for Management*, 2nd edn. New York: John Wiley.

Zimdahl R (1999) *Fundamentals of Weed Science*, 2nd edn. San Diego: Academic Press.

Weed Technology and Control

R L Zimdahl, Colorado State University, Fort Collins, CO, USA

Introduction

Developers of weed technology are concerned with finding economically efficient (profitable) methods of weed control in agronomic and horticultural crops, on rangeland or turf, and in noncrop environments. Weed technology is the application of scientific principles and knowledge to the development of methods or processes for dealing with the presence of weeds. Other articles in this encyclopedia deal with the definition of weeds, the characteristics that weeds share, and the detrimental aspects of weeds. These things will not be addressed in this article.

Concepts of Prevention, Control, Eradication, and Management

Weed technology usually begins with the control but it ought to and, in fact, often does begin with prevention. Prevention includes things done to avoid the necessity of control. It is stopping weeds from invading an area. It is a practical means of weed management but it takes time, careful attention to many details, and success is difficult to achieve. In weed science, as in human medicine, prevention is better than control, but control is required because too often the weeds arrive before they can be prevented.

Weed control includes many techniques designed to limit weed infestations and minimize competition. These techniques attempt to achieve a balance between cost of control and crop yield loss. Weed control techniques have been adopted readily because control is easier than prevention or eradication. Control is compatible with short-term economic goals.

Eradication is complete elimination of all live weeds, weed parts, and weed seed. Eradication is the best program for small populations of perennial weeds, but present technology does not make it easy.

Weed management is the combination of the techniques of prevention, eradication, and control to deal with weeds in a crop, cropping system, or environment. Weed managers recognize that cropping history, management objectives, available technology, financial resources, and many other factors must be combined to make good management decisions. Complete weed control in a crop may be the best decision in some, but not all, cases. Maintenance of a weed population at a low level in a cropping system may be the most easily achievable and financially wise goal.

Weed Prevention

People want to be and stay healthy. When we become ill, we are pleased to have competent physicians, hospitals, and medical services. We would rather remain healthy than have to cure an illness. The same thinking applies to weed management. Weed control treats, it doesn't prevent weeds.

A good weed management program includes vigilance. The good weed manager can identify weed seeds, seedlings, and mature plants and has a management program for each crop and field and appropriate follow-up programs. The good manager is ever watchful for new weeds that may become problems and, whenever possible, emphasizes prevention. Several preventive techniques are available including the best method, planting clean seed. Other preventive strategies include:

- preventing weed seed production especially by new weeds
- preventing vegetative spread of perennials
- isolating introduced livestock to prevent introduction of weed seeds
- use of clean farm equipment and clean itinerant equipment, including combines, cultivators, and grain trucks
- cleaning irrigation water
- using mowing and other appropriate weed control practices on irrigation ditch banks
- inspecting imported nursery stock and soil for weeds, seeds, and vegetative reproductive organs
- paying special attention to fence lines, field edges, rights of way, railroads, etc., to prevent introduction of new weeds
- preventing deterioration of range and pasture
- cleaning crop seed and disposing of cleanings properly by seed and grain handlers.

Mechanical Control

No weed control method has ever been abandoned completely. New techniques have been added but old ones are still used, especially in small-scale agriculture. Mechanical methods have a long history and are the primary control technique in many crops.

Manual Labor

Hand-pulling is practical and efficient, especially in gardens. It is very effective for annual weeds, but not for perennials capable of vegetative reproduction. A disadvantage is that hand-pulling doesn't get the job done when it's most needed. Most of us are too busy or too lazy to go out and weed before weeds become obvious. By the time they become obvious, easy to grab, and pull, yield reduction due to weed competition will have occurred.

Hand-hoeing has been used for weed control for many years. It is the method of choice for most gardens and ornamental plantings and is used in vegetable crops, and on small farms in developing countries. Hand-hoeing controls the most persistent perennials if it is done often enough, but it takes a lot of time and human energy. If human labor is abundant, and labor cost is low, manual labor is an acceptable method of weed control. If human labor is not abundant and it is expensive, hand methods are expensive and not efficient.

Tillage

When most people think of mechanical control, the first thing that comes to mind is tillage with an implement to disturb, cultivate, or mix the soil. On arable land, tillage alone or in combination with other methods may be the most economical system of control. Tillage buries crop residue, conditions soil, and facilitates drainage. It controls weeds by burying them, separating shoots from roots, stimulating germination of dormant seeds and buds (to be controlled by another tillage), desiccating shoots, and exhausting carbohydrate reserves of perennial weeds.

Other reasons for tillage include breaking up compacted soil, soil aeration, seed bed preparation, trash incorporation, and crop cultivation. All of these are important, but the main accomplishment of most tillage in the world's developed countries is weed control. The advent of no-till farming and minimum-till farming have shown that tillage is not essential to grow crops and may do no more than control weeds. Too frequent tillage can increase soil compaction; a disadvantage. Other disadvantages include exposure of soil to erosion, moisture loss, and stimulation of weed growth by encouraging germination of dormant seeds and vegetative buds. Decisions about the role of tillage must be made for each soil type and farming system.

Primary tillage is initial soil breaking or disturbance. Depth varies from at least 15 cm (except where primitive tools are used with limited animal power) to as much as 60 cm. The implements include moldboard and chisel plows, which cut and invert soil and bury plant and other surface residue. This is often the first step of seedbed preparation. Secondary tillage may be subsequent to primary tillage or it may be the first tillage operation. Soil is disturbed, often vigorously, but upper layers are usually not inverted. Secondary tillage is fast, inexpensive, and its tools are appropriate for large areas. Tools include the double disk, several kinds of harrows, field cultivators, spring-tooth harrows, rototillers, rod-weeders, and the cultipacker. This diverse group of implements tills soil from a few cm up to 12 to 15 cm deep. Secondary tillage implements break clods, firm soil, and remove weeds. The tools of secondary tillage are effective for weed control and seedbed preparation.

Primary and secondary tillage is followed, in many row-crops, by selective interrow cultivation or intertillage. Tractor-mounted cultivators or animal-drawn implements move soil between crop rows to loosen it and control weeds. In general, interrow tillage is just that; it works between crop rows. Some implements prepare interrow areas for furrow irrigation (where water runs down furrows between crop rows). Implements used for interrow cultivation include a wide range of tine (long finger-like rods) and flared or straight steel shovel-like tools at the end of solid or flexible (flat, steel) shanks that travel through soil at shallow depths (2 to 5 cm). They break soil crusts, and facilitate irrigation but the main purpose is weed control.

Intrarow tillage implements to control small weeds in the crop row are available. Research has shown that, in row-crops intrarow cultivation must be done early in the growing season for best results. Without herbicides, weeds in row-crops are always controlled better by an in-row cultivator than by the standard interrow cultivator when each operation was performed at the right time. In-row cultivators have special tools that disturb soil around crop plants and uproot weeds in rows. The tools include spyders (toothed disks that move soil toward or away from crop rows) and torsion weeders and spring hoe weeders that flex vertically and horizontally to uproot weeds in crop rows. Spinners displace weeds in crop rows. Standard interrow crop cultivators are most effective on weeds 15 cm or smaller. These cultivators do not work well in row-crops when weed density is high.

There are situations where plowing and subsequent tillage do not adequately prepare land for planting. These include land heavily infested with perennial sod-forming grasses. This may be found in developing country agriculture.

To maximize tillage benefits, uniform spacing of crop rows, straight rows achieved by precision planting, gauge wheels, and depth guides are needed. The success of tillage for weed control is also determined by biological factors:

1. Weeds that share a crop's growth habit and time of emergence may be the most difficult to control, especially when they grow in crop rows. Weeds that emerge earlier or later than the crop are often easier to control.
2. If a weed's seeds have a short, specific period of germination, it is easier to control them with tillage as opposed to those whose seeds germinate over a long time.
3. Perennial weeds that reproduce vegetatively are difficult to control with tillage alone.

Successful mechanical control of weeds is also determined by human factors. Human cultural knowledge tells one when and how to do something on a given soil and farm. Tillage is a cultural practice and therefore, by definition, it requires cultural knowledge. It requires the mind of a good farmer who knows the land. Successful mechanical control requires managerial skill that cannot be acquired from science. Such knowledge is acquired by doing and by observing those who have done things well. Cultural knowledge is the art of farming whereby one knows how to select and apply scientific knowledge to solve problems. Successful mechanical control of weeds, regardless of implement, is always related to the timeliness of the operation. Research can determine when to do something but knowing when to act on a particular farm is part of the cultural knowledge good farmers have.

Growers need to be aware of the effect of tillage type and timing on weed populations and, whenever possible, choose a tillage method and time that contribute to weed control. That is good management and the integration of techniques will follow. Reduced cultivation encourages establishment of wind-disseminated species and annual broadleaved species decline.

Minimum or no-tillage agriculture is practiced for many reasons, including economic ones, and a desire to reduce soil erosion. Tillage, including minimum or no-tillage, affects the weed population. Any method of weed control that minimizes tillage may benefit soil structure. Minimum tillage crops often yield as well and may have lower production costs than more intensive tillage systems.

Mowing

Mowing to remove shoot growth prevents seed production and may deplete root reserves of some upright perennials. If repeated often enough, it can be used to control upright perennials in turf. Prostrate perennials such as field bindweed (*Convolvulus arvensis*) survive mowing.

To maximize mowing's benefits, it must be done before viable seeds have been produced. Weeds should be cut in the bud stage or earlier. Mowing is a useful technique but rarely accomplishes much weed control because it is done late. It removes unsightly growth and, if done at the right time, can prevent seed production, which is important in control of annuals and biennials.

The foregoing deals with mowing as an operation performed to control weeds or clean up an area.

Mowing is a normal cultural operation for some crops (e.g., turfgrass and hay) and is properly regarded as a potential weed management technique rather than just a necessary part of producing the crop.

Flooding, Draining, and Chaining

These techniques cause ecological change. If a normally dry area is flooded or a normally wet area is drained, the ecological system is changed and weed species will change. These techniques are effective only when an area is immersed or drained for 3 to 8 weeks. Immersion, an anaerobic treatment, is not equally effective on all weeds; lowland or paddy rice (*Oryza sativa*) fields have weeds such as barnyardgrass (*Echinochloa crus-galli*) and junglerice (*E. colona*) that survive the flooded conditions of the rice paddy as well as rice does. Flooding eliminates some but not all weeds because it creates an environment where some weeds succeed. Flooding controls some established perennials but the expense of creating dikes and obtaining water make the practice economically unfeasible.

Draining is an excellent control for cattails (*Typha* spp.), bulrushes (*Scirpus* spp.), and reed canarygrass (*Phalaris arundinacea* L.) that grow best in wet areas. Draining and flooding are not applicable to most agronomic or horticultural environments, but they should not be forgotten when considering weed control on some sites.

Chaining is employed on rangelands to destroy emerged vegetation. A large chain similar to a ship's anchor chain is dragged between two bulldozers and uproots sagebrush (*Artemisia tridentata*), rabbitbrush (*Chrysothamnus nauseous*), and other range weeds. Chaining removes emerged growth and controls most annuals, but not perennials that reproduce vegetatively. The technique is not suited for cropland. Chains are also used to stop weeds in irrigation canals. Removing collected weeds from the impoundment created by the chain is a labor-intensive, unpleasant operation.

Nonmechanical Methods

Flame

Many plant processes are susceptible to high-temperature disruption attributed to denaturation of protein, increasing membrane permeability, and enzyme inactivation. Initial thermal disruption of cellular membranes is followed by dehydration. Heat, short of setting fire to an area, usually does not kill by combustion. Thermal death points for most plant tissue are between 45 and 55°C after prolonged exposure. Flame temperature approaches 1100°C but flamers may be used selectively when distance from the crop and speed are controlled. A flamer directs an ignited, pressurized petroleum-based fuel toward weeds and at the base of crop plants. Plant size at treatment influences efficacy much more than plant density.

Flaming kills only emerged, green shoots. It can be used where tillage is impractical, such as along a railroad or irrigation ditch. Buried weed seeds or perennial plant parts are not affected. Dry seeds withstand high temperatures and rather long exposures because soil protects and insulates. Burning can destroy weed seeds but only if they are on the soil surface. Even a small layer of soil from shallow burial will protect most seeds.

Burning mature weeds destroys debris but doesn't prevent crop losses from competition. Flaming has no residue, a problem with chemical methods of control. Other than rain, flaming is not affected by environmental conditions. It may increase erosion by eliminating vegetation that holds soil. Heat could induce germination of dormant seeds or create conditions favorable for their germination by eliminating emerged, competing plants. This is especially true when brush is burned.

Steam and Hot Water

Steam has been used for several years by nursery and greenhouse growers to sterilize or pasteurize soil. It is quite effective against common soil pathogens and some weed seeds. It has also been used, to a limited extent, in orchards and on railroad and highway rights of way. Steam and hot water (hot means near boiling) have the advantage over herbicides of not leaving any chemical residue but they also have the important disadvantage of rapid, contact action with no control persistence over time. The efficacy of steam is, in most ways, similar to that of flame weeding, with the notable exception that steam does not cause fire.

If only annual weeds are present, steam and hot water are effective for control of most, but not all, seedling plants. For example, it has been reported that downy brome (*Bromus tectorum*) was controlled best during mature stages because of its dense pubescence when young. When perennial plants are present, steam is still effective but retreatment is required. The required treatment interval for complete control may be as short as 25 days (similar to that for flame weeding) but if higher energy is used (890 kJ m^{-2} as opposed to 340 kJ m^{-2}) intervals as long as 50 days may be possible. Steam may control weed seeds up to 10 cm deep in soil but usually seed death does not result from steam or hot water application.

Steam has been most appropriate for urban hard surfaces and railroad rights of way. Its success is affected by the speed of movement of the applicator, the energy equivalent, the plants present and their stage of growth.

Solarization and Mulching

The fact that mulches raise soil temperature makes weed control possible using the heat of the sun, a process called solarization. Weed seed germination is suppressed by high soil temperatures and seedlings are killed. Transparent and opaque polyethylene sheets trap solar energy and raise soil temperature above the thermal death point for most seedlings and some seeds.

Its effectiveness for weed control is dependent on a warm, moist climate and intense radiation with long days to raise soil temperature enough to kill weed seeds and seedlings. Moisture increases soil's ability to conduct heat and sensitizes seeds to high temperatures. Solarization has potential to improve weed management but costs, compared to other methods, preclude widespread adoption in other than high-value crops.

Mulching excludes light and prevents shoot growth. Thick wide mulches are required to control perennials that creep to the edge of a mulch and emerge. Mulches increase soil temperature and may promote better plant growth. Several different materials have been used for mulch, including straw, hay, manure, paper, and plastic. It is common to see mulches in greenhouses where plants are grown on a soil floor. Mulches are used most in high-value crops grown on small areas and in crops where laying the mulch can be mechanized.

A synthetic, woven, black cloth is available for mulching. It is sold commercially in rolls about 1.8 m wide and can be applied by machine when trees are planted. It is easy to spread and prevents emergence of most annual weed seedlings.

Cultural Weed Control

Crop Competition

The techniques of cultural weed control are well known to farmers and weed scientists. In fact, they are employed regularly but often are not conscious attempts to manage weeds. Planting a crop is a sure way to reduce growth because the crop interferes with the weeds. It is a method of weed management but most often cultural weed control just happens, rather than occurring as planned weed management. Methods of cultural weed management include conscious use of crop interference, use of cropping pattern, intercropping and no-till or minimum-tillage.

Weed scientists have investigated the relative competitiveness of crop cultivars. Several crops exhibit genotypic differences in competitiveness. The quest to develop integrated weed management systems has encouraged research on the competitiveness of crop cultivars. More vigorous, taller, faster-growing cultivars are likely to be better competitors but too little is known about what makes a cultivar competitive and whether it is a trait that plant breeders can select for and develop.

Alfalfa (*Medicago sativa*, lucerne) and other hay crops are often regarded as smother crops. Land is not plowed when these crops are growing, making it harder for some annuals to succeed but perennial weeds do well. Sudangrass (*Sorghum sudanense*), planted in dense stands, can compete effectively against many, but not all, weeds.

Crops can be favored by knowing and using the effect of row width and crop seeding rate. Varying row width is using the principles of plant population biology to achieve competitive interactions that favor the crop.

Intercropping is a common, farming system in the developing world. The main reasons for mixing crops or planting in close sequence are to maximize land use and reduce risk of crop failure. Intercropping maintains soil fertility, reduces erosion, may reduce insect problems, gives greater stability to yield over seasons, and may provide yield advantages over single crop agriculture. The effectiveness of intercropping depends on the species combined and their relative proportions. Intercropping will not always provide adequate weed control. It is undoubtedly true that plants that are not crops are classified by most farmers in the developed world as weeds. Other farmers classify noncrop plants in a way that judges their potential use or their effects on soil and weeds.

Planting Date and Population

The trend in crop production is toward early planting to optimize yield. Yield is increased because crops have a longer growing season and photosynthesize for more days. Early planting provides a competitive edge to adapted crop cultivars because of an early-season competitive advantage that may be due to the weed's light requirement for growth and to shading by the crop. Choice of planting date is part of integrated weed management.

Planting date is often dictated by considerations other than weed management and by agronomic studies that define the population that gives the best yield. Populations are also determined by row spacings required by planting, cultivating, and harvesting machines. Increasing crop plant populations can often decrease weed density and growth.

Companion Cropping

Cover crops or living mulches can be used as intercrops or companion plants to suppress weeds. Appropriate weed control practices must also consider the need to maintain soil fertility and prevent erosion and open row-crops are inimical to these needs. Integrated low or no-tillage weed management systems, compatible with more than one plant in a field that reduced herbicide use, fertilizer requirement, and soil erosion are practical weed management techniques. Yields are often no different and live mulch plants do not reduce yield; they are complementary, not competitive.

Companion cropping can be a good weed control technique but research is needed to determine how appropriate it may be in specific situations. Limited evidence supports the contention that it can provide weed competition, build soil organic matter, reduce soil erosion, and improve water penetration. In some climates when spring soil moisture is limiting, cover or companion crops can deplete moisture and be detrimental to crops in spite of weed control advantages. Companion crops may also have to be killed before a crop is planted or they become competitors.

Crop Rotations

Crop rotation is done for economic, market, and agronomic reasons. Some weeds associate with certain crops more than with others. Barnyardgrass and junglerice are common in rice. Wild oat (*Avena fatua*) is common in spring wheat (*Triticum aestivum*) and barley (*Hordeum vulgare*) but almost never occurs in rice. Jointed goatgrass (*Aegilops cylindrica*) is common in winter wheat. Nightshades (*Solanum* spp.) are common in potatoes (*Solanum tuberosum*), tomatoes (*Lycopersicon esculentum*), and beans (*Phaseolus* spp.), and kochia (*Kochia scoparia*) and common lambsquarters (*Chenopodium album*) are frequent in sugar beets (*Beta vulgaris*). Dandelions (*Taraxacum officinale*) are common in turf but not in row-crops.

These associations occur because of similarity of crop and weed phenology, adaptation to cultural practices, similar growth habits, and perhaps of most importance, resistance or adaptation to imposed weed control methods. When one crop is grown for many years, some weeds, if they are present in the soil seed bank, will be favored and their populations will increase. Weed–crop associations can be changed by changing crop, time of planting, or weed control method.

A good rotation includes crops that reduce weeds that are especially troublesome in succeeding crops. Removal is accomplished by competition or through use of different weed control techniques in different crops.

Crop rotation regularly changes the crop in each field, soil preparation practices in one field, subsequent soil tillage, and weed control techniques. All of these affect weed populations and while crops are not commonly rotated to control weeds, rotations affect weed problems.

Fertility Manipulation

Manipulation of soil fertility solely to manage weed populations is virtually unknown. Farmers fertilize to maximize yield and assure success, not to manipulate weed populations. However, as is true of most soil manipulations, fertility affects weeds. When weed density is low, added fertilizer, particularly nitrogen, increases crop yield and makes the crop a more vigorous competitor with weeds. But when weed density is high, added nutrients favor weed over crop growth.

A good illustration of the potential of fertility manipulation as an instrument for changing plant populations is the Park Grass Experiment at Rothamsted in England. In unlimed plots amended with a complete fertilizer with nitrogen primarily as ammonium sulfate, a pure stand of common velvetgrass (*Holcus lanatus*) has developed. It was selected out of the original mixture solely by fertility manipulation and lack of lime. With complete fertilizer and lime, plots have one of the heaviest hay yields and a very diverse flora, including orchardgrass (*Dactylis glomerata*, cocksfoot) and meadow foxtail (*Alopecurus pratensis*). In unlimed plots amended with ammonium sulfate and no phosphorus, the vegetation is completely different from either of the above. If potassium is absent, dandelions are absent because they flourish only with potassium and a pH above 5.6.

Biological Weed Control

Biological weed control is the action of parasites, predators, or pathogens to maintain another organisms's population at a lower average density than would occur in their absence. Biological control is usually thought of as intentional introduction of parasites, predators, or pathogens to achieve control but it is also natural. Scientists can discover the control potential of natural parasites, predators, or pathogens and exploit it to achieve human ends. The aim is to maintain a population at a lower average density, not to eradicate it. Biological control will never be the solution to every weed problem. It will be properly employed as one weed management practice among many.

Advantages

In its classical form, biological control can be permanent weed management because once an organism is released, it may be self-perpetuating and control will continue without further intervention. This is true when fungi are released in an inundative manner. Afterwards, the weed doesn't have to be controlled by humans, it just happens. This ideal is not always achieved. Self-perpetuation is an advantage other weed control techniques do not have. There are no chemical environmental residues from biological control other than the organism, which may be an ecological pollutant because it is foreign to the environment in which it is released. In the classical or idealized version of biological control this doesn't happen because extensive research before release establishes, one hopes, that the organism is environmentally benign. There is no mammalian toxicity from biocontrol organisms. In most cases, initial costs are nonrecurring and usually, once established, no further inputs are needed. Development costs have been much less than for herbicides. While all of these advantages do not accrue to all biological control organisms, they are cited to justify research and greater employment of biological control.

Disadvantages

There are some situations where biological control is not appropriate. If a plant is a weed in one place and valued in another place, biological control is inappropriate. Spread of a biological control organism, once introduced, cannot be controlled. The control organism cannot always distinguish valuable native plants from weedy relatives. The absolute demand for specificity in a biological control agent means development is research intensive, requires a large budget, and several years.

Biological control is inherently slow and results are not guaranteed. In many crops, weeds must be controlled during a brief, critical period, often within days or weeks, to prevent yield reduction. Biological control is not rapid. In addition, because eradication is not an appropriate goal for biological control; weeds that should be eradicated on some sites (e.g., larkspur (*Delphinium* spp.) on rangeland) may be better controlled with other techniques. There are some species that are geographically localized minor weeds and development of a biological control for them would be too expensive. Cropland weeds exist in an ecologically unstable habitat that is often a poor environment for successful introduction, survival, and growth of biological control organisms. Cropland weeds exist in a weed complex, rarely as a single species. Because biological control is necessarily directed at a single species, it is often an inappropriate choice for the weed complex found in most crops. Release of a biological control organism can suppress or extinguish native biological control organisms and other organisms. A corollary is that other harmful or beneficial species may increase in abundance. Such events could lead to loss of biological diversity, loss of existing biological control, release of species from competitive regulation, disruption of plant community structure, suppression of essential organisms, and disruption of food chains and nutrient cycling.

In summary, biological control is slow, often less effective, and commonly uncertain. Biological control, particularly in disturbed cropping situations, will not control many different weeds. It won't eradicate weed problems, but neither will most other techniques.

Chemical Control: Herbicides

A herbicide is a chemical that disrupts the physiology of a plant for long enough to kill it or severely reduce its growth. Herbicides change the chemical environment of plants, which can be more readily manipulated than the climatic, edaphic, or biotic environment.

Herbicides reduce or eliminate labor and machine requirements and modify crop production techniques. Herbicides, when used appropriately, are production tools that increase farm efficiency, reduce the need for tillage, and may reduce energy requirements. Herbicides do not eliminate energy requirements because they are petroleum-based.

Understanding the nature, properties, effects, and uses of herbicides is essential if one is to be conversant with modern weed management. Weed management is dominated by herbicides in the developed world. Whether one likes them or deplores them, they cannot be ignored. To ignore them is to be unaware of the opportunities and problems of modern weed management. Ignoring or dismissing herbicides may lead to an inability to solve weed problems in many agricultural systems and may delay development of better weed management systems.

Advantages

An important criterion for a grower is profit or return on investment in technology. Herbicides are profitable where labor is scarce or expensive. They may be advantageous also where labor is plentiful and cheap. Herbicides control weeds in crop rows where cultivation is not possible. They can be used in places where other methods don't work. Preemergence herbicides provide early-season weed control when competition results in the greatest yield reduction and when other methods are less efficient or impossible to use (e.g., it is impossible to mechanically cultivate when soil is wet).

Selective herbicides reduce the need for tillage and control of weeds in crop rows where tillage is not effective and may be harmful to roots. Herbicides reduce destruction of soil structure by decreasing the need for tillage. Erosion in orchards and other perennial crops can be prevented by maintenance of a sod cover with selective herbicides. Many perennial species cannot be controlled effectively with hand labor and herbicides are often the only option.

Herbicides save labor and energy by reducing the need for hand labor and mechanical tillage. They can reduce fertilizer and irrigation requirements by eliminating competing weeds. They reduce harvest costs by eliminating interfering weeds and can reduce grain drying costs because green, weedy plant material is absent. Other methods of weed control will, of course, also accomplish these things but not as efficiently and often not as cheaply.

Disadvantages

It is often suggested that herbicides reduce crop production costs. Many disagree and suggest herbicides are a net cost because they are expensive and the equipment for applying them is an added cost. The argument cannot be concluded in general terms; it must be decided for each specific crop production system.

One of the major concerns about herbicides is their undeniable mammalian toxicity. All have some degree of toxicity to humans and other plant and animal species. Some are no more toxic in terms of their LD_{50} than many common chemicals (e.g., aspirin, gasoline, or table salt). Many worry because, in their opinion, they have no choice about exposure to toxicity.

Some herbicides persist in the environment. None persists forever, but all have a measurable environmental life. In some cases, a herbicide can carry over from one crop season to the next to restrict rotational possibilities and injure succeeding crops. Plants that are not targets may be affected by drift or inappropriate application. Their use in many crops eliminates all plants except the crop and may increase soil erosion by reducing vegetative cover. Herbicides are often inconsistent in control because they are affected by environmental conditions and results are not always predictable.

For a long time, it was assumed that because weeds have relatively long life cycles and the same herbicide is not used repetitively on the same land, weeds would not develop resistance. Herbicide resistance has now been reported in one or more of 15 herbicide chemical families.

Selective herbicides may encourage monoculture and discourage diversity. Unquestioned expansion of herbicide technology into developing countries is not always wise because of the risk to agricultural diversity in many of these countries.

Precision is required when herbicides are used. One must think carefully about what herbicide to use, when to use it, how much to use, and how surplus chemical will be disposed of. They cannot be used casually; intelligence is required in their use and in disposal of surplus chemicals and empty containers.

Finally, because herbicides are so good at what they do, they may actually create problems after their use. Herbicides control certain weeds while leaving a crop unscathed. Natural plant communities are usually a polyculture (not an absolute generalization). Diversity is the rule. When all plants are eliminated save the crop, other plants (weeds) will move into the environment created and they may be more difficult to control than the ones just eliminated.

See also: **Weeds**: Herbicide Resistance; Weed Biology; Weed Competition; Weed Seed Biology.

Further Reading

Ahrens WH (ed.) (1994) *Herbicide Handbook*, 7th edn. Lawrence: Weed Science Society of America.

Aldrich RJ and Kremer RJ (1997) *Principles in Weed Management*, 2nd edn. Ames: Iowa State University Press.

Altieri MA and Liebman M (eds) (1998) *Weed Management in Agroecosystems: Ecological Approaches*. Boca Raton: CRC Press.

Anderson WP (1996) *Weed Science: Principles and Applications*, 3rd edn. New York: West Publishing Co.

Anonymous (1985) *Principles of Weed Control in California*. Fresno: Thomson Publications.

Bond W and Grundy AC (2000) Non-chemical weed management in organic farming systems. *Weed Research* 41: 383–405.

Bowman G (ed.) (1997) *Steel in the Field: A Farmer's Guide to Weed Management Tools*. Beltsville: Sustainable Agriculture Network.

Buhler DD (1999) *Expanding the Context of Weed Management*. New York: Food Products Press.

Hansson D and Ascard J (2002) Influence of developmental stage and time of assessment of hot water weed control. *Weed Research* 42: 307–316.

Hatzios K (ed.) (1998) *Herbicide Handbook: Supplement to the 7th edn*. Lawrence: Weed Science Society of America.

Kolberg RL and Wiles LJ (2002) Effect of steam application on cropland weeds. *Weed Technology* 16: 43–49.

Radosevich S, Holt J, and Ghersa C (1997) *Weed Ecology: Implications for Management*, 2nd edn. New York: J. Wiley & Sons Inc.

Ross MA and Lembi CA (1999) *Applied Weed Science*, 2nd edn. Upper Saddle River: Prentice-Hall.

Sheley RL and Petroff JK (1999) *Biology and Management of Rangeland Weeds*. Corvallis: Oregon State University Press.

Smith AE (ed.) (1995) *Handbook of Weed Management Systems*. New York: Marcel Dekker.

Vencill W (ed.) (2002) *Herbicide Handbook*, 8th edn. Lawrence: Weed Science Society of America.

Zimdahl RL (1999) *Fundamentals of Weed Science*, 2nd edn. San Diego: Academic Press.

Herbicide Resistance

R M Twyman, University of York, York, UK

Introduction

Herbicides are used to control weeds, one of the most significant constraints to crop production. The ideal herbicide would not harm animals or organisms living in the soil, would break down into harmless substances in the environment, and would selectively kill weeds, but not crop plants. It is this "selectivity" that is the most difficult property to attain, since all herbicides act by disrupting essential metabolic or physiological processes occurring in plants. Until recently, selectivity was achieved predominantly by exploiting metabolic differences between species. For example, 2,4-D (2,4-dichlorophenoxyacetic acid) is an analog of the plant hormone idoleacetic acid, and it kills plants by interfering with normal growth and development. At the correct dose, 2,4-D can selectively kill broad-leaf weeds growing among grasses, because dicotyledons and monocotyledons show differential uptake and/or detoxification of the herbicide. This strategy becomes more difficult to apply when crops and the weeds infesting them are closely related, since the metabolic processes in related species are similar. Therefore, some of the major weed problems today are caused by grassy weeds in cereal crops and other similar pairings, particularly in monoculture systems.

One way in which the problem of selectivity can be addressed is to introduce, into crop plants, genes that confer resistance to broad-spectrum herbicides. In this manner, selectivity is achieved not by changing the properties of the herbicide to discriminate between crops and weeds, but by changing the properties of the crops to make them more selectable. Herbicide resistance genes have been isolated from a number of bacteria and from plants subject to laboratory selection experiments. They have been introduced into crops to produce transgenic lines with resistance to a number of broad-spectrum herbicides, including glyphosate, glufosinate, and bromoxynil. Similarly, it has proven possible to generate herbicide-resistant crops by introgression and by the identification of mutants produced in tissue culture or via pollen mutagenesis. Unfortunately, the same selective forces that can be applied in the laboratory also operate in nature. The indiscriminate use of herbicides in the past has therefore led to the evolution of herbicide-resistant weed varieties, some resistant to multiple herbicides with different modes of action, which are much more difficult to control. A more recent concern is the possibility of herbicide-resistance genes escaping from transgenic crops into related weed species, either by introgression or horizontal gene transfer.

Herbicides and their Modes of Action

All herbicides work by interfering with processes that are essential for plant growth and survival. Many of the commonly used herbicides target either photosynthesis or essential metabolic pathways. For example, paraquat generates superoxide radicals when exposed to light and causes bleaching by interfering with electron transport in photosystem I. Similarly, triazines, such as atrazine, and nitriles, such as bromoxynil, interfere with electron transport on the reducing side of photosystem II by competing with plastoquinone for the Q_B-binding protein, the product of the *psb*A gene. Glyphosate is widely used because it has low toxicity in animals and is rapidly broken down in the soil. It works by competitively inhibiting the enzyme 5-enolpyruvylshikimate-3-phosphate synthase (EPSPS), which is essential for the synthesis of aromatic amino acids. Sulfonylureas, such as chlorsulfuron, inhibit the enzyme acetolactate synthase (ALS), which is required for the synthesis of branched chain amino acids. Other herbicides interfere with the synthesis of lipids, chlorophyll, or carotenoids. Still others work by inhibiting the polymerization of microtubules and therefore by interfering with cytoskeletal organization and mitosis.

There are two major systems for herbicide classification, both of which are based on cellular target and mode of action (**Table 1**). The Herbicide Resistance Action Committee (HRAC) system is an alphabetical classification that lists herbicides according to their targets, modes of action, and chemical classes. For example, HRAC group C herbicides inhibit photosynthesis at photosystem II and are placed in subcategories C1, C2, and C3 according to chemical classes and the way in which they interact with the Q_B-binding protein. The Weed Science Society of America (WSSA) system is a numerical classification, and while the basis of classification is similar to the HRAC system, there are no subgroups. For example, herbicides in HRAC subgroups C1, C2, and C3 are classified as groups 5, 7, and 6, respectively, in the WSSA system. Herbicides with unknown modes or sites of action are classified in HRAC group Z until further evidence is available, but they may still receive a classification based on their chemical properties in the WSSA system.

Natural Herbicide Resistance

The indiscriminate use of herbicides has two direct consequences on the weed populations in agricultural systems. First, the successful elimination of one weed species leaves an ecological vacuum, which can be exploited by other weed species that are not affected by the herbicide. For example, broad leaf weeds that are effectively eliminated from cereal crops are rapidly replaced by grassy weeds that are more difficult to control. Second, herbicides place "selective pressure" on the weed population, and this is met by evolutionary adaptations that produce herbicide-resistant varieties. Essentially, herbicide resistance can evolve in one of two ways: target site resistance or metabolic resistance.

Target Site Resistance

Most herbicides interact with a known target protein in the plant and block its activity. Protein–ligand interactions are dependent on the complementary structures and/or chemical properties of the interacting partners, and mutations that reduce the affinity of such interactions confer target-site resistance. This type of resistance often results from single point mutations affecting critical residues in the target protein. For example, resistance to sulfonylureas and other herbicides that inhibit ALS activity has evolved in many weed species through point mutations in the gene encoding this enzyme. In natural weed populations (as opposed to those used in laboratory selection experiments), five key residues are involved: Ala 122, Pro 197, Ala 205, Trp 574, and Ser 653 (positions based on the *Arabidopsis thaliana* ALS sequence).

Table 1 Classification of herbicides by cellular target and mode of action, according to the HRAC and WSSA systems

HRAC	*WSSA*	*Mode of action (chemical class)*	*Example*
A	1	Inhibition of acetyl CoA carboxylase	Diclofop-methyl
B	2	Inhibition of acetolactate synthase	Chlorsulfuron
C1	5	Inhibition of photosystem II (triazines and others)	Atrazine
C2	7	Inhibition of photosystem II (ureas and amides)	Diuron
C3	6	Inhibition of photosystem II (nitriles)	Bromoxynil
D	22	Inhibition of photosystem I	Paraquat
E	14	Inhibition of protoporphyrinogen oxidase	Bifenox
F1	12	Bleaching (inhibition of phytoene desaturase)	Norflurazon
F2	28	Bleaching (inhibition of 4-HPPD)	Benzofenap
F3	11, 13	Bleaching (carotenoid biosynthesis, unknown target)	Clomazone
G	9	Inhibition of EPSPS	Glyphosate
H	10	Inhibition of glutamine ligase	Bialaphos
I	18	Inhibition of dihydropteroate synthase	Asualam
K1	3	Inhibition of microtubule assembly	Benefin
K2	23	Inhibition of microtubule organization	Carbetamide
K3	15	Inhibition of cell division	Acetochlor
L	20, 21	Inhibition of cell wall synthesis	Dichlobenil
M	24	Disruption of membranes	Dinoseb
N	8, 26	Inhibition of lipid synthesis (target other than ACCase)	Butylate
O	4	Synthetic auxins	2,4-D
P	19	Inhibition of auxin transport	Naptalam
Z	various	Mode of action unknown	Difenzoquat

Abbreviations: 2,4-D, 2,4-dichlorophenoxyacetic acid; 4-HPPD, 4-hydroxyphenylpyruvate dioxgenase; ACCase, acetyl CoA carboxylase; EPSPS, 5-enolpyruvylshikimate-3-phosphate synthase. Modified with permission from Heap I (2003) *The International Survey of Herbicide Resistant Weeds.* http://www.weedscience.org/.

Metabolic Resistance

Herbicide sensitivity also depends on the rates of uptake, transport to growing areas, detoxification, and elimination. In principle, mutations could arise that affect any of these processes, but detoxification, through catabolic breakdown or conversion into a nontoxic derivative has been the most widely documented. For example, paraquat-resistant weeds have been identified that overproduce the enzyme superoxide dismutase, and bromoxynil-resistant weeds have been identified that overproduce a cytochrome P-450 monooxygenase. Overproduction can occur through point mutations that increase the activity of the genes' regulatory elements or gene amplification. Metabolic resistance is also the basis of natural herbicide resistance in many crops. For example, corn (*Zea mays*; maize) lines resistant to triazine have an enhanced capacity to detoxify the herbicide by 2-hydroxylation or are able to conjugate the herbicide to glutathione.

Table 2 summarizes the number of weed species that have evolved resistance to some of the more commonly used classes of herbicides based on information available up until January 2003. The number of populations affected is actually much higher than shown, because resistance can evolve in the same species at different sites and on different occasions. Furthermore, some of the resistant populations show cross-resistance (single mutations that confer resistance to multiple herbicides) or multiple resistance to several herbicides with different modes of action through the accumulation of mutations. For example, although 73 weed species have evolved resistance to ALS inhibitors, this represents 195 separate reports. Resistance has evolved in Kochia (*Kochia scoparia*), a dicotyledonous weed of the Chenopodiaceae family that infests wheat (*Triticum* spp.), at 21 sites in the USA, Canada, and Europe since 1987. On three of these occasions, cross-resistance to triazines was also demonstrated. Natural herbicide resistance has evolved in nine of the world's ten most weedy species (**Table 3**).

Engineered Herbicide Resistance

The introduction of herbicide resistance genes into crop plants allows the use of broad-spectrum herbicides with little natural selectivity. This provides perhaps the best method for selecting between closely related species. One way in which resistance can be introduced is by introgression from a related species that shows herbicide tolerance. For example, triazine resistance has been introduced into oilseed rape (*Brassica napus*) from weedy populations of *Brassica campestris* by introgressing a mutant *psbA* gene. There remains the problem of naturally evolved herbicide resistance in weeds, and this can be addressed by the direct introduction of transgenes from distant species (e.g., bacteria). The rational is that a transgene that has no counterpart in the plant kingdom would represent a difficult target for weeds to emulate through the adaptation of their existing

Table 3 The 10 most weedy species in the world

Rank	*Species*	*Common name*
1	*Lolium rigidum*	Rigid rye-grass
2	*Avena fatua*	Wild oat
3	*Amaranthus retroflexus*	Redroot pigweed
4	*Chenopodium album*	Common lambsquarters
5	*Setaria viridis*	Green foxtail
6	*Echinocola crus-galli*	Barnyardgrass
7	*Eleusine indica*	Goosegrass
8	*Kochia scoparia*	Kochia
9	*Conyza canadensis*	Horseweed
10	*Amaranthus hybridus*	Smooth pigweed

Data adapted with permission from Heap I (2003) *The International Survey of Herbicide Resistant Weeds.* http://www.weedscience.org/.

Table 2 Prevalence of herbicide-resistant weed species, grouped by herbicide mode of action

HRAC/WSSA	*Mode of action*	*Example*	*Number of resistant weed species*
B/2	ALS inhibitors	Chlorsulfuron	73
C1/5	Photosystem II inhibitors	Atrazine	64
A/1	ACCase inhibitors	Diclofop-methyl	28
D/22	Photosystem I inhibitors	Paraquat	22
O/4	Synthetic auxins	2,4-D	21
C2/7	Photosystem II inhibitors	Diuron	20
K1/3	Microtubule assembly	Benefin	10
N/8	Inhibition of lipid synthesis	Butylate	6
F3/11	Inhibition of carotenoid synthesis	Amitrole	4
G/9	Inhibition of EPSPS	Glyphosate	4

The number of resistant weed species does not take into account the independent evolution of herbicide resistance in the same species on more than one occasion, and may include instances of multiple herbicide resistance. Data adapted with permission from Heap I (2003) *The International Survey of Herbicide Resistant Weeds.* http://www.weedscience.org/.

gene pool. As is the case for natural herbicide resistance, transgenic plants have been engineered either to express target proteins with reduced sensitivity to the herbicide or detoxifying enzymes that accelerate herbicide elimination.

Target Site Resistance

Target site resistance is achieved by introducing into susceptible crop plants transgenes encoding modified versions of the herbicide target. Among the first examples of this approach was the production of glyphosate-resistant transgenic tomatoes (*Lycopersicon esculentum*) by expressing a mutant version of the *Salmonella typhimurium aroA* gene encoding a modified EPSPS enzyme that was no longer inhibited by the herbicide. Glyphosate inhibits aromatic amino acid synthesis in both bacteria and plants, but bacteria provide a more convenient high-throughput selection system for the isolation of mutants. It was important to include a chloroplast targeting sequence in the expression construct, because this is the normal site of EPSPS activity within the cell. Previous experiments, in which glyphosate-tolerant *S. typhimurium* EPSPS was expressed in the cytosol of transgenic tobacco plants, did not significantly increase the level of herbicide resistance. Mutant, glyphosate-resistant *aroA* genes have also been isolated from *Escherichia coli* and *Agrobacterium tumefaciens* and expressed in a wide range of transgenic plants. Glyphosate resistance is the most common trait in commercially released transgenic plants, accounting for up to 75% of all the transgenic plants in the world.

Transgenic plants have also been generated expressing mutant versions of acetyl CoA carboxylase (ACCase) and acetolactate synthase (ALS), which provide resistance against group A/1 and B/2 herbicides, respectively. A mutant *psbA* gene from *Amaranthus hybridus* has been expressed in transgenic tobacco (*Nicotiana tabacum*) to provide resistance against triazine herbicides (group C1/5). Since *psbA* is a chloroplast gene, and at the time this experiment was carried out only nuclear transformation was a routine procedure, the gene was modified to include a nuclear promoter and polyadenylation site plus a chloroplast transit sequence.

There are two major disadvantages to the target site resistance approach:

- Target site resistance often carries a yield or fitness penalty. Herbicide resistant mutants arise after strong selection in laboratory experiments or in wild weed populations exposed to selection pressure. The mutations do not exist at polymorphic levels in unselected populations, which indicates they are not selectively neutral and certainly do not confer a fitness advantage in the absence of the herbicide. This may reflect a decrease in the affinity of the herbicide target for its natural ligand or substrate. This has been shown to be the case for mutant *E. coli* EPSPS enzymes, which do not bind glyphosate but also have a lower affinity for the natural substrate phosphoenolpyruvate.
- The natural target is still present. Therefore the herbicide-resistant plant is functionally heterozygous. This perhaps explains why herbicide-resistant mutants tend to perform better than herbicide-resistant transgenics – they can be selected to homozygosity.

Metabolic Resistance

Herbicide-resistant transgenic plants have been generated by overexpressing the endogenous target protein, e.g., petunia plants overexpressing EPSPS under the control of a constitutive promoter. However, transgenic plants expressing detoxifying enzymes have an advantage over those with modified or overexpressed herbicide targets because the enzyme can be derived from a completely alien source. Therefore, natural mutations in weed populations would not be sufficient to confer herbicide resistance and the resistance trait in the transgenic plant is dominant, because there is no competitive endogenous activity. This reduces the fitness penalty associated with herbicide resistance because the modification required to confer herbicide resistance does not compromise existing metabolic pathways. A number of bacterial genes have been isolated and transferred to plants in order to detoxify herbicides. Perhaps the most widely used is *bar*, from *Streptomyces hygroscopicus*, which detoxifies bialaphos, glufosinate and other herbicides based on phosphinothricin through the phosphinothricin acetyltransferase activity of its product. The only other detoxifying gene to be used commercially is *bnx*, from *Klebsiella ozaenae*, which encodes a nitrilase that detoxifies bromoynil. Other examples include a gene from *Alcaligenes*, encoding a cytochrome P-450 monooxygenase, which is used to detoxify 2,4-D, and a dehalogenase that detoxifies dalapon.

Transfer of Herbicide Resistance Transgenes to Weeds

Until recently, the evolution of herbicide-resistant weeds was constrained by the gene pool of the weed species themselves, and any species with which they can naturally cross. However, there is now a perceived risk that herbicide resistance transgenes could spread to weed populations, generating a population of superweeds that are immune to all herbicides. In the case of target site resistance transgenes, the risks are small because effectively the same mutations could arise naturally and without the associated fitness penalty. The main concern is the spread of detoxifying transgenes, which could cross to the feral and wild relatives of crops by introgression, and could spread to microorganisms by horizontal gene transfer.

Introgression

Introgression is the gradual movement of genes from one species into the gene pool of another, when there is some opportunity for hybridization between them. For most crops, there is no risk of transgene introgression into weeds, because there are no weeds that are so closely related to the crop that hybridization would be possible. However, in the case of wheat (*Triticum aestivum*), rice (*Oryza sativa*), and oilseed rape, the risks of introgression are higher. Wheat has a number of wild relatives and many traits have been moved from wild grasses to wheat by plant breeders. In some cases, specialized techniques have been required, because crosses do not occur in the field, but a few species share homologous chromosome sets with wheat, including the weedy species *Aegilops cylindrica*. Rice also has numerous weedy relatives, including red rice (also *O. sativa*), *O. rufipogon,* and *O. officinalis*. The transfer of the *bar* gene between field rice and red rice has been demonstrated, but the ease with which this could occur in nature is unknown because rice is predominantly a self-pollinating species. Oilseed rape is related to the weed *Brassica campestris* with which it shares an A genome, and can hybridize in the laboratory to a number of other species.

There are many approaches that can be used to minimize the introgression of transgenes into related weeds or to prevent the survival of plants containing those transgenes:

- Management practices. The physical isolation of transgenic crops, such as greenhouse confinement or the use of weed-free zones around crops is one practical solution to the problem of introgression.
- Genetic control of fertility. For example, male sterility, apomixis.
- Chromosomal placement. This involves placing the transgene in a genome that is not shared with a related weed species. For example, transgenes on the wheat D genome could introgress into *A. cylindrica*, but this would be less likely if the transgene were placed on the A genome.
- Transplastomic plants. Many of the pathways affected by herbicides are located in the chloroplast because these are selective for plants and are less harmful to animals. Plastid transformation allows the direct manipulation of the chloroplast genome to express herbicide-resistance genes, and carries the extra advantage that chloroplast genes are maternally inherited in most crops.
- Transient expression. Instead of stably transformed plants, herbicide resistance could be conferred by genes expressed from nonintegrated transgenes or virus vectors.
- Transgenic mitigation. This involves the inclusion of a tightly linked transgene that confers a trait that is selectively neutral to the crop, but disadvantageous to the weed population. Examples might include dwarfing genes or genes that control seed dormancy or shattering.

Horizontal Gene Transfer

Horizontal gene transfer involves the movement of genes between species that are not sexually compatible and may belong to very different taxonomic groups. The process is common in bacteria, resulting in the transfer of plasmid-borne antibiotic resistance traits from harmless to pathogenic species or strains, and it is envisaged that gene transfer from plants to bacteria, and then from bacteria to other plants, could result in the transfer of herbicide resistance traits from crops to unrelated weeds. Gene transfer from bacteria to plants has been widely demonstrated and indeed exploited in the cases of *A. tumefaciens* and *Agrobacterium rhizogenes*. However, there has been no reported incidence of natural herbicide resistance traits being transferred from crops to weeds via *Agrobacterium*, so it seems unlikely that these species are capable of taking up DNA from plants under natural conditions. Recently, horizontal transgene mobility from the chloroplast genome of transplastomic tobacco plants to the opportunistic bacterium *Acinetobacter* spp. strain BD413 was demonstrated in laboratory experiments, but the conditions were adapted for gene transfer by introducing into the bacteria sequences that were homologous to the plant's transgene. In the absence of such sequences, gene transfer from the plant to the bacterium could not be demonstrated. These results suggest that gene transfer from the plant genome to certain bacteria is possible under ideal circumstances, and therefore it is remotely possible that interactions between *Acinetobacter* spp. and *Agrobacterium* spp. could occur (either directly or indirectly) resulting in the creation of *Agrobacterium* strains capable of transferring herbicide resistance traits to weeds. The data also suggest, however, that endogenous plant genes are not efficient passengers in this transfer process, so the risks of horizontal gene transfer can be effectively eliminated by avoiding bacterial transgenes and using those derived from plants.

List of Technical Nomenclature

Apomixis	Reproduction without fertilization.
Broad-spectrum herbicide	A herbicide that kills all plants.
Cross-resistance	Resistance to several herbicides brought about by a single mutation.
Fitness penalty	A mutation that confers an advantage in one environment may be disadvantageous in another. Target site resistance often carries a fitness penalty because the target protein may show lower affinity for its natural substrate, as well as for the herbicide.
Functional heterozygosity	The introduction of a transgene that encodes a modified version of an endogenous protein, such that both are produced concurrently.
Herbicide	A chemical that kills plants.
Horizontal gene transfer	The movement of genes between unrelated species.
Introgression	The gradual movement of genes from the genepool of one species into another, where there is some scope for hybridization between them.
Male sterility	The absence of functional pollen.
Metabolic resistance	Herbicide resistance brought about by the ability of the resistant plant to eliminate or detoxify the herbicide.
Mode of action	How a herbicide works, in most cases by inhibiting a particular target protein in the plant.
Multiple resistance	Resistance to several herbicides brought about by the accumulation of mutations.
Selectivity	The ability of a herbicide to kill target plants (weeds) but not nontarget plants (crops, garden plants).
Target protein/site	The protein with which a herbicide interacts.
Target site resistance	Herbicide resistance conferred by modification of the target protein so that the herbicide no longer inhibits its activity.
Transgenic mitigation	The cotransfer of a gene, linked to a herbicide resistance gene, that confers a selective disadvantage to weeds.
Transplastomic	Transgenic plants in which the foreign gene is inserted into the chloroplast genome.
Weed	A plant growing where it is not wanted.

See also: **Weeds**: Weed Biology; Weed Technology and Control.

Further Reading

Cole DJ and Rodgers MW (2000) Plant molecular biology for herbicide tolerant crops and discovery of new herbicide targets. In: Cobb AH and Kirkwood RC (eds) *Mechanisms of Herbicide Action*, pp. 239–278. Sheffield: Sheffield Academic Press.

Daniell H, Datta R, Varma S, Gray S, and Lee SB (1998) Containment of herbicide resistance through genetic engineering of the chloroplast genome. *Nature Biotechnology* 16: 345–348.

Gressel J (2000) Molecular biology of weed control. *Transgenic Research* 9: 355–382.

Gressel J (2002) Transgenic herbicide-resistant crops – advantages, drawbacks and failsafes. In: Oksman-Caldentey K-M and Barz WH (eds) *Plant Biotechnology and Transgenic Plants*, pp. 597–634. New York: Marcel Dekker.

Heap I (2003) *The International Survey of Herbicide Resistant Weeds*. http://www.weedscience.org/.

Holt JS, Powles SB, and Holtum JAM (1993) Mechanisms and agronomic aspects of herbicide tolerance. *Annual Review of Plant Physiology and Plant Molecular Biology* 44: 203–229.

Kay E, Vogel TM, Bertolla F, Nalin R, and Simonet P (2002) *In situ* transfer of antibiotic resistance genes from transgenic (transplastomic) tobacco plants to bacteria. *Applied and Environmental Microbiology* 68: 3345–3351.

Oxtoby E and Hughes MA (1990) Engineering herbicide tolerance into crops. *Trends in Biotechnology* 8: 61–65.

Tranel PJ and Wright TR (2002) Resistance of weeds to ALS-inhibiting herbicides: What have we learned? *Weed Science* 50: 700–712.

APPENDIX: List of Systematic and Common Plant Names

Systematic name	Common name
Abies spp.	fir
Acacia melanoxylon	blackwood acacia
Acacia spp.	acacia
Acer platanoides	Norway maple
Acer pseudoplatanus	sycamore
Actinidia deliciosa	kiwifruit
Actinidia chinensis	kiwifruit
Aegilops cylindrica	jointed goatgrass
Aegilops spp.	goatgrass
Aesculus hippocastanum	horse chestnut
Agave fourcroydes	henequen
Agave sisalana	sisal
Allium cepa	onion
Allium porrum	leek
Allium sativum	garlic
Allium ursinum	ramsoms
Alnus spp.	alder
Alopecurus myosuroides	blackgrass
Alopecurus pratensis	meadow foxtail
Alstroemeria	alstroemeria
Amaranthus hybridus	smooth pigweed
Amaranthus retroflexus	redroot pigweed
Amelanchier alnifolia	saskatoon fruit
Amona cherimola	cherimoya
Anacardium occidentale	cashew
Ananas comosus	pineapples
Annona muricata	soursop
Anthriscus sylvestris	cow parsley
Anthurium	anthurium
Antirrhinum spp.	snap-dragon
Apium graveolens	celery
Arachis hypogaea	groundnut, peanut
Areca catechu	areca nut, betel nut
Artemisia tridentata	sagebrush
Artocarpus spp.	breadfruit, jakfruit
Arundinaria	bamboo
Asclepias spp.	milkweed
Asparagus officinalis	asparagus
Atriplex canescens	four-wing saltbush
Atriplex polycarpa	desert saltbush
Avena fatua	wild oat
Avena sativa	oats
Avicennia germinans	black mangrove
Avicennia marina	mangrove
Avverrhoe carambola	carambola
Azadirachta indica	neem tree
Bellis perennis	daisy
Berberis	barberry
Bertholletia excelsa	Brazil nut
Beta vulgaris	beet, sugar beet
Beta vulgaris subsp. *cicla*	Swiss chard
Beta vulgaris subsp. *maritima*	sea beet
Betula spp.	birch
Bidens spp.	burr-marigolds
Brassica campestris	rapini
Brassica napus	canola, colza, oilseed rape
Brassica nigra	black mustard
Brassica oleracea, var *acephala*	kale
Brassica oleracea, var *botrytis*	cauliflower
Brassica oleracea, var *capitata*	cabbage
Brassica oleracea, var *italica*	broccoli
Brassica rapa	Chinese cabbage, turnip rape
Bromus tectorum	brome, downy brome
Brosimum guianense	snake wood
Buehmeria nivea	ramie
Buxus sempervirens	box
Cajanus cajan	pigeonpea
Camellia japonica	cemellia
Camellia sinensis	tea
Cannabis sativa	hemp
Capsella bursa-pastoris	shepherd's purse
Capsicum annuum	bell pepper, cayenne pepper, red pepper
Cardamine hirsuta	hairy bitter-cress
Cariandum sativuum	coriander
Carica papaya	papaya
Castanea spp.	chestnut
Catharanthus roseus	Madagascar periwinkle, periwinkle

Systematic name	Common name
Chamaecyparis nootkatensis	yellow cedar
Chenopodium album	common lambsquarters, fat hen
Chenopodium pallidicaule	canihua
Chenopodium quinoa	inca wheat, quinoa
Chrysanthemum	chrysanthemum
Chrysothamnus nauseous	rabbitbrush
Cicer arietinum	chickpea
Cichorium intybus	chicory
Cirsium arvense	creeping thistle
Citrullus vulgaris	seedless watermelon, watermelon
Citrus aurantium	sour orange (Seville orange)
Citrus limon	lemon
Citrus sinensis	orange
Cocos nucifera	coconut
Coffea spp.	coffee
Colocasia esculenta	taro
Convolvulus arvensis	bindweed, field bindweed
Conyza canadinsis	horseweed
Copernicia cerifera	Brazilian palm
Corchorus spp.	jute
Corylus avellana	hazel
Crocus spp.	crocus
Cucumis melo	muskmelon
Cucumis melo cantalupensis	cantaloupe melon
Cucumis sativus	cucumber
Cucurbita argyrosperma	squash
Cucurbita melo	melon
Cucurbita pepo	pumpkin, zucchini
Cunninghamia lanceolata	Chinese fir
Curcurbita maxima	giant pumpkin
Cuscuta spp.	dodder
Cydonia oblonga	quinces
Cynodon dactylon	bermuda grass
Dactylis glomerata	orchard grass
Dalbergia nigra	Brazilian rosewood
Datura stratonium	jimson weed
Daucus carota	carrot
Dianthus caryophyllus	carnation
Digitalis purpurea	foxglove
Dimocarpus longan	longan
Dioscorea spp.	yams
Diospyros spp.	persimmon
Echinochloa colona	junglerice
Echinochloa crus-galli	barnyardgrass

Systematic name	Common name
Eichhornica crassipes	water hyacinth
Elaeis guineensis	oil palm
Elettaria cardamomum	cardamom
Eleusine coranana	finger miller
Eleusine indica	goosegrass
Elodea canadensis	Canadian pondweed
Elytrigia repens	couch
Eragrotis tef	teff
Eucalyptus spp.	eucalyptus
Euphorbia pulcherrima	poinsettia
Eustroma grandiflora	lisianthus
Fagopyrum esculentum	buckwheat
Fagus sylvatica	beech
Ficus carica	fig
Ficus elastica	rubber tree
Ficus spp.	ornamental ficus
Fragaria spp.	strawberries
Galium aparine	cleavers
Garsinia mangostana	mangosteen
Gentiana spp.	gentian
Glycine max	soybean
Gossypium hirsutum	cotton
Gypsophila paniculata	gypsophila
Hedera hexix	English ivy
Helianthus annuus	sunflower
Helianthus tuberosus	Jerusalem artichoke
Hemerocallis fulva	day lily
Hevea brasiliensis	rubber
Hibiscus cannabinus	kenaf
Holcus lanatus	common velvetgrass
Hordeum vulgare	barley
Humulus lupulus	hops
Hyacinthus orientalis	hyacinth
Hyoscyamus niger	henbane
Hypericum perforatum	St Johns wort
Ipomoea batatas	sweet potato
Iris spp.	iris
Isatis tinctoria	woad
Jasminum officinale	jasmine
Juglans nigra	black walnut tree
Juglans regia	walnut
Kalanchoe spp.	kalanchoe
Kochia scoparia	kochia
Laburnum anagyroides	garden laburnum tree
Lactuca sativa	lettuce
Laguncularia racemosa	white mangrove
Lavendula spp.	lavender
Lemna gibba	duckweed
Leptochloa fusca	kallar grass
Liatris spicata	liatris
Lilium longiflorum	easter lily
Limnanthes alba	meadowfoam
Limonium spp.	statice
Linobium stoloniferum	Amazon frogbire

Systematic name	Common name
Linum uritatissimum	flax
Litchi chinensis	lychee
Lolium multiflorum	Italian ryegrass
Lolium rigidum	rigid ryegrass
Lonicera spp.	honeysuckle
Lupinus spp.	lupins
Lycopersicon esculentum	tomato
Lycopersicon pimpinellifolium	wild tomato fruit
Malus pumila	apple
Mangifera indica	mangoes
Manihot esculenta	cassava
Matricaria discoidea	pineapple weed
Medicago sativa	alfalfa, lucerne
Medicago spp.	medicago
Mentha arvensis	spiderwort
Mentha piperitu	peppermint
Mentha spicata	spearmint
Monstera deliciosa	Swiss cheeseplant
Morus spp.	mulberry
Musa sapientum	bananas
Musa textilis	obaca
Narcissus pseudo-narcissus	daffodil
Nelumbo nucifera	sacred lotus
Nicotiana tabacum	tobacco
Olea europaea	olive
Orchidaceae	orchid
Orobanche spp.	broomrapes
Oryza sativa	rice
Osmia spp.	orchard bees
Otiorhynchus	vine weevil
Oxalis tuberosa	oca
Panax ginseng	ginseng
Papaver rhoeas	corn poppy, field poppy
Papaver somniferum	opium poppy
Parthenium argentatum	guayule
Passiflora edulis f. flavicarpa	passionfruit
Pastinaca sativa	parsnip
Pelargonium spp.	geranium
Pennisetum glaucum	millet, pearl millet
Persea americana	avocado
Petasites hybridus	butterbur
Petunia spp.	petunia
Phalaris arundinacea	canary grass
Pharbitis nil	Japanese morning glory
Phaseolus aureus	mung bean
Phaseolus coccineus	runner bean
Phaseolus vulgaris	common bean, field bean, french bean, kidney bean
Phleum pratense	timothy grass
Phoenix dactylifera	date palm
Phormium tenax	New Zealand flax
Picea abies	Norway spruce
Picea glauca	white spruce
Picea mariana	black spruce
Picea pungens	Colorado blue spruce
Picea sitchensis	Sitka spruce
Pinus caribaea	Caribbean pine
Pinus monticola	western white pine
Pinus taeda	loblolly pine
Pistia stratioles	water lettuce
Pisum sativum	pea
Popophyllum peltatum	American mandrake
Populus spp.	poplar
Populus tremula	quaking aspen
Primula obconica	poison primrose
Primula spp.	primroses
Prunius avium	cherry
Prunus armenaica	apricot
Prunus domestica	plum
Prunus persica	nectarine, peach
Pseudotsuga menziesii	Douglas fir
Psidium guajaba	guava
Pteris vittata	brake fern
Pyrus communis	pear
Pyrus pyrifolia	Japanese pear
Quercus spp.	oak
Quercus rubra	red oak
Ranunculus flabellaris	amphibious buttercup, aquatic buttercup
Ranunculus repens	creeping buttercup
Raphanus raphanistrum	wild radish
Raphanus sativus	radish
Raphia ruffia	raffia
Rheum rhapoponticum	rhubarb
Rhizophora mangle	red mangrove
Rhododendron	azalea, rhododendron
Ribes nigrum	blackcurrant
Ricinus communis	castor bean
Rosa spp.	rose
Rubia tinctorum	madder
Rubus fructicosus	blackberry
Rubus strigosus	raspberry
Rumex crispus	curled dock
Saccharum officianarum	sugar cane
Saintpaulia ionantha	African violet
Salicornia bigelovii	pickleweed
Salix spp.	willow
Sanguisorba canadensis	bloodroot
Santalum album	sandalwood
Scirpus spp.	bulrushes

Systematic name	Common name
Secale cereale	rye
Senecio jacobaea	ragwort
Senecio vulgaris	groundsel
Sequoia sempervirens	coast redwood
Setaria italica	foxtail millet
Setaria viridis	green foxtail
Sinapis alba	white mustard
Solanum spp.	nightshade
Solanum nigrum	black nightshade
Solanum tuberosum	potato
Sorghum bicolor	sorghum
Sorghum halepense	johnsongrass
Spergula arvensis	spurrey
Stellaria media	chickweed
Striga spp.	witchweeds
Tagetes spp.	mexican marigold
Tamarix chinensis	salt cedar
Taraxacum officinale	dandelion
Taxus baccata	European yew
Taxus brevifolia	Pacific yew
Tectona grandis	teak
Theobroma cacao	cocoa
Thinopyrum bessarabicum	sand couch
Tilia spp.	lime, basswood
Toxicodendron diversilobum	poison oak
Toxicodendron radicans	poison ivy
Toxicodendron vernicifluum	Japanese lacquer tree
Trifolium pratense	red clover
Trifolium spp	clover
Trigonella foenum-graecum	fenugreek
Triticum aestivum	wheat
Triticum durum	durum wheat
Triticum turgidum	emmer wheat
Tropaeolum majus	common nasturtium
Tulipa spp.	tulip
Typha spp.	cattails
Ulex europaeus	gorse
Urtica dioica	nettles, stinging nettle
Vaccinium oxycoccus	cranberry
Vaccinium spp.	blueberry, bilberry, cranberry
Veronica hederifolia	ivy-leaved speedwell
Vicia faba	broad bean, faba bean
Vicia narbonensis	narbonne bean
Vigna mungo	blackgram/mung bean
Vigna unguiculata	cowpea
Vitis vinifera	grapevine, seedless grape
Yucca spp.	yucca
Zangibar officinale	ginger
Zea mays	corn, Indian corn, maize

INDEX

The index is in letter-by-letter order, whereby hyphens and spaces within index headings are ignored in the alphabetization (e.g. acids, organic precedes acid-tolerance). Prefixes and terms in parentheses are excluded from the initial alphabetization.

The index is arranged in set-out style, with a maximum of three levels of heading. Major discussion of a subject is indicated by bold page numbers. Page numbers suffixed by *T* and *F* refer to Tables and Figures respectively.

Cross-reference terms in italics are general cross-references, or refer to subentry terms within the main entry (the main entry is not repeated to save space). Readers are also advised to refer to the end of each article for additional cross-references—not all of these cross-references have been included in the index cross-references.

NOTE:

Entries in this index refer to plants unless otherwise stated, e.g. protein(s) refers to plant proteins.
Subentries with the same page number(s) are present to indicate the depth of information available.
vs. indicates a comparison.

Abbreviations used in this index include:

AFLP – amplified fragment length polymorphism
CAM – crassulacean acid metabolism
ESTs – expressed sequence tags
EU – European Union
GM – Genetically Modified
SQLs – sesquiterpene lactones

A

B

C

D

F

G

H

J

K

M

N

P

Q

R

S

T

U

V

W

X

ISBN 0-12-227050-9

9 780122 270505